# VERTEBRATES

## COMPARATIVE

ANATOMY

FUNCTION

EVOLUTION

# VERTEBRATES

## COMPARATIVE

ANATOMY

FUNCTION

EVOLUTION

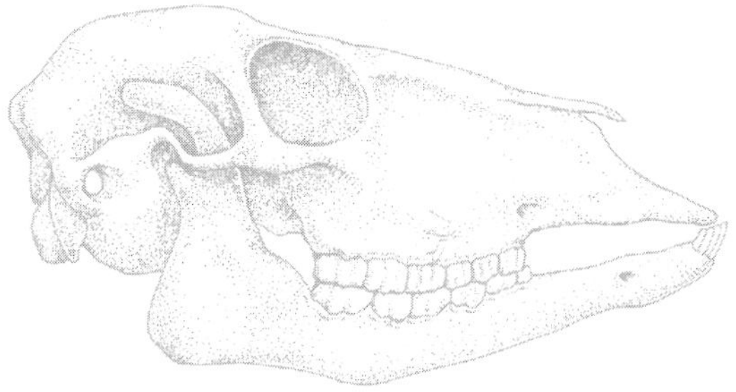

Kenneth V. Kardong, Ph. D.
*Washington State University*



WCB Wm. C. Brown Publishers
Dubuque, Iowa • Melbourne, Australia • Oxford, England

## Book Team

Project Editor *Margaret J. Kemp*
Production Editor *Sue Dillon*
Designer *Jeff Storm*
Art Editor *Mary E. Powers*
Photo Editor *Rose Deluhery*
Permissions Coordinator *Mavis M. Oeth*
Visuals/Design Developmental Consultant *Donna Slade*

**Wm. C. Brown Publishers**
A Division of Wm. C. Brown Communications, Inc.

Vice President and General Manager *Beverly Kolz*
Vice President, Publisher *Kevin Kane*
Vice President, Director of Sales and Marketing *Virginia S. Moffat*
National Sales Manager *Douglas J. DiNardo*
Marketing Manager *Craig Marty*
Advertising Manager *Janelle Keeffer*
Director of Production *Colleen A. Yonda*
Publishing Services Manager *Karen J. Slaght*
Permissions/Records Manager *Connie Allendorf*

**Wm. C. Brown Communications, Inc.**

President and Chief Executive Officer *G. Franklin Lewis*
Corporate Senior Vice President, President of WCB Manufacturing *Roger Meyer*
Corporate Senior Vice President and Chief Financial Officer *Robert Chesterman*

Cover illustration © David Uhl.

Composition by The Clarinda Company

Copyedited by Kathy Massimini

The credits section for this book begins on page 749 and is considered an extension of the copyright page.



A Times Mirror Company

Library of Congress Catalog Card Number 93–070713

ISBN 0–069–21991–7



Printed in the United States of America by Wm. C. Brown Communications, Inc.
2460 Kerper Boulevard, Dubuque, IA 52001

10 9 8 7 6 5 4 3 2 1

Dedicated with pleasure and thanks to
Richard C. Snyder
and
T. H. Frazzetta

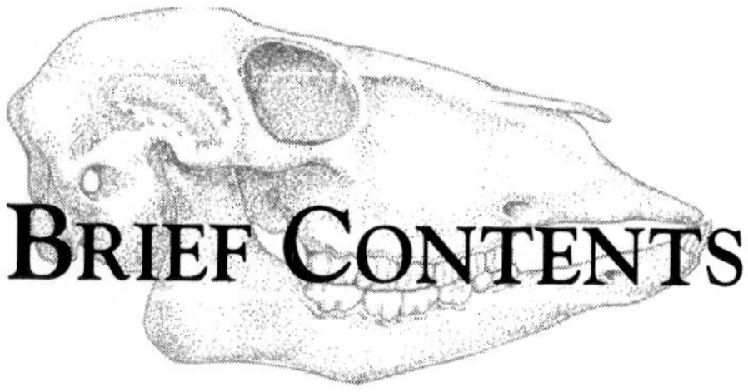

# BRIEF CONTENTS

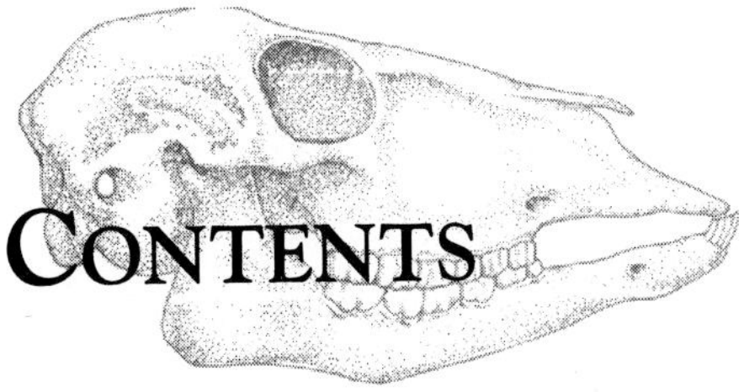

# CONTENTS

## Chapter 13 The Digestive System 491



## Chapter 14 The Urogenital System 533



## Chapter 15 The Endocrine System 584

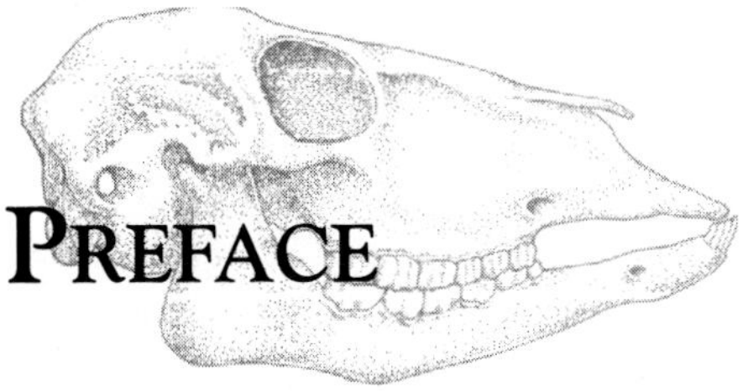

# PREFACE

Comparative morphology has always been a rich source of ideas fundamental to the life sciences. The experimental work that has characterized vertebrate morphology during the last two decades, together with its holistic and integrative approach, has made the discipline of morphology a common meeting ground of scientific research and explains its central place in a modern educational curriculum. A part's structure alone, its function, or the environment in which it serves cannot be studied in isolation if the basic principles of life and its evolution are to be understood.

My intentions in this textbook are (1) to set forth the basic structure of vertebrates, (2) to discuss the functional morphology of anatomical systems, and thus (3) to discuss the basis of major adaptive changes in vertebrate evolution. Students are encouraged to think critically about how organisms work and how vertebrates might have evolved their remarkable and complex adaptations.

## ORGANIZATION

Morphology is foremost, but I have tried to develop and integrate an understanding of function and evolution into the discussion of the anatomy of the various systems. Vertebrate morphology is discussed system by system beginning with the integument in chapter 6. The first five chapters prepare the way.

Chapter 1 introduces the discipline, evaluates the intellectual predecessors to modern morphology, defines central concepts, and alerts students to misunderstandings they may unknowingly bring with them to the study of evolutionary processes. For example, vertebrate morphology is in large part a historical science, and I frequently discuss fossils throughout the book. But most students know little about the techniques used to study the past. Therefore, in the opening chapter, the difficulties inherent in studying past events, the techniques used to chart the chronology of life on Earth, and the basis for interpreting and restoring fossils are discussed. This prepares students for discussions of fossil evidence later in the book. Finally, the experimental basis for the recent renaissance in vertebrate morphology is discussed. By the end of the first chapter, I invite students to think critically and systematically about vertebrates and their evolution.

Chordates and their origins are covered in the second chapter. Considerable attention is given to the neglected protochordates and their evolution. Evolutionarily, their appearance prepared the way for the vertebrates. Giving them their due here prepares the way for the vertebrates, which occupy the remainder of the book.

Chapter 3 discusses vertebrates, their origins, and basic taxonomy.

Chapter 4 addresses a problem many students may forget. Organisms must not only survive the onslaught of biological challenges from predators and competitors, they must also address problems posed by basic physics. Biomechanics and its significance to animal design are fast becoming a large part of the study of vertebrate function and evolution. This chapter discusses basic physical phenomena to prepare students for the selective use of these principles later. This chapter also addresses an old complaint made by J.B.S. Haldane almost 60 years ago. The complaint was that biologists often fail to notice the major differences among animals, namely differences in size. Therefore, scaling is discussed here.

Chapter 5 includes a summary of descriptive embryology and concludes with a discussion of the role embryonic processes play in vertebrate evolutionary events.

Chapter 6, the integument, begins a systematic treatment of vertebrate morphology. Because the skeletal system is complex and illustrates many basic principles of morphology, function, and evolution, it is treated in three chapters—cranial skeleton (7), axial skeleton (8), and appendicular skeleton (9). Muscles are discussed in chapter 10, followed by chapters covering respiration, circulation, digestion, and the urogenital, endocrine, nervous, and sensory systems. Chapter 18 concludes the book and provides an overview of the science of morphology.

## Pedagogy

A number of features are included to enhance the textbook. It is richly illustrated with figures that include new information and provide a fresh perspective. Each chapter opens with an outline. Important concepts and anatomical terms are boldfaced. Cross references direct students to other areas of the text where they can refresh their understanding or clarify an unfamiliar subject. Important literature is cited at the end of each chapter. These references make it easier for instructors to encourage their students to research a particular idea or topic. A glossary of definitions is included at the end of the book, preceded by four appendices: (1) Vector Algebra, (2) International System of Units, (3) Common Greek and Latin Combining Forms, and (4) Classification of Chordates.

Boxed essays are included in most chapters. Their purpose is to present subjects or historical events that students should find interesting, and perhaps from time to time, fun.

## Special Features

I made a special effort to include new subjects and give emphasis to topics central to the vertebrate story. Fossils are prominently featured. Some of the most remarkable vertebrates ever to grace the Earth are now extinct. Their anatomical adaptations and their passing raise some of the most intriguing issues about vertebrate evolution. Their treatment here enriches the study of vertebrate history and brings continuity to vertebrate evolution.

The chapter on embryology presents a perspective that some consider the logical introduction to vertebrates. Biomechanics, biophysics, and the consequences of size are discussed to help students better understand vertebrate design and function.

The mechanics of gill and lung ventilation are also emphasized. Certainly, the route to bird and mammal hearts leads through lower vertebrates. However, hearts and aortic arches of lungfishes, amphibians, and reptiles represent important adaptations in their own right, sophisticated designs exhibiting complex functional adjustments to the changing stresses imposed by the distinct environments in which these animals live.

The basic physical environments, water or air, profoundly affect the designs of feeding structures. Therefore environmental demands are discussed along with basic jaw mechanics when examining vertebrate feeding adaptations. Locomotion, including swimming, running, flying, and digging adaptations, is discussed. Much new data are still being reported on the subject of neural crest and its contribution to vertebrate structures. This has resulted in a complete reevaluation of the neural crest and has made views held until just a few years ago obsolete. This new information on neural crest and the reinterpretation of structures it requires is included. Research on the neural crest has also prompted a new look at protochordate and chordate origins. The chapters on chordates (2) and the introduction to vertebrates (3) should be timely.

New or neglected ideas are also included. Many of these ideas come from Europe where they have been known for a long time. Personally, I find many of these ideas compelling, even elegant, but frankly I find others thin, and unconvincing. However, these unconventional ideas enliven the discussion and encourage students to think about these unresolved issues of vertebrate structure, function, and evolution.

E. Jarvik's ideas on the evolution of girdles and jaws are discussed. The philosophical views of P. Dullemeijer, giving emphasis to the functional coupling of parts, are mentioned. The contrary views of W. Gutmann on chordate origins are also mentioned, primarily to remind students that function must be considered equally in any phylogenetic hypotheses.

Recent work on the evolution of synapsid jaws and the mammalian middle ear bones has been of great interest. Not only is this a remarkable story revealed through the rather complete fossil record, but it also suggests ways that incipient structures might evolve as critical intermediates in a phylogenetic lineage. Here, and elsewhere, the importance of preadaptation as a major phenomenon in evolutionary processes is illustrated.

I point out how ideas of function have been tested and identify questions about structure that remain (e.g., fenestrae). Competing theories that are experimentally resolvable (e.g., serial versus composite theories of jaw origin) are presented. Many new advances in vertebrate morphology (e.g., neural crest, snyapsid jaw evolution, cranial kinesis, and cardiovascular circulation in lower vertebrates) are discussed. Despite my own skepticism, a few contrary ideas are included (e.g., calcichordate origin of vertebrates; much of the evidence for hot-blooded dinosaurs).

I encourage students to do more than just memorize the names of parts. I invite students to actively participate in the discipline of morphological study and think critically about the processes that shape and influence vertebrate design.

## Acknowledgments

In various ways I received help from many people while preparing this textbook. Early in the preparation of the manuscript discussions with many individuals were especially helpful. In this regard I recognize and thank Edgar F. Allin, Dennis M. Bramble, Richard R. Fay, Brian K. Hall, James A. Hopson, George V. Lauder, and R. Erik Lombard.

Chapters of the manuscript, in various stages of preparation, were reviewed by many people whose contribution is appreciated. For special, extensive help during this review, I recognize in particular Walter Bock, T. H.

Frazzetta, G. E. Goslow, Jr., Leonard B. Kirschner, John A. Ruben, Samuel Tarsitano, and Philip S. Ulinski.

Over the years, conversations and friendships with colleagues have helped clarify thoughts and inspire fresh ideas. I extend my thanks for these moments to Neil F. Anderson, Vincent L. Bels, Susan W. Herring, Dominique G. Homberger, John A. Larsen, Jr., Jon Mallatt, Anthony P. Russell, Bruce A. Young, Edward J. Zalisko, and Gart A. Zweers.

I am indebted to the staff of Wm. C. Brown Publishers who saw this book through to publication. Developing the illustrations, photographs, and tables for this book has been a major effort. It was a joy to work with L. Laszlo Meszoly in developing the many new concepts incorporated into the book. For the final rendering, I especially acknowledge the work of Jack Reuter. My thanks goes to Mary Huggins and especially David K. Brake, who were there in the beginning.

Mostly I am grateful to friends and family, especially Willemina J. Kardong, who helped and encouraged throughout all stages of the process.

## Reviewers

**Samuel F. Tarsitano**
*Southwest Texas State Univ.*

**G. E. Goslow, Jr.**
*Brown University*

**Leonard B. Kirschner**
*Washington State University*

**Malcolm T. Jollie**

**Michael R. Weil**
*Univ. of WI-Eau Claire*

**Charles Meszoely**
*Northeastern University*

**Walter Bock**
*Columbia University*

**John Ruben**
*Oregon State University*

**Bedford M. Vestal**
*University of Oklahoma*

**Jay M. Templin**
*Delaware County Community College*

**Richard C. Snyder**
*University of Washington*

**Richard A. Cloney**
*University of Washington*

**T. H. Frazzetta**
*University of Illinois*

**Michael LaBarbera**
*University of Chicago*

**William A. Gern**
*University of Wyoming*

**Ingrith Deyrup-Olsen**
*University of Washington*

**Christopher A. Loretz**
*State University of New York at Buffalo*

**Carl W. Helms**
*Clemson University*

**William P. Wall**
*Georgia College*

**Frank E. Fish**
*West Chester University*

**A. C. Allen**
*Ursinus College*

**Donald O. Straney**
*Michigan State University*

**George V. Lauder**
*University of California-Irvine*

**Charles T. Robbins**
*Washington State University*

**Edward J. Zalisko**

**Robert I. Bowman**

**Farish A. Jenkins, Jr.**
*Harvard University*

**Karel F. Liem**
*Harvard University*

**William Presch**
*California State University-Fullerton*

**Philip S. Ulinski**
*University of Chicago*

**Donald Christian**
*University of Minnesota-Duluth*

**Jon Mallatt**
*Washington State University*

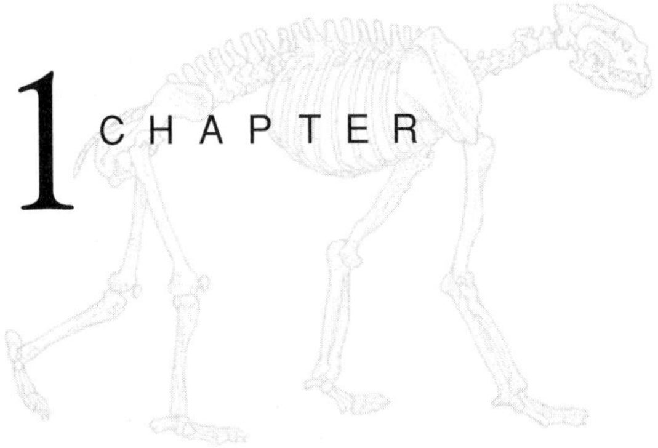

# 1 CHAPTER Introduction



## Comparative Vertebrate Morphology

Comparative morphology deals with anatomy and its significance. We focus on animals, in particular vertebrate animals, and the significance these organisms and their structure may hold. The use of "comparison" in comparative morphology is not just a convenience. It is a tool. Comparison of structures throws similarities and differences into better relief. Comparison emphasizes the functional and evolutionary themes vertebrates carry within their structures. Comparison also helps formulate the questions we might ask of structure.

For example, different fishes have different tail shapes. In the **homocercal** tail, both lobes are equal in size, making the tail symmetrical (figure 1.1a). In the **heterocercal** tail, found in sharks and a few other groups, the upper lobe is elongated (figure 1.1b). Why this difference? The homocercal tail is found in teleost fishes—salmon, tuna, trout, and the like. These fishes have a swim bladder, an air-filled sac that gives their dense bodies neutral buoyancy. They neither sink nor bob to the surface, so they need not struggle to keep their vertical position in the water. Sharks, however, lack swim bladders, so they tend to sink. The extended lobe of their heterocercal tail provides lift during swimming to help counteract this sinking tendency. The differences in structure, homocercal versus heterocercal, are related then to differences in function. Why an animal is constructed in a particular way is related to the functional requirements the part serves. Form and function are coupled. Comparison of parts highlights these differences and helps us pose a question. Functional analysis

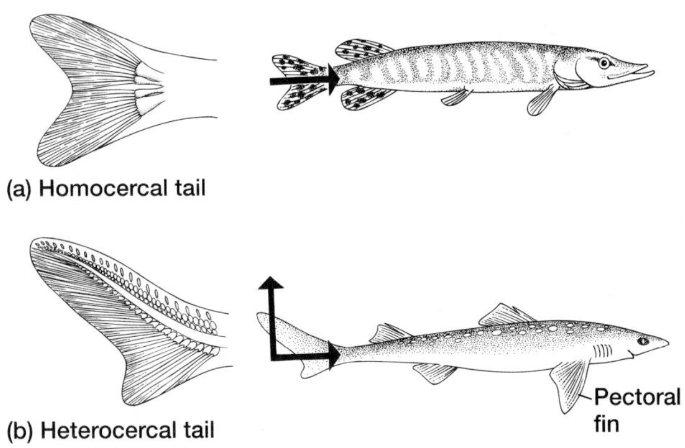


**Figure 1.1** Homocercal and heterocercal fish tails. Form differs because function differs. (a) Sweeping side to side movements of the homocercal tail, common in fishes with neutral buoyancy, drive the body forward. (b) Swimming strokes of the heterocercal tail not only propel the fish forward, but motion of the long extended upper lobe also imparts an upward lift to the posterior end of the fish. Sharks, which are a good deal denser than water, need the upward forces provided by the extended lobe of the tail and anteriorly by the pectoral fins near the head to counteract this tendency to sink.

helps answer our question and gives us a better understanding of animal design. **Functional morphology** is the discipline that relates a structure to its function.

Comparative analysis thus deploys various methods to address different biological questions. Generally, comparative analysis is used either in a historical or a nonhistorical context. When we address historical questions, we examine evolutionary events to work out the history of life. For example, on the basis of the comparison of characters, we may attempt to construct classifications of organisms and the evolutionary phylogeny of the group. Often such historical comparisons are not restricted to classification alone but center on the process of evolution behind morphological units, such as jaws, or limbs, or eyes.

When we make nonhistorical comparisons, as is frequently the case, we look outside an evolutionary context, with no intention of concluding with a classification or elucidation of an evolutionary process. Usually nonhistorical comparisions are extrapolative. For example, by testing a few vertebrate muscles, we may demonstrate that they produce a force of 15 N (newtons) per square centimeter of muscle fiber cross section. Rather than testing all vertebrate muscles, a time-consuming process, we usually assume that other muscles of similar cross section produce a similar force (other things being equal). The discovery of force production in some muscles is extrapolated to others. In medicine the comparative effects of drugs on rabbits or mice are extrapolated to tentative use in humans. Of course, the assumed similarities upon which an extrapolation is based often do not hold in our analysis. Insight into the human female reproductive cycle is best obtained if we compare the human cycle with those in higher primates because primate reproductive cycles, including the human one, differ significantly from those of other mammals.

Extrapolation allows us to make testable predictions. Where tests do not support an extrapolation, science is well served because this forces us to reflect on the assumptions behind the comparison, perhaps to reexamine the initial analysis of structures and to return with improved hypotheses about the animals or systems of interest. Comparison itself is not just a quick and easy device. The point to emphasize is this: Comparison is a tool of insight that guides our analysis and helps us set up hypotheses about the basis of animal design.

## Designs of Students

Such philosophical niceties usually do not entice students into their first course in morphology, however. Most students first venture into a course in vertebrate morphology on their way into some other profession. Customarily, morphology courses prepare students headed into technical fields such as human medicine, dentistry, or veterinary medicine. In addition, morphology is important to taxonomists who use the structure of animals to define characters. In turn, these characters are used as the basis for establishing relationships between species.

Morphology is central to evolutionary biology as well. Many scientists, in fact, would like to see a discipline devoted to the combined subject, namely, **evolutionary morphology.** Evidence of past evolutionary changes is inscribed in animal structure. Within the amphibian limb are the structural reminders of its fish fin ancestry; within the wing of a bird are the evidences of its derivation from the reptilian forelimb. Each modern group living today carries forward mementos of the evolutionary course traveled by its ancestors. For many biologists, a study of the morphological products of the past gives insight into the processes that produced them, insight into the natural forces that drove evolutionary changes, and insight into the limitations of evolutionary change.

## Vertebrate Design—Form and Function

Morphology offers more than charitable assistance to other disciplines. The study of morphology provides its own pleasure. It raises unique questions about structure and offers a method to address these questions. In brief, vertebrate morphology seeks to explain vertebrate design by elucidating the reasons for and processes that produce the basic structural plan of an organism. For most scientists today, evolutionary processes explain form and function. We might hear it said that the wings of birds, tails of fishes, or hair of mammals arose for the adaptive advantages each structure provided, and so they were favored by natural selection. Certainly this is true but it is only a partial explanation for the presence of these respective features in bird, fish, and mammal designs. The external environment in which an animal design must serve certainly brings to bear evolutionary pressures on its survival, and thus on those anatomical features of its design that convey adaptive benefits.

Internal structure itself also affects the kinds of designs that do or do not appear in animals. No terrestrial vertebrate rolls along on wheels. No aerial vertebrate flies through the air powered by a rotary propeller. Natural selection alone cannot explain the absence of wheels in vertebrates. It is quite possible to imagine that wheels, were they to appear in certain terrestrial vertebrates, would provide considerable adaptive advantages and be strongly favored by natural selection. In part the explanation lies in the internal limitations of the structure itself. Rotating wheels could not be nourished through blood vessels nor innervated with nerves without quickly twisting these cords into knots. Wheels and propellers fall outside the range of structural possibility in vertebrates. Structure itself contributes to design by the possibilities it creates; evolution contributes to design by the favored structures it preserves. We must consult both structure and evolution to understand overall design. That is why we turn to the discipline of morphology. It is one of the few modern sciences that addresses the natural unity of both structure (form and function) and evolution (adaptation and natural selection). By wrapping these together in an integrated approach, morphology contributes a holistic analysis of the larger issues before contemporary biology. Morphology is concerned centrally with the emergent properties of organisms that make them much more than the reduced molecules of their parts.

### Grand Design

Vertebrate design is complex, often elegant, and sometimes remarkably precise. To many early day morphologists, this complexity, this elegance, this precision implied the direct intervention of a divine hand in guiding the production of such sophisticated designs. However, not everyone was convinced. After all, towering mountain ranges also offer spectacular vistas but do not require recourse to divine intervention to explain them. Plate tectonics offers a natural explanation. Under pressure from colliding tectonic plates, the Earth's crust crumples to produce these ranges. With knowledge, scientific explanations uncover the mysteries that shroud geological events.

Similarly, biology has found satisfying natural explanations to replace what were once assumed to be direct divine causes. Modern principles of evolution and structural biology offer a fresh approach to vertebrate design and an insight into the processes responsible for producing that design. Just as processes of plate tectonics help geologists understand the origin of the Earth's surface features, structural and evolutionary processes help biologists understand the origin of plant and animal life. Life on Earth is a product of these natural processes. Humans are not exempt nor are we given special dispensation from these processes. Like our fellow vertebrates, humans too are products of our evolutionary past and basic structural plan. The study of morphology, therefore, brings us an understanding of the integrated processes that forged us. To understand the processes behind our design is to understand the product, namely, humans themselves, both what we are and what we can become.

But, I am getting ahead of the story. We have not had an easy intellectual journey in reaching the clarity of morphological concepts we seem to enjoy at the moment. The principles were not always so obvious, the evidence not always so clear. In fact, some issues prevalent over 100 years ago remain unresolved. The significance of underlying structure to the evolution of design, central to much of biology early in the nineteenth century, is only recently being reexamined for its potential contribution to modern morphology. Morphology has often been internally beset by unhappy contentions between those scientists centered on structure and those centered on evolution. To some extent, the fundamental principles of both structure and evolution have grown from different intellectual sources and different intellectual outlooks. To understand this, we need to examine the historical development of morphology. Later in this chapter, we examine the intellectual roots of theories about structure. But first let's look to the intellectual roots of theories about evolution.

## Historical Predecessors—Evolution

The concept of evolution is tied with the name Charles Darwin (figure 1.2). Yet most persons are surprised to learn that Darwin was not first, nor was he ever foremost, in proposing that organisms evolve. In fact, the idea of change

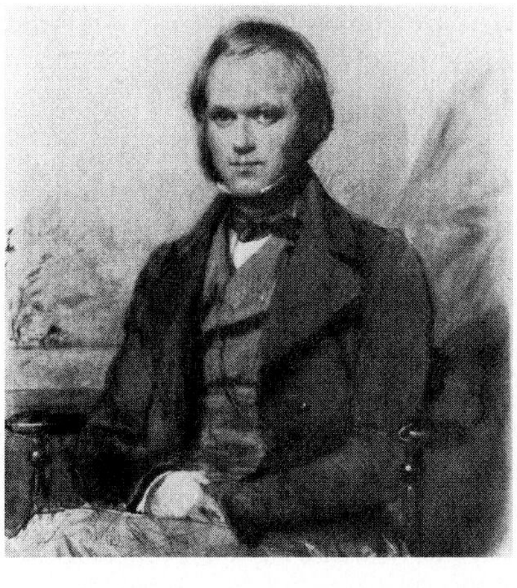

**Figure 1.2** Charles Darwin (1809–1882) at about 30 years old and three years back from his voyage aboard H.M.S. *Beagle*. Although *The Origin of Species* was still just a few notebooks in length and several decades away from publication, Darwin had several accomplishments behind him, including his account of *The Voyage of the Beagle*, a collection of scientific observations. At this time, he was also engaged to his cousin Emma Wedgwood, with whom he would live a happy married life.

through time in animals and plants dates back to ancient schools of Greek philosophy. Over 2,500 years ago, Anaximander developed ideas about the course of change from fishlike and scaly animals to land forms. Empedocles saw original creatures come together in oddly assembled ways—humans with heads of cattle, animals with branches like trees. He argued that most perished, but only those creatures who came together in practical ways survived. Even at their best, these armchair views are more poetic than scientific, so it would be an exaggeration to characterize this Greek philosophical thought as a practical predecessor of modern evolutionary science. Nevertheless, the idea of evolution existed long before Darwin thanks to these Greek philosophers.

## The Process behind the Change

What the Englishman Charles Darwin contributed was not the idea that species evolve. Rather, Darwin proposed the conditions for and mechanism of this evolutionary change. He proposed three conditions:

First, if left unchecked, members of any species increase naturally in number because all possess a *high reproductive potential*. Even slow-breeding elephants, Darwin pointed out, could increase from a pair to many millions in a few hundred years. We are not up to our rooftops in elephants, however, because as numbers increase, resources are consumed at an accelerating rate and become scarce. This brings about condition two, *competition* for the declining resources. In turn, competition leads to condition three, *survival of the few*. The mechanism now determining which organisms survive and which do not Darwin termed **natural selection,** nature's way of weeding out the less fit. In this struggle for existence, those with superior adaptations would, on average, fare better and survive to pass on their successful adaptations. Thus, descent with modification resulted from the preservation by natural selection of favorable characteristics.

As simple as this sounds today, Darwin's insight was profound. He performed no decisive experiment, mixed no chemicals in test tubes, ground no tissue in a blender. Rather, Darwin's insight arose from observation and reflection. The controversy over evolutionary processes emerges at one of three levels—fact, course, mechanism—and asks a different question at each level. The first level addresses the *fact* of evolution and asks if organisms change through time. Did evolution occur? The fact that evolution has occurred is today well established by many lines of evidence, from gene changes to the fossil record. But this does not mean that all controversies over evolution are comfortably settled. At the next level, we might ask what *course* did evolution then take? For example, anthropologists who study human evolution usually agree on the fact that humans did evolve, but they often disagree, sometimes violently, over the course of that evolution. Finally we can ask: What *mechanism* produced this evolution? At this third level in the evolutionary debate, Darwin made his major contribution. For Darwin, natural selection was the mechanism of evolutionary change.

Verbal scuffles over the fact, course, and mechanism of evolution often become prolonged and steamy because opponents ask questions at different levels and end up arguing at cross-purposes. Each of these questions had to be settled historically as well to bring us to an understanding of the evolutionary process. Historians have taken much notice of the violent public reaction to Darwin's ideas on evolution, a reaction spurred by its challenge to religious convention. But what of the scientific climate at that time? Even in scientific circles, opinion was strongly divided on the issue of "transmutation" of species, as evolution was termed then. The issue initially centered around the fact of evolution. Do species change?

## Linnaeus

Foremost among the scientists who felt that species were fixed and unchangeable was Carl von Linné (1707–1778), a Swedish biologist who followed the custom of the day by latinizing his name to Carolus Linnaeus, by which he is most recognized today (figure 1.3). Linnaeus devised a system for naming plants and animals, which is still the basis of modern taxonomy. Philosophically he argued that species were unchangeable, created originally as we find them today. For several thousand years Western thought had kept company with the biblical view, namely, that all species resulted from a single and special act of divine creation, as described in Genesis, and thereafter species remained unchanged.

Although most scientists during the 1700s sought to avoid strictly religious explanations, the biblical view of

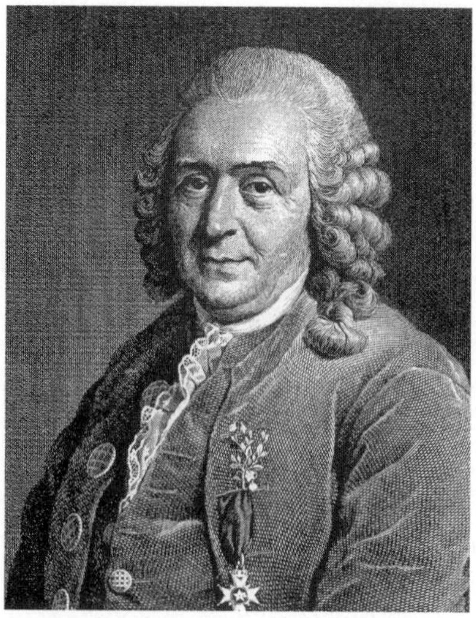

**Figure 1.3** Carolus Linnaeus (1707–1778). This Swedish biologist devised a system still used today for naming organisms. He also firmly abided by and promoted the view that species do not change.

creation was a strong presence in Western intellectual circles because it was conveniently at hand and meshed comfortably with the philosophical arguments put forth by Linnaeus and those who argued that species were immutable (unchanging). However, it was more than just the compatibility of Genesis with secular philosophy that made the idea of immutable species so appealing. At the time, evidence for evolution was not assembled easily, and the evidence available was ambiguous in that it could be interpreted both ways, for or against evolution.

## Naturalists

Today we understand the perfected adaptations of animals—the trunks of elephants, the long necks of giraffes, the wings of birds—as natural products of evolutionary change. Diversity of species results. To scientists of an earlier time, however, species adaptations reflected the care exercised by the Creator. Diversity of plant and animal species was proof of God's almighty power. Animated by this conviction, many sought to learn about the Creator by turning to the study of what He had created. One of the earliest to do so was the Reverend John Ray (1627–1705), who summed up his beliefs along with his natural history in a book entitled *The Wisdom of God Manifested in the Works of the Creation* (1691). William Paley (1743–1805), archdeacon of Carlisle, also articulated the common belief of his day in his book *Natural Theology; or Evidences of the Existence and Attributes of the Deity Collected from the Appearances of Nature* (1802). Louis Agassiz (1807–1873), curator of the Museum of Comparative Zoology at Harvard University, found much public support for his successful work to build and stock a museum that collected the remarkable creatures of this world's manifestations of the divine mind that produced them (figure 1.4). For most scientists, philosophers, and laypeople, there was, in the biological world of species, no change, thus no evolution. Even in secular circles of the mid-nineteenth century, intellectual obstacles to the idea of evolution were formidable.

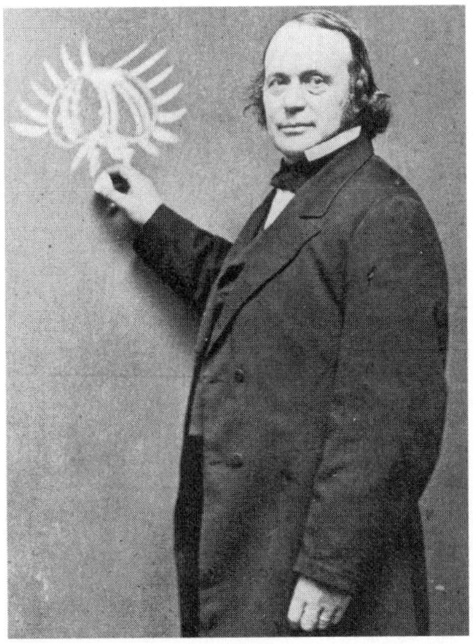

**Figure 1.4** Louis Agassiz (1807–1873) was born in Switzerland but came to his second and permanent home in the United States when he was 39. He studied fossil fishes and was first to recognize evidence of the ice ages, episodes of glaciation in Earth's history. He founded the Museum of Comparative Zoology at Harvard University. Although brilliant and entertaining in public and in anatomical research, Agassiz remained unconvinced of Darwinian evolution to the end of his life.

## J-B. de Lamarck

Among those taking the side of evolution, few were as uneven in their reputation as Jean-Baptiste de Lamarck (figure 1.5a). Most of his life, Lamarck lived on the border of poverty. He did not even hold the equivalent of a professorship at the Jardin du Roi in Paris (later the Muséum d'Histoire Naturelle; figure 1.5b). Abrupt speech, inclination to argument, and strong views did little to endear Lamarck to his colleagues. Yet his *Philosophie Zoologique*, generally dismissed when published in 1809 as the amusing ruminations of a "poet," eventually established the theory of evolutionary descent as a respectable scientific generalization.

Lamarck's ideas spoke to the three issues of evolution—fact, course, and mechanism. As to the fact of evolution, Lamarck argued that species changed through time. Curiously, he thought that the simplest forms of life arose by spontaneous generation; that is, they sprang ready made in muck from inanimate matter but thereafter evolved onward and upward into higher forms. As to the course of evolution, he proposed a progressive change in species along an ascending scale, from the lowest on one end to the most complex and "perfect" (meaning humans) on the other. As to the mechanism of evolution, Lamarck proposed that *need* itself produced heritable evolutionary change. When environments or behaviors changed, an animal developed new needs to meet the demands the environment placed upon it. Needs altered metabolism, changed the internal physiology of the organism, and triggered the appearance of a new part to address these needs. Continued use of a part tended to develop that part further; disuse led to its withering. As environments changed, a need arose, metabolism adjusted, and new organs were created. Once acquired, these new characteristics were passed on to offspring. This, in summary, was Lamarck's view. It has been called evolution by means of the *inheritance of acquired characteristics*. Characters were "acquired" to meet new needs and then "inherited" by future generations.

While a debt is owed Lamarck for championing evolutionary change and so easing the route to Darwin, he also created obstacles. Central to his philosophy was an inadvertent confusion between physiology and evolution. Any person who begins and stays with a weight-lifting program on a

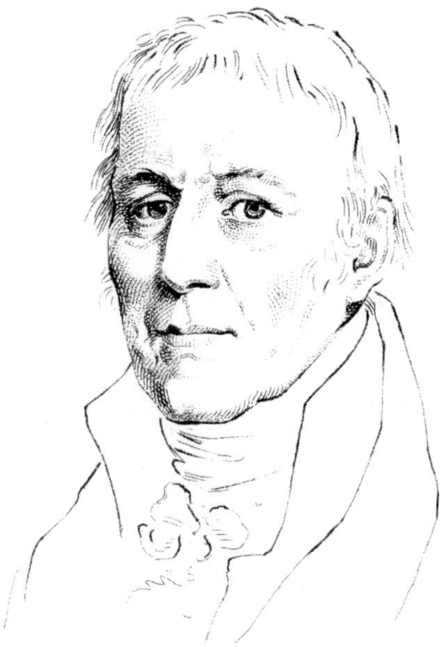

(a)

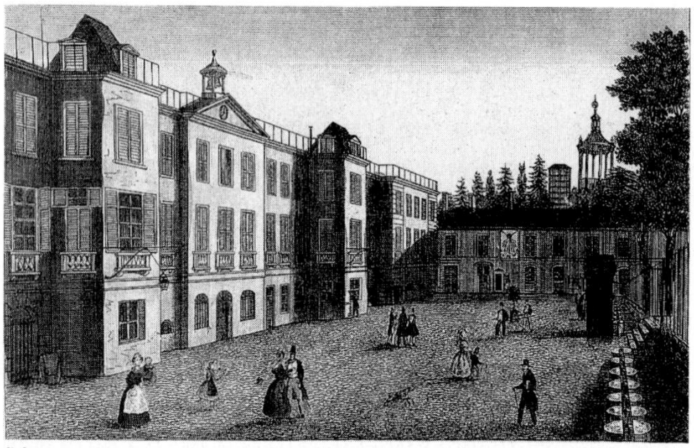

(b)

**Figure 1.5** (a) J-B. de Lamarck (1744–1829) worked most of his scientific life at the Muséum d'Histoire Naturelle (b). His academic position gave him a chance to promote the idea that species change.

regular basis can expect to see strength increase and muscles enlarge. With added weight, use (need) increases; therefore, big muscles appear. This physiological response is limited to the exercising individual because big muscles are not passed genetically to offspring. Charles Atlas, Arnold Schwarzenegger, and other bodybuilders do not pass newly acquired muscle tissue to their children. If their children seek large muscles, they too must start from scratch with their own training program. Somatic characteristics acquired through use cannot be inherited. Lamarck, however, would have thought otherwise.

Unlike such physiological responses, evolutionary responses involve changes in an organism that are inherited from one generation to the next. We know today that such characteristics are genetically based. They arise from gene mutation, not from somatic alterations due to exercise or metabolic need.

## *Acquired Characteristics*

Lamarck's proposed mechanism of inheritance of acquired characteristics failed because it confused immediate physiological response with long-term evolutionary change. Yet most laypeople today still inadvertently think in Lamarckian terms. They mistakenly view somatic parts arising to meet immediate needs. Recently, an actor/moderator of a television nature program on giraffes spoke what was probably on the minds of most viewers when he said that the origin of the long neck helped giraffes meet the "needs" of reaching treetop vegetation. Environmental demands do not reach into genetic material and directly produce heritable improvements to address new needs or new opportunities. Bodybuilding changes muscles, not DNA. That route of inheritable modification does not exist in any organism's physiology.

The other side of the Lamarckian coin is disuse, loss of a part following loss of a need. Some fishes and salamanders live in deep caves not reached by daylight. These species lack eyes. Even if they return to the light, eyes do not form. Evolutionarily, the eyes are lost. It is tempting to attribute this evolutionary loss of eyes to disuse in a dark environment. That of course would be invoking a Lamarckian mechanism. Contrary to Lamarck's theory, somatic traits are not inherited.

Because it comes easily, it is difficult to purge a Lamarckian explanation from our own reasoning. We fall automatically and too comfortably into the convenient habit of thinking of parts as rising to meet "needs," one creating the other. For Darwin, and for students coming to evolution fresh today, Lamarck's theory of acquired characteristics impedes clear reasoning. Unfortunately, Lamarck helped popularize an erroneous outlook that current culture perpetuates.

## *Upward to Perfection*

The proposed course of evolution championed by Lamarck also remains an intellectual distraction. The concept of the "scale of nature" (Latin, *scala naturae*) goes back to Aristotle and is stated in various ways by various philosophers. Its central theme holds that evolving life has a direction beginning with the lowest and evolving to the highest organisms, progressively upward toward perfection. Evolutionists, like Lamarck, viewed life metaphorically as ascending a ladder one rung at a time, up toward the complex and the perfected. After a spontaneous origin, organisms progressed up this metaphorical ladder or scale of nature through the course of many generations.

The concept of a ladder of progress was misleading because it viewed animal evolution as internally driven in a particular direction from the early imperfect soft-bodied forms up toward perfected humans. As water runs naturally downhill, descent of animals was expected to run naturally to the perfected. Simple animals were not seen as adapted in their own right but rather as springboards to a better future. The scale of nature concept encouraged scientists to view animals as progressive improvements driven by anticipation of a better tomorrow. Unfortunately, remnants of this idea still linger in modern society. Certainly humans are perfected in the sense of being designed to meet demands, but no more so than any other organism. Moles and mosquitoes, bats and birds, earthworms and anteaters all achieve an equally perfect match of parts-to-performance-to-environmental demands. It is not the benefits of a distant future that drive evolutionary change. Instead, the immediate demands of the current environment shape animal design.

The idea of perfection rooted in Western culture is perpetuated by continued technological improvements. We bring it unnoticed, like excess intellectual baggage, into biology where it clutters our interpretation of evolutionary change. When we use the terms *lower* and *higher*, we risk perpetuating this discredited idea of perfection. Lower animals and higher animals are not poorly designed and better designed, respectively. Lower and higher refer only to order of evolutionary appearance. Lower animals evolved first; higher animals arose after them. Thus, to avoid any suggestion of increasing perfection, many scientists prefer to replace the terms *lower* and *higher* with the terms ***primitive*** and ***derived*** to emphasize only evolutionary sequence of appearance, early and later, respectively.

To Lamarck and other evolutionists of his day, nature got better and animals improved as they evolved "up" the evolutionary scale. Thus, Lamarck's historical contribution to evolutionary concepts was double sided. On the one hand, his ideas presented intellectual obstacles. His proposed mechanism of change—inheritance of acquired characteristics—confused physiological response with evolutionary adaptation. By championing a flawed scale of nature, he diverted attention to what supposedly drove animals to a better future rather than to what actually shaped them in their present environment. On the other hand, Lamarck vigorously defended the view that animals evolved. For many years, textbooks have been harsh in their treatment of Lamarck, probably to ensure that his mistakes are not acquired by modern students. However, it is also important to give him his place in the history of evolutionary ideas. By arguing for change in species, Lamarck helped blunt the sharp antievolutionary dissent of contemporaries like Linnaeus, gave respectability to the idea of evolution, and helped prepare the intellectual environment for those who would solve the question of the origin of species.

## Natural Selection

The mechanism of evolution by means of natural selection was unveiled publically by two persons in 1858, although it was conceived independently by both. One was Charles Darwin, the other was Alfred Wallace. Both were part of the respected naturalist tradition in Victorian England that encouraged physicians, clergymen, and persons of leisure to devote time to observations of plants and animals in the countryside. Such interests were not seen as a way to pass idle time in harmless pursuits. On the contrary, observation of nature was respectable because it encouraged intercourse with the Creator's handiwork. Despite the reason, the result was thoughtful attention to the natural world.

## A. R. Wallace

Alfred Russel Wallace, born in 1823, was 14 years younger than Darwin (figure 1.6). Although following the life of a naturalist, Wallace lacked the comfortable economic circumstances of most gentlemen of his day; therefore, he turned to a trade for a livelihood. First he surveyed land for railroads in his native England and eventually, following his interest in nature, he took up the collection of biological specimens in foreign lands to sell to museums back home. His search for rare plants and animals in exotic lands took him to the Amazon jungles and later to the Malay archipelago in the Far East. We know from his diaries that he was impressed by the great variety and number of species to which his travels introduced him. In early 1858, Wallace fell ill while on one of the Spice Islands (Moluccas) between New Guinea and Borneo. During a fitful night of fever, his mind recalled a book he had read earlier by the Reverend Thomas Malthus entitled *An Essay on the Principle of Population, as It Affects the Future Improvement of Society*. Malthus, writing of human populations, observed that unchecked breeding causes populations to grow geometrically, whereas the supply of food grows more slowly. The simple, if cruel, result is that people increase faster than food. If there is not enough food to go around, some people survive but most die. The idea flashed to Wallace that the same principle applied to all species. In his own words:

> **It occurred to me to ask the question, Why do some die and some live? And the answer was clearly, that on the whole the best fitted lived. From the effects of disease the most healthy escaped; from enemies, the strongest, the swiftest,**

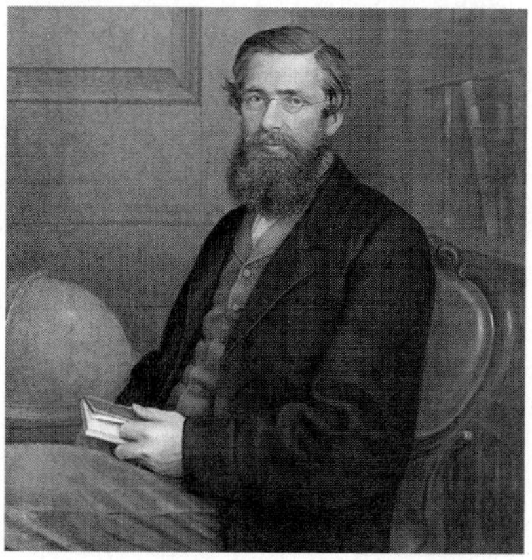

**Figure 1.6** Alfred Russel Wallace (1823–1913) in his 30s.

**or the most cunning; from famine, the best hunters or those with the best digestion; and so on.**

**Then I at once saw, that the ever present variability of all living things would furnish the material from which, by the mere weeding out of those less adapted to the actual conditions, the fittest alone would continue the race.**

**There suddenly flashed upon me the idea of the survival of the fittest.**

**The more I thought over it, the more I became convinced that I had at length found the long-sought-for law of nature that solved the problem of the Origin of Species.**

**(Wallace, 1905)**

Wallace began writing that same evening and within two days had his idea sketched out in a paper. Knowing that Darwin was interested in the subject, but unaware of how far Darwin's own thinking had progressed, he mailed the manuscript to Darwin for an opinion. The post was slow, so the journey took four months. When Wallace's paper arrived out of the blue with its stunning coincidence to his own ideas, Darwin was taken by complete surprise.

## Charles Darwin

Unlike Wallace, Charles Darwin (1809–1882) was born into economic security. His father was a successful physician, and his mother part of the Wedgwood (pottery) fortune. He tried medicine at Edinburgh but became squeamish during operations. Fearing creeping idleness, Darwin's father redirected him to Cambridge and a career in the church, but Darwin proved uninterested. At formal education, he seemed a mediocre student. While at Cambridge, however, his long-standing interest in natural history was encouraged by John Henslow, a professor of botany. Darwin was invited on geological excursions and collected biological specimens. Upon graduation, he joined as *de facto* naturalist of the government's H.M.S. *Beagle* over the objections of his father, who wished him to get on with a more conventional career in the ministry.

He spent nearly five years on the ship and explored the coastal lands it visited. The experience intellectually transformed him. Darwin's belief in the special creation of species, with which he began the voyage, was shaken by the vast array of species and adaptations the voyage introduced to him. The issue came especially to focus on the Galápagos Islands off the west coast of South America. Each island contained its own assortment of species, some found only on that particular island. Local experts could tell at sight from which of the several islands a particular tortoise came. The same was true of many of the bird and plant species that Darwin collected.

Darwin arrived back in England in October 1836 and set to work sorting his collection, obviously impressed by the diversity he had seen but still wedded to misconceptions about the Galápagos collection in particular. He had, for instance, thought that the Galápagos tortoise was introduced from other areas by mariners stashing reptilian livestock on islands to harvest during a later visit. Apparently Darwin dismissed reports of differences among the tortoises of each island, attributing these differences to changes that attended the animals' recent introductions to new and dissimilar habitats. However, in March of 1837, almost a year and a half after departing the Galápagos, Darwin met in London with John Gould, respected specialist in ornithology. Gould insisted that the mockingbirds Darwin had collected on the three different Galápagos Islands were actually distinct species. In fact, Gould emphasized that the birds were endemic to the Galápagos—distinct species not just varieties—although clearly each was related to species on the South American mainland. It seemed to have suddenly dawned on Darwin that not only birds but plant and tortoise varieties were distinct as well. These tortoises geographically isolated on the Galápagos were not only derivatives of ancestral stocks but they were now distinct island species.

Here then was the issue. Was each of these species of tortoise or bird or plant an act of special creation? Although distinct, each species also was clearly related to those on the other islands and to those on the nearby South American mainland. To account for these species, Darwin had two serious choices. Either they were products of a special creation, one act for each species, or they were the natural result of evolutionary adaptation to the different islands. If these related species were acts of special divine creation, then each of the many hundreds of species would represent a distinct act of creation. But if this were so, it seemed odd that they would all be similar to each other, the tortoises to other tortoises, the birds to other birds, and the plants to other plants on the various islands, almost as if the Creator ran out of new ideas. If, however, these species were the natural result of evolutionary processes, then similarity and diversity would be expected. The first animal or plant washed or blown to these oceanic islands would constitute the common stock from which similar but eventually distinct species evolved. Darwin sided with a natural evolution.

But Darwin needed a mechanism by which such evolutionary diversification might proceed, and at first he had none to suggest. Not until his return to England did Darwin's experiences from the Galápagos Islands and throughout his voyage crystalize. Two years after his return, and while in the midst of writing up his results of other studies from the *Beagle*, Darwin read for amusement the essay on population by Malthus, the same essay Wallace would discover years later. The significance struck Darwin immediately. If animals, like humans, outstripped food resources, then competition for scarce resources would result. Those with favorable adaptations would fare best, and new species incorporating these favored adaptations would arise. "Here then I had at last got a theory by which to work" wrote Darwin. In a moment of insight, he had solved the species problem. That was 1838, and you would think the excitement would have set him to work on papers and lecturing.

Nothing of the sort happened. In fact, four years lapsed before he wrote a first draft, which consisted of 35 pages in pencil. Two years later, he expanded the draft to over 200 pages in ink, but he shoved it quietly into a drawer with a sum of money and a sealed letter instructing his wife to have it published if he met an untimely death. A few close friends knew what he had proposed but most did not, including his wife with whom he otherwise enjoyed a close and loving marriage. This was Victorian England. Science and religion fit hand and glove.

Darwin's delay testifies to how profoundly he understood the larger significance of what he had discovered. He wanted more time to gather evidence and write the volumes he thought it would take to make a compelling case. Then in June 1858, 20 years after he had first come upon the mechanism of evolution, Wallace's manuscript arrived. Darwin was dumbfounded. By coincidence, Wallace had even hit upon some of the same terminology, specifically, natural selection. Mutual friends intervened, and much to the credit of both Wallace and Darwin, a joint paper was read in the absence of both before the Linnaean Society in London the following month, July, 1858. Wallace was, as Darwin described him, "generous and noble." Wallace, in "deep admiration," later dedicated his book on the Malay archipelago to Darwin as a token of "personal esteem and friendship." Oddly, this joint paper made no stir. But Darwin's hand was now forced.

## Critics and Controversy

Darwin still intended a thick discourse on the subject of natural selection but agreed to a shorter version of "only" 500 pages. This was *The Origin of Species*, published at the end of 1859. By then word was out, and the first edition was completely sold as soon as it appeared.

Largely because he produced the expanded case for evolution in *The Origin of Species*, and because of a continued series of related work, Darwin is remembered more than Wallace for formulating the basic concept. Darwin brought a scientific consistency and cohesiveness to the concept of evolution, and that is why it bears the name Darwinism.

Science and religion, especially in England, had been tightly coupled. For centuries, a ready answer was at hand for the question of life's origin, a divine explanation, as described in Genesis. Darwinism challenged with a natural explanation. Controversy was immediate, and in some remnant backwaters, it still lingers today. Darwin himself retired from the fray, leaving to others the task of public defense of the ideas of evolution.

Sides quickly formed. Speaking before the English Parliament, the future prime minister Benjamin Disraeli safely chose his friends: "The question is this—Is man an ape or an angel? My lord, I am on the side of the angels."

Despite the sometimes misguided reactions, two criticisms stuck and Darwin knew it. One was the question of variation, the other the question of time. As to time, there seemed not to be enough. If the evolutionary events Darwin envisioned were to unfold, then the Earth must be very old to allow time for life to diversify. In the early part of the nineteenth century, Dr. John Lightfoot, vice chancellor at Cambridge University, calculated from his biblical studies of who begot whom that humans were created in 4004 B.C. on October 23 at 9:00 A.M., presumably Greenwich mean time. Many took this date as literally accurate, or at least as indicative of the recent origin of humans, leaving no time for evolution from apes or angels. A more scientific effort to age the Earth was made by Lord Kelvin, who used temperatures taken in deep mine shafts. Reasoning that the Earth would cool from its primitive molten state to present temperatures at a constant rate, Kelvin extrapolated backward to calculate that the Earth was no more than 24 million years old. He did not know that natural radioactivity in the Earth's crust keeps the surface hot. This fact deceptively makes it seem close in temperature and thus in age to its molten temperature at first formation. The true age of the Earth is actually several billion years, but unfortunately for Darwin, this was not known until long after his death.

Critics also pointed to inheritance of variation as a weak spot in his theory of evolution. The basis of heredity was unknown in Darwin's day. The popular view held that inheritance was blending. Like mixing two paints, offspring received a blend of characteristics from both parents. This view, although mistaken, was taken seriously by many. It created two problems for Darwin. From where did variation come? How was it passed from generation to generation? If natural selection favored individuals with superior characteristics, what ensured that these superior characteristics were not blended and diluted out of existence in the offspring? If favored characters were blended, they would effectively be lost from view and natural selection would not work. Darwin could see this criticism coming and devoted much space in *The Origin of Species* to discussing sources of variation.

Today we know the answers to this paradox. Mutations in genes produce new variations. Genes carry characteristics unaltered and without dilution from generation to generation. This mechanism of inheritance was unknown and unavailable to Darwin and Wallace when they first sought answers to the origin of species. It was probably no coincidence that the intellectual breakthroughs of both were fostered by voyages of separation from the conventional scientific climate of their day. Certainly, study of nature was encouraged, but a ready interpretation of the diversity and order they observed awaited such naturalists. Although the biblical story of creation in Genesis was conveniently at hand and taken literally by some to supply explanations for the presence of species, there were scientific obstacles as well. Confusion between physiological and evolutionary adaptation (Lamarck), the notion of a scale of nature, the idea of fixity of species (Linnaeus and others), the young age of Earth (Kelvin), and the mistaken views of variation and heredity (blending inheritance) all differed from predictions of evolutionary events or confused the picture. It is testimony to their intellectual insight that Darwin and Wallace could see through the obstacles that defeated others.

## Historical Predecessors—Morphology

We might expect that the study of structure and the study of evolution historically shared a cozy relationship, each supporting the other. After all, the story of evolution is written in the anatomy of its products, in the plants and animals that tangibly represent the unfolding of successive changes through time. For the most part, direct evidence of past life and its history can be read in the morphology of fossils. By degrees, living animals preserve evidence of their phylogenetic background. It might seem then that animal anatomy would have fostered early evolutionary concepts. For some nineteenth century anatomists, this was true. T. H. Huxley (1825–1895), remembered for many scientific contributions including monographs on comparative anatomy, remarked upon first hearing Darwin's ideas of natural selection words to the effect, "How truthfully simple. I should have thought of it." Huxley was won over (figure 1.7). Although Darwin retired from public controversy following the publication of *The Origin of Species,* Huxley pitched in with great vigor, becoming "Darwin's Bulldog" to friend and foe alike.

Not all anatomists joined the evolutionary bandwagon so easily, however. Some simply misread morphology as giving evidence of only stasis, not change. On the other hand, many raised solid objections to Darwinian evolution, some of which still have not been addressed even today by evolutionary biologists. To understand the contribution of morphology to intellectual thought, we need to backtrack a bit to the anatomists who preceded Darwin. Foremost among these was the French comparative anatomist, Georges Cuvier.

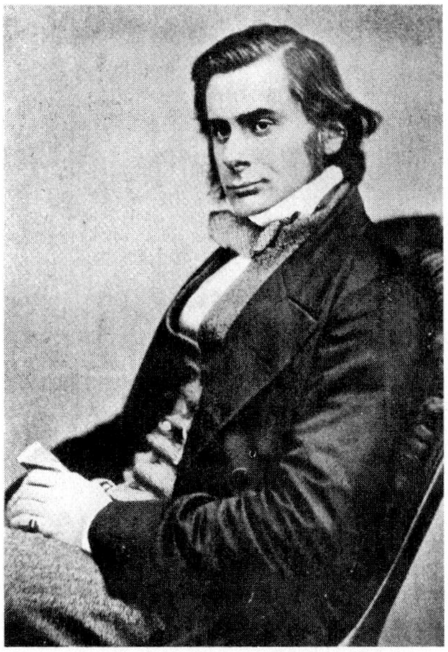

**Figure 1.7** Thomas H. Huxley (1825–1895) at age 32.

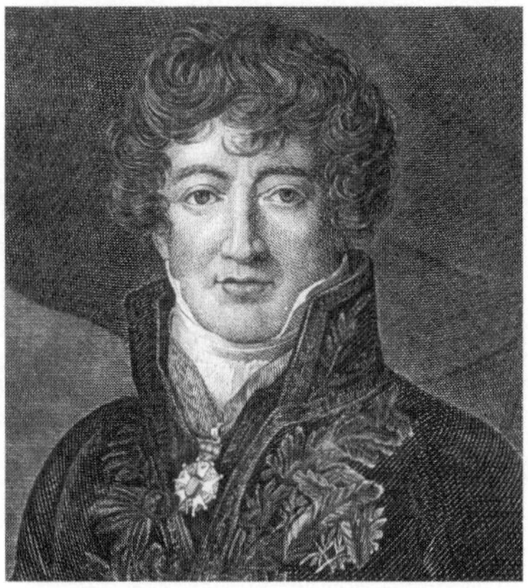

**Figure 1.8** Georges Cuvier (1769–1832). His life spanned the French Revolution, which at first won his sympathies, but as lawlessness and bloodshed became more of its character, he became disgusted. His life also overlapped with Napoleon's rule. Cuvier came to Paris in 1795 to take a post at the Muséum d'Histoire Naturelle, where he pursued administrative duties and studies in paleontology, geology, and morphology for most of his remaining life.

### Georges Cuvier

Georges Cuvier (1769–1832) brought attention to the function that parts performed (figure 1.8). Because parts and the function they served were tightly coupled, Cuvier argued that organisms must be understood as functional wholes. Parts had dominant and subordinate ranking as well as compatibility with each other. Certain parts necessarily went together but others were mutually exclusive. Possible combinations were thus limited to parts that meshed harmoniously and met necessary conditions for existence; therefore, the number of ways parts could be assembled into a workable organism was predictable. Given one part of an organism, Cuvier once boasted, he could deduce the rest of the organism. Parts of organisms, like parts of a machine, serve some purpose. Consequently, for the entire organism (or machine) to perform properly, the parts must harmonize. If one part is altered, function of connected parts is disrupted and performance fails. From this, Cuvier reasoned that evolution simply could not occur because if an animal were altered beyond a certain limit, harmony among the parts would be destroyed, function would fail, and the animal would no longer be viable. Change (evolution) would cease before it began. Cuvier's functional morphology put him in intellectual company with Linnaeus but in opposition to Lamarck's evolutionary ideas.

Cuvier took comfort as well from the known fossil record of his day. Gaps existed between major groups, as would be expected if species were immutable and evolution did not occur. During his time, ancient Egyptian mummies of humans and animals were being pilfered by Napoleon's armies and sent to European museums. Dissection proved that these ancient animal mummies were structurally identical to modern species. Again, this was evidence of no change, at least to Cuvier. Today, with a more complete fossil record at our disposal and a realization that evolution occurred over millions of years, not just within the few millennia since the the time of the pharoahs, we could enlighten Cuvier. In his day, however, the mummies were for Cuvier sweet pieces of evidence confirming what his view of morphology required. Parts were adapted to perform specific functions. If a part was changed, function failed and an animal perished. Thus, there was no change and no evolution of species.

## Richard Owen

English anatomist Richard Owen (1804–1892) believed like Cuvier that species were immutable, but unlike Cuvier, he felt that the correspondence between parts (homologies) could not be left without explanation (figure 1.9a). Virtually the same bones and pattern are present in the flipper of a dugong, the forelimb of a mole, and the wing of a bat (figure 1.9b). Each possesses the same bones. Why?

From our twentieth century perspective, the answer is clear. Out of a common ancestry, evolution passes along similar structures to perform new adaptive functions. But Owen, opposed to evolutionary ideas, was determined to find an alternative explanation. His answer centered around **archetypes.** An archetype was a kind of biological blueprint, a supposed underlying plan upon which an organism was built. All parts arose from it. Members of each major animal group were constructed from the same essential, basic plan. All vertebrates, for instance, were thought to share the same archetype, which explained why all possessed the same fundamental parts. Specific differences were forced on this underlying plan by particular functional needs. Owen was fuzzy about why he ruled out an evolutionary explanation, but he was vigorous in promoting his idea of archetypes.

He even carried this idea to repeated parts within the same individual (figure 1.10a). For example, he envisioned

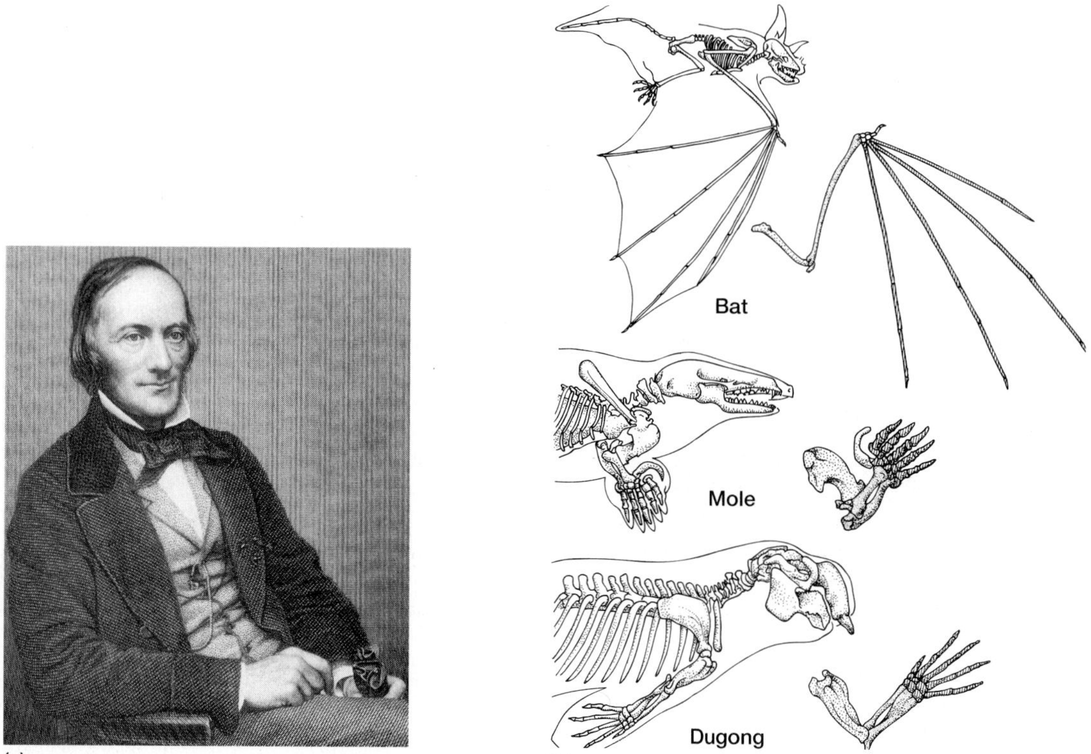


(a) (b)

**Figure 1.9** Richard Owen (1804–1892). (a) Although admired for his anatomical research, Owen was a difficult man from the accounts of those who worked or tangled with him. He agreed with Cuvier's emphasis on adaptation; however, he felt some explanation for homologies was required and, therefore, introduced the idea of archetypes. (b) Forelimbs of bat, mole, and dugong. Owen noted that each limb performs a different function—flight, digging, and swimming, respectively—and each is superficially different, but he could trace all three to an underlying common plan he called the archetype. Today we recognize that common ancestry accounts for these underlying similarities, although we would join Owen in crediting adaptation for the superficial differences among these homologous parts.

## Box Essay 1.1 Querks and QWERTY

Animals function as units, so they survive or perish as units. Georges Cuvier argued that parts of each animal were assembled in ranking order, dominant and subordinate. This unified the organism. Change a part, the harmony fails, and the animal's very existence is in question. If structures linked up into a unified whole, how then could they change without loss of the very functional roles that ensured their existence? Cuvier was no evolutionist.

Hindsight is 20-20. We know today that Cuvier's antievolutionary views were mistaken, but his emphasis upon unity of structure was not. Structure itself is a factor in evolution, either restraining or enhancing possibilities. Structural unity is not just a factor in biological change. Structure can be an obstacle in technological change.

Consider for example the primitive typewriter, or its derived descendant the PC computer (box figure 1). From one brand to another, function keys on the periphery change, but the computer's alphabet keys are fixed in positions relative to each other. The almost universal arrangement is the QWERTY system, named for the first six keys on the upper left row. Most people learned on this system. Once trained, a typist moves with a practiced familiarity from one typewriter (or PC keyboard) to another. QWERTY is the standard, keys are placed predictably in the same position, the structure remains the same from machine to machine. However, change a key's position, and function is disrupted. No standard exists yet for the position of the "break" key or even the "return" key on computer boards. Various manufacturers of different brands take the liberty of putting them in positions of choice, not in positions of familiarity. As a result, if you are using a new computer, these keys and other peripheral keys must be searched out, and typing is slowed.

But, the QWERTY alphabet keys are the same. What surprises most is that the QWERTY arrangement of keys is not the most efficient for typing speed. For example, the A key, frequently used in English, is presided over by the weak fifth little finger of your left (and for most, the nondominant) hand. The QWERTY layout of keys was devised in the early days of typewriters when mechanical levers and linkages moved keys. If keys were typed in close sequence, adjacent levers were susceptible to locking with each other and jamming. To avoid mechanical interference, keys often struck in sequence were physically separated on the keyboard. T and H, for example, are even in different rows entirely.

Electric typewriters made this system unnecessary. Keys more commonly used could be placed under the more convenient guidance of major fingers. Working this out, a different positioning of keys was devised, affectionately termed AZERTY. These more efficiently placed keys increase typing speed at least 20%, other things being equal. But other things are not equal. Few people know the AZERTY system, most learn on QWERTY, and relearning is a problem. For most of us, restructuring the keyboard destroys the utility or the convenience of the typewriter. The technological shift to AZERTY has not occurred. Cuvier would have understood.

Unlike keyboards, the parts of organisms come unlabeled. It is then a more difficult task to reveal the underlying structural unity of an organism. A simple network diagramming the functional influences between parts quickly becomes complicated (box figure 2). P. Dullemeijer, at

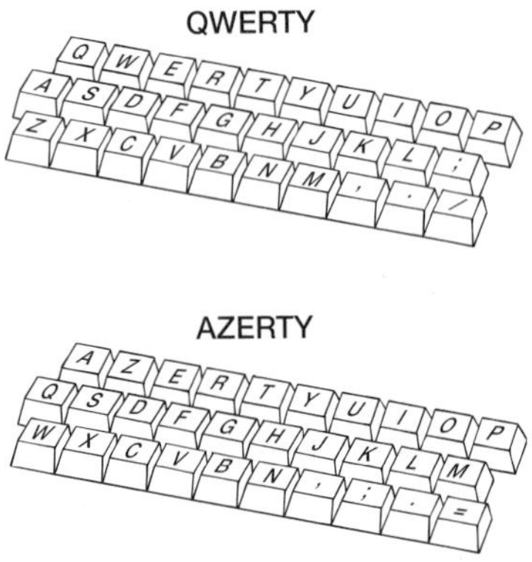


**Box figure 1** QWERTY versus AZERTY. What most people see when they sit down to a typewriter or computer keyboard is an arrangement of keys like that shown at the top. The QWERTY system is named for the six keys in the upper alphabetic row. Although it is the most common, the placement of keys is not the most efficient. On the bottom, the AZERTY arrangement places keys used commonly in English in more convenient positions presided over by more dominant fingers. However, most persons learn on QWERTY and most use systems with a QWERTY arrangement of keys. Changing requires relearning and new machines, so the shift to AZERTY has not widely occurred.

that the vertebrate skeleton consisted of a series of idealized segments he termed vertebrae (figure 1.10b). Not all available parts of these serially repeated vertebrae were expressed at each segment, but all were available if demanded. Taken together, this idealized series of vertebrae constituted the archetype of the vertebrate skeleton. Johann Wolfgang von Goethe (1749–1832), although perhaps best remembered as a German poet, also dabbled in morphology and was the first to suggest that the vertebrate skull was created from modified and fused vertebrae. His idea was expanded by others, such as Lorenz Oken (1779–1851), so by Owen's time, the concept was well known. Owen considered the skull to be formed of vertebrae extended forward into the head. He held that all four vertebrae contributed the University of Leiden in Holland, mapped interactions of parts in the snake skull, attempting to show how dominant and subordinate parts are interrelated. G. Lauder, at the University of California, has attempted to do the same for fishes, showing that of the many possible connections only a few have been exploited in fishes for jaw opening.

Evolution, like a designer of typewriters, is not a perfectionist and can work only with parts available in useful functional combinations. If evolution produced ideal designs, vertebrates might roll along on wheels. But wheels are not a structural option, legs must do. Structural constraints and functional unity are factors in the origin of new designs, typewriters or animals. (An interesting discussion of this topic can be found in Stephen Gould's [1987] article "The Panda's Thumb of Technology.")

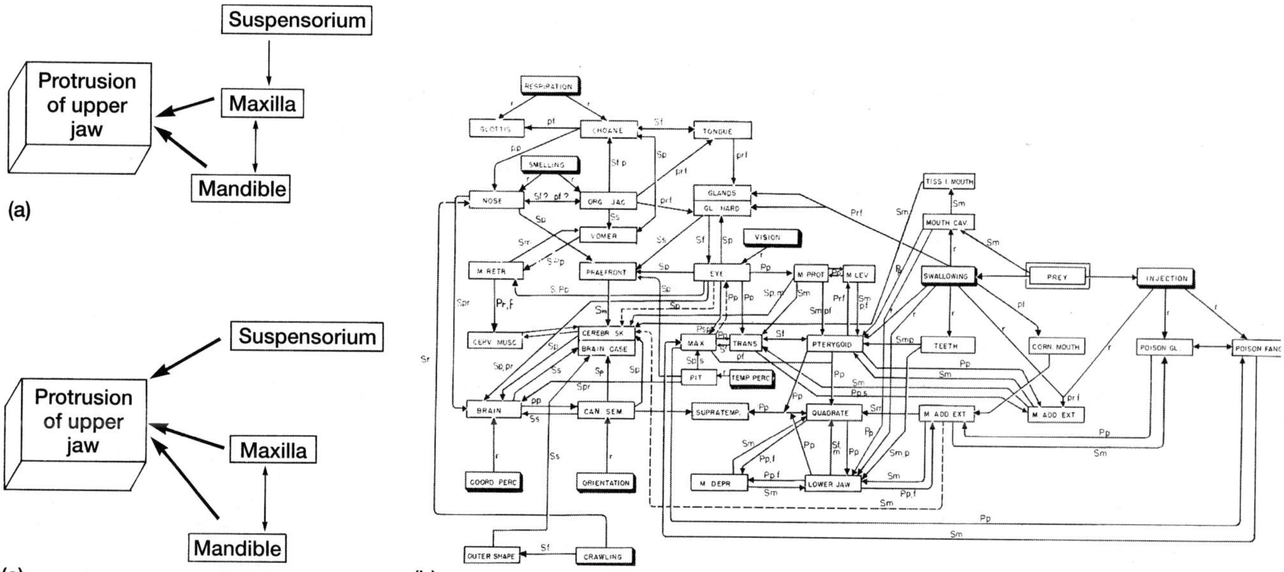


(a) (b)

**Box figure 2** Structural networks. (a) Abstract network representing major anatomical components of a fish head. In some fishes, forward movement of the upper jaw (protrusion) is mediated by different elements that are coupled functionally and structurally. In perches and allied species (top), movements of the maxillary and mandibular bones contribute to this protrusion. The large bony suspensorium exerts an effect on jaw protrusion only indirectly via its effects on the maxilla. In some cichlid fishes (bottom), motions of the maxilla and mandible still mediate jaw protrusion. But as a result of changes in connections of the suspensorium that decouple it from the maxilla, the suspensorium can independently affect upper jaw protrusion. This change in structural design has greatly increased the functional versatility and diversity of jaw morphology in cichlids. (Other changes not included also occur.) (b) Network of structural and functional interrelationships between head elements of a venomous snake. The bewildering complexity of this pattern alone illustrates how tightly coupled and structurally interwoven parts become. This suggests why phylogenetic modification (evolution) occurs within some morphological restrictions. Rectangles are anatomical elements; boxes with shading represent activities of dominant significance; arrows run in direction of influence due to function (f), mechanics (m), position (p), or size/shape (s).

and even went so far as to derive human hands and arms from parts of the fourth contributing vertebra, "the occipital segment of the skull."

T. H. Huxley, in a public lecture (published in 1857–1859), took to task the "vertebral theory of the skull," as it had become known. Bone by bone, he traced homologies and developmental appearances of each skull component. He reached two major conclusions. First, all vertebrate skulls are constructed on the same plan. Second, this developmental plan is *not* identical to the developmental pattern of the vertebrae that follow. The skull is *not* an extension of vertebrae, at least according to Huxley. Ostensibly, the subject of Huxley's public lecture was the skull, but his target was Owen and the archetype. The archetype is, wrote Huxley, "fundamentally opposed to the spirit of modern science."

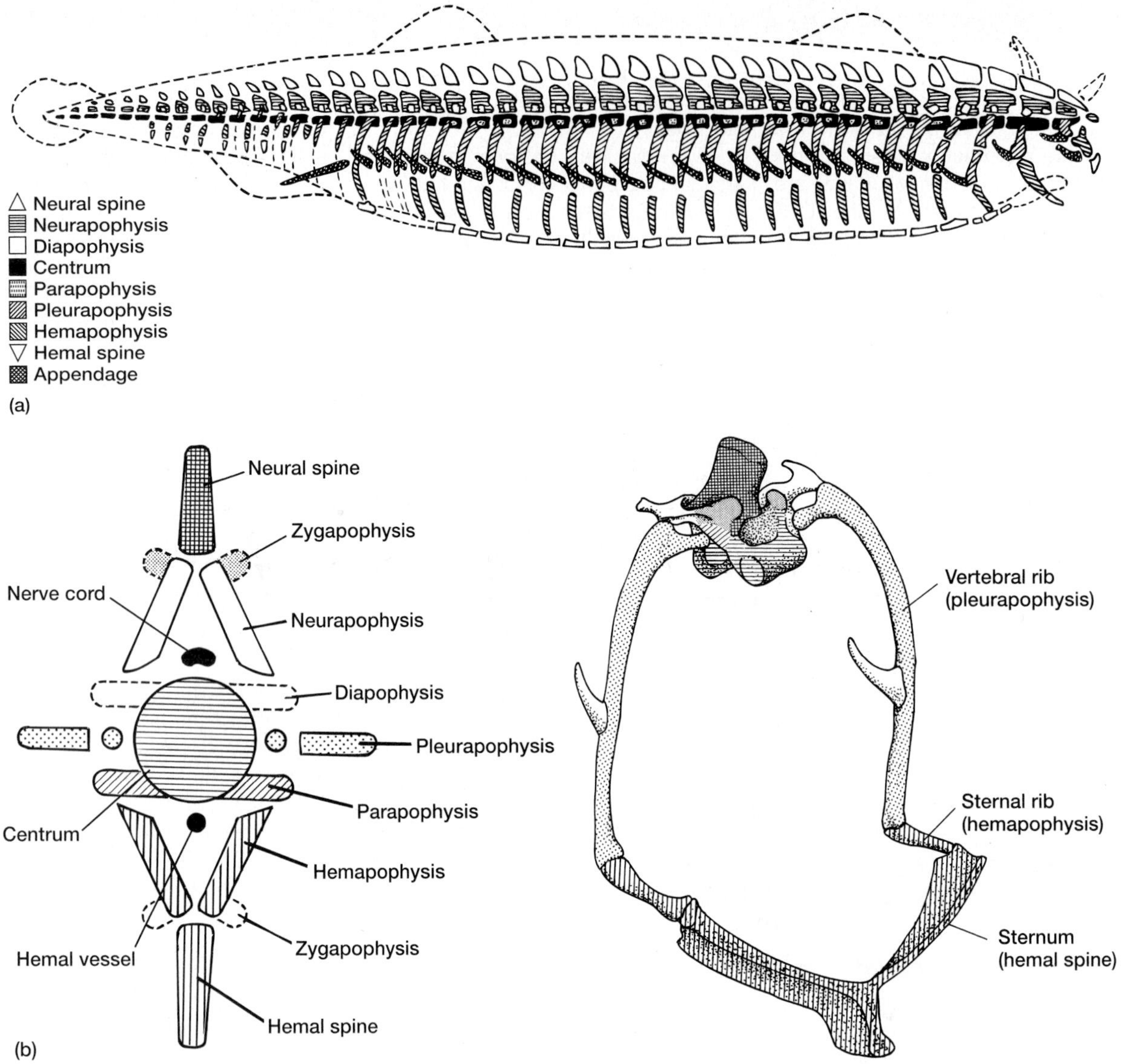


**Figure 1.10** Vertebrate archetype. Richard Owen saw the underlying pattern of the vertebrate body as a repeating series of vertebral units, collectively the vertebrate archetype (a). Owen supported the view that these vertebral units, carried forward into the head, even produced the basic elements of the skull. (b) Ideal vertebra. Potentially each vertebra included numerous elements, although not all were expressed in each segment. An actual section from a bird's skeleton indicates how this underlying plan might be realized.

Certainly Owen was the leader of those morphologists who idealized structure and pushed the vertebral theory of the skull too far and too literally. On the other hand, Huxley succeeded too well in discrediting the concept of archetypes. The two men clashed over archetypes and came down on opposite sides of evolution as well (Huxley for, Owen against). With the eventual triumph of Darwinian evolution in the twentieth century, the issues raised by morphologists such as Owen and Cuvier also tended to be forgotten. In a sense, the baby got thrown out with the bath water; that is, serious morphological issues were forgotten as evolutionary concepts triumphed.

Further contributing to the displacement of morphology has been the rise of molecular biology in recent times. Molecular biology has won a deserved place in modern science, with its successes in medicine and insights into the molecular machinery of the cell. Unfortunately, in some circles, all significant biological issues that humans face have been reduced to the chemical laws that govern molecules. In its extreme, such a reductionist view sees an organism as nothing more than the simple sum of its parts—know the molecules to know the person.

Certainly this is naive. A long distance separates the molecules of DNA from the final product we recognize as a fish or a bird or a human. Furthermore, as obvious as it might sound, the action of DNA does not reach upward to affect the agency of natural selection, but rather natural selection acts downward on DNA to affect the genetic structure of populations. A great deal of what we need to understand about ourselves comes from the world around us, not just from the DNA within.

Practitioners of morphology have begun to bring these issues that occupied Cuvier and Owen a century ago forward in a modern context. Cuvier's emphasis on adaptation has been given new life because of the clarity it brings to our appreciation of biological design. The idea of a pattern underlying the process of design has also been revisited. The result of this has been quite surprising. To explain biological design, we need more than Darwinism. Morphology, too, must be seen as a cause of design.

## Why Are There No Flying Elephants?

Not all animal designs are equally likely. Some imaginable animal concoctions simply do not work mechanically, so they never arise. Their bulk is too great or their design unwieldy. An elephant with wings would literally never fly; that is obvious. Yet many modern evolutionary biologists tend to forget about physical limitations when discussing animal design. Most resort solely to evolutionary explanations. It is tempting to be satisfied with such comfortable explanations of animal design—the long necks of giraffes give them reach to treetop vegetation, the hair of mammals insulates their warm-blooded bodies, the fins of fishes control their swimming, the venom of vipers improves their hunting success.

These and other examples of animal design were favored by natural selection, presumably for the adaptive advantages each conferred. This is reasonable, as far as it goes, but it is only half an explanation. Figuratively, natural selection is an external architect that chooses designs to fit current purposes. But the raw materials or morphology of each animal is itself a factor in design. To build a house with doors, walls, and roof, the architect lays out a scheme, but the materials available affect the character of the house. Use of brick, wood, or straw will place limits or constraints on the design of the house. Straw cannot bear several stories of weight like bricks, but it can be bent into rounded shapes. Wood makes for economical construction but is susceptible to rot. In each material lie opportunities and limitations for design.

To explain form and design, we must certainly consider the environment in which an animal resides. Among bird groups, there are no truly burrowing species that are counterparts to mammalian moles. So-called burrowing owls exist, but these are hardly equal to moles in exploiting a subterranean existence. Most amphibians occur near water because of their moisture requirements. Gliding fishes exist, but truly flying forms with strong wings do not. Elephants are large and ponderous in construction, which precludes a flying form on the elephant plan no matter how strongly natural selection favors it.

To understand form and to explain design, we must evaluate both external and internal factors. The external environment assaults an organism with a wrath of predators, challenges of climate, and competition from others. Natural selection is a manifestation of these factors. Internal factors play a part as well. Parts are integrated into a functionally whole individual. If design changes, it must do so without serious disruption of the organism. Because parts are interlocked into a coherent whole, there exist limits to change before the organism's machinery will fail. The internal construction of an organism sets boundaries to allowable change. It establishes possibilities engendered by natural selection. As new species appear, further possibilities open. But natural selection does not initiate evolutionary changes in design. Like a jury, natural selection acts only on the possibilities brought before it. If natural selection is strong and possibilities are few, then extinction occurs or diversification along that particular evolutionary course is curtailed. As a result, the avian design for delicacy of flight offers few possibilities for evolution of robust design and powerful forelimbs for digging. On the other hand, the avian design allows for the further evolution of airborne vertebrate species. Not all evolutionary changes are equally probable in large part because not all morphologies (combinations of parts) are equally available to natural selection.

Morphology embraces the study of form and function, of how a structure and its function become an integrated part of an interconnected design (the organism), and of how this design itself becomes a factor in the evolution of new forms. The term *morphology* is not just a synonym for the word *anatomy*. It has always meant much more. For Cuvier, it meant the study of structure with function, for Owen it meant the study of archetypes behind the structure, and for Huxley it meant a study of structural change over time (evolution). Today, diverse schools of morphology in North America, Europe, and Asia all generally share an interest in the structural integration of parts, the significance of this for the functioning of the organism, and the resulting limitations and possibilities for evolutionary processes. Morphology does not reduce explanations of biological design to molecules alone. Morphological analysis focuses on higher levels of biological organization—at the level of the organism, its parts, and its position within the ecological community.

## Morphological Concepts

To analyze design, concepts of form, function, and evolution have developed. Some of the most useful of these address similarity, symmetry, and segmentation.

## Similarities

In different organisms, corresponding parts may be considered similar to each other by three criteria—ancestry, function, or appearance. The term ***homology*** applies to two or more features that share a common ancestry, the term ***analogy*** to features with a similar function, and the term ***homoplasy*** to features that simply look alike (figure 1.11). These terms date back to the nineteenth century but gained their current meanings after Darwin established the theory of common descent.

More formally, features in two or more species are homologous when they can be traced back in time to the same feature in a common ancestor. The bird's wing and the mole's arm are homologous forelimbs, tracing their common ancestry to reptiles. Homology recognizes similarity based upon common origin. A special case of homology is **serial homology,** which means similarity between successively repeated parts in the *same* organism. The chain of vertebrae in the backbone, the several gill arches, or the successive muscle segments along the body are examples.

Analogous structures perform similar functions, but they may or may not have similar ancestry. Wings of bats and bees function in flight but neither structure can be traced to a similar part in a common ancestor. On the other hand, turtle and dolphin forelimbs function as paddles (analogy) and can be traced historically back to a common source (homology). Analogy recognizes similarity based upon similar function.

Homoplastic structures look alike and may or may not be homologous or analogous. In addition to sharing common origin (homology) and function (analogy), turtle and dolphin flippers also look superficially similar; they are homoplastic. The most obvious examples of homoplasty come from mimicry or camouflage, where an organism is in part designed to conceal its presence by resembling something unattractive. Some insects have wings shaped and sculptured like leaves. Such wings function in flight, not in photosynthesis (they are not analogous to leaves), and certainly such parts share no common ancestor (they are not homologous to leaves), but outwardly they have a similar appearance to leaves; they are homoplastic.

Such simple definitions of similarities have not been easily won. Historically, morphology has struggled to clarify the basis of structural similarities. Before Darwin, biology was under the influence of idealistic morphology, the view that each organism and each part of an organism outwardly expressed an underlying plan. Morphologists looked for the essence or ideal type behind the structure. The explanation offered for this ideal was the unity of plan. Owen proposed that archetypes were the underlying source for an animal's features. Homology for Owen meant comparison to the archetype, not to other adjacent body parts and not to common ancestors. Serial homology meant something different too, based again on this invisible archetype. But Darwinian evolution changed this by bringing an explanation for similarities, namely common descent.

Analogy, homology, and homoplasy are each separate contributors to biological design. Dolphins and bats live quite different lives, yet within their designs we can find fundamental likenesses—hair (at least some), mammary glands, similarities of teeth and skeleton. These features are shared by both because both are mammals with a distinct but common ancestry. Dolphins and ichthyosaurs belong to quite different vertebrate ancestries, yet they share certain likenesses—flippers in place of arms and legs and streamlined bodies. These features appear in both because both are designed to meet the common hydrodynamic demands of life in open marine waters. In this example, convergence of design to meet common environmental demands helps account for likenesses of some locomotor features (figure 1.12). On the other hand, the webbed hindfeet of gliding frogs and penguins have little to do with common ancestry (they are not closely related) or with common environmental demands (the frog glides in air, the penguin swims in water). Thus, structural similarity can arise in several ways. Similar function in similar habitats can produce convergence of form (analogy); common historical ancestry can carry forward shared and similar structure to descendants (homology); occasionally, accidents or incidental events can lead to parts that simply look alike (homoplasy). In explaining design, we can invoke one, two, or all three factors in combination. To understand design, we need to recognize the possible contribution of each factor separately.

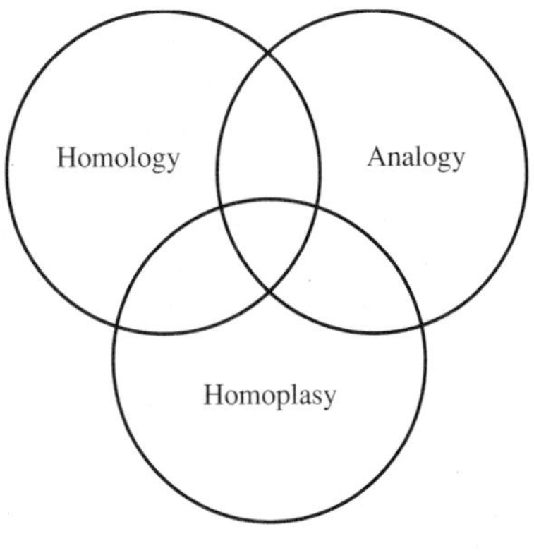


**Figure 1.11** Similarities. Parts may be similar in ancestry, function, and/or appearance. Respectively, these are defined as homology, analogy, or homoplasy. None of these types of similarities are mutually exclusive. Parts may simultaneously be homologous and analogous and homoplastic.

## Symmetry

Symmetry describes the way in which an animal's body meets the surrounding environment. **Radial symmetry** refers to a body that is laid out equally from a central axis, so that any of several planes passing through the center divides the animal into equal or mirrored halves (figure 1.13a). Invertebrates such as jellyfishes, sea urchins, and sea anemones provide examples. With **bilateral symmetry,** only the **midsagittal plane** divides the body into two mirrored images, left and right (figure 1.13b).

Body regions are described by several terms (figure 1.13c). **Anterior** refers to the head end **(cranial), posterior** to the tail **(caudal), dorsal** to the back, and **ventral** to the belly or front. The midline of the body is **medial,** the sides are **lateral.** An attached appendage has a region **distal** (farthest) and **proximal** (closest) to the body. The **pectoral region** or chest supports the forelimbs; the **pelvic region** refers to hips supporting the hindlimbs. A **frontal plane** divides a bilateral body into dorsal and ventral sections, a **sagittal plane** splits it into left and right portions, and a **transverse plane** (coronal plane) separates it into anterior and posterior portions.

Because humans carry the body upright and walk with the belly forward, the terms ***superior*** and ***inferior*** generally replace the terms *anterior* and *posterior*, respectively, in medical anatomy. Like many terms used only in the descriptive anatomy of humans, *superior* and *inferior* are poor ones to employ in general comparative research because few animals other than humans walk upright. If you venture into

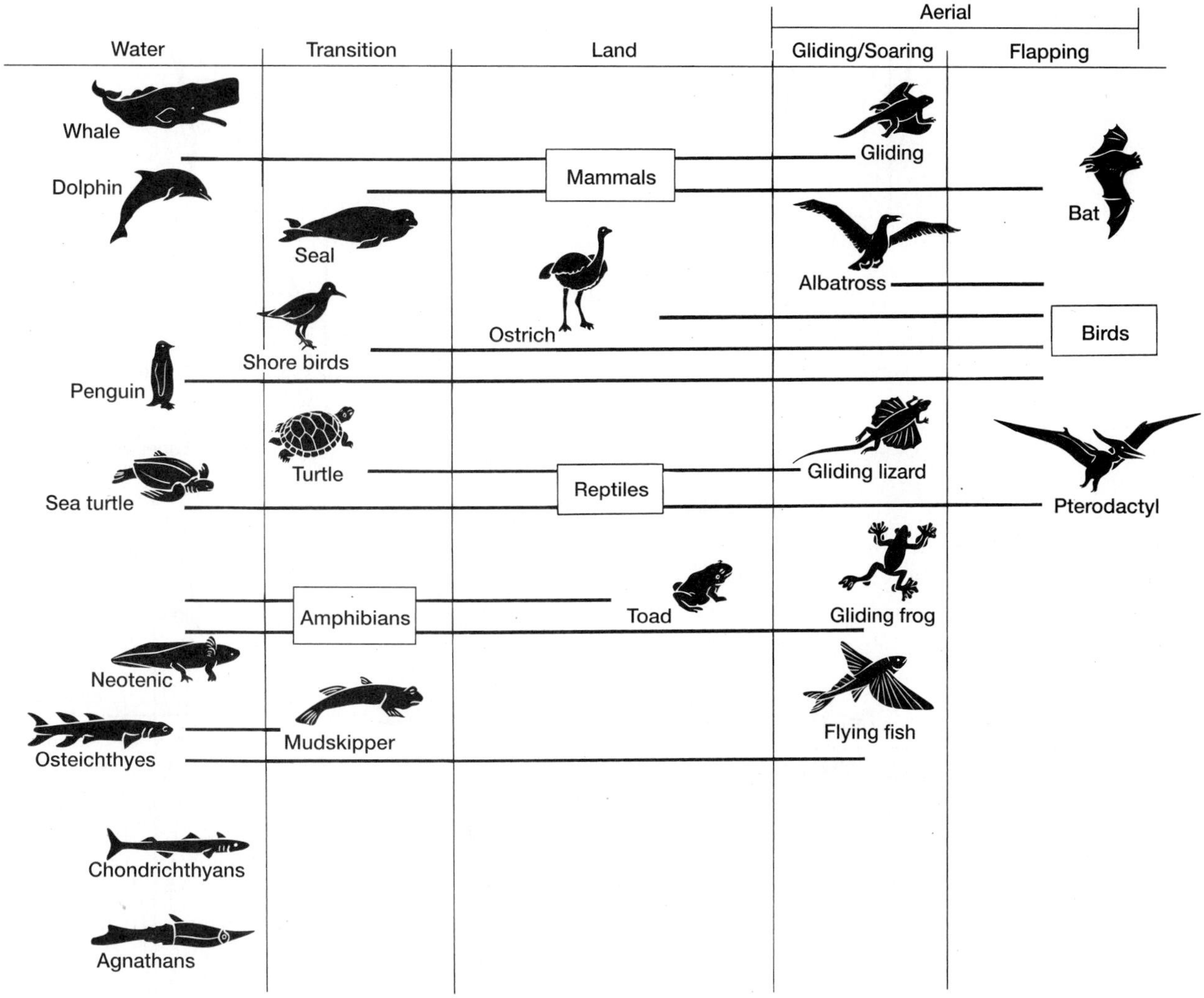


**Figure 1.12** Convergence of design. Groups of animals often evolve in habitats that differ from those of most other members of their group. Most birds fly, but some, such as ostriches, cannot and live exclusively on land; others, such as penguins, live much of their lives in water. Many, perhaps most, mammals are terrestrial, but some fly (bats) and others live exclusively in water (whales, dolphins). "Flying" fishes take to the air. As species from different groups enter similar habitats, they experience similar biological demands. Convergence to similar habitats in part accounts for the sleek bodies and fins or flippers of tuna and dolphins because similar functions (analogy) are served by similar parts under similar conditions. Yet tuna and dolphins come from different ancestries and are still fish and mammal, respectively. Common function alone is insufficient to explain all aspects of design. Each design carries historical differences that persist despite similar habitat.

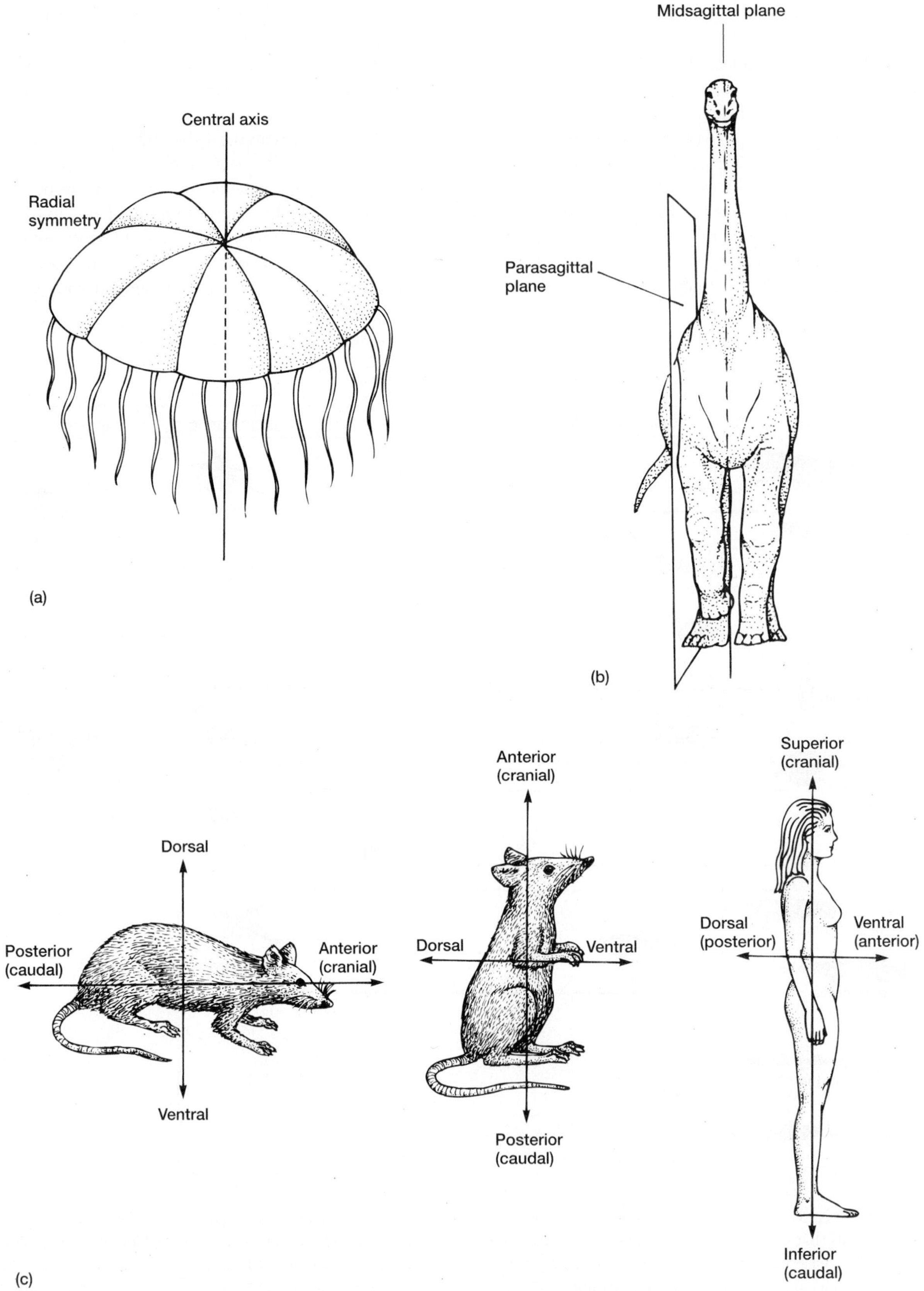


**Figure 1.13** Body symmetry. Radial and bilateral are the two most common body symmetries. (a) Radially symmetrical bodies are laid out regularly around a central axis. (b) Bilaterally symmetrical bodies can be divided into mirror images only through the midsagittal plane. (c) Dorsal and ventral refer to back and belly, respectively, anterior and posterior to cranial and caudal ends, respectively. In animals that move in an upright position (e.g., humans), superior and inferior apply to head (cranial) and posterior (caudal) ends.

the study of human anatomy, you can expect to meet such specialized terms.

## Segmentation

A body or structure built of repeating or duplicated sections is segmented. Each repeated section is referred to as a **segment** (or **metamere**), and the process that divides a body into duplicated sections is called **segmentation** (or **metamerism**). The backbone, composed of repeating vertebrae, is a segmental structure; so is the lateral body musculature of fish that is built from repeating sections of muscle.

Not all body segmentation is the same. To understand design based upon segmentation, we need to turn our attention to invertebrates. Among some invertebrates, segmentation is the basis for amplifying reproductive output. In tapeworms, for example, the body begins with a head (the scolex) followed by duplicated sections called proglottids (figure 1.14). Each section is a self-contained reproductive "factory" housing complete male and female reproductive organs. The more sections, the more factories, and the more eggs and sperm produced. Some overall body unity is established by simple but continuous nerve cords and excretory canals that run from segment to segment. Other than this, each segment is semiautonomous, a way to replicate sex organs and boost overall reproductive output, which is quite unlike segmentation found in other animals.

Annelids, such as earthworms and leeches, have segmented bodies that provide support and locomotion rather than reproduction. Annelid segmentation differs from that of tapeworms because its function is different involving the fluid-filled body coelom that forms a hydrostatic skeleton. The hydrostatic skeleton is one of two basic types of supportive systems found in animals.

The other supportive system we see in animals is a rigid skeleton. We are familiar with a rigid skeleton because our bones and cartilage constitute such a system. Another example is the chitinous outer skeletons of arthropods, such as crabs, lobsters, and insects. Rigid skeletons are efficient systems of levers that allow selective muscle use to produce movement.

Although hydrostatic skeletons are perhaps less familiar to you, they are common among animals nevertheless. As the term *hydro* suggests, this supportive system includes a fluid-filled cavity enclosed within a membrane. A hydrostatic skeleton usually is further encased within a muscular coat. At its simplest, the muscular coat is composed of circular and longitudinal bands of muscle fibers (figure 1.15). Movement is accomplished by controlled muscle deformation of the hydrostatic skeleton. In burrowing or crawling animals, movement is usually based on peristaltic waves produced in the body wall. Swimming motions are based on sinusoidal waves of the body.

The advantage of a hydrostatic skeleton is the relatively simple coordination. Only two sets of muscles, circular and longitudinal, are required. Consequently, the nervous system of animals with hydrostatic systems is usually simple as well. The disadvantage is that any local movement necessarily involves the entire body. Because the fluid-filled cavity extends through the entire body, muscle forces developed in one region are transmitted through the fluid to the entire animal. Thus, even when movement is localized, muscles throughout the body must be deployed to control the hydrostatic skeleton.

In truly segmented animals, **septa** sequentially subdivide the hydrostatic skeleton into a series of internal compartments. As a consequence of compartmentalization, the body musculature is also segmented, and in turn the nerve and blood supply to the musculature are segmentally arranged as well. The locomotor advantage is that such segmentation allows for more localized muscle

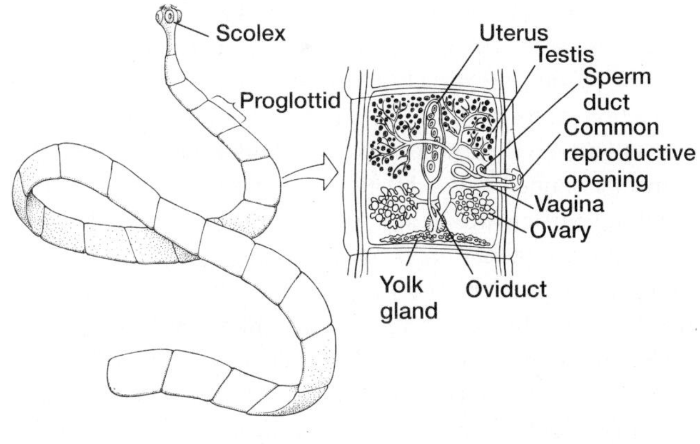


**Figure 1.14** Segmented tapeworm. Each section, or proglottid, is a reproductive factory producing eggs and sperm.

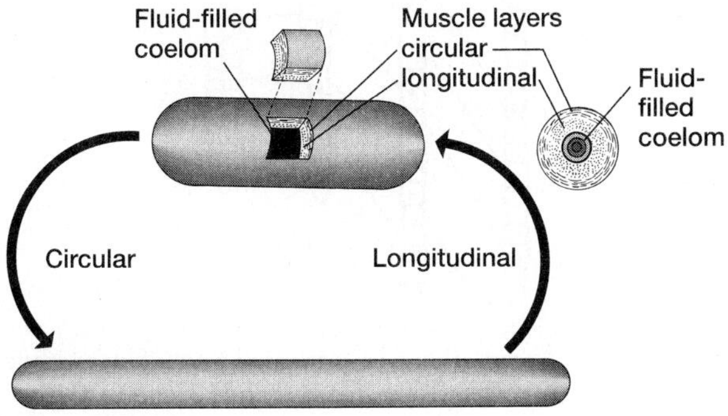


**Figure 1.15** Hydrostatic skeleton. At its simplest, changes in shape and movement involve two mechanical units, the muscle layers of the body wall (longitudinal and circular) and the fluid-filled body coelom within. Contraction of the circular muscles lengthens the shape; contraction of longitudinal muscles shortens the body. The fluid within is incompressible so that muscular forces are spread throughout the body to bring about changes in shape.

control and localized changes in shape (figure 1.16). For instance, the segmented body of an earthworm is capable of localized movement.

Segmentation among vertebrates is less extensive than segmentation among invertebrates. Lateral body musculature is laid out in segmental blocks, and nerves and blood vessels supplying it follow this segmental pattern. But segmentation goes no deeper. The viscera are not repeated units, and the body cavity is not serially compartmentalized. Locomotion is provided by a rigid skeleton, and the vertebral column (or notochord) is served by segmental body musculature; however, segmentation of the outer body musculature does not extend inward to the coelom and viscera.

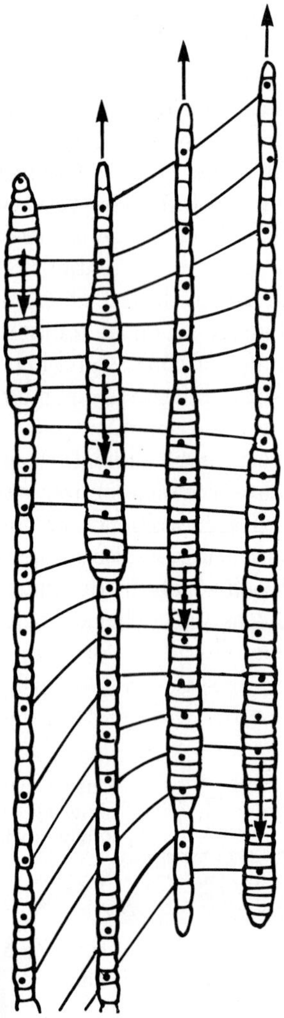

**Figure 1.16** Locomotion of a segmented worm. Fluid within the body cavity flows into selected compartments, filling and expanding each. This ballooning of the body is controlled selectively by each body segment and coordinated overall by the worm's nervous system. As the fluid passes backward from one compartment to the next, each expanded segment pushes against the surrounding soil in turn and establishes a firm hold on the walls of the worm's tunnel-shaped body. Extension of the anterior body pushes the head forward in order for the worm to make progress through the soil.

Although the vertebrate body is not composed of a hydrostatic skeleton, selected organs are based on the principle of hydrostatic support. The notochord, for instance, contains a core of fluid-engorged cells tightly wrapped in a sheath of fibrous connective tissue. This incompressible but flexible rod is a hydrostatic organ that functions to keep the body at a constant length. The penis is another example of a hydrostatic organ. When properly stimulated, cavities within it fill firmly with fluid, in this case with blood, to give the penis an erect rigidity of some functional significance.

# Evolutionary Morphology

As mentioned previously, evolution and morphology have not always been happy companions. On the brighter side, the more recent cooperation between scientists in both diciplines has clarified our understanding of animal design. With this cooperation, concepts of design and change in design have come into better relief.

## Function and Biological Role

For most of us, the concept of function is rather broad and used loosely to cover both how a part works in an organism and how it serves adaptively in the environment. The cheek muscles in some small mice act to close their jaws and chew food. In so doing, these muscles perform the adaptive role of processing food. The same structure works both within an organism (chewing) and in the role of meeting environmental demands (resource processing). To recognize both services, two terms are employed. The term ***function*** is restricted to mean the action or property of a part as it works *in an organism*. The term ***biological role*** (or just role) refers to how the part is used *in the environment* during the course of the organism's life history.

In this context, the cheek muscles of mice function to close the jaws and serve the biological role of food processing. Notice that a part may have several biological roles. Not only do jaws serve a role in food processing, but they might also serve the biological role of protection or defense if used to bite an attacking predator. One part may also serve several functions. The quadrate bone in reptiles functions to attach the lower jaw to the skull. It also functions to transmit sound waves to the ear. This means that the quadrate participates in at least two biological roles: feeding (food procurement) and hearing (detection of enemies or prey). Body feathers in birds provide another example (figure 1.17a–c). In most birds, feathers function to cover the body. In the environment, the biological roles of feathers include insulation (thermoregulation), aerodynamic contouring of body shape (flight), and, in some, display during courtship (reproduction).

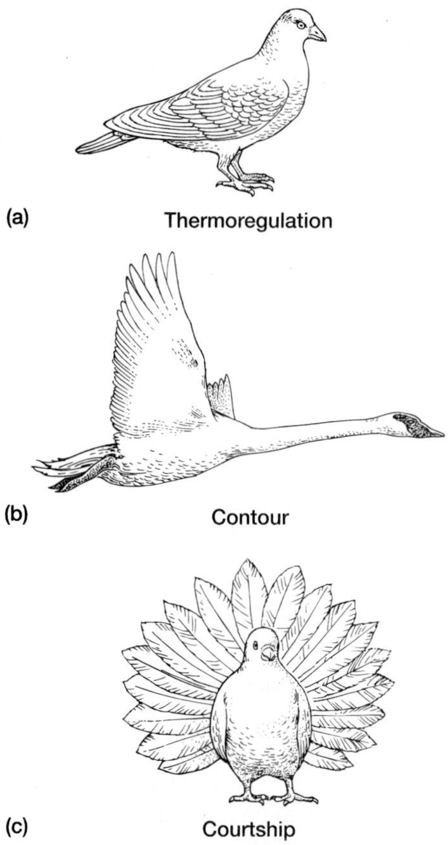


**Figure 1.17** Biological roles. The same structure may serve several biological roles. For example, in addition to producing lift for flight, feathers play a part in (a) thermoregulation (insulation), to prevent heat loss to a cold environment, (b) aerodynamic contouring (flight), to streamline the body, and (c) reproduction (courtship) to display colors to rivals or mates.

Functions of a part are determined largely in laboratory studies; biological roles are observed in field studies. Inferring biological roles only from laboratory studies can be misleading. For example, some harmless snakes produce oral secretions in which laboratory biologists discovered toxic properties. Many leaped to the conclusion that the biological role of such toxic oral secretions must be to kill prey rapidly, but field studies proved that this was not the case. Humans also produce a saliva that is mildly toxic (function), but certainly we do not use it to envenomate prey (biological role). Saliva serves the biological role of processing food by initiating digestion and lubrication of food. Toxicity is an inadvertent by-product of human saliva, without any adaptive role in the environment.

## Preadaptation

For many scientists, the word ***preadaptation*** is chilling because it seems to invite a misunderstanding. Alternative terms have been proposed (protoadaptation, exaptation), but these really do not help and only congest the literature with redundant jargon. If we keep in mind what preadaptation does not mean as well as what it signifies, then the term should present no special difficulty. Preadaptation means that a structure or behavior possesses the necessary form and function *before* (hence pre-) the biological role arises that it eventually serves. In other words, a preadapted part can do the job before the job arrives. The concept of preadaptation does not imply that a trait arises in anticipation of filling a biological role sometime in the future. Adaptive traits serve roles of the moment. If there is no immediate role, selection eliminates the trait.

For example, feathers likely evolved initially in birds (or in their immediate ancestors) as insulation to conserve body heat. Like hair in mammals, feathers formed a surface barrier to retard the loss of body heat. For warm-blooded birds, feathers were an indispensable energy-conserving feature. Today, feathers still play a role in thermoregulation; however, for modern birds, flight is their most conspicuous role. Flight came later in avian evolution. Immediate ancestors to birds were ground- or tree-dwelling reptilelike animals. As flight became a more important lifestyle in this evolving group, feathers already present for insulation became adapted into aerodynamic surfaces in order to serve flight. In this example, we can say that insulating feathers were a preadaptation for flight. They were ready to serve as aerodynamic surfaces before that biological role actually arose.

Similarly, the wings of diving birds are preadapted as paddles. In pelicans and auks, they are used to swim while the bird is submerged. If, as now seems likely, primitive lungs for respiration arose early in fishes, then they were preadapted to become swim bladders, buoyancy devices of later fishes. Fish fins were preadapted to become tetrapod limbs.

One hypothetical scheme of preadaption traces the origin of birds from reptiles through a series of five preflight stages (figure 1.18). Beginning with reptiles that lived in or frequented trees, the sequence shows that some leaped from branch to branch in order to escape pursuing predators or get to adjacent trees without making a long journey down one tree and back up the other. Such behavior established the animal's practice of taking to the air temporarily. Next came parachuting in which the animal spread its limbs and flattened its body to increase resistance and slow descent during the vertical drop, softening the impact on landing. Gliding was next. The animal deflected from the line of fall, so horizontal travel increased. Flailing, an early stage of active flight, further increased the horizontal distance. Flapping flight gave access to habitats unavailable or to terrestrial species. In fact, a new mode of life was achieved and modern birds are the result.

Such a view, although hypothetical, presents a plausible sequence by which flight in birds might have arisen. It helps address several criticisms leveled at morphological processes of evolutionary change. One long-standing

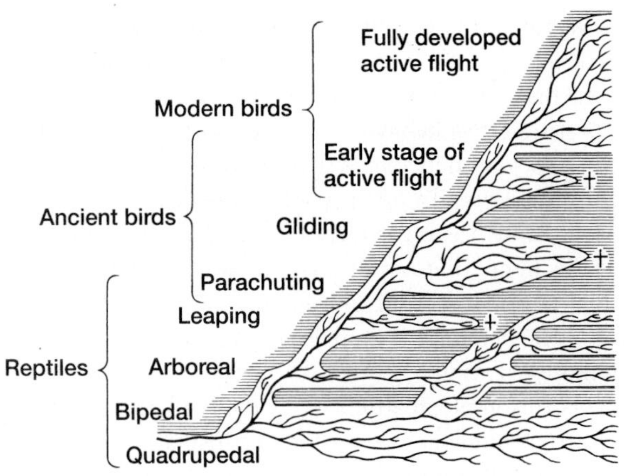


**Figure 1.18** Evolution of bird flight modeled as a series of successive stages, each preadapted to the next, that trace the evolution of birds from reptiles. Each stage is adaptive in its own right, but after having been achieved, each sets the stage for the next.

Based on the research of W. J. Bock.

complaint against evolutionary change is that many structures, such as large complicated wings and feathers, could not possibly have had any selective value when they first appeared. Such **incipient structures** would be small and formative when they first made their evolutionary debut. The argument goes like this: Incipient structures would not enjoy selective favor until they were large and elaborate enough to perform the role that brought an adaptive advantage, such as flapping flight. However, this example shows that large, complicated structures need not have evolved all at once in one large evolutionary binge. In the hypothesized five-stage evolution of bird flight, no preceding stage anticipated the next. There was no drive in the stages themselves propelling them necessarily to the next stage. Each stage was adaptive in its own right, for the immediate advantages enjoyed. If conditions changed, organisms may have evolved further, but there were no guarantees.

Some mammals, like "flying" squirrels, are still gliders. They are well adapted to conifer forests. Others, such as bats, are full-fledged, powered fliers. In an evolutionary sense, gliding squirrels are not necessarily "on their way" to becoming powered fliers like bats. Gliding is sufficient to meet demands the squirrels face when moving through the canopy of northern conifer forests. Gliding in these squirrels serves the environmental demands of the present. It does not anticipate powered flight in the distant future.

The example of bird flight also reminds us that a new biological role usually precedes the emergence of a new structure. With a shift in roles, the organism experiences new selective pressures in a slightly new niche. The shift from leaping to parachuting, or from parachuting to gliding, or from gliding to early flailing flight initially placed old structures in the service of new biological roles. This initial shift in roles exposed the structure to new selection pressures favoring those mutations that solidify a structure in its new role. First comes the new behavior, and then the new biological role follows. Finally a change in structure becomes established to serve the new activity.

## Evolution as Remodeling

The scheme that traces the evolution of bird flight also tells us that evolutionary change usually involves renovation, not new construction. Old parts are altered, but seldom are brand new parts added. Almost always, a new structure is just an old part made over for present purposes. In fact, if a complete novelty made a sudden appearance, it would probably disrupt the organism's smooth, functional harmony and would be selected against.

Because evolution proceeds largely through the process of remodeling, descendant organisms bear the traces of ancestral structures. Preadaptation does not cause change but is only an interpretation of evolutionary outcomes after they occur. Preadaptation is hindsight, a look backward to see out of what ancestral parts present structures arose. In hindsight, we might see that leaping preceded parachuting, parachuting preceded gliding, and gliding preceded flailing. Each preceding step preadapted to the next. The conceptual mistake would be to interpret these steps as internally driven inevitably from grounded reptiles to flying birds. Nothing of the sort is intended. We do not know ahead of time the future course of evolution, so we cannot tell which structures are preadapted until after they have evolved into new roles.

# Phylogeny

The course of evolution, known as **phylogeny,** can be summarized in graphic schemes or **dendrograms** that depict treelike branched connections between groups. Ideally the representation is a faithful expression of the relationships between groups. But the choice of dendrogram is based on intellectual bent and practical outcome. Dendrograms summarize evolution's course. This brevity gives them their attractiveness. All have risks, all flirt with oversimplification, and all take shortcuts to make a point. Let us look at the advantages and disadvantages of several types of dendrograms.

## Of Bean Stalks and Bushes

In 1896, Ernst Haeckel wrote *The Evolution of Man* in which he depicted the human pedigree or human phylogeny (figure 1.19). The book is a useful summary of his thoughts on the subject. Some today might wish to correct points in Haeckel's explicit phylogeny, but what does not stick out so readily is the assumption behind his dendrogram, namely,

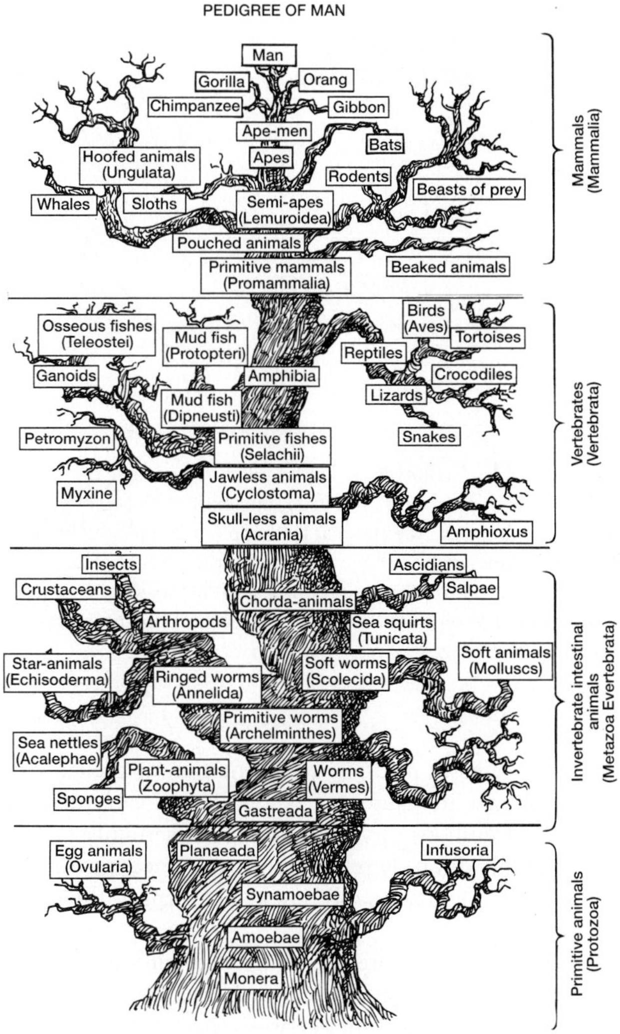


**Figure 1.19** Haeckel's phylogeny. Like a tree, this phylogeny displays the proposed branching of species. Although many lines of evolution are shown, Haeckel chose to call it the "Pedigree of Man," subtle evidence of the common view that humans represent the culmination of evolution's efforts.
Ernst Haeckel.

that humans are the pinnacle of evolution. Neither then (nineteenth century) nor now (twentieth century) was Haeckel alone in assuming that nature climbed from one species to the next like rungs on a ladder, from primitive to perfected, from lower forms to humans at the top of the scale of nature. What such a dendrogram subtly promotes is the mistaken view that humans stand alone as the sole possessor of the top rung of the evolutionary ladder.

In reality, the human species is just one of thousands of recent evolutionary products. Evolution does not proceed up a single ladder but bushes outward along several simultaneous courses. Although mammals continued to prosper largely on land, birds evolved concurrently and teleost fishes diversified in all waters of the world. Birds, mammals, fishes, and all species surviving today represent pinnacles within their groups. No single species is a Mount Everest among the rest. Humans share the current evolutionary moment with millions of other species, all with long histories of their own and all adapted in their own ways to their own environments.

To reflect this diverse pattern of evolution faithfully, dendrograms should look like bushes, not like bean stalks or ladders (figure 1.20a,b). After birds evolved from reptiles,

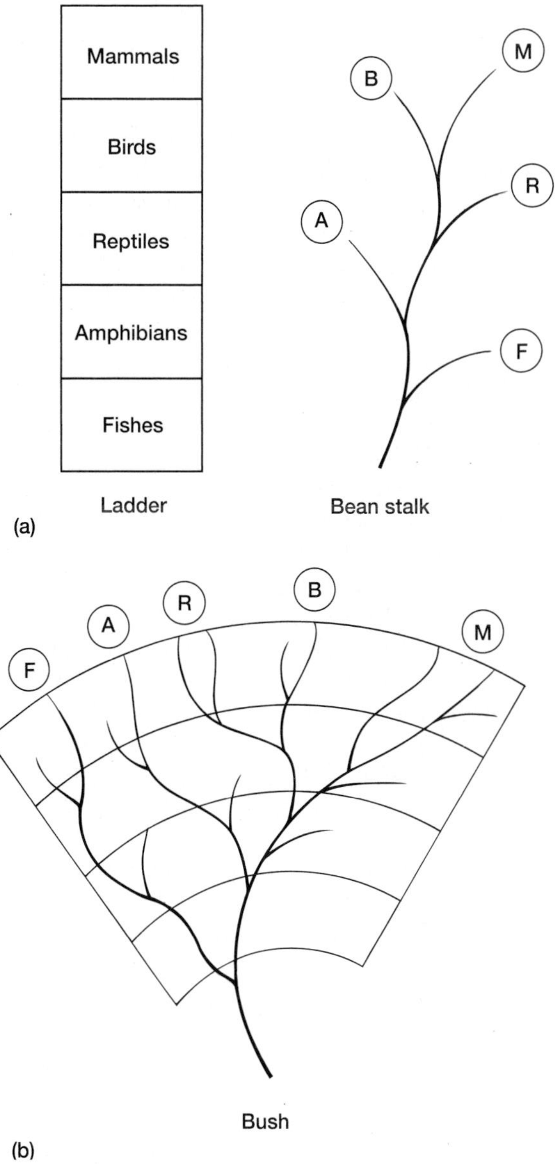


**Figure 1.20** Bean stalks and bushes. (a) The "ladder of creation" is a misleading metaphor. Evolution proceeds not in stately fashion up a ladder of species, one to the next, but along parallel lines that branch outward. Dendrograms shaped like bean stalks illustrate the order in which a group appeared but nurture the misleading view that species evolved in linear sequence up to the present time. (b) The diversity of unfolding evolution is better represented by a dendrogram shaped like a bush.

reptiles not only persisted but they actually diversified and continued to evolve and prosper. The same holds for amphibians that gave rise to reptiles and for fishes that gave rise to amphibians. Certainly modern amphibians have carried forward primitive features from their early ancestors; however, they have also continued to evolve independently of reptiles since the two lineages parted company over 300 million years ago. Frogs are structurally quite different, for instance, from the earliest amphibians.

Dendrograms that look like bean stalks or ladders are quick, uncomplicated summaries of the course of evolution (figure 1.20a). This is their strength, but they can also mislead because they imply that the most significant achievement of an earlier group is to serve as the source for a derivative group—fish for amphibians, amphibians for reptiles, and so on. Dendrograms in the shape of ladders warp our view in that more recent groups are somehow depicted as better perfected than earlier groups. Dendrograms that look like bushes not only track the course of new groups but they also show us that after one group gives rise to another, both may continue to evolve concurrently and adapt to their own environments (figure 1.20b). Once a new group is produced, evolution among ancestors does not stop nor does a derived group necessarily replace its ancestors.

The evolution of life is a continuous and connected process from one moment to the next. New species may evolve gradually or suddenly, but there is no point of discontinuity, no break in the lineage. If a break occurs in the evolving lineage, the consequence is extinction, a finality not redeemed. When taxonomists study current living species, they examine an evolutionary cross section of time in that they view only the most recent but continuing species with a long diverging history behind them. The apparent discreteness of species or groups at the current moment is partly due to their previous divergence. When followed back into their past, the connectedness of species can be determined. A dendrogram showing lineages in three dimensions (figure 1.21) emphasizes this continuity. If reduced to a two-dimensional branching dendrogram, the relationships stand out better but imply an instant distinctiveness of species at branch points. The sudden branches are a taxonomic convention but may not faithfully represent the gradual separation and divergence of species and new groups.

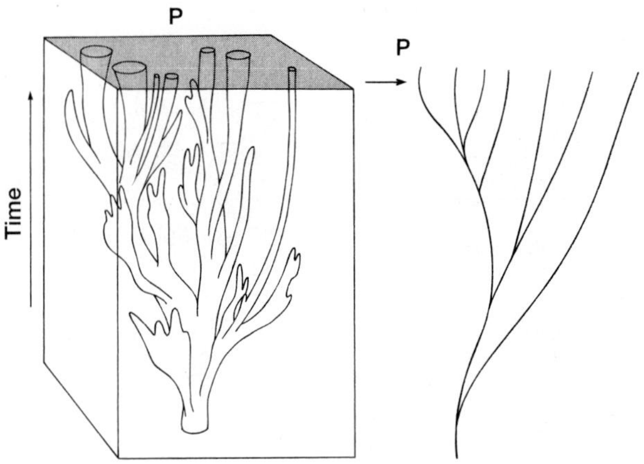


**Figure 1.21** Evolution of dendrograms. The course of evolution, with some branches becoming extinct, is depicted by the left dendrogram. We stand at the time horizon plane (P) to observe the lineages that have persisted to the present. The illustration on the right is one possible two-dimensional dendrogram that represents only the major surviving lines of descent.

## Simplification

Most dendrograms intend to make a point and are simplified accordingly. For example, the evolution of vertebrates is depicted in figure 1.22a to make a point about steps along the way. Although this representation is considerably simplified, it is a convenient summary, but if taken literally, the dendrogram is quite implausible. The first four species are living, so they are unlikely ancestral species in the steps. A more plausible representation of their evolution is shown in figure 1.22b. Species at each division point lived millions of years ago and are certainly extinct by now. Only distantly related derivative ancestors survive to the present and are used to represent steps in the origin of vertebrates.

A more complicated dendrogram of birds is shown in figure 1.23. Many groups are included, their likely evolution traced, and the relationships between them proposed. Thus, their phylogeny is more faithfully represented, although the complexity of the diagram makes major trends less apparent. In choosing a dendrogram, we should strike some compromise between simple (but perhaps misleading) and complex (but perhaps overwhelming).

## Patterns of Phylogeny

Dendrograms can be used to express relative abundance and diversity. The swollen and narrowed shapes of the "balloons" in figure 1.24 roughly represent the relative numbers of vertebrates that existed in each group during various geological times. The first mammals and birds arise within the Mesozoic but do not become abundant and prominant components in terrestrial faunas until much later, in fact, not until after the decline of the contemporaneous reptiles at the end of the Cretaceous. Shapes of branches within a dendrogram convey this additional information.

Rates at which new species appear can also be represented by the sharpness of branching within a dendrogram. One dendrogram shows smooth branches, implying the

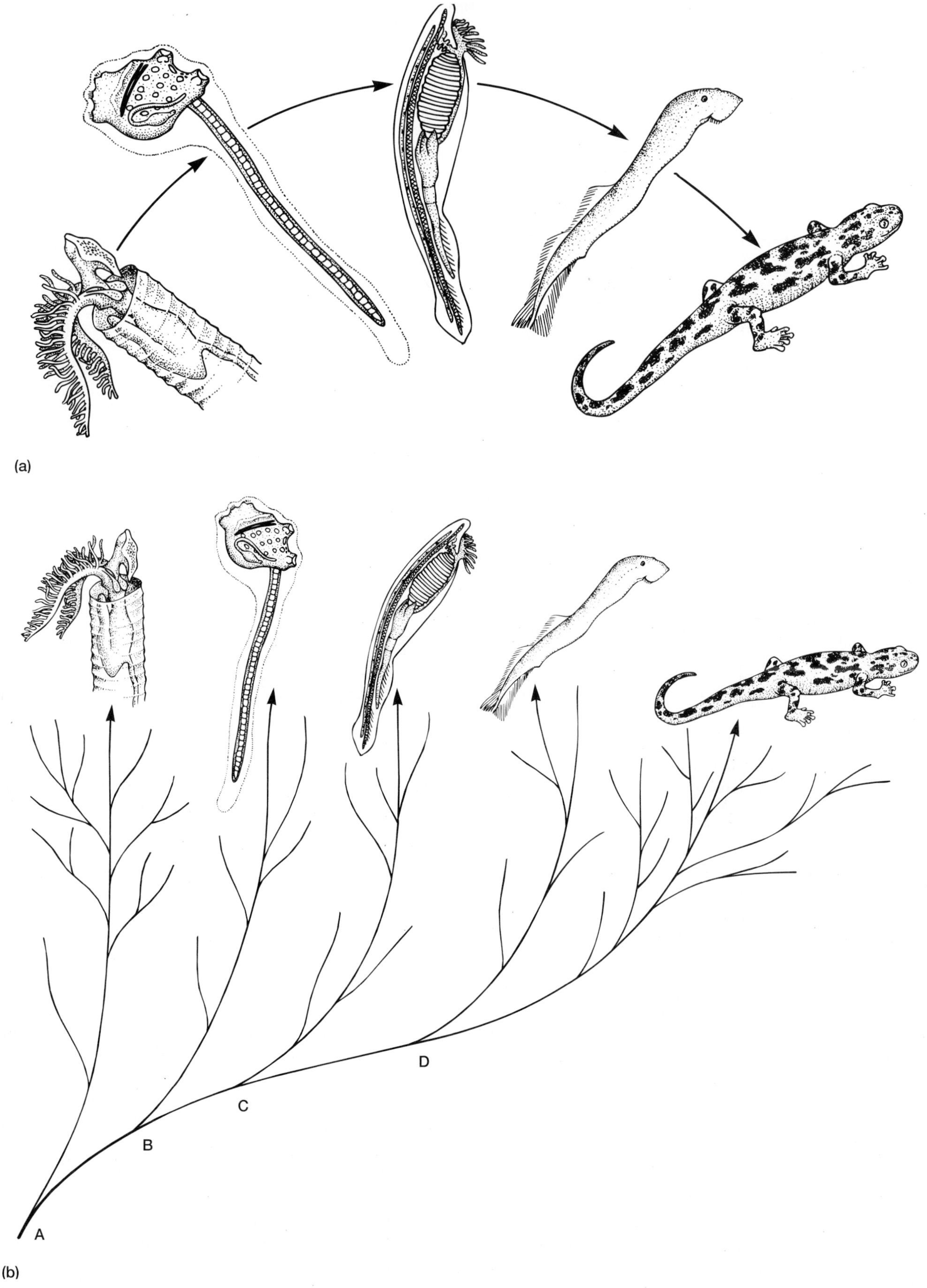


**Figure 1.22** Steps in vertebrate evolution. (a) Examples of a hemichordate, a urochordate larva, a cephalochordate, a lamprey, and a salamander (from left to right). All are living species, so they are not likely the immediate ancestors of each succeeding group. (b) Their actual ancestors (from A to D, respectively) lived millions of years ago and are now extinct. Modified descendants that represent these species today carried forward some of the primitive traits of their extinct ancestors, but they also evolved additional modifications.

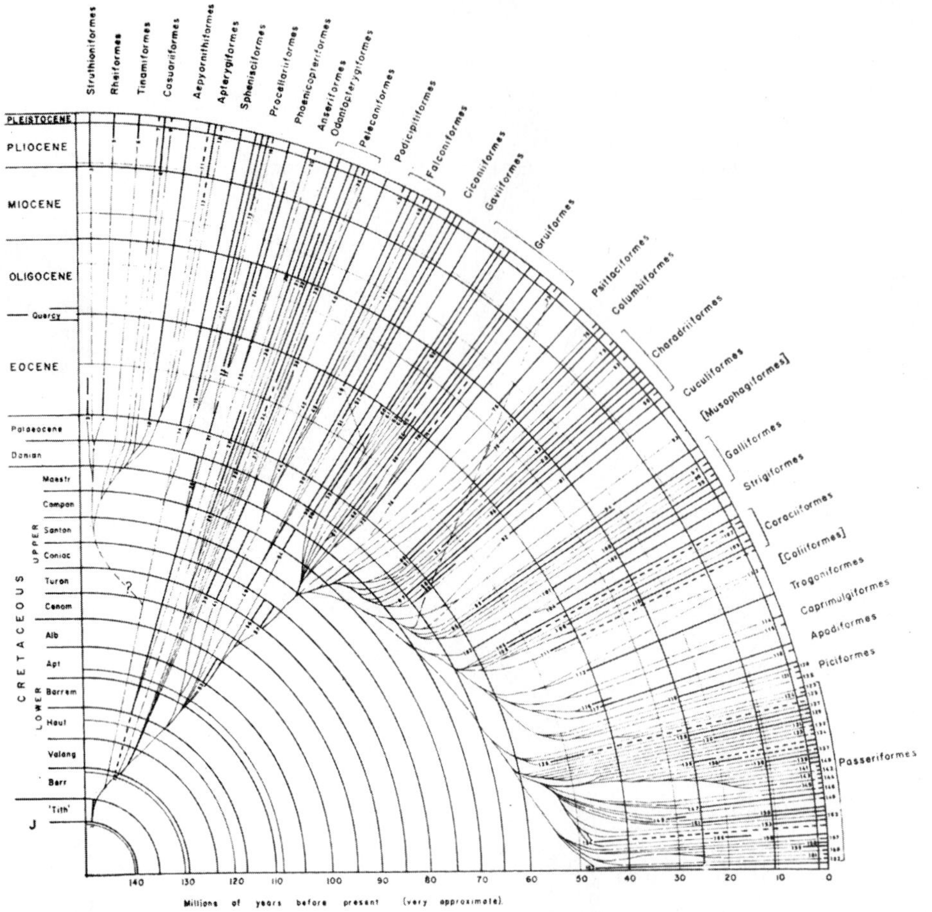


**Figure 1.23** Phylogeny of birds. This dendrogram attempts to detail the relationships and the time of origin of each group of modern birds. Although it expresses the hypotheses of these relationships in detail, the diagram is too complex and difficult to view easily. General trends are less evident as well.

gradual appearance of new species (figure 1.25a). The other is sharply angular, which implies rapid change and relatively sudden appearance of new species (figure 1.25b). Behind these two types of dendrograms stand different sets of assumptions about the process of evolution. One sees evolution working gradually to produce new species. The other sees the process as an event in which species persist for long stretches with relatively little change followed by a rather abrupt appearance of a new species. G. G. Simpson in the 1940s termed such long intervals of unchanged evolution occasionally interrupted by short bouts of rapid change as **quantum evolution.** Efforts to celebrate this in dendrograms have recently found favor again, termed **punctuated equilibrium** by those sharing Simpson's view.

Students should recognize dendrograms as summaries of information about the course of vertebrate evolution but realize that they also contain, even if inadvertently, hidden expressions of intellectual preference and personal bias. Dendrograms are practical devices designed to illustrate a point. Sometimes this requires complex sketches, and other times just a few simple branches on a phylogenetic tree serve our purposes.

## Paleontology

The late paleontologist Alfred Romer once poetically referred to the grandeur and sweep of vertebrate evolution as the "vertebrate story." And in a sense it is exactly that, a story with twists and turns that could not have been known beforehand—the debut of new groups, the loss of old ones, the mysteries of sudden disappearances, the evolutionary tales told by the parade of characters. Like a good story, when we finish it we will know the characters better, and because we ourselves are part of this story, we will come to know ourselves a little better as well. The vertebrate story unfolds over a span of 590 million years, a depth of time almost unimaginable (figure 1.26). To help us fathom this vastness of time, we consult paleontology, the discipline devoted to events of the distant past.

The vertebrate story is a narrative spoken partially from the grave because of all species ever to exist most are now extinct. The evolutionary biologist and paleontologist G. G. Simpson once estimated that of all animal species ever to evolve, roughly 99.9% are extinct today. So in this story of life on Earth, most of the cast of characters are dead.

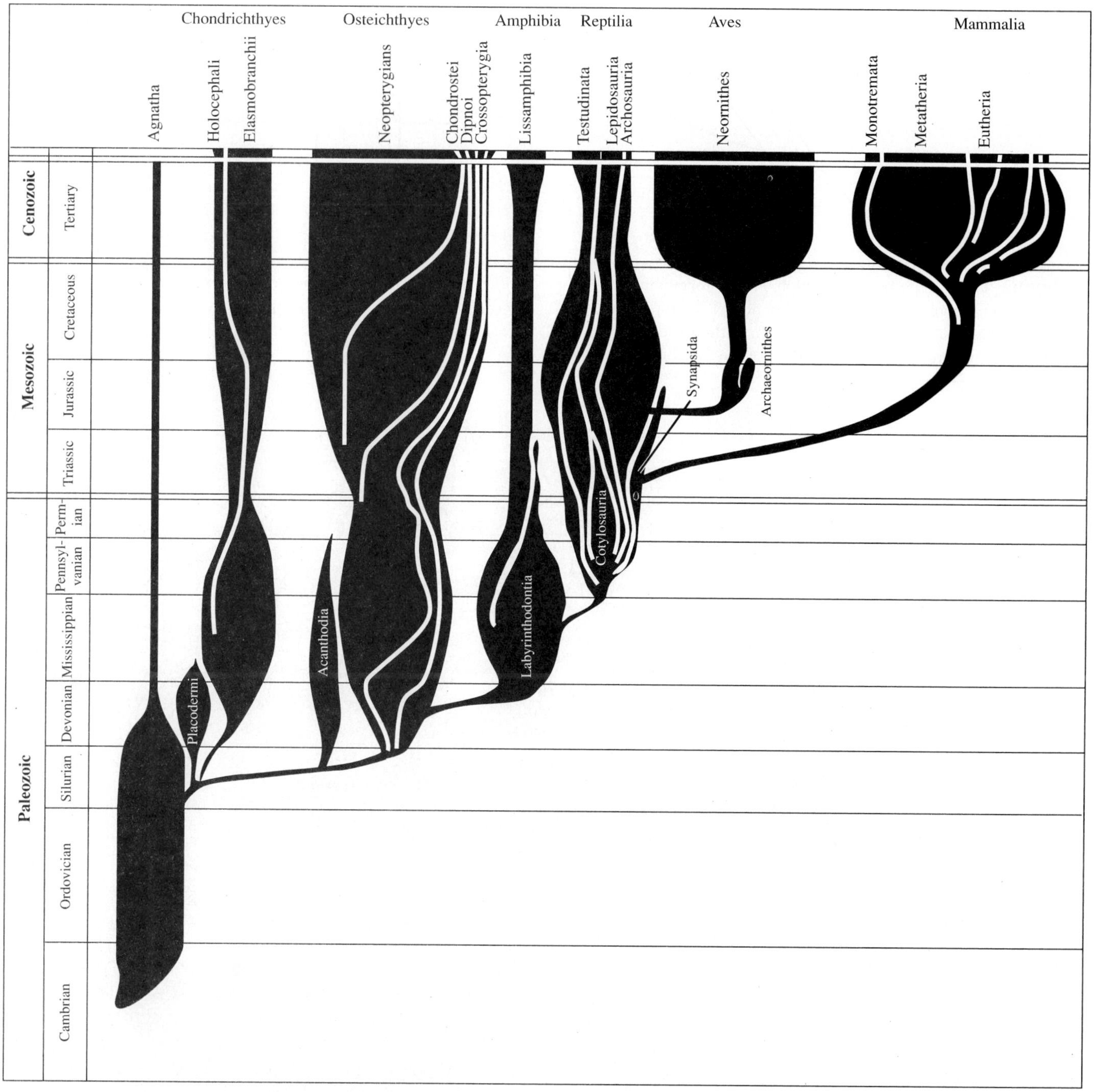


**Figure 1.24** Abundance phylogeny. This dendrogram attempts to represent the first time each vertebrate group appeared, the relationship of one class to the others, and the relative abundance of each group (depicted by the size of each balloon).

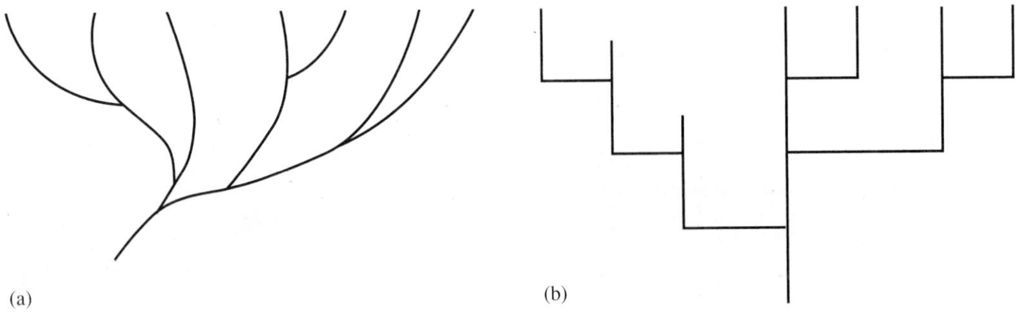


**Figure 1.25** Patterns of evolution. A dendrogram may be intended to represent the gradual (a) or abrupt (b) appearance of new species represented by a new branch. Although the two dendrograms agree on the relationships of species, they depict two different processes behind their evolution, namely, a gradual evolutionary process (a) or a rapid process of quantum evolution (b).

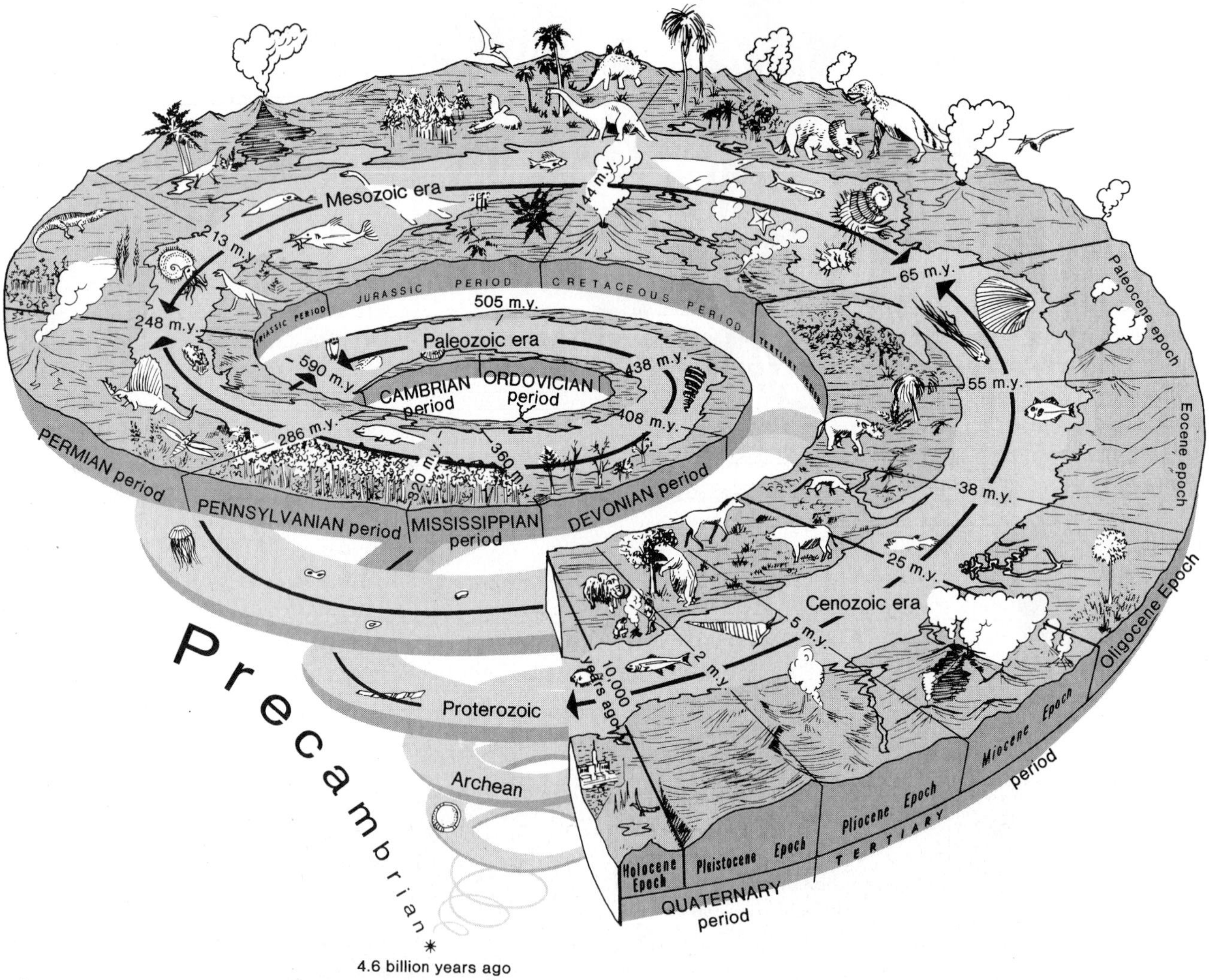


**Figure 1.26** Geologic time. The gathering of cosmic gases under gravity's pull created Earth some 4.6 billion years ago. Yet life became neither abundant nor complicated until the Cambrian period, about 590 million years ago, when the first complex invertebrates and early vertebrates appeared.

Source: After U.S. Geological Survey publication, *Geologic Time*.

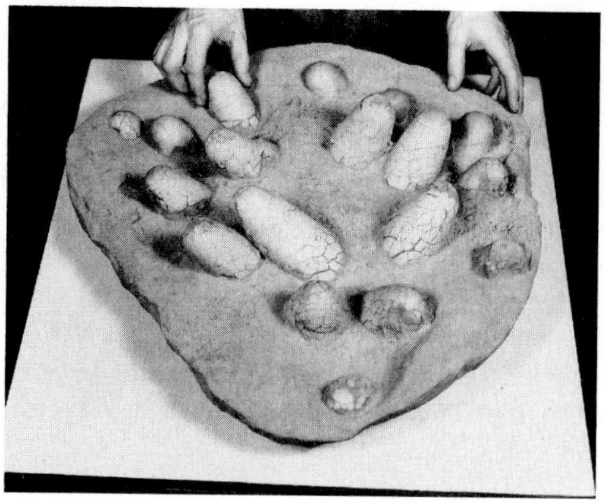

**Figure 1.27** Fossil eggs. Examination of the fetal bones within these eggs reveals that these are of *Protoceratops*, a Cretaceous dinosaur that lived in what is today Mongolia.

What survives are their remnants, the fossils and the sketchy vignettes these fossils tell of the structure and early history of vertebrates.

## Fossilization and Fossils

When we think of fossil vertebrates, we probably picture bones and teeth, the hard parts of a body that more readily resist the destructive processes following death and burial. Certainly most fossil vertebrates are known from their skeletons and dentition. In fact, some extinct species of mammals are named on the basis of a few distinct teeth, the only remnants to survive. The calcium phosphate compound composing bones and teeth is a mineral usually preserved indefinitely, with little change in structure or composition. If groundwater seeps through bones lying in soil or rock, over time other minerals such as calcite or silica may soak into the tiny spaces of bone to add further minerals and harden it.

Fossils are more than bones and teeth, however. Occasionally products of vertebrates, such as eggs, will fossilize. If tiny young bones are preserved inside, we can identify them and the group to which they belong (figure 1.27). This tells us more than just the structure of this species; it also tells us something about their reproductive biology. The recent discovery in Montana of fossilized clumps of eggs belonging to duck-billed dinosaurs testified to the reproductive style of this species, but there was accompanying circumstantial evidence to imply even more. The clumps or clutches of eggs were near each other, about two adult body lengths apart, suggesting that the area was a breeding colony. Analysis of the rock sediments in which they were found indicates that the colony was on an island in the middle of a runoff stream from the nearby Rocky Mountains. At the same site, bones from duck-billed dinosaurs of different sizes, and thus different ages, were present. This could happen only if young stayed around the nest until they were fully grown. Perhaps the parents even gathered food and brought it back to nourish the newly hatched young. For this species of duck-billed dinosaurs, the emerging picture is not one of a dispassionate reptile that laid its eggs and departed. Instead, this reptile appears to have had sophisticated parental care and supportive social behavior. Gathering of food, protecting and teaching of young, and bonding of pairs are implied by the fossils.

A marine fossil of an ichthyosaur, a dolphinlike reptile, was recovered from limestone rocks dating to 175 million years ago (figure 1.28). This adult specimen appears to be a female fossilized in the act of giving birth. Several small (young) skeletons remain with her body, one apparently emerging through the birth canal and another already born lying beside her (figure 1.28). If this represents a "fossilized birth," then unlike most reptiles, ichthyosaurs bore live young who were fully functional like young dolphins today.

Occasionally, fossils preserve more than just their hard parts. If a full animal skeleton is discovered, microscopic analysis of the region occupied in life by the stomach might reveal the types of foods eaten shortly before its death. Dung is sometimes fossilized. Although we might not

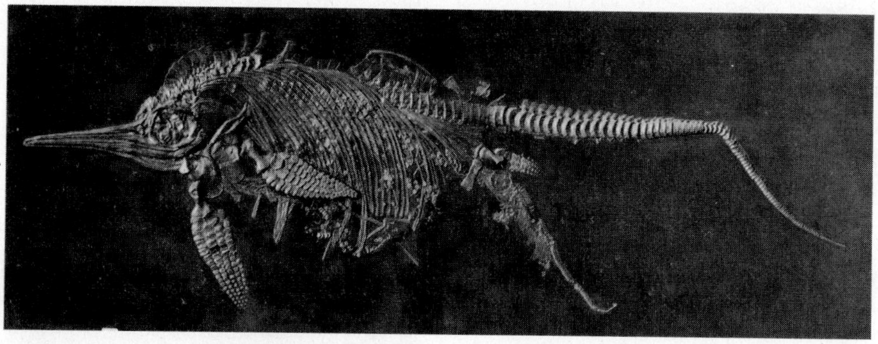

**Figure 1.28** Fossil ichthyosaur. Small skeletons are seen within the adult's body and next to it. This may be a fossilized birth, with one young already born (outside), one in the birth canal, and several more still in the uterus. Such special preservations suggest the reproductive pattern and live birth process in this species.

know which animal dropped it, we can gain some notion about the types of foods eaten. Soft parts usually decay quickly after death and seldom fossilize. A dramatic exception to this has been the discovery of woolly mammoths, distant relatives to elephants, frozen whole and preserved in the arctic deep freeze of Alaska and Siberia. When thawed these mammoths yielded hair, muscles, viscera, and digested food, exceptional finds indeed. Rarely are paleontologists so lucky. Occasionally soft parts leave an impression in the terrain in which they are buried. Impressions of feathers in the rock around the skeleton of *Archaeopteryx* demonstrate that this animal was a bird (figure 1.29). Similar impressions of skin tell us about the surface textures of other animals—scaly or smooth, plated or fine beaded (figure 1.30a,b).

The past behavior of now extinct animals is sometimes implied by their fossilized skeletons. Nearly complete skeletons of fossilized snakes have been found in lifelike positions in rocks dating to 32 million years ago. These natural aggregations seem to represent, as in many modern species of temperate snakes, a social event to prepare for hibernation during the cold winter season. Other vertebrate behaviors, or at least their locomotor patterns, are implied in fossilized footprints (figure 1.31). Size and shape of footprints, together with our knowledge of animal assemblages of the time, give us a good idea of who made them. With dinosaur tracks, it has been possible to estimate the velocity of the animal at the time the tracks were made. Three-million-year-old volcanic ash, now hardened to stone, holds the footprints of ancestral humans. Discovered in present-day Kenya by Mary Leaky, the two-by-two footprints are those of a large individual, presumably an adult male, and of a small individual, possibly an adult female. These human footprints confirm what had been deciphered from skeletons, namely, that our ancestors of 3 million years ago walked upright on two hindlegs.

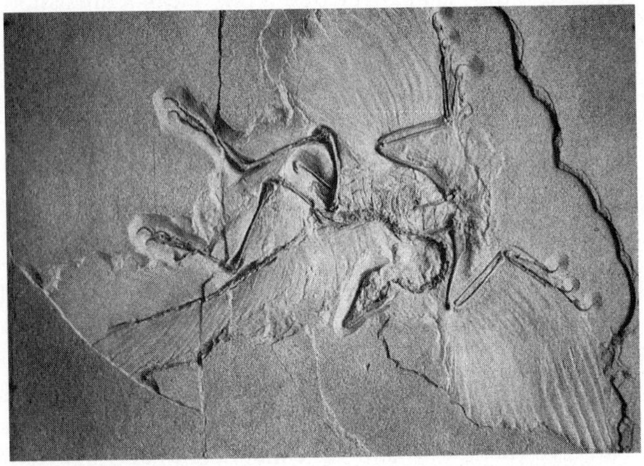

**Figure 1.29** *Archaeopteryx*. The original feathers have long since disintegrated, but their impressions left in the surrounding rock confirm that the associated bones are those of a bird.

## Recovery and Restoration

Paleontologist and artist combine talents to recreate the extinct animal as it might have looked in life. Remnants of long-dead animals provide source material from which basic anatomy is reassembled. After such a length of time in the ground, even mineral-impregnated bone becomes brittle. If the original silty sediments around bone have hardened to stone, they must be chipped or cut away to expose the fossilized bone encased within. Picks and chisels help to partially expose the upper surface and sides of the bone, which are wrapped in protective plaster and allowed to harden (figure 1.32). Following this procedure, the remainder of the bone is exposed and the plaster wrap extended to encase it completely. The brittle bones are shipped to laboratories within their plaster support. Once specimens reach the lab,

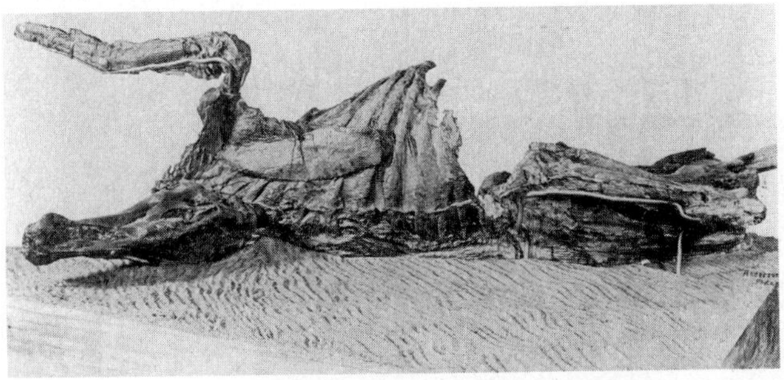

(a)

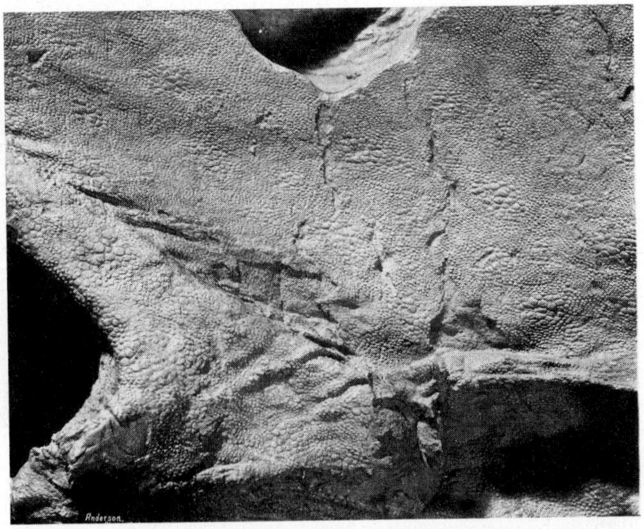

(b)

**Figure 1.30** Mummification. (a) Fossil mummified carcass of the duck-billed dinosaur *Anatosaurus*. (b) Detail shows the surface texture of the skin.

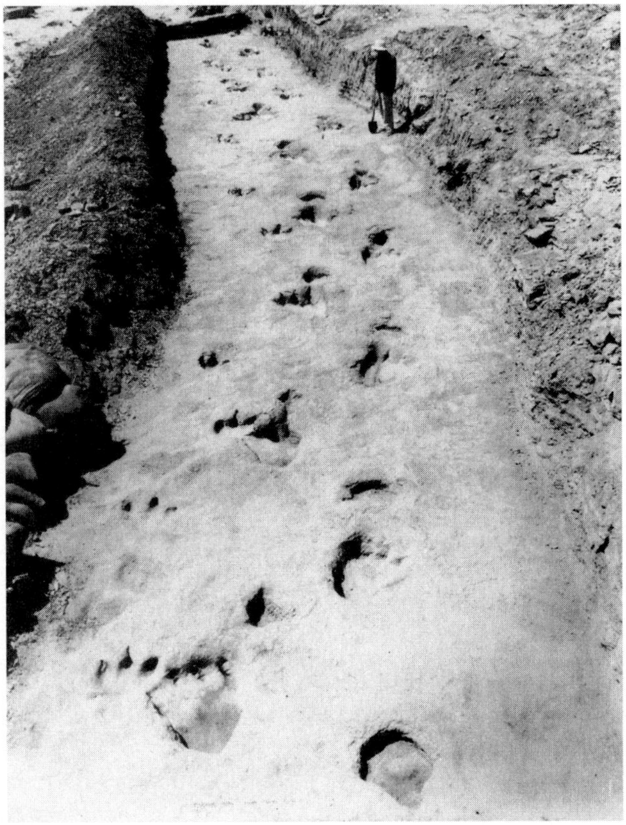

(a)

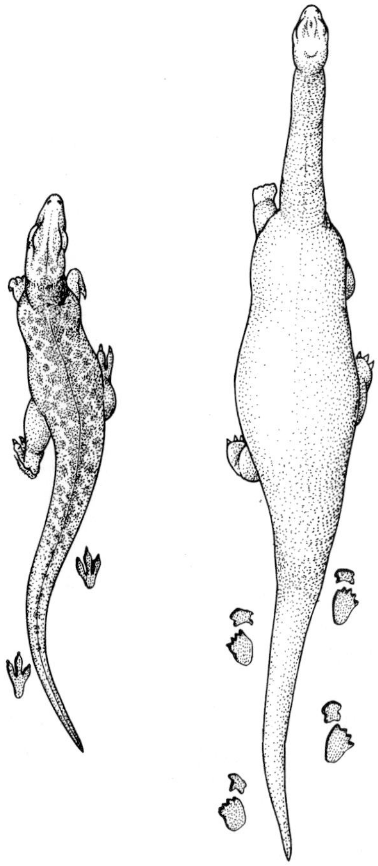

(b)

**Figure 1.31** Dinosaur tracks. (a) Tracks from the late Jurassic were made in soft sand that later hardened to form rock. Two sets are present, the large tracks of (b) a sauropod and the three-toed tracks of a smaller carnosaur, a bipedal carnivorous dinosaur.

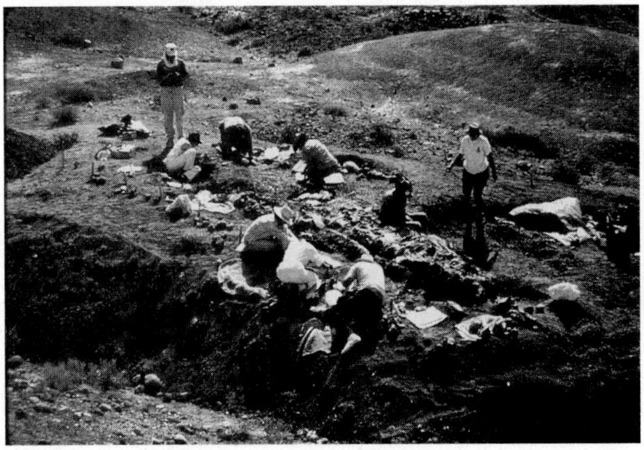

(a)

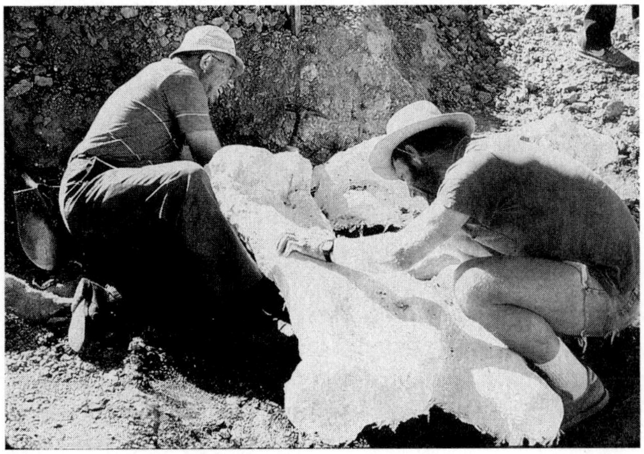

(b)

**Figure 1.32** Fossil dig. (a) Partially exposed dinosaur bones. The work crew prepares the site and notes the location of each excavated part. (b) This *Triceratops* femur is wrapped in a plastic jacket to prevent disintegration or damage during transport back to the museum.

the plaster along with any further rock is removed. Tiny needles were once used to pick away the rock. Today, a stream of fine sand from a pencil-sized nozzle is used to sandblast or carve away rock to free the fossil.

Confidence in a restored version of a fossil rests largely on direct fossil evidence and knowledge of modern, living counterparts, which indirectly supply the likely biology of the fossil (figure 1.33). Size and body proportions are readily determined from the skeleton. Muscle scars on bones help determine how muscles might have run. When added to the skeleton, these give us an idea of body shape. General feeding type, herbivore or carnivore, is implied by the type of teeth; and lifestyle—aquatic, terrestrial, or aerial—is determined by the presence of specialized features such as claws, hooves, wings, or fins. The type of rock from which the fossil was recovered—marine or terrestrial deposits, swamp or dryland—further testifies to its lifestyle. Comparison with related and similarly structured living vertebrates helps fill in locomotor style and environmental requirements (figure 1.33a–c).

The presence or absence of ears, proboscis (trunk), nose, hair, and other soft parts must be guessed at. Living relatives help in this process. For instance, all living rodents have vibrissae, long hairs on the snout, so these might be included in restorations of extinct rodents. Except for some burrowing or armored forms, most mammals have a coat of fur, so it is fair to cover a restored mammal with hair. All living birds have feathers and reptiles have scales, both of which can logically be added to restored avian or reptilian fossils, although the length or size must be guessed. Surface colors or patterns, such as stripes or spots, are never preserved in an extinct vertebrate. In living animals, colored patterns camouflage appearance or emphasize courtship and territorial behaviors. Reasonably, surface patterns had similar functions among extinct animals, but specific colors and patterns chosen for a restoration must be produced from the artist's imagination. A dynamic mural showing dinosaurs at battle may satisfy our curiosity for what these great animals might have looked like in life, but we should remember that in any such restorations human interpretation stands between the actual bones and the fully colored reconstruction.

New fossil finds, especially of more complete skeletons, improve the evidence upon which we build a view of extinct vertebrates. Often, however, new insights into old bones arise from an inspired reassessment of the assumptions upon which original restorations were based. Such is largely the case with our recent reassessment of dinosaurs. Their structures, size, and success now seem to make them warm-blooded, active vertebrates living a lifestyle less like reptilian lizards and turtles of today and more like mammals or birds. New fossil discoveries got us thinking, but the major change in the way artists and paleontologists restore dinosaurs reflects new courage in interpreting them as predominantly active land vertebrates of the Mesozoic.

Reconstruction of human fossils has followed fashion as well as new discoveries. When first unearthed in the late nineteenth century, Neanderthal bones were thought to be those of a single individual, a Cossack soldier from the Napoleonic wars fought a few decades earlier. In the beginning of the twentieth century, this view gave way to a stoop-shouldered, beetle-browed, and dim-witted image. Neanderthals were reassessed to be a breed apart from modern *Homo sapiens*, and restoration reflected this demoted image. Today Neanderthals are classified again as a human species, *Homo sapiens neanderthalensis*. Shaven and suited, the claim goes, a Neanderthal could walk the streets of New

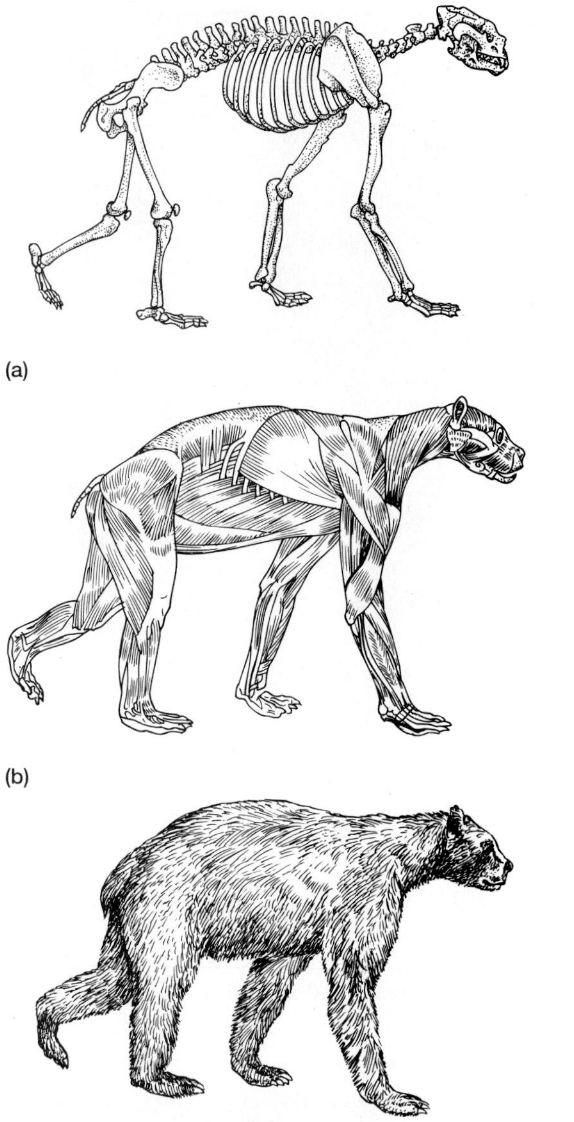

(a)

(b)

(c)

**Figure 1.33** Restoration of a fossil. (a) The skeleton of the extinct short-faced bear, *Arctodus simus*, is positioned in its likely posture in life. (b) Scars on the bones from muscular attachments and knowledge of general muscle anatomy from living bears allow paleontologists to restore muscles and hence to create the basic body shape. (c) Hair added to the surface completes the picture and gives us an idea of what this bear might have looked like in its Alaskan habitat 20,000 years ago.

York without drawing a second glance or a raised eyebrow. In New York, perhaps, but this "new" elevation of Neanderthal to modern status has been inspired by current artists' restorations that make the species look human.

The point is not to smirk at those who err or follow fashion but to recognize that any restoration of a fossil is several steps of interpretation away from the direct evidence of the bones themselves. Reconstructing the history of life on Earth improves with new fossil discoveries as well as with improved knowledge of basic animal biology. The better we understand the function and physiology of animals, the better our assumptions will be when we restore life to the bones of dead fossils. It is worth the risks and pitfalls to recreate the creatures of the past because in so doing we recover the unfolding story they have to tell us about life on Earth.

## From Animal to Fossil

The chance is extremely remote that an animal, upon death, will eventually fossilize. Too many carrion eaters await within the food chain (figure 1.34). Disease or age or hunger may weaken an animal, but a harsh winter or successful predator is often the immediate instrument of death. Its flesh is consumed by carnivores and its bones broken and picked over by marauding scavengers that follow. On a smaller scale, insect larvae and then bacteria feed on what remains. By stages the deceased animal is broken down to its chemical components, which reenter and recycle through the food chain. In a small forest, hundreds of animals die each year, yet as any hiker or hunter can attest, it is rare to find an animal that has been dead for any length of time. Scavengers and decomposers go quickly to work. Even rodents, whose customary food is seeds or foliage, will gnaw on bones of dead animals to obtain calcium. To escape this onslaught, something unusual must intervene before all trace of the dead animal is literally eaten up.

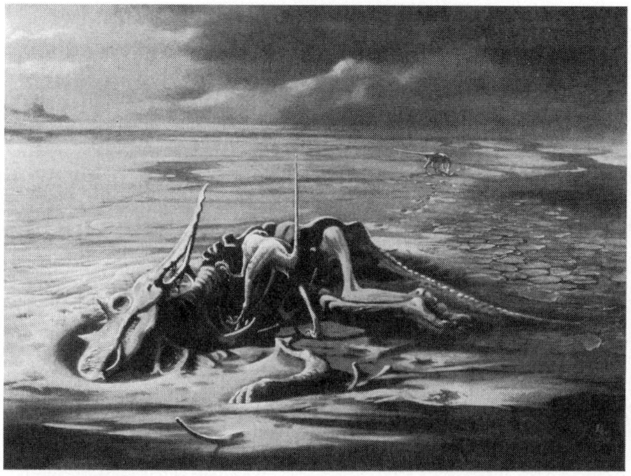

**Figure 1.34** Almost fossils. Upon death, few animals escape the keen eyes of scavengers looking for a meal. Bacteria and bugs descend upon the flesh that is left. Small animals seeking calcium chew up bones. Little if anything is left to fossilize.

Animals living in water or near the shore are more likely to be covered by mud or sand when they die (figure 1.35). Upland animals die on ground exposed to scavengers and decay; thus, most fossil-bearing rocks (i.e., sedimentary rock) are formed in water. Even if successfully buried, bones are still in peril. Under pressure and heat, silt turns to rock. Shifting and churning and settling of rock layers can pulverize fossils within. The longer a fossil lies buried, the greater the chance these tectonic events will obliterate it. This is why older rock is less likely to harbor fossils. Finally, the fossil must be discovered. Theoretically, one could begin to dig straight down anywhere through the Earth's crust at any site and eventually hit fossil rocks. Excavations for roads or buildings occasionally unearth fossils in the process. Usually such a free-lance approach to fossil discovery is too chancy and expensive. Instead, paleontologists visit natural **exposures** where sheets of crustal rock have fractured and slipped apart or been cut through by rivers, revealing the edges of rock layers perhaps for the first time in millions of years. In these layers, or **strata,** the search begins for surviving fossils.

## Aging Fossils

To discover a fossil is not enough. Its position in time with regard to other species must be determined as well because this will help place its morphology in an evolutionary sequence. Techniques for aging fossils vary, and preferably several are used to verify age.

### *Stratigraphy*

One such technique is **stratigraphy,** a method of placing fossils in a relative sequence to each other. It occurred to Giovanni Arduino as early as 1760 that rocks could be arranged from oldest (deepest) to youngest (surface). By the time the British geologist Charles Lyell published his great three-volume classic, *Principles of Geology,* during 1830–1833, a system of relative dating of rock layers was well established. The principle is simple. Similar strata, layered one on top of another, are built in chronological order (figure 1.36). As in construction of a tower, the oldest rocks are at the bottom, with later rocks in ascending sequence to the top where the most recent rocks reside. Each layer of rock is called a **time horizon** because it contains the remains of organisms from one slice in time. Any fossils contained within separate layers can be ordered from the oldest to the most recent, bottom to top. Although this gives no absolute age, it does produce a chronological sequence of fossil species *relative* to each other. By placing fossils in their stratigraphic sequence, we can determine which arose first and which later, relative to other fossils in the same overall rock exposure.

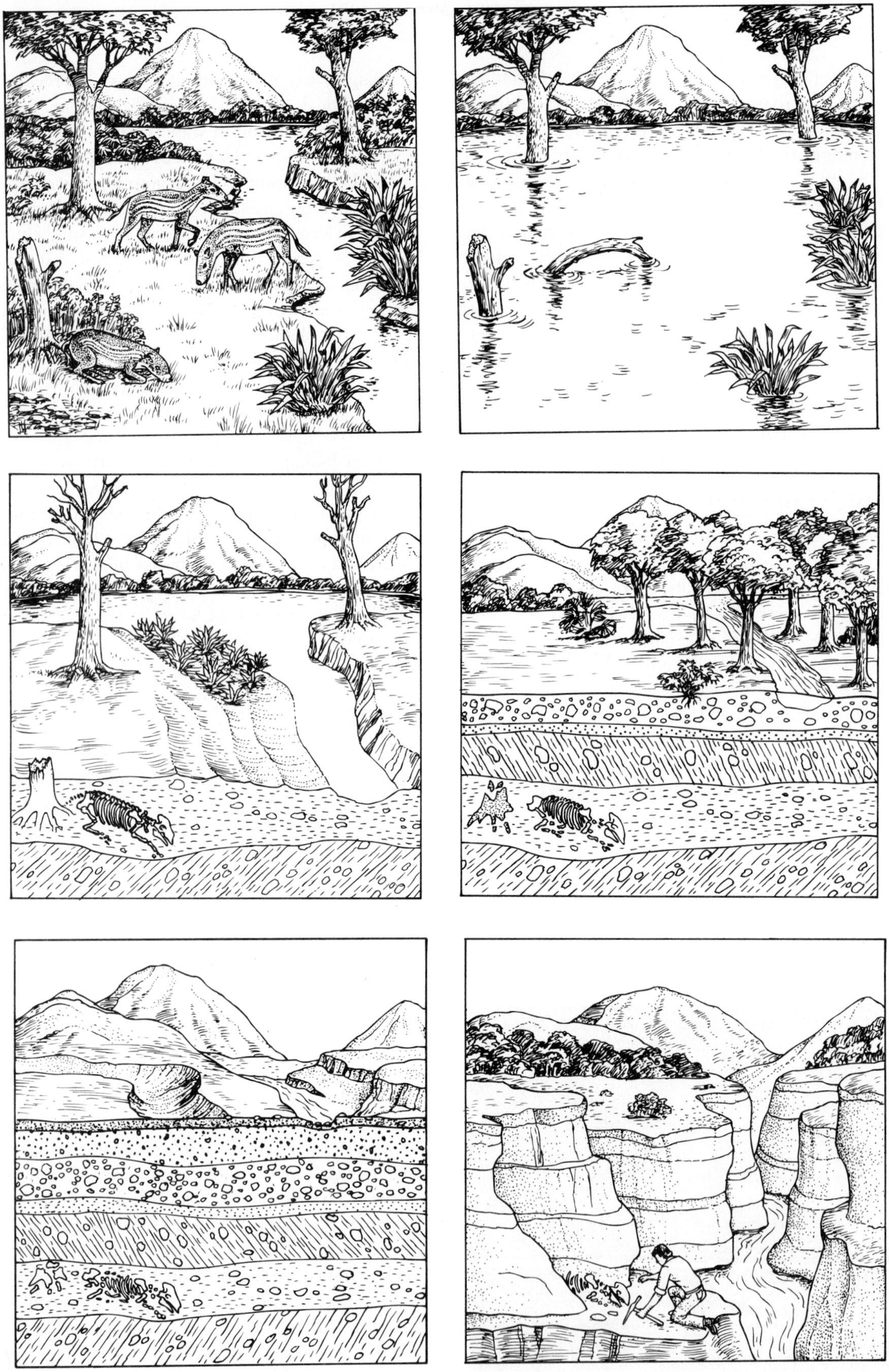

**Figure 1.35** Making fossils. The remains of extinct animals that persist have escaped the appetites of scavengers, decomposers, and later tectonic shifting of the Earth's crustal plates in which they reside. Usually water covers a dead animal so that it escapes the notice of marauding scavengers. As more and more silt is deposited over time, the fossil becomes even more deeply buried in soil compacted into hardened rock. To be discovered, the Earth must open either by fracture or by the knifing action of a river to expose the fossil held in the rock.

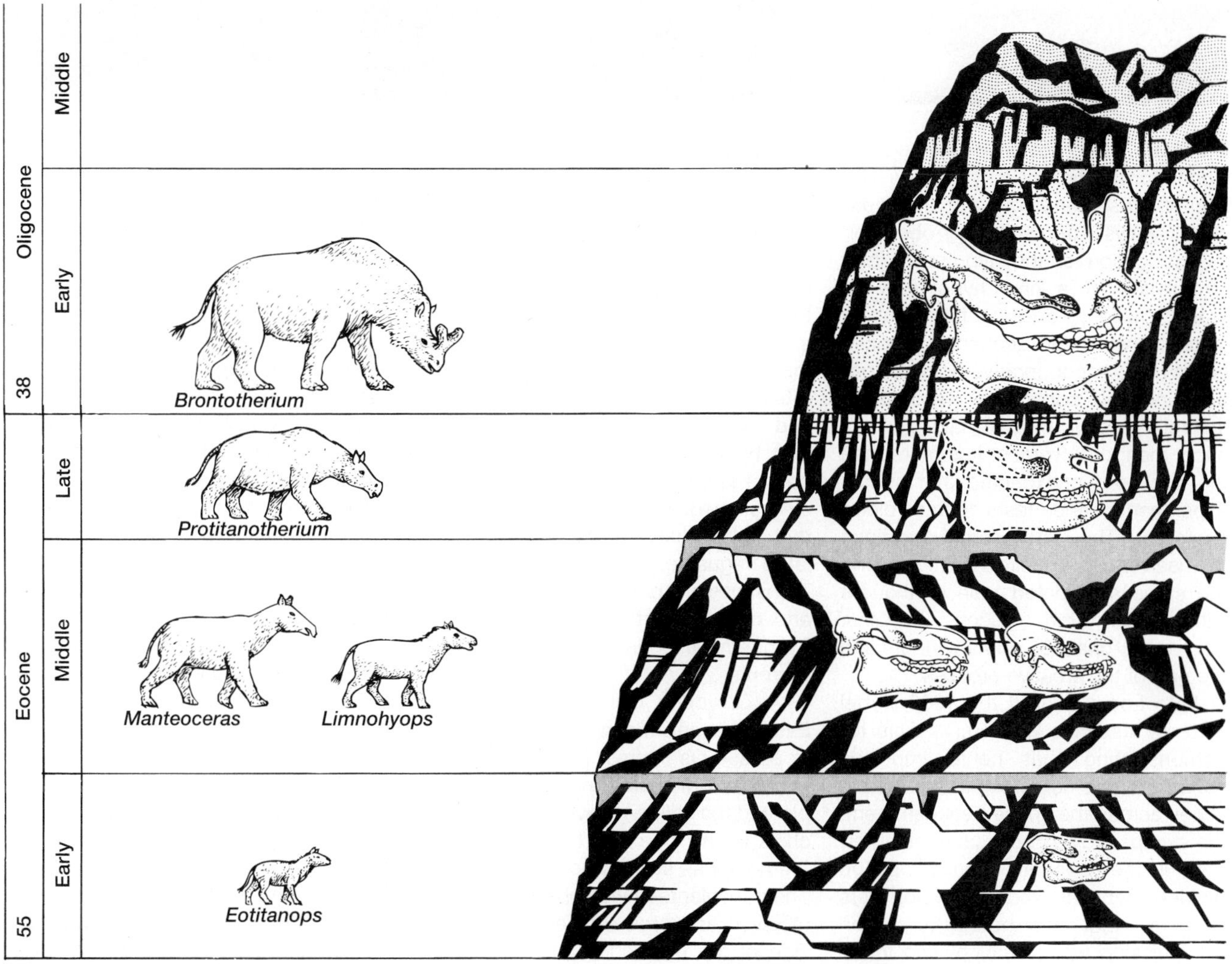


**Figure 1.36** Stratigraphy. Sediment settling out of water collects at the bottoms of lakes. As more sediment collects, the deeper layers are compacted by the ones above until they harden and become rock. Animal remains become embedded in these various layers. Deeper rock forms first and is older than rock near the surface. Logically, fossils in deeper rock are older than those above, and their position within these rock layers gives them a chronological age relative to older (deeper) or younger (surface) fossils.

## *Index Fossils*

By matching rock strata in one location to comparable rocks in another exposed location, we can build up an overlapping chronological sequence longer than that represented at any single location by itself (figure 1.37). The actual correlation of rock strata between two distantly located sites is done by comparison of mineral content and structure. Index fossils are distinctive markers that can facilitate matching of rock strata. These are species of animals, usually hard-shelled invertebrates, that we know from previous work occur only within one specific time horizon. Thus, the presence of an index fossil confirms that the stratigraphic layer is equivalent in age to a similar layer containing the same fossil species elsewhere (figure 1.38).

## *Radiometric Dating*

Relative stratigraphic position is useful, but to assign an age to a fossil, a different technique is used. This is **radiometric dating,** a technique that takes advantage of the natural transformation of an unstable elemental isotope to a more stable form over time (figure 1.39a). Such radioactive decay of an element from one isotope state to another occurs at a constant rate, expressed as the characteristic half-life of an isotope. The **half-life** is the length of time that must pass before half the atoms in the original sample transform into product atoms (figure 1.39b). Common examples include "decay" of uranium-235 to lead-207 (half-life of 713 million years) and potassium-40 to argon-40 (half-life of 1.3 billion years). When rocks form, these radioactive isotopes

## Box Essay 1.2 The Scientific Method—What They Tell You and What They Don't

Formally, the scientific method includes formulation of a hypothesis, design of a test, carrying out of an experiment, analysis of results, corroboration or falsification of the hypothesis, and formulation of a new hypothesis. In practice, science does not follow such a stately and linear sequence. Broken equipment, uncooperative animals, paperwork, and committee meetings all conspire against the well-laid plans of mice, men, and women. It is more than the expected unexpected that affects experiments and tests one's blood pressure. The intellectual questions themselves do not always yield along the same anticipated train of thought with which the scientist begins. Accidents, chance, and even dreams are part of the creative process.

Otto Loewi shared the 1936 Nobel Prize in medicine with Henry Dale for demonstrating that nerve impulses pass from one nerve cell to the next in series across the space between them, the synapse, by a chemical transmitter. Early in the twentieth century, opinion divided between those physiologists who felt that this neuron-to-neuron transmission was chemical and those who felt that it was electrical. A definitive experiment settling the issue was needed. One night, when he was deep in sleep, the definitive experiment came to Loewi and woke him. Relieved and satisfied, he went back to sleep looking forward to the next day. When awaking the next morning, he remembered dreaming the experiment but had forgotten what it was. Several frustrating weeks passed until, once again deep in sleep, Loewi dreamt the same dream, and the experimental design came back. Leaving nothing to chance this time, he got up, dressed, and in the middle of the night went to his laboratory to begin the experiment that would settle the issue of transmission and years later win him a share of the Nobel Prize.

Loewi's experiment was as simple as it was elegant. He removed the heart and associated vagus nerve from the body of a frog and isolated them in a beaker of saline. Next he stimulated the free vagus nerve, causing the heart rate to slow. Loewi then took this saline and poured it over another isolated frog heart from which the vagus had been removed. The rate of this heart also slowed, providing clear evidence that a chemical produced by the stimulated vagus nerve controlled heart rate. Transmission between nerve (vagus) and organ (heart) was brought about by chemical agents, not by electrical currents.

As a young cell biologist, Herbert Eastlick began a series of experiments to pursue his interest in embryonic development of young muscle. He transplanted the still formative hindlimbs of a chick to the side of a host chick while the host was still developing in its egg. The transplanted hindlimbs were usually received and grew well enough on the side of the host chick to allow study. One day, when a local supplier was temporarily out of the white leghorn eggs Eastlick had used, he substituted brown leghorns, a breed with brown feathers. After three days of incubation, one egg was opened and both leg-forming areas of a brown leghorn were transplanted to a white leghorn host. Results were puzzling. The right transplanted leg from the brown leghorn developed brown feathers, the left transplanted leg from the same brown leghorn developed white feathers. What caused these contrary results?

Eastlick checked his notes, repeated his experiments, and used great care in performing more transplants. Still some legs were brown and some were white. It then dawned on him that the stump of the transplanted limb might in some instances include nearby neural crest cells, but not in all instances. Neural crest cells form first on top of the nerve tube and then normally disperse about the embryo. He tried limbs with and without accompanying neural crest cells. That was it. Those brown leghorn limbs with neural crest cells produced brown feathers. Those without lacked pigment cells and were white. Eastlick, who started out working on muscles, confirmed what a few had guessed, namely, that one derivative of neural crest cells are pigment cells that give feathers their color.

Alexander Fleming (1881–1955), while studying bacteria, noticed that when molds occasionally contaminated cultures, the bacteria next to the molds failed to grow. Hundreds of students and fellow bacteriologists before Fleming had seen molds and likely noticed the stunted growth of bacteria. But it was Fleming's curiosity that precipitated the serious question, "What causes this reaction?" In answering it, he discovered that molds produced penicillin, a bacterial inhibitor. Fleming's question opened the way for development of a new branch of pharmacology and a new industry. His answer established the basis of disease control through antibiotics.

Testing of a well-crafted hypothesis forms the center of the scientific method. But where the next hypothesis comes from cannot always be predicted. A thought in the middle of the night, an experiment gone wrong, a close observation of the ordinary, these too may inspire a new scientific hypothesis and are part of the method of science.

are often incorporated. If we compare the ratios of product to original and if we know the rate at which this transformation occurs, then the age of the rock and hence the age of fossils it holds can be calculated. If, for instance, our sample of rock showed lots of argon relative to potassium, then the rock would be quite old and our estimated age quite high (figure 1.39c). Most of the potassium would have decayed to argon, its product. Conversely, if there were little argon compared with potassium, then only a little time would have passed and our calculated age would be young.

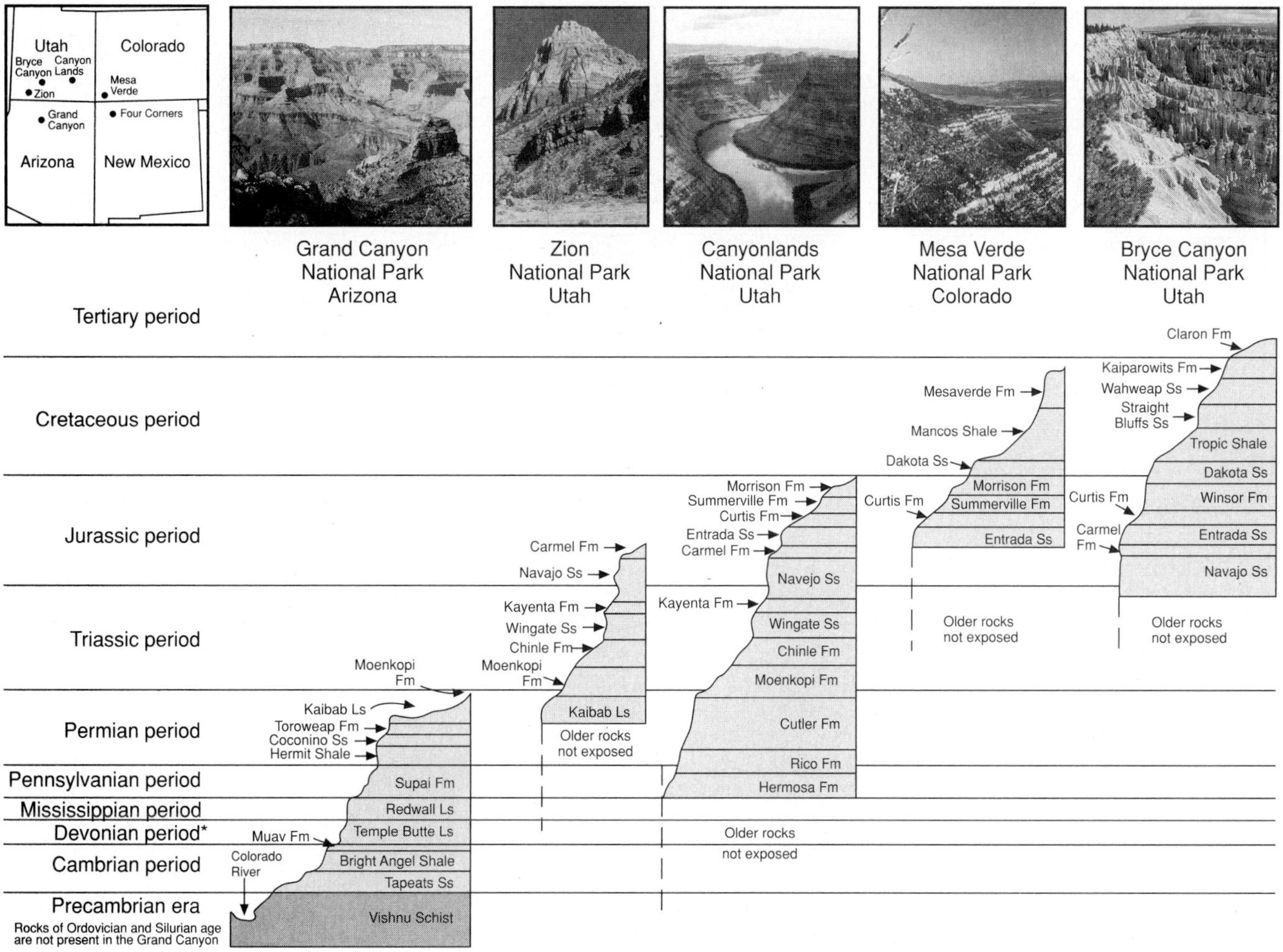


**Figure 1.37** Building a chronology of fossils. Each exposure of rocks can be of a different age from the rest. To build up an overall sequence of fossils, various exposures can be matched where they share similar sedimentary layers (same ages). From five sites in the southwest United States, overlapping time intervals allow paleontologists to build a chronology of fossils greater than that at any site taken by itself.

Because of the sometimes capricious uptake of isotopes when rocks form, not all rocks can be dated by radiometric techniques. But when available and cross-checked, radiometric dating yields the absolute ages of rocks and the fossils these rocks contain.

## *Geological Ages*

Geological time is divided and subdivided in turn into eons, eras, periods, and epochs (figure 1.40). The oldest rocks on Earth, with an age of 3.6 billion years, are found in Tanzania. However, radiometric dates of meteorite fragments fallen to Earth give age estimates of 4.6 billion years. Since astronomers assume that our solar system and everything within it—planets, sun, comets, meteors—formed at about the same time, most geologists take this figure as the Earth's age. The span of Earth history, 4.6 billion years to the present, is divided into two unequal eons, the Phanerozoic (visible life) and Cryptozoic (hidden life). The Cryptozoic eon is the first and by far the longer, stretching from the birth of the planet to 590 million years ago. Understandably, rocks from this eon are rare, and those surviving rocks contain traces only of microscopic organisms, the first primeval forms to appear as life on Earth gained momentum. At about 590 million years ago or, as we now know, slightly earlier, complicated multicellular organisms made a sudden appearance, which is why we start the Phanerozoic eon at this point in time. Attempts to divide the Cryptozoic eon into exact eras are at present premature because some stretches of time, especially the earliest, are unrepresented by rocks. The single era of the Cryptozoic is the Precambrian, sometimes divided into early, middle, and late. Because the Precambrian spans the same length of time as the Cryptozoic eon, the two names are often used interchangeably for this early period in Earth's history.

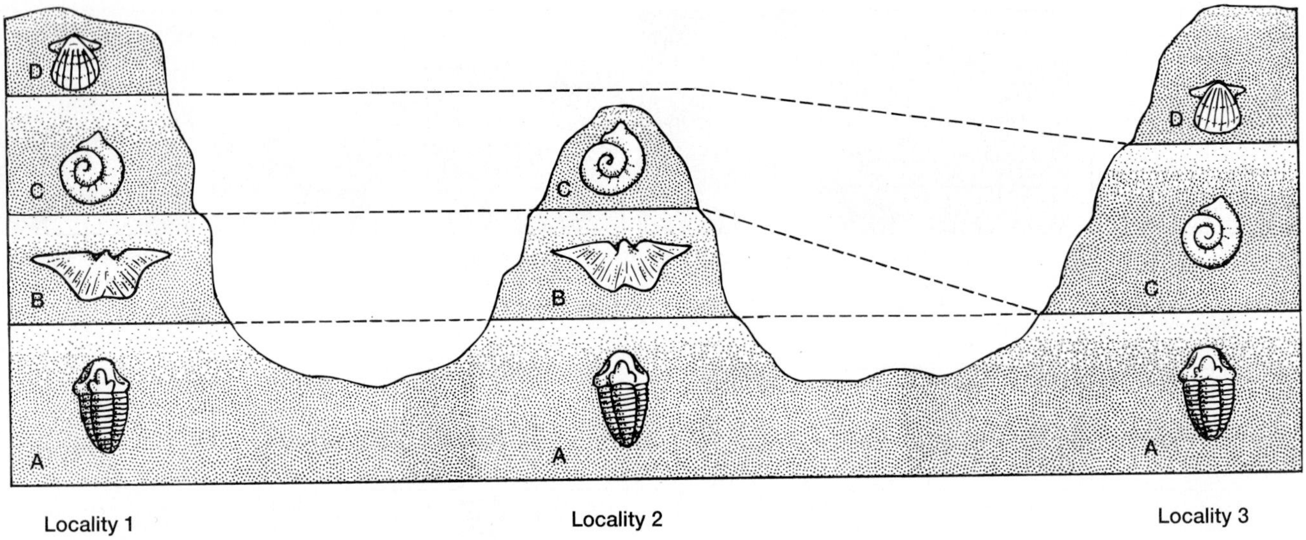


**Figure 1.38** Index fossils. After careful study at many well-dated sites, paleontologists can confirm that certain fossils occur only at restricted time horizons (specific rock layers). These distinctive index fossils are diagnostic fossil species used to age rocks in new exposures. In this example, the absence of index fossils confirms that layer B does not exist at the third location. Perhaps rock-forming processes never reached the area during this time period or the layer was eroded away before layer C formed.

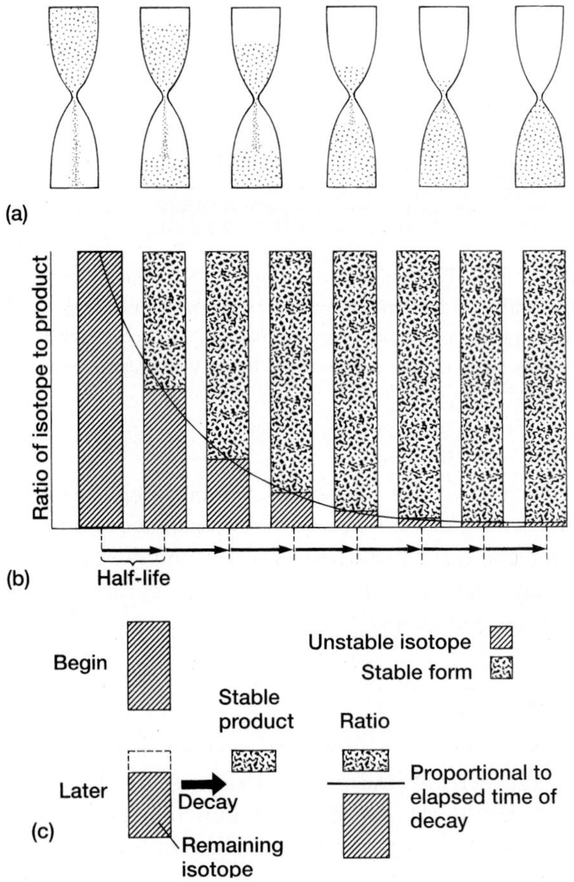


**Figure 1.39** Radiometric dating. (a) Sand flows regularly from one state (upper portion) to another (lower portion) in an hourglass. The more sand in the bottom, the more time has passed. By comparing the amount of sand in the bottom with that remaining in top and by knowing the rate of flow, we can calculate the amount of time that has elapsed since the flow in an hourglass was initiated. Similarly, knowing the rate of transformation and the ratios of product to original isotope, we can calculate the time that has passed for the radioactive material in rock to be transformed into its more stable product. (b) Half-life. It is convenient to visualize the rate of radioactive decay in terms of half-life, the amount of time it takes an unstable isotope to lose half its original material. Shown in this graph are successive half-lives. The amount remaining in each interval is half the amount present during the preceding interval. (c) A radioactive material undergoes decay, or loss of mass, at a regular rate that is unaffected by most external influences, such as heat and pressure. When new rock is formed, traces of radioactive materials are captured within the new rock and held along with the product into which it is transformed over the subsequent course of time. By measuring the ratio of product to remaining isotope, paleontologists can age the rock and thus age the fossils they contain.

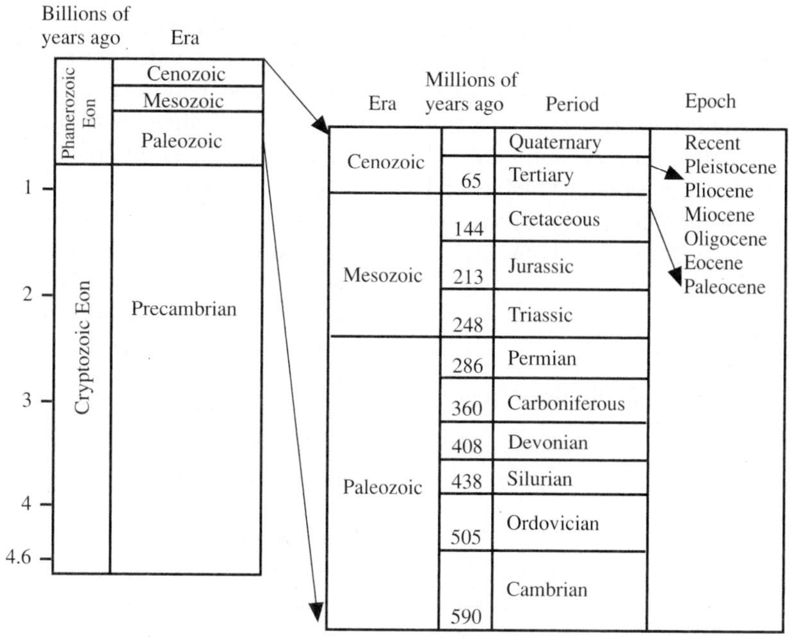


**Figure 1.40** Geological time intervals. The Earth's history, from its beginnings 4.6 billion years ago, is divided into major eons, the Cryptozoic and the Phanerozoic. These eons are divided into four eras of unequal length—Precambrian, Paleozoic, Mesozoic, and Cenozoic. Each era is divided into periods, and periods into epochs. Only epochs of the Cenozoic are listed in this figure.

The Phanerozoic eon is divided into three eras, Paleozoic (old animal life), Mesozoic (middle animal life), and Cenozoic (recent animal life). Invertebrates predominated during the Paleozoic era as they still do today. Among the vertebrate groups of that time, fishes were the most conspicuous and diverse; therefore, the Paleozoic might be called the Age of Fishes. The first tetrapods appeared in the Paleozoic, and by late in this era, they radiated extensively across the planet. During the Mesozoic, the extraordinary diversity of reptiles took them into nearly every conceivable environment. So extensive was this radiation that the Mesozoic is often called the Age of Reptiles. The following Cenozoic era is sometimes called the Age of Mammals. Until then, most mammals were small in number and small in size. At the end of the Mesozoic, widespread extinctions eliminated the dinosaurs and many allied groups of reptiles. Disappearance of these predominant reptiles seems to have opened evolutionary opportunities for mammals, who thereafter enjoyed an expansive radiation of their own during the ensuing Cenozoic. Mammalian radiation must be kept in perspective, however. If the Cenozoic were to be named for the vertebrate group with the most species, it would properly be termed the Age of Teleost Fishes, or secondly, the Age of Birds, or thirdly still, the Age of Reptiles. Despite the previous Mesozoic extinctions that depleted reptilian ranks, reptiles today still outnumber mammals in terms of numbers of species. But in the Cenozoic, mammals displayed a radiation unequaled in their history, and they occupied dominant positions within most terrestrial ecosystems. Because we are mammals and it is our taxonomic class on the rise, the Cenozoic to some, not unexpectedly, is the Age of Mammals.

Eras are divided into periods, names for which originated in Europe. The Cambrian, Ordovician, and Silurian originated from British geologists working in Wales. Respectively, Cambria was the Roman name for Wales, and the Ordovices and Silures were Celtic tribes existing before the Roman conquest. The Devonian period was named for rocks near Devonshire, also on British soil. And the Carboniferous ("coal-bearing") period celebrates the British coal beds on which so much of Britian's participation in the Industrial Revolution depended. In North America, the Lower and Upper Carboniferous are called Mississippian and Pennsylvanian periods, after rocks in the Mississippi valley and the state of Pennsylvania. The Permian, although named by a Scot, is based on rocks in the province of Perm in Asia.

The Triassic takes its name from rocks in Germany, the Jurassic from the Jura Mountains between France and Switzerland, and Cretaceous from the Latin word for chalk *(creta)*, a reference to the white chalky cliffs along the English Channel.

It was once thought that geological eras could be divided into four parts—Primary, Secondary, Tertiary, and Quaternary—from the oldest to the youngest, respectively. This proved untenable, but two names, Tertiary and Quaternary, survive as the two periods of the Cenozoic. Throughout the geological time scale, periods divide into epochs usually named after a characteristic geographic site of that age. Sometimes boundaries between epochs are marked by changes in characteristic fauna. For example, in

North America the late part of the Pliocene epoch is recognized by the presence of particular species of fossil deer, voles, and gophers. The early part of the succeeding Pleistocene is recognized by the appearance of mammoths. The boundary or transitional time between both of these epochs is defined by a fauna that includes extinct species of jackrabbits and muskrats, but not mammoths. Most names of epochs are not in general use and will not be referred to in this book.

## Tools of the Trade

Analysis of vertebrate design proceeds in three general steps, each enhancing the other.

### The Question

A specific question about design is formulated first in any analysis. This is not so trivial or simple as it may sound. A well-formed question focuses thought, suggests the appropriate experiment or line of research to pursue, and promises a productive answer. Physicists of the late nineteenth century believed that space contained a kind of fixed, invisible substance called "ether," which accounted for how light traveled through space. Like sound in air, light in ether was thought to propagate by setting it in motion. As planets circled the sun, they sped through this ether's wind like a person sitting in the open bed of a truck rushing through air. Physicists asked the question, "How might light be affected as it passes with or against the ether wind?" After a series of experiments with light, they found no effect of the ether. For a time, they and other scientists were stumped. As it turned out, they had asked the wrong question. Ether, as an invisible occupant of space, does not exist. No ether, no wind. They should have asked first if ether existed! Our opinion of these physicists should not be harsh, however, because even mistakes inspire better questions and an eventual sounder answer.

In morphology, several practical tools can be used to help define the question. One is dissection, the careful anatomical description of an animal's structural design. Another is taxonomy, the proposed relationships of the animal (and its parts) to other species. From these techniques, we gain insight into the morphological design and can place this in a comparative relationship with other organisms. The specific questions we then formulate about the structure of the organism might be about its function or evolution.

### The Function

To determine how a structure performs within an organism, various techniques are used to inspect the functioning organism or its parts directly. Radiography, X-ray analysis, allows direct inspection of hard parts or marked parts during performance (figure 1.41). Expense or accessibility, however, often makes radiography of a living organism impractical. High-speed video tape or cinematographic film can sometimes be used instead. The event, feeding or running, for example, is filmed with the camera set at a high rate of speed so that the event unfolds in slow motion when played back at normal projection speed. The tape or film preserves a record of the event, and slow motion playback permits careful inspection of motions at a speed where sudden displacements are obvious. Natural markers, for instance, bulging muscles or visible hard parts such as teeth or hooves, allow inferences about the functioning of attached or underlying bones and muscles. Inferences can also be made from gentle manipulation of parts in a relaxed or anesthetized animal. Thus, with radiography or high-speed tape/film, displacement of individual points on the animal can be followed, measured, and plotted frame by frame. From this careful record of displacements, velocity and acceleration of parts can be calculated to describe the motion of parts quantitatively. Together with information on simultaneous muscle activity, this produces a description of the part and an explanation of how its bone and muscle components achieve a characteristic level of performance.

Visceral functions can be addressed in other ways. Thin tubes (cannulae) inserted into blood vessels and connected to calibrated and responsive instruments (transducers) allow us to study an animal's circulatory system (figure 1.42). Similar approaches to kidney and gland function have been used. Radiopaque fluids, those visible on radiographs, can be fed to animals and mechanical events of the digestive tract followed. Muscles, when active, generate low

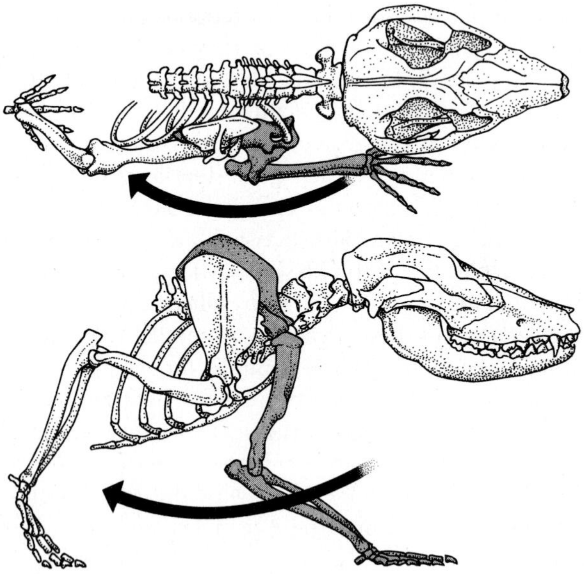

**Figure 1.41** Walking stride of an oppossum. The propulsive phase is depicted in these tracings of motion radiographs from overhead and side views. Change in the position of the shoulder blade is evident.

Based on the research of F. A. Jenkins and W. A. Weijs.

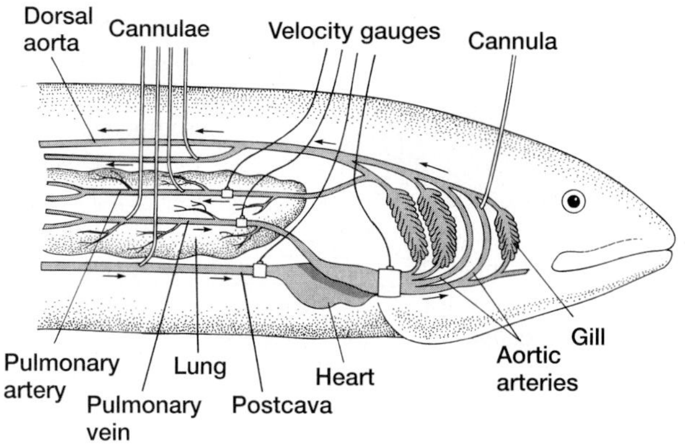


**Figure 1.42** Analysis of viscera in a lungfish. To monitor blood pressure, cannulae (small) tubes are inserted into blood vessels. To monitor rate of blood flow, velocity gauges are placed around selected vessels. From such information, it is possible to determine changes in the rate of blood flow in this lungfish when it breathes in water or gulps air into its lung.

levels of stray electric charges. Electrodes inserted into muscles can detect these on monitors, allowing the investigator to determine when a particular muscle is active during performance of some event. This activity can be compared with activity of other muscles. This is the technique of electromyography (EMG), and the electric record of the muscle an **electromyogram.**

Figure 1.43 depicts an experimental setup combining several techniques simultaneously to analyze the feeding strike of a venomous snake. With the snake under anesthesia and with proper surgical technique, four pairs of bipolar, insulated, fine wire electrodes are inserted into four lateral jaw muscles to record electromyograms of each during the strike. A strain gauge is affixed with glue to a suitable location on top of the snake's head where it can detect motion of underlying skull bones. The wires, termed leads, from the four bipolar electrodes and strain gauge are sutured to its skin, carefully bound into a cable, and connected to preamplifiers that boost

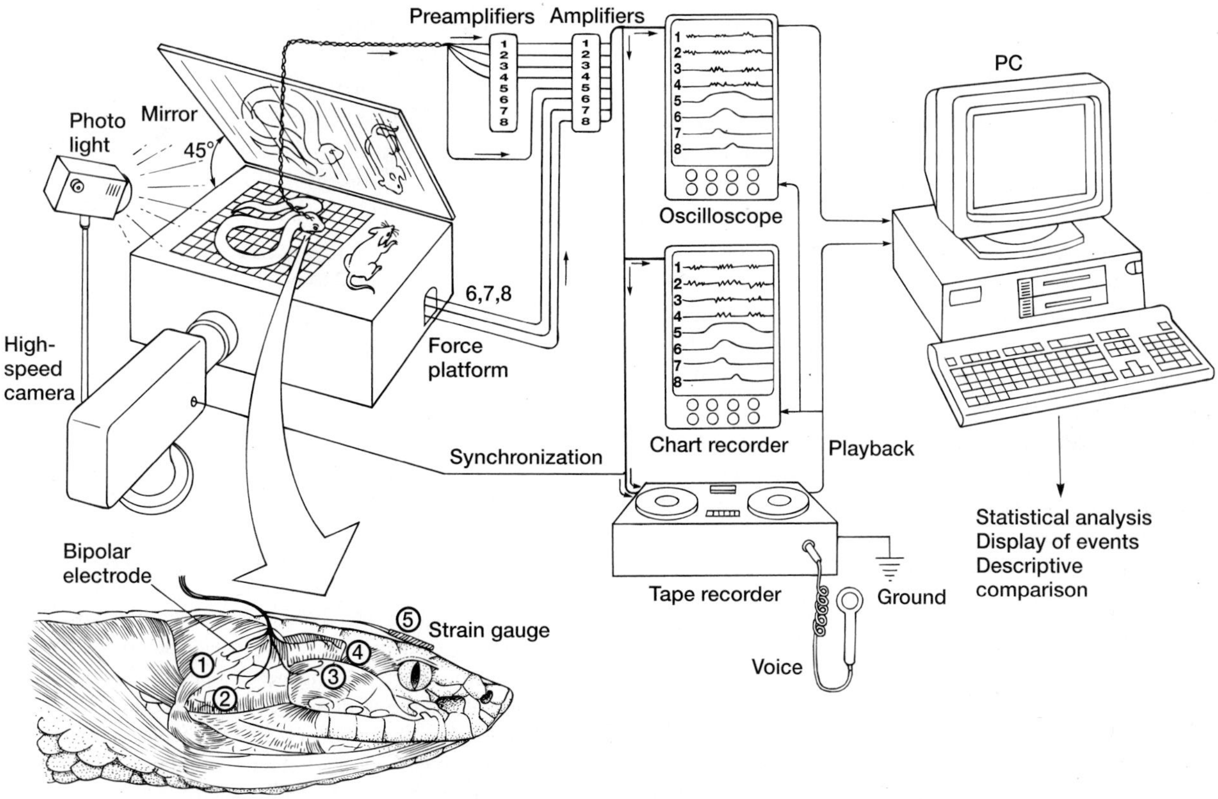


**Figure 1.43** Experimental analysis of function. Careful surgery allows insertion of bipolar electrodes into four selected jaw muscles on the right side of the snake. A strain gauge is fixed over a movable point in the snake's skull. Leads from these electrodes are connected to preamplifiers and then to amplifiers to boost and filter the signals. Channels from the force platform join these four electrodes and carry responses in the three planes of space. The electrical output is displayed on an oscilloscope, recorded on a chart, and saved on magnetic tape. The snake strike is filmed by a high-speed camera or video that is pulse-synchronized with the other electrical outputs. Voice comments may be added on the tape. Electrical "noise" in the room can be reduced by placing the snake (but not recording instruments) in a shielded Faraday cage (not shown). Later, slow playback from tape to oscilloscope and chart recorder permit manual analysis of data. Or playback can be directed into a computer for analysis. Comparison of separate events is done easiest if all are recorded simultaneously, but parts can be done in separate runs and then later matched.

the very low signals from the jaw muscles. Interference with these signals from stray electrical "noise" in the room can be reduced if the snake and apparatus are placed in an electrically shielded cage, a Faraday cage (not shown in figure 1.43). From preamplifiers, each circuit, called a channel, is run next to an amplifier. The strain gauge enters the amplifier at this junction as well (channel 5); special electrical balancing of its signal may be necessary.

The snake is centered on a force platform that records forces produced in the three planes of space (forward/back, up/down, left/right), and the leads enter the last three channels to fill this eight-channel system. A permanent record of the rapid strike is made by a high-speed camera or video system. The camera produces a pulsed output that is simultaneously combined with the rest of the electrical outputs to permit matching of film events with EMG, strain gauge, and force platform data. A background mirror tilted at 45° allows a carefully placed camera to record dorsal and lateral views of the strike simultaneously. Notes on temperature, time, and other environmental data are recorded by hand.

Outputs are displayed on an oscilloscope and chart recorder for immediate viewing, and they are saved on a tape recorder as a permanent record. Voice input to the tape recorder allows comments to be added. Later, the tape recorder can be played back slowly, and data can be redisplayed on the oscilloscope or chart recorder. With appropriate software, a computer allows quantitative description of events, matching of film/video with electrial events and so on.

A partial analysis of feeding data obtained in this way is illustrated in figure 1.44a–c. Three instants during the snake's strike are shown—just before, at the beginning of, and during venom injection. Its head positions at these three points are traced from the film record, and below each position are the outputs from the first five channels (electromyograms 1–4, strain gauge 5). The snake's instantaneous movement unfolds at the beginning (left) of each record and travels across each trace from left to right. From prior dissections, structural components hypothesized to be important in strike performance are set forth in a proposed morphological model, to which these functional data are now added. Before onset of the strike, all muscle channels are silent because no contraction is occurring, and the strain gauge trace indicates that the snake's mouth is closed (figure 1.44a). As the strike begins, the lower jaw starts to open. This is initiated by contraction of muscle 1 and indicated by activity on the electrical trace for the first time (figure 1.44b). The initial rotation of the fang is detected by the strain gauge. At the third point in the strike, the snake closes its jaws firmly on the prey, and all the jaw-closing muscles, including the first, show high levels of activity (figure 1.44c). The strain gauge indicates changes in the jaw positions during this bite, from fully open at first to jaw closure on the prey. Thus, the first muscle opens the lower jaw, but its high electrical activity slightly later during the bite indicates that it continues to play a role. The other three muscles are powerful jaw-closing muscles, adductors, and act primarily during the bite.

This form-function analysis is far from complete. Many more muscles are involved and events on both sides

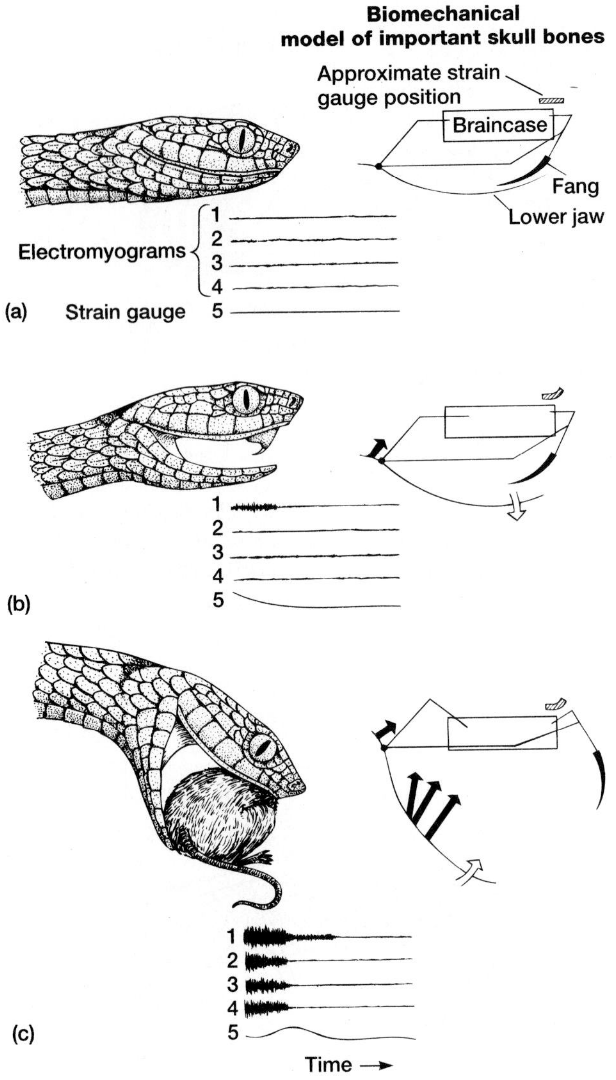


**Figure 1.44** Initial analysis of morphological and functional data. Three points in the feeding strike of a venomous snake are illustrated: (a) just before the strike, (b) at the onset of the strike, and (c) during the bite. Electrical traces from the four muscles (channels 1–4) and strain gauge (channel 5) are shown below each. The biomechanical models (right) of the snake's skull during each stage are based on prior anatomical analysis. (a) No myograms are evident prior to the strike, and no bone or fang displacement occurs. (b) The muscle opening the snake's jaw (channel 1) and the strain gauge records (channel 5) are first to show changes on the myograms. The model incorporates these changes by showing the start of fang erection. (c) The snake's jaws close firmly to embed its fully erected fang within its prey. Electromyograms show that all jaw muscles are active, and the strain gauge indicates that the snake's mouth is closing on the prey. These events are incorporated into the model (right) where solid arrows represent the onset and directional contraction.

of the animal need to be followed. Presentations of different sizes of prey might result in modifications of jaw function and so on. Anatomical analysis produces a knowledge of basic structure. From this, a set of testable questions about design can be formulated. Which structural elements are critical to performance? How do they function? Functional data address these questions.

It is best if motion and muscle events are recorded simultaneously to make comparisons between them easier. Often this is not feasible, however. Equipment may be unavailable or the animal uncooperative. Thus, it is not uncommon and certainly acceptable to perform parts of the functional analysis separately, then later match up bone displacements and muscle activity. It is becoming common now to include analysis of the nervous system along with simultaneous muscle and bone events. This produces a more complete explanation of performance. Not only is the immediate basis of motion described but the basis for neural control of these displacements and for initiating muscle activity are described as well. Activity of muscles at appropriate moments can be seen also.

## The Biological Role

To discover the adaptive role of a part, scientists eventually venture into the field to document how the animal actually deploys the morphological design in the environment. Careful observation of the organism in its environment must be incorporated with techniques of population biology to assess overall ecological performance of a part's form and function. **Ecomorphology** is the term that has been coined to recognize the importance of ecological analysis in the examination of a morphological system.

By this late point in an analysis, one usually has a good idea of how a structure might be used under natural conditions. Occasionally there are surprises. For example, unlike other finches, the "woodpecker" finch of the Galápagos uses its beak to break off a sharp needle or twig, which it deploys as a spear or probe to jab insect grubs hidden under the bark of trees. Deer mice chew rough seeds and grasses but also grab an occasional insect to eat as well. The jaws of deer mice consequently function as more than just a grinding mill of tough seeds. The pronghorn, a deerlike animal of the North American plains, can attain speeds in excess of 96 km/hr, but no natural predator today or in the past existed with a comparable ability. High speed itself, therefore, is not just an adaptation for escape from predators. Instead, pronghorns cruise at 30 to 50 km/hr in order to move between scattered resources. This, not just escape from predators, seems to be the most important aspect of pronghorn speed and design.

Thus, laboratory studies determine the form and function of a design. Field studies assess the biological role of the feature, that is, how the form and function of the feature serve the animal under natural conditions. A feature's biological role, in turn, suggests the kinds of selection pressures brought to bear on the organism and how the feature might be an adaptation that addresses these evolutionary forces. Carrying this a step further, comparison of homologous features from one group to another, or from one class to another, provides insight into how change in animal design might reflect changes in selection pressures.

The story of vertebrate evolution is the story of transition and adaptive change—transition from water to land (from fish to amphibians), from land to air (from reptiles to birds), and in some cases, the reinvasion of water (dolphins, whales) or return to a terrestrial mode of life (e.g., ostriches). In the study of vertebrate evolution, it is useful to think of how a particular design adapts the organism to the particular demands of its present environment, and how structure itself places limitations on or opens opportunities for the kinds of adaptations that might eventually arise.

## Box Essay 1.3 Living Fossils

Taken literally, a "living fossil" is a contradiction in terms because, of course, fossils are dead. But, occasionally a species survives up to the present having changed little in external appearance since the inception of its lineage. In these living fossils, evolution is arrested. Because they retain in their bodies ancient characteristics, and because they are living, they carry forward the physiology and behavior missing in preserved fossils. All living animals, not just a privileged few, retain at least a smattering of characteristics that are throwbacks to an earlier time in their evolution. The duckbill platypus, a furry mammal of Australia, still lays eggs, a holdover from its reptilian ancestors. Even humans retain ancient features. We have hair, for example, that comes down from the most ancient of mammals. I suppose we could even count our backbone as a retained feature of fishes!

However, what most scientists mean by a living fossil is an unspecialized species, alive today, that is built from the same ancient features that first appeared in the early days of the lineage. In terms of head and body shape, crocodiles have been labeled as living fossils, as have sturgeons and *Amia,* the bowfin. Along the coasts of New Zealand persists a lizardlike reptile, *Sphenodon.* Four-legged and scaled, it looks like a squat but otherwise average lizard. Under the skin, however, the skeletal system, especially the skull, is quite ancient. One of the most surprising living fossils is the surviving sarcopterygian, *Latimeria,* a coelacanth. This crossopterygian fish is a distant relative of the group giving rise to the first amphibians. And until 1939, *Latimeria* was thought to have been extinct for millions of years.

*Latimeria* retains many ancient crossopterygian creations: well-developed notochord, unique snout, fleshy appendages, divided tail. Its discovery excited great interest because the last members of this line had apparently expired 75 million years ago. In 1938, Goosen, a commercial fishing captain working the marine waters off the southern tip of Africa, decided, on an impulse, to fish the waters near the mouth of the Chalumna River. He was

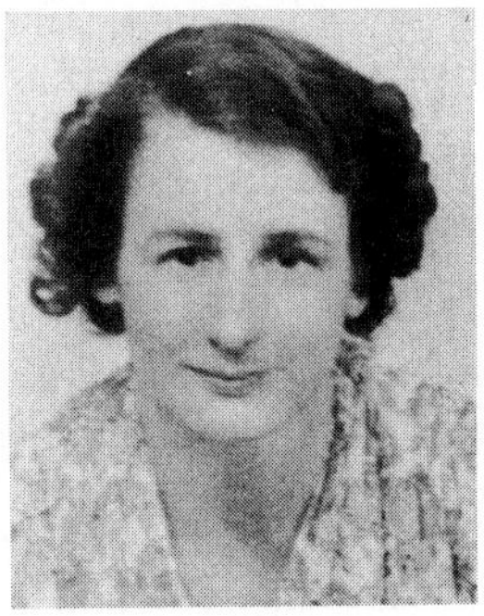

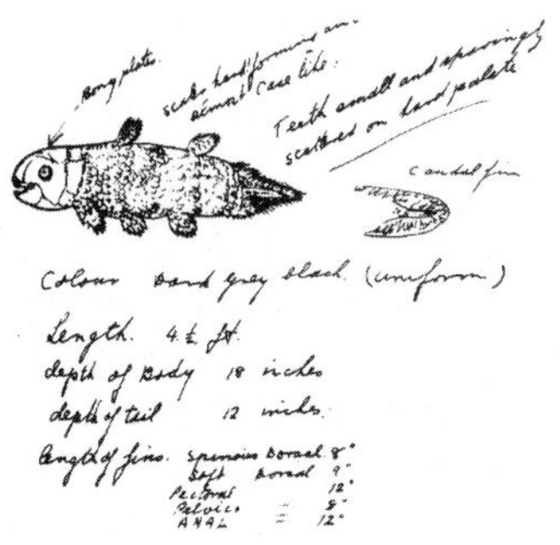

**Box figure 1** M. Courtenay-Latimer, while curator of the East London Museum in South Africa. Her quick sketch and notes of the coelacanth sent to J.L.B. Smith for his opinion are shown next to her.

## Selected References


Alexander, R. McN. 1976. Estimates of speeds of dinosaurs. *Nature* 261:129–30.

Alexander, R. McN. 1989. *Dynamics of dinosaurs and other extinct giants*. New York: Columbia University Press, p. 167.

Appel, T. A. 1987. *The Cuvier-Geoffroy debate, French biology in the decades before Darwin*. New York: Oxford University Press, p. 305.

Appleby, R. 1979. Fossil aches and pains. *New Scientist*, 16 August:516–17.

Bock, W. J. 1959. Preadaptation and multiple evolutionary pathways. *Evolution* 13:194–211.

Bock, W. J. 1979. The synthetic explanation of macroevolutionary change—a reductionistic approach. Bull. Carnegie Mus. *Nat. Hist*. No. 13:20–69.

Bock, W. J. 1988. The nature of explanations in morphology. *Amer. Zool.* 28:205–15.

Bock, W. J. 1989. Principles of biological comparison. *Acta Morphol. Neerl-Scand.* 27:17–32.

Bock, W. J. 1991. The homology concept. In *Philosophical foundation and practical methodology. Zoologisches Jahrb.* 14: 23–64.

Breithaup, B. H., and D. Duvall. 1986. The oldest record of serpent aggregation. *Lethaia* 19:181–85.

Chambers, R. 1884. *Vestiges of the natural history of creation*. Reprint 1969. New York: Leicester University Press.

Coleman, W. 1964. *Georges Cuvier, zoologist*. Cambridge, Mass.: Harvard University Press.

Cuvier, G. 1805. *Leçons d'anatomie comparée*. 5 vols. Paris: Baudouin.

Darwin, C. 1859. *On the origin of species by means of natural selection, or the preservation of favoured races in the struggle of life*. London: John Murray.

Davis, D. D. 1949. Comparative anatomy and the evolution of vertebrates. In *Genetics, paleontology, and evolution*, edited by G. L. Jepsen, E. Mayr, and G. G. Simpson. Princeton: Princeton University Press, pp. 64–89.

Desmond, A. 1982. *Archetypes and ancestors, paleontology in Victorian London 1850–1875*. Chicago: University of Chicago Press.

about 5 km offshore, over the submarine shelf, when he lowered his trawling nets into 40 fathoms (240 ft, about 73 m) of water. An hour or so later, the nets were retrieved and opened to spill onto the deck a ton and a half of edible fish, two tons of sharks, and one coelacanth. None of these old salts had ever seen such a fish, and they had little idea about what it was except to recognize its uniqueness. As was the custom, the crew saved the fish for the curator of the museum in East London, Africa, the then tiny museum in their port city. (Although this was in South Africa, a British heritage inspired local names, hence East London for this museum situated in Africa.)

The curator was Ms. M. Courtenay-Latimer (box figure 1). The museum's budget was thin, to say the least, so to build local enthusiasm and support she had emphasized exhibits representing local sea life. She encouraged crews of fishing trawlers to watch for unusual specimens. If any were caught, they were included in the pile of inedible rubbish fish at the end of the day, and Courtenay-Latimer was called to come pick what specimens she could use. On this particular day while sorting through fish, she spotted the heavy-scaled, blue coelacanth with fins like arms. It was 1.6 m in length and weighed 60 kg. When caught it had snapped at the fishermen, but it was now dead and beginning to decompose in the hot sun. By training, Courtenay-Latimer was not an ichthyologist nor was she blessed with a staff of experts. Besides curator, she was also treasurer and secretary of the museum. Although she did not recognize the coelacanth for exactly what it was, she was keen enough to realize that it was special and convinced a reluctant taxi driver to deliver her, her assistant, and the rather smelly fish back to the museum. Thin budgets again plagued her as there were no freezers or equipment to preserve such a large fish. It was then taken to a taxidermist who was instructed to save even the parts not needed for the job. But, after 3 days in the hot weather and no return word from the nearest fish expert whom Courtenay-Latimer contacted, the taxidermist discarded the soft parts. When she told the chairman of the museum's board of trustees what she suspected, he scoffed, suggesting that "all her geese were swans." Apparently he entertained the idea of discarding it but eventually relented and authorized the stuffing and mounting of the fish.

Unfortunately, her letter to the closest fish expert took 11 days to reach him because East London was still a rather remote area of South Africa and it was the holiday season. The expert whom she contacted was J.L.B. Smith, an instructor in chemistry by profession, an ichthyologist by determination. The letter included a description and rough sketch of the fish, which was enough to tell Smith that this could be the scientific find of the decade. As anxious as he was to see and confirm the fish, however, he could not leave to make the 560-km (350 mile) journey to East London. He had examinations to administer and score. Eventually, his excitement and hopes were realized when he finally did visit the museum and peered on the fish for the first time. It was a coelacanth until then known to science only from Mesozoic fossils. In honor of the person (Courtenay-Latimer) and the place (Chalumna River), Smith named it *Latimeria chalumna*.

Since then, other *Latimeria* have been discovered off the coast of eastern Africa. They seem to be predators living at depths of 40 to 80 fathoms. Thanks largely to a captain, a curator, and a chemist, *Latimeria* is a living fossil again today.


Dullemeijer, P. 1974. *Concepts and approaches in animal morphology*. Assen, Netherlands: Van Gorcum & Comp. B.V.

Felsenstein, J. 1985. Phylogenies and the comparative method. *Amer. Nat.* 125:1–15.

Gould, S. J. Not necessarily a wing. *Nat. Hist.* 94(10):19–25.

Gould, S. J. 1987. The Panda's thumb of technology. *Nat. Hist.* 96(1):15–23.

Gould, S. J., and N. Eldredge. 1977. Punctuated equilibria: The tempo and mode of evolution reconsidered. *Paleobiology* 3:115–51.

Jenkins, F. A., Jr., and W. A. Weijs. 1979. The functional anatomy in the Virginia opossum *(Didelphis virginiana)*. *J. Zool. (London)* 188:379–410.

Liem, K. F., and D. B. Wake. 1985. Morphology: Current approaches and concepts. In *Functional vertebrate morphology*, edited by M. Hildebrand, D. M. Bramble, K. F. Liem, and D. B. Wake. Cambridge, Mass.: Harvard University Press, pp. 366–77.

Mayr, E. 1963. *Animal species and evolution*. Cambridge, Mass.: Belknap Press.

Mayr, E. 1982. *The growth of biological thought*. Cambridge, Mass.: Belknap Press.

Owen, R. 1866–1868. *On the anatomy of vertebrates*. 3 vols. London: Longmans, Green. Reprint 1982. New York: AMS Press.

Rensch, B. 1960. *Evolution above the species level*. New York: Columbia University Press.

Ruse, M. 1979. *The Darwinian revolution*. Chicago: University of Chicago Press.

Simpson, G. G. 1944. *Mode and tempo in evolution*. New York: Columbia University Press.

Simpson, G. G. 1953. *The major features of evolution*. New York: Columbia University Press.

Wake, D. B. 1982. Functional and evolutionary morphology. *Persp. Biol. Med.* 25:603–20.

Wallace, A. R. 1905. *My life: A record of events and opinions*. London: Sonnenschein.

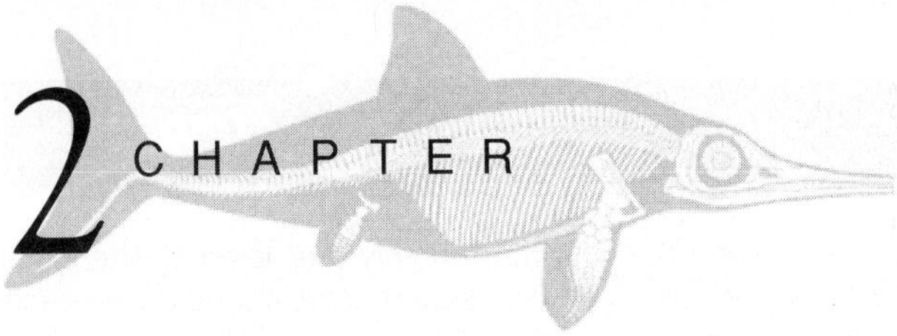

# 2 CHAPTER Origin of Chordates



Chordates are neither the most diverse nor the largest of the animal phyla, although in terms of the number of species, they come in a respectable fourth behind arthropods, nematodes, and molluscs (figure 2.1). Tucked away within this phylum is a small family, the hominids, that includes humans. In part, our interest in chordates derives from the fact that humans belong to this phylum, so studying chordates brings topics concerning us close to home. But we have more than just a vested interest in chordates. Many chordates are constructed of hard parts that survive to yield a respectable history in the fossil record, which has made them especially useful in defining ideas about evolutionary processes. Advanced chordates are also some of the most intricate animals ever to appear. They therefore introduce us to questions about the complexity of biological organization and about the special mechanisms important in evolution.

## CHORDATE PHYLOGENY

Chordates have a fluid-filled internal body cavity termed a **coelom.** Along with other animals possessing such a coelom, chordates are grouped together as the **coelomates.** Among these coelomate animals, two apparently distinct and independent evolutionary lines are present. One line of coelomates, called the **protostomes,** includes molluscs, annelids, and arthropods together with assorted smaller phyla (figure 2.2). The other line, the **deuterostomes,** includes several other small phyla as well as echinoderms, protochordates, and chordates. The distinction between protostomes and deuterostomes rests upon certain embryological characteristics (table 2.1). More will be said later about embryonic development, but here some general introductory features can help clarify the differences between protostomes and deuterostomes.

**Embryonic development; details of early cleavage (p. 153)**

In both coelomate groups, the egg begins to divide repeatedly after fertilization, a process termed **cleavage,** until the very young embryo is made up of many cells formed from the original single-celled egg. In some animals, dividing cells of the embryo are offset from each other, a pattern known as **spiral cleavage.** In others, the dividing cells are aligned, a pattern termed **radial cleavage.** At this point, the embryo is little more than a clump of dividing cells that soon become arranged into a round, hollow ball, with cells forming the outer wall around a fluid-filled cavity within. One wall of this ball of cells begins to indent and grow in-

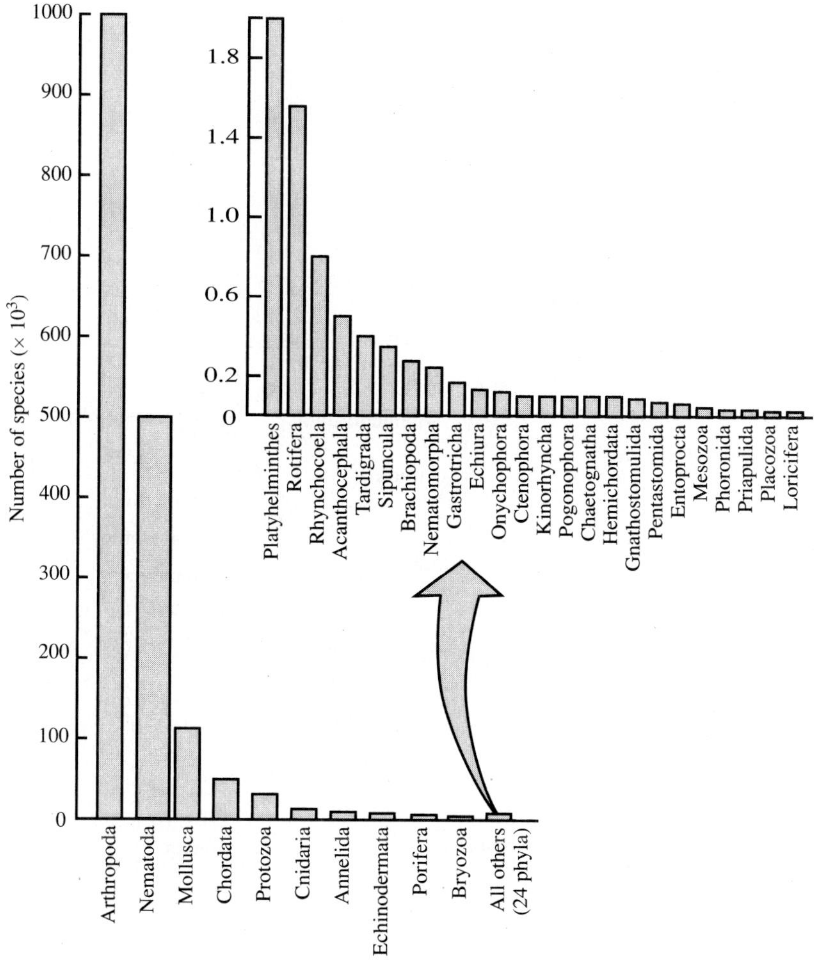


**Figure 2.1** Relative abundance of species within the animal phyla.

ward. The opening into this indentation is the **blastopore,** and the indentated cells themselves are destined to become the gut of the adult. Indentation continues until cells reach the opposite wall where they usually break through, forming a second opening into the primitive gut (the blastopore being the first). At about this point in embryonic development, the now multicellular embryo is composed of three basic tissue layers, an outer **ectoderm,** an inner **endoderm** that forms the wall of the gut, and a **mesoderm** that forms the layer between the two. If the solid sheet of mesodermal cells splits to form the body cavity within them, the result is a **schizocoelom** (figure 2.3a). If, instead, sheets of mesoderm form outpocketings of the gut that pinch off to form the body cavity, the result is an **enterocoelom** (figure 2.3b).

Protostomes, literally meaning "first mouth," are animals in which the mouth arises from or near the blastopore. Additionally, they tend to have spiral cleavage, a schizocoelom, and a skeleton derived from the surface layer of cells (figure 2.3a). Deuterostomes, literally meaning "second mouth," are animals in which the mouth arises not from the blastopore but secondarily at the opposite end of the gut (the blastopore itself becomes the anus). Additionally, embryonic development of deuterostomes includes radial cleavage, an enterocoelom, and a calcified skeleton, when present, derived generally from ectodermal tissues (figure 2.3b). These embryological characteristics shared by deuterostomes testify that they are more closely related to each other in an evolutionary sense than to any of the protostomes. Further, because these distinctions seem so fundamental and involve basic differences in body organization, they are thought by some to imply an ancient division among coelomates that took place far back in the evolutionary past.

On the basis of embryonic criteria, chordates are grouped with the deuterostomes. Their mouth forms opposite to the blastopore, their cleavage generally is radial, their coelom is an enterocoelom, and their skeleton arises generally from mesodermal tissues of the embryo. But we should be clear from the beginning about the character of the chordate phylum itself. It is easy to forget that two of the three chordate subphyla are technically *in*vertebrates. The phylum Chordata is divided into three subphyla of

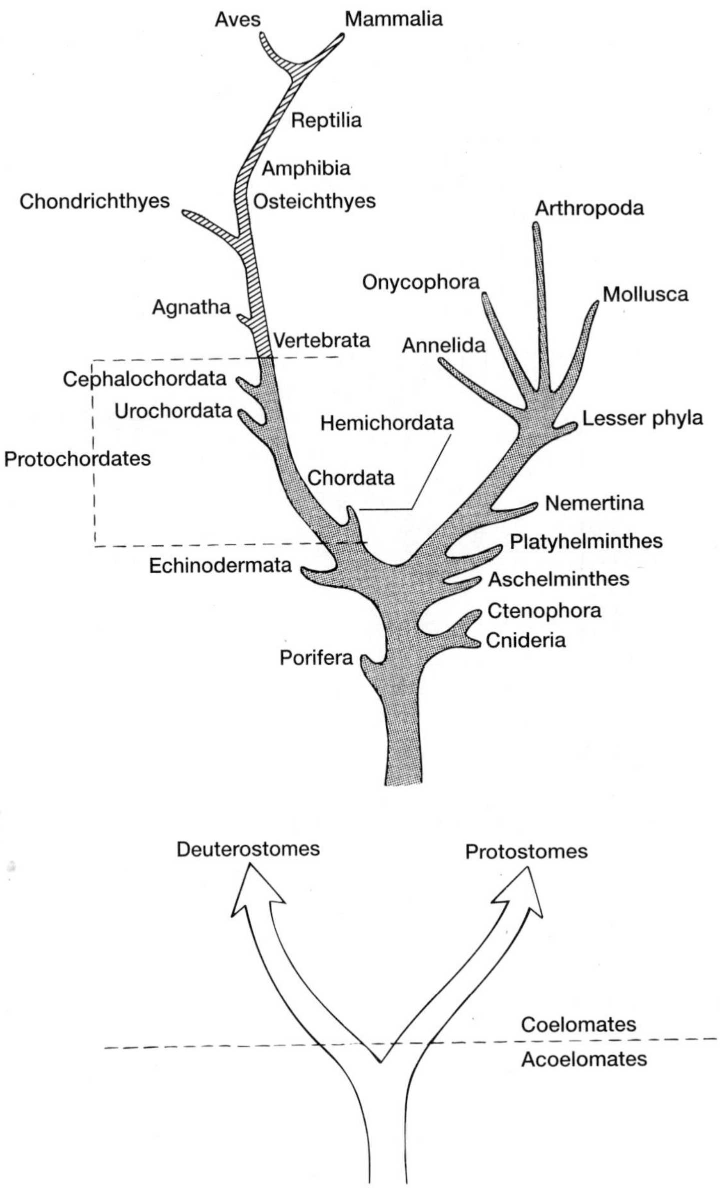


**Figure 2.2** Phylogenetic tree of metazoa. The relationships of various vertebrate (crosshatched) and invertebrate (stippled) groups are diagramed. In this phylogeny, protochordates fall between Echinodermata and Vertebrata. In the upper figure, the left branch includes deuterostomes, the right branch protostomes.

unequal size: the Cephalochordata (amphioxi), the Urochordata (tunicates), and the largest group, the Vertebrata (vertebrates). Strictly speaking, the invertebrates include all animals except members of this last subphylum, the vertebrates.

The earliest chordate fossils appear in the Cambrian period, about 590 million years ago. Although later chordates evolved hard bones and well-preserved teeth that left a substantial fossil testimony to their existence, ancestors to the first chordates likely had soft bodies and left almost no fossil trace of the evolutionary pathway taken from prechor-

**TABLE 2.1 Fundamental Patterns in Coelomate Development**

| Protostomes | Deuterostomes |
|---|---|
| Blastopore (mouth) | Blastopore (anus) |
| Spiral cleavage | Radial cleavage |
| Schizocoelic coelom | Enterocoelic coelom |
| Ectodermal skeleton | Mesodermal skeleton |

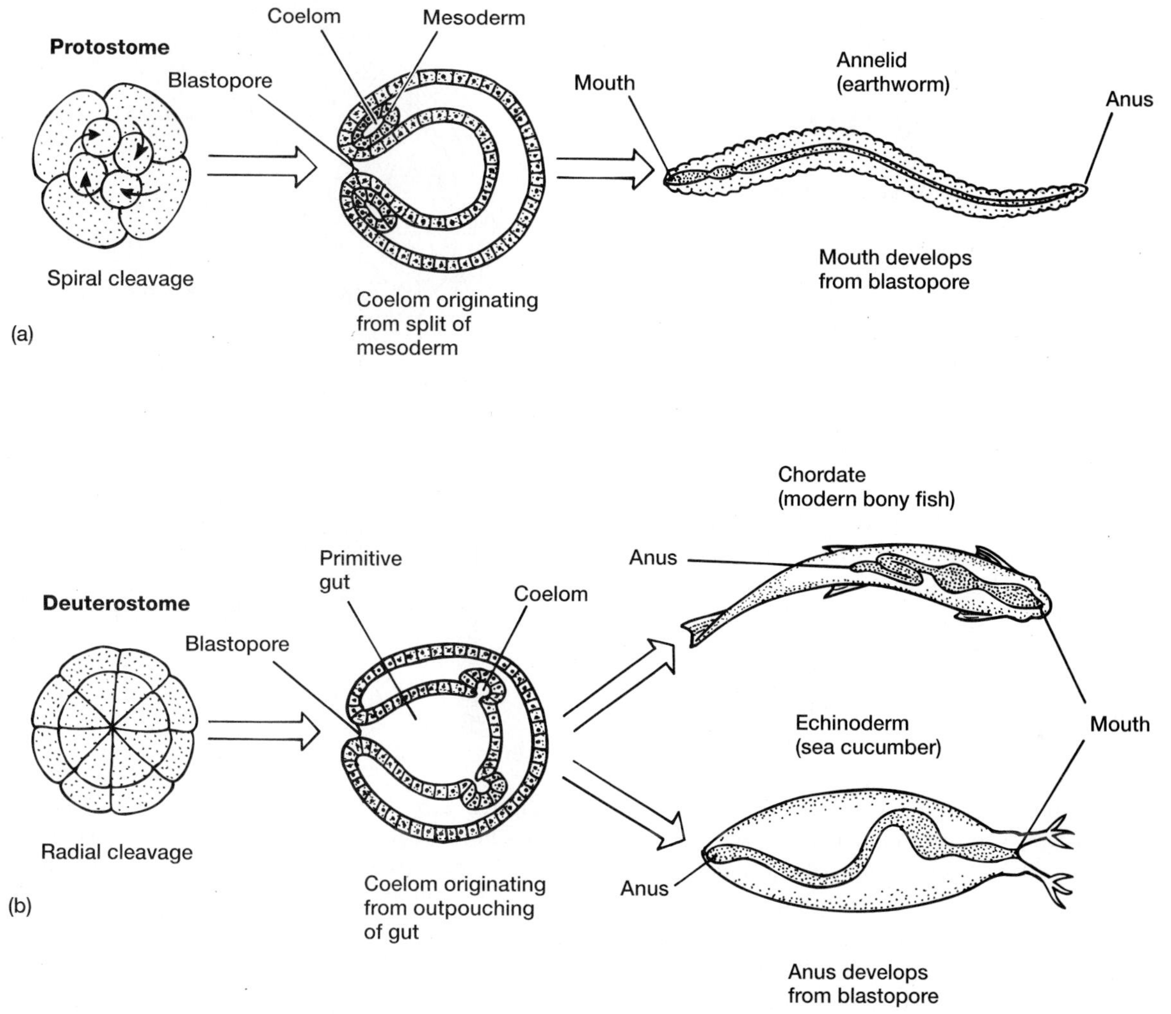


**Figure 2.3** Protostomes and deuterostomes. Coelomates are divided into two major groups on the basis of embryonic characteristics. (a) Protostomes show spiral cleavage, coelom formation by splitting of the mesoderm, and derivation of the mouth from the blastopore. (b) Deuterostomes exhibit radial cleavage, coelom formation by outpocketing of the gut, and derivation of the anus from or in the vicinity of the blastopore.

date to chordate. Thus, to decipher chordate origins, we derive evidence from anatomical clues carried in the bodies of living forms. In order to evaluate the success of our attempts at tracing chordate origins, we first need to decide what defines a chordate. We will then attempt to discover the animal group that is the most likely evolutionary source for chordates.

## Chordate Characteristics

At first glance, the differences among the three chordate subphyla are more apparent than the similarities that unite them. Most vertebrates have an endoskeleton, a system of rigid internal elements of bone or cartilage beneath the skin. The endoskeleton participates in locomotion, support, and protection of delicate organs. Some chordates are terrestrial, and most use jaws to feed. But cephalochordates and urochordates are all marine animals, none are terrestrial, and all lack a bony or cartilaginous skeleton. Their support system may involve rods of collagenous material. Cephalochordates and urochordates are suspension feeders, having a sticky sheet of mucus that strains food from streams of water passing over a filtering apparatus. All three subphyla, despite these superficial differences, share a common body design similar in four fundamental features—notochord, pharyngeal slits, dorsal hollow nerve tube, and postanal tail (figure 2.4). These four features unite all chordates and, taken together, distinguish them from all other phyla. We look next at each characteristic separately.

### Notochord

The notochord is a slender rod that arises out of the dorsal wall of the embryonic gut in primitive chordates. It lies dorsal to the coelom but beneath and parallel to the central nervous system. The phylum takes the name Chordata from

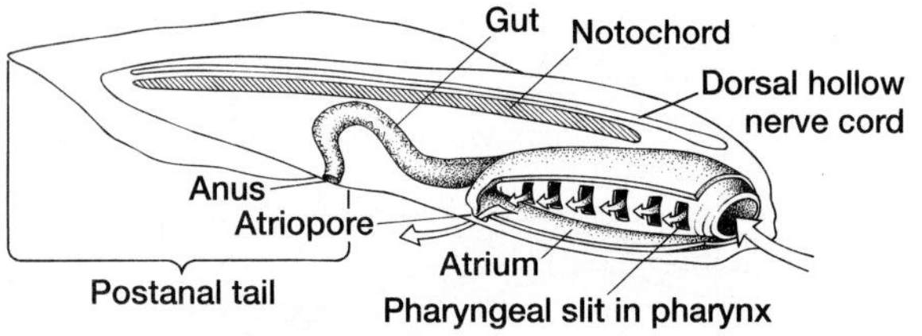


**Figure 2.4** Generalized chordate characteristics. A single stream of water enters the chordate mouth, flows into the pharynx, and then exits through several pharyngeal slits. In many lower chordates, water exiting through the slits enters the atrium, a common enclosing chamber, before returning to the environment via the single atriopore.

this structure. Typically, the notochord is composed of a core of cells and fluid encased in a tough sheath of fibrous tissue (figure 2.5a). Sometimes the fluid is held within swollen cells called vacuolated cells; other times it resides between core cells of the notochord. In having the mechanical properties of an elastic rod, the notochord can flex laterally from side to side (figure 2.5c), but cannot collapse along its length like a telescope (figure 2.5b). This mechanical property results from the cooperative action of the outer fibrous sheath and the fluid core it encloses. If the fluid were drained, then like letting air from a balloon, the outer skin would collapse and form no useful mechanical device. The fluid that normally fills the fibrous sheath remains static and does not flow. Such mechanical structures, in which the outer wall encloses a fluid core, are called **hydrostatic organs.** The notochord is a hydrostatic organ with elastic properties that resist axial compression. It lies along the body axis to allow lateral flexion but prevents collapse of the body during locomotion (figure 2.5d).

To understand its mechanics, imagine what would occur when one block of muscle contracts on one side of an animal without a notochord. As the muscle shortens, it shortens the body wall of which it is part and telescopes the body. In a body with a notochord, the longitudinally incompressible cord resists the tendency of a contracting muscle to shorten the body. Instead of shortening the body, the contraction of the muscle sweeps the tail to the side. Thus, upon contraction, the body's segmentally arranged musculature acts upon the notochord to initiate swimming motions that produce lateral pressure against the surrounding substrate. Upon muscle relaxation, the springy notochord straightens the body. Thus, the notochord prevents the collapse or telescoping of the body and acts as the muscles' antagonist in order to straighten the body. As a result, alternating side to side muscle contractions in partnership with the notochord generate lateral waves of body undulation. This form of locomotion may have been the initial condition that first favored the evolution of the notochord.

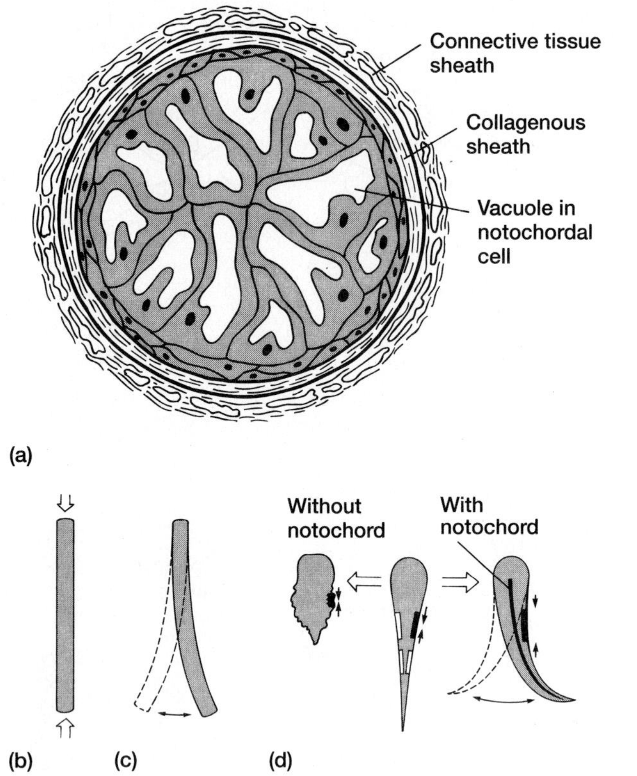


**Figure 2.5** Notochord. (a) Cross section of the notochord of a frog tadpole. (b) The notochord lies above the body cavity and is axially incompressible; that is, it resists shortening in length. (c) The notochord is flexible laterally, however. (d) As seen from above, the consequences of muscle contraction in a body with and without a notochord. Without a notochord, lateral muscle contraction telescopes the body. A notochord prevents collapse of the body, and muscle contractions on alternating sides efficiently flex the body in swimming strokes.

**Gutmann's view (p. 76)**

The notochord continues to be an important functional member throughout most groups of chordates. Only in later forms, such as in bony fishes and terrestrial vertebrates, is it largely replaced by an alternative functional member, the vertebral column. When replaced by the vertebral column, it still appears as an embryonic structure, serving early in ontogeny as a scaffold for the growing embryonic body. In adult mammals with a full vertebral column, the notochord is reduced to a remnant, the **nucleus pulposus.** This is a small core of gel-like material within each intervertebral disk that forms a circular pad lying between successive vertebrae.

**Structure and embryonic development of the notochord (pp. 160, 284)**

## Pharyngeal Slits

Another chordate feature is the **pharyngeal slits** (figure 2.4). The **pharynx** is a part of the digestive tract located immediately posterior to the mouth. During some point in the lifetime of all chordates, the walls of the pharynx are pierced, or nearly pierced, by a longitudinal series of openings, the pharyngeal slits (or pharyngotremy, literally meaning "pharyngeal holes"). The term *gill slits* is often used in place of pharyngeal slits for each of these openings, but a "gill" proper is a specialized structure composed of tiny plates or folds that harbor capillary beds for respiration in water. In vertebrates, gills form adjacent to these pharyngeal slits. The slits are openings only, often with no significant role in respiration. In many primitive chordates, these openings serve primarily in feeding, but in embryos they play no respiratory role; therefore *gill slits* is a misleading term.

Pharyngeal slits may appear early in embryonic development and persist into the adult stage, or they may be overgrown and disappear before the young chordate is born or hatched. Whatever their eventual embryonic or adult fate, all chordates show evidence of pharyngeal slits at some time in their lives.

When slits first evolved, they likely aided in feeding. As openings in the pharynx, they allowed the one-way flow of a water current—in at the mouth and out through the pharyngeal slits (figure 2.4). Secondarily, when the walls defining the slits became associated with gills, the passing stream of water also participated in respiratory exchange with the blood circulating through the capillary beds of these gills. Water entering the mouth could bring suspended food and oxygen to the animal. As it exited through the slits and across the vascularized gills, carbon dioxide was given up to the departing water and carried away. Therefore the current of water passing through pharyngeal slits can simultaneously support feeding and respiratory activity.

In primitive chordates, the pharynx itself is often expanded into a **pharyngeal** or **branchial basket,** and the slits on its walls are multiplied in number, increasing the surface area exposed to the passing current of water. Sticky mucus lining the pharynx snatches food particles from suspension. Sets of cilia, also lining the pharynx, produce the water current. Other cilia gather the food-laden mucus and pass it into the esophagus. This mucus and cilia system is especially efficient in small, **suspension-feeding** organisms, those that extract food floating in water. Such a feeding system is prevalent in primitive chordates and in groups that preceded them.

In vertebrates that feed on large food, mucus and cilia serve less well, and a pharyngeal pump worked by muscles takes the place of cilia to create the water flow. Slits still serve as convenient exit portals for excess or spent water, while adjacent gill structures function in respiration. In fishes and aquatic amphibians, the pharyngeal slits that appear during embryonic development usually persist into the adult and form the exit channel through which water associated with feeding and respiration flows. In vertebrates that reside on land, however, the pharyngeal slits normally never open and thus give rise to no adult derivative.

**Why cilia are replaced by muscles as body size increases (p. 120)**

## Dorsal and Tubular Nerve Cord

The third chordate characteristic is a dorsal hollow nerve cord derived from ectoderm (figure 2.6a). The central nervous system of all animals is ectodermal in embryonic origin, but only in chordates does the nerve tube typically form by a distinctive embryonic process, namely, by **invagination.** Future nerve tube cells of the early chordate embryo gather dorsally into a thickened **neural plate** within the surface ectoderm. This neural plate of cells folds or rolls up and sinks inward from the surface (invaginates) to take up residence internally within the embryo. In most nonchordate embryos, ectodermal cells destined to form the central nervous system do not usually amass as surface placodes; instead, they individually move inward to assemble into the basic nervous system. Further, the major nerve cord in most invertebrates is ventral in position, that is, below the gut, and solid. In chordates, however, the nerve cord lies above the gut and is hollow along its entire length, or more accurately, it surrounds the **neurocoel,** a fluid-filled central canal (figure 2.6b). The advantage, if any, of a tubular rather than a solid nerve cord is not known, but this distinctive feature is found only among chordates.

**Nerve tube formation (p. 158)**

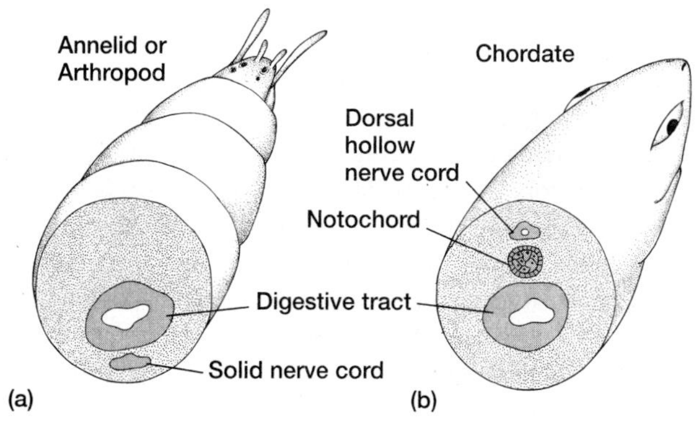


**Figure 2.6** Dorsal hollow nerve cord. (a) Basic body plan of an annelid or arthropod. In such animals, a definitive nerve cord, when present, is ventral in position and lies below the digestive tract. (b) Basic chordate body plan. The nerve cord of chordates lies in a dorsal position above the digestive tract and notochord. Its core is hollow, or more correctly, it has a fluid-filled central canal.

## Postanal Tail

Fourth, chordates possess a postanal tail that represents a posterior elongation of the body extending beyond the anus. The tail is primarily an extension of the chordate locomotor apparatus, the segmental musculature and notochord. More will be said later about the role of the tail in swimming.

**Swimming in fishes (p. 297)**

## Chordate Body Plan

What is common to all chordates are these four primary features—notochord, pharyngeal slits, dorsal hollow nerve cord, and postanal tail. These characteristics may be present only briefly during embryonic development or they may persist into the adult stage, but all chordates exhibit them at some point during their lifetimes. Taken together, they are a suite of characters found only among chordates. Chordates are also bilaterally symmetrical and show segmentation. Blocks of muscle, or **myomeres,** are arranged sequentially along the body and tail as part of the outer body wall.

Now that we have an idea about the basic and secondary characteristics of chordates, let us turn our attention to the ancestry of this group. Biologists interested in such questions often consult an assortment of primitive animals that in some ways seem in their structure and design to have preceded chordates phylogenetically and may represent chordates at their most primitive. These animals are the protochordates.

# PROTOCHORDATES

Between vertebrates and other invertebrates lie the protochordates. Although they are an informal assemblage of animals and are not a proper taxonomic group, protochordates share some or all four features of the fundamental chordate body plan. Because the fossil record reveals little about chordate ancestors, living protochordates have been scrutinized for clues to chordate origins. Living protochordates are themselves, of course, products of a long evolutionary history independent from other phyla. Their anatomy is simple, and their phylogenetic position ancient. The hope is that within living members of the protochordates we can find traces of the evolutionary steps from prechordate to chordate. A closer look at protochordates will indicate why they provide such tantalizing clues to the origin of the chordates.

All protochordates are marine animals that feed by means of cilia and mucus. But they often live quite different lives as young larvae than they do as adults. As larvae, they may be **pelagic,** residing in open water between the surface and the bottom. Although unattached, most free-floating larvae have limited locomotor capability and are therefore **planktonic,** riding from place to place primarily in currents and tides rather than by their own efforts of long distance swimming. As adults, they are usually **benthic,** living on or within a bottom marine substrate. Some **burrow** into the substrate or are **sessile** and attached to it. Some adults are **solitary,** living alone, others are **colonial** and live together in associated groups. Some are **dioecious** (literally, two houses), with male and female gonads in separate individuals, others are **monoecious** (one house), with both male and female gonads in one individual.

Juvenile and adult stages often are anatomically distinct. Each stage is a separate and distinct form known as a **morph.** Thus, juvenile morphs and adult morphs differ in appearance. The term **paedomorphosis** (or, **neoteny** as it is known in some older literature) refers to the retention of some larval characteristics in the adult, that is, the retention of the juvenile morph by the breeding adult. By definition, an adult is the sexually mature stage that reproduces. Hence, a paedomorphic species is one in which the sexually mature stage looks like the juvenile. This may seem odd at first; however, among many animals, the reproductive stage is anatomically similar in appearance to the juvenile morph. More will be said of paedomorphosis in chapter 5.

**Paedomorphosis (p. 188)**

This informal category of protochordates usually includes three groups—hemichordates, urochordates, and cephalochordates. We look next at each.

## Phylum: Hemichordata

Members of the hemichordates are marine "worms" with apparent links to chordates on the one hand and to echinoderms on the other. They share with chordates unmistakable pharyngeal slits (figure 2.7) and embryonic invagination of the nerve cord. Although usually solid, parts of the nerve cord may actually be tubular in some species. Hemichordates lack a notochord and postanal tail, hence the name *hemi-* or *half*-chordates. As larvae, some of these worms pass through a small planktonic stage called the **tornaria larva** (figure 2.8). This planktonic larva is equipped with ciliated bands on its surface and a simple gut. In its ciliated structure, simple digestive system, and planktonic lifestyle, the tornaria larva resembles the **auricularia larva** of echinoderms. In fact, some persons argue that the similarities testify to a phylogenetic link between hemichordates (tornaria larva) and echinoderms (auricularia larva).

Further, hemichordates, like both echinoderms and chordates, are deuterostomes. Their mouth forms opposite to the embryonic blastopore, and they exhibit the characteristic deuterostome patterns of embryonic cleavage and coelom formation. The similarities of hemichordates to the larval design of echinoderms, on the one hand, and adult

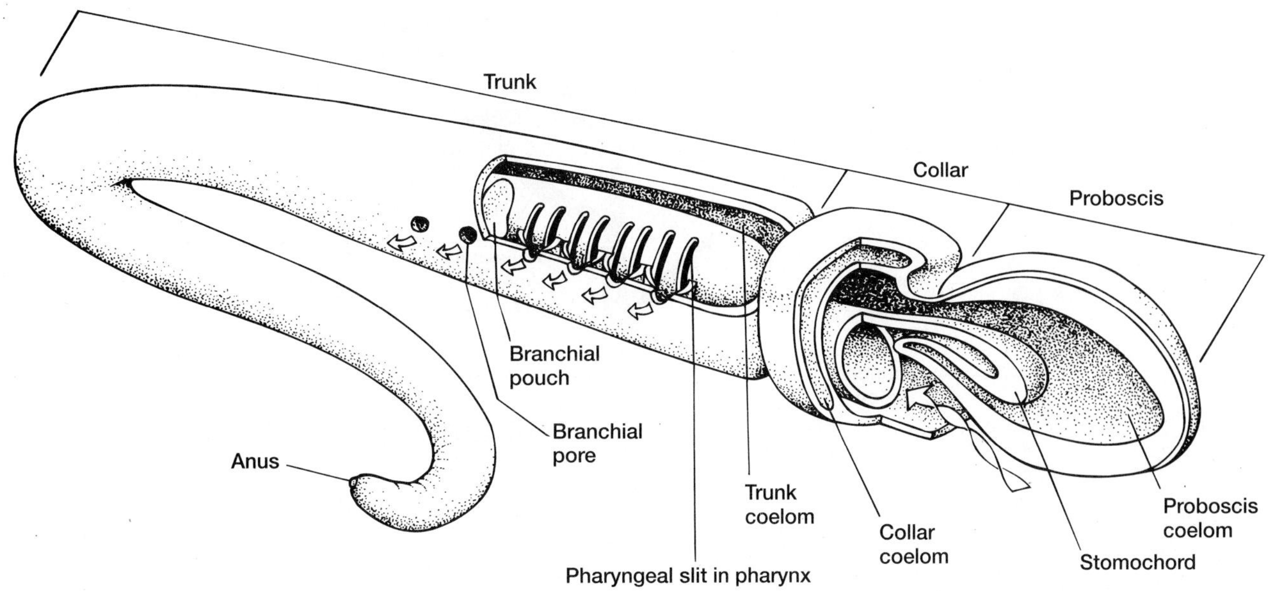


**Figure 2.7** Generalized acorn worm. Proboscis, collar, and trunk regions are shown in partial cutaway view, revealing the coelom in each and the associated internal anatomy of the worm. Within the proboscis is the stomochord, an extension of the digestive tract. The food-laden cord of mucus (spiral arrow) enters the mouth together with water. Food is directed through the pharynx into the gut. Excess water exits via the pharyngeal slits. Several slits open into the branchial pouch, a common compartment with a pharyngeal pore that opens to the outside environment.

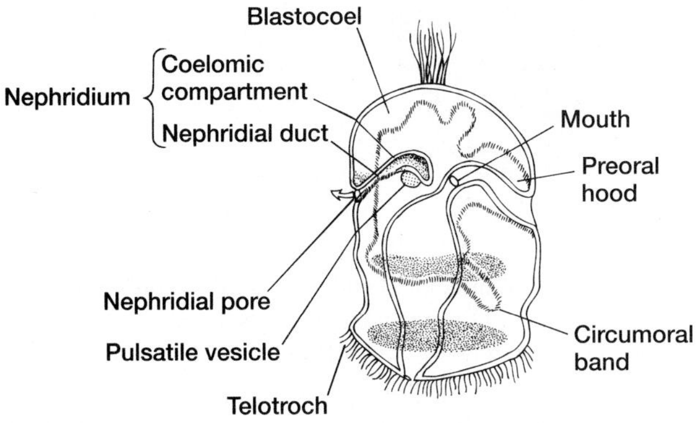


**Figure 2.8** Hemichordate, generalized tornaria larva. The simple gut begins at the mouth under a preoral hood and passes through the body of the larva. On the surface, a meandering circumoral band of cilia runs along each side of the larva. A tuft of cilia projects from the anterior end and the telotroch, an apron of cilia, runs along the posterior end. The excretory organ is a nephridium consisting of a coelomic compartment of podocytes that extends toward the exterior via a ciliated nephridial duct and opens through a nephridial pore.

chordates, on the other, are tantalizing. Perhaps they stand close to the evolutionary route taken by early chordates and still hold clues to the origin of the chordate body plan. But remember that living hemichordates are themselves millions of years departed from the actual ancestors they might share with early prechordates. Their own evolution has dealt them specialized structures serving their sedentary habits. Within the hemichordates are two taxonomic groups, the **enteropneusts,** burrowing forms, and the **pterobranchs,** usually sessile forms.

### *Class: Enteropneusta—"Acorn Worms"*

The enteropneusts or acorn worms are marine animals of shallow waters. Some species reach over a meter in length, but most are shorter than this. They live in mucus-lined burrows and have a body with three regions—**proboscis, collar, trunk**—each with its own coelom (figure 2.9a–c). The proboscis, used in both locomotion and feeding, includes a muscular outer wall that encloses a fluid-filled coelomic space. Muscular control over the shape of the proboscis gives the animal a useful probe to shape a tunnel or inflate itself against the walls of the burrow to anchor its body in place (figure 2.9b). Tucked away in their burrows, many species ingest loosened sediment, extract the organic material it contains, pass the spent sediment through their simple gut, and deposit a casting (fecal waste) on the surface of the substrate where changing tides flush it away.

Other species are suspension feeders, extracting tiny bits of organic material and plankton directly from the water. In these forms, the synchronous beating of cilia on the outer surface of the proboscis sets up water currents that flow across the animal's mucous surface (figure 2.10).

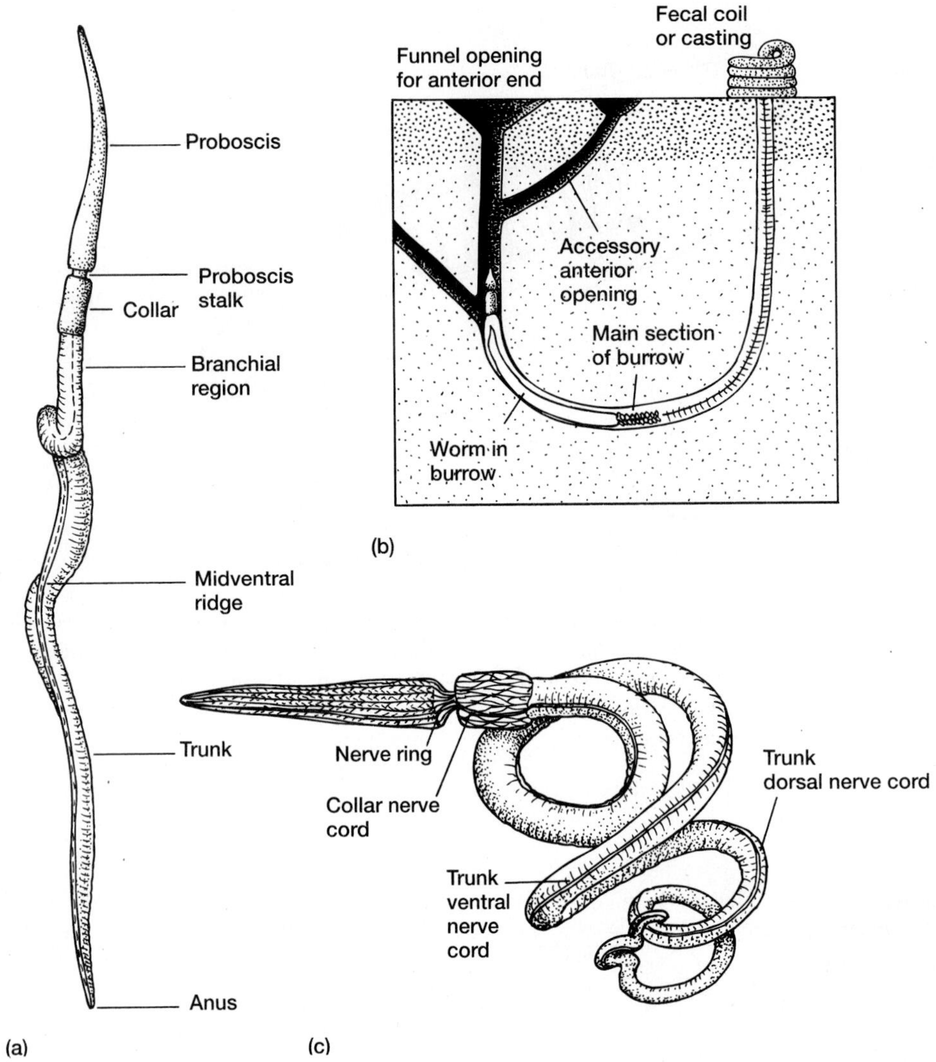


**Figure 2.9** Hemichordata, class Enteropneusta. The hemichordates depicted in this figure are enteropneusts, known informally as acorn worms. (a) External features and body regions of an adult worm. (b) Acorn worm *Balanoglossus* in burrow. (c) Nervous system of the acorn worm *Saccoglossus*. The nervous system is organized into dorsal and ventral nerve cords from which nerves spread to all parts of the body. In some species, the section of the dorsal nerve cord within the collar arises embryologically by invagination of surface ectoderm and often develops a tubular shape.

Suspended materials adhering to the mucus on the proboscis are swept along ciliary tracks to the mouth. The muscular lip of the collar can be drawn over the mouth to reject or sort larger food particles. Excess water that enters the mouth exits through numerous pharyngeal slits located along the lateral walls of the pharynx. Sets of adjacent slits open into a common chamber, the **branchial pouch,** that in turn pierces the outer body wall to form the **branchial pore,** an undivided opening to the outside environment (figure 2.7). Excess water departing from the pharynx thus passes first through a slit, then through one of the several branchial pouches, and finally through the branchial pore to the outside.

During ontogeny, perforations developing in the lateral walls of the pharynx form the original pharyngeal slits (figure 2.11a). However, each such slit next becomes partially subdivided by the **tongue bar,** a downward growth from the top rim of the opening (figure 2.11b). The fleshy bars between the original slits are referred to as the **primary pharyngeal bars** (or septa) and the tongue bars that come to divide them as the **secondary pharyngeal bars.** Secondary, but not primary, pharyngeal bars have a coelomic canal derived from the trunk coelom. The **lateral cilia** lining the edges of both primary and secondary pharyngeal bars move water currents through the pharynx. The **frontal cilia** move mucus and occur in mucous-secreting epithelium along the medial edges of tongue bars and elsewhere within the lining of the pharynx (figure 2.11c). A network of afferent and efferent branchial vessels supplies the tongue bars, possibly participating in respiratory

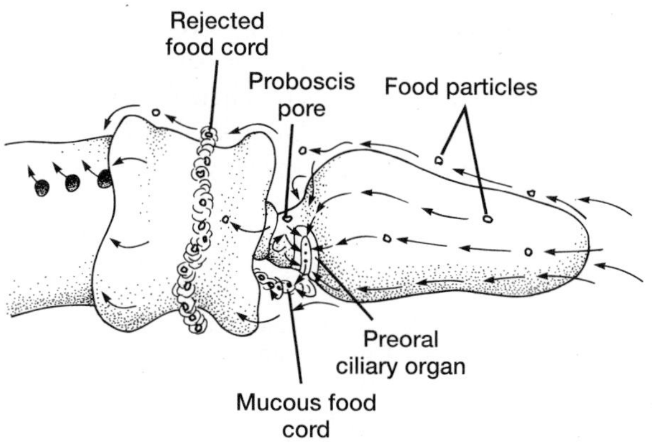


**Figure 2.10** Suspension mucus feeding. Direction and movement of food and mucus are indicated by arrows. Food material, carried along in the water current generated by surface cilia, travels across the proboscis and into the mouth where it is captured in mucus and swallowed. Rejected food material collects in a band around the collar and is shed.

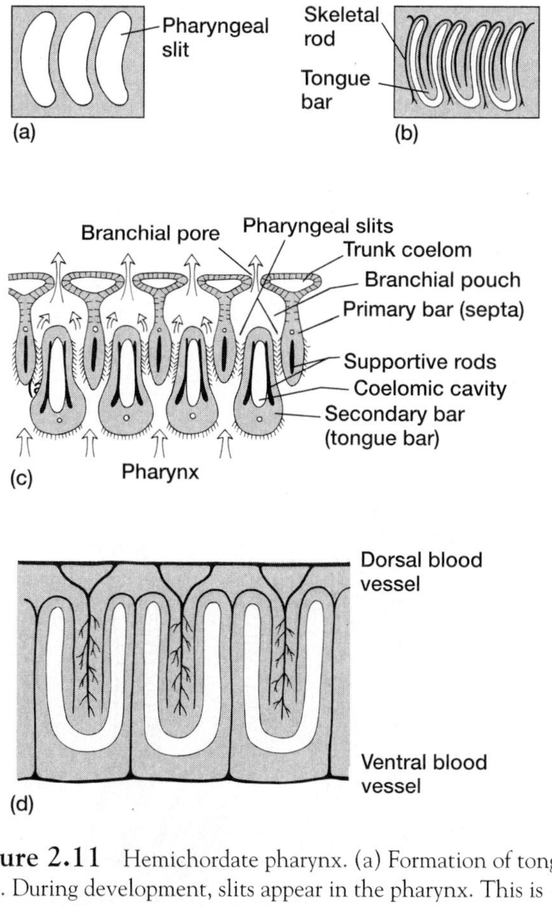


**Figure 2.11** Hemichordate pharynx. (a) Formation of tongue bars. During development, slits appear in the pharynx. This is followed by the partial subdivision of each by growth of a downward process, the tongue bar. (b) M-shaped supportive rods within these tongue bars appear. (c) Cross section through branchial bars. Cilia lining these bars move water from the pharynx past the edges of each tongue bar, past each primary bar, into the common branchial pouch and then out through a branchial pore. (d) Vascular supply to the tongue bars. Branches from the dorsal and ventral blood vessels supply each tongue bar, suggesting that respiratory exchange also occurs in the pharyngeal slits of the hemichordate.

(c) Modified from Pardos, 1988.

exchange with the passing stream of departing water (figure 2.11d).

The **stomochord** (figure 2.7) arises in the embryo as an outpocketing from the roof of the embryonic gut anterior to the pharynx. In the adult, the stomochord retains a narrow connection to what becomes the buccal cavity, but it usually enlarges as it projects forward into the cavity of the proboscis to form a preoral diverticulum. The surface of the stomochord is associated with components of the vascular and excretory systems. Its walls consist of epithelial cells, like those of the buccal cavity, as well as ciliated and glandular cells. Its hollow interior communicates with the buccal cavity.

Excretion in acorn worms probably occurs partly through the skin, but they also possess a **glomerulus,** a dense network of blood vessels within the proboscis. Vascular fluid entering the glomerulus from the dorsal blood vessel is presumably filtered, yielding "urine" that is released into the proboscis coelom and eventually eliminated through the proboscis pore. Within the collar, a pair of ciliated collar ducts that extend from the collar coelom to the exterior via the first pharyngeal pore are also thought to be excretory in function.

The circulatory system is represented by two principal vessels, a **dorsal** and a **ventral blood vessel.** The blood, which contains few cells and lacks pigment, is propelled by muscular pulsations in these major vessels. From the dorsal vessel, blood passes forward into a **central venous sinus** at the base of the proboscis. Riding on top of this sinus is the **heart vesicle** that exhibits muscular pulsations and provides additional motive force to drive blood from the venous sinus forward into the glomerulus. From the glomerulus, blood flows to the ventral blood vessel and posteriorly beneath the digestive tract, which the ventral vessel supplies.

The nervous system in acorn worms consists mainly of a diffuse network of nerve fibers at the base of the epidermis (figure 2.9c). Dorsally and ventrally, the nerve network is consolidated into longitudinal nerve cords joined by nerve interconnections. In some species, the section of dorsal nerve cord in the collar invaginates from the surface ectoderm, sinks downward, and pinches itself off from the ectoderm. This **collar nerve cord** often is partially tubular in nature. Its embryonic origin from dorsal ectoderm and its retention of a tubular structure in a few species suggest its homology to the dorsal hollow nerve cord of chordates.

Enteropneust gonads are housed in the trunk, the sexes are dioecious, fertilization is external. Early cleavage is radial, and formation of the body cavities is usually enterocoelic. In some species, development proceeds directly from egg to young adult. In most, there is a tricoelomic tornaria larval stage in that the three body cavities include an anterior **protocoel,** a middle **mesocoel,** and a posterior **metacoel,** each destined to become the coelom of the proboscis, collar, and trunk, respectively (figure 2.12). The tornaria feeds and may remain a planktonic larva for several months before undergoing metamorphosis into the benthic adult.

The tornaria has a **nephridium,** an excretory organ through which the larva regulates its internal ionic environment and rids itself of metabolic wastes. It consists of a blind-ended tube within the anterior region of the larva. During metamorphosis, the base of the nephridium enlarges into the protocoel of the adult, but in the larva, the ciliated **nephridial duct** (pore canal) conveys waste to the surface

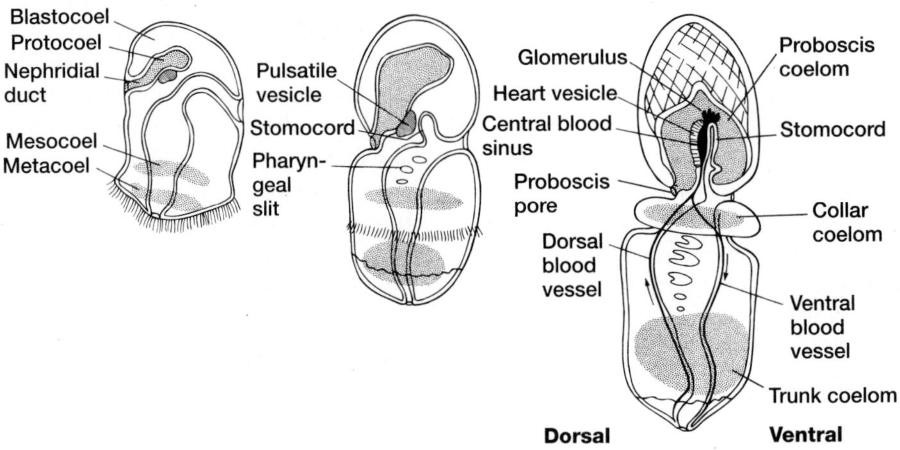


**Figure 2.12** Metamorphosis of hemichordate larva. Transformation of larva into juvenile, from left to right. The three coeloms of the larva give rise to the three respective body cavities of the adult—proboscis, collar, and trunk.

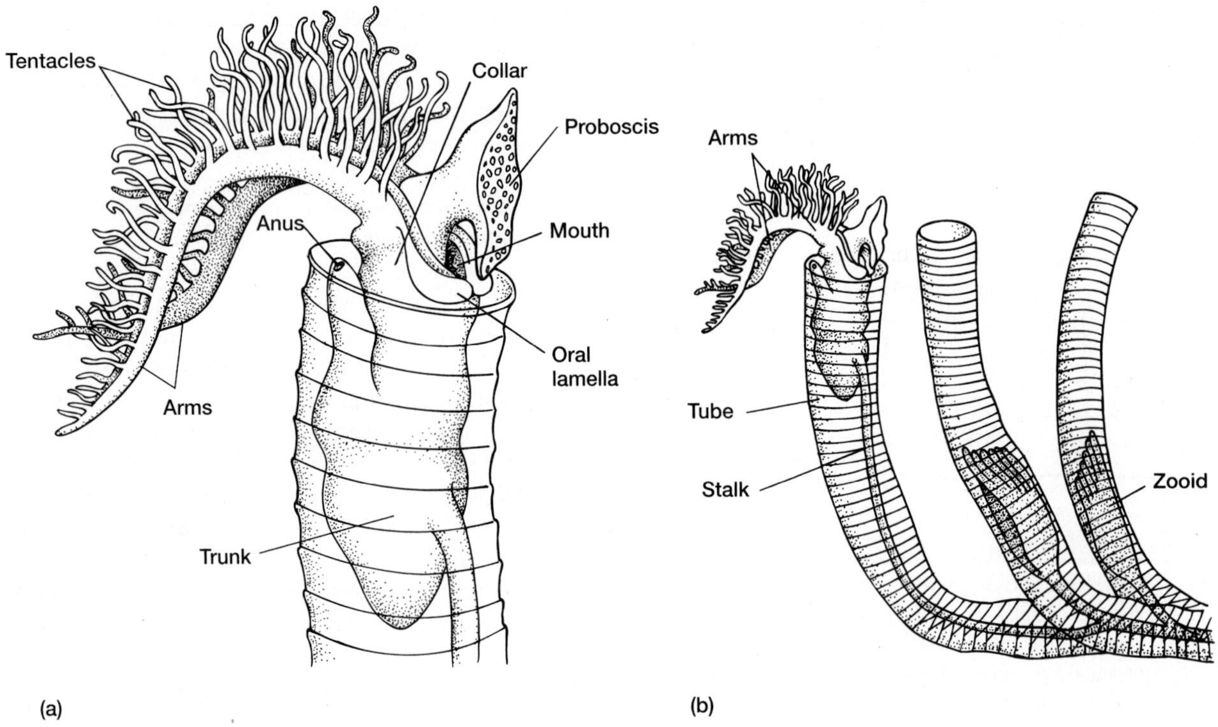


**Figure 2.13** Hemichordata, class Pterobranchia. (a) The sessile pterobranch *Rhabdopleura*. Notice that this pterobranch has the same body plan as an acorn worm—proboscis, collar, trunk—but these three features are modified and the whole animal lives in a tube. (b) Pterobranchs in tubes. When disturbed, the stalk shortens to pull a pterobranch to safety inside the tube. When merged into a colony, each contributing pterobranch individual is often called a zooid.

and opens to the outside via the **nephridial pore** (hydropore; figure 2.8). In addition to ciliated cells, the walls of the nephridium are lined by **podocytes,** specialized excretory cells that form a porous boundary between the lumen of the nephridium and the blastocoel, the larval cavity in which it resides. It is thought that the beat of cilia draws excess fluid from the blastocoel across the porous layer of podocytes, into the lumen of the nephridium, and out the nephridial pore. Lying next to the nephridium is a small contractile **pulsatile vesicle.**

**Vertebrate kidney (p. 534)**

Upon the larva's metamorphosis, the base of the nephridium expands to become the proboscis coelom, its canal becomes the proboscis duct, and much of its lining becomes the muscle and connective tissue of the proboscis. The pulsatile vesicle becomes the contractile tissue, or heart vesicle, that settles on top of the forming central venous sinus. Podocytes associated with local blood vessels gather to form the glomerulus.

### *Class: Pterobranchia*

Most pterobranchs, of which there are only three genera, live in secreted tubes usually in deep oceanic waters (figure 2.13a). Proboscis, collar, and trunk are present, although they may be quite modified. The collar, for instance, is drawn out into two or more elaborate tentacles, part of the animal's suspension-feeding apparatus (figure 2.13a). The trunk is U shaped, with the anus bending back to open at the top of the rigid tube in which the animal resides. An extension of the body, the stalk, attaches it to its tube and jerks the animal safely inside when it is disturbed (figure 2.13b). Most species are colonial. Because individual identity is often lost, each contributing individual to the colony is commonly referred to as a **zooid.**

Excretory organs of pterobranchs include a glomerulus in the proboscis and perhaps a ciliated pair of collar ducts. A stomochord is usually present. The nervous system is even simpler than that of acorn worms. A tubular nerve cord is absent. Next to the epidermis in the dorsal region of the collar lies the **collar ganglion,** which is the closest a pterobranch comes to possessing a central nervous system. Nerve branches emanate forward from the collar ganglion to the tentacles and posteriorly into the trunk. A few pharyngeal slits are present in most species.

### *Hemichordate Phylogenetic Affinities*

With links to chordates on the one hand and echinoderms on the other, hemichordates hold out promise of connecting chordates to their ancestral source among the invertebrates. This was recognized early in the twentieth century, but enthusiasm, perhaps overenthusiasm, led to overinterpretation of hemichordate structure. The stomochord within the proboscis was originally deemed to be a notochord and championed as a further structural link with chordates. But, such a claim is unfounded. Unlike a true notochord, the hemichordate stomochord is hollow, originates anterior to the pharynx, and lacks the fibrous sheath necessary to give it the structural integrity of a rigid notochord. In fact, hemichordates share only two anatomical features with chordates—pharyngeal slits and a short length of tubular nerve cord (in some). The proboscis, collar, and trunk composing their body plan is quite unlike the body plan of any other protochordate. Although most will admit that hemichordates exhibit affinities to chordates and to echinoderms in particular, exact ancestral connections remain debatable.

## Subphylum: Urochordata

At some point in their life histories, urochordates generally show all four defining chordate characteristics—notochord, pharyngeal slits, tubular nerve cord, and postanal tail (figure 2.14a). Consequently, they are proper chordates classified as a subphylum within the Chordata. Urochordates are specialists at feeding on suspended matter, especially very tiny particulate plankton. In some, the pharynx is expanded into a complex straining apparatus, the branchial basket. In other species, the filtering apparatus is secreted by the epidermis and surrounds the animal. All species are marine. Urochordates are divided into three taxonomic classes. Adult Ascidiacea are sessile, having larvae that attach them to firm rock surfaces, piles, or floats, whereas the classes Larvacea and Thaliacea are permanently pelagic and drift unattached to any fixed substrate.

Urochordate literally means "tail backstring," a reference to the notochord. The familiar name, **tunicates,** is inspired by the flexible outer body cover, the **tunic.** It is secreted by the underlying epidermis with contributions from scattered cells within the tunic. This tunic, sometimes referred to by the more general name **test,** characterizes the ascidians, the largest class of urochordates.

### *Class: Ascidiacea—"Sea Squirts"*

Ascidians or sea squirts are marine animals that are often brightly colored. Some species are solitary, others colonial. Adults are sessile, but larvae are planktonic.

***Larva*** The larva, sometimes called the **ascidian tadpole,** does not feed during its short sojourn of a few days as a free-living member of the plankton, but it disperses to and selects the site at which it will take up permanent residence as an adult. Only the larval stage exhibits all four chordate characteristics simultaneously. The small pharynx bears slits in colonial species. The tubular nerve cord extends into a tail supported internally by a turgid notochord. Vacuolated cells are absent from the ascidian notochord. Instead, in

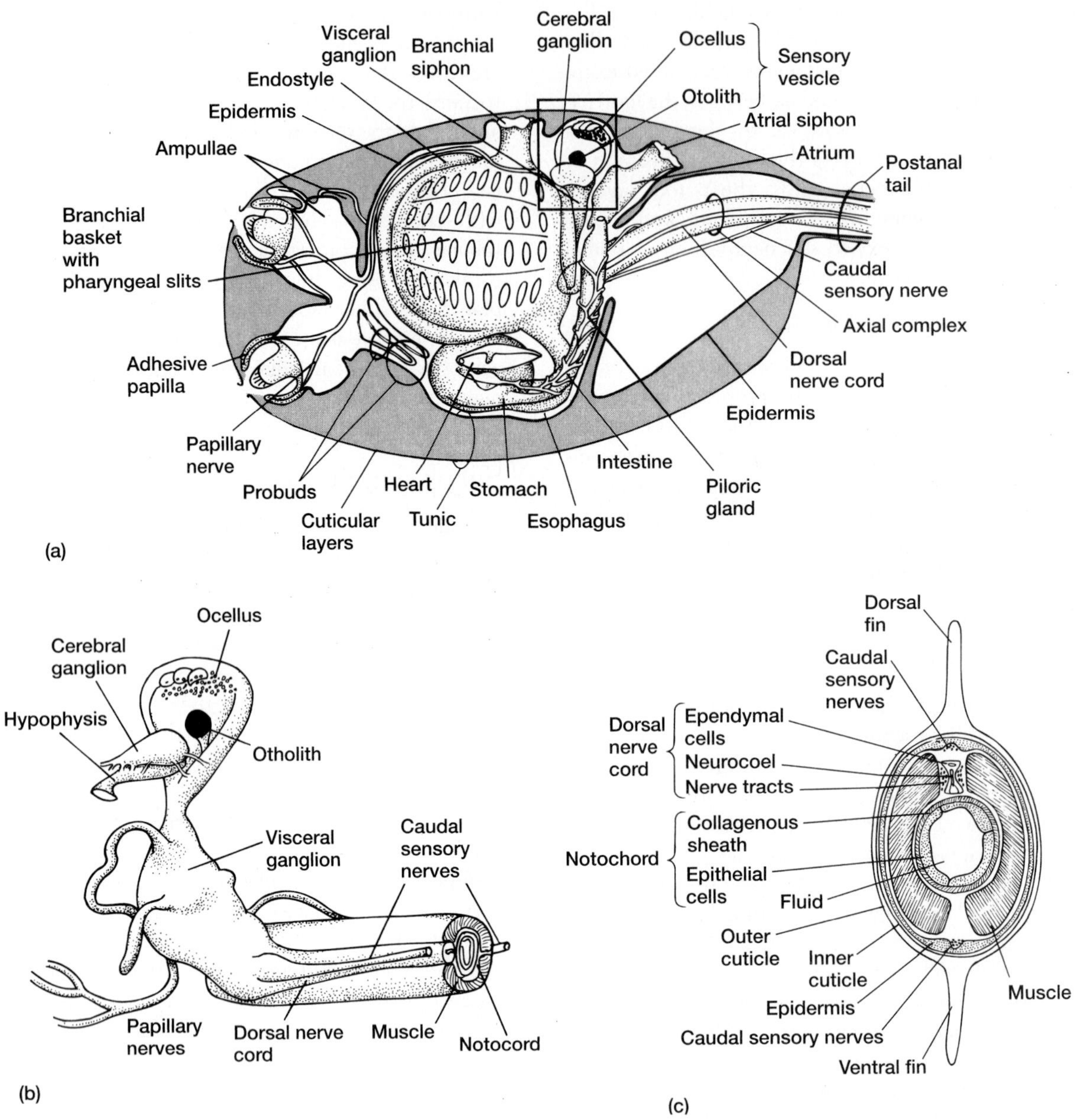


**Figure 2.14** Urochordata, class Ascidiacea. (a) Larva of the ascidian *Distaplia occidentalis*. (b) Enlarged view of the anterior larval nervous system of *Diplosoma*. (c) Cross section of the larval tail of *Diplosoma*. During development, the tail twists the dorsal fin to the left side of the body, but in this figure, the tail is rotated 90° and drawn upright. Notice that the ventral and dorsal fins are formed from the outer layer of the tunic and that the central notochord is bounded by sheets of muscle. The dorsal nerve cord is composed of ependymal cells around a neurocoel, with axons of sensory nerves coursing along its side.

Based on the research of R. A. Cloney.

most solitary and colonial species, the notochord is tubular. Its walls are composed of a single layer of epithelial cells covered externally by a circumferential sheath of collagen fibers. The epithelial layer encloses an extracellular gel- or fluid-filled lumen (figure 2.14b). Therefore, the ascidian notochord is a turgid, tubular rod closed at both ends.

In solitary ascidian species, the gut does not fully differentiate in the nonfeeding larva. An anus is not present to mark the point beyond which the tail continues. In many colonial species, the gut may be fully differentiated, including an anus that opens into the atrial chamber, and feeding may begin within 30 minutes after settlement. The "postanal" tail is sometimes twisted or rotated about 90° from the body so that the tubular nerve cord is not strictly "dorsal" relative to the body. Striated muscle cells lie in sheets along the sides of the tail rather than in segmental blocks. Special **myomuscular** and gap junctions join these muscle cells together so that all cells on one side act as a unit, contracting together to bend the tail. An acellular tunic secreted mostly by the underlying epidermis covers the ascidian larva. The surface of the tunic is covered by thin inner and outer cuticular layers. The outer cuticular layer forms the larval tail fins but is cast off at metamorphosis. The inner cuticular layer remains after metamorphosis to form the outermost surface of the juvenile. Beneath the tunic, the epidermis at the anterior end of the body forms **adhesive papillae** that serve to attach the larva to a substrate at the end of its planktonic existence.

A **sensory vesicle** (figure 2.14a) located next to the rudimentary pharynx contains navigational equipment thought to be involved in orientating the larva during its planktonic existence. Within the sensory vesicle is a light-sensitive **ocellus** and a gravity-sensitive **otolith** (figure 2.14b). A rudimentary **cerebral ganglion,** functional only after metamorphosis, and a **visceral ganglion** are nearby and send nerves to various parts of the body. The nerve cord includes ciliated **ependymal cells** around the neurocoel and **nerve tracks** that arise from the visceral ganglion and pass lateral to the ependymal cells in order to supply the tail muscles (figure 2.14c). Sensory nerves return from the tail and adhesive papillae to the visceral ganglion. Blood stem cells and a rudimentary heart are present (figure 2.14a). In a few colonial species, the blood cells become mature and the heart beats, but like the adult hearts, such larval hearts periodically reverse the direction of pumping.

***Metamorphosis*** At the end of its short planktonic stage, the ascidian larva makes contact with the substrate of choice, usually in a dark or shaded location, adhesive papillae take hold to attach it, and metamorphosis to a young adult begins almost immediately (figures 2.15 and 2.16). Within a few minutes of attachment, the epithelial notochord cells contract and separate from each other, the extracelluar fluid leaks from the central lumen, and the notochord becomes limp. The **axial complex,** consisting of notochord, dorsal nerve cord, and tail muscles, is actively drawn into the body and resorbed. In some ascidian species, this active resorption of the axial complex results from contraction of surface epidermal cells. In other species, active contraction of the notochordal cells causes its resorption. Once inside the body, the axial complex is phagocytized over the next several days and its constituents redistributed to support the young growing adult. Lost too are the outer tunic layer, sensory vesicle, and visceral ganglion; however, the

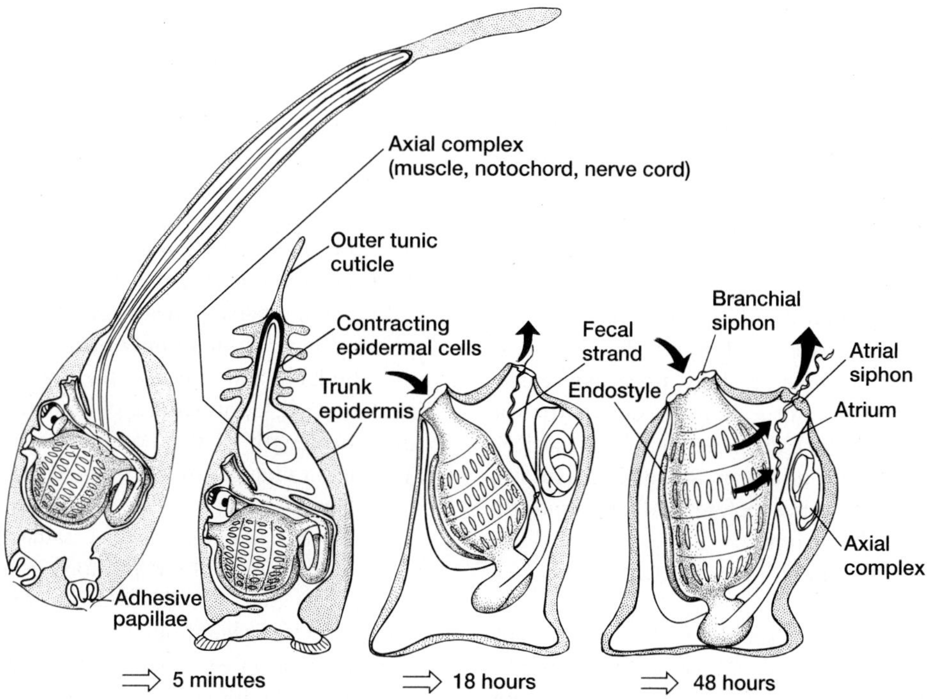


**Figure 2.15** Metamorphosis of the ascidian larva *Distaplia*, from left to right. The planktonic nonfeeding larva settles and attaches to a substrate. Adhesive papillae hold the larva in place, contraction of tail epidermis pulls the axial complex into the body, and the larva sheds its outer cuticle following attachment. By 18 hours, the branchial basket rotates to reposition the siphons, and the appearance of a fecal strand testifies that active feeding has begun. By 48 hours, most of the axial complex is resorbed, rotation is complete, and attachment to the substrate is firm. At this point, the juvenile is clearly differentiated.
Based on the research of R. A. Cloney.

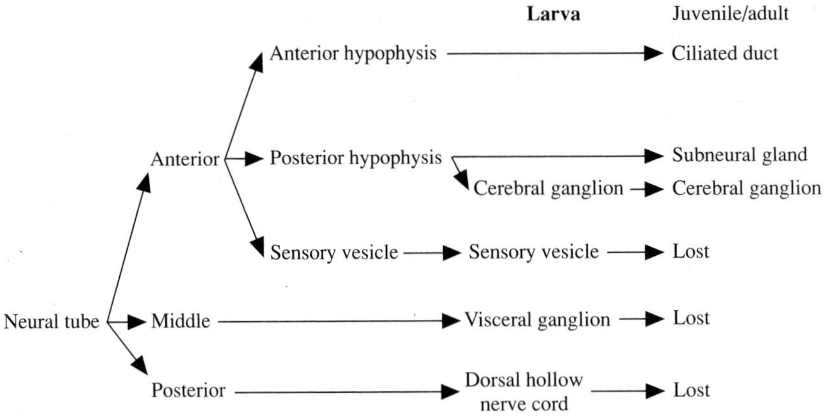


**Figure 2.16** Developmental fate of the nervous system of the urochordate larva *Diplosoma* at metamorphosis. The larval neural tube generally has anterior, middle, and posterior regions that form the basic components of the larval nervous system. At metamorphosis, they are modified into juvenile structures or lost.

pharynx enlarges, slits in its walls increase in number, and the attached individual begins feeding for the first time.

Most of the chordate features that made their debut in the larva, namely, notochord, tail, and dorsal nerve tube disappear as well in the forming adult. Although the pharynx persists and even expands, it becomes highly modified. The slits in its walls proliferate and each subdivides repeatedly, producing smaller openings called **stigmata.** This radically remodeled pharynx forms the barrel-shaped branchial basket (expanded pharynx plus numerous stigmata) of the adult sea squirt (figure 2.17a).

***Adult*** The tunic, composed of a unique protein, **tunicin,** and a polysaccharide similar to plant cellulose, forms the body wall of an ascidian adult. Within the walls formed by the tunic are enclosed the branchial basket, a large atrial cavity, and the viscera. The tunic attaches the base of the animal to a secure substrate (figure 2.17a). Incurrent (branchial) and excurrent (atrial) siphons form entrance and exit portals for the stream of water that circulates through the body of the tunicate. Tiny, fingerlike sensory tentacles encircle the incurrent siphon to examine the water entering and perhaps exclude excessively large particles before water flows next into the branchial basket. The stigmata sieve the passing water before it flows from the branchial basket into the **atrium,** the space between basket and tunic (figure 2.17b). From here, the current exits via the excurrent siphon.

Rows of cilia line the branchial basket. The mucus-producing **endostyle,** a ventral groove, is connected by **peripharyngeal bands** around the inside to the **dorsal lamina.** Particulate matter is extracted from the passing stream of water by a netlike sheet of mucus lining the branchial basket. The rows of cilia collect the food-laden mucus and deliver it to the dorsal lamina, which in turn conveys it to the gut.

The sea squirt's heart is tubular, with a single layer of myoepithelial cells lining its inner wall (figure 2.17c). The surrounding **pericardial cavity** is the only remnant of the coelom. Contraction of the heart pushes blood out to the organs and tunic. After a few minutes, the flow reverses to return blood along the same vessels to the heart. Unlike the vertebrate circulatory system, there is no continuity between the heart myoepithelium and the blood vessels. Ascidian blood vessels are not lined by an endothelium. Instead, they are true hemocoels, that is, connective tissue spaces. The blood contains a fluid plasma with many kinds of specialized cells, including **amoebocytes** that resemble vertebrate lymphocytes. They are phagocytic, and some accumulate waste materials. No specialized excretory organ has been found in tunicates.

The nervous system consists of a **cerebral ganglion** located between the siphons (figure 2.17a). From each end of the ganglion arise nerves that pass to the siphons, gills, and visceral organs. Beneath the ganglion lies the **subneural gland,** a structure of unknown function that is left over from the larva and joined to the branchial basket via a **ciliated funnel.** Contractions of smooth muscle bands running the length of the body and encircling the siphons bring about changes in the shape and size of the adult.

Tunicates reproduce sexually and asexually (figure 2.18). Both sexes occur in the same individual (monoecious) although self-fertilization occurs only in some species. Asexual reproduction, known solely in colonial species, involves **budding.** The rootlike **stolons** at the base of the body may fragment into pieces that produce more individuals, or buds may arise along blood vessels or viscera. In some species, buds even appear in the larva before metamorphosis. Such budding gives the tunicate a method to propagate rapidly when conditions improve. In some species, buds seem especially hardy and are adept at surviving temporary adversity.

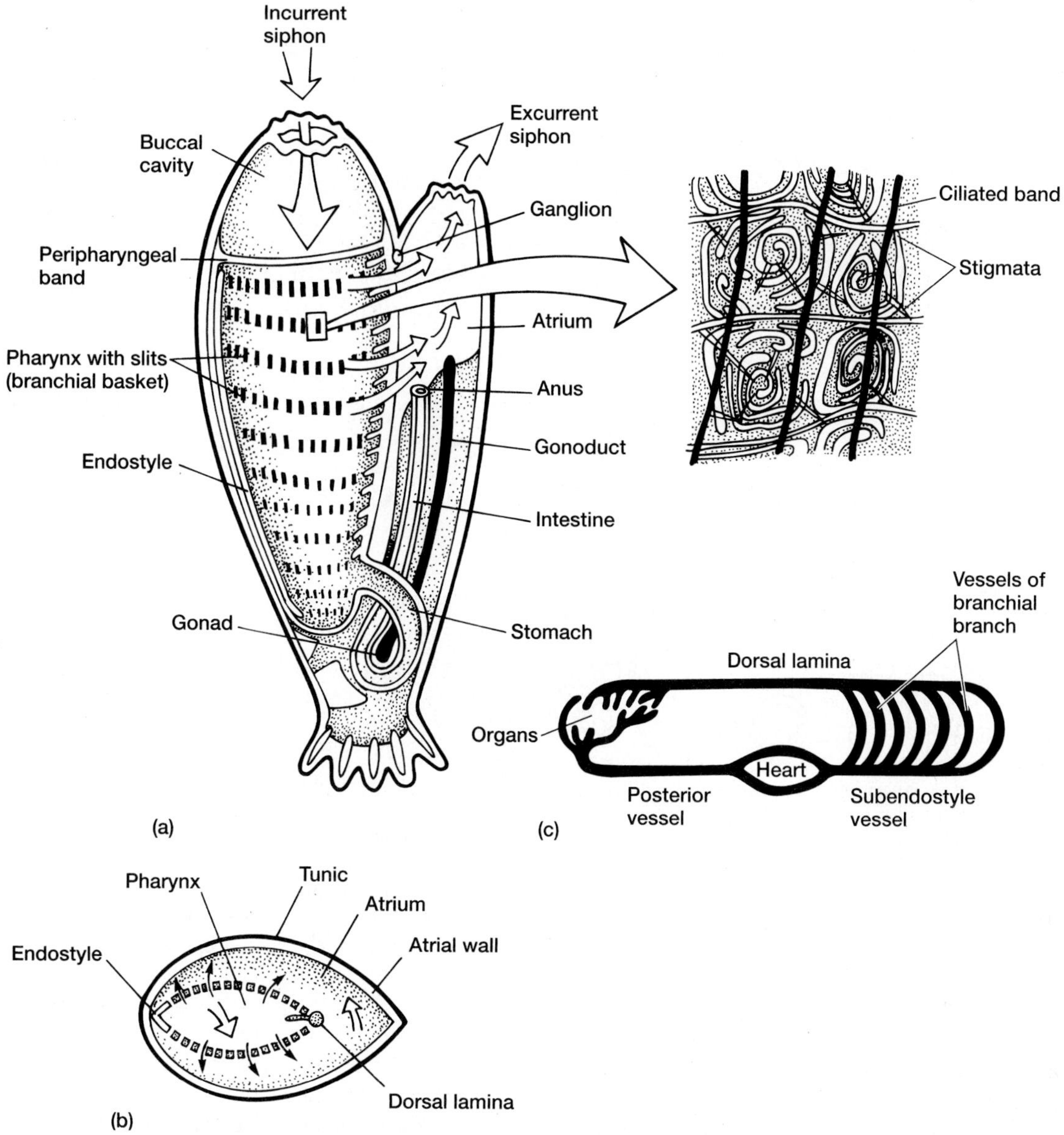


**Figure 2.17** Adult urochordate. (a) The flow of water is shown as it enters the incurrent siphon (branchial siphon) and passes through modified pharyngeal slits (stigmata) into the atrium and out through the excurrent siphon (atrial siphon). On the right, the structure of several highly subdivided pharyngeal slits are depicted. (b) Cross section through the branchial basket and surrounding atrium. (c) Diagram of urochordate circulation. Blood flows in one direction and then reverses itself rather than maintaining a single direction of flow.

## *Class: Larvacea (Appendicularia)*

Members of this worldwide class are tiny marine animals reaching only a few millimeters in length, that reside within the planktonic community. Larvacea received their name because the adults retain larval characteristics similar in some ways to the ascidian tadpole with a tail and trunk (figure 2.19a–c). In fact, the general resemblance of adult larvaceans to ascidian tadpoles suggests that larvaceans may be phylogenetic derivatives of the larval stage of ascidians.

Larvaceans produce a most remarkable feeding apparatus that consists of three components—**screens, filters,** and expanded **gelatinous matrix.** Because the larvacean lives within the gelatinous matrix it constructs, this gelatinous matrix is termed a "house." This house also holds the feeding screens and filters, and forms the channels through which streams of water carry suspended food particles. Houses and feeding styles differ among the various species, but generally the undulating tail of the larvacean creates a feeding current that draws water into

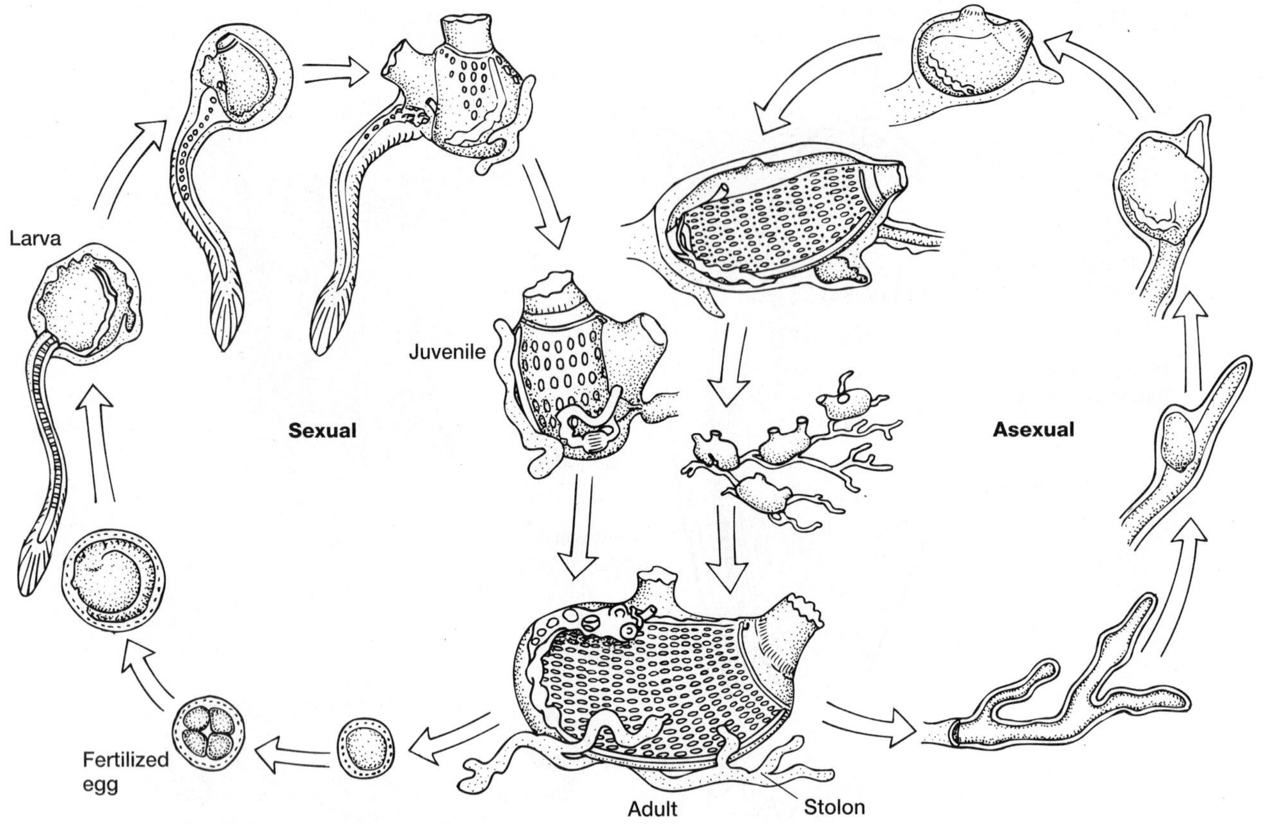


**Figure 2.18** Urochordate—ascidian life cycle. The life cycle of colonial ascidians includes a sexual (left) and an asexual (right) phase. In the sexual phase, the tunicate larva develops from a fertilized egg. This larva is planktonic and persists for a few hours or few days at most. It soon settles on a solid substrate and undergoes metamorphosis into a sessile juvenile that grows into adulthood. The asexual phase begins with external buds from the rootlike stolon or internal buds from organs within the body, depending on the species. These buds grow and differentiate into adults, often forming a colony of tunicates.

the house. Incoming water passes first through the meshwork of screens that exclude large particles; consequently, they serve as an initial sorting device. Water continues its flow through internal passages and then up the sides and through the mucous feeding filters, where tiny suspended food particles are removed. When the current of water is emptied of its suspended material, it leaves the house through an excurrent opening. The larvacean takes advantage of its convenient central position at the base of the feeding filters to gather all intercepted food particles. By means of ciliary action, the animal sucks trapped particles from the filters into its pharynx every few seconds. Mucus, secreted by an endostyle, gathers the food. Excess water exits from the pharynx via a pair of slits and joins the current departing through the excurrent opening.

If the filters become clogged with food, a reverse flow may clear them. If that fails, the house is abandoned and a new one secreted (figure 2.20). Actively feeding larvaceans might abandon and build new houses every few hours. Disturbance of captured larvaceans, perhaps simulating predator attack, can prompt an even more frequent cycle of abandon and build.

The rudiment of a new house, which is secreted by the epithelium, is already present while the animal still occupies its old one. Some houses split to release the larvacean; others have special escape hatches. Almost immediately upon exiting its house, the animal initiates a vigorous series of motions that enlarge the house rudiment to a size it can enter. Once inside, expansion of the house continues with the addition of feeding screens and filters. Sometimes within the space of only a few minutes, the new house is complete and the larvacean is once again feeding actively.

All species, except one, are monoecious, and most of these are **protandrous,** that is, sperm and eggs are produced by the same gonad (of the same individual) but at different times during its life. Maturation is so rapid that within 24 to 48 hours of fertilization miniature larvaceans secrete a house and are set up for feeding.

Their rapid reproductive growth and special feeding apparatus give larvaceans a competitive advantage over

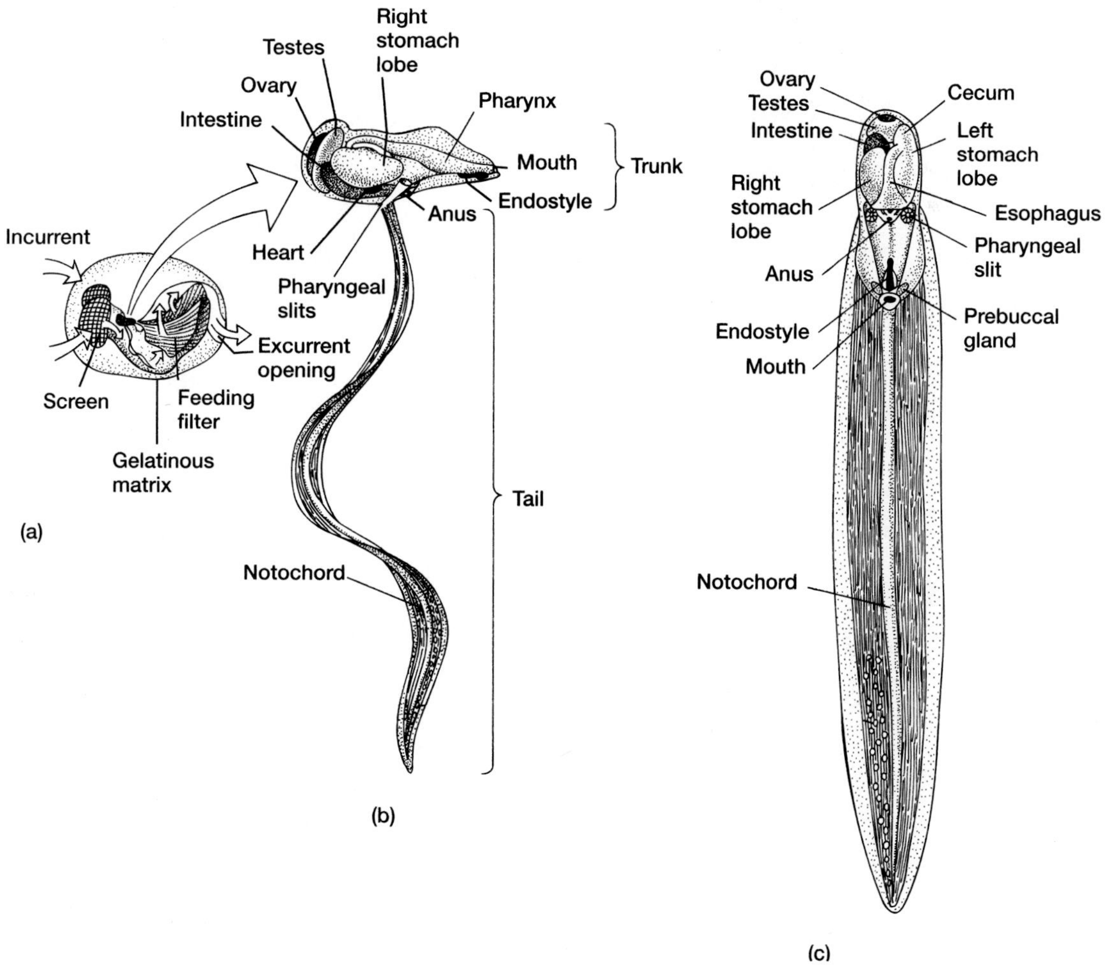


**Figure 2.19** Urochordata, class Larvacea (Appendicularia), *Oikopleura albicans*. (a) *Oikopleura* is shown within its gelatinous house. The animal's feeding filter obtains food from the current of circulating water (small arrows). This larvacean resides at the base of the screen where it sucks food off these screens. (b and c) Enlarged lateral and ventral views, respectively, of the isolated larvacean. The tail, supported by a notochord, is active in producing the current of food-bearing water that moves through the internal channels of the house and through the feeding filter.

other aquatic suspension feeders. Larvaceans are especially adept at gathering ultraplankton, very minute bacteria-sized organisms. Collectively, ultraplankton are the major producers in most open oceans, but they are generally too small to be captured by the filtering traps of most suspension feeders. These tiny organisms that escape the clutches of other suspension feeders fall prey to the efficient filtering gear of larvaceans. Larvacea are able to sift through large volumes of water, ingest a wide range of plankton sizes, including the very tiniest, and proliferate rapidly in response to local blooms in food supply.

The trunk of the larvacean holds its major body organs, although which organs are present varies among the three families of larvacea. Members of the smallest family, the Kowalevskiidae, lack endostyle, heart, and spiracles. In the Fritillaridae, the stomach consists of only a few cells. In the Oikopleuridae, the family best studied, the digestive system includes a U-shaped digestive tube, a pharynx with a pair of pharyngeal slits, and an endostyle that manufactures mucus. The blood, which is mostly devoid of cells, circulates through a system of simple sinuses driven by the pumping action of a single heart and the movement of the tail.

The tail is thin and flat. Muscle bands act on a notochord to produce movement. A tubular nerve cord is present.

### *Class: Thaliacea*

Like larvaceans, the thaliaceans are free-living pelagic urochordates, but unlike them, thaliaceans are apparently derivatives of the adult ascidian rather than of the tadpole morph (figure 2.21a–c). A few pharyngeal slits are present. Details of feeding are unresolved, although cilia, mucus, and a branchial basket certainly participate.

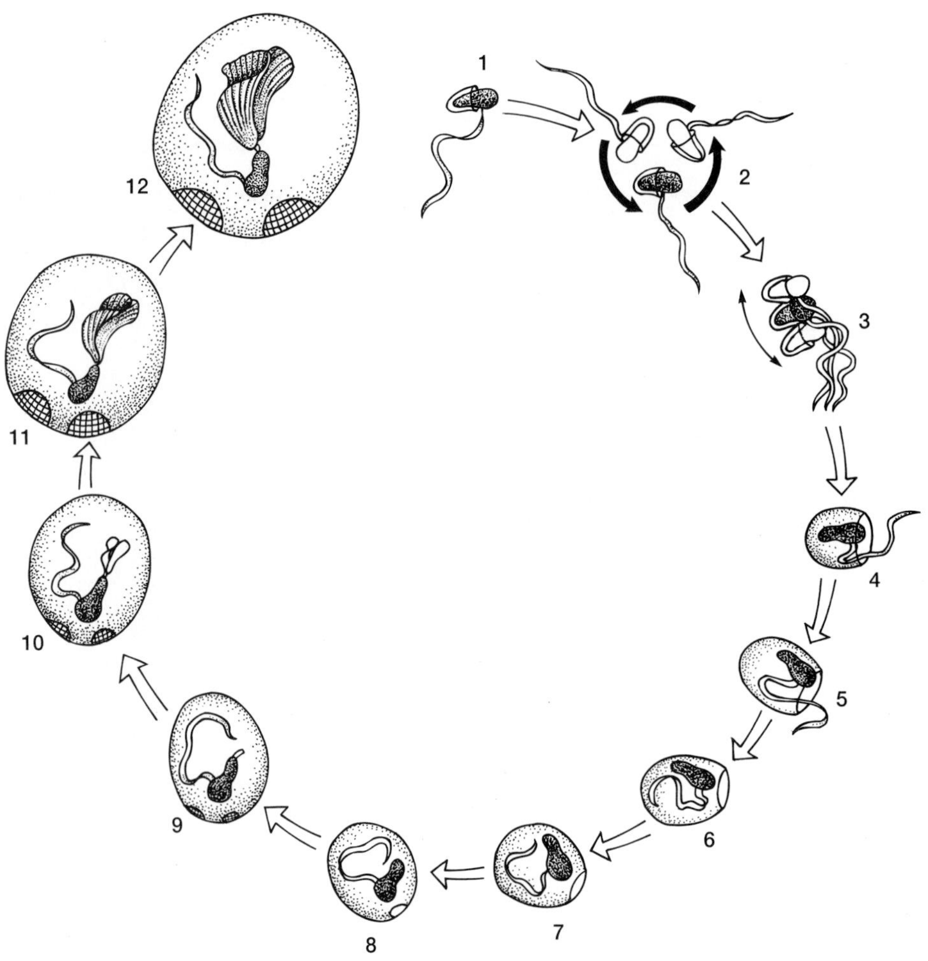


**Figure 2.20** House building by the appendicularian *Oikopleura*. Clogged filters apparently trigger an appendicularian to abandon its house (1). Vigorous movements enlarge the rudiment of a house (2 and 3) until it has room enough for the animal to enter (4). Thereafter, the house is further enlarged, filters are secreted, and feeding begins again (12).

The most primitive species of thaliaceans are built like colonial ascidians except incurrent and excurrent siphons lie at opposite ends of the body (figure 2.21c). The outer body, or test, encloses a water-filled chamber. Most thaliaceans possess encircling circumferential bands of muscle within the walls of the test. Slow contraction of these muscle bands constricts the test and squeezes the water in the chamber out through the posterior aperture. When muscles relax, the elastic test expands, drawing in water through the anterior aperture to refill the chamber. Repeated cycles of muscle contraction and test expansion produce a one-way flow of water through the thaliacean, creating a jet propulsion system for locomotion.

## Subphylum: Cephalochordata

Cephalochordates occur worldwide in warm temperate and tropical seas. They are built upon the characteristic chordate pattern that includes pharyngeal slits, tubular nerve cord, notochord, and postanal tail (figure 2.22a–c). These animals are anatomically simple, with an approach to food gathering we have seen in other protochordates, namely, suspension feeding based on a pharyngeal filtering apparatus surrounded by an atrium. Slits open the walls of the expanded pharynx to allow exit of a one-way feeding current driven by cilia. Supporting edges of each slit constitute the primary pharyngeal bars (figure 2.23). During embryonic development, a tongue bar grows downward from the upper rim of each slit and joins the ventral rim, thereby completely dividing each original pharyngeal slit into two. This dividing support, derived from the tongue bar, constitutes a secondary pharyngeal bar. **Supportive rods** of fibrous connective tissue support the pharyngeal bars internally. Short transverse connecting **synapticules** cross-link these pharyngeal bars.

Major ciliated food corridors line the pharynx. The ventral channel is the endostyle, the dorsal channel is the **epibranchial groove,** and the inside edges of the primary and secondary pharyngeal bars carry ciliary tracts. An **oral hood** encloses the anterior entrance to the pharynx and

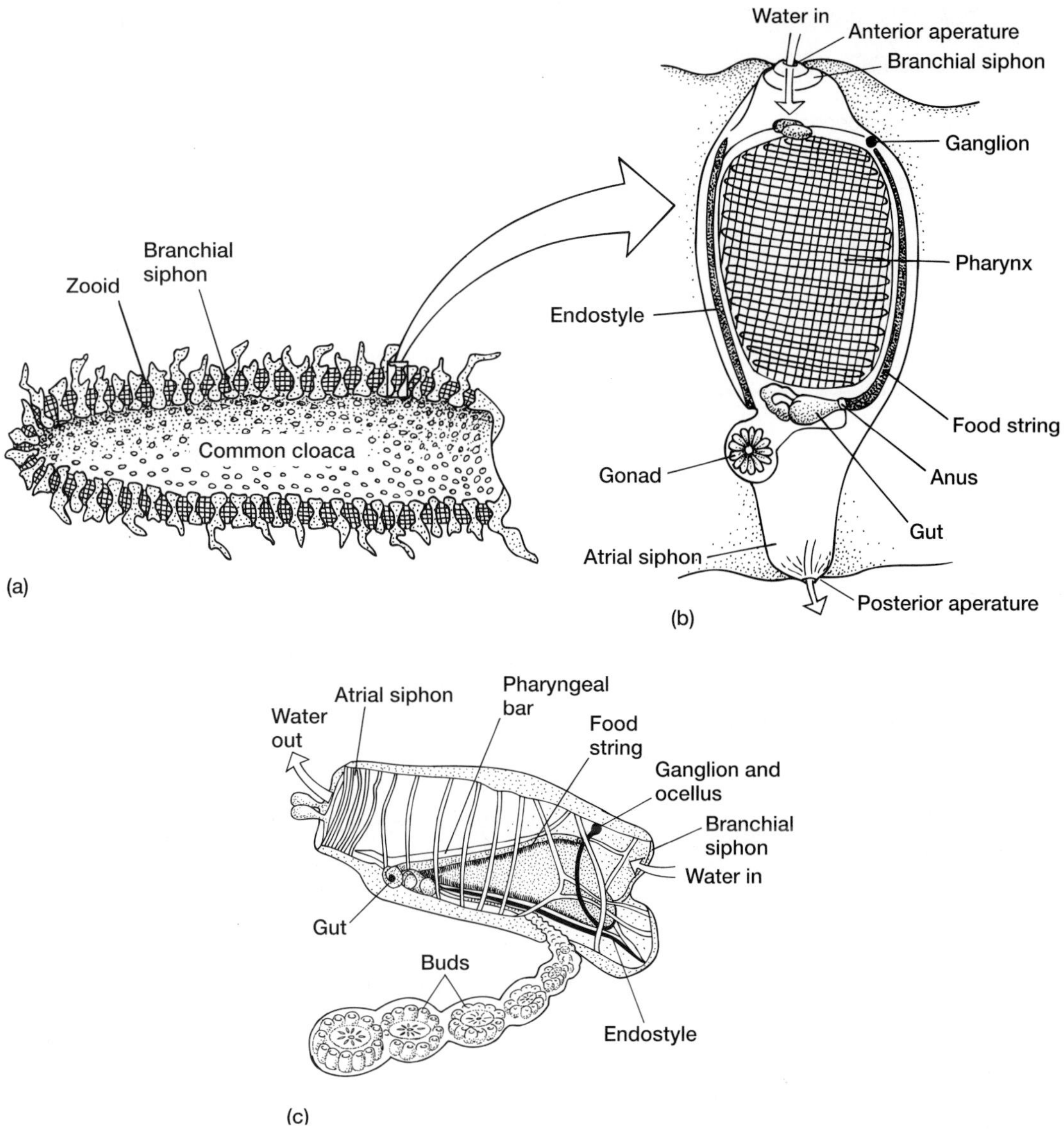


**Figure 2.21** Urochordata, class Thaliacea. (a) Colony of thaliaceans. (b) Isolated zooid. Longitudinal section of the body of this individual member is removed from its "house." Small arrows indicate direction of water flow. (c) An order of thaliacean known as salps. Branchial and atrial siphons are at opposite ends, turning the feeding current of water into a modest propulsive jet. Asexual buds are produced.

supports an assortment of food-processing equipment. Projecting from the free edge of the oral hood are **buccal cirri** that prevent entrance of large particles. The inside walls of the oral hood hold ciliated tracts that sweep food particles into the mouth. The coordinated motion of these cilia gives the impression of rotation and inspired the name **wheel organ** for these tracts (figure 2.22c). One of these dorsal tracts, usually located below the right side of the notochord, bears a ciliated invagination that secretes mucus to help collect food particles and is known as **Hatschek's pit** or **groove.** Hatschek's pit occurs in the roof of the oral cavity, a similarity shared with the vertebrate pituitary gland, part of which also forms by invagination from the roof of the buccal cavity. This has led some to propose that Hatschek's pit may have an endocrine function.

The posterior wall of the oral hood is defined by the **velum,** a partial diaphragm that supports short sensory **velar tentacles.** Suspended material faces a gauntlet of testing, sifting, and sorting devices before passing through the central opening in the velum and entering the pharynx. Mucus, secreted by the endostyle and secretory cells of the pharyngeal bars, is driven up the walls of the pharynx by cilia. Food particles adhere and are gathered dorsally into a thread in the epibranchial groove, from which they are conveyed to the gut. Excess water passes out through

## Box Essay 2.1 *Amphioxus* or *Branchiostoma?*

From their early discovery, cephalochordates seemed destined to be a lesson in taxonomic etiquette. Demeaned in 1774 by the first attempt to classify them, they were thought to be slugs and dubbed *Limax lanceolatus* by the German zoologist P. S. Pallas (although to be fair he had only a scruffy, ill-preserved specimen from which to work). In 1836 William Yarrell recognized the special nature of these animals and named them *Amphioxus* (meaning two ended) *lanceolatus.* Alas, this name stuck too well because much later it was discovered that O. G. Costa, actually two years before Yarrell, had christened them *Branchiostoma* and by rules of taxonomic priority the species should, and now does, carry this official generic name. Amphioxus, however, is a familiar name ingrained in common usage. It is not quite the mouthful of *Branchiostoma,* so we shall keep amphioxus (without italics or capitalization) as one common name, along with "lancelet" as another.

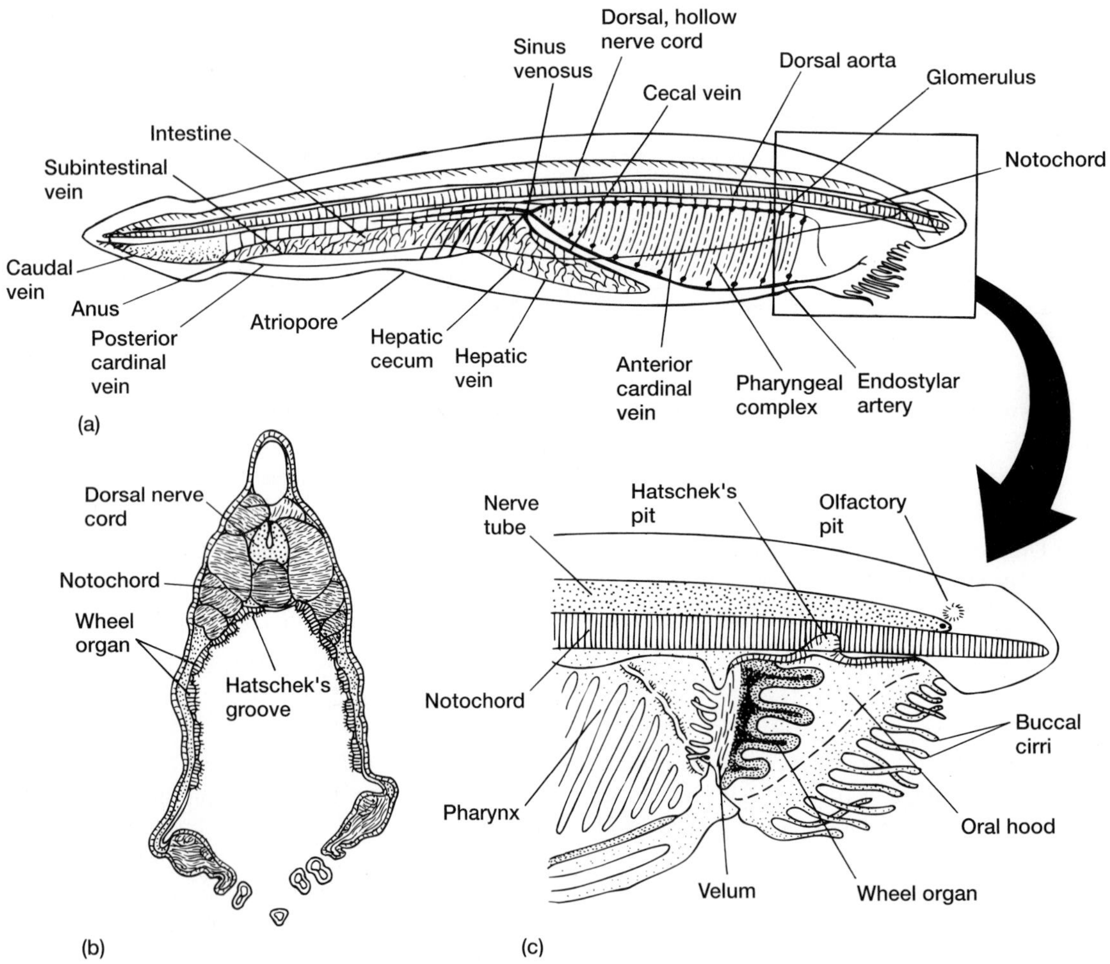


**Figure 2.22** Cephalochordate *Branchiostoma lanceolatum*, known by its more simple name of amphioxus. (a) Lateral view. (b) Cross section through oral hood. (c) Enlargement of mouth.

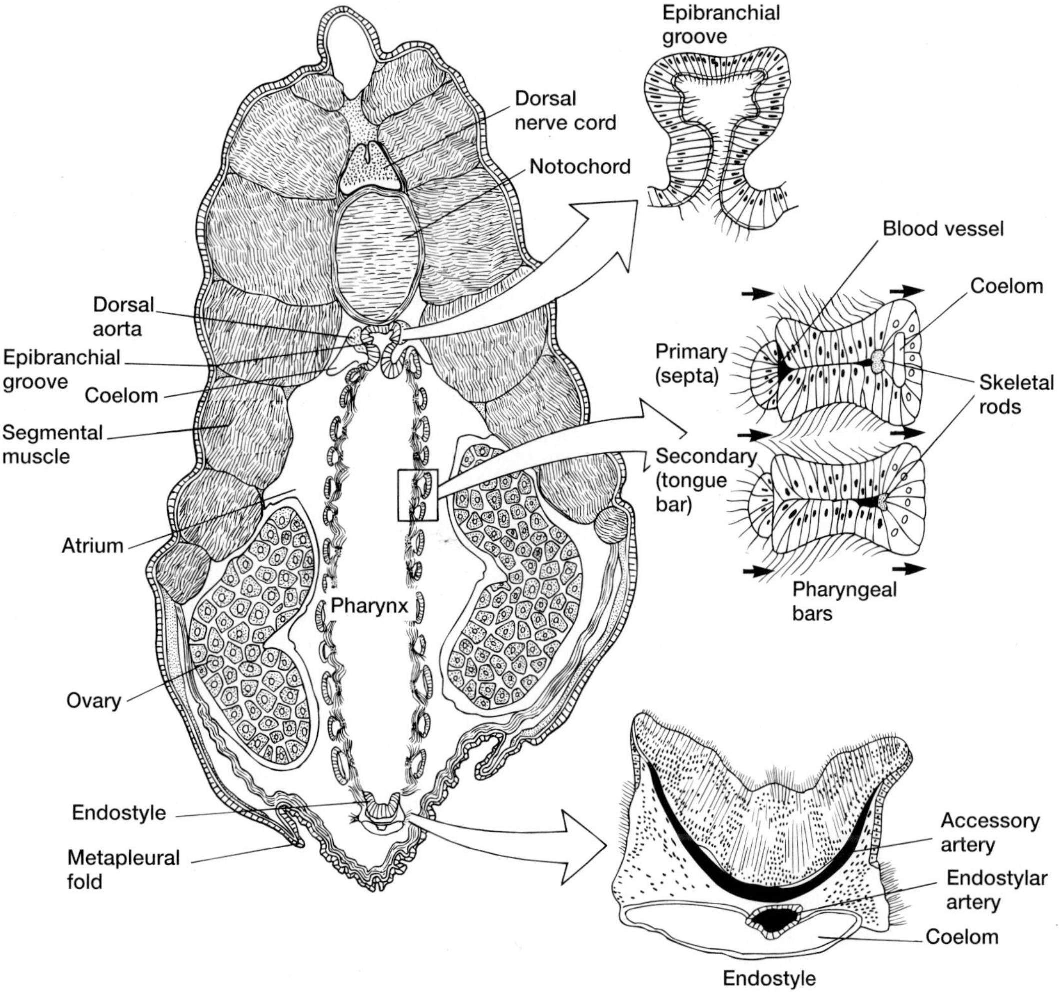


**Figure 2.23** Cross section of amphioxus. The slanted pharyngeal bars encircle the pharynx. On the right, individual pharyngeal bars are enlarged. Notice that they are cut transversely at right angles to their long axis. The coelom continues into the primary branchial bars but is absent from the secondary branchial bars that form as down growths to subdivide each pharyngeal slit.

pharyngeal slits to the atrium and finally departs posteriorly via the single **atriopore.**

Parts of the cephalochordate digestive system may be possible precursors of vertebrate organs. For instance, the endostyle of amphioxus collects iodine as does the thyroid gland, the pharyngeal endocrine gland of vertebrates. The midgut cecum, a forward extension of the gut, is thought by some to be a forerunner of the liver (because of its position and blood supply) and of the pancreas (because cells in its walls secrete digestive enzymes). Whatever their phylogenetic fate, these and other parts of amphioxus are structurally distinct from vertebrate organs, a reflection of the specialized demands of suspension feeding.

The blood of amphioxus is a colorless plasma lacking formed elements, such as corpuscles or amoebocytes, and lacking respiratory pigments. Paired **anterior** and **posterior cardinal veins** return blood from the body, joining in paired **common cardinal veins** (ducts of Cuvier). The paired common cardinal veins and the single **hepatic vein** meet ventrally in the swollen **sinus venosus**(figure 2.24). Blood flows anteriorly from the sinus venosus into the **endostylar artery** (ventral aorta). Below each primary pharyngeal bar, the endostylar artery branches into a set of three vessels—visceral, skeletal, coelomic—to supply the primary bar. At their departure from the endostylar artery, some of these vessels form swellings termed bulbilli. The secondary pharyngeal bars are not directly supplied from the endostylar artery. Instead, blood flows from the primary to the secondary bars through small vessels in the cross-connecting synapticules. Within the secondary bars, blood travels dorsally in the visceral and skeletal vessels. Dorsal to the pharyngeal slits, some vessels from the primary and secondary bars anastomose to form saclike renal glomeruli. Dorsal to these glomeruli, all pharyngeal bar vessels join the paired **dorsal aortae.** The anterior end of amphioxus is supplied by forward extensions of the dorsal aortae. Posterior to the

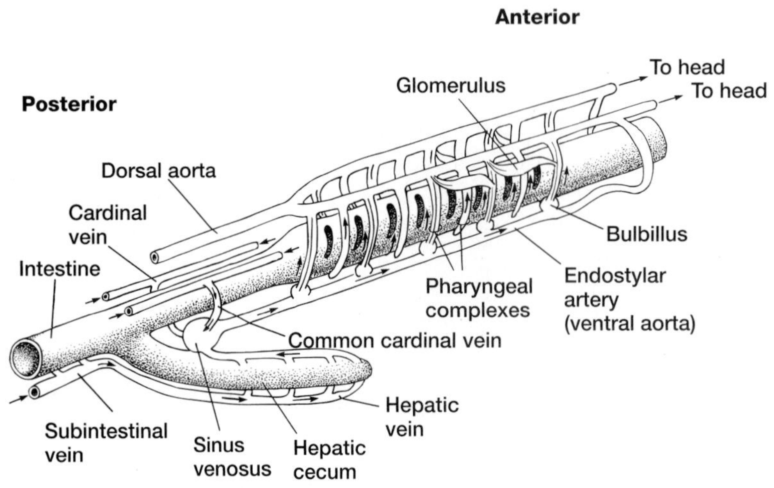


**Figure 2.24** Circulatory system of amphioxus.

pharynx, left and right aortae fuse into an unpaired aorta that supplies the rest of the body (figure 2.24).

Thus, blood circulation in amphioxus is laid out on the same general pattern as that seen in vertebrates. Blood courses forward into the ventral aorta (endostylar artery) and flows posteriorly into the dorsal aorta. Afferent and efferent vessels move blood to and from the cecum, respectively. Blood flow does not reverse in an ebb-and-flow pattern. As in vertebrates, capillary-like vascular networks in the major organs of amphioxus connect afferent and efferent vessels. However, in two important respects, circulation in amphioxus departs from that in vertebrates. First, amphioxus has no heart. A swelling at the confluence of the returning veins is, because of its location in the circulatory system, generally termed a sinus venosus, but it lacks pulsations. Instead, the job of contraction is distributed among other vessels, the hepatic vein, ventral aorta (endostylar artery), bulbulli, and others. These pump blood. Smooth or striated muscles are absent from their walls, but specialized, contractile myoepithelial cells are present. These cells are the presumed source of the pumping forces that move the blood.

In addition to the absence of a heart, the second difference between the amphioxus and the vertebrate circulatory systems occurs in the pharynx. In amphioxus, several parallel vessels travel through each pharyngeal arch from the ventral to the dorsal aorta, rather than the single aortic arch typical of vertebrates. In amphioxus, each primary pharyngeal bar carries at its center a narrow extension of the body cavity, the coelomic space, and each primary bar contains three branches from the ventral aorta—coelomic, skeletal, and visceral vessels. Secondary pharyngeal bars lack a coelomic space and carry only skeletal and visceral vessels. These two vessels within secondary bars form loops connected at their bend, with adjacent vessels in the primary branchial bar. These sets of three (primary) or two (secondary) blood vessels are referred to collectively as a **pharyngeal arch complex.** Although structurally distinct, a pharyngeal arch complex is perhaps analogous to an aortic arch of vertebrates.

Circulatory system (p. 448)

The excretory system of amphioxus was once thought to be its most peculiar feature, a throwback to platyhelminthes, annelids, and molluscs. To some extent this is true because amphioxus lacks a kidney. Instead, its excretory system consists of connected clusters of **solenocytes** (figure 2.25a,b). Each solenocyte is a single cell with cytoplasmic **pedicels,** projections that make contact with the surface of the nearby glomerulus. From the other side of the solenocyte, a long circular stand of microvilli, with a single long flagellum down the center, projects across the coelomic space to enter the **nephridial tubule.** Each nephridial tubule receives a cluster of solenocytes and in turn opens into the atrium. Solenocytes occur commonly among invertebrates; however, the solenocytes of amphioxus, with pedicels that embrace nearby glomerular blood vessels, are very similar to the foot processes of podocytes, cells found in the vertebrate kidney. The exact excretory function solenocytes play in amphioxus is unclear, but their arrangement between blood vessel and atrium suggests a role in eliminating metabolic wastes removed from the blood and flushed away by the stream of water passing through the atrium.

Vertebrate kidney (p. 541)

The cephalochordate larval stage is planktonic, lasting from 75 to over 200 days. The young larval amphioxus is markedly asymmetrical in its head region (figure 2.26). The first pair of coelomic pouches gives rise to two different structures: the left to Hatschek's pit, the right to the lining of the head coelom. Left and right series of pharyngeal slits appear at different times as well. The left series of slits appears first near the ventral midline and proliferates to perhaps as many as 14. The last slits in this series degenerate leaving eight slits on the left side. Some think that the resulting asymmetry of the head might be related to amphioxus spiral body movements during feeding. Next, the remaining slits on the left migrate up the side of the pharynx to a lateral position. At the same time, the right pharyngeal slits make their first appearance, symmetrically positioned with those on the left. More slits are now added on both sides together with the appearance of tongue bars that divide them as they form.

The larva lacks an atrium. During metamorphosis the atrium is added from metapleural folds. These folds appear on either side, grow down over the pharyngeal slits, meet at the ventral midline beneath the pharynx, and fuse to complete the surrounding atrium. Oral cirri, wheel organ, and velum with tentacles are now added to the mouth. During this metamorphosis, the larva sinks out of the plankton to a substrate in which it will take up a burrowing residence as an adult.

Although adults can swim, they usually live buried in coarse sediments with their oral hood protruding. Amphioxus prefers coastal waters and lagoons well aerated by tides but not churned by heavy wave action. Its locomotor system is based on segmental muscles of the body wall, and its hydrostatic notochord serves amphioxus in such habitats. Muscle cells of the segmental body musculature are in contact with the nerve cord via cytoplasmic extensions. Unlike notochords of other chordates, the notochord of amphioxus consists of a series of striated muscle cells arranged transversely (figure 2.27). Fluid-filled spaces separate muscle cells, and both cells and spaces are enclosed in a dense connective tissue sheath. In the larva, a single row of tightly packed, highly vacuolated cells forms the notochord, whereas in the adult, most cellular vacuoles disappear and extracellular fluid-filled spaces emerge between these cells. Notochordal muscles connect to the central nerve cord through cytoplasmic extensions that run dorsally through the connective tissue sheath to the surface of the spinal cord, at which point they meet nerve endings within the cord. The cephalochordate notochord originates from the roof of the gut during development, as it does in most other chordates. But the presence of muscle cells in the cephalochordate notochord is

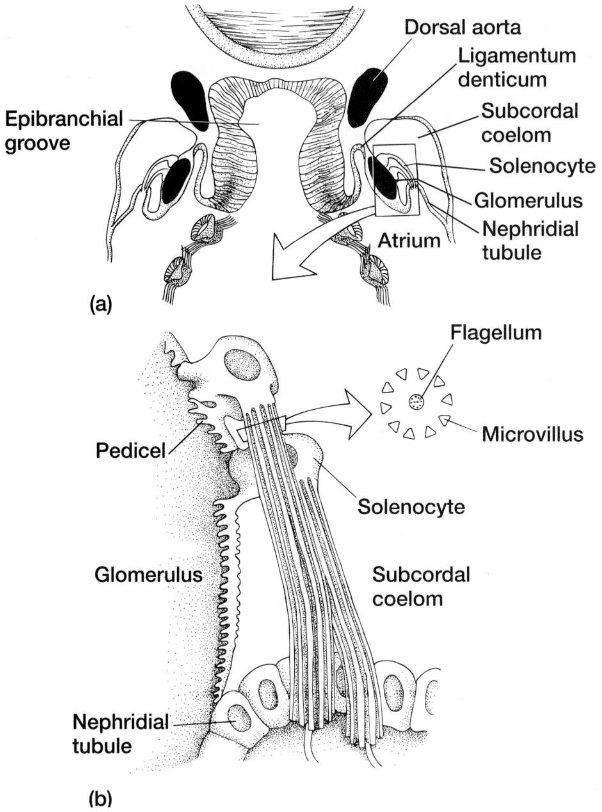


**Figure 2.25** Solenocytes of amphioxus. (a) Dorsal region of the pharynx showing the relationship of solenocytes to the vascular glomerulus on the one end and the atrium on the other. (b) Solenocyte structure enlarged. Solenocytes embrace the walls of the glomerulus through cytoplasmic pedicels and reach the nephridial tubule through microvilli that have a central flagellum.

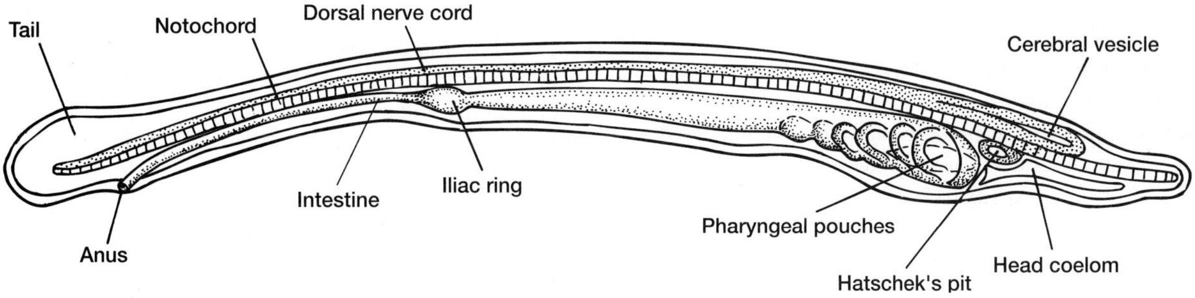


**Figure 2.26** Larval amphioxus. Pharyngeal slits appear only on the left side of the body during this early stage of development, but the basic chordate pattern is evident from the notochord, dorsal nerve cord, and short postanal tail. The atrium does not appear until metamorphosis.

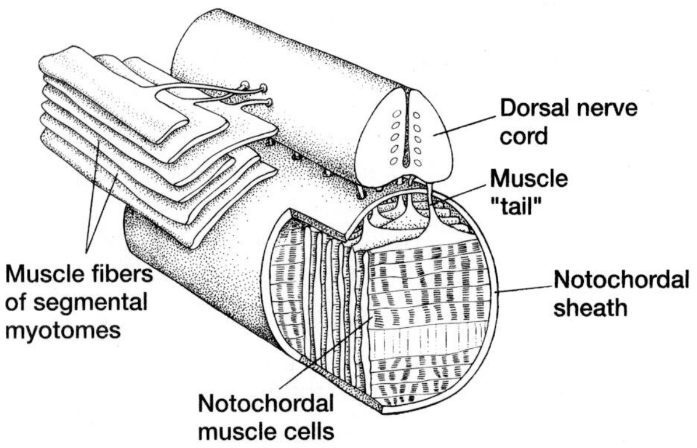


**Figure 2.27** Specialized notochord of amphioxus. Plates of slow contracting muscles are packed within the notochordal sheath. Each plate is a single or sometimes double muscle cell containing contractile fibers arranged transversely. Cytoplasmic extensions of these plates, called "tails," pass upward through holes in the notochordal sheath and synapse with the surface of the dorsal nerve cord. Fluid-filled spaces occur between these muscle cells, although a few vacuoles lie within these specialized cells. Muscle cells comprising the lateral segmental body musculature also send "tails" to the surface of the adjacent nerve cord where they synapse. Presumably the adjacent nerve cord directly stimulates these muscle cells through these synapses.

unique and sets it apart from all other protochordate and vertebrate notochords.

When these muscle cells contract, the tough notochordal sheath prevents ballooning, internal pressure rises, and the notochord stiffens. Stiffening may strengthen burrowing or increase the intrinsic vibration rate of amphioxus to aid it in swimming fast.

The tubular nerve cord of amphioxus does not enlarge anteriorly into a differentiated brain. The segmental body muscles make contact with the spinal cord not by motor nerves reaching out peripherally to the muscles, but by thin processes of the muscles themselves that reach centrally to the surface of the spinal cord.

## Overview of Protochordates

Protochordates possess some (hemichordates) or all (urochordates, cephalochordates) four characteristics that define a chordate—notochord, pharyngeal slits, dorsal hollow nerve tube, postanal tail—although these characteristics may be present at one stage in the life history and not at another. Adults are usually benthic and their larval stage planktonic. Consequently, larva and adult live quite different lifestyles and are structurally quite different in design. Their food consists of suspended particles extracted from a stream of water propelled by cilia. Food particles are collected on sheets of mucus and directed to the gut. Excess water flowing in with food is diverted outside through lateral pharyngeal slits in order to prevent turbulence that might disrupt the carefully gathered mucous cords laden with food. When present, the notochord, along with tail muscles, is usually part of the locomotor apparatus, giving the animal more mobility than afforded by cilia alone.

Protochordates have a phylogenetic history that precedes the vertebrates. They have enjoyed a long and independent evolution of their own dating back perhaps over 600 million years. Their relationships to each other and the sequence of their evolutionary emergence has received attention from biologists for a long time. With this introductory knowledge of protochordates, let us turn to the question of their evolutionary origins.

# Chordate Origins

The fossil record provides little help in charting the phylogenetic route through which the chordate body plan arose from its invertebrate ancestors. Looking for ancestors among the living groups of invertebrates presents problems as well. Most living invertebrate groups diverged from each other millions of years ago and have since gone their own evolutionary ways. It is debatable whether we can ever expect to find anatomical evidence within these living groups that will allow us to derive one from another. Whichever invertebrate group we favor, it cannot be converted in its modern form into a chordate without drastic reorganization. Although living groups are consulted for possible clues they retain of ancestral relationships, most biologists realize that the actual ancestors to chordates are now extinct.

Faced with these intrinsic obstacles and with little evidence from the fossil record to help, it is hardly surprising that disagreement over the origin of chordates is widespread. At one time or another, almost every group of invertebrates has been cited as the evolutionary source of chordates. Although preposterous, even protozoa have been suggested as more or less direct ancestors of chordates! Less extreme but also tenuous are chordate origins among the nemertines, or ribbon worms, with the evertable proboscis of these worms giving rise to the chordate notochord, the pharynx to the branchial region, and so forth.

The more plausible attempts to find the group ancestral to chordates have looked to similarities in anatomy and in embryology. One such view traces chordate origins back to annelids and arthropods.

## Cephalochordates from Annelids and Arthropods

Separate theories put forth in the nineteenth century held that annelids or arthropods might be chordate ancestors. In the early twentieth century, biologist W. Gaskill and shortly thereafter William Patten presented the closely argued case in support of annelids and/or arthropods.

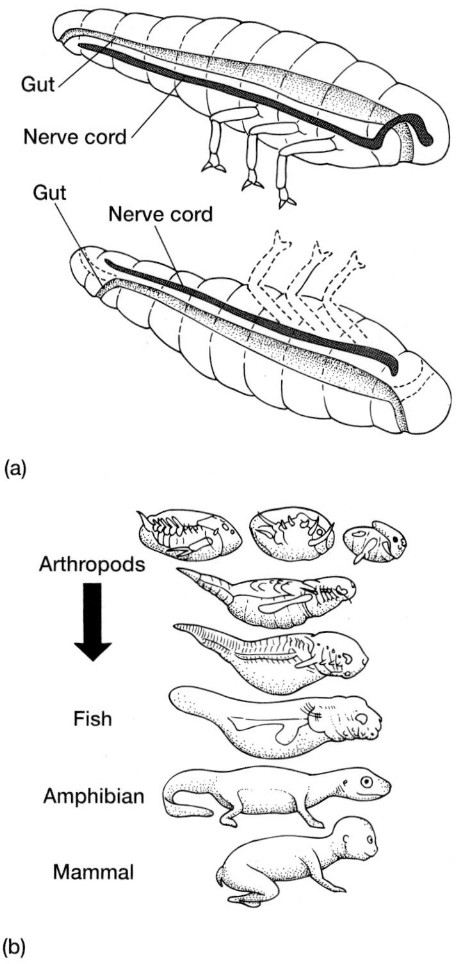


**Figure 2.28** Proposed evolution from annelid/arthropod to chordate. (a) If details are ignored, the basic annelid/arthropod body turned on its back produces the basic chordate body with the nerve cord now dorsally positioned above, rather than below, the gut. (b) Elaboration of this upside-down theory begins with the nauplius larva of crustaceans and other arthropods that swim with their legs up and back down. Through transitional forms, these changes give rise to an "inverted" vertebrate.

Source: W. Patten, *The Evolution of the Vertebrates and Their Kin*, 1912. Philadelphia: P. Blakeston's Son & Co.

Their reasoning, taken together, went as follows. Annelids and arthropods share with chordates similarities of basic body design. All three groups are segmented. All exhibit similarities in gross brain regionalization, with forebrain and hindbrain. Finally, the basic chordate body plan is present in annelids and arthropods, although upside down (figure 2.28a,b). In annelids and arthropods, the nerve cord occupies a ventral position below the gut along with a major blood vessel. If an annelid or an arthropod is flipped over, this brings the nerve cord into a dorsal position, along with the major blood vessel, which becomes the dorsal aorta. In this reversed position, the inverted annelid or arthropod body becomes the fundamental chordate body.

This argument has been embellished since by the imaginative work of others, but it suffers from some major weaknesses. For example, many of the supposed linking similarities between chordates and annelids or arthropods result from analogy rather than homology. The segmentation and jointed appendages that are part of an arthropod exoskeleton are quite unlike the chordate myotomal segmentation. The main nerve cord of annelids and arthropods is solid, not hollow as in chordates. Furthermore, the usual positions for a chordate's mouth and anus are ventral, whereas an annelid or an arthropod rolled on its back would turn both mouth and anus up, pointing skyward. Inverting an annelid or arthropod to produce a chordate body plan would require migration of the mouth and anus back ventrally or formation of new ones ventrally. Unfortunately for this theory, the embryology of chordates preserves no hint of such an event.

The embryonic history of chordates also is fundamentally different in method of coelom formation, derivation of mesoderm, and in basic pattern of early cleavage. Even the body axis is different. In protostomes, such as annelids and arthropods, the anterior end forms on the side with the embryonic blastopore. In deuterostomes, such as chordates, anterior is pointed in the opposite direction, away from the blastopore. Collectively, these difficulties with theories of an annelid or arthropod ancestry for chordates encouraged alternative proposals.

## Cephalochordates from Echinoderms

Echinoderms, like chordates, are deuterostomes. It is perhaps just such underlying similarity that inspired W. Garstang, a biologist of the late nineteenth and early twentieth centuries, to put forth an alternative theory outlining the origin of chordates. Garstang reasoned that, because of these embryonic affinities, echinoderms or a group very similar to echinoderms were the likely chordate ancestors.

At first this seemed farfetched. Adult echinoderms, such as starfish, sea urchins, sea cucumbers, and crinoids, offer little to suggest a phylogenetic affinity with chordates. They have tube feet, calcium carbonate plates in their skin, and pentaradial (five-armed) body symmetry. However, Garstang cleverly looked beyond the adult, to the echinoderm larva as the phylogenetic source of the basic chordate plan (figure 2.29).

### *The Evidence*

Since Garstang's time, the idea that chordates arose from larval echinoderms has been expanded. In its present form, the case for such a chordate ancestry might be stated as follows:

Both echinoderms and chordates are deuterostomes that share embryonic similarities of cleavage and mesodermal and coelomic formation. Echinoderm larvae, like chordates generally, are bilaterally symmetrical. Perhaps most striking of all, auricularia larvae of echinoderms are remarkably similar to tornaria larvae of hemichordates. In fact, for a number of years, tornaria larvae were considered to be echinoderm larvae until some were followed through metamorphosis into their adult and decidedly hemichordate form.

The likely relationships begin to emerge between echinoderms and hemichordates, and between hemichordates and chordates. Shared characteristics link hemichordates back to echinoderms. Both exhibit similarities of early embryonic development, and both pass through a similar appearing pelagic larval stage. In turn, other suites of shared characteristics link hemichordates forward to chordates. For example, like chordates, hemichordates possess pharyngeal slits. Further, some see additional linking evidence in the dorsal tubular nerve cord in the collar region of some hemichordates. Thus, shared pharyngeal slits and a dorsal nervous system link hemichordates to chordates, while tornaria larvae link hemichordates, and through them chordates, to echinoderms.

## *Appearance of Chordate Characteristics*

Just as larval similarities seem to link echinoderms and chordates, Garstang proposed that in fact chordate characteristics first debuted in the echinoderm larva (figure 2.29). Its larva has bilateral symmetry and possesses a simple one-way gut. Near the mouth lies an **adoral band** of feeding cilia; across its lateral body surface meanders a long row of cilia, the **circumoral band,** by which the larva is propelled. From such modest structural ingredients in the echinoderm larva, the basic chordate plan emerges, or so argued Garstang. In particular, he envisioned that the larval body elongated, becoming increasingly muscular and forming a tail that could generate lateral undulations as a means of locomotion. Body elongation drew out the circumoral ciliated band and

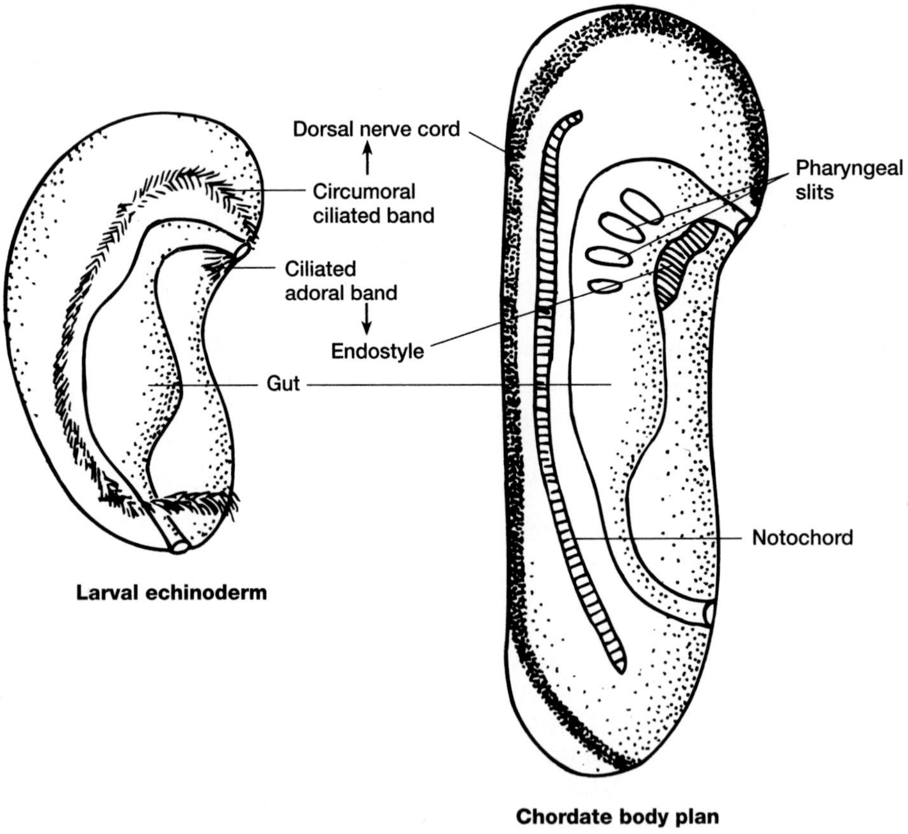


**Figure 2.29** Garstang's theory of the origin of the chordate body plan. The proposed common ancestor of protochordates (left) was bilaterally symmetrical and had the external appearance of a young auricularian (echinoderm) larva. The ancestor's circumoral ciliated bands and their associated underlying nerve tracts moved dorsally to meet and fuse at the midline, forming a dorsal nerve cord in the chordate body plan. The adoral ciliated band gave rise to the endostyle and ciliated tracts within the pharynx of the chordate. Scientists other than Garstang noted that the appearance of pharyngeal slits improved efficiency by providing a one-way flow for the food-bearing stream of water. A notochord is a locomotor advantage in the larger organism.

brought its left and right halves dorsally, where they met at the midline together with the underlying nerve tract, the antecedent of the nerve tube (figure 2.29). Garstang pointed to the rolling up of the neural tube during vertebrate embryology as an embryonic remnant of this phylogenetic event. Lengthening of the adoral band near the mouth and along the gut provided the beginnings of an endostyle. The larva now lacked only pharyngeal slits and a notochord to complete its transformation into a full-fledged chordate.

If evolutionary changes in the larva progressed this far, pharyngeal slits to support suspension feeding and a notochord for muscular locomotion might subsequently have found favor by selection, yielding an echinoderm larva endowed with a full set of chordate characters—notochord, dorsal nerve cord, pharyngeal slits, and postanal tail. However, this series of hypothetical steps leaves unanswered two specific questions. First, what selective advantages might have driven these changes? Second, how might the larva cease being an echinoderm? Remember, these changes presumably arose within the body of an echinoderm larva. Although possessing the chordate anatomy, this larva would still be echinoderm in lifestyle. At the end of its free-living larval stage, it would metamorphose, lose these chordate features, and become a respectable sessile adult echinoderm. How could a distinct and separate chordate group arise from a remodeled echinoderm larva? No certain answers as yet exist to these questions, but several plausible theories have been put forth.

## *Selective Advantage*

Once the full complement of chordate characteristics are in place, they compose a workable body plan. But starting from an echinoderm larva, evolution does not look hopefully ahead to the distant advantages of a chordate lifestyle (see chapter 3). Changes in the echinoderm larva must have been driven by some immediate adaptive advantage at the time of their first appearance. What were these conditions? Suppose, for instance, that the larva of this ancestral echinoderm lasted longer in its planktonic stage, fed, and thereby grew in size. Larger size is an advantage in escaping predation and in becoming established on a substrate once metamorphosis begins. If the larva's size increased for these or other reasons, this change in size alone would require compensating changes in two systems, locomotion and feeding, for the same reasons.

The reason is geometry. As an object gets larger, surface and mass increase unevenly relative to each other. Body mass increases proportionately to the cube of the linear dimensions, but surface area increases only by the square of linear dimensions. In a larva that increased in size, surface cilia propelling the larva would not increase fast enough to keep up with the expanding mass. The locomotor surface would fall behind as the larva got bigger. As a result, there would be relatively fewer surface cilia to move a relatively greater bulk. This, the argument goes, favored the development of an alternative locomotor system. Segmental musculature, elongated body, and stiffened bar (notochord) are the supposed solutions, replacements for the faltering ciliary system.

**Consequences of size on surface and volume ratios (p. 122)**

Similarly, the mode of feeding must change, and for the same reason, namely, a geometric mismatch between surface area bearing cilia and mass requiring nutritional support. The surface around the mouth supports feeding cilia that sweep suspended particles into the mouth. But as larval size increases, body mass outstrips the ability of these surface cilia to meet nutritional needs. An adoral ciliary band expanded into an endostyle would improve food transport. Perforations (slits) within the pharynx would allow one-way flow of a feeding current. Both changes would increase the efficiency of the feeding mechanism. These feeding structures may have been favored by just such selective pressures.

## *Larval Echinoderm to Chordate Tadpole*

However, the problem of larval metamorphosis to adult echinoderm still remains. Sooner or later, the planktonic echinoderm must transform into a benthic adult. But how might this echinoderm larva, now endowed with chordate characters, achieve a separate evolutionary destiny from the adult it is fated to become at metamorphosis?

Garstang's answer to this was clever again. He suggested that the adult stage was eliminated and the larval stage enhanced. The pelagic larva is adapted to a free-living lifestyle, the adult to its benthic lifestyle. If the modified larva enjoyed success, and the adult did not, then time spent in this larval stage might be extended at the expense of time spent as an adult. If gonads ripened early, the larva would become sexually mature and capable of reproduction, an adult function. Attainment of sexual maturity in the larva is paedomorphosis.

Through paedomorphism the echinoderm larva would escape from a life cycle tied to a benthic adult. An echinoderm larva equipped with chordate features might enjoy the adaptive advantages of greater pelagic mobility and, in the process, depart along an independent evolutionary course. Those scientists favoring an echinoderm ancestry for chordates have been quick to invoke paedomorphosis within phylogenetic schemes. Garstang, for instance, suggested that vertebrates might have evolved from echinoderms first through hemichordate-like and then through urochordate-like ancestors via paedomorphosis (figure 2.30).

**Paedomorphosis (p. 188)**

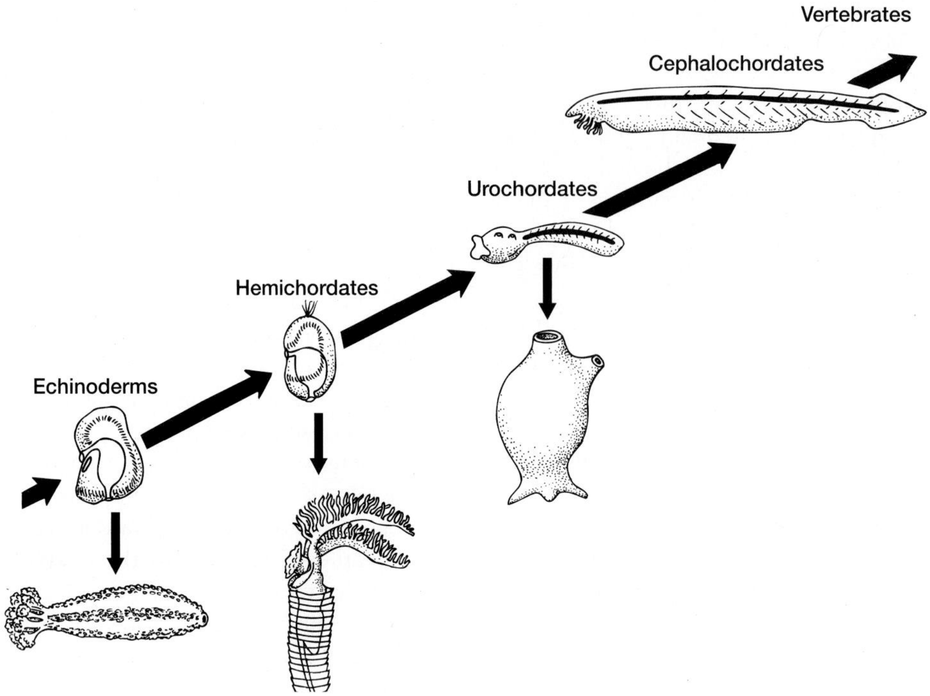


**Figure 2.30** Summary of Garstang's view of vertebrate origins. Beginning with an echinoderm larva, Garstang proposed a series of literal evolutionary steps through the larval stages that involved paedomorphosis and eventually produced chordates.

## *The Critics*

This echinoderm-to-chordate scheme via paedomorphosis has generated some skepticism. Critics have emphasized that existing protochordates should not be taken literally as actual relics of these proposed evolutionary steps but merely as suggestive of the transformations involved. The earliest chordates arose in late Precambrian or early Cambrian times, a vast 590 million or so years ago. Certainly the ancestral chordates are long since extinct. Only distant relatives survive, the living protochordates. In the meantime, protochordates have evolved specialized structures and lifestyles of their own. Larvacea, pterobranchs, and other living protochordates must be a far cry from Precambrian ancestors that produced chordates. Living protochordates perhaps retain clues to early chordate origins, but they cannot be taken as the literal ancestors.

If tunicates are taken as one step in chordate evolution, then prolonged ascidian tadpole life could be seen as part of the evolutionary process. But some have quibbled with this view that an ascidian larva, specialized for relatively short planktonic life and site selection, could experience a reverse selection pressure of prolonged pelagic life and production of a feeding stage.

Structurally, the ascidian larva also seems an unlikely ancestor to later chordates. The so-called intestine that opens into the atrium of an ascidian tadpole is not homologous to the intestine of other chordates such as amphioxus. The ascidian intestine has been lost, and the part of the ascidian gut that opens into the atrium is homologous with the hepatic cecum of amphioxus. Similarly, the ascidian tail musculature is not segmental but a sheet of musculature. The ascidian larva seems to be evolving away from the acquisition of chordate features rather than toward chordates.

Other scientists have more fundamental objections to Garstang's initial theory. Zoologist N. Berrill, for example, takes the dissenting view that the anatomical similarity of larval echinoderms and hemichordates reflects a convergence of larval types to similar pelagic marine environments rather than to phylogenetic closeness. Berrill views the echinoderm and hemichordate larvae that figure in Garstang's hypothesis as highly specialized *convergent* adaptations for prolonging the pelagic larval phase, which in no way represent phylogenetic connections.

## *The Dipleuruloid Theory*

More recently, the comparative morphologist Malcolm Jollie has set forth a theory of chordate origins that addresses some of these objections. He begins, like Garstang, with echinoderms. Jollie finds the similarities in embryonic cleavage, mesoderm formation, and larvae between echinoderms and hemichordates too close to be accidental or the result of ecological convergence. Instead, he argues that such similarities result from retention of characteristics derived from an ancestor common to both. His hypothesized common ancestor is termed a **dipleurula,** envisioned to be a small, ciliated, and bilateral form that was either benthic or planktonic. Although now extinct, its primitive features were incorporated within the larvae of echinoderms and of hemichordates. Pharyngeal pores arose, at least later among derivative hemichordates, as aids in the ciliary and mucus system of feeding.

The other chordate characteristics—notochord, tail, dorsal hollow nerve cord—serve active locomotion, not the planktonic existence of a small filter-feeding form. Thus, Jollie has suggested that a prechordate evolved next with the ability to swim using active undulations of the body. Increase in body size or availability of larger prey may have favored such a change (figure 2.31). Accompanying these changes in active locomotion was muscle segmentation along with an elastic, but anticompressive rod (notochord) to prevent telescoping of the body. The nervous tissue serving these dorsal segmental muscles became consolidated into the adjacent hollow nerve cord.

This prechordate was an incipient predator with a better differentiated head, a simple pharynx with a few slits, and a large mouth. Jollie argued that this active prechordate gave rise to two opposite evolutionary pathways (figure 2.31). One direction was to retreat from this emerging active lifestyle and secondarily move back toward a more highly specialized filter-feeding system. Cephalochordates and urochordates are the surviving results of this reversed trend both taking up benthic and filter-feeding lifestyles after metamorphosis. This would also account for the apparent tendency among ascidians to lose basic chordate features at metamorphosis. The other evolutionary direction was for enhancement of the active lifestyle begun in the prechordate. This trend included increased emphasis on predation rather than on filter feeding and eventually led to the first vertebrates. Complex brain, cranial nerves, and paired nasal, optic, and otic sensory structures concentrated now in a distinct head accompanied these changes and testify to the active nature of even the earliest vertebrates.

## Overview

Although unsettled and controversial in its specifics, the origin of chordates lies certainly somewhere among the invertebrates, a transition occurring in remote Precambrian times. Within the chordates arose the vertebrates, a subphylum of vast diversity that includes some of the most remarkable species of animals ever to grace the land, air, and waters of Earth. Within the early chordates the basic body plan was established, namely, pharyngeal slits, notochord, dorsal hollow nerve cord, and postanal tail. Feeding depended upon the separation of suspended food particles

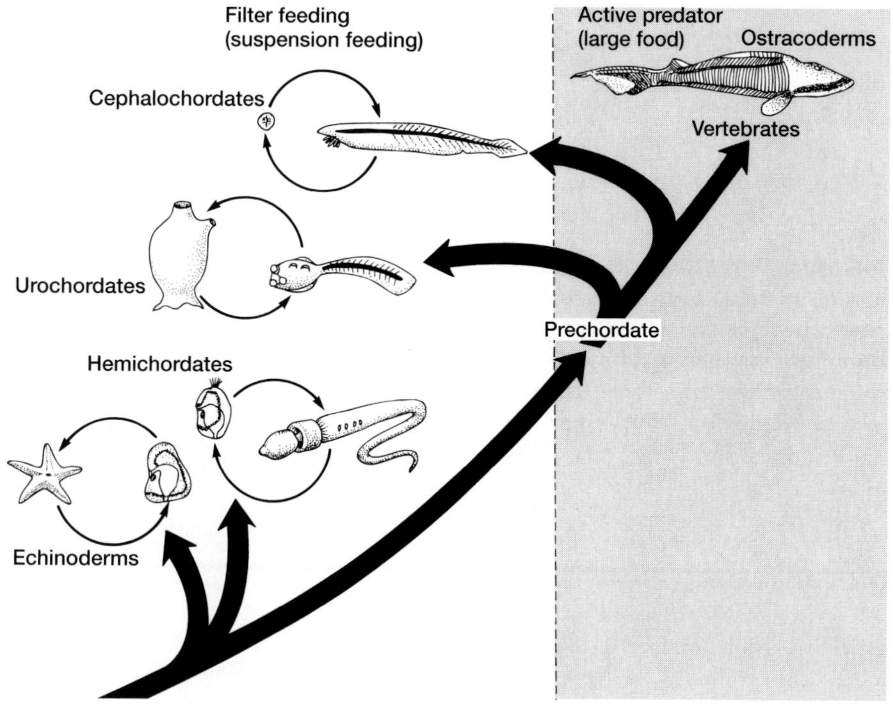


**Figure 2.31** Dipleuruloid theory of chordate evolution. The tendency to abandon a more sedentary lifestyle of filter feeding in favor of a more active predaceous life initially favored development of a prechordate with notochord, muscular tail, and dorsal hollow nerve cord. Continuation of this trend gave rise to vertebrates, but urochordates and cephalochordates reversed the trend and returned secondarily to suspension-feeding habits.

## Box Essay 2.2 In Reverse: Chordates to Echinoderms

Some scientists dissent from the echinoderm-to-chordate hypothesis. In fact, W. Gutmann stands this sequence on its head, proposing that tunicates, hemichordates, and echinoderms are actually advanced deuterostomes, later derivatives of earlier cephalochordates, which in turn derive from still earlier acoelomate worms.

As I read it, Gutmann's evolutionary scenario proceeds through four major stages, each of which envisions the appearance of one or more structural innovations. During the first stage, a wormlike acoelomate animal developed without a fluid-filled body coelom (box figure 1a). In order to maintain its shape and support its body frame, this worm relied upon a viscous, internal, gellike connective tissue, perhaps something like that of jellyfish today. During the second stage, a segmental coelom, a series of fluid-filled spaces enclosed in the body wall, appeared (box figure 1b). The notochord made its debut in stage three and pharyngeal slits in stage four (box figure 1c). At each stage, the innovation enjoyed the adaptive advantages of improved mechanics. For example, the segmental coelom at stage two provided a hydrostatic skeleton. The fluid-filled coelomic spaces reduced body rigidity in comparison with the viscous gel of the acoelomate. Lateral flexions of the body were easier, as was the production of traveling waves of peristalsis in the supple body. The result was to increase locomotory performance in active swimming, via lateral flexions, and in active burrowing, via peristalsis.

The appearance of a notochord in stage three fundamentally changed the mechanical basis of body support and locomotion, and brought with it changes in the arrangement of the nervous system. The notochord, itself a hydrostatic mechanism (gel-filled cells encased in a rigid coat of fibrous connective tissue), was smaller and occupied less space than the extensive coelom it partly replaced. The action of the entire body musculature was no longer required to maintain the shape of the body. Transverse and circular muscles were no longer required to maintain the integrity of the hydrostatic skeleton or produce peristaltic movements. Loss of these muscles eliminated peristaltic movements together with the burrowing lifestyle based upon them. However, the longitudinal muscles adjacent to the notochord enlarged to become the major locomotor muscles. Septa connected these muscles to the notochord on a segmental pattern. A postanal tail lengthened the locomotor apparatus. Greater mechanical protection and decreased distance from peripheral nerves to segmental muscles favored a shift of the central nervous system to a dorsal position, nestled along the notochord.

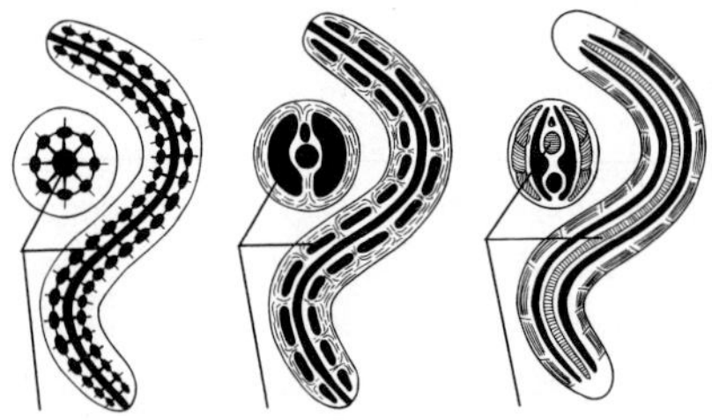


**Box figure 1** Evolution of supporting structures suggested by Gutmann. (a) Acoelomates are predecessors of the deuterostomes. The acoelomate body derived support from a viscous, gellike connective tissue. Although it lacked coelomic cavities, extensive branches of the gut formed a system of canals that distributed food to most parts of the body. (b) As parts of the canal system enlarged, they furnished a series of connected, fluid-filled coelomic chambers between the muscular grid. The fluid-filled metameric coelom that replaced the gel reduced body stiffness; therefore, the body could bend more easily, making efficient burrowing possible as well. (c) During the last major step, the hydrostatic notochord evolved. The notochord is smaller than the coelom, occupies less space, and does not require the action of the whole body musculature to maintain its shape and mechanical integrity. Swimming occurs by means of lateral flexures of the body kept at constant length by the notochordal rod. But, movement by peristalsis is no longer possible because the body cannot be significantly shortened.

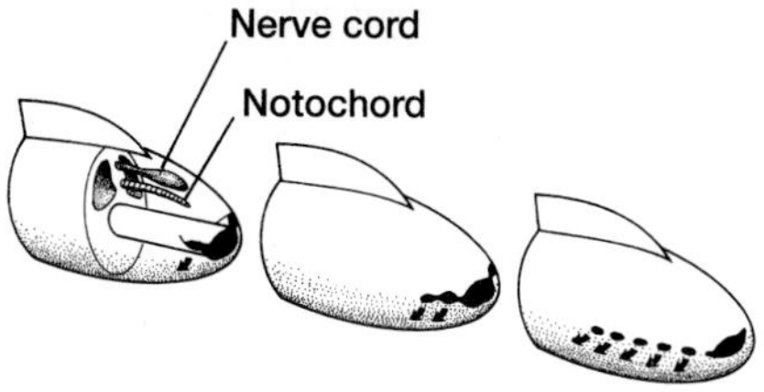


**Box figure 2** Evolution of pharyngeal slits suggested by Gutmann. Pharyngeal slits evolved after the notochord. They could not have arisen before because open slits in the lateral body wall would have disrupted the hydrostatic skeleton by allowing fluid to escape. Slits appear first at the corners of the mouth (left), where they allow filtered water to exit. As they enlarge (middle and right), the body flanks weaken, favoring the evolution of vertical supportive rods between successive pharyngeal slits.

During the fourth stage, the last chordate feature appeared, pharyngeal slits (box figure 2). In fact, Gutmann argues that pharyngeal slits could not have appeared before the notochord was in place, otherwise pharyngeal slits would have punctured the lateral body wall, opened the coelomic spaces, and allowed fluid to escape, destroying the functional integrity of the hydrostatic skeleton upon which earlier, prenotochordal locomotion was based.

According to this argument, all other chordate features had to be in place before pharyngeal slits evolved. The result was a primitive deuterostome, perhaps a cephalochordate much like amphioxus, that could not use peristaltic action to burrow because the notochord resisted this. The resulting deuterostome was designed for an active life based on lateral undulations of a muscular body acting on a notochordal bar. Vertebrates with an endoskeleton are seen as a later derivative of primitive cephalochordates.

If Gutmann's reasoning is substantially correct, echinoderms might have evolved after the first chordates, not before. If pharyngeal slits are necessarily a late innovation, then hemichordates, equipped with them, must be a late-appearing group among the protochordates (box figure 3). As hemichordates reverted to a burrowing existence, structures associated with an active swimming life, such as the stomochord (former notochord?) and nerve cord of the collar (former tubular dorsal nerve cord?), reverted to remnants. In turn, a specialized group of hemichordates, the pterobranchs, is seen as the group that eventually gave rise to echinoderms. Pterobranch tentacles, formed from the collar, gave rise to the echinoderm water vascular system and so on. In this view, tunicates also arose from primitive cephalochordates, the ascidian tadpole testifying to this relationship and the adult tunicate representing a sessile departure.

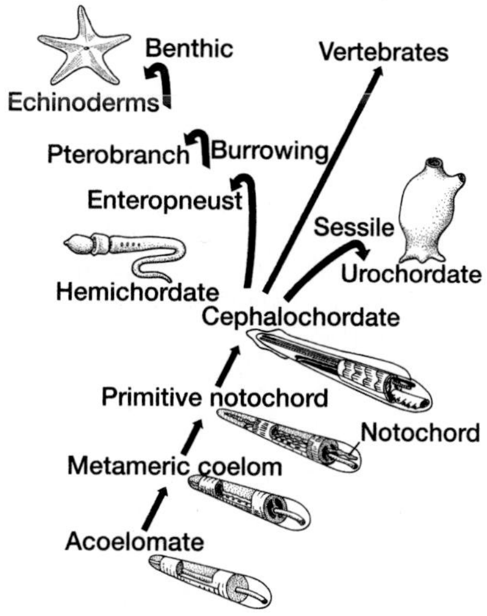


**Box figure 3** Origin of chordates proposed by Gutmann. The earliest chordate looked like amphioxus. Sessile urochordates derived from it as well as burrowing hemichordates and, eventually, benthic echinoderms.

Although few subscribe to Gutmann's view of protochordate evolution, he includes in his evolutionary scenario plausible functional changes that accompany each step. Alternative views that envision, for example, the pharyngeal slits appearing before the notochord ignore the functional disruption this would bring to the hydrostatic skeleton. Thus if Gutmann's contrary view is to be invalidated, alternative evolutionary scenarios must be consistent with both likely anatomical changes and functional changes that accompany them. These changes must also be understood in the context of a realistic ecological setting in which they might have served.

from the water and involved the pharynx, a specialized area of the gut with walls lined by cilia to conduct the flow of food-bearing water. Mucus coated the pharynx walls to snatch suspended particles. Pharyngeal slits allowed a one-way flow of water. Locomotor equipment included a notochord and segmentally arranged muscles extending from the body into a postanal tail.

Feeding and locomotion were activities that favored these novel and specialized structures in early chordates. Subsequent evolutionary modifications would center around feeding and locomotion and continue to characterize the wealth of adaptations found within the later vertebrates.

## Selected References

Alexander, R. Mc.N. 1981. *The chordates*. 2d ed. London: Cambridge University Press.

Alldredge, A. 1976. Appendicularians. *Sci. Amer.* 235 (1):95–102.

Alldredge, A. 1977. House morphology and mechanisms of *feeding in the Oikopleuridae (Tunicata, Appendicularia)*. *J. Zool. (London)* 181:175–88.

Alldredge, A., and L. P. Madin. 1982. Pelagic tunicates: Unique herbivores in the marine plankton. *BioScience* 32(8):655–63.

Andersson, E., and R. Olsson. 1989. The oral papilla of the lancelet larva *(Branchiostoma lanceolatum)* (Cephalochordata). *Acta Zool.* 70:53–56.

Balser, E. J., and E. E. Ruppert. 1990. Structure, ultrastructure, and function of the preoral heart-kidney in *Saccoglossus kowalevskii* (Hemichordata, Enteropneusta) including new data on the stomochord. *Acta Zool.* 71:235–49.

Barrington, E.J.W. 1965. *The biology of the Hemichordata and Protochordata*. San Francisco: W. H. Freeman.

Baskin, D. G., and P. A. Detmers. 1976. Electron microscopic study on the gill bars of amphioxus *(Branchiostoma californiense)* with special reference to neurociliary control. *Cell Tissue Res.* 166:167–78.

Bone, Q., J-C. Braconnot, and K. P. Ryan. 1991. On the pharyngeal feeding filter of the salp *Pegea confoederata* (Tunicata: Thaliacea). *Acta Zool.* 72:55–60.

Bone, Q., and K. P. Ryan. 1979. The Langerhans receptor of *Oikopleura* (Tunicate: Larvacea). *J. Mar. Biol. Assn.* 59:69–75.

Cavey, M. J., and R. A. Cloney. 1972. Fine structure and differentiation of ascidian muscle. I. Differentiated caudal muscle of *Distaplia occidentalis* tadpoles. *J. Morph.* 138:349–73.

Cavey, M. J., and R. A. Cloney. 1974. Fine structure and differentiation of ascidian muscle. II. Morphometrics and differentiation of the caudal muscle cells of *Distaplia occidentalis* tadpoles. *J. Morph.* 144:23–78.

Cloney, R. A. 1969. Cytoplasmic filaments and morphogenesis: The role of the notochord in ascidian metamorphosis. *Z. Zellforsch.* 100:31–53.

Cloney, R. A. 1977. Larval adhesive organs and metamorphosis in ascidians. I. Fine structure of the everting papillae of *Distaplia occidentalis*. *Cell Tissue Res.* 183:423–44.

Cloney, R. A. 1978. Ascidian metamorphosis: Review and analysis. In *Settlement and metamorphosis of marine invertebrate larvae*, edited by F-S Chia and M. Rice. New York: Elsevier, pp. 255–82.

Cloney, R. A. 1979. Larval adhesive organs and metamorphosis in ascidians. II. The mechanism of eversion of the papillae of *Distaplia occidentalis*. *Cell Tissue Res.* 200:453–73.

Cloney, R. A. 1982. Ascidian larvae and the events of metamorphosis. *Amer. Zool.* 22:817–26.

Cloney, R. A., and M. J. Cavey. 1982. Ascidian larval tunic: Extraembryonic structures influence morphogenesis. *Cell Tissue Res.* 222:547–62.

Cloney, R. A., and L. Grimm. 1970. Transcellular emigration of blood cells during ascidian metamorphosis. *Z. Zellforsch.* 107:157–73.

Dawydoff, C. 1948. Conception morphologique du stomocorde. In *Traite de zoologie: Anatomie, systematique*, edited by P. P. Grass. *Biologie* 11:511–57.

Fiala-Medioni, A. 1978. A scanning electron microscope study of the branchial sac of benthic filter-feeding invertebrates (ascidians). *Acta Zool.* 59:1–9.

Flood, P. R. 1978. Filter characteristics of appendicularian food catching nets. *Experientia* 34:173–75.

Flood, P. R. 1981. On the ultrastructure of mucus. *Biomed. Res.* 2(Suppl):49–53.

Fredriksson, G., and R. Olsson. 1991. The subcordal cells of *Oikopleura dioica* and *O. albicans* (Appendicularia, Chordata). *Acta Zool.* 72:251–56.

Garstang, W. 1928. The morphology of Tunicata. *Quart. J. Microsc. Sci.* 72:51–87.

Gilmour, T.H.J. 1979. Feeding in pterobranch hemichordates and the evolution of gill slits. *Can. J. Zool.* 57:1136–42.

Gilmour, T.H.J. 1982. Feeding in tornaria larvae and the development of gill slits in enteropneust hemichordates. *Can. J. Zool.* 60:3010–20.

Goddard, C. K. 1973. Vascular physiology of the ascidian *Pyura praeputialis*. *J. Zool. (London)* 170:271–98.

Gutmann, W. F. 1981. Relationships between invertebrate phyla based on functional-mechanical analysis of the hydrostatic skeleton. *Amer. Zool.* 21:63–81.

Hirakow, R., and N. Kajita. 1990. An electron microscopic study of the development of amphioxus, *Branchiostoma belcheri tsingtauense*: Cleavage. *J. Morph.* 201:331–44.

Holland, N. D., and L. Z. Holland. 1990. Fine structure of the mesothelia and extracellular materials in the coelomic fluid of fin boxes, myocoels and sclerocoels of a lancelet, *Branchiostoma floridae* (Cephalochordata-Acrania). *Acta Zool.* 71:225–34.

Holland, N. D., and L. Z. Holland. 1991. The histochemistry and fine structure of the nutritional reserves in the fin rays of a lancelet, *Branchiostoma lanceolatum* (Cephalochordata-Acrania). *Acta Zool.* 72:203–7.

Jollie, M. 1973. The origin of the chordates. *Acta Zool.* 54:81–100.

Jollie, M. 1982. What are the "Calcichordata"? and the

larger question of the origin of chordates. *Zool. J. Linn. Soc.* 75:167–88.

Katz, M. 1983. Comparative anatomy of the tunicate tadpole, *Ciona intestinalis*. *Biol. Bull.* 164:1–27.

Lester, S. M. 1988. Settlement and metamorphosis of *Rhabdopleura normani* (Hemichordata: Pterobranchia). *Acta Zool.* 69:111–20.

Madin, L. P. 1974. Field observations on the feeding behavior of salps (Tunicata: Thaliacea). *Mar. Biol.* 25:143–47.

Moller, P. C., and C. W. Philpott. 1973. The circulatory system of *Amphioxus (Branchiostoma floridae)*. *J. Morph.* 139:389–406.

Miyamoto, D. M., and R. J. Crowther. 1985. Formation of the notochord in living ascidian embryos. *J. Embr. Exp. Morph.* 86:1–17.

Nakao, T. 1965. The excretory organ of *Amphioxus (Branchiostoma) belcheri*. *J. Ultrastr. Res.* 12:1–12.

Nishida, H. 1986. Cell division pattern during gastrulation of the ascidian, *Halocynthia roretzi*. *Dev. Growth Diff.* 28:191–201.

Nishida, H. 1987. Cell lineage analysis of ascidian embryos by intracellular injection of a tracer enzyme. III. Up to the tissue restricted stage. *Dev. Biol.* 121:526–41.

Nishida, H., and N. Satoh. 1983. Cell lineage analysis in ascidian embryos by intracellular injection of a tracer enzyme. I. Up to the eight-cell stage. *Dev. Biol.* 99:382–94.

Nishida, H., and N. Satoh. 1985. Cell lineage analysis in ascidian embryos by intracellular injection of a tracer enzyme. II. The 16- and 32-cell stages. *Dev. Biol.* 110:440–54.

Olsson, R., K. Holmberg, and Y. Lilliemarck. 1990. Fine structure of the brain and brain nerves of *Oikopleura dioica* (Urochordata, Appendicularia). *Zoomorphology* 110:1–7.

Pardos, F. 1988. Fine structure and function of pharynx cilia in *Glossobalanus minutus* Kowalewsky (Enteropneusta). *Acta Zool.* 69:1–12.

Pardos, F., and J. Benito. 1988. Blood vessels and related structures in the gill bars of *Glossobalanus minutus* (Enteropneusta). *Acta Zool.* 69:87–94.

Patten, W. 1912. *The evolution of the vertebrates and their kin*. Philadelphia: P. Blakiston's Son.

Pechenik, J. A. 1991. *Biology of the invertebrates*. 2d ed. Dubuque, Iowa: Wm. C. Brown.

Rähr, H. 1979. The circulatory system of amphioxus (*Branchiostoma lanceolatum* [Pallas]). *Acta Zool.* 60:1–18.

Ruppert, E. E., and E. J. Blaser. 1986. Nephridia in the larvae of hemichordates and echinoderms. *Biol. Bull.* 171:188–96.

Ruppert, E. E., and P. R. Smith. 1988. The functional organization of filtration nephridia. *Biol. Rev.* 63:231–58.

Sahlin, K., and R. Olsson. 1986. The wheel organ and Hatschek's groove in the lancelet, *Branchiostoma lanceolatum* (Cephalochordata). *Acta Zool.* 67:201–9.

Sugino, Y. M., and M. Nakauchi. 1987. Budding, life-span, regeneration, and colonial regulation in the ascidian, *Symplegma reptans*. *J. Exp. Zool.* 244:117–24.

Torrence, S. A. 1986. Sensory ending of the ascidian static organ (Chordata, Ascidiacea). *Zoomorphology* 106:61–66.

Torrence, S. A., and R. A. Cloney. 1981. Rhythmic contractions of the ampullar epidermis during metamorphosis of the ascidian *Molgula occidentalis*. *Cell Tissue Res.* 216:293–312.

Torrence, S. A., and R. A. Cloney. 1982. Nervous system of ascidian larvae: Caudal primary sensory neurons. *Zoomorphology* 99:103–15.

Torrence, S. A., and R. A. Cloney. 1983. Ascidian larval nervous system: Primary sensory neurons in adhesive papillae. *Zoomorphology* 102:111–23.

3 CHAPTER

# The Vertebrate Story

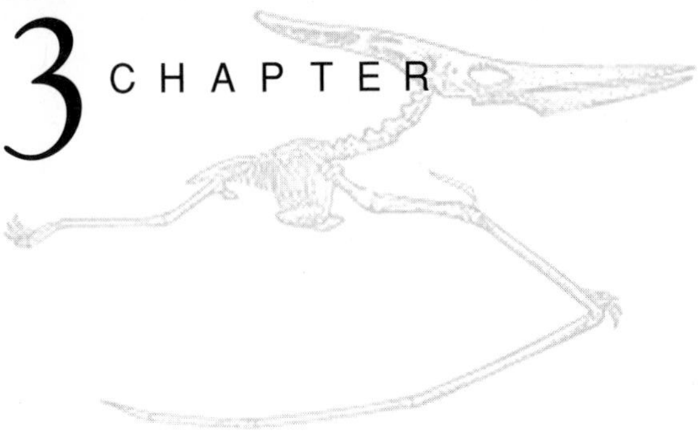



## Introduction

The vertebrate story unfolds over a span of almost 590 million years, an unimaginable depth of time (figure 3.1). During this time, some of the largest and most complex animals ever known evolved among the vertebrates. Vertebrates occupy marine, freshwater, terrestrial, and aerial environments and exhibit a vast array of lifestyles. Like tunicates and amphioxus, vertebrates are proper chordates and possess at some time during their lives all four defining chordate characteristics—notochord, pharyngeal slits, tubular and dorsal nerve tube, and postanal tail. The diversity vertebrates enjoy possibly can be attributed to opportunity. They arose at a time when few large predators existed. Their success may be due to their great variety of innovations as well. Two of these innovations—the vertebral column and the cranium—provide names for this subphylum.

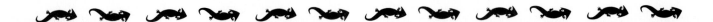

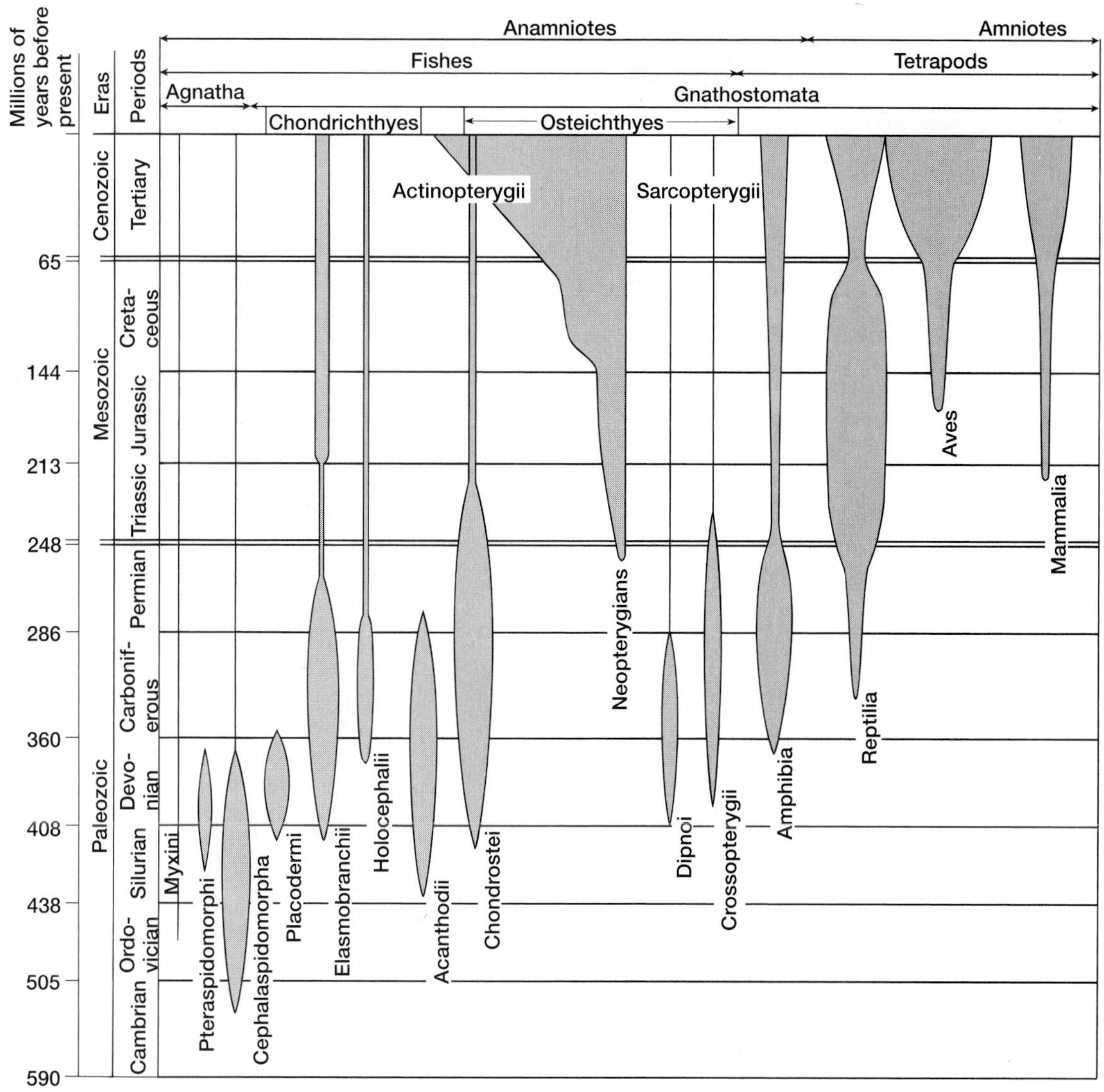


**Figure 3.1** Vertebrate diversity. The vertical scale on the left represents geological time in millions of years before the present. Names of geological eras and periods are listed in conjunction with geological time. Each column of the graph begins with the first known fossil traces of the specific group. The varied widths of each column expresses subjective estimates of the related abundance and diversity of that particular group through time. Among the three agnathan classes, the cephalaspidomorphs are the oldest and include the lampreys living today. The Chondrichthyes are represented by two subclasses, the elasmobranchs and the holocephalans. The Osteichthyes are represented by two subclasses (divided into respective superorders), the actinopterygians (chondrosteans, neopterygians) and the sarcopterygians (dipnoans, crossopterygians). The broad groups of vertebrates, indicated across the top of the graph, include Agnatha and Gnathostomata, fishes and tetrapods, and anamniotes and amniotes, which encompass the classes below them.

## Innovations

### *Vertebral Column*

The **vertebral column** inspired the name *vertebrates* and is composed of **vertebrae,** a series of separate bones or cartilage blocks firmly joined as a backbone that defines the major body axis. Squeezed between successive vertebrae are thin compression pads, the **intervertebral disks.** A typical vertebra consists of a solid cylindrical body, or **centrum,** that often encloses the notochord, a dorsal **neural arch** enclosing the spinal cord, and a ventral **hemal arch** enclosing blood vessels. Extensions of these arches are **neural** and **hemal spines,** respectively (figure 3.2). Ironically, the earliest vertebrates were without well-developed vertebrae and relied upon a strengthened notochord to meet mechanical demands of body support and locomotion. When vertebrae

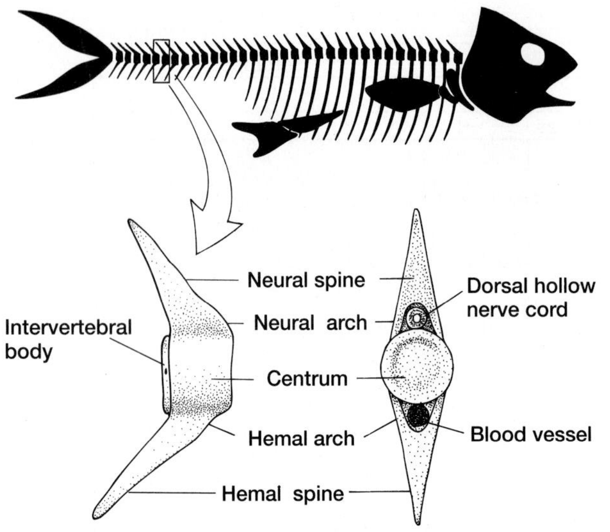


**Figure 3.2** Basic vertebrate. Vertebrae replace the notochord as the predominant means of body support in advanced fishes and tetrapods. A typical vertebra usually consists of a single centrum, with a neural arch and a neural spine dorsally and a hemal arch and a hemal spine ventrally. The notochord becomes enclosed in the centrum. Intervertebral bodies are cartilaginous or fibrous pads that separate vertebrae. In adult mammals, these bodies are called intervertebral disks, which retain gellike cores that are remnants of the embryonic notochord.

first appeared in later fishes, the vertebral elements initially rode upon or surrounded a notochord that continued to serve as the major structural component of the animal's body. In later fishes and terrestrial vertebrates successive vertebrae take over the functions of body support and movement. As the role of the vertebral column enlarged, that of the notochord declined. In adults of most advanced vertebrates, the embryonic notochord disappears, but in mammals, it persists only as a small gellike core called the **nucleus pulposus** within each intervertebral disk.

## *Head*

The other major innovation that evolved in vertebrates is the **cranium.** Eyes, ears, nose, and other sensory organs of the head become more prominent in comparison with those in protochordates. That part of the neural tube with which protochordate sense organs are associated enlarged to form a distinct anterior brain. The cranium is a composite structure of bone or cartilage that supports these sensory organs in the head and encases or partially encases the brain.

Vertebrate evolution has been characterized by a fresh and vast array of cranial structures that collectively form the head. During ontogeny the head develops from **neural crest cells,** a cluster of embryonic cells found only in vertebrates. Because these cells are embryonic, they are transient and seldom spring to mind when we think of vertebrate characteristics. But, these special neural crest cells are the source of most adult structures that distinguish vertebrates from the other chordates.

**Neural crest cells (p. 177); formation of the vertebrate head (p. 226).**

Some vertebrates lack vertebrae but all have a cranium. Therefore, some scientists prefer the term *craniates* in place of vertebrates, subphylum Craniata in place of Vertebrata, to recognize this distinctive feature. However, we use the familiar term of *Vertebrata* for this subphylum, which consists of 11 classes.

## Origin of Vertebrates

The origin and early evolution of vertebrates took place in marine waters; however, at one time, fossil and physiological evidence seemed to point to a freshwater origin. Many early vertebrate fossils were recovered from what appeared to be freshwater or delta deposits. These earliest fish fossils consisted of fragments of bony armor worn smooth, as if upon death the bodies were washed and tumbled down freshwater streams. Eventually they came to rest in the silt and sand that collect in deltas at the mouths of rivers. In the 1930s, the physiologist Homer Smith argued that the vertebrate kidney worked well to rid the body of any osmotic influx of excess water, a problem among freshwater but not among marine animals. However, a more recent survey of early Paleozoic deposits pointed to an extensive assembly of marine vertebrates. It was shown from this find that the vertebrate kidney, while good at maintaining water balance, need not be interpreted as an innovation of freshwater forms. The kidneys of lobsters and squid work in similar ways, yet these invertebrates and their ancestors have always been marine. Today, few scientists insist that the very first vertebrates were products of freshwater environments.

**Kidney physiology and early vertebrate evolution (p. 550)**

Evolution of early vertebrates was characterized by increasingly active lifestyles hypothesized to proceed in three major steps. Step 1 comprised a suspension-feeding *prevertebrate*. The prevertebrate deployed only cilia to produce the food-bearing current. Step 2 comprised an *agnathan*, an early vertebrate lacking jaws but possessing a muscular pump to produce a food-bearing water current. Step 3 comprised a *gnathostome*, a vertebrate with jaws. Food collection was less random. This gnathostome fed on larger food items with a muscularized mouth that rapidly snatched and selected prey from the water. These three steps possibly unfolded as follows.

### *Step 1: Prevertebrate*

Suspension feeding based on ciliary pumps is common to hemichordates, urochordates, and cephalochordates. The first prechordate probably deployed a similar method of suspension feeding (figure 3.3). As we have just dis-

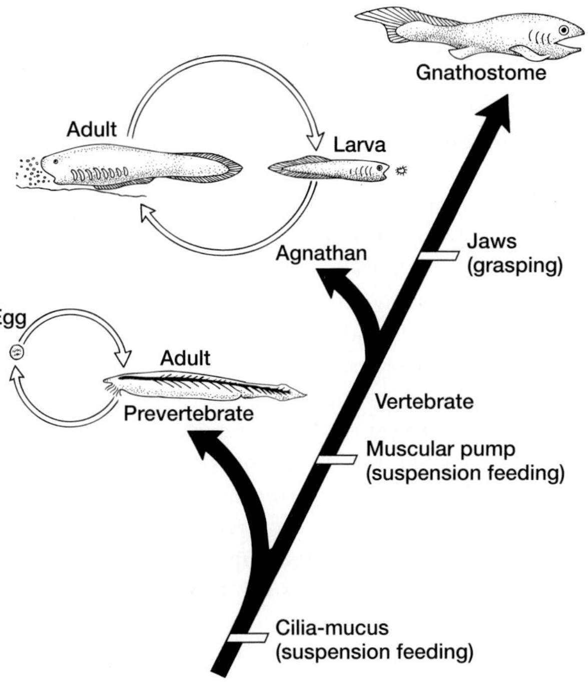


**Figure 3.3** Origin of vertebrates. A more active, predaceous lifestyle characterized vertebrate evolution, leading vertebrates away from suspension feeding that typified their ancestors. Prevertebrates are envisioned as suspension feeders, perhaps something like amphioxus, but they came to depend on a muscularized pharynx to produce feeding currents of water. Following prevertebrates, an agnathan stage developed in which adults might have been benthic feeders but larvae continued the trend toward a more active lifestyle. Selection and capture of specific prey may have next led to gnathostomes. Thus, the early trend in vertebrate evolution was from ciliary to muscular mechanisms of moving feeding currents, and then to jaws that directly snatched prey from water.

cussed, this would have been a marine form, perhaps very similar to amphioxus, but better able to tolerate estuary environments in which rivers enter and mix with the sea. The shift from such a prevertebrate to a vertebrate condition involved two mechanical changes in the pharynx that together produced a muscular pump. First, the pharynx developed an encircling band of muscles. Second, cartilage replaced the collagen of pharyngeal bars. Contraction of the muscle bands constricted the pharynx, squeezing water out the pharyngeal slits. Upon muscle relaxation, the cartilaginous supports sprang back to expand the pharynx, restore its original shape, and draw in new water. Initially, this new muscular pump merely supplemented the existing ciliary pumps in moving water through the pharynx. But in larger animals, surface ciliary pumps became less effective in supplying their greater mass. Increased mass favored prominence of the muscular pump and loss of the ciliary mechanism of moving water. The appearance of an active muscular (and cartilaginous bar) pump now removed the limits to size imposed by a ciliary pump.

**Functional significance of changes in surface-to-volume ratios with increasing size of the organism (p. 121)**

### *Step 2: Agnathan*

Appearance of a muscular pharyngeal pump brought early chordate evolution to the agnathan stage. The ensuing diversification of these jawless fishes was extensive in its own right and exploited an expanded pharyngeal pump. These agnathans were deposit feeders, mud grubbers that pushed their mouths into loose organic or silty mud and drew in sediment rich in organic particles and microorganisms. Although cilia and mucus of the branchial basket still served to collect these passing particles from suspension and transport them to the esophagus, the new muscularized pharynx, not cilia, forced the stream of rich organic material through the mouth. Some ostracoderms possibly created thick suspensions of food by using roughened structures around their mouth to scrape growths of algae from rock surfaces into free suspension that could then be sucked into the mouth by action of the same muscular pharynx.

### *Step 3: Gnathostome*

Prevertebrates, with their ciliary pump, and probably early jawless vertebrates, with their muscular pump, fed on suspended particles. Food-bearing currents carried enough organic material past mucus, where some collided with it and was gathered up. The transition from agnathan to gnathostome involved a switch in feeding method. Transitional species became raptorial feeders that plucked individual food particles selectively from suspension or off surfaces. Some chosen food items would have been wary zooplankton ready to dash off when approached. Other items would have been particles with significant inertia that required forceful effort to be ingested. Raptorial feeding favored a sudden and forceful expansion of the pharyneal pump followed by certain mouth closure to prevent escape of captured food. Elastic recoil of the springy cartilaginous bars allowed early jawless vertebrates to produce a suction drawing food into the mouth; however, this system was too weak to allow forceful capture and ingestion. With the advent of jaws powered by quick muscle action, limits to prey size were also removed. Active predation became a common lifestyle in subsequent vertebrate radiation.

## Vertebrate Classification

Vertebrate pretenders have recently been put forward. Their inclusion within the vertebrate group is still uncertain, and their significance to the vertebrate story remains unclear. For present purposes, they are mentioned only briefly. The **calcichordates,** divided into cornutes and

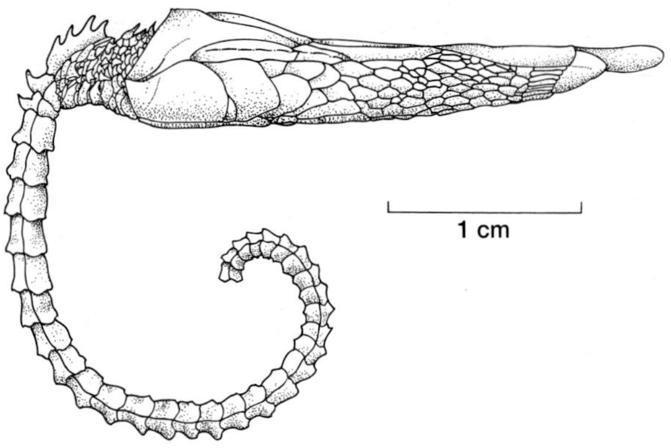


**Figure 3.4** Calcichordate (lateral view). Small imbricated plates of calcium carbonate cover the surface of the animal's body.

mitrates, are a curious group known only from fossils in marine rocks dated 600 to 400 million years ago (figure 3.4). These animals had an echinoderm-like skeleton in which each plate was a single crystal of calcium carbonate, as in starfish, and they resembled flattened crinoids (sea lilies). Proponents see vertebrate-like features as well. Supposed vertebrate- or at least chordate-like characteristics include a row of surface pores (gill slits?), an internal head chamber (branchial basket?), a whiplike stalk (postanal tail?), and an expanded anterior nervous system (brain?).

Arguments over the phylogenetic position of this group arise from disputes over interpretation of the fossils themselves. For example, indentations occur on the sides of some calcichordate fossils, but there is no particular reason to see these as pharyngeal slits. A large chamber is present, presumably at the anterior end, but this may simply be an enlarged stomach rather than a pharyngeal chamber. Although a posterior body projection is present, it lacks evidence of internal notochord, nerve tube, or segmental muscle, and so it does not qualify as a "postanal tail." Possibly it may have been a stalk upon which the body rested. Most likely, calcichordates were a peculiar group of fossil echinoderms, not early vertebrates. The early Cambrian period was a time of experimentation with body plans. Many species died out during this time, leaving no survivors. Whether calcichordates were such an experimental dead end or echinoderm ancestors is not yet clear. Certainly they do not seem to offer any special help in elucidating vertebrate ancestry.

There is even less to say about the second curious group, the **conodonts,** because they are known only from small teeth and a few impressions pressed into marine rock (figure 3.5a–c). All were small, about 4 cm long, thin, and wormlike. Like the bones of vertebrates, their teeth contained traces of calcium phosphate. The body was bilaterally symmetrical, and a few had chevron-shaped blocks of segmented muscle along the body. The tiny rows of teeth resemble those of hagfishes, a primitive group of jawless vertebrates. These teeth seem to establish a place for conodonts close to vertebrates, although the exact relationship is not yet decided. They appeared early (in the Cambrian) and lasted a long time (until the Triassic). One day these carnivores may upset our view about ancestral vertebrates being suspension feeders, but for the moment, they are not known well enough to include them confidently within the vertebrates.

The taxonomy selected for use throughout this book divides vertebrates into 11 classes. These classes can be merged into convenient vertebrate groups with distinctive features (table 3.1). Amphibians, reptiles, birds, and mammals are collectively termed **tetrapods.** Tetrapod literally means four-footed, but the group is understood to include descendants of four-footed ancestors, such as snakes, legless lizards, legless amphibians, flippered marine mammals, and birds, as well as proper **quadrupeds** (four-footed) vertebrates. All other vertebrates are **fishes.** Vertebrates with

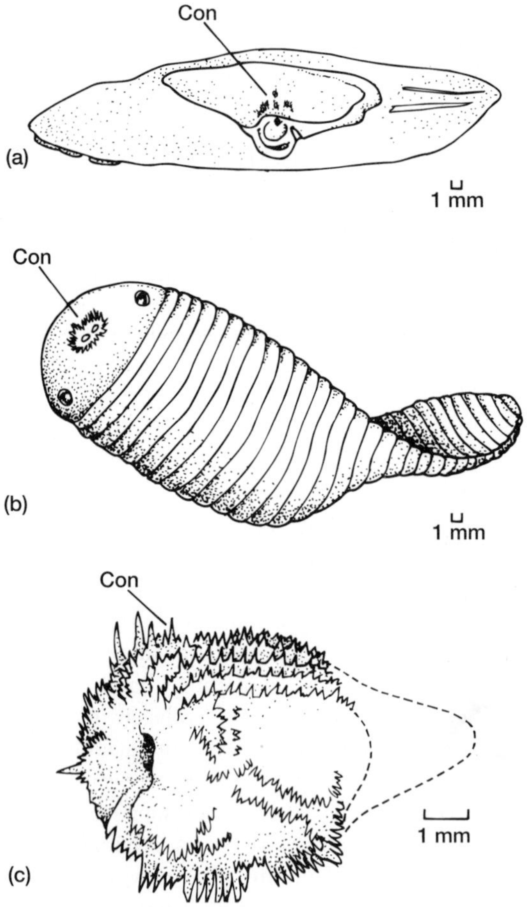


**Figure 3.5** Conodont animals. (a) North American fossil from the Carboniferous period. (b) Cambrian fossil from western Canada. The teethlike points are conodonts (con), from which these animals derive their name. (c) Hypothetical animal with conodonts projecting outward.

## Box Essay 3.1 Ranking the Vertebrates

Except for calcichordates and conodonts, biologists generally agree on which animals belong in the vertebrate subphylum, although taxonomists differ over how to assemble vertebrates into hierarchies that best reflect their relationships. Occasionally, a group ranked as a class in one taxonomic scheme is demoted to a subclass in another. For example, before me as I write are several textbooks that provide a taxonomy of vertebrates. Metatherians, pouched mammals such as kangaroos, are variously ranked as a subclass in some texts and as an infraclass in others. Primitive crossopterygian fishes are a class in one textbook and an order in another. Taxonomy improves certainly as knowledge improves, but differences in personal preference also intervene. Students faced with this variety of taxonomies should recognize that differences do not always reflect the impact of new information but rather express the many ways of shifting and distributing the same animals among available categories.

**Table 3.1 Divisions of the Vertebrates**

| Classes | Groups | | |
|---|---|---|---|
| Myxine | Agnathans | Fishes | Anamniotes |
| Pteraspidomorphi | Agnathans | Fishes | Anamniotes |
| Cephalaspidomorphi | Agnathans | Fishes | Anamniotes |
| Placodermi | Gnathostomes | Fishes | Anamniotes |
| Chondrichthyes | Gnathostomes | Fishes | Anamniotes |
| Acanthodii | Gnathostomes | Fishes | Anamniotes |
| Osteichthyes | Gnathostomes | Fishes | Anamniotes |
| Amphibia | Gnathostomes | Tetrapods | Anamniotes |
| Reptilia | Gnathostomes | Tetrapods | Amniotes |
| Aves | Gnathostomes | Tetrapods | Amniotes |
| Mammalia | Gnathostomes | Tetrapods | Amniotes |

jaws are **gnathostomes** (meaning jaw and mouth); fishes without jaws are **agnathans** (meaning without jaws) that lack such rigid hinged elements supporting the borders of the mouth. Embryos of amniotes possess a delicate, transparent, saclike membrane, the **amnion,** that encases the embryo in a protective water compartment. Vertebrates producing embryos wrapped in such an amnion are **amniotes** (reptiles, birds, and mammals); those without an amnion are **anamniotes** (fishes and amphibians).

**Embryonic amnion (p. 180)**

# Fishes

There are more species of fishes alive today than of all other vertebrates combined. Fishes display a great ecological diversity that is reflected in their diverse morphology (figure 3.6). Considering their numbers and diversity, it is little wonder that our knowledge of most fishes is far from complete and that fish taxonomy constantly undergoes refinement and change. Furthermore, an informal terminology has grown up along with the formal classification schemes. Both are used, so students should become comfortable with each.

## Agnathans

The vertebrate story begins with agnathans. These "jawless" fishes lack a biting apparatus derived from branchial arches. Agnathans first appeared in the late Cambrian and enjoyed their greatest radiation in the Silurian and early Devonian. They were also the first vertebrates in which bone occurs, although it is almost exclusively located in the outer **exoskeleton** that encases the body in bony armor just beneath the epidermis. The **endoskeleton** includes fibrous

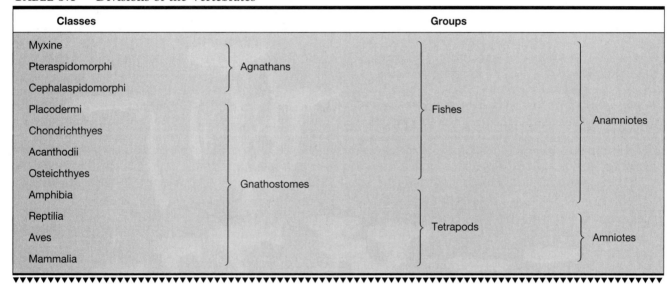

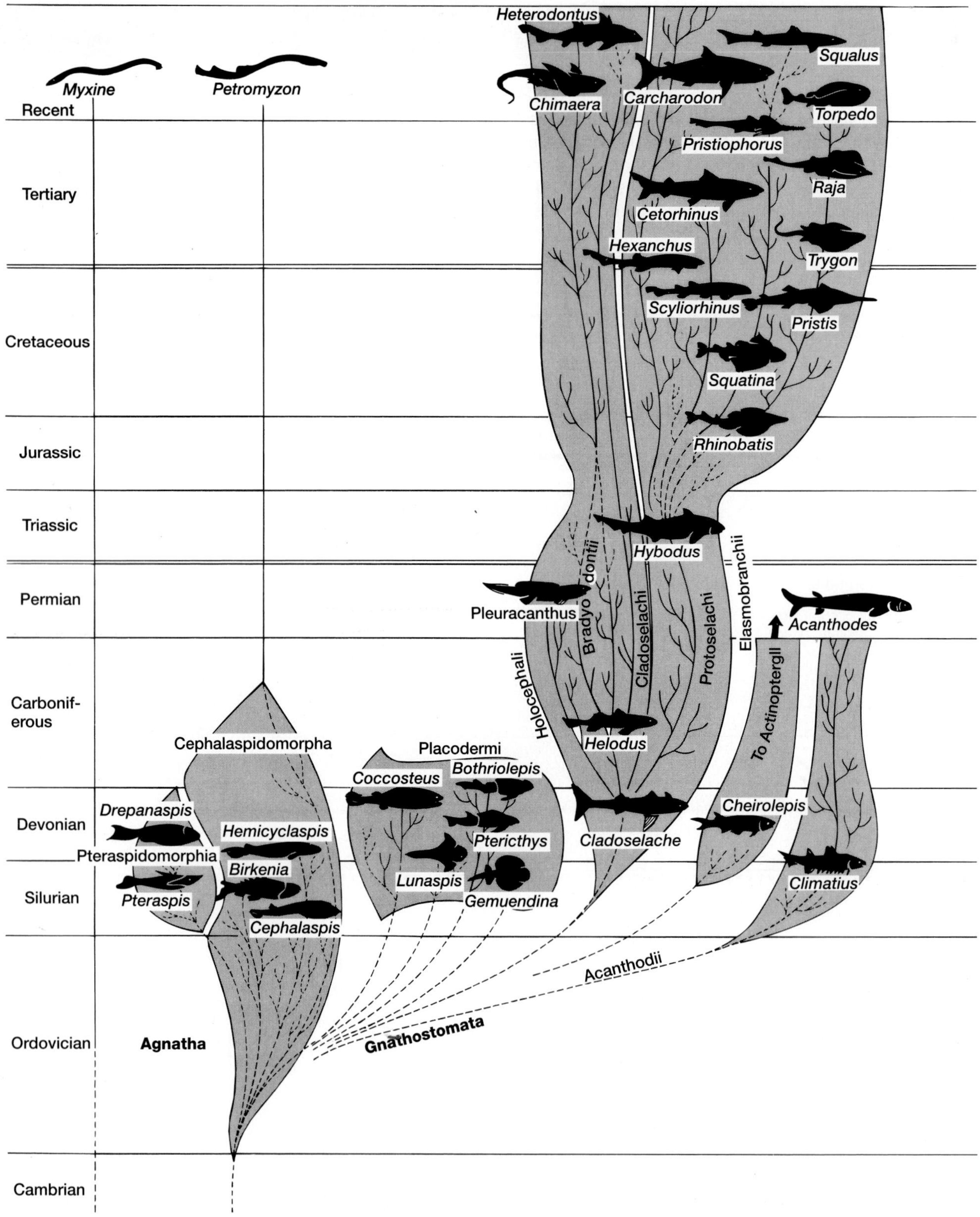


**Figure 3.6** Phylogeny of agnathans and early gnathostomes. Dashed lines represent hypothesized phylogenetic associations where fossil intermediates are unknown.

connective tissue, bone, and cartilaginous elements that lie inside the body. The endoskeleton of the first agnathans was not well developed and, when present, was usually of cartilage.

**Comparison of exoskeleton and endoskeleton (chapter 7)**

The informal terms **ostracoderms** and **cyclostomes** denote extinct and living jawless fishes, respectively. Ostracoderm (meaning bone and skin) applies loosely to the now extinct but earliest agnathans of the Cambrian that were characterized by an extensive bony exoskeleton. Cyclostome (meaning round and mouth) applies to all living agnathans—the lampreys and hagfishes. These fishes lack bone altogether and both have a single median nostril.

Most early ostracoderms were minnow sized, not more than a few centimeters in length. Bony plates of the head were often large and fused into a composite **head shield.** Plates on the trunk were typically smaller, allowing lateral flexibility for swimming. Beneath the superficial bony plates, an endoskeleton of bone is seldom evident, which suggests that a notochord rather than an ossified vertebral column defined the body axis. Spines and lobes projecting from the bodies of many ostracoderms probably offered them some protection from predators and perhaps contributed to their stability as they moved through the water. In some anapsid ostracoderms (e.g., *Phlebolepis*), a lateral pair of ribbonlike fins were present. In osteostracans (e.g., *Hemicyclaspis*), paired muscular fins were evident in the shoulder region. They resemble the paired fins of gnathostomes in position and musculature, but their internal anatomy is poorly known so that homology between ostracoderm and gnathostome paired fins cannot be firmly established.

Small body size, absent or slight fins, small tail, heavy dermal armor, dorsoventral flattening, and of course absence of jaws have led to the view that these early agnathans were bottom dwellers that may have extracted suspended food from organic sediments.

As convenient as the terms *ostracoderm* and *cyclostome* are, neither seems to be a natural evolutionary category. Ostracoderms are not a single taxonomic group, but they comprise at least two classes. Extant (living) lampreys are most likely related to one of these classes, so they should be classified with their extinct cousins. Hagfishes are distinct from all other jawless fishes and may be the most primitive of all agnathans. These relationships are perhaps best recognized formally by dividing agnathans into the following three classes.

## *Class: Myxini*

The **hagfishes,** eellike scavengers that feed on the insides of dead or dying invertebrates and other fishes (figure 3.7a, b), are included in the class Myxini. They use teethlike processes on their muscular tongue to rasp flesh from prey. Slime glands beneath the skin release mucus through surface pores. This mucus, or "slime" as it is called, may serve to slip them from the grip of a predator or clog its gills. In addition, hagfishes can knot their bodies to escape capture or give them force to tear off food (figure 3.7c).

Ovaries and testes occur in the same individual, but only one is functional so hagfishes are not practicing hermaphrodites. The eggs are large and yolky, with up to 30 per individual. No larval stage has ever been found, so development from yolk-filled eggs is thought to be direct, that is, without metamorphosis.

Like lampreys, hagfishes lack bone or surface scales, which was once taken as evidence of close affinity between them. However, lack of bone in lampreys is now thought to be a secondary loss, but in hagfishes this may represent their primitive condition.

**Absence of bone in hagfishes and lampreys (p. 193)**

Hagfishes possess a single median nasal opening rather than paired lateral openings. Water enters this single nasal opening and passes over an unpaired nasal sac on its way to the pharynx and gills. Body fluid of hagfishes is also unique. In other vertebrates, seawater is roughly two-thirds saltier than body fluid. Water moves osmotically with this gradient so that marine vertebrates must regulate their salt and water levels constantly to stay in balance with the surrounding environment. Because salt concentrations in hagfish tissues are similar to surrounding seawater, there is no net flow of water in or out of the body. In having high salt concentrations, hagfishes are physiologically like invertebrates. With physiological similarity to invertebrates and distinctiveness from other vertebrates, hagfishes have been considered as more primitive animals than other vertebrates.

## *Class: Pteraspidomorphi (Diplorhina)*

Although on physiological grounds the myxini are now viewed as the earliest vertebrates, the pteraspidomorphs have the oldest documented fossil record. They appear in the late Cambrian, although they are represented at first only by splinters of primitive bone lacking true bone cells. These bone fragments have been recovered from benthic sediments associated with marine invertebrates. The class extends into the late Devonian, where more complete fossils have been found. Although some species are incomplete, the presence of paired nasal openings seems to characterize pteraspidomorphs.

Most pteraspidomorphs had head shields formed by fusion of several large bony plates (figure 3.8a–c). Behind the head shield, the exoskeleton was composed of small plates and scales. Occasionally, lateral and dorsal spines projected from this shield. These spines may have offered these fishes some protection from attack or anchored their head in mud or sediment to prevent them from being dislodged.

## *Class: Cephalaspidomorpha (Monorhina)*

Body shape of the cephalaspidomorphs is quite varied, suggesting varied lifestyles, and even the living lampreys are included. The single nasal opening merges with a single opening of the hypophysis (endocrine gland) on top of the head into a common keyhole-shaped opening the nasohypophyseal opening. Such a distinctive, dorsally placed nasopharyngeal opening unites all members of this diverse group. The fossil record of this group extends from the early Silurian to the late Devonian.

**Nasal sacs (p. 668)**

Fossil cephalaspidomorphs were heavily armored with bony plates (figure 3.9a–c) that formed a head shield and smaller scales that covered the rest of the body. Bodies were either **fusiform** (spindle shaped) or flattened. In some extinct cephalaspidomorphs, anterior lobes projected from the edges of the head shield and may have conferred some degree of stability during active swimming. Anaspids, a late group of cephalaspidomorphs, showed further reduction of the head shield, increased flexibility of the body armor, and a lobed tail, all of which suggest a trend toward more open-water swimming (figure 3.9c).

Living **lampreys** (figure 3.9d) are now placed among the cephalaspidomorphs. A lamprey uses its oval mouth to grasp a stone and hold its position in a current. In parasitic forms, the mouth clings to live prey so that the rough tongue can rasp away flesh or clear skin, allowing the fish to open blood vessels below and drink of the fluid within. Some species are marine, but all spawn in fresh water. The marine forms often migrate long distances to reach spawning grounds. During spawning, fertilized eggs are deposited in a prepared nest in loose pebbles. An **ammocoete larva** hatches from an egg. Unlike its parents, the ammocoete is a suspension feeder that lies buried in loose sediment with its mouth protruded. Upon metamorphosis, the ammo-

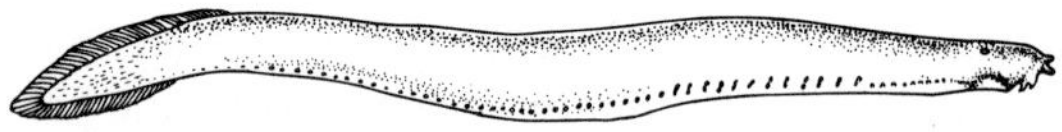

(a) *Bdellostoma*

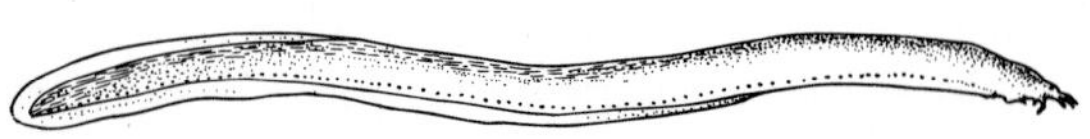

(b) *Myxine*

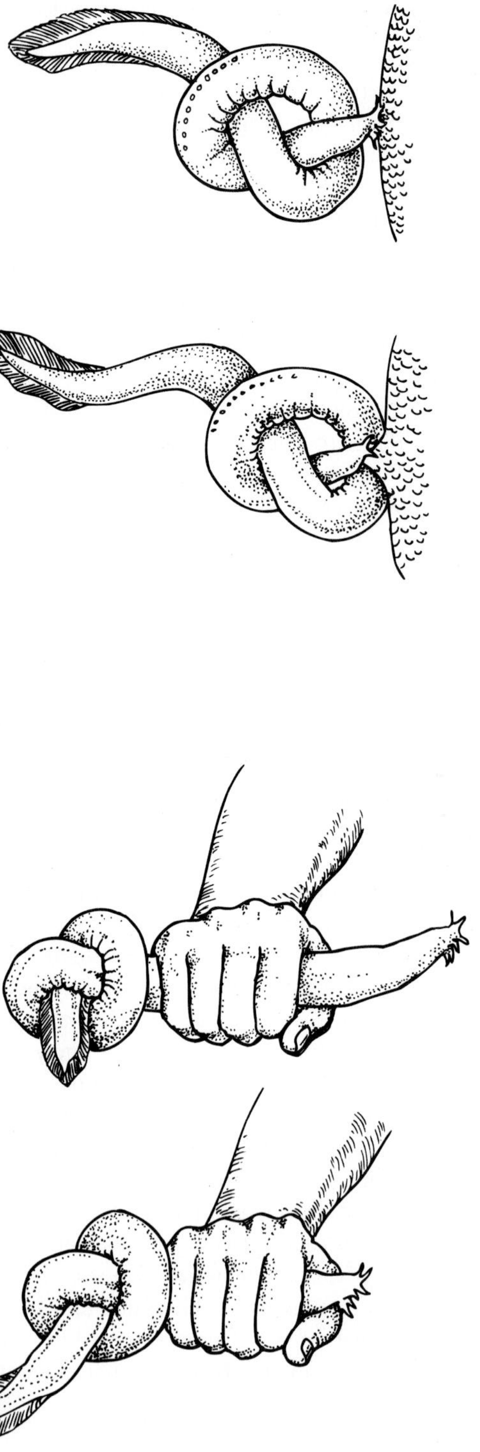

(c) "Knotting" behavior

**Figure 3.7** Myxini, the hagfishes. (a) The slime hag *Bdellostoma*. (b) The hagfish *Myxine*. (c) "Knotting" behavior. Hagfishes are scavengers. When pulling pieces of food off dead prey, they can twist their bodies into a "knot" that slips forward to help tear pieces free. Knotting, together with mucus secreted by skin glands, also helps hagfishes slip free of an unfriendly grip.

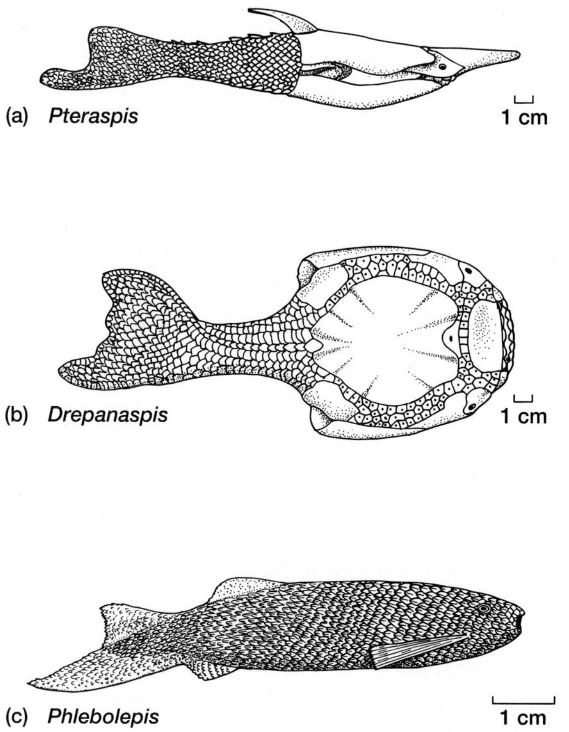


**Figure 3.8** Pteraspidomorphs. All are extinct fishes of the early Paleozoic, with plates of bony armor that developed in the head. (a) The heterostracan *Pteraspis*. (b) The heterostracan *Drepanaspis*. (c) The thelodontid *Phlebolepis*.

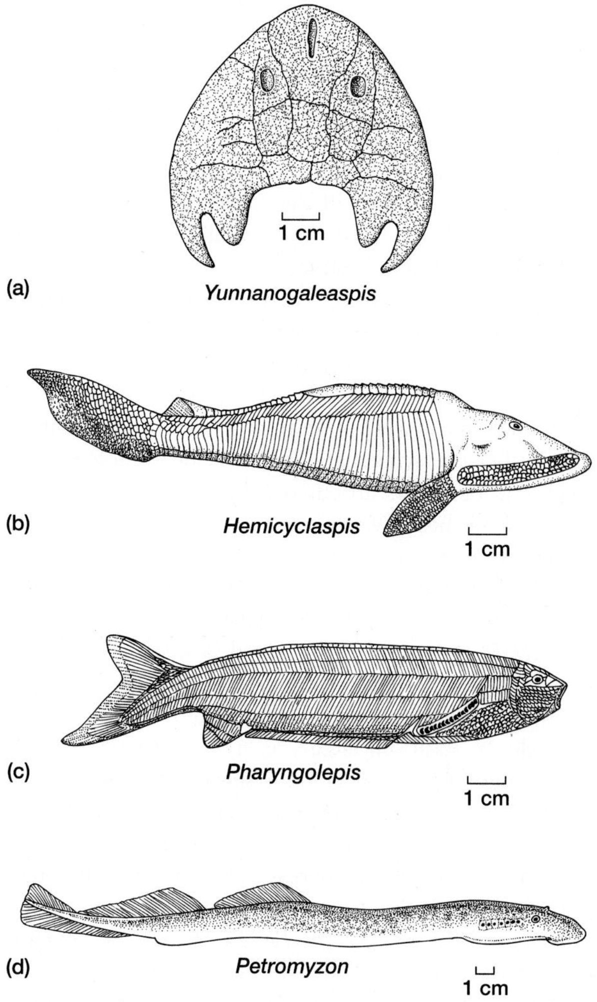


**Figure 3.9** Cephalaspidomorphs. (a) The galeaspid *Yunnanogaleaspis*, for which only the head shield is known. (b) The osteostracean *Hemicyclaspis*. (c) The anaspid *Pharyngolepis*. (d) Living members include the lamprey *Petromyzon*.

coete transforms into a parasitic adult. In some species, the larval stage may last up to seven years, at which time metamorphosis yields a nonfeeding adult that reproduces and soon dies.

Medial fins are present, but paired fins or lobes are absent. Individual blocks of cartilage ride atop the lamprey's prominent notochord. Like their fossil relatives, lampreys are jawless, but they lack a bony exoskeleton that the fossils exhibit. In fact, they lack bone entirely. However, the single medial nasal opening is a distinctive feature in both lampreys and fossil cephalaspidomorphs. Further, the brain and cranial nerves are strikingly similar as well. Nasal opening and central nervous system are the primary basis for uniting lampreys with similar ostracoderms within the cephalaspidomorphs.

## Gnathostomes

One of the most significant changes in early vertebrate evolution was the development of jaws in primitive fishes. These were biting devices derived from anterior pharyngeal arches. Two early groups of jawed fishes are known. The acanthodians appeared first in the early Silurian about 100 million years after the appearance of the first ostracoderms. A second group, the placoderms, is known first from the very late Silurian. Jaws could bite or crush prey, allowing these fishes to process larger food. This adaptation opened up an expanded predatory way of life.

Early gnathostomes also possessed two sets of **paired fins.** One set, **pectoral fins,** was usually anteriorly placed; the other set, **pelvic fins,** was located posteriorly. Both pairs articulated with supportive bony or cartilaginous girdles within the body wall. Supported on girdles and controlled by specialized musculature, paired fins conferred stability and control, allowing a swimming animal to maneuver within and prowl its marine environment actively. Compared to the ostracoderms that preceded them, early gnathostomes probably enjoyed more active lives, venturing into new habitats in search of food, breeding sites, retreats, and unexploited resources.

## Class: Placodermi

Fossil **placoderms** (meaning plate and skin) date from very late in the Silurian, but they flourished in the Devonian. Primitive placoderms were similar to earlier ostracoderms in many ways. Most were still encased in a heavy bony armor, the tail was small, and the head shield was composed of large fused plates of bone (figure 3.10a–g). But unlike ostracoderms, all placoderms had jaws. Paired pectoral and pelvic fins were present. A prominent notochord that supplied longitudinal support to the body was often accompanied by ossified neural and hemal arches. Although true centra were absent, neural and hemal arches were often fused into a sturdy composite bone called the synarcual. This provided a fulcrum with which the braincase articulated and may have facilitated raising of the head. Many species were large, some reaching several meters in length.

Placoderms are usually depicted as bottom-living detritus feeders. Most had a flattened body form. Together with heavy armor and slight paired fins, such a body form suggests a benthic life. Although most placoderms were benthic, some had reduced and lightened body armor along the body. In addition, large size, strong jaws, sleek bodies, and strengthened axial column suggest that some placoderms had an active and predaceous lifestyle.

Placoderms radiated along several lines. Some were adapted to open water, whereas others spread from marine environments in which the class arose to fresh water. Some were specialized bottom dwellers such as the rhenanids, which were skate or raylike forms. The more robust arthrodires and antiarchs enjoyed pelagic lives, cruising in pursuit of food. Some tapered forms, such as the ptyctodontids, resembled chimaeras. Ptyctodontid males usually possessed a set of **claspers,** which were specialized pelvic fins probably associated with the practice of internal fertilization.

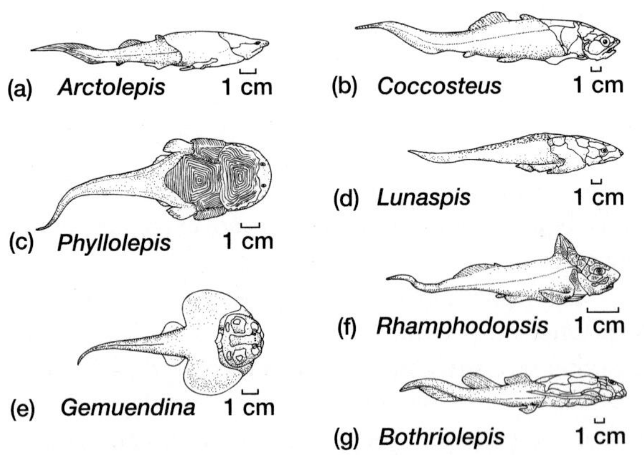


**Figure 3.10** Placoderms. Most placoderms possessed a dermal armor composed of bony plates that were broken up into small scales on the midbody and tail. Many placoderms were large and most were active predators. (a) The arthrodire *Arctolepis*. (b) The arthrodire *Coccosteus*. (c) The phyllolepid *Phyllolepis*. (d) The petalichthyid *Lunaspis*. (e) The rhenanid *Gemuendina*. (f) The ptyctodontid *Rhamphodopsis*. (g) The antiarch *Bothriolepis*.
After Stensiö, 1969.

Whatever their varied lifestyles, all placoderms had jaws, unlike the agnathans before them. They were not limited to a diet of suspended organic particles, but they could exploit larger food or bite big chunks out of unwary victims. Yet, they are difficult to place in a phylogenetic sequence. In fact, their diversity has led some to question whether placoderms even constitute a unified group. They appear at about the time when intermediates between ostracoderms and modern groups might be expected, but placoderms are too specialized to be such direct intermediates. They dominated the Devonian seas but were rather abruptly replaced in the early Carboniferous by the more advanced jawed fishes, the Chondrichthyes (cartilaginous fishes) and Osteichthyes (bony fishes). No living fishes carry extensive plates of external bony armor similar to that of placoderms, so it is difficult even to understand the mechanical or physiological advantages such armored bodies might have enjoyed. Today, most view placoderms as a more or less natural but specialized group that underwent extensive diversification, especially during the Devonian. However, they are without any living descendants and are not even closely related to the cartilaginous or bony fishes that replaced them.

## Class: Chondrichthyes

The modern chondrichthyan consists of two groups, the sharks and rays (elasmobranchs) and the chimaeras (holocephalans) (figure 3.11a,b). Some systematists suggest that each group arose independently, but anatomical evidence argues otherwise. For example, both groups have distinctive placoid scales, cartilaginous skeleton, and pelvic claspers (in males); primitive members show similarities in serial replacement of teeth.

The placoid scales of chondrichthyans are distinctive in that they are usually pointed or cone shaped and show no signs of growth. Initially they form beneath the skin and erupt to the surface. Such scales first appeared in the late Silurian, thus placing Chondrichthyes in this period. However, the first of two major episodes of chondrichthyan radiation began later, in the Devonian, and extended throughout the rest of the Paleozoic. Although one group was common in fresh water, radiation occurred predominantly in marine seas. Most of these chondrichthyans had sleek, fusiform bodies, suggesting that they were active swimmers. The second major episode of radiation began in the Jurassic and extends to the present.

No evidence indicates that chondrichthyans descended from fishes with bony armor, such as ostracoderms or placoderms. As the class name Chondrichthyes (meaning cartilage and fish) suggests, members of this group have skeletons composed predominantly of cartilage impregnated

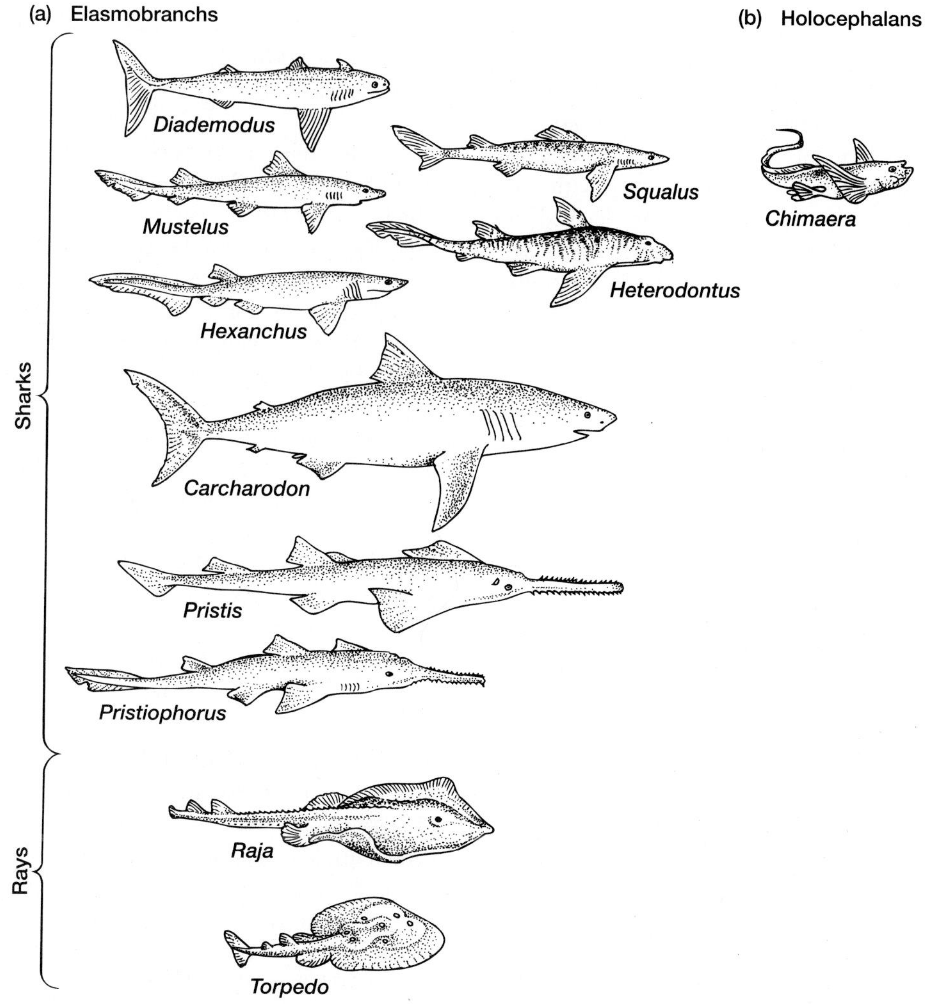


**Figure 3.11** Chondrichthyans. (a) Elasmobranchs, including various sharks and rays. (b) Holocephalans.

with calcium. However, as we have seen, bone was present in the very earliest agnathan vertebrates; therefore, its near absence in later chondrichthyans must represent a secondary loss. Such a view is supported by traces of bone found in the placoid scales and in the teeth. Bone is also found as a thin veneer on the vertebrae of some modern sharks. A fossil shark from the Permian even had a thick layer of bone around its lower jaw.

Like most fishes, chondrichthyans are denser than water, so they tend to sink. For bottom-dwelling rays, this is not a problem, but open-water swimmers must expend extra effort to counteract this tendency. Large livers with buoyant oils, pectoral fins that act like hydrofoils, and heterocercal tails supply lift to help them maintain their depth within the vertical water column.

Unlike most bony fishes, cartilaginous fishes produce relatively small numbers of young. Some females lay eggs usually enclosed in a tough, leathery case; others retain their young in the reproductive tract until they are fully developed. Gestation may last a long time. In dogfish sharks, embryos are carried within the uterus for nearly two years, and young are nourished directly from the yolk. However, for some shark embryos retained in the reproductive tract, yolk is supplemented by a nutrient-rich material secreted from the walls of the uterus. In others, a placenta-like association, complete with an umbilical cord, develops between embryo and mother. The pelvic fin of males is often modified into a clasper used to engage the female and aid internal fertilization.

The braincase of chondrichthyans is usually extensive, but without sutures between elements. In the earliest species, the notochord reigned as the major structural member, although a few cartilaginous neural spines were strung in a series along its dorsal surface. Modern chondrichthyans have a vertebral column composed mostly of cartilage that largely replaces the notochord as the functional support of the body. In many, the first gill slit is reduced to a small rounded opening termed a **spiracle.**

***Subclass: Elasmobranchii*** Among the cartilaginous fishes, *sharks* occupy the spotlight. Modern sharks occur in most oceans of the world. Some species frequent great depths along deep oceanic trenches. The sleeper shark has been photographed with remote cameras at depths of over 1,600 meters. In most sharks, the mouth is armed with serrated, pointed teeth. The functional teeth are backed by rows of replacement teeth, each ready to rotate into position to take the place of a broken or lost functional tooth. This turnover can be rapid. In young sharks, each forward tooth can be replaced weekly.

**Tooth replacement (p. 497)**

*Cladoselache,* a 2-m chondrichthyan was an early shark of the Devonian. As in its modern counterparts, tooth replacement in *Cladoselache* was continuous. But unlike modern sharks, a prominent notochord instead of a vertebral column provided body support. Fins were supported by paired girdles, but these girdle halves were not yet united as a single bar reaching across the midline. The dogfish shark, a basking shark, and the whale shark are examples of modern elasmobranchs. The dogfish shark, a delicacy in restaurants when fresh and a frequent companion of many zoology students in comparative anatomy classes, seldom exceeds 1 m in length. The basking shark and whale shark reach 10 m and 20 m, respectively, making them, after the baleen whales, the largest living vertebrates. Neither shark, however, is a slashing predator. Instead, both strain food from the water. The basking shark feeds by swimming forward with the mouth agape. In this way, it strains up to hundreds of pounds of zooplankton, mainly copepods, daily from the water. During winter months plankton stocks decline in the subpolar and temperate waters. The basking shark is thought to recline on the bottom in deep water during this slow season. Whale sharks feed on plankton all year with gill bars that are modified into great sieves. When feeding, they approach plankton, usually a school of shrimp-like krill, from below and sweep rapidly upward, engulfing both krill and water at once. Excess water exits through their gill slits, and krill are strained out and swallowed.

All *rays* belong to the superorder Batidoidimorpha. Modern rays are bottom-dwelling specialists with a fossil record from the early Jurassic. The pectoral fins are greatly enlarged and fused to the head to give the flattened body an overall disk-shaped appearance. The tail is reduced, and flapping of the pectoral fins provides propulsion. The teeth are designed to crush prey, mostly molluscs, crustaceans, and small fishes discovered buried in sand. On their whip-shaped tail, stingrays carry a jagged-edged sharp spine that they can lash at attackers. Electric rays can even administer severe shocks that are generated by modified blocks of muscle to thwart enemies or stun prey. Manta rays and devil rays, some of the largest members of this group, measure up to 7 m across from fin tip to fin tip. They are often pelagic and gracefully cruise tropical waters in search of plankton that they strain by means of modified gill bars.

**Electric organs (p. 354)**

Rays have a rounded spiracle located dorsally and behind the eyes. The spiracle is the primary means by which some rays get water into the mouth and across the gills. Rays should not be confused with flatfishes (flounders, sole, halibut). Rays have full gill slits ventrally placed and eyes dorsally placed on the body. Flatfishes are bony fish that have gill slits *and* eyes twisted around to the "top" side of the body, with the opposite side flattened against the substrate.

The terms *skate* and *ray* are used loosely and even interchangeably because there is no natural biological difference between them. Generally, skates have a **rostrum,** a pointed noselike extension of the braincase, and produce eggs encased in a leathery case. They are classified in the family Rajidae. Most rays lack a rostrum and give birth to live young, but they are placed in several different families. Taxonomically, however, skates are a type of ray and members of the Batidoidimorpha.

***Subclass: Holocephali—Chimaeras*** Chimaeras (or ratfishes) are modern representatives of the holocephalans. In this group, the body does not end in an enlarged propulsive caudal fin; instead it is long and tapered to a point, inspiring the name *ratfish.* Fossil holocephalans with this shape are known from the Carboniferous, but the relationship of modern forms to these fossils is still undecided.

Chimaeras differ from sharks in many ways. The upper jaws of chimaeras are firmly fused to the braincase. Their gill openings are not exposed to the surface but covered exteriorly by an **operculum.** In adult chimaeras, the small, circular spiracle, derived from the first gill slit, is absent and appears only as a transitory embryonic structure. Their diet includes seaweed and molluscs that the grinding or crushing plates of their teeth can accommodate. Scales are absent. In addition to pelvic claspers, males sport a single median hook, the **cephalic clasper,** on their head, which is thought to clench the female during mating.

Today, there are only about 25 species of chimaeras. They spend most of their time in deep waters over 80 m, and they have no commercial value. Because these factors have discouraged study of this group, chimaeras remain poorly known.

## *Class: Acanthodii*

Acanthodians are represented by spines in the early Silurian; thus, they are the first jawed fishes to appear in the fossil record, even earlier than placoderms. They persisted well into the Permian, long after the placoderms had become extinct. The largest acanthodian was over 2 m, but most were minnow sized (under 20 cm) with streamlined bodies. Early acanthodians were marine but later ones tended to occupy fresh water.

Acanthodii means "spiny forms," a reference to rows of spines along the top and sides of the body. Each fin, except the caudal fin, was defined on its leading edge by a prominent spine that probably supported a thin web of skin (figure 3.12a, b). In some fossils, only these spines are visible. In others, reduced but unmistakable true fin elements were tucked away at the base of the spine. Although an ossified series of neural and hemal arches was present, the prominent notochord extended well into the long dorsal lobe of the tail and served as the major mechanical support for the body. Compared with ostracoderms, the dermal armor was considerably reduced to many small scales across the surface of the body. Dermal armor persisted on the head, but these bony plates were small and formed no composite unit such as a head shield.

Acanthodians have been bounced around within gnathostome taxonomy, a reflection of their still uncertain relationship to other primitive jawed fishes. Their early fossil debut and partial exoskeleton invite their comparison with placoderms. On the other hand, their subterminal mouth below the snout (in contrast with the terminal mouth of bony fishes), caudal fin with projecting dorsal lobe, nonoverlapping scales, and basic jaw structure suggest a relationship with chondrichthyans. In addition, the sleek shape and partially ossified internal skeleton of acanthodians point to a relationship with osteichthyes, the advanced bony fishes.

Their specialized spines indicate that acanthodians represent a distinct offshoot of early gnathostome evolution. They are usually placed in a separate class between cartilaginous and bony fishes. Furthermore, this group is unique in other ways. Acanthodian teeth, unlike those of other fishes, had no enamel surface and were not replaced regularly. Although the body shape was fusiform, acanthodians had large eyes and a short snout, which gave them a shape rather unlike advanced fishes (figure 3.12).

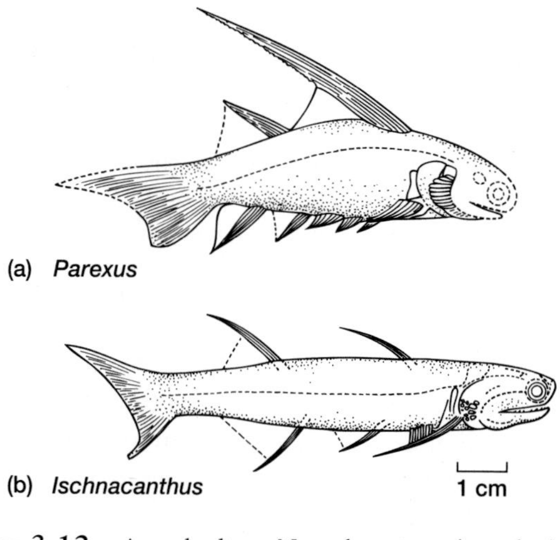


**Figure 3.12** Acanthodians. Note the spines along the body of each that in life supported a web of skin. (a) *Parexus*, Lower Devonian. (b) *Ischnacanthus*, Lower-Middle Devonian.

## *Class: Osteichthyes*

Most living vertebrates are bony fishes, members of the class Osteichthyes. Small overlapping scales from the late Silurian are the first fossil remains known of this group. Osteichthyans are not the only fishes to contain bone in their skeletons, but the taxonomic term *Osteichthyes* (meaning bone and fish) recognizes the pervasive presence of bone, especially throughout the endoskeleton, among members of this class. Even in early bony fishes, the internal skeleton was highly ossified. In most later descendants, ossification persisted or progressed. This trend is reversed only in a few groups, such as sturgeons, paddlefishes, and some later lungfishes, in which endoskeletons are primarily cartilaginous. Whereas cartilaginous fishes address problems of buoyancy with oily livers and hydrofoil fins, most bony fishes possess an adjustable, gas-filled **swim bladder** that provides neutral buoyancy, so they need not struggle to keep from sinking or bobbing to the surface.

**Swim bladder and its distribution within fishes (p. 403)**

No single feature alone distinguishes them from other fishes. Rather, bony fishes have a suite of characteristics including a swim bladder, possibly modified from lungs, and an extensive ossification of the endoskeleton. Dermal bones may cover the body, especially in primitive groups, but these are never large and platelike as in ostracoderms or placoderms. Instead, the body is usually covered by overlapping scales. Fins are often strengthened by **lepidotrichia,** slender rods or "rays" that provide a fanlike internal support.

Bony fish consist of two groups of quite unequal size. The **actinopterygians** compose the vast majority of bony fishes and have been the dominant group of fishes since the mid-Paleozoic (figure 3.13). The other group of bony fishes are the **sarcopterygians**. Although small in numbers, this group is important to the vertebrate story because it gave rise to the tetrapods, all land vertebrates, and their descendants.

***Subclass: Actinopterygii*** Actinopterygians are called "ray-finned" fishes because of their distinctive fins, which are internally supported by numerous slender, endoskeletal lepidotrichia (rays). Muscles that control fin movements are located *within* the body wall, in contrast to the muscles of sarcopterygians that are located *outside* the body wall along the projecting fin.

For many years, actinopterygians were divided into **chondrosteans, holosteans,** and **teleosts,** each intended to

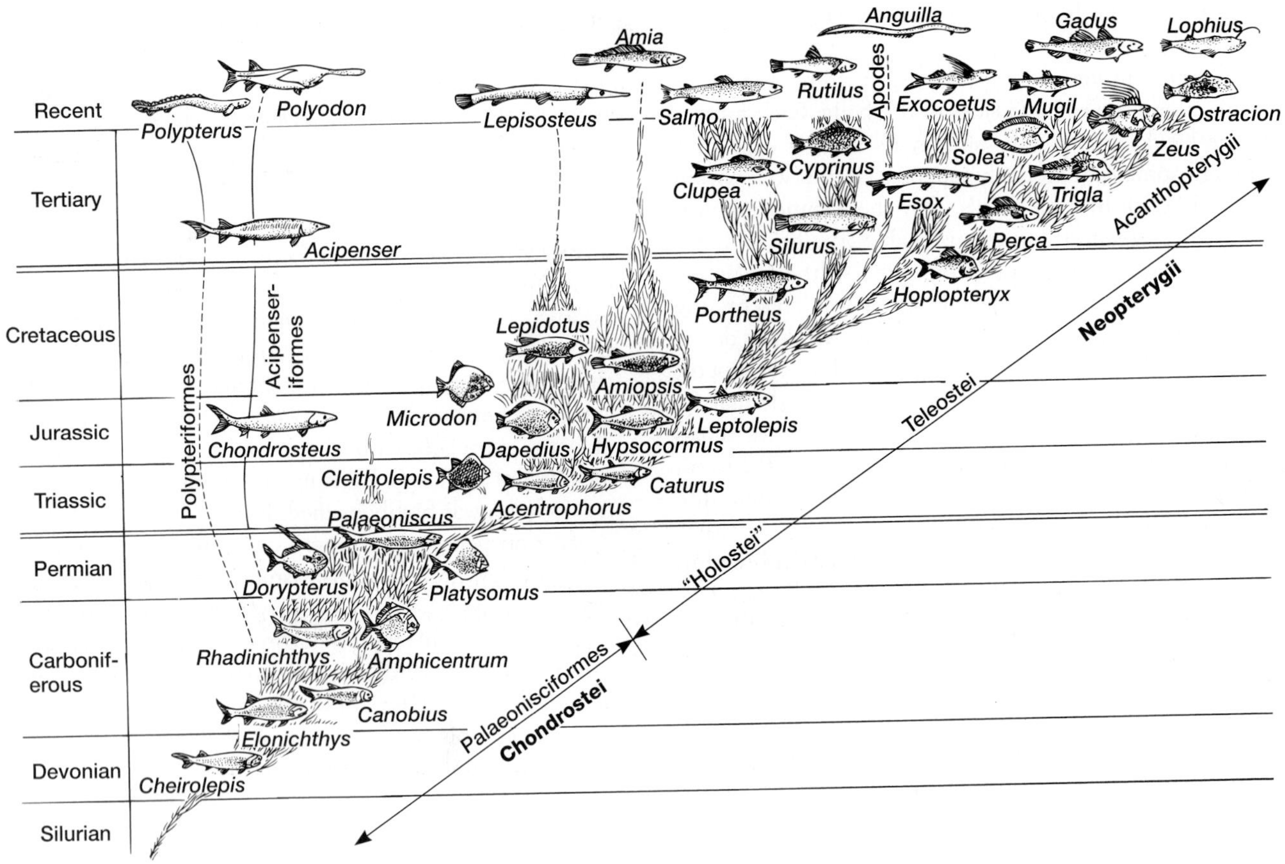


**Figure 3.13** Actinopterygian phylogeny.

represent primitive, intermediate, and advanced groups of ray-finned fishes, respectively (figure 3.14). Although these terms are often recycled in new proposed hierarchies of classification or still used by some to express "grades" of organization, it is evident that the three are unnatural groups as originally envisioned. In our classification scheme, we use two divisions current at the moment, the **Chondrostei** encompassing primitive ray-finned fishes and the **Neopterygii** encompassing advanced ones. These two groups are further divided into lower categories (see appendix A).

*Chondrostei.* Best known of the primitive chondrosteans and probably the earliest bony fishes are the extinct palaeoniscids (Palaeonisciformes). One species reached half a meter in length, but most were smaller. The notochord provided axial support, although ossified neural and hemal arches accompanied it as the notochord reached well into the extended tail. The fusiform palaeoniscid body, suggesting an active life, was covered by small, overlapping rhomboidal scales arranged in parallel rows set closely to one another. The base of each scale consisted of bone, the middle was composed of dentin, and the surface was covered with ganoine, an enamel-like substance that gave them their name of **ganoid scales.** Many find the head of sharks and acanthodians similar to that of palaeoniscids. This may reflect a phylogenetic relationship or an early convergence of a successful feeding style based on quick snatching of prey. Palaeoniscids occupied marine as well as freshwater habitats. They reached their greatest diversity during the late Paleozoic but were replaced in the early Mesozoic by neopterygians.

**Scale types in fishes (p. 202)**

Surviving chondrosteans include acipenserids, paddlefish and sturgeon species placed in the order Acipenseriformes, and bichers placed in the order Polypteriformes. In most acipenserids, the first gill slit is reduced to a spiracle; longitudinal support of the body comes from a prominent notochord. In a departure from palaeoniscids and other primitive bony fishes, acipenserids usually lack ganoid scales except for a few enlarged scales arranged in separate rows. Reversing a trend toward ossification, the skeleton is almost entirely cartilaginous. Paddlefishes occur in fresh waters in North America and China. They are open-water filter feeders of plankton. Sturgeons, the largest species of fresh water fishes, can reach up to 8 m and 1,400 kg. Some migrate between fresh- and marine waters, making treks of over 2,500 kilometers. These toothless bottom feeders eat buried invertebrates, dead fishes, and young fingerlings of other fish species. Some may live to be 100 and do not reach sexual maturity until they are almost 20 years

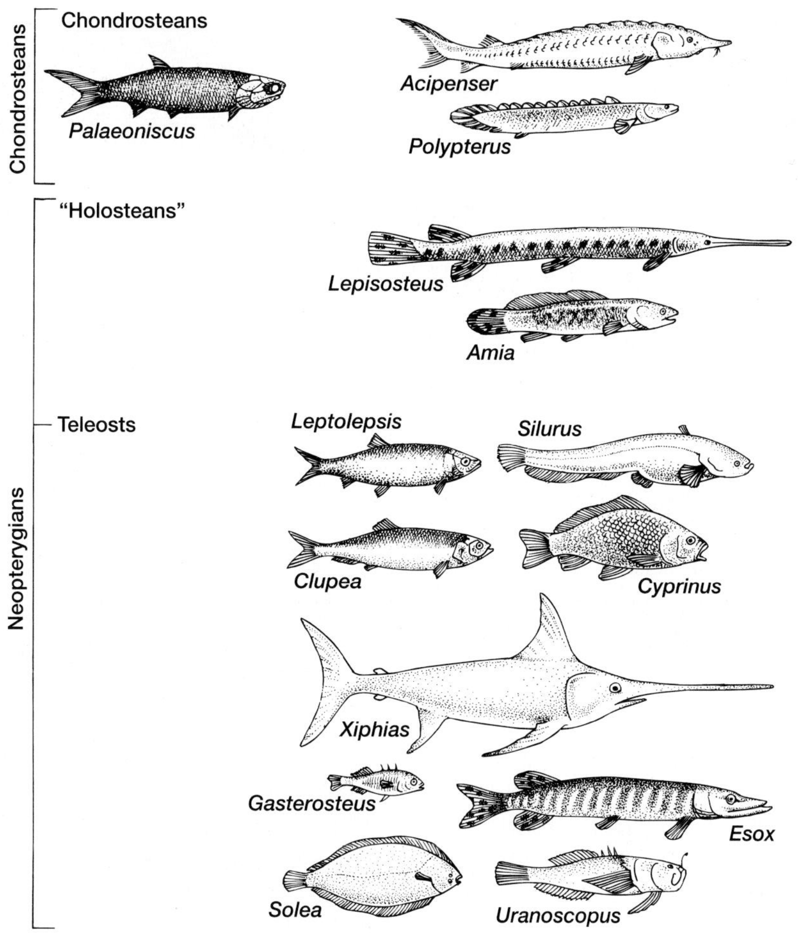


**Figure 3.14** Representative actinopterygians.

old. Their **roe** (eggs) are sold commercially as Russian caviar. Although once considered a nuisance species, sturgeons are a favored food today, especially smoked. Over 50,000 are harvested annually from the Columbia River in North America alone.

Bichers inhabit swamps and streams of Africa and include the living genera *Polypterus* and *Erpetoichthys*. They possess a swim bladder that is more like a paired, ventral lung. Species of *Polypterus* will drown if they cannot occasionally inhale fresh gulps of air to replenish the air in their lungs. Their pectoral fins are "fleshy" as well. Because of their paired lungs and fleshy fins, bichers were formerly classified with lungfishes as sarcopterygians. But today, most see the fleshy pectoral fins as a distinctive feature that evolved independently from the fleshy fins of sarcopterygian. Some taxonomists classify bichers within the bony fish group but elevate them to equal rank with Actinopterygii as the subclass Brachiopterygii (Cladistia). In this text, we keep their classification within the actinopterygians, in the order Polypteriformes, because bichers share rhomboidal ganoid scales, similar patterns of skull bones, and a spiracle with other primitive chondrosteans.

*Neopterygii.* In the early Mesozoic, neopterygians replaced chondrosteans as the most dominant group of fishes and have flourished ever since. They display a great range of morphologies and have adapted to a variety of habitats in all parts of the world. In the course of neopterygian evolution, changes in the skull accommodated increased jaw mobility during feeding and offered attachment sites for associated feeding musculature. Scales became rounder and thinner. The thick, overlapping scales of palaeoniscids afforded protection but restricted flexibility. Reduction of surface scalation probably accompanied development of more active swimming. The notochord was replaced by increasingly ossified vertebrae that also promoted efficient swimming. The asymmetrical heterocercal tail of palaeoniscids was generally replaced by a symmetrical homocercal tail.

Although primitive neopterygians (called "holosteans" by some) had a homocercal tail, internal vestiges remain of a heterocercal ancestry, a spiracle is absent, and scales are reduced. These primitive living neopterygians include gars (Lepisosteiformes), which still retain large, rhomboidal ganoid scales, and bowfins (Amiiformes). Both have more flexible jaws than palaeoniscids, but less flexible than advanced neopterygians.

The most recent group of ray-finned fishes is the advanced neopterygians, or Teleostei (meaning terminal and bony fish). This very diverse group encompasses close to 20,000 living species that enjoy extensive geographic distribution. Representatives occur from pole to pole and at elevations ranging from alpine lakes to deep-ocean trenches. Teleosts have a long history dating back over 225 million years to the late Triassic. Nevertheless, it seems to be a monophyletic group, originating from a common ancestry. Generally, teleosts share a suite of characteristics including homocercal tail, circular scales without ganoine, ossified vertebrae, swim bladder to control buoyancy, and skull with complex jaw mobility allowing for rapid capture and manipulation of food.

Some of the more familiar groups of living teleosts include the clupeomorphs (herrings, eels), salmonids (salmon, trout, whitefishes, pikes, smelts), percomorphs (perches, basses, seahorses, sticklebacks, sculpins, halibut), cyprinids (minnows, carp, suckers, squawfishes), siluroids (catfishes), and atherinomorphs (flying fishes, silversides, grunion).

***Subclass: Sarcopterygii*** Sarcopterygians are the other subclass of bony fishes. Unlike the ray-finned actinopterygians the paired fins of sarcopterygians rest at the ends of short projecting appendages with internal bony elements and soft muscles, hence the name "fleshy-finned fish." Although sarcopterygians never were a diverse group, they are significant because they gave rise to the very first terrestrial vertebrates, the amphibians. Amphibian limbs evolved from sarcopterygian fins; however, these fins do not support the sarcopterygian body, nor do they serve the fish on land. Instead, fleshy fins are aquatic devices that sarcopterygians seem to use for pivoting or maneuvering in shallow waters or for working bottom habitats in deeper waters.

For the first time among fishes, external nostrils open internally to the mouth through holes termed ***choanae,*** which inspired an alternative name for this group, the Choanichthyes. Differences in embryonic development have raised doubts about the homology of choanae among fishes and dampened enthusiasm for this name; thus, we use the term *sarcopterygian*. The two principal groups within the subclass include the **crossopterygians** (lobe-finned fishes) and the **dipnoans** (lungfishes). Sarcopterygians were common in fresh water during most of the Paleozoic, but today the only surviving sarcopterygians are three genera of lungfishes living in tropical streams and the rare coelacanth, *Latimeria*, found in deep oceanic waters.

**Choanae or internal nares (p. 246)**

Other than fleshy fins, primitive sarcopterygians differ from other bony fishes in having scales covered with **cosmine.** These **cosmoid scales,** initially rhomboidal in shape, tend to be reduced to thin circular disks without cosmine in later sarcopterygians. Early dipnoans and crossopterygians had double dorsal fins and heterocercal tails (figure 3.15 a,b). In later species, the dorsal fins usually were reduced, and the caudal fin became symmetrical and **diphycercal,** with the vertebral column extending straight to the end of the tail with equal areas of fin above and below it (figure 3.15c).

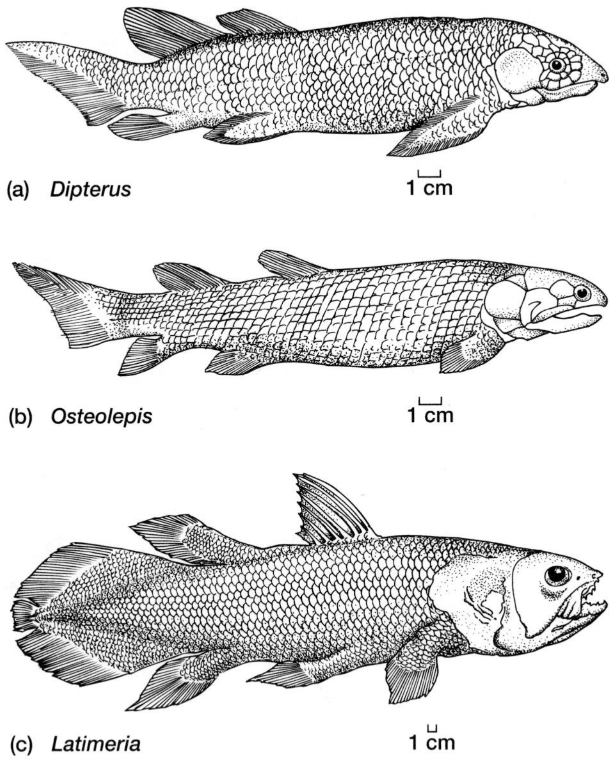


**Figure 3.15** Sarcopterygians. (a) *Dipterus*, fossil lungfish of the Devonian. Note the heterocercal tail. (b) *Osteolepis*, a crossopterygian (rhipidistian) of the Devonian that also had a heterocercal tail. (c) *Latimeria* is a living crossopterygian (coelacanthiformes) exhibiting a diphycercal tail.

**Fish tail types (p. 292)**

*Dipnoi*. The fossil record of lungfishes extends back to the Devonian. *Diabolichthyes* (Lower Devonian), the earliest know lungfish, shared some characteristics with crossopterygians as well, suggesting that it might be a transitional species between crossopterygians and modern lungfishes. Some lungfishes were marine but most occupied fresh water. Three surviving genera occur in continental streams and swamps (figure 3.16a–c). With paired lungs, dipnoans can breathe during periods when oxygen levels in the water fall or when pools of water evaporate during dry seasons. Modern lungfishes lack cosmine, have a skeleton composed mostly of cartilage, and exhibit a prominent notochord.

*Crossopterygii*. Crossopterygians, like lungfishes, date back to the early Devonian. This group includes coelacanths and rhipidistians. Although the notochord is still predominant, it is now accompanied by ossified neural and

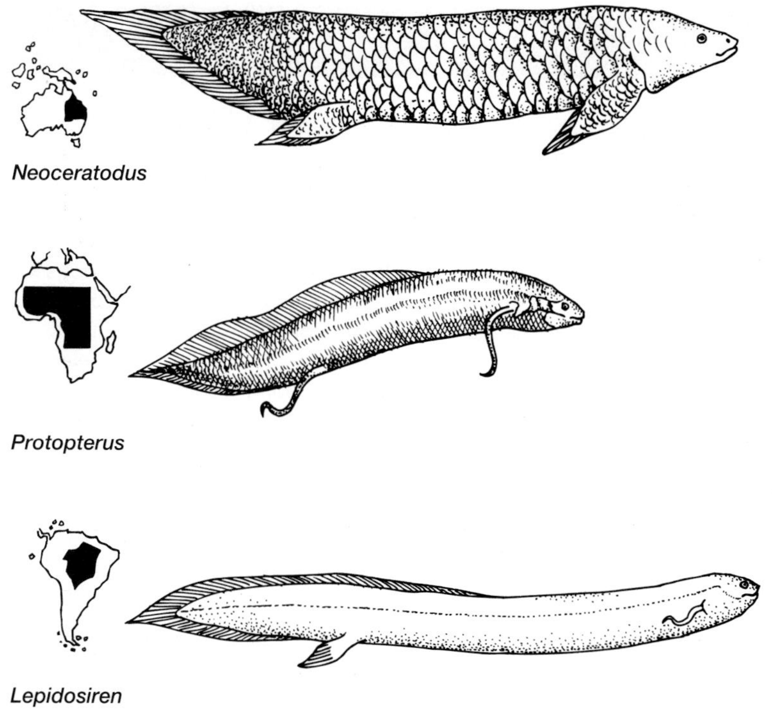


**Figure 3.16** Living lungfishes. (a) Australian lungfish, *Neoceratodus*. (b) African lungfish, *Protopterus*. (c) South American lungfish, *Lepidosiren*.

hemal arches as well as by concentric centra that tend to constrict it and supplement its function. During the late Paleozoic, **rhipidistians** were the dominant freshwater predators among bony fishes. The rhipidistian braincase had a hingelike joint running transversely across its middle so that the front of the braincase swiveled on the back of the braincase. This ability, together with modifications in skull bones and jaw musculature, represents design changes accompanying a specialized feeding style thought to involve a powerful bite. Rhipidistian jaws carried **labyrinthodont teeth** characterized by complex infolding of a tooth wall around a central pulp cavity. Rhipidistians gave rise to amphibians during the Devonian but themselves became extinct in the early Permian.

**Cranial kinesis (p. 245); labyrinthodont teeth(p. 498)**

Coelacanths, the sole surviving crossopterygians, first appeared in the middle Devonian. Throughout this group, the braincase is divided as in rhipidistians, but the vertebral centra are tiny and the notochord especially prominent. Most coelacanths were marine. The only extant genus, *Latimeria*, inhabits deep oceanic shelves. Its lung does not serve in respiration but is filled with fat.

Radically different views of dipnoan relationships that have been put forth are not discussed at length in this chapter. Because the skull of lungfishes is a single undivided unit with other differences in bone patterns and reduced ossification, some taxonomists have suggested that they be placed within the class Chondrichthyes, close to holocephalans. Others keep them within the class Osteichthyes but place them within their own subclass equal in rank with the sarcopterygians. However, differences in the skull and related structures between rhipidistians and lungfishes seem to result from different feeding styles rather than from phylogenetic distance. Lungfishes and early rhipidistians share scales with cosmine, similar skeletons, and a common ancestor, *Youngolepis*, which has intermediate characteristics. For these reasons, we keep rhipidistians and lungfishes together within the subclass Sarcopterygii.

Tetrapods arose from sarcopterygian ancestors, most likely from a group of rhipidistians called the **osteolepiforms.** Erik Jarvik, a Swedish paleontologist, has argued that amphibians had two separate origins among the rhipidistians: salamanders evolved from **porolepiform** rhipidistians and all other amphibians from osteolepiform rhipidistians. Other scientists have been even more radical, suggesting that lungfishes rather than crossopterygians were tetrapod ancestors. However, because of the remarkable anatomical similarities in limbs, vertebral column, skull, and teeth between early amphibians and osteolepiform rhipidistians, these crossopterygians are most likely the ancestors of amphibians.

## Box Essay 3.2 Lungfishes: Dealing with Drought

Many lungfishes live in swamps that dry out on an annual basis. As the water level begins to fall, the lungfish burrows into the still soft mud, forming a bottle-shaped hole into which it curls up (box figure 1a). When the mud dries, mucus secreted by the skin hardens to form the cocoon, a thin lining that resists further water loss within the burrow holding the lungfish (box figure 1b). Usually, the fish's metabolic rate drops as well, thereby curtailing its caloric and oxygen needs. Such a reduced physiological state in response to heat or drought is termed **estivation** (box figure 1c). As long as there is standing water above the burrow, the lungfish occasionally comes to the surface to breathe air through the neck of the burrow. After the surface dries completely, the neck of the burrow remains open to allow direct breathing of air.

Estivation enjoys a long history. Burrows of lungfish from the early Permian and Carboniferous have been discovered. The African lungfish normally estivates for four to six months, the length of the dry summer season, but it can sustain longer periods of estivation if forced to do so. The South American lungfish estivates as well, but it does not form a mucous cocoon nor does it fall into such a deep metabolic torpor. Although the Australian lungfish does not estivate, it can use its lungs to breathe air when oxygen levels drop in the water it frequents.

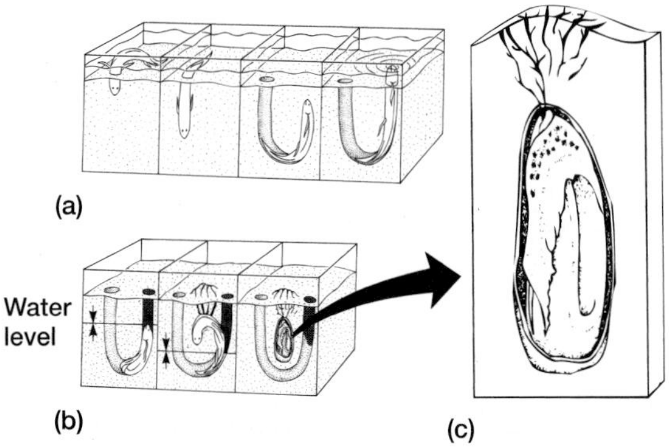


**Box figure 1** African lungfish during estivation within its burrow. Reduced basal metabolism requires only infrequent breathing. The lungfish draws in fresh air through the neck of the burrow that maintains continuity with the environment above. (a) While declining water still covers the swamp, the lungfish burrows into the soft mud, establishes the basic U-shaped burrow, and reaches to the surface to breathe. (b) As the water level drops further, the lungfish moves into a cocoon lined with mucus and maintains contact with the air through breathing holes. (c) In the cocoon, the rolled up lungfish enters an estivative stage during which its metabolic rate drops and its respiratory requirements decrease.

### Overview of Fish Phylogeny

All modern fishes except cyclostomes belong to either the Chondrichthyes or the Osteichthyes. Fishes are diverse in morphology and worldwide in distribution. They outnumber all other vertebrates combined and are one of the most successful groups of animals.

In the earliest fishes, the ostracoderms, bone was already a major part of their external design. In many later groups, there was a tendency for ossification to extend to the internal skeleton, but bone was secondarily reduced or lost in lampreys, chondrichthyans, and some bony fishes, such as chondrosteans and lungfishes.

Fishes are major players in the vertebrate story. Within the fish group, jaws and fins first appeared. Ray-finned fishes have been the dominant aquatic vertebrates since the mid-Paleozoic. Fleshy-finned fishes gave rise to land vertebrates, the tetrapods. In a sense then, the story of tetrapods is a continuation of what began with fishes. Tetrapods inherited paired appendages, jaws, and backbones (vertebrae) from fishes. Some fanciful taxonomic schemes celebrate this close relationship by placing land vertebrates as a subgroup of the sarcopterygians. However, despite the debit of tetrapods to their fish ancestors, the demands of terrestrial life and the new opportunities available led to a rather extensive remodeling of the fish design as tetrapods diversified into terrestrial and eventually aerial modes of life. Tetrapod design is the part of the vertebrate story to which we next turn.

## Tetrapods

### Class: Amphibia

The class Amphibia contains three subclasses, two of which, the **labyrinthodonts** (named for their complex tooth structure) and the **lepospondyls** (named for the structure of their vertebrae) are known only from fossils. Living amphibians—frogs, salamanders, caecilians—are usually lumped together in the third group as **lissamphibians.** Early attempts to classify primitive amphibians depended heavily upon the structure of the vertebrae. More recent fossil discoveries and consideration of additional characteristics have cast doubts on the status of the lepospondyls and even of the labyrinthodonts as natural groups. Similarly, combining all modern amphibians into the common subclass Lissamphibia simplifies their classification, but we cannot safely assume that together they all represent a natural group either. Presently, we maintain the subclasses Labyrinthodontia, Lepospondyli, and Lissamphibia within the class Amphibia, but this scheme is likely to be challenged as amphibian relationships are further clarified.

## *Primitive Amphibians*

***Subclass: Labyrinthodontia*** Ancient amphibians retained bony scales, although these were generally restricted to the abdominal region. Many were surprisingly large. *Eogyrinus*, a Carboniferous species, reached 5 m. Grooves etched in the skulls of some juvenile amphibians carried the **lateral line system,** a strictly aquatic sensory system found in fossils of young amphibians but absent in adults of the same species. At metamorphosis, living terrestrial amphibians also lose the lateral line system of their aquatic larvae. Thus, ancient amphibians, like their modern counterparts, were probably aquatic as juveniles and terrestrial as adults.

The earliest group of labyrinthodont amphibians with a fossil record are the **Ichthyostegalia,** dating from the late Devonian. *Ichthyostega* was a genus described aptly as a "four-footed fish" because of its close relationship to the crossopterygians from which it evolved (figure 3.17a, b). In addition to inheriting the distinctive crossopterygian vertebra, *Ichthyostega,* like its fish ancestors, also possessed radial fin rays supporting a tail fin, a lateral line system, a large notochord, and labyrinthodont teeth. Yet, *Ichthyostega* was clearly an amphibian with a characteristic amphibian pattern of dermal skull bones, limbs with digits, and weight-bearing girdles.

Two independent lineages of labyrinthodonts followed—the Temnospondyli and the Anthracosauria. Anthracosaurs gave rise to reptiles, who are ancestors to both birds and mammals.

Unlike their rhipidistian ancestors, amphibians were adapted for sorties onto land. The limbs replaced fins, the supportive girdles were generally more ossified and stronger, and the vertebral column tended to increase in prominence. Primitive amphibians probably inherited lungs as well as their aquatic mode of reproduction from their rhipidistian ancestors. Fertilization was likely external, with large numbers of small eggs laid in water. Like modern salamanders, fossil larval stages of Paleozoic amphibians exhibit external gills. Utilization of land was almost certainly an adult occupation following metamorphosis from an aquatic larval stage.

***Subclass: Lepospondyli*** The other group of Paleozoic amphibians is the lepospondyls. They appeared quite early in the Carboniferous but were never as abundant as the labyrinthodonts. Most lepospondyls were small. What unites these amphibians is the presence of a solid vertebra in which all three elements are fused on to a single, spool-shaped centrum.

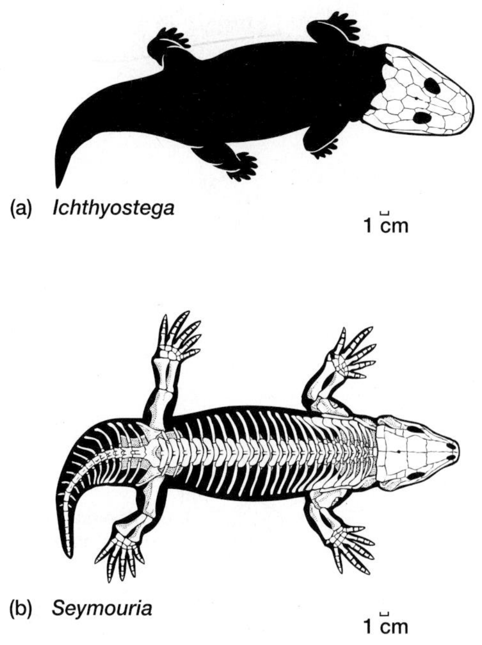


**Figure 3.17** Labyrinthodont amphibians. (a) *Ichthyostega*, from the late Devonian, is a member of the ichthyostegid group and is the oldest known amphibian. The animal was about 1 m long. (b) Skeleton of *Seymouria*, a later anthracosaur from the early Permian. About 50 cm long.

Vertebrae types (p. 298)

The distinctive nectridean lepospondyls were apparently entirely aquatic, reversing a trend in most other early amphibians. Their paired limbs were small and ossification was reduced, but the tail of some species was often quite long. Skulls of "horned" nectrideans of the early Permian were flattened and drawn out into distinctive long winglike processes (figure 3.18a, b).

Microsaurs (meaning small and lizard) were lepospondyl amphibians despite their misleading name. Most were small, around 10 cm, but varied in design. Some seemed to be burrowing forms.

## *Modern Amphibians*

Living amphibians date back over 200 million years to the Jurassic. The three living orders—**frogs, salamanders,** and **caecilians**—include almost 4,000 species displaying a wide range of life histories (figure 3.19a–c). Except for an absence on some oceanic islands, they occur throughout the tropical and temperate regions of the world. Amphibian eggs, which lack shells and extraembryonic membranes, are laid in

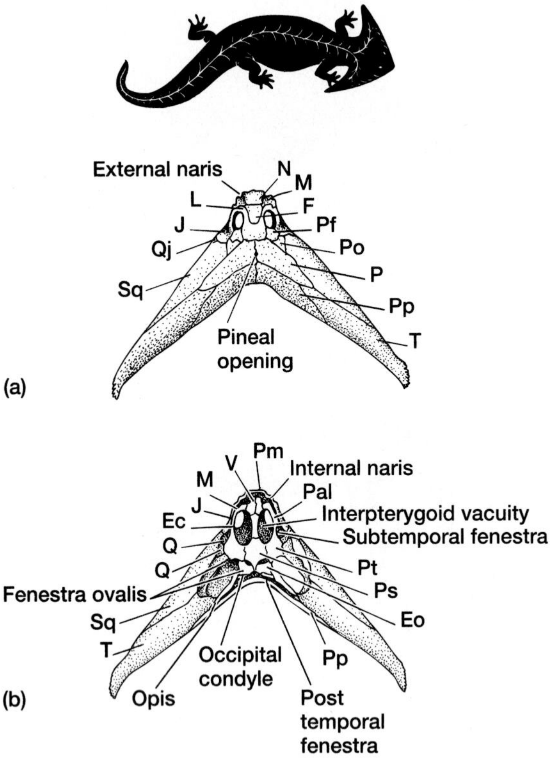


**Figure 3.18** *Diploceraspis*, a lepospondyl amphibian, was a "horned" nectridean of the early Permian. Overall body length was about 60 cm. Dorsal (a) and ventral (b) views of the skull. The various bones of the skull are ectopterygoid (Ec), exoccipital (Eo), frontal (F), jugal (J), lacrimal (L), maxilla (M), nasal (N), parietal (P), palatine (Pal), postfrontal (Pf), premaxilla (Pm), postorbital (Po), postparietal (Pp), parasphenoid (Ps), pterygoid (Pt), quadrate (Q), quadratojugal (Qj), squamosal (Sq), tabular (T), and vomer (V).

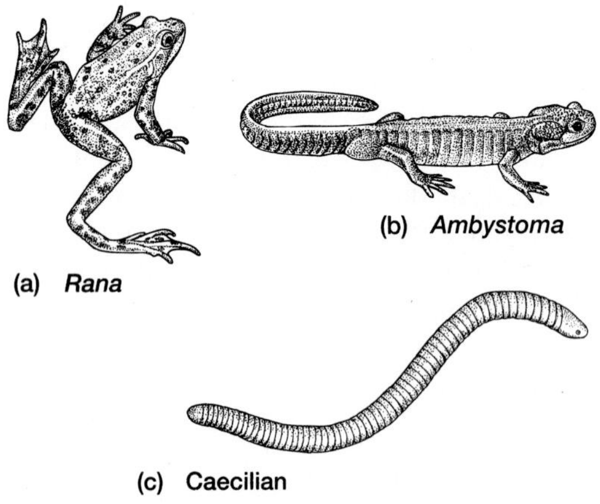


**Figure 3.19** Lissamphibia. (a) Frog *(Rana)*. (b) Salamander *(Ambystoma)*. (c) Gymnophiona (caecilian).

water or moist locations. External fertilization characterizes frogs, whereas internal fertilization characterizes most salamanders and probably all caecilians. Typically, paired lungs are present, although they may be reduced or even absent entirely in some families of salamanders. Mucous glands of the skin keep amphibians moist, and granular (poison) skin glands produce chemicals unpleasant or toxic to predators.

Modern amphibians in some ways stand between fishes and later tetrapods; therefore, they supply us with living intermediates in the vertebrate transition from water to land. In their own right, however, living amphibians are specialized and represent a considerable departure in morphology, ecology, and behavior from the ancient amphibians (figure 3.20). Many bones of the ancient skull and pectoral girdle are lost. Scales are absent, except in caecilians, which allows respiration to occur through the moist skin. Most living amphibians are small. The broken fossil record prevents us from confidently connecting them with either the lepospondyls or the labyrinthodonts. When frogs first appeared in the Lower Jurassic and salamanders in the

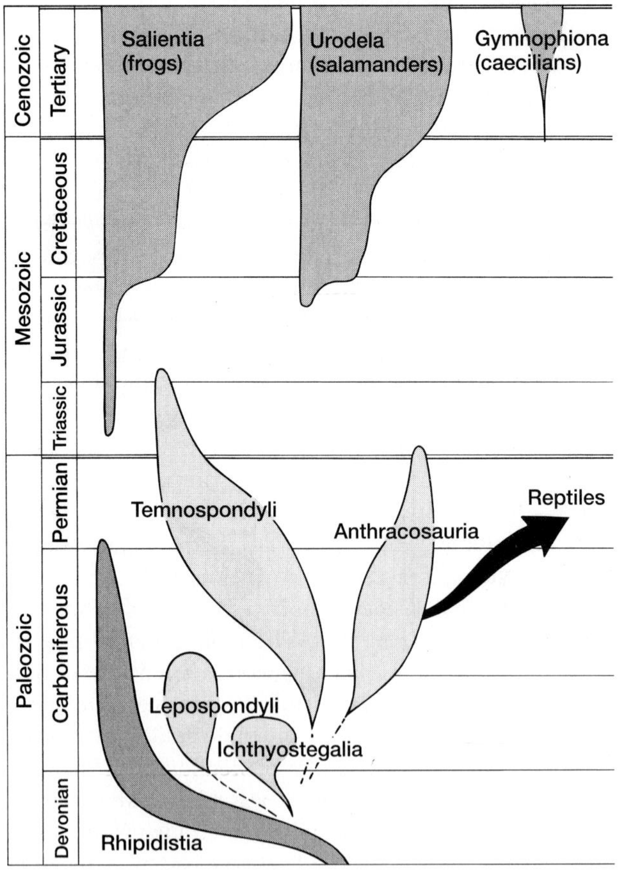


**Figure 3.20** Phylogenetic appearances of amphibian groups. The three orders of labyrinthodonts (ichthyosteigs, temnospondyls, anthracosaurs) and the three orders of lissamphibia (Salientia, Urodela, Gymnophiona) are shown separately. Rhipidistian fishes, from which ancient amphibians arose, are included as well.

Upper Jurassic, they were essentially modern in their skeletal design.

Living amphibians share some common characteristics. Most modern forms are small, respire through their skin, have unique **pedicellate** teeth with a suture dividing the tooth base from the tip, and possess an extra bone associated with the ear, the **auricular operculum.** Because of these reasons and because of the break in the fossil record between extant and extinct amphibians, most taxonomists treat all living amphibians as members of their own group, the subclass Lissamphibia.

***Order: Urodela (Caudata)*** Urodela, or Caudata, contains the salamanders. Informally, "newts" are aquatic salamanders belonging to the family Salamandridae. In general body form, salamanders resemble Paleozoic amphibians, having paired limbs and a long tail. Terrestrial salamanders usually protrude their tongue to feed, but aquatic forms part their jaws rapidly to create a suction that gulps in the food. Compared to the ancestral amphibian skull, the urodele skull is more broad and open, with many bones lost or fused. Salamanders have no "eardrum," or tympanum, nor do they have an **otic notch,** an indentation in the skull to hold a tympanum. Among primitive salamanders, fertilization is external, but in advanced groups, the male produces the **spermatophore,** a package of sperm that is collected by the female and opens in her reproductive tract. Strictly speaking, fertilization is internal in modern salamanders.

***Order: Salientia (Anura)*** Frogs and toads make up the order Salientia. Adult frogs are without a tail, hence the name anurans (no tail). Their long hindlegs are part of their leaping equipment, inspiring their alternative name of salientians (jumpers). In most frogs and toads, fertilization is external, except in the genus *Ascaphus*, and eggs are laid in water or moist locations. The larva or tadpole is a striking specialization of frogs. Tadpoles usually feed by scraping algae from rock surfaces. During this stage, salientians are especially suited to exploit temporary food resources, such as spring algae blooms in drying ponds. Typically, after a brief existence, the tadpole undergoes a rapid and radical change, or **metamorphosis** into an adult with quite a different design. The adult has a stout body and usually protrudes its tongue to feed. A **tympanum** is usually present and is especially well developed in males that produce territorial and courtship vocalizations.

The terms *frog* and *toad* are imprecise. In a strict sense, toads are frogs belonging to the family Bufonidae. More informally, the term *toad* is used for any frog having "warty" skin and **parotoid glands,** large raised glandular masses behind the eyes. "Warts" consist of clumps of skin glands scattered across the body surface. Other frogs have smooth skin without warts and lack parotoid glands.

***Order: Gymnophiona (Apoda)*** Gymnophionians, or caecilians, show no trace of limbs or girdles; hence, they are sometimes called apodans (no feet). All are restricted to damp tropical habitats where they live a burrowing lifestyle. Unlike the open skull of frogs and salamanders, the caecilian skull is solid and compact. Although their life histories are not well known, males possess a copulatory organ; thus, fertilization is internal. Primitive caecilians lay eggs that hatch into aquatic larvae; more advanced species produce live terrestrial young.

## Class: Reptilia

The phylogenetic route from amphibians to birds and mammals leads through reptiles. During the peak of their radiation in the Mesozoic, reptiles produced some of the most extraordinary animals ever to grace the land, sea, and air, including dinosaurs, airborne pterodactyls, aquatic plesiosaurs, and ancestors of modern birds and mammals. But mass extinctions reduced some groups and eliminated others entirely, ending reptile domination by the end of the Mesozoic. Terrestrial survivors found opportunities that might not otherwise have been available to them. Our world today would be a quite different place had the basic stocks of reptiles not evolved initially and then been eliminated by later extinctions.

Reptiles, together with birds and mammals, are amniotes, a monophyletic group evolving from common ancestors within the early Carboniferous. The term *Cotylosauria* was coined for this basal group, the common ancestors to all later amniotes. The **cotylosaurs,** meaning the "stem reptiles," have included various groups and the term has been used loosely, prompting some taxonomists to abandon it. The alternative name, Captorhinida, is a proposed replacement for these most primitive reptiles. New fossils and reinterpretation of temporal regions suggest that the stem reptiles might be found among the diadectomorphs. This primitive reptilian group arose in the late Carboniferous and shows affinities to the amphibian group the Seymouriamorpha. Although promising, the relationships of early reptiles are still being resolved. Because of its familiarity and because the term *Cotylosauria* was originally coined for a diadectomorph, we use the term *cotylosaur* to denote the most primitive reptilian stock that gave rise to later reptile groups and birds, and to mammals.

Although the reptile *Sphenodon*, living on islands near New Zealand, may be unfamiliar, most of you know of snakes and lizards and turtles and crocodiles. From these living forms, we have some composite image of what constitutes a "reptile." Further, the ingredients are available for telling a relatively complete story of reptilian history because of the complete series of fossil intermediates leading from early ancestors to modern descendants. Thus, you may find it odd that taxonomists still quibble over what constitutes a reptile (figure 3.21). However, reptiles turn out to be a rather diverse group, with specializations associated with different diets, patterns of locomotion, and body size. Among modern groups, for example, crocodiles have more

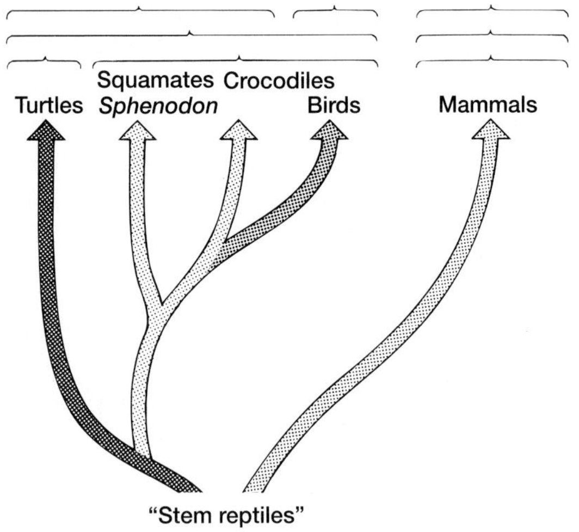


**Figure 3.21** Alternative classification schemes of reptiles. From the time that they first appeared, reptiles evolved along three major lines of descent—one to turtles, one to mammals, and one to squamates, including *Sphenodon*, crocodiles, and birds. The ways in which the three descendant groups are nested together vary, depending on what aspect of these relationships taxonomists want to emphasize. We follow the view that reptiles are amniotes that are *not* birds or mammals (top set of brackets). Other, but less favored, nesting schemes are suggested by the middle and bottom sets of brackets.

features in common with birds than they do with lizards, snakes, or turtles. Among fossil groups, ancestral tetrapods preceding mammals are classified as reptiles, but they, in fact, share more features with mammals than they do with reptiles, their taxonomic companions. As a consequence of this diversity and specialization, no single suite of characteristics unites all fossil and modern reptiles.

Some despair and see "reptiles" as a grade, a level of organization rather than as a recognizable taxonomic category. From this perspective, as figure 3.21 illustrates, primitive amniotes have yielded three major divergent lineages: one that gave rise to mammals, a second to turtles, and a third to other reptilian groups and birds. Alternatively, we could divide amniotes into just two groups: the first mammals and their ancestors, and the second all other amniotes (reptiles and birds). Although relationships of present-day groups are perhaps more accurately highlighted by such a view, the extraordinary radiation of Mesozoic reptiles would be poorly served. One way around this is to define members of Reptilia in a backhanded fashion, that is, as amniotes that are not birds or mammals.

Embryos of living reptiles, like those of all other amniotes, are enveloped in extraembryonic membranes, all of which are usually packaged in a calcareous or leathery shelled egg. Reptiles have scales (but no hair or feathers) composed partly of surface epidermis. They reach preferred body temperature usually by absorbing heat from the environment. Respiration is primarily through the lungs, with very little occurring through the skin. Although these and other soft body characteristics are known from study of living reptiles, they are not necessarily unique to reptiles nor do such features always fossilize. In the search for more distinguishing and preservable characteristics, early taxonomists turned to skeletal features. Among amniotes in general and reptiles in particular, the structure of the temporal region of the skull, the area behind each eye, seemed to be a reliable indicator of evolutionary lineages. As a consequence, formal terminology grew up to describe the temporal region of the amniote skull.

The temporal region in amniotes varies in two ways, in the number of openings, termed **temporal fenestrae,** and in the position of the **temporal arches,** or **bars,** made up of defining skull bones. From these two criteria, we recognize four skull types. In primitive amniotes, as well as in their amphibian ancestors, the temporal region is covered completely by bone that is not pierced by temporal openings (figure 3.22a). This **anapsid skull** is characteristic of the first reptiles and the later turtles. The **synapsid skull** found in mammalian ancestors represents an early divergence from the anapsid. The synapsid skull has a single pair of temporal openings bordered above by a temporal bar formed by squamosal and postorbital bones (figure 3.22b). In another group that diverged from anapsids, we see a **diapsid skull** characterized by two pairs of temporal openings separated by this temporal bar. As points of formal anatomical reference, this squamosal-postorbital bar is designated as the **upper temporal bar.** The **lower temporal bar,** formed by jugal and quadratojugal bones, defines the lower rim of the lower temporal fenestra (figure 3.22c). Diapsids, including pterosaurs and dinosaurs, were predominant during the Mesozoic and gave rise to birds and all

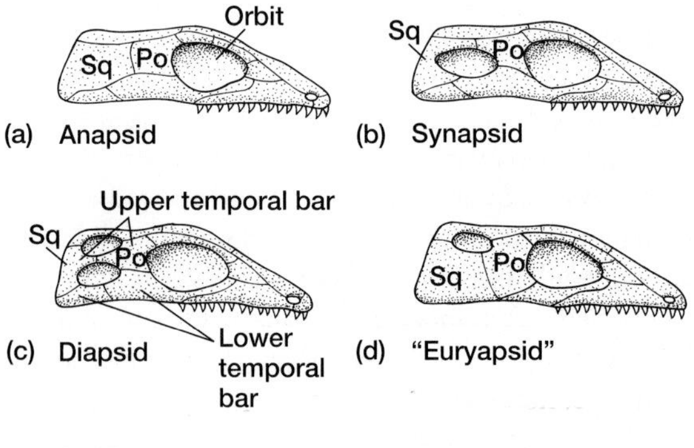


**Figure 3.22** Amniote skull types. Differences among the skulls occur in the temporal region behind the orbit. Two, one, or no fenestrae may be present, and the position of the arch formed by postorbital (Po) and squamosal (Sq) bones varies. (a) The anapsid skull has no temporal fenestrae. (b) The synapsid skull has a bar above its single temporal fenestra. (c) The diapsid skull has a bar between its two temporal fenestrae. (d) The "euryapsid" skull has a bar below its single temporal fenestra. Rather than being a separate skull type, the euryapsid skull is thought to be derived from a diapsid skull that lost its lower temporal bar.

living reptiles (except turtles). Two groups of Mesozoic marine reptiles, the plesiosaurs and ichthyosaurs, possess a fourth skull type, the **euryapsid skull.** A single pair of temporal openings is present, but the squamosal-postorbital arch forms the lower border of the paired opening. Once thought to be a separate derivation from anapsids, the euryapsid skull seems instead to be derived from diapsid ancestors by loss of the lower temporal opening. Further, the euryapsid skull, which two groups of marine reptiles possess, is now thought to result from convergence, not from common ancestry, so both groups are classified separately within the diapsids.

Anapsids, diapsids, and synapsids are directly translated into taxonomic subclasses. Turtles lack temporal openings, so strictly speaking, they have anapsid skulls. Consequently, many taxonomists prefer to treat them as members of the subclass Anapsida. However, the current fashion is to elevate them to their own subclass, Testudinata, in recognition of their specialized morphology. Birds inherit a diapsid skull and modern mammals a synapsid skull, but both usually are placed in their own separate class apart from their immediate reptilian ancestors. Formally, this leaves us four subclasses within the class Reptilia—Anapsida, Testudinata, Synapsida, and Diapsida (figure 3.23).

## *Subclass: Anapsida*

From anapsids sprang all later subclasses (figure 3.23). One of the early groups of anapsids was the captorhinids. Captorhinid reptiles were small, about 20 cm in length, and generally similar to modern lizards in that they had a well-ossified skeleton (figure 3.24). Rows of tiny, sharp teeth along the margins of the jaws and across the roof of the mouth as well as an agile body suggest that insects might have been a major part of their diet, as they are in the similarly designed small, modern lizards. Captorhinids are broadly similar to anthracosaur amphibians, but captorhinids possess reptilian features, such as strong jaw musculature, and reptilian structural details in their skull, limbs, and vertebral column.

Like all anapsids, these early reptiles had skulls that lacked temporal openings. There was no notch or indentation that might betray the presence of a tympanum. Thus, in later reptiles, birds, and mammals, the tympanum is a new structure, not inherited from amphibians through these early anapsid reptiles.

The first captorhinid reptiles date from the Carboniferous (early Pennsylvanian) and occupied tree stumps away from standing bodies of water, providing additional testimony that they exploited the land further than their amphibian ancestors. The fossil record preserves no hint of soft embryonic membranes that enveloped developing eggs, but it seems likely that extraembryonic membranes debuted at about this time as well and contributed to the independence of reptiles from water.

When the last of the Anapsida became extinct at the end of the Triassic, the derivative reptilian groups were established and the Mesozoic radiation of their ancestors was well underway.

## *Subclass: Testudinata*

When turtles first appeared in the late Triassic, they already possessed a distinctive shell made up of a dorsal **carapace** of expanded ribs and surface skin plates (scutes) and a connected ventral **plastron** of fused bony pieces (figure 3.25a). Their anapsid skull suggests a captorhinid origin (figure 3.25b), which is why some taxonomists classify them within Anapsida. But because the shell represents a considerable departure from any previous anapsid, turtles are now more commonly placed in their own subclass, Testudinata.

Modern turtles belong either to the suborder Pleurodira or to the suborder Cryptodira, depending on the method they employ to retract their head into their shell. Pleurodires flex their neck laterally to retract the head, whereas cryptodires flex their neck vertically. These two groups seem to share a common ancestor, *Proganochelys*, from the late Triassic. The term *tortoise* is sometimes applied to turtles restricted to land, but no formal taxonomic distinction is made between turtle and tortoise.

## *Subclass: Diapsida*

The diapsid assemblage is remarkable. Except for turtles, all other reptiles living today are diapsids. Snakes, lizards, *Sphenodon*, and crocodiles trace their ancestry back to diapsids of the Mesozoic. Birds, too, arose from diapsid ancestors. Dinosaurs were diapsids, so were pterosaurs, and so were many large marine reptiles of the Mesozoic.

Fossils of the earliest known diapsid, *Petrolacosaurus*, come from the late Carboniferous in what is today Kansas. The body was about 20 cm long, with slightly elongated neck and limbs, and the tail added another 20 cm to the overall length. The skull was typically diapsid, with a pair of temporal openings defined by complete temporal bars. Other primitive diapsid species became quite specialized. *Coelurosaurus* had greatly elongated ribs that in life likely supported a gliding membrane. *Askeptosaurus* was about 2 m in length, slender, and probably aquatic in habits.

***Ichthyopterygia*** During the Mesozoic, several major diapsid lineages became specialized for aquatic existence, among them were the **ichthyosaurs** (figure 3.26a). They were formerly allied with other marine diapsids, such as the plesiosaurs in the Euryapsida (or Parapsida), on the basis of skull similarities. But the similarities proved to be convergent. Now ichthyosaurs are classified as the Ichthyopterygia and the other marine diapsids as the Sauropterygia. Euryapsida is no longer used as a taxonomic category. From deposits of the early Triassic, the first ichthyosaurs already appeared to be aquatic specialists. Advanced ichthyosaurs

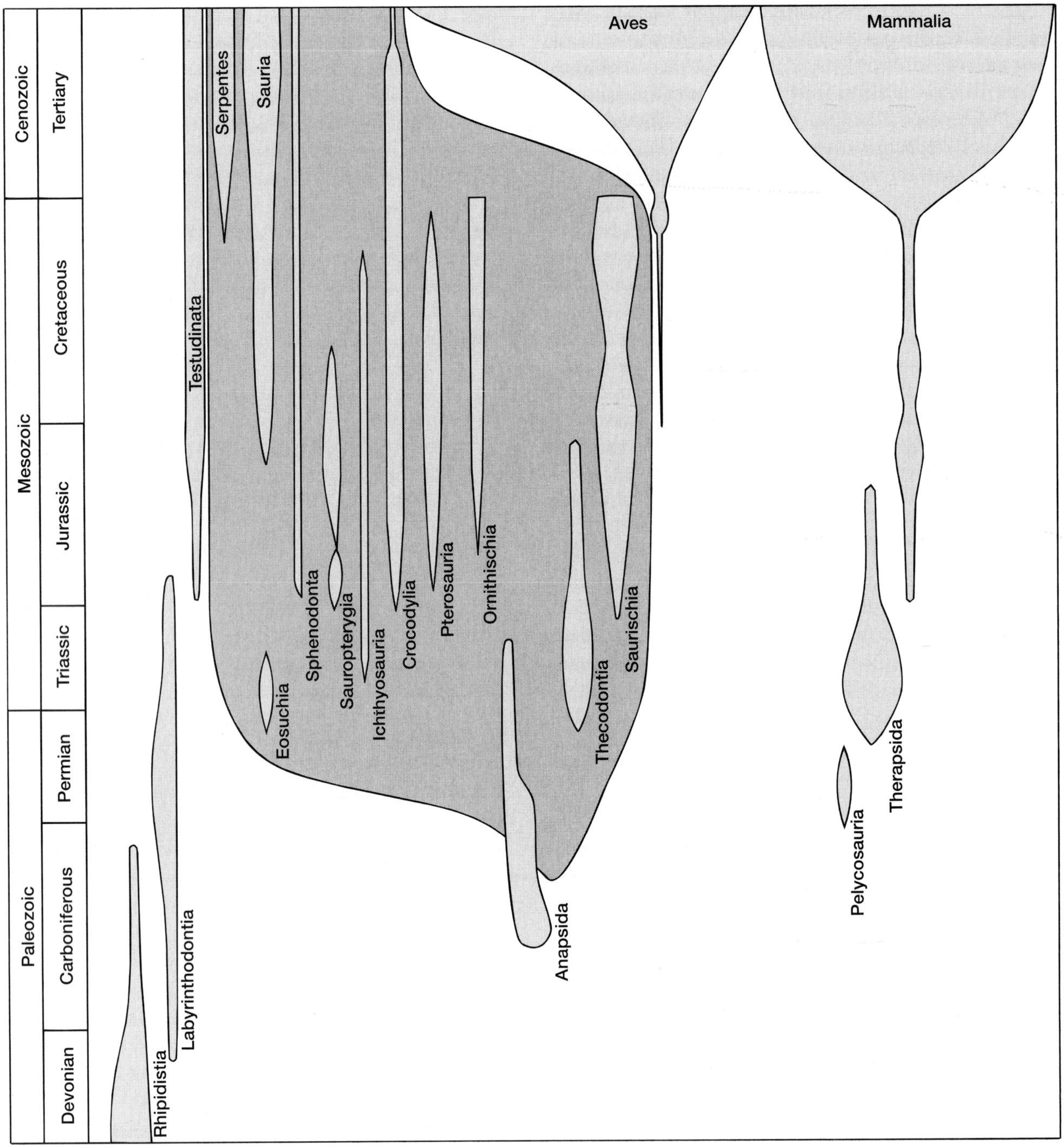


**Figure 3.23** Phylogeny and classification of reptiles and derivative amniotes. The major groups are indicated by balloons that show geological duration by their height and relative abundance by their width. Anapsid reptiles appeared in the middle Carboniferous, and from them arose all the later reptilian groups—testudines, synapsids, and diapsids. The synapsids, including first pelycosaurs and later therapsids, gave rise to mammals. The diapsids include the lepidosaurs, some marine reptiles, and the large archosaur group. *Sphenodon* (order Sphenodontia), snakes (order Serpentes), and lizards (order Sauria) are all lepidosaurs. The diapsid marine groups include Ichthyosauria and Sauropterygia. Archosaurs comprise thecodonts, pterosaurs, crocodiles, dinosaurs (Saurischia and Ornithischia), and birds.

had a porpoise-like body design, but their tail swept from side to side to provide propulsion, unlike the porpoise tail that moves in a dorsoventral direction. Sleek bodies, paddlelike limbs, and teeth around the rim of a beaklike mouth suggest an active predaceous lifestyle.

***Lepidosauromorpha*** The lepidosauromorph group of diapsids encompasses the marine **sauropterygians** and terrestrial **lepidosaurs.**

*Sauropterygia.* Sauropterygians were, along with ichthyosaurs, the other Mesozoic lineage of diapsids spe-

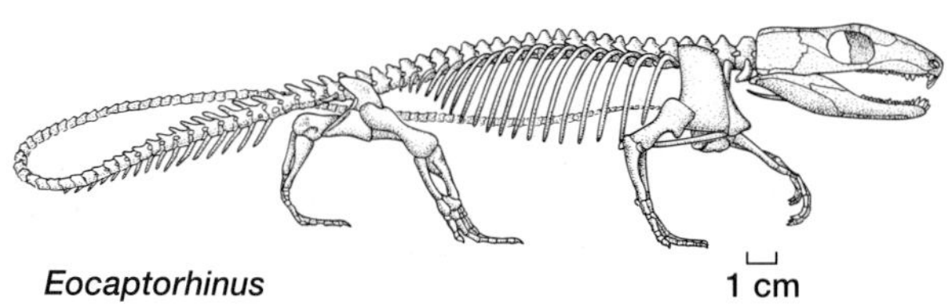


**Figure 3.24** Anapsida. Skeleton of *Eocaptorhinus*, from the Permian. An early North American anapsid of the captorhinid family, *Eocaptorhinus* shows absence of fenestrae in the temporal region of the skull behind the eye.

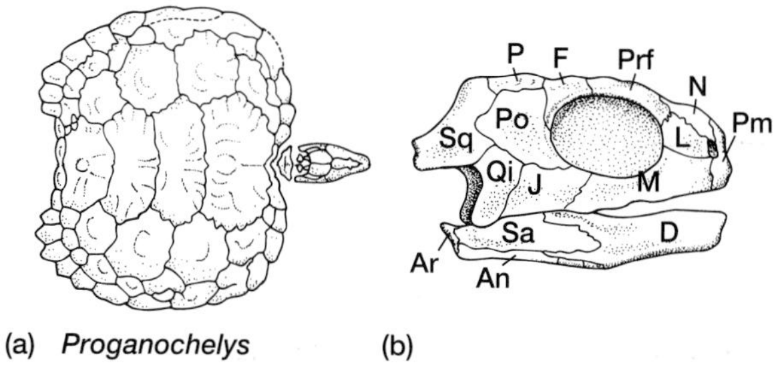


**Figure 3.25** Testudinata. (a) *Proganochelys*, a turtle of the Triassic exhibiting pattern of skin scutes that overlay the carapace. (b) Fossil *Proganochelys* skull showing absence of temporal fenestrae. Bones of the skull include the angular (An), articular (Ar), dentary (D), frontal (F), jugal (J), lacrimal (L), maxilla (M), nasal (N), parietal (P), postorbital (Po), prefrontal (Prf), premaxilla (Pm), quadratojugal (Qj), surangular (Sa), squamosal (Sq).

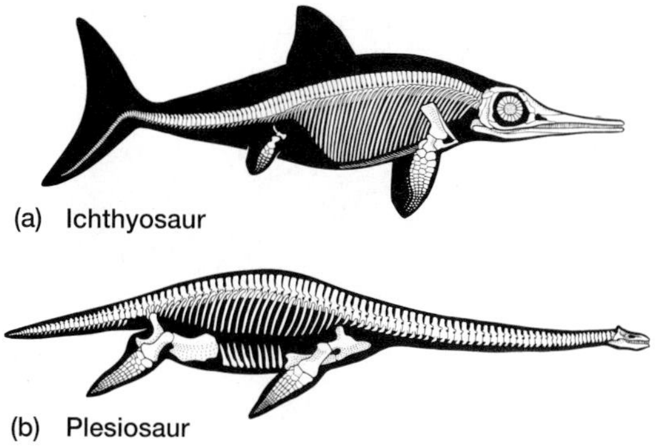


**Figure 3.26** Marine reptiles of the Mesozoic. (a) Ichthyosaur, a porpoise-like reptile about 1 m long. (b) Sauropterygian, a plesiosaur, about 7 m in length.

cialized to an aquatic mode of life. This group includes the early **nothosaurs** and the later **plesiosaurs** that evolved from them. The plesiosaur body, especially the neck, was long, and the limbs were modified into paddles that acted like oars or hydrofoils to propel the animal in water (figure 3.26b).

*Lepidosauria*. Modern snakes, lizards, *Sphenodon*, and their ancestors constitute the lepidosaurs. A late Permian/early Triassic group of lepidosaurs, the **eosuchia,** are most likely the ancestors of all modern lepidosaurs. *Sphenodon*, the tuatura, occurs today only on parts of New Zealand and nearby islands (figure 3.27a). This genus carries forward the primitive eosuchian skull with complete temporal bars defining upper and lower temporal openings. In lizards, the lower temporal bar is absent. Snakes lack both upper and lower temporal bars. Because these connecting constraints are deleted in the skulls of lizards and snakes, both vertebrates, especially snakes, have increased jaw mobility that enhances their ability to capture and swallow prey.

**Functional consequences of the loss of temporal arches (p. 254)**

The **squamates** include snakes, lizards, and a group of tropical or subtropical reptiles, the amphisbaenids. Some taxonomists place amphisbaenids with lizards, others treat them as a distinct group. All amphisbaenids are burrowers; most are limbless and prey upon arthropods (figure 3.27b). The majority of living squamates are lizards or snakes (figure 3.27c,d). Many persons are surprised to learn that some species of lizards (other than amphisbaenids) are limbless, like snakes; therefore, the presence or absence of limbs alone does not distinguish snakes from true lizards. Instead, differences in internal skeletal anatomy, especially in the skull, are used to define the two groups. Further, lizards have movable eyelids and most have an external auditory meatus (opening). Snakes lack both structures.

***Archosauromorpha*** Encompassed within the archosauromorph group are several small, primitive assemblages of diapsids that are poorly known and a very large group, the archosauria, that includes crocodiles, flying pterosaurs, dinosaurs, and their immediate ancestors and descendants. Archosauromorphs display a trend toward increasing **bipedalism,** or two-footed locomotion. The forelimbs tend to be reduced, whereas the hindlimbs are drawn under the body to become the major weight-bearing and locomotor appendages.

The term *archosaur*, meaning "ruling reptile," recognizes the extraordinary radiation and preeminence of this group during the Mesozoic. Surviving descendants of the archosaurs include all crocodiles and birds, although of course birds are separated taxonomically into their own class. Formally then, archosaurs include thecodonts, the most primitive of the group, crocodiles, pterosaurs, and two large orders, the Saurischia and Ornithischia. Taken together, Saurischia and Ornithischia constitute what laypeople know as the "dinosaurs."

The first archosaurs, the **thecodonts,** arose late in the Permian and prospered during the Triassic. Before becoming extinct by the end of the Jurassic, they gave rise to the other orders of archosaurs. Thecodonts take their name from teeth set in deep, individual sockets (thecodont

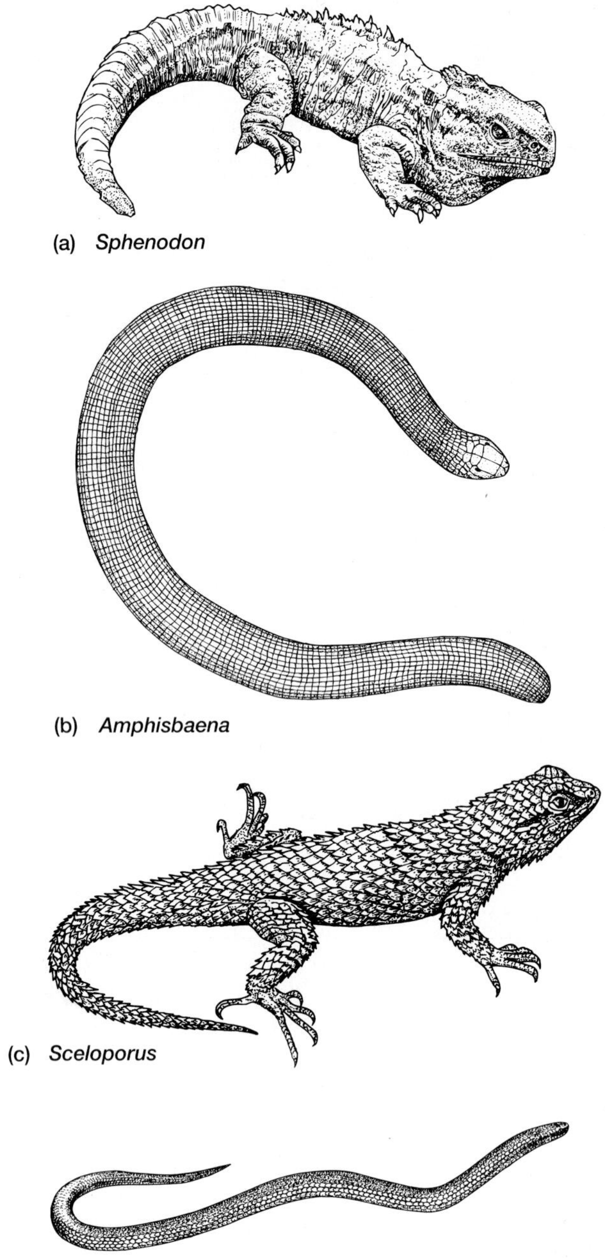

(a) *Sphenodon*

(b) *Amphisbaena*

(c) *Sceloporus*

(d) *Contia*

**Figure 3.27** Lepidosaurs. (a) *Sphenodon*. (b) Amphisbaenian, a burrowing lepidosaur. (c) Lizard (*Sceloporus*). (d) Snake (*Contia*).

condition) rather than in a common groove. The **antorbital fenestra** is a large opening in the skull in front of the eye that distinguishes thecodonts from diapsid contemporaries belonging to other groups. Within the hindlimb, a unique ankle design appeared in some thecodonts along with a tendency to bipedal, upright posture.

**Ankle types (p. 334)**

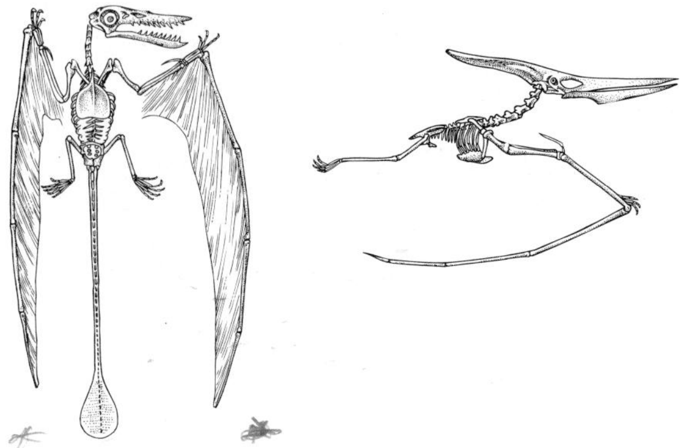

(a) *Rhamphorhynchus* (b) *Pteranodon*

**Figure 3.28** Pterosaurs. The lengthened forelimb of pterosaurs supported a membrane derived from skin to form the wing. (a) *Rhamphorhynchus*. Wingspan was about 1.5 m. (b) Skeleton of *Pteranodon*. Wingspan was about 8 m.

The **pterosaurs,** often called pterodactyls after members of one suborder, could glide and soar but also were capable of powered flight. Pterosaurs, birds, and bats are the only three vertebrate groups to achieve active aerial locomotion. Because of their antorbital fenestra, limb posture, and specialized ankle joints, pterosaurs appear to have a phylogenetic affinity to thecodonts. The first known pterosaur, from the late Triassic, was already specialized for flight with membranous wings. Many were sparrow to hawk sized, but the late Cretaceous *Quetzalcoatlus*, found in fossil beds in Texas, had an estimated wing span of 12 m. Pterosaur teeth suggest a diet of insects in some species and strained plankton in other species. Fossilized stomach contents confirm that one species ate fish.

The early **rhamphorhynchoids** are a suborder of pterosaurs distinguished by long tails and short necks (figure 3.28a), and the later **pterodactyloids,** the second pterosaur suborder, lacked a tail but possessed a long neck and often had a projecting crest at the back of the head (figure 3.28b).

Dinosaurs include the two remaining orders of archosaurs, the Saurischia and Ornithischia. Both dinosaur orders apparently arose independently from thecodont predecessors and thereafter evolved independently, although it is possible that they share a common ancestor among the early archosaurs. The two orders differ in the pelvic structure. In saurischians, the three bones of the pelvis—ilium, ischium, pubis—radiate outward from the center of the pelvis (figure 3.29a). In ornithischians, the ischium and part of the pubis lie parallel and project backward toward the tail (figure 3.29b). All dinosaurs have either a saurischian or an ornithischian pelvis.

Within the Saurischia, there are two independent lines of evolution. The **theropods** include mostly carnivorous species and are divided into two groups, the **carnosaurs** and **coelurosaurs.** The best-known theropods are

*Tyrannosaurus* and *Allosaurus*. The mostly herbivorous **sauropodomorphs** constitute the other saurischian line, including **prosauropods** and **sauropods.** Familiar members of this group are *Apatosaurus* (formerly *Brontosaurus)*, *Diplodocus*, and *Brachiosaurus*.

Within the exclusively herbivorous ornithischians, there are four evolutionary lines—**stegosaurs, ceratopsians** (*Triceratops* being the most familiar), **ornithopods,** and **ankylosaurs** (figure 3.30).

Crocodiles, alligators, and their close allies **(gavials, caimans)** are the only reptilian members of the archosaurs to survive the Mesozoic and live in modern times. In many features, especially the skull and ankle joint, alligators and crocodiles are not far removed from primitive thecodonts. Modern crocodilian families are known from the late Cretaceous.

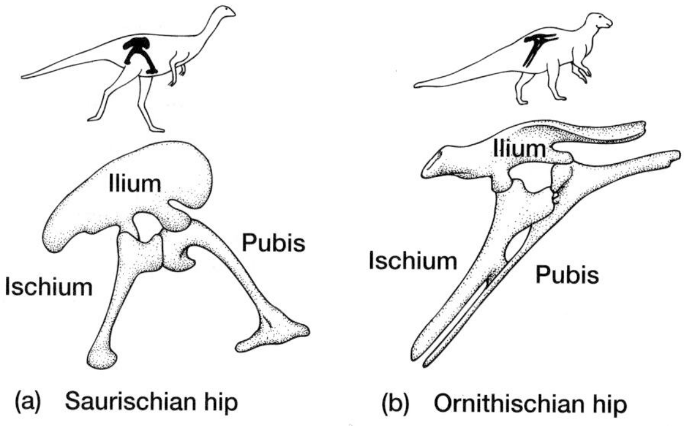


**Figure 3.29** Dinosaur hips. Two types of hip structures define each group of dinosaurs. (a) Saurischians all possessed a pelvic girdle with three radiating bones. (b) Ornithischians had a hip with pubis and ischium bones lying parallel and next to each other.

## *Subclass: Synapsida*

Synapsids are reptiles with a single temporal fenestra bounded above by the upper temporal bar (squamosal-postorbital bones). They are the ancestors of mammals and possess some characteristics of body posture and tooth formation that became elaborated later within the mammals. Anticipating this, paleontologists often refer to some synapsids as mammallike reptiles, an unfortunate designation, because it unfairly encourages one to glance over them in order to emphasize the mammals they produced. Yet the very transitional position of synapsids between ectothermic reptiles and endothermic mammals gives them special significance (figure 3.31). As research on synapsids proceeds, we are coming to better understand what immediate selective factors favored the very changes that eventually were central to mammalian structure and physiology. Taxonomists recognize two groups of synapsids, **pelycosaurs** (sail-backed reptiles) and **therapsids** (mammallike reptiles).

***Pelycosauria*** Pelycosaurs arose in the mid-Carboniferous from cotylosaurs and soon enjoyed an extensive radiation through the early Permian, coming to constitute about half of the known reptilian genera of their time. Some, like *Edaphosaurus*, were herbivorous. Most, however, were carnivores and preyed on fish and aquatic amphibians. Different species of pelycosaurs differed in size, but they were not very diverse in design, perhaps because of their specialized lifestyle. The most notable specialization in some species was a broad "sail" along the back consisting of an extensive flap of skin supported internally by a row of fixed neural spines projecting from successive vertebrae (figure 3.32a, b). If the sail was brightly colored in life, it might have been deployed in courtship or in bluff displays with rivals, like elaborate

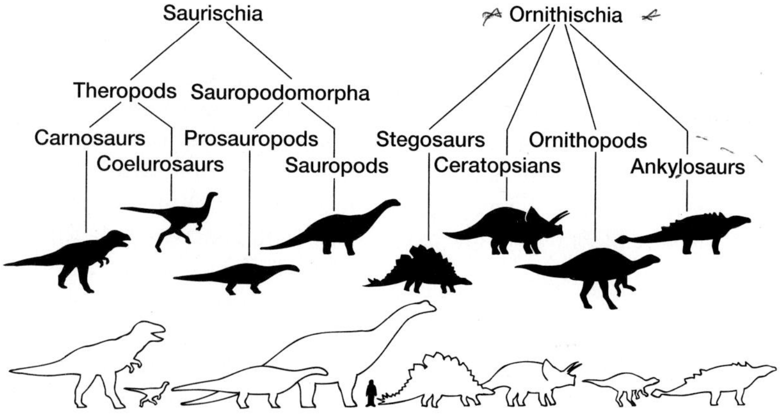


**Figure 3.30** Phylogeny of dinosaurs. Both Saurischia and Ornithischia are divided into several subgroups of dinosaurs. Relative sizes of adults are shown at the bottom of the figure. A 2-m human in dark silhouette is diminutive by comparison.

## Box Essay 3.3 Dinosaurs: Heresies and Heresay—The Heated Debate

The claim is that dinosaurs were warm-blooded, like birds and mammals, not cold and sluggish like lizards and snakes. To be specific, the issue is not really whether the blood of dinosaurs ran hot or cold. After all, on a hot day with the sun beating down, even a so-called "cold-blooded" lizard can bask, heat its body, and, strictly speaking, have warm blood circulating in its arteries and veins. The issue is not blood temperature, hot or cold, but is the source of the heat internal or external? To clarify this question, two useful terms need to be defined, *ectotherm* and *endotherm.* Animals that depend largely upon sunlight or radiation from the surrounding environment to heat their bodies are cold-blooded, or more accurately, **ectotherms** (heat from outside). Turtles, lizards, and snakes are examples. "Warm-blooded" animals produce heat inside their body by metabolizing proteins, fats, and carbohydrates. To be more accurate, warm-blooded animals are **endotherms** (heat from within). Birds and mammals are obvious examples.

Were dinosaurs ectotherms or endotherms? The source of their body heat is in dispute, not their blood temperature.

Heat for *ecto*therms is cheaply won. They need only bask in the sun. The trouble with such a lifestyle is that the sun is not available at night nor is it always available in cold temperate climates. By contrast, heat for *endo*therms is expensive. A digested meal, often caught with great effort, produces fats, proteins, and carbohydrates necessarily spent in part to generate heat to keep the endotherm body warm. Where endotherms have an advantage is that their activity need not be tied to heat available from the environment. These different physiologies are accompanied by different lifestyles. Ectotherms bask. On cold nights they become sluggish and in freezing winters they hibernate. Endotherms remain metabolically active throughout each day and each season, despite the cold or inclement weather. Certainly there are exceptions—bears and some small mammals hibernate—but endothermy requires continuous activity in most cases. Thus, the issue of warm-bloodedness in dinosaurs is not just an issue of physiology but one also about the type of accompanying lifestyle they enjoyed.

Because dinosaurs are reptiles, they were for many years envisioned to be ectotherms just like their living counterparts—lizards, snakes, turtles, and crocodiles. Occasional dissenters from this view objected, but the first person to assemble a case for endothermic dinosaurs was Robert Bakker. He and a few others proposed four principal lines of evidence. Let's look at the arguments.

*Insulation.* First, some mid- to late Mesozoic reptiles had surface insulation, or at least they seemed to. For ectotherms, a surface insulation would only block absorption of the sun's rays through the skin and interfere with efficient basking. But, for endotherms, a surface layer holding in their internally manufactured heat might be an expected adaptation. Unfortunately, soft insulation is rarely preserved, but in a few fossils of the Mesozoic, impressions in the surrounding rock indicate the presence of an insulating layer that resembles hair in some (pterodactyls, therapsids) and feathers in others *(Archaeopteryx).* In fact, feathers likely first arose as thermal insulation and only later evolved into aerodynamic surfaces. Apparently, then, some Mesozoic reptiles had insulation like endotherms rather than bare skin like ectotherms.

*Large and Temperate.* Second, large Mesozoic reptiles are found in temperate regions. Today, large reptiles such as great land tortoises and crocodiles do not occur in temperate regions. They

ornamentations of birds are today. The sail might also have been a solar collector. When turned broadside to the sun, blood circulating through the sail was warmed and then carried to the rest of the body.

Rather suddenly, pelycosaurs declined in numbers and were extinct by the end of the Permian. Therapsids evolved from them and largely replaced them for a time as the dominant terrestrial vertebrates.

***Therapsida*** Therapsids appeared in the early Permian, prospered during the Triassic, a period with relatively mild climate, and became extinct during the Jurassic. They apparently exploited terrestrial habitats more extensively than the pelycosaurs before them; consequently, they exhibit greater diversity of body design. Some trends in therapsids were conservative, however. Their stance was quadrupedal and their feet had five digits, but their limb posture was less sprawled and their legs more directly positioned under the weight of the body (figure 3.33a, b). This reflects a more efficient and active mode of locomotion. Tooth and skull design also departed from other reptiles. Teeth were differentiated into distinct types, perhaps with specialized chewing functions. The skull, especially the lower jaw, became simplified. Some herbivorous therapsids became specialized for rooting or grubbing, some for digging, some for browsing. The overall selection for more efficient terrestrial locomotion and feeding specializations resulted in great diversity within therapsids (figure 3.31).

There is even some evidence from hairlike specializations, bone histology, and latitudinal distribution that therapsids were becoming endothermic in parallel with live in warm tropical or subtropical climates. The only modern reptilian inhabitants of temperate regions are small or slender lizards and snakes. The reason is easy to understand. When winter arrives in temperate regions and freezing cold settles in, these small ectothermic reptiles squeeze themselves into deep crevices where they safely hibernate until spring and escape the freezing temperatures of winter. On the other hand, for a large and bulky animal, there are no suitably sized cracks or crannies into which they can retreat to avoid the winter cold. Large animals must be endothermic to survive in temperate climates. Thus, the presence of large reptiles in temperate climates of the Mesozoic suggest that they were warm-blooded. Like wolves, coyotes, elk, deer, moose, bison, and other large temperate mammals today, the large Mesozoic reptiles depended on heat produced physiologically to see them through.

*Predator-to-Prey Ratios.* Third, the ratio of predators to prey argues for endothermic dinosaurs. Endotherms, in a sense, have their metabolic furnaces turned up all the time, day in and day out to maintain a high body temperature. A single endothermic predator, therefore, requires more "fuel," in the form of prey, to keep the metabolic furnaces stoked than an ectothermic predator of similar size. Bakker reasoned that there should be few predators but lots of prey (lots of fuel to feed the few predators) in ecosystems dominated by endothermic reptiles. But, if ectothermic reptiles dominated, then proportionately more predators should be present. By selecting strata that stepped through the rise of dinosaurs, Bakker compiled the ratios. If Mesozoic archosaurs were becoming endothermic, then the ratio of predator to prey should drop. That happens. As this ratio was followed from early reptiles, to predinosaurs, and to dinosaurs, it dropped. There were proportionately fewer predators and more prey.

*Bone Histology.* Fourth, the microarchitecture of dinosaur bone is similar to that of endothermic mammals, not to that of ectothermic reptiles. Bones of ectothermic reptiles show growth rings, like those of trees, and for much the same reason, they grow in seasonal spurts. Endothermic mammals, with constant body temperature year round, lack such growth rings in their bones. When various groups of dinosaurs were examined, the microarchitecture of their bones told a clear story—no growth rings.

Critics of this case for warm-blooded dinosaurs challenge the various lines of evidence. Insulating hair or feathers, even if present in a very few Mesozoic reptiles, seem absent in all others, including dinosaurs. Fossils from so-called temperate regions may actually come from subtropical deposits or represent dinosaurs that migrated to mild climates during winter months. Calculations of predator-to-prey ratios are not so straightforward and may reflect capricious fossil collection. A nearly constant body temperature may result from large size alone; therefore, evidence from the microarchitecture of bone need not reflect an endothermic physiology. The debate continues.

The important point to keep before us is that dinosaurs were in their own right an extraordinary group. These active animals occupied almost every conceivable habitat—land, water, sea, air. Their social systems were complex, and the adults of some species were enormous. If dinosaurs were endotherms, their complete demise at the end of the Mesozoic can only be more mysterious and the loss of the awesome splendor of this group all the more intriguing.

their archosaur contemporaries. However, by the Jurassic, therapsids went into a precipitous decline and were replaced as dominant terrestrial vertebrates by the archosaurs (figure 3.31).

## Class: Aves

Birds outnumber all vertebrates except fishes. They can be found virtually everywhere, from the edge of the polar ice to tropical forests. As mentioned previously (see page 102), some taxonomists combine birds and dinosaurs into a single heroic taxonomic category. Although this is seldom done, the idea is not as farfetched as it might at first sound. Both lay eggs encased in shells and have similar bone and muscle structures. Over a century ago, these features led T. H. Huxley to call birds "glorified reptiles." Nevertheless, birds depart markedly from their reptilian ancestors among the archosaurs.

### *Flight*

Only birds, bats, and pterosaurs evolved the capacity for flight, but not all bird flight is the same. Some birds soar, some hover, some are sprint fliers, others long-distance fliers, and some don't fly at all. The wings of penguins serve as flippers. Ostriches have lost use of their wings altogether and depend entirely on running for locomotion. In fact, some of the largest birds ever to evolve were flightless. *Diatryma*, a 2-meter flightless bird, cruised the plains of North America 60 million years ago (figure 3.34b). *Phororhacos*, a similar

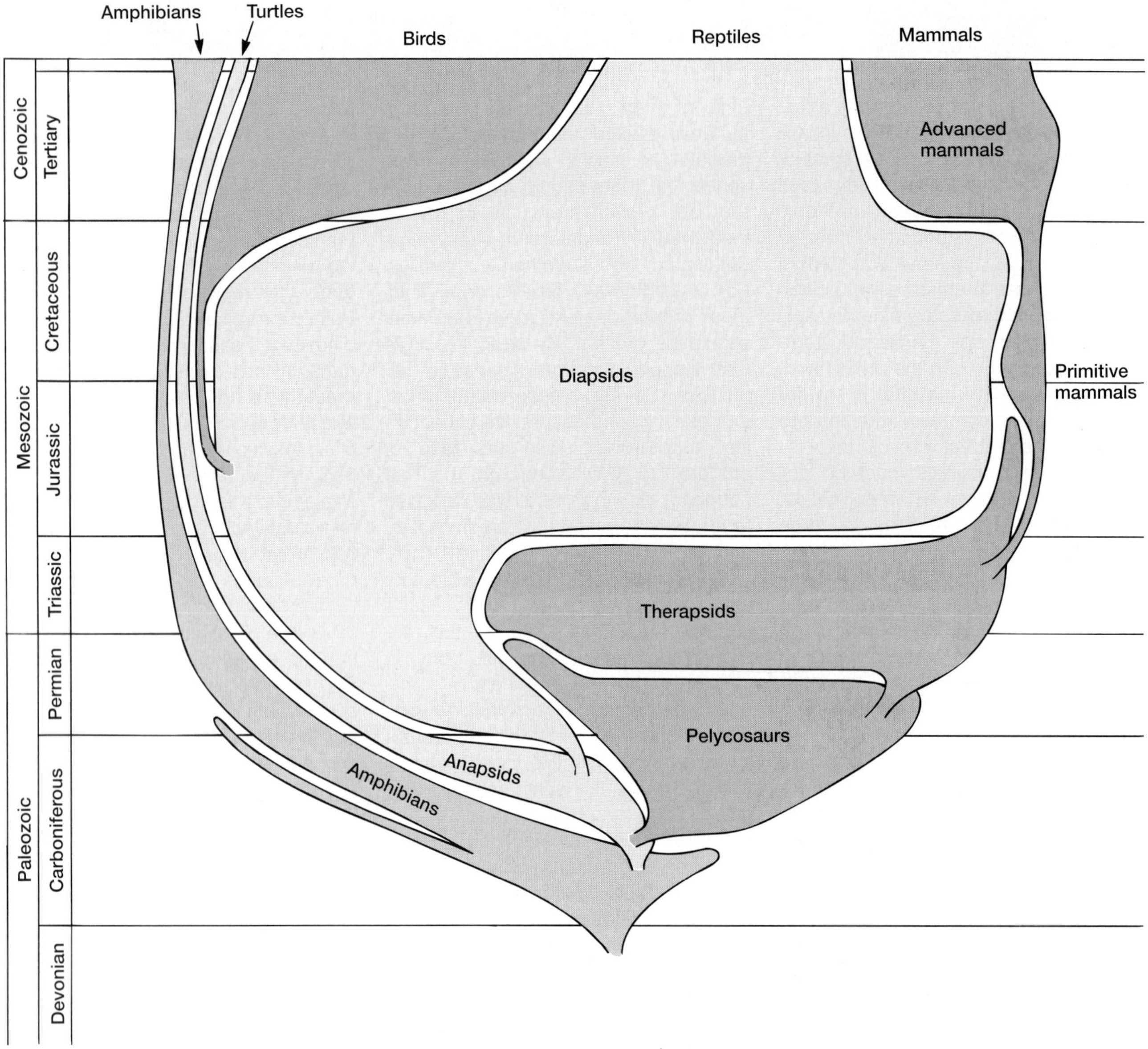


**Figure 3.31** Relative diversity of terrestrial vertebrates. Geological time is represented on the vertical axis, vertebrate diversity on the horizontal. Synapsids and the mammals that arose from them are darkly shaded. Note the diversity of the therapsid reptiles that abruptly give way to the diapsid reptiles during the mid-Mesozoic.

flightless bird, lived in South America 30 million years ago (figure 3.34c). Both were large terrestiral predators. Although they left no descendants, other large flightless birds evolved and even survived into recent times along with primitive humans. Examples are the elephant bird *(Aepyornis)* of Madagascar and the 3-m tall moas *(Dinornis)* of New Zealand. Moas belonged to a family of large ground birds that were plant eaters in New Zealand when no native land mammals resided there. Unfortunately for moas and modern scientists, the Polynesians who arrived in about A.D.1300 hunted moas for food and colorful feathers. By the time Western explorers visited, moas were extinct. Only fossils remained to tell the story.

### *Feathers*

Size, flight, or anatomy alone do not distinguish birds from other vertebrates. What makes birds unique are feathers, specializations of the skin. Were it not for impressions of feathers in rock, the avian fossil *Archaeopteryx* might well have been mistaken for a reptile from its skeletal anatomy alone. This bird of the Jurassic was a comtemporary with dinosaurs, and for a long time, it represented the earliest known bird fossil. Recently, *Protoavis*, an earlier but less complete bird skeleton, has been unearthed in 225-million-year-old deposits. Both species represent the subclass Archaeornithes, or "ancestral birds." All others,

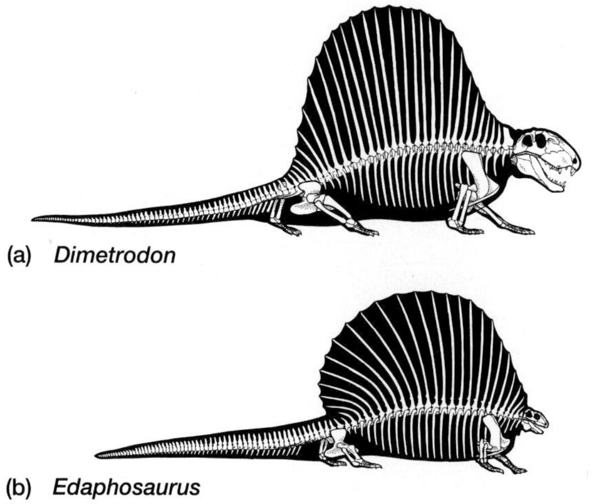


**Figure 3.32** Pelycosaurs. (a) *Dimetrodon*, a predator, reached 3 m in length (Lower Permian of Texas). (b) *Edaphosaurus*, a herbivore (late Carboniferous and early Permian) was about 3 m long.

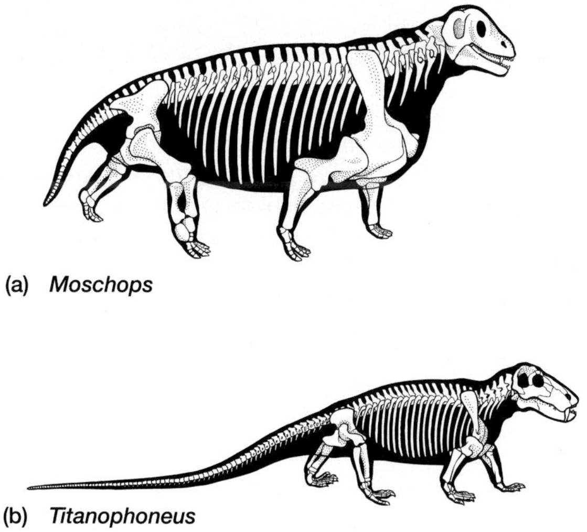


**Figure 3.33** Therapsids. (a) *Moschops*, about 5 m in length. (b) *Titanophoneus*, about 2 m.

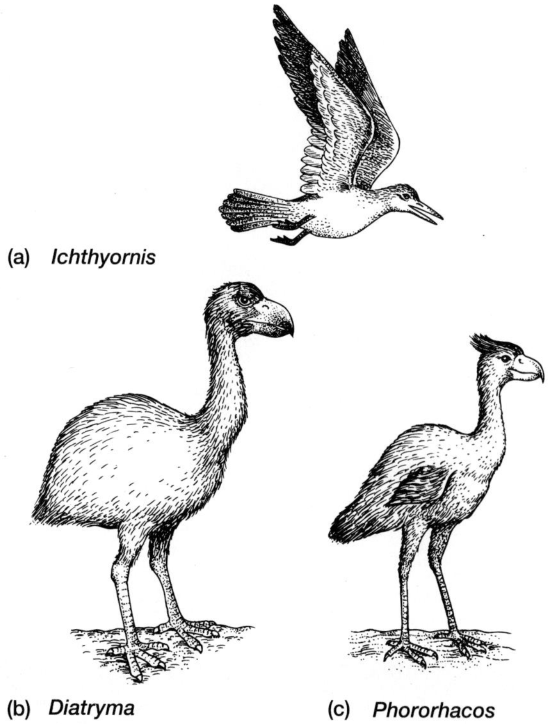


**Figure 3.34** Extinct birds. (a) *Ichthyornis* was pigeon sized, likely sought fish for food, and lived in North America about 100 million years ago. (b) *Diatryma* lived 60 million years ago. It was a flightless bird that stood over 2 m tall and likely ran down small prey much as the diminutive roadrunner does today. (c) *Phororhacos*, another flightless predator, lived in South America some 30 million years ago.

including modern birds, belong to the subclass Neornithes. Although the lineage of birds is old, it was not especially diverse for millions of years. Only following the demise of dinosaurs did birds begin their extraordinary radiation.

**Feather types, development, and function (p. 206)**

## *Diversity*

The basic avian design has proved highly adaptable and birds have undergone extensive diversification. For instance, the early neornithes included a primitive flamingo and *Hesperornis*, a toothed diving bird with such small wings that it was certainly flightless. *Ichthyornis* was a small, ternlike seabird recovered from Cretaceous rock of Kansas (figure 3.34a). By the late Mesozoic, water birds had already diverged widely.

Birds have continued to be successful in exploiting aquatic resources (figure 3.35). Some species dive deep beneath the surface and use their wings to propel themselves in pursuit of fishes. Others are specialized for plunging and use their diving speed to carry them to fishes below. Many species feed at the water's surface, either skimming it from the air or dipping for resources as they float. A few species prowl the air above the water, surprising other birds and pirating their catch.

**Raptors** are birds with **talons,** specialized feet used to stun or grasp prey. Hawks, eagles, and owls are examples. Many hunt prey on the ground. Other raptors, such as the prairie falcon, strike their quarry, usually a dove or a slow migrating duck, in the air and then follow it to the ground to dispatch the injured prey (figure 3.36).

Feet reflect functions performed. Paddling birds have webbed feet and raptors have talons. Feet of running species are robust and waddlers' are broad. Birds that soar on strong winds usually have long and narrow wings like the wings of

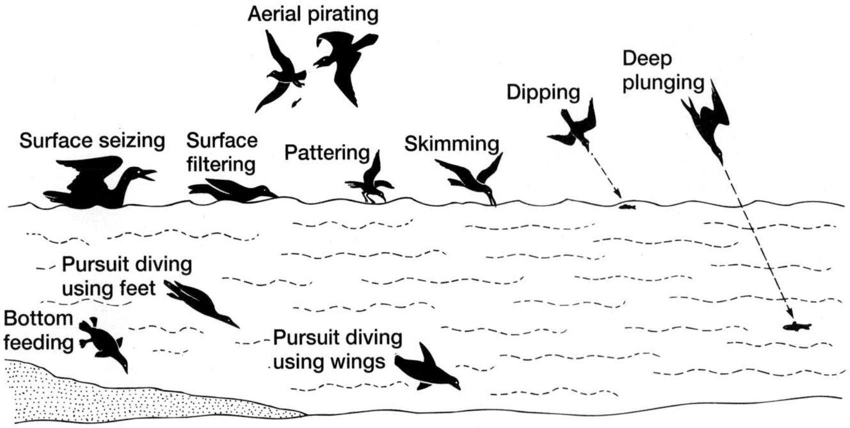


**Figure 3.35** Lifestyles of water birds.

## Box Essay 3.4 Discovery of the First Bird

The discovery of *Archaeopteryx* was especially timely. In 1861 the fossil was quarried from a site in Bavaria in what is present-day Germany. Only two years before, Charles Darwin had published *The Origin of Species,* which immediately ignited public debate. These were the early days of paleontology, with relatively few recovered fossils and even fewer serious scientists to dig them up. Critics of Darwin were quick to point out the absence of fossil intermediates between groups, which his theory of evolution anticipated. If one group gave rise to another, as Darwin's ideas suggested, then transitional forms should occur. *Archaeopteryx* helped address this objection. It was such an intermediate fossil because it possessed features of both birds (feathers) and reptiles (skeleton, teeth).

The discovery of *Archaeopteryx* prompted interest in the possibility of other ancient avian fossils that might further narrow the gap between reptiles and birds. Reptiles have teeth but modern birds do not. Somewhere between the two, evolutionary intermediates developed a bill and lost teeth. Thus, uncovering a fossil bird with reptilian teeth would be of considerable significance and help supply details about this evolutionary transition. O. C. Marsh, an American paleontologist of the mid-1800s, discovered just such birds with teeth, although they were later than *Archaeopteryx* (box figure 1).

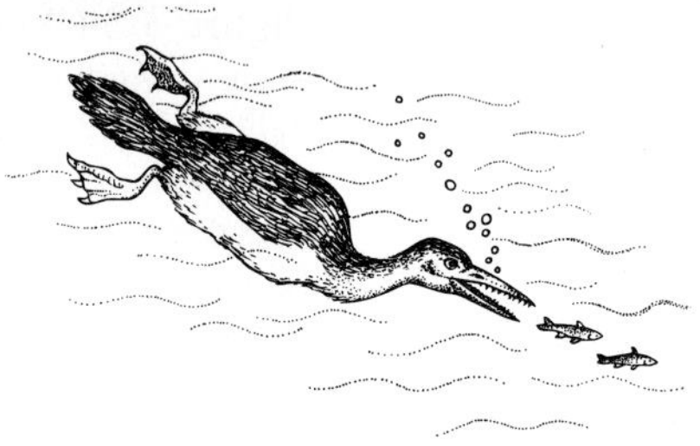

**Box figure 1** Bird with teeth. *Hesperornis* lived 100 million years ago in inland seas of North America. Although larger (almost 1 m overall) in shape, its features and probable lifestyle resembled the modern loon. This bird also retained teeth, a characteristic held over from its reptilian ancestors.

Despite the significance of Marsh's discoveries, enemies of evolution in the U.S. Congress protested the use of taxpayer's money to search out fossils with bird teeth, which everyone knew did not exist (until Marsh discovered them, of course). Today, as in the nineteenth century, science is a predominant feature of our culture, which may be good or bad. Nonetheless, the pervasiveness of science is a modern fact, so it should be dealt with thoughtfully. Most politicians who govern today have no better training in biology or in any science than politicians in Marsh's time. Law schools and businesses still supply most of our public figures. A background heavy in business and light in science gives only lopsided preparation to persons who guide the destiny of science in society.

*Archaeopteryx* occasionally still makes the news. Recently, a well-known astronomer dabbling in paleontology claimed that the Bavarian fossils of *Archaeopteryx* were forgeries. Fossil forgeries have occasionally turned up, but *Archaeopteryx* is decidedly not one of them. Regrettably, this astronomer's cavalier opinion cast undeserved doubt upon these fossils. Although the popular media picked up and spread the premature rumors of a forgery gleefully, they failed to report equally the results of an extensive reinvestigation that showed these forgery claims to be completely groundless. Charitably said, this astronomer could have saved everyone lots of wasted time had he simply made an effort to bring his naive ideas before someone familiar with the pitfalls into which he stumbled.

Stay tuned. *Archaeopteryx* seems to have a public life of its own.

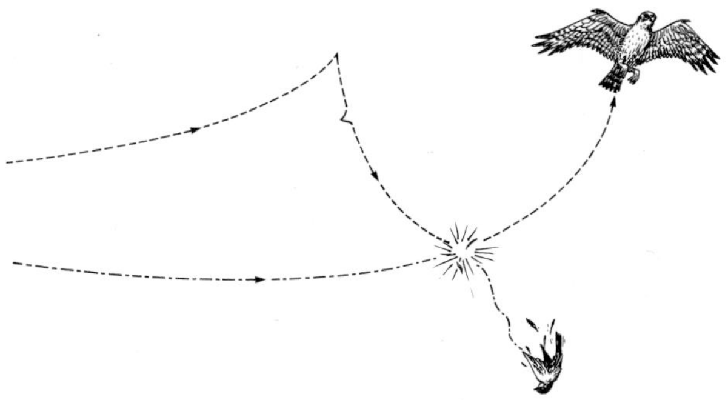

**Figure 3.36** Falcon attack. The falcon's midair blow delivered with the talons is a "stoop" intended to stun and knock the prey from the air. The prey is finally controlled and killed on the ground.

glider planes. High-speed or migratory birds have narrow, often swept-back wings. Pheasants and other birds that deploy short bursts of flight in enclosed bushy or forest habitats have broad elliptical wings for maneuverability. Slotted high-lift wings are seen in birds that soar on warm air updrafts over inland areas.

**Aerodynamics and wing designs serving flight (p. 346)**

## Class: Mammalia

The most primitive mammals are the **Prototheria,** a subclass that includes fossil forms and living monotremes, the duckbill platypuses and spiny anteaters. Most other mammals fall into the subclass **Theria,** made up of several extinct groups and the two modern infraclasses, **metatherians** (pouched marsupials such as kangaroos and opossums) and **eutherians** (placental mammals). Thus, the three living forms of mammals are monotremes, marsupials, and placentals.

### *Characteristics of Mammals*

The two primary characters that define a mammal are hair and mammary glands. In general, mammals are generally endothermic furry animals nourished from birth with milk secreted by their mothers. Hair is unique to mammals, although in whales, armadillos, and some other mammals it is reduced considerably. A thick coat of hair, the **pelage,** primarily insulates the mammalian body to hold in heat although hair also has a sensory function. The bases of sensory hairs stimulate associated nerves when the hair is moved. "Whiskers" around the faces of carnivores and rodents are specialized long hairs called sensory **vibrissae.**

Sebaceous and sweat glands of the mammalian skin are associated with hair. Their products condition the skin and allow evaporative loss of excess body heat. The embryonic similarity between skin glands and mammary glands suggests that milk glands were derived from these specialized skin gland. In addition, mammalian red blood cells that transport oxygen lose their nuclei and most other cell organelles when they mature and enter the general circulation.

Hair, mammary glands, sweat and sebaceous glands, and anucleate red blood cells are unique to mammals. Other characteristics that are not necessarily restricted to this class include large brain in relation to body size, maintenance of high body temperature (except in some young and during resting periods of torpor), and modifications of the circulatory system from that of reptiles.

Hair and mammary glands rarely are preserved in fossils, so they are of little practical value in tracking the early evolution of mammals. Alternatively, fossil mammals exhibit three distinct skeletal characteristics. The first is a chain of three tiny middle ear bones that conduct sound from the tympanic membrane to the sensory apparatus of the inner ear. Reptiles have only one or sometimes two middle ear bones, but never three. Second, the lower jaw of mammals is composed only of the dentary, a single bone, whereas several bones make up the lower jaw of reptiles. The third skeletal feature is a joint between the dentary and squamosal bones of the jaws. In reptiles, other bones form the jaw joint. Even these three features are not always preserved in fossils, so paleontologists often resort to other backup features such as tooth structure. For instance, most teeth in mammals are replaced just once in a lifetime, not continuously, and occlusion of teeth is more precisely controlled than in reptiles.

**Mammalian teeth, their development, and functions(p. 497)**

Note that the characteristics we most associate with mammals (hair and mammary glands) are unavailable to paleontologists. Wherever the line between reptiles and mammals is drawn, it will be somewhat arbitrary. We cannot be certain that a boundary fossil with a mammalian skull or tooth pattern also possessed hair and milk glands in life.

Mammals inherited the basic synapsid skull design from therapsids before them who, in turn, inherited it from the earlier pelycosaurs. Reasonably then we could include all three groups with a synapsid skull—mammals, therapsids, pelycosaurs—in a single all-embracing taxonomic group. But this is usually not done. Most taxonomists like to recognize the new structural and physiological differences that distinguish mammals from therapsids and pelycosaurs.

### *Extinct Mammals*

Mammals are late Triassic derivatives of therapsids, making early mammals contemporaries of Mesozoic reptiles, such as pterosaurs, crocodiles, turtles, and dinosaurs (figure 3.37). Mammals did not become the predominant terrestrial group

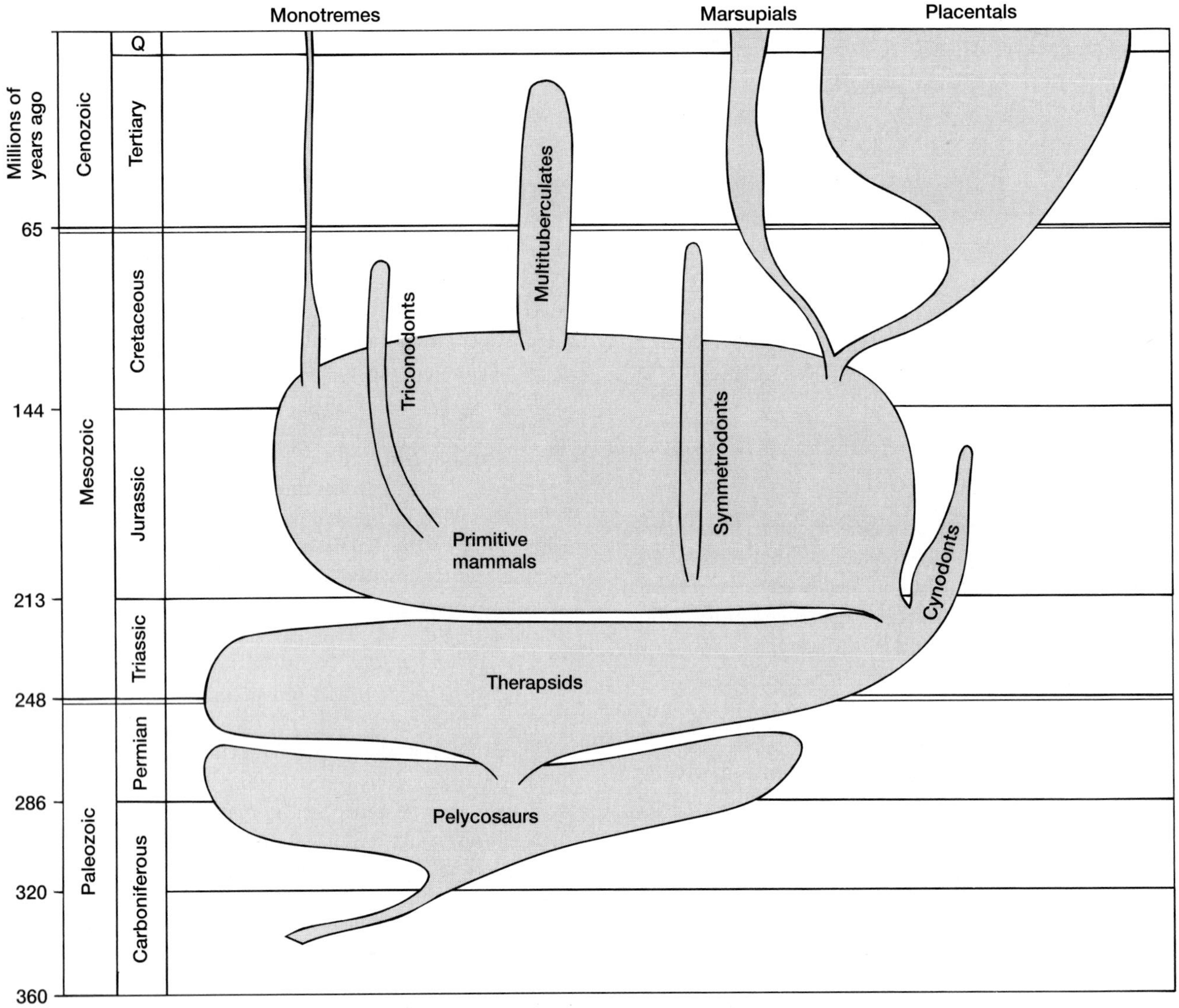


**Figure 3.37** Evolution of mammals. Primitive mammals evolved from therapsid reptiles, specifically from cynodonts. From these primitive mammals emanate several independent lines of descent. Some are now extinct (triconodonts, multituberculates, symmetrodonts). Others survive giving us the modern groups, the prototheres (monotremes) and the therians (marsupials and placentals).

until the Cenozoic, after the great dinosaur extinctions at the end of the preceding Mesozoic.

Extinct mammals include several groups with elaborate names—kuehneotherids, haramiyoids, sinoconodontids, multituberculates, and morganucodonts, to mention a few. Generally, these early mammals were the size of a mouse. They were probably nocturnal and endothermic, with teeth much like modern moles and shrews (insectivores). Brain size was larger, for a given body size, than in their reptilian contemporaries. Teeth in primitive mammals did more than just snag prey or clip vegetation. They had specialized functions—incisors at the front of the mouth, canines, premolars, and molars along the sides of the mouth. This permits division of labor, allowing some teeth to tear or clip food, others to break it up mechanically and prepare it for rapid digestion. Specialized tooth function implies, but does not prove, that primitive mammals were endothermic. If they were, they probably had a coat of insulating fur. Early mammals presumably hatched from eggs and nursed from mammary glands like the monotremes, the most primitive mammals living today.

## *Living Mammals*

All living mammals fall into one of three groups—monotremes, marsupials, and placentals (figure 3.38). Marsupial and placental mammals trace their ancestry to a common group in the early Cretaceous and so are

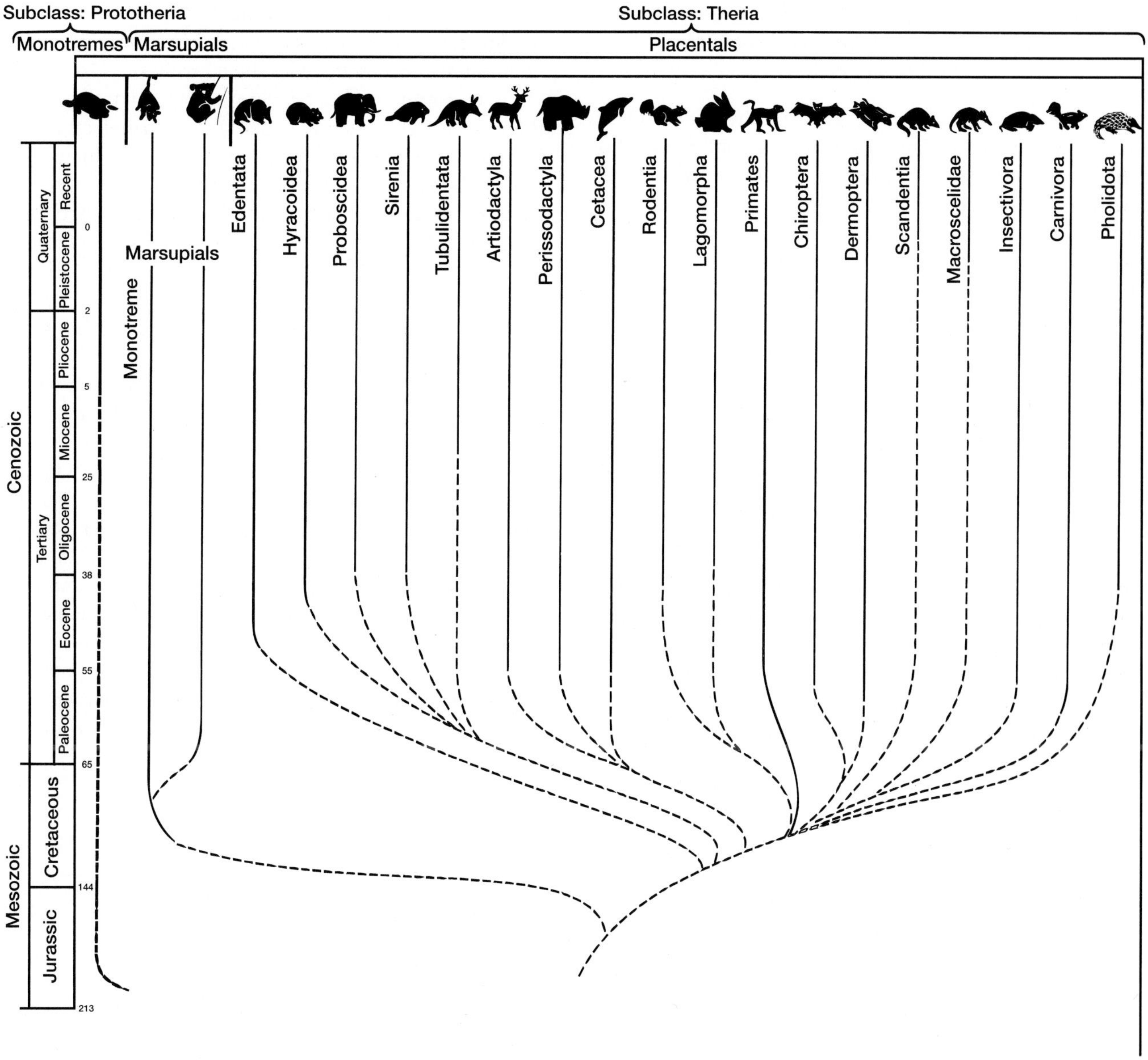


**Figure 3.38** Living mammals. Monotremes, marsupials, and placentals are the three groups of mammals living today, the placentals being the largest group.

placed together within the subclass Theria. Monotremes diverged early from the Theria, in the Lower Jurassic, and have been much on their own course ever since.

The three species of living monotremes include the duckbill platypuses that inhabit Australia and the two species of spiny anteaters that inhabit Australia and New Guinea. Like therian mammals, monotremes have hair, suckle their young, and are endotherms. However, unlike other mammals, monotreme embryos develop in shelled eggs, a primitive feature retained from reptilian ancestors.

**Monotreme embryology (p. 157)**

The earliest fossil marsupials are found in North America and to a lesser extent in South America, and so their first phylogenetic appearance is thought to be somewhere in the Western Hemisphere. Some, such as opossums, still live in the Americas, as did many other marsupial groups through most of the Cenozoic. Today, most marsupials are restricted to Australia and surrounding regions. Kangaroos are a familiar example. Tiny young are born at an early developmental stage, pull themselves into their mother's pouch, and suckle there until they grow considerably larger. Specialized forms still present in Australia, such as a burrowing marsupial (marsupial "mole") and a species that glides through the air (marsupial "flying squirrel"), suggest that marsupials once enjoyed great diversity.

**Marsupials (chapter 5–embryology, chapter 14–reproductive organs, and chapter 15–hormonal control of the breeding cycle)**

Placental mammals are today by far the most numerous and widespread of any mammalian group. They take their name from their mode of reproduction in which the nutritional and respiratory needs of the young are provided through a **placenta,** a vascular organ connecting the fetus and the female uterus. Such a vascular association between fetus and mother is not unique to placental mammals. A temporary "placenta" forms between the early embryo and the female uterus in some marsupials. In fact, nutritional and respiratory support of the embryo is found to varying degrees in some reptiles, fishes, and even a few amphibians. What distinguishes placental mammals is that reproduction in *all* species is based on a placenta.

**Vertebrate placentae (p. 181)**

Some of the most primitive living placentals are moles and shrews that belong to the order **Insectivora,** the group from which most other placental orders probably arose. Bats (order **Chiroptera**) are the only mammals with powered flight, although gliding placental mammals arose twice—as flying lemurs (order **Dermoptera**) in Asia and within the rodents ("flying" squirrels). Two placental groups are fully aquatic—the order Cetacea, which includes toothed whales **(odontocetes)** and the baleen whales **(mysticetes),** and the order Sirenia, which includes the manatees.

The term *ungulate* refers to hoofed animals, including the orders **Perissodactyla** (horses) and **Artiodactyla** (pigs, camels, cattle, deer, etc.). Subungulates loosely includes the orders **Proboscidea** (elephants), **Sirenia** (sea cows), and **Cetacea** (whales and porpoises). The rumen is a specialized part of the digestive tract from which **ruminants** of the suborder **Ruminantia** derive their common name. Giraffes, deer, cattle, bison, sheep, goats, antelopes, and their allies in the order Artiodactyla are all ruminants. Within the order **Carnivora,** the term *fissiped* is used informally for land carnivores (cats, dogs, bear, skunks), and the term *pinniped* refers to semiaquatic carnivores (seals and walruses).

**Rodentia** is the largest of the placental orders and is often divided informally into the **sciuromorphs** (squirrel-like), the **myomorphs** (mouselike), and the **hystricomorphs** (porcupinelike). Mammals of the order **Primates** can swing through trees **(brachiation),** or had ancestors who could, and they possess grasping fingers and toes tipped by nails. The **lower primates** or prosimians, include lemurs, lorises, and tarsiers. The **higher primates** or anthropoids encompass those with a tail (infraorder **platyrrhines**) and those without one (infraorder **catarrhines**). New World and Old World monkeys are found in

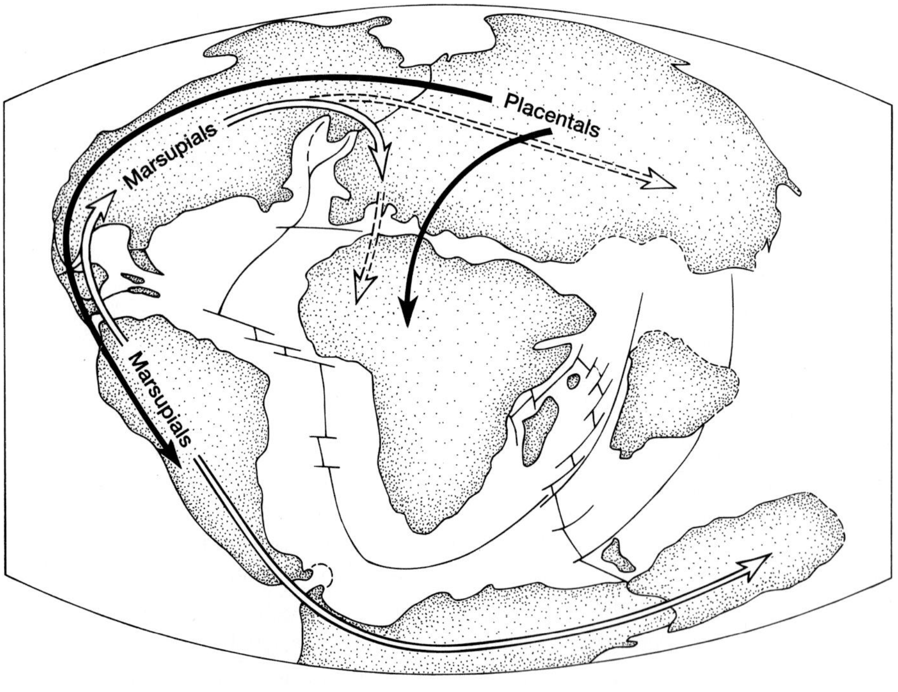


**Figure 3.39** Therian radiation. Position of the continents during the late Mesozoic is shown. Although today most marsupials live in Australia, their center of origin was apparently the New World of the late Cretaceous. From there they spread in two directions. One direction during the Eocene was to Europe and North Africa, although they subsequently became extinct on both those continents (dashed arrows). The other direction in which marsupials spread was through South America and Antarctica to Australia before these continents separated. Placentals originated in the Old World and spread to the New World via land connections that existed between the continents during the Mesozoic.

these different infraorders, but the word *monkey* is a general term that has no formal taxonomic definition. The term *apes* refers to members of the family **Pongidae,** and the term *hominids* to humans and their immediate ancestors of the family **Hominidae.**

The place of origin and routes of dispersal of therian mammals are still debated, although the known fossil record indicates that the earliest marsupials arose in the New World and the earliest placentals in the Old World late in the Cretaceous. Although continental drift was underway then and the Atlantic Ocean was present but small, most of the continents were still in contact (figure 3.39) and the late Cretaceous climate, even in polar regions, was mild. During this time, marsupials dispersed to Asia, Antarctica, and Australia while placentals migrated into Africa and to the New World. As the continents fragmented further, these stocks of mammals were carried into semi-isolation and served as the founding stocks for the distinctive mammalian groups that subsequently evolved on the separating continents.

## Selected References


Alexander, R. McN. 1974. *Functional design in fishes*. 3d ed. London: Hutchinson and Company.

Berman, D., S. S. Sumida, and R. E. Lombard. 1992. Reinterpretation of the temporal and occipital regions in *Diadectes* and the relationships of Diadectomorphs. *J. Paleont.* 66:481–99.

Bolt, J. R., and R. E. Lombard. 1992. Nature and quality of the fossil evidence for otic evolution in early tetrapods. In *The evolutionary biology of hearing,* edited by D. B. Webster, A. N. Popper, and R. R. Fay. New York: Springer-Verlag, pp. 377–403.

Carroll, R. L. 1982. Early evolution of reptiles. *Ann. Rev. Ecol. Syst.* 13:87–109.

Carroll, R. L. 1988. *Vertebrate paleontology and evolution*. New York: W. H. Freeman.

Chatterjee, S. 1991. Cranial anatomy and relationships of a new Triassic bird from Texas. *Phil. Trans. Royal Soc. (London)* 332:277–342.

Clack, J. A. 1988. New material of the early tetrapod *Acanthostega* from the Upper Devonian of east Greenland. *Palaeontology* 31:699–724.

Colbert, E. H. 1980. *Evolution of the vertebrates*. 3d ed. New York: John Wiley and Sons.

Feduccia, A. 1993. Evidence from claw geometry indicating arboreal habits of *Archaeopteryx*. *Science* 259:790–93.

Forey, P., and P. Janvier. 1993. Agnathans and the origin of jawed vertebrates. *Nature* 361:129–34.

Goodrich, E. S. 1958. Studies on the structure and development of vertebrates. New York: Dover Publications.

Hall, B. K. 1982. Bone in cartilaginous fishes. *Nature* 298:324.

Heaton, M. J., and R. R. Reisz. 1986. Phylogenetic relationships of captorhinomorph reptiles. *Can. J. Earth Sci.* 23:402–18.

Janvier, P. 1981. The phylogeny of the craniata, with particular reference to the significance of fossil "Agnathans." *J. Vert. Paleont.* 1:121–59.

Janvier, P. 1984. The relationships of the osteostraci and galeaspids. *J. Vert. Paleont.* 4:344–58.

Janvier, P., and A. Blieck. 1979. New data on the internal anatomy of the Heterostraci (Agnatha), with general remarks on the phylogeny of the Craniota. *Zool. Scripta* 8:287–96.

Jefferies, R.P.S. 1986. *The ancestry of vertebrates*. London: British Museum.

Kemp, T. S. 1982. *Mammal-like reptile and the origin of mammals*. London and New York: Academic Press.

Krejsa, R. J., P. Bringas, Jr., and H. C. Slavkin. 1990. A neontological interpretation of conodont elements based on agnathan cyclostome tooth structure, function, and development. *Lethaia* 23:359–78.

Lombard, R. E., and S. S. Sumida. 1992. Recent progress in understanding early tetrapods. *Amer. Zool.* 32:609–22.

Mallatt, J. 1984. Early vertebrate evolution: Pharyngeal structures and the origin of gnathostomes. *J. Zool. (London)* 204:169–81.

Milinkovitch, M. C. , G. Orti, and A. Meyer. 1993. Revised phylogeny of whales suggested by mitochondrial ribosomal DNA sequences. *Nature* 361:3346–48.

Milner, A. R. 1988. The relationships and origin of living amphibians. In *The phylogeny and classification of the tetrapods. Amphibians, reptiles, birds,* edited by M. J. Benton. Oxford, Clarendon Press. Systematics Assoc. Special Vol. 35A, 1:59–102.

Panchen, A. L. 1977. The origin and early evolution of tetrapod vertebrae. In *Problems in vertebrate evolution,* edited by S. M. Andrews, R. S. Miles, and A. D. Walker. New York: Academic Press, pp. 289–318.

Panchen, A. L. 1980. The origin and relationships of the anthracosaur amphibia from the Late Paleozoic. In *The terrestrial environment and the origin of land vertebrates,* edited by A. L. Panchen. New York: Academic Press, pp. 319–50.

Radinsky, L. B. 1987. *The evolution of vertebrate design*. Chicago: University of Chicago Press.

Romer, A. S. 1966. *Vertebrate paleontology*. 3d ed. Chicago: University of Chicago Press.

Romer, A. S. 1971. *The vertebrate story*. Rev. ed. Chicago: University of Chicago Press.

Stahl, B. J. 1974. *Vertebrate history: Problems in evolution*. New York: McGraw-Hill.

Stock, D. W., and G. S. Whitt. 1992. Evidence from 18S ribosomal RNA sequences that lampreys and hagfishes form a natural group. *Science* 257:787–89.

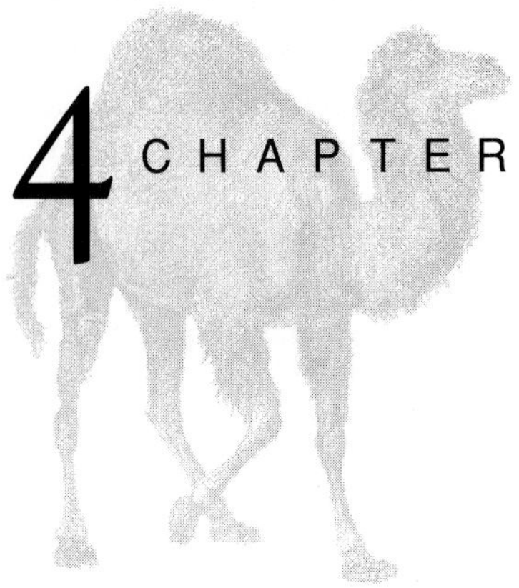

# 4 CHAPTER Biological Design



## Introduction: Size and Shape

Bodies, like buildings, obey laws of physics. Gravity will bring down an ill-designed dinosaur just as certainly as it will fell a faulty drawbridge. Animals must be equipped to address biological demands. The long neck of a giraffe gives it access to treetop vegetation; the claws of cats hook prey; a thick coat of fur gives the bison protection from the cold of winter. In order for animals to catch food, flee from enemies, or endure harsh climates, structures have evolved that serve them against these challenges to survival. But there is more to an animal's environment than predators and prey, climate and cold. An animal's design must address physical demands. Gravity acts on all structures within its reach. Heavy terrestrial vertebrates must exert much effort to move a massive body from one place to another. Bones and cartilage must be strong enough to bear the weight. If these skeletal structures fail, so does the organism, and its survival is in doubt. Animals at rest or in motion experience forces that their structural systems must withstand. As the British biologist J.B.S. Haldane put it,

> **"It is easy to show that a hare could not be as large as a hippopotamus, or a whale as small as a herring. For every type of animal there is a most convenient size, and a large change in size inevitably carries with it a change in form."**
> **(Haldane, 1956, p. 952)**

In this chapter we examine how structures built by humans and those evolved by natural selection have design features that incorporate and address common problems posed by basic physical forces. For example, living organisms come in a great variety of sizes (figure 4.1); however, not all designs work equally well for all sizes (figure 4.2).

A grasshopper can jump a hundred or more times its own body length. From time to time, this feat has tempted some people to proclaim that if we were grasshoppers we could leap tall buildings in a single bound. The implication is that grasshoppers possess special leaping devices absent in humans. Certainly grasshoppers have

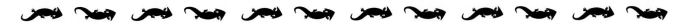

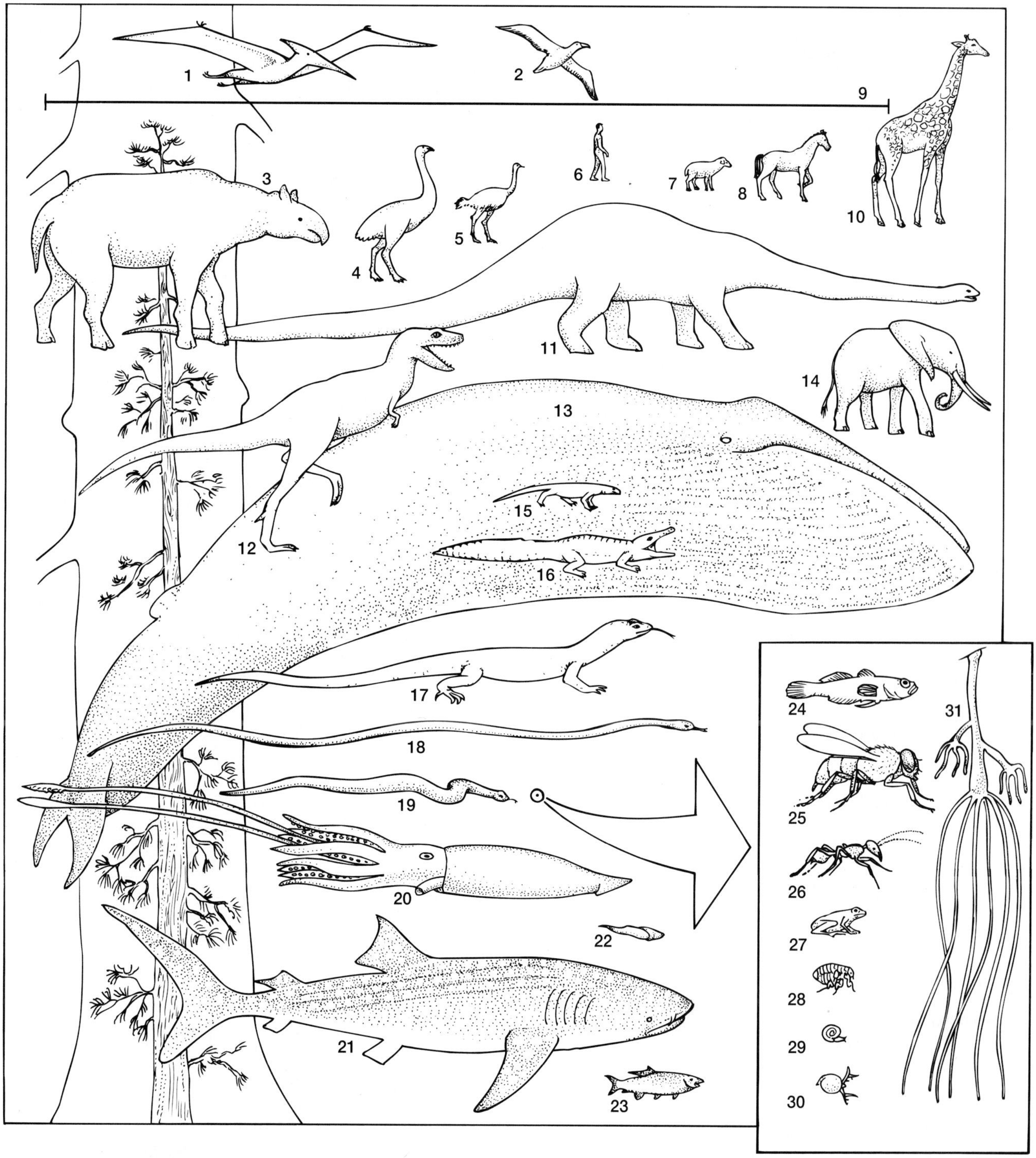


**Figure 4.1** Animal sizes range over many orders of magnitude. The largest animal is the blue whale, the smallest adult vertebrate a tropical frog. All organisms are drawn to the same scale and are numbered as follows: (1) the pterosaur *Quetzalcoatlus* is the largest aerial reptile; (2) the albatross is the largest flying bird; (3) *Baluchitherium* is the largest extinct land mammal; (4) *Aepyornis* is the largest extinct bird; (5) ostrich; (6) a human figure represented by this scale is 6 feet tall; (7) sheep; (8) horse; (9) this line designates the length of the largest tapeworm found in humans; (10) the giraffe is the tallest living land animal; (11) *Diplodocus;* (12) *Tyrannosaurus;* (13) the blue whale is the largest known living animal; (14) African elephant; (15) the Komodo dragon is the largest living lizard; (16) the saltwater crocodile is the largest living reptile; (17) the largest extinct lizard; (18) *Gigantophis* is the largest extinct snake; (19) the reticulated python is the longest living snake; (20) *Architheuthis*, a deep-water squid is the largest living mollusc; (21) the whale shark is the largest fish; (22) an arthrodire is the largest placoderm; (23) large tarpon; (24) *Trimmatom nanus* is one of the smallest fishes; (25) housefly; (26) medium-sized ant; (27) this tropical frog is the smallest vertebrate; (28) cheese mite; (29) smallest land snail; (30) *Daphnia*, is a common water flea; (31) a common brown hydra. The lower section of a giant sequoia is shown in the background on the left of the figure with a 100-foot larch superimposed.

suitably long legs that launch them great distances. But the more important reason why grasshoppers and humans differ in their relative jumping abilities is a matter of size, not a matter of long legs. If a grasshopper was enlarged to the size of a human, it too would be unable to leap a hundred times its new body length, despite its long legs. Differences in size necessarily bring differences in performance and in design.

To illustrate this point, let us look at two examples, one from music and one from architecture. A small violin, although shaped generally like a bass, encloses a smaller resonance chamber; therefore, its frequency range is higher (figure 4.3). The larger bass has a lower frequency range. A Gothic cathedral, because it is large, encloses relatively more space than a small brick and mortar neighborhood church. Large cathedrals include devices to increase surfaces through which light may pass in order to illuminate the congregation within (figure 4.4). The end and side walls of cathedrals are designed with outpocketings that architects call apses and transepts. The side walls are pierced by slotted openings, clerestories, and tall windows. Together, apses, transepts, clerestories, and windows allow more light to enter, so they compensate for the proportionately larger volume enclosed within. Later in this chapter, we will see that this principle applies to animal bodies as well.

Shipbuilders often resort to a scale model to test ideas for hull design. But the model, because it is many times smaller than the ship it represents, responds differently to the wave and surface action of water in a testing tank. Thus, a model may not reliably mimic the performance of a larger ship.

Size and shape are functionally linked whether we look inside or outside of biology. The study of size and its consequences is known as **scaling.** Mammals, from shrews to elephants, fundamentally share the same skeletal architecture, organs, biochemical pathways, and body temperature. But an elephant is not just a very large shrew. Scaling requires more than just making parts larger or smaller. As body size changes, the demands on various body parts change disproportionately. Even metabolism scales with

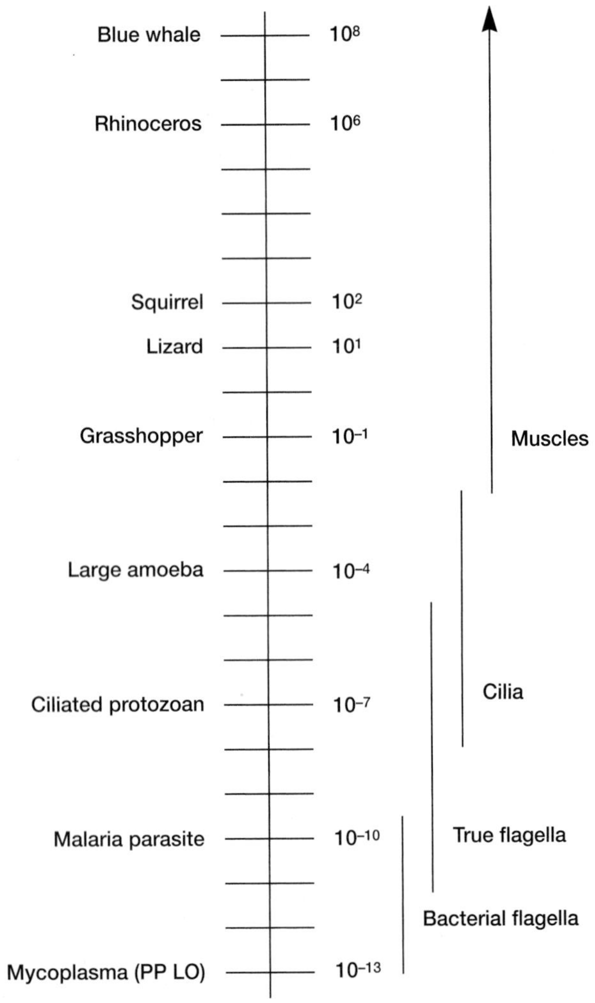


**Figure 4.2** Body size and locomotion. The masses of living organisms are given on a logarithmic scale. The blue whale tops the scale. Mycoplasma, a prokaryotic, bacterium-like organism, is at the bottom. The locomotor mechanism ranges from bacterial flagella to muscle as size increases. Size imposes constraints. Cilia and flagella that move a small mass well become less suitable for locomotion of larger masses. Movement in bigger animals is based on muscles.

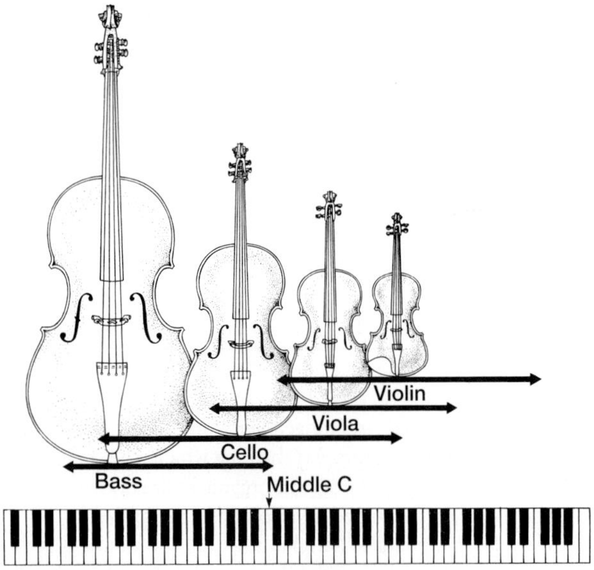


**Figure 4.3** Influence of size on performance. The four members of the violin family are similarly shaped but they differ in size. Size differences alone produce different resonances and account for differences in performance. The bass is low, the violin high, and the middle-sized cello and viola produce intermediate frequencies.

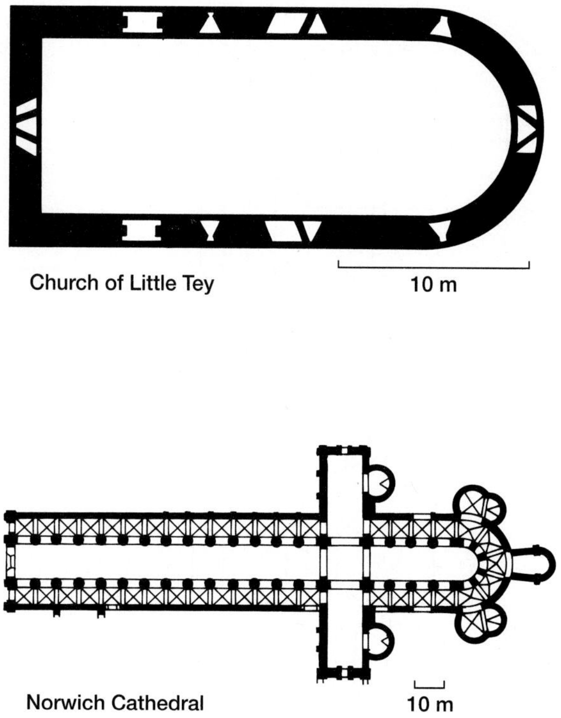


**Figure 4.4** Influence of size on design. The floor plans of a small medieval church (top) and a large Gothic cathedral (bottom) are drawn to about the same size. The medieval church is about 16 m in length, the Gothic cathedral about 139 m. Because the Gothic cathedral is larger in life, however, it encloses relatively greater space. Transept, chapels, and slotted windows of the side walls of the cathedral must let in more light to compensate for the larger volume and to brighten the interior.
For an extended account of the consequences of size on design see Gould, 1977.

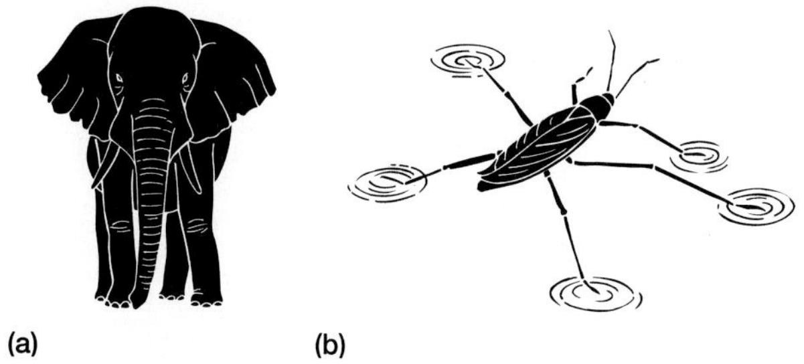


**Figure 4.5** Consequences of being large or small. Gravity exerts an important force on a large mass. Surface tension is more important for smaller masses. (a) The large elephant has stout, robust legs to support its great weight. (b) The small water strider is less bothered by gravity. In its diminutive world, surface forces become more significant as it stands on water supported by surface tension.

size. Oxygen consumption per kilogram of body mass is much higher in smaller bodies. Size and shape are necessarily linked, and the consequences affect everything from metabolism to body design. To understand why, we look first to matters of size.

## Size

Because they differ in size, the world of an ant or a water strider and the world of a human or an elephant offer quite different physical challenges (figure 4.5a,b). A human coming out of his or her bath easily breaks the water's surface tension and, dripping wet, probably carries without much inconvenience 250 g (about half a pound) of water clinging to the skin. But if a person slips in the bath, however, he or she has to contend with the force of gravity and risks breaking a bone. For an ant, surface tension in even a drop of water could hold the insect prisoner if not for properties of its chitinous exoskeleton that make it water repellent. On the other hand, gravity poses little danger. An ant can lift 10 times its own weight, scamper upside down effortlessly across the ceiling, or fall long distances without injury. Generally, the larger an animal, the greater is the significance of gravity. The smaller an animal, the more it is ruled by surface forces. The reason for this has little to do with biology. Instead, the consequences of size arise from geometry and the relationships among length, surface, and volume. Let us consider these.

### Relationships among Length, Area, and Volume

If shape remains constant but body size changes, the relationships among length, surface area, volume, and mass change. A cube, for instance, that is doubled in length and then doubled again is accompanied by larger proportional changes in surface and volume. Thus, as its length doubles, its edge length increases by increments of 1, 2, and 4 cm. However, total surface area of its faces does not simply double; surface area increases by increments of 1, 4, and 16 cm. The cube's volume increases in even faster steps, of 1, 8, and 64 cc (figure 4.6a). The shape of the cube stays constant, but because, and *only* because, it is larger, the biggest cube encloses relatively more volume per unit of surface area than does the smallest cube. In other words, the biggest cube has relatively less surface area per unit of volume than the smallest cube (figure 4.6b).

It is certainly no surprise that a large cube has, in *absolute* terms, more total surface area and more total volume than a smaller cube. But notice the emphasis on *relative* changes between volume and surface area, and between surface area and length. These are a direct consequence of changes in size. These relative changes in surface area in relation to volume have profound consequences for the design of bodies or buildings. Because of them, a change in size

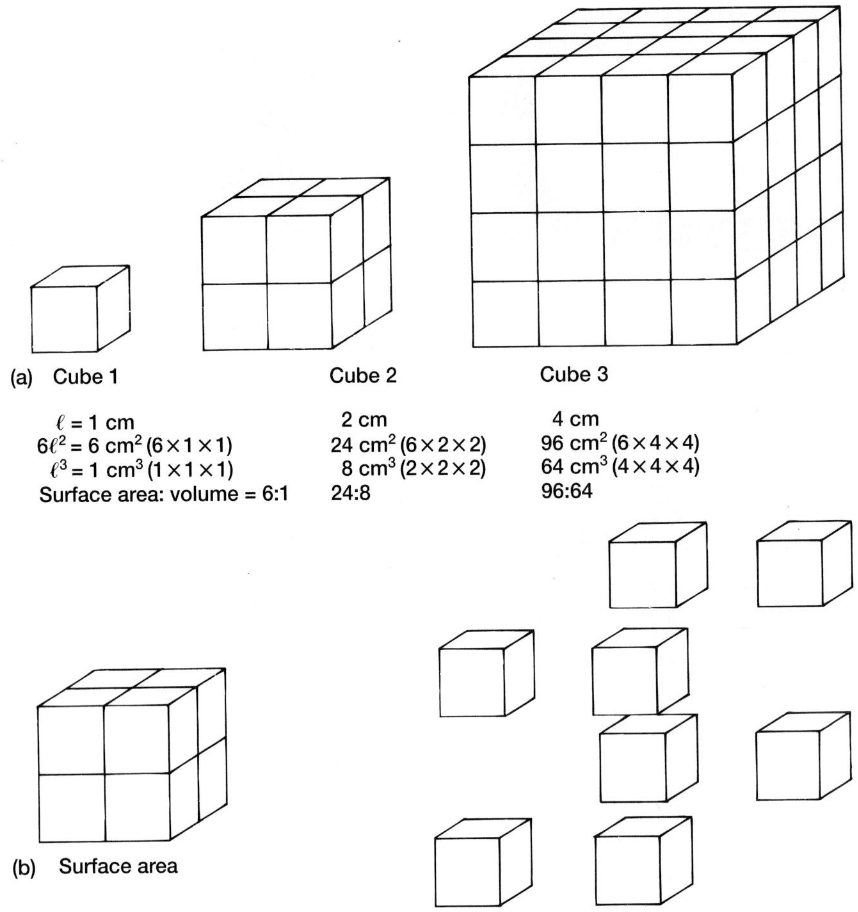


**Figure 4.6** Length, surface, and volume. (a) Even if shape remains the same, a size increase alone changes the proportions between length, surface, and volume. The length of each edge of the cube quadruples from the smallest to the largest size shown. Cubes 1, 2, and 3 are 1, 2, and 4 cm in length (l) on a side, respectively. The length (l) of a side increases by a factor of 2 as we go from cube 1 to cube 2 and from cube 2 to cube 3. The surface area jumps by a factor of 4 ($2^2$) with each doubling of length, and the volume increases by a factor of 8 ($2^3$). A large object has relatively more volume per unit of surface than a smaller object of the same shape. (b) Surface area. By dividing an object into separate parts, the exposed surface area increases. The cube shown on the left has a surface area of 24 $cm^2$, but when it is broken into its constituents, the surface area increases to 48 $cm^2$ ($8 \times 6$ $cm^2$). Similarly, chewing food breaks it into many pieces and so exposes more surface area to the action of digestive enzymes in the digestive tract.

inevitably requires a change in design to maintain overall performance.

More formally stated, the surface area ($S$) of an object increases in proportion to ($\propto$) the square of its linear dimensions ($l$):

$$S \propto l^2.$$

But volume ($V$) increases even faster in proportion to the cube of its linear dimensions ($l$):

$$V \propto l^3.$$

This proportional relationship holds for any geometric shape expanded (or reduced) in size. If we enlarge a sphere, for example, from marble size to soccer ball size, its diameter increases 10 times, its surface increases $10^2$ or 100 times, and its volume increases $10^3$ or 1,000 times. Any object obeys these relative relationships imposed by its own geometry. A tenfold increase in the length of an organism, as can occur during growth, brings a hundredfold increase in its surface area and a thousandfold increase in its volume (if no changes in body shape occur during growth). Consequently, the same organism is necessarily different when large and must accordingly be designed differently to accommodate different relationships among its length, surface, and volume. With this in mind, let us next turn to surface area and volume as factors in design.

## Surface Area

To start a fire, a single log is splintered into many small pieces of kindling. Because the surface area is increased, the fire can start more easily. Similarly, many bodily processes and functions depend on relative surface area. Chewing food breaks it into smaller pieces and increases the surface area available for digestion. The efficient exchange of gases, oxygen and carbon dioxide, for instance, depends in part upon available surface area as well. In gills or lungs, large blood vessels branch into many thousands of tiny vessels, the capillaries, thereby increasing surface area and facilitating gas exchange with the blood. Folds in the lining of the digestive tract increase surface area available for absorption. Drag on a fast-swimming fish is proportional to the resistance of its skin to the water flowing across its body surface; hence, the body shapes of fishes minimize surface area. Bone strength and muscle force are proportional to the cross-sectional areas of parts that particular bones and muscles support or move. Vast numbers of bodily processes and functions depend on relative surface area. These examples show that some designs maximize surface area, while others minimize it. Organs (lungs, gills, stomach, blood vessels) that are adapted to promote exchange of materials typically have large surface areas. But where surface friction resists movement, such as through air or water, efficient locomotion is promoted by having small surface area.

Because, as we have seen, surface and volume scale differently with changing size, processes based on relative surface area must change with increasing size. For example, in a tiny aquatic organism, surface cilia stroke in coordinated beats to propel the animal. As the animal gets larger, surface cilia have to move proportionately more volume, so they become a less effective means of locomotion. It is no surprise that large aquatic organisms depend more on muscle power than on ciliary power to meet their locomotor needs. The circulatory, respiratory, and digestive systems rely particularly on surfaces to support the metabolic needs required by the mass of an animal. Large animals must have large digestive areas to ensure adequate surface for assimilation of food in order to sustain the bulk of the organism. Large animals can compensate and maintain adequate rates of absorption if the digestive tract increases in length and develops folds and convolutions. Rate of oxygen uptake by lungs or gills, diffusion of oxygen from blood to tissues, and gain or loss of body heat are all physiological processes that rely on surface area. As J.B.S. Haldane once said: "Comparative anatomy is largely the story of the struggle to increase surface in proportion to volume" (Haldane, 1956, p. 954). We will not be surprised then, when in later chapters we discover that organs and whole bodies are designed to address the relative needs of volume in relation to surface area.

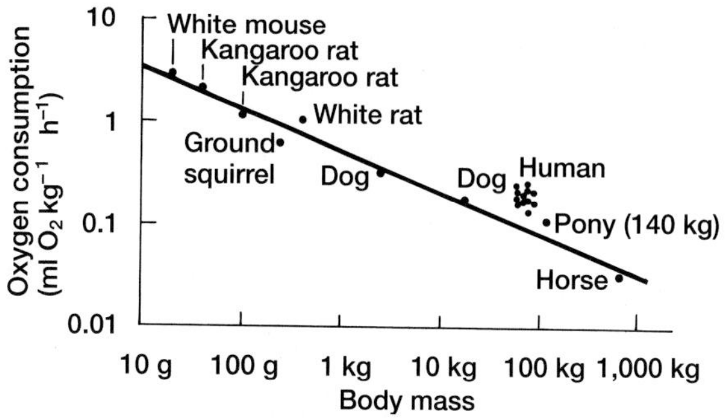


**Figure 4.7** Relationship between metabolism and body size. Physiological processes, like anatomical parts, scale with size. The graph shows how oxygen consumption decreases per unit of mass as size increases. This is a log-log plot showing body mass along the horizontal scale and oxygen consumption along the vertical. Oxygen consumption is expressed as the volume (ml) of oxygen ($O_2$) per unit of body mass (kg) during one hour (h).

As body size increases, oxygen consumption per unit of body mass decreases (figure 4.7). In absolute terms, a large animal, of course, takes in more total food per day than does a small animal to meet its metabolic needs. Certainly, an elephant eats more each day than a mouse does. A cougar may consume several kilograms of food per day, a shrew only several grams. But in relative terms, metabolism per gram is less for the larger animal. The several grams the shrew consumes each day may represent an amount equivalent to several times its body weight; the cougar's daily food intake is a small part of its body mass. Small animals operate at higher metabolic rates; therefore, they must consume more oxygen to meet their energy demands and maintain necessary levels of body heat. This is partly due to the fact that heat loss is proportional to surface area, whereas heat generation is proportional to volume. A small animal has more surface area in relation to its volume than a larger animal does. If a shrew were forced to slow its weight-specific metabolic rate to that of a human, then to keep warm the tiny shrew would need an insulation of fur at least 25 cm thick.

## Volume and Mass

When an object increases in volume, its mass increases proportionately. Because body mass is directly proportional to volume, mass (like volume) increases in proportion to the cube of a body's linear dimensions.

In terrestrial vertebrates, the mass of the body is borne by the limbs, and the strength of the limbs is proportional to their cross-sectional area. Change in body size, however, sets up a potential mismatch between body mass and cross-sectional limb area. As we learned earlier in this section, mass increases faster than surface

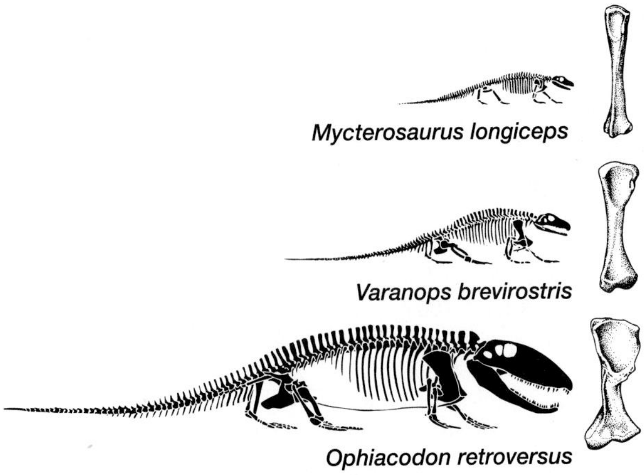


**Figure 4.8** Body size and limb design in pelycosaurs. Relative sizes of three pelycosaur species are illustrated. The femurs of each, drawn to the same length, are shown to the right of each species. The larger pelycosaur carries a relatively larger mass, and its more robust femur reflects this supportive demand.

area when size increases. A tenfold increase in diameter produces a thousandfold increase in mass but only a hundredfold increase in cross-sectional area of the supporting limbs. Without compensatory adjustments, weight-bearing bones fall behind the mass they must carry. In large animals, bones are scaled in proportion to the mass they must support in order to prevent their breaking. For this reason, bones of large animals are relatively more massive and robust than the bones of small animals (figure 4.8). This disproportionate increase in mass compared with surface area is the reason why gravity is more significant for large animals than for small ones.

Whether we look at violins, Gothic cathedrals, or animals, the consequences of geometry reign when it comes to size. Objects of similar shape but different size must differ in performance.

## Shape

To remain functionally balanced, an animal must have a design that can be altered as its length and area and mass grow at different rates. As a result, an organism must have different shapes at different ages (sizes).

### Allometry

As a young animal grows its proportions may also change. Young children, too, change in proportion as they grow; children are not simply miniature adults. Relative to adult proportions, the young child has a large head and short arms and legs. This change in shape in correlation with a change in size is called **allometry** (figure 4.9).

Allometry rests on comparisons, usually of different parts as an animal grows. For instance, during growth, the bill of the godwit, a shorebird, increases in length faster than its head. The bill becomes relatively long compared to the skull (figure 4.10). Generally, the relative sizes of two parts, $x$ and $y$, can be expressed mathematically in the allometric equation

$$y = bx^a$$

where $b$ and $a$ are constants. When the equation is graphed on log-log paper, a straight line results (figure 4.11a,b).

Allometric relationships describe changes in shape that accompany changes in size. Size changes do not occur only during ontogeny. Occasionally, a phylogenetic trend within a group of organisms includes a relative change in size and proportion through time. Allometric plots describe these trends as well. Titanotheres are an extinct group of mammals comprised of ten known species from the Cenozoic. A plot of skull length versus horn height for each species shows an allometric relationship (figure 4.12). In this example, we track evolutionary changes in the relationship between parts followed through several species.

Compared with a reference part, the growing feature may exhibit positive or negative allometry, depending on whether it grows faster than (positive) or slower than (negative) the reference part. For example, compared

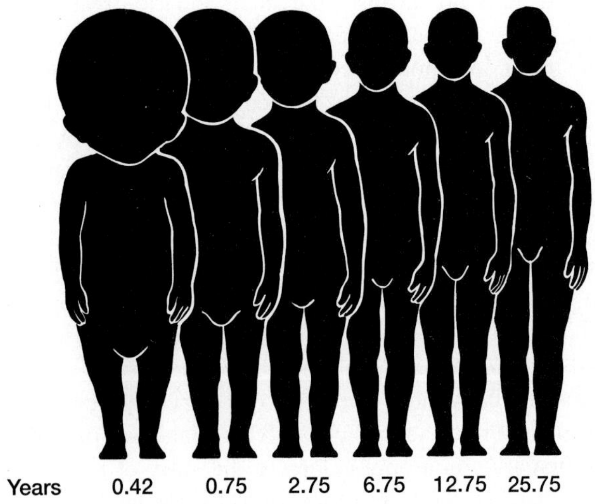


**Figure 4.9** Allometry in human development. During growth, a person changes shape as well as size. As an infant grows, its head makes up less and its limbs make up more of its overall height. Ages, in years, are indicated beneath each figure.

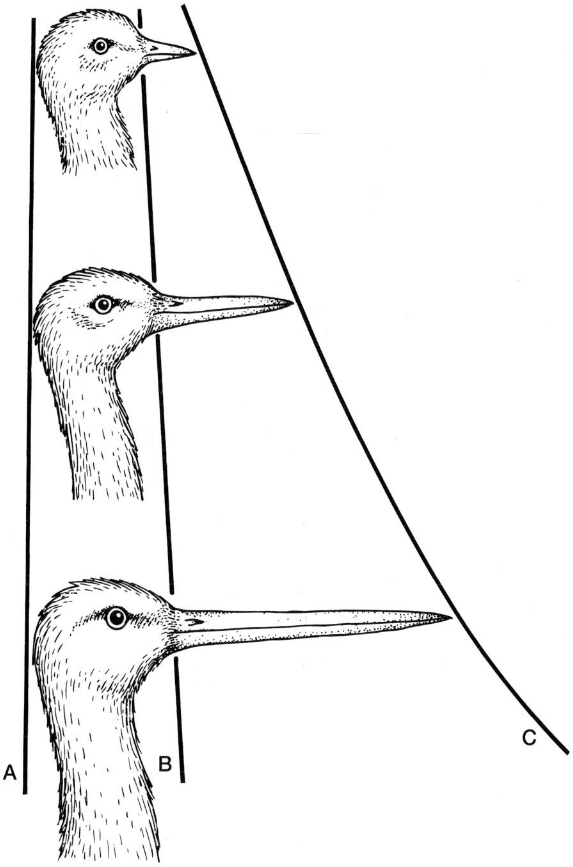


**Figure 4.10** Allometry in the head of a black-tailed godwit. Difference in relative growth between skull length (lines A and B) and bill length (lines B and C) are compared. Notice that for each increase in skull length the bill grows in length as well, but at a faster rate. As a result, the bill is shorter than the skull in the chick (top) but longer than the skull in the adult (bottom).

with skull length, the bill of the godwit shows positive allometry. The term **isometry** describes growth in which the proportions remain constant, and neither positive nor negative allometry occurs. The cubes shown in figure 4.6 exemplify isometry as do the salamanders illustrated in figure 4.13.

## Transformation Grids

D'Arcy Thompson popularized a system of transformation grids that express overall changes in shape. The technique compares a reference structure to a derived structure. For instance, if the skull of a human fetus is taken as a reference structure, a rectilinear transformation grid can be used to define reference points at the intersections of the horizontal and vertical grid lines (figure 4.14). These reference points on the fetal skull are then relocated on the adult skull. Next these reference points are connected again to redraw the grid, but because the shape of the skull has changed with growth, the redrawn grid too is differently shaped. Thus the grid graphically depicts shape changes. Similarly, transformation grids can be used to emphasize graphically phylogenetic differences in shape between species, such as the fishes shown in (figure 4.15).

Transformation grids and allometric equations do not explain changes in shape, they only describe them. However, in describing changes in proportions, they focus our attention on how tightly shape couples with size.

# On the Consequences of Being the Right Size

Animals, large or small, enjoy different advantages because of their sizes. The larger an animal, the fewer are the predators that pose a serious threat. Adult rhinoceroses and elephants are simply too big for most would be predators to handle. Large size is also advantageous in species in which physical aggression between competing males is part of reproductive behavior. On the other hand, small size has its advantages as well. In fluctuating environments struck by temporary drought, the sparse grass or seeds that remain may sustain a few small rodents. Because they are small, they require only a few handfuls of food to see them through. When the drought slackens and food resources return, the surviving rodents, with their short reproductive and generation times, respond in short order and their population recovers. In contrast, a large animal needs large quantities of food on a regular basis. During a drought, a large animal must migrate or perish. Typically, large animals also have long generation times and prolonged juvenile periods. Thus, populations of large animals may take years to recover after a severe drought or other devastating environmental trauma.

The larger an animal, the more its design must be modified to carry its relatively greater weight, a consequence of the increasing effects of gravity. It is no coincidence that the blue whale, the largest animal on Earth, evolved in an aquatic environment in which its great weight received support from the buoyancy of the surrounding water. For terrestrial vertebrates, an upper size limit occurs when supportive limbs become so massive that locomotion becomes impractical. The movie creators of Godzilla were certainly unaware of the impracticality of their design as this great beast crashed about stomping buildings. For lots of reasons, not the least of which is his size, Godzilla is an impossibility.

Body parts used for display or defense often show allometry, as the adult ram horns in figure 4.16 illustrate. As a male lobster grows, its defensive claw grows too, but much more rapidly than the rest of its body. When the lobster attains a respectable size, its claw has grown into a formidable weapon (figure 4.17). The claw exhibits **geometric growth,** that is, its length is *multiplied* by a constant in each time interval. The

| Specimen | Skull dimensions (mm) | |
|---|---|---|
| | x | y |
| A | 1 | 3.5 |
| B | 2 | 11.8 |
| C | 3 | 23.9 |
| D | 4 | 39.6 |
| E | 5 | 58.5 |
| F | 6 | 80.5 |
| G | 7 | 105.4 |
| H | 8 | 133.0 |
| I | 9 | 163.7 |
| J | 10 | 196.8 |
| K | 20 | 662.0 |
| L | 30 | 1,345.9 |
| M | 40 | 2,226.8 |
| N | 50 | 3,290.5 |
| O | 60 | 4,527.2 |
| P | 70 | 5,929.1 |
| Q | 80 | 7,489.9 |
| R | 90 | 9,204.3 |

(a)

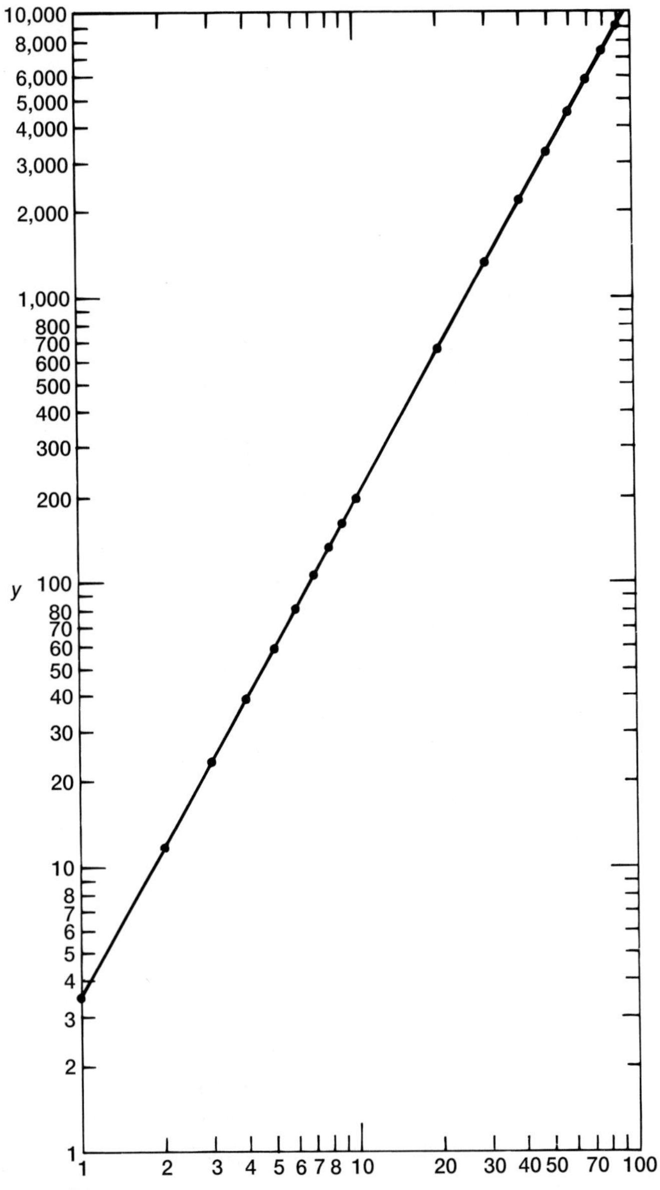


(b)

**Figure 4.11** Graphing allometric growth. (a) If we organize a range of skulls from the same species in order of size (A–R), we can measure two homologous parts on each skull and collect these data points in a table. (b) If we plot one skull dimension ($y$) against the other ($x$) on log-log paper, a line connecting these points describes the allometric relationship between the points during growth in size of the members of this species. This can be expressed with the general allometric equation, $y = bx^a$, wherein $y$ and $x$ are the pair of measurements and $b$ and $a$ are constants, $b$ being the y-intercept and $a$ being the slope of the line. In this example, the slope of the line ($a$) is 1.75. The y-intercept ($b$) is 3.5 observed on the graph or calculated by placing the value of $x$ equal to 1 and solving for $y$. The equation describing the data is $y = 3.5x^{1.75}$.

(a) from *On Size and Life* by Thomas A. McMahon and John Tyler Bonner. Reprinted with permission of W. H. Freeman and Company.

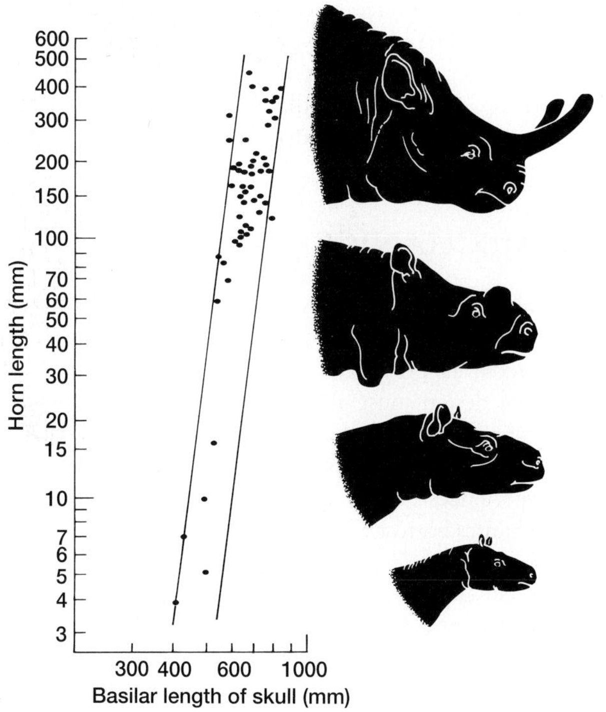


**Figure 4.12** Allometric trends in phylogeny. The skull and horn lengths of titanotheres, an extinct family of mammals, are plotted. The horn length increases allometrically with increasing size of the skull of each species.

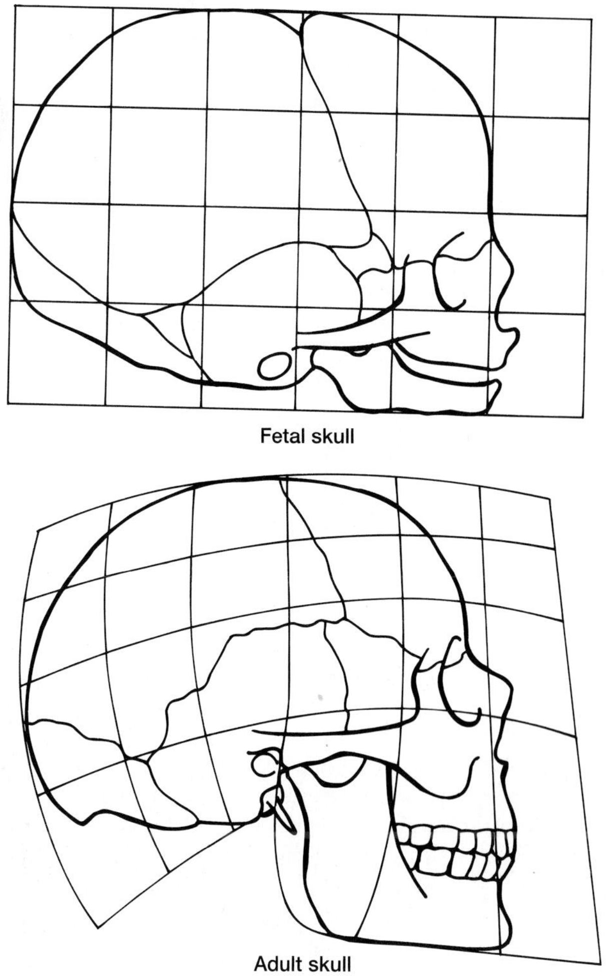


**Figure 4.14** Transformation grids in ontogeny. Shape changes in the human skull can be visualized more easily with correlated transformation grids. Horizontal and vertical lines spaced at regular intervals can be laid over a fetal skull. The intersections of these lines define points of reference on the fetal skull that can be relocated on the adult skull (bottom) and used to redraw the grid. Because the adult skull has a different shape, the reference points from the fetal skull must be reoriented. A reconstructed grid helps to emphasize this shape change.

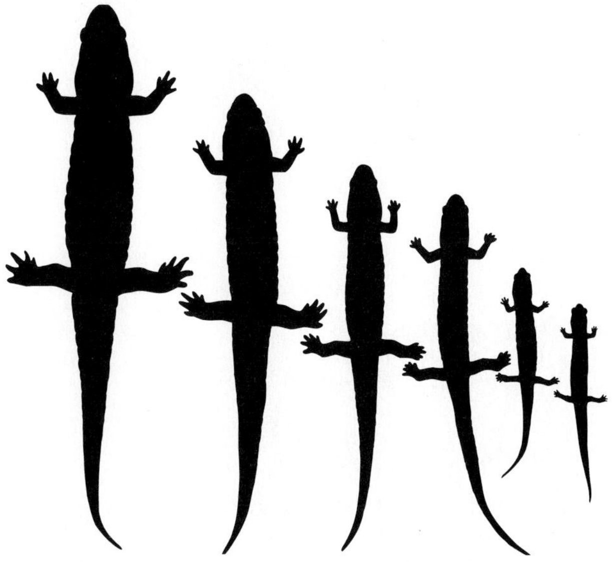

**Figure 4.13** Isometry. These six species of salamanders differ in size; yet, the smallest is almost the same shape as the largest because body proportions within this genus *(Desmognathus)* remain almost constant from species to species.
Kindly supplied by Samuel S. Sweet.

rest of the body shows **arithmetic growth** because a constant is *added* to its length in each time interval. To be effective in defense, the claw must be large, but a young lobster cannot yet wield so heavy a weapon because of its small size. Only after attaining substantial body size can such a claw be effectively deployed in defense. The accelerated growth of the claw brings it in later life up to fighting size. Before that, the small lobster's major defensive tactic is to dash for cover under a rock.

This example shows that size and shape are sometimes linked because of biological function, as with the lobster and its claw. More often, however, design is concerned with the consequences of geometry. Changes in the relationship

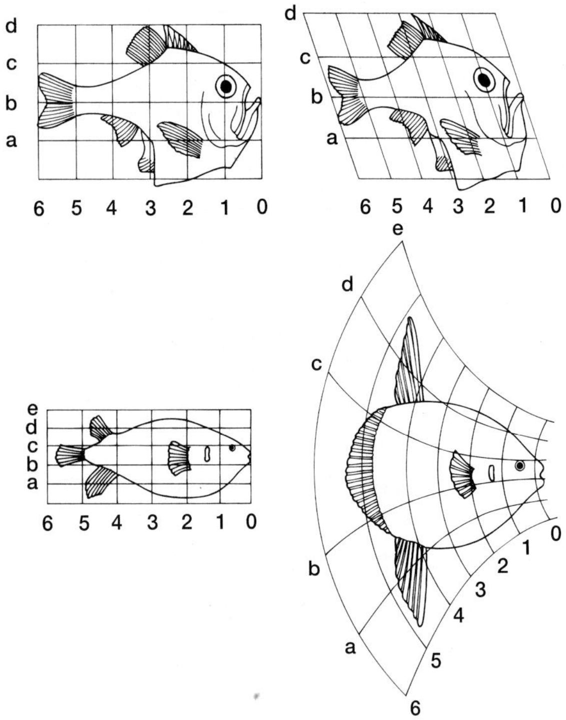


**Figure 4.15** Transformation grids in phylogeny. Changes in shape between two or more species, usually closely related, can also be visualized with transformation grids. One species is taken as the reference (left), and the reference points are relocated in the derived species (right) to reconstruct the transformed grid.

between length, surface, and volume as an object increases in size (figure 4.6) are the major reason why change in size is necessarily accompanied by change in shape. As we see time and again, throughout the book, size itself is a factor in vertebrate design and performance.

## Biomechanics

Physical forces are a permanent part of an animal's environment. Much of the design of an animal serves to catch prey, elude predators, process food, and meet up with mates. But biological design must also address the physical demands placed upon the organism. In part, analysis of biological design requires an understanding of the physical forces an animal experiences. Those in the field of **bioengineering** or **biomechanics** borrow concepts from engineering mechanics to address these questions.

Mechanics is the oldest of the physical sciences, with a successful history dating back at least 5,000 years to the ancient pyramid builders of Egypt. It continues up to the present with the engineers that send spaceships to the planets. Through the course of its history, engineers of this discipline have developed principles that describe the physical properties of objects from bodies to buildings. Ironically, engineers and biologists usually work in reverse directions. An engineer starts with a problem, for instance, a river to span, and then designs a product, a

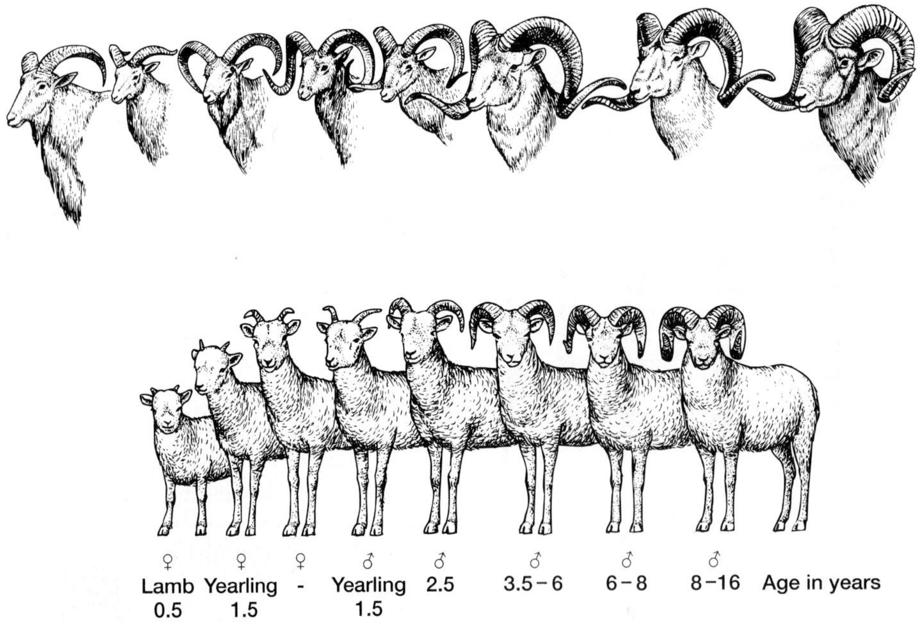


**Figure 4.16** Changes in horn shape. These species and subspecies of Asiatic sheep show changes in horn shape across their geographic distribution (top). The first in the lineup is the Barbary sheep *(Ammotragus)* from North Africa. The others belong to species or subspecies of *Ovis*, Asiatic sheep of the argali group that extend into central Asia. The last sheep on the right is the Siberian argali *(Ovis ammon ammon)*. As a young male bighorn grows in size (bottom illustration), its horns change shape as well. In the adult ram, these horns are used in social displays and in combat with male rivals.

Modified from Geist, 1971.

bridge, to solve the problem. A biologist, however, starts with the product, for instance, a bird wing, and works back to the physical problem it solves, namely, flight. Nonetheless, reducing animals to engineering analogies simplifies our task of understanding animal designs.

## Fundamental Principles

Certainly animals are more than just machines. But the perspective of biomechanics gives a clarity to biological design that we might not otherwise expect. What follows is an introduction to a few basic biomechanical principles.

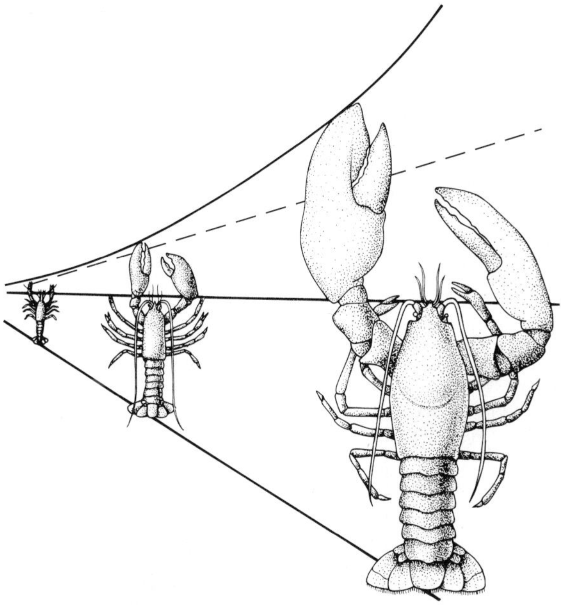

**Figure 4.17** Lobster allometry. Although the defensive claw is small at first, it grows geometrically while body length increases only arithmetically. Because of this, when the body is large enough to use the claw, the claw has increased dramatically in size to become an effective weapon. The dotted line indicates the size the claw would reach if it did not show allometry, that is, if it grew arithmetically instead of geometrically.

### *Basic Quantities—Length, Time, Force, and Mass*

Most of the physical concepts we deal with in biomechanics are familiar. **Length** is a concept of distance, **time** is a concept of the flow of events, and **force** describes the effects of one body acting on another.

When it comes to the concept of **mass,** however, our intuition not only fails but actually interferes because what most people call "weight" is not equivalent to "mass." Mass is a property of matter, weight a measure of force. One way to think of the difference is to consider two objects in outer space, say, a pen and a refrigerator. Both would be weightless and neither would exert a force on a scale. However, both still have mass, although the mass of each is different. To toss the pen to a companion astronaut would require little effort, but to move the massive but weightless refrigerator would require a mighty heave even in the weightlessness of space. Contrary to intuition, therefore, weight and mass are not the same concepts.

### *Units*

Units are not concepts but conventions. They are standards of measurement that when attached to length, time, and mass give them concrete values. A photograph of a building alone gives no necessary indication of its size (figure 4.18); therefore, a friend is often pressed into service to stand in the picture to give a sense of scale to the building. Units serve similarly as a familiar scale. But different systems of units have grown up in engineering, so a choice must be made.

In many English-speaking countries, the "English system" of measurement—pounds, feet, seconds—has been preferred. In engineering schools, these units are usually used. Initially, these units grew up from familiar objects such as body parts. The "inch" was originally associated with the thumb's width, the "palm" was the breadth of the hand, about 3 inches, the foot equaled 4 palms, and so on.

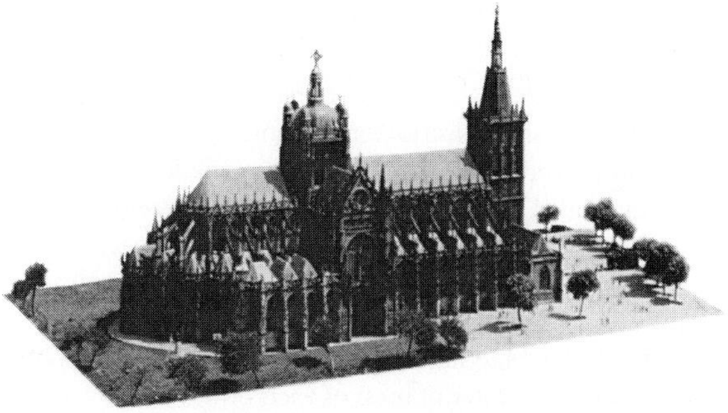

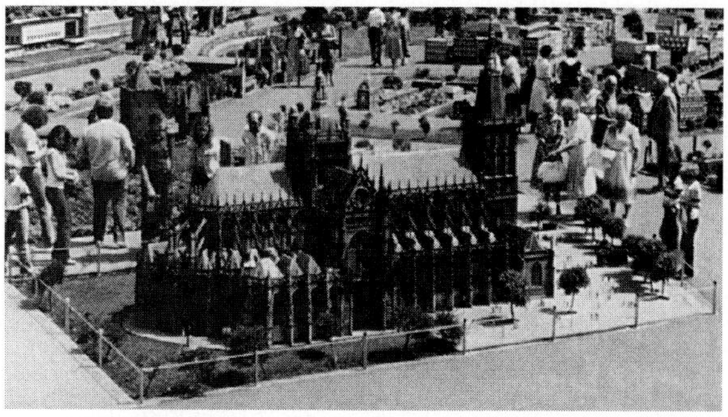

**Figure 4.18** Units as reference. We use familiar objects as references of size. If denied familiar references, such as fellow humans (left), we can easily underestimate the true size of the cathedral (right). Units of measurement such as inches, feet, and pounds are conventions attached to quantities in order for us to set standard references for expressing distances and weight.

Although poetic, the English system can be cumbersome when converting units. For instance, to change miles to yards requires multiplying by 1,760. To convert yards to feet, we must multiply by 3, and to convert feet to inches, we multiply by 12. During the French Revolution, a simpler system based on the meter was introduced. Changing kilometers to meters, meters to centimeters, or centimeters to millimeters requires only moving the decimal point. The **Système Internationale,** or **SI** system, is an extended version of the older metric system. Primary units of the SI system include meter (m), kilogram (kg), and second (s) for dimensions of length, mass, and time, respectively. In this book, as throughout physics and biology, SI units are used. Table 4.1 lists the common units of measurement in both the English and the SI systems.

TABLE 4.1 Common Fundamental Units of Measurement

| English System | Physical Quantity | Système Internationale (SI) |
|---|---|---|
| Slug or pound mass | Mass | Kilogram (kg) |
| Foot (ft) | Length | Meter (m) |
| Second (s) | Time | Second (s) |
| Feet/second (fps) | Velocity | Meters/second (m s) |
| Feet/second$^2$ (ft sec$^2$) | Acceleration | Meters/second$^2$ (m s$^{-2}$) |
| Pound (lb) | Force | Newtons (N or kg m s$^{-2}$) |
| Foot-pound (ft-lb) | Moment (torque) | Newtons meters (Nm) |

## *Derived Quantities—Velocity and Acceleration*

**Velocity** and **acceleration** describe the motion of bodies. Velocity is the rate of change in an object's position, and acceleration in turn is the rate of change in its velocity. In part, our intuition helps our understanding of these two concepts. When traveling east by car on an interstate highway, we may change our position at the rate of 88 km per hour (velocity) (about 55 mph if you are still thinking in the English system). Step on the gas and we accelerate, hit the brake and we decelerate or better stated, we experience negative acceleration. With mathematical calculations, negative acceleration is a better term to use than deceleration because we can keep positive and negative signs in a more straightforward way. The sensation of acceleration is familiar to most, but in common conversation, units are seldom mentioned. When they are properly applied, units may sound strange. For instance, suddenly braking a car may produce a negative acceleration of −290 km/h/h (about −180 mph/h). The units may be unfamiliar, but the concept of acceleration, like that of velocity, is an everyday experience.

## *Reference Systems*

When preparing to record events, a conventional frame of reference is selected that can be overlaid on an animal and its range of activity. Purists use an "inertial reference," a coordinate system defined relative to a fixed star in space. So long as we do not concern ourselves with animals rocketing into outer space, such a definition is more exotic than we need. For our purposes, and for most engineering applications as well, the coordinate system is defined relative to the surface of the Earth.

For reference systems, there are several choices, including the polar and cylindrical systems. The most common, however, is the **rectangular Cartesian reference system** (figure 4.19). For an animal moving in three-

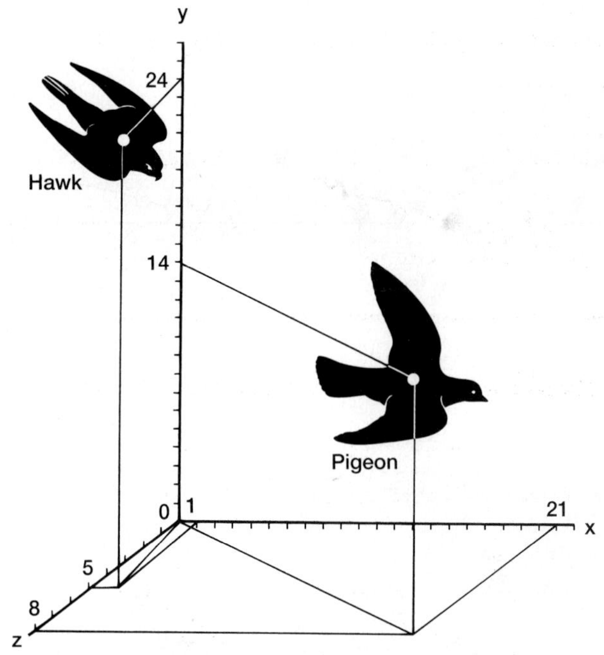


**Figure 4.19** A three-axis Cartesian coordinate reference system defines the position of any object. Customarily the horizontal, vertical, and axis at right angle to these two are identified as x, y, and z, respectively. The three meet at the origin (0). The direct projection line of an object to each axis defines its position at that instant along each axis. Thus, the three projections fix an object's position in space—1, 24, 5 for the hawk and 21, 14, 8 for the pigeon. The white dot graphically represents the center of mass of each bird.

dimensional space, its position at any moment can be described exactly on three axes at right angles to each other. The horizontal axis is $x$, the vertical axis is $y$, and the axis at right angles to these is $z$. Once defined, the orientation of these reference systems cannot be changed, at least not during the episode during which we are taking a series of measurements.

### *Center of Mass*

If we are interested in the motion of a whole organism rather than the separate motion of its parts, we can think of the mass of the animal as being concentrated at a single point called its **center of mass.** The center of mass, in laypersons' terms the center of gravity, is the point about which an animal is evenly balanced. As a moving animal changes the configuration of its parts, the position of its center of mass changes from one instant to the next (figure 4.20).

### *Vectors*

Vectors describe measurements with a magnitude and a direction. Force and velocity are examples because they have magnitude (N in the SI system, mph in the English system) and direction (e.g., northwesterly direction). A measurement with only magnitude and no direction is a **scalar quantity.** Time and temperature have magnitude but no direction, so they are scalar, not vector, quantities. A force applied to an object can also be represented along a rectangular Cartesian reference system. When we use such a reference system, trigonometry helps us to calculate vector values. For example, we can measure the force applied to a dragged object (F in figure 4.21), but the portion of that force acting horizontally against surface friction ($F_x$) is more difficult to measure directly. However, given the force (F) and angle ($\theta$), we can calculate both horizontal and vertical components ($F_x$ and $F_y$). And, of course, conversely, if we know the component forces ($F_x$ and $F_y$), we can calculate the combined resultant force (F).

## Basic Force Laws

Much of engineering is based on laws that were formulated by Isaac Newton (1642–1727). Three of his laws are fundamental:

1. *First law of inertia.* Because of its inertia, every body continues in a state of rest or in a uniform path of motion until a new force acts on it to set it in motion or change its direction. **Inertia** is the tendency of a body to resist a change in its state of motion. If the body is at rest, it will resist being moved, and if it is in motion, it will resist being diverted or stopped.
2. *Second law of motion.* Simply stated, the change in an object's motion is proportional to the force acting on it (figure 4.22). Or, a force *(F)* is equal to the mass *(m)* of an object times its experienced acceleration *(a)*:

$$F = ma$$

   Units of this force are expressed in newtons (N), kg m $s^{-2}$.
3. *Third law of action, reaction.* Between two objects in contact, there is for each action an opposite and equal reaction.

Albert Einstein's (1879–1954) theories of relativity placed limits on these Newtonian laws. But these limitations become mathematically significant only when the speed of an object approaches the speed of light (186,000 miles/s). Newtonian laws serve space travel well enough to get vehicles to the moon and back, and so they will serve us here on Earth as well.

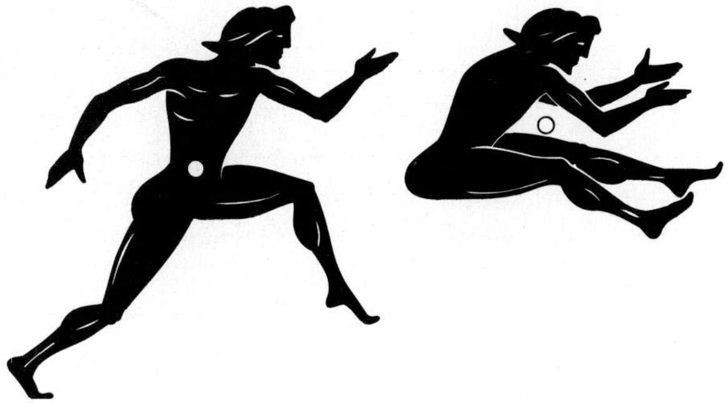

**Figure 4.20** Center of mass. The single point at which the mass of a body can be thought to be concentrated is the center of mass. As the configuration of this jumper's body parts changes from takeoff (left) to midair (right), the instantaneous location of the center of mass (white dot) changes as well.

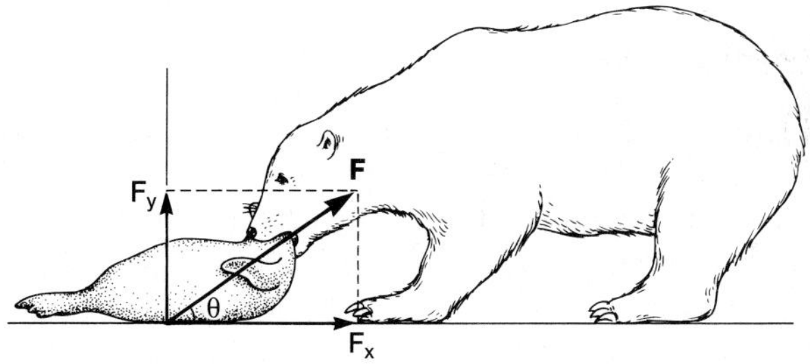


**Figure 4.21** Vectors. When dragging the seal, the polar bear produces a main force (F) that can be represented by two small component forces acting vertically ($F_y$) and horizontally ($F_x$). The horizontal force acts against surface friction. If we know the main force (F) and its angle ( $\theta$) with the surface, we can calculate the component forces using graphic or trigonometric techniques.

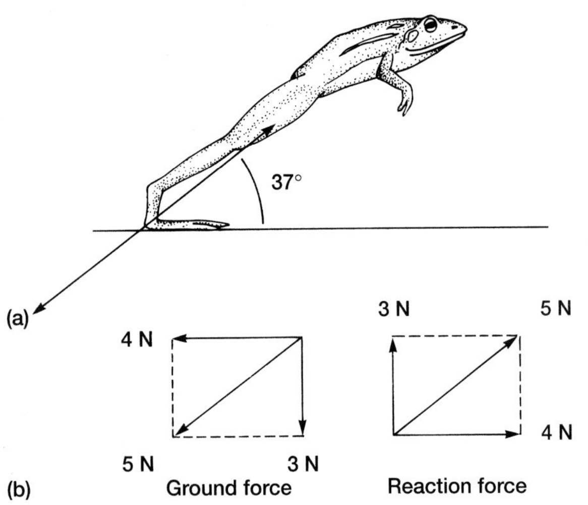


**Figure 4.22** Forces of motion. (a) The force a frog produces at lift-off is the result of its mass and acceleration at that instant ($F = ma$). (b) Forces produced collectively by both feet of a hefty frog and the ground are opposite but equal. The vector parallelograms represent the components of each force. If a frog of 50 g (.05 kg) accelerates 100 m $s^{-2}$, a force of 5 N (100 × .05) is generated along the line of travel. By using trigonometric relations, we can calculate the component forces. If lift-off is at 37°, then 4 N (cos 37° × 5 N) and 3 N (sin 37° × 5 N).

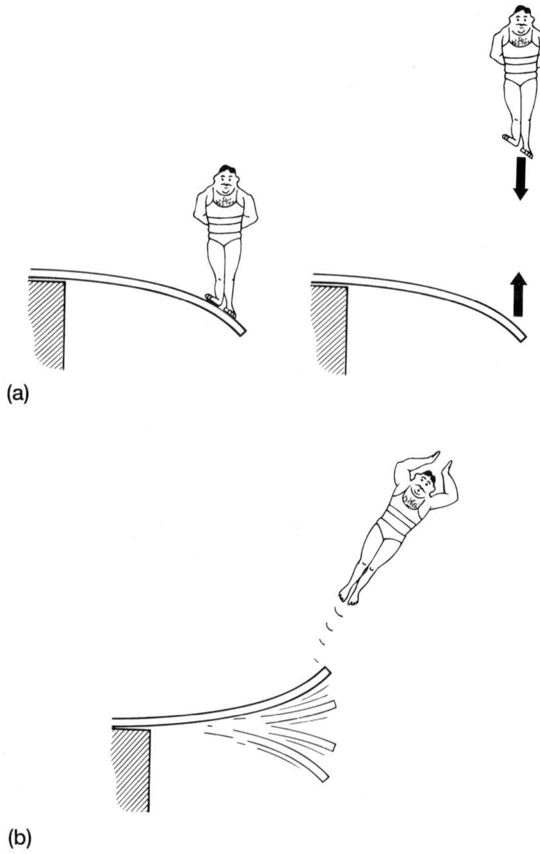


**Figure 4.23** Free-body diagrams. (a) Two physical bodies, the board and the diver, each exert a force on the other. If the forces of the two bodies are equal, opposite, and in line with each other, then no linear or rotational motion results. Although the forces are present, the two bodies are in equilibrium (left). To depict these forces graphically (right), the two bodies are separated in free-body diagrams, and the forces acting on each body are represented by vectors (arrows). (b) If forces are unequal, motion is imparted.

In biomechanics, Newton's second law, or its modifications, are most often used because the separate quantities can be measured directly. In addition, knowing the forces experienced by an animal often gives us the best understanding of its particular design.

## Free Bodies and Forces

When forces are calculated, it often helps to isolate each part from the rest in order to look at the forces acting on it. A **free-body diagram** graphically depicts the isolated part with its forces (figure 4.23a,b).

When you walk across a floor, you exert a force upon it. The floor gives ever so slightly and imperceptibly until it returns a force equal to yours, which exemplifies the action and reaction principle described by Newton's third law. If the floor did not push back equally, you would fall through. Think of a swimmer perched at the end of a diving board. The board bends until it pushes back with a force equal to the force exerted on it by the swimmer. Swimmer and board are separated in the free-body diagram, and the forces on each are shown in figure 4.23a. If both forces are equal and opposite, they cancel and the two are in equilibrium. If not, motion is produced (figure 4.23b).

As a practical matter, mechanics is divided between these two conditions. Where all forces acting on an object balance, we are dealing with that part of mechanics known as **statics.** Where acting forces are unbalanced, we are dealing with **dynamics.**

## Torques and Levers

In the vertebrates, muscles generate forces and skeletal elements apply these forces. There are several ways to represent this mechanically. Perhaps the most intuitive representation is with torques and levers. The mechanics of torques and levers are familiar because most persons have firsthand experience with a simple lever system, the teeter-totter or seesaw of childhood. Action of the seesaw depends on the opposing weights seated on opposite ends and on the distances of these weights from the pivot point, or **fulcrum.** This distance from weight to fulcrum is the **lever arm.** The lever arm is measured as the perpendicular distance from force to fulcrum. Shorten the lever arm and more weight must be added to keep the board in balance (figure 4.24a). Lengthen it sufficiently, and a little sister can keep several big brothers balanced on the opposite end.

A force acting at a distance (the lever arm) from the fulcrum tends to turn the seesaw about this point of rotation, or more formally, it is said to produce **torque.** When levers are used to perform a task, we also recognize an **in-torque** and **out-torque.** If more output force is required, shortening the "out" and lengthening the "in" lever arms increases the out-torque. Conversely, if out-torque speed is required, then lengthening the out and shortening the in lever arms favors greater velocity in the out-torque (figure 4.24c). Of course, this increased speed is achieved at the expense of force in the out-torque. In engineering terms, torque is more commonly described as the **moment** about a point and the lever arm as the **moment arm.**

The mechanics of levers mean that output force and output speed are opposites. Long output lever arms favor speed, short ones favor force. Regardless of how desirable it would be to have both in the design of, say, an animal's limb, simple mechanics do not permit it. For a given input, both output force and output speed cannot be maximized. Compromises in design must be made.

Consider the forelimbs of two mammals, one a runner specialized for speed, the other a digger specialized to generate large output forces. In figure 4.25a, the relatively long elbow process and short forearm of the digger favor large force output. In the runner (figure 4.25b), the elbow is short, the forearm long. Lever arms are less favorable to force output in the foot of the runner but more favorable to speed. The speed of the elbow is magnified by the relatively greater output lever arm, but this is accomplished at the expense of output force.

More formally, we can express the mechanics of input and output forces at different velocities with simple ratios. The ratio of $F_2/F_1$, the output to input force, is the **mechanical advantage.** The ratio of the output to input lever arms, $l_o/l_i$, is the **velocity ratio.**

As we might expect, diggers enjoy a mechanical advantage in their forearm but runners enjoy a velocity advantage in their forearm. There are, of course, other ways of producing output force or speed. Increased size, and hence force, of input muscles and emphasis of fast-contracting muscle cells both affect output. Insofar as the

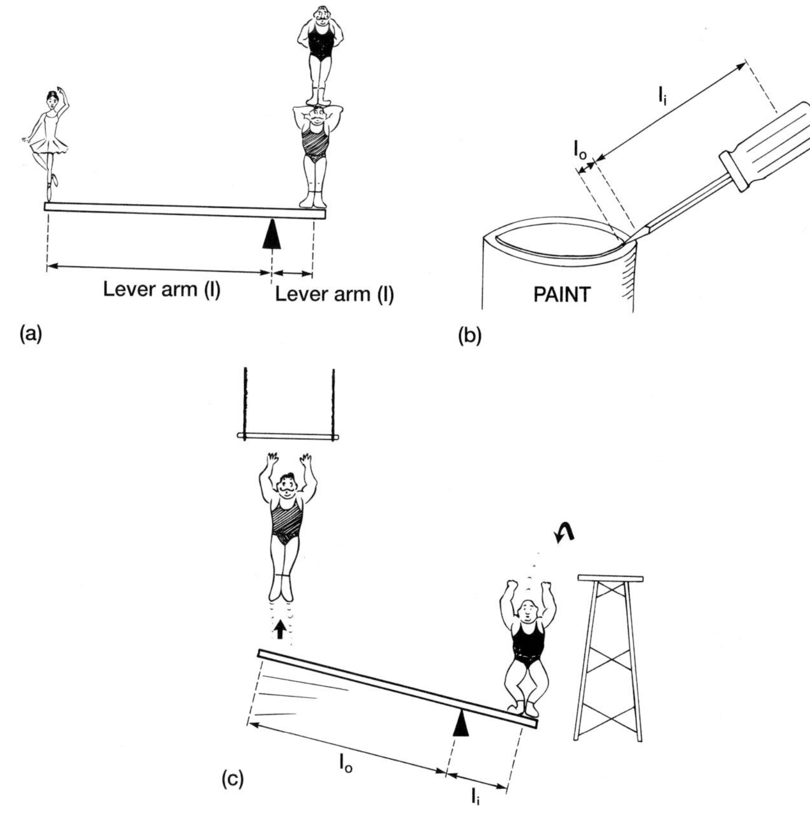


**Figure 4.24** Principles of lever systems. (a) The balance of forces about a point of pivot (fulcrum) depends on the forces times their distance to the point of pivot, their lever arms (l). (b) To get more output force, the point of pivot is moved closer to the output and farther from the input force. In this diagram, the short output lever arm ($l_o$) and long input lever arm ($l_i$) work in favor of more output force. (c) To produce high output speed, the pivot point is moved closer to the input force ($l_i$). Other things being equal, speed is achieved at the expense of output force.

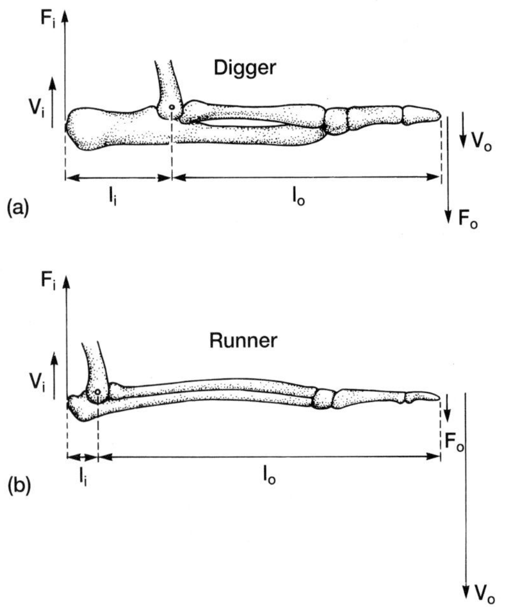


**Figure 4.25** Strength versus speed. The forearms of a digger (a) and a runner (b) are drawn to the same overall length. Input forces ($F_i$) and input velocities ($V_i$) are the same, but output forces ($F_o$) and velocities ($V_o$) differ. The differences result from the differences between the lever arm ratios of the two forearms. Output force is greater in the digger than in the runner, but output velocity of the digger is less. Formally, these differences can be expressed as differences in mechanical advantages and in velocity ratios.

lever system itself is concerned, a compromise between force and speed results from the lever mechanics.

Artiodactyls, such as deer, have limbs designed for both force and speed (figure 4.26). Two muscles, the medial gluteus and the semimembranosus, with different mechanical advantages, make different contributions to force or to speed output. The medial gluteus enjoys a higher velocity ratio ($l_o/l_i$ = 44 compared with $l_o/l_i$ = 11 for the semimembranosus), a leverage that favors speed. If we compare these muscles with the gears of a car, the medial gluteus would be a "high" gear muscle. On the other hand, the semimembranosus has a mechanical advantage favoring force and would be a "low" gear muscle. During rapid locomotion, both are active, but the low gear muscle is most effective mechanically during acceleration and the high gear muscle is more effective in sustaining the velocity of the limb.

The mechanics of levers limit a muscle to maximizing its lever advantage by producing either output force or output speed. The presence of two muscles, one specialized for large forces, the other for speed, represents a feature of biological design that takes advantage of the mechanics of torques and levers to provide the limb of a running animal with some degree of both. Just as

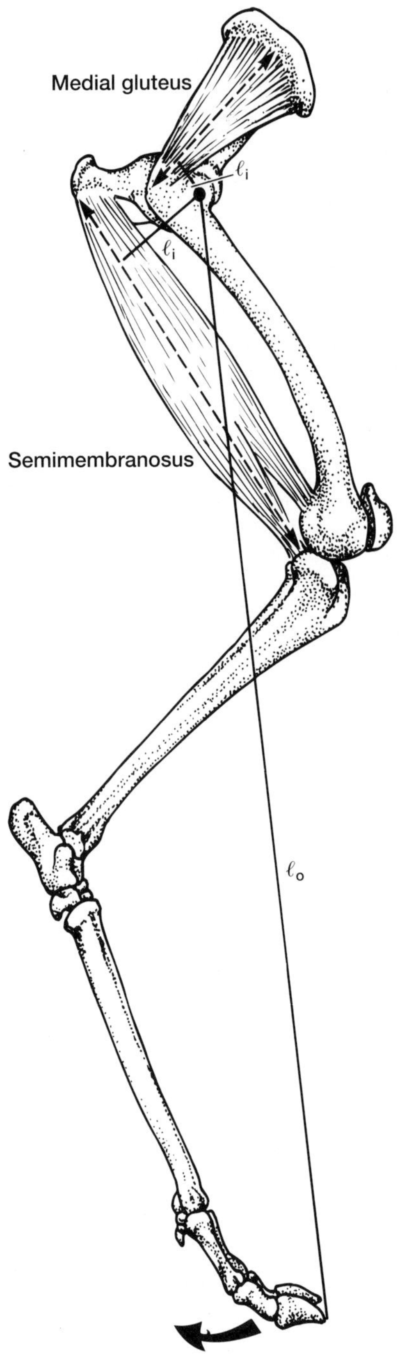


**Figure 4.26** High and low gear muscles. Both the medial gluteus and the semimembranosus muscles turn the limb in the same direction, but they possess different mechanical advantages in doing so. A muscle's lever arm is the perpendicular distance to the point of rotation or pivot point (black dot) from the line of muscle action (dotted line). The velocity ratio is higher in the medial gluteus, which can move the limb faster. But the semimembranosus moves the limb with greater output force because of its longer lever arm. Lever arms in all the muscles ($l_i$) and the common lever arm out ($l_o$) are indicated.

a seesaw does not have a single fulcrum that can maximize output force and output speed simultaneously, similarly one muscle cannot maximize both. A muscle has leverage that can maximize either its force output or its speed output, a limitation that arises from the nature of mechanics, not from any necessity of biology. However, biological design must abide by the laws and limits of mechanics when mechanical problems of animal function arise.

## Land and Fluid

For most terrestrial vertebrates, the external forces they experience arise ultimately from the effects of gravity. Vertebrates in fluids, such as fishes in water or birds in flight, experience additional forces from the water or air around them. Because the forces are different, the designs that address them differ as well.

### *Life on Land: Gravity*

Gravity acts on an object to accelerate it. On the surface of the Earth, the average acceleration of gravity is about 9.81 m $s^{-2}$ acting toward the Earth's center. Newton's second law *($F = ma$)* tells us that an animal with a mass of 90 kg produces a total force of 882.9 N (90 kg × 9.81 m $s^{-2}$) against the Earth upon which it stands. An object held in your hand exerts a force against your hand, which results from the object's mass and gravity's pull. Release the object, and the acceleration from gravity's effects becomes apparent as the object picks up speed as it falls to Earth (figure 4.27). Gravity's persistent attempt to accelerate a terrestrial animal and pull it down constitutes the animal's weight. In tetrapods, this is resisted by the limbs.

The weight of a quadrupedal animal is distributed among its four legs. The force borne by fore- and hindlimbs depends on the distance of each from the center of the animal's mass. Thus, a large *Diplodocus* might have distributed its 18 metric tons (39,600 pounds) with a ratio of 4 tons to its forelimbs and 14 to its hindlimbs (figure 4.28).

When we explored the consequences of size and mass at the beginning of this chapter, we noted that large animals have relatively more mass to contend with than small animals. A small lizard scampers safely across tree limbs and vertical walls; a large lizard is Earth bound. Gravity, like other forces, is a part of an animal's environment and affects performance in proportion to body size. Size is also a factor for animals that live in fluids, although forces other than gravity tend to be predominant.

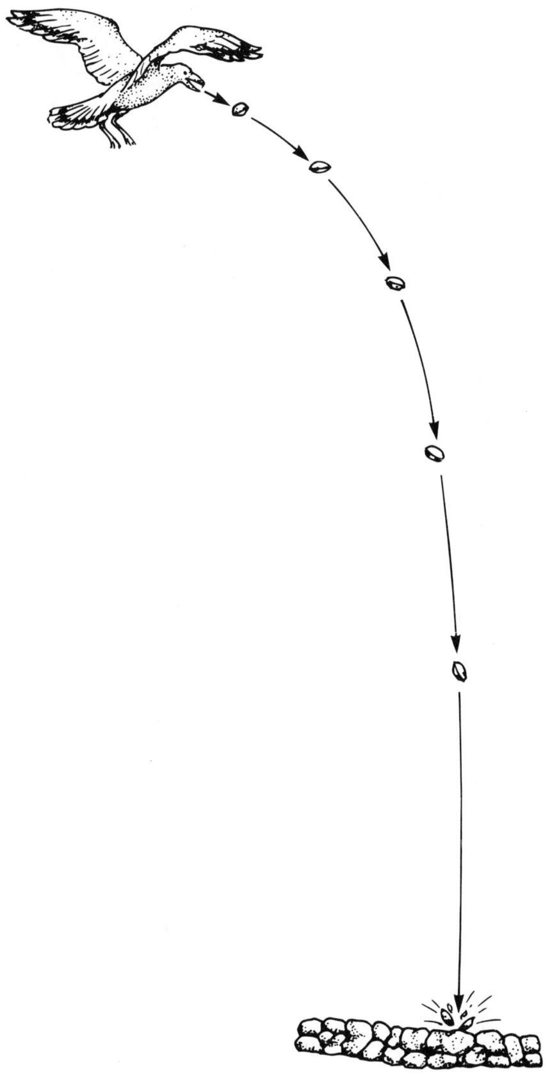

**Figure 4.27** Gravity. The clam released by the sea gull accelerates under gravity's pull and picks up speed as it falls to the rocks. With equal intervals of time designated by each of the six arrows, note the accelerating positions of the clam.

### *Life in Fluids*

***Dynamic Fluids*** Water and air are fluids. Certainly air is thinner and less viscous than water, but it is a fluid nonetheless. The physical phenomena that act on fishes in water generally apply to birds in air. Air and water differ primarily in viscosity, but they place similar physical demands on animal designs. When a body moves through a fluid, the fluid exerts a resisting force in the opposite direction to the body's motion. This resisting force, termed **drag,** may arise from various physical phenomena, but forces caused by **friction drag** (or skin friction) and by **pressure drag** are usually the most important. As an animal moves through a fluid, the fluid flows

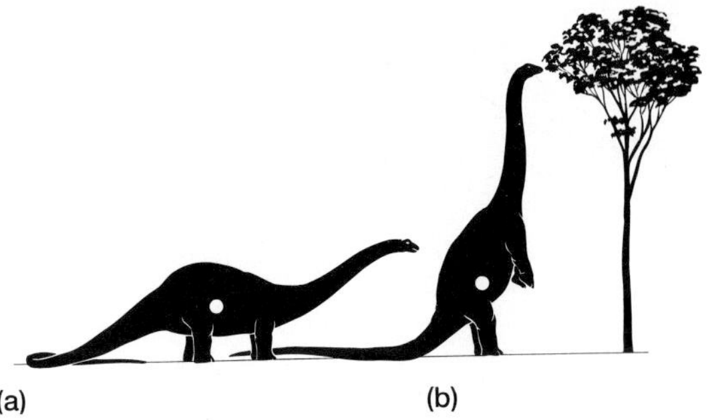


**Figure 4.28** Weight distribution. (a) The estimated center of mass of this dinosaur lies closer to the hindlimbs than to its forelimbs, so the hindlimbs bear most of the animal's weight. For *Diplodocus*, its 18 metric tons (39,600 pounds) might have been carried by a ratio of 4 tonnes to its forelimbs and 14 tonnes to its hindlimbs. (b) If *Diplodocus* lifted its head and forefeet up to reach high vegetation, then all 18 tonnes would have been carried by hindlimbs and tail, which formed three points of support giving each limb and the tail 6 tonnes to carry.

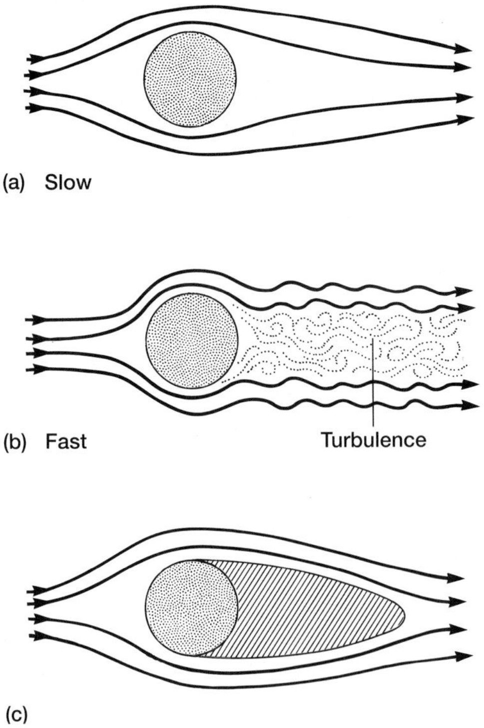


**Figure 4.29** Streamlining. (a) A fluid passing around a circular object flows in smooth streamlines that spread when they meet an object and come together again behind the object once they have passed. (b) As speed increases, turbulence develops behind the object because streamlines begin to separate and fail to return to a smooth laminar flow. (c) Extension and tapering of the object partially into the area of turbulence helps to maintain the smooth laminar flow of the fluid and prevents or reduces turbulence. This extension of the body produces a streamlined shape.

along the sides of its body. As fluid and body surface move past one another, the fluid exerts a resisting force (drag) on the surface of the animal where they make contact. This force creates friction drag and depends among other things on the viscosity of the fluid, the area of the surface, the surface texture, and the relative speed of fluid and surface.

If particles in a passing fluid travel around a body along smooth, layered paths, the flow is said to be **laminar,** with each layer gliding smoothly over adjacent layers. The paths these particles travel describe **streamlines** and can be visualized as smooth parallel lines at representative layers in the laminar flow around a body (figure 4.29a). Fluid particles in **turbulent** (nonlinear) **flow** move along irregular paths (figure 4.29b). Adverse effects of fluid viscosity, of speed, or of body surface properties themselves may change the flow from laminar to turbulent. This change most often begins in the **boundary layer,** the fluid layer closest to and flowing against the surface of the body. Disruption of the laminar flow of the boundary layer produces adverse backflow downstream, which is seen as a wake of disturbed fluid behind the animal. Physically, the turbulence results from a substantial pressure differential (pressure drag) between the front and the back of the animal. An extended, tapering body fills in the area of potential turbulence, encourages streamlines to close smoothly behind it, and thereby reduces pressure drag (figure 4.29c). The result is a streamlined shape common to all bodies that must pass rapidly and efficiently through a fluid. An active fish, a fast-flying bird, and a supersonic aircraft are all streamlined for much the same reason—to reduce pressure drag (figure 4.30a–d).

Engineers examine the physical problems associated with motion through fluids within the disciplines of hydrodynamics (water) or aerodynamics (air). Applied to animal designs that move through fluids, these disciplines reveal how size and shape affect the way the physical forces of a fluid act on a moving body.

In general, four physical characteristics affect how the fluid and body dynamically interact. One of these is the *density*, or mass per unit volume of the fluid. A second is the *size and shape* of the body as it meets the fluid. The resistance a rowboat oar experiences when the blade is pulled broadside-on is of course quite different than when it is pulled edge-on. The third physical characteristic of a fluid is its *velocity*. Finally, the *viscosity*, or elasticity, of a fluid refers to its resistance to flow. These four characteristics are brought together in a ratio known as the Reynolds number:

$$R_e = \frac{plU}{\mu}$$

where $p$ is the density of the fluid and $\mu$ is a measure of its viscosity; $l$ is an expression of the body's charac-

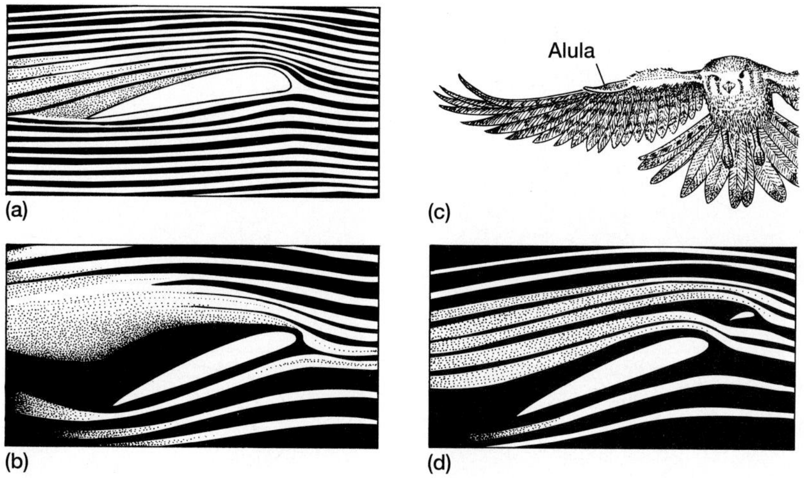


**Figure 4.30** Life in fluids. (a) The airplane wing, shown in cross section, encourages smooth flow of the passing airstream. (b) As its angle of attack to the direction of flow increases, turbulence suddenly forms and lift is lost. (c) Birds, such as this falcon have a small feather (alula) that can be lifted to smooth the airflow at high angels of attack. (d) When turbulence forms, this small airfoil can be lifted to form a slot that accelerates air over the top of the wing and prevents stalling.

teristic shape and size; and *U* is its velocity through the fluid.

Perhaps because we ourselves are large land vertebrates, we have some intuition about the importance of gravity but no special feeling for all that the Reynolds number has to tell us about life in fluids. The units of all variables of the ratio cancel each other, leaving the Reynolds number without units. No feet per second, no kilograms per meter, nothing. It is dimensionless, a further factor obscuring its message; yet, it is one of the most important expressions that summarizes the physical demands placed upon a body in a fluid. The Reynolds number was developed during the last century to describe the nature of fluid flow, in particular, how different circumstances might result in fluid flows that are dynamically similar. The Reynolds number tells us how properties of an animal affect fluid flow around it. In general, at low Reynolds numbers, skin friction is of great importance; at high Reynolds numbers, pressure drag might predominate. Perhaps most importantly, at least for a biologist, the Reynolds number tells us how changes in size and shape might affect the physical performance of an animal traveling in a fluid. It draws our attention to the features of the fluid (viscosity) and the features of the body (size, shape, velocity) that are most likely to affect performance.

For scientists performing experiments, the Reynolds number helps them to build a scale model that is dynamically similar to the original. For example, several biologists wished to examine air ventilation through prairie dog burrows but lacked the convenient space to build a life-sized tunnel system in the laboratory. Instead, they built a tunnel system ten times smaller but compensated by running winds ten times faster through it. The biologists were confident that the scale model duplicated conditions in the full-sized original because a similar Reynolds number for each verified that both were dynamically similar even though their sizes differed.

***Static Fluids*** The expression "as light as air" betrays the common misconception that air has almost no weight. In fact, air weighs about 101,000 Pa (14.7 psi or pounds per square inch) at sea level, which is equivalent to one **atmosphere** (atm) of pressure. The envelope of air surrounding Earth extends up to several hundred kilometers. Although not dense, the column of air above the surface of Earth is quite high, so the additive weight at its base produces a substantial pressure at Earth's surface. We and other terrestrial animals are usually unaware of this weight because terrestrial vertebrates evolved in its presence and possess bodies designed to comfortably withstand the weight of the overhead atmosphere.

If we drive from low elevation to high elevation in a short period of time, we might notice the unbalanced pressure in that builds uncomfortably in our ears until a yawn or stretch of our jaw "pops" and equilibrates the inside and outside pressures to relieve the mismatch. Most of us have experienced increasing pressure as we dive deeper in water. At a given depth, the pressure surrounding an animal in water is the same from all sides. The deeper the animal, the greater is the pressure. In fresh water, with each meter of depth, atmospheric pressure increases by about $9.8 \times 10^3$ Pa. At 5 meters, atmospheric pressure would be about $49 \times 10^3$ Pa. Scaled for a human, that would be like trying to breathe with a 90-kg slab placed upon the chest. A fully submerged sauropod would experience $49 \times 10^3$ N on each square meter of its entire chest (figure 4.31). It is not likely that even the massive chest muscles of this dinosaur could overcome so much pressure when it drew in a breath. Therefore, *Branchiosaurus* and other long-necked animals probably did

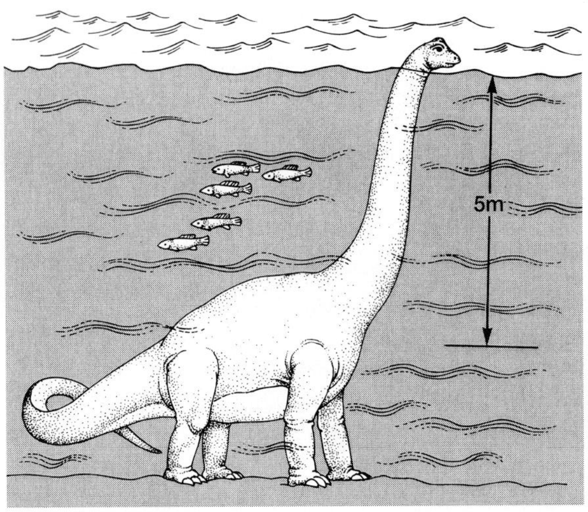


**Figure 4.31** Water pressure. Water pressure increases with depth, but at any given depth, pressure is equal from all directions. For each meter below the surface, the pressure in fresh water increases by about 9,800 Pa. A large sauropod submerged up to its chin would experience water pressure of about 49,000 Pa (5 m × 9,800 Pa) around its chest, too much pressure to allow chest expansion against this force. Breathing would be impossible. Sauropods such as *Brachiosaurus* shown here were probably not completely aquatic.

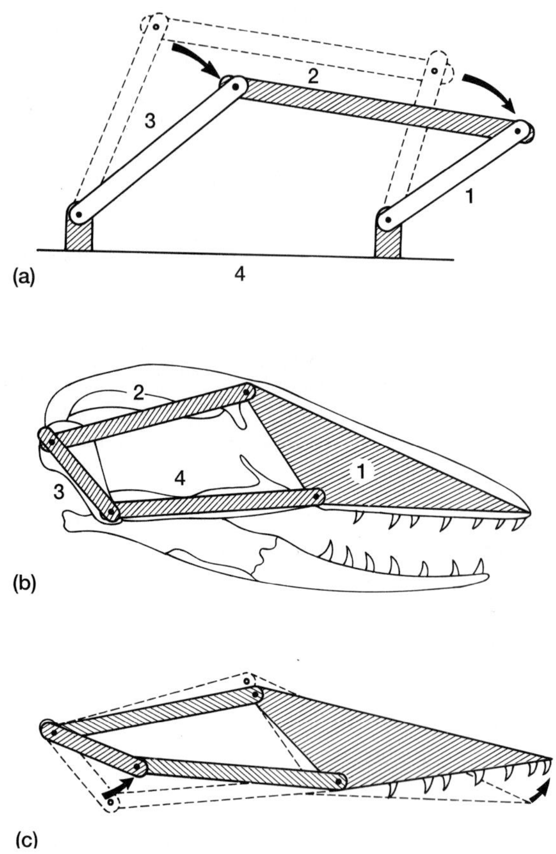


**Figure 4.32** Kinematic chain. (a) This four-linkage mechanism is joined by pin connections so that the motion of link 3 imparts a specific motion to the three other links. (b) The four-linkage chain of a lizard skull (ignoring the lower jaw) is constrained. (c) Again motion of link 3 imparts a specific motion to each of the other links.

not live aquatic lives with their body deeply submerged and their head reaching far above to the surface to snorkel air.

## Machines

When we are interested in the motions of parts of the same animal, it is customary to represent each movable part with a link. A joined series of links is a **kinematic chain** representing the main elements of an animal. If these linkages are floppy and without control, then the chain is said to be unconstrained. A kinematic chain restricted in motion is constrained and formally constitutes a **mechanism.** The motion of one link will impart a definite and predictable motion in all other links of the same mechanism (figure 4.32a).

A kinematic mechanism simulates the relative motions of the parts of the animal it represents, so it helps identify the role of each element. For example, several bony elements on both sides of a lizard's skull are involved when it lifts its snout during feeding. These elements can be represented by a kinematic chain that constitutes the jaw mechanism of the lizard (figure 4.32b,c).

Often we are interested in more then just the motion of a mechanism. We might want to know something about the transfer of actual forces. Such devices that transfer forces are **machines.** Formally defined, a machine is a mechanism for transferring or applying forces. In a car's engine, the pistons transfer the explosive forces of gasoline combustion to the connecting rod, the rod to the crankshaft, and the crankshaft in turn to the gears, axles, and wheels. Pistons to wheels collectively form a "machine" that transfers energy from the ignited gasoline to the road. Levers that transfer forces qualify as machines too. The input force brought into a machine by a lever arm is applied elsewhere as an output force by the opposite lever arm. In this engineering sense, the jaws of a herbivore are a machine whereby the input force produced by the jaw muscles is transferred along the mandible as an output force to the crushing molar teeth (figure 4.33).

## Strength of Materials

A weight-bearing structure carries or resists the forces applied to it. These forces may be experienced in three general ways. Forces pressing down on an object to compact it are **compressive forces,** those that stretch it are **tensile forces,** and those that slide its sections are **shear forces** (figure 4.34a–c). Surprisingly, the same structure is not able to withstand the three types of force applications

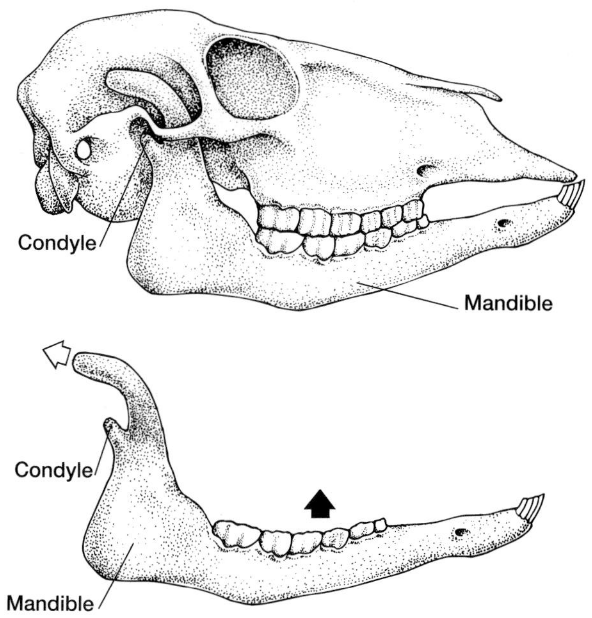


**Figure 4.33** Jaws as machines. A machine transfers forces. Here the lower jaw of a herbivore transfers the force of the temporalis muscle (open arrow) to the tooth row (solid arrow) where food is chewed. Rotation occurs about the condyle (dot).

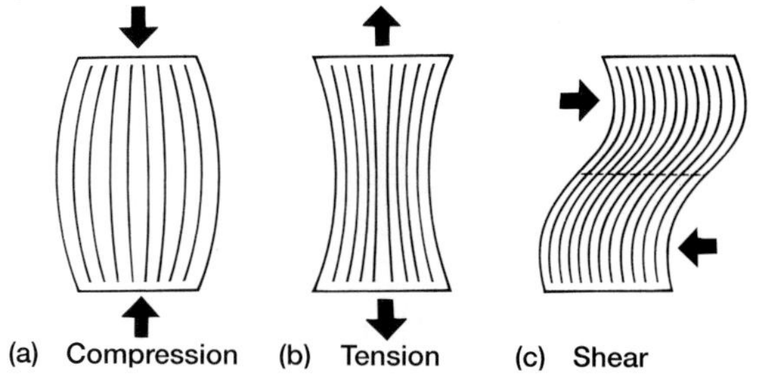


**Figure 4.34** Direction of force application. The susceptibility of a material to breaking depends on the direction in which the force is applied (arrows). Most materials withstand compression best (a) and are weaker when placed in tension (b) and shear (c).

equally. For any structure, the maximum force a structure sustains in compression before breaking is its **compressive strength;** in tension, is its **tensile strength;** and in shear, is its **shear strength.** Table 4.2 lists the strengths of several materials when they are exposed to compressive, tensile, and shear forces.

Notice from this table that most materials are strongest in resisting compressive forces and weakest in their ability to withstand tension or shear. This is very significant in design. Ordinarily, supportive columns of buildings bear the load in a compressive fashion, their strongest weight-bearing orientation. However, if the column bends slightly, tensile forces, to which they are more susceptible, appear.

TABLE 4.2 Strength of Different Materials Exposed to Compressive, Tensile, and Shear Forces

| Material | Compressive Strength (Pa) | Tensile Strength (Pa) | Shear Strength (Pa) |
|---|---|---|---|
| Bone | $165 \times 10^6$ | $110 \times 10^6$ | ? |
| Cartilage | $27.6 \times 10^6$ | $3.0 \times 10^6$ | ? |
| Concrete | $41.4 \times 10^6$ | $4.0 \times 10^6$ | ? |
| Cast iron | $620.5 \times 10^6$ | $1.17 \times 10^6$ | $124 \times 10^6$ |
| Granite | $103 \times 10^6$ | $10 \times 10^6$ | $13.8 \times 10^6$ |

*Source:* Adapted from J. E. Gordon, 1978. *Structures, or why things don't fall down*, New York: DaCapo Press. Other sources have been used as well.

When any object bends, compressive forces build up on the inside and tensile forces on the outside of the bend. Opposite sides experience different force applications. The column may be strong enough to withstand compressive forces, but the appearance of tensile forces introduces forces it is intrinsically weaker in resisting. If bending persists, breaks may originate on the side experiencing tension, propagate through the material, and cause the column to fail. Flying buttresses, side braces on the main support piers of Gothic cathedrals, were used to prevent the piers from bending, thus keeping them in compression and allowing them to better carry the weight of the cathedral's arched roof (figure 4.35).

### *Loads*

How a load is positioned upon a supportive column affects its tendency to bend (figure 4.36a–c). When the load is arranged evenly above the column's main axis, no buckling is induced and the column primarily experiences the load as compressive force (figure 4.36b). The same load placed asymmetrically off center causes the column to bend (figure 4.36c). Tensile forces now appear. Tensile and compressive forces are greatest at the surfaces of the column, least at its center. Development of surface tensile forces is especially ominous because of the intrinsic susceptibility of supportive elements to such forces.

### *Biological Design and Biological Failure*

***Fatigue Fracture*** With prolonged or heavy use, bones, like machines, can fatigue and break. When initially designed, the working parts of a machine are built with materials strong enough to withstand the calculated stresses they will experience. However, with use over time, these parts often fail, a condition known to engineers as **fatigue fracture.** Not long after the Industrial Revolution, engineers noticed that moving parts of machinery occasionally broke at loads within safe limits. Axles of trains, in use for some time, suddenly broke for no apparent reason. Cranks or cams that had withstood peak loads many times before sometimes suddenly broke under routine operation. Eventually engineers appreciated that one of the factors leading to these failures was fatigue fracture. Although a moving part might be strong enough initially to bear up easily to peak loads, over time tiny microfractures form in the material. These are insignificant individually, but cumulatively they can add up to a major fracture that exceeds the strength of the material, and breakage follows.

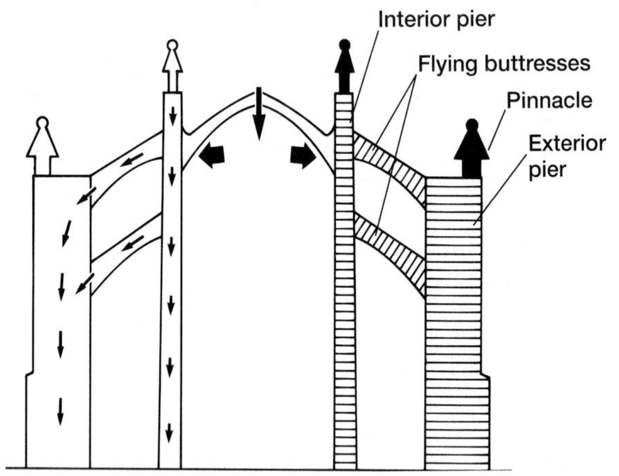


**Figure 4.35** Gothic cathedral. The right side of the cathedral shows its structural elements. The left side illustrates how these structures carry the thrust lines of forces. At its simplest, the cathedral includes the exterior pier topped by a pinnacle, the main interior pier, and the flying buttresses between the interior and exterior piers. The weight of the vault (roof) produces an oblique thrust against the interior piers. Wind pressure or snow load accentuates this lateral pressure, which tends to bend the main interior piers. Flying buttresses act in an opposite direction to resist this bending and help carry lateral thrust from the roof to the ground (small arrows).

***Load Fracture*** In vertebrates, bones are loaded symmetrically or, where that is not possible, muscles and tendons act as braces to reduce the tendency for a load to induce bending of a bone (figure 4.37). The greatest stresses develop at the surface of bone, while forces are almost negligible at its center. Consequently, the core of a bone can be hollow without much loss of its effective strength. Probably for the same reason, cattail reeds, bamboos, and fishing poles are hollow as well. This economizes on material without a great loss in strength.

Most fractures likely begin on the side of the bone experiencing tensile forces. A fracture propagates through the bone matrix, causing failure. Bone, however, is composed of a composite material consisting of several substances with different mechanical properties. Together these substances resist the propagation of a fracture better than either

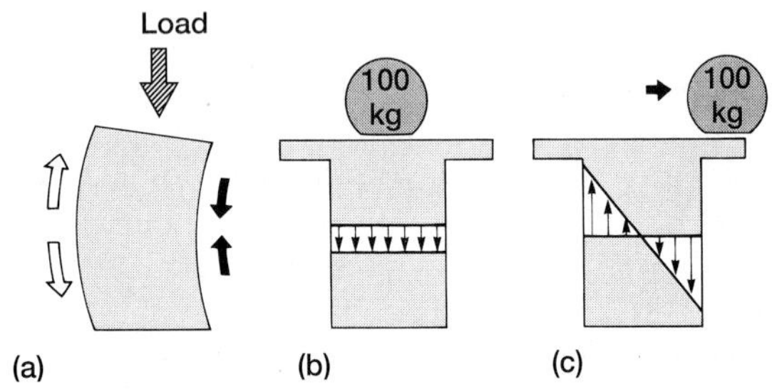


**Figure 4.36** Loading. (a) When a material bends under a load, compressive forces (solid arrows) develop along the concave side, tensile forces (open arrows) along the convex side. (b) When a supportive column is loaded symmetrically (with the weight centered), the only type of force experienced is compressive force. The distribution of the 100-kg force within a representative section is depicted graphically. The lengths of the down arrows show equal distribution of compressive forces within this representative section. (c) Asymmetrical loading of the same mass causes the column to bend. The column experiences compressive forces (down arrows) and tensile forces (up arrows). Both compressive and tensile forces are greatest near the surface and least toward the center of the column.

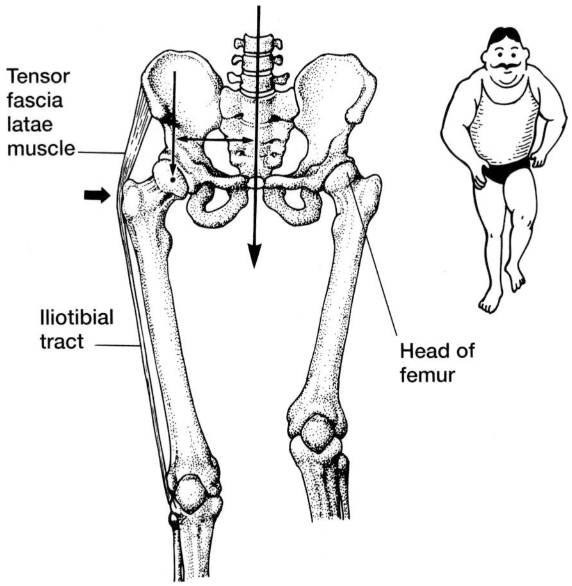


**Figure 4.37** Braces. The weight of the upper body is carried on the heads of the femurs (left). This means that during the striding gait (right), the head of one femur carries all the upper body weight; and consequently, the long shaft of the femur is loaded asymmetrically, increasing its tendency to bend. The iliotibial tract, the long tendon of the tensor fascia latae muscle that runs laterally across the femur, partially counteracts this tendency to bend and thereby reduces the tensile forces that would otherwise develop within the femur.

constitutent alone (figure 4.38a–c). This same principle of composites gives fiberglass its resistance to breaking. Fiberglass consists of glass fibers embedded in a plastic resin. Glass is brittle, resin weak, but together they are strong because they blunt small cracks and prevent their spread. As a crack approaches the boundary between the two fiberglass materials, the resin gives slightly. A small space opens to receive the tip of the crack, absorb the stresses concentrated there, and interrupt its propagation.

Collagen fibers and hydroxyapatite crystals are the composite materials of bone matrix. They are thought to act in a manner analogous to the glass and resin of fiberglass to blunt small fractures. Further, the orientation of collagen fibers alternates in successive layers so that they better receive tensile and compressive forces.

**Bone structure (p. 172)**

## Tissue Response to Mechanical Stress

Tissues can change in response to mechanical stress. If living tissue is unstressed, it tends to decrease in prominence, a condition termed **atrophy** (figure 4.39a). If it experiences increased stress, tissue tends to increase in prominence, a condition termed **hypertrophy** (figure 4.39b). Cell division and proliferation under stress is termed **hyperplasia.** Thus, in response to exercise, the muscles of an athlete will increase in size. This overall increase is primarily due to an increase in the size of existing muscle cells, not to an increase in cell number (hypertrophy but not much hyperplasia). During pregnancy, smooth muscles of the uterus increase both in size and number (hypertrophy and hyperplasia).

**Muscle response to chronic exercise (p. 363)**

Tissues can, under some circumstances, change from one type to another, a transformation called **metaplasia.** Metaplastic transformations are often pathological. For example, the normal ciliated pseudostratified columnar epithelium of the trachea may become stratified squamous epithelium in tobacco smokers. But some metaplastic changes seem to be part of normal growth and repair processes as well. For example, reptiles exhibit metaplastic bone formation during growth of long bones. Chondrocytes become osteoblasts and cartilaginous matrix becomes osseous as cartilage undergoes direct transformation into ossified bone. During bone repair in reptiles, amphibians, and fishes, the cartilaginous callus appears to arise from connective tissue through metaplasia.

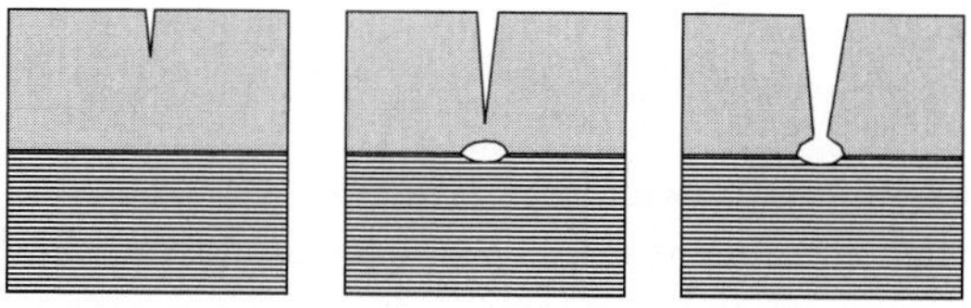


**Figure 4.38** Fracture propagation. (a) Failure of a structure begins with the appearance of a microfracture that spreads rapidly. (b) In composite materials, such as bone, the advancing fracture is preceded by stress waves that may open a small space at the boundary between the composite materials where they give slightly. (c) As the fracture line meets this space, its sharp tip is blunted and its progress curtailed.

**Tissue types (p. 168)**

All tissues retain some physiological ability to adjust to new demands, even after embryonic development is complete. Weight training causes an athlete's existing muscles to increase and his or her tendons to strengthen. Regular long-distance running enhances circulation, increases blood volume, improves oxygen delivery to tissues, and metabolizes stored lipids more efficiently. Although the number of nerve cells does not usually increase in response to the physiological stress of exercise, coordination of muscle performance often does. Tissues continue to adapt physiologically to changes in demand throughout the life of the individual. One of the best examples is bone because it illustrates the complexity of tissue response.

### *Responsiveness of Bone*

While performing in a protective or supportive role, bone cannot significantly deform or change shape. Leg bones that telescope or bend like reeds would certainly be ineffective as supports for the body. Bones must be firm. But because living bone is dynamic and responsive, it gradually changes during the life of an individual. The genetic program of a person sets forth the basic form a bone takes, but immediate environmental factors contribute to ultimate bone form as well. Some peoples of the New World developed the practice of wrapping a baby's head against a cradle board (figure 4.40a, left). As a result, the normal skull shape of the baby was altered so that the side pressed to the board was flattened. In parts of Africa and Peru, prolonged bandaging of the back of the skull caused elongation of the cranium (figure 4.40a, right). Until recent times, young Chinese girls who looked forward to a leisured life had their feet permanently folded and tightly bound to produce tiny feet in adulthood. The toes were crowded and the arch exaggerated (figure 4.40b, right). The normal and, by comparison, large foot was considered ugly in women (figure 4.40b, left). Because foot-binding impaired biomechanical performance, this also had what was considered the proper social consequence of keeping women literally "in their place."

***Environmental Influences*** Four types of environmental influences alter or enhance the basic shape of bone set down by the genetic program of all species in general. One is infectious disease. A pathogenic organism can act

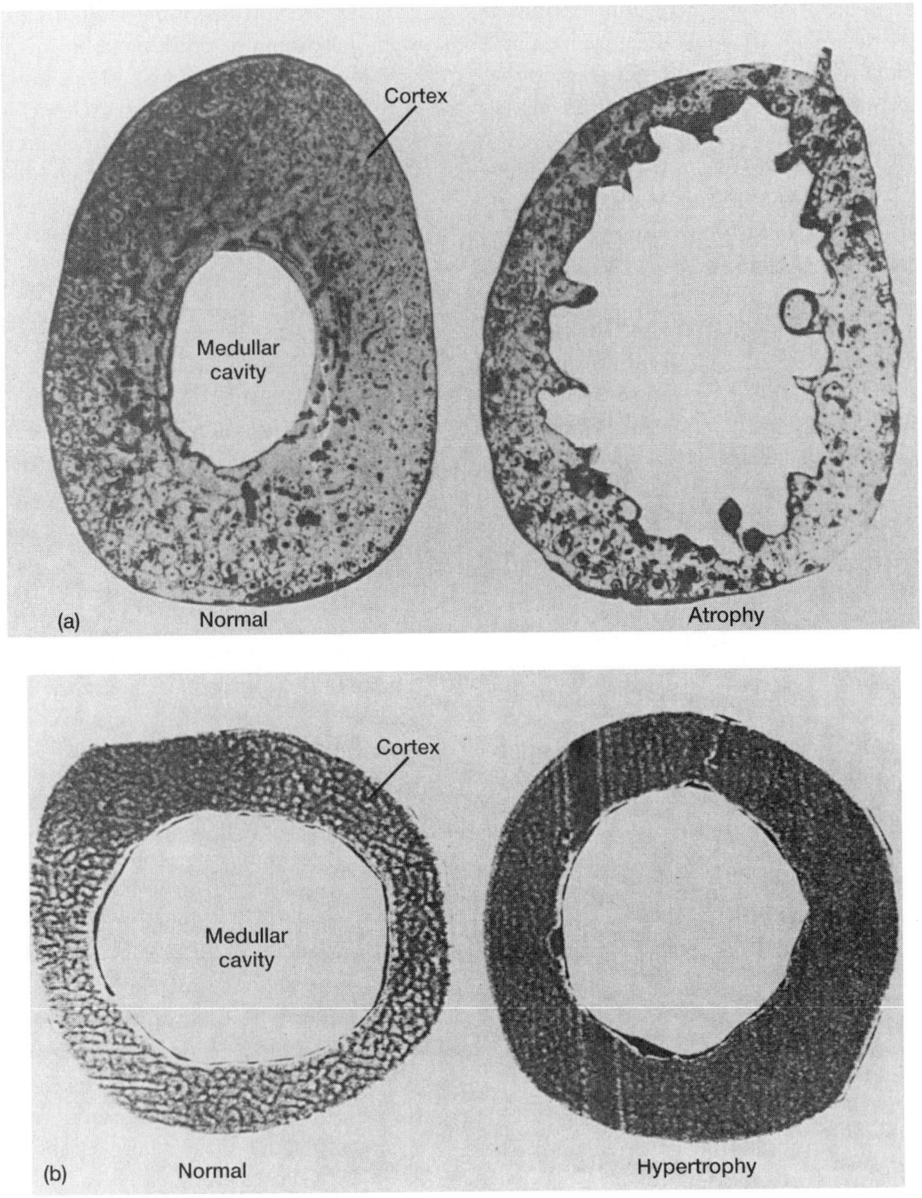


**Figure 4.39** Loss (atrophy) and increase (hypertrophy) of bone. (a) Cross section of a normal foot bone from a dog is illustrated on the left. Cross section of the same bone from the opposite foot (right) that was immobilized in a cast for 40 weeks reveals significant atrophy. (b) Cross section of a normal femur from a pig is depicted on the left. Cross section of a femur from a pig that had been vigorously exercised on a regular basis for over a year shows increased bone mass (right). Hypertrophy is evident from the thickening and greater density of the cortex.

directly to alter the pattern of bone deposition and change its overall appearance. Or, the pathogen can physically destroy regions of a bone. A second environmental influence is nutrition. With adequate diet, normal bone formation is usually taken for granted. If the diet is deficient, bones can suffer considerable abnormalities. Rickets, for example, caused by a deficiency of calcium in humans, results in buckling of weight-bearing bones (figure 4.40c). Ultraviolet radiation transforms dehydrocholesterol into vitamin D, which the human body needs to incorporate calcium into bones. Sunshine and supplements of milk are usually enough to prevent rickets. Hormones are the third factor that can affect bone form. Bone is a calcium reservoir, which is perhaps its oldest function. When demanded, some calcium is removed from bone matrix. Calcium drains occur during lactation when the female produces calcium-rich milk, during pregnancy when the fetal skeleton begins to ossify, during egg laying when hard shell is added, and during antler growth when the bony rack of antlers are being developed.

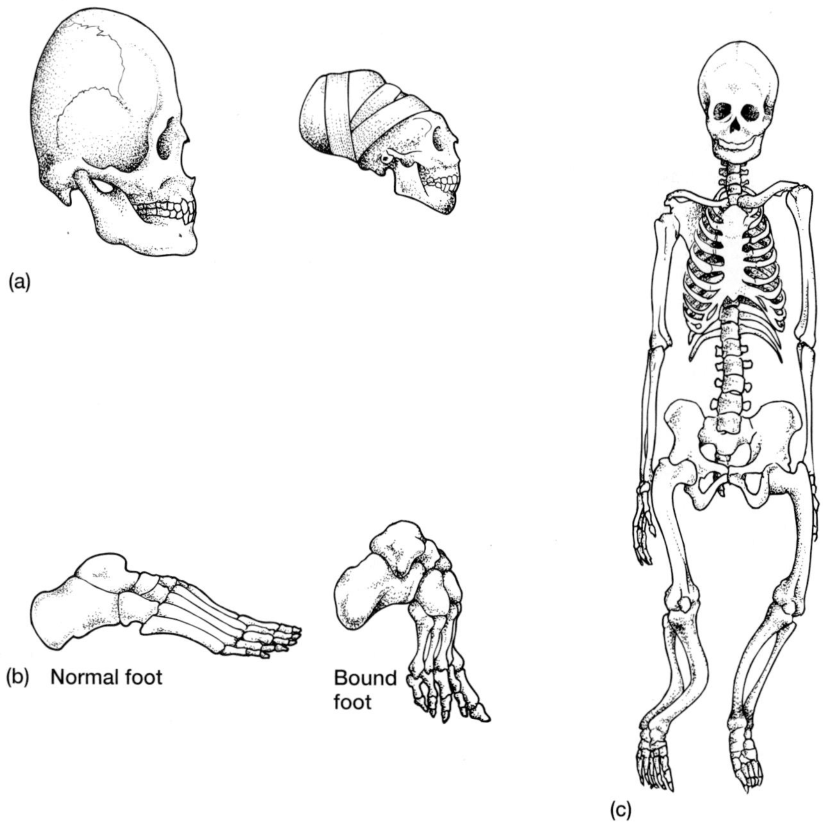


**Figure 4.40** Responsiveness of bone to mechanical stress. (a) The continuous mechanical pressure of a cradle board flattened the back of the Navajo Indian skull (left), and wrapping of the Peruvian native skull (right) caused it to elongate. (b) Historically, many Chinese followed the practice of binding the feet of young girls with tight wrappings. The small, deformed foot shown on the right was considered socially "attractive." (c) A nutritional calcium deficiency during infancy led to rickets, which weakened this woman's skeleton, shown here at age 70. Her bones bent under the normal load of her body.

**Endocrine control of bone calcium (p. 589)**

The fourth environmental influence on bone form is mechanical stress (figure 4.40b,c). Each weight-bearing bone experiences gravity, and muscles tug on most bones. Forces produced by gravity and muscle contraction place bone in an environment of stresses that determine the final shape of bone. Throughout an individual's life, these stresses upon bone change. As a young animal gains its footing and daring, it becomes more active. As an adult, it might migrate, battle for territory, or increase its foraging to support offspring of its own. As the animal grows bigger, scaling becomes a factor. Geometric increase in the mass of a growing animal places greater mechanical demands upon the supportive elements of its body. Human athletes on a continuous training program intentionally push their strength and stamina to stimulate physiological adaptation to the heightened activity. Conversely, age or inclination can lead to declining activity and reduced stress on bones. Teeth might decay, and this changes the stress pattern experienced by the jaws. An injury can lead to favoring one limb over another. For a variety of reasons then, the forces experienced by bones change.

***Atrophy and Hypertrophy*** The response of bone to mechanical stresses depends upon force duration. If bone experiences continuous pressure, bone tissue is lost and atrophy occurs. Continuous pressure against bone arises occasionally with abnormal growths, such as brain tumors that bulge from the surface of the brain and press on the underside of the bony skull. If this continuous pressure is prolonged, the bone erodes, forming a shallow depression along the surface of contact. Aneurysms, balloonings of blood vessels at weak spots in the vessel wall, can exert continuous pressure against nearby bone and cause it to atrophy. Orthodontic braces cinched to teeth by a dentist force teeth up against the sides of the bony sockets in which they sit. Resorption of continuously stressed bone opens the way for teeth to migrate slowly but steadily into new and presumably better positions within the jaws.

Thus, bone that experiences a continuous force atrophies; however, so does bone that experiences no force. When forces are absent, bone density actually thins. People restricted to prolonged bed rest without exercise show signs of osteoporosis. This has been studied experimentally in dogs on whom a cast has been applied to one leg. The immobilizing cast eliminates or considerably reduces the normal loads carried on a leg bone. Bones so immobilized exhibit significant signs of resorption, which can actually occur rather quickly. Experiments with immobilized wings of roosters show that within a few weeks wing bones become extensively osteoporetic. Rarification of bone matrix occurs in astronauts during extended periods of weightlessness. Calcium salts leave bones, circulate in the blood, and this excess is actually excreted. When astronauts return to Earth's gravitational forces, their skeletons gradually recover their former density. Even over long voyages, the ossified skeleton is not likely to disappear all together, but it may fall to a genetically determined minimum. And, of course, muscle contractions maintain some regime of forces on bone. But during deep space travel lasting many months, bone atrophy can progress far enough to make return to Earth's gravity hazardous. Prevention of bone atrophy in space travel remains an unsolved problem.

Between continuously stressed and unstressed bone is the third type of force application, *intermittent* stress. Intermittent stress stimulates bone deposition, or hypertrophy. The importance of intermittent forces on bone growth and form has long been suspected from the fact that bone atrophies when intermittent forces are removed. Conversely when rabbit bones were intermittently stressed by a special mechanical apparatus, hypertrophy occurred. More recently, bones of the rooster wing were stressed once daily with compressive loads but otherwise left immobilized. After a month, the artificially stressed bones did not exhibit osteoporosis but did show growth of new bone, clearly an appropriate physiological response to the artificially induced intermittent stresses.

***Internal Design*** The overall shape of a bone reflects its role as part of the skeletal system. Internal bone tissue consists of areas of **compact** and **spongy bone.** Distribution of compact and spongy bone is also thought to be directed by mechanical factors, although there is little hard evidence to support this correlation. According to an engineering theory called the trajectorial theory, when a load is placed on an object, the material within the object carries the resulting internal stress along stress trajectories or paths that pass these forces from molecule to molecule within the object (figure 4.41a). A beam embedded at its base in the wall will bend under its own weight. The lower surface of the beam experiences compressive forces as the material is pushed together, and the upper surface of the beam experiences tension as material here is pulled apart. The resulting compressive and tensile stresses are carried along stress trajectories that cross at right angles to each other and bunch under the beam's surface.

Culmann, a nineteenth century engineer, applied this trajectorial engineering theory to the internal architecture of bone. Because the femur carries the load or weight of the upper body, he reasoned that similar stress trajectories must arise within this bone. In order for the body to build a strong structure yet be economical with material, bony tissue should be laid down along these stress trajectories, the lines along which the load is actually carried. After looking at sections of bone, Culmann suggested that nature arranged bone spicules **(trabeculae)** into a lattice of spongy bone at the ends of long bones (figure 4.41b). Because these lines of stress move to the surface near the middle of the bone, the trabeculae follow suit and the overall result is a tubular bone. If the trabeculae of bone follow internal lines of stress, these trabeculae might be expected to form a lattice of spongy bone after birth when functional loads are first experienced. This is borne out. Trabeculae of young fetuses display random honeycomb architecture. Only later do they become arranged along presumed lines of internal stress.

***Wolff's Law*** As applied mechanical forces change, bone responds dynamically to adapt physiologically to changing stresses. Wolff's Law, named for a nineteenth century scientist who emphasized the relationship between bone form and function, states that remodeling of bone occurs in proportion to the mechanical demands placed upon it.

When bone experiences new loads, the result is often a greater tendency to buckle. When buckling occurs, tensile forces appear. Bones are less able to withstand tensile forces than they can compressive forces. In order to compensate, bone undergoes a physiological remodeling to better adapt to the new load (figure 4.42a–c). Initially, adaptive remodeling entails thickening along the wall experiencing compression. Eventually overall remodeling restores the even, tubular shape of the bone. How are cells along the compressive side selectively stimulated to deposit new bone? Because nerves penetrate throughout bone, they might be one way of promoting and coordinating the physiological response of osteocytes to changes in loading. However, bones to which nerves have been cut still abide by Wolff's Law and adjust to changes in mechanical demand.

Muscles pulling on bone affect the shape of vascular channels near their points of attachment to bone, which alters the blood pressure in vessels supplying bone cells. Increased muscle activity accompanying increased load might, via such blood pressure changes, stimulate bone cells to remodel. However, muscle action on bone, even if sufficient to change blood pressure, seems too global a mechanism to lead to the specific remodeling responses actually observed in bone.

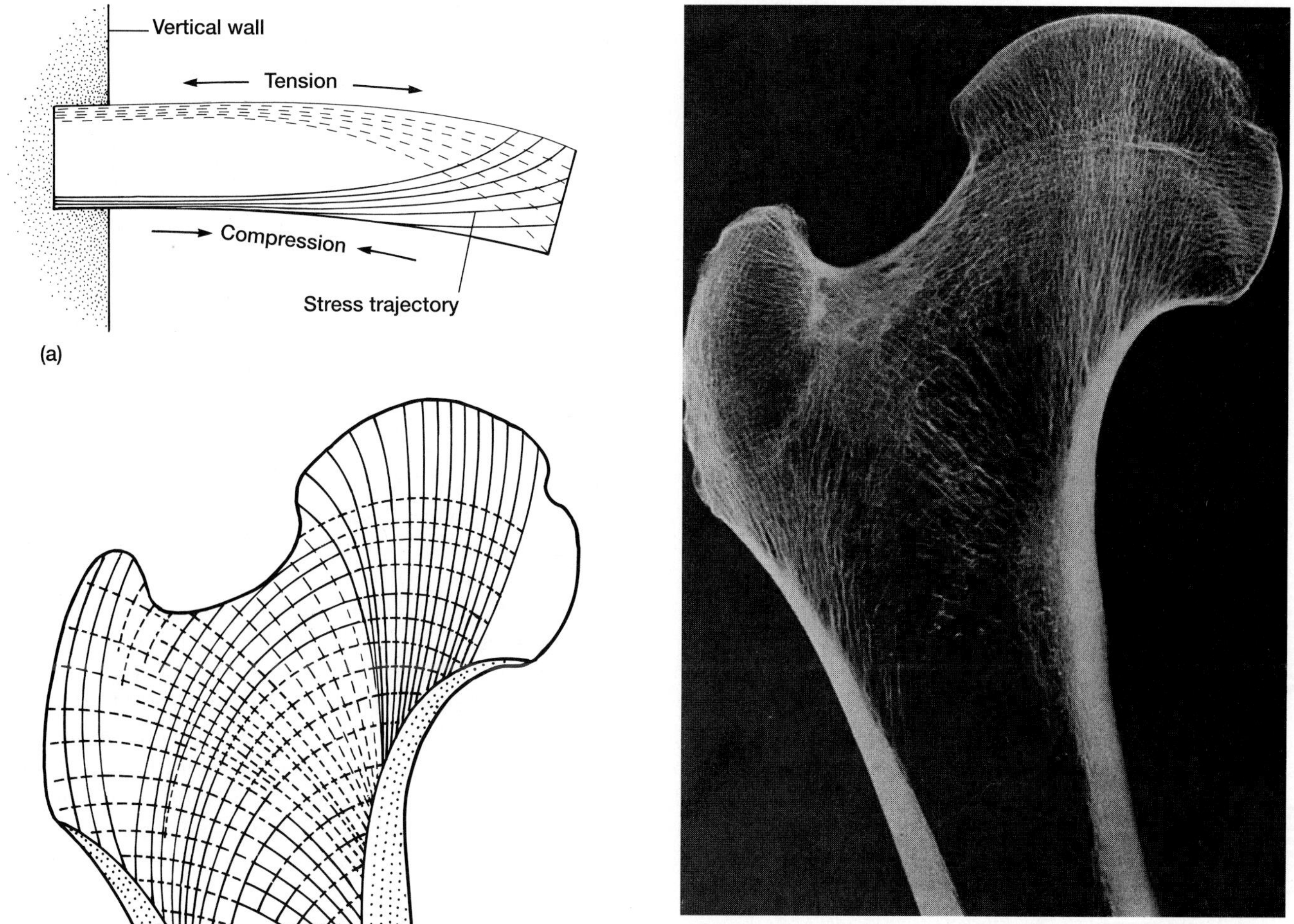


**Figure 4.41** Stress trajectories. (a) A beam projecting from a wall tends to bend under its own weight, placing internal stresses on the material from which it is made. Engineers visualize these internal stresses as being carried along lines called stress trajectories. Compressive forces become concentrated along the bottom of the beam, tensile forces along the top. Both forces are greatest at the surface of the beam. (b) Stress trajectories in living bone. When this theory is applied to living bone, the matrix of bone appears to be arranged along the lines of internal stress. The result is an economical latticework of bone, with material concentrated at the surface of a tubular bone. A cross section through the proximal end of a femur reveals the lattice of bone spicules within the head that become concentrated and compacted along the wall within the shaft of the femur.

Bone cells occupy small lacunae, spaces within the calcium matrix of a bone. Slight configurational changes in the lacunae occupied by bone cells offer a more promising mechanism. Under compression, lacunae tend to flatten; under tension they tend to become round. If these configurational changes produced under load could be read by the bone cells occupying the lacunae, then bone cells might initiate a remodeling matched to the type of stress experienced.

Another mechanism might involve **piezoelectricity,** or low-level electric charges. These are surface charges that arise within any crystalline material under stress—positive

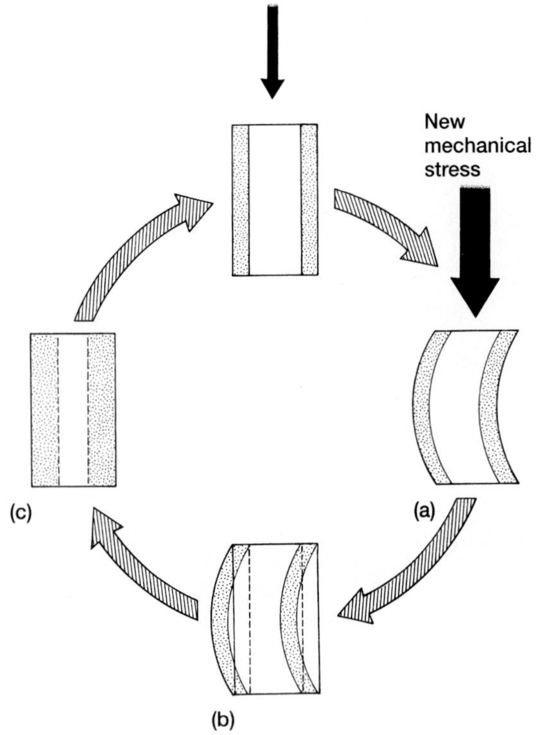


**Figure 4.42** Bone remodeling. When tubular bone experiences a new and more distorting stress (a), it undergoes a physiological response that both thickens and straightens it. New bone forms along the concave surface (b), remodeling the bone, and the straight shape is restored (c). Further remodeling returns the bone to its original shape (top) although the walls would now be thicker to withstand the new, increased load.

charges appear on the surface under compression and negative charges on the surface in tension. Bone, with its structure of apatite crystals, experiences piezoelectric charges when it is loaded. It can be easily imagined that under a new load, a new environment of piezoelectric charges would appear within the tissue of stressed bone. If individual bone cells could key off these localized piezoelectric charges, then a specific remodeling response might follow.

Although promising, each of these proposed mechanisms by itself seems insufficient to account for the physiologically adaptive remodeling that occurs during bone response to the functional demands placed upon it. This is a challenging area for further research.

# Biophysics

Biophysics is concerned with principles of energy exchange and the significance of these principles for living organisms. The use of light, the exchange of heat, and the diffusion of molecules are fundamental to the survival of an organism. Biological design and its limits are determined by the physical principles governing energy exchange between an organism and its environment, and internally between active tissues within the organism. One of the most important of these physical principles applies to the exchange of gases.

## Diffusion and Exchange

### *Pressures and Partial Pressures*

Air pressure varies slightly with weather conditions, such as low- and high-pressure fronts, and with temperature. When animals ascend in altitude, air pressure drops significantly as the air thins (becomes less dense) and breathing becomes more labored. This drop in pressure of the gases, especially oxygen, creates the difficulty. Air is a mixture of nitrogen (about 78% by volume), oxygen (about 21% by volume), carbon dioxide, and trace elements. Each gas in air acts independently to produce its own pressure irrespective of the other gases in the mixture. Of the total 101,000 Pa (pressure of air) at sea level, oxygen contributes 21,210 Pa (101,000 Pa × 21%) to the total, nitrogen 78,780 Pa (101,000 Pa × 78%), and the remaining gases 1,010 Pa. Because each gas contributes only a part of the total pressure, its contribution is its **partial pressure.** The rate at which oxygen can be inhaled depends on its partial pressure. At 5,300 m (18,000 ft), air pressure drops to about 0.5 atm, or 50,500 Pa. Oxygen still comprises about 21% of the air, but because the air is thinner, there is less total oxygen present. Its partial pressure falls to 10,605 Pa (50,500 Pa × 21%). With a drop in the partial pressure of oxygen, the respiratory system picks up less and breathing becomes more labored. Animals living in the high mountains, and especially high-flying birds, must be designed to address this change in atmospheric pressure.

Because water weighs much more than air per unit of volume, an animal descending through water experiences pressure changes much more quickly than one descending through air. With each descent of about 10.3 m (33.8 ft), water pressure increases by about 1 atm (atmosphere). Thus, a seal at a depth of 20.6 m experiences almost two additional atmospheres of pressure more than it experiences when basking on the beach. The effect of this pressure change on body fluids and solids is probably inconsequential, but gas in the lungs or in the gas bladders of fishes is compressed significantly. A 1-m descent in water adds 9,800 Pa of pressure, or about 1.5 lb of pressure per square inch of chest wall. Compressing the lungs or the gas bladder reduces their volume and thus affects buoyancy. The movement of gases into and out of the bloodstream is affected by the difference in the partial pressure of oxygen breathed in at the surface and its different partial pressure when it is diffused into the blood once the animal is submerged. We look specifically at these properties of gases and the way in which the vertebrate body is designed to accommodate them when we examine the respiratory and circulatory systems in chapters 11 and 12, respectively.

### *Countercurrent, Concurrent, and Crosscurrent Exchange*

Exchange is a large part of life. Oxygen and carbon dioxide pass from the environment into the organism or from the organism into the environment. Chilled animals bask to pick up heat from their surroundings; active animals lose heat to their surroundings to prevent overheating. Ions are exchanged between the organism and its environment. This process of exchange, whether it involves gases, or heat, or ions is sometimes supplemented by air or water currents passing one another. Efficiency of exchange depends on whether the currents pass in opposite or equivalent directions.

Imagine two parallel but separate tubes carrying streams of water. Water entering one tube is hot, and water entering the other is cold. If the tubes are made of conducting material and contact each other, heat will pass from one to the other (figure 4.43a,b). Water flow may be in the same direction, as in **concurrent exchange,** or in opposite directions, as in **countercurrent exchange.** The efficiency of heat exchange between the tubes is affected by the directions of flow.

If the streams are concurrent, as the two tubes come in contact, the temperature difference will be at its maximum but will drop as heat is transferred from the hotter to the colder tube. The cold stream of water will warm, the hot stream will cool so at their point of departure, both streams of water approach the average of their two initial temperatures (figure 4.43a). If we take the same tubes and same starting temperatures but run the currents in opposite directions we have a countercurrent exchange; heat transfer becomes much more efficient than if both currents flowed in the same direction (figure 4.43b). A countercurrent flow keeps a differential between the two passing streams throughout their entire course, not just at the initial point of contact. The result is a much more complete transfer of heat from the hot stream to the cold stream. When the tubes are separated, the cold stream is nearly as warm as the adjacent hot stream. Conversely, the hot stream gives up most of its heat in this countercurrent exchange so that its temperature has fallen to almost that of the entering cold stream.

This physical principle of countercurrent exchange can be incorporated into the design of many living organisms. For example, endothermic birds that wade in cold water could lose much of their critical body heat to the icy water if warm blood circulated through their feet and was exposed to the cold water. Replacing this lost heat could be expensive. A countercurrent heat exchange between outgoing warm blood in the arteries supplying the feet and returning cold blood in the veins prevents heat loss in wading birds. In the upper legs of such birds, small arteries come in contact with small veins, forming a **rete,** a network of intertwining vessels. Because arterial blood in these vessels passes in opposite directions to venous blood, a countercurrent system of heat exchange is established. By the time the blood in the arteries reaches the feet, it has given up almost all of its heat to the blood in the veins returning to the body. Thus, there is little heat lost through the foot into the cold water. The countercurrent system of the rete forms a **heat block,** preventing the loss of body heat to the surroundings. Estimates indicate that the rete is so efficient in heat transfer that if boiling water were poured through a wading bird's arteries at one end and ice water through its veins at the other end, blood vessels in the feet would lose less than 1/10,000 of a degree in temperature.

Respiration in many fishes is characterized by a countercurrent exchange also. Water high in oxygen flows across the gills, which contain blood capillaries low in oxygen flowing in the opposite direction. Because water and blood pass in opposite directions, gas exchange between the two fluids is very efficient.

In bird lungs, and perhaps in other animals as well, gas exchange is based on another type of flow, a crosscurrent exchange between blood and air capillaries. Because blood capillaries cross at right angles to the air capillaries in which gas exchange occurs, a crosscurrent is created (figure 4.43c). Blood capillaries run sequentially from an arteriole to supply each air capillary. When blood capillaries cross an air capillary, oxygen passes into the bloodstream and $CO_2$ is given up to the air. Each blood capillary contributes stepwise to the rising level of oxygen in the venule it joins. Partial pressures vary along the length of an air capillary, but the additive effect of these blood capillaries in series is to build up efficient levels of oxygen in venous blood as it leaves the lungs. Unlike a countercurrent system, a crosscurrent system is relatively unaffected if airflow is reversed.

## Optics

Light carries information about the environment. Color, brightness, and direction all arrive coded in light. Decoding this information is the business of light-sensitive organs. However, the ability to take advantage of this information is affected by whether the animal sees in water or in air, and it is affected by how much the two eyes share overlapping fields of view.

### *Depth Perception*

The position of the eyes on the head represents a trade-off between panoramic vision and depth perception. If the eyes are positioned laterally, each scans separate halves of the surrounding world, and the total field of view at any

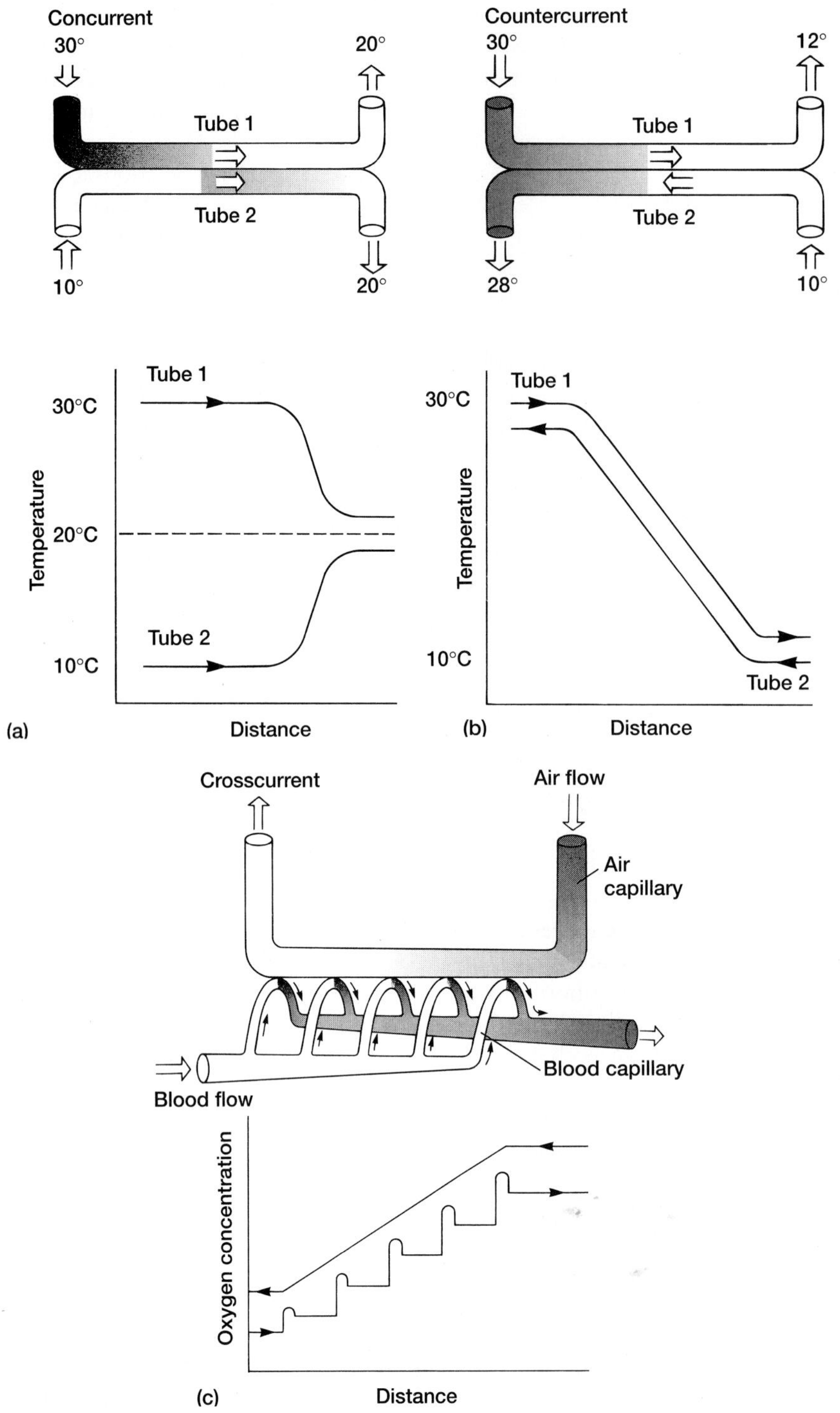


**Figure 4.43** Systems of exchange. Direction and design of exchange tubes affect the efficiency of transfer regardless of whether the exchange involves heat, gases, ions, or other substances. The first two examples (a and b) illustrate heat transfer. The third example (c) depicts gas exchange. (a) Concurrent exchange describes the condition in which separated fluids flow in the same direction. Because the temperature gradient between the fluids is high when they enter the tubes and rather low when they exit, the average difference in temperature exchanged between the two fluids is relatively low. The fluid in tube 2 is at 10° when it enters and at 20° when it exits. (b) In countercurrent exchange, the fluids pass in opposite directions within the two tubes so that the temperature difference between them remains relatively high all along their lengths. The fluid in tube 2 is at 10° when it enters and at 28° when it exits. Thus, more heat is transferred with countercurrent exchange than with concurrent exchange. (c) In crosscurrent exchange, each blood capillary branch passes across an air capillary at about right angles to it and picks up oxygen. The levels of oxygen rise serially in the departing blood. Arrows indicate the direction of flow.

moment is extensive. Where visual fields do not overlap, an animal has **monocular vision.** It is common in animals preyed upon and gives the individual a large visual sweep of its environment to detect the approach of potential threats from most surrounding directions. Strict monocular vision, in which the visual fields of the two eyes are totally separate, is relatively rare. Cyclostomes, some sharks, salamanders, penguins, and whales have strict monocular vision.

Where visual fields overlap, vision is **binocular vision.** Extensive overlap of visual fields characterizes humans. We have as much as 140° of binocular vision, with 30° of monocular vision on a side. Binocular vision is important in birds (up to 70°), reptiles (up to 45°), and some fishes (as much as 40°). Within the area of overlap, the two visual fields merge into a single **stereoscopic image** (figure 4.44). The advantage of stereoscopic vision is that it gives a sense of depth perception. Closing one eye and maneuvering about a room demonstrates how much sense of depth is lost when the visual field of only one eye is used.

Depth perception results from how the brain processes visual information. With binocular vision, the visual field seen by each eye is divided in the brain. In most mammals, half goes to the same side, and the other half crosses via the **optic chiasma** to the opposite side of the brain. For a given part of the visual field, inputs from both eyes are brought together on the same side of the brain. Within the brain, the **parallax** of the two images is compared. Parallax is the slightly different views one gets of a distant object when it is viewed from two different points. Look at a distant lamp post from one position, and then step a few feet laterally and look at it again from this new position. Slightly more of one side of the lamp post can be seen, less of the opposite side, and the position of the post relative to background reference points changes as well. The nervous system takes advantage of parallax resulting from eye position. Each visual image gathered by each eye is slightly offset from the other because of the distance between the eyes. Although this distance is slight, it is enough for the nervous system to produce a sense of depth resulting from the differences in parallax.

**Depth perception and stereoscopic vision (p. 682)**

## *Accommodation*

Sharp focusing of a visual image upon the retina is termed **accommodation** (figure 4.45a). Light rays from a distant object strike the eye at a slightly different angle than rays from a nearby object. As a vertebrate alters its gaze from close to distant objects of interest, the eye must adjust, or accommodate to keep the image focused. If the image falls behind the retina, **hyperopia,** or farsightedness, results. An image focused in front of the retina produces **myopia,** or nearsightedness (figure 4.45b,c).

The lens and the cornea are especially important in focusing entering light. Their job is considerably affected by the **refractive index** of light, a measure of the bending effects on light passing from one medium to another. The refractive index of water is similar to the refractive index of the cornea; therefore, when light passes through water to the cornea in aquatic vertebrates, there is little change in the amount it bends as it converges on the retina. But when light passes through air to the liquid medium of the cornea in terrestrial vertebrates, it bends considerably. Similarly,

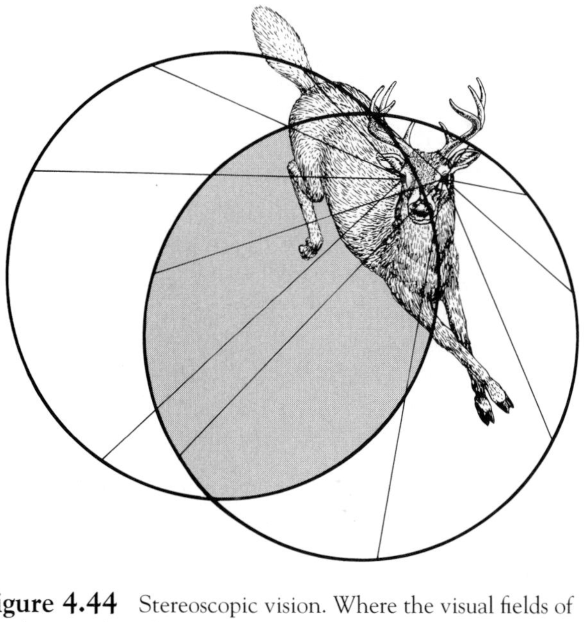

**Figure 4.44** Stereoscopic vision. Where the visual fields of the deer overlap, they produce stereoscopic vision.

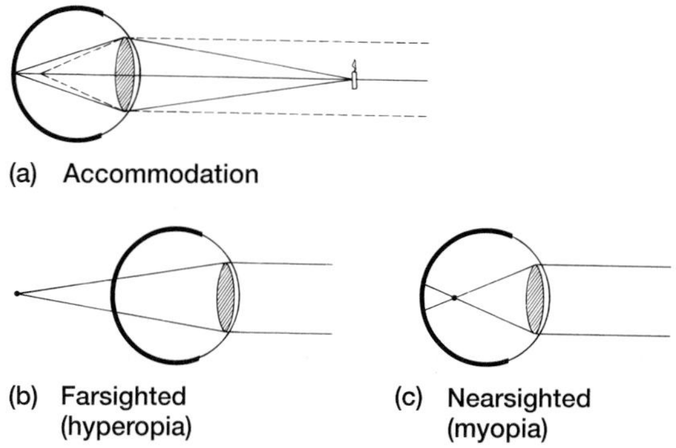


**Figure 4.45** Accommodation. (a) Normal vision in which the image is in sharp focus on the retina of the eye. (b) Farsighted condition (hyperopia) in which the lens brings the light rays to focus behind the retina. (c) Nearsighted condition (myopia) in which the sharpest focus falls in front of the retina.

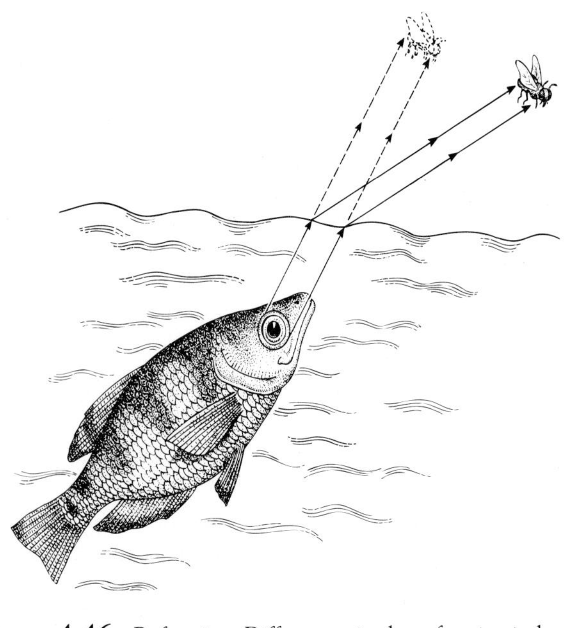

**Figure 4.46** Refraction. Differences in the refractive indexes of water and air bend light rays that enter the water from the insect. The result makes the insect seem to be in a different position than it really is, indicated by the dashed lines. The archer fish must compensate in order to shoot a squirt of water accurately and hit the real, not the imaginary, image.

aquatic animals viewing an object in air must compensate for the distortion produced by differences in the refractive indexes of air and water (figure 4.46). As a consequence of these basic optical differences, eyes are designed to work either in water or in air. Underwater vision is not necessarily out of focus. It just looks that way to our air-adapted eyes when we jump into a clear stream and attempt to focus our eyes. If we place a pocket of air in front of our eyes (e.g., a diving face mask), the refractive index our eyes are designed to accommodate returns and things become clear.

Accommodation can be accomplished by mechanisms that change the lens or the cornea. Cyclostomes have a corneal muscle that changes the shape of the cornea to focus on entering light. In elasmobranchs, a special protractor muscle changes the position of the lens within the eye. The elasmobranch eye is focused for distant vision. For near objects, the protractor muscle moves the lens forward. In most amniotes, the curvature of the lens changes to accommodate the eye's focus on objects near or far. Ciliary muscles act on the lens to change its shape and thus alter its ability to focus passing light.

**Eyes and mechanisms of accommodation (chapter 17)**

## Selected References


Alexander, R. McN. 1981. Factors of safety in the structure of animals. *Sci. Prog. (Oxford)* 67:109–30.

Alexander, R. McN. 1982. *Locomotion of animals*. London: Chapman Hall.

Alexander, R. McN. 1983. *Animal mechanics*. London: Blackwell Scientific Publications.

Behari, J. 1991. Electrostimulation and bone fracture healing. *Biomed. Engineering* 18:235–54.

Bookstein, F., B. Chernoff, R. Elder, J. Humphries, G. Smith, and R. Strauss. 1985. *Morphometrics in evolutionary biology*. Special pub. 15. Philadelphia: Academy of Natural Sciences.

Calder, W. A. 1984. *Size, function, and life history*. Cambridge, Mass.: Harvard University Press.

Cowin, S. C., R. T. Hart, J. R. Balser, and D. H. Kohn. 1985. Functional adaptation in long bones: Establishing *in vivo* values for surface remodeling rate coefficients. *J. Biomech.* 18:665–84.

Dimery, N.J., R. McN. Alexander, and R. F. Kerr. 1986. Elastic extensions of leg tendons in the locomotion of horses *(Equus caballus)*. *J. Zool.* 210:415–25.

Frost, H. M. 1990. Skeletal structural adaptations to mechanical usage (SATMU): 1. Redefining Wolff's Law: The bond modeling problem. *Anat. Rec.* 226:403–13.

Gans, C. 1974. *Biomechanics. An approach to vertebrate biology*. Philadelphia: J. B. Lippincott. Reprint 1980. Ann Arbor: University of Michigan Press.

Gordon, J. E. 1968. *The new science of strong materials*. London: Penguin Books.

Gordon, J. E. 1978. *Structures, or why things don't fall down*. New York: DaCapo Press.

Gould, S. J. 1977. *Ever since Darwin. Reflections in natural history*. New York: W. W. Norton.

Haldane, J.B.S. 1956. On being the right size. In *The world of mathematics*, edited by James R. Newman. New York: Simon and Schuster.

Lanyon, L. E. 1974. Experimental support for the trajectorial theory of bone structure. *J. Bone Joint Surg.* 56B:160–66.

Lanyon, L. E., and C. T. Rubin. 1985. Functional adaptation in skeletal structures. In *Functional vertebrate morphology*, edited by M. Hildebrand, D. M. Bramble, K. F. Liem, and D. B. Wake. Cambridge, Mass.: Harvard University Press, pp. 1–25.

McMahon, T. A., and J. T. Bonner. 1983. *On size and life*. New York: Scientific American Library.

Peter, R. H. 1983. *The ecological implications of body size*. Cambridge: Cambridge University Press.

Ruben, C. T., and L. E. Lanyon. 1987. Osteoregulatory nature of mechanical stimuli: Function as a determinant for adaptive remodeling in bone. *J. Orthop. Res.* 5:300–310.

Sadegh, A. M., G. M. Luo, and S. C. Cowin. 1993. Bone ingrowth: An application of the boundary element method to bone remodeling at the implant interface. *J. Biomech.* 26:167–82.

Schmidt-Nielsen, K. 1984. *Scaling: Why is animal size so important?* Cambridge: Cambridge University Press.

Thompson, D'A. W. 1942. *On growth and form*. 2d ed. Cambridge: Cambridge University Press.

Vogel, S. 1981. *Life in moving fluids*. Princeton: Princeton University Press.

Vogel, S. 1988. *Life's devices. The physical world of animals and plants*. Princeton: Princeton University Press.

Vogel, S., C. P. Ellington, Jr., and D. C. Kilgore, Jr. 1973. Wind-induced ventilation of the burrow of the prairie dog, *Cynomys ludovicianus*. *J. Comp. Physiol.* 85:1–14.

Wainwright, S. A., W. D. Biggs, J. D. Currey, and J. Gosline. 1982. *Mechanical design in organisms*. Princeton: Princeton University Press.

Woo, S. L-Y., S. C. Kuei, D. Amiel, M. A. Gomex, W. C. Hayes, F. C. White, and W. H. Akeson. 1981. The effect of prolonged physical training on the properties of long bone: A study of Wolff's law. *J. Bone Joint Surg.* 63A:780–87.

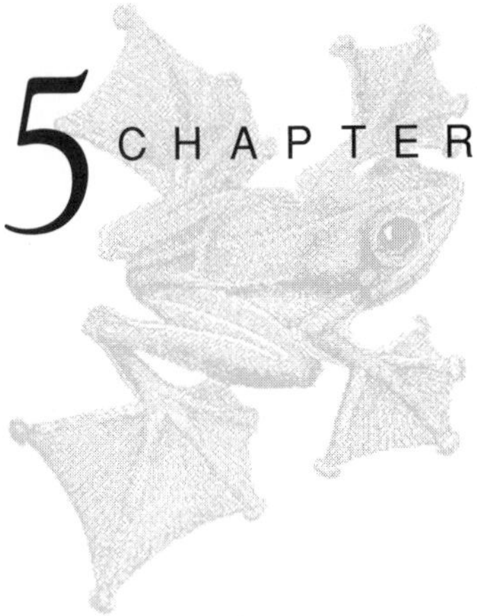

# 5 CHAPTER Life History



## Introduction

The English politician Benjamin Disraeli put it as follows, "Youth is a blunder; manhood a struggle; old age a regret!" The unfolding course of normal events from embryo to death constitutes an individual's life history. Whether it falls into blunder, struggle, and regret, as Disraeli proposed, is a matter for poets to debate. For biologists, life history begins with fertilization followed by embryonic development, maturation, and in some cases senescence, each stage being a prelude to the next. Embryonic development extends from fertilization to birth or hatching. During this time, a single cell, the egg, is fertilized and divides into billions of cells from which the basic structural organization of the individual takes shape. Maturation includes the time from birth to the point of reproductive ability. Maturation usually involves growth in size and acquisition of learned skills as well as appearance of anatomical features that distinguish the reproductive-ready adult. Prereproductive individuals are called juveniles or immatures. If juvenile and adult are strikingly different in form and the change from the one to the other occurs abruptly, the transformation is termed **metamorphosis.**

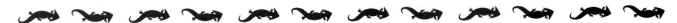

Loss of physical vigor and reproductive ability accompany **senescence,** or **aging.** This phenomenon is apparent in humans but rare in wild animals. In fact, senescent animals usually provide an unwilling but easy meal for ready predators. Most examples among animals other than humans come from zoos, because zoo animals are spared the natural fate of their free-ranging colleagues. Only a few examples of senescence in the wild are known. Some species of salmon senesce quickly after spawning and die within a few hours. Aging individuals occasionally survive in social species, such as canids and higher primates. But humans are unusual among vertebrates in that aging individuals commonly enjoy an extended postreproductive life. Even before life-extending medicines and health care, senior citizens characterized ancient human societies. The value of the elderly to human societies does not stem from their services as warriors or hunters or tillers of the soil because physical vigor has declined, nor can it be found in their procreative capacity. Perhaps aging humans were valued because of the child care service they could perform. Or perhaps aging individuals were living libraries, repositories of knowledge gained from a lifetime of experiences. Whatever the reasons, most human societies are unusual in protecting senescent individuals within the safety of society and not casting them to the wolves.

## Early Embryology

At the other end of an individual's life history are the events of early embryology. This is a complex, fascinating study in its own right. Embryonic development has profoundly contributed to evolutionary biology and to morphology. Early in development, the cells of the embryo become sorted into three primary germ layers—**ectoderm, endoderm,** and **mesoderm.** In turn, each layer gives rise to specific regions that form body organs. Structures of two species that pass through closely similar steps of embryonic development can be taken as evidence of homology between these structures. Close homology testifies to the phylogenetic relationship of both species.

Although embryonic development is an unbroken continuous process, we recognize stages in this progress in order to follow events and compare the developmental processes between groups. The youngest stage of the embryo is the fertilized egg, or **zygote,** which develops subsequently through the **morula, blastula, gastrula,** and **neurula** stages. During these early stages, the **embryonic area** becomes delineated from the **extraembryonic area** that supports the embryo or delivers nutrients but does not become a part of the embryo itself. The delineated embryo first becomes organized into basic germ layers and then passes through **organogenesis** (meaning organ and formation) during which the well-established germ layers differentiate into specific organs.

### Fertilization

Union of two mature sex cells, or **gametes,** constitutes fertilization. The male gamete is the **sperm** and the female the **ovum,** or egg. The sperm and egg carry genetic material from each parent. Both are **haploid** at maturity, with each containing half the chromosomes of each parent. The sperm's passage through the outer layers of the ovum sets in motion or **activates** embryonic development.

Although an egg can be very large, as is a chicken egg, it is but a single cell with a nucleus, cytoplasm, and cell membrane, or **plasma membrane.** While still in the ovary, the ovum accumulates **vitellogenin,** a transport form of yolk formed in the liver of the female and carried in her blood. Once in the ovum, vitellogenin is transformed into **yolk platelets** consisting of storage packets of nutrients that help support the growing needs of the developing embryo. The quantity of yolk that collects in the ovum is specific to each species. Eggs with slight, moderate, or enormous amounts of yolk are **microlecithal, mesolecithal,** or **macrolecithal,** respectively. Further, the yolk can be evenly distributed **(isolecithal)** or concentrated at one pole **(telolecithal)** of the spherical ovum. When yolk and other constituents are unevenly arranged, the ovum shows a **polarity** defined by a **vegetal pole,** where most yolk resides, and an opposite **animal pole,** where the prominent haploid nucleus resides.

The region immediately beneath the plasma membrane of the ovum is referred to as the **cortex** of the ovum. It often contains specialized **cortical granules** activated at fertilization. Outside the plasma membrane, three envelopes surround the ovum. The first, the **primary egg envelope,** lies between the plasma membrane and the surrounding cells of the ovary. The most consistent component of this primary layer is the **vitelline membrane,** a transparent jacket of fibrous protein. In mammals, the homologous structure is called the **zona pellucida** (figure 5.1). When the zona pellucida is viewed with a light microscope, a thin, striated line that was once called the "zona radiata" seems to constitute another discrete component of this primary layer. However, the high resolution electron microscope reveals that the zona radiata is not a separate layer but an effect produced by a dense stand of microvilli projecting from the surface of the ovum. These microvilli intermingle with microvilli reaching inward from the surrounding cells of the ovary. This stand of microvilli increases surface contact between the ovum and its environment within the ovary. Often after fertilization, a **perivitelline space** opens between the vitelline membrane and the plasma membrane.

The **secondary egg envelope** is composed of ovarian or **follicle cells** that immediately surround and help transfer nutrients to the ovum. In most vertebrates, follicle cells fall away from the ovum as it departs from the ovary. However, in placental mammals, some follicle cells cling to the ovum, becoming the **corona radiata** that accompanies the ovum on its journey to the uterus. The successful sperm must penetrate all three layers—follicle cells (in placental mammals), vitelline membrane, and plasma membrane.

The **tertiary egg envelope,** the exterior wrapping around the egg, forms in the oviducts. In some sharks, it consists of an egg case. In birds, reptiles, and monotremes, it includes the shell, shell membranes, and albumen enveloping the ovum. The tertiary layer is added after fertilization when the ovum travels down the uterine tubes. Vertebrates laying eggs encapsulated in such shells or other tertiary egg envelopes are **oviparous** (meaning egg and birth). If parents nestle over the eggs to add warmth, the eggs are **incubated.** Those vertebrates giving birth to embryos without such

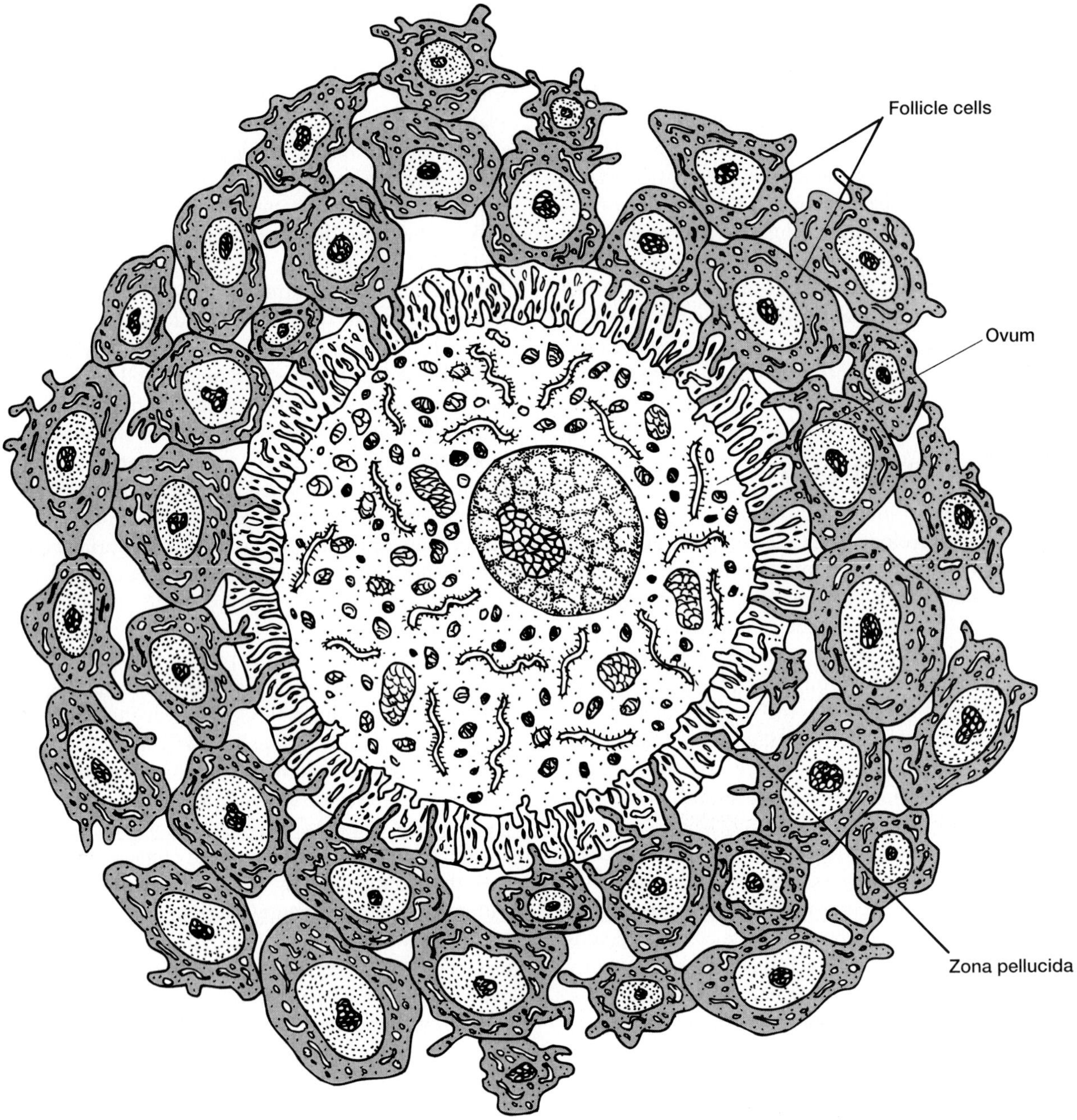


**Figure 5.1** Ovum of a placental mammal. The ovum is a large cell with organelles. The plasma membrane projects as microvilli into the surrounding protein membrane, the zona pellucida. Follicle cells arise from the ovary in which the ovum initially resides and, depending on the species, they may or may not accompany the ovum at ovulation. Processes of the follicle cells project to the surface of the ovum through the zona pellucida.

shells are **viviparous** (meaning live and birth). The **gestation** period includes the time the embryo develops within the female.

Viviparity has independently evolved over a hundred times in vertebrates. Many of these occasions occur in fishes, but most are in squamates. Oddly, no instance of viviparity is known in turtles, crocodiles, or birds, perhaps because they use the eggshell as a calcium reservoir on which the embryo draws when its own skeleton is undergoing ossification. In squamates, calcium is stored in the embryonic yolk so that evolutionary loss of the eggshell creates no loss of access to calcium stores. Viviparity has evolved repeatedly in squamates.

In some species, shelled eggs are retained within the oviducts of the female until they hatch or until the shells fall away. Shortly thereafter, the young are released to the world from the oviduct. Such reproductive patterns make it clear we must distinguish between the act of giving birth and the mode of supplying the fetus with nutrition. Specifically, **parturition** is the act of giving birth via viviparity, and **oviposition** is the act of laying eggs. The general term **parition** includes parturition and oviposition. Two general terms describe patterns of fetal nutrition. Embryos that draw nutrients from the yolk of the ovum are **lecithotrophic.** Lecithotrophic nutrition occurs through direct transfer of yolk to the connecting part of the digestive tract, as in some fishes, or through the vitelline arteries and veins that provide a vascular connection between the embryo and its yolk reserves. If nutrients are drawn from alternative sources, the embryos are **matrotrophic.** Vascular placentae or secretions of the oviduct that deliver nutrients to embryos are examples of matrotrophy. If offspring are provisioned with nutrients after birth or hatching, matrotrophy can be continued. In mammals, nutrient delivery changes from preparitive matrotrophy (placenta) to postparitive matrotrophy (lactation).

Release of the ovum from the ovary is **ovulation.** Fertilization usually takes place soon thereafter. With the fusion of egg and sperm, the **diploid** chromosome number is restored. Activation of development, initiated by sperm penetration, ushers in the next process, **cleavage.**

## Cleavage

During cleavage there is repeated mitotic cell division of the zygote. The embryo experiences little or no growth in size, but the zygote is transformed from a single cell into a solid mass of cells called the morula. Eventually the multicelled and hollow blastula forms (figure 5.2a–c). The **blastomeres** are the cells resulting from these early cleavage divisions of the ovum.

The first cleavage furrows appear at the animal pole and progress toward the vegetal pole. Where yolk is sparse, as in the microlecithal eggs of amphioxus and placental mammals, cleavage is **holoblastic**—mitotic furrows pass successfully through the entire zygote from animal to vegetal pole. After the first few furrows pass from the animal to the vegetal pole, subsequent furrows perpendicular to these develop until a hollow ball of cells forms around an internal fluid-filled cavity. Structurally, the blastula is the hollow ball of cells around the internal **blastocoel** cavity. In embryos where yolk is plentiful, cell division is impeded, mitotic furrowing is slowed, only a portion of the cytoplasm is cleaved, and cleavage is said to be **meroblastic.** In extreme cases, such as in the eggs of many fishes, reptiles, birds, and monotremes, meroblastic cleavage becomes **discoidal** because extensive yolk material at the vegetal pole remains undivided by mitotic furrows and cleavage is restricted to a cap of dividing cells at the animal pole.

In all chordate groups, cleavage converts a single-celled zygote into a multicellular, hollow blastula. Variations in the fundamental cleavage process result from characteristic differences in the amount of accumulated yolk reserves. The simplest pattern occurs in amphioxus, where little yolk is present. Eggs of amphibians possess slightly more yolk than amphioxus. In most fishes, reptiles, birds, and monotremes, great stores of yolk are packed into the egg. Placental mammals have little yolk present (figure 5.2a–d; table 5.1).

### *Amphioxus*

Eggs of amphioxus are microlecithal. The first cleavage plane passes from animal to vegetal pole, forming two blastomeres. The second cleavage plane is at right angles to the first and also passes from animal to vegetal poles, producing an embryo of four cells that resembles an orange with four wedges. The third cleavage plane is at right angles to the first two and lies between poles just above the equator, producing the eight-celled morula stage (figure 5.2a). Subsequent divisions of the blastomeres, now less and less in synchrony with each other, yield the 32-celled blastula surrounding the fluid-filled blastocoel.

### *Fishes*

In gars and bowfins, cleavage is holoblastic, although cleavage furrows of the vegetal pole are slowed. Most cell division is restricted to the animal pole (figure 5.3). Blastomeres in the vegetal pole are relatively large and hold most of the yolk reserves; those in the animal pole are relatively small and form the **blastoderm,** a cap of cells arched over a small blastocoel. The blastula produced is very much like that of amphibians.

In hagfishes, chondrichthyans, and most teleosts, cleavage is strongly discoidal, leaving most of the yolky cytoplasm of the vegetal pole undivided. Cleavage in teleosts produces two cell populations in the blastula. One of these is the blastoderm, also called the **blastodisc** because it is a

**TABLE 5.1 Comparison of Cleavage Patterns and Yolk Accumulation in Representative Vertebrates**

| Cleavage Pattern | Yolk Accumulation | Representative Animals |
|---|---|---|
| Holoblastic | Microlecithal | Amphioxus, placental mammals |
| | Mesolecithal | Lampreys, bowfin, gars, amphibians |
| Meroblastic | Macrolecithal | Elasmobranchs, teleost fishes |
| Discoidal[a] | Macrolecithal | Reptiles, birds, monotremes |

[a]Discoidial cleavage is an extreme case of meroblastic cleavage.

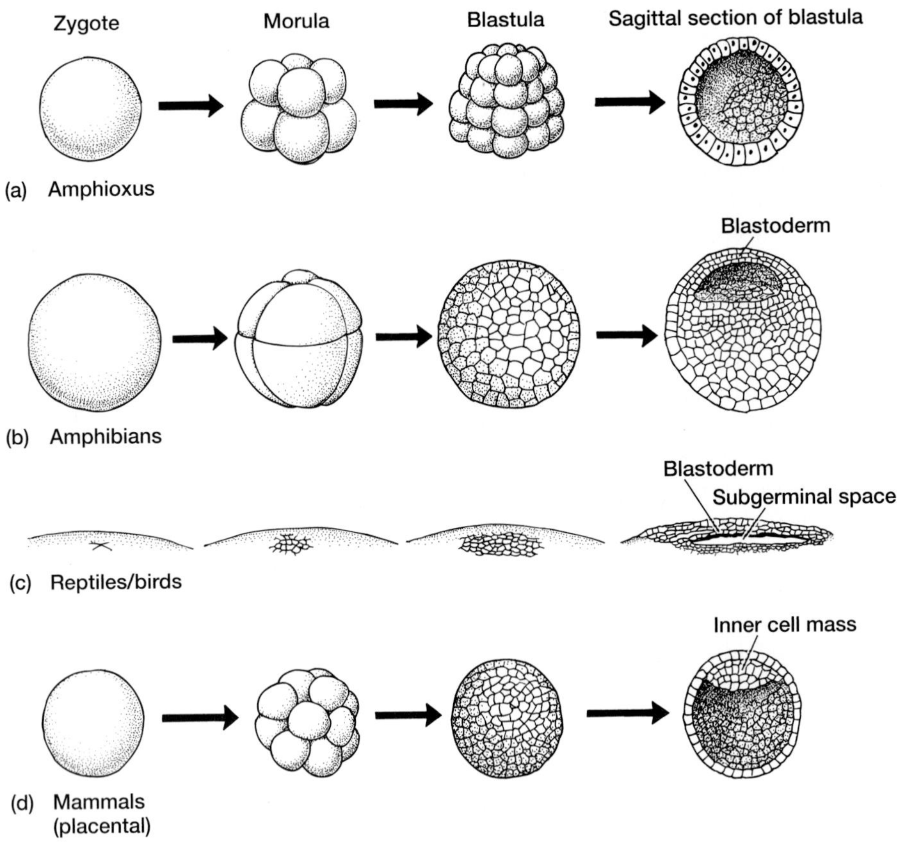


**Figure 5.2** Cleavage stages in five chordate groups. Relative sizes are not to scale. (a) Amphioxus. (b) Amphibian. (c) Reptiles and birds. (d) Placental mammal.

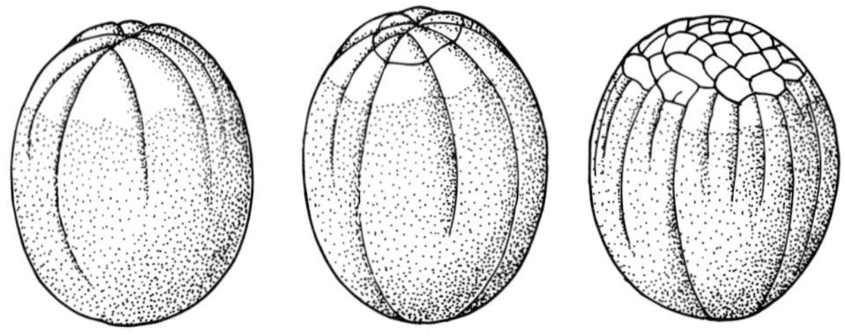

**Figure 5.3** Holoblastic cleavage in the bowfin, *Amia*.

discrete patch of embryonic tissue, or **embryonic disc** because it is destined to form the embryonic body (figure 5.4a–d). The other cell population formed is the **periblast,** a syncytial layer closely adhering to the uncleaved yolk. The periblast helps to mobilize this yolk so that it can be used by the growing embryo (see figure 5.9).

## *Amphibians*

As in the gars and bowfins, blastomeres of the animal pole divide more often than those of the vegetal pole, in which cell division is presumably slowed by abundant yolk platelets. Consequently, cells of the vegetal pole, having undergone fewer divisions, are larger than those of the more active animal pole. When the blastula stage is reached, the small blastomeres of the animal pole constitute the blastoderm and form a roof over the emerging blastocoel.

## *Reptiles and Birds*

In reptiles and birds, yolk is so prevalent within the vegetal pole that cleavage furrows do not pass through it at all; thus, cleavage is discoidal. Blastomeres resulting from successful cleavage clump at the animal pole forming the blastoderm, (descriptively termed a blastodisc in reptiles and birds) that rests atop the undivided yolk (figure 5.2c). The term ***subgerminal space*** applies to the fluid-filled cavity between blastoderm and yolk at this point in development.

The blastoderm becomes **bilaminar** (two layered). Cells at its border migrate forward, beneath the blastoderm, toward the future anterior end of the embryo. Along the way, these cells are joined by cells dropping from the blastoderm, an event known as **ingression.** Migrating cells together with ingressing cells form the new **hypoblast.** Cells remaining in the depleted blastoderm now properly constitute the **epiblast.** The space between the newly formed hypoblast and epiblast is the compressed blastocoel.

## *Mammals*

In mammals, the blastula stage is termed a **blastocyst.** The three living groups of mammals differ in their modes of reproduction. The most primitive living mammals, the monotremes, retain the reptilian mode of reproduction and lay shelled eggs. Marsupials are viviparous, but the neonate is born at a very early stage in its development. Placental mammals retain the embryo within the uterus until a later stage in development and supply most of its nutritional and respiratory needs through the specialized placenta. Because of such differences, embryonic development in these three groups will be treated separately.

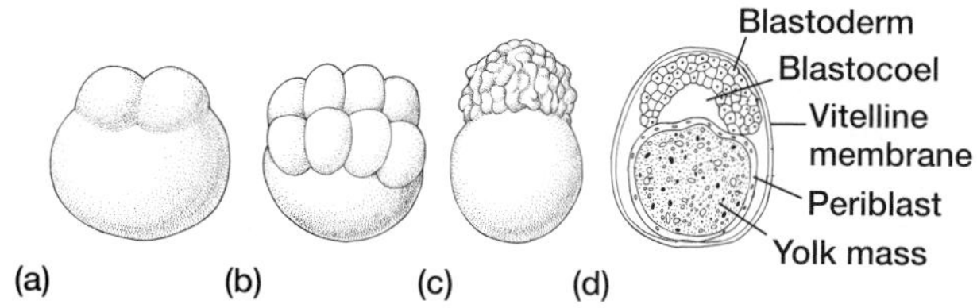


**Figure 5.4** Discoidal cleavage in a teleost (zebra fish). (a) Cleavage begins with the appearance of the first mitotic furrow. After successive mitotic divisions (b), the blastula (c) results. (d) Cross section of the blastula. A cap of blastoderm rests on the uncleaved yolk mass, and a vitelline membrane is still present around the entire blastula.

***Monotremes*** In monotremes, yolk platelets collect in the ovum to produce a macrolecithal egg. When the ovum is released from the ovary, the follicle cells are left behind. Fertilization occurs in the oviduct. The walls of the oviduct secrete first an "albumenlike" layer and then a leathery shell before the egg is laid. Cleavage, which is discoidal, begins during this passage of the embryo down the oviduct and gives rise to the blastoderm, a cap of cells that rests atop the undivided yolk. The blastoderm grows around the sides of the yolk and envelopes it almost completely (figure 5.5a).

***Marsupials*** In marsupials, the ovum accumulates only modest amounts of yolk. Upon ovulation, it is surrounded by a zona pellucida but lacks follicle cells (lacks the corona radiata; figure 5.5b). Once the ovum is fertilized, the oviduct adds a **mucoid coat** and then an outer, thin **shell membrane.** The shell membrane is not calcified,

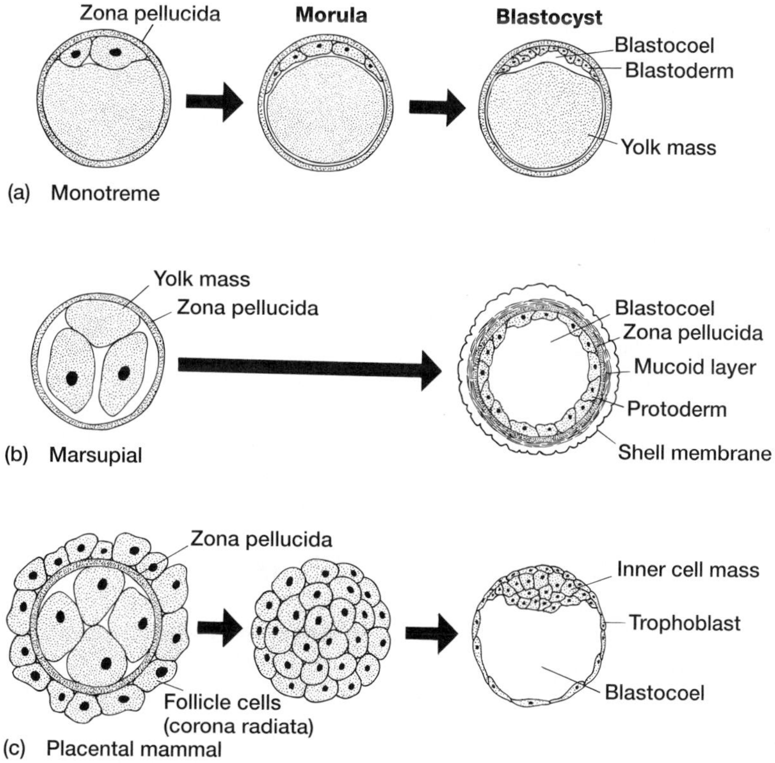


**Figure 5.5** Cleavage in three groups of living mammals. (a) Monotremes exhibit discoidal cleavage with a blastocyst composed of a cap of blastoderm atop uncleaved yolk. (b) In marsupials, cleavage does not result in a distinct morula stage composed of a mass of solid cells. Instead, cells produced during cleavage spread along the inside of the zona pellucida and directly become the protoderm. The oviduct produces a mucoid coat and thin shell membrane. (c) Placental mammals pass from a morula to a blastocyst, where cells are set aside as an inner cell mass and an outer wall (the trophoblast). Very little yolk is present. A mucoid coat and zona pellucida are present around the morula but are not shown here.

but it is generally similar in mode of formation, chemical composition, and structure to the shell of monotremes and some oviparous reptiles. It remains around the embryo until near the end of gestation, when it is finally shed.

Early cleavage in marsupials does not result in formation of a morula. Instead, blastomeres spread around the inner surface of the zona pellucida, forming a single-layered **protoderm** around a fluid-filled core. Initially the blastocyst is a **unilaminar** (single-layered) protoderm around a blastocoel. Through uptake of uterine fluids, the blastocyst and its enveloping mucoid and shell membranes expand in size. Cells at one pole of the blastocyst give rise to the embryo and to its amnion, whereas the remaining cells give rise to a **trophoblast.** Trophoblastic cells help establish the embryo during its brief residence within the uterus, after which they participate in physiological exchange between maternal and fetal tissues, contribute to extraembryonic membranes, and possibly protect against the female's premature immunological rejection of the embryo before birth.

***Placentals*** In placentals, the ovum contains very little yolk when it is released from the ovary. It is surrounded by the zona pellucida and clinging follicle cells, which form the corona radiata. After fertilization, cleavage results in the morula, a compact ball of blastomeres still within the zona pellucida and with an added exterior mucoid coat. The appearance of fluid-filled cavities within the morula ushers in the blastocoel. Cells organize around the blastocoel to form the blastocyst. The zona pellucida prevents the blastocyst from prematurely attaching to the oviduct until it reaches the uterus. Upon arrival in the uterus, the blastocyst lyses a small hole in the zona pellucida and squeezes out. At this point, the blastocyst consists of an outer sphere of trophoblastic cells and an **inner cell mass** clumped against one wall (figure 5.5c). The trophoblast contributes to the extraembryonic membranes that will establish a nutritive and respiratory association with the uterine wall. The inner cell mass contributes additional membranes around the embryo and eventually forms the body of the embryo itself.

Doubt has recently been raised about the homology of the trophoblastic layers in placentals and marsupials. Marsupials lack a morula and an inner cell mass and differ from placentals in other aspects of cleavage as well. The terms choriovitelline or chorioallantoic membrane have been suggested as replacements for the term trophoblast. The issue is not just a struggle over names. If the trophoblast proves to be unique to placentals, this implies that it arose as a new embryonic structure in the Cretaceous when placental mammals emerged. This new trophoblast would have been a vital component in the emerging reproductive style of placentals, allowing prolonged interuterine exchange between fetal and maternal tissues. But the trophoblast of marsupials accomplishes most of the same functions as the trophoblast of placentals. Until the evidence is more persuasive, we will follow the conventional view of a homologous trophoblast in both placental mammals and marsupials.

**Overview of mammalian evolution (p. 113)**

### *Overview of Cleavage*

During cleavage, repeated cell divisions produce a multicellular blastula, each cell of which is a parcel containing within its walls some of the original cytoplasm of the egg. Because ingredients within the original polarized ovum were unevenly distributed, each cell holds a slightly different cytoplasmic composition that it carries during migration to new positions within the embryo. In some species, the blastula imbibes uterine fluids to swell in size, but it does not grow by incorporating new cells. During gastrulation, the stage following blastula formation, most cells arrive at their final destinations. Some of the initial ability of these cells to differentiate along many pathways has been narrowed, however, so that most cells at this stage are fated to contribute to just one part of the embryo. During subsequent embryonic stages, cell fate narrows further until each cell eventually differentiates into a terminal cell type.

## Gastrulation and Neurulation

Cells of the blastula undergo major rearrangements within the embryo to reach the gastrula and neurula stages. **Gastrulation** (meaning gut and formation) is the process by which the embryo forms a distinct endodermal tube that constitutes the early gut. The space enclosed within the gut is the **gastrocoel,** or **archenteron. Neurulation** (meaning nerve and formation) is the process of forming an ectodermal tube, the **neural tube.** This tube is a forerunner of the central nervous system and encloses the **neurocoel.** Gastrulation and neurulation occur simultaneously in some species and include other embryonic events with far-reaching consequences. During this time, the three germ layers come to occupy their characteristic positions—ectoderm on the outside, endoderm lining the primitive gut, and mesoderm between the other two (figure 5.6a). Sheets of mesoderm become tubular, and the resulting body cavity enclosed within the mesoderm is the **coelom** (figure 5.6b).

Cleavage is characterized by cell division; gastrulation is characterized by major rearrangements of cells. By the end of gastrulation, large populations of cells, originally on the surface of the blastula, divide and spread toward the inside of the embryo, a process that is much more than simple cell shuffling. As a result of this reorganization, tissue layers and cell associations are established strategically within the embryo. How they are positioned will largely determine their subsequent interactions with each other. Tissue-to-

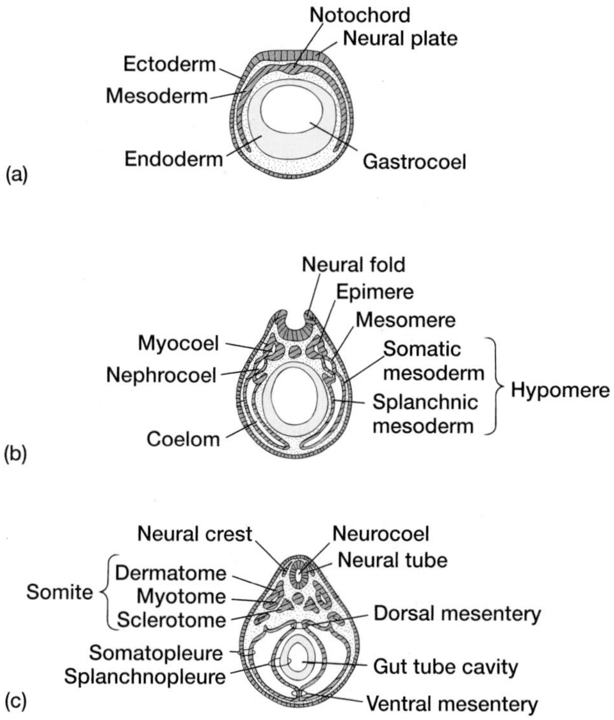


**Figure 5.6** General steps in successive differentiation of the mesoderm and neural tube. Mesoderm initially comes to lie between the other two germ layers (a), and differentiates into three major regions, the epimere, the mesomere, and the hypomere (b). Each of these gives rise to specific layers and groups of mesodermally derived cell populations (c). Neurulation begins with a dorsal thickening of the ectoderm into a neural plate (a). This plate folds (b), and its folds fuse into a hollow neural tube (c). Note the formation and separation of the neural crest (c) from the edges of the original neural plate.

tissue interaction is one of the major determinants of later organ formation.

Although the pattern of gastrulation varies considerably among chordate groups, it is usually based on a few methods of cell movement in various combinations. Cells may spread across the outer surface as a unit **(epiboly);** cells may turn inward and then spread over the internal surface **(involution);** a wall of cells may indent or simply fold inward **(invagination);** sheets of cells may split into parallel layers **(delamination);** or individual surface cells may migrate to the interior of the embryo **(ingression).**

By whatever method, cells moving to the interior leave behind the sheet of surface cells that constitute the ectoderm. This surface ectoderm thickens into a strip of tissue that forms the **neural plate** along what is to be the dorsal side and anterior-posterior axis of the embryo (figure 5.6a). Next the margins of the neural plate grow upward into parallel ridges that constitute the **neural folds** (figure 5.6b). The neural folds eventually meet and fuse at the midline, forming the neural tube that encloses the neurocoel (figure 5.6c). The neural tube is destined to differentiate into the brain and spinal cord. Just before or just as the neural folds fuse, some cells within these ectodermal folds separate out and establish a distinct population of **neural crest cells.** In the embryo's trunk, these neural crest cells are organized initially into cords, but in the head, they usually form into sheets. From their initial position next to the forming neural tube, neural crest cells migrate out along defined routes to contribute to various organs. Such cells are unique to vertebrates and are discussed at length later in this chapter.

The endoderm is derived from cells moving inward from the outer surface of the blastula. At first the endoderm forms the walls of a simple gut extending from anterior to posterior within the embryo. But as development proceeds, outpocketings from the gut and its interactions with other germ layers produce associated glands and their derivatives.

The mesoderm also is derived from cells entering from the outer surface of the blastula. Mesodermal cells proliferate as they expand into a tissue sheet around the insides of the body between outer ectoderm and inner endoderm. Occasionally, rather than forming a sheet, mesodermal cells become dispersed and intermingle with neural crest cells to produce a network of loosely connected cells called **mesenchyme.** Some prefer the term **ectomesenchyme** for this loose confederation of cells in recognition of its double embryonic origin. The notochord arises from the dorsal midline between lateral sheets of mesoderm. Each lateral sheet of mesoderm becomes differentiated into three regions—a dorsal **epimere,** a middle **mesomere** or **intermediate mesoderm,** and a ventral **hypomere** or **lateral plate mesoderm** (figure 5.6b). The central cavity within the mesoderm is the **primary** or **embryonic coelom.** Parts of the embryonic coelom often become enclosed in the mesoderm, forming a **myocoel** within the epimere, a **nephrocoel** within the mesomere, and simple **coelom** (body cavity) within the lateral plate mesoderm.

Two processes can produce these cavities within the mesoderm. In **enterocoely,** the most primitive method of coelom formation among chordates, the interior cavity is contained within the mesoderm when it first pinches off from other tissue layers. In **schizocoely,** the mesoderm forms first as a solid sheet and splits later to open the cavity within. If you remember that vertebrates are deuterostomes, which are characterized by enterocoely, you may be surprised to learn that schizocoely predominates in this group as a whole. In fact, cephalochordates and lampreys are the only chordates in which the coelom is formed by strict enterocoely. This has led many to conclude that the method of coelom formation is not a useful criterion for characterizing superphyletic groups. Others hold that the absence of enterocoely in most vertebrates is likely a secondary condition derived from enterocoelous ancestors. Coelom formation via mesodermal splitting can be attributed to developmental modifications, perhaps to accommodate enlarged yolk stores. In this view, schizocoely evolved independently in vertebrates and protostomes. Until this dilemma is resolved, we will follow the view that schizocoelic coelom formation in vertebrates is derived from enterocoelous ancestors.

**Phylogenetic significance of coelom formation (p. 47)**

The epimere of the trunk becomes organized into segmental clumps of cells called **somites** that in turn split into three separate populations. These populations of somite cells contribute to the skin musculature **(dermatome),** the body musculature **(myotome),** and the vertebrae **(sclerotome).** The mesomere gives rise to portions of the kidney. As the coelom expands within the hypomere, inner and outer mesodermal sheets of cells are defined. The inner wall of the hypomere is the **splanchnic mesoderm,** and the outer wall the **somatic mesoderm** (figure 5.6b). These sheets of mesoderm come into association with endoderm and ectoderm, with which they interact later to produce specific organs. Collectively, the paired sheet of splanchnic mesoderm and the adjacent sheet of endoderm form the **splanchnopleure;** the somatic mesoderm and the adjacent ectoderm form the **somatopleure** (figure 5.6c).

## *Amphioxus*

Gastrulation in amphioxus occurs by invagination of the vegetal wall (figure 5.7a). As vegetal cells grow inward, they obliterate the blastocoel. Cells on the inside next separate into endoderm and mesoderm. Some investigators prefer to emphasize the potential of this single layer, calling it future endoderm and mesoderm. Others refer to it as **endomesoderm** in recognition of its present unity. The endomesoderm eventually moves up against the inside wall of the ectoderm and forms the primitive gut. Th.e gastrocoel communicates to the exterior through the **blastopore** (figure 5.7a). The embryo is consequently transformed during early gastrulation from a single layer of blastomeres to a double layer of cell sheets consisting of the ectoderm and the endomesoderm. Each layer will give rise to specific adult tissues and organs.

Delineation of the mesoderm from ectoderm occurs during neurulation in the amphioxus embryo. A series of paired outpocketings form and pinch off from the mesoderm. These cavities merge to become the coelom (figure 5.7b). As the paired mesodermal outpocketings take shape, the mesoderm at the dorsal midline between them differentiates into the **chordamesoderm.** In addition to giving rise to the notochord, the chordamesoderm stimulates differentiation of the overlying ectoderm into the central nervous system. The mesoderm then becomes delineated into epimere, mesomere, and hypomere.

## *Fishes*

Gastrulation, like cleavage, is modified in proportion to the amount of yolk present. The amount of yolk varies considerably from one group of fishes to the next, so the patterns of gastrulation are quite varied in fishes as well.

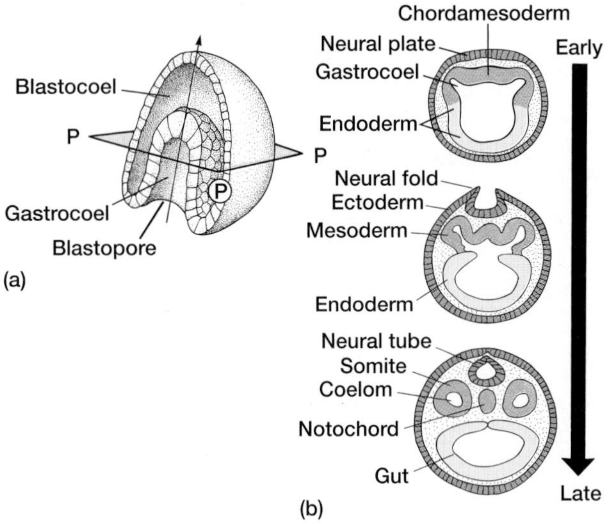


**Figure 5.7** Amphioxus gastrulation and neurulation.
(a) Invagination at the vegetal pole pushes cells into the interior of the blastula. The blastocoel is eventually obliterated, and the new space these ingrowing cells define becomes the gastrocoel. The arrow indicates the anteroposterior axis of the embryo.
(b) Successively older cross sections taken along the plane (P) defined in the illustration on the left (a). As development proceeds, mesodermal outpocketings appear and pinch off to form somites, leaving the endoderm to form the lining of the gut.

In lampreys and primitive bony fishes, the onset of gastrulation is marked by the appearance of an indentation, the dorsal edge of which is the **dorsal lip of the blastopore** (figure 5.8a–j). The dorsal lip of the blastopore is an important organizing site within the embryo. Surface cells flow to the blastopore by epiboly, slip over its lip and turn inward, and then begin to spread along the internal ceiling of the embryo. These entering surface cells constitute the endomesoderm, the name reminding us again of the two germ layers (endoderm and mesoderm) into which it will separate. The endomesoderm surrounds a gastrocoel and obliterates the blastocoel as it grows.

During gastrulation in sharks and teleost fishes, the blastoderm grows over the surface of the yolk, eventually engulfing it completely to form the extraembryonic **yolk sac.** While this is occurring, the endomesoderm arises under and at the edges of the spreading blastoderm (figure 5.9a,b). The endomesoderm is continuous with the surface layer of the blastoderm, but its source is disputed. Some claim that it is formed by cells flowing around the edge of the blastoderm and inward. Others claim that deep cells already in place become rearranged to produce the endomesoderm. Whatever its embryonic source, the endomesoderm tends to be thickest at the posterior edge of the blastoderm where it becomes concentrated into the **embryonic shield** that produces the body of the embryo (figure 5.10a–d).

Separation of the endomesoderm into endoderm and mesoderm occurs next. When finally separate, the endoderm is a flat sheet of cells stretched over the adjacent yolk,

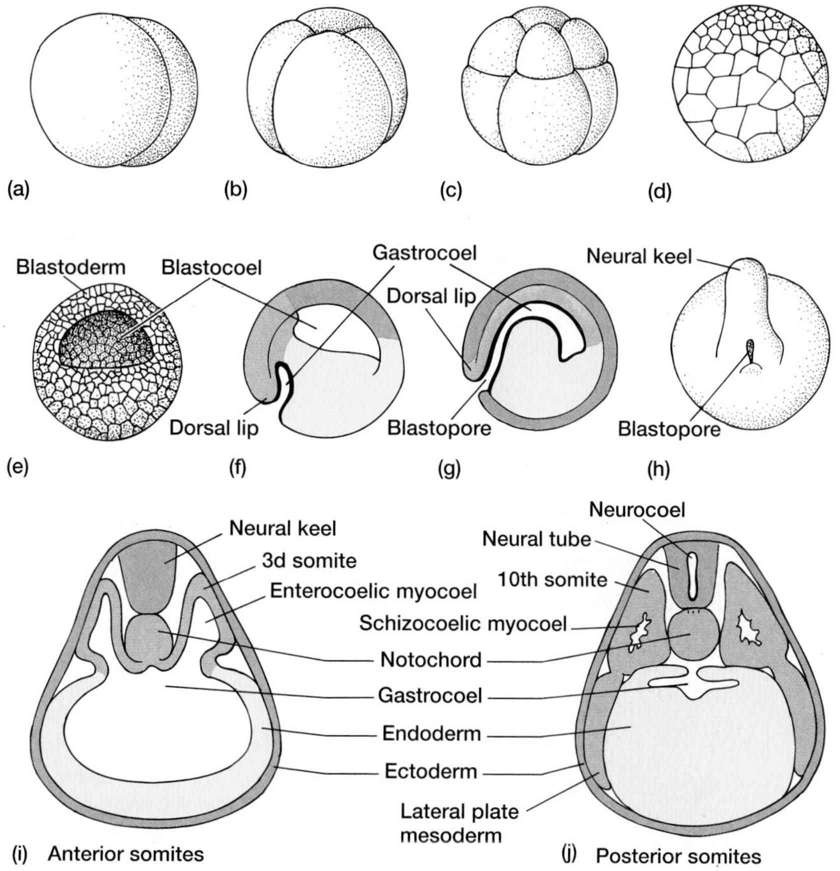


**Figure 5.8** Lamprey early embryonic development. (a–d) Cleavage stages leading to a blastula. (e) Cross section of the blastula. (f,g) Cross section of successive stages in gastrulation. (h) Exterior view of the entire gastrula. Formation of the myocoel within the somites is different in the anterior (i) compared with the posterior (j) region. In the anterior region, the myocoel is enterocoelous; posteriorly, it is schizocoelous. No open neural plate is formed. Instead, a solid cord of ectodermal cells sinks to the interior from the dorsal midline, forming the solid neural keel. This solid cord of cells becomes secondarily hollowed out to form the characteristic dorsal tubular nerve cord.

but it does not grow around the entire yolk mass. A recognizable gastrocoel has not yet appeared. The mesoderm forms chordamesoderm at the midline. Chordamesodermal cells give rise to the notochord and lateral plates of mesoderm that grow around the yolk. Thus, the yolk is eventually enclosed by a membrane consisting of periblast, mesoderm, and ectoderm, but no endoderm.

Gastrulation is based on several noteworthy differences within different groups of fishes. In lampreys, the coelom is enterocoelic, forming as the mesoderm pinches off from the rest of the endomesoderm. This is similar to the enterocoelic process in amphioxus and suggests that enterocoely represents the primitive method of coelom formation. We do not know how coelom formation occurs in hagfishes, but in all other vertebrates, the coelom forms by schizocoely in which the solid sheet of mesoderm splits to open spaces that become the body cavity.

In tetrapods, sharks, lungfishes, and some protochordates, the thickened neural plate rolls up into folds that join to produce a dorsal tubular nerve tube. However, in lampreys and teleosts, no such open neural plate is formed. Instead, a solid rod of ectodermal cells, the **neural keel,** grows along the dorsal midline and sinks inward from the surface (figure 5.8d,e). Later, a neurocoel appears within the core of the formerly solid neural keel to produce the characteristic dorsal tubular nerve tube (figure 5.8f).

## *Amphibians*

In amphibians, a superficial indentation marks the beginning of gastrulation and establishes the dorsal lip of the blastopore. Three major and simultaneous cell movements occur. First, the movement of surface cells by epiboly creates a stream of cells flowing toward the blastopore from all directions (figure 5.11a). Second, these cells involute over the lips of the blastopore. Third, entering cells move to and take up specific sites of residence within the embryo. Cells entering by such migratory routes become part of the endomesoderm surrounding the gastrocoel. The chordamesoderm, forerunner to the notochord, arises middorsally within the endomesoderm. Separation of the endomesoderm into distinct germ layers begins with the appearance of

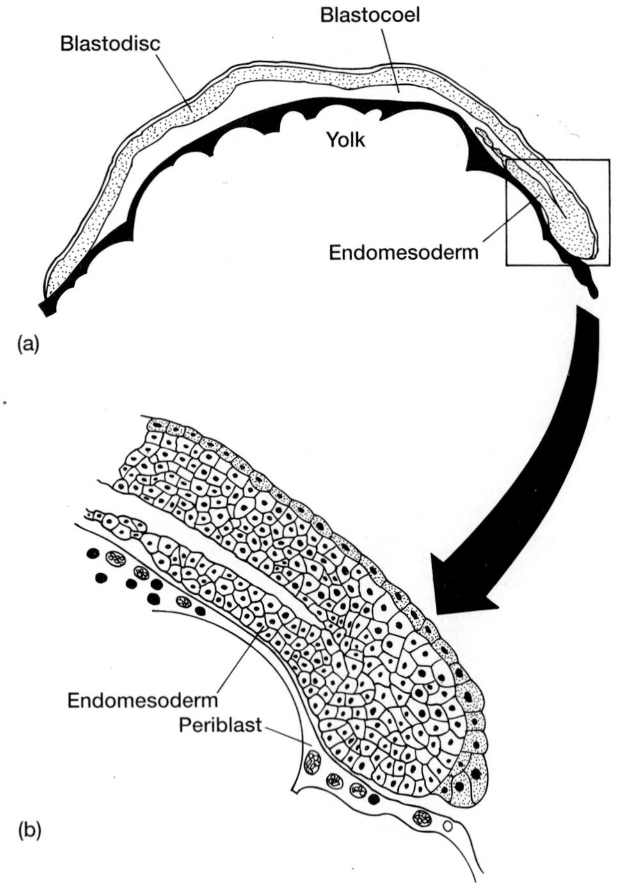


**Figure 5.9** Gastrulation at an early stage in a teleost fish (trout). (a) Cross section of the (blastoderm) arched over a compressed blastocoel. (b) Enlarged view of the posterior region of the blastoderm when the second layer, the endomesoderm, first appears.

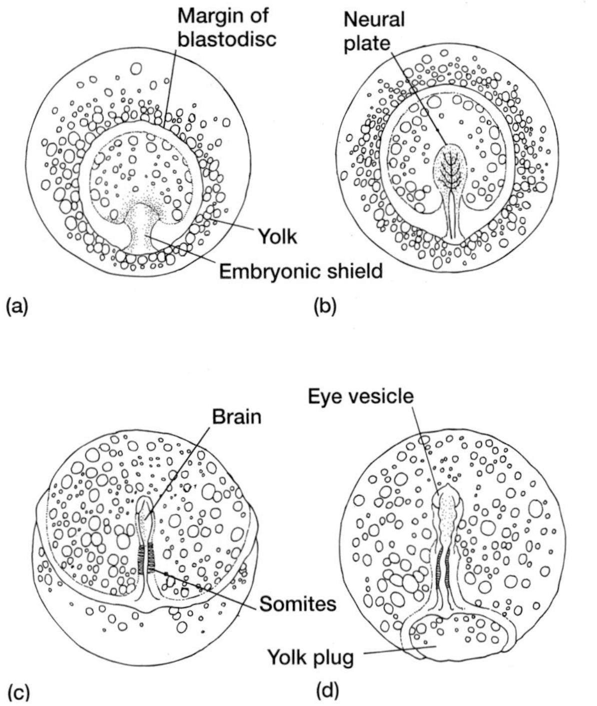


**Figure 5.10** Differentiation within the embryonic shield of a teleost fish (trout). (a) Early gastrulation. (b) Later gastrulation. (c) Formation of anterior regions of the embryo within the embryonic shield. (d) Blastoderm has nearly completed its overgrowth of the yolk.

paired projections of tissue growing out from the endomesoderm's lateral inner wall and upward to meet beneath the forming notochord. These paired projections of tissue together with the ventral region of yolk-laden endomesoderm separate into the endoderm proper. The rest of the endomesoderm becomes the mesoderm proper. From a solid sheet of cells, the mesoderm grows downward between the newly delineated endoderm and outer ectoderm. Distinct epimere, mesomere, and hypomere become evident in the mesoderm, and by schizocoely the solid mesodermal layer splits to produce the coelom within (figure 5.11b).

Gastrulation establishes the ectoderm and endoderm, obliterates the blastocoel, forms the new gastrocoel, and leaves a blastopore that is partially plugged by yolk-laden cells not completely drawn into the interior of the embryo.

Neurulation in amphibians usually begins before the endomesoderm has separated into its distinct germ layers. As in all tetrapods, neurulation proceeds by thickening of the neural plate that rolls up into the hollow neural tube (figure 5.11b). An external view of anuran development from fertilization to growth of operculum and forelimbs is illustrated in figure 5.12.

## *Reptiles and Birds*

In birds and reptiles, enormous accumulations of yolk alter embryonic processes. The flattened blastula includes the superficial epiblast, the hypoblast beneath, and the blastocoel between them. The onset of gastrulation is marked in the epiblast by the appearance of a thickened area at what will eventually be the posterior region of the embryo. This thickened area constitutes the **primitive streak** (figure 5.13a,d) and originates as a raised clump of cells called the **primitive node** (Hensen's node). The **primitive groove** is a narrow gully that runs down the middle of the primitive streak. Cells spread across the surface of the epiblast through epiboly and reach the primitive streak, where they involute at the edges of the streak and enter the embryo. Cells entering via the primitive streak either contribute to the mesoderm by spreading between epiblast and hypoblast or form the endoderm by sinking further to the level of the hypoblast. At this level, they displace hypoblastic cells by pushing them to the periphery (figure 5.13b).

By the end of gastrulation, many surface cells originally belonging to the epiblast have migrated to new positions within the embryo. Cells remaining on the surface

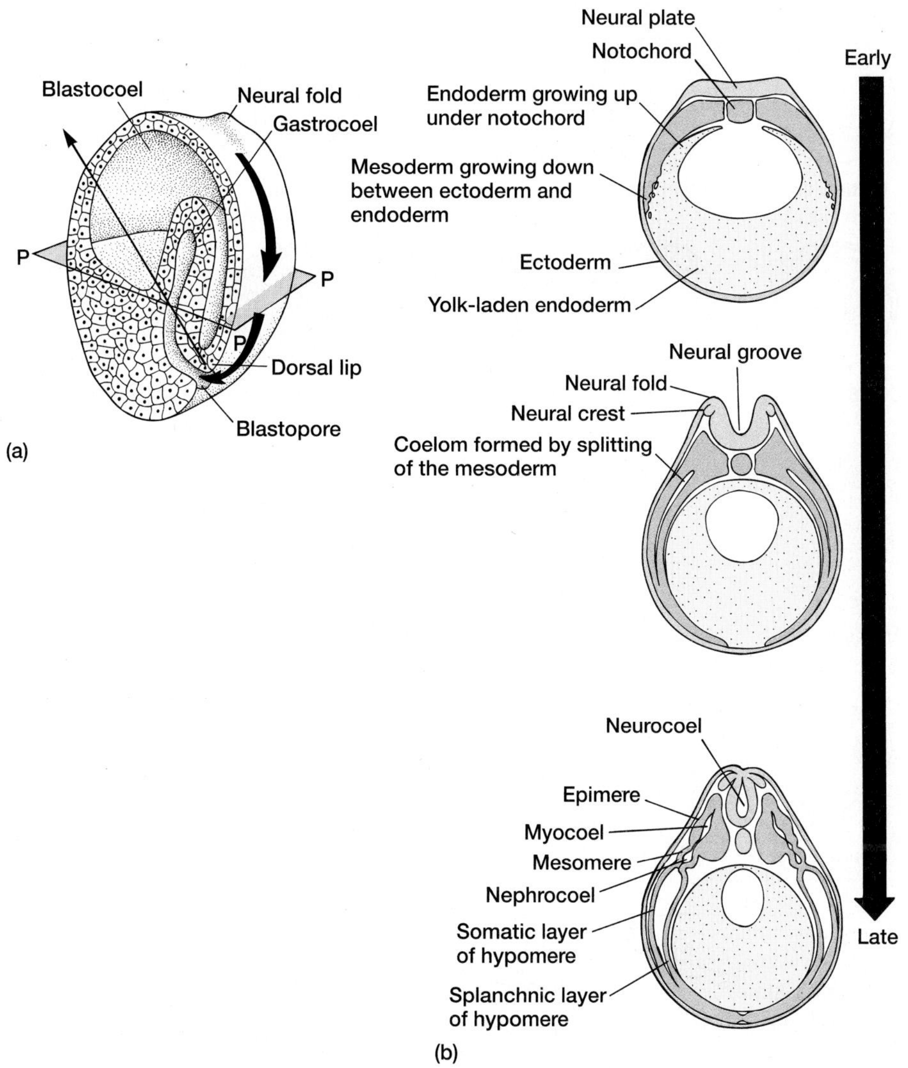


**Figure 5.11** Amphibian gastrulation and neurulation. (a) Sagittal section of an amphibian gastrula. Cells move along the surface (epiboly) and turn inward at the blastopore to form the enlarging gastrocoel. Solid arrows indicate the surface movements of cells. The long arrow indicates the anteroposterior axis of the embryo. (b) Successively older cross sections taken through the plane (P) illustrated in the sagittal section (a). As development proceeds, wings of endoderm grow, fuse, and become distinct from the mesoderm. The mesoderm grows downward and differentiates into various body regions. Notice that the coelom forms within the mesoderm by a splitting of this mesodermal layer.

now constitute a proper ectoderm. Involuting cells have pushed cells of the hypoblast to the extraembryonic area. In their place over the yolk are the newly arrived cells of the embryonic endoderm. Between ectoderm and endoderm is the mesoderm, also composed of cells arriving via involution through the primitive streak. Along the midline, a notochord differentiates within the mesoderm (figure 5.13c).

Neurulation involves formation of a neural tube from a neural plate precursor. At the onset of neurulation, the three germ layers have already been delineated (figure 5.14a), and reorganization of the lateral mesoderm begins. Initially, the mesoderm is a plate of solid tissue that lies lateral to the notochord with recognizable epimere, mesomere, and hypomere. The hypomere splits, forming the splanchnic and somatic layers of mesoderm and the schizocoelic coelom between. Association of these mesodermal layers with adjacent endoderm and skin ectoderm produces the composite splanchnopleure and somatopleure (figure 5.14b). Although the primitive streak does not have an opening like a blastopore, it functions like a blastopore as the site through which superficial cells enter the embryo.

## *Mammals*

***Monotremes*** Gastrulation, like cleavage, is quite different in the three living groups of mammals. In monotremes, as in reptiles, gastrulation involves a blastodisc atop a large yolk mass. At the end of cleavage, the blastocyst is unilaminar. The blastoderm is five to seven cells thick at its center

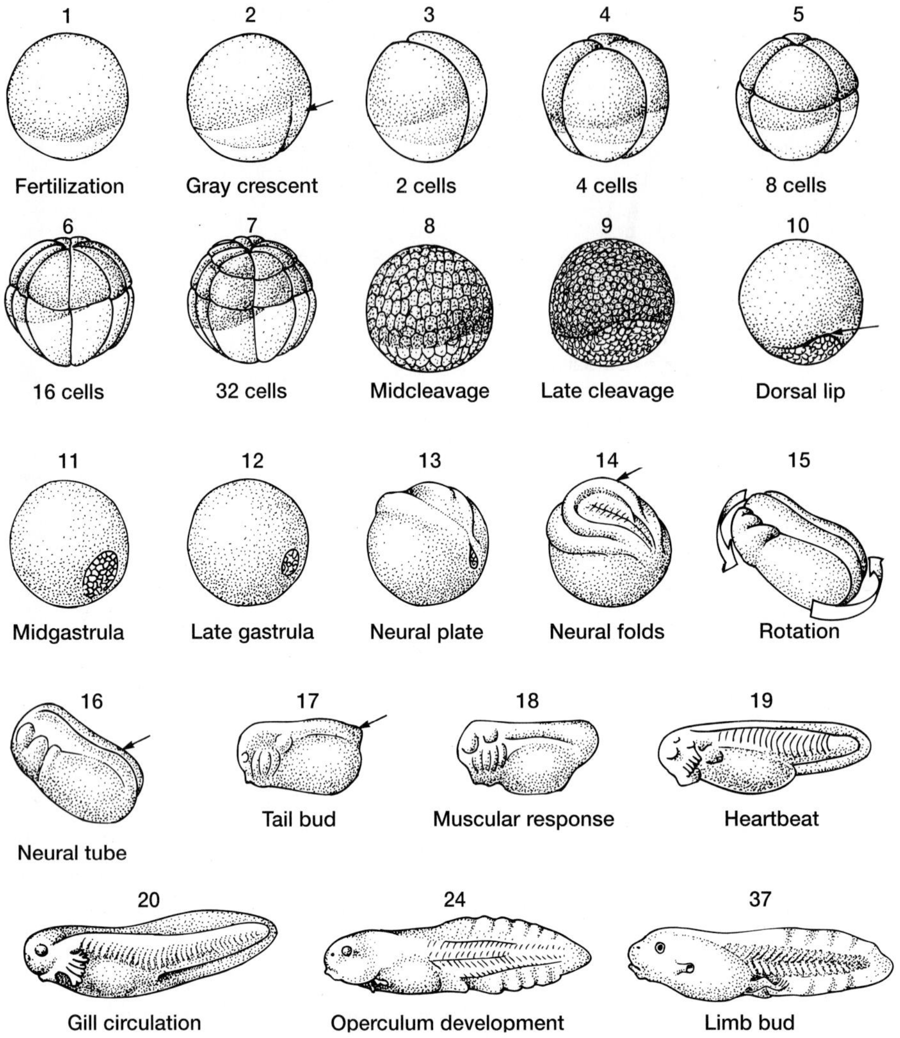


**Figure 5.12** External view of anuran development. Beginning with fertilization (1), morula (6–8), blastula (9 and 10), gastrula (11 and 12), and neurula (13–16) stages follow successively. In later development, the tail bud forms (17), muscular twitches begin (18), heartbeat commences (19), functional external gills develop (20), and blood circulation occurs through the caudal fin. Subsequent events include the formation of an operculum (24), a flap of head skin that grows over and covers the gills. The hindlimbs develop first and then the forelimbs. Eventually the embryo undergoes metamorphosis into a juvenile frog. Stages between 20–24 and 24–37 are not illustrated.

but thinned at its margins. This sheet of blastoderm grows through mitotic cell division and spreads around the yolk. During pregastrulation, the monotreme blastocyst becomes bilaminar. As the blastoderm grows around the yolk, it sheds cells inward, forming distinct endodermal and ectodermal layers. The endoderm is formed from these inward moving cells; the ectoderm is composed of cells that remain behind on the surface. These two layers grow into the vegetal pole so that the yolk becomes completely enclosed within the embryo.

As in reptiles and birds, gastrulation in monotremes begins with the appearance of the primitive streak. From its initial development as a thickened area in the ectoderm to which surface cells converge, the primitive streak becomes a major elongate axis around which the embryonic body is organized. The term ***medullary plate*** is sometimes used to describe the early thickening of ectodermal cells before a distinctive primitive streak can be distinguished. Details of cell movements are not known, but presumably epiboly and involution bring surface cells around the primitive streak into the interior of the embryo.

The second event of gastrulation is the appearance of a mesodermal sheet (figure 5.15a). It probably arises from cells entering via the primitive streak and becomes interposed in its customary position between existing ectoderm and endoderm.

In the echidna, one species of monotreme, the embryo within the uterus increases in size because it takes up fluid secreted by the uterus before it is enveloped in a leathery outer shell. This absorbed fluid is thought to provide

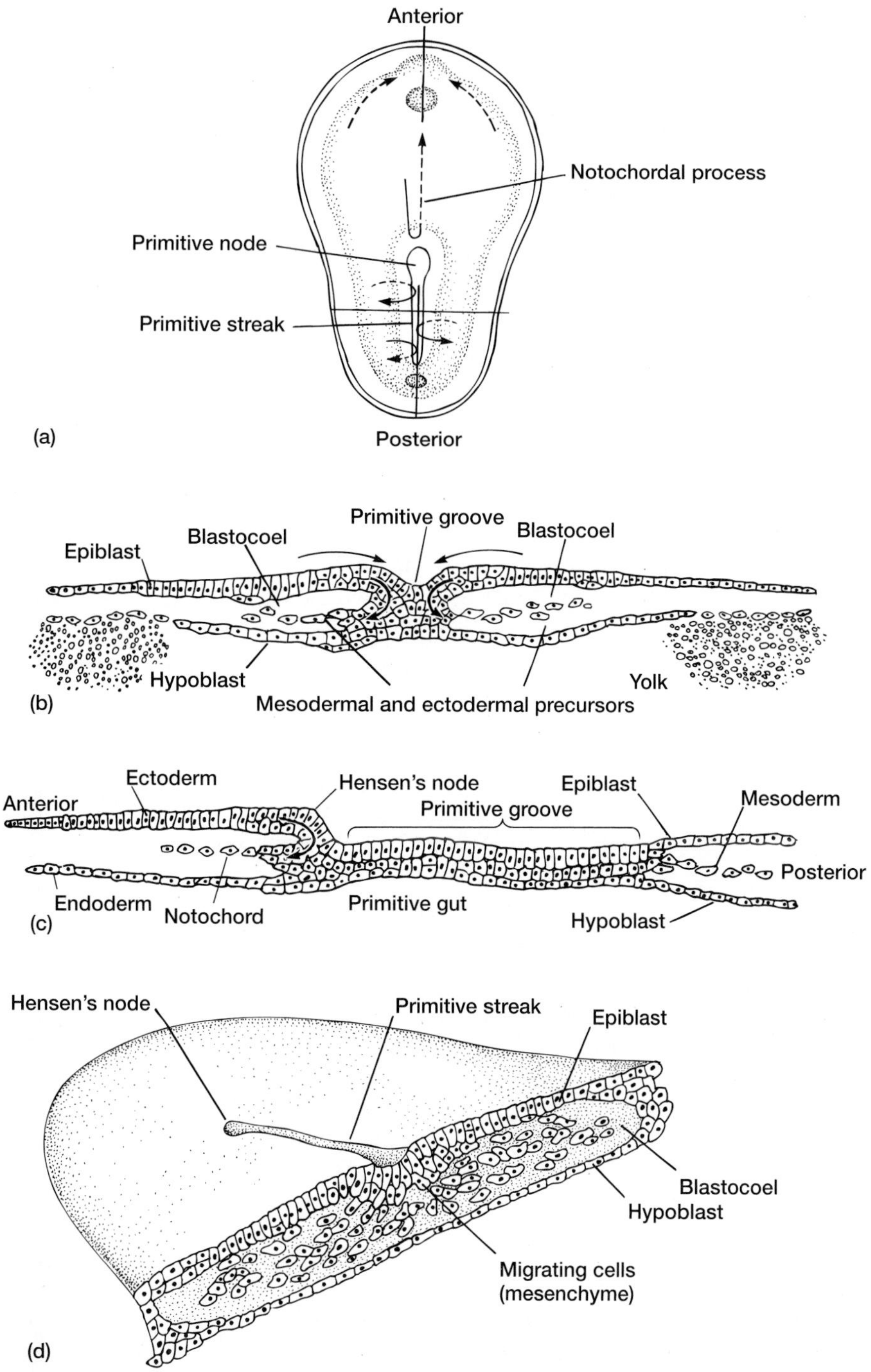


**Figure 5.13** Bird gastrulation. (a) Dorsal view of the primitive streak. Arrows indicate the direction of major cell movements from the surface through the primitive streak to the interior. (b) A cross section through the embryo illustrates the inward flow of cells. Some of these cells contribute to the mesoderm, others displace the hypoblast to form the endoderm. (c) A longitudinal medial section through the embryo shows the forward migration of a separate stream of cells that produce the notochord. (d) Three-dimensional view of the primitive streak during early gastrulation.

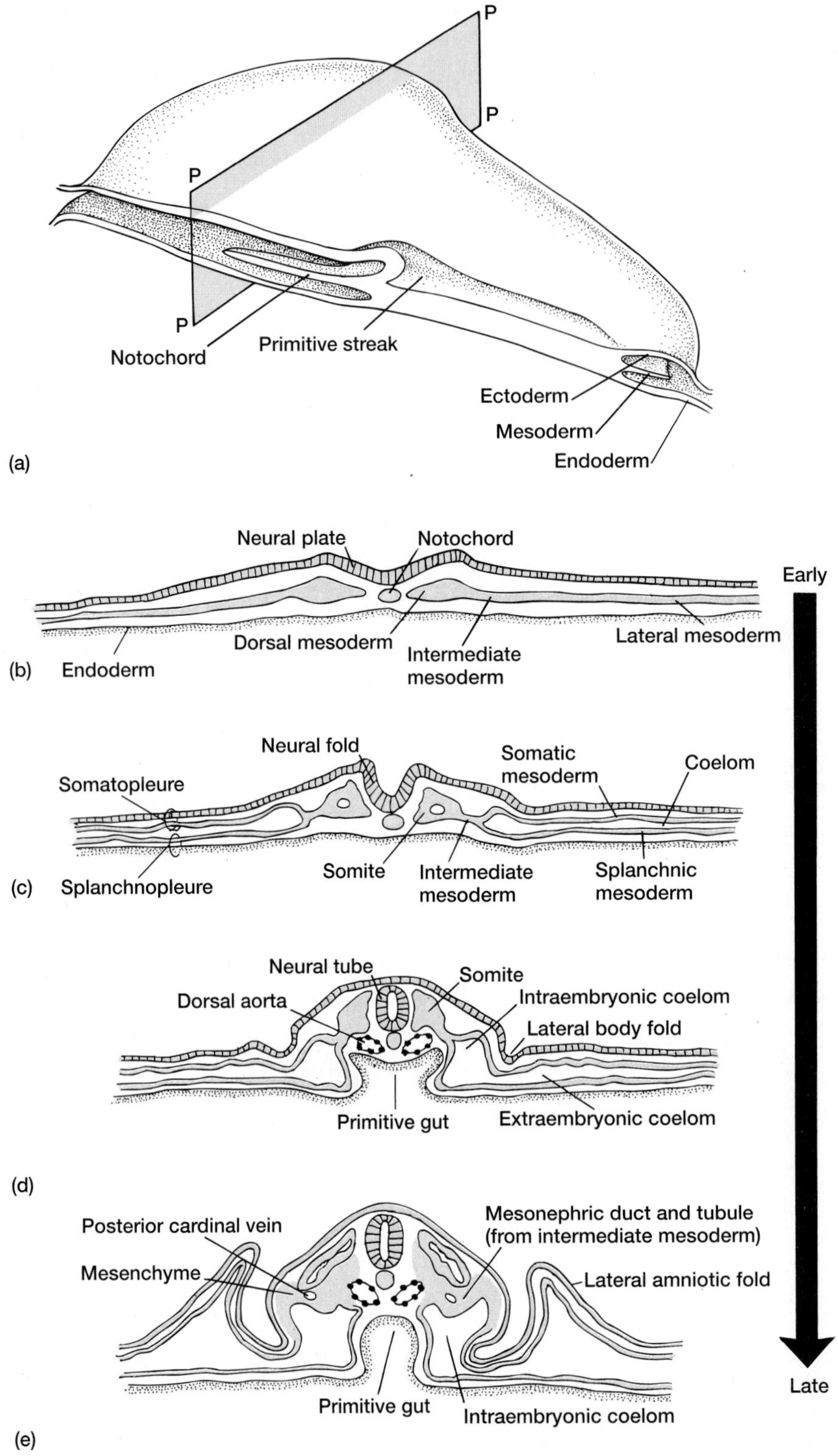


**Figure 5.14** Bird gastrulation and neurulation. (a) Sagittal section of the embryonic disc showing the primitive streak and the extent of the three primary germ layers. (b–e) Successively older cross sections through the plane (P) indicated in the top figure (a). As gastrulation proceeds, cells entering through the primitive streak form the mesoderm and the endoderm. The mesoderm becomes further differentiated into specific regions, and the endoderm displaces the former hypoblast to the periphery. Successive cross sections show neurulation proceeding from neural plate to neural folds to hollow nerve tube. Note also the regionalization of the mesoderm and the appearance of extraembryonic membranes (lateral amniotic fold).

nutrition for embryonic growth during the last days of gestation and the ten-day incubation period.

Neurulation in monotremes appears to involve the rolling up of a neural plate into a hollow neural tube.

***Marsupials*** In marsupials, the blastocyst is composed of a single layer of protodermal cells spread around the inside wall of the zona pellucida. The marsupial blastocyst is distinct among mammals, forming neither a blastodisc like monotremes nor an inner cell mass like placentals. Strictly speaking, it is unilaminar at the end of cleavage. During pregastrulation, this unilaminar blastocyst is transformed into a bilaminar embryo with an ectoderm and an endoderm. Cells of the protoderm proliferate near the animal pole and migrate around their own inner surface, forming a deeper endodermal layer (figure 5.15b). The superficial protoderm layer at this point is now called the ectoderm. As the two germ layers become delineated, the primitive streak appears in the ectoderm, marking the beginning of gastrulation.

Surface cells stream to the primitive streak and involute to the interior of the embryo. Once inside, they contribute to the mesoderm that spreads between outer ectoderm and deeper endoderm (figure 5.15b).

As in other vertebrates, the neural plate rolls into a neural tube during neurulation.

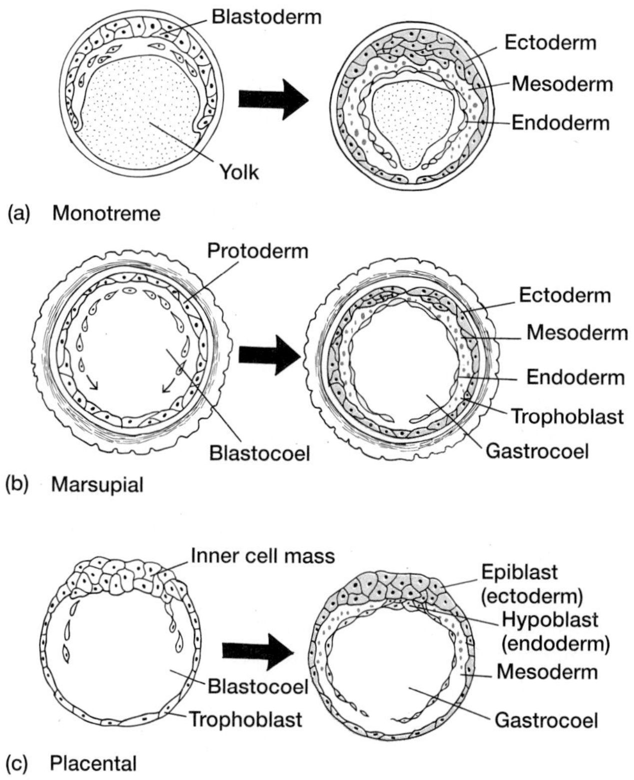


**Figure 5.15** Gastrulation in mammals. In all three mammalian groups, a primitive streak is formed through which cells enter to contribute to the mesoderm. (a) Monotreme. (b) Marsupial. (c) Placental.

***Placentals*** In placentals, the blastocyst is composed of two distinct populations of cells at the end of cleavage, an outer trophoblast and an inner cell mass. During pregastrulation, reorganization of the inner cell mass produces a bilaminar embryonic disc composed of epiblast (future ectoderm and mesoderm) and hypoblast (future extraembryonic tissue; figure 5.15c). This occurs when some cells depart from the inner cell mass and migrate into and around the periphery of the blastocoel, forming a thin hypoblastic layer, which is sometimes referred to as the endoderm at this point. The remaining population of cells of the depleted inner cell mass is the epiblast. The now flattened and circular epiblast together with the adjacent and underlying cells of the hypoblast constitute the embryonic disc. At this point, the epiblast contains all cells that will produce the actual embryo. The **exocoelomic membrane** is a term occasionally applied to endodermal cells outside the embryonic disc; this is based on the unproven theory that endodermal cells arise from the trophoblast rather than from the inner cell mass as hypoblastic cells do.

In placental mammals, as in other amniotes, appearance of a primitive streak marks the beginning of gastrulation (figure 5.16a). Surface cells of the epiblast stream toward the primitive streak (epiboly) and over its edges (involution) to reach the inside. As in reptilian and avian embryos, some entering cells move deep into the embryo, displacing the hypoblast to the periphery where its cells contribute to extraembryonic tissues. Other entering cells become organized into a middle mesoderm. These mesodermal cells grow outward between the deep hypoblast (now more correctly termed the endoderm) and the superficial epiblast (now termed the ectoderm) that has been depleted of cells. The notochord arises from these entering cells (figure 5.16b). The laterally placed mesoderm is at first a solid sheet of tissue but subsequently becomes differentiated into epimere, mesomere, and hypomere. Splitting of the solid mesodermal layer produces the coelom by schizocoely and defines somatic and splanchnic mesodermal sheets.

As regionalization of the mesoderm takes place, neurulation results in development of a tubular nerve tube from a neural plate (figure 5.16b).

## Organogenesis

By the end of neurulation, several major reorganizations of the embryo have been accomplished. First, polarity based on the animal-vegetal pole axis of the egg has been superseded by bilateral symmetry based on an anterior-posterior axis of the emerging embryonic body. Second, the three primary germ layers have been delineated—ectoderm, endoderm, mesoderm. In all vertebrates, ectoderm gives rise to nervous tissue and epidermis; endoderm to digestive and respiratory tubes; mesoderm to skeletal, muscular, and circulatory systems and to connective tissues (figure 5.17). There are exceptions, but generally across vertebrate groups, the same major adult tissue has as its source the

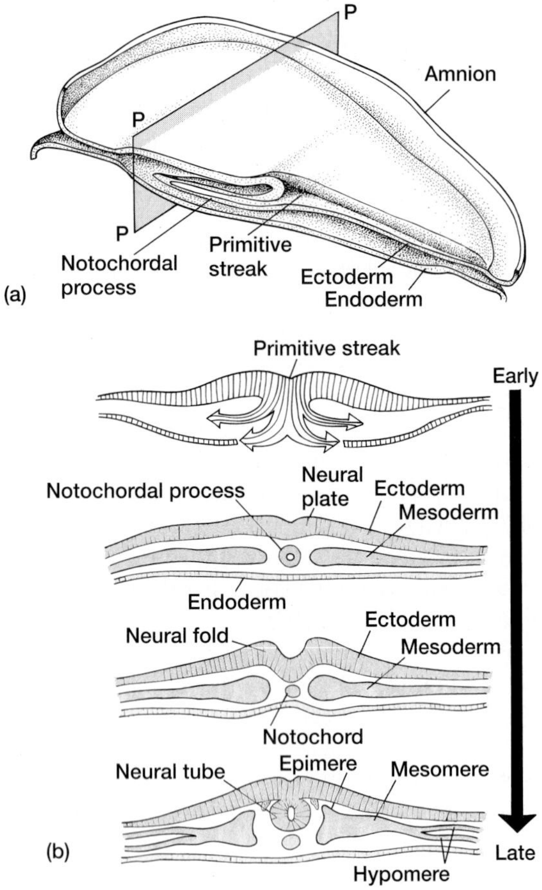


**Figure 5.16** Placental mammal gastrulation and neurulation. (a) Sagittal section of the embryonic disc. (b) Successively older cross sections through the plane (P) indicated in the top illustration (a). As gastrulation proceeds, cells entering through the primitive streak form the mesoderm that differentiates into various body regions (bottom cross section).

same specific germ layer of the embryo. Third, the three germ layers become strategically positioned next to one another so that they can mutually interact during **organogenesis,** the differentiation of organs from tissues. Mesoderm is especially important in organogenesis because of its cooperative associations with both ectoderm and endoderm. It is partially supported in its own differentiation by the other two layers, but in turn it stimulates or induces the other layers to form parts of organs.

## Histogenesis

The environment immediately around the cell is the **extracellular matrix,** meaning outside the cell, or the **interstitial space (interstitium),** meaning around the cell. But separate cells functioning in isolation are seldom found within the body. Instead, like cells are usually associated into sheets or confederations of cells. Where these aggregations of similar cells are specialized to perform a common function, they constitute a **tissue.** One early accomplishment of development is to place cells produced during cleavage into one of the cellular germ layers—ectoderm, mesoderm, endoderm. In turn, these formative germ layers differentiate into proper tissues through the process of **histogenesis** (meaning tissue formation). There are are four primary categories of adult tissues—**epithelium, connective tissue, muscle tissue,** and **nervous tissue.** Muscle and nervous tissues are discussed more fully in chapters 10 and 16, respectively. Because we meet epithelia and connective tissues repeatedly, they are introduced next and aspects of their embryonic development are discussed.

## Epithelium

Epithelial tissues are formed of closely adjoined cells with very little extracellular matrix between them. Usually one side of the epithelium rests upon a **basal layer.** For many years, **basement membrane** was the term used to describe this basal layer, but the electron microscope revealed that the basement membrane is a blend of two structures with separate origins, the **basal lamina** (derived from epithelium) and the **reticular lamina** (derived from connective tissue). By convention, the choice of terms depends on what can be resolved by the microscope, a basement membrane (light microscope) or basal and reticular laminae (electron microscope). Opposite to the basal layer is the **free surface** that faces a **lumen** (cavity) or the exterior environment. This free surface is the usual site at which secretory products are released from the cell **(exocytosis)** or materials are taken into the cell **(endocytosis).** The free surface is most likely to form tiny fingerlike processes, such as microvilli and cilia. The **stereocilia** are very long microvilli. Epithelia are divided into two categories, sheets (membranes) and glands (secretory; figure 5.18).

### *Covering and Lining Epithelium*

Epithelial membranes cover surfaces or line body cavities, ducts, and lumina of vessels. Arranged in sheets, epithelia can be either (1) **simple,** composed of a single layer of cells or (2) **stratified,** composed of more than one layer of cells. The cells themselves can be **squamous** (flat), **cuboidal** (cube shaped), or **columnar** (tall) in form. Epithelial names take advantage of these features of arrangement and cell shape. For example, simple squamous epithelium is made up of a single layer (hence simple) of flat cells (hence squamous). Simple squamous epithelium most commonly lines body cavities and vessels. The tissue lining blood and lymph vessels is called **endothelium,** and that lining body cavities is **mesothelium.** Simple cuboidal epithelium appears in many ducts. Simple columnar epithelium lines the digestive tract and some other tubular structures (figure 5.18).

In stratified squamous epithelium, characteristic of the skin, mouth, and esophagus, cells occur stacked in layers (stratified) and surface cells are flat (squamous). Stratified

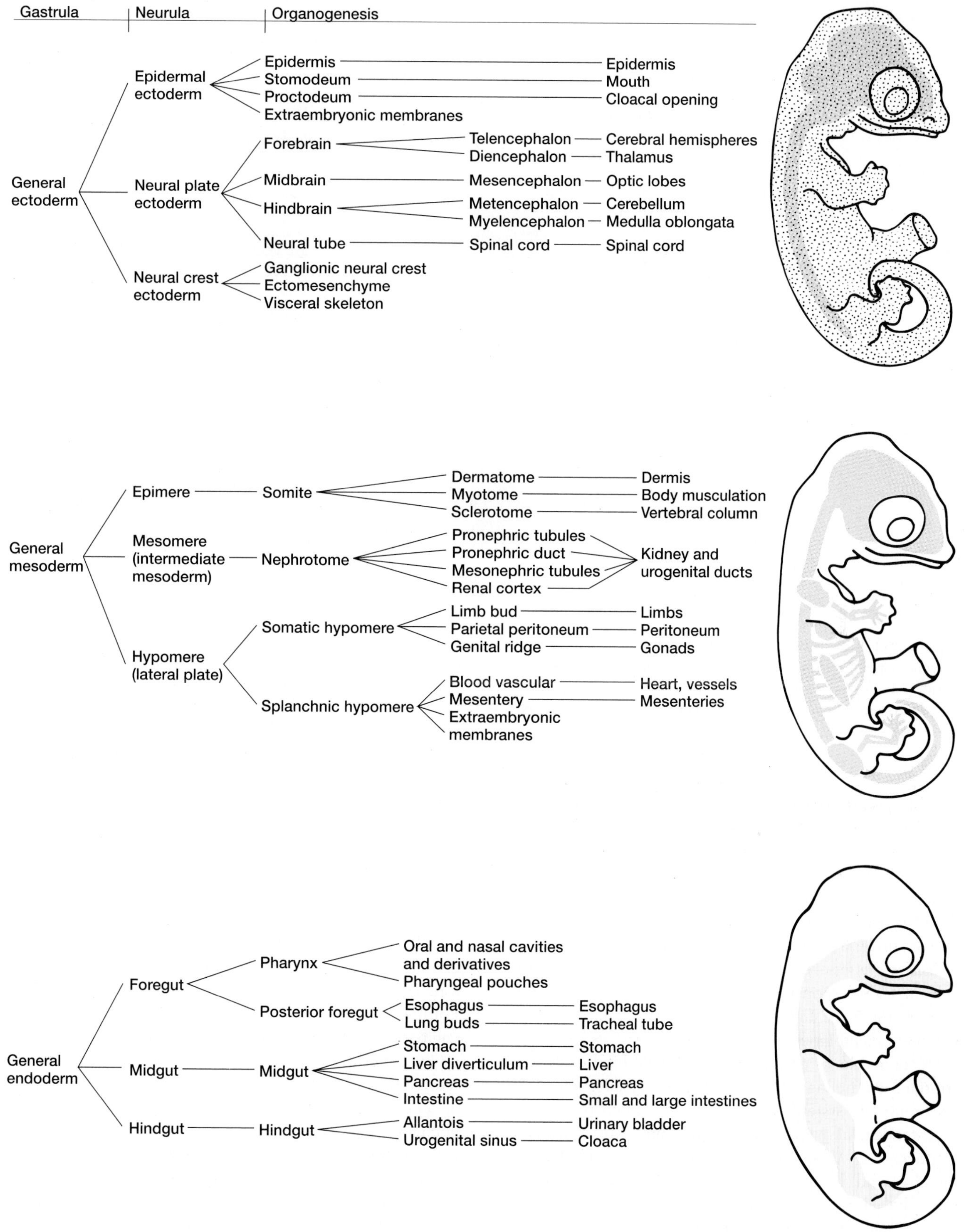


**Figure 5.17** Organogenesis. The three primary germ layers are delineated during gastrulation and neurulation. Thereafter, they become differentiated into various body regions, and these regions produce the major organs of the vertebrate body. The embryonic origin of each organ can be traced back to these specific germ layers. In general, ectoderm produces the skin and nervous system; mesoderm the skeleton, muscle, and viscera; and endoderm the digestive tract and its derivatives.

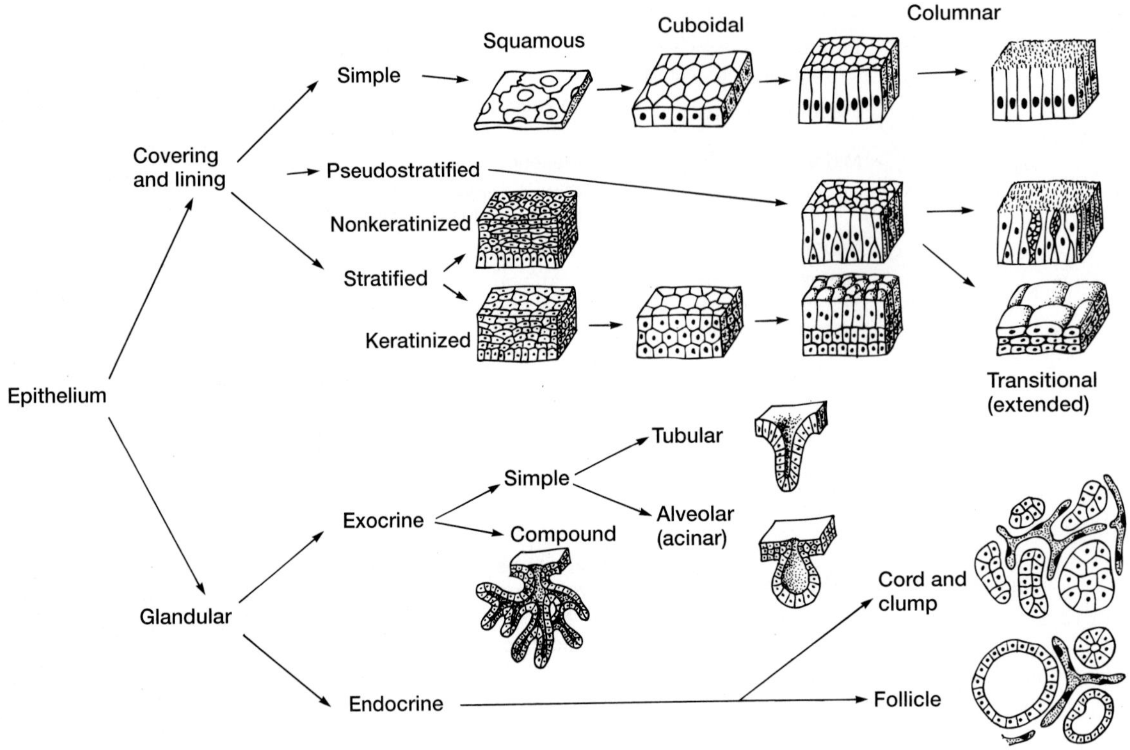


**Figure 5.18** Classification of epithelia. Epithelia fall into one of two groups: (1) membranes that line or cover cavities or (2) glands that secrete products that act elsewhere in the body. Membranes are single-layered (simple) or multilayered (stratified) sheets of cells. Cells in the sheets can be squamous, cuboidal, or columnar in shape. Exocrine glands release their products (secretions) into ducts that are single (simple) or branched (compound). Endocrine glands release their products into blood vessels; they are arranged into clusters (cord and clump) or into tiny balls (follicles).

cuboidal and stratified columnar epithelia are rare. In mammals, cells of the male urethra and cells of the Graafian follicles of the ovary are examples.

In addition to simple and stratified epithelia, the third type of lining epithelium is **pseudostratified epithelium** found in the trachea. Cells appear stacked when first inspected under a microscope, but a more careful look reveals that they are falsely layered. The staggered arrangement of cell nuclei is responsible for this false (pseudo-) stratification. Actually all cells, even those at the top, rest on the basement membrane.

**Transitional epithelium** is a special kind of pseudostratified epithelium found only in the bladder and ducts of the urinary system. The cells stretch when the bladder is distended, allowing them to accommodate changes in bladder size. When relaxed, transitional cells become bunched and deceptively appear to constitute a multilayered epithelium. Recent study indicates that even when relaxed each cell touches the basement membrane, so properly the tissue is a pseudostratified epithelium. The name *transitional epithelium* is a misnomer held over from the time when this tissue was erroneously thought to be intermediate (hence transitional) between other types of epithelia.

### *Glandular Epithelium*

Cells specialized to secrete a product are called **glands.** Glands with ducts that collect and carry away the product are **exocrine glands;** if the product is carried away by the circulatory system, the glands are **endocrine glands.** Glands usually arise from **glandular epithelium.** The ectoderm and endoderm of the early embryo are lining epithelia; therefore, adult organs derived from them are epithelial organs. Epithelial glands arise as tubes or solid cords through invagination and outgrowths from these two epithelial germ layers. Strictly speaking, however, not all cells that produce secretions are epithelial glands derived from ectoderm or endoderm. Some connective tissue cells derived from mesenchyme secrete products that are carried away by ducts or blood vessels; or their products simply collect in the extracellular matrix around the secreting cell. Thus, most but not all glands of the vertebrate body are epithelial in origin.

A **multicellular gland** is composed of many secretory cells in aggregation, and a **unicellular gland** has only a single secretory cell. Exocrine glands can be **tubular** (cylindrical) or **alveolar** (acinar; rounded in shape). Glands can be **simple,** drained by a single duct, or **compound,** drained by multiple branching ducts. The **myoepithelial cells** are derived from ectoderm (hence they are epithelial), but they possess contractile properties (hence myo-). They are associated with the basal regions of secretory cells and mechanically assist with release of products from exocrine glands. Endocrine glands are composed of cells aggregated into **cords** and **clumps** (sheets and solid masses) or **follicles** (tiny, hollow spheres; figure 5.18).

## Connective Tissues

Connective tissues generally include bone, cartilage, fibrous connective tissue, adipose tissue, and blood (figure 5.19). At first glance, connective tissues seem to be the misfits of histology—the leftovers after all other tissues have been categorized. Connective tissues have a variety of functions and occur in diverse contexts. Adipose tissue stores lipids; bone and cartilage support the body; blood transports respiratory gases; dense connective tissue packs organs. Bone cells reside in a hard casing of calcium phosphate; blood cells occur in liquid plasma. To complicate matters, schemes for classifying connective tissues vary among different textbooks. Elegant but futile efforts have been made to find a common denominator for all connective tissues. Some physiologists define them functionally on the basis of their mechanical role in support. Bone, cartilage, and perhaps fibrous connective tissues qualify as supportive tissue, but blood certainly does not. Others define connective tissues as developing from mesenchyme. Certainly, many connective tissues arise from mesoderm, but there are exceptions to this as well. For instance, connective tissues of jaw muscles arise from neural crest cells, not from mesoderm.

Rather than search for an overly restrictive definition, perhaps it is best to view connective tissues as bringing a convenient order to what otherwise would be a jumble of tissue types. Generally, each type of connective tissue includes a distinctive *cell type* that is isolated from other cells and surrounded by or embedded in a relatively abundant *extracellular matrix*. Of course, adipose tissue is the exception that sticks out because almost no matrix surrounds individual adipose cells.

The consistency of the extracellular matrix surrounding connective tissues determines the physical properties of the tissue and hence its functional role. In bone, the matrix is hard; in loose connective tissue it is gellike; in blood it is fluid. The matrix is made up of **protein fibers** and a surrounding **ground substance.** Consistency of the ground substance varies from liquid to solid, depending on the tissue type.

Connective tissues can be categorized as general or special as well.

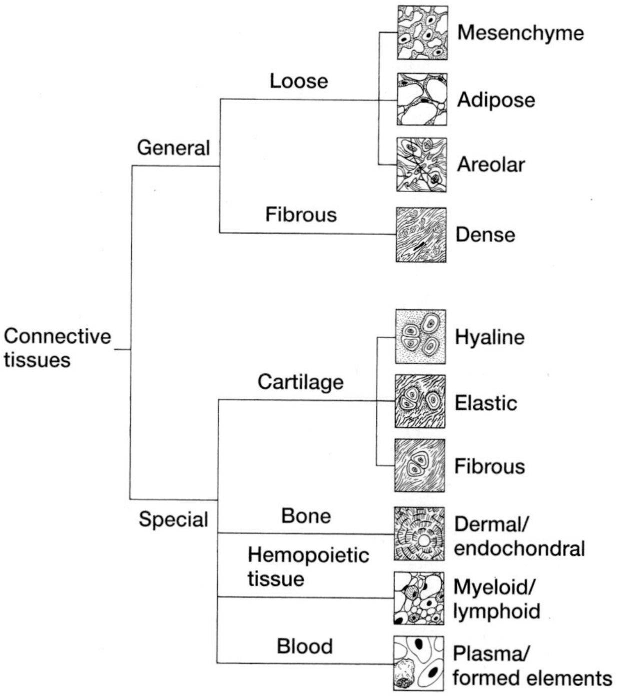


**Figure 5.19** Categories of connective tissue. Bone, cartilage, fibrous tissue, adipose tissue, and blood are some of the body's connective tissues. Each type of connective tissue includes a distinctive cell type surrounded by an extracellular matrix.

### *General Connective Tissues*

General connective tissues are dispersed widely throughout the body. The most common is fibrous connective tissue that forms tendons and ligaments as well as much of the dermis of the skin and the outer capsules of organs. The distinctive cell is a **fibroblast,** and the extracellular matrix secreted by fibroblasts is principally a network of protein fibers in a ground substance of polysaccharide gel.

### *Special Connective Tissues*

Examples of special connective tissues are bone, cartilage, blood, and hemopoietic tissues. The two types of **hemopoietic tissues** form blood cells; **myeloid tissue** is located inside cavities of bone, and **lymphoid tissue** occurs in the spleen, lymph nodes, and elsewhere. It was once thought that myeloid and lymphoid tissues each produced only one type of circulating blood cell, **myelocytes** and **lymphocytes,** respectively. Today we realize that both types of hemopoietic tissue are capable of manufacturing either of these blood cell types.

Cartilage and bone are specialized connective tissues in which inorganic salts and protein fibers have been deposited in the matrix. They differ in cell type (chondrocytes in cartilage, osteocytes in bone), in composition of the matrix (chondroitin sulfate in cartilage, calcium phosphate in bone), and in vascularization (cartilage is avascular, bone vascular). They also differ in their microarchitecture, bone can be highly ordered into osteons, and cartilage is usually less organized. On their surfaces, both are covered by a similar coat of fibrous connective tissue. Although virtually identical, these fibrous sheaths are logically termed the **perichondrium** around cartilage and the **periosteum** around bone.

***Cartilage*** Cartilage is a firm but flexible special connective tissue. The matrix primarily consists of chondroitin sulfate (ground substance) and collagenous or elastic proteins (fibers). Spaces within the matrix are called **lacunae** that house cartilage cells, or **chondrocytes.** The physical properties of cartilage and hence its functional roles are determined largely by the type and abundance of protein fibers in the matrix. There are three types of cartilaginous tissue.

The most widespread is **hyaline cartilage.** In the embryo, hyaline cartilage makes up many bones before they undergo **ossification** (bone formation). In the adult, hyaline cartilage persists at the articular ends of long bones, at the tips of ribs, in tracheal rings, and in many parts of the skull. Collagen fibers are present in the matrix but not in sufficient abundance to be easily seen with a light microscope. The name hyaline, meaning "glassy," refers to the homogenous appearance of the matrix (figure 5.20a).

Where cartilage is subjected to tensile or to warping loads, the ground substance is liberally reinforced with collagen fibers, which are obvious under microscopic examination. Such cartilage is **fibrocartilage** (figure 5.20b). The solid ground substance is especially effective in resisting compressive forces, and the embedded collagen fibers are better at addressing tensile forces. Fibrocartilage occurs in intervertebral disks, the pubic symphysis, disks within the knee, and selectively in other sites.

As the name suggests, **elastic cartilage** is flexible and springy, a property due to the presence of elastic fibers in the matrix (figure 5.20c). The internal support for your ear and epiglottis is a good example of elastic cartilage.

Cartilage does not receive blood supply directly. Blood vessels reside only within the perichondrium on its surface. Thus, nutrients and gases must pass between blood and chondrocytes by long-range diffusion through the intervening matrix. Similarly, no nerves directly penetrate cartilage. Cartilage may be heavily invested with calcium salts, as, for example, in the skeletons of chondrichthyan fishes, but cartilage is never as highly organized as bone.

***Bone*** Bone is a specialized connective tissue in which calcium phosphate and other organic salts are deposited in the matrix. In many vertebrates, these organic salts are arranged in a regular and highly ordered unit known as an **osteon** (also called the **Haversian system;** figure 5.21). Each osteon is a series of concentric rings made up of bone cells and layers of bone matrix around a central canal through which blood vessels, lymphatic vessels, and nerves travel.

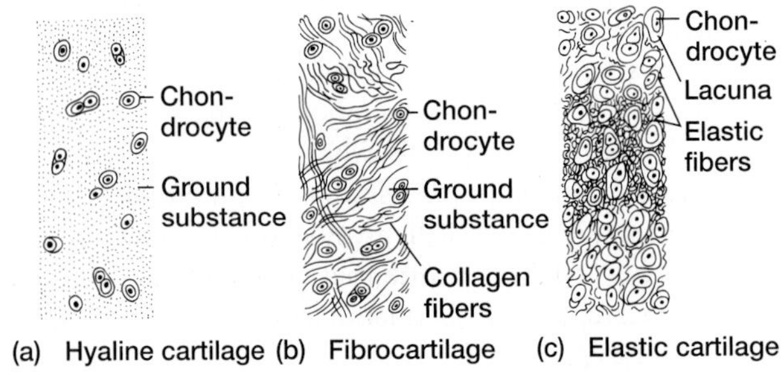


**Figure 5.20** Types of cartilage. The cartilage cell, or chondrocyte, is surrounded by a matrix composed of a ground substance and protein fibers. (a) Fibers are not apparent in the matrix of hyaline cartilage when it is viewed with a light microscope. (b) Collagen fibers are abundant in fibrocartilage, giving it mechanical resistance to tensile forces. (c) Elastin, the predominant protein fiber in elastic cartilage, makes it springy and flexible.

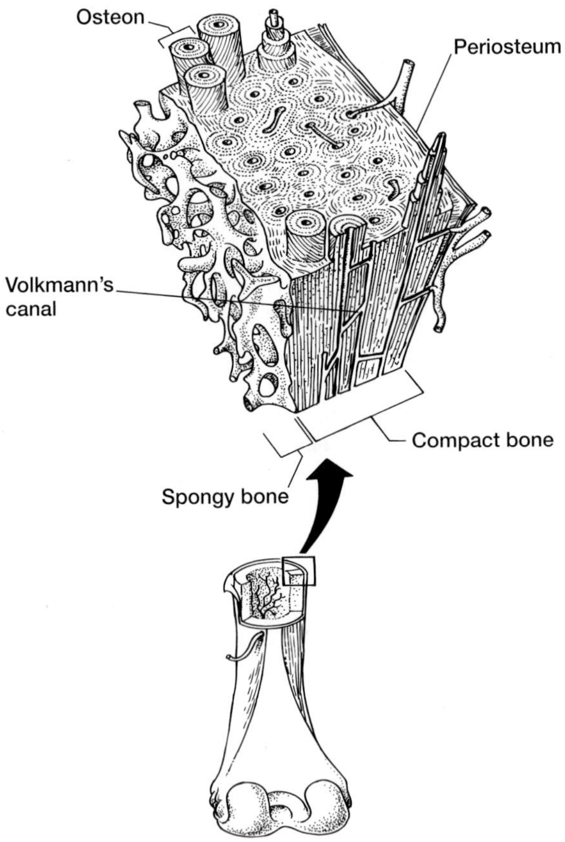


**Figure 5.21** Bone architecture. Osteons make up compact bone. Each osteon is a series of concentric rings of osteocytes and their matrix. Nerves and blood vessels pass through a central canal within each osteon. Diagonal connections, known as Volkmann's canals, allow blood vessels to interconnect between osteons. As new osteons form, they usually override existing, older osteons as part of the ongoing dynamic process of bone remodeling.

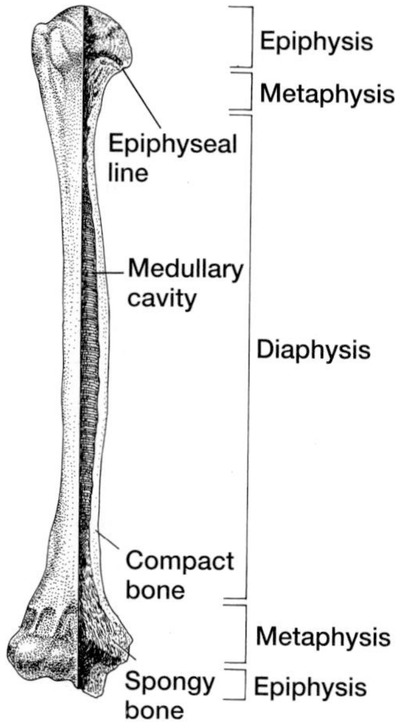


**Figure 5.22** Regions of a long bone. The middle section of bone is the diaphysis (shaft) containing the medullary cavity. In mammals, the ends of the bone are secondary centers of ossification, or epiphyses, although this term is sometimes used loosely to mean simply the end of a bone. Between the diaphysis and the epiphysis is the metaphysis, the actively growing region of bone. Compact bone is dense; spongy (or cancellous) bone is porous. The medullary cavity and all spaces in spongy bone are filled with blood-forming hemopoietic tissues.

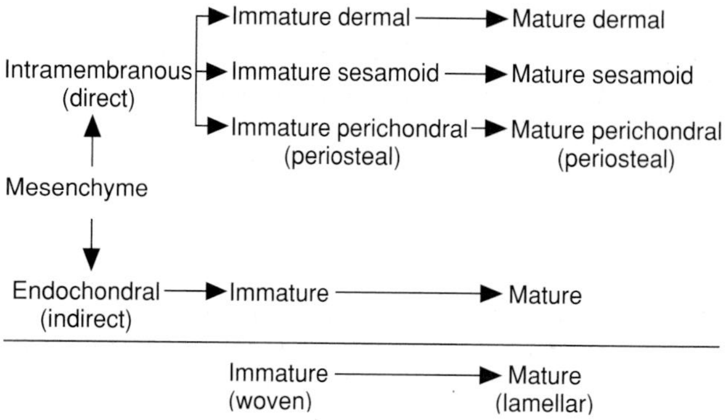


**Figure 5.23** Bone development. All bone develops from initial condensations of mesenchyme; thereafter, events proceed by one of two developmental sequences. In endochondral (indirect) bone development, mesenchyme forms a cartilage analog that subsequently is replaced by bone. In intramembranous (direct) bone development, mesenchyme proceeds directly to bone without an intermediate cartilaginous step. Intramembranous development may occur in dermis (dermal bone), tendon (sesamoid bone), or under the perichondrium (perichondral bone). Whether endochondral or intramembranous, when bone first appears it is termed "immature" bone (woven) because of the formative, irregular arrangement of matrix and osteocytes. As lamellae become more regularly organized, bone is referred to as "mature" bone (lamellar bone).

Volkmann's canals, running diagonally through this system, interconnect blood vessels between osteons. The layers of bone matrix are termed **lamellae.** Bone cells are identified on the basis of their activity: **osteoblasts** engage in osteogenesis (i.e., producing new bone); **osteoclasts** remove existing bone; and **osteocytes** maintain equilibrium in fully formed bone. Unlike cartilage, bone has a direct vascular and nerve supply.

There are several criteria by which we classify bone. From its visual appearance, we see two types of bone—**cancellous** or **spongy bone,** which is porous, and **compact bone,** which appears dense to the naked eye (figure 5.22). From its position, we recognize **cortical bone** in the outer boundary or cortex of a bone and **medullary bone** that lies within the core. From their pattern of embryonic development, we know there are two types of bone—**endochondral** and **intramembranous bone.** In the following section and subsections on bone, we trace these two types of bone development.

## Bone Development and Growth

Both endochondral and intramembranous development begin with local aggregations of loosely arranged mesenchymal cells. Thereafter, the two processes differ. In intramembranous development, bone is formed directly; in endochondral development, cartilage is formed initially and subsequently replaced by bone (figure 5.23). Once bone matrix appears, the two developmental routes again proceed through the same steps. During the first step, formative **immature (woven) bone** arises with lots of cells and irregularly strewn bundles of collagen. This proceeds to **mature (lamellar) bone** that is sparsely populated with bone cells but possesses orderly arranged layers of matrix. It is impossible to tell from the gross visual appearance of mature bone whether it was produced by endochondral or intramembranous development. We look at these two types of development in more detail.

### *Endochondral Bone Development*

Endochondral means within or from cartilage, and bones resulting from this developmental process are sometimes referred to as **cartilage** or **replacement bones.** During endochondral development, we can recognize three regions of a bone. The middle shaft is the **diaphysis;** each end is an **epiphysis;** and the region between is the **metaphysis** or **epiphyseal plate** (figure 5.22). Endochondral bone development involves the formation of a cartilage model of the future bone from mesenchymal tissue and the subsequent replacement of this cartilage model by bone tissue (figure 5.23). Replacement of cartilage continues throughout most of an individual's early life.

The steps of this process are illustrated in figure 5.24a–g. First, loose collections of mesenchymal cells condense to form a hyaline cartilage model surrounded by a perichondrium (figure 5.24a). Second, the periosteal bone collar forms in the region of the diaphysis (figure 5.24b). Cells on the inner surface of the diaphyseal perichondrium become osteoblasts and deposit the bone collar. As the bone collar is being formed, inorganic calcium salts accumulate in the matrix to calcify the cartilage in the core of the diaphysis (figure 5.24c). Calcium salts also seal off chondrocytes from nutritional and gas exchange with blood vessels on the surface of the cartilage. The entombed chondrocytes die as calcification proceeds. Next, the vascular system invades this calcified cartilage. These proliferating blood vessels erode away the cartilage debris to thus form the initial spaces of the marrow cavity.

Finally, osteoblasts appear within the core of the bone, and the primary center of ossification is established (figure 5.24d). Within this center of ossification, old bits of calcified cartilage become overlaid by new bone. Spikelike **trabeculae** are transitional composites of new bone and resorbing calcified cartilage. Later, when an ossified matrix predominates, trabeculae are called **bone spicules.** Additional osteoblasts circulating in blood are brought in by the invading vascular tissue. At about the same time, osteoclasts appear as well, signaling the active nature of bone remodeling through matrix deposition (osteoblasts) and removal (osteoclasts). Cartilage replacement, begun in the diaphysis, continues in the metaphyses. The epiphyseal plate is the active area of cartilage growth, calcification, cartilage removal, and new bone deposition. As ossification process approaches, the chondrocytes proliferate and hypertrophy while the surrounding matrix calcifies (figure 5.24, bottom inset). Blood vessels invade and erode the calcified cartilage. Ossification is the last process to overtake a region and finally replace the cartilage remnants.

Proliferation of cartilage in the epiphyses lengthens bone. Continued deposition of bone under the diaphyseal periosteum contributes to increased growth in bone girth. Bones in fishes, amphibians, and reptiles grow thoughout their lifetimes, although growth slows in later life. Thus, some fishes, turtles, and lizards can reach quite large sizes. In birds and mammals, however, bone growth ceases when adult size is attained.

In mammals, secondary centers of ossification arise in the epiphyses (figure 5.24e,f). The events that occur are similar to those that occurred during primary ossification in the bone shaft, namely, cartilage calcifies, blood vessels invade the epiphyses, osteoblasts appear, and new bone is deposited. In humans, these secondary centers of ossification appear at two to three years of age.

At or shortly after mammals reach sexual maturity, the epiphyseal plates and the metaphyseal regions they occupy ossify completely (figure 5.24g). Stated another way, the zone of ossification overtakes cartilage proliferation. At this point, the mammal's major growth phase is over. However, active remodeling and reorganization of the bone matrix continues throughout the remainder of life.

### *Intramembranous Bone Development*

In intramembranous bone development, bone forms directly from mesenchyme without a cartilage precursor (figure 5.23). Initially the mesenchyme is compacted into sheets or membranes, hence the resulting bones occasionally are referred to as "membrane bones."

As mesenchymal cells condense, they quickly become richly supplied with blood vessels. Between these compacted cells there appears a gellike ground substance. Dense bars of bone matrix are deposited within this ground substance, and osteoblasts simultaneously become evident for the first time. The dense bars of matrix become more numerous, eventually replacing the gellike ground substance. Subsequent growth proceeds by application of successive layers of new bone to the surface of these existing bone matrix bars (figure 5.25a–c). There are three types of specialized intramembranous bone development—dermal bone, sesamoid bone, and perichondral bone.

**Dermal bone** forms directly through ossification of mesenchyme. Many bones of the skull, pectoral girdle, and integument are examples. They are called dermal bones because the mesenchymal source of these bones lies within the dermis of the skin.

**Sesamoid bones** form directly within tendons, which are themselves derived from connective tissue. The patella of the knee and the pisiform bone of the wrist are examples. Sesamoid growth seems to be a response of tendons to mechanical stresses.

**Perichondral** and **periosteal bone** are formed from the deep cell layer of the fibrous connective tissue covering cartilage (perichondrium) or bone (periosteum). This type of bone develops early and retains the ability to form bone directly in the adult. Osteoblasts differentiate within this inner layer of the perichondrium or periosteum to produce bone without a cartilage precursor.

After ossification occurs and bone is formed, breakage or trauma to this bone may be followed by the appearance of cartilage. Because this cartilage forms after initial bone formation, it is called **secondary cartilage.** Following breakage, cartilage holds the ends of broken bone together and is replaced soon thereafter through endochondral bone ossification. Repair of a fracture involving cartilage is widespread throughout vertebrates. Some embryologists prefer a more restrictive definition, recognizing as secondary cartilage only the cartilage that arises on the margins of intramembranous bones from periosteal cells in response to mechanical stresses. Once formed, this cartilage may ossify or remain as cartilage throughout life. In this restricted sense, secondary cartilage is known only in birds and mammals.

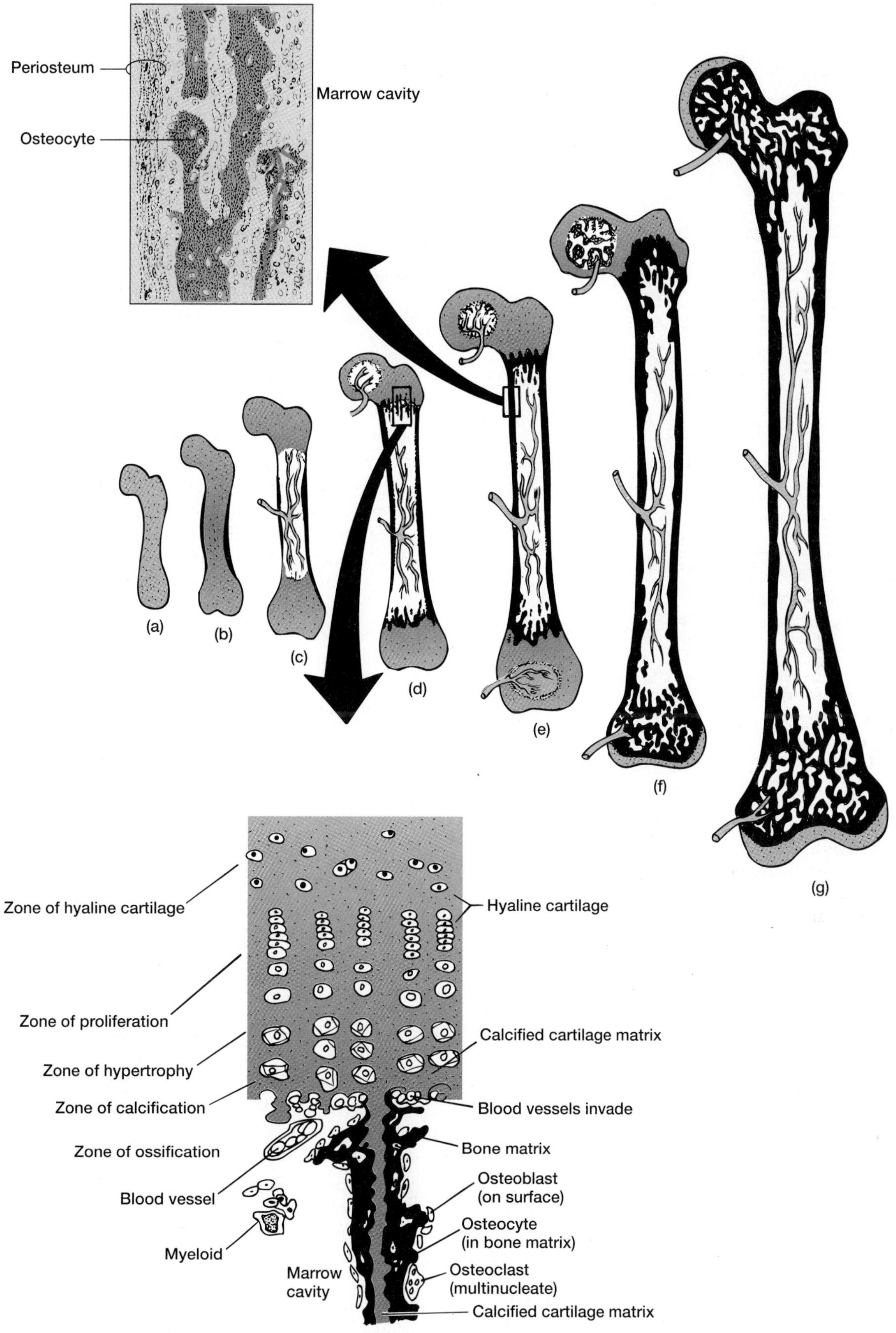


**Figure 5.24** Steps in endochondral bone growth. (a) Hyaline cartilage model. (b) Appearance of a bone collar. (c) Calcification of cartilage in the diaphysis followed by invasion of blood vessels. (d) Onset of ossification. (d,e,f) Appearance of secondary centers of ossification (epiphyses). (g) At maturity, the growth center (metaphysis) disappears. The top inset illustrates a portion of the wall of the diaphysis in which perichondral bone appears under the periosteum. The bottom inset is a section through the diaphysis showing successive proliferation of new cartilage, calcification, and replacement by the advancing line of ossification.

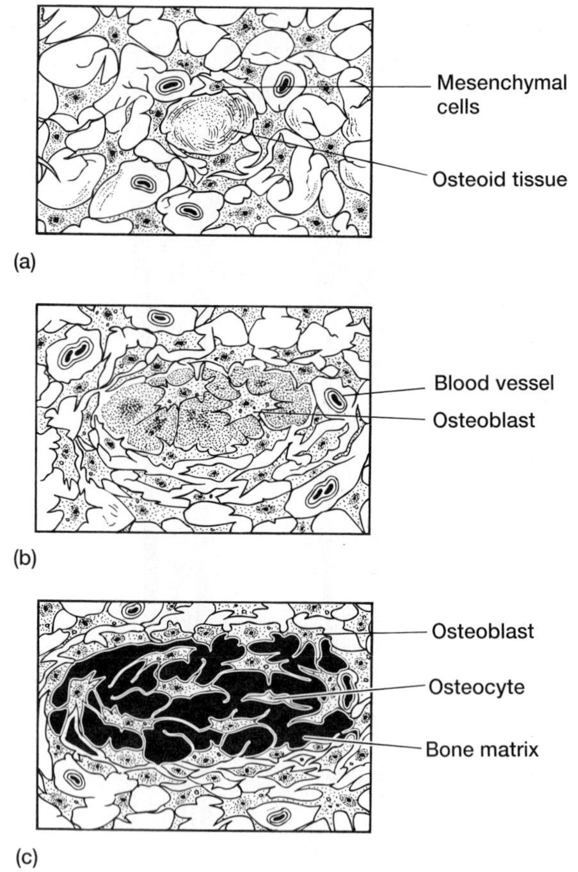


**Figure 5.25** Intramembranous bone formation. (a) Mesenchymal cells converge and produce osteoid tissue, a precursor of bone matrix. (b) Blood vessels invade, osteoblasts appear, and initial osteoid tissue becomes enriched with calcium, forming the matrix of immature bone. (c) After more and denser matrix forms, the cells within are more properly called osteocytes. Those on the surface still actively produce more bone matrix and so are osteoblasts.

## *Comparative Bone Histology*

Bone comprised of osteons is found throughout vertebrates, but it is not the only histological pattern of bone nor is it even the most common. In many teleost fishes, bone is **acellular,** entirely lacking osteocytes within the calcium phosphate matrix. During growth, osteoblasts at the surface secrete new matrix. However, these cells remain at the surface of the bone and do not become encased in their own secretions, so the bone they produce is acellular. Ostracoderms as well as some other groups of fishes have both acellular and cellular bone. In amphibians and reptiles, bone is often lamellar and cellular, with osteocytes present. Occasionally, osteons are present as well, being formed secondarily during continued growth and remodeling. More frequently, however, new bone is formed on a seasonal basis, producing growth rings in the cortex.

The view of bone as composed of an extensive osteon system comes from human bone and may generally apply to higher primates, but even among mammals this pattern shows differences. In many nonprimate mammals, large areas of acellular and even nonvascular bone may be found within the same individual. Bone from rats exhibits few osteons. In many marsupials, insectivores, artiodactyls, and carnivores, osteons may be absent from bone or from large regions of bone.

## *Bone Remodeling and Repair*

Microfractures accumulate in the mineralized matrix of bone over time. If left unattended, these microfractures might coalesce into a major fracture and the bone fail at a critical moment. To repair damage before it weakens bone significantly, new bone must replace older bone on a regular basis. An advancing front of osteoclasts in partnership erode channels through existing bone. In the wake of these osteoclasts, a large population of osteoblasts gathers to line the newly eroded channel and deposit new bone in characteristic concentric rings, forming a new osteon that often overrides lamellae of older osteons (figure 5.26).

This bone repair process is not only an important part of preventative maintenance, it is also a continuous remodeling process through which bone adapts to new functional demands throughout an individual's lifetime. In spite of preventative maintenance, however, an unexpected blow or twist might break a bone.

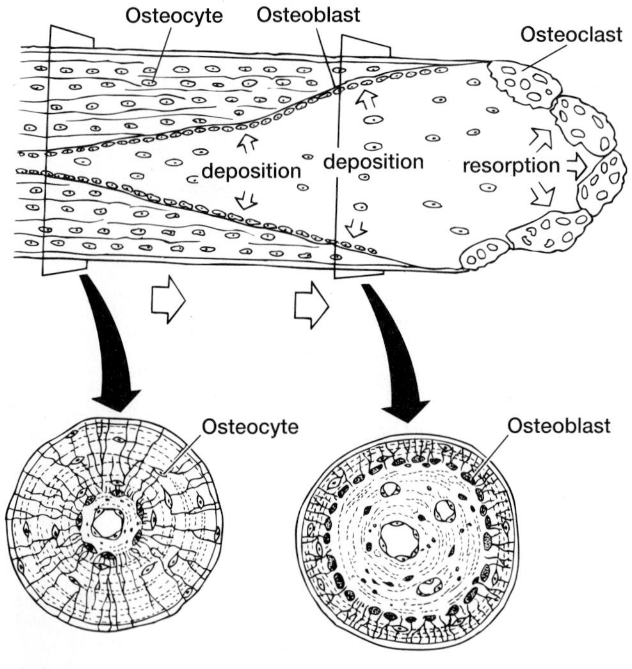


**Figure 5.26** Formation of a new osteon. An advancing line of osteoclasts removes bone cells by eroding through existing bone matrix to open a channel. Osteoblasts appear along the perimeter of the channel and immediately begin to form concentric rings of new matrix. As they themselves become surrounded by the matrix, they become proper osteocytes.

## Box Essay 5.1 Evolution of Bone

Bone is found only in vertebrates. Why it should make an evolutionary debut here and not in some other animal group is not known. One theory holds that bone arose first not as a supportive tissue but as a stored form of calcium or phosphate. Because salts of calcium and phosphate and other minerals occur in greater concentration in seawater than in the tissues of marine organisms, they tend to invade an animal's body, seeking an equilibrium. Excess salts and minerals can be excreted by the kidney or deposited out of the way, perhaps in the skin. Calcium and phosphate ions participate usefully in cellular metabolic pathways, so if they are stored rather than excreted, they would be easily accessible during times of increased metabolic demand. Large calcium and phosphate stores, if located superficially, would also form a hard surface protecting vertebrates from physical assault by predators. This secondary protective role might then favor development of the more extensive bony armor characteristic of early fishes. The appearance of a bony internal skeleton came still later, under selection for enhanced mechanical support.

As plausible as this hypothesis might be, it does not account for the particular form calcium takes in vertebrate bone. The hard, inorganic fraction of vertebrate bone is calcium phosphate in the crystalline form of hydroxyapatite, rather than the calcium carbonate in the crystalline form of calcite or aragonite that characterizes most invertebrate skeletons. Perhaps, as recently suggested, calcium in vertebrate bone is more stable under conditions of physiological stress associated with active lifestyles. In contrast to most invertebrates, vertebrates show an ancient and unusual lifestyle characterized by intense bursts of activity. Bursts of activity lead to lactic acid formation followed by marked fluctuations of blood pH accompanied by prolonged acidosis (more acidic) before resting pH levels return. Under conditions of acidosis, calcium carbonate of invertebrates tends to literally dissolve, whereas calcium phosphate of vertebrates is more stable. A skeleton that tended to dissolve following extended activity would obviously weaken. This would also flood the circulating blood with excess calcium, perhaps further complicating normal metabolism of internal organs.

Thus, a skeleton of calcium would afford some mechanical protection, but one of calcium phosphate in particular (but not of calcium carbonate) would make bone matrix more stable. It would also reduce the physiological disadvantages bone dissolution otherwise might create for an animal that depended on bursts of activity. This hypothesis for the evolution of vertebrate bone also fits well with the views of those who see early vertebrates or prevertebrates as animals abandoning sedentary lifestyles of their ancestors in favor of more active ones (see also Ruben and Bennett, 1987).

Early chordate evolution (p. 70)

A break initiates a four-step repair process. First, a blood clot forms between the broken ends of bone (figure 5.27a). Smooth muscle contraction and normal clotting seal the severed ends of blood vessels that run through the bone. Second, a callus develops between the ends of the break, mostly from the activity of cells within the periosteum (figure 5.27b). The callus is composed of hyaline and fibrocartilage, often with remaining bits of the blood clot. A few new bony spicules appear at this time as well. Third, the cartilaginous callus is replaced by bone, largely through a process reminiscent of endochondral bone formation. Cartilage calcifies, chondrocytes die, vascular tissue invades, osteoblasts and osteoclasts arrive, and bone matrix appears (figure 5.27c). After the cartilage is replaced, the two broken bone ends are usually knitted together by irregular spicules of bone (figure 5.27d). Finally, osteoclasts and osteoblasts participate in the remodeling of this rough mend to finish the repair process. This final remodeling step can continue for months. If the original break was severe, the area of repair can remain rough and uneven for many years (figure 5.27e).

In 1843 Dr. David Livingstone (of "Dr. Livingstone, I presume"), the famous Scot who explored Africa in the early nineteenth century, was badly mauled by a lion. He sustained a severe fracture of his upper arm but survived to carry on a prolonged missionary campaign. After his death 30 years later, his remains were returned to England and positively identified, in part by the distinctive fracture callus still evident.

## Neural Crest and Ectodermal Placodes

Neural crest cells, ectodermal placodes, and their many derivatives have been known from the nineteenth century, but their extraordinary significance to vertebrate evolution has only recently received the attention deserved. These cells and placodes are embryonic structures found solely in vertebrates and nowhere else in the animal kingdom, not even among the other subphyla of chordates.

Long wedges of neural crest cells are set aside early in vertebrate development. Before complete closure of the neural folds, neural crest cells break loose from the surface epithelium to assemble temporarily into distinct cords above the forming neural tube. From here they migrate out along defined routes within the embryo to permanent sites at which they differentiate into a great variety of structures, including ganglia of spinal and cranial nerves, Schwann cells that form the insulating sheath around peripheral nerves, chromaffin cells of the adrenal medulla, pigment cells of the body (except in retina and central

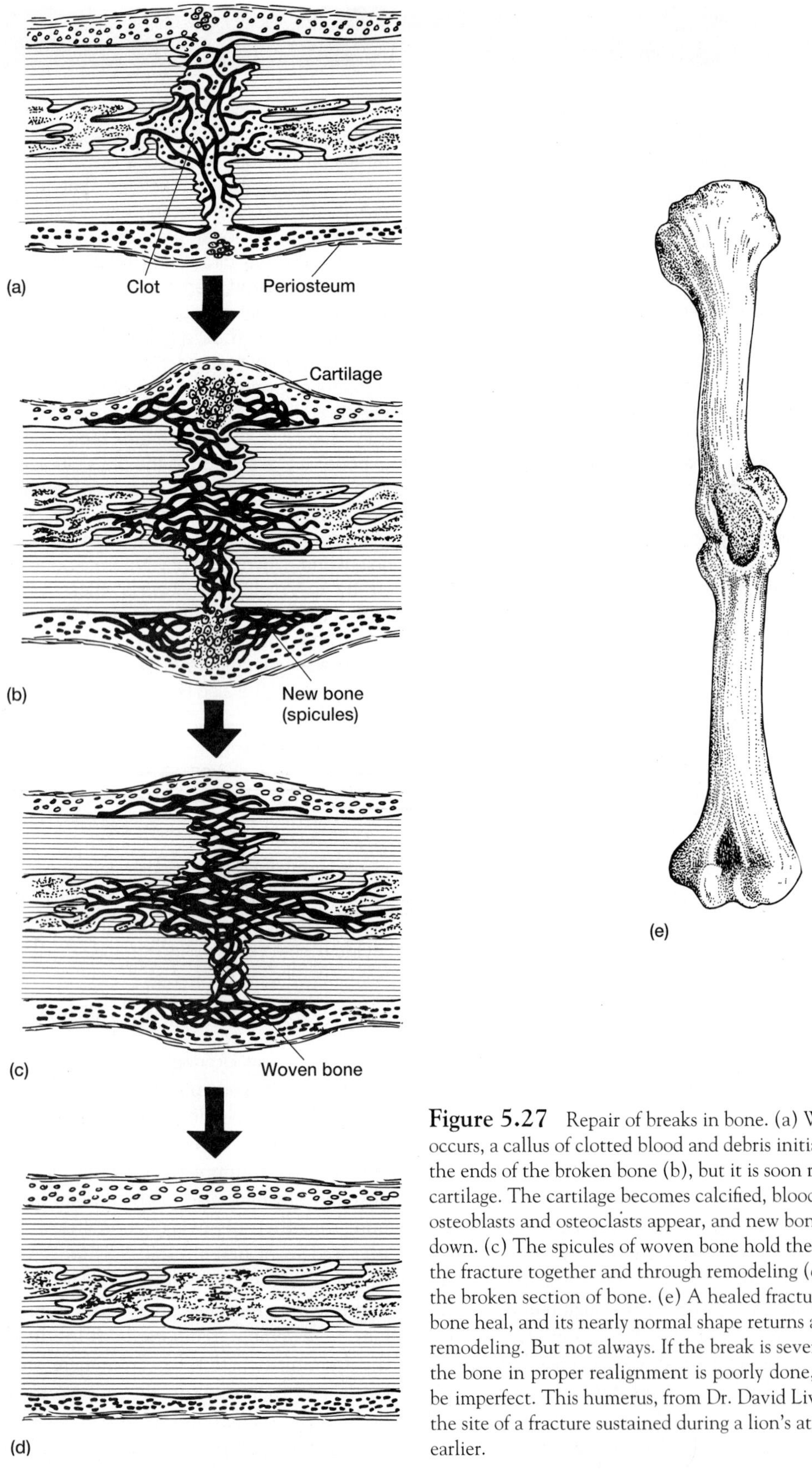


**Figure 5.27** Repair of breaks in bone. (a) When a fracture occurs, a callus of clotted blood and debris initially forms between the ends of the broken bone (b), but it is soon replaced by cartilage. The cartilage becomes calcified, blood vessels invade, osteoblasts and osteoclasts appear, and new bone matrix is laid down. (c) The spicules of woven bone hold the broken ends of the fracture together and through remodeling (d) come to replace the broken section of bone. (e) A healed fracture. Most breaks in bone heal, and its nearly normal shape returns after a period of remodeling. But not always. If the break is severe and "setting" the bone in proper realignment is poorly done, then repair may be imperfect. This humerus, from Dr. David Livingstone, shows the site of a fracture sustained during a lion's attack 30 years earlier.

TABLE 5.2 Neural Crest Derivatives

| |
|---|
| Dorsal root ganglion neurons |
| Sympathetic and parasympathetic ganglia |
| Some neurons |
| Hormone-producing cells |
| Chromaffin cells of the adrenal medulla |
| Calcitonin cells |
| Schwann cells |
| Parts of meninges |
| Branchial cartilage cells |
| Pigment cells (except in retina and central nervous system) |
| Odontoblasts |
| Dermis of facial region |
| Vasoreceptors |
| Sensory capsules and parts of neurocranium |
| Cephalic armor and derivatives |

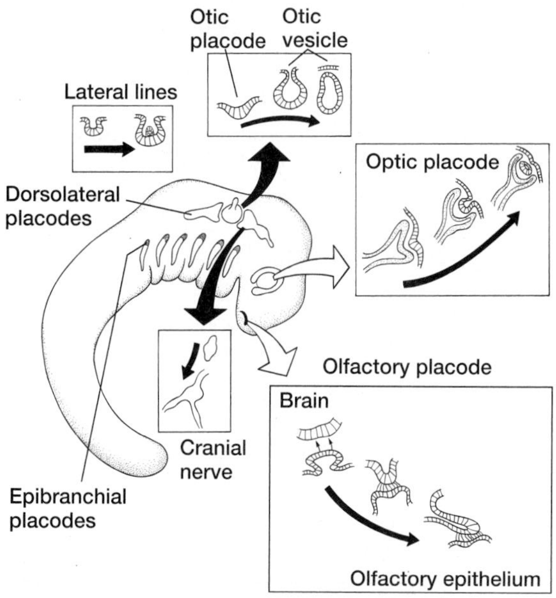


**Figure 5.28** Ectodermal placodes in a representative vertebrate. There are two paired sets of ectodermal placodes, the dorsolateral and the epibranchial placodes, as well as the olfactory and the optic placodes. All form sensory organs or receptors.

nervous system), and several types of widely dispersed hormone-producing cells. In the head, neural crest cells give rise to most cartilage and bone of the lower jaw and to most connective tissue of the voluntary muscles. Within the cores of teeth, **odontoblasts** that secrete the inner layer of dentin also arise from neural crest cells. The derivatives of neural crest cells are summarized in table 5.2.

**Ectodermal placodes** are not related to neural crest cells. Placodes are thickenings of the surface ectoderm that sink inward to form specific sensory receptors. Sensory (but not motor) fibers of the spinal nerves departing from along the length of the spinal cord arise embryologically from neural crest cells. Cranial nerves that depart from the brain arise from neural crest cells and ectodermal placodes in the embryo. In fishes and amphibians, placodes contributing to cranial nerves are located in two rows within the head. The upper row of **dorsolateral placodes** and the lower row of **epibranchial placodes** lie sequentially just above the gill slits (figure 5.28). Some cells of the dorsolateral placodes also contribute to other sensory systems as well. They migrate to positions over the head and along the body where they differentiate into receptor cells and associated sensory nerves of the lateral line sensory system. An especially prominent member of the dorsolateral series of placodes is the **otic placode,** which sinks inward from the surface as a unit to form the vestibular apparatus concerned with balance and hearing.

**Cranial and spinal nerves (chapter 16); sensory organs derived from placodes (chapter 17)**

TABLE 5.3 Placodes and their Derivatives

| Placode | Derivative |
|---|---|
| Dorsolateral | |
| Lateral line | Lateral line mechanoreceptors and electroreceptors |
| Otic | Vestibular apparatus |
| Cranial nerve | Sensory nerve ganglia |
| Epibranchial | |
| Cranial nerve | Sensory nerve ganglia |
| Olfactory | Sensory epithelium |
| Optic | Lens of eye |

The paired **olfactory placodes** form at the tip of the head and differentiate into sensory receptors of smell that grow to and connect with the brain. The paired **optic placodes** form laterally to produce the lens of the eye. Placodes can interact with the neural crest but do not arise from it. All vertebrate placodes, except the optic placode, differentiate into sensory nerves. The derivatives of ectodermal placodes are summarized in table 5.3.

The vertebrate body, especially the head, is in large measure a collection of structures of neural crest or placode origin. Although integrated harmoniously in the adult, these unique derivatives distinguish vertebrates from all other chordates.

# Extraembryonic Membranes

While the embryo is in the ovary or during its passage down the oviduct, it gains extrinsic secondary and tertiary egg envelopes. Intrinsic membranes should not be confused with these wrappings added by the oviducts. Intrinsic membranes that arise from the embryonic germ layers and grow to surround the developing embryo are **extraembryonic membranes** (figure 5.29a–d). They function in sequestering waste products, transporting nutrients, and exchanging respiratory gases. They create a tiny aquatic environment, enveloping the embryo in a self-contained, fluid-filled capsule. The embryo effectively floats in an almost weightless environment, with gravity having only a slight effect upon its delicate and growing tissues. Extraembryonic membranes also protect the young embryo within its own moist environment so that an external body of water is not needed.

Vertebrates whose embryos possess extraembryonic membranes are **amniotes,** the **amnion** being one of the several extraembryonic membranes. Amniotes include reptiles, birds, and mammals. **Anamniotes,** meaning without an amnion, include fishes and amphibians. Fishes lay their eggs in water, and amphibians seek moist spots or return to water to deposit their eggs. Embryos of fishes and amphibians lack extraembryonic membranes.

Extraembryonic membranes appear early and continue to enlarge throughout development, keeping pace with the enlarging metabolic needs of the growing embryo. At birth or hatching, the young individual breaks free of these membranes and must depend on its own internal organs to meet its nutritional (digestive tract) and respiratory (lungs) needs. The four extraembryonic membranes and their origins in reptiles, birds, and mammals are summarized in table 5.4 and discussed in detail in the following subsections.

## Reptiles and Birds

In birds and generally in reptiles, the extraembryonic membranes form soon after the basic germ layers are established. The germ layers that contribute to the extraembryonic membrane are continuous with the germ layers from the body of the embryo but they spread outward, extending away from the embryo. The bilaminar splanchnopleure of endoderm and splanchnic mesoderm form one membrane sheet that spreads around the yolk, eventually enclosing it as the **yolk sac.** Blood vessels develop in the mesodermal component of the spreading splanchnopleure and form a network of **vitelline vessels.** This vascularization network is important in mobilizing the energy and nutrients of the yolk during embryonic growth. The somatopleure of surface ectoderm and the somatic mesoderm forms the other bilaminar sheet that spreads outward from the embryonic body (figure 5.30). The somatopleural sheet grows upward over the embryo as **amniotic folds,** that eventually meet and fuse at the midline. Two membranes are produced from the amniotic folds. One is the amnion that immediately surrounds the embryo and encloses it in a fluid-filled **amniotic cavity.** The other is the more peripheral **chorion** (figure 5.29c).

As the amniotic folds develop, the **allantois,** a diverticulum of the hindgut endoderm, grows outward, carrying splanchnic mesoderm with it. The endoderm and splanchnic mesoderm of the allantois continue to expand, slipping

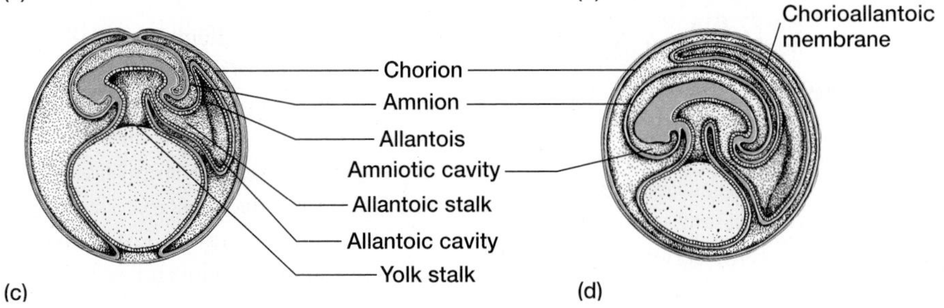

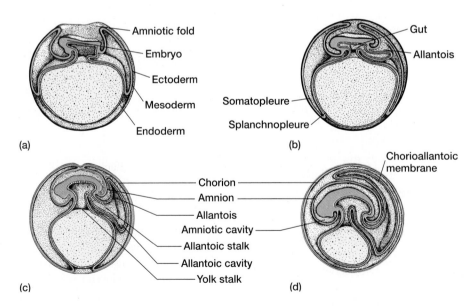

**Figure 5.29** Extraembryonic membrane formation in a bird (sagittal sections). Somatopleure lifts upward (a), forming amniotic folds that join (b) and fuse (c) above the embryo to produce the chorion and the amnion. The expanding allantois comes into association with the chorion to produce the chorioallantoic membrane (d). An extensive vascular network forms within the mesoderm and serves as a site of respiratory exchange for gases passing through the porous shell (not shown).

TABLE 5.4 Sources of the Four Extraembryonic Membranes in Most Reptiles, Birds, and Mammals

| | Extraembryonic Membrane | | | | |
|---|---|---|---|---|---|
| *Vertebrate Group* | *Amnion* | *Chorion* | *Allantois* | *Yolk Sac* | *Respiratory Membrane* |
| | Germ Layer Sources | | | | |
| Birds | Ectoderm, somatic mesoderm | Ectoderm, somatic mesoderm | Endoderm, splanchnic mesoderm | Endoderm, splanchnic mesoderm | Chorion, allantois |
| Reptiles | Ectoderm, somatic mesoderm | Ectoderm, somatic mesoderm | Endoderm, splanchnic mesoderm | Endoderm, splanchnic mesoderm | Chorion, allantois |
| Monotremes | Ectoderm, somatic mesoderm | Ectoderm, somatic mesoderm | Endoderm, splanchnic mesoderm | Endoderm, splanchnic mesoderm | Chorion, allantois |
| Marsupials | Ectoderm, somatic mesoderm | Ectoderm, somatic mesoderm | Endoderm, splanchnic mesoderm | Endoderm, splanchnic mesoderm | Chorion, splanchnopleure |
| Placental Mammals | Ectoderm (trophoblast), somatic mesoderm | Ectoderm (trophoblast), somatic mesoderm | Endoderm, splanchnic mesoderm | Endoderm, splanchnic mesoderm | Chorion, allantois |

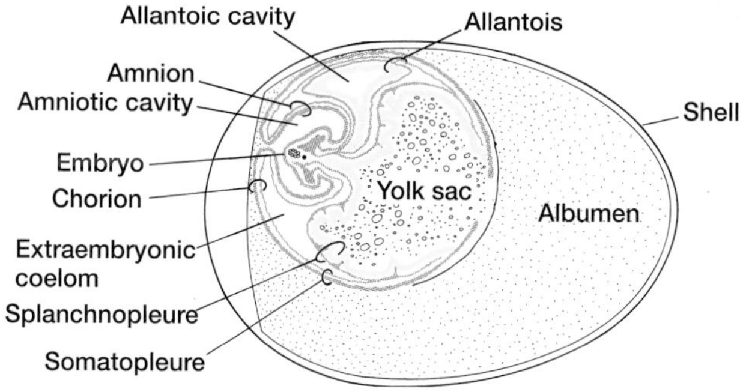


**Figure 5.30** Cross section of a bird embryo within the shelled egg after about eight hours of incubation. Note the early formation of the allantois and the amnion.

between the amnion and the chorion. Eventually, the outer allantois and the chorion fuse to form a single composite membrane, the **chorioallantoic membrane** (figure 5.29c,d). The mesoderm sandwiched within this membrane forms an extensive network of **allantoic vessels** that function in respiratory exchange through the porous shell. The **allantoic cavity** bounded by the allantois becomes a repository for the embryo's excretory wastes.

## Mammals

Structures homologous to the four extraembryonic membranes of reptiles and birds appear in mammals—amnion, chorion, yolk sac, and allantois. In placental mammals, the amnion forms over the embryo. In some placental mammals, such as dogs and pigs, it forms from amniotic folds in the somatopleure after the embryo is well organized. In other placental mammals, such as humans, fluid-filled spaces appear within the inner cell mass prior to the establishment of germ layers. These spaces coalesce to form the initial amniotic cavity.

In monotremes, the extraembryonic membranes are formed in much the same way in which they are formed in reptiles and birds (table 5.4). In therian mammals, marsupials and placentals, a structure homologous to the yolk sac is present, but it contains no yolk platelets. Instead, it is filled with fluid. The embryonic disc is suspended between the amniotic cavity and the yolk sac. As in other amniotes, the allantois begins as an outgrowth of the hindgut that expands outward, becoming surrounded by a layer of mesoderm as it grows. The chorion of placental mammals is bilaminar, as in reptiles and birds, but it forms from the trophoblast and includes the adjacent mesodermal layer. The expanding allantois grows into contact and fuses with much of the internal wall of the chorion, producing the chorioallantoic membrane. The allantoic vessels or, the **umbilical vessels,** as they are more often called in mammals, develop within the mesodermal core of the chorioallantoic membrane. These vessels function in respiration and nutritional exchange with the uterus of the mother.

### *Eutherian Placenta*

In eutherian mammals, the chorioallantoic membrane of the fetus establishes intimate contact with the adjacent vascular wall of the mother's uterus to produce the **placenta,** a composite structure formed in part from tissues of the fetus and in part from tissues of the mother (figure 5.31a–d). It is often called the **allantoic placenta** because the allantois (chorioallantoic membrane) serves as the fetal contribution to the placenta. Blood from the mother does not pass into the fetus. Rather, the placenta brings capillary beds of both

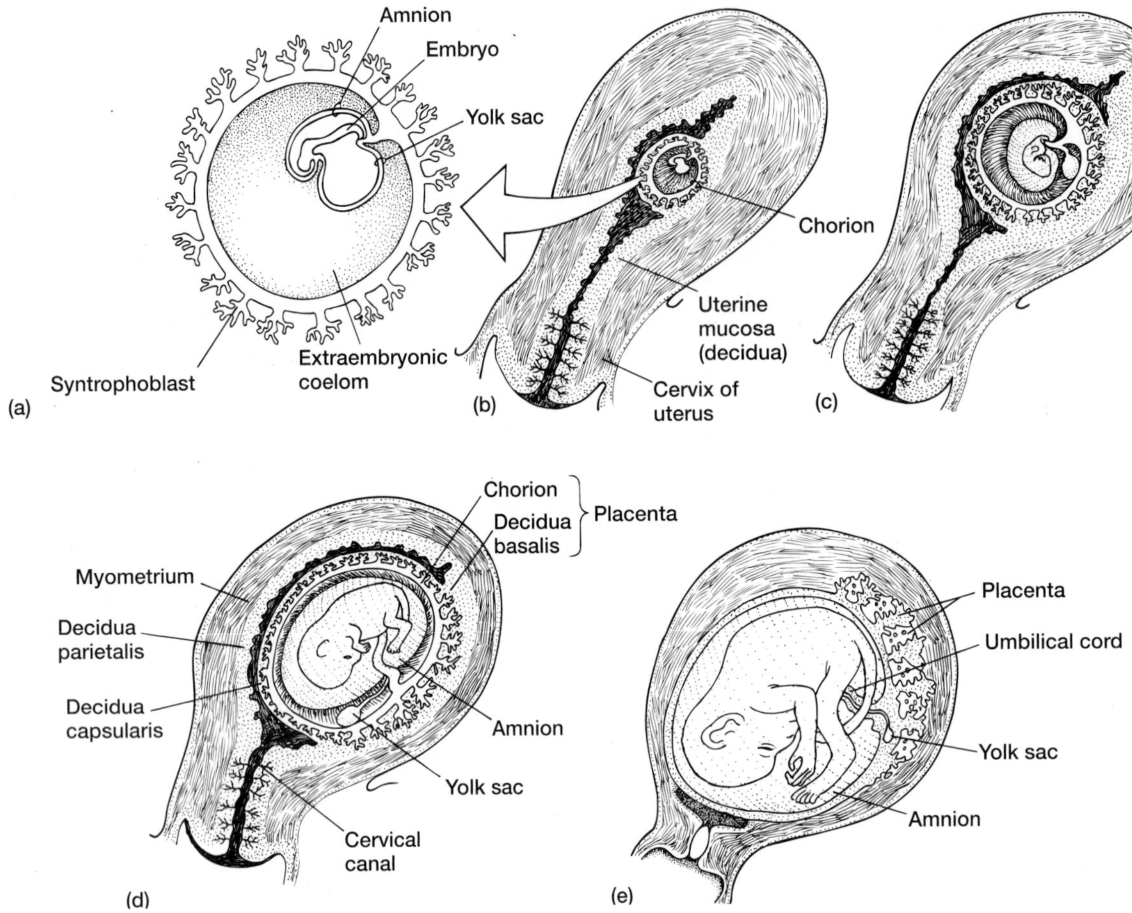


**Figure 5.31** Uterus during pregnancy. (a–e) Primate embryo and its membranes are shown at successive stages of development. The decidua is the inner lining of the uterus, the myometrium is the outer muscular wall. That part of the decidua associated with the fetal chorion is the decidua basalis. Together the maternal decidua basalis and fetal chorion form the placenta. The decidua parietalis and the decidua capsularis compose the remainder of the decidua. Once the placenta is formed, the umbilical cord carries the paired umbilical arteries, unpaired umbilical vein, and stalk of the yolk sac from the placenta to the embryo. The amniotic cavity continues to grow with the embryo until term, at which time it contains liquid (the so-called bag of water).

fetus and female into close association, but not into direct union, to allow transfer of nutrients and oxygen from the mother to the fetus and nitrogenous wastes and carbon dioxide from the fetus to the mother.

Eutherian mammals are also called **placental mammals** because eutherian reproduction is characterized by a placenta. The placenta of eutherian mammals begins to form when the blastocyst first makes contact with the wall of the readied uterus. In humans, implantation of the blastocyst results in its taking up residence in the uterine wall about six days after ovulation (figure 5.32a–d). In some species, implantation is postponed for weeks or months as further development of the blastocyst is temporarily arrested. This postponement, termed **delayed implantation,** extends the length of gestation so as to prevent an inopportune birth of a new individual while the female is still nursing young of a previous litter or while seasonal resources are slight. Badgers, bears, seals, some deer, and camels have delayed implantation.

Upon implantation, cells of the trophoblast proliferate to form two recognizable layers. Cells of the outer **syntrophoblast (syncytiotrophoblast)** layer lose their boundaries to form a multinucleated syncytium. The syntrophoblast helps the embryo enter the uterine wall and establish an association with maternal blood vessels. The second derivative layer of the trophoblast is the **cytotrophoblast,** the cells of which retain their boundaries and contribute to the extraembryonic mesoderm (figure 5.32c).

In summary, the placenta is formed of fetal and maternal tissues. Blood vessels of the fetus grow out into the syntrophoblast where they establish a close association with maternal blood vessels. The placenta supports respiratory and nutritional functions of the fetus. Hormones produced by the placenta stimulate other endocrine organs of the mother and help maintain the uterine wall with which the embryo is associated.

**Placental blood circulation (p. 480)**

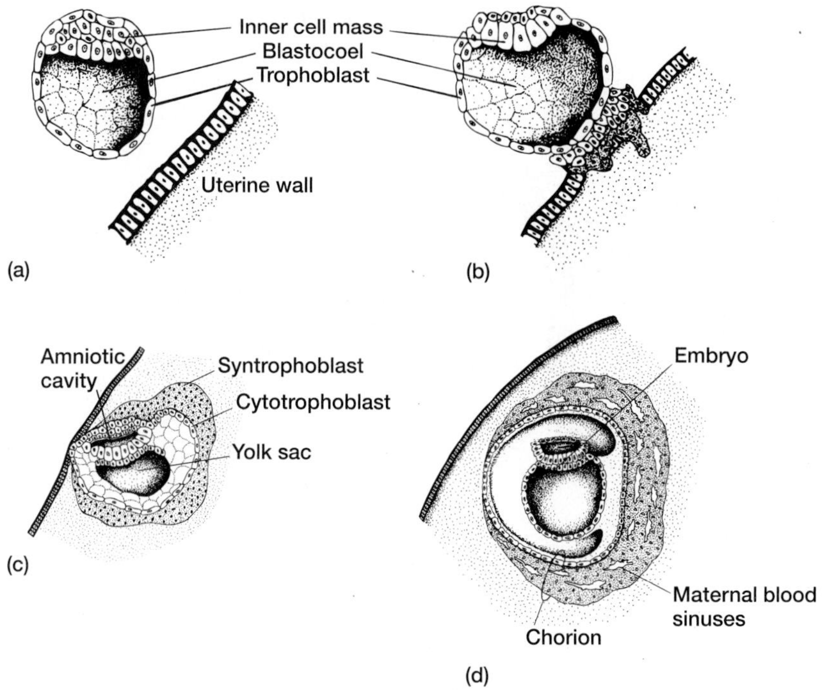


**Figure 5.32** Implantation of a mammalian embryo (human) within the wall of the uterus. (a) The blastocyst is not yet attached to the wall at about five days, but note that the inner cell mass, the trophoblast, and the blastocoel are already present and the zona pellucida has been shed. (b) Initial contact of the blastocyst with the uterine wall. (c) Deeper penetration of the blastocyst into the wall of the uterus. The trophoblast gives rise to an outer syntrophoblast, which is a syncytium, and the inner cytotrophoblast. The amniotic cavity forms by cavitation within the inner cell mass. (d) Blood sinuses of the maternal circulation course through the syntrophoblast to provide nutritional support and respiratory exchange for the embryo.

### *Other Placentae*

Most people are surprised to learn that placentae develop in marsupials and even in some fishes, amphibians, and reptiles. In fact, birds are the only vertebrate class in which no members possess a placenta. Some marsupials have developed placentae to support the fetus during its brief time in the uterus. One of the most widespread placental types among marsupials is the **yolk sac placenta (choriovitelline placenta)** in which the yolk sac serves as the fetal contribution to the placenta (figure 5.33a). In opossums and many other marsupial groups, the expanded yolk sac makes contact with the chorion to form the **choriovitelline membrane.** Part of this choriovitelline membrane is avascular, and part is extensively vascularized. In some species, the choriovitelline membrane may penetrate the uterine wall to form a yolk sac placenta that presides over gas and nutrient exchange between fetus and uterine tissues. In the marsupial bandicoots, two types of placentae are present. One is the yolk sac placenta just described and the other is an **allantoic placenta (chorioallantoic placenta)** structurally similar to that of eutherian mammals (figure 5.33c,d). Implantation in bandicoots is similar to that of placental mammals in that the chorion invades the uterus, bringing fetal and maternal capillary beds into close association.

Even in eutherian mammals, additional placentae sometimes form. For example, shortly after implantation in the dog, rapid outgrowth of the yolk sac brings it into contact with the chorion to produce a composite choriovitelline membrane that serves as a transitory yolk sac placenta (figure 5.34a). Within a few days, the growing allantois catches up and makes contact with the chorion to supplant the yolk sac placenta eventually with the more definitive allantoic placenta (figure 5.34b). During this period of change from one placenta type to another, the dog embryos appear most vulnerable to physiological stress.

## Overview of Early Embryonic Development

Yolk stores affect the pattern of cleavage and subsequent gastrulation. When yolk accumulates in the ovum in large quantities, it mechanically interferes with the formation of mitotic furrows and restricts cleavage to the relatively yolk-free area at the animal pole. In extreme cases, such as in teleost fishes, reptiles, birds, and monotremes, cleavage is discoidal, with the blastodisc confined to a cap of cells on top of the yolk. Subsequent gastrulation involves rearrangement of surface cells that move through an embryonic shield or primitive streak. Like blastopores, both embryonic shields and primitive streaks function as embryonic organizing areas. Both may be homologous to blastopores but flattened to accommodate the large amount of yolk.

Discoidal cleavage evolved independently in teleost fishes on the one hand and in reptiles, birds, and monotremes on the other. We, of course, do not know what cleavage pattern characterized the labyrinthodont amphibians. Modern amphibians have mesolecithal eggs and holoblastic cleavage. If labyrinthodonts shared this cleavage pattern as their modern amphibian descendants, then the discoidal cleavage seen in later reptiles, birds, and monotremes must represent a derived condition that evolved independently of the discoidal cleavage of teleosts.

In eutherian mammals, the yolk sac is almost entirely devoid of yolk, yet cleavage is discoidal and gastrulation oc-

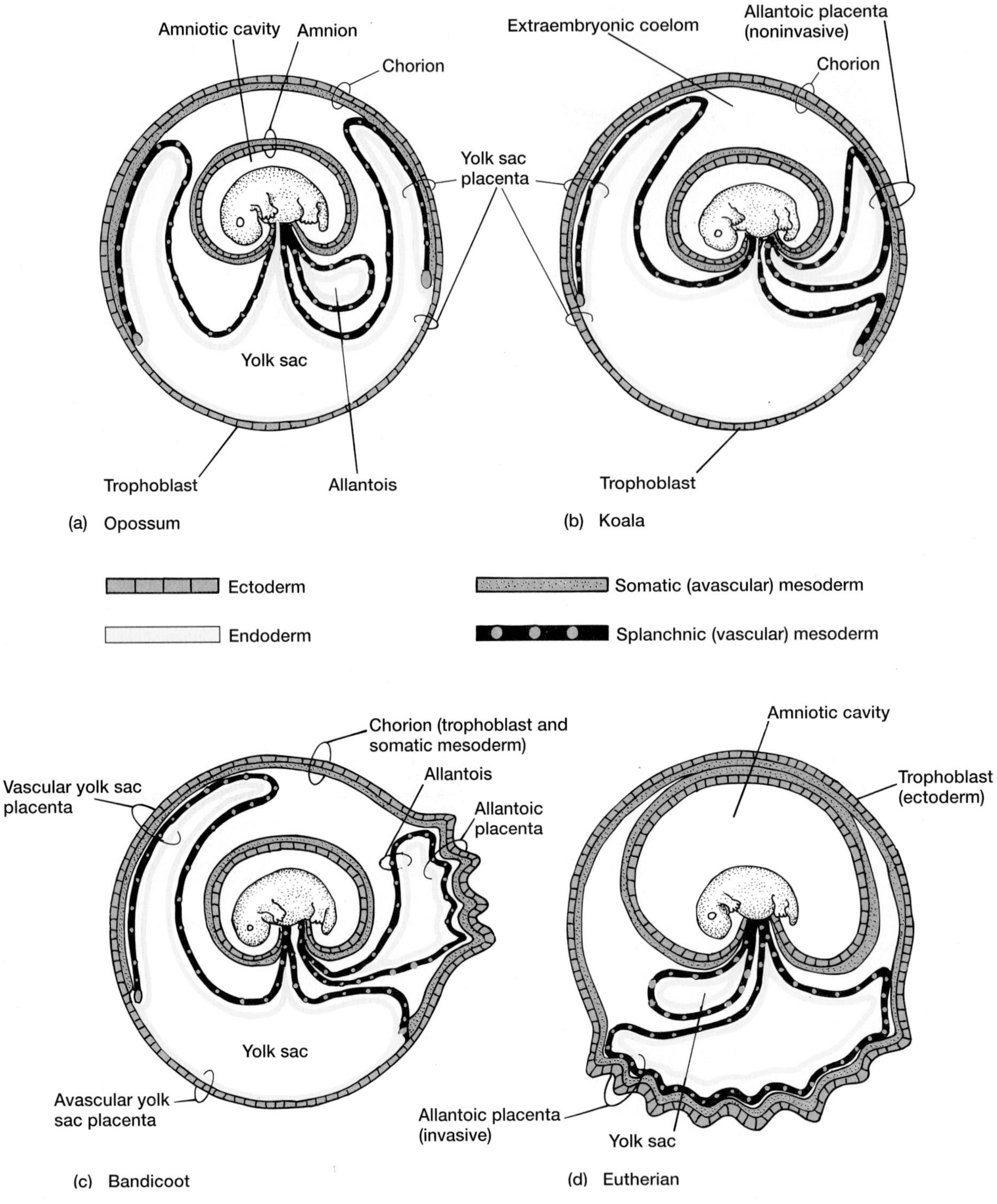


**Figure 5.33** Fetal membranes of some marsupials (a–c) and a placental (d). (a) Yolk sac placenta (choriovitelline placenta) of the opossum *Didelphis*. (b) Yolk sac placenta (choriovitelline placenta) and the allantoic placenta (chorioallantoic placenta) of a koala. (c) Yolk sac placenta and allantoic placenta of a bandicoot. The allantoic placenta invades the uterine wall extensively. (d) The characteristic allantoic placenta of a eutherian mammal is derived from the chorioallantoic membrane and becomes associated with the uterine wall.

curs via a primitive streak just as if large quantities of yolk were present and cells had to move around such an obstruction. This cleavage process likely represents the retention of features inherited from ancestors with yolk-laden eggs. Without reference to the phylogenetic background of eutherian mammals, such a pattern of early embryonic development would be difficult to explain.

Division of vertebrates into amniotes and anamniotes reflects a fundamental difference in mechanism of embryonic support. Appearance of the amnion along with

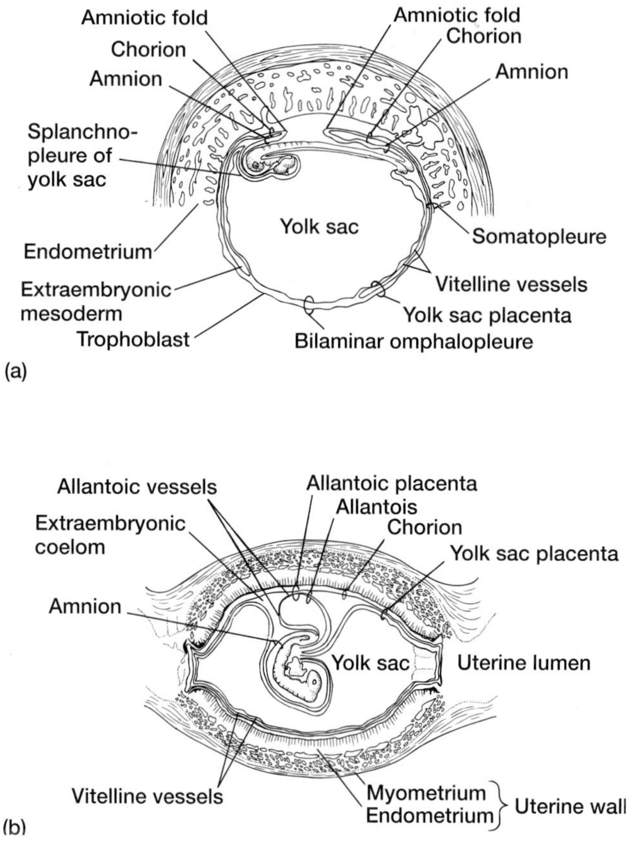


**Figure 5.34** Fetal membranes in the dog. (a) Early in development, the yolk sac forms a simple transient placenta with the uterine wall. (b) Within a few days of its establishment, the transient yolk sac placenta is replaced by the functional allantoic placenta that arises from the rapidly expanding chorioallantoic membrane.

other extraembryonic membranes in the reptiles represents an adaptation to an increasingly important terrestrial mode of life that took advantage of many new possibilities. Most reptiles, birds, and monotremes have shelled or **cleidoic eggs.** Once the cleidoic egg evolved, females no longer needed to trek long distances to bodies of water to lay their eggs in safety. The cleidoic egg is a self-contained little world. The yolk sac holds nutrients to support the developing embryo, the allantois serves as a repository into which nitrogenous waste products can be sequestered safely away from the embryo, and the amnion floats the embryo in a water jacket to prevent desiccation and lessen mechanical shocks. Either the yolk sac or the allantois becomes vascularized to serve a respiratory function.

Among mammals, we see a range of compromises in the pattern of embryonic development. With less yolk, the embryo correspondingly increases its dependence on oviducts and uterus for nutrients. In marsupials, nutrients are absorbed directly from the egg, but a modest placenta develops in some. However, prolonged development within the female poses additional problems. As the embryo gets larger, respiratory demands increase, and oxygen delivery must be improved. In eutherian mammals, a well-developed placenta develops to exchange gases with the maternal blood and address this problem. However, another potential problem arises for the embryo because the placenta keeps it in such close association with maternal tissues. At least half of the embryo is immunologically foreign because half its proteins are produced by the male's genetic contribution. If recognized as foreign, the mother's immune system will try to reject the embryo.

In marsupials, the embryo spends a relatively brief time within the uterus and is born at an early stage in development. Adult kangaroos can reach 70 kg, but the young weigh less than 1 g when they are born. A short gestation period addresses, in part, possible immunological rejection and provides evidence for why a fetus is born early. Further, the marsupial blastocyst is protected initially from immunological recognition by an inert eggshell membrane of strictly maternal origin that is retained throughout most of the brief gestation period. In placental mammals, the outer layer of the trophoblast is thought to promote implantation and prevent rejection of the embryo during its prolonged gestation.

# Development of the Coelom and its Compartments

The coelom produced within the hypomere during early embryonic development is partitioned during later development. In fishes, amphibians, and most reptiles, the coelom is subdivided into a **pericardial cavity** that contains the heart and a **pleuroperitoneal cavity** that houses most other viscera (figure 5.35a–c). The **transverse septum** is a complicated fibrous partition that separates these two compartments of the coelom. Large embryonic veins pass through this septum as they return to the heart. These veins eventually make contact with the **hepatic diverticulum** from the gut, which is destined to become the liver. As the hepatic diverticulum grows into the mesenchymal core of the septum, it meets these large embryonic veins that subdivide into the vascular sinusoids of the liver. As growth continues, the liver bulges from the confines of the transverse septum. The septum's posterior wall becomes the **serosa** covering the liver and a constricted connection to the septum becomes the **coronary ligament.** In reptiles, the transverse septum generally lies oblique within the body rather than dorsoventrally. This results from its posterior shift to a position beneath the pleuroperitoneal cavity that is situated dorsally. Lungs reside in the cranial end of the pleuroperitoneal cavity, but they usually do not become housed separately in their own coelomic compartments.

However, in some reptilian groups, each lung is sequestered into a separate coelomic compartment, the **pleural cavity.** Pleural cavities form in crocodiles, turtles, and some lizards as well as in birds and mammals, although the developmental pattern is different in mammals from that in other groups. In reptiles that have a pleural cavity and

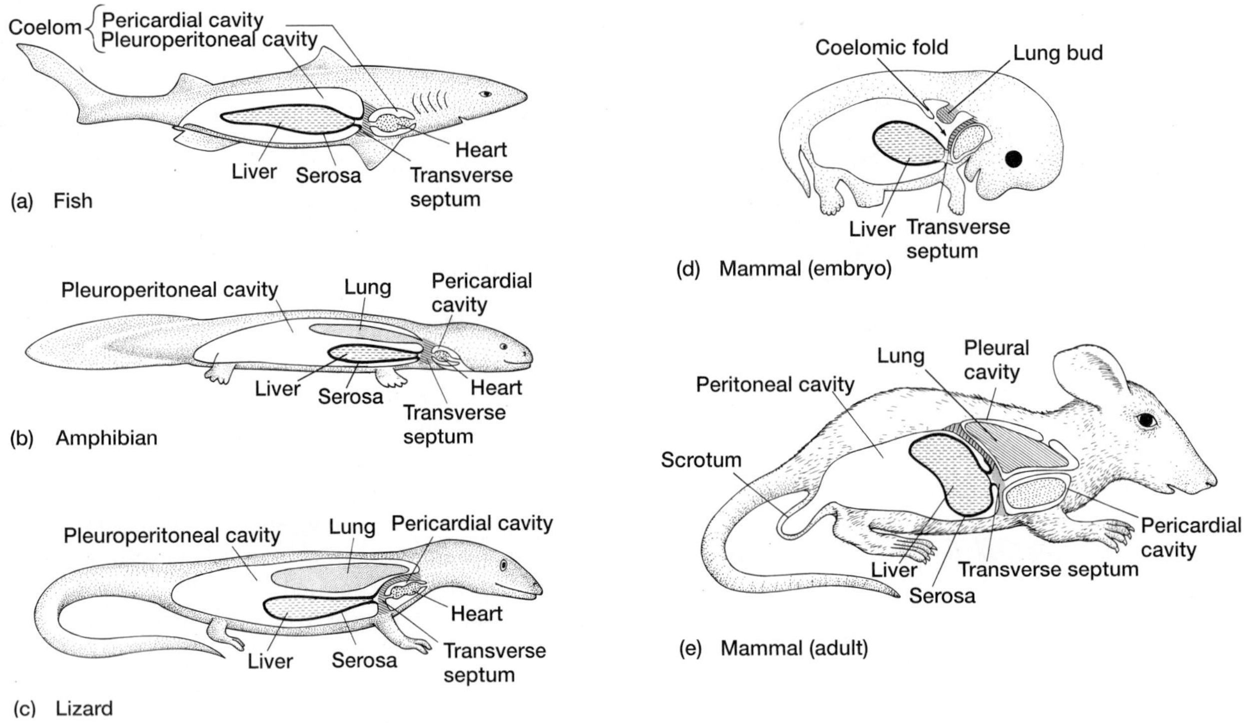


**Figure 5.35** Body cavities. The coelom, arising in the hypomere, becomes divided by a fibrous transverse septum into pericardial and pleuroperitoneal cavities in fishes (a), amphibians (b), and most reptiles (c). In embryonic mammals, a coelomic fold grows past the posterior face of the lung and makes contact with the transverse septum (d), thus separating the pleural cavity from the peritoneal cavity. This fold subsequently becomes invested with muscle primordia and together with the transverse septum becomes the muscular prehepatic diaphragm of the adult (e). In the males of some species, a posterior extension of the coelom through the body wall produces the scrotal pouch (scrotum) that receives the testes.

in all birds, the pleural cavity is cordoned off by a thin, nonmuscular, oblique septum known as the **pulmonary fold** (figure 5.36). This fold grows from the midline toward and into the serosa of the liver. Growth continues until the pulmonary fold joins the body wall. Thus, the pulmonary fold partly suspends the liver and sequesters each lung into its own pleural cavity.

In mammals, a **coelomic fold (pleuroperitoneal membrane)** originating in the dorsal body wall grows ventrally eventually to meet and fuse with the transverse septum. This fusion confines each lung in its own pleural cavity. The coelomic fold becomes a muscularized **diaphragm** so that its contractions directly influence lung ventilation after hatching or birth (figure 5.35d,e). Muscularization of the diaphragm is complex. Some muscle cells populating the diaphragm arise in thoracic myotomes in the adjacent body wall. However, most muscle cells arise in cervical myotomes far anterior to the diaphragm. These cervical muscle primordia migrate posteriorly and enter the margins of the dorsal coelomic fold where they differentiate into striated muscles. The ventral transverse septum remains relatively unmuscularized and forms the **central tendon** of the dome-shaped diaphragm. The **phrenic nerve,** a collection of several nerves, develops in the neck region adjacent to the cervical myotomes. As these myotomes migrate posteriorly into the septum, the phrenic nerve accompanies them serving to innervate the diaphragm. The diaphragm's position anterior to the liver makes it a **prehepatic diaphragm** (figure 5.35d). Only mammals have such a prehepatic diaphragm; however, many vertebrates possess an analogous sheet or sheets of striated muscle located posterior to the liver. These sheets function in lung ventilation and are called **posthepatic diaphragms.** In crocodiles, for instance, **diaphragmatic muscles** function collectively as a posthepatic diaphragm to pull the liver posteriorly, using it as a plunger to help inflate the lungs.

**Vertebrate diaphragms and lung ventilation (p. 421)**

The coelom and its subdivisions are lined by thin cellular sheets of mesothelium called **mesenteries.** The mesenteries secure the integrity of cavities, define spaces in which active organs more freely operate, and help sequester organs with conflicting activities. For example, the pericardial cavity separates the heart from other viscera to allow the transient buildup of favorable pressure around this organ at critical stages in its pumping cycle so that its chambers can be refilled. The pleuroperitoneal cavity accommodates the intestine through which peristaltic waves move food during digestion. The cavity gives the intestine freedom of movement during digestive episodes, yet the digestive tract activity remains controlled by the mesenteries that suspend it. Division of the coelom into compartments also allows more

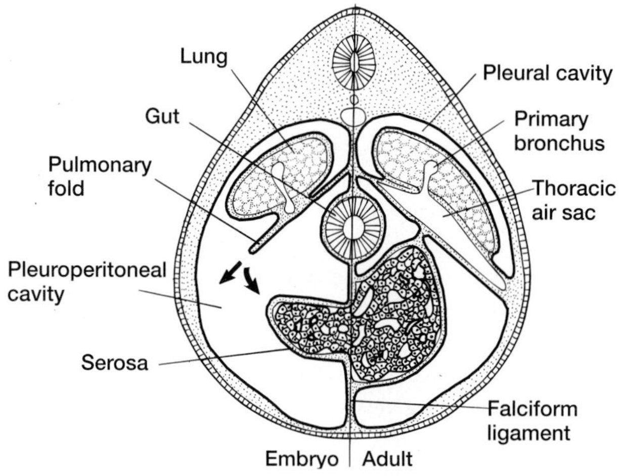


**Figure 5.36** Avian body cavities. Cross section of a bird illustrating the embryonic (left) and adult (right) cavities. In the embryo, the pulmonary fold grows obliquely to establish contact with the liver and the body wall. This confines the lung to its pleural cavity.

localized control of internal organs. For example, within the pleural cavities, lungs are placed directly under the control of muscles that ventilate them. Some mammals possess a scrotum, a coelomic pouch that protrudes outside the body cavity and offers a cooler environment favorable to sperm production and storage (figure 5.35e).

# Maturation

## Metamorphosis

As the events of early development come to a close, the embryo takes shape. If this emerging individual is fundamentally unlike the adult, it is termed a larva and will eventually undergo metamorphosis, a radical and abrupt change in structure to become an adult. Even in vertebrates lacking a distinct metamorphosis, the newborn still undergoes a period of maturation during which it develops from a juvenile to an adult. Strictly speaking, the overall process of **ontogeny** (development) is ongoing throughout the life of the individual and does not end at hatching or birth.

It is not uncommon for larva and adult, or juvenile and adult, to live different lives often in quite different environments. Among marine chordates such as tunicates, larvae are unattached and mobile or freely carried by currents to new locations. Such larvae are dispersal stages. Less restricted than sessile, bottom-bound adults, the tunicate larvae select the specific location that will be their permanent residence as adults. The adult tunicate is a feeding and reproductive stage. In frogs, the young larva, or tadpole, is typically a feeding stage through which the individual takes advantage of fleeting resources in a drying puddle or seasonal algal bloom. The sexually mature adult stage is less confined to bodies of water. If larva and adult live in different environments, they necessarily will have dissimilar designs.

If conditions experienced by the larva are more hospitable than those endured by the adult, the balance of time an individual spends as a larva compared with its stretch as an adult might change adaptively as well. For example, in some species of lampreys, the individual may persist in larval form for several years, metamorphosing into the brief adult form only long enough for a few weeks of breeding before dying. The sole function of the adult is reproduction (figure 5.37a).

In some species of salamanders, the adult form fails to appear during the life cycle. Instead, the larval form becomes sexually mature and breeds. In lowland populations of the northwestern salamander, *Ambystoma gracile*, individuals remain as aquatic larvae for several years and then metamorphose into sexually mature terrestrial adults that breed. In high montane populations of this same species, many individuals stop short of metamorphosis (figure 5.37b). Their larval forms become sexually mature and breed. For these montane individuals, foregoing of metamorphosis means that they avoid becoming a terrestrial form exposed to harsh alpine winters. By remaining larval, they retain their aquatic lifestyle in which they can safely overwinter in the unfrozen depths of ponds. Theoretically, the transformed adult could scamper safely back into ponds with the onset of winter, but the larval form already possesses external gills and feeding jaws that are better suited to pond life.

## Heterochrony

In some salamanders, mature gonads develop before metamorphosis; in others, gonads mature only after metamorphosis (figure 5.37b). The term **heterochrony** describes an evolutionary change in the relative time that a feature, such as gonad maturation, appears in a species' development compared with the time that same feature appeared in the species' ancestors. Sometimes the embryonic appearance of a feature is accelerated relative to other parts so that it appears earlier in descendants than in their ancestors. If, compared to other parts, it is retarded in appearance, the feature then arises relatively later in descendants than in their ancestors. Changes in relative rates of development have profound effects on the design of an organism. The morphological result of heterochrony is **paedomorphosis** (meaning child and form) if embryonic (or juvenile) characteristics of ancestors appear in adult stages of descendants. The reverse morphological result, in which adult characteristics of ancestors appear in larval (or juvenile) stages of descendants, is **recapitulation.**

The term *recapitulation* has been used in various ways by different scientists, and it has been the center of debate over the "biogenetic law." We will examine recapitulation and this law in a larger context following the discussion on paedomorphism. Although paedomorphosis has been a contentious term as well, it is less burdened with historical connotations. We look at its meaning first.

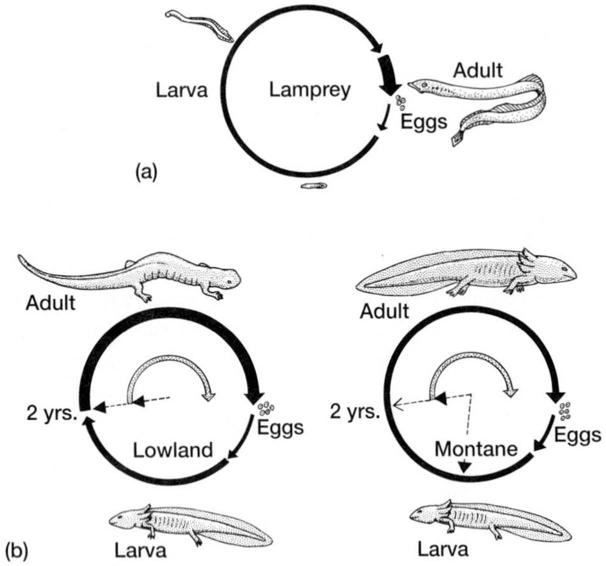


**Figure 5.37** Heterochrony. (a) Life cycle of a lamprey. In many species, the larval stage lasts the longest, perhaps several years, and metamorphosis produces an adult that breeds during a brief few weeks and then dies. (b) The northwestern salamander, *Ambystoma gracile*, ranges from lowlands to high mountainous regions of the Pacific Northwest. The *Ambystoma* larva is primarily aquatic, the adult more terrestrial. In lowland populations, the larva undergoes metamorphosis at about two years of age, becomes an adult, and reproduces. In montane populations, this species usually does not metamorphose, although individuals become sexually mature and reproduce at about two years of age. Thus, in montane populations, two-year-old individuals are anatomically larval in appearance and habits, but they are capable of reproduction. Paedomorphosis describes an individual larval in anatomy, but one that is sexually mature. Neoteny is a special case of paedomorphosis wherein sexual maturity occurs but somatic development slows, allowing juvenile characteristics to persist. Like hands on a clock, both sexual maturity (small hand) and somatic development (large hand) arrive during metamorphosis at about the same time (two years of age) in lowland populations of the northwestern salamander. In montane populations, somatic development lags and metamorphosis does not occur, but sexual maturity arrives at more or less the right time. This is indicated by a slowing of the big hand of the clock (somatic development) relative to the small hand (reproductive development). These are neotenic forms—adults (sexually mature) but in juvenile garb (anatomically larval). The outer series of arrows (solid black) follow somatic changes. The inner arrow (light) shows the onset and extent of sexual maturity during a salamander's life history.

### *Paedomorphism*

Adults are paedomorphic if they resemble juveniles of their ancestors. Stated slightly differently, paedomorphosis results when the larval form becomes reproductively mature. Adaptively, it represents a trade-off between the advantages or disadvantages of larval versus adult morphologies. From an embryological standpoint, paedomorphosis is based on differences in the timing between somatic development and the onset of reproductive maturity, a kind of developmental race between the two. If sexual maturity is accelerated relative to somatic development, paedomorphosis is called **progenesis.**

Amphibians and some insects undergo progenesis. During early development in members of the tropical salamander genus *Bolitoglossa,* hands and feet are webbed and paddlelike. Only late in their development do distinct digits finally become delineated. Unlike other members of this tropical genus, *Bolitoglossa occidentalis* lives in trees. It has webbed feet and a small body, both adaptations to arboreal life. The flat paddlelike feet give it grip on slippery leaves, and the small body reduces the risks of gravity's downward pull. Because growth ceases at a still small juvenile size, a small body results. As a consequence of this early cessation of development, other developmental processes in *B. occidentalis* are also arrested early. Limb development stops before digits become delineated, leaving the animal with webbed, paddle-shaped hands and feet. Other stunted characteristics occur as well, again truncated by the early cessation of development. Not all changes correlated with small body size necessarily have adaptive significance, however; but small size and webbed feet seem to have overriding advantages. Sexual maturity in *B. occidentalis,* when compared with other closely related species, is advanced so it occurs earlier relative to somatic development, giving us an example of paedomorphosis that results from progenesis.

On the other hand, if somatic development slows and is overtaken by normal sexual maturity, paedomorphosis occurs through **neoteny,** the more common method of paedomorphosis in vertebrates (figure 5.37b). The mudpuppy *Necturus maculosus* is permanently neotenic. It lives on the bottoms of lakes and retains its gills throughout life. However, populations of the tiger salamanders *Ambystoma tigrinum* exhibit neoteny in response to immediate environmental conditions. In western North America, some populations are neotenic and reproduce as aquatic, gill-breathing forms; others lose their gills, develop lungs, and metamorphose into sexually mature adults. As mentioned previously, some populations of the northwestern salamander also exhibit neoteny (figure 5.37b).

Each stage in ontogeny is adaptive in its own right. To be a successful adult, the individual must first be a successful infant or juvenile. Larval and juvenile characteristics function not just as predecessors to adult structures to come, but most serve the individual in the environment it currently occupies. The entire ontogeny of an individual is the sum total of adaptive responses to different environments and selective pressures during its entire lifetime. Change in emphasis between larval and adult morphologies reflects this adaptive change in the time an individual spends within each stage of its life history.

## Ontogeny and Phylogeny

### Biogenetic Law

It has long been supposed that ontogeny, especially early events of embryonic development, retains current clues to distant evolutionary events. Ernst Haeckel, a nineteenth century German biologist, stated this boldly in what became known as the **biogenetic law.** Pharyngeal slits, numerous branchial arches, and other fish characteristics even appear in the early embryos of reptiles, birds, and mammals, but they are lost as these tetrapod embryos proceed to term (figure 5.38). Although lost as tetrapod development unfolds, these and many similar structures are remnants of fish features from the evolutionary past. Haeckel argued that from ovum to complete body, the individual passes through a series of developmental stages that are brief, condensed repetitions of stages through which its successive ancestors evolved. The biogenetic law states that ontogeny in abbreviated form recapitulates (repeats) phylogeny.

Haeckel certainly recognized that recapitulation was approximate. Comparing it to an alphabet, he suggested that the ancestry behind each organism might be a sequence of stages: A, B, C, D, E, . . . Z, whereas the embryology of a descendant individual might pass through an apparently defective series: A, B, D, F, H, K, M, etc. In this example, several evolutionary stages have fallen out of the developmental series. Although the ancestry of an organism might include an entire series of steps, Haeckel did not believe that all these would necessarily appear in the ontogeny of a later individual. Evolutionary stages could disappear from the developmental series. Nevertheless, he felt that the basic series of major ancestral stages remained the same, and thus the biogenetic law applied.

Development certainly exhibits a conservatism wherein ancient features persist like heirlooms in modern groups. Ontogeny, however, is not so literally a repeat of

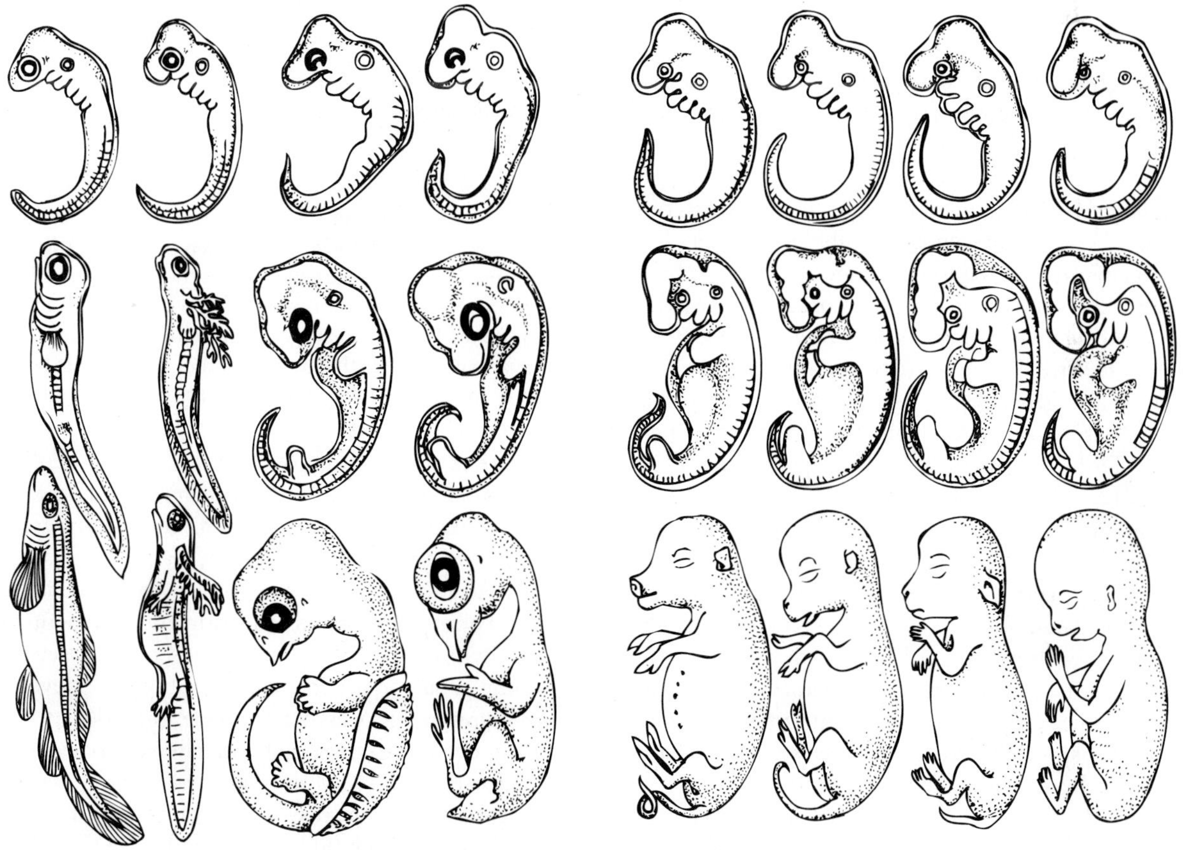

**Figure 5.38** Haeckel's comparison of early embryonic stages across vertebrate groups. Eight species are shown across the figure—fish, salamander, tortoise, chick, pig, calf, rabbit, and human. The youngest developmental stage of each is at the top of the figure followed by two successively older stages below.

phylogeny as Haeckel supposed. A contemporary of his, Karl Ernst von Baer, cited examples from embryos of descendant animals that did not conform to the biogenetic law—chick embryos lack the scales, swim bladders, fin rays, and so forth of adult fishes that evolutionarily preceded them. Furthermore, the order of appearance of ancestral structures is sometimes altered in descendant embryos. Haeckel allowed for exceptions, von Baer did not. Von Baer said that these exceptions and "thousands" more were too much. He proposed alternative laws of development. Foremost was von Baer's proposal that development proceeds from the general to the specific. Development begins with undifferentiated cells of the blastula that become germ layers, then tissues, and finally organs. Young embryos are undifferentiated (general), but as development proceeds, distinguishing features (specific) of the species appear. Each embryo, instead of passing through stages of distant ancestors, departs more and more from them. Thus, the *embryo* of a descendant is never like the *adult* of an ancestor and only generally like the ancestral embryo. Other scientists since von Baer have also dissented from strict application of the biogenetic law. What can be made of all this?

We should recognize that embryos are adapted to their environments. Anamniotes usually make their homes in salt- or fresh water. Amniote embryos grow in an aquatic environment as well, the amniotic fluid. Thus, similarities between embryos might be expected. But, as von Baer pointed out, what we observe at best is correspondence between the *embryos* of descendants and the *embryos* of their ancestors, not between the descendant *embryos* and the ancestral *adults*. Haeckel was certainly mistaken to see a correspondence between descendant embryos and ancestor adults. Organs are adapted to their functions. As a fish embryo approaches hatching, its "limb" buds become fins, a bird's become wings, a mammal's become paws or hooves or hands, and so forth. There is, however, an element of conservatism in ontogeny, even if it is not an exact telescoping of evolutionary events. After all, the young embryos of mammals, birds, and reptiles do develop pharyngeal slits that never become functional as respiratory devices. Is this recapitulation? No. It is better to think of this as preservationism for reasons not too difficult to imagine.

Each adult part is the developmental product of prior embryonic preparation. The zygote divides to form the blastula; gastrulation brings germ layers to their proper positions; mesoderm interacts with endoderm to form organ rudiments; tissues within organ rudiments differentiate into adult organs. Skip a step, and the whole cascade of ensuing developmental events may fail to unfold properly.

In mammals, the notochord of the embryo is replaced almost entirely in the adult by the solid vertebral column (figure 5.39). For the young embryo, the notochord provides an initial axis, a scaffolding along which the delicate body of the embryo is laid out. The notochord also stimulates development of the overlying nerve tube. If the notochord is removed, the nervous system does not develop. The adult

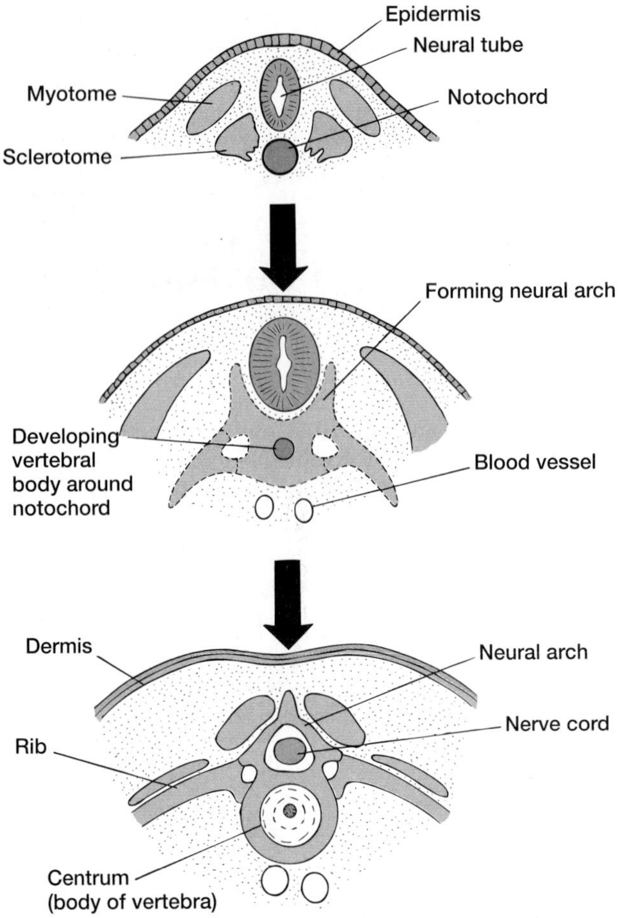


**Figure 5.39** Vertebrae replace the notochord in mammalian embryos. The sclerotomes are segmented clusters of cells that gather around the notochord and differentiate into the segmentally arranged vertebrae, collectively known as the vertebral column. Vertebrae protect the nerve cord and provide sites of attachment for muscles. The vertebral column functionally replaces the notochord that persists only as a small core of the intervertebral disks between successive vertebrae.

supportive role is taken over by the vertebral column, but the notochord performs a vital *embryonic* role before disappearing; namely it serves the young embryo as a central element of embryonic organization.

A notochord that persists in the mammalian embryo should not be interpreted as a sentimental momento of a distant phylogenetic history. Instead, it should be seen as a functioning component of early embryonic development. Structures and processes intertwine to produce the conservatism evident in development. They are not easily eliminated without a broad disruption of ensuing events. Anatomical innovations, new structures brought into service in the adult, are usually added at the end of developmental processes, not at the beginning. A new structure inserted early into the developmental process would require many simultaneous replacements of many disrupted developmental processes thereafter. Evolutionary innovations thus usually arise by remodeling rather than by entirely new construction.

The forelimbs of ancestors that supported the body and allowed the organism to romp over the surface of the land are renovated into the wings that carry bats and birds aloft. We need look no further than our own human bodies to find similar examples of evolutionary remodeling. The backbone and legs that carried our distant ancestors comfortably on all fours hold us upright in a bipedal stance. The arms and hands that can control the delicate strokes of a paintbrush or the writing of a novelist come refashioned from ancient forelegs that carried a hefty trunk and helped our ancestors dash from predators. The past is hard to erase. When parts are already available, renovation is easier than new construction.

## Epigenetics

Genes carry characteristics from generation to generation down through the ages. They also synthesize the material from which the organism is constructed, or so the story goes. But this is too simple a view. Actually genes make only varieties of RNA. Thereafter, RNA varieties assemble different amino acids into proteins. Proteins construct parts of cells that cooperate in making whole cells that join to form tissues, and so on until an organism is finally assembled. As these events move further from the genes, the genes have less and less direct hold on how the organism is eventually formed. The mutual associations established among cells and tissues play a large role in the eventual developmental outcome. These events are **epigenetic,** literally above the genes. Each level of organization—proteins, cells, tissues, organs, etc.—comes under the jurisdiction of additional constraints by which further development proceeds. An example might help.

### *Induction*

During early development, the predecessors to the chain of vertebrae composing the vertebral column appear as a series of paired blocks or segments of tissue, the sclerotomes, nestled along either side of the neural tube (figure 5.39). If development proceeds normally, the sclerotomes give rise to cartilages that eventually ossify into vertebrae, and the neural tube gives rise to the spinal cord. If a section of the neural tube is removed experimentally at this early stage, then of course the affected stretch of the spinal cord fails to develop. Surprisingly, however, the adjacent vertebral column fails as well, even though the sclerotomes are not directly affected. That is because the neural tube, in addition to providing the foundation for the spinal cord, also stimulates proper development of the neighboring sclerotomes. This stimulatory effect between developing parts of the embryo is known as **induction.** Developmental events are coupled and locked into step with each other. In the adult, the vertebral column comes to protect the spinal cord by surrounding it. Nerves that reach out from the cord squeeze between successive vertebrae. To create well-fit structures in the adult, nerves and vertebrae must match and grow together. Induction between neural tube and sclerotomes ensures that they are paced with each other so that neither races ahead prematurely. Not genes but tissue interactions are the most immediate developmental events to promote and shape the outcome.

Between neural tube and sclerotomes, induction is a one-way street—neural tube to sclerotomes. The reverse experiment, removal of the sclerotome, leads to little interruption in the growth of the neural tube. Reciprocal induction between tissues is common, however. The growing embryonic tetrapod limb provides an example. Two pairs of limb buds sprout along the sides of the body, being the first evidence of the future fore- and hindlimbs. As each limb bud lengthens like a sprouting branch, proximal, middle, and distal parts take shape in that order. Within the early limb bud, there is a recognizable mesodermal core and a surface thickening of ectoderm at the tip, the **apical ectodermal ridge (AER).** Both mesoderm and AER must interact to produce limb development. If the AER is removed, limb development ceases immediately. AER promotes outward growth of the limb bud. The core of mesoderm determines whether the limb produced is a forelimb or a hindlimb. Exchange of mesodermal cores between anterior and posterior limbs in birds results in a reversal of the arrangement of wings and legs. The AER stimulates growth of the mesoderm, but it in turn is maintained by the underlying core of mesoderm.

### *Phylogeny*

The tight coupling of AER and mesoderm arises from interactions between the tissues themselves rather than from distant dictates of genes. The course of developmental events arises predominantly out of these mutual inductions between tissues. A tiny alteration of one tissue can have a profound effect on the adult structures produced. Such epigenetic interactions have been central not only in development but in evolution. Legless lizards exemplify this.

Most lizards possess four legs used to great advantage. But in some species in which habitats favor sleakness, legless forms have evolved. These forms use their whole body like snakes to slip through crowded terrain. As in other vertebrates, limbs of lizards also grow as lateral body sprouts along the sides of the young embryo. Additionally, nearby somites, clumps of mesoderm, grow downward to contribute cells to the core of mesoderm in the limb bud and establish an interaction with the AER (figure 5.40a). In limbless lizards, these somites fail to grow downward completely, AER regresses, and limbs do not appear (figure 5.40b). In these specialized lizards, a major adaptive change to limblessness has occurred by simple modification of an early developmental pattern. In this case, the evolution of limblessness did not require the accumulation of hundreds of mutations each closing down one tiny anatomical part of the limb, one for the thumb, one for the second finger, and

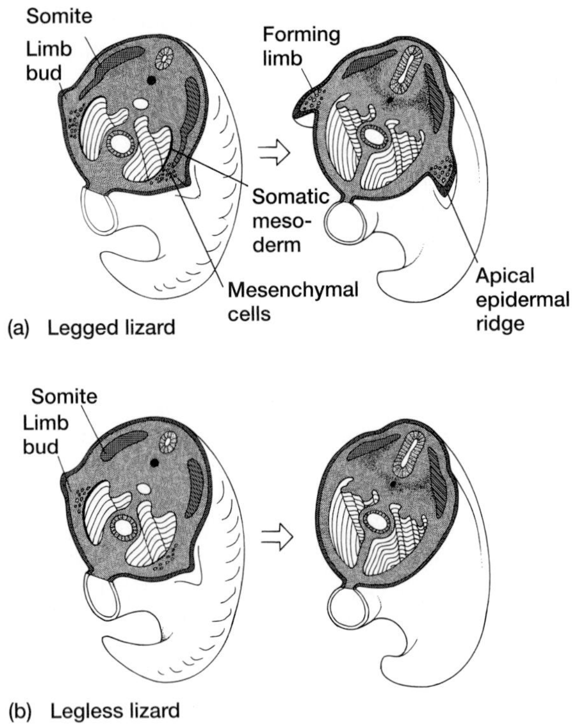


**Figure 5.40** Limb formation in lizards. Cross sections through the posterior end of the embryo are depicted. (a) Mesenchymal cells normally depart from the somatic mesoderm, enter the forming limb bud, and become the core of the growing limb. Ventral processes from the local somites reach into this area of migrating mesenchymal cells. (b) In legless lizards, early rudimentary limb buds form, but somites do not grow into this vicinity. This apparently denies an inductive influence over events. The apical epidermal ridge regresses, the limb bud recedes, and no limbs develop.

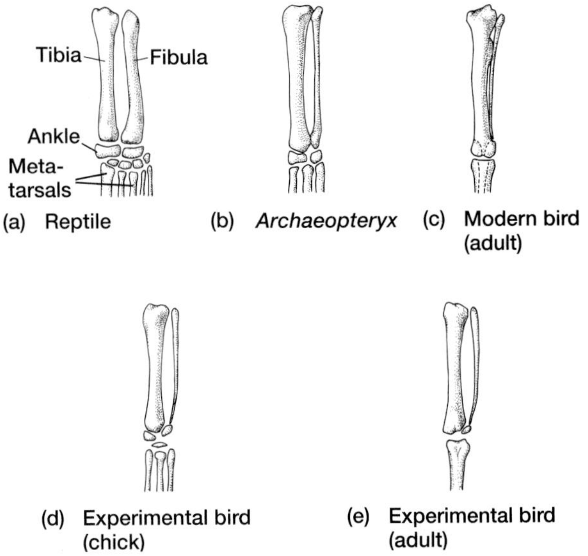


**Figure 5.41** Hindlimbs of a reptile (a), a primitive bird (*Archaeopteryx*) (b), and a modern bird (c,d). The several ankle bones of reptiles and the pair of shank bones, fibula and tibia, are reduced through this series to modern birds. Although Müller did nothing to alter the genotype of experimental chicks, his mechanical barrier separating differentiating regions produced an embryonic (d) and adult (e) hindlimb similar to *Archaeopteryx* and especially similar to reptile limbs (a). Apparently, the underlying developmental program in modern birds was altered very little in the course of evolution. Müller was able to recreate much of the ancestral condition in the foot by making only modest changes in the developmental pattern in modern birds.

so on. Rather, a few changes in limb bud growth during early developmental stage apparently gave rise to the limbless condition that found adaptive favor in the specialized habitat frequented by these lizards.

A similar alteration of a developmental pattern seems to have been the basis for evolution of the specialized foot of modern birds. In reptiles, the tibia and fibula bones of the lower leg are about equal in length and articulate with several small bones of the ankle (figure 5.41a). In *Archaeopteryx*, this feature began to change. Although both tibia and fibula are equal in length, the ankle bones were reduced to two in *Archaeopteryx* (figure 5.41b). In modern birds, the fibula is short and tiny, but the tibia has enlarged to engulf the two ankle bones and form a single composite bone (figure 5.41c).

In an attempt to clarify the evolution of the bird leg, embryologist Armand Hampé performed experiments in which he either separated tibia and fibula or provided additional mesenchyme to the ankle during early development of the limb bud. The limb produced in both cases bore a remarkable resemblance to the limb of *Archaeopteryx*. Tibia and fibula were of equal length; separate ankle bones were again present (figure 5.41d). These experiments were extended by Gerd Müller who used inert barriers inserted into early chick hindlimbs to separate regions differentiating into tibia and fibula. The resulting limb suggested to Müller similarities to reptile limbs in that tibia and fibula were of equal length and were not closely adjacent (figure 5.41e). Further, the musculature of the experimental chick hindlimb reverted to a characteristic reptilian pattern of insertion. Such experimental manipulations could not have affected the genome because only the developmental pattern was altered. Could it be that Hampeé and Müller had experimentally run evolution in reverse and discovered the simple method by which profound changes were initially brought about in birds? In ancient archosaurs, a few mutations affecting the supply of cells to or interaction with the fibula could have had a cascading effect on ankle development and resulted in an extensively altered adult design.

It is tempting to interpret other specialized structures in a similar light. Among modern horses, only a single toe (the middle or third toe) persists on each leg to form the functional digit (figure 5.42a). However, ancestral horses, such as *Hyracotherium*, had four toes on the front foot and three on the back. Occasionally, modern horses develop vestiges of these old second and fourth toes (figure 5.42b–d). When this occurs,

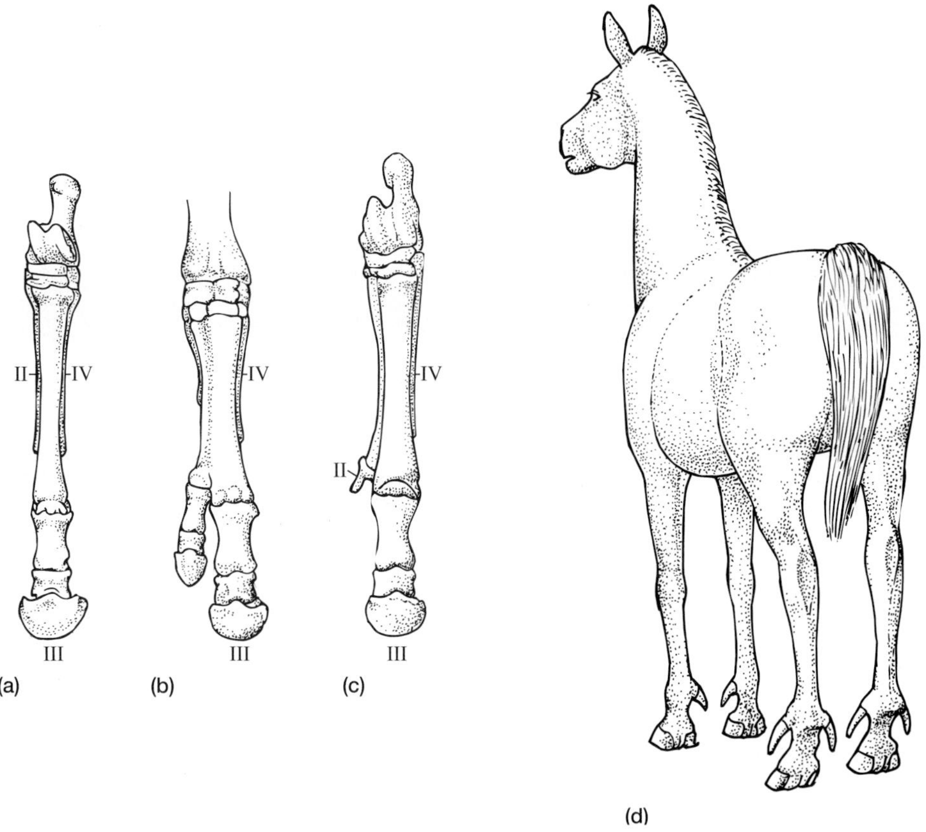


**Figure 5.42** Extra toes in modern horses. (a) Modern horses have only one enlarged digit on each foot, a single toe. The one toe evolved from ancestors with three or four toes. During the course of their evolution, the peripheral toes IV, II, and I were lost and the central toe (III) emphasized. (b,c) On rare occasions, however, these "lost" toes or their remnants reappear, testifying to the lingering presence of the underlying ancestral developmental pattern. (d) On rare occasions, modern horses, such as the one illustrated, exhibit additional toes. Such toe remnants in modern horses apparently represent the partial reemergence of an ancient ancestral pattern.

For more information on extra toes in modern horses, see Gould, S. J. 1983. *Hen's teeth and horse's toes. Further reflections in natural history.* New York: W.W. Norton.

we get a glimpse of the underlying developmental pattern that produces the foot. The reduction of toes in horses enjoyed adaptive favor because it contributed to locomotor performance. Literally hundreds of structural changes occurred in bones, muscles, ligaments, nerves, and blood vessels from the four-(or five-) toed ancestors to the single-toed modern horses. If this, like the evolution of limblessness in some lizards, was based on a narrowing of the underlying developmental pattern, then these hundreds of changes could be accomplished with relatively few gene mutations.

Experimental tests can be undertaken to examine the mechanisms behind the evolutionary alterations of developmental patterns. The skeletons of living lampreys lack bone and are composed principally of uncalcified cartilage and dense connective tissue. But the ability to produce bone represents an ancestral condition found among ostracoderms. In lampreys, therefore, absence of bone represents a secondary loss. The uncalcified skeleton of lampreys is a paedomorphic condition wherein development no longer reaches the point at which calcified tissues debut. If the chondrocytes are removed from the lamprey skeleton and grown in vitro under conditions hospitable for mineralization, the chondrocytes do produce calcified cartilage. This experimental result illustrates that lamprey chondrocytes retain the ancient ability to form a mineralized skeleton even though they do not do so in modern lampreys. Loss of bone in lampreys may confer advantages in economy of material or in producing a supple body suited to their specialized environment. The evolutionary mechanism, however, is based on modification of the embryonic process of bone formation. In lampreys, evolution of an unossified skeleton is achieved by suppression of this ancient developmental potential at a point before a mineralized skeleton is produced.

Evolutionary alteration of developmental patterns offers a simple way to produce profound anatomical changes. But we must keep in mind that we see only the successes in retrospect, not the failures. If appropriate gene mutations fail to appear at a timely moment, there is nothing the organism can do to summon them up to produce a desired part. Needs do not usher in the desired genetic improvements. For horses, birds, legless lizards, and lampreys, the fortuitous but timely appearance of new genes affecting developmental patterns produced renovated adult structures that found adaptive favor at the time. For those that have succeeded and persisted, many have failed and perished.

These patterns of development and evolution teach that no part is an island. All parts are linked and integrated with the rest of the organism. Consequently, there is no one-to-one correspondence between genes and body parts. Some parts are affected by many genes. Evolution does not proceed gene by gene, each bringing a tiny change that over millions of years eventually add up to a new structure. With its influence dispersed, one small genetic change can produce large, integrated structural modifications that are the basis for major and rapid evolutionary changes in design.

## Box Essay 5.2 Panda's Thumb

The giant panda from the forests of China is related to bears. Unlike bears, which are omnivores and eat almost anything, the giant panda feeds almost exclusively on bamboo shoots for about 15 hours per day. Pandas strip leaves by passing the stalks between their thumb and adjacent fingers. In addition to diet, pandas are unique among bears in apparently possessing six digits on the forelimb instead of the customary five. The extra digit is the "thumb," which is actually not a thumb at all but an elongated wrist bone controlled by muscles that work it against the other five digits to strip leaves from bamboo stalks. The true thumb is committed to another function, so it is unavailable to act in opposition to the other fingers. The radial sesamoid bone of the wrist has become remodeled and pressed into service as the effective "thumb" (box figure 1).

The availability of parts narrows or enlarges evolutionary opportunities. Had the wrist bone been locked irretrievably into another function, as the original thumb was, the doors for evolution of bamboo feeding might have been closed and this endearing bear might never have evolved (see also Davis, 1964 and Gould, 1980).

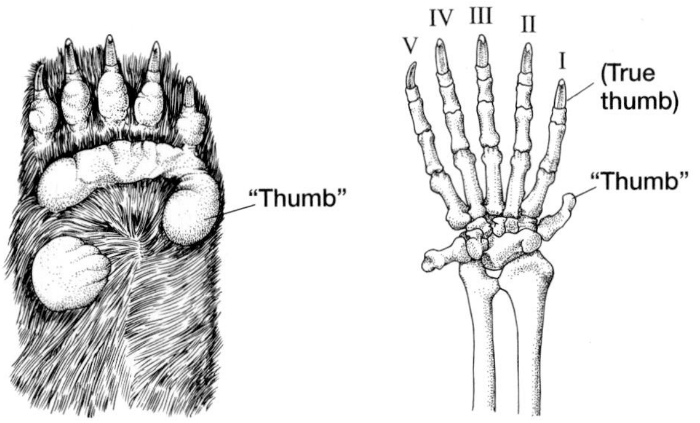


**Box figure 1** The panda's thumb. The panda has five digits like most mammals; however, opposing these is another digit, a "thumb," which is actually not a thumb at all but an elaborated wrist bone.

# Selected References

Arey, L. B. 1974. *Developmental anatomy*. 7th ed. Philadelphia: W. B. Saunders.

Balinsky, B. I. 1981. *An introduction to embryology*. 5th ed. Philadelphia: W. B. Saunders.

Blackburn, D. G. 1982. Evolutionary origins of viviparity in the Reptilia. I. Sauria. *Amphibia-Reptilia* 3:185–205.

Blackburn, D. G. 1985. Evolutionary origins of viviparity in the Reptilia. II. Serpentes, Amphisbaenia, and Ichthyosauria. *Amphibia-Reptilia* 5:259–91.

Blackburn, D. G. 1992. Convergent evolution of viviparity, matrotrophy, and specializations for fetal nutrition in reptiles and other vertebrates. *Amer. Zool.* 32:313–21.

Blackburn, D. G., J. M. Taylor, and H. A. Padykula. 1988. Trophoblast concept as applied to therian mammals. *J. Morph.* 196:127–36.

Blackburn, D. G., L. J. Vitt, and C. A. Beuchat. 1984. Eutherian-like reproductive specializations in a viviparous reptile. *Proc. Natl. Acad. Sci. USA* 81:4860–63.

Bloom, W., and D. Fawcett. 1975. *Textbook of histology*. 10th ed. Philadelphia: W. B. Saunders.

Bockman, D. E., M. E. Redmond, K. Waldo, H. Davis, and M. L. Kirby. 1987. Effect of neural crest ablation on development of the heart and arch arteries in the chick. *Amer. J. Anat.* 180:332–41.

Caplan, A. I. 1984. Cartilage. *Sci. Amer.* 252(10):84–94.

Cowin, S. C., R. T. Hart, J. R. Balser, and D. H. Kohn. 1985. Functional adaptation in long bones: Establishing *in vivo* values for surface remodeling rate coefficients. *J. Biomech.* 18:665–84.

Davis, D. D. 1964. The giant panda: A study in evolutionary mechanisms. Fieldiana. *Zool. Mem.* 3:1–339.

DeBeer, G. R. 1958. *Embryos and ancestors*. 3d ed. London: Oxford University Press.

Ekanayake, S., and B. K. Hall. 1987. The development of acellularity of the vertebral bone of the Japanese medaka, *Oryzias latipes* (Teleostei: Cyprinidontidae). *J. Morph.* 193:253–61.

Frost, H. M. 1990. Skeletal structural adaptations to mechanical usage (SATMU): 1. Redefining Wolff's Law: The bond modeling problem. *Anat. Rec.* 226:403–13.

Gould, S. J. 1977. *Ontogeny and phylogeny*. Cambridge, Mass.: Harvard University Press.

Gould, S. J. 1980. *The panda's thumb. More reflections in natural history*. New York: W.W. Norton.

Gould, S. J. 1983. *Hen's teeth and horse's toes. Further reflections in natural history*. New York: W.W. Norton.

Griffiths, M. 1978. *The biology of monotremes*. New York: Academic Press.

Hall, B. K. 1984a. Developmental processes underlying heterochrony as an evolutionary mechanism. *Can. J. Zool.* 62:1–7.

Hall, B. K. 1984b. Developmental processes underlying the evolution of cartilage and bone. *Symp. Zool. Soc. Lond.* 52:155–76.

Hall, B. K., and J. Hanken. 1985. Repair of fractured lower jaws in the spotted salamander: Do amphibians form secondary cartilage? *J. Exp. Zool.* 233:359–68.

Halstead, B., and J. Middleton. 1972. *Bare bones*. Edinburg: Oliver and Boyd.

Hirakow, R., and N. Kajita. 1990. An electron microscope study of the development of Amphioxus, *Branchiostoma belcheri tsingtauense:* Cleavage. *J. Morph.* 203:331–44.

Krause, W. J., and J. H. Cutts. 1983. Ultrastructural observation on the shell membrane of the North American opossum *(Didelphis virginiana)*. *Anat. Rec.* 207:335–38.

Langilee, R. M., and B. K. Halt. 1993. Calcification of cartilage from the lamprey *Petromyzon marinus* (L.) in vitro. *Acta Zool.* 74:31–41.

Lanyon, L. E. 1974. Experimental support for the trajectorial theory of bone structure. *J. Bone Joint Surg.* 56B:160–66.

Lanyon, L. E., and C. T. Rubin. 1985. Functional adaptation in skeletal structures. In *Functional vertebrate morphology,* edited by M. Hildebrand, D. M. Bramble, K. F. Liem, and D. B. Wake. Cambridge, Mass.: Harvard University Press, pp. 1–25.

Lillegraven, J. A. 1985. Use of the term "trophoblast" for tissues in therian mammals. *J. Morph.* 183:293–99.

Luckett, W. P. 1977. Ontogeny of amniote fetal membranes and their application to phylogeny. In *Major patterns in vertebrate evolution,* edited by M. K. Hecht, P. C. Goody, and B. M. Hecht. New York: Plenum Press, pp. 439–516.

Maderson, P.F.A. (ed). 1987. Developmental and evolutionary aspects of the neural crest. New York: John Wiley and Sons.

Moss, M. L. 1961. Osteogenesis of acellular teleost fish bone. *Amer. J. Anat.* 108:99–109.

Müller, G. B. 1989. Ancestral patterns in bird limb development: A new look at Hampé's experiment. *J. Evol. Biol.* 2:31–47.

Müller, G. B., and P. Alberch. 1990. Ontogeny of the limb skeleton in *Alligator mississippiensis:* Developmental invariance and change in the evolution of archosaur limbs. *J. Morph.* 203:151–64.

Nichols, D. H. 1986. Mesenchyme formation from the trigeminal placodes of the mouse embryo. *Amer. J. Anat.* 176:19–31.

Noden, D. M. 1983. The role of the neural crest in patterning of avian cranial skeletal, connective, and muscle tissues. *Dev. Biol.* 96:144–65.

Patten, B. M., and B. M. Carlson. 1981. *Foundations of embryology*. 4th ed. New York: McGraw-Hill.

Renfree, M. B. 1981. Marsupials: Alternative mammals. *Nature* 293:100–101.

Renfree, M. B. 1983. *Marsupial reproduction: The choice between placentation and lactation. Oxford review of reproductive biology*, vol. 5. Oxford: Clarendon Press, pp. 1–29.

Ruben, C. T., and L. E. Lanyon. 1987. Osteoregulatory nature of mechanical stimuli: Function as a determinant for adaptive remodeling in bone. *J. Orthop. Res.* 5:300–310.

Ruben, J. A. 1989. Activity physiology and evolution of the vertebrate skeleton. *Amer. Zool.* 29:195–203.

Ruben, J. A., and A. A. Bennett. 1987. The evolution of bone. *Evolution* 41:1187–97.

Sadegh, A. M., G. M. Luo, and S. C. Cowin. 1993. Bone ingrowth: An application of the boundary element method to bone remodeling at the implant interface. *J. Biomech.* 26:167–82.

Selwood, L., and J. J. Young. 1983. Cleavage *in vivo* and in culture in the Dasyurid marsupial, *Antechinus stuartii* (Macleay). *J. Morph.* 176:43–60.

Shine, R. 1985. The evolution of viviparity in reptiles: An ecological analysis. In *Biology of the Reptilia,* vol. 15, edited by C. Gans and F. Billett. New York: John Wiley and Sons, pp. 605–94.

Shubin, N. H., and P. Alberch. 1986. A morphogenetic approach to the origin and basic organization of the tetrapod limb. *Evol. Biol.* 20:313–87.

Stewart, J. R. 1985. Placentation in the lizard *Gerrhonotus coeruleus* with a comparison to the extraembryonic membranes of the oviparous *Gerrhonotus multicarinatus* (Sauria, Anguidae). *J. Morph.* 185:101–14.

Stewart, J. R. 1990. Development of the extraembryonic membranes and histology of the placentae in *Virginia striatula* (Squamata:Serpentes). *J. Morph.* 205:33–43.

Stewart, J. R. 1992. Placental structure and nutritional provision to embryos in predominately lecithotrophic viviparous reptiles. *Amer. Zool.* 32:303–12.

Stewart, J. R., and D. G. Blackburn. 1988. Reptilian placentation: Structural diversity and terminology. *Copeia* 1988:839–52.

Tyndale-Biscoe, H., and M. Renfree. 1987. Reproductive physiology of marsupials. Cambridge: Cambridge University Press.

Van Haarlem, R. 1983. Early ontogeny of the annual fish genus *Nothobranchius:* Cleavage plane orientation and epiboly. *J. Morph.* 176:31–42.

Wimsatt, W. A. 1974. Morphogenesis of the fetal membranes and placenta of the black bear, *Ursus americanus* (Pallas). *Amer. J. Anat.* 140:471–96.

Woo, S. L-Y., S. C. Kuei, D. Amiel, M. A. Gomex, W. C. Hayes, F. C. White, and W. H. Akeson. 1981. The effect of prolonged physical training on the properties of long bone: A study of Wolff's law. *J. Bone Joint Surg.* 63A:780–87.

Wourms, J. P., B. D. Grove, and J. Lombardi. 1988. The maternal-embryonic relationship in viviparous fishes. In *Fish physiology,* edited by W. S. Hoar and D. J. Randall. San Diego: Academic Press, pp. 1–134.

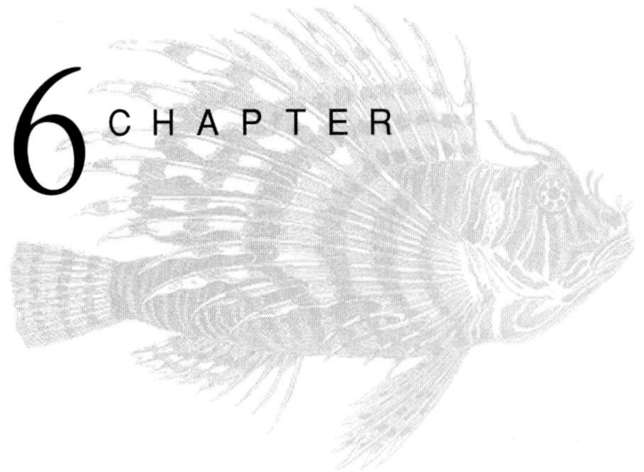

# 6 CHAPTER Integument



The integument (or skin) is a composite organ. On the surface is the **epidermis,** below it is the **dermis,** and between them lies the **basement membrane** (basal lamina and reticular lamina). The epidermis is derived from the ectoderm and produces the basal lamina (figure 6.1a). The dermis develops from mesoderm and mesenchyme and produces the reticular lamina. Between the integument and deep body musculature is a transitional subcutaneous region made up of very loose connective and adipose tissues. In microscopic examination this region is termed the **hypodermis.** In gross anatomical dissection, the hypodermis is referred to as the **superficial fascia** (figure 6.1b).

The integument is one of the largest organs of the body, making up some 15% of the human body weight. Epidermis and dermis together form some of the most varied structures found within vertebrates. The epidermis produces hair, feathers, baleen, claws, nails, horns, beaks, and some types of scales. The dermis gives rise to dermal bones and osteoderms of reptiles. Collectively epidermis and dermis form teeth, denticles, and scales of fish. In fact, the developmental destinies of dermis and epidermis are so closely linked across the basement membrane that in the absence of one, the other by itself is incapable of or inhibited from producing these specialized structures. In terms of embryonic development then, epidermis and dermis are tightly coupled and mutually necessary.

As the critical border between the organism and its environment, the integument has a variety of specialized functions. It forms part of the exoskeleton and thickens to resist mechanical injury. The barrier it establishes prevents the entrance of pathogens. The integument helps hold the shape of an organism as well. Osmotic regulation and movement of gases and ions to and from the circulation are aided by the integument in conjunction with other systems. Skin gathers needed heat or radiates the excess and houses sensory receptors. It holds feathers for locomotion, hair for insulation, and horns for defense. Skin pigments block harmful sunlight and display bright colors during courtship. The list of functions can easily be extended.

The remarkable variety of skin structures and roles makes it difficult to briefly summarize the forms and functions of the integument. Let us begin by examining the embryonic origin and development of the skin.

## Embryonic Origin

By the end of neurulation in the embryo, most skin precursors are delineated. The single-layered surface ectoderm proliferates to give rise to the multilayered epidermis. The deep layer of the epidermis, the **stratum basale (stratum germinativum),** rests upon the basement membrane. Through active cell division, the basement membrane re-

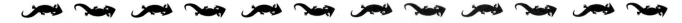

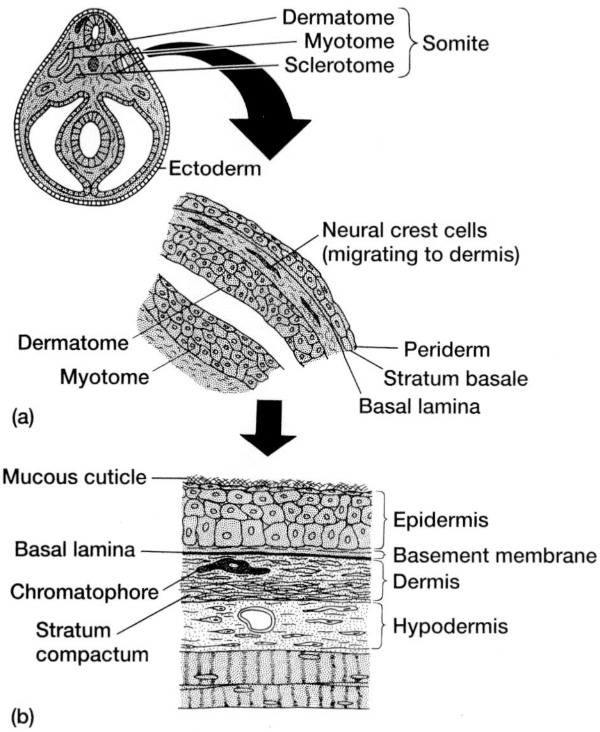


**Figure 6.1** Embryonic development of the skin. (a) Cross section of a representative vertebrate embryo. The ectoderm initially differentiates into a deep stratum basale, which replenishes the outer periderm. The dermatome settles in under the epidermis to differentiate into the connective tissue layer of dermis. As migrating neural crest cells pass between dermis and epidermis, some settle between these layers to become the chromatophores. (b) The epidermis further differentiates into a stratified layer that often has a mucous coat or cuticle on the surface. Within the dermis, collagen forms distinctive plies (layers) that constitute the stratum compactum. The basement membrane lies between the epidermis and the dermis. Beneath the dermis and the deeper layer of musculature is the hypodermis, a collection of loose connective and adipose tissues.

plenishes the single layer of outer cells called the **periderm** (figure 6.1a). Additional skin layers are derived from these two as differentiation proceeds.

The dermis arises from several sources, principally from the dermatome. The segmental **epimeres** (somites) divide, producing the **sclerotome** medially, the embryonic source of the vertebrae, and the **dermomyotome** laterally. Inner cells of the dermomyotome become rearranged into the **myotome,** the major source of skeletal muscle. The outer wall of the dermomyotome spreads out under the ectoderm as a more or less distinct **dermatome** that differentiates into the connective tissue component of the dermis. Connective tissue within the skin is usually diffuse and irregular, although in some species collagen bundles are arranged into a distinct, ordered layer within the dermis. This layer is called the **stratum compactum** (figure 6.1b). Cells of neural crest origin migrate into the region between dermis and epidermis, contributing to bony armor and to skin pigment cells called **chromatophores** (meaning color and bearing). Usually chromatophores reside in the dermis, although in some species they may send pseudopods into the epidermis or take up residence there themselves. Often, chromatophores are scattered within the hypodermis. Nerves and blood vessels invade the integument to round out its structural composition.

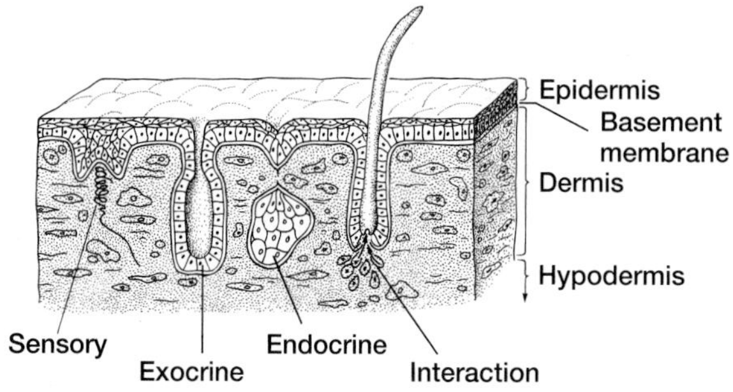


**Figure 6.2** Specializations of the integument. Sensory receptors reside in the skin. Exocrine glands with ducts and ductless endocrine glands form from invaginations of the epidermis. Through a dermal-epidermal interaction, specialized skin structures such as hair, feathers, and teeth arise.

Fundamentally, the integument is composed of two layers, epidermis and dermis, separated by the basement membrane. Vascularization and innervation are added, along with contributions from the neural crest. From such simple structural ingredients, a great variety of integumentary derivatives arise. The integument houses sensory organs that detect arriving stimuli from the external environment. Invagination of the surface epidermis forms skin glands, exocrine if they retain ducts and endocrine if they separate from the surface and release products directly into blood vessels (figure 6.2). Interaction between epidermis and dermis stimulates specializations such as teeth, feathers, hair, and scales of several varieties (figure 6.3a–i).

# General Features of the Integument

## Dermis

The dermis of many vertebrates produces plates of bone directly through intramembranous ossification. Because of their embryonic source and initial position within the dermis, these bones are called **dermal bones.** They are prominent in ostracoderm fishes but appear secondarily even in derived groups, such as in some species of mammals.

**Dermal (intramembranous) bone development (p. 173)**

The most conspicuous component of the dermis is the fibrous connective tissue composed mostly of collagen

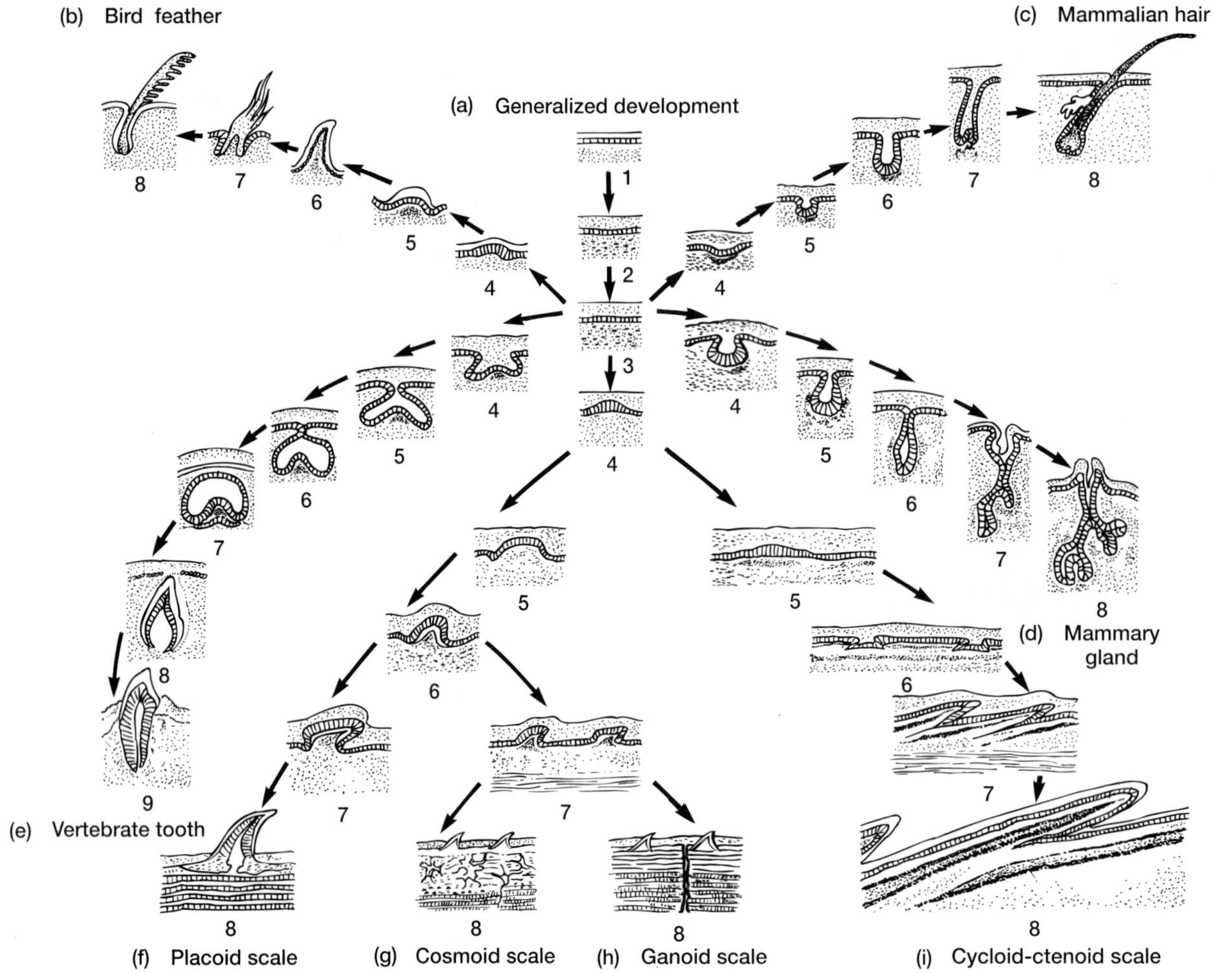


**Figure 6.3** Skin derivatives. (a) Out of the simple arrangement of epidermis and dermis, with a basement membrane between them, a great variety of vertebrate integuments develop. Interaction of epidermis and dermis gives rise to feathers in birds (b), hair and mammary glands in mammals (c and d), teeth in vertebrates (e), placoid scales in chondrichthyans (f), and cosmoid, ganoid, and cycloid-ctenoid scales in bony fishes (g–i). Based on the research of Richard J. Krejsa.

fibers. Collagen fibers may be woven into distinct layers called **plies.** The dermis of the protochordate amphioxus exhibits an especially ordered arrangement of collagen within each ply (figure 6.4). In turn, plies are laminated together in very regular, but alternating orientation. These alternating layers act like warp and weft threads of cloth fabric, giving some shape to the skin and preventing it from sagging. In aquatic vertebrates, such as sharks, the bundles of collagen lie at angles to each other, giving the skin a **bias,** like cloth; that is, the skin stretches when it is pulled at an angle oblique to the direction of the bundles. For example, if you take a piece of cloth, like a handkerchief, and pull it along either warp or weft threads, the cloth extends very little under this parallel tension. But if you pull at opposite corners tension is applied obliquely at a 45° angle to the threads, and the cloth stretches considerably (figure 6.5a,b). This principle seems to govern the tightly woven collagen of shark skin. Its flexible skin bias accommodates lateral bending of the body but simultaneously resists distortions in body shape. As a result, the skin stretches without wrinkling. Because it does not wrinkle, water flows smoothly and without turbulence across the surface of the body (figure 6.5d).

In fishes and aquatic vertebrates, including cetaceans and aquatic squamates, collagen fibers of the dermis are usually arranged in orderly plies that form a recognizable stratum compactum. In terrestrial vertebrates, the stratum compactum is less obvious because locomotion on land depends more on the limbs and less on the trunk. And, of course, any wrinkling of the skin is less disruptive to a terrestrial vertebrate moving through air. Consequently, collagen fibers are present, even abundant, in the skin of terrestrial vertebrates, but they are much less regularly ordered and usually do not form distinct plies.

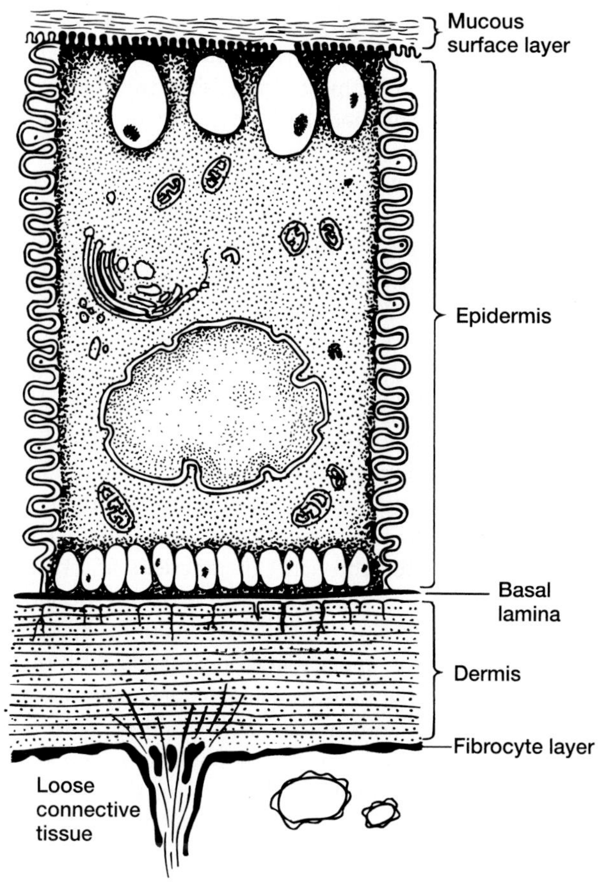


**Figure 6.4** Protochordate, skin of amphioxus. The epidermis is a single layer of cuboidal or columnar cells that secrete a mucus that coats the surface and rests upon a basal lamina. The dermis consists of very highly ordered collagen fibers arranged in alternating plies (layers) to form a "fabric" that brings structural support as well as flexibility to the outer body wall. Pigment is secreted by the epidermal cells themselves.

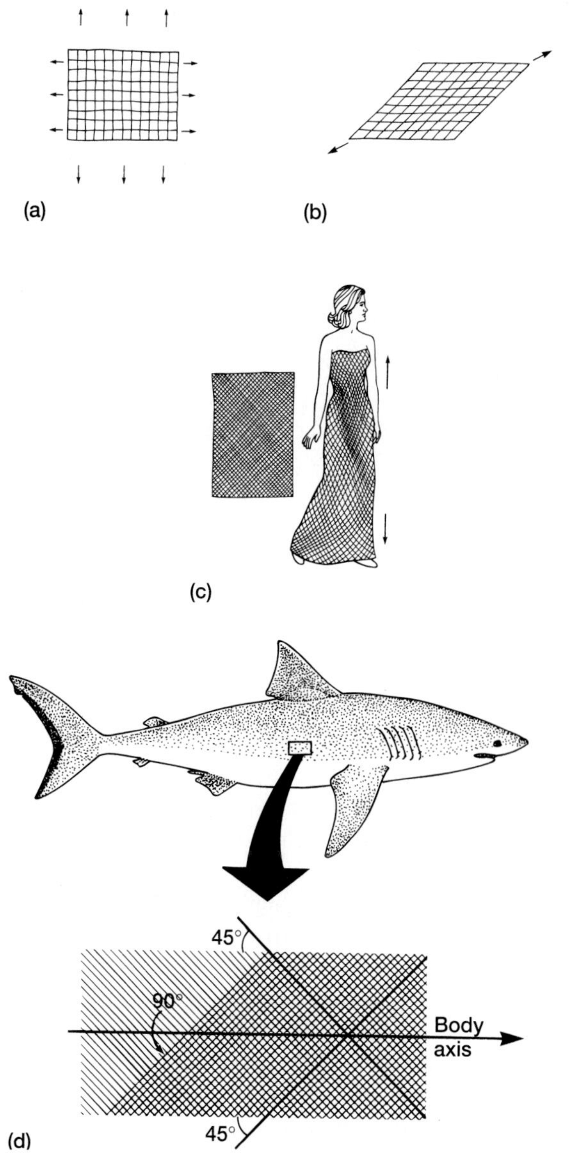


**Figure 6.5** Bias in woven material. (a) Warp (longitudinal threads) and weft (cross threads) compose the fibers of fabric. If the tensile force is parallel with the threads (indicated by the arrows), little distortion of the fabric occurs. (b) However, tension along the bias at 45° to the threads results in a substantial change in shape. (c) Fashion designers take advantage of these features of fabric when designing clothes. In the loose bias direction, the fabric falls into folds and pleats but can hold its shape along the warp-and-weft threads. (d) Plies of collagen of the stratum compactum of fish skin act in a similar way. The flexible bias of the skin is oriented at 45° to the body length, thus accommodating lateral bending during swimming. This arrangement keeps the skin flexible but tight so that surface wrinkling does not occur and turbulence is not induced in the streamlines passing over the body as the fish swims.

## Epidermis

The epidermis of many vertebrates produces mucus to moisten the surface of the skin. In fishes, mucus seems to afford some protection from bacterial infection and helps ensure the laminar flow of water across the body surface. In amphibians, mucus probably serves similar roles and additionally keeps the skin from drying during the animal's sorties onto land.

In terrestrial vertebrates, the epidermis covering the body often forms an outer **keratinized** or **cornified** layer, the **stratum corneum.** New epidermal cells are formed by mitotic division, primarily in the deep stratum basale. These new epidermal cells push more superficial ones toward the surface, where they tend to self-destruct in an orderly fashion. During their demise, various protein products accumulate and collectively form **keratin** in a process called **keratinization.** Thus, keratin is a class of proteins produced during keratinization, and the specific epidermal cells that

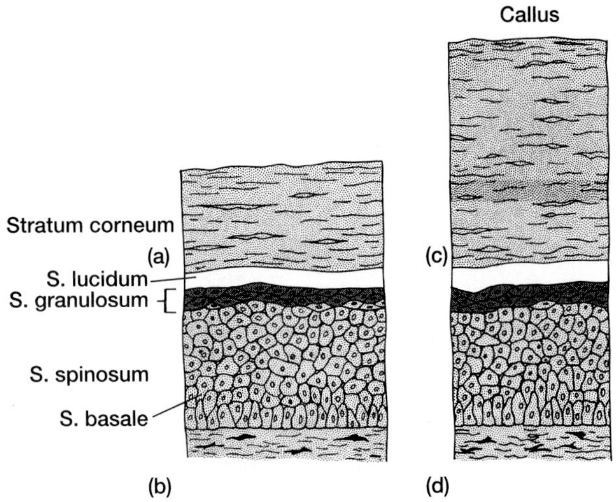


**Figure 6.6** Keratinization. In places where mechanical friction increases, the integument responds by increasing production of a protective keratinized callus, and the stratum corneum thickens as a result.

participate are **keratinocytes.** The resulting superficial stratum corneum is a nonliving layer that serves to reduce water loss through the skin in dry terrestrial environments.

Keratinization and formation of a stratum corneum also occur where friction or direct mechanical abrasion insult the epithelium. For example, the epidermis in the oral cavity of aquatic and terrestrial vertebrates often exhibits a keratinized layer, especially if the food eaten is unusually sharp or abrasive. In areas of the body where friction is common, such as the soles of the feet or palms of the hands, the cornified layer may form a thick protective layer, or **callus,** to prevent mechanical damage (figure 6.6). The stratum corneum may be differentiated into hair, hooves, horn sheathes, or other specialized cornified structures. The term ***keratinizing system*** refers to the elaborate interaction of epidermis and dermis that produces the orderly transformation of keratinocytes into such cornified structures.

Finally, scales form within the integument of many aquatic and terrestrial vertebrates. Scales are basically folds in the integument. If dermal contributions predominate, especially in the form of ossified dermal bone, the fold is termed a **dermal scale.** An epidermal fold, especially in the form of a thickened keratinized layer, produces an **epidermal scale.**

# Phylogeny

## Integument of Fishes

With few exceptions, the skin of most living fishes is nonkeratinized and covered instead by mucus. Exceptions include keratinized specializations in a few groups: the "teeth" lining the oral disk of lampreys, the jaw coverings of some herbivorous minnows, and the friction surface on the belly skin of some semiterrestrial fish are all keratinized derivatives. However, in most living fishes, the epidermis is alive and active on the body surface, and there is no prominent superficial layer of dead, keratinized cells. Surface cells are often patterned with tiny **microridges** that perhaps hold the surface layer of mucus. The mucous layer is formed from various individual cells in the epidermis with contributions from multicellular glands. This mucous coat, termed a **mucous cuticle,** resists penetration by infectious bacteria, probably contributes to laminar flow of water across the surface, makes the fish slippery to predators, and often includes chemicals that are repugnant, alarming, or toxic to enemies.

Two types of cells occur within the epidermis of fishes: **epidermal cells** and specialized **unicellular glands.** In living fishes, including cyclostomes, prevalent epidermal cells make up the stratified epidermis. Superficial epidermal cells are tightly connected through cell junctions and contain numerous secretory vesicles that are released to the surface where they contribute to the mucous cuticle. Epidermal cells of the basal layer are cuboidal or columnar. Mitotic activity is present in but not restricted to the basal layer.

Unicellular glands are single, specialized, and interspersed among the epidermal cell population. There are several types of unicellular glands. The **club cell** is an elongate, sometimes binucleate, unicellular gland (figure 6.7). Some chemicals within club cells excite alarm or fear. They are thought to be released by observant individuals to warn others of imminent danger. The **granular cell** is a diverse cell found in the skin of lampreys and other fishes (figure 6.7). Both granular and club cells contribute to the mucous cuticle, but their other functions are not fully understood. The **goblet cell** is a type of unicellular gland that is absent from lamprey skin but usually found in other bony and cartilaginous fishes. It too contributes to the mucous cuticle and is recognized by its "goblet" shape, namely, a narrowed basal stem and wide apical end holding secretions. The electron microscope has helped distinguish an additional type of unicellular gland in the epider-

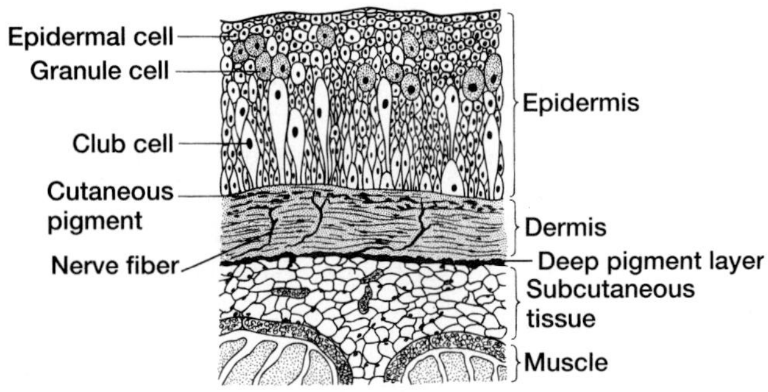


**Figure 6.7** Lamprey skin. Among the numerous epidermal cells are separate unicellular glands, the granular cells and the club cells. Note the absence of keratinization. The dermis consists of regularly arranged collagen and chromatophores.

mis, the **sacciform cell.** It holds a large, membrane-bound secretory product that seems to function as a repellent or toxin against enemies once it is released. As increased attention is given to the study of fish skin, other cell types are being recognized. This growing list of specialized cells within the epidermis reveals a complexity and variety of functions that was not previously appreciated.

Collagen within the stratum compactum is regularly organized into plies that spiral around the body of the fish, allowing the skin to bend without wrinkling. In some fishes, the dermis has elastic properties. When a swimming fish bends its body, the skin on the stretched side stores energy that helps unbend the body and sweeps the tail in the opposite direction.

The fish dermis often gives rise to dermal bone, and dermal bone gives rise to dermal scales. In addition, the surface of fish scales is sometimes coated with a hard, acellular **enamel** of epidermal origin and a deeper **dentin** layer of dermal origin. Until recently, both enamel and dentin were recognized on the basis of appearance, not on their chemical composition. As the surface appearance of scales changed between fish groups, so did the terminology. Enamel was thought to give way phylogenetically to "ganoin" and dentin to "cosmine." These terms were inspired by the superficial appearance of scales, not by their chemical composition nor even by their histological organization. Perhaps it is best to think of ganoin as a different morphological expression of enamel, and cosmine as a different morphological expression of dentin and to be prepared for subtle chemical differences as we meet them.

## *Primitive Fishes*

In ostracoderms and placoderms, the integument produced prominent bony plates of dermal armor that encased their bodies in an exoskeleton. Dermal bones of the cranial region were large, forming the head shields, but more posteriorly along the body, the dermal bones tended to be broken up into smaller pieces, the dermal scales. The surface of these scales was often ornamented with tiny, mushroom-shaped tubercles. These tubercles consisted of a surface layer of enamel or an enamel-like substance over an inner layer of dentin (figure 6.8). One or several radiating pulp cavities resided within each tubercle. The dermal bone supporting these tubercles was **lamellar,** organized in a layered pattern.

The skin of living hagfishes and lampreys departs considerably from that of primitive fossil fishes. Dermal bone is lost, and the skin surface is smooth and without scales. The epidermis is composed of stacked layers of numerous living epidermal cells throughout. Interspersed among them are unicellular glands, namely, the large granular cells and elongate club cells. In addition, the skin of hagfishes includes **thread cells** that discharge thick cords of mucus to the skin surface when the fish is irritated. The dermis is highly organized into regular layers of fibrous connective tissue. Pigment cells occur throughout the dermis. The hypodermis includes adipose tissue. Within the dermis, hagfishes also possess multicellular **slime glands** that release their products via ducts to the surface.

## *Chondrichthyes*

In cartilaginous fishes, dermal bone is absent, but surface denticles, termed **placoid scales,** persist. These scales are what give the rough feel to the surface of the skin (figure 6.9a). Recent evidence suggests that these tiny placoid scales favorably affect the water flowing across the skin as the fish swims forward to reduce friction drag. Numerous secretory cells are present in the epidermis as well as stratified epidermal cells. The dermis is composed of fibrous connective tissue, especially elastic and collagen fibers, whose regular arrangement forms a fabriclike warp and weft in the dermis (figure

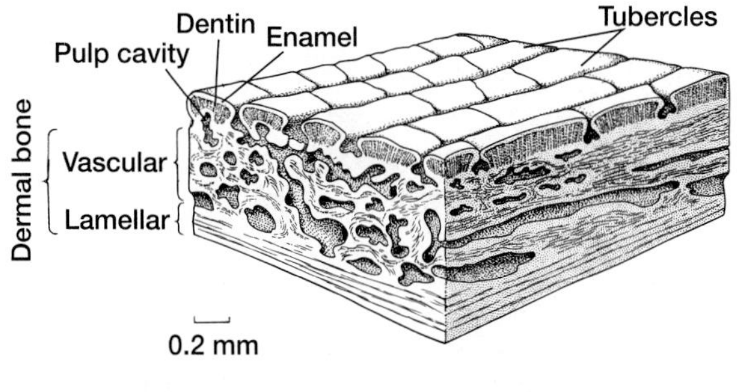


**Figure 6.8** Section through an enlarged ostracoderm scale. The surface consists of raised tubercles capped with dentin and enamel enclosing a pulp cavity within. These tubercles rest upon a foundation of dermal bone, part of the dermal armor covering the body.

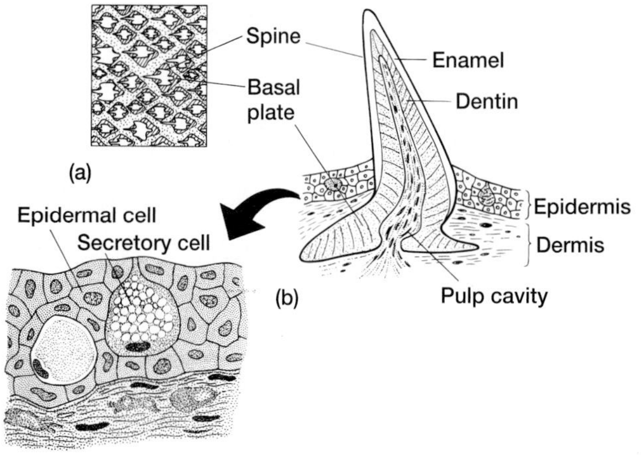


**Figure 6.9** Shark skin. (a) Surface view of the skin showing regular arrangement of projecting placoid scales. (b) Section through a placoid scale of a shark. The projecting scale consists of enamel and dentin around a pulp cavity.

6.5d). This gives the skin strength and prevents wrinkling during swimming.

The placoid scale itself develops in the dermis but projects through the epidermis to reach the surface. A cap of enamel forms the tip, dentin lies beneath, and a pulp cavity resides within (figure 6.9a,b). Chromatophores occur in the lower part of the epidermis and upper regions of the dermis.

### *Bony Fishes*

The dermis of bony fishes is subdivided into a superficial layer of loose connective tissue and a deeper layer of dense fibrous connective tissue. Chromatophores are found within the dermis. The most important structural product of the dermis is the scale. In bony fishes, dermal scales do not actually pierce the epidermis, but they are so close to the surface they give the impression that the skin is hard (figure 6.10a,b). The epidermal covering includes a basal layer of cells. Above this layer are stratified epidermal cells. As they move toward the surface, epidermal cells undergo cytoplasmic transformation, but they do not become keratinized. Within these layered epidermal cells occur single unicellular glands, the secretory and club cells. These unicellular glands, along with epidermal cells, are the source of the mucous cuticle, or surface "slime."

On the basis of their appearance, several types of scales are recognized among bony fishes. The **cosmoid scale,** seen in primitive sarcopterygians, resides upon a double layer of bone, one layer of which is vascular and the other lamellar. On the outer surface of this bone is a layer that is now generally recognized as dentin, and spread superficially on the dentin is a layer now recognized as enamel. The unusual appearance of these enamel and dentin coats inspired, in the older literature, the respective names of ganoin and cosmine, on the mistaken belief that ganoin was fundamentally a different mineral from enamel and cosmine different from dentin. Although the chemical nature of these layers is now clear, the earlier names have stuck to give us the terms for distinctive scale types. In the cosmoid scale, there is a thick, well-developed layer of dentin (cosmine) beneath a thin layer of enamel (figure 6.11a).

The **ganoid scale** is characterized by the prevalence of a thick surface coat of enamel (ganoin), without an underlying layer of dentin (figure 6.11b). Dermal bone forms the foundation of the ganoid scale, appearing as a double layer of vascular and lamellar bone (in palaeoniscoid fishes) or a single layer of lamellar bone (in other primitive actinopterygians). Ganoid scales are shiny (because of the enamel), overlapping, and interlocking. Living polypteriforms and gars retain ganoid scales. However, in most other lines of bony fishes, ganoid scales are reduced through the loss of the vascular layer of bone and loss of

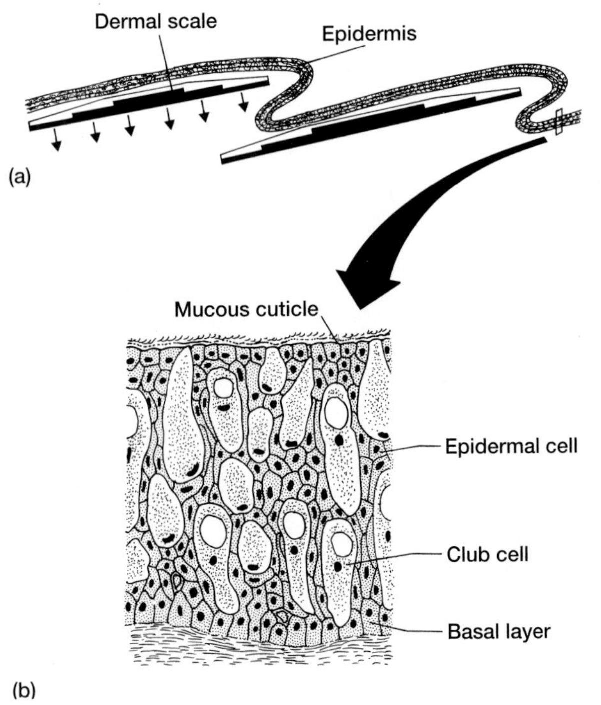


**Figure 6.10** Bony fish skin. (a) Arrangement of dermal scales within the skin of a teleost fish (arrows indicate direction of scale growth). (b) Enlargement of epidermis. Note epidermal cells and club cells.

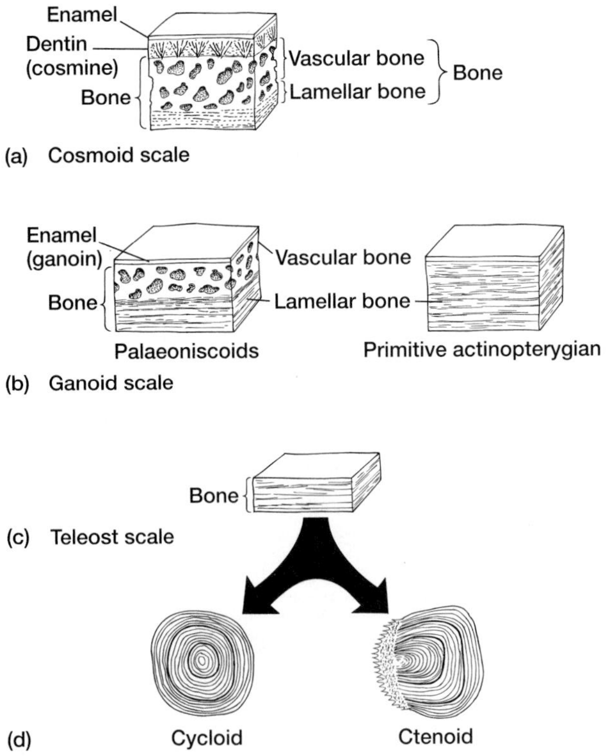


**Figure 6.11** Scale types of bony fishes. Cross section of a cosmoid scale (a), a ganoid scale (b), and a teleost scale (c). Surface views of the two types of the teleost scale, cycloid and ctenoid scales (d).

the enamel surface. This produces, in teleosts, a rather distinctive scale.

The **teleost scale** lacks enamel, dentin, and a vascular bone layer. Only lamellar bone remains, which is acellular and mostly noncalcified (figure 6.11c). Two kinds of teleost scales are recognized. One is the **cycloid scale,** composed of concentric rings, or **circuli.** The other is the **ctenoid scale** with a fringe of projections along its posterior margin (figure 6.11d). New circuli are laid down, like rings in a tree, as a teleost fish grows. Annual cycles are evident in the groupings of these circuli and allow aging of individual fish from this pattern in their scales.

## Integument of Tetrapods

Although keratinization occurs in fishes, among terrestrial vertebrates it becomes a major feature of the integument. Extensive keratinization produces a prominent outer cornified layer, the stratum corneum, that resists mechanical abrasion. Lipids are often added during the process of keratinization or spread across the surface from specialized glands. The cornified layer along with these lipids increases the resistance of the tetrapod skin to desiccation.

Multicellular glands are more common in the skin of tetrapods than in the skin of fishes. In fishes, the mucous cuticle and secretions of the unicellular glands at or near the surface of the skin coat it. In contrast, among tetrapods, multicellular glands usually reside in the dermis and reach the surface through common ducts that pierce the cornified layer. Thus, the stratum corneum that protects the skin and prevents desiccation also controls the release of secretions directly to the surface. If it were not for these openings in the stratum corneum, the surface of the skin could not be coated or lubricated by these secretions.

### *Amphibians*

Amphibians are of special interest because during their lives they usually metamorphose from an aquatic form to a terrestrial form. Phylogenetically, amphibians are also transitional between aquatic and terrestrial vertebrates. In most modern amphibians, the skin is also specialized as a respiratory surface across which gas exchange occurs with the capillary beds in the lower epidermis and deeper dermis. In fact, some salamanders lack lungs and depend entirely on **cutaneous respiration** through the skin to meet their metabolic needs.

Cutaneous respiration (p. 403)

The most primitive amphibians had scales like the fishes from which they arose. Among living forms, dermal scales are present only as vestiges in some species of tropical caecilians (apoda). Frogs and salamanders lack all traces of dermal scales (figure 6.12a). In salamanders, the skin of the aquatic larvae includes a dermis of fibrous connective tissue, consisting of superficial loose tissue over a compact deep layer. Within the epidermis are deep basal cells and surface apical cells. Scattered throughout are large **Leydig cells** thought to secrete substances that resist entry of bacteria or viruses (figure 6.12b). In terrestrial adults, the dermis is similarly composed of fibrous connective tissue. In the epidermis, Leydig cells are now absent, but distinct regions can be recognized, such as the strata basale, spinosum, granulosum, and corneum. Presence of a thin stratum corneum affords some protection from mechanical abrasion and retards loss of moisture from the body without unduly shutting off cutaneous gas exchange. During the breeding season, **nuptial pads** may form on digits or limbs of male salamanders or frogs. Nuptial pads are raised calluses of cornified epidermis that help the male hold the female during mating.

Generally, the skin of frogs and salamanders usually includes two types of multicellular glands, mucous and poison glands. Both are located in the dermis and open to the surface through connecting ducts (figure 6.12b). The **mucous glands** tend to be smaller, each being made up of a little cluster of cells that release their product into a common duct. The **poison glands** (granular glands) tend to be larger and often contain stored secretion within the lumen of each gland. Secretions of poison glands tend to be distasteful or even toxic to predators. However, few persons handling amphibians are bothered by this secretion nor need they be concerned because it is potentially harmful only if eaten or injected into the bloodstream.

Chromatophores may occasionally be found in the amphibian epidermis, but most reside in the dermis. Capillary beds, restricted to the dermis in most vertebrates, reach into the lower part of the epidermis in amphibians, a feature serving cutaneous respiration.

### *Reptiles*

The skin of reptiles reflects their greater commitment to a terrestrial existence. Keratinization is much more extensive, and skin glands are fewer than in amphibians. Scales are present, but these are fundamentally different from the dermal scales of fishes, which are built around bone of dermal origin. The reptilian scale usually lacks the bony undersupport or any significant structural contribution from the dermis. Instead, it is a fold in the surface epidermis, hence, it is an epidermal scale. The junction between adjacent epidermal scales is the flexible **hinge** (figure 6.13a). If the epidermal scale is large and platelike, it is sometimes termed a **scute.** Additionally, epidermal scales may be modified into crests, spines, or hornlike processes.

Although not usually associated with scales, dermal bone is present in many reptiles. The **gastralia,** a collection of bones in the abdominal area, are examples. Where dermal bones support the epidermis, they are called **osteoderms,** plates of dermal bone located under the epidermal scales. Osteoderms are found in crocodilians, some lizards, and some extinct reptiles. Some bones of the turtle shell are probably modified osteoderms.

The dermis of reptilian skin is composed of fibrous connective tissue. The epidermis is generally delineated into three regions—stratum basale, stratum granulosum, and stratum corneum. However, this changes prior to molting in those reptiles that slough large pieces of the cornified skin layer. In turtles and crocodiles, sloughing of skin is modest, comparable to birds and mammals, in whom small flakes fall off at irregular intervals. But in lizards, and especially in snakes, shedding of the cornified layer, termed **molting** or **ecdysis,** results in removal of extensive sections

## Box Essay 6.1 Poison Arrows and Poison Frogs

The skin of most amphibians contains glands that secrete products that are distasteful or even toxic to predators. In tropical regions of the New World lives a group of frogs, the poison arrow frogs, with especially toxic skin secretions (box figure 1). Native peoples of the region will often gather these frogs, hold them on sticks over a fire to stimulate release of these secretions, and then collect the secretions on the tips of their arrows. Game shot with these toxin-laced arrows are quickly tranquilized or killed.

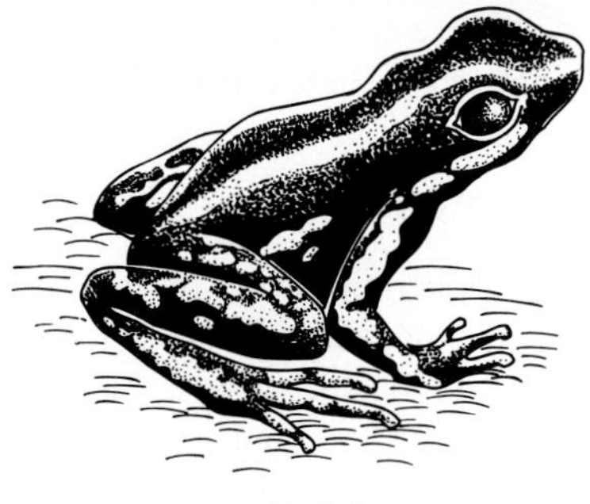


**Box figure 1** Poison arrow frog. Its bright colors advertise toxic skin secretions that are poisonous to most predators.

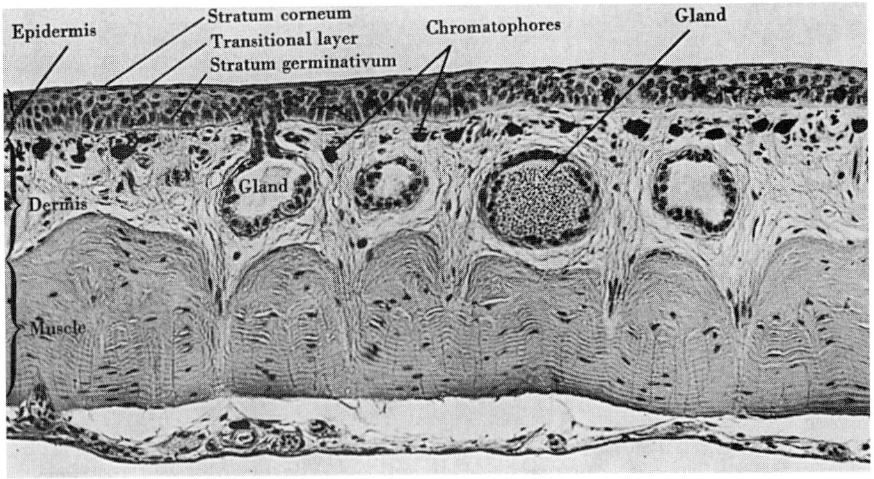


(a)

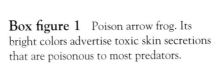

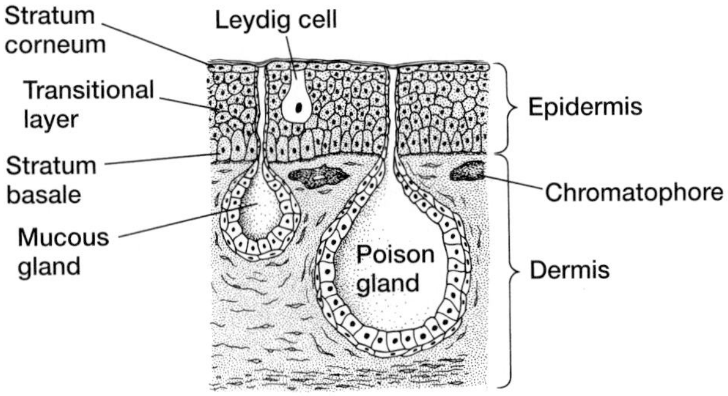


(b)

**Figure 6.12** Amphibian skin. (a) Section through an adult frog skin. A basal stratum basale and a thin, superficial stratum corneum are present. The transitional layer between them includes a stratum spinosum and a stratum granulosum. (b) Diagrammatic view of amphibian skin showing mucous and poison glands that empty their secretions through short ducts to the surface of the epidermis.

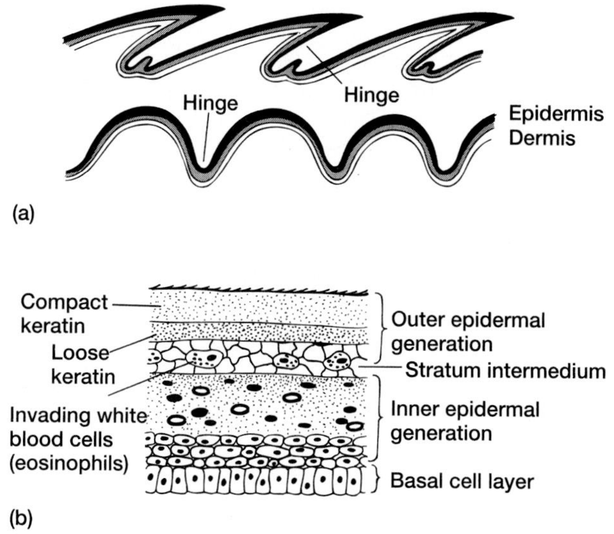


**Figure 6.13** Reptile skin. (a) Epidermal scales. Extent of projection and overlap of epidermal scales varies among reptiles and even along the body of the same individual. Snake body scales (top) and tubercular scales of many lizards (bottom) are illustated. Between scales is a thinned area of epidermis, a "hinge" allowing skin flexibility. (b) Skin shedding. Just before the old outer layer of epidermis is shed, the basal cells produce an inner epidermal generation. White blood cells collect in the splitting zone to promote separation of new from old outer epidermis.

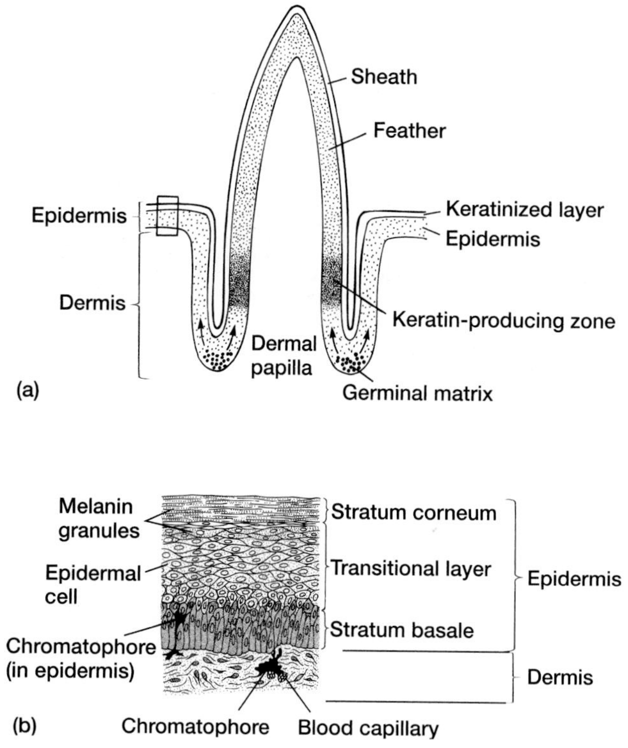


**Figure 6.14** Bird skin. (a) Growth of a feather follicle. The feather itself forms within a sheath that, like the feather, is a keratinized derivative of the epidermis. (b) Section of skin showing the stratum basale and the keratinized surface layer, the stratum corneum. Cells moving out of the basal layer spend time first in the transitional layer before reaching the surface. This middle transitional layer is equivalent to the spinosum and granulosum layers of mammals.

of superficial epidermis. As molting begins, the stratum basale, which has given rise to the strata granulosum and corneum, duplicates the deeper layers of granulosum and corneum, pushing up under the old layers. White blood cells invade the **stratum intermedium,** a temporary layer between old and new skin (figure 6.13b). These white blood cells are thought to promote the separation and loss of the old superficial layer of the skin.

Integumental glands of reptiles are usually restricted to certain areas of the body. Many lizards possess rows of **femoral glands** along the underside of the hindlimb in the thigh region. Crocodiles and some turtles have **scent glands.** In alligators of both sexes, one pair of scent glands opens into the cloaca, another pair opens on the margins of the lower jaw. In some turtles, scent glands can produce quite pungent odors, especially when the animal is alarmed by handling. Most integumental glands of reptiles are thought to play a role in reproductive behavior or to discourage predators, but the glands and their social roles are not well understood.

## *Birds*

***Basic Structure*** The feathers of birds have been called nothing more than elaborate reptilian scales. This oversimplifies the homology but probably not by much. Certainly the presence of epidermal scales along the legs and feet of birds testifies to their debt to reptiles.

The dermis of bird skin, especially near the feather follicles, is richly supplied with blood vessels, sensory nerve endings, and smooth muscles. During the brooding season, the dermis in the breast of some birds becomes increasingly vascularized, forming a **brood patch** in which warm blood can come into close association with incubated eggs.

The epidermis comprises the stratum basale and the stratum corneum. Between them is the transitional layer of cells transformed into the keratinized surface of the corneum (figure 6.14a,b).

Bird skin has few glands. The **uropygial gland,** located at the base of the tail (figure 6.15) secretes a lipid and protein product that birds collect on the sides of their beak and then smear on their feathers. Preening coats the feathers with this secretion, making them water repellent, and probably conditions the keratin of which they are composed. The other gland, located on the heads of some birds, is the **salt gland,** which is well-developed in marine birds. Salt glands excrete excess salt obtained when these birds ingest marine foods and seawater.

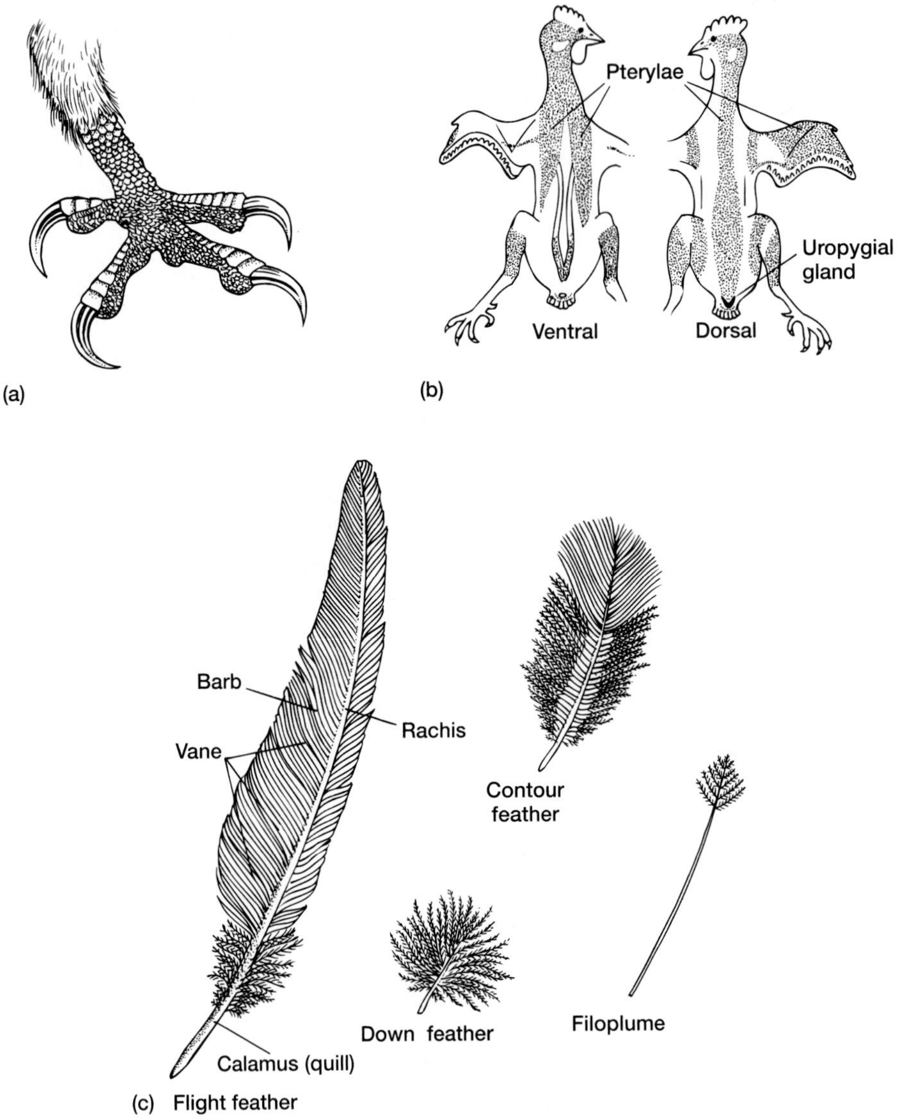


**Figure 6.15** Epidermal derivatives in the bird. (a) Epidermal scales are present on the feet and legs of birds. (b) Feathers arise along specific pterylae tracts. (c) Feather types. Flight feathers constitute the major locomotor surfaces. Contour feathers aerodynamically shape the surface of a bird. Filoplumes are often specialized for display. Down feathers lie close to the skin as thermal insulation.

Salt excretion (p. 548)

Feathers distinguish birds from all other vertebrates. Feathers can be structurally elaborate and come in a variety of forms. Yet feathers are nonvascular and nonnervous products of the skin, principally of the epidermis and the keratinizing system. They are laid out along distinctive tracts, termed **pterylae,** on the surface of the body (figure 6.15b). Via one or several molts, they are replaced each year. Feathers develop embryologically from **feather follicles,** invaginations of the epidermis that dip into the underlying dermis. The root of the feather follicle, in association with the dermal pulp cavity, begins to form the feather. The feather itself grows outward in a sheathed case (figure 6.16a). Within the sheath, the central axis is divided into a distal **rachis** that bears **barbs** with interlocking connections, termed **barbules,** and a proximal **calamus** that attaches to the body (figure 6.16b).

***Functions*** There are several types of feathers (figure 6.15c). Contour feathers aerodynamically shape the surface of the bird. Down feathers lie close to the skin as thermal insulation. Filoplumes are often specialized for display, and flight feathers constitute the major aerodynamic surfaces. Flight feathers of the wings are characterized by a long rachis and prominent **vane** (figure 6.16b). These feathers have some value as insulation, but their primary function is locomotion. Most feathers receive sensory stimuli and carry colors for display or courtship. Chromatophores occur

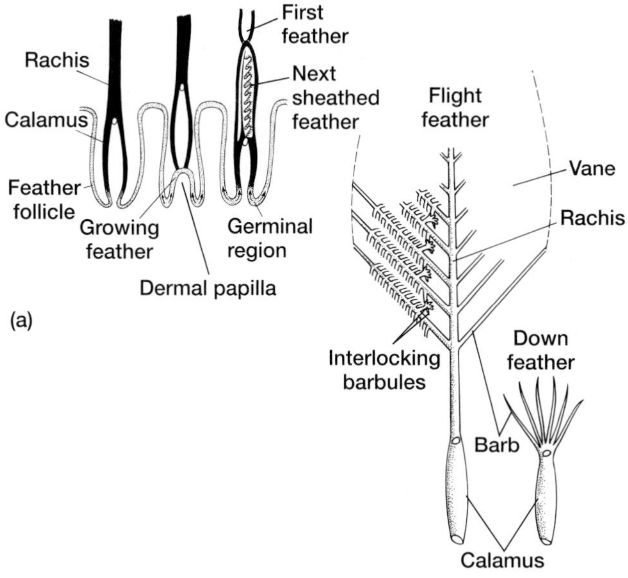


**Figure 6.16** Feather growth and morphology. (a) Successive stages in the growth of new feathers. During molting of old feathers, new ones arise in the feather follicles to replace those shed. (b) General morphology of flight and down feathers.

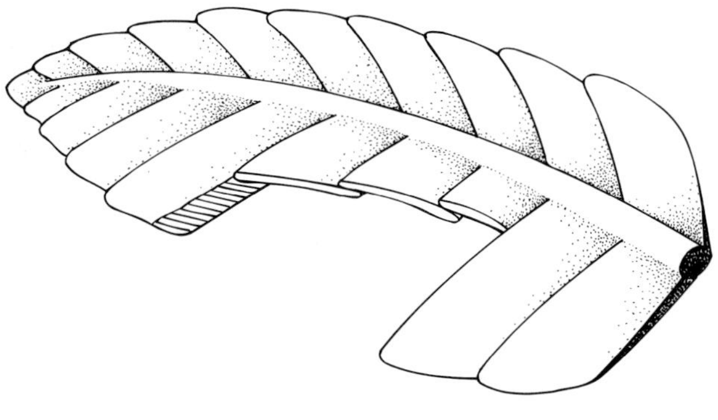

**Figure 6.17** Hypothetical scale, intermediate stage between an enlarged reptile scale and an early bird feather. Some living lizards use such enlarged scales to reflect away excess solar radiation. Subdivision of the scale provides the flexibility required for free movement in an active animal.

within the epidermis, and their pigments are carried into the feathers to give them color. But light refraction on the feather barbs and barbules also creates some of the iridescent colors that feathers display.

At least one bird has feathers and skin lightly coated with a toxin thought to deter predators. The brightly colored bird, called a hooded pitohui, lives in New Guinea and is about the size of a blue jay. The poison works by repelling snakes, hawks, or other predators tasting one of the feathers. The bright plumage of the pitohui may represent a warning coloration to predators.

***Evolution of Feathers*** When we think of feathers, we think of their roles in flight, but they likely had other functions when they first arose. One view is that feathers, or their scaly predecessors, played a role in surface insulation. Surface insulation, of course, holds heat in or shields the body from taking up excess heat. Either may have been the initial advantage of feathers. Surface insulation would have interfered with the absorption of environmental heat, a disadvantage if the ancestors of birds were ectothermic. However, many species of ectothermic lizards have enlarged surface scales. Once the basking lizard is warmed, it turns so that the scales act like many tiny parasols to shade the skin surface and block further uptake of solar radiation (figure 6.17). Once enlarged and shaped for heat exclusion, these protofeathers would be preadapted for heat retention or for flight.

Others argue that the ancestors of birds were endothermic. In this view, protofeathers initially functioned to conserve internally produced body heat. The evolution of aerodynamic devices serving flight came later.

Both views, whether beginning with an ectotherm or an endotherm, suggest that feathers played a role in surface insulation when they first appeared and became secondarily co-opted for a role in flight. A different view entirely stems from the argument that feathers evolved initially as aids to gliding and then to flight. Feathers were selected because of their favorable effect on the airstream passing over the body or limbs of a gliding animal. If the protoavian limb were not streamlined, then pressure drag would result and turbulence would have reduced aerodynamic efficiency. However, surface scales projecting from the trailing edge of the limb would have streamlined the limb, reduced drag, and thus been favored by selection.

**Aerodynamic principles (p. 135)**

Regardless of whether they evolved first for gliding or for insulation, feathers were modified from reptilian scales. In modern birds, feathers that serve flight are highly modified. Interlocking barbs and barbules give some structural integrity to the flexible flight feather. In flight feathers, the rachis is offset, making the vane asymmetrical (figure 6.18a). This design affects the action of the flight feather during wing beats . On the downstroke, the pressure on the underside of each feather acts along its anatomical midline, the **center of pressure.** But because the rachis is offset, the result is to twist the feather slightly about its point of attachment to the limb, forcing feathers of the wing together into a broad surface that presses against the air and drives the bird forward (figure 6.18b). On the upstroke, the center of pressure is now across the topside of the asymmetrical feather and forces it to twist in the opposite direction, opening a channel between the feathers (figure 6.18c). This reduces their resistance to the airstream and allows the wing to recover and prepare for the next power downstroke.

**Bird flight (p. 345)**

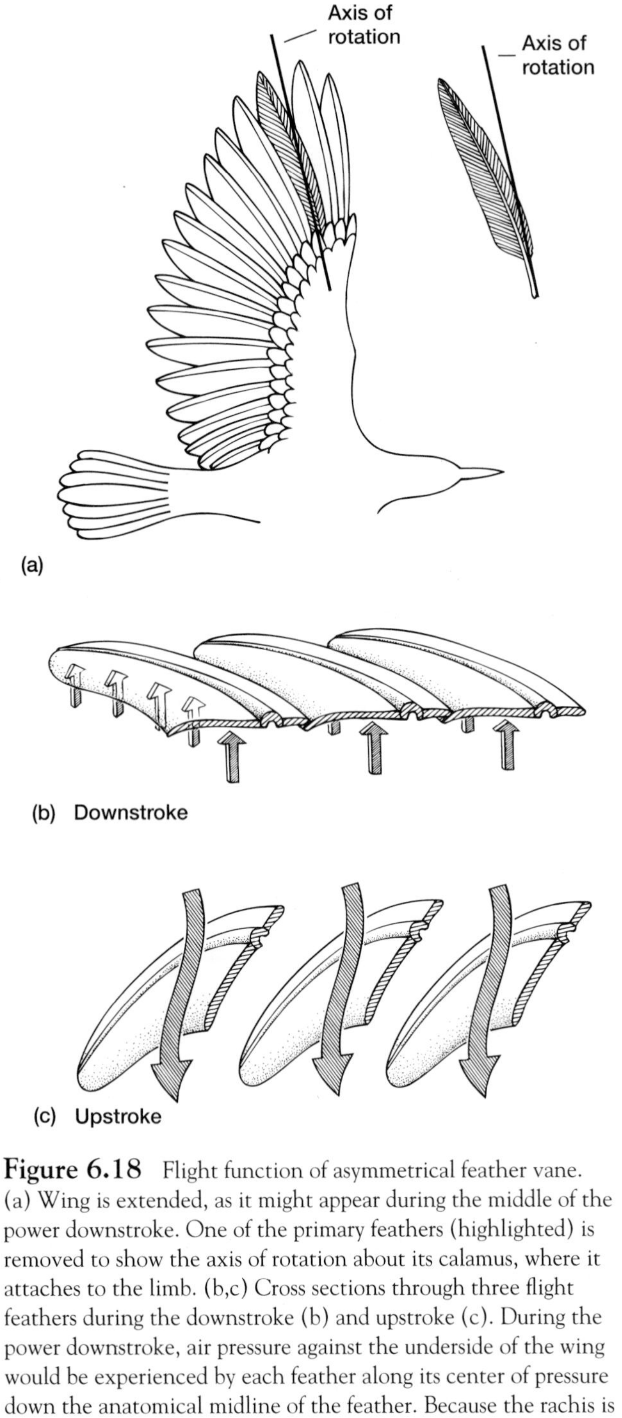


**Figure 6.18** Flight function of asymmetrical feather vane. (a) Wing is extended, as it might appear during the middle of the power downstroke. One of the primary feathers (highlighted) is removed to show the axis of rotation about its calamus, where it attaches to the limb. (b,c) Cross sections through three flight feathers during the downstroke (b) and upstroke (c). During the power downstroke, air pressure against the underside of the wing would be experienced by each feather along its center of pressure down the anatomical midline of the feather. Because the rachis is offset, however, this center of pressure forces the feather to rotate about its axis, and the primary feathers temporarily form a closed uniform surface. During the recovery upstroke, air pressure against the back of the wing forces rotation in the opposite direction, spaces open between feathers, and air slips between the resulting slots, thus reducing resistance to wing recovery.

This controlled twisting of flight feathers passively responding to wing beat depends on the asymmetrical design of the feather and hence on the action of air pressure against it during powered flight. A close look at the wing

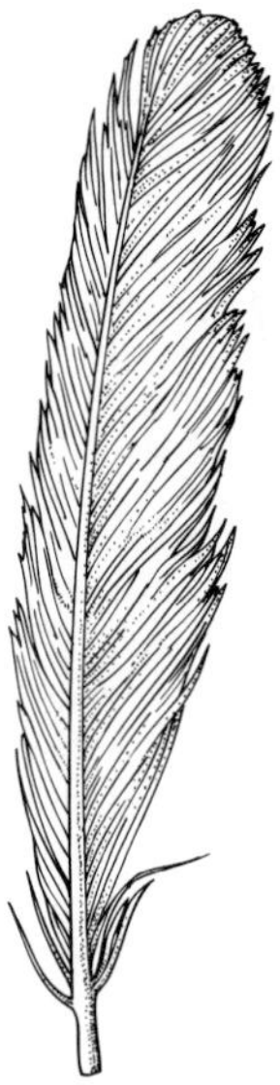

**Figure 6.19** *Archaeopteryx* feather. This feather from the wing of *Archaeopteryx* shows the asymmetrical design of the vane, suggesting that it may have been used during powered flight as in modern birds.

feathers of *Archaeopteryx* also reveals an offset rachis and an asymmetrical vane (figure 6.19). This suggests that by the time of *Archaeopteryx*, powered flight had already evolved.

## *Mammals*

As in other vertebrates, the two main layers of the mammalian skin are epidermis and dermis, which join and interface through the basement membrane. Beneath lies the hypodermis, or superficial fascia, composed of connective tissue and fat.

***Epidermis*** The epidermis may be locally specialized as hair, nails, or glands. Epithelial cells of the epidermis are keratinocytes and belong to the keratinizing system that forms the dead, superficial cornified layer of the skin. The surface keratinized cells are continually exfoliated and replaced by cells arising primarily from the deepest layer of the epidermis, the stratum basale. Cells within the basale divide mitotically, producing some that remain to maintain the population of stem cells and others that are pushed outward. As they are displaced to higher levels, they pass through keratinization stages exhibited as distinct, successive layers toward the surface—**stratum spinosum, stratum granulosum,** often a **stratum lucidum,** and a **stratum corneum** (figure 6.20). The process of keratinization is most distinct in regions of the body where the skin is thickest, as on the soles of the feet. Elsewhere, these layers, especially the lucidum, may be less apparent.

Keratinocytes are the most prominent cell type of the epidermis, but others are recognized although their func-

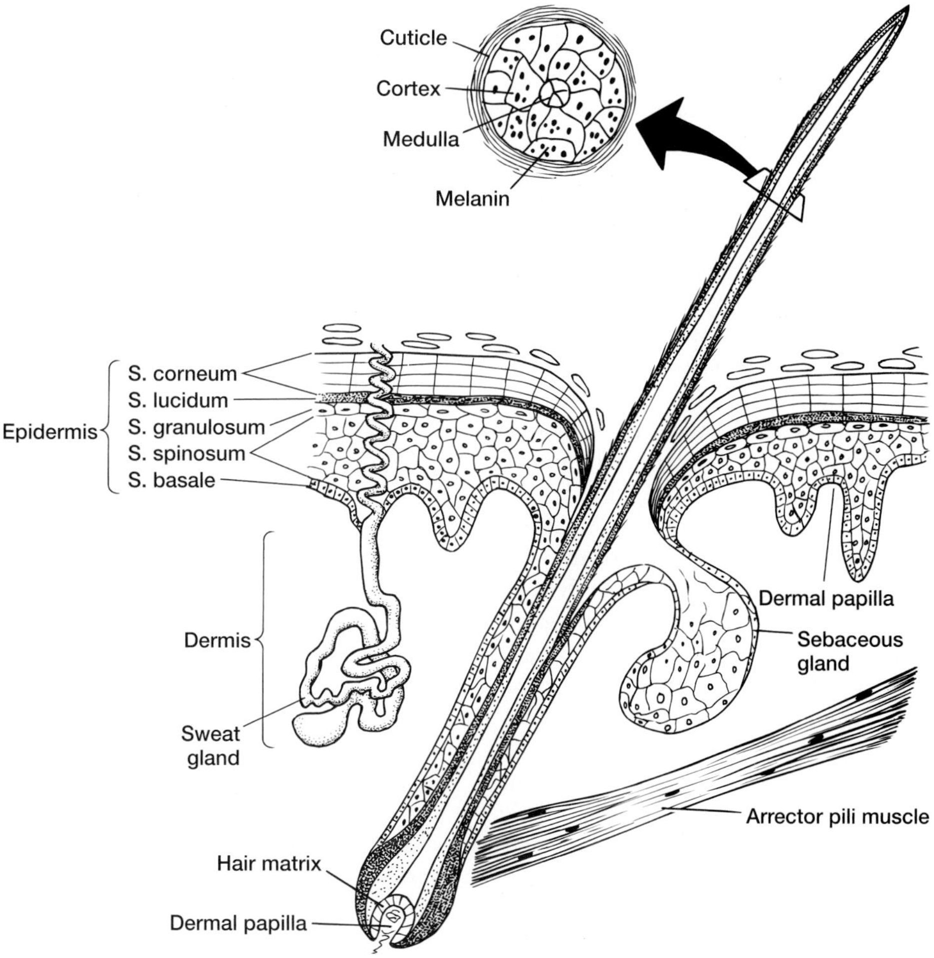


**Figure 6.20** Mammalian skin. The epidermis is differentiated into distinct layers. As in other vertebrates, the deepest is the stratum basale, which through mitotic division produces cells that as they age become successively part of the stratum spinosum, the stratum granulosum, often the stratum lucidum, and finally the surface stratum corneum. The dermis pokes up dermal papillae that give the overlying epidermis an undulating appearance. Sweat glands, hair follicles, and sensory receptors lie within the dermis. Notice that the sweat ducts pass through the overlying epidermis to release their watery secretions on the surface of the skin.

tions are less clearly known. The **Langerhans cells** are stellate cells dispersed singly throughout the upper parts of the stratum spinosum. Current evidence suggests that they play a role in cell-mediated actions of the immune system. The **Merkel cells,** originating from neural crest and associated with nearby sensory nerves, are thought to respond to tactile stimulation (mechanoreceptors).

In addition to these epithelial cell types, the other prominent cell type that becomes secondarily associated with the epidermis is the chromatophore. Chromatophores arise from embryonic neural crest cells and may be found almost anywhere within the body. Those that reach the skin occupy sites within the deeper parts of the epidermis itself. They secrete granules of the pigment **melanin,** which are passed directly to epithelial cells and eventually carried into the stratum corneum or into the shafts of hair. Skin color results from a combination of the yellow stratum corneum, the red underlying blood vessels, and the dark pigment granules secreted by chromatophores.

***Dermis*** The mammalian dermis is double layered. The outer **papillary layer** pushes fingerlike projections, termed **dermal papillae,** into the overlying epidermis. The deeper **reticular layer** includes irregularly arranged fibrous connective tissue and anchors the dermis to the underlying fascia. Blood vessels, nerves, and smooth muscle occupy the dermis but do not reach the epidermis. The mammalian dermis produces dermal bones, but these contribute to the skull and pectoral girdle and only rarely form dermal scales in the skin. One exception is *Glyptodon,* a fossil mammal whose epidermis was underlaid by dermal bone. A similar situation exists in the living armadillo. These species represent secondary developments of dermal bone in the mammalian integument.

Blood vessels and nerves enter the dermis. Hair follicles and glands project inward from the epidermis (figure 6.20). The dermis is usually composed of irregularly arranged fibrous connective tissue that is often impregnated with elastic fibers to give it some stretch but return it to its original shape. As a person ages, this elasticity is lost, and the skin sags.

***Hair*** Hairs are slender, keratinous filaments. The base of a hair is the **root.** Its remaining length constitutes the **shaft.** The outer surface of the shaft often forms a scaly **cuticle.** Beneath this is the **hair cortex,** and at its core is the **hair medulla** (figure 6.20).

The hair shaft projects above the surface of the skin, but it is produced within an epidermal **hair follicle** rooted in the dermis. The surface of the epidermis dips down into the dermis to form the hair follicle. At its expanded base, the follicle receives a small tuft of the dermis, the **hair papilla.** This papilla seems to be involved in stimulating activity of the **matrix cells** of the epidermis but itself does not directly contribute to the hair shaft. The tiny clump of living matrix cells, like the rest of the stratum basale, is the germinal region that starts the process of keratinization to produce hair within the follicle. Unlike keratinization within the epidermis, which is general and continuous, keratinization within the hair follicle is localized and intermittent.

Chromatophores in the follicle contribute pigment granules to the hair shaft to give it further color. The **arrector pili** muscle, a thin band of smooth muscle anchored in the dermis, is attached to the follicle and makes the hair stand erect in response to cold, fear, or anger.

A thick covering of hair is **fur,** or **pelage,** generally composed of guard hairs and underfur. The **guard hairs,** the larger, coarse hairs are the most apparent on the outer surface of the fur. The **underfur** is stationed beneath the guard hairs and is usually much finer and shorter. Both guard hairs and underfur function largely as insulators. In most marine mammals, the underfur is reduced or lost entirely, and only a few guard hairs are evident. Hair has grain; that is, it is laid out in a particular direction. Strokes against the grain are resisted (figure 6.21). An exception occurs in moles, which lack turnaround space and must back up in their tunnels. The hair of moles can be combed forward or backward without much difference in grain.

Some hairs are specialized. Sensitive nerves are associated with the roots of **vibrissae,** or "whiskers," around the snouts of many mammals. Not surprisingly, these are common in nocturnal mammals and in mammals that live in burrows with limited light. The **quills** of porcupines are stiff, coarse hairs specialized for defense.

***Evolution of Hair*** The phylogenetic origin of hair remains speculative. One view holds that hair arose initially as surface insulation, retaining body heat in primitive mammalian endotherms. An alternative view is that hair

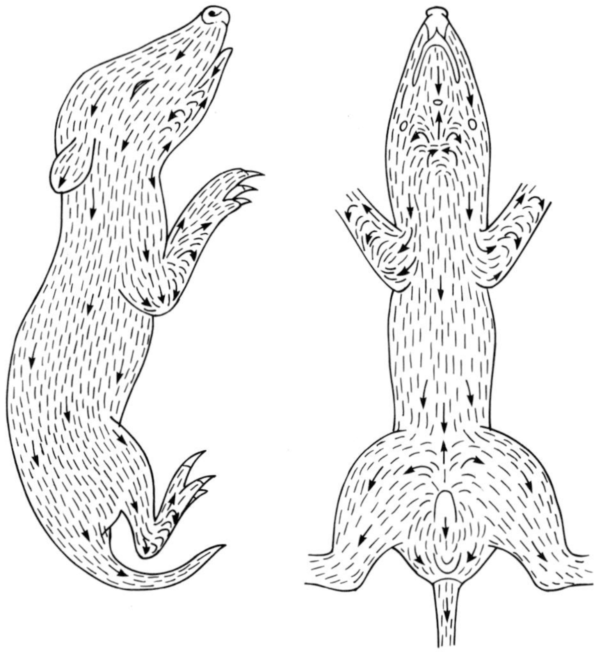

**Figure 6.21** Hair tracks. Hair grows with a grain, a particular direction in which it slants. Notice the various growth directions (arrows) in which the hairs of the marsupial bandicoot lie.

evolved first as tiny projecting rods in the hinges between scales and served as tactile devices. These "protohairs" could help monitor surface sensory data when an animal was hiding from an enemy or retreating from the weather. If such a role increased in importance, it would have favored longer shafts and perhaps the evolution of structures resembling vibrissae. This sensory protohair might then have evolved secondarily into an insulative pelage as mammals became endothermic. Although insulative in modern mammals, hair still retains a sensory function.

Being soft and decomposable, hair does not leave a reliable trace in the fossil record. Some therapsids, ancestors to mammals, have tiny pits in the facial region of their skulls. These pits resemble pits on skulls associated with sensory vibrissae in modern mammals. Some have interpreted these pits as indirect evidence of hair in therapsids. But the skulls of some modern lizards with scales have similar pits and, of course, lizards have no hair. Thus, such pits are not conclusive evidence of the presence of hair. Further, one especially well-preserved skin impression of *Estemmenosuchus*, a therapsid from the Upper Permian, shows no evidence of hair. The epidermis was smooth, without scales, and undifferentiated, although it was supplied with glands. In life, its skin was probably soft and pliable. Thus, we still do not know when hair first arose in primitive mammals or in their therapsid ancestors.

***Glands*** Principally there are two main types of glands in mammals—sebaceous and sweat glands. Derived from them are scent and mammary glands.

The **sebaceous glands** produce an oily secretion, **sebum,** that is released into hair follicles in order to condition and help waterproof fur. Sebaceous glands are absent from the palms of hands and soles of feet, but they are present, without associated hair, at the angle of the mouth, on the penis, near the vagina, and next to the mammary nipples. At these sites, their secretion lubricates the skin surface. The **wax glands** of the outer ear canal, which secrete earwax, and **Meibomian glands** of the eyelid, which secrete an oily film over the surface of the eyeball, probably are derived from sebaceous glands.

The **sweat glands** produce a watery product called perspiration or **sweat.** Two types are usually recognized by the viscosity of their sweat (viscous or thin), by their associations (with or without hair follicles), and by their functional onset (at puberty or before). One type produces thin sweat, is not associated with hair follicles, and functions before puberty. Its products function in regulation of body temperature. The other produces viscous sweat, is associated with hair follicles, and begins functioning at puberty. It is responsible for "body odor." Both types of sweat glands are long, coiled invaginations of the epidermis that reach deep within the dermis but maintain continuity through the surface of the skin even through the cornified stratum corneum.

Sweat glands are not found in all mammals, and their distribution varies. Chimpanzees and humans have the greatest numbers of sweat glands, including some on the palms and soles. In the duckbill platypus, sweat glands are limited to the snout. In deer they are present at the base of the tail. In mice, rats, and cats, they are present on the paws and in rabbits they appear around the lips. In elephants, sweat glands and sebaceous glands are absent entirely.

Surface evaporation of the watery product of sweat glands helps dissipate heat, perhaps the main function of these glands. However, sweat also contains waste products; therefore, the integument represents one avenue for elimination of metabolic by-products.

The **scent glands** are derived from sweat glands and produce secretions that play a part in social communication. These glands may be located almost anywhere on the body, as on the chin (some deer, rabbits), face (deer, antelope, bats), temporal region (elephants), chest and arms (many carnivores), anal region (rodents, dogs, cats, mustelids), belly (musk deer), back (kangaroo rats, peccaries, camels, ground squirrels), or legs and feet (many ungulates). Secretions of these glands are used to mark territory, identify the individual, and communicate during courtship.

The **mammary glands** are also thought to be derived from sweat glands or perhaps from sebaceous glands. Functional only in the female, they produce **milk,** a watery mixture of fats, carbohydrates, and proteins that nourishes the young. Ectodermal mammary ridges, within which mammary glands form, are located along the ventrolateral side of the embryo. The number of mammary glands varies among species. Release of milk to a suckling is **lactation.**

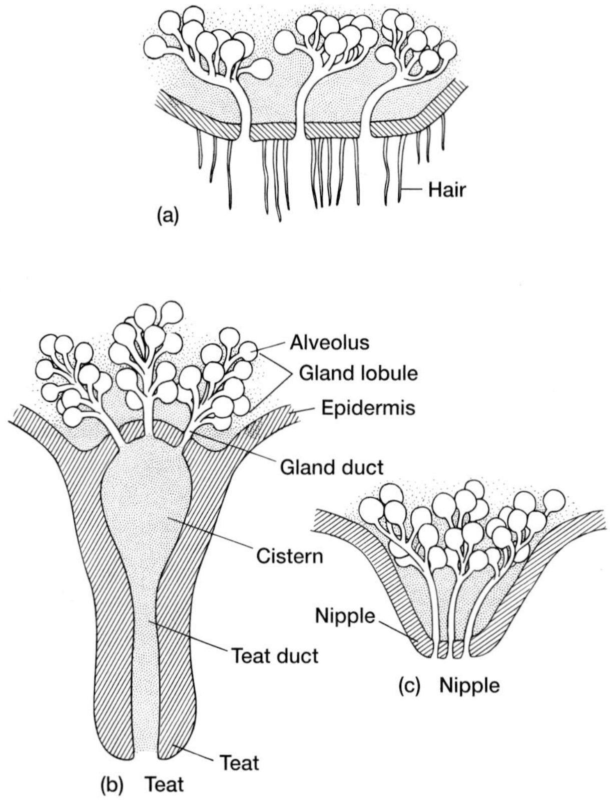


**Figure 6.22** Mammary glands. Glandular mammary tissue derived from the integument lies in the dermis, but ducts reach through the epidermis to the surface. Mammary glands are arranged in lobules, each lobule being a collection of alveoli and their immediate ducts. (a) In monotremes, the mammary glands open directly to the unspecialized skin surface, and the young press their shaped snout to the patch of skin where these glands open. (b) In some marsupials and in many placental mammals, the mammary ducts open through specializations of the integument. The teat is a tubular specialization of the epidermis that is expanded at its base into the cistern, a chamber that receives milk from the mammary glands before passing it along the common teat duct to the suckling infant. (c) The nipple is a raised epidermal papilla around which the supple lips of the infant fit directly to drink the released milk.

Mammary glands consist of numerous **lobules.** Each lobule is a cluster of secretory alveoli in which milk is produced. The alveoli can open into a common duct that, in turn, can open directly to the surface through a raised epidermal papilla, or **nipple.** The nipple is usually surrounded by a circular pigmented area of skin called the **areola.** Alveolar ducts also can open into a common chamber, or **cistern,** within a long collar of epidermis, called the **teat.** The teat forms a secondary duct carrying milk from the cistern to the surface (figure 6.22a–c). Adipose tissue can build up beneath the mammary glands to produce **breasts.**

In monotremes, nipples and teats are absent, and breasts do not form. Milk is released from ducts onto the flattened milk patch, or areola, on the surface of the skin (figure 6.22a). The front of the infant's snout is shaped to fit the surface, permitting vigorous suckling. In short 20 to 30 minute bursts of suckling, a young echidna can take in milk equivalent to about 10% of its body weight. In marsupials and placental mammals, either teats or nipples are present (figure 6.22b,c). At sexual maturity, adipose tissue builds up under the mammary gland to produce the breast. Enlargement of the mammary glands occurs under hormonal stimulation shortly before the birth of suckling young. Suckling stimulates a neural response to the nervous system that results in release of **oxytocin,** the hormone that stimulates contraction of myoepithelial cells enveloping the alveoli, and hence milk is released. In common language, this active release of milk is termed *letdown*.

**Milk release (p. 602)**

# Specializations of the Integument

## Nails, Claws, Hooves

Nails are plates of tightly compacted, cornified epithelial cells on the surface of fingers and toes; thus, they are products of the keratinizing system of the skin. The **nail matrix** forms new nail at the nail base by pushing the existing nail forward to replace that worn or broken at the free edge. Nails protect the tips of digits from inadvertent mechanical injury. They also help stabilize the skin at the tips of the fingers and toes, so that on the opposite side the skin can establish a secure friction grip on objects grasped.

Only primates have nails (figure 6.23a). In other vertebrates, the keratinizing system at the terminus of each digit produces claws or hooves (figure 6.23b,c). **Claws,** or **talons,** are curved, laterally compressed keratinized projections from the tips of digits. They are seen in some amphibians and in most birds, reptiles, and mammals. **Hooves** are enlarged keratinized plates on the tips of the ungulate digits.

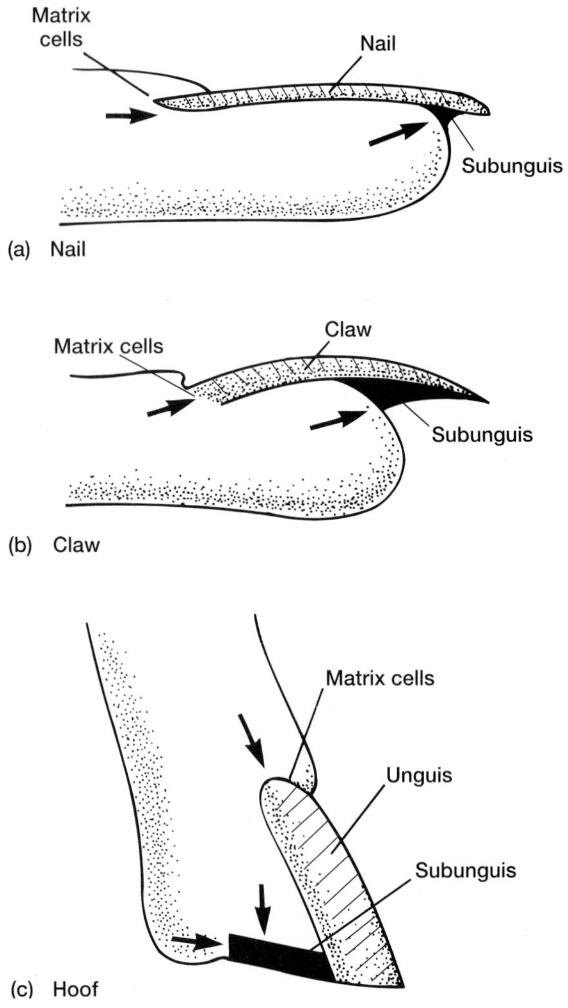


**Figure 6.23** Epidermal derivatives. (a) The nail is a plate of cornified epithelium growing outward (arrows) from proliferating matrix cells at its base and from the subunguis. (b) Claw. (c) Hoof.

## Horns and Antlers

"Horned" lizards have processes extending from behind the head that look like horns but are specialized, pointed epidermal scales. Mammals are the only vertebrates with true horns or antlers.

The skin, together with the underlying bone, contributes to both true horns and antlers. As these structures take shape, the underlying bone rises up, carrying the overlying integument with it. In **horns,** the associated integument produces a tough, cornified sheath that fits over the bony core (figure 6.24a). In **antlers,** the overlying living skin (called "velvet") apparently shapes and provides vascular supply to the growing bone. Eventually the velvet falls away to unsheath the bare bone, the actual material of the finished antlers (figure 6.24b).

True antlers occur only in members of the Cervidae (e.g., deer, elk, moose). Typically, only males have antlers, which are branched and shed annually. There are notable exceptions. Among caribou, both sexes have seasonal antlers. In deer, the antler usually consists of a main **beam,** from which branch shorter **tines,** or **points.** In yearling bucks, antlers are usually no more than prongs or spikes that may be forked. The number of tines tends to increase with age, although not exactly. In old age, antlers may even be deformed. In caribou and especially in moose, the main antler beam is compressed and **palmate,** or shovellike, with a number of points projecting from the rim.

The annual cycle of antler growth and loss in the white-tail deer, for example, is under hormonal control. In the spring, increasing length of daylight stimulates the pitu-

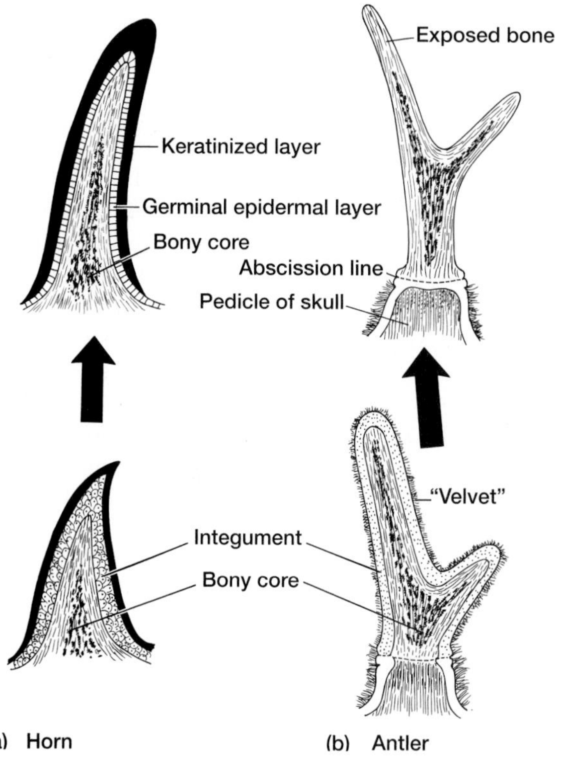


**Figure 6.24** Horns and antlers. (a) Horns appear as outgrowths of the skull beneath the integument, which forms a keratinized sheath. Horns occur in bovids of both sexes and are usually retained year round. (b) Antlers also appear as outgrowths of the skull beneath the overlying integument, which is referred to as "velvet" because of its appearance. Eventually this overlying velvet dries and falls away, leaving the bony antlers. Antlers are restricted to members of the deer family and, except for caribou (reindeer), they are present only in males. Antlers are shed and replaced annually.

itary gland at the base of the brain to release hormones that stimulate antlers to sprout from sites on the skull bones. By late spring, the growing antlers are covered by velvet. By fall, hormones produced by the testes inhibit the pituitary and the velvet dries. By thrashing and rubbing, the deer wipes the velvet off to expose the fully formed, but now dead bone of the antlers (figure 6.25a–e). Males use their antlers during clashes with other males to maintain access to reproductively receptive females. Following this brief mating season, further hormonal changes lead to a weakening of the antler at its base where it attaches to the living bone of the skull. The antlers break off, and for a short time during winter, deer are without antlers.

Among mammals, **true horns** are found among members of the family Bovidae (e.g., cattle, antelope, sheep, goats, bison, wildebeests). Commonly, horns occur in both males and females, are retained year round, and continue to grow throughout the life of the individual. The horn is unbranched and formed of a bony core and a keratinized sheath (figure 6.26). Those of the males are designed to withstand the forces encountered during head-butting combat. In large species, females usually have horns as well, although they are not as large and curved as in males. In small species, females are often hornless.

Unlike the true horns of bovids, horns of the pronghorn, family Antilocapridae, are forked in adult males. The old outer cornified sheath, but not the bony core, is shed annually in early winter (figure 6.27a). The new sheath beneath, already in place, becomes fully grown and forked by summer. Female pronghorns also have horns in which their keratinized sheath is replaced annually, but these are usually much smaller and only slightly forked. The horns of giraffes are different still. They develop from separate, cartilaginous processes that ossify, fuse to the top of the skull, and remain covered

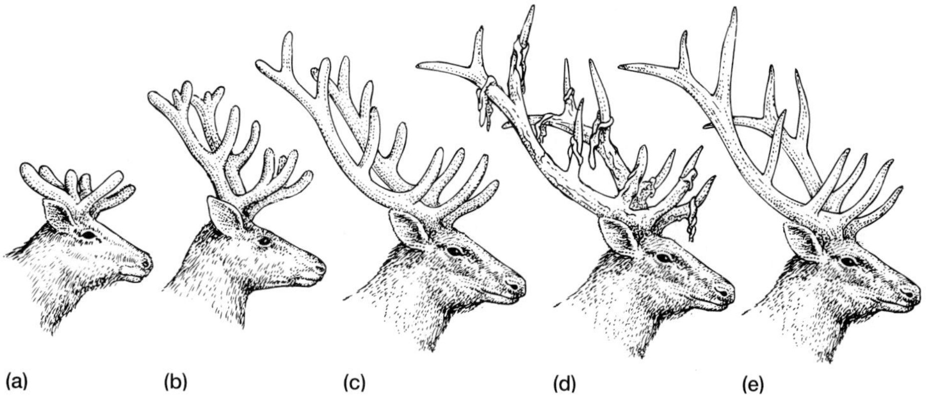


**Figure 6.25** Annual growth of elk antlers. (a,b) New antlers begin to grow in April. (c) By May, antlers are nearly fully formed even though they are still covered by the living integument (velvet). (d) By late summer, the velvet has begun to dry and peel off. (e) Fully formed bony antlers are in place.

with living, noncornified skin (figure 6.27b). The rhinoceros horn does not include a bony core, so it is exclusively a product of the integument. It forms from compacted keratinous fibers (figure 6.27c).

## Baleen

The integument within the mouths of mysticete whales forms plates of **baleen** that act as strainers to extract krill from water gulped in the distended mouth. Although it is sometimes misleadingly referred to as "whalebone," baleen contains no bone. It is a series of keratinized plates that arise from the integument. During its formation, groups of dermal papillae extend and lengthen outward, carrying the overlying epidermis. The epidermis forms a cornified layer over the surface of these projecting papillae. Collectively, these papillae and their covering of epidermis constitute the plates of frilled baleen (figure 6.28).

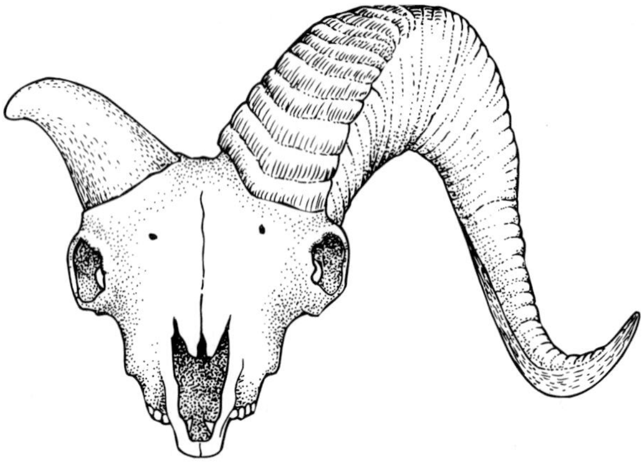

**Figure 6.26** True horns of the mountain sheep (Bovidae). The cornified covering of the horn of the mountain sheep is removed on the right side of the skull to reveal the bony core.

**Feeding by baleen whales (p. 274)**

## Scales

Scales have many functions. Both epidermal and dermal scales are hard, so when they receive mechanical insult and surface abrasion, they prevent damage to soft tissues beneath. The density of scales also makes them a barrier against invasion of foreign pathogens, and they retard water loss from the body. In sharks and other fishes, scales dampen the boundary layer turbulence to increase swimming efficiency. Some reptiles regulate the amount of surface heat they absorb by turning their bodies toward or away from the sun. This determines whether the sun rays are deflected off the full face of the scale or shine under the lifted posterior edge of the scale to reach the thin epidermis beneath.

Epidermal scales compose the major component of the skin of reptiles. They are also present in birds along their legs and in some mammals, such as the beaver, they cover the tail (figure 6.29a).

## Dermal Armor

Dermal bone forms the armor of ostracoderm and placoderm fishes. Being a product of the dermis, dermal bone finds its way into alliances with a great variety of structures. Dermal bone supports the scales of bony fishes but tends to be lost in tetrapods. It is absent in the skin of birds and most mammals. Exceptions have been noted earlier, namely, in the fossil mammal *Glyptodon* (figure 6.29b) and in the skin of the living armadillo. However, selected dermal bones take up residence in the fish skull and pectoral girdle and have persisted into modern groups of vertebrates. Most dermal bones of the skull and shoulder girdle all began phylo-

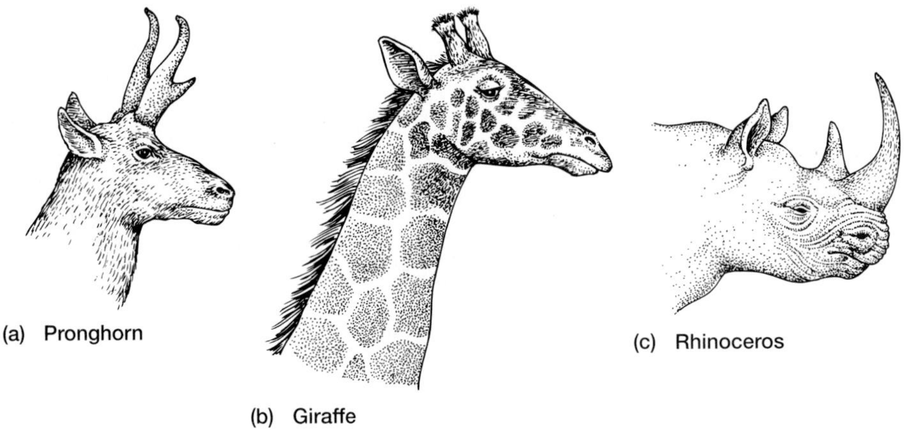


**Figure 6.27** Other types of horns. (a) In pronghorns, the bony core of the horns is unbranched, but the cornified sheath is branched. (b) Giraffe horns are small ossified knobs covered by the integument. (c) Rhinoceroses have several horns that rest on a low knob on the skull, but these horns have no inner core of bone. As outgrowths of the epidermis alone, they are mainly composed of compacted keratinized fibers.

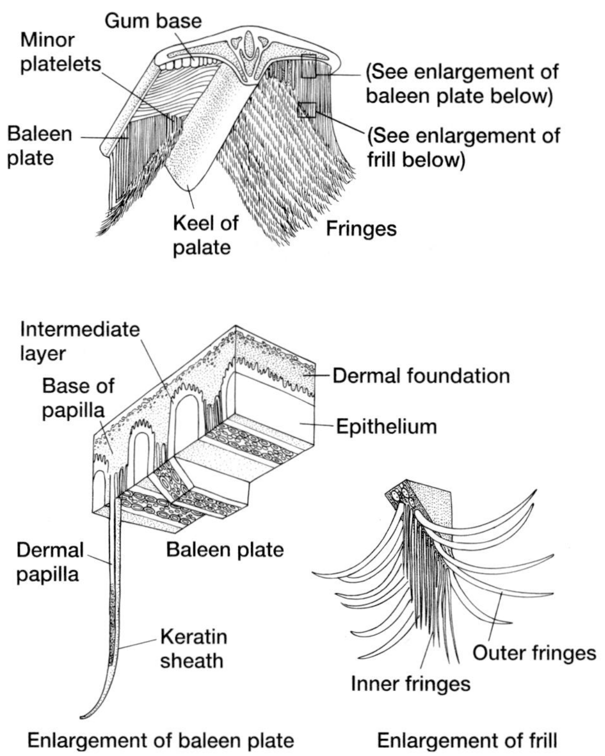


**Figure 6.28** Baleen from a whale. The lining to the mouth includes an epithelium with the ability to form keratinized structures. Groups of outgrowing epithelium become keratinized and frilly to form the baleen. Longitudinal sections of the keratinized layer and the frill are enlarged below the diagram of the baleen.

genetically in the skin and later sank inward to become parts of the skeleton. This sharing of available parts between systems reveals again the remodeling character of evolution.

The shell of turtles is a composite structure. The dorsal half of the shell is the **carapace** formed by fusion of dermal bone with expanded ribs and vertebrae (figure 6.29c). Ventrally, the **plastron** represents fused dermal bones along the belly. On the surface of both carapace and plastron, keratinized plates of epidermis cover this underlying bone.

**Turtle shell (p. 288)**

## Mucus

Mucus produced by the skin serves several functions. In aquatic vertebrates, it inhibits entrance of pathogens and may even have some slight antibacterial action. In terrestrial amphibians, mucus keeps the integument moist, allowing it to function in gas exchange. Although cutaneous respiration is prominent in amphibians, it occurs in other vertebrates as well. For example, many turtles rely on cutaneous gas exchange as they hibernate submerged in ice-covered ponds during the winter. Their shells are too thick, of course, to allow significant gas exchange, but exposed areas of skin around the cloaca offer a suitable opportunity. Sea snakes may depend on cutaneous respiration for up to 30% of their oxygen uptake. Similarly, fishes such as the plaice, European eel, and mudskipper may depend on some cutaneous gas exchange to meet their metabolic requirements.

**Cutaneous respiration (p. 403)**

Mucus is also involved in aquatic locomotion. As a surface coat, it smoothes the irregularities and rough surface features on the epidermis to reduce the friction met by a vertebrate swimming through relatively viscous water.

## Color

Skin color results from complex interactions between physical, chemical, and structural properties of the integument. Changes in blood supply can redden the skin, as in blushing. The **differential scattering** of light, referred to as Tyndall scattering, is the basis for much color in nature. This is the phenomenon that makes the clear day sky appear blue. In birds, air-filled cavities within feather barbs take advantage of this scattering phenomenon to produce the blue feathers of kingfishers, blue jays, bluebirds, and indigo buntings. Many black, brown, red, orange, and yellow colors result from pigments that produce color by selective light reflection. Interference phenomena are responsible for **iridescent colors.** As light is reflected from materials with different refractive indices, interference between different wavelengths of light produces iridescent colors. In many birds, iridescent colors result from interference of light reflected off the tiny barbs and barbules of the feathers.

Many of the pigments producing colors by this variety of physical phenomena are synthesized by and held in specialized chromatophores. Because these are cells, the suffix *-cyte* instead of *-phore* might seem more logical; however, the tradition of using the suffix *-phore* (meaning bearer of) for chromatophores and for all the various types of chromatophores is an entrenched convention, especially applied to pigment cells of ectothermic vertebrates and all invertebrates with chromatophores. We follow the widespread practice in this text. Most chromatophores arise from embryonic neural crest and can take up residence almost anywhere within the body. It is not uncommon to find them associated with the walls of the digestive tract, within the mesenteries, or around the reproductive organs. Their function at these remote sites is not resolved, but they are thought to protect deep cell layers from penetrating solar radiation.

On the basis of form, composition, and function, four groups of chromatophores are currently recognized. The most well known of these is the **melanophore** that

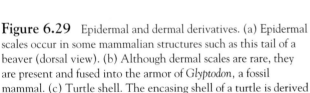

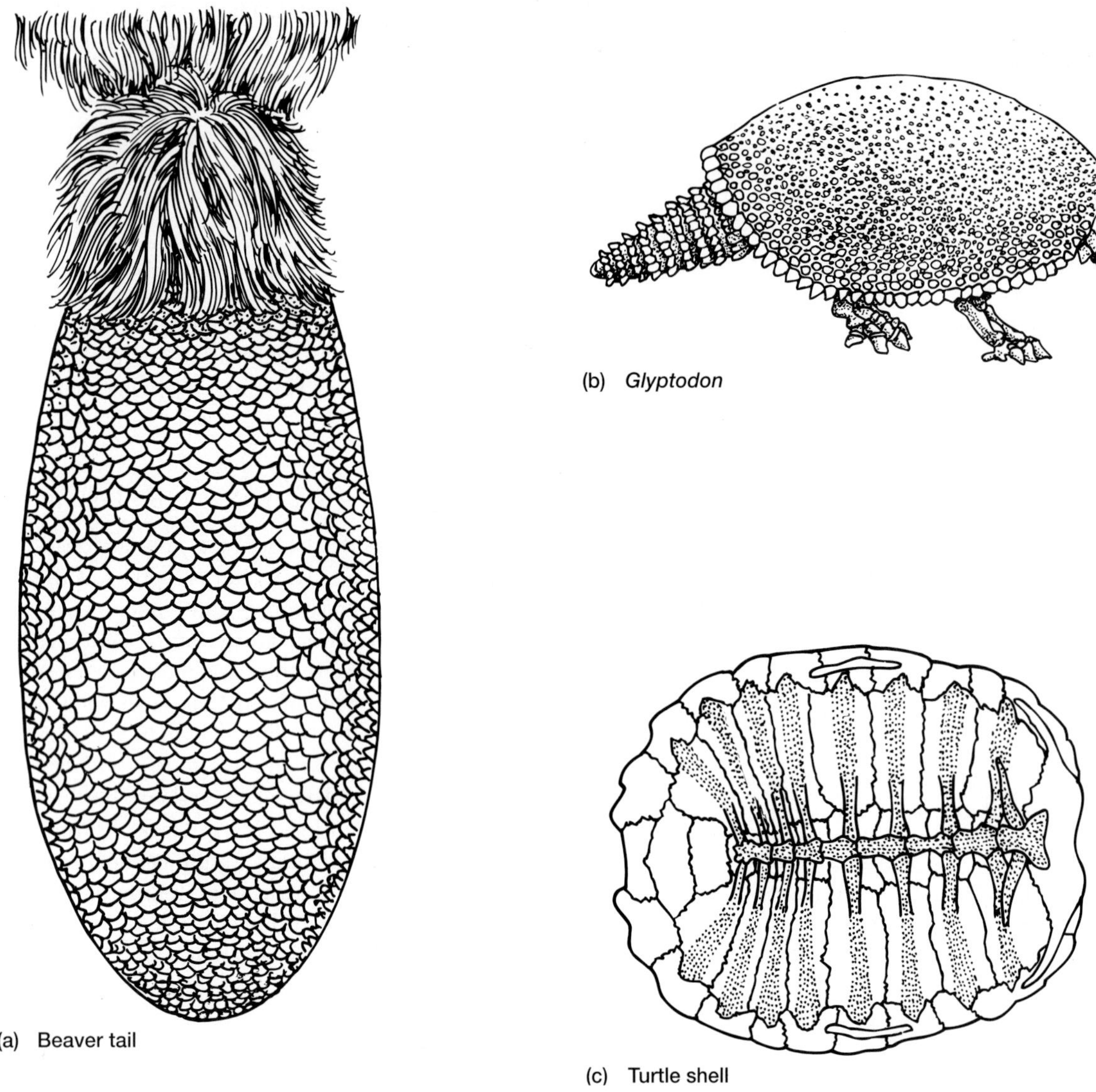


**Figure 6.29** Epidermal and dermal derivatives. (a) Epidermal scales occur in some mammalian structures such as this tail of a beaver (dorsal view). (b) Although dermal scales are rare, they are present and fused into the armor of *Glyptodon*, a fossil mammal. (c) Turtle shell. The encasing shell of a turtle is derived from three sources—the ribs and the vertebrae of the endoskeleton (stippled) and the dermal bone arising in the integument (white). The surface of this bony shell is covered by large, thin epidermal scales (not shown).

contains the pigment melanin. Cellular organelles called **melanosomes** house these melanin granules that intercept sunlight striking the surface of an animal to prevent penetration of harmful radiation. They, of course, also add color to the integument that may camouflage an animal, making it less detectable, or brighten a part that contributes to a behavioral display. There are two types of melanophores. The **dermal melanophore** is a broad, flat cell that changes color rapidly and is found only in ectotherms. The **epidermal melanophore** is a thin, elongated cell prominent in endotherms but present in all vertebrates. By contributing melanosomes, it adds color to keratinocytes, hair, and feathers.

A second type of chromatophore is the **iridophore,** which contains light-reflecting, crystalline guanine platelets. It is found in ectothermic vertebrates and in the iris of the eye of some birds. Two other types of chromatophores are the **xanthophore,** containing yellow pigments, and the **erythrophore,** so called because of its red pigments. In addition, a few chromatophores contain several of these pigments but are not classified. For example, in the iris of the Mexican ground dove, chromatophores contain both reflecting platelets (as expected in iridophores) and melanin (as in melanophores). This suggests that differentiation of chromatophores from neural crest stem cells must be responsive to a variety of developmen-

## Box Essay 6.2 Skin Color

In humans, the conversion of dehydrocholesterol to vitamin D, which is necessary for normal bone metabolism, requires small amounts of ultraviolet radiation. If vitamin D is insufficient, bones become soft and deformed. On the other hand, too much ultraviolet radiation can be very damaging to deep living tissues. Skin alone is not especially effective in reflecting or safely absorbing these wavelengths of solar radiation. This task falls to the chromatophores and the pigment they produce.

Only a few minutes of exposure to sunlight each day is necessary to convert enough precursor (dehydrocholesterol) into vitamin D to meet an individual's metabolic needs. In tropical regions near the equator, sunlight passes directly through the absorbing layers of the atmosphere to strike the surface of the Earth. Terrestrial vertebrates covered with hair, feathers, or scales have some external protection against sun exposure. Humans, who essentially lack a thick coat of hair, do not. Too much ultraviolet radiation can produce harmful amounts of vitamin D, sunburn, and a higher incidence of skin cancer. The evolution of increased numbers of chromatophores in the skin of people in tropical regions protects against too much ultraviolet radiation. In temperate regions away from the equator, the angle of incidence of sunlight is low, passing more diagonally through more atmosphere and thus filtering out much of the ultraviolet radiation. Fewer chromatophores in the skin compensate for decreased availability of ultraviolet radiation, apparently allowing just enough radiation to convert dehydrocholesterol to a sufficient amount of vitamin D. Differences in skin color among the human races are an indirect result of these adaptive compromises.

Thus, the number of chromatophores in the skin is an evolutionary adaptation to the level of exposure to ultraviolet radiation. In addition, the production of pigment granules can change in response to short-term changes in exposure to sunlight. If exposure to sunlight is reduced, the chromatophores decrease their level of synthesis of pigment granules and the skin lightens. If exposure is increased, pigment granule production increases and the skin darkens. Such tanning occurs in all races of humans, but it is most conspicuous in light-skinned Caucasians. Sudden exposure to high levels of sunlight may result in sunburn, or radiation damage to the integument. As it does with a burn from a hot stove, the skin repairs itself and sheds the damaged layers. This is why the skin "peels" several days following a sunburn.

tal cues that produce pigment cells with intermediate properties.

Sunlight can influence physiological changes in chromatophore activity. Increased exposure stimulates increased production of pigment granules, resulting in darker skin over a period of days. In some vertebrates, the response is more immediate. Some fishes and lizards can change their colors almost instantly. The true chameleon, for example, can change colors to match its environment, at least if the background is light brown to dark green. Some fishes, such as the flounder, can change not only their color but also their color pattern to resemble the background (figure 6.30a). This physiological adjustment of color to background is mediated by the endocrine system and involves redistribution of pigment granules within the chromatophores. It was once thought that chromatophores themselves changed shape, sending out cytoplasmic pseudopods. Now it appears that color changes are not based on changes in cell shape. Instead, chromatophores assume a relatively fixed shape, and in response to hormonal stimulation, their pigment granules are either shuttled out into the previously positioned pseudopods or returned to become concentrated centrally within the cell (figure 6.30b).

**Endocrine control of melanophores (p. 603)**

## Skin Evolution: The Epidermal-Dermal Unit

In a general way, it is easy to homologize integumentary structures (figure 6.31). Hair, feathers, and reptilian scales all are products of the epidermis so they are all broadly homologous. But taken separately, controversies persist about homologies. For instance, some would claim that hair is a transformed reptilian scale, originally protective in function. Others argue that hair is a derivative of epidermal bristles, originally sensory in function. Some authors point to the similarity in structure between placoid scales and shark teeth to support the view that vertebrate teeth arose from shark scales. Others dissent, pointing out that teeth were present in early fishes before sharks evolved, so shark scales could not be the forerunners of vertebrate teeth.

When approaching controversies surrounding skin evolution, we need to remember that the skin consists of two layers, an epidermis and a dermis, not a single evolving structure. Interactions between these two layers play a part in their evolution. The dermis helps maintain, regulate, and specify the types and proliferation of epidermal cells. This has been explored with experimental embryology. For example, the epidermis from the leg of a chick embryo destined to form leg scales can be peeled away from its underlying dermis and kept alive in isolation with

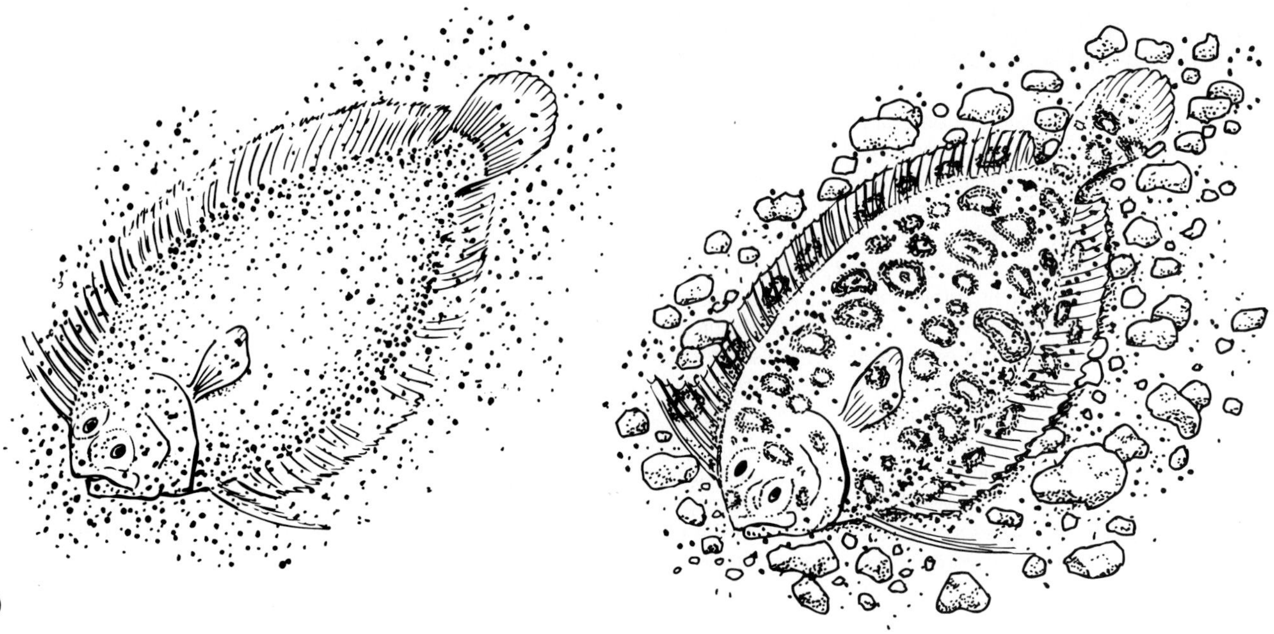


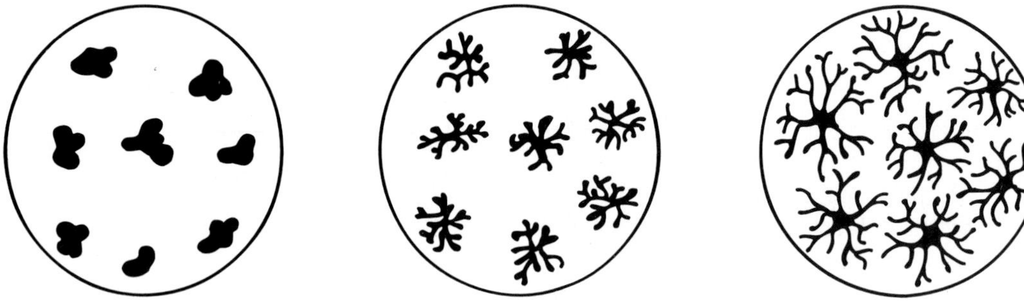


**Figure 6.30** Color changes. (a) The flounder changes its surface color as texture and pattern of the substrate changes. (b) Mediated by the endocrine system, chromatophores in the integument change the position of pigment granules within their cellular processes to change the hue and pattern of skin color.

sufficient nutrients. Cells of such living but isolated epidermis cease to proliferate. If recombined in vitro with embryonic dermis, the epidermal cells resume proliferation and scales form. We know that the stimulus is within the dermis because if any other tissue, such as cartilage or muscle, is substituted, the epidermis fails to respond.

In a few instances, the epidermis acts autonomously from the dermis. When exposed to air, isolated chick epidermis shows the intrinsic capacity to transform itself into a keratinized layer without contact with an underlying dermis. This is not well understood, but this degree of epidermal autonomy seems dependent on the ability of the epidermis to reconstruct the basement membrane or its chemical equivalent.

Despite its occasional independence, activity of the epidermis is largely influenced by the underlying dermis. Its direction of differentiation is also set by the dermis. For example, in the chick embryo, dermis from the leg promotes overlying epidermis to form keratinized scales, and dermis from the trunk induces overlying epidermis to produce feathers. If trunk dermis is experimentally replaced by leg dermis, the overlying trunk epidermis that customarily would produce feathers instead produces scalelike thickenings characteristic of the transplanted dermis. In the guinea

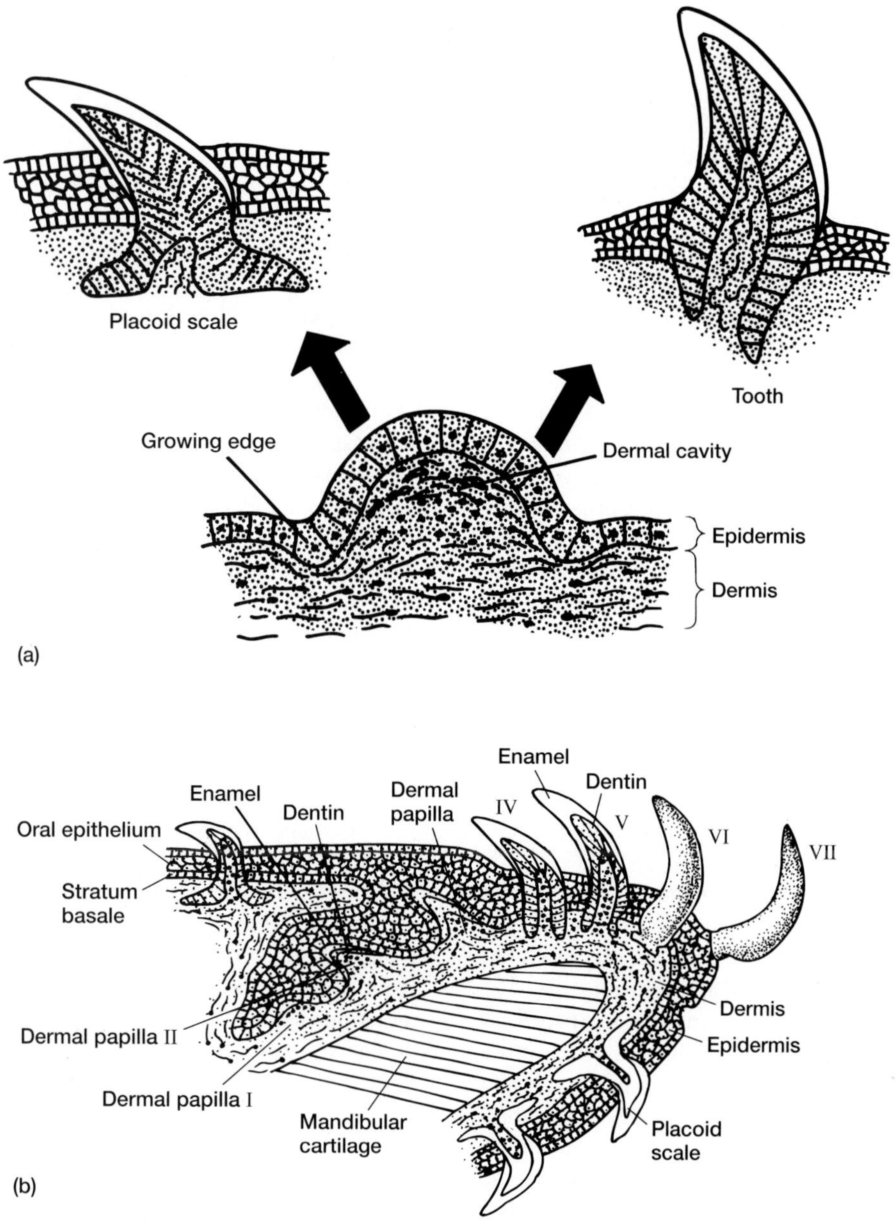


**Figure 6.31** Interaction between epidermis and dermis. (a) Interactions between dermis and epidermis produce a variety of structures, such as teeth and placoid scales. (b) Shark teeth are derivatives of the oral epithelium. The basic similarities of composition (enamel, dentin) and method of formation (epidermis-dermis) suggest that teeth and placoid scales are homologous as integumentary structures. Successive stages in tooth development are indicated by roman numerals.

pig, if dermis from the trunk, ear, or sole of the foot is transplanted beneath the epidermis elsewhere in the body, the epidermis responds by producing epidermal derivatives characteristic of the trunk, ear, or sole of the foot, respectively. In some lizards, the skin forms two types of epidermal scales, one tiny and tubercular in shape, the other large and overlapping. Scale type is determined by the underlying dermis. If embryonic dermis is switched between the two developing scale types, the overlying epidermis differentiates in accordance with its transplanted dermis. In mouse embryos, dermis specifies the type of hair as well as the general pattern of hairs produced. Upper lip dermis promotes the formation of vibrissae; trunk dermis promotes the formation of guard hairs.

In large measure, then, response of the epidermis is specific to the type of underlying dermis. To some extent, age of an experimentally transplanted dermis or epidermis influences this response. Results tend to vary if the source of one is from a young embryo and the source of the other is from an older embryo or adult. Nevertheless, under normal circumstances, the dermis seems to bring a necessary physical substrate and organization along with nutrient supply to the epidermis. This stimulatory effect of dermis upon epidermis is **embryonic induction.** Although the dermis does not directly contribute cells of its own to the epidermal derivative (hair, feather, scale), it induces the type of epidermal specialization. The epidermis responds by altering activity of its germinal layer to produce the specified structure.

This epidermal-dermal interaction is evident even in tissue transplants between species from different classes. However, the dermis usually cannot induce the epidermis to form a specialization that is not typical of its class. Lizard epidermis can be paired with chick or mouse dermis. Likewise chick epidermis can be paired with lizard or mouse dermis, and mouse epidermis can be paired with lizard or chick dermis. In these reciprocal transplants between reptile (lizard) and bird (chick) and between reptile (lizard) and mammal (mouse), the type of skin specialization induced (scale, feather, or hair) conforms to the origin of the epidermis, not to the transplanted dermis. Thus, the lizard epidermis is induced to form a reptilian scale, the chick epidermis to form a feather, the mouse epidermis to form hair, irrespective of the origin of the dermis with which it is paired. Interestingly, if the transplanted dermis does not come from a region producing a skin specialization, it seems to lack the necessary ability to induce the cross-class epidermis to form a specialization. Further, specializations induced by these cross-class transplants of dermis do not develop fully. Lizard scales, bird feathers, and mammal hair form but cease to grow after a certain stage. Apparently, the foreign dermis is sufficient to stimulate epidermal proliferation, but it cannot specify the type of epidermal derivative.

The evolution of the skin, in particular its specializations, has apparently involved changes in the ability of the dermis to induce and the epidermis to respond, as well as in the interactions between them. From experimental embryology on living forms, we realize that if we speak only of the evolution of epidermal structures, we are neglecting the role of the dermis in this process. Although the dermis may not actually contribute cells to specialized skin derivatives, it is indispensable for their normal formation. Remove the dermis from the pulp cavity of a placoid scale, and the enamel and dentin fail to form normally. If the dermis is missing beneath the pulp cavity of a forming tooth, the tooth enamel forms incompletely. And the reverse holds. Remove the epidermis and the dermis alone is unable to form a placoid scale or vertebrate tooth properly. Interaction of epidermis and dermis is necessary to produce a normal skin derivative.

Experimental embryology has extended this insight into evolutionary events. Modern birds, of course, lack teeth. The young chick breaking out of the eggshell uses what is termed an "egg tooth." In reality, this is not a tooth at all but a projection on the cornified bill. That is why recent research has been surprising. Koller and Fisher took tooth-inducing dermis from a mouse jaw, placed it under a bird's beak, and allowed the pair to differentiate. In several successful experiments, rudimentary teeth appeared. Chick epidermis had been induced by mouse dermis to form teeth! Although teeth do not form in modern birds, the epidermis of birds has not entirely lost its tooth-forming potential. This latent potential in birds is not expressed because the inductive interaction between bird dermis and epidermis has been lost. The epidermis is present and the dermis is present, but in birds their interaction has changed.

Perhaps the focus of evolutionary events in the integument has been as much on this interaction as on the layers themselves. Obviously, interactions do not fossilize and they are hard to characterize structurally. It is little wonder that controversies about homology exist. If we think of the epidermis, the dermis, and their interaction as an evolving unit, then their specialized products (hair, feathers, and reptilian scales) are broadly homologous. Shark scales, vertebrate teeth, and bony fish scales can be seen as products of this interacting epidermal-dermal system; therefore, they are homologous integumentary structures.

# Selected References

Bereiter-Hahn, J., A. G. Matoltsy, and K. S. Richards (eds). 1986. *Biology of the integument*, vol. 2. Berlin: Springer-Verlag.

Fritzch, B., and U. Wahnschoffe. 1983. The electroreceptive ampullary organs of urodeles. *Cell Tissue Res.* 229:483–503.

Halata, Z., and B. L. Munger. 1983. The sensory innovation of primate facial skin II. Vermillion border and mucosa of lip. *Brain Res. Rev.* 5:81–107.

Jakobowski, M. 1980. Size and vascularization of the gill and skin respiratory surfaces in the white amur, *Ctenopharyngodon idella* (Val.) (Pisces, Cyprinidae). *Zoologia* 24:93–104.

Koller, E. J., and C. Fisher. 1980. Tooth induction in chick epithelium: Expression of quiescent genes for enamel synthesis. *Science* 207:993–95.

Lethbridge, R. C., and I. C. Potter. 1981. The skin. In *The biology of lampreys*, edited by M. W. Hardsty and I. C. Potter. London: Academic Press, pp. 377–448.

Maderson, P.F.A. 1965. Histological changes in the epidermis of snakes during the sloughing cycle. *Amer. Zool.* 146:98–113.

Maderson, P.F.A. 1972. When? Why? and How?: Some speculations on the evolution of the vertebrate integument. *Amer. Zool.* 12:159–71.

Maderson, P.F.A. 1975. Embryonic tissue interactions as the basis for morphological change in evolution. *Amer. Zool.* 15:315–27.

Maderson, P.F.A. 1985. Some developmental problems of the reptilian integument. *Biology of the Reptilia* 14:523–98.

McGowan, C. 1989. Feather structure in flightless birds and its bearing on the question of the origin of feathers. *J. Zool. (London)* 218:537–47.

Munger, B. L., and Z. Halata. The sensory innovation of primate facial skin I. Hairy skin. *Brain Res. Rev.* 5:45–80.

Norberg, R. A. 1985. Function of vane asymmetry and shaft curvature in bird flight feathers; inferences on flight ability of *Archaeopteryx*. In *The beginnings of birds*, edited by M. K. Hecht, J. H. Ostrom, G. Viohl, and P. Wellnhofer. Eichstätt, Germany: Bronner and Daentler, pp. 303–18.

Regal, P. J. 1975. The evolutionary origin of feathers. *Quart. Rev. Biol.* 50:35–66.

Regal, P. J. 1985. Common sense and reconstructions of the biology of fossils: *Archaeopteryx* and feathers. In *The beginnings of birds*, edited by M. K. Hecht, J. H. Ostrom, G. Viohl, and P. Wellnhofer. Eichstätt, Germany: Bronner and Daentler, pp. 67–74.

Reif, W. E. 1992. Evolution of dermal skeleton and dentition in vertebrates. The Odonatode regulation theory. *Evol. Biol.* 15:287–368.

Sengel, P. 1986. Epidermal-dermal interaction. In *Biology of the integument*, vol. 2, edited by J. Bereiter-Hahn, A. G. Matoltsy, and K. S. Richards. Berlin: Springer-Verlag, pp. 374–408.

Spearman, R.I.C., and P. A. Riley (eds). 1980. The skin of vertebrates. *Linn. Soc. Symp.* Ser no. 9.

Tarsitano, S. F. 1985. The morphological and aerodynamic constraints on the origin of avian flight. In *The beginnings of birds*, edited by M. K. Hecht, J. H. Ostrom, G. Viohl, and P. Wellnhofer. Eichstätt, Germany: Bronner and Daentler, pp. 319–32.

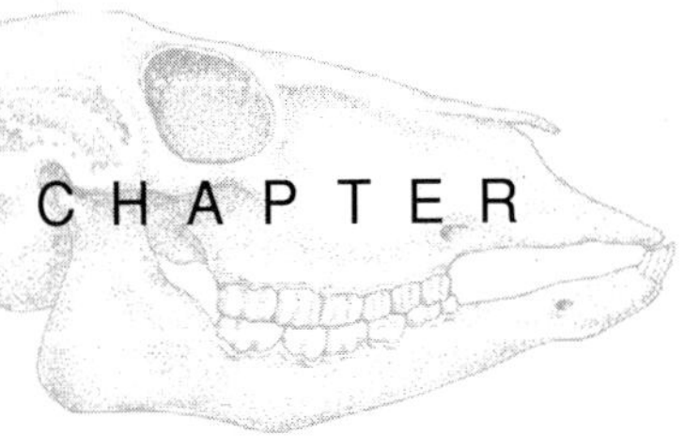

# 7 CHAPTER

# Skeletal System: The Skull



The skeleton gives the vertebrate body shape, supports its weight, offers a system of levers that together with muscles produce movement, and protects soft parts such as nerves, blood vessels, and other viscera. Because it is hard, bits of the skeleton often survive fossilization, so our most direct contact with long-extinct animals is often through their skeletons. In the architecture of the skeleton is written the story of vertebrate function and evolution.

The skeletal system is composed of an exoskeleton and an endoskeleton (figure 7.1a). The **exoskeleton** is formed from or within the integument, the dermis giving rise to bone and the epidermis to keratin. The **endoskeleton** forms deep within the body from mesoderm and other sources, not directly from the integument. Tissues contributing to the endoskeleton include fibrous connective tissue, bone, and cartilage.

During the course of vertebrate evolution, most bones of the exoskeleton stay within the integument and protect surface structures. Dermal armor of ostracoderms and bony scales of fishes are examples. Other bones have sunk inward, merging with deeper bones and cartilaginous elements of the endoskeleton to form composite structures. As a practical matter, this makes it difficult to examine the exoskeleton and the endoskeleton separately. Parts of one are often found in company with the other. Instead, we select composite structural units and follow their evolution. This way of dividing the skeleton for study gives us two units, the skull, or cranial skeleton, and the postcranial skeleton (figure 7.1b). The postcranial skeleton includes the vertebral column, limbs, girdles, and associated structures, such as ribs and shells. In the next two chapters, we examine the postcranial skeleton. Our discussion of the skeleton begins with the skull.

## INTRODUCTION

Although merged into a harmonious unit, the vertebrate skull or **cranium** is actually a composite structure formed of three distinct parts. Each part of the skull arises from a separate phylogenetic source. The most ancient part is the

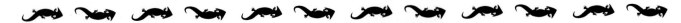

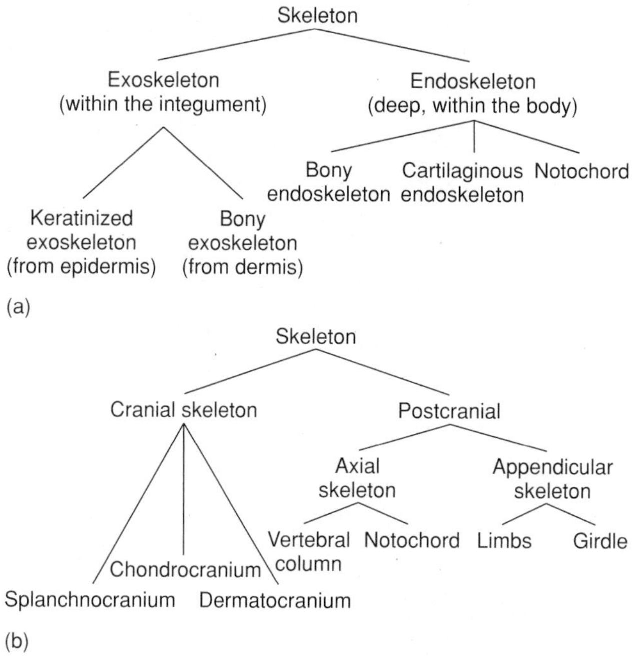


**Figure 7.1** Organization of skeletal tissues in vertebrates. Components of the skeletal system function together as a unit but, as a convenience, they can be divided into manageable parts for closer analysis. (a) As a protective and supportive system, the skeleton can be divided into structures on the outside (exoskeleton) and inside (endoskeleton) of the body. (b) On the basis of position, the skeleton can be treated as two separate components, the cranial skeleton (skull) and the postcranial skeleton. The postcranial skeleton includes the axial and appendicular skeletons.

**splanchnocranium (visceral cranium),** which first arose to support pharyngeal slits in protochordates (figure 7.2a). The second part, the **chondrocranium,** underlies and supports the brain and is formed of endochondral bone or of cartilage (figure 7.2b). The third part of the skull is the **dermatocranium,** a contribution that in later vertebrates forms most of the outer casing of the skull. As its name suggests, the dermatocranium is composed of dermal bones (figure 7.2c).

**Endochondral and dermal bone (p. 173)**

In addition to these formal components, two general terms apply to parts of the cranium as well. The **braincase** is a collective term that refers to the fused cranial components immediately surrounding and encasing the brain. Structures of the dermatocranium, the chondrocranium, and even the splanchnocranium can make up the braincase, depending on the species. The **neurocranium** is used as an equivalent term for the chondrocranium by some morphologists. Others expand the term to include the chondrocranium along with fused or attached sensory capsules—the supportive nasal, optic, and otic capsules. Still others consider the neurocranium to be only the ossified parts of the chondrocranium. Be prepared for slightly different meanings in the literature. Although I use the term *neurocranium* sparingly, I mean neurocranium to include the braincase (ossified or not) plus associated sensory capsules.

# Chondrocranium

Elements of the chondrocranium appear to lie in series with the bases of the vertebrae. This arrangement inspired several morphologists of the nineteenth century to propose that the primitive vertebral column initially extended into the head to produce the skull. By selective enlargement and fusion, these intruding vertebral elements were seen as the evolutionary source of the chondrocranium. Consequently, the idea grew that the head was organized on a segmental plan like the vertebral column that produced it. Today this view is not held as confidently, although many allow that the occipital arch forming the back wall of the skull may represent several ancient vertebral segments that now contribute to the posterior wall of the chondrocranium (table 7.1).

In elasmobranchs, the expanded and enveloping chondrocranium supports and protects the brain within. However, in most vertebrates, the chondrocranium is primarily an embryonic structure serving as a scaffold for the developing brain and as a support for the sensory capsules.

## Embryology

Although the embryonic formation of the chondrocranium is understood, details may differ considerably from one species to another. Generally, condensations of head mesenchyme form elongate cartilages next to the notochord. The anterior pair are the **trabeculae,** the posterior pair the **parachordals,** and in some vertebrates, a pair of **polar cartilages** lies between them (figure 7.3a). Behind the parachordals, several **occipital cartilages** usually appear as well. In addition to these cartilages, the sensory capsules associated with the nose, eyes, and ears develop supporting cartilages—**nasal, optic,** and **otic capsules,** respectively. Two types of embryonic cells differentiate to form the chondrocranium. Neural crest cells contribute to the nasal capsule, trabeculae (possibly only the anterior part), and perhaps to part of the otic capsule (figure 7.4a). Mesenchyme of mesodermal origin contributes to the rest of the chondrocranium (figure 7.4b). As development proceeds, these cartilages fuse. The region between the nasal capsules formed by the fusion of the anterior tips of the trabeculae is the **ethmoid plate.** The parachordals grow together across the midline to form the **basal plate** between the otic capsules. The occipitals grow upward and around the nerve cord to form the **occipital arch**

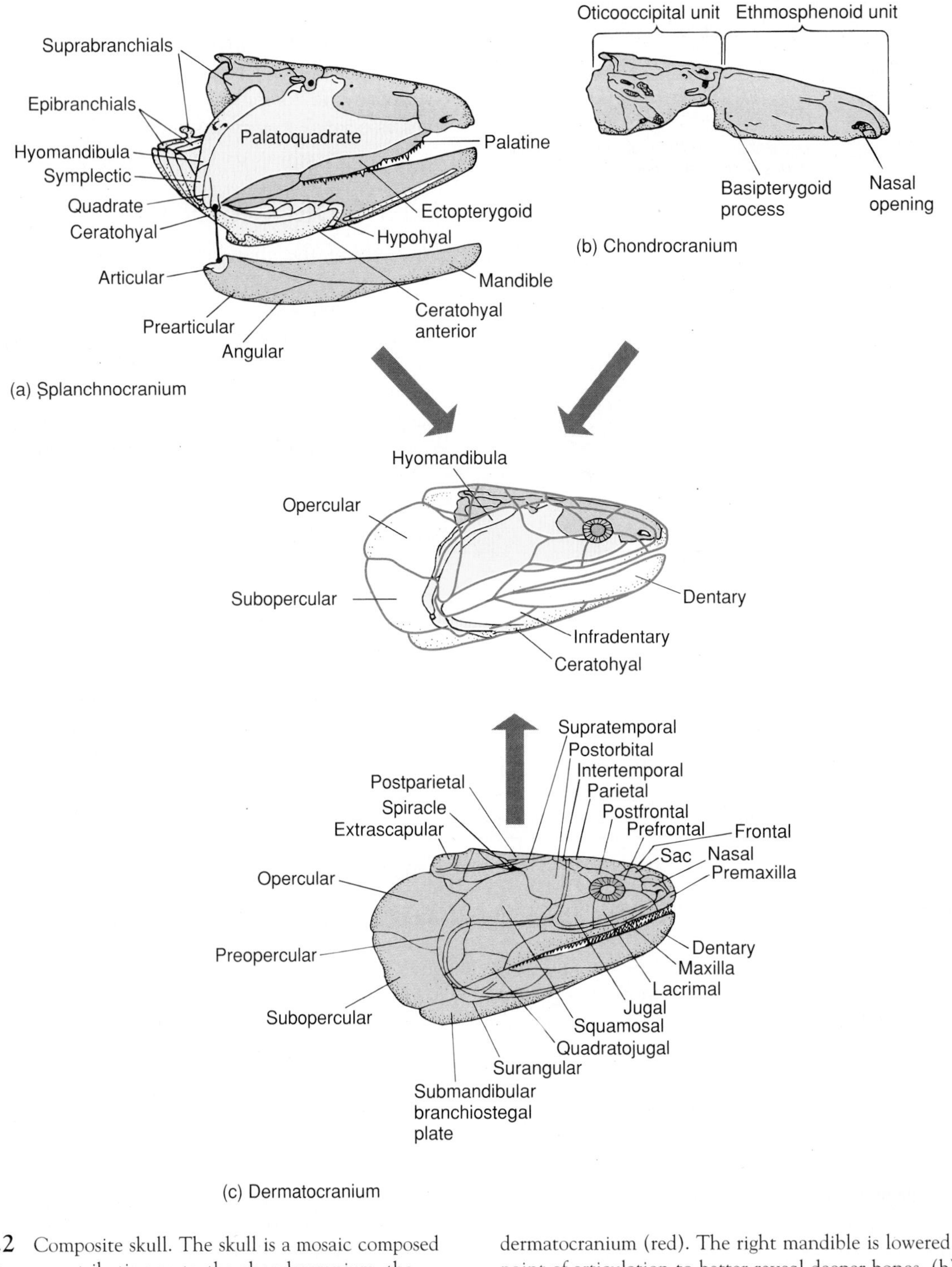


**Figure 7.2** Composite skull. The skull is a mosaic composed of three primary contributing parts, the chondrocranium, the splanchnocranium, and the dermatocranium. Each has a separate evolutionary background. The skull of Eusthenopteron, a Devonian crossopterygian fish, illustrates how parts of all three phylogenetic sources contribute to the unit. (a) The splanchnocranium (yellow) arose first and is shown in association with the chondrocranium (blue) and parts of the dermatocranium (red). The right mandible is lowered from its point of articulation to better reveal deeper bones. (b) The chondrocranium in Eusthenopteron is formed by the union between the anterior ethmosphenoid and the posterior oticooccipital units. (c) The superficial wall of bones compose the dermatocranium. The central figure depicts the relative position of each contributing set of bones brought together in the composite skull. (Sac: nasal series)

TABLE 7.1 Endochondral Contributions to the Chondrocranium

| Endochondral Structure | Fishes (Teleost) | Amphibians | Reptiles/Birds | Mammals |
|---|---|---|---|---|
| Occipital bones | Supraoccipital<br>Exoccipital<br>Basioccipital | Supraoccipital<br>Exoccipital<br>Basioccipital | Supraoccipital<br>Exoccipital<br>Basioccipital | Supraoccipital, Exoccipital, Basioccipital } Occipital bone |
| Mesethmoid bone | Mesethmoid[a] (internasal) | Absent | Absent | Mesethmoid (absent in primitive mammals, ungulates) } Ethmoid |
| Ethmoid region | Ossified | Unossified | Unossified | Turbinals (ethmo-, naso-, maxillo-) } Ethmoid |
| Sphenoid bones | | | | |
| Sphenethmoid | Sphenethmoid | Sphenethmoid | Sphenethmoid | Presphenoid } Sphenoid[c] |
| Orbitosphenoid | Orbitosphenoid | Orbitosphenoid | Orbitosphenoid | Orbitosphenoid } Sphenoid[c] |
| Basisphenoid | [Basisphenoid][b] | Basisphenoid | Basisphenoid | Basisphenoid } Sphenoid[c] |
| Pleurosphenoid | Pleurosphenoid | ? | Pleurosphenoid (crocodilians, amphisbaenians) | Absent |
| Laterosphenoid | | | Laterosphenoid (snakes) | Absent |
| Otic capsule<br>Periotic | Prootic<br>Epiotic<br>Sphenotic | Prootic<br>Opisthotic | Prootic<br>Opisthotic<br>Epiotic (absent in birds) | Petrosal with mastoid process |

[a]This bone is of dermal origin.
[b]This bone is usually absent or reduced in fishes.
[c]Alisphenoid from the splanchnocranium contributes.

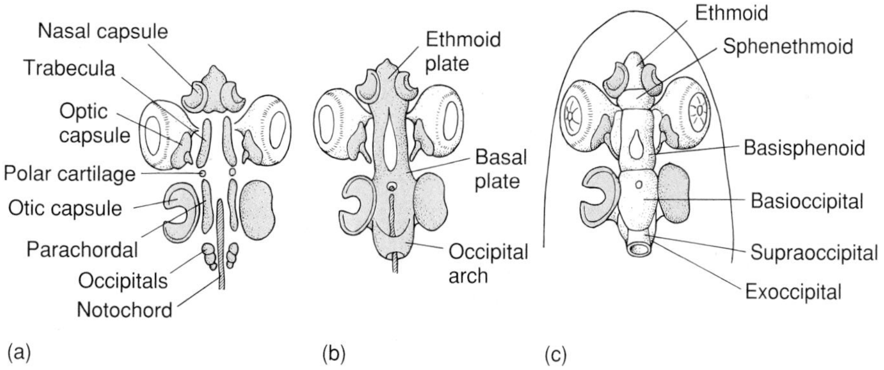


**Figure 7.3** Embryonic development of the chondrocranium. Cartilage (blue) appears first but in most vertebrates is replaced by bone (white) later in development. The chondrocranium includes these cartilaginous elements that form the base and back of the skull together with the supportive capsules around sensory organs. Early condensation of mesenchymal cells differentiates into cartilage (a) that grows and fuses together to produce the basic ethmoid, basal, and occipital regions (b) that later ossify (c), forming basic bones and sensory capsules.

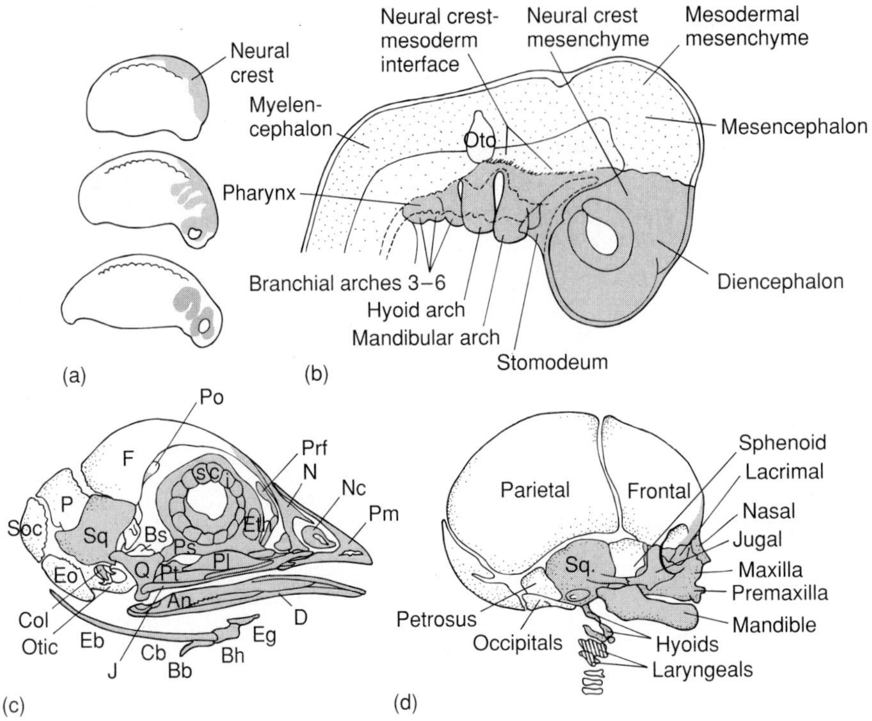


**Figure 7.4** Neural crest contributions to the skull. (a) Salamander embryo illustrating the sequential spread of neural crest cells. During early embryonic development, neural crest cells contribute to the head mesenchyme, which is called the ectomesoderm because of its neural crest origin. (b) Also contributing to the head mesenchyme are cells of mesodermal origin, the mesodermal mesenchyme. The position of the mesodermal (stippled) and the neural crest (shaded) mesenchyme, and the approximate interface between them, are indicated in the chick embryo. Skull of a chick (c) and a human fetus (d) show bones or portions of bones derived from neural crest cells (shaded). Abbreviations: angular (An), basibranchial (Bb), basihyal (Bh), basisphenoid (Bs), ceratobranchial (Cb), columella (Col), dentary (D), epibranchial (Eb), entoglossum (Eg), exoccipital (Eo), ethmoid (Eth), frontal (F), jugal (J), nasal (N), cartilage nasal capsule (Nc), parietal (P), palatine (Pl), premaxilla (Pm), postorbital (Po), prefrontal (Prf), parasphenoid (Ps), pterygoid (Pt), quadrate (Q), scleral ossicle (Sci), supraoccipital (So), squamosal (Sq).

(figure 7.3b). Collectively, all these expanded and fused cartilages constitute the chondrocranium.

In elasmobranchs, the chondrocranium does not ossify. Instead the cartilage grows still farther upward and over the brain to complete the protective walls and roof of the braincase. In most other vertebrates, the chondrocranium becomes partly or entirely ossified (figure 7.3c).

## Splanchnocranium

The splanchnocranium is an ancient chordate structure. In amphioxus, the splanchnocranium, or at least its forerunner, is associated with the filter-feeding surfaces.

Among vertebrates, the splanchnocranium generally supports the gills and offers attachment for the respiratory muscles. Elements of the splanchnocranium contribute to the jaws and hyoid apparatus of gnathostomes.

### Embryology

The mistaken view that the splanchnocranium developed from the same embryonic source as the walls of the digestive tract inspired the name "visceral" cranium, a name that unfortunately has stuck despite the misnomer. Embryologically, the splanchnocranium arises from neural crest cells, *not* from lateral plate mesoderm like the smooth muscle in the walls of the digestive tract. In protochordates, neural crest cells are absent. Pharyngeal bars, composed of fibrous connective tissue, but never bone, arise from mesoderm and form the unjointed branchial basket, the phylogenetic predecessor of the vertebrate splanchnocranium. In vertebrates, cells of the neural crest depart from the sides of the neural tube and move into the walls of the pharynx between successive pharyngeal slits to differentiate into the respective pharyngeal arches. Pharyngeal arches of aquatic vertebrates usually are associated with their respiratory gill system. Because of this association they are referred to as **branchial arches,** or **gill arches.**

Each arch can be comprised of a series of up to five articulated elements per side, beginning with the **pharyngobranchial** element dorsally and then in descending order the **epibranchial, ceratobranchial, hypobranchial,** and **basibranchial** elements (figure 7.5). One or more of these anterior branchial arches may come to border the mouth, support soft tissue, and bear teeth. Branchial arches that support the mouth are called **jaws,** and each contributing arch is numbered sequentially or named. The first fully functional arch of the jaw is the **mandibular arch,** the largest and most anterior of the modified series of arches. The mandibular arch is composed of the **palatoquadrate** dorsally and **Meckel's cartilage** (mandibular cartilage) ventrally. The **hyoid arch,** whose most prominent element is the **hyomandibula,** follows the mandibular arch. A varying number of branchial arches, often designated with roman numerals, follow the hyoid arch (figure 7.5).

## Origin of Jaws

In agnathans, the mouth is neither defined nor supported by jaws. Instead, the splanchnocranium supports the roof of the pharynx and lateral pharyngeal slits. Lacking jaws, ostracoderms would have been restricted to a diet of small, particulate food. The ciliary-mucus feeding surfaces of protochordates probably continued to play a large part in the food-gathering technique of ostracoderms. In some groups, small teethlike structures, derived from surface scales, surrounded the mouth. Perhaps ostracoderms used these rough "teeth" to scrape rock surfaces and dislodge encrusted algae or other organisms. As these food particles became suspended in water, ostracoderms drew them into their mouth with the incurrent flow of water. The mucus-lined walls of the pharynx collected these dislodged food particles from the passing stream.

Jaws appear first in acanthodian and placoderm fishes that used them as food traps to grab whole prey or take bites from large prey. Within some groups, jaws also served as crushing or chewing devices to process food in the mouth. With the advent of jaws, these fishes became more free-ranging predators of open waters.

Jaws arose from one of the anterior pair of gill arches. Evidence supporting this comes from several sources. First, the embryology of sharks suggests that jaws and branchial arches develop similarly in series (figure 7.6) and both arise from neural crest. The spiracle appears to have once been a full-sized gill slit, but in modern sharks it is crowded and much reduced by the enlarged hyoid arch next in series. Furthermore, nerves and blood vessels are distributed in a pattern similar to branchial arches and jaws. Finally, the musculature of the jaws appears to be transformed and modified from branchial arch musculature.

So it seems reasonable to conclude that branchial arches phylogenetically gave rise to jaws. But the specifics remain controversial. For example, we are not sure whether jaws represent derivatives of the first, second, third, or even

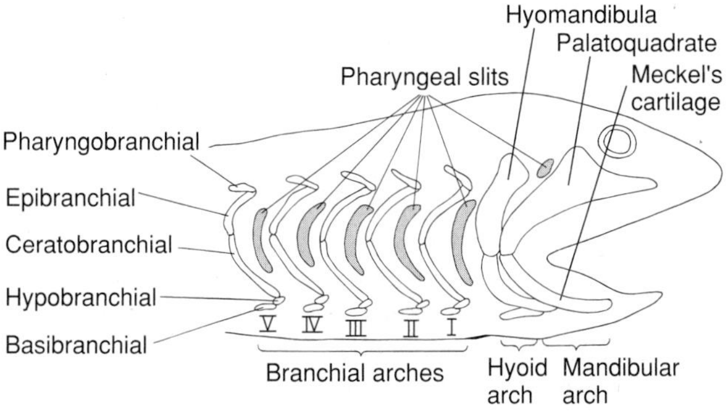


**Figure 7.5** Primitive splanchnocranium. Seven arches are shown. Up to five elements compose an arch on each side beginning with the pharyngobranchial dorsally and in sequence to the basibranchials most ventrally. The first two complete arches are named, mandibular arch for the first and hyoid arch for the second that supports it. The characteristic five-arch elements are reduced to just two in the mandibular arch, the palatoquadrate and Meckel's cartilage. The large hyomandibula, derived from an epibranchial element, is the most prominent component of the next arch, the hyoid arch. Behind the hyoid arch are variable numbers of branchial arches I, II, and so on.

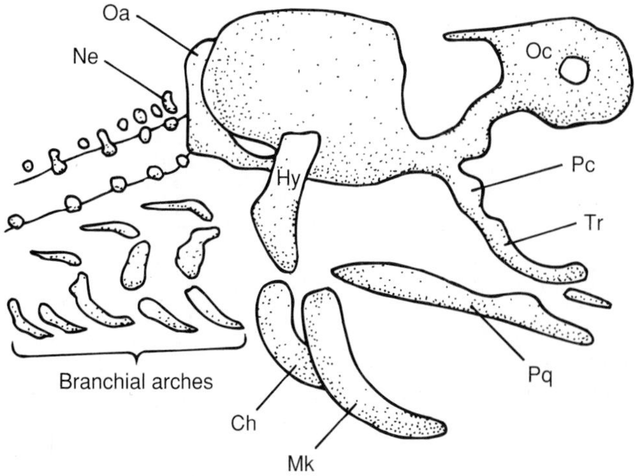


**Figure 7.6** Shark embryo, the dogfish *Scyllium*. Jaws appear to be in series with the branchial arches. The mandibular arch is first, followed by the hyoid and then several branchial arches. Such a position of the jaws, in series with the arches, is taken as evidence that the jaws derive from the most anterior branchial arch. Abbreviations: ceratohyal (Ch), hyomandibula (Hy), Meckel's cartilage (Mk), neural arch (Ne), occipital arch (Oa), orbital cartilage (Oc), polar cartilage (Pc), palatoquadrate (Pq), trabecula (Tr).

fourth branchial arches of primitive ancestors. Derivation of the mandibular arch also excites some controversy. The **serial theory** is the simplest view and holds that the first or perhaps second ancient branchial arch gave rise exclusively to the mandibular arch, the next branchial arch exclusively to the hyoid arch, and the rest of the arches to the branchial arches of gnathostomes (figure 7.7a).

Erik Jarvik, a Swedish paleontologist, proposed the **composite theory,** a more complex view based on his examination of fossil fish skulls and embryology of living forms (figure 7.7b). He hypothesized that ten branchial arches were present in primitive species, the first and following arches being named terminal, premandibular, mandibular, hyoid, and six branchial arches. Rather than the "one arch, one mandible" view, he envisioned a complex series of losses or fusions between selective parts of several arches that came together to produce the single composite mandible. According to his theory, the mandibular arch of gnathostomes is formed by fusion of parts of the premandibular arch and parts of the mandibular arch of jawless ancestors. The palatoquadrate forms from the fusion of the epibranchial of the premandibular arch with the epibranchial and one pharyngobranchial of the mandibular arch. Meckel's cartilage arises from the expanded ceratobranchial element. Next, the hyomandibular arch arises phylogenetically from the ceratobranchial and hypobranchial elements of the third primitive gill arch. The remaining branchial arches persist in serial order. The other elements of the primitive arches are lost or fused to the neurocranium.

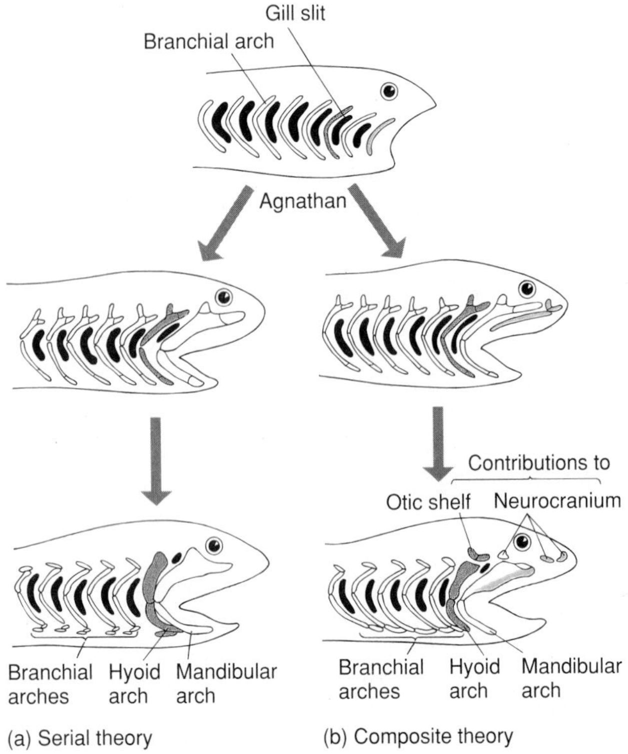


**Figure 7.7** Serial and composite theories of jaw development. (a) The serial theory holds that jaws arise completely from one of the anterior branchial arches. Elements may be lost within it, but other elements from other arches do not contribute. (b) In the composite theory, the mandibular arch is formed from elements of several adjacent arches that also contribute to the neurocranium.

Descriptive embryology provides much of the evidence put forth in these theories. However, descriptive embryology alone cannot trace arch components from embryo to adult structures with complete confidence. We can look forward to the use of more modern techniques to help settle this. For example, populations of cells can be marked with chemical or cellular markers early in embryonic development and followed to eventual sites of residence in the adult. These markers would permit us to detect the contributions of gill arches to jaws or chondrocranium. Nevertheless, even though some argue over details, we know in general that vertebrate jaws are derivatives of ancient gill arches (table 7.2).

## Types of Jaw Attachments

Because of the mandible's prominence, evolution of the jaws is often traced through how the mandible is attached (i.e., its **suspensorium**) to the skull (figure 7.8). Agnathans represent the earliest **paleostylic** stage in which none of the arches attach themselves directly to the skull. The earliest jawed condition is **euautostylic,** found in placoderms and acanthodians. The mandibular arch is suspended from the skull by itself (hence "auto"), without help from the hyoid arch. In early sharks, some osteichthyans, and crossopterygians, jaw suspension is **amphistylic;** that is, the jaws are attached to the braincase through two primary articulations, anteriorly by a ligament connecting the palatoquadrate to the skull and posteriorly by the hyomandibula. Many, perhaps most, modern sharks exhibit a variation of amphistylic jaw suspension. In most modern bony fishes, jaw suspension is **hyostylic** because the mandibular arch is attached to the braincase primarily through the hyomandibula. Often a new dermal element, the **symplectic bone,** aids in jaw suspension. The visceral cranium remains cartilaginous in elasmobranchs, but within bony fishes and later tetrapods, ossification centers appear, forming distinctive bony contributions to the skull. In most amphibians, reptiles, and birds, jaw suspension is **metautostylic.** Jaws are attached to the braincase directly through the quadrate, a bone formed in the posterior part of the palatoquadrate (figure 7.8). The hyomandibula plays no part in supporting the jaws; instead, it gives rise to the slender **columella** or **stapes,** involved in hearing. Other elements of the second arch and parts of the third contribute to the **hyoid** or **hyoid apparatus** that supports the tongue and the floor of the mouth. In mammals, jaw suspension is **craniostylic.** The entire upper jaw is incorporated into the braincase, but the lower

TABLE 7.2 Derivatives of Branchial Arches in Sharks, Teleosts, and Tetrapods

| Arch | Sharks | Teleosts | Amphibians | Reptiles/Birds | Mammals |
|---|---|---|---|---|---|
| I | Meckel's cartilage | Articular[a] | Articular | Articular | Malleus |
| | Pterygoquadrate | Quadrate<br>Epipterygoid | Quadrate<br>Epipterygoid | Quadrate<br>Epipterygoid | Incus<br>Alisphenoid |
| II | Hyomandibula | Hyomandibula<br>Symplectic<br>Interhyal | { Columella<br>Extracolumella | Columella<br>Extracolumella | Stapes |
| | Ceratohyal | Ceratohyal<br>Hypohyal | Ceratohyal<br>Hypohyal | Ceratohyal | Anterior horn hyoid |
| | Basihyal | Basihyal | | Body of hyoid | Body of hyoid |
| III | Pharyngobranchial<br>Epibranchial<br>Ceratobranchial<br>Hypobranchial | Pharyngobranchial<br>Epibranchial<br>Ceratobranchial<br>Hypobranchial } | Body of hyoid | Second horn of hyoid | Second horn of hyoid |
| IV | Branchial arch | | Last horn and body of hyoid<br>Laryngeal cartilages (?) | Last horn and body of hyoid<br>Laryngeal cartilages (?) | Thyroid cartilages (?) |
| V | Branchial arch | Branchial arch | Laryngeal cartilages (?) | Laryngeal cartilages (?) | Laryngeal cartilages |
| VI | Branchial arch | Branchial arch | Not present | Not present | Not present |
| VII | Branchial arch | Branchial arch | | | |

[a]Sometimes dermal bone contributes.

jaw is suspended from the dermal **squamosal** bone of the braincase. The lower jaw of mammals consists entirely of the **dentary** bone, which is also of dermal origin. The palatoquadrate and Meckel's cartilages still develop, but they remain cartilaginous except at their posterior ends, which give rise to the **incus** and **malleus** of the middle ear, respectively (figure 7.9). Thus, in mammals, the splanchnocranium does not contribute to the adult jaws or to their suspension. Instead, the splanchnocranium forms the hyoid apparatus, styloid, and three middle ear bones—malleus, incus, and stapes.

# DERMATOCRANIUM

Dermal bones that contribute to the skull belong to the dermatocranium. Phylogenetically, these bones arise from the bony armor of the integument of early fishes and sink inward to become applied to the chondrocranium and splanchnocranium. Bony elements of the armor also become associated with the endochondral elements of the pectoral girdle to give rise to the dermal components of this girdle.

**Dermal girdle (p. 325)**

Dermal bones first become associated with the skull in ostracoderms. In later groups, additional dermal bones of the overlying integument also contribute. The dermatocranium forms the sides and roof of the skull to complete the protective bony case around the brain; it forms most of the bony lining of the roof of the mouth, and encases much of the splanchnocranium. Teeth that arise within the mouth usually rest on dermal bones.

As the name suggests, bones of the dermatocranium arise directly from mesenchymal and ectomesenchymal tissues of the dermis. Through the process of intramembranous ossification, these tissues form dermatocranial bones.

## Parts of the Dermatocranium

Dermal elements in modern fishes and living amphibians have tended to be lost or fused so that the number of bones present is reduced and the skull simplified. In amniotes, bones of the dermatocranium predominate, forming most of the braincase and lower jaw. The dermal skull may contain a considerable series of bones joined firmly at sutures in order to box in the brain and other skull elements. As a convenience, we can group these series and recognize the most common bones in each (figure 7.10; table 7.3).

### *Dermal Bone Series*

***Facial Series*** The facial series encircle the external naris and collectively form the snout. The **maxilla** and **premaxilla** (incisive) define the margins of the snout and usually bear teeth. The **nasal** lies medial to the naris. The **septomaxilla** is a small dermal bone of the facial series that is

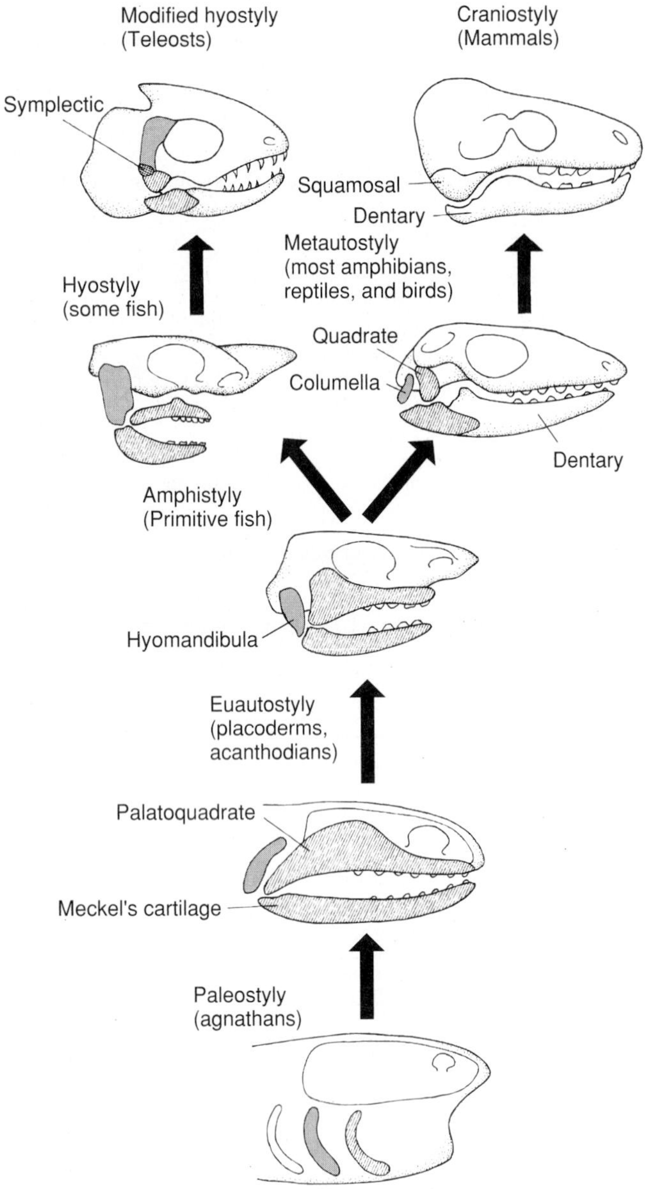


**Figure 7.8** Jaw suspension. The points at which the jaws attach to the rest of the skull define the type of jaw suspension. Note the mandibular arches (crosshatched areas) and hyoid arches (dark areas). The dermal bone (white areas) of the lower jaw is the dentary.

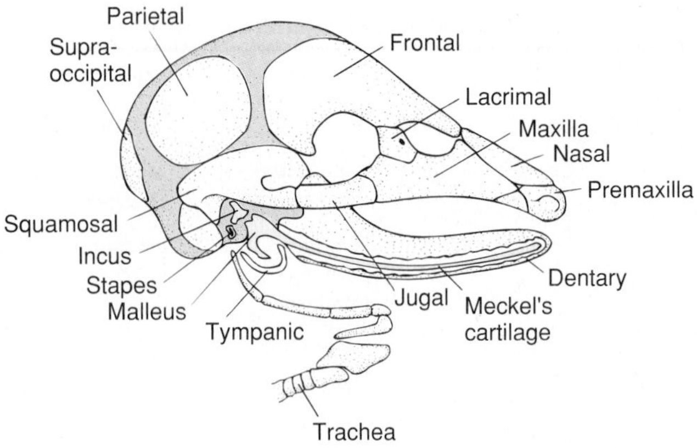


**Figure 7.9** Skull of armadillo embryo. During embryonic formation of the three middle ear ossicles (incus, stapes, malleus), the incus and stapes arise from the mandibular arch, testifying to the phylogenetic derivation of these bones from this arch. The dermal dentary is cut away to reveal Meckel's cartilage, which ossifies at its posterior end to form the malleus.

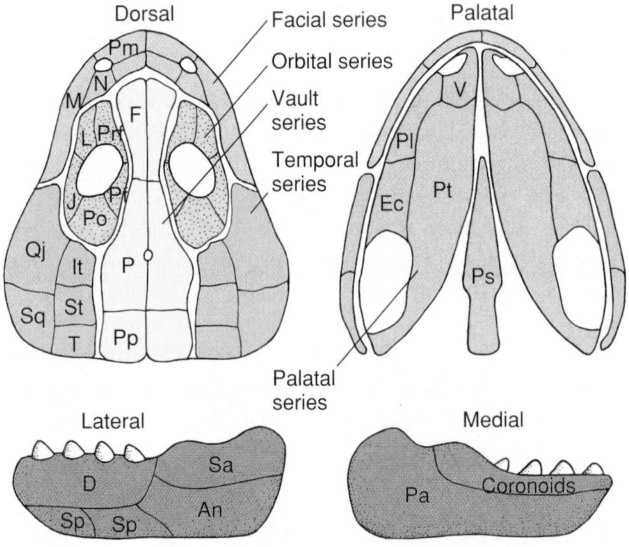


**Figure 7.10** Major bones of the dermatocranium. Sets of dermal bones form the facial series surrounding the nostril. The orbital series encircles the eye, and the temporal series composes the lateral wall behind the eye. The vault series, the roofing bones, run across the top of the skull above the brain. Covering the top of the mouth is the palatal series of bones. Meckel's cartilage (not shown) is encased in the mandibular series of the lower jaw. Abbreviations: angular (An), dentary (D), ectopterygoid (Ec), frontal (F), intertemporal (It), jugal (J), lacrimal (L), maxilla (M), nasal (N), parietal (P), prearticular (Pa), palatine (Pl), premaxilla (Pm), postorbital (Po), postparietal (Pp), prefrontal (Prf), parasphenoid (Ps), pterygoid (Pt), quadratojugal (Qj), surangular (Sa), splenial (Sp), squamosal (Sq), supratemporal (St), tabular (T), vomer (V).

often absent. When present, it is usually sunken below the surface bones and aids in forming the nasal cavity.

***Orbital Series*** The dermal bones encircle the eye to define the orbit superficially. The **lacrimal** takes its name from the nasolacrimal (tear) duct of tetrapods that passes through or near this bone. The **prefrontal, postfrontal,** and **postorbital** continue the ring of bones above and behind the orbit. The **jugal** usually completes the lower rim of the orbit. Not to be confused with these dermal bones are the **scleral ossi-**

TABLE 7.3 Major Dermal Bones of the Skull

| Braincase | | | | | Mandible |
|---|---|---|---|---|---|
| *Facial Series* | *Orbital Series* | *Temporal Series* | *Vault Series* | *Palatal Series* | *Mandibular Series* |
| Premaxilla | Lacrimal | Intertemporal | Frontal | Vomer | Lateral |
| Maxilla | Prefrontal | Supratemporal | Parietal | Palatine | Dentary (teeth) |
| Nasals (septomaxilla) | Postfrontal | Tabular | Postparietal | Ectopterygoid | Splenials (2) |
| | Jugal | Squamosal | | Pterygoid | Angular |
| | | Quadratojugal | | Parasphenoid (unpaired) | Surangular |
| | | | | | Medial prearticular |
| | | | | | Coronoids |

**cles** of neural crest origin that, when present, reside within the orbit defined by the ring of dermal bones.

***Temporal Series*** The temporal series lies behind the orbit, completing the posterior wall of the braincase. In many primitive tetrapods, an indentation in this series, the **otic notch (temporal notch)**, in life suspends the eardrum (tympanic membrane). Openings called **fenestrae** (sing., fenestra) arise within this region of the outer braincase in many tetrapods in association with the jaw musculature. A row of bones, the **intertemporal, supratemporal,** and **tabular,** make up the medial part of the temporal series. This row is reduced in early tetrapods and usually lost in later species. Laterally, the **squamosal** and **quadratojugal** complete the temporal series and form the "cheek."

***Vault Series*** The vault or **roofing bones** run across the top of the skull and cover the brain beneath. These include the **frontal** anteriorly and the **postparietal** (interparietal) posteriorly. Between them is the large **parietal,** occupying the center of the roof and defining the small **parietal foramen** if it is present. The parietal foramen is a tiny skylight in the skull roof that exposes the pineal gland, an endocrine gland, to direct sunlight.

***Palatal Series*** The dermal bones of the **primary palate** cover much of the roof of the mouth. The largest and most medial is the **pterygoid.** Lateral to it are the **vomer, palatine,** and **ectopterygoid.** Teeth may be present on any or all four of these palatal bones. In fishes and lower tetrapods, there also is an unpaired medial dermal bone, the **parasphenoid.**

***Mandibular Series*** Meckel's cartilage is usually encased in dermal bones of the mandibular series. Laterally, the wall of this series includes the tooth-bearing dentary and one or two **splenials,** the **angular** at the posterior corner of the mandible and the **surangular** above. Many of these bones wrap around the medial side of the mandible and meet the **prearticular** and one or several **coronoids** to complete the medial mandibular wall. Left and right mandibles usually meet anteriorly at the midline in a **mandibular symphysis.** If firm, the mandibular symphysis unites them into an arched unit. Most notably in snakes, the mandibular symphysis is composed of soft tissues, permitting independent movement of each mandible.

# Overview of Skull Morphology

## Braincase

In chondrichthyan fishes, the braincase is an elaborate cartilaginous case around the brain. The dermatocranium is absent, reflecting the elimination of almost all bone from the skeleton. However, in most bony fishes and tetrapods, the braincase is extensively ossified with contributions from several sources. For descriptive purposes, it is useful to think of the braincase as a box with a platform of endoskeletal elements supporting the brain, all encased in exoskeletal bones (figure 7.11). The endoskeletal platform is assembled from a series of **sphenoid** bones. The **occipital** bones, which apparently are derived from anterior vertebrae, form the end of this sphenoid platform. These occipital bones, up to four in number **(basioccipital, supraoccipital,** and paired **exoccipitals),** close the posterior wall of the braincase except for a large hole they define, the **foramen magnum,** through which the spinal cord runs. Articulation of the skull with the vertebral column is established through the **occipital condyle,** a single or double surface produced primarily within the basioccipital but with contributions from the exoccipitals in some species.

# Box Essay 7.1 Getting a Head

The idea that the skull is derived from serial compacted vertebrae dates to the eighteenth century. The German naturalist and poet, W. Goethe (1749–1832), was apparently the first to think of but not the first to publish this idea. Goethe gave us the word *morphology,* which meant to him the search for underlying meaning in organic design or form. Among his discoveries was the observation that plant flowers are modified stem petals compacted together. His venture into vertebrates and vertebrate skulls in particular occurred in 1790 whilst he was strolling in an old cemetery in Venice. He spied a dried ram's skull disintegrated at its bony sutures but held in sequence by the soil. The separated bones of the ram's skull seemed to be the foreshortened anterior vertebrae of the backbone, but Goethe did not publish this idea until about 1817. Public credit for this idea and for elaborating it goes to another German naturalist, L. Oken (1779–1851). In 1806, Oken was strolling in a forest and came upon a dried sheep skull. He was similarly struck by its serial homology with the vertebrae, and shortly thereafter published the idea (box figure 1a).

Next, the vertebral theory of skull origin fell into the hands of Richard Owen and became part of his much embellished theoretical view on animal archetypes (box figure 1b). Because of Owen's prominence in early nineteenth-century science, the idea of skull from vertebrae became a central issue within European scientific communities. One of the most persuasive dissenters from this view of a vertebral source for the skull was T. H. Huxley, who based his critique upon a detailed comparative study of vertebrate skulls and their development. This came to a head (no pun intended) in an invited lecture, the Croonian lecture of 1858, in which Huxley argued that the development of the skull showed that it was not composed of vertebrae. He suggested that the "skull was no more derived from vertebrae, than vertebrae are derived from the skull." The skull, Huxley argued, arose in much the same way in most vertebrates, by fusing into a unit, not as a jointed series. Skull ossification showed no similarity with ossification

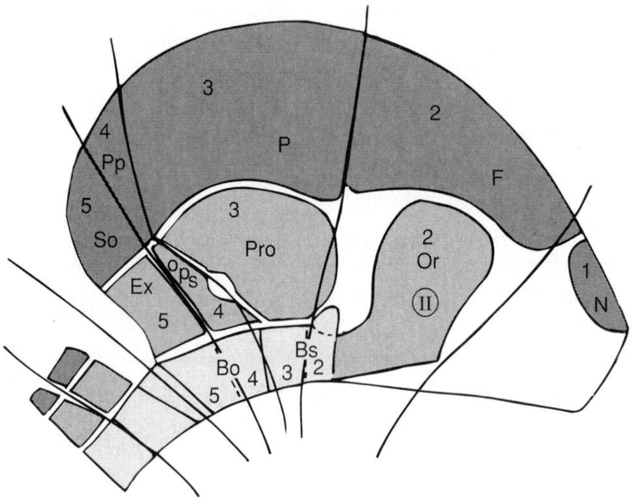


(a)

**Box figure 1** Getting a head. Derivation of the head from anterior vertebrae was proposed separately by Goethe and Oken. Owen expanded on their ideas. (a) Ram's skull showing how its presumed segmental pattern might be interpreted as being derived from parts of anterior vertebrae that expanded. (b) Richard Owen's elaborated view of head segmentation from vertebrae. Owen proposed that anterior vertebrae within the body moved forward to contribute to skeletal elements to the head. Therefore, Owen believed, the bony elements of the head could be homologized to the parts of a fundamental vertebral pattern. (c) Taking several vertebrates, he indicated how named parts of the skull might represent respective parts of this underlying vertebral pattern from which they derive. (d) Rather than being derived from vertebrae moved forward into the head, T. H. Huxley proposed alternatively that the components of the head were derived from a basic segmentation unrelated to the vertebral segmentation behind the skull. These basic segments (roman numerals) are laid out across a generalized vertebrate skull to show the respective contributions to specific parts. Abbreviations: basioccipital (Bo), basisphenoid (Bs), exoccipital (Eo), frontal (F), nasal (N), opisthootic (Ops), orbitosphenoid (Or), parietal (P), postparietal (Pp), prootic (Pro).

of the following vertebrae. Although Huxley was probably right about this for most of the skull, the occipital region does ossify in a manner similar to vertebrae. Perhaps the real weakness of the vertebral theory was, as others would point out later, that vertebrae debuted phylogenetically *after* the head, so they are unlikely its ancestral source!

While disposing of the vertebral theory, Huxley substituted a segmental theory, tracing the segmentation to somites, not to vertebrae (box figure 1c). He took the otic capsule housing the ear as a "fixed" landmark and envisioned four somites (preotic) in front and five somites (postotic) behind it as segmental sources for segmental adult derivatives of the head.

Today, some would argue that the head is a unique developmental system without any tie to the segmental somites (somitomeres). The neural crest cells that also contribute to parts of the skull show no segmental pattern in the head. However, at least in fishes, the branchial arches are segmental, as is the head paraxial mesoderm (somitomeres), and segmentation apparently can be carried into the accompanying neurocranium.

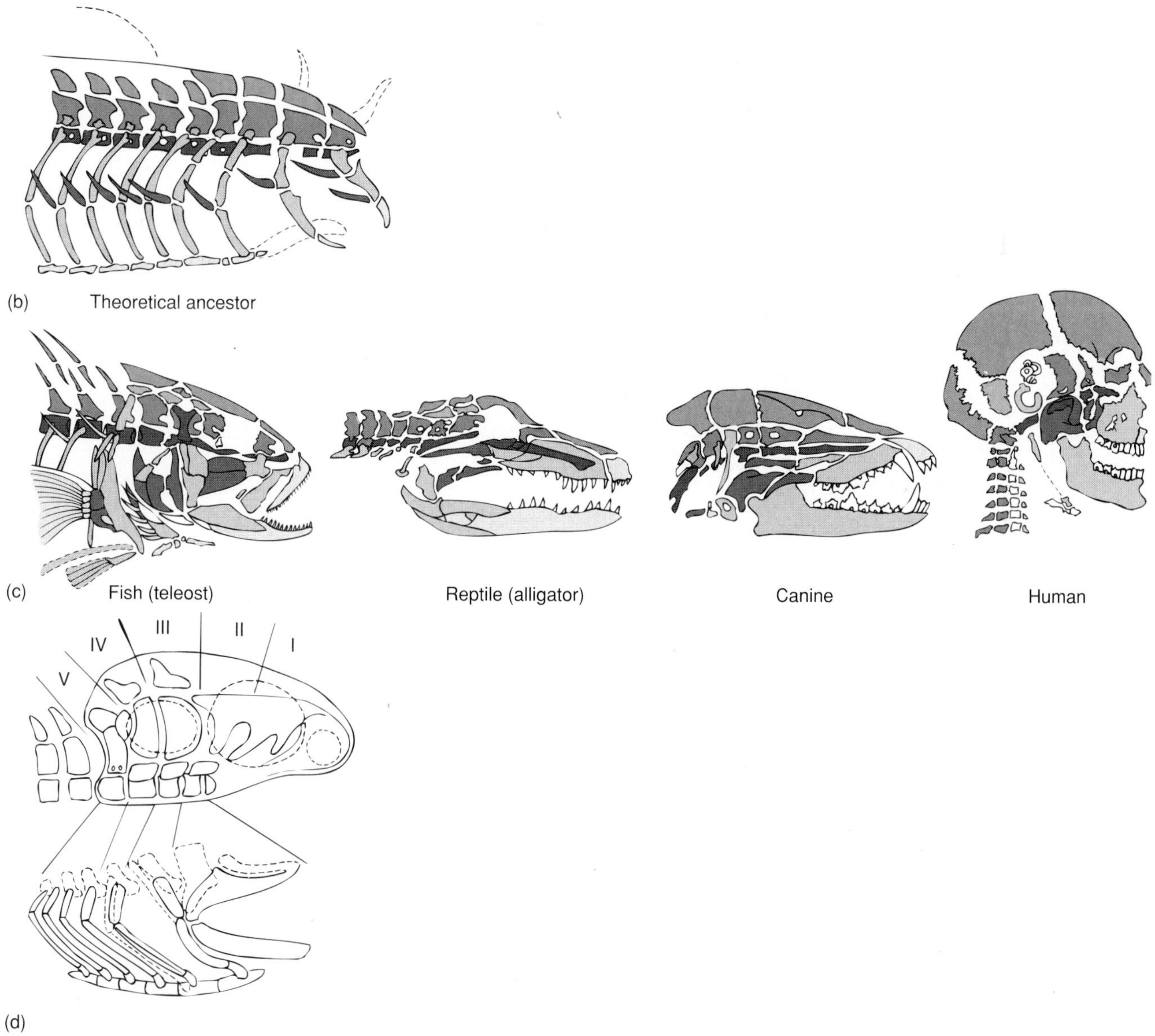

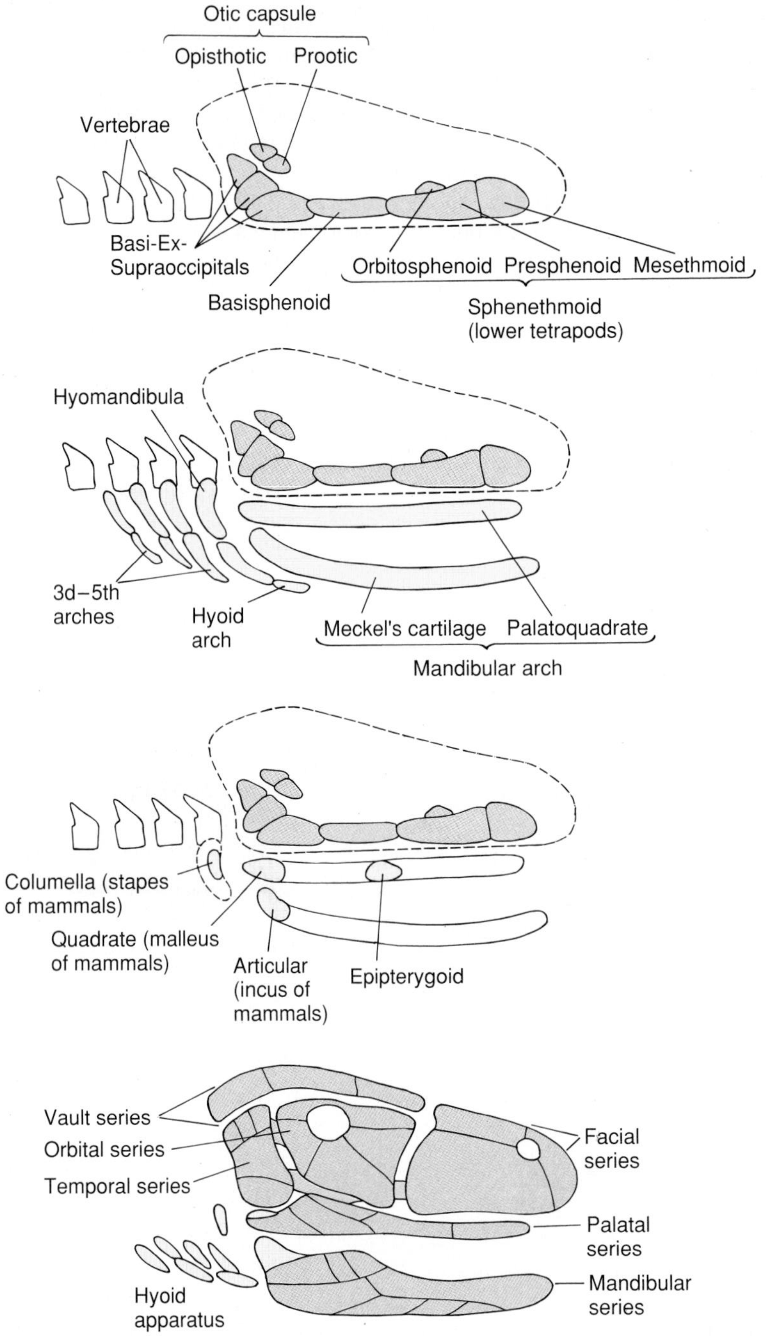


**Figure 7.11** Contributions to the skull. The chondrocranium (blue) establishes a supportive platform that is joined by contributions from the splanchnocranium (yellow), in particular the epipterygoid. Other parts of the splanchnocranium give rise to the articular, quadrate, and hyomandibula, as well as to the hyoid apparatus. The dermatocranium (red) encases most of the chondrocranium together with contributions from the splanchnocranium.

The otic capsule rests on the posterior part of the endoskeletal platform and encloses the sensory organs of the ear. The splanchnocranium contributes the **epipterygoid** (**alisphenoid** of mammals) to the endoskeletal platform and gives rise to one (columella/stapes) or more (malleus and incus of mammals) of the middle ear bones housed in the otic capsule.

In most vertebrates, these endoskeletal elements, along with the brain and sensory organs they support, are enclosed by the exoskeletal elements, derivatives of the dermis, to complete the braincase.

### Jaws

The **upper jaw** consists of the endoskeletal palatoquadrate in primitive vertebrates. The palatoquadrate is fully functional in the jaws of chondrichthyans and primitive fishes, but in bony fishes and tetrapods, the palatoquadrate usually makes limited contributions to the skull through its two derivatives, the epipterygoid, which fuses to the neurocranium, and the **quadrate,** which suspends the lower jaw except in mammals. The dermal maxilla and premaxilla replace the palatoquadrate as the upper jaw.

The **lower jaw,** or **mandible,** consists only of Meckel's cartilage in chondrichthyans. In most fishes and tetrapods, Meckel's cartilage persists but is enclosed in exoskeletal bone of the dermatocranium, which also supports teeth. Meckel's cartilage, encased in dermal bone, usually remains unossified, except in some tetrapods where its anterior end ossifies as the **mental** bone. In most fishes and tetrapods (except mammals), the posterior end of Meckel's cartilage can protrude from the exoskeletal case as an ossified **articular** bone.

In mammals, the lower jaw consists of a single bone, the dermal dentary. The anterior tooth-bearing part of the dentary is its **ramus.** Jaw-closing muscles are inserted on the **coronoid process,** an upward extension of the dentary. Posteriorly, the dentary forms the transversely expanded **mandibular condyle,** a rounded process that articulates with the **glenoid fossa,** a depression within the squamosal of the braincase. Thus, in mammals, the mandibular condyle of the dentary replaces the articular bone as the surface of the lower jaw through which mandibular articulation with the braincase is established.

### Hyoid Apparatus

The hyoid or hyoid apparatus is a ventral derivative of the splanchnocranium behind the jaws. In fishes, it supports the floor of the mouth. Elements of the hyoid apparatus are derived from the ventral parts of the hyoid arch and from parts of the first few branchial arches. In larval and paedomorphic amphibians, the branchial bars persist but form a reduced hyoid apparatus that supports the floor of the mouth and functional gills. In adults, the gills and the associated part of the hyoid apparatus are lost, although elements persist within the floor of the mouth usually to support the tongue. Typically, the hyoid apparatus includes a main body, the **corpus,** and extensions, the **cornua** ("horns"). In many mammals, including humans, the distal end of the hyoid horn fuses with the otic region of the braincase to form the **styloid process.**

## Cranial Kinesis

Kinesis means movement. Cranial kinesis refers literally then to movement within the skull. But if left this general, it becomes too broad to provide a useful context in which to discuss skull function. Some authors restrict the term to skulls with a transverse hingelike joint across the skull roof. This is usually accompanied by a basal joint in the braincase as well. But this restricted definition precludes most teleost fishes, despite their highly mobile skull elements. Here, we use cranial kinesis to mean movement between the upper jaw and the braincase about joints between them (figure 7.12a). Such **kinetic skulls** characterize most vertebrates. They are found in ancient fishes (crossopterygians and probably palaeoniscoids), bony fishes (especially teleosts), very early amphibians, most reptiles (including most Mesozoic forms), birds, and reptilian ancestors to mammals. Kinetic skulls are not present in modern amphibians, turtles, crocodiles, and mammals (with the possible exception of rabbits). The widespread presence of cranial kinesis among vertebrates, but its essential absence among mammals, seems to create a problem for humans. Because we, like most other mammals, have **akinetic skulls** with no such movement between upper jaw and braincase, we tend to underestimate its importance (figure 7.12b).

Kinesis and akinesis each have advantages. Cranial kinesis provides a way to change the size and configuration of the mouth rapidly. In fishes and other vertebrates that feed in water, rapid kinesis creates a sudden reduction of pressure in the buccal cavity so that the animal can suck in a surprised prey. This method of prey capture, which takes advantage of a sudden vacuum to gulp in water carrying the intended food, is known as **suction feeding.** Cranial kinesis also allows tooth-bearing bones to move quickly into strategic positions during rapid feeding. Some teleost fishes, for instance, swing their anterior teeth-bearing bones forward at the last moment to reach out quickly at the intended prey. In many venomous snakes, linked bones along the sides of the skull can rotate forward. The snake erects the maxillary bone bearing the fang and swings it from a folded position along its upper lip to the front of the mouth, where it can more easily deliver venom into prey. In many reptiles with kinetic skulls, teeth on the upper jaw can be

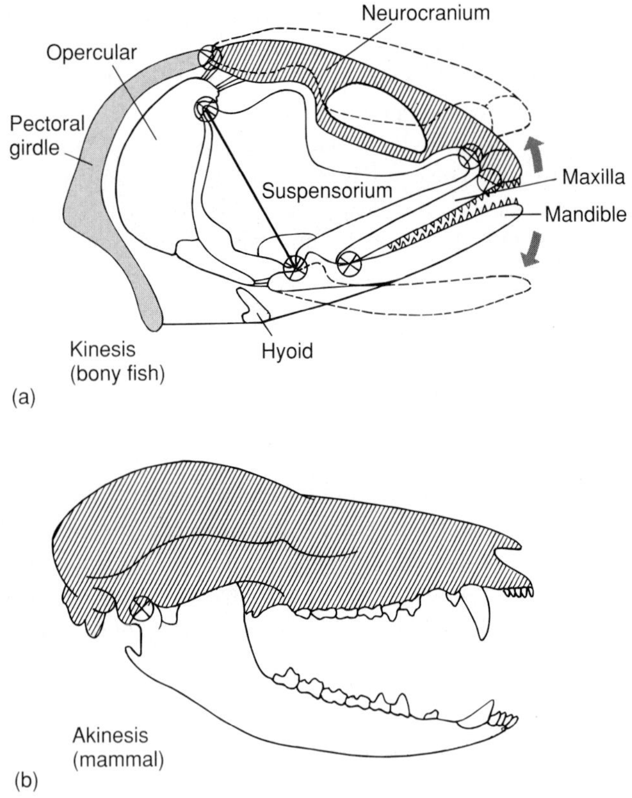


**Figure 7.12** Mobility of skull bones. (a) The fish skull is kinetic. The upper jaw and other lateral skull bones rotate upon each other in a linked series, resulting in displacements of these bones (dashed outline) during feeding. Circles represent points of relative rotation between articulated elements. (b) The mammal skull is akinetic because no relative movement occurs between the upper jaw and the braincase. In fact, the upper jaw is incorporated into and fused with the braincase. There are no hinge joints through the braincase nor any movable linkages of lateral skull bones.

reoriented with respect to the prey in order to assume a more favorable position during prey capture or to align crushing surfaces better during swallowing. On the other hand, loss of kinesis in mammalian skulls allows infants to suckle easily. Juveniles and adult mammals can chew firmly with sets of specialized teeth that work accurately.

**Tooth structure and occlusion (p. 495)**

# Phylogeny of the Skull

The skull is a composite structure derived from the splanchnocranium, dermatocranium, and chondrocranium. Each component of the skull comes from a separate phylogenetic source. The subsequent course of skull evolution is complex, reflecting complex feeding styles. With a general view of skull structure now in mind, we can turn to a more specific look at the course of this evolution.

## Agnathans

### *Ostracoderms*

Osteostracans were one of the more common groups of ostracoderms. They possessed a head shield formed from a single piece of arched dermal bone, two close-set eyes dorsally placed with a single pineal opening between them, and a median nostril in front of the pineal opening. Along the sides of the head shield ran what are believed to be sensory fields, perhaps electrical field receptors or an early lateral line system sensitive to currents of water.

The broad, flattened head shield lowered the profile of ostracoderms, perhaps allowing them to hug the bottom surface, and their slight body suggests that they were benthic-dwelling fishes. The head shield formed the roof over the pharynx and held the sequential branchial arches that stretched like beams across the roof of the pharynx. Paired gill **lamellae** supported on **interbranchial septa** were stationed between these bars. Reconstructions of the head of *Hemicyclaspis,* a cephalaspidomorph, indicate that a plate, presumably of cartilage, stretched across the floor of the pharynx (figure 7.13a). Muscle action is thought to have raised and lowered this plate to draw a stream of water first into the mouth, and then over the gills, and finally out the branchial pores along the ventral side of the head. Suspended particles held in the stream of water could be captured within the pharynx before the water was expelled (figure 7.13b).

Anaspids were another group of early ostracoderms. Instead of a single bony shield, many small bony scales covered the head (figure 7.14a–c). The eyes were lateral, with a pineal opening between them and a single nostril in front. The body was streamlined, suggesting a slightly more active life than other ostracoderms enjoyed.

Heterostracans had flat to bullet-shaped heads composed of several fused bony plates (figure 7.15a). Their eyes were small and laterally placed, with a median pineal opening but no median nostril. Presumably water flowed through the mouth, across the gill slits of the large pharynx, into a common tunnel, and out a single exit pore. The mouth of some was rimmed with sharp, pointed oral scales that could have been used to dislodge food from rocks, allowing it to join the stream of water that entered the mouth (figure 7.15b).

Some scientists think that a few ostracoderms were predaceous, using the buccal cavity to gather up large prey, but because ostracoderms lacked jaws, feeding could not be based on powerful biting or crushing. The heavily plated heads and slight bodies of most ostracoderms argue for a relatively inactive lifestyle spent feeding on detritus and organic debris stirred up and drawn into the pharynx.

## Box Essay 7.2 Cranial Kinesis in Hares?

In hares or "jackrabbits" (but not in distantly related pikas or in their fossil ancestors), a suture between regions of the fetal braincase remains open in the adult, forming an intracranial joint (box figure 1). This intracranial joint runs along the sides and base of the adult braincase and hinges across the top via the postparietal. The joint permits relative motion between anterior and posterior parts of the braincase. It has been hypothesized that this joint helps absorb the impact forces sustained as the forelimbs strike the ground when a rabbit runs. Upon impact, mechanical deformation of the joint would absorb some kinetic energy as the hinge is strained. This deformation and absorption would reduce the shock sustained by the anterior part of the braincase. Additionally, the impact forces would tend to drive blood from intracranial sinuses into a complex association of venous channels and spaces within the skull. This would help dissipate these kinetic forces further as they acted against resistance offered by the walls of the blood vascular system.

The external ears (pinnae) of hares radiate heat generated during strenuous activity, but apparently only after locomotor exercise ceases. During locomotion, the ears are usually held erect by strong muscles at their bases. It has been hypothesized that these erect ears help reopen the intracranial joint as the hare pushes off on another leap to accelerate again, thus in a sense "resetting" this cranial mechanism and preparing it to act as a shock-absorbing device when the forelimbs again strike the ground (box figure 1c).

The functional significance of the intracranial joint is still debated. However, if such hypotheses are confirmed, this specialized joint in hares together with their projecting ears might also serve to reduce jarring of the eyes carried in the anterior braincase. Rabbit kinesis represents an independent and apparently unique condition among mammals that did not evolve from therapsid kinesis. Further, it evolved not for its advantages during feeding but rather for its advantages during rapid locomotion.

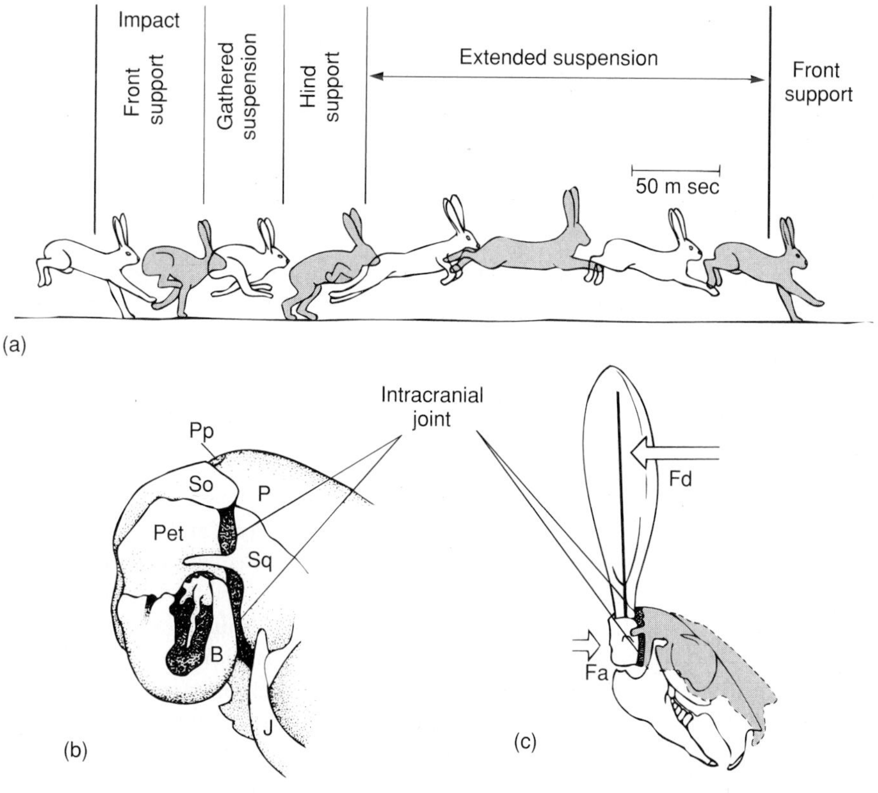


**Box figure 1** Possible cranial kinesis in hares. (a) Phases during a running stride are illustrated. Note that the forelimbs receive the initial impact upon landing. (b) Posterior regions of the skull of the jackrabbit *Lepus*. The intracranial joint extends along the sides of the skull between squamosal (Sq) and otic regions and then along the base of the skull. The interparietal bone forms the hinge across the top of the skull. (c) External ears held erect and attached to the posterior part of the skull may help to reposition the posterior part of the skull relative to the anterior part during the extended suspension phase of running. The presumed motion (slightly exaggerated) of the anterior braincase relative to the posterior braincase is indicated. Fa is the force vector due to acceleration resulting from thrust, and Fd is the force vector due to drag of the ears in the oncoming wind. Abbreviations: bulla (B), postparietal (Pp), jugal (J), parietal (P), petrosal (Pet), supraoccipital (So), squamosal (Sq).

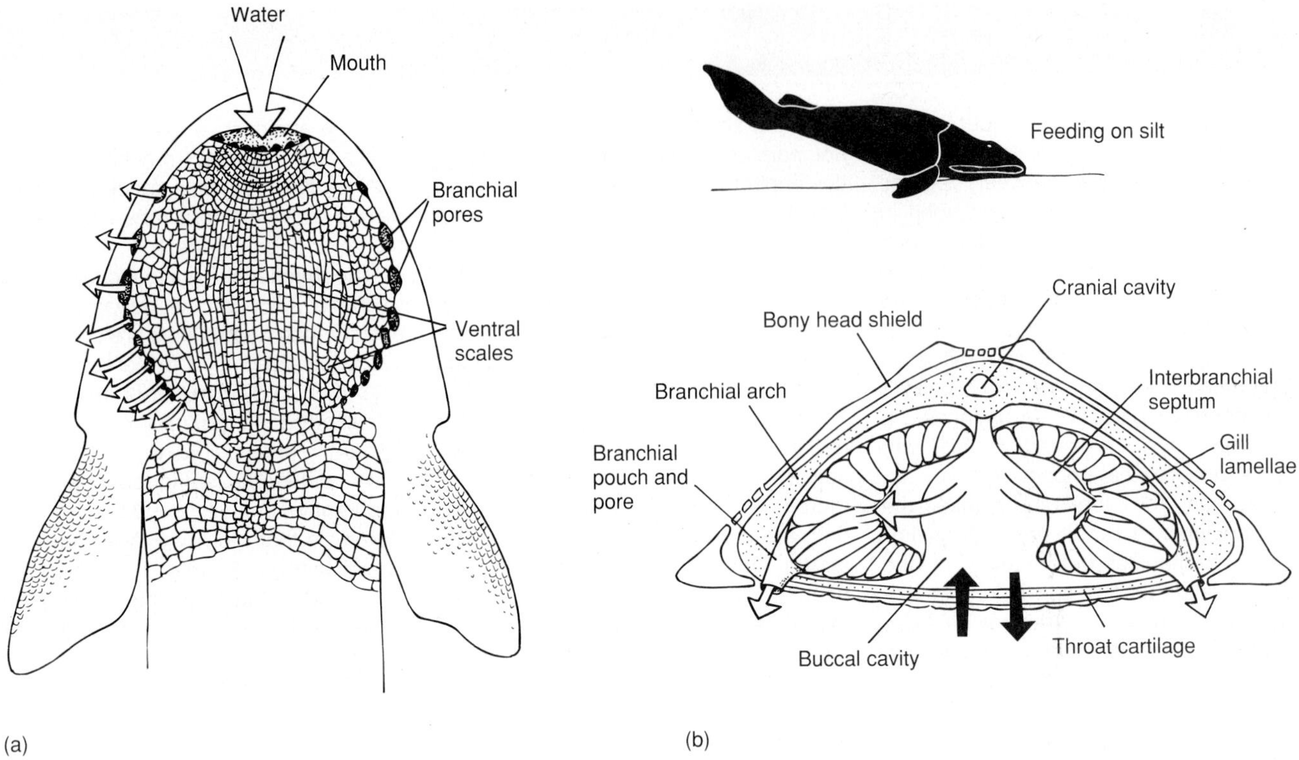


**Figure 7.13** Ostracoderm *Hemicyclaspis*, a cephalaspidomorph. (a) Ventral view showing branchial pores, the presumed sites of exit for water moving through the pharynx. (b) Cross section through the pharynx illustrating respiratory gill lamellae and supporting branchial arch. Presumably, the floor of the pharynx could be raised and lowered to actively draw water into the mouth and drive it out through the several branchial pores. The current crossed the respiratory gills before exiting. Suspended food may have been collected in the pharynx and then passed to the esophagus.

### *Cyclostomes*

Lampreys and hagfishes are the only surviving agnathans and heirs of the ostracoderms. However, subsequent specializations have left cyclostomes with anatomies quite unlike those of the early ostracoderms. Cyclostomes lack bone entirely and are specialized for parasitic or scavenging lives that depend on a rasping tongue to scrape up tissue for a meal. Lampreys have a single medial nostril and a pineal opening. Branchial pouches are present. The braincase is cartilaginous. Branchial arches, although present, form an unjointed branchial basket. Hagfishes have a median nostril but no external pineal opening.

## Gnathostomes

All vertebrates, except agnathans, have jaws and form the embracing group gnathostomes (jaw mouth). Some biologists mark the advent of vertebrate jaws as one of the most important transitions in their evolution. Powerful closing muscles, derivatives of the branchial arch musculature, make the jaws strong biting or grasping devices. It is not surprising then that with the advent of jaws, gnathostomes experience a dietary shift away from suspension feeding of the ostracoderms to larger food items. With a change in diet also comes a more active lifestyle.

### *Fishes*

***Placoderms*** As much as a third to a half of the anterior placoderm body was composed of heavy plates of dermal bone that also enclosed the pharynx and braincase. The rest of the body was covered with small bony scales. The dermal plates of the head were thick and tightly joined into a unit termed the **cranial shield** (figure 7.16a,b). Although the pattern of these dermal plates has been compared to scales of bony fishes, their arrangement was sufficiently different that it seems best to follow the convention of using different names until some agreement is reached on their homologies. The braincase was heavily ossified, and the upper jaws attached to it. In most, a well-defined joint existed between the braincase and the first vertebra. A spiracle was apparently absent. Water departing from the mouth exited posteriorly at the open junction between cranial and trunk shields. Most placoderms were 1 m in length, although one species possessing strong jaws reached nearly 6 m overall.

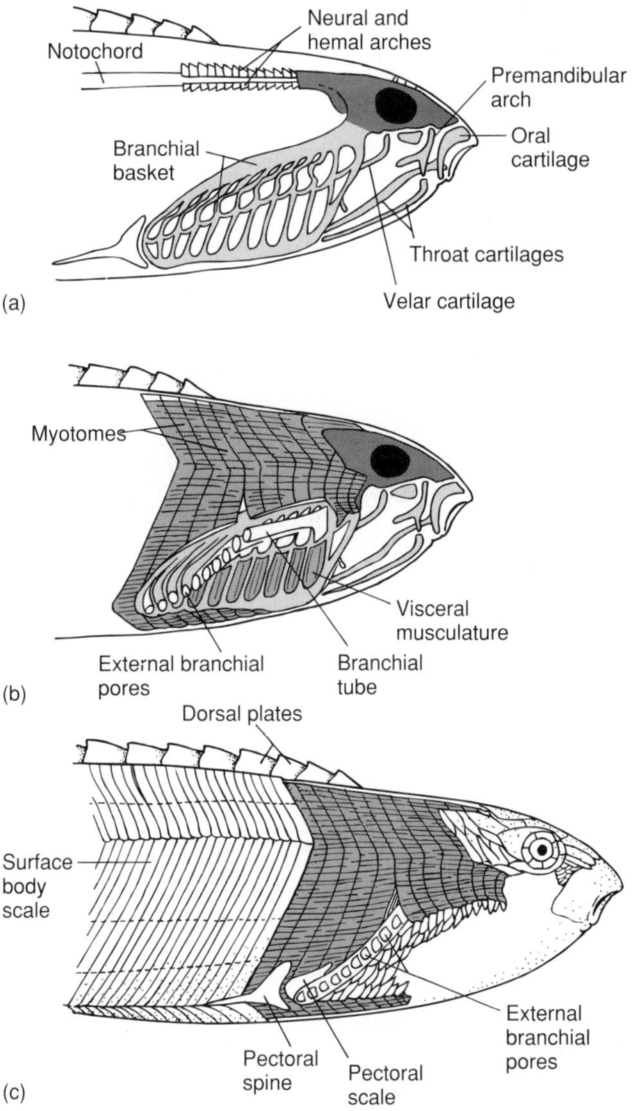


**Figure 7.14** Ostracoderm *Pterolepis,* an anaspid. (a) Exposed skull. The splanchnocranium included a few elements around the mouth, and the chondrocranium held the eye. A notochord was present and vertebral elements rested on it. (b,c) Restoration of muscles and some of the surface scales. The throat cartilages supported the floor of the buccal cavity, which might have been part of a pump to draw water into the mouth and then force it across the gills and out through the external branchial pores.

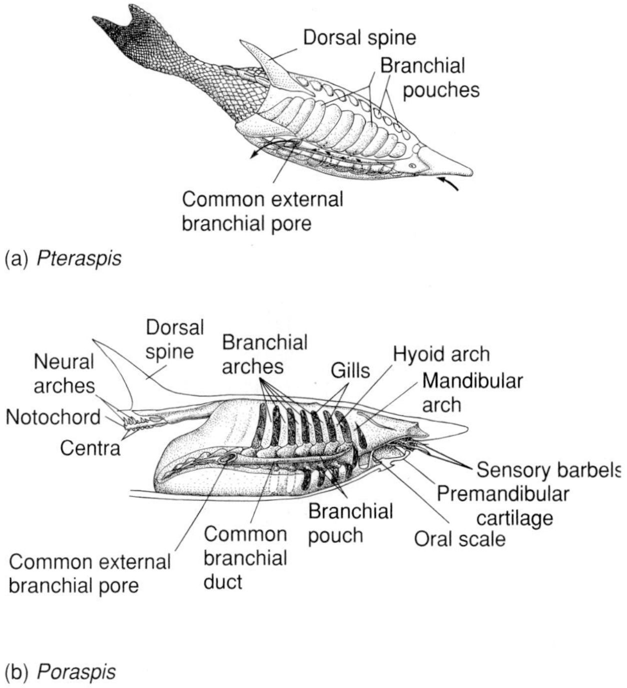


**Figure 7.15** Ostracoderm feeding. (a) Lateral view of *Pteraspis,* a heterostracan. Water flowed through the mouth, over the gills, suspended in branchial pouches, and into a common chamber before finally exiting via the branchial pore. Large, fused bony plates formed the head shield. Throughout the tail, the bony scales were small to accommodate the lateral bending of the tail. (b) Schematic reconstruction of the head of a heterostracan. Pointed, rough oral scales rimmed the mouth and might have been used to scrape or dislodge food from rock surfaces. This reconstruction of a heterostracan is based primarily on *Poraspis*.

***Acanthodians*** The gnathostomes with the earliest surviving fossil record are the acanthodians. Most were small, several centimeters in length, with streamlined bodies, suggesting an active swimming lifestyle. Their bodies were covered with nonoverlapping, diamond-shaped, dermal bony scales. The bony scales of the head region were enlarged into small plates. The pattern of cranial dermal scales resembled bony fishes, but as with placoderms, these are usually given their own names. Some species had an **operculum,** a bony flap that covered the exit gill slits. Eyes were large, suggesting that visual information was especially important to these fishes. *Acanthodes* (early Permian) possessed a **lateral cranial fissure,** a gap that partially divided the posterior braincase. This fissure is an important fixture in actinopterygian fishes, where it allows exit of the tenth cranial nerve. The mandibular arch that formed the jaws was much like that of sharks and bony fishes. Three centers of ossification appear within the palatoquadrate: the **metapterygoid** and **autopalatine** both articulated with parts of the braincase and the posterior quadrate articulated with the ossified Meckel's cartilage (figure 7.17a). A dermal bone, the **mandibular,** reinforced the ventral edge of the lower jaw. A hyoid arch and five successive branchial arches were present in *Acanthodes* (figure 7.17b).

***Chondrichthyans*** Cartilaginous fishes possess almost no bone. Denticles are present, vestiges of scales made up of the minerals enamel and dentin. A dermatocranium is absent. Instead, the chondrocranium has been expanded upward and over the top of the head to form the braincase. As a consequence, the chondrocranium is a much more

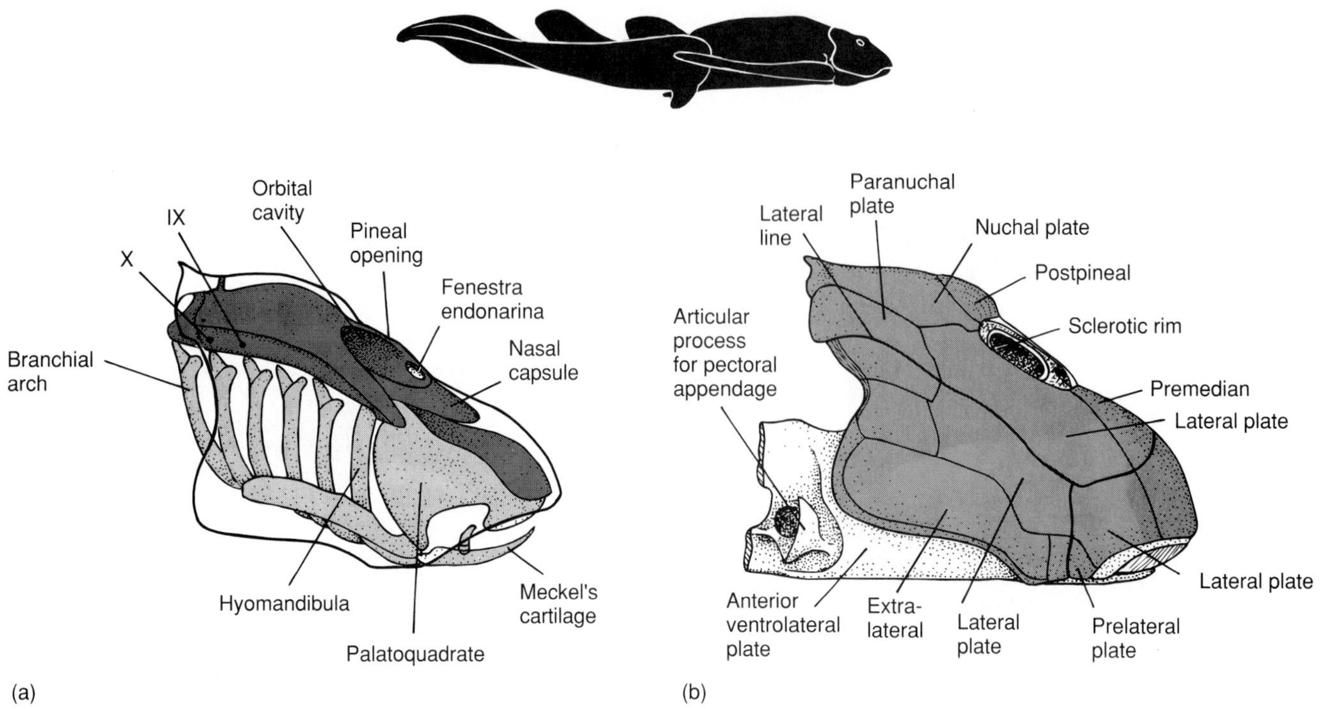


**Figure 7.16** Placoderm skull. *Pterichthyodes* was about 15 cm long and lived in the middle Devonian. (a) Lateral view of splanchnocranium and chondrocranium. (b) Skull with overlying dermatocranium in place. Note the dermal plates.
After Stensiö, 1969.

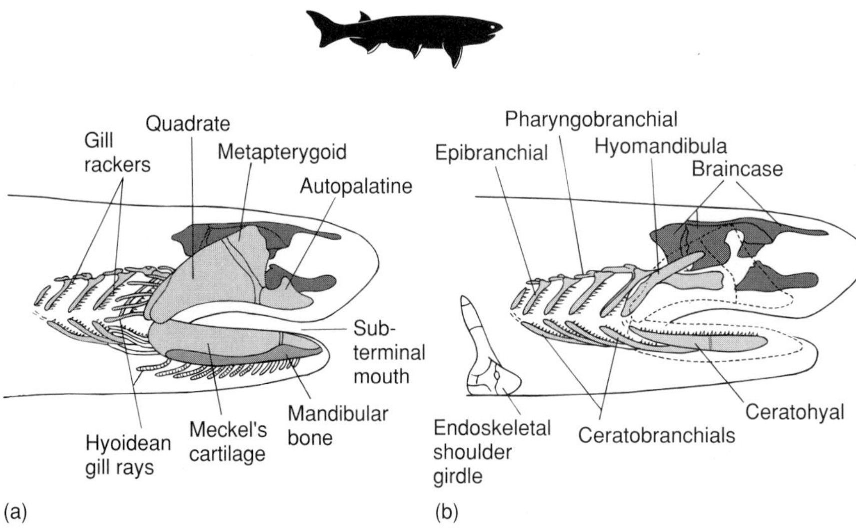


**Figure 7.17** Acanthodian skull, *Acanthodes*. (a) Lateral view with mandibular arch shown in its natural position. (b) Mandibular arch is removed to reveal the chondrocranium, hyoid arch, and five successive branchial arches.

prominent component of the skull than it is in most other vertebrates. The **ethmoid** and **orbital** anterior regions and posterior **oticooccipital** region are merged into an undivided braincase. The splanchnocranium is present. In primitive chondrichthyans, six gill arches trailed the mandibles (figure 7.18a,b). The upper jaw (palatoquadrate) of primitive sharks was supported by the braincase and probably by the hyomandibula.

Modern sharks usually lack a strong, direct attachment between hyomandibula and palatoquadrate. Instead, the jaws are suspended at two other sites, by the ceratohyal and Meckel's cartilage and by a strong, ligamentous

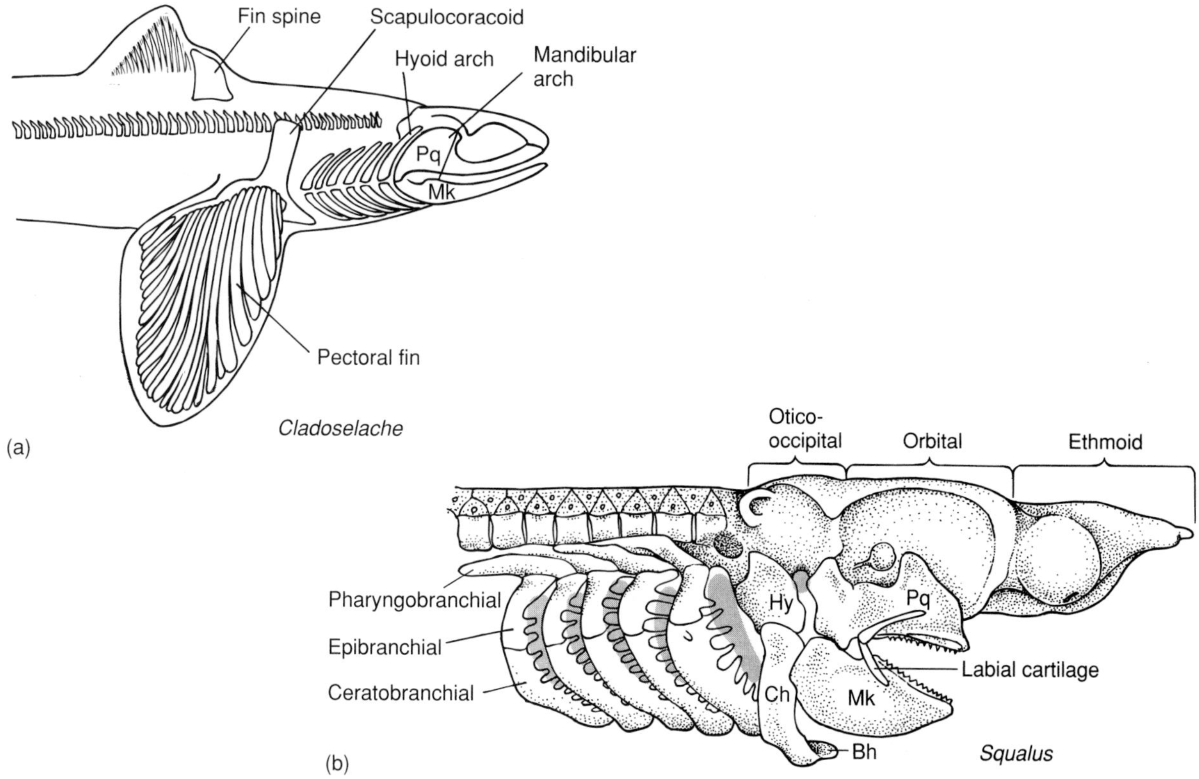


**Figure 7.18** Shark skull. (a) Primitive shark *Cladoselache*, a late Devonian shark that reached perhaps 55 cm in length. Mandibles were followed by a complete hyoid arch and five branchial arches. Full gill slits were present between each arch. (b) Modern shark *Squalus*, the dogfish shark. The hyoid arch, second in series, is modified to support the back of the mandibular arch. As the hyoid moves forward to help suspend the jaw, the gill slit in front is crowded and reduced to the small spiracle. Although fused into one unit, the three basic regions of the chondrocranium are ethmoid, orbital, and oticooccipital. Abbreviations: basihyal (Bh), ceratohyal (Ch), hyomandibula (Hy), Meckel's cartilage (Mk), palatoquadrate (Pq).

connection running from the base of the nasal capsule to the orbital process of the palatoquadrate. As the ceratohyal, and to some extent the hyomandibula, have moved in to aid in supporting the jaws, the gill slit in front has become crowded, leaving only a small opening, the **spiracle.** In some bony fishes, the spiracle has vanished altogether. In chondrichthyans, such as holocephalians, the jaws mechanically crush hard shells of prey, but in active chondrichthyans, such as predaceous sharks, the jaws capture prey.

Sharks may use suction to draw small prey toward or into the mouth, but more commonly, they attack prey directly, approaching it head-on. As sharks raise their head, the lower jaw descends (figure 7.19a). Upper and lower jaws articulate with each other, and both in turn are suspended like a pendulum from the hyoid arch. The hyoid arch swings about its attachment to the braincase, which permits the jaws to descend and shift downward and forward over the prey (figure 7.19b). Teeth along the upper (palatoquadrate) and lower (Meckel's cartilage) jaws are often oriented with their points in an erect position to engage the surface of the prey. Occasionally the nictitating membrane, a movable flap of opaque skin, is drawn protectively across each eye.

Jaw protrusion may also assist the synchronized meeting of upper and lower jaws on the prey. If the lower jaw alone was responsible for closing the mouth, it might prematurely strike the prey before the upper jaw was suitably positioned to assist. Protracting the mandibles away from the head allows the jaws to assume a more favorable geometric configuration so that they meet the prey simultaneously and avoid deflecting it when they close. As the jaws clamp on the prey, the mandibular arch often is protracted near the end of closure. If the prey is large, the shark may violently shake its head to cut free a section of the prey and swallow it.

When protracted, the jaws disrupt the streamlined body silhouette characteristic of an active, open-water fish. Retraction of the jaws following feeding restores the hydrodynamic, streamlined shape of the fish and tucks them back up against the chondrocranium.

***Actinopterygians*** Early actinopterygians had relatively large eyes and small nasal capsules. The jaws were long, extending to the front of the head. The jaws carried

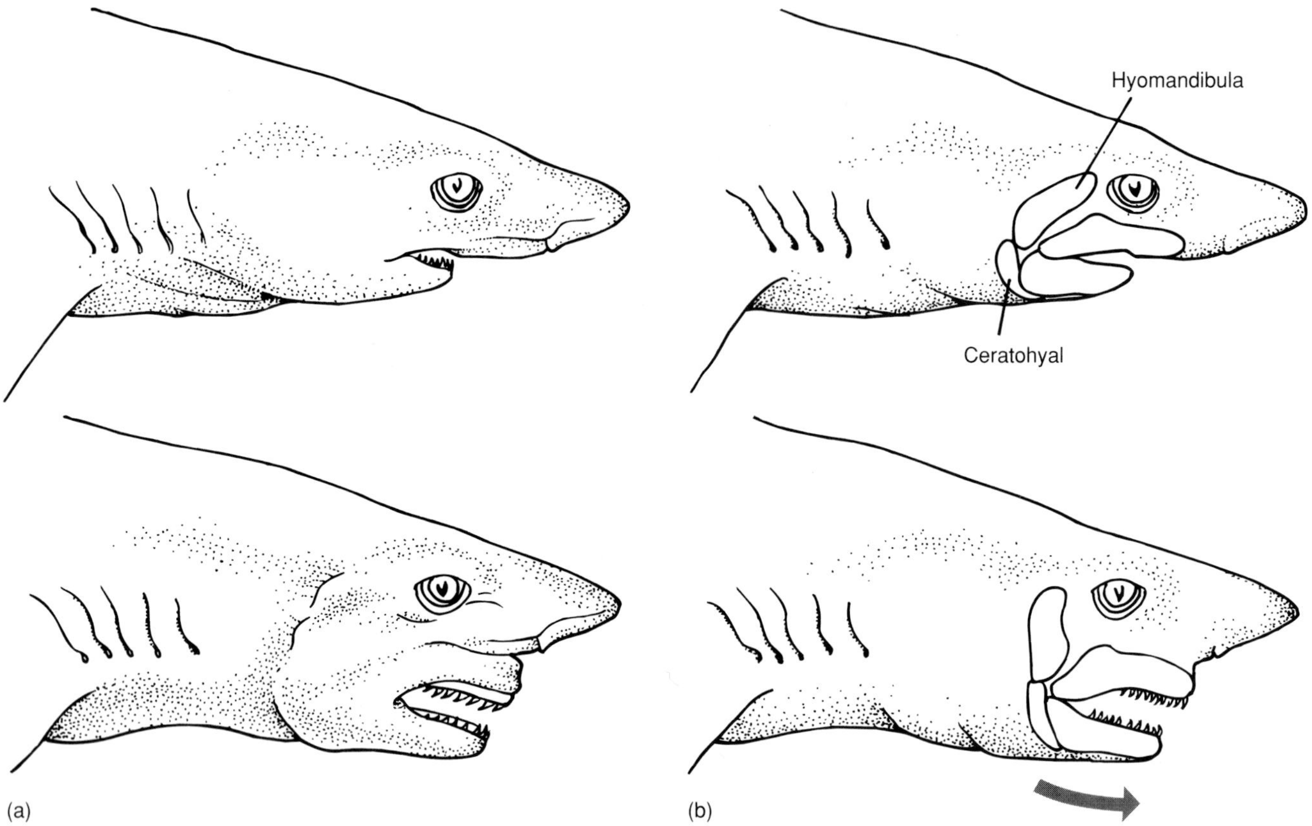


**Figure 7.19** Feeding in sharks. (a) Sketches of shark with jaws retracted (top) and manually protracted (bottom). (b) Interpreted positional changes in the mandibular arch as it rides forward on its suspension from the ceratohyal. Position depicted is near the completion of jaw closure on the prey. Arrow indicates ventral and forward shift of the jaws.
Based on the research of T. H. Frazzetta.

numerous teeth, and an operculum covered the gill arches. The hyoid arch increased its support of the mandibles. Homologies of dermal bones in some groups have been difficult to assign, partly because of the proliferation of extra bones, especially facial bones. Around the external naris, there may be many tiny bones variously ascribed by position to nasals, rostral, antorbitals, and others. One common scheme is shown in figure 7.20a,b, but several varieties occur as well. Notice in particular the set of **opercular bones** covering the gills and the set of **extrascapulars** at the dorsal, posterior rim of the skull. These are major dermal bones in actinopterygians that are lost in tetrapods (figure 7.21a–d).

Within actinopterygians, an extraordinary radiation occurred that continues to the present. It is difficult to generalize about trends within the skull because so many varied specializations of modern bony fishes are part of this radiation. If a common trend exists, it is for increased liberation of bony elements to serve diversified functions in food procurement.

Most actinopterygians employ rapid suction feeding, with prey capture completed within 1/40 of a second. The almost explosive expansion of the buccal cavity creates a vacuum to accomplish swift capture. Negative pressure sucks a pulse of water carrying the prey into the mouth. Once captured, teeth hold the prey. Compression of the buccal cavity expels excess water posteriorly out the gill slits. Fishes that feed by suction take in larger chunks of food than suspension feeders. Larger food particles have more inertia and require a stronger feeding device. Suction feeders consequently possess a well-muscularized buccal cavity and powerful, kinetic jaws.

In primitive actinopterygians, such as the fossil *Cheirolepis* and living *Amia* (figures 7.21a–d and 7.22a,b), the feeding apparatus includes several units. One is the neurocranium, to which the premaxilla and maxilla are usually fused. The posterior part of the neurocranium articulates with and is free to rotate on the anterior vertebra. The opercular bones form a unit along the side of the head. The **suspensorium** is formed from the fusion of various bones in different species but usually includes the hyomandibula, palatine, and quadrate. The suspensorium is shaped like an inverted triangle, its two upper corners articulating with the snout and braincase, its third lower corner articulating with the mandible. During jaw open-

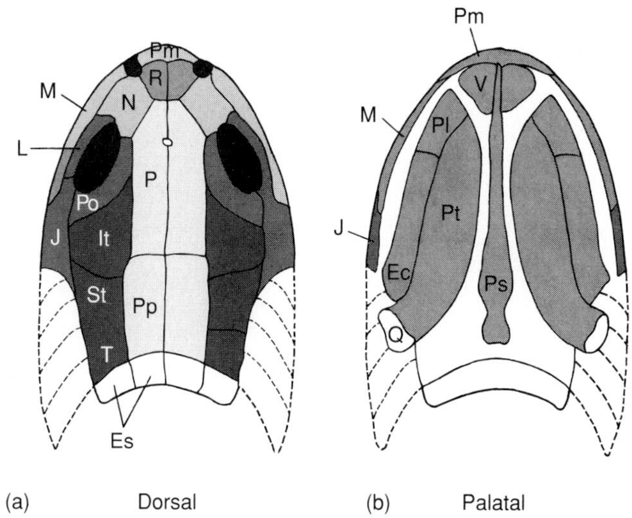


**Figure 7.20** Major skull bones of an actinopterygian fish. (a) Dorsal view. (b) Palatal (ventral) views. Opercular bones are represented by dashed lines. Abbreviations: extrascapulars (Es), intertemporal (It), jugal (J), lacrimal (L), maxilla (M), nasal (N), parietal (P), palatine (Pa), premaxilla (Pm), postorbital (Po), postparietal (Pp), parasphenoid (Ps), pterygoid (Pt), quadrate (Q), rostral (R), squamosal (Sq), supratemporal (St), tabular (T), vomer (V).

ing, epaxial muscles of the trunk raise the neurocranium and the attached upper jaw. Sternohyoideus muscles in the throat move the hyoid apparatus to lower the mandible (figure 7.23a,b). Strong adductor muscles of the jaws run from the suspensorium directly to the mandible to close the lower jaw.

In advanced actinopterygians, the teleosts, there is usually even greater freedom of skull bone movement (figure 7.24a–e). The premaxilla and maxilla are now usually freely articulated with each other and with the neurocranium (figure 7.25). During jaw opening, the neurocranium is raised, and the mandible is lowered. In addition, the geometric arrangement of the jaws allows it to move forward. The hyoid apparatus forms struts within the floor of the buccal cavity. When pulled backward by the throat musculature, these hyoid struts help push the lateral walls of the buccal cavity apart and so contribute to its sudden enlargement and creation of suction within.

***Sarcopterygians*** In early lungfishes, the upper jaw (palatoquadrate) was fused to the ossified braincase, which was a single unit with teeth flattened into plates. This suggests that the earliest lungfishes fed on hard

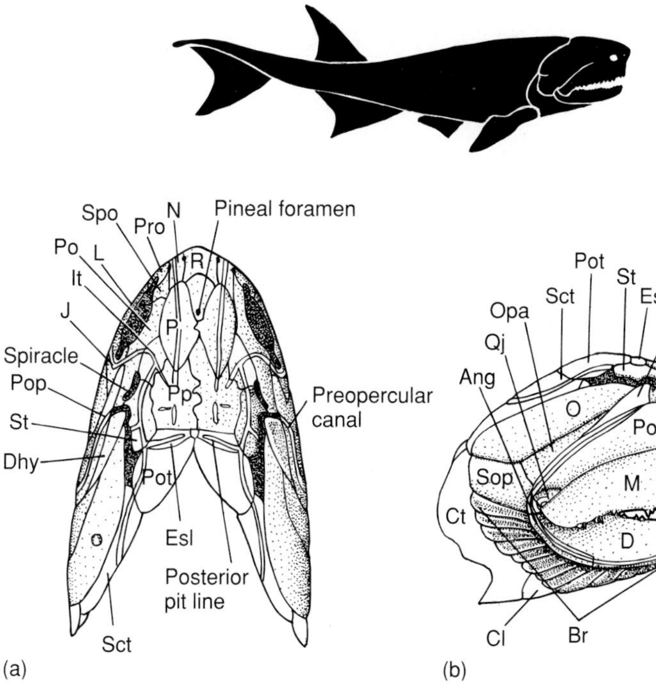

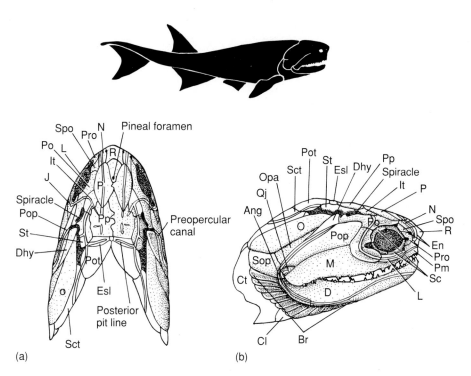

**Figure 7.21** Skull of the primitive palaeoniscoid fish *Cheirolepis*, from the late Devonian. Overall length of the fish was about 24 cm. (a,b) Dorsal and lateral views of the skull, respectively. Bones of the pectoral girdle (red) are tightly connected to the posterior wall of the skull. Abbreviations: angular (Ang), branchiostegals (Br), clavicle (Cl), cleithrum (Ct), dentary (D), dermohyal (Dhy), external naris (En), lateral extrascapular (Esl), frontal (F), gulars (G), intertemporal (It), jugal (J), lacrimal (L), maxilla (M), nasal (N), opercular (O), accessory opercular (Opa), parietal (Pa), premaxilla (Pm), postorbital (Po), preopercular (Pop), postrostral (Por), posttemporal (Pot), postparietal (Pp), preorbital (Pro), quadratojugal (Qj), rostral (R), sclerotic ring (Sc), supracleithrum (Sct), suborbital (Sb), supraoccipital (So), subopercular (Sop), supraorbital (Spo), supratemporal (St).

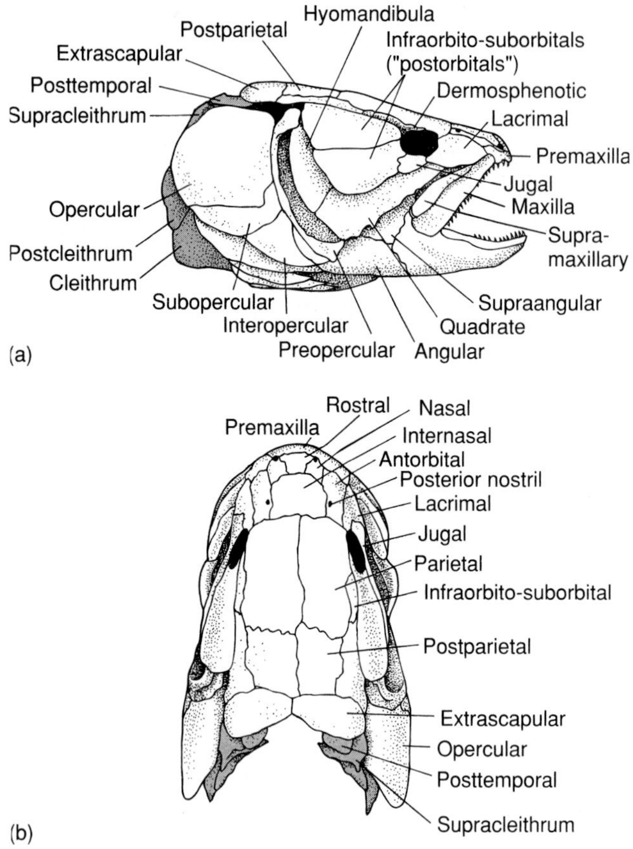


**Figure 7.22** Skull of the bowfin, *Amia*, a chondrostean. Lateral (a) and dorsal (b) views.

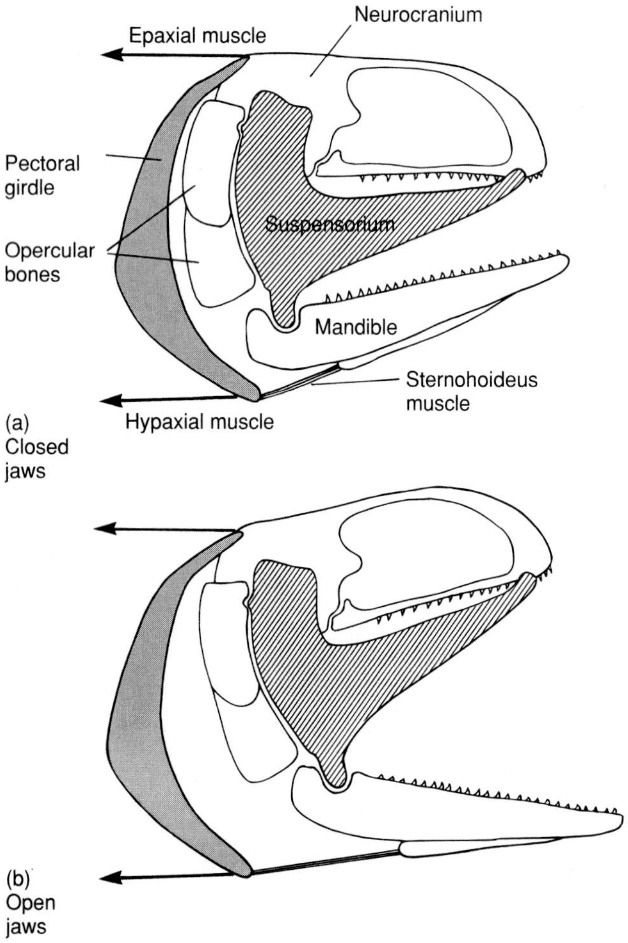


**Figure 7.23** Cranial kinesis in a primitive actinopterygian fish. (a) Jaws are closed. (b) Jaws are open. The mandible rotates on its articulation with the suspensorium, which in turn is articulated with the opercular bones. The pectoral girdle remains relatively fixed in position, but the neurocranium rotates on it to lift the head. Lines of action of major muscles are shown by arrows.

UNIVERSITY OF WOLVERHAMPTON
Harrison Learning Centre

ITEMS ISSUED:

**Customer ID: WPP61191450**

Title: Vertebrates : comparative anatomy, function, evolution
ID: 7608972079
**Due: 05/05/2015 23:59**

Total items: 1
14/04/2015 16:40
Issued: 4
Overdue: 0

Thank you for using Self Service.
Please keep your receipt.

Overdue books are fined at 40p per day for 1 week loans, 10p per day for long loans.

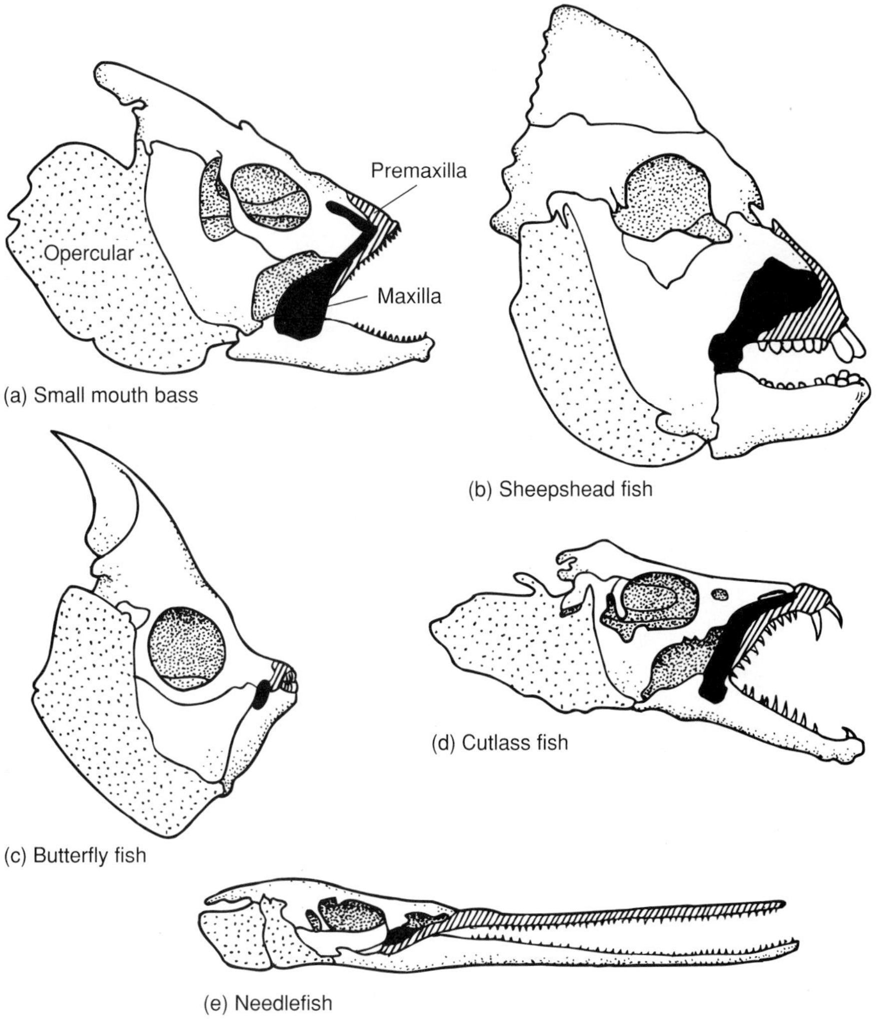


**Figure 7.24** Teleost skulls. Despite the great diversification of teleosts in many habits, the basic pattern of skull bones is preserved. (a) Small mouth bass *(Micropterus dolomieu)*. (b) Sheepshead fish *(Archosargus probatocephalus)*. (c) Butterfly fish *(Chaetodon ocellatus)*. (d) Cutlass fish *(Trichiurus lepturus)*. (e) Needlefish *(Tylosurus marinus)*.

foods, like their living counterparts that have similar tooth plates and jaws for feeding on shellfishes, snails, and crustaceans. The other group of sarcopterygians, the crossopterygians, had strong jaws with small, pointed teeth. However, unlike teeth of other fishes, the walls of crossopterygian teeth were extensively infolded, producing distinct **labyrinthodont teeth.** Large teeth were carried on the dentary of the lower jaw and along the lateral bones of the palate—vomer, palatine, ectopterygoid. Bones of the dermatocranium resembled those of actinopterygians, and like actinopterygians, the palatoquadrate articulated anteriorly with the nasal capsule and laterally with the maxilla. Unlike actinopterygians and lungfishes, the braincase of crossopterygians typically ossified into two articulated units, an anterior **ethmoid unit** (ethmosphenoid unit) and a posterior oticooccipital unit with a flexible joint between them. In the dermal roofing bones above this joint, a hinge formed between the parietal and postparietal. Consequently, the snout could rotate upward about the rest of the skull, a displacement thought to be important during feeding (figure 7.26). The functional notochord also extended well forward into the head, passing through a tunnel in the oticooccipital segment, eventually abutting the back of the ethmoid unit and perhaps bringing added support into this region of the skull.

**Labyrinthodont teeth (p. 498)**

*Nasal Capsules*. From fishes to tetrapods, the nasal capsules have had a complex history. The nasal capsules hold the olfactory epithelium in the form of a paired **nasal**

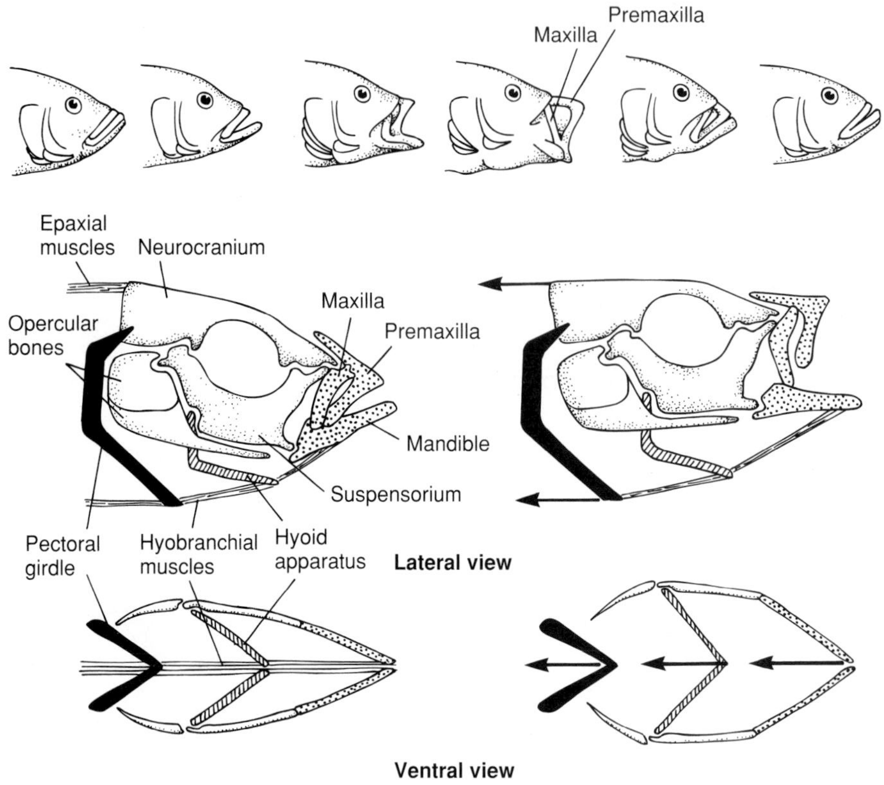


**Figure 7.25** Suction feeding of a teleost fish. Top series are traces from a high-speed film of jaw opening (food not shown). Note changes in position of the jaws. Lateral and ventral views, respectively, of the major kinetic bones of the skull are shown when jaws are closed (left) and when they are open (right). Note the forward movement of the jaws (stippled areas) and outward expansion of the buccal cavity. Lines of muscle action are shown by arrows.

**sac** (figure 7.27a). In actinopterygians, the nasal sac typically does not open directly into the mouth. Instead, its anterior (incurrent) and posterior (excurrent) narial openings establish a route for one-way water flow across the olfactory epithelium, delivering to it fresh chemical odors. By contrast, each nasal sac of tetrapods opens directly into the mouth via an **internal naris,** or **choana** (figure 7.27b). Each nasal sac also opens to the exterior by way of an **external naris** (nostril), thus establishing a respiratory route for airflow in and out of the lungs. In addition to internal and external nares, a third opening within the nasal sac begins as a tube, the **nasolacrimal duct,** that runs toward the orbit in order to drain away excess secretions of the adjoining lacrimal gland after helping to moisten the surface of the eye.

**Olfactory organs (p. 668)**

Among sarcopterygians, the nasal capsules of rhipidistians are similar to those of tetrapods. In rhipidistians, the nasolacrimal duct is an adaptation that benefits surface fishes that poke their eyes and nostrils out of the water. The lacrimal gland moistens exposed sensory organs that are subjected to drying. The nasolacrimal duct is probably homologous to the posterior (excurrent) naris of actinopterygian fishes. Rhipidistians (but not coelacanths) also possess internal nares, apparently representing a new derivative of the nasal sac connecting it with the mouth. However, lungfishes probably lack internal nares, although this is still debated. In lungfishes, the posterior (excurrent) naris opens near the margin of the mouth but does not pierce the palatal series of dermal bones as does the true internal naris of rhipidistians and tetrapods.

## *Amphibians*

The earliest amphibians arose from crossopterygian ancestors and retained many of their skull features, including most of the bones of the dermatocranium. Numerous bones in the snout were reduced, leaving a distinct nasal bone occupying a position medial to the external naris (figure 7.28a,b). The opercular series of bones covering the gills are typically lost. Extrascapulars across the back of the fish skull also disappear in primitive amphibians. Along with this, the amphibian pectoral girdle loses its attachment to the back of the skull. Roofing bones and chondrocranium become more tightly associated, reducing the neurocranial mobility of the snout in comparison with rhipidistians.

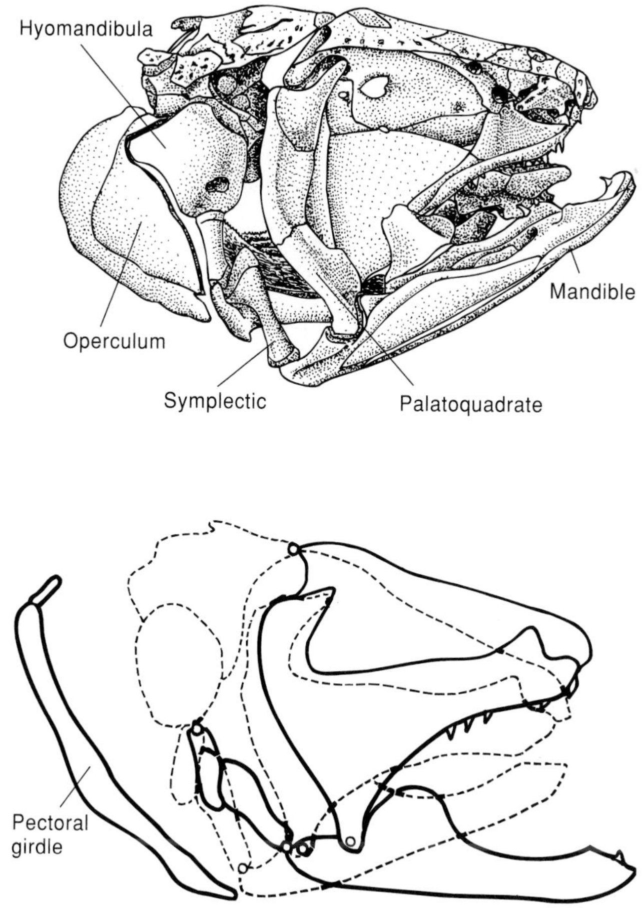


**Figure 7.26** Cranial kinesis of a crossopterygian *Latimeria*. (a) Lateral view of the skull. (b) Biomechanical model of major functional elements showing displacement pattern during jaw opening (solid lines) compared with closed position (dashed lines).

The lateral line system, an aquatic sensory system, is evident in skulls of the earliest amphibians, at least among the juveniles that were presumably aquatic stages (figure 7.29a,b). The skull is flattened, and in some an otic notch at the back of the skull marks the presumed location of an eardrum (tympanic membrane). The columella conveys sound vibrations from the tympanic membrane to the inner ear. But the columella in early amphibians is still a robust bone that also seems to be a buttress between the braincase and the palatoquadrate. Teeth were conical in labyrinthodonts, with the enamel folded into complex patterns. Teeth of lepospondyls lacked the highly folded enamel and the otic notch was absent.

The skulls of modern amphibians are greatly simplified compared with those of their fossil ancestors, with many of the dermal bones being lost or fused into composite bones. Caecilian skulls are compact and firmly ossified, although the pattern of dermal bones can be quite varied. In salamanders, the chondrocranium consists primarily of **orbitosphenoid** and **prootic** bones, with exoccipitals closing

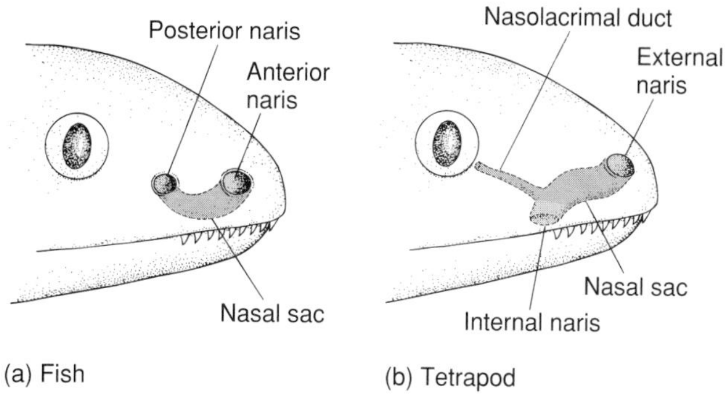


**Figure 7.27** Openings of the nasal sac. (a) In an actinopterygian fish, the nasal sac typically has an anterior naris through which water enters and a posterior naris through which water exits, but the nasal sac does not open into the mouth. (b) In a tetrapod, the nasal sac has an external naris (homologous with the anterior naris of the fish) and a nasolacrimal duct to the orbit (presumed homologue of the posterior naris of the fish). In addition to these, a third extension of the nasal sac, the internal naris, opens into the buccal cavity through the roof of the mouth.

the posterior wall of the braincase (figure 7.30). Nasal bones are usually present. Up to four pairs of roofing bones contribute to the skull: frontals and parietals are present in all, but prefrontals and lacrimals vary among groups. In anurans (figure 7.31), ossification of the chondrocranium is highly variable, usually with just five bones present, a single sphenethmoid and paired prootics and exoccipitals. A nasal bone is present, but only a paired, composite **frontoparietal** remains of the roofing bones. In both frogs and salamanders, the single parasphenoid has expanded to form a large plate that has crowded other palatal bones.

The splanchnocranium, a major component of the fish skull, is reduced in amphibians. In modern amphibians, the hyomandibula plays no role in jaw suspension. This task is taken over exclusively by the articular and quadrate bones through which the mandible articulates with the skull. The branchial arches composing the hyobranchial apparatus support external respiratory gills in the larvae, but when the larvae metamorphose into the adult, these arches are reduced to the hyoid apparatus that supports the action of the tongue.

Salamanders commonly use suction feeding in water. The floor of the throat is rapidly expanded and the jaws parted enough so that the pulse of water carrying the intended prey enters (figure 7.32). Excess water gulped in with the prey exits at the back of the mouth through the gill slits. In salamanders, as in fishes, there is a **unidirectional** flow of food and water into the mouth and out the gill slits. In metamorphosed salamanders and in adult frogs, gill slits are absent, so excess water entering the mouth during feeding must reverse its flow to exit via the mouth. Such flow is said to be **bidirectional.** On land, amphibians commonly use a sticky, projectile tongue. At close range, muscles catapult the tongue over the parted mandibles and into contact with the prey. At longer range, muscle action

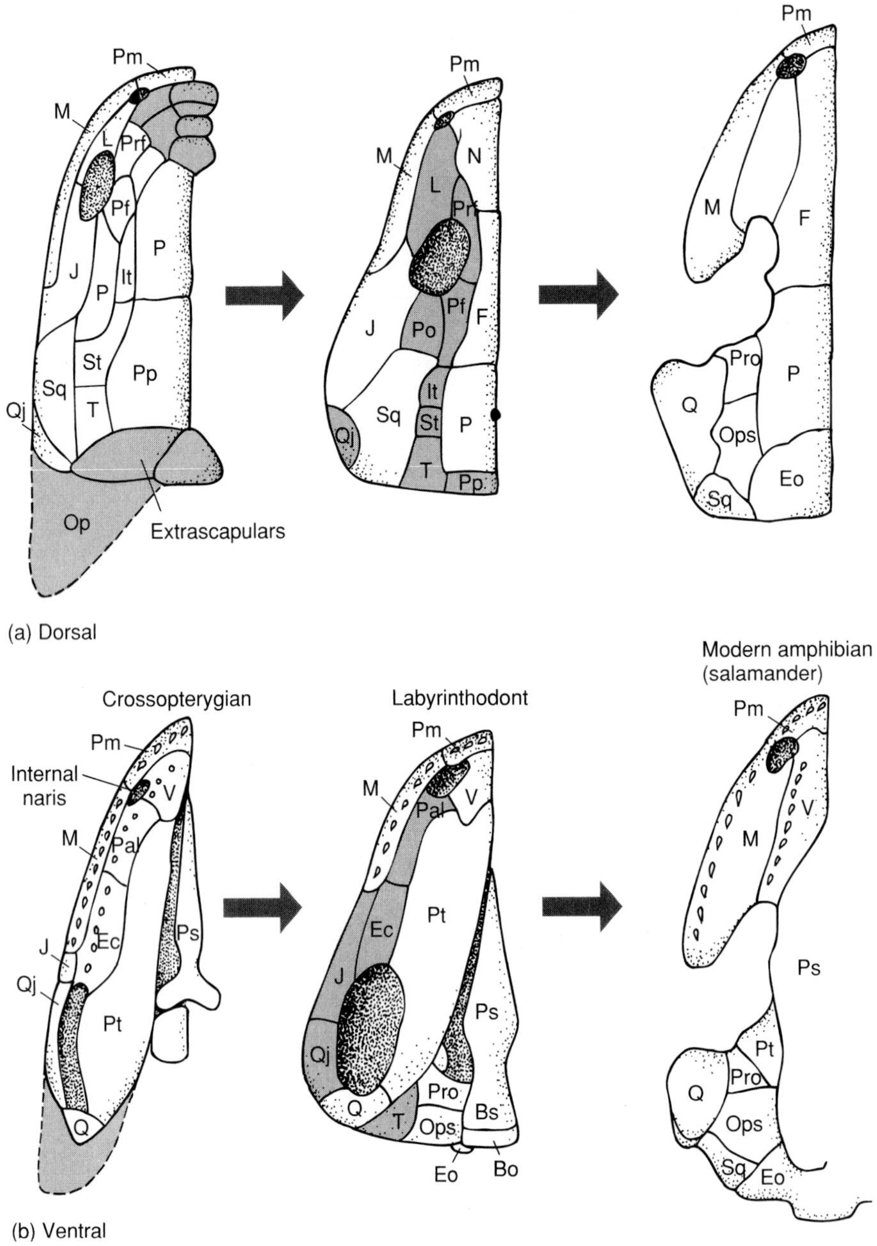


**Figure 7.28** Diagrammatic views of skull modifications from crossopterygian to labyrinthodont to modern amphibian (salamander). (a) Dorsal views. (b) Ventral (palatal) views. Skull bones lost in the derived group are shaded in the skull of the preceding group. Abbreviations: basioccipital (Bo), basisphenoid (Bs), ectopterygoid (Ec), exoccipital (Eo), frontal (F), intertemporal (It), jugal (J), lacrimal (L), maxilla (M), nasal (N), opercular (Op), opisthotic (Ops), parietal (P), palatine (Pal), postfrontal (Pf), premaxilla (Pm), postorbital (Po), prefrontal (Prf), prootic (Pro), postparietal (Pp), parasphenoid (Ps), pterygoid (Pt), quadrate (Q), quadratojugal (Qj), supratemporal (St), squamosal (Sq), tabular (T), vomer (V).

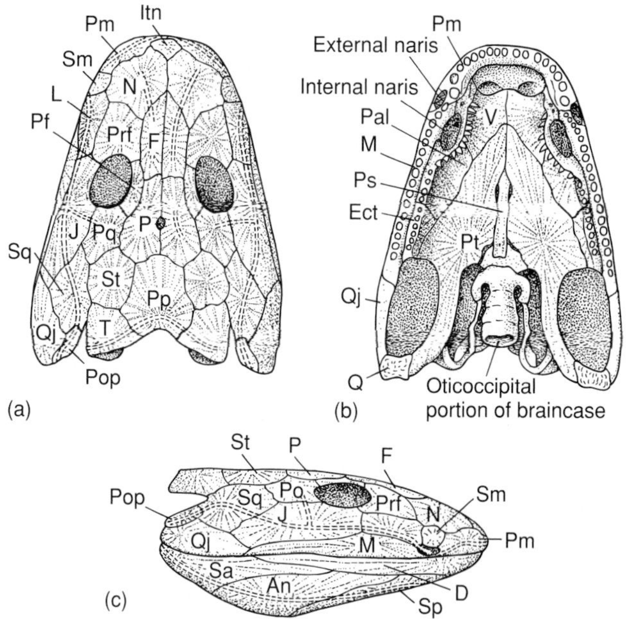


**Figure 7.29** Skull of *Ichthyostega*, a primitive labyrinthodont amphibian of the late Devonian. Dorsal (a), ventral (b), and lateral (c) views. Parallel tracks of dashed lines indicate course of the aquatic lateral line system on the skull bones. Abbreviations: angular (An), dentary (D), ectopterygoid (Ect), frontal (F), internasal (Itn), jugal (J), lacrimal (L), maxilla (M), nasal (N), parietal (P), palatine (Pal), postfrontal (Pf), premaxilla (Pm), postorbital (Po), preopercular (Pop), postparietal (Pp), prefrontal (Prf), parasphenoid (Ps), pterygoid (Pt), quadrate (Q), quadratojugal (Qj), surangular (Sa), septomaxilla (Sm), splenial (Sp), supratemporal (St), squamosal (Sq), tabular (T), vomer (V).

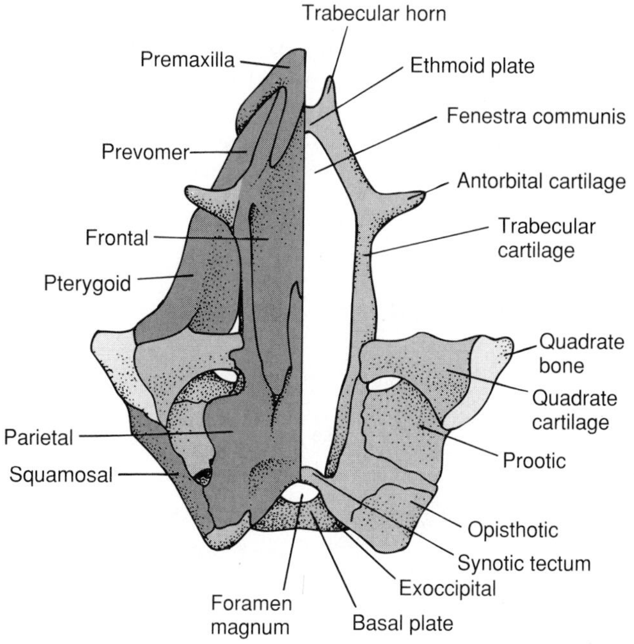


**Figure 7.30** Skull of *Necturus*, a modern amphibian. Superficial skull bones are indicated on the left. These bones have been removed to reveal the chondrocranium and derivatives of the splanchnocranium on the right.

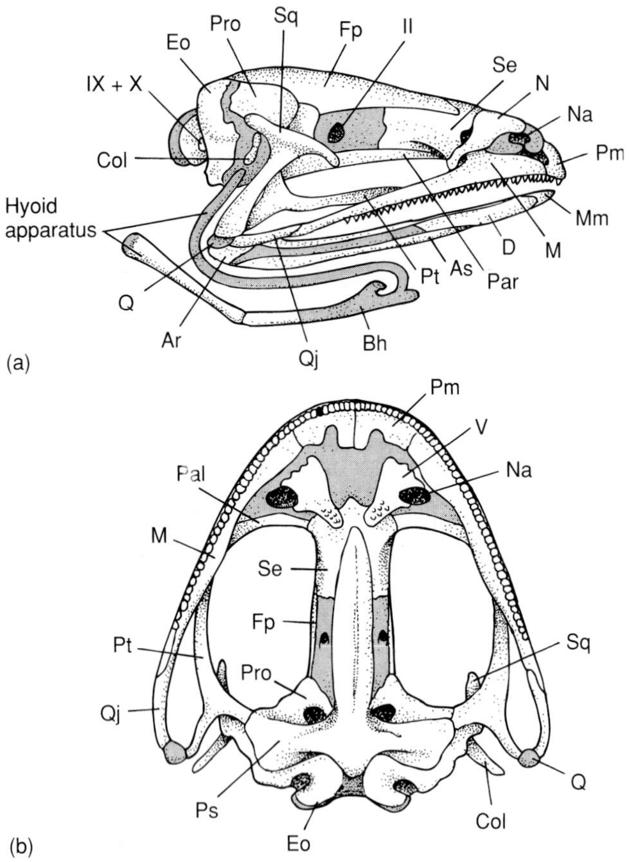


**Figure 7.31** Frog skull. Lateral (a) and ventral (b) views. Abbreviations: articular (Ar), angulosphenoid (As), basihyal (Bh), columella (Col), dentary (D), exoccipital (Eo), frontoparietal (Fp), maxilla (M), mento-Meckelian (Mm), nasal (N), naris (Na), palatine (Pal), premaxilla (Pm), prootic (Pro), parasphenoid (Ps), pterygoid (Pt), quadrate (Q), quadratojugal (Qj), sphenethmoid (Se), squamosal (Sq), vomer (V). Roman numerals indicate foramina serving specific cranial nerves.

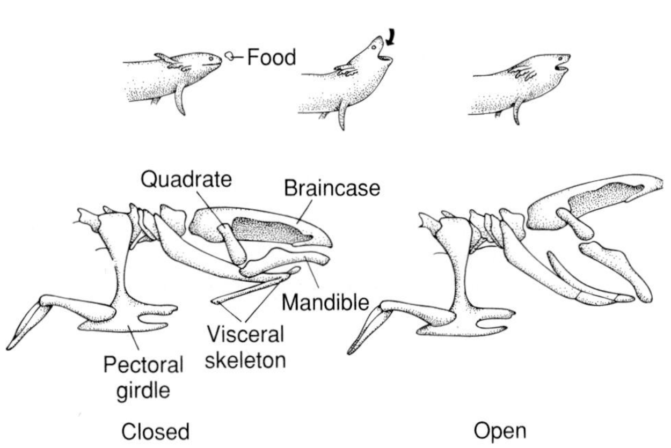


**Figure 7.32** Suction feeding by an aquatic salamander. Before, during, and after suction feeding traced from a high-speed film feeding sequence (top series). Note the interpreted positions of the skull elements when the jaws are closed (bottom left) and open (bottom right).

works in cooperation with fluid-filled spaces within the tongue to accelerate it along the hyoid apparatus. Retraction of the tongue returns the attached food to the mouth, and teeth close on it to control the struggling prey.

## *Primitive Reptiles*

The first reptiles were small and would probably remind us of lizards in general appearance. The skull roof, like that of early amphibians, was formed from the dermatocranium with openings for eyes, pineal organ, and nostrils (figure 7.33a–d). Robust attachment flanges and processes are evidence of strong jaw-closing muscles. The palatoquadrate of the mandibular arch was reduced to the small epipterygoid and separate quadrate. The hyoid arch produced a columella, a stout bone that braced the back of the dermatocranium against the chondrocranium. These early reptiles lacked an otic notch, so presumably the tympanic membrane was absent. Sound transmission to the inner ear may have occurred along bones of the lower jaw.

***Skull Fenestrae*** As mentioned earlier, the temporal region of the dermatocranium contains features particularly revealing of amniote lineages (figure 7.34). Fenestrae are

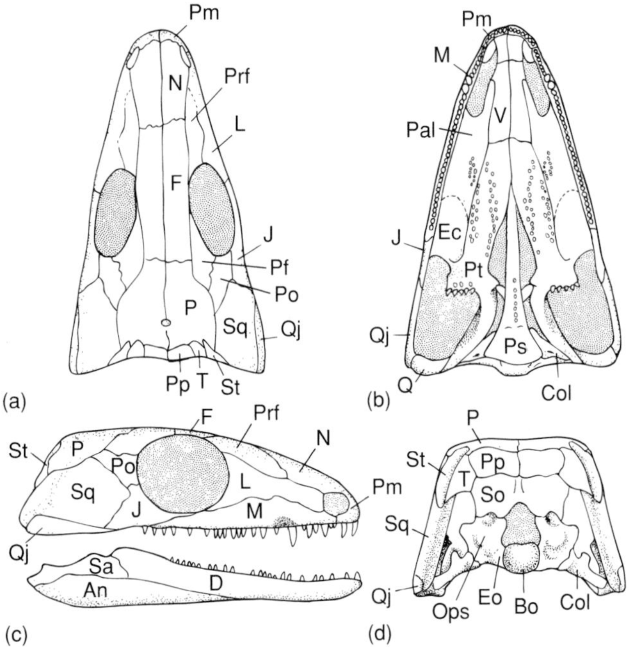


**Figure 7.33** Skull of *Paleothyris*, a captorhinomorph reptile from the Carboniferous. Dorsal (a), ventral (b), lateral (c), and posterior (d) views. Abbreviations: angular (An), basioccipital (Bo), columella (Col), dentary (D), ectopterygoid (Ec), exoccipital (Eo), frontal (F), jugal (J), lacrimal (L), maxilla (M), nasal (N), opisthootic (Ops), parietal (P), palatine (Pal), postfrontal (Pf), premaxilla (Pm), postorbital (Po), postparietal (Pp), prefrontal (Prf), parasphenoid (Ps), pterygoid (Pt), quadrate (Q), quadratojugal (Qj), surangular (Sa), supraoccipital (So), supratemporal (St), squamosal (Sq), tabular (T), vomer (V).

openings in the outer dermatocranium. The anapsid skull lacks temporal fenestrae. In recent turtles, **emarginations** often encroach upon the posterior margin of the skull roof. These emarginations are large notches that function like fenestrae, but they are independent phylogenetic derivatives. The diapsid skull includes two temporal fenestrae, a condition carried forward in *Sphenodon* and in crocodiles and their allies. However, the lower and upper temporal bars are often lost in other modern forms. This gives us several contemporary varieties of a modified diapsid skull in which the diapsid condition is substantially altered, such as birds, lizards, and especially snakes.

The synapsid skull of pelycosaurs, therapsids, and modern mammals contains a single temporal opening. Loss or reduction of the postorbital bone in many modern mammals allows merging of the temporal fenestra with the orbit.

#### Taxonomic implications of temporal fenestrae (p. 102)

Although used by taxonomists to delineate phylogenetic lineages within tetrapods, the functional significance of fenestrae is not clear. With few exceptions, most notably lepospondyl amphibians, fenestrae are absent in amphibians and primitive reptiles. Because fenestrae are associated with strong jaw adductor muscles, it has been suggested that they open space in the skull for these muscles to bulge during contraction (figure 7.35a–d). But it is difficult to see how such a function could have afforded some initial advantage favoring their evolution. Initially fenestra would have been too small to provide space for bulging muscles

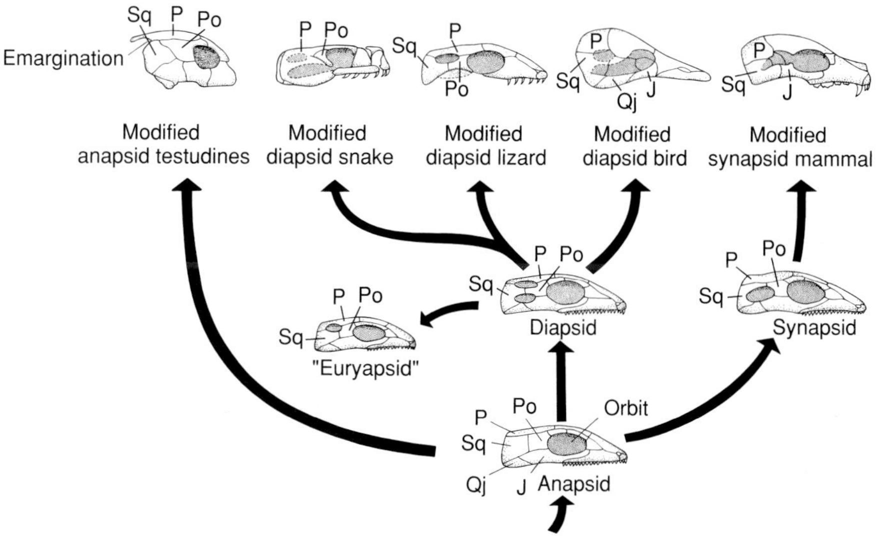


**Figure 7.34** Major lineages of dermatocranium evolution within amniotes. The anapsid skull occurs in cotylosaurs and their modern descendants, turtles and tortoises. Two major groups, the diapsids and synapsids, independently evolved from the anapsids. *Sphenodon* and crocodilians retain the primitive diapsid skull, but it has been modified in diapsid derivatives such as snakes, lizards, and birds. Shading indicates positions of temporal fenestrae and orbit. Abbreviations: jugal (J), parietal (P), postorbital (Po), quadratojugal (Qj), squamosal (Sq).

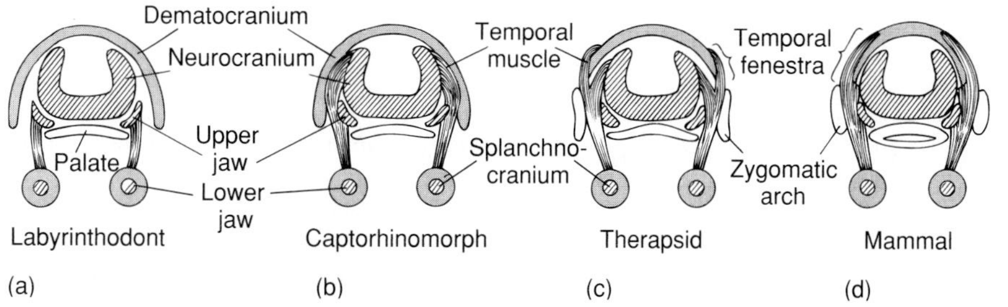


**Figure 7.35** Temporal fenestrae. The shift in jaw muscle attachment to the skull is shown. (a) The temporal muscles of the jaw run from upper to lower jaw in primitive tetrapods. As these muscles become more prominent (b,c), their site of origin from the skull expands (d). In amniotes, these accompanying fenestrae become important taxonomic characteristics helping to identify lines along which advanced tetrapods evolved.

that presumably favored their appearance. Alternatively, some have suggested that unstressed bone of the dermatocranium might have little selective value if it did not contribute to muscle attachment. Its loss would have been expected, leading to the initial appearance of fenestrae in these areas. More positively, it has been proposed that the rims of open fenestrae offer a more secure attachment site for muscles than does a flat surface. Muscle tendons merge with the periosteum, spread the tensile forces around the rim, and distribute them across the extended surface of the bone. This might render the attachment site less susceptible to being torn loose from the bone.

Whatever the function of fenestrae, their presence would be possible only if holes would not unduly weaken the ability of the skull to withstand stresses. Their absence in labyrinthodonts and primitive reptiles, their presence in later tetrapods, and the appearance of emarginations by a different route in turtles imply a complex and not fully understood interaction between function and design in early tetrapods.

***Cranial Kinesis in Reptiles*** Skull elements of reptiles exhibit varying degrees of mobility. The most extensive motions are found in the skulls of lizards and especially snakes. In these two groups, a transverse hinge extends across the skull roof. Depending on the position of this hinge, three names apply. Where a hinge passes across the back of the skull, permitting rotation between the neurocranium and outer dermatocranium, the skull is said to exhibit **metakinesis** (figure 7.36a). If a joint passes through the dermatocranium behind the eye, the skull exhibits **mesokinesis.** If a joint in the dermatocranium passes in front of the orbits, the skull exhibits **prokinesis.** Depending on the number of hinges, the skull may be **monokinetic,** having one joint, or **dikinetic** (amphikinetic), having two joints. Metakinesis is found in some eosuchian reptiles. Although rare, mesokinesis is possibly present in amphisbaenians and some burrowing lizards. Prokinesis is typical in snakes and birds. Most modern lizards are dikinetic, with both meta- and mesokinetic joints across their skull roofs.

The term ***streptostyly*** applies not to the skull roof but to the quadrate and describes the condition in which the quadrate is free to undergo some degree of independent rotation about its dorsal connection with the braincase (figure 7.36b). Most lizards, snakes, and birds are streptostylic.

## *Modern Reptiles*

Modern turtles have anapsid skulls, but emarginations that develop from the posterior region forward often result in the opening of large regions within outer bones of the dermatocranium (figure 7.37a–e). Large jaw-closing muscles occupy this space. Although turtles lack teeth, the opposing surfaces of upper and lower jaws are usually covered with keratinized "tooth" plates that deliver powerful biting forces to food.

Several modern reptiles are surviving members of the diapsid radiation. In *Sphenodon*, complete upper and lower temporal bars firmly join the front and the back of the lateral skull wall (figure 7.38a–d). Consequently, no significant mobility is permitted within the dermatocranium. However, the mandible slides back and forth on the fixed quadrate from which it is suspended. The single row of teeth of the mandible moves between a double row of teeth on the upper jaw, an action that seems to be important in slicing through some types of prey.

Loss of the lower temporal bar produces the modified diapsid skull of lizards (figure 7.39). Loss of this lower bony strut laterally liberates the snout from the posterior part of the skull. Lizard ancestors, the eosuchians, apparently possessed a single metakinetic joint across the back of the skull. A second kinetic joint, the mesokinetic joint, has been added to this in most modern lizards, making the skulls of most lizards dikinetic. Although skulls of some specialized lizards, such as bur-

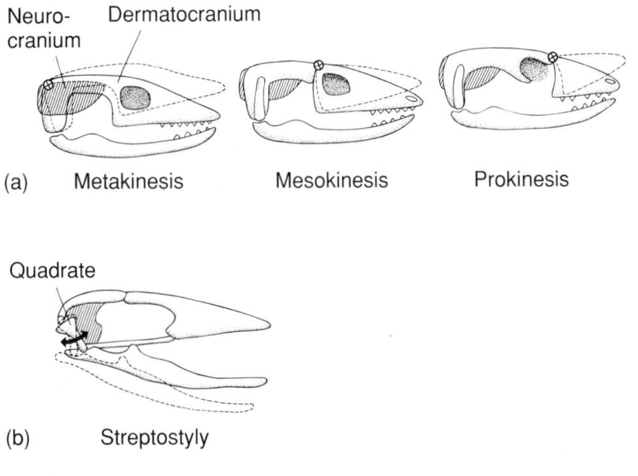


**Figure 7.36** Cranial kinesis in squamates. (a) There are three types of cranial kinesis based largely on the position at which the hinge (X) lies across the top of the skull. The hinge may run across the back of the skull roof (metakinesis), behind the orbit (mesokinesis), or in front of the orbit where the snout articulates (prokinesis). (b) The ubility of the quadrate to rotate about its dorsal end is called streptostyly.

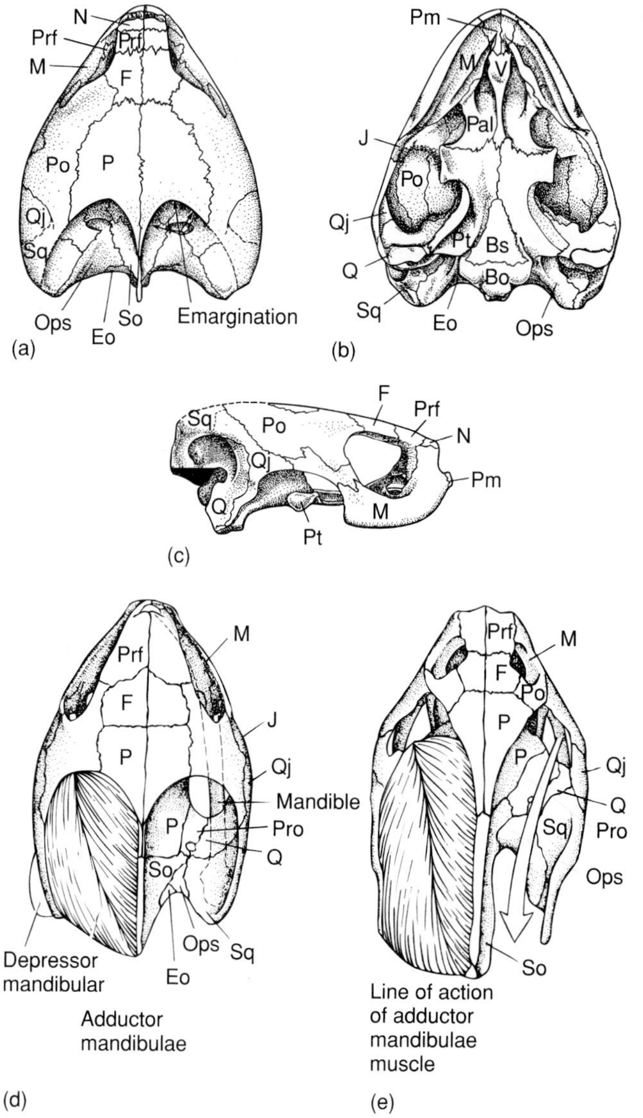


**Figure 7.37** Turtle skulls. (a–c) Skull of *Pleisochelys*, from the late Jurassic. *Pleisochelys* is the earliest known member of the cryptodires. Note the absence of any temporal fenestrae but the presence of emarginations etched in the dorsal, posterior rim of the skull. Dorsal (a), ventral (b), and lateral (c) views. (d) Modern box turtle *Emys*, showing site of residence of jaw opening (depressor mandibulae) and closing (adductor mandibulae) muscles in relation to emargination. (e) Modern softshell turtle *Trionyx*, showing line of action of adductor mandibulae from lower jaw to skull within enlarged emargination. Abbreviations: basioccipital (Bo), basisphenoid (Bs), exoccipital (Eo), frontal (F), jugal (J), maxilla (M), nasal (N), opisthotic (Ops), parietal (P), palatine (Pal), prefrontal (Prf), premaxilla (Pm), prootic (Pro), postorbital (Po), pterygoid (Pt), quadrate (Q), quadratojugal (Qj), supraoccipital (So), squamosal (Sq), vomer (V).

rowers, anteaters, and some herbivores, seem monokinetic, this is likely a secondary condition. This kinetic machinery of lizard jaws has been modeled as a four-bar linkage system (figure 7.40a,b). One unit is the triangular-shaped snout. Its posterior wall forms one of the four linkages. The dorsal corner of the snout participates in the mesokinetic joint and forms a second mechanical link with the dorsal end of the quadrate through the top of the skull. The quadrate represents the third link. The fourth mechanical link connects the quadrate's lower end (where it meets the pterygoid) forward to the posterior lower corner of the snout to complete and close the four-bar kinematic chain.

**Biomechanical mechanisms (p. 137)**

Without such a kinematic series of linkages, jaw closure would be scissorslike, and jaw-closing forces on the prey would have a forward component that might deflect or squirt the prey out of the mouth, increasing the chance of prey loss (figure 7.40c). However, in the skull of many lizards, rotation of the four linkages permits changes in geometric configuration. As a consequence, these lizards can alter the angle of the tooth row borne by the snout as it closes on the prey. Upper and lower jaws close and meet the prey nearly simultaneously, delivering forces directed at the prey, and thus the lizards are less likely to experience prey loss.

The metakinetic joint is not directly part of this linkage train of bones. It permits the dermatocranium, to which the linkage chain is joined, to move relative to the deeper neurocranium. The axis of the metakinetic joint is almost coincident with the superficial joint between the dorsal end of the quadrate and the braincase, but it is not part of this outer set of linkages. Thus, rotation about this metakinetic joint lifts the whole dermatocranium along with the entire set of linkages relative to the neurocranium.

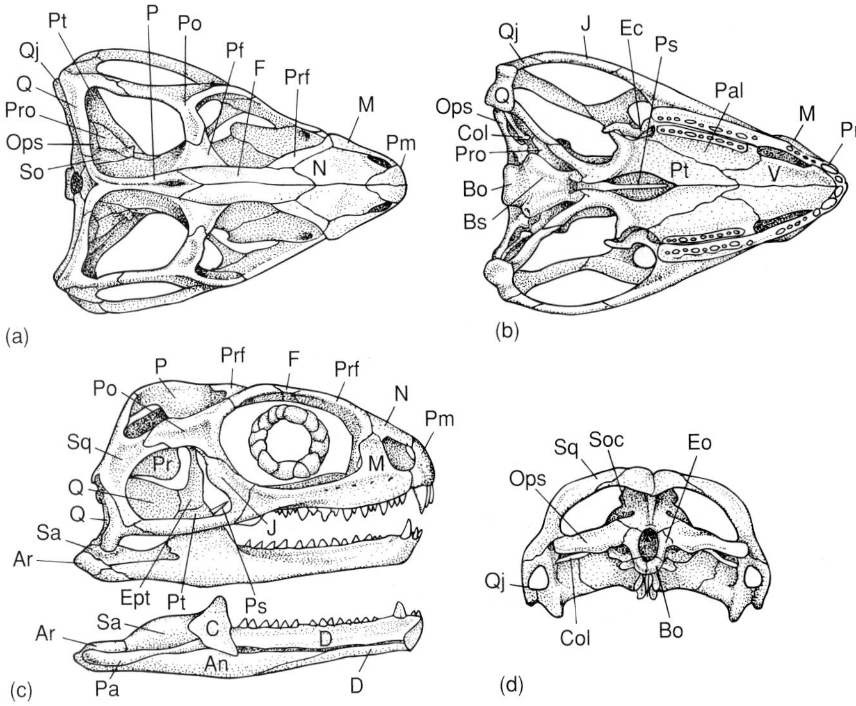


**Figure 7.38** Living rhynchocephalian. The two temporal fenestrae are still bounded by bone in *Sphenodon*, a living diapsid. Dorsal (a), ventral (b), lateral (c), and posterior (d) views. Abbreviations: angular (An), articular (Ar), basioccipital (Bo), basisphenoid (Bs), coronoid (C), columella (Col), dentary (D), ectopterygoid (Ec), exoccipital (Eo), epipterygoid (Ept), frontal (F), jugal (J), maxilla (M), nasal (N), opisthotic (Ops), parietal (P), prearticular (Pa), palatine (Pal), postfrontal (Pf), premaxilla (Pm), postorbital (Po), prefrontal (Prf), prootic (Pro), parasphenoid (Ps), pterygoid (Pt), quadrate (Q), quadratojugal (Qj), surangular (Sa), supraoccipital (Soc), squamosal (Sq), supratemporal (St), vomer (V).

Some lizards, like many terrestrial salamanders, project their tongues during feeding. When the tongue is prominently deployed, a lizard engages in **lingual feeding** (figure 7.41a). The jaws part, and the sticky tongue is projected at the prey. In chameleons, a circular **accelerator muscle** wraps around the **lingual process** of the hyoid apparatus (figure 7.41b,c). Upon contraction, the accelerator muscle squeezes the lingual process, picks up speed as it slides down the tapered process, perhaps like squeezing a slippery bar of soap, and carries along the glandular tip of the tongue (figure 7.41d). The gathered momentum of the tip of the tongue launches it out of the mouth toward the prey. Upon impact, the fleshy glandular tip of the tongue flattens against the target, establishing firm adhesion. Retraction of the tongue back into the mouth retrieves the prey. The jaws then close to hold the captured prey.

In snakes, the frontal and parietal roofing bones have grown down around the sides of the skull to form most of the walls of the braincase as well (figure 7.42). Their enlargement results in crowding or loss of many of the other dermal bones. Snake skulls are prokinetic. A joint across the skull forms in front of the orbit between frontal and nasal regions. However, most of the extensive mobility of the snake jaw results from changes in skull design in the lateral bones. Both upper and lower temporal bars are lost, thus removing struts that in other diapsid skulls form restrictive braces across the temporal region. The kinematic machinery of the snake skull includes more elements than the linkage system of lizards (figure 7.43a–c). The quadrate, as in lizards, is streptostylic but more loosely articulated with the pterygoid. Forces imparted to the pterygoid are transmitted to the tooth-bearing maxilla via the linking ectopterygoid. The maxilla rotates upon the prefrontal, from which it is suspended from the braincase. In many snakes, especially in advanced venomous snakes such as vipers, the prefrontal and supratemporal also enjoy some degree of rotation upon the braincase. Thus, the kinetic system can be modeled on a linkage chain with up to six links (supratemporal, quadrate, pterygoid, ectopterygoid, maxilla, prefrontal) suspended at either end from a seventh link, the braincase (figure 7.43d).

The mandible of snakes, suspended from the quadrate, includes a tooth-bearing dentary that articulates with a posterior **compound bone** derived from the fused surangular and angular. A tiny splenial is usually present on the medial

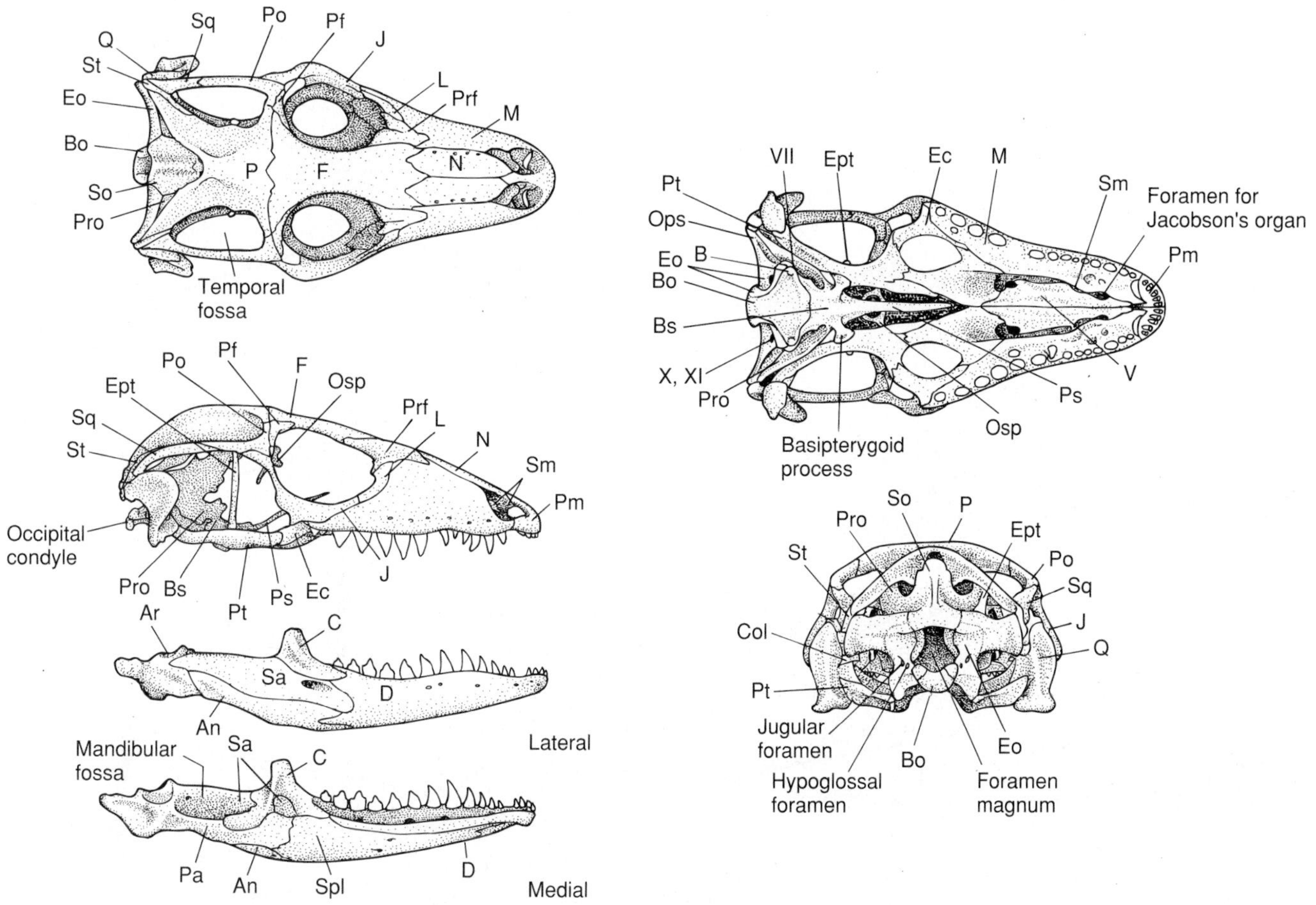


**Figure 7.39** Lizard skull. Lizards are modified diapsids. Two fenestrae are present, but the ventral bony border of the lower fenestra is absent, a result of changes serving increased cranial kinesis. Abbreviations: angular (An), articular (Ar), basioccipital (Bo), basisphenoid (Bs), coronoid (C), dentary (D), ectopterygoid (Ec), exoccipital (Eo), epipterygoid (Ept), frontal (F), jugal (J), lacrimal (L), maxilla (M), nasal (N), opisthotic (Ops), orbitosphenoid (Osp), parietal (P), prearticular (Pa), palatine (Pal), postfrontal (Pf), premaxilla (Pm), postorbital (Po), prefrontal (Prf), parasphenoid (Ps), prootic (Pro), pterygoid (Pt), quadrate (Q), surangular (Sa), septomaxilla (Sm), splenial (Sp), supraoccipital (So), squamosal (Sq), supratemporal (St), vomer (V).

side. Both halves of the lower jaw are joined at the mandible symphysis, not by bony fusion but by flexible soft tissues that unite the tips of the mandible. Mandibular tips thus enjoy independent movement. Because there are no bony cross connections between chains of movable bones on left and right sides, each kinematic set of linkages can spread and move independently of the other on the opposite side. This is particularly important during swallowing when alternating left and right sets of bones are walked over the prey (see figure 7.63). It is a mistaken view that snakes "unhinge" their jaws when they swallow. Instead, the great freedom of rotation between elements of the kinematic chains, the independent movement of each, and the ability of flaring the flexible jaws outward to accommodate bulky prey all account for the suppleness of snake jaws. These processes, not disarticulation, permit snakes to swallow (although slowly) relatively large, whole prey.

Crocodilians, together with *Sphenodon* and squamates, represent the last of the surviving reptiles with a primitive diapsid skull. The crocodilian skull is a composite of chondro-, dermato-, and splanchnocranial elements, although the dermatocranium tends to predominate (figure 7.44). Both temporal bars are present and the skull is firm, without any evidence of cranial kinesis. Thus, crocodilians carry forward a rather primitive skull design, departing very little from that of early fossil diapsids. The major difference is within the roof of the mouth, where marginal bones (premaxilla, maxilla, palatine) grow inward to meet at the midline beneath the sphenoid region. Together with the pterygoid, these marginal bones produce the bony secondary palate that separates the nasal passageway from the mouth.

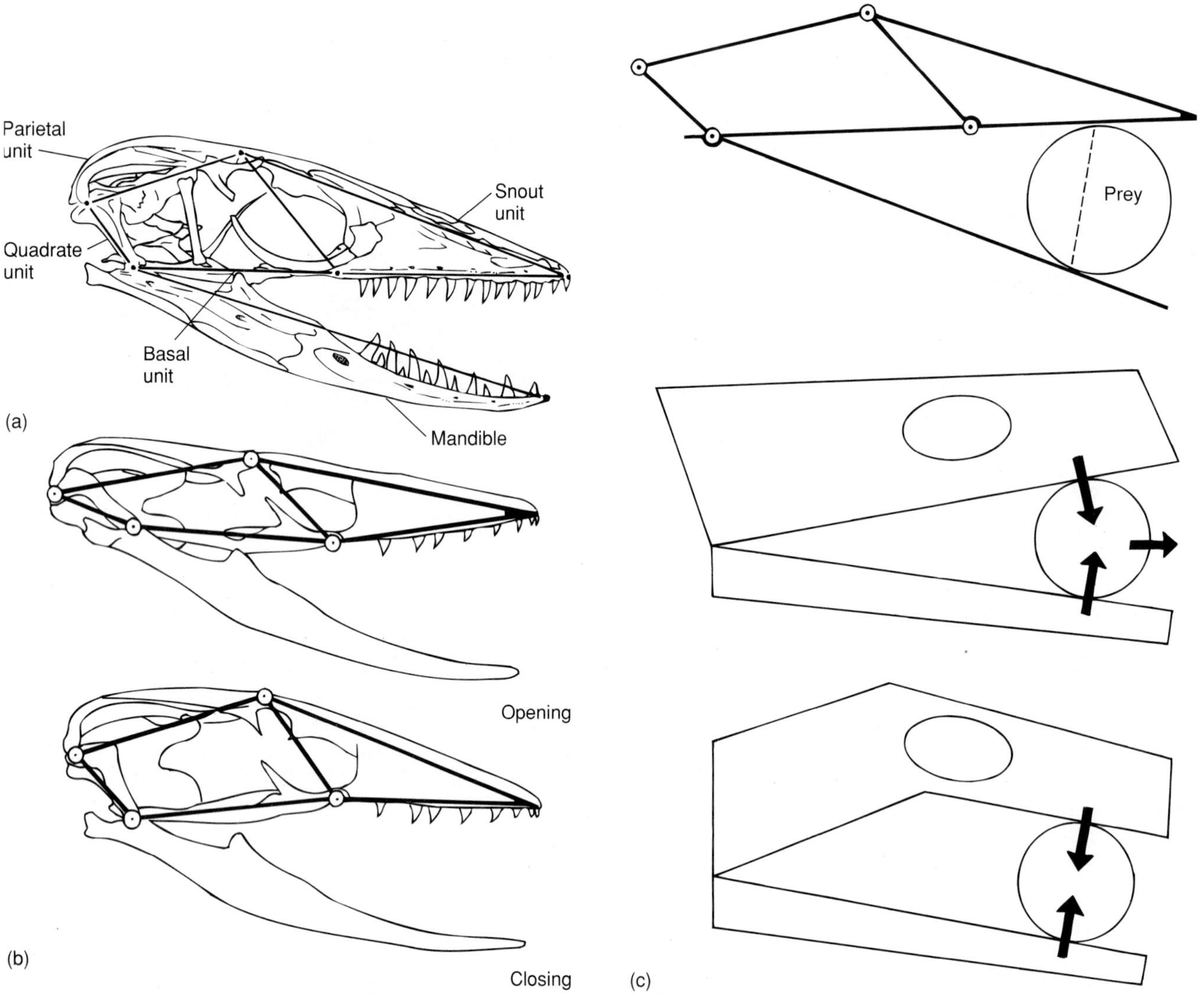


**Figure 7.40** Kinesis of a lizard skull. (a) Joints within the skull allow the snout to lift upward or bend downward about its mesokinetic articulation with the rest of the braincase. This results in a change in the angle of closure of the teeth when the animal grabs its prey. (b) These movable units of the lizard skull can be represented as a kinematic mechanism by linkages (heavy lines) and points of rotation (circles). Compared with the rest position of these linkages (a), geometric changes are shown during opening (middle) and closing (bottom) on the prey. (c) The functional significance of cranial kinesis in lizards is related to the resulting change in angle of tooth rows. Kinesis bends the snout so that both rows close directly on the prey (bottom). Were this not the case (middle and top), jaw closure would be more of a scissors action, tending to squirt the prey back out of the mouth.
Based on the research of T. H. Frazzetta.

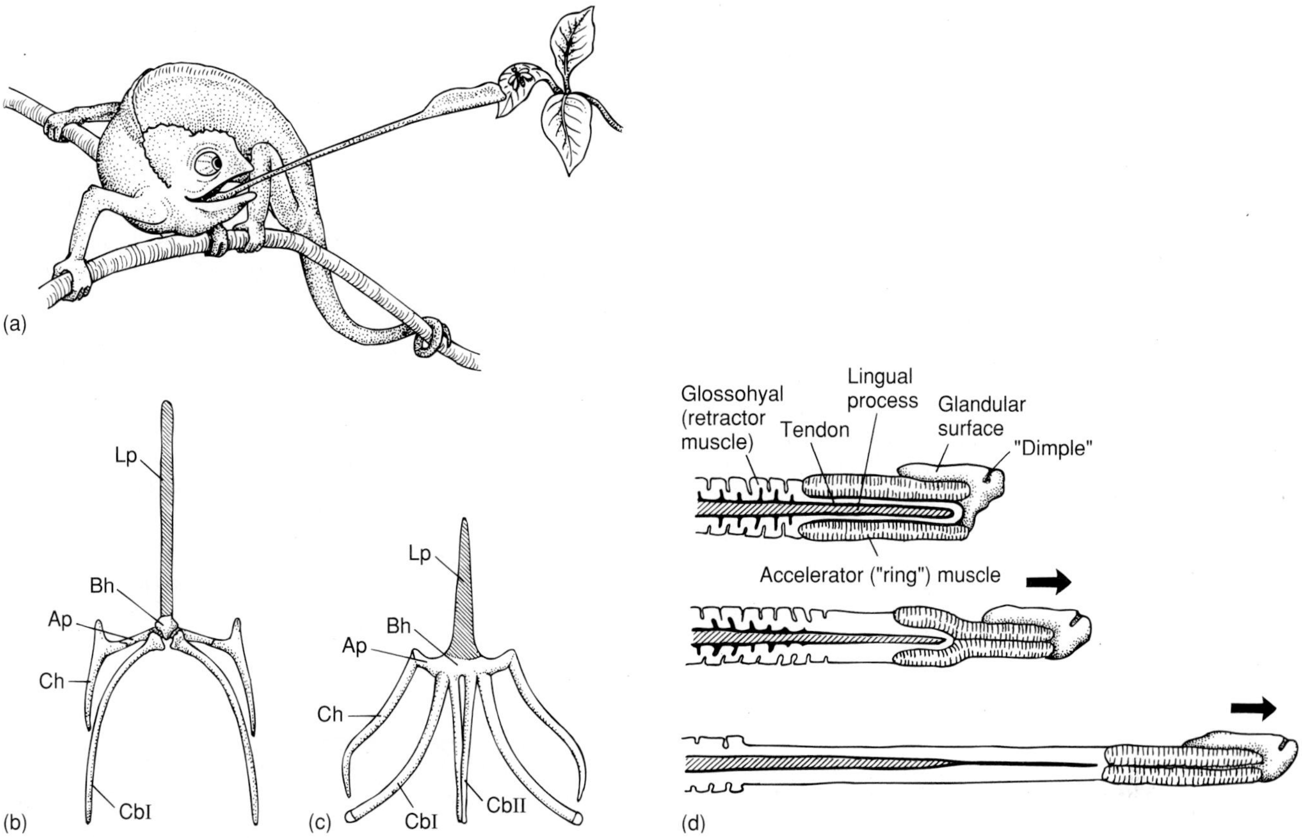


**Figure 7.41** Lingual feeding in lizards. (a) Jackson's chameleon uses its projectile tongue to "shoot" long distances at prey. (b) Hyoid apparatus of the chameleon includes an elongated lingual process (Lp) along which the tongue slides during launch. (c) Hyoid apparatus of a lizard without a projectile tongue. (d) Mechanical basis of tongue projection. The accelerator muscle, a circular band around the lingual process, contracts to squeeze the lingual process. The squeeze of accelerator muscles causes the muscle to slide rapidly toward the tip of the lingual process, carrying with it the glandular surface of the tongue. With gathered momentum, the tongue is launched from the lingual process toward the prey. The folded glossohyal muscle attached to the tongue's tip is carried out as well and eventually is responsible for retrieving the tongue and the adhering prey. Abbreviations: anterior process (Ap), basihyal (Bh), ceratobranchials I and II (CbI and CbII), ceratohyal (Ch), lingual process (Lp).

## *Birds*

Birds also arise from a diapsid ancestry, but like squamates, they show considerable modification of this skull pattern (figure 7.45). The braincase is much inflated and ossified in birds, accommodating a relatively expanded brain within. Sutures between bones are usually overgrown in the adult so that boundaries are not easily delineated. The palatal bones are quite varied, but generally all show some degree of reduction and lightening. Vomers and ectopterygoids are small, pterygoids are short struts articulating with the quadrate, and epipterygoids are usually lost (figure 7.46a–d).

Like turtles and some dinosaurs, birds are toothless, and their jaws are covered by keratinized sheaths. Birds that feed on slippery prey, such as fish-eating shorebirds, have serrated keratin to improve the friction grip. The jaws are drawn out into a **beak.** The upper temporal bar is absent, and the lower temporal bar is a slender rod called the **jugal bar** (quadratojugal-jugal bar), which extends from the beak posteriorly to the side of the movable (streptostylic) quadrate. The skull is prokinetic. A strong **postorbital ligament** extends from behind the eye to the lower jaw. When the lower jaw opens, this strong postorbital ligament becomes taut, and the posterior mandible along with the distal end of the quadrate swings forward and up (figure 7.47a,b). Through the jugal bar, this motion is imparted to the upper jaw, forcing it to lift by rotating about its prokinetic joint with the frontal bone.

Many birds use their beak like a probe to reach buried grubs or insects embedded in tree bark or soft soil. The jaws need not be parted far to seize the food. Other birds have beaks that open tough seeds and short, stout jaws that concentrate closing forces at the base of their beak.

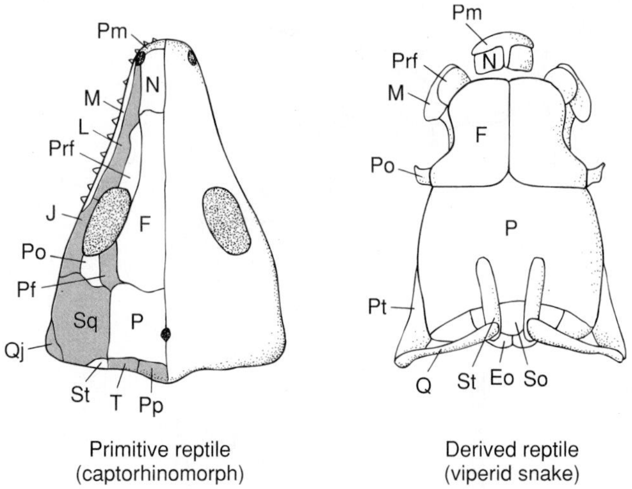


**Figure 7.42** Diagrammatic comparison of a derived modern snake skull with a captorhinomorph skull, a primitive reptile. Bones lost in the modern snake are indicated by shading in the primitive captorhinomorph. Abbreviations: exoccipital (Eo), frontal (F), jugal (J), lacrimal (L), maxilla (M), nasal (N), parietal (P), postfrontal (Pf), premaxilla (Pm), postorbital (Po), postparietal (Pp), prefrontal (Prf), pterygoid (Pt), quadrate (Q), quadratojugal (Qj), supraoccipital (So), squamosal (Sq), supratemporal (St), tabular (T).

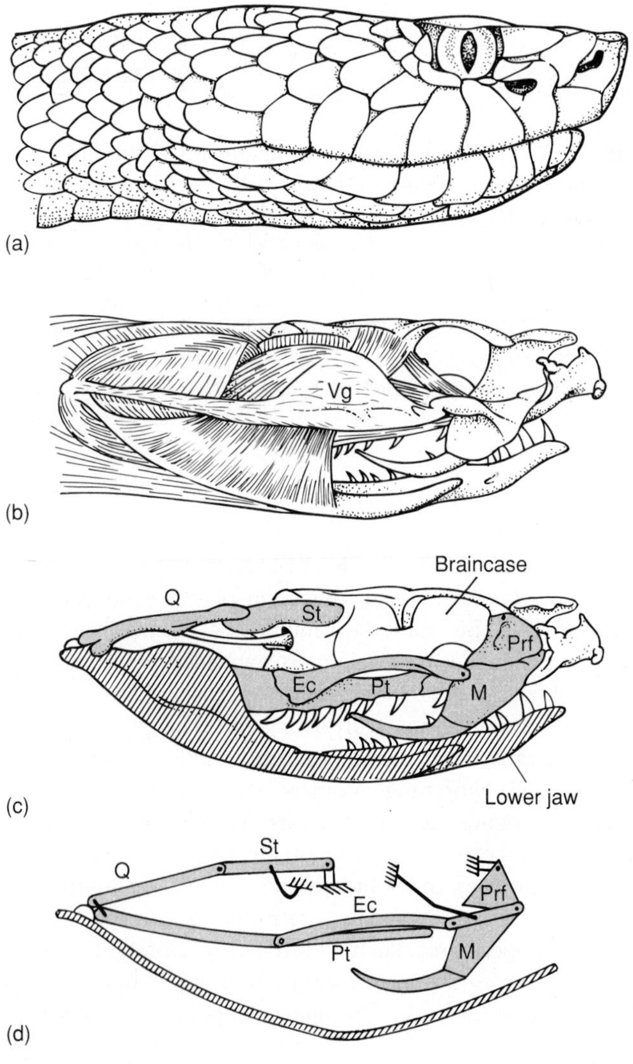


**Figure 7.43** Kinematic model of movable skull bones in a venomous snake, the water moccasin. Whole head (a) with successive removal of skin and muscles (b) reveals bones of the skull (c). Bones movable relative to the braincase are in color; lower jaw is crosshatched. (d) Biomechanical model of movable bones rotatable about pin connections. Movable bones include the ectopterygoid (Ec), maxilla (M), pterygoid (Pt), prefrontal (Prf), quadrate (Q), supratemporal (St). Location of the main venom gland (Vg) is shown as well.

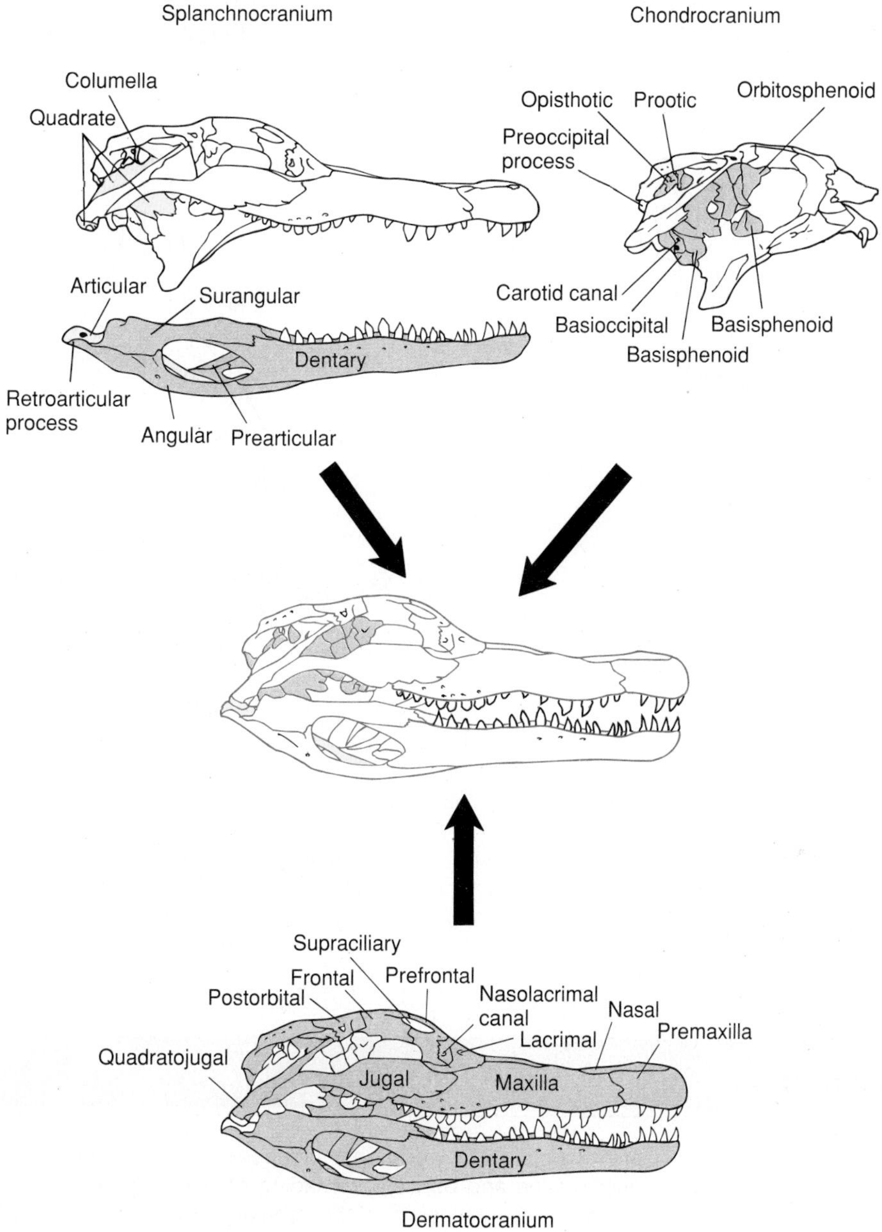


**Figure 7.44** Alligator skull. A composite skull design characteristic of vertebrates. The skull is a combination of elements receiving contributions from the chondrocranium, the splanchnocranium, and the dermatocranium.

## Synapsids

***Primitive Synapsids*** *Dimetrodon* represents a primitive synapsid. Therapsids continue the synapsid line and exhibit considerable diversity (figure 7.48). For a time in the Permian and early Triassic, they were fairly abundant. Some were herbivores, most were carnivores. Most skull bones of cotylosaurs persist, but, characteristic of synapsids, the temporal region develops a single opening bound horizontally along its lower border by a bony connection between jugal and squamosal bones. In advanced therapsids and primitive mammals, the vertical bar dividing the orbit from the single temporal fenestra is lost, leaving the narrow bony connection between jugal and squamosal bowing outward in the cheek region. This bony squamosal-jugal bar is now commonly called the

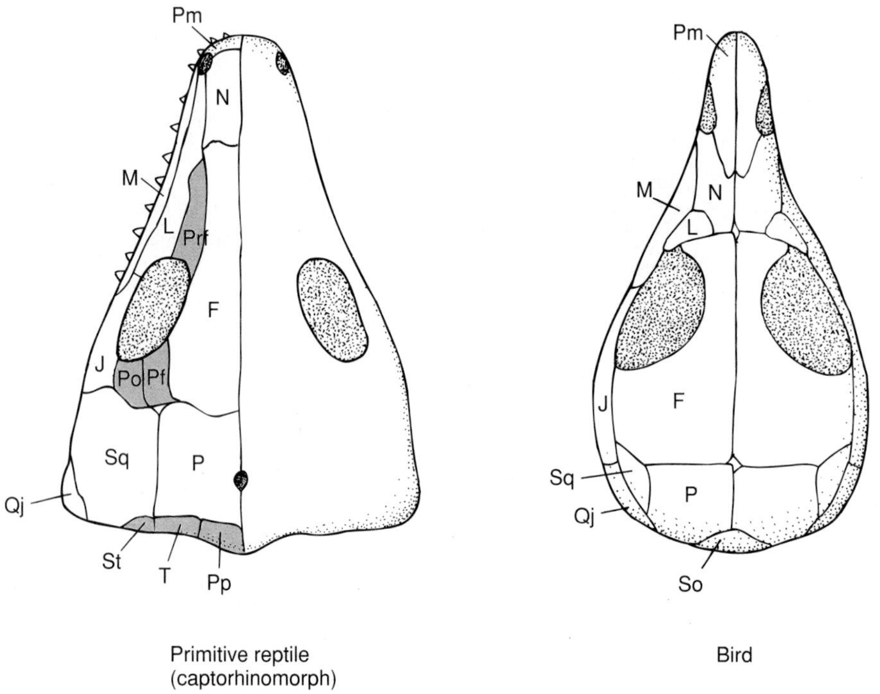


**Figure 7.45** Diagrammatic comparison of a derived bird skull with a primitive reptile skull, a captorhinomorph. Bones lost in birds are shaded in the primitive reptile. Abbreviations: frontal (F), jugal (J), lacrimal (L), maxilla (M), nasal (N), parietal (P), postfrontal (Pf), premaxilla (Pm), postorbital (Po), postparietal (Pp), prefrontal (Prf), quadratojugal (Qj), supraoccipital (So), squamosal (Sq), supratemporal (St), tabular (T).

## Box Essay 7.3 Striking Features of Snakes

The jaws of snakes are highly kinetic, with great freedom of motion. Skull bones that in other reptiles are fixed to the braincase or have restricted movement, are joined in snakes into linked chains with extensive motion relative to the braincase. Further, the series of linked bones on left and right sides are not joined directly, so they experience independent displacement, a feature allowing alternating left and right reciprocating motion of jaw bones over the prey being swallowed. This independent motion and outward spreading of the jaws (not "unhinging" of the jaws) allows most snakes to swallow large prey. Little by little, the distended jaws are walked in alternating steps over the prey until it is completely engulfed.

During the rattlesnake strike, the forward swing of these linked bones erects the maxilla and the fang it carries into position to inject venom into the prey. Snake fangs are modified teeth with hollow cores so that venom flows from their base into the prey. The fangs of most venomous snakes are longer than other teeth in the mouth, and the fangs of vipers and pit vipers are especially long. Extensive rotation of the fang-bearing maxilla in such snakes allows this long fang to be folded up and out of the way along the upper lip when it is not in use.

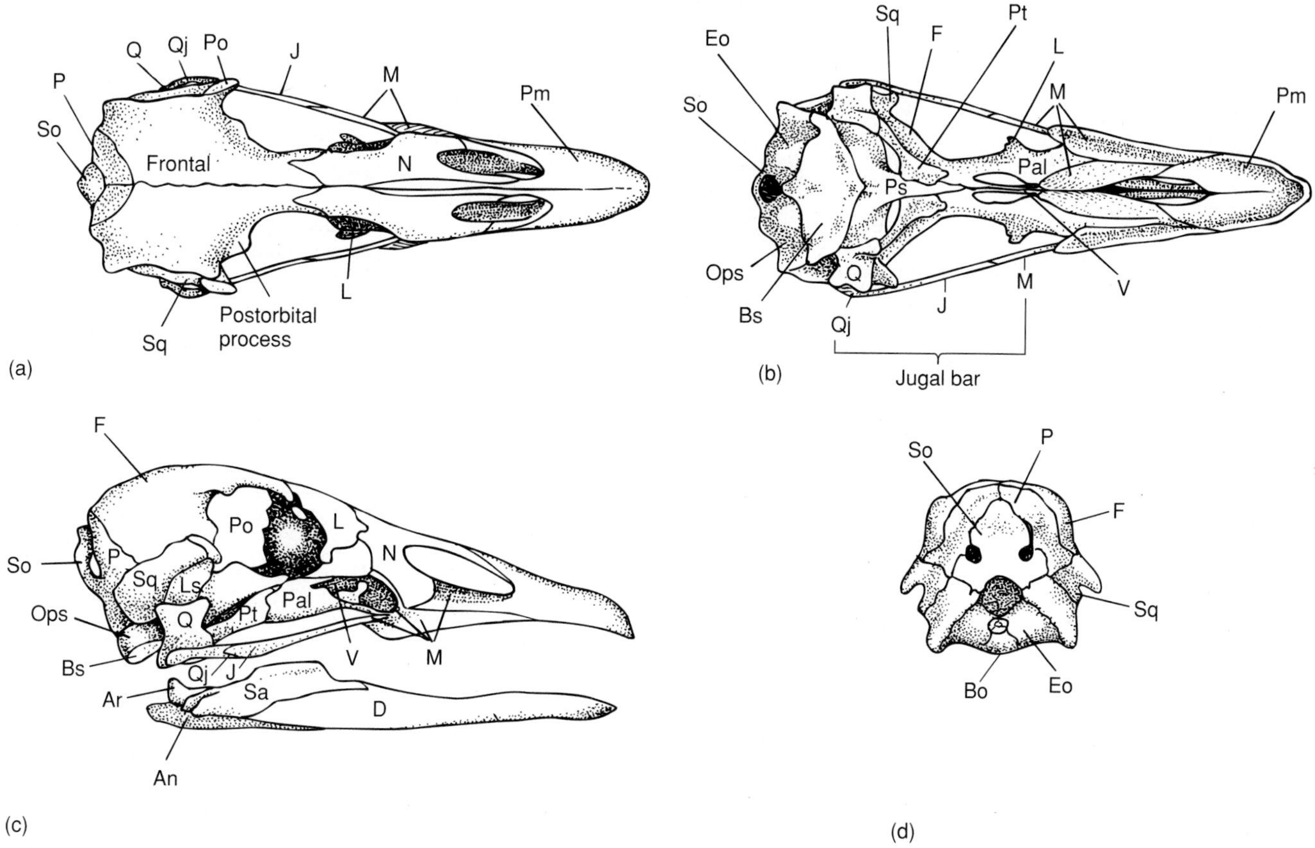


**Figure 7.46** Bird skull. In the adult bird, sutures between skull bones fuse to obliterate identifiable borders. Dorsal (a), ventral (b), lateral (c), and posterior (d) views of the skull of a young gosling *(Anser)* before bones fuse. Abbreviations: angular (An), articular (Ar), basioccipital (Bo), basisphenoid (Bs), dentary (D), exoccipital (Eo), frontal (F), jugal (J), lacrimal (L), laterosphenoid (Ls), maxilla (M), nasal (N), opisthootic (Ops), parietal (P), prearticular (Pa), palatine (Pal), premaxilla (Pm), postorbital (Po), parasphenoid (Ps), pterygoid (Pt), quadrate (Q), quadratojugal (Qj), surangular (Sa), supraoccipital (So), squamosal (Sq), vomer (V).

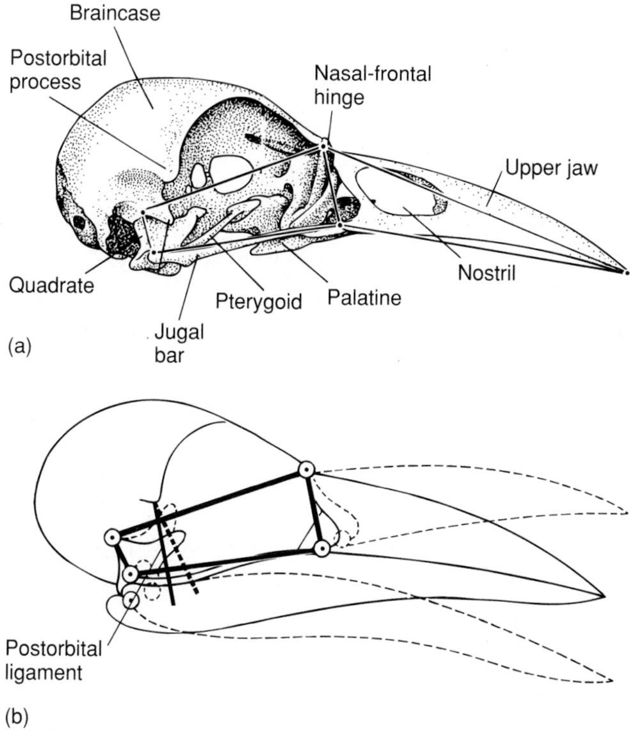


**Figure 7.47** Cranial kinesis in the crow skull *(Corvus)*. (a) Skull with lower jaw removed. Flexion of the beak occurs at its base with the skull. (b) Kinematic model that represents the major mechanical elements of the skull with linkages. The lower jaw is in place showing the postorbital ligament and points of rotation. The strong postorbital ligament helps control kinesis and opening of the lower jaw by forcing the lower jaw to swing forward and up. The connected jugal imparts this motion to the beak, resulting in its upward rotation about the prokinetic joint (nasofrontal hinge).

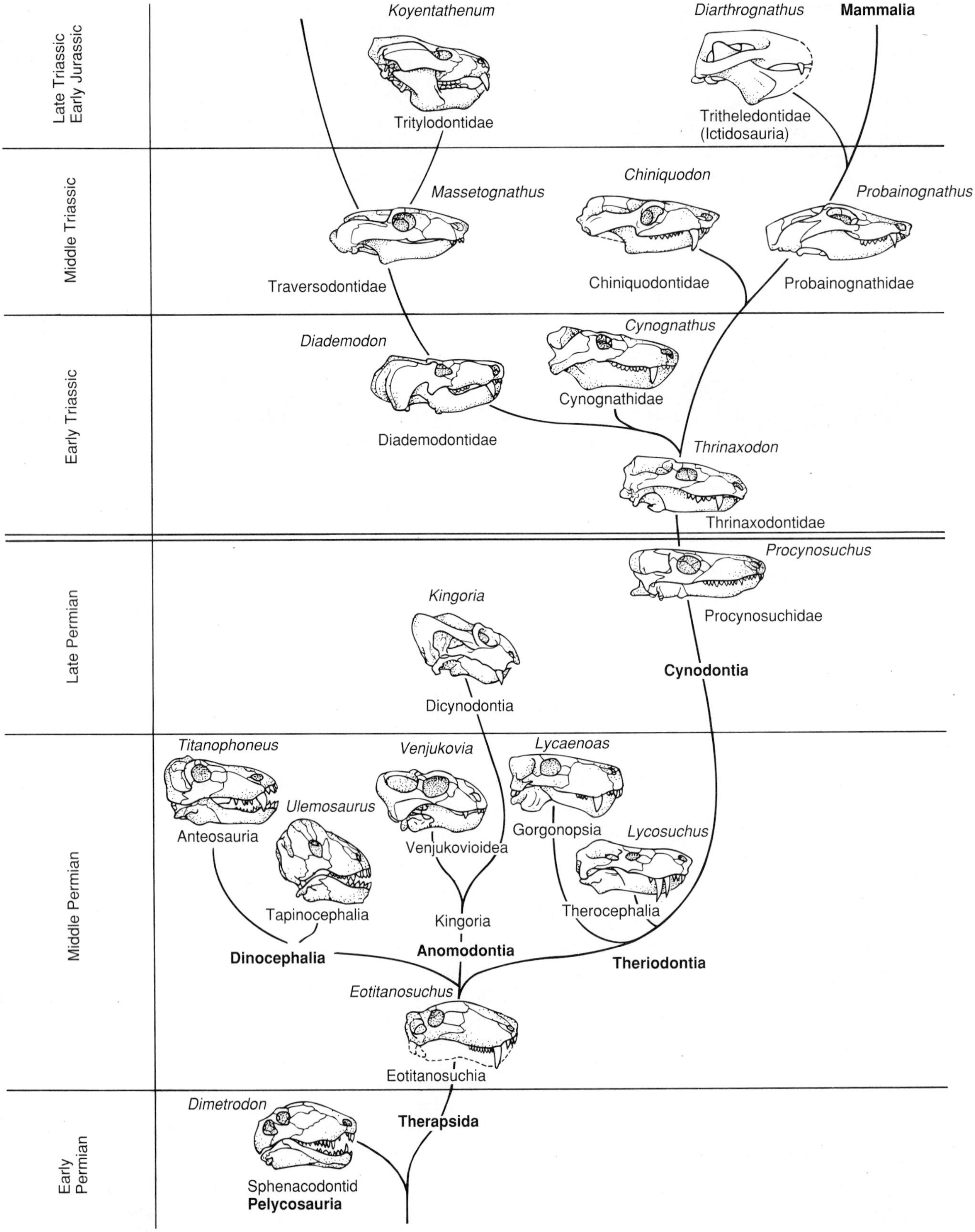


**Figure 7.48** Radiation of therapsids. Therapsids continue the synapsid line and exhibit considerable diversity. For a time in the Permian and early Triassic, they were fairly abundant. Some were herbivorous, most were carnivorous. Based on the research of James A. Hopson.

**zygomatic arch.** Within therapsids, there is a tendency for the temporal opening to move dorsally and for the zygomatic arch to become delineated.

***Mammals*** The skull of mammals represents a highly modified synapsid pattern. Various dermal elements are lost in therian mammals, including the septomaxilla, prefrontal, postorbital, postfrontal, quadratojugal, and supratemporal (figure 7.49). The postparietals, typically paired in reptiles, fuse into a single, medial **interparietal** in therapsids, which in mammals may incorporate the tabular and fuse with the occipital bones. Monotremes retain several reptilian skull features including prefrontal, postfrontal, and pleurosphenoid bones together with unfused occipitals. Monotremes are somewhat specialized as well. The lacrimal of therians is absent in them, and the jugal bones are small (figure 7.50a–d). A tympanic ring encircles the middle ear bones of monotremes, but in most therians, this ring has expanded into a large swollen capsule, the **auditory bulla,** that houses the middle ear ossicle (figure 7.51a–c).

***Placental Mammals*** Fusions between separate centers of ossification produce composite bones in the skull of placental mammals. The single occipital bone represents the fusion of basioccipital, paired exoccipitals, supraoccipital, and interparietal (and perhaps tabular); (figure 7.52a). The occipital bone defines the foramen magnum and closes the posterior wall of the braincase. Ventrally, there is a bilobed occipital condyle that articulates with the **atlas,** the first vertebra of the cervical region. Dorsally, a raised **nuchal crest** may form across the back of the occipital region, offering a secure attachment site for neck muscles and ligaments that support the head.

Several embryonic centers contribute to the sphenoid bone, representing the orbitosphenoid, presphenoid, basisphenoid, and a large alisphenoid (the epipterygoid of lower vertebrates; figure 7.52b).

On the side of the braincase behind the orbit, a large **temporal** bone is formed by the fusion of contributions from all three parts of the skull (figures 7.52c and 7.53). The dermatocranium contributes the squamosal and the **tympanic bulla** (a derivative of the angular) in many mammals. The chondrocranium contributes the **petrosal,** itself a derivative of prootic and opisthootic bones (figure 7.52c). The petrosal often bears a ventrally directed projection, the **mastoid process.** The splanchnocranium contributes three tiny middle ear bones (malleus, incus, stapes) and the styloid (figure 7.54).

In most tetrapods, the nasal capsule remains unossified. However, in mammals, the ethmoid portion ossifies to form scroll-like **turbinates** (turbinals, conchae). There are usually three sets of turbinates attached to respective neighboring bones, the **nasoturbinate, maxilloturbinate,** and **ethmoturbinate.** The coiled walls of the turbinates support

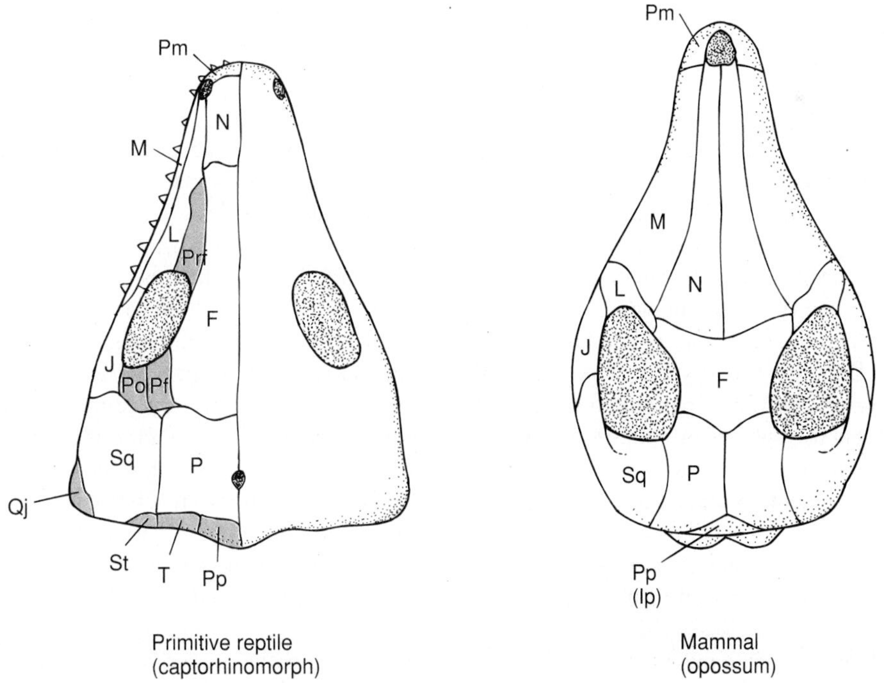


**Figure 7.49** Diagrammatic comparison of a derived mammal skull with a primitive reptile skull, a captorhinomorph. Bones lost in the derived mammal are shaded in the primitive reptile. Abbreviations: frontal (F), jugal (J), interparietal (Ip), lacrimal (L), maxilla (M), nasal (N), parietal (P), postfrontal (Pf), premaxilla (Pm), postorbital (Po), postparietal (Pp), prefrontal (Prf), quadratojugal (Qj), squamosal (Sq), supratemporal (St), tabular (T), vomer (V).

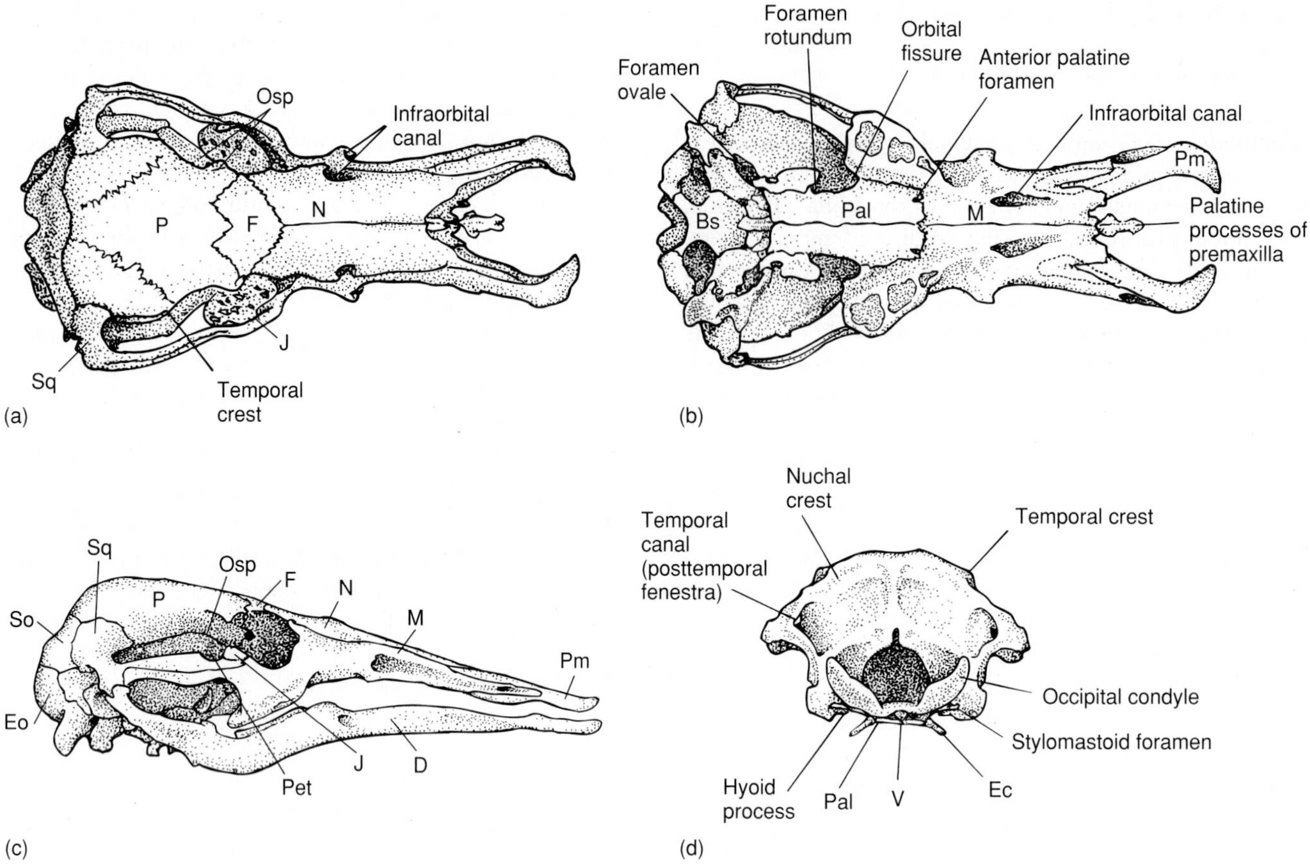


**Figure 7.50** Monotreme, skull of the duckbill platypus *Ornithorhynchus*. Dorsal (a), ventral (b), lateral (c), and posterior (d) views. Abbreviations: basisphenoid (Bs), dentary (D), ectopterygoid (Ec), exoccipital (Eo), frontal (F), jugal (J), maxilla (M), nasal (N), orbitosphenoid (Osp), parietal (P), palatine (Pal), premaxilla (Pm), petrosal (Pet), supraoccipital (So), squamosal (Sq), vomer (V).

the mucous membrane within the nasal passage. Air entering these passages is warmed and moistened before reaching the lungs, functions that are especially important in endotherms. Absent in ungulates but present in most other orders, such as rodents, carnivores, and primates, is another region of the nasal capsule, the **mesethmoid.** This element forms the septum between the nasal capsules and usually remains cartilaginous. Between the nasal area and cranial cavity stands the transverse and finely perforated **cribriform plate** (figure 7.54). Olfactory nerves originating in the olfactory epithelium of the nasal capsule pass through this plate to reach the olfactory bulb of the brain.

***Middle Ear Bones*** Two profound changes in the lower jaw mark the transition from therapsid to mammal (figure 7.55). Both changes go hand in glove. They result in such an alteration in skull design that some anatomists doubted them until the surprisingly good fossil record made the evolutionary transition undeniable. One of these changes is the loss of the postdentary bones of the lower jaw. The other is the presence of three middle ear bones. In vertebrates, the inner ear is embedded deep within the otic capsule and holds the sensory apparatus responsive to sounds. The hyomandibula or its derivatives deliver sound vibrations to the sensitive inner ear. In all tetrapods except mammals, the hyomandibula tends to become reduced to a slender, light bone called the columella. Sometimes there is a second hyomandibula-derived bone, the **extracolumella.** The columella is usually suspended in the middle ear cavity where damping by restrictive attachments is minimized. As sounds set the tympanic membrane into motion, these vibrations are imparted to the small, responsive columella. Its opposite end often expands to reach the sensitive inner ear apparatus that responds to the vibrations the columella delivers.

In mammals, two tiny additional bones join the columella in the middle ear. Together these bones transmit sound to the inner ear. Specifically, these three middle ear bones are malleus (derived from the articular), incus (derived from the quadrate), and stapes (derived from the columella). So distinctive is the presence of three middle ear bones, that many anatomists mark the fossil transition to mammals at the point of their acquisition.

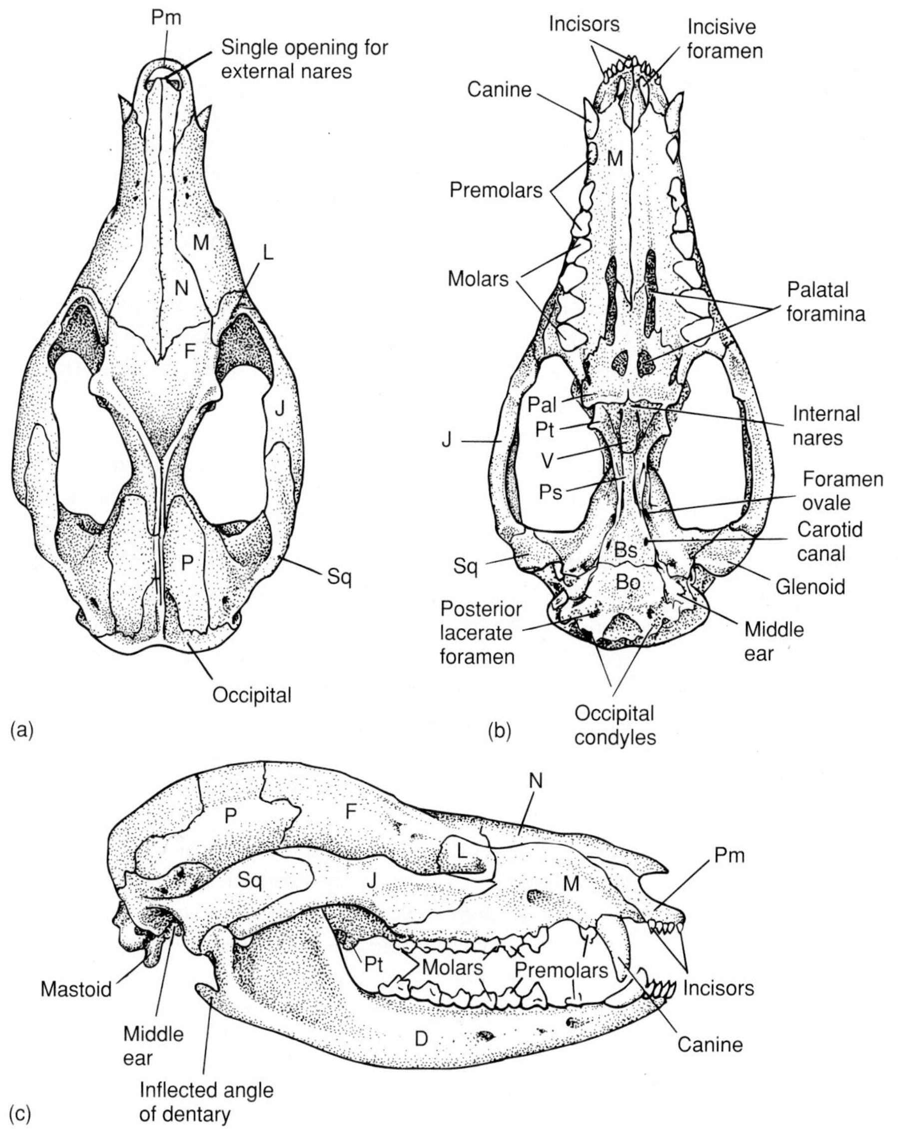


**Figure 7.51** Marsupial, skull of the opossum *Didelphis*. Dorsal (a), palatal (b), and lateral (c) views. Abbreviations: basioccipital (Bo), basisphenoid (Bs), dentary (D), frontal (F), jugal (J), lacrimal (L), maxilla (M), nasal (N), parietal (P), palatine (Pal), premaxilla (Pm), parasphenoid (Ps), pterygoid (Pt), squamosal (Sq).

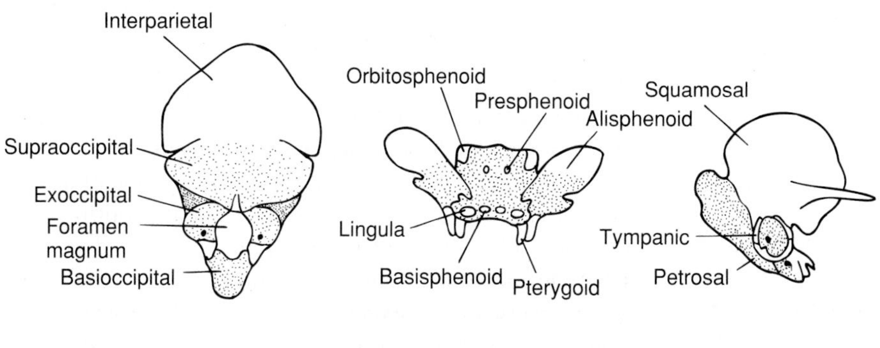


**Figure 7.52** Composite bones of the placental mammal skull during embryonic development, *Homo sapiens*. (a) Occipital bone has centers of ossification that include the interparietal (postparietal), supraoccipital, paired exoccipital, and the basioccipital. (b) The sphenoid bone is a fusion of the orbitosphenoid, presphenoid, basisphenoid, pterygoid, and alisphenoid (epipterygoid). In many mammals, these fused bones are joined by parts of the pterygoid and lingula. (c) The temporal bone results primarily from the merger of the squamosal, petrosal, and tympanic.

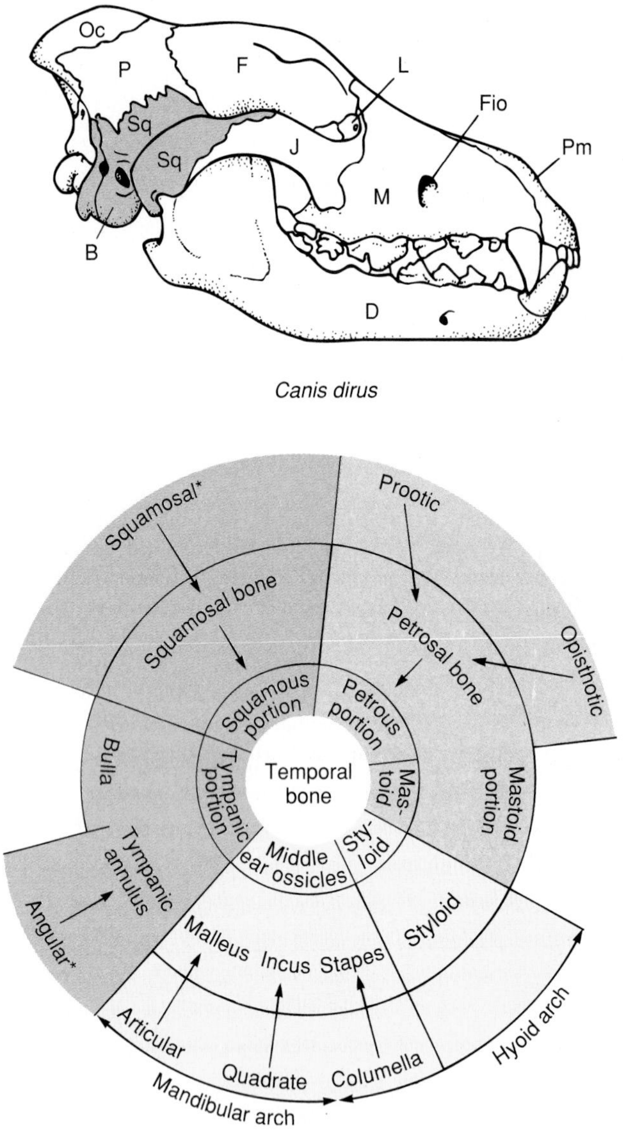


**Figure 7.53** Mammalian temporal bone. The temporal bone forms phylogenetically from the dermatocranium (angular, squamosal), the chondrocranium (prootic, opisthootic), and the splanchnocranium (articular, quadrate, columella, styloid). The separate bony elements in reptiles (outer circle) contribute to the composite temporal bone of mammals (middle and inner circle). Some of these contributions are dermal bones (*). The otic capsule is buried beneath the surface of the skull, leaving the exposed and often elongated mastoid process. The bulla or auditory bulla forms, at least in part, from the tympanic annulus, itself a phylogenetic derivative of the angular bone. The exposed squamous portion of the temporal bone is illustrated in color on the skull of the Pleistocene wolf *Canis dirus*. Abbreviations: auditory bulla (B), dentary (D), frontal (F), infraorbital foramen (Fio), jugal (J), lacrimal (L), maxilla (M), occipital (Oc), parietal (P), premaxilla (Pm), squamosal (Sq).

Anatomy and function of the ear (p. 692)

Coupled with the derivation of the three middle ear bones are changes in the posterior bones of the mandible. In reptiles, the lower jaw includes the tooth-bearing dentary in addition to several postdentary bones (angular, articular, coronoid, prearticular, splenial, surangular; figure 7.56). In mammals, this set of postdentary bones has been entirely lost from the lower jaw, and the dentary has enlarged to assume the exclusive role of lower jaw function. From pelycosaur to therapsid to mammal, the anatomical details of these changes are well documented in an ordered time sequence by the fossil record. In pelycosaurs, the articular (future malleus) resides at the back of the mandible and establishes lower jaw articulation with the quadrate (future incus). In early to later therapsids, these two bones become reduced, along with the postdentary bones, eventually moving out of the lower jaw and taking up a position in the middle ear. The functional reason for these changes is thought to be related to improved hearing. The phylogenetic reduction in size of these bones would reduce their mass and thus increase their oscillatory responsiveness to airborne vibrations. Their removal from the jaw joint permits their more specialized role in transmitting sound to the inner ear. Alternatively, or along with such changes related to hearing, some morphologists have proposed that changes in feeding style led to changes in the preferred site of insertion of the jaw-closing muscles, specifically a shift forward on the dentary and closer to the teeth. Larger jaw muscles acting close to the tooth row lessen the stresses at the back of the jaw where it articulates with the skull. Loss of postdentary bones then might reflect this shift in forces forward to the tooth row and away from the joint these bones formed.

These changes in the lower jaw were accompanied by changes in the method of food preparation prior to swallowing. Most reptiles bolt their food, swallowing it whole or in large chunks. Mammals typically chew their food before swallowing it, a process termed **mastication.** Mastication also occurs in a few groups of fishes and lizards. But, it is within mammals that feeding strategy is based on mastication of food. If mastication became a more characteristic part of food preparation, then changes in jaw-closing muscles might be expected, with greater emphasis shifting to the dentary.

***Secondary Palate and Akinesis*** In addition to changes in the mammalian lower jaw, the presence of a secondary palate is also related to mastication. The secondary palate includes a **hard palate** of bone and a posterior continuation of fleshy tissue, the **soft palate** (figure 7.57a,b). The hard palate is formed from the inward growth of bony processes of the premaxilla, maxilla, and palatine that meet at the midline as a bony platform (figure 7.58a–c). This hard palate and its fleshy continuation effectively separate the food chamber below from the respiratory passage above. Some turtles and also crocodilians have a secondary palate, and they benefit from the advantages of separation of routes

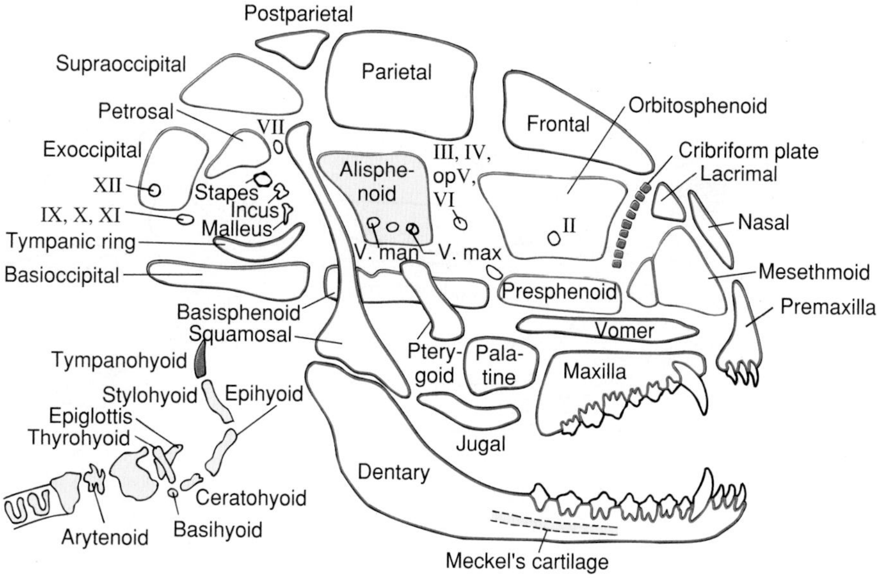


**Figure 7.54** Diagram of a dog's skull. Sources of the various bones are outlined: dermatocranium (red), chondrocranium (blue), and splanchnocranium (yellow).

for food and air. But chewing requires that food stays in the mouth for an extended period of time in mammals; therefore, separation of respiratory and oral passages is especially important. Mastication can proceed without impeding regular breathing. Similarly, the secondary palate completes the firm roof of the food chamber, so that the pumping action of the throat of a suckling infant creates effective negative pressure within the mouth without interfering with the respiratory passage.

**Mastication (p. 525)**

Mastication in mammals has been accompanied by very precise tooth occlusion to serve the mechanical breakdown of food. Precise, strong occlusion requires a firm skull, so mammals have lost the cranial kinesis of reptiles, leaving them an akinetic skull. The mammalian mandibular condyle fits into a very restrictive articulation with the squamosal bone. When jaws close about this joint, upper and lower rows of teeth are placed in very precise alignment. This allows specialized teeth to function properly. As a further consequence of precise occlusion, the pattern of tooth eruption in mammals differs from most other vertebrates. In lower vertebrates, teeth wear and are replaced continuously **(polyphyodonty);** therefore, the tooth row is always changing. If teeth function primarily to snag prey, this causes little difficulty. However, continuous replacement means that at some location in the jaws, worn teeth are missing or new ones are moving into position. To avoid disruption of occlusion, teeth in most mammals are not continuously replaced. Mammals exhibit **diphyodonty.** Only two sets of teeth erupt during the lifetime of a mammal, the "milk teeth" of the young and the "permanent" teeth of the adults.

**Tooth types and their development (p. 497)**

The chain of events leading from mastication to akinesis and to diphyodonty should not be viewed as inevitable. Some fishes chew their food but retain kinetic skulls and polyphyodonty. But evolutionary events that produced the mammalian skull underscore the importance of examining anatomical changes in partnership with functional changes that must accompany phylogenetic modification of vertebrate design. Form and function necessarily go together, an issue to which we next turn.

## Overview of Skull Function and Design

The skull performs a variety of functions. It protects and supports the brain and its sensory receptors. It may function to cool the brain during sustained activity or during a rise in environmental temperature. In many active terrestrial mammals, the nasal epithelium lining the nasal passages dissipates excess heat by evaporation as air moves across this moist lining. A similar function has been proposed for the elaborate air passageway in some groups of hadrosaurs, the duck-billed dinosaurs (figure 7.59). Air entering their nostrils would have coursed through intricate passageways formed within premaxilla and nasal bones to provide evaporative cooling. The skull of many animals also supports the voice box and occasionally serves as a sound resonator to deepen or amplify an animal's call. The Weddell seal takes advantage of its jaws to open and maintain its breathing holes in surface ice (figure 7.60).

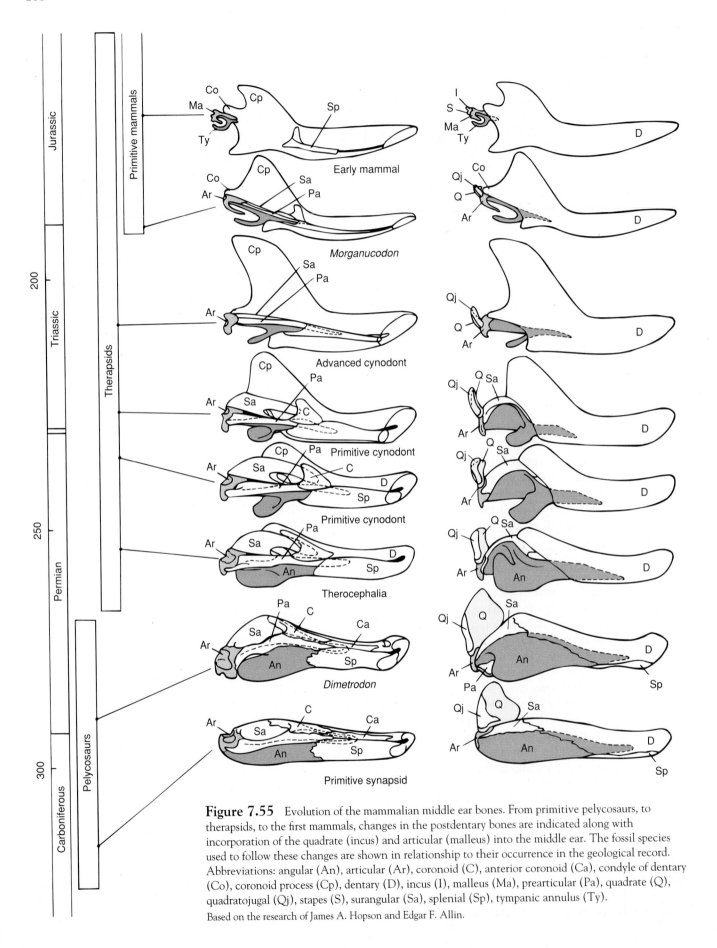

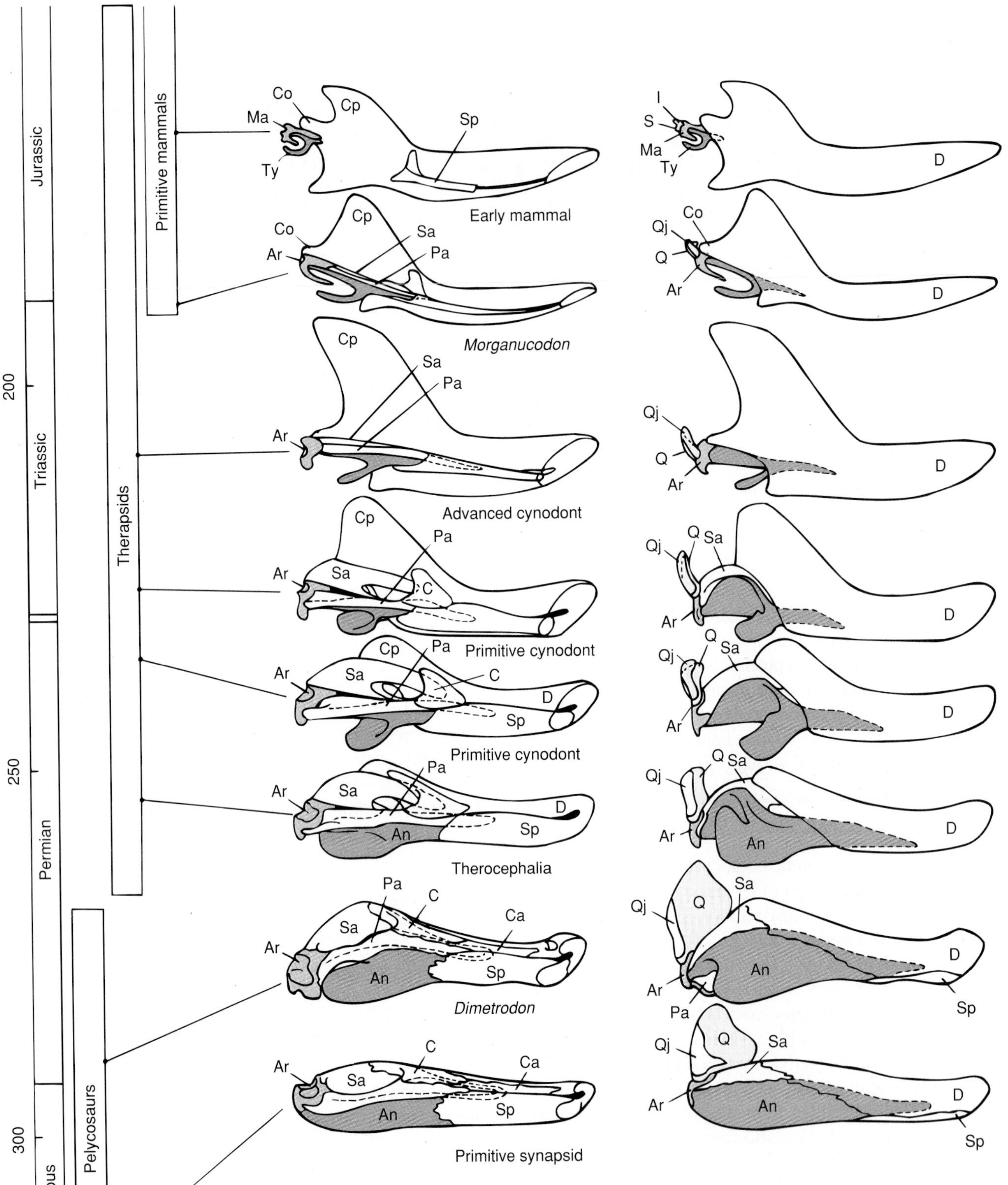


**Figure 7.55** Evolution of the mammalian middle ear bones. From primitive pelycosaurs, to therapsids, to the first mammals, changes in the postdentary bones are indicated along with incorporation of the quadrate (incus) and articular (malleus) into the middle ear. The fossil species used to follow these changes are shown in relationship to their occurrence in the geological record. Abbreviations: angular (An), articular (Ar), coronoid (C), anterior coronoid (Ca), condyle of dentary (Co), coronoid process (Cp), dentary (D), incus (I), malleus (Ma), prearticular (Pa), quadrate (Q), quadratojugal (Qj), stapes (S), surangular (Sa), splenial (Sp), tympanic annulus (Ty).

Based on the research of James A. Hopson and Edgar F. Allin.

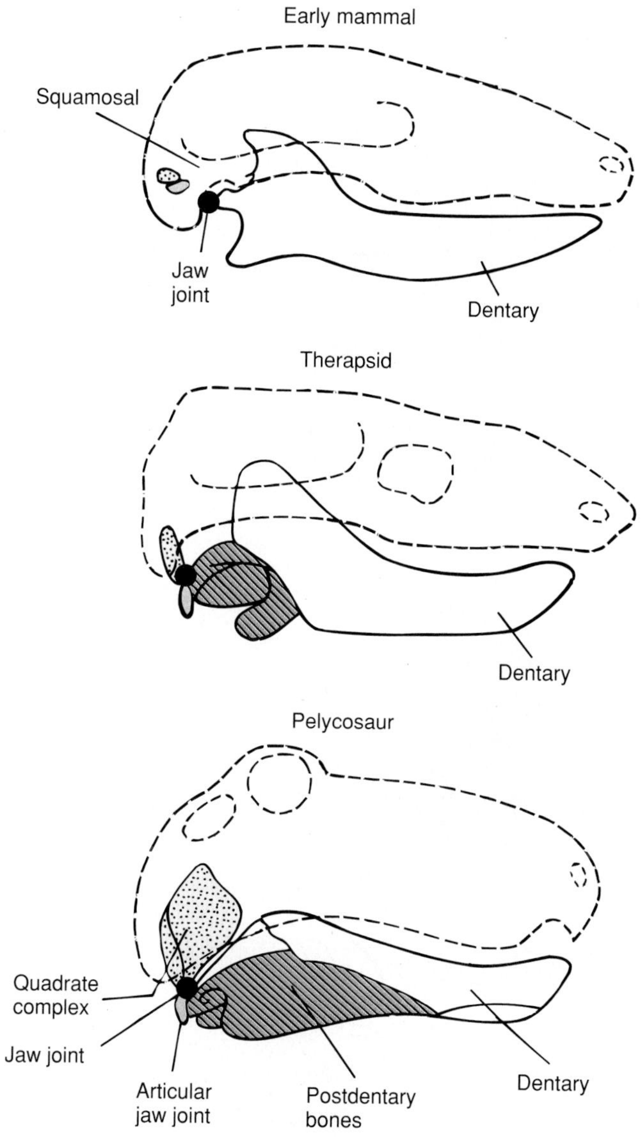


**Figure 7.56** Changes in jaw articulation during transition from reptiles to mammals. In mammals, the postdentary bones of the lower jaw are lost and the dentary enlarges. Bones involved in jaw articulation in reptiles, the articular and the quadrate complex (quadrate and quadratojugal) become reduced and move in to contribute to the inner ear ossicles of mammals. Jaw articulation in mammals is taken over by the dentary and squamosal. The columella is not shown.

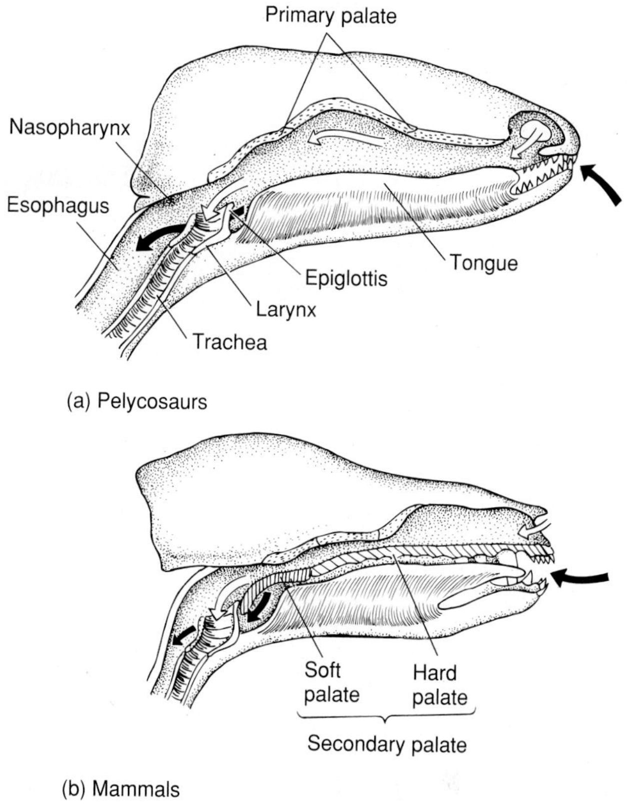


**Figure 7.57** Secondary palate. (a) In most reptiles, air entering the nostrils and food entering the front of the mouth share a common passage, although each exits differently at the back. Air moves down the trachea and food travels down the esophagus. (b) In mammals, the routes of air and food are separated from the beginning of their travel by the secondary palate, a structure of bone (hard palate) and soft tissue (soft palate). White arrows indicate the path of air; dark arrows indicate the path of food.

These examples remind us that the skull is a multipurpose "tool" involved in a great variety of functions. Its design reflects and incorporates these multiple roles. Generalizations about skull design and function can be misleading if we ignore its multiple functions. However, if we are cautious, we can understand how skull design reflects fundamental functional problems. The skull primarily functions as part of the feeding system of vertebrates. How it addresses problems of feeding depends largely upon whether feeding occurs in air or in water. Each medium presents different limitations and opportunities. The viscosity of water and the buoyancy of tiny organisms within it mean that water, much more than thin air, contains a richer community of floating planktonic organisms. Suspension feeding and harvesting of these tiny organisms becomes economical, and filter-feeding devices enjoy some adaptive favor. Generally, feeding proceeds in two steps, food capture and then swallowing. We look at each of these in turn.

## Prey Capture

### *Feeding in Water*

The first step in feeding is food capture, which depends generally on the medium in which feeding takes place. The higher viscosity of water presents both problems and opportunities for the animal feeding in water. Feeding in water poses a disadvantage in that water easily carries

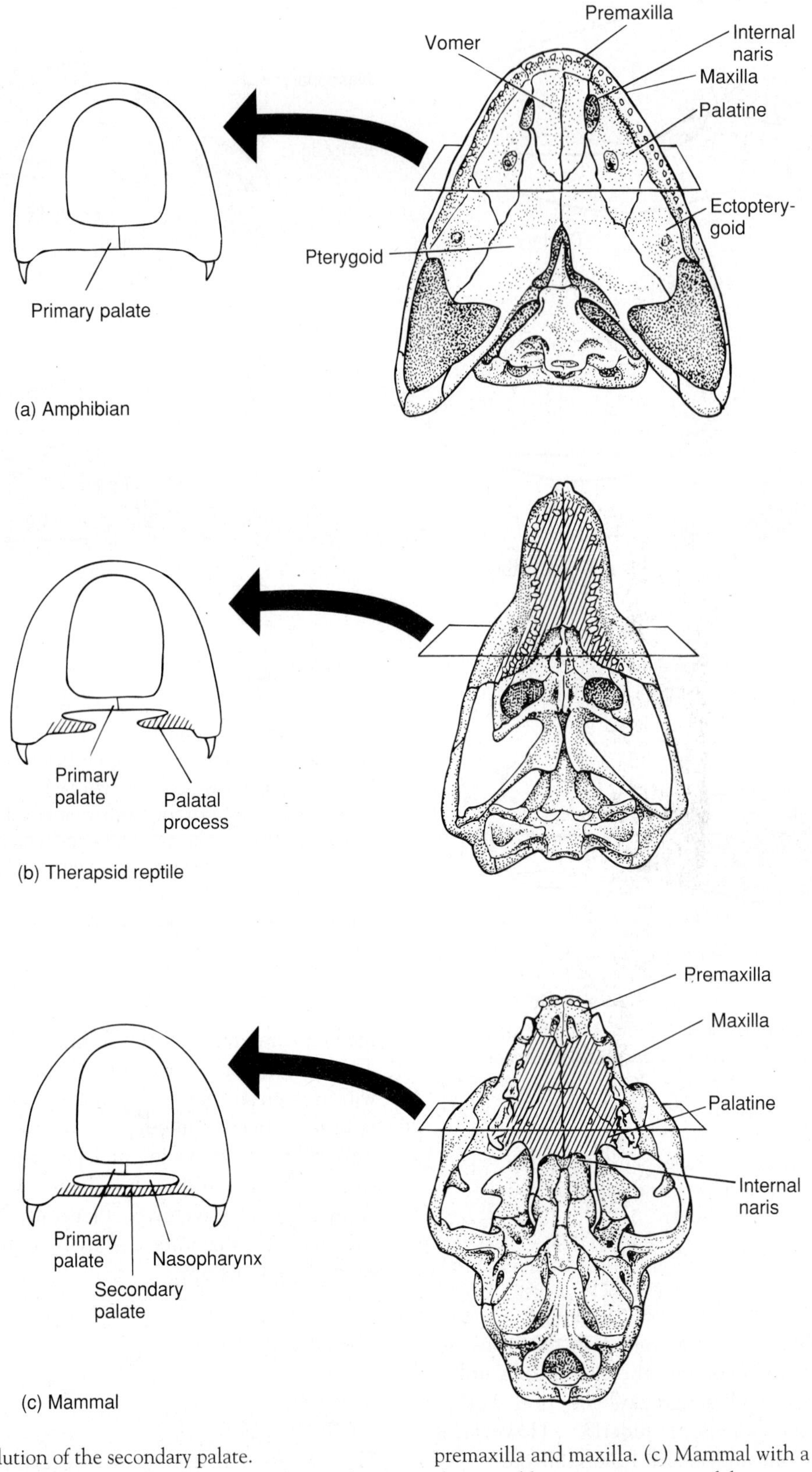


**Figure 7.58** Evolution of the secondary palate. (a) Amphibian with a primary palate in cross section (left) and ventral (right) views. (b) Therapsid-reptile with a partial secondary palate formed by the medial extension of the premaxilla and maxilla. (c) Mammal with a secondary palate that, in addition to extensions of the premaxilla and maxilla, includes part of the palatine bone.

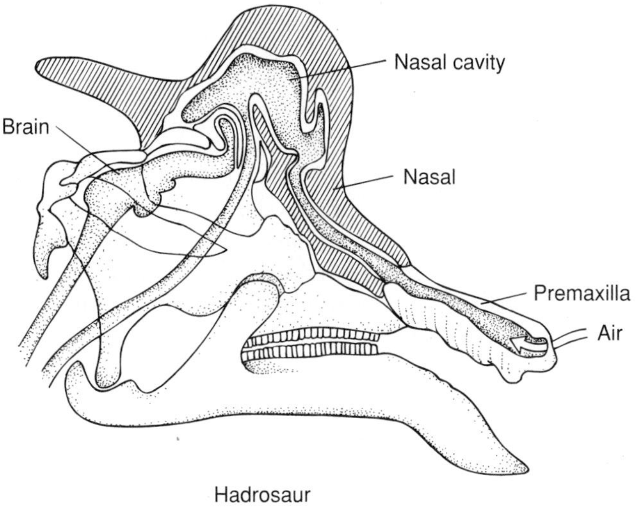


**Figure 7.59** Air passage of duck-billed dinosaurs. The air passageway formed by the premaxillae and nasal bones of the hadrosaur is crosshatched. Air that flowed through the nasal cavity on its way to the lungs cooled the nasal epithelial lining and hence the blood flow through it. Although the vascular system of the hadrosaur is not known, if it were similar to some mammals, then this cooled blood might have circulated in such a fashion as to precool the blood flowing to the brain. In this way, the brain was protected from elevated temperatures. Alternatively or additionally, such an expanded air passage might have been a resonating chamber to amplify vocalizations.

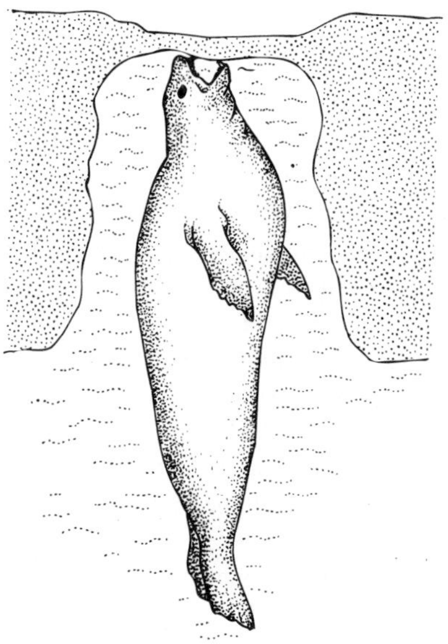

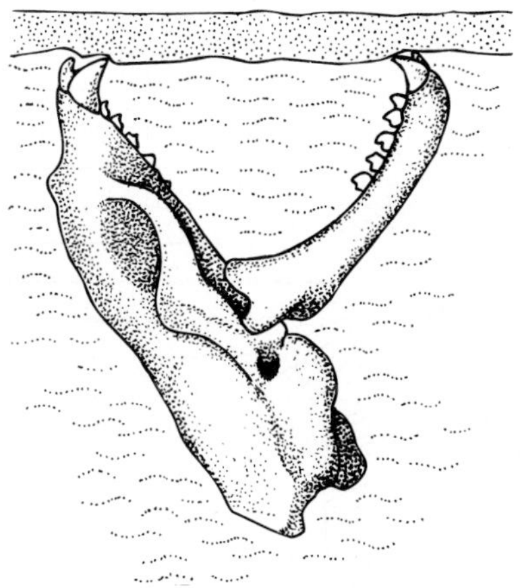

**Figure 7.60** Weddell seal. In addition to feeding, the jaws of this seal are used to ream away ice to open or reopen a frozen breathing hole.

shock or pressure waves ("bow waves") immediately in front of the predator approaching its food. These pressure waves can arrive an instant before the advancing predator and alert or deflect the intended prey. On the other hand, when a vertebrate quickly gulps water into its mouth, the viscosity of the water drags along the prey as well. This viscosity makes **suction feeding,** used with relatively large prey, possible.

To capture small foods, aquatic animals use **suspension feeding.** The density of water gives it the viscosity to retard the fall of particulate material out of suspension. Compared to air, water holds a floating bounty of tiny organic particles and microorganisms, a rich potential nutritional resource for an organism with the equipment to harvest it. Cilia move and control currents of water (and transport captured food) and sticky mucus snatches suspended food from the current of water as it glides by.

***Suspension Feeding*** Suspension feeding is a feeding strategy confined largely, perhaps exclusively, to animals living in water. Some, making an esoteric point, argue that bats "filter" insects "suspended" in air, but this misses the point. Air is too thin to hold suspended food for long. Bats catch or grasp prey, but they do not really use a filtering apparatus nor face the same mechanical problems as aquatic organisms in a viscous medium, so they are not suspension feeders. Most suspension feeders are benthic (bottom-dwelling) organisms or are associated with a herbivorous/detritus feeding style. Respiration and feeding are tightly coupled. Often the same water currents support both activities.

Suspension feeders use several methods to intercept and gather nutrients traveling within streams of water. Captured particles are usually smaller than the pores of the filter. They may collide directly with the filter (figure 7.61a) or because of their inertia, they deviate from the streamlines to collide with the mucus-covered surface of the filter (figure 7.61b). Upon impact, the particles cling to the sticky mucus and are rolled up in mucous cords, and then are passed by cilia into the digestive tract.

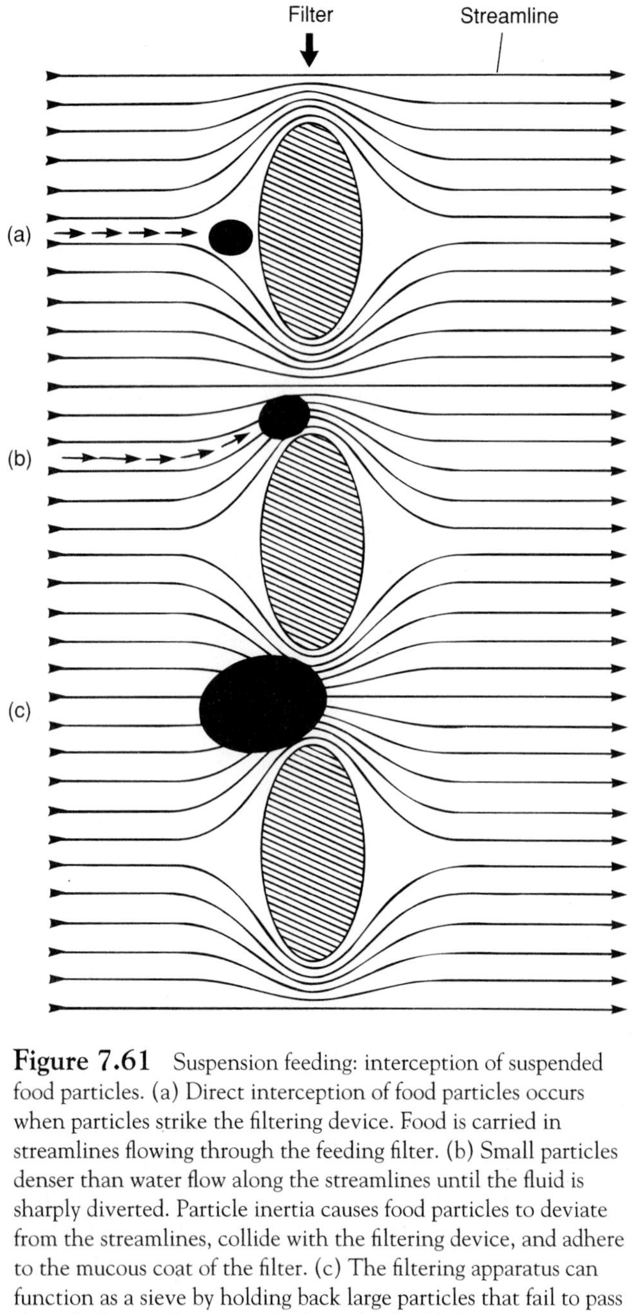


**Figure 7.61** Suspension feeding: interception of suspended food particles. (a) Direct interception of food particles occurs when particles strike the filtering device. Food is carried in streamlines flowing through the feeding filter. (b) Small particles denser than water flow along the streamlines until the fluid is sharply diverted. Particle inertia causes food particles to deviate from the streamlines, collide with the filtering device, and adhere to the mucous coat of the filter. (c) The filtering apparatus can function as a sieve by holding back large particles that fail to pass through the small pores. Cilia drive the food-laden mucus to the digestive tract.

Less commonly, a sieve can be used to strain suspended particles larger than the pores of the sieve. As the stream of water passes through the sieve, the particles are held back and then collected from the face of the selective filter (figure 7.61c). This method is rare among animals, perhaps because the relatively large particles filtered tend to plug and foul the sieve. The buccal cirri of amphioxus intercept large particles, apparently to prevent them from entering the pharynx and clogging the suspension-feeding system. The gill rakers of bony fishes also remove particulate matter. When the filter becomes clogged, these fishes can clear the material by a kind of cough or quick expansion of the gill arches. Larvaceans (urochordates) abandon their filter when it becomes clogged, secrete a new filter, and continue straining microorganisms from the circulating current of water.

In some invertebrates, the mucus is electrically charged. Mild attraction pulls particles out of suspension and into contact with the walls of the filtering device. However, such mechanisms of suspension feeding are unknown in vertebrates and protochordates.

In amphioxus, the endostyle and lining of the pharynx secrete mucus, which is swept upward by the action of cilia also lining the pharynx. The main current, driven by cilia, passes the cirri upon entry into the mouth and pharynx, through the pharyngeal slits, into the atrium, and via the atriopore exits to the outside environment once again. Small suspended particles in the current pass the pharyngeal bars. Some deviate from the stream of water to collide with and become entrapped in the mucous layer. The mucus and its captured particles are gathered dorsally in the epibranchial groove where it is formed into a mucous cord that other cilia sweep into the digestive tract.

In the ammocoetes larvae of lampreys, suspension feeding is similar to amphioxus except that a pair of muscular velar flaps, rather than cilia, beat rhythmically to create the current that flows into the pharynx. Mucus, secreted along the sides of the pharynx, is driven upward by cilia into the epibranchial groove where a row of cilia at the base of this groove form mucus and captured food into a cord that is passed into the digestive tract. The ventral endostyle of ammocoetes adds digestive enzymes to the forming mucous food cord but does not secrete mucus.

Although some envision the ostracoderms as employing new modes of feeding, the absence of jaws would have made this unlikely. They seem to have lacked even the muscular tongue of cyclostomes to break up food and place it into suspension. Thus, ostracoderms likely carried forward a suspension-feeding style similar to that of the protochordates before them. Not until gnathostomes do we see a significant tendency away from suspension feeding.

In gnathostomes, suspension feeding is less common. Some actinopterygians use gill rakers like a sieve to filter larger particles from the stream of passing water. The larvae of anurans employ a buccal pump. They draw in a stream of water containing food particles or scrape rock surfaces to enrich the entering stream with these dislodged materials.

The success and efficiency of suspension feeding depends on the size and speed of passing particles. It is most effective with small food particles that neither foul the filter nor escape the sticky mucous lining. To take advantage of large food items, another feeding style evolved, namely, suction feeding.

***Suction Feeding*** Like most fishes, amphibians living in water typically use suction feeding (see figures 7.25 and 7.32). The buccal cavity expands rapidly, pressure drops, and food is aspirated into the mouth. Geometry and enlargement of the buccal cavity are controlled by the muscularized visceral skeleton. Excess water, gulped in with the food, is accommodated in several ways. In salamanders prior to metamorphosis and in fishes, gill slits at the back of the mouth offer an exit for excess water. Flow is unidirectional. In salamanders after metamorphosis, in frogs, and in all other aquatic vertebrates, gill slits are absent, so excess water entering the mouth reverses its direction of flow to exit via the same route. Flow is bidirectional. Turtles possess an expansive esophagus that receives and temporarily holds this excess water until it can be slowly expelled without loosing the captured prey.

The early stages in vertebrate evolution took place in water, mostly in marine waters, occasionally in fresh water. Adaptations for feeding and respiration took advantage of these conditions. Adaptations for suspension and suction feeding are present in early vertebrates. With the transition of vertebrates to land and air, neither suspension nor suction feeding provided efficient ways to procure or process food. The jaws became specialized for grasping.

### *Feeding in Air*

Terrestrial feeding in most amphibians and many lizards requires a projectile tongue. The term **lingual feeding** recognizes the use of a rapid, projected, and sticky tongue to capture prey (figures 7.62 and 7.41a–d). However, in many other animals, prey is captured by **prehension,** a method by which the animal rapidly grasps the prey with its jaws. In such animals, the jaws are prey traps, designed to snare the unwary.

As a strategy to capture prey, prehension does not always involve the jaws. Birds of prey snatch quarry with their talons, and mammalian predators often use claws to catch and then control intended prey. Jaws are used secondarily to help hold the struggling victim or to deliver a killing bite.

## Swallowing

Once an animal has captured and dispatched its prey, it must swallow the prey in order to digest it. In suspension feeders, the food-laden cords of mucus are swept by synchronized ciliary action into the esophagus. Other animals usually swallow prey whole or in large pieces. Suction feeders rapidly expand the buccal cavity repeatedly to work the captured prey backward into the esophagus. Terrestrial vertebrates use the tongue to reposition the food bolus and work it toward the back of the mouth. The highly kinetic skull of snakes allows great freedom of jaw movement. A snake swallows a relatively large animal by stepping the tooth-bearing bones over the dispatched prey (figure 7.63a,b).

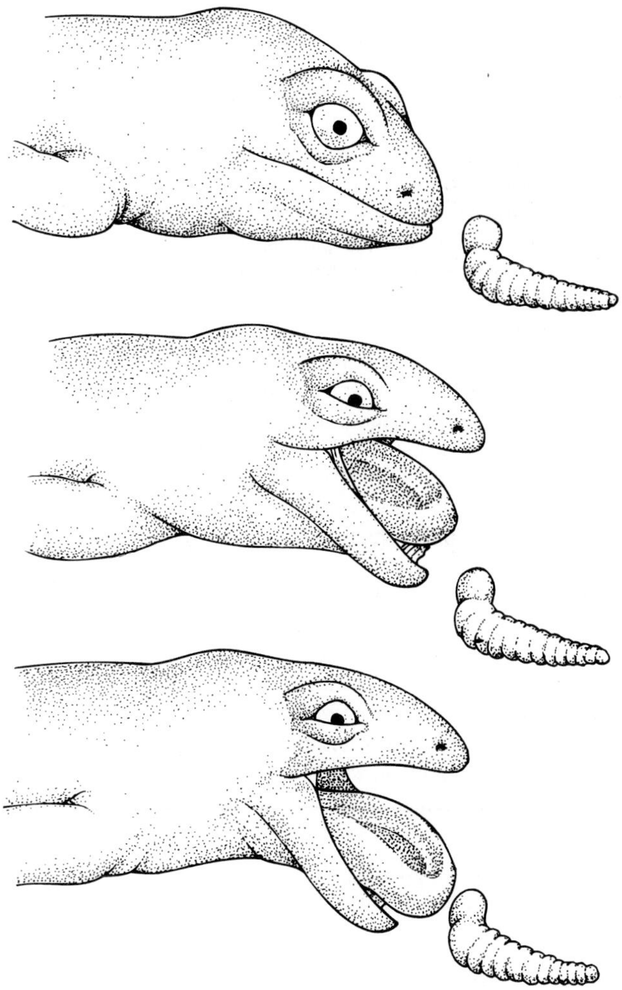

**Figure 7.62** Terrestrial feeding by a salamander. In this filmed sequence, the salamander's jaws open (top), its tongue starts to project (middle), and then makes contact with the prey (bottom).

Based on the research of J. H. Larsen.

## Box Essay 7.4 Whales and Whalebone

The largest animal alive today is the blue whale. It makes its living as a filter feeder. The filtering device is the baleen, a brushlike specialization of the oral epithelium occupying the site where teeth might otherwise be expected in the upper jaw. "Whalebone" is a misnomer for baleen. The term is inaccurate because no bone occurs within the baleen. Because baleen is frayed, it acts like a strainer to hold back food from the stream of water passed through it. Food preference depends a little on species, but most baleen whales strain small fishes or shrimplike crustaceans called "krill," which school or gather in dense swarms. Food collected in the baleen is licked free by the tongue and swallowed.

Blue and humpback whales represent one subgroup of baleen whales called fin whales or rorquals. Right whales are the other subgroup. In both groups, the teeth are absent, the baleen is present, and the skull is long and arched to hold the filter-feeding equipment.

To feed, right whales part their jaws slightly and swim through swarms of krill. The stream of water enters the front of the mouth and passes out the lateral suspended wall of baleen. Here the krill become entangled in the frilly baleen and are licked up and swallowed (box figure 1a). The blue whale feeds differently. As it approaches a school of fishes or krill, it opens its mouth wide to swim over and engulf the concentrated prey and accompanying water. Pleated furrows along its neck and belly allow the throat to inflate like a pouch and fill with this huge mass of water (box figure 1b). Up to 70 tons of water are temporarily held in the expanded throat. The whale then contracts the swollen pouch, forcing water through the baleen where the food is strained, collected up by the tongue, and swallowed.

Humpback whales have been observed to release air bubbles while circling a school of prey swimming above them. As the air bubbles rise, they form a "bubble cloud" that may corral or drive the school up to the surface ahead of the whale. The bubble cloud may also immobilize or confuse schools of prey, causing them to clump together, or it may disguise the whale as it surges upward with its mouth open through the center of the bubble cloud. Some humpbacks begin feeding at the surface by lopping or slapping their tail against the water as they dive. Just as the tail flukes are about to reenter the water, a whale flexes its tail so that the flukes hit the water, leaving a bubbling effervescence at the surface. This is thought to

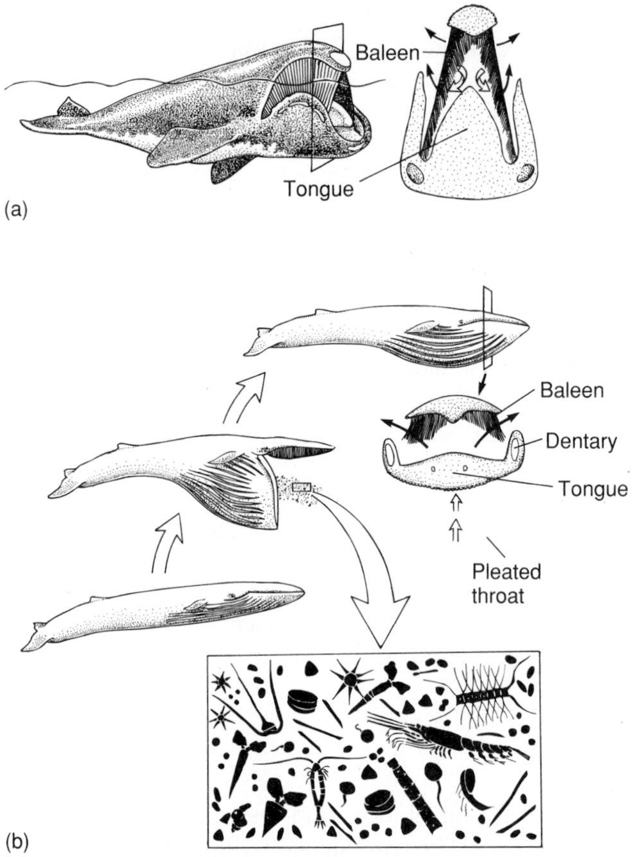


Swallowing mechanisms of terrestrial vertebrates (p. 503)

As we have seen in many vertebrates, swallowing involves mastication (the chewing of food). Mastication occurs in a few groups of fishes and lizards. Within mammals mastication has had a profound influence upon skull design, producing an akinetic skull with precise tooth occlusion and only two replacement sets of teeth, a secondary palate, large jaw-closing musculature, and changes in lower jaw structure.

# Overview of Evolutionary Issues

## Transition from Reptiles to Mammals

Within the skull, several major changes in design occur as mammals evolve from reptilian ancestors. One change already noted is in the lower jaw. In reptiles, the jaws articulate with the braincase via the articular-quadrate joint.

startle prey and stimulate them to clump tightly together in a school. The whale then releases a bubble cloud as it dives, which it follows by a feeding lunge back up through the bubble cloud to collect the prey in its mouth.

The oldest fossil whales come from the Oligocene and bear unmistakable resemblances to primitive terrestrial mammals. Distinct incisor, canine, premolar, and molar teeth were present. From these earliest whales, two major modern lines soon arose. One is the baleen whales, formally called the mysticetes. The other major line of whales is the toothed whales, or the odontocetes, including sperm, killer, and other whales with teeth.

In both baleen and toothed whales, the skull is telescoped. Some bones are pushed together and even overlap, yet a long snout persists (box figure 1c). In the odontocetes, the backward lengthening of the facial bones creates the snout. In the mysticetes, the occipital bones are pushed forward. Although achieved differently, the result is the same—to reposition the nostrils to a more central and dorsal position. When a whale surfaces to breathe, this position of the nostrils allows easy venting of the lungs and drawing in of fresh air without the whale having to tip its entire head out of water.

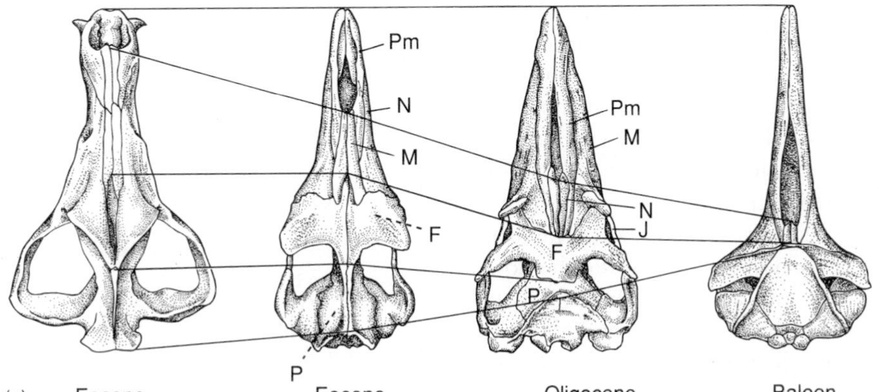


**Box figure 1** Whales feeding. (a) The right whale has long plates of baleen suspended from its upper jaw and it feeds by swimming through plankton with its mouth parted. Water enters, passes along the sides of the tongue, and then departs through the curtain of baleen, leaving the plankton entangled in the baleen. (b) As a fin whale approaches a concentration of planktonic organisms, usually krill, it opens its mouth and engulfs these organisms together with the water in which they reside. Its pleated throat allows for considerable expansion of the mouth in order to accommodate the plankton-filled water. The whale lifts its throat to force water out through its baleen, which holds back the food but allows excess water to filter out. Its tongue licks this food from the baleen and swallows it. (c) Skulls of whales have been highly modified during their evolution, especially the design of the face and the position of the nostrils. *Andresarchus*, a terrestrial carnivore from the Eocene, may have belonged to a group from which early cetaceans arose. For comparison, Eocene (*Prozeuglodon*), Oligocene (*Actiocetus*), and a modern baleen whale are illustrated. Although not on a direct evolutionary line with each other, these comparisons show the changes in cetacean skull design, especially in the facial region. Abbreviations: frontal (F), jugal (J), maxilla (M), nasal (N), parietal (P), premaxilla (Pm), supraoccipital (So), squamosal (Sq).

In mammals, the jaws articulate via the dentary-squamosal joint. Several postdentary bones become lost during this transition, and a few move to the ear. The dentary expands posteriorly to form a new articulation with the skull, namely, via the dentary-squamosal joint. Although the factors favoring these changes are disputed, the reality of these changes is not. Bones located at the back of the reptilian lower jaw were either lost or altered in function from jaw articulation to hearing. But this fact raises a new problem. How could bones involved in jaw suspension change function without disrupting the intermediate species? If postdentary bones moved to the middle ear, how could they abandon jaw suspension without producing an individual with no method of supporting the jaw against the skull? G. Cuvier, nineteenth-century French anatomist, would have understood the dilemma.

He argued that evolution could not occur for just such a reason, because a change in structure would disrupt function and stop evolution in its tracks before it had begun.

*Diarthrognathus*, a late cynodont close to primitive mammals, suggests an answer. Its name means two (di-) sites of articulation (arthro-) of the jaw (gnathus). In addition to the articular-quadrate joint inherited from reptiles, a dentary-squamosal joint was apparently present (figure 7.64). We do not know the feeding style of *Diarthrognathus*, so we cannot be sure of the biological role played by this second articulation.

What do living vertebrates suggest? Some birds, such as the skimmer, for example, feed by holding their lower jaw just below the water's surface and flying swiftly along until they strike a fish. Then the jaws snap shut to snatch the fish. A secondary articulation seems to strengthen the lower jaw and help prevent its dislocation as it collides with the fish. *Diarthrognathus* did not feed on fish, but it may have wrestled with struggling prey or fought with competitors. A second jaw articulation would make the jaw stronger. Whatever its advantages, a dentary-squamosal joint was established before the postdentary bones departed from the lower jaw; therefore, when the quadrate and articular bones departed, an alternative method of lower jaw-skull articulation was already in place. This is significant because loss or movement of these bones to support hearing did not disrupt the function they abandoned, jaw suspension. The existing dentary-squamosal articulation was in a sense "ready to serve," preadapted for a new or expanded function.

**Preadaptation (p. 21)**

*Probainognathus*, another late cynodont, like *Diarthrognathus* exhibits a posterior extension of the dentary to establish a secondary point of jaw articulation with the skull. *Probainognathus*, *Diarthrognathus*, and several other late cynodonts with similar transitional double jaw articulations suggest how a harmonious transition in form and function might have occurred. They remind us again that a series of anatomical changes alone are an incomplete statement about evolutionary events. They must be coupled with hypotheses about the accompanying functional series of changes. Form and function go together and both must receive attention if we are to bring some understanding to the process of evolutionary change.

## Evolutionary Modifications of Immature Forms: Akinesis in Mammals

We sometimes forget that an evolutionary modification can debut in an embryonic or infant stage and later become incorporated or expanded in the adult. Such may have been the case with akinesis in mammals. In all mammals, infants suckle milk from their mothers. Suckling re-

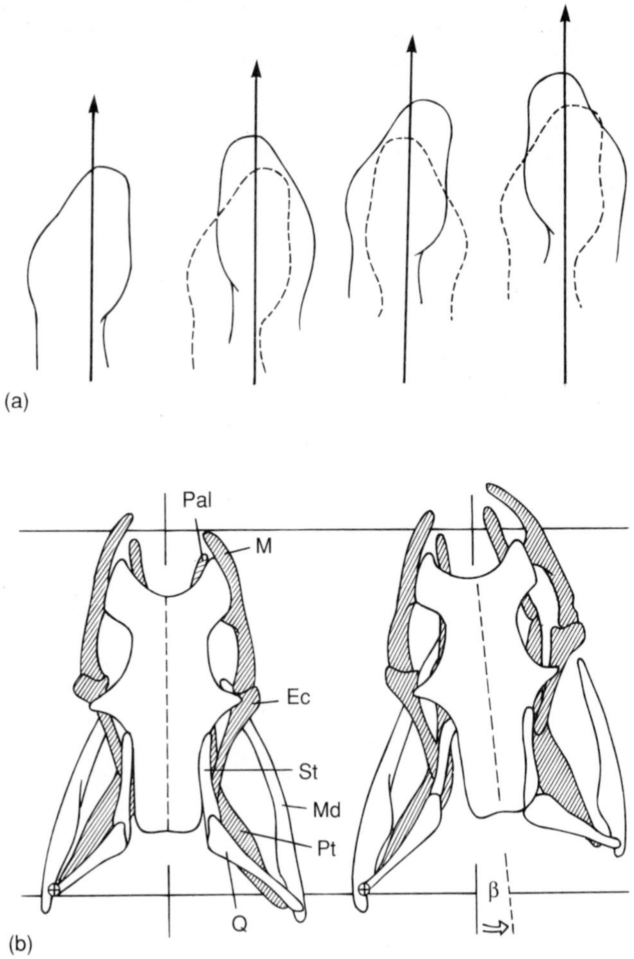


**Figure 7.63** Swallowing by a rat snake, *Elaphe*, as seen in dorsal view. (a) Outline of snake's head during successive swallowing motions, left to right. Previous head position is indicated by the dotted outline. With alternating left and right advances, the jaws walk over the prey along a line of progress, the axis of swallowing, until the jaws pass over the entire prey. These jaw walking displacements place the prey at the back of the throat where contractions of neck muscles move the prey along to the stomach. (b) Movable bones of the skull (shaded) on one side swing outward from the prey and advance farther forward where they come to rest momentarily on the surface of the prey at a new position. Movable bones of the opposite side now take their turn. By such reciprocating motion, jaws walk along the prey. In addition to jaw displacement, the skull itself swings outward from the swallowing axis (arrow) through an angle (β) in the direction of the advancing bones to place them farther along the prey. Abbreviations: braincase (Bc), ectopterygoid (Ec), maxilla (M), mandible (Md), palatine (Pal), pterygoid (Pt), quadrate (Q), supratemporal (St).

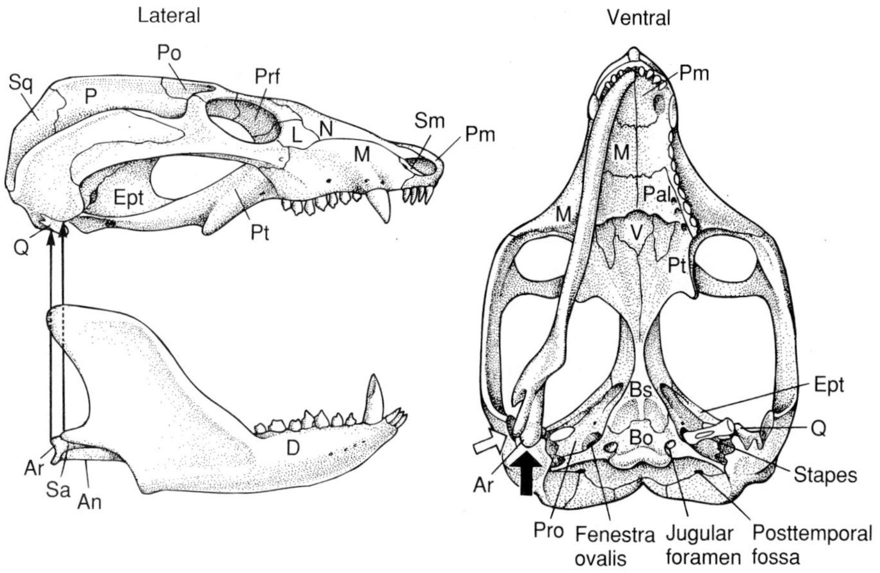


**Figure 7.64** Double jaw articulation. Skull of *Probainognathus*, a late cynodont reptile (therapsid). Double jaw articulation in the reptile occurs between the quadrate and articular (solid arrow), the reptilian condition, and another articulation occurs between dentary and squamosal (open arrow) that came to predominate in later mammals. Abbreviations: angular (An), articular (Ar), basioccipital (Bo), basisphenoid (Bs), dentary (D), epipterygoid (Ept), lacrimal (L), maxilla (M), nasal (N), parietal (P), prefrontal (Prf), premaxilla (Pm), prootic (Pro), postorbital (Po), pterygoid (Pt), quadrate (Q), septomaxilla (Sm), vomer (V).

quires a pump and a seal. Fleshy lips provide the seal around the teat of the mammary gland, the mouth is the chamber that receives the milk, and the up-and-down action of the tongue pumps the milk from the mother to the infant's mouth and esophagus. If respiration and feeding shared a common chamber, as in most reptiles, the infant would have to interrupt nursing and release its attachment to the nipple in order to breathe. A secondary palate makes this inefficient interruption in feeding unnecessary. It separates feeding from breathing by separating the mouth from the nasal chambers. But a secondary palate that separates the mouth from the nasal passages also fuses left and right halves of the skull, thus preventing any movement within or across the braincase. The result is an akinetic skull.

Further changes in the adult evolved later. With loss of kinesis, the skull is firm and ready to serve strong jaw-closing muscles. Mastication, development of specialized teeth to serve chewing (accurately occlusal tooth rows), and a muscular tongue (to move food into position between tooth rows) might then find adaptive favor. Certainly there are other ways to chew food. Some fishes with kinetic skulls and teeth that are continuously replaced chew their food. In mammals, the conditions seem especially favorable for mastication and we find this adaptation in almost all mammalian species. Analysis of evolutionary events often centers on adult stages, yet understanding of these events must come from a knowledge of the entire life history of species.

## Composite Skull

Dermatocranium, chondrocranium, and visceral cranium contribute to the skull. Although their phylogenetic backgrounds are different, parts of each combine into a functional unit, the skull. If species were stamped out one at a time, each being a unique creation, then there would be little reason to expect a composite skull. Yet, clearly the skull is a combination from different phylogenetic sources. In Darwin's day as now, we can point to this as evidence that supports evolution, not special creation. Evolution is behind the origin of new structures and the species that display them.

The skull also illustrates a point made earlier. Evolution proceeds by remodeling, seldom by new construction. Consider the splanchnocranium. From fishes to mammals, it is variously modified to serve the adaptive demands of the organism of the moment, first as support for gill slits, and then as the source of jaws, as support for the tongue, and in mammals as part of the ear (figure 7.65). In this series of transformations seen in retrospect, we have no reason to believe that the splanchnocranium of the first fish anticipates its eventual contribution to the hearing devices of mammals. Evolution does not look ahead. The sequences or trends we see are apparent only after the fact, after evolution has taken place. There is no evolutionary arrow anticipating the future. We have seen that changes in the splanchnocranium, chondrocranium, and dermatocranium serve biological roles of the moment. They are not predestined changes preparing for the future.

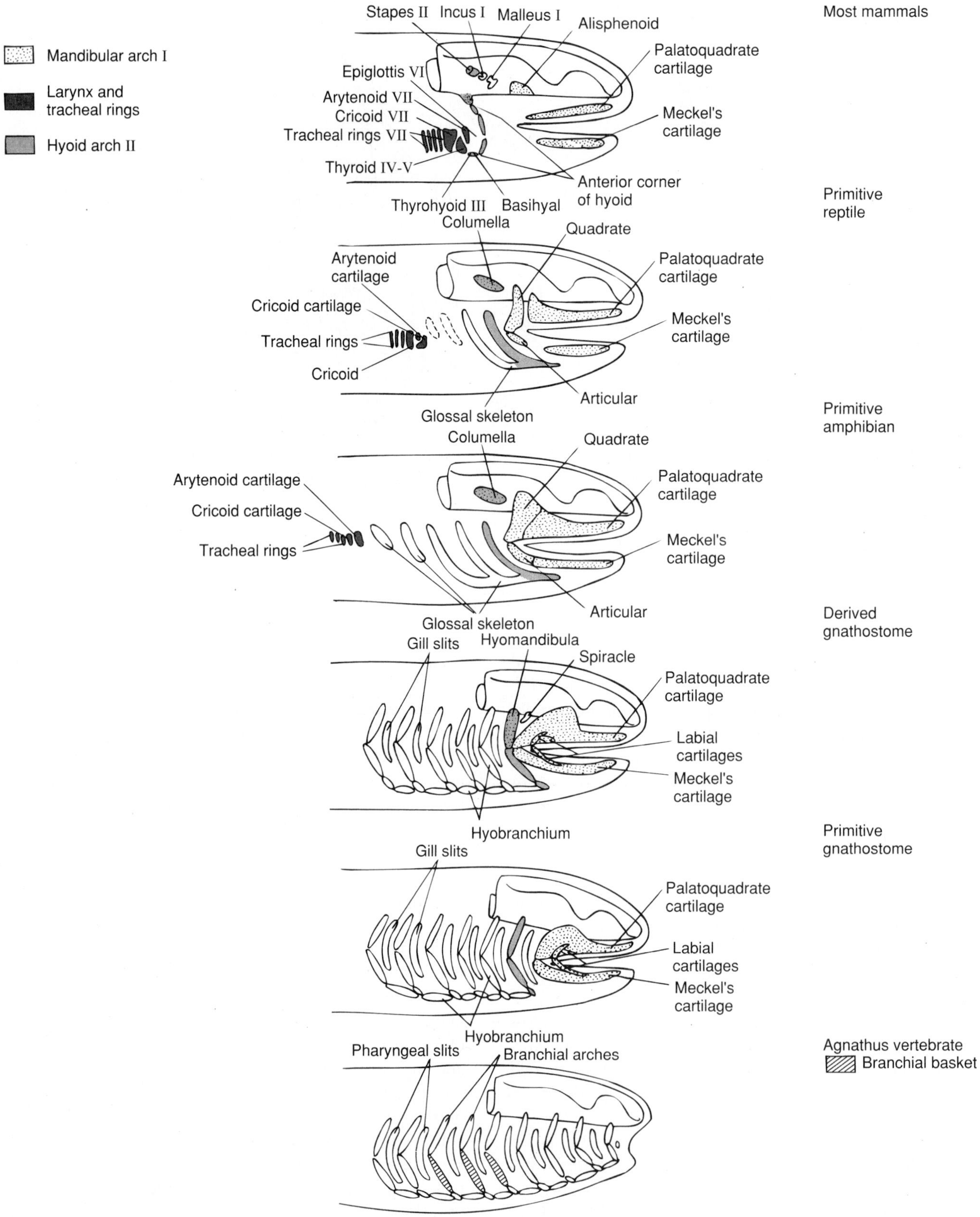


**Figure 7.65** Phylogeny of the splanchnocranium. Notice how the branchial arches are remodeled to serve various functions within each succeeding group. Parts of the branchial basket become the jaws, tracheal cartilages, tongue supports, components of the neurocranium, and ear ossicles. Roman numerals indicate branchial arch number.

# Selected References


Allin, E. F. 1986. The auditory apparatus of advanced mammal-like reptiles and early mammals. In *The ecology and biology of mammal-like reptiles*, edited by N. Hotton, III, P. D. MacLean, J. J. Roth, and E. C. Roth. Washington, D.C.: Smithsonian Institution Press, pp. 283–94.

Allin, E. F., and J. A. Hopson. 1992. Evolution of the auditory system in Synapsida ("mammal-like reptiles") and primitive mammals as seen in the fossil record. In *The evolutionary biology of hearing*, edited by D. B. Webster, R. R. Fay, and A. N. Popper. New York: Springer-Verlag, pp. 587–614.

Bartsch, P. 1993. Development of the snout of the Australian lungfish *Neoceratodus forsteri* (Krefft, 1870), with special reference to cranial nerves. *Acta Zool.* 74:15–29.

Bemis, W. E. 1986. Feeding systems of living Dipnoi: Anatomy and function. *J. Morph.* (suppl.) 1:249–75.

Bjerring, H. C. 1977. A contribution to structural analysis of the head of craniate animals. *Zool. Scripta* 6:127–83.

Bock, W. J. 1968. The avian mandible as a structural girder. *J. Biomech.* 1:89–96.

Bramble, D. M. 1989. Cranial specialization and locomotor habit in the lagomorpha. *Amer. Zool.* 29:303–17.

Cundall, D., and C. Gans. 1979. Feeding in water snakes: An electromyographic study. *J. Exp. Zool.* 209:189–208.

Erdman, S., and D. Cundall. 1984. The feeding apparatus of the salamander *Amphiuma tridactylum:* Morphology and behavior. *J. Morph.* 181:175–204.

Frazzetta, T. H. 1968. Adaptive problems and possibilities in the temporal fenestration of tetrapod skulls. *J. Morph.* 125:145–57.

Frazzetta, T. H. 1986. The origin of amphikinesis in lizards. *Evol. Biol.* 20:419–61.

Gorniak, G. C., H. I. Rosenberg, and C. Gans. 1982. Mastication in the tuatara, *Sphenodon punctatus* (Reptilia: Rhynchocephalia): Structure and activity of the motor system. *J. Morph.* 171:321–54.

Homberger, D. G., and R. A. Meyers. 1989. Morphology of the lingual apparatus of the domestic chicken, *Gallus gallus*, with special attention to the structure of the fasciae. *Amer. J. Anat.* 186:217–57.

Hopson, J. A., and H. R. Barghusen. 1986. An analysis of therapsid relationships. In *The ecology and biology of mammal-like reptiles*, edited by N. Hotton, III, P. D. MacLean, J. J. Roth, and E. C. Roth. Washington, D.C.: Smithsonian Institution Press, pp. 83–106.

Jarvik, E. 1980. *Basic structure and evolution of vertebrates*. 2 vols. New York and London: Academic Press. Primarily a detailed study of certain extinct or primitive fishes.

Jollie, M. 1977. Segmentation of the vertebrate head. *Amer. Zool.* 17:323–33.

Jollie, M. 1986. A primer of bone names for the understanding of the actinopterygian head and pectoral girdle skeletons. *Can. J. Zool.* 64:365–79.

Kardong, K. V. 1977. Kinesis of the jaw apparatus during swallowing in the cottonmouth snake, *Agkistrodon piscivorus*. *Copeia* 1977:338–48.

Kardong, K. V. 1986. Kinematics of swallowing in the yellow rat snake, *Elaphe obsoleta quadrivittata:* A reappraisal. *Jpn. J. Herpet.* 11:96–109.

Kardong, K. V., P. Dullemeijer, and J.A.M. Fransen. 1986. Feeding mechanism in the rattlesnake *Crotalus durissus*. *Amph.-Rept.* 7:271–302.

Keynes, R., and A. Lumsden. 1990. Segmentation and the origin of regional diversity in the vertebrate central nervous system. *Neuron* 4:1–9.

Langille, R. M., and B. Hall. 1987. Development of the head skeleton of the Japanese Medaka, *Oryzias latipes* (Teleostei). *J. Morph.* 193:135–58.

Larsen, J. H., Jr., J. T. Beneski, Jr., and D. B. Wake. 1989. Hyolingual feeding systems of the Plethodontidae: Comparative kinematics of prey capture by salamanders with free and attached tongues. *J. Exp. Zool.* 252:25–33.

Larsen, J. H., Jr., and D. J. Guthrie. 1975. The feeding system of terrestrial tiger salamanders (*Ambystoma tigrinum melanostictum* Baird). *J. Morph.* 147:137–54.

Lauder, G. V. 1979. Feeding mechanics in primitive teleosts and in the halecomorph fish *Amia calva*. *J. Zool. (London)* 187:543–78.

Lauder, G. V. 1980. Evolution of the feeding mechanism in primitive actinopterygian fishes: A functional anatomical analysis of *Polypterus*, *Lepisosteus*, and *Amia*. *J. Morph.* 163:283–317.

Lauder, G. V. 1982. Patterns of evolution in the feeding mechanism of actinopterygian fishes. *Amer. Zool.* 22:275–85.

Lauder, G. V. 1985. Aquatic feeding in lower vertebrates. In *Functional vertebrate morphology*, edited by M. Hildebrand, D. M. Bramble, K. F. Liem, and D. B. Wake. Cambridge, Mass.: Harvard University Press, pp. 210–29, 397–99.

Liem, K. F. 1980. Acquisition of energy by teleosts: Adaptive mechanisms and evolutionary patterns. In *Environmental physiology of fishes*, edited by M. A. Ali. New York: Plenum Press, pp. 299–334.

Liem, K. F. 1990. Aquatic *versus* terrestrial feeding modes: Possible impacts on the trophic ecology of vertebrates. *Amer. Zool.* 30:209–21.

Lombard, R. E., and D. B. Wake. 1976. Tongue evolution in the lungless salamanders, family Plethodontidae. I. Introduction, theory and a general model of dynamics. *J. Morph.* 148:265–86.

Lombard, R. E., and D. B. Wake. 1977. Tongue evolution in the lungless salamanders, family Plethodontidae. *J. Morph.* 153:39–80.

Miller, B. T., and J. H. Larsen, Jr. 1990. Comparative kinematics of terrestrial prey capture in salamanders and newts (Amphibia: Urodela: Salamandridae). *J. Exp. Zool.* 256:135–53.

Nishikawa, K. C., and D. C. Cannatella. 1991. Kinematics of prey capture in the tailed frog *Ascaphus truei* (Anura: Ascaphidae). *Zool. J. Linn. Soc.* 103:289–307.

Noden, D. M. 1984. Craniofacial development: New views on old problems. *Anat. Rec.* 208:1–13.

Panchen, A. L. 1967. The nostrils of choanate fishes and early tetrapods. *Biol. Rev.* 42:374–420.

Rieppel, O. 1980. The evolution of the ophidian feeding system. *Zool. J. Anat.* 103:551–64.

Sanderson, S. L., J. J. Cech, Jr., and M. R. Patterson. 1991. Fluid dynamics n suspension-feeding blackfish. *Science* 251:1346–48.

Sanderson, S. L., and R. Wassersug. 1990. Suspension-feeding vertebrates. *Sci. Amer.* (March):96–101.

Sibbing, F. A., J.W.M. Osse, and A. Terlouw. 1986. Food handling in the carp *(Cyprinus carpio)*: Its movement patterns, mechanisms and limitations. *J. Zool. (London)* 210:161–203.

So, K-K. J., P. C. Wainwright, and A. F. Bennett. 1992. Kinematics of prey processing in *Chamaeleo jacksonii*: Conservation of function with morphological specialization. *J. Zool. (London)* 226:47–64.

Thexton, A. J., D. B. Wake, and M. H. Wake. 1977. Tongue function in the salamander *Bolitoglossa occidentalis*. *Arch. Oral Biol.* 22:361–66.

Wainwright, P. C., D. M. Kraklau, and A. F. Bennett. 1991. Kinematics of tongue projection in *Chamaeleo oustaleti*. *J. Exp. Biol.* 159:109–33.

Wainwright, P. C., and A. F. Bennett. 1992. The mechanism of tongue projection in chameleons. I. Electromyographic tests of functional hypotheses. II. Role of shape change in a muscular hydrostat. *J. Exp. Biol.* 168:1–21; 23–40.

Wassersug, R. J., and K. Hoff. 1982. Developmental changes in the orientation of the anuran jaw suspension. *Evol. Biol.* 15:223–46.

Weinrich, M. T., M. R. Schilling, and C. R. Belt. 1992. Evidence for acquisition of a novel feeding behaviour: Lobtail feeding in humpback whales, *Megaptera novaeangliae*. *Anim. Behav.* 44:1059–72.

# 8 CHAPTER

# Skeletal System: The Axial Skeleton



## Introduction

Two structural components combine to define the long axis of the vertebrate body, offer sites for muscle attachment, prevent telescoping of the body, and support much of the weight. One is the notochord, the other, the vertebral column. The **notochord** is a long, continuous rod of fibrous connective tissue wrapping a core of fluid or fluid-filled cells. The **vertebral column** consists of a discrete but repeating series of cartilaginous or bony elements. The notochord is phylogenetically the oldest of the two structural components, but it tends to give way to the vertebral column, which assumes the role of body support in most later vertebrates.

Evolution of the vertebral column is not entirely clear, especially during its phylogenetic inception. Consulting the earliest vertebrate fossils, the ostracoderms, we find only a few with even a hint of a vertebral column. This may be because vertebrae were present but unossified or poorly preserved. Or, more probably, vertebrae were absent or uncommon. Instead, the central body axis was formed by the prominent notochord in ostracoderms. The subsequent evolution of the vertebral column in fishes and tetrapods is also complicated, partly because some components became enlarged, others were lost, and some evolved independently several times. Events of early embryology, which often can be consulted to help clarify such phylogenetic uncertainties, fail us in this instance because opinions differ about even basic comparative details of embryonic events in living forms.

Perhaps it is best to begin with a generalized view of the structure of the vertebral column and the companion terminology. Next we look to each vertebrate class for actual examples. The original function of the vertebrae was to protect the spinal cord and dorsal aorta. Later, vertebrae became important as attachment sites for body musculature. In tetrapods, their roles expanded to include suspension of the body and locomotion on land.

# Basic Components

## Vertebrae

The first components of the vertebra to appear were the dorsal and ventral arches that rested upon the notochord (figure 8.1). The dorsal arches, **neural** and **interneural** (intercalary) **arches,** protected the neural tube. The ventral arches, **hemal** and **interhemal arches,** enclosed blood vessels. The next stage in the evolution of the basic elements of a vertebra was the formation of two **centra,** an **intercentrum** (hypocentrum) and a **pleurocentrum.** The bases of the ventral arches expand to form these centra where they meet the notochord. The centra served to anchor and support these arches.

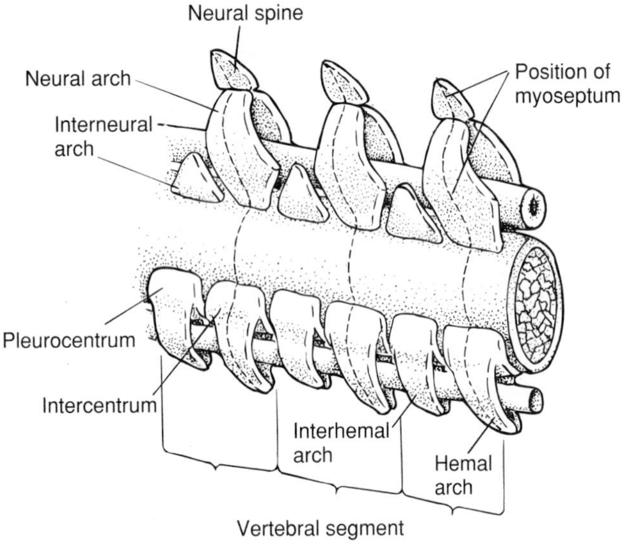


**Figure 8.1** The axial skeleton of a generalized primitive vertebrate. Three vertebral segments from the tail region are illustrated. The notochord is prominent. Vertebral elements are represented by a pair of centra (intercentrum, pleurocentrum), their associated ventral arches (hemal, interhemal), and by dorsal arches (neural, interneural) that often support a neural spine. Dashed lines indicate the location of myosepta, connective tissue sheets that bound each section of body musculature.

### *Regions of the Vertebral Column*

Each **vertebral segment** consists of arches and centra, up to two dorsal arches (neural and interneural), as many as two ventral arches (hemal and interhemal arches), and often two centra (intercentrum and pleurocentrum). The evolution of these basic vertebral elements is characterized first of all by enlargement of some elements at the expense of others. Second, the vertebral components generally come to displace the notochord as the primary mechanical axis of the body. Third, the vertebral segments comprising the axial column tend to become regionally differentiated within the vertebral column they collectively define (figure 8.2). In most fishes, the vertebral column is differentiated into two regions, an anterior **trunk region** and a posterior **caudal region.** In tetrapods, the trunk becomes further differentiated anteriorly into the neck, or **cervical region,** and posteriorly into the hip, or **sacral region.** In some tetrapods, there is further differentiation of the trunk into the chest, or the **thoracic region,** and into the area between the thorax and the hips, the **lumbar region.**

### *Centra*

Among vertebrates there is great variation in the structure of the centra, in the relative importance of the pleurocentrum compared with the intercentrum, in the extent of ossification, and in the degree to which centra supplement or replace the notochord as mechanical elements of the axial column. Each centrum constitutes the **body** of the vertebra. In some vertebrates, centra may be absent **(aspondyly).** Others exhibit one **(monospondyly)** or two **(diplospondyly)** centra per segment. In many anamniotes, the caudal vertebrae may be diplospondylous, the trunk vertebrae, monospondylous. In some groups, the single centrum in trunk vertebrae is derived from the pleurocentrum; in others, it is derived from the intercentrum. In Holocephali and Dipnoi, the number of centra may secon-

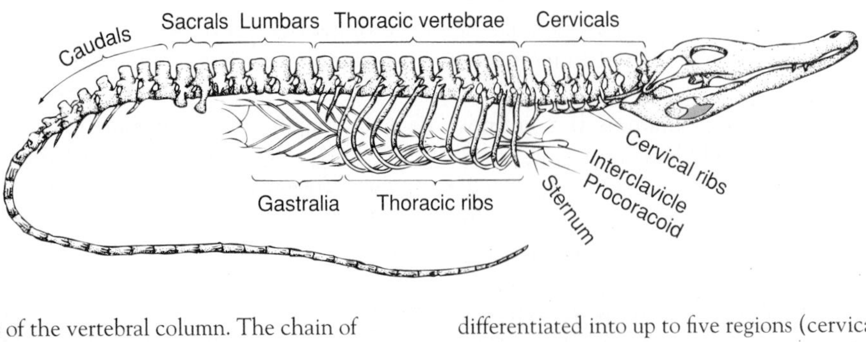


**Figure 8.2** Regions of the vertebral column. The chain of vertebrae comprising the axial column differentiates into two regions (trunk and caudal) in all fishes, but becomes differentiated into up to five regions (cervical, thoracic, lumbar, sacral, and caudal) in amniotes. The vertebral column of an alligator depicted in the figure illustrates these five regions.

darily increase to five or six per segment **(polyspondyly).** In amniotes, the pleurocentrum predominates and becomes the body of each vertebral segment. Remnants of the intercentrum contribute to cervical vertebrae and perhaps in a minor way to other vertebral elements, but for the most part, the intercentrum is not a significant mechanical element of the amniote vertebral column.

In tetrapods, a descriptive terminology has grown up to characterize the two general anatomical relationships between centra and their neural arches (figure 8.3a,b). In one condition, termed **aspidospondyly,** all elements are separate. Specifically, the three arch elements (intercentrum, pleurocentrum, and neural arch), either paired or single, remain as separate ossified elements. The **rhachitomous vertebra,** found in some crossopterygians and some early amphibians, is a specialized type of aspidospondylous vertebra. The term means "cut up" spine, a reference to the numerous separate parts that constitute each vertebral segment (figure 8.3a). Both large intercentrum and small pleurocentrum are usually paired and a separate neural arch is associated with them. Several derived aspidospondylous vertebral types are recognized. In the **embolomerous vertebra,** separate but approximately equal-sized centra are present. In the **stereospondylous vertebra,** the vertebra consists of a single body derived entirely from the intercentrum.

In the other general vertebral condition of tetrapods, termed **holospondyly,** all vertebral elements in a segment are fused into a single piece. In the **lepospondylous vertebra,** a specialized type of holospondylous vertebra, the centrum of the solid vertebra is husk shaped and usually perforated by a notochordal canal (figure 8.3b).

At one time, the vertebral type was used as the major criterion to define tetrapod taxa, and each type was thought

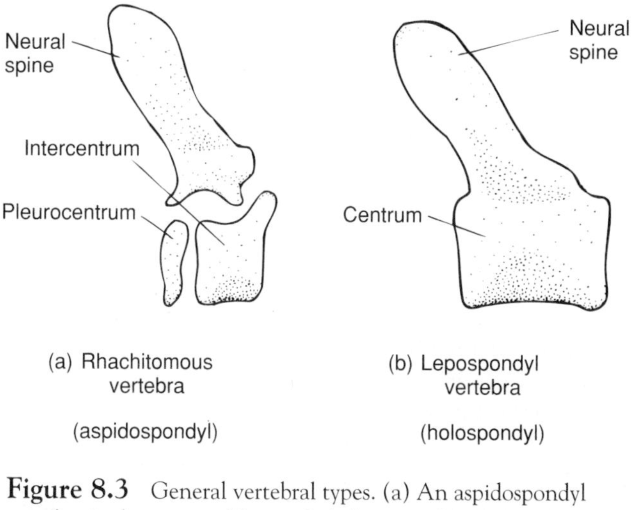


**Figure 8.3** General vertebral types. (a) An aspidospondyl vertebra is characterized by ossified elements that remain separate. The specific type illustrated is a rhachitomous vertebra that has three discrete parts, pleurocentrum, intercentrum, and neural spine. (b) A holospondyl vertebra is characterized by fused construction of all components. The specific type shown is a lepospondylous vertebra, a holospondyl vertebra with a husklike centrum.

to characterize a separate phylogenetic trend. With this taxonomic emphasis came a proliferation of descriptive terminology to track supposed vertebral phylogeny, especially in amphibians; however, problems became evident with this approach. Many early tetrapods evolved from an aquatic ancestry into new terrestrial habitats, and their vertebrae became modified to accommodate life on land where walking predominated. But other derived amphibians reinvaded or secondarily returned to aquatic habitats where swimming received renewed emphasis. The vertebrae of these later, but secondarily aquatic amphibians are similar to the vertebrae of the most primitive and predominantly aquatic amphibians. Thus, the morphologically similar types of vertebrae represent multiple evolutionary trends, testimony to functional convergence but not evidence of close phylogenetic unity. Consequently, much of the elaborate terminology, based on the mistaken assumption of close phylogenetic affinity, has been largely abandoned, although a few terms linger on in taxonomic usage. For example, a temnospondylous vertebra formally designated a vertebra of several parts with a separate arch, but in descriptive usage, this meaning has now been assigned to other terms. Further, other surviving terms, although developed in reference to tetrapods, are now often applied to vertebrae of fishes as a descriptive convenience.

The centra are linked successively into a chain of vertebrae, the **axial column.** The shapes of surfaces at the articular ends of the centra affect the properties of the vertebral column and the way in which forces are distributed between vertebrae. A functional scheme for classifying centra might be desirable, but analysis of their complicated mechanical functions has proved difficult and remains incomplete. Thus, the more traditional anatomical criteria employing articular shape are used more often, yielding several types of centra.

Centra with flat ends are **acoelous (amphiplatyan)** and seem especially suited to receive and distribute compressive forces within the vertebral column (figure 8.4a). If each surface is concave, the centrum is **amphicoelous,** a design that seems to allow limited motion in most directions (figure 8.4b). Centra that are concave anteriorly and convex posteriorly are **procoelous** (figure 8.4c). The reverse shape, concave posteriorly and convex anteriorly, characterizes centra that are **opisthocoelous** (figure 8.4d). Centra that are **heterocoelous** bear saddle-shaped articular surfaces at both ends (figure 8.4e). In both procoelous and opisthocoelous centra, the convex articular surface of one centrum fits into the concave surface of the next to form a kind of ball-and-socket joint, permitting extensive motion in most directions without stretching the nerve cord that their neural arches protect. By comparison, if acoelous or amphicoelous vertebral series are flexed, adjacent centra hinge about their edges. If rotation is extensive, then like opening a door, the space will tend to widen between centra and stretch the central nerve cord running dorsally above them (figure 8.5a). However, in procoelous and opisthocoelous

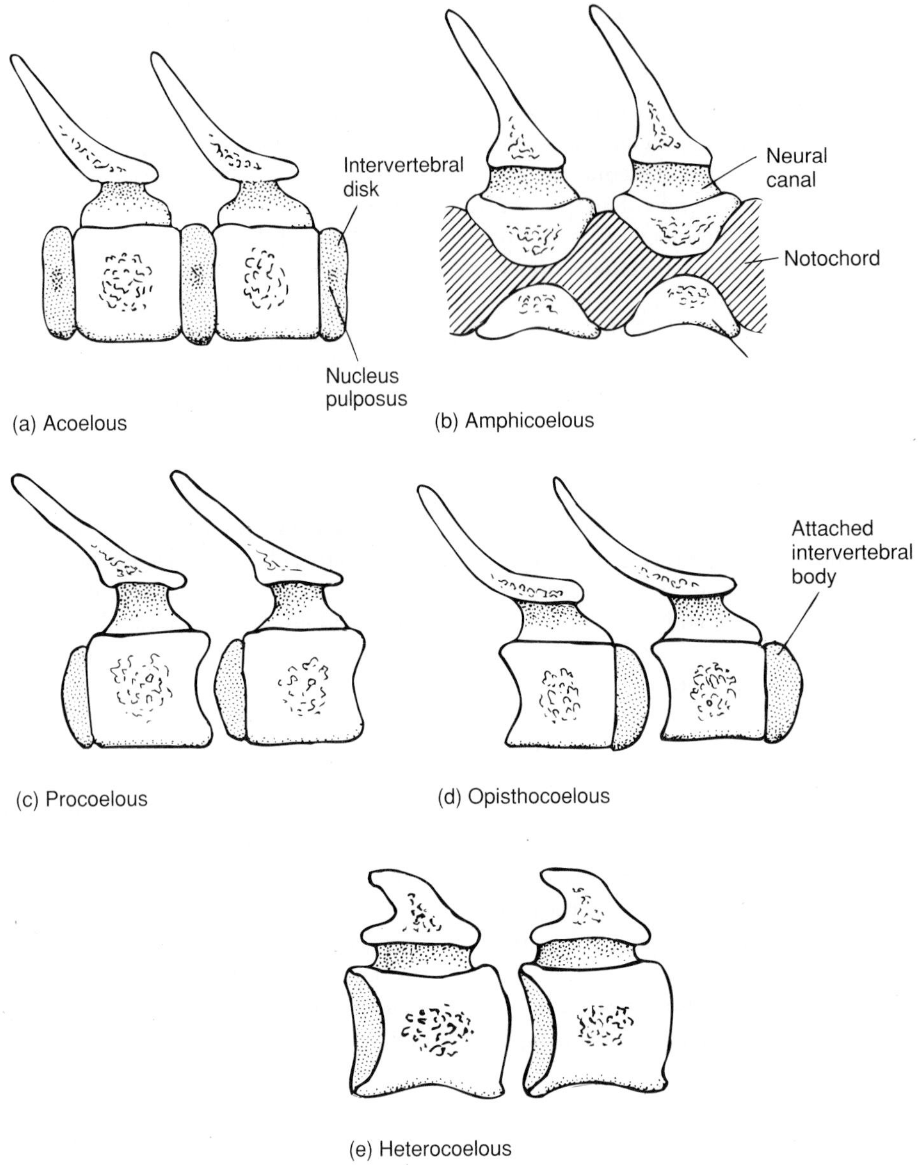


**Figure 8.4** General centra shapes. The shapes of articulating centra ends define specific anatomical types: (a) acoelous, both ends are flat; (b) amphicoelous, both ends are concave; (c) procoelous, anterior end is concaved; (d) opisthocoelous, posterior end is concaved; (e) heterocoelous, saddlelike articulating ends.

centra, with ball-and-socket articulation, the point of rotation is not at the edge but at the center of the convex surface of the centrum. Flexion of the vertebral series does not open a space between them, and the central nerve cord is not unduly stretched (figure 8.5b). Heterocoelous centra allow great lateral and vertical flexion, but they prevent wringing or rotation of the vertebral column about its long axis (figure 8.5c). Heterocoelous centra are most common in turtles that retract their necks and in cervical vertebrae of birds.

This anatomical classification includes only the criterion of centrum shape, but soft tissues are often associated and are usually extremely important in affecting function. The notochord or its adult derivatives often run through and fill the concavities at the articular ends of centra, which are capped by cartilaginous pads. The term ***intervertebral disk*** has been used broadly to designate any pad of tissue between articular surfaces of centra. However, strictly speaking, an intervertebral disk in the adult is a pad of fibrocartilage whose gellike core, the **nucleus pulposus,** is derived from the embryonic notochord. By this strict definition, intervertebral disks are found only in mammals in whom they reside between successive surfaces of adjacent centra. In other groups, the pad between centra is called an **intervertebral cartilage** or **body.** Joining the rims of adjacent centra is the **intervertebral ligament** that

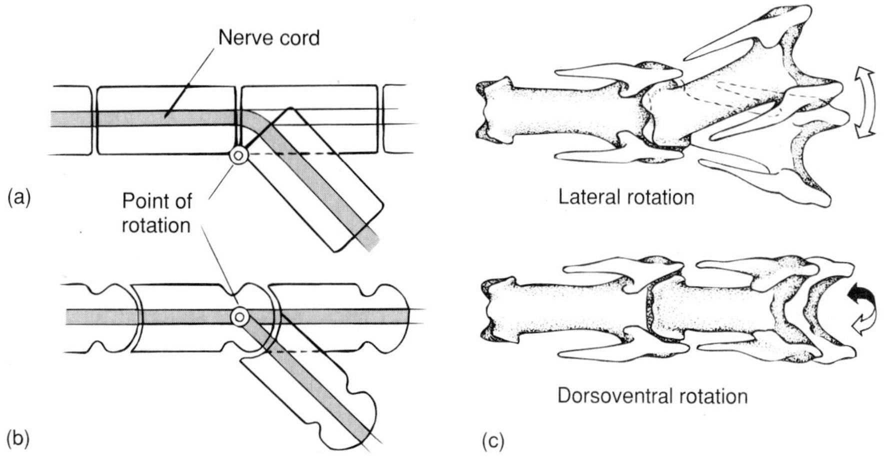


**Figure 8.5** Centra functions. (a) Amphicoelous or acoelous centra flex about a point on their rims, tending to stretch the centrally located dorsal nerve cord. (b) Opisthocoelous and procoelous centra eliminate this potentially damaging stretching tendency with ball-and-socket ends that establish a centrally located point of rotation instead of one at the rims. (c) In heterocoelous centra, opposite saddle-shaped surfaces fit together, allowing extensive lateral and dorsoventral rotation. Ventral view of two vertebrae of the ostrich *Struthino*.

is important in controlling the stiffness of the vertebral column when it flexes.

Projecting from centra and from their arches are **apophyses,** processes that will be described more fully as we meet them during the survey of the axial column later in the chapter. In general, apophyses include **diapophyses** and **parapophyses,** both of which articulate with ribs. Apophyses also form antitwist, interlocking processes, or **zygapophyses,** between successive vertebrae. The term **transverse process** generally applies to any process extending from the centrum or the neural arch, but historically it has been used so loosely that it retains no exact morphological meaning.

## Ribs

Ribs are struts that sometimes fuse with vertebrae or articulate with them. Ribs provide sites for secure muscle attachment, help suspend the body, form a protective case around viscera (rib cage), and sometimes serve as accessory breathing devices. Embryologically, ribs preform in cartilage within **myosepta** (myocommata), that is, within the dorsoventral sheets of connective tissue that partition successive blocks of segmental body musculature (figure 8.6a–c). Caudal vertebrae never develop ribs.

In many fishes, there are two sets of ribs with each vertebral segment, a dorsal and a ventral set. The **dorsal ribs** form at the intersection of each myoseptum with the **horizontal septum** (horizontal skeletogenous septum), a longitudinal sheet of connective tissue (figure 8.6a). The **ventral ribs** form at points where the myosepta meet the walls of the coelomic cavity. They are serially homologous with the hemal arches of the caudal vertebrae (figure 8.6c). In tetrapods, one of these sets of ribs is lost and the other, apparently the dorsal ribs, persists to become the ribs of terrestrial vertebrates. Ribs of primitive tetrapods are **bicipital,** having two heads that articulate with the vertebrae. The ventral rib head, or **capitulum,** articulates with the **parapophysis,** a ventral process on the intercentrum. The dorsal head, or **tuberculum,** articulates with the **diapophysis,** a process on the neural arch (figure 8.7). If these vertebral processes fail to develop, the articular surface persists, forming a small concavity, the **facet,** to receive the rib. In amniotes, the intercentrum is lost or incorporated into other elements, so the capitulum must shift its articulation to the pleurocentrum (in most reptiles and birds) or between centra (in mammals).

Although ribs function in locomotion in tetrapods, they become an increasingly important part of the respiratory system to move air through the lungs. Classification of tetrapod ribs is based on the type of association they establish with the sternum. Ribs that meet ventrally with the sternum are **true ribs.** Those that articulate with each other but not with the sternum are **false ribs.** Those articulating with nothing ventrally are **floating ribs.** True ribs consist of two jointed segments, the **vertebral (costal) rib,** a proximal segment articulated with the vertebrae, and the **sternal rib,** a distal segment that is usually cartilaginous and meets the sternum. The joint between vertebral and sternal segments accommodates changes in chest shape during respiratory expansion and compression.

In birds, cervical ribs are reduced and fused to the vertebrae. In the thoracic region, the first several ribs are floating ribs followed by true ribs that articulate with the sternum. Some floating and most true ribs bear **uncinate processes,** projections that extend posteriorly from proximal rib segments. Uncinate processes offer attachment to muscles supporting the scapula of the shoulder, but they are

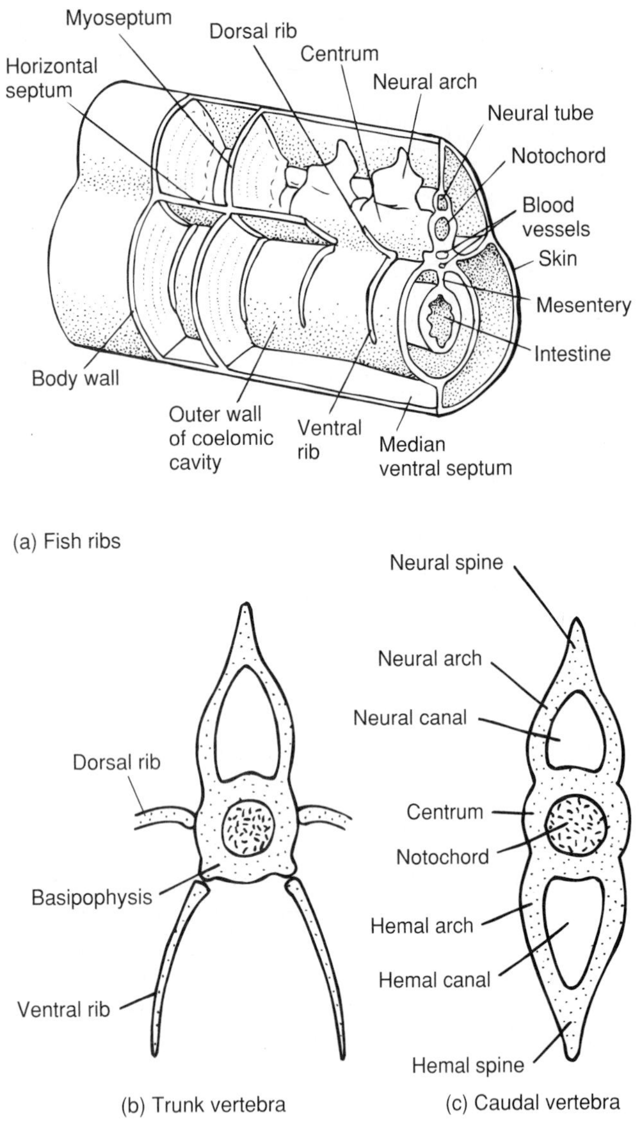


**Figure 8.6** Ribs. (a) In fishes, dorsal ribs develop where myosepta intersect with the horizontal septum, and ventral ribs develop where myosepta meet the wall of the body cavity. (b) Cross section of trunk vertebra of a fish. (c) Cross section of caudal vertebra of a fish. Trunk ribs are serially homologous with caudal hemal arches.

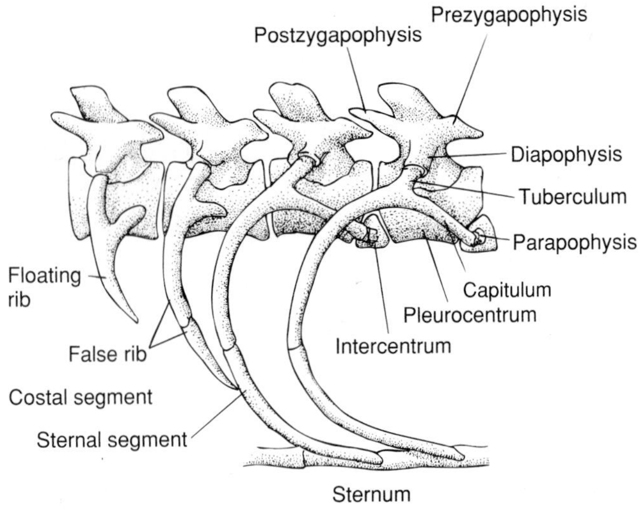


**Figure 8.7** Amniote ribs. Ribs are named on the basis of their articulation with the sternum (true ribs), with each other (false ribs), or with nothing ventrally (floating ribs). Primitively, ribs are bicipital, having two heads, a capitulum and a tuberculum, that articulate respectively with the parapophysis on the intercentrum or the diapophysis on the neural arch. The body of the rib may differentiate into a dorsal part, the vertebral rib or segment, and a ventral part, the sternal rib or segment that articulates with the sternum.

also found in some living and fossil reptiles as well as in the early amphibian, *Ichthyostega,* where they project posteriorly to overlap with the next adjacent rib. This overlap between successive ribs may introduce some overall firmness into the thoracic ribs, giving them the functional integrity to act as a unit during lung ventilation.

In mammals, ribs are present on all thoracic vertebrae and define this region. Some are floating and others are false. Most are true ribs, however, and they meet the sternum through cartilaginous sternal rib segments. Within cervical and lumbar regions, ribs exist only as remnants fused with transverse processes, forming what should properly be termed **pleurapophyses** (transverse process plus rib remnant).

## Sternum

The sternum is a midventral skeletal structure that is endochondral in embryonic origin and arises within the ventral connective tissue septum and adjacent myosepta (figure 8.8a–f). The sternum offers a site of origin for chest muscles. As noted, it also secures the ventral tips of true ribs to complete the protective chondrified or ossified rib cage. The **rib cage** consists of ribs and sternal elements that embrace the visceral. Size and shape changes in the rib cage also act to compress or expand the lungs, promoting ventilation. The sternum may consist of a single bony plate or several elements in series.

Fishes lack a sternum. When it first appears in tetrapods, it is apparently not a phylogenetic derivative of either the ribs or the pectoral girdle, although in many groups it has become secondarily associated with each. A sternum is absent in the first fossil amphibians, but it is present in modern forms. In many urodeles, the sternum is a single midventral **sternal plate** grooved along its anterior borders to receive the ventral elements of the shoulder girdle, the **coracoid plate** (figure 8.8a). In anurans, a single element, the **xiphisternum,** often tipped with the **xiphoid cartilage,** lies posterior to the pectoral girdle, and in some, a second element, the **omosternum** capped by the **episternal cartilage,** lies anterior to the girdle (figure 8.8b). The sternum is absent in turtles, snakes, and many limbless lizards but is common in other reptiles, where it consists of a single, midventral element associated with the shoulder girdle (figure 8.8c). During locomotion, the reptilian sternum confers

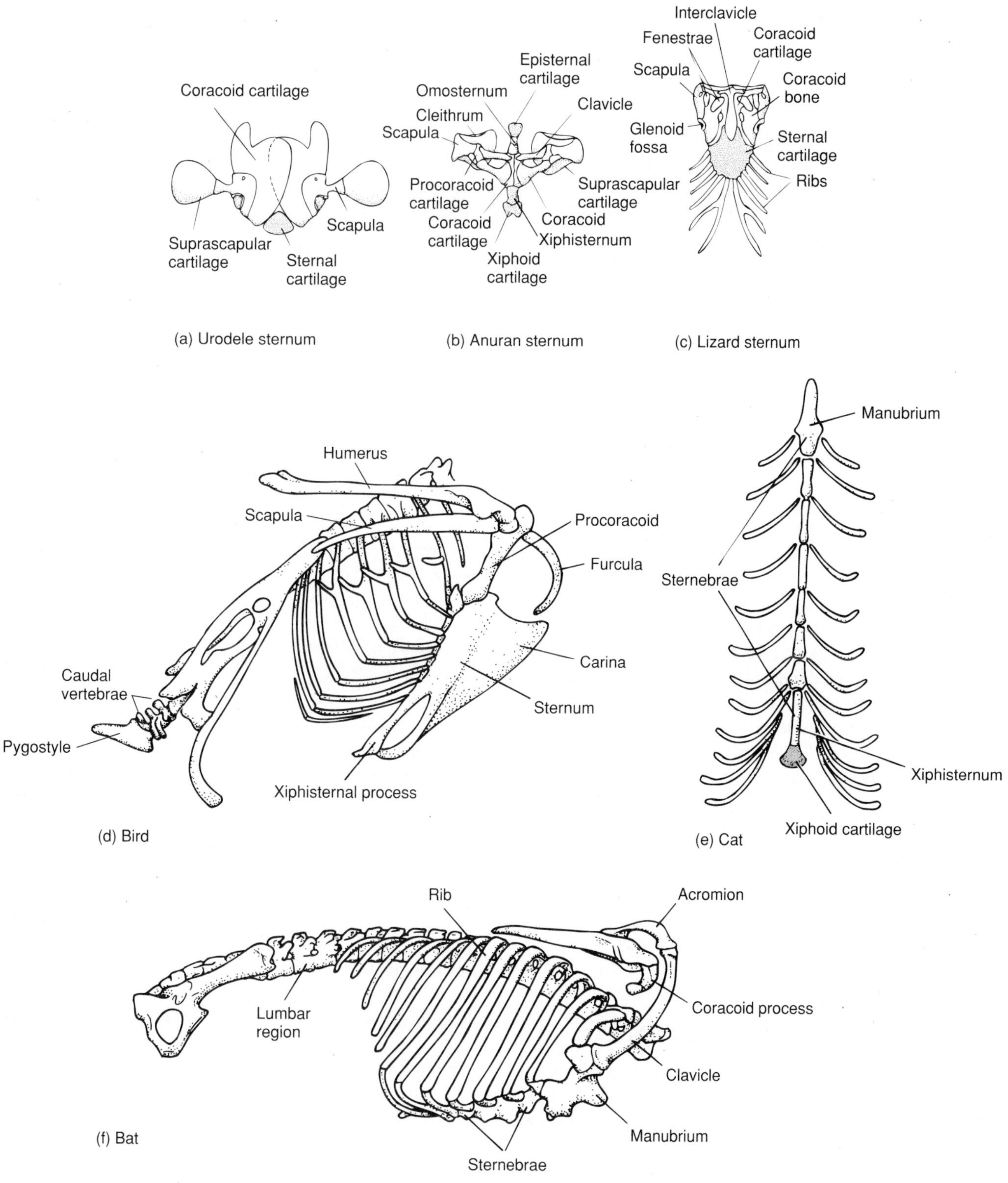


**Figure 8.8** Tetrapod sterna. (a) Urodele, ventral view. (b) Anuran, ventral view. (c) Lizard sternum, ventral view. (d) Bird sternum, lateral view. In birds, the sternum is deeply keeled, forming a carina that offers increased attachment area for enlarged flight muscles. Within the axial column, the tail is short, ending in a specialized pygostyle that supports a fan of tail feathers; the pelvic bones and many of the vertebrae are fused; and the shoulder is braced by the large procoracoid. In bats, the sternebrae are robust and fused. Within the axial column, the lumbar region and neural spines are short; the ribs are broad; and the coracoid process and clavicle are large, reflecting enhanced roles in flight. (e) Mammal (cat) sternum, ventral view. (f) Bat sternum, lateral view.

stability on weight-bearing girdle elements. In flying birds, the massive flight muscles arise from a large sternum that bears a prominent ventral keel, the **carina.** The carina provides additional surface for muscle attachment (figure 8.8d). In most mammals, the sternum consists of a chain of ossified elements in series, the **sternebrae** (figure 8.8e,f). The first and last of these are often modified and called the **manubrium** and **xiphisternum,** respectively.

Thus, a sternum occurs in some modern amphibians, birds, mammals, and archosaurs. However, its absence in the common ancestors to these groups means that it has arisen independently several times within the field of the midventral connective tissue.

## Gastralia

Posterior to the sternum in some vertebrates is a separately derived set of skeletal elements, the **gastralia,** or abdominal ribs (figure 8.2). Unlike the sternum and unlike ribs, the gastralia are of dermal origin. Gastralia are restricted to the sides of the ventral body wall between sternum and pelvis and do not articulate with the vertebrae. They are common in some lizards, crocodiles, and *Sphenodon,* serving as an accessory skeletal system that provides sites for muscle attachment and support for the abdomen.

Within turtles, the **plastron** is a composite bony plate forming the floor of the shell (figure 8.9a–c). It consists of a fused group of ventral dermal elements, including contributions from the clavicles (epiplastrons) and interclavicle (entoplastron) as well as dermal elements from the abdominal region (possibly the gastralia). Such ventral dermal bones are usually absent in birds and mammals, but in many fishes, bones form within the dermis of the belly region. In fishes and other vertebrates, the dermis exhibits a ready potential for producing skeletal derivatives like the gastralia independently in different phylogenetic lineages. Because of such multiple but independent derivations from the dermis, it is perhaps best to restrict the term *gastralia* to rib-shaped elements in the abdominal region rather than to apply this term to all abdominal dermal bones.

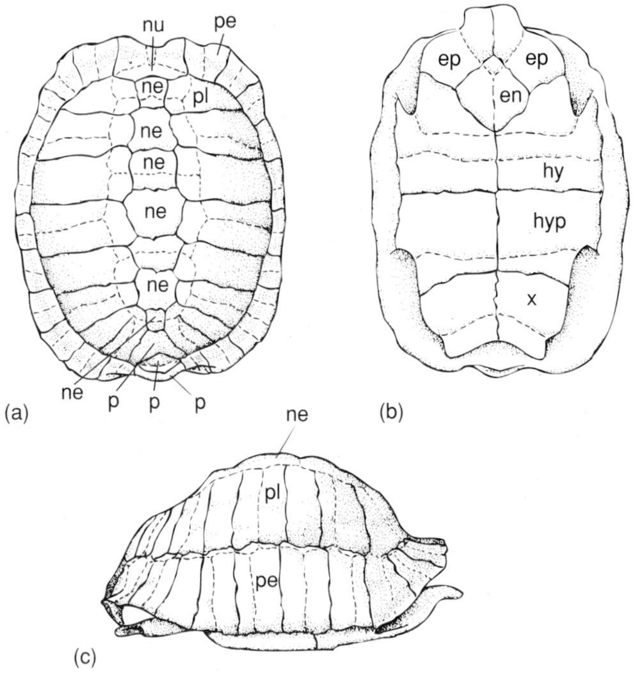


**Figure 8.9** Shell of the tortoise *Testudo.* Solid lines indicate sutures between bony plates; dashed lines represent outlines of more superficial epidermal scales. (a) Dorsal view of convex plastron. (b) Ventral view of flat plastron. (c) Lateral view of whole shell. Carapace consists of numerous peripheral plates (pe) along the margin, eight paired pleural plates (pl), and a single nuchal plate (nu) followed by a series of neural plates (ne) down the dorsal midline, ending with three pygal plates (p). The three plates at the anterior margin of the plastron represent the epiplastron (ep), or paired clavicles, and the entoplastron (en), or the single interclavicle. The remaining plastral plates, the hypoplastrons (hy, hyp) and the xiphiplastron (x) may represent elements of the sternum or gastralia that have become incorporated within the shell.

# Embryonic Development

For the most part, vertebrae arise embryologically from mesenchyme. In some groups, the steps in development have been abbreviated, amended, or deleted, and this has complicated the interpretation of developmental events. Homologous structures and parallel developmental events cannot always be easily determined. Consequently, interpretation and accompanying terminology vary greatly. Without becoming embroiled in the esoteric parts of these controversies, let us see what we can state with some confidence.

In the trunk and tail of tetrapods, the segmental chains of somites each become internally subdivided into discrete layers of cells. Laterally, somites form the dermatome, beneath it, the myotome, and medial to both, the sclerotome. In most tetrapods, vertebrae arise from streams of cells that depart from this inner sclerotome (figure 8.10a). These cells migrate inward toward the midline and cluster along the sides of the notochord. Here, next to and around the notochord, they form a coat or **perichordal tube** (around and notochord). Although more or less continuous, this perichordal tube varies in thickness. It is thin-walled at some sites and thick at other sites. Enlarged sections of the perichordal tube are called **perichordal rings,** arranged serially but slightly out of phase with the segmental musculature derived from the neighboring myotome (figure 8.10b). As a consequence, the deep perichordal rings come to lie between the adjacent muscle segments forming in the outer body wall. Vertebrae arise within these perichordal rings (figure 8.10c). Because these perichordal rings are offset from the myotomes, the vertebrae that differentiate within the rings are also offset from the muscles. Thus,

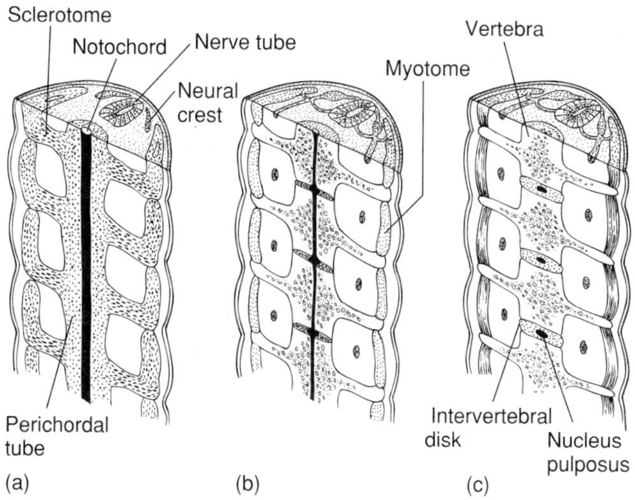


**Figure 8.10** Development of vertebrae in a generalized mammal. (a) Cells in the segmentally arranged sclerotome (itself derived from the somite) stream inward to surround the embryonic notochord where they collect as the perichordal tube of uneven thickness around the notochord. (b) Within the perichordal tube, massed cells differentiate into cartilaginous elements known as perichordal rings. (c) Ossification produces individual vertebrae with intervertebral disks between them. The notochord persists only as the fluid center, or nucleus pulposus, within each intervertebral disk. Muscles that differentiate from the myotome (also derived from the somite) run between successive vertebrae.

the musculature forms *across* adjacent vertebrae, rather than within the same vertebra. In this way, muscles act on adjacent vertebrae in a suitable functional position. If a muscle is attached to only one vertebra, the muscle would, of course, have no functionally significant role. Although this is the general vertebrate pattern, there are some departures from or developmental additions to these steps.

Somite differentiation (p. 160)

Among elasmobranchs and many primitive bony fishes, the inward streams of cells arriving from the sclerotomes first congregate in discrete clusters and differentiate into paired cartilages rather than into ossified vertebrae directly. Up to four pairs of cartilages are formed per segment. The embryologist Hans Gadow called these paired cartilages **arcualia.** The developmental fate of each arcualium could be charted from embryo to adult, and its specific contribution to the adult vertebra identified (figure 8.11). Although such developmental steps commonly occur in elasmobranchs and many primitive bony fishes, discrete cartilages do *not* always appear in later groups and certainly are always absent in tetrapods. Although absent in these derived groups, Gadow nevertheless proposed that arcualia were the underlying pattern in all later groups. He attributed their absence in amniotes to developmental shortcuts, reductions, and elimination of intervening developmental steps. However, this seems to force an interpretation on vertebral formation that does not fit tetrapods and is even inaccurate in teleosts. Let us look at the events of vertebral formation in teleosts and tetrapods to see the departures from Gadow's theory.

In most teleosts, embryonic formation of vertebrae proceeds in three steps. First, the sheath of the notochord itself differentiates into a chain of cartilaginous elements, the **chordal centers** (or chordacentra; figure 8.12a). Between successive chordal centers, the undifferentiated notochordal sheath is destined to become the intervertebral ligament between vertebrae of the adult. Second, local mesenchyme condenses at the level of the myosepta. These condensations become cartilaginous anlagen called **arch centers** (arcualia by some accounts) that give rise to the dorsal and ventral arches. Third, cells of sclerotomal origin condense on the surface of the notochordal sheath, forming the perichordal tube that becomes ossified without first passing through a cartilaginous stage (figure 8.12b). As vertebral formation proceeds, the deep chordal centers fuse with their respective perichordal centers on their surfaces. The arch centers often, but not always, fuse with the perichordal tube and ossify (figure 8.12c). Therefore, although arcualia precede and then contribute to vertebrae in elasmobranchs and some primitive bony fishes, this pattern is not strictly followed in derived fishes such as teleosts. In teleosts, cartilaginous anlagen are the source of the arch centers, but the perichordal tube and notochordal sheath, not the arcualia, are the sources of the centra.

In tetrapods, vertebrae do not develop from modified arcualia, not even in part. Tetrapod centra arise from a perichordal tube of mesenchymal origin, not from discrete blocks of cartilage (arcualia). Consequently, Gadow's comprehensive view of a common arcualial pattern underlying vertebral development in all vertebrates is generally not accepted today.

Another view of vertebral development is being challenged as well. It has been claimed that as cells stream out of the segmental **primary sclerotomes** on their way to form arcualia or the perichordal tube, they first regroup (figure 8.13a,b). This cellular regrouping is accomplished by the caudal half of one sclerotome fusing with the cranial half of the next to form resegmented blocks of cells, the **secondary sclerotomes.** These regrouped cells then supposedly move to the notochord to form the vertebral elements (figure 8.13c,d).

Such a view is appealing because it provides a developmental mechanism by which the sclerotome and its myotome, initially in register with one another, become offset or staggered before differentiating into respective vertebrae and muscles (figure 8.13c,d). Experimental embryology has attempted to address this issue of resegmentation. Sclerotomal cells from a chick and a quail were used because each is microscopically distinctive and thus can be recognized. Every other sclerotome was surgically replaced

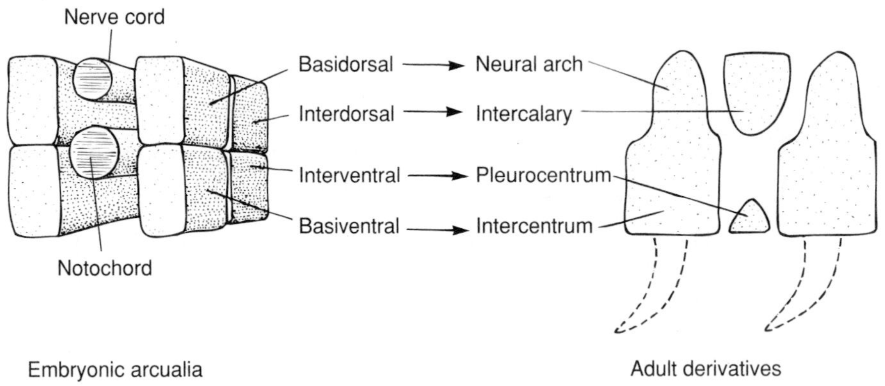


**Figure 8.11** Arcualia. During embryonic development in some primitive fishes, mesenchymal cells that gather around the notochord first form discrete blocks of cartilage, up to four pairs per segment. These blocks are called arcualia. In such fishes, each arcualium can be followed through its subsequent embryology to the part of the vertebra it forms in the adult.

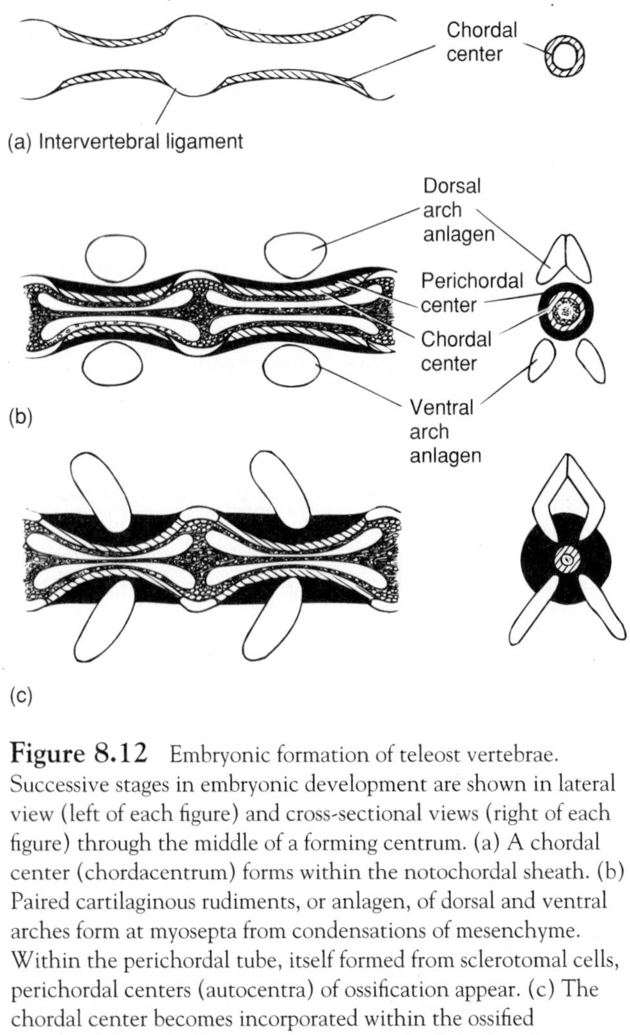


**Figure 8.12** Embryonic formation of teleost vertebrae. Successive stages in embryonic development are shown in lateral view (left of each figure) and cross-sectional views (right of each figure) through the middle of a forming centrum. (a) A chordal center (chordacentrum) forms within the notochordal sheath. (b) Paired cartilaginous rudiments, or anlagen, of dorsal and ventral arches form at myosepta from condensations of mesenchyme. Within the perichordal tube, itself formed from sclerotomal cells, perichordal centers (autocentra) of ossification appear. (c) The chordal center becomes incorporated within the ossified perichordal center, forming the centrum. Arches often, but not always, fuse with the centrum to ossify along with it. The notochord may persist as intervertebral cartilaginous pads, as intervertebral ligaments, and as a constricted notochord running through the centers of the centra.

in chick hosts with a sclerotome from a quail donor before resegmentation. Development was then allowed to proceed normally. The visible differences between chick and quail cells made it possible to determine the contribution of each to the resulting vertebrae. In these experiments, individual vertebrae contained both chick and quail cells, suggesting that initially alternate chick-quail sclerotomes did resegment prior to differentiation into vertebrae.

On the other hand, other scientists working with serial sections of developing vertebrae from mammals claim that no major resegmentation occurs and that cells migrating from the sclerotome move obliquely to congregate and take up residence between myotomes without any prior regrouping. The differences in results may reflect differences in techniques or in species, but for the moment, we cannot be certain if significant resegmentation occurs and if it does how prevalent this pattern might be.

# Phylogeny

## Fishes

### *Agnathans*

A notochord, rather large and prominent, is usually present in ostracoderms. Vertebral elements, however, are rare. Only among heterostracans, osteostracans, and galaeaspids have trace impressions of vertebral elements been observed in fossil specimens. These elements probably are small unossified pieces of vertebrae surrounding a prominent notochord. Thus, among ostracoderms, a strong notochord provided the central mechanical axis for the body.

Among living hagfishes and lampreys, the situation is similar. Hagfishes possess a prominent notochord but lack any hint of vertebral elements. Lampreys possess vertebral elements, but these are small, cartilaginous elements resting

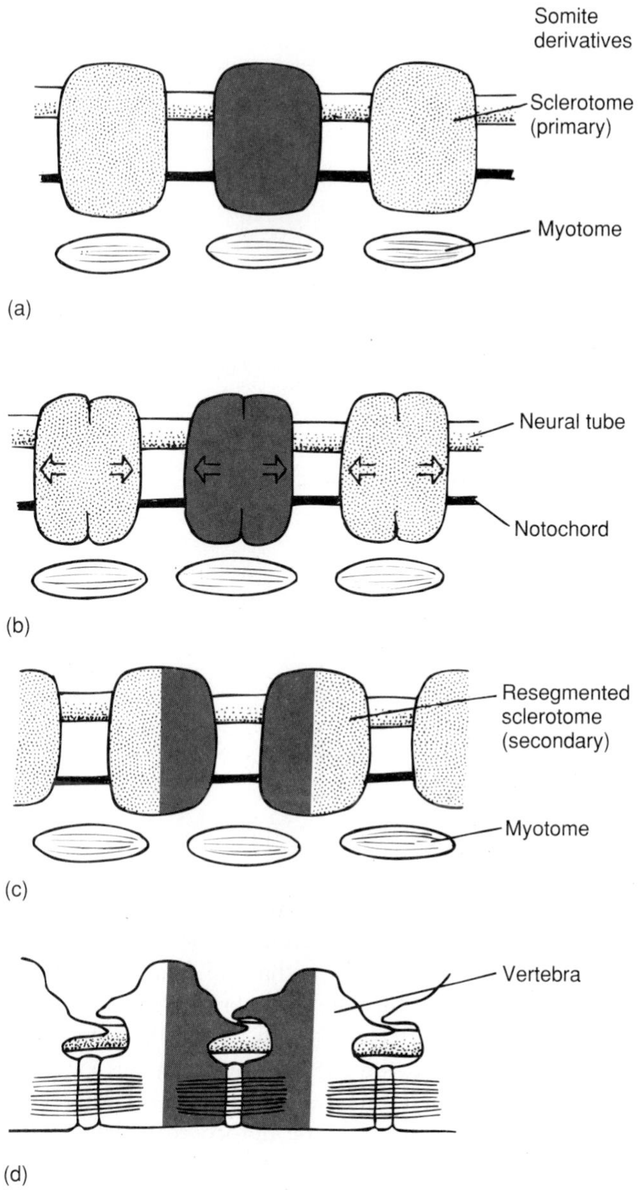


**Figure 8.13** Sclerotomal resegmentation hypothesis. Lateral views of developing vertebral column. (a) Sclerotomes and accompanying myotomes are segmental derivatives of somites. (b) Before differentiating into vertebrae, sclerotomes split and adjacent halves join to form resegmented sclerotomes. (c) These resegmented sclerotomes now are offset from the myotomes. (d) Resegmented sclerotomes differentiate into vertebrae, and myotomes differentiate into associated muscles that run between adjacent vertebrae.

dorsally upon a very prominent notochord that gives axial support for the body (figure 8.14a,b).

## *Gnathostomes*

***Primitive Fishes*** In most primitive fishes, the axial column consisted of a prominent notochord. There is no evidence of vertebral centra, although dorsal and ventral arches were usually present. Primitive chondrichthyans exhibited such a pattern. A prominent notochord provided axial support (figure 8.15a–c), but a vertebral column was represented by only cartilaginous neural and hemal arches. In advanced sharks, these vertebral elements enlarge to become the predominant structural element of the body axis, although the constricted notochord enclosed within the vertebral centra persists (figure 8.15d).

Some placoderms preserve evidence of a prominent notochord supporting ossified neural and hemal arches (figure 8.16a). Fossil impressions of most acanthodians show clear evidence of an ossified series of neural and hemal arches (figure 8.16b). These arches rode upon a prominent notochord. Among palaeoniscoids, the notochord was unconstricted and reached from the skull nearly to the tip of the tail. A series of neural spines lay dorsally along the notochord, and ventral hemal arches accompanied it into the trunk and tail regions.

Among living primitive bony fishes, such as sturgeons and paddlefishes, the vertebral column is unossified, presumably a secondary condition, but several elements of the

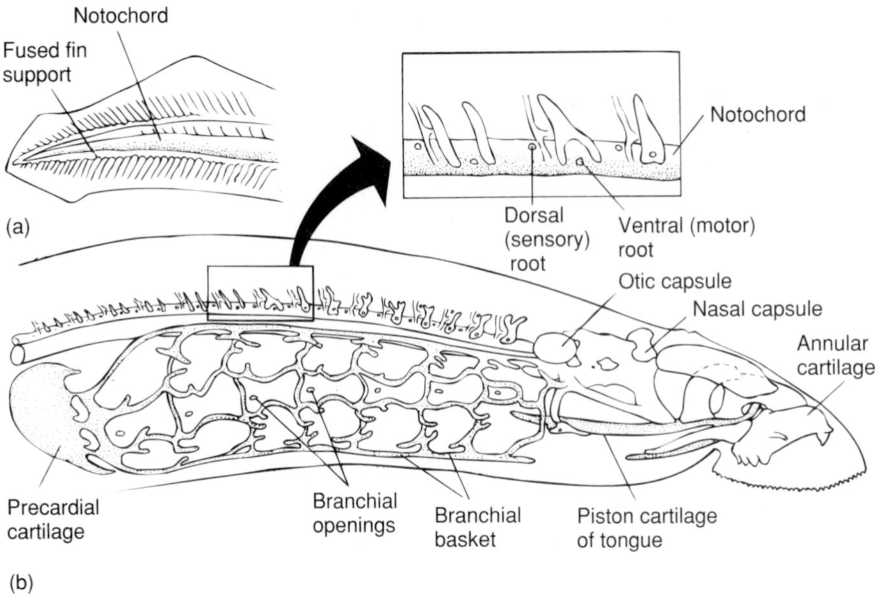


**Figure 8.14** Lamprey skeleton. (a) Enlargement of caudal body section of the lamprey. (b) Anterior end of the lamprey with enlargement of the axial skeleton illustrating the prominent notochord. Note that only a few cartilaginous vertebral elements are present.

vertebrae are present in each segment (figure 8.17a). In more derived bony fishes, such as the bowfin (figure 8.18a–d) and teleosts (figure 8.17b,c), the vertebral column typically is ossified and its centra more prominent to replace the notochord as the major mechanical support for the body. Neural spines and ribs become more developed, as do accessory bony elements that help internally stabilize some of the unpaired fins.

Mechanically, the axial column of fishes represents an elastic beam. Lateral bending movements produced by the body musculature place the column in compression (figure 8.19a,b). Even during peak bursts of speed, the fish's notochord or ossified vertebrae experience stresses well within their capacity to withstand without breaking or collapsing (figure 8.19c). However, when laterally flexed, the vertebral column is in danger of buckling, and its separate vertebrae could become disarticulated if they were too loosely joined (figure 8.19d). The intervertebral ligaments resist this and return stiffness to the vertebral column. Thus, centra seem to function as compression members, and the stiffness that resists buckling is controlled by the degree of lateral flexure permitted by these ligaments between centra (figure 8.19e).

Although compression seems to be the most prevalent force, the axial column in some fishes must be able to resist torsion, the tendency to twist or "wring" the axial column. Torsional forces are especially acute in fishes with asymmetrical tails, where one lobe is quite long. In these fishes, oscillation of the asymmetrical tail produces desirable lift but also tends to twist the axial column, possibly even affecting the trunk vertebrae. In these fishes and in later tetrapods in whom twisting, or **torsion,** places the integrity of the axial column at risk, several features of axial column design seem to address the mechanical demands of torsion. Consolidation of separate vertebral elements into a holospondylous vertebral column of solid vertebrae helps withstand torsional forces. Long neural spines extending over several segments functionally tie together adjacent regions to resist twisting. If the notochord remains prominent, its sheath is often quite thick and invested with bands of fibrous connective tissue oriented in such a way as to resist excessive torsion.

***Caudal Skeleton and Fins*** In most fishes, the axial skeleton continues into the tail, where it can take several forms. In many fishes, the tail is asymmetrical, with a long dorsal and a small ventral lobe separated by a notch. If the posterior end of the vertebral column turns upward and into this dorsal lobe forming its central axis, a **heterocercal tail** forms (figure 8.20). In the **diphycercal tail,** the vertebral column extends straight back, with the fin itself developed symmetrically above and below it. Living lungfishes and bichirs are examples. The **homocercal tail,** characteristic of teleosts, has equal lobes and appears to be symmetrical, but the narrowed vertebral column that runs to its base slants upward to form the support for the dorsal edge of the fin. The hemal arches below expand into supportive struts, known as **hypurals,** to which the rest of the fin attaches (figure 8.20a–c). Among the earliest vertebrates, the tail was not commonly

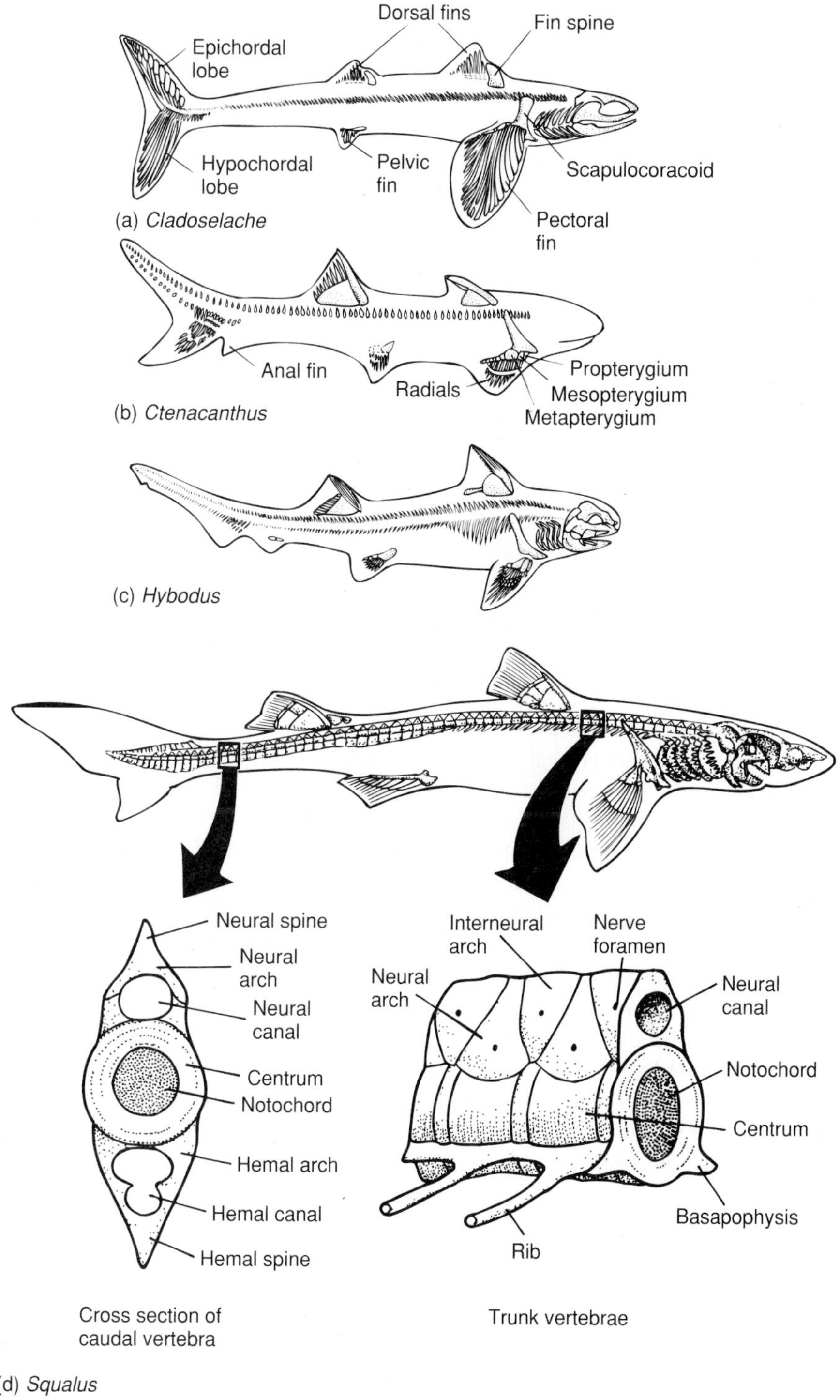


**Figure 8.15** The axial skeleton in sharks and their ancestors. (a) Paleozoic shark *Cladoselache* with a chain of neural arches presumably riding upon a notochord that extended into the tail. (b) *Ctenacanthus* from the late Paleozoic. (c) *Hybodus* from the Mesozoic. (d) Modern shark *Squalus*. The vertebral elements tend to enlarge in elasmobranchs, surpassing the notochord as the major mechanical support for the body in modern forms.

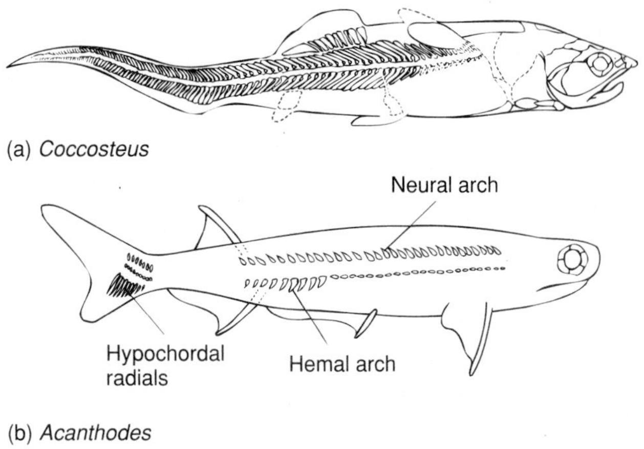


**Figure 8.16** Axial skeleton of primitive fishes. (a) Placoderm *Coccosteus*, with prominent notochord supporting dorsal and ventral vertebral elements. (b) Acanthodian *Acanthodes*, with neural and hemal arches that presumably rode upon a notochord.

symmetrical. Rather, most ostracoderms show the heterocercal condition (figure 8.20a) or even a "reversed" heterocercal condition, termed a **hypocercal tail,** in which the vertebral axis enters the tail and turns down into an extended ventral lobe. The symmetrical diphycercal and homocercal tails (figure 8.20b,c) are usually derived from ancestors with asymmetrical heterocercal tails. They are common among fishes with lungs or air bladders that give their dense bodies neutral buoyancy. In sharks, which lack lungs or air bladders, lift is apparently provided by the extended dorsal lobe of the heterocercal tail.

When the heterocercal tails of sharks are removed and tested separately in experimental tanks, they have a tendency to push downward against the water, resulting in an upward reaction force on the tail, which produces lift (figure 8.21). In fact, removal of the dorsal or the ventral lobe alone reveals that within the tail, lift produced by the two lobes differs in magnitude and direction (figure 8.21b). In general, as the tail sweeps back and forth, the small ventral lobe deflects water upward, causing a small downward component of force, whereas the large dorsal lobe deflects water downward, resulting in an opposite large upward force (figure 8.21b). The overall effect is for the tail to produce a resultant force directed forward and upward. Although at first it might seem strange that the ventral lobe produces forces contrary to the overall upward lift generated by the tail, this action of the ventral lobe might represent a method for fine tuning the lift. In sharks that have just eaten a large meal or in gravid females, the center of body mass might shift unfavorably, tilting or angling the body out of its line of travel. The ventral lobe might help level the shark in a more direct body orientation. In nautical terminology applied to submarines, adjustment for vertical tilt is called "trimming." Small radial muscles reside in the ventral lobe of the shark's tail. Their contraction might alter stiffness, change the forces produced in the tail, help trim the body, and adjust the shark about its center of gravity.

If this interpretation of the function of a heterocercal tail is correct, then the reversed heterocercal tail, the hypocercal tail of ostracoderms, would have produced forces that tilted the body down and drove the mouth into the substrate. This would aid the animal in feeding on foods buried in soft sediments.

***Sarcopterygians*** The notochord continues to serve as the major supportive element within the axial skeleton of sarcopterygians, including the crossopterygian ancestors of early amphibians. In living sarcopterygians, the vertebral column can be rudimentary and cartilaginous (figure 8.22a,b). However, in many early species, such as the rhipidistians, vertebral elements were usually ossified and exhibited a rhachitomous type of aspidospondyly in which each vertebra consisted of three separate vertebral elements, a neural arch, a hoop or crescent-shaped intercentrum, and paired pleurocentrum (figure 8.23a). In the tail, the intercentrum expanded into the continuous hemal arch and spine. Although differing in some details, an aspidospondylous condition occurred in many early rhipidistians, including *Eusthenopteron* (from late in the Devonian) and *Osteolepis* (from the mid-Devonian). In the tail of these early rhipidistians, each aspidospondylous vertebra included the small, paired pleurocentrum and dorsal (neural) and ventral (hemal) arches. In the trunk region, the hemal arch became reduced, and its base expanded into the prominent intercentrum (figure 8.23b,c). The segmental myosepta mark the borders of the earlier embryonic segments and attach to the neural arch and intercentrum (hemal arch) medially in the adult. In later and usually larger rhipidistians, fusion of central elements produced a derived aspidospondylous condition in which each segment consisted of a single centrum of ringlike bone to which the neural arch may or may not have fused.

## Tetrapods

### *Amphibians*

The vertebrate transition to land brought to bear considerable changes in the selection pressures acting on design. As animals evolved from water to air, their bodies went from a buoyant support design to a design in which bodies were suspended between limbs. All systems, including respiration, excretion, and body support, were affected. Changes in the axial skeleton are especially indicative of these new mechanical demands.

The taxonomic division of the earliest amphibians into two groups, lepospondyls and labyrinthodonts, is

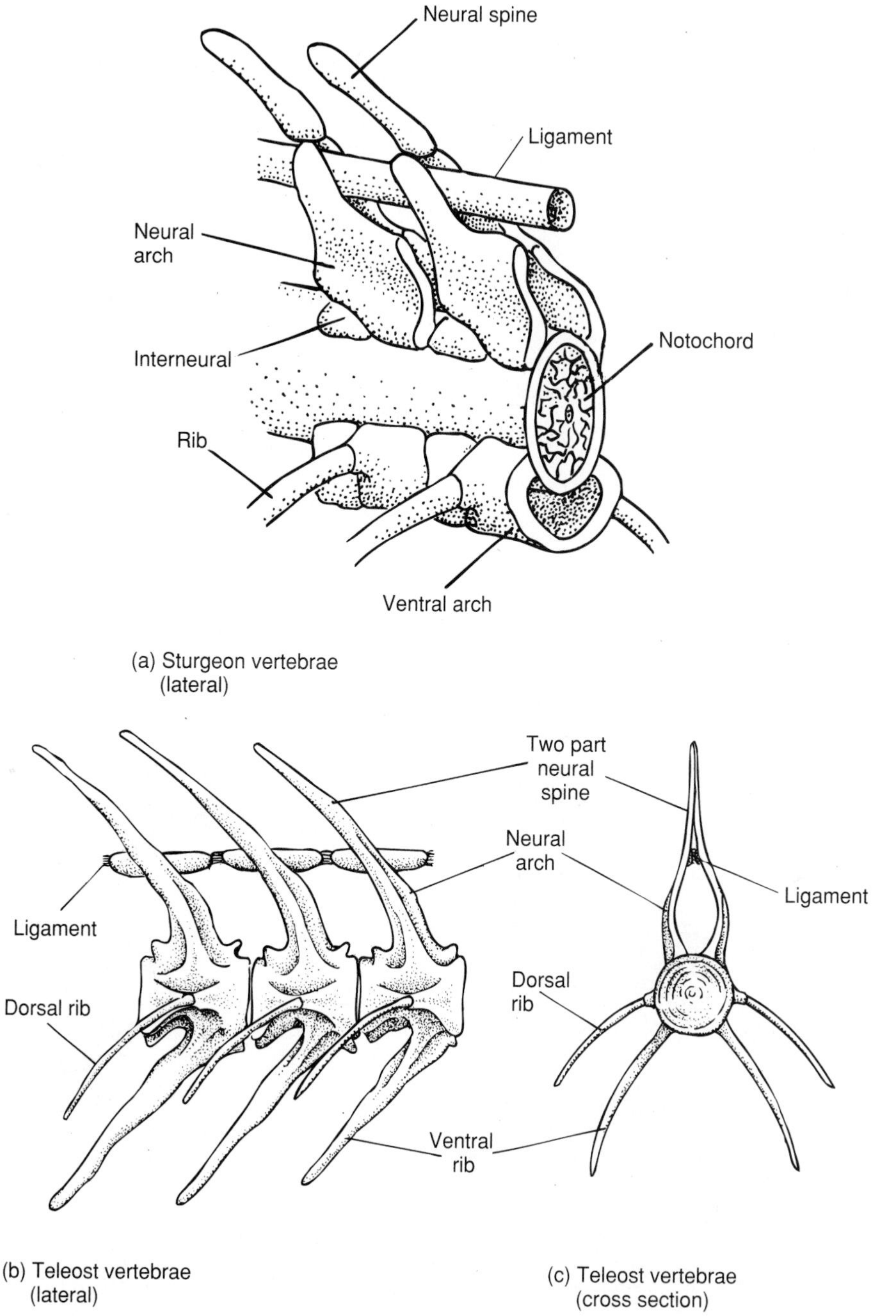


**Figure 8.17** Actinopterygian vertebrae. (a) Sturgeon vertebrae, lateral view. (b) Teleost vertebrae, lateral view. (c) Teleost vertebra, cross section.

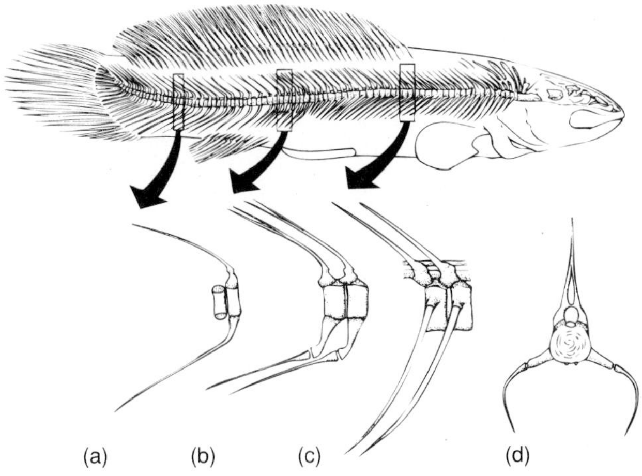


**Figure 8.18** Axial skeleton of the bowfin *Amia calva*. (a–c) Representative lateral sections of the vertebral column. (d) Cross section of a trunk vertebra. Note the predominance of ossified vertebrae.

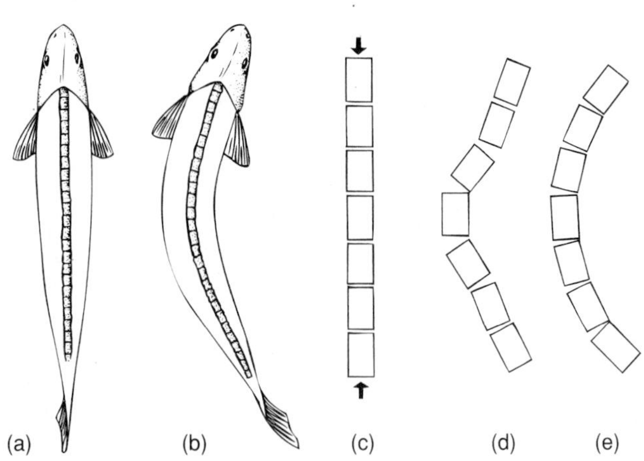


**Figure 8.19** Function of amphicoelous vertebrae in a teleost. (a,b) Swimming involves development of lateral flexion of the vertebral column induced by contractions of the body musculature. (c) Chain of vertebrae shown under axial loads. Even during maximum bursts of speed, the ossified vertebrae are strong enough to withstand maximum compressive loads. (d) When flexed, the chain of vertebrae might buckle and fail. (e) Firm intervertebral ligaments that resist buckling return stiffness to the vertebral column.

Based on the research of J. Laerm, 1976.

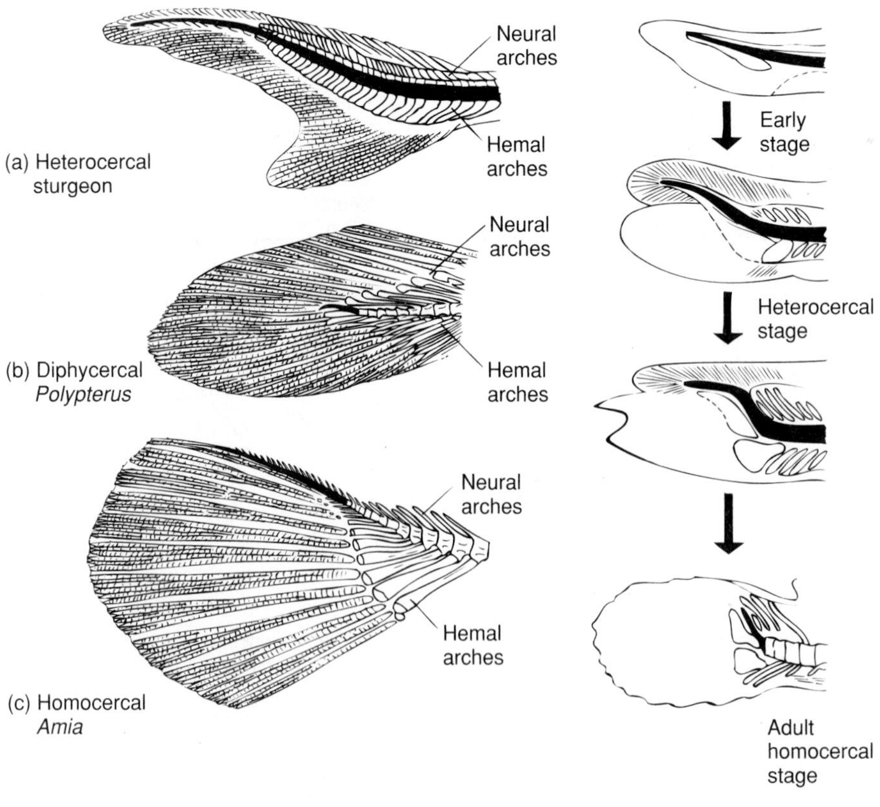


**Figure 8.20** Caudal fins of fish. (a) Sturgeon. (b) Bichir *Polypterus*. (c) Bowfin *Amia*. Note the positions of the vertebral column and the conditions of the remaining notochord. Sequence leading to the heterocercal tail is shown to the right of each figure.

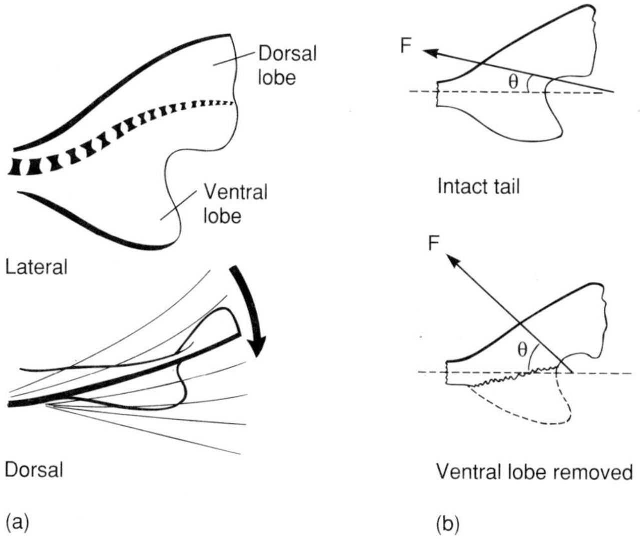


**Figure 8.21** (a) Tail of a Port Jackson shark as seen from the side and from above, moving in the direction of the arrow. The vertebral column extends into the dorsal lobe. The thick black lines indicate the stiff edges that lead the more flexible parts of the lobes that lag behind. Because of this bend, the dorsal lobe produces a large upward force and its ventral lobe produces a small downward component. (b) Without the ventral lobe, the thrust is inclined at a greater angle (Θ) with the body axis. Resultant force of intact tail (top) and tail with the ventral lobe removed (bottom).

Modified from J. R. Simons, 1970.

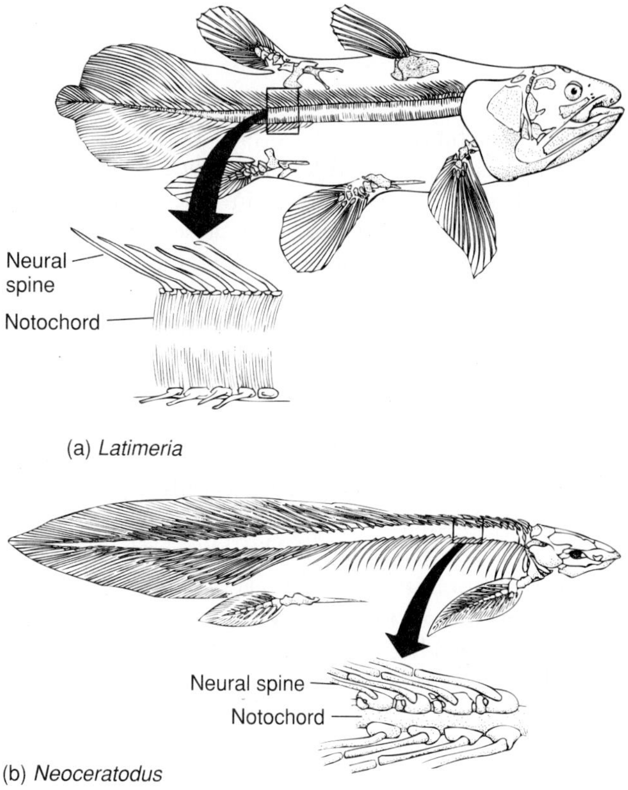


**Figure 8.22** Axial skeletons of living sarcopterygians. (a) Enlarged lateral view of posterior axial skeleton of the coelacanth *Latimeria*. (b) Enlarged lateral view of trunk vertebrae and notochord of the lungfish *Neoceratodus*.

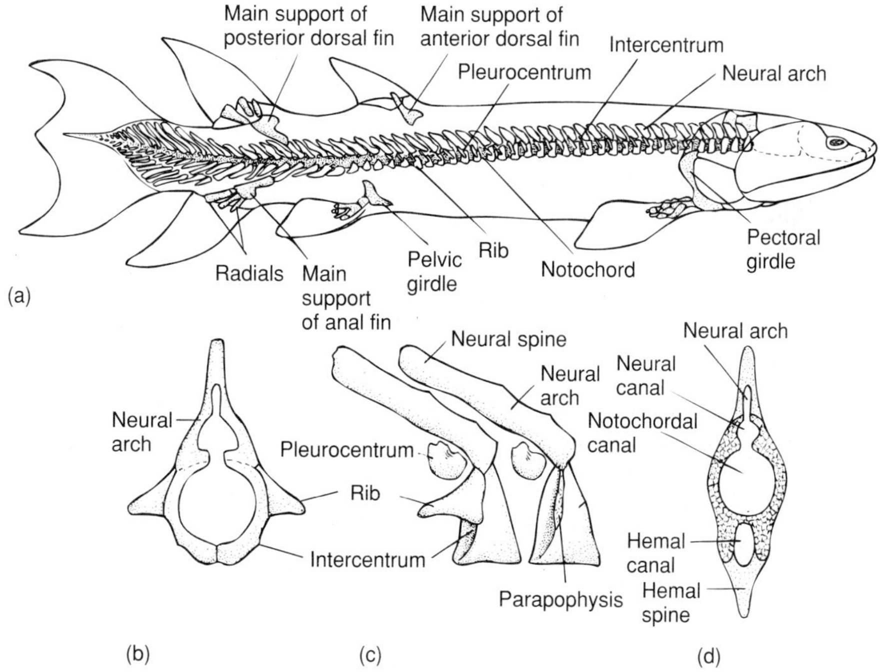


**Figure 8.23** Axial skeleton of the fossil crossopterygian *Eusthenopteron*. (a) Restored axial skeleton. Cross section (b) and lateral views (c) of trunk vertebrae. (d) Cross section of a caudal vertebra.

largely inspired by the two vertebral types found in Paleozoic amphibians (figure 8.24). Lepospondyls are named for their distinctive type of holospondylous vertebra, termed a lepospondylous vertebra, in which the vertebral elements are fused. They presumably arose from ancestors with aspidospondylous vertebrae. Thus, the single, solid vertebra typical of lepospondyls represents the fusion of vertebral elements that were separate originally.

Most lepospondyls had long, deep tails, suggesting that they, like modern salamanders, were swimmers. Modern amphibians also have a vertebral column composed of single, solid vertebrae at each segment, suggesting that they might have evolved from these early lepospondyls. However, a silent gap in the fossil record extends from the last lepospondyls (in the Permian) to the first frogs (in the early Jurassic) or first salamanders (late Jurassic), almost 40 million years without fossils to connect modern amphibians and late lepospondyls confidently. Their similar vertebrae may reflect convergence of morphological design to parallel functional roles in swimming. Consequently, vertebrae of solid construction may have been derived independently in one or all groups of modern amphibians.

Labyrinthodonts evolved directly from rhipidistians, taking over their aspidospondylous type of vertebra as well. The characteristic mode of fish progression in which locomotion depends on lateral waves of undulation in the vertebral column has been retained in modern salamanders and was probably present in early amphibians as well (figure 8.25a–c). Swimming in most fishes depends on the production of lateral body bends that sweep posteriorly as traveling waves that push the sides of the fish against the surrounding water (figure 8.25a,b). These traveling waves produce lateral undulations of the fish body and are also the basis for terrestrial locomotion in salamanders and even most reptiles. Synchronized with these lateral body swings are limb movements that lift and plant the foot to establish points of rotation about which the tetrapod vertebral column undulates (figure 8.25c).

Along with the vertebrae, such lateral body undulations were carried forward from fish ancestors into early labyrinthodont amphibians as well, constituting the basic mode of locomotion of early labyrinthodont amphibians walking on land. What was mechanically new in this early mode of terrestrial locomotion was a tendency to twist the vertebral column, placing it in **torque.** Without surrounding water to support the body and with planted feet establishing pivot points, walking on land placed new torsional

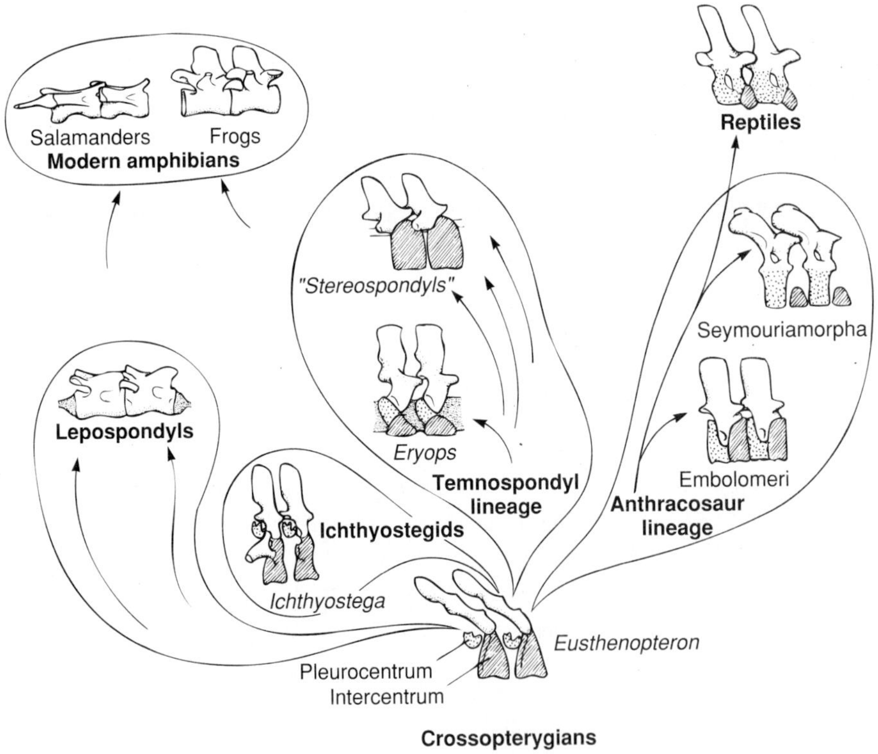


**Figure 8.24** Evolution of tetrapod vertebrae. The lepospondyl condition appeared early. It may have given rise to modern amphibians, but more likely it was restricted to the lepospondyls, the taxonomic group named for this vertebral type. The solid vertebrae of modern amphibians probably arose independently. The rhachitomous vertebra, inherited from crossopterygian fishes, evolved along two major lines, temnospondyl and anthracosaur. In the temnospondyl line, the intercentrum enlarged at the expense of the pleurocentrum. In the anthracosaur line, however, the pleurocentrum came to predominate.

**Figure 8.25** Lateral undulatory locomotion from fishes to tetrapods. Lateral swimming motions of fishes are incorporated into the basic pattern of terrestrial locomotion of primitive tetrapods. (a) Side to side sweeps of the body of an eel exert a force against the surrounding water as the fish travels forward. (b) Similar lateral undulations of a shark's body push against the water and drive the fish forward. (c) Lateral undulations of a salamander do not press the body against its terrestrial surroundings. But these undulations serve to advance each foot forward, plant it, and then rotate the body about this point of pivot for locomotion on land.

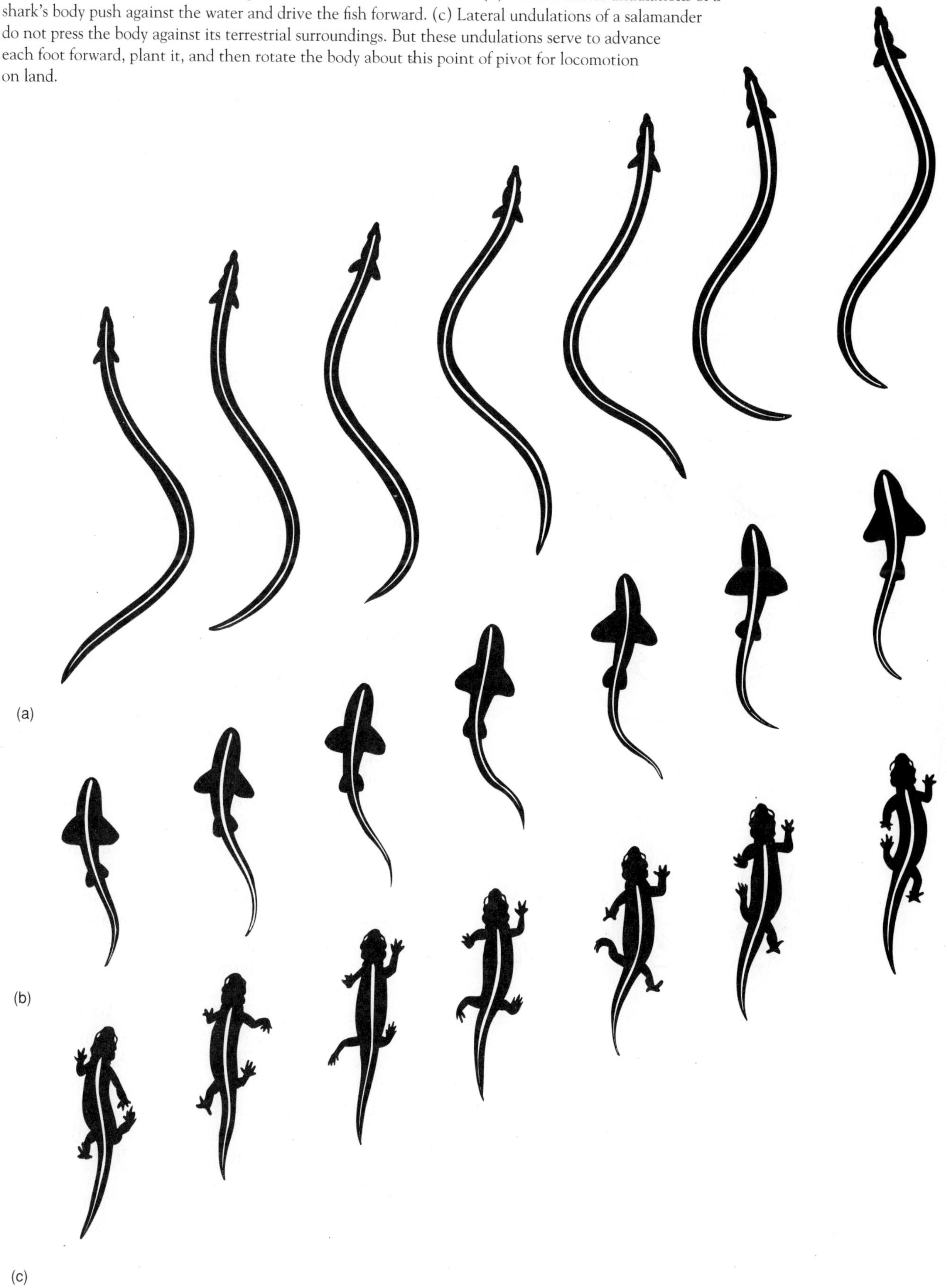

stresses on the vertebrae. Several features of the design of early amphibian vertebrae can be interpreted as functional modifications addressing these new stresses.

As in *Ichthyostega,* most early labyrinthodont vertebrae were aspidospondylous. Although this tended to give way to derived conditions in later species, the vertebrae of these early species consisted of separate components applied to a still prominent notochord. Such a loose confederation of bony elements might at first seem ill suited to address the torsional forces introduced in the axial column when amphibians ventured onto land. However, a functional hypothesis incorporating the fibrous nature of the notochord and solid structure of the vertebral elements suggests otherwise. If fibrous bands within the outer sheet of the notochord were wound in opposite spirals crossing at about 45°, they would create a kind of geodetic or warp-and-weft framework resistant to torsional forces (figure 8.26a). Rigid bony pieces of the vertebra might have been so placed as to occupy the spaces between these fibrous bands. Lateral flexions of the body during terrestrial locomotion would bring the edges of these bony vertebral pieces into contact, checking further torque. Up to that point, however, the elastic sheath of the notochord would allow the flexibility required to produce these lateral body bends during locomotion. This functional model depicts the axial column of early labyrinthodonts as consisting of two mechanical components, the notochordal sheath introducing limited flexibility and the hard vertebral elements preventing excessive torque (figure 8.26b,c).

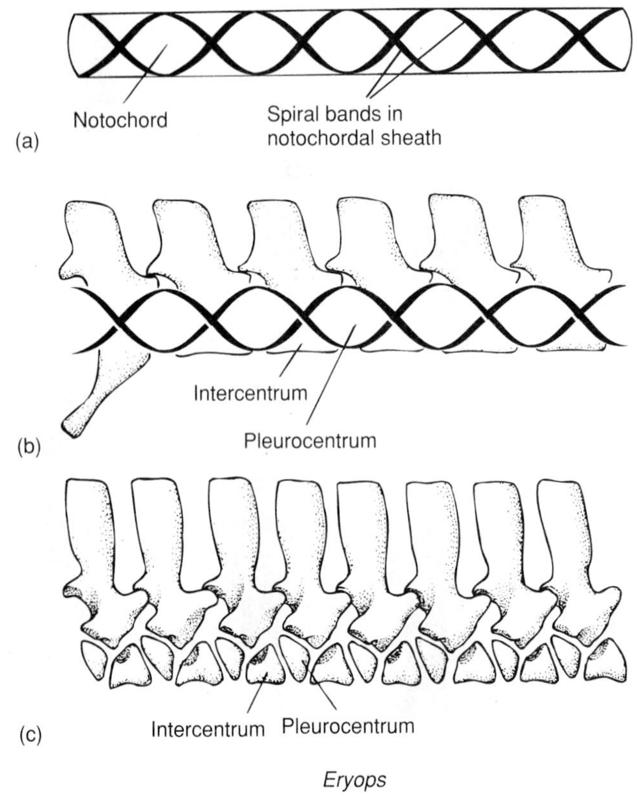


**Figure 8.26** Geodesic model of the axial skeleton of an early tetrapod. (a) Bands of connective tissue wound in opposite directions, each at 45° to the long axis of the notochord, form a geodetic framework that resists both flexion and torque. (b) A rhachitomous vertebral column has much the same pattern, shown by dashed outline and overlaid on the geodetic framework. (c) Vertebral column of *Eryops* wherein the spaces of the geodetic framework are occupied by rigid bone around the notochord, permitting controlled flexion that resists torque.

Two subsequent lineages of labyrinthodont radiation occurred and were characterized by differences in the relative prominence of each vertebral centrum. In the **temnospondyl lineage,** the intercentrum became predominant. In the **anthracosaur lineage,** the pleurocentrum became predominant. In both, the notochord became reduced as respective centra enlarged to assume the central role in axial support.

In the early temnospondyls, the vertebrae were rhachitomous, a type of specialized aspidospondylous vertebra. This type consisted of a neural arch, a crescent-shaped intercentrum below the notochord, and a pair of bony pleurocentra above the notochord. However, in later temnospondyls, the intercentra became preeminent, large, and completely ossified cylinders upon which the neural arch rested. Conversely, the pleurocentrum became much reduced or was lost entirely. The term *stereospondyl* has in the past been applied to these later temnospondylous amphibians. This was based on the view that all shared a common ancestry because all shared a common vertebral design (prominent intercentrum). However, it now seems more likely that these late temnospondyls evolved independently from separate earlier groups and that the similar vertebral designs are instead a consequence of convergent evolution.

Within the anthracosaur lineage, the opposite centrum, the pleurocentrum, enlarged. Initially, in aquatic anthracosaurs, the pleurocentrum was about the same size as the intercentrum; subsequently, in terrestrial anthracosaurs, it came to predominate.

Vertebral evolution within these labyrinthodont lineages poses several questions. Why, for instance, does the intercentrum predominate in one lineage (temnospondyls) and the pleurocentrum in the other (anthracosaurs)? Or we might ask, why in *both* lineages does one centrum become predominant at all? Unfortunately, the relationship between vertebral structure and function remains poorly understood, so let us start with what is known.

With continued commitment by early amphibians to life on land, consolidation of the separate vertebral elements occurred in both lineages, leading to vertebrae composed of a single predominant centrum. In the two lines of labyrinthodont evolution, different elements are reduced, but the functional advantages are equivalent, namely increased strength. Locomotion on land imposed significantly greater weight-bearing stresses on the axial column. Thus, terrestrial locomotion required a vertebral column characterized by firmness and strength to suspend and support the body. Enlargement of the ossified centra at the expense of the notochord brought firm support to the body. Enlargement of one centrum at the expense of the other has the overall effect of reducing the number of centra per segment from two to one. This reduces flexibility, firms the axial column, and therefore increases its ability to support the weight of the body on land. Conversely, the more centra per segment, the greater the flexibility of the vertebral column, an advantageous design for an aquatic organism employing lateral flexions of its vertebral column during swimming.

Radiations in both temnospondyls and anthracosaurs included reinvasion of aquatic habitats as well as entry into semiaquatic and terrestrial habitats. The vertebral column, central to locomotion, was consequently as varied as the emerging amphibian lifestyles. If swimming was favored, the trunk and tail regions were usually elongated and the number of vertebrae increased. This was especially true in the later embolomeres, a group of anthracosaurs that apparently returned secondarily to swimming as the primary mode of locomotion, and in the early lepospondyls, a group that seems to have been swimming specialists from their first appearance. However, in early temnospondyls, such as *Eryops*, and later anthracosaurs, such as a *Seymouria*, emphasis was on terrestrial locomotion. This was accompanied by a reduction in the number of vertebrae, extensive vertebral ossification, centra enlargement, reduction of the notochord, and greater overall firmness of the vertebral column.

It is not known why the intercentrum (temnospondyls) or the pleurocentrum (anthracosaurs) came to predominate. Simple chance events in the independent pathways of evolution might have tipped the advantage different ways on two occasions. However, it is more likely that the differences reflect functional differences in the two lines of labyrinthodont evolution. Increasing prominence of the intercentrum in temnospondyls may have been favored by their emphasis upon aquatic locomotion. Of the two centra, the intercentrum was more closely associated with axial muscles and ribs that served swimming. Enlargement of the intercentrum might then have accompanied the increased functional demands of aquatic locomotion. On the other hand, increasing prominence of the pleurocentra in anthracosaurs and later reptiles might have been favored by an opposite trend toward terrestrial locomotion. Pleurocentra supported neural arches, successively interlocked through their zygapophyses, that became more important with increased load-bearing function. Thus, enlarged pleurocentra might then have accompanied enlargement of their associated neural spines and zygapophyses as these came to play more prominent mechanical roles during terrestrial locomotion.

Certainly one of the vertebral innovations of tetrapods were these zygapophyses seen first in amphibians. Terrestrial vertebrates faced a new mechanical problem, a tendency for excessive twisting of the vertebral column. In fishes, the axial skeleton receives more or less continuous and even support along its entire length, whereas in tetrapods, only two pairs of points, the fore- and hindlimbs, provide support. As opposite feet plant themselves on a surface to establish points of support during locomotion, the intervening vertebral column is wrung or twisted, placing shearing stress on the fibrous connections between successive vertebrae. The bony zygapophyses reach across these vertebral joints to interlock in gliding articulations. They are oriented to allow bending in a horizontal or vertical plane, but they resist twisting. In snakes, where twisting forces might be even greater because they are legless, additional sets of zygapophyses, anterior **zygosphene** and posterior **zygantrum,** provide additional checks on torsion but do not significantly restrict normal lateral bending of the vertebral column (figure 8.27a,b).

The other new feature of the axial skeleton that also appears first in amphibians is delineation of a sacral region, the site of attachment of the pelvic girdle to the vertebral column. The earliest amphibians show such a region. Presence of a sacral region joining pelvic girdle and vertebral column is taken as evidence that direct transfer of propulsive forces in the hindlimbs to the axial skeleton became an important component of the terrestrial locomotor system very early in tetrapod evolution.

Other changes in the axial skeleton, related to extended exploitation of land, are evident for the first time in amphibians as well. Connection between the pectoral girdle and back of the skull was lost. This occurred in *Ichthyostega*, for example. Accompanying this loss was the redesign of the first vertebra that became a cervical vertebra, allowing greater freedom of head rotation upon it. For early amphibians, life on land meant that the lower jaw rested on the ground. Opening the jaws required lifting of the head because the lower jaw could not be dropped. Uncoupled from the pectoral girdle, the head could be lifted without restraint or interference from the shoulder. This uncoupling, along with the appearance of a cervical vertebra, allowed the amphibian to turn its head to one side without reorienting the rest of its entire body. Further, when the head was uncoupled from the pectoral girdle, it experienced less jarring as the feet struck the ground during terrestrial locomotion. This was advantageous because the head carried most of the sensory organs.

### *Amniotes*

Amniotes phylogenetically receive their vertebra from the anthracosaur line, so their major centrum is a pleurocentrum and the small centrum, an intercentrum. In many reptiles and birds and in all mammals, the intercentrum is usually lost being remembered only by the rib's capitulum that still articulates between vertebrae where the intercentrum would occur. In some amniotes, the intercentrum contributes to parts of the cervical vertebrae. Most specialized textbooks, such as those on human anatomy, trouble

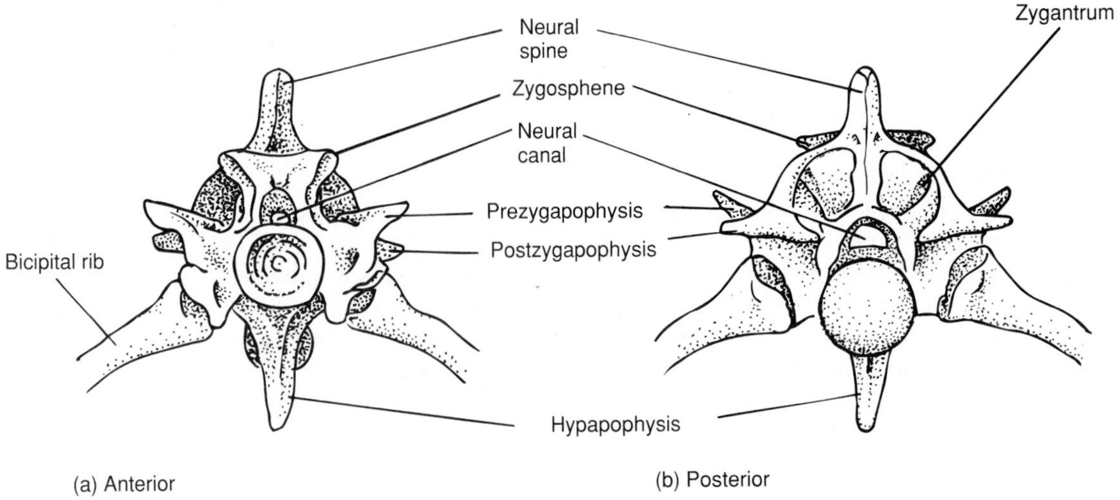


**Figure 8.27** Trunk vertebrae from a snake in anterior (a) and posterior (b) views. In addition to interlocking pre- and postzygapophyses, snakes have an additional set of processes, the zygosphene and zygantrum, that engage to further prevent wringing of the long serpentine vertebral column.

themselves only to call the surviving pleurocentrum just the "centrum" or sometimes just the "body" of the vertebra, a reference to its fused unity with the neural spine. After this long, thought-provoking, and intriguing evolutionary history, such a bland name is unequal to the pleurocentrum's phylogenetic service.

In amniotes, two cervical vertebrae develop, an apparent answer to the problem of maintaining bony strength while retaining cranial mobility (figure 8.28a–g). The first cervical vertebra is the **atlas,** the second, the **axis** (figure 8.28f,g). Vertical (nodding) and horizontal (tilting) movements of the head are largely limited to the skull-atlas joint, whereas twisting movements occur largely within the atlantoaxial joint. This divides the labor between two joints yet maintains bony strength in the neck.

In turtles, the shell into which the limbs and head retreat is a composite unit made of expanded ribs, vertebrae, and dermal bones of the integument that fuse into a protective bony box that harbors the soft viscera (figure 8.29a–c). Turtles are unique in that the appendicular skeleton lies *within* the rib cage rather than on the outside as in all other vertebrates (figure 8.30a,b).

The vertebral column of amniotes is often specialized. In birds, numerous cervical vertebrae have highly mobile heterocoelous articulations between them, giving the skull that rides on this flexible chain of vertebrae great freedom of movement and reach (figure 8.31). At the other end of the vertebral column, the posterior thoracic, lumbar, and sacral vertebrae fuse into a unit, the **synsacrum.** Similarly, adjacent bones of the pelvic girdle fuse into the **innominate** bone, which in turn fuses with the synsacrum (figure 8.32). The overall result is the union of pelvic and vertebral bones into a sturdy but light structure supporting the body during flight.

In mammals, the vertebral column is differentiated into distinct regions. Typically mammals have seven cervical vertebrae, beginning with an atlas and axis that permit the head great freedom of movement. Even the long-necked giraffe and "neckless" whale have seven cervical vertebrae, although exceptions occur in sloths (with six to nine) and sirenians (with six). In armadillos and many jumping mammals such as kangaroo rats, the seven cervical vertebrae may fuse. The number of vertebrae within the thorax and lumbar regions ranges from about fifteen to twenty, and there are usually two or three sacral vertebrae, although humans have five. The caudal vertebrae are quite variable in number. The mammalian tail is much less massive than the reptilian tail. Arches, zygapophyses, and transverse processes diminish toward the posterior tip of the tail so that most caudal vertebrae near the end of the series consist only of centra.

## Form and Function

Most phylogenetic changes in form of the vertebral column address new functions. Transition from water to land was one significant change in vertebrate lifestyle, and it was accompanied by considerable change in the mechanical demands experienced by the axial skeleton. To understand these mechanical forces and their impact on design, we should first compare the general problems faced by aquatic and terrestrial vertebrates.

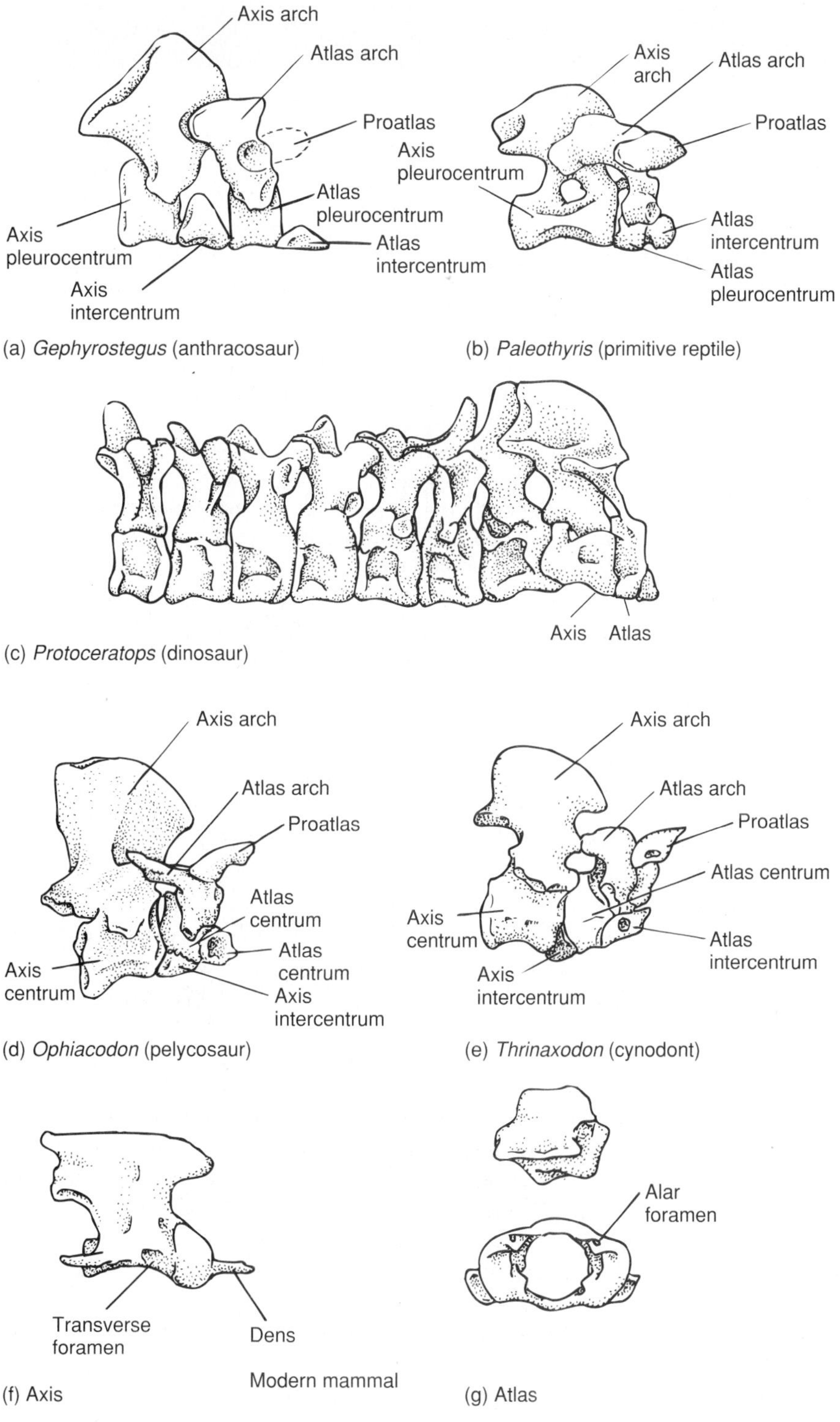


**Figure 8.28** Cervical vertebrae. Fusions and reductions in the first few vertebrae produce the distinctive cervical vertebrae. (a) Anthracosaur *Gephyrostegus*. (b) Primitive reptile *Paleothyris*. (c) Ornithischian *Protoceratops*. (d) Synapsid pelycosaur *Ophiacodon*. (e) Therapsid cynodont *Thrinaxodon*. (f) Axis of a modern mammal. (g) Atlas of a modern mammal.

## Box Essay 8.1 Human Engineering

A vertebrate with one of the most interesting pieces of personal body engineering are humans. Our posture is upright when we walk. In other words, we are bipedal. We depend upon two legs rather than four like our quadrupedal ancestors. Our unusual posture has required some reengineering to restabilize our upright carriage. Few other mammals are built to stand and walk comfortably on two legs. Prairie dogs sit up on their hindlegs, some deer lift up their forelegs, other primates become bipedal for short distances, but humans are built to be comfortable bipeds.

Upright instability comes from two features of our bipedal posture. First, we use half the number of support posts, two limbs, compared with the four of quadrupeds. Second, upright posture places the thorax and much of the rest of our body well above our center of gravity. For these reasons, unusual adaptations to upright posture have been incorporated into our design.

We might imagine that our upright posture is built up of three changes, each elevating the torso in increments of about 30° (box figure 1a–e). First, the front of the body is lifted about 30° (box figure 1a). This change is seen in some primates and can be accomplished without much redesign of the leg muscles. Second, the upper part of the pelvis is tilted back, rotating the vertebral column an additional 30° (box figure 1b). Third, the vertebral column in the lumbar region curves the last 30° to bring the upper body fully upright (box figure 1c).

To accommodate birth of the infant with a relatively large head, the birth canal through the pelvis is expanded by shifting the sacral region further back (box figure 1d). In reaching this upright posture, the lines of muscle action from hips to femur are altered. A broad pelvis is required to accomplish the same spread of muscle orientations and favorable mechanical angles on the femur. Shortening and broadening of the pelvis restores a more advantageous muscle leverage of the gluteal muscles during a striding gait (box figure 1e).

**Birth canal (p. 337); human striding gait (p. 369)**

The column of vertebrae projecting above the hips, like the mast of a ship

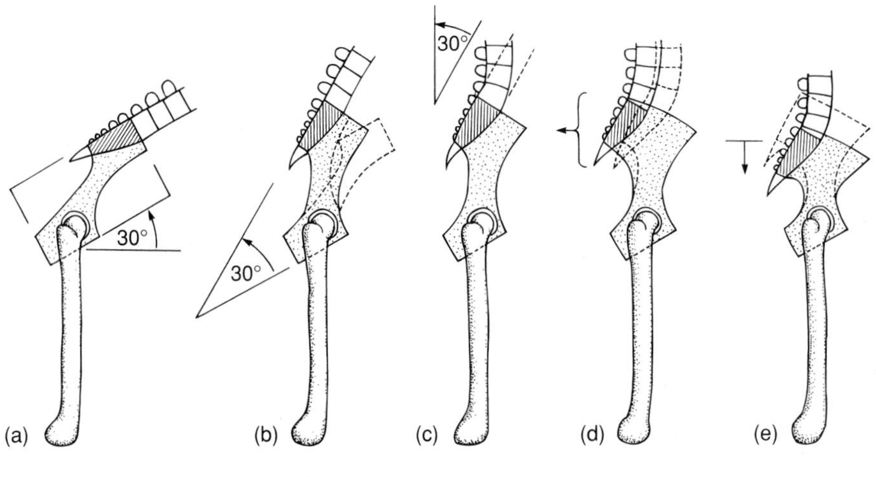


**Box figure 1** Human bipedal posture. Compared with quadrupedal primates (a), human upright posture is engineered by changes in the slant of the pelvic girdle (b), increased curvature in the lower back (c), broadening (d), and then shortening of the hips (e).

projecting above the hull, is stabilized by a system of ligaments and muscles that act like rigging to support the column (box figure 2a). Flaring and shortening the pelvis also broadens the base of support (box figure 2b). If the pelvis had remained tall, the upper body would have been placed well above its balanced support on the femur (box figure 2c). By shortening the distance between the hips and the head of the femur, the weight of the upper body is brought close to and directly above the femur upon which it is balanced and situated in a less precarious position.

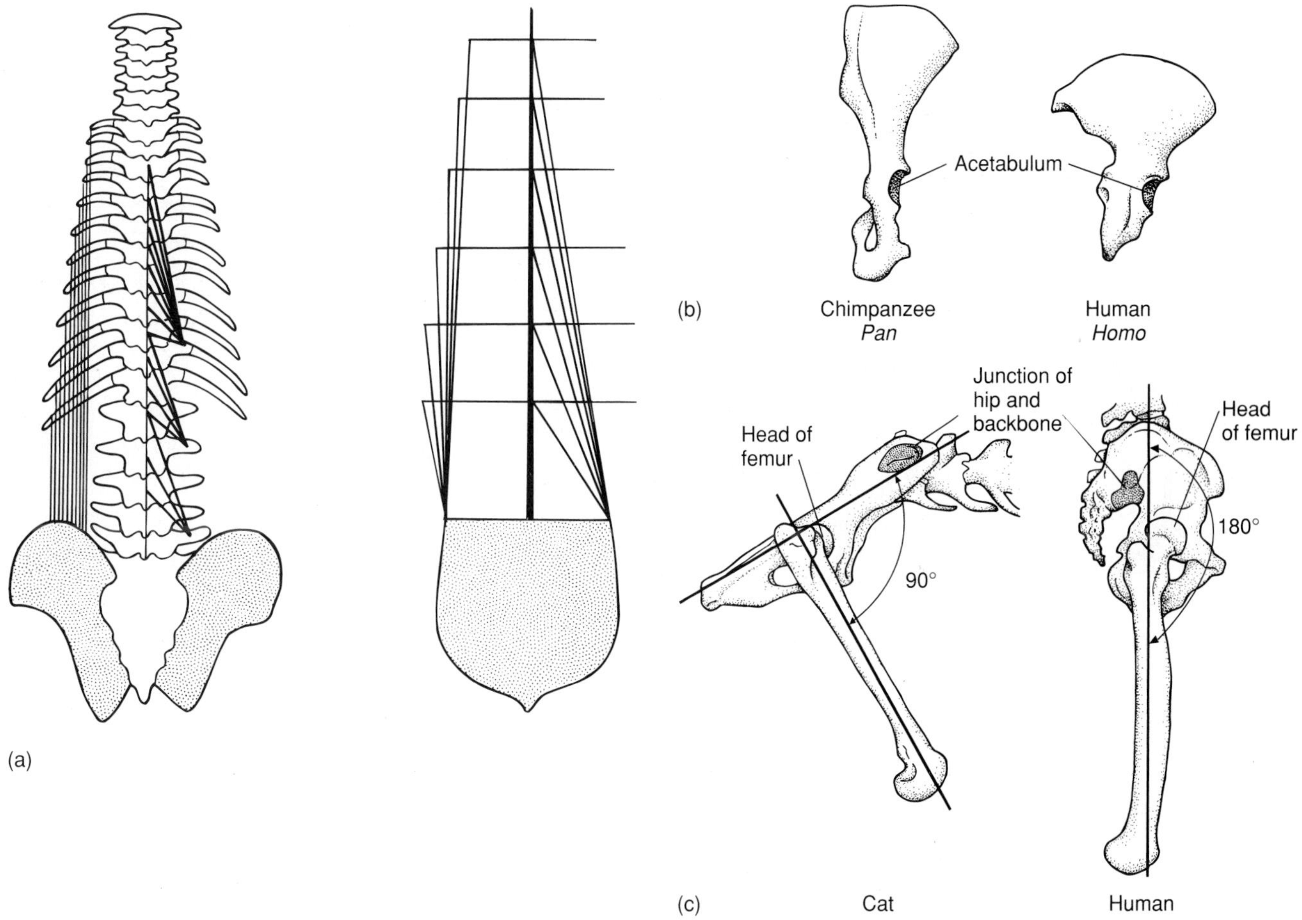


**Box figure 2** Upright posture of humans brings with it some instability as the weight of the upper body is brought above the hips. Several features of the human skeleton are remodeled to return some stability. (a) Muscles and ligaments, like the rigging of a sailing ship, span the distance between vertebrae and up from the hips to the lower vertebrae and ribs to stabilize the rising vertebral column. (b) The blade of the ilium is widened to broaden the base of support at the hips. (c) The distance between the sacrum and the head of the femur is shortened in humans, thus bringing the base of the vertebral column closer to its eventual support by the acetabulum (socket of the hip), and offering the same spread of hip muscle attachments to maintain a favorable line of action on the femur.

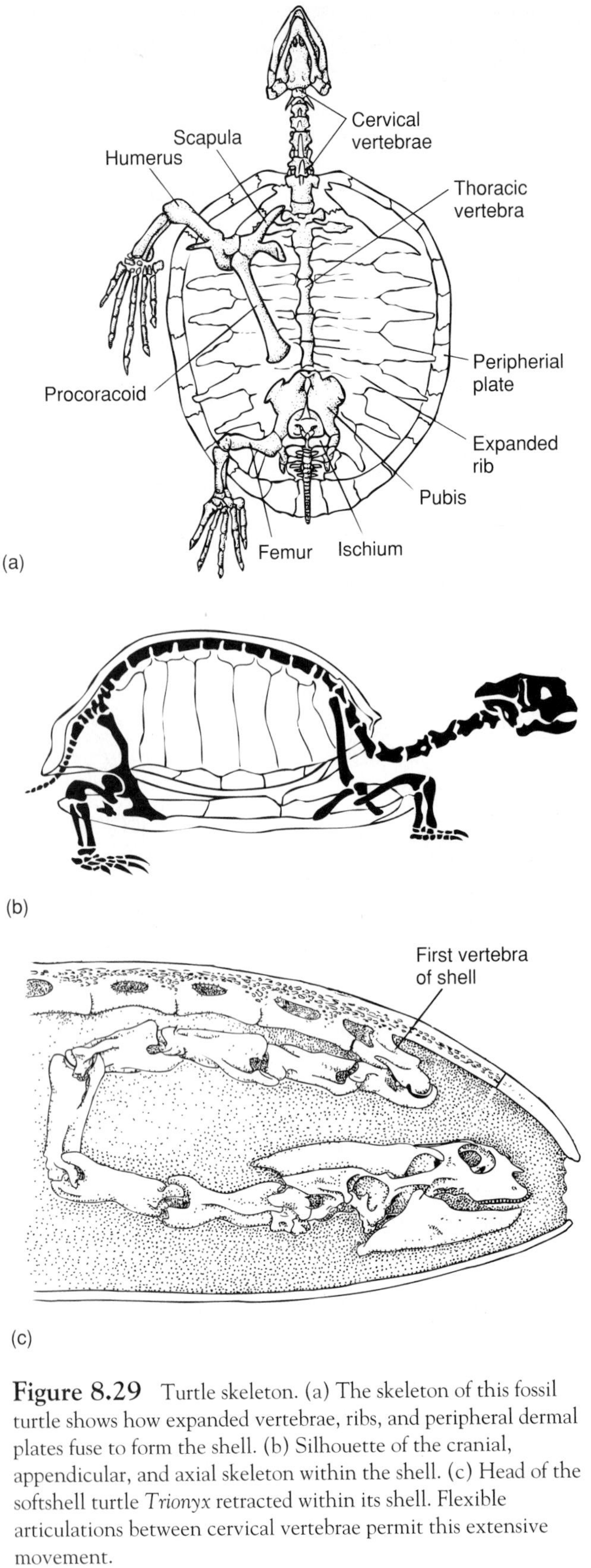


**Figure 8.29** Turtle skeleton. (a) The skeleton of this fossil turtle shows how expanded vertebrae, ribs, and peripheral dermal plates fuse to form the shell. (b) Silhouette of the cranial, appendicular, and axial skeleton within the shell. (c) Head of the softshell turtle *Trionyx* retracted within its shell. Flexible articulations between cervical vertebrae permit this extensive movement.

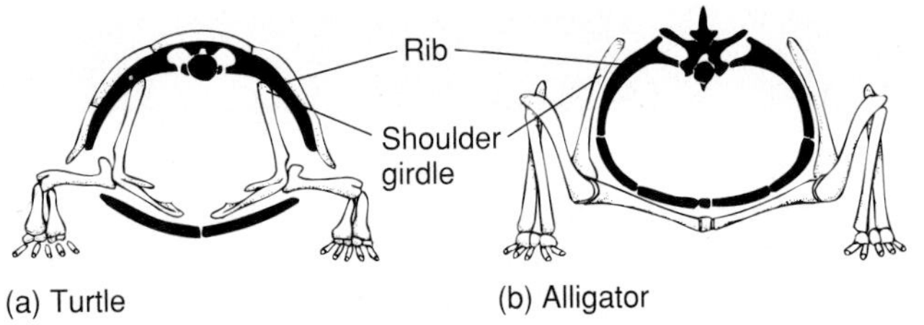


**Figure 8.30** Cross section of turtle body (a) showing the unusual position of the appendicular skeleton inside the rib cage (dark) compared with the external skeleton of other vertebrates (b), illustrated by this cross section of an alligator.

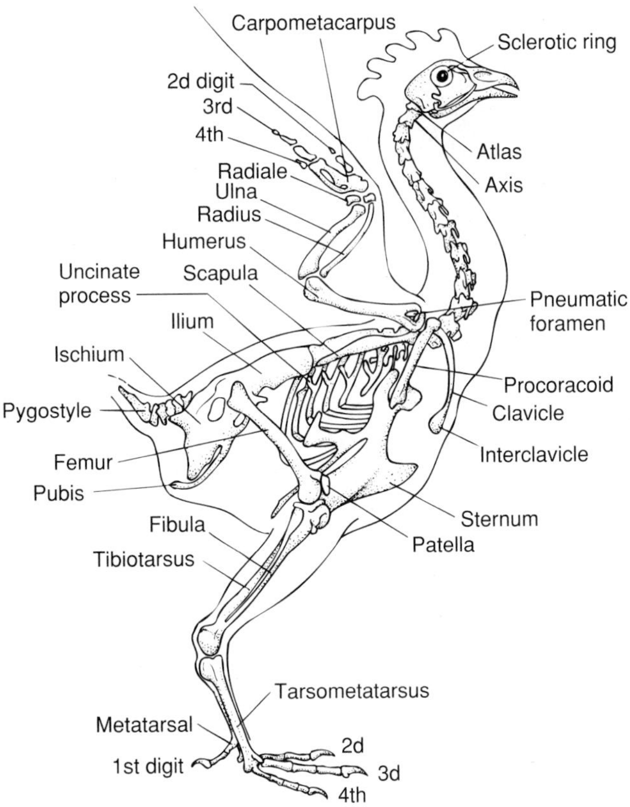


**Figure 8.31** Skeleton of a chicken.

## Fluid Environment

In an aqueous medium, such as fresh water or marine seas, an organism does not depend for support primarily on the endoskeletal framework. Instead, the body takes advantage of its buoyancy in the surrounding water (figure 8.33a). For an active aquatic organism, two problems are uppermost. The first is drag on the body as it slips through a relatively dense medium, water. The answer is streamlining, contouring of the body to reduce drag forces. It is no accident that the general body shapes of fast-swimming fishes and supersonic aircraft are both streamlined. This shape improves the

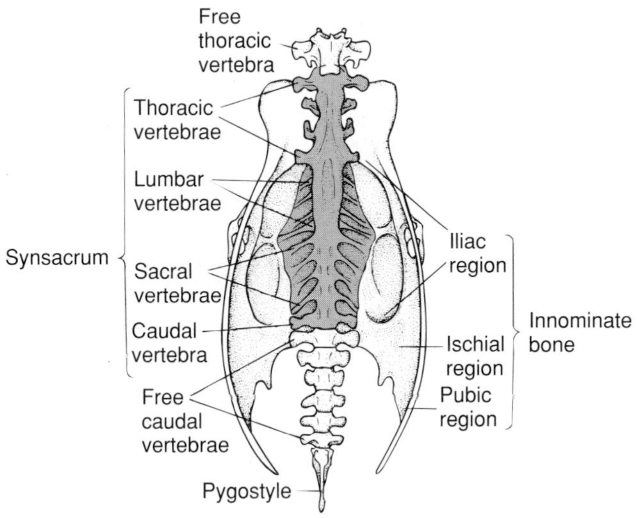


**Figure 8.32** Synsacrum of a pigeon, ventral view. Notice how the synsacrum (shaded) is fused to joined elements of the pelvis, the innominate bone.

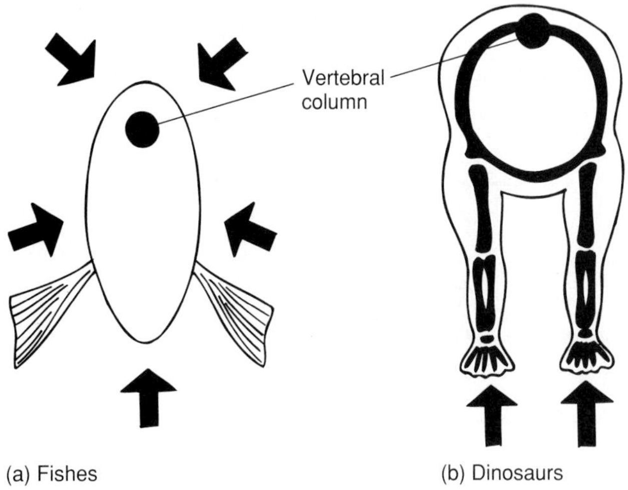


**Figure 8.33** Body support. (a) In fishes, the surrounding water (arrows) supports and buoys the weight of the body. (b) In tetrapods, the limbs support and the vertebral column suspends the weight of the body.

performance of both fishes and aircraft as they meet common physical demands while traveling through a medium that resists their passage.

**Streamlining (p. 135)**

The second problem for an active aquatic organism is orientation in three-dimensional space. Any streamlined body has a tendency to tip and deviate from its line of travel, rotating about its center of mass. In fishes, these perturbations are countered by stabilizing fins appropriately positioned along the body.

**Three-dimensional stability (p. 315)**

## Terrestrial Environment

Land generally presents a two-dimensional surface across which to maneuver. Because tetrapods live on land without the buoyancy of a dense medium such as water, gravity presents a problem. When remaining in place, the tetrapod's body either rests on the ground between sprawled legs, or it is suspended between the pairs of legs, as in most mammals and some dinosaurs. The pairs of legs function as abutments that support the body between them. The vertebral column serves as a bridge between the support posts, the legs, and suspends the body from it (figure 8.33b). A convenient mechanical analogy has been drawn between this posture and engineered structures such as bridges.

What bridge engineers call the Forth bridge is a two-armed, bridge in which both extensions are balanced against each other, or cantilevered, and carry the weight of the roadbed to the pier (figure 8.34a). Compressive forces are borne by solid structural members, tensile forces by cables. A pair of such bridges carries the extended roadbed between them. The weight of each section of the roadbed is transferred to the nearest pier. The point between piers where weight transfer changes is the **nodal** (figure 8.34b).

**Mechanics of loading (p. 138)**

The mammalian vertebral column, if viewed in engineering terms, might be represented by two Forth bridges, with the body suspended from them. The spines and centra represent the compression members; the ligaments and muscles, the tension members; the two pairs of legs, the piers. The point of the nodal depends on the relative weight distribution between the two piers, the two pairs of legs (figure 8.34c). Where the nodal occurs, force distribution within the vertebral column changes, and structural members that receive these forces become modified as well. Such an engineering analogy helps explain the reverse orientation of the neural spines midway along the length of the vertebral column between the two pairs of legs. The point at which the neural spines reverse might correspond to the biological nodal, and thus structurally reflect the underlying mechanical forces that the vertebral column must address.

If the body is heavy, one region is often cantilevered against another. In the iguanodon, a bipedal dinosaur, the heavy tail helps balance the weight of the thorax and anterior body across the hindlimbs (figure 8.35).

Other engineering analogies help clarify the ways in which some features of biological form might represent solutions to problems of mechanical stress. For example, to carry weight, any arch must maintain its bowed shape and avoid flattening out. Arch suspension bridges suspend the weight of the roadbed (figure 8.36a). The same mechanical principle seems to be incorporated into the design of mammals. Between the pairs of limbs, the abdominal muscles and sternum keep the arched vertebral

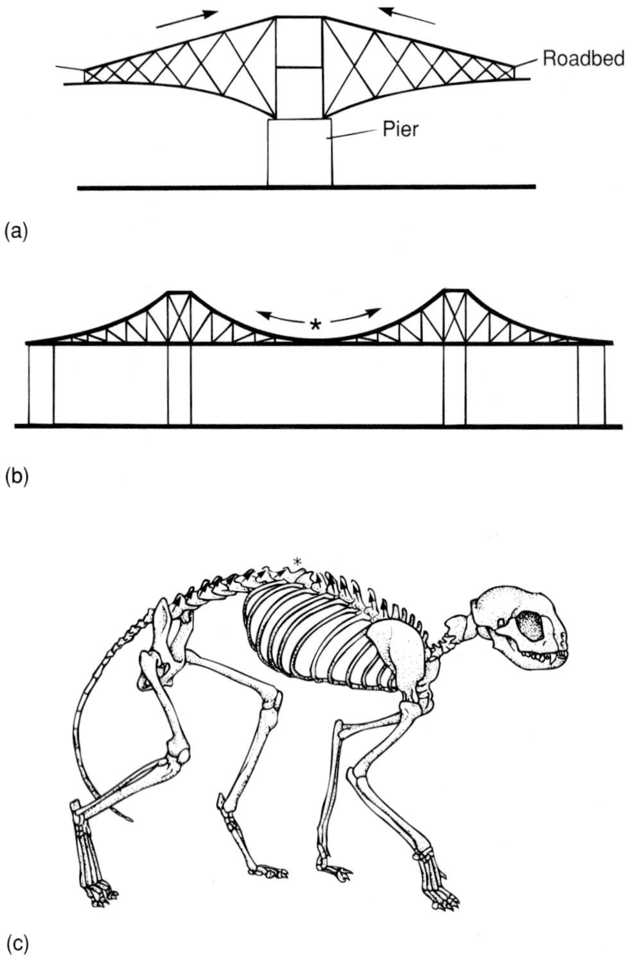


**Figure 8.34** Engineering analogies and design of the vertebral column. (a) The Forth bridge works by resisting compression in its rigid members and tension in its flexible members. Each section of the bridge rests on piers. (b) If sections and piers are combined, the weight of the roadbed can span the distance between the nearest piers. The "nodal" (*) marks the point of trade-off in weight distribution between two piers. (c) By analogy, the vertebral column might be viewed as serving roughly the same function, spanning the distance between fore- and hindlimbs. The bones resist compression; the muscles and ligaments resist tensile forces. Change in orientation of the neural spines marks the point of the nodal.

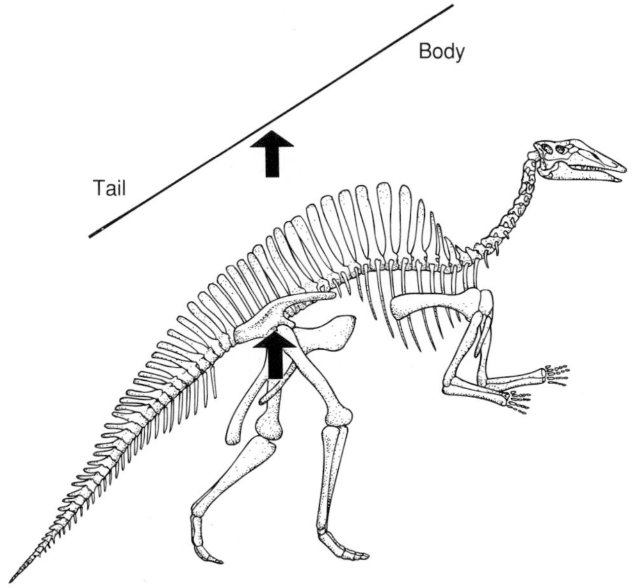


**Figure 8.35** In bipedal animals, such as in this iguanodontid dinosaur *Ouranosaurus*, the weight of the heavy tail and the upper body are balanced like a simple seesaw on the fulcrum of the hips. Firmness of the vertebral column was apparently maintained by networks of strong ligaments that tied together the tall neural spines.

column from sagging and so effectively maintain its structural and thus functional integrity. The neck forms a reversed arch, with ligaments and muscles holding the head (figure 8.36b–d).

## Design of Vertebrae

Not all vertebrae are morphologically alike even within the same vertebral column. Differences in design reflect different mechanical demands within parts of the column as well.

### *Direction of the Neural Spine*

The angle that the neural spine makes with its centrum often varies from vertebra to vertebra. This angle may represent a structural way to orient the spine so that it receives the suite of mechanical forces in the least stressful direction. Local mechanical forces on the spine arise largely from contraction of the axial musculature. The complex axial musculature originates at distant sites along the vertebral column and reaches to the ends of the neural spines, applying forces on these spines. Rostral muscles that are inserted on the neural spine bend it forward; more caudal muscles bend it backward. If these groups contract together, then the spine experiences the single resultant force of both acting together, not one or the other force separately. Recall that bones, like most structures, are weakest in tension and sheer, but strongest when loaded in compression. If this resultant force bent the spine, it would place parts of the spine in tension or shear, which is worse, and so expose it to forces it is least able to withstand. Thus, the neural spine appears to be oriented in such a way that its long axis is in parallel with the resultant forces imposed collectively by all axial muscles inserted on it. This orientation means that the spine experiences these forces as a compressive force, the direction of stress loading in which it is strongest (figure 8.37a).

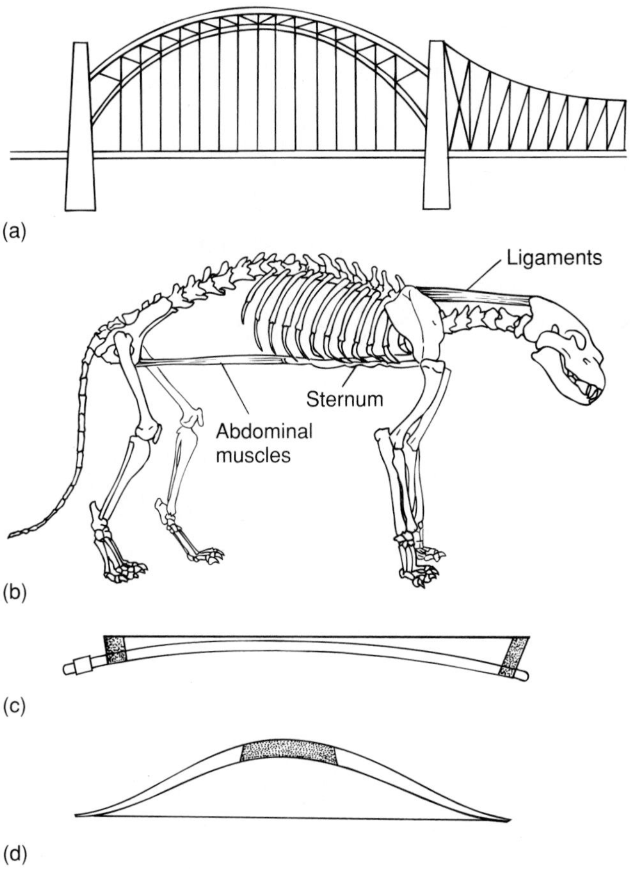


**Figure 8.36** (a) The roadbed and piers of an arch suspension bridge hold the span between them from an arch. As long as arch integrity is maintained, it supports the weight of the bridge. (b) Similarly, muscles and ligaments hold the vertebral column in arches. (c) The arch of the cervical vertebrae is like the reverse arch of a violin bow. (d) The other arch, formed by the trunk vertebrae, resembles an archer's bow.

### *Height of the Neural Spine*

The height of a neural spine is apparently proportional to the mechanical leverage the muscles must exert to move or stabilize the vertebral column. In a sense, the neural spines are levers that transmit the force of muscle contraction to centra (figure 8.37b). This force is proportional to the physiological cross section of the muscle and to its lever arm, its perpendicular distance to the centrum. To increase this force, the muscle could be enlarged or the neural spine could be lengthened. Increasing spine length increases the lever arm from centrum to line of muscle action, and thus effectively increases the mechanical advantage of the muscle.

Vertebral designs incorporate modifications to meet these mechanical problems. For instance, in many reptiles, the spines on trunk vertebrae are about equal in height and similar in orientation (figure 8.38a). Compared with mammals, the axial musculature of reptiles is less specialized for rapid locomotion. In many mammals, height and direction of the spines vary within the same vertebral column, indicating the specialized functions performed by different sections of the vertebral column (figure 8.38b).

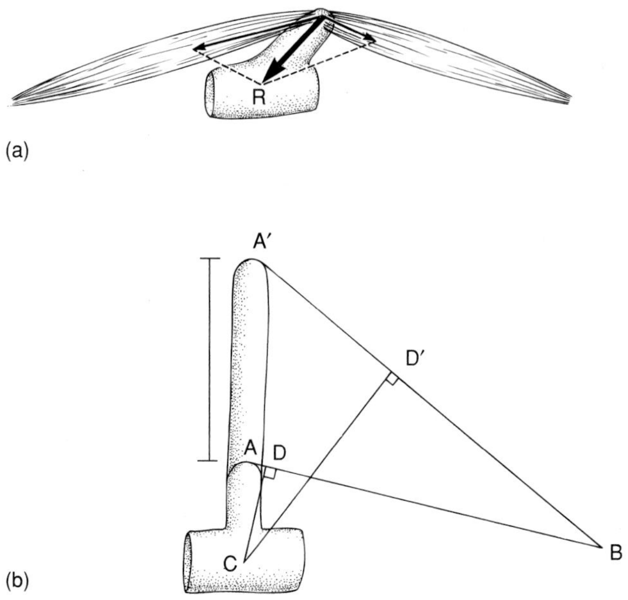


**Figure 8.37** Orientation and height of the neural spine reflect mechanical forces acting on or through it. (a) Axial muscles develop forces, the resultant of which is coincident with the length of the neural spine. If the resultant produced a shearing or bending of the spine, it would introduce forces that the neural spine, like most supportive structures, is least able to withstand. (b) Muscles that are inserted in the neural spine act on it like a lever to bring a force to the centrum. One way to increase the mechanical leverage of this force is to increase the length of the neural spine. Increasing the length from A to A′ changes the length of the lever arm, the perpendicular distance from the line of action to the centrum. In this example, the lever arm increases from CD to CD′ and so it increases the effective force on the centrum.

## Regionalization of the Vertebral Column

We can now step back and take an overall look at the vertebral column with a view to summarizing how the transition from water to land has changed the mechanical demands placed on vertebrae and how the changes in design have followed.

In fishes, the vertebral column is differentiated into two regions, the caudal and trunk regions (figure 8.39a). Zygapophyses and similar interlocking projections are generally absent. The centra are unspecialized except if they receive ribs or hemal or neural arches. The relatively undifferentiated vertebral column of fishes reflects the fact that it is not used to support the body. Support comes generally from the buoyancy of the surrounding water. The vertebral column mainly offers sites of attachment for the swimming musculature. It serves as a mechanical replacement of the

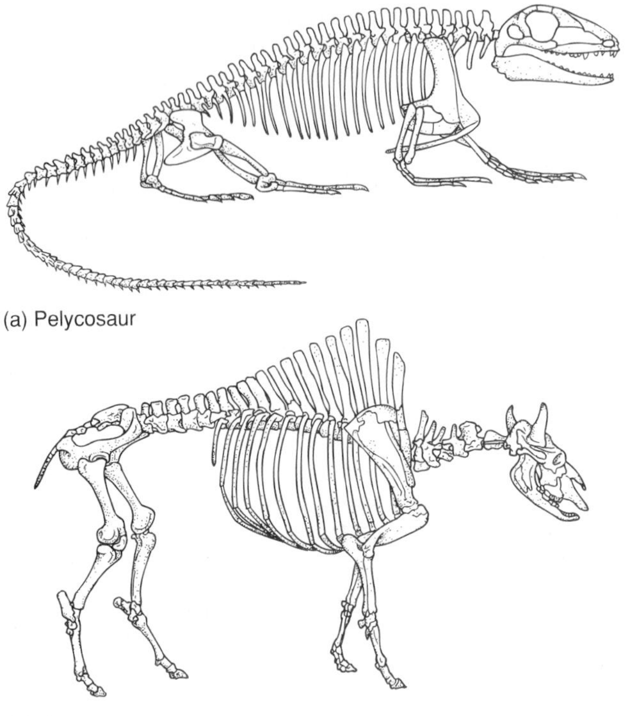


**Figure 8.38** Variation in the height of neural spines can be seen in vertebrates in which relatively heavy weights must be supported by the vertebral column. (a) Skeleton of a pelycosaur, with most of the neural spines of similar height and orientation. (b) Skeleton of a bison, illustrating the tall neural spines in the shoulder. Through ligaments to the skull and cervical vertebrae, these neural spines help support the weight of the heavy head.

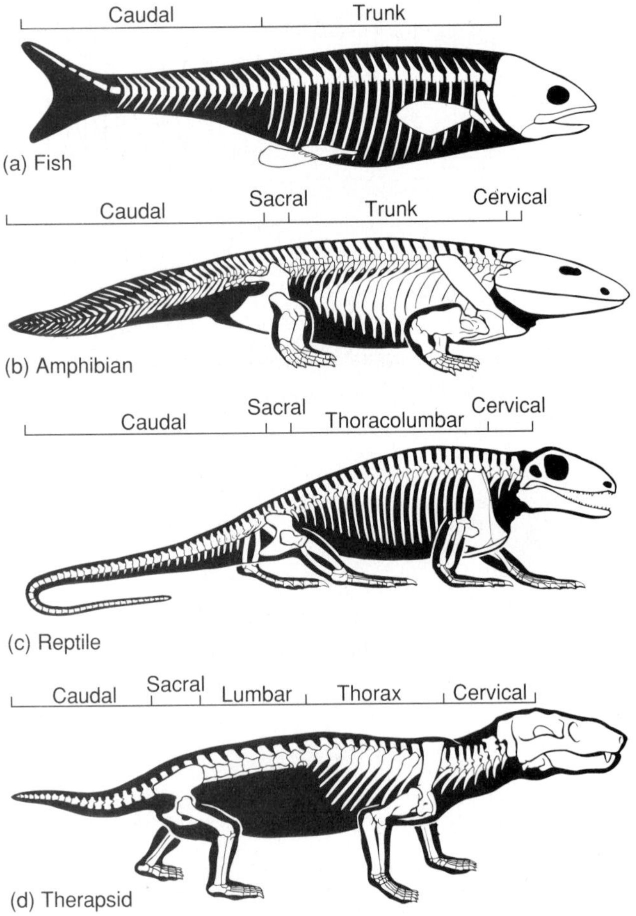


**Figure 8.39** Regionalization of the vertebral column. (a) Fish. (b) Amphibian. (c) Reptile. (d) Therapsid.

notochord, resisting the tendency of the body to telescope, yet it allows lateral flexibility for swimming.

In tetrapods, however, the vertebral column supports the body against gravity and receives and transmits the propulsive forces limbs generate during locomotion. Diverse functional demands are placed on the vertebral column, so we might expect to find delineation of specialized regions.

In amphibians, caudal, sacral, trunk, and modest cervical regions of the vertebral column are delineated (figure 8.39b). Most amphibians are not strictly terrestrial and return frequently to water. Much of the musculature and axial skeleton still retain similarities to their fish ancestors. For instance, the long tail often supports a broad fin and the trunk region is relatively undifferentiated, as in fishes. However, locomotion on land is important especially among adults. Through the pelvic girdle, the hindlimbs are directly attached to the adjacent region of the vertebral column to define the sacral region. The cervical region also is differentiated, allowing some freedom for the skull to turn independently of the body.

In reptiles, cervical, thoracolumbar, sacral, and caudal regions are present (figure 8.39c). The sacral region is stronger than in amphibians, designed to support more habitual existence on land. Most reptiles retain a trunk (**dorsal, thoracolumbar** region). Ribs on the vertebrae immediately in front of the hindlimbs may be shortened, and in some reptiles, the trunk may differentiate into two regions, the thorax with ribs and the lumbar region without ribs. The appearance of a lumbar region within the posterior thorax deserves some notice because it reflects an increase in locomotor performance. As hindlimbs swing forward to take long strides during rapid locomotion, the vertebral column usually flexes laterally on itself. This can cause ribs on adjacent vertebrae to crowd one another. Loss of ribs in the area of greatest flexion addresses this crowding, producing a presacral section of the vertebral column without ribs, the lumbar region. Consequently, the appearance of a lumbar region marks a point at which tetrapods begin to experiment with more rapid forms of locomotion. This occurs in many archosaurs, some modern reptiles, synapsids, and mammals. Locomotor behavior cannot be easily determined from fossils directly. But behavior can be inferred from morphology. In the next chapter, we see that the morphology of the appendicular skeleton confirms this interpretation. Presumably more active lifestyles accompanied these more rapid forms of locomotion in later tetrapods.

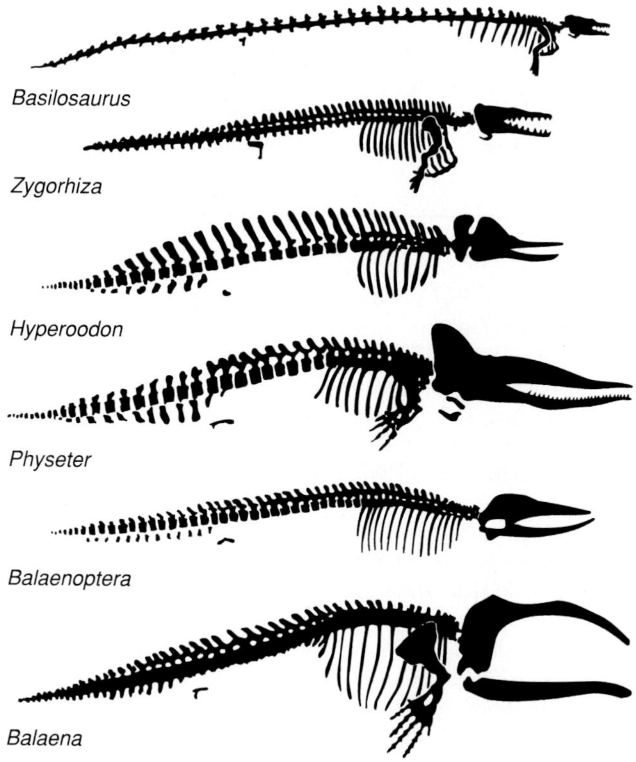


**Figure 8.40** Axial skeletons of whales in lateral silhouettes. Shown are two fossil species *(Basilosaurus, Zygorthiza)*, two odontocetes *(Hyperoodon, Physeter)*, and two mysticetes *(Balaenoptera, Balaena)*. Notice the reduction in size of the limbs and girdles and the proportionate increase in size of the vertebral column compared with the size of limbs, girdles, and vertebral columns of quadrupeds.

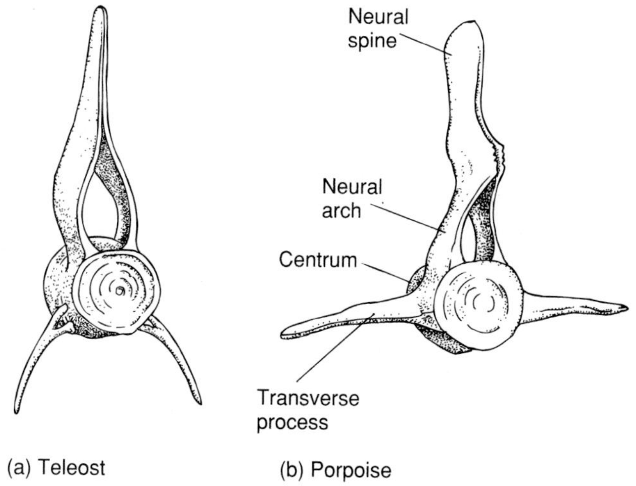


**Figure 8.41** Vertebrae of aquatic vertebrates. (a) Vertebra of a teleost fish. (b) Vertebra of a porpoise. Notice the reduction in zygapophyses in the vertebra of the aquatic porpoise.

Five distinct regions are differentiated within the vertebral column of mammals—cervical, thoracic, lumbar, sacral, and caudal (figure 8.39d). The musculature is attached to the vertebral column in complex ways, corresponding to the demands that active locomotion places upon the individual vertebrae. In mammals and in other tetrapods that secondarily resume an aquatic lifestyle, the axial column returns, at least in part, to the compression girder of fishes. Hindlimbs are often reduced, and forelimbs form paddles (figure 8.40). In the porpoise, for example, antitwist zygapophyses are absent from the centra, although interlocking notches of successive neural spines resist torsion (figure 8.41a,b).

Birds are an interesting example, exhibiting a close match of form and function within the vertebral column. Cervical vertebrae are flexibly articulated to give the head great freedom of movement and reach when a bird preens its feathers or probes for food. On the other hand, most of the vertebrae in the middle and posterior part of the column are fused to each other and to the pelvic girdle (figure 8.42).

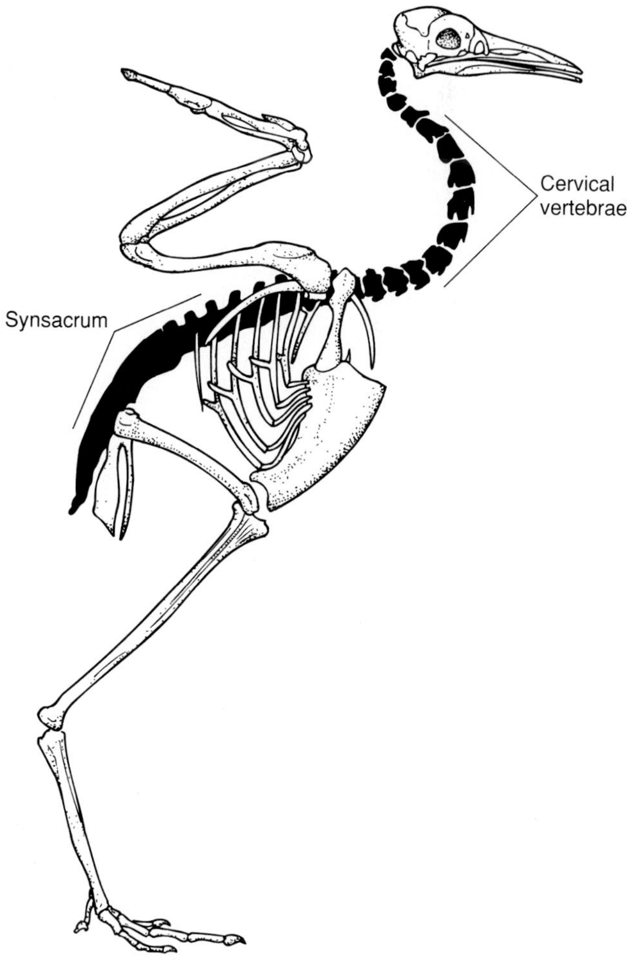


**Figure 8.42** Bird vertebral column. Regions of extensive vertebrae fusion are indicated posteriorly within the synsacrum. The numerous heterocoelous cervical vertebrae allow for great mobility of the head.

This brings rigidity to the vertebral column and establishes a firm and stable axis for control while a bird is in flight. Indirectly, this fusion of elements decreases the weight of the body because less muscle is required to control individual vertebrae. Muscles otherwise required to bring stability can be reduced, economy of design realized, and weight of the bird lightened. The vertebral columns of most birds show much uniformity, a likely indication of the overriding significance of flight and its demands on biological design of these vertebrates that have mastered life in the air.

## Selected References

Bagnall, K. M., S. J. Higgins, and E. J. Sanders. 1988. The contribution made by a single somite to the vertebral column: Experimental evidence in support of resegmentation using the chick-quail chimaera model. *Development* 103:69–85.

Blake, R. W. 1983. *Fish locomotion*. London: Cambridge University Press.

Francois, Y. 1966. Structure et développment de la vertèbre de *Salmo* et des Téléostéens. *Arch. Zool. Exp. Gen.* 107:287–328.

Janvier, P., and A. Blieck. 1979. New data on the internal anatomy of the Heterostraci (Agnatha), with general remarks on the phylogeny of the craniota. *Zool. Scripta* 8:287–96.

Jenkins, F. A., Jr. 1969. The evolution and development of the dens of the mammalian axis. *Anat. Rec.* 164:173–84.

Jenkins, F. A., Jr., and G. E. Goslow, Jr. 1983. The functional anatomy of the shoulder of the Savannah monitor lizard *(Varanus exanthematicus) J. Morph.* 175:195–216.

Krantz, G. S. 1981. *The process of human evolution*. Cambridge, Mass.: Schenkman Publishing Co.

Laerm, J. 1976. The development, function, and design of amphicoelous vertebrae in teleost fishes. *Zool. J. Linn. Soc.* 58:237–54.

Laerm, J. 1979. On the origin of rhipidistian vertebrae. *J. Paleont.* 53:175–86.

Laerm, J. 1982. The origin and homology of the neopterygian vertebral centrum. *J. Paleont.* 56:191–202.

Panchen, A. L. 1977. The origin and early evolution of the tetrapod vertebrae. *Linn. Soc. Symp.* 4:289–318.

Parrington, F. R. 1967. The vertebrae of early tetrapods. *Colloq. Int. Cent. Nat. Rech. Sci.* 163:267–79.

Simons, J. R. 1970. The direction of the thrust produced by the heterocercal tails of two dissimilar elasmobranchs: The Port Jackson shark, *Heterodontus portusjacksoni* (Meyer), and the piked dogfish, *Squalus megalops* (Macleay). *J. Exp. Biol.* 52:95–107.

Verbout, A. J. 1976. A critical review of "Neugliederung" concept in relation to the development of the vertebral column. *Acta Biotheoretica* 25:219–58.

Verbout, A. J. 1985. The development of the vertebral 5 column. *Advances Anat. Embry. Cell Biol.* 90:1–120.

Wassersug, R. J. 1989. Locomotion in amphibian larvae (or "Why aren't tadpoles built like fishes?"). *Amer. Zool.* 29:65–84.

9 CHAPTER

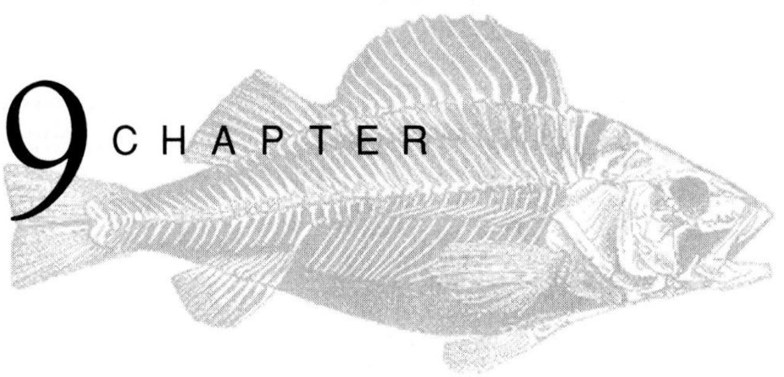

# Skeletal System: The Appendicular Skeleton



## Introduction

From components of the appendicular skeleton, evolution has fashioned some of the most elegant and specialized locomotor devices from the fins of fishes to the limbs of tetrapods. Like the rest of the skeletal system, the appendicular system is represented well in the fossil record. This brings us into direct contact with the structural details of extinct animals and helps us to track the general course of phylogenetic modifications of these skeletal elements. Within the appendicular skeleton, the relationship between structure and biological role is direct, at least in a general way. We need not be aerodynamic engineers to understand that the wings of birds give them access to the air and the special lifestyles that follow, that the limbs of tetrapods serve them on land, and that the fins of fishes are suited to water. Transitions from water to land and from land to air have had an impact upon the design and redesign of the appendicular system.

What we see in this chapter, however, is that form and function can be closely matched. Not all birds use the air in equal ways. Some, in fact, such as penguins and ostriches do not fly at all. Tetrapods use the land differently. Some lumber, some dash, some dig, some climb trees. For some fishes, fins provide lift for cruising through the water; for others, fins become specialized for maneuvering in tight places. Form and function are slightly different for each, and biological design reflects these differences.

## Basic Components

The appendicular skeleton includes the **paired fins** or **limbs** and the **girdles,** the braces within the body that support them. The anterior girdle is the **shoulder** or **pectoral girdle,** to which dermal and endochondral skeletal elements contribute, and that support a pectoral fin or limb. The posterior girdle is the **hip** or **pelvic girdle,** consisting of endochondral skeletal elements that support the pelvic fin or limb.

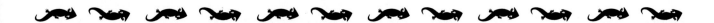

## Fins

Particularly in primitive fishes, the body is apt to carry projecting spines, lobes, or processes. Unlike these projections, fins are membranous or webbed processes internally strengthened by radiating and thin **dermal fin rays.** In elasmobranchs, these dermal fin rays, termed **ceratotrichia,** are slender keratinized rods (figure 9.1a). Dermal fin rays, or **lepidotrichia,** in bony fishes are usually an ossified or chondrified series of tiny elements that strengthen this web (figure 9.1b). In some bony fishes, the tip of the fin may be additionally stiffened by keratinized rods, the **actinotrichia.** The proximal part of the fin close to the body is supported by **pterygiophores** of two general types, the enlarged **basals** within the proximal part of the fin and the slender **radials** that extend support from the basals into the middle region of the fin (figure 9.2a).

Fins occur singly, except for a pair near the head and a second pair posterior to this, the pectoral and pelvic fins, respectively. The basal pterygiophores of these projecting paired fins articulate with and are braced by girdles inside the body wall. These paired fins will receive our greatest attention because they are the phylogenetic source of the tetrapod limbs.

## Limbs

The forelimbs and hindlimbs of tetrapods are all built on the same pattern comprised of three recognized regions. The **autopodium,** the distal end of the limb, consists of numerous elements comprising the wrist and ankle that in turn support their respective digits (figure 9.2b). The special term *hand* implies a structure modified for grasping, and *foot* suggests a role in standing. But these two terms do not apply logically across tetrapods. For instance, the terminal part of a horse forelimb is not a hand, nor is the terminal part of the hindlimb of a whale a foot. Therefore, the terms **manus** and **pes,** respectively, are preferred for the autopodium of fore- and hindlimbs. The middle limb region is the **zeugopodium,** with two internal supportive elements—ulna and radius of the forearm, tibia and fibula of the shank. The limb region closest to the body is the **stylopodium,** with a single element—humerus of the upper arm, femur of the thigh.

A depression within the pectoral girdle, the **glenoid fossa,** articulates with the humerus. A deep socket in the pelvis, the **acetabulum,** receives the femur.

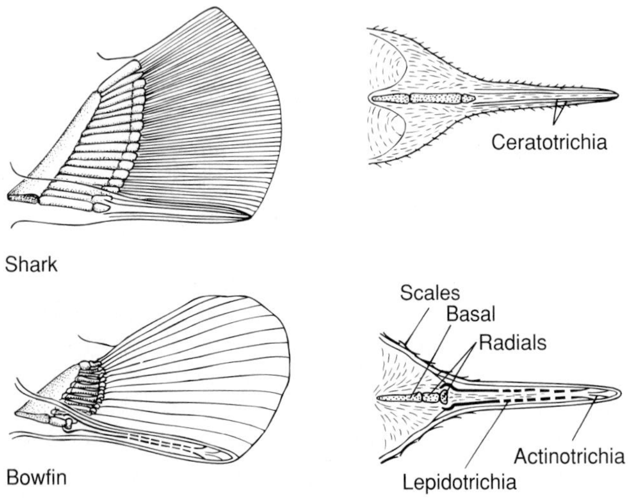


**Figure 9.1** Dermal fin rays. (a) Ceratotrichia are keratinized rods that radiate out like vanes in a fan to internally support the fins of chondrichthyan fishes. (b) Lepidotrichia are cartilaginous or ossified supports within fins of bony fishes.

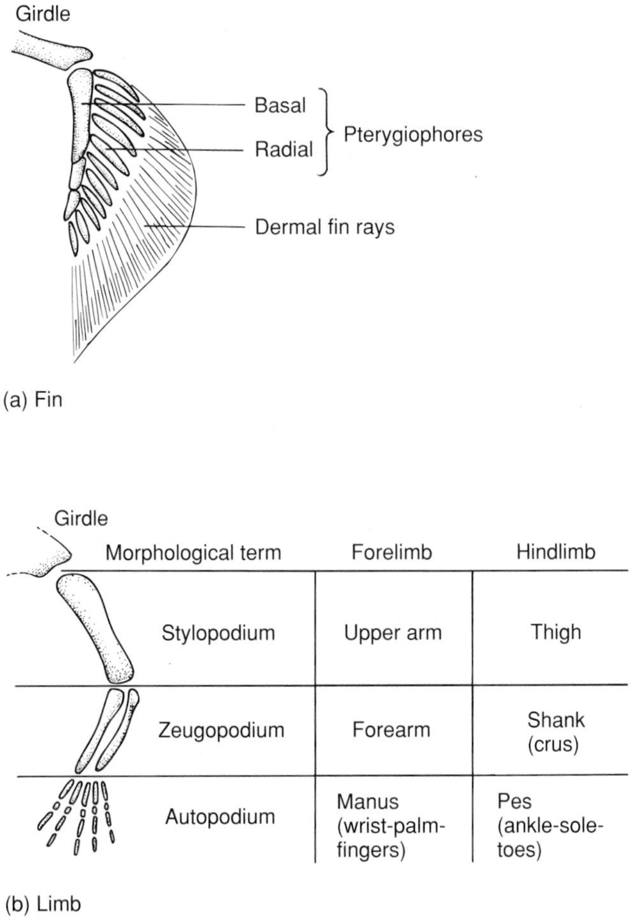


**Figure 9.2** Basic components of the fin and the limb. (a) The fin is composed of pterygiophores, basals and radials, and dermal fin rays. Fin rays are called lepidotrichia in bony fishes and ceratotrichia in elasmobranchs. (b) The limb, either fore- or hindlimb, includes three regions—stylopodium (upper arm/thigh), zeugopodium (forearm/shank), and autopodium (manus/pes).

# Origin of Paired Fins

Like any object traveling in three-dimensional space, the body of a fish is susceptible to deflections from its line of travel about its center of mass. It may swing from side to side **(yaw),** rock about its long axis **(roll),** or buck forward and back **(pitch;** figure 9.3a). Wind tunnel tests using shark models with selected fins removed have helped clarify how fins bring stability to a streamlined body. It appears that dorsal and lateral fins control the body by resisting perturbations of the body about its center of mass. In sharks, at least, the pectoral fins act like hydrofoils to lift the anterior part of the body. The heterocercal tail contributes to lift posteriorly.

Dynamics of heterocercal tails (p. 294)

As early fishes became more active, they would have experienced instability while in motion. Presumably, just such conditions favored any body projection that resisted pitch, roll, or yaw, and led to the evolution of the first paired fins. The associated girdles stabilized the fins, served as sites for muscle attachment, and transmitted propulsive forces to the body.

In gnathostome fishes, two fundamental types of fins developed from two different arrangements of the **metapterygial stem** or **axis,** a chain of endoskeletal basals. One fin type is the **archipterygial fin** in which the metapterygial stem runs down the middle of the fin. From this central stem, endoskeletal radials project outward to support the **preaxial** (anterior) and **postaxial** (posterior) sides of the fin evenly. Slender dermal fin rays extend to the edges of the fin to complete this support. Externally, the archipterygial fin looks leaf-shaped and narrowed at its base. The second basic fin type is the **metapterygial fin** in which the metapterygial stem of basals is located posteriorly. Most radials project from this posterior axis into the preaxial side of the fin, and dermal fin rays extend from the ends of the radials to the edges of the fin (figure 9.3b). These two fin types have influenced theoretical work on the origin of paired fins. To track the phylogenetic source of early fish fins, the gill-arch and the fin-fold theories have been put forth.

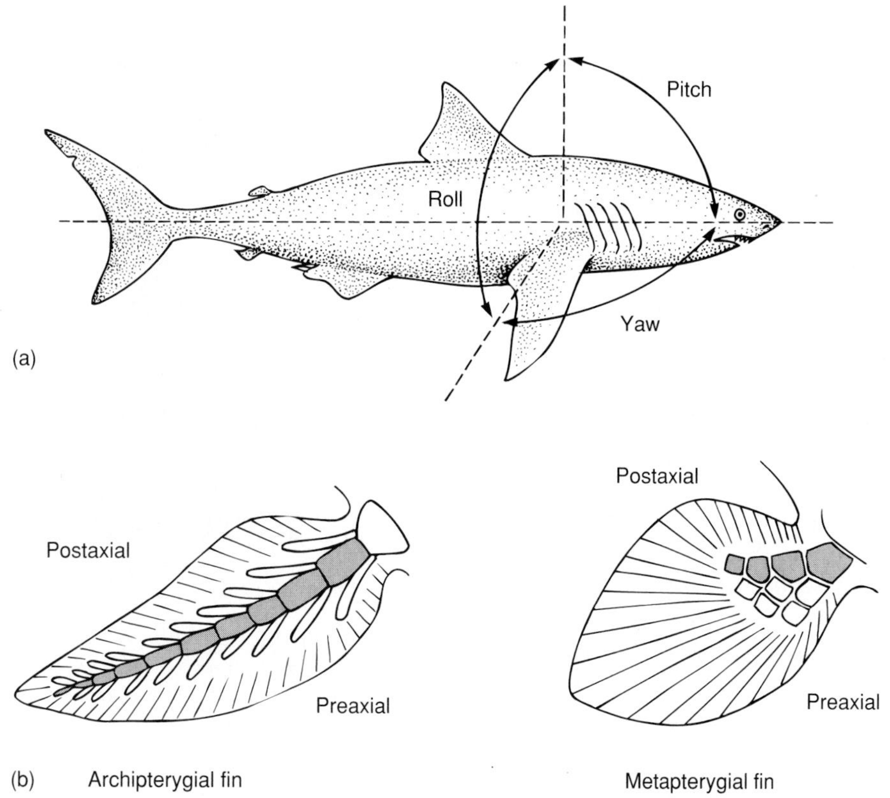


**Figure 9.3** Fins as stabilizers. (a) The body of a fish can deviate from its intended line of travel in three ways. A roll rocks the fish about its long axis, a yaw swings it from side to side, and a pitch bucks it up and down about its center of mass. (b) There are two types of fin structure, the archipterygial fin with a symmetrical central axis (left) and the metapterygial fin with the asymmetrical axis displaced toward the postaxial side (right). The metapterygial axis or stem present in both is shaded.

## Gill-arch Theory

During the second half of the nineteenth century, morphologist C. Gegenbaur proposed that paired fins and their girdles arose from gill arches (figure 9.4). Specifically, the endoskeletal girdle arose from the gill arch and the primitive archipterygial fin arose from the gill rays. Initially, Gegenbaur based his theory on fin anatomy in sharks. However, the discovery (1872) of the Australian lungfish, *Neoceratodus*, convinced him that the primitive fins were archipterygial fins similar to the paired fins in *Neoceratodus*—a central stem supporting a series of radials. This central stem articulated with the lungfish's endoskeletal shoulder girdle, the future scapulocoracoid.

Nonetheless, the gill-arch theory left much unexplained. Although it accounted for the evolution of the pectoral girdle, it does not explain (1) the appearance of a posterior pelvic girdle distantly placed from the gill arches, nor (2) the presence of dermal bone in the pectoral girdle, nor (3) the different embryologies of pectoral girdle and gill arches.

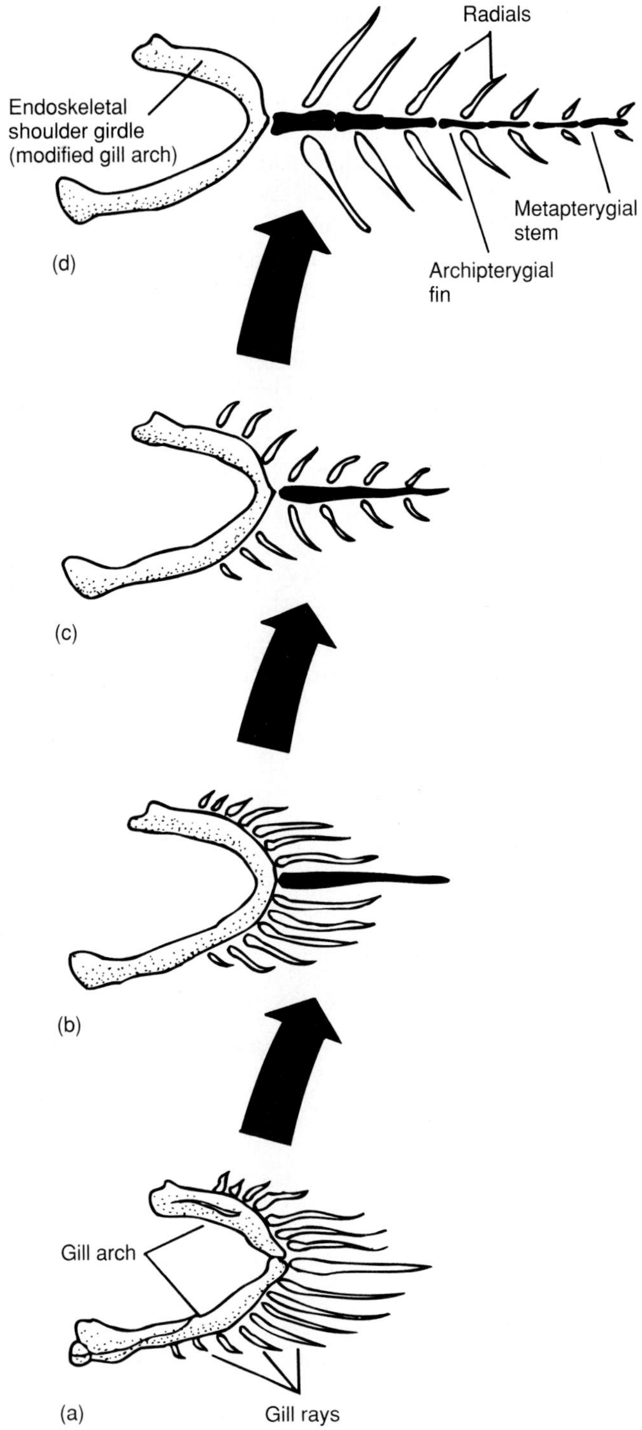


**Figure 9.4** Gill-arch theory proposed by Gegenbaur for the origin of paired fins. Gill rays expand (a,b) and proliferate (c), forming a long central support for an external fin, not unlike the archipterygial condition (d) found in some present-day lungfishes.

## Fin-fold Theory

At about the same time, the second half of the nineteenth century, morphologists J. K. Thacher and F. M. Balfour independently put forth the fin-fold theory, an alternative view expanded by later scientists. With this view, paired fins arose within a paired but continuous set of ventrolateral folds in the body wall that were stiffened by a transverse series of endoskeletal pterygiophores, proximally the basals and distally the radials (figure 9.5a–c). Additional stability came from the inward extension of basals and their eventual fusion across the midline to produce the supportive girdles. Dermal bone, a contribution of the overlying bony armor, was later added to the pectoral girdle to strengthen the paired fins further.

In support of the fin-fold theory, several indirect pieces of evidence are usually cited. The pregnathostomes with the first fin folds left no fossil record themselves (figure 9.6a,b). However, many surviving fossils of early fishes carry hints or presumed remnants of these earlier fin folds. For instance, some primitive ostracoderms possessed lateral continuous folds along the ventral body wall. Acanthodians possessed a paired row of spines to mark where a paired fin fold presumably resided in their ancestors (figure 9.6c). Furthermore, if pectoral and pelvic fins arose from a fin fold, then they likely arose at the same time (figure 9.6d). In this context, it is significant that the paired fins of embryonic sharks develop together from a continuous thickening of ectoderm along the lateral body wall. This has been interpreted as an embryonic recapitulation of the phylogenetic transition from fin folds to paired fins (figure 9.6e).

More recently, the fin-fold theory has been given greater detail. For example, Erik Jarvik has emphasized the segmental contributions to the archipterygial fins in some living fishes. If fins initially were steering keels, contributions from the adjacent segmental myotomes muscularize the fin folds, making them movable. Jarvik further suggested that the endoskeletal basals and radials developed from the mesenchyme within the core of the fold, supporting the

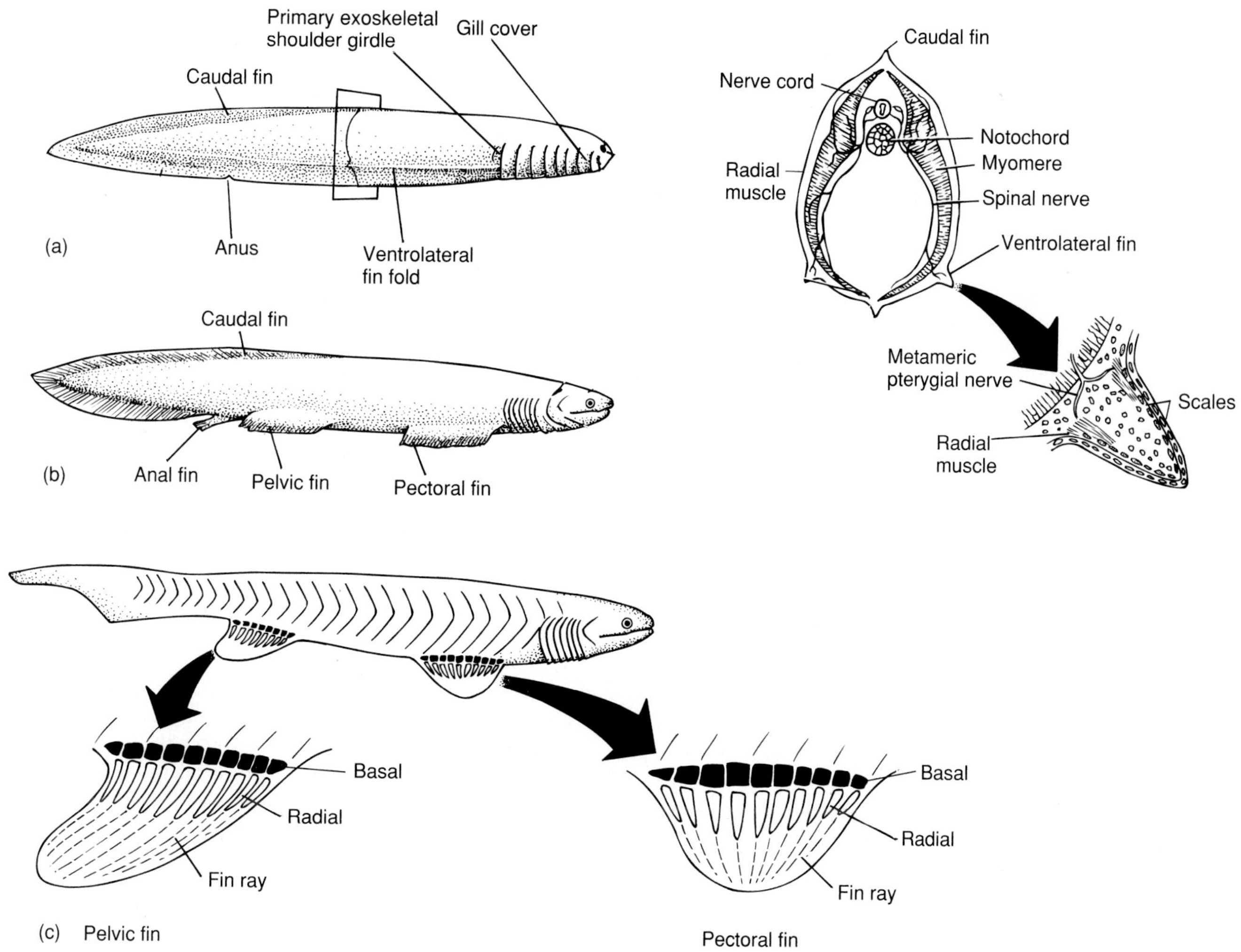


**Figure 9.5** Fin-fold theory proposed by Balfour and Thacher for the origin of paired fins. Lateral stabilizer fins (a) become divided into specialized pectoral and pelvic fins (b). Basals and radials enlarge to support the fins (c). The metapterygial stem is solid black.

projecting fin and offering attachments to muscles. Within the fin, the supportive dermal fin rays developed from modified rows of scales, an event that seems to be repeated during the embryonic development of dermal fin rays in many living fishes.

Jarvik has also taken exception to the view that dermal bone is added initially to the shoulder girdle because of selection forces favoring fin stability. He notes that the fish shoulder girdle lies at the transition from trunk to head. At this point, the axial musculature is interrupted by the pharyngeal slits. Consolidation of small dermal skin bones into a composite dermal girdle may have been initially advantageous because it offered an anterior site for attachment of the interrupted axial musculature at this point of transition. This dermal girdle would also form the posterior wall of the buccal cavity, protect the heart, and be the attachment site of some sets of jaw and gill-arch muscles. For one or all of these reasons, an anteriorly placed dermal girdle may have arisen and only secondarily joined with endoskeletal elements in support of the fin. Of course, no similar selection forces would be acting posteriorly, where the axial musculature runs uninterrupted from the trunk to the tip of the tail. This would help account for the presence of dermal contributions to the pectoral girdle and their absence from the pelvic girdles.

## Embryonic Development of Tetrapod Limbs

Although Gegenbaur considered archipterygial fins the most ancient type of fin, this now seems unlikely. The archipterygial fins of lungfishes are more likely modified from metapterygial fins. Metapterygial fins are common in gnathostome fishes, and neither type shows derivation from embryonic gill arches as Gegenbaur envisioned. The most

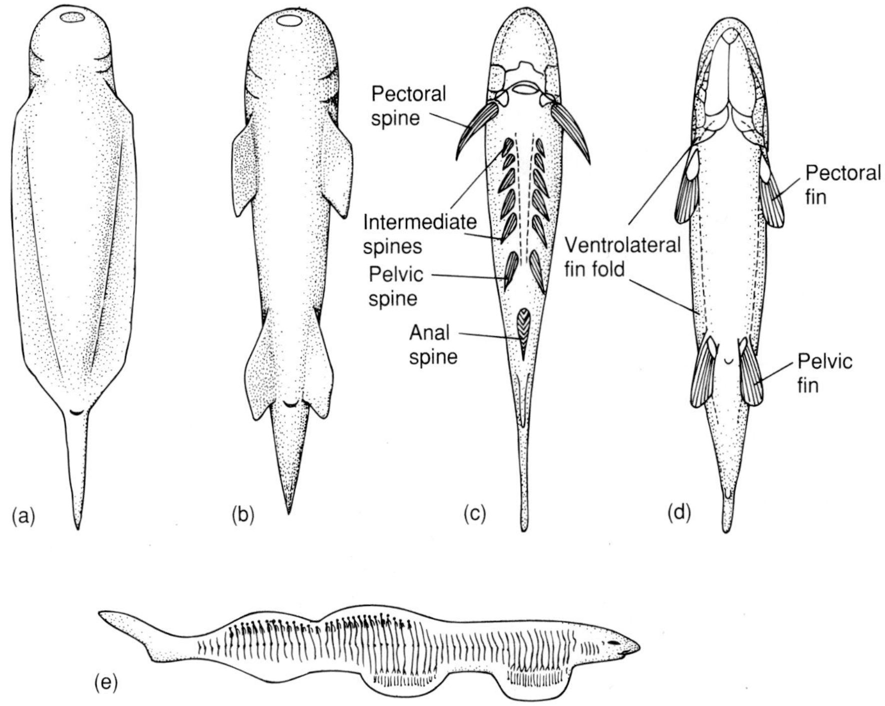


**Figure 9.6** Fin-fold theory. Indirect evidence pointing to the fin-fold theory can be seen from the position of rows of ventral spines in acanthodians (c), presumably remnants of the lost fin fold in pregnathostome ancestors (a,b). Ventral views (a–c) can be compared with the ventral view of a bony fish (d). Further evidence can be drawn from shark embryology (e), in which the discrete paired fins develop from a continuous thickening along the lateral body wall of the shark, a developmental event taken to be reminiscent of phylogenetic events.

constant and perhaps most ancient part of the fin is the posteriorly placed metapterygial stem, which we recognize in the paired fins of gnathostome fishes as well as in the tetrapod limb.

In fact, recent embryonic studies detect what seems to be a common developmental pattern underlying most tetrapod fore- and hindlimbs. The predominance of elements arising posteriorly on the postaxial side of the limb characterize this pattern. First, a stylopodium appears, which consists of a single proximal element (humerus or femur). The stylopodium subsequently branches to yield an adjacent set of preaxial (radius or tibia) and postaxial (ulna or fibula) elements that represent the intermediate section of the limb, the zeugopodium. Through subdivision and budding of embryonic primordia, the postaxial element alone extends the emerging embryonic pattern further distally, forming most of the primordia of the autopodium (manus or pes), which includes all the digits. The preaxial element contributes only a few wrist or ankle elements but none of the digits to the emerging embryonic autopodium (figure 9.7a).

This embryonic pattern of limb formation is asymmetrical. The postaxial series of skeletal elements branch to produce most of the limb and all its digits. The preaxial series never branch. Limb asymmetry in tetrapods seems to result from retention of the postaxial metapterygial stem in fishes (figure 9.7b). In sharks, primitive actinopterygians, and fossil rhipidistians, the branching metapterygial stem forms the series of postaxial elements in their fins. Elements of the metapterygial stem become incorporated into the tetrapod limb along with the basic asymmetry; therefore, the metapterygial axis of the tetrapod limb is asymmetrical, running through the postaxial elements and digits. Variation occurs from group to group in adult limbs, but this variation can be traced back to developmental modifications within the underlying embryonic pattern. Modifications of the basic pattern include fusion or loss of its fundamental elements, expansion of existing elements, and occasional appearance of new skeletal components, or **neomorphs.**

The basic configuration of supportive elements is generally the same in pectoral and pelvic appendages; however, some differences arise from changes in or to the pelvic girdle. For example, in elasmobranchs and in many placoderms, the pelvic fins of males are often endowed with **claspers,** modifications of the pterygiophores used during mating to engage the female. In many teleost fishes, the pelvic girdle and associated fins move forward to reside with the pectoral fins within the shoulder region.

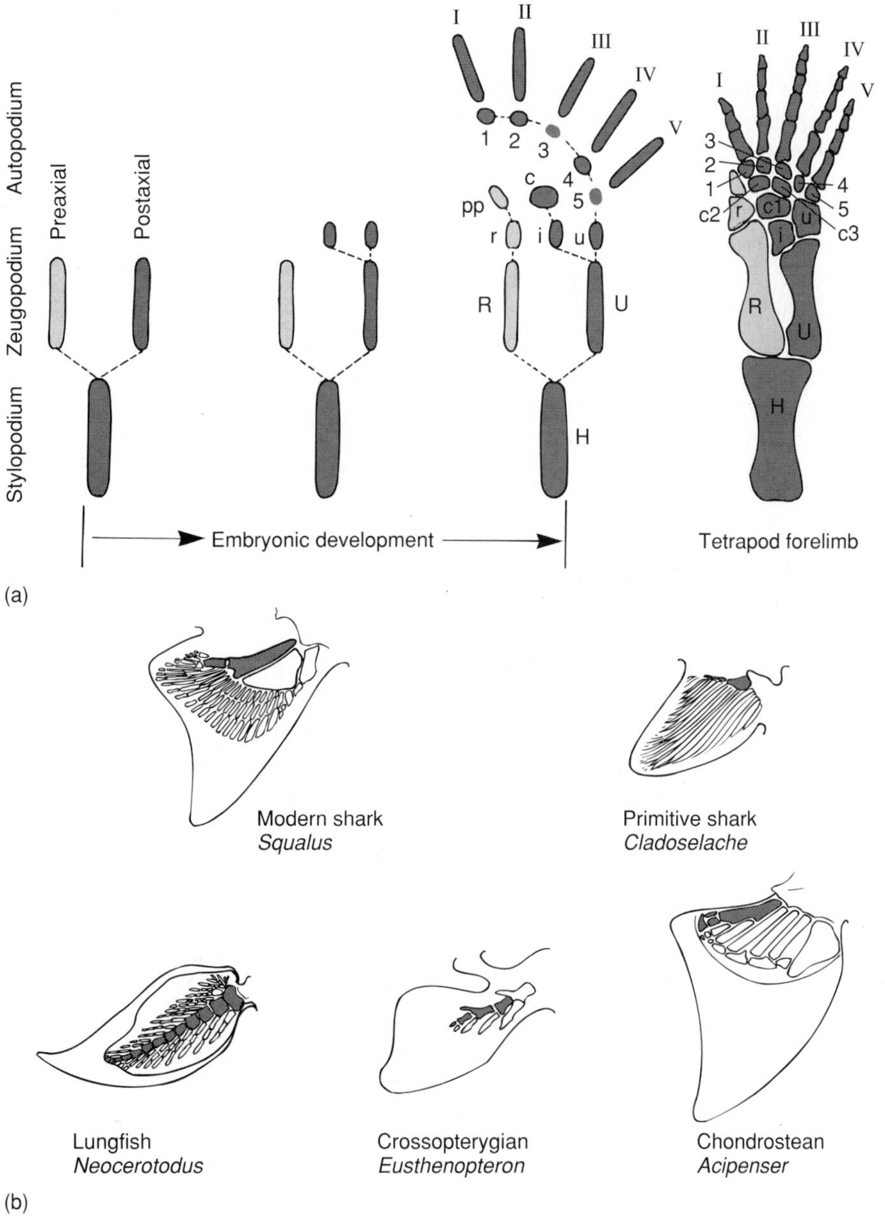


**Figure 9.7** Hypothesized developmental pattern underlying most limb development. (a) The stylopodial element appears and then divides into a preaxial and a postaxial element in the zeugopodium. The preaxial element (radius/tibia) does not branch but gives rise to more distal elements contributing to the autopodium. The postaxial element (ulna/fibula) branches to form the carpals or tarsals and the digital arch, which yields the fingers and toes. This postaxial side of the limb is thought to be derived from the metapterygial stem of fishes. (b) Position of the metapterygial stem in representative fishes. Abbreviations: humerus (H), radius (R), ulna (U), radiale (r), prepollux (pp), intermedium (i), ulnare (u), centrales (C1–C5), carpals (1–5), digits (I–V).

# Phylogeny

## Fishes

### *Agnathans*

Ostracoderms had unpaired medial fins, a caudal fin on the tail and often unpaired anal and dorsal fins. Most lacked true paired pectoral and pelvic fins. Anaspids usually possessed a paired sharp spine in the shoulder region, and some genera had long and stabilizing lateral fin folds running the length of their bodies. Heterostraci and Galeaspida fossils lacked all trace of paired fins. Similarly, living cyclostomes clearly lack paired fins. Only among some osteostracans were paired fins present and only in the pectoral region. The corners of the head shield bore indented fossa into which lobe-shaped pectoral fins fitted, and the margins of the fossa offered attachment sites for associated fin musculature. Details of the fins themselves are incomplete, but there is evidence of an endoskeleton and associated muscles. Thus, all ostracoderms lacked pelvic fins, and most lacked even rudimentary pectoral fins.

Like acanthodians, sharks, and most placoderms, ostracoderms lacked lungs or air bladders as well. Their bony armor surface gave them a density greater than the surrounding water, so they tended to sink to the bottom when they stopped swimming. Pectoral fins or spines provided anterior lift as did the flattened head shield. Both gave ostracoderms some devices to generate modest lift when they swam. However, absence or slight development of pectoral fins, small body musculature, and absence of jaws suggest that these fishes were bottom dwellers and feeders, only occasionally becoming active swimmers in open waters.

### *Placoderms*

Placoderm fishes first appear in the Silurian and enjoyed quite an extensive radiation, apparently taking advantage of powerful jaws and active lifestyles. Both pectoral and pelvic girdles were present. The pelvic girdle seems to have been a single endoskeletal element. The more complex pectoral girdle consisted of various fused dermal elements that contributed to the walls of the thoracic bony armor and braced the endoskeletal scapulocoracoid. Indented within the scapulocoracoid was an articular fossa that received the basal pterygiophores of the fin. In some placoderms, such as the antiarchs, the pectoral "fin" was quite specialized, forming a tapered appendage of endochondral elements encased in dermal bone (figure 9.8a,b).

### *Chondrichthyans*

Primitive chondrichthyans, such as early sharks, possessed pectoral and pelvic fins that were primarily stabilizers. They consisted of basal elements and tightly packed radials supporting the fin; the girdle was a single, enlarged basal element (figure 9.9a). In later sharks, the paired basal components of the pectoral and pelvic girdles became extended across the midline of the body to fuse into U-shaped **scapulocoracoid** and **pubioischiac bars,** respectively (figure 9.9b,c). Even the earliest chondrichthyans show no evidence of dermal contributions to the shoulder girdle. Modern sharks possess three enlarged pterygiophores at the

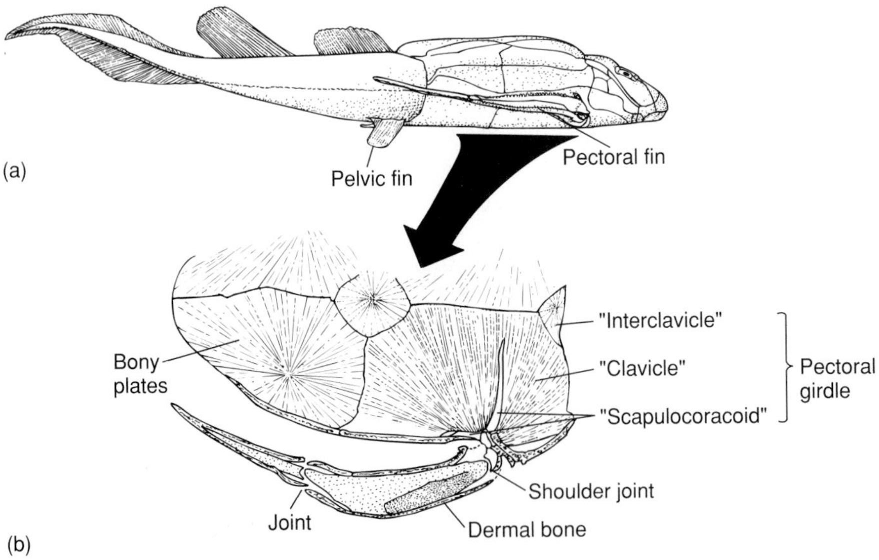


**Figure 9.8** The antiarch *Bothriolepis*, a placoderm from the late Devonian. (a) Lateral view. (b) Ventral view detailing the pectoral fin. The pelvic fin was only slightly developed and the pectoral fin, although distinct, was more of a specialized spine articulated with the girdle. Endochondral elements were encased in dermal bone. Pectoral fin is sectioned to show endochondral elements within the dermal exoskeleton. Homologies of bones contributing to the pectoral girdle are uncertain, so they are placed in quotes.
Source: After Stensiö, 1969.

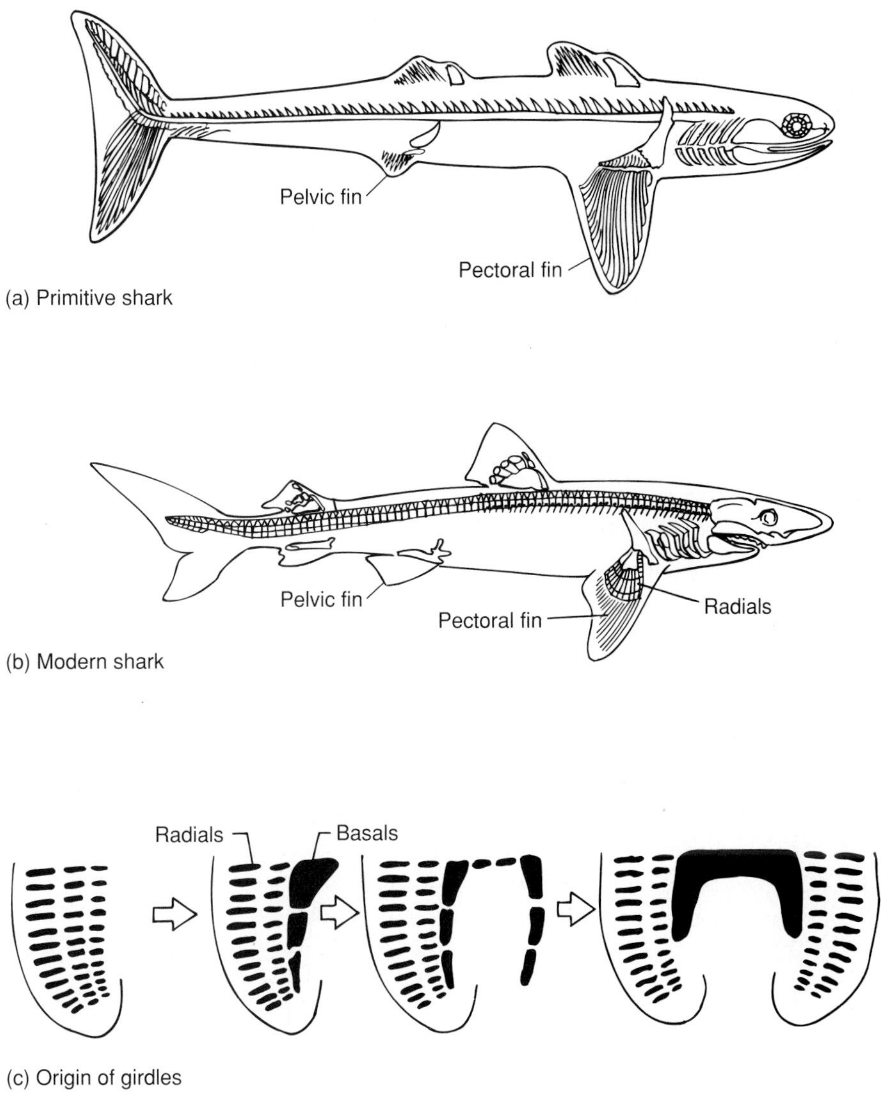


**Figure 9.9** Primitive (a) and modern (b) sharks. One trend in the evolution of the shark appendicular skeleton was the fusion of separate basal girdle elements across the midline (c). These fusions of pterygiophores produced the pubioischiac and scapulocoracoid bars.

base of the pectoral fin. The most posterior of these three is the **metapterygium,** arising within the series of basals of the metapterygial stem, followed by the **mesopterygium** and **propterygium,** enlarged derivatives of the radials. The metapterygial stem within the pelvic fin consists of a postaxial series usually with one long element supporting a stand of radials (figure 9.10).

## *Acanthodians*

In acanthodians, large spines formed the leading edge of dorsal, anal, and paired fins. Often, additional spines ran in rows between the pectoral and pelvic fins. In life, a web of skin usually stretched across these spines and was covered by rows of delicate scales (figure 9.11a–c). If present at all, basal and radial elements tended to be quite small. In some acanthodians, the pectoral spine articulated with a scapulocoracoid, but the pelvic spine is not known to articulate with an endoskeletal girdle.

## *Bony Fishes*

***Actinopterygians*** The pectoral girdle of actinopterygians is partly endochondral but mostly dermal. An air bladder is common throughout the group, so most members are neutrally buoyant. Large hydrofoil fins are rare. The fins function mainly as small oars for close maneuvering or slight adjustments of body position.

The dermal shoulder girdle, well established even in primitive bony fishes, forms a U-shaped collar of bone around the posterior border of the gill chamber and braces the small endoskeletal scapulocoracoid. The largest element of the dermal shoulder girdle is the **cleithrum,** upon which

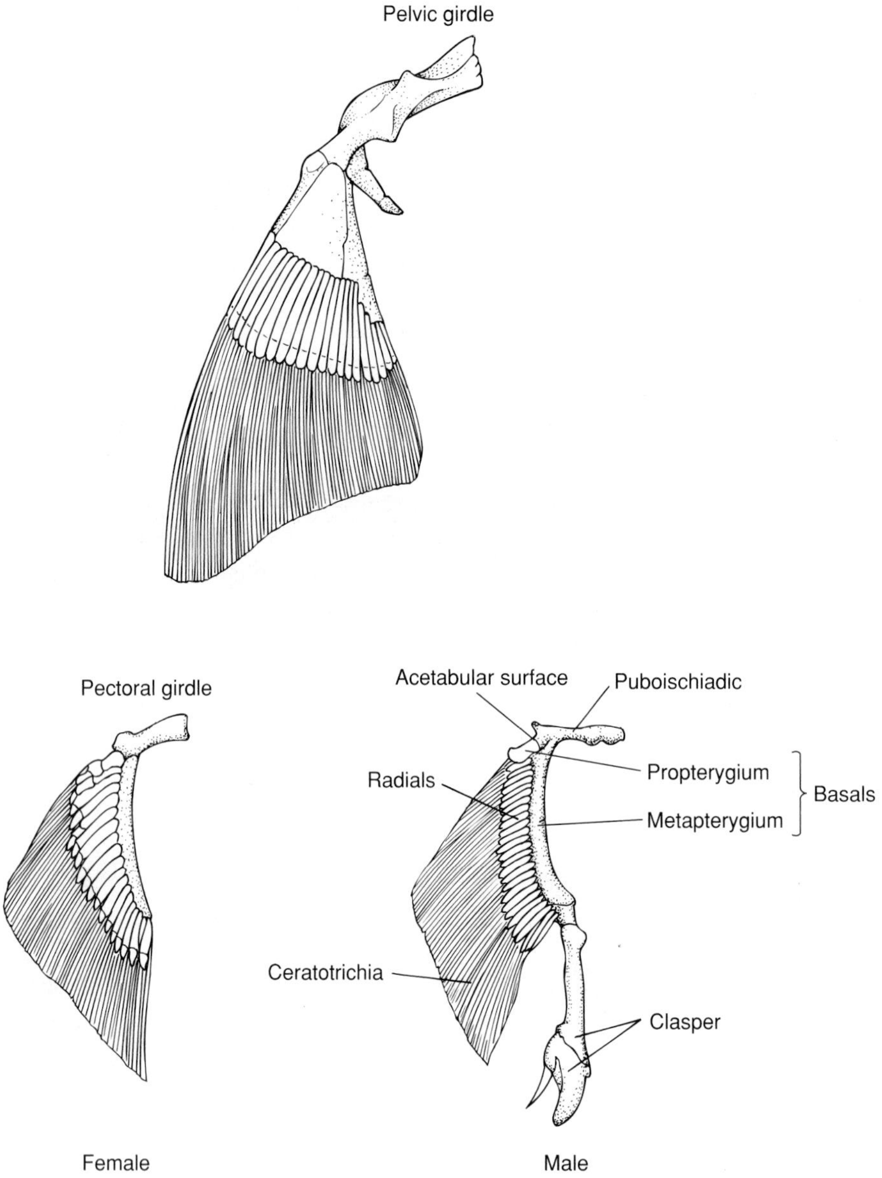


**Figure 9.10** Appendicular elements in the pectoral and pelvic fins and girdles of the modern shark *Squalus*.

the scapulocoracoid usually resides (figure 9.12). Ventrally, the cleithrum meets the **clavicle,** which bends medially to meet the opposite clavicle at the midline beneath the gill chamber. Where they meet they form a **symphysis.** Dorsally, the cleithrum supports a **supracleithrum** and, in turn, a **posttemporal** through which the dermal girdle is attached to the back of the skull. In some actinopterygians, additional dermal bones may join this girdle (e.g., the anocleithrum), whereas in others, dermal bones may be lost (e.g., the clavicle is usually lost in teleosts). However, this basic set of dermal bones is common in the shoulder girdle of actinopterygians.

***Sarcopterygians*** The sarcopterygians are sometimes referred to as the lobe-finned fishes, a reference to the muscles and internal supportive elements that project from the body to form the fleshy base of the dermal fin. Among the sarcopterygians, the crossopterygians are of particular interest because these primitive fishes possess certain fin features that approach limb features in early tetrapods.

Surviving sarcopterygians include three living genera of lungfishes and a single crossopterygian species, *Latimeria*. Lungfishes dating from the early Devonian show some skeletal specializations that characterize their subsequent evolution. Among the living genera, the fins are considerably reduced; only the skeletal elements of the Australian lungfish *(Neoceratodus)* exhibit features that can be homologized to those of other fishes. The dermal shoulder girdle includes a cleithrum and clavicle together with a dorsal anocleithrum (figure 9.13a). The endoskeletal girdle includes

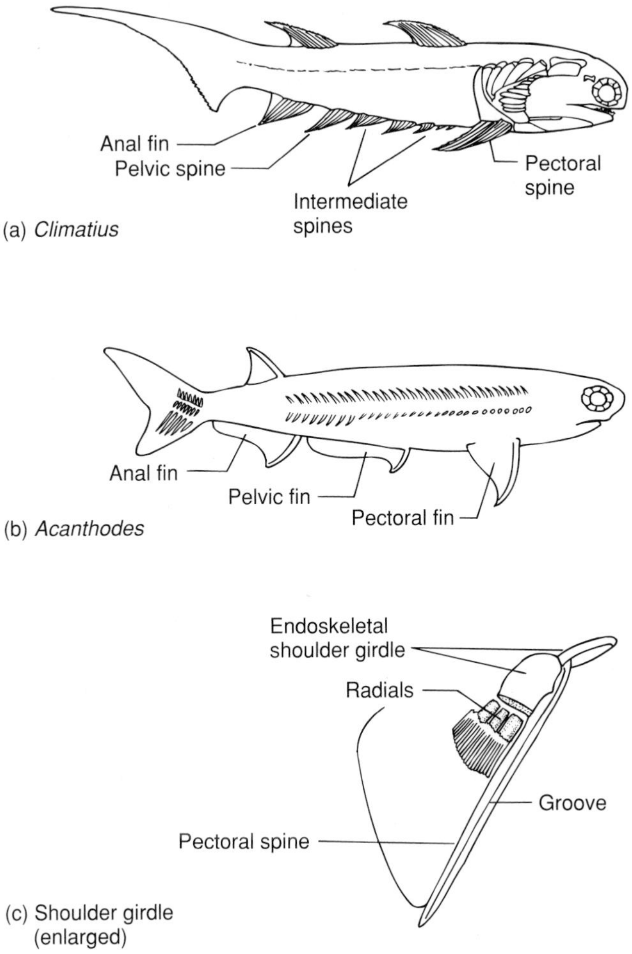


**Figure 9.11** Acanthodian fishes showing the row of spines between pectoral and pelvic elements. (a) *Climatius*. (b) *Acanthodes*. (c) Enlargement of restored endoskeletal shoulder girdle. Note the pectoral spine with the small radials and fused elements at its base.

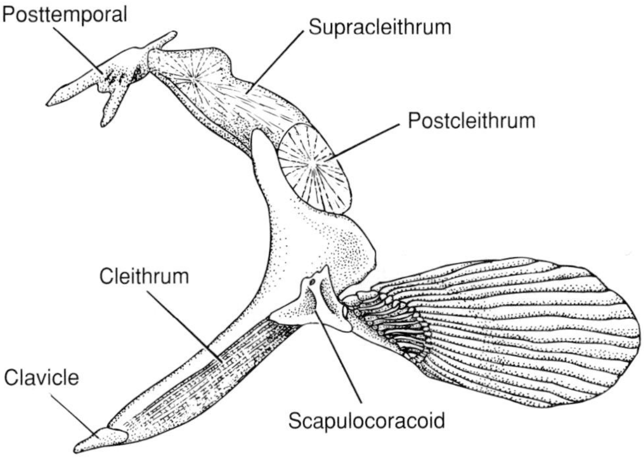


**Figure 9.12** Pectoral girdle of *Amia*, a primitive actinopterygian.

a scapulocoracoid that supports the projecting series of fin elements in an archipterygial pattern. Paired pelvic fins are also archipterygial and rest on a single cartilaginous girdle element.

Despite their endoskeletal structure, these fins do not carry lungfishes on trips across land. The thin, threadlike fins of the South American and African lungfishes are unsuited for terrestrial transport of the body. The Australian lungfish has slightly stronger fins, but it is, in fact, the most aquatic of the three genera. Fins serve these lungfish to maneuver in shallow water or crawl about aquatic vegetation and obstacles on the bottom.

The only surviving crossopterygian is the coelacanth, *Latimeria*. This group first appeared during the middle Devonian. Skeletal elements within the fins of *Latimeria* form a long, unbranched axis. The dermal shoulder girdle lacks an interclavicle but includes a crescent of four bones, a ventral clavicle, a cleithrum and extracleithrum that support the scapulocoracoid, and a dorsal bone believed to be

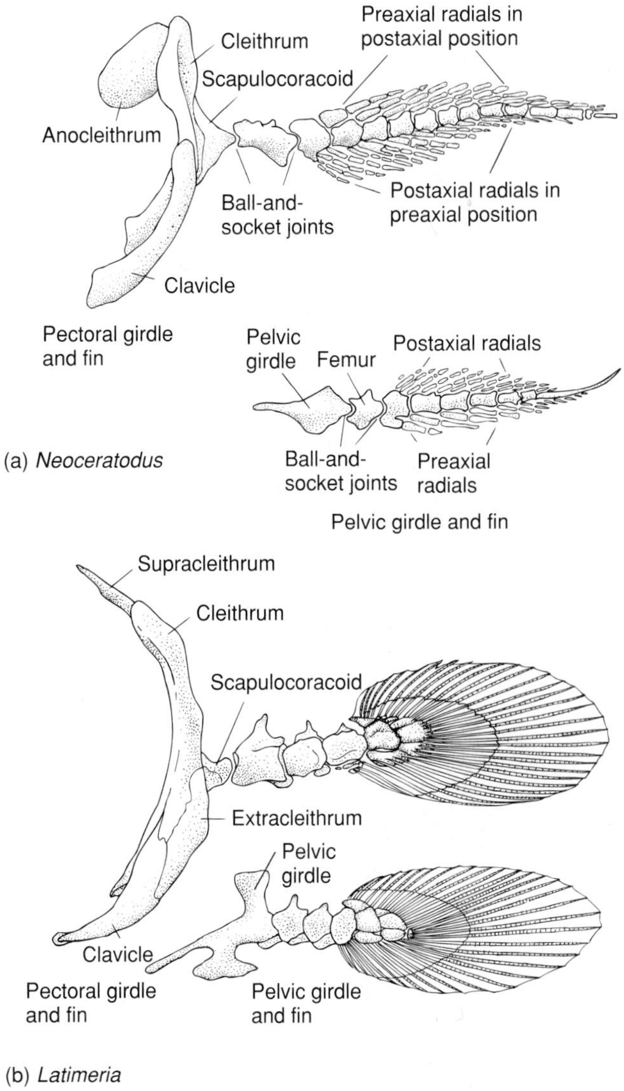


**Figure 9.13** Appendicular skeletons of living sarcopterygians. (a) The dipnoan *Neoceratodus*. (b) The crossopterygian *Latimeria*.

the supracleithrum (figure 9.13b). The pelvic girdle consists of a single element bearing several processes. Direct observation of living *Latimeria* in their natural habitats at depths of about 150 m reveals that their fins play no significant role as they crawl along the bottom. Instead, the paired fins are used to stabilize and control their position in underwater currents in which they drift slowly or remain suspended.

Some fossil rhipidistians have left a remarkably detailed record of the structure of their lobed fins. One of the best studied of these is *Eusthenopteron,* a late Devonian rhipidistian. Pectoral and pelvic appendages support dermal fins but internally possess bones homologous to those of early tetrapod limbs (figure 9.14a,b). The pectoral fin articulates with a scapulocoracoid and a series of supporting paired dermal elements of the girdle—clavicle, cleithrum, anocleithrum, supracleithrum, posttemporal. In addition to these, a single unpaired dermal element is present midventrally, which overlaps both lower tips of the two halves of the girdle (figure 9.14b). This oval bone, essentially an enlarged oval scale, is the **interclavicle.** The interclavicle is a new member of the fish girdle, joining it first here in crossopterygians. It is retained in the dermal girdle of later tetrapods as well.

The pelvic fin articulates with a single endoskeletal girdle bone. Left and right members of this paired girdle do not meet at the midline nor do they articulate with the axial column. Instead, they are embedded within the body wall, offering a bony base from which the fleshy fin projects outward from the sides of the fish (figure 9.14a).

## Tetrapods

Early amphibians, being the first tetrapods, quickly displayed changes in the appendicular skeleton correlated with locomotion on land and exploitation of the terrestrial environment. One of these adaptations has already been mentioned. The pectoral girdle tends to lose its attachment to the skull, a feature allowing increased cranial mobility and perhaps reduced jarring of the head. Girdles and limbs became stronger, more robust, and more completely ossified (figure 9.15a,b). One of the earliest amphibians was *Ichthyostega.* Its pelvic girdle was composed of three fused bones—**pubis, ischium,** and **ilium.** Through the ilium, the pelvic girdle became attached to the vertebral column and so defined the sacral region (figure 9.16). The pectoral girdle lost its attachment to the skull. The dermal fins of fish ancestors were replaced by digits.

*Eogyrinus,* a primitive labyrinthodont of the Carboniferous, reached over 2 m in length. The limbs and girdles, although strengthened, were relatively small for its size and mostly cartilaginous, so they were unlikely used to romp across the land. Instead, like modern salamanders and some fish, *Eogyrinus* must have used its limbs as pivots around which it moved (figure 9.17a). Limbs and girdles of amphibians became progressively larger and strengthened, reflecting an increased terrestrial habit and exploitation of the land. *Eryops,* a temnospondyl of the Carboniferous and Permian, also reached lengths of almost 2 m. But the limbs and girdles of *Eryops* were robust, extensively ossified, and strong supportive structures of an amphibian committed to life on land (figure 9.17b,c).

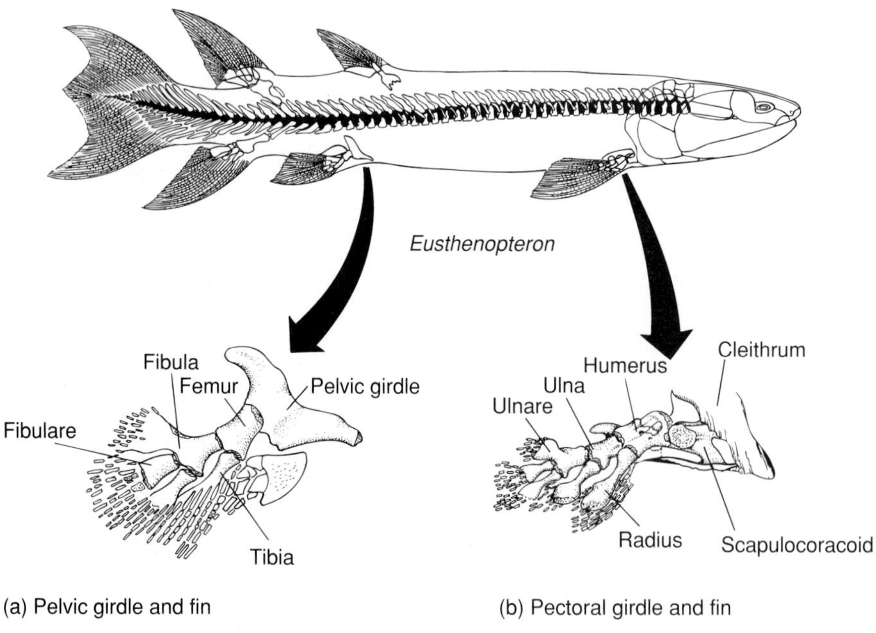


**Figure 9.14** Appendicular skeleton of the fossil crossopterygian *Eusthenopteron.* (a) Pelvic girdle and fin. (b) Pectoral girdle and fin.

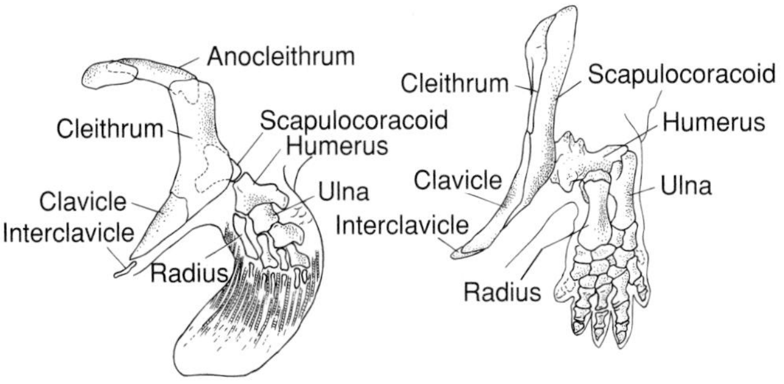


(a) Crossopterygian *Eusthenopteron*

(b) Labyrinthodont *Eryops*

**Figure 9.15** Appendicular skeleton of a crossopterygian and early amphibian, showing left girdle and appendage. (a) *Eusthenopteron*, crossopterygian fish from the late Devonian. (b) *Eryops*, a temnospondyl amphibian from the Carboniferous.

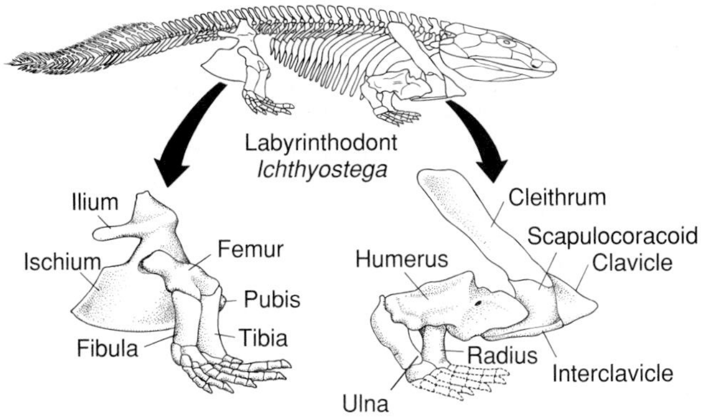


**Figure 9.16** Primitive amphibian *Ichthyostega*, showing components of the appendicular skeleton. The number of digits carried by the forelimb is not known. Six digits were present on the hindlimb.

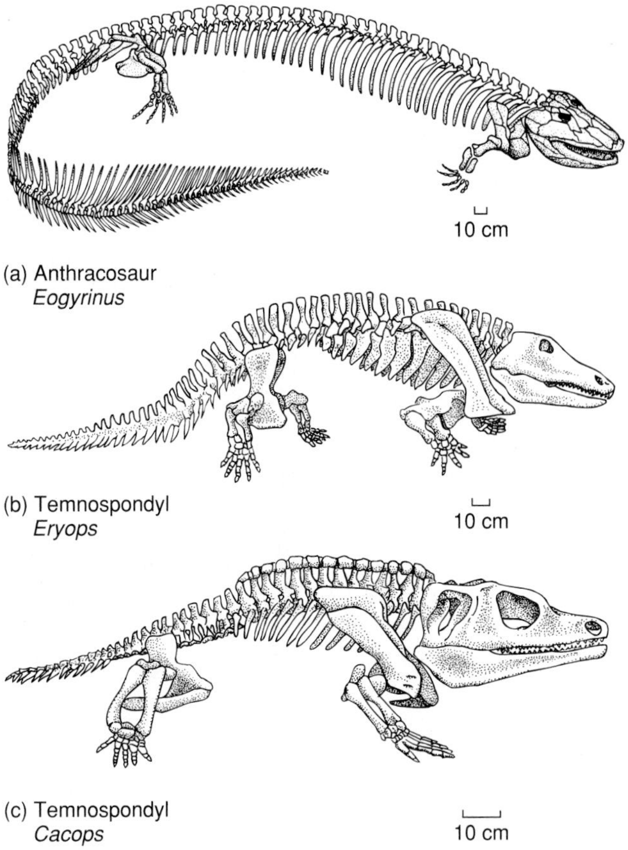


**Figure 9.17** Primitive amphibians. (a) *Eogyrinus*, from the Carboniferous, was over 2 m in length. Although limbs and girdles were present, they were built slightly for an animal of this size, perhaps serving as points of pivot rather than actual weight-bearing structures. (b,c) By comparison, the more robust limbs and girdles of *Eryops* and *Cacops*, both from the Permian, reflect increased use of the limbs for terrestrial locomotion.

Skulls of young *Eryops*, unlike those of adults, show evidence of a lateral line system, an aquatic sensory system. This suggests that adults were predominantly terrestrial, whereas young were predominantly aquatic, a life cycle not unlike that of many modern amphibians.

**Lateral line system (p. 689)**

## *Pectoral Girdle*

Tetrapods carry over from crossopterygians a shoulder girdle consisting of dermal and endoskeletal elements; however, unlike their fish ancestors, tetrapods have a shoulder girdle that is structurally and functionally detached from the skull. Because the pectoral girdle is no longer connected to the back of the skull, the dorsal series of dermal bones, which previously were involved in establishing this connection in fish, are lost as well. Thus, in early amphibians, the post-temporal, supracleithrum, anocleithrum, and postcleithrum are absent, leaving a dermal shoulder girdle composed of the remaining ventral elements—the paired cleithrum and clavicle and an unpaired midventral interclavicle that joins both halves of the girdle across the midline (figure 9.18). In modern amphibians, the dermal bones are usually lost entirely, as in salamanders, or reduced in prominence, as in anurans. The endoskeletal scapulocoracoid becomes the predominant girdle element yet retains its fidelity to the cleithrum (figure 9.18). In fishes, the scapulocoracoid, as its composite name suggests, tends to be a single element. However, in early tetrapods, it actually arises from two distinct embryonic centers of endochondral ossification and produces two distinct bones, the **scapula** and the **coracoid** (figure 9.19).

In primitive reptiles, the clavicle and interclavicle persist, but the cleithrum is usually absent. The clavicle is lost in some modern reptiles, but it is retained in many, and in turtles it is incorporated into the plastron as the **entoplastron.** In birds, the paired clavicle usually fuses with the unpaired interclavicle, producing the composite wishbone,

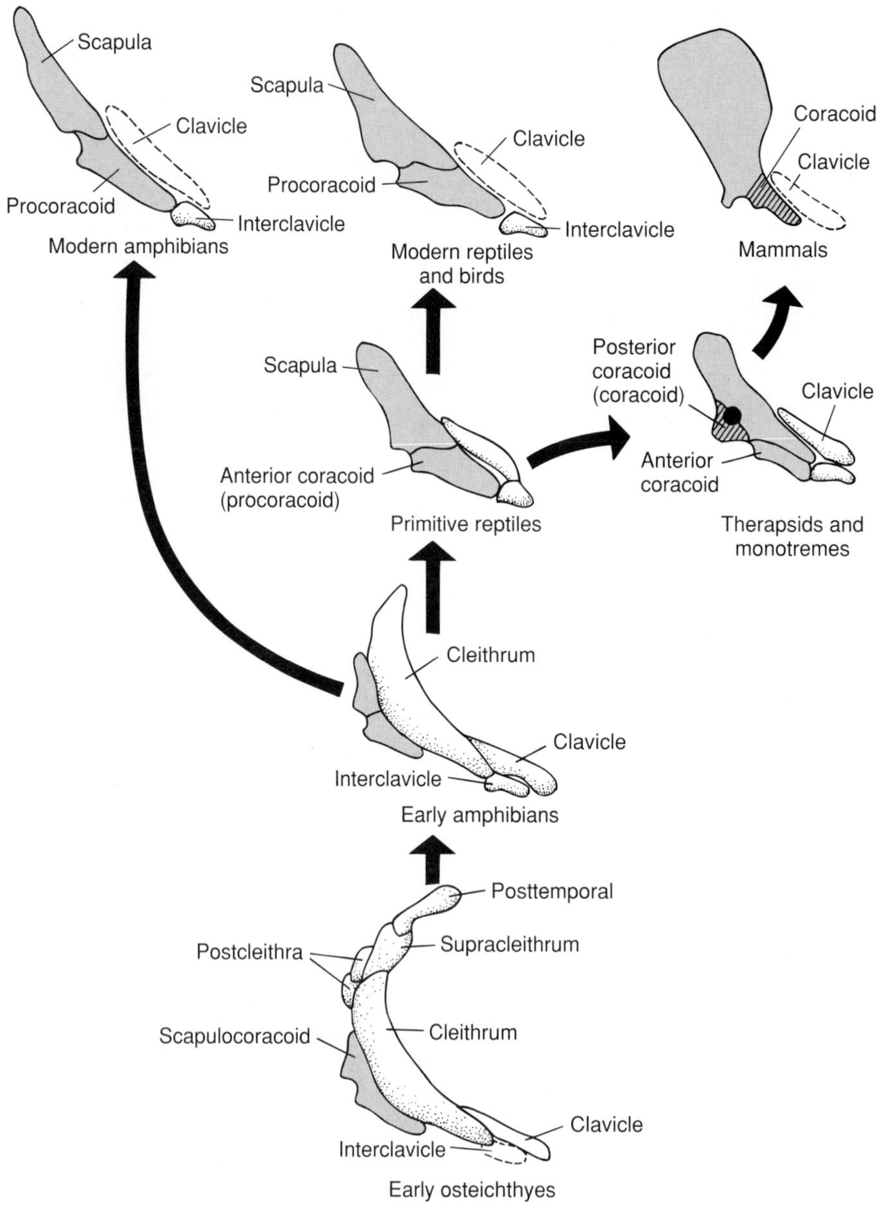


**Figure 9.18** Summary of pectoral girdle evolution. Notice that dermal elements of the girdle tend to be lost and endochondral elements tend to assume a greater role. In primitive therapsids, a third endochondral bone appears, the posterior coracoid, to join with the phylogenetically older scapula and anterior coracoid bones. The three persist into primitive mammals. In marsupials and placental mammals, only the scapula and posterior coracoid (called just coracoid) persist. In modern reptiles and birds, the scapula and anterior coracoid (or procoracoid) persist.

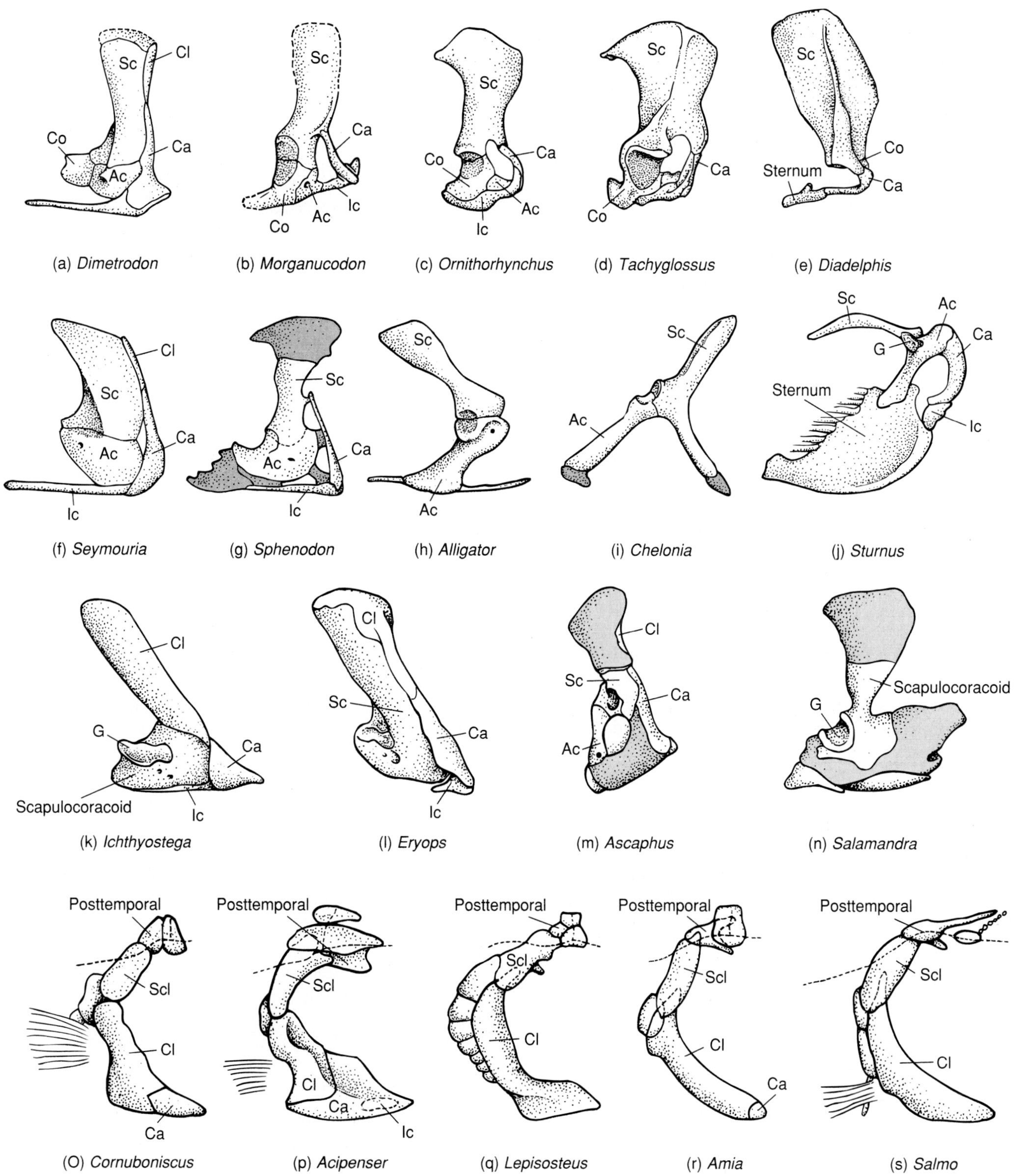


**Figure 9.19** Pectoral girdles of selected vertebrates. (a) Pelycosaur. (b) Primitive extinct mammal. (c) Duckbill platypus, a monotreme. (d) Echidna, a monotreme. (e) Opossum, a marsupial. (f) Anthracosaur, an amphibian. (g) *Sphenodon*, a squamate. (h) Alligator, a reptile. (i) Turtle, a reptile. (j) Bird. (k) Ichthyostegid, an amphibian. (l) Temnospondyl, an amphibian. (m) Frog. (n) Salamander. (o–s) Actinopterygians. Abbreviations: procoracoid (Ac), clavicle (Ca), cleithrum (Cl), coracoid (Co), glenoid (G), interclavicle (Ic), scapula (Sc), supracleithrum (Scl).

or **furcula.** Both scapula and coracoid of the endochondral girdle persist. In fact, they now become a more prominent part of the shoulder girdle in birds and modern reptiles.

As mentioned previously, endochondral elements of the early tetrapod shoulder develop from two centers of ossification, giving rise to a scapula and a "coracoid." However, in primitive synapsid reptiles, three centers of ossification develop. The dorsal center gives rise to the scapula, and the two ventral centers produce *two* coracoids. To avoid confusion and to track the fate of each, separate names are given to these two coracoids. The anterior of these synapsid coracoids is homologous to the coracoid of fishes, amphibians, and other reptiles we have followed so far. This **anterior coracoid** is more often called the **procoracoid** (precoracoid). The posterior of these synapsid coracoids is a new center of ossification, the **posterior coracoid,** or more often it is called just the **coracoid.**

Both coracoids are present in pelycosaurs, therapsids, and monotremes, but only the coracoid (posterior coracoid) persists into marsupial and placental mammals. The "coracoid" in therian mammals, then, is really a different coracoid from that found in other amniotes. Thus, using the terminology to work on our behalf when following phylogenetic events, the coracoid element in birds, reptiles, amphibians, and fishes (where applicable) should be called procoracoid. The term *coracoid* should be reserved for the new coracoid element in synapsids and therian mammals. We shall do so here. Unfortunately, in practice, others apply the single term *coracoid* loosely and inappropriately to all ventral endochondral elements in the shoulder girdle. The student should be prepared for this more general usage elsewhere.

Several dermal elements of the shoulder persist in early synapsids. The clavicle and interclavicle are present in therapsids and monotremes, but in marsupials and placentals, the interclavicle is absent and the clavicle often is reduced in size, whereas the scapula becomes the predominant shoulder element. On the other hand, the coracoid (posterior coracoid) is reduced and fused to the scapula as the **coracoid process** (figure 9.19).

## *Pelvic Girdle*

The pelvic girdle is never joined by contributions of dermal bone. From its first appearance in placoderms, the pelvic girdle is exclusively endoskeletal. It arose from pterygiophores, perhaps several times, in support of the fin. In crossopterygians, as in most fishes, it is formed of a single element, but in tetrapods, three endochondral bones contribute—ilium, ischium, and pubis (figures 9.20 and 9.21). Through the ilium, the pelvic girdle is attached to the vertebral column first in amphibians, establishing and therefore defining the sacral region. Throughout later amniotes, these three bones of the pelvic girdle persist, although their general pattern varies.

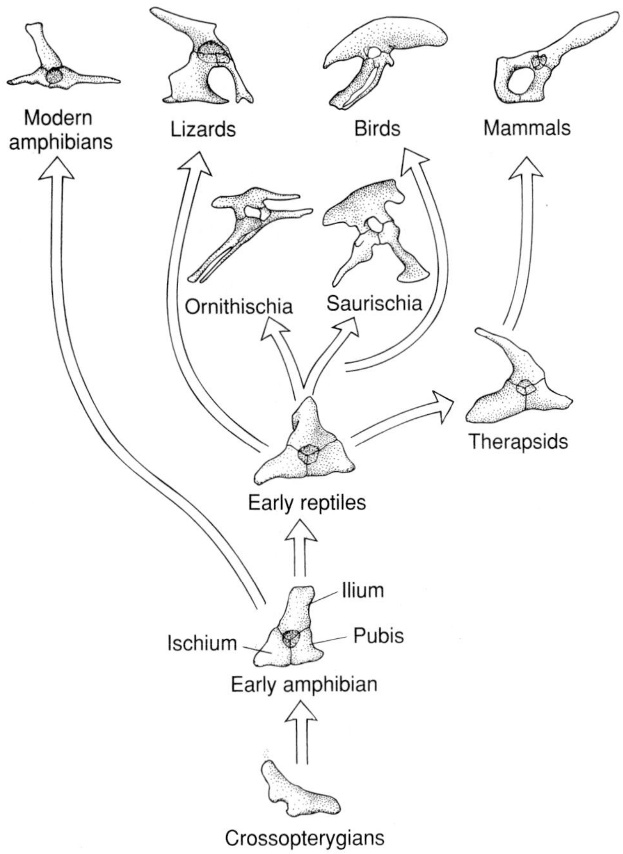


**Figure 9.20** Summary of pelvic girdle evolution. Three endochondral elements—ilium, ischium, pubis—characterize the pelvic girdle in early tetrapods. This basic pattern persists into later tetrapods.

For example, two distinctive patterns, the saurischian and ornithischian pelvic girdles, define two respective groups of dinosaurs. In birds, all three bones appear embryologically as distinct centers of ossification, but then they fuse to form the composite **innominate bone,** usually with no trace of sutures between them. Further fusion between the composite innominate and composite synsacrum introduces considerable firmness in the posterior avian skeleton.

## *Manus and Pes*

The autopodium at the end of each tetrapod limb has undergone a complex evolution of its own. Tracking this evolution has been difficult, largely because of the numerous elements that participate. Their are several **digits,** each beginning proximally with a **metapodial element** (**metacarpals** on the forelimb, **metatarsals** on the hindlimb) followed by a chain of **phalanges** (sing., phalanx). The digits rest upon several separate bones, collectively known as the **carpals** in the wrist and the **tarsals** in the ankle. In some marine vertebrates (ichthyosaurs, plesiosaurs, cetaceans,

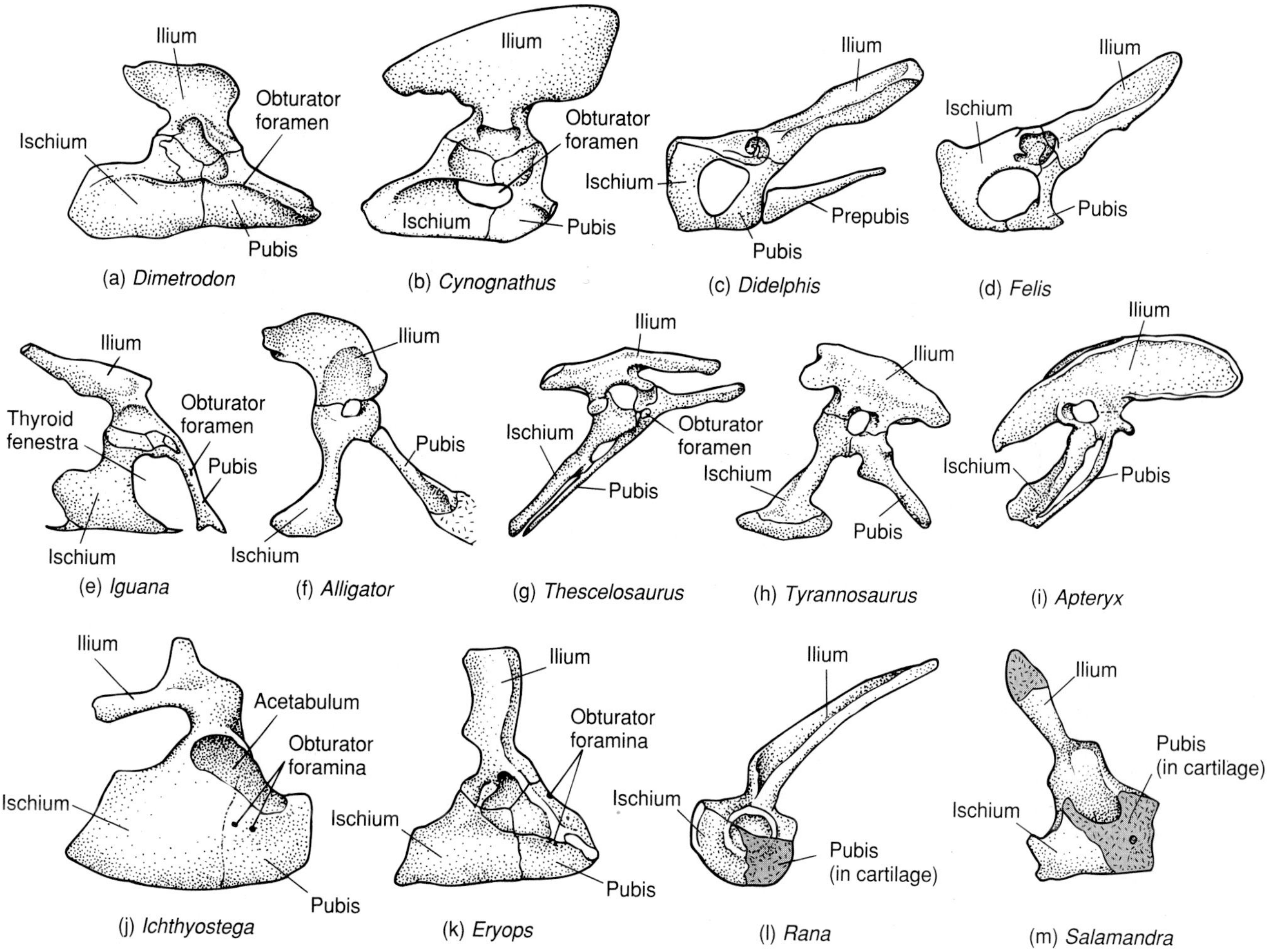


**Figure 9.21** Pelvic girdles of selected vertebrates. Stippling represents cartilaginous areas. (a) Pelycosaur. (b) Therapsid reptile. (c) Opossum, a marsupial. (d) Cat, a placental. (e) Lizard. (f) Alligator, a reptile. (g) Ornithischian, a dinosaur. (h) Saurischian, a dinosaur. (i) Bird. (j) Labyrinthodont, an amphibian. (k) Temnospondyl, an amphibian. (l) Frog. (m) Salamander.

sirenians, and marine carnivores, for example), the major trend has been toward **polyphalangy,** a proliferation in the number of phalanges. It is very unusual to find species with more than five digits, a condition known as **polydactyly.** But in many groups, such as ungulates and some terrestrial carnivores, the opposite trend has occurred, namely, toward reduction in the number of phalanges and loss or fusion of associated carpals and tarsals.

Traditionally, the basic tetrapod limb was thought to consist of five digits that were named and numbered (roman numerals) by their **pentidactylous** pattern (figure 9.22). Some were lost in specialized lines, but this underlying five-digit pattern is a reasonable hypothesis because the limbs of most primitive amniotes commonly carry five digits. Unfortunately, fossil specimens of the very earliest tetrapods, such as *Ichthyostega,* did not preserve enough of the digits to permit a confident count, at least not until recently. New discoveries of early tetrapod specimens test this five-digit hypothesis. The hindlimb of *Ichthyostega* had seven digits (its manus remains unknown still), the manus of *Acanthostega* (an ichthyostegalian) included eight digits (the pes number is unknown), and the fore- and hindlimbs of *Tulerpeton* (an anthracosaur) had six digits (figure 9.23). These late Devonian fossils are the earliest tetrapod remains available. Collectively, they indicate that the primitive tetrapod pattern was polydactylous and the five-digit pattern is a later stabilization.

Recent speculation also suggests that two independent tetrapod limb designs may have emerged from the polydactyl condition in primitive tetrapods. One was the amniote lineage in which digit number stabilized at five on each limb. The other lineage leads to modern amphibians, with five toes on the hindlimbs but only four on the forelimbs. The older view is that this four-digit pattern is derived from a five-digit ancestor. If we trace modern amphibian ancestry back to these earliest tetrapods, as some now suggest, then amphibians

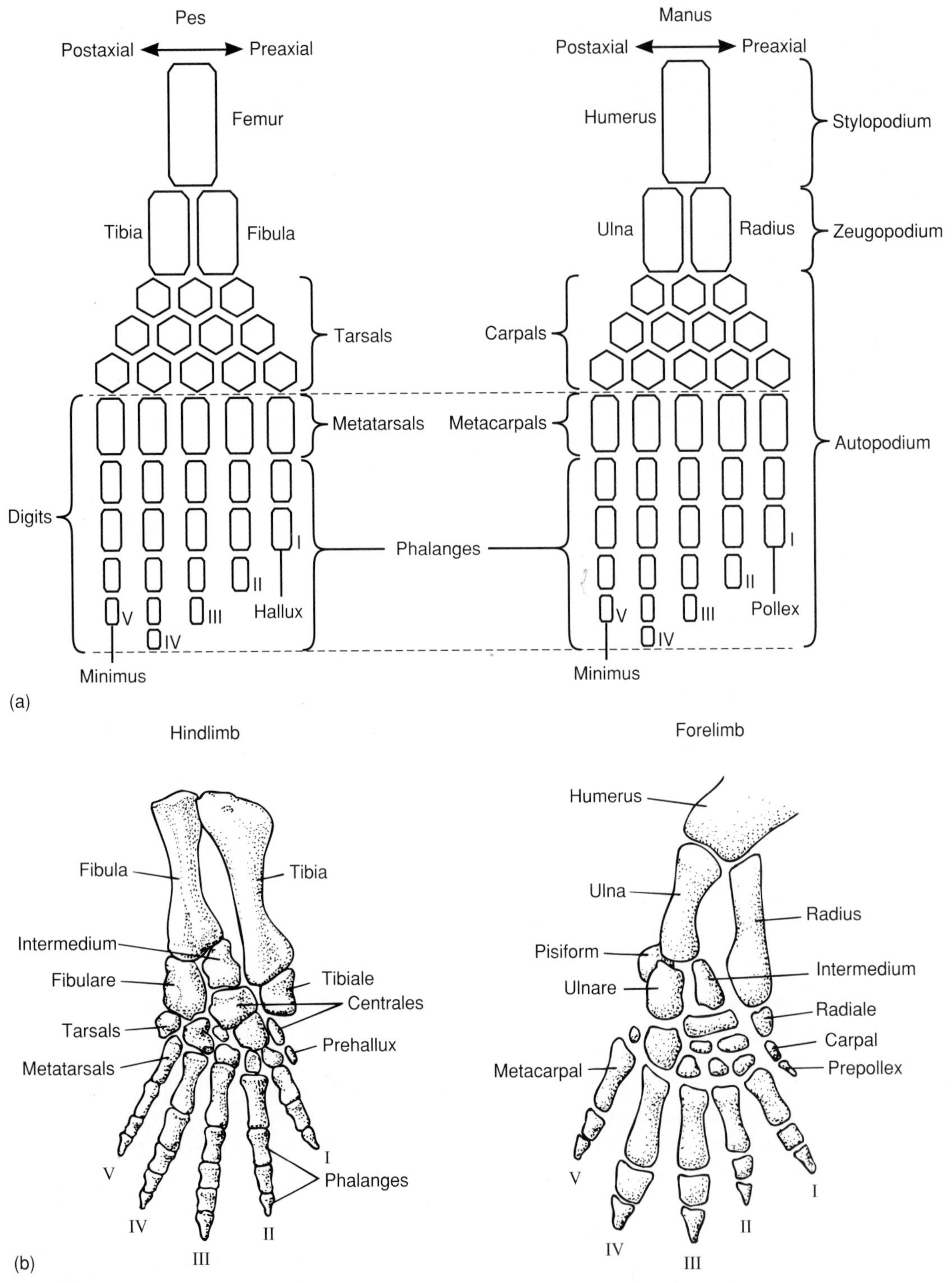


**Figure 9.22** Basic organization of the fore- and hindlimbs. (a) Manus and pes have five digits; each digit includes its metacarpal or metatarsal and chain of phalanges. These digits in turn articulate with various wrist and ankle bones. (b) Fore- and hindlimbs of primitive tetrapod.

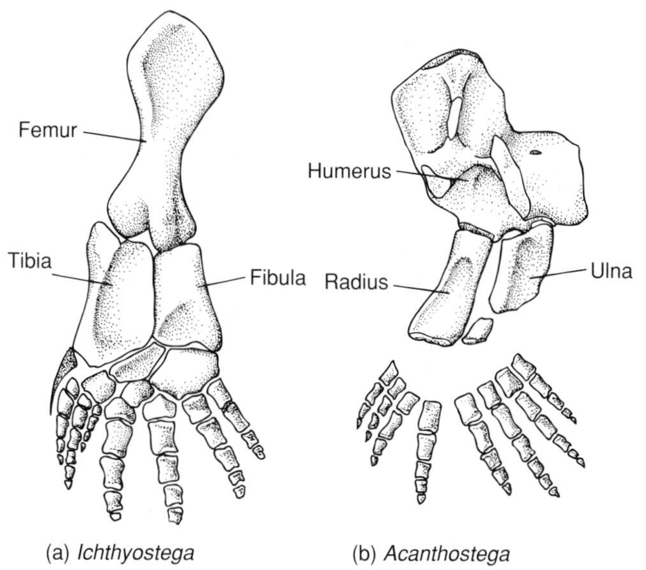


**Figure 9.23** Polydactylous limbs of the earliest tetrapods (dorsal views). (a) Hindlimb of *Ichthyostega* with seven digits. (b) Forelimb of *Acanthostega* (an ichthyostegalian) with eight digits.

Source: After Coates and Clack, 1990.

independently derived their reduced digit number directly from polydactylous ancestors.

Such speculations are tantilizing and refreshing but still quite tentative. The earliest tetrapods were clearly polydactylous. But the significance of this to later tetrapod evoluton awaits further study. For our purposes, the five-digit pattern represents a useful basis for discussing limb evolution and changes in functional design.

In the manus, a digit consists of several phalanges with a metacarpal at its base. Each of the five metacarpals articulates in turn with a carpal. The wrist bones that articulate with the radius and ulna are, respectively, the **radiale** and the **ulnare.** The **intermedium** lies between these two wrist bones. Within the middle of the wrist are one to three **centrales.** In the pes, the primitive number of digits is also taken as five, each with a metatarsal at its base. In turn, each digit articulates proximally with the following sequence of bones: tarsal, centrale, tibiale, intermedium, and fibulare, the last three meeting the tibia and fibula of the shank.

Although this stately pattern of expected manus and pes elements gives us a starting point when we look at distal limb anatomy, the actual morphology is often considerably modified by fusions, elongations, eliminations, and additions of apparently new elements to this pattern (figures 9.24a–h and 9.25a–g). For example, the **pisiform** is a sesamoid bone that can lie to the outside of the carpus, especially in reptiles and mammals. In birds, fusion of the forelimb elements produces an autopodium with three digits (II, III, IV). The ulnare regresses during ontogeny and in its space a neomorph arises from a new embryonic condensation (figure 9.24e). This new avian ankle bone has not been named. Some still call it an ulnare, others say "ulnare" in quotes. Some number it; others call it a neomorph. All will be found in new references in the avian literature for a while.

In the hindlimb, lateral digits tend to be lost in mammals, and the medial metatarsals (III and IV) fuse into a composite ankle bone commonly called the **cannon bone** (figure 9.25e). In birds, fusion of elements in the hindlimb produces a composite bone, the **tarsometatarsus,** named for the elements contributing (figure 9.25c). In mammals, the fibulare is the specific tarsal that articulates with the fibula, but it is more commonly called the **calcaneus,** and the element resulting from fusion of the intermedium and tibulare is called the **astragalus,** which articulates with the tibia (figure 9.25f). Although originating in mammals, these names have been carried over to ankles of reptiles and birds; however, differences in embryonic contributions to these two bones in reptiles, birds, and mammals perhaps betray differences in homology and, strictly speaking, call for different names. But some uncertainties regarding embryonic contributions remain, and the convenience of the terms *calcaneum* and *astragalus* for tarsal bones of similar function justifies using these terms for all groups, at least for the present.

Within archosaurs, two types of ankle joints occur. An **intratarsal joint** (crurotarsal joint) forms such that the line of ankle flexion passes between the calcaneus and the astragalus (figure 9.26a). Or both calcaneus and astragalus can be interlocked and the line of flexion passes distal to them, a **mesotarsal joint** (figure 9.26b). Such differences in ankles may be used to discern lineages within archosaurs. The intratarsal joint occurs in crocodilians and in most advanced thecodonts. The mesotarsal joint characterizes dinosaur ankles and likely represents an adaptation to bipedal locomotion. A tight interlocking union of astragalus and calcaneum ensures that the metatarsal joint will restrict hingelike flexion in the same plane of motion as the overall limb. Further, the metatarsal joint brings the calcaneus back into tight contact with the fibula, where it becomes part of the shank. In this instance, the calcaneus assumes a greater weight-bearing function within the hindlimbs that are themselves now brought under the body to help support the animal.

Changes in the primitive pattern of manus and pes correlate with changes in functional demand arising from the biological roles in which the limbs participate. We look at these changes in function shortly, but first let us reflect on the overall significance of the morphological patterns we have encountered so far in the appendicular system.

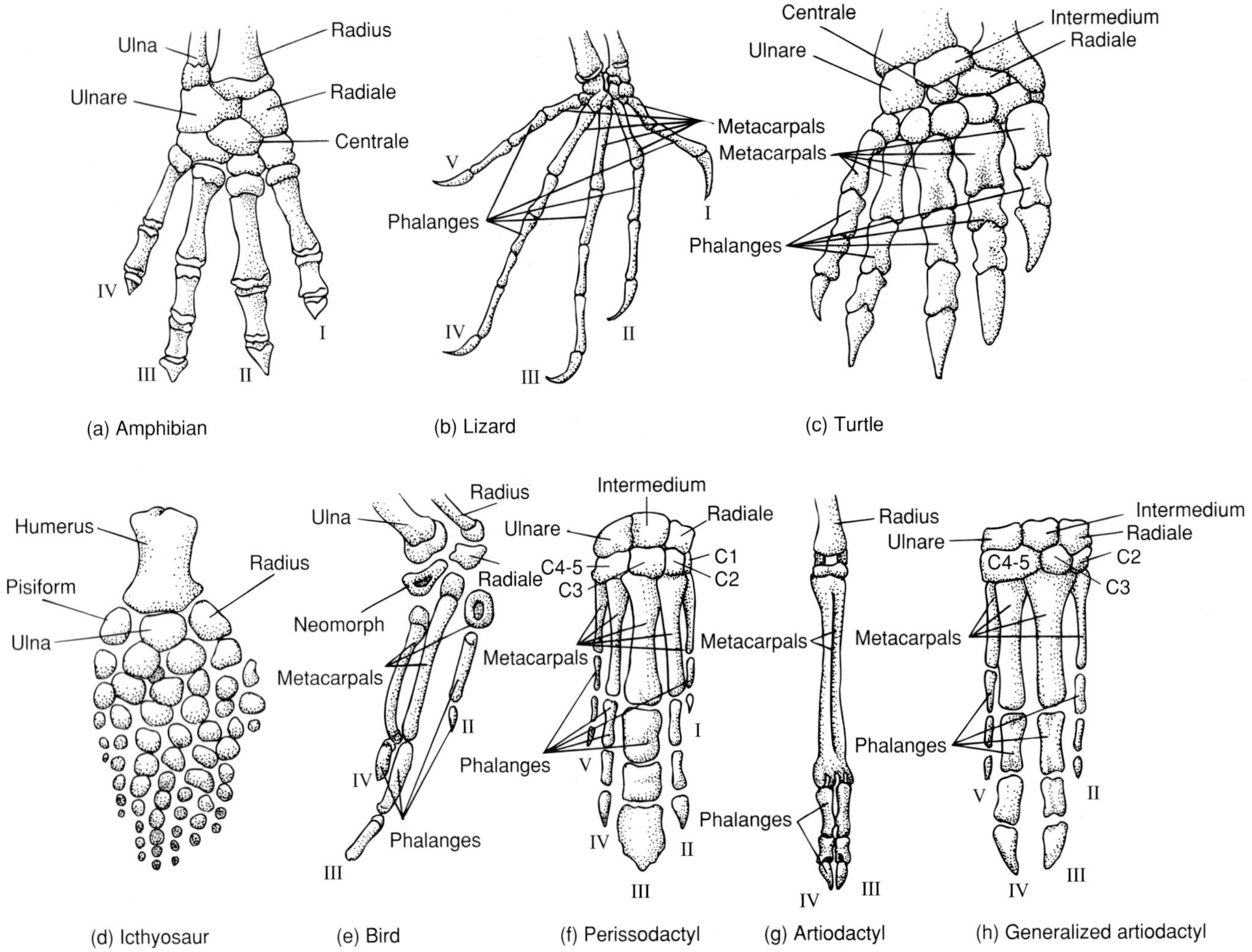


**Figure 9.24** Variations of the tetrapod manus.
(a) Amphibian *(Necturus)*. (b) Lizard. (c) Turtle *(Pseudemys)*.
(d) Ichthyosaur. (e) Bird. (f) Perissodactyl. (g) Artiodactyl.
(h) Generalized artiodactyl.

# Evolution of the Appendicular System

## Dual Origin of the Pectoral Girdle

Throughout its evolution, the pelvic girdle is endochondral. In early tetrapods, it comprises three distinct bones, the ilium, ischium, and pubis, instead of the single element characteristic of fishes.

However, the pectoral girdle is clearly of dual origin, composed of dermal as well as endochondral bones. The endochondral component, the scapulocoracoid, evolved by fusion or enlargement of several basal fin elements. It functions as the articular surface of the fin and later of the limb. The appendicular musculature of the forelimb is attached securely to it. The dermal component of the shoulder girdle evolved from dermal bones of the body's surface. In ostracoderm fishes, these bones composed the outer protective armor. As the early fish girdle evolved, some dermal armor may have sunk inward to join the existing endochondral components of the fish girdle. Alternatively, before becoming associated with the pectoral fins, these dermal bones may have developed the additional role of securing the anterior point of transition between axial musculature and the branchial chamber. Whichever their evolutionary route, dermal bones became an important brace for the endoskeletal pectoral girdle. Like endochondral bones, these dermal bones were passed along to tetrapods, where a new dermal element, the interclavicle, debuted in some crossopterygians. The interclavicle was incorporated at the midventral symphysis between both halves of the shoulder girdle. In general, the dermal girdle offered additional area for muscle attachment and protected the heart, but it functioned primarily as a firm brace for the endochondral elements of the shoulder.

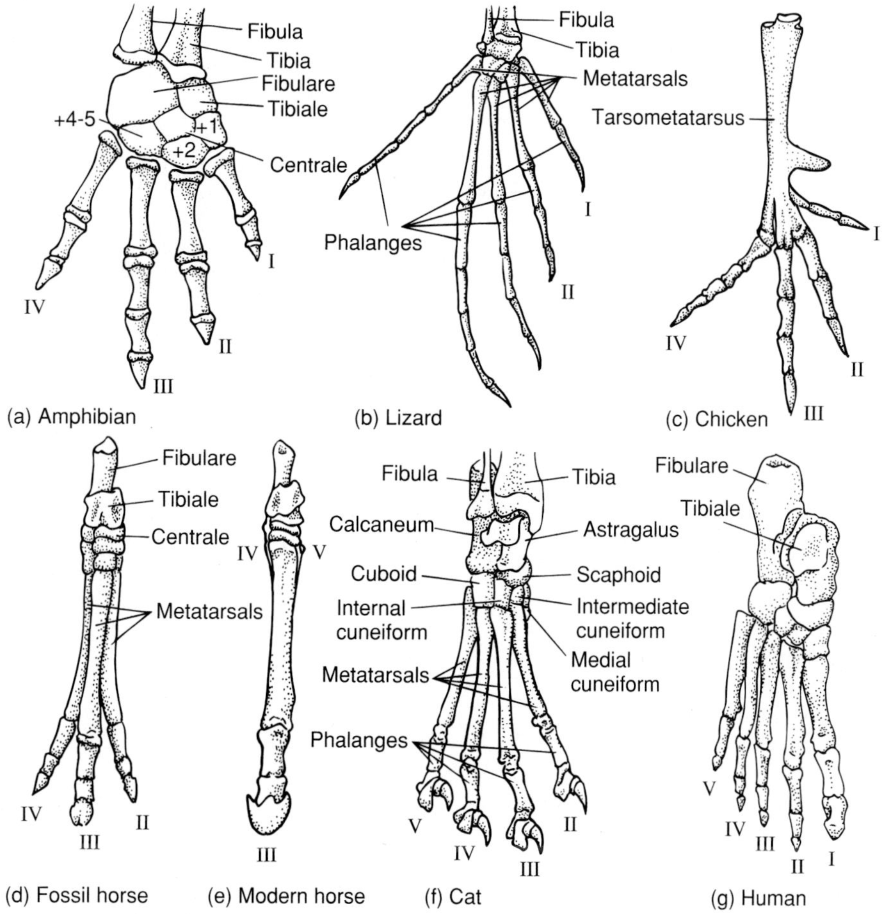


**Figure 9.25** Variations of the tetrapod pes. (a) Amphibian (*Necturus*). (b) Lizard. (c) Bird (chicken). (d) Fossil horse. (e) Modern horse. (f) Cat. (g) Human.

## Adaptive Advantage of Lobe Fins

The transition from water to land led to significant changes in the appendicular system. Fortuitously, the crossopterygian predecessors to the amphibians possessed lobe fins preadapted to become tetrapod limbs. But, lobe fins were present in crossopterygians not in anticipation of future roles on land but to serve these fishes in the aquatic environments in which they lived. What then might have been the immediate biological role of lobe fins in crossopterygian fishes?

We know from the types of geological deposits in which their bones are found that many early crossopterygians lived in fresh water, as their dipnoan relatives do today. Living lungfishes use their rudimentary lobe fins to "walk" along the bottom of slow moving streams or backwaters; that is, they use their fins as points of pivot about which their buoyant body moves. If the freshwater pools in which they live dry during the hot, rainless seasons of the year, lungfishes apparently do not venture a short distance over mud to remaining pools of water. Instead they burrow into the mud, form around them a protective cocoon, and estivate. Their metabolic rate drops, their breathing slows, and they slumber until rains return to replenish the dried pools.

**Lungfish estivation (p. 98)**

It would be helpful to consult living crossopterygians directly to see how lobe fins are actually used. But, of course, no freshwater crossopterygian survives to the present. (*Latimeria* lives in deep marine waters.) Possibly early crossopterygians, like their lungfish cousins today, used their maneuverable lobe fins as points of pivot to move them through aquatic vegetation near the edges of ponds and streams. In shallow waters, aquatic plants and fallen debris from surrounding forests would produce a complex underwater environment offering retreat when a crossopterygian was threatened or prey when it was foraging. Lobe fins offer one solution to maneuvering within such a freshwater "jungle." With this view, lobe fins are aquatic adaptations that serve a fish in shallow waters. Terrestrial environments present different challenges. How well lobe fins might serve there is another issue.

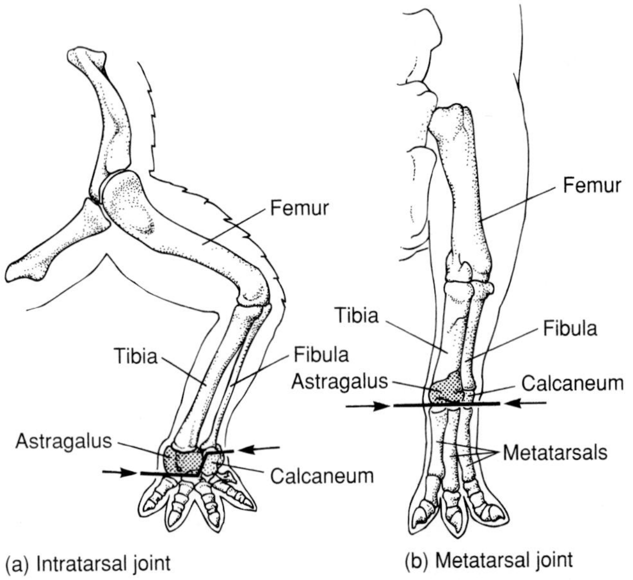


**Figure 9.26** Ankle types among archosaurs. (a) Intratarsal joint. The line of flexion, indicated by heavy black line and arrows occurs between (hence intra-) the astragalus and the calcaneum. Crocodiles, shown here, and most advanced thecodonts have an intratarsal joint. (b) Mesotarsal joint. The joint occurs directly between proximal tarsals (astragalus and calcaneum) and distal metatarsals (hence metatarsal). Dinosaurs and several other groups of Mesozoic reptiles had metatarsal joints.

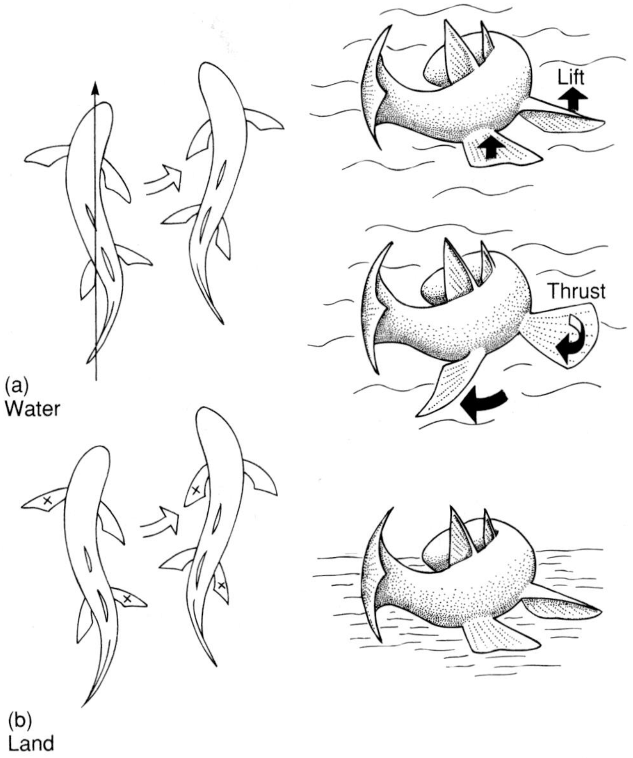


**Figure 9.27** Transition from water to land. The same swimming motions used in water could have served crossopterygians venturing onto land. (a) In water, typical lateral undulations of the fish body provides propulsion for swimming. Horizontally held fins may have functioned like hydrofoils to produce lift. Fins rotated vertically and drawn backward would have served as oars, adding to forward propulsion. (b) On land, these fishes could have used the same lateral body undulations to place fins as points of pivot (x) about which the body "swam." Limbs would not have needed the strength of fully developed tetrapod limbs because they were not used to carry weight or produce locomotor force. They were needed only as pegs about which the strong body musculature could pivot.

## On to the Land

The musculature associated with the fins of early crossopterygians (rhipidistians specifically) was probably too weak to have supplied propulsive thrust directly for transport on land or to have borne the weight of the body out of water. However, the slight musculature was sufficient to fix the fins on the body like pegs. This allowed the well-developed axial musculature to produce lateral undulation and the peglike fins to act as pivots around which the body could rotate. Thus, the same trunk undulations used in swimming could be used with little modification for short journeys across land (figure 9.27a,b). Existing morphology and swimming behavior provided the basis for gradual transition to land. The Devonian amphibian *Ichthyostega,* even though it had tetrapod limbs, still seems to have had aquatic habits. Its tail bore a caudal fin, its lateral line system was present, even in adults, and its vertebrae had not yet replaced the notochord as the predominant basis for axial support. Similarly, most lepospondyls and early temnospondyls appear to have been predominantly aquatic. Not until the Permian, 50 million years after vertebrates first invaded land, did communities of more fully terrestrial tetrapods appear.

But why leave the water at all? What advantages might there have been for crossopterygians that left their aquatic world and ventured onto land? Several ideas have been proposed. One hypothesizes that overland travel developed, ironically, to keep these fishes in water. The Devonian was a time of occasional droughts and floods, which suggests that crossopterygians might have used strengthened fins/limbs to move from a small drying pool to larger permanent ones. However, this presupposes that the limb was already strong enough to be used during sojourns across land before such droughts occurred. Further, modern lungfishes respond to drought by estivating rather than by leaving their aquatic environment.

An alternative hypothesis stresses that movement to the land was favored by predation in the water. To escape predation from other species or from cannibalistic adults, young fishes may have frequented shallow waters where predators could not follow. Because the young had to maneuver in shallow water among thick shoreline vegetation, their lobe fins may have evolved into more supportive appendages. Thereafter, movement onto land would have been an option. In this view, the first movement to land did not involve long treks to alternative pools but a short step up onto the nearby beach. Lobe fins would have needed only to be strong enough (pegs?) to participate mechanically in such a first tentative exploitation of the land.

Others have suggested that limbs developed to enable fishes to get out of the water and breathe. But because of their lungs, these fishes could breathe, if that was required, by simply coming to the surface and gulping fresh air above. Food has been proposed as another enticement for movement onto land. Of course, no other vertebrates were on land, and crossopterygian teeth were not suited for feeding on plants. But arthropods were abundant, having radiated into terrestrial environments much earlier (Silurian). They may have offered an alternative food source for crossopterygians that scampered up onto beaches or shores to seek them out.

Because no one was there to record events, we cannot be sure of the selective pressures that favored the transition to land. But, the fossil record clearly demonstrates that just such a transition in vertebrate evolution occurred during the Devonian. As a matter of interest, the transition from water to land has happened several times, although with a less lasting phylogenetic impact. In some teleost fishes today, for instance, there are species, such as the mudskipper, that use a strengthened skeleton to venture temporarily onto land to search for food and perhaps gain respite from predators left behind in the water.

# Form and Function

Changes in the skeletal system, like many other systems, have been extensive in the transition from aquatic environments to life on land. On land, the main contributors to locomotion are the limbs, not the tail. Consequently, the limbs undergo extensive and significant morphological alterations. In addition, the shoulder and hip generally establish different structural associations with the axial column as a result of the transition to land. In tetrapods, the axial column is slung by muscles from the shoulder girdle; however, the hip is attached directly to the column (figure 9.28a,b). The shoulder moves on the thorax via these muscles so that the forelimb's impact with the ground is softened and these sudden forces are not transmitted to the skull. The hip is firmly associated with the sacrum via a bony connection (figure 9.28b). The powerful hindlimbs transmit their propulsive force directly to the bony axial column.

Different modes of locomotion on land place different mechanical demands on the appendicular skeleton. Tetrapods specialized for running are **cursorial.** Those that dig or burrow are **fossorial,** and those that hop are **saltatorial** (ricochetal). The fastest modes of locomotion occur among flying, or **aerial** (volant), specialists. **Arboreal** locomotion refers generally to animals that live in trees. One form is **scansorial** locomotion that applies to climbing in

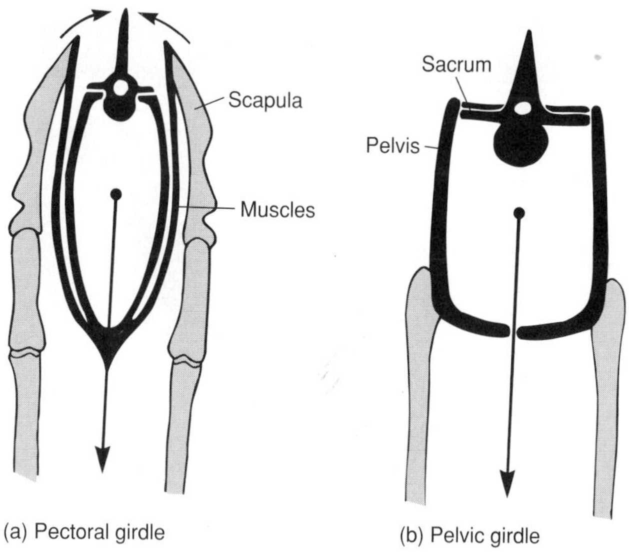


**Figure 9.28** Appendicular girdles of tetrapods. (a) Muscles of the pectoral girdle support the anterior part of the tetrapod body in a muscular sling. (b) The pelvic girdle is attached directly to the vertebral column via the sacrum.

trees with the use of claws. Squirrels are an example. The other form of arboreal locomotion is **brachiation,** which applies to locomotion by arm swings under branches with the use of gripping hands. Monkeys, chimpanzees, and occasionally gorillas are examples.

Each of these terrestrial specialties is accompanied by morphological modifications of the basic limb and girdle structure. To understand the different functional demands these specialized modes of terrestrial locomotion place on the appendicular skeleton, we need first to step back and view at what point the basic terrestrial structure and behavior began, namely, with fishes swimming in water.

## Swimming

As we have already seen, the body of an active fish coursing through a viscous medium such as water creates turbulence in its wake. This turbulence results in drag, which slows the fish's forward progress. Streamlining prevents turbulence, reduces drag, and improves performance. Lateral undulations passing along the body move the fish through the aqueous medium, producing backthrust against the water and providing forward force. The basic primitive tetrapod locomotion evolved from this characteristic lateral undulation fishes use to swim.

This same mode of progress still serves most modern amphibians and reptiles quite well, giving them access to a great variety of habitats. In tetrapods that secondarily become aquatic, as in cetaceans, for example, the limbs again may become secondary to the tail and lose their prominence in aquatic locomotion (figure 9.29a). However, not all secondarily aquatic vertebrates have reduced limbs. For

# Box Essay 9.1 Human Engineering: Arms and Hands, Legs and Feet

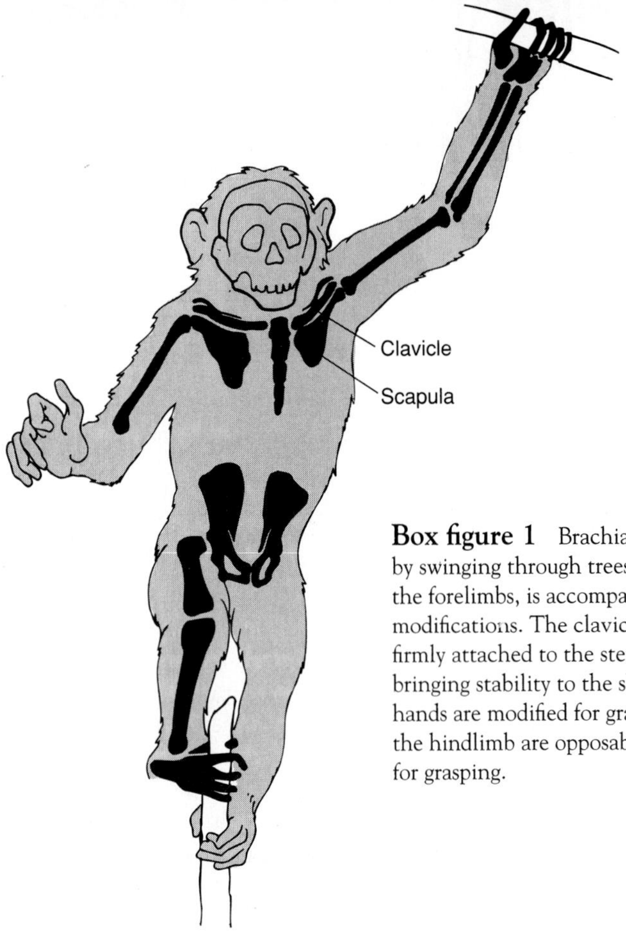


**Box figure 1** Brachiation, locomotion by swinging through trees suspended from the forelimbs, is accompanied by structural modifications. The clavicle is large and firmly attached to the sternum, thus bringing stability to the shoulder. The hands are modified for grasping. The toes of the hindlimb are opposable and also used for grasping.

Although a long 5 to 10 million years have passed since distant human ancestors swung through trees, we still retain evidence of this brachiating mode of locomotion.

In the forelimb, for example, brachiators are characterized by long arms with grasping hands. Although our arms are shorter than those of primates that still depend on locomotion through trees, nevertheless our arms are relatively long compared to those of other vertebrates. If we stand comfortably upright with arms to our sides, our fingers reach below our hips. By contrast, the forelimbs of a nonbrachiating animal such as a dog or cat, if pressed back, do not reach so far. In the manus of a brachiator, digits II through V form a hook, with which they grasp overhead branches. Without thinking it anything special, we use this same comfortable design to grip the handle of a suitcase carried at our side. The arm position changes from the overhead to the side, but the grip used is the same. In cursorial vertebrates such as cats, the clavicle is reduced. But, in brachiators like monkeys, the clavicle is a prominent structural element of the shoulder serving to transfer the weight of the body to the arm (box figure 1). *Homo sapiens* retains this prominent clavicle.

example, as birds become more aquatic, their wings often take on greater roles in swimming. Their forelimb bones become stouter and more robust, reflecting the increased strength required to provide the bird with flippers to propel them while swimming after food in the water. In penguins, the wings are flightless and used exclusively like flippers to enable the animal to swim underwater (figure 9.29b). The hindlimbs of swimming birds may become partially or completely webbed feet to increase pressure against the water when these birds paddle (figure 9.29c).

## Terrestrial Locomotion

### *Early Modes of Locomotion*

In early tetrapods, limbs were placed laterally in a sprawled stance, establishing points of pivot (figure 9.30a). Locomotion was accomplished, as in fishes, by alternating lateral undulations of the vertebral column about these pivots. In modern amphibians and reptiles, the characteristic mode of progression still depends on this pattern of lateral swings by which the vertebral column moves about points of rotation established by the feet. However, in some terrestrial birds, in many species of Mesozoic reptiles, and in

The design of our hindlimbs and pelvic girdle accommodates compromises to our upright bipedal posture (box figure 2a,b). The birth canal, the opening enclosed by left and right pelvic girdles through which the infant passes during birth, is wide, especially in human females. This canal accommodates the large size of the infant's cranium (box figure 2b). But widening the hips to accommodate an adequate birth canal places the heads of the femurs far apart and outside the center line of body weight. A bend in the femur just above the knee allows the limbs to swing directly beneath the body.

Our bipedal posture and pendulum-like leg motions result in changes in foot design. Apes retain a grasping hindfoot with a projecting large toe. In humans, the toe is aligned with the other digits of the foot so that as the limbs are swung beneath the body, they can be placed close to the line of travel without catching the projecting toe on the opposite leg. The human foot forms an arch, a way of broadening the base of support upon which the upper body stands. The arch also changes the geometry of the foot: As the foot pushes off during walking, it extends the ankle farther than if the arch were absent.

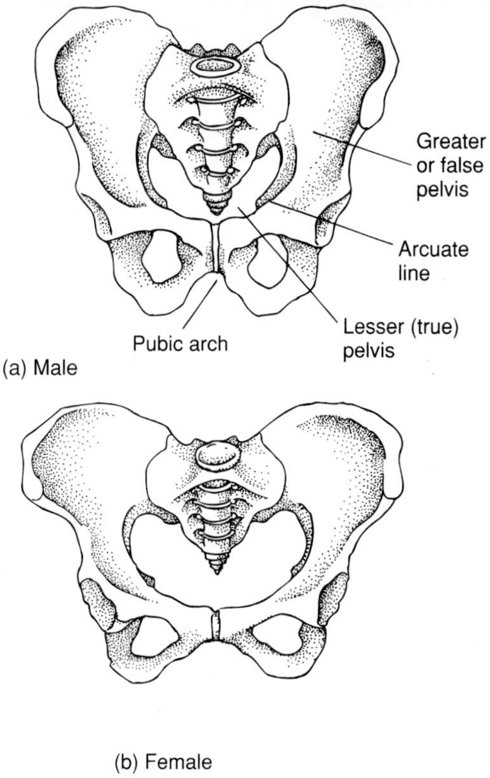


**Box figure 2** Pelvises of human male (a) and female (b). In humans, the birth of a baby with a relatively large head requires a relatively large birth canal. The hips of females are correspondingly wider than those of males to accommodate the infant's head.

many groups of mammals, the trend has been toward cursorial locomotion.

#### Fish swimming to early tetrapod walking (p. 298)

From the characteristic sprawled posture of early tetrapods, many later tetrapods have developed limbs drawn under their body, a change in posture that increases the ease and efficiency of limb swing during rapid locomotion (figure 9.30b). Crocodiles and alligators use sprawled postures when resting on shore, but they can change their limb position when they move. If making a quick dash to water, they can draw their limbs under the body more directly beneath their weight. This allows the limbs to swing more easily beneath the raised body. In several lines of therapsids, in most placental mammals, and in many Mesozoic reptiles, this change in limb posture results from a structural change in limb design. The femur and especially the humerus of the limbs show torsion of their distal ends, which rotates the digits carried at the ends of the limbs forward and more in line with the direction of travel (figure 9.31).

Accompanying this change in posture was a tendency to restrict limb movement to one plane, the sagittal plane. Amphibians with a sprawled posture must use an overarm swing after each propulsive stroke to establish a new forward point of pivot (figure 9.32a). However, with legs positioned beneath the body, limb recovery after propulsive strokes can be accomplished efficiently by swinging its limbs forward beneath its body in an easy pendulum fashion (figure

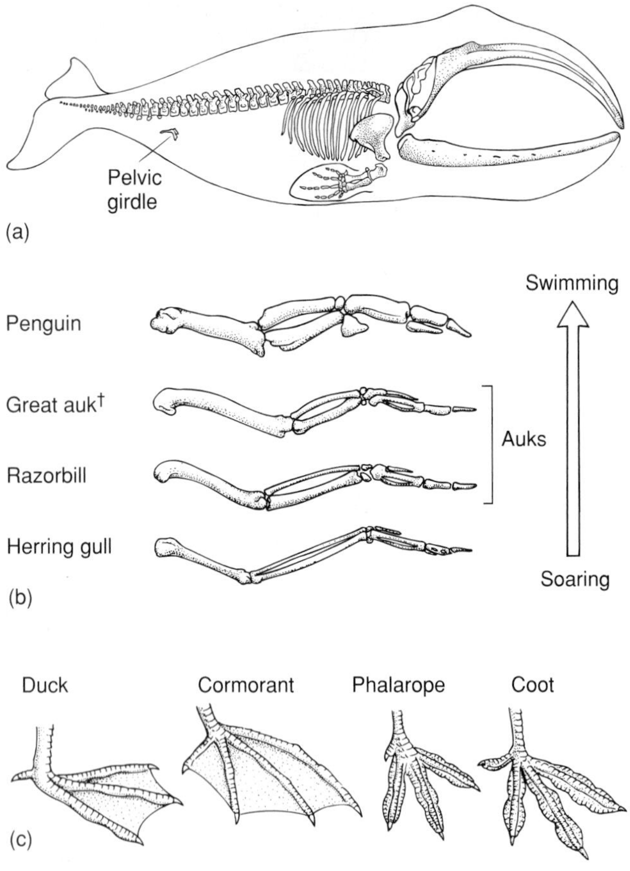


**Figure 9.29** Adaptations of the appendicular skeleton in secondarily aquatic tetrapods. (a) Skeleton of a right whale, showing reduction of the appendicular skeleton, especially the pelvic girdle and fin, in this aquatic mammal. (b) The gracile bones of the seagull contrast with the robust forelimb of the penguin. Forelimbs of several species of auks are shown between. Some are extinct (†). These changes reflect an increasing role in underwater swimming. (c) Swimming birds typically have webbed feet.

9.32b). In tetrapods with a sprawled posture, the adductor muscles, which run from girdle to limb, are massive in order to lift and hold the body in a push-up position. As the limbs move more directly under the body, the adductor musculature is reduced.

In therapsids, the acetabulum and glenoid fossa shifted ventrally to follow the inward shift in limb posture. Most noticeably in the shoulder girdle, placement of the forelimbs directly beneath the scapula shifted mechanical forces away from the midline to the scapula (figure 9.33a,b). This gave the scapula a greater role in locomotion and weight bearing. Conversely, the medial elements—clavicle, interclavicle, coracoid, and procoracoid—with reduced supportive roles became reduced in prominence. The hindlimbs were also drawn under the body, accompanied by a reduction in the adductor muscles. In turn, the pubis and ischium, sites of origin of these adductor muscles, were reduced as well. A shift in orientation of the pelvic girdle allowed for a forward thrust more aligned with the forward direction of travel (figure 9.34a,b).

A remarkable change in the functional way in which the vertebral column participates in locomotion appears first in therapsids. This change is characterized by a shift from lateral to vertical flexions. In animals with sprawled postures, lateral flexion of the vertebral column contributes to the sweeping overhand recovery of the limbs. With limbs carried under the body, lateral undulations contribute little to limb oscillations. Consequently, the structural changes first seen in therapsids were accompanied by a shift from lateral to vertical flexion of the vertebral column, coordinating it with the limbs that are swung in the same plane. Loss of ribs from the posterior trunk, which produces a more distinct lumbar region in some therapsids, represents a structural specialization that allowed greater flexibility of the axial column in a vertical plane.

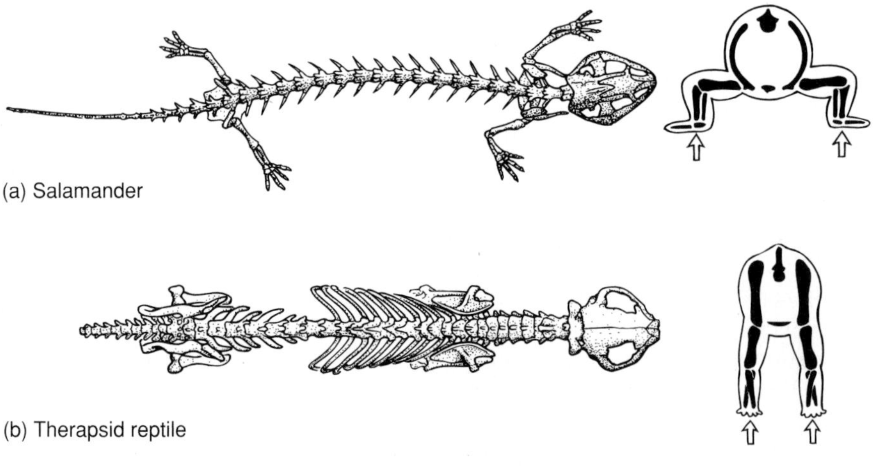


**Figure 9.30** Change in limb posture. (a) The sprawled posture exhibited by this salamander was typical of fossil amphibians as well as of most reptiles. (b) Placental mammal. This posture began to change in synapsids so that in late therapsid reptiles the limbs were thought to be carried more under the body, a reflection of increased efficiency in rapid locomotion.

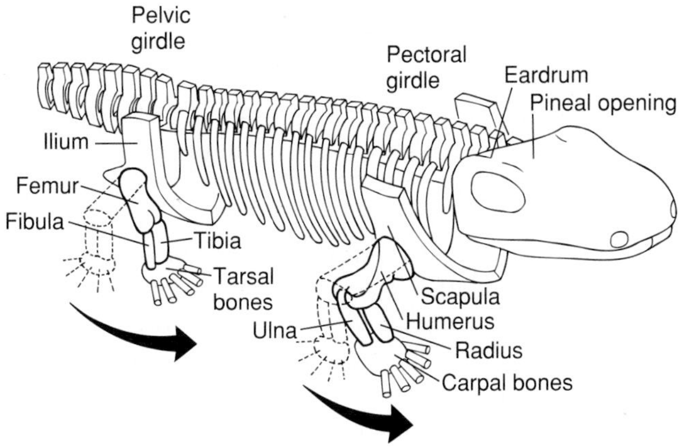


**Figure 9.31** Digit orientation. Toes tended to point laterally in early tetrapods (dashed lines). However, accompanying more efficient terrestrial locomotion, the direction of digits changed along with limb position. Torsion of the humerus and femur brought the toes forward and more in line with the direction of travel. Note in particular how opposite ends of the humerus are rotated to bring the toes forward.

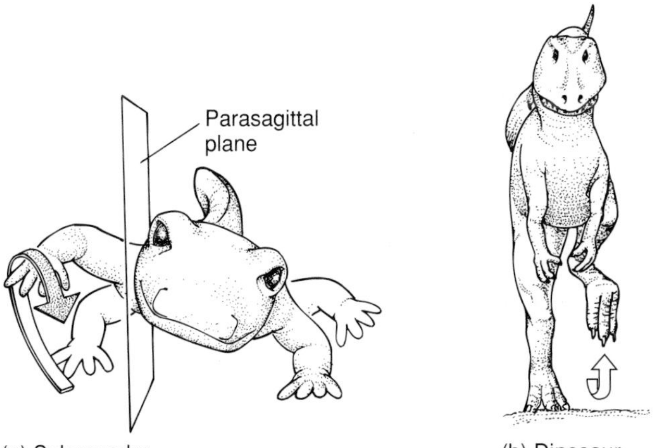


**Figure 9.32** Terrestrial locomotion. (a) Terrestrial but noncursorial salamanders achieve limb recovery by an overhand swing of the arm outside the parasagittal plane. (b) Cursorial dinosaurs achieved limb recovery by a pendulumlike swing in a parasagittal plane, which keeps the limbs directly below the body so that they support the body weight. The pendulumlike swing improves the ease and efficiency of limb recovery.

Generally then, as locomotion was used by therapsids for more sustained, efficient, and rapid transport on land, a variety of structural modifications were incorporated into the appendicular skeleton. Torsion brought the digits forward and more in line with the direction of travel. Sprawled limbs were brought under the body. Vertical flexion of the vertebral column added its motions to the limb displacements. Together, these changes increase the ease and efficiency of limb oscillation and contribute to active lifestyles.

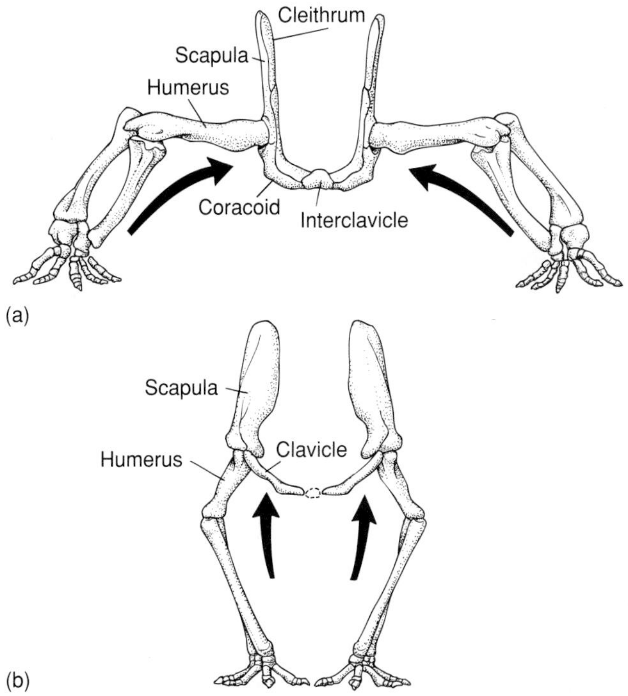


**Figure 9.33** Change in the role of the shoulder girdle with change in limb posture. (a) Sprawled posture brings a medially directed force toward the shoulder girdle, conferring on medial elements a major role in resisting these forces. (b) As limbs are brought under the body, these forces are directed less toward the midline and more in a vertical direction. This position of the limbs might account for loss of some pectoral elements in phylogenetic lines in which limb posture shifted.

Similar changes in the appendicular skeleton appear in archosaurs as well. Limbs were positioned under the body to carry the body weight more efficiently as these animals moved or migrated in search of resources. However, locomotion was usually based on a bipedal posture with trunk and tail balanced across the hindlimbs.

## *Cursorial Locomotion*

Beyond an increase in the ease and efficiency of limb oscillation, many later tetrapods became specialists at rapid locomotion accompanied by further modifications that serve such a specialized mode of transport. Rapid locomotion has evolved in both predators and their prey, two sides to the evolutionary coin. It also provides an animal with the means to cruise from areas of locally depleted resources to new pastures and to locate dispersed resources in sparse lands.

The speed or velocity attained by a vertebrate is a product of its *stride length* and *stride rate*. Other things being equal, vertebrates with longer strides can cover more ground than those with short legs, so they attain greater speeds. The faster the rate of limb oscillation, the faster the animal travels. Let us consider adaptations that serve either

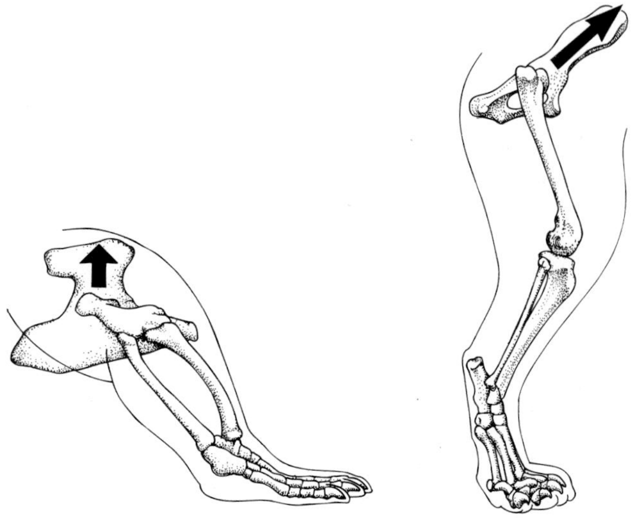


**Figure 9.34** Changes in the pelvic girdle. (a) When limbs are sprawled, propulsive forces are transferred more vertically through the sacrum. (b) In mammals, in whom rapid locomotion becomes common, orientation of the pelvic girdle changes so that the forward thrust of the hindlimbs is brought more into alignment with the line of travel and transferred to the vertebral column.

stride length or stride rate, and hence contribute to cursorial locomotion.

***Stride Length*** One way to increase stride length is to lengthen the limb. Highly cursorial vertebrates exhibit a marked lengthening of their distal limb elements. A related modification is a change in foot posture. Humans walk with the entire sole of their foot in contact with the ground, exhibiting a **plantigrade** posture. Cats walk with a **digitigrade** posture in which only the digits bear the weight. Deer use a **unguligrade** posture, traveling on the tips of the toes (figure 9.35). The change from plantigrade to digitigrade to unguligrade postures effectively lengthens the limb and increases the length of stride.

Another way to increase stride length is to increase the distance through which the limbs move while they are off the ground. For example, the cheetah, when increasing its speed from 50 to 100 km/hr (about 30 to 60 mph), does not appreciably change its rate of limb oscillation but instead it increases its length of stride. With greater spring in each forward leap and with extreme flexion and extension in the vertebral column, the cheetah limbs extend their reach during each stride to increase speed (figure 9.39a).

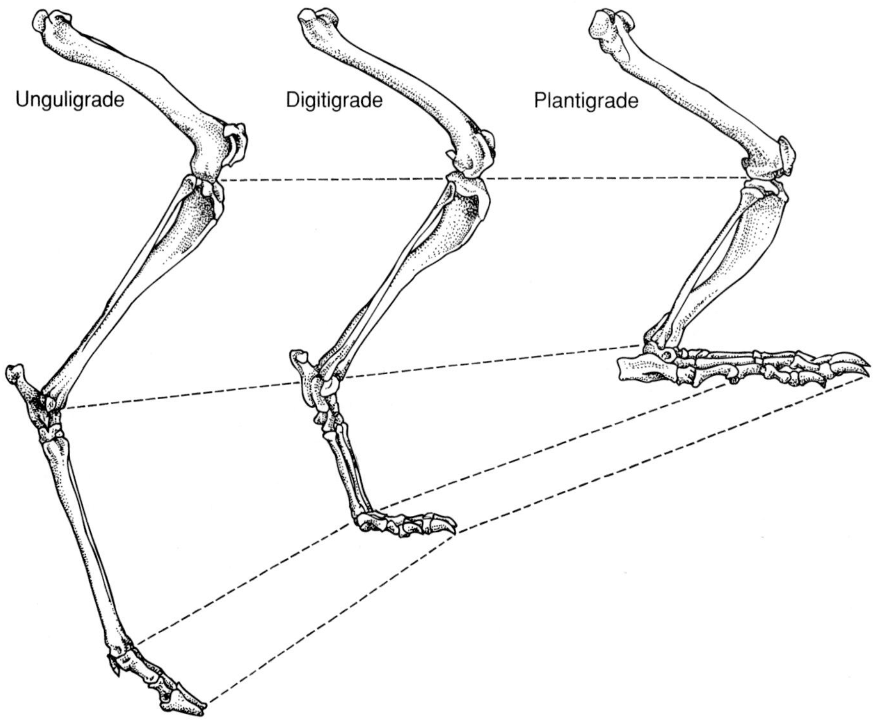


**Figure 9.35** Foot postures. Unguligrade, digitigrade, and plantigrade designs for feet. Note how changes in foot posture produce relatively longer limbs.

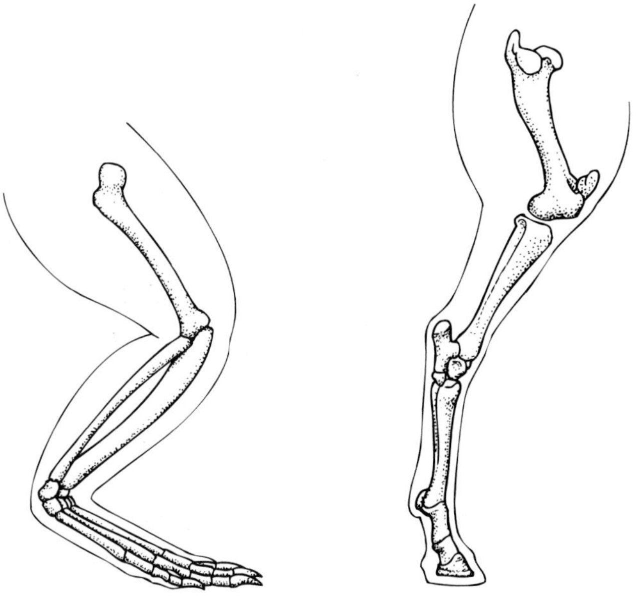


**Figure 9.36** Location of limb muscles in a lizard (left) and a horse (right). In cursorial animals, such as horses, muscles acting along the leg tend to be bunched close to the body and exert their forces along the limb through long, light tendons. This design reduces the mass of the lower limb and thus reduces the inertia that must be overcome during rapid limb oscillation.

***Stride Rate*** Velocity of travel also depends on the rate at which the limbs are moved. Larger, more mechanically efficient muscles increase the rate of limb movement. Shortening the limb would certainly make limb oscillation easier and increase stride rate, but it would also shorten the length of the stride and compromise speed. However, flexion of the limb during recovery effectively shortens it, increasing the rate of forward oscillation.

Another way to promote stride rate is to lighten the distal end of the limb in order to reduce the mass and thus the inertia that must be overcome due to mass. If the bulk of the powerful limb muscles are located close to the body and carry their force distally through light tendons to the point of force application at the end of the limb, inertia is reduced at the limb's distal end. By so lightening the limb, it can be more easily and efficiently moved with less energy. The bunched limb muscles in the shoulders and hips of deer, horses, and other fast animals are examples (figure 9.36). Another adaptation that increases stride rate is reduction in the number of digits. In highly cursorial mammals, one or two of the central digits are strengthened to receive the forces of impact with the ground. But the more peripheral digits tend to be reduced or lost (figure 9.37a,b). Overall, the result is to lighten the end of the limb and allow it to oscillate more rapidly. Birds specialized for rapid terrestrial locomotion, such as ostriches, show similar cursorial adaptations such as lengthening of hindlimbs and loss of digits (figure 9.37c).

***Gait*** The pattern of foot contacts, or **footfalls,** with the ground during locomotion constitutes an animal's **gait.** The gait selected depends on the rate of travel, obstructions in the terrain, maneuverability sought, and body size of the animal (figure 9.38).

In the **amble,** the animal swings its fore- and hindfeet on the same side more or less in unison. Long-legged animals amble to avoid tangling their limbs, which might otherwise occur in a running cycle. Camels and long-legged cheetahs traveling at slow speeds amble. A fast amble is termed a **pace.** Some harness horses are trained to use this type of gait in racing.

An animal that **trots** moves its diagonally opposite feet together. The trot is advantageous at slow speeds because the connecting line of support between diagonally opposite limbs runs directly under the center of mass. This provides stability to animals with sprawled postures, such as amphibians, or animals with broad bodies, such as hippopotamuses.

In the **bound,** called the **pronk** in artiodactyls, all four feet strike the ground in unison. Although this gait abruptly jars and decelerates the animal, it gives great four-footed stability each time the feet contact the ground. Conversely, all four feet are off the ground during most of each locomotor cycle, perhaps an advantage to an animal that must clear low brush.

The half bound and gallop are more complex and are used at high rates of speed. When a pair of feet approaches the ground, the **leading foot** strikes second but is in front of the **trailing foot** that makes contact first. In the **half bound,** the hindfeet make contact more or less in unison, but the forefeet make contact with a distinct leading and trailing pattern. In the **gallop,** both fore- and hindfeet display distinct lead and trailing pattern. At slow speeds, the gallop is called a **canter.** Galloping and half bounding are said to be **asymmetrical gaits** because footfalls of a pair, forefeet or hindfeet, are unevenly spaced during a cycle. These gaits may be less stable than **symmetrical gaits,** such as the trot or pace, but they have the advantage of introducing a **suspension phase** in the cycle, an interval during which all four feet are off the ground. This arises from the greater reach of the limbs and results in increased stride length.

***Uses of Cursorial Locomotion*** Cursorial locomotion is widespread among vertebrates, especially among mammals, but it is deployed in many ways. The lion and cheetah can use their speeds to accelerate quickly for short bursts, whereas the horse and pronghorn more often use their cursorial abilities to cruise open plains in search of dispersed resources or to put distance between themselves and potential predators. But it is even more complicated than this. Body size is a factor as well (table 9.1).

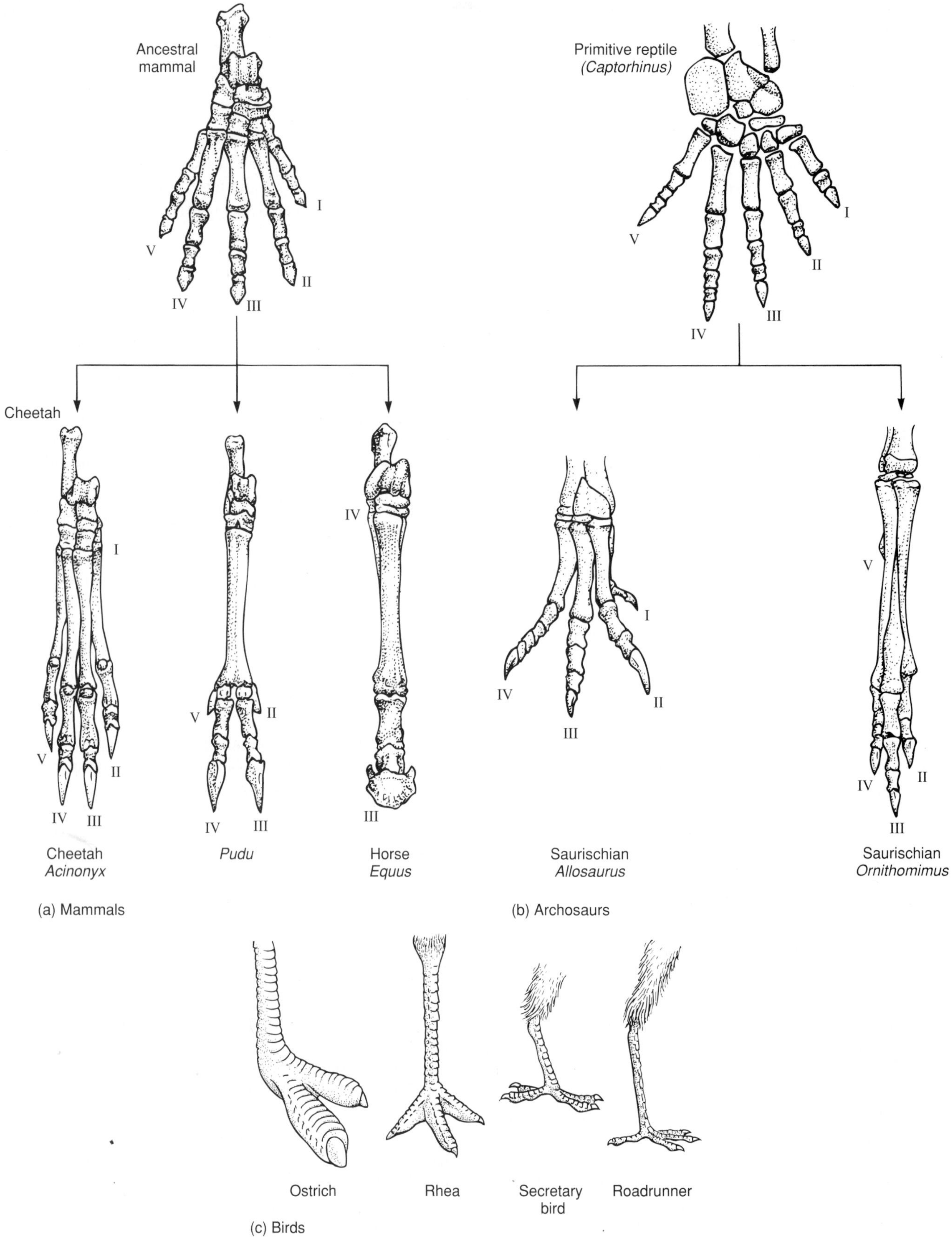


**Figure 9.37** Reduction of digits in cursorial animals. Central digits at the ends of limbs tend to be strengthened while more peripheral digits are lost. The overall result is to lighten the distal portion of the limb. (a) The hindfoot of a cheetah, artiodactyl (pudu), and perissodactyl (horse) are shown. Note the varying degrees of digit reduction in comparison with a more general ancestral mammal. (b) A similar trend, apparently related to cursorial locomotion, occurs in archosaurs, although no archosaur foot is reduced so extremely as the horse foot, in which only a single digit remains. (c) Cursorial birds have slender hindlimbs and sometimes show a loss of peripheral digits, as in the ostrich.

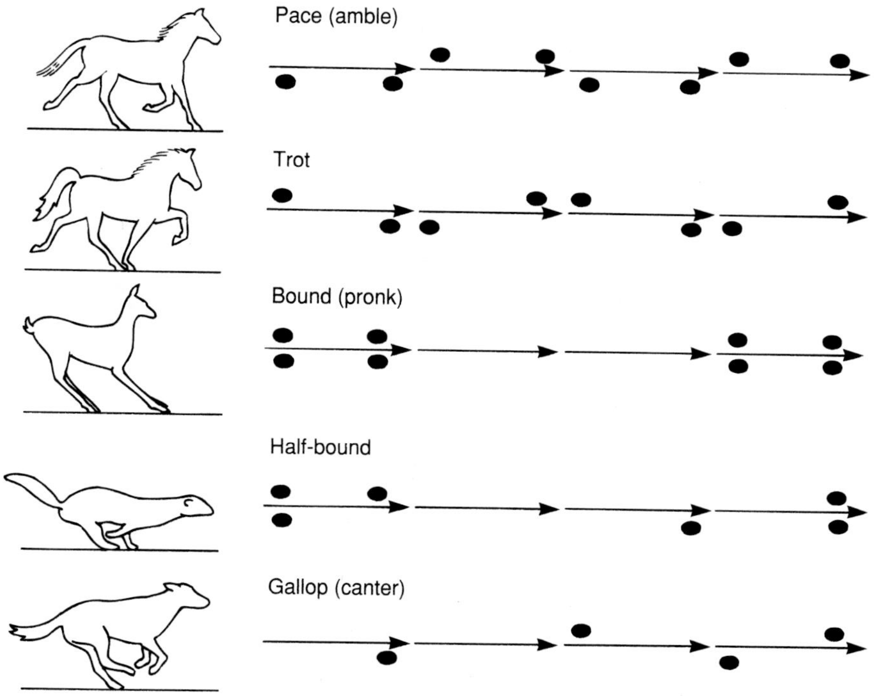


**Figure 9.38** Gaits or footfall patterns of various mammals. The particular gait selected depends on speed of travel, size of the animal, and structure of the terrain. The footfall patterns produced by each gait when the feet strike the ground are indicated.

For example, both the horse and cheetah are adapted for cursorial modes of locomotion (figure 9.39a,b). The cheetah, in addition to using its speed differently, is also smaller than the horse. The horse can maintain a speed of 30 km/hr for over 30 km. But if it were built like a fox or cheetah, it could not maintain even moderate speeds for more than a few kilometers. The cheetah is not an endurance machine; rather, it is designed for quick bursts over short distances. Were it larger with increased mass to carry, then its morphological design would also have to scale up accordingly. In the cheetah, the great flexion of the vertebral column extends and then gathers the limbs during suspension phases. The result is to increase the effective stride length. It is estimated that this extensive bending of the vertebral column alone, by increasing stride length, adds almost 10 km/hr (6 mph) to the speed of the animal. But such flexions also mean that much of the body mass is displaced vertically rather than in the direction of travel. The cheetah must expend significant energy to lift this mass during each set of strides. For a heavy animal like a horse, this design consumes too much energy to sustain long distance travel. Consequently, the vertebral column of the horse bends very little even during full gallop. Vertical displacement of the vertebral column in the hips may be less than 10 cm and in the shoulder less than 5 cm. Proportionately less energy must be used to lift this mass and more of the weight of the animal is carried along the linear path of forward travel.

**TABLE 9.1 Maximum Speeds and Sizes of Cursorial Animals**

| Animal | Maximum Speed (km/hr) | Weight (kg) |
|---|---|---|
| Fox | 60 | 4.5 |
| Horse | 60 | 540 |
| Lion | 80 | 180 |
| Cheetah | 110 | 60 |
| Pronghorn | 95 | 90 |
| Human | 35 | 85 |
| Coyote | 65 | 10 |

The horse and the cheetah represent cursorial animals that use their speed differently in different biological roles—endurance versus quick bursts, respectively. But their designs also represent different compromises with body size—large versus small size, respectively. If the larger horse were built like the smaller cheetah, it could not contend so well with its greater mass and be able to

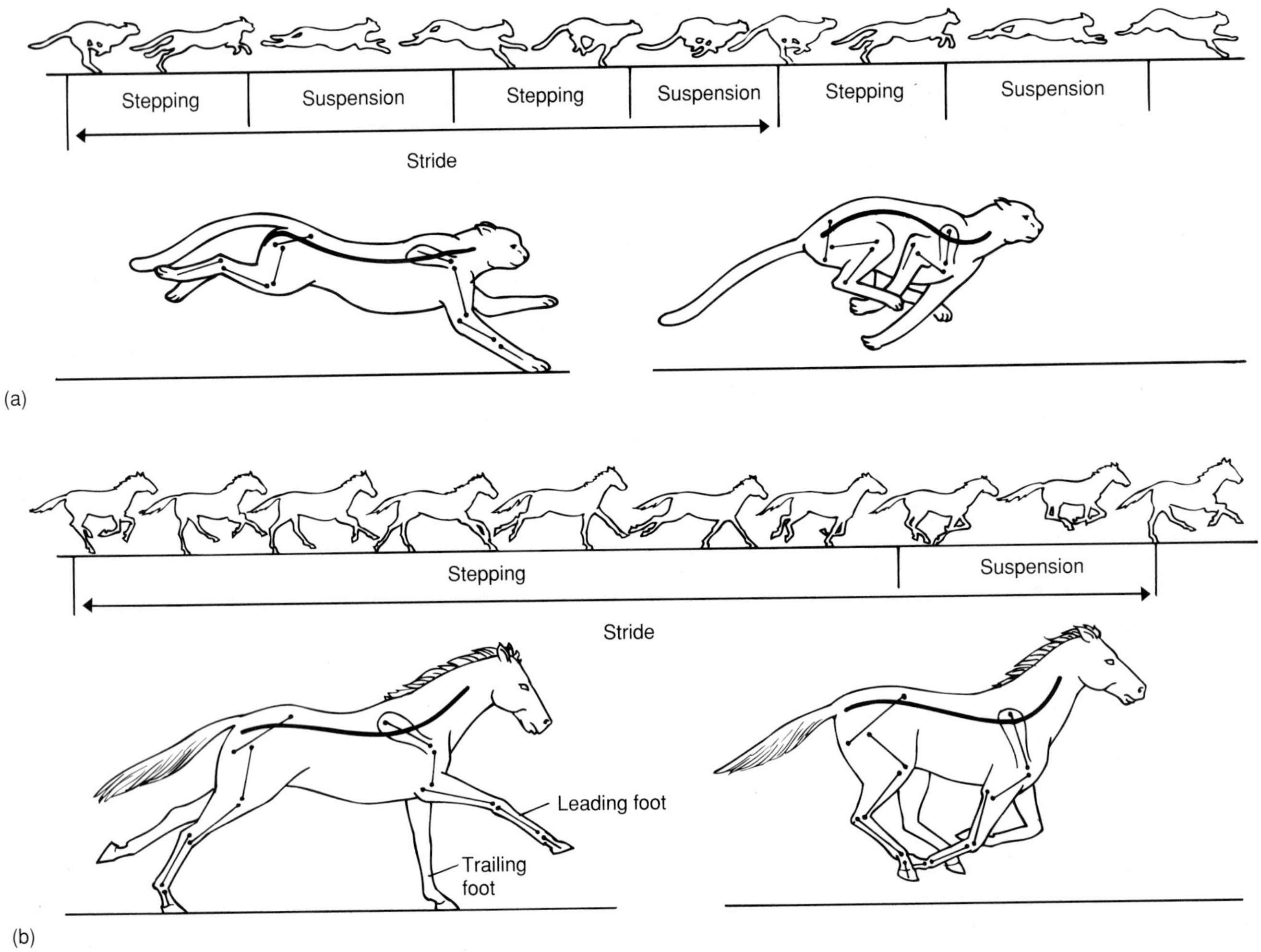


**Figure 9.39** Comparison of two cursorial mammals, a horse and a cheetah. (a) The cheetah depends on quick bursts of speed to overtake prey. Notice the extensive flexion of the vertebral column that increases stride length and adds about 10 km/h (6 mph) to its overall speed. (b) The horse uses its speed for sustained locomotion; therefore, the vertebral column flexes much less to avoid the exhausting vertical rise and fall of the body mass characteristic of the cheetah. A less flexible vertebral column keeps the mass of the horse more linear along its line of travel. The lead foot and trailing foot change during sustained bouts of rapid running.

sustain its locomotor endurance on which it greatly depends. The dynamic demands of locomotion placed on the skeletal system depend both on the biological roles served by locomotion and on the demands imposed by body size.

## *Aerial Locomotion*

***Gliding and Parachuting*** Travel through the air occurs in at least a few species in all classes of vertebrates. "Flying" fish spread especially wide pectoral fins during short flights in the air above water (figure 9.40a). A species of tropical frog spreads its long, webbed toes to slow its airborne fall (figure 9.40b). Lizards with special flaps of skin and squirrels with loose skin between fore- and hindlimbs spread these membranes to slow their drop through air or extend the distance of their horizontal travel (figure 9.40c–f). These tentative fliers are not really fliers at all, however. Instead they are gliders or parachutists. True powered flight occurs in just three groups—bats, pterosaurs, and most birds (figure 9.41). In each group, the forelimbs are modified into wings that both generate the force driving them forward through the air and provide lift against gravity.

***Flight*** Most functional analyses of powered flight have centered on birds, taking advantage of the sophisticated aerodynamic equations engineers use to design aircraft. But borrowing directly from engineers has been especially difficult because the wings of birds have all the characteristics a designer of aircraft seeks to eliminate. Bird wings flap

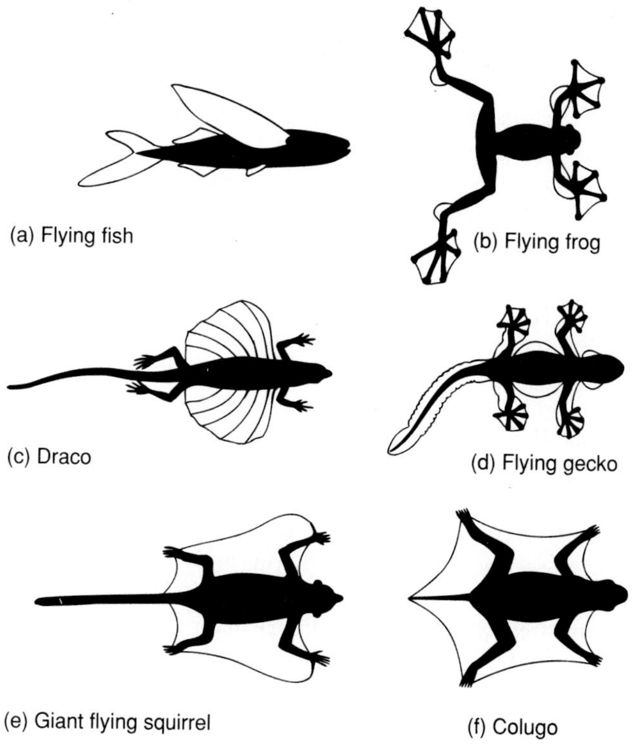


**Figure 9.40** Gliding and parachuting. In all classes of vertebrates, at least a few species can be found that occasionally take to the air. (a) Fish. (b) Amphibian. (c,d) Reptiles. (e,f) Mammals.

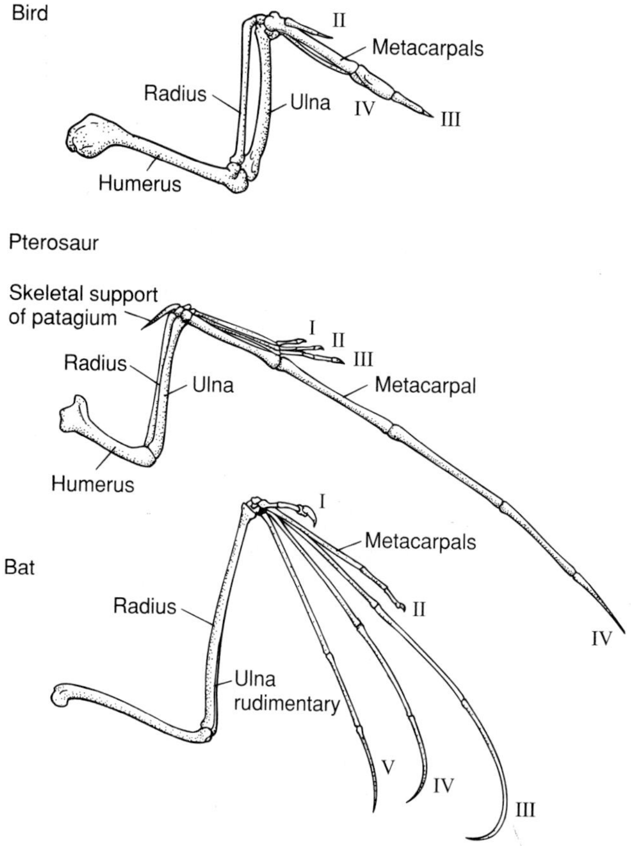


**Figure 9.41** Modifications of the forelimb in bird, pterosaur, and bat to support the aerodynamic surface. Generally, the participating digits are lengthened and bones lightened.

(whereas airplane wings are fixed), are porous (rather than solid), and yield to air pressure (rather than resisting it like airplanes). Although simplifying assumptions must usually be made, such analyses have yielded an understanding of several adaptations for flight in birds.

*Feathers*. Contour feathers give the body of a bird its streamlined shape to help it cut efficiently through air. By filling the body out into a streamlined aerodynamic silhouette, contour feathers help maintain a laminar airflow across the body and reduce friction drag. It has even been suggested that the body shape, which is similar to that of an airplane wing, also produces **lift.** However, most of the lift is produced by the wing. The primary feathers, attached to the manus, are responsible for providing forward thrust. Secondary feathers, attached to the forearm, provide lift (figure 9.42). Thus, the functions of flight are divided between these two types of flight feathers. Primaries act like propellers, providing forward thrust, and secondaries act like airplane wings, providing lift.

**Aerodynamics (p. 135); feathers (p. 207)**

*Skeleton*. High-speed and radiographic films of birds in flight give a detailed view of wing motions and the role played by the pectoral girdle and rib cage. One wing beat cycle can be divided into four phases: (1) upstroke-downstroke transition, (2) downstroke, (3) downstroke-upstroke transition, and (4) upstroke. During **upstroke-downstroke transition,** the leading edge of the wing is elevated above the body and lies nearly within a sagittal plane. The elbow and wrist joints are fully extended. From this position, the wing is forcefully brought downward (depressed) and forward (protracted) during the **downstroke,** producing thrust and lift. The wrist and elbow remain extended during downstroke. The wing continues downward and forward until its tip extends in front of the body. During the **downstroke-upstroke transition,** wing motion reverses, leading into the upstroke. The **upstroke** is complicated and apparently generates little lift but repositions the wing for the next downstroke. During the upstroke, the wing is folded and lifted upward (elevated) and backward (retracted), while the elbow and wrist are fully flexed (figure 9.43a–c).

Synchronous changes in the rib cage and shoulder girdle occur as well when these wings beat. During the downstroke, the U-shaped and flexible furcula, accompanied by the procoracoids, bends laterally. The sternum shifts upward and backward. During the upstroke, these motions are reversed. The furcula springs back and the sternum drops

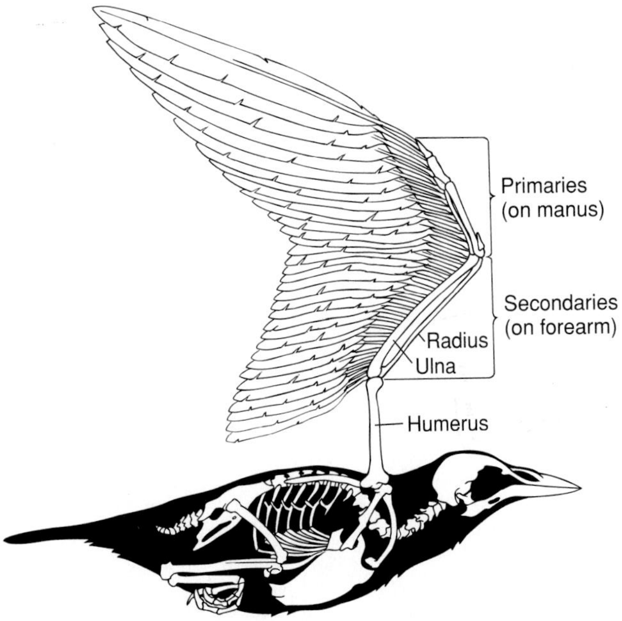


**Figure 9.42** Feathers along the wing divide the functions of flight among them. Those at the tip, the primaries, attach to the manus and are primarily responsible for producing thrust; those more proximal, the secondaries, attach to the forearm and are primarily involved in producing lift.

downward and forward (figure 9.43d). These configurational changes within the rib cage alter the size of the thoracic cavity. In addition to whatever contributions such changes make to flight, it is hypothesized that they also are part of the respiratory mechanism ventilating the lungs. This coupling of locomotor and respiratory systems takes advantage of the muscle forces produced during flight to ventilate the lungs and air sacs at the same time.

**Airflow through avian lungs (p. 426)**

In most birds, the furcula probably functions like a spring when it bends and then recoils during flight. Energy is stored as elastic energy in bent bones during one part of the stroke and then recovered during recoil later in the cycle. But in birds such as parrots and toucans, the clavicles are not fused, and in flightless birds, they are typically vestigial or absent entirely. In some soaring birds, the furculae are quite rigid and probably resist bending. Although the functional significance of this structural diversity is not understood, the bird furcula might be expected to play additional roles in flight other than just a mechanism to store and return energy.

We have already mentioned that the innominate bone and the synsacrum fuse, which stabilizes the body in flight. The flexibility of the cervical vertebrae allows a bird to reach all parts of its body. These two design features, fusion and flexibility, are nearly uniform throughout birds, testimony to the influence of flying on biological design. The skeleton also exhibits other modifications for flight. The bones of birds and pterosaurs, but not bats, are hollow rather than filled with blood-forming or fatty tissues like the bones of other vertebrates (figure 9.44). Absence of these tissues from bird bones results in overall lightening of the skeleton and reduces the weight that must be launched into the air. In pterosaurs and in birds, an expanded sternum serves as the origin for the powerful **pectoralis** flight muscles.

**Sternum (p. 288)**

*Types of Flight.* Although flight is the common denominator in most birds, not all flight is the same. For hovering birds and strong fliers, emphasis is on maximum propulsive force and thus on the primary feathers attached to the manus. In such birds, the manus is proportionately the largest section of the forelimb (figure 9.45a). For soaring birds, emphasis is on lift and thus upon the secondary feathers attached to the forearm. The forearm is proportionally the longest section of the wing in soaring birds (figure 9.45b,c).

Hummingbirds, swifts, and swallows depend on strong, frequent wing beats. Soaring birds take advantage of air in motion to gain altitude and stay aloft (figure 9.46a–d). Those that soar over open oceans take advantage of strong prevailing winds and have long narrow wings like those of glider aircraft (figure 9.47a). But the aerodynamic character of moving air can be different, so the character of soaring flight is different as well. Birds that soar over open country ride thermals, fountains of warm air rising upward. As the sun warms the Earth, the nearby air is warmed and begins to rise. Vultures, eagles, and large hawks find these rising thermals, circle to stay within them, and ride them to gain easy altitude. These birds have slotted wings (figure 9.47b). For flight in enclosed habitats such as woodlands and shrubby forests, elliptical wings give birds such as pheasants a quick, explosive takeoff and maneuverability within tight spaces (figure 9.47c). Birds of prey, migratory waterfowl, swallows, and others that depend on fast flight possess swept-back wings (figure 9.47d). To understand these general wing designs, we need to examine the aerodynamic basis of flight itself and the problems different wing designs address.

***Aerodynamics*** During horizontal flapping flight, four forces act on a bird at equilibrium. The upward lift (L) force is opposed by the force of gravity (mg), tending to pull the bird down. Various drag (D) forces act in the direction opposite to the direction of travel and wings generate thrust (T), a forward force component (figure 9.48a). The angle at which the wing meets the airstream is its **angle of attack.** Increasing this angle increases lift, but only up to a point. As the angle of attack increases, drag forces increase as well because of the change in the wing profile meeting the airflow (from edge-on to broadside) and because of the increased turbulence across the wing (disruption of laminar flow). Thus, at some extreme angle of attack that depends on air speed and particular wing shape, drag becomes predominate and the wing no longer produces net lift. When this happens, the wings **stall** (figure 9.48b). Stalling can be

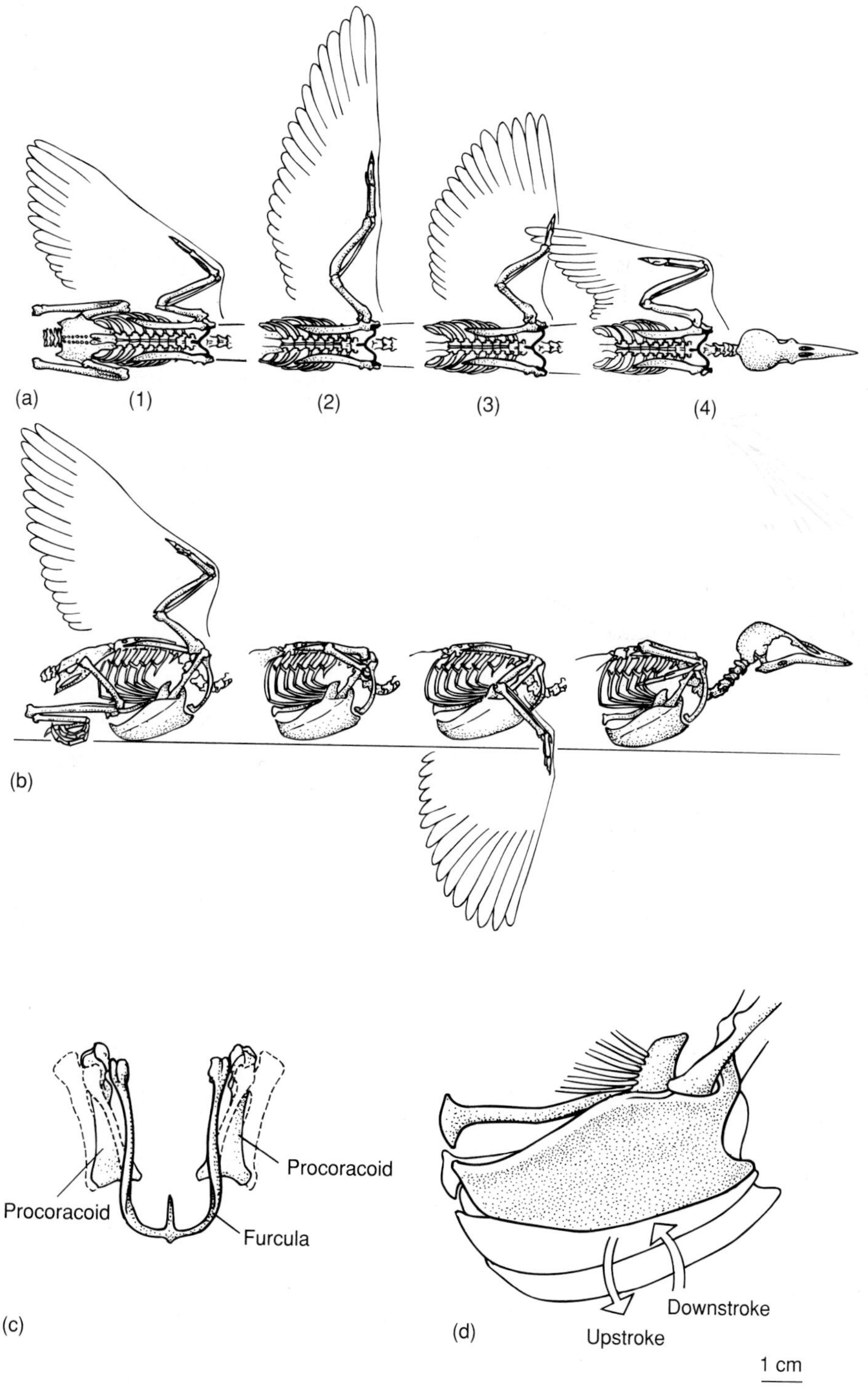


**Figure 9.43** Wing beat cycle of a European starling. Positions of shoulder girdle and wing are illustrated in dorsal (a) and lateral (b) views. (c) Anterior view of furcula and procoracoids that bend laterally on the downstroke (dashed lines) and recoil medially on the upstroke (solid lines). (d) Lateral view of the excursion of the sternum, which moves in a posterodorsal direction during the downstroke and reverses this motion to an anteroventral direction during the upstroke.

From F. A. Jenkins, Jr., et al., "A Cineradiographic Analysis of Bird Flight: The Wishbone in Starlings is a Spring." in *Science*, 16 September 1988, Vol. 241, pp. 1495–1498. Copyright © 1988 American Association for the Advancement of Science, Washington DC. Reprinted by permission.

delayed if the layers of air in the laminar flow are prevented from separating. In birds, the small **alula** controls the laminar airstream passing over the wing and prevents its early separation as the angle of attack initially increases. Thus, greater angles of attack can be reached before stalling, so greater lift can be produced.

**Aerodynamics (p. 135); feathers (p. 207)**

An **airfoil** is any object placed in a moving stream of air that produces a useful reaction because of its shape.

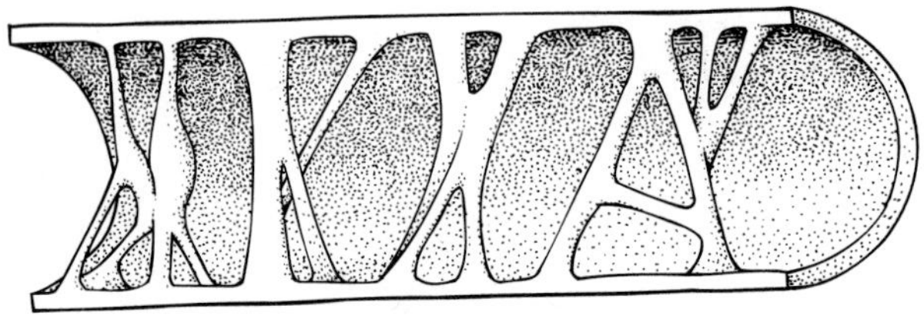

**Figure 9.44** Long bones of birds. Many tissues that contribute to weight are reduced in birds. No tissue fills the marrow cavities of the long bones, and the bone walls are thinned. Thin struts stiffen the bone and prevent it from buckling. In life, the spaces are filled with extensions of the air sacs.

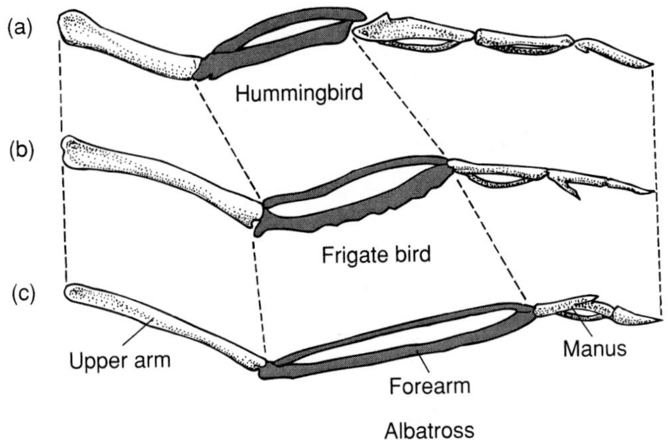


**Figure 9.45** Differences in flight are reflected in differences in wing design. In a hovering bird, such as the hummingbird (a), emphasis is on the primary feathers and the distal part of the forelimb, offering attachment for the primaries, and consequently the manus is relatively lengthened. In soaring birds, such as the frigate (b) and especially the albatross (c), emphasis is on the secondary feathers, and the part of the forearm that supports these feathers is relatively lengthened.

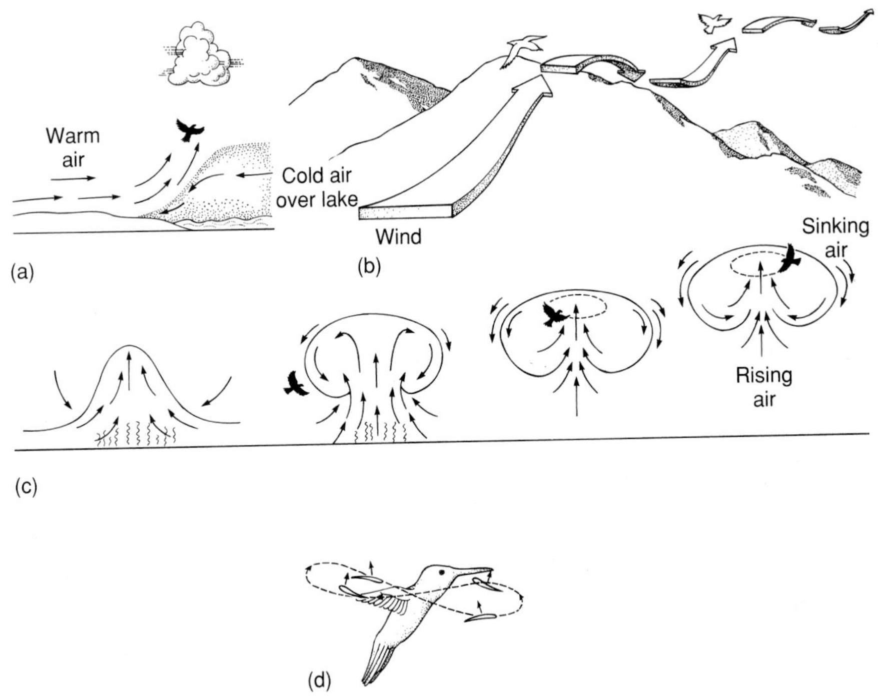


**Figure 9.46** Soaring and hovering flight. (a) Soaring birds take advantage of updrafts from wind currents. Cold air moving off water slides under the lighter air warmed over land. Rising air creates updrafts birds use to gain altitude. (b) Ridge soaring. Uplift is created when wind is forced upward by a low mountain. The wind rebounds again behind the mountain, creating repeated opportunities for ridge soaring. (c) Thermals. Local areas of the ground warmed by sunlight heat the adjacent air, which begins to rise. This bubble of rising warm air is a thermal with internal circulation as shown by the arrows. As the thermal rises, soaring birds enter and circle within the core to rise to a higher altitude. (d) Hovering birds, such as the hummingbird, must depend entirely upon the strength of their wing muscles to generate lift. Wings are rapidly swept along the dashed path producing lift (arrows) on the backstroke as well as during the forward downstroke.

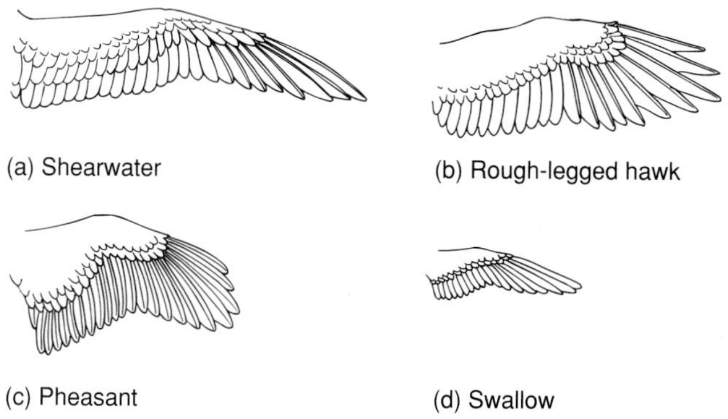


**Figure 9.47** Wing shape differs with the type of flight. (a) Soaring birds have long, narrow wings like glider planes. (b) Birds that soar over land, as hawks do, have slotted wings, with primaries slightly spaced at their tips. (c) For birds that must maneuver in close quarters, such as pheasants, the wings are elliptical in shape to allow quick bursts of flight in enclosed habitats such as forests. (d) Fast-flying birds, such as falcons and swallows, have swept-back wings.

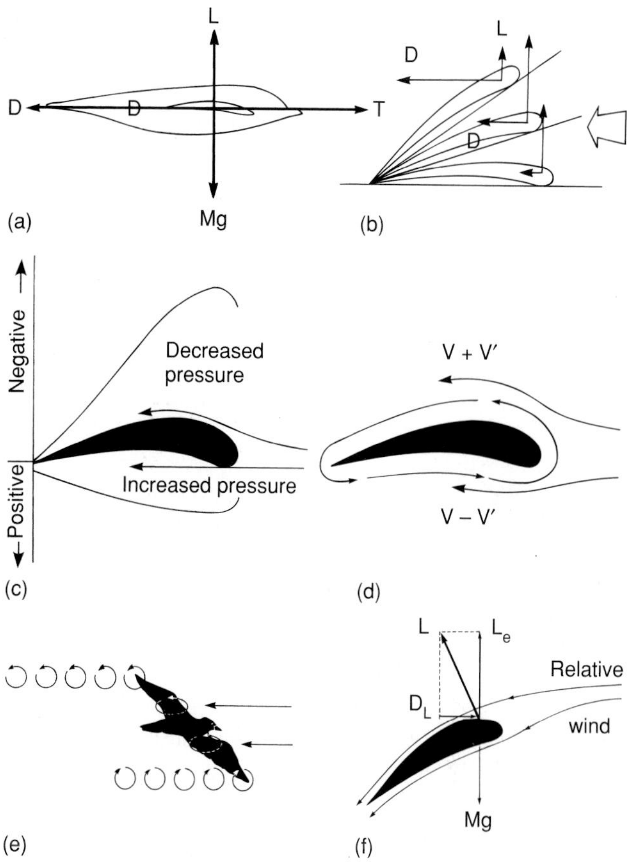


**Figure 9.48** Aerodynamics of flight. (a) During stable, flapping flight, four basic forces act on the animal. Active wing beats produce a forward thrust (T), and the cambered wing shape produces lift (L). In opposition are the downward force of gravity (mg) and the drag force (D) acting in the opposite direction to the bird's line of travel. (b) Both lift and drag are affected by the angle of attack the wing makes with the oncoming airflow. The trade-off between them in part determines how well the animal remains airborne. At some critical angle of attack, drag predominates and the animal's wing stalls. If it does not adjust the position of its wing, the bird begins to drop. (c) A useful cambered airfoil placed in an airstream forces air to travel faster along the foil's upper rather than its lower surface. As a result, the lower surface of the airfoil experiences relatively greater pressure than the upper surface. This difference is measured as lift. (d) Theoretically, this pressure difference (upper to lower) further results in a local circulation of air about the airfoil, depicted by the velocity (V) of the airstream meeting the wing and the speed of circulation (V′). This local circulation may contribute to drag. (e) Long, tapered wing tips may represent a way of spreading, dissipating, and thus minimizing the effects of this drag. (f) Induced drag. Total lift produced by the wing acts perpendicular to the surface. The vertical component of lift ($L_e$) acts directly opposite to gravity (mg). The horizontal component of lift is the induced drag ($D_L$) acting opposite to the direction of travel.

Wings are cambered airfoils that produce lift. Whether the airfoil is the wing of a bird, pterosaur, bat, or airplane, it generates lift by dividing the onrushing stream of air, directing some across upper and some across lower surfaces. The airfoil shape forces the divided airstream to move more rapidly across its long, curved upper surface rather than along its shorter, lower surface. As D. Bernoulli (1700–1782) described long ago, and the principle now bears his name, the faster a fluid moves, the lower the pressure it exerts on adjacent structures over which it moves. Thus, the air pressure exerted upon the upper surface of an airfoil by the faster moving air is less than on the lower surface. This difference in upper and lower pressures results in lift (figure 9.48c).

But the pressure differential is also hypothesized to generate a local circulation of air that follows this pressure differential encircling the airfoil. The effective velocity of air moving over the upper surface results from the speed of the airstream plus the speed of the local circulation traveling in the same direction. The velocity of air moving over the lower surface results from the speed of the airstream minus the local circulation moving in the opposite direction. This reduction in velocity induces resistance drag opposed to forward travel (figure 9.48d). It is thought that the tapered wing tips of birds dissipate these vortices to help mitigate their effects (figure 9.48e).

Total drag is the overall force that resists movement of an animal through a fluid. Two general categories of drag contribute, parasitic drag and induced drag. **Parasitic drag** is the resistance to passage of an animal through a fluid. In turn, several types of resistance contribute to parasitic drag: profile drag is the portion of this resistance caused by the shape of the animal moving through fluid; friction drag is

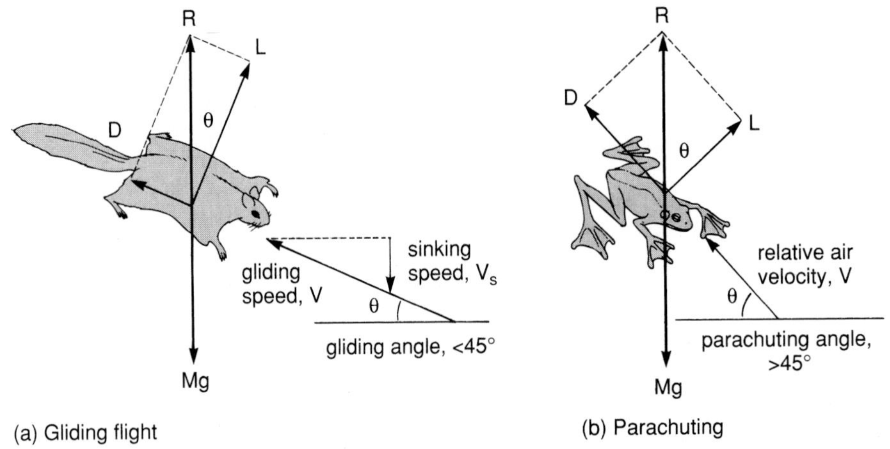


**Figure 9.49** Aerodynamics of nonflapping aerial locomotion. (a) Gliding flight. (b) Parachuting. The basic difference between the two types of air travel is that the area supporting the weight of the parachuting animal is relatively smaller. Consequently, its overall lift to drag ratio (L/D) is lower, resulting in a steeper gliding path and in a steeper angle with the ground (θ). The force due to gravity (mg) is acting in a direction opposite to the resultant (R) of drag and lift.

caused by shear stress at the boundary layer; pressure drag is caused by adverse backflow in the wake. **Induced drag** is associated with the production of lift (figure 9.48f). A wing meeting the onrushing relative wind produces lift at right angles to the wing's surface. The useful part of this lift acts vertically, directly opposite to gravity. The vector difference between lift and its effective vertical component represents the induced drag. Therefore, induced drag is the vector component of the lift force acting opposite to the direction of travel. Ironically, in producing lift, the bird wing generates a retarding force component that contributes to overall drag.

**Friction and pressure drag (p. 135)**

Animals that glide and animals that parachute depend on the same aerodynamics as those that fly, except thrust is not produced. Performance differences have to do with relative differences between lift and drag forces. A gliding animal, such as the flying squirrel, spreads its body when airborne to present a broad surface to the air. If its glide is steady, several forces act upon it. Resistance of the outstretched body against the airstream produces lift (L). Drag (D) in the direction opposite to travel is also present, and the effect of gravity (mg) acts as well. The descending glide path makes an angle (θ) with the ground. Relative to this angle, the component of force producing lift is mg cos θ and drag is mg sin θ. Their L/D ratio is greater than one during a steady glide (figure 9.49a). For a parachuting animal, force relationships are the same except that the area supporting weight is small, lift is small, the L/D ratio is lower; thus, the glide path is steeper (figure 9.49b).

## *Fossorial Locomotion*

Animals that spend part or all of their lives underground are said to be **subterranean.** With such a lifestyle, an animal takes advantage of existing tunnels or holes into which it retreats. Snakes, lizards, turtles, and many amphibians escape down burrows or deep natural recesses in the earth to find relief from the harsh, cold winter or the excess heat of midsummer. Many fishes find relief from predators in tunnels, whereas predators often use tunnels to conceal themselves until they can pounce on unsuspecting prey. Sleek predators such as snakes or weasels follow their prey beneath ground into subterranean chambers. Some subterranean animals store food in underground caches.

However, many subterranean vertebrates excavate their own tunnels by active digging. Such active earth movers are fossorial. The fossorial habit has evolved in every vertebrate class. Prairie dogs and rabbits excavate extensive, interconnecting tunnels; the subterranean living quarters of rabbits are **warrens.** These can include a maze of passageways with escape routes and snug nesting chambers in which young can be raised. Thus, digging may produce underground microhabitats with safer conditions, moderate climate, and more abundant food than offered on the surface.

***Ways of Digging*** A lungfish seeking temporary retreat as a pool of water dries digs into the soft mud using its body and fins. Flattened flounders wave pectoral fins to stir loose sand into suspension. As it settles, the sand covers the fish to conceal its body. Frogs back into a shallow burrow scooped out by their hindlimbs. Among reptiles, amphisbaenians use their pointed heads to penetrate soft soil. Body pressure against the walls of the tunnel compacts the substrate so the walls will hold and not immediately collapse on the animal. Some snakes push their way through loose sand

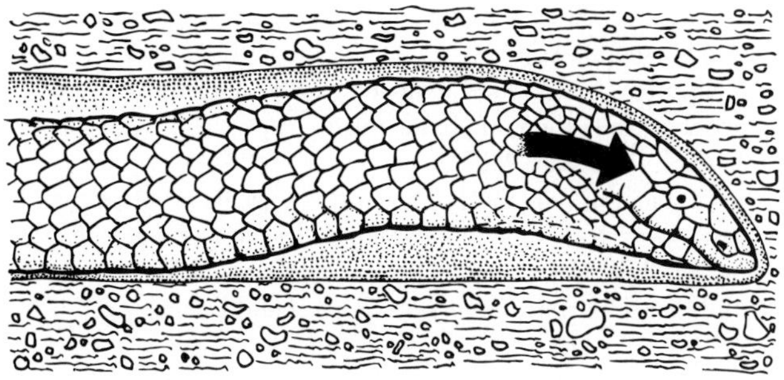

**Figure 9.50** While buried, fossorial animals face special problems, not the least of which is obtaining sufficient air. The sand snake uses its head to push a hole slightly larger than its body in loose sand, thereby creating a sand-free space to facilitate breathing.

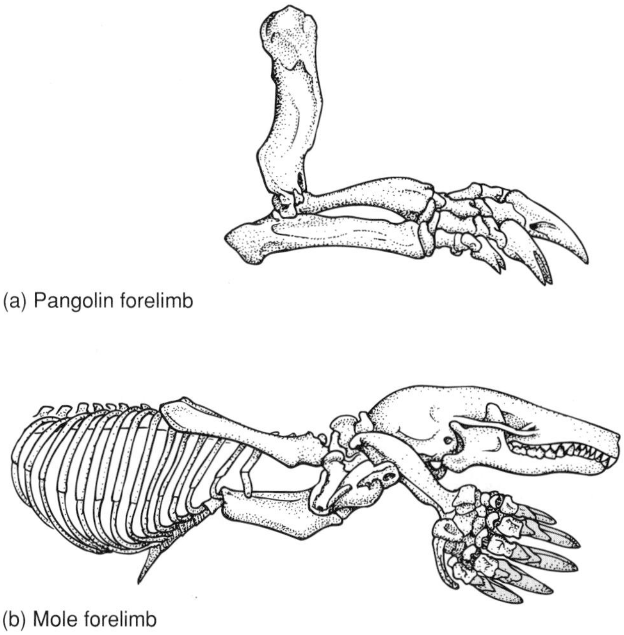


**Figure 9.51** Skeletal adaptations for digging. (a) The forelimb of a digging mammal, the pangolin, is short and robust, giving it a power advantage to move earth. (b) The mole is similarly designed, with powerful forelimbs and a broad shovellike manus.

to descend several inches beneath the surface away from the desert heat above (figure 9.50). Many rodents gnaw into soil to loosen it with their powerful incisor teeth before excavating it with their limbs.

***Fossorial Adaptations*** The appendicular skeleton, especially the forelimbs, can apply great force to move earth. Several structural modifications usually are involved. First, the limb bones of fossorial animals are especially stout and robust, and the muscles that are attached to them are relatively large (figure 9.51a,b). This produces a short and forceful bone-muscle system, unlike the long or delicate limbs of cursorial or aerial specialists. Second, the limb as a lever system is adapted for high-force output. The forearm and hand of fossorial vertebrates, which deliver the outforce, are relatively short; the elbow, which delivers the in-force, is lengthened to increase the lever input of muscle contraction. Third, the hand is usually broad and wide, like a shovel, and extended with stout claws that scoop soil with each stroke.

**Lever mechanics (p. 132)**

## Selected References

Balfour, F. M. 1876. The development of elasmobranch fishes. *J. Anat. Physiol. (London)* 11:128–72.

Balfour, F. M. 1881. On the development of the skeleton of the paired fins of Elasmobranchii, considered in relation to its bearings on the nature of the limbs of the Vertebrata. *Proc. Zool. Soc. (London)* 1881:656–71.

Coates, M. I., and J. A. Clack. Polydactyly in the earliest known tetrapod limbs. *Nature* 347:66–69.

Edwards, J. L. 1977. The evolution of terrestrial locomotion. In *Major patterns in vertebrate evolution*, edited by M. K. Hecht, P. C. Goody, and B. M. Hecht. New York: Plenum Press, pp. 553–77.

Edwards, J. L. 1989. Two perspectives on the evolution of the tetrapod limb. *Amer. Zool.* 29:235–54.

Fricke, H., O. Reinicke, H. Hofer, and W. Nachtigall. 1987. Locomotion of the coelacanth *Latimeria calumnae* in its natural environment. *Science* 329:331–33.

Goslow, G. E., Jr., K. P. Dial, and F. A. Jenkins, Jr. 1989. The avian shoulder: An experimental approach. *Amer. Zool.* 29:287–301.

Gould, S. J. 1991. Eight (or fewer) little piggies. *Nat. Hist.* 100:22–29.

Hildebrand, M. 1987. The mechanics of horse legs. *Amer. Sci.* 75:594–601.

Hinchliffe, J. R., and M. K. Hecht. 1984. Homology of the bird wing skeleton, embryological versus paleontological evidence. *Evol. Biol.* 18:21–39.

Hom, M. H., and R. N. Gibson. 1988. Intertidal fishes. *Sci. Amer.* 251(1):64–70.

Jarvik, E. 1965. On the origin of girdles and paired fins. *Israel J. Zool.* 14:141–72.

Jenkins, F. A., Jr., K. P. Dial, and G. E. Goslow, Jr. 1988. A cineradiographic analysis of bird flight: The wishbone in starlings is a spring. *Science* 242:1495–98.

Jenkins, F. A., Jr., and G. E. Goslow, Jr. 1983. The functional anatomy of the shoulder of the Savannah monitor lizard *(Varanus exanthematicus)*. *J. Morph.* 175:195–216.

Krantz, G. 1981. *The process of human evolution*. Cambridge, Mass.: Schenkman Publishing Co.

Lovejoy, C. O. 1988. Evolution of human walking. *Sci. Amer.* 251(1):118–25.

Norberg, U. M. 1985. Flying, gliding, and soaring. In *Functional vertebrate morphology*, edited by M. Hildebrand, D. M. Bramble, K. F. Liem, and D. B. Wake. Cambridge, Mass.: Harvard University Press, pp. 129–58.

Norberg, U. M. 1990. Vertebrate flight. Mechanics, physiology, morphology, ecology and evolution. *Zoophysiology* 27:1–291.

Padin, K. 1988. The flight of pterosaurs. *Nat. Hist.* 97:58–66.

Rackoff, J. S. 1980. The origin of the tetrapod limb and the ancestry of tetrapods. In *The terrestrial environment and the origin of land vertebrates*, edited by A. L. Panchen. New York: Academic Press, pp. 255–92.

Schultze, H-P. 1986. Dipnoans as sarcopterygians. *J. Morph.* (suppl.) 1:39–74.

Shubin, N. H., and P. Alberch. 1986. A morphogenetic approach to the origin and basic organization of the tetrapod limb. *Evol. Biol.* 20:319–87.

Tarsitano, S. F. 1985. The morphological and aerodynamic constraints on the origin of avian flight. In *The beginnings of birds*, edited by M. K. Hecht, J. H. Ostrom, G. Viohl, and P. Wellnhofer. Eichstätt, Germany: Bronner and Daentler, pp. 319–32.

Thacher, J. K. 1876. Medial and paired fins, a contribution to the history of vertebrate limbs. *Trans. Conn. Acad. Arts Sci.* 3:281–310.

10 CHAPTER

# The Muscular System



## INTRODUCTION

Muscles make things happen. They supply force for movement and together with the skeletal system are the movers and levers that make an animal act. Just as importantly, muscles restrain motion. When we stand comfortably or sit reflectively, muscles hold our body in position to keep it from toppling over. Muscles also act upon the viscera—blood vessels, respiratory channels, glands, organs—to affect their activity. For instance, muscles wrapping the tubular digestive tract contract in peristaltic waves that mix and move the food within. Muscles form sphincters, gatekeepers, that control the passage of materials out of tubular ducts. Sheets of muscle within the walls of the respiratory track affect the flow of air to and from the lungs. Muscles lining the walls of blood vessels affect circulation.

Secondarily, muscles play a role in heat production. As any exercising athlete knows, a by-product of muscle contraction is heat. Ordinarily, the human body generates enough heat, but if its core temperature should drop in cold weather, large muscles throughout the body strongly contract to produce shivering. Shivering muscles do no extra work but they give off extra heat, and the core temperature comes up to normal. In some species of fishes, the extrinsic eye muscles that rotate the eyeball take on the additional specialized function of producing heat. These enlarged muscles include biochemical pathways that generate heat. From these muscles, this heat is carried by blood vessels directly to the brain to warm it.

Two by-products of muscle contraction that usually go unnoticed are noise and very low voltage of electricity. However, many sharks and some other predacious fishes have sensory receptors that detect such stray noise and electric signals at close range. Even when prey are hidden or buried, their muscles contract to pump water across their gills during regular breathing. The electric noise from these contracting muscles can betray their position to predators.

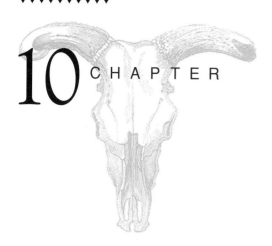

In some species of fishes, these stray electric by-products have become a major function of specialized muscles. Blocks of these specialized muscles produce high levels of voltage, not force. Such blocks of muscle are **electric organs** and occur to varying degrees in over 500 species of fishes (figure 10.1). Electric organs have arisen in several species of chondrichthyans as well as in teleost fishes belonging to different families. Such an independent appearance of electric organs represents an example of convergent evolution.

Electroreceptors (p. 704)

Electric organs generate bursts of energy to paralyze prey. Other fishes, such as the torpedo fish, employ jolts of voltage to protect themselves from predators. Still others use the electric organs to generate an electric field around their bodies. As they move through murky or dark waters, objects in the environment that come close disturb this surrounding electrical field and alert the fish to objects in its path. Thus, the specialized muscles of electric organs play a biological role in food capture, defense, and navigation.

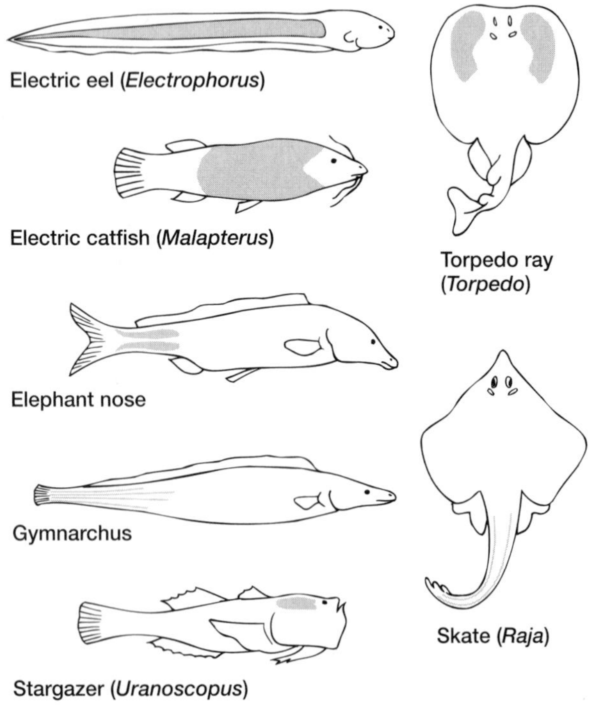


**Figure 10.1** Electric organs. The electric organs are specialized blocks of muscle derived, for example, from branchial muscles in the torpedo ray and from axial muscles in the skate. In the torpedo, skate, electric eel, and electric catfish, these electric organs can produce a jolt of voltage sufficient to stun prey or discourage a predator's attack. In other fishes, the electric organs produce a weak electric field around the fish's body, allowing it to detect any object that disrupts the field. In this way, fishes with electric fields can navigate and find food in dark or silty waters. Each fish illustrated in the figure belongs to a different family. The torpedo and skate are elasmobranchs, the others bony fishes. Thus, electric organs have arisen independently several times within different groups.

However, in most vertebrates, muscles produce forces that control motion. Motion can drive the organism through its environment or control the actions of its internal body processes.

# Organization of Muscles

## Classification of Muscles

Because muscles have many functions and many scientists from diverse fields study them, it is not surprising to find different criteria used to classify muscles. The criterion picked depends on which property of muscles is of personal interest. The following criteria are most commonly used as the basis for distinguishing muscles:

1. Muscles are classified according to their color. There are **red** and **white** muscles. This classification has fallen out of favor because such a color distinction alone underestimates the complexity of muscles.
2. Muscles are classified according to their location. **Skeletal** or **somatic muscles** move bones (or cartilages) and **visceral muscles** control the activity of organs, vessels, and ducts.
3. Muscles are classified by the way in which they are controlled by the nervous system. **Voluntary muscles** are under immediate conscious control, but **involuntary muscles** are not.
4. Muscles are classified according to their embryonic origin. This is discussed more fully later in this chapter.
5. Muscles are classified by their general microscopic appearance. There are **striated, cardiac,** and **smooth muscles.** Let us look at this general microscopic appearance next.

### *Striated Muscle*

Viewed with a microscope, striated muscle appears to have cross bands or striations that result from its underlying structure. Striated muscle is also under voluntary control and is usually associated with the skeletal system. Each striated muscle cell is multinucleate, with many nuclei distributed throughout its cytoplasm. Individual cells are usually less than 5 cm long, but they can be attached end to end to form longer composite fibers. Internally, each striated muscle cell is packed with long units called **myofibrils.** Each myofibril is a chain of repeating units, or **sarcomeres.** In turn, each sarcomere itself is composed of underlying **myofilaments** of two kinds, **thick** and **thin filaments.** Under the electron microscope, thick and thin filaments appear as a highly ordered and repeated design within each sarcomere, giving each sarcomere a distinct banding pattern (figure 10.2). This ordered molecular arrangement is too small to be seen directly with the light microscope. However, because myofibrils within a muscle cell tend to be aligned in register with each other, the overall effect produces a "striated" pattern on the muscle fiber that is visible even under a light microscope.

## *Cardiac Muscle*

Cardiac muscle occurs only in the heart. Like striated muscle, it is characterized by a banding pattern. Unlike striated muscle, however, individual cardiac muscle cells are short, mononucleate, often branched, and joined to each other by distinct **intercalated disks** into sheets (figure 10.3). Cardiac muscle cells are involuntary. Waves of contraction spread through cells and across the intercalated disks that conduct electric impulses.These waves of contraction can be initiated by nerves or contraction arising intrinsically within the muscle tissue itself. Cardiac muscle tissue kept healthy and active outside the body can contract spontaneously and rhythmically without stimulation from the outside.

## *Smooth Muscle*

Viewed through a light microscope, smooth muscle lacks striations, so they have been labeled as "smooth." Smooth muscle is almost entirely concerned with visceral functions—digestive tract, blood vessels, lungs—so it is called

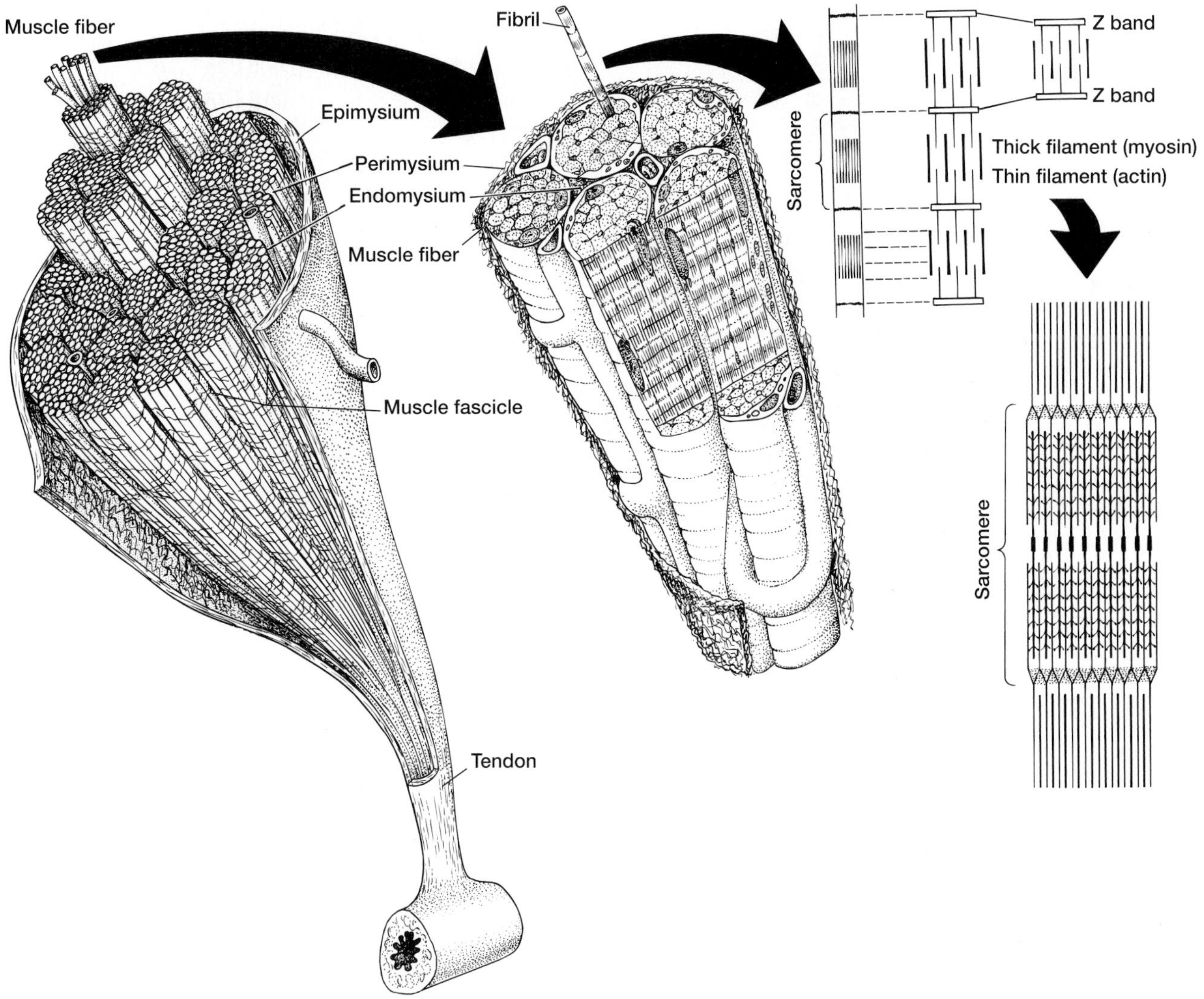


**Figure 10.2** Striated muscle. Each muscle fiber is internally composed of myofibrils, each myofibril is a chain of sarcomeres, and each sarcomere is composed at the molecular level of myofilaments specifically overlapping myosin (thick) and actin (thin) myofilaments. The underlying molecular arrangement of these filaments produces the pattern of striations within a myofibril. Because bundles of myofibrils are aligned in register within the muscle cell, they produce a visible striated pattern superficially on the muscle cell. Individual muscle cells are wrapped in fibrous connective tissue (endomysium), groups of cells are bundled by more wrappings (perimysium), and the entire muscle is covered by an outer sheet of fibrous connective tissue (epimysium). These layers of fibrous connective tissue extend beyond the ends of the muscle cells to form tendons that connect muscles to bones.

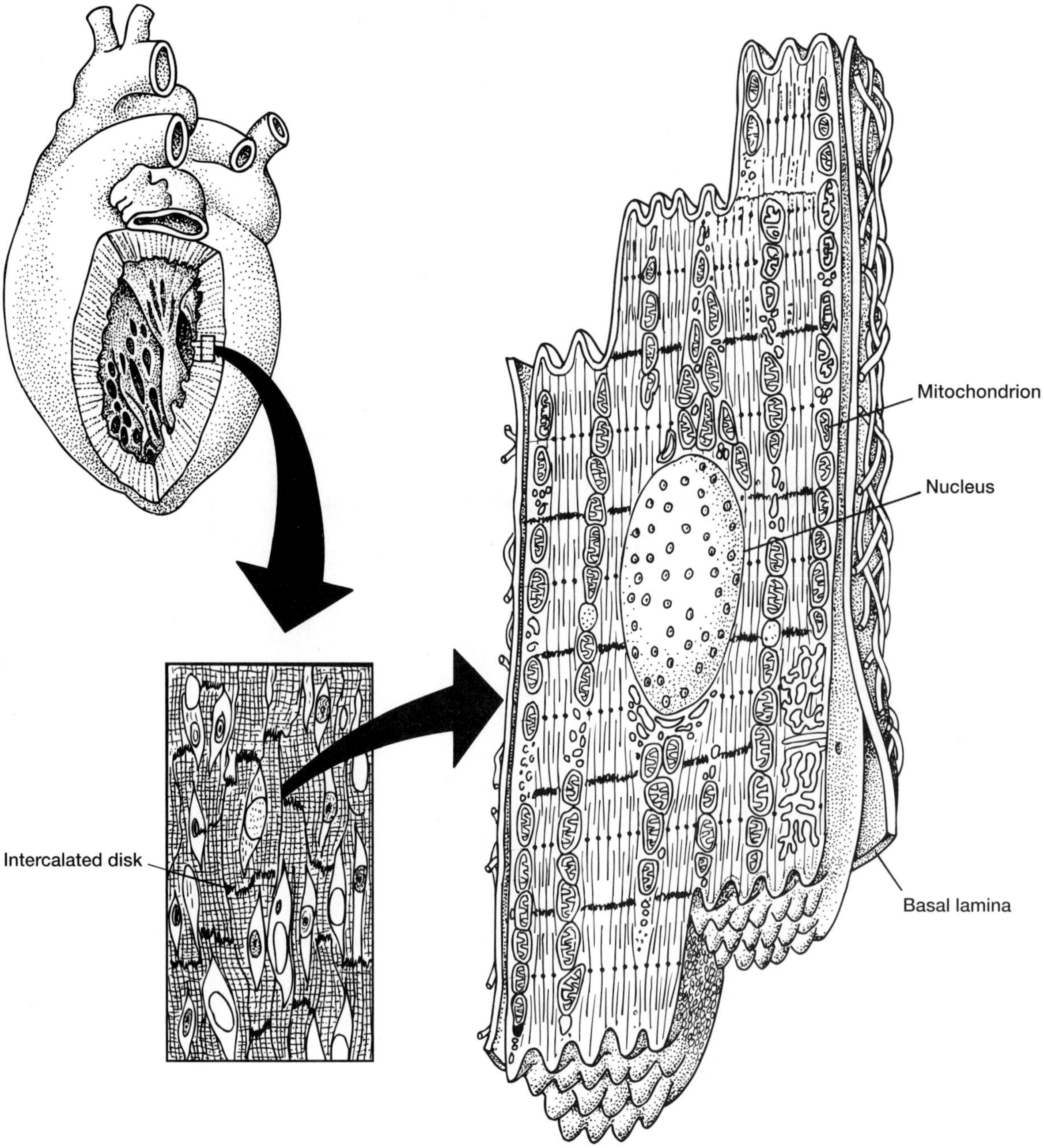


**Figure 10.3** Cardiac muscle. Found only in the heart, cardiac muscle cells are short and joined one to another by intercalated disks, specialized attachment sites. Cardiac muscle cells form sheets that constitute the thick pumping walls of the heart.

visceral muscle as well. Activity of smooth muscle is outside voluntary control. Typically, contractions are slow and sustained compared with the rapid contractions characteristic of striated muscle. Consequently, smooth muscle is suited for sphincters where fatigue might mean untimely loss of control.

Each smooth muscle cell is mononucleate, short, and fusiform in shape (figure 10.4). All are about the same size. Smooth muscle cells are joined to each other at junctions to form sheets. These sheets are wrapped around the organs on which they exert mechanical control. The molecular mechanism of contraction is not as well understood as it is in striated muscle, but it is generally assumed to be based on a sliding filament mechanism.

In our discussion of muscles in this chapter, striated muscles receive our center of attention. Striated muscles supply the force that moves the skeleton. Cardiac and smooth muscles are discussed with the viscera in later chapters.

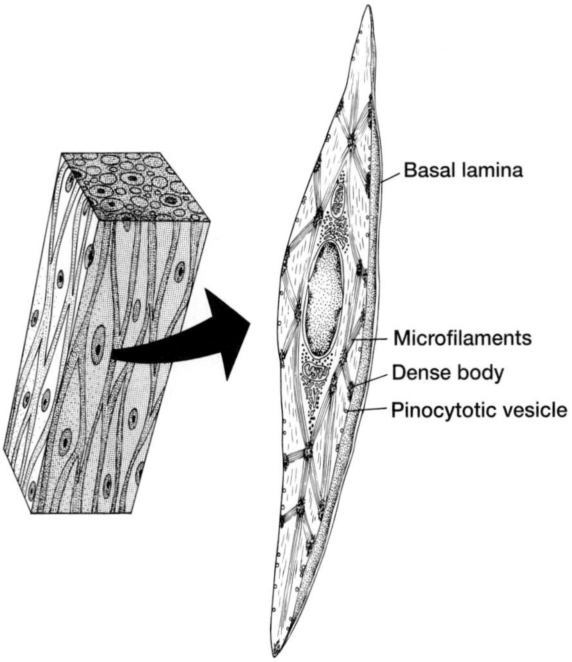


**Figure 10.4** Smooth muscle. On the right, a single smooth muscle cell is enlarged and isolated from the block of smooth muscle. Although smooth muscle lacks striations, the underlying contractile mechanism is thought to be based on sliding filaments of actin and myosin.

**Heart (p. 462); digestive system (p. 505)**

## Structure of Skeletal Muscles

The term *muscle* has at least two meanings. Sometimes muscle refers to muscle tissue (muscle cells only); other times muscle refers to the whole organ (muscle cells plus associated connective tissue, nerves, blood supply). The meaning intended must sometimes be decided from the context in which the term *muscle* is used. We use the specific term **muscle cell** to denote the active contractile component of a **muscle organ.**

Often, in place of the term *muscle cell,* **muscle fiber** is used. To the naked eye or under low microscopic magnification, a muscle organ teased apart looks like frayed rope (figure 10.2). This inspired the term *muscle fiber* for these tiny frayed strands, which in fact are the long individual muscle cells. Because these are actually the striated muscle cells, choice of the term *fiber* is unfortunate. Logically, the term *muscle cell* should be used for them, but this is not the customary practice. Because this usage is firmly established, we follow the convention of anatomists and physiologists and use the term *muscle fiber* to refer to a whole muscle cell.

The fleshy part of a muscle organ is its **belly (gaster),** and its ends that join the skeleton or adjacent organs form **attachments.** The biceps brachii muscle of your arm is composed, like all striated muscles, of packages of muscle cells. Each muscle cell is lightly wrapped by an immediate layer of connective tissue, the **endomysium.** Groups of muscle cells are wrapped in a **perimysium.** The entire muscle organ is surrounded by an outer coat of connective tissue, the **epimysium.** A **fascicle** refers to a bundle of muscle cells defined by its particular perimysium.

## Tendons

The muscle organ is not actually attached to bones by the contractile muscle fibers that comprise it. Instead, the various wrappings of connective tissue extend beyond the ends of the muscle fibers to connect with the periosteum of the bone. These connective components of the muscle organ that establish a cordlike attachment to bone are called **tendons.** Tendons drawn out into thin, flat sheets of tough connective tissue are **aponeuroses** (sing., aponeurosis). Aponeuroses that wrap and bind parts of the body together are considered **fascia.**

Tendons serve various functions. Muscle mass may be located in one convenient location, yet muscle force can be transmitted to a distant point via tendons. For example, the limb muscles of cursorial animals are usually bunched close to the body, but through long tendons, their force is applied at the ends of the legs (figure 10.5). Tendons also permit delicacy of control by distributing forces to digits for precise movements. The long tendons that stretch from the forearm muscles to the tips of the fingers in raccoon or primate hands are examples.

Tendons are metabolically economical and vascular supply is modest. They require little maintenance and consume little energy compared with muscle fibers. Tendons allow the metabolically expensive muscle fibers to be just long enough to produce the required amount of shortening or force. The rest of the length of the muscle between its two sites of attachment is provided by tendons.

## Basis of Muscle Contraction

### *Resting and Active Muscle*

A muscle receiving no nervous stimulation is relaxed, or in a **resting state.** A muscle organ in a resting state is soft, and collagenous fibers surrounding the muscle maintain its shape during this phase. The muscle generates no force, and if tensile force is applied to it, it stretches. Resistance to the applied tensile force arises from the collagenous fibers. When nerves stimulate a muscle to its threshold level, contraction results and generates **tensile force,** constituting the **active state** of a muscle. The bone to which the muscle is attached and the mass that must be moved represent an external resistance called the **load.** Whether or not a muscle actually shortens upon contraction depends on the relative balance between the tensile force of contraction and the load to be moved.

## Box Essay 10.1 Venom Injection

Striated muscles usually run between cartilage or bony elements of the skeletal system, sometimes attaching to the dermis to control the skin, but seldom do they act directly on glands. An exception is found within poisonous snakes, where a prominent compressor muscle originates from the lower jaw, sweeps upward around the venom gland, and is inserted on the gland's surface (box figure 1a,b). When the snake strikes its prey, the sharp, hollow fangs pierce skin to reach the prey's tissues and blood system beneath. The compressor muscle contracts, squeezing the venom gland and raising the pressure within the fang. This pressure forces the premanufactured venom, stored within the lumen of the gland, from the main venom gland through a departing venom duct that includes, in some species, a small accessory gland along its length. The venom flows along the duct to the base of the fang, where enveloping sheets of tissue direct the venom into the fang opening. The venom now courses down the hollow core of the fang to its tip through an exit aperture into the tissues of the prey.

In some nonvenomous snakes, jaw muscles adjacent to the oral glands bulge or press on the glands to facilitate intermittent release of secretion into the mouth. The situation in venomous snakes differs in that striated muscles are inserted directly on the venom gland and are thus dedicated to the function of sudden and timely release of venom to help quickly dispatch the prey.

Snake jaws (p. 254); venom gland (p. 519)

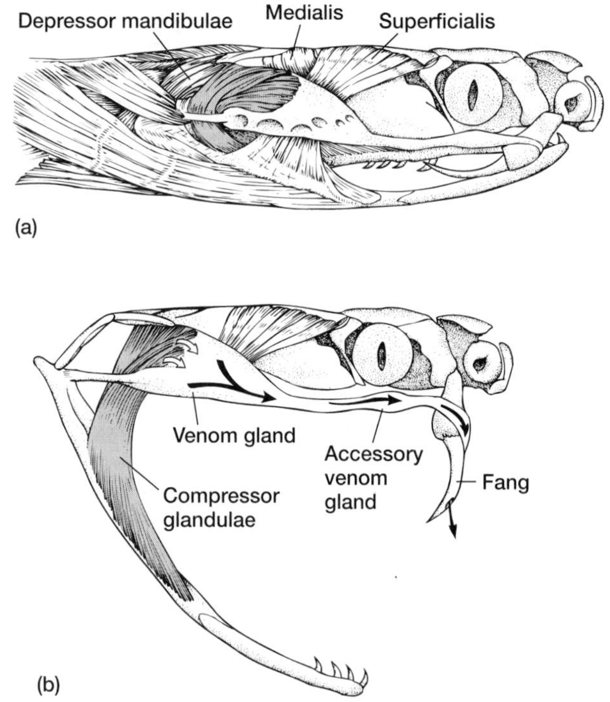


**Box figure 1** Primitive viperid snake *Azemiops*. The skin has been removed to expose the lateral jaw musculature and venom gland. (a) Ligaments tether the triangular main venom gland in position at each of its three corners. The duct runs to the base of the fang from the anterior corner of the venom gland. Notice the accessory venom gland along the duct's length. (b) Cranial muscles are lifted to show the attachment of the compressor muscle (compressor glandulae) to the venom gland. During the strike, contraction of the compressor squeezes stored venom out of the gland, along the duct, through the hollow fang, and into the prey. Solid arrows indicate the route of venom.

### *Molecular Mechanisms of Contraction*

Although muscle contractions are active and produce force, they cannot lengthen to push apart their sites of attachment. The underlying chemistry of muscle contraction is built on sliding filaments of muscle proteins that slip past one another to shorten the muscle. In striated muscle, where the contractile mechanism is best understood, contraction involves chemical cross-bridges that form and reform between thick and thin filaments to ratchet or slide these filaments past one another. We do not need to understand the biochemistry involved, but we must realize that the effect of their sliding is to shorten the sarcomere of which they are part. The sarcolemma (cell membrane) invaginates into the muscle cell at regular intervals and associates with the sacroplasmic reticulum. The terminus of the innervating neuron, known as the motor end plate, initiates the electric wave of depolarization in the sarcolemma. The sarcolemma spreads the propagating stimulus to all parts of the muscle cell. Within the muscle fiber, this electric wave of depolarization stimulates local chemical events, resulting in the sliding of molecular filaments. Because contraction occurs simultaneously throughout all sarcomeres within a muscle cell, the overall result is for the

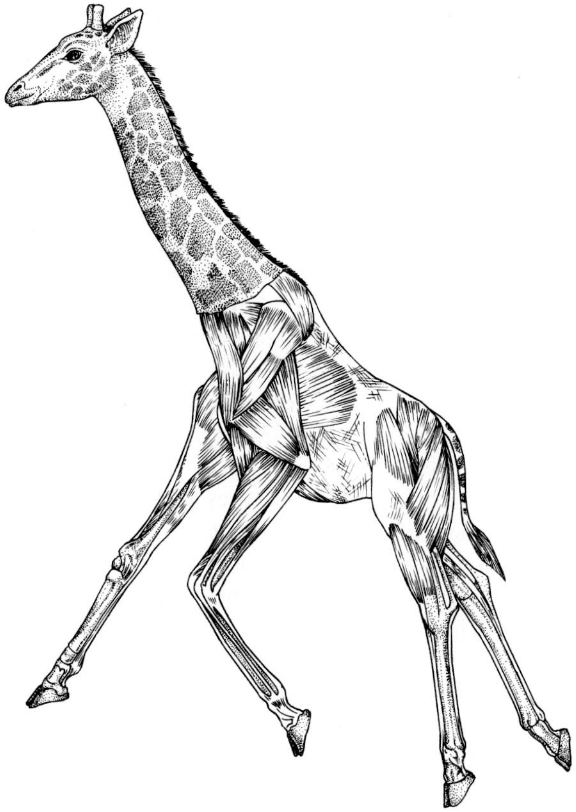

**Figure 10.5** Limb tendons of a giraffe. Tendons distribute the forces of muscle contractions to sites distant from the muscle itself. The limb muscles of a giraffe are located close to the body, but tendons of these muscles extend outward along the leg bones and deliver their forces at the giraffe's hooves.

chains of sarcomeres to shorten together, which shortens the muscle fiber that in turn generates a tensile force.

# Muscle Function

## Muscle Fibers

Some of the major contractile characteristics of a muscle fiber include how rapidly it reaches maximum tension, and how long it can sustain this tension. A variety of properties of the muscle fiber generate tension, beginning with its underlying molecular mechanism of shortening itself, namely, the sliding of thick and thin filaments. The consequences of this underlying contractile mechanism emerge in what are called tension-length curves.

### *Tension-Length Curves for a Single Muscle Fiber*

Tension produced by a particular muscle fiber is not constant, but it depends on the muscle's fixed length when it is stimulated. We can hold the two ends of a muscle fiber at a fixed length and then stimulate it and record the tension produced. If this is done for the same muscle fiber fixed at different lengths, different tensions are produced at different lengths. Tensions and lengths can be plotted in a tension-length curve that peaks at intermediate lengths but drops at both ends (figure 10.6a–c). This curve arises from the limitations of cross-bridging between the underlying muscle myofilaments.

When the muscle fiber is fixed in a lengthened position, filaments overlap very little, few cross-bridges form, and tension is low (figure 10.6b). When the muscle is fixed in the shortest positions, filaments overlap, interfere with cross-bridge formation, and tension is again low (figure 10.6a). Only at intermediate lengths is the number of cross-bridges maximized and tension peaks (figure 10.6c).

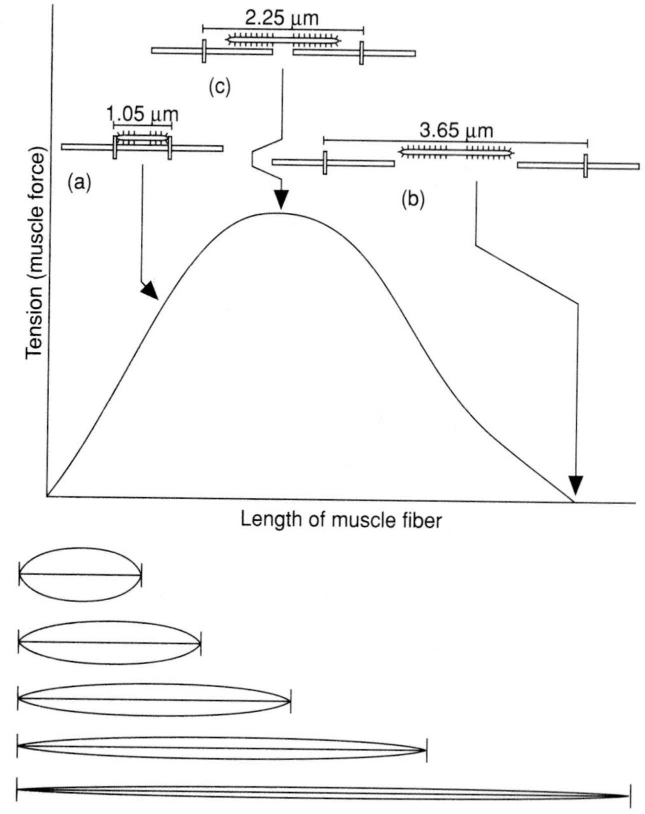


**Figure 10.6** Tension-length curves of a muscle fiber. If a muscle fiber is fixed at set lengths and then stimulated, the force it produces will vary with the length. Somewhere between extreme lengths, its force peaks. (a) When muscle length is short, overlap of thick and thin filaments reduces total force. (b) When muscle length is stretched, filaments establish fewer cross-bridges and thus less force is generated. (c) The optimum occurs at intermediate lengths because the maximum number of cross-bridges is formed to achieve the maximum force. At the bottom of the graph, a muscle fiber is shown at five different fixed lengths below the tension generated respectively above on the curve.

## *Properties of Muscle Fibers*

**Color** Even omnivores like ourselves might notice occasionally during a festive feeding frenzy that meat from the same animal is colored differently. Turkey, for example, contains light and dark meat. In fishes, you may notice that most of the muscle is white, but occasionally in some species a small lateral strip of red muscle is present. The two types of muscle, which early experimenters conveniently termed red and white muscle, also have different physiological properties.

Muscles made up of red fibers tend to be highly vascularized and rich in **myoglobin,** a dark macromolecule that stores oxygen and looks red. Red muscle is resistant to fatigue. Muscles made up of white fibers are less vascularized and low in myoglobin, but they contract rapidly. Game birds, such as turkeys, fly in quick rapid bursts. They do not migrate long, sustained distances, and their pectoralis or "breast" muscles used in flight are white muscle fibers. However, their leg muscles used to scamper along the ground are red. In migratory birds, the same pectoralis muscles are dark, capable of supporting sustained flight.

In fishes such as pike and perch, which make quick darts to catch prey, the lateral body muscles characteristically are white. In migratory fishes and those that swim in swift streams against a sustained current, the same lateral body muscles tend to be red.

If not pushed too far, this match of muscle color with contraction speed, endurance, and physiology helps us understand the basis of animal performance and the muscle types that serve it. However, muscle color alone does not always reveal underlying subtle differences in fiber physiology. For example, another important distinguishing feature of a muscle fiber is its capacity to establish sustained generated force.

***Tonic and Twitch Fibers*** On the basis of a fiber's ability to establish and sustain generated force, muscles can be categorized as tonic or twitch fibers. **Tonic fibers** are relatively slow contracting and produce low force, but they can sustain contraction for prolonged periods of time. Such fibers are involved in postural support, so they compose much of the axial and appendicular musculature. Tonic fibers are common in amphibians and reptiles, less so in fishes and birds, and absent in mammals. **Twitch (phasic) fibers,** by contrast, generally produce fast contraction, so they often make up muscles used for rapid movements. Twitch fibers are found in somatic muscles of all classes of vertebrates.

Twitch fibers have been most extensively studied, and generally, are of two kinds, **slow twitch** and **fast twitch** fibers. As their names suggest, slow twitch fibers take longer to reach maximum force than fast twitch fibers, more than twice as long in some cases. But *fast* and *slow* are relative terms and species specific. For example, fast-contracting muscles in hummingbirds have a twitch time of 8 ms. In guinea pigs, twitch time can be around 21 ms. Within mammals, fast twitch and slow twitch contractions in rats average around 13 and 38 ms, respectively. In cats, twitch times are 40 and 90 ms, depending on the muscles involved. Differences in contraction speeds seem related to differences in myosin types in the fibers, to differences in activation of actin and myosin, and to differences in nervous innervation.

There are subtypes of both tonic and twitch fibers based in part on reactions of fibers to specific histochemical stains for ATP, myoglobin, oxidative enzymes, etc. Translating a fiber's staining properties into a functional profile has proved difficult. However, the staining reaction of a fiber reveals its underlying biochemical character, which suggests its possible physiological function. These techniques reveal that some types of muscle fibers carry large stores of glycogen, other types contain enzymes that support either short bursts of activity or sustained activity. Some fiber types are intermediate (figure 10.7). Muscles that appear red may have one or more of these subtypes present, depending on the species of vertebrate. This mixing of fiber types is why color alone is usually not sufficient to characterize the underlying physiology of the muscle, and most now turn to histochemical characteristics to help identify fiber types.

In addition to contraction speed (fast and slow twitch) and histochemical profiles (subtypes), another distinguishing feature of muscle fibers is their resistance to fatigue during sustained exercise. Slow twitch fibers (S) tend to be resistant to fatigue. For example, some slow twitch fibers studied in cats maintained constant tension over 60 minutes of sustained activity. Fast twitch fibers are more diverse in their contractile properties. In cats, three types are recognized (table 10.1). At one extreme are fast twitch fibers that produce a large force but fatigue quickly, their tension falling to zero after less than 1 minute of continuous stimulation. Such fibers are fast twitch, fatigable (FF). At the other extreme are fast twitch, fatigue-resistant (FR) fibers that produce smaller force but can sustain prolonged contractions. The third type is between these two in contractile properties, fast twitch, intermediate (FI).

The muscle organ is often a mixture of fiber types with different resistances to fatigue. It has been hypothesized that within a muscle organ, the neurons establish a priority in recruiting the fiber types during prolonged exercise. As activity begins, slow twitch fibers are called on first to generate tension. As activity continues, fast twitch fibers resistant to fatigue are recruited. If activity persists, then finally fast twitch fibers that can produce large output but fatigue relatively quickly are recruited. Just when each fiber type is employed apparently depends on the strenuousness of the activity, the species of animal, and the prior conditioning of the muscle.

To summarize, contractile characteristics of muscle fibers depend on molecular properties of thick and thin filaments, fiber types, proportions of fibers within a mus-

(a)

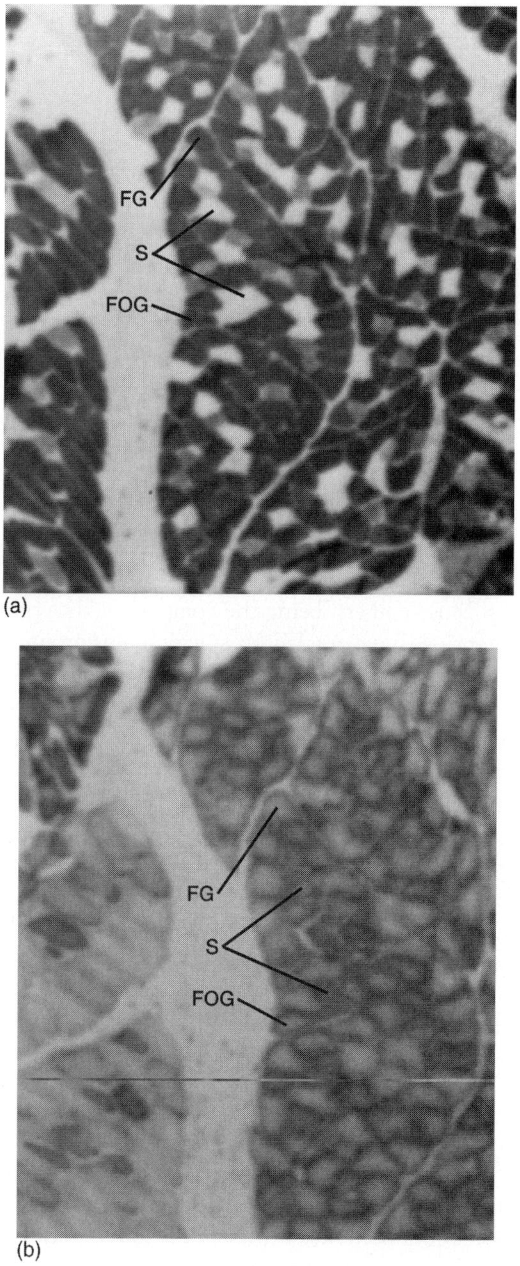


(b)

**Figure 10.7** Histochemical profiles of muscle fibers in cross section. Fresh muscle is removed, serially sectioned, and stained. (a) and (b) are adjacent sections of the same muscle treated with different stains. In (a), slow twitch (S) fibers are light; fast twitch fibers are dark. In (b), the slow twitch fibers are known, leaving only fast twitch to identify. Fast twitch glycolytic (FG) are light and fast twitch oxidative glycolytic (FOG) are dark. (a) Stained for myosin ATPase, pH 10.4 preincubation. (b) Stained for NADH-D (nicotinamide adenine dinucleotide dehydrogenase). Modified from and thanks to Young, Magnon, and Goslow, 1990.

cle, and the pattern of fiber recruitment during muscle activity. No single characteristic alone is diagnostic of the fiber's contractile speed, level of tension produced, or sustained force. Not all fiber types occur in any one vertebrate class. As we see next, the muscle organ itself might perform different functions in different species. Furthermore, differences in performance can result in part from changes in a muscle organ's overall design and internal architecture.

## Muscle Organs and Fibers

Some muscle organs produce strong forces, and others move their loads quickly. Some move loads long distances, others displace loads only short distances. Some muscles produce graded movements, moving heavy loads, but when given a light load, they perform smoothly as well. These differences in muscle performance do not entirely result from modulation of molecular cross-bridging nor from differences in fiber types. All muscles share a similar underlying contractile mechanism, the sliding of filaments. How then is this variety of properties produced out of a common molecular mechanism?

### *Whole Muscle Force Generation*

The total force output of a muscle organ is ultimately based on two functional components. The sliding of molecular filaments is responsible for the **active component** that contributes to total force. The **elastic component** can contribute to overall muscle force as well (figure 10.8). As opposing muscle groups or gravity stretch a muscle, some of this energy is stored within the muscle organ. The elastic components act like a stretched rubberband. When the muscle shortens, this elastically stored energy adds to the active component of contraction to help shorten the muscle. The elastic energy is stored in the connective tissue around muscle fibers and in muscle tendons. This elastic feature of muscles seems prevalent especially in muscles involved in repetitive events such as limb oscillation during running or trunk flexure during sustained swimming. For example, the long tendons in the lower legs of camels act like elastic "springs," storing energy when the limb touches the ground to bear the animal's weight as it walks or runs. When the limb pushes off, over 90% of this stored energy is recovered and contributes to forward momentum. It is suspected, but not yet proved, that the extensive sheets of tendons and aponeuroses in the dolphin tail act similarly to store energy when muscles are stretched during swimming. When the tail stroke is reversed, this energy is returned.

The muscle organ includes the active contractile machinery comprised of the sliding of molecular units and the elastic component residing in connective tissue. Active contraction consumes chemical energy delivered by ATP (adenosine triphosphate), the cell's source of chemical energy. The elastic component depends on mechanical energy resulting from gravity or motion of body parts that load the muscle like a spring, storing energy until it is released. Thus, the total force output generated by a muscle

TABLE 10.1 Muscle Fiber Types and Their Physiological Properties

| Fiber Type | Contraction Time | Force Output | Resistance to Fatigue | Histochemistry |
|---|---|---|---|---|
| Slow twitch | | | | |
| S | Slow | Very small | Very high | Slow oxidative (SO) |
| Fast twitch | | | | |
| FR | Fast | Small | High | Fast oxidative glycolytic (FOG) |
| FI | Fast | Medium | Medium | Fast intermediate (FI) |
| FF | Fast | Large | Low | Fast glycolytic (FG) |

*Note:* Abbreviations of fibers are: slow (S); fast twitch, resistant (FR); fast twitch, intermediate (FI); fast twitch, fatigable (FF).

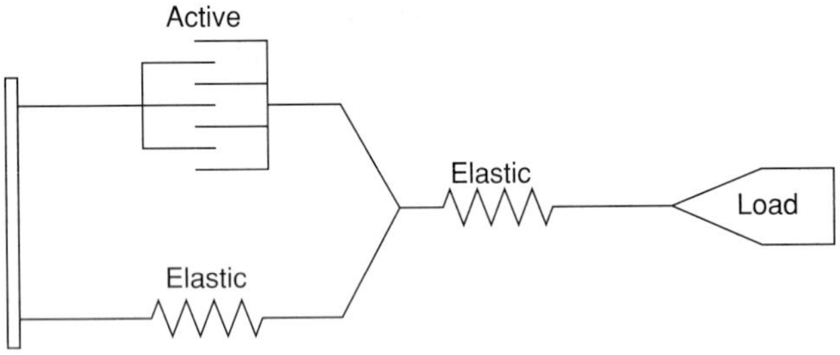


**Figure 10.8** Diagrammatic representation of functional components of a muscle organ. Sliding filaments represent the active component of a muscle. Springs represent the elastic component and may lie next to (parallel) or follow (series) the contractile component. Total output force of the muscle arises from both components.

organ arises from the combined action of active contraction and elastic recoil.

## *Tension-Length Curves for a Whole Muscle*

The tension-length curve for a whole muscle has different properties from the tension-length curve for one isolated muscle fiber. This is because the muscle organ includes packed sheets of connective tissue that add an elastic component to the active component of force. Because a muscle's active and elastic components contribute to the force generated, its tension-length curve is a combination of both. The **resting tension** represents the force required to stretch the relaxed muscle to greater lengths and results from the elastic constituents of the muscle, mainly from its collagenous fibers. The curve for **total tension** is measured at different lengths when the muscle is contracting. The sum of active and elastic components constitute total tension. The **active tension,** the contribution made by the active component alone, is derived from the difference between total tension and resting tension.

The shape of this tension-length curve becomes important when we think about the design of bone and muscle systems. Muscle fibers, being the component that produces active tension, have one length at which tension in them is greatest. Thus, any given muscle has a length at which it produces its maximum active tension (figure 10.9a), implying that a muscle that must be shortened over a relatively long distance cannot produce maximum force throughout its entire range of motion. Therefore, if a skeletal part is moved a long distance, it is common to find several muscles acting to move the bone in the same direction (figure 10.9b). Each muscle reaches its peak tension at a slightly different point during motion, passing the responsibility of generating the maximum force from one muscle to the next as the bone rotates. This may explain why several muscles with duplicate actions might be part of a design where one large muscle might seem sufficient.

## *Graded Force*

The same muscle can produce a graded motion, delivering a large force to a bone when the animal moves a heavy load or a small force when it moves a small load. For example, when you lift a heavy but not impossible weight in your hand, the biceps brachii is equal to the chore and produces the necessary large force. But when you lift a light pencil with your forearm, the same biceps brachii produces a lesser force to match. One way in which graded force is generated is by **rate modulation.** To stimulate contraction, the arriving nerve impulse must be at or above a threshold level, otherwise no muscle twitch occurs. Only electric impulses above this threshold stimulate contraction. Up to a point, force increases as the rate of arriving nerve impulses increases. Increasing force with increasing impulse rate constitutes rate modulation. Eventually, this force peaks and does not increase further even if the impulse rate continues to increase. Within the range of graded response to rate modulation, motor nerves to a muscle fiber can develop a graded force output.

## Box Essay 10.2 Exercising Muscles

When muscles are exercised, they get bigger, or at least they get bigger if the exercise involves increased load and is continued on a regular basis over a period of time. The late humorist Robert Benchley had a couch he named "the track." As he described it, when friends urged him to exercise, he obliged by telling them as he graciously departed the room that he was on his way to spend some time at the track. They were always surprised to see how rested he seemed upon his return.

Exercising means more than this to most of us. But, strictly speaking, when a muscle contracts, even during slow walking, it is being "exercised." Thus, exercise physiologists prefer the term *chronic overload* to describe elevated levels of sustained muscle activity, or just "training" to recognize the elevated muscle demands and their consequences.

Muscles enlarge in response to training. This enlargement results from several changes within the muscle. Capillaries proliferate and fibrous connective tissue increases to add to muscle bulk. However, increased muscle bulk results primarily from the enlargement of existing cells. Each cell adds more myofilaments. This results in an increase in the cross-sectional area of fibers, up to a 50% increase following some schedules of exercise training. Until recently, evidence of an accompanying increase in fiber number was less clear. Individual fibers got bigger, but they did not seem to increase in number. If a weight is applied to a quail wing, the stretched latissimus dorsi muscle shows a dramatic gain in mass through increases in both fiber number and size. But in this avian example, the overload is continuous, not intermittent as in most exercise training. In cats, a small increase in fiber (9% increase) can be induced by exercise. These additional fibers apparently do not arise by the splitting of existing fibers. Instead, new fibers are added from undifferentiated cells within the muscle. Therefore, present studies indicate that large increases in muscle mass induced by training result from some change in the number of fibers, but predominantly from an increase in individual fiber size.

Where humans, or rats, or cats have been exercised to test the physiological responses of their muscles to training, the muscles responding to chronic overload are those actually involved in the increased activity. For example, if a human is trained on an exercise bicycle, muscles of both legs get larger over the extended period of training. If only one leg is strapped to the exercise bicycle, then only muscles of that leg get larger. The physiological response is localized.

How fiber types adapt to overload is complicated. Certainly muscle fibers change during growth. Muscles of the limbs of young cats possess slow twitch fibers. In the adult, some of these same muscles take on the properties of fast twitch fibers. Further, nervous innervation seems to play a part in determining fiber type. Experimentally, if nerves stimulating slow and fast twitch fibers are switched, to some extent the muscle fiber takes on the contractile properties of its new nerve (i.e., slow twitch muscles become fast twitch and vice versa).

However, how muscle fibers respond to training is less clear. The response of muscle to training depends somewhat on the nature of the training, namely, on the loads moved and on the length of training. Pumping a bicycle against easy resistance has different effects from pumping a bicycle against high resistance.

World class athletes have been examined in order to study the response of muscles to training. Small samples of muscle can be biopsied, removed, and examined. When this is done in the quadriceps muscles of elite marathon runners, the muscle shows a large proportion of slow twitch fibers. When the same muscle is biopsied from elite sprinters, the quadriceps show a large proportion of fast twitch fibers. This suggests that the physiological characteristics of the leg muscles are matched to the athletic event. Endurance runners have more slow twitch but fatigue-resistant fibers. Sprinters have more fast twitch fibers.

Are these muscle fiber proportions the result of training or inheritance? To answer this, humans were trained on an endurance program consisting of intense bicycling twice a day. The proportion of slow twitch fibers in the vastus lateralis was biopsied before and after this six-month endurance training. The oxidative potential of the muscle, its ability to utilize oxygen in synthesizing ATP, nearly doubled. This increased oxidative capacity occurred in all fiber types, not just in slow twitch or fast twitch fibers but in both kinds. Further, there was an increase in the concentration of capillaries that served the muscle. Finally, the muscles got bigger with training, as might be expected, but the proportions of slow to fast twitch fibers did not change significantly. Training improved performance but did not change the basic fiber types composing the muscle. Training alone does not seem to be enough to make a world class marathon runner out of a world class sprinter.

This has raised the prospect of future athletes, be they animal athletes such as race horses or greyhounds or human Olympic athletes, being discovered early, by biopsy, and set on a course to greatness. It is not, however, so simple, even if it were desirable. Peak performance of an athlete is more than just muscle. The extent of muscular vascularization, the ability of the respiratory system to deliver oxygen, the rate of conversion of stored energy to available energy, and so forth also affect the performance of an athlete, to say nothing of "motivation," a major factor as well. Muscle physiology and fiber type certainly set boundaries to performance, but in complex ways. We are not yet able to forecast with certainty the future athletic performance a human or a race horse can attain.

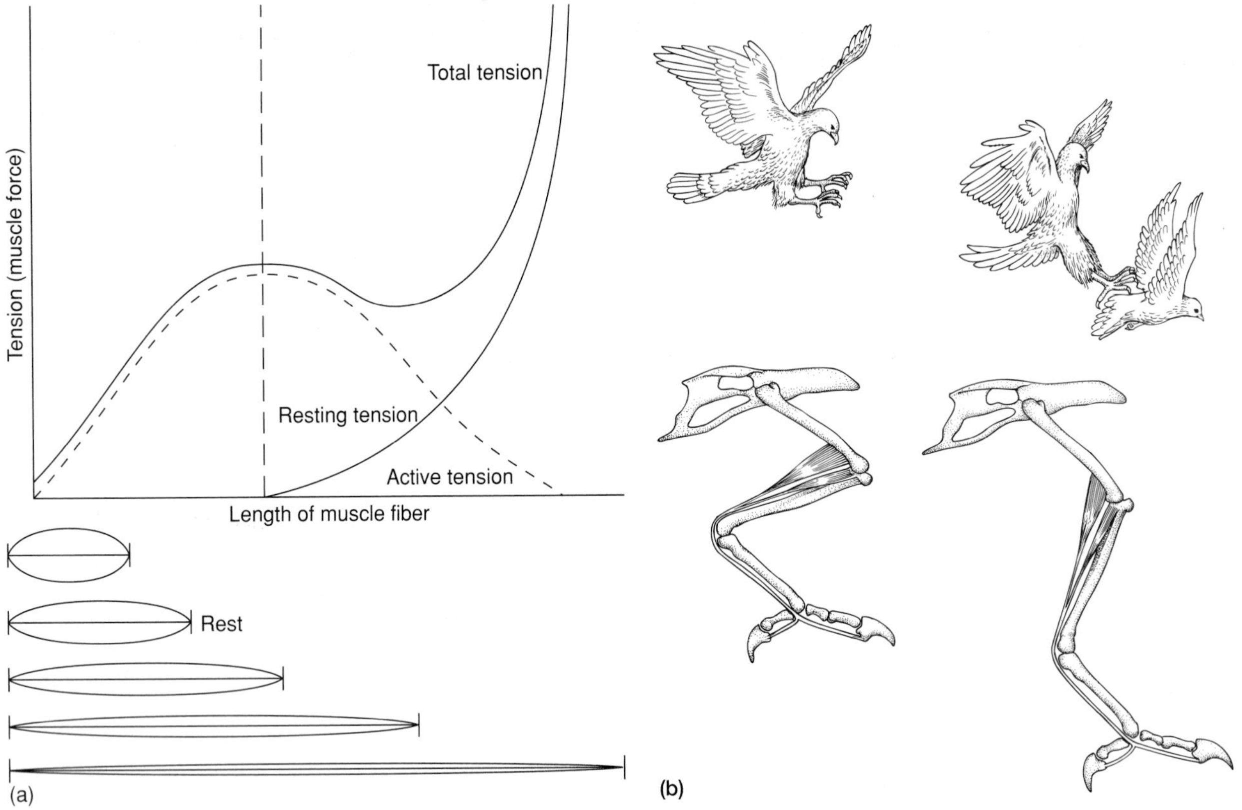


**Figure 10.9** Tension-length curve of a muscle organ. (a) The curve of resting tension represents the force required to stretch a relaxed muscle. Total tension curve is the force measured from an active muscle at various lengths. The difference between the resting and the tension curves is the active tension curve, representing the force only of the active contractile components of the muscle. Beneath the horizontal axis, note that five lengths of the muscle are depicted. (b) When this hawk strikes its prey, its limbs reach out to make contact. As they do, the distance between origin and insertion of leg muscles changes; thus, the tension produced by these muscles changes as well. Because each of the two muscles attains its peak force at different lengths, one of the two muscles is at or close to its peak tension through the range of leg extension.

A second way this graded force can be matched to loads is accomplished by selective contraction of a few, many, or all muscle fibers within a muscle organ. How is this done? One motor neuron exclusively supplies a group of the muscle fibers. Another neuron does the same but innervates a different set of muscle fibers within the muscle organ. A single motor neuron together with the unique set of muscle fibers it innervates is called a **motor unit.** By recruiting additional motor neurons, the central nervous system can selectively increase the total output force a muscle generates until it matches the load (figure 10.10). Not surprisingly, if delicate movements are required, there are fewer muscle fibers for each innervating neuron. Motor units of laryngeal muscles that control vocalizations or extrinsic eye muscles that move the eyes can contain as few as ten muscle fibers, whereas a motor unit like the large gastrocnemius muscle of the leg can have several thousand muscle cells per motor neuron.

## *Cross-Sectional Area*

The maximum force produced by a muscle is proportional to the total cross-sectional area of all its sarcomeres (or total area of all its fibers). Two terms express this relation between tension and muscle fibers. The cross-sectional area of a *muscle* perpendicular to its longitudinal axis at its thickest part constitutes its **morphological cross section.** A muscle's **physiological cross section** represents the cross-sectional area of all muscle *fibers* perpendicular to their longitudinal axes. In muscles in which all fibers run parallel to each other and to the long axis of the muscle, morphological and physiological cross sections are equal. If fibers are oblique to the long axis of the muscle, its physiological cross section represents a more accurate index of the muscle's ability to generate tension. The physiological cross section is an expression of the number of muscle fibers present. Reasonably, the more fibers present, the greater is the tension and the maximum force produced.

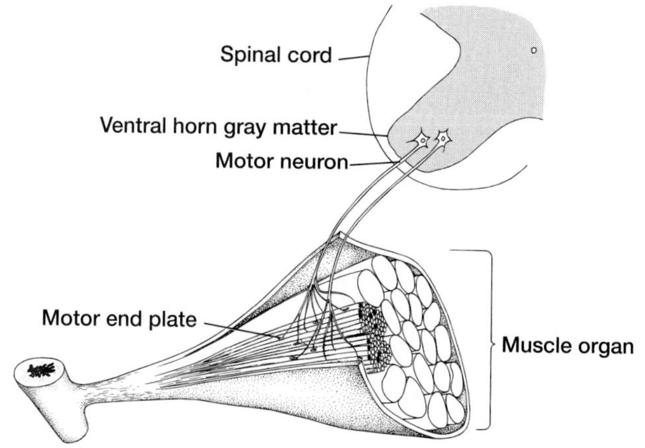


**Figure 10.10** A motor unit. Many motor neurons supply a muscle, but each neuron innervates only a few muscle fibers. Selective recruitment of additional motor units can increase muscle force. Thus, the number of contracting muscle fibers can be increased in graded amounts to equal the force required to accomplish the job.

Thus, contrary to what you might at first expect, a long muscle and a short muscle, of equal physiological cross sections, generate the *same* forces, not different forces (figure 10.11a–c).

This is somewhat analogous to the properties of a chain. A chain is no stronger than its weakest link. Increasing a chain's length does not make it stronger. To increase strength, more parallel sections of chain are added adjacent to each other. Similarly, a muscle cell might be thought of as a package containing chains of sarcomeres, with its tension limited by the weakest sarcomere. Therefore, increasing length does not increase tension. To increase tension, the number of adjacent chains of sarcomeres must be increased by increasing the number of parallel muscle fibers (figure 10.11d).

### *Fiber Orientation*

Other factors being equal, tension generated by a muscle varies with the orientation of fibers within it. Muscle fibers may be arranged in one of two general ways, each conferring different mechanical properties. A **parallel muscle,** in which all fibers lie along the line of tension generated, has as the name implies, fibers parallel to each other. A **pinnate muscle** has fibers that lie oblique to the line of force generated, and it is inserted on a common tendon that receives the inclined muscle fibers (figure 10.12a). Each type of muscle has mechanical advantages and disadvantages.

Parallel muscles are best at moving a light load through a long distance. The sternomastoid that turns the head or the long sartorius that adducts the hindleg are examples. Pinnate muscles are best suited for moving a heavy load through a short distance (figure 10.12b). The strong gastrocnemius muscle of the calf is an example. It is inserted on the calcaneus and exerts considerable force to extend the foot and lift the weight of the body, but it can be shortened only through a small distance.

For both parallel and pinnate muscles, generation of tension is based on the contracting mechanism of the sliding filaments. Their mechanical properties arise from differences in fiber arrangement. Pinnate muscle permits the packing of more muscle fibers in the same space. Consider two muscles of equal size and shape but with different fiber arrangements. Within a given volume, pinnate muscles have shorter fibers but more of them than parallel muscles. Being shorter and inclined at an angle to the line of muscle action, pinnate muscles shorten less, so their insertion tendon moves a shorter distance. However, because more fibers are packed in the same space, the usable force produced along the line of action is greater (figure 10.12b).

More formally, the physiological cross-sectional area of a pinnate muscle is greater than a comparable parallel muscle. In a pinnate muscle, the force produced by individual fibers can be separated into its vector components, one a useful component aligned with the tendon, the other at right angles contributing no useful force (figure 10.12c). The useful force component calculated from the vector trigonometry equals the fiber force ($F_f$) times the cosine of the angle of pinnation ($\theta$), or $F_f \cos \theta$. So long as this angle of pinnation does not become too large, most of the force of fiber contraction will yield a large useful component vector along the line of tendon action. This useful force component is a bit less than the fiber's tension, but this is compensated for by having more fibers than a comparable sized parallel muscle.

In practice, most muscles are compromises between the two specialized extremes, parallel and pinnate. What should be emphasized, however, is that the different properties of parallel and pinnate muscles do not arise from differences in their mechanism of contraction at the molecular level. It is at the tissue level of organization that different properties of overall performance arise.

### *Velocity of Shortening*

Other factors being equal (such as fiber physiology and angle of pinnation), the velocity of shortening is greater in a long muscle than in a short muscle (figure 10.13a,b). Assume that we have two muscles identical in all properties and dimensions except that one is long and the other short. The time it takes for each to contract to half its resting length is equal, but the velocity traveled by the point of insertion is greater in the longer muscle. The longer muscle has more sarcomeres in series, and their velocities are additive. Thus, the longer a muscle, the greater is the velocity of its insertion.

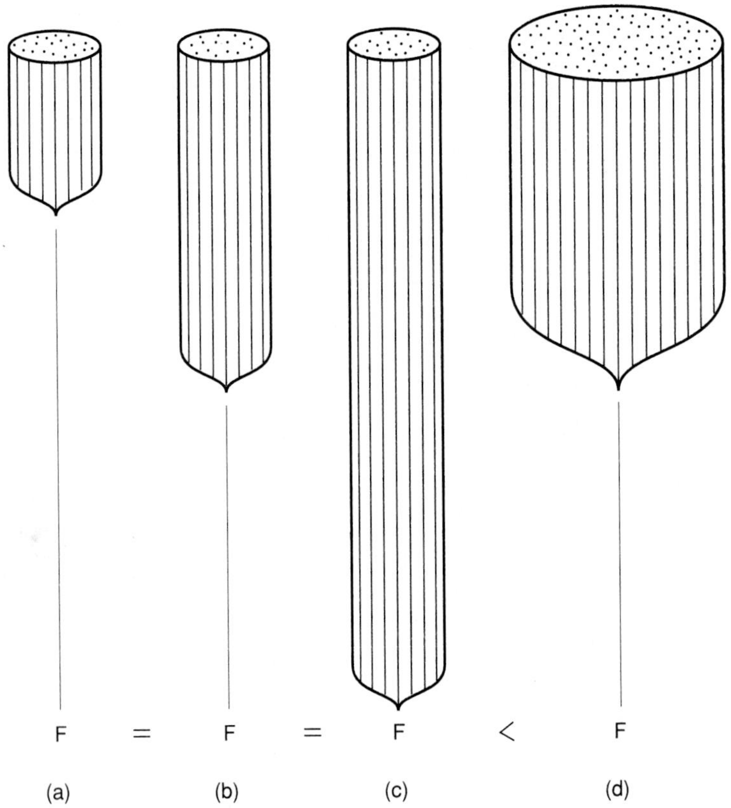


**Figure 10.11** Muscle force is proportional to cross-sectional area. (a–c) Muscles that differ in lengths but have equal cross-sectional areas therefore produce the same force. (d) A muscle with greater cross-sectional area, and hence with more muscle fibers, produces a greater tension than the other muscles, other things being equal.

### *Distance of Shortening*

Other factors being equal, the absolute distance through which a muscle contracts is greater for a long muscle than a short muscle. This property, like velocity, is a consequence of the additive effect of chains of sarcomeres. As individual sarcomeres shorten, their distance of travel is added to that of adjoining sarcomeres in series. Because there are more sarcomeres in each chain of a long muscle, the additive effect is greater in a long muscle than in a short muscle. As a result, the insertion of a long muscle is displaced a greater distance than the insertion of a short muscle.

## Bone-Muscle Lever Systems

The action of a muscle is more than just a property of its underlying physiology or fiber arrangement. Often the performance of a muscle depends on how it is attached to bones of the lever system. This aspect of muscle performance has been difficult to confirm experimentally, but theoretical work provides some insight. For example, a muscle crossing a single joint can be attached close to (proximal) or away from (distal) the point of rotation at the joint. Each site of insertion, proximal or distal, results in different mechanical properties. Inserted distally, the muscle is best suited for strong movements; inserted proximally, it is best suited for fast movements. If you wish to swing a heavy gate or one that is stuck at the latch, you apply force farthest from the hinges to get more leverage. If you wish to open a light gate rapidly, applying force close to the hinges yields best results. These differences arise from differences in simple mechanical advantage, but one is at the expense of the other. Rapid motion (proximal insertion) comes at the expense of strong motion (distal insertion), and vice versa.

The site of insertion also affects the distance through which a moving part swings. For example, the biceps brachii muscle generates a long sweep of the end of the forearm if it is inserted close to the elbow joint. If the biceps is inserted distally and shortened the same length, it provides a much shorter sweep of the forearm (figure 10.14a,b).

A hypothetical analysis of limbs indicates that tradeoffs must inevitably be struck between designs that favor strength and those that favor speed of limb displacement. For example, the teres major, which runs from the scapula to the humerus, is inserted more distally in the forelimb of a badger (strong digger) than in the forelimb of a cheetah (rapid runner; figure 10.15a,b). This proximal and distal relationship can be expressed as the **lever advantage,** the ratio of lever arm *in* to lever arm *out*. The perpendicular distance from the point of bone rotation to the line of muscle action

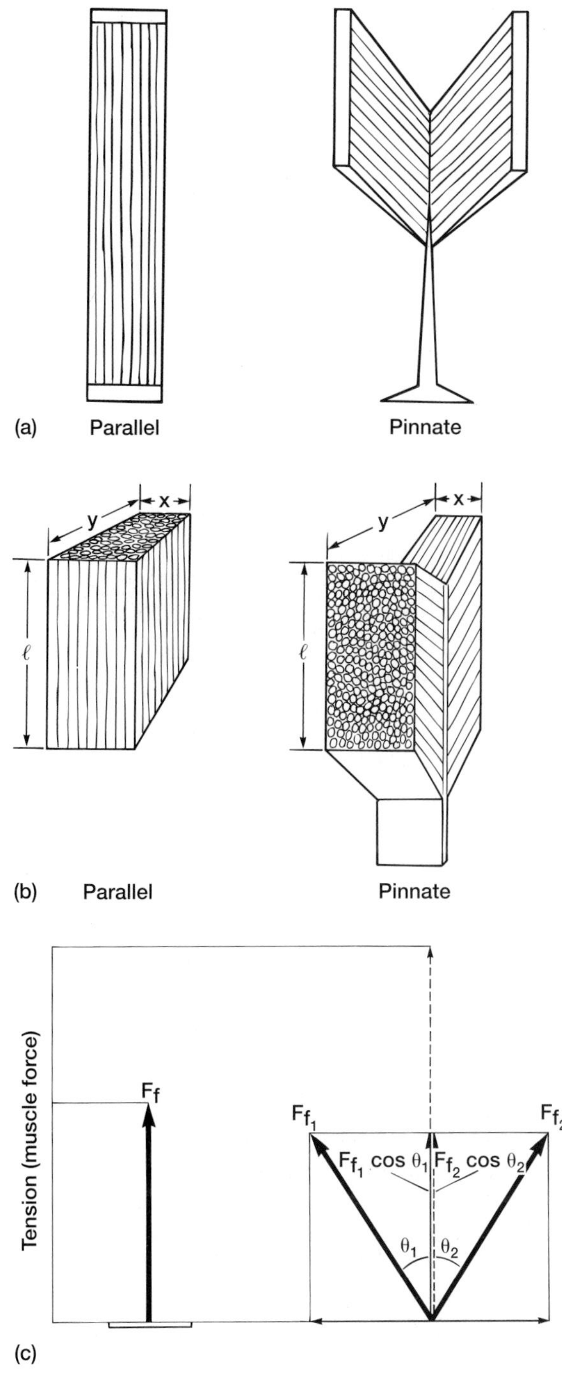


**Figure 10.12** Parallel and pinnate muscles. (a) Muscles with fibers aligned along the line of action are parallel. Those with fibers oblique to the line of action are pinnate. (b) The pinnate orientation permits the packing of more fibers in the same volume than the parallel arrangement does. The oblique orientation of fibers in pinnate muscles reduces the effective distance through which the insertion can be moved and slightly reduces the force that each fiber can direct along the line of action. However, the greater overall number of fibers compensates and makes pinnate muscles especially suited for moving heavy loads short distances. Both blocks of muscles are of equal three-dimensional size, xyl. The force the parallel muscle can produce is proportional to its cross section, xy. The force of the pinnate muscle is greater, being proportional to its physiological cross section, xl of one side plus xl of the other. (c) The force a fiber produces along the line of action in a parallel muscle is equal to the force of that fiber. In a pinnate muscle, the useful force of a fiber lies along the line of action of the muscle organ. This useful force is the trigonometric component of the force of the fiber ($F_f$) times the cosine of the angle the fiber makes with the line of muscle action ($\theta$). Because more fibers can be packed within the same volume of a pinnate muscle, the useful force from two fibers is additive, giving a total force greater than in a similar sized parallel muscle.

on this point of rotation is the lever arm in ($L_i$). From the point of rotation to the point at which motion is applied is the lever arm out ($L_o$). For the badger, this ratio is about 1:5, and for the cheetah it is about 1:9. The badger has a greater strength advantage in its forearm system than the cheetah does. In the badger, the point of muscle insertion is more distally placed and the leg is relatively shorter to give the teres major a strength advantage. With respect to velocity, however, the ratios tell us that the cheetah's forearm muscles have the speed advantage. The 1:9 ratio indicates that the toes move nine times faster than the point of muscle attachment. A higher rate of limb oscillation is achieved but at the expense of strength. These hypothesized changes in performance with changes in ratios still await experimental verification in a variety of animals. Such analyses provide provisional insight into the mechanical advantages and disadvantages of alternative limb designs.

Lever arms (p. 132)

Even within the same limb of the same individual, different muscles can be inserted in bone in such a way as to enjoy different lever advantages and make different

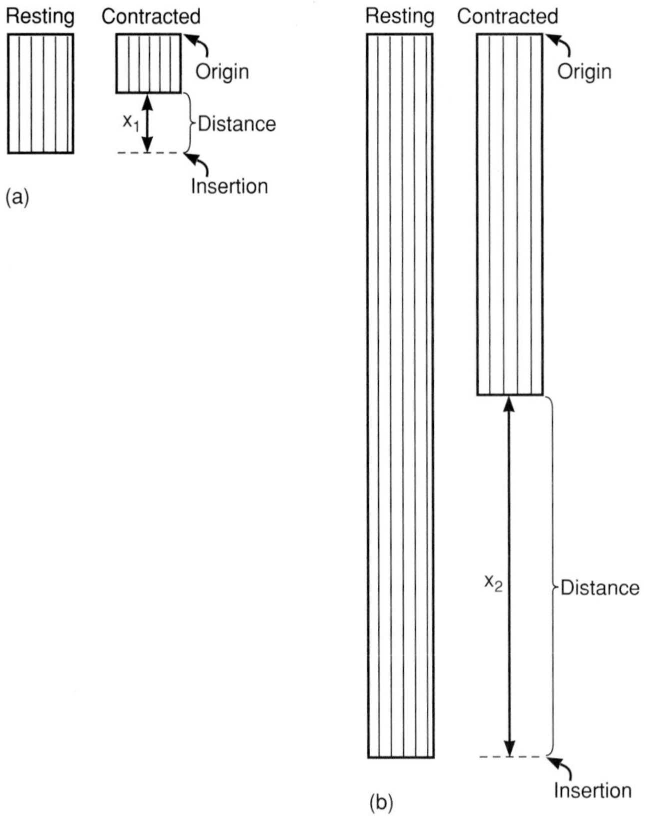


**Figure 10.13** The velocity and distance that the point of insertion moves in a long muscle is greater than in a short muscle. (a) A short muscle that contracts to half its resting length shortens by the distance $x_1$. (b) A long muscle contracting to half its length will do so in the same amount of time if it is made up of the same type of sarcomeres. But because contraction speed of chains of sacromeres is additive, this produces a faster speed over a longer distance ($x_2$). When fast displacement is required, long muscles are often involved.

contributions to strength or speed during limb oscillations. Many cursorial animals have **low** and **high gear muscles.** Low gear muscles, such as the hamstrings, enjoy a mechanical strength advantage to help overcome inertia when accelerating or moving the mass of the limb. High gear muscles enjoy a speed advantage and produce rapid limb oscillation.

In addition to effects on strength or speed, the proximal insertion of a muscle has further consequences on performance. The muscle force moving a bone must be maintained at a peak level throughout the entire arc of the bone's rotation to maintain peak speed. As we saw earlier, the force generated by a muscle falls off at the ends of its tension-length curve; thus, several muscles often move a bone, each in turn reaching its peak force to maintain bone speed. Proximal insertion of a muscle offers an alternative mechanical design. If a muscle is inserted near the point of rotation, it can produce a long excursion of the distal end of a bone while shortening very little itself around the peak force of its tension-length curve. If inserted distally, the muscle would have to be shortened much more to produce the same distal displacement of the bone. Generally, the more a muscle is actually shortened, the more energy it consumes even if the force remains the same. Therefore, proximally inserted muscles use less energy

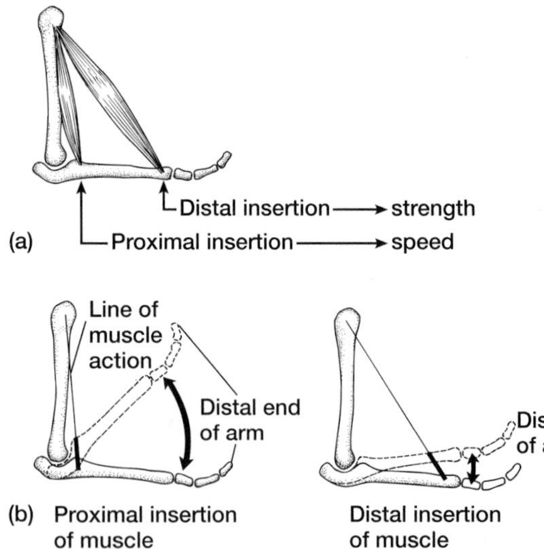


**Figure 10.14** Strength versus speed. A muscle inserted on different points in a lever system produces different mechanical advantages. (a) If inserted near (proximal) the point of rotation, the muscle favors speed. If inserted distal to its point of rotation, it favors strength. (b) Proximal insertion also favors greater excursion of the distal end of the part rotated. The thick solid bar represents the distance of muscle shortening, which is equal in both.

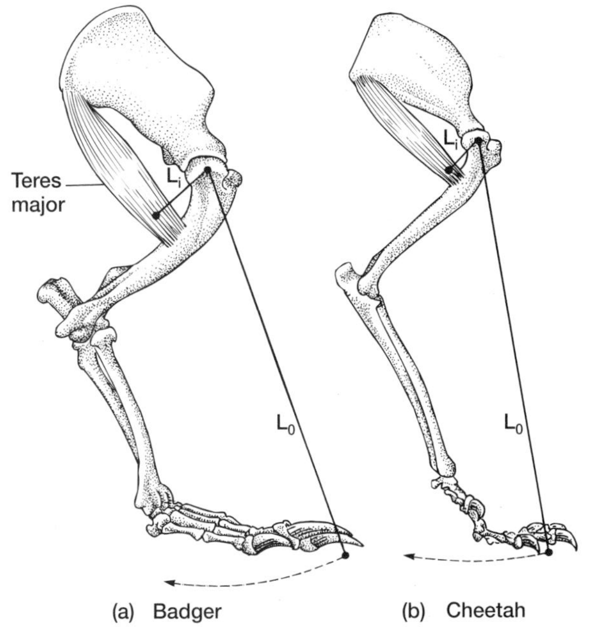


**Figure 10.15** Strength versus speed in the design of forelimbs. (a) In the badger, the teres major is inserted distally on the humerus. (b) In the cheetah, the teres major is inserted closer to the point of rotation. The resulting change in the lever arm in ($L_i$) and lever arm out ($L_o$) changes the mechanical advantage from one favorable to strength (the badger) to speed (the cheetah). Both forelimbs viewed medially are drawn to the same overall length.

and provide a more economical design for rotating limb segments during locomotion.

Some precautions should be highlighted. First, most of these principles of performance and design are based on theoretical arguments taken from the presumed mechanical consequences of muscular attachment sites and lever arm advantages or disadvantages. It has proved difficult to confirm many of these by direct experimental tests. Second, most of the arguments assume that changes in the lever arm will not become too short for the job, or else the muscle would not be up to the task. Obviously, even a big muscle working with a very small lever arm could not move a part effectively. Third, we have assumed that the muscle and its mechanical advantage are comfortably matched to its external load. For example, we noted that the longer the muscle fibers, the faster is the velocity and the greater is the distance of shortening. However, if a *heavy* load is to be moved, then the reverse may be true. A short but strong muscle with many fibers provides more force to address the large load than a long but weaker muscle with fewer fibers. We are simply stating the obvious. Both the speed of shortening and the distance of shortening depend not only on the length of muscle fibers but also on the relationship of the force generated and the size of the external load. From cautious generalizations about force velocity and distance of shortening, we can recognize the trade-offs between internal muscle design and size of external load.

## Sequencing of Muscle Actions

Muscles of course do not act in isolation. Any movement involves various muscles and each reaches its peak force at different times during movement. For example, the striding gait of humans is composed of two movement phases, the **stance phase,** when the heel strikes the ground until the toe is lifted from it, and the **swing phase,** when the toe is lifted from the ground until the heel strikes it again (figure 10.16a–e). Muscles that move the limb become active at different points during these phases. As the heel strikes the ground, the hamstrings and pretibial muscles reach peak activity; thereafter, the quadriceps increase in activity as the torso is carried forward over the limb. With the heel off the ground, the calf group (gastrocnemius, soleus) increases in activity. During the swing phase, most of these muscles are electrically silent or show low activity as gravity swings the bent leg forward like a pendulum beneath the body. Selective use of muscles greatly reduces muscle contractions and overall energy expenditure.

During normal locomotion, the central nervous system coordinates this elaborate pattern of selective muscle deployment. Part of the reason it is difficult to design

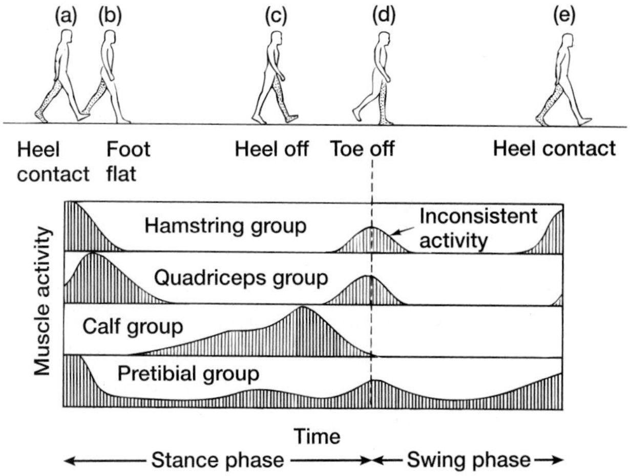


**Figure 10.16** Sequential action of muscles. During activity, muscles reach peak activity at slightly different times to distribute and spread the forces producing motion. The human striding gait includes a stance phase (a–c, right leg), during which the foot is in contact with the ground, and a swing phase (d,e, right leg), during which the limb freely swings forward. Groups of leg muscles reach their peak forces at different times during each phase. Most are inactive during the swing phase because gravity pulls the limb forward in pendulum fashion. Greatest activity of hamstring, quadriceps, and pretibial muscles occurs early in the stance phase. The calf group reaches peak activity and presumably peak force just before the end of the stance phase.
Modified from J.V. Basmajian.

artificial human limbs to replace amputations is because of this complexity of muscle sequence and action. Not only must the prosthetic limb produce the necessary forces, but these forces must be generated in the proper order if they are to simulate normal locomotion faithfully.

Recruitment of muscle types is also selective. For example, in elasmobranchs and some primitive teleosts, red fibers are fatigue resistant, whereas white fibers fatigue more easily but contract faster. Electromyographic studies of sharks show that the block of red muscles along the body produces swimming undulations at slow speeds. As speed increases, the white blocks of axial musculature are recruited (figure 10.17a). Carp and a few other teleosts have "pink" muscle, a third type of fiber that is intermediate in physiological character between red and white muscle. As swimming speed increases in carp and some teleosts, there is an orderly recruitment of blocks of axial musculature, first red, and then pink, and finally white muscles at high rates of speed (figure 10.17b).

## Overview of Muscle Mechanics

All muscle contraction is fundamentally based on the mechanism of sliding filaments of macromolecules, the formation of cross-bridges between the molecules of primarily actin and myosin. As the cross-bridges form, they slide past one another to shorten the sarcomere, which collectively shortens the muscle fiber. Despite this universal mechanism, muscles perform a variety of tasks and are involved in quite a range of functions. Changes in performance and function are brought about by alterations at higher levels of organization, not by changes in the basic sliding filament mechanism. For instance, the properties of a muscle are influenced by the length or orientation of a muscle fiber, by physiology, by the way muscles attach to the lever system they move, and by the sequence of muscle actions relative to each other.

By reducing the question to underlying actin and myosin molecules, we can only partially understand muscle design and function. The particulars of organization at the cell, tissue, and organ level are also required to explain the basis of muscle performance.

## Muscle Actions

A **motor pattern** generally means any repetitive movement activated by the nervous system. Muscles can act independently, concurrently, or in sequence to produce complex motor patterns that control the skeletal system. Even seemingly simple motor patterns may involve many muscles. During a strong cough to clear the throat, 253 named muscles contract. Muscles that act together to produce motion in the same general direction are **synergists.** The biceps brachii and brachialis muscles of the upper human arm that flex the forearm are synergists. Muscles that produce opposing motions are **antagonists.** The biceps brachii on one side of the upper arm and the triceps brachii on the other are antagonists. Both contract in opposite directions during rapid rotation of the forearm, not to frustrate each other but to balance each other and control and coordinate rapid or powerful movements.

Under different conditions, the same muscle can have several actions. The primary action a muscle produces is its **prime motion.** Its two points of attachment to the skeletal system are defined accordingly. A muscle's **origin** is its relatively fixed point of attachment and its **insertion** is its relatively movable point of attachment. Each site of origin of a muscle is a **head,** and each site of insertion is a **slip.** Occasionally a muscle may act synergistically and have a secondary motion as well as its prime motion. For example, the geniohyoid of mammals runs between the hyoid and chin. Its prime motion is to move the hyoid, which in turn moves the larynx forward. Secondarily, when the larynx is held fixed in position, the geniohyoid acts in support of the digastric muscle to lower the mandible and help open the jaws. Muscles may also act as **fixators** to stabilize a joint or lever system. If you make a gentle fist, only your forearm muscles contract. Your biceps and triceps remain relaxed. However, if you clench your fist vigorously, your biceps and triceps of the upper arm involuntarily contract as well, not to help close your fingers directly but to stabilize the elbow joint while the nearby forearm muscles tighten your fist.

Other terms describe muscle actions as well (figure 10.18). Flexion and extension apply chiefly to limbs. **Flexors** bend one part relative to another about a joint. **Extensors** straighten a part (e.g., "pointing the toe" during ballet). Adduction and abduction are most often used to describe motion of the limb relative to the body. **Adductors** draw the limb toward and **abductors** move the limb away from the ventral surface of the body. Applied to jaw action, **levators** (a special kind of adductor) close and **depressors** (a special kind of abductor) open the jaws. Contraction of **protractors** result in projection of a part, such as the tongue of a frog, away from its base, whereas **retractors** bring it back. A limb can be turned by **rotators,** specifically **supinators** if they rotate the palm or sole up or **pronators** if they rotate it down. Rotation is sometimes used in a general way to mean overall limb oscillation or swing as well. The **constrictor** or **sphincter** muscles surround openings (e.g., gill constrictors around the mouth, intestinal sphincters around the anus) and tend to close them; **dilators** act antagonistically to open the orifice. As we meet them later in this chapter, we define other muscle actions.

## Muscle Homologies

During their evolution, some muscles have fused with one another, others have split into distinct new muscles, some have become reduced in prominence, and others have changed their points of attachment and hence their function. Unlike the evolutionary history of bones, muscles leave no direct trace in the fossil record. Their positions

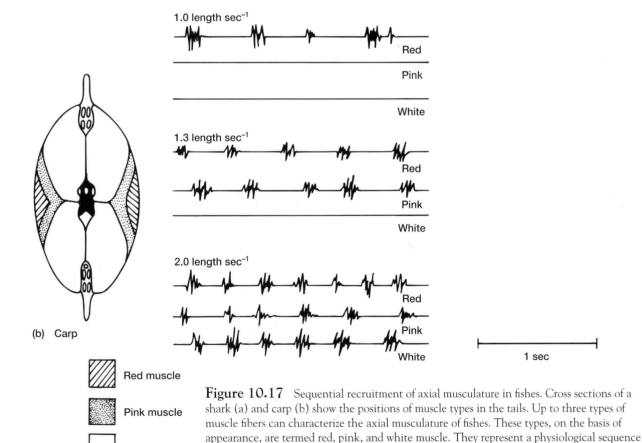

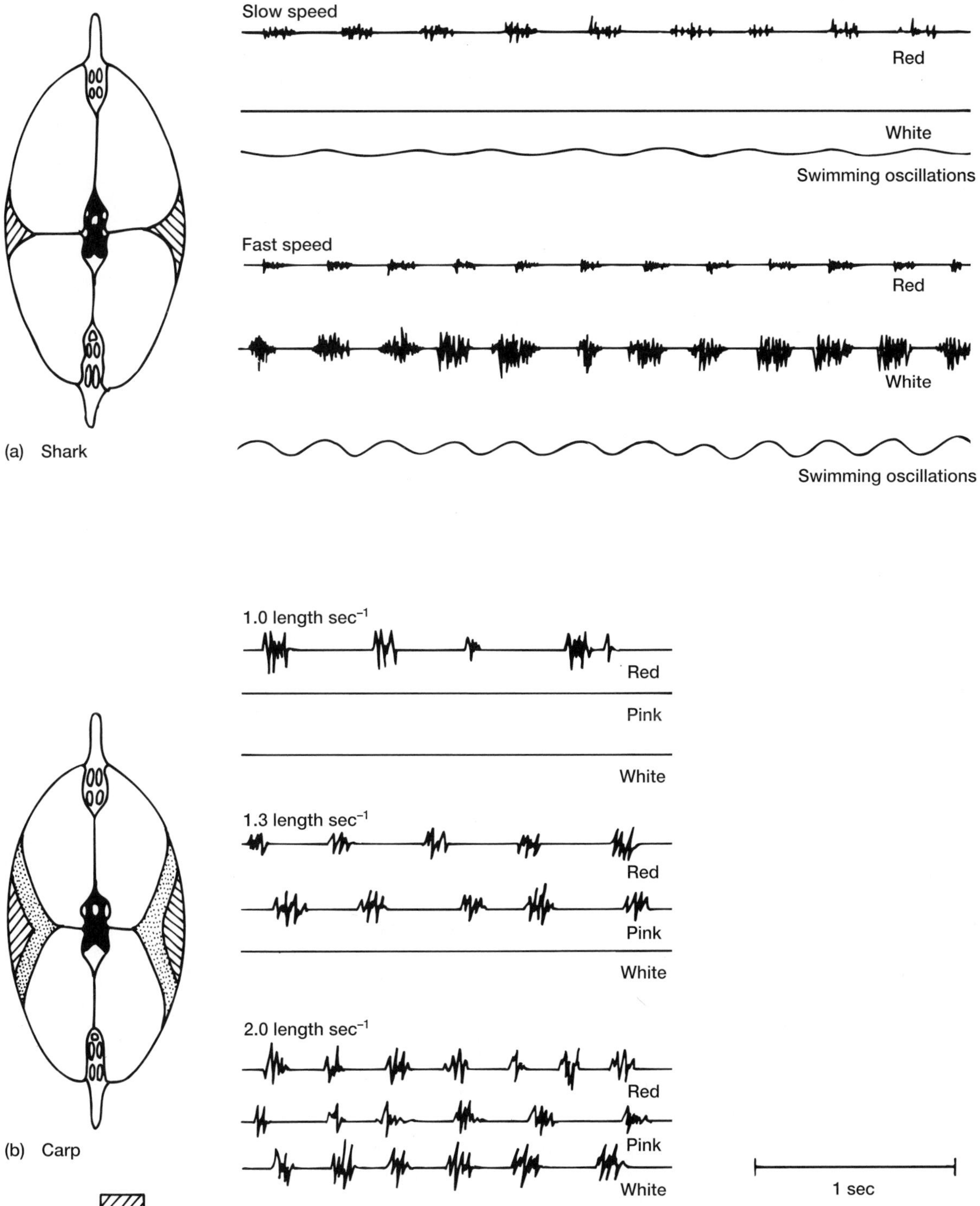


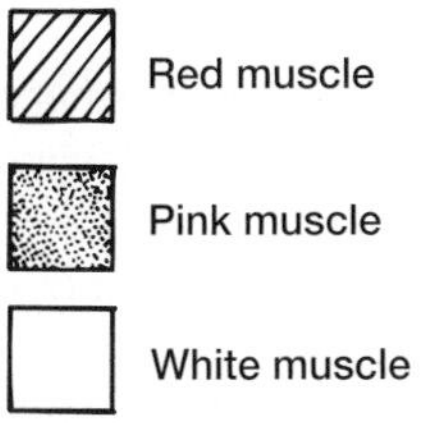


**Figure 10.17** Sequential recruitment of axial musculature in fishes. Cross sections of a shark (a) and carp (b) show the positions of muscle types in the tails. Up to three types of muscle fibers can characterize the axial musculature of fishes. These types, on the basis of appearance, are termed red, pink, and white muscle. They represent a physiological sequence from slow twitch (endurance) to fast twitch (fatigable) fibers. Because these fibers are organized into discrete regions within a body, electrodes can be inserted to record the swimming speed at which each set of fibers is recruited and begins contributing to swimming undulations. (a) In sharks, only red and white muscles are present. At slow speed, red fibers contract, and at higher speeds, white fibers join in. The wavy tracings under each set of electromyograms represent the shark's swimming oscillations. (b) In some teleost fishes, like carp, all three types of fibers are present. Red, and then pink, and then white muscle fibers are recruited sequentially as swimming speed increases. Each myogram represents electrical activity and hence reveals muscle contraction at that point. Body lengths per second are used to express swimming speed.
Modified from Johnson, et al., 1977.

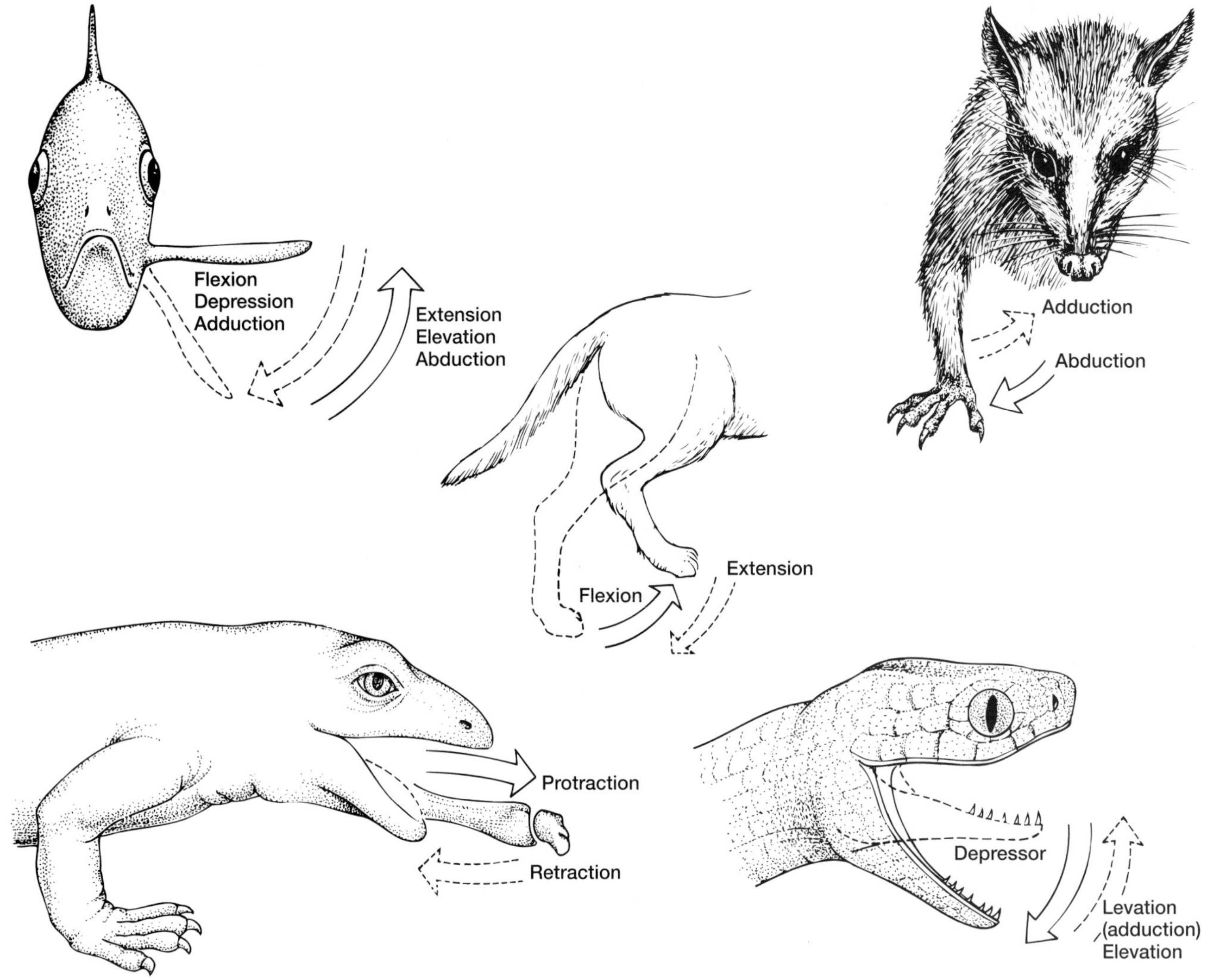


**Figure 10.18** Muscle actions. Muscle adduction draws an appendage toward the ventral midline, and muscle abduction moves it away. Although these terms apply to tetrapod limbs and fish fins, the terms *depression* and *flexion* are sometimes used synonymously with adduction in fishes; *extension* and *elevation* are synonymous with abduction. In tetrapods, flexion means bending a part, extension means straightening it. Protraction sends a part outward from its base, retraction pulls it back. Opening the jaws is depression or abduction and closing them is elevation or levation or adduction.

must be inferred from attachment scars on fossilized bones. Tracking these changes in the fossil record to establish homologies is difficult. Consequently, alternative criteria are often used. One such criterion is attachment similarity. Similar attachments in different muscles are assumed to attest to their homology. However, sites of attachment for the same muscle can vary in different groups (figure 10.19a). In mammals, the gastrocnemius on the posterior side of the shank is inserted in the calcaneus of the heel. In frogs, the tendon of the gastrocnemius instead stretches across the bottom of the foot as the plantar aponeurosis.

Another criterion is functional similarity. Similar function of two muscles is assumed to represent retention of a common ancestral pattern. This criterion can be misleading as well. For example, the depressor mandibulae opening the lower jaw in reptiles has a single belly. In mammals, the digastric serves roughly the same function, but it has two bellies (hence di- and gastric), and its parts arise from separate embryonic sources (figure 10.19b).

Another criterion often used is nervous innervation because there seems to be some phylogenetic permanence between a muscle and its nerve supply. In mammals, the diaphragm muscle of the posterior thorax might be expected to be innervated by nearby posterior thoracic nerves. Instead, it is innervated by a cervical nerve arising well anterior to it near the head. During embryonic development, the muscular predecessor to the diaphragm arises in the cervical region and migrates posteriorly. The cervical nerve that comes to innervate it, the phrenic nerve, grows from the cervical region as well and accompanies

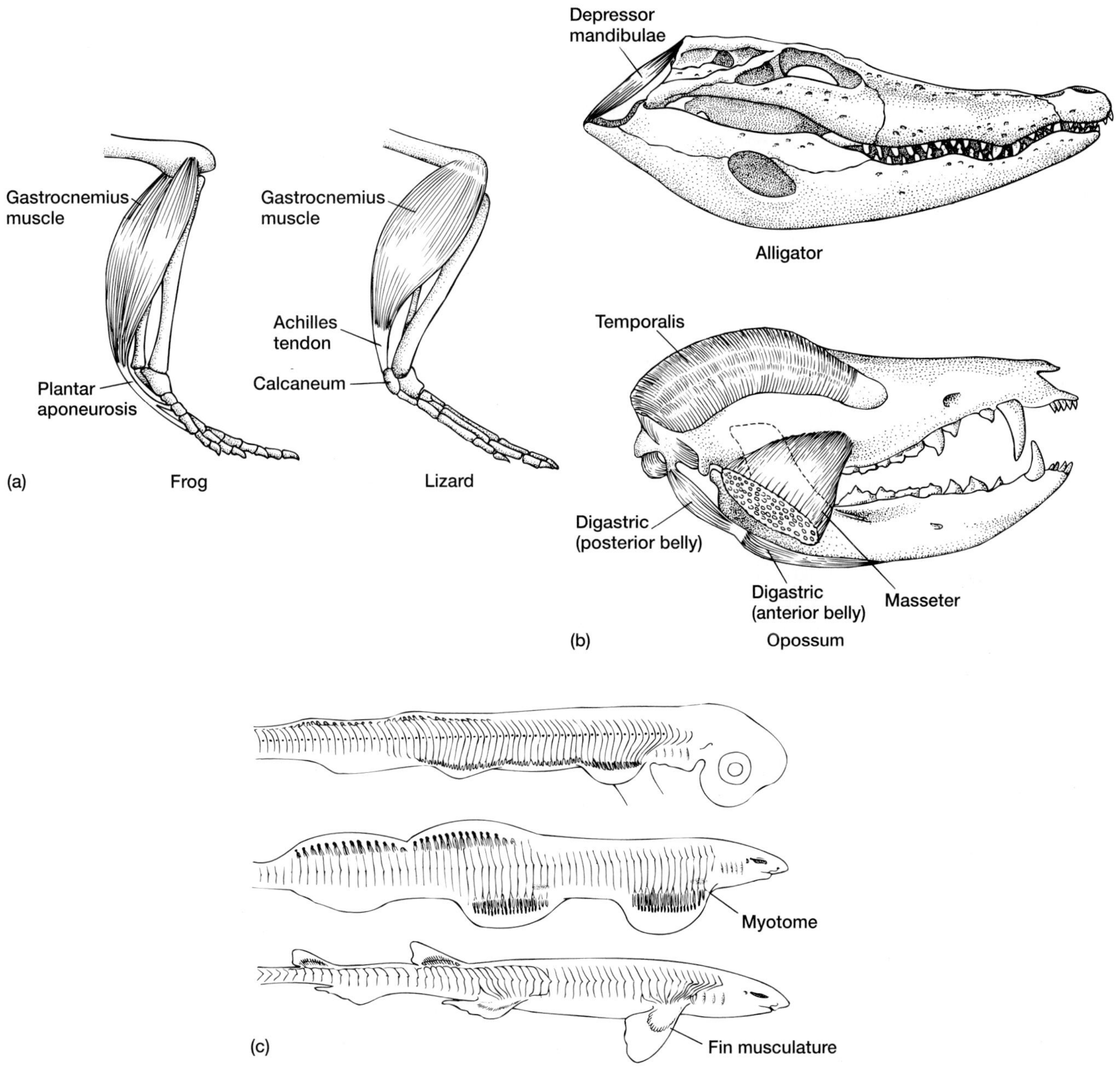


**Figure 10.19** Criteria for muscle homology. Muscle homologies can be based on several criteria, although each may have its uncertainties. (a) Similar function between muscles suggests homology. The posterior shank muscles of the frog and reptile extend the foot. Although this muscle is thought to be the gastrocnemius, the insertions of each are slightly different in each animal. (b) The depressor mandibulae and the digastric muscles depress the lower jaws of reptiles (alligator) and mammals (opossum), but the nerve supply to these muscles of each is different, suggesting that the two are not strictly homologous. (c) A common embryonic pattern of development is often used to establish muscle homologies, but this criterion may be difficult to interpret. For example, during the embryonic development of fin muscles in a shark, the ventral tips of myotomes grow downward along the body, forming a low ridge and eventually entering the fin. In tetrapods, limb muscles arise from mesenchyme in the limb bud, not directly from myotomes. Although fin and limb musculature are likely homologous in a general sense, embryonic development has apparently become modified in tetrapods, leading to a different pattern of formation.

the diaphragm muscle back to its eventual posterior site of residence.

Embryonic origin is often used to establish muscle homologies. As with other structures, similar embryonic development is suggestive of similar phylogenetic ancestry. However, even this criterion can present difficulties. For example, in many fishes, the ventral tips of nearby myotomes grow into the fin rudiment, pinch off, and differentiate into the fin musculature (figure 10.19c). In tetrapods, the myotomes grow ventrally, but their cells do not actually pinch off nor do they contribute to limb musculature directly. Instead, limb musculature seems to form from mesenchyme in the limb bud itself. Apparently, embryonic events have been cut short in tetrapods, but this makes the use of embryonic criteria less useful for assigning homologies between fish fins and tetrapod limbs.

Whichever criteria are used, they are most reliable when closely related species are compared. Given the evolutionary continuity of the skeletal system between fishes and tetrapods, it is unlikely that the muscles that work the limbs evolved as brand new structures. It seems more likely that the muscles of tetrapods are broadly homologous with those of their fish ancestors. The mistake would be to take "homologous" muscles literally in derived mammals and birds as exact modifications of comparable muscles in fish ancestors. In general, it is probably more reasonable to think in terms of modifications of muscle groups through phylogenetic changes rather than to think of modifications of individual muscles among distantly derived groups.

To identify these major groups of muscles and to track their evolution, we turn to events of early embryonic development when these formative blocks of muscle are established.

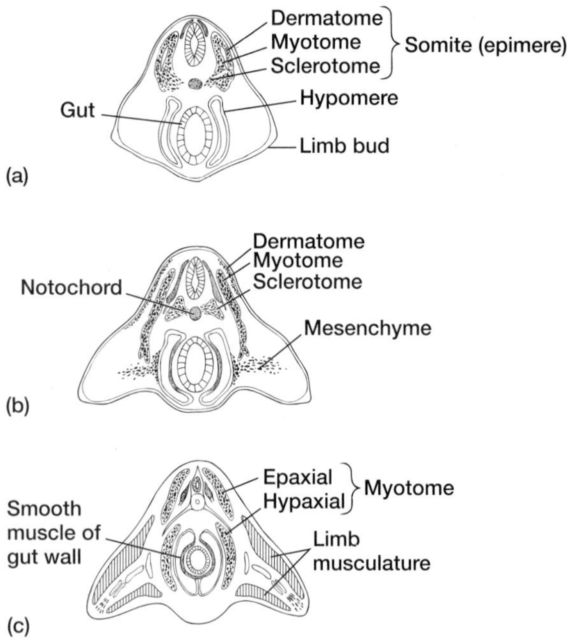


**Figure 10.20** Embryonic origin of postcranial muscles in a generalized tetrapod. (a) Cross section illustrating each of the three regions of the mesoderm—epimere, mesomere, and hypomere—as they differentiate during embryonic development. The epimere becomes segmented to form the somite, which in turn forms the dermatome, myotome, and sclerotome. (b) Cells of the dermatome move beneath the skin and there differentiate into dermal muscles. The ventral tip of the myotome grows downward to meet mesenchymal cells migrating out of the hypomere. Cells of the sclerotome grow medially to form around the notochord and differentiate into the vertebrae. (c) Interactions between myotomal cells and migrating mesenchymal cells from the hypomere promote the development of limb muscles. Longitudinal division of the myotome produces epaxial and hypaxial muscles of the body (shown in cross section).

# Embryonic Origin of Muscles

In general, muscles arise from three embryonic sources. One source is the **mesenchyme,** a loose confederation of cells that become dispersed throughout the embryonic body. Smooth muscles within the walls of blood vessels and some viscera arise from mesenchyme. The second source of muscles is the paired **hypomere.** As the hypomere becomes distinct from the rest of the body mesoderm, its medial walls (splanchnic) embrace the gut and differentiate into the smooth muscle layers of the alimentary tract and its derivatives (figure 10.20a). Cells of the hypomere also form the cardiac muscle of the tubular heart. The third embryonic source is the **paraxial mesoderm** from which most striated muscles develop (figure 10.21). During or shortly after neurulation, the paraxial mesoderm, as its name suggests, forms next to the neural tube along the axis of the embryonic body (figure 10.22a). In the trunk, the paraxial mesoderm becomes segmentally arranged into anatomically separate **somites** (epimeres). In the head, the paraxial mesoderm does not differentiate into discrete somites but it forms clusters of mesoderm, called **somitomeres,** in series with the separate somites that follow behind them (figure 10.22a).

In amniotes, there are usually seven pairs of somitomeres in the head, but sometimes there are fewer. The somites of the body divide into populations of cells that contribute to the skin (dermatome), vertebral column (sclerotome), and body musculature (myotome; figure 10.20b). The somitomeres of the head make similar contributions to the regions of the head but generally do not divide into such easily recognizable populations of cells as do the somites.

**Differentiation of mesoderm (p. 169)**

## Postcranial Musculature

### *Appendicular Musculature*

In many fishes, the ventral tips of adjacent myotomes grow downward into the emerging fin bud and differentiate directly into the fin musculature. In tetrapods, the myotomes neighboring the limb bud also grow downward but stop before actually entering the limb bud itself (figure 10.20b).

Experimentally, myotomes can be removed early before the full limb develops. But doing so interrupts subsequent limb growth and differentiation. Consequently, myotomes seem to be required for normal limb development. However, it is still unclear how many myotomal cells actually contribute to the growing limb. Perhaps the myotome indirectly induces limb development by mobilizing or stimulating nearby tissues that develop into limbs. When the ventral tips of the myotomes are present, some cells of the hypomere (somatic mesoderm) depart and migrate into the adjacent limb, contributing to and comingling with the intrinsic mesenchyme already in place within the limb bud (figure 10.20c). Next, the mesenchyme within the tetrapod limb bud differentiates into skeletal elements and associated limb musculature.

At present, opinion differs on how many distinct embryonic sources directly contribute to the appendicular musculature. In general, the myotomes certainly contribute together with mesenchyme to the appendicular musculature in fishes. In tetrapods, the intrinsic mesenchyme is involved.

### *Axial Musculature*

The axial musculature arises from myotomes that differentiate from somites. These myotomes grow and expand along the sides of the body, forming the musculature associated with the vertebral column (or notochord), ribs, and lateral body wall (figure 10.21a,b). In gnathostomes, a longitudinal sheet of continuous connective tissue, the **horizontal septum,** divides the myotomes into dorsal and ventral regions, each destined to become, respectively, the **epaxial** and **hypaxial musculature** (figures 10.20c and 10.21b).

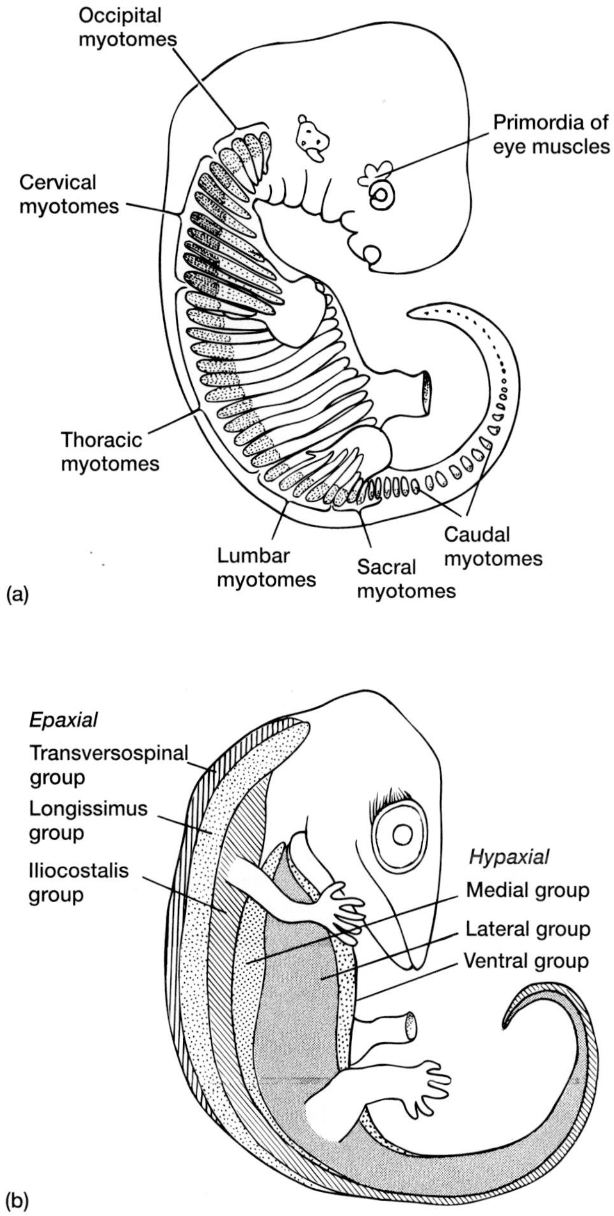


**Figure 10.21** Muscles derived from embryonic myotomes in a reptile (lizard). (a) During embryonic development, the myotomes expand into respective areas of the body. (b) Differentiation of muscle groups of the superficial trunk and tail are shown.

## Cranial Musculature

### *Jaw Musculature*

The jaw musculature arises from two distinct embryonic sources, each with a different nerve supply. Although functionally integrated to work the jaws cooperatively, there are two separate sets of head muscles that move the jaw. One set is the **hypobranchial musculature** (figure 10.22b). It arises from myotomes of trunk somites whose ventral tips grow downward and forward into the throat to form the musculature along the ventral side of the branchial arches, hence hypo- (beneath) and -branchial (arches). Although they grow forward into the throat, these myotomes are accompanied by nerves emanating from the spinal column adjacent to the original trunk somites. Consequently, the hypobranchial musculature is supplied by spinal nerves. The hypobranchial musculature runs between the ventral elements of the gill arches and between the gill arches and the pectoral girdle. It also contributes to the tongue.

The other set of jaw muscles, the **branchiomeric musculature,** is derived from somitomeres in the head and supplied by cranial nerves. Notice that the branchiomeric muscles originate from somitomeres, negating an older view. Because the gill arches and their branchial muscles are located within the wall of the pharynx, they were once thought to be serially homologous with the smooth muscles that are also located within the wall of the digestive tract. Like the smooth muscle differentiating in the gut wall, it seemed reasonable to conclude that head structures also arose embryologically from the

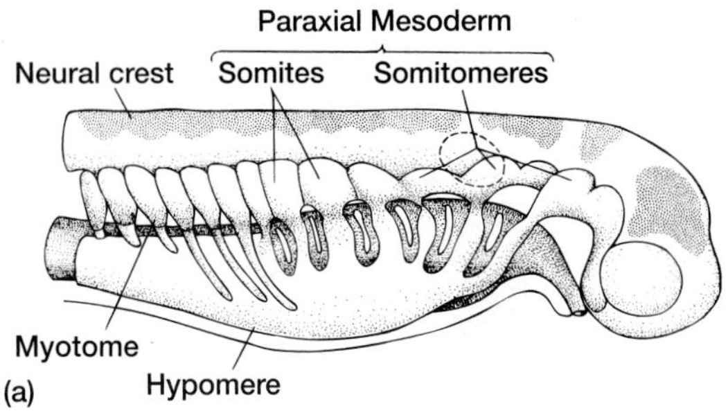


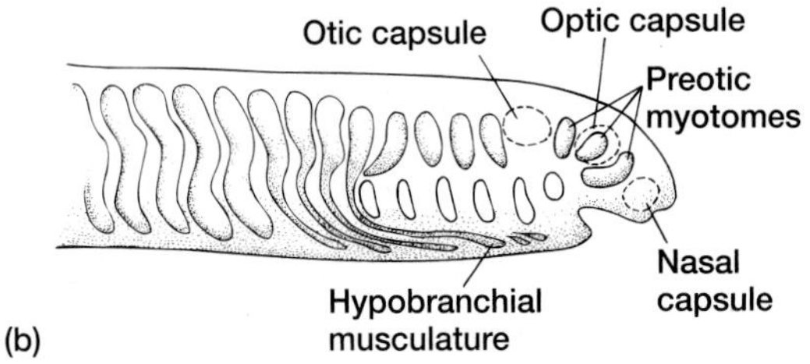


**Figure 10.22** Embryonic origin of cranial muscles in a shark embryo. (a) Paraxial mesoderm, the dorsal mesoderm next to the embryonic notochord, divides into discrete somites in the trunk but forms somitomeres, localized swellings that remain connected in the head. Somitomeres contribute to much of the head musculature, including that which is associated with the branchial arches. The trunk somites differentiate into axial musculature and contribute to the fin musculature. (b) At a slightly later stage in embryonic development, cervical myotomes derived from the somites grow ventrally into the throat to give rise to the hypobranchial muscles beneath the gill arches.

visceral or splanchnic part of the hypomere. Cells of the hypomere, to the extent they could be followed in microscopic sections of the embryo, seemed to confirm this view. The term *visceral skeleton*, based on that view, survives in use today.

This now abandoned view, that much of the head is derived from the hypomere, fueled debate over the organization of the head itself. One hypothesis suggested that all cranial components of the head were serially homologous with the segmental plan of the trunk. The opposite hypothesis stressed the degree to which cranial tissues were distinct and fundamentally different from the trunk. With the advantage of recent techniques for marking individual cells and groups of cells, the embryonic sources of the head can be identified with greater confidence. Not all groups of vertebrates have been sampled and analyzed with these new techniques. However, we now have enough evidence to abandon the idea that striated jaw muscles arise from the hypomere. Instead, we realize that the two sets of jaw muscles arise from serial parts of the paraxial mesoderm—the hypobranchial muscles develop from somites, the branchiomeric muscles from somitomeres. Neither of course is part of the hypomere.

### *Extrinsic Eye Muscles*

Tiny muscles that move or shape the lens to focus light on the retina are intrinsic, within the eyeball, and are discussed in chapter 17 with sensory organs. The extrinsic muscles on the outside of the eyeball rotate it within the ocular orbit to direct the eye's gaze at objects of interest (figure 10.23). The six extrinsic eye muscles originate from the walls of the orbit and are inserted on the outer surface of the eyeball. Their attachments allow rotation of the eye to desired positions. These six muscles arise from three (or perhaps four) different somitomeres. The most anterior somitomere of the three gives rise to the superior, inferior, and medial rectus and to the inferior oblique muscles, all supplied by the third (III) cranial nerve. The next somitomere gives rise to the superior oblique, supplied by the fourth (IV) cranial nerve. A third somitomere gives rise to the external rectus, supplied by the sixth (VI) cranial nerve. The contribution, if any, of a fourth somitomere to the extrinsic eye musculature is still under study.

The number of somitomeres that contribute to the extrinsic eye musculature has always been contentious, perhaps because it is now clear that the number arising within the head varies between groups of vertebrates. The more general term **preotic myotomes** is sometimes used to recognize, but without commitment to number, the somitomeres contributing to the extrinsic eye musculature.

# Comparative Anatomy

Generally cranial and postcranial somatic musculature arise from paraxial mesoderm. Thus, they are serially homologous with each other. This section looks at this system within fishes and then examines the complex remodeling of the muscular system in tetrapods as they meet the quite different set of functional demands on land.

## Postcranial Musculature

### *Axial Musculature*

***Fishes*** In fishes, the axial musculature arises directly from the embryonic and segmental myotomes. Once fully differentiated into the adult musculature, the blocks of axial musculature retain their segmentation, but they are termed **myomeres** to distinguish them from the formative embryonic myotomes from which they arose. Successive myomeres are separated from each other by connective tissue sheets, the **myosepta** (myocommata). Myosepta extend inward, become attached to the axial column (vertebral column or notochord), and join successive myomeres into muscle masses. The **horizontal septum** of the skeleton is absent in cyclostomes (figure 10.24a) but present in all gnathostome fishes, where it divides the myomeres into epaxial and hypaxial muscle masses (figure 10.24b). Each spinal nerve that

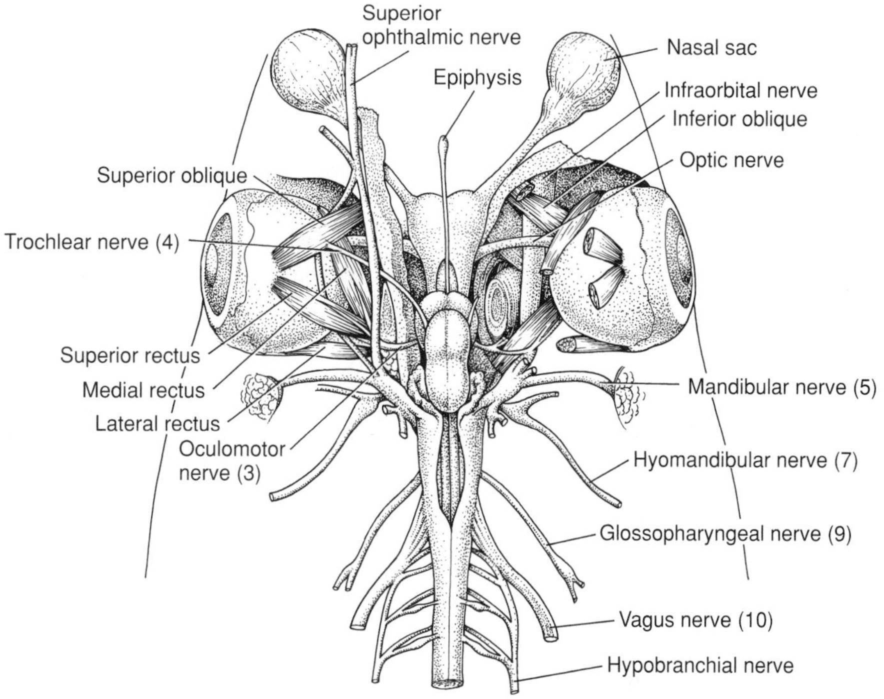


**Figure 10.23** Extrinsic eye musculature of a shark (dorsal view). The extrinsic eye muscles are derived from somitomeres and rotate the eyeball within the orbit in order to direct the gaze. The roof of the chondrocranium over the eyeball has been removed to expose several extrinsic muscles (left). The superior oblique and superior rectus muscles have been cut to expose the deeper extrinsic muscles (right).

supplies a myomere bifurcates. The first branch, the **dorsal ramus,** supplies the epaxial division, and the second branch, the **ventral ramus,** supplies the hypaxial division. Dorsal ribs, when present, develop at the intersection of the horizontal septum with successive myosepta.

The axial musculature of fishes supplies the major propulsive forces for locomotion and, not surprisingly, constitutes the bulk of the body's musculature. Viewed from the lateral surface, the myomeres are folded into zigzag blocks that often look V- or W-shaped (figure 10.24b). Muscle fibers that comprise the myomeres are short, but this folded shape of each myomere extends over several axial segments, giving it and its short fibers control over an extended length of the body. A contraction spreading within the axial musculature alternates from side to side, developing characteristic waves of lateral undulation. These powerful bends produced by the axial musculature are responsible for developing the body's lateral thrusts against the water and driving the fish forward. The axial column, be it a jointed vertebral column or a flexible notochord, receives the attachments of these muscles and acts as a compression girder, resisting telescoping of the body that might otherwise result.

The propulsive force of lateral undulation is perpendicular to the surface of the section of fish generating the force. Because the undulating body increasingly bends toward the tail, the direction of the force relative to the line of travel becomes inclined more posteriorly. Acceleration of the tail is also greater than that of the body sections near the head; hence, the force is greater in the tail. This increase in force and its more posteriorly directed inclination help to explain why the tail is important in generating swimming forces.

More formally, this can be explained by propulsive and reaction forces. The reaction or normal force that is returned by the water is equal and opposite to the propulsive force of the fish's body against the water (figure 10.25a). The normal force is resolved into two vector components, one laterally directed and the other at right angles forwardly directed. The lateral force vector adds nothing to forward progression, but the forward force vector drives the fish (or at least that section of the fish) forward. The size of this forward force vector represents the size of the forward thrust generated at that point on the fish. Because the normal force is larger in the tail and its forward directed component vector is larger than that in the trunk, the tail is the most important part of the body generating useful swimming forces (figure 10.25b,c).

Posteriorly, the axial musculature continues from the trunk into the tail. Anteriorly, the axial musculature is attached to the skull and pectoral girdle. Some fishes use these attachments during feeding to lift the neurocranium or stabilize the girdle from which active jaw muscles originate.

***Tetrapods*** In tetrapods, the appendicular muscles generally take on more responsibility for locomotion and accordingly account for more muscle bulk. Although the axial

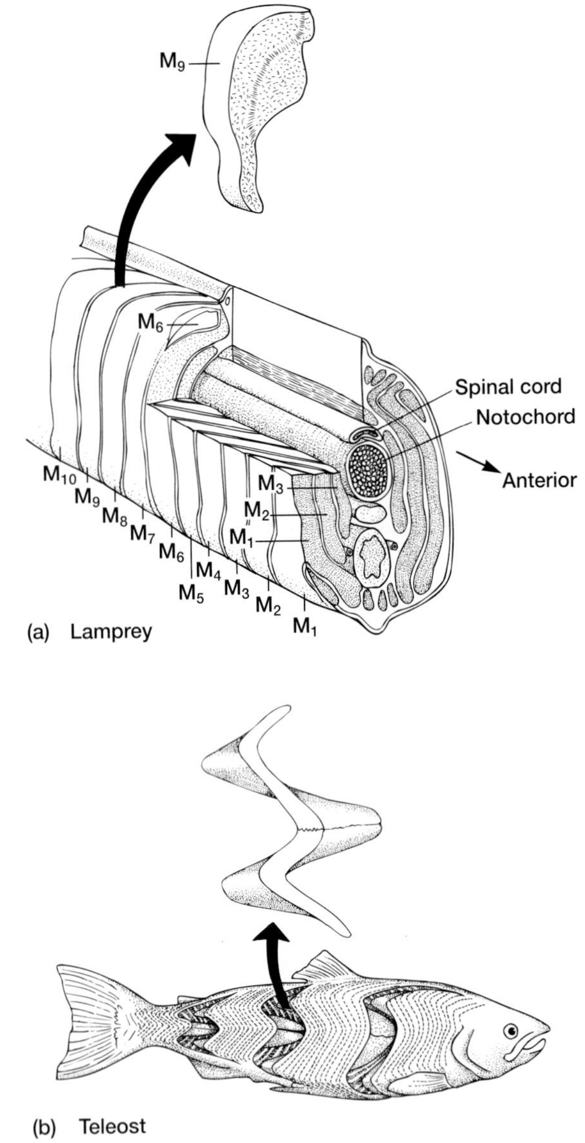


**Figure 10.24** Axial musculature of fishes. (a) Lamprey trunk in cutaway view showing the arrangement of segmental and numbered myomeres. (b) Lateral view of teleost showing the arrangement of myomeres that form the extensive trunk musculature. Sections of the trunk musculature have been removed to reveal the arrangement of the folded myomeres. A block of segmental muscle is enlarged and shown in isolation.
(a) Modified from Hardisty, 1979.

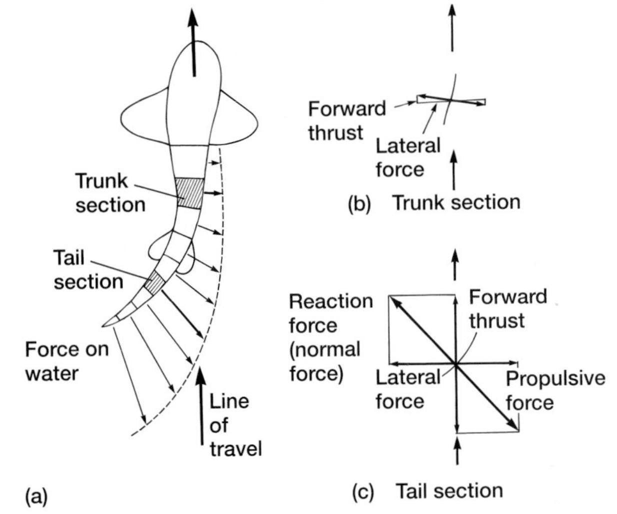


**Figure 10.25** Swimming forces in a shark. (a) Traveling waves of undulation pass backward along the body, throwing it into curves. These press against the water to produce a propulsive force. The water returns a reaction, or normal, force. The body is arbitrarily divided into sections. The level and direction of the propulsive force generated by each section are indicated by the vectors (arrows). Notice the change in angle of the propulsive force along the body relative to its line of travel. (b) Diagram of trunk forces. The propulsive and normal forces from an anterior trunk section are shown. Notice that the forward component of the normal force is much reduced compared with the normal force that the tail experiences. (c) Diagram of tail forces. The lateral and forward component vector forces of the normal force are shown in this tail section. The forward vector along the line of travel drives the fish forward.

musculature tends to be reduced, that which remains differentiates into specialized muscles, a reflection of the more complicated control exerted over flexion of the vertebral column and movement of the rib cage (figure 10.26a–c).

In salamanders, epaxial muscles are still essentially one muscle mass, the **dorsalis trunci** (figure 10.26b). The hypaxial musculature has differentiated into a few muscles, but compared with other tetrapods, the axial musculature is still quite simple and constitutes a large proportion of the overall body musculature. Continued prominence of the axial musculature in salamanders is thought to reflect the continued central role of the axial column in locomotion and the limbs' modest contribution to propulsion. In frogs, in whom the hindlegs serve the specialized saltatorial locomotion, the appendicular musculature of the hindlimbs is large and the axial musculature is reduced in prominence.

In reptiles, the horizontal septum is lost or indistinct, although supply by the dorsal and ventral rami of the spinal nerve still betrays which muscles are of epaxial and hypaxial origin (figure 10.26c). Even though lateral undulations of the vertebral column contribute to locomotion, the limbs become much more important in producing the propulsive forces central to locomotion. The epaxial musculature associated with the vertebral column is reduced. The hypaxial musculature forms much of the body wall and is associated with breathing because these muscles are attached to the rib cage. Because the rib cage in turtles is rigid, the hypaxial muscles are reduced or lost. In snakes, which are of course without limbs, the axial column figures significantly in lateral undulation and the axial musculature is prominently developed.

The axial muscles in reptiles tend to split into several layers, forming many differentiated muscles that span several segments (figure 10.27a–d). There are three general divisions of the epaxial musculature, the **transversospinalis,** the **longissimus,** and the **iliocostalis** muscle groups. Muscles of these three groups usually attach to ver-

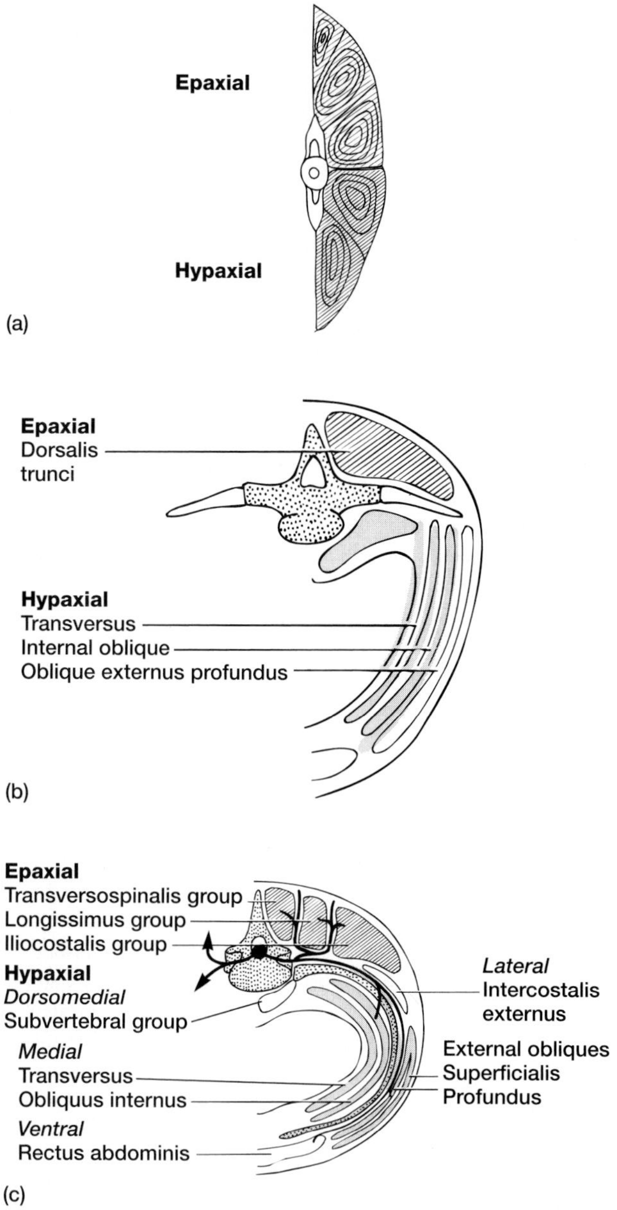


**Figure 10.26** Organization of axial musculature (cross sections). (a) Teleost fishes have regions of relatively undifferentiated epaxial and hypaxial muscle masses. (b) Salamanders have an epaxial musculature that is a relatively undifferentiated muscle mass, the dorsalis trunci. Hypaxial muscles differentiate into several discrete muscles. (c) Lizards have both epaxial and hypaxial muscle masses that have differentiated into several specialized groups of muscles. The horizontal septum is not easily recognized, but distribution of the spinal nerve branches permits identification of derivatives of epaxial (dorsal branch of spinal nerve) and hypaxial (ventral branch) musculature.

tebrae, and in some species, they further split into additional muscle groups.

The hypaxial musculature attached to the rib cage controls breathing and also aids in moving the trunk. As mentioned, this can be a prominent role in snakes. Most descriptive work on reptile hypaxial musculature recognizes three embryonic precursors giving rise to three direct and a fourth composite group of muscles. One group is the **dorsomedial** musculature that runs beneath the vertebral column as the subvertebralis and extends anteriorly as the longus collis that aids in moving the neck. The second group is the **medial** musculature, which is distributed along the inside of the rib cage, and includes the transversus abdominis and internal obliques. The third group is the **lateral** musculature spanning the outside of the rib cage; it includes the external obliques and external intercostals. Apparently derivatives of medial and lateral musculature contribute to the **ventral** musculature along the belly, which includes the rectus abdominis that extends from the sternum and ribs to the pelvis. It is divided midventrally along its length by the linea alba and is crossed at regular intervals by connective tissue inscriptions, a pattern suggesting a basic segmental nature.

In birds, the same divisions of the axial musculature are represented, but they are reduced, especially in regions where associated vertebrae are fused.

In mammals, the three reptilian divisions of the epaxial and four divisions of the hypaxial musculature are present, although they tend to form numerous splits that yield additional muscles.

## *Appendicular Musculature*

***Fishes*** In fishes, two opposing muscle masses extend over the dorsal and ventral surfaces of the fins from girdle to pterygiophores. Embryologically, these arise from myotomes that grow out into the fins and differentiate into these dorsal and ventral muscle masses. Dorsal muscles elevate the fin, ventral muscles depress or adduct the fin. Occasionally, these muscles produce distinct muscle slips that aid in fin rotation. Compared with the massive axial musculature, the fin musculature of fishes is relatively slight.

***Tetrapods*** In tetrapods, these dorsal and ventral appendicular muscles tend to be more prominent as the limbs assume more of the task of producing locomotor forces and the axial musculature becomes less involved. In addition to becoming more prominent, these muscle masses also tend to split and divide, forming many distinct muscles that increase the complexity of the adult limb musculature considerably. This story is further complicated by the fact that the tetrapod limb musculature receives phylogenetic contributions from other regions. The axial musculature along the body and the branchiomeric musculature of the gill arches also contribute to the tetrapod limb muscles especially to the shoulder muscles. Further, the hip and

## Box Essay 10.3 Careful Turns, Fast Starts, and Cruising in Open Waters

To best generate swimming thrust, the body of a fish should be deep (tall from dorsal to ventral surface) to present a broad, oarlike surface to the water. The sculpin is an example (box figure 1a). Not only is the body deep, but the lateral silhouette is enlarged further by expansive fins along the dorsal and ventral surfaces. Thus, the total lateral area of the fish pressed against the water during swimming is extensive and helps generate thrust. But not all fish are designed in this way because swimming is deployed significantly differently among fishes.

For fishes that must perform careful maneuvers, the body is often disk shaped (box figure 1b). Angelfish, popular in tropical fish shops, slip carefully through crowded vegetation and position their bodies to find food along blades of aquatic plants. Butterfly fish, inhabitants of coral reefs, maneuver at slow speed among the coral shelves, poking their mouths into tiny crevices for food. Their disk-shaped bodies keep the anteroposterior body axis short so it can be turned in tight spaces. The edge of the body is often rimmed by a fringe of fins that control precise and specific adjustments.

Other fishes are designed for quick bursts in which they accelerate to surprise prey or dart from sudden approach of danger. The pike is an example (box figure 1b). Deploying swimming for quick dashes requires that the fish overcome inertia. Consequently, almost 60% of a pike's mass is axial muscle. This gives it a powerhouse of contractile units to generate sudden large forces. Further, its body is relatively flexible, at least through the tail, so it can bend into large amplitude curvatures to produce a normal force more in line with its intended line of travel.

Although less specialized than the pike, the trout also has a relatively large mass of axial musculature and a flexible body. When threatened, it can quickly form a C-shaped bend and use its large mass of axial musculature to "push off" suddenly, accelerating in a quick burst in some other direction to make its escape (box figure 1c).

Still other fishes, such as tuna, are designed for cruising (box figure 1b). The trailing edge of the tail is expanded to deliver the tail thrust to the surrounding water. But the **peduncle** that connects the tail to the body is very narrow. Consequently, the overall mass of the tail and its peduncle is quite low, and thus the inertia that must be oscillated during swimming is low as well. On the other hand, most of the axial musculature is bunched more anteriorly in the trunk, increasing its inertia. There is less tendency for the tail to impart its oscillations on the more massive body and cause wasteful lateral body swings. Overall, these changes in location of muscle mass together with the streamlined shape make the swimming design of the tuna especially efficient for sustained cruising, an advantage in a fish that covers great stretches of open ocean in search of schools of small fishes, its principal prey.

Of course, most fishes are not such specialists and must compromise between these specialized extremes. Further, the design of a fish requires more than just attention to the mass of the axial musculature. For example, the sculpin has a relatively large head in order to dislodge and pick up benthic animals with its powerful suction feeding. Consequently, the head is a relatively large part of the sculpin; thus, its overall optimum design is a compromise between the requirements of feeding and swimming.

**Box figure 1** Swimming specializations. (a) Broad lateral profile of a sculpin that undulates against the water to produce thrust. The tail and prominent dorsal and ventral fins enlarge the thrusting surface. (b) Three body shapes illustrating three specialized roles for swimming—top, maneuvering (butterfly fish); middle, quick lunges (pike); bottom, sustained cruising (tuna). (c) Trout seen from above. In response to a threat, it sharply flexes its body and then straightens the tail in order to accelerate quickly and change direction.

shoulder muscles transmit locomotor forces to the vertebral column differently. The pelvic girdle is attached directly to the sacral region of the vertebral column, but the pectoral girdle of terrestrial vertebrates is hung in place by a **muscular "sling"** (figure 10.28). This is a set of muscles that run from thorax to shoulder to suspend the anterior part of the body through muscular ties from the blades of the pectoral girdles. Finally, many specialized tetrapods depart from the general trend. Frogs, for example, are specialized to leap and have complicated hindlimb musculature; birds are specialized for flight, and their limb musculature serves the special demands of aerial locomotion.

*Pectoral girdle and forelimb.* Contributions to tetrapod shoulder and forelimb muscles come from four sources—branchiomeric, axial, dorsal limb, and ventral limb muscles (figure 10.29a–c).

1. **Branchiomeric muscles.** The branchiomeric muscles contribute the **trapezius** and **mastoid** groups. In mammals, the trapezius group includes the clavotrapezius, acromiotrapezius, and spinotrapezius; the mastoid

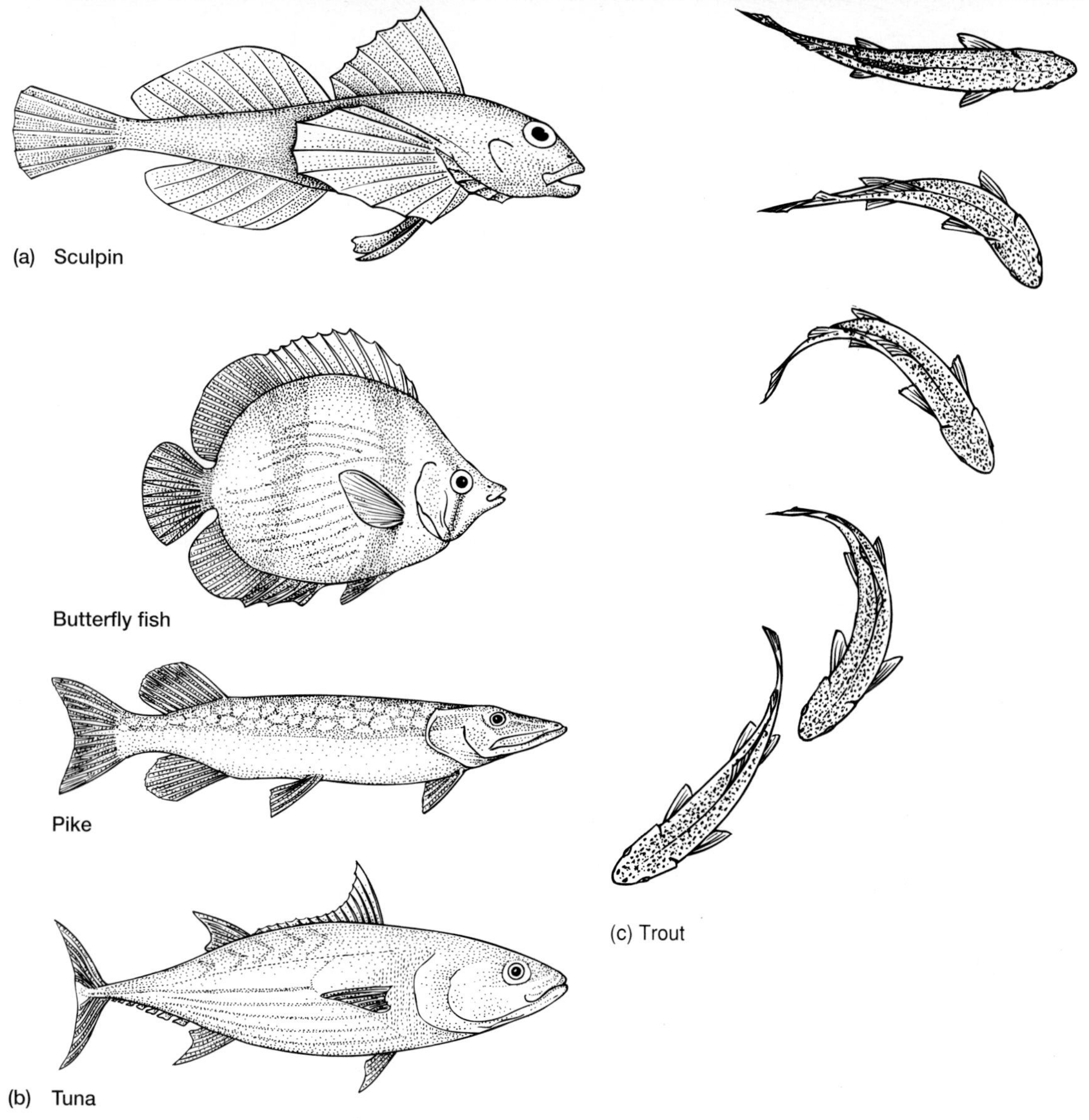


group includes the cleidomastoid and sternomastoid muscles (table 10.2).

2. **Axial musculature.** The axial musculature contributes the **levator scapulae, rhomboideus complex,** and **serratus** muscles. These three derivatives of the axial musculature, together with the trapezius of branchiomeric origin, form the muscular sling that suspends the body between the two scapular blades. The pectoral girdle of turtles is an exception because it is directly attached to the shell. In some tetrapods, such as pterosaurs, birds, and bats, the pectoral girdle rests on the sternum. In fishes, the pectoral girdle is usually attached to the back of the skull, but in most tetrapods it is not.

Freeing of the shoulder from the skull is established in early amphibians during the transition to land and in part seems related to increased cranial mobility. As the shoulder girdle became freed from the skull, the nearby branchiomeric and axial muscles were pressed into serving as part of the muscular sling through which the forelimbs are attached to the body. Most of the remaining pectoral and forelimb muscles of tetrapods arise from the dorsal and ventral muscle masses (table 10.2).

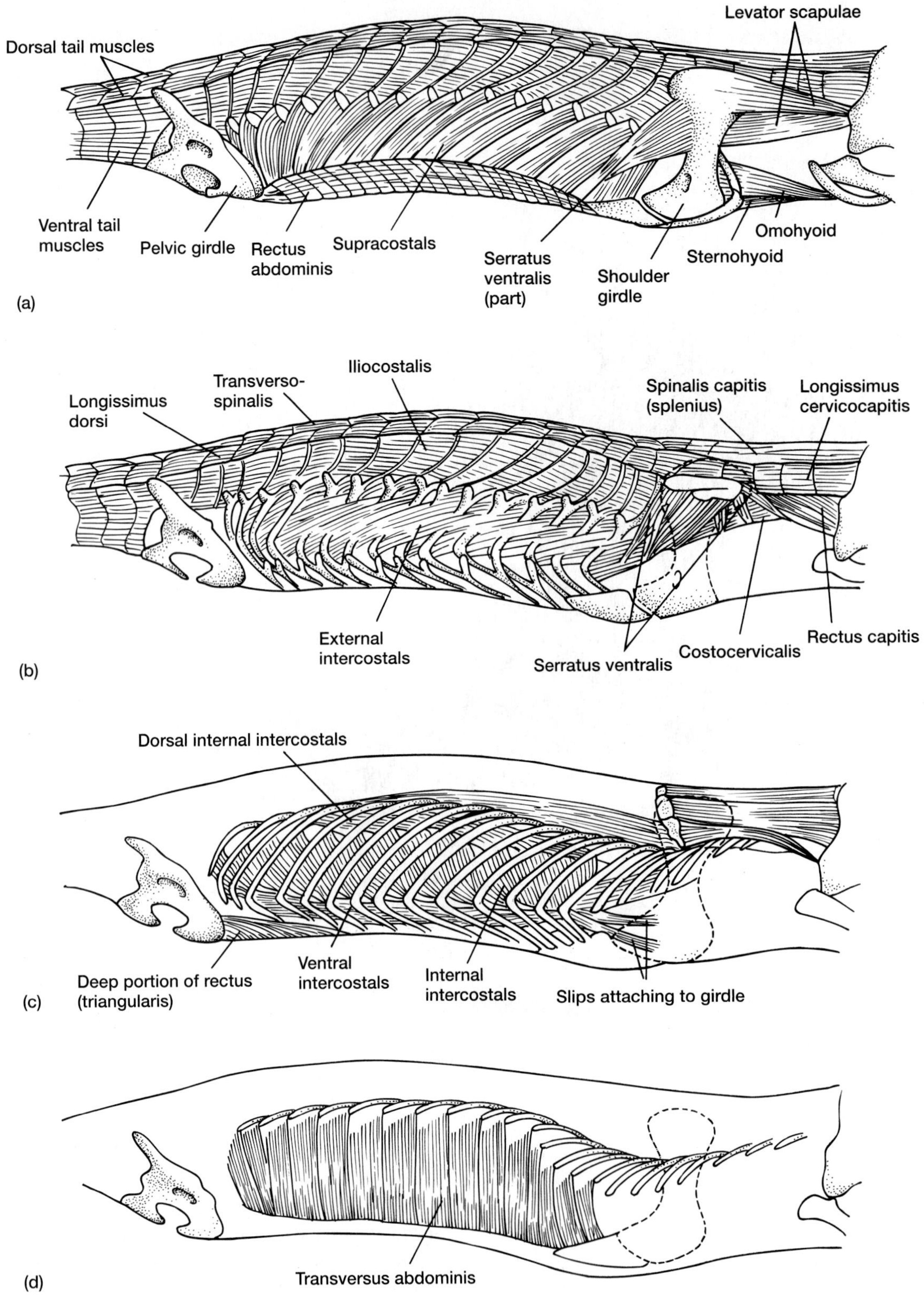


**Figure 10.27** Lateral views of the axial musculature of the reptile *Sphenodon*. (a–d) Muscle layers are successively removed to reveal deeper muscles beneath.

TABLE 10.2 Homologies of Axial and Appendicular Musculature

| Muscle Groups | Salamander | Reptiles | Mammals |
|---|---|---|---|
| | | *Pectoral Girdle and Forelimbs* | |
| Branchiomeric | Cucullaris | Trapezius | Clavotrapezius<br>Acromiotrapezius<br>Spinotrapezius |
| | Levatores arcuum | Sternomastoid | Cleidomastoid<br>Sternomastoid |
| Axial | Levator scapulae<br>Thoraciscapularis | Levator scapulae<br>Serratus ventralis | Levator scapulae<br>Rhomboideus<br>Serratus ventralis |
| Dorsal | Latissimus dorsi | Latissimus dorsi | Latissimus dorsi<br>Teres major |
| | Subcoracoscapularis | Subcoracoscapularis | Subscapularis |
| | Dorsalis scapulae<br>Procoracohumeralis longus | Dorsalis scapulae<br>Deltoideus clavicutans | Deltoideus (Acromiodeltoid and Scapulodeltoid) |
| | Triceps | Triceps | Triceps |
| | Forearm extensors | Forearm extensors | Forearm extensors |
| Ventral | Pectoralis | Pectoralis group | Pectoralis (4) |
| | Supracoracoideus | Supracoracoideus | Supraspinatus<br>Infraspinatus |
| | Coracoradialis<br>Humeroantebrachialis | Biceps brachii<br>Brachialis inferior | Biceps brachii (part)<br>Biceps brachii (part) |
| | Coracobrachialis | Coracobrachialis | Coracobrachialis |
| | Forearm flexors | Forearm flexors | Forearm flexors |
| | | *Pelvic Girdle and Hindlimbs* | |
| Axial | | | |
| | Subvertebralis | Subvertebralis | Psoas minor |
| Dorsal | Puboischiofemoralis internus | Puboischiofemoralis internus | Psoas<br>Iliacus<br>Pectineus |
| | Ilioextensorius<br>Puboischiofemoralis externus | Iliotibialis<br>Femorotibialis | Rectus femoris<br>Vasti |
| | Iliotibialis | Ambiens | Sartorius |
| | Iliofemoralis | Iliofemoralis | Tensor fascia latae<br>Gluteus minimus<br>Gluteus medius<br>Pyriformis |
| | Tibialis anterior | Tibialis anterior | Tibialis anterior |
| | Extensor digitorum communis | Extensor digitorum communis | Extensor digitorum longus<br>Extensor hallucis longus<br>Peroneus tertius |
| | Peroneus longus | Peroneus longus | Peroneus longus |
| | Peroneus longus and brevis | Peroneus brevis | Peroneus brevis |
| | Extensor digitorum brevis | Extensor digitorum brevis | Extensor digitorum brevis |
| Ventral | Puboischiofemoralis externus | Puboischiofemoralis externus | Obturator externus<br>Quadratus femoris |

*continued*

**TABLE 10.2 Homologies of Axial and Appendicular Musculature, continued**

| Muscle Groups | Salamander | Reptiles | Mammals |
|---|---|---|---|
| Ventral (continued) | Adductor femoris<br>Pubotibialis | Adductor femoris<br>Pubotibialis | Adductor femoris brevis<br>Adductor femoris longus |
| | Caudofemoralis<br>Ischioflexorius | Caudofemoralis<br>{ Flexor tibialis externus<br>Flexor tibialis internus II<br>Flexor tibialis internus I } | Caudofemoralis<br>Dorsal semitendinosus<br>{ Ventral semitendinosus<br>Biceps femoris }<br>Semimembranous |
| | Puboischiotibialis | Puboischiotibialis | Gracilis |
| | Flexor digitorum sublimis and longus | Gastrocnemius internus | { Gastrocnemius medialis<br>Flexor hallucis longus } |
| | { Popliteus<br>Fibulotarsalis } | Gastrocnemius externus | { Gastrocnemius lateralis<br>Soleus<br>Plantaris } |

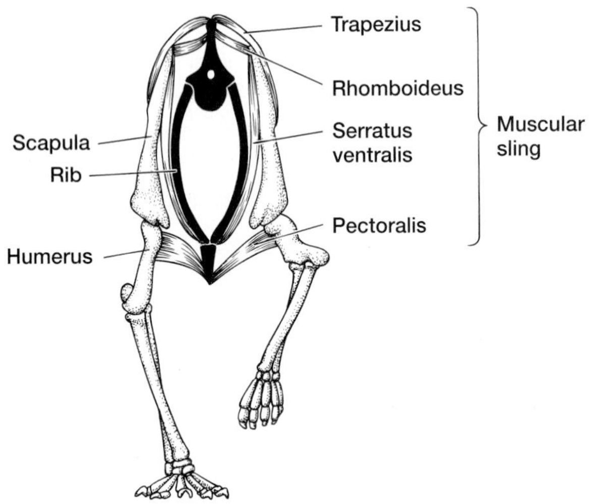


**Figure 10.28** The "muscular sling" of tetrapods. Appendicular muscles of the forelimbs suspend the anterior body of tetrapods from the shoulders. Some of these muscles arise from axial muscles (rhomboideus, serratus ventralis), some from branchial muscles (trapezius), and some from the forelimb musculature itself (pectoralis).

3. **Dorsal muscles.** The dorsal muscles of the shoulder are inserted on the humerus and function to oscillate it during movement or fix it in position while an animal stands. Of these muscles, only the **latissimus dorsi** originates outside the limb, from the body wall. In mammals, a tiny slip of the latissimus within the scapula separates as the **teres major.** The other dorsal muscles that act on the humerus are the **teres minor, subscapularis,** and **deltoideus,** which may form two distinct muscles. The prominent **triceps,** often showing several heads, is also a derivative of the dorsal musculature, but it acts to extend the forearm. Dorsal muscles of the forearm form most of the extensor musculature, which extend or straighten the digits via tendons (table 10.2).

4. **Ventral muscles.** The **pectoralis** is a very prominent ventral muscle of the chest. From a long origin along the sternum, its fibers converge on the humerus (figure 10.30a–c). This muscle tends to split into four more or less distinct derivatives in mammals—the **pectoantebrachialis,** the deeper **pectoralis major,** the deeper still **pectoralis minor,** and the deepest of all, the **xiphihumeralis.** The ventrally positioned **supracoracoideus** of reptiles extends from its origin on the coracoid laterally to its insertion on the humerus. However, in mammals, the supracoracoideus originates dorsally on the lateral face of the scapula. The bony scapular spine divides this muscle into the **supraspinatus** and **infraspinatus** muscles, which are inserted on the humerus as well. The **coracobrachialis** from the coracoid runs along the underside of the humerus. In mammals, the **biceps brachii** has two heads, representing the apparent fusion of two muscles that have their insertions on the forearm and flex it in lower vertebrates. Forearm flexors from ventral muscles act through tendons on the digits (table 10.2).

*Pelvic girdle and hindlimb.* Unlike the shoulder, the tetrapod hip has no muscular sling in which it "floats." Instead, the pelvic girdle is fused to the vertebral column. Consequently, few extrinsic muscles control the hindlimbs. The **psoas minor** from the axial musculature is an exception. However, most hindlimb musculature derives from dorsal and ventral muscles that differentiate into the complex assortment of hip, thigh, and shank muscles (table 10.2).

1. **Dorsal muscles.** The **puboischiofemoralis internus** of lower tetrapods is a dorsal muscle that runs from the lumbar region and girdle to the femur, making it an important limb rotator. Three muscles differentiate from it in mammals. All three are inserted on the femur but originate

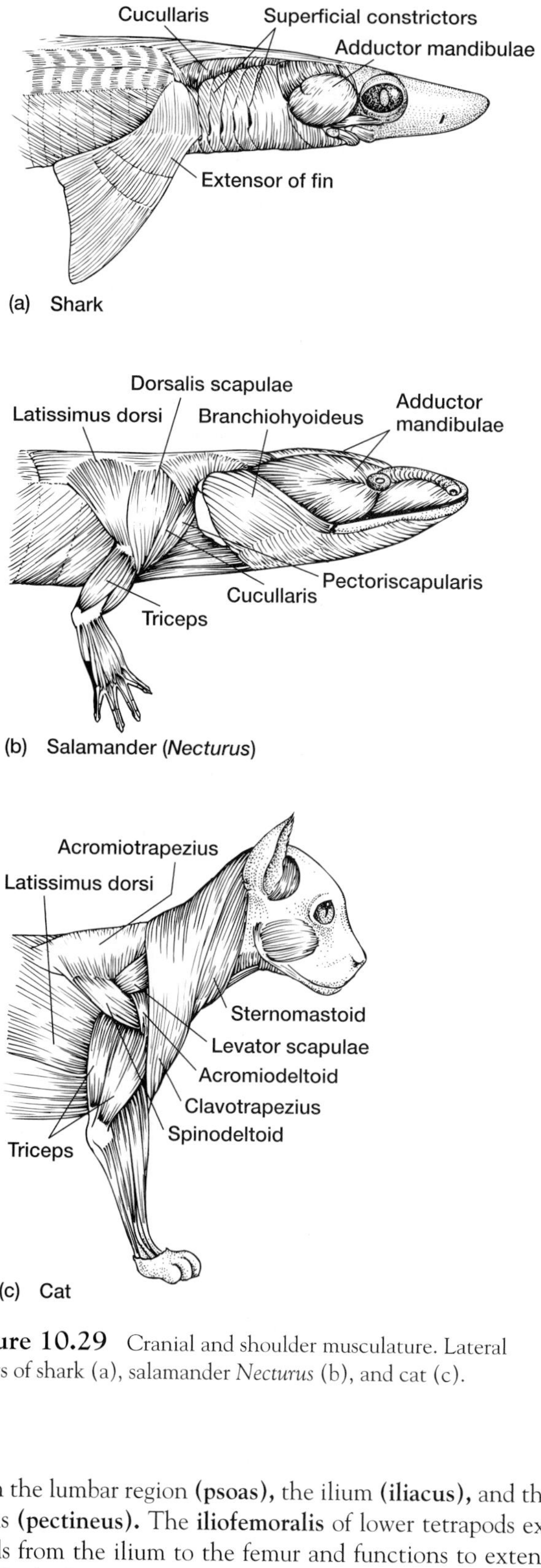


**Figure 10.29** Cranial and shoulder musculature. Lateral views of shark (a), salamander *Necturus* (b), and cat (c).

from the lumbar region **(psoas),** the ilium **(iliacus),** and the pubis **(pectineus).** The **iliofemoralis** of lower tetrapods extends from the ilium to the femur and functions to extend the limb. In mammals, it divides into the **tensor fascia latae, pyriformis,** and **gluteus complex.** The *quadriceps* is a collective term for the **rectus femoris** and the three heads of the **vastus** (lateralis, medialis, intermedius). These muscles lie along the anterior margin of the femur and are usually quite prominent. Through their common insertion on the patella, they are very powerful shank extensors. The long **sartorius** originates on the ilium but crosses two joints, the acetabulum and the knee, before being inserted on the tibia. The **ambiens** of reptiles and the **iliotibialis** of amphibians are likely homologues of the sartorius. The **tibialis anterior** and various other dorsal muscles of the shank constitute the shank extensors that flex the ankle via long tendons (table 10.2).

2. **Ventral muscles.** In lower tetrapods, the **puboischiofemoralis externus** is a ventral muscle that extends from the pubis and ischium to the femur (figure 10.31a–c). In mammals, the **obturator externus** and **quadratus femoris** are derivatives. In lower vertebrates, the **caudofemoralis,** extending from the base of the tail to the femur, is a powerful muscle that retracts the hindlimb. When the hindlimb is fixed, the caudofemoralis has the opposite action of swinging the tail. In mammals, it is reduced in prominence. Similarly, the **obturator internus** and **gemelli muscles** in mammals are relatively reduced compared to their homologue, the **ischiotrochantericus** in reptiles. The **adductor femoris,** which is large in most tetrapods, straightens the hindlimb. The name *hamstrings* is a collective term for three muscles, the **semimembranosus, semitendinosus,** and **biceps femoris.** All arise from the pelvis, run along the posterior margin of the femur, and have insertions on the shank or nearby on the distal end of the femur. Together these prominent muscles flex the shank. The **puboischiotibialis** of lower tetrapods covers much of the ventral surface of the thigh and retracts it. Its mammalian homologue is the **gracilis.**

The most prominent ventral muscle of the shank is the **gastrocnemius,** the "calf" muscle. In mammals, it has two heads, resulting from the fusion of two different phylogenetic predecessors. The mammalian **gastrocnemius medialis** and the **flexor hallucis longus** arise from the reptilian **gastrocnemius internus.** The **gastrocnemius lateralis** along with the **soleus** and **plantaris** arise from the reptilian **gastrocnemius externus** (table 10.2).

*Specializations among tetrapods*. Locomotion among tetrapods is based on alternating limb displacements at moderate rates. Departure from this generalized mode of progression usually depends on modifications of the musculature that powers it. In anurans, for example, locomotion is saltatorial and both limbs are activated simultaneously by contraction of powerful hindlimb extensors. At the end of a leap, the pectoral girdle and forelimbs of anurans absorb the impact of landing. The muscular sling of the pectoral girdle suspending the body in other tetrapods probably functions to absorb the jolts and jars during locomotion. But this role in anurans is clearly accentuated and the mode of progression is a marked departure from the alternating limb swings of other tetrapods. The muscles of the anuran forelimb are stout to help during landing, and the extensor muscles of the hindlimb are prominent to launch the animal. This specialized mode of locomotion

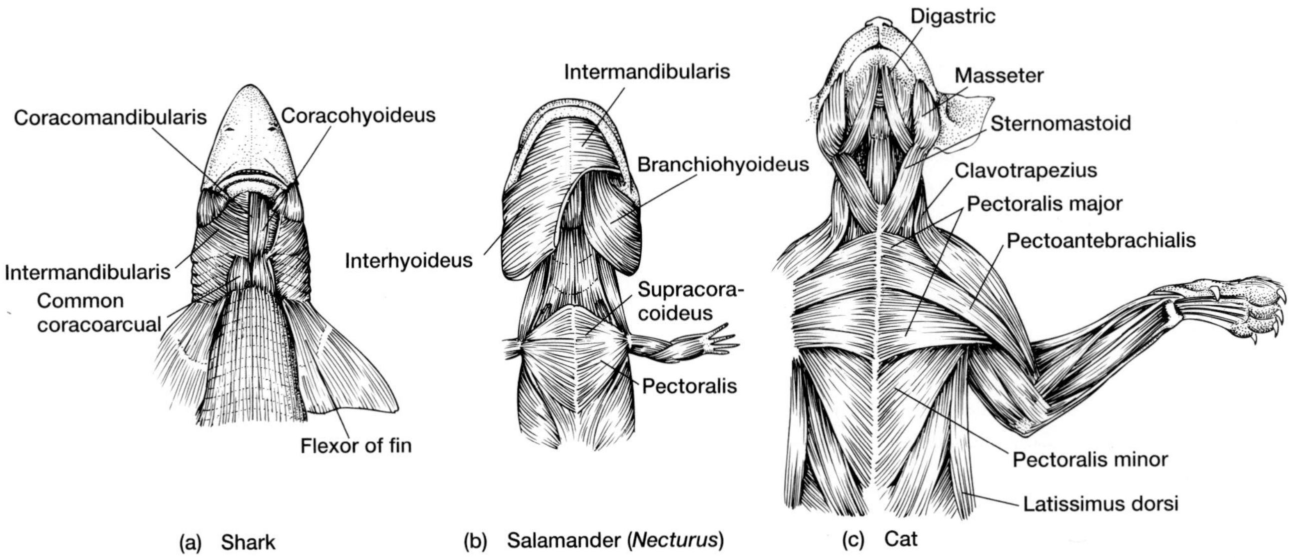


**Figure 10.30** Cranial, hypobranchial, and shoulder muscles. Ventral views of shark (a), salamander *Necturus* (b), and cat (c).

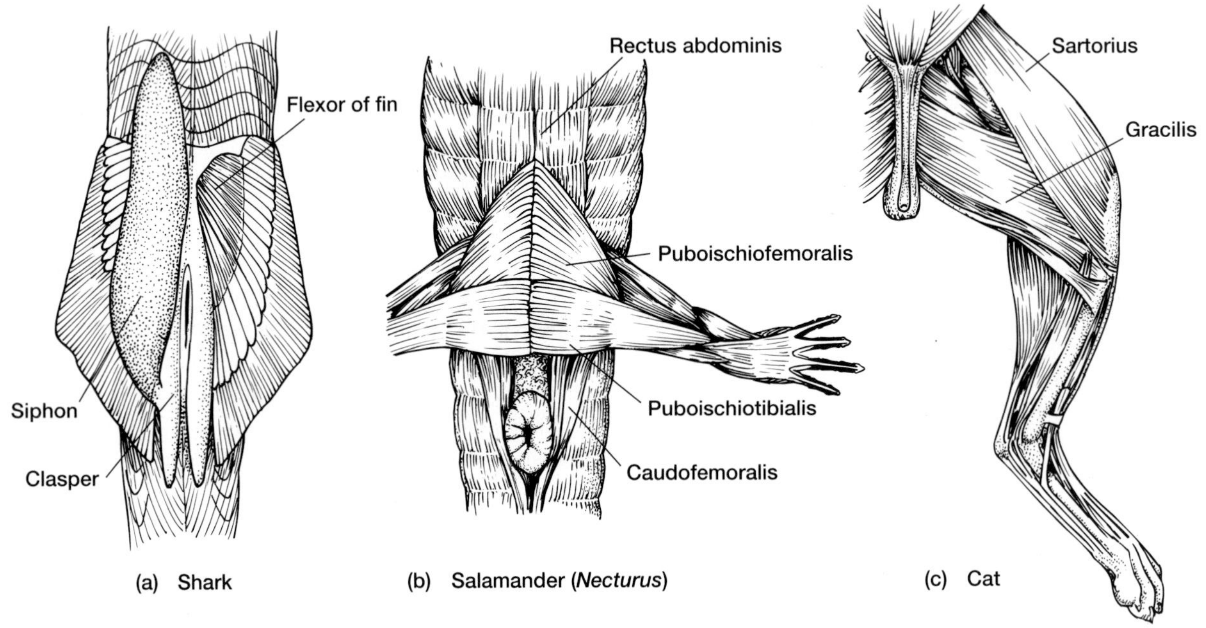


**Figure 10.31** Pelvic musculature. Ventral views of shark (a), salamander *Necturus* (b), and cat (c).

might account for the relatively complex and differentiated musculature of anurans compared with salamanders. Figure 10.32a,b illustrates the superficial musculature of the frog.

In tetrapods specialized for cursorial locomotion, the appendicular muscles tend to be bunched proximally near the body and their forces distributed distally through long tendons to the ends of limbs. This design reduces the mass carried by the limb itself, which in turn reduces the inertia that needs to be overcome during reciprocating limb oscillation. Among mammals, the perissodactyls (figure 10.33) and artiodactyls exhibit the most well-developed proximal repositioning of limb muscle mass, but similar trends can be found in most other mammalian groups that depend on rapid locomotion. Parallel trends apparently arose in reptiles of the Mesozoic, especially within fleet archosaurs. The appendicular muscles of the hindlimbs of these bipedal reptiles show evidence of proximal bunching, presumably with long tendons that extended to the ends of the limbs.

Cursorial trends (p. 339)

In birds, the general trends of muscle evolution characteristic of tetrapods are evident (figure 10.34). The axial musculature tends to decline in prominence, whereas the appendicular musculature increases. Muscle masses become differentiated into a more complex set of discrete muscles. Birds also show changes in musculature design related to the specialized demands of powered flight and secure landing. Fusion of the posterior vertebral column with elements of the pelvic girdle into a rigid bony support reduces the requirement for large masses of axial muscles to firm up the vertebral column. Consequently, there is a reduction in the

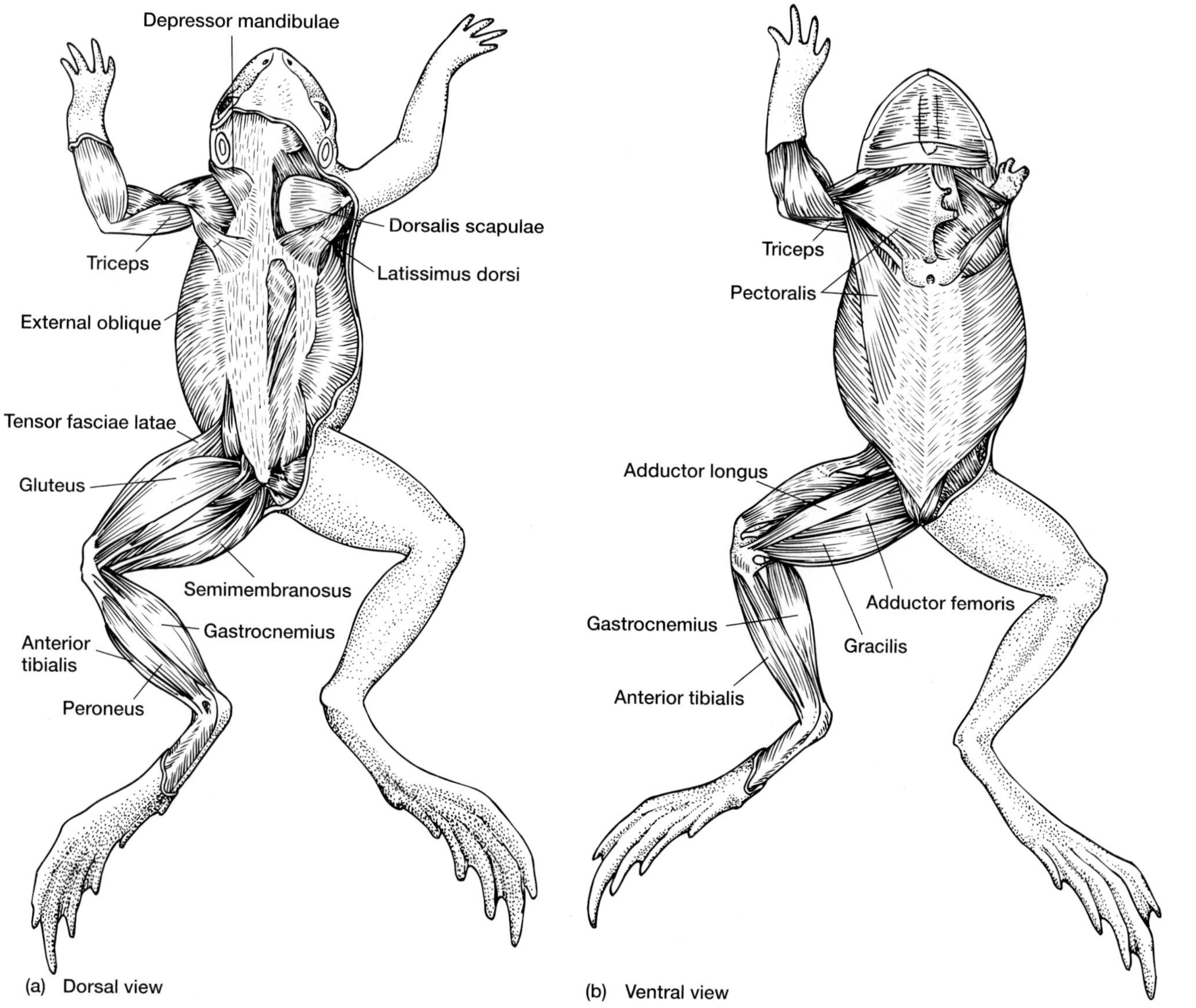


**Figure 10.32** Superficial musculature of a frog. Dorsal (a) and ventral (b) views.

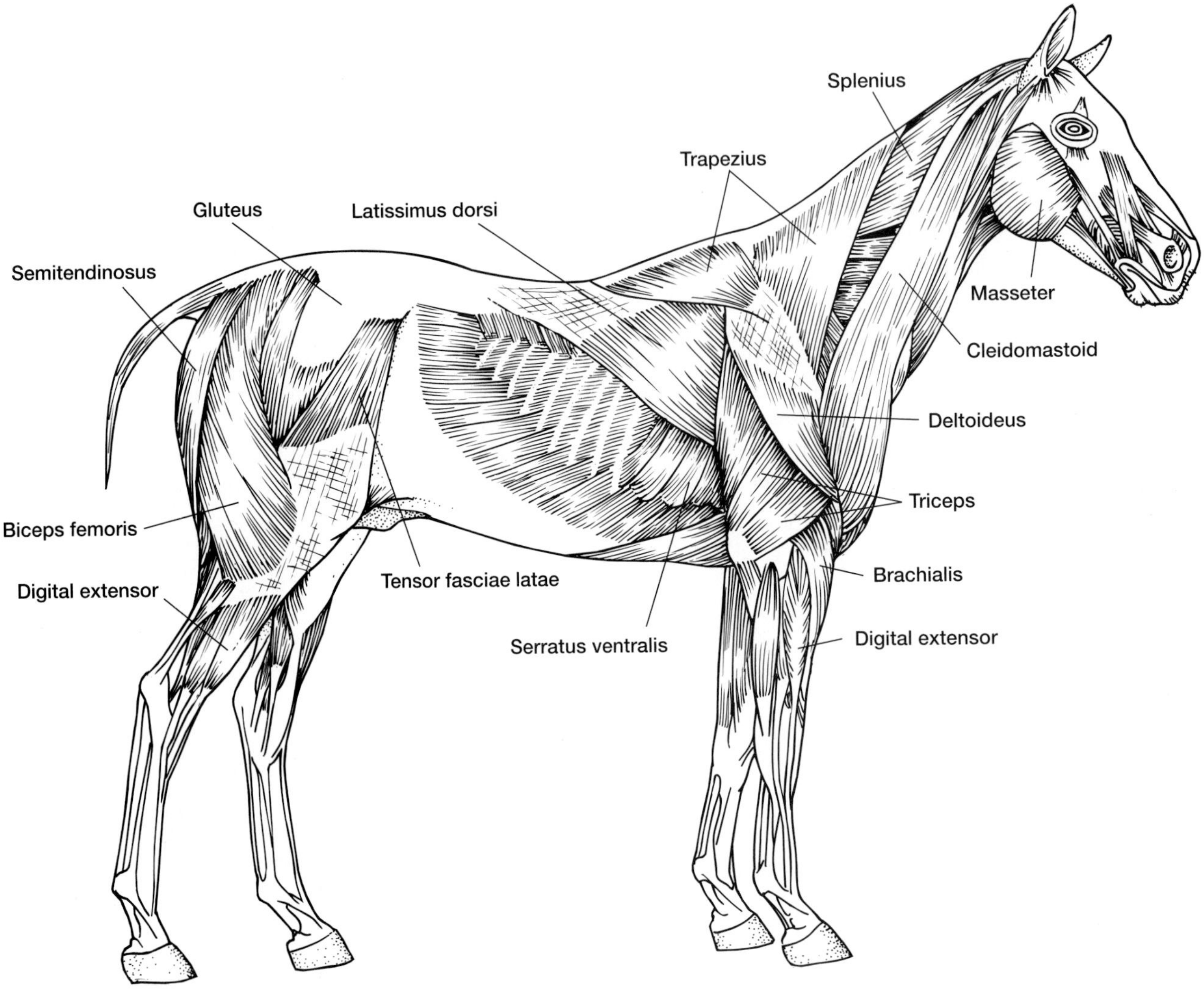


**Figure 10.33** Superficial musculature of a horse.

posterior axial musculature. At the anterior end of the axial column, the chain of cervical vertebrae controlled by a complex set of cervical muscles provide flexibility and very precise control of head movement. Musculature of the pelvic girdle and hindlimb is differentiated into numerous muscles that form a prominent muscle mass. When a bird lands, the hindlimb muscles catch and balance the mass of the body on impact as it comes to rest on the ground or a branch. Most muscles are bunched proximally and extend to the toes through long tendons. The proximal bunching of muscles keeps the mass close to the midline of the body, a feature important during flight, but the long tendons reaching to the toes give more precision to toe placement, a feature especially significant in perching birds and raptors.

**Avian skeletal adaptations (p. 302)**

Muscles of the pectoral girdle and forelimb (wing) are particularly well developed and specialized for powered flight. Most wing muscles are bunched proximally, especially the massive pectoralis that resides near the midline of the sternum from which it originates. The pectoralis is inserted on the humerus and provides a powerful downstroke during flight. Beneath the pectoralis is the supracoracoideus. Recall that this muscle runs from its origin on the pectoral girdle to its insertion on the humerus in reptiles (figure 10.35), a course that makes the supracoracoideus a limb adductor (depressor). However, in birds, the strong tendon of the supracoracoideus runs over the pulleylike end of the coracoid and is inserted on the dorsal surface of the humerus (figure 10.36a–c). This reorientation of the point of insertion allows the supracoracoideus to lift the wing, thus turning this muscle into a wing elevator. Consequently, the depressor (pectoralis) and the elevator (supracoracoideus) that are responsible for producing opposite motions during wing downstroke and upstroke both lie on and originate from the sternum (figure 10.36c). Because of the phylogenetic shifts in position of points of insertion, their actions in birds are quite different.

When a bird is flying, especially gliding, its forearm is extended to draw the overlying skin taut. The anterior region

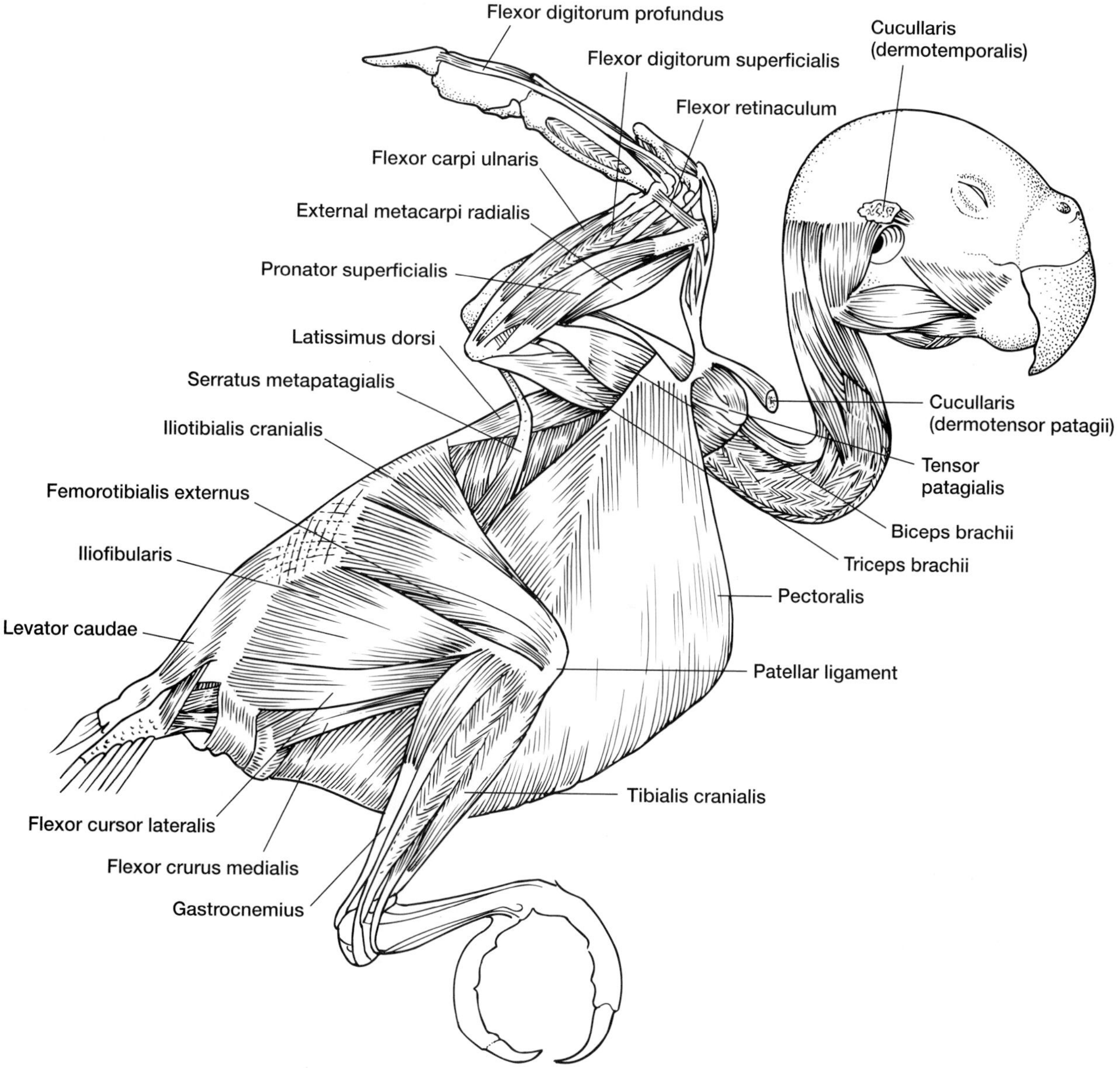


**Figure 10.34** Superficial musculature of a parakeet, with wing elevated.

of skin between the shoulder and wrist is the **patagium,** and within its leading edge is the **patagialis** muscle (figure 10.34). The patagialis, which may form several small slips, arises from the clavicle and extends via a long tendon to the metacarpals of the wrist. Like a clothesline, the leading edge of the patagium is hung from this cordlike muscle, which forms the anterior edge of the flight surface of the wing. If the patagialis or its long tendon is cut, the patagium loses its aerodynamic shape and the wing becomes almost useless for flight.

The forearm muscles of a bird are small but numerous. They probably act to refine the positions of wing feathers, forming and controlling the aerodynamic surface that the wing presents to the airstream.

**Flight (p. 345); aerodynamics (p. 135)**

## Cranial Musculature

Until recently, the striated muscles associated with the jaws and branchial arches were thought to arise embryologically from two quite different sources, the branchiomeric musculature and the hypobranchial musculature. However, as discussed in the section on postcranial muscles (figure 10.22a,b), the current view envisions both branchiomeric and hypobranchial jaw musculature arising from a common source, namely, from the paraxial mesoderm. The branchiomeric musculature arises from the cranial paraxial mesoderm (somitomeres) and the hypobranchial musculature from the trunk paraxial mesoderm (somites). Having emphasized the unity of the jaw musculature rather than its embryonic distinctions, we nonetheless follow this

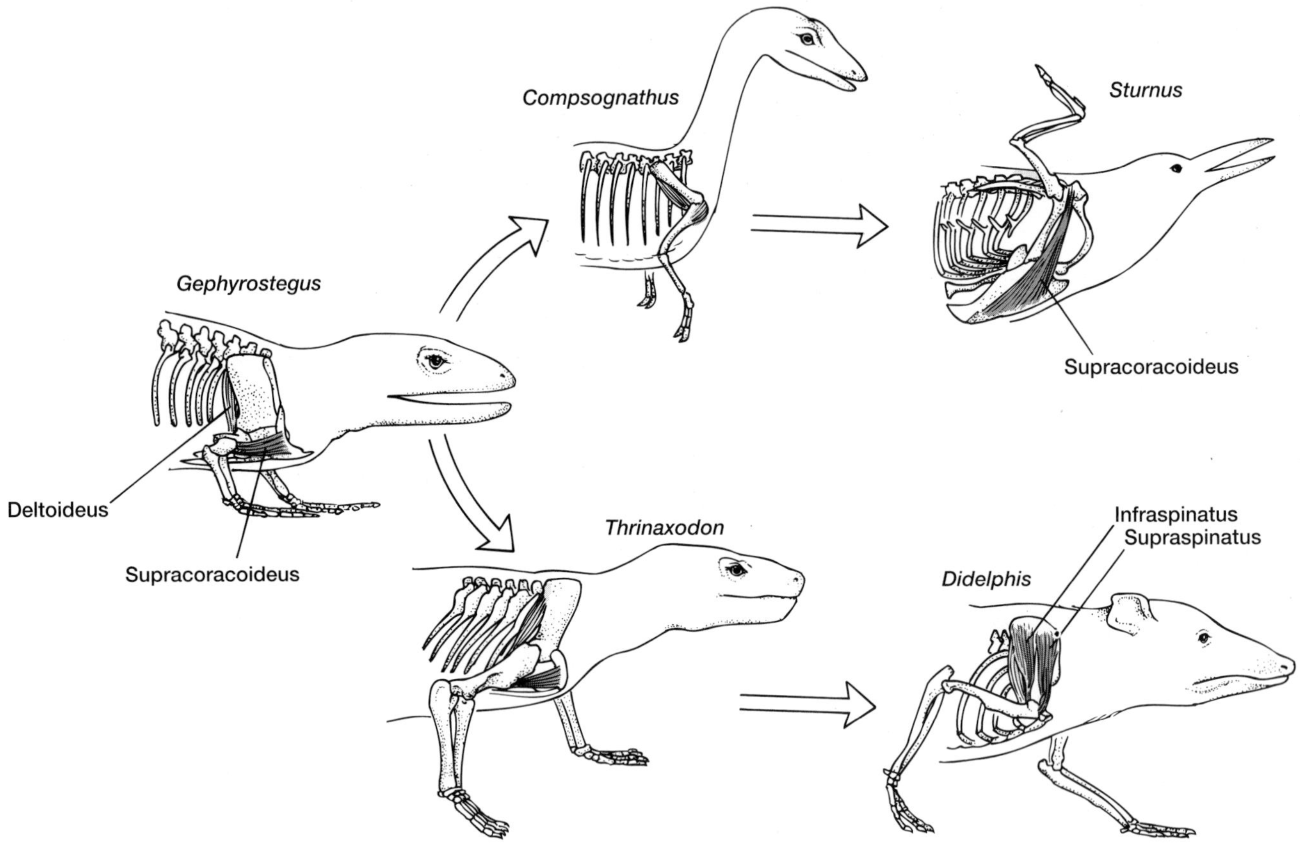


**Figure 10.35** Evolution of the supracoracoideus muscles. In primitive reptiles such as *Gephyrostegus*, this muscle likely originated on the coracoid and was inserted on the proximal head of the humerus to adduct the limb. This muscle probably ran in a similar course in therapsids *(Thrinaxodon)*. In mammals, such as the opossum *(Didelphis)*, the supracoracoideus originates on the scapula and becomes divided by the scapular spine into the supraspinatus and infraspinatus. In the theropod *Compsognathus*, an archosaur, the supracoracoideus probably followed a course similar to that of primitive reptiles, having an origin on the coracoid and an insertion on the proximal end of the humerus. In modern birds, the supracoracoideus originates at the sternum, runs over the coracoid, and is inserted on the dorsal, proximal end of the humerus. The deltoideus and its derivatives are also shown.

historical division for convenience in describing comparative aspects of cranial musculature (table 10.3).

## *Branchiomeric Musculature*

Cranial nerves supply the branchiomeric musculature associated with the sides of the branchial arches (figure 10.37a–d). In fishes, the branchial arches together with their branchiomeric muscles function as a pumping device to move water across the gills, replacing the ciliary system of protochordates. As the anterior branchial arch (or arches) evolved into the jaws, the associated musculature accompanied the bony or cartilaginous elements to become part of the jaw-closing and -opening system of gnathostome fishes.

In general, each branchial arch is endowed with its own set of branchiomeric musculature, a set of **levator muscles** that extend from the arch to the skull and raise the arch, and two sets of **constrictor muscles** that run between arch elements to bend or straighten them (figure 10.37d). One set of constrictor muscles is superficial (near the surface) and the other is deep (sunken) beneath the surface. The **superficial constrictor muscles** bow as a thin sheet across the surface of the arch and compress the arches (draw them inward). The **deep constrictor muscles** are divided into dorsal and ventral constrictors above and below the level of the gills. The dorsal deep constrictor muscles include the **adductors,** which span the middle of the arch and bend it. They are especially well developed on the mandibular (first) arch where they close the jaws. The **interarcuals** are dorsal constrictors at the upper ends of the arches that bend the branchial elements there. The ventral deep constrictor muscles run between the lower ends of the paired branchial arches. Within the mandibular arch, this muscle is the **intermandibularis,** and within the hyoid arch, it is the **interhyoideus** (figure 10.38). Within the remaining branchial arches, ventral constrictors contribute to various oblique and transverse slips of muscle.

Phylogenetically there is great fidelity between a cranial nerve and its branchiomeric muscles, and in turn

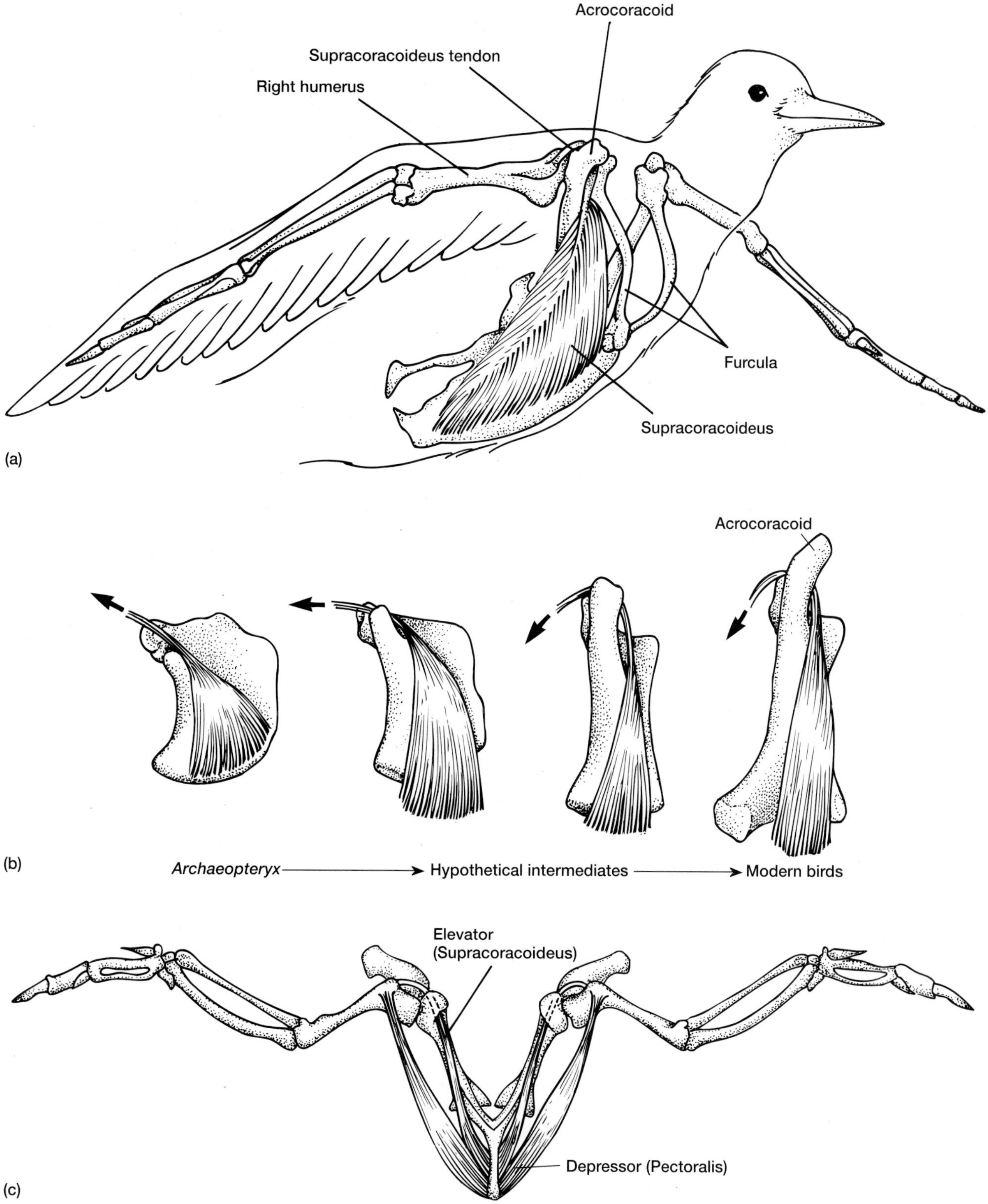


**Figure 10.36** Supracoracoideus in modern birds. (a) The supracoracoideus is shown as it originates at the sternum beneath the pectoralis, which has been removed. Its tendon passes over the coracoid and is inserted at the dorsal surface of the humerus; therefore, this muscle elevates the wing. (b) Proposed changes in the supracoracoideus from *Archaeopteryx* through hypothetical intermediates to modern birds. (c) Separate sets of pectoralis muscles originating at the sternum produce both the powerful downstroke of the wing and the return recovery. Recovery muscles originate at the sternum, pass up through an opening defined by scapula, furcula, and coracoid, and are inserted on the dorsal side of the humerus. As the pectoralis muscles contract, they lift the humerus and elevate the wing. Deltoideus muscles arising from the scapula may also aid in wing elevation. Powerful depressor muscles inserting on the ventral side of the humerus pull the wing downward.

TABLE 10.3 Homologies of Cranial Musculature

| Arch | Cranial Nerve Supply | Shark | *Necturus* | Alligator, Lizard | Cat, Mink |
|---|---|---|---|---|---|
| | | | ***Branchiomeric Musculature*** | | |
| 1 | V | Levator palatoquadrati<br>Spiracularis<br>Adductor mandibulae<br>Preorbitalis | Adductor mandibulae<br>(levator mandibulae) | Adductor mandibulae<br>Pterygoideus<br>(4 present) | Masseter<br>Temporalis<br>Pterygoids<br>Tensor veli palati<br>Tensor tympani |
| | | Intermandibularis | Intermandibularis | Intermandibularis | Mylohyoid<br>Anterior digastric |
| 2 | VII | Levator hyomandibulae<br>(Epihyoidean) | Depressor mandibulae | Depressor mandibulae | Stapedius |
| | | Interhyoideus | Interhyoideus<br>Constrictor colli | Interhyoideus<br>Constrictor (sphincter) colli (Gularis) in part | Platysma and facial muscles (part)<br>Posterior digastric<br>Stylohyoid |
| 3 | IX | Superficial constrictor | Brachiohyoideus | | Intrinsic muscles of the larynx and certain muscles of the pharynx |
| 4 | XI<br>X | Cucullaris | Cucullaris<br>Levatores arcuum | Trapezius<br>Sternomastoid | Trapezius complex<br>Sternocleidomastoid complex |
| | | Interarcuals | | | |
| | | Superficial constrictors and interbranchials | Dilatator laryngis<br>Subarcuals<br>Transversi ventrales<br>Depressores arcuum | | |
| | | | ***Hypobranchial Musculature*** | | |
| | | Coracoarcuals<br>Coracohyoideus<br>Coracomandibularis<br>Coracobranchialis | Rectus cervicis<br>Geniohyoid<br>Pectoriscapularis | Rectus cervicis<br>Sternohyoid<br>Omohyoid | Geniohyoid, others of tongue and larynx<br>Rectus cervicis sternohyoid, omohyoid, thyrohyoid |

fidelity between a set of branchiomeric muscles and its respective arch. Consequently, by tracking gill arches, we can track muscle homologies as well to discover what each muscle or its derivatives becomes in different groups.

***Mandibular Arch*** In sharks, the superficial constrictor within the branchial arches is a thin sheet of muscle within a flap of skin representing septa that cover gill slits (figure 10.38). In the mandibular arch, the dorsal part of the deep constrictor forms the largest of the jaw muscles, the **adductor mandibulae,** located at the angle of the jaws where it provides powerful closing forces. Part of the adductor is separated in sharks as the **preorbitalis,** arising near the orbit and tapering as it passes posteriorly to its insertion on the adductor mandibulae. In bony fishes, the adductor is composed of several derivative muscles that act on selected parts of the highly kinetic skull. In tetrapods, the adductor mandibulae persists as a strong jaw adductor. It often has several prominent and distinctive heads that converge as a pinnate muscle on a common tendon. In mammals, the **masseter** and **temporalis** are jaw-closing muscles with different lines of action, and both are derived from the adductor mandibulae.

The ventral part of the deep constrictor differentiates into the intermandibularis, a transverse sheet of muscle, extending between the ventral tips of the paired mandibles. In tetrapods, the intermandibularis persists as a transverse muscle sheet, usually located between the tips of the lower jaws. In mammals, the anterior part of the digastric muscle, involved in opening the jaws, is derived from the intermandibularis.

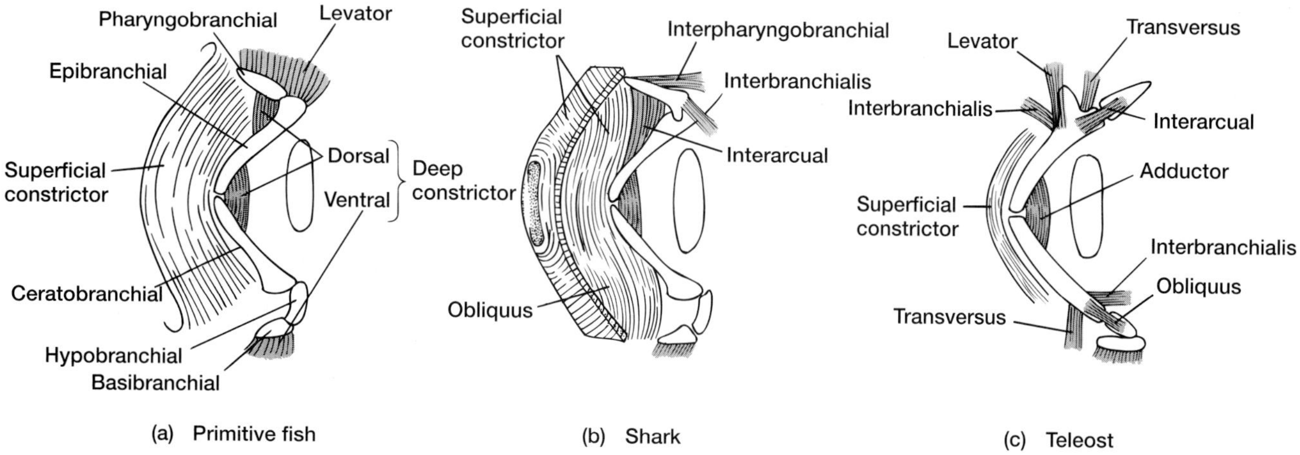


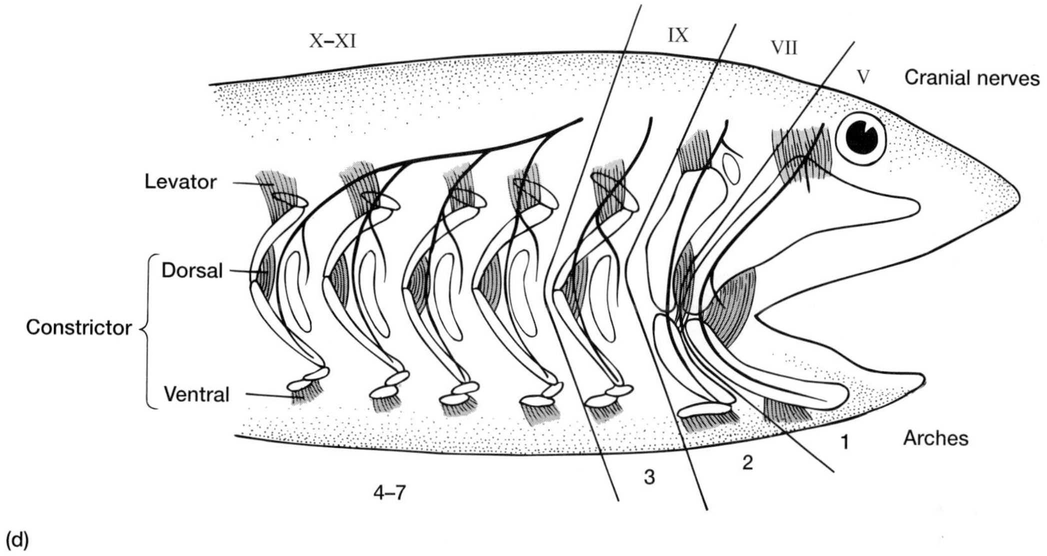


**Figure 10.37** Branchiomeric musculature. (a) Lateral view of a branchial primitive arch illustrating the basic sets of levator and constrictor muscles together with their skeletal structures. (b) Shark branchial arch. (c) Teleost branchial arch. (d) Jaw muscles and their cranial nerve supply tend to stay with their respective branchial arch during the course of subsequent evolution. Each arch has levator and constrictor muscles that, respectively, elevate and close the articulated elements. Cranial nerves V, VII, IX, and X–XI supply muscles of arches 1, 2, 3, and 4–7, respectively. Fidelity of muscles, nerves, and arches generally was maintained as the branchial arches evolved and subsequently became modified into components of the jaws.

In sharks, the muscle derivative of the mandibular levator is the **levator palatoquadrati,** running from the chondrocranium to the palatoquadrate cartilage. In some fishes, such as the chimaera, and in tetrapods, the palatoquadrate becomes fused to the braincase, becoming part of it, and the levator palatoquadrati muscle is absent.

***Hyoid Arch*** The hyoid arch begins as a separate gill arch in primitive fishes, but elements of the hyoid arch become secondarily involved in suspension of the mandibles in some vertebrates and separate as the hyoid apparatus in others. Associated muscles shift their positions as well. Most of the deep constrictor muscles are reduced or lost. In sharks, the largest of the hyoid muscles are derivatives of the hyoid levator. One of these is the **levator hyomandibulae,** which reaches from the chondrocranium to the hyomandibular cartilage. The second of these, often closely fused with the first, is the **epihyoidean,** which is inserted on connective tissue behind the angle of the jaw. In bony fishes, the equivalent of the epihyoidean is the **levator operculi,** with its insertion on the operculum. The **depressor mandibulae** of tetrapods, which opens the jaws, is the homologue of the levator operculi and epihyoidean. In mammals, the depressor mandibulae evolves

| Arch | Cranial nerve supply | Muscles | Muscle regions |
|---|---|---|---|
| 1st | V | Levators<br>Levator palatoquadrati<br>Spiracularis<br><br>Constrictors<br>(Dorsal)<br>Adductor mandibulae<br>Preorbitalis<br><br>(Ventral)<br>Intermandibularis | 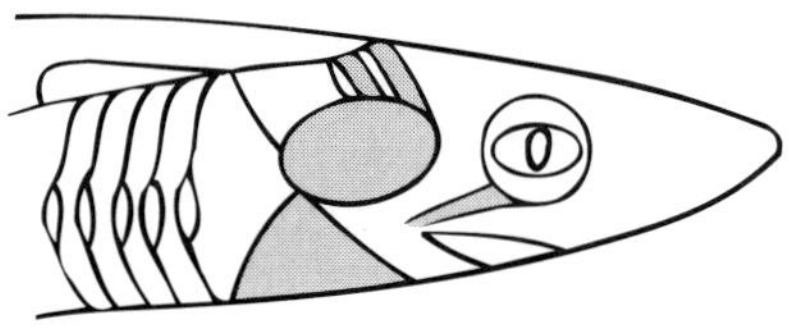 |
| 2d | VII | Levators<br>Levator hyomandibulae<br><br>Constrictors<br>(Dorsal)<br>Superficial constrictors, dorsal<br><br>(Ventral)<br>Interhyoideus | 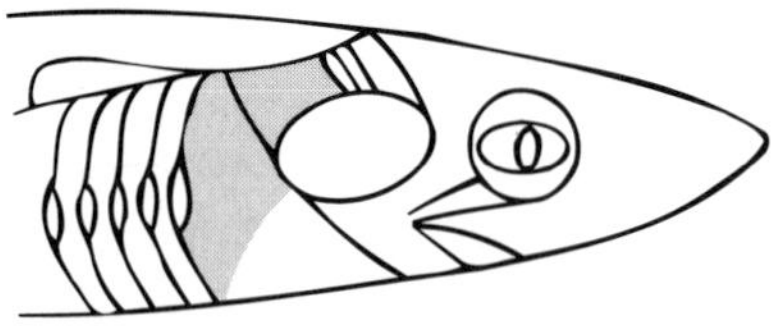 |
| 3d | IX | Levators<br><br>Constrictors<br>(Dorsal)<br>Superficial constrictors, dorsal<br><br>(Ventral)<br>Superficial constrictors, ventral | 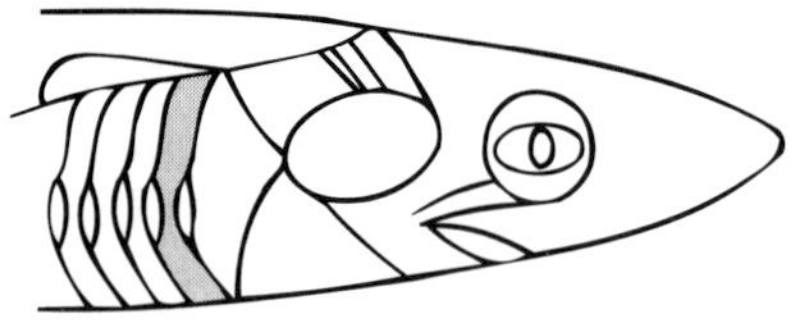 |
| 4–7th | X | Levators<br>Cucularis<br>Interarcuals<br><br>Constrictors<br>(Dorsal)<br>Superficial constrictors, dorsal<br><br>(Ventral)<br>Superficial constrictors, ventral | 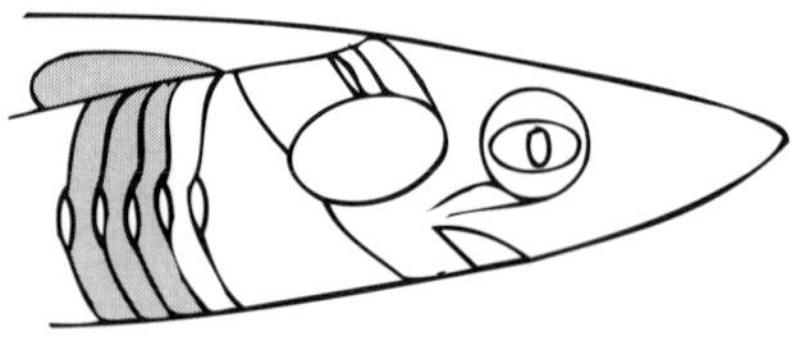 |

**Figure 10.38** Branchiomeric musculature of sharks. Specific derivatives of the levator and constrictor series of muscles in each arch are shown together with their cranial nerve supply.

into the **stapedius,** but the **digastric** functions in opening the jaw, not the stapedius. The posterior section of the digastric is derived from the ventral hyoid musculature, the interhyoideus.

The **interhyoideus** forms in the ventral part of the deep constrictor. Like its serial counterpart in the mandibles, the interhyoideus runs transversely between the lower tips of the paired hyoid. In tetrapods, it forms additional thin sheets of muscles, the **constrictor colli,** that become rather extensive layers of facial muscles in mammals. The **platysma** is an unspecialized muscle derived from the hyoid arch. Generally, it is a thin subcutaneous muscle layer spanning the throat. Other muscles derived from the hyoid arch have more specialized functions, including control of facial expression and lips during feeding. Facial control during feeding is especially important in herbivores because the lips are used to help grasp and break off parts of plants. In the rhinoceros (figure 10.39), the **levator labi superioris,** the **levator nasolabialis,** and the **depressor labi mandibularis,** respectively, part the upper and lower lips. The **zygomaticus** controls the corner of the mouth. The **orbicularis oris** closes the lips. Contraction of the **caninus** flares the nostril. The **buccinator** flattens the cheeks, thus pressing food between the tooth rows.

***Branchial Arches*** Within fishes, the levator and constrictor muscles form many discrete small muscles within the arches (figure 10.40a–c). These muscles run between branchial arch elements to control local movements during gill ventilation. The most prominent of these is the **cucullaris,** formed by the fusion of several levators on adjacent branchial arches. It is inserted along the dorsal margin of the branchial arches and extends to the shoulder girdle. In tetrapods, it extends from the axial musculature to the scapula, usually forming a pair of muscle complexes, the trapezius and mastoid groups.

The branchial arches are important structural components of the pumping and feeding apparatus in fishes. They become reduced in tetrapods and contribute only to the larynx and other parts of the throat. Associated constrictor muscles similarly become reduced to laryngeal muscles; however, some levators take on enlarged roles, contributing as the trapezius and mastoid muscles to the muscular sling supporting the shoulder girdle.

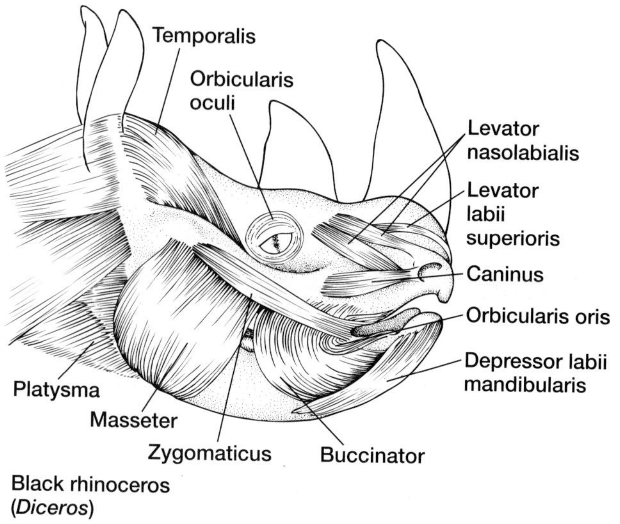


**Figure 10.39** Head of black rhinoceros illustrates facial muscles that act on the borders of the mouth and nostril.

### *Hypobranchial Musculature*

The hypobranchial musculature is supplied with spinal nerves and runs between the lower ends of the branchial arches in an anteroposterior course. In fishes, the hypobranchial musculature consists primarily of the **coracoarcuals,** so named because they originate from the coracoid region of the shoulder girdle and through various slips are inserted on the gill arches. They are prominent jaw openers and expanders of the buccal cavity. In tetrapods, they accompany the branchial arches with derivatives in the throat, muscles associated with the hyoid apparatus, larynx, and tongue.

## Overview of Muscle Evolution

Within fishes, the axial musculature predominates and is represented by segmentally arranged blocks of muscle. In tetrapods, the axial musculature is relatively reduced, and the appendicular muscles increase in prominence. Tetrapods also exhibit greater complexity within the muscle masses, as illustrated by the several hundred different muscles that differentiate from these masses.

Phylogenetic differentiation of muscles occurs in many ways. The **migration** of muscle primordia, termed **anlagen,** to other regions of the body is one way. Muscularization of the diaphragm in mammals begins with appearance of muscle anlagen in the neck that migrate posteriorly during embryonic development into the body septum behind the liver. There they differentiate into the striated muscles of the diaphragm. The hyoid musculature spreads throughout the neck and head in some tetrapods to form the platysma and other muscles of the face (figure 10.41). Another way muscles differentiate phylogenetically is by **fusion.** The rectus abdominis extending along the belly of tetrapods is formed by fusion of the ventral regions of several myotomes that grow into the abdominal region during early embryonic development. In this example, a single muscle arises from fusion of parts of successive myotomal segments. Muscles can also differentiate by **splitting.** The pectoralis muscle in the chest is a large fan-shaped muscle in lower tetrapods, but in mammals it divides into as many as four discrete muscles that act on the forelimb.

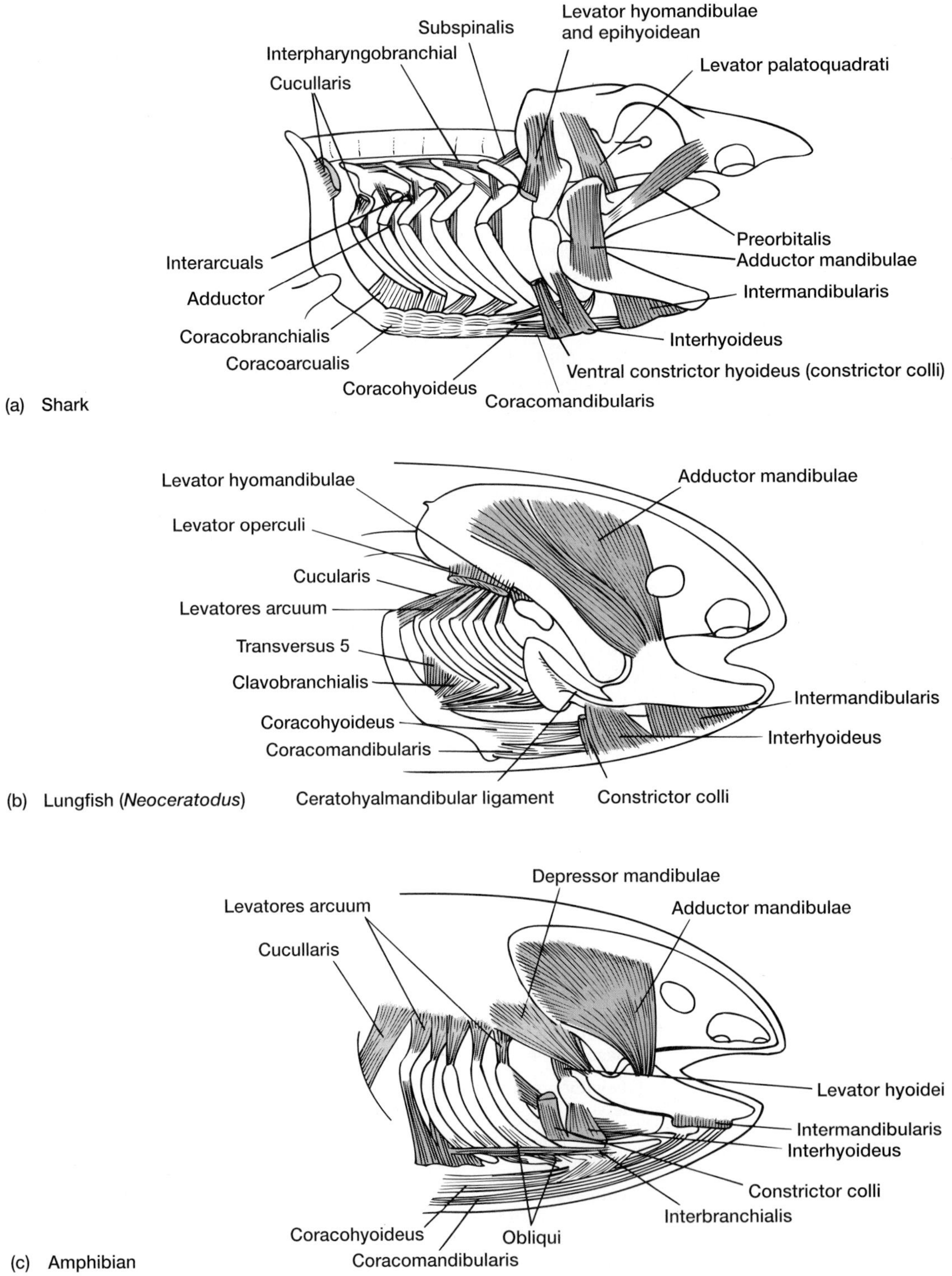


**Figure 10.40** Lateral view of head musculature. (a) The shark *Squalus*. (b) The lungfish *Neoceratodus*. (c) Amphibian (composite of adult muscles of an anuran and urodele).

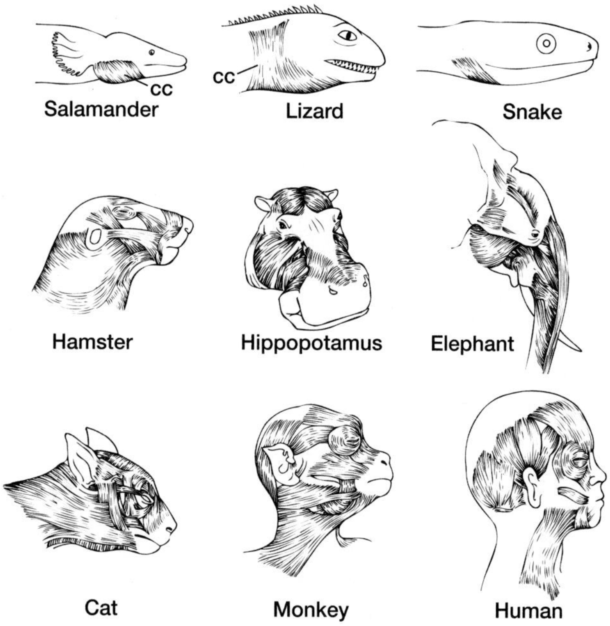


**Figure 10.41** Evolution of facial muscles. In tetrapods, the hyoid musculature expands over and partially encircles the neck as a thin sheet, the constrictor colli (cc). This muscle also tends to adhere to the dermis of the skin. In mammals, the musculature derived from the hyoid arch expands radically over the head and differentiates into a suite of facial muscles. They are best differentiated around the eyes, lips, and ears, where they serve to accent facial expression.

**Diaphragm (p. 186)**

The greater differentiation of the tetrapod musculature compared with the fish musculature reflects the changing demands of terrestrial support and locomotion. Running and flying and other tetrapod activities involve more than just the mechanics of swinging limbs or flapping wings. These are complex motor patterns that require precise control. A cheetah dashing across an irregular landscape not only must oscillate its limbs rapidly, but its foot placement must quickly accommodate slight surface irregularities each time the foot touches the ground. As the changing airstream and wind gusts suddenly alter the aerodynamic flow across a bird's wing, the bird must quickly adjust its wings and flight feathers. The more differentiated musculature of tetrapods is an indirect indication of this greater variety and precision of movements they can and must perform.

Muscle phylogeny also illustrates the remodeling character of evolution. Muscles arising initially in the jaws (e.g., trapezius, mastoid) and in the axial musculature (e.g., serratus) become incorporated into the muscular system of the shoulder and forelimb. On the other hand, we also see within muscle evolution a remarkable fidelity between muscles and their nerve supply. The phrenic nerve to the diaphragm arises anteriorly, like the muscle anlage, in the cervical region and accompanies this muscle to its final destination, posteriorly within the body. Sets of muscles associated with the branchial arches are, in general, supplied faithfully throughout vertebrates by the same cranial nerve, despite the fact that these muscles of the branchial arch are often remodeled to serve new roles in tetrapods.

## Selected References


Alway, S. E., P. Winchester, E. Davis, and W. J. Gonyea. 1989. Regionalized adaptations and muscle fiber proliferation in stretch-induced enlargement. *J. Appl. Physiol.* 66:771–81.

Bock, W. J. 1968. The mechanics of one- and two-joint muscles. *Amer. Mus. Novitates* 2319:1–45.

Bock, W. J. 1974. The avian skeletomuscular system. *Avian Biology* 4:119–257.

Bone, Q. 1966. On the function of the two fiber types of myotomal muscle fiber in elasmobranch fish. *J. Marine Biol.* (U.K.) 46:321–49.

Bone, Q. 1989. Evolutionary patterns of axial muscle systems in some invertebrates and fish. *Amer. Zool.* 29:5–18.

Dial, K. P., G. E. Goslow, Jr., and F. A. Jenkins, Jr. 1991. The functional anatomy of the shoulder in the European starling (*Sturnus vulgaris*). *J. Morph.* 207:327–44.

Gans, C., and W. J. Bock. 1965. The functional significance of muscle architecture—A theoretical analysis. *Ergeb. Anat. Entwicklgesch* 38:115–42.

Gasc, J.-P. 1981. Axial musculature. In *Biology of the Reptilia*, edited by C. Gans and T. Parsons. New York: Academic Press, pp. 355–435.

Giddings, C. J., and W. J. Gonyea. 1992. Morphological observations supporting muscle fiber hyperplasia following weight-lifting exercise in cats. *Anat. Rec.* 233:178–95.

Goslow, G. E., Jr. 1971. The attack and strike of some North American raptors. *Auk* 88:815–27.

Goslow, G. E., Jr. 1985. Neural control of locomotion. In *Functional vertebrate morphology*, edited by M. Hildebrand, D. M. Bramble, K. F. Liem, and D. B. Wake. Cambridge, Mass.: Harvard University Press, pp. 338–65.

Goslow, G. E., Jr., K. P. Dial, and F. A. Jenkins, Jr. 1989. The avian shoulder: An experimental approach. *Amer. Zool.* 29:287–301.

Hardisty, M. W. 1979. *Biology of the cyclostomes*. London: Chapman and Hall.

Herring, S. W., L. E. Wineski, and F. C. Anapol. 1989. Neural organization of the masseter muscle in the pig. *J. Comp. Neurol.* 280:563–76.

Jacobson, A. G., and S. Meier. 1984. Morphogenesis of the head of a newt: Mesodermal segments, neuromeres, and distribution of neural crest. *Dev. Biol.* 106:181–93.

Jenkins, F. A., Jr., and W. A. Weijs. 1979. The functional anatomy of the shoulder in the Virginia opossum *(Didelphis virginiana)*. *J. Zool. (London)* 188:379–410.

Jensen, D. 1966. The hagfish. *Sci. Amer.* 214(2):82–90.

Johnson, I. A. 1985. Sustained force development: Specializations and variation among the vertebrates. *J. Exp. Biol.* 115:239–51.

Johnson, I. A., W. Davison, and G. Goldspink. 1977. Energy metabolism of carp swimming muscles. *J. Comp. Physiol.* 114:203–16.

Martindale, M. Q., S. Meier, and A. G. Jacobson. 1987. Mesodermal metamerism in the teleost, *Oryzias latipes* (the Medaka). *J. Morph.* 193:241–52.

McCormick, K. M., and E. Schultz. 1992. Mechanisms of nascent fiber formation during avian skeletal muscle hypertrophy. *Dev. Biol.* 150:319–34.

Meier, S., and P.P.L. Tam. 1982. Metameric pattern development in the embryonic axis of the mouse. I. Differentiation of the cranial segments. *Differentiation* 21:95–108.

Noden, D. M. 1983. The embryonic origins of avian cephalic and cervical muscles and associated connective tissues. *Amer. J. Anat.* 168:257–76.

Noden, D. M. 1984. Craniofacial development: New views on old problems. *Anat. Rec.* 208:1–13.

Sacks, R. D., and R. R. Roy. 1982. Architecture of the hindlimb muscles of cats: Functional significance. *J. Morph.* 173:185–95.

Saltin, B., and P. D. Gollnick. 1983. Skeletal muscle adaptability: Significance for metabolism and performance. In *Handbook of physiology-skeletal muscle*, edited by L. D. Peachey, R. H. Adrian, and S. R. Geiger. Baltimore: Waverly Press, pp. 555–631.

Tam, P.P.L., S. Meier, and A. G. Jacobson. 1982. Differentiation of the metameric pattern in the embryonic axis of the mouse. II. Somitomeric organization of the presomitic mesoderm. *Differentiation* 21:109–22.

Trotter, J. A., J. D. Salgado, R. Ozbaysal, and A. S. Gaunt. 1992. The composite structure of quail pectoralis muscle. *J. Morph.* 212:27–35.

Young, B. A., D. Magnon, and G. E. Goslow, Jr. 1990. Length-tension and histochemical properties of select shoulder muscles of the Savannah monitor lizard *(Varanus exanthematicus)*: Implications for function and evolution. *J. Exp. Zool.* 256:63–74.

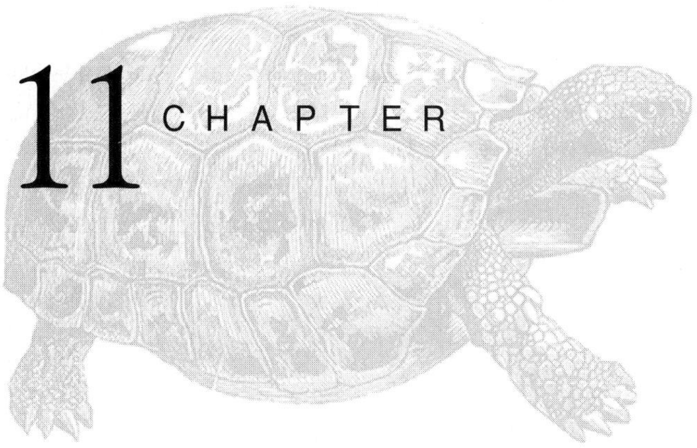

# 11 CHAPTER The Respiratory System



## INTRODUCTION

To metabolize effectively and survive, cells within the body of a vertebrate must replenish the oxygen used and rid themselves of the by-products accumulated during metabolism. These chores fall primarily to two transport systems, the circulatory and respiratory systems. The circulatory system basically connects cells deep within the body with the environment and is discussed in chapter 12. The respiratory system, our focus in this chapter, involves gas exchange between the surface of an organism and its environment. At their simplest, these two systems aid in the process of **passive diffusion,** the random movement of molecules from an area of high to an area of low partial pressure (figure 11.1a). Oxygen is usually (but not always) at high partial pressure in the environment and tends to diffuse into the organism. Carbon dioxide collects in tissues and tends to diffuse out.

Yet, if unaided, passive diffusion alone is insufficient to meet the needs of large multicellular organisms. For example, a hypothetical spherical aquatic organism could have a radius no larger than 0.5 mm for its core tissues to receive adequate oxygen by diffusion alone, even if the surrounding water were saturated with oxygen. For oxygen to diffuse passively from your lungs to your extremities, the journey would require several years! Obviously this would not do. In large multicellular organisms, transport systems aid passive diffusion. The circulatory and respiratory systems speed the process.

Major modifications in the design of respiratory organs have occurred during animal evolution to optimize the diffusion of important gases. The rate of passive diffusion between an organism and its environment depends on several factors. One is surface area. The greater the available surface area, the greater is the opportunity for molecules to

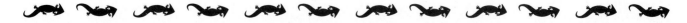

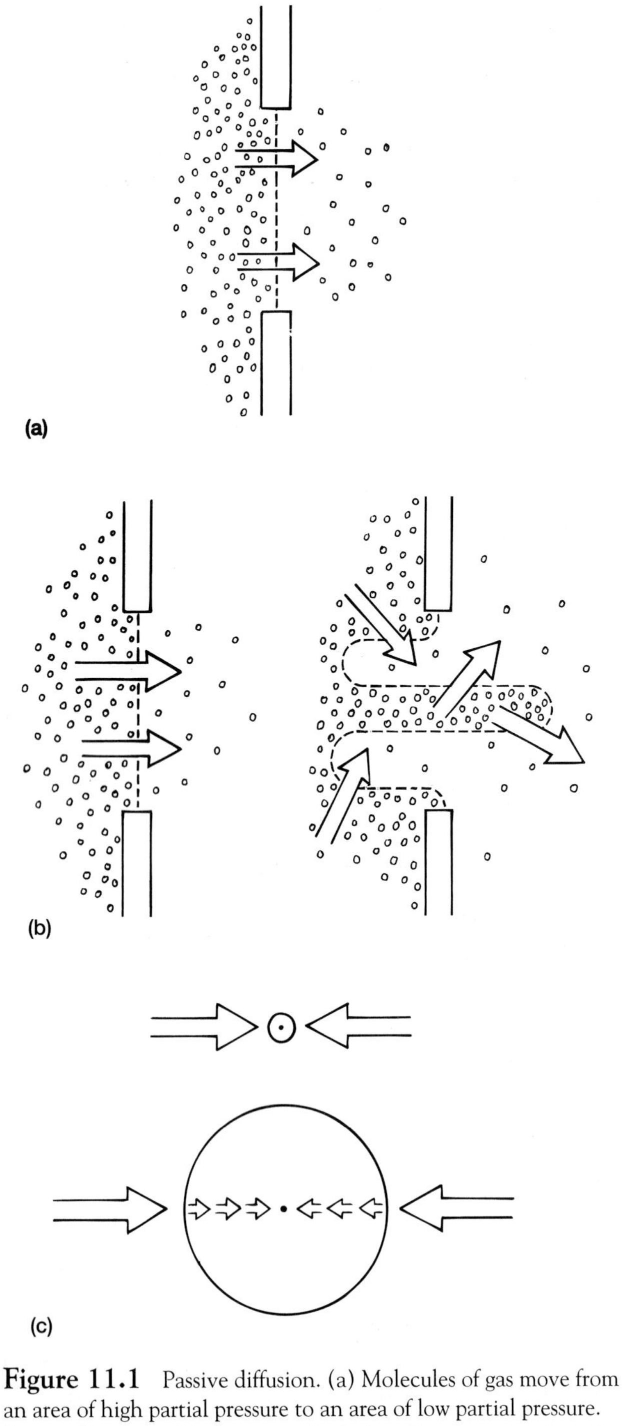

**Figure 11.1** Passive diffusion. (a) Molecules of gas move from an area of high partial pressure to an area of low partial pressure. Eventually an equilibrium is reached when the concentration of molecules becomes equal on both sides of the surface across which diffusion occurs. (b) The rate of movement of diffusing molecules depends upon available surface area. Increasing the surface area increases the rate at which diffusion occurs, although the final equilibrium concentration will eventually be the same regardless of the surface area available. (c) The time it takes for molecules to reach deep tissues depends upon the distance they must travel. Molecules moving into cells at the center of the small circle reach the core much sooner than molecules that must traverse thicker tissues to reach those cells within the center.

move across an epithelial surface (figure 11.1b). For instance, the gas exchange organs of vertebrates are highly subdivided to increase the surface available to transfer gases between air and blood. Another factor is distance. The greater the distance, the longer it will take for molecules to reach their destinations (figure 11.1c). Thick tissues slow diffusion, and thin barriers aid the process. The thin walls of the respiratory organs reduce the distance between the environment and the blood. A third factor is the resistance to diffusion by the tissue barrier itself. The moist skin of living amphibians facilitates gas transfer. In contrast to this situation, the skin of most mammals is cornified and thick, a feature that slows gas diffusion with the environment.

One of the most important factors affecting diffusion rate is the difference in partial pressures across the exchange surface. The gills of most fishes experience a high partial pressure of oxygen relative to the blood; therefore, oxygen diffuses across the gills into the blood. Occasionally fishes living in warm stagnant water may encounter a partial pressure of oxygen in the water below that of its blood. Under these unusual conditions, oxygen may actually diffuse in the reverse direction, and the fish is in danger of losing oxygen to the water!

Both respiratory and circulatory systems have "pumps" that move fluids, such as air or water (respiration) or blood (circulatory). The heart is a pump that circulates blood. In fishes, the predominant respiratory pump is the branchial apparatus that drives water across the gills (figure 11.2a). In tetrapods, one familiar pump is the rib cage, sometimes assisted by a diaphragm, that moves air through the lungs (figure 11.2b). As we will see, many types of supplementary pumping devices in vertebrates are also part of the respiratory mechanism. By moving fluids that contain gases, these pumps function to maintain high partial pressure gradients across exchange surfaces.

The respiratory and circulatory systems, although anatomically distinct, are functionally coupled in the process of **respiration,**[1] the delivery of oxygen to tissues and the removal of waste products, principally carbon dioxide. **External respiration** refers to gas exchange between the environment and blood via the respiratory surface. **Internal respiration** refers to gas exchange between the blood and the deep body tissues.

During external respiration, gases diffuse between the environment and the organism—oxygen enters, carbon dioxide departs. **Ventilation** is the active process of moving the respiratory medium, water or air, across the exchange surface. Pumping of blood through an organ via capillaries is known as **perfusion.** The **respiratory organs** specialize in ventilation to deliver oxygen and remove carbon dioxide accumulated during perfusion. The demands on the respiratory organ vary, depending on whether the medium is water or air. This is partly due to differences in density. Water, be-

[1]Biochemists have usurped the term and use it to refer to something quite different, namely, to *chemical* respiration, the aerobic degradation of substrates in biochemical pathways.

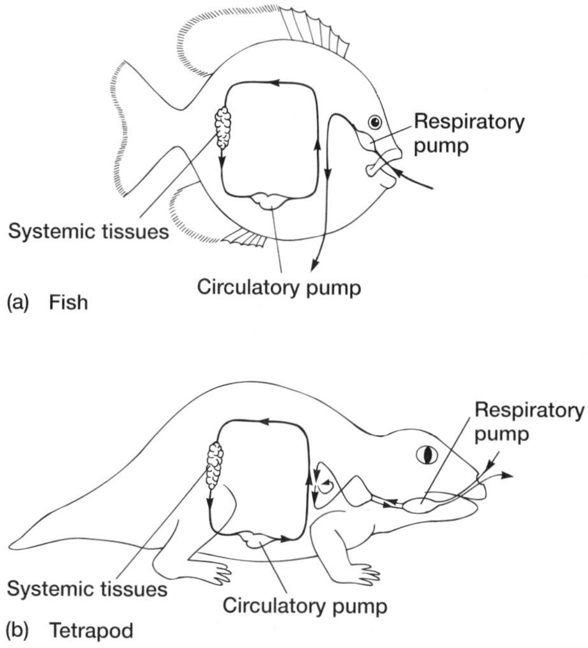


**Figure 11.2** Respiratory and circulatory systems cooperate to deliver oxygen to deep tissues and carry away carbon dioxide. Both systems are diagramed in the figure. During external respiration, air or water is inhaled and transported to the exchange capillaries of the blood. Thereafter, blood circulates oxygen to all systemic (body) tissues, represented here by a small patch of tissue, where internal respiration occurs. Oxygen is delivered to these tissues and carbon dioxide is carried away. (a) In fishes, the respiratory pump usually includes the branchial arches and their musculature. External respiration occurs in the gill capillaries. The heart, being the primary circulatory pump, drives blood through the gills and then to the systemic tissues. (b) In tetrapods, this respiratory pump can include the buccal cavity, which forces air into elastic lungs against resistance, and a rib cage around the lungs. External respiration occurs in the lungs. The circulatory pump, or heart, drives blood through vessels. Internal respiratory exchange occurs between blood and systemic tissues.

ing denser than air, requires more energy to set it in motion. Other things being equal, ventilation that involves moving water is energetically more expensive than ventilation that involves air. In addition, because water is more dense, structures are more buoyant in water than in air. Gills supported by water tend to collapse in air and therefore fail as respiratory organs on land. Lungs are structurally reinforced to work better in air.

But it is not just differences in the physical properties of air and water that affect ventilation and the devices that serve ventilation. The partial pressure of a gas in air often differs from its partial pressure in water. This means that the availability of gases to respiratory organs differs in air and water. Atmospheric air is composed of oxygen (about 21%), nitrogen (about 78%), and carbon dioxide (less than 0.03%). Trace elements make up the rest. In microenvironments, such as animal burrows, the composition may change slightly. But, in general, the partial pressure of the physiologically important gases at sea level are extremely constant worldwide. Although partial pressure varies with altitude, the composition of gases in air is relatively unchanged up to over 100 km, thanks to mixing by winds and air currents. However, in water, the situation is quite different. When brought into contact with water, these gases go into solution. The amount of gas that dissolves in water depends on the chemistry of the gas itself, its partial pressure in air, the temperature of water, and the presence of other dissolved substances. As a result, the amount of oxygen in water can be quite variable; furthermore, it is never as concentrated as it is in air.

In most fish gills, ventilation is **unidirectional.** Water enters the buccal cavity through the mouth, passes across the row of gills known as the **gill curtain,** and exits flowing in one direction only (figure 11.3a). In active fishes, ventilation is almost continuous to keep a more or less steady stream of new water washing across the exchange surfaces of the gills. Lung ventilation, however, is usually **bidirectional (tidal),** with air entering and exiting through the same channels (figure 11.3b). A fresh breath of air that is **inhaled** into the lungs mixes with spent air and is **exhaled.** The exchange capillaries of the lung are replenished intermittently, not continuously, with air.

Vertebrates that live in aqueous environments most often encounter too little oxygen, termed **hypoxia,** partly because water is already low in dissolved oxygen. For this reason, most organs that supplement respiration are found among aquatic rather than strictly terrestrial animals.

One of the major transitions in vertebrate evolution was the change from water breathing to air breathing. This major evolutionary event, together with the physiology of the respiratory system, has made respiration the focus of

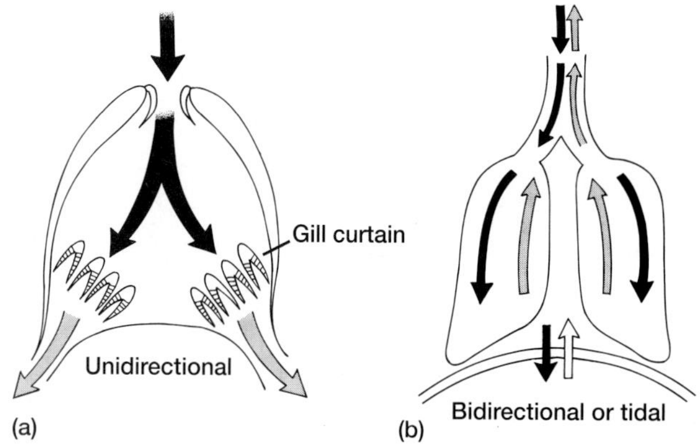


**Figure 11.3** Unidirectional and bidirectional flow. (a) In fishes and many aquatic amphibians, water movement is unidirectional because water flows through the mouth, across the gill curtain, and out the lateral gill chamber. (b) In many air-breathing vertebrates, air flows into the respiratory organ and then reverses its direction to exit along the same route, creating a bidirectional or tidal flow.

much research. Let us begin by looking at the various organs that have arisen to facilitate respiration. They have something to tell us about the evolutionary forces at work in designing the respiratory system in water, in air, and in between.

# Respiratory Organs

## Gills

Vertebrate gills are designed for water breathing. Specifically, they are dense capillary beds in the branchial region that serve external respiration. They are supported by skeletal elements, the branchial arches. The mechanism of gill ventilation depends on whether they are located internally or externally. **Internal gills** are associated with **pharyngeal slits** and **pouches.** Often they are covered and protected laterally by soft skinfolds, such as the **interbranchial septum** in chondrichthyan fishes, or by a firm **operculum,** as in many osteichthyan fishes (figure 11.4a–c). Ventilation usually involves the muscular pump of the buccal cavity actively driving water across the internal gills. **External gills** arise in the branchial region as filamentous capillary beds that protrude into the surrounding water (figure 11.4d). They are found in the larvae of many vertebrates including lungfishes, some actinopterygians, and amphibians. Water currents flow across their projecting surfaces, or in still water specialized muscles sweep external gills back and forth to ventilate them.

## Lungs

Vertebrate lungs are designed for air breathing. Lungs are elastic bags that lie within the body. Their volume expands when air is inhaled and decreases when air is exhaled. Embryologically, lungs arise as endodermal outpocketings from the gut. In primitive fishes and most tetrapods, the lungs of adults are usually paired. They lie ventral to the digestive tract and are connected to the outside environment through the **trachea.** Entrance into the trachea is gained through the **glottis,** which is guarded by tiny sets of muscles that open and close it. Usually the trachea branches into two **bronchi,** one to each lung. In some species, each bronchus branches into successively smaller **bronchioles** that eventually supply air to the respiratory surfaces within the lung. In tetrapods with slender bodies,

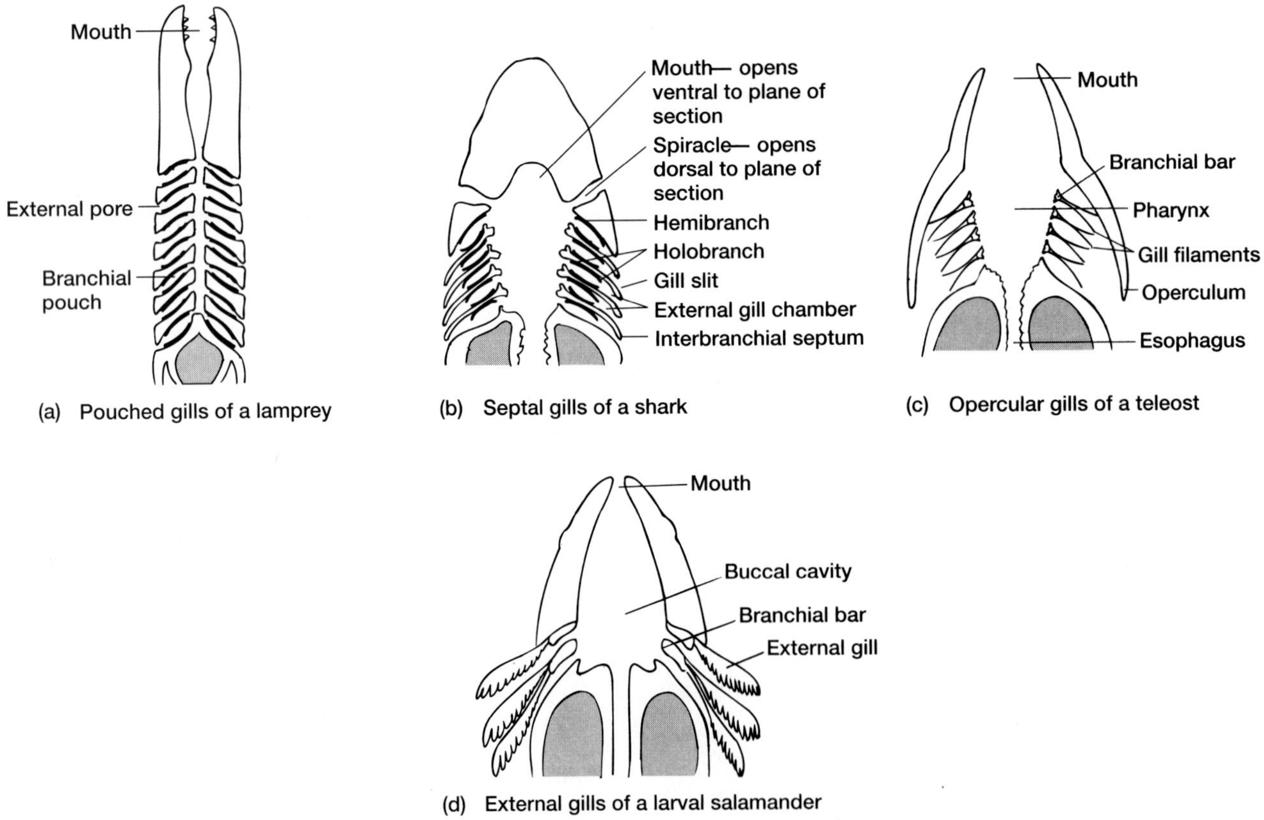


**Figure 11.4** Gill coverings. (a) Branchial pouch in lampreys. No cover protects the lateral opening of the gill chamber. (b) Septal gills in sharks. Individual flap valves formed from individual gill septa guard each gill chamber. (c) In most teleosts and some other species, a common operculum covers their several gills. (d) In larval salamanders, the branchial arches support vascular external gills that project into the surrounding water.

one lung may be reduced in size, and in some amphisbaenids and most advanced snakes, only a single lung is present.

The trachea, bronchi, and bronchioles can hold a significant volume of air. Although exhalation forces most of the spent air from the lungs, some remains in these passageways. Upon inhalation this "spent" air is drawn back into the lungs before fresh air from outside reaches the lungs to mix with the used air. This volume of used air within the respiratory passageways is called the **dead space.** The total volume inhaled in a single breath is referred to as the **tidal volume.** In a chicken, the dead space may represent up to 34% of the total tidal volume. Normal tidal volume of a human at rest is about 500 ml. Because the dead space is about 150 ml (30%), 350 ml (500 ml − 150 ml) of fresh air actually reaches the lungs.

## Gas Bladders

Many actinopterygian fishes possess a **gas bladder,** a single elongated sac located dorsal to the digestive tract. This bladder usually retains a **pneumatic duct,** its tubular connection to the digestive tract. Gas bladders are filled with air that enters via the pneumatic duct or with gas secreted into the bladder from the blood. If used to control the buoyancy of the fish in the vertical water column, they are referred to as **swim bladders.** Occasionally they may be heavily vascularized to participate in supplementary respiration and are called **respiratory gas bladders.** The internal vascular walls of respiratory gas bladders are subdivided into many partitions that increase the surface area available for external respiratory exchange.

Gas bladders differ from lungs in two ways. First, gas bladders are usually situated dorsal to the digestive tract whereas lungs are ventral, and second, gas bladders are single whereas lungs are usually paired. *Neoceratodus*, the Australian lungfish, is an exception because as an adult it has a single lung dorsal to the digestive tract; however, its trachea originates ventrally from the digestive tract. Its embryonic lung arises initially as a paired primordium, suggesting that the single lung of *Neoceratodus* is a derived condition.

Despite their differences, gas bladders and lungs share many basic similarities of development and anatomy. Both are outpocketings from the gut or pharynx and have roughly equivalent nerve and muscle supplies. Some morphologists take these similarities as evidence that lungs and gas bladders are homologous. Even if they are homologous, it is not clear which function came first, gas transfer or buoyancy control. The two functions are not mutually exclusive. A filled gas bladder that aids gas transfer also makes the fish more buoyant, and a gas bladder used for buoyancy can also be tapped as a temporary source of oxygen. Among fishes, evolutionary reversals between respiratory and buoyancy functions have occurred repeatedly. Lungs have evolved into nonrespiratory swim bladders that in subsequent evolutionary processes have reverted to respiratory gas bladders (figure 11.5).

Phylogenetically, neither lungs nor gas bladders are present in agnathans, elasmobranchs, or placoderms (figure 11.5), except in the fossil placoderm *Bothriolepis* of the late Devonian. *Bothriolepis* possessed lungs or lunglike structures, apparently derived from the posterior set of branchial pouches, comprised of a pair of ventral sacs that were connected by a common duct to the floor of the digestive tract. Whether this was unique to *Bothriolepis* or shared by other placoderms is unknown. We also do not know whether the lungs of later vertebrates arose from placoderm lungs or represent a separate, independent derivation. If they arose from placoderms, then *Bothriolepis* represents the primitive condition, and the absence of lungs in elasmobranchs represents a loss and their presence in bony fishes indicates a derived condition. Gas bladders of actinopterygians, if homologous, would be later derivatives of lungs (figure 11.5).

## Cutaneous Respiratory Organs

Although lungs and gills are the primary respiratory organs, the skin can supplement breathing. Respiration through the skin, referred to as **cutaneous respiration,** can take place in air, in water, or in both. In the European eel and plaice, oxygen uptake through the skin may account for up to 30% of total gas exchange (figure 11.6). Amphibians rely heavily on cutaneous respiration, often developing accessory skin structures to increase the surface area available for gas exchange. In fact, in salamanders of the family Plethodontidae, adults lack lungs and gills and depend entirely on cutaneous respiration to meet their metabolic needs. Like most mammals, humans respire very little cutaneously, although our skin is permeable to some chemicals applied topically (spread on the surface). In fact, many medicinal ointments are absorbed through the skin. Bats take advantage of cutaneous respiration across their well-vascularized wing membranes to eliminate as much as 12% of their total carbon dioxide waste, but they take up only 1 or 2% of their total oxygen requirement through this cutaneous route (figure 11.6). Feathers and poorly vascularized skin of birds preclude cutaneous respiration. Similarly, in reptiles, the surface covering of scales limits cutaneous respiration. However, in areas between scales (at the hinges of scales) and in areas with reduced scales (e.g., around the cloaca), the skin is heavily vascularized to allow some cutaneous respiration. Sea snakes can supplement up to 30% of their oxygen intake via cutaneous respiration across the skin on their sides and back. Many turtles pass the cold winter in hibernation safely at the bottom of ponds where the limited respiration around their cloaca is sufficient to meet their reduced metabolic needs.

The newly hatched larva of the teleost fish *Monopterus albus*, an inhabitant of southeast Asia, uses predominantly cutaneous respiration during its early life. At hatching, the large and heavily vascularized pectoral fins

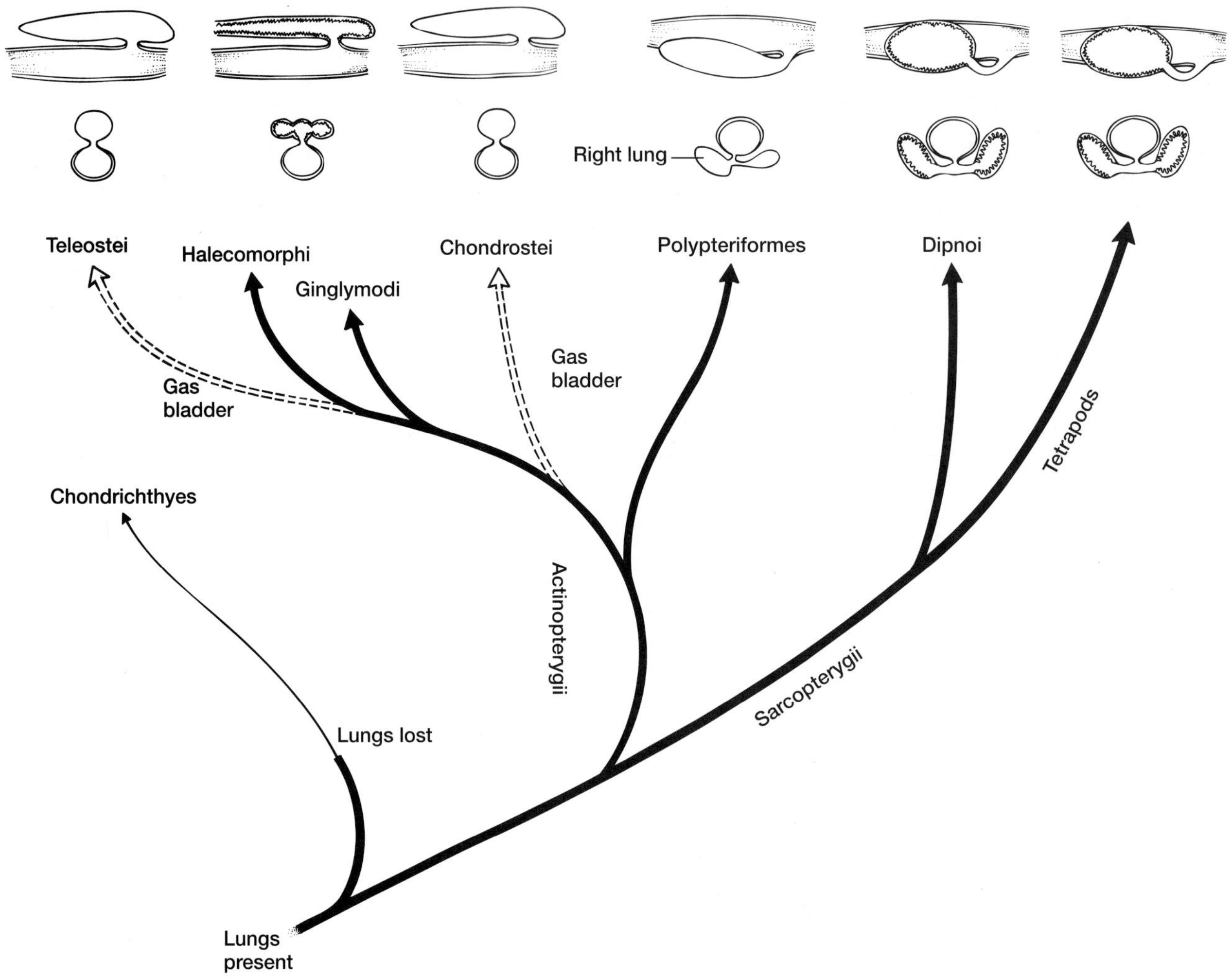


**Figure 11.5** Evolution of lungs and gas bladders. If lungs were present early in gnathostomes, their absence in chondrichthyans would be secondary and their presence in actinopterygians, sarcopterygians, and tetrapods would be a retention of these primitive lungs. Gas bladders in actinopterygians may have evolved independently, or they may have been modified from earlier lungs. Some gas bladders are respiratory in function, a feature apparently representing a secondary condition. Above the dendrogram outlining the evolutionary rise of each group, there are sagittal (top) and cross-sectional (bottom) views of the lung and its connection to the digestive tract. In Polypteriformes (*Polypterus*), paired lungs open through a common muscular glottis into the right floor of the pharynx. The left lung is reduced, the right one long, but the epithelial lining of both is smooth. Gas bladders of Chondrostei originate from the stomach and those of primitive teleosts from the esophagus, suggesting that these nonrespiratory gas bladders may be of independent origin in these two groups.

beat in such a fashion as to drive a stream of water backward across the surface of the larva and its yolk sac. Blood in superficial skin vessels courses forward. This establishes a countercurrent exchange between water and blood to increase the efficiency of cutaneous respiration in this larva (figure 11.7a). Such a respiratory organ allows the larva to inhabit the thin layer of surface water into which nearby oxygen from the air has dissolved. Similarly, in many amphibians, increased surface area allows for increased cutaneous gas exchange (figure 11.7b,c).

Countercurrent exchange (p. 146)

## Accessory Air-breathing Organs

Lungs and skin are not the only organs that tap sources of oxygen in the air. Many fishes have specialized regions that take up oxygen from the air. *Hoplosternum*, a tropical carp-like fish found in fresh waters in South America, gulps air and swallows it into its digestive tract (figure 11.8a). Oxygen in the gulped air diffuses across the wall of the

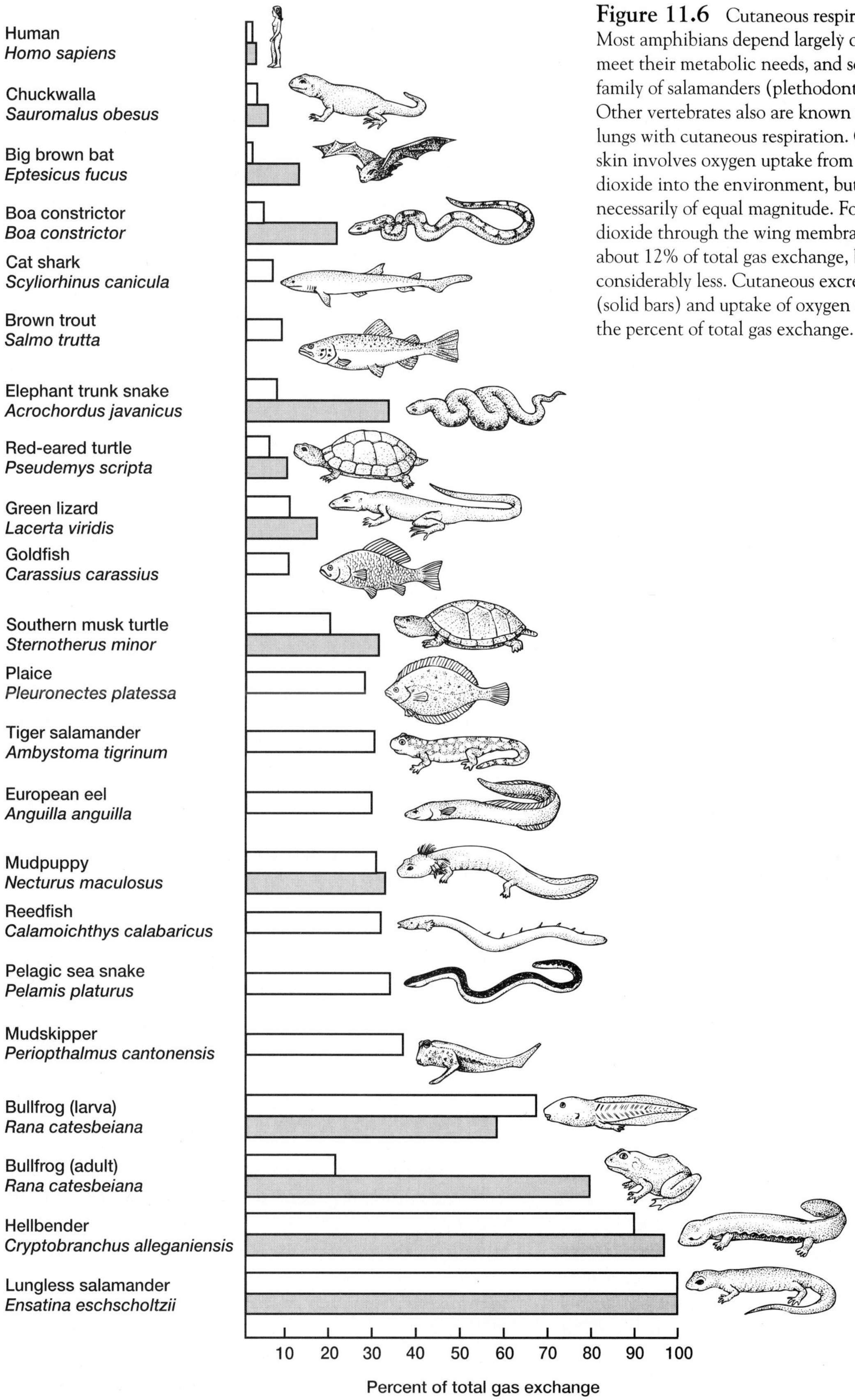


**Figure 11.6** Cutaneous respiration among vertebrates. Most amphibians depend largely on cutaneous respiration to meet their metabolic needs, and some, such as the lungless family of salamanders (plethodontids), use it exclusively. Other vertebrates also are known to supplement gills or lungs with cutaneous respiration. Gas exchange through the skin involves oxygen uptake from and release of carbon dioxide into the environment, but these exchanges are not necessarily of equal magnitude. For example, loss of carbon dioxide through the wing membranes of bats accounts for about 12% of total gas exchange, but oxygen uptake is considerably less. Cutaneous excretion of carbon dioxide (solid bars) and uptake of oxygen (open bars) is indicated as the percent of total gas exchange.

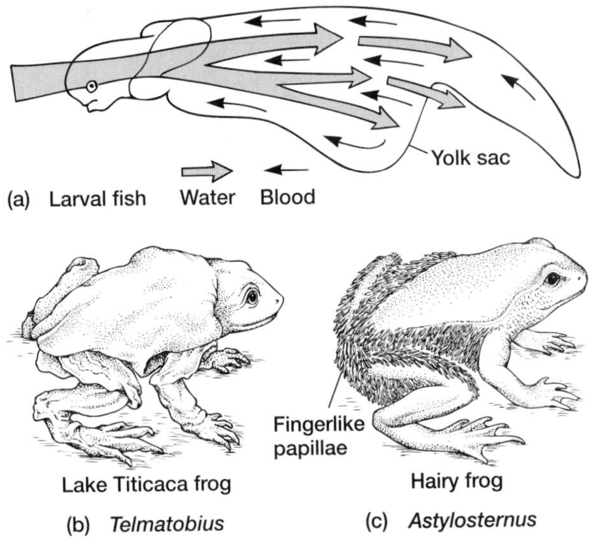


**Figure 11.7** Adaptations for cutaneous respiration. Many vertebrates exhibit complex or elaborate specializations that enhance the efficiency of gas exchange through the skin. (a) While still small, this fish larva, *Monopterus albus*, occupies the thin layer of water adjacent to the surface where oxygen levels are relatively high. Its pectoral fins beat, forcing water to flow across its body surface. Blood circulating through the skin flows in the opposite direction from the water, establishing a countercurrent exchange between blood and water. (b) In the Lake Titicaca frog, *Telmatobius culeus*, prominent loose skin folds on its back and limbs provide extensive surface area for cutaneous respiration. (c) In the male hairy frog, *Astylosternus robustus*, numerous papillae appear during the breeding season, forming a ruffled supplementary respiratory organ on its sides and hindlimbs.

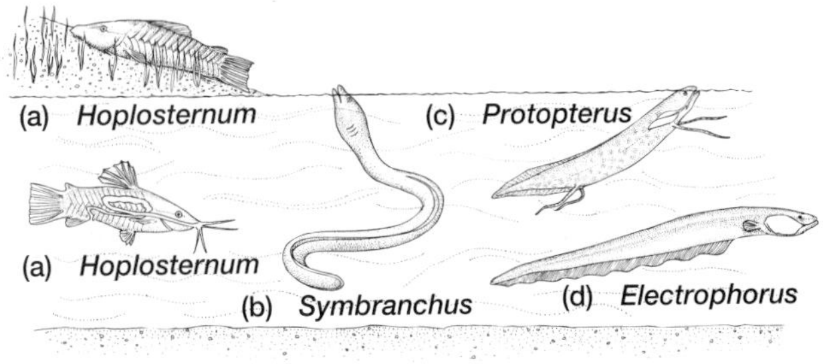


**Figure 11.8** Air-breathing fishes. Fishes that temporarily breathe air usually live in waters where oxygen depletion occurs seasonally or frequently. Gulping air supplements depressed oxygen uptake through gills and helps a fish through short periods of hypoxia. (a) *Hoplosternum*, a carplike fish, swallows air into its intestine where extra capillary beds take up this supplemental oxygen. (b) *Symbranchus* holds a gulped air bubble against its reinforced gills to take up extra oxygen. (c) *Protopterus*, a lungfish, has well-developed lungs for breathing air. (d) *Electrophorus*, an electric eel, gulps air into its mouth and takes up oxygen through the wall of its mouth.

digestive tract into the bloodstream. The digestive tract is richly supplied with blood vessels that supplement gill respiration. The electric eel *Electrophorus* gulps and holds air in its mouth to expose capillary networks of the mouth to oxygen (figure 11.8d).

Gills ordinarily are unsuitable organs for air breathing. The moist, leaflike exchange surfaces stick together in air and collapse without the buoyant support of water. However, in some fishes, gills are used in air breathing (figure 11.8b,c). The rockskipper *Mnierpes*, an inhabitant of wave-swept rocky shores of the tropical Pacific coast of Central and South America, occasionally makes brief sorties onto land to scrounge for food, to evade aquatic predators, and to avoid periods of intense wave action. During these sojourns, it holds gulped air against its gills to extract oxygen. Its gills are reinforced to prevent their collapse during these bouts of air breathing.

### Breathing and Embryos

Among anamniotes, respiration generally takes place directly between the surrounding environment and the embryo across the skin. In birds and most reptiles, the embryo is wrapped in extraembryonic membranes and enclosed in a shell. One of these membranes, the chorioallantois, lies directly beneath the shell and acts as the respiratory organ. The porous shell allows oxygen to be picked up by the blood circulating within the chorioallantois and carbon dioxide to be eliminated from it. The chorioallantois sustains the respiratory needs of the chicken embryo for most of its time in the egg (figure 11.9a,b). About six hours before hatching, the chick pokes through the inner shell membrane to push its beak into a small air space within the egg. This allows its lungs to fill for the first time and begin to participate, along with the chorioallantois, in air breathing. When the chick breaks through the outer shell several hours later, its lungs breathe atmospheric air directly and the chorioallantois quickly shuts down (figure 11.9c).

## Ventilatory Mechanisms

Whatever the exchange organ—lungs, gills, skin, or accessory structures—water or air moves actively across the respiratory surfaces to increase the rate of diffusion. A few ventilatory mechanisms are based on cilia, but most involve the action of muscles.

### Cilia

If an animal is small and its metabolic demands modest, microscopic cilia are sufficient to move water across the respiratory surfaces and support the exchange of gases between tissues and environment. Cilia line the routes along which

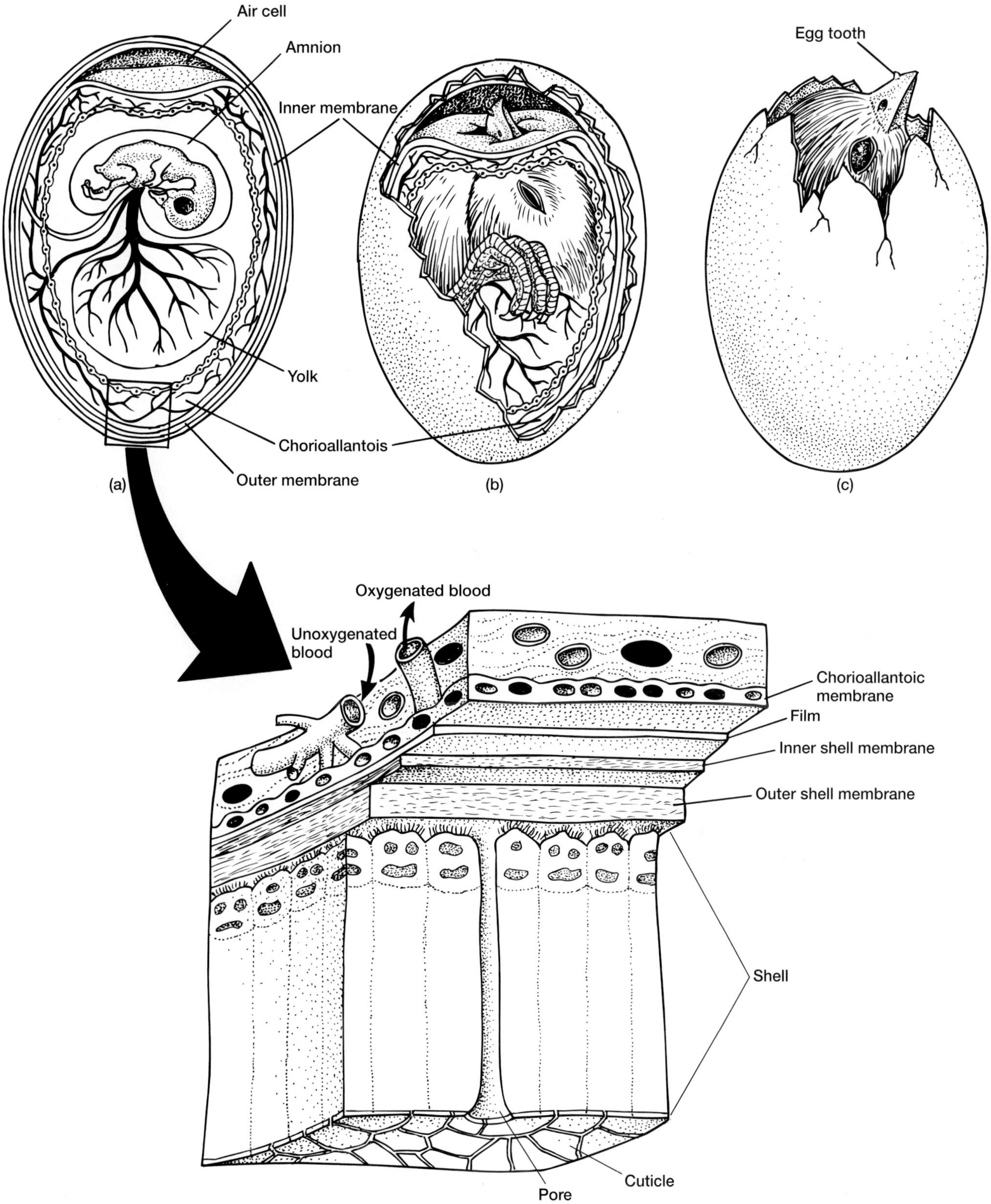


**Figure 11.9** Respiration in the chicken embryo. (a) While the chick embryo is enclosed in its shell, it respires through this porous shell. The chorioallantois carries blood to the inner surface of the shell to exchange gases at this interface. The shell proper is made up of calcite crystals pierced by tiny pores. Inner and outer shell membranes separate the shell from the vascularized chorioallantois. The chick embryo meets all its respiratory needs, up to day 19 of incubation, as air passes through the porous shell and exchanges gases with blood in the chorioallantois. (b) On day 19, the embryo pokes its beak through the inner shell membrane into the air space between both membranes. Its lungs inflate, and the chick breathes air in addition to continued respiration via the chorioallantois. (c) Six hours later, the chick pecks through the shell proper, a process termed *pipping,* to breathe atmospheric air directly. Thereafter, chorioallantoic respiration declines and the chick further cracks the shell and soon steps out.

the water current flows. Their coordinated sweeps drive water, a relatively viscous medium, through the pharynx and across the gills. Cilia, like oars, are ineffective against a relatively thin medium such as air. Furthermore, cilia are surface structures, so they are limited by available surface area. As an animal's size increases, mass increases faster than surface area and cilia become less suitable as a mechanism for moving the ventilatory current that delivers oxygen to the organism. Thus, cilia, as part of the ventilation system, are found in small aquatic organisms with low metabolic demands, such as the protochordates.

In large vertebrates, the respiratory channels often retain cilia, but they are involved in clearing surface debris that can foul the breathing device. Although "inside" the body, lungs are continuously exposed to fresh air from the outside environment. Ciliated and mucous cells are specialized to remove impurities from this air. They are interspersed throughout the lining of the lungs and secrete mucus over the lining to trap dust and particulate material. Cilia beat in coordinated patterns to move this mucous blanket laden with foreign material up the airways and into the pharynx where it is swallowed unnoticed.

Another secretion that lines the lungs (and gas bladders) is **surfactant.** Surfactant reduces surface tension at the water-air interface. This becomes an increasingly important function where there is partitioning of the interior respiratory surface. Surface tension can collapse the resulting microscopic compartments in which gas exchange takes place. Surfactant lowers this surface tension, helps stabilize these compartments, and maintains their structural integrity as elaborated surfaces for respiratory exchange.

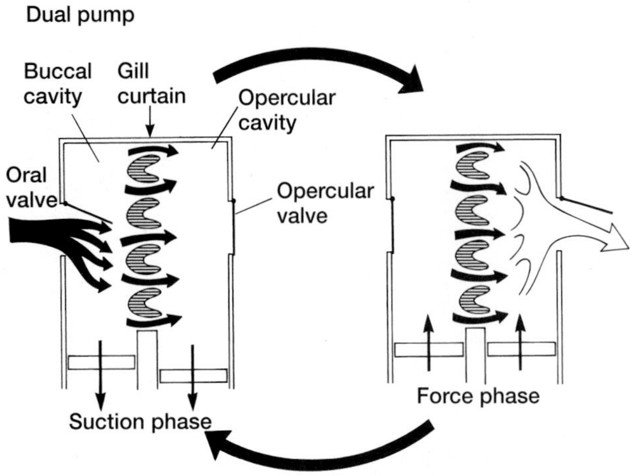


**Figure 11.10** Water-breathing fishes: dual pump. In most fishes, the buccal and opercular cavities form dual pumps on opposite sides of the gill curtain. Muscle action expands both cavities, represented by the falling pistons (downward dark arrows, left) in the suction phase. During the force phase, muscles contract to compress the cavities, represented by the rising pistons (upward dark arrows, right). As pressure within each cavity falls and rises, more water (suction phase) is drawn in and expelled (force phase). Because of the slight difference in pressure between buccal and opercular cavities, water is almost continuously moving from buccal to opercular cavity. The valves of the mouth and operculum prevent reverse flow of water. Thus, a one-way and more or less continuous flow of water across the gills is established.

## Muscular Mechanisms

Ventilation in vertebrates usually depends on muscle action. Water moving across gills ventilates them. In amphibians with external gills, muscles within or associated with the bases of the projecting gills contract to wave the gills back and forth through the water. Some swimming fishes take advantage of their forward progress through water. They open their mouths slightly, allowing water to enter and irrigate the gills. This technique by which a fish's own forward locomotion contributes to gill ventilation is known as **ram ventilation.** It is characteristic of many large, active pelagic fishes such as tuna and some sharks. More commonly, muscular pumps actively drive water or air through the respiratory organ. There are three principal types of pumps, one common in water-breathing and two found among air-breathing vertebrates.

### *Water Ventilation: Dual Pump*

In water-breathing fishes, the most common pump is a **dual pump** (figure 11.10). As the name suggests, it is two pumps in tandem, buccal and opercular, that work in a synchronous pattern to drive water in a nearly continuous unidirectional flow across the gill curtain between them. This mechanism of gill irrigation can be viewed as a two-stroke pump. The first stroke, or *suction phase,* begins with compressed buccal and opercular cavities and closed oral and opercular valves. As the buccal cavity expands, creating a low intraoral pressure, the oral valves open and outside water rushes in following the pressure gradient. The simultaneous expansion of the more posterior opercular cavity with its closed valve also creates a pressure that is even lower than in the adjoining buccal cavity. Consequently, water that first enters the buccal cavity is encouraged by the pressure differential to continue on across the gill curtain and into the opercular cavity.

During the second stroke, or *force phase,* the oral valves close and the opercular valves open. Simultaneous muscle compression of the buccal and opercular cavities raises pressure in both, but because of the open opercular valve, pressure in the opercular cavity is slightly lower. Consequently, water flows from the buccal cavity across the gill curtain and exits via the open opercular valve. The timing of the suction and force phases together with the pressure differentials between them results in a unidirectional, nearly continuous flow of new water across the gills.

## *Air Ventilation: Pulse Pump*

Fishes that occasionally gulp atmospheric air, such as lungfishes, are no different from other fishes when they are actively *water* breathing. They use the same dual pump mechanism to irrigate their gills. However, when the lungfish breathes air, the dual pump is modified into a **pulse pump** to move air in and out of the lungs. The buccal cavity pumps air in several phases, but it can be summarized in two, an exhalation phase and an inhalation phase. The **exhalation phase** begins with the *transfer* of spent air from the lungs into the buccal cavity. In some fishes, relaxation of a sphincter around the glottis permits this transfer from the lungs to the buccal cavity. Exhalation concludes with *expulsion* of air from the buccal cavity to the outside either through the mouth or under the operculum. As the fish rises and its snout breaks the surface, its mouth opens to *intake* atmospheric air, the first step in the **inhalation phase.** Inhalation concludes with *compression* which forces a bubble of fresh air from the buccal cavity into the lungs (figure 11.11).

Theoretically, this bidirectional or tidal exchange of air to and from the lungs of air-breathing fishes could be aided by the hydrostatic pressure of the water column surrounding the fish. Because surrounding hydrostatic pressure increases with depth, a fish rising to the surface with its head tipped upward experiences slightly greater pressure on its deeper body than on the buccal cavity near the surface. During exhalation, this could help force air from the lungs into the buccal cavity and out the mouth. In reverse, after the fish has gulped atmospheric air and turned downward, air in the deeper buccal cavity would be under slightly greater pressure than air in the slightly shallower lung. This could help move the bubble of freshly gulped air into the lung.

In practice, some fishes do take advantage of the hydrostatic differential in water pressure on their bodies when transferring or expelling air during exhalation. Usually this is augmented by muscle contractions within the buccal cavity and with striated muscles around the lung. However, inhalation seems to be based primarily on active contractions of the branchial musculature. First, muscle expansion of the buccal cavity helps draw air into the mouth, and then active compression generated by alternative sets of muscles forces it into the lungs.

In addition to air-breathing fishes, amphibians use a pulse pump to ventilate their lungs. Airflow is bidirectional. When an amphibian is in water, the hydrostatic pressure against the sides of its partially submerged body presses on the lung to produce a pressure that is higher than atmospheric pressure. When the amphibian exhales air, this hydrostatic pressure aids in venting the lungs. During inhalation, however, the buccal cavity must work against this force to refill the lungs. A strong, muscular buccal cavity addresses this problem of taking breaths of air while the animal is immersed in water. On the other hand, the buccal cavity on which this pulse pump is centered is also involved in feeding. As we will see later in the chapter, the dual roles

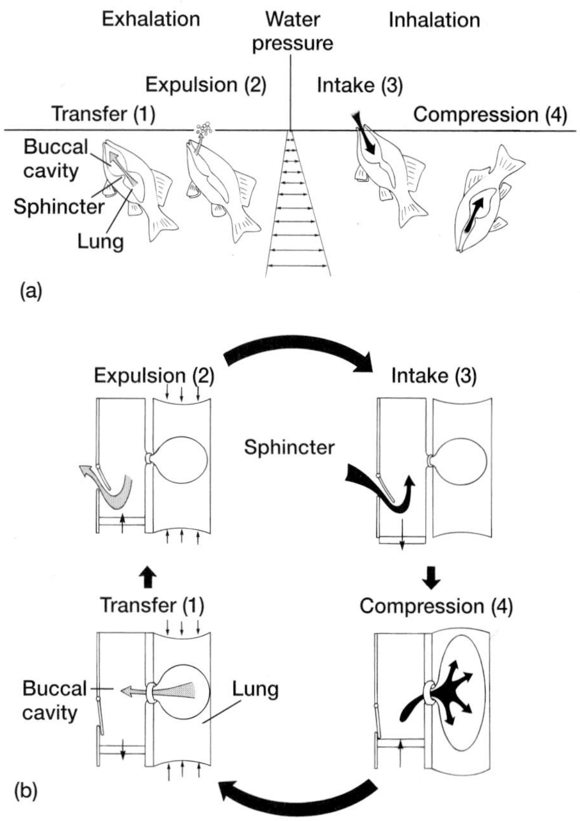


**Figure 11.11** Air-breathing fishes: pulse pump. When breathing air to supplement gill ventilation, most fishes employ a pulse pump to replenish air in their lungs. Two phases are involved—exhalation and inhalation. During exhalation, surrounding water pressure on the body walls forces air from the lungs and out the open mouth. During inhalation, the fish's head breaks the surface and its buccal cavity expands, drawing in air. Muscle compression of the buccal cavity forces the valve in the mouth closed, and positive pressure moves air into the lungs. A sphincter between the mouth and lungs closes to prevent escape of air from the lungs. (a) The pulse pump in action. During exhalation, spent air is transferred (1) from the lungs to the buccal cavity and then expelled (2). Inhalation begins with intake of air (3) and then compression (4) forces fresh air into the lungs. The increasing water pressure with depth is depicted in the center. (b) The mechanical events during exhalation and inhalation phases are diagramed. Large, solid arrows show path of airflow. Tiny arrows against the body wall illustrate the effects of water pressure on the lungs.

of the buccal cavity in feeding and ventilation can lead to conflicting demands and compromises in design.

## *Air Ventilation: Aspiration Pump*

The **aspiration pump** is a third type that does not push air into the lung against a resisting force. Rather air is sucked in, or aspirated, by the low pressure created around the lungs (figure 11.12). The lungs are located

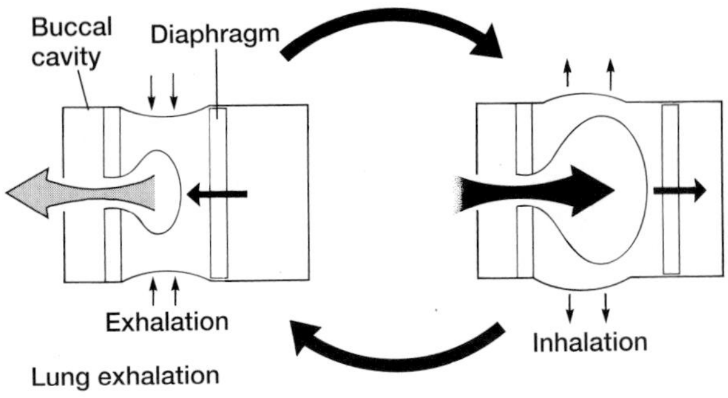


**Figure 11.12** Air-breathing amniotes: aspiration pump. In most amniotes, the buccal cavity has little to do with forcing air in or out of the lungs. Instead, a rib cage expands and compresses and/or a diaphragm moves forward and back within the body cavity to create a positive pressure that expels air or negative pressure that draws air into the lungs.

*within* the pump so that the force required to ventilate them is applied directly. The "pump" includes the rib cage and often a muscular diaphragm. A movable diaphragm in the thorax causes pressure changes rather than the action of the buccal cavity. The diaphragm, like a plunger, alters the pressure on the lungs to favor entry or exit of air.

The aspiration pump is bidirectional and moves air tidally. It is found in terrestrial amniotes—reptiles, mammals, and birds. In birds especially, the aspiration pump is highly modified. The buccal cavity is no longer part of the amniote pumping mechanism. Unlike the pulse pump, feeding and ventilation are decoupled in vertebrates using an aspiration pump. This functional decoupling increases the opportunities for independent diversification of the feeding and ventilation mechanisms.

# Phylogeny

## Agnathans

As in cephalochordates, the ammocoete larva of the lamprey depends on cilia-lined channels to gather food collected. However, unlike cephalochordates, the feeding-ventilation current of water is produced by pumps composed of muscular **velar folds,** or **velum,** and by compression and expansion of the branchial apparatus (figure 11.13a). Closure of the velum and muscle compression of the branchial apparatus drive water across the gills and out the pharyngeal slits. Relaxation of these same muscles allows the elastic branchial apparatus to spring back into its expanded shape, thus drawing in exterior water past the open velum. The pharyngeal openings are small and round, not long slits like those in amphioxus. There are usually seven pairs of slits. Flaps of skin cover these openings, which act as valves. Although water can exit through them, inward water movement forces them closed; thus reverse flow is prevented (figure 11.13b).

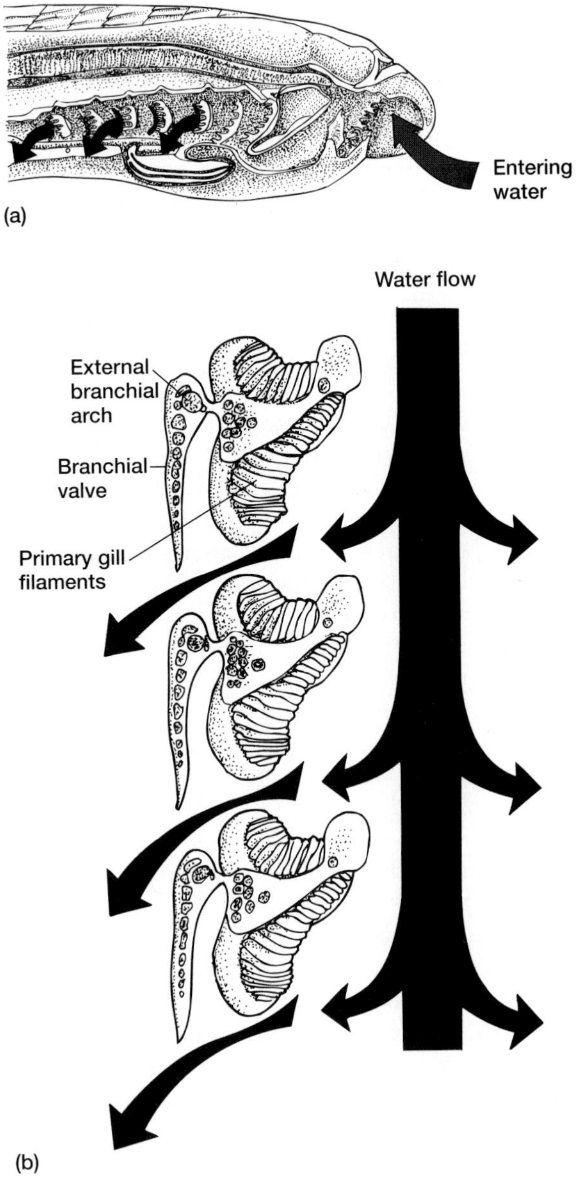


**Figure 11.13** Ventilation in the ammocoete larva. (a) The muscular velum draws water into the mouth and forces it through the pharyngeal slits and across the gills before exiting to the outside. (b) Frontal section through three branchial arches showing position of gills and direction of water flow.

Unlike the lateral gills of gnathostomes, the gills of ammocoetes lie medial to the branchial arches. Each gill includes a central partition, the interbranchial septum, that supports a set of **primary lamellae (gill filaments)** on its anterior and posterior sides. Each filament is extensively subdivided into numerous tiny platelike **secondary lamellae** that contain the respiratory capillary beds. The current of water is directed across the sides of these secondary lamellae. Blood flowing within the capillary beds of the lamellae courses in the opposite direction. Thus, water and blood passing in opposite directions establish a countercurrent system between them to improve gas diffusion.

In many species, the adult lamprey is a short-lived reproductive stage that does not feed and dies soon after breeding. In species with a prolonged adult stage, the adult feeds by attaching its circular mouth to the sides of living prey. The tongue is used to scrape flesh. In such species, the mouth grips the prey, making it unavailable for entry of water to ventilate the gills. Instead, water exits *and* enters through the pharyngeal slits (figure 11.14a,b). Muscle compression and relaxation of the branchial apparatus drive this water, which moves tidally in and out of the branchial pouch via the associated slits, unlike in most fishes. A partition that divides the pharynx into a dorsal esophagus connected to the digestive tract and a ventral water channel that furnishes the branchial pouches prevents the mixing of food and water of respiration.

In hagfishes, no major expansions and contractions of the branchial apparatus occur. Instead, scrolling and unscrolling of the velum, one on each side, together with synchronized contractions and relaxations of the branchial pouches produce a current of water that enters via the nostril and nasopharyngeal ducts, flows in one direction across the gills, and exits (figure 11.15a). In cross section, the velum is shaped like an inverted T (figure 11.15b,c). Its sides furl and unfurl to produce the current of water that enters through the nostril and courses posteriorly through the

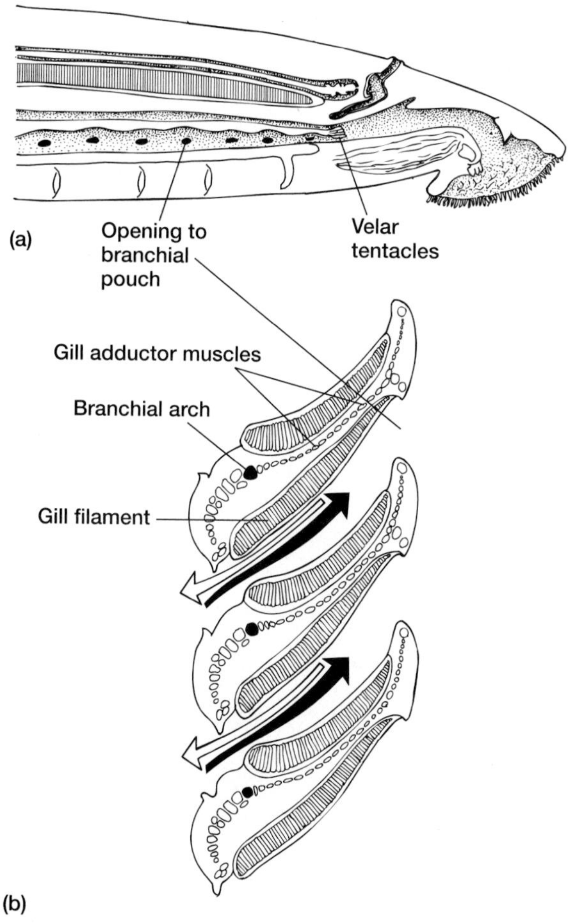


**Figure 11.14** Ventilation in the adult lamprey. (a) Longitudinal section. Because the adult lamprey's mouth often is attached to prey, water must alternatively enter as well as exit via pharyngeal slits. Thus, unlike most fishes, gill ventilation in the lamprey is tidal. (b) Frontal section of three gill arches. Double-headed open arrows indicate tidal flow of water.

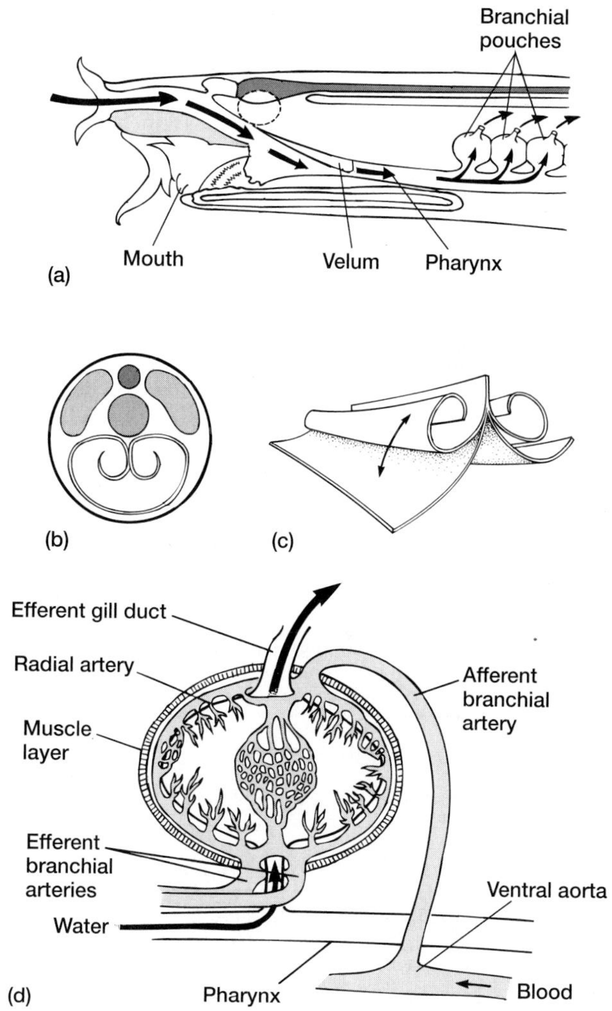


**Figure 11.15** Ventilation in the hagfish. (a) Longitudinal section. Water (indicated by arrows) enters via the nostril, not the mouth, to reach the pharynx. The scroll-shaped velum, rolls up and down as the branchial pouches contract to drive this current across the gills and out the gill pores. (b) Cross section of the scroll-shaped velum. (c) Lateral view of velum scrolling and unscrolling to move water through the pharynx. (d) An individual branchial pouch showing the sites of entry and exit of water and the position of the capillary beds within. The muscular walls of these pouches are compressed by contraction but expanded by elastic recoil.

branchial pouches. The branchial pouches are defined by an outer muscular wall that encloses the gill lamellae. Afferent blood vessels supply the lamellae and efferent branchial vessels drain them (figure 11.15d). The current of water propelled by the velum and by the pumping action of these branchial pouches flows across the gill lamellae and out the common branchial duct.

## Elasmobranchs

As in all gnathostome gills, elasmobranch gills lie lateral to the branchial arch. Each gill consists of a central partition, the interbranchial septum, covered on each face by primary lamellae (gill filaments). The primary lamellae are composed of standing rows of secondary lamellae. Across their sides flows water that irrigates the gills. Like ribs of a fan, **gill rays** within the septum give it support. The term ***holobranch*** refers to a branchial arch and the lamellae on both anterior and posterior faces of its septum. A gill arch with lamellae on only one face is a **hemibranch.** Facing plates of lamellae on adjacent gills constitute a **respiratory unit** (figure 11.16a,b).

Among elasmobranchs, the respiratory mechanisms of sharks has been studied most. Ventilation is based on a dual pump mechanism that creates alternating negative (suction) and positive pressures to draw water in and then drive it across the gill curtain. Pressures recorded on either side of the gill curtain within the buccal and **parabranchial** compartments reveal the efficiency of this dual pump. Although pressures rise and fall in each cavity, the pressure is always relatively lower in the parabranchial cavity, located lateral to the gills, than in the buccal cavity, located medial to the gills. In addition to bringing new pulses of water into the mouth, the dual pump mechanism of the shark also maintains a nearly constant pressure difference between buccal and parabranchial compartments. As a result, the pressure oscillations of the dual pump are converted into a smoother, almost continuous unidirectional irrigation of the gills (figure 11.17a–d). Blood coursing within the capillaries of the secondary lamellae sets up a countercurrent or perhaps a crosscurrent pattern, promoting efficient gas exchange.

**Countercurrent and crosscurrent exchange (p. 146)**

In sharks swimming in open waters, ram ventilation can add to gill irrigation and nearly replace the dual pump during such times.

What is embryologically the first gill slit is reduced to a small oval opening, the **spiracle,** which carries a much reduced hemibranch, sometimes referred to as a **spiracular pseudobranch.** In bottom-dwelling skates and rays, the ventral mouth may be partially buried, leaving the dorsally

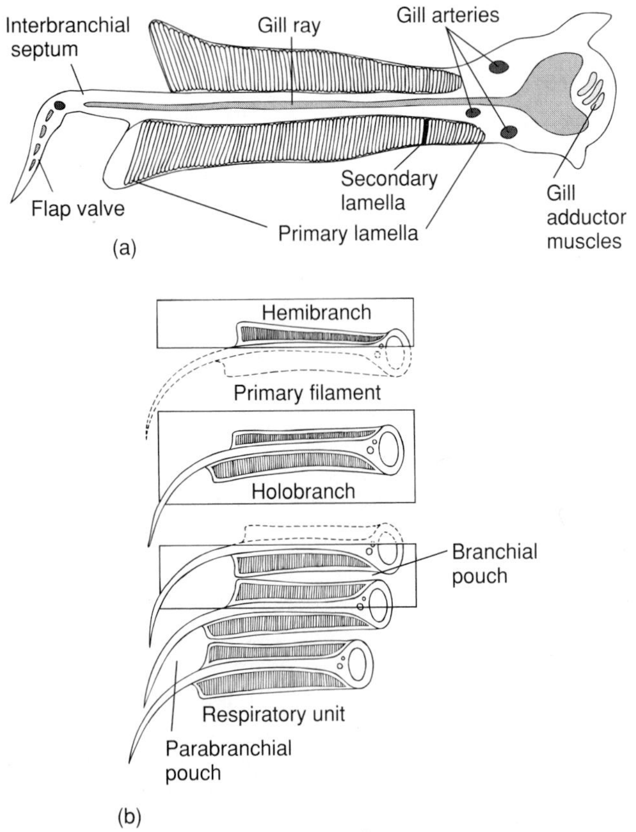


**Figure 11.16** Shark gill. (a) The interbranchial septum has banks of lamellae supported by gill rays and a medial branchial arch. (b) Structural units include a hemibranch and a holobranch as well as a functional respiratory unit.

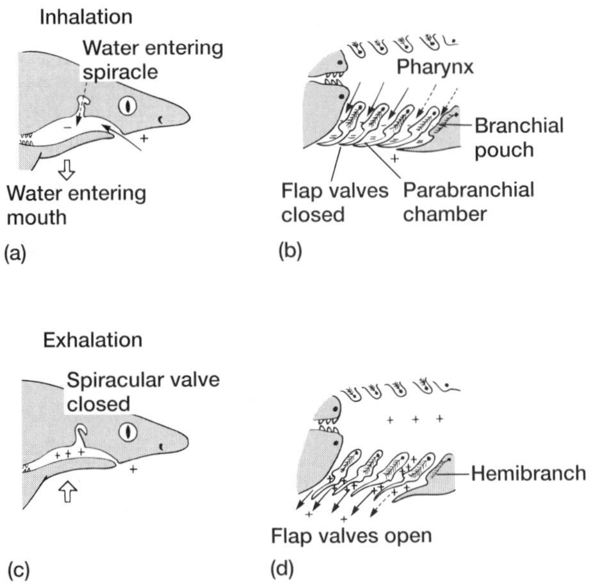


**Figure 11.17** Gill ventilation in a shark. Lateral (a,c) and frontal (b,d) views. Relative positive and negative pressures are indicated by + and −, respectively. The ventilation mechanism consists of a buccal pump that draws water in and forces it across the gill curtain and out. Notice that the flap valves close during inhalation and that relative pressures are always lower in the parabranchial chamber than in the pharynx. Thus, water moves unidirectionally across the gills in a pulsing but continuous flow.

placed spiracle in an unobstructed position where it allows water for gill irrigation to enter. The spiracle possibly may play a role in chemical sampling of the passing stream of water as well. For most other elasmobranchs, the function of the spiracular pseudobranch is unknown. In sharks, it probably does not have a respiratory function because blood supplying the pseudobranch comes from an adjacent fully functional gill and is already oxygenated.

Holocephalians (ratfishes) lack spiracles altogether. They also differ from other elasmobranchs in having a single extensive flap of skin, or operculum, covering all the branchial arches rather than individual flap valves over each pharyngeal slit.

## Bony Fishes

The operculum of osteichthyans is bony or cartilaginous. It provides a protective cover over the branchial arches and gills they support. In addition, the operculum is part of the dual pump used to ventilate the gills.

In cross section, each gill is V shaped and composed of primary lamellae (gill filaments) that are subdivided into secondary lamellae and supported on a branchial arch. Tiny adductor muscles cross between filaments to control the arrangement of adjacent gills that govern the flow of water across the secondary lamellae (figure 11.18a). As in most other fish gills, the blood in the secondary lamellae flows one direction and water flows in the opposite direction to establish a countercurrent exchange (figure 11.18b).

Fishes that ventilate a gas bladder do so by gulping and forcing fresh air through the pneumatic duct. Usually, a fish expels spent air as it approaches the water's surface, captures and swallows a new gulp of fresh air, and descends again. In the jeju, a freshwater fish of the Amazonian region, the anterior muscular compartment of the gas bladder is connected to a posterior compartment through a sphincter. As the jeju breaks the surface, the fresh air it gulped into the buccal cavity is forced along the pneumatic duct and preferentially enters the anterior chamber of the gas bladder (figure 11.19a,b). The sphincter closes and spent air in the posterior chamber exits. Finally the sphincter opens, and the muscular walls of the anterior chamber contract, forcing the new air into the vascularized posterior chamber (figure 11.19c,d).

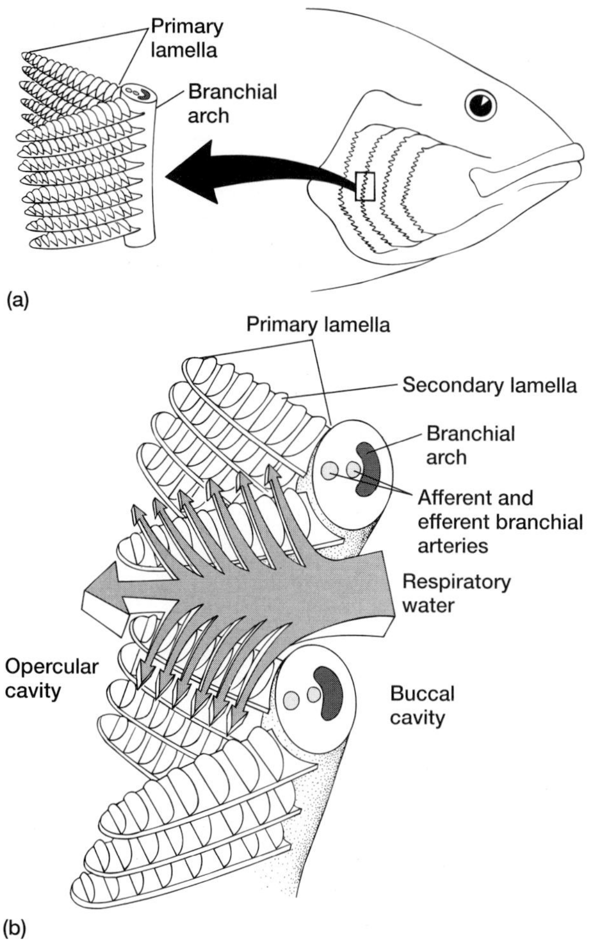


**Figure 11.18** Gill ventilation in teleost. (a) One gill bar is removed, showing the stack of gill lamellae. (b) Water flow is directed across the secondary lamellae opposite to that of blood flowing within each secondary lamella, establishing a countercurrent exchange between them.

## Overview of Fish Respiration

### *Gills*

In water-breathing fishes, different devices have arisen to serve a common function, driving a stream of water across vascularized gills. Compression and expansion of the branchial apparatus irrigates lamprey gills tidally. Scrolling of a velum moves water across the gills of hagfishes. Ram ventilation occurs in active fishes swimming forward through the water. They open their mouth, permitting oncoming water to enter and pass across the gills. In gnathostomes, the most common device serving gill irrigation is the dual pump. The branchial arches and their associated muscles are the central components of this pump. Because they are also involved in feeding, the design of the branchial apparatus represents a compromise between the demands of feeding and ventilation.

### *Lungs and Gas Bladders*

Air-filled sacs arise early in fish evolution and serve respiratory and hydrostatic functions. In lungfishes and tetrapods, the respiratory function predominates. In the African lungfish, *Protopterus*, the trachea arises from the floor of the esophagus, bends around the right side of the esophagus and joins the lungs in their dorsal position within the body cavity, a location that is more favorable to buoyancy control. The lungs are subdivided into faveoli (figure 11.20a,b). Air forced into these lungs exchanges with capillary blood circulating in the walls of the faveoli.

## Box Essay 11.1 Mouth in the Sand

In most bony fishes, gill irrigation is based on a dual pump that draws water into the mouth, across the gill curtain, and out under the operculum. However, some fishes with specialized feeding habits exhibit a modified ventilation mechanism, similar to that of the parasitic lamprey. An example is the sturgeon *Acipenser,* whose mouth is used as a protrusible suction tube for probing and feeding in muddy bottom sediments. When the sturgeon is not feeding, gill ventilation occurs as it does for a bony fish—water enters the mouth, moves across the gill curtain, and out the opercular opening (box figure 1a). However, when feeding, the sturgeon's mouth is buried in bottom sediments so it cannot breathe. Under these circumstances, water enters the buccal cavity not through the feeding mouth, but through a permanent aperture at the upper margin of the operculum. The water then turns and passes across the gill curtain in the normal direction to exit out the customary opercular opening (box figure 1b). Curiously, although it is present and theoretically available, the spiracle accounts for very little of the water that enters during these alternative ventilatory movements during feeding.

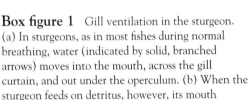

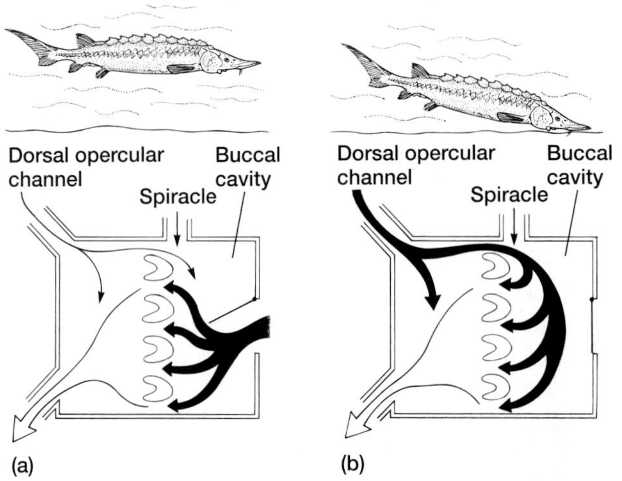


**Box figure 1** Gill ventilation in the sturgeon. (a) In sturgeons, as in most fishes during normal breathing, water (indicated by solid, branched arrows) moves into the mouth, across the gill curtain, and out under the operculum. (b) When the sturgeon feeds on detritus, however, its mouth cannot serve as an entrance portal for water. During these times, water instead enters along a dorsal opercular channel to sweep across the gill curtain (solid, branched arrows) and then out the normal ventral channel under the operculum.

In actinopterygian fishes, the hydrostatic function became more pronounced as these fishes entered new adaptive zones of the marine environment and encountered a new array of selective forces. To understand this, we need to examine why a fish might require a hydrostatic organ.

Most fishes are denser than the water in which they live, so they tend to sink. If their skeletons are highly ossified, as in bony fishes, the high density of bone makes this sinking tendency more pronounced. It is not surprising that almost all osteichthyans possess some form of gas bladder (or lung). Air-filled gas bladders give buoyancy to the fish body and help resist its tendency to sink. Gas bladders are usually absent among bottom-dwelling bony fishes and fishes of open water, such as tuna and mackerel, that swim continuously.

A gas bladder that serves primarily in controlling buoyancy is a swim bladder. In primitive teleosts, the swim bladder is **physostomous,** retaining its connection to the digestive tract via the pneumatic duct through which air is released and taken in (figure 11.21a). In most advanced teleost fishes, this connection is lost and the swim bladder is a closed bag of gases called a **physocleistous** swim bladder (figure 11.21b). Both types adjust the buoyancy of the fish to varying water depth.

The volume occupied by the swim bladder determines its buoyancy and its ability to compensate for the greater density of the fish body. Because water pressure increases with depth, the thin-walled swim bladder tends to be compressed when a fish descends and expanded when it rises. Thus, if the swim bladder is to maintain a constant

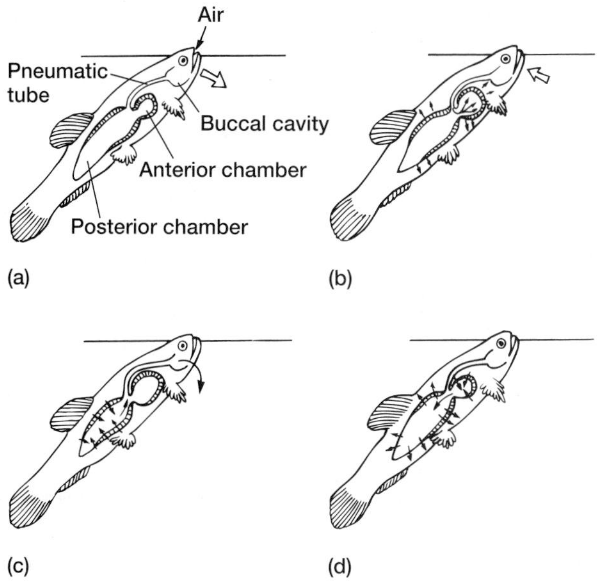


**Figure 11.19** Air-breathing fishes. Most air-breathing fishes use a pulse pump to fill their air bladders or lungs, which are able to separate spent and incoming air during ventilation. The mouth breaks the surface (a) so that air drawn in along the pneumatic tube preferentially enters the anterior air chamber (Ac) (b). Spent air in the posterior chamber is forced out through the pneumatic tube and exits under the operculum (c). The sphincter between the anterior and posterior chambers opens, allowing air to replenish the posterior chamber as well (d).

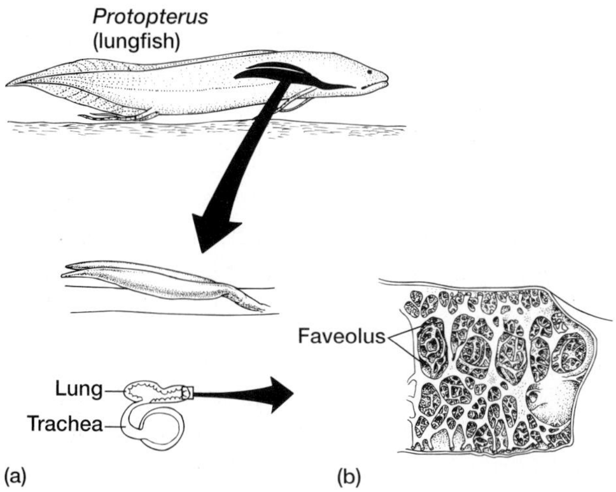


**Figure 11.20** Lungs of the lungfish *Protopterus*. (a) View of the lungs from the right side and in cross section. (b) Enlargement of the internal wall of the lung. The lung is subdivided internally, forming small compartments or faveoli. Faveoli are most numerous in the anterior part of the lung. Approximate location of the lungs is indicated by the darkened area (top) in the lateral view of the fish's body.

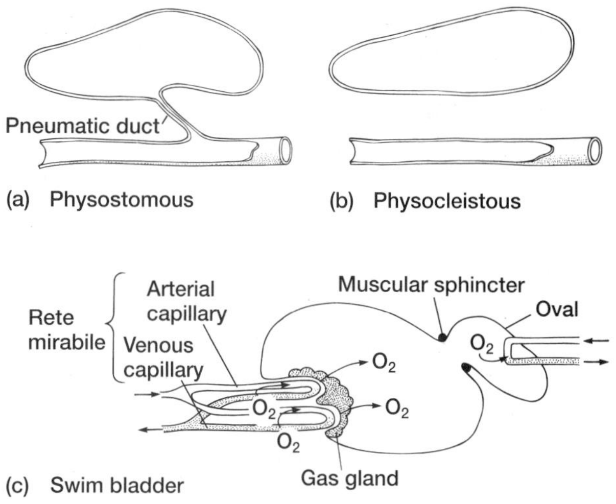


**Figure 11.21** Swim bladders. (a) Physostomous swim bladders retain their connection to the pharynx via the pneumatic duct. Air volume can be controlled if a fish gulps in more air or releases extra through the pneumatic duct. (b) In the physocleistous swim bladder, the connecting pneumatic duct has been lost. Air volume, and hence buoyancy, is controlled if more gas is released into the bladder at the rete mirabile or if some is removed at the oval. (c) The rete mirabile is a knot of capillaries. As blood leaves the gas gland of the swim bladder via the venous capillaries of the rete, lactic acid is added. This reduces hemoglobin's affinity for oxygen. Oxygen, therefore, tends to diffuse out and enter adjacent arterial capillaries passing blood to the rete. Consequently, the oxygen concentration builds in the arterial blood as it approaches the gas gland so that the partial pressure of oxygen in the arterial capillaries of the rete is high when it reaches the gas gland. This encourages oxygen release into the swim bladder.

volume, gas must be added when a fish dives and removed when it surfaces. Fishes with physostomous swim bladders can do this by gulping extra air or releasing spent air via the pneumatic duct. More commonly, gas secretion occurs directly across the walls of the bladder. Some swim bladders have special **gas glands** from which gas from the blood is released into the bladder. In the gas gland, blood vessels form a countercurrent capillary arrangement, the **rete mirabile** (figure 11.21c). Incoming arterial and outgoing venous capillaries within this rete lie next to one another in the gas gland. Experiments on gas secretion into the swim bladder suggest that the mechanism involves lactic acid. During passage through the gas gland, lactic acid is added to the blood leaving the gland, increasing this blood's acidity. Increased acidity reduces the solubility of gases and the affinity of hemoglobin for oxygen. As a result, the partial pressure of oxygen in the venous capillaries is higher than the partial pressure of oxygen arriving in the adjacent arterial capillaries. Oxygen diffuses into the arterial capillaries, raising its partial pressure before the arterial blood flows into the gas gland. As the process is repeated, the partial pressure of oxygen in the arterial capillaries of the rete builds until it exceeds the partial pressure of oxygen in the swim bladder; therefore oxygen is released into the bladder (figure 11.21c).

Resorption of gas often involves specialized regions. In many advanced teleosts, there is an **oval,** a pocket at one end in which gas is absorbed back into the blood. During resorption, blood vessels to the oval dilate and the smooth muscle sphincter dividing it from the rest of the bladder opens. Gas at high partial pressure in the bladder can now come into contact with the vascular walls of the oval, be taken up by the blood, and removed from the bladder.

Generally, the gases in swim bladders (78% nitrogen, 21% oxygen) are similar in composition to the gases in air, at least when the bladder is filled initially with gulped air. Among fishes with physocleistous swim bladders, which do not gulp air, the gas composition varies. In fishes living at great depths, the gas in the swim bladder is mainly oxygen. In trout and other salmonids, nitrogen is at very high proportions in the swim bladders regardless of the depth at which they live.

Swim bladders also have secondary functions. In some fishes, the bladder is connected to the hearing apparatus and aids in sound detection. Some fishes produce sounds within the swim bladder or use it as a resonator. Releasing air by belching is one source of sound. Grinding of the teeth is another. Sounds can cause the swim bladder to vibrate, or it may amplify or resonate them. Other fishes have specialized muscles that strum the bladder itself to produce a sound. Because males have specialized muscles that females do not, it is thought that the resulting sounds are part of territorial or courtship displays.

**Sound detection by the swim bladder (p. 696)**

No elasmobranch has a swim bladder. The tendency of these fishes to sink is addressed in a different fashion. A cartilaginous skeleton avoids the added density of extensive ossification. In addition, two other sources counteract the tendency to sink. One is the fins. Elasmobranchs have broad pectoral fins and can change their angle to the flow of water in order to steer their body up or down. The heterocercal tail, as it sweeps back and forth during swimming, produces lift and compensates for the fish's density along with the pectoral fins. A second source of lift is generated by an oil **(squalene)** consisting of lipids and hydrocarbons. Oils are lighter than water, so they reduce the density of the elasmobranch. As any student who has dissected a shark knows, the copious oil permeates the large liver. In some sharks, liver oil alone can constitute 16 to 24% of the body weight. Squalene, by reducing the density of the elasmobranch, reduces the energy needed for swimming because the pectoral fins and heterocercal tail need not devote as much of their efforts to compensating lift.

**Heterocercal tails (p. 292)**

# Amphibians

In modern amphibians, the skin is a major respiratory organ, and in some species, it is the exclusive respiratory organ. The skin is moist and the layer of keratin relatively thin, allowing easy diffusion of gases between the environment and the rich supply of capillaries within the integument.

The significance of cutaneous respiration in modern amphibians is almost certainly greater than it was in early amphibians. Many early amphibians, like the crossopterygian fishes from which they arose, had scales, that would have obstructed gas exchange through the skin. Ancient amphibians likely depended on lungs for respiration. Many, including *Ichthyostega,* had prominent ribs encircling the thorax, which suggests, but does not prove, that these early amphibians moved their ribs to ventilate their lungs. However, in modern amphibians, ventilation depends not on ribs but on pumping movements of the throat to irrigate gills or fill lungs.

In aquatic amphibians, pharyngeal slits often persist with internal gills. Feathery external gills are often present as well, especially among larval amphibians. Most, but not all, amphibians have lungs for breathing air. The respiratory surface within the lungs is usually developed best anteriorly, and it decreases posteriorly along the inner walls. This surface is **septal,** meaning that partitions form and subdivide to increase the surface area exposed to incoming air. The interconnecting septa divide the internal wall into compartments, **faveoli,** that open into the central chamber within each lung. Faveoli differ from the alveoli of mammalian lungs in that they are not found at the end of a highly branched tracheal system. Faveoli are internal subdivisions of the lung wall that open into a common central chamber. Inspired air travels along the trachea into the central lumen of the lung and from here diffuses into the surrounding faveoli. Capillaries located within the thin septal walls of the faveoli take up oxygen and give up carbon dioxide.

## *Amphibian Larvae*

Salamander larvae typically have both internal and external gills. Pumping action of the throat irrigates the internal gills with a unidirectional stream of water across their surfaces. Feathery external gills are held out in the passing current, allowing water to flow across them. If there is no current or if water is stagnant, the larvae can wave their gills back and forth through the water to irrigate the capillary beds they carry.

Larvae of anurans employ buccal and pharyngeal force pumps to produce a unidirectional flow of water across the gills and generate a food-bearing current. The "piston" for the buccal part of this pump includes enlarged elements of the splanchnocranium (ceratohyal, copula, hypobranchial plate). These elements articulate with the palatoquadrate, which acts as a fulcrum about which they rotate to expand

## Box Essay 11.2 Blowholes and Breathing

***"It [spermaceti] had cooled and crystallized to such a degree, that when, with several others, I sat down before a large Constantine's bath of it, I found it strangely concreted into lumps, here and there rolling about in the liquid part. It was our business to squeeze these lumps back into fluid. A sweet and unctuous duty! No wonder that in old times this sperm was such a favorite cosmetic. Such a clearer! such a sweetener! such a softener! such a delicious mollifier! After having my hands in it for only a few minutes, my fingers felt like eels, and began, as it were, to serpentine and spiralize."***

Herman Melville, *Moby Dick*

The sperm whale was the special quarry of whaling ships. Besides its blubber, the large spermaceti organ in its head was a special prize because of the large quantity of high-quality oil it yielded, up to 4 tons in some large males. This organ helps the sperm whale alter its buoyancy during deep dives.

When the whale is at the surface, air in its lungs is replenished by venting and then refilling them through long nostrils that open at the blowhole. When the whale exhales, the exiting warm air condenses, giving the appearance of spouting streams of water. When it dives, the sperm whale may reach a depth of a mile or more in search of its favored quarry, large squid. Deep diving gives it access to a resource unavailable to most other large predators. By warming or cooling its spermaceti organ, the oil inside can be changed from a liquid to a crystalline state and back again and thereby change the buoyancy of the whale.

The spermaceti organ is richly invested with large arteries supplying dense capillary networks. It is hypothesized that a whale at great depth might augment its ascent by shunting warm blood through this organ. If the oil begins to melt, it turns from a crystalline state to a slightly expanded liquid state, thereby increasing positive buoyancy and contributing to lift. When the whale descends from the surface, cold seawater might be flushed through the nasal passageway lying within the spermaceti organ to cool and hence crystallize the oil. This would increase density and help descent.

The two nasal passages are different. The left seems to be a direct route from blowhole to throat by which to fill and vent the lungs. However, the right nasal passage runs forward from the blowhole into a vestibular sac at the front of the snout beneath the blubber. From this sac, this nasal passageway broadens into a wide tube as it courses posteriorly through the spermaceti organ, to reach the back of the throat where it joins the left nasal passage. Before joining its partner, the right nasal passage expands into a second sac, the nasofrontal sac (box figure 1). Perhaps by shunting or flushing seawater between vestibular and nasofrontal sacs at either end of the spermaceti organ, the right nasal passage might allow selective cooling of the oil within the spermaceti organ and thus fine-tune the sperm whale's descent or establish its neutral buoyancy at a desired depth.

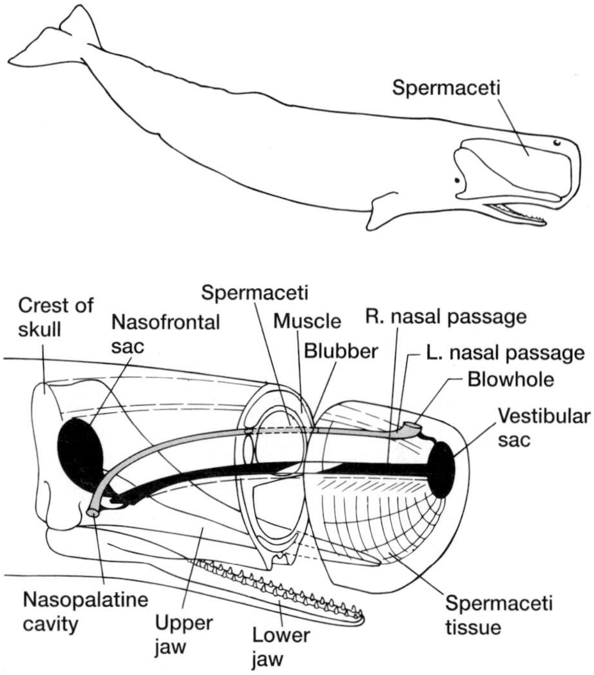


**Box figure 1** Sperm whales. The large head of sperm whales holds the spermaceti tissue, which is impregnated with oil. The two nasal passages are asymmetrical. The left nasal passage runs from the nasopalatine cavity to the blowhole. The right runs through the spermaceti organ. Cool water channeled through the spermaceti organ causes surrounding oil to crystallize and become more dense. Warming of the organ causes the reverse, reformation of the oil into a liquid and increased buoyancy. By such changes, the whale can descend or rise to the surface and float there comfortably.

and compress the buccal cavity (figure 11.22a,b). The action of muscles on the pharyngeal pump is not yet understood but seems to involve compression and expansion of this cavity.

The basic mechanism of amphibian gill ventilation includes a buccal cavity and a pharyngeal cavity separated from each other by a valve, the velum. The buccal cavity is separated from the mouth by the **oral valve** and from the nares by an **internal narial valve.** Inhalation depresses the floor of the buccal cavity, which lowers the pressure within it. The velum closes temporarily to prevent entry of water into the pharyngeal cavity, but water fills the buccal cavity through the mouth and nares (figure 11.23a). Near the end of the inhalation stage, pharyngeal constriction causes a rise in pressure within the pharyngeal cavity relative to the buccal cavity. This keeps the velum closed and pushes water across the gill curtain. The exhalation stage begins with elevation of the floor of the buccal cavity, raising the pressure within it and forcing the oral and narial valves closed. Nearly simultaneous expansion of the pharyngeal cavity drops the pressure within it relative to the buccal cavity. Consequently, water in the buccal cavity pushes open the velum and refills the pharyngeal cavity, displacing the water within it (figure 11.23b). As with water-breathing fishes, the gills of frog tadpoles see an almost continuous unidirectional stream of water across their surfaces.

In some tadpoles, such as those of the tailed frog *Ascaphus truei*, the prominent oral sucker around the mouth is used to grip the surface of rocks in the fast-flowing streams in which these tadpoles live. A sucker that is firmly attached prevents entry of water through the mouth. However, action of the floor of the buccal cavity draws water in via the nares and then forces it across the gills before exiting (figure 11.24a). This same action of the buccal cavity, together with valves guarding the mouth, removes water from the area of the oral sucker to produce the low pressure that helps hold the tadpole to the rock (figure 11.24b).

## *Amphibian Adults*

When the amphibian larva undergoes metamprphosis into an adult, gills are lost. Cutaneous respiration continues to play an important role in meeting respiratory demands after metamorphosis, and lungs, if present, are ventilated by a buccal pump.

The four stages of lung ventilation in frogs are best understood. In the first stage, the buccal cavity expands to draw fresh air in through the open nares (figure 11.25a). In the second stage, the glottis opens rapidly, releasing spent air from the elastic lungs. This air streams across the buccal cavity with little mixing and is vented through the open nares (figure 11.25b). In the third stage, the nares close, and the floor of the buccal cavity rises, forcing the fresh air held in this cavity into the lung through the open glottis (figure 11.25c). In the fourth stage, the glottis closes, retaining the

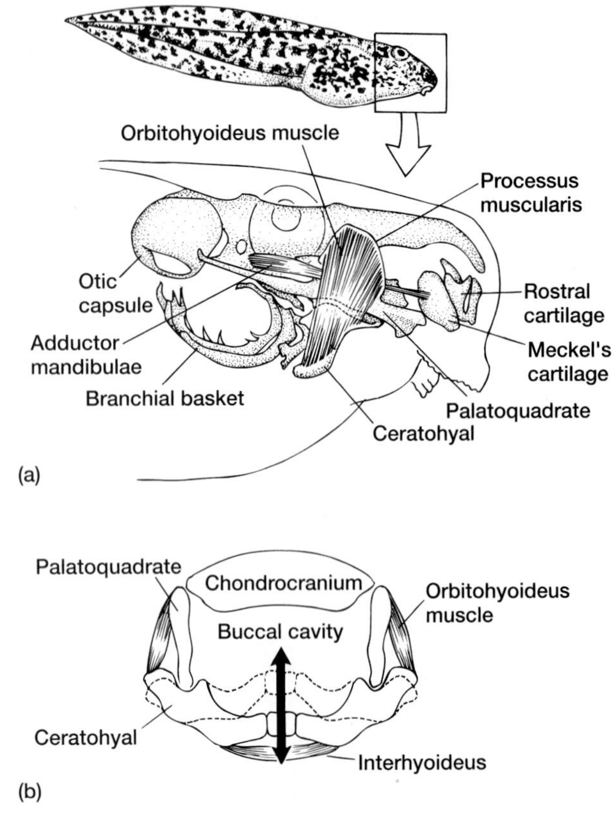


**Figure 11.22** Ventilation of tadpole gills. (a) The chondrocranium and major components of the visceral cranium are illustrated. (b) The floor of the buccal cavity is raised and lowered (double-headed arrow) to produce the movement of water. Two sets of muscles are primarily responsible. The orbitohyoideus depresses the floor, and the interhyoideus elevates it.

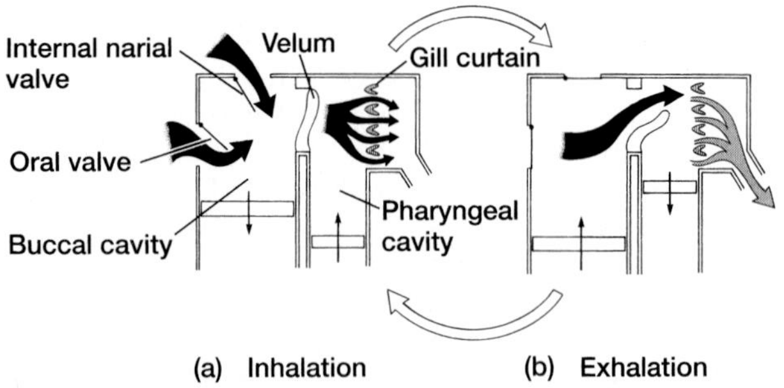


**Figure 11.23** Schematic diagram of ventilation in a frog tadpole. Between buccal cavity and pharyngeal cavity stands the velum, a flap that permits only one-way flow of water from buccal to pharyngeal cavities (but not the reverse). (a) During inhalation, the floor of the buccal cavity drops (down arrow), which brings water in via the mouth and nares. Elevation of the floor (up arrow) of the pharyngeal cavity closes the velum and forces water across the gill curtain. (b) During exhalation, elevation of the buccal floor (up arrow) forces the oral and narial valves closed and pushes water past the open velum into the pharyngeal cavity.

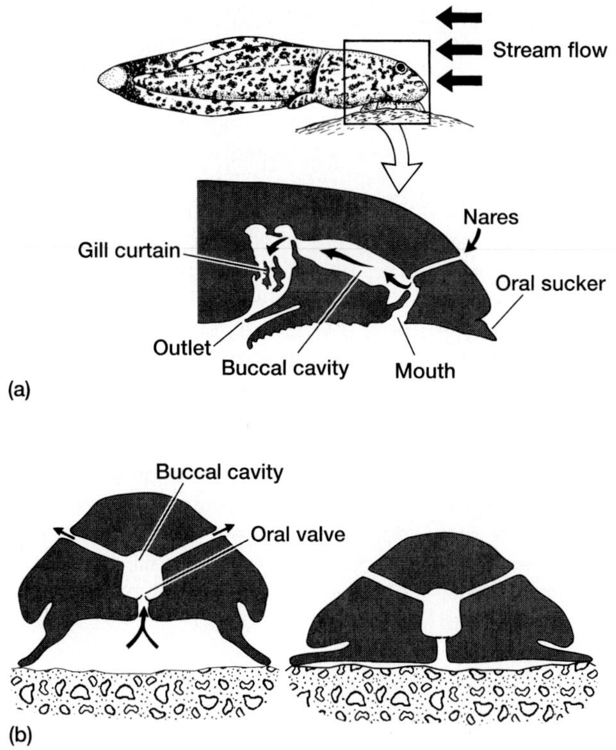


**Figure 11.24** Gill ventilation in the tailed frog larva. The tadpole uses its extensive oral sucker around its mouth to establish a secure attachment to the undersurface of a rock in a fast-moving stream (solid arrows). (a) When the oral sucker is attached, water (solid arrows) to irrigate the gills enters through the nares, passes through the buccal cavity across the gill curtain, and then exits. (b) Water removed from the area to which the oral suction was attached creates a vacuum that helps the sucker hold the rock. The oral valve prevents a break in this seal.

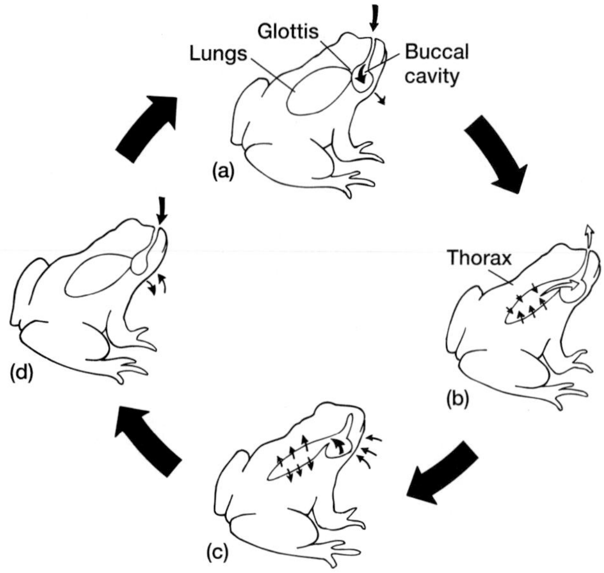


**Figure 11.25** Lung ventilation in the frog. (a) The frog's throat drops to replenish the air in the buccal cavity. (b) As the glottis opens, the thorax is compressed, forcing spent air from the lungs past that held in the buccal cavity and expelling it (open arrows). (c) Elevation of the throat and closure of the nares forces fresh air from the buccal cavity into the lungs. (d) Pumping of the throat flushes the buccal cavity.

air that has just filled the lungs and the nares open again. Between cycles, the buccal cavity may oscillate repeatedly (figure 11.25d). This rapid oscillation was once thought to turn the lining of the mouth temporarily into an accessory breathing organ. However, experimental evidence refutes this. The capillaries lining the mouth do not serve in gas exchange. Instead, such buccal oscillations between lung fillings serve mainly to flush the buccal cavity of any stray residue of expired air in the mouth following each ventilatory cycle.

The pulse pump, and hence the buccal cavity, in frogs is also deployed in producing vocalizations that play a key role in the social organization and breeding success of frogs. Evolutionary modifications of the buccal cavity consequently affect three significantly different functions.

Opinions differ about how close in function the buccal pump of frogs is to the pulse pump of lungfishes. Certainly they differ in subtleties. For example, exchange of spent air in the lungs and fresh air held in the mouth seems to be more efficient in frogs. However, the similarities are striking. In both frogs and lungfishes, movement of the hyoid apparatus aids in filling of the buccal cavity, and spent air expelled from the lungs crosses this same chamber. In both groups, fresh air is pushed into the lungs against pressure. To some extent then, frogs have retained the basic pattern of lung filling deployed by lungfishes. However, all this changes in reptiles, birds, and mammals. The mechanism of ventilation in these groups is the aspiration pump, a departure from that of amphibians and earlier air-breathing fishes.

## Reptiles

Pharyngeal furrows and occasionally pharyngeal slits appear during the early embryonic development of reptiles, but they never become functional after birth. In some groups, supplemental cutaneous respiration is significant, but for the most part, paired lungs meet their respiratory needs.

The lungs of snakes and most lizards typically include a single central air chamber into which faveoli open (figure 11.26 a,b). Like purse strings, cords of smooth muscle define and encircle the opening into each faveolus. The thin walls of each carry capillary beds and may be subdivided by even smaller internal septa. Sometimes the faveoli are reduced in the posterior part of the lung, leaving it as a nonexchange region. In monitor lizards, turtles, and crocodiles, the single central air chamber itself is subdivided into numerous internal chambers that receive air from the trachea. These internal chambers are ventilated by respiratory movements,

## Box Essay 11.3 Frog Songs

In addition to the buccal cavity and lungs, frog vocalizations involve a third compartment, the vocal sac, a chamber opening off the floor of the buccal cavity. Access to it is gained through a muscle-controlled slit. Contractions of the body wall force air from the lungs, through the larynx, into the buccal cavity, and through the open slit into the vocal sac, inflating it. Next, contractions of muscles on the floor of the buccal cavity reverse the path of air so it flows back from the vocal sac to the buccal cavity, through the larynx, and to the lungs to reinflate them (box figure 1a–c).

In the toad *Bufo valliceps,* the larynx consists of a pair of arytenoid cartilages embraced by the circular cricoid cartilage. The arytenoid cartilages form a unit between the cornua of the hyoid. The laryngeal constrictor muscle originates from the hyoid cornua and is inserted on the arytenoid cartilage near the glottal opening. Upon contraction, it spreads the arytenoids to widen the opening. Anterior and posterior laryngeal muscles form a strap across the front and back of the arytenoids. When they both contract, they slip across the arytenoids toward their middle, having the greatest mechanical advantage at this spot to close these cartilages. Cooperative action of this dilator and these constrictor muscles affects airflow and modulates sound production.

As air is forcefully shifted back and forth between lungs and vocal sac, the nares are closed to prevent temporary escape of air. If the vocal sac is large, as in some species, then several pulses of filling are often used to inflate it fully. The paired vocal cords are two thin strips of tissue within the larynx, each held by an arytenoid cartilage and stretched across the airflow. As air rushes out of the lungs across the vocal cords, the cords and often nearby margins of the larynx are vibrated. The inflated vocal sac serves as a resonating chamber to modulate the sound produced. In a few species, sound is produced as the lungs fill, but in most species, sound is produced when air exits from the lungs.

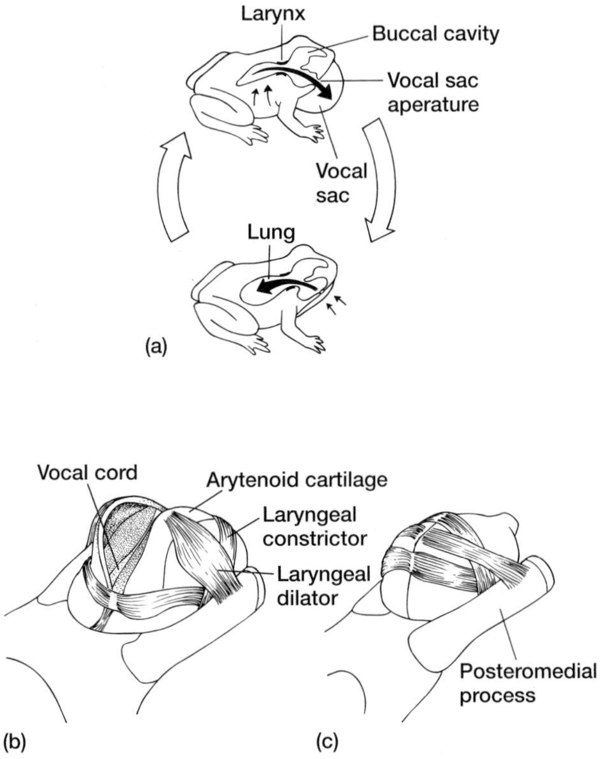


**Box figure 1** Frog songs.
(a) Musculature in the body wall forces air out of the lungs through the larynx and into the buccal cavity. From the buccal cavity, air enters the vocal sac via an aperture. Compression of the throat forces this air back along the reverse route into the lungs. (b) Larynx opened. (c) Larynx closed.

whereas the exchange of gas between the faveoli and these chambers appears to occur by diffusion.

Filling of the lungs in all reptiles is based on an aspiration pump mechanism, although the anatomical parts that actually participate may differ. The aspiration pump acts on the walls of the lung to change its shape and induce airflow in or out. Ribs alter the shape of the body walls around the lungs, and intercostal muscles running between these ribs move them. In lizards, for instance, sets of intercostal muscles actively move the ribs forward and outward during inhalation. The result is to enlarge the cavity around the lungs, decrease pressure within them, and draw air into the lungs. During active exhalation, different sets of intercostal muscles contract to fold the ribs back and inward, thus com-

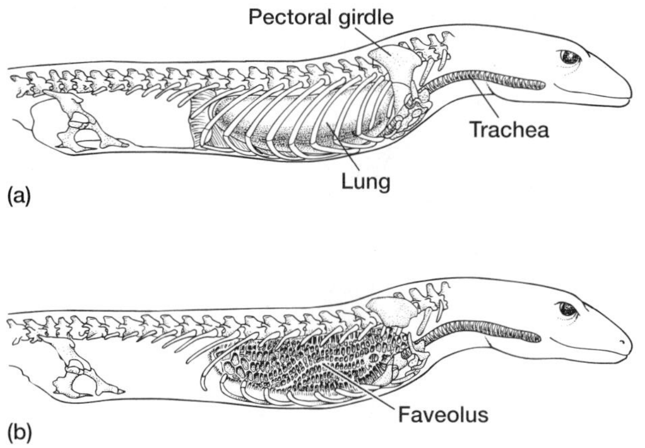


**Figure 11.26** Lung ventilation in a lizard. (a) The lungs are located in the thorax surrounded by ribs, and connected to the trachea. Compression and expansion of the rib cage forces air in or out of the lungs. (b) Cutaway view of the internal lining of the lungs showing numerous faveoli that collectively give the lining a honeycomb appearance. The internal faveoli of the lungs increase their respiratory surface area and function in gas exchange with capillaries lining their walls.

pressing the lungs within their cavity and expelling air. Occasionally, exhalation is passive. In this instance, muscle contraction is minimal, and gravity (and some elastic recoil) acts on the ribs, causing them to compress the lung cavity. Between breaths, the glottis is closed to prevent premature escape of air.

In snakes, the long, narrow lungs extend through most of the length of the body. In primitive snakes, as in other reptiles, the lungs are paired, but in many advanced snakes, the left lung is reduced and often lost entirely. In most snakes, faveoli are prominent anteriorly, but they decrease gradually and become absent posteriorly, producing two regions of the lung, an anterior respiratory portion (faveoli) and a posterior saccular portion (avascular) (figure 11.27a–c). Ribs and associated muscles run the entire length of the thorax so that regional compression and expansion of the body wall expand or deflate the lung. Opening and closing of the glottis are synchronized with these movements. Gas exchange occurs in the respiratory portion of the lung. The saccular portion of the lung acts as a bellows when the anterior body is occupied with different functions and unavailable to compress or expand the lung. For instance, when a snake swallows prey, the body becomes distended as food passes slowly through the esophagus, yet ventilation of the lungs must continue. Although the trachea, reinforced with semicircular rings of cartilage, stays open, the anterior body cannot act as the aspiration pump. Instead, the posterior body behind the prey expands and contracts, causing the saccular lung to fill and empty the lungs.

In caimans and other crocodiles, the liver assists the aspiration pump by acting like a "piston" to ventilate the lungs. During inhalation, the ribs rotate forward and outward, expanding the cavity around the lungs. In addition, the liver, located immediately behind the lungs, is pulled posteriorly by the action of **diaphragmatic muscles.** These muscles are derived from internal abdominal musculature. They extend forward from the pelvis and gastralia to the **posthepatic septum,** a thin sheet connected to the posterior side of the liver. Contraction of the diaphragmatic muscles draws the liver back, increasing the volume of the lung cavity and dropping the pressure within the lungs. This draws in atmospheric air. Exhalation reverses these movements. The ribs fold back into position, and the liver moves forward against the lung as a result of the contraction of **abdominal muscles.** Because pressure on the walls of the lung increases, air is expelled (figure 11.28).

Ventilation in turtles represents a special problem in design. The shell around the lungs prevents changes in shape and precludes aspiration pumping using the ribs. In soft-shelled turtles, movements of the hyoid apparatus draws water in and out of the pharynx. Oxygen is absorbed in the pharynx to sustain the turtle while it is submerged. In snapping turtles, the carapace is reduced, permitting deformations of the shell that contribute to lung ventilation. More commonly, in-and-out movements of the limbs alters pressure on the lungs, and special sheets of muscles within the shell change pulmonary pressure (figure 11.29a). Turtle lungs and other viscera reside in a single fixed cavity, so any change in volume alters pressure on the lungs. A limb extended from or pulled into the shell affects pressure in this cavity and aids the aspiration pump (figure 11.29b). In addition the posterior visceral cavity is closed by a **limiting membrane,** connective tissue to which the **transversus abdominis** and **obliquus abdominis** muscles are attached. Contraction or relaxation of these muscles alters the volume of the cavity within the shell and contributes to the inhalation or exhalation of air (figure 11.29c). The diaphragmaticus muscle, although absent in tortoises, is present in most other turtles. The diaphragmaticus together with the transversus abdominus compress the visceral cavity to act as exhalation muscles. The glottis opens and the obliquus abdominis expands the visceral cavity to act as an inhalation muscle.

## Mammals

An aspiration pump ventilates the lungs of mammals. Changes in the shape of the rib cage and pistonlike action of a muscularized **diaphragm** contribute to this pumping mechanism. The diaphragm consists of **crural, costal,** and **sternal** parts, all of which converge on a **central tendon.** Unlike the diaphragmatic muscles of crocodiles, which are located posterior to the liver, the diaphragm of mammals lies anterior to the liver, and acts directly on the **pleural cavities** in which the lungs reside (figure 11.30a,b). Intercostal muscles run between the ribs. The transversus abdominis, serratus, and rectus abdominis that are inserted on the ribs and originate outside the rib cage (figure 11.30c,d) all aid in mammalian lung ventilation.

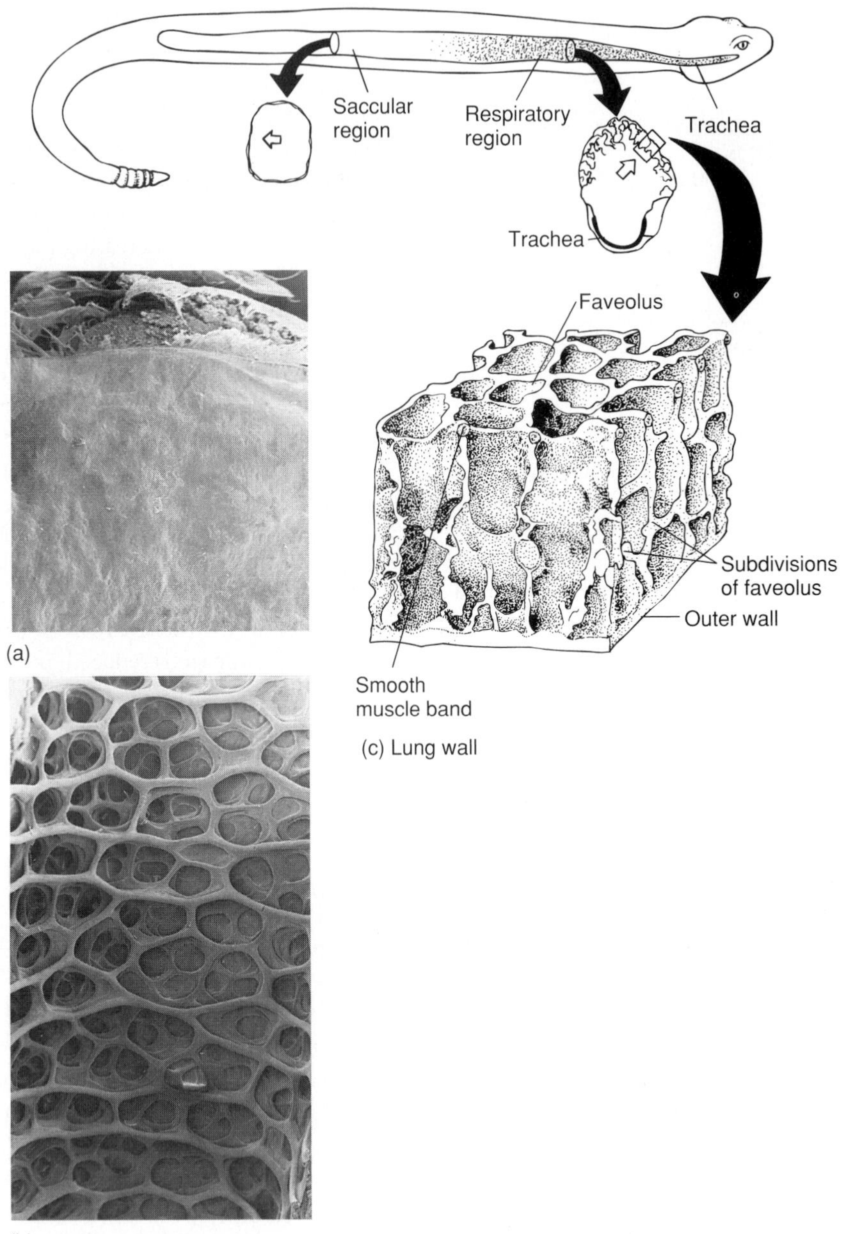


**Figure 11.27** Snake lung, rattlesnake. Like the snake body, the rattlesnake's one lung is long and attenuated. Air travels down the long trachea to the lung. Most snakes have two lungs of unequal length, but in many venomous snakes, the left lung is lost. The trachea of the rattlesnake lung becomes an open trough where it meets the lung. The anterior lung is heavily vascularized and functions in respiratory exchange. The posterior part of the lung basically is a saccular, avascular region. Ribs along the sides of the body compress and expand to empty or fill the lungs. As the snake swallows prey, the tip of the trachea is pushed in front of the prey, so breathing continues. As the prey moves along the esophagus, which parallels the trachea, the anterior ribs expand to allow passage. At this time, they cannot compress and expand the anterior lung. Therefore, the posterior ribs act upon the saccular region of the lung, working like a bellows to move air across the respiratory surfaces. Representative cross sections of saccular and respiratory regions are illustrated at the top of the figures. (a) Luminal view of the surface of the saccular region. (b) Luminal view of the respiratory region showing the faveoli. The entrance to each faveolus is defined by a honeycombed network of smooth muscles. (c) A section of wall from the respiratory region showing further subdivisions within the faveoli.

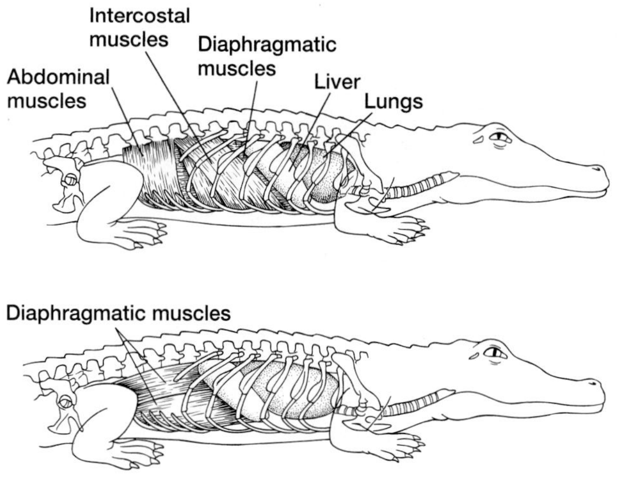


**Figure 11.28** Ventilation in the crocodile. In addition to a rib cage, the aspiration pump in the crocodile uses back-and-forth movements of the liver like a piston to act on the lungs. During inhalation, the rib cage expands and the liver is pulled back while the crocodile aspirates fresh air into its lungs. During exhalation, the rib cage and forward-moving liver compress the lungs and the crocodile expels spent air.

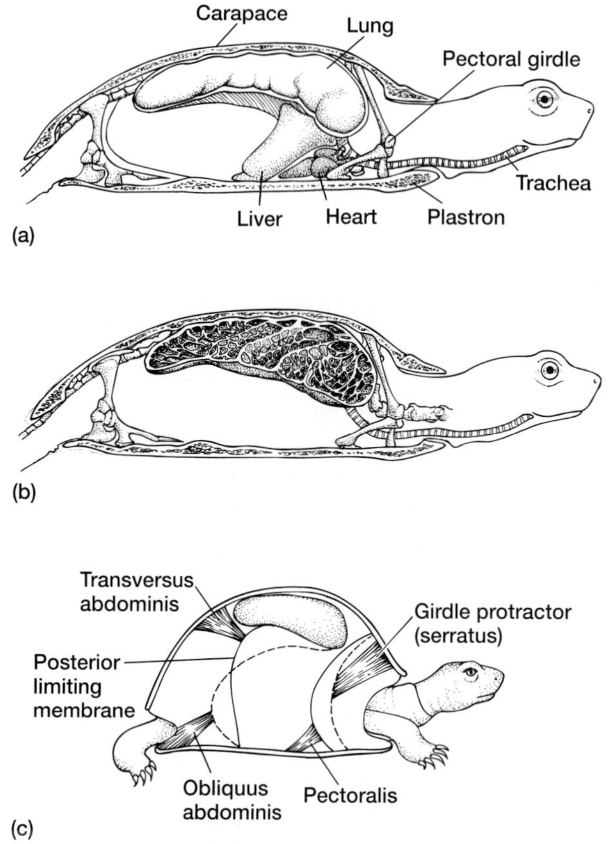


**Figure 11.29** Ventilation in the turtle. (a)Location of the lung inside the turtle shell. (b) Cutaway view of the lung showing its internal structure. Turtle lungs lie within a protective, rigid shell. Consequently, the fixed rib cage cannot act in ventilating the lungs. Instead, turtles have sheets of muscles within the shell that contract and relax to force air in and out of the lungs. Turtles also have the ability to alter air pressure within the lungs by moving their limbs in and out of the shell. (c) In the specialized tortoise, a diaphragmatic muscle is absent but other respiratory muscles take its place. Within the rigid shell, the viscera are enclosed by limiting membranes that under muscle action alter their position during exhalation (solid line) and inhalation (dashed line). During active exhalation, contraction of the transversus abdominis pulls the posterior limiting membrane up against the lung and contraction of the pectoralis draws the shoulder girdle back into the shell, further compressing the viscera. During active inhalation, exhalation muscles relax and contraction of the obliquus abdominis and girdle protractor expand the visceral cavity by pulling the posterior limiting membrane outward and the shoulder girdle forward, respectively.

### *Ventilation*

Mammalian ventilation is bidirectional and involves the rib cage and diaphragm. Upon inhalation, the external intercostal muscles contract to rotate the adjacent ribs and medial sternum forward. Because the ribs are bowed in shape, this rotation includes an outward as well as a forward swing of each arched rib. The result is to expand the space that the rib cage encloses around the lungs. Contraction of the dome-shaped diaphragm causes it to flatten, further enlarging the thoracic cavity. The elastic lungs expand to fill the enlarged thoracic cavity and air is drawn in (figure 11.31a,b).

During active exhalation, internal intercostal muscles slant in the opposite direction of the relaxed external intercostals and pull the ribs back. Relaxation of the diaphragm causes it to recoil and resume its arched, dome shape. Rib retraction and diaphragm relaxation decrease chest volume, forcing air from the lungs. Elastic energy stored in the lung and gravity acting to fold or collapse the rib cage may aid exhalation (figure 11.31c).

Although scientists agree on the muscles that control mammalian breathing, their precise functions have proved elusive, partly because of the surprisingly complex pattern of rib movement and partly because the rib cage and diaphragm are not equally involved in ventilation at all times. For example, during quiet breathing, only inhalatory muscles may show activity. At such times, exhalation muscles may not contract, and compression of the rib cage results from elastic and gravitational forces. As you can confirm for yourself, it is even possible to ventilate your lungs moving only the diaphragm and not the rib cage. When supporting vigorous ventilation during exercise, the rib cage, diaphragm, and most muscles are involved. To complicate the matter further, there appears to be a coupling of breathing cycles with locomotor cycles so that both are synchronized.

The mammalian diaphragm lies immediately posterior to the lungs and separates the thoracic cavity containing the lungs from the abdominal cavity containing other major viscera. When an animal is at rest, the muscularized

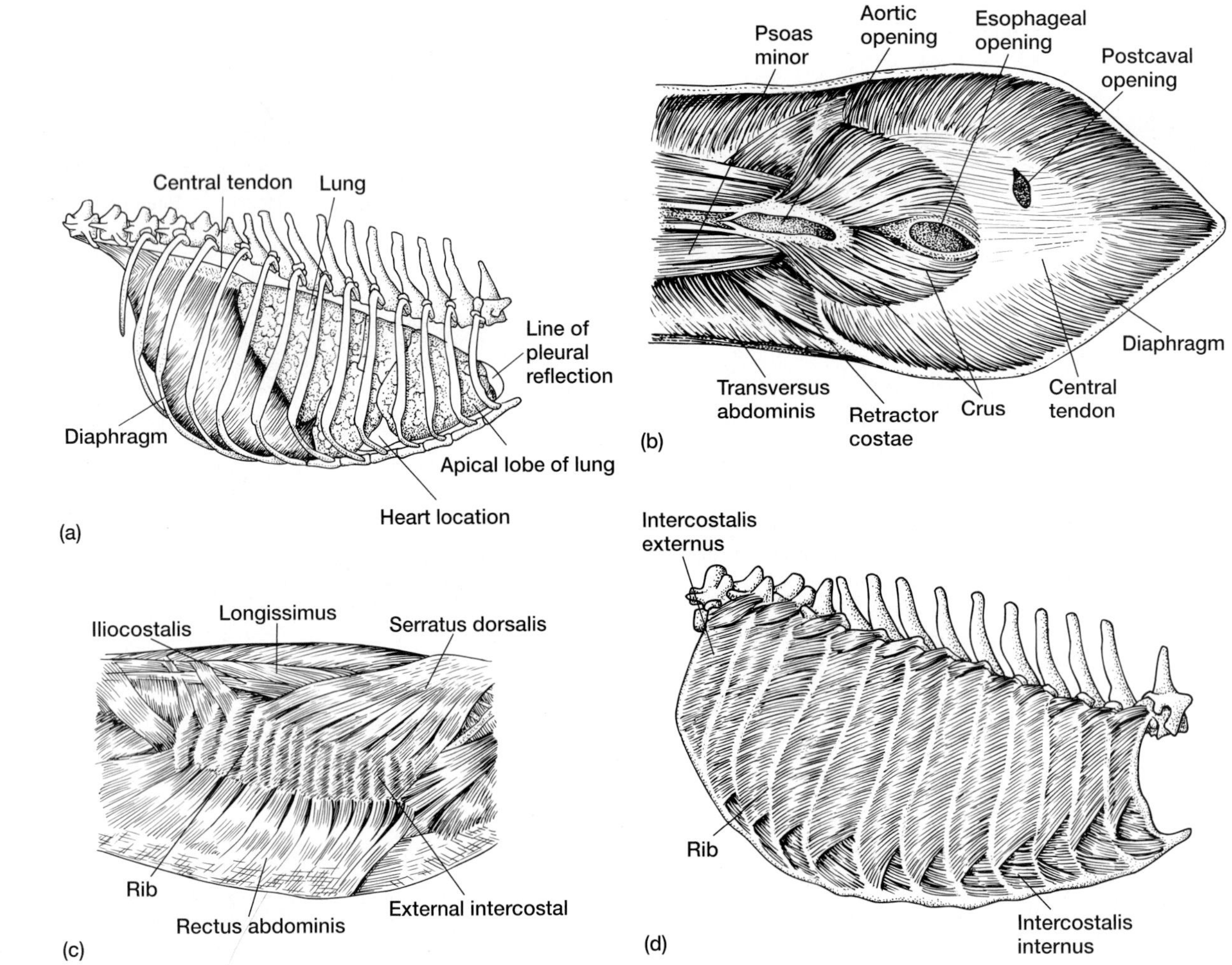


**Figure 11.30** Ventilation in the dog. Generally, ventilation of mammalian lungs involves expansions and contractions of the rib cage along with depression and elevation of the diaphragm. The details are remarkably complex. (a) Location of lungs and diaphragm within the rib cage of the dog (lateral view). (b) Ventral view of the diaphragm, which lies behind the lungs and has a dome shape. Notice the openings that allow anterior-posterior passage of the aorta, esophagus, and postcava. Superficial (c) and deep (d) muscles of the rib cage.

diaphragm is the principal component in mammalian lung ventilation. However, during locomotion in quadrupedal mammals, the rib cage may receive ground reaction forces through the forelimbs that slightly change its shape. Further, the abdominal viscera, somewhat free to move within the body cavity, slide forward and backward in synchrony with the rhythm imposed on the body by the pattern of limb oscillation. The abdominal viscera act as a kind of "piston," first pressing anteriorly on the thoracic cavity and then sliding posteriorly, releasing pressure on the lungs. A running mammal takes advantage of this rhythmic movement of the viscera, expelling air when the viscera press against the thorax and inhaling when they move away. Thus, in cursorial mammals, breathing patterns and locomotor gait are often coupled (figure 11.32a–c).

## *Gas Exchange*

As we have seen in lower tetrapods, faveoli along the interior walls of the lungs form the respiratory exchange surface. Air is drawn into the center core of the lung and diffuses outward into the faveoli. However, in mammals, the sites of respiratory exchange are reached via a different route. The respiratory passageway (including trachea, bonchi, bronchioles) repeatedly divides, producing smaller and smaller branches until they finally terminate in blind-ended compartments, the **alveoli,** in which gas exchange occurs (figure 11.33a–c). The trachea, bronchi, and bronchioles that transport gas to and from the alveoli are called the **respiratory tree** in recognition of their branching pattern. No gas exchange occurs along the conducting passageway of the respiratory tree until air reaches the alveoli. In mammals, the total alveolar area is extensive, perhaps over ten times that of amphibians of similar mass. Such a large exchange

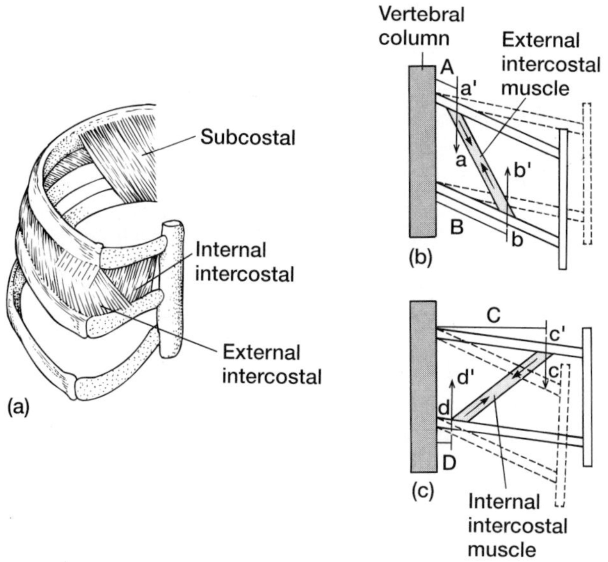


**Figure 11.31** Rib cage movements in mammals. (a) Various muscles run between adjacent ribs at slanted angles. (b) During inhalation, external intercostals contract, causing adjacent ribs to be drawn forward, expanding the pleural cavities around the lungs, and aspirating air into them. (c) Exhalation is often passive. Gravity pulls the ribs down, compressing the lungs and expelling air. During vigorous respiration, exhalation may be active. When this occurs, internal intercostals, slanted in an opposite direction, contract to compress the rib cage.

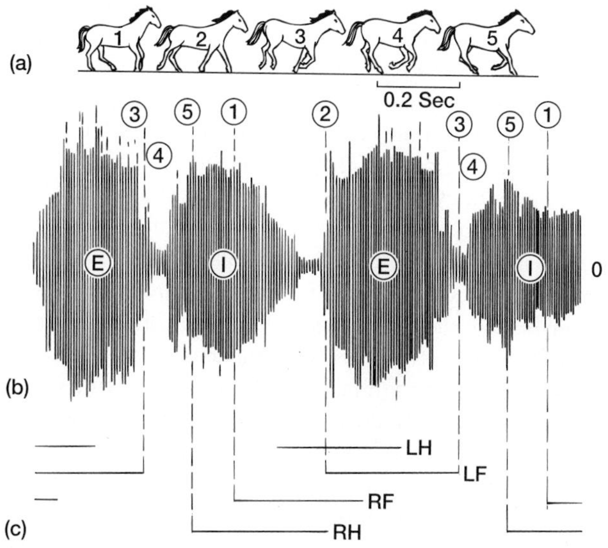


**Figure 11.32** Locomotor and ventilatory cycles in mammals. During rapid locomotion, cycles of inhalation and exhalation are often synchronized with phases of the locomotor cycle. (a) Body positions of a horse at five successive points in a canter, indicated by circled numbers below. (b) Bursts of sound recorded at the nostrils reveal points of inhalation (circled I) and exhalation (circled E). (c) The footfall pattern indicates times of foot contact: left forelimb (LF), right forelimb (RF), left hindlimb (LH), right hindlimb (RH).

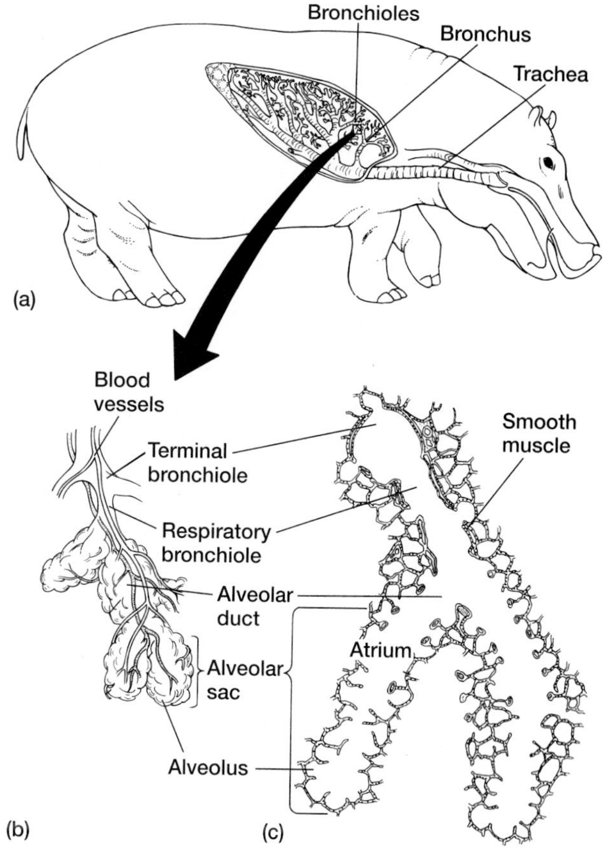


**Figure 11.33** Mammalian lung. The lungs of mammals are blind-ended, terminating in small alveoli through which respiratory exchange occurs. (a) The trachea leads to the pleural cavities and branches into bronchi to supply left and right lungs. Repeated bronchial branchings produce smaller and smaller bronchioles that eventually lead to alveolar sacs. (b) Enlarged alveolar sac. Arteries and veins supply the alveoli to accommodate gas exchange within them. (c) Internal subdivisions of the alveolar sacs are shown. Each small compartment is an alveolus where actual respiratory exchange between blood and air occurs. Note the smooth muscle bands at the openings to the alveolar sacs.

area is essential in mammals to sustain the high rate of oxygen uptake required by an active endotherm. The nasal passages not only form part of this conducting system but serve to warm and moisten the entering air.

## Birds

Cutaneous respiration is insignificant in birds. The almost exclusive respiratory organ is the lungs. Like mammals, birds have two lungs connected to a trachea and ventilated by an aspiration pump. Beyond this, however, the structural similarities are few. For example, there are no blind-ended alveoli in and out of which air moves. Instead, the conducting passages branch repeatedly and eventually form numerous tiny, one-way passageways, the **parabronchi,** that permit air to flow through the lungs.

Small **air capillaries** open off the walls of each parabronchus, and gas exchange with the blood actually occurs in the air capillaries. Further, nine avascular **air sacs** are connected to the lungs, although they are tucked in among the viscera and extend into the cores of most large bones (figure 11.34a,b). Thus, the bones of birds contain air, not marrow. The anterior air sacs include the single **interclavicular sac** and the paired **cervical** and **anterior thoracic air sacs.** The posterior air sacs include the paired **posterior thoracic** and paired **abdominal air sacs** (figure 11.34a).

The trachea is divided into two primary bronchi, termed **mesobronchi,** that do not enter the lung but extend posteriorly to reach the posterior air sacs. Along the way, the mesobronchi give rise to numerous branches, the most prominent of which include **latero-, ventro-,** and **dorsobronchi** as well as **secondary bronchi.** These lead to the parabronchi (figure 11.35a–c). During passage through the parabronchus, gases diffuse between the lumen of the parabronchus and the connecting, blind-ended air capillaries. Oxygen diffuses in turn from the air capillaries into the adjacent blood capillaries that give up carbon dioxide to the air capillaries. Thus, the walls of air and blood capillaries constitute the sites of gas exchange.

Within this vast system of connecting passageways, there are no valves to suggest what the pattern of airflow might be. This has led to much speculation about what roles the different parts of the respiratory system play. Without giving it much thought, some have proposed that air sacs function to lighten the bird like helium balloons. But because air in the sacs has the same density as air outside the bird, the air sacs provide no lift. Adding air sacs does not make the bird lighter. Others propose that air sacs serve to cool hot testes, but female birds have similar sacs. Certainly, air sacs are not a prerequisite for flight because bats, who have typical mammalian lungs, are good fliers and can even, on occasion, migrate long distances.

Recent research suggests another possibility—air sacs act as bellows. Debate still rages about the details of this mechanism, but some aspects are understood. If we follow a single breath, its passage through the sacs and lungs includes two complete cycles of inhalation and exhalation. During the first inhalation, air enters through the trachea and flows mainly into the posterior air sacs, filling them. During the first exhalation, this air moves forward into the lungs and through the parabronchi. Gas exchange with the blood occurs now in the air capillaries, but air does not leave the lungs yet. With the second inhalation, the air inhaled with the first breath, which is now depleted of oxygen and saturated with carbon dioxide, enters the anterior air sacs. The second exhalation forces it to pass from the trachea to the outside (figure 11.36a,b). In addition, fresh air moves into the posterior sacs with inhalation and begins a new cycle. Thus, this pattern of ventilation produces a nearly continuous unidirectional flow of fresh air across the lungs. Speculating further, such a unidirectional flow may also establish a crosscurrent exchange within the lung, with air flowing from posterior to anterior air sacs as circulating blood flows next to it in the opposite direction (figure 11.37).

This continuous one-way flow, perhaps in a crosscurrent pattern, gives birds an efficient respiratory system to support their high metabolic requirements. Mammals have

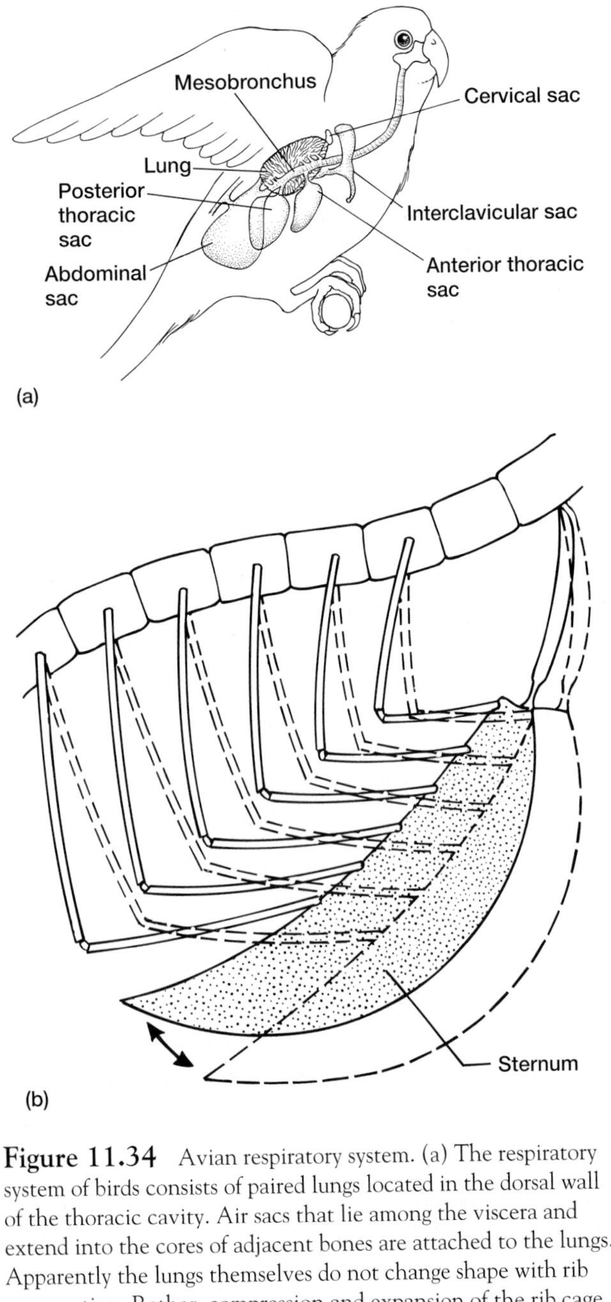


**Figure 11.34** Avian respiratory system. (a) The respiratory system of birds consists of paired lungs located in the dorsal wall of the thoracic cavity. Air sacs that lie among the viscera and extend into the cores of adjacent bones are attached to the lungs. Apparently the lungs themselves do not change shape with rib cage motion. Rather, compression and expansion of the rib cage acts on the air sacs, drawing air through them and then into the lungs. (b) Ventilation of the avian lung. Ribs are hinged to each other and to the sternum in such a way that lowering of the sternum results in expansion of the rib cage and inhalation. Elevation of the sternum compresses the lungs and air is expelled.

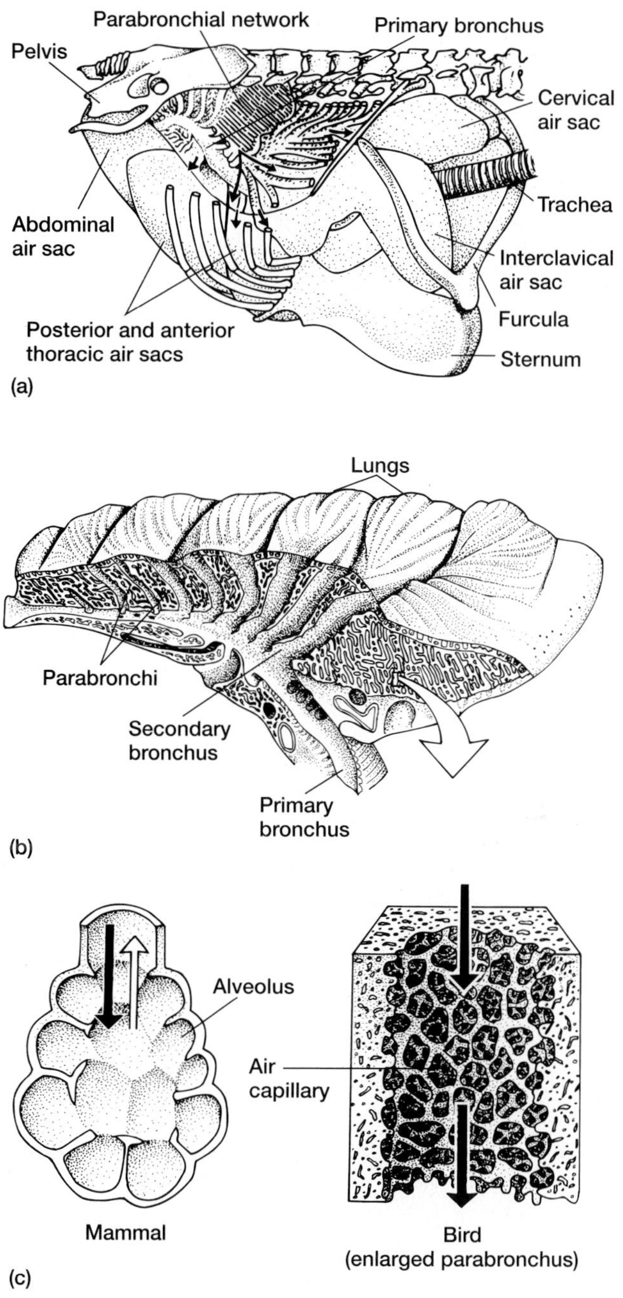


**Figure 11.35** Avian lung. (a) Lungs and air sacs are located within the body cavity between the sternum and the axial column. The lung is cut away to show the primary bronchus and parabronchial network inside. Inflated air sacs are indicated. (b) The isolated lung is sectioned. The small pores in the exposed lung are parabronchi. The trachea branches into two primary bronchi (mesobronchi) that extend to the posterior air sacs. Along the way, they open into secondary bronchi. These lead to parabronchi that open into the highly subdivided respiratory tissue, the air capillaries. In the bird lung, flow through the parabronchi is one way, unlike the mammalian airflow that ends in blind aveoli. (c) Comparison of avian and mammalian respiratory surfaces. In the avian lung, air passes one way (solid arrows) through the parabronchi, replenishing the air capillaries that surround and open into the parabronchi. In the mammalian lung, the alveoli are blind-ended. In order for air exchange to take place, the lung must move tidally (open and solid arrows).

a ventilatory system that supports a high metabolic rate as well. But the especially efficient bird lung, with its ability to take up oxygen even in thin air, gives them a special advantage when they fly at high altitudes. Birds fly, apparently comfortably, over high mountain ranges while human climbers below struggle along, often requiring supplements of bottled oxygen.

**Countercurrent and crosscurrent exchange (p. 146)**

# Form and Function

## Patterns of Gas Transfer

In a general sense, a respiratory organ couples blood flow with ventilation. One function of the respiratory organ is to orient blood flow in relation to ventilation. The orientation is important because it affects the efficiency of gas exchange. One common pattern is countercurrent flow, illustrated in the gills of some fishes, in which water flows across the secondary lamellae in one direction and blood flows through capillaries in the opposite direction (figure 11.38a).

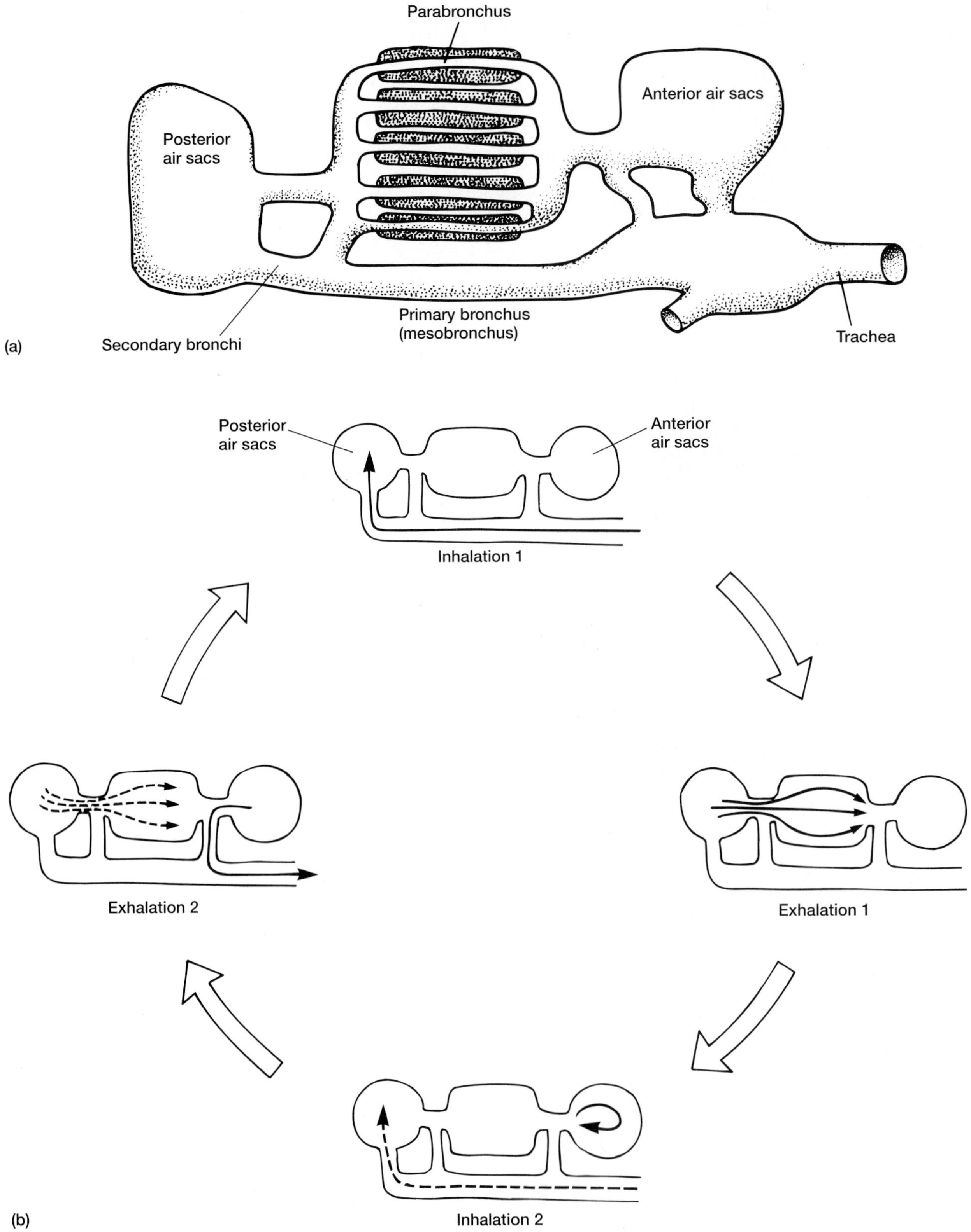


**Figure 11.36** Schematic representation of lung ventilation in birds. (a) The avian respiratory system includes anterior and posterior air sacs that connect to the parabronchi and hence to the respiratory tissue. (b) Route of air movement during breathing. Movement of one draft of air in, through, and out the lungs requires two cycles of inhalation/exhalation. During the first inhalation (1), the draft of air (solid arrow) is carried to the posterior air sacs so that upon exhalation (1), this air passes into the lung. Inhalation (2) of the next breath forces the air from the first breath out of the lung into the anterior air sacs so that upon exhalation (2), this spent air exits. During the second set of breaths, the next draft of air (dashed arrows) moves through the avian respiratory system.

This arrangement maintains high partial pressure gradients of gases while water and blood stream pass each other. As noted, crosscurrent flow is thought to occur between air and blood capillaries in avian lungs. Airflow and blood flow cross each other obliquely rather than lying in parallel. Blood capillaries are in series with each other as they cross a gas gradient of air capillaries. Oxygen is efficiently loaded into the blood before it departs this exchange system. The gills of some fishes may operate on a crosscurrent pattern as well (figure 11.38b). Mammalian lungs illustrate gas exchange involving a **uniform pool.** Lung ventilation tends to keep the partial pressures of gases within the alveolar spaces uniform thanks to frequent breathing, mixing of gases, and absence of significant barriers to diffusion. The circulating blood in the alveolar capillaries encounters more or less uniform partial pressures (figure 11.38c).

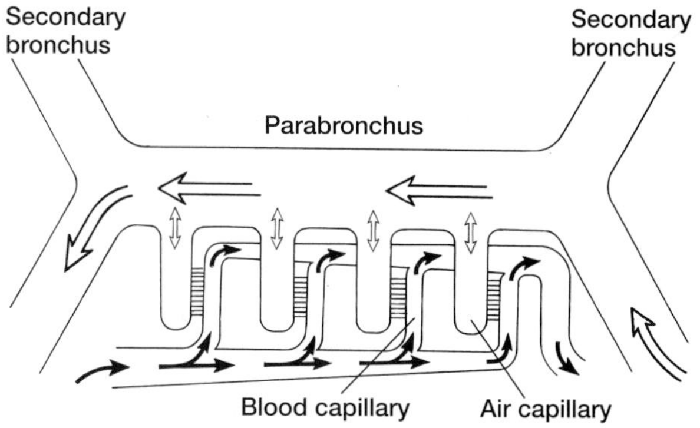


**Figure 11.37** Crosscurrent gas exchange in the avian lung. Diffusion of gases between the air capillaries and the parabronchus (open arrows) replenishes the gases available for exchange between the lungs and the blood capillaries (solid arrows). It is hypothesized that oxygen is progressively loaded into the blood (and carbon dioxide given up) based on an efficient crosscurrent system.

The respiratory area within vertebrate lungs has often been described as *alveolar,* a term inspired by the structure of mammalian lungs. However, the term is inappropriate for other groups. The respiratory compartments of most nonmammalian vertebrates do not form at the terminus of a bronchial tree. Instead, most compartments are subdivided by secondary and tertiary septa and should be called faveoli. This pattern should be called **flaviform** to distinguish it from the mammalian alveolar pattern. In birds, this type of subdivision yields a third structural pattern in which blind sacs surround and open to a central parabronchus. The ways in which these differences in lung structure serve different metabolic demands in the various groups of vertebrates are not well understood at present.

## Rates of Gas Transfer

The respiratory organs must also be designed to match the *rate* at which air or water passes the respiratory surface (ventilation) with the *rate* at which blood moves through the respiratory organ (perfusion). When the lungs are functioning efficiently, the rates of ventilation and perfusion are balanced so that the amount of oxygen available to diffuse across the respiratory surface from one side is exactly matched by the ability of the blood perfusing the opposite side to carry this oxygen away (figure 11.39a). For carbon dioxide, the reverse is true. The amount of carbon dioxide carried by the blood must be matched to the ability of the respiratory medium to carry it away. If perfusion is too rapid relative to ventilation rate, blood moves through the organ too quickly and departs before it is fully saturated with oxygen (figure 11.39c). On the other hand, if perfusion is too slow, blood lingers too long in the organ after it has become saturated and can no longer take up additional oxygen (figure 11.39b). In either case, the metabolic cost of oxygen extraction will be higher than optimum, and respiration will be inefficient.

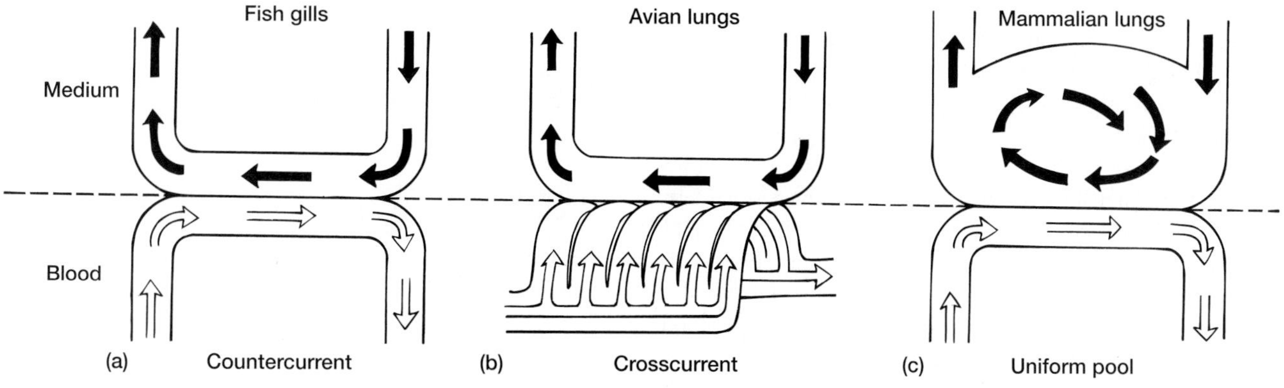


**Figure 11.38** Patterns of gas transfer. Orientation of ventilation (solid arrows) to blood flow (open arrows) is established by the respiratory organ. (a) Countercurrent. (b) Crosscurrent. (c) Uniform pool.

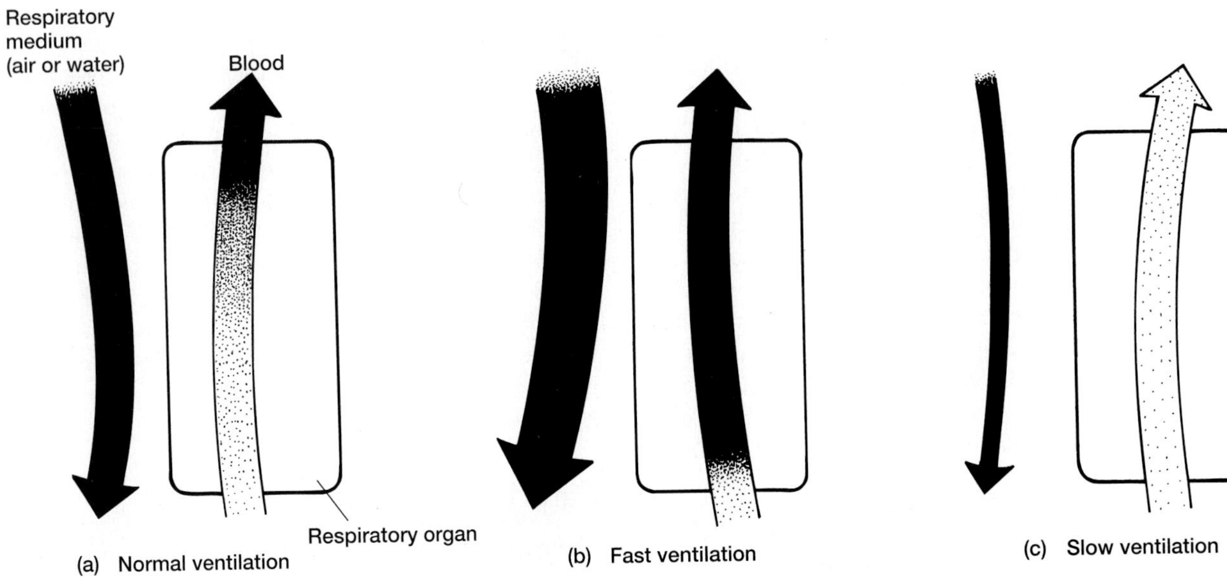


**Figure 11.39** Ventilation: perfusion ratios. The air- or water-breathing organ balances blood flow (perfusion) with movement of the respiratory medium (ventilation). (a) If perfusion and ventilation are matched properly, blood departs from the respiratory organ just as it becomes saturated with oxygen. (b) If ventilation is too fast, blood lingers longer than necessary in the respiratory organ and becomes saturated early but takes up no more oxygen. (c) If ventilation is too slow, blood is only partially oxygenated when it departs from the exchange organ. Breathing that is too fast or too slow is inefficient. Widths of arrows are proportional to flow rates. Shading on arrows passing through the respiratory organ indicates degree of oxygen saturation.

The ratio of ventilation to perfusion depends on the species. Within a species the ratio changes with activity levels and with availability of oxygen in the environment. In mammals, the ratio can be 1:1; in some reptiles, it is as much as 5:1. Some fishes have shown a ratio of 35:1. As a relative measure of the interaction of the respiratory and circulatory systems within a species, gas transfer ratios give us insight into the problems a species faces as well as its physiological response.

For example, water, even when it is saturated with dissolved air, still contains considerably less dissolved oxygen than an equal volume of air. Further, water is 1,000 times denser and more viscous than air, so gases diffuse much more slowly. Consequently, relatively large volumes of water must be moved across gill surfaces to match the high oxygen affinity of the perfusing blood; therefore, ventilation-perfusion ratios are generally large in fishes. Water flow may be up to 35 times blood flow. Water flows almost continuously and in a countercurrent fashion. In contrast to this, inactive reptiles with low metabolic demands may take a breath only every minute or so. In mammals, with high metabolic demands and tidal ventilation, breathing is more or less continuous, so the blood flowing from the alveoli is saturated. In exercising humans, metabolic demands of active tissues increase further. Both ventilation and perfusion increase in step at such times. You breathe faster (ventilation), and your heart rate accelerates (perfusion).

Many subtle adjustments help to optimize gas exchange. For example, fishes respond to a drop in available oxygen in the water in several ways. As might be expected, gill ventilation increases, as does output of blood by the heart. Other adjustments occur as well. Gill filaments are repositioned to allow more secondary lamellae to participate in respiration (figure 11.40a–c). The transit time of water moving through the gills increases, and the distance of diffusion across the lamellae probably decreases. Acting in concert, these collective changes maintain oxygen uptake by maintaining, as well as possible, favorable exchange ratios during times of low oxygen availability.

As we have seen, the respiratory organs involved in breathing water and air are necessarily different in design because of the different problems faced with gas exchange in the two media. We explore these differences in the next two sections.

## Breathing in Water

At 15°C, water holds about 1/30 as much oxygen as air. Furthermore, water is considerably denser than air. Nevertheless, water-breathing fishes can usually maintain sufficient delivery of oxygen to their tissues. In part, this is possible because of their high ventilation rate. Water flow is usually over ten times that of blood flow. But it is also due to the dual pump that keeps a nearly continuous stream of new water washing across the gills and to the efficient counter-

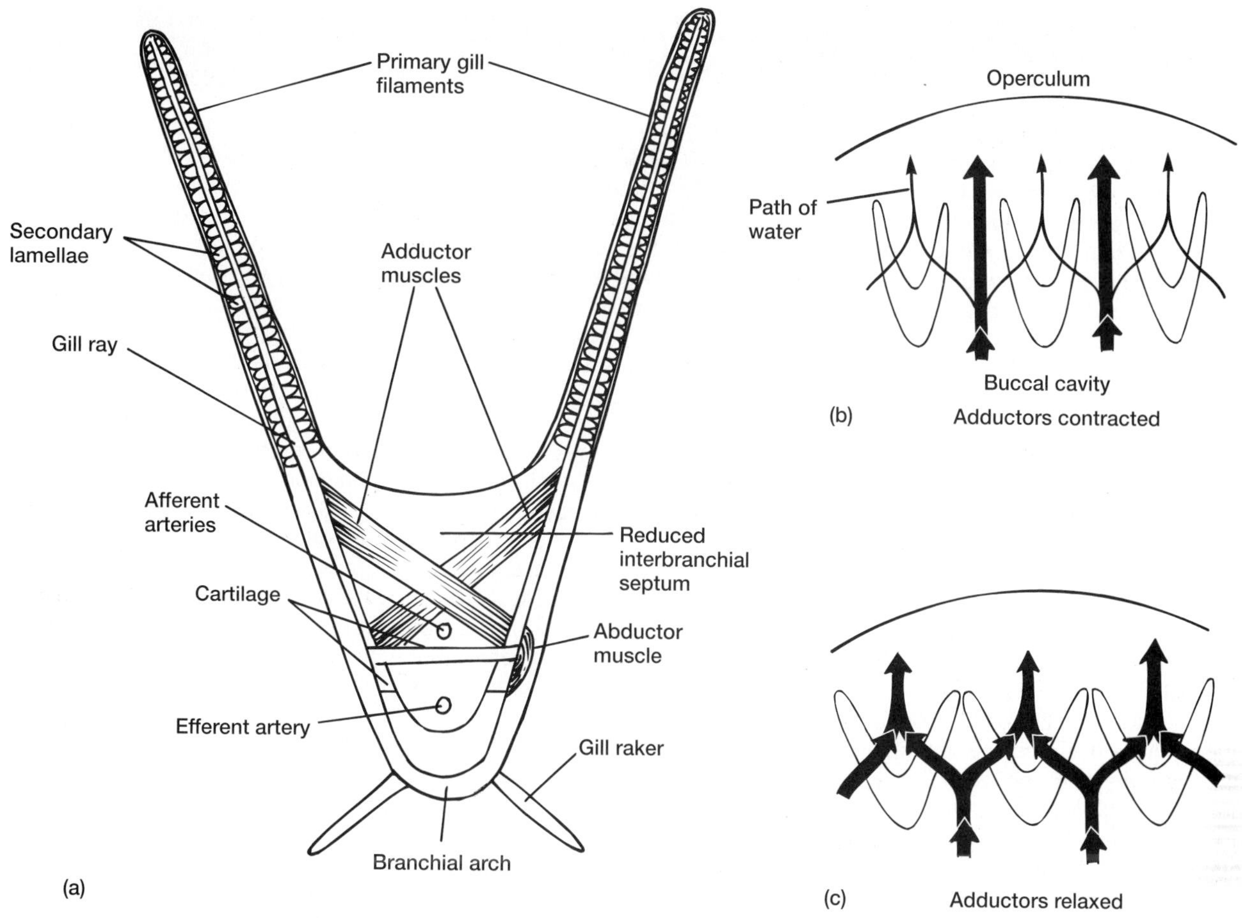


**Figure 11.40** Fish gills. (a) Positions of gill filaments are controlled by crossing adductor muscles. (b) When resting, primary gill filaments may part to allow excess water to bypass the respiratory surfaces. (c) When active, the primary filaments are moved more directly into the water flow to increase irrigation of secondary gill filaments and increase the opportunity for gas exchange.

current pattern of flow. As a consequence of this design, water leaving the gills may have given up 80 to 90% of its oxygen. This is a rather high oxygen extraction. In mammals, for example, only about 25% of the oxygen in the lung is taken up before it is exhaled. Although fish ventilation extracts more oxygen, the metabolic cost of moving the denser water is higher, so the high level of oxygen extraction is won only at a cost.

## Breathing in Air

Water is dense, so its tidal movement would be a relatively expensive method of respiration as the water mass was accelerated in first one direction and then in the other. On the other hand, air is light, so its tidal movement requires relatively less energy. However, exchange surfaces are exposed to evaporation in vertebrates that breathe in air. For this reason air-breathing organs, such as lungs, are usually recessed in cavities, precluding unidirectional flow or countercurrent exchange and requiring a tidal method of ventilation. The exception, of course, is the unusual arrangement that evolved in birds in which air-breathing organs are ventilated by unidirectional flow and gas exchange involves a crosscurrent pattern.

# Evolution of Respiratory Organs

## Acid-Base Regulation

The evolution of respiratory organs is in large part related to problems involving the *extraction* of oxygen from water or air to fuel metabolism, but not entirely. Often the design of a respiratory organ depends on its opposite role, namely, *elimination* of metabolic waste products. Mechanisms that regulate the acid-base balance of the blood illustrate how the body handles by-products of metabolism.

## Box Essay 11.4 Getting in Over Your Head—SCUBA Diving

The lure to explore or exploit underwater worlds directly has tempted humans for centuries. The easiest route has been to hold your breath and dive in. An obvious limitation is that you can stay under only until your breath runs out. To extend the downtime, various devices have been used to pump air to divers wearing hoods or hard hats. The limitation to these, of course, is that the dive is restricted by the air line to the surface.

Scuba gear solved this. Scuba, or more correctly S.C.U.B.A., means self-contained underwater breathing apparatus. A large volume of air, compressed into a small cylinder, and a device to deliver it on demand, the regulator, allow the diver great freedom of movement while underwater (box figure 1a).

A diving suit with compressed air was invented in 1825 and a demand regulator in 1866, but no one seemed to make the connection between the two at least for purposes of underwater exploration, except for Jules Verne who brought these together in the world of fiction, *20,000 Leagues Under the Sea.* Credit for bringing these devices together in the real world goes to Jacques Cousteau of France and Emile Gagnan of Canada. They fitted a demand regulator to a compressed air tank and tested it during the summer of 1943. It worked. Diving was revolutionized, but with some added risks.

Most risks come from the effects of increased partial pressures of gases. At the surface of Earth, the column of air above weights down on a person at sea level, producing 1 atmosphere of pressure, or 101,000 Pa (14.7 psi). In seawater, every 10 m of descent increases pressure on a diver by about 1 more atmosphere. Thus at 20 m, pressure is 3 atmospheres, or 303,000 Pa (1 atmosphere from the column of air at sea level plus 2 more for the additional pressure of seawater at 20 m). And this is what creates problems. For the compressed air to fill the lungs at 20 m, the regulator must match the pressure on the lungs. Thus, air enters the lungs at a higher pressure than it would on dry land at sea level. The high pressure of air in the lungs forces high levels of gases into the blood. When the blood is saturated, it holds more gas than it does at lower pressures. As a result of these elevated saturation levels, problems can result if a diver goes too deep or comes up too fast.

On the one hand, if the diver continues to descend, nitrogen (about 78% of air) reaches unusually high levels in the blood. At depths of over 30 m, elevated nitrogen levels can cause dizziness, loss of judgment, and impaired simple motor functions (box figure 1b). This condition is known as **nitrogen narcosis.** Surprisingly, oxygen (about 21% of air) can become toxic if high partial pressures force it to reach high levels in the blood. Excess oxygen can cause lung injury and permanent damage to the central nervous system.

On the other hand, if the diver ascends too quickly, problems from nitrogen can arise as well. When dives are deep and prolonged, the elevated partial pressures push high levels of nitrogen into the blood. When ascending, the diver must take adequate time to allow the excess nitrogen to escape through the lungs as air is expelled.

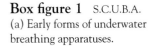

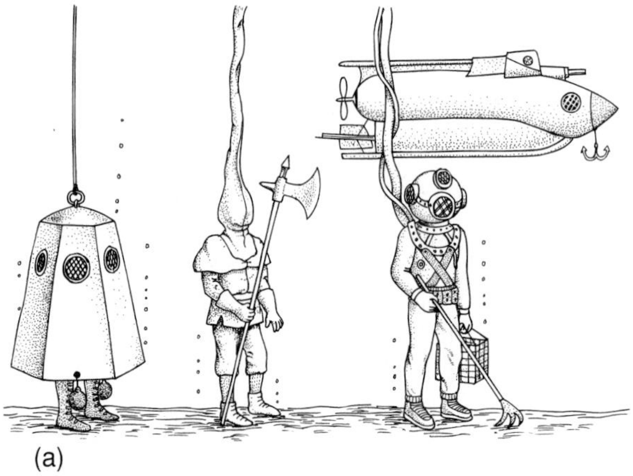

(a)

**Box figure 1** S.C.U.B.A. (a) Early forms of underwater breathing apparatuses.

During respiration, oxygen is transported to active tissues of the body. Simultaneously, by-products of metabolism are removed. Ammonia ($NH_3$), a toxic by-product, is excreted via the gills or the kidneys in the form of less harmful urea and uric acid. But carbon dioxide excretion is another matter. Surprisingly, carbon dioxide itself is not very toxic, although the hydrogen ions ($H^+$) it generates can be a problem. Elimination of carbon dioxide from the body is related to its effects on acid-base levels of the blood, or **pH balance.**

When you quickly pop a soft drink lid, gas under pressure is suddenly returned to atmospheric pressure, and it comes out of solution to fizz and form bubbles. The same thing happens in blood when a diver comes to the surface too quickly. Nitrogen comes out of solution too quickly and forms bubbles in the blood. These bubbles can lodge anywhere—lungs, joints, muscles, stomach, brain—and cause pain or death. The condition is known as **caisson disease** or the "bends." Treatment calls for quickly placing the afflicted diver in a recompression chamber to once again elevate the pressure on his or her body and force the nitrogen bubbles back into the blood. Then, slowly, the individual is brought back to normal atmospheric pressures, allowing time for the excess nitrogen to escape by diffusion through the lungs.

Deep-diving marine mammals, such as dolphins and seals, are not immune to the bends, but they seem to depend on devices that minimize the problem. Most obviously, their lungs are not filled with air under pressure. They take a breath at the surface and then dive. No more air is added during the deep dive. In fact, excess air retained in the lungs is usually expelled. Thus, as an animal descends and pressure on its rib cage and lungs increases, this pressure does not force residual air into the blood at a high pressure. High nitrogen levels in the blood do not develop, so there is less risk of nitrogen coming out of the blood upon return to the surface. Further, the bronchial tree is supported by cartilaginous rings only up to the level of the respiratory bronchioles. Absence of supporting rings beyond this point in the bronchial tree permits the alveoli to collapse under high pressure. Consequently, not much air is trapped in the lungs, where it might make prolonged contact with capillary exchange surfaces during a prolonged dive at great pressure. Finally, although the mechanism is not yet well understood, it seems that the tissues of marine mammals are resistant to the bends. Their fat, in particular, seems to be able to absorb excess nitrogen safely.

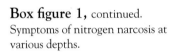

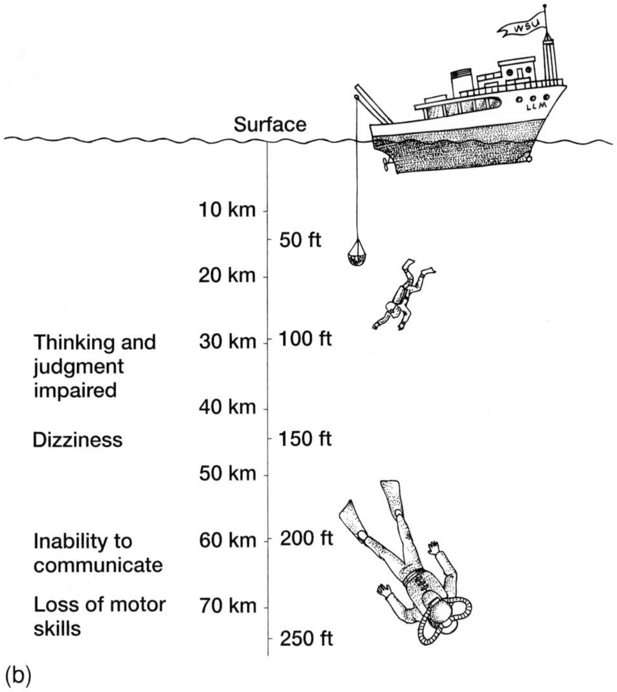


**Box figure 1,** continued.
Symptoms of nitrogen narcosis at various depths.

When carbon dioxide enters the blood, it combines with water to dissociate reversibly into carbonic acid, which in turn forms a hydrogen ion and a bicarbonate ion (figure 11.41). An increase in hydrogen ions in the blood causes a decrease in blood pH. As more hydrogen ions accumulate, the blood becomes more acidic. As hydrogen ions are removed, blood becomes less acidic (more basic). This is critical. The affinity of hemoglobin for oxygen decreases with decreasing pH. More fundamentally, protein enzymes, that control essential cell metabolism operate

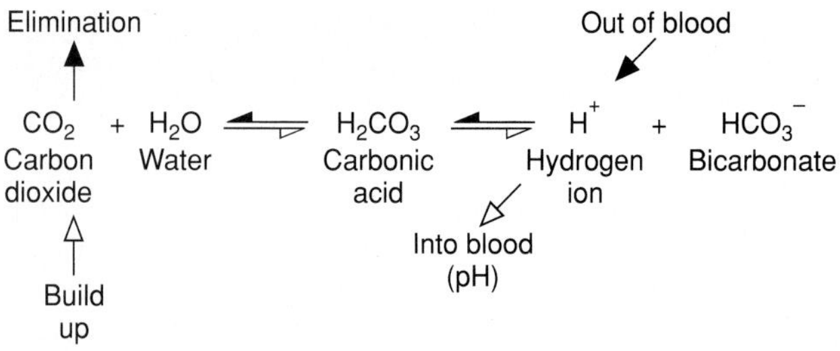


**Figure 11.41** Dissociation equations for carbon dioxide and its effects on pH balance. Buildup or elimination of carbon dioxide ($CO_2$) affects blood pH. When $CO_2$ blood levels are low, hydrogen ions ($H^+$) combine with bicarbonate ions ($HCO_3^-$) to form carbonic acid ($H_2CO_3$), which dissociates into water ($H_2O$) and $CO_2$. The equation shifts to the left. When $CO_2$ accumulates in the blood, the equation shifts to the right, resulting in the buildup of $H^+$ in the blood and a more acidic pH.

within a narrow pH range. If blood pH is too high or too low, they become nonfunctional. Control of pH centers on control of hydrogen ions, and this in turn is affected by carbon dioxide levels in the blood. Elimination of carbon dioxide shifts the equation shown in figure 11.41 to the left. A hydrogen ion recombines with a bicarbonate ion so the hydrogen ion concentration in the blood is reduced, and acid levels fall. The buildup of carbon dioxide in the blood has the reverse effect—blood becomes more acidic.

When a vertebrate exercises, lactic acid accumulates in the blood as a by-product of protein metabolism. Eventually it is broken down by chemical degradation, but not immediately. Thus, the buildup of lactic acid threatens to change unfavorably the blood pH. A compensatory increase in carbon dioxide elimination counters this lactic acid-induced change in blood pH. Hydrogen diffuses out of the blood and pH returns to normal levels. Carbon dioxide is part of a complex buffering system that prevents drastic swings in pH levels. In mammals, breathing rate increases with exercise; therefore, more oxygen is taken into the lungs to support aerobic metabolism and more carbon dioxide is eliminated to buffer blood pH.

In fishes, some carbon dioxide is eliminated via the skin, but most exits across the gills. In adult amphibians, oxygen is taken in through the lungs, but carbon dioxide is almost exclusively eliminated through the skin. In addition, the amphibian kidney participates in regulating acid balance by secreting hydrogen ions, but it does so by employing a secretory mechanism that depends on an immediate supply of water. Because amphibians usually frequent sources of fresh water, this is simple and easy. Thus acid balance in amphibians is maintained indirectly by carbon dioxide elimination via the skin (which affects the dissociation equation and hence the concentration of hydrogen ions) or directly by hydrogen ion secretion via the kidneys.

Amphibians, however, pay a price for this simple system of elimination. Because acid balance via the kidneys is based on a secretory mechanism requiring large influxes of water, they must have immediate access to a supply of water. Water for elimination, and not the demands of oxygen uptake, is one of the major reasons why amphibians are so closely linked to aquatic environments.

The loss of fish gills by the first tetrapods meant the loss of one major route used in regulating pH balance. As we have just seen, the kidneys and skin of modern amphibians take over this function. It is not known how *Ichthyostega*, who had some thick skin scales, and other fossil amphibians coped with this problem. In amniotes, the lungs take on the function of regulating pH balance by eliminating carbon dioxide. In the mammalian kidney, elimination is based on a different mechanism from that of amphibians, one that conserves water. In chapter 15, we discuss the kidney's role in water balance and in acid-base regulation. This short journey into blood chemistry reminds us that the evolution of life on land required more than the appearance of limbs. New physiological problems had to be solved as well.

## Ventilation

The evolution of respiratory organs is also a story of mechanical devices that move water or air. Some respiratory pumps depend on cilia; however, most are based on muscle contraction.

### *Ciliary Pumps*

Cutaneous gas exchange may have played an important part in early vertebrate respiration, and in some groups, such as amphibians, still does. Direct exchange of gases between tissues and the environment through the skin is a simple and direct way to meet modest metabolic needs of small organisms. The small larvae of some fishes still depend on cutaneous respiration. In some cases, such as the Australian lungfish larva, surface cilia are used to develop respiratory currents across the organism's surface.

Among protochordates, cilia move the current of water that brings food to mucous traps within the branchial basket. Ventilated by this "ciliary pump," the branchial basket, with its large surface area and extensive blood supply, also assumes a large portion of the respiratory chores of the skin. If such a condition existed in ancestors to the first fishes, it would have had consequences for their subsequent evolution.

First, such a branchial basket, based on an active ciliary pump specialized for ventilation and feeding, would allow the evolution of larger and more active species than would be possible with cutaneous respiration alone. Second, by reducing a vertebrate's dependence on cutaneous respiration, ciliary pumps made possible the evolution of the thick bony armor that prevents cutaneous respiration. Presence of dermal armor in ostracoderm fishes might reflect this evolutionary opportunity.

### *Muscular Pumps*

As mentioned earlier, if an animal is large or active, the ventilatory abilities of cilia fall behind the increase in metabolic requirements. Muscular pumps that replace cilia as the mechanism to move currents of water are an answer to this problem. For example, the ammocoete larva of lampreys employs a muscular velar to pump water across its gills. In the adult, the branchial basket participates in the muscular movement of water across exchange surfaces.

The appearance of muscular pumps in early vertebrates was probably a prerequisite to the attainment of large size and active lifestyles. Without such respiratory mechanisms, the kinds of vertebrates that evolved would have been considerably restricted.

## Water-to-Land Transition

No single change in lifestyle had more important effects on vertebrate design than the transition from life in water to life on land. In regard to the respiratory system, this transition included a change from water- to air-breathing organs and eventually a change in the type of ventilatory pump. Air-breathing lungs arose before this transition was underway, and the aspiration pump that efficiently fills the lungs arose much later after terrestrial vertebrates had become established.

### *Air-breathing Organs*

One prerequisite to life on land is the presence of a respiratory organ that can serve in gas exchange with air. The evolution of air-breathing organs occurred several times within different lines of bony fishes. These organs include vascularized swim bladders, parts of the digestive tract, specialized compartments to the gill chamber, and in dipnoans, lungs. One feature common to most fishes with air-breathing organs is that they live in fresh water susceptible to seasonal hypoxia. As a result of high temperatures, drought, decay of organic material, or stagnant water, the oxygen levels in water occasionally plummet. Hypoxia can be a time of intense stress for fishes, yet, ironically, oxygen is within easy reach in the atmosphere above them. Presumably it was just such conditions of seasonal hypoxia that favored the evolution of accessory organs capable of extracting oxygen from gulps of atmospheric air.

It is instructive to compare this situation faced by many bony fishes with that of elasmobranchs who never developed an air-breathing capability. Elasmobranchs approach neutral buoyancy because their skeletons have been reduced to cartilage and their oils reduce their density. They do not have gas bladders. They frequent well-aerated marine waters, and some sharks cruise deep open water distant from the water-air interface.

The evolutionary transition from water to land occurred between crossopterygians and ancient amphibians. Today, all crossopterygians, except for *Latimeria*, are extinct. Unfortunately, *Latimeria* occurs in deep marine waters and apparently is specialized. Its lung, inundated with fat, is a nonrespiratory organ. Among the lungfishes, the Australian *(Neoceratodus)* and the South American *(Lepidosiren)* lungfishes live in shallow freshwater streams, and the African lungfish *(Protopterus)* primarily occupies lakes. When their aqueous environment becomes hypoxic or dries entirely, they use their lungs to tap atmospheric oxygen.

The lives of these lungfishes suggest that lungs evolved not in anticipation of life on land but for the immediate adaptive advantage they conferred, namely, as supplements to gill respiration when dissolved oxygen in water became inadequate. Lungs were preadapted. Their biological role was supplementary, allowing fishes to tap into an alternative source of oxygen in the atmospheric air above their aquatic world. When the first tetrapods began exploiting the terrestrial world, lungs were ready to serve in their new role as the primary respiratory organs. Terrestrial life came after the appearance of lungs, not before.

### *Advantages of Movement to Land*

What conditions might have favored the movement to land? One suggestion has been that the seasonal drying of freshwater pools favored movement of stranded fishes across land to pools that persisted. Perhaps. But modern lungfishes faced with similar conditions do not normally migrate across land to new water. Instead, they estivate by burying themselves in mud, where their metabolic rate drops. Encased in mud cocoons, they can survive several years until rains return to replenish their pools.

Another suggestion has been that low oxygen levels in water prompted fishes onto land in search of alternatives. But, as we have seen, hypoxia stimulates air breathing but not necessarily migration.

Fishes today that venture onto land, such as the teleost mudskipper, apparently do so to search for food and to leave behind water-bound predators. Similarly, these advantages may have favored the movements of the first crossopterygians onto land, thereby beginning the terrestrial phase of vertebrate evolution.

**Skeletal modifications for land (p. 334)**

### *Air-breathing Mechanisms*

Although air breathing itself evolved before vertebrates moved to land, the air-breathing mechanisms carried onto land by early tetrapods were modifications of the dual pump, a water-breathing mechanism of fishes (figure 11.42). We

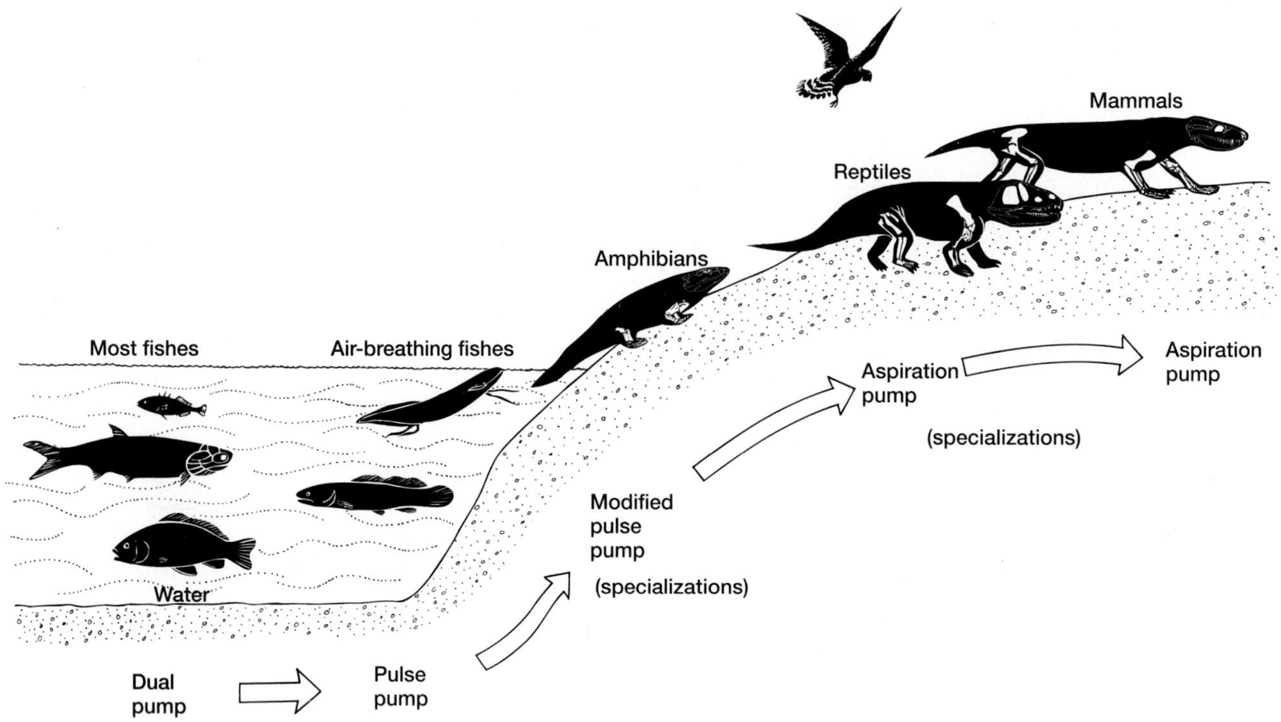


**Figure 11.42** Evolution of ventilatory mechanisms. Water-breathing fishes irrigate their gills with a dual pump mechanism in which buccal and opercular cavities operate in tandem. Air-breathing fishes use a pulse pump mechanism, a modification of the dual pump in which the buccal cavity is the major mechanical component. In adult amphibians, lung ventilation is based on a modified pulse pump in which the opercular pump is lost entirely. However, the amphibian skull design is compromised because the buccal cavity must function in feeding and in lung ventilation. One solution is found in plethodontid salamanders. Gas exchange is taken over entirely by cutaneous respiration and lungs are lost; therefore, the buccal cavity serves only feeding. In amniotes, the aspiration pump completely separates feeding from lung ventilation, uncoupling demands on the jaws for service in both activities. The basic dual and pulse pumps are specialized in many fishes. Bird respiration represents a specialization of the aspiration pump.

have seen the evolutionary stages involved. The dual pump is modified into the pulse pump of air-breathing fishes to force air into their lungs or gas bladders. This same pulse pump mechanism is the basis on which living adult amphibians fill their lungs, but with modifications. Because amphibians do not have gills, the tandem opercular component of the pulse pump becomes redundant in adult amphibians, and in adult frogs and salamanders it is lost. The job of ventilating the lungs now falls mainly to the other component of the pulse pump, the buccal cavity, which becomes enlarged and broadened. This means that the buccal cavity of adult amphibians must serve two major functions, feeding and breathing, often with contradictory demands on design.

On the one hand, the amphibian buccal cavity moves a large volume of air to ventilate the lungs. To move a large tidal volume, the jaws should be light and the skull broad. On the other hand, to serve feeding, we would expect the jaws to be short and robust. It has been hypothesized that the plethodontid salamanders addressed these competing demands on the buccal cavity by losing their lungs. In this salamander family, respiration occurs entirely through the skin; therefore, the buccal cavity exclusively serves feeding and is accordingly narrow and robust.

The aspiration pump separates the respiratory apparatus (rib cage and diaphragm) from the feeding apparatus (jaws), offering another way to address opposing demands on the buccal cavity. The evolutionary consequences of this separation of functions are most evident in reptiles. Reptiles possess small heads and strong jaws. Feeding styles are varied and specialized. Unlike modern amphibians, they eliminate carbon dioxide through their lungs, not through their armored or thickened skin that resists water loss. These combined changes allowed reptiles to travel farther from sources of water and become more committed to terrestrial lifestyles.

# SELECTED REFERENCES

Brainerd, E. L., K. F. Liem, and C. T. Samper. 1989. Air ventilation by recoil aspiration in polypterid fishes. *Science* 246:1593–95.

Bramble, D. M., and D. R. Carrier. 1983. Running and breathing in mammals. *Science* 219:251–56.

Burggren, W. W. 1978. Gill ventilation in the sturgeon, *Acipenser transmontanus:* Unusual adaptations for bottom dwelling. *Resp. Physiol.* 34:153–70.

Burggren, W. W., K. Johansen, and B. McMahon. 1985. Respiration in phyletically ancient fishes. In *Evolutionary biology of primitive fishes*, edited by R. E. Foreman, A. Gorbman, J. M. Dodd, and R. Olsson. New York: Plenum Press, pp. 217–52.

Carrier, D. R. 1987. The evolution of locomotor stamina in tetrapods: Circumventing a mechanical constraint. *Paleobiology* 13:326–41.

Carrier, D. R. 1990. Activity of the hypaxial muscles during walking in the lizard *Iguana iguana*. *J. Exp. Biol.* 152:453–70.

Carrier, D. R. 1991. Conflict in the hypaxial musculoskeletal system: Documenting an evolutionary constraint. *Amer. Zool.* 31:644–54.

Duncker, H.-R. 1978. General morphological principles of amniotic lungs. In *Respiratory function in birds, adult and embryonic,* edited by J. Piiper. Berlin: Springer-Verlag, pp. 2–15.

Feder, M. E., and W. W. Burggren. 1985. Skin breathing in vertebrates. *Sci. Amer.* 253(5):126–42.

Gans, C. 1970. Strategy and sequence in the evolution of the external gas exchanger of ectothermal vertebrates. *Forma et Functio* 3:61–104.

Gans, C. 1973. Sound production in the Salientia: Mechanism and evolution of the emitter. *Amer. Zool.* 13:1179–94.

Gans, C., and B. Clark. 1976. Studies on ventilation of *Caiman crocodilus* (Crocodilia, Reptilia). *Resp. Physiol.* 26:285–301.

Gans, C., H. J. DeJongh, and J. Faber. 1969. Bullfrog (*Rana catesbeiana)* ventilation: How does the frog breathe? *Science* 163:1222–25.

Gans, C., and G. M. Hughes. 1967. The mechanism of lung ventilation in the tortoise *Testudo graeca* (Linné). *J. Exp. Biol.* 47:1–20.

Gaunt, A. S., and C. Gans. 1969. Mechanics of respiration in the snapping turtle, *Chelydra serpentina* (Linné). *J. Morph.* 128:195–228

Gradwell, N. 1972. Gill irrigation in *Rana catesbeiana.* Part II. On the musculoskeletal mechanism. *Can. J. Zool.* 50:501–21.

Hughes, G. M., and C. M. Ballintijn. 1965. The muscular basis of the respiratory pumps in the dogfish (*Scyliorhinus canicula*). *J. Exp. Biol.* 43:363–83.

Johansen, K. 1968. Air-breathing fishes. *Sci. Amer.* 219:102–11.

Liem, K. F. 1981. Larvae of air-breathing fishes as countercurrent flow devices in hypoxic environments. *Science* 211:1177–79.

Liem, K. F. 1985. Ventilation. In *Functional vertebrate morphology*, edited by M. Hildebrand, D. M. Bramble, K. F. Liem, and D. B. Wake. Cambridge, Mass.: Harvard University Press, pp. 185–209.

Liem, K. F. 1987. Functional design of the air ventilation apparatus and overland excursions by teleosts. *Fieldiana: Zoology* 37:1–29.

Liem, K. F. 1988. Form and function of lungs: The evolution of air-breathing mechanisms. *Amer. Zool.* 28:739–59.

Liem, K. F. 1989. Functional design and diversity in the feeding morphology and ecology of air-breathing teleosts. In *Trends in vertebrate morphology*, edited by H. Splechtna and H. Hilgers. New York: Gustav Fischer Verlag, 35:487–500.

Luchtel, D. L., and K. V. Kardong. 1981. Ultrastructure of the lung of the rattlesnake, *Crotalus viridis oreganus*. *J. Morph.* 169:29–47.

Martin, L. D., and B. M. Rothschild. 1989. Paleopathology and diving mosasaurs. *Amer. Sci.* 77:460–67.

Martin, W. F., and C. Gans. 1972. Muscular control of the vocal tract during release signaling in the toad *Bufo valliceps*. *J. Morph.* 137:1–27.

Munshi, J.S.D., K. R. Olson, T. K. Ghosh, and J. Ojha. 1990. Vasculature of the head and respiratory organs in an obligate air-breathing fish, the swamp eel *Monopterus (Amphipnous) cuchia*. *J. Morph.* 203:181–201.

Perry, S. F. 1988. Functional morphology of the lungs of the Nile crocodile, *Crocodylus niloticus:* Nonrespiratory parameters. *J. Exp. Biol.* 134:99–117.

Perry, S. F., A. M. Bauer, A. P. Russell, J. T. Alston, and J. E. Maloney. 1989. Lungs of the geckoo *Rhacodactylus leachianus* (Reptilia: Gekkonidae): A correlative gross anatomical and light and electron microscopic study. *J. Morph.* 199:23–40.

Piiper, J. (ed.) 1978. *Respiratory function in birds, adult and embryonic*. Berlin: Springer-Verlag.

Piiper, J., and P. Scheid. 1977. Comparative physiology of respiration: Functional analysis of gas exchange organs in vertebrates. *Internat. Rev. Physiol.* 14:219–53.

Rahn, H.A. Jr., and C. V. Paganelli. 1979. How bird eggs breathe. *Sci. Amer.* 240 (2):46–55.

Rahn, H., K. B. Rahn, B. J. Howell, C. Gans, and S. M. Tenney. 1971. Air breathing of the garfish *(Lepisosteus osseus)*. *Resp. Physiol.* 11:285–307.

Randall, D. J., W. W. Burggren, A. P. Farrell, and M. S. Haswell. 1981. *The evolution of air breathing in vertebrates*. Cambridge: Cambridge University Press.

Ruben, J. A., and A. J. Boucot. 1989. The origin of the lungless salamanders (Amphibia: Plethodontidae). *Amer. Nat.* 134:161–69.

Scheid, P., and J. Piiper. 1989. Respiratory mechanics and airflow in birds. In *Form and Function in Birds*, edited by A. S. King and J. McLeeland. New York: Academic Press, 4:369–91.

Schmidt-Nielsen, K. 1971. How birds breathe. *Sci. Amer.* 225(December):72–79.

Weibel, E. R., and C. R. Taylor. (eds.) 1981. Design of the mammalian respiratory system. *Resp. Physiol.* 44:1–164.

# 12 CHAPTER The Circulatory System



## INTRODUCTION

Twelve miles per hour is faster than many people can run and a speed few can sustain. Yet, in the next Olympic marathon, the top finishers will average this speed for over 26 miles lasting several hours! Whales can dive from the surface to depths of over 2,000 m and feed there for up to an hour. During that time, they experience immense pressures on their bodies, over 16 million Pa (about 2,300 psi) per square meter of body surface. That is about equal to a column of lead 150 m high pressing on each square meter of the body. Animals such as the oryx, an African antelope, can be exposed to searing ambient temperatures during the day, and their body temperatures can exceed 45°C (113°F). In large measure, the ability of human and animal athletes to adjust to changes in activity and physical stress is possible because of adjustments orchestrated by the circulatory system.

Along with the respiratory system, the circulatory system transports gases between the sites of external and internal respiration. But the circulatory system also has many other important functions. It adjusts to changes in pressure on or within the body. Blood transports excess heat produced within the body to the skin to dissipate it. Conversely, a cool reptile basking in the sun gathers surface heat to warm its blood, which is then circulated to the rest of the body. Glucose and other end products of digestion are carried to active organs for metabolic use or to other organs for temporary storage. The circulatory system transports hormones to target organs and waste products to the kidneys. Blood also carries cells and chemicals of the immune system to defend the body from invasion by foreign organisms.

The circulatory system is basically a set of connecting tubes that transport fluid. The ability of the organism to adjust to immediate physiological changes in physical and metabolic activity depends on the rapid response of this system. The circulatory system includes the blood and lymph vascular systems. Lymphatic vessels and **lymph,** the fluid they circulate, collectively constitute the **lymphatic system,** discussed later in this chapter. The vascular system includes the blood vessels that carry blood pumped by the heart. Together blood, vessels, and heart constitute the **cardiovascular system,** which we discuss first.

# Cardiovascular System

## Blood

Cells produced by hemopoietic tissues usually enter the circulation to become the **peripheral** or **circulating blood.** Circulating blood comprises plasma and formed elements. The **plasma** is the fluid component and can be thought of as the ground substance of blood, a special connective tissue. The **formed elements** are the cellular components of blood. **Red blood cells,** or **erythrocytes,** are one cell type of the formed elements. All erythrocytes have nuclei, except those in mammals. Mature red blood cells in mammals lack nuclei. **Hemoglobin,** the major oxygen transport molecule, is excreted by the kidneys if it is left free in the plasma. Red blood cells function as containers for hemoglobin, preventing its elimination. Red blood cells vary in individual size, from 8 μm in humans, to 9 μm in elephants, to 80 μm in some salamanders. Most live three to four months in the circulating blood before being broken down and replaced.

**White blood cells,** or **leucocytes,** are a second major cellular constituent of the formed elements. Leucocytes defend the body from infection and disease. The **platelets** are a third formed element in the blood. They release factors that produce a cascade of chemical events leading to the formation of a **clot,** or **thrombus,** at sites of tissue damage.

Plasma and formed elements give blood a wide variety of roles in body processes. In addition to functioning in respiration and disease protection, blood also plays a part in nutrition (carries carbohydrates, fats, proteins), excretion (carries spent metabolites), regulation of body temperature (carries and distributes heat), maintenance of water balance, and transport of hormones.

## Arteries, Veins, and Capillaries

Although they vary in size, there are three principal types of blood vessels—arteries, veins, and capillaries. **Arteries** carry blood away from the heart, **veins** carry blood toward the heart, and **capillaries** are the tiny vessels that lie between them. Although arteries usually carry blood high in oxygen and veins carry blood low in oxygen, this is not always true. For example, the pulmonary artery carries blood low in oxygen from the heart to the lung to be replenished, and the pulmonary vein usually returns blood high in oxygen. Thus, the direction of blood flow with reference to the heart defines the type of vessel, not the oxygen content of blood it carries.

Arteries and veins have tubular walls organized into three layers that enclose a central lumen (figure 12.1). The innermost layer, the **tunica intima,** includes the lining of endothelial cells that face the lumen. On the outside is the **tunica adventitia,** composed mostly of fibrous connective tissue. Between these two layers is the **tunica media,** which differs the most in arteries and veins. Some smooth muscle contributes to the tunica media of large arteries, but elastic fibers predominate. In large veins, this middle layer contains mostly smooth muscle with almost no elastic fibers. Veins usually have one-way valves within their walls, and arteries lack such valves. Very small arteries and veins are called **arterioles** and **venules,** respectively. In these small vessels, the tunica adventitia is thin, and the tunica media is composed mostly of smooth muscle; thus, arterioles and venules are quite similar in structure (figure 12.1).

Smooth muscles form sheets that surround the walls of arteries or veins. Smooth muscle cells respond to nervous and hormonal stimuli. When they contract, the caliber of a vessel narrows, a response termed **vasoconstriction.** When circular muscle contraction ceases, resident blood pressure forces the vessel to open and restore or enlarge the size of the lumen, a response called **vasodilation.**

Gases, nutrients, water, ions, and heat are transported across the walls of blood capillaries. To facilitate efficient exchange, capillaries are small and have thin walls. Capillaries lack a tunica media and a tunica adventitia. Only the endothelial wall of the tunica intima remains. Sets of capillaries serving one area of tissue constitute a **capillary bed.** Each tissue is invested with multiple sets of overlapping capillary beds. As activity of tissues increases or decreases, more or less of these overlapping capillary beds open or close to regulate blood supply.

### *Arteries*

The structure of arteries varies with their size. Large arteries have considerable amounts of elastic fibers in their walls; small arteries have almost none. Structural differences occur because of functional differences between large and small arteries. Arteries function primarily as a supply system that carries blood away from the heart and out to body tissues. They also absorb and distribute the sudden surge of blood through them when the heart contracts. Rhythmic contractions of the heart send spurts of blood into the large arteries. With their elastic walls, these large arteries expand to receive the sudden injection of blood, which can be felt in your wrist or neck arteries as a "pulse." Between contractions, the stretched arterial walls elastically recoil, driving this volume of blood smoothly along through the smaller arteries and into the arterioles. Arterioles direct this blood to local tissues. Humans are one of the few species prone to arterial disease characterized by hardening of the arterial walls and loss of elastic recoil. As a consequence, afflicted arteries do not expand to blunt the sudden pulse of blood, nor do they move the column of blood along between heartbeats. The heart must work harder, and the smaller arteries and arterioles experience higher surges of blood pressures. Not designed for such pressures, these small vessels may rupture. If this occurs in a critical organ, death may follow.

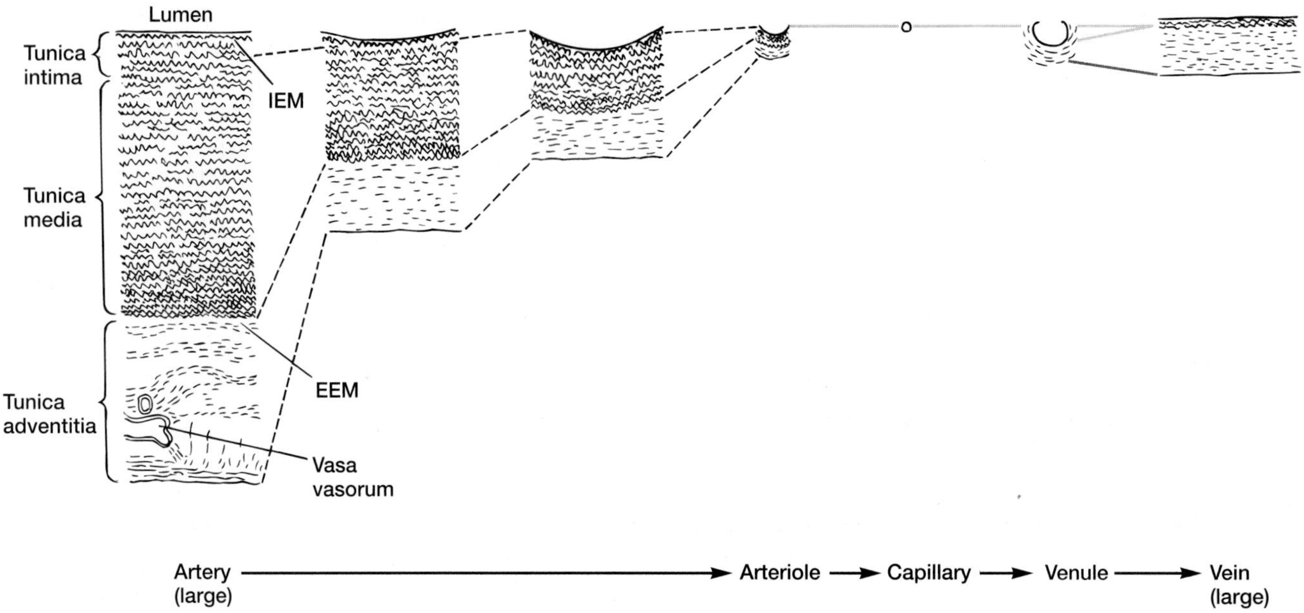


**Figure 12.1** Blood vessels. The three layers of blood vessel walls change in thickness and relative size from large arteries to small arterioles, capillaries, venules, and veins. In large arteries, the tunica media is especially well impregnated with elastic fibers, including belts termed *internal* (IEM) and *external* (EEM) *elastic membranes*. Their walls allow arteries to stretch and receive the pulse of blood delivered suddenly from the heart. Large blood vessels receive their own blood supply from small vessels to their walls, the vasa vasorum.

## *Hemodynamics of Circulation*

The pressures and flow patterns of the blood circulating through vessels constitute the **hemodynamics** of circulation. When the ventricles of the heart contract, the peak force produced is the **systolic pressure.** The **diastolic pressure** constitutes the lowest pressure within the blood vessels, which is reached between heartbeats. Diastolic pressure results from the force sustained by the elastic recoil of arteries. Blood pressure usually is expressed in a shorthand fashion, with the systolic pressure read and recorded first. For example, in most young adult humans, 120/80 are normal systolic and diastolic values, respectively, obtained from the vessels of the arm (figure 12.2). If the arteries start to show signs of disease, blood pressure will rise, a telltale indication that the major arteries are beginning to fail in absorbing cardiac pumping forces as a result of structural changes in their walls.

As blood flows away from the heart, pressure declines (figure 12.2). This drop in pressure results from two factors, friction as the blood encounters resistance from the luminal walls of vessels and increase in the total cross-sectional area of blood vessels. The flow of any fluid in tubes is resisted by the friction of the liquid against the walls. For blood to circulate, a force must be used to overcome this frictional resistance to blood flow; nevertheless, as a result of this resistance, blood pressure falls as circulation proceeds. Additionally, as blood flows from large to small arteries, arterioles, and capillaries, the total cross-sectional area of vessels increases, especially in capillaries. Like a fast stream entering a large lake, pressure declines as the larger volume is filled. As a result, blood reaching the venous side of the circulatory system retains very little pressure. In fact, in some of the large veins, forces moving the blood may fall to zero or even become negative. When this happens, blood will tend to flow in a reverse or retrograde direction. Dealing with these unfavorable pressures is a task that falls to the veins.

## *Veins*

Because veins return blood to the heart, they are collecting tubes. At any one moment, up to 70% of the circulating blood within the body may reside in veins. During times of stress, slight vasoconstriction of strategic veins effectively decreases available volume and moves some blood from this reservoir to the arterial side of the circulatory system.

Veins are also designed to address low blood pressures. One-way valves that prevent retrograde blood flow are common within their walls. If veins pass between active muscles or through parts of the body subjected to pressure changes (e.g., the pleural cavities containing the lungs), external forces impinge upon and squeeze their walls. These supplemental forces affect venous flow, and because of the one-way valves, blood moves only one way back to the heart (figure 12.3). Understandably, in veins that pass through body organs and tissues that offer no induced forces, such as those within the bones or the brain, one-way valves are ab-

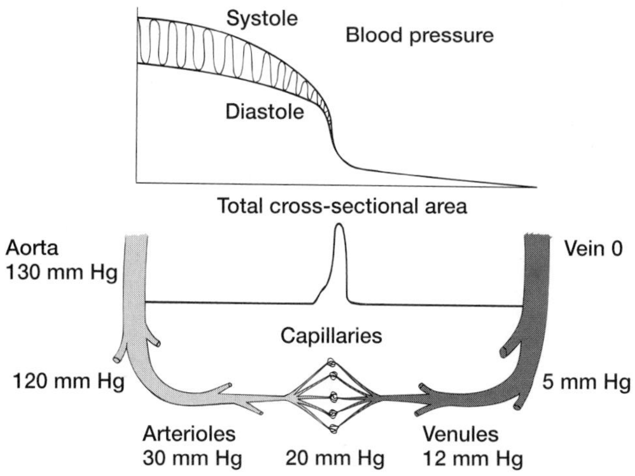


**Figure 12.2** Hemodynamics of blood flow. The systemic flow of blood is depicted graphically at the bottom of the figure. Above the respective vessels is their total cross-sectional area. Note the blood pressure in the different vessels. As blood flows from large arteries, such as the aorta, to capillaries and veins, the initial pressure imparted by the force of heart contraction falls. This is due to the frictional resistance from the walls of the vessels and from the increasing total cross-sectional area. Capillaries have an especially large cross-sectional area. The difference between systolic and diastolic pressures declines as blood approaches the capillaries and is usually minimal thereafter in the venous flow. Normal blood pressure for an adult human is indicated in mm Hg.

sent, and return of the blood to the heart depends on any remaining intrinsic pressure and gravity.

## *Microcirculation*

The specific component of the cardiovascular system that regulates and supports cell metabolism intimately is the **microcirculation.** Capillary beds plus the arterioles that supply them and the venules that drain them form the microcirculation. Blood flow to the capillary beds is controlled by smooth muscles. The **precapillary sphincters** are little rings of smooth muscle restricting the entrance to the capillary beds. The walls of both arterioles and venules include thin sheets of smooth muscles. Selective nervous and hormonal control of these smooth muscles regulates the flow of blood to the capillaries and hence to the tissues they supply in order to equal the needs of local cell activity. Blood can be diverted through shunts that bypass some regions entirely (figure 12.4).

As an animal lowers its head to drink from a stream, blood pressure within its tissues changes quickly (figure 12.5). Quick adjustments in the microcirculation serve to equalize and distribute these temporary pressure fluctuations to prevent undue stress on especially sensitive organs such as the brain and spinal cord. Heat distribution within the body is also influenced by the microcirculation. When an animal is active, excess heat transported by the blood

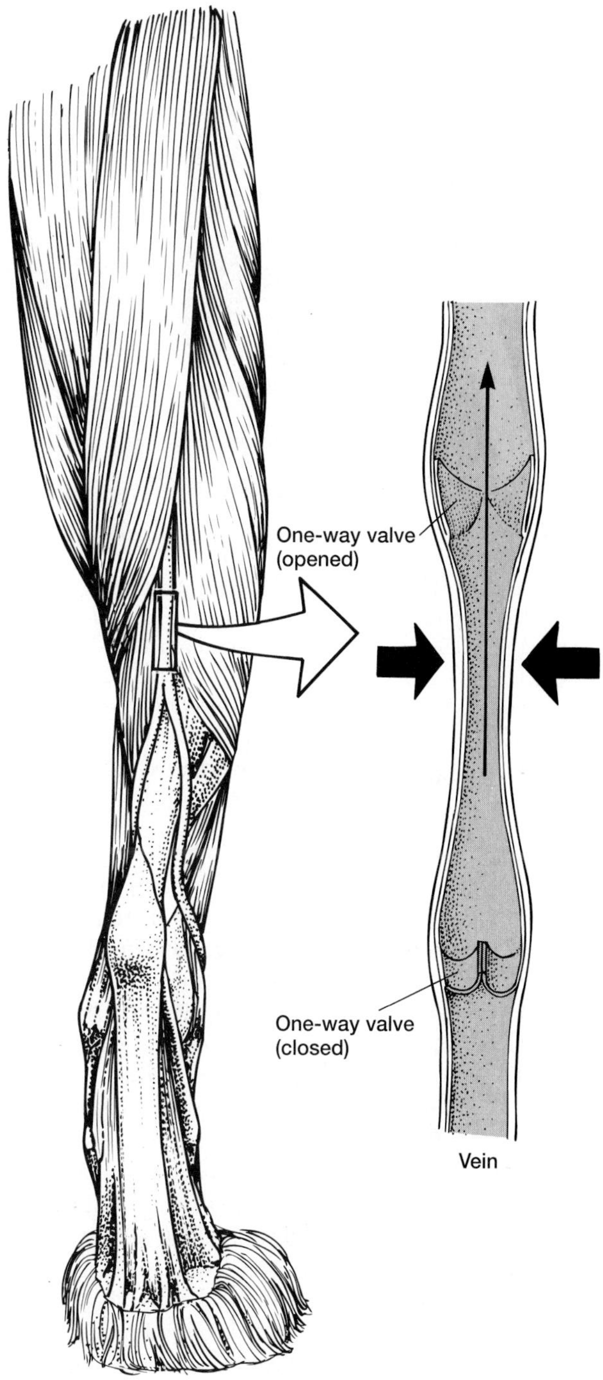


**Figure 12.3** One-way valves in veins. The one-way valves within the lumina of veins prevent retrograde movement of blood and ensure its return flow toward the heart (vertical arrow). The pressure that drives blood flow (horizontal solid arrows) comes from surrounding organs, usually muscles, that impinge upon and squeeze veins.

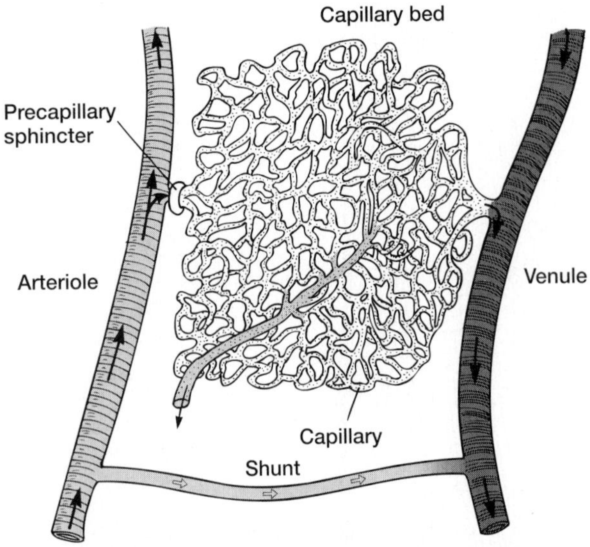


**Figure 12.4** Microcirculation. The microcirculation includes the capillary beds as well as the arterioles supplying and venules draining them. The usual flow of blood to, through, and from a capillary bed is diagrammed (solid arrows). Smooth muscles of the walls of the arterioles form small bands, the precapillary sphincters, that control blood flow to the capillary bed. A direct shunt running from the arterial to the venous side of circulation allows for major diversions of blood (open arrows).

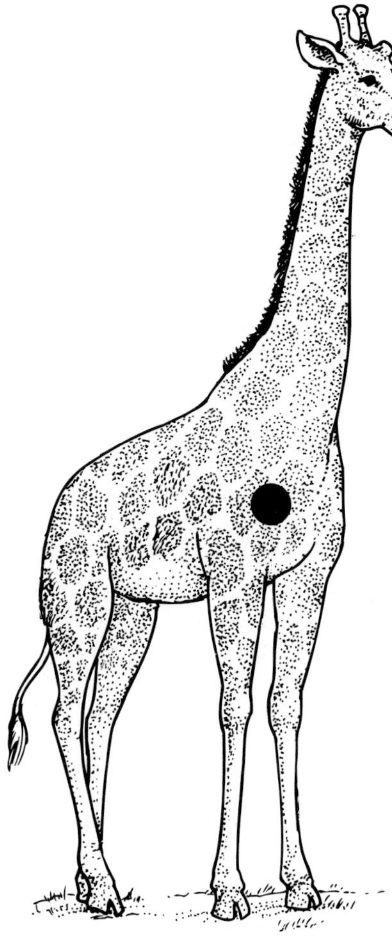

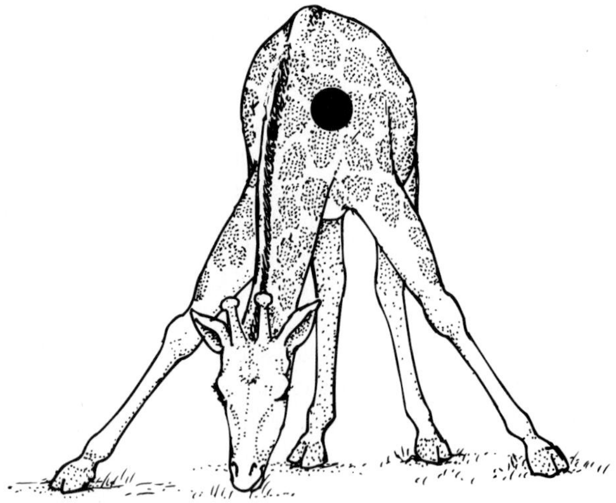

**Figure 12.5** Blood pressure changes. As posture changes, the head and extremities are raised or lowered relative to the heart. A giraffe lowering its head to drink quickly experiences increased pressure within its brain and cranial tissues. Adjustments in the microcirculation prevent such pressures from creating a problem.

reaches the body surfaces. Capillary beds of the skin open to increase blood flow, bringing more heat to the body's surface where it can be dissipated. Humans with fair skin redden during exercise, indicating this increased peripheral blood flow. In cold weather, the opposite occurs. As body temperature drops, peripheral blood supply decreases, reducing heat loss and helping to maintain the core body temperature.

The microcirculation is also involved in allocating blood to active organs. Capillaries are so small that it would take an hour for just a few drops of blood to pass through a single one. Yet, collectively, capillary beds represent an extensive volume, their linear extent being somewhere around 96,500 km (about 60,000 miles) of microtubing in all. No animal has enough ready blood volume to fill all its capillaries at once. If all the capillaries of a body opened simultaneously, all major blood vessels would be quickly emptied, and the circulatory system would fail. This does not happen because blood is selectively directed to open capillary beds only in active organs.

Ordinarily, not all body tissues are active simultaneously, so the supply of blood to those active tissues is sufficient. By selective deployment, the volume of blood required at any moment can be kept low. However, under some circumstances, the microcirculation fails to deliver sufficient blood to meet tissue needs. For example, if there are more active organs than available blood, then the microcirculation gives preference to some and not to others.

If strenuous exercise is undertaken soon after a large meal, then the digestive system and skeletal muscles compete for blood to support their activities. Preference is given to the skeletal muscles as more capillary beds open in them and the stomach receives less blood. The "side cramp" humans who exercise might complain of results from **ischemia,** a localized lack of sufficient blood to the stomach to meet metabolic expectations. Following severe injury or trauma, the microcirculation may fail to regulate blood distribution. When this happens, a condition called shock, properly **hypotensive shock,** results from a cascade of events. Too many vessels open, not enough blood is available, pressure drops, and circulation fails. If shock is not quickly reversed, death can soon follow. The chemical arsenal in some snake venoms takes advantage of this general physiological feature of the cardiovascular system. When injected into prey, the venom induces shock, helping the snake to dispatch its meal quickly.

## Single and Double Circulation

Blood travels in one of two general patterns. Most fishes have a **single circulation** pattern in which blood passes only once through the heart during each complete circuit. With this design, blood moves from the heart to the gills to the systemic tissues and back to the heart (figure 12.6a). Amniotes have a **double circulation** pattern in which blood passes through the heart twice during each circuit, traveling from the heart to the lungs, back to the heart, out to the systemic tissues, and back to the heart a second time (figure 12.6b). The rise of this double circulation involving the addition of a pulmonary circuit was a major evolutionary event. Between those vertebrates with a single and those with a double circulation stand functional intermediates with characteristics of both conditions. The intermediates include lungfishes, amphibians, and reptiles. They suggest the adaptive advantages of transitional forms venturing onto land and highlight the evolution of the circulatory system design. Let us begin by examining the basic embryonic derivation of the circulatory system.

## Embryonic Development of the Cardiovascular System

Most blood vessels arise within embryonic mesoderm (or from mesenchyme) almost as soon as this germ layer becomes established. Small clusters of mesodermal cells called **blood islands** mark the embryonic debut of the cardiovascular system (figure 12.7a–d). Embryonic blood islands yield both blood vessels and blood cells, so they are involved in **angiogenesis** (blood vessel formation) and **hemopoiesis** (blood cell formation). Blood islands merge, forming a connected vascular network that eventually links parts of the embryo to each other and connects it to its nutrient supply and respiratory organs. The embryonic heart is tubular.

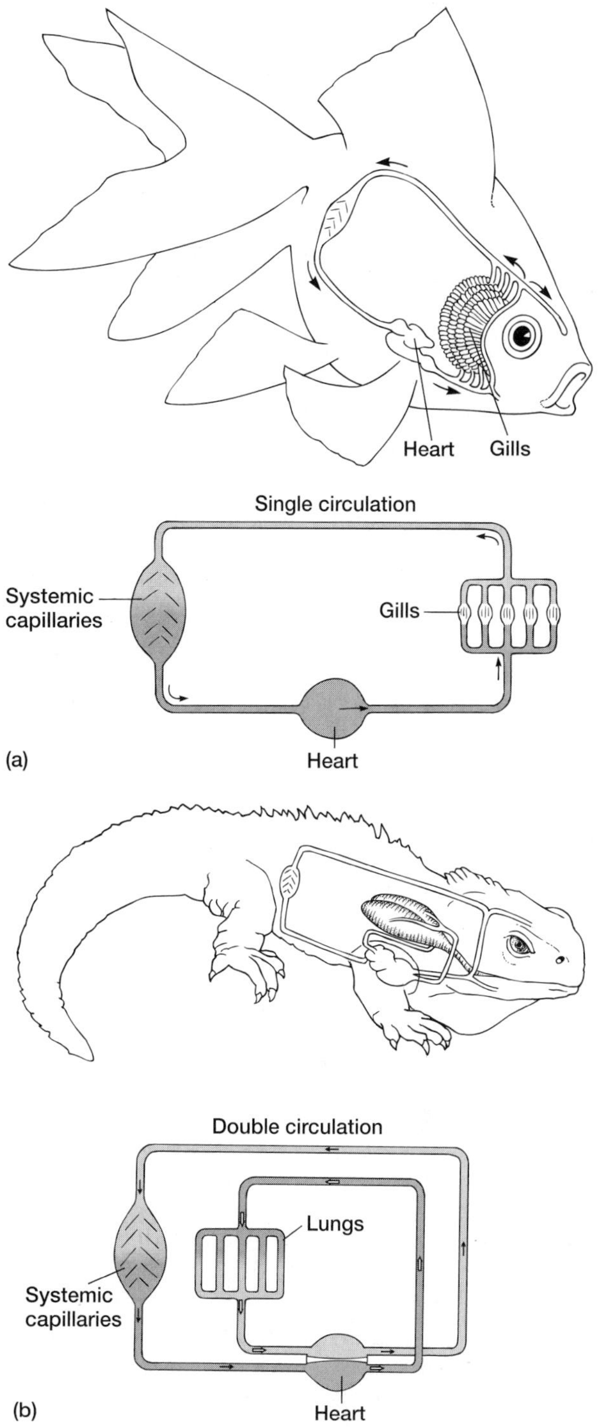


**Figure 12.6** Single and double circulation. (a) The single circulation of fishes includes heart, gills, and systemic capillaries in series with one another (arrows indicate path of the blood flow). (b) The double circulation of most amniotes includes heart, lungs, and systemic capillaries. Blood passes twice through the heart before completing one route. This places lungs and systemic tissues in separate circuits that parallel each other. Solid arrows indicate systemic circulation to and from systemic capillaries. Open arrows indicate pulmonary circulation to and from lungs.

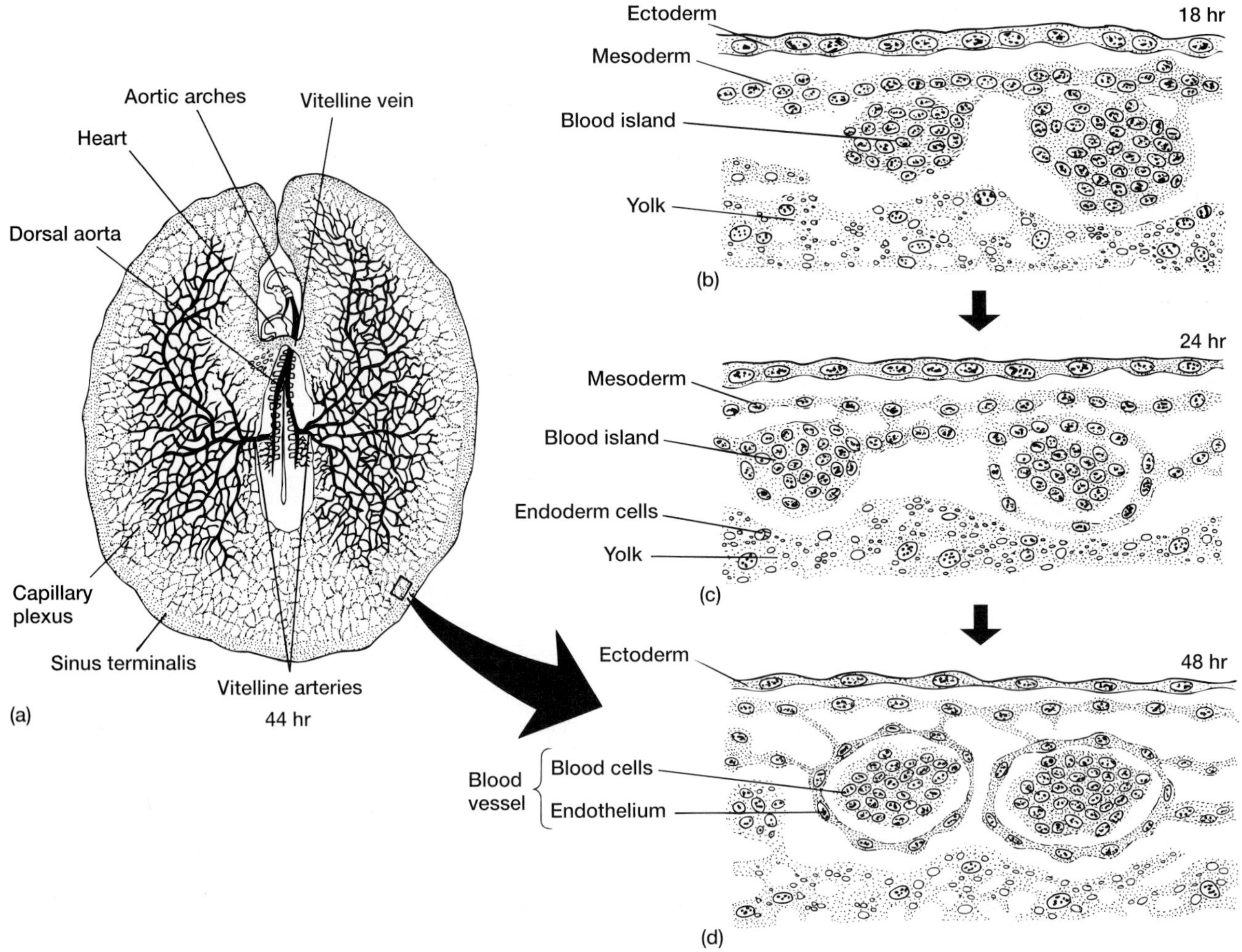


**Figure 12.7** Embryonic blood cell formation. (a) Chick embryo after about 44 hr of incubation (ventral view). The circulatory system is already well established. Peripheral blood islands have coalesced into major vitelline vessels. (b–d) Detailed developmental sequence of blood vessel formation. (b) Local clusters of mesoderm are organized into blood islands after 18 hr of incubation. (c) Development at 24 hr of incubation. (d) At about 44 to 48 hrs of incubation, distinct blood vessels and blood cells that are part of the vitelline vascular network have been formed.

Early on, it has autonomous, rhythmic beats that drive blood through the developing vascular network. As in the adult, the cardiovascular system of the embryo assumes an active and essential role in respiration, metabolism, excretion, and growth.

When first formed, the embryonic vertebrate heart is already contractile and includes four major adjoining chambers. The **sinus venosus** is the first chamber to receive returning blood. Blood flows next into the **atrium,** then into the **ventricle,** and finally into the fourth chamber, the **bulbus cordis.** From the bulbus cordis, blood leaves the heart to enter arteries departing for the body of the embryo. In most tetrapods, splanchnic mesoderm forms the basic four-chambered, tubular heart. Development of the heart begins when cells leave the splanchnic mesoderm to form a medial pair of **endocardial tubes** (figure 12.8a,b). Cells remaining in the splanchnic mesoderm proliferate, producing a thickened lateral region, the paired **epimyocardium.** Cells of the endocardial tube and epimyocardium grow toward the midline and fuse into the single, centrally located, tubular heart. Specifically, the fused endocardial tubes form the endothelial lining of the heart, called the **endocardium,** and the epimyocardium gives rise to the extensive cardiac muscle of the heart wall, the **myocardium,** together with the thin visceral peritoneum covering the heart's surface. With these fusions, the basic four-chambered embryonic heart is established (figure 12.8c).

Flexions and expansions of the tubular heart twist the heart into different configurations, but the internal path of blood flow remains the same (figure 12.9). In most fishes, adults retain this basic four-chambered embryonic heart. However, in lungfishes and tetrapods, varying degrees of internal subdivisions cordon off additional compartments within the heart, and some of the original chambers may

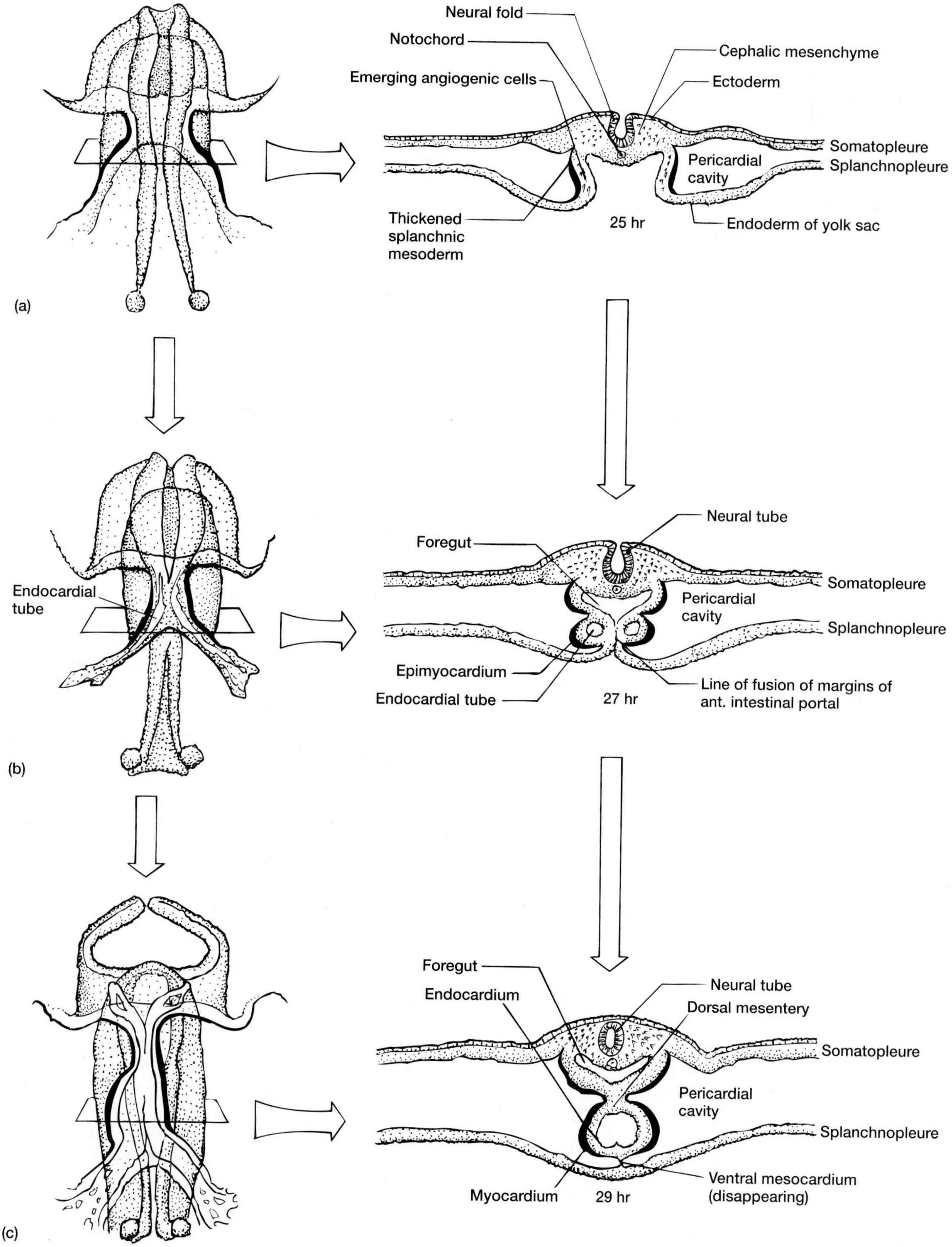


**Figure 12.8** Embryonic heart formation. Chick embryo in successive stages of incubation (25, 27, 29 hr, respectively). Ventral (left) and corresponding cross-sectional (right) views of heart formation are illustrated. (a) Angiogenic cells emerge from the epimyocardium, a thickened splanchnic mesoderm. (b) Angiogenic cells differentiate into a pair of primordial endocardial tubes. (c) This pair of endocardial tubes fuses medially into the single endocardial tube, the future lining of the heart. The thickened epimyocardium forms the thin peritoneum on the surface of the heart and the extensive myocardium, the muscular wall of the heart.

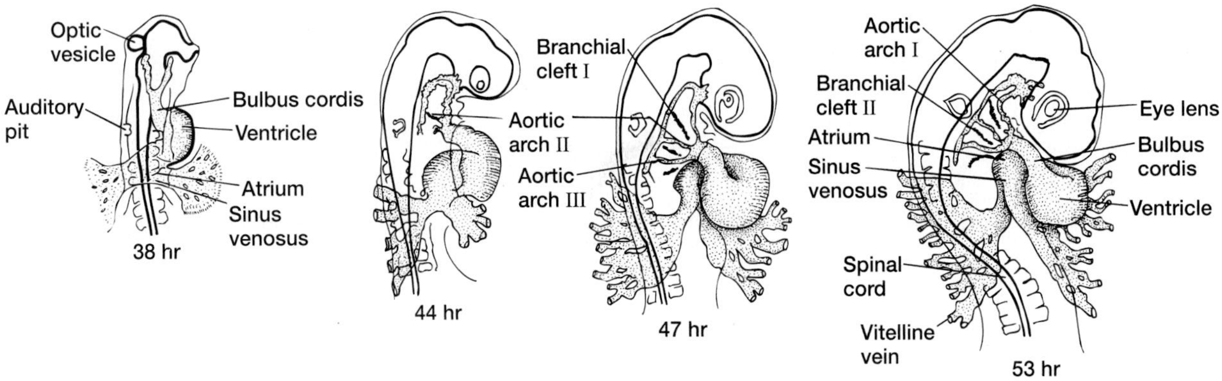


**Figure 12.9** Growth of the chick heart. The four-chambered heart consists of sinus venosus, atrium, ventricle, and bulbus cordis. Once it forms, subsequent folding and enlargement shift the relative positions of these chambers. This process does not alter the route of blood flow through the functioning embryonic heart.

become reduced or appropriated by other parts of the adult vascular system. We examine these anatomical modifications and their functional significance as we meet them in this chapter. First, let us review the basic structure of the major arteries and veins, the blood distribution system the heart serves (figure 12.10).

## Phylogeny of the Cardiovascular System

The vessels of the cardiovascular system are as varied as the diverse organs they supply. However, these variations are based on modifications of a fundamental plan of organization common to vertebrates. Because it is usually highly modified in advance forms, this fundamental organization of the cardiovascular system is most evident in primitive vertebrates. Blood leaving the heart first enters an unpaired **ventral aorta** and courses forward below the pharynx. Anteriorly, the ventral aorta divides into the **external carotids** that carry blood into the ventral region of the head. Before producing these external carotids however, the ventral aorta gives off a series of **aortic arches,** which pass dorsally within the branchial arches between pharyngeal slits. Above the pharynx, these aortic arches meet a paired **dorsal aorta.** Sprouting from the anterior end of the dorsal aorta are the **internal carotids,** which carry blood forward into the head and usually penetrate the braincase to supply the brain. The dorsal aorta itself, however, carries blood posteriorly (figures 12.10 and 12.11).

At about the level of the liver, the paired vessels of the dorsal aorta unite to form the unpaired **aorta,** which distributes blood to the posterior part of the body and eventually extends into the tail as the **caudal artery.** Along the way, the dorsal aorta gives off numerous small **parietal arteries** to the local body wall as well as several major arteries, usually paired, to somatic tissues. Paired **subclavian arteries** supply the anterior appendages (fins or limbs) and usually branch from the dorsal aorta, as do the caudal **iliac arteries** that supply the posterior appendages. The gonads receive blood from paired **genital arteries (ovarian** or **spermatic).** Paired **renal arteries** to the kidneys are large, major branches from the dorsal aorta. This ensures that the kidneys receive blood early in the arterial circuit while blood pressure is still relatively high, a feature of the hemodynamics that aids renal filtration. Typically in vertebrates, three unpaired arteries depart from the dorsal aorta to supply the viscera. These are the **celiac,** supplying the liver, spleen, stomach, and part of the intestines; the **anterior mesenteric,** supplying most of the small intestine; and the **posterior mesenteric,** supplying the large intestine.

In primitive vertebrates, blood return to the heart includes several prominent veins. The **common cardinal vein** or **sinus** is the major vein that receives blood returning from the **anterior cardinal vein (precardinal)** and the **posterior cardinal vein (postcardinal).** These drain the anterior and posterior regions of the body, respectively. Tributaries from the anterior appendage usually empty into the common cardinal vein via the **subclavian vein.** Veins from the lateral body wall and posterior appendage also empty into the common cardinal vein via the **lateral abdominal vein.**

A **portal system** is a vascular channel that begins in one set of capillaries and runs to another without coursing through the heart in between. There are two major portal systems of the venous circulation in vertebrates. The hepatic portal system begins in capillaries within the wall of the digestive tract and runs as the **hepatic portal vein** to the liver, where it empties into the capillaries and blood sinuses of the liver. This hepatic portal vein transports absorbed nutrients directly from the digestive tract to the liver for storage or processing of many end products of digestion. The renal portal system transports blood returning from capillary beds within the tail or hindlimbs through the paired **renal portal veins** that empty into capillaries within the kidneys.

The function of the renal portal system is not well understood. Because it carries caudal blood to the kidneys, some have suggested that it provides a direct route for

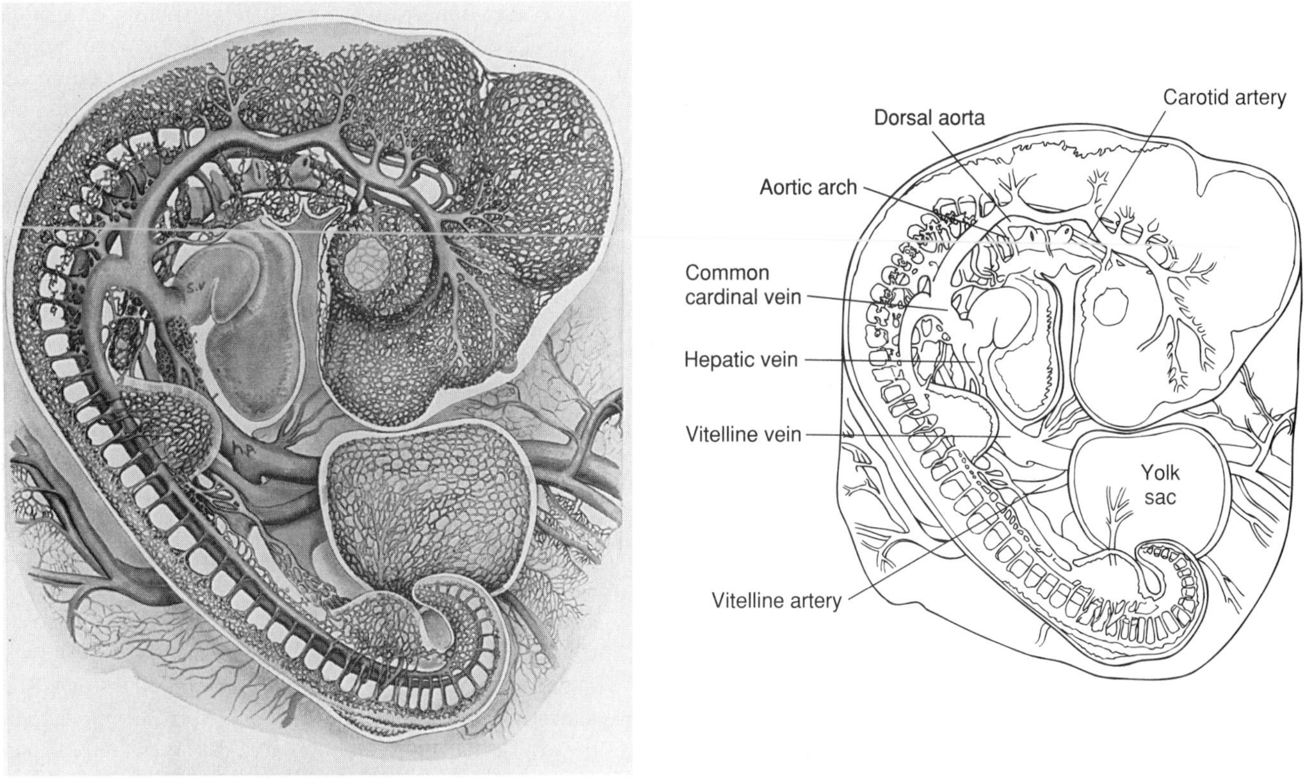


**Figure 12.10** Cardiovascular system of a four-day-old chick. Venous circulation shows anterior and posterior cardinal veins draining into the common cardinal vein that enters the sinus venosus. Vitelline veins return through the forming postcava and travel through the liver sinusoids to enter the heart via the hepatic vein. Arterial circulation is also well established. Aortic arches pass around the pharynx to join above in the dorsal aorta, which supplies blood to the head via the carotid arteries. The dorsal aorta continues posteriorly, eventually forming the vitelline arteries to the yolk.

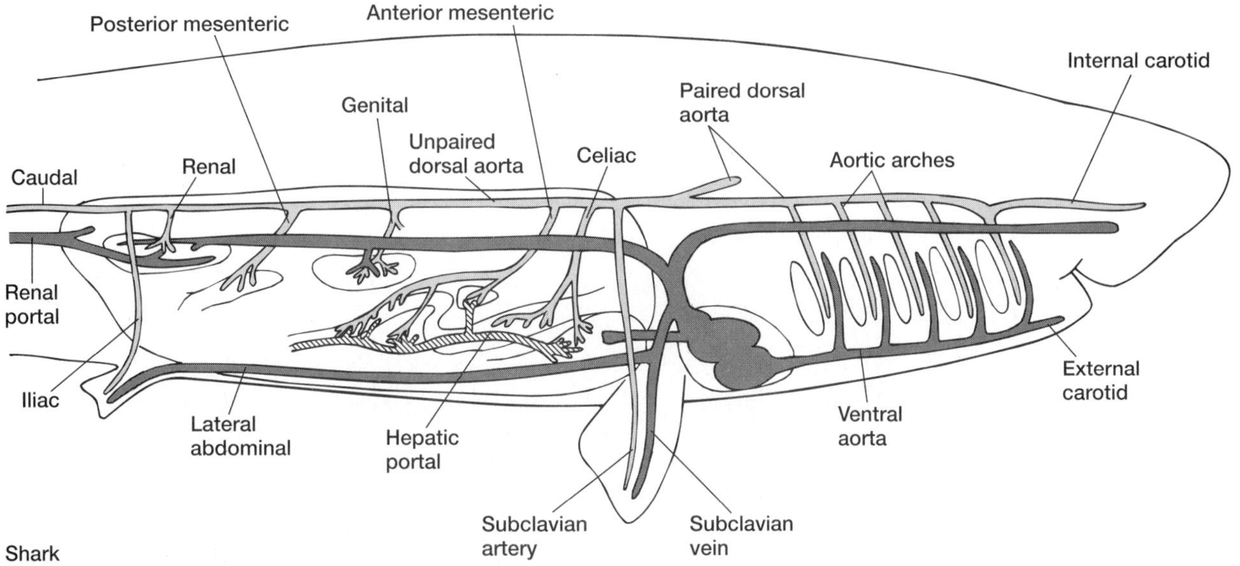


**Figure 12.11** Basic vertebrate circulatory pattern illustrated in a shark. The heart pumps blood to the ventral aorta, from which it is distributed to the paired aortic arches and then to the single dorsal aorta. From the dorsal aorta, blood flows forward to the head and posteriorly to the body, where major branches carry it to visceral and somatic tissues.

delivering metabolic by-products to the kidneys that result from active locomotion involving the caudal musculature. Others suggest that it may represent a way of improving kidney filtration. Arterial blood entering the renal arteries directly from the dorsal aorta has high pressure; venous blood of the renal portal system has low pressure. Kidney filtration depends in part on an initial high pressure to move fluid out of the blood and into the kidney tubules, but low pressure in the renal portal veins aids in the recovery of water and other usable solutes, returning these fluids to the general circulation. The renal portal system is present in all classes of vertebrates except mammals. Even though mammals lack a renal portal system, the mammalian kidney nevertheless has a low-pressure vascular network that may be its counterpart and function similarly to recover fluids from the urine.

Phylogenetic modifications within this basic pattern of arteries and veins are largely correlated with functional changes. In the transition from water to land, gills gave way to lungs, accompanied by the establishment of a pulmonary circulation. In some fishes and certainly in tetrapods, the cardinal veins become less involved in blood return. Instead, the composite, prominent **postcava (posterior vena cava)** arose to drain the posterior part of the body and the **precava (anterior vena cava)** developed to drain the anterior part of the body. Beginning first with arterial vessels, let us now follow the major phylogenetic modifications within the cardiovascular system in detail.

## *Arterial Vessels*

***Aortic Arches*** The number of primitive aortic arches and branchial arches through which they run are still debated. Some ostracoderms had as many as ten pairs of branchial arches and presumably ten pairs of aortic arches. The number of pairs of aortic arches varies in living forms. Lampreys have eight (figure 12.12), hagfishes fifteen. Some species of sharks possess ten or twelve pairs. Nevertheless, only up to six pairs customarily appear during embryonic development in most gnathostome fishes and in all tetrapods. Accordingly, six is the number of aortic arches usually taken as the basic embryonic pattern, and they are designated by roman numerals (I–VI; figure 12.13). The phylogenetic variation within the aortic arches can be intricate (figure 12.14a–e; see also figure 12.17a–c). Thus, reference to a basic six-arch pattern brings a simplifying theme to a complex anatomy. However, you should be prepared for personal preferences of other authors who may use different numbers for aortic arches. Some insist on using numbers up to ten in recognition of the presumed primitive number. Some abandon the effort to assign homologies, ignore those arches lost phylogenetically, and simply number the arches as they find them in the adult, 1, 2, and 3, for example. In this book, I use roman numerals to track the presumed phylogenetic fate of the arches, taking six pairs to represent the basic embryonic pattern.

*Fishes*. Soon after branching from the ventral aorta, the aortic arches divide into capillary beds within the gills. The section of the aortic arch delivering blood to the gills is the **afferent artery,** and the dorsal section carrying it away is the **efferent artery.** The capillary beds between them partially or completely encircle the gills and empty first into the **collecting loop** that joins the efferent artery.

In chondrichthyans, the first pharyngeal slit becomes reduced but not lost, forming the small **spiracle.** During embryonic development, the ventral section of the first aortic arch, which is ordinarily expected to supply the first pharyngeal slit, does not appear. Instead, a vascular sprout from the adjacent collector loop grows to the spiracle, feeding a small capillary bed in its wall. This vessel constitutes the **afferent spiracular artery.** The dorsal section of the first arch forms the **efferent spiracular artery,** which drains this small capillary bed. Because of its small size and the fact that it receives oxygenated blood via the afferent spiracular artery, the spiracle is thought to play little role in respiratory ex-

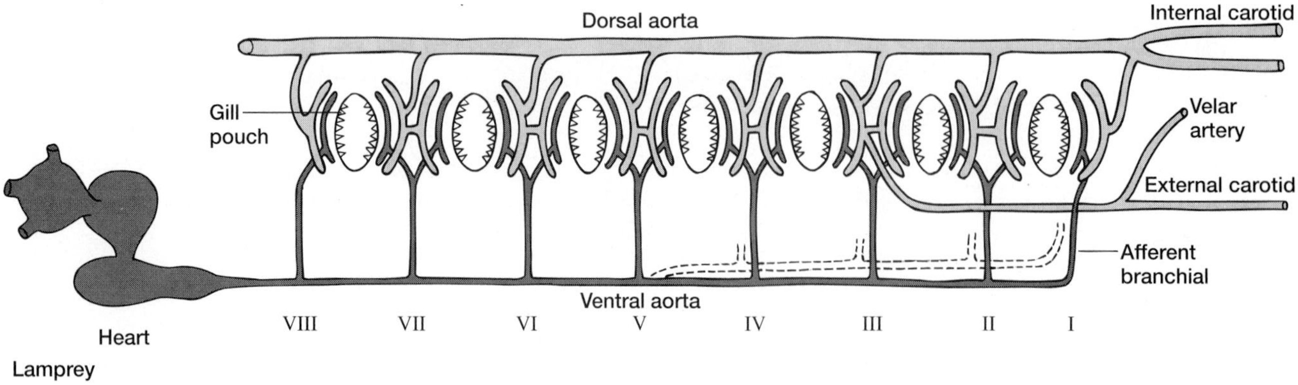


**Figure 12.12** Aortic arches, gills, and anterior arteries of a lamprey.

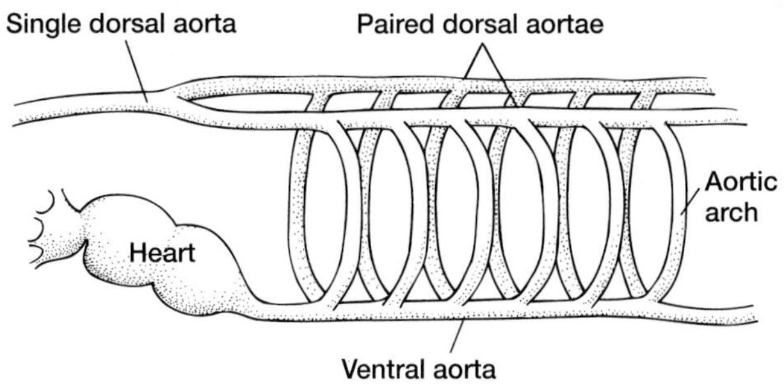


**Figure 12.13** Primitive pattern of aortic arches. Diagram of the basic six-arch pattern.

change. Instead it may develop as part of a secretory or sensory organ.

The remaining aortic arches (II–VI) form small sprouts halfway along their lengths. These merge and cross connect as the collecting loops serving the vascular capillary beds within the gills that form adjacent to the enlarged pharyngeal slits. The anterior and posterior halves of each collecting loop are its **pretrematic** and **posttrematic branches,** respectively (figure 12.15a). Although the external carotid artery arises embryologically from the anterior end of the ventral aorta, it becomes associated with the collecting loop, an understandable change if it is to carry oxygenated blood to the lower jaw. The internal carotid artery supplying the brain receives oxygenated blood from the first fully functional collecting loop (pharyngeal slit II) via the efferent branchial artery (II) (figure 12.15b).

In most actinopterygian fishes, four pairs of aortic arches (III–VI) arise from the ventral aorta. They service the gills associated with five pharyngeal slits. In sturgeons and a few other species, the first pharyngeal slit persists as a small spiracle, but in most fishes, even this modest slit is absent in the adult. The first aortic arch is lost along with this first slit (figure 12.14a).

In most fishes with supplementary air-breathing organs, oxygenated blood departing from these organs enters the general venous circulation, boosting the overall oxygen level of the blood returning to the heart. However, in lungfishes, blood leaving the highly vascularized lungs directly reenters the heart via a separate pulmonary vein (figure 12.14b). Lungfishes are the only fishes that have a separate pulmonary circuit to and from the lung.

In lungfishes, as in other bony fishes, the first pharyngeal slit is reduced to a spiracle that has no respiratory function. Its associated aortic arch (I) is reduced as well. In the Australian lungfish, *Neoceratodus*, the remaining five pharyngeal slits open to fully functional gills supplied by four aortic arches (III–VI). In the African lungfish, *Protopterus*, the functional gills are reduced further. The third and fourth gills are absent entirely, but their aortic arches (III–IV) persist (figure 12.14b). In all lungfishes, the efferent vessel of the most posterior aortic arch (VI) gives rise to the **pulmonary artery** but maintains its connection to the dorsal aorta via the short **ductus arteriosus.**

If you think carefully about the blood flow implied by this anatomical pattern, you will understand the misconceptions it invites. For example, notice that if blood low in oxygen in the ventral aorta flows along its presumed route through arches II, V, and VI, it passes through the capillary beds of the gills, is replenished with oxygen, and enters the dorsal aorta as oxygenated blood. But notice that in the African lungfish, blood low in oxygen in the ventral aorta seems to have an alternative route through arches III and IV, which lack gills. Theoretically, blood could reach the dorsal aorta unaltered, still lacking oxygen. If that happened, as the anatomy alone might suggest, then both oxygenated and deoxygenated blood would mix in the dorsal aorta, reduce the overall oxygen tension in blood flowing to the systemic tissues, and apparently defeat most advantages that air-breathing lungs might bring.

Because of this, the aortic arch pattern seemed inefficient to early anatomists, but they excused it, perhaps even expected such a pattern, because they thought that lungfishes were neither complete water breathers nor air breathers. Lungfishes held a middle ground capable of a bit of both, but doing neither particularly well. This mistaken view, that lungfishes were imperfectly designed compared to more advanced vertebrates, was tested when careful studies of their circulatory physiology were completed. In fact, very little mixing of oxygenated and deoxygenated blood actually occurs, thanks largely, as we see later in this chapter, to the role played by the heart.

*Amphibians*. The same mistaken view that the cardiovascular system was imperfectly designed was also held about amphibians and for much the same reasons. The anatomical arrangement of their aortic arches suggests that some mixing occurs between oxygenated blood from the gills and deoxygenated blood returning from the body.

In amphibians, the first two aortic arches (I, II) disappear early in development. The pattern of the remaining arches differs between larvae and metamorphosed adults. In most larval salamanders, the next three aortic arches (III–V) carry external gills, and the last aortic arch (VI) sprouts the pulmonary artery to the developing lung. A notable exception is the neotenic salamander *Necturus* in which part of the sixth arch disappears and only its dorsal section persists, forming the base of the pulmonary artery (figures 12.16 and 12.19). In most species of salamanders, the external gills are lost following the larva's transformation into the adult, but the aortic arches are retained as major systemic vessels.

The short section of dorsal aorta between aortic arches III and IV, termed the **carotid duct,** usually closes at metamorphosis. This forces the carotids to fill with blood from a derivative of the ventral aorta. The section of ventral aorta between arches III and IV becomes the **common carotid artery,** which feeds the external carotid (from the anterior ventral aorta) and the internal carotid (the

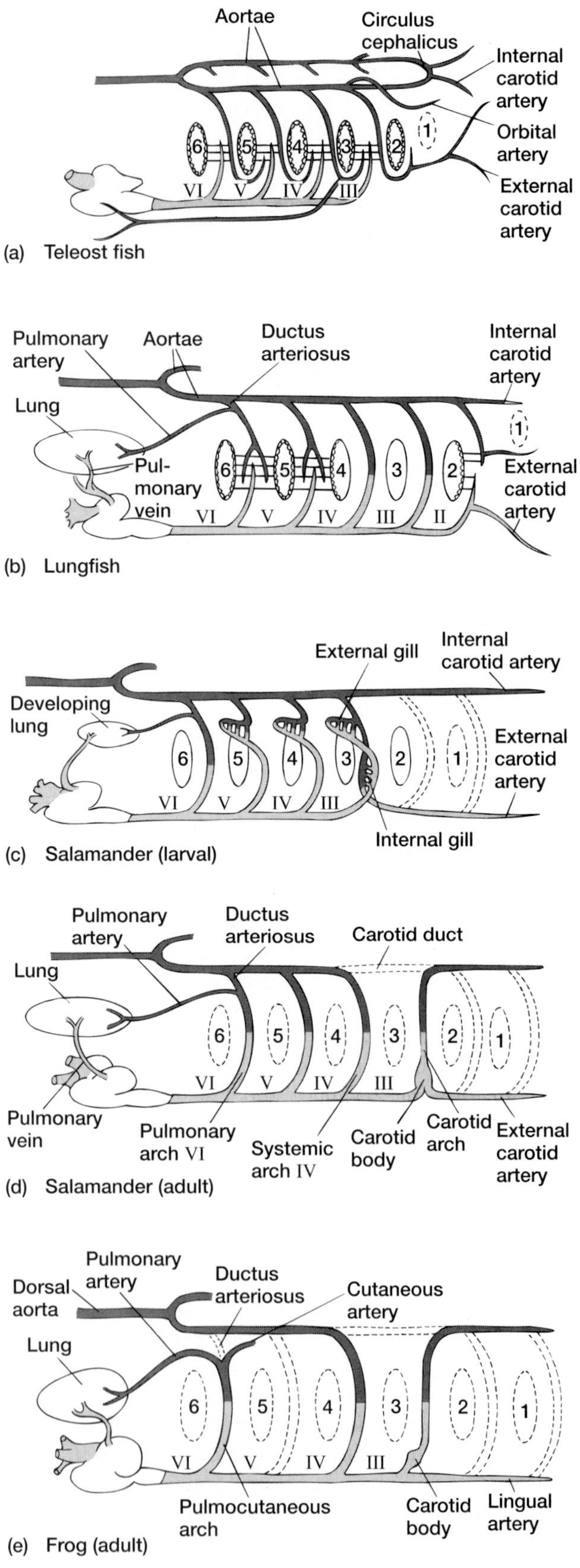


**Figure 12.14** Aortic arches of anamniotes and some derivatives. Diagrams of the basic six-arch pattern. (a) Teleost fish. (b) Lungfish *(Protopterus)*. (c) Neotenic and larval salamander. (d) Adult salamander. (e) Adult frog.

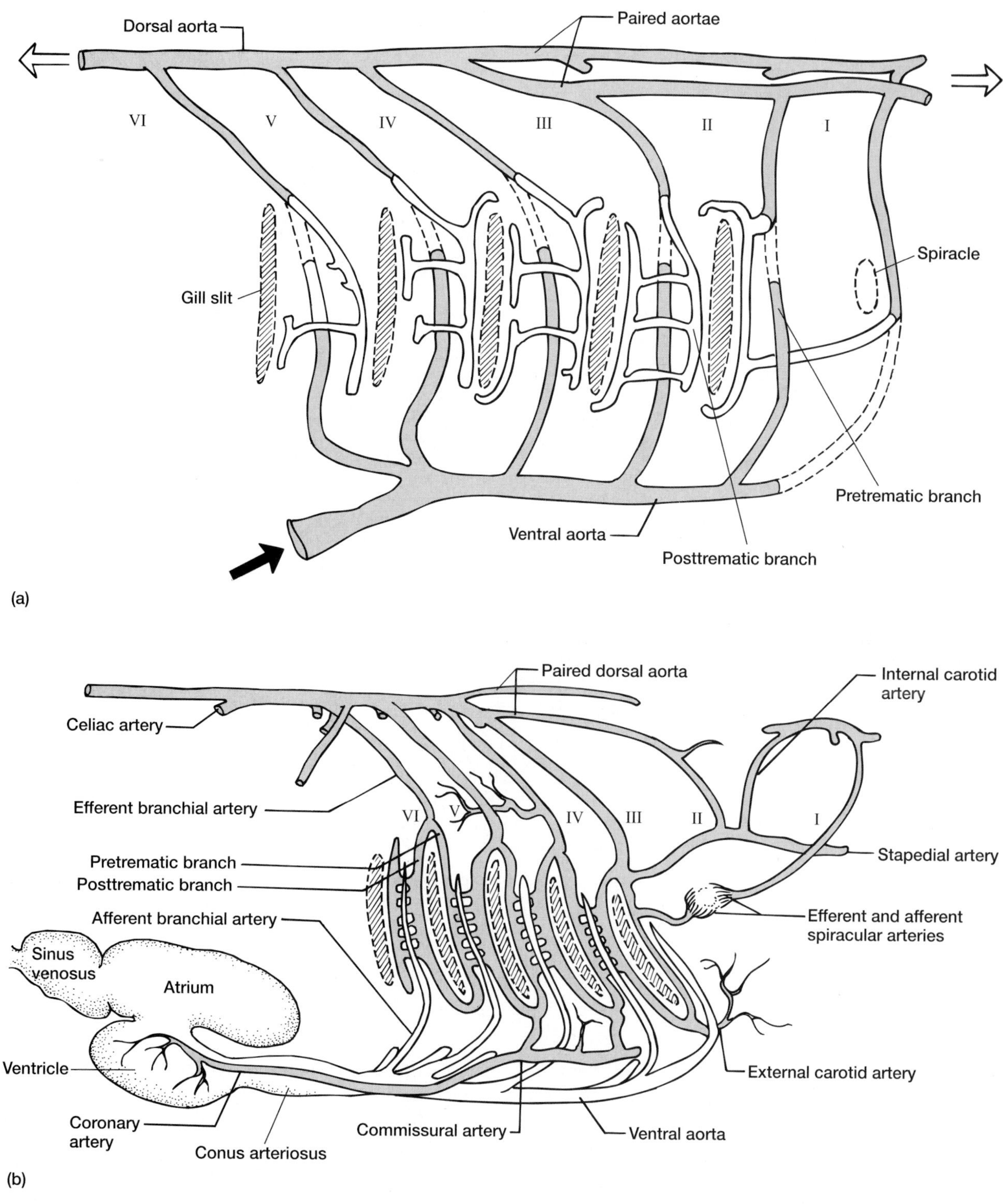


**Figure 12.15** Aortic arches of a shark. (a) Embryonic modifications of aortic arches. New additions (white) to the arches establish the pretrematic and posttrematic parts of the collecting loops that receive afferent and supply efferent branchial arteries, derivatives of ventral and dorsal sections of the aortic arches, respectively. (b) Adult shark aortic arch derivatives. Roman numerals indicate aortic arches.

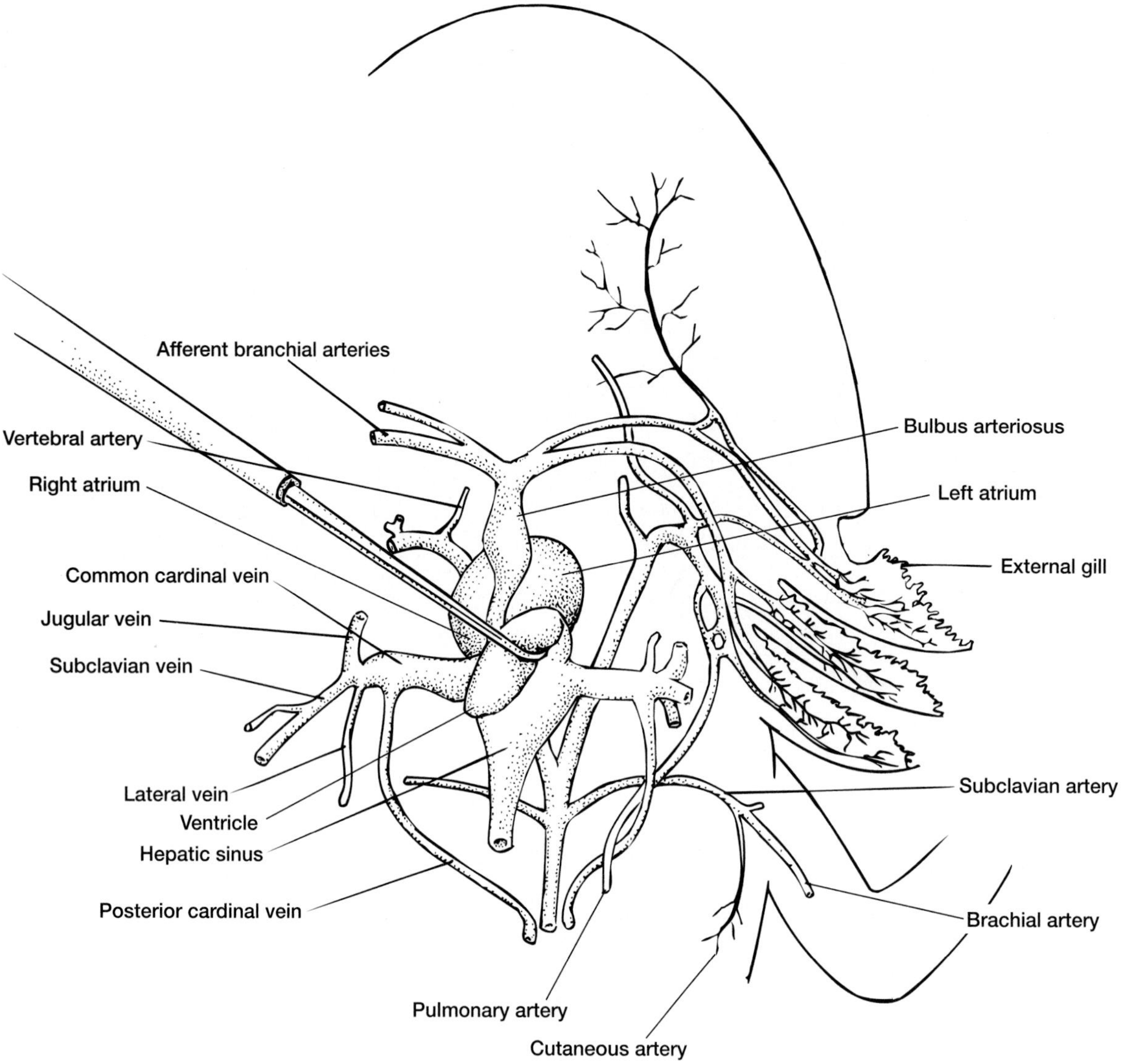


**Figure 12.16** Aortic arches of the salamander *Necturus* (ventral view).

anterior section of the dorsal aorta together with the third aortic arch). The **carotid body** is an enlarged portion of the carotid arteries that usually forms near the point at which the common carotids branch. Its functions are not completely known. Certainly the carotid body plays a role in sensing the gas content or pressure of the blood as well as having some endocrine functions.

The next two arches (IV, V) constitute major systemic vessels that join the dorsal aorta. The final aortic arch (VI) also joins the dorsal aorta, its last short section forming the ductus arteriosus. Shortly before joining the dorsal aorta, the sixth aortic arch gives off the pulmonary artery, which itself divides into small branches to the floor of the mouth, pharynx, and esophagus before actually entering the lungs. In lungless salamanders, the pulmonary artery supplies the skin of the neck and back.

In frogs, the larva usually has internal gills that reside on the last four aortic arches (III–VI), and the embryonic pulmonary artery buds from arch VI. At metamorphosis, these gills are lost together with the carotid duct and all of arch V. The aortic arches that persist (III, IV, and VI) expand to supply blood to the head, body, and pulmonary circuits, respectively. The third arch and associated section of anterior dorsal aorta become the internal carotid. The anterior extension of the ventral aorta is the external carotid. Internal and external carotids both branch from the common carotid, the section of ventral aorta between arches III and IV. A carotid body can usually be found at the root of the internal carotid. The next enlarged aortic arch (IV) joins with the dorsal aorta, the major systemic artery supplying the body. The last arch (VI) loses its connection to the dorsal aorta because the ductus arteriosus closes and becomes the **pulmocutaneous artery.** One branch of the pulmocutaneous artery is the now well-developed pulmonary artery that enters the lung. The other branch is the **cutaneous artery,** which delivers blood to the skin along the dorsal and lateral body wall.

Some early morphologists concluded that inefficiencies in amphibian blood flow would result from these anatomical patterns. For example, oxygenated blood returning to the heart was thought to mix with deoxygenated blood returning from systemic tissues. Suffice it to say that this does not happen. In fact, little mixing occurs. But this mistaken view gained popularity and to some extent still prevails in some scientists' attitudes about the physiology of these lower vertebrates.

*Reptiles*. Beginning in reptiles, but carried into birds and mammals, the symmetrical aortic arches of the embryo tend to become asymmetrical in the adult. Aortic arches III, IV, and VI persist in reptiles, but most of the changes center on enhancements and modification of the fourth arch. Perhaps the most significant anatomical modification of the arterial system in reptiles is the subdivision of the ventral aorta. During embryonic development, the ventral aorta splits to form the bases of three separate arteries leaving the heart—the left aortic arch, the right aortic arch, and the pulmonary trunk (figure 12.17a).

The pulmonary trunk incorporates the bases of the paired sixth arch and their branches as part of the **pulmonary arch** to the lungs. The base of the left aortic arch, the left aortic arch (IV) itself, and the curved section of the left dorsal aorta into which it continues constitute the **left systemic arch.** The **right systemic arch** includes the same components on the right side of the body—the base of the right aortic arch, the right aortic arch itself, and the arched section of the right dorsal aorta. The two systemic arches unite behind the heart to form the common dorsal aorta. The right systemic arch tends to be the most prominent of the two, primarily because of the additional vessels that it supplies. For example, the carotid arteries, originating from the ventral aorta in more primitive vertebrates, arise in reptiles from the right systemic arch. Blood passing through the right systemic arch might flow to the body or enter the carotid arteries to supply the head. In most reptiles, the subclavian arteries branch from the dorsal aorta, but in some reptiles, they branch from the systemic arches. These modifications of the aortic arches in reptiles produce one pulmonary circuit and two systemic circuits, each of which arises independently from the heart.

*Birds*. In birds, the right systemic arch becomes predominant (figure 12.17b). The bases of the aortic arch, the right aortic arch (IV), and the adjoining section of the right dorsal aorta form the right systemic arch during embryonic

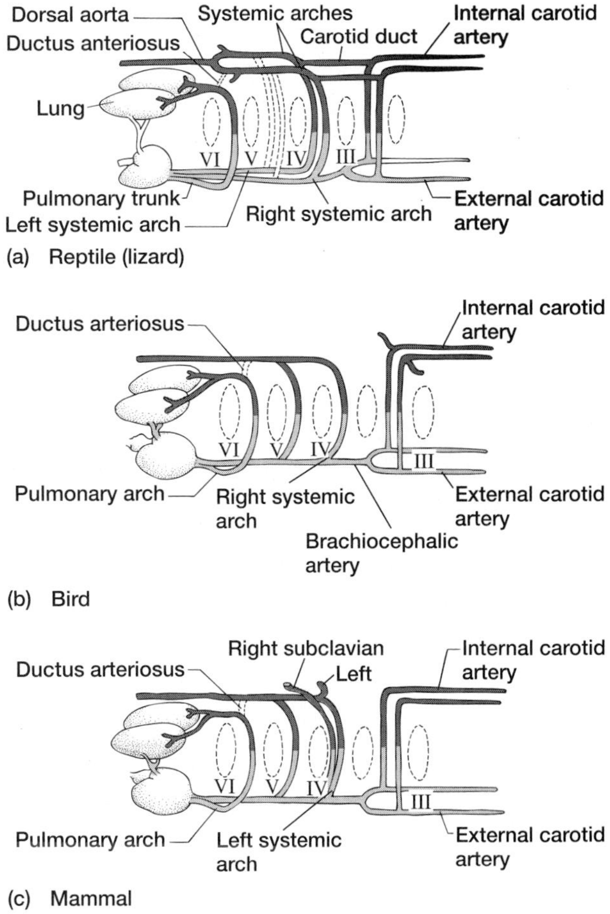


**Figure 12.17** Aortic arches of amniotes. Diagrams of derivatives of the basic six-arch pattern. (a) Reptile. (b) Bird. (c) Mammal.

development. Its opposite member, the left systemic arch, never fully develops. The carotids arise generally from the same components of the aortic arches as in reptiles (aortic arch III and parts of the ventral and dorsal aortae), and they branch from the right systemic arch. However, the paired subclavians to the wings arise from the internal carotids and not from the dorsal aorta. The common carotids and subclavians supply the head and forelimbs, respectively. The common carotids can branch from the right systemic arch separately, or both can join to form a single carotid (figure 12.18a–c). A short but major vessel, the **brachiocephalic artery,** is present in a few reptiles but serves as the major anterior vessel in many birds. It too branches from the right systemic arch. Beyond this junction of the brachiocephalic artery, the systemic arch curves posteriorly to supply the rest of the body. In birds as in reptiles, the pulmonary arch forms from the bases of the paired sixth arch and their branches to supply both lungs.

*Mammals*. Up to six aortic arches arise in the mammalian embryo, but only three persist in the adult as the

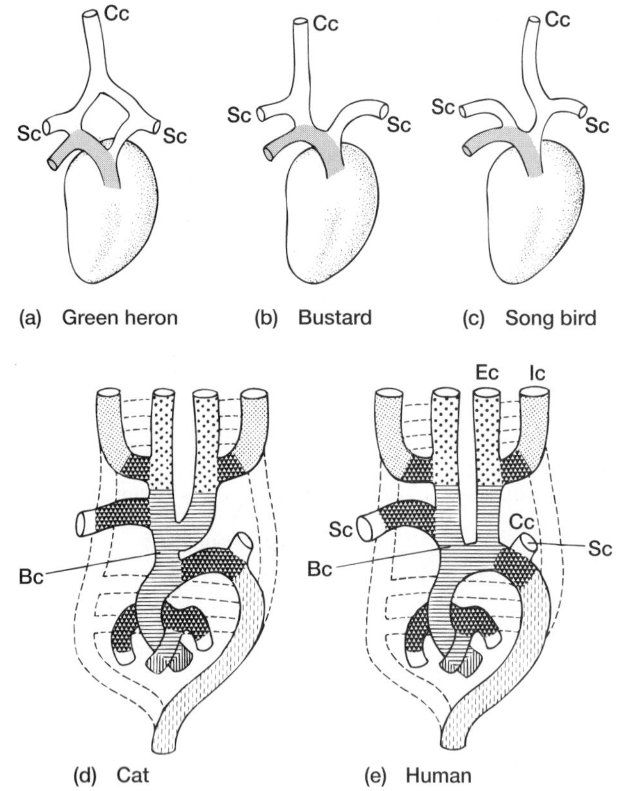


**Figure 12.18** Ventral views of aortic arches. In birds, various alternative configurations of the departing arches can be found. In the green heron (a), two carotid arches unite. A single carotid persists on the right side of the bustard *Eupodotis* (b) and on the left side of the songbird *Passeres* (c). In mammals, formation of major anterior arteries can vary between species, such as the cat (d) and the human (e). Abbreviations: brachiocephalic (Bc), common carotid (Cc), external carotid (Ec), internal carotid (Ic), systemic arch (Sc), subclavian (Sc).

major anterior arteries—the carotid arteries, the pulmonary arch, and the systemic arch (figure 12.17c). The carotid arteries and pulmonary arch are assembled from the same arch components as those of reptiles. The avian carotids arise from the paired aortic arches (III) and parts of the ventral and dorsal aortae. The pulmonary arch forms from the bases of the paired sixth arch and its branches. The systemic arch arises embryonically from the left aortic arch (IV) and left member of the paired dorsal aorta, and therefore is a *left* systemic arch in mammals. The common carotids may share a brachiocephalic origin or branch independently from different points on the aortic arch (figure 12.18d,e). The other notable difference in mammals is in the formation of the subclavian arteries. The left subclavian departs from the left systemic arch in mammals. The right subclavian, however, includes the right aortic arch (IV), part of the adjoining right dorsal aorta, and the arteries that grow from these into the right limb (figure 12.18d,e).

***Overview of Aortic Arch Evolution*** In most fishes, the aortic arches deliver deoxygenated blood to the respiratory surfaces of the gills and then distribute oxygenated blood to tissues of the head (via the carotids) and remainder of the body (via the dorsal aortae). In lungfishes and tetrapods, the aortic arches form the pulmonary arch, the arterial circuit to the lungs, and the systemic arches, the arterial circuits to the rest of the body (figure 12.19). The carotid arteries still bear the primary responsibility for supplying blood to the head in tetrapods, but now they usually branch from one of the major systemic arches. The double systemic arches (left and right) present in amphibians and reptiles (figure 12.20a,b) become reduced to a single systemic arch, the right in birds the left in mammals (figure 12.20c,d). Although birds and mammals share many similarities, including endothermy, active lives, and diverse radiation, they arose out of different reptilian ancestries. Any similarities in their cardiovascular anatomies represent independent evolutionary innovations.

The basic six-arch pattern of aortic arches is a useful concept that allows us to track aortic arch derivatives and organize the diversity of anatomical modifications we encounter. Furthermore, the appearance of six aortic arches during the embryonic development of living gnathostomes suggests that this is the ancestral pattern. However, as we have seen, the actual adult anatomy can be quite varied among different species.

## *Venous Vessels*

The major veins that return blood to the heart are complicated and highly variable. Within each vertebrate group, the veins compose a few main functional systems that arise embryologically from what seems to be a common developmental pattern. Before examining the anatomy of veins in each group, we turn first to these basic systems of venous circulation. In vertebrates with an established double circulation, there are two general functional systems of venous circulation, the systemic system draining the general body tissues and the pulmonary system draining the lungs. Within the systemic system, hepatic portal veins serve the liver, renal portal veins serve the kidneys, and general body veins drain the remaining systemic tissues.

***Systemic System*** Early in development, three major sets of paired veins are present, the **vitelline veins** from the yolk sac, the **cardinal veins** from the body of the embryo itself, and the **lateral abdominal veins** from the pelvic region. The paired vitelline veins are among the first vessels to appear in the embryo. They arise over the yolk and follow the yolk stalk into the body. They then turn anteriorly, continue along the gut, and enter the sinus venosus. The liver primordium grows into the vitelline veins. Proliferation of liver cords breaks up the associated vitelline veins into **hepatic sinusoids.** The remaining short sections of the vitelline

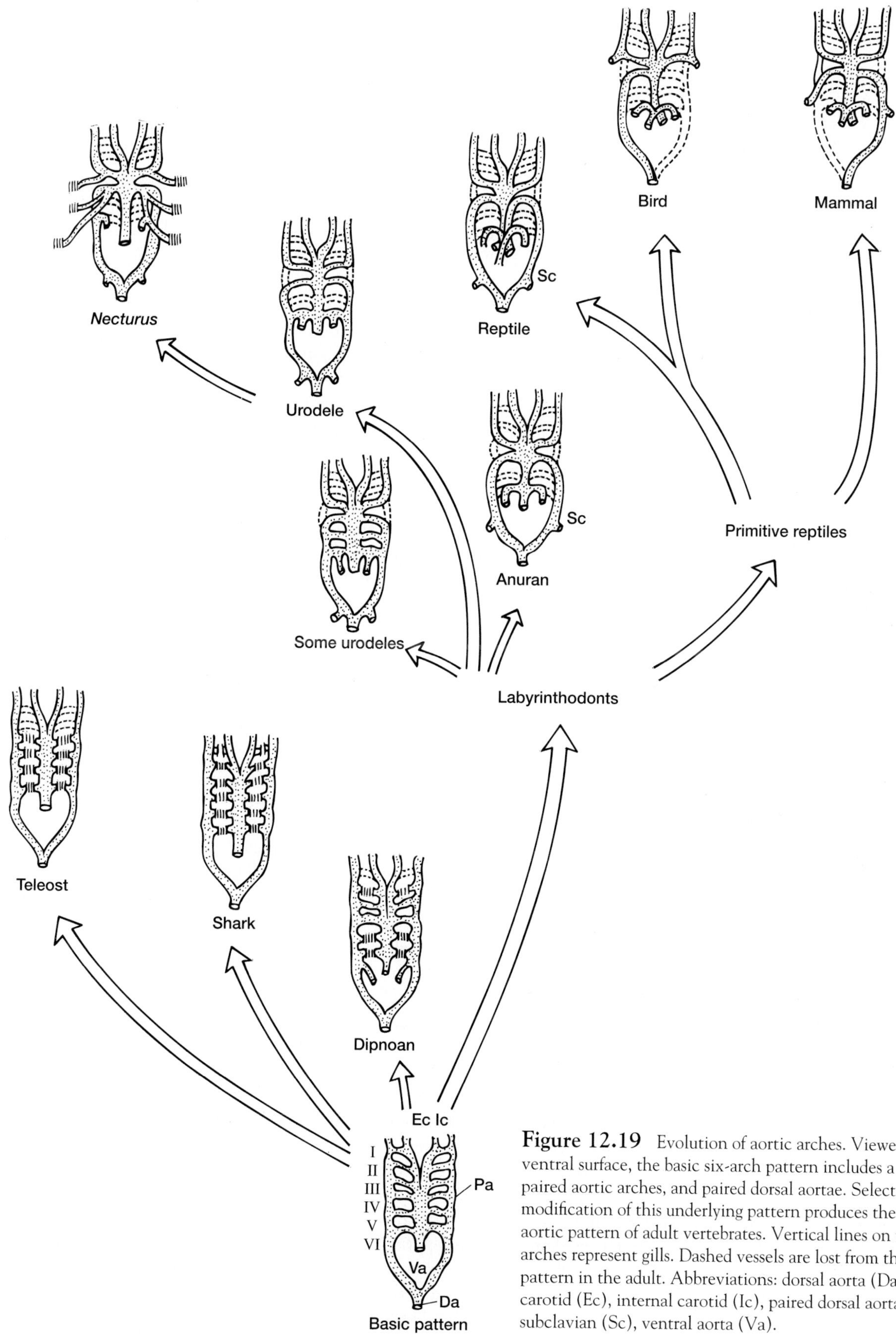


**Figure 12.19** Evolution of aortic arches. Viewed from the ventral surface, the basic six-arch pattern includes a ventral aorta, paired aortic arches, and paired dorsal aortae. Selective loss or modification of this underlying pattern produces the derived aortic pattern of adult vertebrates. Vertical lines on the aortic arches represent gills. Dashed vessels are lost from the basic pattern in the adult. Abbreviations: dorsal aorta (Da), external carotid (Ec), internal carotid (Ic), paired dorsal aorta (Pa), subclavian (Sc), ventral aorta (Va).

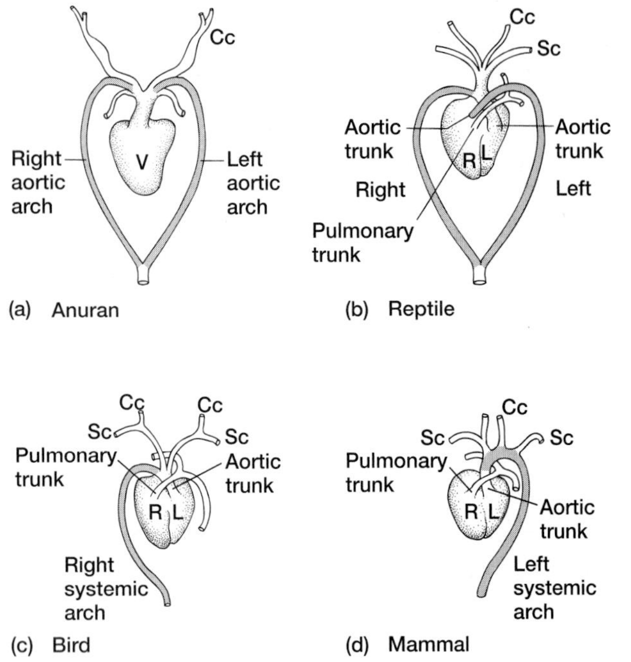


**Figure 12.20** Fate of the systemic arches in tetrapods (ventral views). Systemic arches of both sides persist in the adult in anurans (a) and reptiles (b). The right systemic arch persists in birds (c), and the left in mammals (d). Abbreviations: common carotid (Cc), left ventricle (L), right ventricle (R), subclavian (Sc), ventricle (V).

veins that drain these hepatic sinusoids and enter the sinus venosus are the **hepatic veins.**

The cardinal veins include the **anterior cardinal veins,** which drain blood from the head region, and the **posterior cardinal veins,** which return blood from the embryo's body. Both pairs of anterior and posterior cardinals unite at the level of the heart into short **common carotid veins** that open into the sinus venosus. The anterior cardinals consist of several parts that develop as vessels receiving tributaries from the brain, cranium, and neck. The posterior cardinals develop primarily as vessels of the embryonic kidneys.

The lateral abdominal veins are present in fishes, but they are usually merged or absent in tetrapods. In fishes, each vein joins the **iliac vein** from the pelvic fin and travels forward in the lateral body wall. At the level of the shoulder, the iliac vein joins the **brachial vein** and thereafter becomes the **subclavian vein,** which turns medially to enter the common cardinal vein. However, in tetrapods, the subclavian returns separately to the heart, and the lateral abdominal veins enter the liver. In amphibians, left and right lateral abdominal veins may unite into a single median vein, the **ventral abdominal vein,** that runs along the floor of the body coelom. In birds and mammals, the abdominal vein is absent.

Subsequent venous development involves changes in these early paired vessels accompanied by anastomoses between them, loss of parts by atrophy, and appearance of additional embryonic vessels. The alterations are usually more extensive than those seen among arteries and produce major, often asymmetrical, adult venous routes of blood return to the heart.

*Hepatic portal vein.* The hepatic portal vein runs from the digestive tract to the liver and forms a direct route to transport absorbed end products of digestion immediately to the liver. It is common to all vertebrates and develops mostly from the embryonic **subintestinal vein,** an unpaired vessel originating in the caudal vein (figure 12.21a). The subintestinal vein loops around the anus and extends forward, running along the ventral wall of the intestine from which it collects blood. It passes through the liver from the intestine and finally joins the left vitelline vein. Within the proliferating liver cords, the vitelline veins become broken into a network of small hepatic sinusoids. The anterior end of the subintestinal vein empties blood into these hepatic sinusoids, and its posterior end regresses to lose contact with the caudal vein. This modified subintestinal vein is now properly called the **hepatic portal vein** (figure 12.21b). It collects blood not just from the intestines but also from the stomach, pancreas, and spleen and delivers it to the vascular sinusoids within the liver.

*Renal portal system.* Early in development, blood returning in the caudal vein from the tail flows through the subintestinal vein or through the posterior cardinals, the posterior cardinals being the more usual route (figure 12.21a). The posterior cardinals travel dorsal to the kidneys, drain blood from them, and then continue forward to empty into veins entering the heart. Subsequently, a set of **subcardinal veins** arise ventral to the kidneys, drain them, and run forward to empty into the posterior cardinals. Once this route via the subcardinals becomes established, the short section of posterior cardinal atrophies between its junction with the subcardinal and the kidney. At this point in development, the subintestinal vein has also lost its connection with the caudal vein. As a consequence of these vascular alterations, blood from the tail must now pass through the kidneys. With the routing of caudal blood through the kidneys, the caudal vein becomes the renal portal system. From the kidneys, blood is drained by the newly established subcardinal veins (figure 12.21b).

In general, blood entering the renal portal system arrives from the caudal vein draining the tail. However, alternative renal portal routes occur in some vertebrates. In cyclostomes and some teleosts, blood of the renal portal system enters the kidneys via segmental veins from the body wall. In some lungfishes, additional blood from the pelvic fins and the posterior abdominal region contribute to the renal portal flow entering the kidneys. The caudal vein of these lungfishes does not supply but instead drains the kidneys and then continues forward to join the posterior cardinals or postcava.

*General body veins.* In primitive vertebrates, the basic early embryonic pattern is retained, and blood from anterior

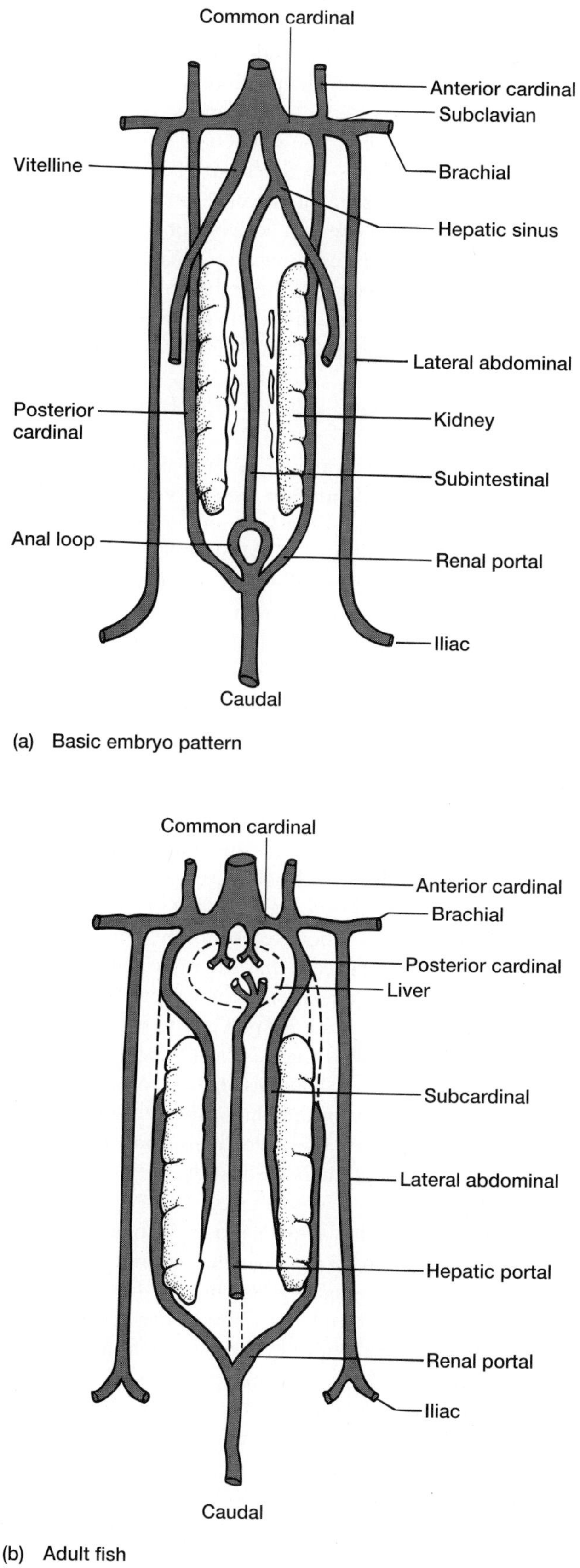


**Figure 12.21** Major veins. The basic embryonic (a) and modified adult (b) pattern of major veins in the shark. The hepatic portal vein forms from the embryonic subintestinal vein. Anterior parts of the embryonic vitelline veins give rise to the short hepatic veins draining the liver. Addition of a subcardinal vein drains the kidney, and the renal portal becomes established from posterior derivatives of the posterior cardinal veins. The lateral abdominal vein drains the pelvic appendage, receives blood from the body wall as it runs anteriorly, and joins with the subclavian vein from the pectoral appendage and the anterior cardinal vein from the head to enter into the heart.

and posterior systemic tissues is returned in anterior and posterior cardinal veins, both pairs of veins uniting in common cardinal veins near the heart. In derived vertebrates, the cardinals appear but usually persist only in the embryo, being functionally replaced by alternative adult vessels, the precava and postcava (anterior and posterior venae cavae).

The embryonic derivation of the precava and postcava from precursor veins reveals the extensive modification on which the adult venous system is based. Formation of the precava is preceded by the early embryonic appearance of the anterior, posterior, and common cardinal veins (figure 12.22a). Formation of the precava itself begins with the enlargement of small intersegmental veins into the subclavian veins that empty into the anterior cardinals (figure 12.22b). Next, an **intercardinal anastomosis** develops between the anterior cardinals (figure 12.22c). With growth of the embryo, these newly established channels become used increasingly, especially on the right side, to return blood from the head. The common cardinal of the right side enlarges to receive this returning blood and becomes the precava of the adult (figure 12.22d). The common cardinal of the left side regresses, persisting only as a small vein from the atrium of the adult heart.

Formation of the postcava is even more elaborate. Initially, the paired posterior cardinal veins return blood from the embryonic body behind the heart. However, subsequent consolidation of parts of three embryonic vessels—hepatics, subcardinals, and supracardinals—and extensive anastomoses between them result in a progressive shift of returning blood away from the posterior cardinals to an emerging single medial channel made up of parts of several veins. Contributing vessels to this return channel eventually merge into a single vein, the postcava. Its developmental history begins with the appearance of the posterior cardinals that drain the early embryonic kidneys (mesonephros). Next, the subcardinal veins arise and connect with each other through the **subcardinal anastomosis.** Finally, the **supracardinal veins** develop and provide supplementary drainage of the posterior body.

Anteriorly, the right vitelline vein (right hepatic vein) joins the right subcardinal, and posteriorly a new connection becomes established with subcardinals and supracardinals (figure 12.22c). As a result of these anastomoses and consolidations between vessels, an unpaired medial channel develops, which offers an alternative return route to the heart as the earlier return route via the posterior cardinals regresses. This channel is modest at first, but it enlarges as more blood seeks this path of return to the heart, and it eventually becomes the adult postcava. Thus, the precava and postcava are mosaics of preceding vessels, parts of which are pirated during embryonic development to produce the definitive adult vessels that drain the anterior and posterior parts of the body, respectively.

***Pulmonary System*** Many fishes have supplementary air-breathing organs, but only fishes with lungs possess a pulmonary system. Among living fishes, only dipnoans have true lungs. If the ancient placoderms had lungs, a possibility mentioned earlier, then the pulmonary system would have evolved early in vertebrate evolution.

*Pulmonary veins.* The pulmonary veins return blood from the paired lungs to the heart. Before entering the heart, they usually unite into a single vein. Embryologically, the pulmonary vein does not arise by conversion of existing vascular channels. Instead, numerous small vessels originate separately within and drain the embryonic lung buds. They then converge into several common vessels that become the pulmonary veins entering the left atrium.

**Lung evolution (p. 404)**

***Fishes*** The head is drained by the paired anterior cardinal veins and small **inferior jugular veins** that join the common cardinal veins just before they empty into the sinus venosus of the heart. The subclavian and iliac veins drain the appendages via the lateral abdominal vein. Both also join the common cardinal. In most fishes, modification of the posterior cardinal diverts all returning blood from the tail so that it flows through the kidneys before emptying into the remaining sections of the posterior cardinal. The hepatic portal vein transports blood from the digestive tract to the capillaries in the liver. From the liver, blood flows to the heart via the short hepatic veins (figure 12.23a,b).

In actinopterygians, the lateral abdominal veins are usually lost and the pelvic fins are drained by the posterior cardinal. Blood from gas bladders empties into the hepatic or common cardinal veins.

In lungfishes, venous return to the heart is similar to that of other fishes except that the right posterior cardinal vein enlarges to assume most of the responsibility for draining blood from the posterior part of the body and, accordingly, is usually called the postcaval vein (figure 12.23c). The paired lateral abdominals fuse to form the unpaired ventral abdominal vein that drains the pelvic fins and empties into the sinus venosus. Blood returning from the lungs enters the atrium of the heart directly.

***Amphibians*** In larval salamanders, such as *Necturus*, the **internal jugular** (derived from the anterior cardinal) and **external jugular veins** together with the small **lingual vein** from the tongue return blood from the head. The large postcava offers one route for blood returning from the kidneys. The **ventral abdominal vein** transports blood primarily from the hindlimbs to the liver sinusoids. Blood from the tail is offered several alternative return routes—through the ventral abdominal, the posterior cardinal, or the postcava via the kidneys. The hepatic portal persists. Numerous hepatic veins that enter the postcava drain the liver (figure 12.23d,e).

The veins of adult salamanders are much the same as the larvae. In adult anurans, the major difference is in the posterior cardinal, part of which is lost between the kidney

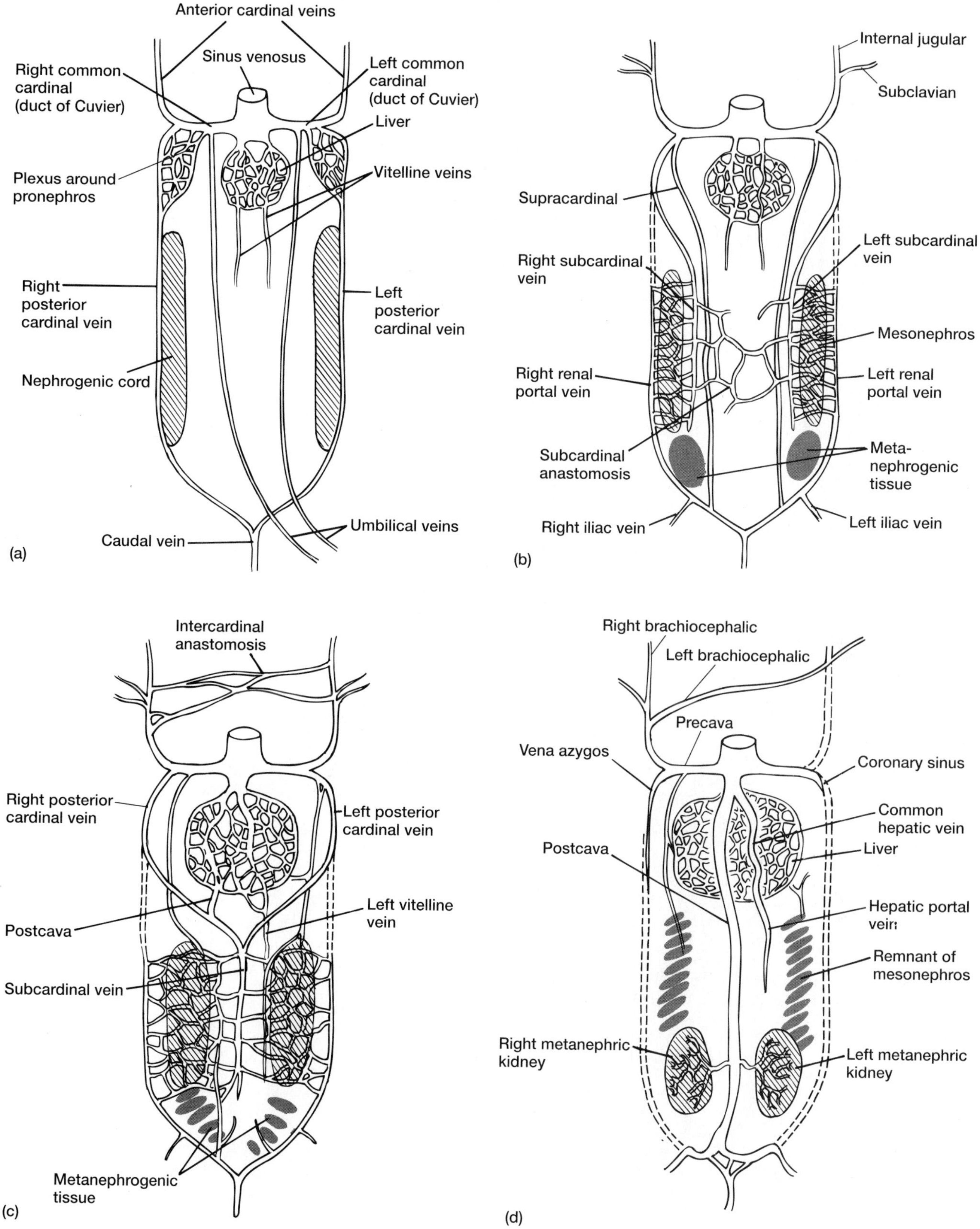


**Figure 12.22** Embryonic development of mammalian veins. (a) Early in development, the anterior, posterior, and common cardinals become established. (b) Intersegmental veins close to the pectoral limbs come to empty into the anterior cardinals. The subcardinals arise between the kidneys and pass forward to enter the posterior cardinals. (c) Intercardinal anastomosis becomes established between the anterior cardinals. Returning blood from the posterior body now includes a route through the liver because part of the right vitelline vein has been incorporated within the right subcardinal. (d) The precava receives blood from left and right branchiocephalic veins (intercardinal anastomosis and right anterior cardinal, respectively). The postcava is the major channel returning blood from the posterior region of the body.

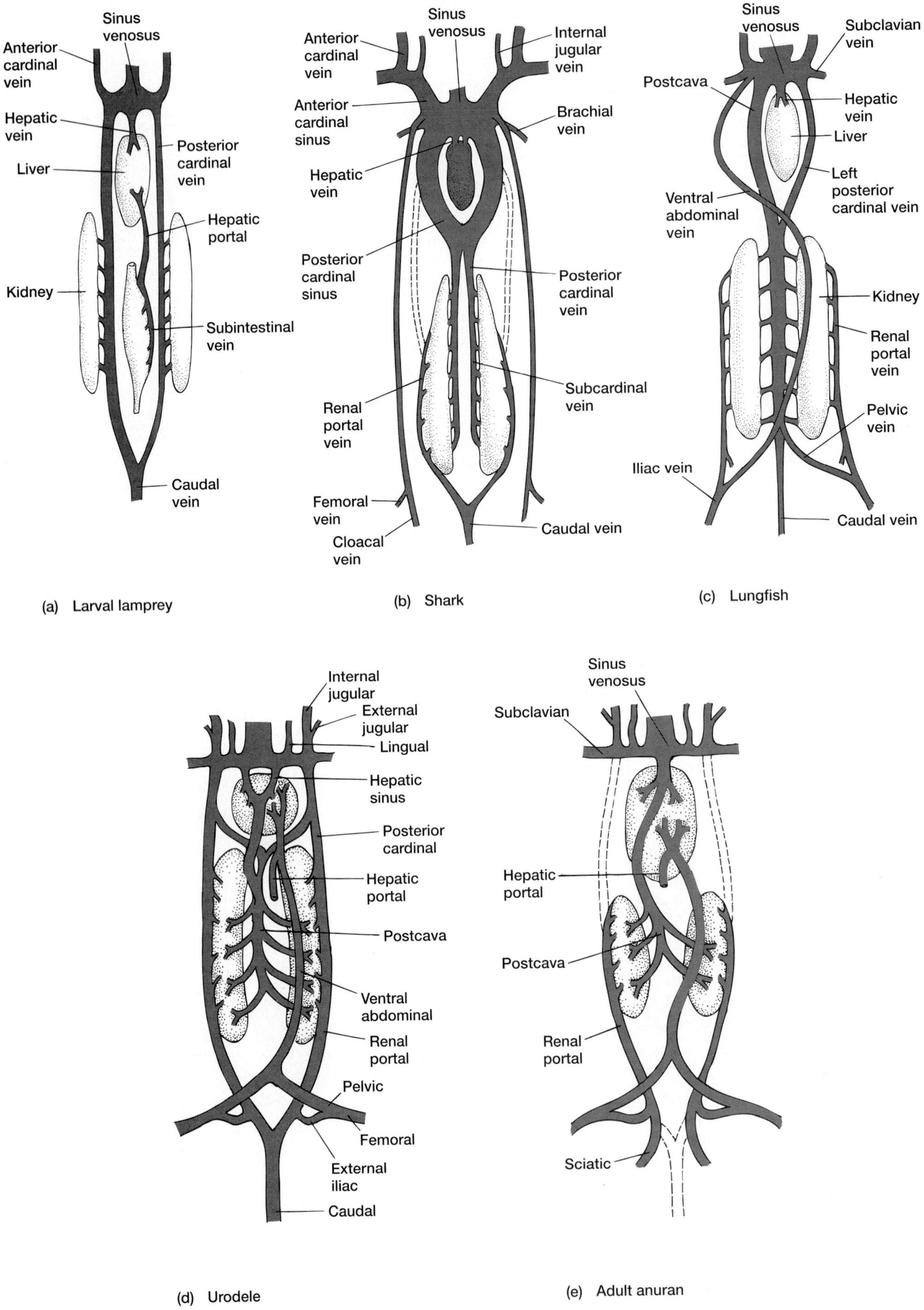


**Figure 12.23** Major venous channels of vertebrates. (a) Larval lamprey. (b) Shark. (c) Lungfish (*Protopterus*). (d) Urodele. (e) Adult anuran.

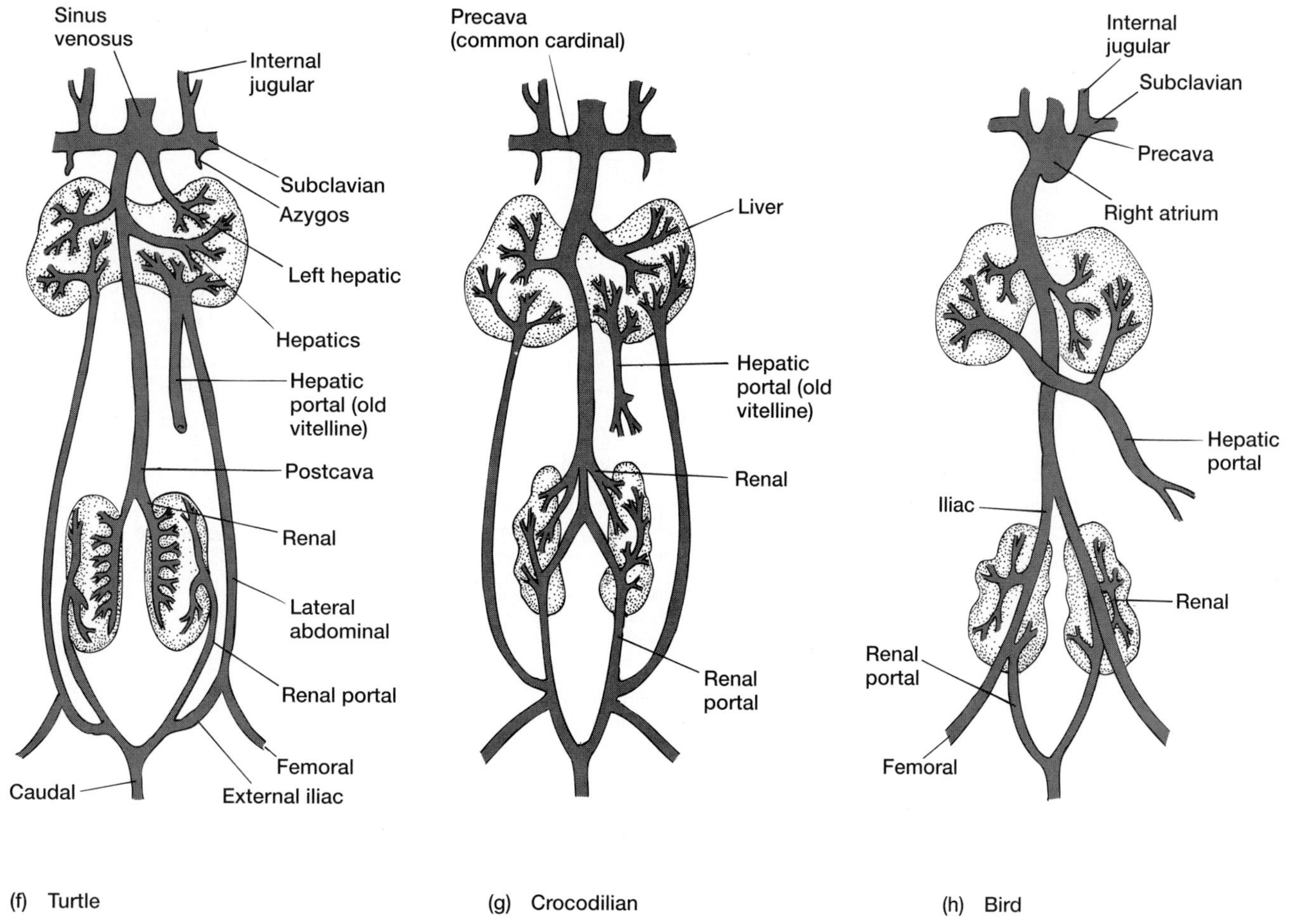


**Figure 12.23,** continued. Major venous channels of vertebrates. (f) Turtle. (g) Crocodilian. (h) Bird.

and the common cardinal; therefore, the posterior cardinal cannot return blood from the kidneys to the heart.

***Reptiles*** The internal jugular (derived from the anterior cardinal), external jugulars, and subclavian from the forearm are tributaries of the paired common cardinals. The enlarged and modified common cardinals are customarily called the precava in reptiles. The posterior cardinal is considerably reduced to small **azygous veins** in the thorax. Paired lateral abdominal veins are present as is a single postcava. The hepatic portal joins the capillaries of the digestive tract with the sinusoids of the liver. Blood from the liver sinusoids returns by short hepatic veins that join the postcava. The precava and postcava enter the much reduced sinus venosus of the heart (figure 12.23f,g).

***Birds*** Short external jugulars join long internal jugulars (anterior cardinals) to return blood to the common cardinals, which are modified into the paired precava. The femoral, caudal, and renal veins are tributaries of the extensive postcava, which also receives hepatic veins before entering the heart. The hepatic portal and renal portals are also present (figure 12.23h).

***Mammals*** Renal portal and abdominal veins are absent in mammalian venous circulation, but a hepatic portal vein is present. The cardinal vessels are substantially modified to produce two major vessels, the single precava (the superior vena cava in humans) and the single postcava (inferior vena cava in humans). These vessels collect blood from the anterior and posterior parts of the body, respectively, and return it to the right atrium of the heart. The posterior vena cava is divided into several sections, including the hepatic, renal, and subcardinal veins (figure 12.23i).

## Hearts

The heart is a pump that moves blood through vessels. In a slow-swimming dogfish shark, the heart can move 7.5 liters of blood per hour; in a resting hen 24 liters per hour; in a human 280 liters (about 75 gallons) per hour. In a giraffe, almost 1,200 liters of blood can circulate throughout the body per hour. If heart rate increases, a response known as

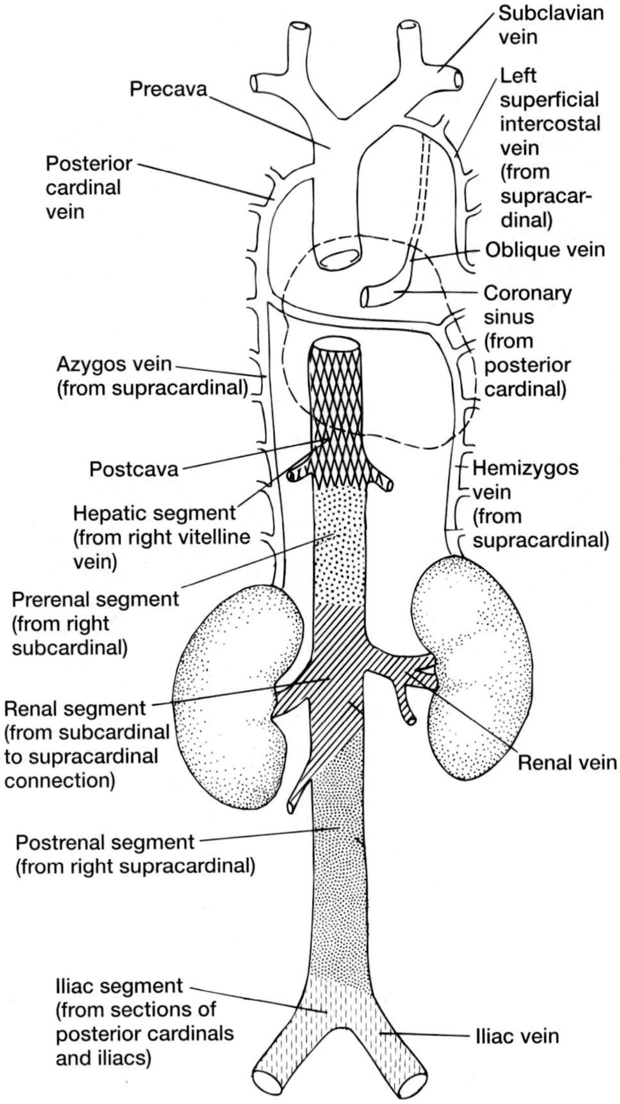


**Figure 12.23,** continued. (i) Mammal.

**tachycardia,** these values can more than triple. If heart rate decreases, **bradycardia** results and these values can fall precipitously. For example, when a turtle dives, its cardiac output can drop to less than a tenth of its prior output. In addition to functioning as a pump, the heart also serves to channel deoxygenated and oxygenated blood to appropriate parts of the circulation, thereby preventing their mixing. Before discussing the special functions of the heart, we first look to its structure in vertebrates.

## *Basic Vertebrate Heart*

Phylogenetically, the heart probably began as a contractile vessel, much like those found within the circulatory system of amphioxus. In most fishes, the heart is part of a single circulation. Vessels serving gas exchange in the gills and systemic capillary beds are in series with each other. The embryonic fish heart consists of four chambers, which are also in series, so that blood flows in sequence from the sinus venosus, to the atrium, to the ventricle, and finally to the fourth and most anterior heart chamber, the bulbus cordis, before entering the ventral aorta. Differences in structure, doubts about homology, and loose use of terms have led to confusion about nomenclature for this fourth chamber. We use the term *bulbus cordis* for this chamber in embryos. In adults, we use the term ***conus arteriosus*** for this anterior chamber if it is composed of cardiac muscle, contractile, and contains various numbers of **conal valves** internally.

A conus arteriosus is generally present in chondrichthyans, holosteans, and dipnoans. Although absent in adult tetrapods, during embryonic development its forerunner, the bulbus cordis, divides into the bases of the major arteries leaving the heart. In some fishes, most notably in teleosts, this anterior chamber is thin walled with smooth muscle and elastic fibers, but it lacks both cardiac muscle and conal valves so it usually is called the **bulbus arterio-**

**sus.** The adult bulbus arteriosus, like the conus arteriosus, arises generally from the embryonic bulbus cordis, but the adult bulbus arteriosus may incorporate part of the adjoining ventral aorta as well in some fishes. Another term often used ambiguously in the older literature is **truncus arteriosus,** which should apply only to the ventral aorta or its immediate derivatives but not to any part of the heart proper. In tetrapods, the ventral aorta often becomes reduced, sometimes persisting only as a small section of vessel at the base of major departing aortic arches. In these cases, the term *truncus arteriosus* is most apt.

Like any active muscle, the heart requires a vascular supply to support its metabolism. The **coronary vessels** perfuse the cardiac tissue of the heart. In fishes, the coronary arteries are derived from the efferent arches or collecting loops of the gills, which carry oxygenated blood. Usually, only the outer part of the myocardium receives coronary arteries. The inner wall of the myocardium, especially of the ventricle, often forms projecting cones of muscle termed **trabeculae** set off by deep recesses. The resulting texture, when viewed from the lumen, looks spongy and is referred to as **trabeculate.** The coronary veins enter the sinus venosus.

In addition to the conal valves, the endocardium develops sets of valves between its chambers: the **sinoatrial (SA) valves** form between the sinus venosus and the atrium and the **atrioventricular (AV) valves** form between the atrium and ventricle. During normal flow, the valves are pushed open, although blood reversal immediately forces them closed, thereby preventing retrograde blood flow. The heart lies within the **pericardial cavity** lined by a thin epithelial membrane, the **pericardium.** In many fishes, the pericardial cavity lies within bone or cartilage, forming a semirigid compartment that holds the heart (figure 12.24a). Sequential contraction of the heart chambers helps move blood from one chamber to the next and finally drives it from the heart into the ventral aorta. Normal muscular movements impinging on nearby veins raises internal pressure and helps drive venous blood back to the heart. But refilling of the sinus venosus and atrium by returning blood is often aided by the low pressure produced within the confines of the semirigid compartment holding the heart. This is termed the **aspiration effect.** As the large muscular ventricle contracts, blood exits through the conus into the ventral aorta to empty the ventricle. This temporarily reduces the volume the ventricle occupies within the pericardial cavity, which lowers the pressure throughout the pericardial cavity surrounding the thin-walled atrium and sinus venosus. Negative pressure surrounding the relaxed sinus venosus and atrium causes them to expand; in turn, they develop a negative pressure that aspirates or sucks in venous blood. Once refilled, the atrium and sinus venosus contract to fill the ventricle (figure 12.24b).

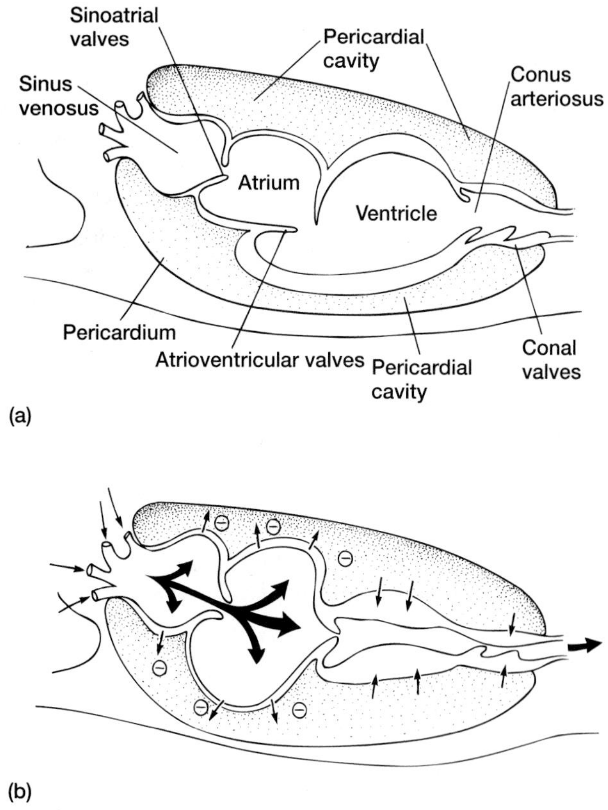


**Figure 12.24** Basic heart structure and aspiration filling.
(a) The four chambers of the fish heart are enclosed within the pericardial cavity. One-way valves between each chamber prevent reverse flow of blood as successive chambers contract.
(b) When the ventricle contracts, the volume occupied by the ventricle within the pericardial cavity is momentarily reduced (this is exaggerated in the diagram). Reduced ventricular volume creates a negative pressure around the other chambers. Because the walls of the sinus venosus and atrium are thin, this low surrounding pressure causes them to expand, creating within their lumina a negative pressure that sucks in or aspirates blood from the returning veins.

Contraction, as mentioned earlier, is an intrinsic property of cardiac muscle. Individual cells even show rhythmic contractions if they are isolated outside the body in a suitable culture medium. Cardiac cells tend to beat in synchrony. Contraction of the entire heart usually begins within a restricted region in the sinus venosus called the **pacemaker** or **sinoatrial node** and then spreads to other regions of the heart. The node consists of **Purkinje fibers,** neuronlike fibers that are apparently modified cardiac muscle cells. The rate at which heartbeats are initiated is under the influence of the nervous and endocrine systems. Heart rate also responds to the rate of venous filling. During exercise, venous return to the heart increases in part due to the increased pressure veins experience from active muscles surrounding them. As returning venous blood fills the heart chambers, they are stretched, and this stretching stimulates the **Starling reflex,** named after the physiologist who first documented it. This reflex self-adjusts the heart rate, increasing it as venous return increases and decreasing it as venous return slows.

Birds and mammals have four-chambered hearts, but of the original four fish chambers, only two persist as major receiving compartments, the atrium and the ventricle, both of which are divided into left and right compartments to produce four chambers. Although the hearts of birds and mammals are both derived from reptiles, they arose independently from different reptilian ancestors. Phylogenetically between these derived tetrapods and fishes stand amphibians and reptiles, a position that seduced many into believing that these intermediate vertebrates possessed hearts that should be evaluated in light of how well they anticipated the hearts of birds or mammals. Certainly the evolutionary route to birds and mammals led through primitive amphibians and reptiles. But living amphibians and reptiles are themselves millions of years removed from these earliest ancestors. Their hearts, like their cardiovascular systems generally, should be examined for the special functional roles they serve in amphibians and reptiles today. To do this, we next examine heart structure and its relationship to the functions it serves in all vertebrates.

### *Fishes*

Like all vertebrate hearts, the hagfish heart lies within the anterior trunk region, is composed of cardiac muscle, and receives blood returning from the general systemic circulation (figure 12.25a). It includes three chambers in series, the sinus venosus, the atrium, and the ventricle (figure 12.25b). Blood returning from the two common cardinals and liver first enters the sinus venosus. It flows through the atrium and then the ventricle, finally it is pumped directly into the ventral aorta and hence to the gills. One-way valves between heart chambers prevent reverse blood flow. No major nerves innervate the hagfish heart to stimulate contraction. Instead, filling of the sinus venosus by returning blood elicits the Starling reflex. This reflex stimulates contraction that originates in the pacemaker and then spreads sequentially to the other chambers.

Occasionally, the hagfish heart is called a **branchial heart** to distinguish it from unique accessory blood pumps elsewhere in its circulation (figure 12.25b). These supplementary circulatory pumps are sometimes called **accessory "hearts,"** in quotes because they contract but usually lack the cardiac muscle of true branchial hearts (figure 12.25c,d). In most fishes, systemic tissues drain via discrete venules and veins. However, in the hagfish, venous drainage of some regions, such as the head and the subcutaneous caudal regions, is provided by large open sinuses. A probable consequence is that venous blood pressure is especially low. Accessory hearts on the venous side of the circulation are an apparent answer to the problem of returning low-pressure venous blood.

The **cardinal hearts** lying within the anterior cardinal veins are like sacs whose pumping action is initiated by skeletal muscles around their outer walls. The paired **caudal hearts** are located in the tail. They are composed of a central cartilaginous rod, skeletal muscles on the side, and veins between. Alternative contraction of these muscles bends the rod back and forth, which presses on the walls of the vessels and pumps blood into the caudal vein (figure 12.25d).

The **portal heart** is a single expanded vascular sac that receives venous blood from one anterior and one posterior cardinal vein, and then it contracts to drive the blood through the liver (figure 12.25c). Only hagfishes have such an accessory heart in the course of the hepatic portal vein, which elevates blood pressure prior to the blood's entry into the liver sinusoids. Furthermore, the portal heart is the only accessory heart to have walls of cardiac muscle like the cardiac muscle of true branchial hearts.

The lamprey heart (branchial heart) is more like the typical fish heart. There are four compartments. Blood flows sequentially into the sinus venosus, the atrium, the ventricle, and the conus arteriosus. One-way valves are present between compartments. The sinoatrial and atrioventricular valves prevent retrograde blood flow. Like the other heart chambers, the conus arteriosus is muscular. Its luminal walls are thrown into folds collectively forming the **semilunar valves,** which prevent reverse blood flow and possibly aid in distributing blood to the aortic arches. From the arches blood flows to the delicate gill capillaries next in line in the circulation (figure 12.26).

Like the lamprey, the hearts of chondrichthyans and bony fishes consist of four basic chambers—sinus venosus, atrium, ventricle, and conus arteriosus (or bulbus arteriosus)—with one-way valves stationed between compartments (figure 12.27a,b). Like the other chambers, the muscular conus arteriosus contracts, acting as an auxiliary pump to help maintain blood flow into the ventral aorta after the onset of ventricular relaxation. Its contraction also brings together the conal valves located on its opposing walls. When these valves meet, they prevent the backflow of departing blood. In teleosts, this fourth chamber is an elastic, noncontractile bulbus arteriosus. A single pair of **bulbar valves** at the juncture of the bulbus arteriosus and the ventricle prevents retrograde flow. When receiving blood following ventricular contraction, the bulbus arteriosus stretches and then gently undergoes elastic recoil to maintain blood flow into the ventral aorta. The result is a **depulsation** or dampening of the large oscillations in blood flow and pressure introduced by ventricular contractions. This has been proposed as a means of protecting the delicate gill capillaries from exposure to sudden spurts of blood at high pressure that would otherwise occur.

The S-shaped arrangement of chambers in the fish heart places the thin-walled sinus venosus and atrium dorsal to the ventricle, so that gravity and atrial contraction assist ventricular filling. Blood flows from posterior chambers to anterior chambers in the following sequence. First, venous blood fills the sinus venosus and pushes open the sinoatrial

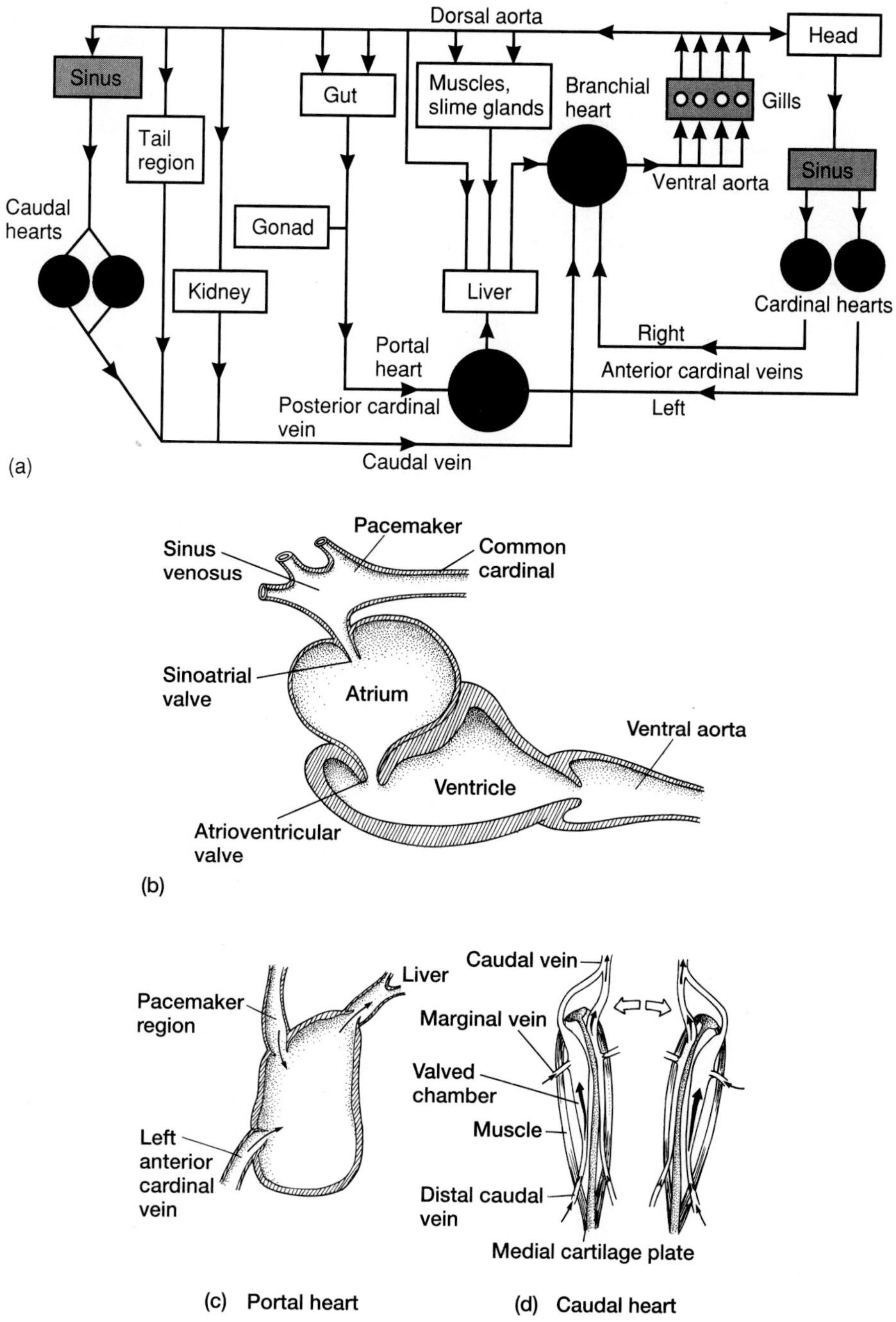


**Figure 12.25** Hagfish circulation. (a) Diagram of the cardiovascular system. (b) Branchial heart illustrating the three chambers. Accessory hearts. The walls of the cardinal (not shown) and portal hearts (c) pulsate to help drive blood. The caudal hearts (d) have paired striated muscles and a central flexible support. They lie almost at the end of the hagfish's tail. Contraction of the left muscle bends the medial cartilage plate, expelling blood from the compressed right chamber and allowing the left to fill. Contraction of the right muscle has the opposite effect. Alternating contractions of the caudal heart musculature enlarge and compress the veins, causing them to fill and then empty.

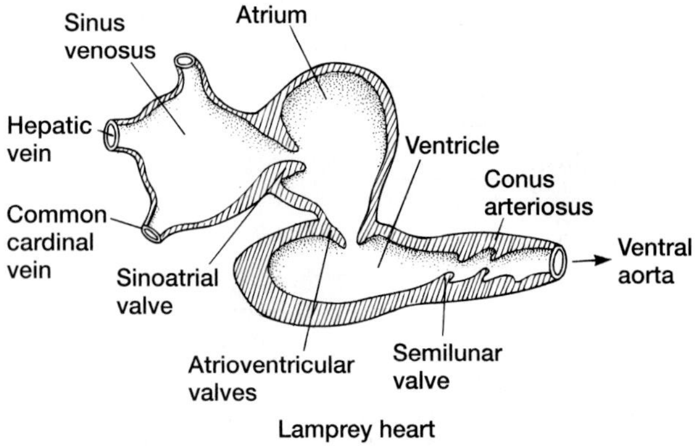


**Figure 12.26** Lamprey heart. The four chambers characteristic of most fishes are present in the lamprey.

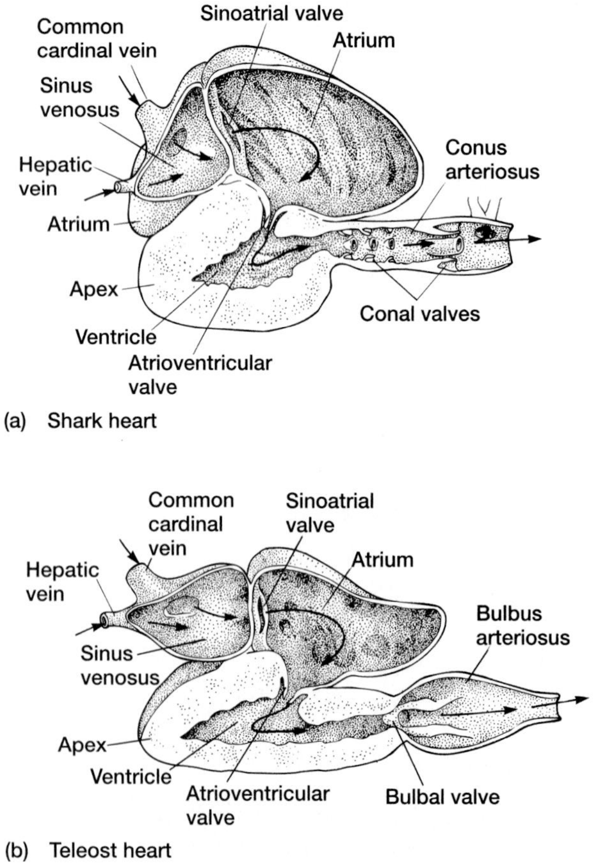


**Figure 12.27** Fish hearts. (a) Shark. (b) Teleost. Blood leaves the shark heart through the muscular conus arteriosus, a chamber that is absent in teleost fishes. Instead, in the teleost heart, the base of the ventral aorta is swollen, creating the elastic bulbus arteriosus.

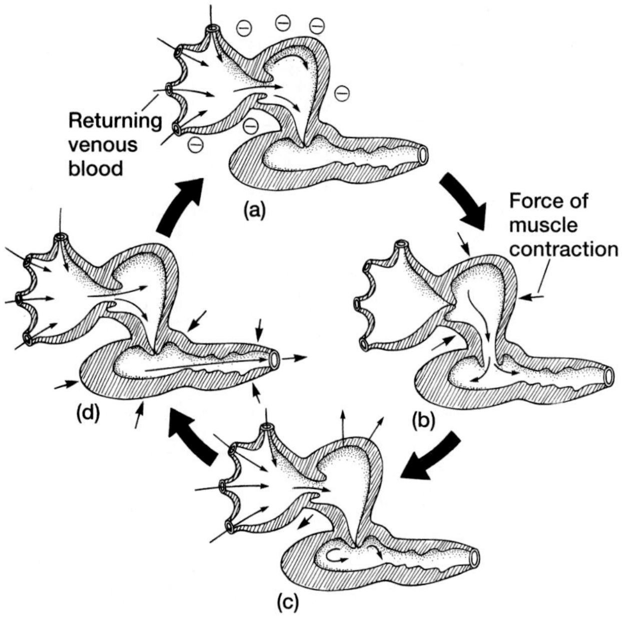


**Figure 12.28** Contraction cycle of the teleost heart. (a) Relaxation of the sinus venosus and atrium draws in blood via the hepatic and common cardinal veins. (b) Contraction of the atrium closes the sinoatrial valve and forces blood into the ventricle. (c) As the atrial walls relax again, blood enters the atrium. (d) Contraction of the ventricle forces blood through the bulbus arteriosus and distributes it to the aortic arches. The aspiration effect (negative signs) contributes to and completes the refilling of the sinus venosus and atrium to start the cycle again.

valve to fill the atrium. The aspiration effect drives this movement of venous blood and encourages the initial filling of the sinus venosus and atrium (figure 12.28a). Second, the atrium contracts, boosting pressure within its lumen. Atrial contraction forces the sinoatrial valve closed and opens the atrioventricular valve, thus allowing blood to flow into and fill the ventricle (figure 12.28b). Third, the atrium relaxes, dropping the pressure in it and in the sinus venosus. Consequently, blood is drawn in by the aspiration effect and begins to fill both chambers again (figure 12.28c). Fourth,the ventricle contracts to drive the blood it holds forward into and through the conus arteriosus, which now begins its contraction (figure 12.28d).

***Lungfishes*** The lungfish heart is modified from that of other bony fishes. The first chamber to receive returning blood is still the sinus venosus. In all three lungfish genera, the single atrium is partially divided internally by an **interatrial septum** (pulmonalis fold) that defines a larger **right** and smaller **left atrial chamber** (figure 12.29a). Pulmonary veins conveying blood from the lungs empty into the sinus venosus (Australian lungfish, *Neoceratodus*) or directly into the left atrial chamber (South American lungfish, *Lepidosiren* and African lungfish, *Protopterus*). The sinus venosus conveying systemic venous blood opens into the right atrial chamber (figure 12.29a). In place of the atrioventricular valves is the **atrioventricular plug,** a raised cushion in the wall of the ventricle. It moves into and out of the opening from the atrium, like the AV valves, to prevent

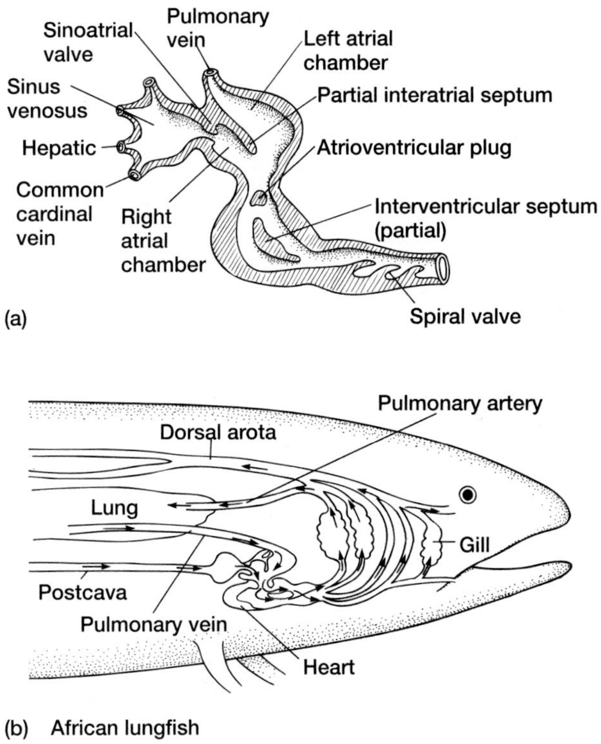


**Figure 12.29** Heart of the African lungfish *Protopterus*. (a) Internal structure of the heart. (b) Path of blood. When the lungfish breathes air, venous blood returning from systemic tissues flows through the heart and tends to be directed to the last aortic arch. The pulmonary artery carries most of the deoxygenated blood to the lung. Blood high in oxygen returning from the lung passes through the heart and then tends to enter the aortic arches without gills. In this manner, blood is shunted directly to the general circulation. Thus, when lungfishes breathe air, they display the beginnings of a double circulation system. The five aortic arches phylogenetically represent the second through the sixth (roman numerals). The first (II) and last two (V, VI) of these carry gills.

retrograde flow of blood into the atrium. The ventricle is also divided internally, but only partially by an **interventricular septum.** Within dipnoans, the South American lungfish shows the greatest degree of both ventricular and atrial internal subdivision. The Australian lungfish shows the least. Alignment of the interventricular septum, atrioventricular plug, and interatrial septum establishes internal channels within and through the heart. When the lungfish breathes air, the left channel tends to receive oxygenated blood returning from the lungs. The right channel tends to carry deoxygenated systemic blood (figure 12.29b). Thus, despite the anatomically incomplete internal septation of the lungfish heart, blood entering from the sinus venosus does not tend to mix with blood returning from the lungs.

The spiral valve within the conus arteriosus aids in separating oxygenated and deoxygenated blood. Apparently derived from conal valves, the spiral valve consists of two endocardial folds whose opposing free edges touch but do not fuse. The conus makes a couple of sharp bends and rotates about 270°, thus turning these folds into a spiral within its lumen. Although unfused, these twisting folds internally divide the conus into two spiraling channels. Because the conus is attached directly to the ventricle, oxygenated blood entering the left channel and deoxygenated blood entering the right channel tend to flow through different spiraling channels within the conus and remain separate. As the oxygenated and deoxygenated streams of blood exit from the conus arteriosus, they enter different sets of aortic arches.

When a lungfish surfaces to gulp fresh air into its lungs, pulmonary blood flow to the lungs increases. When this oxygenated blood returns from the lung, it is shunted through arches III and IV, which lack gills, and flows to systemic tissues directly. Venous blood returning from systemic tissues is shunted through the posterior arches, V and VI, and then diverted to the lung. The blood supply to these posterior arches is derived from the spiral channel that itself received deoxygenated blood from the right side of the heart. Oxygenated blood traveling through the left side of the heart is channeled along the opposite spiral of the conus to enter the anterior set of aortic arches.

These cardiovascular adjustments of the lungfish to breathing air are nicely matched to environmental demands. Under most conditions, oxygen tension is high in the rivers and ponds in which lungfishes live. Deoxygenated blood flowing though the arches with gill capillaries picks up sufficient oxygen from the water to meet metabolic demands. However, as a result of seasonal drought, high temperatures, or stagnant waters, oxygen levels in the water can significantly decline, leaving little to diffuse across gills into the blood. During such times, the lungfish comes to the surface to gulp fresh air into its lungs. Under these deteriorating conditions, physiological changes take full advantage of this added source of oxygen. In *Protopterus*, deoxygenated blood returning from systemic tissues tends to be diverted to the lungs (not to the gills), and about 95% of the oxygenated blood from the lungs tends to be directed via anterior aortic arches to the systemic tissues (not through the gills). The fraction of blood that passes from the lungs to the anterior arches steadily declines to about 65% just before the next breath, following which the fraction returns again to 95%.

This air-breathing system has several physiological advantages. First, streams of oxygenated blood (from the lungs) and deoxygenated blood (from the systemic tissues) tend to be kept separate. Thus, the stream of oxygenated blood on its way to active systemic tissues is not diluted by blood depleted of oxygen, and blood passing through the exchange surfaces of the lung is low in oxygen, promoting rapid uptake of oxygen. Second, the adjusted blood flow in an air-breathing fish prevents loss of oxygen to the water. Paradoxically, if oxygenated blood from the lungs passed through the gills, it could actually lose oxygen by diffusion to the oxygen-poor water. However, oxygenated blood is preferentially directed along aortic arches without gill capillaries to flow directly to systemic tissues. In addition to the preferential shunting of oxygenated blood into the anterior arches, a secondary mechanism, involving a shunt at the base of the gill capillaries, prevents oxygenated blood from being exposed to water low in oxygen in the gills. Some lungfishes have thick muscular arteries that connect afferent and efferent aortic arches. When these shunts are opened, blood entering arches with gills can bypass the gill capillaries entirely and avoid exposure to oxygen-poor water irrigating the respiratory beds.

Not all modern lungfishes exhibit the same ability to adjust to air ventilation. The gills of the Australian lungfish are well developed and its lung less so. This lungfish does quite well in oxygenated water, but if forced onto land, it cannot maintain sufficiently high oxygen levels through its lung to sustain itself for long. However, the African lungfish has gills not as well developed, but its lung is more so. If forced onto land, its circulatory system and lung can sustain the lungfish indefinitely. Thus, the degree of physiological response to breathing water or air depends on the species of lungfish.

## *Amphibians*

Amphibians rely on cutaneous gas exchange (plethodontid salamanders), on gills (many larval forms), on lungs (some species of toads), or on all three modes (most amphibians). Because the sources of oxygenated and deoxygenated blood vary, heart structure varies as well. Generally, in amphibians with functional lungs, the heart includes a sinus venosus, right and left atria divided by an anatomically complete interatrial septum, a ventricle lacking any internal subdivision, and a conus arteriosus with a spiral valve (figure 12.30a). Except for the salamander of the species *Siren*, which has a partial ventricular septum, amphibians are unique among air-breathing vertebrates in lacking any internal division within the ventricle.

The cardiovascular system is perhaps best studied in frogs. The conus arteriosus of the frog heart arises from a single trabeculate ventricle (figure 12.30b). Semilunar valves lie at the base of the conus and prevent retrograde flow of blood back into the ventricle. Internally, a spiral valve twisting nearly through a complete rotation establishes channels within the conus that target blood to specific sets of systemic and pulmocutaneous arches. The systemic and pulmocutaneous arches both arise from the truncus arteriosus, a remnant of the ventral aorta, but the two sets of arches receive blood from different sides of the spiral valve.

In lungless salamanders or those with reduced lung function, the interatrial septum and spiral valve may be much reduced or absent entirely. Unlike frogs, in which the pulmocutaneous artery branches to give rise to the cutaneous artery, salamanders lack a cutaneous artery. Instead,

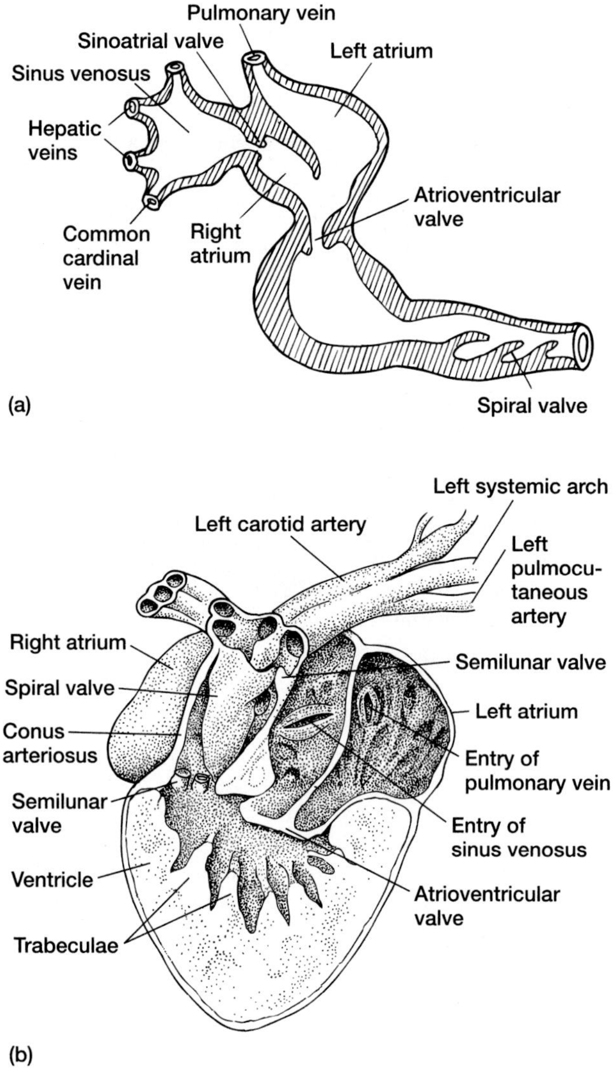


**Figure 12.30** Amphibian hearts. (a) Diagram of typical amphibian heart. Notice that the atrium is divided into left and right chambers but the ventricle lacks an internal septum. (b) Bullfrog *(Rana catesbeiana)* heart. Although lacking internal septa, the wall of the ventricle folds into numerous trabeculae. The small compartments between these trabeculae are thought to aid in separating bloodstreams that pass through the heart.

branches from vessels supplying the systemic circulation carry blood to the salamander skin. The pulmonary artery and the systemic arches in salamanders arise from the truncus arteriosus (figure 12.31a,b).

The two different streams of blood returning from the systemic and pulmonary circuits of amphibians are kept separate as they pass through the heart (figure 12.31c). As in lungfishes, deoxygenated blood is selectively directed to the lung via the pulmonary artery, and oxygenated blood is directed to systemic tissues via the

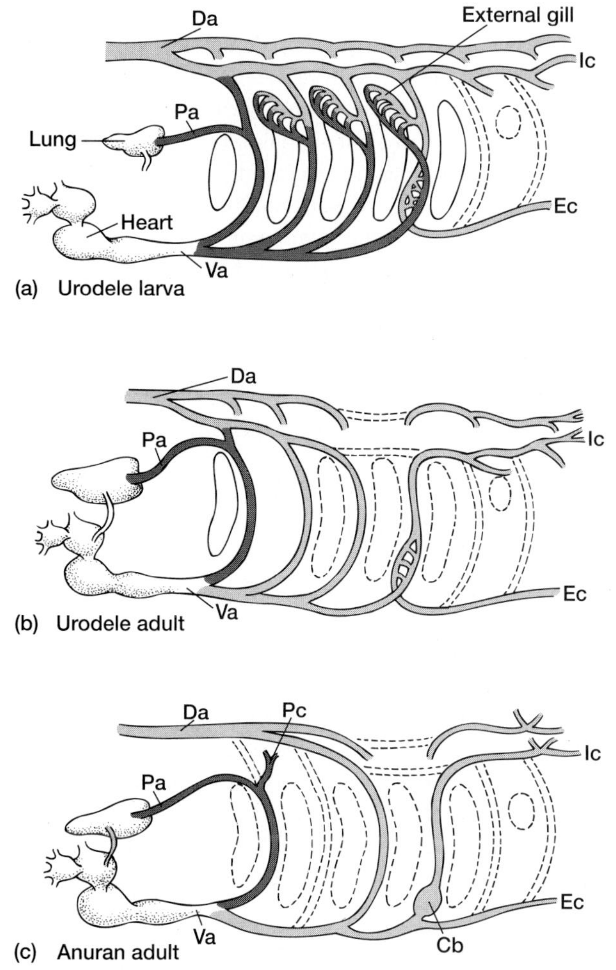


**Figure 12.31** Blood flow to aortic arches in amphibians. (a) Larval salamander. (b) Adult salamander. (c) Anuran. Notice the pulmocutaneous (Pc) branch to the skin. In frogs, a sphincter prevents blood flow to the lung during diving, thus diverting blood flow to the skin to increase cutaneous respiration. Abbreviations: carotid body (Cb), dorsal aorta (Da), external carotid (Ec), internal carotid (Ic), pulmonary artery (Pa), pulmocutaneous artery (Pc), ventral aorta (Va).

aortic arches. In frogs breathing air, oxygenated and deoxygenated blood are separated and distributed by the heart. What is somewhat surprising about this ability is that the ventricle of the frog heart, like that of other amphibians, lacks even a partial internal septum. The trabeculate topography produces deep recesses in the walls of the ventricle that are thought to separate streams of blood that differ in oxygen tension. It is hypothesized that as one stream enters the ventricle, it preferentially fills the recesses between trabeculae, whereas the second stream occupies the center of the ventricle. Because of their different positions, the oxygenated and deoxygenated streams depart by different exits to reach appropriate sets of arteries. Thus, the trabeculae apparently are the structures in the frog ventricle that separate pulmonary and systemic venous streams of blood flowing through the heart (figure 12.30b).

When a frog dives, the lung collapses from water pressure on the body wall. In the genus *Rana*, a sphincter at the base of the pulmonary artery constricts, resulting in reduced blood flow to the lung and increased flow to the skin. Thus, while a frog is submerged, loss of pulmonary respiration is somewhat offset by increased cutaneous respiration (figure 12.31c).

In adult salamanders, pulmonary and systemic circuits are similarly separated in the heart. In specialized species, heart design is modified. For example, in the lungless plethodontids, in which 90% of the respiratory needs are met through the skin and 10% through the buccal cavity, the heart lacks a left atrium, the compartment that would receive blood returning from the lungs. Where gills predominate over lungs as respiratory organs (e.g., *Necturus*), the interatrial septum is reduced or perforated.

### *Reptiles*

Reptiles have entered more fully terrestrial environments and adopted more active lifestyles than the amphibians that preceded them. The cardiovascular system of reptiles supports accompanying higher metabolic rates and elevated levels of oxygen and carbon dioxide transport. It is capable of generating elevated blood pressures, higher cardiac output, and efficient separation of oxygenated and deoxygenated bloodstreams. The diversity of hearts and heart function in reptiles is becoming better understood. It is clear that no single reptile heart can represent all others. Further, looking at reptile hearts as evolutionarily incomplete and imperfect bird or mammal hearts does little justice to the specialized, elegant, and rather effective cardiovascular designs of reptiles that support their specialized and distinctive lifestyles. In general, two basic reptilian heart patterns are recognized. One is found in chelonians and squamates, the other in crocodilians. We shall take them in that order.

***Chelonian/Squamate Hearts*** In these reptiles, the sinus venosus is reduced in comparison to amphibians but it retains the same functions. It is still the first chamber to receive venous blood and contains the pacemaker. The conus arteriosus (or bulbus cordis) appears during early embryonic development but becomes divided in the adult to form the bases (trunks) of three large arteries leaving the ventricle—the **pulmonary trunk** and the **right** and **left aortic trunks.** Usually, the brachiocephalic artery, delivering blood to the subclavians and carotids, emanates directly from the right aortic arch, but in some turtles, it arises directly from the ventricle, crowded in with the trunks of the three aortic arches (figure 12.32a,b). The conus also gives rise to a band of contractile muscle tissue at the base of the pulmonary trunk

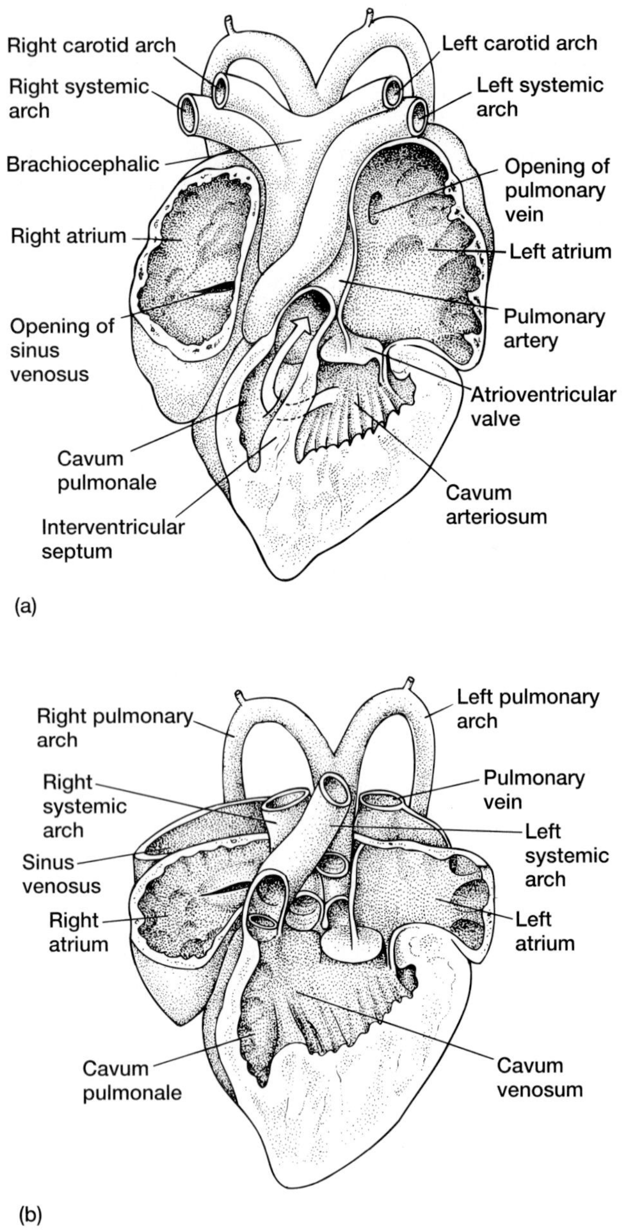


**Figure 12.32** Lizard heart, ventral view. (a) Part of the ventral wall of the heart has been removed. The open arrow indicates the flow of blood from the cavum arteriosum via the interventricular canal into the cavum venosum. From the cavum venosum blood reaches the base of the systemic arches. (b) More of the ventral wall has been removed.

to control the resistance blood meets as it flows to the lungs. The atrium is completely divided into right and left atria. Prominent atrioventricular valves guard the entrance to the ventricles. Strictly speaking, the ventricle is a single chamber functioning as a single fluid pump to drive blood into the major arteries leaving the heart. Internally, however, it has three interconnected compartments, the **cavum venosum** and the **cavum pulmonale** separated from each other by a **muscular ridge** and the **cavum arteriosum** connected to the cavum venosum via an **interventricular canal.** The cavum arteriosum fills with blood from the left atrium but has no direct arterial output. During systole, the blood it receives flows through the interventricular canal to the aortic arches. The cavum pulmonale does not receive blood directly from the atria. Instead, blood from the cavum venosum moving across the muscular ridge fills the cavum pulmonale. In turn, much of the blood filling the cavum venosum is deoxygenated blood from the right atrium. Thus, the heart has five chambers, composed of two atria and three compartments of the ventricle, or six chambers if you count the sinus venosus.

The pattern of blood flow through the hearts of Chelonia and squamates differs depending on whether they breathe air or hold their breath, a condition termed *apnea.* For example, in an air-breathing turtle on land, most deoxygenated blood returning from systemic tissues is directed to the lungs, and most oxygenated blood from the lungs is directed to the systemic tissues via the aortic trunks.

Specifically, from the sinus venosus, the right atrium receives deoxygenated blood returning from the body. The left atrium receives oxygenated blood returning from the lungs. When the atria contract, deoxygenated blood in the right atrium flows into the cavum venosum and then across the muscular ridge to the cavum pulmonale. Additionally, when the right AV valves open, they lie across the opening to the interventricular canal and temporarily close it. Oxygenated blood in the left atrium enters the cavum arteriosum and temporarily remains there while the AV valves occlude the interventricular canal. When the ventricle contracts, the muscular ridge is compressed against the opposite wall to separate the cavum venosum from the cavum pulmonale momentarily. The AV valves close to prevent retrograde backflow into the atria, but in so doing, the right AV valve opens the interventricular canal and allows blood to flow through it. Thus, blood leaves the ventricle via the most accessible routes—deoxygenated blood in the cavum pulmonale exits primarily through the pulmonary artery to the lung, although some also squirts across the muscular ridge to enter the left aortic arch; oxygenated blood in the cavum arteriosum moves through the interventricular canal to reach the bases of the aortic trunks, through which it then exits (figures 12.33a and 12.34a,b).

There is also a slight asynchrony in the timing of contractions in the walls of the ventricle. As a result, deoxygenated blood is driven into the pulmonary artery before oxygenated blood is set in motion. When the adjacent ventricular walls contract, oxygenated blood finds high resistance in the mostly filled pulmonary artery. Consequently, oxygenated blood exits through the systemic arches because they offer the least resistance.

Measurements of oxygen content in the major arteries confirm that the distribution of systemic and pulmonary blood is highly directional—deoxygenated blood flows to the lungs, oxygenated blood to the systemic tissues. In turtles breathing air, 70 to 90% of all blood reaching the left

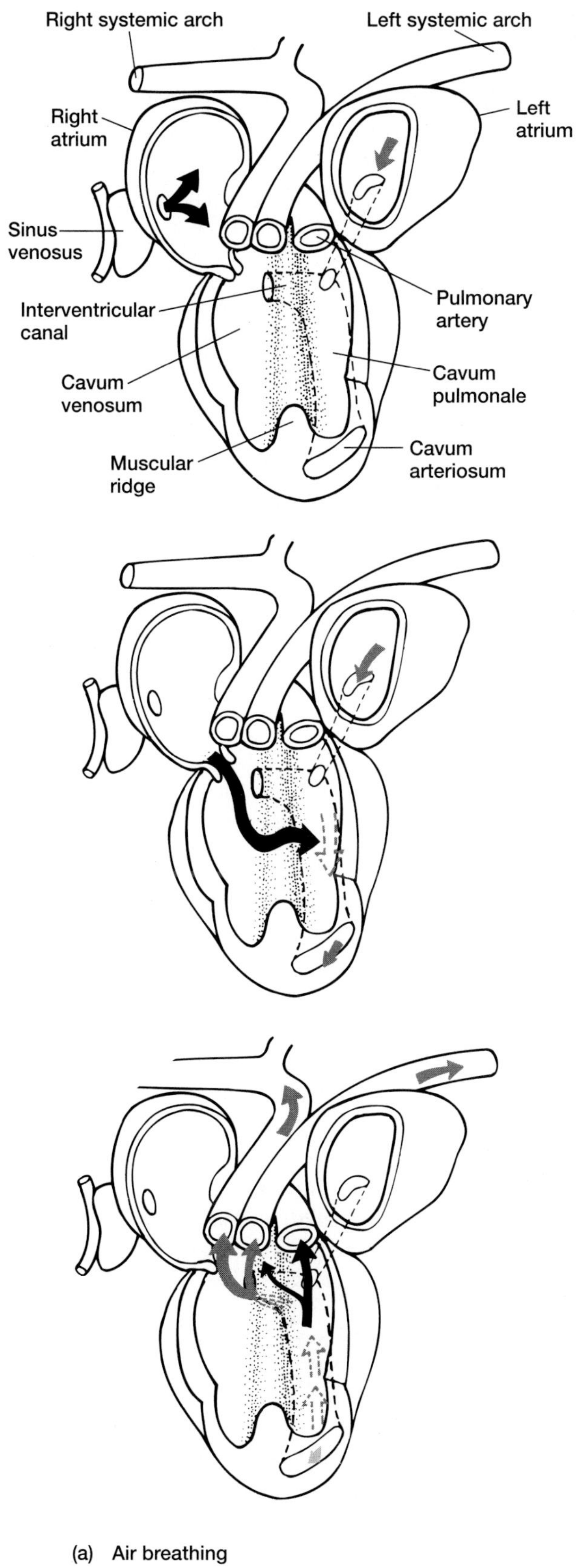

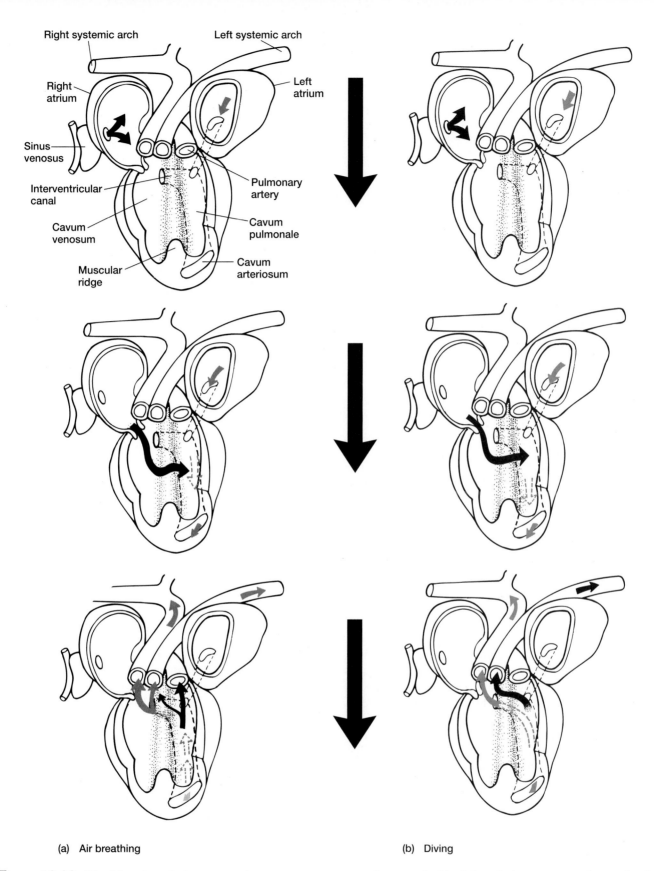

**Figure 12.33** Blood flow through the squamate heart. (a) When squamates breathe air on land, venous blood from the right atrium enters the cavum venosum of the ventricle and crosses a muscular ridge, to fill the cavum pulmonale momentarily. Upon ventricular contraction, most of this blood exits via the pulmonary artery. Simultaneously, blood from the left atrium enters the deep cavum arteriosum. Contraction of the ventricle squirts this blood through the interventricular canal and then the blood departs via the left and right systemic arches. (b) When squamates dive, resistance to pulmonary blood flow encourages blood that would normally exit to the lungs to move instead across the muscular ridge and depart primarily via the left aortic arch.

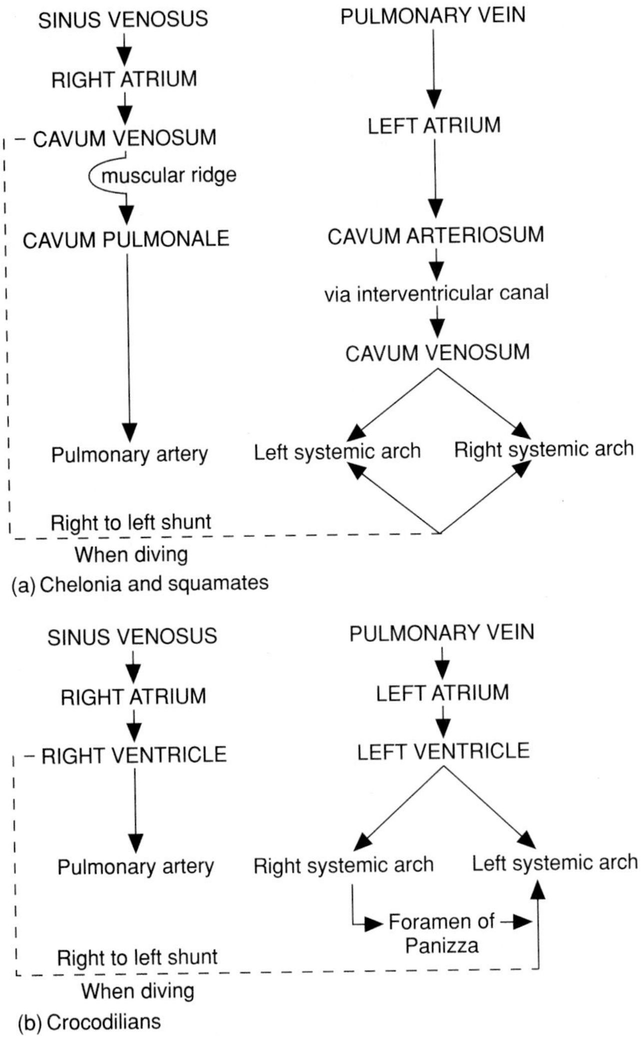


**Figure 12.34** Reptilian hearts. This flowchart compares the path of blood in the hearts of Chelonia and squamates (a) and crocodilians (b). Dashed lines indicate the cardiac diving shunts that divert blood from the pulmonary circuit directly to the systemic circuits.

systemic arch is oxygenated blood coming from the pulmonary circuit; 60 to 90% of deoxygenated blood reaching the lungs comes from systemic tissues. Isolation of oxygenated and deoxygenated streams occurs in spite of the fact that the compartments of the ventricle are not anatomically separate. Notice that this functional separation extends to the aortic trunks. The left aortic arch fills mainly with oxygenated but also with some deoxygenated blood, but when the turtle breathes air, the right aortic arch carries only oxygenated blood to ensure a flow of highly oxygenated blood to the brain through the carotids from the brachiocephalic artery (figure 12.35).

When the turtle dives beneath the water, the physiological problems with which the circulatory system must cope change significantly. The heart responds with a right-to-left **cardiac shunt.** Blood entering the cavum venosum is directed to the opposite side and out the aortae rather than out the pulmonary circuit (figure 12.33b). Differences in the resistances of systemic and pulmonary circuits are believed to control this shunting. A sphincter at the base of the pulmonary artery contracts to increase the pulmonary resistance to blood flow after a turtle dives. Because blood tends to follow the path of least resistance, it flows into the systemic circulation. When a turtle dives, blood that would pass to the lungs during air breathing is shunted instead through the aortic arches to the systemic circuit (figure 12.35).

A diving turtle makes the best of a difficult situation. The air held in its lungs is soon depleted of oxygen, so there is little advantage in circulating large quantities of blood to it. The energy used to pump blood to the lungs would return no physiological benefit. Instead, diverting blood to the systemic circuit increases the blood volume that can remove metabolites or gather oxygen stored in the tissues.

***Crocodilian Hearts*** In many respects, the hearts of alligators and crocodiles are structurally similar to those of other reptiles. The conus arteriosus (bulbus cordis) produces the bases of the trunks of the three departing arteries—pulmonary and left and right aortic trunks. One-way **lunar valves** at the bases of each trunk permit blood to en-

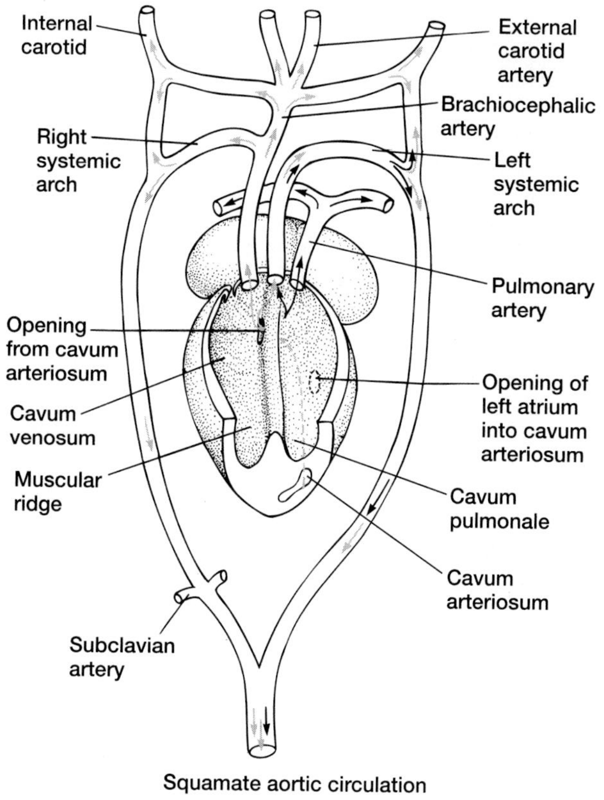


**Figure 12.35** Squamate aortic circulation. Blood flows to major arteries when squamates breathe air. Oxygenated blood is directed to the systemic arches. Most, but apparently not all, of the deoxygenated blood enters the pulmonary artery. The small amount of deoxygenated blood flowing to the systemic circulation enters the left systemic arch. Thus, oxygen tension of blood in this arch is slightly lower than it is in the right systemic arch. It may be significant that carotid vessels supplying the head and brain branch from the right systemic arch.

ter the conus but halt reverse backflow into the ventricle. The sinus venosus is reduced but still functions as the receiving chamber for returning systemic blood. The atrium is completely subdivided into two distinct left and right chambers, and the sinus venosus empties into the right atrium. The pulmonary vein enters the left atrium in adults, but it does not open into the left atrium during embryonic development. What are initially separate pulmonary veins, one from each lung, unite as a single stem, the pulmonary vein, that enters the sinus venosus. However, as embryonic development proceeds, this part of the sinus venosus together with the associated pulmonary vein become incorporated into the developing left atrium (figures 12.34b and 12.36).

In other respects, the crocodilian heart is quite different from what we have seen so far. The ventricle is divided by an anatomically complete interventricular septum into distinct left and right chambers. The pulmonary trunk and *left* aortic arch open off the *right* ventricle. The *right* aortic arch opens off the *left* ventricle. A narrow channel called the **foramen of Panizza** connects the left and right aortic arches shortly after they depart from the ventricle (figure 12.37a).

In a crocodile breathing air, right and left atria fill with deoxygenated systemic and oxygenated pulmonary blood, respectively. Contraction of the atria delivers blood to the respective ventricles. When the ventricles contract, blood flows through the nearest portals of least resistance.

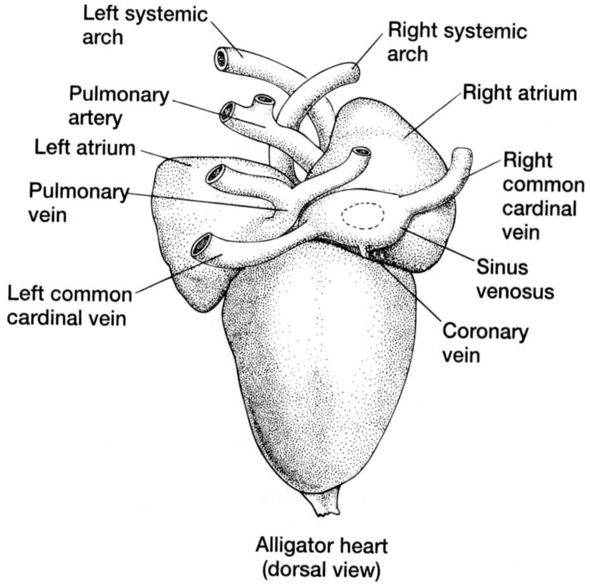


**Figure 12.36** Alligator heart, dorsal view. Notice that the sinus venosus, although reduced, is the first chamber of the heart to receive returning venous blood via the common cardinals. The coronary vein also flows into the sinus venosus.

# Box Essay 12.1 Lizard Hemodynamics

Lizard hearts invite misunderstandings. Two anatomically separate atria are present, but the ventricle is a single chamber with interconnecting compartments. Three major arteries depart directly from the ventricle. Such a design suggested to early anatomists that bloodstreams mixed as oxygenated and deoxygenated blood entered the common ventricle. The subtle assumptions behind this interpretation were as much an obstacle to understanding heart function as was the complex anatomy itself. Lizards were seen as primitive, and anatomists were looking ahead to the cardiovascular systems of advanced endotherms. As stated by one such scientist, the "perfect solution" to the separation of bloodstreams "was not attained until the avian and mammalian stages were reached." Recent experimental research on the blood flow through the lizard heart has shown just how wrong this earlier physiology and philosophy was.

Several techniques have been used to clarify blood flow within the cardiovascular system of living lizards. One technique uses radiology and takes advantage of contrast fluids that are generally nontoxic and compatible with blood. These contrast media are **radiopaque,** that is, visible when viewed by X ray. By the introduction of a radiopaque medium into selected veins, the subsequent route taken by the blood can be followed, usually through a sequential series of photographs or on a video monitor. Because the radiopaque medium is within the blood vessels of a live animal, its course appears to represent the normal circulation of blood. One such experiment was performed by Kjell Johansen on the lizard *Varanus niloticus,* a large member of the family Varanidae (Johansen, 1977). The radiopaque contrast medium was injected into the right jugular vein and postcava.

The right atrium, ventricle, and pulmonary arteries could be seen to fill with contrast medium in successive stages. Although the lizard ventricle shows no complete internal division, none of the contrast medium entered the systemic arches (box figure 1a–c). This is experimental confirmation that the heart of this varanid lizard keeps returning deoxygenated blood separate from oxygenated blood and targets deoxygenated blood to the pulmonary circuit for oxygenation within the lung.

Another technique uses small blood sampling tubes called cannulae. Cannulae have been used in experiments on the savannah monitor lizard, *Varanus exanthematicus,* to clarify the hemodynamics of blood pressure and the flow of oxygenated blood (Burggren and Johansen, 1982). Using anesthesia and surgical procedures, Burggren and Johansen inserted small cannulae connected to pressure transducers and recorders into the lumina of selected arteries of the monitor lizard. Additional cannulae in other arteries permitted the researchers to withdraw tiny samples of blood and measure the oxygen tension. As in other squamates and Chelonia, the varanid lizard heart was able to separate oxygenated and deoxygenated bloodstreams, targeting them to systemic and to pulmonary circuits, respectively. However, the researchers discovered that in the varanid, unlike in other squamates and Chelonia, blood pressure in the systemic circuit reached levels over twice that in the pulmonary circuit during systole (box figure 2a–c). In most other lizards, the systolic pressures in both circuits are quite similar. Thus, the heart of the varanid lizard not only diverts separate streams of oxygenated and deoxygenated blood to systemic and pulmonary circuits, but it also generates separate pressures within each circuit as well.

The cavum venosum within the varanid ventricle is considerably reduced, but otherwise the varanid heart is anatomically similar to that of other squamates and Chelonia. However, generation of high systemic and low pulmonary blood pressures makes it hemodynamically similar to the hearts of crocodilians, birds, and mammals. Why this should be so in varanids is not clear. Once they reach their preferred body temperatures, varanids have a higher metabolic rate than most other lizards. It has been suggested that the high systemic pressure might permit perfusion of a larger number of capillary beds than the systemic pressure in other lizards, without a consequent drop in capillary pressure. High systemic pressure would allow delivery of high levels of oxygenated blood to support the high oxygen requirements of active varanid muscles. However, if lung capillaries experienced such high pressures, they might leak excess fluid that would collect in the lung tissues and interfere with gas exchange. Because the pulmonary capillaries are part of the low-pressure pulmonary circuit, they are protected in varanid lizards. This suggests that by being separated into two pressure pumps, the varanid ventricle protects the pulmonary capillaries from excess pressure while meeting the high-pressure demands of active muscles.

Lizard hearts did not evolve in anticipation of the "perfect" hearts of birds and mammals. The lizard heart proved to be functionally complex and finely adapted to the special demands of a squamate lifestyle. The experimental research has settled much of the physiological function, but it should also invite a reassessment of the philosophy behind interpretation of systems in primitive vertebrates. We should abandon the view that lower vertebrates, because they arose early in vertebrate evolution, are imperfectly adapted. Modern research in functional morphology demonstrates just the opposite. The cardiovascular system like other morphological systems are surprisingly sophisticated both in primitive vertebrates and in their derived descendants.

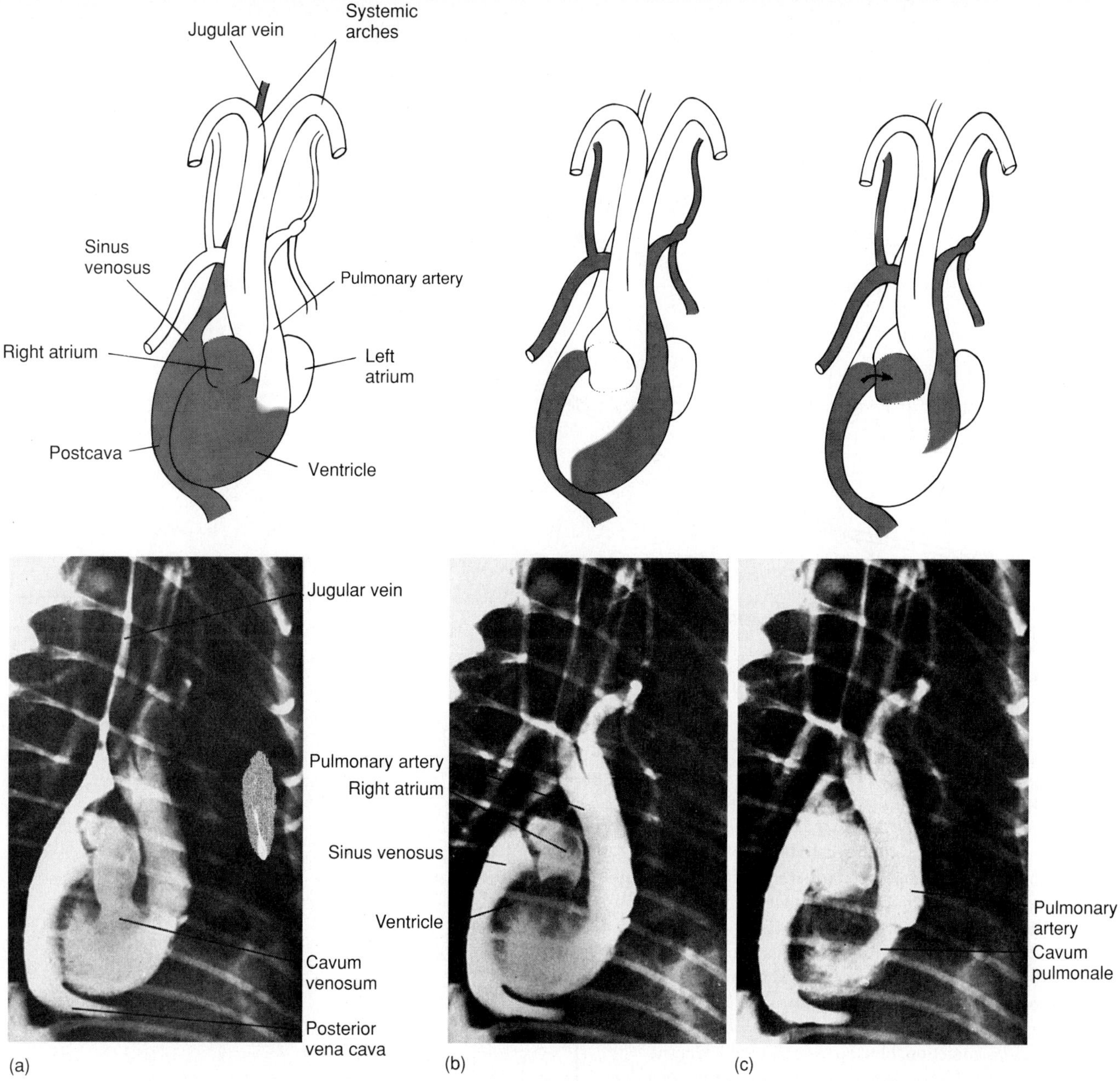


**Box figure 1** Tracing the course of blood through the heart of the lizard *Varanus niloticus*. (a) Radiopaque contrast medium was injected into the right jugular vein and forced back slightly into the postcaval vein. The contrast medium has already entered the right atrium, which has contracted to fill part of the ventricle (the cavum venosum and cavum pulmonale). (b) As the ventricle is undergoing contraction, and the pulmonary artery is filled. (c) Near completion of ventricular contraction, the contrast medium has been almost entirely expelled, and the branches of the pulmonary artery to each lung are clearly filled. Note in the last two stages that closure of the atrioventricular valves prevents retrograde blood flow into the right atrium. Notice also that this deoxygenated medium does not enter the systemic arches; therefore, these arches do not show up on the radiographs.

# Box Essay 12.1, continued

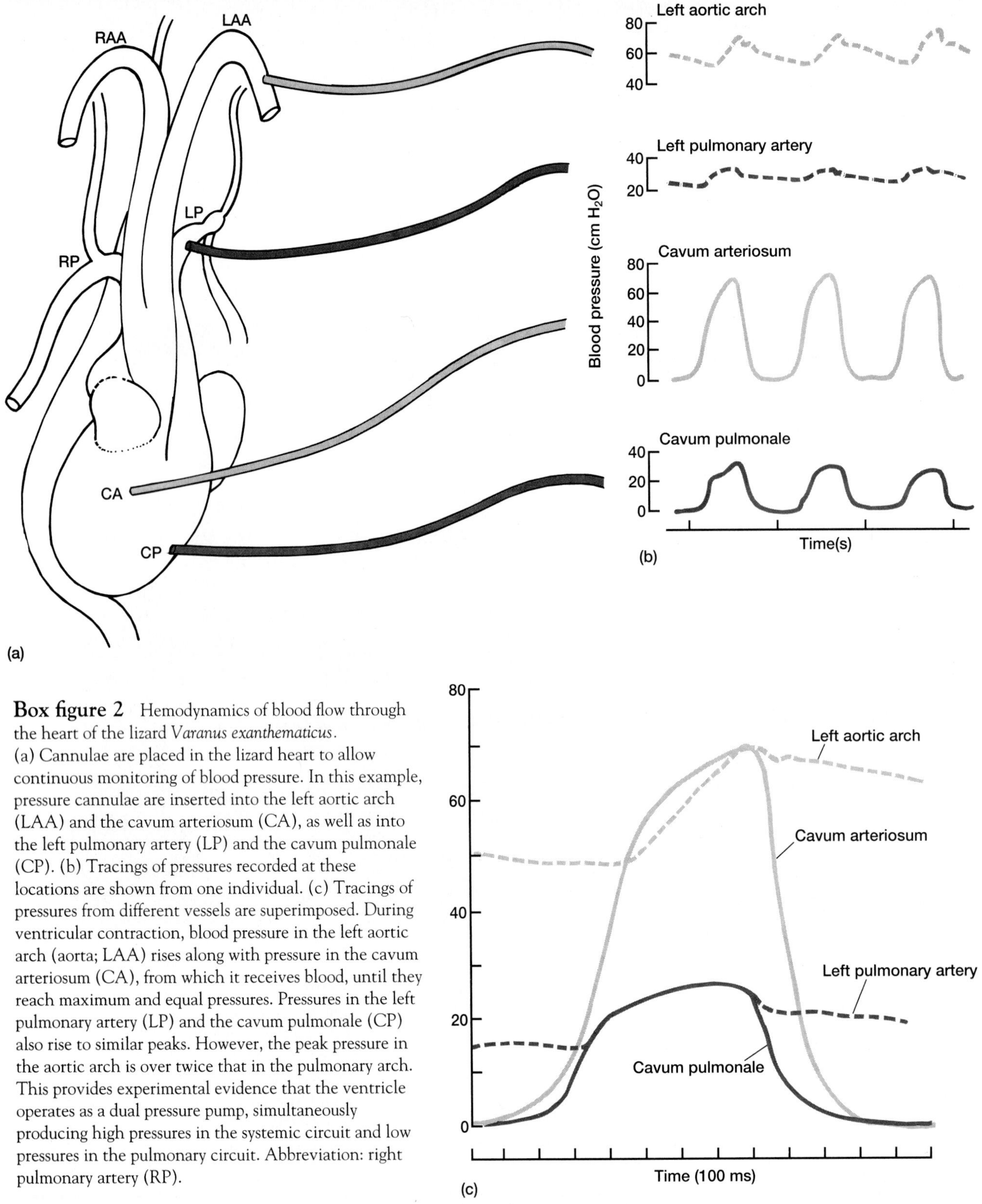


**Box figure 2** Hemodynamics of blood flow through the heart of the lizard *Varanus exanthematicus*.
(a) Cannulae are placed in the lizard heart to allow continuous monitoring of blood pressure. In this example, pressure cannulae are inserted into the left aortic arch (LAA) and the cavum arteriosum (CA), as well as into the left pulmonary artery (LP) and the cavum pulmonale (CP). (b) Tracings of pressures recorded at these locations are shown from one individual. (c) Tracings of pressures from different vessels are superimposed. During ventricular contraction, blood pressure in the left aortic arch (aorta; LAA) rises along with pressure in the cavum arteriosum (CA), from which it receives blood, until they reach maximum and equal pressures. Pressures in the left pulmonary artery (LP) and the cavum pulmonale (CP) also rise to similar peaks. However, the peak pressure in the aortic arch is over twice that in the pulmonary arch. This provides experimental evidence that the ventricle operates as a dual pressure pump, simultaneously producing high pressures in the systemic circuit and low pressures in the pulmonary circuit. Abbreviation: right pulmonary artery (RP).

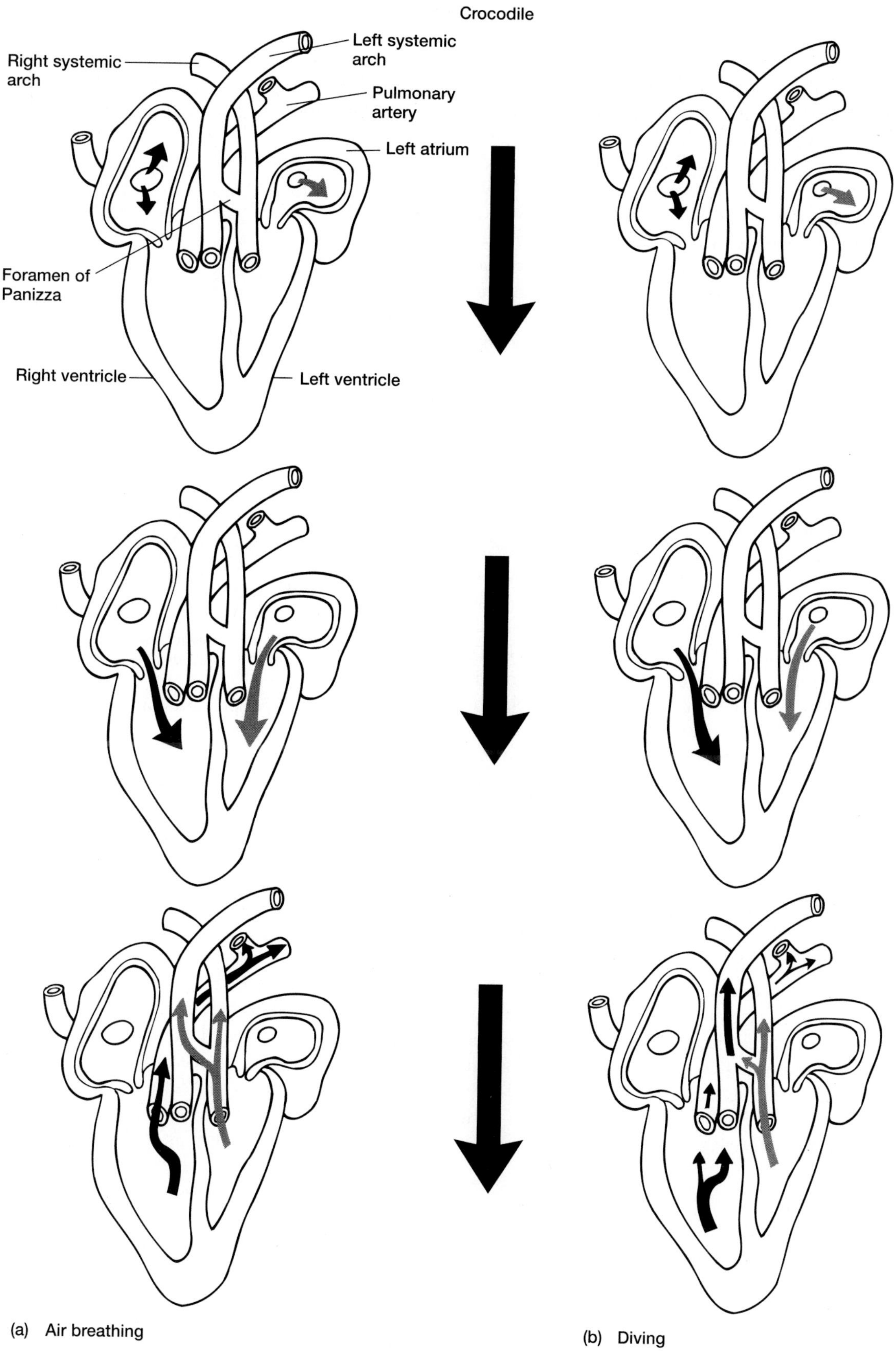


**Figure 12.37** Blood flow through the crocodile heart. (a) Systemic and pulmonary blood flow when the crocodile breathes air. (b) Internal changes that result in decreased pulmonary flow when the crocodile dives.

At the moment of systole, pressure is greatest in the left ventricle. The oxygenated blood it holds enters the base of the right aortic arch, but because of its high pressure, it also enters the left aortic arch via the foramen of Panizza. High pressure in the left aortic arch keeps the lunar valves at its base closed, leaving only the pulmonary route of exit for blood in the right ventricle. As a result, both aortic arches carry oxygenated blood to systemic tissues, and the pulmonary artery carries deoxygenated blood to the lungs (figure 12.37a).

When a crocodile dives, this pattern of cardiac blood flow changes because of a cardiac shunt. Resistance to pulmonary flow increases due to vasoconstriction of the vascular supply to the lungs and partial constriction of a sphincter at the base of the pulmonary artery. As a result, systolic pressure within the right but not the left ventricle rises substantially, matching and somewhat exceeding that within the left aortic arch. Blood in the right ventricle now tends to exit through the left aortic arch rather than through the pulmonary circuit, which presents high resistance to blood flow. Diversion of blood in the right ventricle to the systemic circulation represents a right-to-left cardiac shunt. Blood in the right ventricle, which would flow to the lungs in an air-breathing crocodile, instead travels through the left aortic arch, joining the systemic circulation and bypassing the lungs (figures 12.34b and 12.37b). This lung bypass confers the same physiological advantages we have seen in turtles, namely, an increase in efficiency of blood flow while fresh air is unavailable.

Apnea occurs not only during diving. Most reptiles at rest on land go for long intervals without taking a breath. As apnea continues, oxygen from the lungs becomes depleted, and pulmonary perfusion declines until just before another breath. Thus, for reptiles breathing air but ventilating intermittently, the cardiac shunt allows pulmonary perfusion to match air ventilation. In reptiles in temperate (or desert) regions, the cardiac shunt probably diverts blood during times of hibernation (or estivation) when metabolic needs are reduced, ventilation rate declines, and high levels of pulmonary perfusion would bring few physiological benefits.

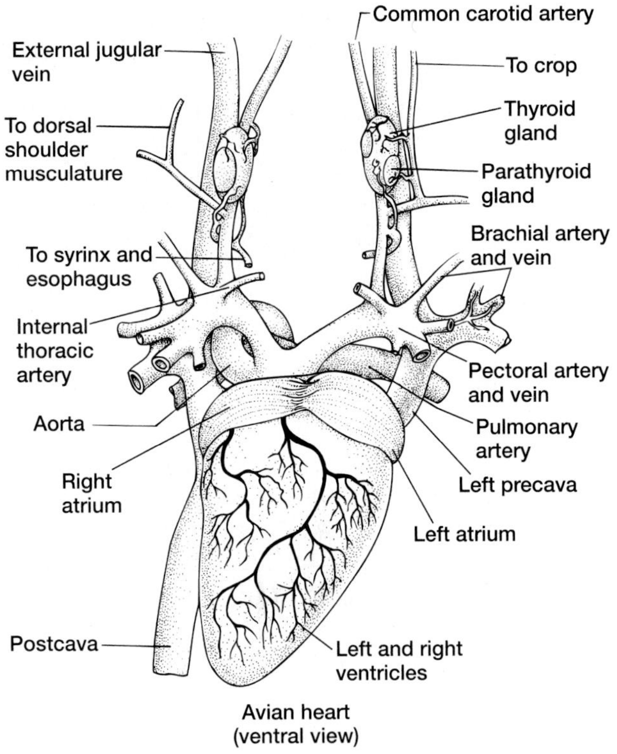


**Figure 12.38** Avian heart, ventral view.

## *Birds and Mammals*

As noted, the hearts of birds and mammals have four chambers that arise from the two chambers (atrium and ventricle) of the fish heart. In birds the sinus venosus is reduced to a small but still anatomically discrete area. The conus arteriosus (bulbus cordis) is only a transient embryonic chamber that gives rise to the pulmonary trunk and a single aortic trunk in the adult (figure 12.38). In mammals the sinus venosus is reduced to a patch of Purkinje fibers, or sinoatrial node, in the wall of the right atrium. The sinoatrial node functions as a pacemaker, initiating the wave of contraction that spreads across the heart like in all other vertebrates. As in birds, the conus arteriosus splits during embryonic development in mammals to produce the pulmonary trunk and single aortic trunk of the adult (figure 12.39).

Although structurally similar, bird and mammal hearts arose independently from different groups of reptilian ancestors. This difference is reflected in their embryonic development. Appearance of the interventricular and interatrial septa that form the paired chambers occurs quite differently in the two groups. Bird and mammal hearts function similarly as well. Both consist of parallel pumps with double circulation circuits. The right side of the heart gathers deoxygenated blood from systemic tissues and pumps it into the pulmonary circuit. The left side of the heart pumps oxygenated blood from the lungs through the systemic circuit. The hearts of birds and mammals are anatomically divided into left and right compartments; thus, there is no cardiac shunting with changing ventilation rates. Therefore, unlike amphibians and reptiles, a cardiac shunt cannot be used in birds and mammals to decouple perfusion of the lung and systemic tissues. Although the reasons are not well understood, some propose that endothermic animals (birds and mammals) may require complete anatomical separation of the cardiac chambers.

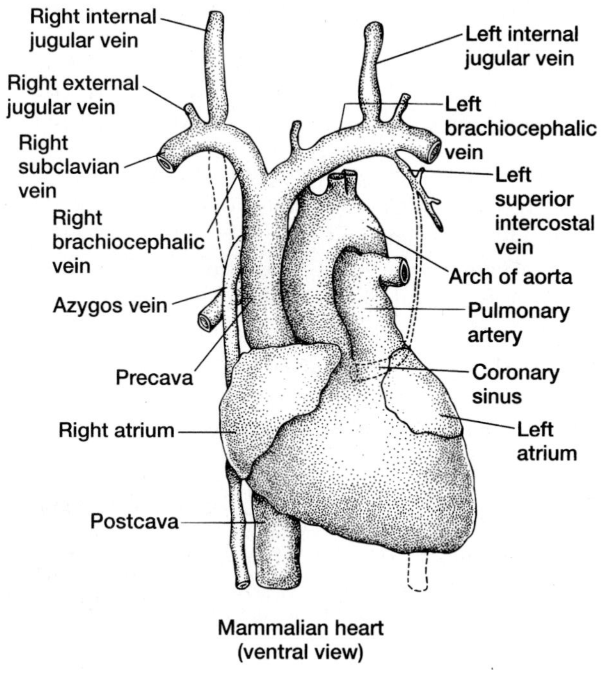


**Figure 12.39** Mammalian heart, ventral view.

## Overview of the Cardiovascular System

It is tempting to measure heart performance of lower vertebrates by how well their cardiovascular systems might serve mammalian needs. Partial internal heart septa have been termed "incomplete" in comparison with the "complete" anatomical divisions in mammalian hearts. Hearts and aortic arches of lower vertebrates have been interpreted as "imperfect" designs because the mammalian design was considered optimal. As we have seen, the lungfish cardiovascular system was mistakenly thought to mix oxygenated and deoxygenated blood.

If we begin with the view that lower vertebrates are designed imperfectly, then such a naive conclusion is bound to follow. In lungfishes, if oxygenated blood from the respiratory gills (II, V, and VI) met with deoxygenated blood from arches without gills (III, IV) in the dorsal aorta, then the two would mix. And if mixing occurred, blood perfusing active systemic tissues would have a lower oxygen tension. Certainly this would be an inefficient design. Experimental work coupled with knowledge of basic vertebrate anatomy now proves this interpretation incorrect.

The "incomplete" internal heart divisions and associated aortic arch arrangements enable lungfishes to adjust their physiological patterns of circulation to changes in the availability of oxygen in their environment. The cardiovascular systems of lower vertebrates are extraordinarily flexible, permitting adjustments to air and water ventilation patterns. Their cardiovascular systems are no less adaptive for the environments in which they serve than the systems of more "advanced" birds and mammals. Evolution of the cardiovascular system represents not a progressive improvement of design but rather an equally adaptive alternative way of addressing the demands that different lifestyles place on the circulatory system.

### *Accessory Air-breathing Organs*

Many fishes do not experience the extremes of hypoxia lungfishes face; nevertheless, they occasionally endure temporary stress from low oxygen availability. Vascularized gas bladders seem to be one answer. Blood is diverted from the general circulation to the gas bladder, where oxygen is taken up and circulated back to the general circulation (figure 12.40a). In *Hoplosternum*, a carplike fish, branches from the dorsal aorta carry blood to areas of the digestive tract enriched with capillary beds. Blood in these capillary beds is exposed to a bubble of air swallowed into the digestive track. Oxygen from the air is taken up and added directly to the systemic circulation to boost overall oxygen tension (figure 12.40b).

In fishes with accessory air-breathing organs, oxygenated blood joins the general circulation before entering

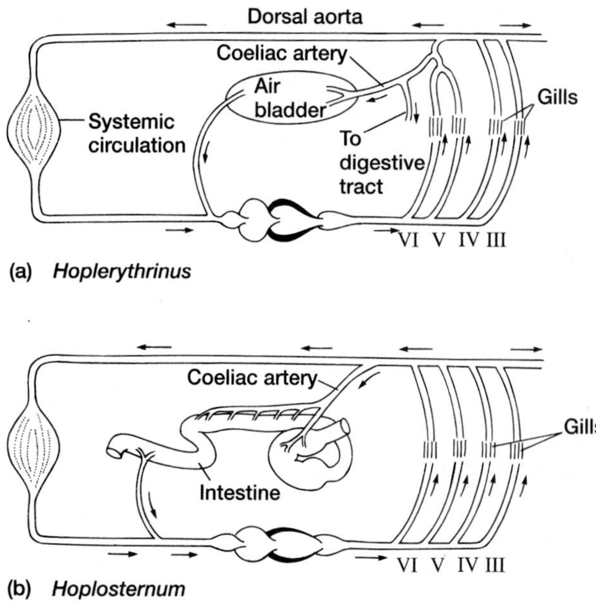


**Figure 12.40** Blood supply to accessory air-breathing organs of fishes. (a) The jeju, *Hoplerythrinus*, has evolved a vascularized swim bladder supplied by a branch of the coeliac artery. When this fish breathes air, blood flow to the swim bladder almost doubles, but there is no anatomical structure to separate the returning oxygenated blood from the venous circulation before both enter the heart. (b) *Hoplosternum* gulps air into its intestine, from which oxygen enters the systemic circulation. As in the jeju, oxygenated blood mixes with deoxygenated blood on its way back to the heart.

the heart. The result is to raise oxygen levels in the blood enough to compensate for low levels in the water and get the fish through temporary periods of hypoxia. This design is sufficient and adaptive for the limited stresses introduced by occasional low levels of oxygen. Lungfishes are unique among living fishes in possessing a discrete pulmonary vein that returns blood directly to the heart. Under more severe conditions of frequent, prolonged hypoxia and during droughts, this cardiovascular design allows dipnoans to survive. Thus, this group can occupy habitats and tolerate conditions for which other fishes are much less suited.

### *Diving Birds and Mammals*

The hearts of diving birds and mammals do not offer the physiological options that amphibians and reptiles employ to adjust to the demands of diving. In a sense, birds and mammals are locked into a design unsuited to aquatic life. When they dive, oxygen in the lungs is quickly depleted. It soon becomes disadvantageous for the heart to pump the customary large volume of blood to the nonfunctioning lungs. Yet, because of the complete internal division of the heart, diversion of blood away from the lung cannot occur within the heart. Adjustments must occur by other means.

When a tetrapod dives, three major physiological adjustments take place within the circulatory system. First, bradycardia occurs. The decreased heart rate reduces the energy spent on pumping blood to lungs depleted of oxygen. Second, anaerobic metabolism in skeletal muscles increases. Third, the microcirculation alters the blood flow to major organs and tissues. For example, blood flow to the brain and adrenal glands is maintained, but blood flow to the lungs, digestive tract, and appendicular muscles (which function under anaerobic conditions) is decreased.

Overall a diving bird or mammal makes the best of a stressful condition. When it is submerged, there is little oxygen available. The lungs are depleted and often collapse under the water pressure. Consequently, energy-consuming activities switch to metabolic pathways that require no immediate oxygen (muscles change to anaerobic metabolism), energy is conserved (bradycardia occurs), and available blood is shunted to priority organs (microcirculation shifts). None of these physiological responses is unique to birds and mammals. All tetrapods show similar responses when they dive in water. But because birds and mammals lack a heart with a cardiac shunt, these are the only major cardiovascular adjustments available.

In contrast to birds and mammals, the hearts of reptiles and amphibians function as two independent pumps. During dives, pulmonary resistance increases, and more blood can be diverted to the systemic circulation. But because the pumps are independent, independent pressures can be produced. It may, for example, be important to keep systemic pressure high so renal filtration of the blood does not decline; this can be accomplished within the systemic circuits without requiring simultaneous elevation of lung perfusion as well.

### *Hemodynamics*

The heart not only produces the initial pressure that moves blood, but it also separates oxygenated and deoxygenated bloodstreams and directs blood into appropriate aortic or pulmonary trunks. Separation of oxygenated and deoxygenated blood depends on many features of heart structure and function, including septation, position of entry and exit portals, dynamics of blood flow, and texture of the heart lining. Internal septa, whether complete or incomplete, aid in separating oxygenated from deoxygenated bloodstreams flowing through the heart. Locations of entrance and exit portals of the heart also assist in maintaining separate arterial and venous blood flow. For instance, in the crocodile heart, if the left aortic arch arose from the left rather than the right ventricle, the cardiac shunt would fail. In the lizard heart, delivery of blood from the left atrium into the cavum arteriosum places oxygenated blood at favorable locations within the heart so that it is strategically positioned to take the proper exit. Further, the balance of resistance between pulmonary and systemic circuits also influences the direction of blood flow from the heart. In the amphibian heart, the dynamics of blood flow through the ventricle partly explain why so little mixing of oxygenated and deoxygenated blood occurs in this common chamber. The recesses of the trabeculate myocardium may provide temporary sites in which blood entering from one stream is momentarily sequestered from another. The heart and its lining, in ways not yet understood, probably produce laminar rather than turbulent flow, further reducing the churning that might induce mixing of oxygenated and deoxygenated blood. For the heart to function properly, many subtleties of design must interact, even though we sometimes do not fully understand their indispensable contributions.

### *Ontogeny of Cardiovascular Function*

Embryo and adult often live in quite different environments. It is not surprising, then, that the circulatory system is different in these two stages of an individual's life history. For example, the heart of an embryonic chick consists of a pump with a single undivided ventricle that has the same hemodynamic demands as an adult fish heart. In both adult fish and chick embryo, gas exchange tissues and systemic tissues are in series. They are served by a single cardiac pump that must generate sufficient pressure to drive blood through both. Similar structural designs serve common functional demands. The embryonic heart no less than the adult serves the needs of the embryo, although it is a transitory stage. For most vertebrates, critical changes in the circulatory system must quickly accommodate sudden changes in physiological demands that occur at birth or hatching. These changes are most extensive and perhaps best known in placental mammals.

## *Fetal Circulation in Placental Mammals*

In eutherian mammals, the fetus depends exclusively on the placenta for its oxygen supply (figure 12.41). A single **umbilical vein** carrying oxygenated blood away from the placenta flows to the liver, where approximately half the blood enters the sinusoids of the liver and the other half bypasses the liver via the **ductus venosus** and enters the hepatic vein. Blood in the hepatic vein joins the large volume of blood returning via the precava and postcava to the right atrium. Pulmonary circulation to the nonfunctional lungs is reduced. About 90% of the blood that reaches the pulmonary artery bypasses the lung via the **ductus arteriosus** and is diverted instead to the dorsal aorta. Within the heart, the interatrial septum is incomplete. The **foramen ovale,** an opening between right and left atria, allows most blood entering the right atrium to flow directly to the left atrium without first passing through the lungs. Thus, the foramen ovale, like the ductus arteriosus, diverts most blood away from the nonfunctional lungs and into the systemic circulation. Blood returns to the placenta via the paired **umbilical arteries** that branch from the iliac artery (figure 12.42a).

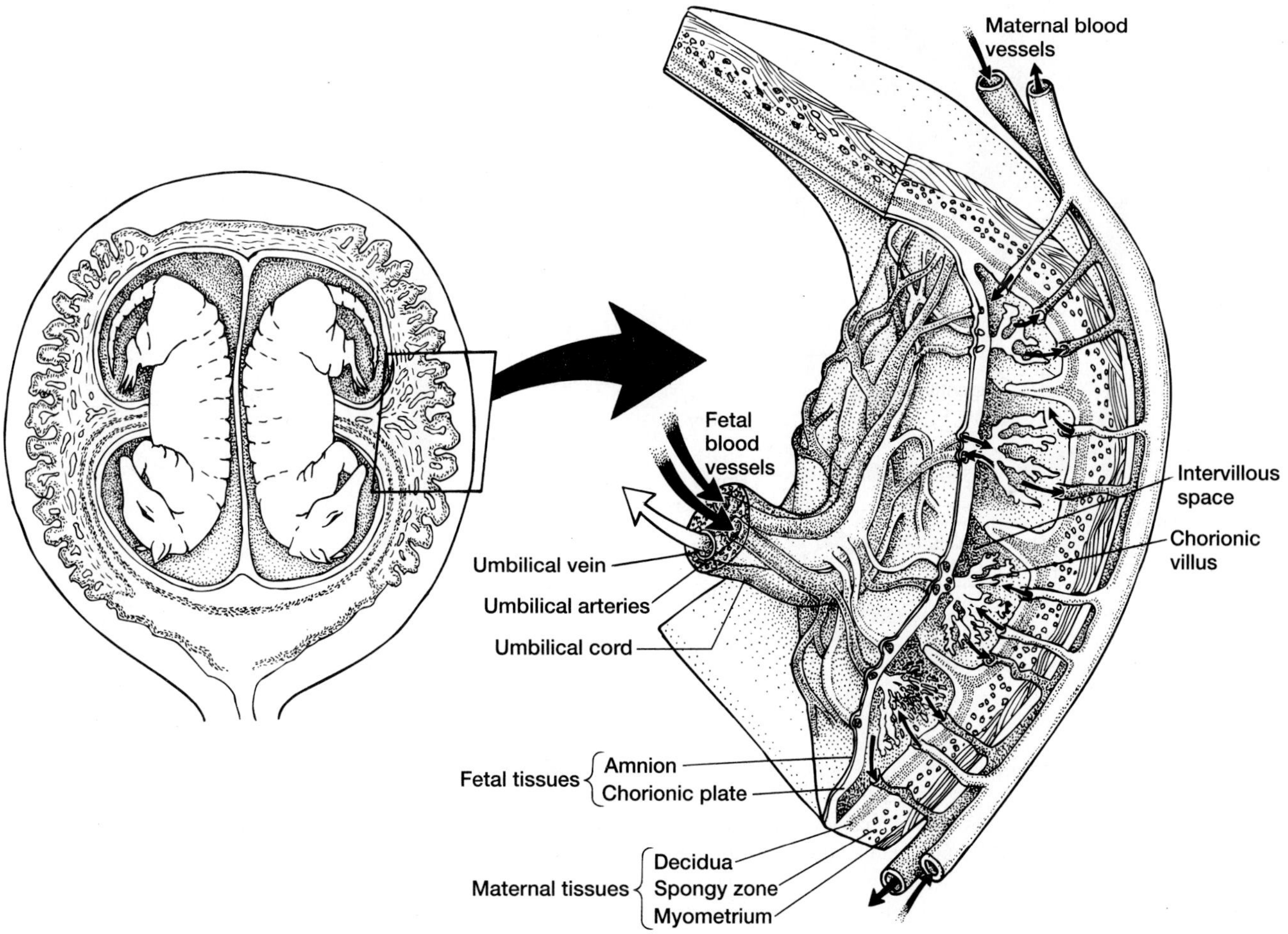


**Figure 12.41** Mammalian placenta. Extraembryonic membranes of the fetus produce the chorionic plate associated with the maternal tissues of the placenta. At parturition, the placenta separates from the uterus at the spongy zone. Fetal blood low in oxygen tension flows through two umbilical arteries into a dense branching network of capillaries in the chorionic villi. Fetal blood takes up oxygen from maternal blood in the chorionic villi. Oxygenated blood flows from these capillaries through the umbilical vein to enter the fetal circulation. Maternal blood flows through the placenta via branches of the uterine artery. It saturates the intervillous spaces and bathes the walls of the chorionic villi, giving up oxygen to the fetal capillaries. Maternal blood flows from these spaces through tributaries to the uterine vein.

After Mossman, 1937.

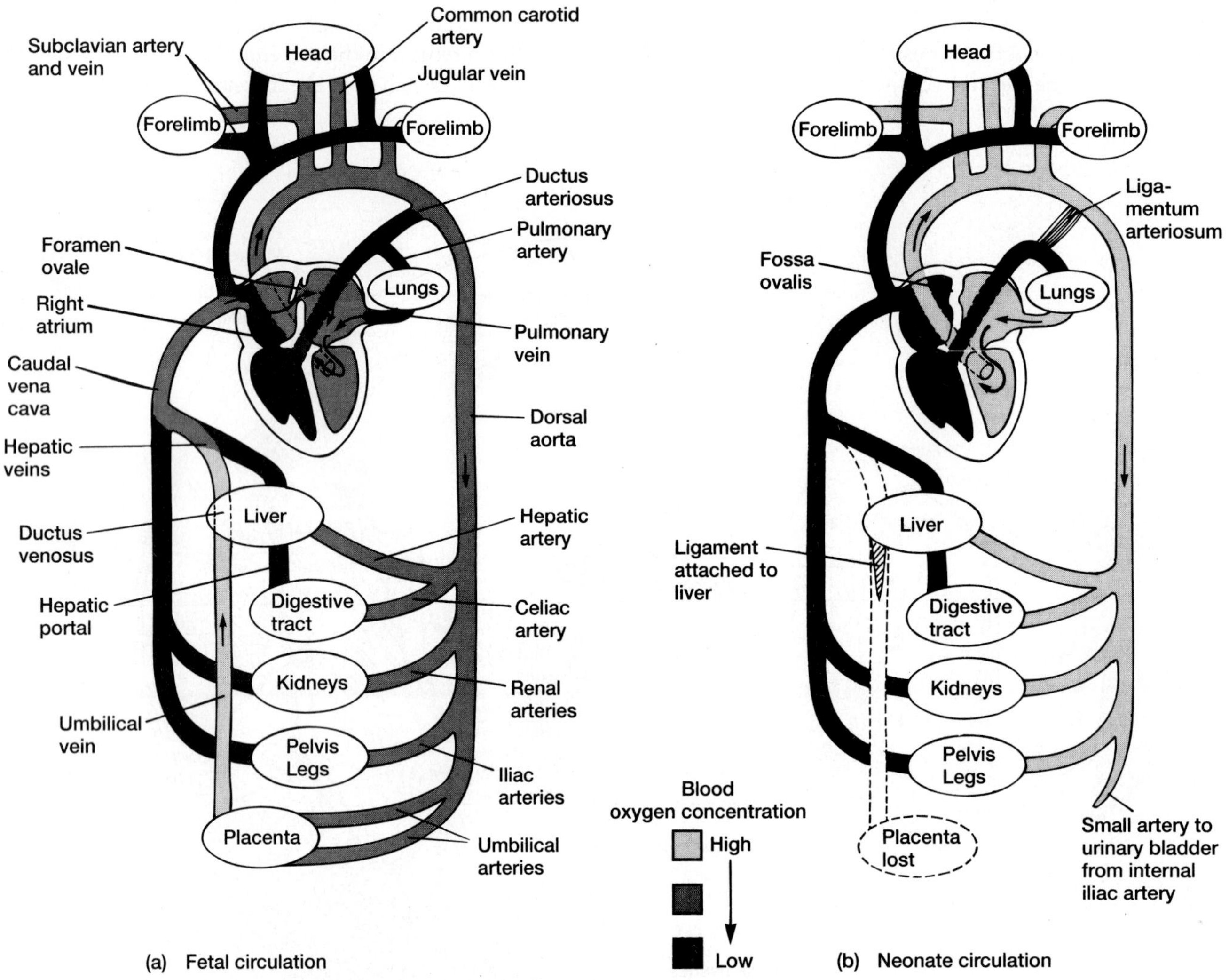


**Figure 12.42** Mammalian circulatory changes at birth. (a) Fetal circulation. Because the lungs are nonfunctional, uptake of oxygen and nutrients occurs through the placenta. The ductus venosus is a liver bypass. The foramen ovale and ductus arteriosus are lung bypasses. (b) Neonatal circulation. Following birth, the lungs become functional, the placenta departs, and the ductus venosus, foramen ovale, and ductus arteriosus close.

**Placentae (p. 181)**

Near the end of gestation, the young mammalian fetus has a specialized and complex circulatory system. Blood entering the right atrium is a mixture of deoxygenated blood (from the liver, precava, postcava, and coronary sinus) and oxygenated blood from the placenta (via the umbilical vein and ductus venosus). However, even with this mixing in a common chamber, oxygenated blood from the placenta tends to be shunted through the foramen ovale to the left atrium. From the left atrium it flows in turn to the left ventricle, the dorsal aorta, the carotids and to the head. Therefore, the fetal brain preferentially receives blood that is higher in oxygen tension compared with blood sent to organs elsewhere in the body.

Because pulmonary resistance is high, pressures are higher on the right side of the heart than on the left side. This pressure differential and the one-way action of the foramen ovale ensure that blood flows only from the right to the left atrium.

***Changes at Birth*** When a young human is born, several changes in the circulatory system occur nearly simultaneously. As maternal and fetal tissues separate in the birth process, placental circulation ceases. The neonatal lungs expand with the first vigorous breaths and become functional for the first time (figure 12.42b). When breathing begins, the ductus arteriosus is closed by muscular contractions of smooth muscles in its walls. Over a period of several weeks, fibrous tissue invades the lumen and obliterates the ductus arteriosus, which becomes a cord of tissue, the *ligamentum arteriosum* (ligament of Botallus). Because more blood enters the functional lungs after birth, more blood returns to the heart, increasing the pressure in the left atrium and forcing the septa of the foramen ovale closed. In most individuals, the septa gradually fuse so that an anatomically complete wall forms between the atria when a human is about a year old.

Smooth muscles within the walls of the umbilical vessels contract and gradually become invaded by fibrous connective tissue. This continues through the first two or three months of postnatal life. Occluded sections of the umbilical arteries become the lateral umbilical ligaments. Other sections of the umbilical arteries contribute to the common and internal iliac arteries. The umbilical vein persists only as a cord of connective tissue, the ligamentum teres. Over a two-month period, the ductus venosus atrophies into a fibrous mass, the ligamentum venosum (figure 12.42b).

As a result of these changes at birth, a double circulation pattern is quickly established and becomes anatomically consolidated over the first few months of neonate life. Failure of one or several of these changes to occur can result in inadequate oxygenation or distribution of blood. As poorly oxygenated blood reaches the peripheral circulation, the infant's skin darkens, a condition known as **cyanosis** (blue-baby syndrome). The severity of the condition and the appropriate medical response depend on which and how many of these changes fail to occur.

## *Heat Transfer*

In addition to delivering gases and metabolites, the circulatory system also functions in heat transfer. For example, reptiles basking in the sun absorb heat in their peripheral blood vessels. As this heated blood circulates throughout the body, it warms deeper tissues. Conversely, blood transports heat produced as a by-product of exercising muscles to the surface of the body (figure 12.43a,b). It can more easily be dissipated through the integument to prevent overheating. Whether blood transports warmth to deep tissues or carries away excess heat to the surface, the appropriate changes in circulation are largely mediated by changes in the microcirculation. During cooling, more capillary beds open in the peripheral circulation of the skin to increase blood flow and transfer heat to the environment. When heat is conserved, peripheral blood flow is reduced.

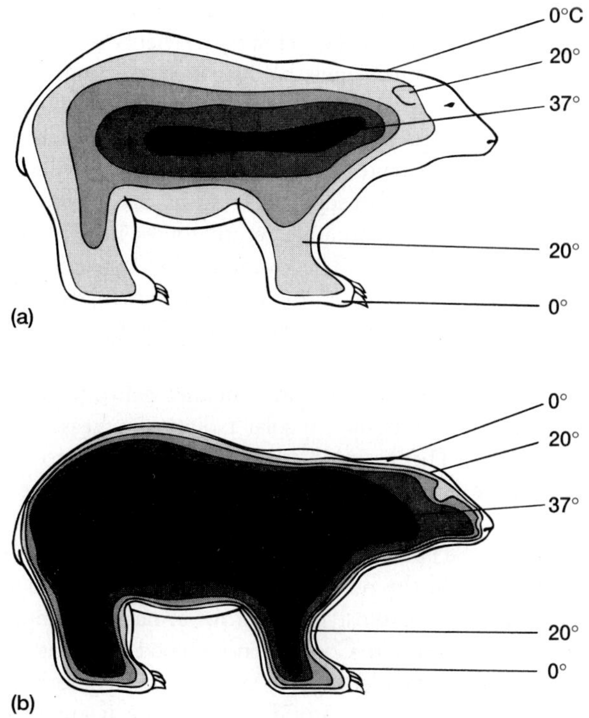


**Figure 12.43** Heat distribution in mammals. (a) Mammals, such as polar bears, preferentially maintain a relatively high core temperature. As blood nears the body's surface, it reaches cooler areas of the animal. (b) When an animal overheats, as during exercise or on hot days, excess heat is circulated to the skin surface where it can be dissipated more easily.

Animals in water often face special problems with respect to controlling heat loss or gain. The flippers or webbed feet of whales, seals, or wading birds are bathed in cold water. Blood circulating to these extremities is warm but the water can be icy; therefore, much heat would be lost to the environment were it not for specialized features of the circulatory system. In the upper regions of flippers or legs, an elaborate intertwining network forms between outgoing arteries and returning veins. These adjacent networks of arteries and veins are termed *retes*. Blood in a rete establishes a countercurrent pattern between outgoing arteries and returning veins. Before reaching the flipper or foot, warm blood passes through the rete. Heat carried in the arteries is transferred almost completely to the returning blood in the veins. By the time blood reaches the extremity, little heat remains to be dissipated to the environment. Such retes function as **heat blocks** to prevent body heat from being lost through the extremities.

**Countercurrent exchange (p. 146)**

In dolphins and whales, an additional mechanism is used to control heat loss. Deep within the core of the fin, the single central artery is surrounded by numerous veins. The outgoing central artery and numerous returning veins form a countercurrent exchange system that functions as a heat block (figure 12.44). However, when the animal is active and excess heat must be dissipated, the same circulatory mechanism participates. Additional blood flows to the fin via the central artery. As the blood dilates this artery, it puts pressure on the surrounding veins causing them to collapse. Because these veins are closed, the warm blood seeks alternative return routes in veins near the surface of the fin. The overall result is to close the deep heat block temporarily and simultaneously divert blood to the surface where excess heat can be transferred to the water.

Pursuing prey or evading enemies generates excess metabolic heat; exposure to solar radiation increases body temperature. The brain is especially susceptible to such temperature extremes. If it overheats even slightly, the result can be lethal. In many animals, a special **carotid rete** at the base of the brain addresses this problem (figure 12.45a). For example, in the nose of a dog, highly folded *turbinate bones* support an extensive area of moist nasal membranes cooled by evaporation. Cooled venous blood returning from this nasal membrane enters the carotid rete to absorb heat from the blood in the carotid artery before it enters the brain (figure 12.45b). Of course, not all heat is blocked to the brain because the brain must be kept warm. But the carotid rete serves as another heat block. In this instance, the heat block provides a mechanism for absorbing excess heat and preventing harmful thermal extremes in the brain (figure 12.45c).

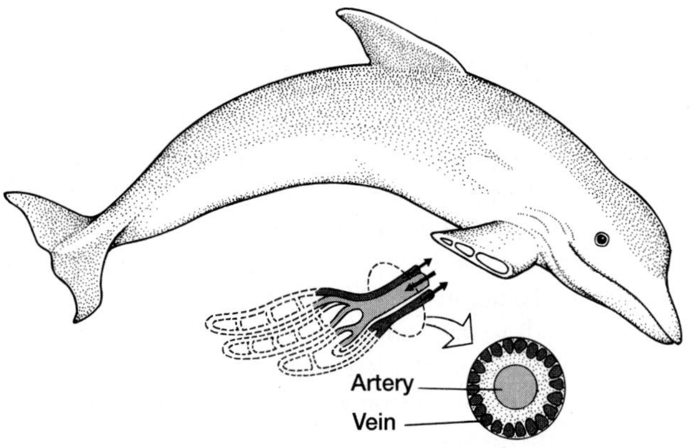


**Figure 12.44** Heat blocks. Birds and mammals that live in cold waters are in danger of losing heat to their surroundings continuously. This is especially true in the extremities that have large surface area in relation to their mass. Warm blood flowing to the extremities loses heat to the environment. Potentially this could be a severe heat drain on the organism. In dolphins, warm arterial blood circulating to the flipper passes returning venous blood, which is cold. Because returning veins of the flipper surround its central artery, a countercurrent system of flow is established. Arterial blood flowing into the flipper gives up its warmth to the returning venous blood, reducing the dolphin's heat loss to its environment. This countercurrent arrangement in the upper flipper serves as a heat block, preventing excess loss of heat to the environment.

# Lymphatic System

The lymphatic system is part of the circulatory system, but because of several special functions in which it is involved, it is treated separately. Structurally, there are two components of the lymphatic system—lymphatic vessels and lymphatic tissue.

## Lymphatic Vessels

Collectively, the lymphatic vessels constitute a blind-ended, tubular system that recirculates fluid from the tissues back to the cardiovascular system. The walls of lymphatic vessels are similar to those of veins and, like veins, they also contain one-way valves.

Pressure within the arterioles of the blood arises from two sources. **Hydrostatic pressure** represents the remaining force generated initially by ventricular contraction. It tends to favor the flow of fluid from the blood into the surrounding tissue. **Osmotic pressure** results from unequal concentrations of proteins within the arteriole and outside in the surrounding tissue fluid, so fluid moves from the surrounding tissue into the blood. As an arteriole approaches a capillary bed, residual hydrostatic pressure is usually higher than osmotic pressure. Consequently, fluid seeps from the blood to bathe the surrounding cells. This fluid that has es-

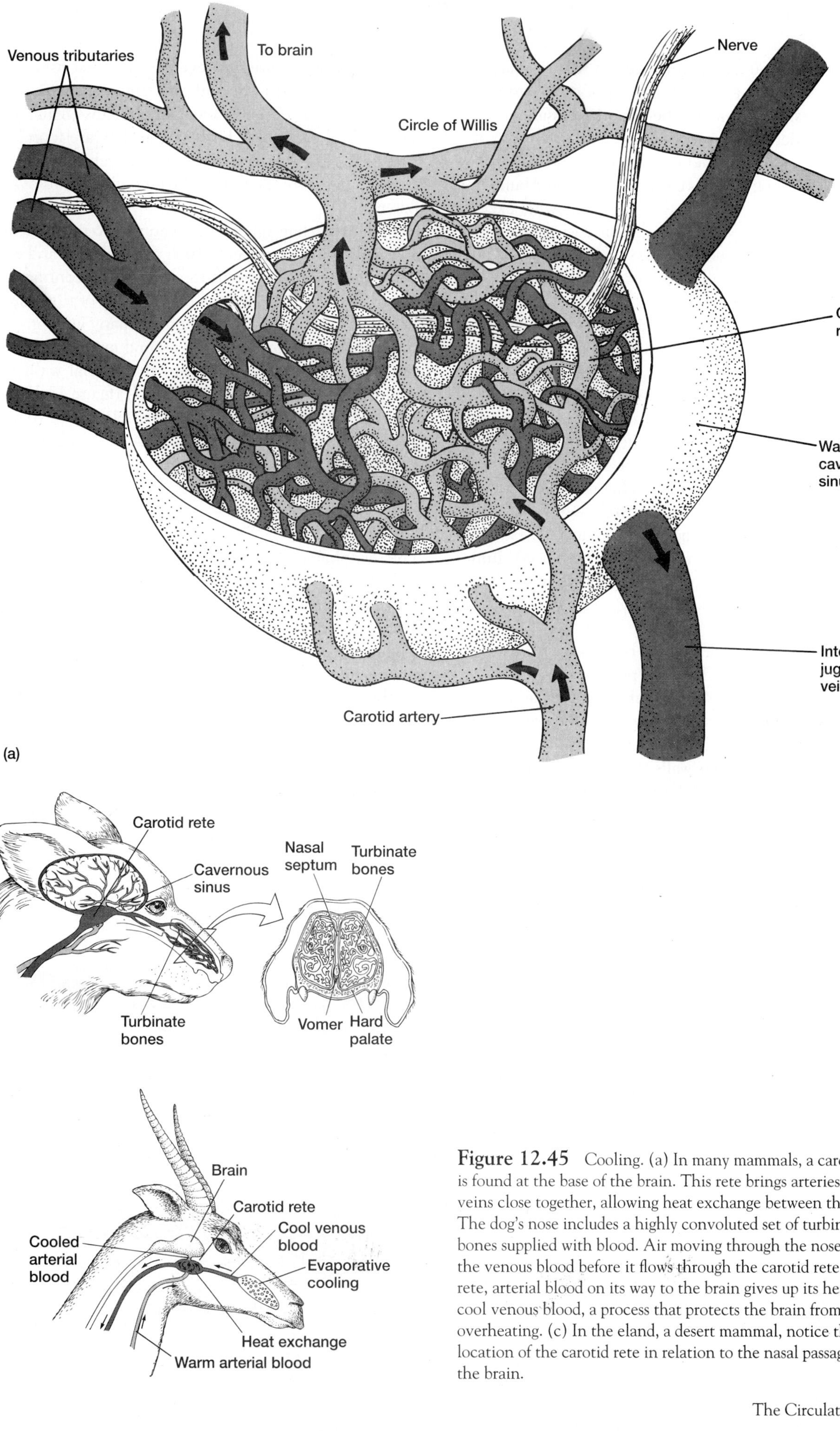


**Figure 12.45** Cooling. (a) In many mammals, a carotid rete is found at the base of the brain. This rete brings arteries and veins close together, allowing heat exchange between them. (b) The dog's nose includes a highly convoluted set of turbinate bones supplied with blood. Air moving through the nose cools the venous blood before it flows through the carotid rete. In the rete, arterial blood on its way to the brain gives up its heat to this cool venous blood, a process that protects the brain from overheating. (c) In the eland, a desert mammal, notice the location of the carotid rete in relation to the nasal passages and the brain.

caped from the blood capillaries is called **tissue fluid.** At the venule side of the capillary bed, most of the hydrostatic pressure has dissipated, leaving the osmotic pressure to predominate. The net inward pressure results in recovery of almost 90% of the original fluid that leaked from the arterial blood. The remaining 10%, if not recovered, would build up in connective tissues, causing them to swell with excess fluid, a condition termed **edema.** Edema does not usually occur because excess tissue fluid is picked up by the lymphatic tubules and eventually returned to the general blood circulation (figure 12.46).

The fluid carried by the lymphatic vessels is **lymph.** It consists mostly of water and a few dissolved substances such as electrolytes and proteins, but it contains no red blood cells. The main vessels of the lymphatic system collect the lymph resorbed by the tiny, blind lymphatic capillaries and return it to the venous circulation near the precaval and postcaval veins (figure 12.47). Lymphatic vessels form a network of anastomosing channels. The major vessels that generally compose the lymphatic network and the parts of the body they drain are the **jugular lymphatics** (head and neck), **subclavian lymphatics** (anterior appendage), **lumbar lymphatics** (posterior appendage), and **thoracic lymphatics** (trunk, viscera of body cavity, tail; figures 12.47a and 12.48a,b).

The low pressure within the lymphatic vessels aids them in taking up tissue fluid but presents a problem in moving lymph along. In some vertebrates, such as teleost fishes, **lymph "hearts"** occur along the route of return. These are not true hearts because they lack cardiac muscle, but striated muscles in their walls slowly develop pulses of pressure to drive the lymph. Spinal nerves supply lymph hearts, although the hearts can also pulse rhythmically on their own if innervation is severed. In teleost fishes, lymph hearts are found in the tail and empty into the caudal vein. They also occur in some amphibians (figure 12.49), reptiles, and embryonic birds. Often they are found where lymphatic vessels enter veins. One-way valves in the lymph hearts help ensure the return of lymph to the cardiovascular system.

The mechanism of lymph return also takes advantage of general body movements, such as inhalation and exhalation pressure differences in the thorax and contractions of nearby muscles that impinge on the walls of the lymphatic vessels to force the flow of lymph. In many vertebrates, the lymphatic vessels form sheaths around major pulsing arteries. Pulse waves traveling within the arterial walls impart their energy to the surrounding lymph (figure 12.50a). The one-way valves within lymphatic vessels ensure that these forces move the lymph back to the blood circulation (figure 12.50b).

## Lymphatic Tissue

The lymphatic system also includes lymphatic tissue, a collection of connective tissue and free cells. The free cells comprise mostly leucocytes, plasma cells, and macrophages, all of which play a part in the body's immune system. Lymphatic tissue can be found almost anywhere in the body, as diffusely distributed tissue, in patches, or encapsulated in lymph nodes. A **lymph node** (figure 12.47b) is a collection of lymphatic tissue wrapped in a capsule of fibrous connective tissue. Lymph nodes are located within channels of the lymphatic vessels along the route of returning lymph. This position in lymphatic vessels ensures that the lymph will percolate

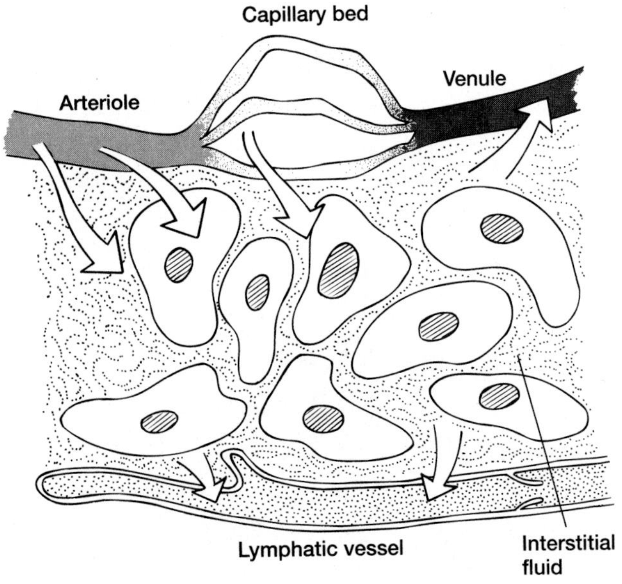


**Figure 12.46** Formation of lymph. The relatively high pressure in capillaries results in fluid from the blood leaking into surrounding tissues. Some of this interstitial fluid returns to the blood on the low-pressure venous side of the circulation. Blind-ended channels called lymphatic vessels collect the excess fluid (lymph) and return it to the general circulation, usually through one of the large thoracic veins (arrows indicate movement of fluid).

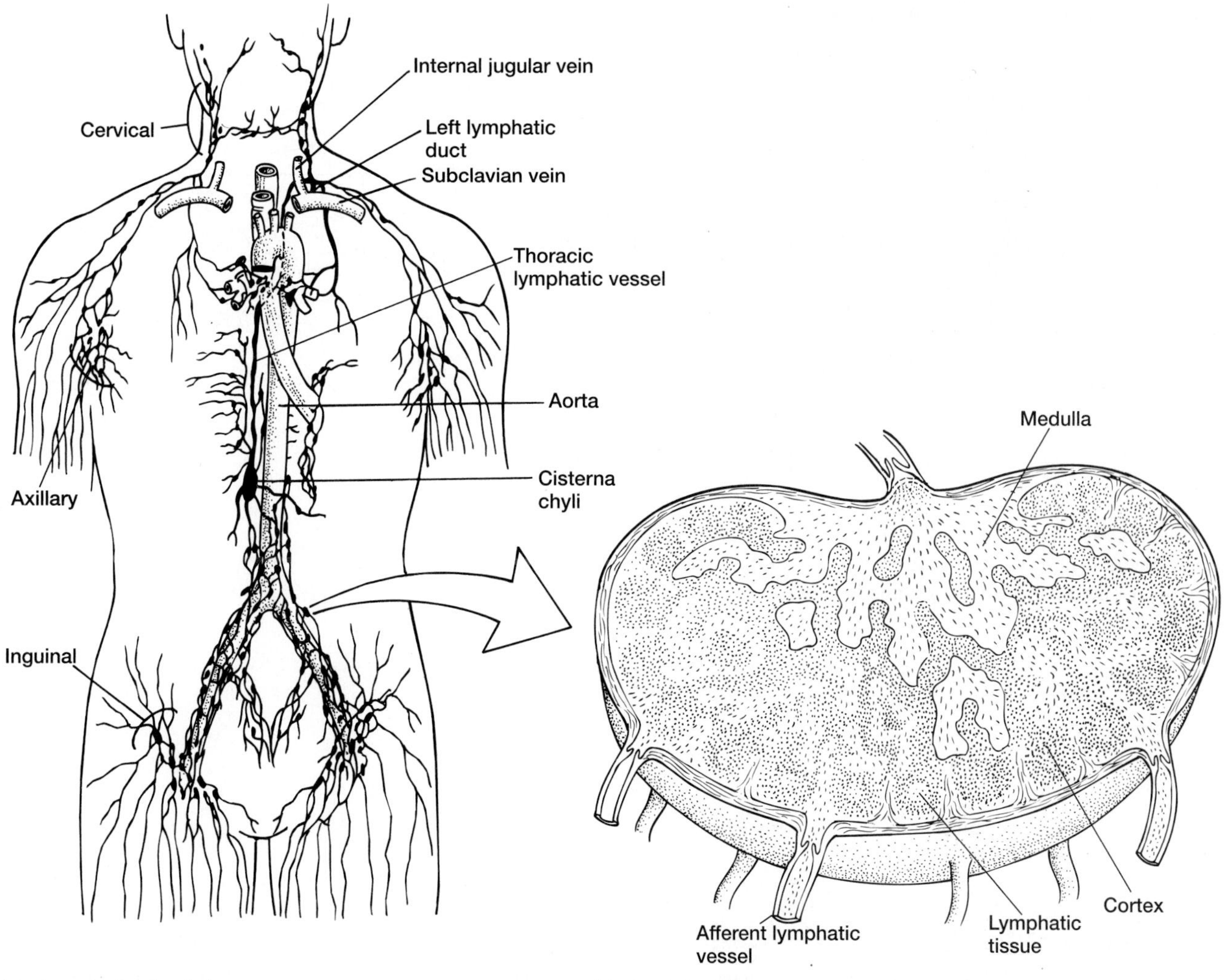


(a) Lymphatic circulation

(b) Lymph node

**Figure 12.47** Lymphatic circulation and lymph nodes. (a) Lymphatic vessels returning from all parts of the body join to form major lymphatic vessels, the largest being the thoracic duct, which empties lymph into the postcaval or subclavian veins. (b) Cross section of a lymph node. In mammals and in a few other species, small swellings or nodes occur along lymphatic vessels. These lymph nodes house lymphatic tissue, which functions to remove foreign materials from the lymph circulating through them. Lymph nodes have a cortex and medulla bounded by a fibrous connective tissue capsule. Notice the entering and departing lymphatic vessels.

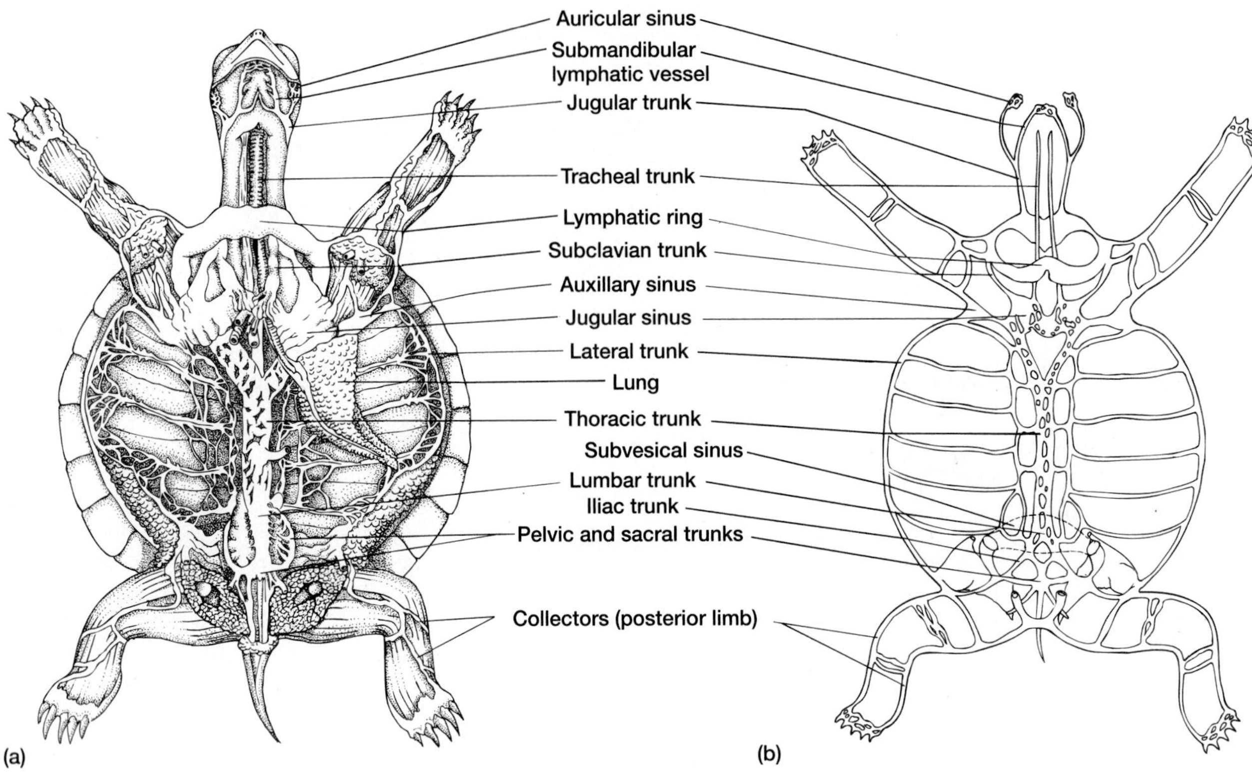


**Figure 12.48** Lymphatic vessels in the turtle *Pseudemys scripta* (ventral view). (a) The plastron and most of the viscera have been removed to show the lymphatic vessels. (b) Diagrammatic view of major lymphatic trunks in turtles.

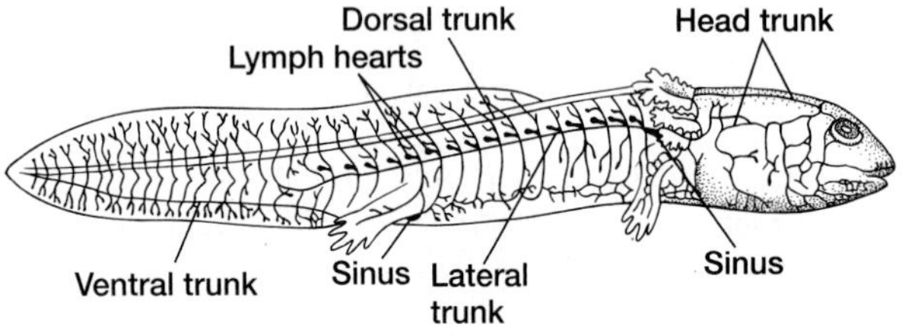


**Figure 12.49** Lymphatic vessels of a salamander. Lymph hearts aid in returning fluid to the blood circulation.

through the lymphatic tissue held in the node and be presented to the free cells. Lymph nodes occur in mammals and some water birds but are absent in other vertebrates. In reptiles, dilation or expansion of lymphatic vessels termed **lymphatic cisterns** or **lymphatic sacs** occur at locations usually occupied by true lymph nodes in birds and mammals.

## Form and Function

The lymphatic vessels function as an accessory venous system, absorbing and returning escaped fluid to the general circulation. They also absorb lipids from the digestive tract. Numerous lymphatic vessels in the digestive tract termed **lacteals,** pick up large chain fatty acids and return them to the blood circulation (figure 12.51).

In addition, lymphatic tissue is involved in the removal and destruction of harmful foreign material, such as bacteria and dust particles. Plasma cells produce some antibodies that circulate in blood. Macrophages cling to leucocytes as they function to destroy bacteria. Lymphatic tissue also intercepts cancer cells migrating through the lymph nodes, although free cells cannot destroy cancer cells. Eventually, lymph nodes are successively overwhelmed by rapidly dividing cancer cells. If a cancer is detected early, surgical intervention usually can cure a patient. Follow-up tests should be performed to detect the extent to which cancer cells have spread through the lymphatic vessels, and then all affected nodes should be removed.

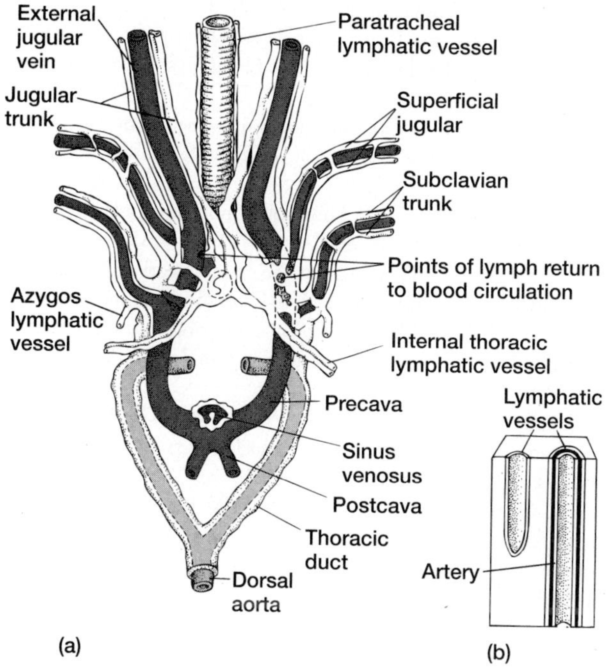


**Figure 12.50** Lymphatic system in reptiles. (a) Anterior lymphatic vessels in the crocodilian *Caiman crocodilus*. The heart has been removed to reveal the major blood vessels and associated lymphatic vessels more clearly. The pressure necessary to move lymph through lymphatic vessels is derived from the action of surrounding organs. Many lymphatic vessels lie next to active muscles as well. The walls of the lymphatic vessels passing through the thoracic cavity are compressed by the rhythmic respiratory movements. One-way valves in these vessels ensure that this pressure drives lymph back into the general blood circulation. (b) In addition, lymphatic vessels often surround major arteries, deriving force from the pulse waves of these arteries. Most lymphatic vessels form an extensive network of interconnected blind channels that gather and return lymph to the systemic blood circulation.

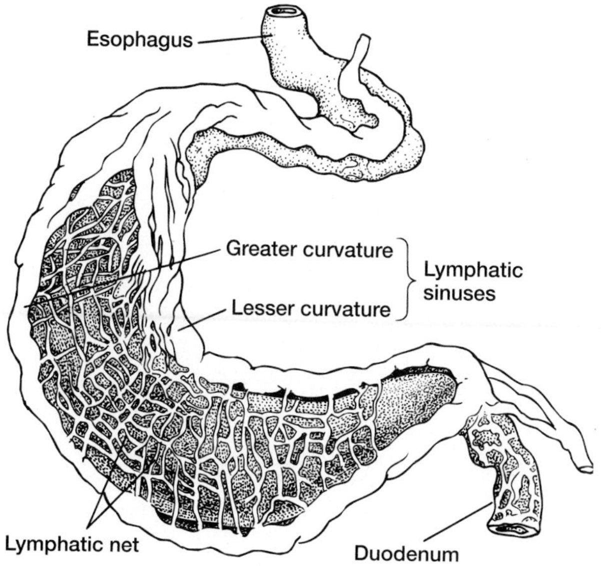


**Figure 12.51** Lymphatic vessels associated with the stomach of the turtle *Pseudemys scripta*. Microscopic lymphatic vessels within the walls of the stomach are lacteals that primarily take up large chain fatty acids absorbed through the wall of the stomach. Lacteals empty into extensive lymphatic nets that drain into large lymphatic sinuses along the lesser and greater curvatures of the stomach. Lymphatic sinuses empty into lymphatic trunks that travel through mesenteries, often receiving tributaries from other visceral lymphatic vessels and entering the thoracic lymphatic duct.

# Selected References

Birch, M. P., C. G. Carre, and G. H. Satchell. 1969. Venous return in the trunk of the Port Jackson shark, *Heterodontus portusjacksoni*. *J. Zool. (London)* 159:31–49.

Block, B. A., and F. G. Carey. 1985. Warm brain and eye temperatures in sharks. *J. Comp. Physiol.* B 156:229–36.

Burggren, W. W. 1988. Cardiac design in lower vertebrates: What can phylogeny reveal about ontogeny? *Experientia* 44:919–30.

Burggren, W. W., and K. Johansen. 1982. Ventricular hemodynamics in the monitor lizard, *Varanus exanthematicus*: Pulmonary and systemic pressure separation. *J. Exp. Biol.* 96:343–54.

Burggren, W. W., and K. Johansen. 1986. Circulation and respiration in lungfishes (Dipnoi). *J. Morph.* (suppl.)1:217–36.

Davie, P. S., M. E. Forster, W. Davison, and G. H. Satchell. 1987. Cardiac function in the New Zealand hagfish, *Eptatretus cirrhatus*. *Physiol. Zool.* 60:233–40.

De Vries, R., and S. De Jager. 1984. The gill in the spiny dogfish, *Squalus acanthias*: Respiratory and nonrespiratory function. *Amer. J. Anat.* 169:1–29.

Feder, M. E., and W. W. Burggren. 1985. Cutaneous gas exchange in vertebrates: Design, patterns, control, and implications. *Biol. Rev.* 60:1–45.

Forster, M. E. 1989. Performance of the heart of the hagfish *Eptatretus cirrhatus*. *Fish Physiol. Biochem.* 6:327–31.

Johansen, K. 1968. Air-breathing fishes. *Sci. Amer.* 219(10):102–11.

Johansen, K. 1977. Respiration and circulation. In *Chordate structure and function*, edited by A. G. Kluge. New York: Macmillan, pp. 306–91.

Johansen, K., and W. W. Burggren. 1980. Cardiovascular function in the lower vertebrates. In *Hearts and heart-like organs*, edited by G. Bourne. New York: Academic Press, pp. 61–117.

Johansen, K., and W. W. Burggren. 1984. Venous return and cardiac filling in varanid lizards. *J. Exp. Biol.* 113:389–99.

Lillywhite, H. B. 1988. Snake, blood circulation and gravity. *Sci. Amer.* 259(6):92–98.

Martin, L. D., and B. M. Rothschild. 1989. Paleopathology and diving mosasaurs. *Amer. Sci.* 77:460–67.

Percy, R., and I. C. Potter. 1986. Description of the heart and associated blood vessels in larval lampreys. *J. Zool. (London)* 208:479–92.

Priede, I. G. 1976. Functional morphology of the bulbus arteriosus of rainbow trout (*Salmo gairdneri* Richardson). *J. Fish Biol.* 9:209–16.

Randall, D. J. 1968. Functional morphology of the heart in fishes. *Amer. Zool.* 8:179–89.

Satchell, G. H. 1984. On the caudal heart of *Myxine* (Myxinoidea: Cyclostomata). *Acta Zool.* (Stockholm) 65:125–33.

Satchell, G. H. 1986. Cardiac function in the hagfish, *Myxine* (Myxinoidea: Cyclostomata). *Acta Zool.* (Stockholm) 67:115–22.

Satchell, G. H., and L. J. Weber. 1987. The caudal heart of the carpet shark, *Cephaloscyllium isabella*. *Physiol. Zool.* 60:692–708.

Satoh, Y-C., and T. Nitatori. 1980. On the fine structure of lymph hearts in amphibia and reptilia. In *Hearts and heart-like organs*, edited by G. Bourne. New York: Academic Press, pp. 149–69.

Vogel, W.O.P. 1985. The caudal heart of fish: Not a lymph heart. *Acta Anat.* 121:41–45.

Yamamoto, K. 1988. Contraction of spleen in exercised freshwater teleost. *Comp. Biochem. Physiol.* 89A:65–66.

Zapol, W. M. 1988. Diving adaptations of the Weddell seal. *Sci. Amer.* 256(6):100–105.

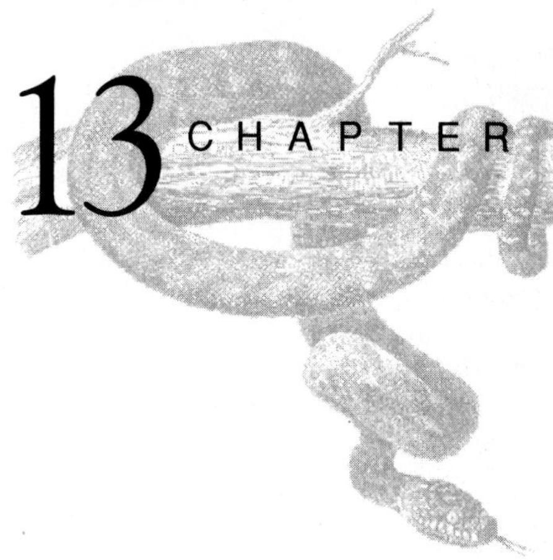

# 13 CHAPTER The Digestive System



## INTRODUCTION

In the nineteenth century Alfred, Lord Tennyson served up the grim description of "Nature, red in tooth and claw," a poetic reminder that animals must procure food to survive, a sometimes harsh but practical necessity. For predators that means another animal; for herbivores that means plants. A quick chase and overpowering kill might characterize prey capture by a carnivore; prolonged browsing or migration to fresh sources of succulent plants might characterize feeding by a herbivore. But such a hard won meal is initially unusable. The process of turning a meal into usable fuel for the body is the business of the digestive system. The digestive system breaks up the large molecules contained in a succulent meal so they can be absorbed and made available for use in the body.

A lump of food in the mouth is called a **bolus.** Both mechanical and chemical processes go to work to digest this bolus. Initially, mechanical chewing with teeth and churning of the digestive tract break up the bolus, reducing it to many smaller pieces and thereby increasing the surface area available for chemical digestion by enzymes. Muscles encircling the walls of the digestive tract produce waves of contraction, termed **peristalsis,** that constrict food in the lumen, forcing it from one section of the tract to the next. As mechanical and chemical action work on the bolus, it soon becomes a pulpy mass of fluid more commonly called **chyme,** or **digesta.**

### Overview

The adult digestive system includes the digestive tract and accessory digestive glands. The **digestive tract** is a tubular passageway that extends through the body from the lips of the mouth to the anus or cloacal opening. Glands embedded in the walls lining the tract release secretions directly into the lumen. On the basis of histological differences among these intrinsic **luminal glands** and differences in size, shape, and embryonic derivation, three regions of the digestive tract are recognized. The **buccal cavity,** or mouth, leads into the **pharynx** and then into the **alimentary canal.** From histological differences in the luminal wall of the alimentary canal, four regions are identified: **esophagus, stomach, small intestine,** and **large intestine** (figure 13.1).

In most vertebrates, the alimentary canal ends in a **cloaca,** a terminal chamber receiving both fecal materials from the intestines and products of the urogenital tract. The exit portal of the cloaca is the **cloacal opening,** or **vent.**

However, in some fishes and most mammals, a cloaca is absent and the intestines and urogenital tract have separate exit portals. A coiled large intestine often straightens into a **rectum** with an anal opening **(anus)** to the outside. The **accessory digestive glands** are extrinsic glands located outside the walls of the digestive tract, but they secrete chemical enzymes of digestion into the lumen via long ducts. Principal glands of the digestive tract are the **salivary glands, liver,** and **pancreas.**

In the embryo, the **gut** is a simple tube of endoderm from which the pharynx and alimentary canal arise along with their associated digestive glands. During embryonic development, invaginations of the surface ectoderm come into contact with the endodermal gut at opposite ends of the body. The anterior invagination, or **stomodeum,** meets the anterior gut, or **foregut.** Between stomodeum and foregut a temporary **buccopharyngeal membrane** forms and eventually ruptures to join the lumina of both. The stomodeum gives rise to the buccal cavity. The posterior invagination of ectoderm, the **proctodeum,** meets the posterior gut, or **hindgut.** Between hindgut and proctodeum the **cloacal membrane** forms and then ruptures, creating the outlet for the hindgut. The proctodeum becomes the adult cloaca (figure 13.2a–d).

# Components of the Digestive System

Terms that describe parts of the digestive tract are often casually used. Some take alimentary canal to mean the entire digestive system, mouth to anus; others apply it in a restricted sense. The term *gastrointestinal tract,* or *GI tract,* literally means stomach and intestines, but most apply it to the whole digestive tract from buccal cavity to anus. Clinical physicians speak of small and large bowels, slang terms for small and large intestines, respectively. There is nothing sinister or sloppy about this, just different professionals seeking terminology to serve their needs. In this text, we use the terminology defined in this section.

## Buccal Cavity

The buccal cavity contains the teeth, tongue, and palate. Oral glands empty into it. Salivary glands help moisten food and secrete enzymes to begin chemical digestion. In some species, mastication begins the mechanical breakdown of food.

### *Boundaries*

The **oral opening,** the margins of which are the lips, creates the buccal cavity entrance. Generally, upper and lower lips follow the line of the tooth rows to meet posteriorly at the angle of the jaws. In mammals, upper and lower lips meet well forward of the jaw angle near the front of the mouth, thus creating a skin-covered **cheek** region. Cheeks prevent loss of food out the sides of the mouth during chewing. In some rodents and Old World monkeys, they expand into **cheek pouches,** small compartments in which gathered food can be temporarily held until it is chewed or carried to caches. Lips are usually pliable, although birds, turtles, some dinosaurs, and a few mammals have rigid beaks with firm margins. In most mammals, lips are fleshy, a feature helping an infant form a seal around

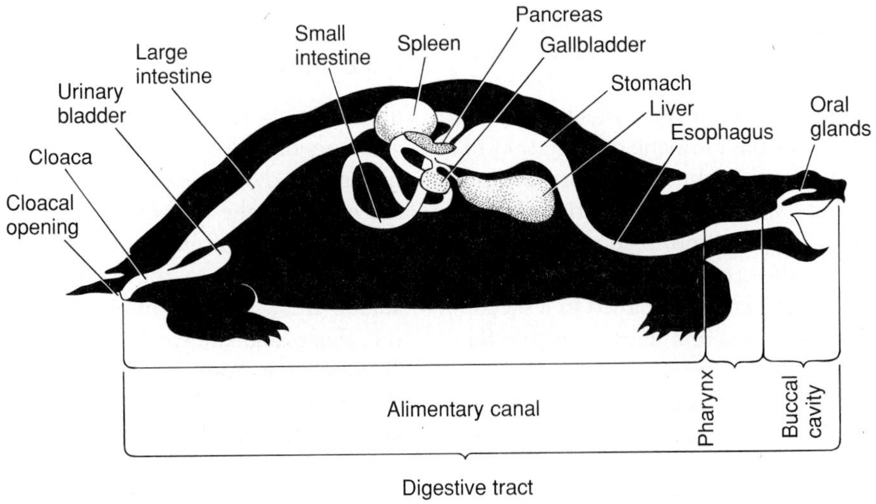


**Figure 13.1** Vertebrate digestive system. The digestive system consists of the digestive tract plus associated glands of digestion. The digestive tract includes the buccal cavity, pharynx, and alimentary canal. The alimentary canal is divided into esophagus, stomach, intestines, and cloaca.

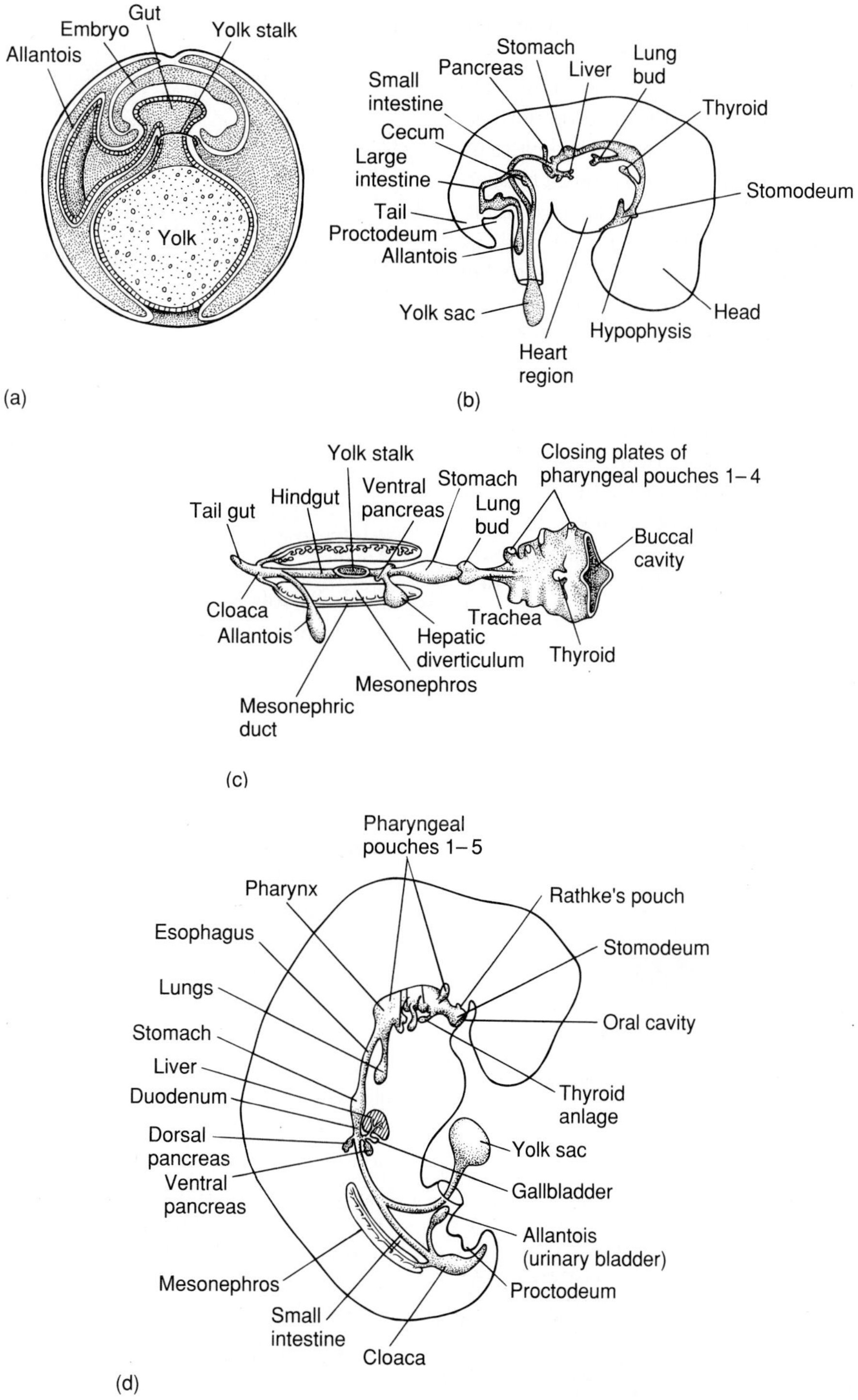


**Figure 13.2** Embryonic formation of the digestive system. (a) Early amniote embryo in sagittal section showing initial position of the gut. Notice the embryo's connections to the yolk via the yolk stalk and to the allantois. (b) Generalized amniote embryo in sagittal section. Note regions of the gut and invaginations destined to form associated glands of the digestive tract. (c) Ventral view of the isolated gut together with the embryonic kidneys. Notice the extensive pouches produced by the pharynx, each of which contributes to specific adult structures. (d) Lateral view of differentiating gut.

the nipple during nursing. Human lips help form the vocalizations of speech.

The lips define the anterior border of the mouth, and the **palatoglossal arch** is a fold that marks the posterior border of the mouth and lies between the mouth and pharynx (figure 13.3a). However, if this or other anatomical markers are absent, the mouth and pharynx form a collective chamber called the **oropharyngeal cavity** (figure 13.3b).

The stomodeum not only forms the buccal cavity, but it also contributes to surface features of the head in some vertebrates. Two good embryonic landmarks present within the stomodeum, the **hypophyseal pouch** and the **nasal placode,** make this evident (figure 13.4a). In cyclostomes, only the posterior part of the stomodeum turns inward contributing to the mouth. The anterior part turns outward, contributing to the outer surface of the head. In sharks and actinopterygians, the hypophyseal pouch lies within the mouth, the nasal placodes differentiate on the outside of the head, and the margins of the mouth form between these anatomical landmarks. In rhipidistians and tetrapods, both landmarks lie within the adult buccal cavity (figure 13.4b).

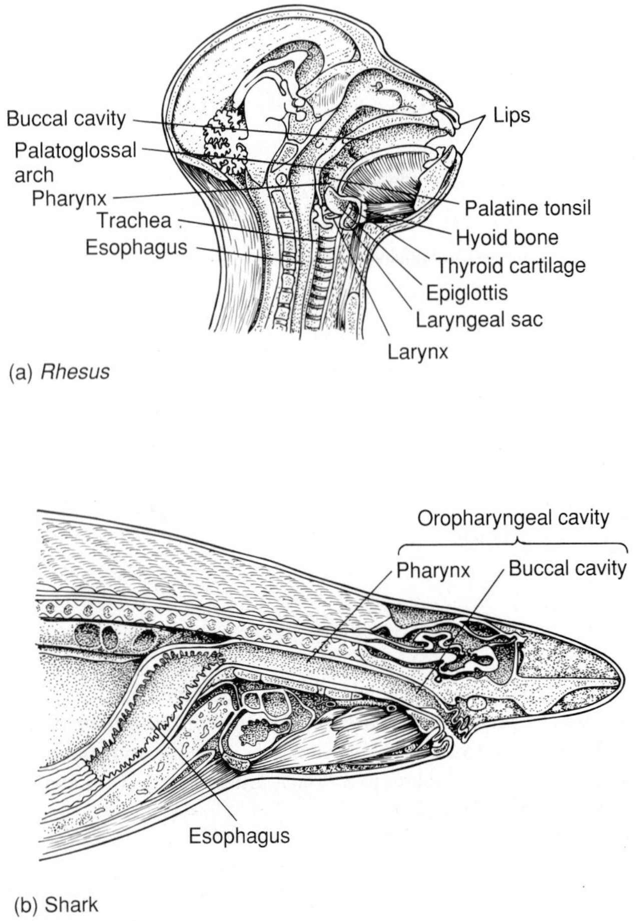


**Figure 13.3** Sagittal view of the buccal cavity, pharynx, and developing esophagus. (a) Head and neck of *Rhesus*. (b) Shark.

## *Palate*

The roof of the buccal cavity is the palate, formed from the fusion of ventral skull bones overlying the mouth. The primitive, or **primary palate,** includes a medial series of bones (vomers, pterygoids, parasphenoid) and a lateral series (palatines, ectopterygoids). In most fishes, the primary palate is a low vault with no openings. In rhipidistians and tetrapods, the nasal passages reach the mouth through paired openings in the primary palate, the **internal nares** or **choanae** (figure 13.5a,b). The **palatal folds** are inward growths of lateral bones that meet at the midline and form a second horizontal roof that separates the nasal passages from the mouth. This new roof, present in mammals and some reptiles, is called the **secondary palate.** The anterior part of the secondary palate is the **hard palate,** comprising paired bony contributions of the premaxilla and maxilla. In some species, the palatine and pterygoids contribute as well. In mammals, the posterior margin of the secondary palate is the fleshy **soft palate,** which extends the position of the internal nares even farther to the back of the buccal cavity (figure 13.5c).

**Secondary palate (p. 266)**

## *Teeth*

Teeth are unique among vertebrate animals. They are usually capped with enamel, a mineralized coat found only in vertebrates. Embryologically, neural crest cells contribute to parts of the forming tooth. Teeth are generally thought to have arisen phylogenetically from the bony armor of primitive fishes, probably from surface structures together with contributions from the underlying dermis. The fact that neural crest cells also contribute to the dermis supports the close relationship between teeth and bony armor.

The presence of teeth, but not dermal armor, in conodonts complicates this picture of tooth evolution. Conodont teeth were mineralized, but it is not certain if the mineral was enamel. If conodonts prove to be very early vertebrates, then teeth apparently preceded exterior armor rather than the other way around. However, at the moment, it seems most likely that teeth evolved from parts of the dermal armor.

**Conodonts (p. 84)**

Teeth help catch and hold prey. They also offer strong opposing surfaces that jaws work to crush hard shells of prey. In mammals and a few other vertebrates, mechanical digestion begins in the mouth. After each bite, the tongue and cheeks collect food and place it between the upper and lower tooth rows and the teeth break down the bolus mechanically, reducing it to smaller chunks to make swallowing easier. By breaking the large bolus into many smaller pieces, chewing also increases the surface exposed to chemical digestion. Even in vertebrates that do not

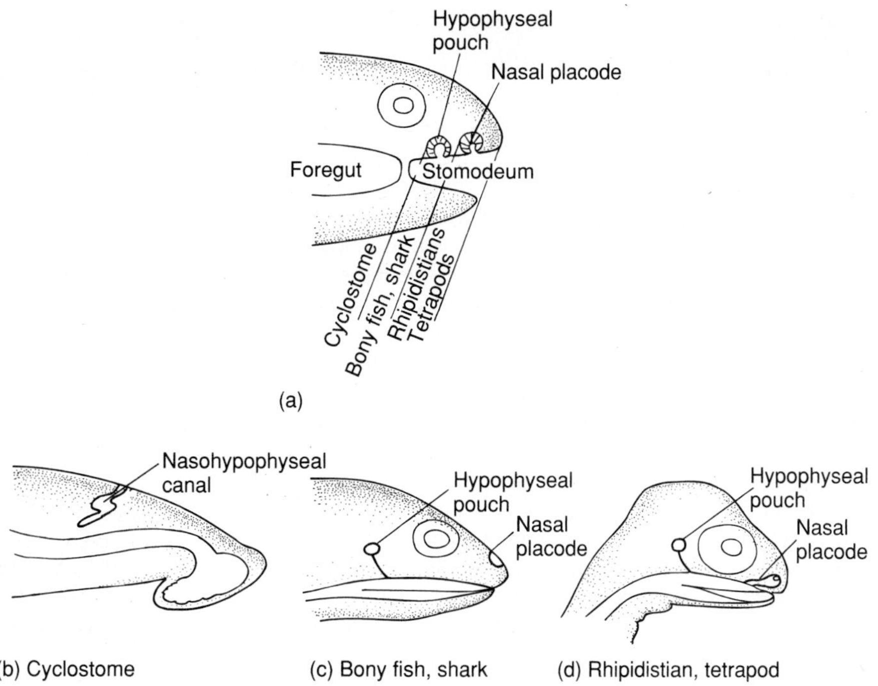


**Figure 13.4** Boundaries of the buccal cavity. The extent to which the embryonic stomodeum contributes to the mouth can be followed by two markers: the nasal placode and hypophyseal pouch. (a) Comparative positions of the anterior margin of the mouth in various groups of vertebrates. (b) Diagrammatic view of each group. These two markers remain outside the mouth in cyclostomes. In sharks and bony fishes, the hypophyseal pouch is pinched off from the mouth. In rhipidistians and tetrapods, both nasal placode and hypophyseal pouch open to or are derived from the mouth.

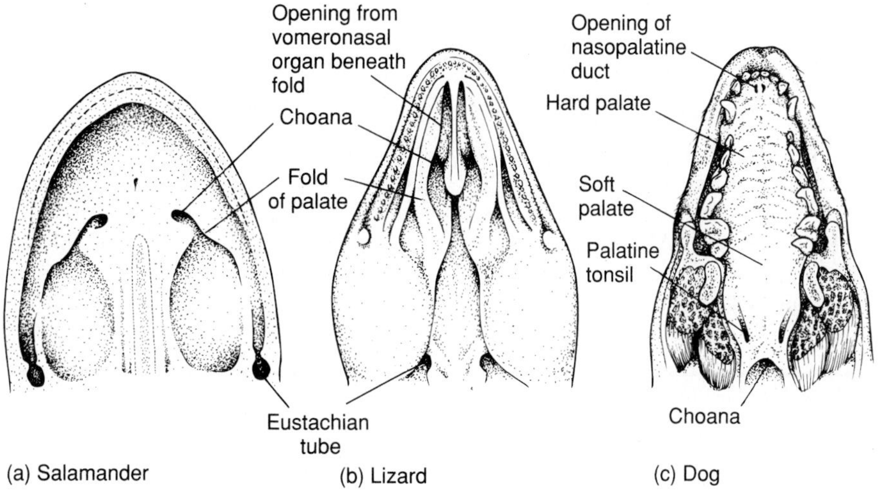


**Figure 13.5** Roof or palate of the mouth in tetrapods. (a) Salamander. (b) Lizard. (c) Mammal (dog). Notice the point of entry of the choana, or internal naris, in each animal.

chew their food, sharp teeth puncture the surface of the prey, creating sites through which digestive enzymes penetrate when food reaches the alimentary canal. For vertebrates that feed on insects and other arthropods, punctures through the chitinous exoskeleton are especially important in giving proteolytic enzymes access to the digestible tissues within.

***Tooth Anatomy*** The part of the tooth projecting above the gum line, or **gingiva,** is the **crown,** the region below is the **base.** If the base fits into a hole, or **socket** (alveolus) within the jaw bone, the base is referred to as a **root.** Within the crown, the **pulp cavity** narrows when it enters the root, forming the **root canal,** and opens at the tip of the root as the **apical foramen.** Mucous connective tissue, or

## Box Essay 13.1 Human Speech

Human speech is much more than loud grunts, at least when done well. Words are built up from carefully formed sounds called **phonemes.** By themselves, sounds have no meaning. Animal communication with sounds is mainly an emotional response to immediate circumstances. But to humans, phonemes in combinations carry ideas and thoughts about past events or future actions. We assign meaning to combinations of sounds rather than to individual sounds themselves. So functionally distinctive is our speech from the vocalizations of other vertebrates that some anthropologists mark the transition to *Homo sapiens* at the point in our ancestry where speech enters.

Not the sounds themselves but the relationships of sounds build words. And words placed in ordered sentences build an idea. But our speech apparatus can produce sounds quickly and shape them carefully only because it is redesigned. Anatomical changes serving speech centered on lengthening of the pharynx, which was accomplished by the separation of soft palate and epiglottis. By such lengthening, air can be effortlessly channeled on a sustained basis through the mouth, where it is shaped into sounds. Apes, with a short pharynx, must "bark" out sounds through short bursts of released air. Wolves can sustain a howl by lifting their heads, stretching their throats, and thereby temporarily lengthening their pharynx. Through control of its muscular walls, the redesigned pharynx of humans became the major vowel-producing chamber.

In nonhuman primates and in most other mammals, the larynx sits high in the neck and fits into the nasopharynx at the back of the nasal passage (box figure 1a). This establishes a direct route of air from nose to lungs without interfering with the food route from mouth to esophagus. In humans, the larynx has dropped down, lengthening the pharynx and forcing food and air routes to cross (box figure 1b). You, as a human, cannot swallow and breathe at the same time. When a mix-up occurs, food headed for the esophagus becomes instead caught in the epiglottis, and you choke. To right matters quickly, residual air in the lungs can be forcibly expelled to shoot out the obstructing food. This is the basis of the Heimlich maneuver done on people who are choking.

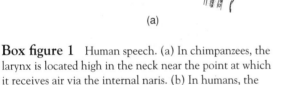

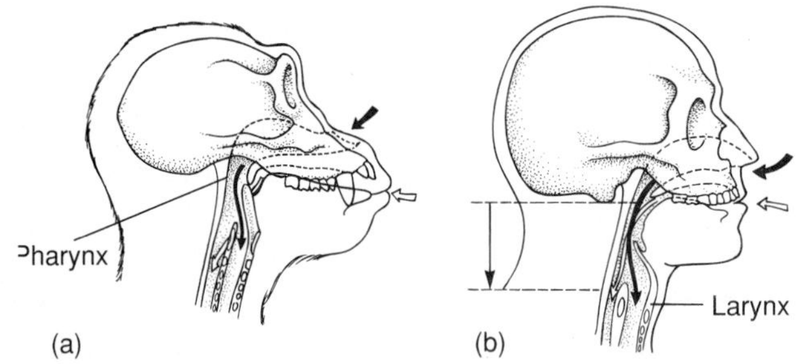


**Box figure 1** Human speech. (a) In chimpanzees, the larynx is located high in the neck near the point at which it receives air via the internal naris. (b) In humans, the larynx is lower, serving to lengthen the pharynx used to produce speech sounds. However, this separates the larynx from easy connection with the air passageway, and air and food routes cross.

**pulp,** fills the pulp cavity and root canal to support blood vessels and nerves that enter the tooth via the apical foramen. The **occlusal surface** of the crown makes contact with opposing teeth. The **cusps** are tiny, raised peaks or ridges on the occlusal surface (figure 13.6a,b).

Three hard tissues compose the tooth—enamel, dentin, and cementum. **Enamel** is the hardest substance in the body and forms the surface of the tooth crown. Concentric rings seen under microscopic examination are believed to result from pulses of calcium salt deposits before tooth eruption. However, with only a few exceptions among mammals, no further enamel is deposited on the crown after the tooth erupts.

**Dentin** resembles bone in chemical composition but it is harder. It lies beneath the surface enamel and cementum and forms the walls of the pulp cavity. Even after tooth eruption, new dentin is laid down slowly throughout the life of an individual. Growth occurs by daily apposition along the walls of the pulp cavity so that in very old animals, dentin may almost fill the entire cavity. The daily layers of dentinal growth are called the **incremental lines of von Ebner.**

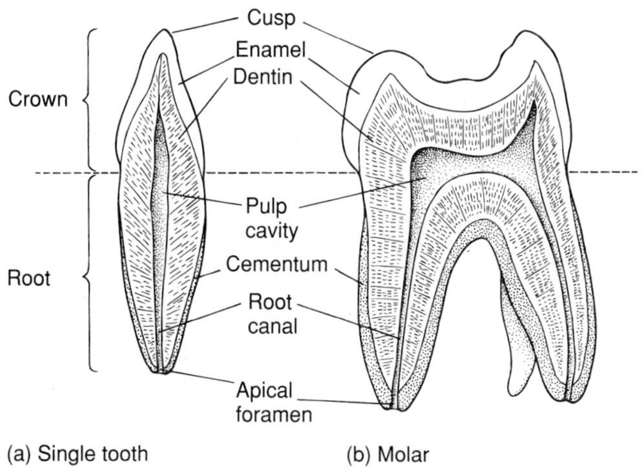


**Figure 13.6** Tooth structure. (a) Tooth with single root. (b) Molar tooth with three roots.

**Cementum,** like bone, has cellular and acellular regions. Cementum rests upon the dentin and grows in layers on the surface of the roots. In many herbivores, cementum can extend up along the crown to between the enamel folds and actually contribute to the occlusal surface. Cells within the cementum, termed **cementocytes,** elaborate the matrix but in seasonally related pulses, so that cementum increases irregularly with age. The result is the production of **cemental annuli,** concentric rings that characterize the cementum layer.

The **periodontal membrane** (periodontal ligament) consists of thick bundles of collagenous fibers that connect the cementum-covered root to the bone of the socket.

In lower vertebrates, teeth are usually **homodont,** similar in general appearance throughout the mouth. Modern turtles and birds lack teeth altogether, but mammals have **heterodont** teeth that differ in general appearance throughout the mouth. Most lower vertebrates have **polyphyodont** dentition, that is, their teeth are continuously replaced. A polyphyodont pattern of replacement ensures rejuvination of teeth if wear or breakage diminish their function. However, most mammals are **diphyodont,** with just two sets of teeth. The first set, the **deciduous dentition** or "milk teeth," appears during early life. It consists of incisors, canines, and premolars, but no molars (figure 13.7a). As a mammal matures, these are shed and replaced by the **permanent dentition,** consisting of a second set of incisors, canines, and premolars and now molars, which have no deciduous predecessors (figure 13.7b).

***Tooth Development*** Teeth are embryonic derivatives of the epidermis and dermis and develop beneath the surface of the skin initially. When mature, fully formed teeth **erupt** through the skin and extend into the buccal cavity. The epidermis produces the **enamel organ,** and mesenchyme cells of neural crest origin collect nearby within the dermis to produce the **dermal papilla** (figure 13.8a). Cells within the enamel organ form a specialized layer of **ameloblasts** that secrete enamel. Cells within the dermal papilla form the **odontoblasts** that secrete dentin (figure 13.8b). Thus, neural crest cells directly contribute to producing dentin through odontoblasts and indirectly induce the overlying ameloblast cells to deposit enamel.

Neural crest (p. 177)

The crown of the tooth forms first, and then shortly before eruption, the root begins to develop (figure 13.8c,d). The cementum and periodontal ligament develop last. In mammals, growth of the permanent tooth begins from a separate primordium of the enamel organ and dermal papilla that is usually adjacent to or deeper than the newly erupted deciduous tooth (figure 13.8e). Through similar steps, expansive growth of the permanent tooth against the roots of the deciduous tooth gradually cuts off its nutrition, causing resorption of the root and eventual loss of the deciduous tooth, which is replaced by the emerging permanent tooth. Finally, appearance of the cementum and periodontal ligament ensure firm attachment of the replacement dentition.

In rodents and lagomorphs, incisors continue to grow from their roots as the tooth crowns are worn down (figure 13.9a). In elephants, molars erupt sequentially over a prolonged period of time. The newest molars erupt at the back of the jaws, and as they slowly emerge, they push older and worn molars to the front of the tooth row (figure 13.9b). Molars moving forward are sequentially worn down completely but are replaced from behind until very late in an elephant's life when the limited number of molars is spent (figure 13.9c). However, for most mammals, once the permanent teeth are in place, they are not replaced nor do they grow in length.

***Specialized Teeth in Lower Vertebrates*** Teeth are attached to supporting bones in three general ways. Mammals and a few lower vertebrates have **thecodont** teeth sunken into sockets within the bone (figure 13.10a). Other vertebrates exhibit an **acrodont** condition, with shallow sockets and teeth attached to the crest of the bone, or a **pleurodont** condition, with teeth attached to the medial side of the bone (figure 13.10b,c).

Among some herbivores and predators, teeth are often broadly flattened into anvillike surfaces for crushing fibrous plant material or hard mollusc shells. The teeth of many teleost fishes form abrasive surfaces used to scrape encrusted algae from rocks and place it into suspension to gather up (figure 13.11a–e).

The oral cavity and its teeth also serve as a prey trap, an apparatus designed to snare unwary prey. Among most carnivores, teeth are simple sharp cones. They puncture the

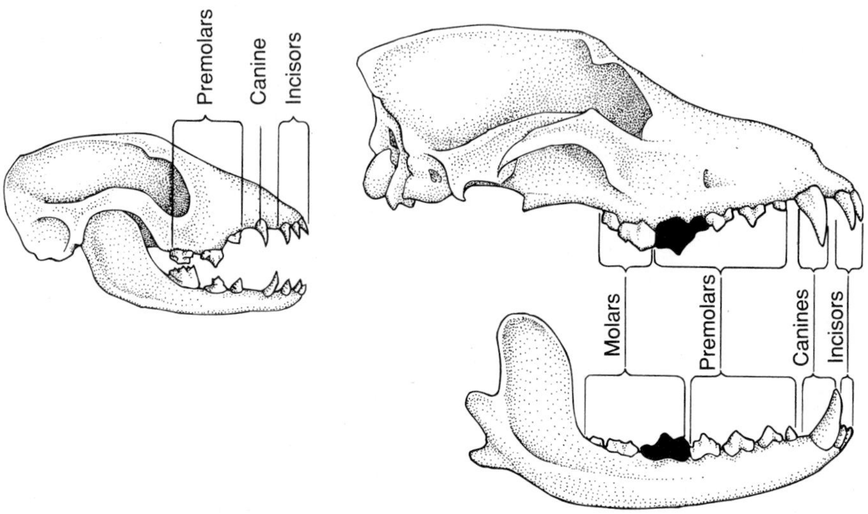


**Figure 13.7** Deciduous (a) and permanent teeth (b) in a dog. The carnassials (shaded teeth) are specialized carnivore teeth derived from last premolar (upper) and first molar (lower).

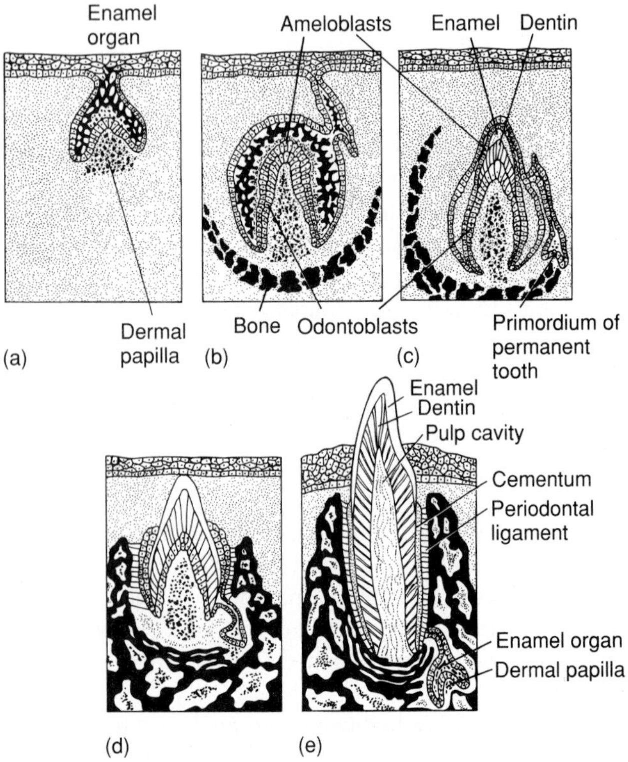


**Figure 13.8** Mammalian tooth development. (a) Enamel organ (from the epidermis) and dermal papilla (from the dermis) appear. (b) Ameloblasts are the source of tooth enamel and form from the enamel organ. Odontoblasts are the source of dentin and form from the dermal papilla. Bone appears and begins to delineate the socket in which the tooth will reside. (c) The primordium of the permanent tooth appears. (d) Tooth growth continues. (e) The deciduous tooth erupts and is anchored in the socket by a well-established periodontal ligament. The enamel organ and dermal papilla of the permanent tooth primordium will not begin to form the tooth until shortly before the deciduous tooth is lost.

skin of the prey to give the jaws a firm grip on the captured and often still struggling animal. Skin is what engineers call a compliant material. Because of its great flexibility, or compliancy, it easily deforms or yields to attempts at puncture. To address this mechanical problem, the teeth of predators have pointed cusps to pierce or cut this compliant material. In addition, the teeth of some predators, such as sharks, have sharp, knifelike cutting edges along the sides of their teeth to help pierce skin. For slicing chunks from flesh, these edges are further serrated, like those on a bread knife, to cut the soft compliant skin (figure 13.12a). The teeth of crossopterygians and some early amphibians have sharp single cusps and their enamel sides are complexly convoluted, inspiring the name labyrinthodont for such teeth. This infolded enamel produces surface ridges that may improve tooth penetration and strengthen the tooth internally (figure 13.12b).

In larval salamanders, most teeth are pointed cones, but the teeth of metamorphosed adults often show specializations. The crowns in some species are **bicuspid,** having two cusps, and the crown itself sits upon a basal **pedicel** to which it is attached by collagenous fibers. When a tooth is replaced, the crown is lost and the pedicel quickly resorbed, leading some to argue that such pedicelated teeth represent a way for rapid tooth replacement. However, the main advantage of this design is that it aids in grasping prey. The "joint" formed between crown and pedicel allows the tooth tip to bend inward but not outward. Thus, when a struggling prey is advanced further into the mouth of a salamander, the teeth tips relax and bend in the same direction, encouraging movement toward the throat. Because these tooth tips are slanted inward, they resist prey escape (figure 13.12c).

Like the teeth of most carnivorous reptiles, snake teeth generally taper to a sharply pointed cusp that penetrates skin and gives it a firm hold on the prey. Some snake

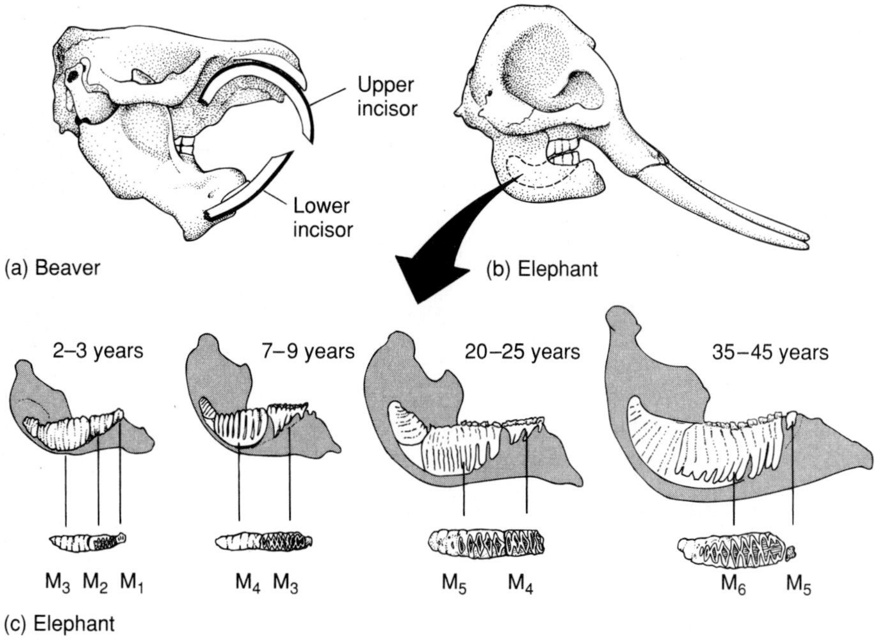


**Figure 13.9** Specialized tooth growth in mammals. In most mammals, molars do not grow nor are they replaced after they erupt. (a) One exception is found in rodents whose incisor teeth continue to grow at their roots as their chisellike crowns are worn away. The superficial bone is cut away to show the roots of upper and lower incisors. (b) In elephants, molar teeth erupt sequentially over a protracted period of time. A new molar that erupts at the back of the tooth row pushes older molars to the front. (c) The superficial bone has been removed over the elephant's teeth to show their eruption and positional changes with age. Corresponding view of molar crowns ($M_{1-6}$) are shown at the bottom.

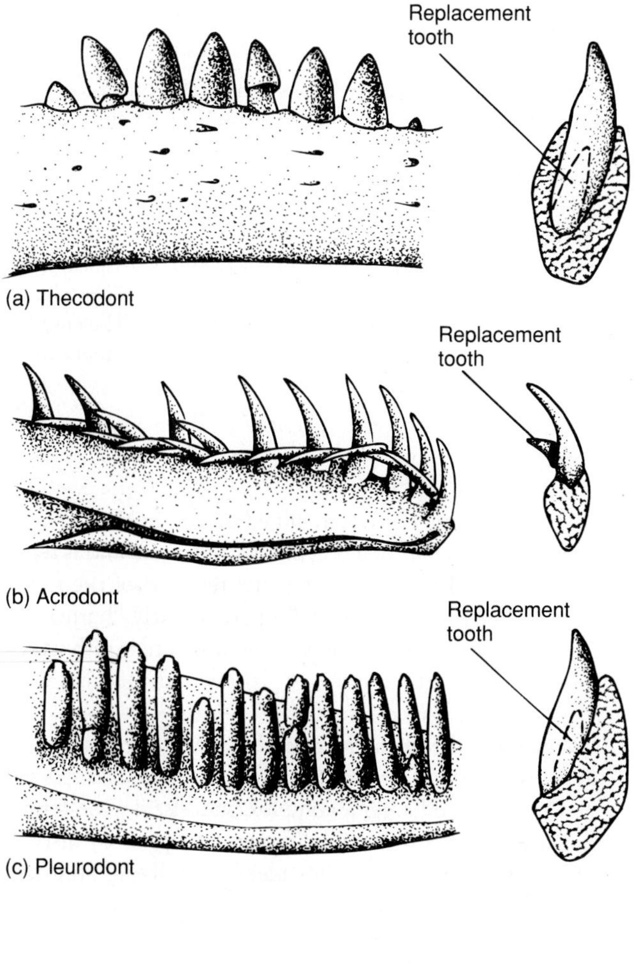


**Figure 13.10** Types of tooth attachment. (a) Thecodont teeth are set in sockets (alligator). (b) Acrodont teeth attach more or less on the occlusal surface of the bone (snake). (c) Pleurodont teeth attach to the side (lizard).

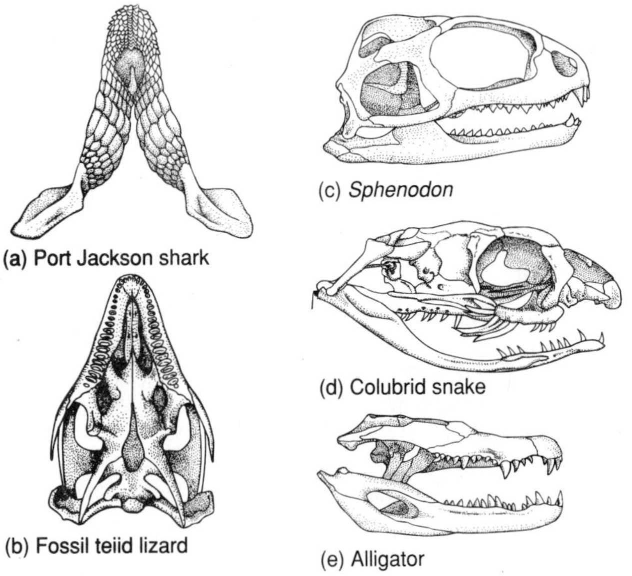


**Figure 13.11** Heterodont dentition. Heterodont dentition is most pronounced among mammals in whom distinct incisors, canines, premolars, and molars are discernible. However, among many ectotherms, teeth differentiation is also evident. (a) Lower jaw of Port Jackson shark. (b) Fossil teiid lizard. (c) *Sphenodon*. (d) Colubrid snake, a boomslang, exhibiting an enlarged set of grooved fangs on the posterior end of the maxilla. (e) Alligator with some large teeth.

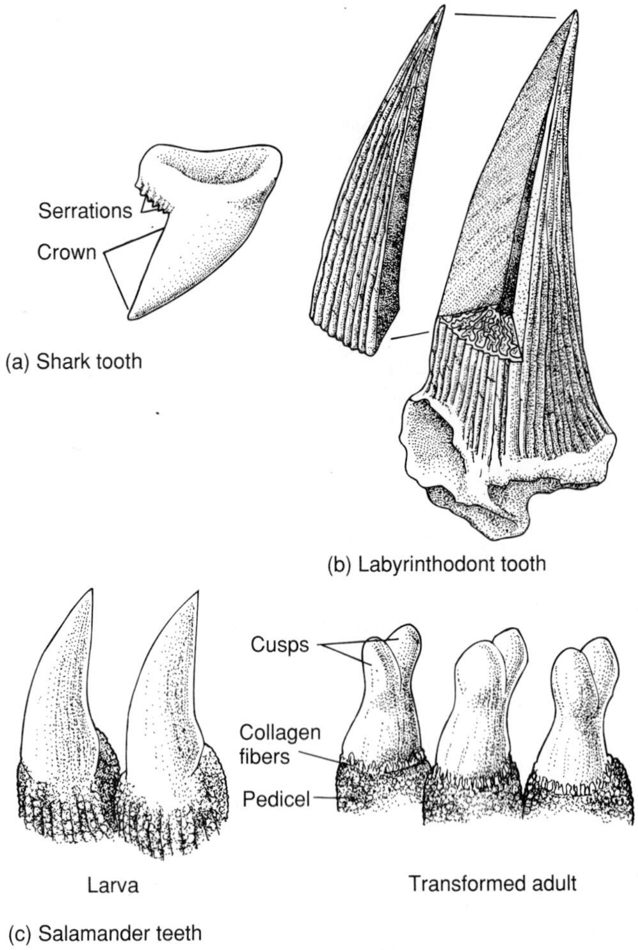


**Figure 13.12** Specializations of teeth. (a) Shark tooth (blacknose shark, *Carcharhinus acronotus*). The pointed crown is nearly smooth edged for piercing prey; the base is serrated for cutting flesh. (b) Lateral view of a labyrinthodont tooth from a fossil amphibian. A wedge of tooth has been removed to show infolded enamel. (c) Teeth before (larval) and after (adult) metamorphosis in northwestern salamander (*Ambystoma gracile*). Larval teeth are pointed. Those of the transformed adult have divided cusps that articulate with a basal pedicel. The cusps are thought to inflect with the struggling prey, thus discouraging its escape from the mouth.

teeth are specialized and bear a bladelike edge or low ridges along their sides thought to aid in tooth penetration. When a snake strikes, it brings its mouth over the prey quickly. The series of needlelike teeth form a prickly surface that easily snag the surface of the prey. Teeth at the front of the snake's mouth are often **recurved,** with the tip inclined forward from the rest of the tooth (figure 13.13a,b). This gives the tooth a major posterior bend at its base and a forward slant at its tip. Forward inclination of the cusp means that during the strike, the sharp tip is brought more in line with the snake's line of approach to the prey. Alignment of teeth tips with the prey facilitates skin puncture upon impact. The posterior-directed bend at the base of the tooth works to hold the prey and facilitate its swallowing. If the prey should pull back in an attempt to escape, the teeth sink more deeply and securely into the skin because of their backward slant. Recurved teeth are found in other vertebrates, such as sharks, and presumably function in a similar fashion. The cusp penetrates on impact and the base holds the struggling prey.

Some maxillary teeth of snakes have open grooves down which oral secretions flow during feeding. In venomous snakes, the edges of these grooves fuse, forming a hollow channel down the core of the tooth through which venom passes from the venom duct into the prey. The term ***fang*** is appropriate for such a hollow tooth modified for venom delivery. Being a modified tooth, the snake fang, like other teeth, is part of a polyphyodont system and is replaced on a regular basis. Thus, artificial removal of the fang will not render a venomous snake permanently "harmless" because within a day or less a replacement tooth takes its place.

***Specialized Teeth in Mammals*** In mammals, the teeth not only capture or clip food but they are also specialized to chew it, producing a complex and distinctive dentition. In fact, the dentition in different groups is so distinctive that it is often the basis for identifying living animals and

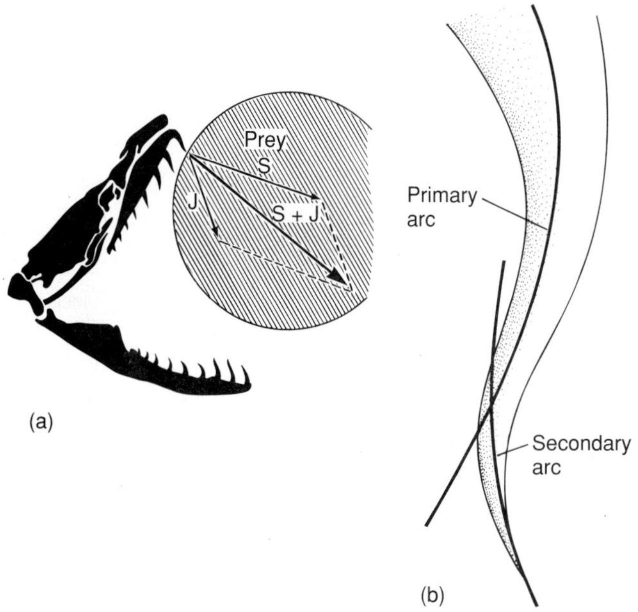


**Figure 13.13** Recurved teeth of snakes. (a) When a snake launches its head at and closes its jaws on prey, two forces are transmitted through the tip of the anterior teeth. These component forces are represented here by vectors. One vector represents the force arising from the forward momentum of the skull (S). The other represents the force of jaw closure (J). The resultant force at impact is S + J. (b) The forward inclination of the tooth's tip (secondary arc) relative to its base (primary arc) may bring the tip into closer coincidence with the line of this resultant force upon impact. The primary arc of the tooth's base orients the tooth posteriorly. When the snake swallows prey, this backward slant resists escape of the prey out of the mouth. During the strike, the reverse secondary arc helps the tooth penetrate the surface of the prey.
Based on the research of T. H. Frazzetta.

fossil species. Not surprisingly, an elaborate terminology has grown up to describe the particular features of mammalian teeth.

The heterodont dentition of mammals includes four types of teeth within the mouth—**incisors** at the front, **canines** next to them, **premolars** along the sides of the mouth, and **molars** at the back. The number of each type differ among groups of mammals. The **dental formula** is a shorthand expression of the number of each kind of tooth on one side of the head for a taxonomic group. For example, the dental formula of the coyote *(Canis latrans)* is:

I 3/3, C 1/1, PM 4/4, M 2/3.

This means that there are three upper and three lower incisors (I), one upper and one lower canine (C), four upper and four lower premolars (PM), and two upper and three lower molars (M), 21 per side or 42 total. Sometimes the dental formula is written as 3–1–4–2/3–1–4–3, the first four numbers indicating the upper teeth and the second four the lower teeth of the coyote. The dental formula for the mule deer *(Odocoileus hemionus)* is 0–0–3–3/3–1–3–3. Notice that the missing upper incisors and canines are indicated by zeros (figure 13.14).

Generally, incisors at the front of the mouth are used for cutting or clipping; canines, for puncturing or holding; premolars and molars, for crushing or grinding food. As a practical matter, it is often hard to distinguish premolars from molars visually. The collective term embracing both is **cheek teeth** or **molariform teeth.** Cheek teeth may be quite diverse, a reflection of their many specialized functions. In humans and pigs, crowns are low, or **brachyodont** (figure 13.15a). In horses, crowns are high, or **hypsodont** (figure 13.15b). If the cusps form peaks, as in omnivores, the teeth are **bunodont** (figure 13.15c). Cusps drawn out into ridges, as in perissodactyls and rodents, produce **lophodont** teeth (figure 13.15d). Crescent-shaped cusps, as in artiodactyls, characterize **selenodont** teeth (figure 13.15e). Hypsodont teeth are typically found in herbivores that grind plant material to break tough cell walls. Their occlusal surface is worn unevenly because the minerals that form the surface—enamel, dentin, and cementum—differ in hardness. Occlusal surfaces are functionally important because they ensure that ridges and depressions persist throughout life, thereby maintaining a rough grinding surface that does not become smooth with continued use (figure 13.15b).

Mammals possess a variety of specialized teeth. In some, **sectorial teeth** are modified so that ridges on opposing teeth slice by one another to cut tissue. In some primates, cutting edges form on the upper canine and lower first premolar, the sectorial teeth. These teeth are deployed in fights between individuals or in defense. In carnivores, the upper premolar and lower molar form **carnassials,** specialized sectorial teeth that slice against each other like scissors to cut sinew and muscle. **Tusks** arise from different teeth in different species. The single 3-m spiral tusk of the narwhal is the left upper incisor (figure 13.16a). In elephants, the tusks are elongate incisors (figure 13.16b), and in walruses, the paired tusks are upper canines that protrude downward (figure 13.16c). In carnivorous mammals, canine teeth together with powerful jaws are used to kill prey. Sometimes these teeth puncture major blood vessels in the neck, causing the victim to bleed profusely and weaken. A practiced carnivore such as an adult lion, is more likely to bite into the neck and collapse the trachea to suffocate its prey. Some mammals, such as anteaters and baleen whales, lack teeth altogether (figure 13.16f,g).

The cusp patterns of mammalian cheek teeth are so distinctive that they are used in species identification. Cusps are termed **cones.** Major cones are identified by adding the prefixes *proto-*, *para-*, *meta-*, *hypo-*, or *ento-*; minor cusps are indicated by the suffix *-ul(e)*. For connoisseurs, the terminology goes on: **cingulum** is used for accessory enamel ridges on the margins of the crown, *-loph* denotes ridges across the crown connecting cusps, *-cone* signifies cusps on upper teeth, and *-conid* signifies cones on lower teeth. For example, proto*cone* and proto*conid* denote

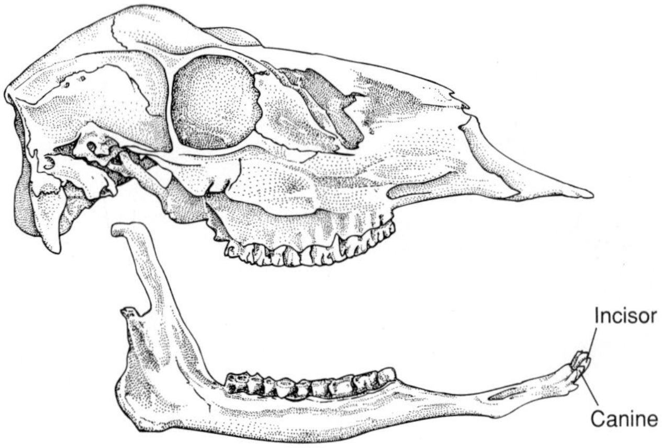


**Figure 13.14** Skull of the mule deer *(Odocoileus hemionus)* with lower jaw lowered. The dental formula for the upper row is 0–0–3–3, and the formula for the lower is 3–1–3–3. The absence of upper incisors and canine is normal, indicated in the dental formula by zeros. The lower incisors and canine are present, and the canine is adjacent to the incisors at the front of the mandible.

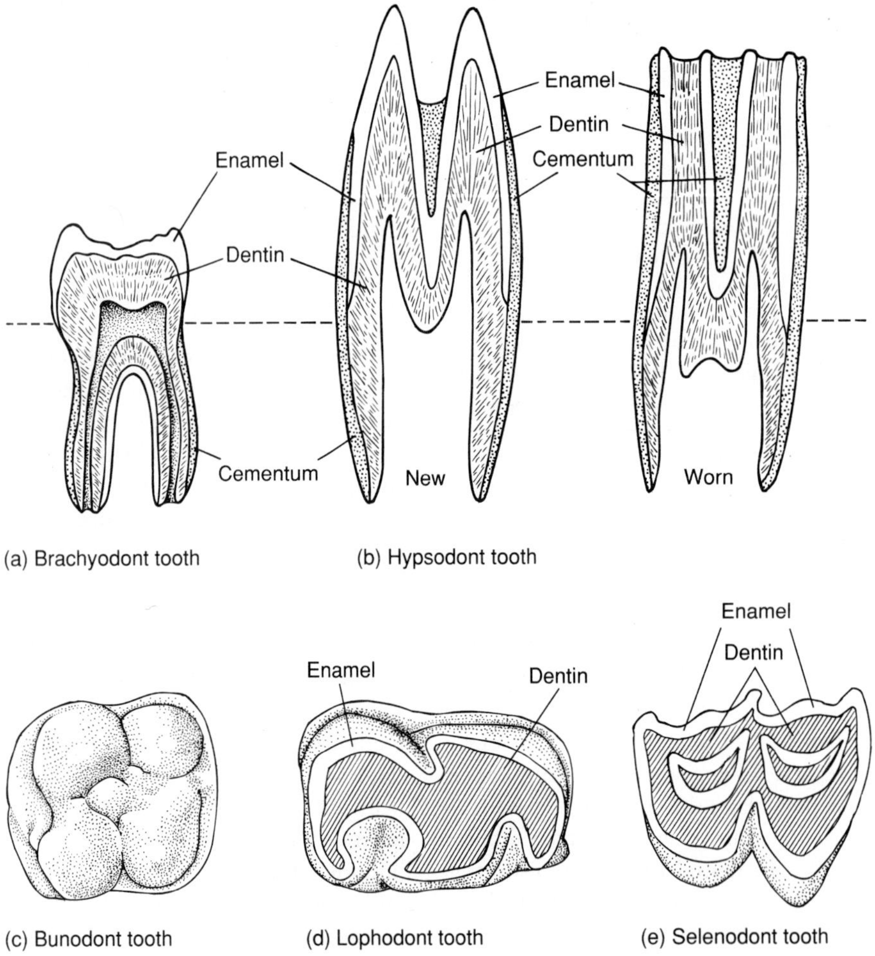


**Figure 13.15** Crown height and occlusal surfaces. Tooth height: (a) Brachyodont tooth. (b) Hypsodont tooth. When the occlusal surface of a newly erupted hypsodont tooth (left) becomes worn, alternating layers of dentin and enamel are exposed (right). The alternating layers of varying hardness ensure that ridges and depressions will form, producing a rough surface that does not become smooth after prolonged use. In mammalian teeth, various occlusal surfaces occur: (c) Bunodont tooth. (d) Lophodont tooth. (e) Selenodont tooth.

the same cusps, but on upper and lower teeth, respectively (figure 13.17a–e).

Much of this terminology was inspired by late nineteenth century paleontologists who proposed that the three cusps on each molar of ancestral reptiles spread across the expanded crown of mammalian descendants to become the paracone (paraconid), protocone (protoconid), and metacone (metaconid) of upper (and lower) molars. Other cones were added later. Expecting such exactness between tiny cusps of reptiles and later mammals was probably overly optimistic. But, it did and continues to provide one practical technique to characterize mammals taxonomically.

### *Tongue*

During feeding, cyclostomes protrude their soft tongue bearing ridges of rasping keratinized "teeth." However, most gnathostome fishes lack a tongue. Occasionally, teeth borne on the lower ends of gill bars can be worked against those of the palate, but a fleshy muscularized tongue is not usually present. A mobile tongue develops first in tetrapods from the hypobranchial musculature attached to and resting on the underlying **hyoid apparatus,** a skeletal derivative of modified lower ends of the hyoid arch and adjacent branchial arches.

The tongue of many tetrapods holds the **taste buds,** sensory organs responsive to chemicals entering the mouth. Lizards and snakes project their foretongue out of the mouth to retrieve air and/or substrate chemicals that their sensory **vomeronasal organ** (Jacobson's organ) evaluates. In many carnivores, the surface of the tongue is roughened, like a file, with numerous keratinous spiny projections, or **filiform papillae,** that help rasp flesh from bones.

**Vomeronasal organ (p. 671)**

Many tetrapods use their tongue in **lingual feeding.** They project their tongue at prey, and its sticky surface holds the catch until tongue retraction brings it back into the mouth. Many terrestrial salamanders and lizards use this technique. In fact, some argue that development of just such a mobile and projectile tongue represents a major feeding innovation in the transition of early amphibians to life on land. Woodpeckers use their long specialized tongues like a probe to obtain insects between cracks in tree bark or in holes they manufacture (figure 13.18a–c).

The tetrapod tongue can also transport captured prey, that is, move prey through the buccal cavity to the back of the pharynx where it is swallowed (figure 13.19a). This process is known as **intraoral transport** and proceeds in several steps. First, the jaws part slowly, and the tongue is advanced forward and beneath the food to fit partially around it (figure 13.19b). Second, the jaws open more rapidly, and the tongue draws the adhering food backward in the mouth (figure 13.19c). Third, the jaws close slowly and grip the more posteriorly positioned food, and the tongue comes away from its adhesive hold on the food. The cycle is repeated, with the tongue working in synchrony with the jaws to move the food in increments to the back of the buccal cavity and into the pharynx (figure 13.19d).

The tongue's grip on food during transport depends partially on its surface irregularities that interlock or physically engage the prey. The ability of the fleshy tongue to shape itself to the food may help it gain a physical grip on the prey. This tongue adhesion also depends on a "wet" adhesion, the sticky effects created by surface tension in air and by capillary action. Where tetrapods have returned to aquatic feeding, these physical phenomena are less effective. This may account for the much simplified tongue devoid of intrinsic musculature found, for example, in crocodilians. Many aquatic tetrapods such as aquatic feeding turtles resort to modified suction feeding in which the tongue has little role to play. In fact, if the tongue of aquatic tetrapods were a large fleshy structure occupying the floor of the buccal cavity, it could interfere with the sudden expansion of the buccal cavity necessary for suction feeding.

## Pharynx

In adults, the pharynx is little more than a corridor for passage of food and air. But, phylogenetically it is the source of many organs, and developmentally its history is complex (table 13.1). During embryonic development, the stomodeum (ectoderm) opens into the pharynx, but the pharynx itself forms from the anterior foregut (endoderm) and is relatively prominent compared with the rest of the still-forming digestive tract. A series of bays, or **pharyngeal pouches,** form on its lateral walls and grow out to meet in-pocketings of the skin ectoderm, termed **branchial grooves.** At their point of contact, these pouches and grooves establish a partition, or **closing plate,** between them. In fishes and larval amphibians, closing plates are perforated to form the functional gill slits. In other vertebrates, these closing plates fail to rupture, or if they do, they are soon sealed over so that functional gill slits do not develop.

The subsequent contribution of the embryonic pharynx to adult structures is staggering. In mammals, the first pharyngeal pouch expands into an elongated **tubotympanic recess** that envelops the middle ear bones, giving rise to the narrow eustachian tube and part of the tympanic cavity. The second pharyngeal pouch gives rise to the palatine tonsil. The third and fourth pouches contribute to the parathyroid. The fifth pharyngeal pouch gives rise to what are termed the ultimobranchial bodies. The ultimobranchial bodies are separate glands in fishes, amphibians, reptiles, and birds, but in mammals, they become part of the thyroid gland and apparently form its internal population of C cells. C cells are involved in controlling blood calcium levels. All pouches contribute to the thymus in fishes, variable numbers contribute in amphibians, and pharyngeal pouches III and IV contribute in mammals. The roof of the pharynx gives rise to the

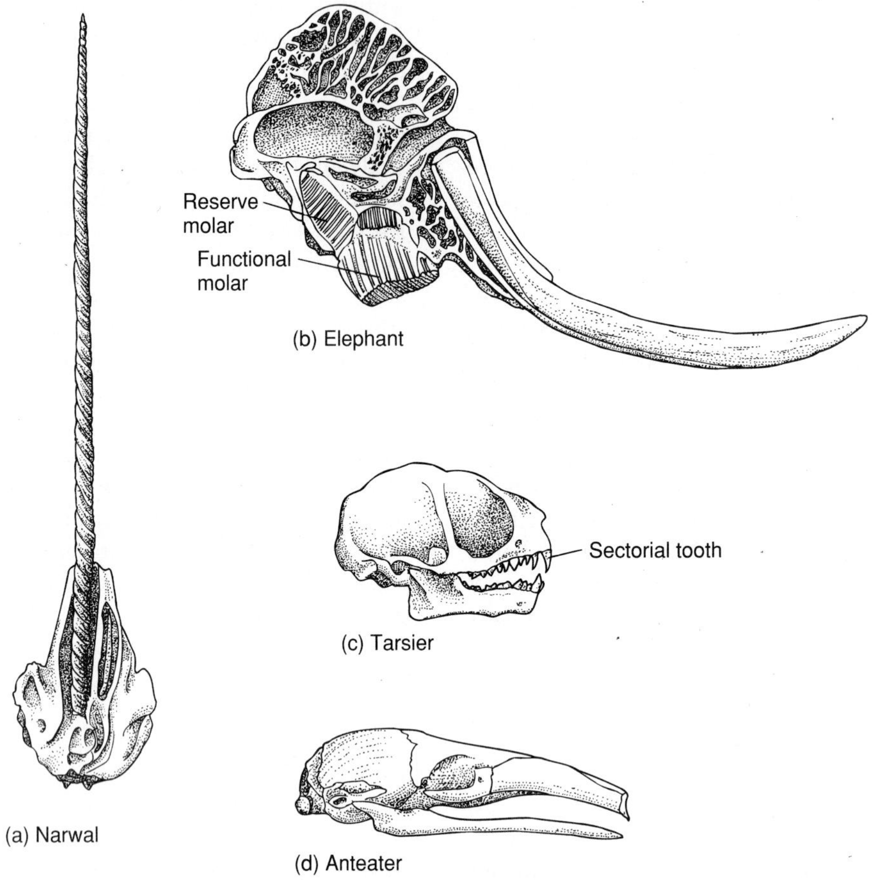


**Figure 13.16** Specialized mammalian teeth. Lateral and frontal views of tusks. Tusks arise from upper left incisor in the narwal (a), from both upper incisors in the elephant (b), and from canines in walruses (e). Sectorial teeth in the primate *Tarsier* (c) and peglike teeth of a porpoise (f) are shown. Teeth are absent in adult anteaters (d) and baleen whales (g).

pharyngeal tonsil; the floor, to the thyroid, part of the tongue, lingual tonsil, and lung primordium.

Swallowing, or **deglutition,** involves the forceful movement of the bolus from the mouth and pharynx into the esophagus and then the stomach. Most vertebrates bolt their food, swallow it whole, and expand their esophaguses to accommodate its size. Seabirds catch fishes in their bills and toss them to the back of the throat. The esophagus becomes distended as food enters it. Contractions of muscles within its walls squeeze the fish along and into the stomach. Snakes work their flexibly articulated jaw bones over the prey, drawing the sides of their mouth over the prey to engulf it. When food enters the esophagus, waves of contraction in its walls and general neck movements force the bolus along to the stomach (figure 13.20).

Three temporary seals form as most mammals chew and swallow their food. The anterior oral seal is formed by the lips. The middle oral seal develops between the soft palate and back of the tongue. The third posterior oral seal occurs between the soft palate and the **epiglottis.** As an animal chews, food tends to gather temporarily in the **vallecula,** the space in front of the epiglottis, and the **pyriform recess,** the passageways around either side of the **larynx.** The epiglottis sits above the larynx, and the trachea sits below it. When the animal swallows, the back and sides of the tongue expand against the soft palate (you can notice this yourself when you are eating), forcing food out of the vallecula, through the pyriform recess, and into the esophagus. The **glottis** is a muscular slit that closes momentarily across the larynx to prevent inadvertent aspiration of food into the trachea and lungs. In most mammals, the posterior seal (soft palate-epiglottis) is in place during swallowing to ensure passage of food into the esophagus without blocking the air passage (figure 13.21a). In human infants, the posterior seal directs milk to the esophagus, but in adults, this seal is lost because the pharynx descends in order to accommodate the onset of speech. Adult humans rely on the middle oral seal (soft palate-back of tongue) to keep food and air passages separate during swallowing (figure 13.21b).

## Alimentary Canal

In some vertebrates, digestion begins in the buccal cavity. But we get down to the serious business of food processing in the alimentary canal, which encompasses further breakdown of the bolus, absorption of its available constituents,

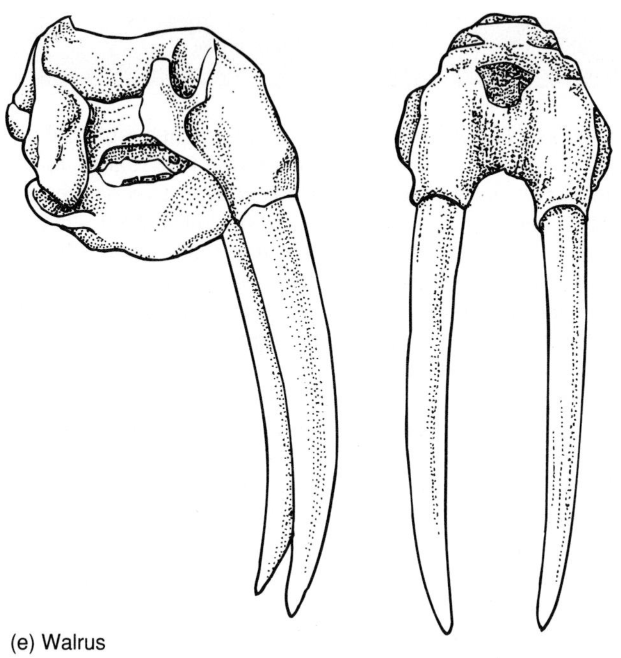

(e) Walrus

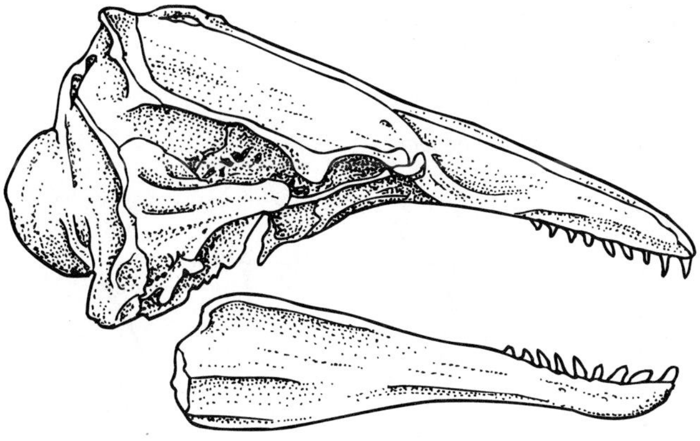

(f) Porpoise

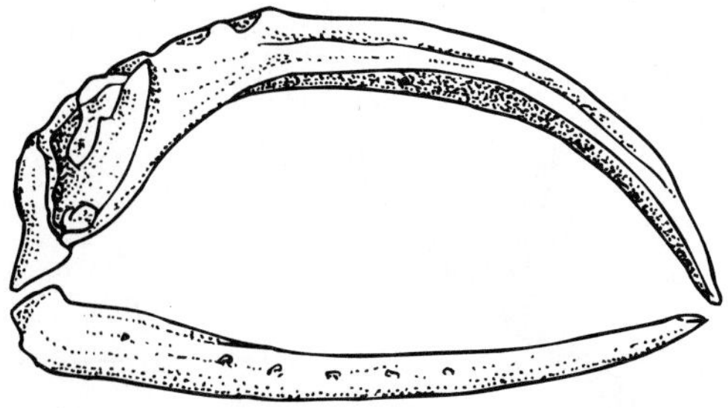

(g) Right whale

**Figure 13.16** (continued).

and elimination of indigestible remnants. The design of the alimentary canal suits the diet of the organism. Because diet can differ even between related groups, alimentary canals can differ significantly between phylogenetically related vertebrates. Most vertebrates have an alimentary canal made up of an esophagus, stomach, intestines, and cloaca. As distinctive as these might appear, all share an underlying unity of design (figure 13.22).

Each region is built on a common plan of organization, namely, a hollow tube with walls composed of four layers. The innermost layer is the **mucosa,** which includes the epithelium that lines the lumen, the thin smooth muscle fibers of the muscularis mucosa, and the region of loose connective tissue, the lamina propria between the epithelial lining and the muscularis mucosa. The **submucosa,** consisting of loose connective tissue and nerve plexes of the autonomic nervous system, forms the second layer of the digestive tract. Outside this layer lies the **muscularis externa,** composed of circular and longitudinal sheets of smooth muscle. The surface layer is the **adventitia,** consisting of fibrous connective tissue. If a mesentery envelopes the alimentary canal, this outer layer of connective tissue plus mesentery is called the **serosa.**

During embryonic development, the endoderm gives rise to the lining of the gut, and the surrounding mesoderm forms smooth muscles, connective tissue, and blood vessels. In most tetrapods, a series of positional changes transform the relatively straight gut of the embryo into the coiled digestive tube of the fetus (figure 13.23a–c). First, the primary gut loop forms (figure 13.23d,e), a large bend in what was up to this point a straight gut. Next, rapid elongation of the gut twists this loop into the first major coil (figure 13.23f). Thereafter, continued elongation and coiling produce the compact digestive tube, and distinctive regions within it become delineated (figure 13.23g).

## *Esophagus*

The esophagus connects the pharynx with the stomach. It is a slender tube that easily becomes distended to accommodate even a large bolus of food. Mucus, to aid in the passage of food, is often secreted, but the esophagus seldom produces enzymes that contribute to chemical digestion. In some vertebrates, the esophageal mucosa is lined with ciliated cells that control the flow of lubricating mucus around the food. The ciliated epithelium may also help gather small crumbs from the meal and move these along to the stomach. In others, the mucosa is stratified epithelium; this may even be keratinized in animals ingesting rough or abrasive foods. In vertebrates that swallow large quantities of food at once, the esophagus serves as a site of temporary storage until the rest of the alimentary canal begins digestion. Anteriorly, muscle coats tend to be composed of striated muscle that becomes replaced posteriorly by smooth muscle.

## Box Essay 13.2 Saber-toothed "Cats"

In some mammals, the upper canines evolved into curved, saberlike teeth. This occurred independently on four occasions, three times within placentals—once in primitive carnivores, the creodonts, and twice in cats, the fossil nimravids and the felids (box figure 1)—and once within a Pliocene marsupial family (Thylacosmilidae). All saber-toothed mammals are extinct, but fossil canines from saber-toothed cats are well known. Their canine teeth were long and curved, and the posterior edge of the blade carried a faint serration. One saber-toothed nimravid bore evidence of a stab wound inflicted by another saber tooth. A fossil dire wolf was found with part of the saber tooth from the felid *Smilodon* embedded in its skull. Modern carnivores feed mostly on herbivores and only rarely dine on each other. Thus, fossil evidence of saber-tooth attacks on other carnivores likely represents a cat defending its kill from scavenging or marauding carnivore competitors seeking to steal the dispatched carcass.

How these canine sabers were deployed during feeding is not well understood. Saber-tooth cats could open their jaws wide, but these jaws were not strong enough to rip out a large mouthful of prey. It seems more likely that the sabers made slashing wounds that bled profusely, but they were not used to rend large sections of flesh from the bodies of the prey.

Why then don't cats today possess similar saber teeth to serve them in dispatching prey? Tigers, lions, cougars, and smaller cats make rapid dashes from ambush and use claws to grasp and control prey while they bite into its neck. Biting inflicts puncture wounds and clamps off the trachea, suffocating prey. It is unknown how saber teeth would have been advantageous if hunting strategies were similar. Alternatively, some suggest that teeth were different because prey were different and required a different hunting strategy. If saber-tooth mammals fed on large ground sloths or other large, slow herbivores, then this type of prey might have presented problems different from the swift herbivores that most large cats prey on today. Without a living saber-tooth cat as reference, the special function of these teeth is difficult to determine. As of yet, there is no consensus. But, if saber teeth represent a specialization for specialized prey, then the absence of large, slow herbivores today might also account for the absence of saber-toothed predators as well.

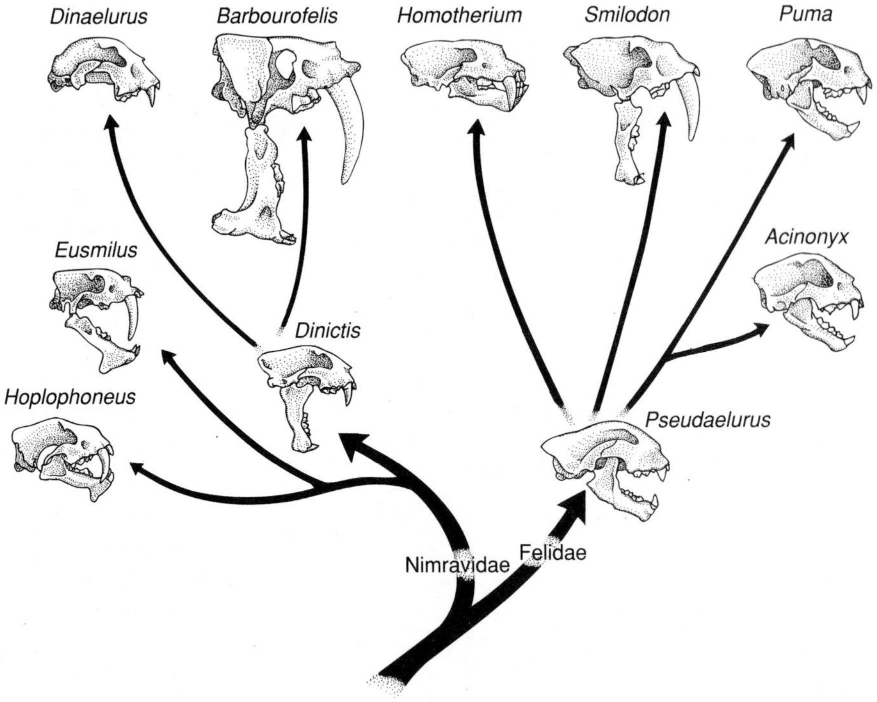


**Box figure 1** Possible phylogeny of cats. One major branch produced the Nimravidae, the other, the Felidae. If this phylogeny is correct, the saber-toothed forms arose independently at least twice within the cat group.

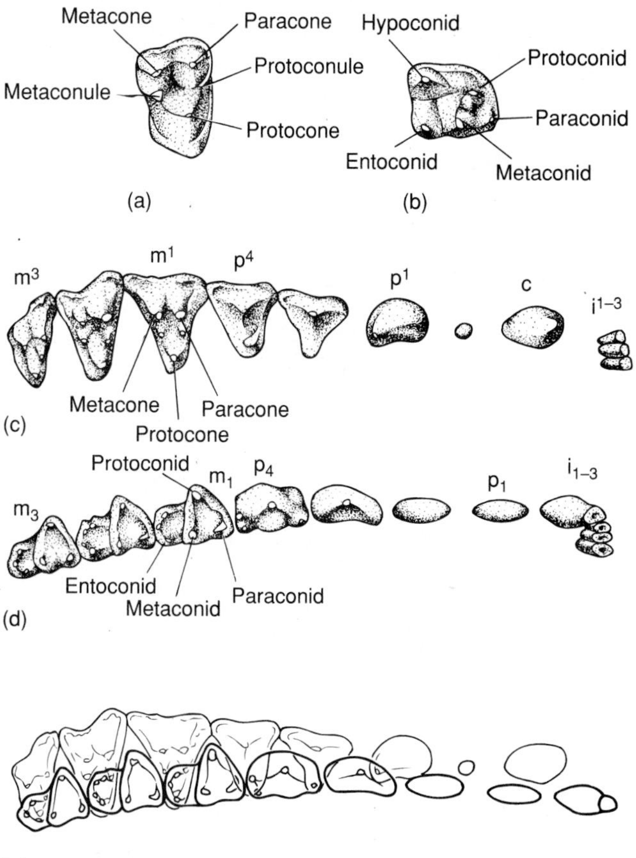


**Figure 13.17** Molar patterns of placental mammals. (a) Upper right molar. (b) Lower left molar. (c) Occlusal view of right upper teeth. (d) Occlusal view of right lower teeth, same species as (c). (e) Upper and lower tooth rows placed in occlusion, with outlines of lower teeth (heavy outlines) superimposed on those of upper. Teeth include canine (c), incisor (i), molar (m), premolar (p) accompanied by their number within the upper (superscript) or lower (subscript) tooth row.

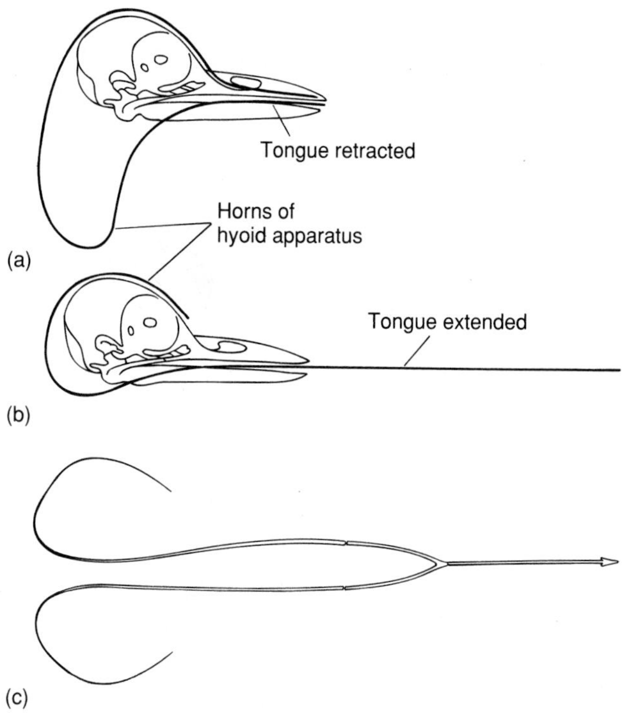


**Figure 13.18** Tongue protrusion in a woodpecker. (a) The flexible and thin hyoid apparatus supports the fleshy tongue. (b) When the woodpecker protrudes its tongue, the hyoid apparatus slips forward and the tongue extends from it. (c) Ventral view of the hyoid apparatus from the woodpecker *Picus*.

## *Stomach*

The esophagus delivers the bolus of food to the stomach, an expanded region of the alimentary canal. Animals that take in large quantities of food on an irregular basis, such as many carnivores, have stomachs that serve as storage compartments until the processes of mechanical and chemical digestion catch up. Such food storage may have been an initial function of the stomach when early vertebrates evolved from suspension feeding to feeding on larger chunks of food. Hydrochloric acid produced by the stomach may have functioned to retard food putrefaction by bacteria, thus preserving it until digestion was underway. In most vertebrates, the stomach performs an expanded role. Some absorption of water, salts, and vitamins occurs in the stomach, but predominately it serves to churn and mix food mechanically and add digestive chemicals collectively called **gastric juice.** Gastric juice includes some enzymes and mucus but is primarily composed of hydrochloric acid released from the mucosal wall of the stomach.

The stomach's expanded size sets it apart from the narrow esophagus that enters and the small intestine into which it empties. When not distended with food, the stomach wall relaxes into folds known as **rugae,** which also help delineate its boundaries (figure 13.24). However, gross external morphology does not always reliably mark internal differences in structure of the mucosal wall. Consequently, the histological character of the mucosal wall is often used to distinguish important functional regions within the stomach.

On the basis of mucosal histology, two regions of the stomach can be distinguished. The stomach's **glandular epithelium** is characterized by the presence of **gastric glands.** These are branched, tubular glands, several of which empty into the bases of surface indentations, or **gastric pits.** There are three divisions of the stomach—cardia, fundus, and pylorus—based on the relative position and type of gastric gland. The **cardia** is a very narrow region found only in mammals, and it marks the transition between the esophagus and the stomach. Its gastric glands, termed **cardiac glands,** are composed predominantly of mucous secreting cells. The **fundus** is usually the largest region of the stomach and contains its most important gastric glands, the **fundic glands.** Mucous cells are present in fundic glands, but these glands are distinguished by their abundance of **parietal cells,** the source of hydrochloric acid, and **chief cells,** the presumed source of several proteolytic enzymes. Before emptying into the intestine, the stomach usually

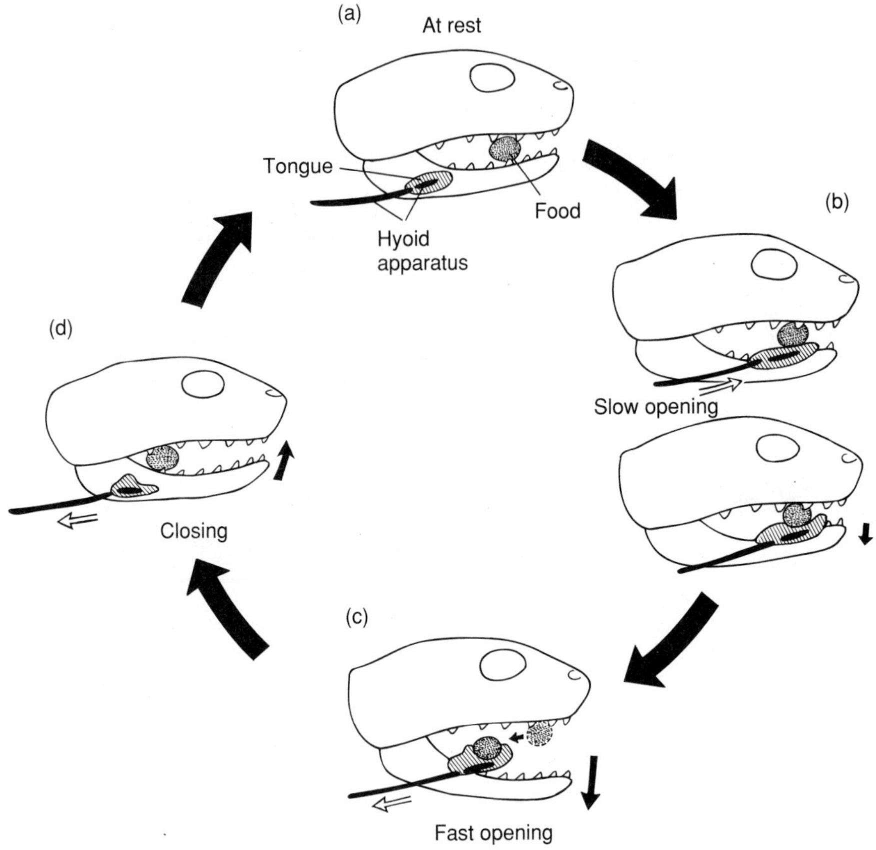


**Figure 13.19** Intraoral transport of food in a generalized tetrapod. (a) At rest. (b) Slow opening. The jaws begin to open slowly and the tongue moves forward to make contact with the food, after which it becomes fitted to the food. (c) Fast opening. The jaws part, and the retracting tongue transports the food to the back of the mouth. (d) Closing. The jaws close to establish a purchase on the food, after which the tongue comes free of the food. If an animal repeats this sequence, food is moved successively to the back of the mouth and into the pharynx, where it is swallowed into the esophagus.

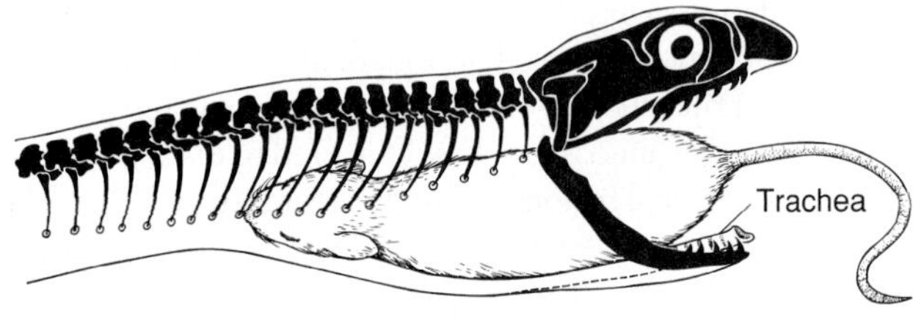


**Figure 13.20** A snake swallowing prey. A snake steps its jaws alternately left and right along its prey, "walking" them along its surface until the prey reaches the back of the snake's throat. Smooth muscles in the wall of the esophagus aided by striated muscles in the lateral body wall move the prey toward the stomach. While the prey is in the snake's mouth, the trachea slips beneath it and forward to maintain an open route through which the snake breathes even while swallowing.

narrows into a **pylorus,** whose mucosal walls hold distinct gastric glands called **pyloric glands.** The pyloric glands are predominantly composed of mucous cells whose secretions help to neutralize the acidic chyme as it moves next into the intestine. Thus, most of the chemical and mechanical process of gastric digestion occur in the fundus. The cardia (when present) and pylorus add mucus. Smooth muscle bands in their walls act as sphincters to prevent the retrograde transfer of food (figure 13.24).

In addition to a region of glandular epithelium, the stomach of some vertebrates also has a second region characterized by **nonglandular epithelium,** devoid of gastric glands. As in some herbivores, the nonglandular region may develop from the base of the esophagus. In other species, such as rodents, loss of gastric glands in the mucosa leaves a nonglandular epithelial stomach in which smooth muscle contractions knead and mix digesta. This nonglandular epithelium in rodents also can be keratinized, perhaps as a result of mechanical abrasion from rough foods such as seeds, grasses, and insect chitinous exoskeletons. Chemical insult from digestive enzymes added in the mouth may also cause a keratinized nonglandular epithelium.

TABLE 13.1 Pharyngeal Pouch Derivatives in Vertebrates

| Pharyngeal Pouch | Position | Lamprey | Elasmobranch | Urodele | Anuran | Reptile | Bird | Mammal |
|---|---|---|---|---|---|---|---|---|
| 1 | Dorsal | Thymus | Spiracle | Tubotympanic recess[a] | Tubotympanic recess[a] | Tubotympanic recess[a] | Tubotympanic recess[a] | Tubotympanic recess[a] |
| | Ventral | Branchial pouch | — | — | | | | |
| 2 | Dorsal | Thymus | Thymus | — | Thymus | Thymus[1] | — | Tonsil (Palatine) |
| | Ventral | Branchial pouch | — | — | — | Parathyroid | — | — |
| 3 | Dorsal | Thymus | Thymus | Thymus | — | Thymus[1,2] | Thymus | Parathyroid |
| | Ventral | Branchial pouch | — | Parathyroid | Parathyroid | Parathyroid | Parathyroid | Thymus |
| 4 | Dorsal | Thymus | Thymus | Thymus | — | Thymus[2,3] | Thymus | Parathyroid |
| | Ventral | Branchial pouch | — | Parathyroid | Parathyroid | Parathyroid | Parathyroid | Thymus |
| 5 | Dorsal | Thymus | Thymus | Thymus | — | Thymus[3] | — | — |
| | Ventral | Branchial pouch | Ultimo-branchial body | Ultimo-branchial body | Ultimo-branchial body | Ultimo-branchial body | Ultimo-branchial body | Ultimo-branchial body |

[a]Auditory cavity of middle ear and eustachian tube.
[1]Lizard.
[2]Turtle.
[3]Snake.

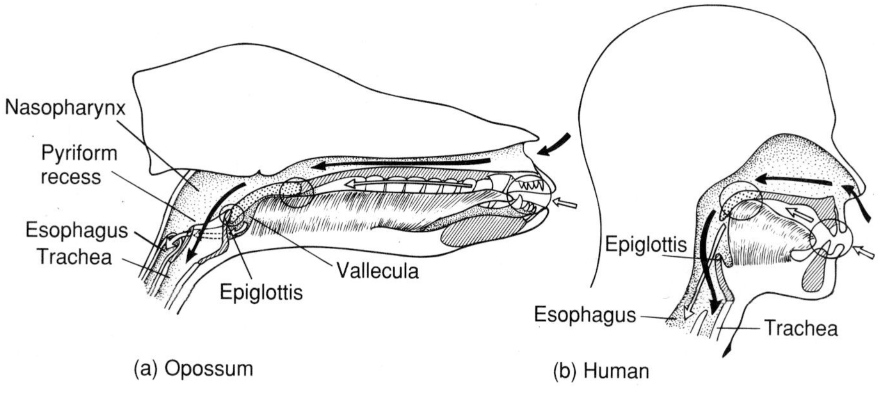


**Figure 13.21** Food and air passages. (a) Sagittal view of an opossum head showing three oral seals: anterior (lips), middle (soft palate and tongue), and posterior (back of tongue and epiglottis). Air (solid arrows) flows directly from the nasal passages into the trachea. Food (open arrows) passes around the sides of the larynx to reach the esophagus. (b) Sagittal view of a human head. The posterior oral seal is absent because the pharynx drops lower in the neck to accommodate sound production for speech. Thus, food and air potentially cross in the lengthened human pharynx.

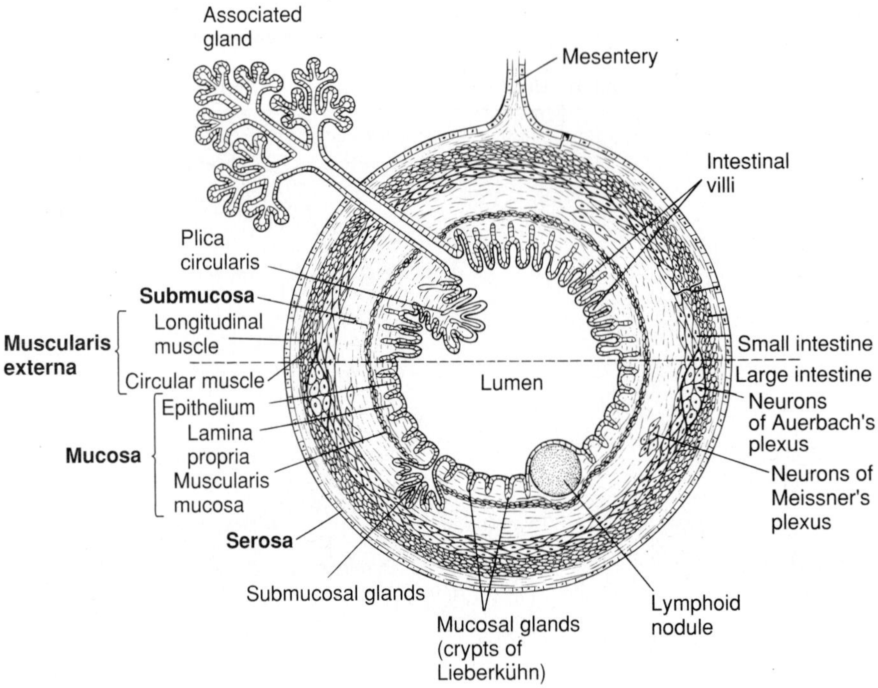


**Figure 13.22** General organization of the alimentary canal. The concentric layers of mucosa, submucosa, muscularis externa, and serosa (or adventitia) are common to all regions of the alimentary canal. Within the mucosa are folds or intestinal villi. Lymphoid tissue occurs throughout, although it may form discrete nodules within the mucosa. Glands may occur within the mucosa, submucosa, or even outside the digestive tube.

### *Intestines*

The mucosa of the intestines is distinctive. First, it contains an epithelium whose free surface facing the lumen has many **microvilli,** perhaps up to several thousand per cell. These tiny, fingerlike projections substantially increase the absorptive surface area of the alimentary canal. Their surfaces also seem to harbor a microenvironment away from the large central lumen, where digestive enzymes can more favorably act upon food. Second, the intestinal mucosa also includes **intestinal glands** (crypts of Lieberkühn), the source of many of the digestive enzymes released into the intestines.

Generally, there are two major regions of the intestines, the small and large intestines. The small intestine may be quite lengthy, but it is usually smaller in diameter than the large intestine. It possesses **villi,** small surface projections that increase the surface area of the mucosa (not to be confused with the much smaller *micro*villi, tiny projections of individual cells; figure 13.25). There can be three successive parts of the small intestine—**duodenum, jejunum,** and **ileum.** The duodenum receives chyme from the stomach and exocrine secretions primarily from the liver and pancreas. The jejunum and ileum are best delineated in mammals on the basis of histological features of the mucosal wall (figure 13.26). The **ileocolic valve** (ileocecal valve) is a sphincter between the ileum of the small intestine and the large intestine. This valve regulates the movement of food into the large intestine.

The large intestine, so named for its large diameter, is usually a straight tube passing to the cloaca or anus. Its mucosa lacks villi. It may be pushed to one side of the body cavity or, as in many mammals, form a large gentle loop called the **colon.** The large intestine often straightens, forming a section termed the rectum. The rectum narrows into the anal canal in which a transition occurs from simple columnar epithelium to stratified epithelium within its mucosal wall. A smooth muscle sphincter within the muscularis of the anal canal controls the release of waste products from the digestive tract.

Generally, the intestines serve several functions. First, peristalsis within the intestinal walls moves food along the digestive tract. Second, the intestines add secretions to food being digested. Mucous secretions protect the epithelial lining from digestive enzymes and lubricate it to facilitate passage of food. The **intestinal juice** produced by the intestinal glands includes enzymes for digestion of proteins, carbohydrates, and lipids. Accessory glands add some secretions as well. For example, the **duodenal glands (Brunner's glands),** located in the submucosa, empty their secretions

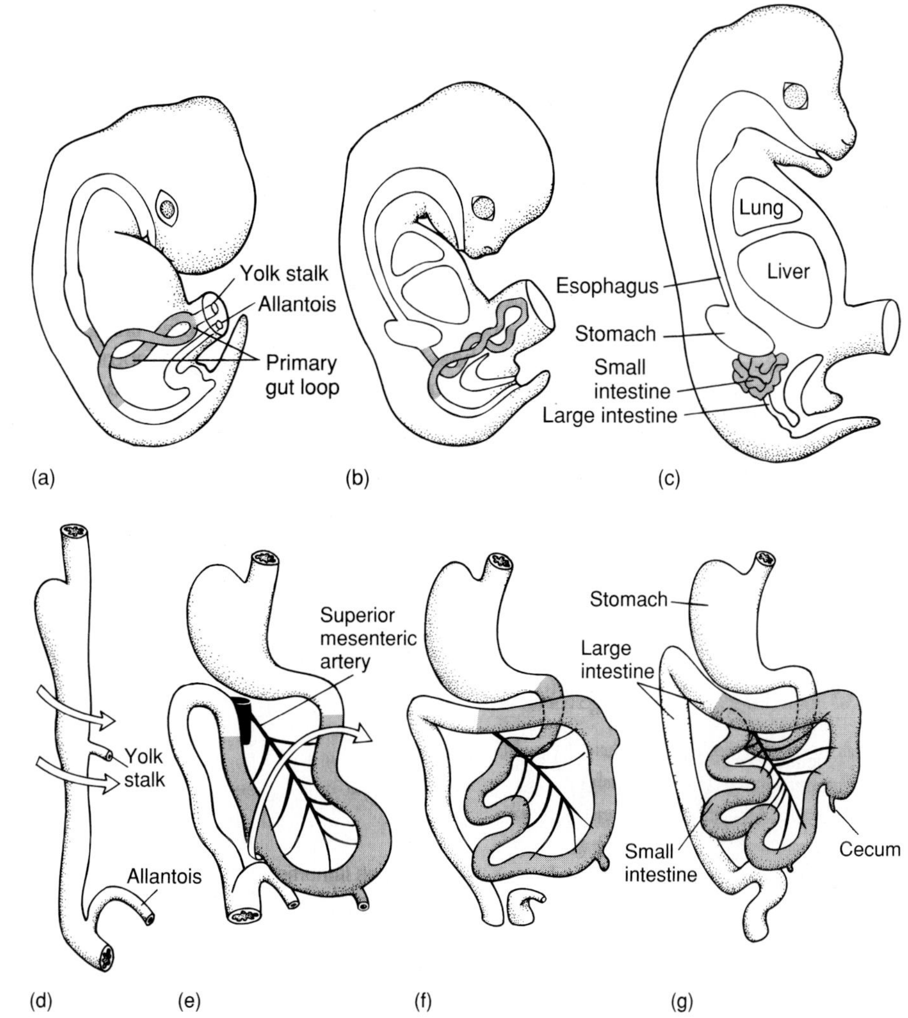


**Figure 13.23** Embryonic differentiation of the mammalian gut. (a–c) The intestines coil (shaded), elongate, and become tucked behind the stomach within the body cavity. (d,e) Differentiation begins with the formation of the primary loop when the gut is still straight. (f,g) Growth, elongation, and coiling of the digestive track ensues. The result is a long but tightly coiled intestine and differentiated stomach.

into the duodenum to help neutralize the acidity of the chyme entering from the stomach. The pancreas releases its proteolytic enzymes into the duodenum also. Third, the intestines selectively absorb the final products of digestion—amino acids, carbohydrates, and fatty acids. Water is also absorbed, especially in the large intestine.

## *Cloaca*

As already mentioned, the proctodeum at the end of the embryonic gut gives rise to the cloaca, a common chamber that receives products from the intestines and urogential tracts. In some fishes and most mammals, the cloaca is absent. Instead, the intestine opens to the outside through the anus, which is a separate opening from that of the urogenital system.

**Cloacae (p. 572)**

## *Specializations of the Alimentary Canal*

Structural modifications accommodating specialized diets occur in several ways within the digestive tract. First, the path that food travels can be lengthened according to the time required for digestion. A **spiral valve** within the lumen of the alimentary canal is one way to increase the length of the route through the digestive tract. This valve creates a helical partition that forces passing food to wind through a spiral channel, thereby increasing the amount of time the bolus spends in the intestines and prolonging digestion (figure 13.27). Another way to increase the path traveled by food is by increasing the length of the alimentary canal. Carnivores have relatively short intestines, but herbivores that must extract nutients from resistant plant cells usually have long intestines (figure 13.28).

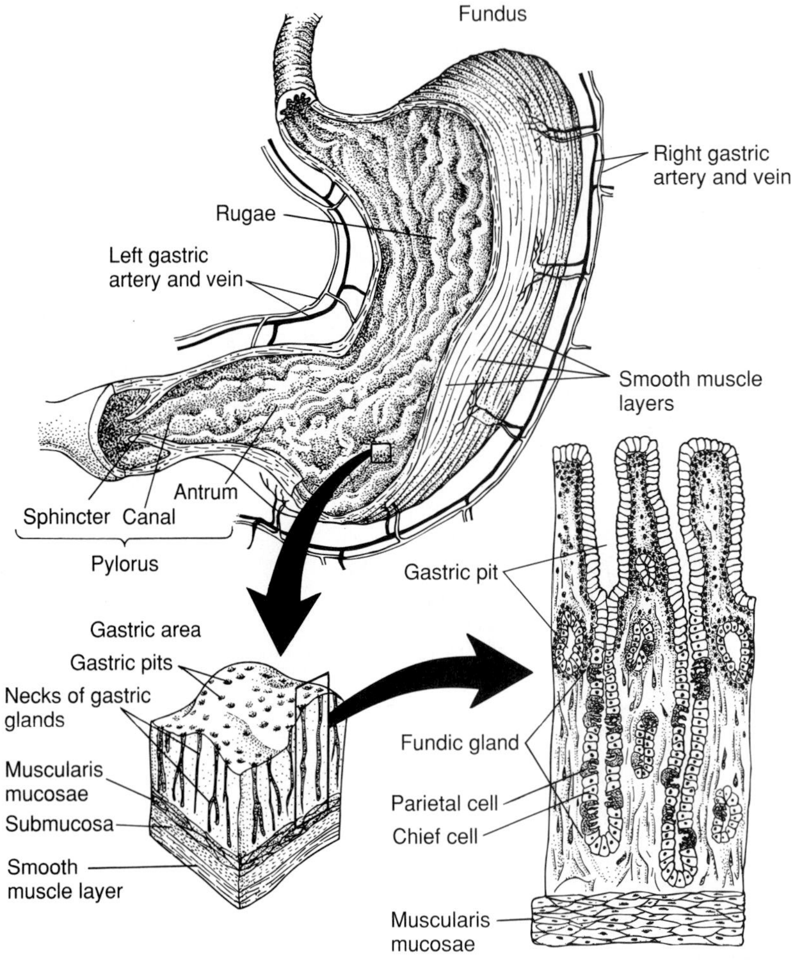


**Figure 13.24** Anatomy of the stomach. Up to three regions are usually discernible in the stomach. The largest of these is the fundic region. Gastric pits open into fundic glands having parietal and chief cells at their bases. The two other regions of the stomach include the cardia and the pylorus, which contain cardiac and pyloric glands, respectively, at the bases of their gastric pits. Various types of mucous cells predominate in these glands.

Second, expansions or extensions of the alimentary canal also may develop to accommodate specialized diets. A **crop** is a baglike expansion of the esophagus often used to store food temporarily during processing. One of the most common extensions is a **cecum,** a blind-ended outpocketing from the intestines through which food circulates as part of the digestive process (figure 13.28).

Third, differentiation of the alimentary canal may occur through regionalization as well. Occasionally, new regions form within the gut. What is a single intestinal tube in some species becomes differentiated into small and large intestines in other species. As we will see, regions are sometimes divided or lost. For example, the cloaca receives the contents of the intestines and urogenital tracks, but as these two systems develop their own separate outlets, the cloaca is lost.

### *Fishes*

In cyclostomes, the alimentary canal is a straight tube leading from mouth to anus without coils, folds, or major bends. The ciliated esophagus runs directly from the pharynx to the intestine. No distinct stomach is present (figure 13.29a). Diet includes small particulate matter, blood and tissue rasped from prey, and detritus. Storage in an expanded stomach before entering the intestine would be of little value, so food passes directly from the esophagus into the intestine. In lampreys (figure 13.29a), metamorphosis from ammocoetes to adults is usually accompanied by the appearance of a "new" esophagus. A cord of cells evaginates from the dorsal surface of the pharynx, acquires a lumen, and offers a new esophageal route for passage of food from the buccal cavity to the intestine. This metamorphic change accommodates changes in adult feeding habits and use of the oral disk for attachment. The pharynx attains an

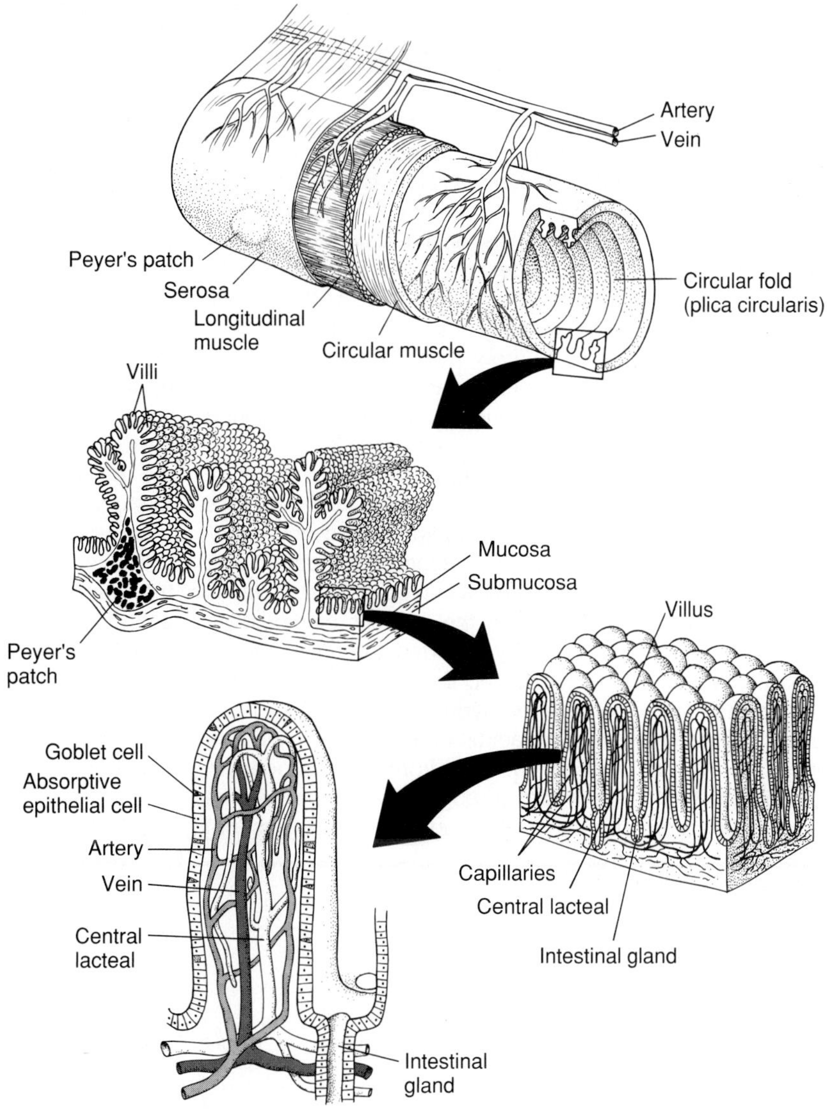


**Figure 13.25** Anatomy of the small intestine. Villi project above the level of the mucosal wall; intestinal glands are sunken within the mucosal wall. In addition to having a blood supply, each villus houses within its core a system of lacteals, specialized lymphatic vessels that absorb long-chain fats.

independent role in tidal ventilation. The new esophagus, often with numerous longitudinal folds, maintains digestive continuity from the mouth to the alimentary canal. Part of the larval esophagus regresses and part becomes incorporated into the anterior intestine of the adult. The cranial end of the intestine bears one or two (depending on species) diverticula near the point of entry of the esophagus. Products from the liver enter the cranial end via a bile duct. The adult intestine is lined with epithelium that contains numerous gland cells dispersed along internal folds that form a modest spiral valve. Digestive enzymes are released into the anterior intestine and mucus is secreted into the posterior intestine. In parasitic lampreys, the anterior intestine is especially important in absorption of fats. In addition, this region of the intestine of marine forms holds swallowed salt water and is important in osmoregulation. The posterior section of the intestine is important in protein absorption and elimination of biliverdin, a pigment of bile (figure 13.30).

Within gnathostome fishes, there is considerable variation in the design of the alimentary canal, perhaps as much as among all the rest of the terrestrial vertebrates (figure 13.29b–f). Generally, an esophagus, stomach, and intestine are present, although a stomach is usually not

## Box Essay 13.3 William Beaumont and Gastric Secretion

At Fort Mackinac, in June 1822, in what was then Michigan Territory, an incident occurred that changed the life of the victim and the course of biology. It started with an accidental gunshot that sent powder and duckshot into Alexis St. Martin, a French-Canadian trapper who was standing but three feet away. What was St. Martin's misfortune was the good fortune of the army surgeon who attended him, William Beaumont. St. Martin, not expected to survive, in fact lived to the next day, and then a week, and eventually recovered his health.

However, the large hole caused by the gunshot did not properly heal. Instead, the edges of his torn stomach and the hole in the rib cage formed an open fistula, an abnormal passage leading from the stomach through the side of the body to the outside. After many months of convalescence, St. Martin was declared a pauper and refused further treatment. William Beaumont took the patient into his own home, dressed his wounds, and continued to nurse him. Beaumont also began what he called his "experiments," taking advantage of the fistula that gave a direct view and access into the stomach. Beaumont drew samples of the gastric juice, tossed in various foods on a string and withdrew them later to see what had happened, and observed the mechanical action of the stomach during digestion.

Physiologists of the day thought of the stomach as a vat or stewpot or as an organ that worked by allowing putrefaction of food. Because Beaumont was able to draw samples of gastric juices and observe the process of digestion, he correctly reported the chemical nature of gastric digestion based on the release of hydrochloric acid and the churning action of the stomach. He also massaged bile from the duodenum backward through the pylorus into the stomach. Although rare, this was not the first gastric fistula opening the process of digestion to direct viewing. But, to his credit, Beaumont was the first to observe digestion carefully and place the physiology of digestion on a sound basis.

As for St. Martin, he lived to the ripe old age of 83, actually outliving Beaumont. But his special stomach had by then become a valuable commodity within scientific circles. On several occasions, he escaped back to the life of the trader he knew best, only to be found and returned to Beaumont. He was pursued (hounded would better describe it) by many physiologists seeking their fame from his special gastric talents. When St. Martin died, his family, by now quite fed up (no pun intended) with his peculiar stomach, refused permission for an autopsy. To make sure he remained unmolested in death, they buried him eight feet deep.

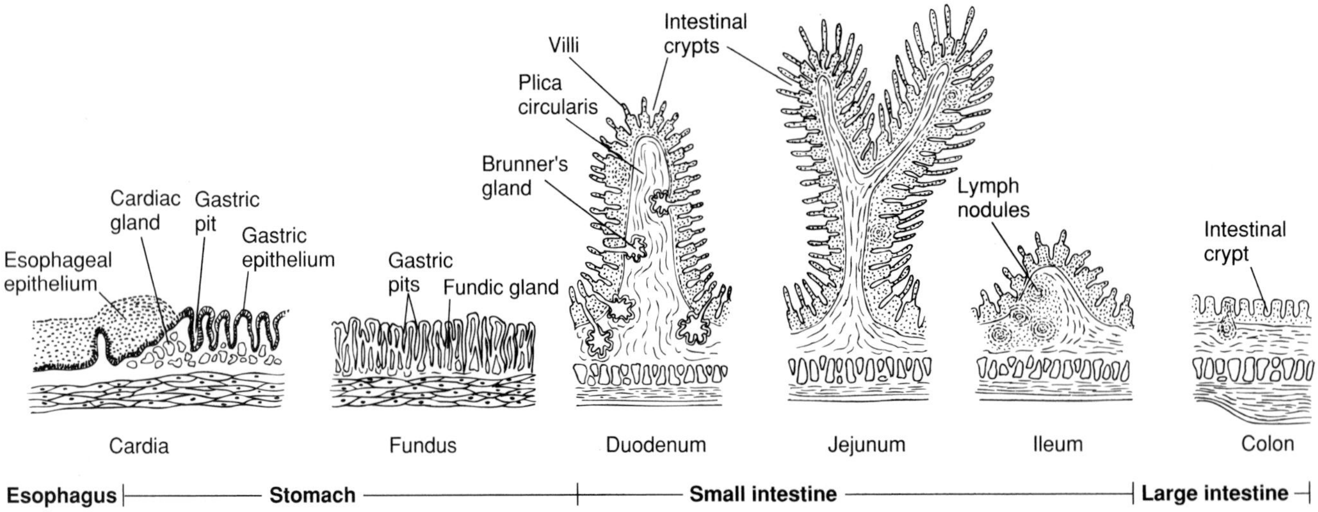


**Figure 13.26** Comparative histology of the mucosa along the alimentary canal of a mammal. Note that the large intestine, like the small intestine, contains intestinal glands, but villi are absent.

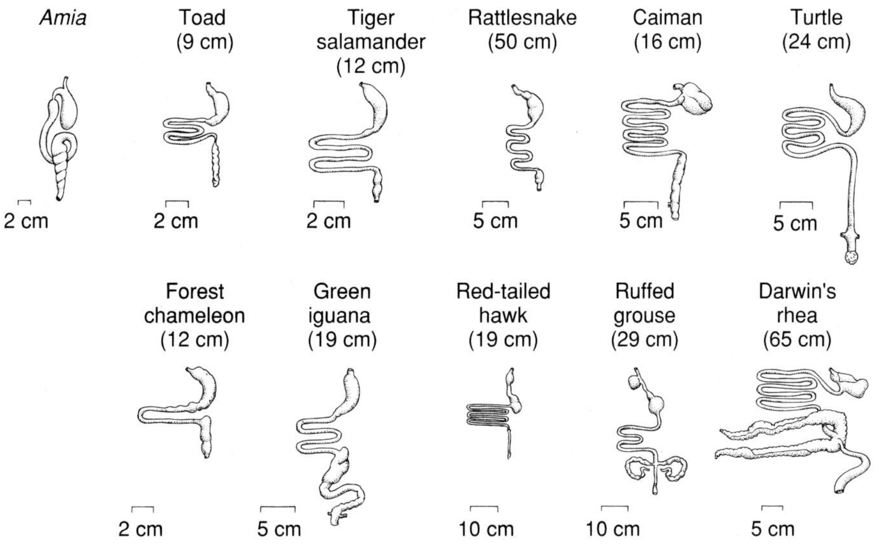


**Figure 13.27** Stomach and intestines of lower vertebrates and birds. Amphibians, snakes, caimans, forest chameleons, and red-tailed hawks are carnivores with relatively short, unspecialized intestines. To prolong passage of digesta, various specializations occur such as the spiral valve of the bowfin, ceca, or the enlarged large intestines of the grouse and rhea.

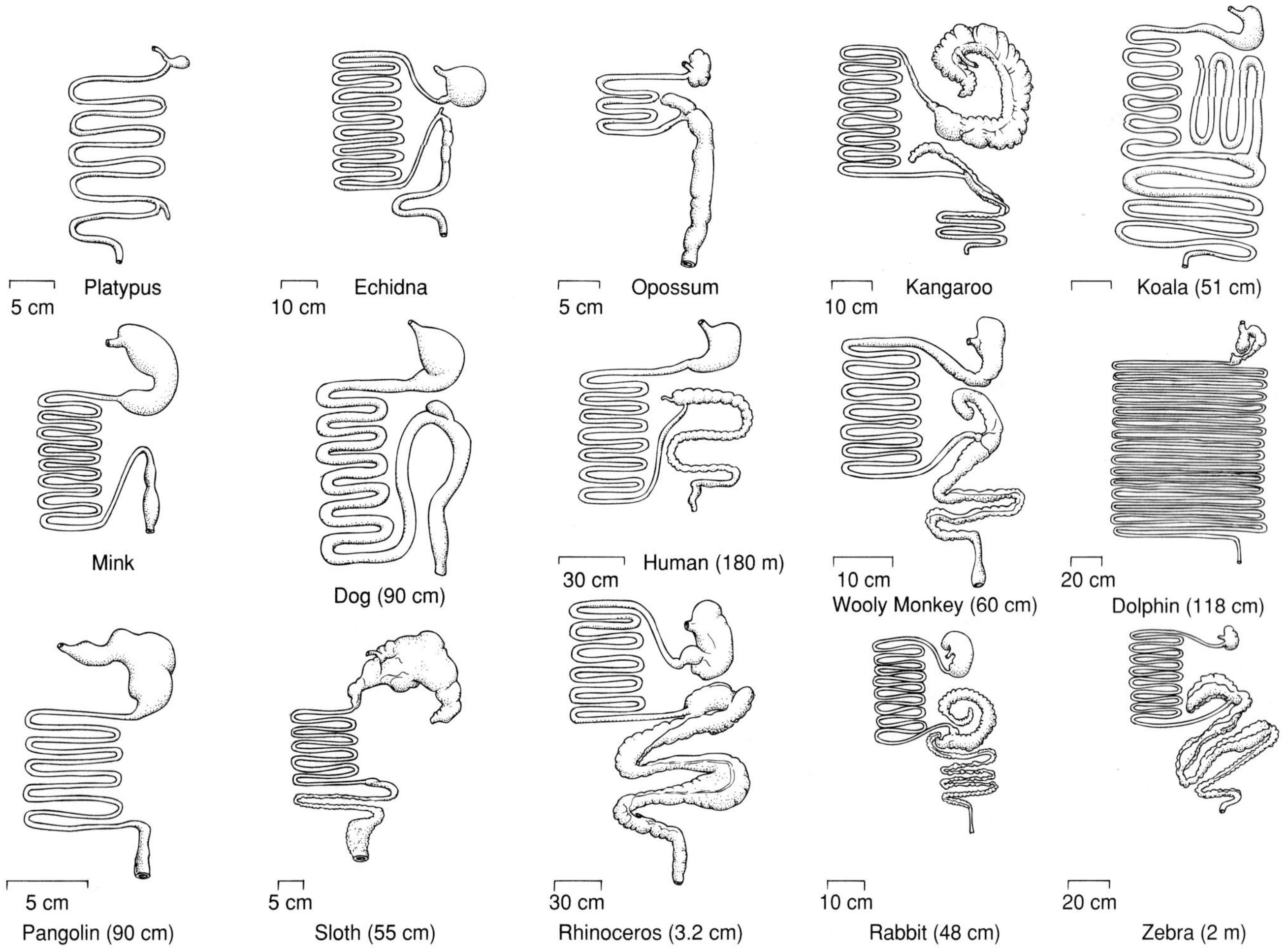


**Figure 13.28** Stomach and intestines, mammals. Relatively long, small intestines are found in the anteating echidna and pangolin, and also in dolphins. Terrestrial mammals that are strict carnivores, such as the mink and dog, have relatively short, unspecialized intestines. Kangaroos, koalas, sloths, rhinoceroses, rabbits, and zebras are herbivores with intestinal specialization that promotes fermentation. Note the relatively simple stomach and intestines of the platypus, whose diet is not well known but is thought to consist of aquatic insects and worms.

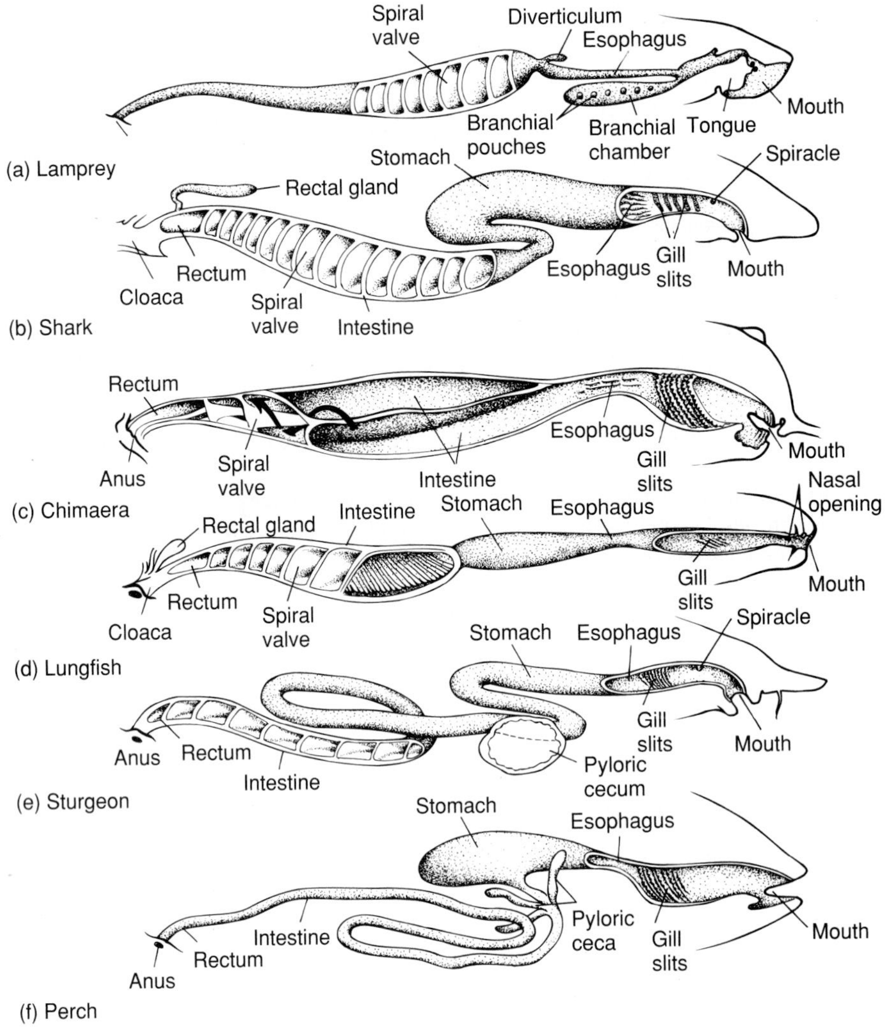


**Figure 13.29** Digestive tracts of selected fishes. (a) Lamprey. (b) Shark. (c) Chimaera. (d) Lungfish. (e) Sturgeon. (f) Perch. When a spiral valve is absent, the intestine is often lengthened, as in the perch.

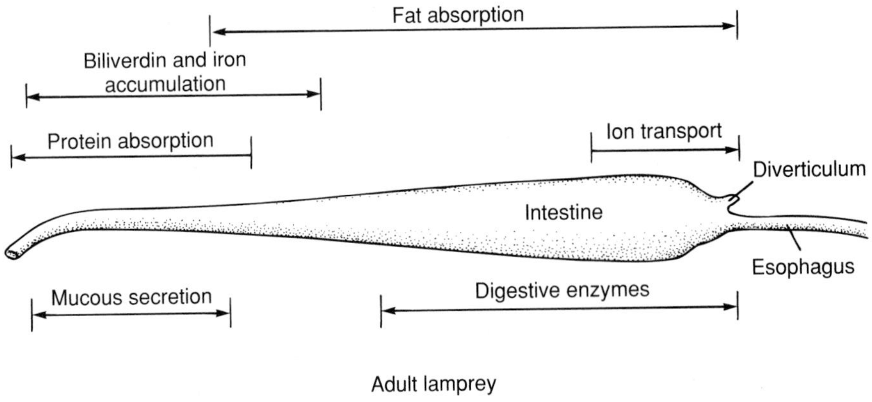


**Figure 13.30** Alimentary canal of adult lamprey. Absorption, elimination, and transport from the various regions of the alimentary canal are at the top of the diagram. Regions releasing digestive enzymes and mucus into the alimentary canal are shown at the bottom.

differentiated in chimaeras, lungfishes, and some teleosts. When present, the stomach is commonly J-shaped, consisting of a wide fundic and a narrow pyloric region (figure 13.31). In sharks, the muscle layers within the wall of the fundus are composed of striated muscle anteriorly and replaced by smooth muscle posteriorly. A spiral valve is found in the intestines of elasmobranchs and many primitive bony fishes, but it is absent in teleosts. In teleosts, it is more common to find elongated intestines folded back on themselves in coils. The terminal section of the intestines is usually slightly widened into a rectum. The rectum leads into a cloaca in sharks or directly to the outside via an anal opening in bony fishes. In elasmobranchs, a **rectal gland** opens into the rectum. Although the rectal gland is not directly involved in digestion, it eliminates excess salt ingested during feeding.

In most bony fishes, **pyloric ceca** that open into the duodenum form at the junction between the stomach and the intestine. These number from several to nearly 200 in some teleosts. They are primary areas for digestion and absorption of food, not fermentation chambers.

### *Tetrapods*

In amphibians, the esophagus is short and its transition to the stomach is gradual, but both regions are discernible. The esophageal epithelium is a single or double layer of mucous (goblet cells) and ciliated cells. The stomach mucosa contains characteristic gastric glands, including fundic glands throughout most of the stomach and pyloric glands at its narrowed approach to the intestine. The intestines are differentiated into a coiled small intestine, the first part of which is the duodenum, and a short straight large intestine that empties into a cloaca (figure 13.32a).

In reptiles, the alimentary canal is similar to that of amphibians, except that in some reptile species it is larger and more elaborate. In many lizards, the stomach is heavy walled and muscular (figure 13.32b). Crocodiles and alligators possess a **gizzard,** a region of the stomach endowed with an especially thick musculature that grinds food against ingested hard objects, usually small stones deliberately swallowed into the stomach (figure 13.32c). The thin-walled glandular region of the crocodilian stomach lies in front of the gizzard where gastric juices are added. A distinct large intestine is usually present in reptiles. In some herbivorous lizards, a cecum is present between the small and large intestines. The cloaca is partially differentiated into the **coprodeum,** a chamber into which the large intestine empties, and the **urodeum,** a chamber into which the urogenital system empties.

In birds, the esophagus produces an inflated crop, in which food is held temporarily before proceeding along the digestive tract or regurgitated as a meal for nestlings. In pigeons, the crop secretes a nutritional fluid called "milk," which is fed to the young for several days after hatching. The esophagus joins the thin-walled glandular section of the stomach, the **proventriculus,** which is connected to the posterior gizzard (figures 13.32d and 13:33a,b). The proventriculus

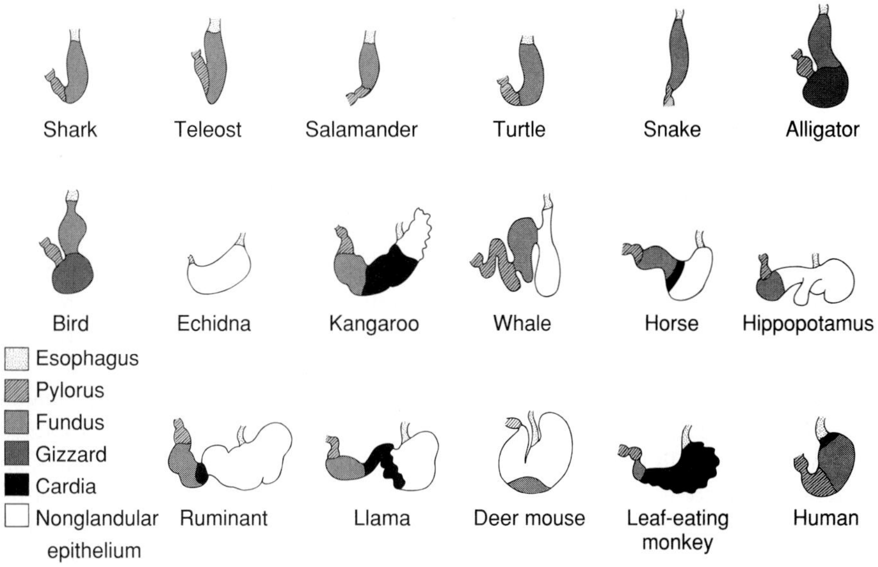


**Figure 13.31** Stomachs of various vertebrates. Two regions of the stomach may be recognized, glandular and nonglandular regions. The glandular region of the stomach includes gastric glands and often exhibits three divisions: cardia, fundus, pylorus. The nonglandular region of the stomach is characterized by a lining epithelium devoid of gastric glands that may be keratinized. The walls of the stomach are composed of layers of smooth muscle, but in some species, these muscular coats are enlarged into a specialized gizzard.

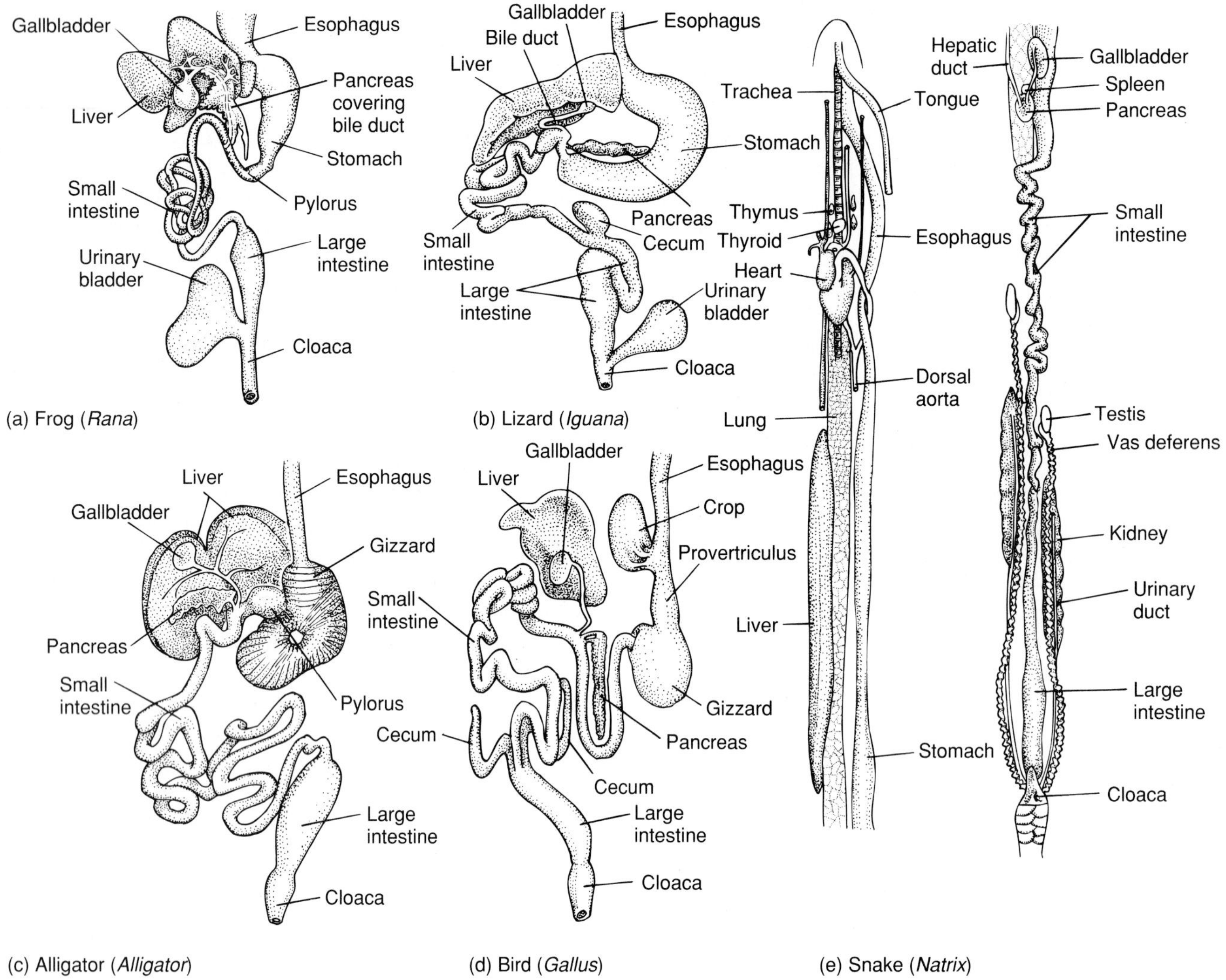


**Figure 13.32** Ventral views of the alimentary canals in tetrapods. (a) Frog. (b) Iguana. (c) Alligator. (d) Bird (chicken). (e) Snake.

secretes gastric juice to help digest the bolus, and the gizzard, together with selected pieces of hard grit and pebbles, grinds large food into smaller pieces. The long, coiled small intestine consists of a duodenum and ileum. A short, straight large intestine empties into the cloaca. In many species, one or several ceca can sprout from the intestine, usually near the junction of large and small intestines.

In mammals, the esophagus usually lacks a crop and the stomach shows no tendency to form a gizzard. In some cetaceans, the stomach or esophagus may expand into a pouch that apparently serves, like an avian crop, to store food temporarily, although some gastric digestion may begin as well in this pouch. The mammalian small intestine is long and coiled and usually can be differentiated histologically into duodenum, jejunum, and ileum. The large intestine is often long, although not as long as the small intestine, and ends in the rectum. In herbivores, a cecum is usually present at the junction between large and small intestines. In humans, this much reduced cecum is called the appendix, or more specifically the **vermiform appendix.** In monotremes and a few marsupials, the large intestine terminates in the cloaca. In placental mammals, it opens directly to the outside through the anal sphincter.

In ruminants and in camels, the stomach is highly specialized. It has four chambers, although the first three—**rumen, reticulum,** and **omasum**—arise from the esophagus and only the fourth—**abomasum**—is an actual derivative of the stomach (figure 13.34a). The large rumen, which gives its name to these mammals, receives the food after it is clipped by teeth and swallowed. The reticulum is a small accessory chamber with a honeycombed texture. Like the first two chambers, the omasum is lined with esophageal epithe-

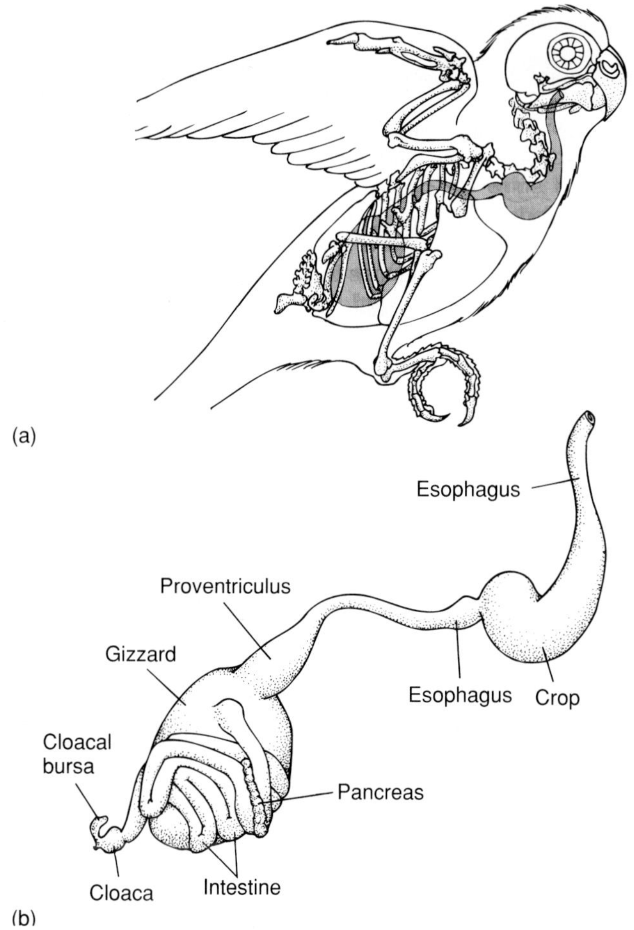


**Figure 13.33** Alimentary canal of a parakeet.
(a) Approximate position of the alimentary canal within the bird.
(b) Alimentary canal enlarged.

lium, although it is folded into overlapping leaves. The three types of mucosa distinctive of the mammalian stomach (cardia, fundus, pylorus) are found only in the abomasum, the presumably "true" stomach.

In many herbivores, digestion of plant cellulose is enhanced by a cecum situated between small and large intestines (figure 13.34b). The cecum contains additional microorganisms effective in cellulose digestion and provides an expanded region prolonging the time available for digestion.

## Associated Glands of Digestion

### *Oral Glands*

The epithelium lining the buccal cavity contains a rich source of cells that secrete mucus and serous fluid. When these secretory cells are gathered together and emptied by a common duct, they constitute an **oral gland.** Such discrete glands are rare in fishes.

In tetrapods, oral glands are more prevalent, perhaps reflecting the absence of a watery medium to moisten food. The most common are the **salivary glands,** a term loosely applied to most major oral glands in tetrapods. These glands are not all homologous, however. In salamanders, mucous glands are found on the tongue, and a large intermaxillary gland is located within the palate. Reptiles also possess oral glands. Usually, strips of glandular tissue, called **supralabial** and **infralabial glands,** are present along the upper and lower lips. In addition, glands may occur within the tongue **(lingual glands)** or below it **(sublingual glands),** in association with the snout **(premaxillary** and **nasal glands),** and along the roof of the mouth **(palatine gland).** These glands release mucus to lubricate the prey during intraoral and esophageal transport. The **lacrimal** and **Harderian glands** release secretions that bathe the eye and vomeronasal organ. **Duvernoy's gland,** situated along the posterior upper lip, is found in many nonvenomous snakes and releases its serous secretion via a duct adjacent to the posterior maxillary teeth (figure 13.35).

Secretions from one or most of these glands, in addition to lubrication of food, may also help maintain healthy oral membranes, neutralize toxins carried by prey, and perhaps initiate the chemical stages of digestion. In poisonous snakes, the **venom gland,** a homologue of Duvernoy's gland, secretes a cocktail of different chemicals with various functions, some toxic, some digestive (figure 13.36). Thus, the injected secretion of the poison gland not only functions to dispatch prey quickly, but it also contains a suite of digestive enzymes introduced along with the toxins during the strike. These digestive enzymes are injected deep into the prey to initiate the breakdown of the prey's tissues from within. Gastric and intestinal juices released later from the walls of the alimentary canal act on the surface of the prey. The overall result is to provide a very speedy digestion from within (venom) and from without (alimentary tract digestive juices). This efficient system apparently allows the snake to swallow large prey and digest it before putrefaction sets in. Rapid digestion may also allow venomous snakes to take advantage of seasonally abundant prey or to clear the digestive tract of bulky prey that might otherwise interfere with lateral bends of the body during locomotion.

When a venomous snake strikes defensively, these digestive enzymes and toxins are released into the victim. Medical treatment of a snakebite in domestic animals, pets, and humans must not only include neutralizing the toxic components of the venom but also inactivating the proteolytic enzymes. Otherwise, even if a patient recovers, extensive scarring may persist from local tissue damage inflicted by enzymes at the site of the bite.

Most birds, particularly those feeding in water, lack oral glands, but there are exceptions. Some passerine birds use mucus from oral secretions to help bind together materials composing their nests.

The most common oral glands in mammals are the salivary glands. There are usually three primary pairs of salivary glands, named for their approximate positions—**mandibular** (submandibular or submaxillary), **sublingual,**

## Box Essay 13.4 The "Spare" Appendix

The human appendix receives bad press. Like an old sock, it is thought of as being without further function, so it is expendable. Its full name is vermiform appendix. Certainly, a person can get along without an appendix. For reasons not particularly clear, it occasionally becomes infected and inflamed. When this happens, it may rupture, spilling out intestinal contents (digestive juices, bacteria, partially digested food, and pus from the inflammation) into the surrounding viscera and there create a life-threatening condition. So if the appendix becomes infected, a reasonable surgeon will quickly relieve you of it. But what of its function?

The human appendix, at the junction of small and large intestines, is a much reduced cecum. Its small size reflects its small, in fact negligible role in cellulose fermentation. No longer does our cecum harbor large-scale microbial fermentation. But just because the appendix performs no digestive function, this does not mean that it lacks a function. The walls of the appendix are richly endowed with lymphoid tissue, much like the rest of the intestine. Just like lymphoid tissue elsewhere, lymphoid tissue of the appendix monitors the passing food, detecting and responding to harmful foreign materials and potential pathogenic bacteria. In short, the human appendix is part of the immune system.

The fact that you can get along without an appendix does not mean it lacks a function. You can get along without some of your fingers, but that does not mean they lack a function. Many senior and near-senior citizens may recall when it was the fashion to surgically remove a child's chronically inflamed tonsils, thinking this would improve an infant's health. Tonsils are stationed around your throat and are the first members of the lymphatic system to detect the arrival of foreign pathogens entering with food. Children certainly got along without their tonsils, probably because the lymphatic system underwent a compensatory enlargement elsewhere, but tonsils do perform a function. There was no malice in the surgeon's urge to remove a child's tonsils. It was well intended, but perhaps one of the misplaced fashions in medicine.

and **parotid.** They form the **saliva,** which is added to food in the mouth. These three pairs of glands lie at the angle of the jaws, usually at about the juncture between the head and the neck, but they are positioned superficial to the neck musculature. Ducts from the mandibular and sublingual glands run anteriorly and release secretions into the floor of the buccal cavity. The duct from the parotid gland opens into the roof of the buccal cavity. In some species, additional salivary glands may be present. In dogs and cats, but in no other domestic animals, a **zygomatic** (orbital) **gland** is present, usually beneath the zygomatic arch (figure 13.37). Like most digestive secretions, saliva contains mucus, salts, proteins, and a few enzymes, most notably **amylase,** which initiates starch digestion. Saliva also aids swallowing by lubricating food.

### *Liver*

The liver is the second largest organ of the body, exceeded in size only by the skin. It functions in a wide variety of roles. Early in fetal life, the liver is directly involved in the production of red blood cells and later it is involved in the destruction of old blood cells. Throughout life, it detoxifies and removes toxic substances from the blood. Bile is manufactured in the liver and released into the intestine to **emulsify** fats, or breaks them up into smaller droplets. Carbohydrates, proteins, and fats are stored and metabolized in the liver.

The liver is one of the most heavily vascularized organs of the body, being supplied with arterial blood via the hepatic artery. However, unlike most organs, it is also supplied with venous blood via the hepatic portal vein that runs directly from the intestines and spleen to the liver, delivering absorbed products of digestion.

During embryonic development, the liver appears as a ventral evagination, or **hepatic diverticulum,** of the gut floor that grows forward into the surrounding mesenchyme (figure 13.38a). The mesenchyme does not directly contribute to the liver, but it induces the endoderm of the hepatic diverticulum to proliferate, branch, and differentiate into **hepatocytes,** glandular cells of the liver. As growth of the hepatic diverticulum continues, it makes contact with embryonic blood vessels, the vitelline veins. These veins form the **hepatic sinusoids,** blood vessels within the spaces between the sheets of hepatocytes (figure 13.38b).

All vertebrates possess a liver. Among the protochordates, a cecum from the gut in amphioxus can be found in the approximate position in which the embryonic liver forms in vertebrate embryos. This cecum has a venous portal system similar to a hepatic portal system. Consequently, the cecum in amphioxus is sometimes called the hepatic cecum. However, it is a site of enzyme production and food absorption, quite unlike the vertebrate liver, so it is unlikely a literal antecedent of the liver of vertebrates.

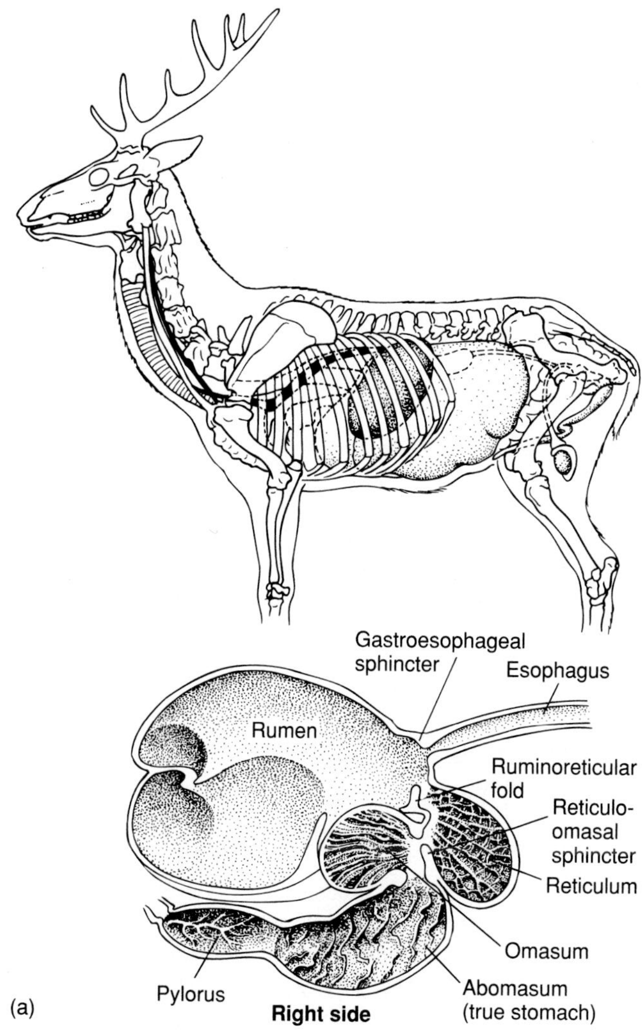


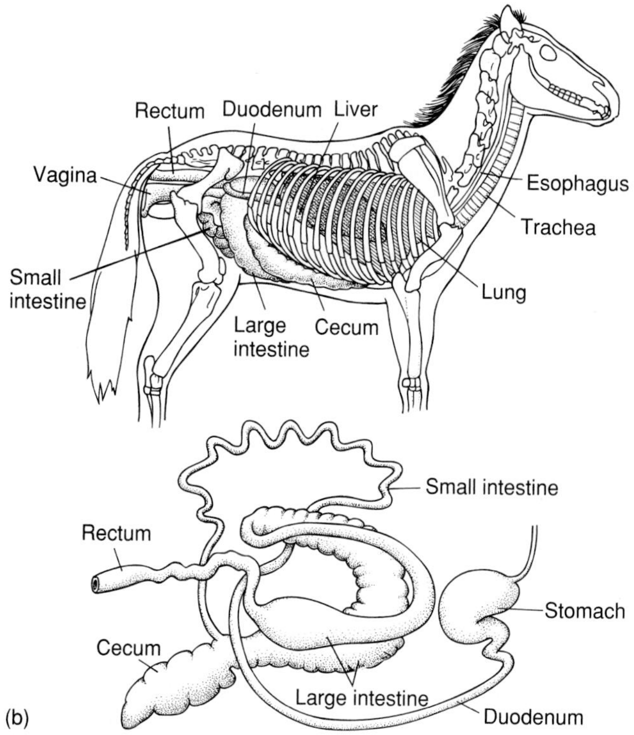


**Figure 13.34** Alimentary canal of foregut and hindgut fermenters. (a) Ruminants ferment food in their foregut. Note the position of the alimentary canal (top) in this deer. The sagittal section of a sheep's "stomach" is illustrated below the deer. Note the series of four chambers. Rumen, reticulum, and omasum are derivates of the esophagus. The fourth compartment, the abomasum, is the actual or true stomach. (b) Hindgut fermenters. Position of the alimentary canal (top) in a horse. Isolated view (bottom) of the large cecum near the juncture of small and large intestines. There is no four-chambered stomach. In hindgut fermenters, the cecum is the major site of fermentation.

At the gross level, the vertebrate liver is bulky and situated within the rib cage, conforming to the available shape of the body cavity. In snakes, it is long and narrow within the tubular body cavity. Although details differ, the microscopic structure of the liver is basically similar throughout vertebrates. It is composed of sheets of hepatocytes separated by blood sinuses, through which courses venous blood returning from the intestines and arterial blood from the hepatic artery (figure 13.39).

The exocrine product of the liver is bile, which is delivered to the intestine where it serves principally to emulsify fats. In most vertebrates, bile is stored in the **gallbladder** and released in sufficient quantities when digesta enter the intestine. The gallbladder is absent in cyclostomes, most birds, and a few mammals, but otherwise is present throughout vertebrates including elasmobranchs and bony fishes, amphibians, reptiles, a few birds, and most mammals. Why it might be absent in some vertebrates and present in most others is not clear. For example, among ungulates, the gallbladder is absent in cervids but present in bovids (except for Duikers, which lack one).

### *Pancreas*

Embryonic development of the pancreas is closely associated with liver development. The pancreas arises from two unpaired diverticula, the **dorsal pancreatic diverticulum,** a bud directly from the gut, and the **ventral pancreatic diverticulum,** a posterior bud of the hepatic diverticulum. These dorsal and ventral pancreatic rudiments may have

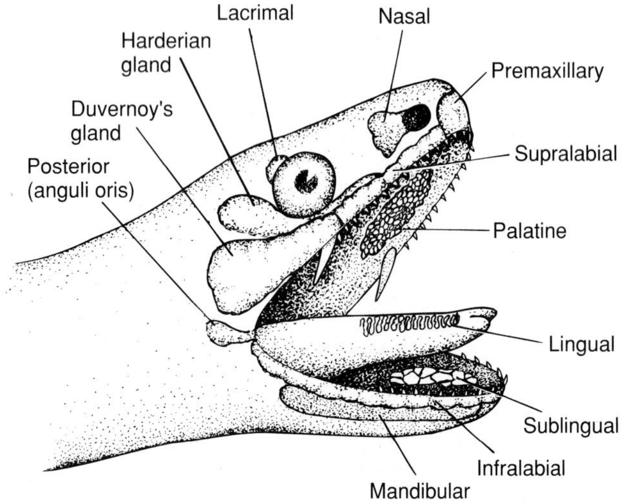


**Figure 13.35** Oral glands of reptiles. Not all oral glands are present in all reptilian species. The venom gland of advanced snakes is a phylogenetic derivative of Duvernoy's gland and is located within the temporal region behind the eye, a position similar to that of Duvernoy's.

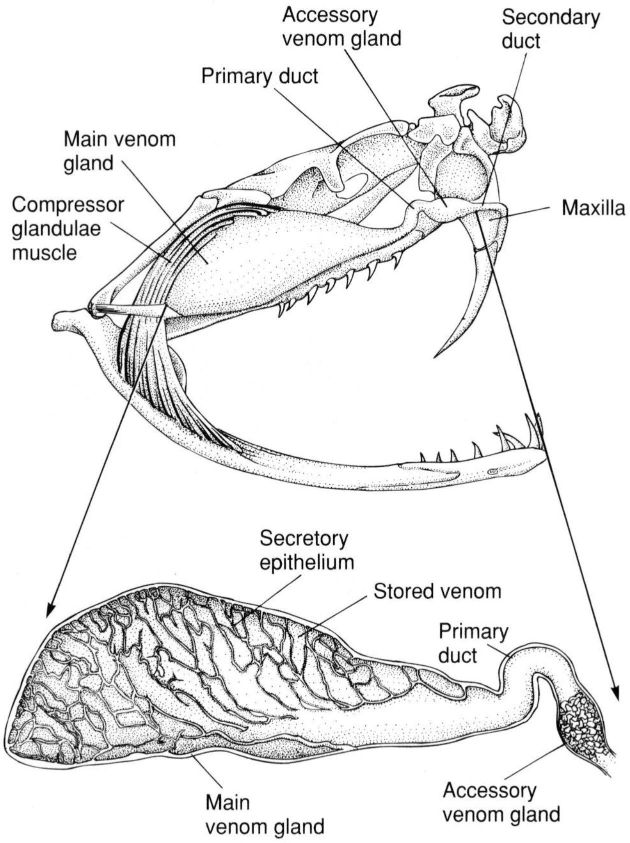


**Figure 13.36** Internal structure of a venom gland in a viperid snake. The secretory epithelium releases venom into the lumen of the gland where large quantities accumulate, ready for a strike. During the strike, contraction of the compressor glandulae muscle applies pressure on the gland, forcing a charge of venom through the ducts and into the prey. During normal hunting of small rodents, the snake usually does not expend all of its venom reserves within the lumen in a single strike. If a snake is artificially forced to expend all its reserves of venom, full replacement of the venom takes about two days.

independent ducts to the intestine, as in some fishes and amphibians, or they may merge as in amniotes, to form a common pancreatic gland. Even if they merge, each rudiment may retain separate ducts to the intestine, as in horses and dogs, or share a single duct, as in humans, pigs, and cows.

Whether one or two, the ducts empty into the duodenal portion of the intestine and release an alkaline exocrine product, **pancreatic juice,** composed primarily of the proteolytic enzyme **trypsin.** Amylases for carbohydrate digestion and lipases for fat digestion are also secreted. Embedded in the pancreas are small **pancreatic islets** (islets of Langerhans) that produce the hormones **insulin** and **glucagon,** both of which regulate the level of glucose in the blood. The pancreas is thus both an exocrine gland producing pancreatic juice and an endocrine gland producing insulin and glucagon (figure 13.40). Both the exocrine and endocrine epithelium of the pancreas arise embryologically from endoderm induced by surrounding mesenchyme.

The pancreas is present throughout vertebrates, both as an exocrine (pancreatic cells) and endocrine (pancreatic islets) gland, although it is not always organized into a discrete organ. In cyclostomes, the exocrine pancreas is dispersed throughout the submucosa of the intestine and on the liver also. In larval cyclostomes, the endocrine pancreas (islets) apparently has ductless follicles located in the submucosa at the anterior part of the intestine. In adult hagfishes, the endocrine follicles develop as discrete encapsulated clumps near the opening of the bile duct into the intestine. They receive their own rich vascular supply. In adult lampreys, the endocrine pancreas is a distinct patch of tissue close to the bile duct and separate from the dispersed exocrine pancreas along the intestine itself. In elasmobranchs, the pancreas is sometimes dispersed along the course of blood vessels within the liver or, as in sharks, it forms a discrete gland with associated exocrine and endocrine components. In bony fishes, a distinct exocrine and endocrine pancreas is present, and the pancreatic islets are clearly delineated. In tetrapods, an exocrine and endocrine pancreas is always present as a discrete organ located near the duodenum.

**Endocrine pancreas (p. 592)**

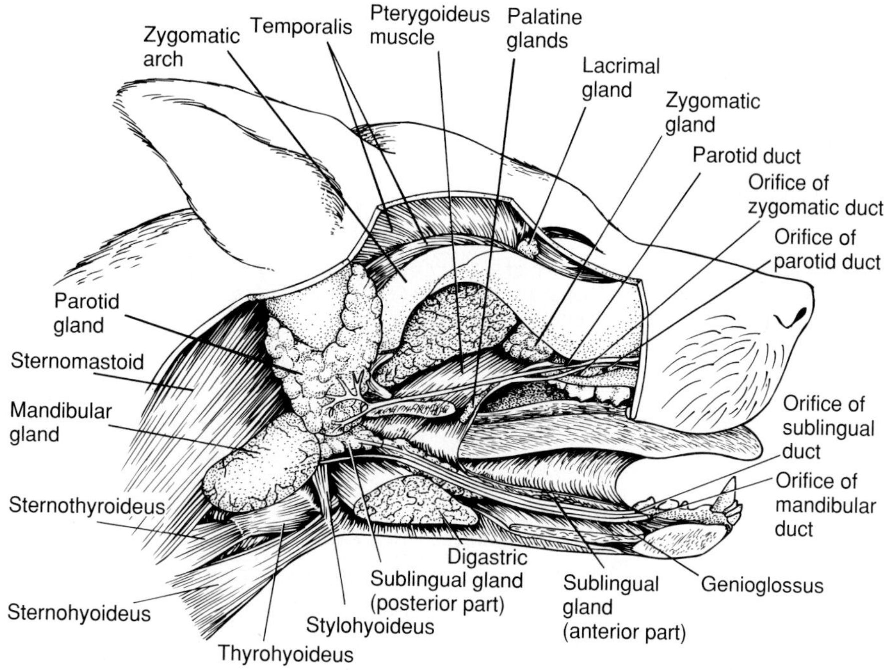


**Figure 13.37** Salivary glands of a mammal, dog. Note the locations of the main salivary glands (sublingual, mandibular, and parotid) along with their ducts leading to the buccal cavity. All mammals possess these three salivary glands. In dogs and cats, a zygomatic gland is also present.

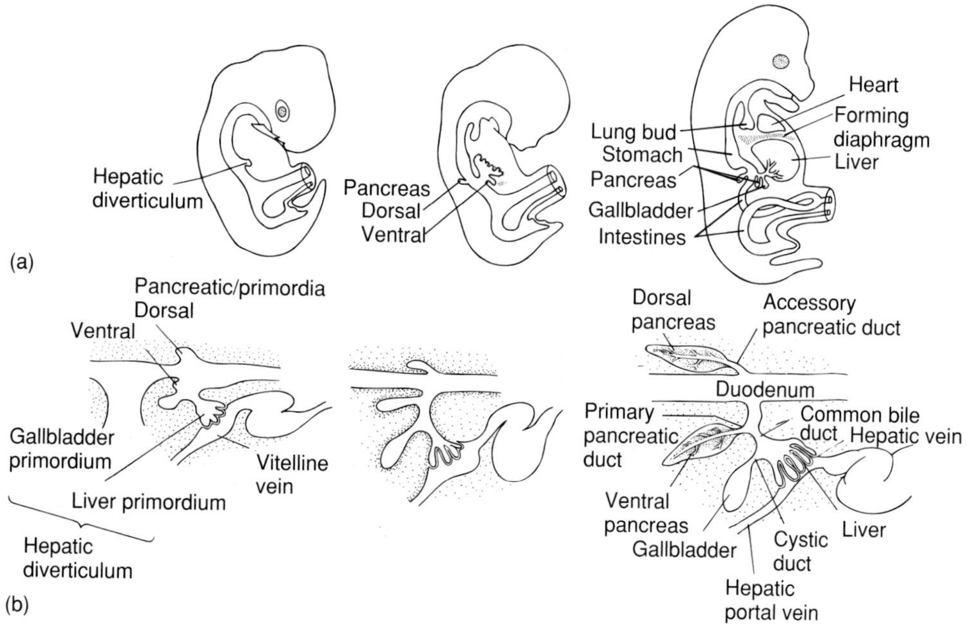


**Figure 13.38** Embryonic formation of the liver. (a) Growth of the liver in a mammalian embryo. (b) Dorsal and ventral pancreatic primordia appear about the same time as the liver primordium. As the liver bud grows, it comes in contact with the vitelline vein, out of which the lining to the liver sinusoids arises.

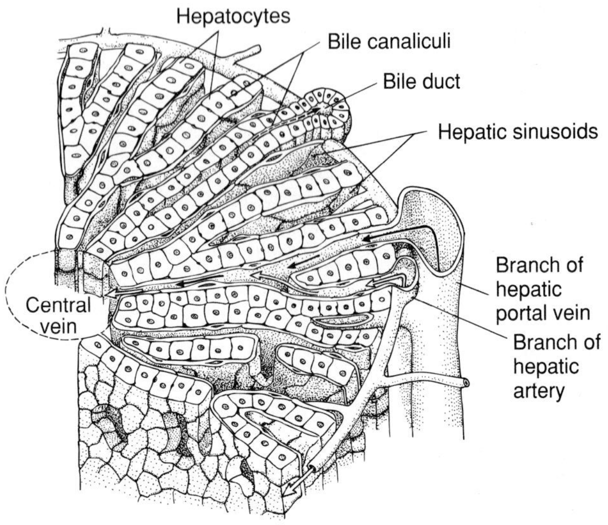


**Figure 13.39** Blood and bile flow in the liver. About three quarters of the blood that reaches the periphery of each liver lobule comes from the hepatic portal vein. The other quarter comes from the hepatic artery. Blood empties into the sinusoids between cords or stacks of hepatocytes (liver cells) and eventually reaches the central vein. From the central vein, it enters the postcaval vein. Solid and open arrows indicate the flow of blood through the liver. Bile is manufactured by hepatocytes, collected in the bile ducts, stored in the gallbladder, and emptied into the duodenum via the common bile duct as it is needed to emulsify fats.

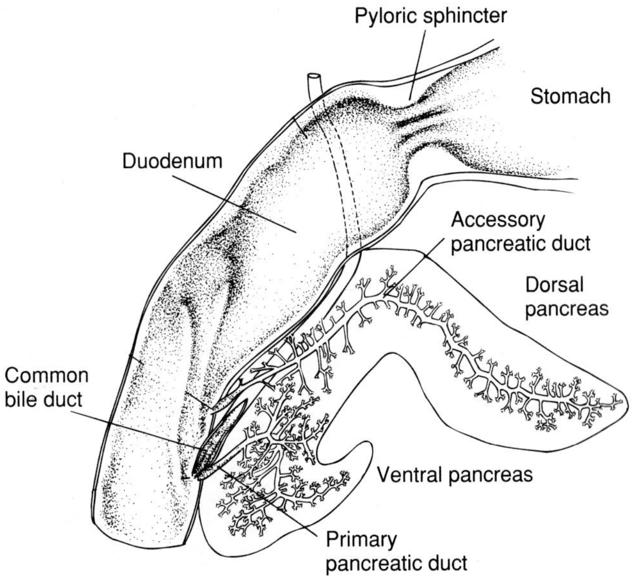


**Figure 13.40** Pancreatic ducts of the giant panda. Both dorsal and ventral pancreas join as a common organ but retain their separate ducts that enter the duodenum. The accessory pancreatic duct drains the dorsal pancreas. The primary pancreatic duct drains the ventral pancreas and enters the duodenum with the common bile duct.

# Function and Evolution of the Digestive System

## Absorption

Absorption of food begins in the stomach. Water, salts, and simple sugars often cross the mucosa and are absorbed in blood capillaries. However, in most vertebrates, the end products of digestion are usually formed and absorbed in the intestine.

Absorption of food depends on the area available and time spent in the alimentary canal. Both microscopic and gross anatomical features can increase surface area. The numerous microscopic microvilli on the epithelial lining of the intestine increase manyfold the area available for absorption. At the gross level, the spiral valve found in the intestines of many fishes serves to force food through the winding channel and increases the time it is exposed to digestion. In herbivorous vertebrates, the intestines may be quite long, and the ceca numerous. These modifications prolong the time food takes to traverse the intestines and allow microbial fermentation to digest cellulose more completely.

Long intestines present a packing problem. In leaf-eating monkeys and other herbivores, the abdominal space expands after a large meal, resulting in an enlarged belly. Occasionally, structural reorganization is required. In ornithischian dinosaurs, for example, the pubic bone is rotated backward, enlarging the abdominal area.

Appearance of a long, distinct large intestine in terrestrial vertebrates correlates with greater requirements to conserve water. The mucosa of the large intestine contains mainly mucous glands so that digestion is brought about by enzymes the digesta carry down from the small intestine and by action of resident microorganisms. The large intestine retains digesta so that the electrolytes and water secreted in the upper digestive tract can be reabsorbed by the body. In lower vertebrates, the large intestine resorbs electrolytes and water produced by the kidneys. The kidneys of amphibians, reptiles, and birds are limited in their ability to concentrate urine. Much of the urinary sodium and water are resorbed in the cloaca into which ducts from the kidneys empty. In addition, retrograde peristaltic waves can reflux material from the cloaca back into the large intestine and ceca, providing a further opportunity to resorb these byproducts.

Retrograde peristalsis essentially prolongs the time digesta spend in the digestive tract. In some warblers, retrograde peristalsis forces intestinal contents back into the gizzard. This seems especially characteristic of birds feeding on fruits with waxy coatings of saturated fats. When the waxy digesta reach the duodenum, high levels of bile salts and pancreatic lipases are added. This mix is refluxed back into the efficient emulsification mill, the gizzard, for further processing. Saturated fatty acids in the wax can be more efficiently broken down and assimilated.

## Feces

For some animals, feces are a resource. Rabbits, hares, many rodents, and even gorillas eat their feces, a behavior termed **coprophagy,** or refection. But the feces eaten usually come only from the cecum. The cecum is emptied in the early morning, and only these droppings are consumed. Within the alimentary canal, there is also a selection process at the cecal-intestinal junction. Liquids and fine particles are diverted into the cecum for extended fermentation and coarser fibers are excluded. Thus, the coarse fiber that bypasses the cecum is not reingested, and only a small percentage of the digesta from the cecum is consumed a second time. Coprophagy allows reingestion, an additional opportunity for the full length of the alimentary canal to capture the products of fermentation, namely, vitamins (K and all B vitamins), amino acids, and volatile fatty acids. If normal coprophagy is prevented, the animal may require vitamin supplements to remain healthy. Coprophagy has been reported in captive apes, cervids, and some other animals, but it is not known if such behavior is important in the wild among these groups.

For the marsupial koala of Australia, feces are eaten by the growing neonate as a transitional food between milk and leaves. The koala mother feeds her own droppings to the six-month-old neonate to begin the process of weaning it from milk to feces to eucalyptus leaves.

The odor from feces can alert a predator to the presence of vulnerable young. Among many herbivores that hide from predators, the young animal does not pass feces until it is licked by its mother. Licking stimulates elimination of feces, which the mother eats so that feces do not collect at sites where young hide and leave a telltale odor. Many young birds bundle their feces. As feces move into the cloaca, its walls secrete a mucous bag that holds the digesta. Parents carry off these bundles of dirty feces, both contributing to good housekeeping (the nest is not fouled) and removing any smelly feces that might attract the attention of a predator.

## Mechanical Breakdown of Food

The purpose of mechanical manipulation of food is to improve the access of digestive enzymes. Biting teeth can puncture an otherwise impermeable exoskeleton (arthropods) or protective armor (bony armor) of prey and allow digestive enzymes to invade the tissue. Some fishes and aquatic salamanders often spit out captured prey only to snatch it with the jaws again. As they repeat this process, their tiny teeth tear down the tough outer skin of the prey.

### *Mastication*

**Mastication,** or chewing, occurs in some fishes and lizards but is characteristic of mammals. The mastication process reduces a large bolus to smaller particles so that digestive enzymes can work on more surface area.

The physical properties of food govern the mastication process. Soft but sinewy foods, such as muscle and skin, are best cut up by the blades of specialized carnassial teeth of carnivores, for example. As upper and lower sets of carnassials close, they tightly slide by each other, like scissors, slicing the food into smaller pieces (figure 13.41a). Fibrous foods, such as grasses and other plant material, are best broken down by grinding. The molar teeth of ungulates, subungulates, and rodents are corrugated on their working surfaces. As the jaws move from side to side, these tooth surfaces slide past one another to tear plant fibers. Chewing mechanically shreds tough plant fibers and breaks down cell walls, thereby exposing the cytoplasm within to digestive enzymes (figure 13.41b). Hard brittle foods, such as nuts and seeds, yield best to compression, like that of a mortar and pestle. Molar teeth that roll over each other pulverize this type of food into smaller pieces (figure 13.41c).

### *Gizzards*

Reduction of food by mechanical action is not restricted to teeth. The churning action of the stomach and intestines also contributes, and the gizzard represents a specialized region dedicated to this function. Hard stones are selected, swallowed, and held in the gizzard, where repeated use over time smoothes them down. The muscularized gizzard works these swallowed stones against the bolus and grinds it into smaller pieces. Eventually, more gritty stones are swallowed to replace those ground up along with the food. The gizzard is especially important in animals that process plant materials with tough cellulose walls. "Gizzard stones" have been found within the abdominal regions of some fossil herbivorous dinosaurs. Smooth and polished, these stones provide indirect evidence that some large dinosaurs had specialized gizzards. Crocodiles and alligators also possess gizzards. The gizzard is well developed in birds, especially in those that feed on seeds.

## Chemical Breakdown of Food

Enzymes from the liver, pancreas, and mucosal wall of the intestine all digest food as it passes through the digestive tract. In some species, chemical digestion begins in the mouth and usually involves amylase digestion of carbohydrates.

The end products of digestion are amino acids, sugars, and fatty acids as well as vitamins and trace minerals indispensable to fuel the organism for growth and maintenance. Most of these end products come from the breakdown of three classes of macromolecules—proteins, carbohydrates,

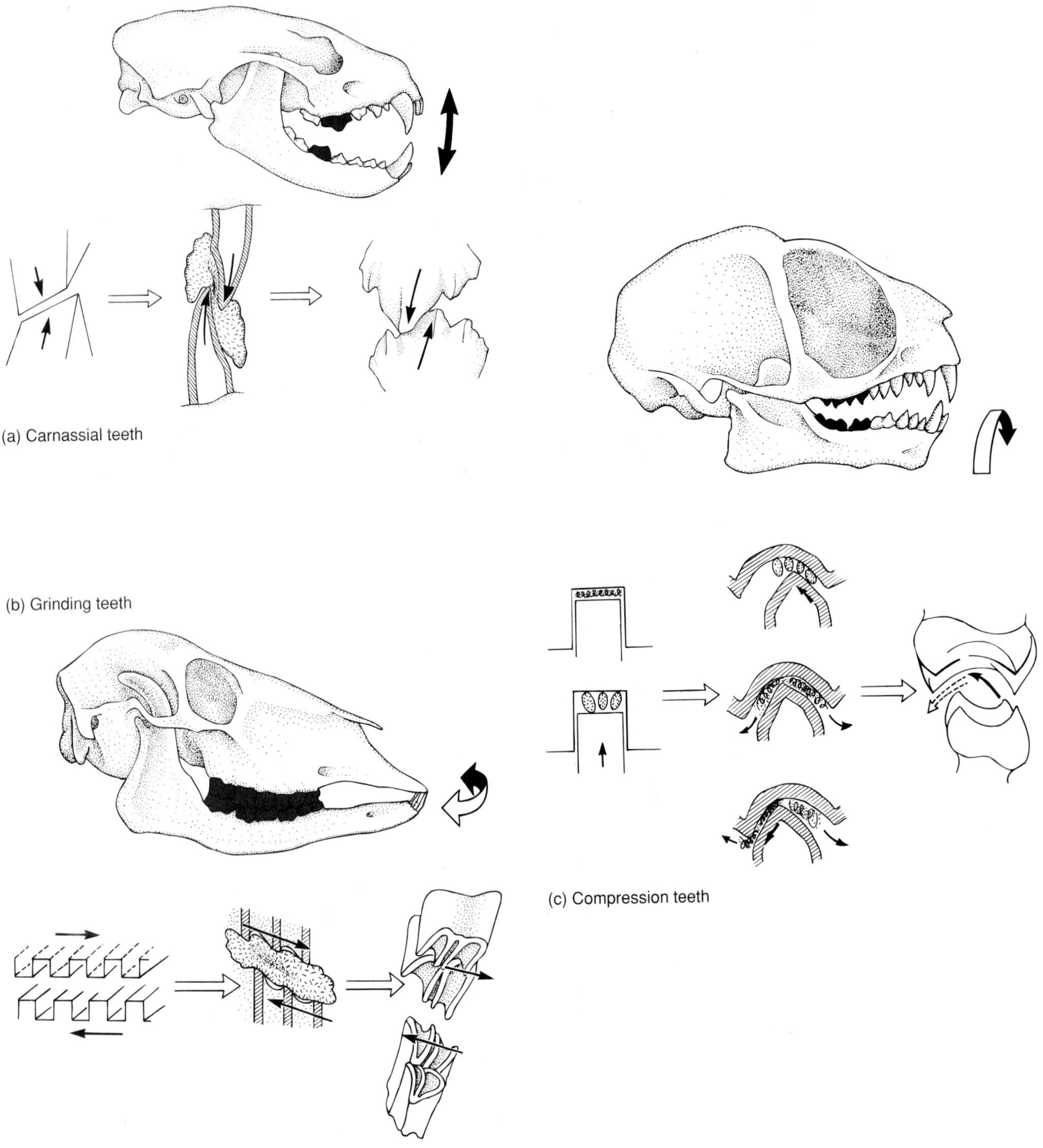


**Figure 13.41** Mastication in mammals. (a) Carnivore skull showing position of carnassials (shaded). Carnassials function like scissors to slice through soft but sinewy foods. (b) Artiodactyl skull showing position of grinding teeth (shaded). Corrugated occlusal surfaces of these teeth grind fibrous foods. (c) Primate skull showing position of compression teeth (shaded) that pulverize hard foods.

## Box Essay 13.5 A Gift Horse in the Mouth

The old adage, "Don't look a gift horse in the mouth," means that if you are kindly given a free horse, don't insult the donor by looking for defects in the gift. It is based on the fact that as horses grind away at fibrous food, their teeth wear down. Eventually, even hypsodont teeth are reduced to small stubs. Because wear increases progressively with age, tooth height is proportional to age. Looking at a horse's mouth then is a not too subtle way of assessing its age and thus value of the gift horse.

Aging animals is of interest to others besides the owners of horses. Generally, larger animals live longer than smaller animals, but not always. Some bats may live as long as bears. Small vipers may live almost as long as pythons. The ages of members of a population are important to wildlife biologists. A scarcity of young individuals may imply a decline in potential new breeding members of the population. On the other hand, if there are low numbers of older individuals, there may be too few young produced to breed and sustain the population. Management decisions are based on such information, but how do biologists age members of a population?

One way, although only a rough way, is to examine tooth height. The lower the crown, the older is the individual. Another way is to examine the width of the pulp canal. Inside the tooth, odontoblasts persist and continue to add layers of dentin slowly to the inner walls of the pulp cavity throughout the life of the individual. Therefore, the pulp cavity progressively narrows with advancing age. On the outer surface of the root, additional cementum is added usually on a seasonal basis to produce annuli, or rings, within the cementum (box figure 1a,b). These three features of teeth—wear, narrowing of the pulp cavity, and deposition of cemental annuli—are roughly proportional to age. Techniques for estimating the ages of individuals are most reliable when they are compared against similar characteristics from a sample containing individuals of known age. In lower vertebrates, where new teeth replace worn teeth, these techniques do not work of course. But in mammals, who have only one set of permanent teeth to last a lifetime, these three dental features are rough indicators of age.

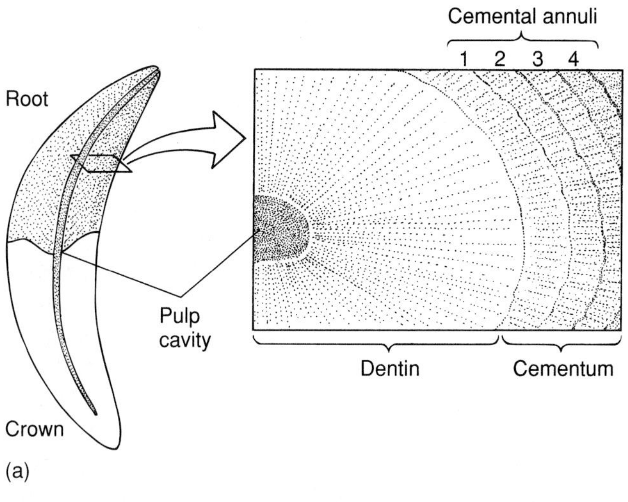


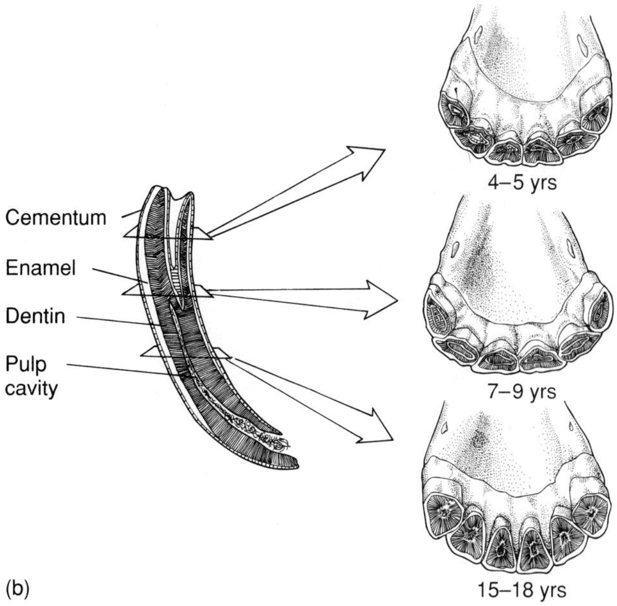


**Box figure 1** Techniques used to age mammalian teeth. (a) Upper canine tooth from a carnivore. A cross section through the root reveals cemental annuli in the cementum superficial to the dentin. The rate of rhythmic formation of these annuli can be calibrated if biologists first study canines from known age groups in the same species. Then, by counting annuli in teeth from free-ranging individuals, they can determine the age of individuals in the population and estimate the age structure of the population. (b) Horse teeth. Progressive use with age wears horse teeth down through successive levels, exposing different layers within the teeth. By examining these teeth and comparing them with teeth of known age, biologists can identify the distinctive pattern on the crowns and determine an approximate age of the horse.

and fats. Digestive enzymes are themselves proteins that are usually sensitive to pH and temperature. Most are inactivated at temperatures above 45°C. Many enzymes are named for the substrate on which they act to which the suffix *-ase* is added.

**Proteases** digest proteins by splitting their peptide bonds. Fat digestion begins with emulsification of large globules into many smaller ones. (Household detergents act in this way to break up grease into small droplets.) Bile, produced by the liver, is one of the body's major emulsifying agents. Emulsification is a physical process, not a chemical one, because it does not break chemical bonds within fats. This is done by **lipases** that chemically break down long-chain fat molecules into smaller chain fatty acids. Emulsification increases the surface area of fats exposed to these lipases.

Digestion of carbohydrates produces simple sugars. One of the most important carbohydrates is cellulose, a structural component of all plants. Cellulose is insoluble and extremely resistant to chemical attack. Many herbivores depend on it as a major energy source, yet surprisingly, no vertebrate is able to manufacture **cellulases,** the enzymes that can digest cellulose. Symbiotic microorganisms, bacteria and protozoans, that live in the digestive tract of the host vertebrate produce cellulases to break down cellulose from ingested plants. The process of breaking down cellulose is known as **fermentation,** which yeilds organic acids that are absorbed and utilized in oxidative metabolism. Carbon dioxide and methane ($CH_4$) are unusable by-products released by belching.

In many vertebrate herbivores, parts of the digestive tract are specialized as fermentation chambers in which symbiotic microorganisms eventually digest cellulose. Because microbial fermentation is relatively slow and cellulose relatively resistant, these chambers are often quite extensive and lengthy. Microbial fermentation may occur in specialized stomachs or pouches that open off the intestine, respectively, known as gastric and intestinal fermentation.

### *Gastric Fermentation*

When the digestion of cellulose is centered in or near a specialized stomach, it is called **gastric fermentation.** More generally, animals in which microbial fermentation arises within the esophagus as well as the stomach are categorized as **foregut fermenters** (figure 13.42a,b). Ruminants illustrate such a digestive specialization. When a ruminant feeds, food initially collects in the saclike rumen, the first of four chambers. The rumen is thin walled and lined with numerous projecting papillae that increase its absorptive surface area. It serves as a large holding and fermentation vat. Later, food in the rumen is regurgitated back into the mouth, remasticated, and swallowed again. This process is repeated until there has been thorough mechanical breakdown of plant material (mastication) and chemical attack on cellulose (fermentation).

Rumination involves complicated waves of contraction that sweep through the rumen and that are synchronized with remastication and with passage of food along the digestive tract. Initially, ruminant animals clip plant material, mix it with saliva, roll it into boli, and swallow it into the rumen (figure 13.43a). Cycles of contraction pass through the rumen and reticulum to circulate and mix the ingested food with microorganisms. This also results in physical separation of coarse and fine food particles. Small particles sink into the fluid that accumulates ventrally within the rumen. Large, undigested plant fibers float on top of this fluid. Methane gas that forms during fermentation collects above this fluid and plant fiber (figure 13.43b). The ruminant belches up the gas, which is a substantial byproduct of foregut digestion. Worldwide, ruminants contribute up to 60 tons of methane per year, about 15% of the

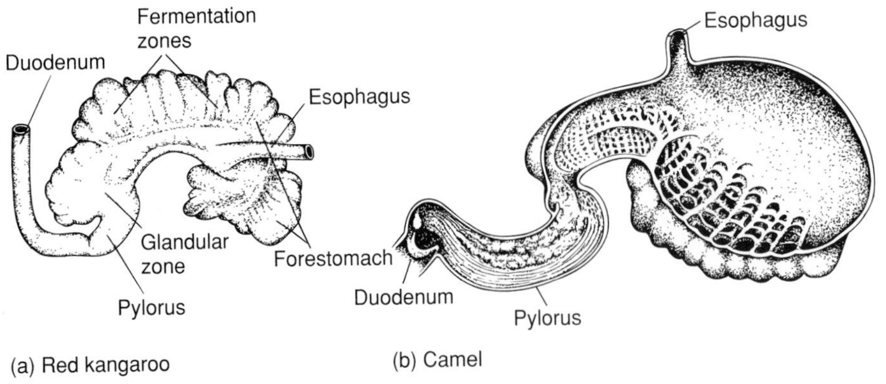


**Figure 13.42** Fermentation in the stomach. (a) Kangaroo stomach. (b) One-hump camel stomach. Bacteria in the stomach release cellulases that break down cellulose, making it available for absorption. Both the marsupial kangaroo and the placental camel independently evolved stomachs that harbor bacteria to ferment the fibrous plants that constitute a large part of their diets.

total atmospheric methane, making ruminants the second major source of atmospheric methane, after natural plant fermentation.

Poorly masticated food is regurgitated for remastication in the mouth (figure 13.43c). Three steps are involved in regurgitation. First, the ruminant contracts its diaphragm as if taking in a breath but keeps the glottis (entrance into the trachea) closed. This produces a negative pressure in the thorax around the esophagus. Second, the gastroesophageal sphincter is relaxed and digesta are aspirated from the rumen into the esophagus. Third, peristaltic contractions sweep the digesta up the esophagus into the mouth, so the animal can rechew the undigested plant material. The process of regurgitation and remastication, termed **ruminating,** occurs repeatedly until most of the material is broken down mechanically. The amount of time an animal spends ruminating depends proportionately on the fiber content of the food. In grazing cattle, this may occupy up to eight hours per day and involve rumination of each bolus 40 to 50 times.

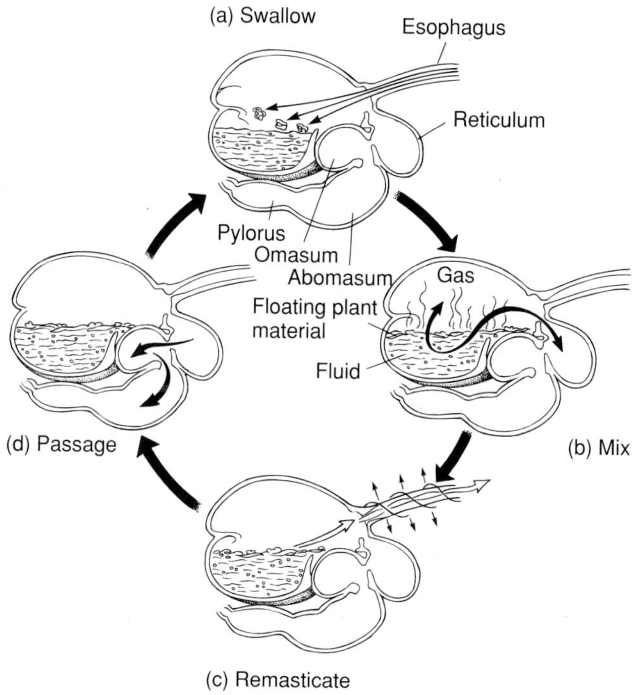


**Figure 13.43** Foregut fermentation in the bovine stomach. (a) In ruminants, food is clipped, rolled into a bolus, mixed with saliva, and swallowed. (b) Contractions spread through rumen and reticulum in cycles that circulate and mix the digesta. Contents separate into fluid and particulate material. Floating, fibrous plant material and a pocket of gas forms during fermentation. (c) Poorly masticated boli of plant material are regurgitated and rechewed later to break down fibrous cell walls mechanically and expose further plant tissue to cellulases. Respiratory inhalation, without opening the trachea, produces negative pressure around the esophagus to draw some of this poorly masticated material into the esophagus through the gastroesophageal sphincter. Peristaltic waves moving forward in the wall of the esophagus carry the bolus into the mouth for rechewing. (d) The omasum transports reduced digesta from the reticulum to the abomasum in two phases. First, relaxation of omasal walls produces negative pressure that draws fine particulate material from the reticulum into its own lumen. Next, contraction of the omasum forces these particulates into the abomasum, the stomach region rich in gastric glands. Thus, the abomasum is the first true part of the stomach.

The reticulum contracts to slosh digesta between itself and the rumen. Possibly it also separates coarse from fine plant material, making the fine plant material available for further transit. The omasum operates like a two-phase pump to transfer digesta from the reticulum to the abomasum (figure 13.43d). First, relaxation of the muscular walls of the omasum aspirates fluid and fine particles from the reticulum into the lumen of the omasum. Second, the omasum contracts to force this digesta into the abomasum. The abomasum is the fundic part of the stomach in which further digestion occurs before digesta pass to the intestines.

In animals that feed upon fibrous plants, the combination of remastication and fermentation process is very efficient. In cattle, organic acids produced in the rumen alone make up 70% of their total energy requirements. Eventually, rechewed food travels through the reticulum into the omasum. The omasum absorbs volatile fatty acids and ammonia and water and at the same time separates the fermenting contents of the rumen and reticulum from the highly acidic contents of the abomasum. The omasum moves smaller food particles into the abomasum, the true stomach in which enzymatic and acidic hydrolysis takes place. Finally, digesta enter the intestine (figure 13.43c).

In the suckling ruminant neonate, the abomasum and intestine digest milk, so fermentation in the rumen is unnecessary. Milk bypasses the rumen of a neonate through a **reticular groove** that reflexively closes when the neonate swallows milk. Thus, milk passes directly from the esophagus to the abomasum via the reticular groove.

Foregut fermentation has arisen independently in groups other than ruminants. Only slightly less elaborate is the multichambered stomach of some nonruminants, including leaf-eating sloths, langur monkeys, peccaries, hippopotamuses, many rodents, and rock hyraxes. Among marsupials, wallabies and kangaroos depend on microbial fermentation in a specialized region of the stomach to digest plants on which they feed (figure 13.42a). However, only ruminants, camels, and some marsupials regurgitate stomach contents and remasticate food. Such remastication together with microbial fermentation is called **rumination** in ruminants and **merycism** in all nonruminants.

### *Intestinal Fermentation*

Microbial digestion of cellulose centered in the intestine is **intestinal fermentation.** Extensive elongation of the intestine and large ceca extend the volume available for fermentation. More generally, herbivores in which fermentation occurs primarily in or associated with the intestine and its ceca are **hindgut fermenters** or **cecal fermenters.** Rabbits, pigs, horses, and koalas are examples of hindgut fermenters (figures 13.28 and 13.44).

## Gastric versus Intestinal Fermenters

In both gastric and intestinal fermenters, microorganisms of the digestive tract release enzymes that digest the plant cellulose (figure 13.45a,b). However, the physiological advantages of such fermentation differs between gastric (foregut) and intestinal (hindgut) fermenters. At first glance, it might seem that gastric fermenters, such as ruminants, camels, kangaroos, and wallabies, enjoy all the advantages when it comes to efficient digestion. First, fermentation takes place in the anterior part of the alimentary canal, yielding end products of digestion early in the digestive process so they are ready for uptake next in the intestine (figure 13.45a). Second, the ruminant system allows rechewing and more complete mechanical breakdown of the cell walls. By shuttling food between mouth and rumen via the esophagus, the ruminant can keep grinding away at plant fibers. The distant ceca of intestinal fermenters make such shuttling impossible. Third, the ruminant system turns nitrogen, which in most vertebrates is a waste product, into a resource. Nitrogen in the form of ammonia is a by-product of cellulose digestion. Microorganisms take up ammonia and use it to make their own cell proteins. Periodically, the rumen contracts, flushing these microorgansisms into the abomasum and intestine where, like any food, the microorganisms themselves are digested and their high-quality proteins are absorbed. Finally, the large storage capacity of the rumen means that a large meal can be gathered quickly in an exposed site, and the animal can retreat to a safe spot in order to digest it.

Gastric fermenters are also able to turn urea, another waste product, into a resource. For example, a camel fed nearly protein-free foods excretes almost no urea in its urine. Certainly urea forms during metabolism, but it reenters the rumen, partly by direct transfer across the rumen wall and partly in the camel's saliva. In the rumen, urea is broken down into carbon dioxide and ammonia from cellulose digestion. Microorganisms take up this ammonia as well and make cell proteins.

Thus, gastric fermentation is especially efficient at extracting the most even from food of poor quality. Ruminant and ruminant-like animals have been especially successful in habitats where the forage is scarce, fibrous, and poor for at least part of the year, such as in alpine regions (goats), deserts (camels), and harsh winter areas (bison).

However, intestinal, or hindgut, fermentation has some advantages. For the intestinal fermenter, the bolus passes through the major absorptive regions of the alimentary canal *before* reaching the principal sites of fermentation, usually the ceca (figure 13.45b). Soluble nutrients such as carbohydrates, glucose, and proteins can be safely absorbed before fermentation begins. By contrast, among foregut fermenters, fermentation occurs early and many of these necessary nutrients are disposed of before they can be absorbed. To compensate for premature digestion, foregut fermenters must rely on occasional flushing of microorganisms into the intestine where they are digested to replace nutrients lost in fermentation.

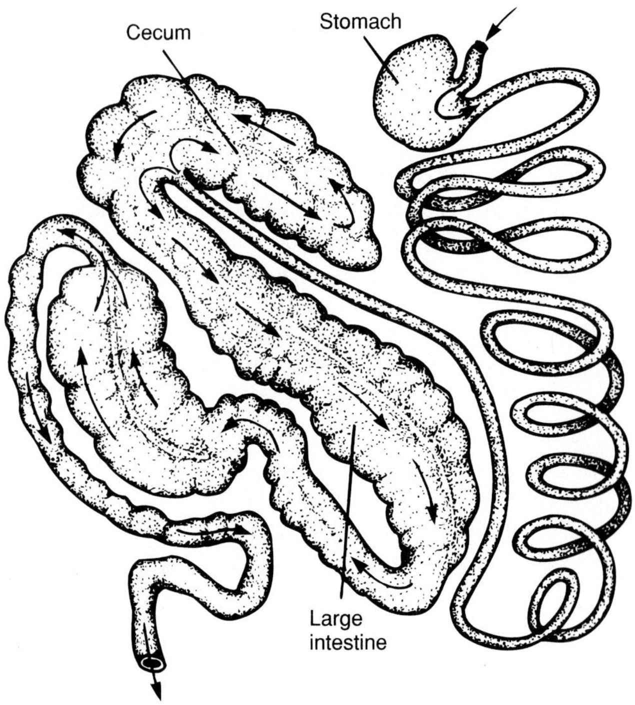


**Figure 13.44** Hindgut fermentation in the alimentary canal of a horse. The cecum acts as a diverticulum at the junction between small and large intestines. In the cecum, microbial fermentation of the digesta helps break down plant cell walls. The large intestine continues the fermentation process.

Further, although foregut fermentation is thorough, it is also slow. To ferment fibrous plants, food must occupy the rumen for extended cellulose processing. Where forage is abundant, the intestinal fermenter can move large quantities of food through the digestive tract, process the most easily digestible component of the forage, excrete the low quality component, and replace that which has been excreted with fresh forage. Horses are hindgut fermenters that depend on a rapid transit and a high food intake to meet their nutritional needs. In addition, coprophagy allows rabbits and some rodents to reingest feces containing fermentation products produced on the first pass through the digestive tract. This gives these intestinal fermenters a second chance to extract some of the undigested material.

***Size and Fermentation*** The body size of a herbivore affects the relative advantages of gastric versus intestinal fermentation. Small animals have relatively high

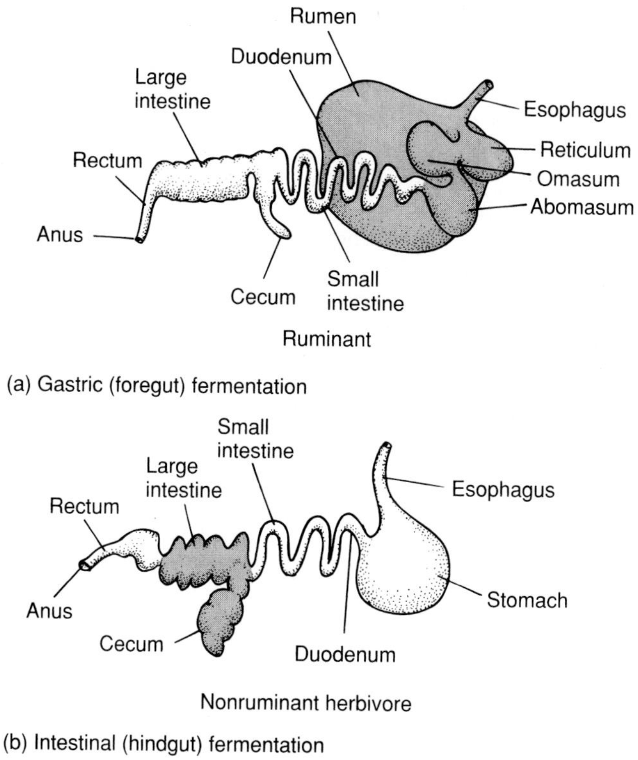


**Figure 13.45** Gastric (foregut) versus intestinal (hindgut) fermentation. (a) Ruminants and other foregut fermenters depend upon microbial activity early in the digestive process as the ruminant "stomachs" chemically attack cellulose in plant cell walls. (b) In intestinal fermenters, fermentation takes place in the long intestines and extended ceca.

metabolic rates, and large animals have relatively low metabolic rates. In general, a small herbivore must digest food rapidly to meet the demands of its high metabolic rate. Thus, most small herbivores are intestinal fermenters with digestion based on the rapid passage of abundant and relatively high-quality foliage. By contrast, large herbivores have low metabolic rates and proportionately more volume in which to process food. For large ruminants, therefore, the slower transit of food does not cause digestion to lag behind metabolic demand because metabolic rate is also lower and the rumen volume devoted to digestion is relatively greater. But similarly, large nonruminants can also enjoy some advantages. In fact, if body size is large enough, intestinal fermenters can attain relatively complete digestion, approaching that of ruminants. Thus, large herbivores, be they ruminants or nonruminants, can extract more energy from plant material than small herbivores can. Only ruminants of intermediate body sizes seem to enjoy an advantage over nonruminants of intermediate body sizes, but only if forage quality is low.

The earliest ruminants for which we have a fossil record were small. But as we have just seen, slow digestion in the rumen of small herbivores is a disadvantage compared with rapid food passage of nonruminants. This has led to the suggestion that the rumen probably evolved for other functions initially, such as detoxification or synthesis of proteins. Later, when grasslands expanded, the foregut was preadapted for medium-sized herbivores requiring more efficient processing of this fibrous forage.

## Digesting Toxins

The digestive tract has evolved to do much more than reduce food to its end products making them available to the organism. Many animals have digestive tracts involved in the detoxification of potentially poisonous chemicals in food. The koala exists on a diet exclusively of eucalyptus leaves. There are over 500 species of eucalyptus in Australia, and the koala favors perhaps a dozen of these as food. Because plants cannot run from herbivorous animals, many produce chemical defenses to make their tissues unpalatable or toxic to herbivores. Examples include tannins in grapes, caffeine in coffee plants, and cannabis in marijuana leaves. These chemicals are bitter or distasteful causing animals to avoid them. Some produce altered states of alertness, leading perhaps to a herbivore that is less attentive to its own safety and thus an easy mark for a predator not on drugs. Such unpalatable or toxic antiherbivore compounds are called **secondary plant compounds** because they are not part of the plant's primary metabolic activity. Similarly, many prey animals have evolved toxins of their own, analogous to secondary plant toxins, that discourage predators. The toxic skin glands of most amphibians are examples.

Eucalyptus produces volatile oils that are toxic to most animals if eaten. Thanks to these secondary compounds, few herbivores can safely eat eucalyptus. The koala is an exception. Its digestive system can detoxify the harmful oils and exploit a resource that is for the most part unavailable to other herbivore competitors.

For many ruminants that browse on foods high in tannins, the ruminant saliva plays a central role in neutralizing the detrimental effects. In addition to being toxic, tannins bind with proteins to reduce their absorption, thus reducing the digestibility of plants. However, ruminants that consume plants rich in tannins produce salivary proteins that bind firmly with the tannins when food first enters the mouth to reduce the toxicity of tannins immediately. But salivary proteins further neutralize such secondary compounds to reduce their later detrimental effects upon digestibility. Not surprisingly, the salivary glands of such ruminants are usually enlarged. In particular, their parotid gland is three times the size of ruminants not grazing on tannin-laden plants.

# Selected References


Beneski, J. T., Jr., and J. H. Larsen, Jr. 1989. Interspecific, ontogenetic, and life history variation in the tooth morphology of mole salamanders (Amphibia, Urodela, and Ambystomidae). *J. Morph.* 199:53–69.

Degabriele, R. 1979. The physiology of the koala. *Sci. Amer.* 241(1):110–17.

Demment, M. W., and P. J. VanSoest. 1985. A nutritional explanation for body-size patterns of ruminant and nonruminant herbivores. *Amer. Nat.* 125:641–72.

Eakin, R. M. 1988. Gastric digestion. *Amer. Zool.* 28:665–70.

Frazzetta, T. H. 1988. The mechanics of cutting and the form of shark teeth (Chondrichthyes, Elasmobranchii). *Zoomorphology* 108:93–107.

Gilmore, B. 1992. Scroll coprolites from the Silurian of Ireland and the feeding of early vertebrates. *Palaeontology* 35:319–33.

Goldblatt, P. J., J. A. Hampton, L. N. Didio, K. A. Skeel, and J. E. Klaunig. 1987. Morphologic and histochemical analysis of the newt *(Notophthalmus viridescens)* liver. *Anat. Rec.* 217:328–38.

Hofmann, R. R. 1989. Evolutionary steps of ecophysiological adaptation and diversification of ruminants: A comparative review of their digestive system. *Oecologia* 78:443–57.

Janis, C. 1976. The evolutionary strategy of the equidae and the origins of rumen and cecal digestion. *Evolution* 30:757–74.

Johnston, D. H., D. G. Joachim, P. Bachmann, K. V. Kardong, R. A. Stewart, L. M. Dix, J. A. Strickland, and I. D. Watt. 1988. Aging furbearers using tooth structure and biomarkers. In *Wild furbearer management and conservation in North America,* edited by M. Novak, J. A. Baker, M. E. Obbard, and B. Malloch. Ottawa: Ministry of Natural Resources, pp. 228–43.

Kardong, K. V. 1980. Evolutionary patterns in advanced snakes. *Amer. Zool.* 20:269–82.

Laitman, J. T. 1984. The anatomy of human speech. *Nat. Hist.* 93(8):20–27.

Langer, P. 1988. *The mammalian herbivore stomach.* New York: Gustav Fischer.

Martin, P. J. 1982. Digestive and grazing strategies of animals in the arctic steppe. In *Paleoecology of Beringia,* edited by D. M. Hopkins, J. V. Mathews, C. E. Schweger, and S. B. Young. New York: Academic Press, pp. 259–66.

Moog, F. 1981. The lining of the small intestine. *Sci. Amer.* 245(5):154–76.

Place, A. R., and E. W. Stiles. 1992. Living off the wax of the land: Bayberries and yellow-rumped warblers. *Auk* 109(2):334–45.

Robbins, C. T. 1983. *Wildlife feeding and nutrition.* New York: Academic Press.

Robbins, C. T., S. Mole, A. E. Hagerman, and T. A. Hanley. 1987. Role of tannins in defending plants against ruminants: Reduction in dry matter digestion? *Ecology* 68:1606–15.

Schwenk, K. 1986. Morphology of the tongue in the tuatara, *Sphenodon punctatus* (Reptilia: Lepidosauria), with comments on function and phylogeny. *J. Morph.* 188:129–56.

Uvnas-Moberg, K. 1989. The gastrointestinal tract in growth and reproduction. *Sci. Amer.* 261(1):78–83.

Wright, D. L., K. V. Kardong, and D. L. Bentley. 1979. The functional anatomy of the teeth of the western terrestrial garter snake, *Thamnophis elegans*. *Herpetologica* 35:223–28.

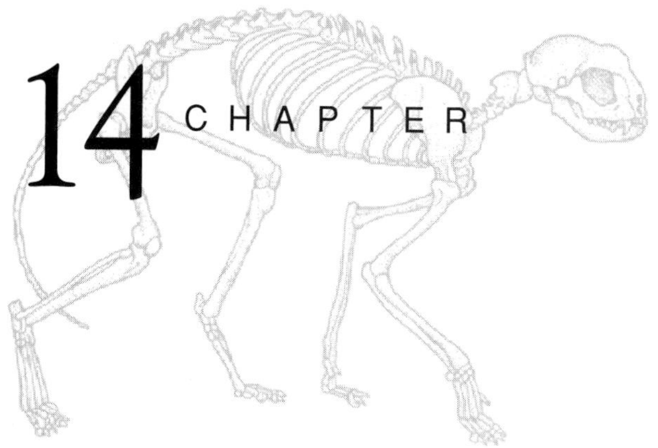

# 14 CHAPTER The Urogenital System



## Introduction

Evolutionary survival depends on doing many things successfully, escaping from predators, procuring food, adjusting to the environment, and so on. All this comes down to reproducing successfully, which is the primary biological role of the genital system. On the other hand, the urinary system is devoted to quite different functions, namely, to the elimination of waste products, primarily ammonia, and to the regulation of water and electrolyte balance. Although urinary and reproductive functions are quite different, we treat both systems together as the urogenital system because both share many of the same ducts.

Anatomically, the urinary system includes the kidneys and the ducts that carry away their product, **urine.** The genital system includes the gonads and their ducts that carry away the products they form, **sperm** or **eggs.** Embryologically, urinary and reproductive organs arise from the same or adjacent tissues and maintain close anatomical association throughout the organism's life.

## Urinary System

### Structure of the Mammalian Kidney

The vertebrate kidneys are a pair of compact masses of tubules situated dorsal to the abdominal cavity. Urine produced by the tubules is ultimately released into the **cloaca** or its derivative, the **urogenital sinus.** We discuss the urinary ducts in some detail later in the chapter when we consider the reproductive system. In this section, we examine the kidney using the mammalian kidney to introduce the terminology that describes the anatomical complexity of this organ.

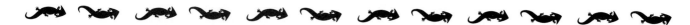

A cutaway view of the mammalian kidney reveals the two regions, an outer **cortex** surrounding a deeper **medulla** (figure 14.1a). Urine produced by the kidney enters the **minor** and then the **major calyx,** which joins the **renal pelvis,** a common chamber leading to the **urinary bladder** via the **ureter.** Elimination of urine from the body occurs through the **urethra.** Within the kidney, the functional unit that forms urine is the microscopic **uriniferous tubule** (figures 14.1b and 14.2). The uriniferous tubule consists of two parts, the **nephron (nephric tubule)** and the **collecting tubule** into which the nephron empties. The number of uriniferous tubules varies from only a few hundred in the kidneys of cyclostomes to over a million per kidney in mammals, in whom the tubules of both kidneys combined constitute over 120 km of tubing. The nephron forms urine. The collecting tubule affects the concentration of urine and conveys it to the minor calyx, the beginning of the excretory duct.

The renal artery, one of the major branches from the dorsal aorta, delivers blood to the kidneys. Through a series of subsequent branches, it eventually forms tiny capillary beds known as **glomeruli,** each being associated with a **renal capsule (Bowman's capsule)** constituting the first part of the nephron. Collectively, the glomerulus and renal capsule form the **renal corpuscle.** An ultrafiltrate without blood cells and proteins is forced through the capillary walls and collects in the renal capsule before it passes through the **proximal** convoluted tubule, **intermediate** tubule, and **distal** convoluted tubule of the nephron, eventually entering the collecting tubules. During transit, the composition of the fluid is altered and water is removed. After circulating through the glomerulus, blood flows through an extended capillary network entwined about the rest of the uriniferous tubule (figure 14.1b). Thereafter, blood is collected in progressively larger veins that join the common renal vein leaving the kidneys.

## Embryonic Development

### *Nephrotome to Nephric Tubules*

The kidneys form within the intermediate mesoderm located in the dorsal and posterior body wall of the embryo. At the onset of its differentiation, this posterior region of the intermediate mesoderm expands, forming a **nephric ridge** that protrudes slightly from the dorsal wall of the body cavity (figure 14.3a). The next structure to appear usually is the paired **nephrotome,** the embryonic forerunner of the nephric tubule (figure 14.3b). The nephrotome is often segmental and contains the **nephrocoel,** a coelomic chamber that may open via a ciliated **peritoneal funnel** to the coelom. Next, the medial end of the nephrotome widens into a thin-walled renal capsule into which grows the glomerulus, a tuft of arterial capillaries. The lateral end of the nephrotome grows outward. This outgrowth fuses with similar outgrowths from successive nephrotomes to form the common **nephric duct** (figure 14.3c). From this point in embryonic development, the modified nephrotome is more properly called a nephric tubule. It may retain its connection with the coelom via the persistent peritoneal funnel.

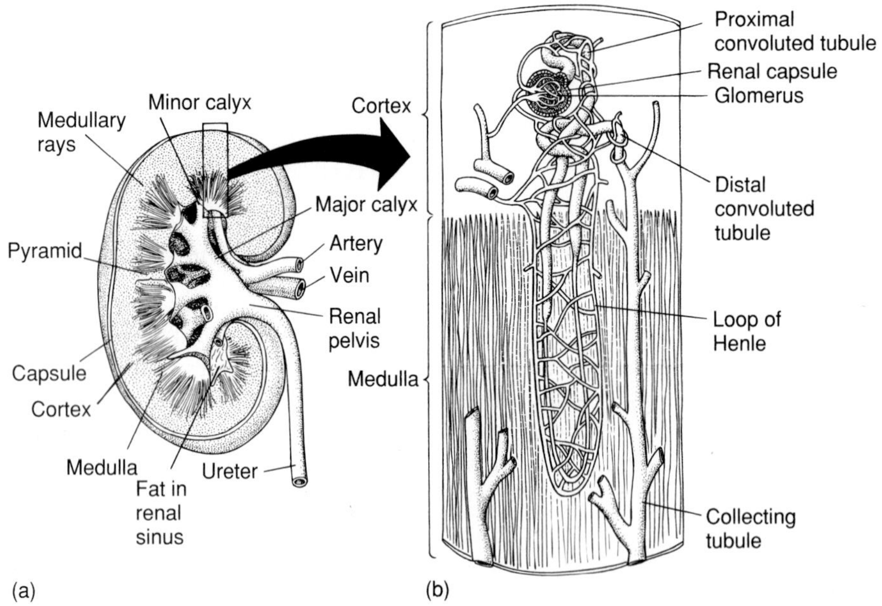


**Figure 14.1** Structure of the mammalian kidney. (a) Section of kidney showing cortex, medulla, and departure of ureter. (b) The uriniferous tubule begins in the cortex, loops through the medulla, and then returns to the cortex where it joins with the collecting tubule.

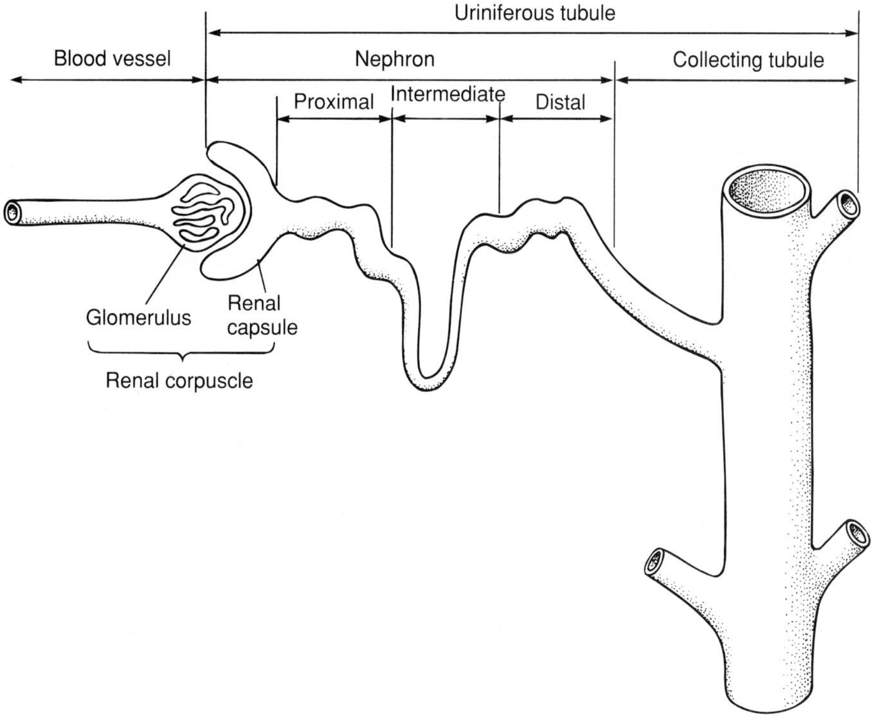


**Figure 14.2** Uriniferous tubule. The nephron (nephric tubule) and collecting tubule make up the uriniferous tubule. In turn, the nephron is comprised of the renal (Bowman's) capsule and the proximal, intermediate, and distal tubules. The glomerulus is the capillary bed associated with the renal capsule. The excretory duct carries away waste products from several uriniferous tubules.

Thus, the fundamental plan underlying the excretory system consists of paired and segmented nephric tubules that open on one end to the coelom and on the other end to the nephric duct, with a glomerulus in between. The ciliated peritoneal funnel seems to drive fluid from the coelom into the tubule, the associated glomerulus adds fluids from the blood, and the tubule itself modifies this collected fluid before it flows into the nephric duct. Although this structure represents the primitive or fundamental plan of excretory tubule organization, tubules opening to the coelom are rarely found in the kidneys of adult vertebrates. Even during embryonic development, the nephric tubule usually develops directly in the nephric ridge without ever establishing a direct opening to the coelom via a peritoneal funnel (figure 14.3a).

## *Tripartite Concept of Kidney Organization*

Developmental and structural differences in the nephric tubules that arise within the nephric ridge inspired a view of kidney formation known as the **tripartite concept.** This concept envisions formation of nephric tubules in one of three locations within the nephric ridge. Subsequent loss, merger, or replacement of these tubules constitutes the developmental basis for the definitive adult kidneys. Specifically, nephric tubules may arise within the anterior, middle, or posterior region of the nephric ridge, giving rise to a *pronephros*, *mesonephros*, or *metanephros*, respectively (figure 14.4a–d). In addition to positional differences, the three regions vary with respect to connections to the coelom. In the pronephros, tubules retain their connections to the coelom through the peritoneal funnel; however, tubules arising within the middle or posterior regions are not connected to the coelom.

***Pronephros*** The anterior pronephros is usually only a transient embryonic developmental stage in all vertebrates. Tubules that appear within the anterior part of the nephric ridge are called **pronephric tubules.** These tubules join to form a common **pronephric duct.** This duct grows posteriorly in the nephric ridge, eventually reaching and opening into the cloaca (figure 14.4a). Glomeruli may protrude into the roof of the body coelom, and fluid filters from them into the body cavity. Pronephric tubules then take up this coelomic fluid through ciliated peritoneal funnels, act on it, and eventually excrete the fluid as urine. However, in most pronephric kidneys, glomeruli make direct contact with pronephric tubules.

Pronephric tubules become associated with glomeruli to form functional kidneys in larval cyclostomes, some adult fishes, and embryos of most lower vertebrates. Fluid filtered from the blood enters the tubules directly, and the

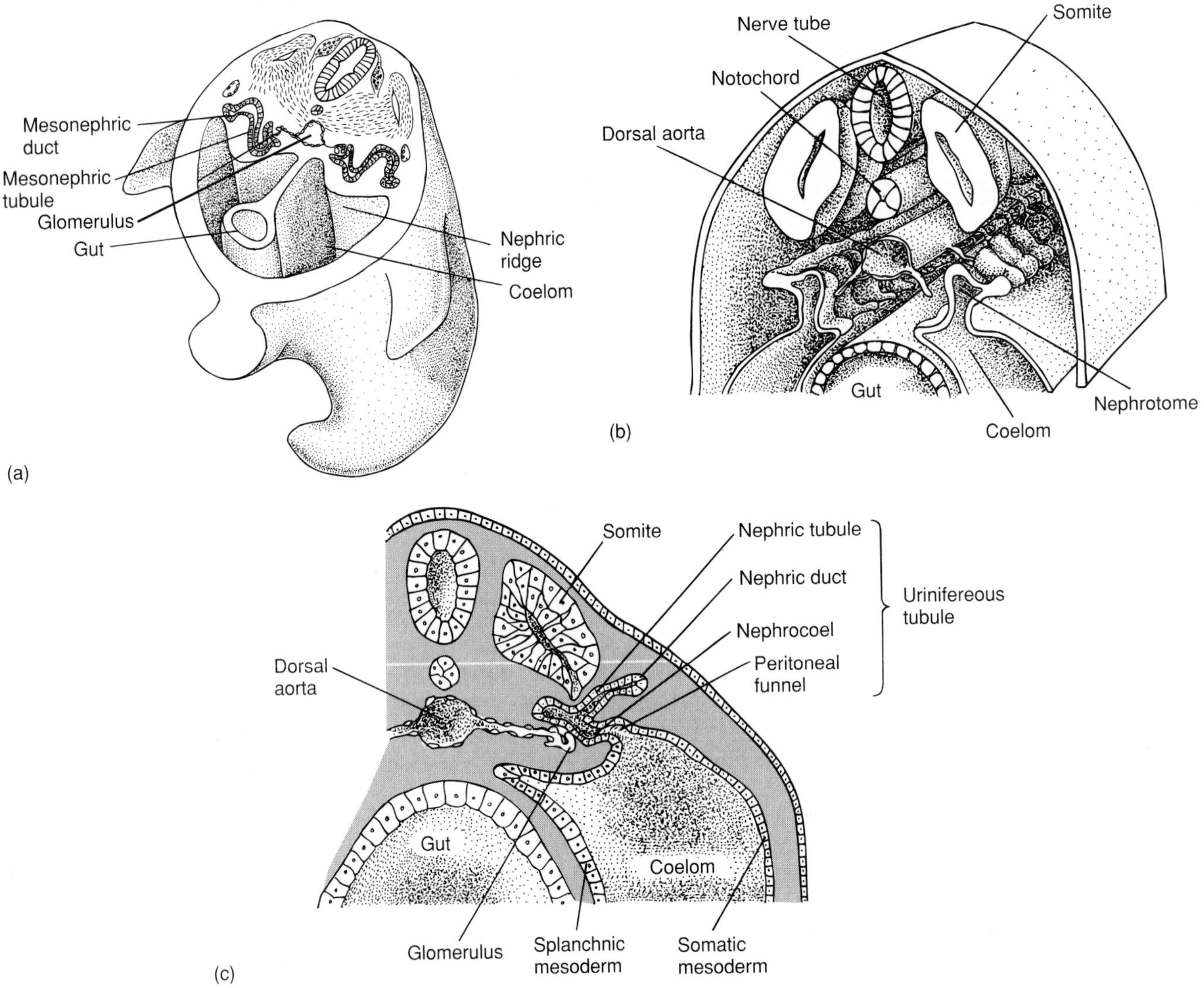


**Figure 14.3** Embryonic appearance of nephric tubules. (a) Nephric tubules develop within the nephric ridge. (b) Preceding this, segmental nephrotomes appear in the posterior part of the intermediate mesoderm. (c) The medial end of nephrotomes differentiates into the first part of the nephric tubule, the renal capsule, into which the glomerulus grows. Arterial sprouts from the dorsal aorta form the glomerulus. The lateral ends of nephrotomes grow outward and fuse with each other into the nephric duct. Sometimes the nephrotome remains connected to the coelom via the ciliated peritoneal funnel.

peritoneal funnels may or may not remain open, depending on the species. In a few amniotes, usually only several pronephric tubules appear during embryonic development. They are not connected to the coelom and do not become functional. In most vertebrates, the embryonic pronephros regresses, and as it does, it is replaced by a second type of embryonic kidney, the mesonephros.

***Mesonephros*** Tubules of the mesonephric kidney arise in the middle of the nephric ridge. These **mesonephric tubules** do not produce a new duct but instead tap into the preexisting pronephric duct. To be consistent, the pronephric duct is now properly renamed the **mesonephric duct** (figure 14.4b).

The mesonephros usually becomes functional in the embryo, but if it persists into the adult, it is modified by incorporation of additional tubules arising within the posterior nephric ridge. This extended mesonephric kidney with additional posterior tubules is termed the **opisthonephros** (figures 14.4d and 14.5). The opisthonephros is found in

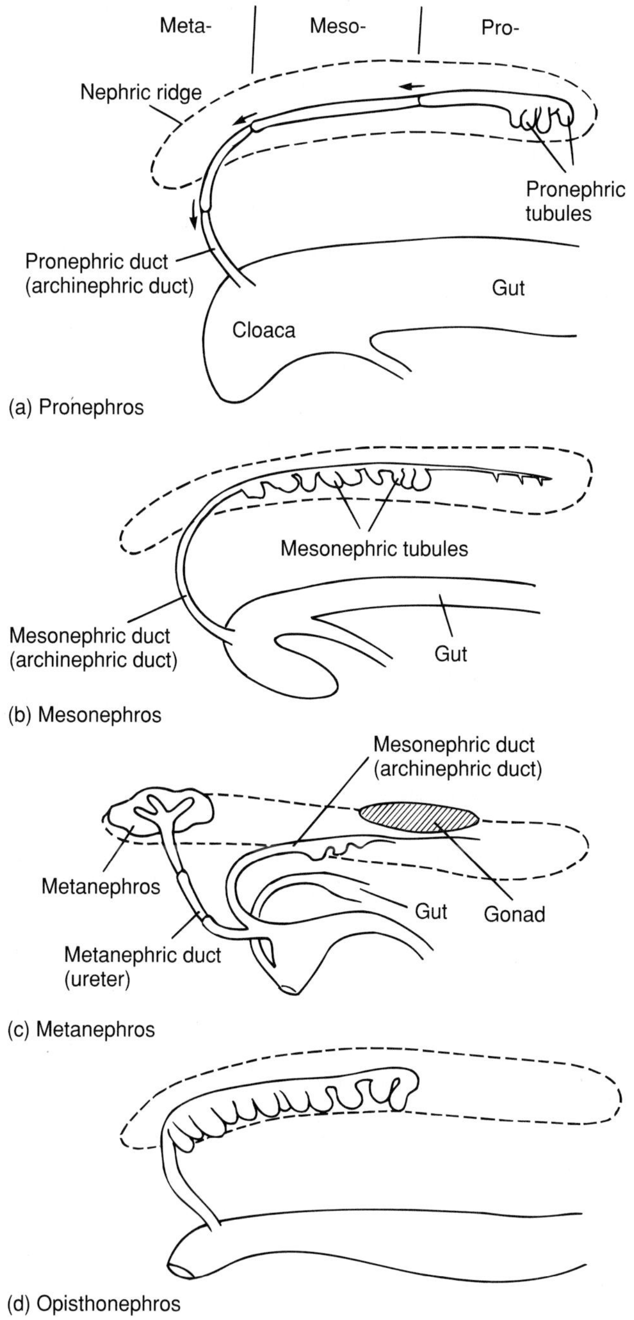


**Figure 14.4** Embryonic origin of the kidneys. Tubules forming the kidney arise in one of three regions of the nephric ridge—anterior (pro-), middle (meso-), or posterior (meta-). (a) Pronephros. Tubules arise in the anterior part of the nephric ridge. They produce a pronephric duct that grows posteriorly in the nephric ridge and empties into the cloaca. Of the three types of kidneys, the pronephros is the first to arise during embryonic development. It becomes the adult kidney in a few fishes but is usually replaced during embryonic development by the mesonephros. (b) Mesonephros. Tubules arise in the middle of the nephric ridge and tap into the existing pronephric duct, now appropriately renamed the mesonephric duct. The mesonephros is usually embryonic and transient. (c) Metanephros. Sprouting from the mesonephric duct, the ureteric diverticulum (later the ureter) grows into the posterior section of the nephric ridge where it stimulates differentiation of tubules that form the metanephros. In males, the mesonephric duct usually takes over the task of sperm transport and is called the vas deferens. In the female, the mesonephric duct degenerates. (d) Opisthonephros. Tubules arising from the middle and posterior nephric ridge form an extended kidney, the opisthonephros, that may develop into the adult kidney of fishes and amphibians.

most adult fishes and amphibians. In amniotes, the mesonephros is replaced in later development by a third type of embryonic kidney, the metanephros.

***Metanephros*** The first embryonic hint of a metanephros is the formation of the metanephric duct that appears as a **ureteric diverticulum** arising at the base of the preexisting mesonephric duct. The ureteric diverticulum grows dorsally into the posterior region of the nephric ridge. Here it enlarges and stimulates the growth of **metanephric tubules** that come to make up the metanephric kidney. The metanephros becomes the adult kidney of amniotes and the metanephric duct is usually called the ureter (figures 14.4c and 14.6).

***Overview*** The nephric ridge is a **nephrogenic region,** meaning that it is the embryonic source of the kidneys and their ducts. Anterior, middle, or posterior parts of the nephric ridge may contribute to kidneys and ducts. Transient stages often yield to later urinary structures. The

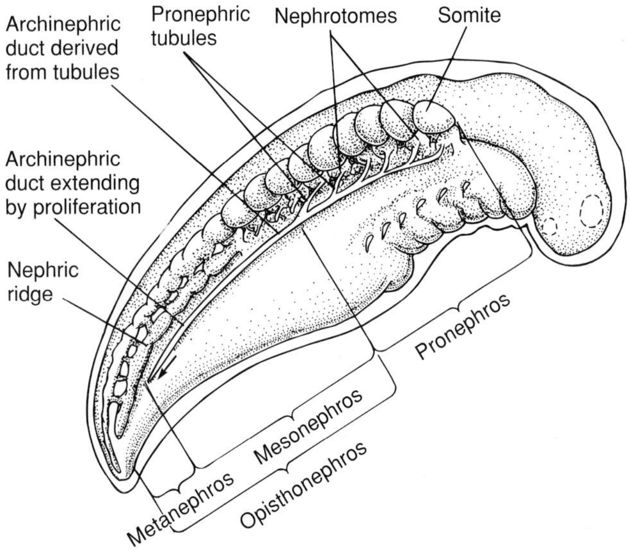


**Figure 14.5** Three-part kidney. Within the nephric ridge, which is derived from intermediate mesoderm, up to three sets of tubules may arise. An extended adult kidney, drained by the mesonephric duct and composed of mesonephric and metanephric posterior tubules, is called the opisthonephros.

tripartite concept that we have used as the conceptual framework in which to discuss these events envisions development of the adult vertebrate kidney as stemming from one of the three regions of the nephric ridge. These three regions are treated as anatomically discrete, and the kidneys they yield as distinct types—pronephros, mesonephros, or metanephros. Additionally, the ontogenetic appearance of such kidneys seems to retrace their phylogenetic origins.

On the other hand, anatomical demarcations between these three regions of the nephric ridge are not always apparent, and the entire nephric ridge may be more a unit than comprised of three parts. Consequently, many morphologists prefer to use an alternative conceptual framework to interpret kidney development and evolution. This alternative view stresses the unity of the entire nephric ridge and is termed the **holonephric concept.** Morphologists who take such a view emphasize that the three types of kidneys arise as parts of one organ, the **holonephros,** which produces tubules in anterior to posterior succession during development. There is no anatomical discontinuity marking separate kidney types. Thus, the holonephros is that part of the nephric ridge that produces the kidney.

Experimental embryology is provocative. For example, transplantation of mesonephros-forming or metanephric-forming mesoderm to the "pronephros" region of mesoderm results in differentiation of these transplanted tissues into pronephric tubules and not into what they would have become if left in place. This indicates that tissues within the nephric ridge are flexible and not committed to one type of kidney or another. Differentiation of the nephric ridge into pronephric, mesonephric, or metanephric tubules is induced by tissue location or by interactions with adjacent tissues and not by intrinsic regionalization within the intermediate mesoderm itself. Because the nephric ridge is nonspecific and developmentally pliable, it has the capacity to form different types of nephrons; therefore, some morphologists argue that the term *holonephros* should be used to describe the unity of the nephric ridge. A holonephric kidney seems to characterize the early development of some hagfishes, elasmobranchs, and caecilians. However, no adult vertebrate retains a holonephros. Absence of examples from adults seems contrary to what we would predict from the holonephric concept and leads other morphologists to retain the tripartite concept.

We seem to be a long way from resolving the disparity between these competing concepts of kidney development and structure. Unfortunately, even basic descriptions of kidney anatomy vary between morphologists from the two camps. Our understanding of kidney morphology is profoundly affected by our intellectual framework. In comparative anatomy, as elsewhere in biology, the concepts we use to interpret what we see can themselves influence how we understand the world around us. Our consideration of kidney structure reminds us that the preconceived views we bring to a subject can affect our interpretation of our anatomical observations. As a practical matter, we use the descriptive richness of the tripartite concept to examine the evolution of vertebrate kidneys. To characterize the kidney, we use terms that indicate which sections of the nephric ridge contribute to its formation. Tubules from all regions contribute to the holonephros. Middle and posterior regions form the opisthonephros. If the kidney forms just from the anterior region, it is a pronephros; if it forms just from the middle, it is a mesonephros; and if it is formed just from the posterior region, it is a metanephros.

## Kidney Phylogeny

### *Fishes*

The most primitive vertebrate kidneys are found among cyclostomes. In the hagfish *Bdellostoma,* pronephric tubules arise in the anterior (cranial) part of the nephric ridge during embryonic development. These tubules unite successively with one another, forming the urinary or pronephric duct (figure 14.7a). Anterior tubules lack glomeruli but open to the coelom via peritoneal funnels, whereas posterior tubules are associated with glomeruli but lack connection to the coelom. In the adult, anterior aglomerular tubules together with several persisting posterior glomerular tubules become the compact pronephros. Although the

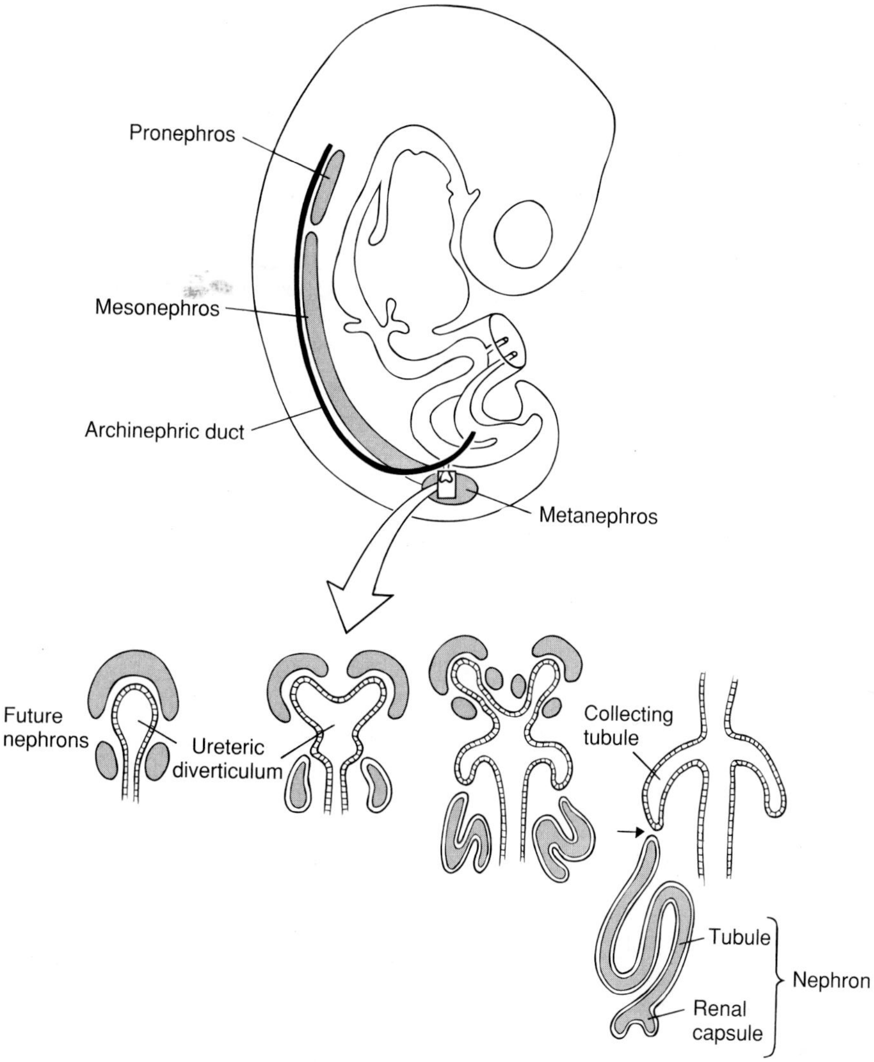


**Figure 14.6** Stages in formation of the amniote kidney. Cross section of the forming metanephros shows that the ureteric diverticulum stimulates surrounding tissue in the nephric ridge to differentiate into nephrons. The ends of the ureteric diverticulum form the collecting tubules.

adult pronephros may contribute to formation of coelomic fluid, the mesonephros is considered to be the functional adult kidney of hagfishes. Each paired mesonephros consists of 30 to 35 large glomerular tubules arranged segmentally along the excretory duct (pronephric duct) and connected to it by short tubules.

In lampreys, the early larval (ammocoetes) kidneys are pronephric, consisting of three to eight coiled tubules served by a single compacted bundle of capillaries called a **glomus.** A glomus differs from a glomerulus in that each vascular glomus services several tubules. Each pronephric tubule opens to the coelom through a peritoneal funnel and empties into a pronephric duct. The pronephros is the sole excretory organ of the young larva. Later in larval life, it is joined by additional mesonephric tubules posteriorly. Upon metamorphosis, additional tubules are recruited from the hindmost part of the nephric ridge, yielding an opisthonephros that becomes the functional adult kidney. The pronephros degenerates, although a few tubules appear to persist into the adult in some lamprey species (figure 14.7b).

In larval fishes, the pronephros usually develops and may for a time become functional, but it is usually supplemented by a mesonephros. In a few teleost species, the pronephros persists as the functional adult kidney; however, in most fishes, the pronephros degenerates and tubules are added caudal to the mesonephros to form a functional opisthonephric kidney in the adult.

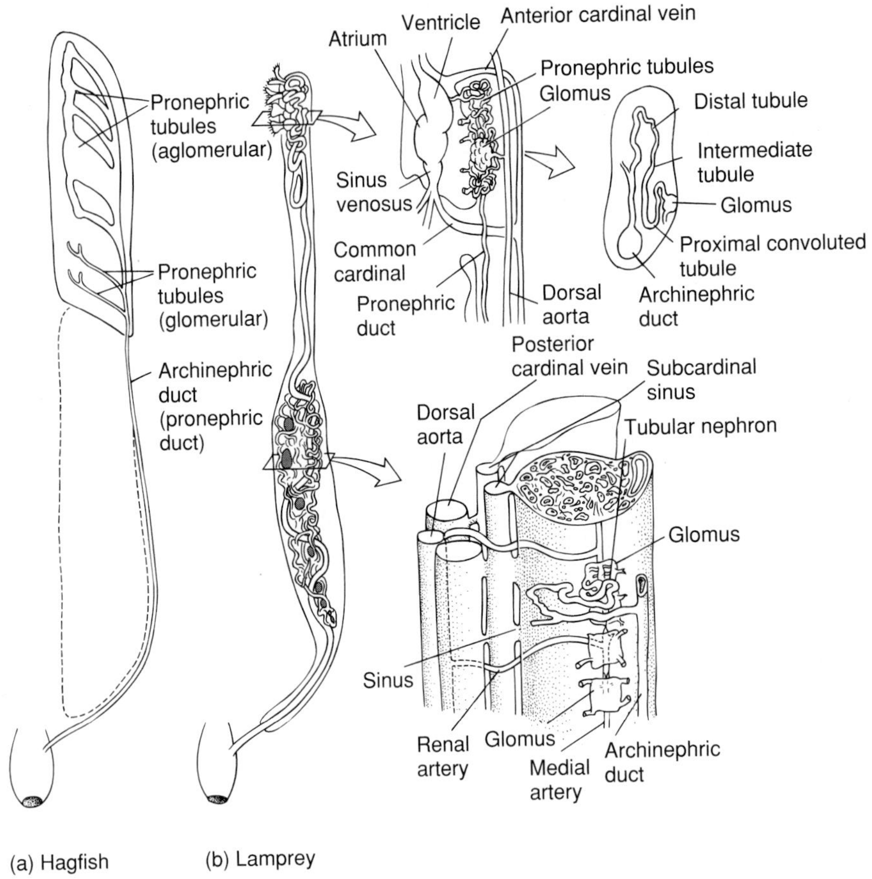


**Figure 14.7** Cyclostome kidneys. (a) Hagfish. The adult kidney includes aglomerular anterior tubules and a few posterior glomerular tubules. (b) Lamprey. The adult kidney includes a posterior opisthonephros. In some species, a few anterior pronephric tubules with peritoneal funnels may persist. Several pronephric tubules share a glomus, and each can be composed of proximal, intermediate, and distal sections.

### *Tetrapods*

Among amphibians having active, free-living larvae, a pronephros may develop and become functional for a time. One or two pronephric tubules may contribute to the adult kidney as well. In caecilians, as many as a dozen pronephric tubules have been reported in the adult kidney. However, the early embryonic pronephros is usually succeeded by the larval mesonephros, which upon metamorphosis is replaced by an opisthonephros in most amphibians. Nephrons within the opisthonephros tend to differentiate into proximal and distal regions before joining the urinary ducts. In amphibians, as in many sharks and teleosts with opisthonephric kidneys, the anterior kidney tubules transport sperm, illustrating again the dual use of ducts that serve both genital and urinary systems (figure 14.8a–c).

In amniotes, the anterior end of the nephric ridge rarely produces pronephric tubules. When present, these are few in number and without excretory function. The predominant embryonic kidney is a mesonephros, but in all amniotes, it is supplemented in late development and completely replaced in the adult by the metanephros drained by a new urinary duct, the ureter. Metanephric tubules tend to be long with well-differentiated proximal, intermediate, and distal regions. In mammals, in particular, the intermediate section of the tubules is especially elongated, constituting the major part of the **loop of Henle.** This term refers to both a positional and a structural feature of the nephron. Positionally, the loop includes the part of the nephron that departs from the cortex and dips into the medulla (the descending limb), makes a sharp turn, and returns to the cortex (the ascending limb). Structurally, three regions contribute—the straight portion of the proximal tubule, the thin-walled intermediate region, and the straight portion of the distal tubule (figure 14.1b). Notice that the terms *descending* and *ascending limbs* refer to the parts of the loop that are departing or entering the cortex, respectively. The terms *thick* and *thin* refer to the height of the epithelial cells forming the loop. Cuboidal cells are thick and squamous cells are thin.

Loops of the nephron occur only in groups capable of producing concentrated urine. Among vertebrates, only the

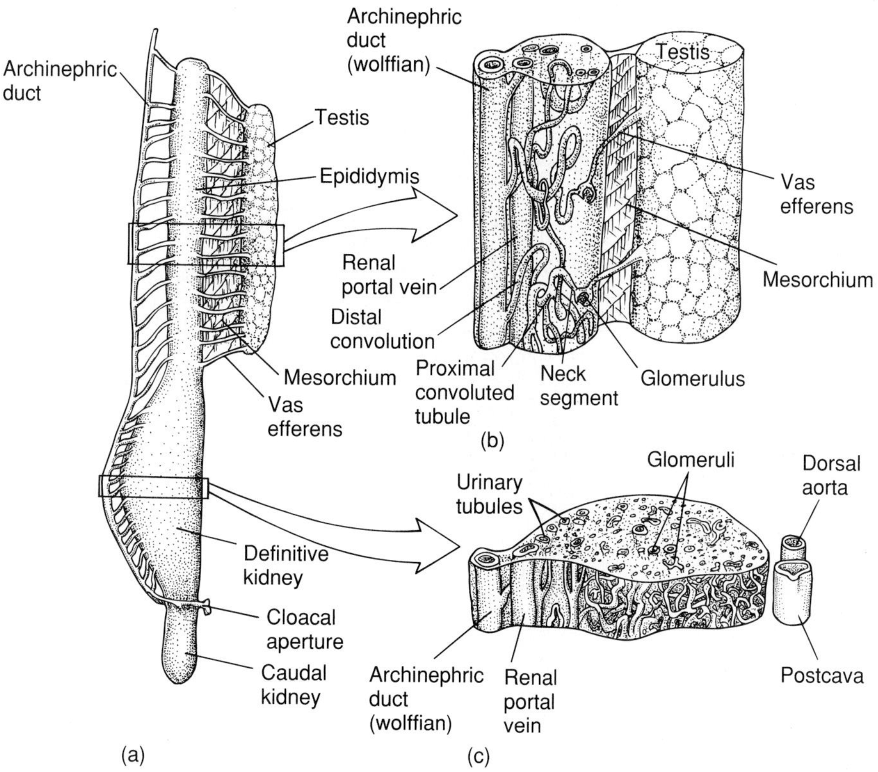


**Figure 14.8** Urogenital organs of a male siren salamander. (a) Whole kidney and testis with associated ducts. (b) The anterior kidney contains tubules that drain the testis in addition to excretory nephrons. Both reproductive and urinary tubules enter the archinephric duct. (c) The posterior kidney is involved in urine formation and drained by the archinephric duct.

kidneys of mammals and some birds can produce urine in which solutes in the urine are more concentrated than in the blood, and only these two groups possess nephrons with loops. All mammalian nephrons have loops, specifically loops of Henle. Mammalian kidneys produce urine 2 to 25 times more concentrated than blood. Further, the ability to concentrate urine is correlated with loop length, and loop length is correlated with availability of water. The beaver has short loops and excretes urine only about twice the osmotic concentration of its blood plasma, but some desert rodents have long loops and can produce urine that is about 25 times as concentrated as their blood.

In a few species of birds, the kidneys contain some nephrons with short, distinct loop segments (figure 14.9), analogous to the loops of Henle in mammals, but these short avian loops evolved independently. These avian kidneys exhibit a modest ability to produce concentrated urine. Their product is about 2 to 4 times more concentrated than their blood. However, the nephrons of most birds do not have loops. In the absence of a loop, the avian nephron is similar to the nephron of reptiles.

## Kidney Function and Structure

Nephron structure can be quite different from one taxonomic group to the next and may appear at first to have no obvious correlation with the phylogenetic position of the taxon. In hagfishes, the nephron is quite simple. A short tubule connects the renal capsule to the excretory duct (figure 14.10a). In lampreys and freshwater bony fishes, the nephron is more differentiated. It includes a renal capsule, proximal and distal tubules usually joined by an intermediate segment, and a collecting tubule (figure 14.10c). However, the nephron of saltwater teleosts is usually reduced because the distal tubule is lost and in some the renal capsule is lost (figure 14. 10a). In amniotes, the nephron is again quite differentiated, and the intermediate segment that contributes to the loop of Henle in mammals is often elaborated (figure 14.10b).

To understand kidney design, the adaptive basis of its excretory and regulatory functions, and the evolution of the nephron, we must look at the demands placed on the kidneys. In general, the vertebrate kidney contributes to the maintenance of a constant, or nearly constant, internal environment, termed **homeostasis,** so that active cells (e.g., striated muscle, cardiac muscle, neurons) are not stressed by radical departure from optimum operating conditions. To

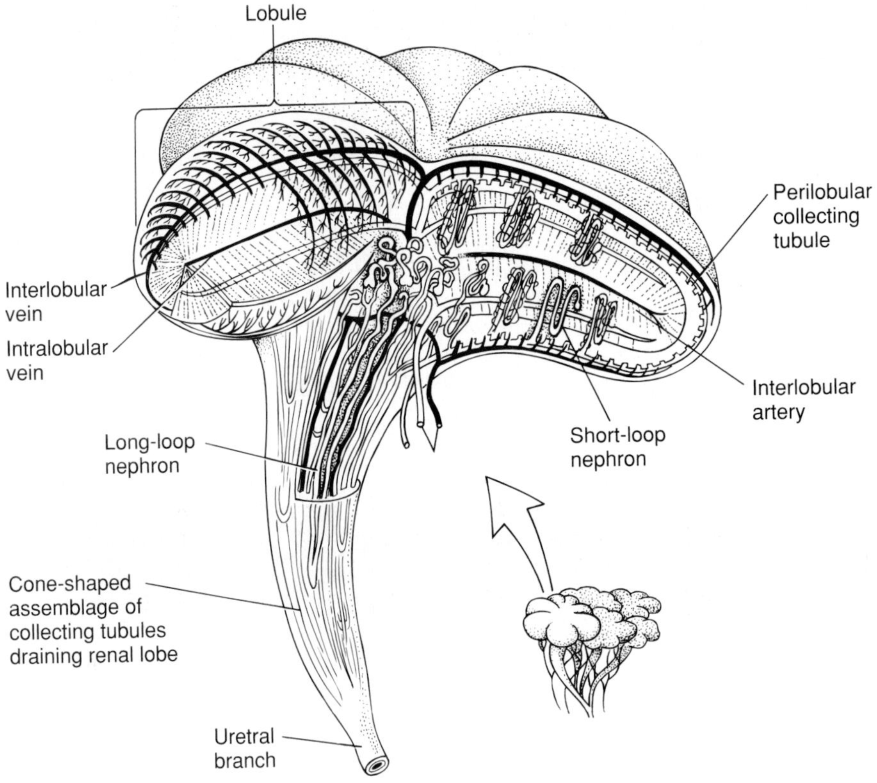


**Figure 14.9** Avian kidney. A section of the kidney is enlarged and cut away to reveal the arrangement of nephrons within and the blood supply to the nephrons.

accomplish this, the kidney performs two fundamental physiological functions, **excretion** and **osmoregulation.** Both are related to maintaining a constant internal environment in the face of accumulating metabolic by-products and perturbations in salt and water concentrations.

## *Excretion: Removing the Products of Nitrogen Metabolism*

Most excreted components in the urine are metabolic by-products that collect within the organism and must be voided so they will not interfere with the organism's physiological balance.

Energy to support growth and activity comes from the metabolism of food. Carbon dioxide and water are end products of carbohydrate and fat metabolism, and both are easily eliminated. But metabolism of proteins and nucleic acids produces nitrogen, usually in the reduced form of ammonia ($NH_3$). Because ammonia is highly toxic, it must be removed from the body quickly, sequestered, or converted into a nontoxic form to prevent accumulation in tissues. Three routes of eliminating ammonia, sometimes in combination, exist in vertebrates. Direct excretion of ammonia is **ammonotelism.** Excretion of nitrogen in the form of uric acid is called **uricotelism.** The third route is **ureotelism,** excretion of nitrogen in the form of urea (figure 14.11). Ammonotelism is common in animals living in water. Ammonia is soluble in water, and a great deal of water is required to flush it from body tissues. For vertebrates living in an aqueous medium, water is plentiful. Thus, ammonia is eliminated through the gill epithelium, skin, or other permeable membranes bathed by water. However, in terrestrial vertebrates, water is often scarce, so water conservation becomes more critical. Because amniotes have lost gills, the gill epithelium is no longer a major route for ammonia excretion. Given these terrestrial constraints, ammonia is converted into urea or uric acid, both being nontoxic forms that address the immediate problem of ammonia toxicity. Furthermore, less water is required to excrete urea or uric acid, so water is conserved as well.

In advanced tetrapods, two evolutionary routes have been followed in addressing the related problems of water economy and nitrogen elimination. Birds and most living reptiles primarily depend on uricotelism. Uric acid, only slightly soluble in water, is formed in the kidneys and transported via the ureters to the cloaca. In the cloaca, uric acid joins with ions and forms a precipitate of sodium, potassium, and ammonium salts. The water not used diffuses through the walls of the cloaca back into the blood.

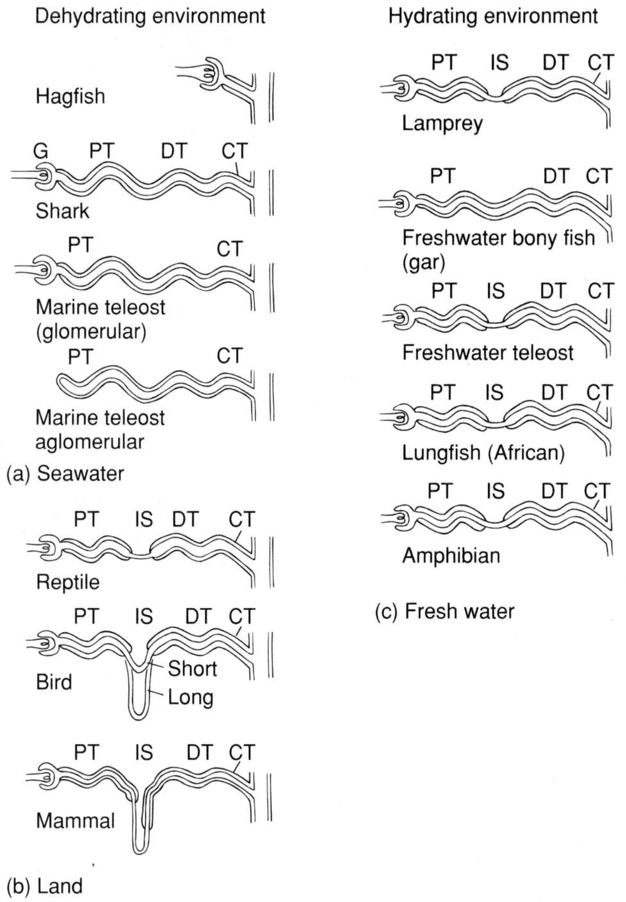


**Figure 14.10** Nephrons from major groups of vertebrates. The segments contributing to the vertebrate nephron depend in large part on whether the animal lives in a dehydrating environment, such as seawater or land (a,b), or in a hydrating environment such as fresh water (c). Nephrons are diagramed and not rendered to scale among groups. Abbreviations: collecting tubule (CT), distal tubule (DT), glomerulus (G), intermediate segment (IS), proximal tubule (PT).

A concentrated, nearly solid uric acid "sludge" forms, allowing nitrogen elimination with little accompanying loss of water.

It is hypothesized that the synthesis of uric acid arose first as an embryonic adaptation, but because of its advantages in water conservation, it was carried over into the adult physiology. The cleidoic egg that first evolved in reptiles is usually laid in dry sites, making water conservation a factor in the embryo's survival. Embryonic adaptations that conserve water include (1) the eggshell, which retards water loss, (2) internal production of water through metabolism of stored yolk, and (3) uricotelism. Because uric acid precipitates out of solution, it does not exert osmotic pressure within the embryo; therefore, it is safely sequestered within the egg without requiring large volumes of water to remove it.

Mammals have followed a different evolutionary route in dealing with nitrogen elimination. They depend largely on ureotelism, the conversion of ammonia into urea. Mammalian kidneys accumulate urea and excrete it as a concentrated urine, thus also detoxifying ammonia and conserving water.

Within an individual, routes of nitrogen excretion can vary in relation to the availability of water. For example, the African lungfish excretes ammonia when it swims in rivers and ponds. But, during droughts, when ponds dry and the lungfish estivates, ammonia is transformed into urea, which can accumulate safely in the body during times of scarce water. With the return of rain, the lungfish rapidly takes up water and excretes the accumulated urea. Similarly, many amphibians eliminate ammonia in water and then excrete urea when they emerge onto land after metamorphosis. In alligators, both ammonia and uric acid are excreted. Turtles excrete primarily ammonia in aquatic habitats but eliminate urea or uric acid when on land (figure 14.11).

## *Osmoregulation: Regulating Water and Salt Balance*

The second major physiological function of the kidneys is osmoregulation. Osmoregulation involves the maintenance of water and salt levels. The external world may vary considerably for an active vertebrate, but cells within see a relatively constant environment. A steady-state intracellular environment is maintained largely by exchange of solutes between the body fluids and the blood and lymph. In turn, the kidneys in large part regulate the constant volume and composition of blood and lymph in terrestrial vertebrates. In aquatic vertebrates, the gill epithelium and digestive tract are as important as the kidneys in addressing problems of salt balance.

***Water Balance*** Most vertebrates require physiological vigilance to maintain internal balance because the external world constantly intrudes. This is particularly true for water, which may be drawn from an organism and dehydrate it or seep inward across permeable surfaces and dilute body fluids. For example, a terrestrial vertebrate is usually in danger of losing water from its body. To counter dehydration, drinking can help replace lost water (figure 14.12a). Some groups, such as reptiles, control water loss with a thick integument that reduces the permeability of their skin to water. In addition, the kidneys, cloaca, and even urinary bladder are **water conservers,** meaning that they recover water before nitrogen is eliminated from the body.

On the other hand, an aquatic life presents other challenges in the management of **water fluxes.** Water may move in or out of the body. In freshwater fishes, the osmotic problem results from a net tendency for an *inward* flux of water. Relative to fresh water, the body of the fish is **hyperosmotic,** meaning that its body fluids are osmotically more

## Box Essay 14.1 Mammals in Deserts, Frogs in the Sea

Dehydration threatens all vertebrates that venture onto land, but it is especially severe for animals living in hot, dry deserts. To manage dehydration, the kangaroo rat *(Dipodomys spectabilis)* has developed several physiological adaptations that allow it to inhabit desert habitats. Even rainwater is scarce for the kangaroo rat, so it doesn't depend on drinking water to replace water that is evaporated during the day. During lush springs, the vegetation it eats contains some water, but late in summer when the diet consists largely of dry seeds, food is not an important source of water. Instead, kangaroo rats depend on water produced as a by-product of carbohydrate and fat metabolism. When metabolized, these foods yield carbon dioxide and water. In fact, up to 90% of the kangaroo rat's water budget may come from the oxidation of food. On the other hand, less water is excreted in the urine than in most other mammals. The loops of Henle are elaborated in the kidneys of kangaroo rats. The long loops allow the production of a concentrated urine, up to four times as concentrated as that of humans. Thus, the kangaroo rat recovers some water from metabolism of its food and loses little in its urine. These adaptations allow it to maintain water balance even under desert conditions.

For amphibians moving from land to seawater, dehydration is also a problem. They are hyposmotic to the salty medium; therefore, water is drawn from their bodies. If water loss is not regulated, they will dehydrate and die. Most amphibians live in fresh water or on land. One of the few exceptions is the Southeast Asian frog, *Rana cancrivora.* At low tide, it ventures into saltwater pools to feed on crabs and crustaceans, a habit that has led to its common name, the crab-eating frog. Yet, it tolerates these salty conditions. It does so through an increase in blood concentrations of sodium and chloride ions and, especially, of urea, as in sharks. At least for short periods in tide pools, it is able to keep its blood levels hyperosmotic to seawater, thus preventing severe water loss and dehydration.

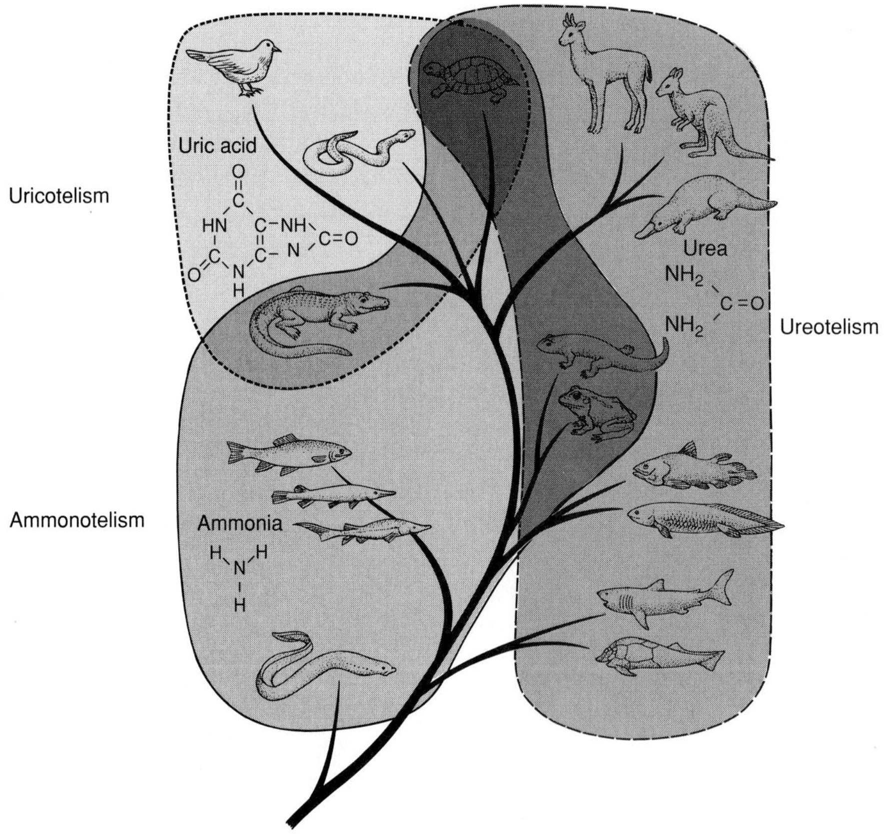


**Figure 14.11** Mechanisms of eliminating nitrogenous wastes. Among many fishes, amphibians, and some reptiles, nitrogen is excreted in the form of ammonia (ammonotelism). Excretion of nitrogen as uric acid (uricotelism) occurs in some reptiles and all birds. In mammals and some amphibians and fishes, nitrogen is eliminated as urea (ureotelism).

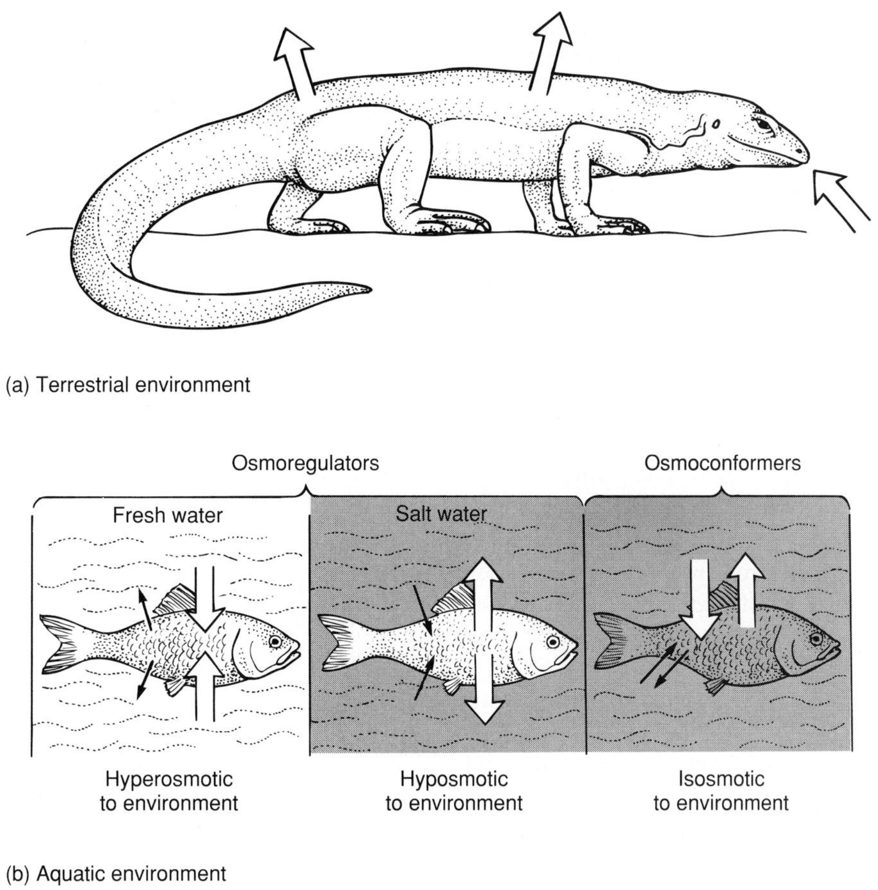


**Figure 14.12** Water balance. (a) In terrestrial vertebrates, the relatively dry surrounding environment tends to draw water from the body, posing the problem of dehydration. (b) In aquatic vertebrates, the tendency to gain, lose, or be in balance with the surrounding water depends on the relative concentration of solutes in the animal compared with those of the surrounding water. Osmoregulators control the concentrations of salt and water in their bodies. In fresh water, an animal is usually hyperosmotic to the medium, and the osmotic gradient leads to an influx of excess water. In salt water, most vertebrates are hyposmotic; therefore, water tends to flow from their bodies into the surrounding environment. As with terrestrial vertebrates, dehydration is the result. In both aquatic situations, the vertebrate must make physiological adjustments to eliminate or take up water in order to maintain homeostasis. In a third aquatic situation in which the level of solutes in body tissues rises to meet that of the surrounding salt water, no significant osmotic gradient develops. Such vertebrates are called osmoconformers because they are isosmotic to seawater and no net flux of water occurs. Open arrows represent net direction of water fluxes; solid arrows represent net direction of solute movement. Shading indicates relatively high concentration of solutes in water.

concentrated (hence hyper-) than the surrounding water. Because fresh water is relatively dilute and the body is relatively salty, water flows into the body (figure 14.12b). If allowed to continue, the net influx of water would substantially dilute body fluids and thus create an imbalance in the extracellular environment. For freshwater fishes, the major homeostatic problem is ridding the body of this excess water. To address this problem, the kidneys are designed to excrete large quantities of dilute urine, about ten times the volume excreted by their marine counterparts.

For most saltwater fishes, the osmotic problem is just the reverse. There is a tendency for a net *outward* flux of water from the body tissues, dehydrating them. Relative to salt water, the bodies of most marine fishes are **hyposmotic,** meaning that the body is osmotically less concentrated (hence hypo-) than seawater. Water tends to be drawn from the body, and dehydration of the body will result if this condition is not controlled physiologically. In this respect, a fish in salt water faces a physiological problem much like that of a tetrapod on land, loss of body water to the environment (figure 14.12a). For marine fishes, osmoregulation is complex. They can drink to recover water, but if they do, they must excrete the excess salt ingested along with the seawater. To aid in water conservation, the kidneys are designed to excrete very little water, thus reducing water loss. To address the problem of excess salt, the gills and sometimes special glands become partners with the kidneys in the business of osmoregulation.

The body of some animals is **isosmotic,** meaning that the osmotic concentrations of the internal environment and surrounding seawater are approximately equal (hence iso-). Because of this balance, there is no net tendency for water to move in or out of the body, so the animal faces no special problems from excess water or dehydration (figure 14.12b). Dissolved molecules and ions, known as **solutes,** in the body increase in concentration until the osmotic concentration in the body equals that of the surrounding seawater. Such an animal is called an **osmoconformer.** Among vertebrates, hagfishes are osmoconformers. Concentrations of sodium and other ions are close to those of the surrounding seawater. Elasmobranchs and coelacanths (*Latimeria*) also have tissue fluids osmotically close to seawater, but this is due to elevated levels of the organic compound urea circulating in the blood. As a result, the osmotic concentration of the blood approaches that of seawater. Although this reduces the physiological problems of dealing with water fluxes, it requires that cells of hagfishes, elasmobranchs, and coelacanths operate efficiently in a fluid environment that is higher in osmotic concentration than that of other vertebrates. It is believed that such elevated concentrations may incur energetic costs in osmoregulation.

Except for hagfishes, elasmobranchs, coelacanths, and some amphibians, all other vertebrates are **osmoregulators.** Despite fluctuations in external environmental osmotic levels, they maintain body fluids at constant osmotic levels through active physiological adjustments. Adjustments may involve conservation or elimination of body water to compensate for osmotically driven water loss or uptake relative to the external environment. Solutes also are regulated through excretion and uptake to maintain homeostasis of body fluids. Thus, osmoregulation involves adjustments of water *and* solutes. Next we meet structures designed to move both. Let us begin by looking at kidney adaptations that serve water balance. There are two problems that the environment imposes—water elimination and water conservation (figure 14.13).

*Water elimination.* Water elimination is a problem for hyperosmotic vertebrates living in fresh water. The vertebrate mechanism of urine formation seems especially well suited to address such a problem. The kidneys of most insects and some other invertebrate animals are **secretion kidneys.** Urine is formed by secretion of constituents into the tubules along their length. However, vertebrate kidneys, like the kidneys of most crustaceans, annelids, and molluscs, are **filtration kidneys.** Large quantities of fluid and solutes pass immediately from the glomerulus into the the renal capsule to form a **glomerular filtrate.** As this filtrate moves along the tubule, selective secretion adds constituents, but most of the initially filtered water and solutes are absorbed back into the capillaries entwined about the tubules. In humans, for example, each day the kidneys form about 170 liters (45 gallons) of glomerular filtrate in their 2 million renal capsules. This is four to five times the total volume of water in the body. If this volume were voided

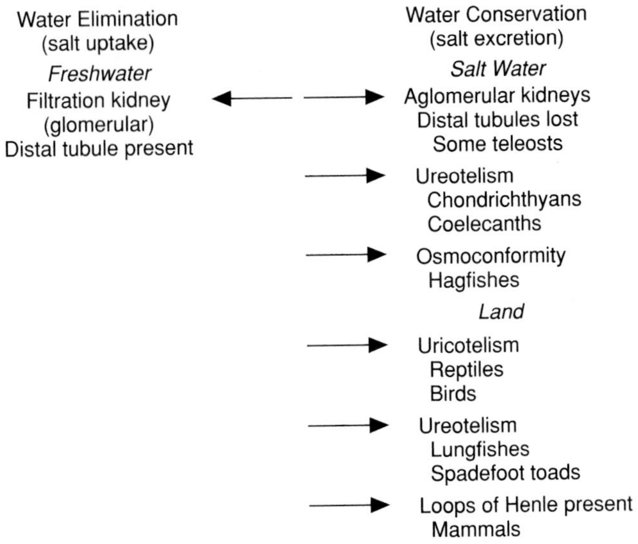


**Figure 14.13** Summary of kidney adaptations to two environmentally imposed homeostatic problems, water elimination and water conservation. In fresh water, most vertebrates need to eliminate excess water. A filtration kidney with fully developed glomerular apparatus and distal tubules can produce copious amounts of dilute urine and rid the body of excess water. In salt water and in terrestrial environments, water-conserving kidneys are advantageous. In marine fishes, aglomerular kidneys that lack distal tubules, ureotelism resulting in elevated levels of blood solutes, and osmoconformity represent three different adaptive routes to water conservation. Vertebrates in terrestrial environments conserve water through structural changes in the nephron (loop of Henle) that promote recovery of water or through more economical means of ridding the body of nitrogen, such as uricotelism or ureotelism, that require less water than ammonotelism.

each day, there would be little time for anything else, to say nothing of the large volumes of water we would need to drink to replace water that was excreted. In fact, all but about 1 liter of filtrate is reabsorbed back into the blood along the uriniferous tubules.

In freshwater fishes and aquatic amphibians, the kidneys characteristically have large, well-developed glomeruli. Consequently, relatively large volumes of glomerular filtrate are produced. The prominent distal tubule absorbs solutes (salts, amino acids, etc.) from the filtrate to retain these in the body, but it absorbs only a third to a half of the filtered water. In this instance, a large proportion of the water is eliminated in the urine. Thus, the kidney is designed to produce large amounts of dilute urine and address the main osmotic problem of excess water in freshwater vertebrates.

*Water conservation.* As emphasized, water conservation is a problem not just for terrestrial vertebrates facing a hot, drying environment but also for vertebrates in salt water. A variety of structural and physiological adaptations have arisen to address the problems of desiccation in salt water and in terrestrial environments.

The filtration kidney is disadvantageous for hyposmotic fishes in seawater because it is designed to form large volumes of urine. These fishes must conserve body water, not eliminate it. Consequently, in many species of marine teleosts, parts of the nephron that contribute to water loss are absent, specifically the glomerulus and the distal tubule. Absence of the glomerulus and associated renal capsule reduces the quantity of tubular fluid that initially forms. These marine teleosts have **aglomerular kidneys** that, by not producing copious amounts of glomerular filtrate, never face the problem of resorbing it later. Essentially, aglomerular kidneys conserve water by eliminating the filtration process in the renal capsule.

Loss of the distal tubule also contributes to water conservation. The distal tubule absorbs salt from urine but allows water to be excreted. Loss of the distal tubule therefore favors water retention by the fishes. Without glomeruli and distal tubules, these teleosts depend largely on selective secretion of solutes into the aglomerular tubules to form a concentrated urine.

Terrestrial vertebrates have alternative adaptations to conserve water. In mammals, and to a lesser extent in birds, water conservation is based on modification of the loop of Henle. The loop creates an environment around the tubules that encourages the absorption of water before it can be excreted from the body. Consequently, urine becomes concentrated, and kidney design serves water conservation.

In the mammalian kidney, the relationship between tubule design and water conservation is complex. The first step in urine formation is formation of glomerular filtrate. Circulating blood cells, fat droplets, and large plasma proteins do not flow into the nephron, but most water and solutes from the blood plasma pass from the capillaries of the glomerulus into the renal capsule. Second, most of the sodium ions, nutrients, and water are absorbed in the proximal tubule. Absorption is facilitated by the large surface area of proximal tubule cells and depends on active transport of sodium. Usable proteins that were part of the glomerular filtrate are also absorbed in the proximal tubule. Third, the filtrate enters the intermediate tubule of the loop of Henle. Contrary to earlier theories, the loop of Henle is not an additional site in which water is extracted from the filtrate. Instead, the loop actively pumps sodium ions from the filtrate out into the interstitial space to create hyperosmotic interstitial fluid around the collecting ducts. Fourth, as collecting ducts carry the modified filtrate to the renal pelvis, they pass through a region that, thanks to the loops of Henle, is hyperosmotic to the filtrate. The osmotic gradient between surrounding tissue fluid and dilute urine entering the collecting ducts provides the driving force that moves water out of the collecting ducts and into the surrounding fluid. When the body is dehydrated, the permeability of cells of the collecting duct changes under hormonal influence and water is drawn from the tubular fluid into the surrounding interstitial fluid. Here blood capillaries, collectively termed the **vasa recta,** take up water together with some solutes and return them to the circulation. The urine that remains in the collecting ducts thus becomes concentrated before it flows into the renal pelvis and ureter (figure 14.14).

Blood flow to the uriniferous tubules is necessary for filtration and reabsorption to occur. The glomeruli sprout from the renal arteries, which branch directly from the dorsal aorta. Blood pressure is still high in the renal arteries; therefore, blood pressure in the glomeruli is high and promotes the flow of fluid into the renal capsules. On the other hand, pressure in the vasa recta is low, as these vessels arise from the arterioles beyond the glomeruli and pressure drops as blood flows through the glomeruli. The lower pressure in the vasa recta encourages uptake of the water that collects around the loops of Henle.

Notice that unlike the water-conserving kidney of aglomerular teleosts, the distal tubule is retained in the water-conserving kidney of mammals. In mammals, part of the distal tubule is incorporated into the loop of Henle, where its ability to absorb salts contributes to the production of a hyperosmotic interstitial environment around the collecting ducts. Thus, in aglomerular teleosts, water conservation is accomplished by elimination of parts of the uriniferous tubule that allow water loss, whereas in mammals, those homologous parts of the uriniferous tubule are retained but become incorporated into a totally different mechanism of concentrating urine.

*Osmoconformers*. In a sense, one way to address the problem of water fluxes is to avoid the problem in the first place. This is the strategy of osmoconformers, whose body fluids have the same osmotic concentration as that of the surrounding medium. Isosmotically balanced with their environments, osmoconformers do not have to cope with the problems of water entry or loss. Osmoconformer vertebrates are all marine. In hagfishes, unlike the hyposmotic body fluids of most marine fishes, concentrations of $Na^+$ and $Cl^-$ in blood and extracellular fluid are elevated, so they are close to those of seawater. Hagfish tissues tolerate these relatively high levels of solutes. Because the hagfish is in osmotic equilibrium with its environment, the nephron does not need to secrete large volumes of urine. Consequently, the nephron is reduced to little more than a renal capsule connected to the archinephric duct by a short, thin-walled duct (figure 14.10a). Surprisingly, the renal corpuscles are quite large. Because water elimination is not a problem for the hagfish, the well-developed renal corpuscle probably functions in regulating divalent ions such as $Ca^{++}$ and $SO_4^{--}$.

Elasmobranchs and the coelacanth *Latimeria* are also approximate osmoconformers, but this is achieved via ureotelism. Urea accumulates at high concentrations in the blood and elevates the blood osmolarity up to that of seawater. For these fishes, large fluxes of water do not occur, and maintenance of water balance presents no special problem. Excess salts that enter the body fluids from seawater are

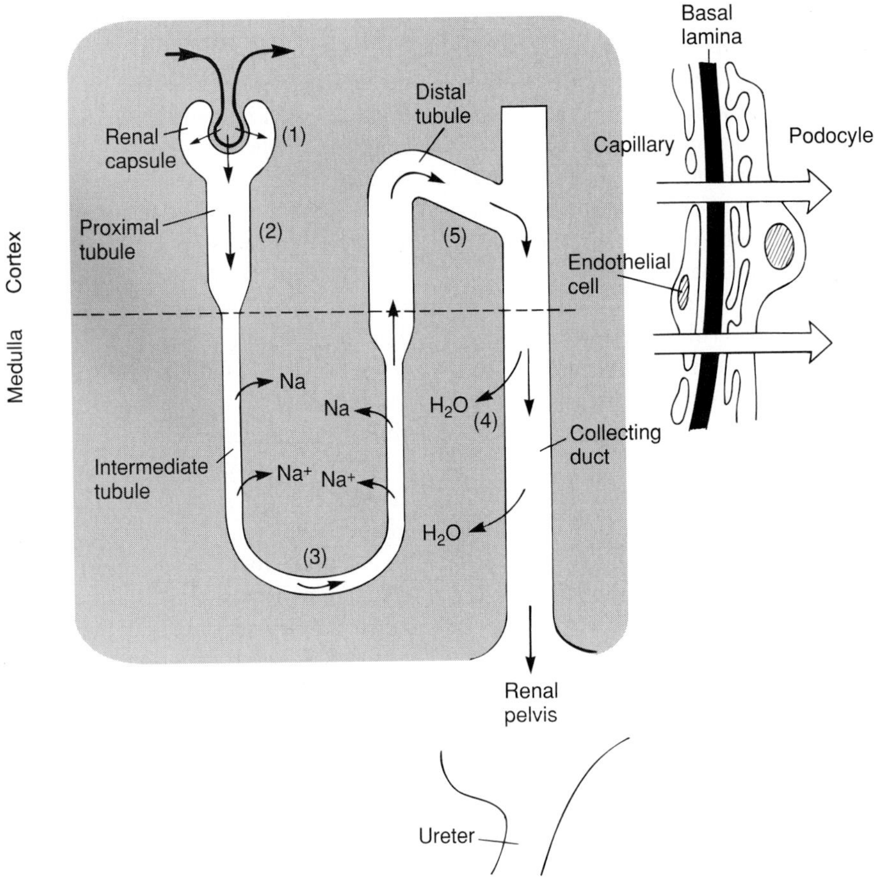


**Figure 14.14** Mammalian kidney function. At the start of urine formation, high pressure in the glomerulus encourages fluid in the blood to flow from the capillaries into the renal capsule, forming a glomerular filtrate. (1) As the glomerular filtrate passes through the rest of the nephron, some constituents are added, but most of the water is absorbed back into the capillaries. In mammals (and birds), this absorption occurs primarily in the proximal tubule (2) and collecting ducts (4). The intermediate tubule of the loop of Henle produces a salty environment (3) in the kidney medulla. As urine flows in the collecting duct through the medulla (4), it is carried through this hyperosmotic region, and water follows the osmotic gradient out of the tubule into the surrounding tissue. Blood vessels of the vasa recta (not shown) take up this water and return it to the systemic circulation. This produces a concentrated urine in the collecting ducts that is excreted from the kidney via the ureter (5). The insert (upper right) is an enlarged view of the renal corpuscle showing the endothelial wall of the glomerulus, the specialized endothelial cell of the renal capsule (podocyte), and the thick basal lamina between these endothelial layers. Arrows indicate the direction of flow of fluid from the blood into the renal capsule to form glomerular filtrate.

eliminated through special glands, such as the *rectal gland* of sharks, or through the gills.

*Tolerance of fluctuations*. Changes in kidney structure and osmoconformity are not the only ways of dealing with osmotic stress. Some aquatic vertebrates can tolerate wide variations in salinity. Those that are osmotically tolerant are **euryhaline** (eury-, wide; haline, salt) animals. Some euryhaline animals can move from marine to brackish water and even to fresh water. Other vertebrates that can withstand only a narrow range of environmental salinities are **stenohaline** (steno-, narrow; haline, salt) animals.

***Salt Balance*** Although we have focused on renal mechanisms that eliminate or conserve water, striking an osmotic balance involves moving salts as well as water. A variety of structures are devoted to the task of regulating salt balance. As mentioned, the distal tubule within the kidney recovers salts from the urine. Gills address ion imbalances by pumping salts out of (marine bony fishes) or into (freshwater fishes) the body. The rectal gland of elasmobranchs also collects, concentrates, and rids the body of salts (figure 14.15a).

Marine reptiles and birds that eat salty foods or drink seawater to replace lost fluids also ingest high levels of salt. Because their kidneys cannot handle this excess salt, it is excreted by special **salt glands.** In response to a salt load, salt glands intermittently produce a highly concentrated secretion containing $Na^+$ and $Cl^-$ primarily. In reptiles, these salt glands can be specialized nasal glands (in some marine lizards), orbital glands (in some marine turtles), sublingual glands (in sea snakes), or glands on the tongue's surface (in saltwater crocodiles and North American crocodiles).

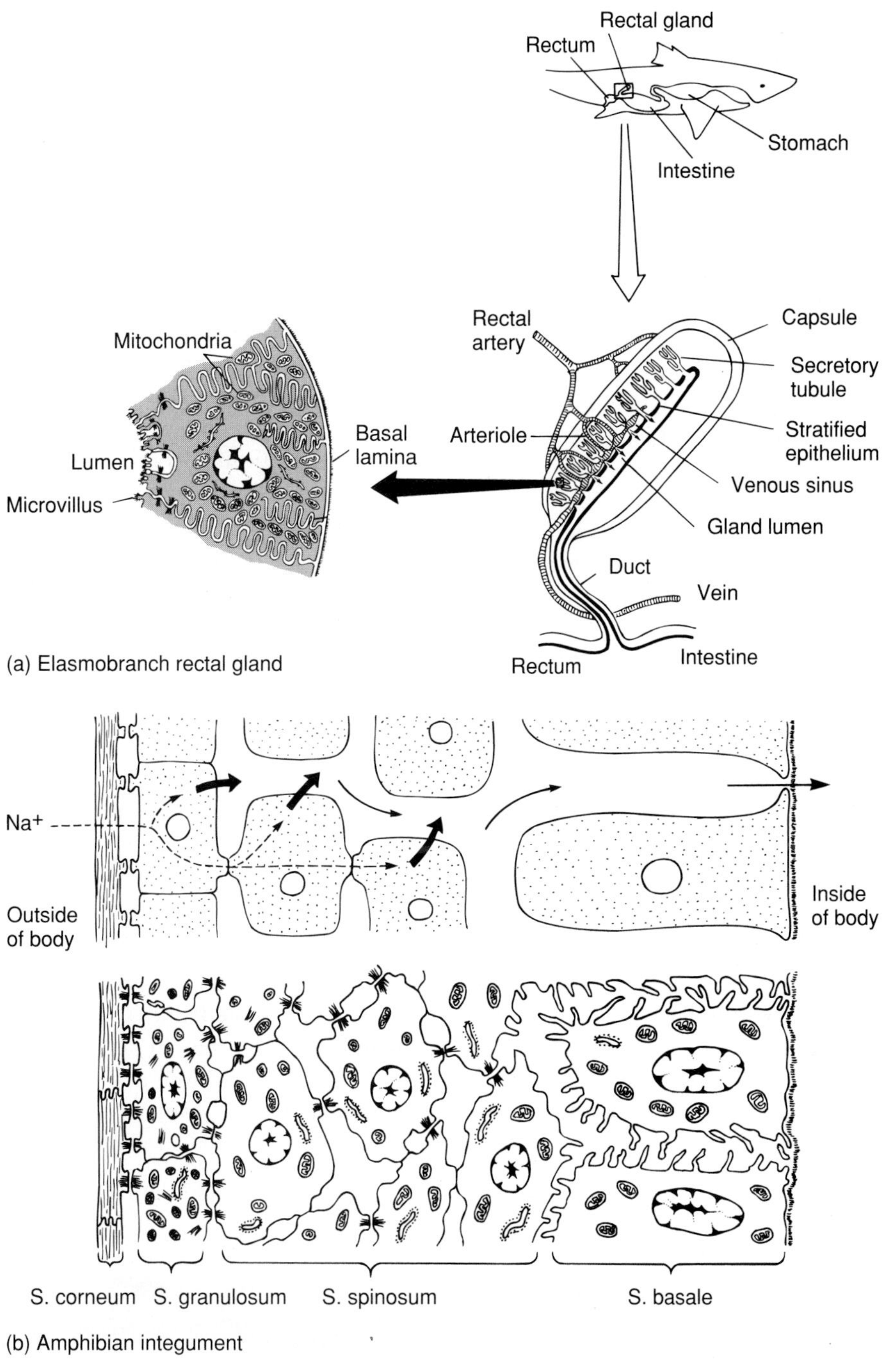


**Figure 14.15** Regulating salt levels. (a) Rectal glands of sharks and other elasmobranchs. These glands have evolved to eliminate salts efficiently from the body without the expenditure of large volumes of water. The outer capsule of the elasmobranch rectal gland consists of connective tissue and smooth muscle. Blood enters via the rectal artery, circulates around the secretory tubules, enters a venous sinus, and then flows into the renal vein. Salt collected by the secretory tubules passes into the lumen of the rectal gland and is then forced into the intestine to be eliminated with the feces. (b) Diagrammatic cross section of amphibian integument. Salt tends to diffuse from amphibians into fresh water. They have evolved the ability to take up replacement salts, especially sodium ions, through the skin by means of active transport. Sodium is taken up across the stratum granulosum and moved by active transport into the spaces between cells. Eventually it makes its way into capillaries within the dermis.

In marine birds, paired nasal salt glands are present. These large, specialized glands are usually located within shallow depressions on the dorsal surface of the skull and release their concentrated secretion into the nasal cavity. Marine mammals lack specialized salt glands. Their kidneys produce urine that is much more concentrated than seawater, so most salt is eliminated through the kidneys. Many terrestrial mammals have sweat glands in the integument primarily serving thermoregulation, but they also eliminate some salt.

In fresh water, the problem is totally different. Salt tends to be lost to the environment. Freshwater fishes absorb salts through their gills. In aquatic amphibians, the skin aids in the regulation of salt balance (figure 14.15b).

***Balancing Competing Demands*** The cloaca, urinary bladder, and large intestine also aid in the regulation of both salt and water balances. Managing salt and water balance must be compromised with other demands. We have already seen that the demands of nitrogen excretion must sometimes be balanced with the need for water conservation. Furthermore, amniotes often incur a heat load if they live in hot climates or lead active lives. Birds pant and mammals sweat to help dissipate heat through the process of evaporative cooling. Water also is lost in this process. Although reptiles lack sweat glands, they possess a thick water-resistant skin and exhibit only a modest panting mechanism, so they cannot regulate their body temperature through evaporative cooling. Instead, they move out of the sun (into the shade or into burrows) or become active at night. Behavioral thermoregulation and lower metabolic rates reduce evaporative water loss and contribute to water conservation in reptiles.

## Evolution

Vertebrate kidneys illustrate preadaptation, a theme we have seen within other systems. But the preadaptation of the urinary system raises an issue we have not addressed—the freshwater origins of vertebrates.

### *Preadaptation*

Excretion of urea or uric acid conserves water and is adaptive for life on land; however, conversion of ammonia into urea or uric acid probably arose well before vertebrates actually ventured onto land. In chondrichthyans and coelacanths, the formation of urea answers the problem of water balance by turning these fishes into osmoconformers. Detoxification of ammonia, by converting it to urea, allows lungfishes to address the immediate problem of surviving droughts. The amniote embryo, confined to a cleidoic egg, converts ammonia into uric acid so it can safely sequester nitrogenous wastes without requiring large quantities of water to flush them away. One or more of these conditions may have preceded life on land and been preadaptive. When vertebrates eventually ventured onto land, they entered an environment in which water was scarce, making water conservation especially important. But as such a transition occurred, the metabolic means of conserving water may already have been in place.

### *Origin of Vertebrates*

Homer Smith, a physiologist, was the first person to notice that the vertebrate kidneys appeared to be better suited to life in fresh water than in salt water. In fact, he argued that the kidneys were so well designed for fresh water that vertebrates must have evolved in fresh water and only later entered salt water. His reasoning went like this: The kid-

## Box Essay 14.2 Between Fresh and Salt Water

Most fishes are stenohaline; they can tolerate only a narrow range of salinities. A few fishes are euryhaline; they tolerate wide swings in salinity and may in fact migrate between fresh and salt water. **Anadromous** fishes hatch in fresh water, migrate to salt water where they mature, and then return to fresh water to spawn. Salmon are an example. Depending on species, anadromous fishes spend one to several years at sea feeding and growing, then return to their natal stream where they breed. **Catadromous** fishes migrate in the opposite direction, from salt to fresh water. European and American eels, *Anguilla,* are examples. They mature in streams and migrate to the ocean to breed.

Although euryhaline fishes pass part of their lives in fresh water and part in salt water, the transition from one to the other cannot be abrupt. A period of adjustment, usually involving several weeks in brackish water, is often required to allow acclimation. When these fishes swim into fresh water, the major physiological challenge is coping with salt loss across the gills. Marine stenohaline fishes placed in fresh water cannot compensate for the high permeability of their gills to salt. Salt continuously leaks out, and the fishes die. Euryhaline fishes develop reduced permeability to salt and survive.

## Box Essay 14.3 Philosophers to Finals

The kidney seems to bring out the philosopher in all of us. Homer W. Smith, who spent a lifetime in the study of kidney physiology, produced the reflective book, *Man and His Gods,* examining the effects of religious and secular myths on human thought and human destiny. No less than Albert Einstein found it an intensely interesting book and wrote the foreword. Isak Dinesen, perhaps today best known for her book turned into a movie, *Out of Africa,* similarly reflected upon the kidney. In a 1934 collection of writing, *Seven Gothic Tales,* her character, an Arab sailor on deck of his ship cruising off the African coast, philosophizes as follows,

> ***What is man when you come to think upon him, but a minutely set, ingenious machine for turning, with infinite artfulness, the red wine of Shiraz into urine?***

Urine comes from the Latin word, *urina,* entering into English usage about the fourteenth century. Before then the French word, "pissier" gave the English "piss," used comfortably by Geoffrey Chaucer (fourteenth century) and even by proper Elizabethan ladies and gentlemen. Not until Oliver Cromwell and puritanism (seventeenth century) did the term fall out of favor. Only recently has it enjoyed rediscovery and use once again in mixed company.

Urine has been put to a variety of household uses—as a hairdressing, as a fermentator of bread, to flavor cheese, and to macerate tobacco leaves. Wealthy French ladies of the seventeenth century could often be found in urine-enriched baths to beautify the skin. In various cultures it has been tried as a mouthwash and gargle. For centuries, it was considered proper and humane to wash battle wounds by urinating on the wounds of comrades (no more sterile nor antiseptic elixir was available). In the early nineteenth century, uroscopy or "water casting" was in great vogue in the medical profession throughout North America and Europe. This involved inspection of the "piss pot" as Elizabethans called it, or urinal, and these medical devices were often elaborately decorated with flowers in middle class households and by gold and silver in finer families. So prominent was nineteenth century uroscopy that the urinal became an emblem of the medical profession.

Students too have come to appreciate their kidneys. When loading up with coffee and great thoughts the night before a final exam, or when celebrating with drink and great excuses after the exam, we are reminded of our kidneys and the volumes of urine they can produce when required. Although evolved for their water-conserving abilities, our kidneys possess the physiological flexibility to rid our bodies of excess water when we overindulge. The collecting ducts become impermeable to the egress of water (ADH, a pituitary hormone, changes its permeability), less moves from collecting ducts into the interstitial space, less is available to be absorbed by the vasa recta, and more fluid is left behind to be excreted in copious amounts. Drinking establishments around the world serve different beers, wines, coffees, and soft drinks, but all have restrooms.

Such inspiration between physiology and philosophy has made public watering holes the site where we celebrate our kidneys. Perhaps it was just such a homage to the human kidney that prompted Samuel Johnson, himself a legendary raconteur and heavy user of his kidneys, to observe,

> ***There is nothing which has yet been contrived by man by which so much happiness is produced as by a good tavern or inn.***
>
> —*Life of Dr. Johnson,* James Boswell

neys of vertebrates are filtration kidneys that can produce large volumes of glomerular filtrate. Such a design would be a liability in marine environments in which water must be conserved, but it would be an asset in freshwater environments in which fishes must rid their bodies of influxes of excess water.

Marine invertebrates are osmoconformers. The levels of salts in their blood are close to those of seawater, making them isosmotic. They are in no danger of dehydration; however, this is not true for marine vertebrates. Compared with marine invertebrates, the levels of salt in the blood of marine vertebrates is almost two thirds lower. Consequently, vertebrates are hyposmotic to seawater and can become dehydrated. To make matters worse, vertebrates have a filtration kidney capable of producing large volumes of water, not conserving it.

Such disadvantageous features of marine vertebrates can be explained, Smith reasoned, if vertebrates originated in fresh water. If vertebrate ancestors lived in fresh water, evolution of filtration kidneys and low solute levels would be adaptive to cope with water influxes an animal experienced in such an environment. However, when these vertebrates later radiated from fresh to salt water, their filtration kidney was disadvantageous and modifications were required. In chondrichthyans and coelacanths, solute levels rose in the blood to address this problem. Other fishes developed adaptations, such as drinking seawater, that recovered water, and salt glands and gills, that eliminated excess salt along with loss of glomeruli and distal tubules. Smith felt that the fossil record available in 1931 also supported a freshwater origin for the earliest vertebrates.

Others have taken issue with Smith's hypothesis and favor instead a marine origin for vertebrates. First, the filtration kidneys of vertebrates are characteristically high-pressure kidneys that produce large volumes of glomerular filtrate. Large volumes of fluid moving from blood to kidney

## Box Essay 14.4 Water, Water, Everywhere, Nor Any Drop to Drink

Adrift on the ocean, seamen that have survived the loss of their ships face an irony. Exposed to heat, they dehydrate. They are surrounded by water, yet to drink it would only make matters worse. The reason is that seawater is hyperosmotic to body fluids. If a person drinks seawater, the salt is absorbed, and blood osmotic levels rise. But to flush the excess salt from the body, the kidney must spend as much or more water than was originally gulped in by the thirsty castaway. The net result is to make the body even more dehydrated. Furthermore, there is another problem. Seawater also contains magnesium sulfate, an ingredient used in laxatives. It stimulates diarrhea, and hence even more fluid is lost via the digestive tract.

Many marine animals address this problem differently. They drink seawater but excrete the excess salt by active transport in special salt glands rather than by flushing it through the kidneys with water. This allows them to use the seawater but not fall behind like humans in their water balance.

tubules give the kidneys a greater chance to act on the constituents within the circulating fluids of the body. A high-pressure system produces a high volume of filtrate, which aids in processing nitrogenous wastes. Thus, the filtration kidney could represent an efficient system for eliminating nitrogenous and other wastes by moving large volumes of filtrate through the kidney. Second, the filtration kidney is not unique to vertebrates. Crustaceans and many other invertebrates possess filtration kidneys, yet clearly they evolved from marine ancestors. Moreover, many are marine osmoconformers today. Finally, reexamination of early vertebrate fossil deposits suggests that they came from marine seas and not from freshwater habitats as Smith supposed. Contrary to Smith's views, the filtration kidney of the marine vertebrates was preadapted to fresh water, but it did not arise there.

In this debate, the hagfish poses a problem for everyone. Hagfishes are osmoconformers like most marine *in*vertebrates but unlike most vertebrates. They are members of the oldest surviving vertebrate group, the cyclostomes, and possess a filtration kidney, yet they live in salt water. If Smith were correct, then these primitive vertebrates would live in fresh water. They do not, of course. If the marine origin of vertebrates is to hold, then hagfishes should be osmoregulators like other vertebrates. Of course, they are not. Perhaps it is best to recognize that the hagfish, although a representative of the earliest group of vertebrates, is very ancient and may have diverged significantly in its physiology from an ancestral condition.

# Reproductive System

The reproductive system includes the gonads, their products, hormones and gametes, and the ducts that transport gametes. Reproductive hormones facilitate sexual behavior and parental care, prepare the reproductive ducts to receive the gametes, support the zygote, and perform other functions that we turn our attention to in the next chapter on the endocrine system. Now we look at the gametes and the ducts that provide a home for and convey gametes during reproduction. The placental mammal is used again to introduce the terminology applied to the reproductive system.

## Structure of the Mammalian Reproductive System

In mammals, each ovary consists of an outer connective tissue capsule, the **tunica albuginea,** that encloses a thick **cortex** and deeper **medulla.** The **ova,** or eggs, occupy the cortex and are wrapped in layers of **follicle cells** derived from connective tissue. An ovum plus its associated follicle cells is termed a **follicle.** Some follicles remain rudimentary, never change, and never release their ova. Others pass through a series of growth or **maturation** stages at the end of which the ovum and some of its clinging follicle cells are cast out of the ovary in the process of **ovulation,** and become ready for fertilization. If fertilization occurs, the ovum continues down the **oviduct** and becomes **implanted** in the wall of the prepared **uterus,** where subsequent growth of the embryo occurs. If fertilization does not occur, the undeveloped ovum continues down the oviduct and is flushed out of the uterus during the next menstruation (figure 14.16).

**Embryonic implantation (p. 181)**

Each mammalian testis also consists of an outer tunica albuginea, which encloses the **seminiferous tubules** that produce sperm. Within the walls of the seminiferous tubules, stem cells multiply and grow into sperm that eventually are released into the lumen. The coiled seminiferous tubules straighten, forming **tubuli recti** just before they join the **rete testis.** Via the **efferent ductules,** the rete testis joins the **epididymis** where sperm are temporarily stored. Upon ejaculation, sperm travel along the **vas deferens (ductus deferens)** into the urethra. Along the way, three accessory sex glands, the **seminal vesicle, prostate,** and **bulbourethral** (Cowper's) **gland,** respectively, add their secre-

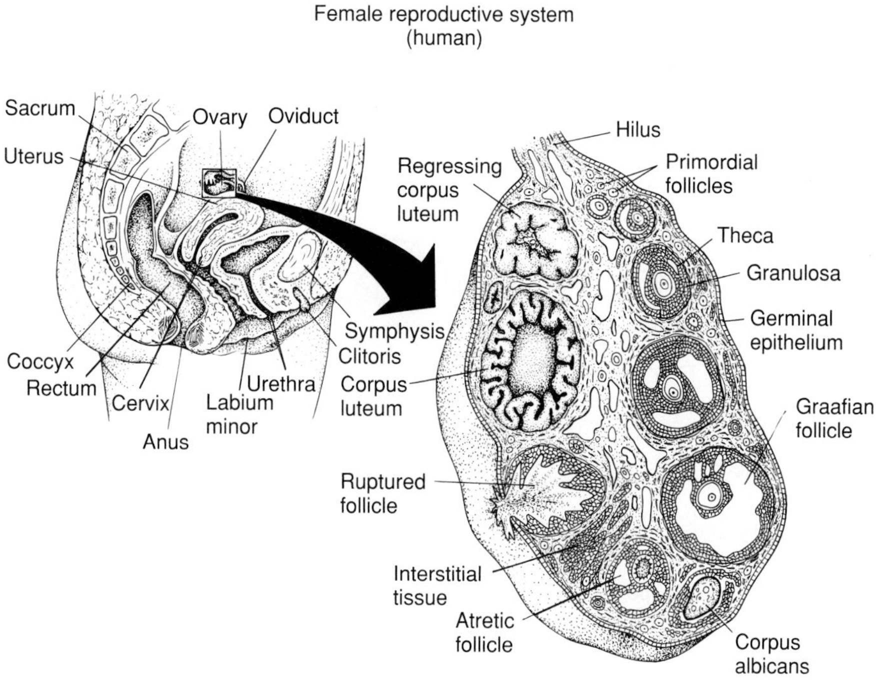


**Figure 14.16** Female reproductive system (human). This sagittal section of the female pelvis shows the reproductive organs and their relationships to the urinary and digestive systems. The ovary is enlarged and sectioned at the right. Summarized within the representative ovary are the successive stages in follicle maturation, beginning with the primordial follicles and then clockwise to the Graafian follicle and corpus luteum. Atretic follicles and other regressing stages are included.

tions as sperm move from the testes to the urethra. This fluid and the sperm it contains constitute **seminal fluid,** or **semen** (figure 14.17).

# Embryonic Development

## *Gonads and Gametes*

The paired gonads arise from the **genital ridge,** initially a thickening in the splanchnic mesoderm to which adjacent mesenchyme cells contribute (figure 14.18). The early gonad is little more than a swelling on the dorsal wall of the coelom with a thick outer cortex around a deeper medulla (figure 14.19a,b). Because the gonad shows neither unique male or female characters at this early stage, it is termed an **indifferent gonad.** The gonads of both sexes initially contain **germ cells,** the future sperm or eggs. Surprisingly, germ cells themselves do not arise in the genital ridge nor even in the adjacent mesoderm. In fact, they do not arise in the embryo at all. They first debut in remote sites outside the embryo in the extraembryonic endoderm. From the extraembryonic endoderm, they undergo a journey that takes them eventually to the indifferent gonad where they take up a permanent residence. In females, germ cells establish residence in the cortex. In males, arriving germ cells establish residence in the medulla, which develops into the seminiferous tubule (figure 14.19c,d).

## *Reproductive Tracts*

Parts of the embryonic urinary system are salvaged by or shared with the genital system. In female mammals, the mesonephric duct **(wolffian duct)** drains the embryonic mesonephros, but it regresses later in development when the metanephros and its ureter become the kidney of the adult. However, a second parallel **müllerian duct** arises next to the embryonic mesonephric duct before it regresses. The müllerian duct, rather than the wolffian, forms the oviduct, uterus, and vagina (figure 14.20). A few mesonephric tubules may persist as the **paroöphoron** and **epoöphoron.** In male mammals, the mesonephric duct becomes the vas deferens. Mesonephric tubules and some of the associated ducts contribute to the epididymis. A rudimentary müllerian duct occasionally arises in embryonic males but never assumes a significant role in the adult male (figure 14.20).

## *Overview*

The urogenital system of vertebrates certainly does not heed the Shakespearean warning, "Neither a borrower nor lender be." Parts evolved first to serve the kidneys (e.g., pronephric duct), later ended up serving the testis in males (e.g., vas deferens). In some species, a given duct is shared between the urinary and reproductive systems. In others, the same duct functions in only one of these systems. Even within the

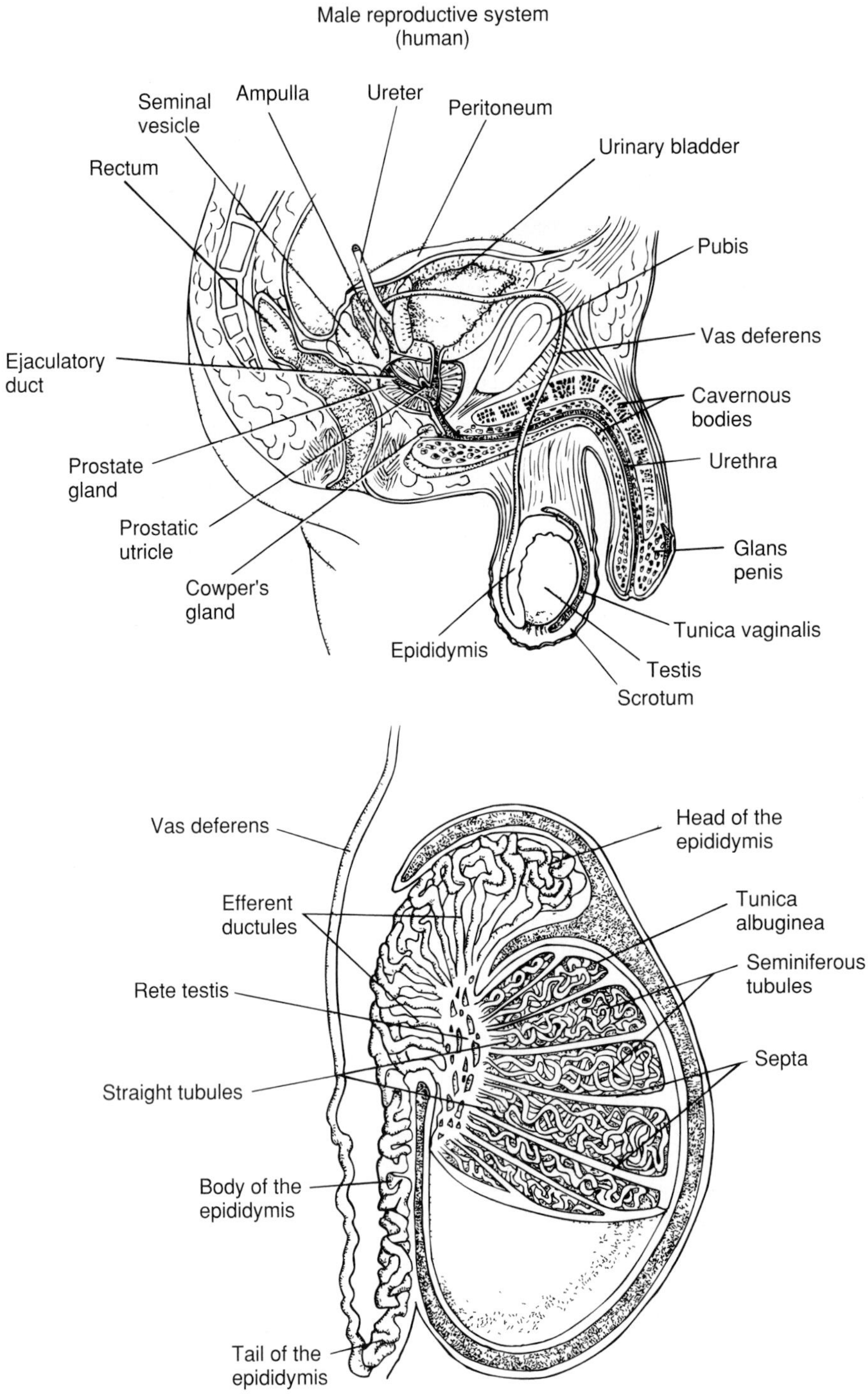


**Figure 14.17** Male reproductive system. This sagittal section of the male pelvis shows the reproductive organs and their relationships to the urinary and digestive systems. The enlarged and cutaway view of the testis and its duct system is shown at the bottom. Spermatozoa produced in the seminiferous tubules eventually pass through the straight tubules into the rete testis and enter the epididymis. Fluid is added as spermatozoa are moved through the vas deferens by contractions of sheets of smooth muscle in its walls.

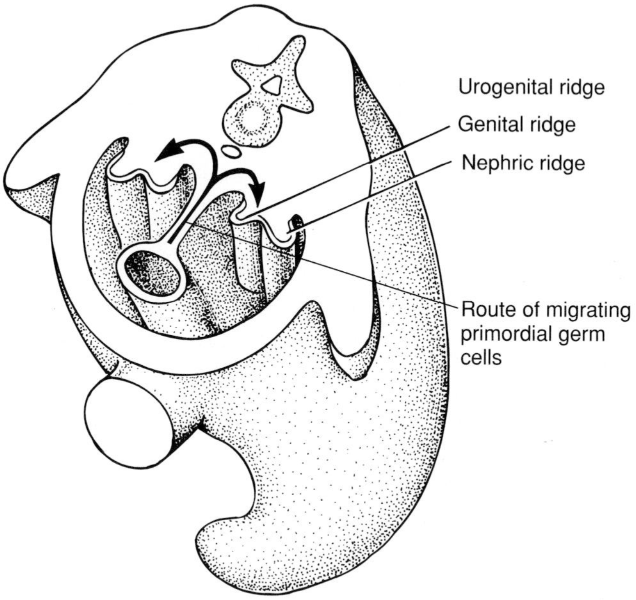


**Figure 14.18** Urogenital ridge. In the posterior part of the developing embryo, paired urogenital ridges arise within the roof of the coelom. The medial ridges are the genital ridges and give rise to the gonads. The lateral nephric ridges give rise to the kidney and its ducts. Primordial germ cells that develop into eggs or sperm arise outside the gonads, migrate to them, and colonize the early gonad rudiments.

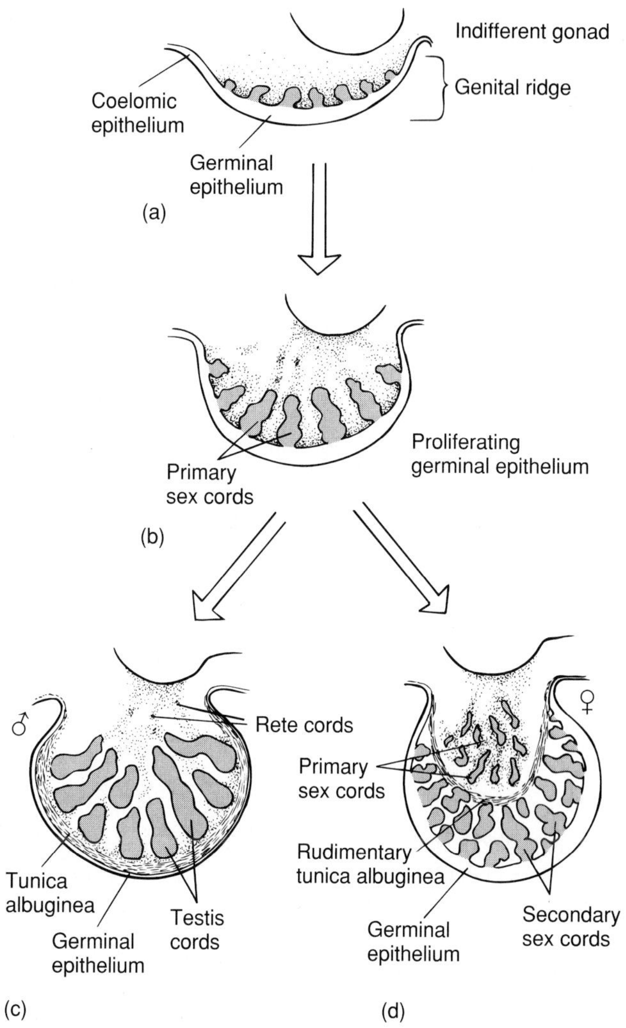


**Figure 14.19** Embryonic formation of the gonad. (a,b) Thickening of the genital ridge and inward movement of adjacent mesenchymal cells gives rise to a swelling, the genital ridge, from the roof of the coelom. Because this early developmental stage is similar in both sexes, it is referred to as the indifferent gonad that includes cortex and medulla. Primordial germ cells arriving from distant locations outside the embryo usually take up residence in the indifferent gonad. (c) In males, the medulla enlarges to become the testis cords that will form the seminiferous tubules. (d) In females, the cortex expands, forming secondary sex cords that house the follicles.

same species, homologous parts perform different roles in opposite sexes. Keeping track of these anatomical differences is no simple matter. A prolific terminology developed to track these anatomical and functional differences can obscure the underlying unity of the system. We select a set of terms applicable throughout the vertebrate urogenital system (noting synonyms) and apply it consistently (figure 14.21). When we examine phylogeny, we use terminology that applies to the homology of reproductive parts throughout vertebrates and place the more common or functional term for a given species and sex in parentheses.

As we saw earlier in our discussion of the kidney, the pronephric duct usually persists and drains the mesonephros or extended opisthonephros. It is renamed the mesonephric duct or the opisthonephric duct, respectively. In some males, this duct transports sperm and is called the vas deferens. In females, it is known embryologically as the wolffian duct. Because this duct plays different roles in different groups, the more general term **archinephric duct** is preferred. The **metanephric duct** is commonly called the ureter. In some males, the kidney divides its services between reproductive and excretory roles. To recognize this, it is common to speak of the **reproductive kidney** and the **uriniferous kidney.**

In females, the archinephric (mesonephric) ducts tend to function only within the urinary system. The müllerian duct arises embryologically next to the archinephric (wolffian) duct. In males, the müllerian duct regresses if it appears at all, but in females, the müllerian ducts become the oviducts of the reproductive system. Released ova enter the oviduct through the **ostium,** which typically flares into a **funnel** (infundibulum) in many vertebrates. The fringed margins of the ostium are the **fimbria** that embrace the ovary. The ovary and ostium are sometimes enclosed in a common peritoneal sac, but usually the oviducts are not connected to the ovaries directly. Instead, the ciliated fimbria and infundibulum gather released ova and move them into the oviduct. Fertilization, if it is internal, usually

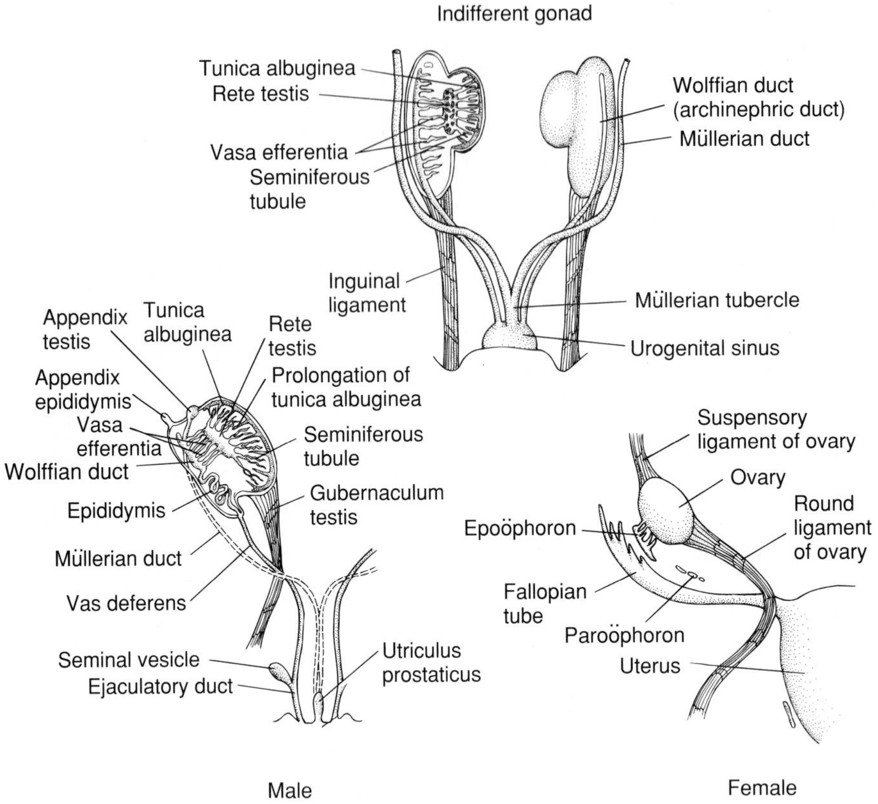


**Figure 14.20** Embryonic formation of the male and female genital system in mammals.

Ducts of the urogenital system

| *General term* | *Alternative term* |
|---|---|
| Archinephric duct | Pronephric duct/mesonephric duct<br>Wolffian duct<br>Opisthonephric duct<br>Ductus deferens<br>(vas deferens) |
| Müllerian duct | Oviduct |
| Metanephric duct | Ureter |

**Figure 14.21** Terminology of the urogenital system. Associations of ducts change during evolution and development. Sometimes the same duct performs different roles in males and females. The result has been a proliferation of synonyms, which are summarized in this figure. The duct serving the early pronephros is the pronephric duct, but when the mesonephros replaces the pronephric kidney, this duct now serves the new mesonephros and is called the mesonephric duct. With the advent of the metanephros, this duct degenerates in the female but becomes the vas deferens of the testis in the male. Some authorities prefer the term archinephric duct or wolffian duct for this structure. Although the term *metanephric* duct might parallel the terms *pronephric* and *mesonephric* ducts, more often the term *ureter* is used for the metanephric duct.

occurs soon after the ovum enters the oviduct. Shortly before their terminus, the oviducts may expand into the uterus, the organ in which an embryo is housed and nourished. If the fertilized egg is wrapped in a shell, **shell glands** or **shell-secreting regions** may be evident in the oviduct.

# Female Reproductive System

## Ovary

The ovary produces both hormones and mature ova (sing., ovum). **Oogenesis** is the process of egg maturation, which occurs from the time of its appearance in the ovary until it completes meiosis. Oogenesis is a complex process involving **mitotic** as well as **meiotic** cell division, growth in egg size, and changes in cytoplasmic composition (figure 14.22). Once germ cells take up residence in the ovary, they are called **oogonia.** The diploid oogonia undergo mitotic division, yielding diploid cells. At the end of this phase of development, they are **primary oocytes.** Primary oocytes then begin meiotic cell division for the first time. As a result of the first meiotic division, each ovum yields a first **polar body** and a **secondary oocyte.** Although the first polar body may divide again, its role in helping to reduce the chromosome number is complete; thereafter, it is of little importance.

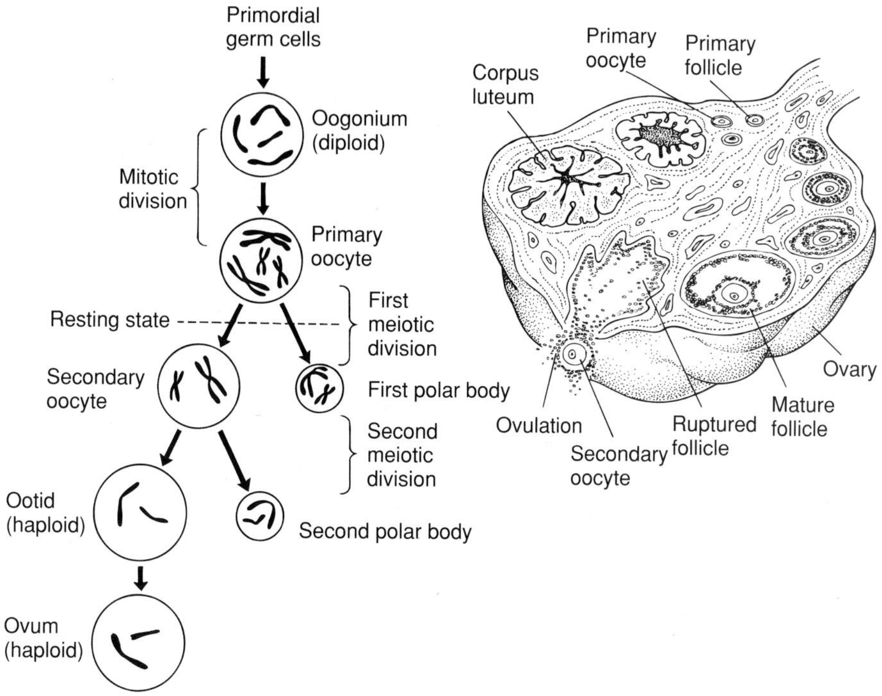


**Figure 14.22** Oogenesis. Diploid primordial germ cells colonize the ovary of the embryonic female. When they arrive in the ovary, these germ cells are called primary oocytes. They gather around themselves a layer of connective tissue cells to form an ovarian follicle. Most oocytes begin meiosis but do not complete this process until ovulation or later, depending upon the species. Of the hundreds or thousands of oocytes residing in follicles within the ovary, only a few will ever mature, be released at ovulation (after which they are usually termed ova), and become fertilized.

The secondary oocyte undergoes a second meiotic division, yielding a second polar body and a haploid ovum.

A capsule of supportive connective tissue cells termed follicle cells form around the primary oocyte. Follicle cells and the oocyte they embrace form an **ovarian follicle.** The follicle cells contribute to nutritional support and help build up yolk within the ovum. During the breeding season, selected follicles and the oocytes they contain resume maturation under hormonal stimulation. As meiosis is completed, a **secondary oocyte** is formed. Release of the oocyte from the ovary is termed ovulation.

Much variation occurs during the time before meiosis takes place. These events of oogenesis may occur largely before or after sexual maturity, depending on the species. At the birth of a mammalian female, the primordial germ cells have already migrated into the ovary and started to undergo meiosis, but further oogenesis is usually arrested until the onset of sexual maturity. In fact, not all the primary oocytes mature. For example, the human female is born with half a million primary oocytes in her ovaries, but perhaps only several hundred of these complete oogenesis. The rest eventually degenerate. In some mammalian species, meiosis occurs before ovulation. In other species, it does not occur until after fertilization.

The ovary is suspended from the dorsal wall of the coelom by a mesentery, the **mesovarium** (figure 14.23). Except for cyclostomes, in which eggs escape through secondary pores in the body wall, vertebrate eggs travel through genital ducts after they are released from the ovaries. In most vertebrates, the ovaries are paired; however, in cyclostomes, some reptiles, most birds, the duckbill platypus, and some bats, only a single ovary is functional (table 14.1).

**Ovipary, vivipary (p. 153)**

## *Genital Ducts*

***Fishes*** In cyclostomes, the single large ovary is suspended from the middorsal wall. In lampreys, as many as 24,000 to over 200,000 ovarian follicles can develop in synchrony, and are ovulated during a single breeding season. Most lampreys spawn once and die shortly thereafter. Few follicles are present in hagfishes and little is known about their reproductive behavior. Cyclostome ovaries lack ducts. Instead, eggs are shed into the coelom. From the coelom, they reach the cloaca (in lampreys) or anus (in hagfishes) through secondary pores. The archinephric ducts drain the kidneys exclusively.

In elasmobranchs, the ovaries are initially paired, but in some species, only one may develop. The müllerian duct, or oviduct, differentiates into four regions—funnel, shell

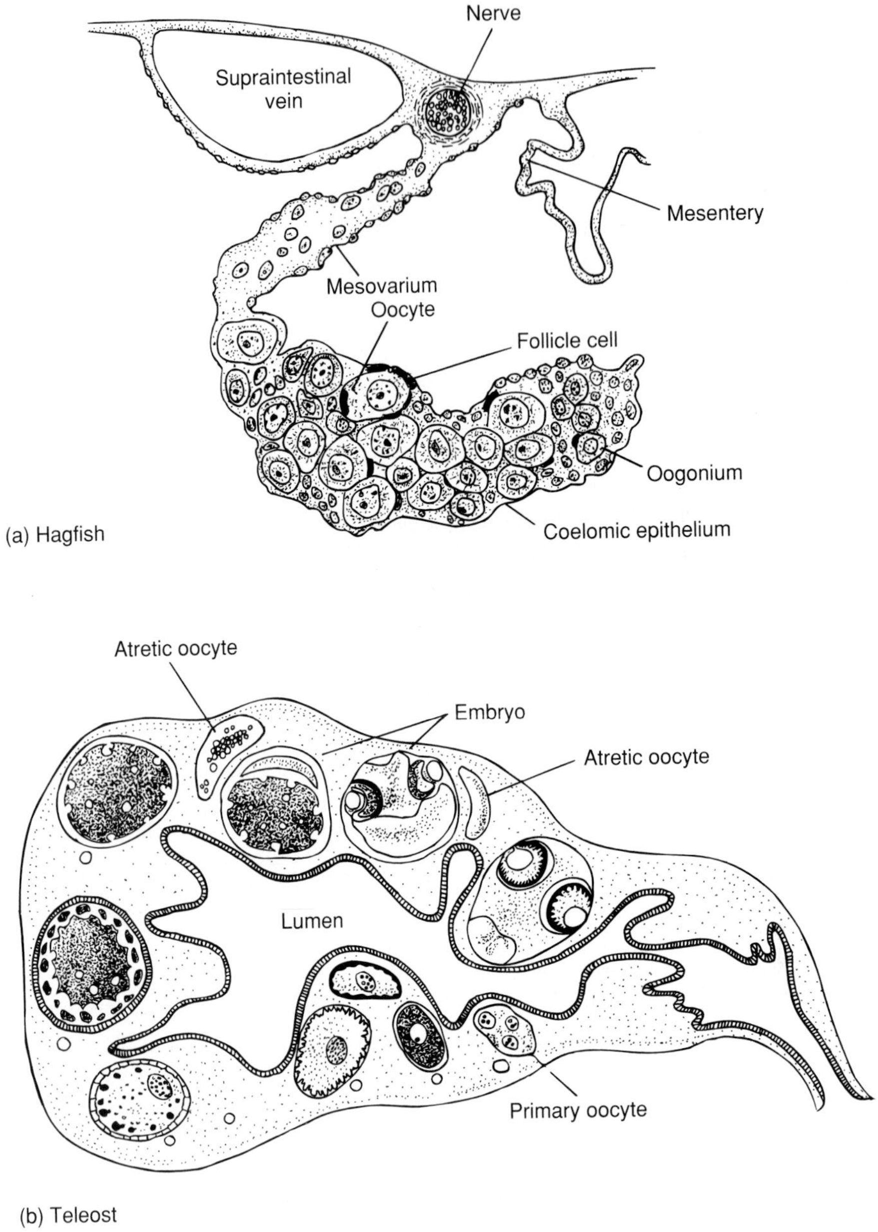


**Figure 14.23** Fish ovaries. (a) Hagfish. Oocytes and surrounding follicle cells are held within the ovary. (b) Teleost. Section of the ovary of the guppy *Poecilia reticulata*. Ova are fertilized while they are within the ovary, and they are retained well into embryonic development. There can be one to seven oocytes in progressive stages of development. Atretic oocytes that fail to develop and developing embryos are illustrated.

gland, **isthmus,** and uterus (figure 14.24a). The funnel collects the eggs shed from the ovary. Anterior ends of the paired oviduct may fuse into a single funnel, or asymmetric development may leave only one primary funnel. In some species, the shell gland (nidamental gland) stores sperm, but in most elasmobranchs, it secretes albumen and mucus. In oviparous species, the shell gland produces the egg case as well. In viviparous species, in particular, the shell gland may be indistinguishable. The isthmus connects the shell gland to the uterus. The uterus nutritionally supports embryos if they are held in the oviduct for an extended period. Oviducts may join before they enter the cloaca, or they may enter separately. The genital ducts of chimaeras are similar to those of sharks except the oviducts always share a common funnel and each oviduct opens separately into the cloaca. The archinephric duct drains the female opisthonephric kidney.

In female bony fishes, like most other anamniote females, the archinephric ducts serve the kidneys, and the paired oviducts (müllerian ducts) serve the paired ovaries

**Table 14.1 Vertebrates with One Functional Ovary**

| Species | Explanation for One-Ovary Condition |
|---|---|
| Agnatha | |
| Lampreys | Fusion of two gonads |
| Hagfishes | One gonad fails to develop |
| Osteichthyes | |
| Perches, *Perca* | Fusion of two gonads |
| Pike perch, *Lucia-Stizostedion* sp. | Fusion of two gonads |
| Stone loach, *Noemacheilus* sp. | Fusion of two gonads |
| European bitterling, *Rhodeus ararus* | Fusion of two gonads |
| Japanese ricefish, *Oryzias latipes* | One gonad fails to develop |
| Guppy, *Poecilia reticulata* | One gonad fails to develop |
| Chondrichthyes | |
| Sharks | |
| *Scyliorhinus* | Left ovary becomes atrophic |
| *Pristiophorus* | Left ovary becomes atrophic |
| *Carcharhinus* | Left ovary becomes atrophic |
| *Galeus* | Left ovary becomes atrophic |
| *Mustelus* | Left ovary becomes atrophic |
| *Sphyrna* | Left ovary becomes atrophic |
| Rays | |
| *Urolophus* | Left ovary functional |
| *Dasyatis* | Right ovary absent |
| Reptilia | |
| Blind worm snakes, *Typhlops* | Left ovary and oviduct absent |
| Aves | |
| Birds | Left ovary functional in most species; right ovary regresses in embryos |
| Mammalia | |
| Duckbill platypus, *Ornithorhynchus anatinus* | Left ovary functional |
| Bats | |
| *Miniopterus natalensis* | Left ovary functional |
| *Miniopterus schreibersi* | Right ovary functional |
| *Rhinolophus* | Right ovary functional |
| *Tadarida cyanocephala* | Right ovary functional |
| *Molossus ater* | Right ovary functional |
| Mountain viscacha, *Lagidium peruanum* | Right ovary functional |
| Water buck, *Kobus defassa* | Left ovary functional |

(figures 14.24b–d and 14.25a–c). In some teleosts, such as salmonids, eggs released from the ovaries fill the body cavity. Eventually they reach short funnellike remnants of the oviducts situated at the posterior part of the coelom. However, in many teleosts, the oviducts regress entirely, leaving egg transport to new **ovarian ducts** (figure 14.26a–c). These ovarian ducts are not homologous to the oviducts (müllerian ducts) of other vertebrates. Instead, they are derived from peritoneal folds that embrace each ovary and have grown posteriorly to form new ducts.

Most teleost fishes lay eggs, but some bear live young. Among these viviparous teleosts, maternal tissues may nourish the embryo. One extreme case is found in the family of teleosts that includes the guppy. In this group, fertilization occurs while the ova are still in the ovarian follicles. The ovary continues to hold the embryos during subsequent development until they are released as tiny fry. Oocytes that fail to reach a point in maturation where they can be fertilized usually undergo involution and are called **atretic oocytes** (figure 14.23b). Recycling of atretic tissue provides nutrition for the surviving oocytes.

***Tetrapods*** Amphibian ovaries are paired, hollow structures that usually show a prominent cortex covered by germinal epithelium. The genital ducts of female amphibians are usually simple and consistent. The archinephric ducts serve the opisthonephric kidneys, the oviducts (müllerian ducts) serve the ovaries.

In amniotes, remnants of the mesonephros may persist in larval stages, but adults have metanephric kidneys drained exclusively by new paired ducts, the ureters (metanephric ducts). In females, the archinephric ducts are rudimentary. The oviducts (müllerian ducts) persist in their roles of transporting ova from the ovaries and supporting the embryo while it is in transit. The tubular oviducts (müllerian ducts) of amniotes often have prominent sheets of smooth muscle within their walls and a lumen lined by a secretory mucosa. In oviparous amniotes, a shell gland may be prominent; in viviparous amniotes the uterus may be distinct (figures 14.27a–c and 14.28a–d).

## *Oviduct*

After ovulation, the fimbria move the ovum into the oviduct. If fertilization is internal, the ovum and sperm meet almost immediately in the upper reaches of the oviduct. If fertilization is external, the smooth muscle and cilia lining the oviduct drive the ovum to the outside where it is fertilized.

In addition to transporting the ovum, the oviduct in some vertebrates may add layers of membrane or a shell. In many species, parts of the oviduct are specialized as distinct shell glands that add these coats. Because membranes and shells are impervious to sperm, they are added after fertilization. In birds and egg-laying reptiles, a layer of albumen, and then a shell membrane, and finally a calcareous outer shell are added as the fertilized ovum slides along the oviduct (figure 14.29). The encapsulated egg is then held within the oviduct until a suitable environmental site in which to lay it is prepared.

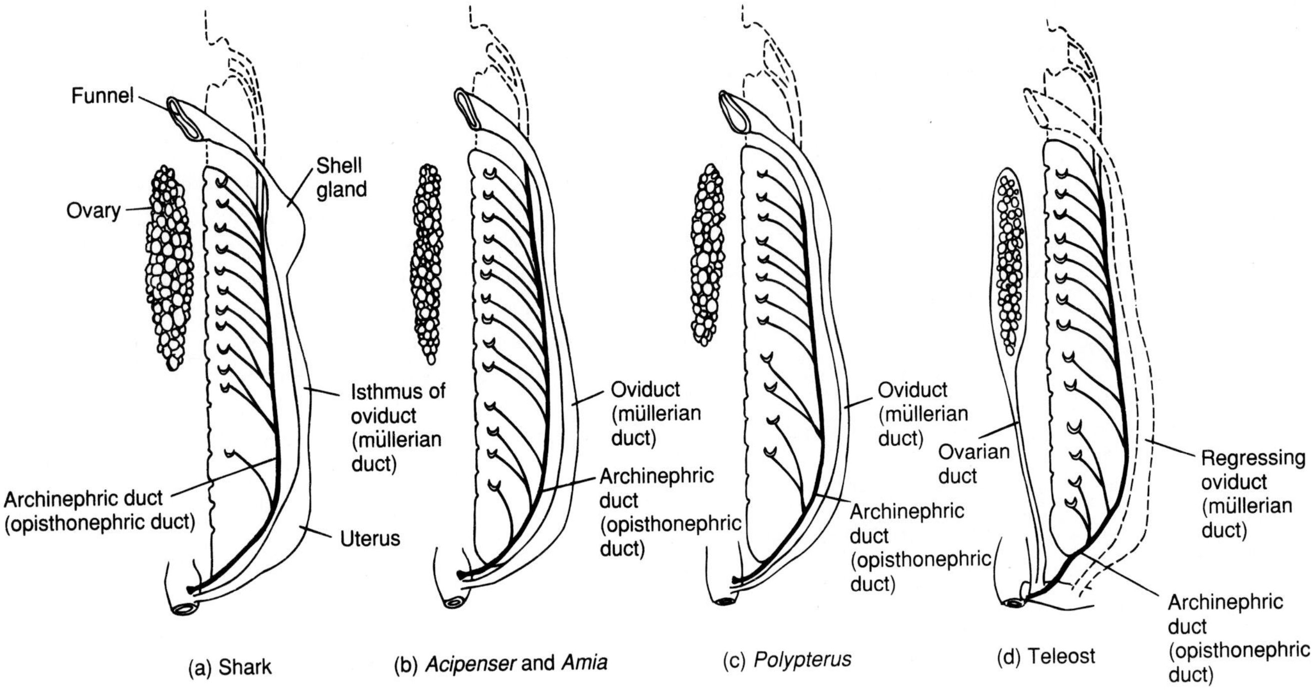


**Figure 14.24** Oviducts of female fishes. (a) Shark. (b) Sturgeon and bowfin. (c) Bichir. (d) Teleost. The oviduct (müllerian duct) arises adjacent to and parallel with the archinephric duct in most fishes. In teleosts, the oviduct is usually replaced by an ovarian duct that is derived separately.

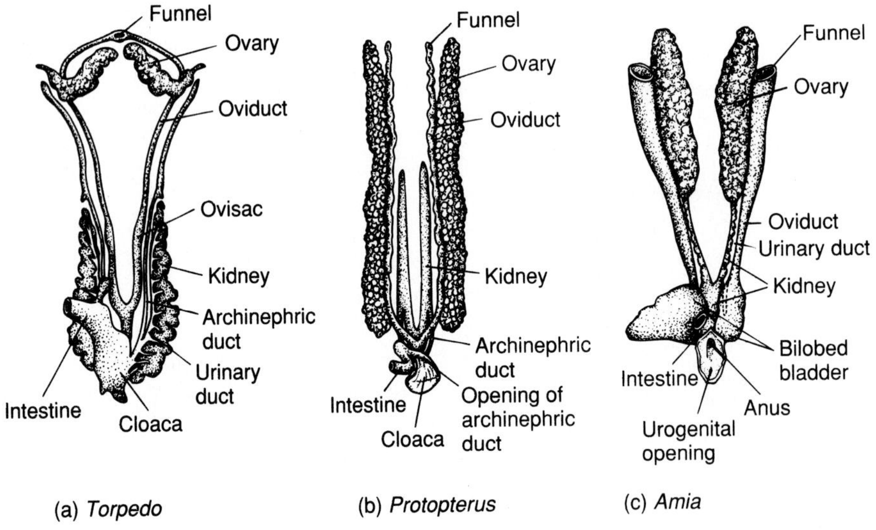


**Figure 14.25** Urogenital systems of female fishes. (a) Ray *Torpedo*. (b) Lungfish *Protopterus*. (c) Bowfin *Amia*.

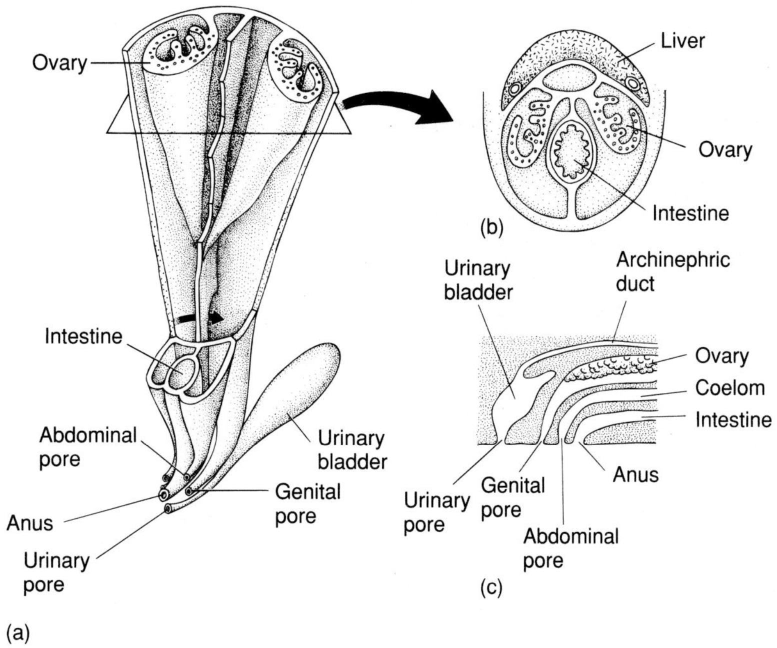


**Figure 14.26** Urogenital system of a teleost female. (a) Ventral view, partially cut away, of the urogenital system in a generalized teleost fish. Ovaries are suspended from the dorsal wall, and release ova into the genital funnels formed from folds in the peritoneal wall. The coelom connects with the outside through abdominal pores. Feces are eliminated via the anus, and urine via the urinary pore of the bladder. (b) Cross section at the level of the ovaries. (c) Sagittal section.

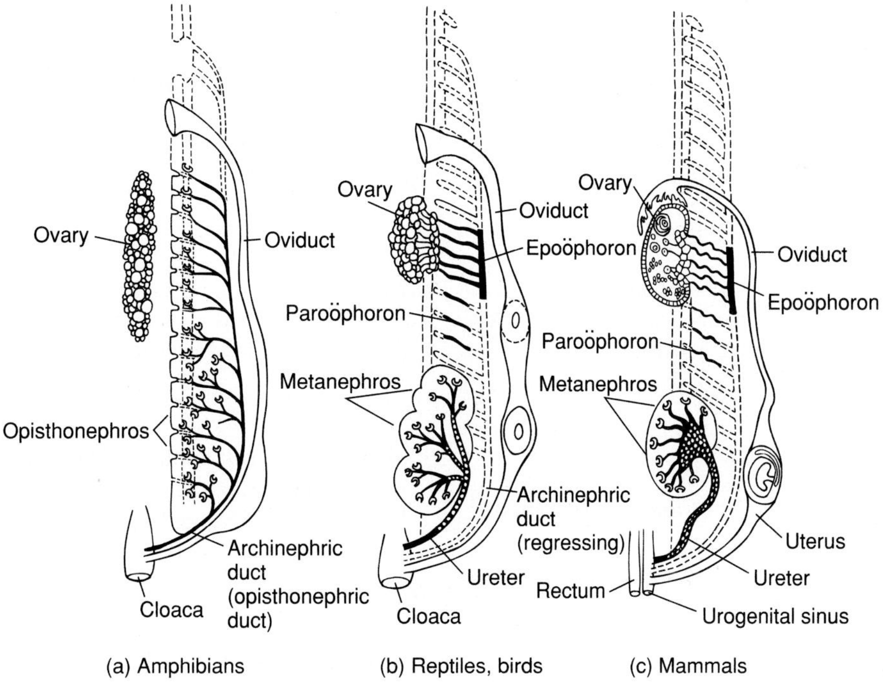


**Figure 14.27** Urogenital anatomy of tetrapod females. (a) Amphibians. (b) Reptiles and birds. (c) Mammals.

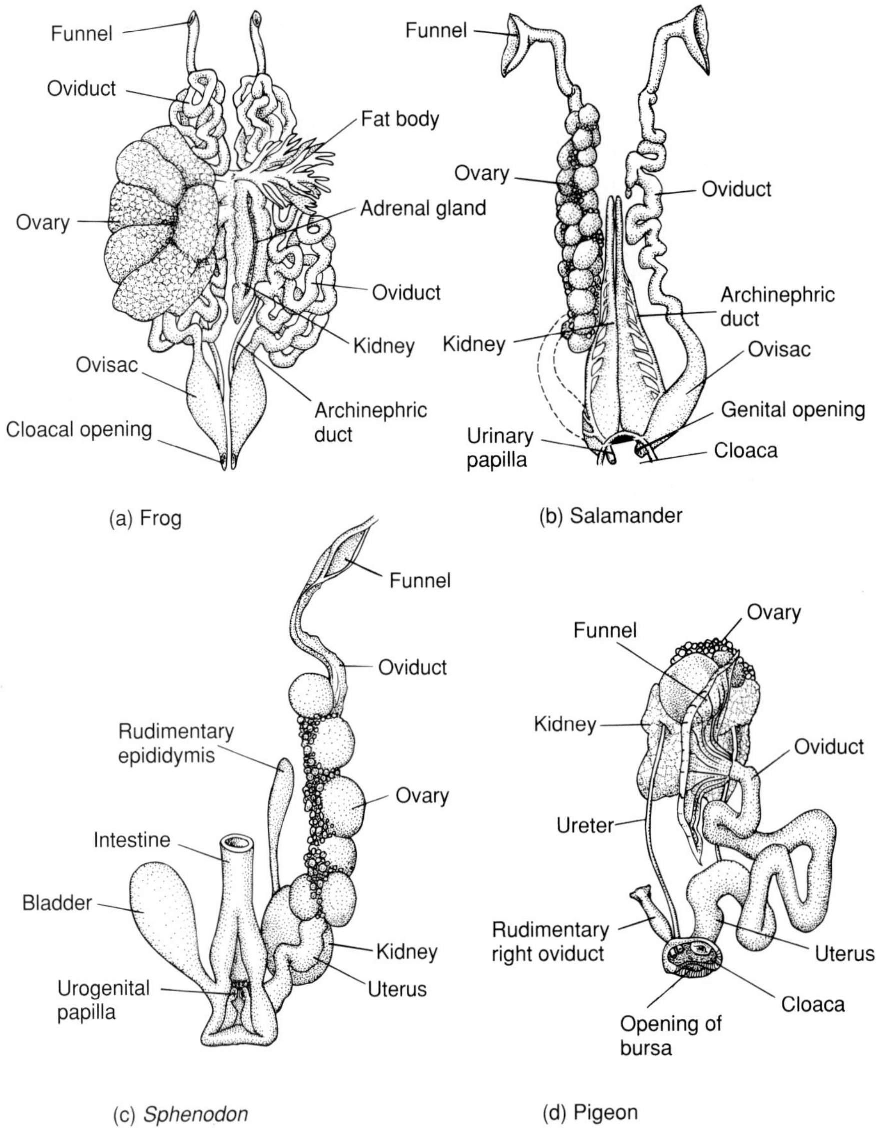


**Figure 14.28** Urogenital systems of amphibian females, ventral views. (a) Frog *Rana*. The intestine, urinary bladder, and left ovary have been removed to reveal underlying structures. Urinary ducts of the right side are pulled away from the kidney to show their course. (b) Salamander *Salamandra*. (c) Reptile *Sphenodon*. (d) Bird *Columba*.

### *Uterus*

The uterus is the terminal portion of the oviduct. Shelled eggs waiting to be laid or embryos completing their development are held within the uterus. In placental mammals and a few other vertebrates, the walls of the uterus and extraembryonic membranes of the embryo establish a close vascular association through a **placenta.** Nutrients and oxygen are transported to the developing embryo and carbon dioxide is given up to the maternal circulation via the placenta.

**Placentae (p. 181)**

In higher mammals, the terminal ends of the oviduct tend to merge into a single uterus and **vagina** that lie along the body's midline. The vagina receives the male penis or intromittent organ during copulation. The female homologue of the male penis is the **clitoris.** Unlike the penis, the clitoris participates neither in sperm transfer nor in urination.

## Male Reproductive System

### *Testis*

Except in cyclostomes and some teleosts, testes are paired and each is suspended from the dorsal wall of the coelom by a mesentery, the **mesorchium.** The testes of vertebrates have two functions—sperm production and hormonal se-

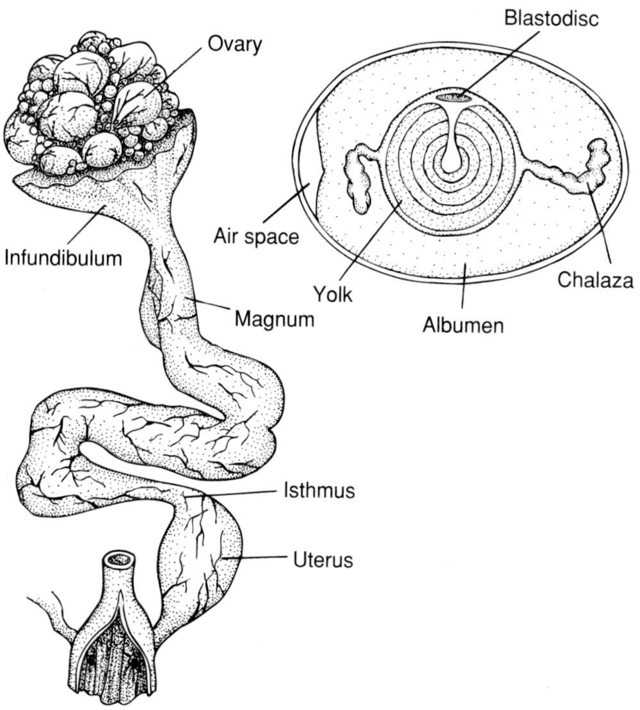


**Figure 14.29** Oviduct of a hen. After ovulation, ova are gathered by the infundibulum of the oviduct. If egg membranes are to be added, fertilization occurs here in the upper reaches of the oviduct. The oviduct adds a coat of albumen, a shell membrane, and eventually a calcareous shell.

cretion. The hormones of the testes are steroids collectively called **androgens.** The principal androgen is **testosterone,** secreted primarily by the **interstitial cells** (Leydig cells) of the testes. Testosterone controls the development and maintenance of secondary sexual characteristics, elevates the sex impulse (or libido), and helps maintain the genital ducts and accessory sex organs. More will be said about the endocrine role of the testes in the next chapter.

During the breeding season, primordial germ cells in the testes begin the process termed **spermatogenesis,** whereby selected germ cells eventually become spermatozoa. Spermatogenesis (like oogenesis) involves both mitotic and meiotic divisions, as well as cytoplasmic reorganization (figure 14.30). In vertebrates, there are two general patterns of spermatogenesis, one in anamniotes and the other in amniotes.

***Amniotes*** In reptiles, birds, and mammals, sperm form within the luminal wall of the seminiferous tubules, which lack subcompartments. Resident primordial germ cells, more commonly called **spermatogonia** at this stage, divide by mitosis. One member of the resulting pair of cells stays within the wall of the seminiferous tubule to propagate further spermatogonia, while the other grows in size. At the end of this growth, the diploid spermatogonium is called a **primary spermatocyte** and begins meiotic division. During meiosis, it briefly becomes a **secondary spermatocyte** and then a haploid **spermatid;** thereafter, it undergoes no further division. However, spermatids undergo cellular reorganization in which nuclear DNA condenses and excess cytoplasm and organelles are jettisoned to form sleek **spermatozoa,** or sperm.

For a time, **Sertoli cells** embrace and nutritionally support spermatids, perhaps promoting further maturation. Most sperm are stored in the lumina of the seminiferous tubules and in the connecting epididymis. At orgasm, sheets of smooth muscle in the walls of the ducts rhythmically contract, forcibly expelling sperm in the process of **ejaculation.** Sperm are transported in a thick, composite fluid secreted by accessory sex glands. In mammals, there are three such glands. The bulbourethral gland discharges mucus during erection and ejaculation. The prostate gland secretes an alkaline substance during ejaculation to protect the sperm from the acidity of any urine remaining in the male urethra. Finally, the seminal vesicle adds a thick secretion rich in the sugar fructose as a source of nutritional support for the sperm.

***Anamniotes*** In fishes and amphibians, sperm are produced in clones, each located within a cyst or follicle, all of which are housed in separate tubular compartments within the testes (figure 14.31a,b). Generally, a spermatogonium is engulfed by one or several connective tissue cells, called (as in females) follicle cells, that become functional Sertoli cells as maturation proceeds. Proliferation of a spermatogonium within the follicle (Sertoli) cells produces a nested clone of many spermatogonia, sometimes called a spermatocyst. Cells within this **spermatocyst** undergo spermatogenesis in unison, eventually producing mature sperm.

### *Genital Ducts*

***Fishes*** In cyclostomes, the large unpaired testes are not served by any genital ducts. Sperm are shed into the coelom and exit via abdominal pores. The archinephric ducts drain the kidneys exclusively (figure 14.32a). In elasmobranchs, the prominent müllerian ducts of the female are rudimentary in the adult male (figure 14.32b). The **accessory urinary ducts,** distinct from the archinephric ducts, are usually present to service the posterior uriniferous kidney (figure 14.33a). Each anterior reproductive kidney has short tubules that join the testis to the archinephric duct, which because of its role in sperm storage and transport may be termed a vas deferens (figure 14.32b). These tubules within the anterior part of the kidney function as an epididymis, connecting the rete testis to the vas deferens and perhaps storing sperm. Adjacent Leydig cells in this cranial region secrete seminal fluid into the genital ducts.

In bony fishes, the archinephric ducts drain the kidneys and may receive sperm from the testes. However, the testes tend to develop separate sperm ducts and routes of

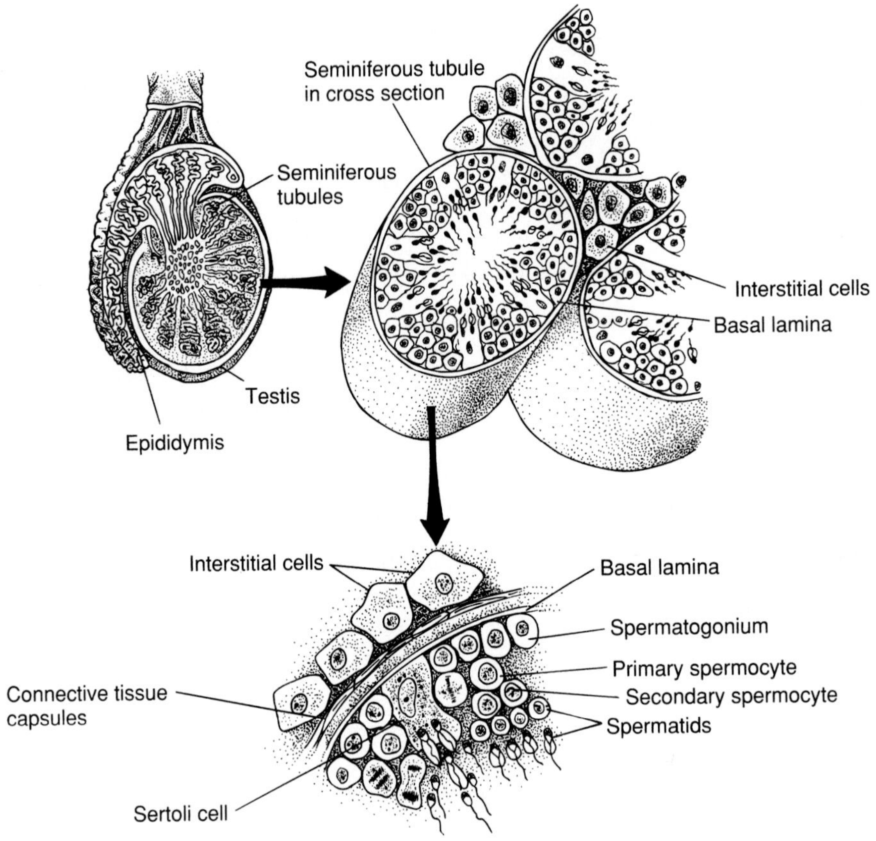


**Figure 14.30** Spermatogenesis. Within the walls of the seminiferous tubules, spermatogonia divide, giving rise to cells that stay in place and preserve the population of spermatogonia as well as to cells that undergo meiotic and cytological reorganization. These become first primary and then secondary spermatocytes. Secondary spermatocytes undergo changes that transform them into spermatozoa. Sertoli cells hold spermatozoa and then release them into the lumen of the seminiferous tubules and connecting epididymis. Interstitial cells (cells of Leydig) lying between the seminiferous tubules secrete male hormones.

exit (figure 14.33b,c). In most teleosts, this separate duct system forms a **testicular duct,** which is not homologous to the archinephric duct and may even establish its own opening to the exterior (figure 14.33d). Some teleosts, such as salmonids, lack sperm ducts entirely. Sperm are released into the body cavity and exit the body through pores near the posterior part of the coelom.

***Tetrapods*** In male amphibians, several genital duct configurations can occur (figure 14.34a,b). In *Necturus* and a few other species, the archinephric ducts transport both sperm from the testes and urine from the uriniferous kidneys. However, this is likely a specialized condition of the paedomorphic *Necturus*. In general, this condition occurs only in larval salamanders. In some salamander families, new accessory urinary ducts service the caudal kidneys, and sperm is transported from the testes through tiny ducts in the cranial kidneys to the archinephric ducts (vas deferens) to be stored. In all frogs and a few species of salamanders, tiny ducts that reach directly from the testes to the archinephric ducts bypass the anterior part of the kidneys. Elimination from the uriniferous kidneys occurs exclusively via the accessory urinary ducts. Thus, in some adult amphibians, the archinephric ducts may have both reproductive and excretory roles, whereas in other species, these ducts may be involved exclusively in sperm transport and new accessory urinary ducts may drain the opisthonephros (figures 14.35a,b and 14.36a–c).

In male amniotes, the archinephric duct (vas deferens) transports sperm exclusively (figures 14.35c,d and 14.36c,d). Several mesonephric tubules of the embryonic kidney may contribute to the epididymis that connects each testis to a vas deferens (figure 14.37). Each amniote kidney is drained by a new duct, the ureter (metanephric duct).

In most vertebrate males, the testes reside within the abdomen; however, the testes of some mammals descend into the **scrotum,** a coelomic pouch suspended outside the body but connected to the abdominal coelom via an **inguinal canal** (figure 14.35d). In other mammals, the testes either remain in the body cavity (e.g., monotremes, some primitive insectivores, sirenians, elephants, sloths, cetaceans, armadillos) or descend into a muscular pouch but

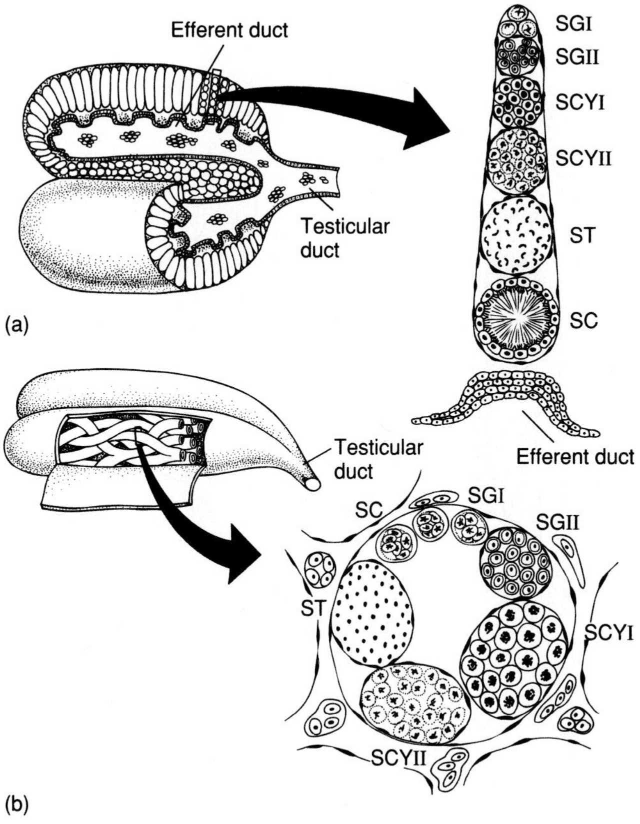


**Figure 14.31** Sperm production in the teleost testis. Sperm may develop within compartments (a) or tubules (b). During copulation, mature sperm pass into the testicular duct. Primary spermatogonia (SGI) become in succession secondary spermatogonia (SGII), primary spermatocytes (SCYI), secondary spermatocytes (SCYII), spermatids (ST), and finally mature spermatozoa (S). Sertoli cells (SC) form part of the epithelium lining the compartments or tubules.

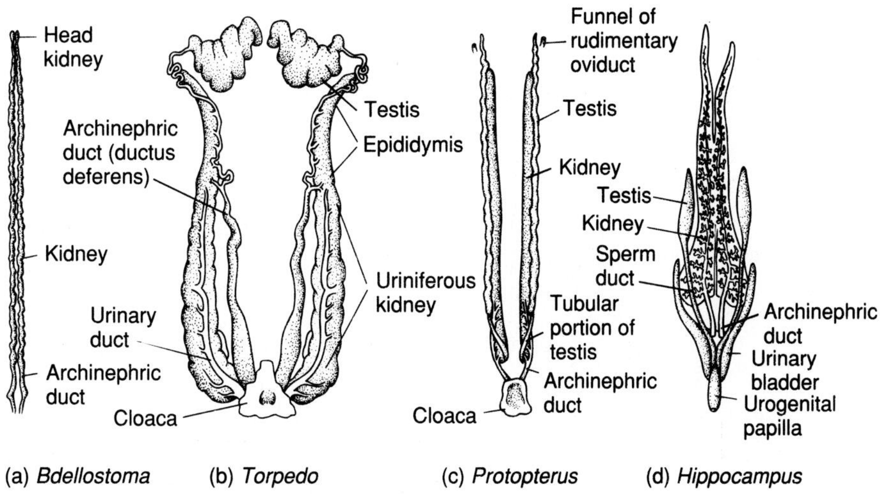


**Figure 14.32** Urogenital systems of male fishes. (a) Hagfish *Bdellostoma*. The single testis of the hagfish hangs in the dorsal body wall between the kidneys. (b) Elasmobranch *Torpedo*. (c) Lungfish *Protopterus*. (d) Teleost, sea horse *Hippocampus*.

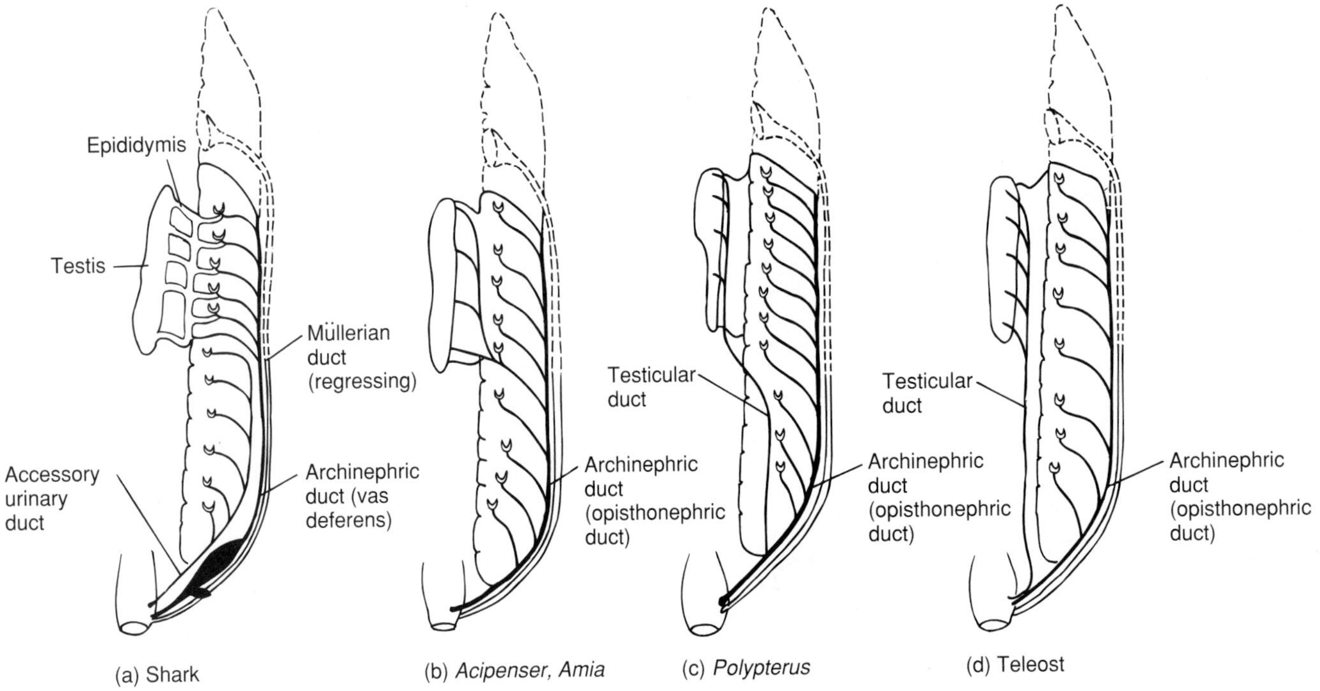


**Figure 14.33** Urogenital ducts of male fishes. (a) Shark. (b) Sturgeon and bowfin. (c) Bichir. (d) Teleost. In sharks, an accessory urinary duct develops to drain the kidney, and the archinephric duct is concerned with sperm transport. In other groups, additional ducts that develop to drain the testis sometimes join the archinephric duct. In teleosts, these exit independently.

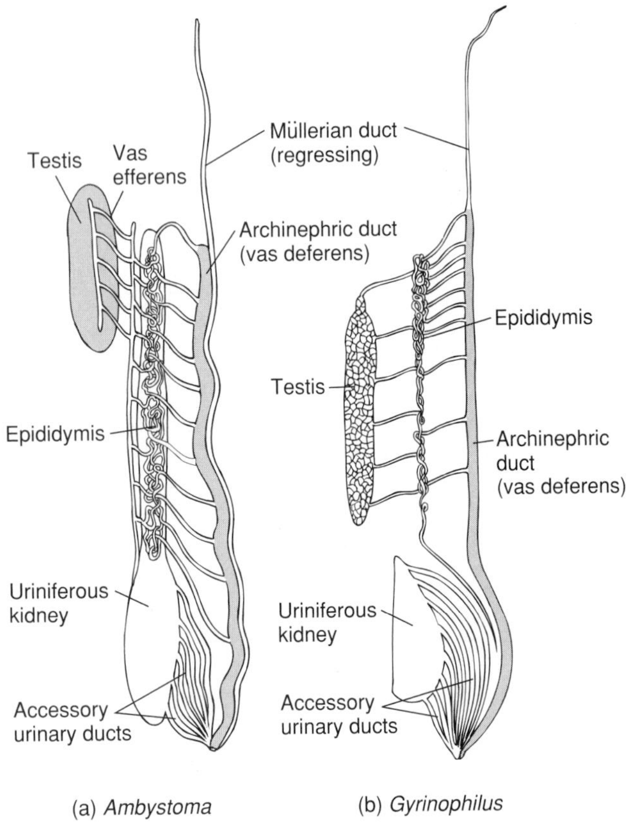


**Figure 14.34** Urogenital systems of male amphibians. (a) Salamander *Ambystoma*. (b) Salamander *Gyrinophilus*.

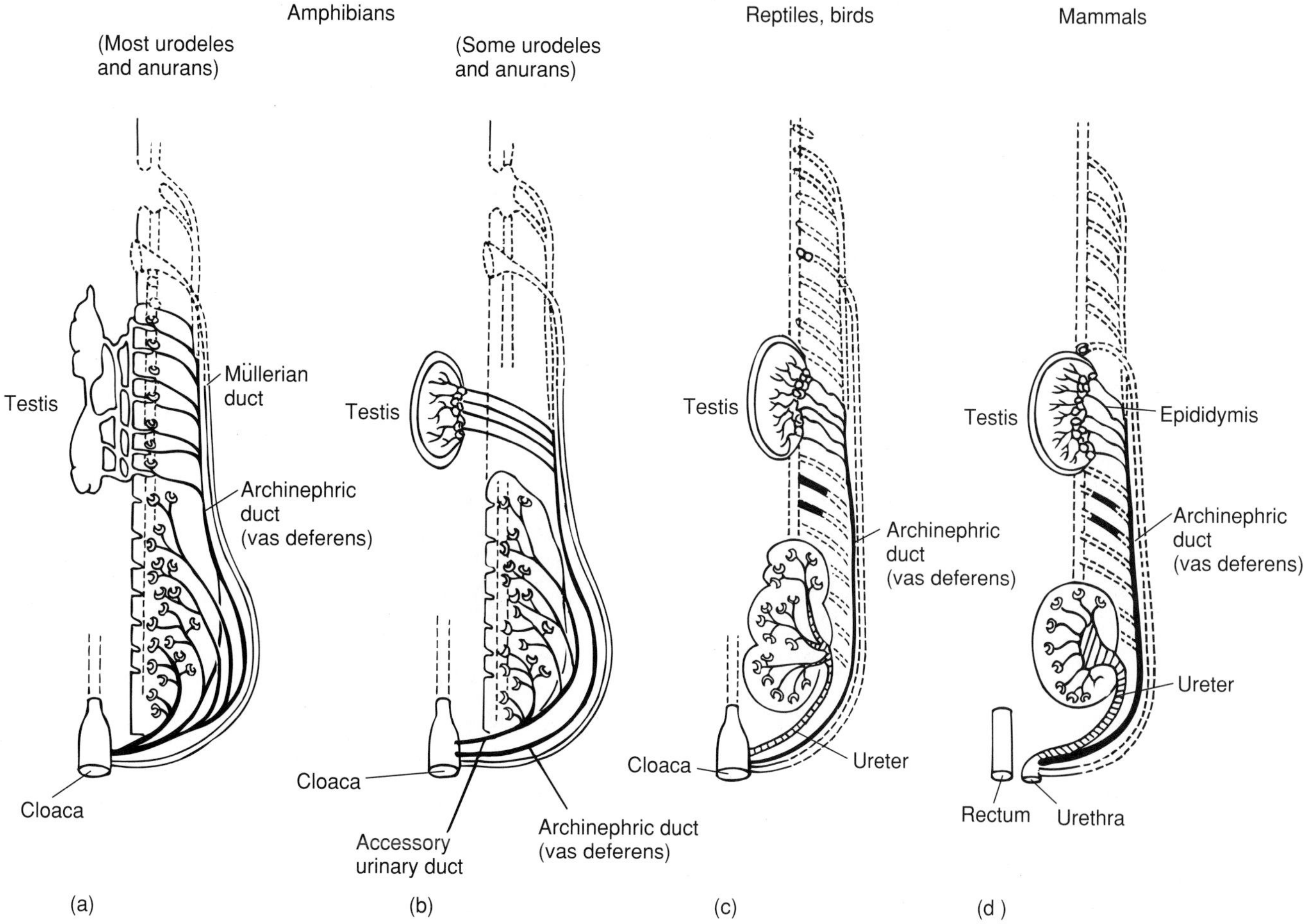


**Figure 14.35** Urogenital ducts of tetrapod males. (a) Most urodeles and most anurans (adults). (b) Some urodeles and some anurans (adults). (c) Reptiles and birds. (d) Mammals.

not a true coelomic scrotum (e.g., moles, shrews, many rodents, lagomorphs, pinnipeds, hyaenas). Some mammals have testes that descend temporarily into the scrotum during the breeding season (e.g., chipmunks and squirrels, some bats, some primates). Most other mammals have testes that descend permanently, which typically occurs during embryonic development. The testes migrate from the body cavity, through the abdominal wall via the inguinal canal, and into the scrotum where the temperature is cooler, often up to 8°C cooler than in the abdomen. The **external cremaster** muscles lift the testes closer to the body under cool conditions and allow them to descend under warm conditions, thus warming or cooling the testes as required. Further, arteries and veins entering and leaving the testes intermingle in a **pampiniform plexus,** a countercurrent exchange mechanism that serves as a heat block to the testes. If the testes fail to descend (a condition called **cryptorchidism**) or are artificially warmed in the scrotum, sperm production falls or even ceases in these species. Thus, the testes in mammals with scrotums seem to have lost the capacity to function at body temperature. However, why some mammals have evolved a scrotum and others have not is still not understood.

## *Copulatory Organs*

In most water-dwelling vertebrates, fertilization is external. Eggs and sperm are shed simultaneously from the body into the water where fertilization occurs. However, if the female uterus houses the embryo or if a shell seals an egg, sperm must fertilize the egg before it descends from the oviduct. In these instances, fertilization is internal. Sperm deposited within the female genital tract journey to the upper reaches of the oviduct to fertilize the egg. In many vertebrates, **copulation (coitus)** involves the direct, momentary apposition of the male and female cloacae to transfer sperm. Often, however, the male possesses external **intromittent organs** specialized to deliver sperm during coitus. In salamanders, sperm transfer is external and involves a **spermatophore,** but fertilization is internal.

In male sharks, rays, chimaeras, and some placoderms, the pelvic fins are specialized as **claspers** (figure 14.38a–c). During copulation, one clasper is inserted into the female cloaca and its terminal cartilages spread by muscle action to help hold the clasper in place. Sperm leave the male cloaca, enter a groove on the clasper, and are flushed by water squirted from siphon sacs within the body wall of the male

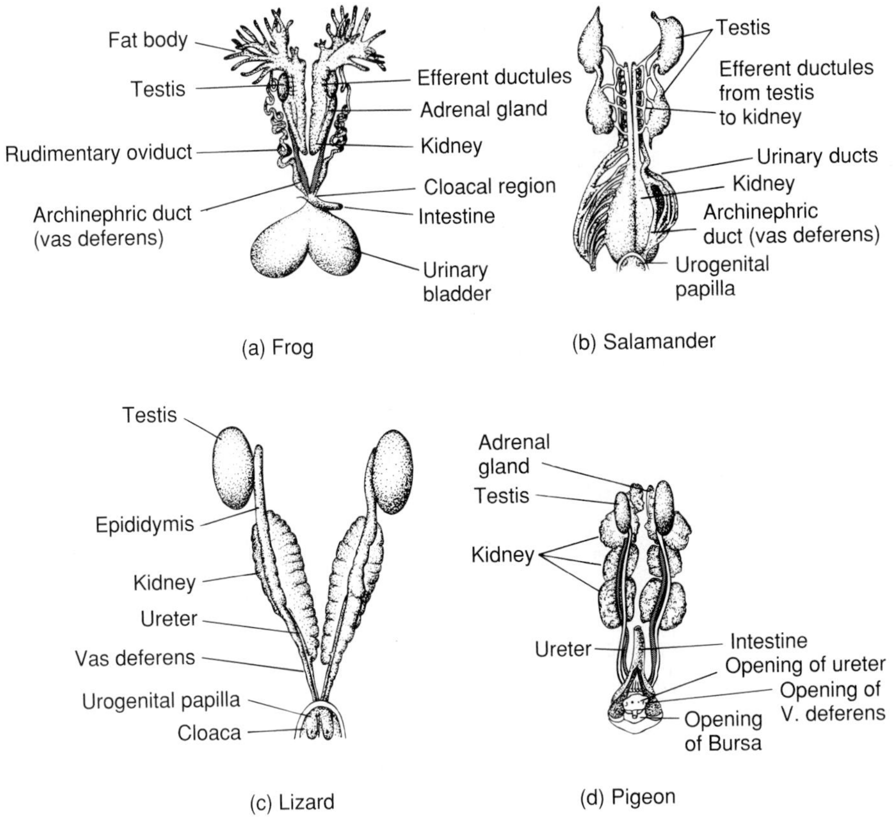


**Figure 14.36** Urogenital systems of tetrapod males, ventral view. (a) Frog *Rana*. (b) Salamander *Salamandra*. (c) Lizard *Varanus*. (d) Bird *Columba*.

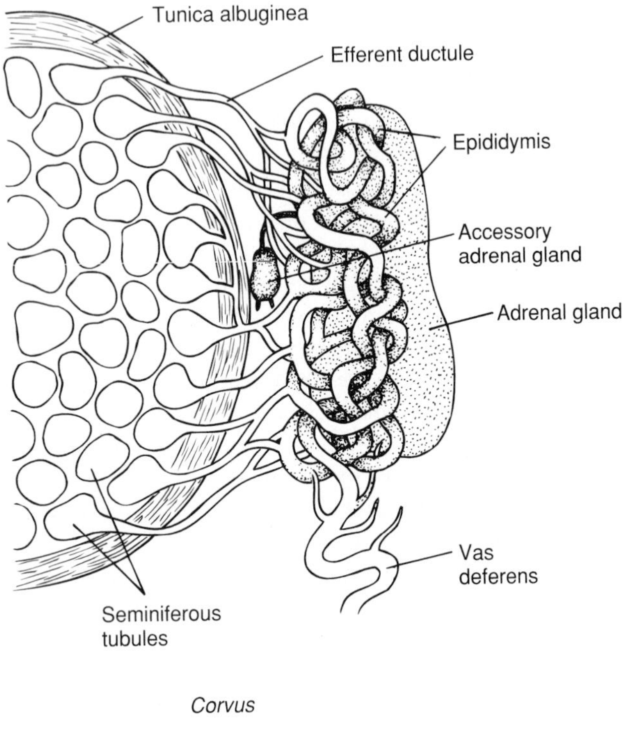


**Figure 14.37** Avian testis and epididymis in the jackdaw *Corvus*.

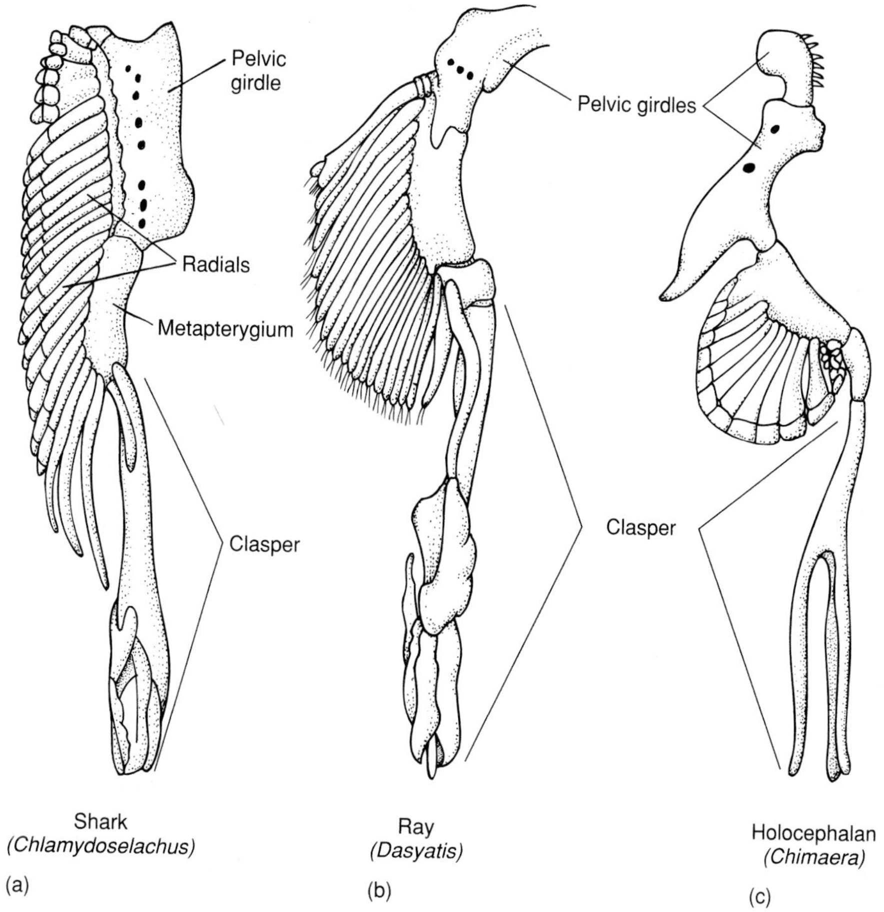


**Figure 14.38** Intromittent organs of chondrichthyans. (a) Shark *Chlamydoselachus*. (b) Ray *Dasyatis*. (c) Holocephalan *Chimaera*.

into the female cloaca. In the killfish *(Fundulus)*, a teleost, pelvic and anal fins interlock during spawning, holding male and female cloacae close together as gametes are released (figure 14.39a). In a few species of teleosts, the anal fin is fashioned into a grooved intromittent organ, termed a **gonopodium,** that deposits sperm into the female during copulation (figure 14.39b).

Fertilization in almost all frogs is external. The male grasps the female from above in a behavior called **amplexus** and releases sperm from his cloaca as eggs leave the female's cloaca. An exception among frogs is the tailed frog, *Ascaphus*. The male possesses a short grooved, taillike extension of the cloaca used to transfer sperm directly into the cloaca of the female. The males of most salamander species produce a spermatophore, which consists of a cap of sperm on top of a gelatinous pedestal (figure 14.40a–c). The spermatophore is deposited in front of the female at the culmination of a stylized courtship. The female nips off the sperm cap with the lips of her cloaca to gather in the sperm (figure 14.41). Females of some species collect only a portion of each spermatophore sperm cap, but they sample from as many as 20 or 30 different spermatophores. Sperm are stored in a dorsal pocket of the cloaca, the **spermatheca,** until they are released to fertilize eggs internally as they travel from the oviducts and out the cloaca. This method of reproduction decouples sperm transfer from fertilization. Thus, sperm transfer may occur at a time and place favorable to courtship but not to egg deposition. In caecilians, the male everts the posterior part of his cloaca through the vent and fits it into the female cloaca to aid sperm transfer.

Intromittent organs are absent in *Sphenodon*. The cloacae are pressed together during courtship and sperm are transferred directly. Some male birds and male turtles, crocodiles, and mammals have a single **penis,** an intromittent organ down the midline of the body (figure 14.42a–c). The evolutionary origin of the penis is unknown, but it seems to be a derivative of the cloaca. When not in use, the penis is usually flaccid and may be retracted into a sheath or returned to the cloacal chamber. It becomes engorged with blood or lymph that fills its specialized compartments and makes it erect. When the penis is erect, it

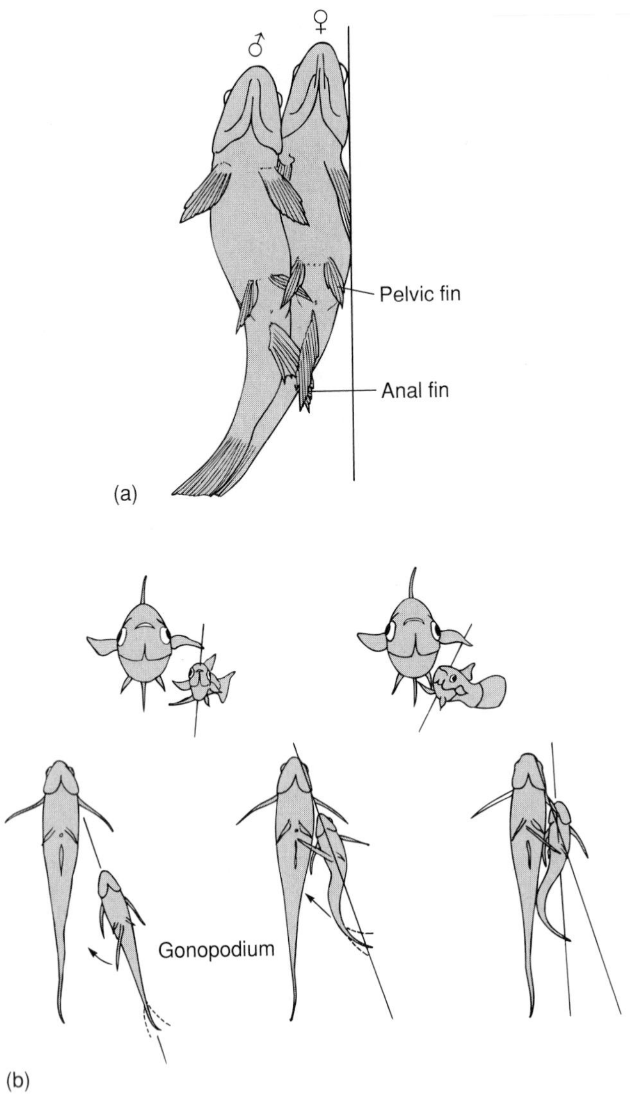


**Figure 14.39** Spawning in teleost. (a) Ventral view of interlocking anal and pelvic fins of *Fundulus*. (b) Gonopodium of male is inserted into anal region of female.

penetrates the female and holds the channel open to ejaculate sperm. Erection achieved by blood infiltration is termed **hemotumescence.** In turtles, the midventral penis consists of two parallel bands of sinusoidal tissue, the **corpora cavernosa.** Between them lies a groove, the **sulcus spermaticus** (figure 14.42a,b). When engorged with blood, the corpora cavernosa enlarge, protrude the penis from the cloacal wall through the vent, and shape the sulcus spermaticus into a duct that receives and transfers sperm from each vas deferens. The females of some turtle species possess a homologue to the penis. Although this structure may be functionless, it possibly completes the other half of the male's sperm groove and therefore contributes to the sperm transfer channel.

The penis of male crocodilians is similar to that of turtles, except that it is relatively longer and the whole organ projects farther from the cloaca (figure 14.42c–e). Although the mechanism of erection is not clear, hemotumescence that defines a sulcus spermaticus seems to be involved. Female crocodilians also possess a rudimentary homologue of the male penis, but it remains within the cloaca and does not protrude.

In lizards and snakes, males possess a pair of intromittent organs, the **hemipenes.** Each hemipenis is usually grooved to allow for sperm transport. It is rough or spinous at its tip to ensure secure engagement when the male inserts it into the female's cloaca. A retractor muscle returns each hemipenis to the body by turning it outside in, a process called invagination. The retractor pulls it into a pocket located at the base of the tail, behind the vent. During erection, muscle action and hemotumescence force each hemipenis through the cloaca and balloon it out through the vent, turning it inside out—this is evagination (figure 14.43a,b). A sulcus spermaticus is defined in each hemipenis, which is sometimes Y shaped. During copulation, only one hemipenis is inserted in the cloaca of the female (figure 14.43c).

In birds, two types of intromittent organs are found. In the domestic turkey, little more than the edges of the cloaca swell during copulation (figure 14.44a). Male and female cloaca are pressed together at coitus. Semen flows between the lateral penile swellings of the male and is ejaculated into the female cloaca. Ostriches and some other groups have another intromittent organ. It is a true penis with an erectile shaft that the male inserts into the female cloaca. In the male ostrich, the erect penis is conical and widened at its base. It bears a sulcus spermaticus along its length (figure 14.44b). In ducks, the erect penis may be quite elaborate, with the sulcus spermaticus spiraling along the tapering shaft. When relaxed, the penis is coiled and tucked within the cloaca along the ventral wall. Lymphatic channels within the penis connect to expanded chambers. The mechanism of erection is thought to involve filling of these internal chambers. As a result, the penis projects from the cloaca and bends forward (figure 14.44c,d).

All mammals copulate with a penis. In addition to the paired corpora cavernosa, a third sinusoidal tissue is present, the **corpus spongiosum** that surrounds the closed sulcus, or **cavernous urethra** (figure 14.45a). These spongy sinuses in the penis become engorged with blood and stiffen. In addition, insectivores, bats, rodents, carnivores, and most primates except humans have a **baculum** (os penis), a permanent bone located within the connective tissue of the penis to stiffen it. In these animals, the already stiffened penis becomes engorged with blood into a fully erect position (figure 14.45b,c). The sensitive tip of the penis is the **glans penis.** The male penis is single in mammals, although in marsupials the tip is forked to fit into the two lateral vaginas of the female. As a result,

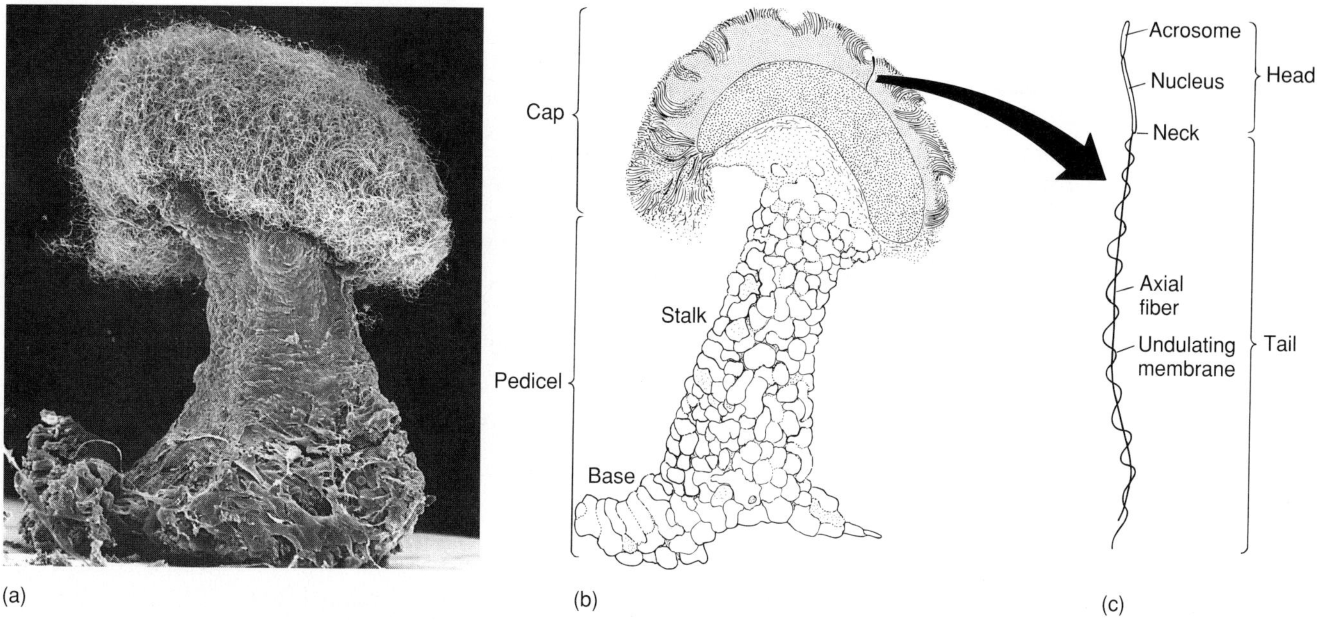


**Figure 14.40** Spermatophores of amphibians. (a) Whole spermatophore deposited by male *Ambystoma macrodactylum*. (b) Longitudinal section of a spermatophore from *Ambystoma texanum*. Generally, sperm heads point outward, tails point inward. (c) An enlarged spermatozoon.

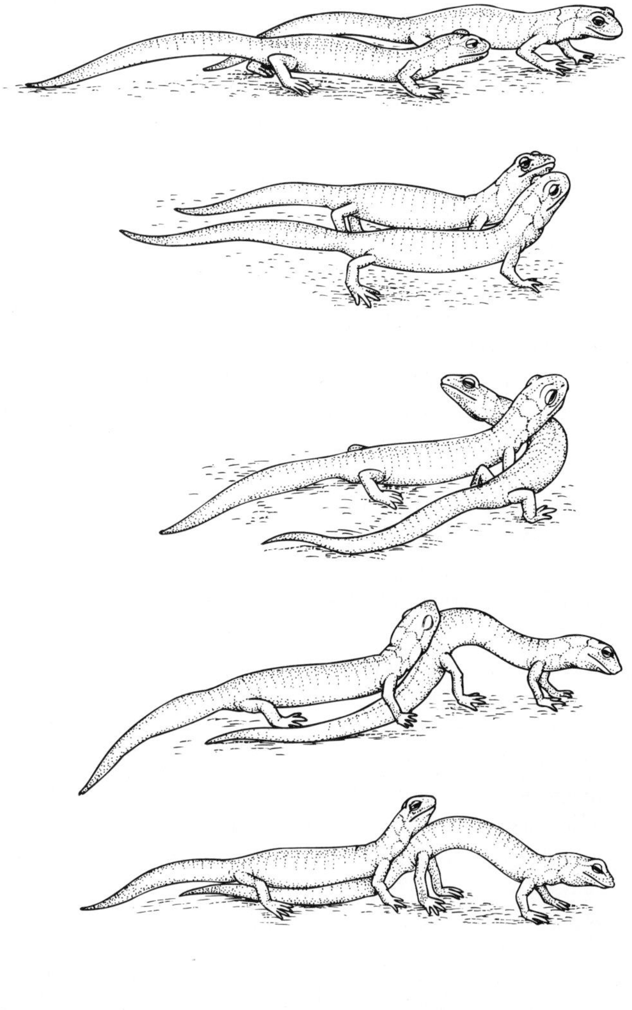

**Figure 14.41** Courtship in salamanders. Stroking and rubbing characterize courtship in the salamander *Ensatina eschscholtzi*. Once the male has the female's attention, he walks before her and deposits clumps of packaged sperm, called spermatophores. Following this, she walks over the spermatophores and collects them with her cloaca. Sperm migrate up her urogenital tracts to fertilize ova internally.

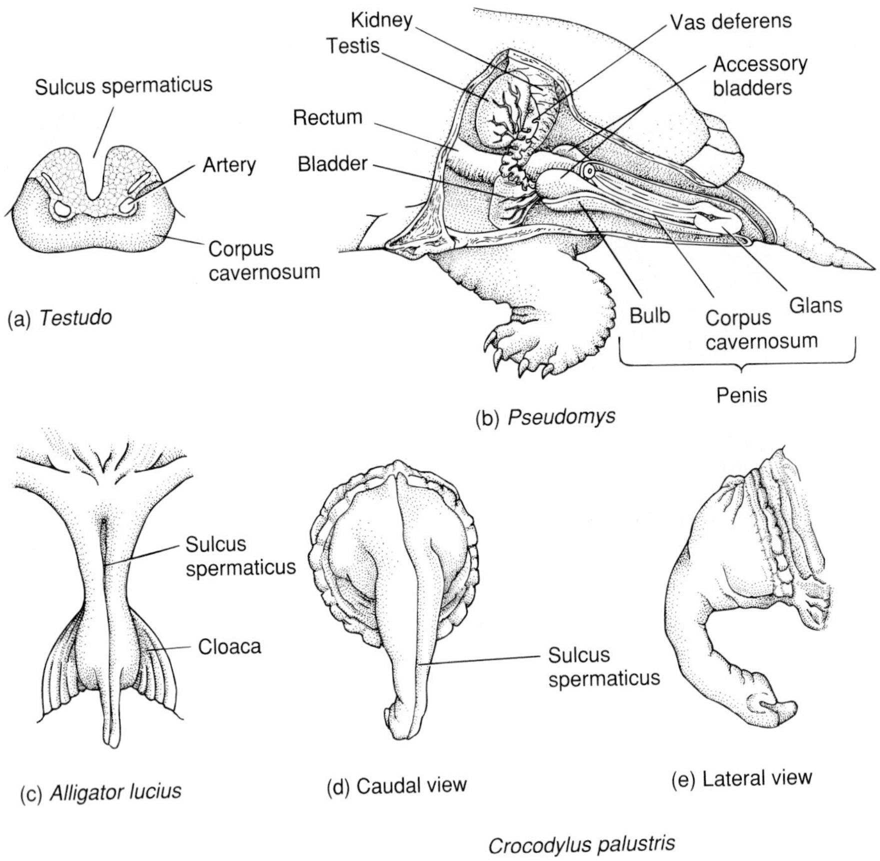


**Figure 14.42** Penises of reptiles. (a) Turtle *Testudo*: cross section of the penis within the cloaca. (b) Turtle *Pseudomys*: sagittal section of the penis. (c) Alligator *Alligator lucius* penis. Caudal (d) and lateral (e) views of the penis of the crocodile *Crocodylus palustris*.

ejaculated sperm move into each lateral vagina and then into the vaginal sinus, a chamber that receives both uteri (figure 14.51).

## Cloaca

The cloaca has already been defined as a common chamber receiving products from kidneys, intestines, and often gonads. It opens to the outside through a cloacal opening or vent. (It is customary to point out that in Latin cloaca means sewer.) The cloaca arises at some point during embryonic development in all vertebrates, but in many it becomes subdivided, lost, or incorporated into other adult structures (figure 14.46a–f). A well-developed cloaca occurs in adult sharks and lungfishes (figure 14.46b,d). But in teleosts, distinct urinary, anal, and genital openings are present replacing the cloaca (figure 14.46f). Among tetrapods, a cloaca is present in amphibians, reptiles, birds, and monotremes. A shallow cloaca persists even in marsupials (figure 14.47a–k).

A cloaca is apparently a primitive vertebrate feature because it occurs in most primitive gnathostomes and persists in the embryos of almost all veretebrates. Its absence in chimaeras (Holocephali), ray-finned bony fishes (Actinopterygii), *Latimeria* (Crossopterygii), and most placental mammals (Eutheria) may represent independent losses.

Embryologically, the cloaca arises from hindgut endoderm and proctodeal ectoderm. Structurally, three functions influence it—defecation, urination, and copulation. Each function tends to be associated with a compartment and each compartment is controlled by muscles that regulate the entry and departure of products from the intestines, kidneys, and gonads. The most proximal compartment is the **coprodeum** into which the intestine empties. The **urodeum** receives products from urinary and genital ducts. The most distal compartment is the **proctodeum,** which functions in copulation and in many amniotes develops a penis (figure 14.47a). Many urogenital ducts, upon approach to the cloaca, inflate slightly to form an expanded

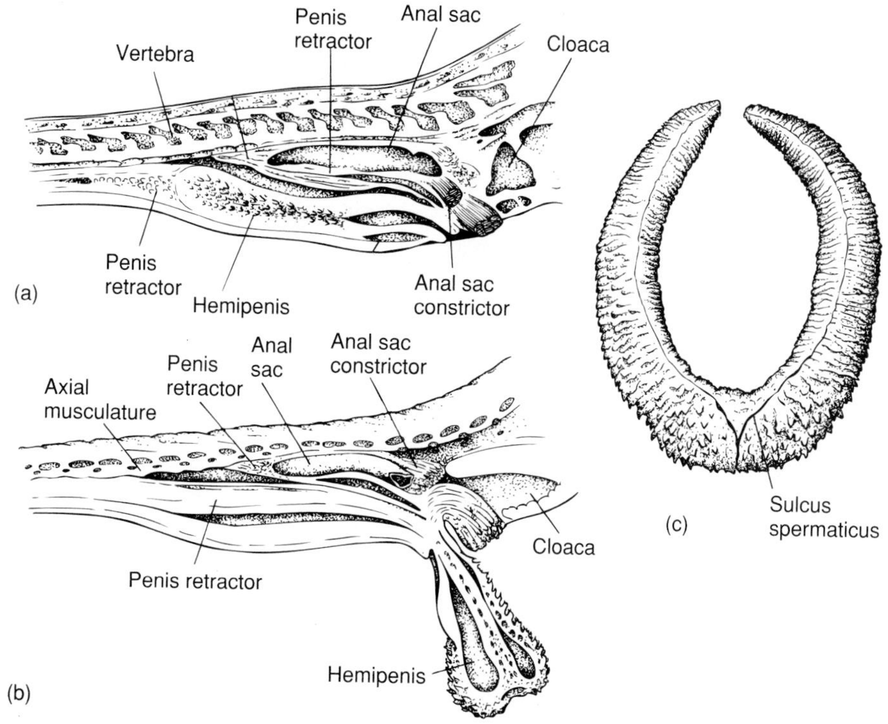


**Figure 14.43** Hemipenis of a snake. Lizards and snakes have paired hemipenes, but usually only one is used during copulation. (a) The hemipenis is pulled back into the body by the retractor muscle (sagittal view). (b) When erect, the hemipenis's internal sinuses become engorged with blood and it pops through the vent (sagittal view). During copulation, the male inserts its hemipenis into the cloaca of the female. Sperm travel down the sulcus spermaticus into the female. (c) One of the two hemipenes from the rattlesnake *Crotalus atrox* is shown everted. This single hemipenis is divided, which gives it a horseshoe shape. Note the divided sulcus spermaticus that runs along each arched branch of the hemipenis.

Modified from Dowling and Savage, 1960.

**urogenital sinus.** These ducts often open into the cloaca via a small projection called the **urogenital papilla.**

Late in the nineteenth century, Hans Gadow suggested that each of the three cloacal compartments was separated from the other by folds in the mucosal wall—the **rectocoprodeal fold** between intestine and coprodeum, the **coprourodeal fold** between coprodeum and urodeum, and the **uroproctodeal fold** between urodeum and proctodeum. Although such folds occur in many vertebrates, they are low or absent in some, making it difficult to delineate boundaries between compartments of the cloaca. Gadow's terminology describing compartments and folds was based on tetrapods, but it is now applied to fishes as well. Unfortunately, there have not been any comparative studies of fishes in which a large sample of species were examined, so it is difficult to generalize about the presence or absence of these cloacal compartments within groups of fishes.

The cloaca of most amphibians is simple. Folds usually delineate the coprodeum and urodeum, but in the absence of an intromittent organ or a uroproctodeal fold, the proctodeum is not anatomically demarcated from the rest of the cloaca (figure 14.47b). Among reptiles, the cloaca of *Sphenodon* is subdivided by folds into three compartments; the proctodeum is simplified and lacks a penis. The cloaca of snakes and lizards also has three compartments, but the proctodeum is usually reduced (figures 14.47c–e and 14.48a,b). The internal subdivision of the cloaca is much less distinct in turtles (figure 14.47f), and in crocodilians the coprodeum and urodeum and, to a lesser extent, the proctodeum are more or less united into a single large chamber (figure 14.47g). In birds, the cloacal folds are quite variable. The ostrich cloaca has a rectocoprodeal fold (figures 14.47h and 14.49a,b), but this is apparently lacking in other groups (figure 14.47i). In birds, the proctodeum is associated with a **cloacal bursa** (bursa of Fabricius) that has an immune function.

The cloaca persists in monotremes, where distinct coprourodeal and uroproctodeal folds demarcate the urodeum from other compartments (figure 14.47j). The ureter and vas deferens open to a urogenital sinus, but urine flows directly into the urodeum, and semen flows through a sperm duct within the penis. Marsupials possess a reduced cloaca that is represented primarily by the proctodeum (figure 14.47k). The ectodermal part of the cloaca persists in some rodents and insectivores; however, in all other placental mammals, the cloaca divides in the sexually indifferent stage and forms separate orifices from the coprodeum and urodeum (figure 14.50a–c). Generally, the coprodeum

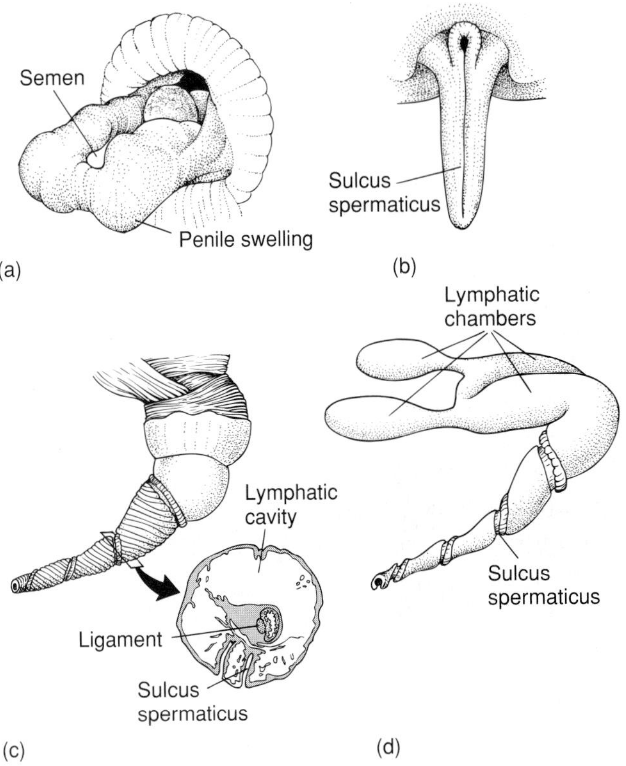


**Figure 14.44** Intromittent organs of birds. (a) Domestic turkey with penile swellings. The margins of the cloaca form the central gully down which sperm flows during copulation. (b) Erect ostrich penis. (c) Erect duck penis with bird in standing position. Cross section shows lymphatic cavities thought to be responsible for eversion of the penis from the cloaca. (d) Diagrammatic lateral view of lymphatic chambers whose filling is thought to be responsible for penile erection.

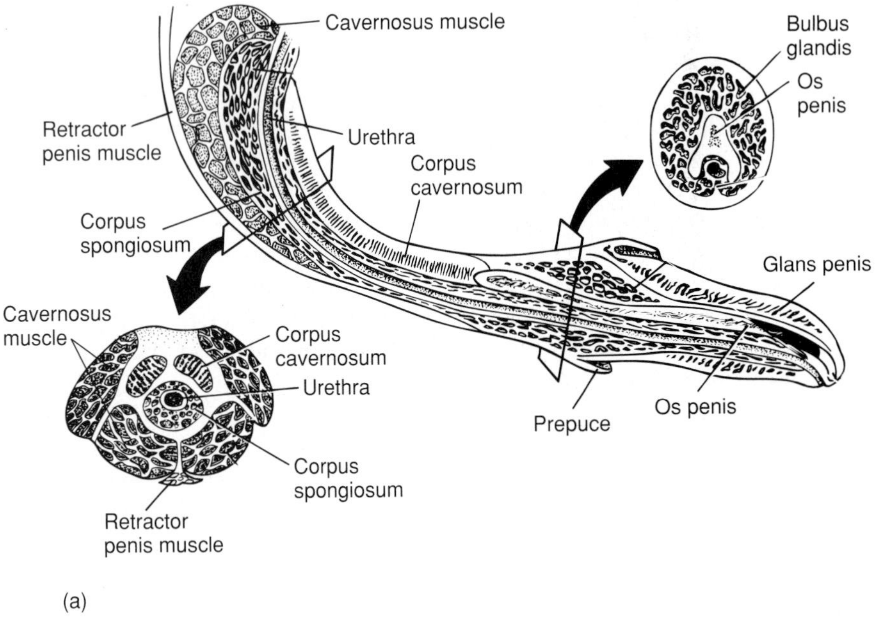


**Figure 14.45** Penile erection in the dog. (a) Sagittal view and cross sections of the penis. (b) Flaccid penis. Arterial blood enters the internal pudendal artery, circulates through capillaries of the penis, and flows from the penis through the pudendal vein. (c) Erect penis. Stimulation of the nerves of erection causes increased blood flow to the penis (1). In addition, partial inhibition of venous drainage (open arrows at 2) results in diversion of blood into the cavernous bodies (corpus cavernosum and bulbus glandis), which fill, stiffen the penis, and result in erection. The os penis (baculum) also helps firm the penis.

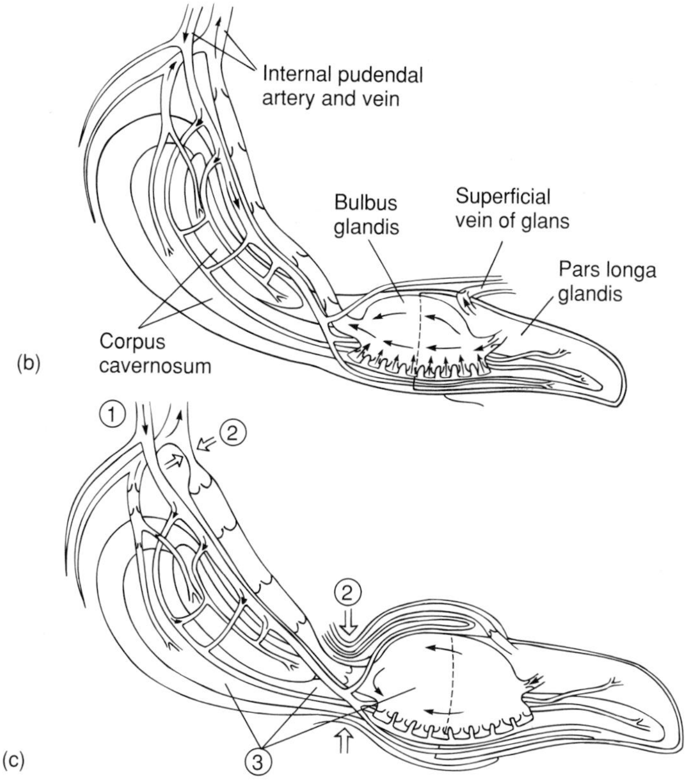


**Figure 14.45** (continued)

becomes the rectal region of the digestive tract with an anal opening. The urodeum yields separate structures, depending on the sex. In the male, the urogenital sinus becomes the urethra that transports sperm and urinary products (figure 14.50d). In the female, the urogenital sinus divides again to produce a urethral opening for the urinary system and a vaginal opening for the reproductive system (figure 14.50c).

Two patterns are evident in the reproductive organs of marsupial females. In opossums, the oviducts enter a vaginal sinus that loops symmetrically around the viscera to form lateral vaginas (figure 14.51a). In kangaroos, the vaginal sinus, via an unpaired central vaginal canal, joins the lateral vaginal loops in the common urogenital sinus (figure 14.51b). In placental females, one end of each oviduct narrows into a slender fallopian tube, which receives the egg released from the ovary. At the other end, the oviducts expand into the uterus to support the young during their embryonic development. In some species, the oviducts join the vagina separately, forming a **duplex uterus.** In **bipartite** and **bicornuate uteri,** the uteri partially fuse. If the uteri fuse entirely, a **simplex uterus** is formed (figure 14.52).

## Urinary Bladder

Before being excreted, urine is usually stored in specialized regions of the urogenital system. In this way, the vertebrate can void urine at opportune times rather than continuously as it is formed. If water conservation is important, the bladder sequesters the concentrated urine so that it does not create osmotic pressure that draws water out of the tissues of the animal.

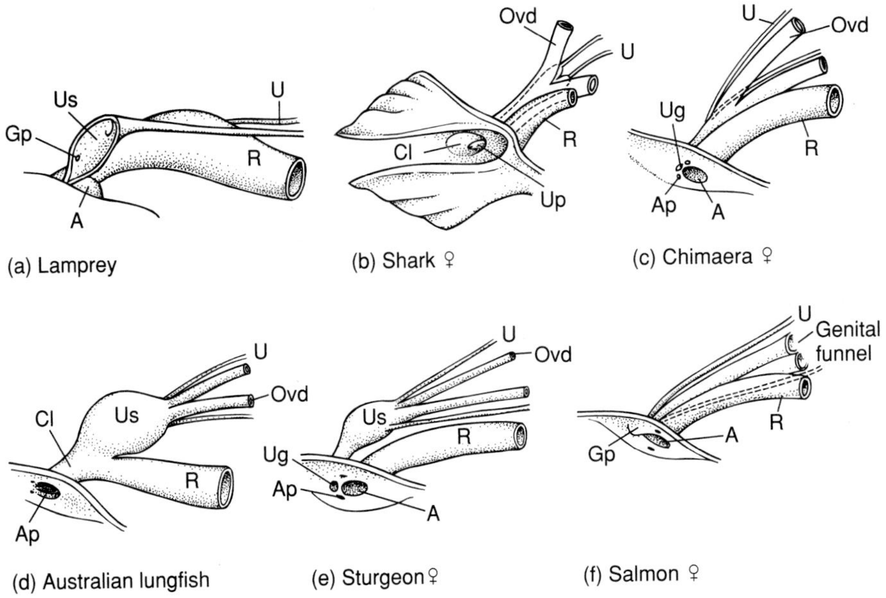


**Figure 14.46** Cloacal and anal regions of fishes. (a) Lamprey. (b) Female shark. (c) Female chimaera. (d) Australian lungfish. (e) Female sturgeon. (f) Female salmon. Structures of the urogenital system include the anus (A), abdominal pore (Ap), cloaca (Cl), genital opening (G), genital pore (Gp), oviduct (Ovd), rectum (R), urinary ducts (U), urogenital opening (Ug), urogenital sinus (Us), urinary papilla (Up).

In fishes, urine is usually stored within the ends of the urinary ducts where they join the cloaca or open to the outside. A urinary bladder of this type is mesodermal and noncloacal in origin. It is found among elasmobranchs, holocephalians, and most teleost fishes (figure 14.53a).

In tetrapods, the urinary bladder arises as an outpocketing of the cloaca. Urine flowing from the urinary ducts usually empties into the cloaca first and then fills the baglike urinary bladder (figure 14.53b). In mammals, the urinary ducts (ureters) empty directly into the urinary bladder (figure 14.53c). The tetrapod urinary bladder appears first among amphibians and is present in *Sphenodon*, turtles, most lizards, ostriches among birds, and all mammals. The urinary bladder has been lost in snakes, some lizards, crocodilians, and all birds except the ostrich.

# Function and Evolution

In most vertebrates, reproduction is seasonal. Courtship and copulation are usually restricted to a brief annual breeding season. During the breeding season, hormone-readied genital ducts receive and transport released eggs and sperm. Onset of reproductive readiness is called **recrudescence.** Only among humans is breeding a year-round affair.

## *Potency and Fertility*

**Fertility** refers to the ability of the female to produce fertilizable eggs or of the male to produce sperm in sufficient numbers to achieve fertilization. A male producing insufficient numbers of sperm is **infertile** or **sterile.** In a human male, ejaculated semen can contain 200 million sperm. And although it just takes one sperm to fertilize an egg, a drop in the sperm count to 50 million may result in sterility. Although millions of sperm may be ejaculated into the vagina, the number of sperm that eventually survive the journey to the upper reaches of the oviduct rarely exceeds a few hundred. Considering that the spermatozoon is small in comparison to the volume of the oviduct, it is not surprising that only a very modest number of sperm arrive at the site of fertilization. Finally, many sperm interact to break through follicle cells or surface mucus clinging to the egg so that one sperm can penetrate the egg cell membrane. Thus, fertilization is performed by the fusion of a single spermatozoon with a single egg, but this comes after much attrition and cooperation among many sperm to promote penetration of the egg.

**Potency** refers to the ability of the male to engage in copulation. **Impotence** results from the failure to achieve an erection. Impotence is different from sterility. Castrated males are sterile because they lack testes and produce no sperm. However, if the testes are removed after puberty, there has often been enough time for androgens to masculinize the individual so that some secondary sexual characteristics, sex drive, and ability to engage in sexual intercourse (potency) are retained.

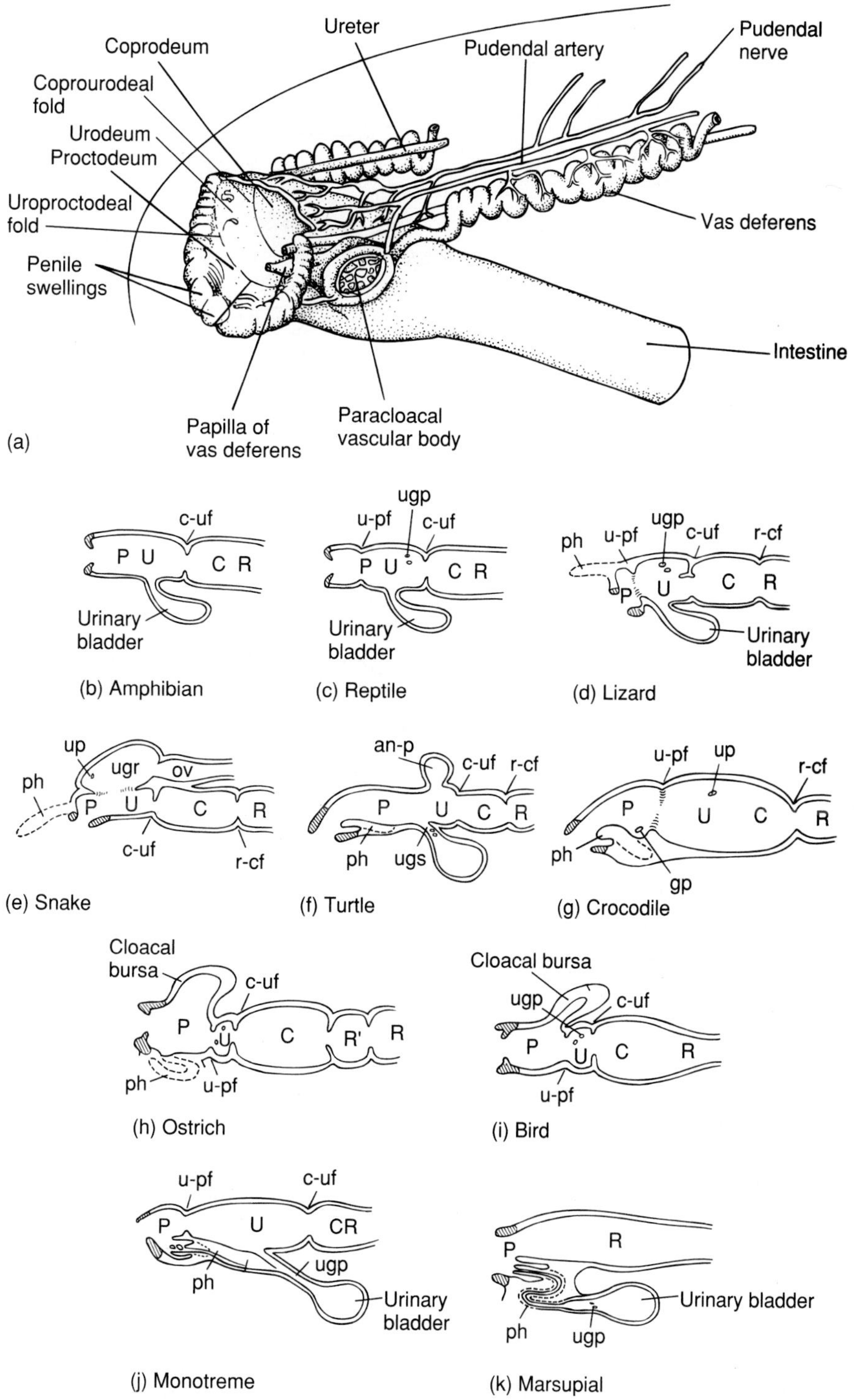


**Figure 14.47** Diagrams of sagittal sections of tetrapod cloacae. (a) Bird cloaca with ducts and organs that open into each of the three chambers. (b) Amphibian. (c) Reptile *Sphenodon*. (d) Lizard *Lacerta*. (e) Snake *Tropidonotus*. (f) Turtle *Pseudomys*. (g) Crocodile. (h) Ostrich. (i) Bird. (j) Monotreme. (k) Marsupial. Parts of the cloacae include the coprodeum (C), rectum (R), urodeum (U). Other abbreviations: anal gland (angl), coprourodeal fold (c-uf), genital pore (gp), oviduct (ov), penis (ph), rectocoprodeal fold (r-cf), urinary pore (up), urogenital pores (ugp), uroproctodeal fold (u-pf), urogenital sinus (ugs), urogenital reservoir (ugr).

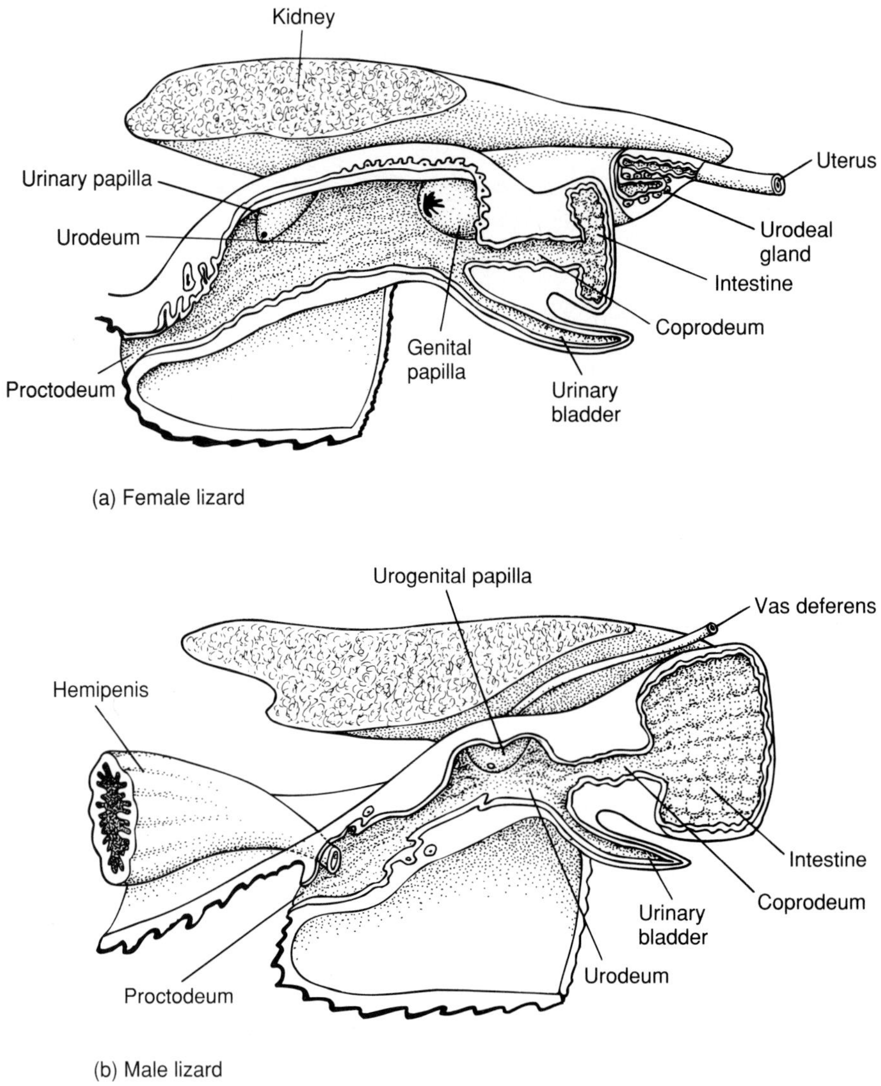


**Figure 14.48** Cloaca of the lizard *Coleonyx.* (a) Female. (b) Male.

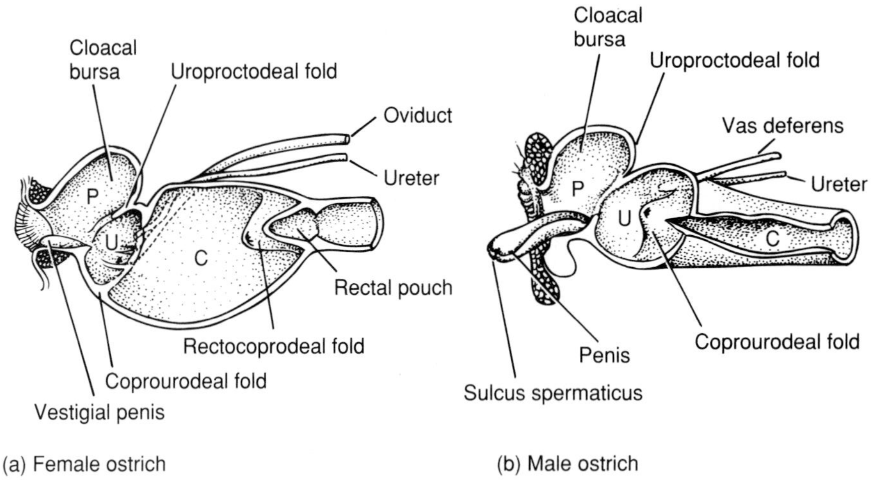


**Figure 14.49** Bird cloacae. (a) Female ostrich cloaca, longitudinal view. (b) Male ostrich cloaca.

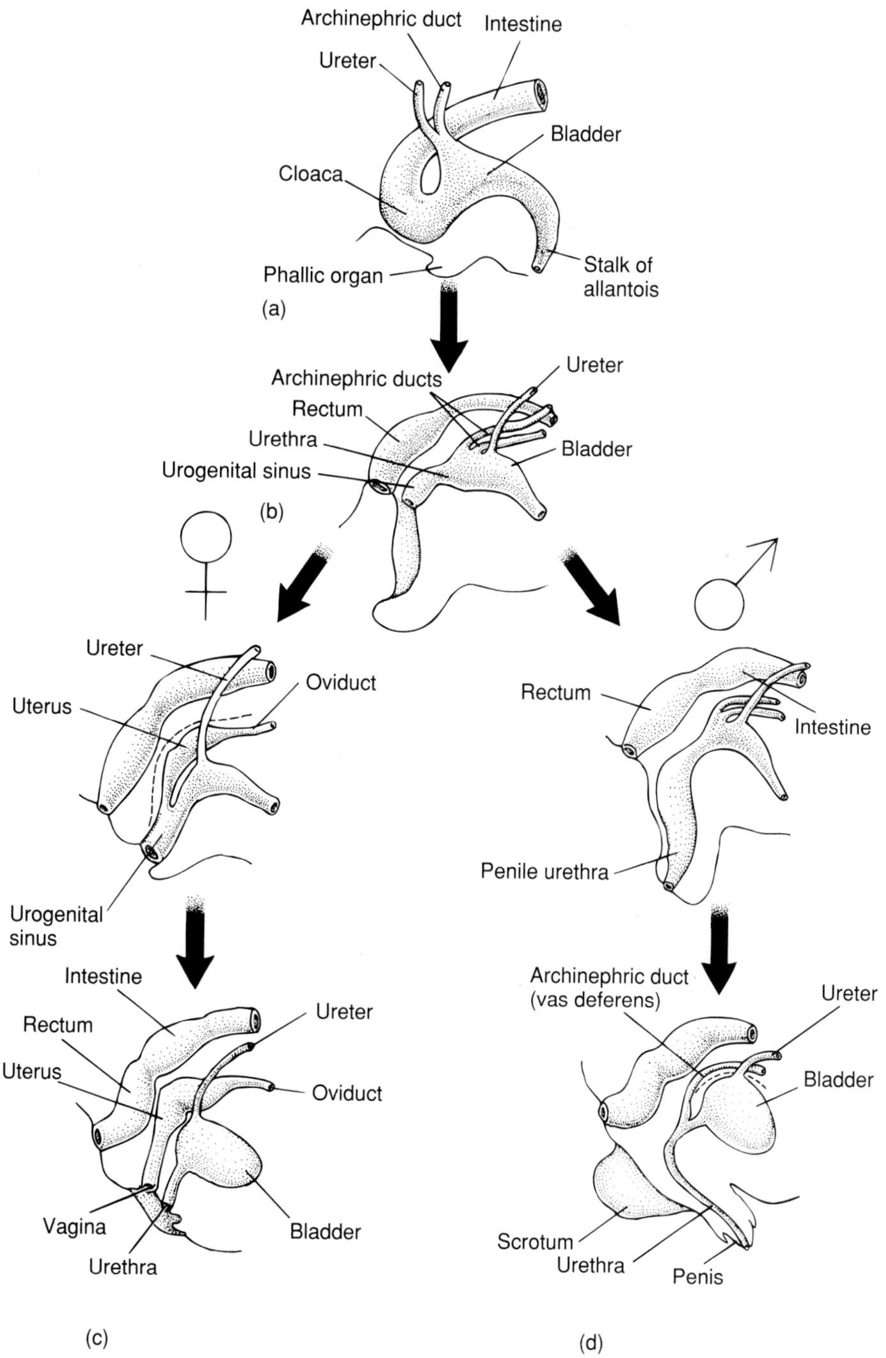


**Figure 14.50** Embryonic derivatives of the urogenital sinus in placental mammals. (a) In the indifferent stage, the cloaca is undivided. (b) The first step toward differentiation is separation of the urogenital sinus from the rectum. (c) In the female, the urogenital sinus divides to form the urethra and the vagina, both with separate external openings. (d) In the male, the urogenital sinus becomes the urethra of the penis and transports both sperm and urine.

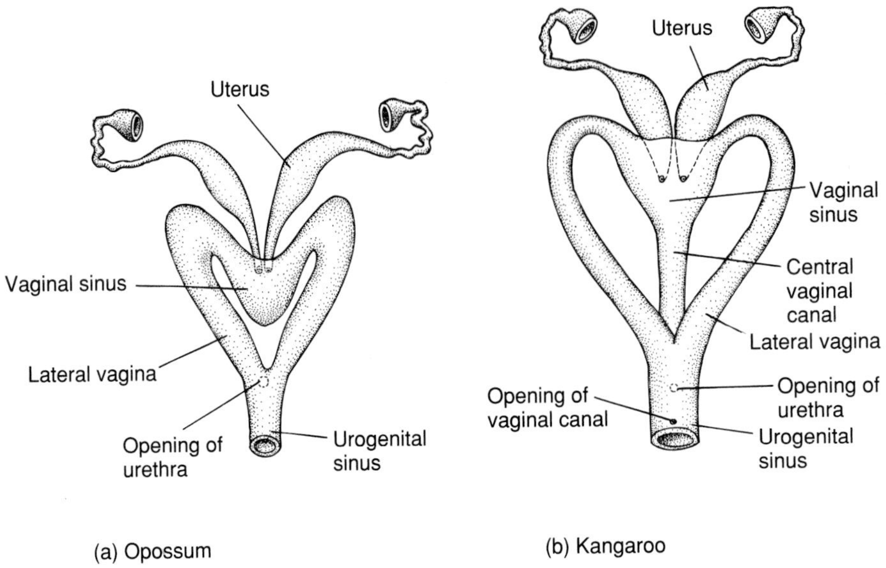


**Figure 14.51** Reproductive organs of female marsupials. (a) Opossum. (b) Kangaroo.

Spermatogenesis is under hormonal control. In a seasonal breeder, sperm are produced only certain times of the year. **Follicle-stimulating hormone (FSH),** a pituitary gonadotrophic hormone stimulates the multiplication of spermatogonia in the seminiferous tubules as the breeding season approaches. With advancing age, there may be a slow decline in the ability of the seminiferous tubules to produce mature sperm, but there is no abrupt cessation comparable to the female **menopause** that occurs in some mammals.

## *External and Internal Fertilization*

External fertilization is common among invertebrates and primitive vertebrates. Eggs and sperm meet outside the body. However, many vertebrates live in environments in which external fertilization is disadvantageous. The tailed frog *Ascaphus*, for instance, lives and mates in fast-moving streams where swift currents might wash away eggs and sperm released into the environment. Internal fertilization via an intromittent organ increases the success of sperm transfer under these conditions.

But internal fertilization offers a further adaptive advantage. The events of courtship and fertilization can be separated from the events of egg deposition. Fertilization does not always occur in an environment that is also suitable for egg deposition. For example, some salamanders mate on land where courtship displays are visible, but dry land offers few favorable sites for the development of salamander eggs. In most salamanders, a spermatophore is taken up by the female during courtship, but eggs are not released at that time. Instead, the sperm are held in the spermatheca until she has found a suitable location for deposition. The eggs are fertilized as they are laid (figure 14.54a–c).

Physiological constraints can restrict the evolution of vivipary in some groups. Among amniotes, calcium for ossification of the embryonic skeleton can be stored in the yolk (e.g., squamates) or in the eggshell (e.g., turtles, crocodiles, and birds). In vivipary, the calcareous eggshell is lost, allowing for efficient exchange between fetal and maternal tissues. However, the shell's calcium reservoirs are lost as well. This may help explain why vivipary is absent among turtles, crocodiles, and birds, groups in which the eggshell is used for calcium storage. Vivipary is common among lizards and snakes that do not use the shell as a calcium reservoir.

In both ovipary and vivipary, the young are carried internally, extending the time between courtship and birth or egg deposition and giving the female a chance to seek safe sites for young to be born or hatch. In vertebrates that regulate their temperature internally or behaviorally, the females retain their embryos, allowing them to develop at a stable temperature. If an ectothermic reptile deposits her eggs under a rock, the eggs will be subjected to environmental fluctuation in temperature. But if she retains them in her body, she can shuttle between sites and on cool days bask in whatever warmth is available to elevate the temperature of the developing embryos within her body.

## *Delays in Gestation*

**Gestation** lasts from conception to hatching or birth. It includes fertilization, implantation (in some species), and development. In some species of mammals, the onset of each stage may be prolonged or delayed. For example, **delayed fertilization** occurs in some bats. Copulation occurs in autumn just before hibernation, but females do not ovulate at that time. Instead, sperm are stored either in the uterus or

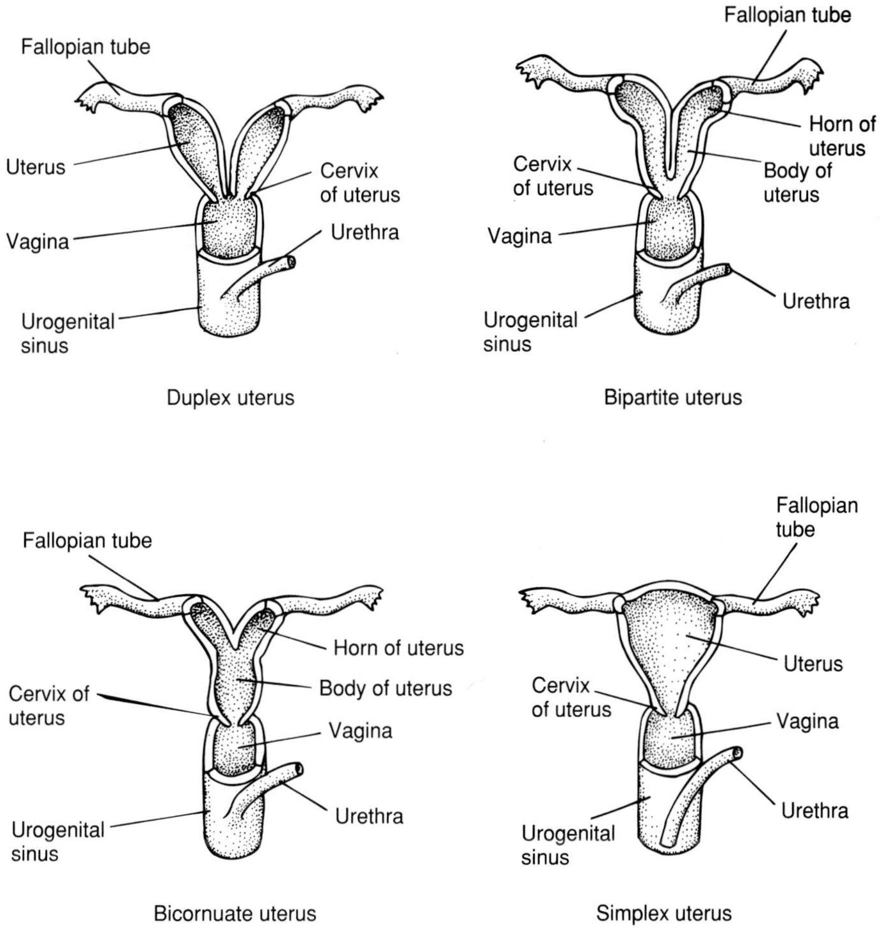


**Figure 14.52** Reproductive organs of female placental mammals. The uterus is characterized by the degree of fusion of the paired uteri.

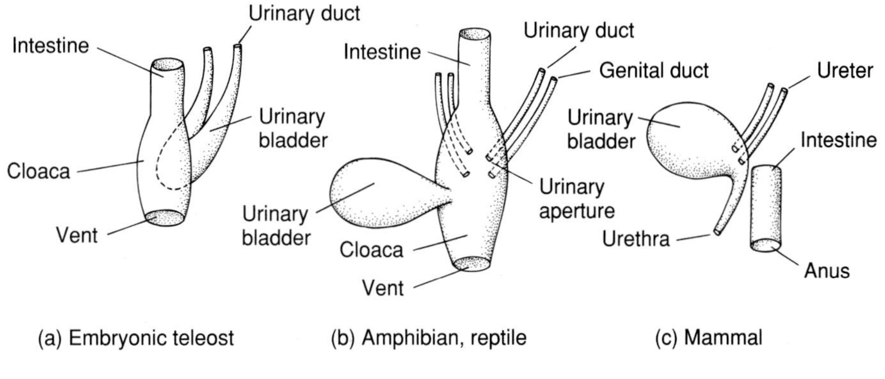


**Figure 14.53** Evolution of the urinary bladder. (a) In teleosts, the intestine and urinary ducts establish separate exits, anus and urinary pores, respectively. As a consequence, the embryonic cloaca is lost in the adult. The teleost urinary bladder, when present, is formed from the expanded ends of the urinary ducts. (b,c) In tetrapods, the urinary bladder is an outgrowth of the cloaca. It empties into the cloaca in amphibians and reptiles (b) but exits via the urethra in mammals (c).

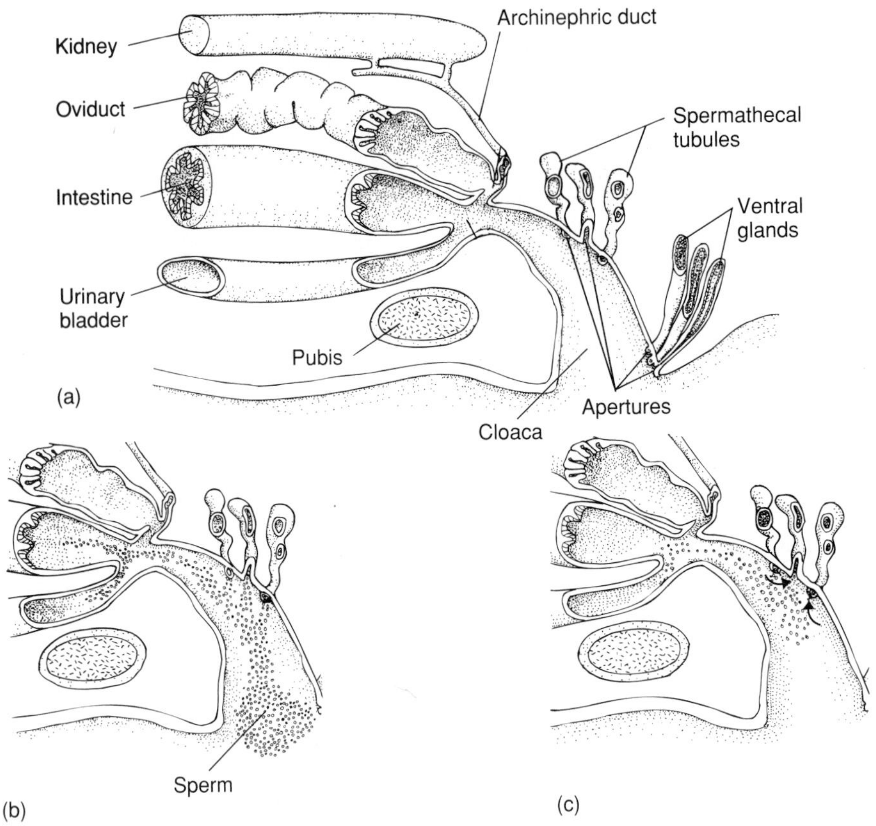

Kidney
Oviduct
Intestine
Urinary bladder
Pubis
(a)
Archinephric duct
Spermathecal tubules
Ventral glands
Apertures
Cloaca
Sperm
(b)
(c)


**Figure 14.54** Sperm storage within the spermatheca of the salamander *Notophthalmus*. (a) Diagram of the urogenital system. A few hours after sperm enter the cloaca (b), they move into the spermathecal tubules (c) where they are stored. In this species, the ova are not released for several months. When they are released, stored sperm are discharged into the cloaca to fertilize the passing eggs.

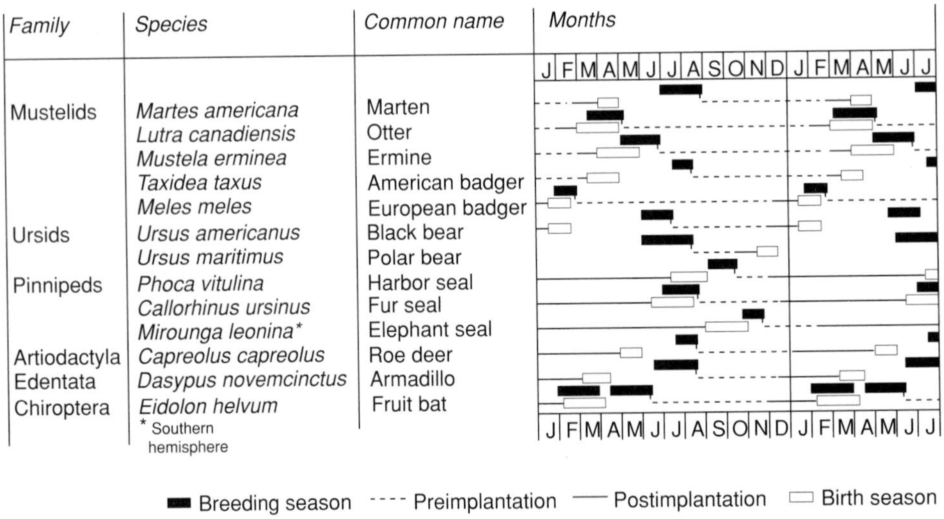

Family
Species
Common name
Months
J F M A M J J A S O N D J F M A M J J
Mustelids
Martes americana
Lutra canadiensis
Mustela erminea
Taxidea taxus
Meles meles
Ursids
Ursus americanus
Ursus maritimus
Pinnipeds
Phoca vitulina
Callorhinus ursinus
Mirounga leonina*
Artiodactyla
Capreolus capreolus
Edentata
Dasypus novemcinctus
Chiroptera
Eidolon helvum
* Southern hemisphere
Marten
Otter
Ermine
American badger
European badger
Black bear
Polar bear
Harbor seal
Fur seal
Elephant seal
Roe deer
Armadillo
Fruit bat
Breeding season
Preimplantation
Postimplantation
Birth season


**Figure 14.55** Delayed implantation. Seasons of breeding, preimplantation, postimplantation, and birth for several species of mammals are shown. To ensure that young are born when resources are most likely available, many mammals have evolved methods to lengthen gestation beyond harsh seasons or times of migration so that birth occurs when conditions are favorable. Delayed implantation occurs following fertilization when the embryo does not immediately implant itself in the uterine wall. Instead, the embryo goes into a stage in which further development is slowed or arrested. Later, after implantation occurs, the pace of embryonic development picks up. Note that a few species give birth during the winter.

the upper vagina. When bats emerge from hibernation several months later, eggs are released, sperm become active, and fertilization finally occurs. Young are born in early summer, a season that is usually characterized by an abundance of insects for food.

In **delayed implantation,** known only in mammals, fertilization and early development occur, but the embryo fails to implant in the uterus. Development is arrested for an extended period, until implantation finally occurs and gestation resumes. Delayed implantation occurs in many members of the weasel family (Mustelidae), bears (Ursidae), and a few other groups (figure 14.55). In most cases, delayed implantation is tied to the annual seasonal cycle. In some marsupials, such as kangaroos and wallabies, however, delayed implantation of the **blastocyst** is tied to the presence of a young kangaroo in the pouch, termed a joey. Suckling by an older joey in the pouch inhibits implantation of the next blastocyst, a type of delay referred to as **embryonic diapause.** In **delayed development,** known from several species of bats, fertilization and implantation occur on schedule, but subsequent growth of the embryo is slow.

Delay in fertilization, implantation, or development increases the time between mating and birth to ensure that young will not be born at an inopportune time (for example, during migration) or when food is scarce (as in the middle of winter). Female caribou give birth immediately after their migration from winter forests to summer tundra. Many species of whales give birth after they migrate from polar seas and arrive in temperate or tropical oceans. Seals give birth when they reach their breeding beaches after an extended migration at sea.

# Selected References


Bentley, P. J. 1966. Adaptations of amphibia to arid environments. *Science* 152:619–23.

Braun, E. J., and W. H. Dantzler. 1972. Function of mammalian-type and reptilian-type nephrons in kidney of desert quail. *Amer. J. Physiol.* 222:617–29.

del Pino, E. M. 1989. Marsupial frogs. *Sci. Amer.* 260(5):110–18.

Dodd, J. M., and M.H.I. Dodd. 1985. Evolutionary aspects of reproduction in cyclostomes and cartilaginous fishes. In *Evolutionary biology of primitive fishes,* edited by R. E. Foreman, A. Gorbman, J. M. Dodd, and R. Olsson. New York: Plenum Press, pp. 295–319.

Dowling, H. G., and J. M. Savage. 1960. A guide to the snake hemipenis: A survey of basic structure and systematic characteristics. *Zoologica* 45:17–28.

Gilkey, J. C. 1981. Mechanisms of fertilization in fishes. *Amer. Zool.* 21:359–75.

Gordon, M. S., B. Schmidt-Nielsen, and H. M. Kelly. 1961. Osmotic regulation in the crab-eating frog *(Rana cancrivora)*. *J. Exp. Biol.* 38:659–78.

Grier, H. J. 1981. Cellular organization of the testis and spermatogenesis in fishes. *Amer. Zool.* 21:345–57.

Griffith, R. W. 1985. Habitat, phylogeny, and the evolution of osmoregulatory strategies in primitive fishes. In *Evolutionary biology of primitive fishes,* edited by R. E. Foreman, A. Gorbman, J. M. Dodd, and R. Olsson. New York: Plenum Press, pp. 69–80.

Guraya, S. S. 1989. *Ovarian follicles in reptiles and birds.* New York: Springer-Verlag.

Halstead, L. B. 1973. The heterostracan fishes. *Biol. Rev.* 48:279–332.

Hardy, M. P., and J. N. Dent. 1986. Transport of sperm within the cloaca of the female red-spotted newt. *J. Morph.* 190:259–70.

Johnson, O. W. 1979. Urinary organs. In *Form and function in birds,* edited by A. S. King and J. McLelland. New York: Academic Press, pp. 183–235.

King, A. S. 1981. Phallus. In *Form and function in birds,* edited by A. S. King and J. McLelland. New York: Academic Press, pp. 107–47.

Lambert, J.G.D. 1970. The ovary of the guppy, *Poecilia reticulata*. *Gen. Comp. Endocrin.* 15:464–76.

Packard, G. C. et al. 1989. How are reproductive systems integrated and how has viviparity evolved? In *Complex organismal functions: Integration and evolution in vertebrates,* edited by D. B. Wake and G. Roth. New York: John Wiley and Sons, pp. 281–93.

Russell, D., R. A. Brandon, E. J. Zalisko, and J. Martan. 1981. Spermatophores of the salamander *Ambystoma texanum*. *Tissue and Cell* 13:609–21.

Schmidt-Nielsen, B., and R. O'Dell. 1961. Structure and concentrating mechanism in the mammalian kidney. *Amer. J. Physiol.* 200:1119–29.

Schroeder, P. C., and P. Talbot. 1985. Ovulation in the animal kingdom: A review with an emphasis on the role of contractile processes. *Gamete Res.* 11:191–221.

Sever, D. M. 1991. Comparative anatomy and phylogeny of the cloacae of salamanders (Amphibia: Caudata). I. Evolution at the family level. *Herpetologica* 47:165–93.

Smith, H. W. 1932. Water regulation and its evolution in the fishes. *Quart. Rev. Biol.* 7:1–26.

Townsend, D. S., M. M. Stewart, F. Harvey Pough, and P. F. Brussard. 1981. Internal fertilization of an oviparous frog. *Science* 212:469–70.

van Tienhoven, A. 1983. *Reproductive physiology of vertebrates.* Ithaca: Cornell University Press.

Wake, M. H. 1980. Fetal tooth development and adult replacement in *Dermophis mexicanus* (Amphibia: Gymnophiona): Fields versus clones. *J. Morph.* 166:203–16.

Wake, M. H. 1986. Urogenital morphology of dipnoans, with comparisons to other fishes and to amphibians. *J. Morph.* (suppl.)1:199–216.

Wallace, R. A., and K. Selman. 1981. Cellular and dynamic aspects of oocyte growth in teleosts. *Amer. Zool.* 21:325–43.

Zalisko, E. J., R. A. Brandon, and J. Martan. 1984. Microstructure and histochemistry of salamander spermatophores (Ambystomatidae, Salamandridae, and Plethodontidae). *Copeia* (3):739–47.

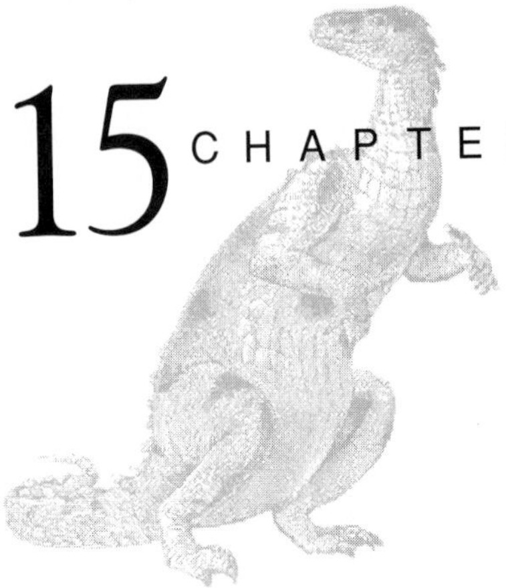

# 15 CHAPTER The Endocrine System



## Survey of Endocrine Organs

Two major control systems preside over activities within the body. One is the nervous system covered in the next chapter, the other is the endocrine system. These control systems are responsible for coordinating activities between organs, increasing organ activity in response to increased physiological needs, and maintaining steady-state conditions.

The endocrine system includes the **endocrine glands,** the chemical messengers, or **hormones** they produce, and the **target tissues** they affect. Endocrine glands are located throughout the body. Hormones are not transported in ducts; instead they are carried by the blood. Although endocrine hormones circulate throughout the body, each one usually affects selected target tissues, so its influence is localized.

Endocrine glands are as varied as the target tissues they control. They preside over reproduction, metabolism, osmoregulation, embryonic development, growth, metamorphosis, and digestion. We begin by looking at the distribution of endocrine organs and the hormones they produce among vertebrate groups.

### Thyroid Gland

#### *Structure and Phylogeny*

The thyroid gland produces, stores, and releases two separate **thyroid hormones** that regulate metabolic rate, metamorphosis, growth, and reproduction. The thyroid hormones are said to be **permissive,** meaning that they "permit" target tissues to be more responsive to stimulation by other hormones, by the nervous system, or possibly by environmental stimuli (such as light or temperature). The thyroid secretes hormones containing iodine. In 1915, **thyroxine,** the first thyroid hormone, was isolated and identified. Another name for this hormone is **tetraiodothyronine,** or $T_4$ for short. A second thyroid hormone identified in 1952 is **triiodothyronine,** or $T_3$ (table 15.1). Initially isolated in mammals, both $T_3$ and $T_4$ are now known to be synthesized in all vertebrates. In cyclostomes, these hormones are stored *intra*cellularly. However, in gnathostomes, the thyroid stores large quantities of hormones *extra*cellularly within the lumina of

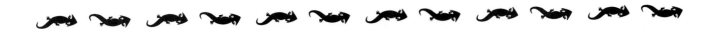

TABLE 15.1 Endocrine Tissues and Secretions in Mammals

| Hormone | Source of Hormone |
|---|---|
| *Adenohypophysis* | |
| Growth hormone (STH) | Pars distalis |
| Prolactin (PRL), Luteotropin hormone (LTH) | Pars distalis |
| Thyrotropin (TSH) | Pars distalis |
| Follicle-stimulating hormone (FSH) | Pars distalis |
| Luteinizing hormone (LH) or Interstitial Cell-stimulating hormone (ICSH) | Pars distalis |
| Adrenocorticotropin (ACTH) | Pars distalis |
| Melanophore-stimulating hormone (MSH) | Pars intermedia |
| *Neurohypophysis* | |
| Antidiuretic hormone (ADH) | Neurons projecting to neurohypophysis from paraventricular and supraoptic nuclei of hypothalamus |
| Oxytocin (OXY) | |
| *Parathyroid* | |
| Parathormone | Chief cells |
| *Thyroid* | |
| Thyroxine (tetraiodothyronine) ($T_4$) | Principal cells |
| Triiodothyronine ($T_3$) | Principal cells |
| Thyrocalcitonin | Parafollicular cells |
| *Adrenal gland* | |
| *Cortex* | |
| Aldosterone | Zona glomerulosa |
| Glucocorticoids | Zona fasciculata, zona reticularis |
| Androgens | Zona reticularis |
| *Medulla* | |
| Norepinephrine | Chromaffin cells |
| Epinephrine | Chromaffin cells |
| *Pancreatic islets* | |
| Insulin | B cells |
| Glucagon | A cells |
| Somatostatin | D cells |
| Pancreatic polypeptide | PP cells |
| *Duodenum* | |
| Cholecystokinin | Intestinal mucosa |
| Pancreozymin | Intestinal mucosa |
| Secretin | Intestinal mucosa |
| *Testis* | |
| Testosterone | Interstitial cells |
| *Ovary* | |
| Estradiol | Theca interna, interstitial cells, granulosa cells (?) |
| Progesterone | Corpus luteum, theca interna |
| Other follicular estrogenic steroids in smaller amounts | |
| *Placenta* | |
| Chorionic gonadotropin | Syntrophoblast |
| Estradiol | Syntrophoblast |
| Estriol | Syntrophoblast |
| Adrenal corticoids | Syntrophoblast |
| Chorionic growth hormone | Syntrophoblast |
| Prolactin | Syntrophoblast |
| ACTH-like substances | Syntrophoblast |

hundreds of tiny irregular spheres, or **follicles.** This condition is unique compared with all other vertebrate endocrine glands. A single layer of epithelial cells called **principal cells** (follicle cells) forms the walls of these follicles (figure 15.1a–c). Principal cells produce a gelatinous **colloid** in which these hormones are stored within the follicles. Principal cells also mobilize thyroid hormones on demand (figure 15.1b). In all vertebrates, the thyroid arises as an outgrowth from the floor of the pharynx. This outgrowth may initially be solid or hollow but soon breaks free of the pharynx (figure 15.2a–d). In some species, it fragments into dispersed masses of follicles. In most, it forms a discrete organ in the throat enclosed in a connective tissue capsule.

In amphioxus and larval ammocoetes, the endostyle is similar in function to a thyroid gland in that it secretes iodine-rich products, but it releases these directly into the digestive tract. During metamorphosis of the ammocoetes, the endostyle is converted into a thyroid gland that releases its hormones into the circulatory system.

In adult cyclostomes and most teleost fishes, the thyroid is fragmented as clumps of follicles scattered widely throughout the pharyngeal region (figure 15.3a). In tetrapods and most other fishes, however, connective tissue encapsulates the thyroid as a single- or double-lobed gland located anterior to the heart (figure 15.3b–l).

### *Function*

Thyroid hormones are attached to the colloid **thyroglobulin** within the follicular lumen and stored in this colloid (figure 15.4a). To manufacture thyroglobulin, principal cells synthesize glucoproteins and unite them with amino acids absorbed from the blood. Just before or as the thyroglobulin enters the follicular lumen, principal cells take up inorganic iodine from the blood and produce iodinated

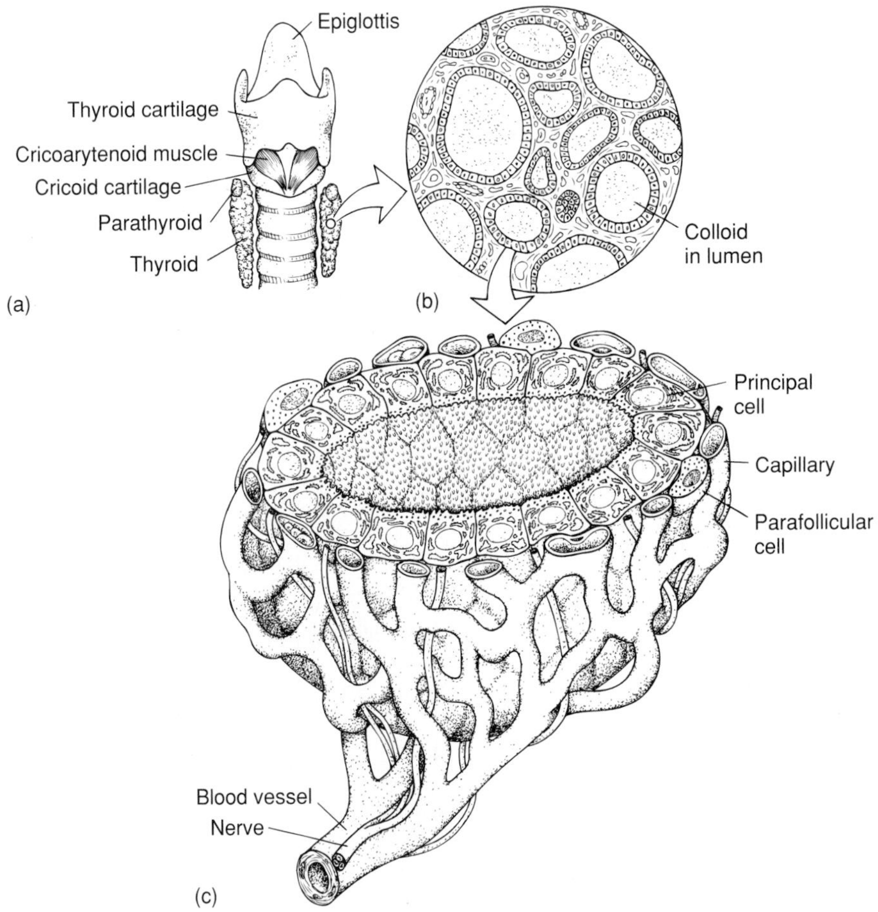


**Figure 15.1** Mammalian thyroid gland. The thyroid gland is composed of numerous spherical follicles. Principal cells in the wall of each follicle produce thyroid hormones and secrete them on demand into capillaries. (a) Ventral view of larynx and trachea of a dog showing paired thyroid and parathyroid glands. (b) Enlarged histological section of the thyroid illustrates follicles and colloid that fills the lumina. (c) Cutaway view of a single thyroid follicle showing the arrangement of principal cells and parafollicular cells (C cells) composing the follicular wall. Note the nerve supply and capillaries embracing the basal regions of these cells.

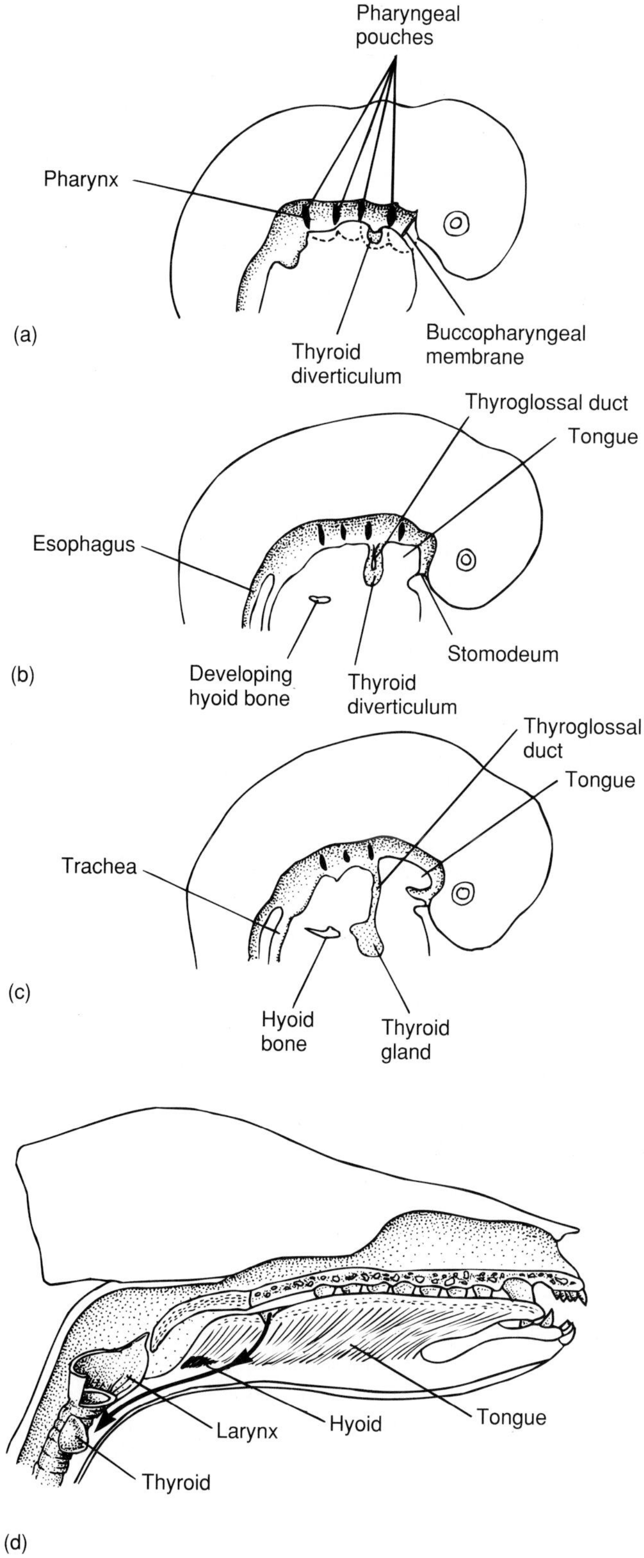


**Figure 15.2** Embryonic development of the mammalian thyroid. (a) Sagittal section through the embryonic pharynx. (b,c) Successive stages in the appearance and growth of the thyroid diverticulum. (d) Location of the thyroid in an adult mammal.

thyroglobulin, the storage form of $T_3$ and $T_4$. Next, synthesis of these thyroid hormones occurs, but the exact method is not understood. Apparently, the coupling or joining of iodinated sections of the folded thyroglobulin molecule produces hormones or their precursors, which remain attached to the thyroglobulin while they are stored in the lumen.

The pituitary hormone **thyrotropin,** or **thyroid-stimulating hormone** (TSH), stimulates principal cells to mobilize these stored thyroid hormones. Principal cells become taller and form apical extensions that envelope the thyroglobulin-hormones complex, allowing these cells to phagocytize and then hydrolize the colloid in lysosomes. Eventually, thyroid hormones are liberated into the nearby capillaries (figure 15.4b).

Thyroid hormones are present in cyclostomes, but their function is not known. The effects of thyroid hormones on target tissues are best known in mammals and birds.

***Metabolism*** In endotherms, thyroid hormones elevate oxygen consumption and heat production by tissues. Injections of thyroid hormones can increase basal metabolic rate several fold. There is little evidence to indicate that thyroid hormones play a similar role in ectotherms, which of course have no "basal" metabolic rate and are not designed to produce heat. The only unequivocal evidence of an effect the thyroid has on ectotherm metabolism can be seen in reptiles when their temperature becomes environmentally elevated. At low temperatures (20°C), lizard tissues are unresponsive to thyroid hormones; however, at preferred temperatures (30°C), tissues respond to thyroid hormones by increasing their oxygen consumption.

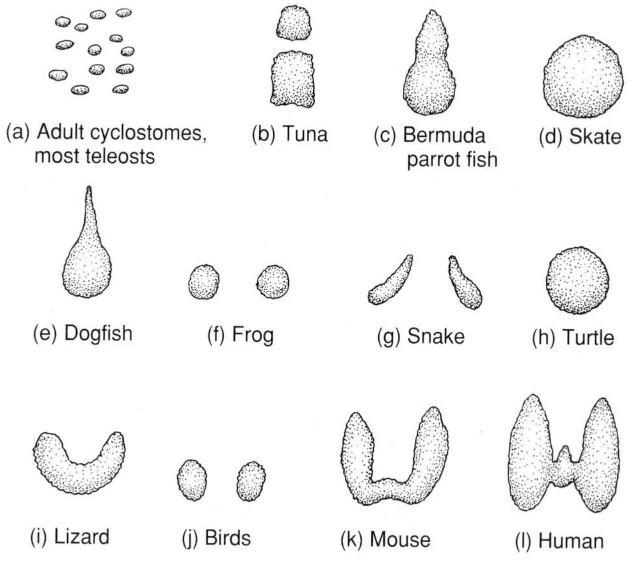


**Figure 15.3** Vertebrate thyroid glands. (a–e) Fishes. (f) Amphibians. (g–i) Reptiles. (j) Bird. (k,l) Mammals.

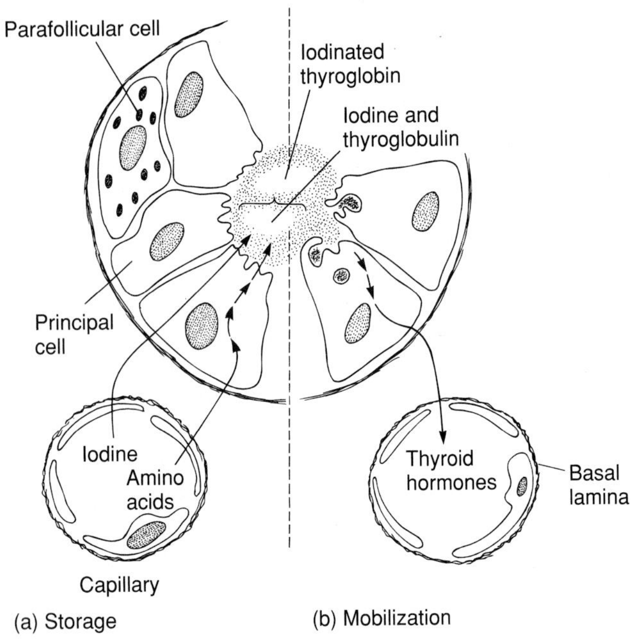


**Figure 15.4** Thyroid secretion and mobilization.
(a) Principal cells take up iodine and amino acids, combine them with thyroglobulin, and secrete the resulting colloid into the lumen. (b) Stimulated by thyroid-stimulating hormone from the pituitary, principal cells mobilize hormones stored in the colloid and release them into adjacent capillaries. The capillary walls are usually fenestrated, but the basal lamina is complete.

***Growth and Metamorphosis*** In birds and mammals, normal growth depends on normal levels of thyroid hormones. **Hypothyroidism,** underproduction of these hormones, results in stunted growth and mental retardation in infants, a syndrome known as **cretinism.** In adults, hypothyroidism results in lethargy and slow mental ability. **Hyperthyroidism,** overproduction of thyroid hormones, results in heightened activity, nervousness, bulging eyes, and rapid weight loss, a medical condition called **Graves' disease.**

The growth of reptiles and perhaps fishes similarly depends on thyroid hormones. For example, enlargement of the thyroid gland occurs when a young salmon (termed a **parr**) is transformed into a **smolt,** its migratory stage in which it travels downstream to the sea. Amphibians differ from most vertebrates in that their thyroid hormones arrest growth but promote metamorphosis.

***Molt*** Thyroid hormones affect loss and subsequent replacement of hair or feathers when an animal **molts.** Thyroxine promotes sloughing or shedding of the skin, which suggests a general effect of thyroid hormones on the vertebrate integument. If thyroid hormones are deficient in birds or mammals, hair or feather growth is impaired, pigment deposition is reduced, and the skin tends to thin. The skin of fishes, amphibians, and reptiles is also adversely affected by thyroid hormone deficiencies.

***Reproduction*** In most vertebrates, elevated levels of thyroid hormones are correlated with gonad maturation and oogenesis or spermatogenesis. Again amphibians seem to be an exception because their thyroid hormones apparently arrest physiological processes that promote reproduction. Surgical removal of the amphibian thyroid glands is followed by accelerated gonadal development.

## Ultimobranchial Body and Parathyroid Gland

The ultimobranchial body and parathyroid gland release hormones with opposite, or antagonistic, effects. The ultimobranchial body secretes **calcitonin** (thyrocalcitonin), which lowers blood levels of calcium and phosphorus. The parathyroid gland secretes **parathormone,** which elevates levels of blood calcium and phosphorus. Because their roles center on the same physiological function, both glands are treated together.

### *Ultimobranchial Body*

Embryonic primordia from the fifth pharyngeal pouches form the **ultimobranchial bodies** (figure 15.5). These bodies are separate, usually paired cell masses located in the neck region of fishes, amphibians, reptiles, and birds. Cyclostomes do not appear to have ultimobranchial bodies. In mammals, their distribution is unique, the primordia being incorporated directly into the thyroid to form a small, dispersed population of **parafollicular cells** (pale cells or C cells) scattered among the principal cells in the walls of the thyroid follicles (figure 15.1c).

The neural crest is the embryonic source of ultimobranchial cells. It is not yet clear whether neural crest cells enter the pharyngeal primordium before it migrates to its site of differentiation or whether neural crest cells colonize the primordium later during differentiation.

### *Parathyroid Gland*

The ventral edges of the embryonic pharyngeal pouches are the source of the **parathyroid glands.** The pouches that contribute vary between species (figure 15.5). The term *parathyroid* describes the close association of this gland with the thyroid gland in mammals, which is embedded in (e.g., mouse, cat, human) or near (e.g., goats, rabbits) the thyroid gland. One or two pairs may be present. However, in amphibians, reptiles, and birds, the parathyroid may be located either on the thyroid or dispersed along the major veins in the neck (figure 15.6a–c). In fishes, the parathyroid is absent. Because it is absent in fishes and does not appear in amphibians until after the gills are lost at metamorphosis, it has been suggested that the role of the parathyroid is preceded phylogenetically by cells in the gills.

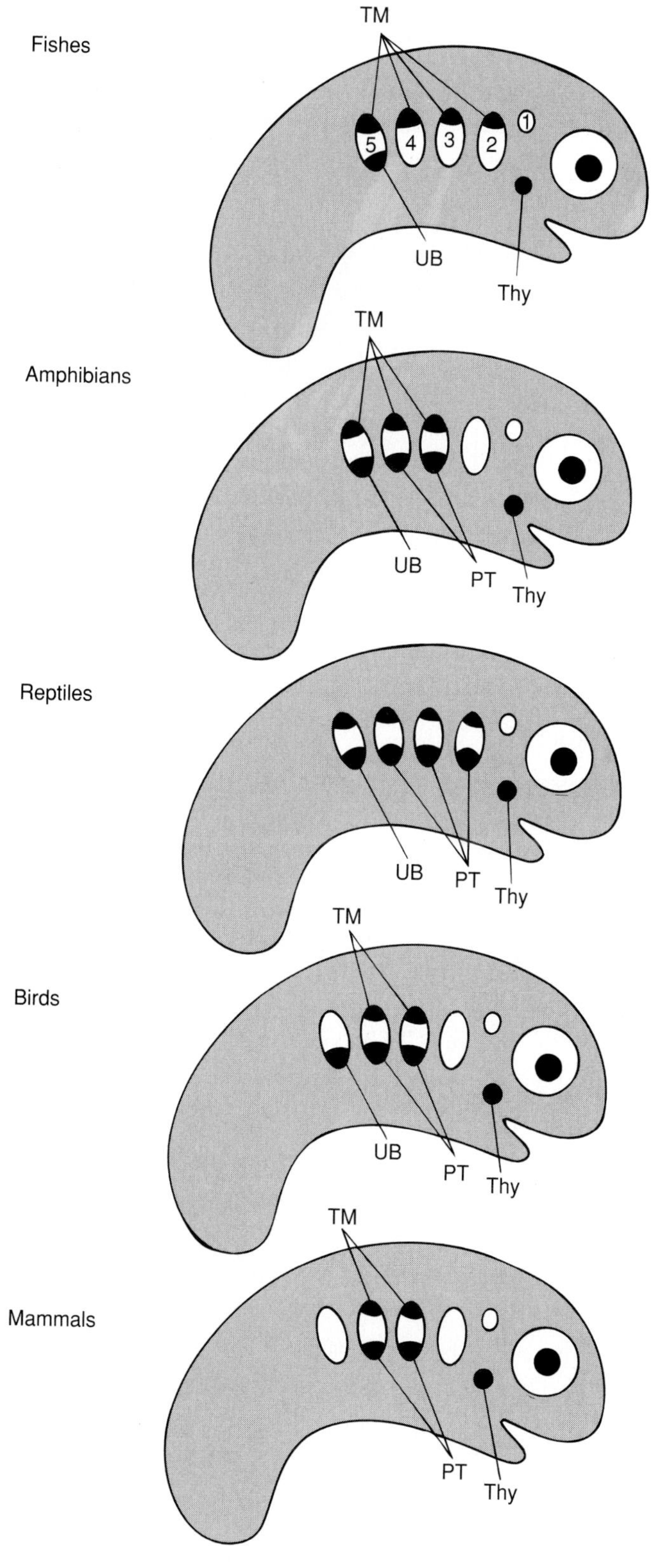


**Figure 15.5** Embryonic contributions of vertebrate pharyngeal pouches to thyroid (Thy), parathyroid (PT), thymus (TM), and ultimobranchial bodies (UB). The reptilian thymus develops from pouches 2 and 3 in lizards, pouches 3 and 4 in turtles, and pouches 4 and 5 in snakes. The ultimobranchial bodies in mammals settle into the thyroid gland as the parafollicular cells (C cells). Pharyngeal pouches are numbered, the first usually being reduced in embryonic development.

Within the parathyroid gland, cells fall into a cord and clump arrangement. **Chief cells,** the most abundant cell type, are probably the source of parathormone. In humans and a few other mammalian species, **oxyphil cells** of unknown function are present as well.

### *Form and Function*

Immediate access to calcium is important in most vertebrates. When birds secrete calcified eggshells or deer grow a new rack of antlers, large amounts of calcium must be rapidly mobilized and transported from one site to another. Maintenance of normal bone strength depends on calcium levels. If levels of calcium in the blood fall too low, skeletal muscles can go into uncontrolled spasms. If blood levels rise too high, osteogenic cells cannot retain calcium in the bone matrix to maintain bone density and strength.

Parathormone secreted by the parathyroid acts to raise blood levels of calcium by promoting kidney retention of calcium, encouraging its absorption across the walls of the digestive tract, and affecting bone deposition. The competing processes of bone deposition and bone removal occur simultaneously and continuously, but they are usually dynamically balanced. Parathormone tips the balance toward net bone removal. As a result, more bone matrix is removed than is deposited; therefore, calcium is liberated from the matrix and taken up by the circulation, causing blood levels of calcium to rise. Calcitonin from the parafollicular cells of the ultimobranchial bodies has the opposite effect. It shifts the balance toward net bone deposition. Calcitonin causes calcium to be extracted from the blood and used to build new bone matrix.

Details of the mechanism controlling calcium levels are still debated, but generally three organs are involved—intestines, kidneys, and bones. The interaction of these organs is depicted in the diagram in figure 15.7. Soft tissues, such as muscles, also require calcium, but their net effect on blood levels of calcium is usually minimal. Calcium in food is absorbed by the intestines. The kidneys can recover all calcium from glomerular filtrate and return it to the extracellular fluid. Control of calcium levels in bone is more complicated. Calcium is incorporated into bone in a crystalline form. The calcium saturation level in bone is lower than it is in the blood, so the net flux of calcium is from blood to bone. Formation of new crystals of bone is a passive process. Although elevated levels of calcitonin correlate with falling blood levels of calcium, the details of how this is accomplished are not yet clear. Parathormone promotes the opposite reaction—the efflux of calcium back into the blood—by promoting osteoclasts that resorb bone. Removal of calcium from bone is an active process. It is not known if the two hormones interact directly or indirectly to inhibit each other's actions.

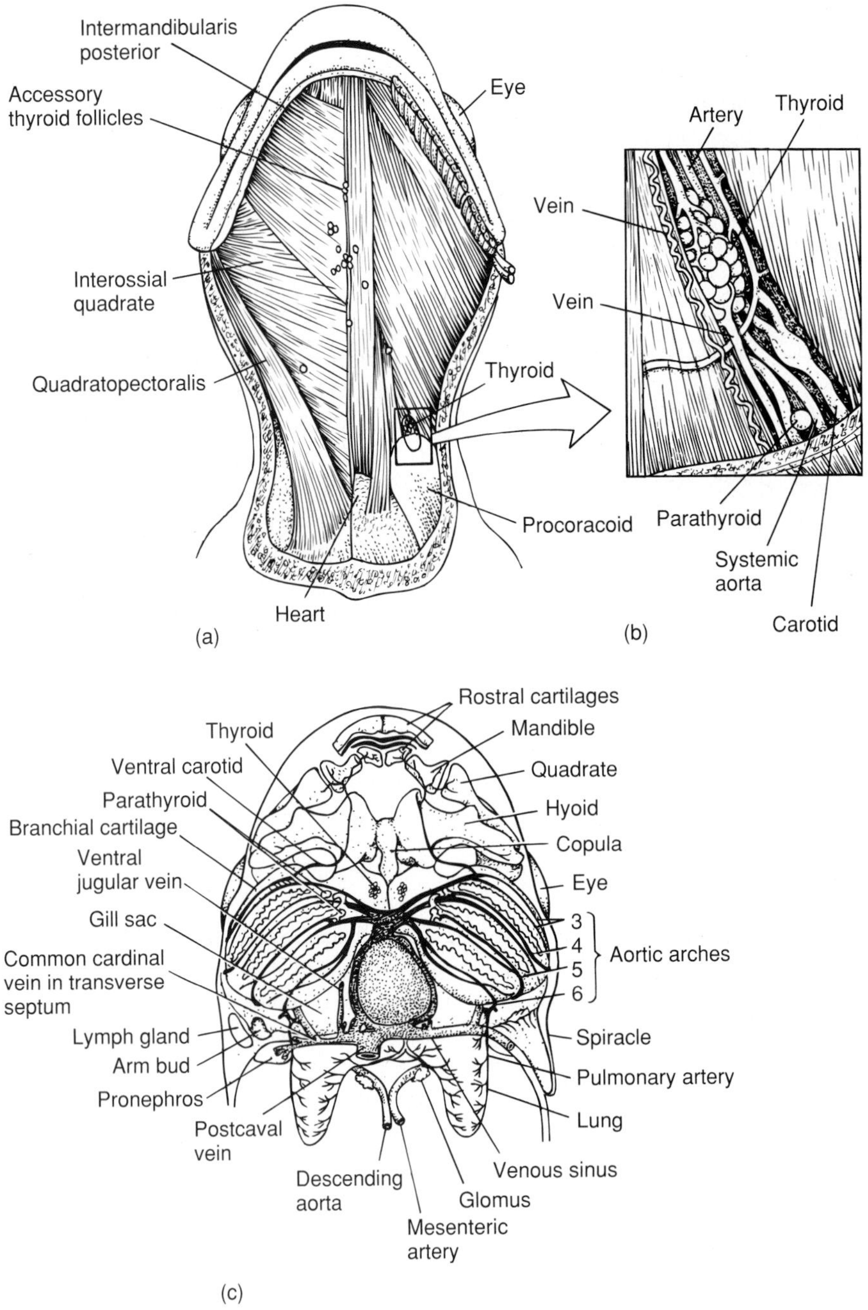


**Figure 15.6** Locations of thyroid and parathyroid glands in amphibians. (a) Exposed ventral view of throat of the salamander *Triturus viridescens*. The left intermandibularis, interossial quadrate, and quadratopectoralis muscles are removed to show deeper muscles and glands. (b) Enlarged area showing thyroid and parathyroid glands with surrounding arteries and veins. (c) Ventral view of bullfrog *(Rana catesbeiana)* throat. Note paired thyroid glands anterior to the heart and paired sets of parathyroids at the bases of the branchial (aortic) arches.

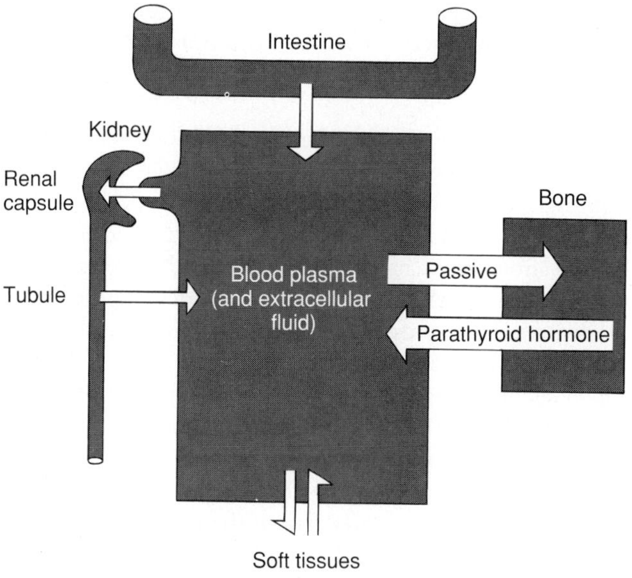


**Figure 15.7** Homeostasis of calcium. Arrows indicate major routes by which calcium is withdrawn or added to the blood plasma and extracellular fluid. Calcium from food is absorbed in the intestines. In the kidney it initially enters the ultrafiltrate that forms in the renal capsule of the kidney, but all calcium ions are recovered and returned to the blood. Calcium moves passively out of the supersaturated blood and crystallizes to form bone. Active bone resorption, under parathyroid hormone stimulation, returns some calcium to the blood.

# Adrenal Gland

## *Structure and Phylogeny*

The **adrenal gland** is a composite organ derived from two separate phylogenetic sources. One is the **interrenal tissue,** or **interrenal bodies** (adrenocortical tissue), that produce **corticosteroid hormones.** Corticosteroids belong to a class of organic compounds called **steroids.** There are three categories of steroids: those involved in (1) water reabsorption and sodium transport by the kidney **(mineralocorticoids),** (2) metabolism of carbohydrates **(glucocorticoids),** and (3) reproduction **(estrogens** and **androgens).** The other phylogenetic source of the adrenal gland is **chromaffin tissue,** or **chromaffin bodies,** that produce **catecholamines.** Catecholamines are **medullary hormones** such as **epinephrine** (adrenaline) and **norepinephrine.** The embryonic origins of these tissues, like their phylogenetic origins, are distinct (figure 15.8). Interrenal tissue arises from splanchnic mesoderm in the region adjacent to the urogenital ridge. Chromaffin tissue arises from neural crest cells.

In adult cyclostomes and teleosts, the interrenal bodies remain separate from the chromaffin bodies. In cyclostomes, the interrenal tissue is scattered along the posterior cardinal veins in the vicinity of the pronephros. Chromaffin cells reside in clusters near but not in contact with the interrenal tissue. In teleosts, the interrenal tissue occurs within the pronephros in scattered clusters or in a strip of tissue around the posterior cardinal veins. In

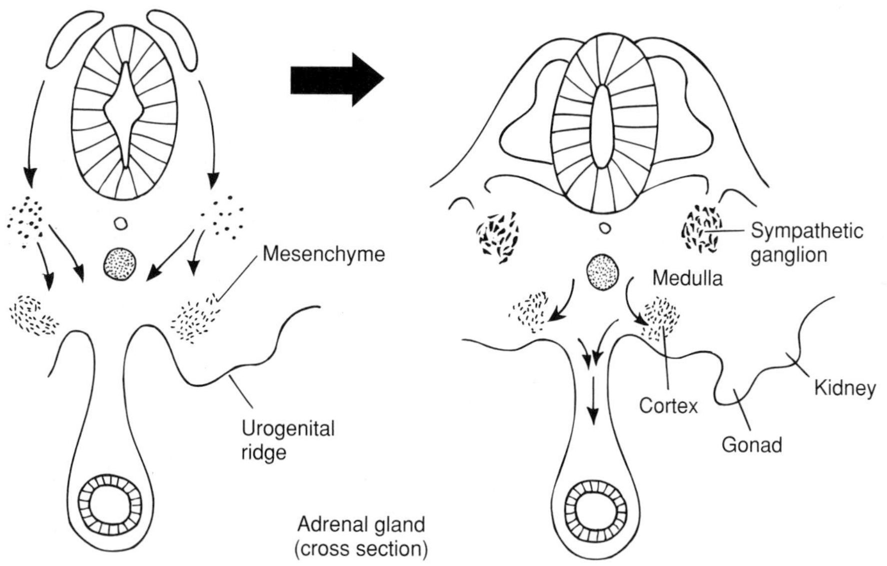


**Figure 15.8** Development of the adrenal gland in a mammalian embryo (cross-sectional views). Mesenchyme adjacent to the urogenital ridge forms the adrenal cortex. Arriving neural crest cells take up residence within the cortex to form the adrenal medulla.

elasmobranchs, the interrenal tissue forms distinct glands along the edges of the kidneys, but the chromaffin tissue is still separate, consisting of arrays of cell clusters between and anterior to the kidneys (figure 15.9a,b). In amphibians, interrenal and chromaffin tissues mingle or reside adjacent to each other and form strands or rows of adrenal tissue (figure 15.9c). The two tissues also mingle in reptiles and birds, although the adrenal glands in amniotes tend to be distinct structures located on or close to the kidneys (figure 15.9d–f). In mammals, for the first time, interrenal and chromaffin tissues form a **cortex** (from interrenal tissue) and a **medulla** (from chromaffin tissue) to create the composite adrenal gland (**suprarenal;** figure 15.9g,h).

### *Function*

In mammals, the adrenal cortex produces corticosteroids. Histological studies have shown three zones within the adult adrenal cortex (figure 15.10). Cells of the outermost **zona glomerulosa** region are small and compact. The kidney releases the hormone **renin,** which leads to a series of events that ultimately stimulate cells of the zona glomerulosa to release mineralocorticoids (e.g., **aldosterone**). Mineralocorticoids in turn affect the kidney's ability to retain water. Cells of the middle **zona fasciculata** region of the adrenal cortex are arranged in rows or cords with blood sinuses between them. **Adrenocorticotropic hormone** (ACTH) released by the pituitary stimulates cells of the zona fasciculata to secrete glucocorticoids. Cells of the third and innermost cortical region, the **zona reticularis,** are small and compact. They are controlled by the pituitary to secrete androgens and additional glucocorticoids.

In many mammals (e.g., primates), an extensive **fetal zone** occupies the periphery of the adrenal cortex prior to birth. This zone is responsible for producing circulating steroids that are chemical precursors of the estrogens synthesized in the placenta. Failure of the fetal zone to function terminates gestation and results in premature birth. Normally, the fetal zone of the adrenal gland ceases to function at birth and dramatically declines in size thereafter.

Thirty or more corticosteroids have been isolated from the mammalian cortex, but most of these are not secreted. Those not secreted seem to be intermediates in the synthesis of definitive hormones released into the blood. In nonmammalian vertebrates, zonation of the interrenal tissue is less conspicuous. Distinct histological regions have been found in anurans, reptiles, and birds, but these may be seasonal.

In vertebrates other than mammals, the cortical hormones primarily regulate sodium transport. In addition to transport of sodium through the walls of kidney tubules, cortical hormones are thought to control sodium transport through the rectal glands in chondrichthyans, the gills and digestive tract in teleosts, the skin and urinary bladder in amphibians, and the salt glands in reptiles and birds.

The medulla, composed of chromaffin tissue, forms the core of the mammalian adrenal gland (figure 15.10). Unlike the cortex, no distinct histological regions are recognized in the adrenal medulla. Catecholamines produced in this region prepare the organism to meet threats or short-term challenges.

Blood supply reaches the cortex through the connective tissue capsule. Blood percolates through the sinuses, bathing cords of cortical tissue, and enters the veins within the medulla. In addition, the medulla is supplied by blood vessels from the capsule that pass without branching through the cortex but break up into a rich capillary network around the cords and clumps of chromaffin cells in the medulla. Thus, the mammalian medulla receives a dual blood supply, one directly from the capsule and one from the cortical sinuses. This second vascular supply via the cortical sinuses places cells within the medulla downstream from the cortex; therefore, cortical hormones released into the blood sinuses are carried first to the medulla and act on it before they leave the adrenal gland. The advantages of this chemical support of chromaffin cell function are not entirely clear. In higher vertebrates, interrenal tissue produces glucocorticoids that control carbohydrate metabolism. In this capacity, interrenal tissue, like chromaffin tissue, affects metabolic activity. Thus, the close association of interrenal and chromaffin tissues established by blood vessels that serve both might be of some advantage in synchronizing their activities.

## Pancreatic Islets

### *Structure and Phylogeny*

The **pancreas** is a composite gland consisting of exocrine and endocrine portions (figure 15.11a). The **exocrine** portion consists of **acini** that secrete digestive enzymes into ducts. The endocrine portion, the **pancreatic islets** (islets of Langerhans), constists of masses of endocrine cells associated with the exocrine pancreas (figure 15.11b,c). In cyclostomes and most teleost fishes, exocrine and endocrine portions of the pancreas are adjacent to each other although they are separate groups of glandular tissue (figure 15.12). In hagfishes, islets are found at the base of the common bile duct, and in lampreys, they are embedded within the mucosal wall of the intestine and even within the liver. In chondrichthyans and crossopterygians *(Latimeria)*, islets occur around the ducts of and within the exocrine pancreas. In most bony fishes, isolated masses of pancreatic islet tissue known as **Brockman bodies** are scattered along the liver, gallbladder, bile ducts, abdominal blood vessels, and surface of the intestines. In a few bony fish species, islets accumulate in the exocrine pancreas. In most tetrapods, the endocrine islets are typically distributed evenly in small clumps. In many birds and in the toad *Bufo*, they form lobes embedded within the exocrine portion of the pancreas.

Both exocrine acini and endocrine islets differentiate within the **pancreatic diverticulum,** which grows out from

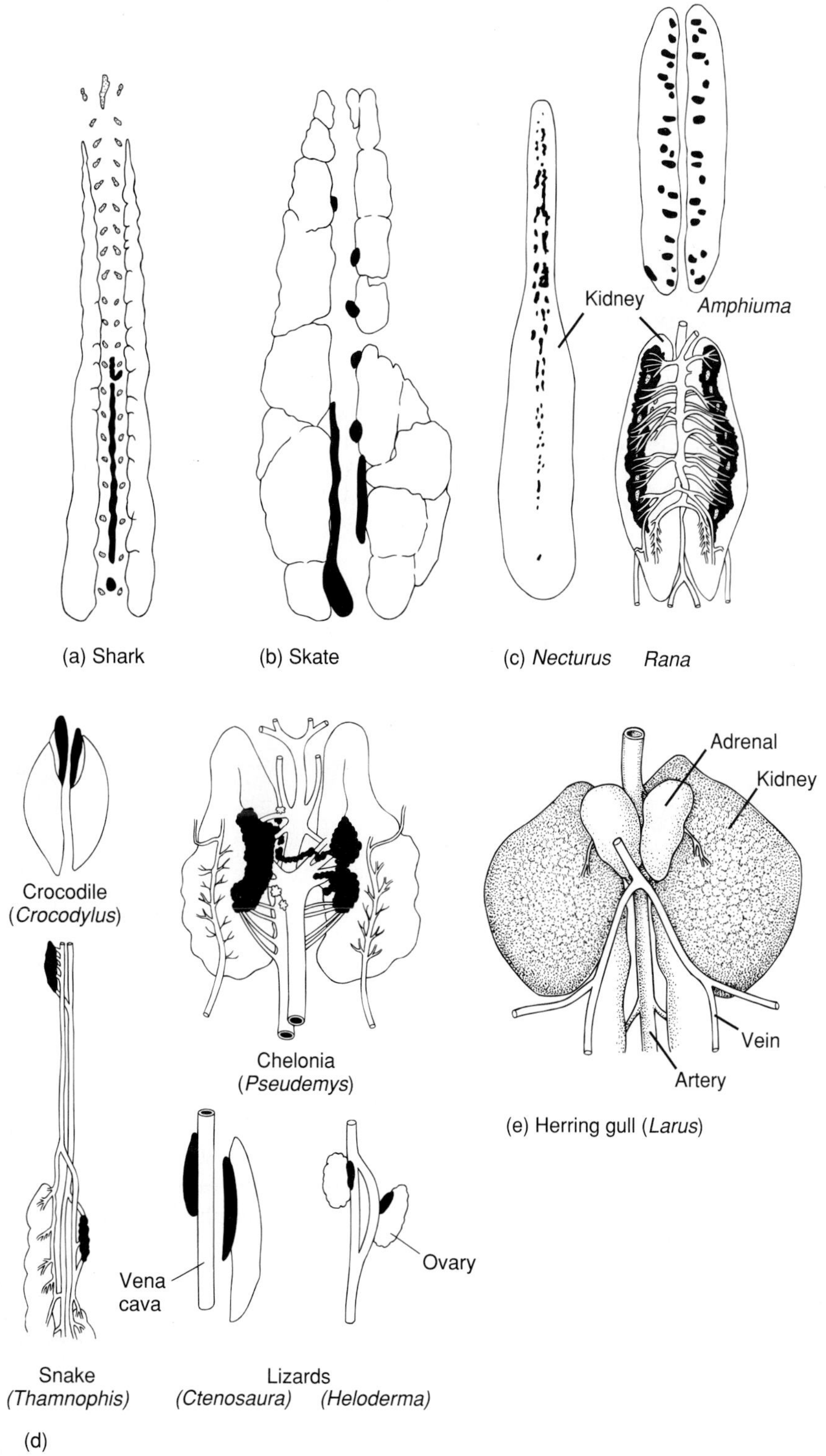


**Figure 15.9** Vertebrate adrenal tissues. Interrenal tissues of elasmobranchs (a,b), amphibians (c), reptiles (d). Location of adrenals in relationship to kidneys in a bird (e). Bird adrenal glands in cross section (f). Locations of mammalian adrenal glands (g). Cross section of mammalian adrenal glands (h). *(continued)*

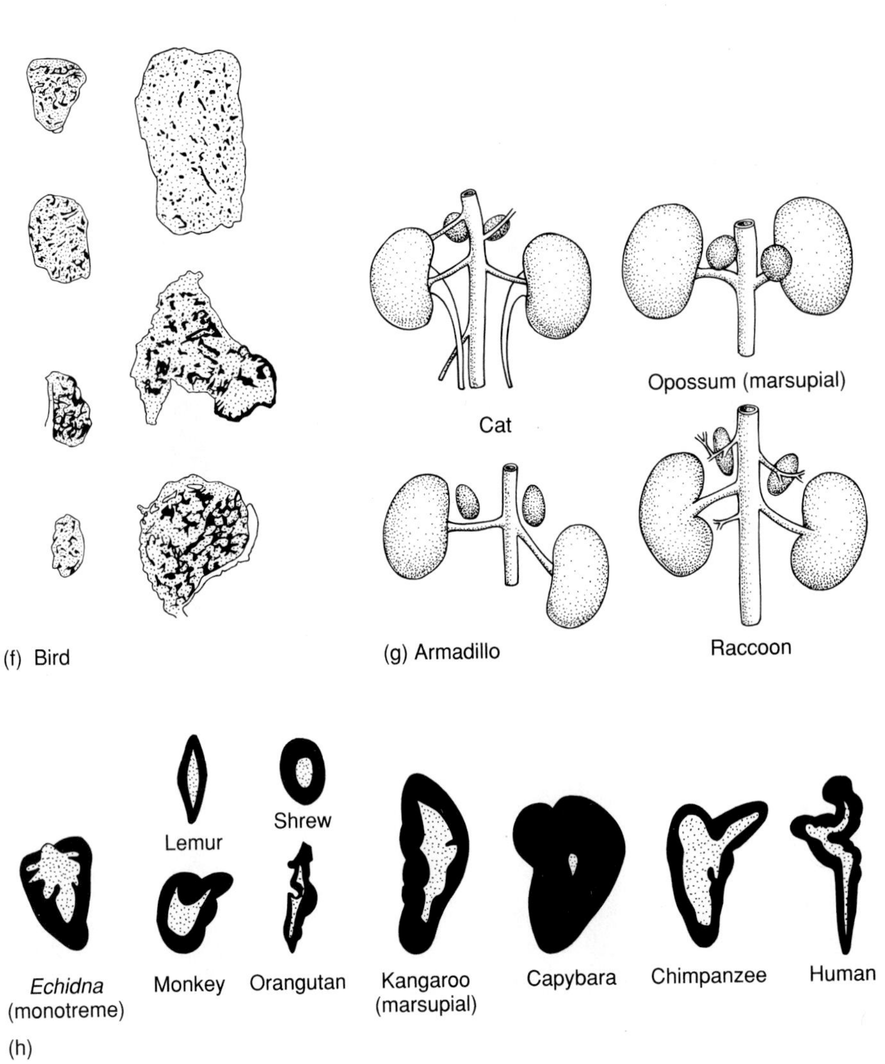


**Figure 15.9** (continued)

the embryonic gut and pushes its way through surrounding mesenchyme. Transplants of marked neural crest cells from quail into early chick embryos reveal that these transplanted neural crest cells give rise to parasympathetic ganglia in the chick pancreas, but apparently the neural crest cells make *no* contribution to the pancreatic islets.

## *Function*

With special stains, up to four cell types can be distinguished within the pancreatic islets of most vertebrates (table 15.2). **Insulin** is produced by **B cells** of the islets. Insulin augments intracellular metabolism of glucose and inhibits the breakdown of glycogen in the liver, but its most important function is to bind to cell membranes and promote the entry of glucose into cells. Thus, blood levels of glucose fall as intracellular levels of glucose rise. If insulin production is too low, glucose is unable to enter cells, builds up in the blood, and is excreted in the urine. This condition is termed **diabetes mellitus,** meaning "sweet urine" disease. In earlier days, physicians used their taste buds to diagnose this disorder.

**Glucagon** is produced by **A cells** of the pancreatic islets. Its effects are opposite to those of insulin because it stimulates glucose levels to rise in the blood. This action appears to be mediated through glucagon's effect on the liver. Glucagon promotes the conversion of stored glycogen into glucose, which enters the blood.

**Somatostatin** is produced by **D cells** of the islets to promote the growth of skeletal and soft tissues in young individuals. It also inhibits secretion of both insulin and glucagon, but the physiological significance of this is not known. **Pancreatic polypeptide** (PP) is secreted by **PP cells** in the islets and is usually released into the blood following a meal. This hormone aids in controlling such gastrointestinal activities as promoting the flow of gastric juice, especially hydrochloric acid, in the stomach.

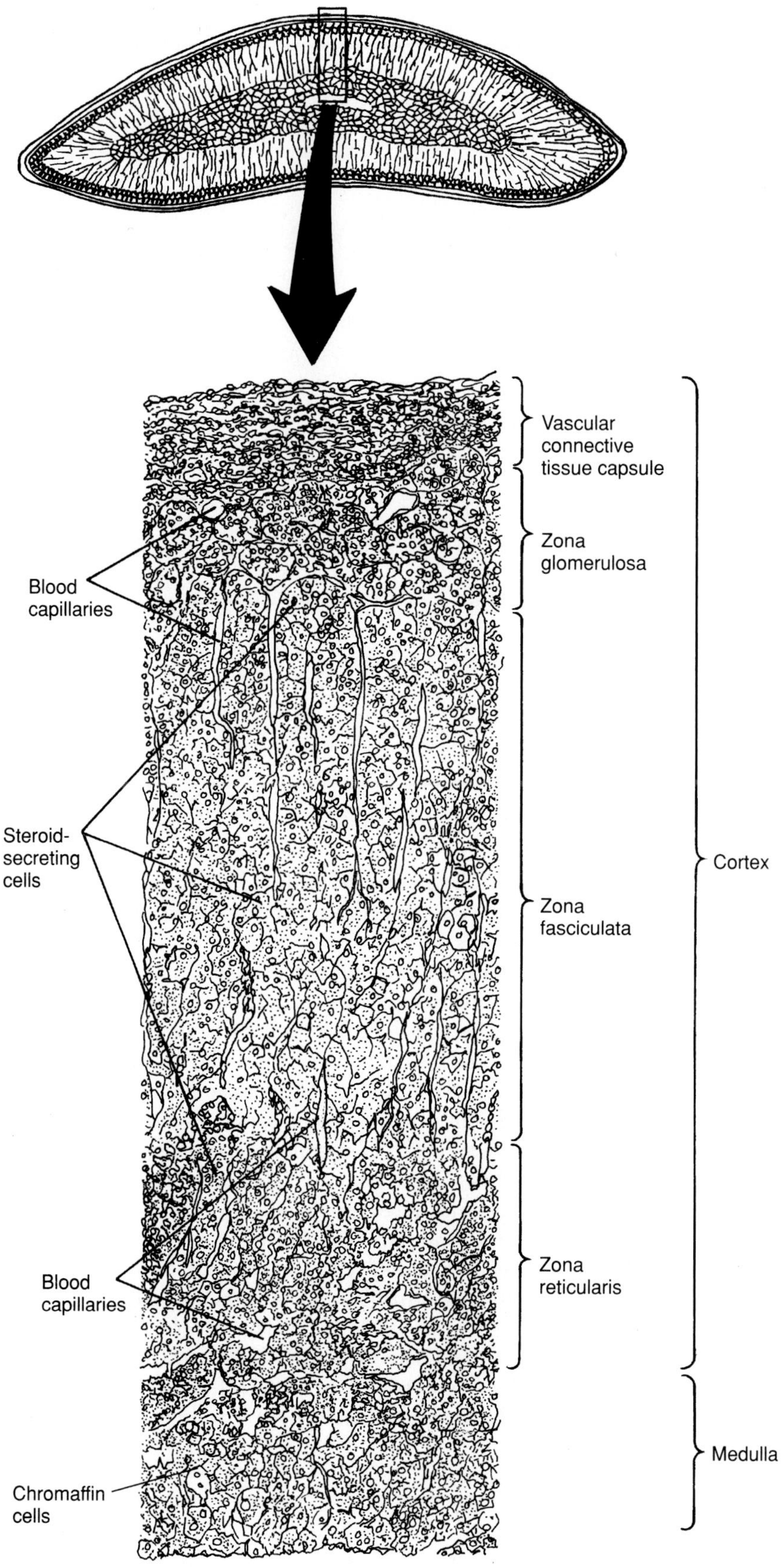


Adult mammal

**Figure 15.10** Zones within the adrenal glands of adult mammals. Three zones are recognized within the adrenal cortex—zona glomerulosa, zona fasciculata, and zona reticularis. The medulla is composed of nonregionalized chromaffin tissue of neural crest origin.

TABLE 15.2 Distribution of the Endocrine Pancreas Cell Types Among Vertebrate Groups

| Class | A Cells | B Cells | D Cells | PP Cells |
|---|---|---|---|---|
| *Agnatha* | | | | |
| Hagfishes | – | ++++ | + | ? |
| *Chondrichthyes* | | | | |
| Elasmobranchs | +++ | ++++ | + | + |
| *Osteichthyes* | | | | |
| Teleosts | +++ | ++++ | + | + |
| *Amphibia* | | | | |
| Anura and Urodela | +++ | ++++ | + | + |
| *Reptilia* | | | | |
| Lepidosauria | ++++ | + | + | + |
| Crocodilia | +++ | +++ | + | ? |
| *Aves* | +++++ | +++ | + | + |
| *Mammalia* | ++ | +++++ | + | + |

*Note:* The number of plus signs (+) represents the relative abundance of each cell type in a group and should not be construed as a precise ratio. The minus sign (–) indicates no cells present.

# Pituitary Gland

## Structure

The **pituitary gland,** or **hypophysis,** is found in all vertebrates. The name *hypophysis* is a recent term inspired by its position beneath the brain (*hypo-* means under and *-physis* refers to growth). The name *pituitary* is centuries old and refers to the mistaken view that this gland produces slime or viscous mucus called *pituita* (phlegm). Although small, this gland has pervasive effects over most of the body's activities. The pituitary has two embryonic sources. One source is the **infundibulum,** a ventral outgrowth from the **diencephalon** of the brain. The other is Rathke's pouch, a diverticulum from the **stomodeum,** which grows dorsally and becomes associated with the infundibulum (figure 15.13a,b). The infundibulum retains its connection to the brain and becomes the **neurohypophysis.** Rathke's pouch is pinched off from its connection to the stomodeum, and becomes the **adenohypophysis** (figure 15.13b–d).

The adenohypophysis and neurohypophysis in turn differentiate into regions that we recognize by their tissue arrangements (cords and clumps), staining properties (acidophils, basophils, and chromophobes), or anatomical position. Three distinct regions subdivide the adenohypophysis—the **pars distalis,** the **pars tuberalis,** and the **pars intermedia** (figure 15.13e). In all vertebrates, the pars distalis is the major portion of the adenohypophysis and the source of a variety of hormones. Often it is differentiated into lobes (cephalic and caudal) or subregions (proximal and rostral). The pars tuberalis is located anterior to the pars distalis. Its function is not well understood but it is found only in tetrapods and is implicated in reproductive processes. The pars intermedia adjoins the neurohypophysis and often retains a cleft, a remnant of the embryonic lumen of Rathke's pouch.

The neurohypophysis consists of up to two subdivisions—the **pars nervosa** and the more anterior **median eminence.** Each of these regions has its own vascular supply. A short portal system between them places the adenohypophysis downstream from the median eminence. The pars nervosa has an extensive blood supply from the general body circulation, which is separate from the supply to the adenohypophysis of the pituitary. The descriptive terms **anterior** and **posterior lobes** are avoided in this book because they are *not* synonymous with the embryonic divisions of the pituitary. Instead they refer to anatomical divisions. The term *posterior lobe* actually includes parts derived from both embryonic sources (table 15.3). The preferred terms adenohypophysis and neurohypophysis divide the pituitary according to its embryonic origin from Rathke's pouch and the infundibulum, respectively.

## Phylogeny

***Fishes*** The size and organization of the pituitary are quite variable even among vertebrates of the same class. In hagfishes, the embryonic sources of the pituitary differ from those of other vertebrates. As in other vertebrates, the

**TABLE 15.3 Divisions of the Pituitary Gland**

| Embryonic Source | Embryonic Divisions | | Anatomical Divisions |
|---|---|---|---|
| Rathke's pouch | Adenohypophysis | Pars tuberalis | Anterior lobe |
| | | Pars distalis | |
| | | Pars intermedia | Posterior lobe |
| Infundibulum | Neurohypophysis | Pars nervosa | |
| | | Infundibular stalk and median eminence | |

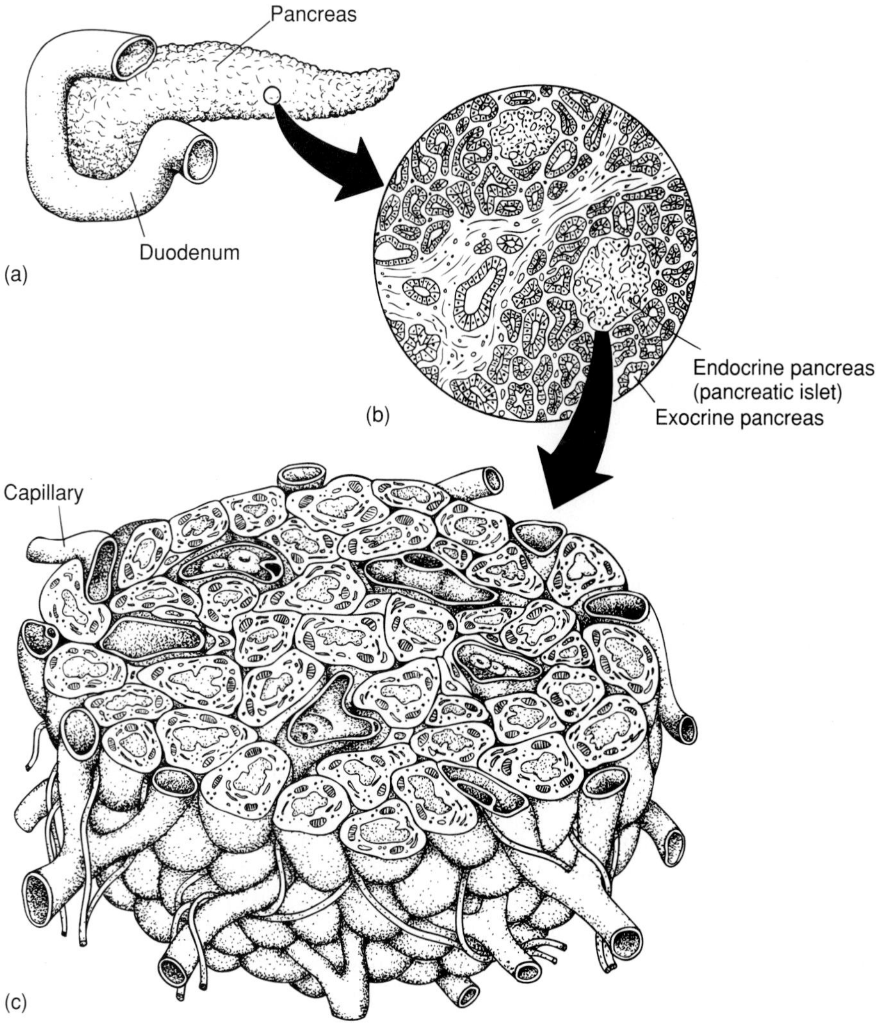


**Figure 15.11** Mammalian pancreatic islets. (a) The pancreas is composed of an exocrine gland and an endocrine gland. (b) Patches of heavily vascularized tissue called pancreatic islets form the endocrine pancreas. (c) Enlarged section of endocrine pancreas.

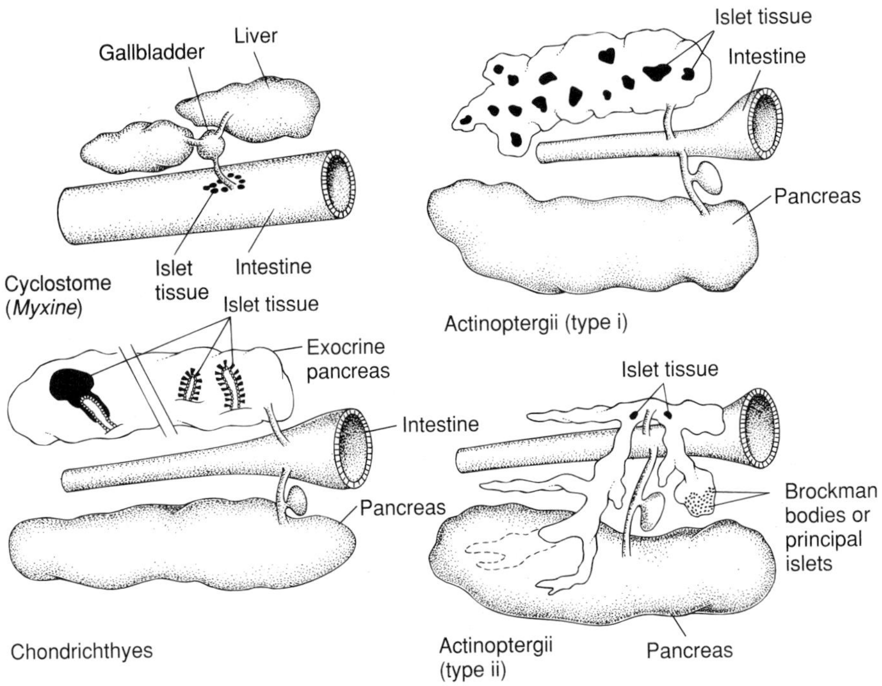


**Figure 15.12** Distribution of pancreatic islets among vertebrates. Shading indicates endocrine tissue (pancreatic islets), and dotted areas indicate exocrine tissue that secretes digestive enzymes. Two general arrangements of pancreatic tissue are known in actinopterygians, type i and type ii.

hagfish neurohypophysis is a hollow elongated sac that extends from the diencephalon of the brain, but a median eminence is absent. The hagfish adenohypophysis appears to arise from endoderm rather than from stomodeal ectoderm. It consists of patches of cells embedded in a dense connective tissue layer but undifferentiated into regions. Thus, the hagfish adenohypophysis may not be homologous with other vertebrate pituitaries.

In lampreys, although a median eminence is absent, in most other respects the pituitary closely resembles that of other fishes (figure 15.14). The neurohypophysis of lampreys extends from the ventral part of the brain and contacts the adenohypophysis. The adenohypophysis arises as an ectodermal pocket but usually retains its connection with the olfactory organ until metamorphosis. Both a pars intermedia and a pars distalis are present. The pars distalis is subdivided further into a **rostral** and a **proximal pars distalis** (table 15.4).

In the pituitary of gnathostome fishes, at least two regions are recognized typically in the adenohypophysis (pars intermedia and pars distalis) and two regions in the neurohypophysis (pars nervosa and median eminence; figure 15. 14). The elasmobranch pituitary exhibits additional features. Unique to elasmobranchs is a forward projection from the pars distalis termed the **ventral lobe,** which some endocrinologists call the **pars ventralis.** The function of the ventral lobe is unknown, although some investigators argue that it is homologous to the pars tuberalis of tetrapods because it secretes similar hormones that promote reproductive processes. The **saccus vasculosus** of the elasmobranch pituitary is a structural specialization derived from the hypothalamus and located above the neurohypophysis, but its function is still unknown. A vascular portal system is present between the median eminence and the pars distalis. As in lampreys, the pars distalis of elasmobranchs is subdivided into a rostral pars distalis and a proximal pars distalis. In osteichthyans other than dipnoans, rostral and proximal subdivisions are recognized within the pars distalis, the saccus vasculosus is present, and the ventral lobe is absent (figure 15.14 and table 15.4).

***Tetrapods*** The pars tuberalis appears in amphibians and persists in most later tetrapods (figure 15.14). Thus, the tetrapod pituitary characteristically consists of an adenohypophysis with three subdivisions (pars intermedia, pars distalis, and pars tuberalis) and a neurohypophysis that retains two subdivisions (pars nervosa and median eminence). In amphibians, the pars distalis shows no regionalization and its function has not yet been demonstrated. Reptilian pituitaries conform to the tetrapod pattern but are remarkably varied in size and shape. The adenohypophysis of snakes is lobed and a cleft may be present in some reptiles. Within

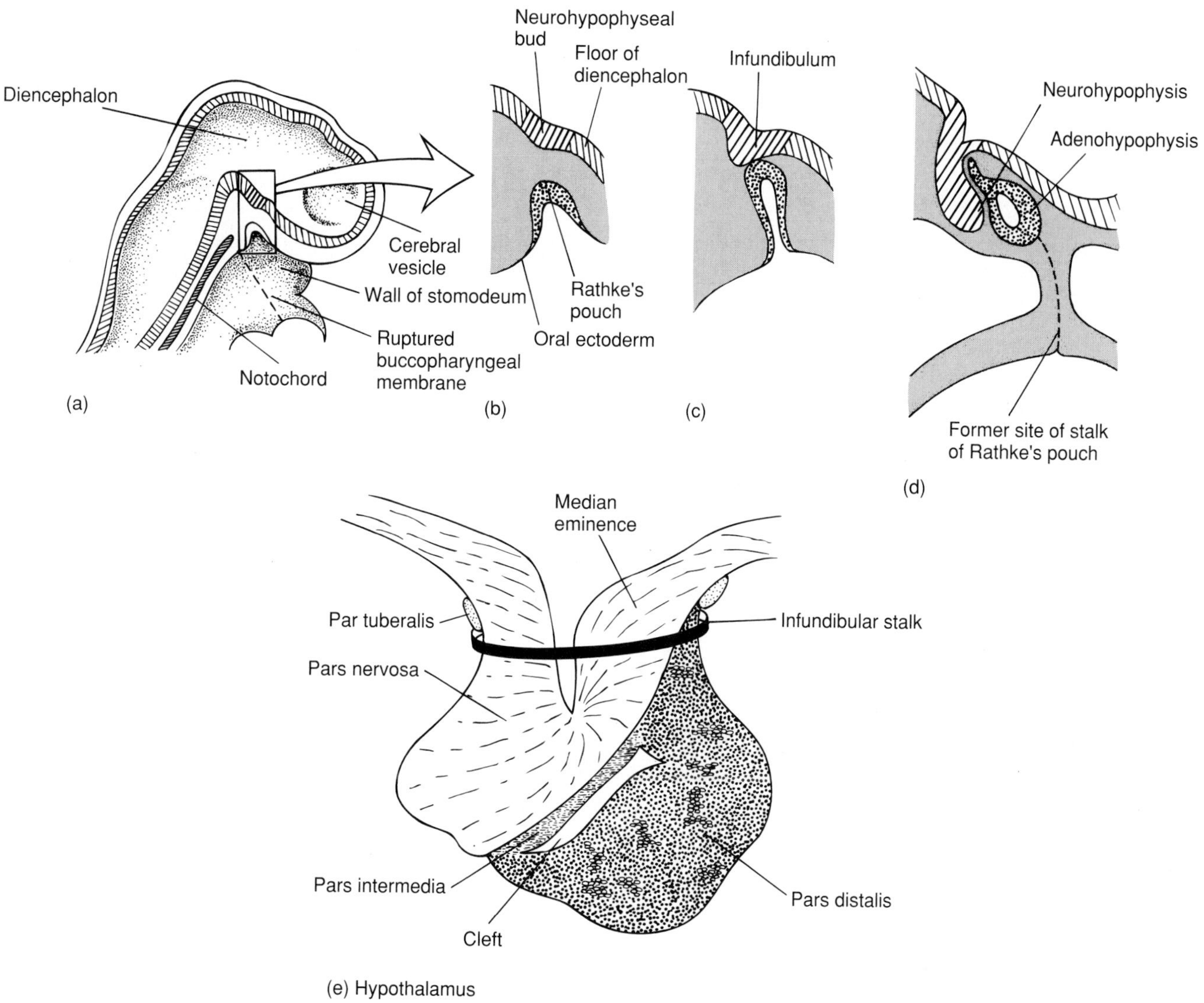


**Figure 15.13** Development of the vertebrate pituitary. (a) Sagittal section of young embryo showing formation of Rathke's pouch and rudimentary infundibulum. (b–d) The two diverticula make contact during embryonic development, and Rathke's pouch breaks free from its source in the stomodeum. (e) Anatomy of the adult pituitary gland. Note how the two embryonic sources are combined.

the reptilian pars distalis, **cephalic** and **caudal lobes** are recognized. The pars tuberalis is well developed in most reptiles but reduced in lizards and absent in snakes. The bird pituitary is similar to that of other tetrapods but the pars intermedia is absent. The pars distalis again consists of cephalic and caudal lobes. The well-developed median eminence is sometimes divided into anterior and posterior regions. The structure of the mammalian pituitary varies, but the basic pattern of an adenohypophysis with pars tuberalis, pars intermedia, and pars distalis and a neurohypophysis with pars nervosa and median eminence is retained in both prototherians and therians (figure 15.14 and table 15.4).

## *Function*

Strictly speaking, cells within the neurohypophysis do not produce pituitary hormones. Instead, axons of **neurosecretory neurons** of the hypothalamus dorsal to it project into the neurohypophysis, where their secretions are released into blood vessels or temporarily stored. In addition to these axons, **pituicytes** within the neurohypophysis are thought to support neurosecretory neurons, but they do not synthesize or secrete hormones.

In contrast to the cells of the neurohypophysis, cells of the adenohypophysis synthesize pituitary hormones. The hypothalamus indirectly influences their activity. Neurosecretory neurons from the hypothalamus project into the region of the median eminence and there secrete their **neurohormones** into capillaries. Through a tiny

TABLE 15.4 Summary of Anatomical Features of the Vertebrate Pituitary

| Group | PT | Adenohypophysis RPD | PPD | PI | PV | Neurohypophysis ME | PN | SV |
|---|---|---|---|---|---|---|---|---|
| *Agnatha* | | | | | | | | |
| Hagfishes | | × | × | | | | + | |
| Lampreys | | + | + | + | | + | + | |
| *Chondrichthyes* | | + | + | + | + | + | + | + |
| *Osteichthyes* | | | | | | | | |
| *Polypterus* | | + | + | + | | + | + | + |
| Teleosts | | + | + | + | | + | + | + |
| *Latimeria* | | + | + | + | | + | + | + |
| Dipnoans | | | | + | | + | + | |
| *Amphibia* | + | × | × | + | | + | + | |
| *Reptilia* | + | × | × | + | | + | + | |
| *Mammalia* | + | × | × | + | | + | + | |

*Note:* Plus signs (+) indicate that the part is present. Similar regions whose homologies are unclear are designated by an ×. Abbreviations: pars tuberalis (PT), rostral pars distalis (RPD), proximal pars distalis (PPD), pars intermedia (PI), median eminence (ME), pars nervosa (PN), saccus vasculosus (SV), pars ventralis (PV).

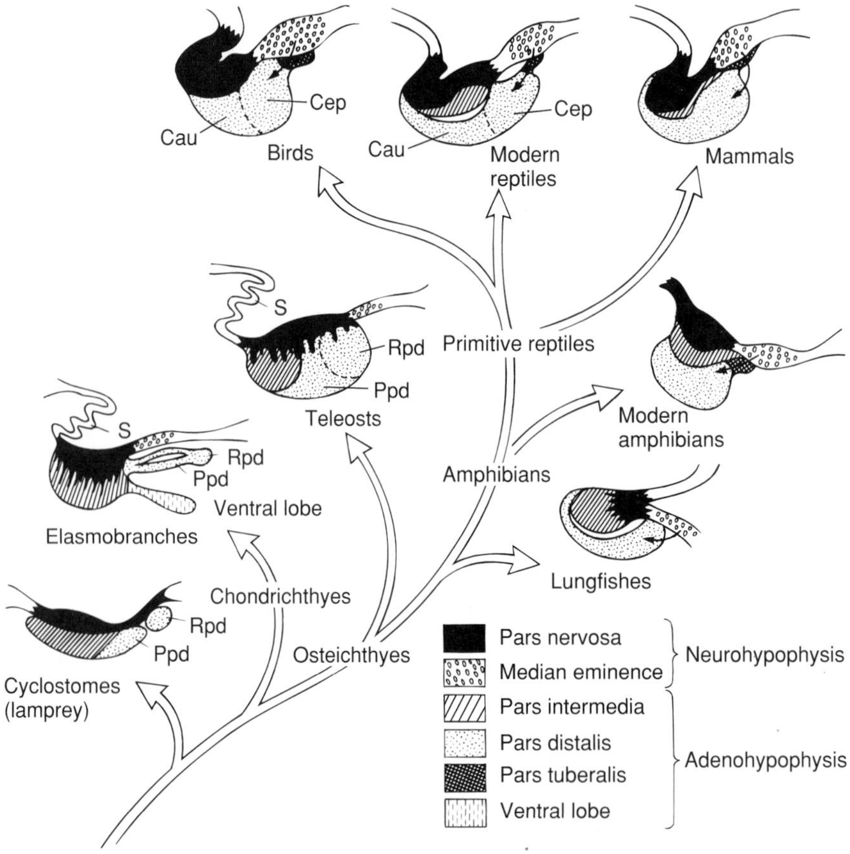


**Figure 15.14** Phylogeny of the vertebrate pituitary. Thin solid arrows within the pituitaries designate the vascular portal connection from the median eminence to the pars distalis. The pars distalis often exhibits anterior and posterior regions, the rostral pars distalis (Rpd) and the proximal pars distalis (Ppd) or the cephalic pars distalis (Cep) and caudal pars distalis (Cau). In mammals, the pars distalis is not subdivided. The ventral lobe, a projection of the adenohypophysis, is unique to elasmobranchs. In some lower vertebrates, the saccus vasculosus (Sv) is present and is derived from the hypothalamus of the brain.

vascular portal link, these neurohormones are transported a short distance through a capillary plexus and then diffuse into the adenohypophysis (figure 15.15). These neurohormones are **releasing hormones** or **release-inhibiting hormones** depending on whether they stimulate or inhibit cells of the adenohypophysis.

From early staining methods, cell types were identified on the basis of their reactions with dyes. **Acidophils** and **basophils** have affinities for acidic and basic dyes, respectively. **Chromophobes** do not react with dyes. Although these terms are still useful for descriptive purposes, new stains and better techniques for identifying hormones have shown that these cell types do not correspond to specific hormones.

***Neurohypophysis*** Two hormones synthesized by neurosecretory cells of the hypothalamus have been identified in the neurohypophysis. One hormone is **vasopressin,** which acts on smooth muscle in the walls of peripheral arterioles causing them to constrict. Resistance to blood flow increases and brings about a rise in blood pressure. If an organism sustains a fair amount of blood loss, pressure sensors in the carotid artery detect declines in blood pressure and stimulate increased secretion of vasopressin through reflex control of the hypothalamus.

Vasopressin is also called **antidiuretic hormone** (ADH) because it promotes water conservation within the kidneys (figure 15.16a). If an organism becomes dehydrated, neurosecretory neurons of the hypothalamus release ADH

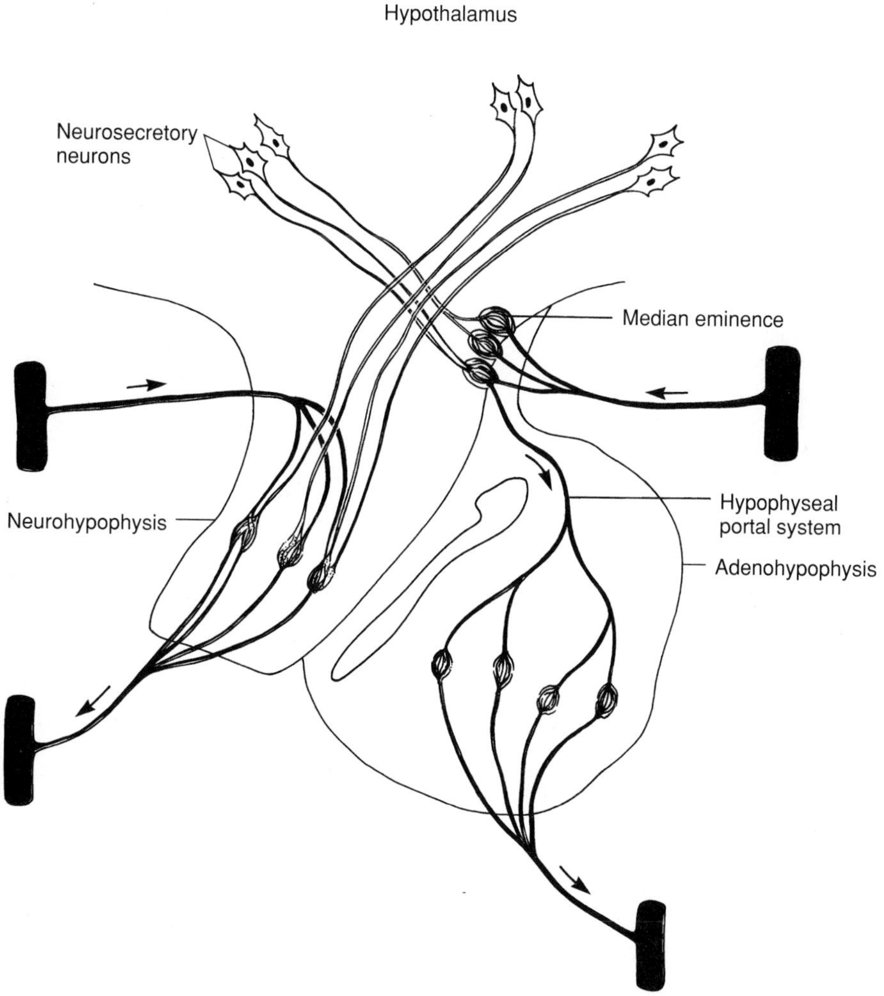


**Figure 15.15** Vascular supply and circulation within the pituitary gland. Note the short hypophyseal portal shunt between the median eminence and the adenohypophysis. A separate capillary supply to the neurohypophysis arises from the general circulation. Neurosecretory neurons release neurohormones into both these capillary networks. Neurohormones entering the median eminence are carried to cells within the adenohypophysis. Neurohormones released in the neurohypophysis enter the general body circulation.

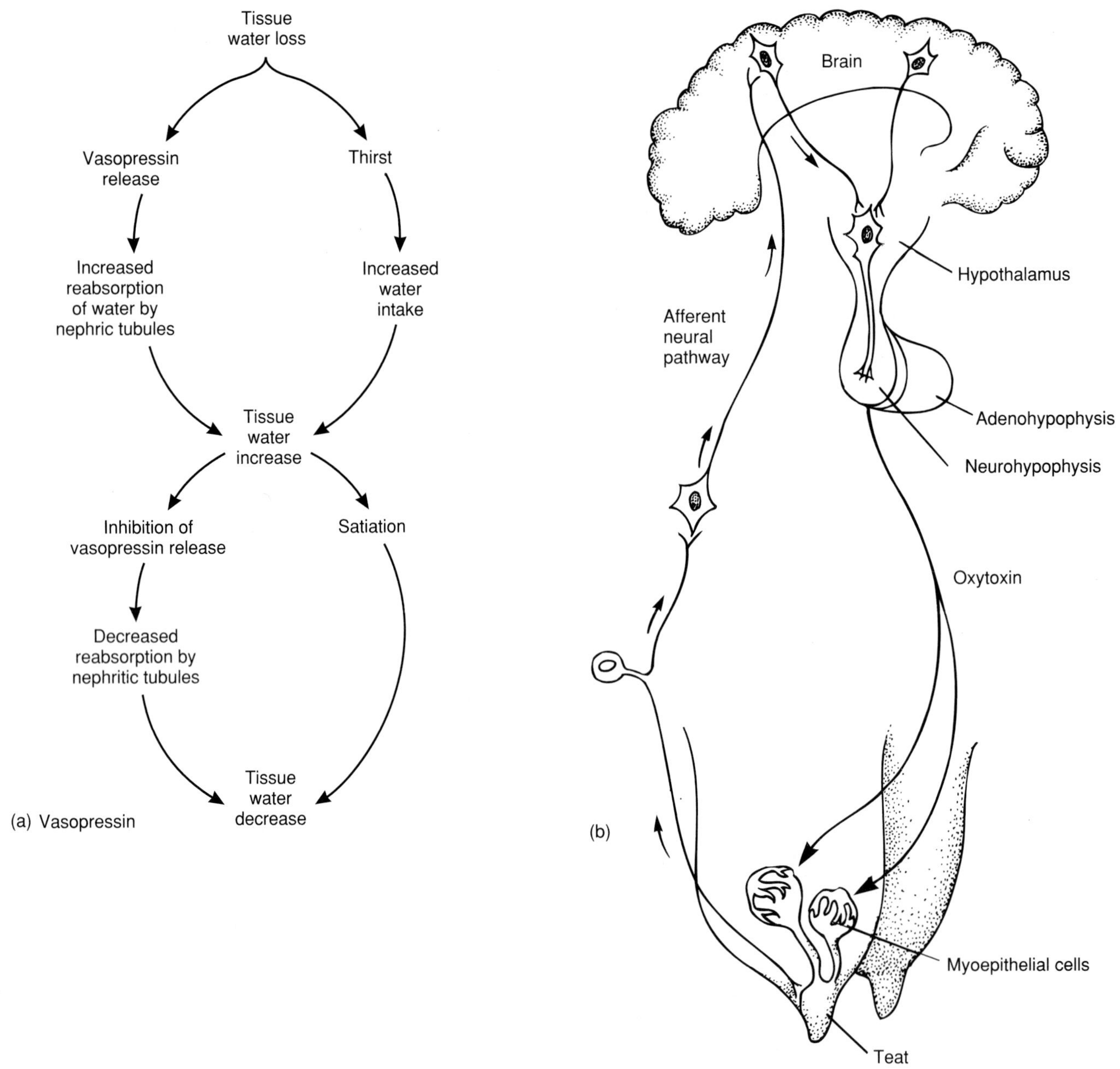


**Figure 15.16** Hormones found in the neurohypophysis. (a) Vasopressin restores water balance through a complex series of steps. (b) Oxytocin promotes the release of milk during suckling. Impulses from tactile stimulation of the nipple are transmitted by afferent nerves first to the brain and then to the hypothalamus. As the neurosecretory cells are activated by these afferent nerves, they manufacture and release oxytocin. Oxytocin secreted by the neurohypophysis is transported by the blood to the mammary gland where it stimulates contractile myoepithelial cells that causes the release of milk.

into the neurohypophysis, where it is picked up in the blood and carried to the kidneys. ADH acts on the walls of the renal collecting ducts, making them highly permeable to water; therefore, water flows from the tubules into the hyperosmotic interstitial fluid and creates a concentrated urine. In the absence of ADH, walls of the collecting ducts remain impermeable to water. Less water is reabsorbed, and the urine is copious and dilute. Under pathological conditions in which disease or tumors prevent sufficient release of ADH, large volumes of urine are passed. As a result, the individual experiences constant thirst and drinks large amounts of water, a medical condition known as **diabetes insipidus.**

The second hormone found in the neurohypophysis is **oxytocin.** Its target tissues are the **myometrium,** the smooth muscle layer of the uterus, and the contractile **myoepithelial cells** of the mammary gland. Late in pregnancy, the level of oxytocin in the blood increases, which is thought to play a role in uterine contractions during parturition. A suckling neonate initiates a reflex through sensory nerves

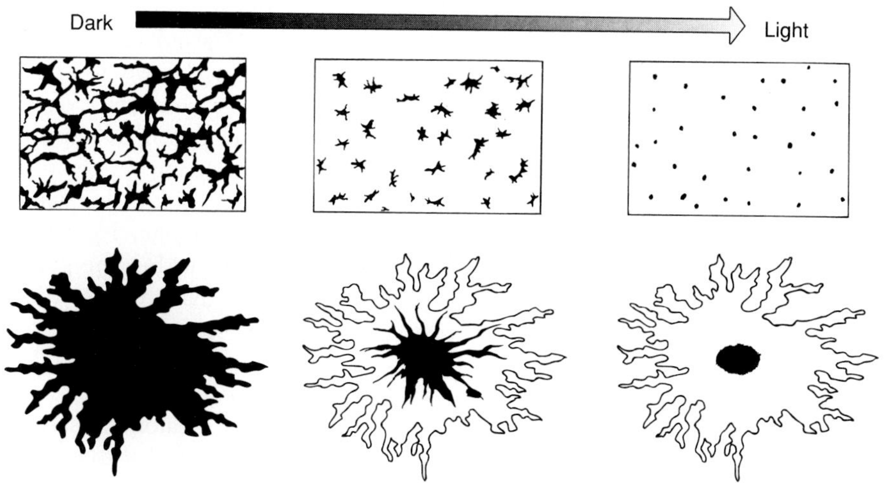


**Figure 15.17** Melanophores of frog skin. Melanophores, located in the integument, respond to melanophore-stimulating hormone by dispersing pigment granules to darken skin color (left) or by concentrating them to lighten skin color (right).

that eventually stimulates neurosecretory neurons of the hypothalamus to release oxytocin into the neurohypophysis. The bloodstream transports the hormone to the mammary gland, where it promotes contractions of the myoepithelial cells on the walls of the exocrine milk glands. About one minute after the onset of suckling, milk begins to flow from the nipple or teat (figure 15.16b).

***Adenohypophysis*** Six major hormones have been identified within the adenohypophysis. **Growth hormone** (GH), or **somatotropic hormone** (STH), unlike other hormones of the pituitary, does not have a specific target tissue. It produces effects throughout the body, including increased protein synthesis, increased mobilization of fatty acids, and decreased glucose utilization. In young animals, deficient levels of growth hormone lead to **pituitary dwarfism,** and excess levels lead to **pituitary gigantism. Acromegaly** is a condition that occurs in adults in which a disproportionate thickening of bones results from an excess of growth hormone released after puberty.

In mammals, **prolactin,** or **luteotropin** (LTH), promotes development of the mammary glands and lactation during pregnancy. In birds, prolactin stimulates lipid synthesis during premigratory fattening and supports brooding behavior. In some species, prolactin stimulates the appearance of a **brood patch,** a defeathered highly vascularized region of the breast skin placed against the incubating eggs to warm them. In pigeons and related birds, prolactin promotes the secretion of **crop milk,** a nutritional fluid produced in the crop and fed to fledglings. In lizards, prolactin affects tail regeneration, and in amphibians it affects growth.

Thyrotropin, or thyroid-stimulating hormone (TSH), stimulates the thyroid gland to synthesize and release $T_3$ and $T_4$ into the blood.

The adenohypophysis releases **gonadotropins,** a group of hormones that affect the gonads and reproductive tracts. The principal gonadotropins produced by the adenohypophysis are follicle-stimulating hormone and luteinizing hormone. Rising levels of **follicle-stimulating hormone** (FSH) induce the development of selected ovarian follicles. In males, FSH initiates and helps maintain spermatogenesis, although the term may seem illogical for this situation. **Luteinizing hormone** (LH) acts in females to finalize maturation of the ovarian follicles. A rise in LH level promotes ovulation. Following ovulation, it promotes reorganization of the follicle cells into the **corpus luteum.** In males, luteinizing hormone, more aptly termed **interstitial cell-stimulating hormone** (ICSH), stimulates the interstitial cells of the testis to secrete testosterone.

Adrenocorticotropic hormone (ACTH) stimulates the cortex of the adrenal gland to release glucocorticoids.

**Melanophore-stimulating hormone** (MSH) is located in the pars intermedia. Its targets are the **melanophores,** pigment cells of the skin. Within a few minutes, MSH affects melanin distribution within melanophores, changing the darkness of the skin in lower vertebrates. Stimulation causes the pigment **melanin** to disperse into fixed cytoplasmic **pseudopods** of the melanophores, which darkens the skin. In the absence of MSH, pigment granules gather at the center of the cell. The overall effect is to lighten the skin (figure 15.17). In birds and mammals, skin pigmentation results from the release of melanin granules into skin, feathers, and hair. MSH may act to increase the production of pigment over the long term or on a seasonal basis.

**Chromatophores (p. 215)**

At one time, the term melano*cyte* was used for pigment cells in which MSH caused increased melanin synthesis but no pigment movement within the cell. Melano*phore* designated another cell type in which melanin moved about within the cell in response to MSH. However, discovery of melanocytes in which both synthesis and movement occur casts doubts on the usefulness of such a sharp distinction.

## Gonads

In addition to producing gametes, the gonads produce hormones that support secondary sex characteristics. In humans, these include pubic hair, male facial hair, female mammary glands, preparation of the sexual ducts for reproduction, and maintenance of sex drive. In males, the **interstitial cells** (Leydig cells) that cluster between seminiferous tubules produce androgens. The principal androgen is **testosterone.** In females, the endocrine tissues of the ovary include the follicles, corpus luteum, and interstitial tissue. The principal hormones produced are estrogens (e.g., estradiol), and **progesterone.** Endocrine coordination of reproduction is discussed in more detail later in the chapter.

## Pineal Gland

The unpaired **pineal gland,** or **epiphysis,** is a dorsal evagination of the midbrain. It is part of a complex of evaginations from the roof of the midbrain that we meet in more detail in chapter 17 when we examine photoreceptive organs. In some vertebrates, the pineal gland affects perception of photoradiation. For example, in some fossil vertebrates, the pineal gland was inserted into an opening in the bony cranium, known as the **pineal foramen,** and covered only by a thin layer of integument. This may have allowed the pineal to respond to changes in photoperiod. In some living vertebrates, this gland is still located just under the skin, but more often it resides beneath the bony cranium. Nevertheless, the presence of light-sensitive cells within the pineal gland of lower vertebrates indicates that this organ may be involved in detecting seasonal or daily light schedules. The pineal has also been shown to regulate reproductive cycles in a variety of vertebrates.

Early Greek anatomists speculated that the pineal gland regulated the flow of thoughts. Absence of evidence did not deter later speculation about the pineal being the seat of the soul. The first experimental hint of an endocrine function came in 1927 when an extract prepared from ground pineal gland was placed in an aquarium with frog tadpoles. The tadpoles' skin blanched, suggesting that the extract affected melanophores. Later the hormone responsible for this effect was isolated and called **melatonin.** However, subsequent research has proved frustrating. The pineal seems to modulate activities already in progress rather than to initiate activities. In lower vertebrates, it clearly affects melanophores in the skin, but in birds and mammals, this role is less important. As mentioned, considerable research suggests that the pineal gland regulates seasonal reproductive patterns. In reptiles and birds, the pineal may aid in the organization of daily or **circadian** rhythms. In mammals, experiments in which the pineal was removed or injections of pineal extracts were administered

**TABLE 15.5 Gastrointestinal Hormones**

| Hormone | Tissue of Origin | Action |
|---|---|---|
| Gastrin | Stomach mucosa | Stimulation of gastric acid secretion |
| Secretin | Duodenum | Inhibition of gastric acid secretion |
| Cholecystokinin | Duodenum | Stimulation of gallbladder contraction and pancreatic secretion |
| Chymodenin | Small intestine | Stimulation of pancreatic chymotrypsinogen secretion |
| Duocrinin | Small intestine | Stimulation of Brunner's gland secretion |
| Enterocrinin | Small intestine | Stimulation of intestinal secretion |
| Enterogastrone | Small intestine | Inhibition of gastric acid secretion |
| Enterooxyntin | Small intestine | Stimulation of gastric acid secretion |
| Gastrone | Intestinal wall | Inhibition of gastric acid secretion |
| Gastrozymin | Small intestine | Stimulation of pepsin secretin |
| Incretin | Small intestine | Stimulation of insulin secretion |
| Pancreotone | Small and large intestines | Inhibition of exocrine pancreatic secretion |
| Villikinin | Small intestine | Stimulation of motility of villi |

provide circumstantial evidence that the pineal may be involved in release of ACTH from the adenohypophysis, in heightened vasopressin secretion, in inhibition of thyroid activity, and even in stimulation of components of the immune system. However, some of these effects may be artifacts of the experimental procedure. For example, heightened immune response might have resulted from the introduction of a foreign material, the pineal extract itself, rather than from any endocrine role of the extract. Thus, the role of the pineal is not as well understood as that of most other endocrine organs.

## Secondary Endocrine Organs

Some organs that play a central role in activities other than endocrine regulation also release chemicals carried by the vascular system to responsive tissues. Such organs function secondarily in the endocrine system. Usually the hormones they release help these **secondary endocrine organs** regulate their own primary activities. Two examples are the digestive tract and the kidneys.

### *Gastrointestinal Tract*

The alimentary canal functions primarily, of course, in digestion. However, the walls of the digestive tract produce chemicals that stimulate or inhibit target tissues. These chemicals are secreted directly, rather than being discharged through ducts. Thus, the digestive tract functions secondarily as an endocrine organ (table 15.5).

When food enters the stomach of amniotes, the gastric mucosa releases the hormone **gastrin** (figure 15.18). Gastrin enters the blood and is transported to the stomach, where it stimulates the secretion of gastric juice. When the stomach empties the churned and acidified food into the duodenum, the intestinal mucosa releases **secretin.** Secretin stimulates the pancreas to release highly alkaline pancreatic juice that buffers the acidic **chyme** arriving from the stomach. **Enterogastrone,** also released by the intestinal mucosa, inhibits further gastric secretion and mobility. Fats, proteins, and acids stimulate the secretion of **cholecystokinin** (CCK), or **cholecystokinin-pancreozymin,** from the intestinal mucosa. Originally, cholecystokinin was thought to be

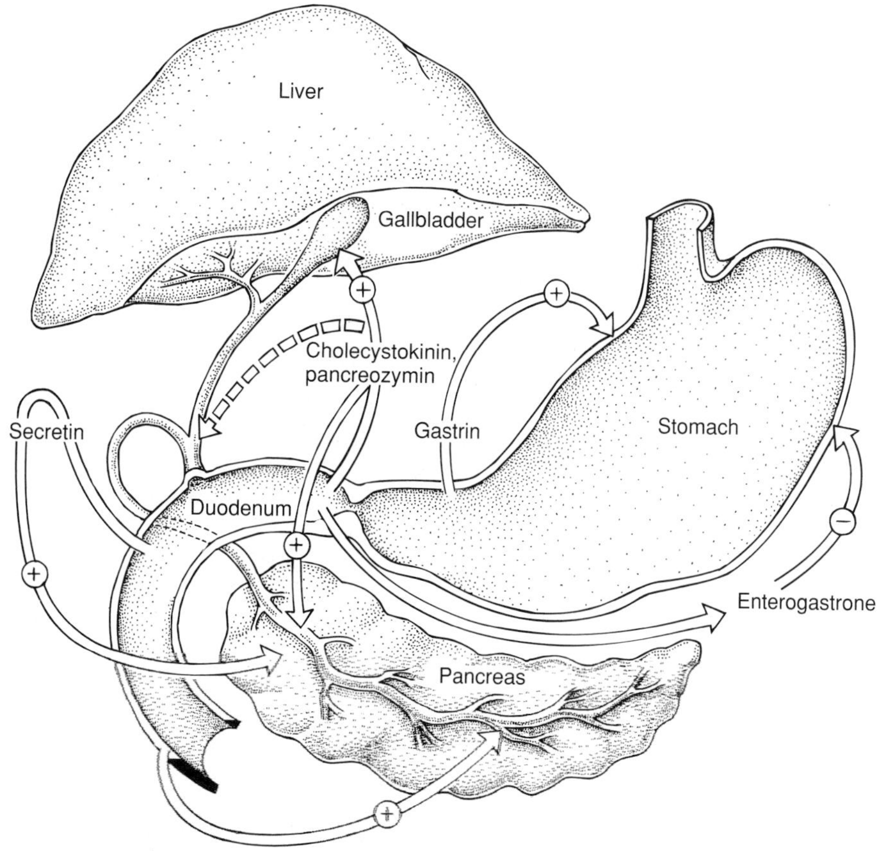


**Figure 15.18** Some mammalian gastrointestinal hormones. Sites of release and effects on target tissues are indicated. Hormonal promotion or inhibition of secretory activity is indicated by plus (+) and minus (−) signs, respectively.

two hormones, hence its hyphenated alternative name, because it has two functions. It stimulates the relaxation of the sphincter at the base of the bile duct and the ejection of bile, allowing bile to flow into the duodenum and act on fats. Cholecystokinin also stimulates the pancreas to secrete pancreatic juice containing digestive enzymes (figure 15.19).

Since the discovery of these gastrointestinal hormones, others with more restrictive actions have been discovered. For example, **enterocrinin,** released by the intestinal mucosa, increases the production of intestinal juice. We examine the endocrine function of the digestive organs in more detail when we consider the evolution of endocrine regulation later in this chapter.

### *Kidneys*

Primarily the kidneys excrete nitrogenous wastes and function in osmoregulation, but they act as an endocrine organ as well (figure 15.20). When blood pressure drops, the **juxtaglomerular cells** wrapped around renal arterioles release the hormone renin. Renin sets in motion a cascade of changes that eventually result in elevating the blood pressure. It catalyzes the transformation of **angiotensinogen** in blood to **angiotensin I,** which is converted into **angiotensin II** in the lungs and in other organs also. Angiotensin II is a vasoconstrictor that increases blood volume by stimulating the release of aldosterone from the adrenal gland. Aldosterone causes the distal tubules of the kidneys to reabsorb more sodium, which causes a rise in blood volume.

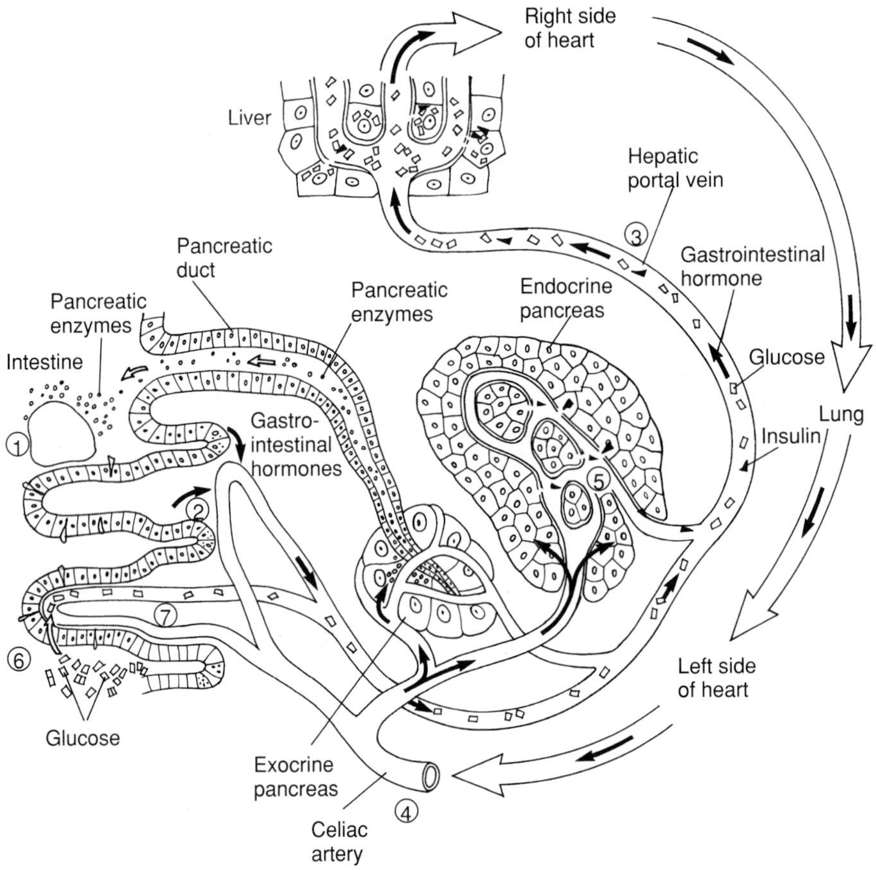


**Figure 15.19** Endocrine control of digestion. As food enters the alimentary canal (1), the release of gastrointestinal hormones is stimulated (2). These hormones enter the circulatory system, travel through the hepatic portal vein (3) to the liver and then to the heart. From the heart, they are transported back to the alimentary canal via the celiac artery (4). When arriving in the pancreas, these gastrointestinal hormones stimulate the release of pancreatic enzymes. Foremost among these gastrointestinal hormones is cholecystokinin, which triggers the release of digestive enzymes from the exocrine pancreas and bile from the gallbladder. The gastrointestinal hormone secretin causes the pancreas to release bicarbonate to help neutralize the acidic chyme that enters the duodenum from the stomach. Both cholecystokinin and secretin also stimulate the endocrine pancreas to release insulin, indicated by solid arrowheads (5). One end product of digestion is glucose (6), which is absorbed across the intestinal wall (7) and transported to the liver via the hepatic portal vein (3).

## Box Essay 15.1 The Pill

In large part, medical successes in developing oral contraceptives for females but not for males can be attributed to intrinsic differences in the natural hormonal control of reproductive functions in the two sexes.

Contraception, mediated by sex hormones, is a *normal* monthly phenomenon in the human female. Progesterone, produced by the corpus luteum, suppresses further ovulation by inhibiting the release of new follicle-stimulating hormone (FSH). Thus, while progesterone is secreted, no FSH is produced, no more follicles mature, no ova are released, and no fertilization occurs. Oral contraceptives seek to mimic this natural series of events.

Oral contraceptives for women contain progesterone, which prevents ovulation by suppressing FSH secretion. But this means that women taking an oral contraceptive do not form corpora lutea; therefore, estrogen produced by the corpus luteum is absent. One modification of the original recipe for oral contraceptives was to add estrogen, to compensate for the missing corpus luteum. Another modification has been to modify the levels of progesterone administered. Usually they are lower in oral contraceptives to match the woman's individual needs.

Unlike women, sperm production in the human male does not follow monthly rhythms. Sperm are produced more or less continuously, so that cyclic, hormone-mediated contraception does not occur in the human male as it does in females. This precludes any mimicking in the male of a natural contraceptive process. Instead, strategies for producing a male oral contraceptive have sought to inhibit FSH secretion directly. Of course, this cannot be accomplished with progesterone. If it could, the side effects would produce a sterile but feminized male. Other hormonal disruptions of testicular function sufficient to stop sperm production have usually had similar side effects.

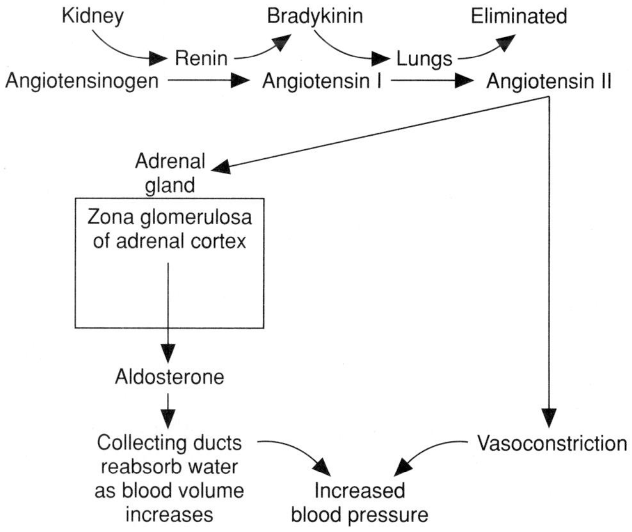


**Figure 15.20** The kidney as an endocrine organ. A drop in blood pressure in blood vessels serving the kidney results in release of renin, the hormone that catalyzes the transformation of angiotensinogen to angiotensin I. In the lungs and elsewhere, angiotensin I is converted to angiotensin II, which indirectly causes the kidney to retain more water. Angiotensin II also acts as a vasoconstrictor. The collective result is restoration of blood pressure.

Together, vasoconstriction and increased blood volume elevate the blood pressure.

# Endocrine Coordination

So far we have surveyed endocrine organs, their hormones, and their target tissues. Next we examine two ways in which endocrine organs interact to coordinate activities. Let us begin by considering reproduction in mammals.

## Mammalian Reproduction

### *Male*

In males, the adenohypophysis releases two gonadotropic hormones that have immediate effects on the testes (figure 15.21). Follicle-stimulating hormone (FSH) plays a prominent role in controlling spermatogenesis. Luteinizing hormone (LH), also called interstitial cell-stimulating hormone (ICSH), acts on the interstitial cells in the testis to promote the production of androgens, especially testosterone. First, testosterone regulates the development and maintenance of the secondary sexual characteristics (including antlers and brightly colored plumage), the sex impulse, and the accessory sex glands. Second, it promotes spermatogenesis. Third, testosterone has a negative feedback effect on the adenohypophysis to limit the production of ICSH and hence prevents overproduction of this gonadotropic hormone (figure 15.21).

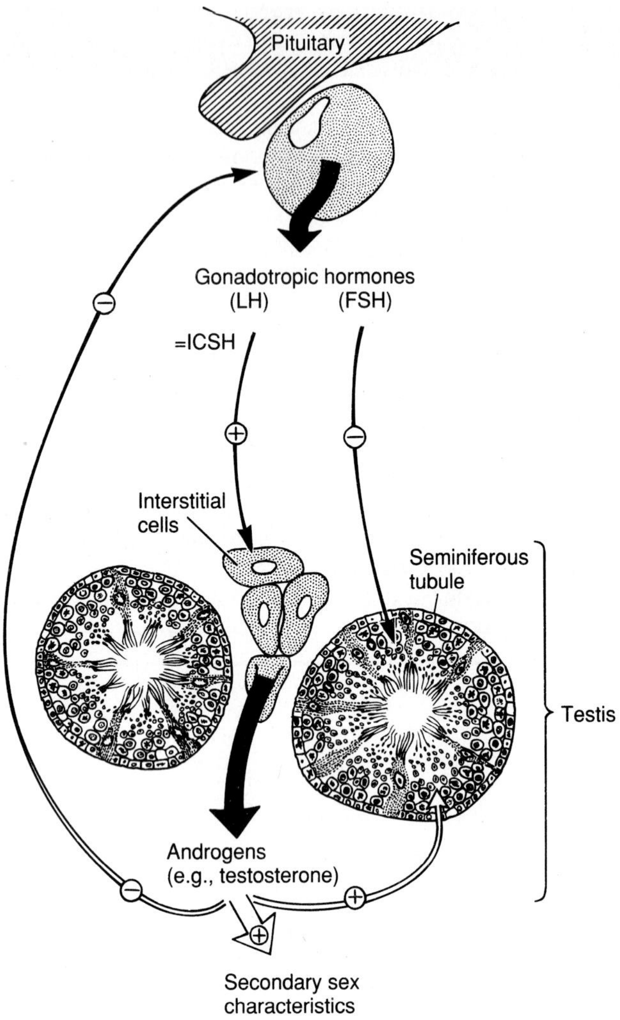


**Figure 15.21** Gonadotropic hormones in male mammals. Follicle-stimulating hormone (FSH) promotes spermatogenesis. Luteinizing hormone (LH) stimulates interstitial cells of the testes to release testosterone. Testosterone in turn helps maintain secondary sex characteristics in males and inhibits secretion of gonadotropic hormones. Hormonal promotion and inhibition of activity are indicated by plus (+) and minus (–) signs, respectively.

### *Female*

In females, oocytes within the chordate ovary are coated with follicle cells derived from ovarian epithelium. In most tetrapods, each ovary houses hundreds or thousands of oocytes wrapped in follicle cells. However, only a few follicles actually undergo **maturation** to release their ova during **ovulation,** making **fertilization** possible. As maturation of an ovum progresses, the enveloping single layer of follicle cells proliferates, becoming the thickened, multilayered **granulosa.** Later in maturation, spaces filled with fluid appear within the granulosa and coalesce into the **antrum,** a single fluid-filled space. In addition, connective tissue cells within the ovary form an outer coat, called the **theca,** around the follicle. Following ovulation, the follicle becomes the **corpus luteum.** Cells of the granulosa become **granulosa lutein cells,** which constitute most of the corpus luteum, and thecal cells persist as **theca lutein cells,** which form the outer capsule of the corpus luteum. Eventual regression of the corpus luteum yields the **corpus albicans** in progressive stages of degeneration. Regression of follicles before ovulation yields **atretic follicles** (figure 15.22; see chapter 14).

The events of follicle maturation are best understood in mammals, especially in humans. Hormones promote follicle maturation and simultaneously prepare the uterus to receive a fertilized ovum (figure 15.23a). There are four major steps involved. First, falling levels of progesterone are accompanied by rising levels of FSH. As levels of FSH rise, selected follicles begin to mature. Why some and not other follicles in the ovary respond is not known. In those that respond, the thin layer of follicle cells divides to produce a thickened coat of cells. Fluid-filled spaces within, the forerunners of the antrum, appear as well.

Second, as follicles grow under continued FSH stimulation, the inner layer of cells secrete increased amounts of estrogen. At this point, estrogen has two actions. It stimulates the endometrium of the uterus to proliferate, and it stimulates secretion of luteinizing hormone (LH).

Third, LH release causes ovulation. A mature follicle ruptures and releases its ovum. Thereafter, LH promotes consolidation of the ruptured follicle into the corpus luteum.

Fourth, the corpus luteum takes over the function of secreting estrogen that was initiated by the follicles, although secretion is now at lower levels. In addition, the corpus luteum produces progesterone. Progesterone is an "optimistic" hormone, promoting the final stages of preparing the uterus for a fertilized ovum. Further, progesterone inhibits the secretion of FSH from the pituitary; therefore, no more follicles mature at this time.

In humans, if pregnancy does not occur, hormonal support for growth of the corpus luteum drops after 10 to 12 days and it deteriorates. When this happens, it involutes, becoming a patch of scar tissue, the corpus albicans. With the decline of the corpus luteum, progesterone levels drop, FSH secretion rises, and the cycle begins again.

If pregnancy occurs, **chorionic gonadotropin** (CG) hormone stimulates the growth of the corpus luteum. CG is produced by the rudimentary placenta established by the embryo implanted in the uterine wall (figure 15.23b). Placental CG functions to maintain the corpus luteum, which in turn produces progesterone to maintain the uterus housing the implanted embryo and its rudimentary placenta. In humans, the corpus luteum, the placenta, and the growing embryo are mutually maintained in this reciprocal fashion until about the third month of pregnancy. Thereafter, the corpus luteum undergoes a slow involution. At this point in the pregnancy, however, involution of the corpus luteum and consequent drop in its output of progesterone do not bring about menstruation and loss of the

## Box Essay 15.2 The Rabbit Died: A Word on Pregnancy Tests

It was and occasionally still is a fixture of many comedy routines to imply positive results for pregnancy if "the rabbit dies." The test is an old one no longer used. In fact, the rabbit never died from a positive test!

Most pregnancy tests are tests for chorionic gonadotropin (CG), a hormone produced by the trophoblast almost as soon as the embryo is implanted in the uterine wall when it is about six days old. Like all hormones, CG circulates in the blood, and some is excreted in the urine. The urine can be tested for the presence (implying pregnancy) or absence (implying no pregnancy) of CG. Early pregnancy tests used a rabbit, hence the source of the anecdote.

The test took advantage of the fact that ovulation in the rabbit does not occur until luteinizing hormone (LH) is secreted. As it turns out, CG mimics LH and produces ovulation in the rabbit. Thus, urine from a woman was injected into a rabbit's blood. If the woman was pregnant, then CG would be present in her urine and stimulate ovulation when injected into the rabbit. The rabbit's ovaries were inspected following injection of the woman's urine. If evidence of ovulation was found, the woman was pregnant. Technically, the rabbit did die when its ovaries were examined, but death was not the indicator of pregnancy.

Frogs were used in another early pregnancy test. CG mimics hormones that cause female frogs to lay eggs. Contemporary tests are simpler. Antibodies produced against CG are mixed with urine from the woman. If agglutination occurs, CG is present, and the test is positive, indicating pregnancy. Nowadays, test kits can be purchased in pharmacies, and a rabbit or frog need not be directly involved.

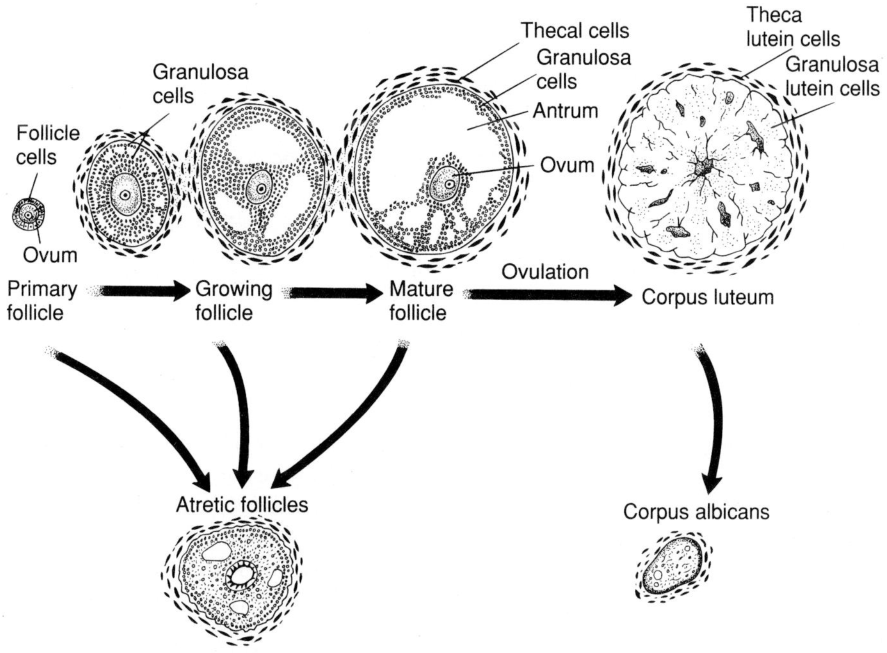


**Figure 15.22** Maturation of an ovarian follicle in mammals. In a female entering her first reproductive season, the ovary is populated by oocytes wrapped in follicle cells that have remained quiescent since primordial germ cells first colonized the ovary during her embryonic development. At the onset of the breeding season, hormones stimulate maturation of some of these follicles. The follicle cells proliferate, mature, and enclose the antrum, a fluid-filled space in the follicle. Upon ovulation, the mature follicle ruptures, releasing the ovum and some clinging follicle cells. After the ovum is released, the walls of the follicle form the corpus luteum, which continues to play an endocrine role for awhile. After completing its endocrine role, the corpus luteum then undergoes involution and remains as a remnant patch of connective tissue, the corpus albicans. If follicles regress before ovulation, the involuted follicles form atretic follicles.

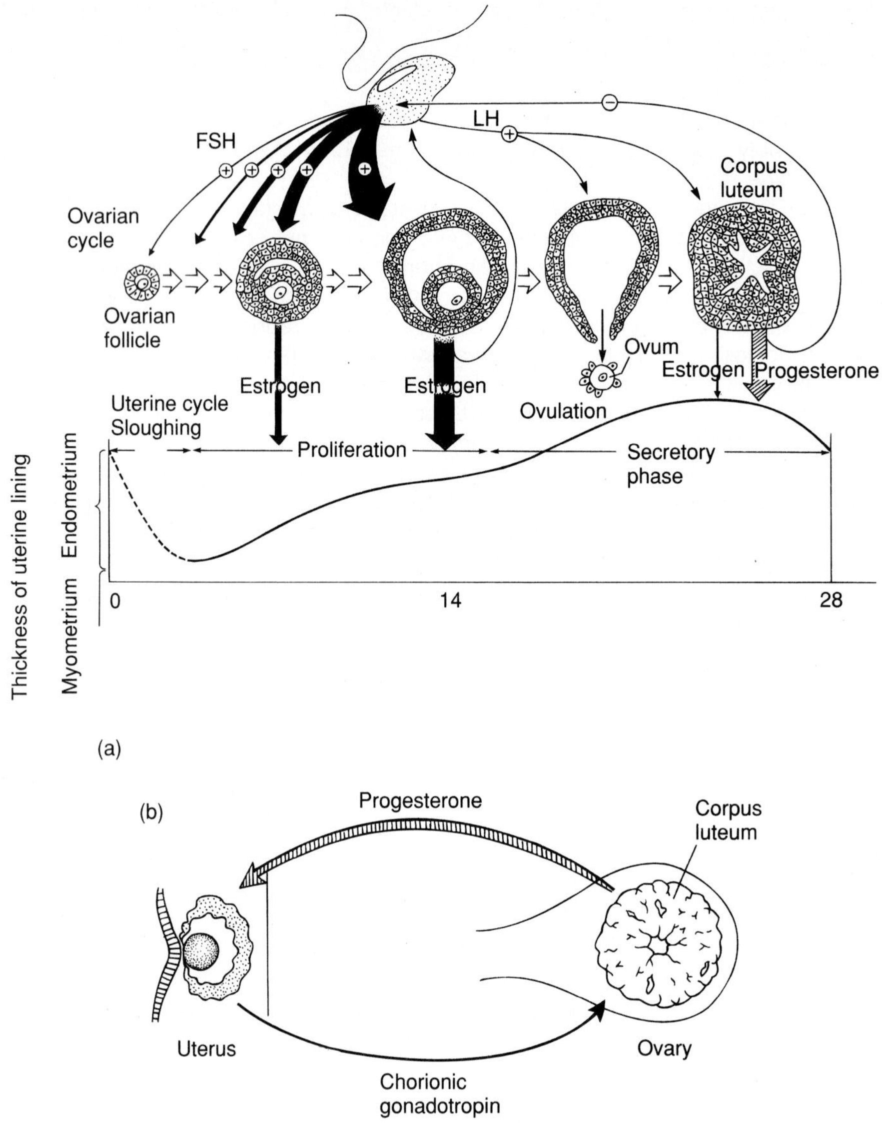


**Figure 15.23** Ovarian and uterine cycles of human females. (a) Follicle maturation and accompanying thickening of the uterus. The endometrium of the uterus thickens and then enters a secretory phase in anticipation of receiving a fertilized ovum. The follicle is an endocrine organ that releases increasing levels of estrogens in response to rising levels of FSH from the adenohypophysis. Following ovulation, the ruptured follicle persists as a corpus luteum. This modified endocrine tissue still produces some estrogens; however, it primarily produces progesterone, which inhibits FSH production and temporarily prevents the maturation of additional follicles. Progesterone also stimulates the uterus to maintain an environment hospitable to the implanted embryo. If implantation fails, the corpus luteum disintegrates at about day 28 of the cycle, progesterone levels fall, and FSH is released, allowing the cycle of follicular development to begin again. (b) If pregnancy occurs, chorionic gonadotropin released by the placenta initially supports the corpus luteum, which secretes progesterone to maintain the uterine wall during and following implantation. This mutual hormonal support lasts until about the third month of pregnancy, when the placenta begins to secrete progesterone. The corpus luteum undergoes slow involution at this time.

implanted embryo because by now the placenta itself is producing progesterone to maintain itself.

The reproductive cycle of the red kangaroo (*Megaleia rufa*) illustrates how the endocrine and nervous systems coordinate reproductive processes (figure 15.24a–c). Like most marsupials, the red kangaroo has a short gestation period. The female kangaroo may support up to three young at staggered stages of development. Her reproductive tract is designed to accommodate embryos at different stages of maturation. Ovulation alternates between the two ovaries. The blastocyst enters the central vaginal canal where it develops during its brief gestation. Sperm from a subsequent mating travel along the lateral vaginal canals without encountering the embryo. After birth, the infant, now called a **joey,** migrates into the pouch and begins suckling from a teat. Via afferent nerves to the pituitary, suckling

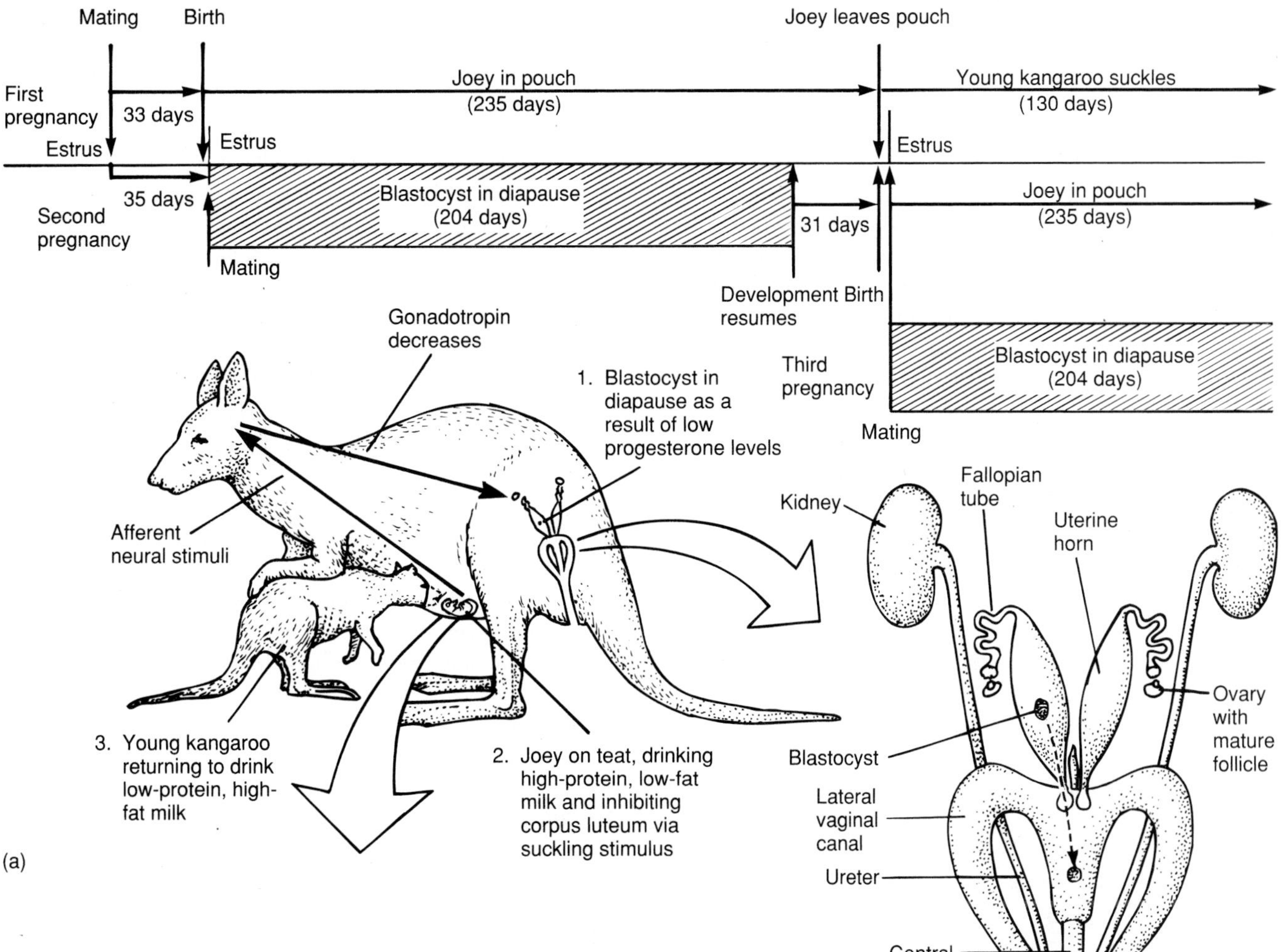


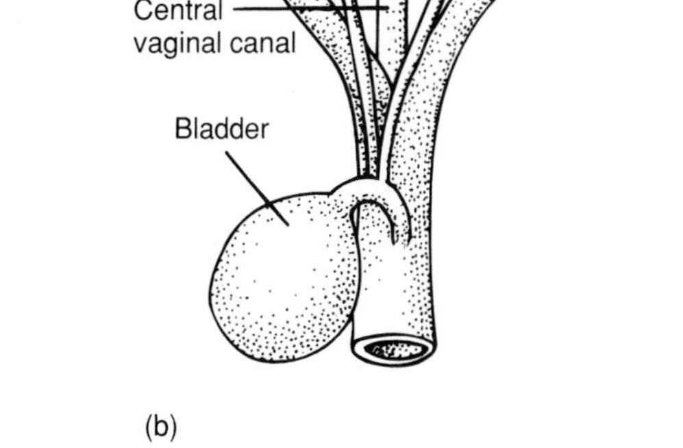

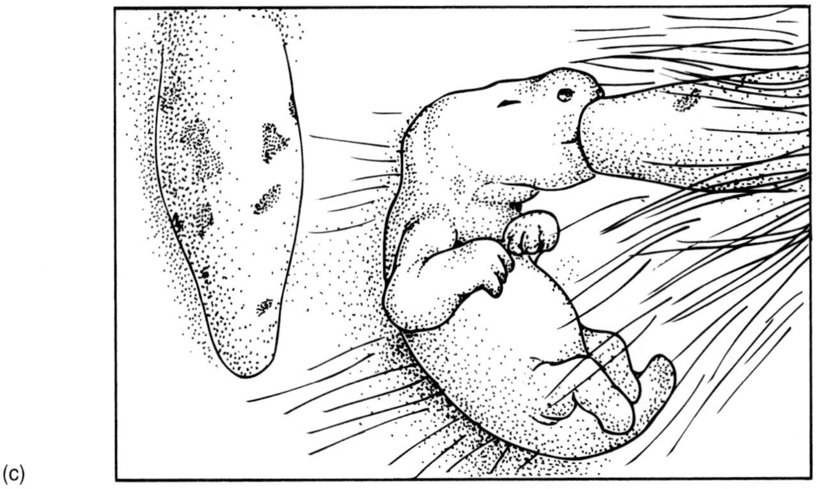


**Figure 15.24** Hormonal and neural control of reproduction in the red kangaroo, *Megaleia rufa*. (a) At any one time, the female may support up to three young at various stages of development: a blastocyst, a joey, and a young kangaroo. This staggering of three pregnancies is diagramed. At the end of the first 33-day gestation period, the first birth occurs. The female enters estrous again and mates, resulting in pregnancy. However, the first young, now termed a joey, enters the pouch and attaches its mouth to a teat. Its suckling stimulates (via afferent neural stimulation to the pituitary) increased levels of prolactin and decreased levels of gonadotropic hormones. Consequently, progesterone secretion by the corpus luteum decreases, so the uterus is no longer able to support development of the second embryo. Now the second embryo enters embryonic diapause. As the first joey becomes more independent and begins to make forays from the pouch, the suckling stimulus wanes, the levels of prolactin decline, and the levels of gonadotropic hormones rise. High gonadotropic levels stimulate the corpus luteum again. High progesterone levels return, and the second embryo is reactivated to complete gestation and take up residence in the pouch. At this time, the female may enter estrous, mate, and become pregnant again. Development of this third pregnancy is controlled by the suckling stimulus of the second joey in the pouch. If a joey is prematurely removed from the pouch or dies, the blastocyst resumes development and the female enters estrous again. If she becomes pregnant, the process is repeated. The first young is not fully weaned until about four months after it begins its forays from the pouch. (b) Reproductive tract of female red kangaroo. Following mating, sperm migrate down the lateral vaginal canal. The central vaginal canal usually holds the blastocyst. (c) Young joey attached to the teat. As the joey grows, the mammary gland enlarges to supply greater volumes of milk. The larger teat is still being suckled by a young kangaroo. Note the large forearms of the joey, which it uses to pull its way from the uterus to the pouch.

(a) Modified from Short, 1972. (b) Modified from G. G. Sharman, 1967.

stimulates the release of prolactin and causes a decrease in gonadotropin. As a result of these hormonal changes, the ovarian corpus luteum is inhibited and its progesterone output declines. Without progesterone, the uterus no longer promotes development of the next blastocyst. Its development is temporarily arrested, and the blastocyst enters **embryonic diapause** (see chapter 14). When the growing joey begins to make tentative forays away from the female pouch, the intensity of its suckling stimulus decreases, allowing the corpus luteum to become activated again. The female comes into estrous and mates. The blastocyst in diapause resumes its development and completes gestation. The newborn moves to the pouch and attaches itself to an available teat. Again the suckling stimulus arrests the development of the new blastocyst, and it enters embryonic diapause. During suckling, the composition of the milk also changes. It increases in fat content as the joey grows.

Death or premature removal of a joey results in decreased prolactin and increased gonadotropic hormonal secretion by the pituitary. Consequently, the corpus luteum is reactivated, progesterone secretion is increased, and development of the blastocyst is resumed. The female enters estrous and usually mates. However, environmental events, such as short photoperiods, can have an effect similar to suckling young. If a young joey is removed in the fall, the blastocyst may not resume development until the spring.

## Metamorphosis in Frogs

Metamorphosis in frogs is an excellent example of the coordination of a complex physiological process involving nervous, secretory, and vascular responses mediated by the endocrine system. The frog tadpole undergoes three developmental stages (figure 15.25a–c). The first **premetamorphosis** stage is characterized by growth in body size. In the second **prometamorphosis** stage, the most conspicuous change is development of the hindlimbs, although some growth in body size still continues. The third stage is **metamorphic climax,** at which time the tadpole is transformed into the young froglet. The forelimbs emerge, the beak is lost, the mouth widens, and the tail is resorbed. Hormones, developmental events, and the nervous system are involved in each stage.

During premetamorphosis, the adenohypophysis produces high levels of prolactin, which stimulates growth but inhibits metamorphosis. The adenohypophysis also produces small amounts of thyroid-stimulating hormone (TSH) autonomously, without any prompting from the hypothalamus. TSH stimulates the thyroid to secrete thyroxine but not in sufficient levels to initiate metamorphosis. During this early stage of development, the median eminence of the pituitary does not respond to thyroxine and remains undeveloped. Thus, during premetamorphosis, the tadpole grows in size but few other changes occur (figure 15.25a).

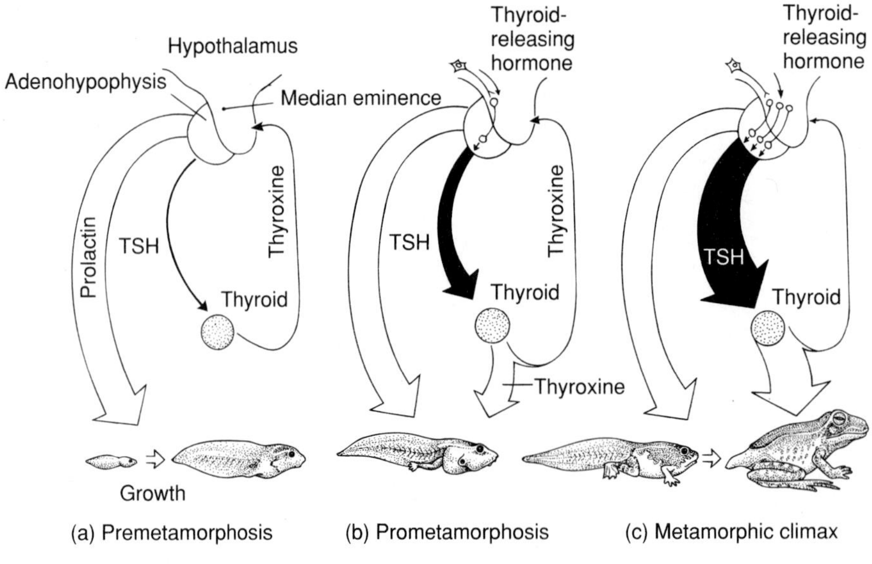


**Figure 15.25** Frog metamorphosis. (a) Premetamorphosis is characterized by high levels of prolactin, which promote growth of the tadpole, and low levels of thyroid-stimulating hormone (TSH) and thyroxine. (b) Prometamorphosis includes elaboration of the median eminence and its portal system, which allows the hypothalamus to influence the adenohypophysis. As a result, levels of TSH rise, which promotes rising levels of thyroxine. Thyroxine stimulates hindlimb development. (c) In metamorphic climax, additional vascular routes via the enlarging median eminence stimulate increased secretion of TSH. The resulting elevated level of thyroxine promotes metamorphosis of the tadpole into a froglet.

During prometamorphosis, the median eminence becomes responsive to thyroxine and begins to develop, establishing a modest but complete portal system that allows neurohormones to be transported from the hypothalamus to the adenohypophysis. The neurohormone **thyrotropin-releasing hormone** (TRH) stimulates the secretion of increasing amounts of TSH. Rising levels of TSH stimulate the thyroid to produce more thyroxine. When circulating levels of thyroxine become high enough, hindleg development is initiated (figure 15.25b).

These events generate a positive feedback system in which rising levels of thyroxine promote the more responsive median eminence to develop a more extensive portal connection so that more TRH is delivered to the adenohypophysis. The arrival of TRH stimulates the secretion of even higher levels of TSH and in turn more thyroxine. As these events snowball, thyroxine levels continue to increase, leading to metamorphic climax. Early models of endocrine control of frog metamorphosis envisioned levels of prolactin falling as levels of thyroxine rise, but this seems not to be true. Levels of prolactin remain high through metamorphic climax, at least in frogs, but its inhibitory effects on metamorphosis are apparently overridden by the rising levels of thyroxine.

# Fundamentals of Hormonal Control

Frog growth and metamorphosis highlight some basic features of hormonal control. First, hormones act not only by exerting a positive influence on target tissues, they also control events by inhibiting target tissues. Second, a target tissue, such as the median eminence, responds to hormones only after earlier stages of development have been completed. Third, endocrine control is exerted not just on the basis of the presence or absence of a hormone but also on changes in its level. Fourth, the endocrine system is also responsive to environmental conditions and can, within limits, extend or shorten metamorphosis. If a tadpole is placed in an environment that is unusually cold or without sufficient nutrients, growth and metamorphosis are retarded.

## *Functional and Structural Linkage*

The endocrine and nervous systems are functionally linked through the hypothalamus within the midbrain. This places the endocrine system under the influence of the central nervous system; thus, through the endocrine system the nervous system indirectly extends its control to target tissues.

The physiological bridge between nervous and endocrine systems is mediated by neurosecretory neurons, so named because they exhibit properties of both nerve cells (they carry electrical impulses) and endocrine cells (they secrete chemicals into blood vessels). Under the influence of higher brain centers, neurosecretory cells in the hypothalamus secrete hormones into the short portal system that begins in the median eminence. When they arrive in the pars distalis, these neurosecretory hormones stimulate or suppress the secretion of other pituitary hormones. Hormones secreted by the pituitary can in turn directly affect target tissues or they may stimulate another endocrine organ to produce a third hormone that is then carried to target tissues. For example, the neurohormone thyrotropin-releasing hormone stimulates the release of thyroid-stimulating hormone, which stimulates the thyroid gland to release thyroxine, which affects target tissues (figure 15.26).

## *Target Tissue Responses*

The actions of a hormone on tissues are usually selective, and the ability of a tissue to respond to a hormone depends on cell **receptors** that recognize it (figure 15.27). These receptors may be located in the cell membrane or in the cytoplasm. Unresponsive tissues lack cell receptors. To produce an effect, a hormone must be bound to receptor chemicals in or on cells that are selective for specific hormones. The

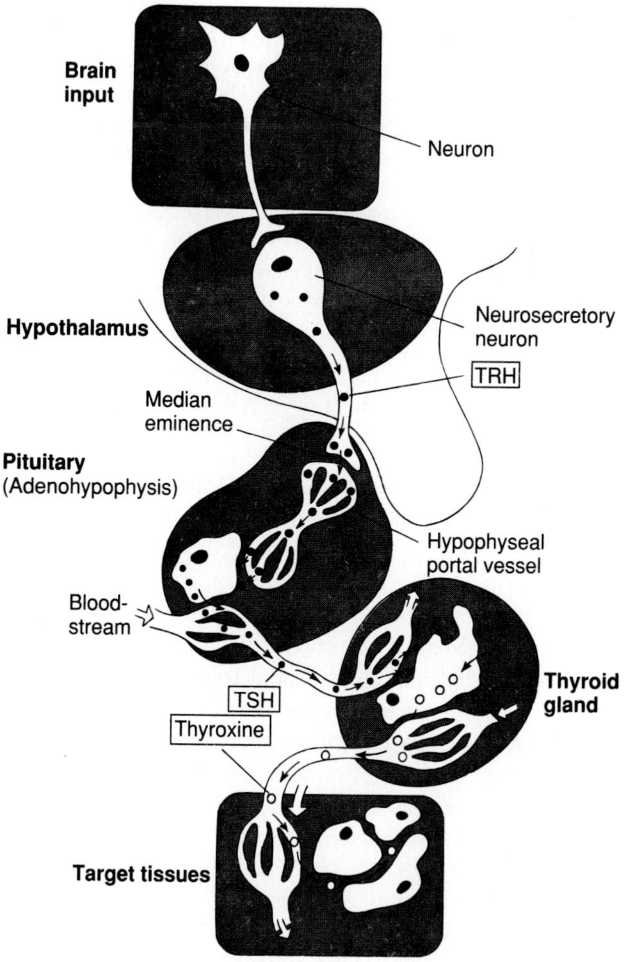


**Figure 15.26** Schematic representation of the paths by which neurons from the brain influence target tissues through intermediate endocrine glands.

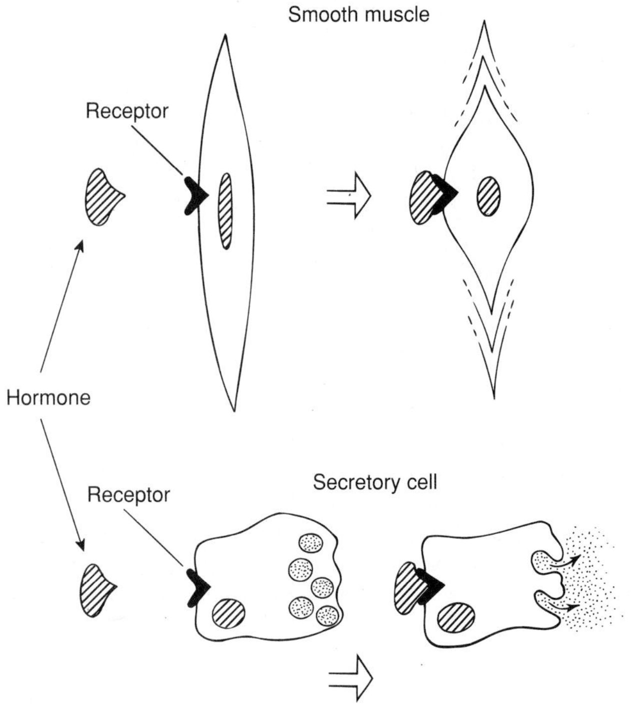


**Figure 15.27** Hormonal receptor sites. Smooth muscle cells and secretory cells possess receptor sites. Receptors for some hormones are located on the plasma membrane. Receptors for others, such as steroids, are located in the cytoplasm. The appropriate hormone links with its receptor in the cell, and through a chain of coupled metabolic pathways, they promote a cellular response. Evolution often involves changes in receptor sites rather than changes in hormones. Evolution of the appropriate receptor site makes a cell responsive and brings it under the influence of the endocrine system. Loss of the receptor removes it from immediate endocrine control.

hormone-receptor complex exerts an influence by promoting synthetic or catabolic reactions. For example, androgen levels rise at puberty in human males, but the selective response to these rising levels depends on the presence of receptors in target tissues that promote cellular differentiation of secondary sexual characteristics. Hair follicles in the axillary, pubic, facial, and chest regions respond with increased hair growth. In human females, mammary gland cells and ducts have receptors that allow them to respond to rising levels of the circulating hormone estradiol, but gland cells elsewhere in the body do not.

Ultimately, hormones influence target tissues by altering rates of cell division or by initiating or inhibiting synthesis of new products. Cell types differ in their responses to a given hormone. For example, a smooth muscle may respond to a hormone by contracting, whereas a gland may respond to the same hormone by releasing a secretory product (figure 15.27). Although the character of hormones is important in endocrine control of metabolism, so is the character of the target tissue itself. LH and ICSH are chemically identical, yet they initiate different processes. LH stimulates ovulation in females, whereas the same hormone, called ICSH in males, promotes interstitial cell growth within the testes. These functional differences primarily result from differences in target tissues, not from differences in the triggering hormone.

## The Endocrine System and the Environment

The endocrine system regulates internal physiology, coordinates embryonic development, balances levels of minerals and nutrients to match demands, stimulates growth and metabolism, and synchronizes activities between distant parts of the organism. Yet the endocrine system itself is influenced by the external environment. Many physiological events such as reproduction, migration, and hibernation are seasonal. The endocrine system acts as an intermediary between the environment and the internal physiology of an organism to coordinate internal changes with external conditions.

For ectotherms, environmental temperature is central to activity. Cooling autumn temperatures may foster reductions in metabolic rate and send temperate reptiles into hibernation. Warming spring temperatures may bring them out of hibernation. Similarly, changing lengths of daylight hours affect the endocrine system, apparently via the eyes or the pineal organ. For many tetrapods, lengthening days can promote the onset of reproduction. Shortening days often result in internal physiological changes that lead to fat deposition and hibernation or migration to warmer climates. The social environment can also affect the endocrine system. For instance, female lizards show signs of accelerated ovarian activity, or **recrudescence,** if they are exposed to a courting male, but recrudescence is delayed if the female sees male-male territorial displays. The endocrine system thus links physiological changes, especially those based on a seasonal cycle, to changes in the surrounding environment. In this way, physiology and behavior respond optimally to environmental conditions.

# Evolution

The evolution of the endocrine system includes phylogenetic changes in hormones, endocrine organs, and target tissues. As you know from the first part of this chapter, the structure of endocrine organs is quite varied. In lower vertebrates, some endocrine organs tend to be distributed in patches and dispersed compared with a more compact arrangement in higher vertebrates. For example, the components of the adrenal gland appear as separate glands containing interrenal and chromaffin tissue in lower vertebrates. In later vertebrates, these components form the cortex and medulla of a composite adrenal gland. Incorporation of parafollicular cells into the thyroid is another example of an evolutionary merging of what in lower vertebrates are separate glands. Location of the pancreatic islets within the pancreas is yet another example of a composite organ in which endocrine and exocrine tissues are combined. Little is known about the functional significance of these mergers. Combining different glands gives them immediate influence over one another and this would seem to make coordination of activities more convenient. However, because many of the specific roles of endocrine organs in vertebrates, especially in lower vertebrates, are still uncertain, it is not possible to say what adaptive advantages might have favored the phylogenetic merging of separate glands.

Adaptive changes in the endocrine system often involve changes in the responsiveness of local tissues to existing hormones rather than changes in the hormones themselves. Consequently, similarities between hormones among different classes do not necessarily imply similar function. For example, prolactin has a wide variety of roles in different classes, including stimulation of milk production in mammals, inhibition of metamorphosis and promotion of growth in amphibians, development of dermal pigmentation in amphibians and reptiles, and modulation of parental care and water balance in fishes. Even within the same class, the role of a hormone may change. For example, in most birds, prolactin initiates behavior that leads to incubation. In addition, in some bird species, the integument of the lower part of the breast develops a brood patch in response to elevated levels of prolactin, an adaptation that does not involve a new hormone but merely a change in the responsiveness of the breast integument to an existing hormone.

Another example of the evolution of tissue responsiveness to an existing hormone can be found in some hormones of the digestive tract. Cholecystokinin (CCK) stimulates the release of digestive enzymes at least as far back phylogenetically as the protochordates; however, later in phylogeny, the gallbladder also became responsive to this ancient hormone.

Interactions between digestive hormones and their target tissues have undergone a complicated sequence of evolutionary changes (figure 15.28a–e). In bony fishes, for example, two hormones are especially important in controlling the secretion of acid from the walls of the stomach. **Bombesin,** a blood-borne hormone, is secreted by endocrine cells residing in the stomach. When food arrives in the stomach, it prompts these cells to secrete bombesin, which promotes the release of gastric acid. As food passes from the stomach to the small intestine, CCK cells of the intestine are stimulated to release CCK. Transported by the blood to the stomach, CCK acts to inhibit the release of additional gastric acid into the empty stomach (figure 15.28a,b).

In amphibians, CCK cells are located in the stomach as well as in the intestine. Cells that produce bombesin are

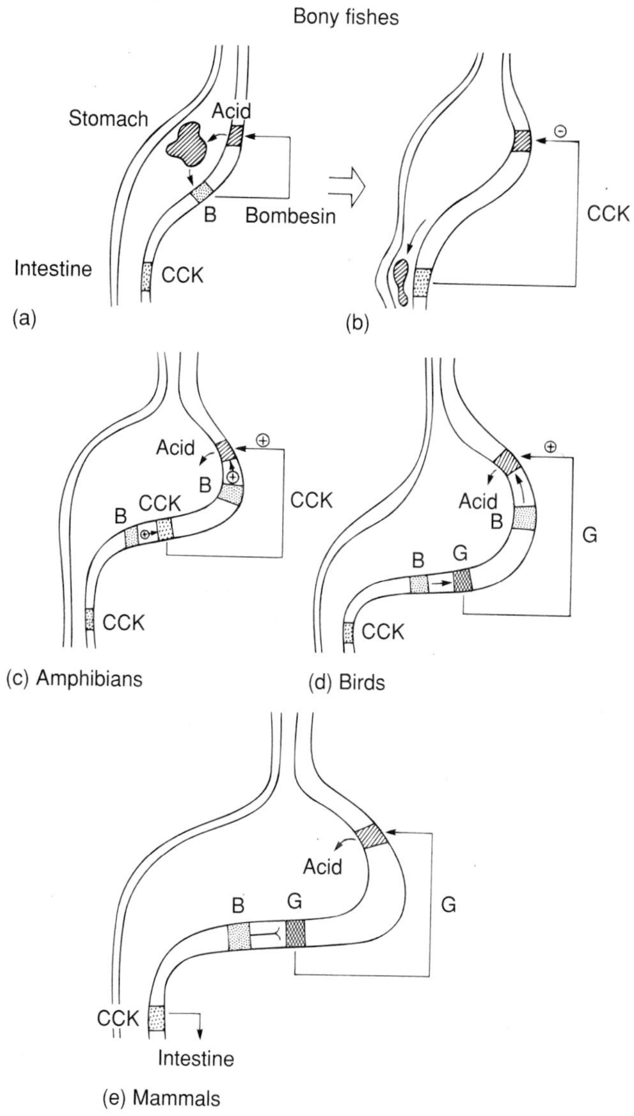


**Figure 15.28** Evolution of gastrointestinal control by the endocrine system. (a) Bony fishes. Food in the stomach of a bony fish stimulates bombesin cells to release blood-borne bombesin (B), which stimulates the secretion of gastric acid. (b) As food moves into the teleost intestine, CCK cells are stimulated to release CCK, which inhibits gastric secretion. (c) Amphibians. Note that CCK cells reside both in the intestine, as in bony fishes, and in the stomach. Bombesin cells release a secretion that diffuses through adjacent epithelium, directly stimulating CCK cells to promote gastric acid secretion. (d) Birds (and probably reptiles). CCK cells are restricted to the intestine. In the stomach, they are replaced by cells that secrete gastrin (G). (e) Mammals. Bombesin cells, or their derivatives, stimulate gastrin cells through direct chemical contact.

located in the stomach, but instead of entering the circulatory system as it does in fishes, bombesin directly stimulates adjacent CCK cells or even gastric cells in the wall of the stomach. Gastric cells respond to CCK or to direct bombesin stimulation by secreting acid into the stomach (figure 15.28c). In mammals, in birds, and probably in reptiles, CCK cells occur only in the intestine. In their place in the stomach are cells that produce the hormone gastrin (figure 15.28d,e). In most amniotes, gastrin rather than CCK stimulates the release of acid from the stomach walls, and stomach cells that secrete bombesin stimulate adjacent gastrin-secreting cells. In mammals, bombesin-secreting cells become neurosecretory in that they contain short axons that extend to the gastrin-secreting cells and stimulate them to produce gastrin.

Several phylogenetic changes are evident in this sequence. First, in fishes CCK inhibits gastric acid secretion, whereas in amphibians CCK promotes the release of gastric acid. In contrast to these two situations, gastrin in amniotes replaces CCK as the hormone that activates the release of stomach acid. Second, bombesin-secreting cells that release their bombesin into the blood in fishes become neurosecretory cells that activate local gastric target tissues in mammals. Bombesin has a wide repertoire of effects in addition to its action on the stomach, including effects on thermoregulation, pituitary activity, and digestive tract mobility. The advantage of this phylogenetic change in bombesin cells from an endocrine to a neurosecretory role is probably related to more localized and precise delivery of stimulation that does not interfere with the other endocrine effects of bombesin.

Replacement of CCK by gastrin as the hormone controlling gastric secretion probably made digestion more efficient. CCK arose first as an intestinal hormone important in processing food. When a distinct stomach appeared, CCK cells were located in both the intestine and the stomach and could be stimulated by food at either site. But the roles of the stomach and intestine in processing food are different, especially in later vertebrates (see chapter 13). With the restriction of CCK cells to the intestine and the appearance of gastrin cells in the stomach, the gastric and intestinal phases of digestion could be controlled separately.

Mammalian CCK is chemically similar to that of fishes, so there has been little evolution of the hormone. However, there have been significant alterations in the endocrine control of gastric and intestinal digestion. Gastrin arose and bombesin-secreting cells changed their routes of action from delivery via the blood to direct neural stimulation.

In some instances, evolution has involved important changes in hormonal structure, or new molecules have been co-opted for hormonal roles. This has been especially true with protein hormones. For example, epinephrine is a neurotransmitter that is released by axons into synaptic spaces. The adrenal cortex releases the hormone epinephrine into the blood. In the endocrine system, chemical messages coordinate internal activities by traveling long distances through the circulatory system. In the nervous system, chemical messages travel short distances across the spaces between neurons and responding cells. Thus, the nervous system, like the endocrine system, regulates activities of the body, and its functional basis is much the same—the release of chemical messages that affect responses. Therefore, we turn next to the nervous system.

## Selected References


Bagnara, J. T., and M. E. Hadley. 1973. *Chromatophores and color*. New Jersey: Prentice-Hall.

Bentley, P. J. 1976. *Comparative vertebrate endocrinology*. Cambridge: Cambridge University Press.

Binkley, S. 1988. *The pineal: Endocrine and non-endocrine function*. New Jersey: Prentice-Hall.

Epple, A., and J. E. Brinn. 1987. *The comparative physiology of the pancreatic islets*. Berlin: Springer-Verlag.

Gorbman, A., W. W. Dickhoff, S. R. Vigna, N. B. Clark, and C. L. Ralph. 1983. *Comparative endocrinology*. New York: John Wiley and Sons.

Jameson, E. W., Jr. 1988. *Vertebrate reproduction*. New York: John Wiley and Sons.

Norris, D. O. 1985. *Vertebrate endocrinology*. 2d ed. Philadelphia: Lea and Febiger.

Nozaki, M., and A. Gorbman. 1992. The question of functional homology of Hatschek's pit of amphioxus (*Branchiostoma belcheri*) and the vertebrate adenohypophysis. *Zool. Sci.* 9:387–95.

Vigna, S. R. 1985. Functional evolution of gastrointestinal hormones. In *Evolutionary biology of primitive fishes*, edited by R. E. Foreman, A. Gorbman, J. M. Dodd, and R. Olsson. New York: Plenum Press, pp. 401–12.

16 CHAPTER

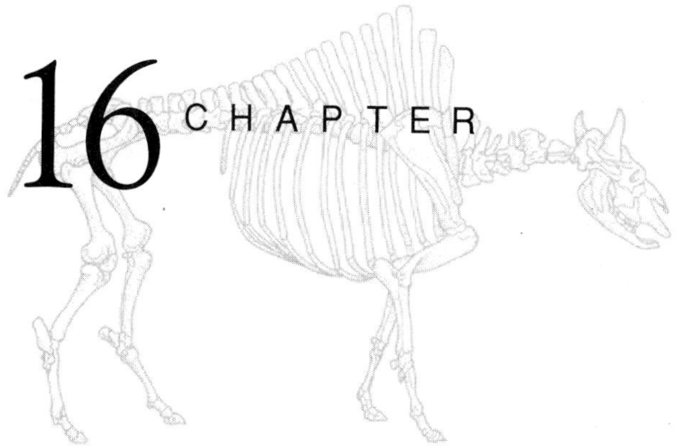

# The Nervous System



## INTRODUCTION

The nervous system is divided into the **central nervous system** (CNS), which includes the brain and spinal cord, and the **peripheral nervous system** (PNS), which consists of all nervous tissue outside the CNS. The nervous system *receives* stimuli from one or more **receptors** and *transmits* information to one or more **effectors** that respond to stimulation. Effectors include **mechanical effectors,** such as muscles, and **chemical effectors,** such as glands. Thus, responses of the nervous system involve the muscle contractions and glandular secretions. The nervous system regulates an animal's performance by integrating immediate incoming sensory information with stored information, the results of past experience, and by then translating past and present information into action through the effectors.

The nervous system includes billions of nerve cells, each of which establishes thousands of contacts with other nerve cells, so the total number of interconnections is astronomical. That is why analysis of the function of the nervous system often includes as much philosophy as it does science. The task is formidable, but not hopeless. Let us begin by looking at the fundamental cellular components of the nervous system.

### Types of Cells within the Nervous System

There are two types of cells within the nervous system—**neurons** and **neuroglial cells,** or **glia.**

#### *Neuroglia*

Neuroglial (nerve and glue) cells do not transmit impulses. They support, nourish, and insulate neurons. All neuroglia bind together nervous tissue, and they can be specialized (figure 16.1). **Astrocytes** pass nutrients between blood capillaries and neurons, **microglia** engulf foreign material and bacteria, **oligodendroglia** and **Schwann cells** insulate neurons, and **ependymal cells** line the central canal of brain and spinal cord.

#### *Neurons*

Neurons are specialized for long-distance transmission of electrical stimuli throughout the body. The neuron is the structural and functional unit of the nervous system. It consists of the **perikaryon,** the **body** or **soma** of the neuron, and thin cell processes called **nerve fibers** if they are long (figure 16.2). The processes are of two types. There usually is one

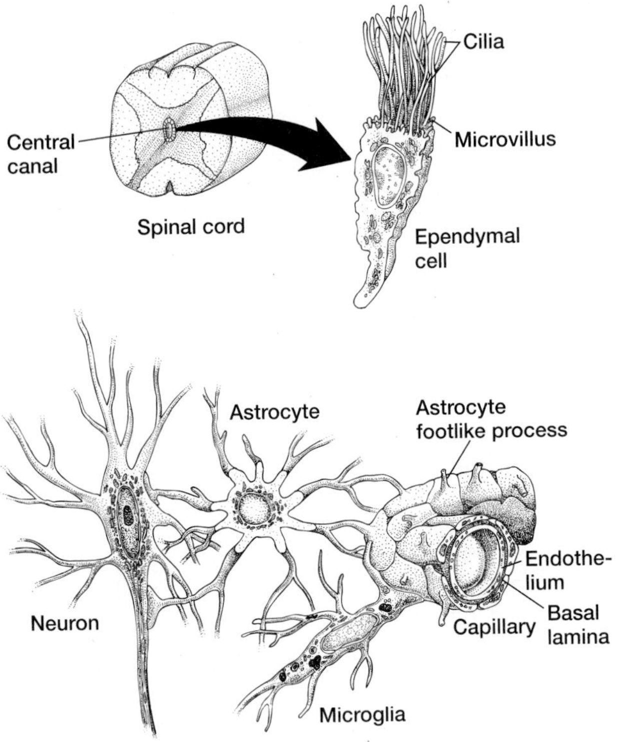


**Figure 16.1** Four types of neuroglia found within the central nervous system. Astrocytes form cytoplasmic connections to transport nutrients between blood capillaries and neurons. Phagocytic microglia engulf stray or foreign materials. Ependymal cells line the central canal of the central nervous system. Oligodendroglia insulate nerves within the central nervous system (not shown).

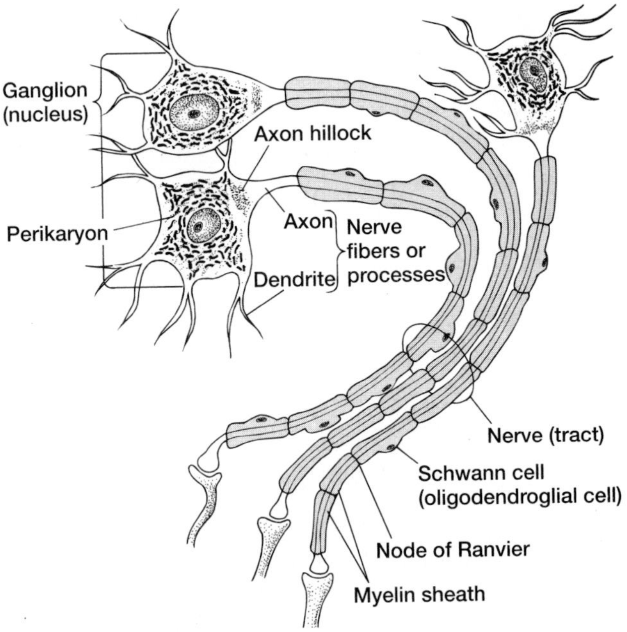


**Figure 16.2** Structure of a neuron. The cell nucleus and surrounding cytoplasm form the body of a neuron (perikaryon). Nerve fibers or processes are cytoplasmic extensions from the perikaryon. Axons carry impulses away from the perikaryon and dendrites carry impulses toward it. The same structures are given different names in the peripheral and the central nervous systems. Central nervous system terms are given in parentheses.

**axon** per neuron and one or many **dendrites.** Dendrites transmit incoming electrical impulses toward the perikaryon. Axons carry impulses away from the perikaryon. Neurons are grouped by the number of their processes. **Unipolar neurons** have a single stem that divides into a dendrite and axon. **Bipolar neurons** have two processes, usually at opposite ends. **Multipolar neurons** have many processes associated with the cell body (figure 16.3a–h).

Neurons and their processes are often known by different terms, depending on whether they occur in the CNS or the PNS. For example, a collection of axons traveling together is a nerve **tract** in the CNS and a **nerve** in the PNS. A collection of nerve cell bodies is a **nucleus** in the CNS and a **ganglion** in the PNS. Neuroglial cells wrap some axons in a thick **myelin** sheath. Such fibers are called **myelinated nerves** and those without sheaths are **unmyelinated nerves** (figure 16.4a,b). A neuroglial cell that produces the myelin sheath is an oligodendroglial cell in the CNS and a Schwann cell in the PNS. The **nodes of Ranvier** are indentations between adjacent neuroglial cells in the myelin sheath.

## Transmission of Information

Information traveling through the nervous system is transmitted in the form of electrical and chemical signals. Electrical signals are **nerve impulses** that travel within the plasma membrane of the neuron and are of two kinds, graded potentials and action potentials. A **graded potential** is a wave of electrical excitation proportional to the magnitude of the stimulus that triggers it. The graded potential declines in magnitude as it travels along a nerve fiber. An **action potential** is an all-or-none phenomenon. Once initiated, it propagates without decrement along a nerve fiber. Action potentials are often used for long-distance signaling in the nervous system. Within the dentrites and the perikaryon, nerve impulses are usually graded potentials, but they become action potentials as they travel out the axon.

Chemical signals are generated at **synapses,** gaps between the junctions of neurons (figure 16.5a). These gaps occur between the processes of one neuron and the next, and between axons and perikarya. Upon arrival at the terminus of an axon, an electrical impulse stimulates the release of stored **neurotransmitters** into the tiny space between processes. Neurotransmitters diffuse across this synaptic junction and settle on the associated cellular process of the next neuron. When collected in sufficient concentration, neurotransmitters initiate an electrical impulse

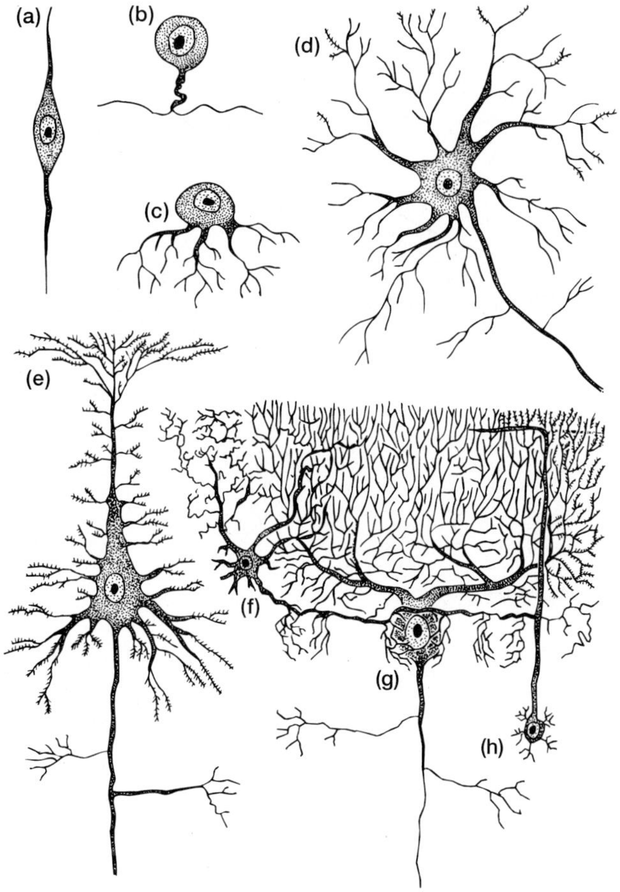


**Figure 16.3** Types of neurons. (a) Bipolar neuron. (b) Unipolar neuron. (c–h) Multipolar neurons.

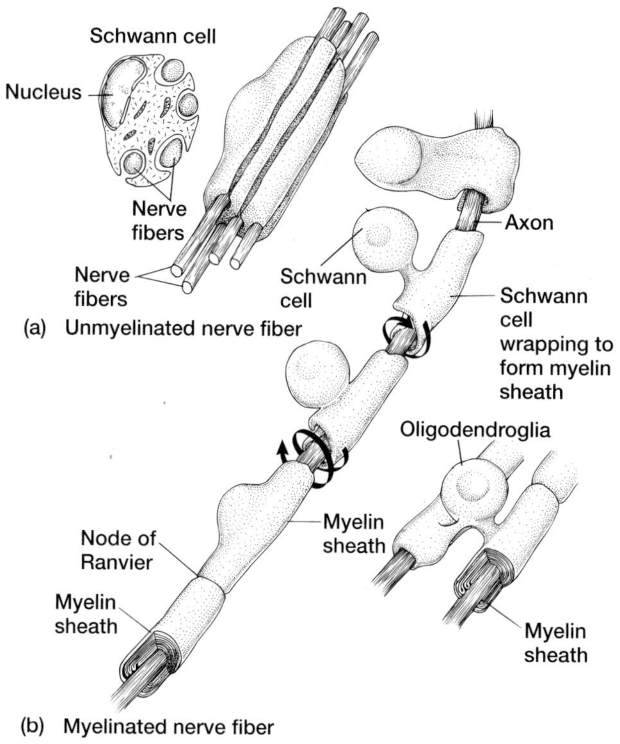


**Figure 16.4** Myelinated and unmyelinated nerve fibers. (a) Despite their name, unmyelinated nerve fibers are associated with neuroglial cells. Usually there are several fibers per neuroglial cell, but these neuroglial cells are not wrapped repeatedly around the fibers as they are in myelinated nerves. (b) The myelin sheath is formed by a neuroglial cell that is repeatedly wrapped around a section of the nerve fiber. In the peripheral nervous system, the neuroglial cell is a Schwann cell. In the central nervous system, it is an oligodendroglial cell. Successive neuroglial cells collectively form the myelin sheath. The boundaries between them are termed nodes of Ranvier.

in the next neuron. Excess and spent neurotransmitter molecules are rapidly inactivated to prevent prolonged effects. Neurotransmitters must quickly reach a threshold level to initiate an electrical impulse in the next neuron. Thus, passage of information through chains of connected neurons includes alternating events of electrical and chemical transmission involving nerve impulses and neurotransmitters, respectively (figure 16.5b).

Synapses introduce control into the processing of information transfer. If synapses were absent and neurons were in direct contact with each other, excitation in one neuron might spread inevitably throughout an entire network of interconnected neurons like ripples on a pond, without any local control. Synapses break up a network of neurons into information-processing units. Whether an impulse is transmitted to the next neuron in a sequence depends on whether or not there is a sufficient concentration of neurotransmitters at the synapse. Where neurons converge, transmission of a single impulse from one nerve might be insufficient. Several impulses might need to arrive simultaneously in order to release enough neurotransmitter molecules to trigger an electrical impulse in the next neuron. Convergence promotes **summation** of information. Conversely, if a neuron sends branches to several circuits, information diverges and is distributed to appropriate areas. Branches of a single axon are called **collateral branches.** Inhibition also affects the flow of information by decreasing the responsiveness of neurons to incoming information. Convergence, divergence, and inhibition are modes of information processing that take advantage of the character of the synapse (figure 16.5). Furthermore, the structure of neurons at a synapse ensures that transmission across the gap occurs in only one direction.

## Neurosecretory Cells

**Neurosecretory cells** are specialized neurons. Most neurons release neurotransmitters at the ends of their axons. Neurosecretory cells also release secretions at the ends of their axons, but these secretions are delivered into a blood

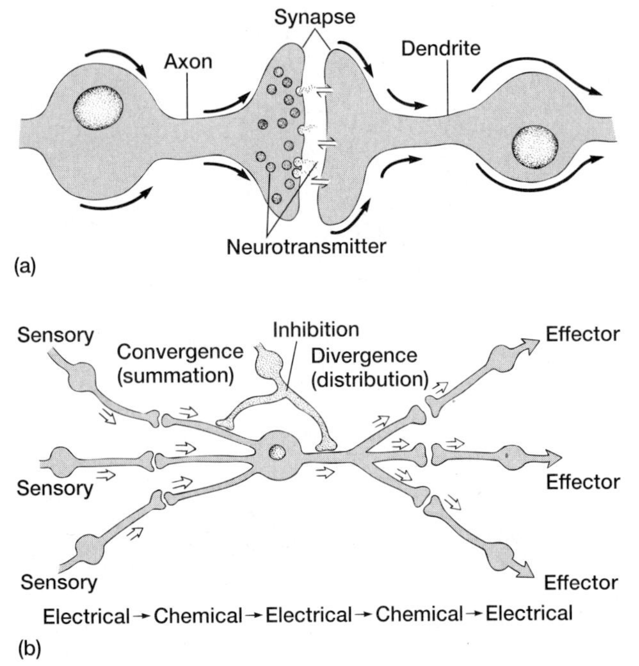


**Figure 16.5** Transmission of information within the nervous system. (a) Neurons transmit and receive stimuli as electrical impulses along their fibers. Synapses are junctions between nerve cells. Axons release chemical messengers (neurotransmitters) that diffuse across the synapse. When they arrive at the dendrite in sufficient concentration, the neurotransmitter molecules initiate an electrical impulse in the next neuron. (b) Synapses assist in processing information. Electrical input can converge or diverge. Inputs of some neurons inhibit or reduce the sensitivity of other neurons.

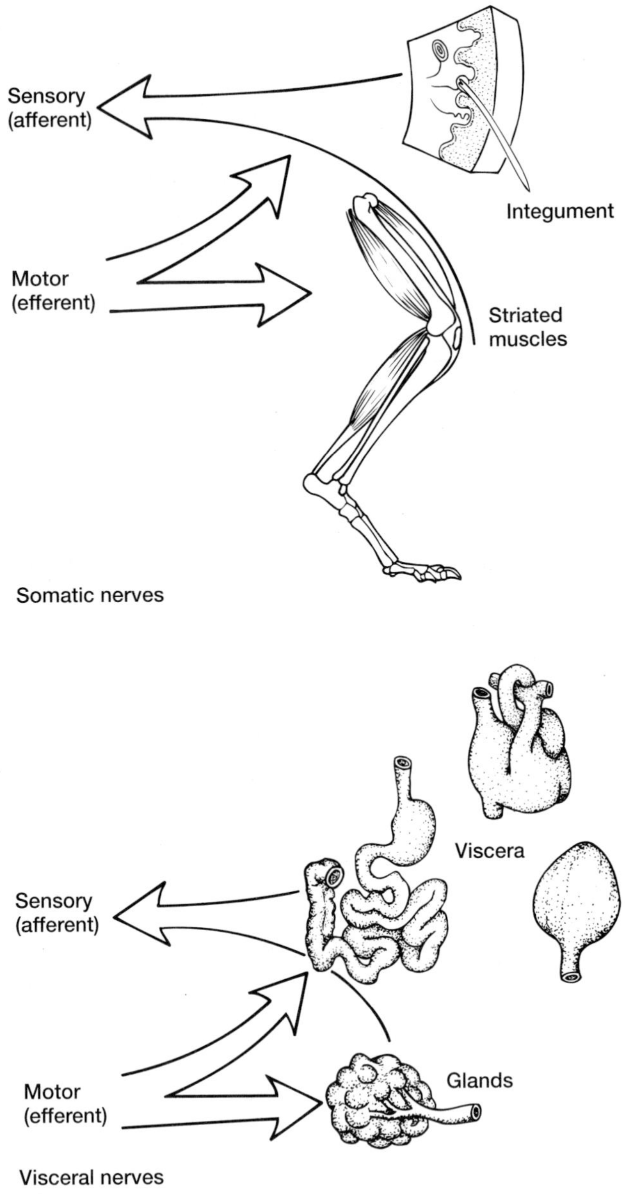


**Figure 16.6** Functional categories of neurons of the peripheral nervous system. Some neurons supply somatic tissues, others visceral tissues. They can be sensory and respond to stimuli from these tissues, or they can be motor and deliver stimuli to these tissues.

capillary and transported to a target tissue. Neurosecretory cells are thus endocrine in function.

# Peripheral Nervous System

The terms used to describe the components of the peripheral nervous system refer to the anatomical or functional properties of the nerves (figure 16.6). Peripheral nerves serve either somatic or visceral tissues and carry sensory or motor information. **Somatic nerves** pass to or from somatic tissues—skeletal muscle, skin, and their derivatives. **Visceral nerves** pass to or from viscera—involuntary muscles and glands. Nerves carrying information from tissues *to* the central nervous system are **afferent,** or **sensory, neurons.** Nerves carrying information *away* from the CNS to effectors are **efferent,** or **motor, neurons.** Thus, a somatic sensory nerve might carry information about touch, pain, or temperature from the skin to the central nervous system. A somatic motor nerve carries impulses from the CNS to a striated muscle to stimulate its contraction. A visceral sensory nerve delivers information about the condition of internal viscera to the CNS. A visceral motor nerve innervates visceral effectors (cardiac muscle, smooth muscle, or glands). The components of the PNS that control visceral activity constitute the **autonomic nervous system** (ANS).

Nerves have two additional properties based on their distribution. Neurons are termed *general* if the innervated tissues are widely distributed or *special* if the tissues are restricted in location. Thus, general somatic neurons innervate sense organs or supply effectors to the integument and most striated muscles. Special somatic neurons are associated with somatic sense organs (e.g., eyes, olfactory organs, inner ears) or effectors (e.g., ciliary eye muscles, extrinsic ocular muscles) that are limited in distribution. General visceral neurons innervate sensory organs or supply effectors in glands or smooth muscles of the digestive tract, heart, and other viscera. Special visceral neurons concerned with sensory input innervate the taste buds. Neurons to the striated branchiomeric muscles are termed special visceral motor neurons.

From anatomical criteria, the peripheral nervous system can be divided into **spinal nerves** emanating from the spinal cord and **cranial nerves** emanating from the brain. We begin by looking at these anatomical divisions of the peripheral nervous system.

## Spinal Nerves

Spinal nerves are sequentially arranged and numbered (C-1, T-1, L-1, S-1) according to their association with regions of the vertebral column (cervical, thoracic, lumbar, sacral). Early anatomists recognized **dorsal** and **ventral roots** of each spinal nerve. Afferent fibers enter the spinal cord via the dorsal root, and efferent fibers leave by way of the ventral root. The **dorsal root ganglion,** a swelling in the dorsal root, is a collection of neuron bodies whose axons contribute to the spinal nerve. Parallel to the spinal cord and attached to each spinal nerve through the **ramus communicans** is the **sympathetic chain** of ganglia (paravertebral ganglia), a paired series of linked ganglia adjacent to the vertebral column or notochord (figure 16.7a,b). Other peripheral ganglia form the **collateral ganglia** (prevertebral ganglia). The paired **cervical, coeliac,** and **mesenteric ganglia** are examples of the collateral ganglia. The **visceral ganglia** occur within the walls of visceral effector organs (figure 16.7b). Thus, there are three types of ganglia—sympathetic, collateral, and visceral.

Peripheral nerves arise during embryonic development from two sources (figure 16.8a–c). One source is the neurons that differentiate within the spinal cord. Axonal processes sprout from these neurons and grow outward to the ganglia or to the effectors they supply (figure 16.8b). The other source is the neural crest. Cells migrate from the neural crest to specific locations and sprout processes that grow back to the central nervous system and out to the tissues they innervate (figure 16.8a). The ventral roots arise from neurons in the spinal cord that send fibers out from the spinal cord. The dorsal root arises from cells of neural crest origin that send fibers into the spinal cord. In higher vertebrates, the two roots usually merge to form the composite spinal nerve and attached sympathetic chain.

The fibers of each spinal nerve innervate restricted structures at that level of the cord. This is especially pronounced with spinal nerve innervation of somatic tissues. Each growing spinal nerve tends to accompany its adjacent embryonic **myotome,** the source of somatic muscles, and its **dermatome,** the source of dermal connective tissue and muscle, as they spread and differentiate during development (figure 16.9a–c). Once it is differentiated, a spinal nerve

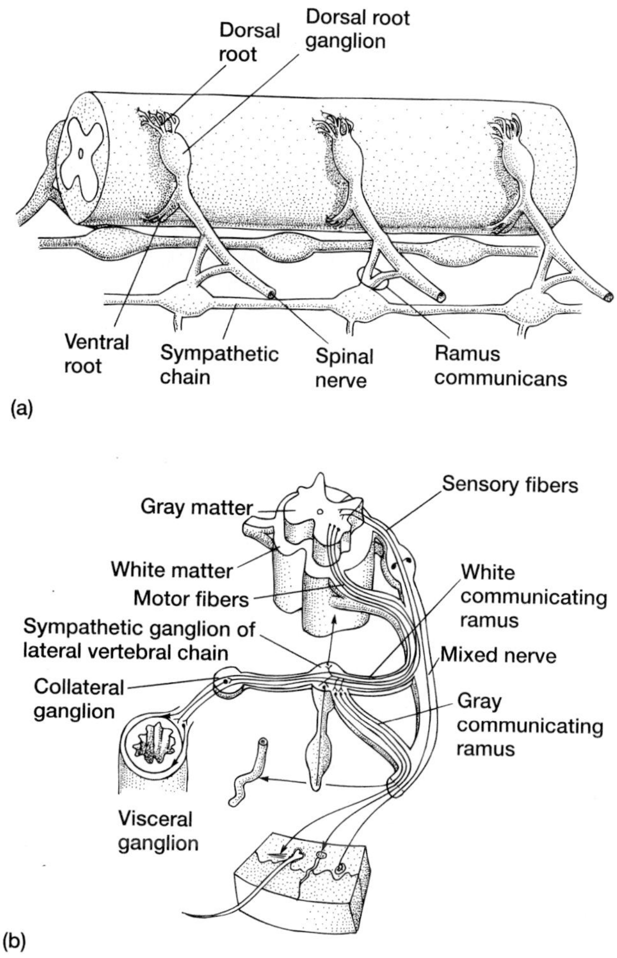


**Figure 16.7** Spinal nerve anatomy. (a) Dorsal and ventral roots connect spinal nerves to the spinal cord. A dorsal root is enlarged into a dorsal root ganglion. Spinal nerves join with the sympathetic chain through communicating rami. (b) Configuration of sensory and motor neuronal routes in an adult mammal.

supplies the skeletal muscles derived from its adjacent myotome and receives somatic sensory input from the restricted area of the body surface differentiated from its dermatome. Strictly speaking, a dermatome refers to an embryonic structure, but the term is often used to denote the region of the adult body derived from it. The fidelity between a dermatome and its spinal nerve permits mapping of the body surface in terms of the corresponding spinal nerves that supply each region. Loss of sensation in a dermatome can be diagnostic for the specific spinal nerve involved.

## Cranial Nerves

Cranial nerves have roots enclosed in the braincase. These are named and numbered by roman numerals from anterior to posterior. The conventional system for numbering these nerves is sometimes inconsistent. For instance, most

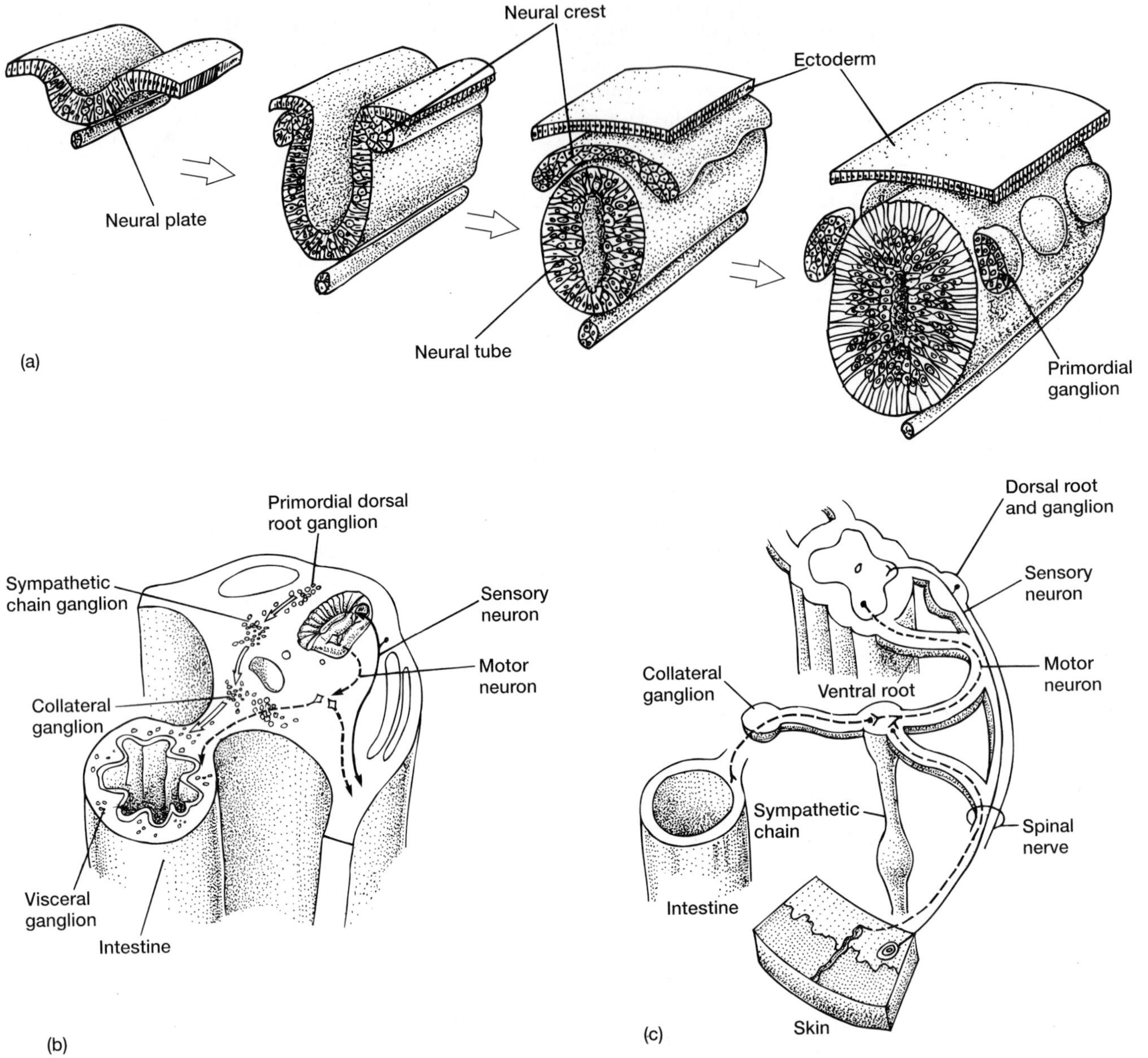


**Figure 16.8** Embryonic development of afferent and efferent spinal nerves. (a) The neural crest forms from ectoderm during neurulation and becomes organized as segmental populations of cells arranged dorsally along the neural tube. (b) From this dorsal location, some cells migrate (open arrows) to specific sites within the body, forming distinct populations of neural cells at these sites. Neurons differentiating within the primordial dorsal root sprout cell processes that grow back to the neural tube and out to somatic tissues. Neuronal bodies that remain in position constitute the dorsal root ganglion. Neurons differentiating within other populations grow cell processes to effectors, and their bodies constitute ganglia. Motor neurons differentiate within the neural tube and grow cell processes to these peripheral ganglia or directly to effectors. (c) Diagrammatic representation of established afferent and efferent neurons within spinal nerves.

anamniotes are said to have ten cranial nerves. A few anamniotes and all amniotes are said to have 12. In fact, there is an additional terminal nerve at the beginning of this series. If counted at all, it is numbered 0 to avoid renumbering the conventionally numbered sequence. Further, the second cranial nerve (II) is not a nerve at all but an extension of the brain. Nevertheless, by convention it is called the optic "nerve." The eleventh cranial nerve (XI) represents the merger of a branch of the tenth cranial nerve (X) with elements of the first two spinal nerves (C-1 and C-2). Despite its composite structure, it is called the spinal accessory nerve and designated by roman numeral XI.

Phylogenetically, the cranial nerves are thought to have evolved from dorsal and ventral roots of a few anterior

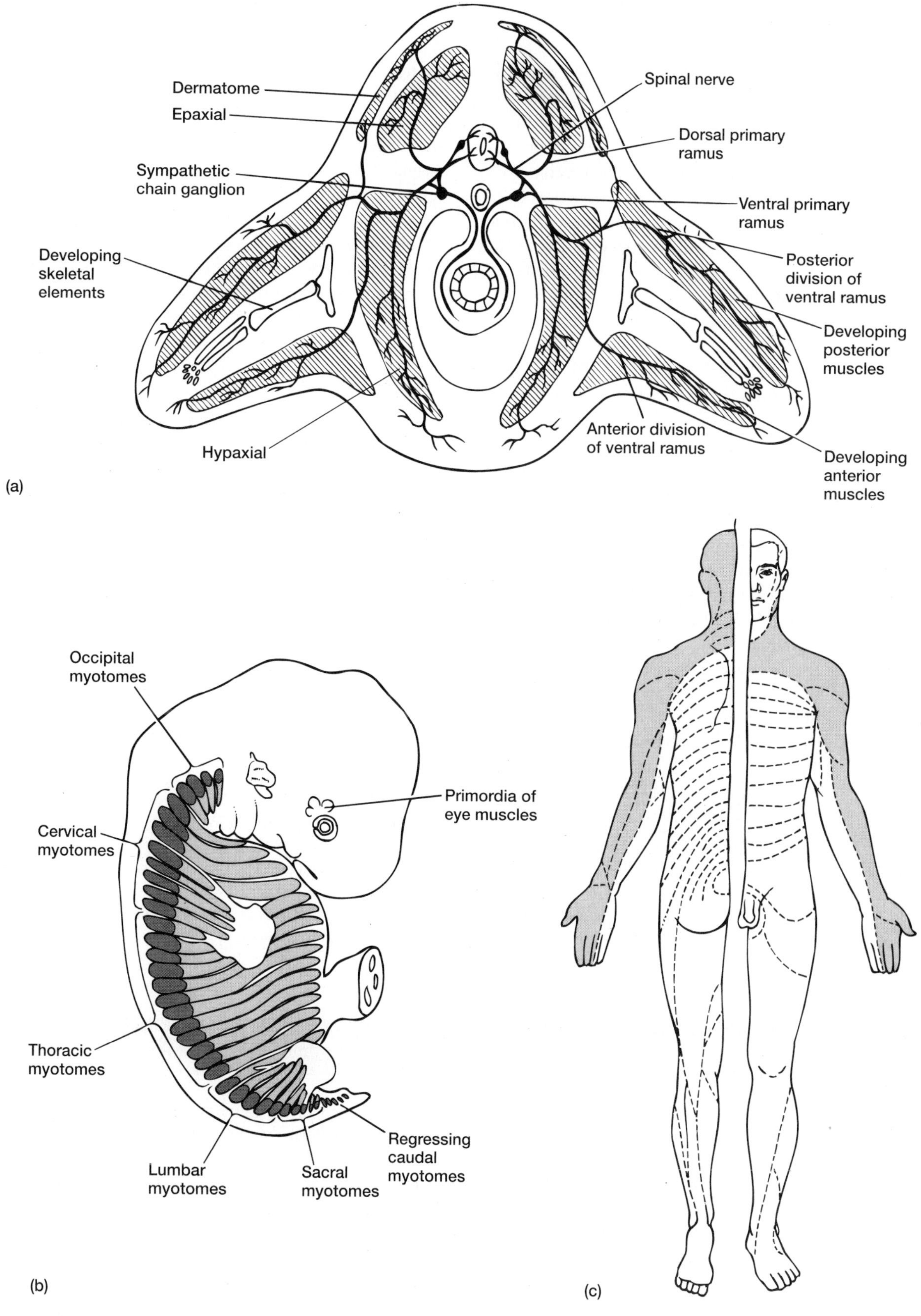


**Figure 16.9** Spinal nerves supply vertebrate limbs and body walls. (a) Cross section of a generalized vertebrate. Note the distribution of spinal nerves to axial and appendicular muscles. (b) Sagittal section of human embryo illustrating the distribution of myotomes served by segmental spinal nerves. (c) Split image showing dorsal and ventral distribution of dermatomes within the human body.

spinal nerves that became incorporated into the braincase. Like spinal nerves, the cranial nerves supply somatic and visceral tissues and carry general sensory and motor information. Some cranial nerves consist of only sensory or only motor fibers. Other nerves are **mixed,** containing both types. Cranial nerves concerned with localized senses (e.g., sight, hearing, olfaction, taste) are called **special cranial nerves** to distinguish them from those concerned with the sensory or motor innervation of the more widely distributed viscera, **general cranial nerves.**

Primitively, all cranial nerves serving the branchial pouches formed three branches per pouch—**pretrematic, posttrematic,** and **pharyngeal** (figure 16.10). In later gnathostomes, these tend to be lost or their homologies become uncertain.

Most anamniotes possess ten cranial nerves. The first few spinal nerves behind the braincase become housed in the skull of later derived groups. But in anamniotes, these anterior spinal nerves are still partially outside the skull, although they have lost their dorsal roots and differ from other spinal nerves. In cyclostomes, these anterior spinal nerves outside the skull are called **occipitospinal nerves.** In other fishes and amphibians, the anterior spinal nerves become partially incorporated into the braincase. They exit via foramina in the occipital region of the skull and are called **occipital nerves.** Occipital nerves unite with the next few cervical spinal nerves to form the composite **hypobranchial nerve** that supplies hypobranchial muscles in the throat (figure 16.11a,b).

Crossopterygians and all tetrapods except modern amphibians possess 12 cranial nerves. Living amphibians secondarily revert to the primitive ten. In amniotes, the occipitospinal nerves are incorporated into the skull and modified. Their roots shift from the spinal cord forward into the medulla. In this way, amniotes derive the eleventh and twelfth cranial nerves. The 12 cranial nerves are illustrated in figures 16.12 through 16.15. They are described in more detail next and their functions are summarized in table 16.1.

**Nervus terminalis (0).** The **terminal nerve** is not numbered. It may be testimony to an ancient anterior head segment that has been lost. This nerve is present in all classes of gnathostomes except birds. It runs to blood vessels of the olfactory epithelium in the olfactory sac and most likely carries visceral sensory and some motor fibers.

**Olfactory nerve (I).** The **olfactory nerve** is a sensory nerve concerned with the sense of smell. Olfactory cells lie

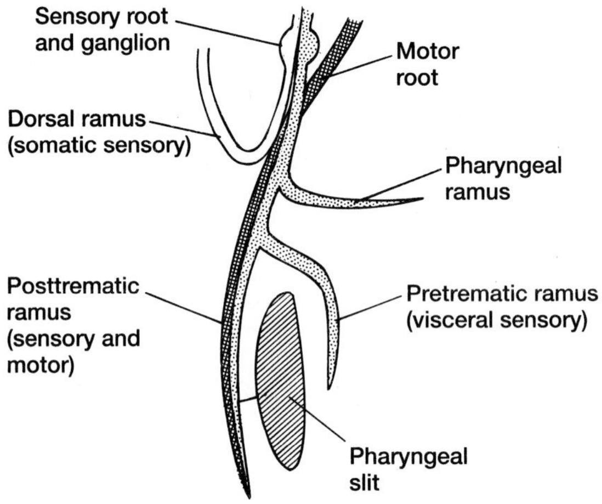


**Figure 16.10** Components of a cranial nerve in a fish. The pharyngeal ramus, to the lining of the pharynx, and the small pretrematic ramus, to the front of the pharyngeal slit, both carry visceral sensory fibers. The dorsal ramus from the skin is comprised of somatic sensory fibers. The posttrematic ramus running down the back of the pharyngeal slit includes both sensory and motor fibers.

## Box Essay 16.1 Shingles

Shingles is the common name for a disease caused by the virus *Herpes zoster,* the same virus that produces chicken pox. Shingles is characterized by a line of blisters that usually radiate out along one side of the body following one of the spinal or cranial nerves to a dermatome.

For most people, chicken pox is a childhood disease that lasts several weeks. It causes itchy blisters over the body and also provides a few days of respite from school. Eventually, the immune system forces the virus into remission. Subsequent events are not well known, but it is thought that the virus retreats to the perikarya of neurons and is held in check there by the immune system. For most people, that is the end of *Herpes zoster.* But in some, the immune system lets down a bit and the virus proliferates, except that during this flare-up its spread is more restricted. The virus migrates along a nerve to the dermatome it supplies (see figure 16.9b,c). The tissue along this pathway reacts by forming the characteristic but very painful blisters. When the immune system responds again, the virus is beaten back, usually for the last time, and the symptoms of the disease abate.

Our knowledge of nerve anatomy and corresponding dermatome associations aids in the diagnosis of shingles. In most cases, it is possible to determine which spinal or cranial nerve the virus spreads along from the pattern of blisters.

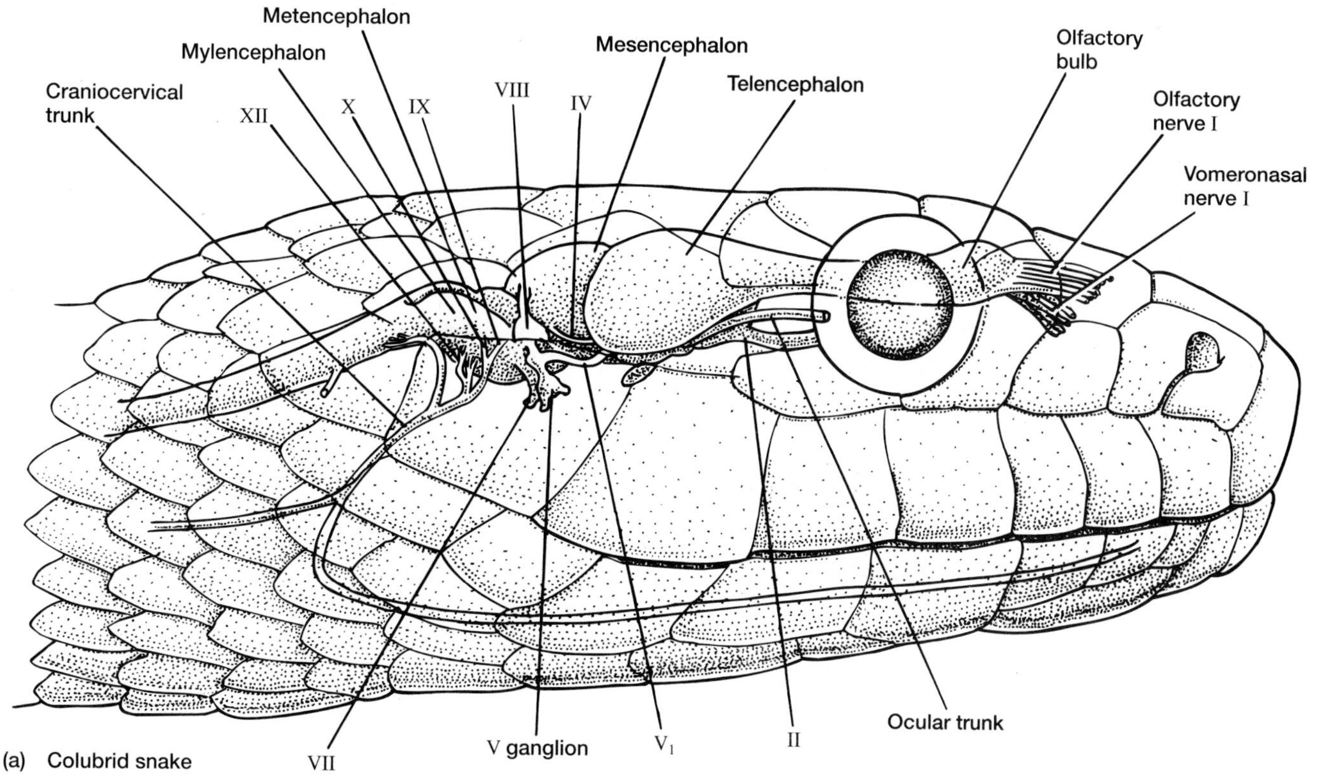


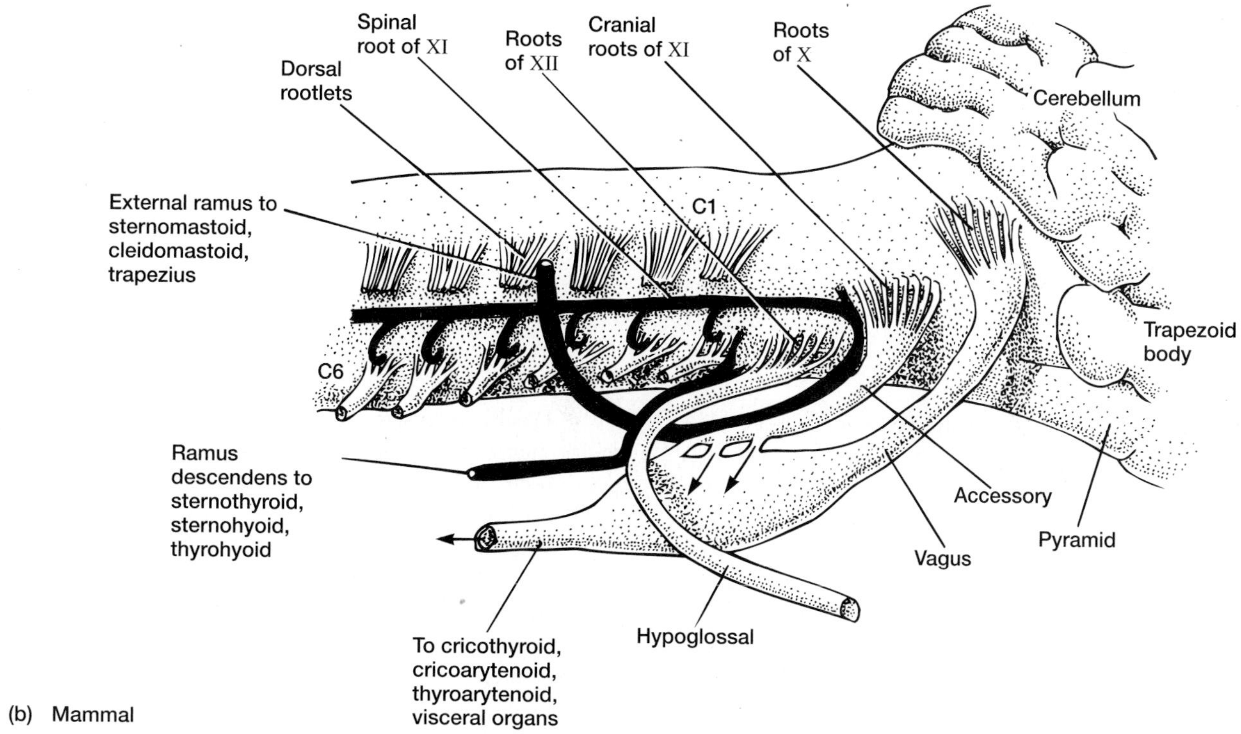


**Figure 16.11** Posterior cranial nerves. (a) Colubrid snake. The glossopharyngeal (IX), vagus (X), hypoglossal (XII), and one of the spinal nerves join to form the craniocervical trunk. Unlike most other amniotes, snakes appear to lack a spinal accessory nerve (XI). (b) Mammal. The roots of the hypoglossal nerve are in series with the ventral roots of the preceding spinal nerves. Spinal nerve contributions to the accessory and hypoglossal nerves are shown in solid black. The vagus receives contributions from the accessory nerves (arrows).

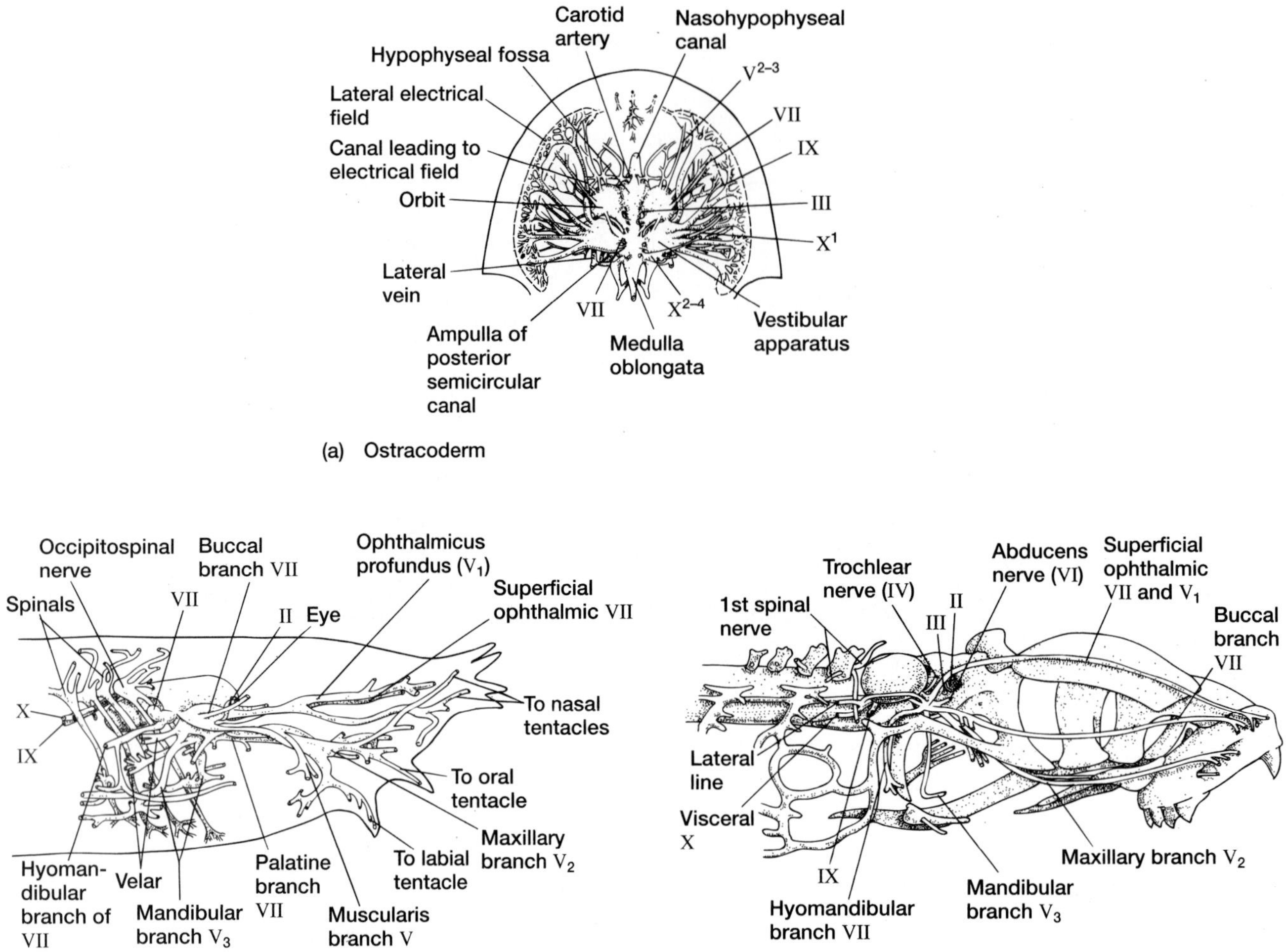


**Figure 16.12** Cranial nerves of vertebrates. (a) Ostracoderm *Kiaeraspis*. (b) Hagfish *Myxine*. (c) Lamprey. (d) Lateral view of cranial nerves in the shark *Squalus*.

in the mucous membrane of the olfactory sac. A short axon leads from each cell to the olfactory bulb. Each axon constitutes an olfactory fiber. Collectively, the olfactory fibers form the short olfactory nerve.

**Optic nerve (II).** Strictly speaking, the **optic nerve** is not a nerve but a sensory tract. That is, it is not a collection of peripheral axons; it is a collection of fibers in the CNS. Embryologically, it develops as an outpocketing of the brain. However, once it is differentiated, it lies outside the brain. Its fibers synapse in the thalamus and midbrain.

**Oculomotor nerve (III).** The **oculomotor nerve** primarily supplies extrinsic eye muscles (superior rectus, medial rectus, inferior rectus, and inferior oblique muscles) derived from preotic myotomes. It is a motor nerve that also carries a few visceral motor fibers to the iris and ciliary body of the eye. Fibers arise in the oculomotor nucleus in the floor of the midbrain.

**Trochlear nerve (IV).** The **trochlear nerve** is a motor nerve that supplies the extrinsic, superior oblique eye muscle. Fibers arise in the trochlear nucleus of the midbrain.

**Trigeminal nerve** or **trigeminus (V).** The **trigeminus** is so named because it is formed of three branches, **ophthalmic** ($V_1$), **maxillary** ($V_2$), and **mandibular** ($V_3$) in amniotes (figures 16.12c and 16.15). The ophthalmic nerve, sometimes called the deep ophthalmic nerve to distinguish it from a more superficial nerve, usually merges with the other two branches. However, in lower vertebrates, the ophthalmic nerve often emerges from the brain separately, possibly representing an ancient condition in which the ophthalmic nerve supplied an anterior arch that has since been lost. The other two branches, the maxillary ramus ($V_2$) to the upper jaw and the mandibular ramus ($V_3$) to the lower jaw, presumably represent pretrematic

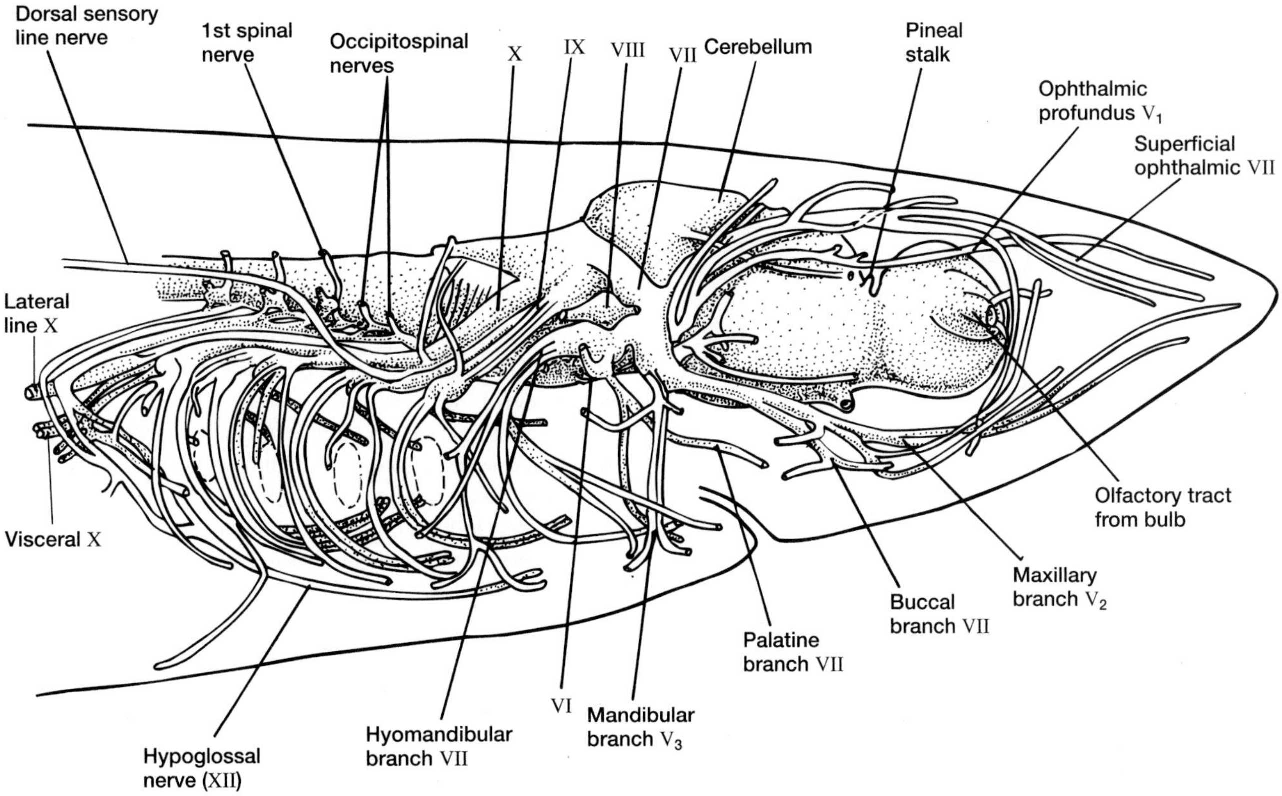


(d) Shark

**Figure 16.12** (continued)

**TABLE 16.1 Functional Components of Cranial Nerves**

| | | Somatic Sensory | | Visceral Sensory | | Visceral Motor | | Somatic Motor |
|---|---|---|---|---|---|---|---|---|
| *Cranial Nerve* | | *Special* | *General* | *General* | *Special* | *Special* | *General* | *General Special* |
| 0 | Terminal | | X | | | | | |
| I | Olfactory | X | | | | | | |
| II | Optic | X | | | | | | |
| III | Oculomotor | | | | | | (X) | X |
| IV | Trochlear | | | | | | | X |
| $V_1$ | Trigeminal | | X | | | | | |
| $V_{2,3}$ | Trigeminal proper | | X | | | X | | |
| VI | Abducens | | | | | | | X |
| VII | Facial | | (X) | X | X | X | X | |
| VIII | Auditory | X | | | | | | |
| IX | Glossopharyngeal | | (X) | X | X | X | X | |
| X | Vagus | | X | X | | X | X | |
| XI | Spinal accessory | | | | | X | | |
| XII | Hypoglossal | | | | | | | X |
| | Lateral line | X | | | | | | |

*Note:* Parentheses indicate variable or negligible function in the category indicated.

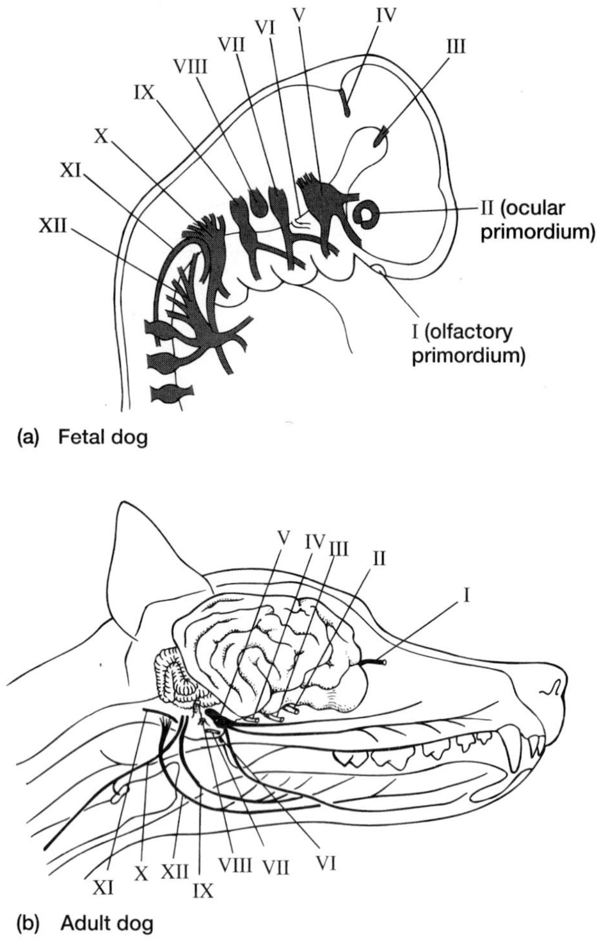


**Figure 16.13** Embryonic development of cranial nerves. (a) Fetus. (b) Adult dog.

and posttrematic rami of a typical branchial nerve to the mandibular arch.

The mixed trigeminus includes sensory fibers from the skin of the head and areas of the mouth and motor fibers to derivatives of the first branchial arch. Sensory fibers of the trigeminus return to the brain from the skin, teeth, and other areas through each of the three branches. The mandibular branch also contains visceral motor fibers to muscles of the mandibular arch.

**Abducens nerve (VI).** The **abducens** is the third of the three cranial nerves that innervates muscles controlling movements of the eyeball. It is a motor nerve that supplies the extrinsic, lateral rectus eye muscle. Fibers arise in the abducens nucleus located in the medulla.

**Facial nerve (VII).** The mixed **facial nerve** includes sensory fibers from the lateral line of the head, ampullae of Lorenzini, and taste buds as well as motor fibers to derivatives of the second (hyoid) arch.

**Auditory nerve (VIII).** The sensory **auditory nerve** (acoustic, vestibulocochlear, statoacoustic) carries sensory fibers from the inner ear, which is concerned with balance and hearing. The nerve synapses in several regions of the medulla.

**Glossopharyngeal nerve (IX).** The mixed **glossopharyngeal nerve** supplies the third branchial arch. It contains sensory fibers from the taste buds, the first gill pouch, and the adjacent pharyngeal lining and lateral line. Motor fibers innervate muscles of the third branchial arch.

**Vagus nerve (X).** The term ***vagus*** is Latin for wandering and aptly applies to this mixed nerve. The vagus meanders widely, serving areas of the mouth, pharynx, and most of the viscera. It is formed by the union of several roots across several head segments.

**Spinal accessory nerve (XI).** In anamniotes, the **spinal accessory nerve** is probably composed of a branch of the vagus nerve and several occipitospinal nerves. In amniotes, especially in birds and mammals, it is a small but distinct motor nerve that supplies derivatives of the cucullaris muscle (cleidomastoid, sternomastoid, trapezius). A few of its fibers accompany the vagus nerve to supply part of the pharynx and larynx and perhaps the heart. Fibers arise from several nuclei within the medulla.

**Hypoglossal nerve (XII).** The **hypoglossal nerve** is a motor nerve that innervates hyoid and tongue muscles. Fibers originate in the hypoglossal nucleus within the medulla.

**Lateral line nerves.** In addition to cranial nerves, fishes possess anterior and posterior **lateral line nerves** that are rooted in the medulla and supply the lateral line system. They were once thought to be components of the facial and vagal nerves, but they are now recognized as independent cranial nerves. Unfortunately, this late recognition as distinct cranial nerves has left them without an identifying roman numeral.

## Evolution

In vertebrate ancestors, each head segment may have been innervated by anatomically separate dorsal and ventral roots in much the same way that spinal nerves supply each trunk segment. Each segment was perhaps innervated by a mixed dorsal root and a motor ventral root. It has been suggested that the cranial nerves are derived from losses or mergers of these separate dorsal and ventral roots. But complex fusions and losses make it difficult to determine the distribution of ancient nerves to their respective head segments. Of the several ancient anterior branchial arches, only the mandibular arch persists. As noted, it incorporates the deep ophthalmic dorsal root into its own dorsal root branches (the maxillary and mandibular branches), forming the composite trigeminal nerve. Other persisting dorsal roots include the terminal, facial, glossopharyngeal, vagus, and accessory nerves. Ventral root derivatives include the oculomotor, trochlear, abducens, and occipital nerves.

Once associated with a branchial arch, each cranial nerve exhibits fidelity to that particular arch and its mus-

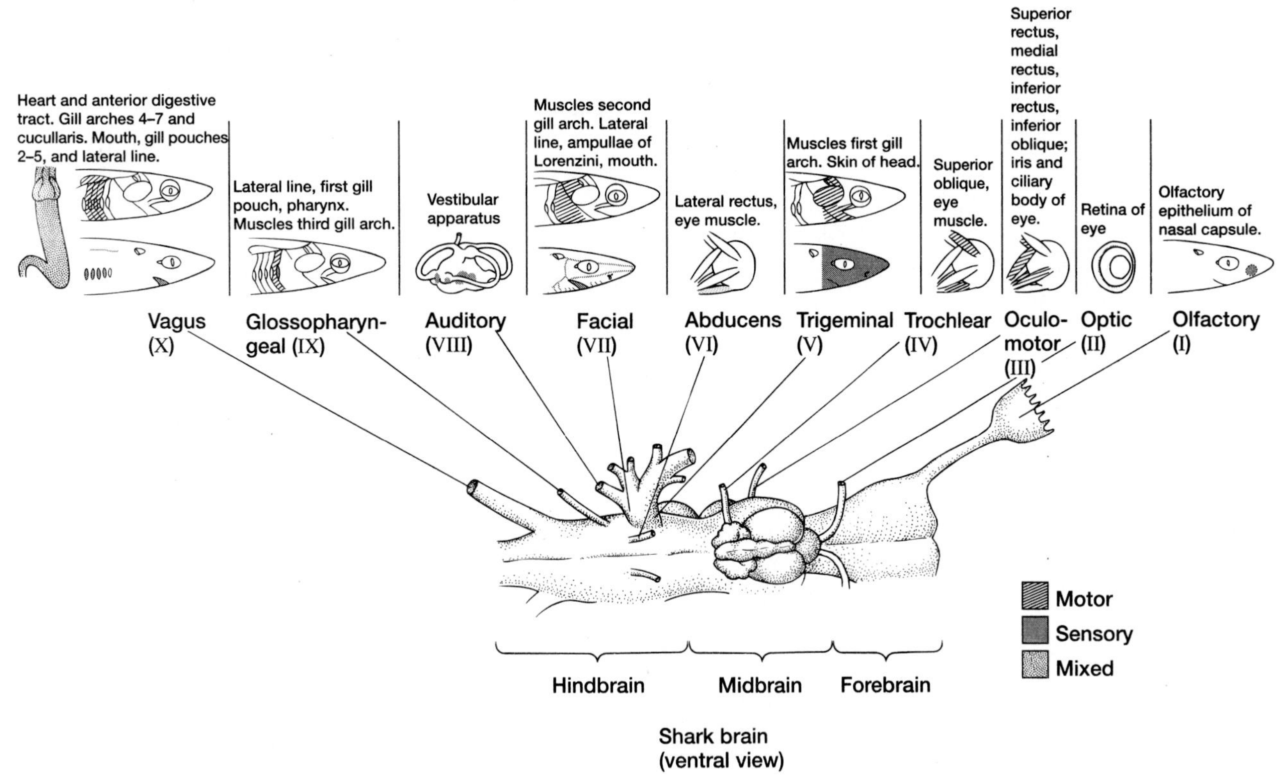


**Figure 16.14** Distribution of cranial nerves in the shark *Squalus*. Enlarged views of the innervated structures of cranial nerves II, III, IV, VI, and X. Sagittal sections of the head indicate the location of the first ten cranial nerves.

cles. Consistently throughout vertebrates, the first arch, the mandibular, is innervated by the trigeminal nerve (V); the second, the hyoid, by the facial nerve (VII); the third, by the glossopharyngeal (IX); and the remaining arches by the vagus (X) and spinal accessory (XI) nerves (table 16.2 and figure 16.16a,b).

The olfactory (I), optic (II), and auditory (VIII) nerves are believed to be derived separately in conjunction with their respective special sense organs rather than in association with ancient head segments.

The shift from aquatic to terrestrial life is reflected in the cranial nerves. The lateral line system, devoted to detecting water currents, is completely lost in terrestrial vertebrates, as are the cranial nerves that served it. Pre- and posttrematic branches associated with the gills are modified as well. The spinal accessory and hypoglossal nerves enlarge or emerge as separate cranial nerves. The spinal accessory separates from the vagus nerve. It supplies the branchiomeric muscles that become more prominent in holding and rotating the head. The hypoglossal nerve to the tongue and hyoid apparatus becomes prominent as the role of these structures in terrestrial feeding and manipulating food in the mouth expands.

## Functions of the Peripheral Nervous System

### *Spinal Reflexes*

Spinal reflexes exhibit the simplest level of control within the nervous system. Although reflexes can disperse information to higher centers, all their necessary and functional components reside or are rooted in the spinal cord. The **spinal reflex** is a circuit of neurons from a receptor to the spinal cord and out to an effector. Incoming sensory and departing motor information travels in circuits laid down by neurons in the spinal nerves. Within the spinal cord, **association neurons (interneurons, internuncial neurons)** connect these sensory and motor neurons to complete the circuit between them. There are two types of spinal reflex arcs, somatic and visceral (figure 16.17). The neuronal circuitry for each type of arc is distinctive, at least in mammals where it has been most widely studied (table 16.3). The role of the central nervous system in modifying spinal reflexes is discussed later in this chapter.

Most **somatic reflex arcs** at the level of the spinal cord include three neurons, somatic sensory and somatic

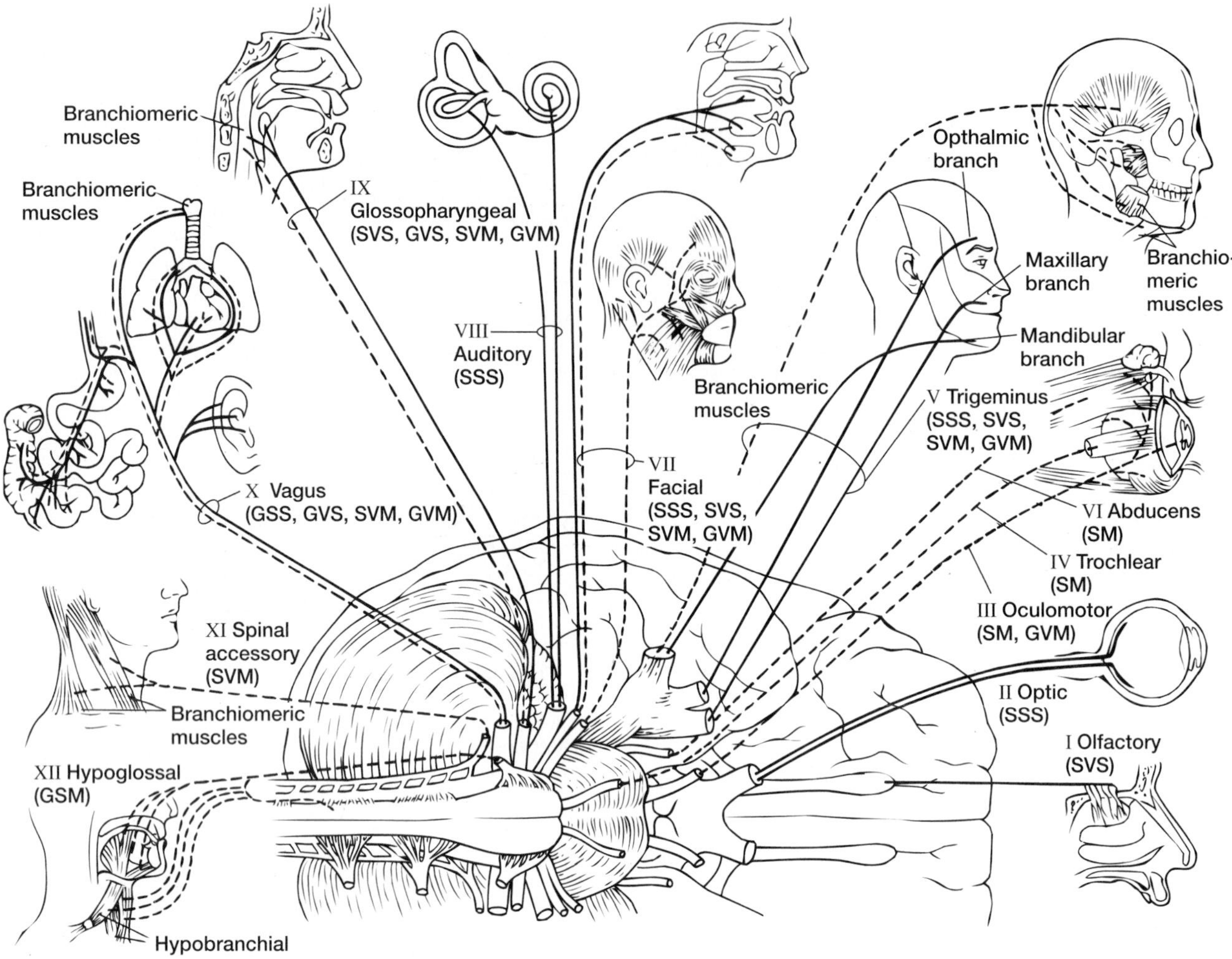


**Figure 16.15** Distribution of cranial nerves in a mammal, *Homo sapiens*. Sensory (solid lines) and motor (dashed lines) nerve fibers are indicated. Enlarged views of innervated structures of cranial nerves are shown around the human brain in ventral view. Abbreviations: general somatic sensory (GSS), general visceral sensory (GVS), general somatic motor (GSM), general visceral motor (GVM), special somatic sensory (SSS), special visceral sensory (SVS), special visceral motor (SVM).

motor neurons with an association neuron connecting them. The body of the somatic sensory neuron is located in the dorsal root. Its nerve fibers travel through the spinal nerve and synapse with an association neuron within the spinal cord. The association neuron may transmit impulses in several possible directions. It may synapse with a somatic motor neuron on the same side of the cord, on the opposite side of the cord, or travel up or down the cord to motor neurons at different levels. The motor neuron then transmits the impulse through the ventral root to a somatic effector. A somatic reflex arc may be even more simple. Spinal reflexes controlling posture involve only two neurons. The sensory neuron synapses directly with the motor neuron. If an animal should start to deviate inadvertently from its normal posture, its muscles are stretched. These stretched muscles elicit a somatic reflex that causes the appropriate muscle to contract and restore the animal to its original posture (figure 16.18a,b).

The **visceral reflex arc** is structurally more complex. The body of a visceral sensory neuron also resides in the dorsal root, but its nerve fibers travel through one or more sympathetic chain ganglia and then through the ramus communicans. Its axons eventually synapse within the spinal cord with an association neuron (figure 16.17). Unlike the somatic arc, the motor output of the visceral reflex arc includes two neurons in sequence. The first is the **preganglionic neuron** that extends out the ventral root and synapses in the sympathetic ganglion, in a collateral ganglion, or in the wall of a visceral organ with a second neuron, the **postganglionic neuron.** The postganglionic neuron travels on to innervate the effector visceral organ. Thus, at its simplest, the visceral arc includes four neurons, one visceral sensory neuron, two visceral

**Table 16.2** Cranial Nerves and Their Associated Branchial Arches

| Ancient Segment | Current Arch | Dorsal Root Representative | Ventral Root Representative |
|---|---|---|---|
| ? | | Terminal (0) | |
| 0 | | Deep opthalmic (V) | Oculomotor |
| — | Mandibular | | |
| 1 | | Superficial opthalmic (skin, V) maxillary (pretrematic; V) mandibularis (posttrematic; V) | Trochlear |
| 2 | Hyoid | Facial (VII) | Abducens |
| 3 | Branchial 3 | Glossopharyngeal (IX) | |
| 4 | Branchial 4 | Vagus (X) | |
| 5 | Branchial 5 | Vagus (X) | Hypoglossal |
| 6 | Branchial 6 | Vagus (X) | Hypoglossal |
| 7 | Branchial 7 | Spinal accessory (XI) | Hypoglossal |

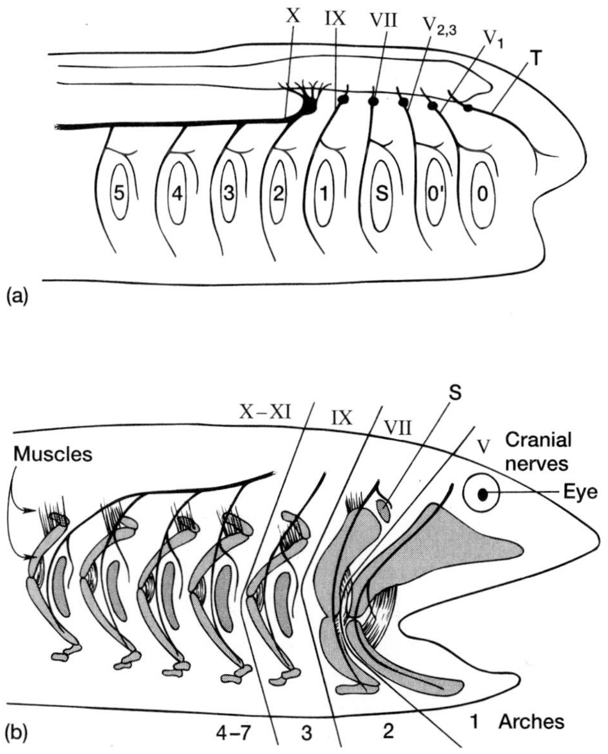


**Figure 16.16** Phylogenetic derivation of cranial nerves. (a) Hypothesized primitive condition. Each pharyngeal slit was supplied by a nerve. The first, or terminal (T), nerve supplied an anterior arch that was lost early in vertebrate evolution. (b) Nerve supply to associated branchial arches. Cranial nerves V, VII, IX, and X–XI supply the following arches: mandibular (1), hyoid (2), third (3), and fourth–seventh (4–7), respectively. These associations between cranial nerves and their derivatives remain stable throughout teleosts and tetrapods. Abbreviations: gill slits lost in gnathostomes (O,O′), gill slits usually present in gnathostomes (1–5), spiracular slit (S).

**Table 16.3** Reflexes in Mammals

| Components of a Reflex Circuit | Somatic Arc | Visceral Arc |
|---|---|---|
| Effector | Skeletal muscle | Cardiac and smooth muscle, glands |
| Number of neurons in circuitry | Three (or two) neurons, sensory (association), and motor | Four neurons, sensory (association), preganglionic motor, postganglionic motor |
| Neurotransmitters | Acetylcholine | Acetylcholine, norepinephrine |

motor neurons in series, and an interconnecting association neuron.

In summary, the somatic arc includes somatic afferent neurons that carry sensory impulses to the CNS from skin, voluntary muscles, and tendons. Somatic efferent neurons deliver motor impulses to somatic effectors. The visceral arc includes visceral afferent neurons that carry sensory impulses to the CNS from the digestive tract and other internal structures. The visceral efferent neurons carry motor impulses to visceral organs; this part of the circuit includes two neurons, preganglionic and postganglionic.

In higher vertebrates, the dorsal root carries predominantly sensory information, which can be somatic or visceral. The ventral root carries almost exclusively motor information, which can be somatic or visceral. In lower vertebrates, there is considerable variation in both the structure of the spinal nerve pathways and the information

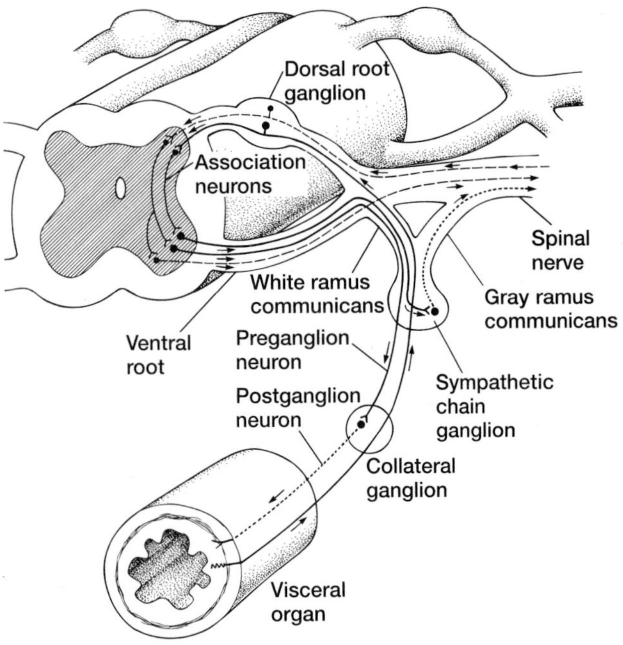


**Figure 16.17** Somatic and visceral reflex arcs of mammals. Sensory input arrives in fibers of the dorsal root that synapse in the spinal cord. Motor output departs in fibers of the ventral root. An association neuron usually connects input and output within the spinal cord. Somatic sensory fibers reach the dorsal root via a spinal nerve. Visceral sensory fibers travel from the visceral organ through one or more ganglia and then through the ramus communicans and the dorsal root finally to synapse in the spinal cord. The somatic motor outflow includes a single neuron that sends its fiber out from the spinal nerve to the effector. The visceral motor outflow includes two neurons in series, a preganglionic neuron (solid line) and a postganglionic neuron (dashed line). The synapse between them can occur in a sympathetic chain ganglion, in a collateral ganglion, or in the wall of the innervated organ. If they synapse in the sympathetic chain, postganglionic fibers usually reach the effector via the spinal nerve.

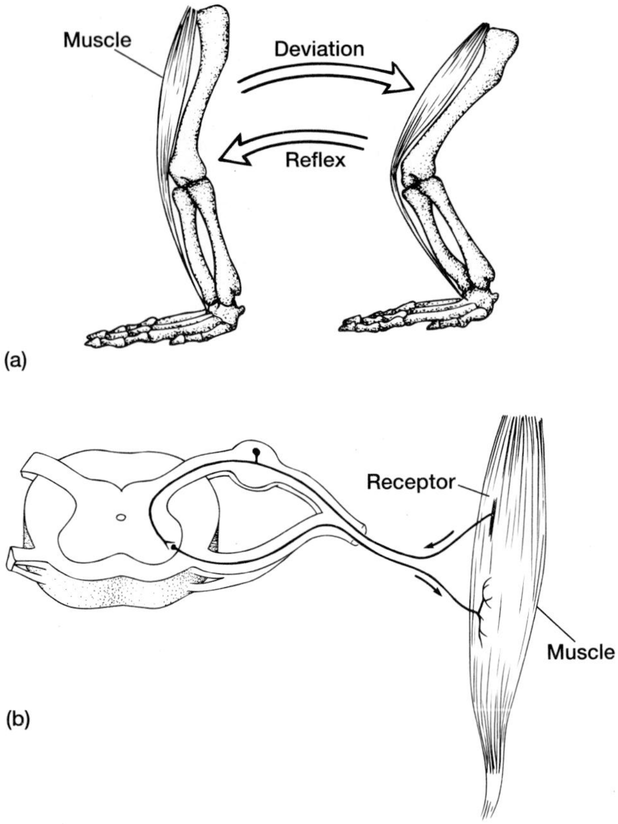


**Figure 16.18** Somatic reflex arc. (a) Posture can be maintained through a spinal reflex involving a single sensory and single motor neuron connected directly within the spinal cord. When a tetrapod begins to deviate from its normal posture, sensory receptors within joints and muscles detect the deviation. (b) Sensory fibers that carry this stimulus to the spinal cord synapse with appropriate motor neurons that stimulate skeletal muscle motor units to contract, straighten the limb, and restore normal posture.

they carry. In lampreys, the dorsal and ventral roots do not join. The ventral root carries only somatic motor information transmitted to striated muscles at that level of the spinal cord. The dorsal root carries somatic and visceral sensory information as in higher vertebrates, but it also carries visceral motor fibers (figure 16.19a). In fishes and amphibians, dorsal and ventral roots are joined, but visceral motor fibers depart via both the dorsal root, as in lampreys, and the ventral root, as in higher vertebrates (figure 16.19b).

## *The Autonomic Nervous System*

Early anatomists noticed that visceral activity did not appear to be under voluntary control. The peripheral nerves and ganglia associated with visceral activity seemed to be autonomous, or independent of the rest of the nervous system. Collectively, they were considered to constitute the autonomic nervous system, a functional division of the peripheral nervous system that presides over visceral activity. Both sensory and motor fibers are included. Autonomic sensory fibers monitor the internal environment of the organism, that is, blood pressure, oxygen and carbon dioxide tension, core and skin temperature, and activity of the viscera. Motor fibers are general visceral motor neurons that innervate cardiac muscle, smooth muscle, and glands. Thus, they control the digestive tract, blood vessels, respiratory tree, bladder, sex organs, and other general body viscera. Because the autonomic nervous system includes the general visceral motor circuit, pre- and postganglionic neurons in series characterize the motor innervation to each organ.

Conscious centers also can affect visceral activity controlled by the autonomic nervous system. For instance,

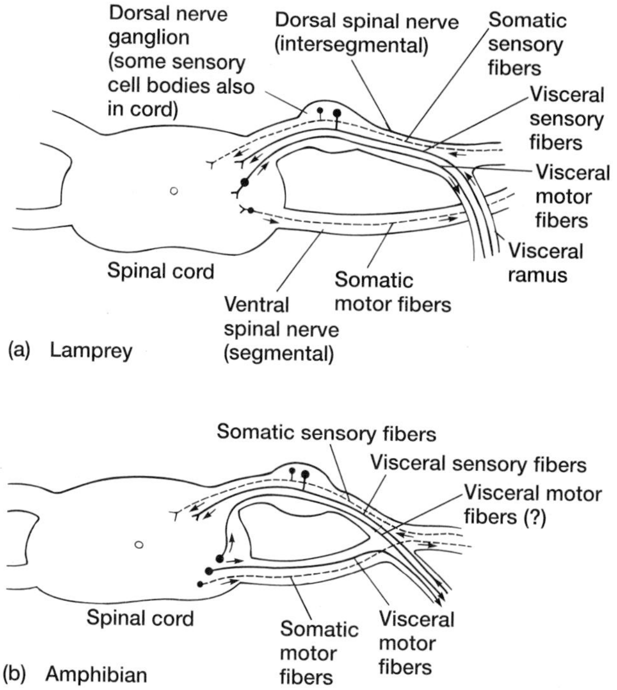


**Figure 16.19** Somatic and visceral circuits in lower vertebrates. (a) Lamprey. (b) Amphibian. It is unclear whether the amphibian dorsal root carries visceral motor output.

through practiced meditation or through deliberate effort that brings a chilling thought to mind, it is possible to affect the heartbeat or the release of sweat. But, for the most part, the autonomic system operates subconsciously and is not under voluntary control. Reflexes control activities maintaining the internal environment. In its simplest form, the neuronal circuity of the autonomic nervous system includes four neurons linked in a reflex loop: a sensory neuron that synapses with an association neuron that synapses with a preganglionic motor neuron in series with a postganglionic motor neuron.

***Functional Divisions of the Autonomic Nervous System*** In mammals, the autonomic nervous system is divided into two contrasting, antagonistic systems of control over visceral activity, the sympathetic system and the parasympathetic system.

The **sympathetic nervous system** prepares the body for strenuous action by increasing activity of the viscera, although it slows digestive processes. Stimulation of the sympathetic system inhibits activity of the alimentary canal but promotes contraction of the spleen (causing it to release extra red blood cells into the general circulation), increases heart rate and blood pressure, dilates coronary blood vessels, and mobilizes glucose from glycogen storage in the liver. It is often said that the sympathetic nervous system prepares the individual to fight or to flee (table 16.4).

The general visceral motor nerves that participate in sympathetic activity depart from the thoracic and lumbar regions of the mammalian spinal cord. This activity is referred to as the **thoracolumbar outflow.** The sympathetic preganglionic neuron is usually short and synapses in the sympathetic chain ganglion or in a ganglion located away from the vertebral column. The postganglionic fiber is usually long (figure 16.20).

The **parasympathetic nervous system** restores the body to a restful or vegetative state by lowering its activity level, although digestion is stimulated. The effects of the parasympathetic system are antagonistic to those of the sympathetic system. It enhances digestion, slows heart rate, drops blood pressure, constricts coronary vessels, and promotes glycogen formation.

Participating visceral motor neurons include cranial nerves V, VII, IX, and X together with spinal nerves departing from the sacral region. This is referred to as **craniosacral outflow.** Parasympathetic preganglionic fibers are long and reach to the wall of the organ they innervate and synapse with very short postganglionic fibers (figure 16.20).

***Adrenergic and Cholinergic Control*** The sympathetic system is said to be **adrenergic** because the neurotransmitters released during stimulation are **adrenaline** or **noradrenaline** (also termed **epinephrine** and **norepinephrine**). The parasympathetic system is said to be **cholinergic** because the neurotransmitter released is **acetylcholine.** Acetylcholine is also released between pre- and postganglionic fibers in both systems (figure 16.20) and at junctions between nerves and skeletal muscles.

In mammals, almost every visceral organ has sympathetic and parasympathetic innervation (figure 16.21 and table 16.4). Exceptions to this double innervation include the adrenal gland, peripheral blood vessels, and sweat glands, all of which receive only sympathetic innervation. Cessation of sympathetic stimulation allows these organs to return to a resting state.

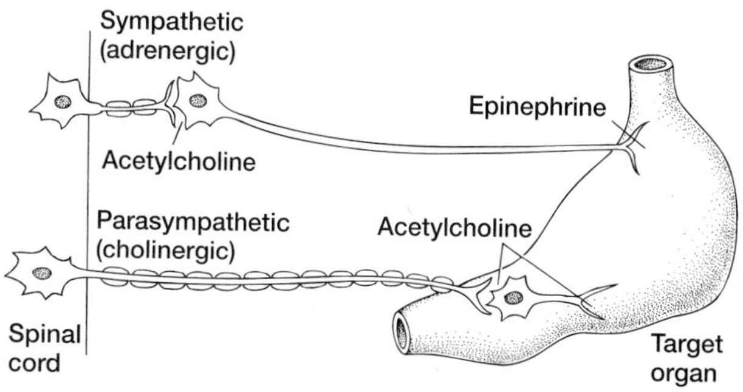


**Figure 16.20** Neurotransmitters of the autonomic nervous system. Adrenergic and cholinergic neurotransmitters are released at the ends of the sympathetic and parasympathetic circuits, respectively. This is the basis for differential organ response.

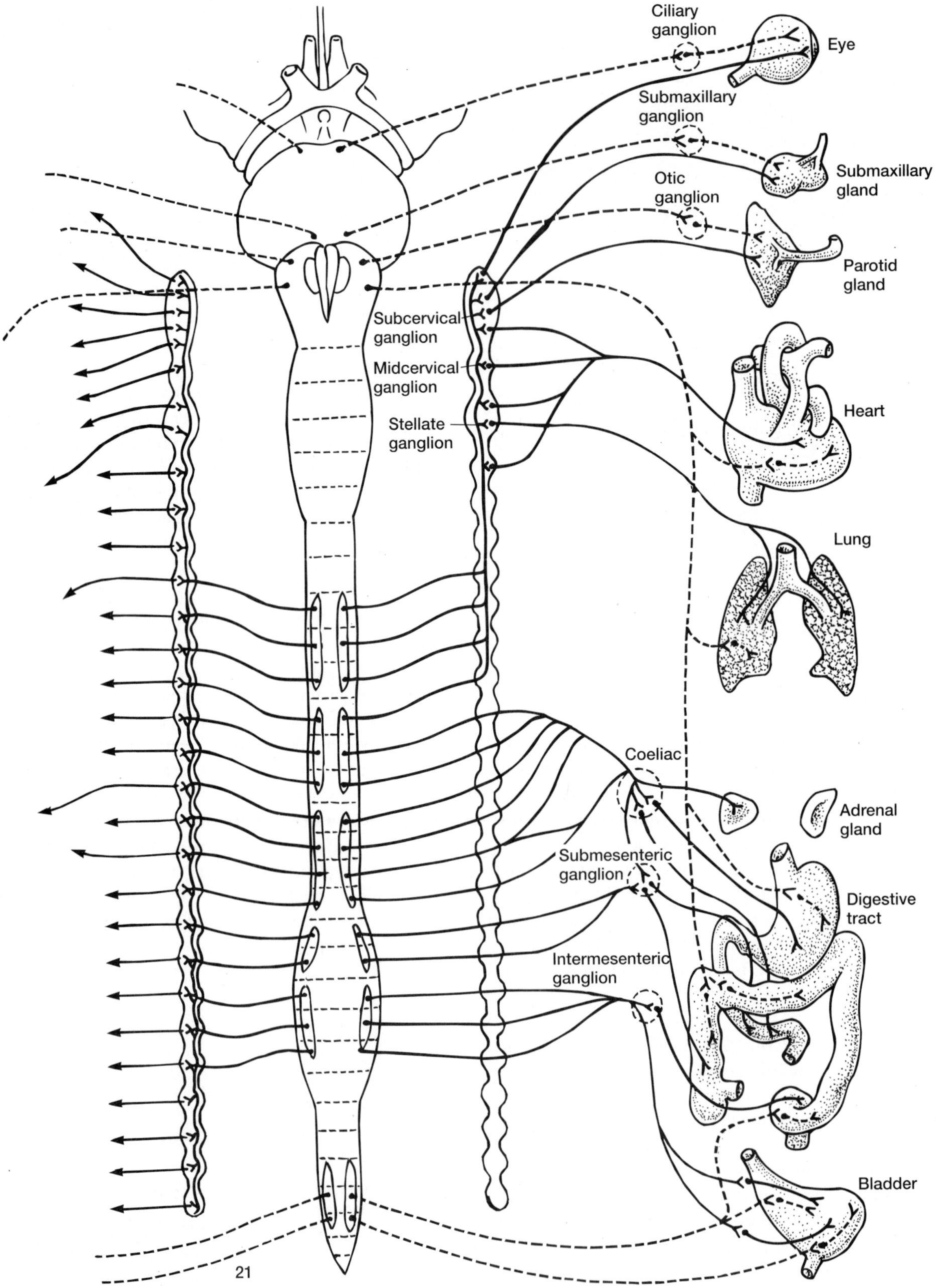


**Figure 16.21** Sympathetic (solid lines) and parasympathetic (dashed lines) systems in a mammal. Note the double innervation of most organs. Preganglionic and postganglionic fibers are indicated.

**TABLE 16.4 Functional Divisions of the Autonomic Nervous System**

| Organ/Activity | Sympathetic Stimulation | Parasympathetic Stimulation |
|---|---|---|
| *Eye* | | |
| Ciliary muscle | Relaxation | Contraction |
| Pupil | Dilation | Constriction |
| *Glands* | | |
| Salivary | Vasoconstriction<br>Slight secretion | Vasodilation<br>Copious secretion |
| Gastric | Inhibition of secretion | Stimulation of secretion |
| Pancreas | Inhibition of secretion | Stimulation of secretion |
| Lacrimal | None | Secretion |
| Sweat | Sweating | None |
| *Digestive tract* | | |
| Sphincters | Increase tone | Decrease tone |
| Walls | Decrease motility | Increase motility |
| *Liver* | Glucose release | None |
| *Gallbladder* | Relaxation | Contraction |
| *Bladder* | | |
| Smooth muscle | Relaxation | Contraction |
| Sphincter | Contraction | Relaxation |
| *Adrenal gland* | Secretion[a] | None |
| *Heart* | | |
| Muscle | Increase rate and force | Slowed rate |
| Coronary arteries | Dilatation | Constriction |
| *Lungs (bronchi)* | Dilatation | Constriction |
| *Spleen* | Contraction | Relaxation |
| *Blood vessels* | | |
| Abdomen | Constriction | None |
| Skin | Constriction | None |
| *Sex organs* | | |
| Penis | Ejaculation | Erection |
| Clitoris | ? | Erection |
| *Metabolism* | Increased | None |

[a]Preganglionic neuron innervation.

The adrenal gland is also exceptional in that it is innervated by the preganglionic fiber only; the postganglionic fiber is absent. Because epinephrine and norepinephrine serve both as adrenergic chemical signals of the sympathetic circuit and as hormones produced by the adrenal gland (see chapter 15), there is a possibility for chemical confusion. But the preganglionic nerve releases acetylcholine rather than adrenaline or similar chemicals, so direct innervation of the adrenal gland by preganglionic fiber removes the possibility of chemical ambiguity between parasympathetic innervation and hormonal stimulation by the gland.

### *Anatomical Divisions of the Autonomic Nervous System*

The division of the autonomic nervous system into sympathetic and parasympathetic functional components holds reasonably well for mammals; however, in other vertebrates, the comparative anatomy of the autonomic nervous system is poorly understood. Most viscera receive contrasting sympathetic and parasympathetic innervation, but these functional divisions do not always correspond to thoracolumbar and craniosacral outflow, respectively. Often, in nonmammalian vertebrates, the autonomic nerves departing from these regions have a mixed function. When examining the autonomic nerves of lower vertebrates, we cannot safely infer function from anatomical position. Therefore, we prefer anatomical distinctions alone, without implied functional significance, when describing the autonomic nervous system of nonmammalian vertebrates.

There are three anatomical divisions of the autonomic nervous system—cranial autonomic, spinal autonomic, and enteric autonomic systems (table 16.5). The **cranial autonomic system** includes the cranial nerves leaving the brain. The **spinal autonomic system** consists of all autonomic fibers departing from the central nervous system in the spinal segments, specifically all thoracic, lumbar, and sacral autonomic fibers.

The **enteric autonomic system** includes intrinsic sensory and motor neurons residing in the wall of the digestive tract. Nerves formed from these neurons interconnect and mingle to form woven patches of nerve processes, termed **plexuses,** within the wall of the digestive tract. The **myenteric plexuses** (Auerbach's plexuses) are situated within the outer wall of smooth muscles, and the **submucosal plexuses** (Meissner's plexuses) are located deep within smooth muscles near their lumen. The enteric autonomic system is responsible for coordination of digestive tract activity. It is independent from but can be modified by the spinal and cranial autonomic systems. Food distending the smooth muscles of the digestive tract mechanically stimulates enteric neurons. These neurons, in turn, activate contractions of the circular and longitudinal smooth muscles in

**TABLE 16.5 Relationship between Functional and Anatomical Divisions of the Autonomic Nervous System**

| Location | Function (Mammals) | Anatomical Designation |
|---|---|---|
| Cranial | Parasympathetic | Cranial autonomic |
| Thoracic | Sympathetic | Spinal autonomic |
| Lumbar | Sympathetic | Spinal autonomic |
| Sacral | Parasympathetic | Spinal autonomic |
| Intrinsic digestive tract | Enteric | Enteric autonomic |

the wall of the digestive tract, resulting in synchronized **peristaltic waves** that propel food through the tract. The enteric autonomic system seems to be present in all classes of vertebrates, although it may be poorly developed in some.

*Fishes*. In cyclostomes, the autonomic nervous system is fragmentary. Sympathetic chains are absent, but collateral ganglia, presumably part of the autonomic system, are scattered throughout the viscera. In hagfishes, cranial autonomic fibers apparently occur only in the vagus (X). However, in lampreys, in addition to the vagus (X), the facial (VII) and glossopharyngeal (IX) nerves include autonomic fibers that mediate events in the gills. In hagfishes, spinal autonomic fibers pass through the ventral roots of spinal nerves, but their subsequent distribution is poorly known. In lampreys, spinal autonomic fibers depart through both dorsal and vental roots of the spinal nerves to supply the kidneys, gonads, blood vessels, posterior digestive tract, cloaca, and other viscera.

In chondrichthyan and osteichthyan fishes, the autonomic nervous system is well represented. A sympathetic chain of ganglia is present within the spinal autonomic system. Nerves of the cranial autonomic system pass to the viscera. However, in elasmobranchs, collateral ganglia are absent and sympathetic chains apparently do not contribute fibers to the cranial nerves. The vagus is well developed with branches to the stomach and heart, but the heart apparently lacks a sympathetic counterpart to the inhibitory vagal innervation (figure 16.22a). Further, the sympathetic ganglia of elasmobranchs are associated with populations of chromaffin cells, neural crest derivatives that in teleosts and most tetrapods (except urodeles) become separated from the ganglia. In most teleosts, collateral ganglia occur, and some spinal nerve fibers are shared with cranial nerves (figure 16.22b).

*Tetrapods*. The autonomic nervous system is well developed in tetrapods. The paired sympathetic chain is present, collateral ganglia are dispersed among the viscera, and cranial and spinal nerves are well delineated. Autonomic outflow in amphibian spinal nerves passes through the ventral roots, but it is still unclear whether motor fibers also occur in the dorsal root (figure 16.23a). In reptiles, birds, and mammals, the autonomic systems are quite similar in basic construction. Spinal autonomic motor fibers depart via the ventral roots of spinal nerves (figure 16.23b,c).

***Overview*** With the exception of cyclostomes, the anatomical organization of the autonomic nervous system is similar in all vertebrate classes. The spinal autonomic outflow includes a paired sympathetic chain (except in cyclostomes and elasmobranchs) with some fibers contributing to cranial nerves. The cranial autonomic outflow includes the trigeminal (V), facial (VII), glossopharyngeal (IX), and vagus (X) cranial nerves, although the facial and glossopharyngeal can be reduced in fishes. In vertebrates with eyes, the oculomotor (III) nerve may send fibers to the iris and ciliary muscles in the eyes.

In mammals, especially in humans, the autonomic nervous system is better known and the circuits for motor outflow can be traced with greater confidence. The human autonomic nervous system includes a sympathetic thoracolumbar outflow and a parasympathetic craniosacral outflow. The existence of a sacral parasympathetic system in other vertebrates is still uncertain. The pelvic nerves of amphibians arising from the posterior end of the spinal cord and supplying the urinary bladder and rectum have traditionally been regarded as parasympathetic sacral outflow. However, even in mammals, sacral parasympathetic fibers mingle with thoracolumbar sympathetic fibers in the pelvic plexus, making it difficult to trace the posterior circuitry of these two systems on their way to visceral effectors. Until the comparative features of the autonomic nervous system are better known in vertebrates, the functional roles of autonomic nerves in nonmammalian classes must be inferred.

To summarize, somatic and visceral effectors receive motor information. Effectors and receptors are linked through the central nervous system. Control of much of the body's activity involves simple reflexes. The somatic reflex arc is primarily involved in controlling skeletal muscles. The visceral reflex arc is the basic component of the autonomic nervous system, which is responsible for monitoring internal visceral activity. We turn next to the central nervous system to examine its role in processing information.

# Central Nervous System

The central nervous system primarily coordinates activities that enable an organism to survive in its environment. In order to do this, the central nervous system must receive incoming information from several sources. Sensory receptors known as **interoceptors** gather information and respond to general sensations of organs within the internal environment. **Proprioceptors** are a type of interoceptor that inform the central nervous system about the position of the limbs and the degree to which joints are bent and muscles are stretched. This information-processing component of the nervous system is referred to as the **somatosensory system,** which includes proprioceptors and surface receptors within the skin. Sensations gathered by the somatosensory system are especially important in coordinating limb and body positions during locomotion. **Exteroceptors** gather information from the external environment. Sensations of touch, pressure, temperature, sight, hearing, smell, taste and other stimuli from the external environment are transmitted via exteroceptors to the brain and spinal cord. A third source of information comes from memory, which allows an organism to adjust its activity on the basis of past experiences.

The central nervous system processes incoming information and returns instructions to the effectors (figure 16.24). These constitute the response of the organism. Entering information **diverges** to inform various areas of the brain and spinal cord about the state of affairs at that point. When a decision is made, instructions **converge** to the

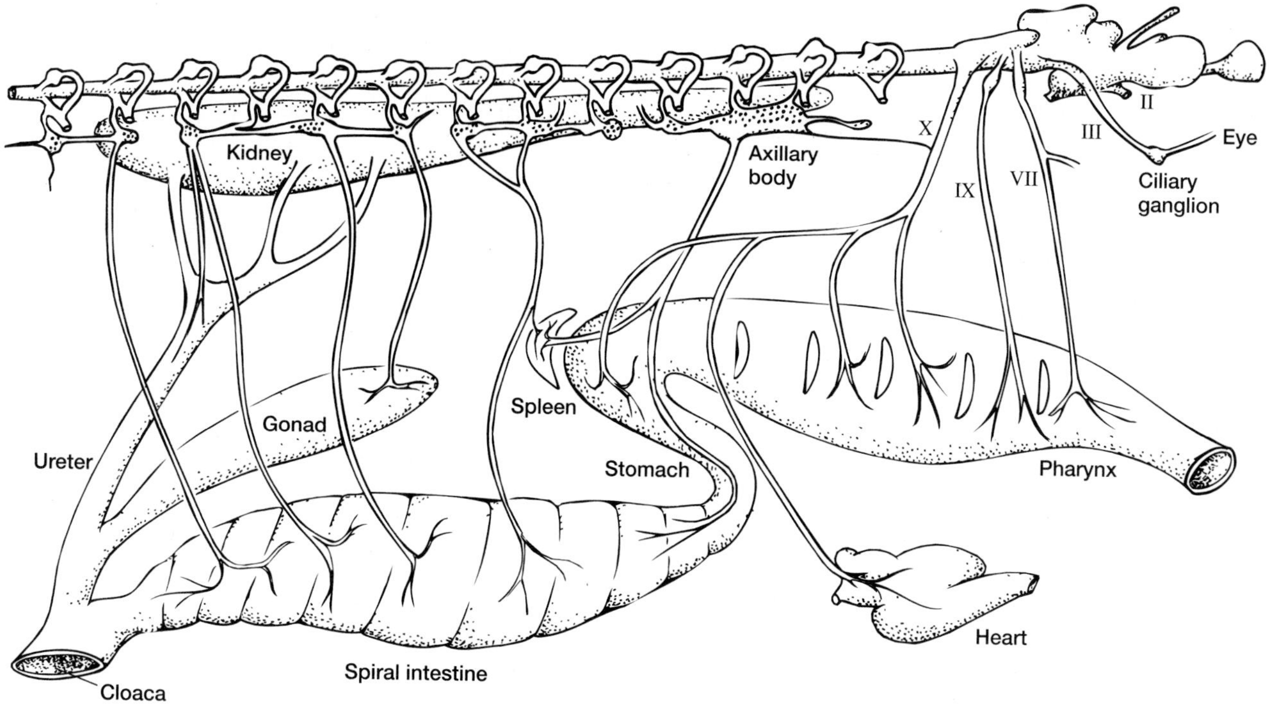


(a) Elasmobranch

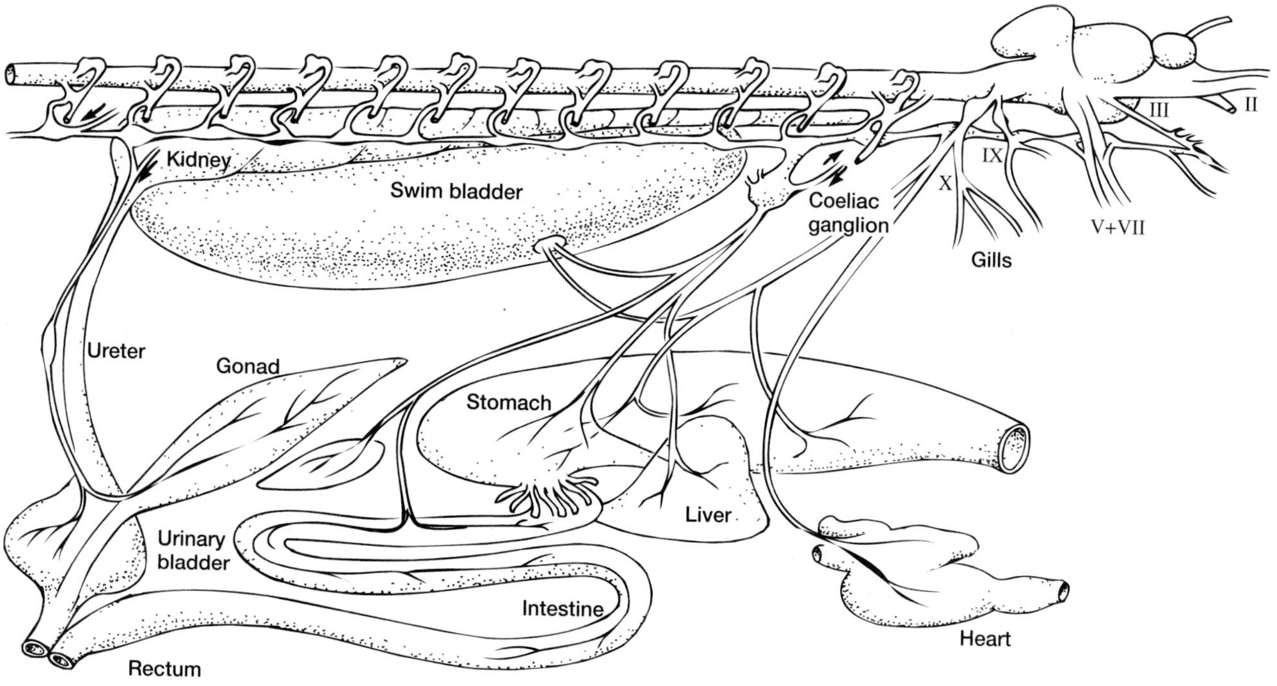


(b) Teleost

**Figure 16.22** Autonomic nervous system of fishes. (a) Elasmobranch (shark). Notice that the vagus nerve (X), which supplies the pharynx, stomach, and heart, does not receive fibers from any of the spinal nerves. (b) Teleost. The vagus (X), which supplies most of the same viscera as in the shark as well as contributes to innervation of the swim bladder, is connected to the sympathetic chain.

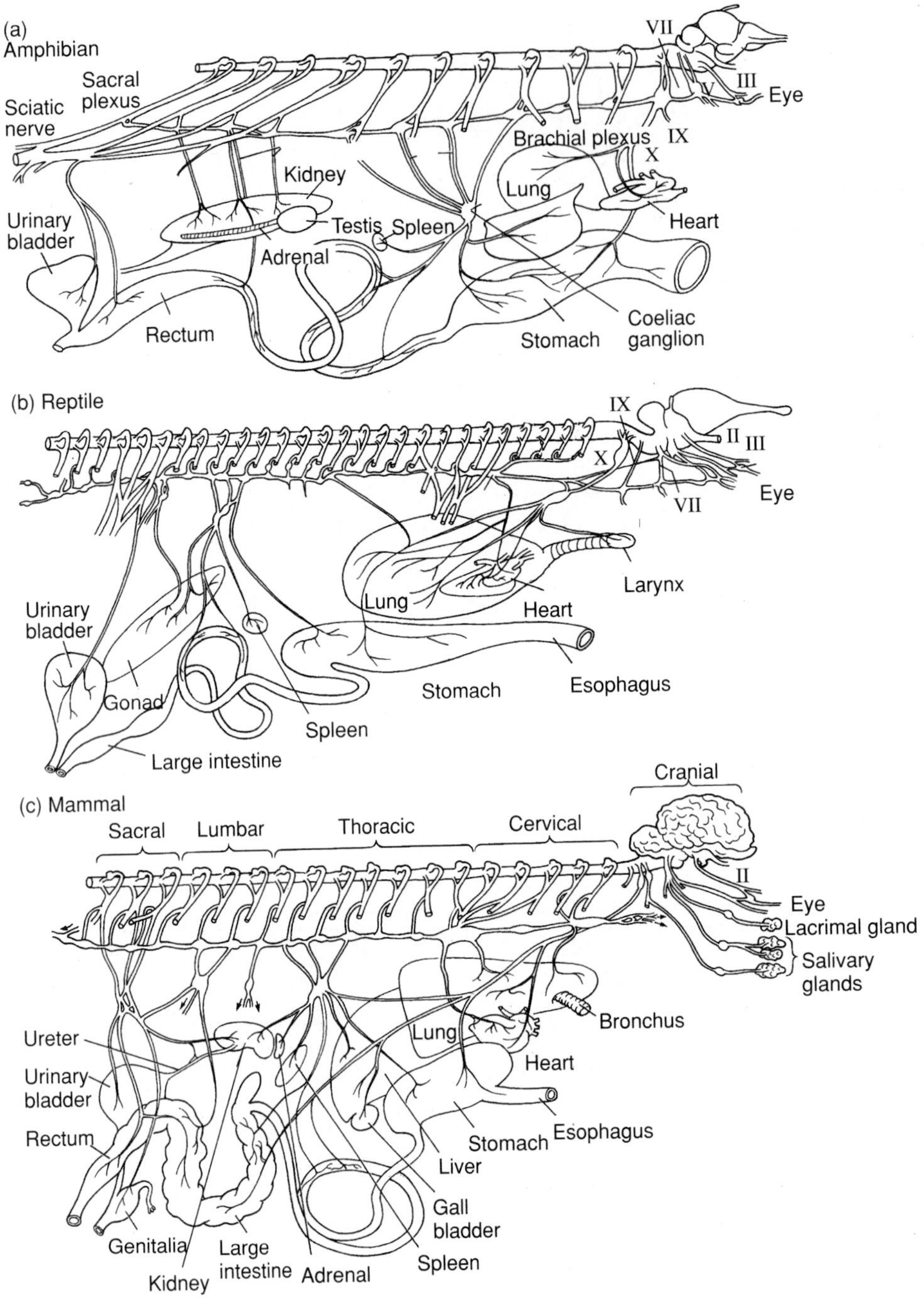


**Figure 16.23** Autonomic nervous system of various tetrapods. (a) Amphibian (anuran). (b) Reptile (lizard). (c) Mammal (placental).

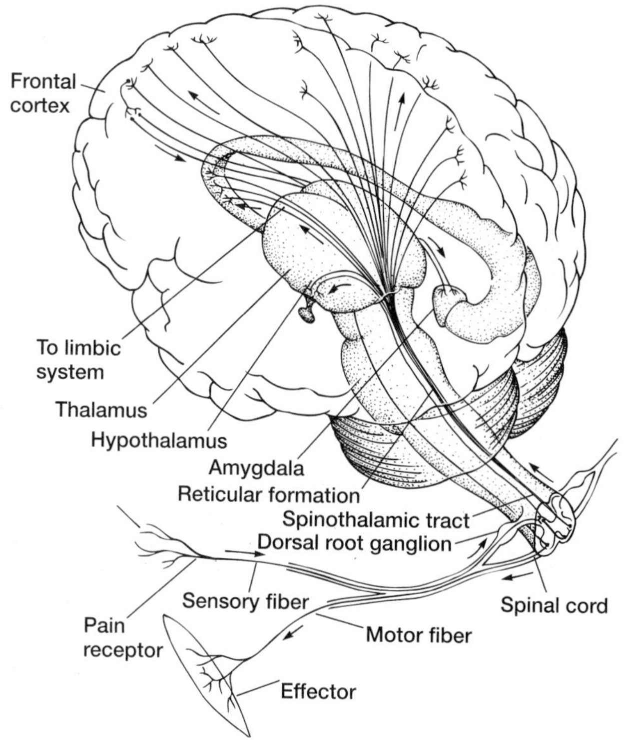


**Figure 16.24** Sensory and motor circuits. Sensory receptors in the skin respond to stimuli by generating an electrical impulse that travels to the spinal cord and synapses in the thalamus. This impulse is relayed by other neurons to other areas of the brain, which produce a response that travels down the spinal cord to a motor neuron and out to an effector.

appropriate effectors. The spinal cord and brain carry the pathways through which this information travels and form the association areas where it is evaluated.

## Embryology

The vertebrate central nervous system is hollow, a result of the fusion of two raised neural folds within the ectoderm. In the brain, the central canal enlarges into fluid-filled **ventricles** that are connected spaces located within the center of the brain. Within the anterior neural tube, three embryonic regions of the brain differentiate into the **prosencephalon, mesencephalon,** and **rhombencephalon** (figure 16.25a–c). These give rise to three regions of the adult brain—**forebrain, midbrain,** and **hindbrain** (figure 16.25f).

The brain and spinal cord are wrapped in **meninges** (sing., meninx) derived in part from neural crest. In mammals, the meninges consist of three layers, the tough outermost **dura mater,** the weblike **arachnoid** in the middle, and the innermost **pia mater** (figure 16.26a). The pia mater contains blood vessels that supply the underlying nervous tissue. **Cerebrospinal fluid** (CSF) is a slightly viscous fluid that flows slowly through the ventricles of the brain, the subarachnoid space beneath the arachnoid, and the central canal. The **choroid plexes,** small tufts of blood vessels associated with ependymal cells, project into the ventricles at specific sites and are the primary sources of cerebrospinal fluid. This fluid is reabsorbed into venous sinuses. Although cerebrospinal fluid is derived from the blood and returns to it, it is devoid of red blood cells or any other large formed elements. When a person is injured and trauma to the central nervous system is suspected, a procedure called a spinal tap is done to sample the cerebrospinal fluid. If it contains red blood cells, then the brain or spinal cord may be damaged. Cerebrospinal fluid forms a cushion of fluid around the brain and spinal cord to support the delicate nervous tissues and absorb shocks from concussions.

In fishes, the meninges consist of a single membrane, the **primitive meninx,** wrapped around the brain and spinal cord (figure 16.26b). With the adoption of terrestrial life, the meninges doubled. In amphibians, reptiles, and birds, the meninges include a thick outer dura mater derived from mesoderm and a thin inner **secondary meninx** (figure 16.26c). Cerebrospinal fluid may circulate more effectively and absorb shocks from the jolts sustained during terrestrial locomotion with a double meningeal layer. In mammals, the dura mater persists, but division of the secondary meninx yields both the arachnoid and the pia mater from ectomesoderm (figure 16.26d).

## Spinal Cord

The vertebrate spinal cord, like the brain, is organized into two regions and named because of their appearance in fresh preparations (figure 16.27a–f). The **gray matter** of the spinal cord includes nerve cell bodies that lie within the core of the spinal cord. Dorsal and ventral extensions of the gray matter are the **dorsal horns** and **ventral horns,** respectively. The dorsal horns contain the bodies of neurons receiving incoming sensory information. The ventral horns contain the bodies of motor neurons (figure 16.17). The **white matter** of the spinal cord surrounds the gray matter. It is predominantly composed of nerve fibers linking different levels of the spinal cord with each other and with the brain. Many of these fibers are myelinated, creating their white color.

The spinal cord functions in two capacities. It establishes simple reflexes and contains pathways of diverging and converging information.

### *Spinal Reflexes*

As you know from the discussion of somatic and visceral reflex arcs, the spinal cord completes the reflex loop between sensory input and motor output. In doing this, the spinal cord selects the effectors to be activated or inhibited. Although the spinal cord operates at the reflex level, it also contains circuits that coordinate different parts of the cord.

Arriving sensory fibers synapse in the dorsal horn of the gray matter with association neurons (figure 16.28a).

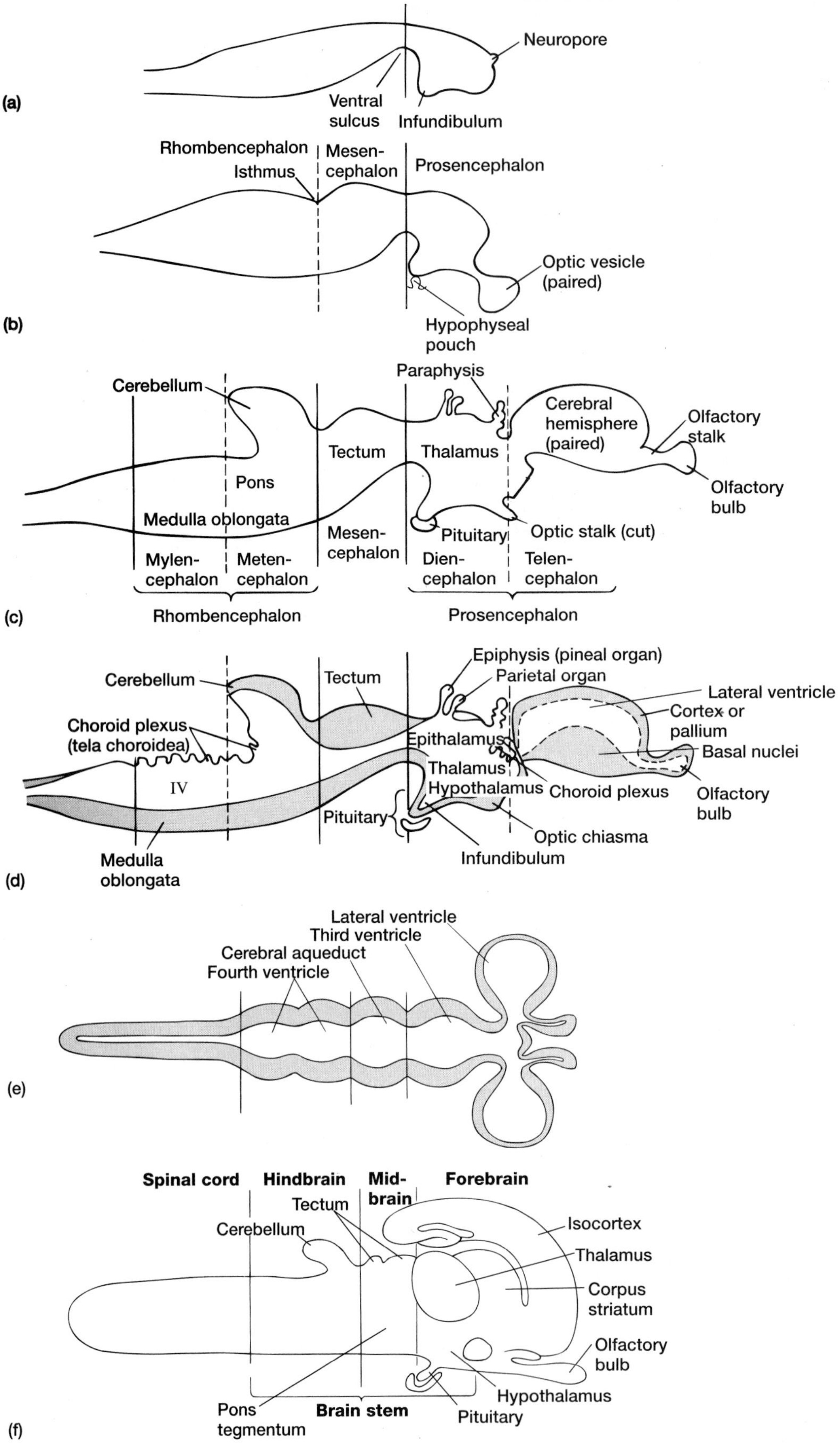


**Figure 16.25** Development of the central nervous system. (a–d) Embryonic development. (e) Fluid-filled ventricles within the central nervous system. (f) Anatomical regions of the adult brain.

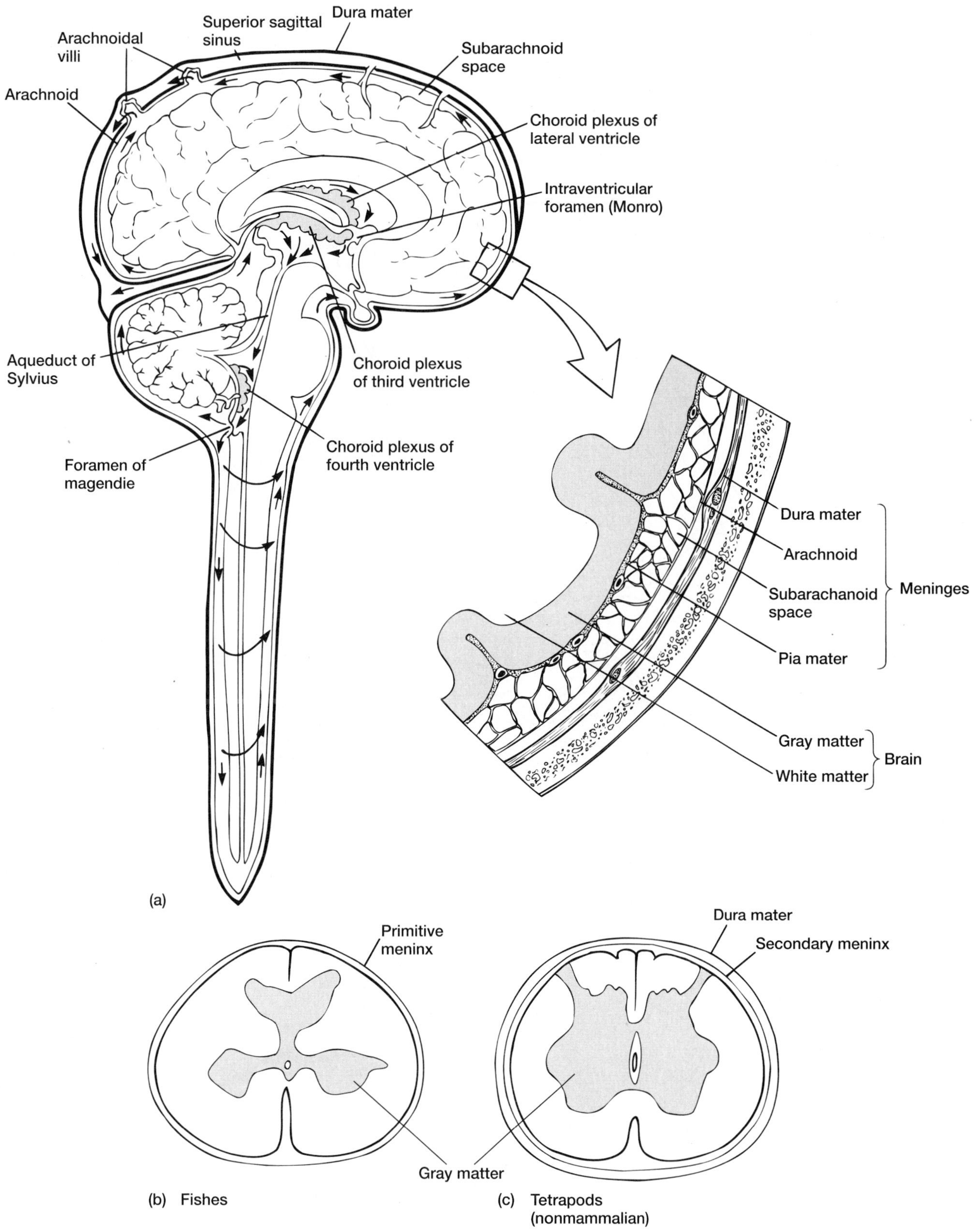


**Figure 16.26** Cerebrospinal fluid and meninges. (a) Arrows trace the circulation of cerebrospinal fluid through the brain and spinal cord of a mammal. The triple-layered meninges are enlarged to the right. (b) The meninges of fishes consist of a single thin layer, the primitive meninx. (c) In all tetrapods except mammals, the meninges are double layered and consist of an outer dura mater and inner secondary meninx. (d) Cutaway section of the spinal cord in a mammal illustrates the three meningeal layers—dura mater, arachnoid, and pia mater. Branches of the spinal nerve are shown along with their connections to the sympathetic chain. *continued.*

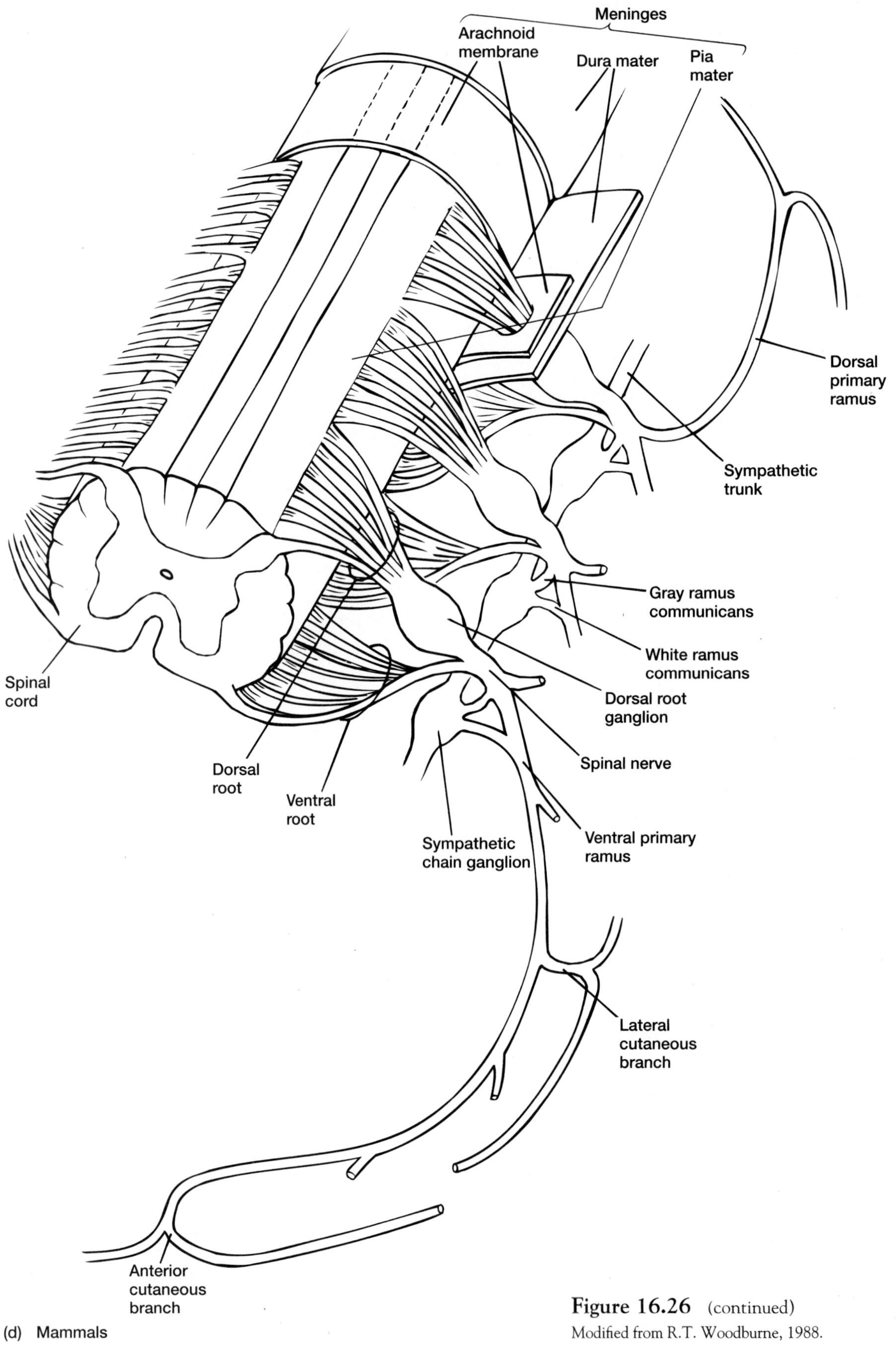


**Figure 16.26** (continued)
Modified from R.T. Woodburne, 1988.

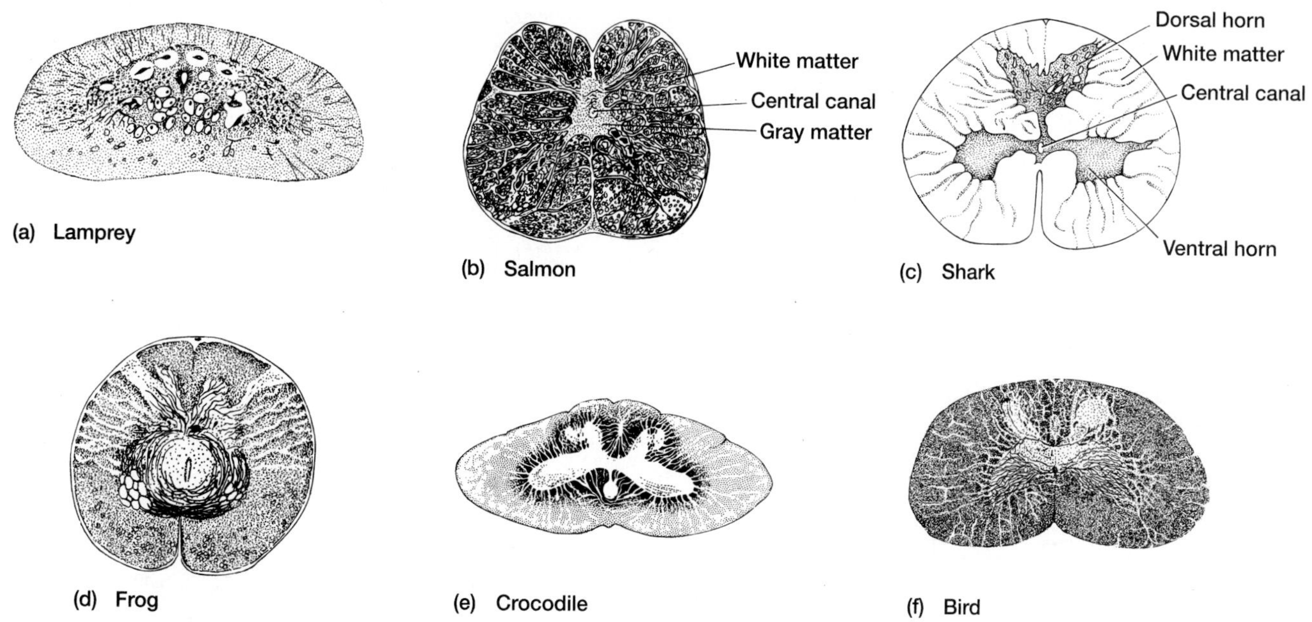


**Figure 16.27** Cross sections of vertebrate spinal cords. (a) Lamprey. (b) Salmon. (c) Shark. (d) Frog. (e) Crocodile. (f) Bird.

Association neurons carry the impulse to the ventral horn on the same side, to the opposite side, or to a different level of the spinal cord or brain. In the ventral horn, the association neuron synapses with a motor neuron whose axon travels out the ventral root to the effector. Dispersion of information within the spinal cord can produce complex responses to stimuli without involving higher centers. For example, if an animal should inadvertently place its foot on a sharp object, the reflex to withdraw it could involve as few as three neurons (figure 16.29). The first, the afferent sensory neuron, carries the painful stimulus to the spinal cord, where it synapses with an association neuron. The association neuron transmits the stimulus to the ventral horn, where it synapses with a motor neuron whose axon conveys the impulse to appropriate retractor muscles that contract and withdraw the foot. Association neurons that connect to appropriate levels on the opposite side of the cord reach motor neurons innervating extensor muscles in the opposite leg. These muscles contract, extend the leg, and prevent the animal's collapse when it lifts its other leg from the sharp object. The circuitry involved demands a connection between painful stimulus and appropriate effectors (retractor and extensor muscles). It need not involve higher brain centers. Usually, the association neurons also convey the painful stimulus to conscious centers of the brain, where it is perceived (figure 16.28b); however, by the time the higher centers become aware of the surprise trauma to the foot, the spinal reflex to retract it is already underway.

### *Spinal Tracts*

Not all information is processed at the level of the spinal cord. Much, perhaps most, information is carried to higher levels of the nervous system for evaluation. The resulting decisions are carried down the spinal cord to appropriate effectors. Nerve fibers carrying similar information tend to travel together in nerve tracts, bundles of similar fibers that occupy a specific region of the spinal cord. Nerve tracts may be ascending or descending tracts, depending on whether they convey information up or down the cord, respectively (figure 16.30). They are usually named for their source and their destination. For example, the spinothalamic tract begins in the spinal cord and extends to the thalamus (table 16.6).

Wars, accidents, and diseases can lead to localized wounds of the spinal cord that sever the ascending or descending flow of information. In humans, such losses of function have been correlated with the specific region in which the wound occurred and used to map the positions of these nerve tracts. More precise information from animal studies has added to our understanding of spinal cord organization. For convenience, these mapped tracts are drawn in discrete locations. In practice, their precise positions may change slightly at different levels of the cord, and there is some overlap of tracts as well.

**Ascending tracts** carry sensory impulses from the spinal cord to the brain. Among the most prominent are the

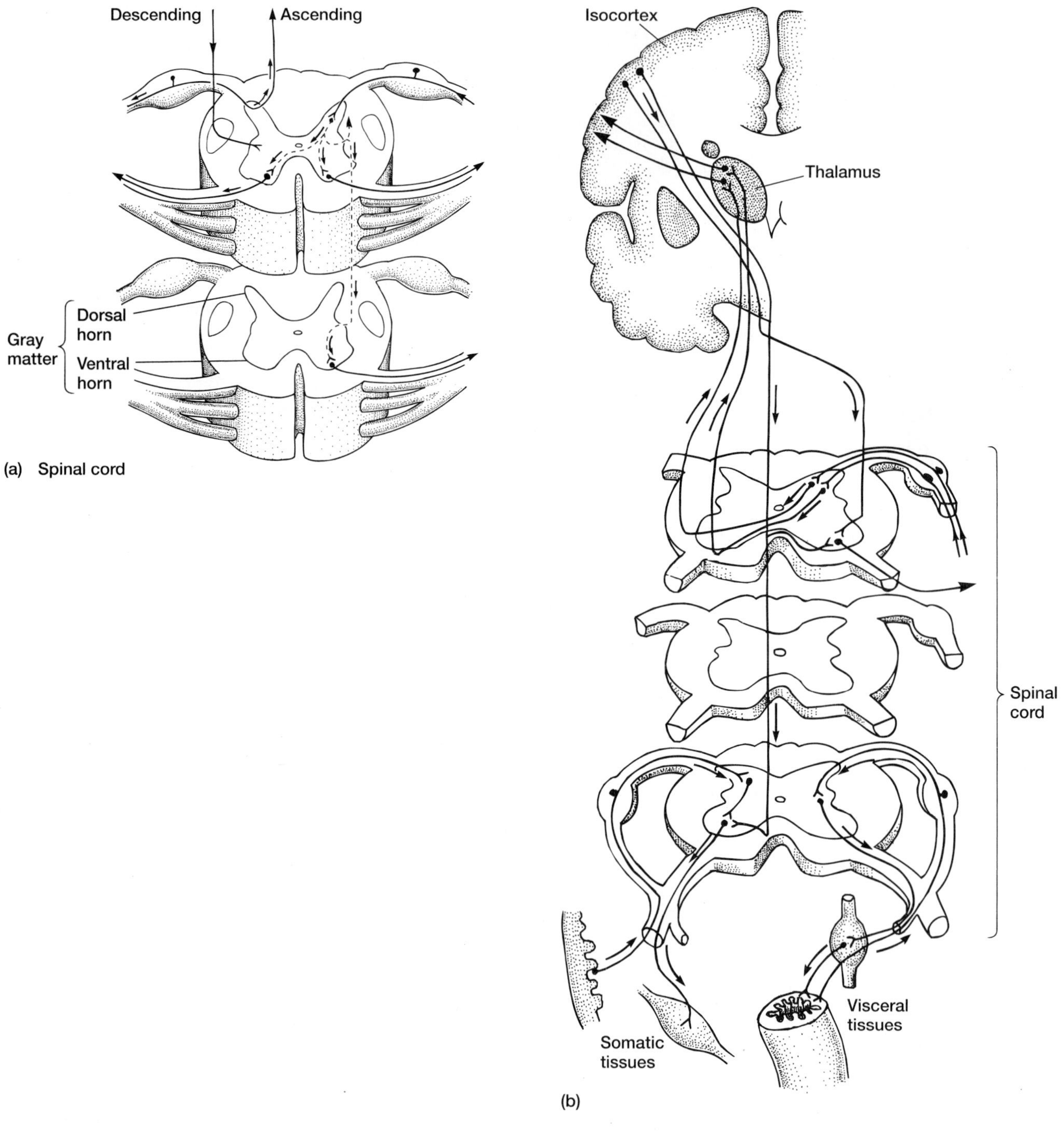


**Figure 16.28** Spinal reflexes. (a) Association neurons (dashed lines) within the gray matter receive afferent signals and relay these across the cord, to the same side of the cord, or to a different level of the spinal cord. (b) Spinal reflexes place somatic and visceral effectors under the immediate control of sensory information. But motor neurons that travel to these tissues are also influenced by descending circuits from conscious centers of the brain.

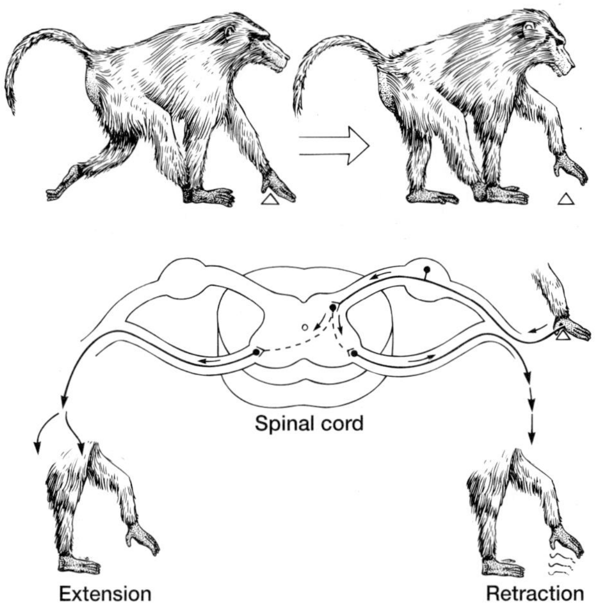


**Figure 16.29** Spinal reflex. Association neurons (dashed lines) within the spinal cord deliver stimuli to motor neurons, causing the retractor muscles of this animal to lift its foot from a harmful object. These stimuli also spread to motor neurons in other areas of the cord that innervate extensor muscles of the opposite limb to contract and support the weight of the body.

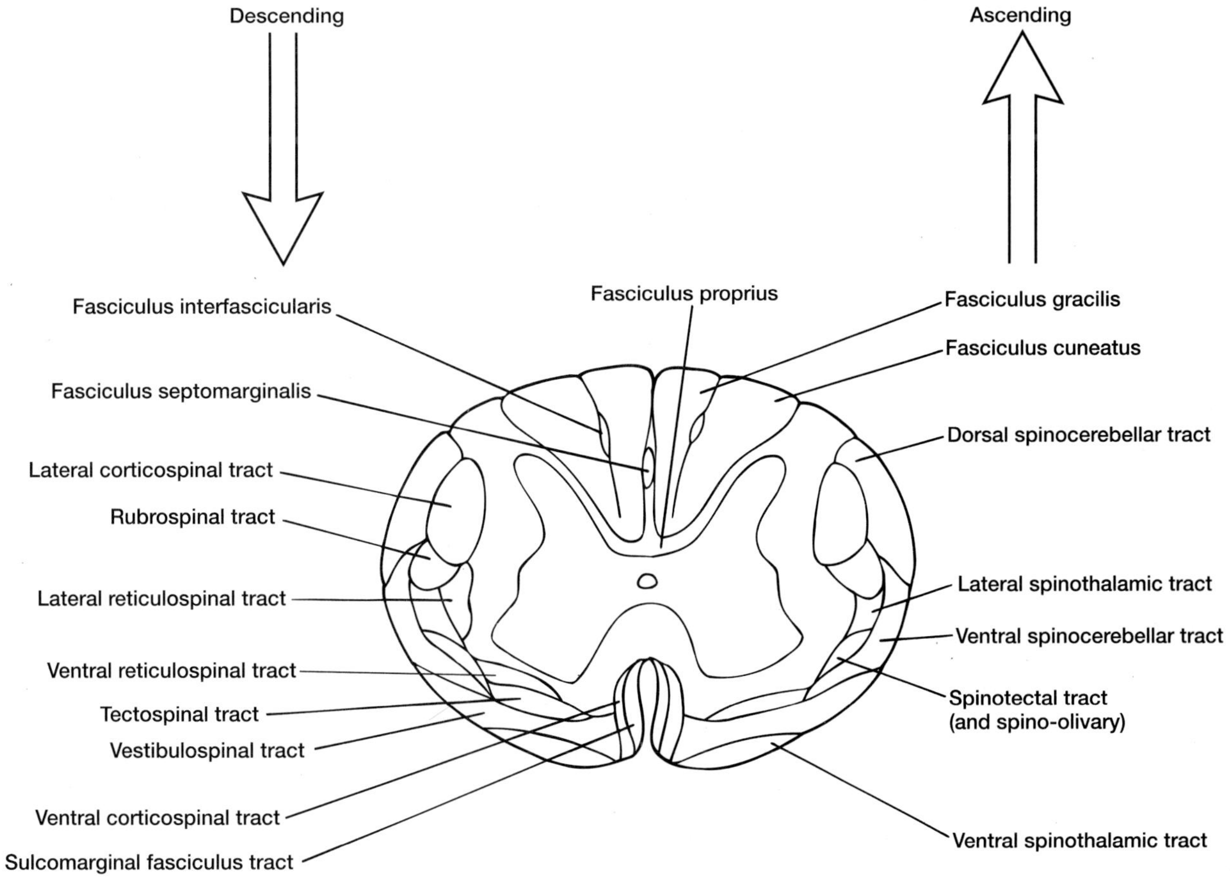


**Figure 16.30** Cross section of the human spinal cord showing approximate locations of ascending (right) and descending (left) nerve tracts.

**Table 16.6 Locations and Functions of Descending and Ascending Nerve Tracts of the Spinal Cord**

| Tract | Source | Destination | Function |
|---|---|---|---|
| | | *Descending* | |
| Lateral and ventral corticospinal tract | Cerebral cortex | Spinal cord | Motor connections direct from cortex to primary motor neurons of arms and legs (places motor neurons under direct voluntary cortical control) |
| Rubrospinal tract | Midbrain (red nucleus of tegmentum) | Spinal cord | Motor connections in spinal cord |
| Lateral and ventral reticulospinal tract | Medulla reticular formation | Spinal cord (dorsal horn) | Postural reflexes |
| Tectospinal tract | Midbrain (colliculus, roof) | Spinal cord | Visual and auditory stimuli to limbs and trunk |
| Vestibulospinal tract | Medulla (vestibular nucleus) | Spinal cord | Postural reflexes accomplished by axial and limb musculature |
| | | *Ascending* | |
| Fasciculus gracilis and fasciculus cuneatus | Spinal cord | Medulla | Sensations of posture and spatial judgments about positions of limbs and body |
| Dorsal and ventral spinocerebellar tract[a] | Spinal cord | Cerebellum via peduncle | Proprioceptive information from muscles to cerebellum |
| Lateral spinothalamic tract | Spinal cord | Thalamus | Pain and temperature sensations to thalamus |
| Ventral spinothalamic tract | Spinal cord | Thalamus | Tactile sensations to thalamus |
| Spinotectal tract | Spinal cord | Midbrain (tectum) | Proprioceptive information from neck and shoulders |
| Spinoreticular tract | Spinal cord | Medulla (reticular formation) | Pain and sensations from internal organs |

[a]May be single tract.

**fasciculus gracilis** and **fasciculus cuneatus,** located in the dorsal region of the spinal cord. Both carry proprioceptive stimuli and sensations associated with posture to the medulla. As each tract ascends, more axons are added. For instance, the gracilis is supplemented laterally to produce the cuneatus (figure 16.31). Thus, at higher levels of the cord, the more medial fasciculus gracilis carries sensations from the lower limb and the more lateral fasciculus cuneatus carries sensations from the upper limb.

The **spinocerebellar tracts** carry proprioceptive information concerning positions of the limbs and body to the cerebellum. This information is not consciously perceived, but it enables the cerebellum to coordinate movements of different parts of the body. The **spinothalamic tracts** carry information to the thalamus. The ventral spinothalamic tract carries sensory information concerned with touch. The lateral spinothalamic tract transmits sensations of pain and temperature.

**Descending tracts** transmit impulses from the brain to the spinal cord. One of the most important is the **corticospinal tract** that runs directly from the cerebral cortex to motor neurons going to the limbs; thus, it places skeletal muscles under cerebral control. The **tectospinal tract** conveys optic and auditory stimuli to the limbs and trunk without going through conscious centers. The **rubrospinal tract** conveys impulses from the midbrain to the spinal cord and is involved in initiating coordinated movements.

## Brain

The brain forms embryologically from the neural tube anterior to the spinal cord. It includes three anatomical regions (figures 16.25a–f and 16.32). The most posterior region is the hindbrain, which includes the **medulla oblongata, pons,** and **cerebellum.** Next is the midbrain, which in-

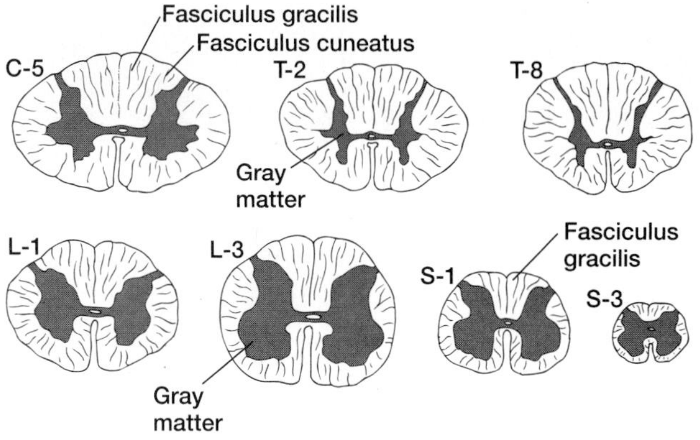


**Figure 16.31** Gray and white matter at various levels of the human spinal cord. Sections are identified by their region—cervical (C), thoracic (T), lumbar (L), sacral (S)—and by specific vertebra numbered (arabic number) within each of these regions from which they came. At the level of the arms (C-5) and legs (L-3), additional sensory and motor fibers enter and leave the spinal cord. This is reflected in the more extensive gray matter compared to other regions of the cord (e.g., T-2, T-8). Note the addition of the fasciculus cuneatus at the highest level on the cord. Primarily, this carries sensory information from the arms.

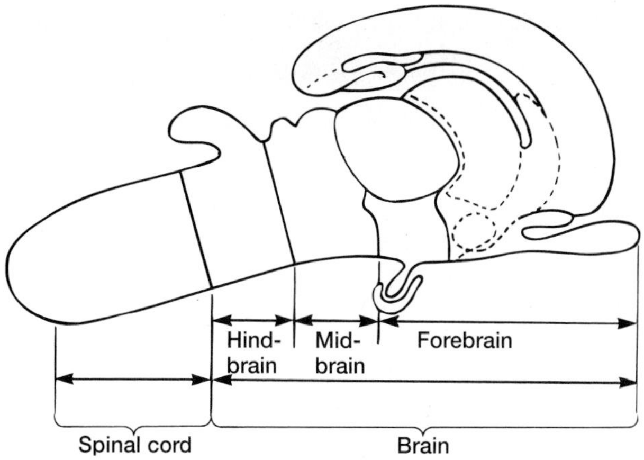


**Figure 16.32** Regions of the vertebrate brain represented diagrammatically.

cludes a sensory **tectum** and a motor **tegmentum.** The **brain stem** includes all regions of the hindbrain and midbrain except for the cerebellum. The most anterior region of the brain, the forebrain, includes the **telencephalon,** or **cerebrum,** and the **diencephalon,** which is the source of the **thalamus.**

## *Phylogeny*

Phylogenetically, there is a tendency for the forebrain to enlarge during vertebrate evolution (figure 16.33). This accompanies increasingly complex behaviors and muscle control. In amniotes, limb posture and body carriage change as terrestrial modes of locomotion become predominant. The limbs move from a sprawled position to one in which they are carried more directly under the weight of the body, increasing the ease and efficiency of limb oscillation (see chapter 8). Coordination of limb oscillation and placement during rapid locomotion become especially complicated in bipedal archosaurs and birds. Increased input of somatosensory information and increased output of motor responses to skeletal muscles requires mediation. The enlargement of the amniote forebrain reflects its increasing role in this mediation within the locomotor system.

In advanced teleost fishes, the midbrain tends to enlarge rather than the forebrain. This seems to be correlated with increasing importance of sensory input from the lateral line system and with greater movement of teleosts in the three-dimensional space of their aquatic environment.

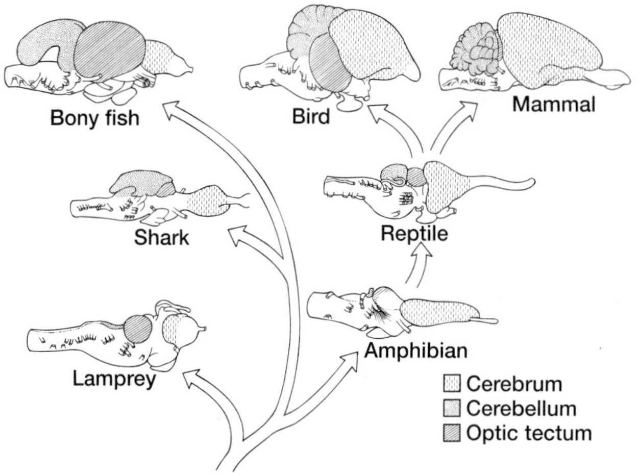


**Figure 16.33** Evolution of the vertebrate brain. Note the phylogenetic enlargement of the cerebrum.

Within these general patterns, the brain of each species reflects the demands of information processing required by its habitat and mode of life (figure 16.34). Cavefish, for instance, have reduced eyes and live in caves, a permanently dark subterranean environment. Correspondingly, the tectum of the midbrain, which normally receives visual input is reduced as well. On the other hand, when visual information constitutes a large part of the brain's sensory input, as in salmon, the tectum is enlarged. Thus, reduction or loss of sensory input from an exteroceptor or interoceptor results in a corresponding reduction or loss of brain nuclei that receive and process this information, whereas increased sensory input leads to increased prominence of the appropriate association.

## Box Essay 16.2 Disease, War, Barroom Brawls—The Early Science of Neurology

Some of the first insights into the functioning of the nervous system came from the results of damage to it. Distribution of information in the nervous system is very orderly in mammals. Sensory information arrives via the dorsal root, motor responses depart via the ventral root, and association neurons intervene between them (see figure 16.29). Anatomical structure and functional activity are closely matched. Because form and function are closely matched in the nervous system, disruption of function can be used to identify the location of an anatomical injury. Because the spinal cord and brain are organized into discrete functional areas, damage to a part results in selective impairment of function. The earliest indication of this came from battle wounds that soldiers survived, but with persistent deficits in function (box figure 1a). Stab or bullet wounds causing restricted damage to the dorsal horn of the gray matter leave patients with more or less normal motor ability but impair their ability to feel sensations from the level of the body where a wound was inflicted (box figure 1b).

Other pathologies affect motor output rather than sensory input. In 1861, the German neurologist H. Broca performed a postmortem examination of the brain of a patient who suffered from a speech defect following injury to his head. While still alive, the patient's lips, tongue, and vocal cords were fully functional, but he could not speak intelligibly. His speech was slow, and many nouns and verbs were deleted. A lesion was found in a restricted area of the forebrain, a region still known as Broca's motor speech area (box figure 1c).

Poliomyelitis, once a common disease primarily afflicting children, struck motor nerves in the ventral horns of the spinal cord. If the disease settled low in the spinal cord, the lesion would likely cause paralysis in the leg on the same side.

In the twentieth century, car accidents have been added to the list of events that inflict this type of damage. Experiments with animals have augmented our knowledge of the functional organization of the central nervous system.

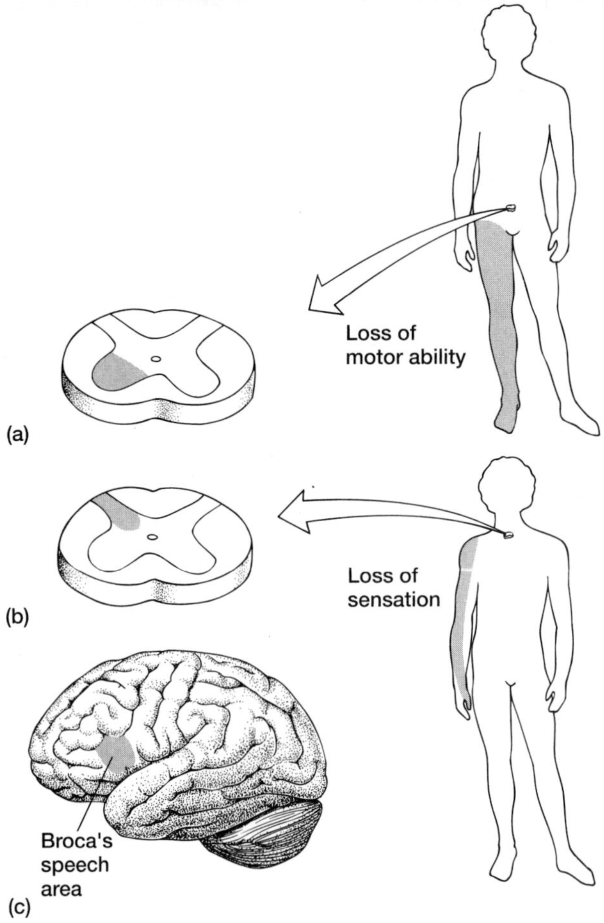


**Box figure 1** Clinical evaluation of injuries to the nervous system. (a) Loss of motor control to right leg muscles can imply selective injury to the ventral horn of the spinal cord at the level where motor neurons to the skeletal muscles of the leg reside. (b) Loss of sensation to the right arm can result from loss or injury to the dorsal horn of the cord at that level. (c) Injury to Broca's area of the brain leaves a person with an understanding of language but results in impaired speech.

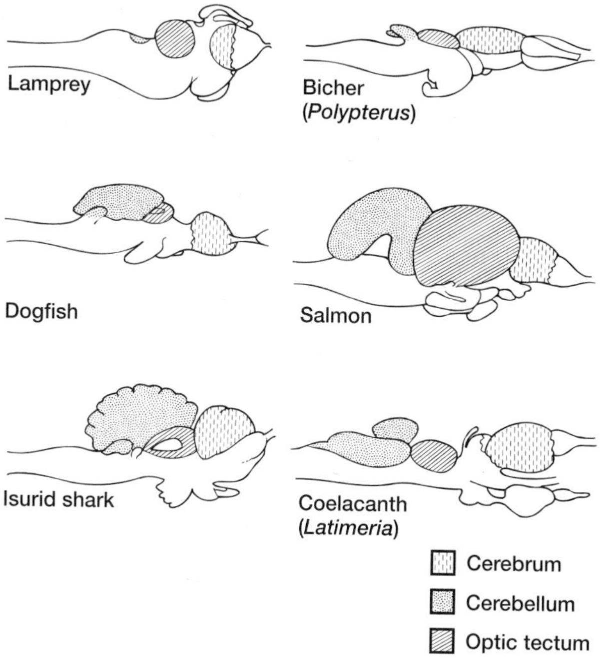


**Figure 16.34** Brains of fishes. Note variations in the sizes of different regions of the brain. These reflect differences in the role that each region plays in processing information important to different species.

## *Form and Function*

Representative vertebrate brains are shown in figures 16.35 through 16.37.

***Hindbrain*** The **medulla oblongata** operates primarily at the reflex level. It has three major functions. First, it houses the primary nuclei of cranial nerves (figure 16.36a–c). In sharks, the primary nuclei or roots of cranial nerves VII through X are contained in the medulla, whereas in mammals, the primary nuclei of cranial nerves VII through XII reside in the medulla. Second, the medulla serves as a major route through which ascending and descending pathways run to and from higher centers of the brain. Third, the medulla contains centers for visceral, auditory, and proprioceptive reflexes, including reflex centers for respiration (figure 16.38), heartbeat, and intestinal motility. Damage to the medulla can be life-threatening because these centers control vital functions.

Medullary nuclei receive afferent signals from sensory nerves at lower levels and descending signals from higher centers such as the hypothalamus. Within these medullary centers, arriving information is processed and efferent output is initiated to adjust visceral activity.

The floor of the midbrain in amniotes becomes a crossroads of increasing importance for the flow of information. In mammals, it develops into a distinct enlargement, the ***pons*** (figure 16.37e). The pons has two basic parts, the pontine tegmentum that contains ascending fibers passing to the thalamus and a basilar portion that contains descending fibers passing from other parts of the brain stem. The trigeminal (V), abducens (VI), and facial (VII) nerves take root in the pons.

The ***cerebellum*** is a dome-shaped extension of the hindbrain. Its surface is often highly convoluted and folded. The cerebellum modifies and monitors but does not initiate motor output. It operates at an involuntary level and has two primary functions. First, it is important in maintaining equilibrium (figure 16.39). Information pertaining to touch, vision, hearing, proprioception, and motor input from higher centers is processed in the cerebellum. Integration of these incoming sensations results in the maintenance of muscle tone and balance. For an organism to run, jump, fly, or swim in a three-dimensional world, it must be able to keep itself upright and orient itself in space relative to gravity. The cerebellum is involved in processing information that results in maintaining positional equilibrium of the organism.

The second primary function of the cerebellum is the refinement of motor action. The cerebellum compares incoming impulses and sends modified signals to motor centers. Direct electrical stimulation of the cerebellum does not produce muscle contractions. Following removal of the cerebellum, an organism can still move in space, but its movement is uncoordinated, exaggerated, or insufficient, and its motion is likely to be uneven. Thus, the role of the cerebellum is to monitor and modify rather than initiate action.

As with other parts of the brain, the size of the cerebellum is proportional to its role. In fishes, the cerebellum is usually relatively large because of extensive input from the lateral line sensory system regarding water currents and electrical stimuli. Furthermore, active aquatic organisms must navigate and orient themselves in a three-dimensional medium. Equilibrium and balance are important; therefore, the cerebellum is well developed. As we would expect, in bottom-dwelling fishes (e.g., flounders) and in fishes that are not active swimmers (e.g., lampreys), the cerebellum has a reduced role and is relatively small (figure 16.35a).

With the advent of terrestrial life, the lateral line system is lost and sensory input to the cerebellum decreases. However, as robust limbs used in terrestrial locomotion develop, proprioceptive information and refinement of muscle action become important and place increased demands on the cerebellum. The cerebellum of terrestrial vertebrates thus remains large and prominent.

***Midbrain*** The roof of the midbrain is the **tectum,** which receives sensory information. In mammals, this region is specialized into **superior** and **inferior colliculi.** The floor of the midbrain is the **tegmentum,** which initiates motor output.

In fishes and amphibians, the midbrain is often the most prominent region of the brain (figure 16.35a–e). The

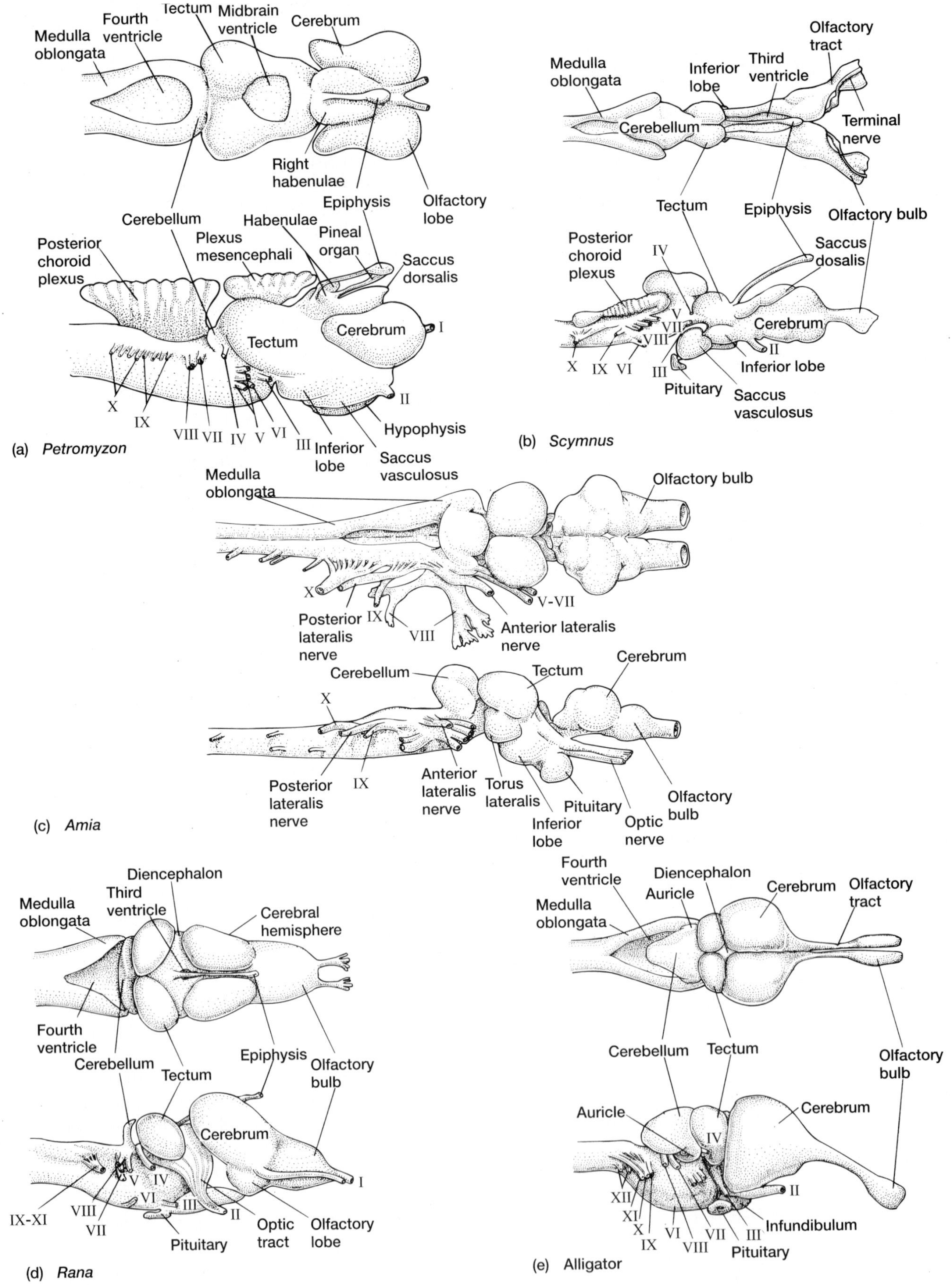


**Figure 16.35** Vertebrate brains. Dorsal views are shown above, lateral views below. (a) Lamprey. (b) Shark. (c) Bowfin. (d) Frog. (e) Alligator. (f) Goose. (g) Insectivore. (h) Horse.

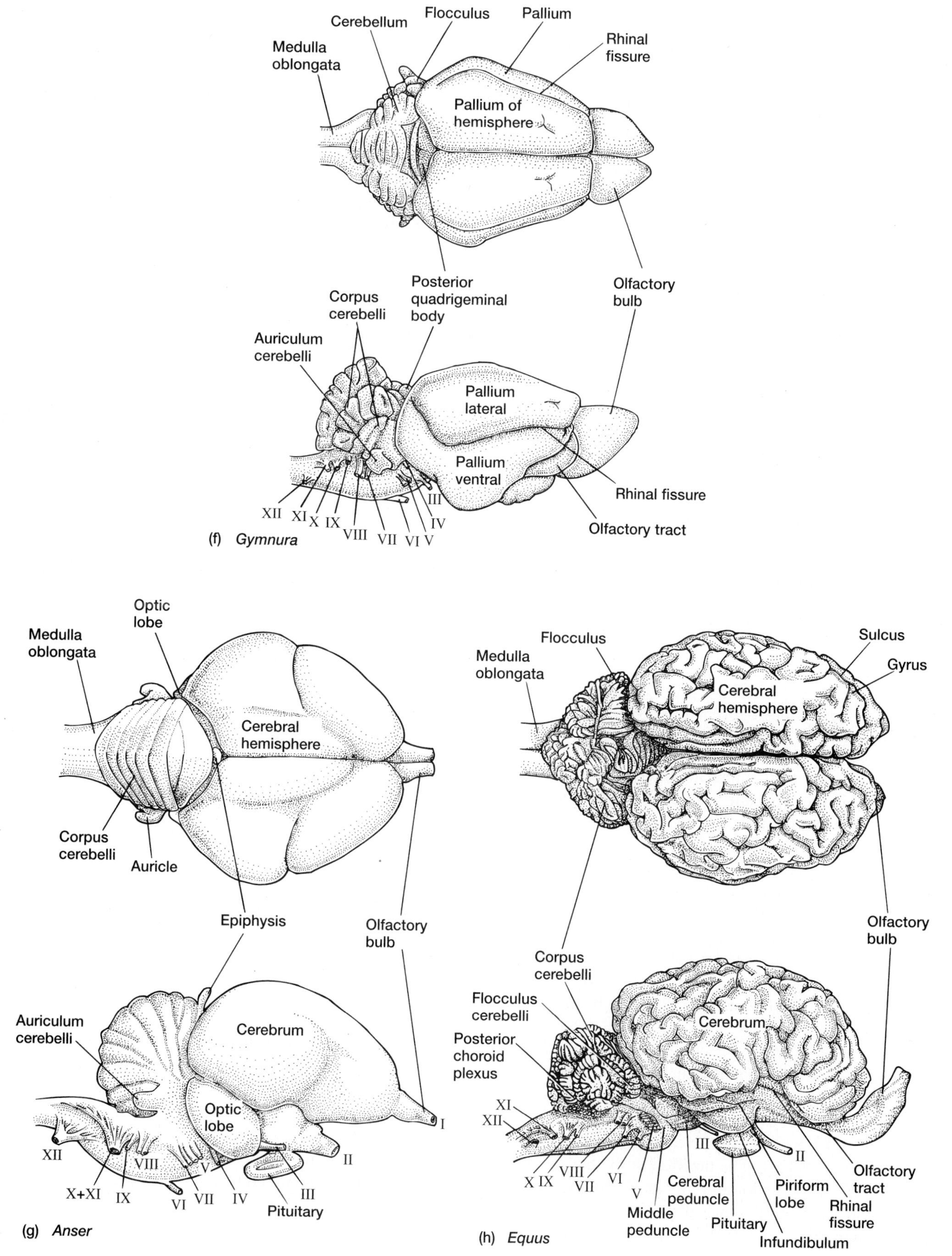


**Figure 16.35** (continued)

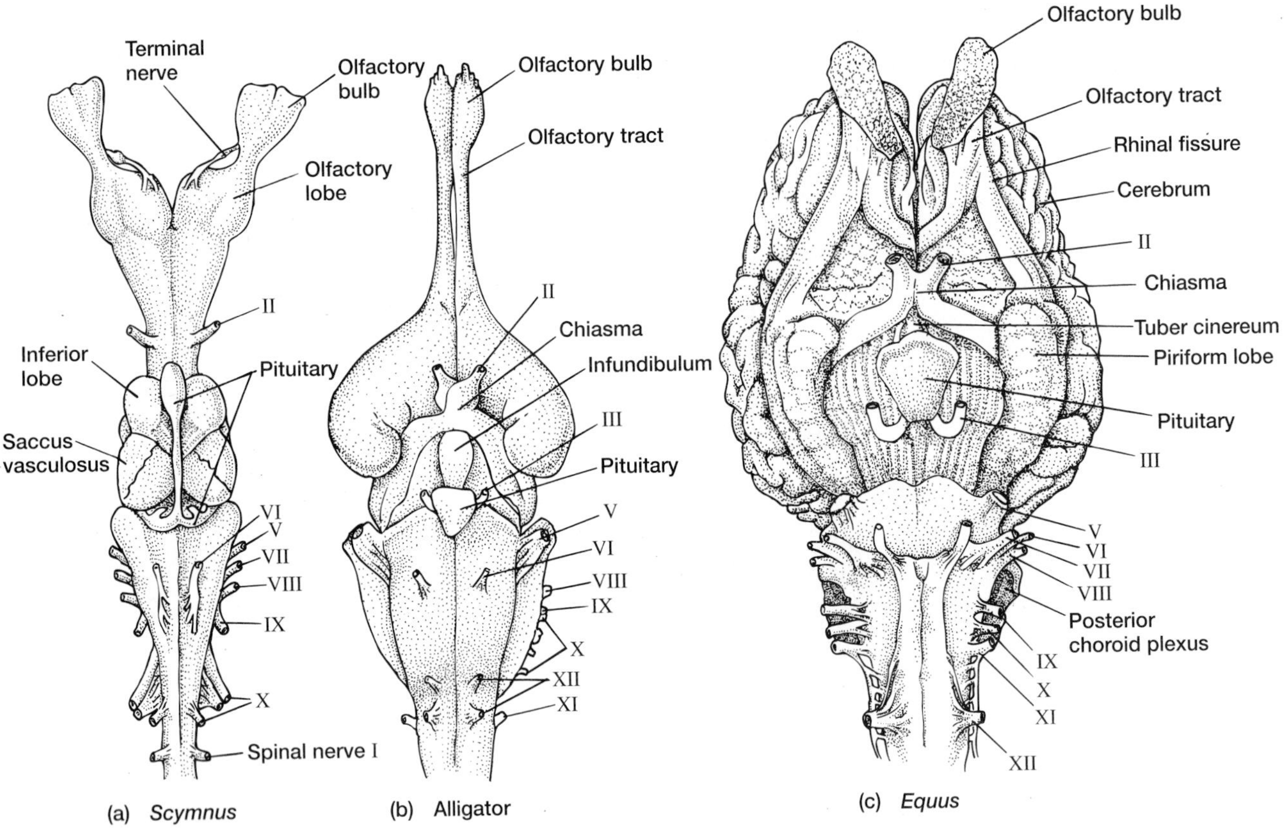


**Figure 16.36** Vertebrate brains, ventral views. (a) Shark. (b) Alligator. (c) Horse.

tectum receives direct input from the eyes. In addition, information from the acoustico-lateralis system, the cerebellum, the olfactory epithelium, and the cutaneous sensors is transmitted indirectly to the tectum. The tegmentum is also prominent in lower vertebrates. In some fishes, it seems to be an important learning center.

In reptiles, birds, and mammals, the tectum continues to receive visual and auditory input, which it relays to the telencephalon through the thalamus. Thus, visual information in all vertebrates reaches the telencephalon via the tectum. A second route by which visual information reaches the telencephalon is through the thalamus of the forebrain, without passing through the tectum (figure 16.40).

***Forebrain*** The ***diencephalon*** includes four regions—**epithalamus, hypothalamus, ventral thalamus,** and **dorsal thalamus.** The roof of the diencephalon produces the epithalamus, which includes the **pineal gland** and the **habenular nucleus** at its base. The function of the habenular nucleus is uncertain. In lower vertebrates, the pineal gland affects skin pigmentation by acting on melanocytes, and it may be important in regulating photoperiod as well. In higher vertebrates, the pineal plays a role in regulating biological rhythms (see chapter 15).

The floor of the diencephalon produces the hypothalamus, and the **mammillary bodies,** which contain nuclei that function in olfaction (figure 16.41). The hypothalamus houses a collection of nuclei that regulate homeostasis to maintain the body's internal physiological balance. Homeostatic mechanisms adjusted by these nuclei pertain to temperature, water balance, appetite, metabolism, blood pressure, sexual behavior, alertness, and some aspects of emotional behavior. The hypothalamus stimulates the pituitary gland situated beneath it to regulate many homeostatic functions. The **limbic** and **reticular systems** influence the functions of the hypothalamus as well. These systems are discussed later in the chapter.

The ventral thalamus is a small area between the midbrain and the rest of the diencephalon. The largest part of the diencephalon is the dorsal thalamus, or sometimes called just thalamus, an area comprised of nuclei that receive sensory input. The thalamus is the major coordinating center of afferent sensory impulses from all parts of the body. Except for the olfactory tracts, which transmit stimuli directly to the cerebral cortex, all sensory tracts including those relaying sensations of touch, temperature, pain, and pressure, as well as all visual and auditory fibers synapse in thalamic nuclei on their way to the cortex. Thus, the thalamus is a relay center for sensory information going to the

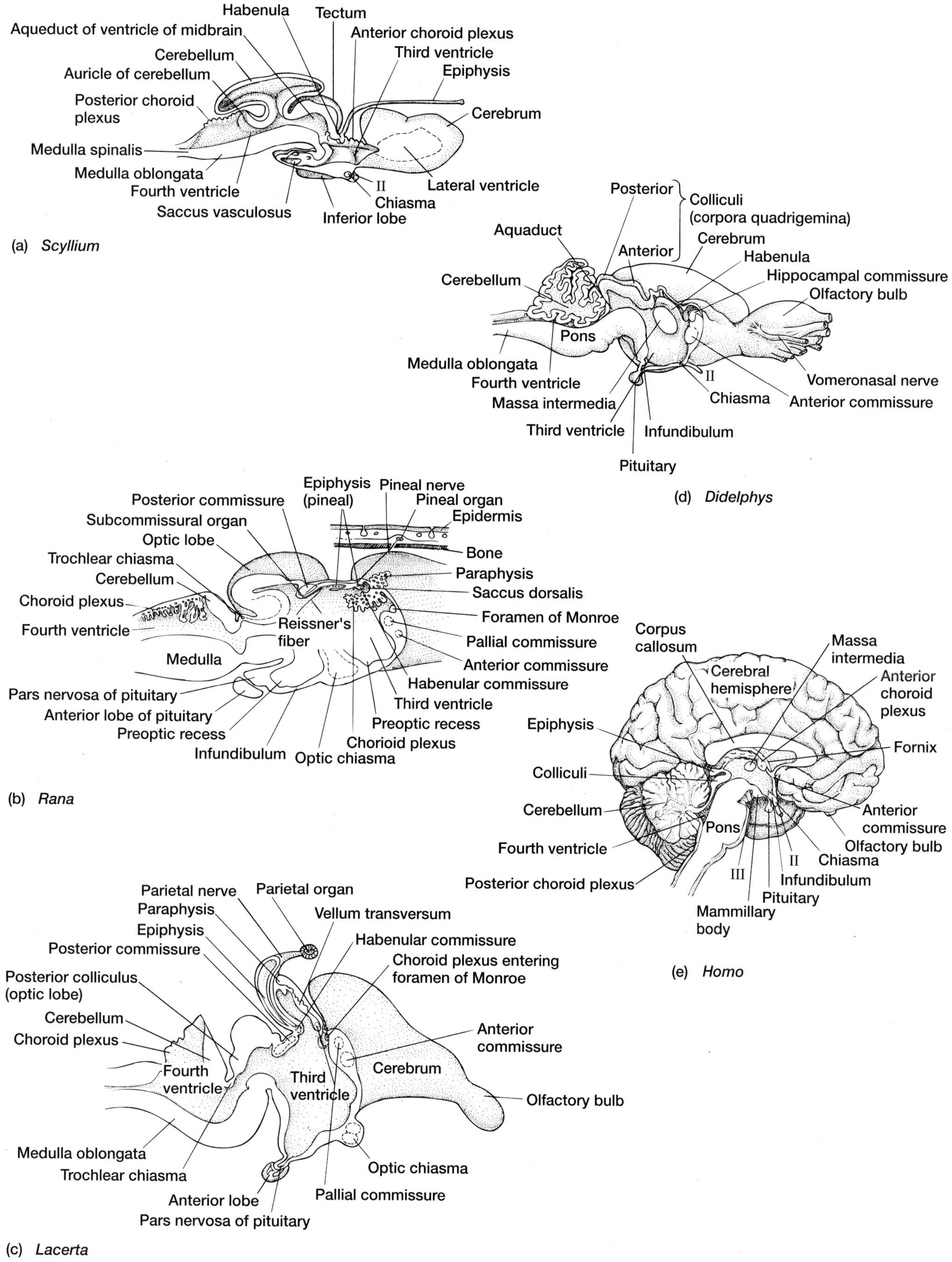


**Figure 16.37** Vertebrate brains, sagittal views. (a) Shark. (b) Frog. (c) Lizard. (d) Opossum. (e) Human.

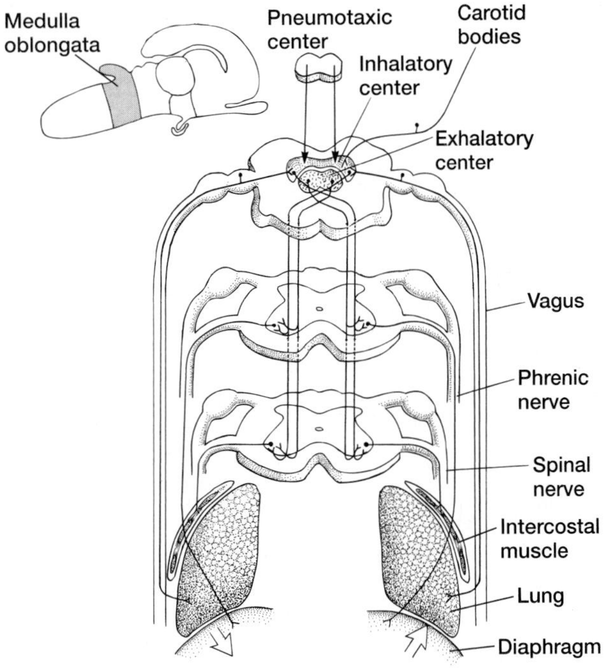


**Figure 16.38** Coordination of breathing by the mammalian medulla. Reflex control of respiration is under the influence of three paired nuclei—the pneumotaxic center in the pons and the dorsal inhalatory center and ventral exhalatory center in the medulla. The inhalatory center receives information about gas composition and blood pH from the carotid bodies and about the degree of lung expansion from the vagus nerve. The inhalatory center excites descending neurons that terminate in motor neurons of the phrenic nerve to the diaphragm. It also stimulates a spinal nerve to the intercostal muscles. When these nerves are excited, inhalation and lung expansion results. The ventral expiratory center, does not seem to function during quiet, normal breathing. This center is connected to motor neurons (not shown) serving antagonistic intercostal and accessory muscles of expiration.

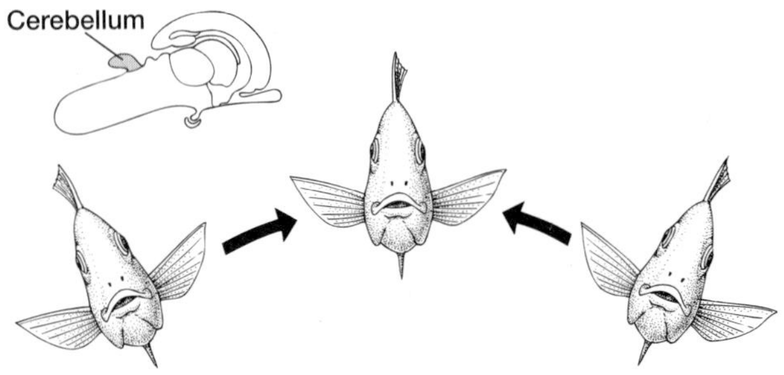


**Figure 16.39** Function of the cerebellum. Balance and orientation are mediated through the cerebellum. As an animal changes its orientation in a gravitational field (left and right sketches), sensory organs that detect its altered position send impulses to the cerebellum. The cerebellum stimulates responses that restore the animal's position (center sketch).

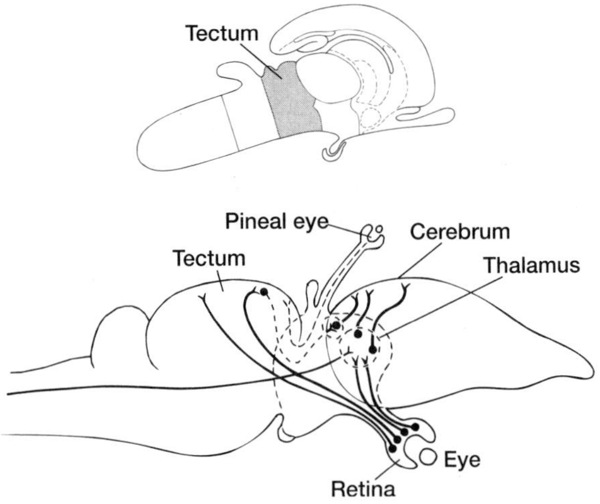


**Figure 16.40** Function of the amniote tectum. The tectum receives visual information directly from the retina of the eye and relays this first to the thalamus and then to the cerebrum. In most vertebrates, visual information from the retina reaches the cerebrum via a second route without passing through the tectum. From the retina, visual information first reaches the thalamus and then is relayed to the cerebrum.

cerebral cortex. The thalamus integrates sensory somatic impulses into a pattern of sensations that is projected to the somatic sensory area of the cerebral cortex.

The ***telencephalon,*** or cerebrum, is a pair of expanded lobes known as **cerebral hemispheres.** The outer wall of these hemispheres forms the **cerebral cortex,** or **cortical region.** The **subcortical region** comprises the remaining cerebral tissue. The hemispheres appear embryologically at the most anterior end of the neural tube. In actinopterygian fishes, the embryonic telencephalon proliferates outward to form the everted adult cerebrum. In all other fishes and tetrapods, the embryonic telencephalon forms lateral swellings, which give rise to the cerebral hemispheres of adults (figure 16.42).

Reception of olfactory information is a major function of the telencephalon. Even in primitive vertebrates, however, ascending fibers arrive from the thalamus, suggesting that the telencephalon has assisted in regulating other sensory integrative functions as well since early in vertebrate evolution. In reptiles and especially in birds and mammals, the cerebral region enlarges five- to twentyfold compared with most anamniotes of similar body size. This phylogenetic enlargement occurs, in part, because the cerebrum must process more sensory information from the thalamus. This is accompanied by an increased number of association centers within the cerebrum. Nevertheless, within any vertebrate class, the size of the telencephalon may vary considerably among species. For example, among chondrichthyan fishes, primitive sharks and rays possess cerebrums comparable in size with those of amphibians, but in advanced sharks

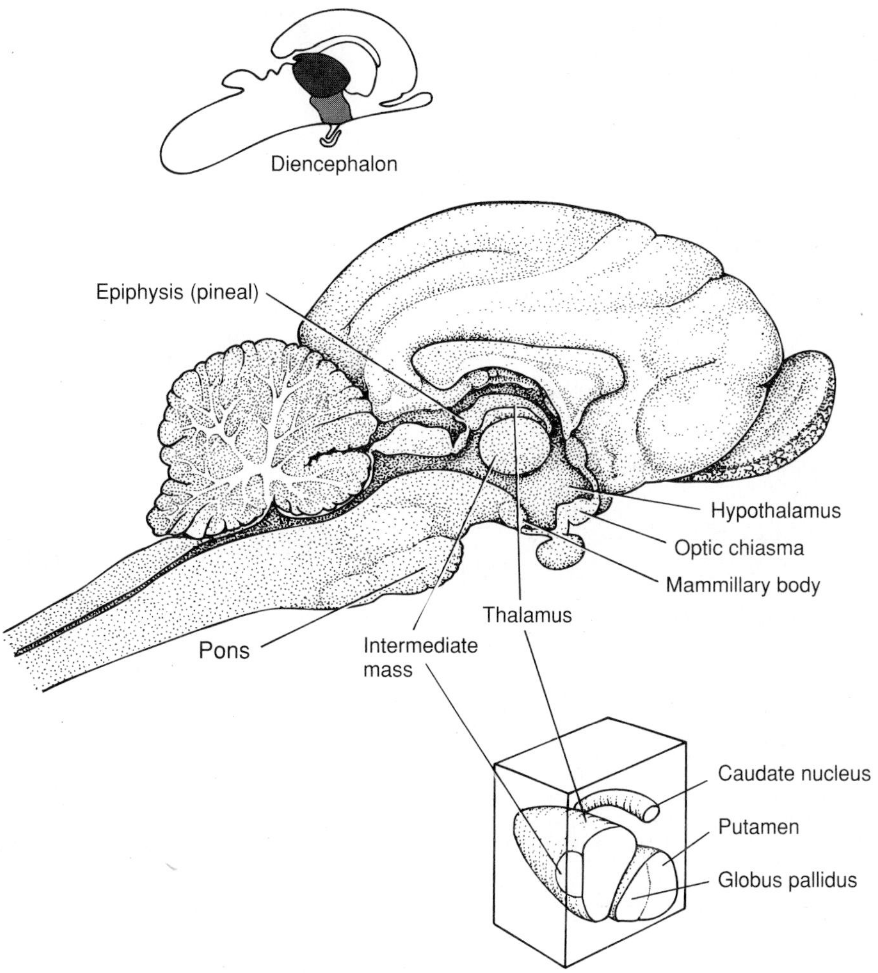


**Figure 16.41** Hypothalamus and its relationship to adjacent regions of the brain. The region of the diencephalon is shaded in the small, top figure. The isolated thalamus in three-dimensional cutaway view is shown in the small block below the brain.

and skates, the relative size of the cerebral hemispheres approaches that of birds and mammals.

In many mammals, the cerebral cortex is folded in a complicated fashion to accommodate its increased volume. The rounded folds are **gyri,** and the intervening grooves are called **sulci.** The term ***fissure*** is often used to note a deep sulcus that separates major surface regions of the cerebrum. Not all mammals show such folding. In the duckbill platypus, opossum, and many rodents, the cerebral cortex is smooth. In the echidna, kangaroos, and most primates, the degree of folding is variable. In all groups of mammals, the extent of folding seems to be more pronounced in large species.

Early theories about evolution of the cerebrum held that new regions progressively emerged out of preexisting regions. A recent "neostructure" was thought to arise from an "archistructure" that evolved from an initial "paleostructure." The morphological terms coined attempted to express these presumed phylogenetic relationships. In addition, much early study of the brain centered on mammals, especially humans, in which descriptive terms were preferred. Instead of recognizing phylogenetic homologies, these terms expressed quaint or fanciful features. For example, hippocampus means horse tail, amygdala means almond, and putamen refers to a fruit pit. Some of these older terms, including *hippocampus* and *amygdala* are still used today. However, within recent years, new experimental techniques have improved our understanding of comparative brain structure, leading to an ongoing reinterpretation of earlier ideas and introducing a new and still formative terminology. These differences in terminology are compared in table 16.7. Proliferation of terms has been especially noticeable in mammalian neuroanatomy and human medicine.

The revised view of cerebral evolution challenges not just the terminology but the assumptions on which the old terminology was based. The current view holds that the

**Table 16.7 Comparison of Recent and Former Terms Designating the Telencephalon**

| Former Terms: Morphological | Former Terms: Descriptive | Recent Terms |
|---|---|---|
| | *Roof of Telencephalon* | |
| Pallium | | Pallium |
| Archipallium | Hippocampus | Medial pallium |
| | | Dorsal pallium |
| | | Dorsomedial cortex (cingulate) |
| Neopallium | Cerebral cortex | Dorsolateral cortex |
| (Neocortex) (Isocortex) | | Lateral pallium |
| | | Dorsal ventricular ridge |
| Paleopallium | Piriform lobe | Lateral cortex |
| | *Floor of Telencephalon* | |
| Corpus striatum | Basal nuclei | Subpallium |
| Paleostriatum | Globus pallidus | Striatum |
| Neostriatum | Caudate nucleus, putamen | Striatum |
| Archistriatum | Amygdala | Striatum |
| Septum | Septal area | Septum |

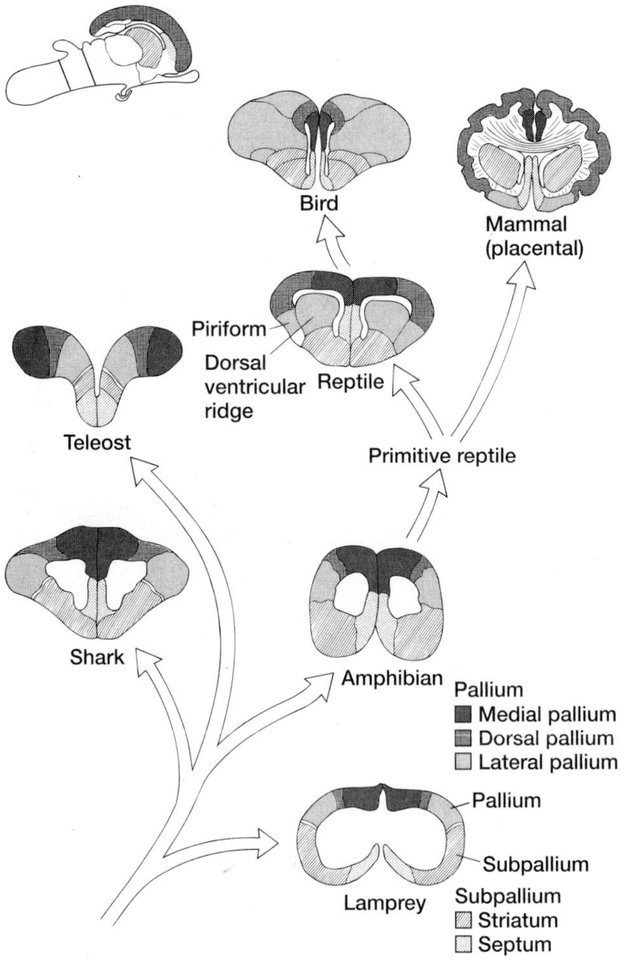


**Figure 16.43** Evolution of vertebrate cerebral hemispheres.

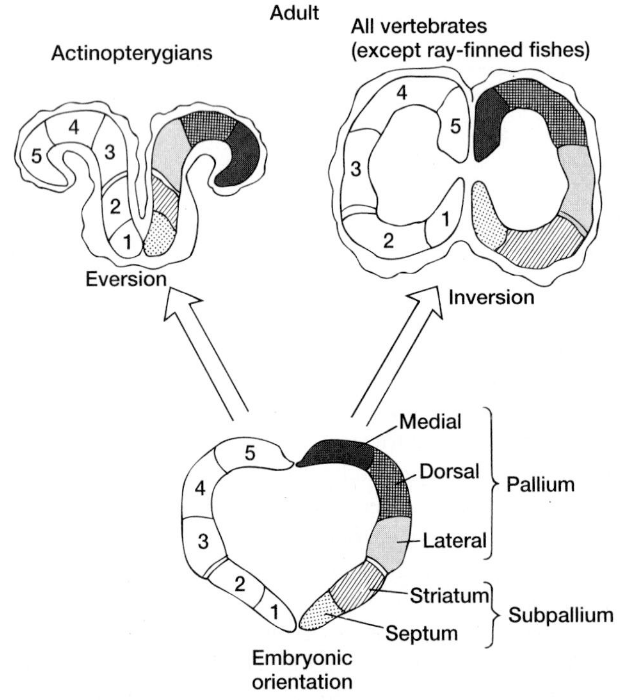


**Figure 16.42** Embryonic development of the telencephalon. In actinopterygians, the telencephalon becomes everted during development. In all other vertebrates, it becomes inverted.

basic regions of the telencephalon did not emerge in a stepwise fashion. The pattern in which these regions are laid out is very ancient and was present in the common ancestor of all vertebrates. From this fundamental pattern, we see that the cerebrum has two regions, a dorsal **pallium** and a ventral **subpallium.** The pallium possesses **medial, dorsal,** and **lateral** divisions. The subpallium consists of a **striatum** and a **septum** (figure 16.42). All vertebrates have a cerebrum based on this basic plan. Major phylogenetic changes in the cerebrum center on loss, fusion, or enlargement of one or more of these regions.

*Pallium*. The *medial pallium* receives secondary olfactory information. The dorsal and lateral pallia receive ascending input, including visual information relayed from the thalamus. Agnathans possess a pallium with the characteristic medial, dorsal, and lateral divisions (figure 16.43). The olfactory bulb is large. In lampreys, it is even larger than the cerebral hemisphere. Processing of olfactory information is an important role of the olfactory bulb, but what additional sensory inputs reach the adjacent telencephalon from ascending tracts is unclear. As mentioned earlier, the telencephalon of ray-finned fishes is everted. The basic pallial and subpallial regions can be recognized, but no consensus exists as to the boundaries between them or the number of cell groups comprising divisions within each (figure 16.42).

The elasmobranch pallium includes lateral, dorsal, and medial divisions, although these may in turn be subdivided. The lateral pallium receives the main olfactory input via the lateral olfactory tract. Parts of the dorsal pallium receive visual, lateral line, thalamic, and possibly auditory stimuli. Less is known about the medial pallia, but exchange of information between hemispheres is likely because they fuse across the midline.

The amphibian pallium is similar to that of primitive sharks but is less complex than that of reptiles. In living amphibians, the pallium consists of three regions—dorsal, lateral, and medial pallial divisions—which receive olfactory input as well as sensory input from the thalamus. The **amygdala** is another region of the amphibian pallium that is primarily concerned with receiving accessory olfactory information from the vomeronasal organ.

The pallium of reptiles includes dorsal, lateral, and medial divisions as well as a hypertrophied region, the **dorsal ventricular ridge** (DVR), that dominates the central region of the cerebral hemisphere. Once thought to be part of the striatum, the DVR is now generally believed to be a derivative of the lateral pallium. In birds, the DVR expands further. It accounts for much of the relative increase in size of the cerebral hemispheres and crowds the lateral ventricle into a slit. The dorsal part of the DVR of birds hypertrophies into a region usually called the **Wulst,** containing highly organized visual information important in stereoscopic vision. The DVR receives visual, auditory, and somatosensory input from several major thalamic nuclei and projects this information to the striatum and to other parts of the pallium. Its size and central position in the flow of information suggest that the DVR may be a major higher association area in both reptiles and in birds. Both the lateral pallium (formerly termed the piriform lobe) and the medial pallium (the former hippocampus) persist as significant cortical areas in reptiles and birds, but the dorsal pallium is usually reduced in prominence, especially in birds.

Mammals also show a dramatic increase in proportionate size of the cerebral hemispheres, but not because of an enlarged DVR as in reptiles and birds. Instead, the dorsal pallium is enlarged in mammals. In the course of this enlargement, the dorsal pallium thickens and differentiates into layers. The resulting mammalian cerebral cortex is an extensive area called the **isocortex.** In primates, approximately 70% of the neurons in the central nervous system are found in the cerebral cortex. The isocortex is devoted to deciphering auditory, visual, and somatosensory information as well as to controlling the function of the brain stem and spinal cord. All sensory areas are channeled or relayed to the cerebral cortex, bringing together sensory and recall information.

The mammalian medial pallium (hippocampus) receives sensory information and seems to initiate inquisitive or investigative behaviors. It is also concerned with memory of recent events. Olfactory information is shunted to the mammalian lateral pallium (piriform).

*Subpallium.* As mentioned, the *subpallium* is divided into two regions, a medial septum and a more extensive lateroventral striatum. Both regions are distinct, even in the earliest fishes. The septum receives information from the medial pallium and is connected to the hypothalamus of the forebrain as well as to the tegmentum of the midbrain. It is an important part of the limbic system. The striatum has a more complicated phylogeny.

The function of the striatum is not well known. The term *striatum,* or **basal ganglia,** originally referred to a collection of nuclear groups at the base of the cerebral hemispheres. This structure appears to be present in all vertebrates and controls the sequence of actions involved in complex movements. It receives sensory input from the pallium and input from a nucleus called the **substantia nigra,** located in the midbrain tegmentum. The striatum has been most thoroughly studied in mammals. The mammalian striatum consists of four parts—the **caudate nucleus,** the **putamen,** the **globus pallidus,** and the **amygdala.** The first three are directly related to motor control, particularly of the limbs. The amygdala is closely affiliated with the limbic system. Disruption of the basal ganglia leads to involuntary and purposeless motions known as dyskinesias. Parkinson's disease, characterized by an involuntary tremor that is often worse when the patient is at rest, is associated with degeneration of the basal nuclei.

In reptiles and birds, the striatum receives information from the DVR and transmits it first to the brain stem and then to the optic region of the tectum. Neurons within the avian striatum are often organized into layers or bands. Expansion of the DVR (reptiles and birds) and the isocortex (mammals) is accompanied by a corresponding expansion of the striatum.

## *Functional Associations of Parts of the Central Nervous System*

***Telencephalon*** The pallium receives direct sensory input, especially auditory, visual, and somatosensory information, from the thalamus, processes this information, and transmits responses to the striatum, hypothalamus, and brain stem. Thus, it indirectly controls locomotion. Major reorganization and expansion of the forebrain are correlated with changes in terrestrial locomotion and posture. In birds, upright posture and complex wing movements are served by the expanded DVR. In mammals, the isocortex enlarges to assume an increasing role in the coordination of complex locomotion.

Especially important sensory signals may be duplicated several times within the telencephalon, giving

# Box Essay 16.3 Phrenology to Neuroanatomy: Attempts to Decipher the Relationship between Brain Anatomy and Personality

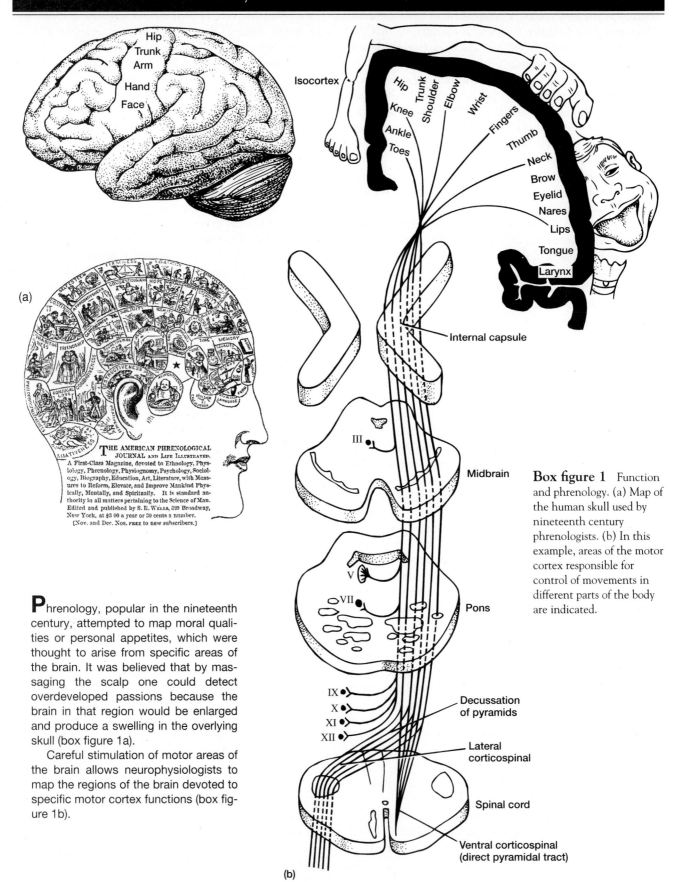

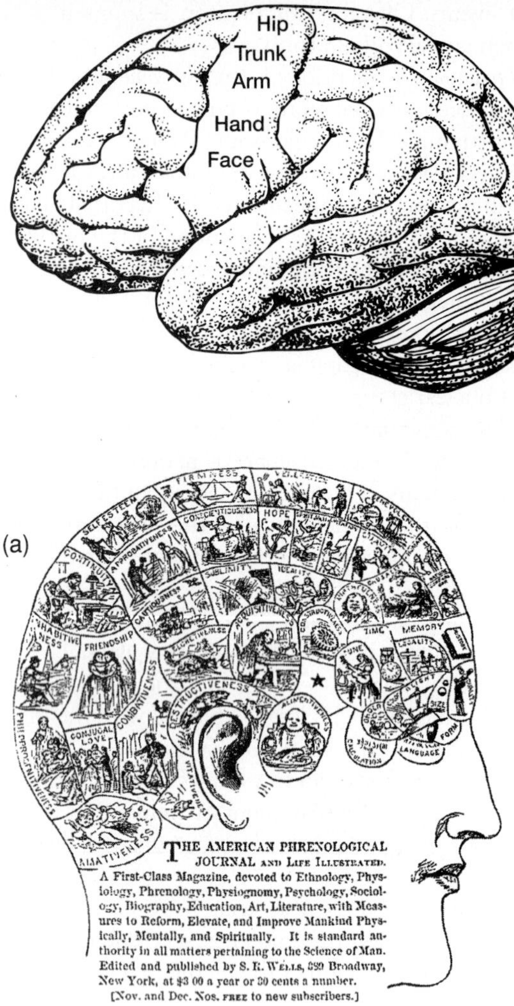


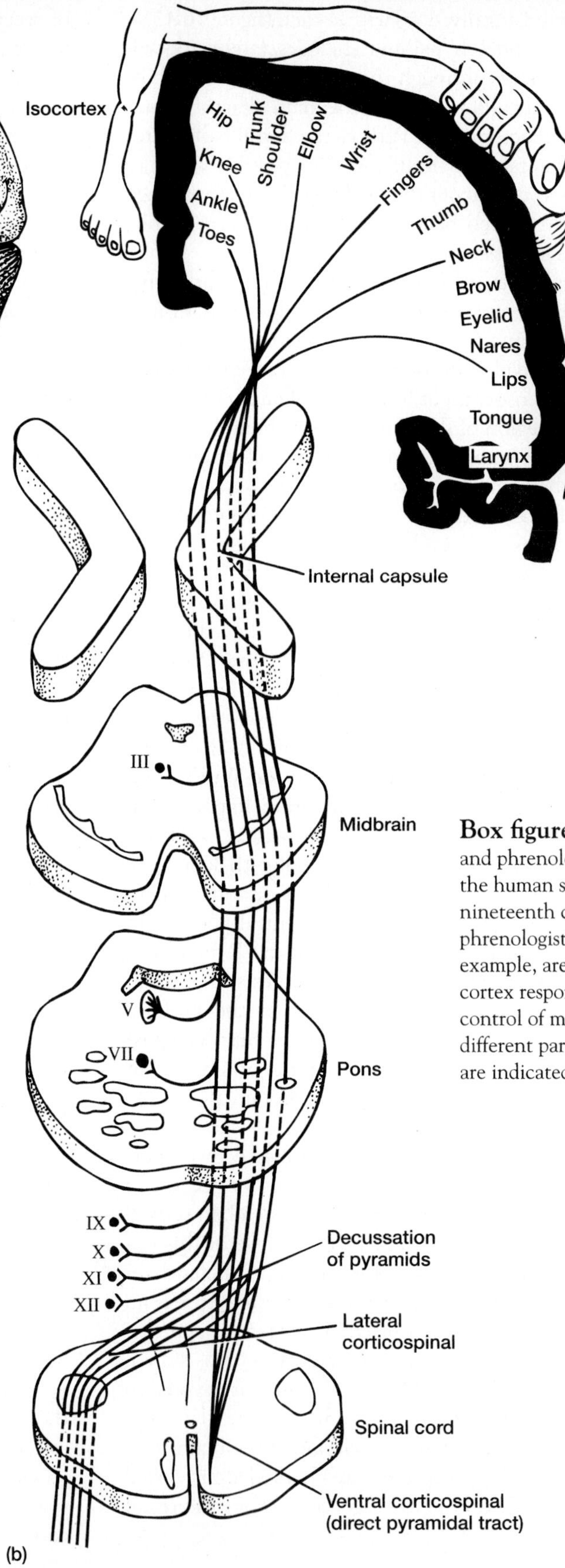


**Box figure 1** Function and phrenology. (a) Map of the human skull used by nineteenth century phrenologists. (b) In this example, areas of the motor cortex responsible for control of movements in different parts of the body are indicated.

Phrenology, popular in the nineteenth century, attempted to map moral qualities or personal appetites, which were thought to arise from specific areas of the brain. It was believed that by massaging the scalp one could detect overdeveloped passions because the brain in that region would be enlarged and produce a swelling in the overlying skull (box figure 1a).

Careful stimulation of motor areas of the brain allows neurophysiologists to map the regions of the brain devoted to specific motor cortex functions (box figure 1b).

multiple representations of the same information. For example, visual input, which is important in almost all vertebrates, has two parallel routes to the telencephalon. One is from the retina to the tectum and then to the telencephalon via a relay in the thalamus. The other is from the retina to the thalamus to the telencephalon. In some placental mammals in which vision is a major source of information, there may be a dozen areas in the telencephalon that decipher visual stimuli. Similarly, multiple visual areas are found in cats, squirrels, bats, and primates. Duplication of centers that process stimuli apparently improves comparison of sensory input within the nervous system and helps extract information it contains. The anatomical consequence is an increase in the size of the brain area to accommodate the reception and processing of multiple sets of similar sensory information.

***Limbic System*** The limbic system is a functional association of brain centers that include nuclei of the thalamus, hypothalamus, amygdala, hippocampus (medial pallium), **cingulate gyrus,** and septum. The **fornix** is a two-way fiber system that connects all nuclei of the limbic system (figure 16.44a,b and table 16.8). The limbic system receives stimuli from the isocortex and returns responses to the isocortex and to the autonomic nervous system. The hypothalamus contains nuclei that affect heart rate, respiration, and general visceral activity through the autonomic nervous system. Changes in these usually accompany strong emotion. The amygdala is active in the production of aggressive behavior and fear. The hippocampus (medial pallium) lies adjacent to the amygdala. Damage to it causes loss of recent memory. The cingulate gyrus and septum are other routes of input to this system.

The limbic system is involved in two functions. First, as mentioned, it regulates the expression of emotions. Experimental or accidental removal of parts of the limbic system leads to emotional passiveness. This function is important to survival. To sustain itself, an animal must actively seek food, be alert to danger, and respond appropriately when threatened. Phylogenetically, the limbic system, or at least many of its centers, arise early in evolution even before there are many direct connections between the thalamus and the cerebral cortex. The limbic system has been called the "visceral brain" because of its substantial influence on visceral functions through the autonomic nervous system.

The second function of the limbic system involves memory. The hippocampus (medial pallium) seems to be essential to sustain recent memory. Damage to the hippocampus does not destroy the memory of events prior to the injury, but subsequent events are recalled only with great difficulty or not at all. Memory is probably resident in the isocortex rather than in the limbic system, but the limbic system is involved in temporarily retaining the memory of a recent experience until the experience becomes established as long-term memory in the isocortex.

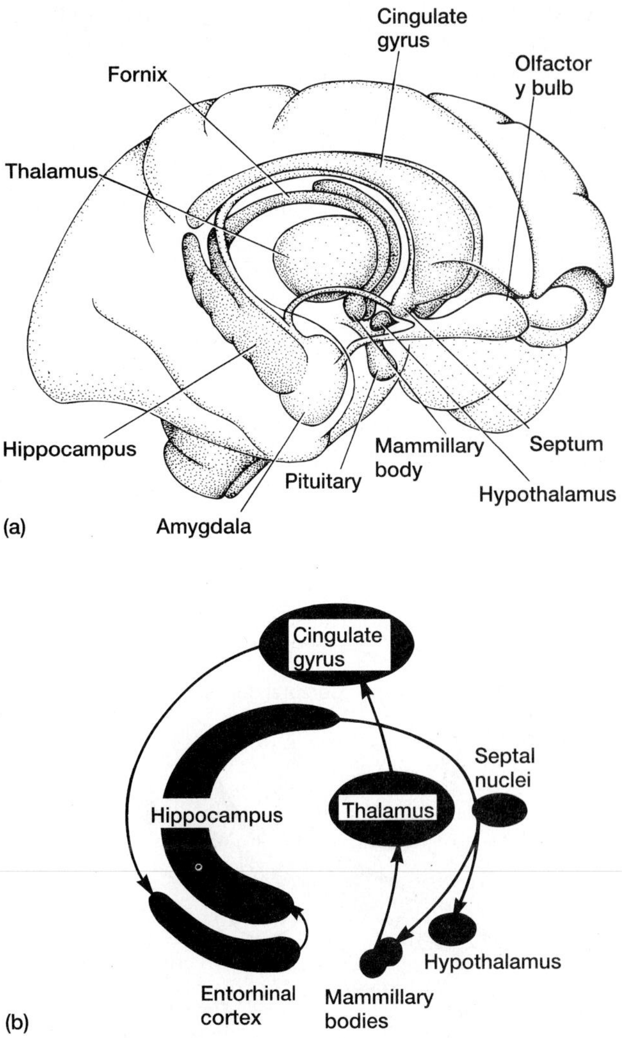


**Figure 16.44** Limbic system. (a) Anatomical components of the limbic system. (b) Flow of information through the limbic system.

## TABLE 16.8 Centers of the Brain Allied as the Limbic System

| Cortical Centers | Subcortical Centers |
|---|---|
| Telencephalon | Telencephalon |
| Pallium | Subpallium |
| Hippocampus | Septum |
| Parahippocampus | Amygdala |
| Cingulate gyrus | Diencephalon |
| Dentate gyrus | Habenular nucleus |
| | Thalamus |
| | Hypothalamus |
| | Mammillary bodies |

***Reticular Formation*** The reticular formation resides in the medulla, midbrain, and thalamus (figure 16.45). This structure is defined in several ways, but it generally consists of enmeshed neurons and their fibers. The term *reticular* and the term *formation* refers to the fact that it lacks delineated tracks or nuclei like a "center" or a "system." This diffuse arrangement of fibers resembles some parts of the nervous systems of lower vertebrates and has inspired the idea that the reticular formation is a phylogenetic retention of an earlier feature.

The reticular formation has two functions. First, it is arousal in action via its awakening or stimulation of the cerebral cortex. An alert animal is more attentive to sensory input. Some anesthetics and tranquilizers act by suppressing transmissions through the reticular formation. Damage to the reticular formation can lead to a prolonged coma. Second, the reticular formation also acts as a filter. It selects information to be relayed to higher centers or down the spinal cord. It tends to pass along information that is novel or persistent.

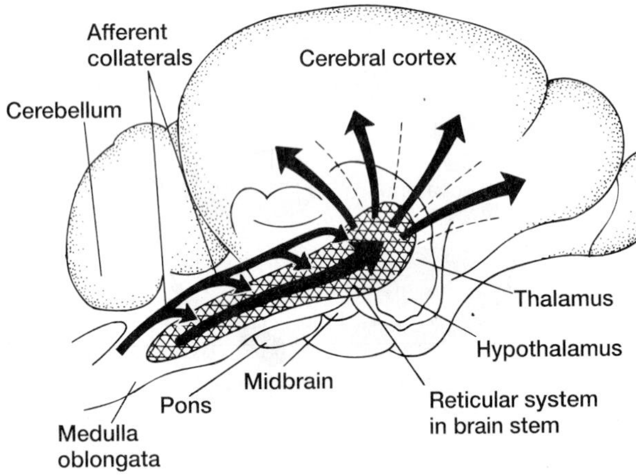


**Figure 16.45** Reticular formation. The reticular formation lies in the medulla, midbrain, and thalamus and projects to the higher centers of the brain. When active, it seems to bring about general alertness. Sensory afferent pathways traveling to higher centers send branches, afferent collaterals, into the reticular system. Through the thalamus, the reticular system is then projected to the cerebral hemispheres resulting in general arousal. Modified from T. E. Stize, et al., 1951.

***Spinocortical Associations*** So far we have looked at regions of the central nervous system that perform local functions—reflexes of the spinal cord, association centers of the brain, and systems of alliance. But the central nervous system shows a high degree of integration. Even reflexes completed at the level of the spinal cord are registered at higher centers, and events in higher centers influence lower levels in the spinal cord. This flow of information tends to occur along distinctive tracts.

Sensory impulses travel up the spinal cord, but before they reach conscious centers in the cerebral cortex, they synapse in the gray matter, in the thalamus, and even in additional nuclei. Thus, information that reaches conscious centers has already been sifted and filtered (figure 16.46a). As discussed previously, all sensory fibers synapse in the thalamus on their way to the cerebral cortex except for olfactory tracts. In the thalamus, sensory impulses are coordinated into an integrated pattern of sensations that is then projected to specialized sensory areas in the cortex. In other words, the cortex receives information that has already been interpreted by subcortical centers. Information traveling down the spinal cord, even if it originated in the cerebral cortex, is modified by the cerebellum, by subcortical centers, and by reflexes at the level of the cord (figure 16.46b).

The subtleties of the central nervous system are profound. We have seen that the nervous system gathers information about the body's internal status, the outside world, and the results of previous experience and turns these into responses that might allow the organism to maintain itself in its environment. But there is more to this process than the mechanical processing of information. Emotions, goals, and conscious participation shape a response, at least in humans. To a large extent, our responses are shaped by our perceptions of physical stimuli, which are mediated by sensory receptors. In the next chapter, we take a closer look at these sensory receptors.

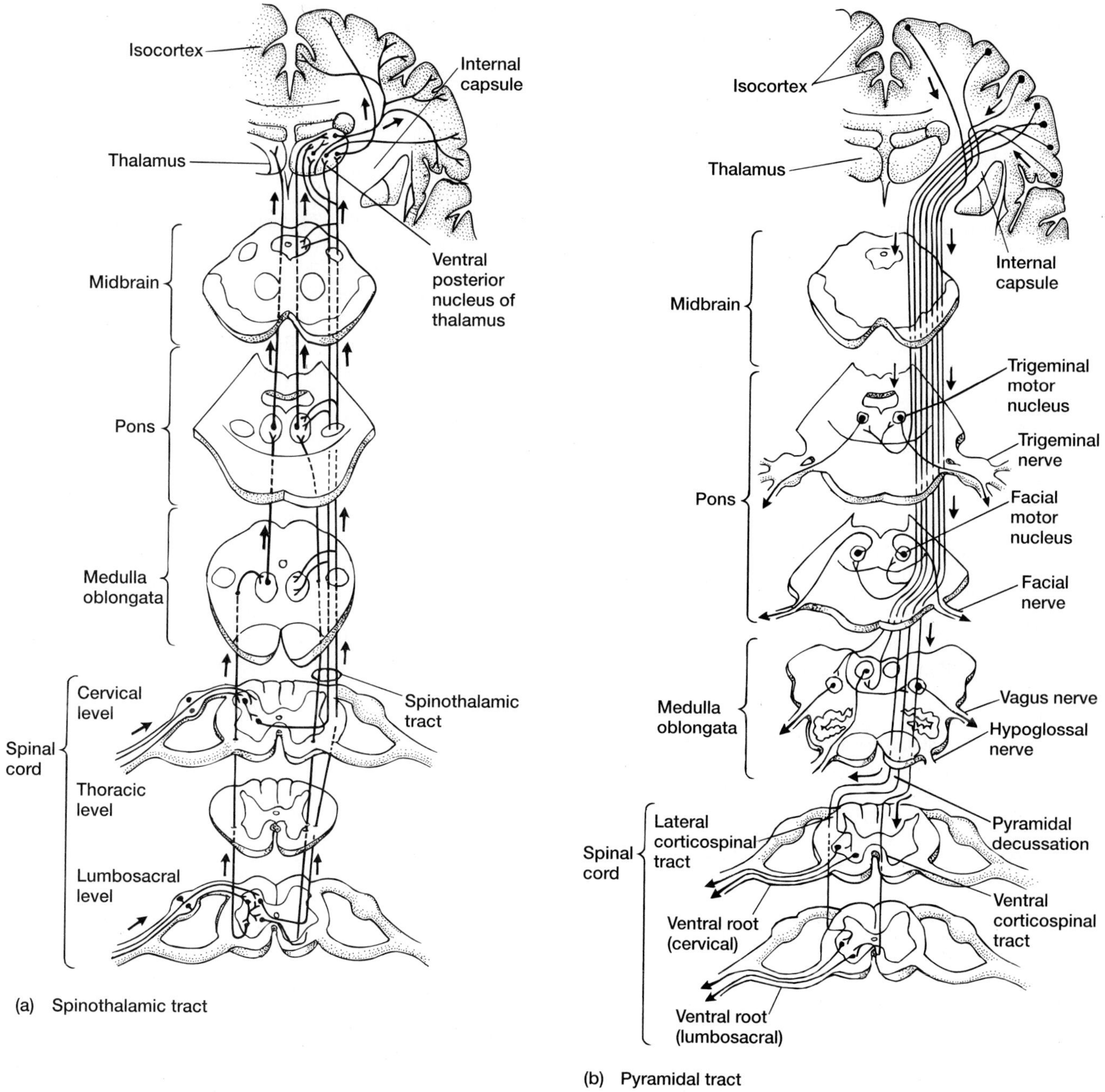


(a) Spinothalamic tract

(b) Pyramidal tract

**Figure 16.46** Processing of sensory and motor information. (a) The spinothalamic tract gathers sensory neurons that carry sensations of pain and courses to the thalamus. Sensations are relayed from the thalamus to higher brain centers. (b) Pyramidal tract. Decisions initiated in the cerebral cortex are conveyed along descending motor neurons, which form the pyramidal tract, to the appropriate level of the spinal cord. From the spinal cord, the response is passed along a motor neuron to the effector.

## Selected References

Halpern, M. 1980. The telencephalon of snakes. In *Comparative neurology of the telencephalon*, edited by S.O.E. Ebberson. New York: Plenum Press.

Kimelberg, H. K., and M. D. Norenberg. 1989. Astrocytes. *Sci. Amer.* 257(4):66–76.

Mongane, P. J., M. S. Jacobs, and A. Galaburda. 1986. Evolutionary aspects of cortical organization in the dolphin brain. In *Research on dolphins*, edited by R. J. Harrison and M. Bryden. Oxford: Oxford University Press.

Nieuwenhuys, R. 1982. An overview of the organization of the brain of actinopterygian fishes. *Amer. Zool.* 22:287–310.

Nilsson, S. 1983. Autonomic nerve function in the vertebrates. In *Zoophysiology*, edited by D. S. Farmer. New York: Springer-Verlag.

Northcutt, R. G. 1977. Elasmobranch central nervous system organization and its possible evolutionary significance. *Amer. Zool.* 17:411–29.

Northcutt, R. G. 1981. Evolution of the telencephalon in nonmammals. *Ann. Rev. Neurosci.* 4:301–50.

Northcutt, R. G., and R. E. Davis (eds.). 1983. *Fish neurobiology*. 2 vols. Ann Arbor: University of Michigan Press.

Olsson, R. 1986. Basic design of the chordate brain. In *Indo-Pacific fish biology: Proceedings of the second international conference on Indo-Pacific fishes*, edited by T. Uyeno, R. Arai, T. Taniuchi, and K. Matsuura. Tokyo: Ichthyological Society of Japan, pp. 86–93.

Parent, A. 1986. *Comparative neurobiology of the basal ganglia*. New York: John Wiley and Sons, Interscience.

Reiner, A., S. E. Brauth, and H. J. Karten. 1984. Evolution of the amniote basal ganglia. *Trends Neurosci.* 7:320–25.

Stevens, C. F. 1979. The neuron. *Sci. Amer.* 241 (3):54–65.

Ulinski, P. S. 1983. *Dorsal ventricular ridge: A treatise on forebrain organization in reptiles and birds*. New York: John Wiley and Sons, Interscience.

Ulinski, P. S. 1984. Thalamic projections to the somatosensory cortex of the echidna, *Tachyglossus aculeatus*. *J. Comp. Neurol.* 229:153–70.

Ulinski, P. S. 1986. Neurobiology of the therapsid-mammal transition. In *The ecology and biology of mammal-like reptiles*, edited by N. Hotton, III, P. D. MacLean, J. J. Roth, and E. C. Roth. Washington, D.C.: Smithsonian Institution Press, pp. 149–71.

Ulinski, P. S. 1990. Cerebral cortex in reptiles. In *Cerebral cortex*, vol. 8, edited by E. G. Jones and A. Peters. New York: Plenum Press.

Ulinski, P. S. 1990. Neuronal organization of the striatum in the alligator, *Alligator mississippiensis*. *Exp. Brain Res.* 19:119–33.

Wake, M. H. 1992. Patterns of peripheral innervation of the tongue and hyobranchial apparatus in caecilians (Amphibia: Gymnophiona). *J. Morph.* 212:37–53.

Wilczynski, W., and R. G. Northcutt. 1983. Connections of the bullfrog striatum: Afferent organization. Efferent projections. *J. Comp. Neurol.* 214:321–42.

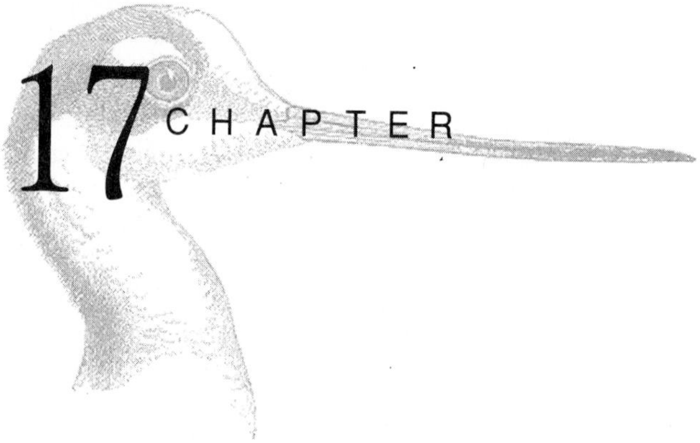

17 CHAPTER

# Sensory Organs



## Introduction

To survive, an organism must react to danger and take advantage of opportunity. Appropriate responses require information about the external environment, the body's internal physiology, and previous experience. The results of previous experience are recorded within the nervous system as memory, but **sensory receptors** monitor the external and internal environment (figure 17.1). Sensory receptors are specialized organs that respond to selected information. Sensory receptors code or translate environmental energies into nerve impulses that are transmitted to the central nervous system (CNS) via afferent fibers. These impulses may or may not be received at conscious levels of the brain.

Sensations we as humans become conscious of are referred to as **perception.** Our view of the world is partially determined by the kinds of information our sensory receptors detect and by how that information is processed. Vertebrates differ in their ability to perceive stimuli. Bats and even dogs hear sounds at frequencies our ears cannot. Hawks can hunt high above the ground and detect tiny rodents scampering below. Rattlesnakes hunt in light too dim for humans to see in. Yet humans see the world in color while most other mammals see it only in black and white. Our perception of the world is limited or enhanced by the availability and sensitivity of our sensory receptors.

It should also be emphasized that conscious sensations are an organism's subjective interpretation of the environment. The environment contains chemicals and photons of light, but the senses of taste and color are interpretations of these phenomena. Similarly, pain does not exist in the environment. You cannot measure pain as you can temperature or force or solar radiation. Taste, color, and pain are perceptions arising out of events in the brain itself.

Nerve impulses carried by sensory nerves are electrical impulses. The optic nerve carries the same kind of electrical impulses as the auditory nerves, olfactory nerves, taste nerves, and so on. Different sensations result from different ways in which the nervous system interprets signals from different sensory receptors. Because the impulses are the same, the central nervous system can be fooled. Artificial stimulation of the auditory nerve is perceived as a sound. Artificial stimulation of the optic nerve is perceived as light. Mechanical pressure on the eyeball stimulates the optic nerve to send electrical impulses to the brain. These impulses are interpreted as what they are expected to be, sensations of light, rather than as mechanical stimuli. This is why a blow to the eye can make one "see stars."

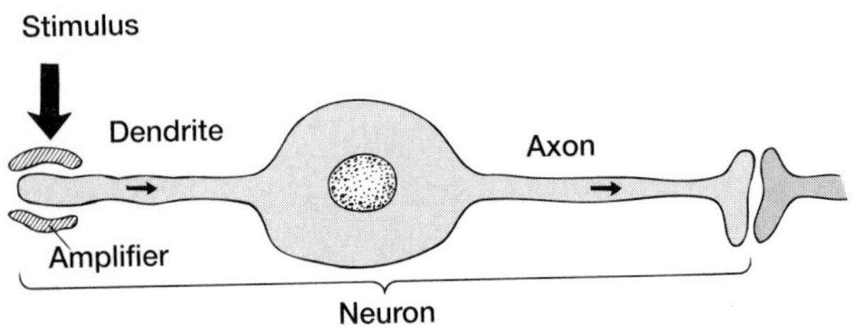


**Figure 17.1** Sensory receptor. A sensory receptor is usually composed of the dendrites of a neuron and can include tissue that amplifies the stimulus. The receptor is a transducer that transforms a stimulus into an electrical impulse or burst of electrical impulses that spread to the cell body and along its axon to other neurons, usually in the central nervous system.

To be clear in discussing perception, we should distinguish an environmental stimulus from the way in which it is interpreted, but this is seldom done. In common usage and even in scientific research, convention rules. We speak of the senses of seeing, hearing, tasting, smelling, and so on as if the stimulus and the perception of the stimulus were the same. The treatment of sensory receptors in this chapter attempts to separate the kind of environmental energy or stimulus monitored from the nervous system's interpretation of that stimulus.

## Components of a Sensory Organ

Sensory neurons are nerve cells specialized for detecting and transmitting information about the external or internal environment. Each sensory neuron sends out slender processes or **nerve fibers.** Sensory receptors usually contain **dendrites,** processes that are responsive to stimuli and carry impulses toward the body of the nerve cell. A sensory neuron usually has an **axon** as well, a nerve process that transmits impulses away from the cell body to other neurons.

Neurons (p. 617)

The sensory receptor acts as a **transducer,** a device that translates energy from one form into another. The microphone of a public address system, which translates sound waves into electrical energy, is another example of a transducer. Most sensory receptors translate light, mechanical, or chemical stimuli into electrical impulses. Often the tip of the sensory nerve fiber is associated with accessory tissues that amplify the stimulus and thereby increase the sensitivity of the receptor. A sensory nerve fiber with its associated tissues is termed a **sensory organ.**

Sensory organs can be classified according to several criteria. **Somatic sensory organs** refer to those of the skin, body surfaces, and skeletal muscles. **Visceral sensory organs** reside in the viscera. **Exteroceptors** receive sensations from the environment and **interoceptors** respond to sensations from organs. A special type of interoceptor is the **proprioceptor,** a sensory organ located in striated muscles, joints, and tendons. A third way to classify sensory organs, and the one used in this chapter, is on the basis of how extensively they are distributed. **General sensory organs** are widely distributed throughout the body and concerned with sensations of touch, temperature, and proprioception. **Special sensory organs** are localized and often specialized.

## General Sensory Organs

General sensory receptors may be placed in one of three anatomical categories—free, encapsulated, or with associated nerve endings. The structure of the nerve ending is designed to increase the effect of the stimulus (figure 17.2).

### Free Sensory Receptors

When the terminus of a sensory process lacks any specialized association, it is termed a **free nerve ending** or **free sensory receptor.** At its terminus, the free nerve ending may **arborize,** or branch, extensively to increase the area monitored. Free sensory receptors are primarily concerned with sensations interpreted as painful, but they can also be stimulated by extremes of heat or cold. Tissue damage can lead to swelling and direct stimulation. A toothache is an example. Tactile sensations—that is, sensations of pressure or touch—are often detected by free nerve endings as well. Free sensory receptors are abundant in areas in which sensitivity is highly developed, such as the skin, cornea, oral cavity, tooth pulp, and intestines.

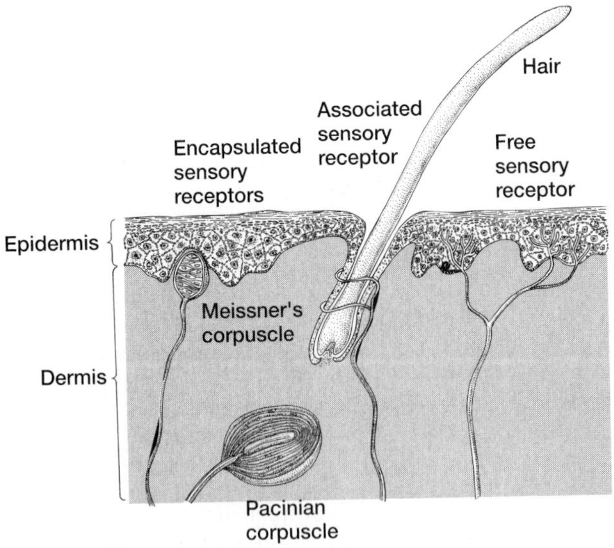


**Figure 17.2** General sensory receptors. A free nerve ending, encapsulated sensory receptors (Meissner's and Pacinian corpuscles), and an associated sensory receptor of a hair follicle.

## Encapsulated Sensory Receptors

When the terminus of a sensory process is enclosed in a specialized structure, it is called an **encapsulated nerve ending** or **encapsulated sensory receptor.** For example, **Meissner's corpuscle** is a sensory ending wrapped in mesodermal cells located in the dermis of the skin just beneath the epidermis (figure 17.2). It responds to touch. The **corpuscle of Ruffini,** responsive to warmth, and the **end-bulb of Krause,** responsive to cold, are other encapsulated receptors located in the dermis. **Pacinian corpuscles** (corpuscles of Vater-Pacini) are located in the skin, joints, and deep tissues of the body. For instance, it is not uncommon to find them associated with the pancreas. They respond to pressure. In encapsulated receptors, the capsules enhance the deformation of the nerve endings, thus assisting in the initiation of the nerve impulse. In Pacinian corpuscles, for example, the nerve ending is enclosed in a series of concentric layers that form an "onion skin" capsule. This capsule acts as a tiny transducer that converts pressure into electrical depolarization of the nerve ending.

## Associated Sensory Receptors

When the terminus of a sensory process is wrapped around another organ, it is called an **associated nerve ending,** or an **associated sensory receptor.** For example, nerve endings are associated with the base of a hair follicle (figure 17.2). When a hair is moved, the entwined nerve endings at the base of the hair are stimulated.

### *Proprioception*

Proprioception is based largely on information gathered by associated sensory receptors located in muscles, tendons, and joints. These receptors monitor the state of limb flexion and the degree of muscle contraction. As a result, the central nervous system is kept informed about limb or body position. If a body part is moved, the muscles involved and the amount of contraction supplied will differ depending on the initial position of the part. Proprioceptive information is indispensable for determining the location of a part before and during its movement. If you are a sighted person, you probably do not take advantage of proprioceptive information relayed to your conscious centers. However, if you are blindfolded and a partner gently swings your extended arm to a new location, you will be aware of the new position to which your arm has been rotated. Nevertheless, most proprioceptive information is processed at subconscious levels of the nervous system to make automatic adjustments of posture or to synchronize body and limb movements.

Some proprioceptive fibers come from encapsulated Pacinian corpuscles located in joint capsules, but most come from two types of associated receptors, muscle spindles and Golgi tendon organs.

***Muscle Spindles*** Within skeletal muscles, the muscle fiber that produces the major force moving a part is the **extrafusal muscle cell** (extrafusal muscle fiber). Such fibers are innervated by **alpha motor neurons,** whose bodies are located in the gray matter of the spinal cord. Interspersed among the extrafusal muscle fibers are fusiform-shaped packages of **muscle spindles** that contain modified striated **intrafusal muscle cells** (intrafusal muscle fibers). Unlike extrafusal fibers that work the lever system, intrafusal fibers are specialized sensory organs.

There are two types of intrafusal muscle fibers. The **nuclear bag intrafusal fiber** has nuclei clustered in a swollen region near the middle of the fiber and is associated with a **primary afferent sensory nerve** (annulospiral nerve). The **nuclear chain intrafusal fiber** has nuclei strung out along the fiber instead of clustered. It is associated with a **secondary afferent sensory nerve** (flower spray nerve). Both types of intrafusal fibers are innervated by **gamma motor neurons** (figure 17.3).

Muscle spindles function to maintain muscle tone. Normal muscle maintains a small amount of tension even when it is relaxed, a state in which the muscle has **tonus.** When a muscle relaxes more than normal, the muscle spindle sags. Primary and secondary afferent nerves wound around the intrafusal fibers sense this sag. Through reflex connections in the spinal cord, these afferent neurons synapse with alpha motor neurons to stimulate contraction of extrafusal fibers that stimulate muscle tension and restore muscle tone.

The stretching of a muscle lends to its reflex contraction. When postural muscles are stretched or load is added, muscle spindles lengthen, initiating the **stretch reflex** (figure 17.4a). Because gamma motor neurons cause the intrafusal fibers to contract, these neurons are thought to increase or decrease the sensitivity of this reflex (figure 17.4b,c).

The sensory function of the intrafusal muscle fibers is to inform the nervous system about the rate of change in the length of the extrafusal muscle fibers with which they are associated. This information can initiate a stretch reflex to adjust tonus. It is also relayed to the cerebellum, which modulates muscle activity.

**Postural reflex (p. 649)**

***Golgi Tendon Organs*** **Golgi tendon organs** are sensory receptors in the tendons that attach muscles to bone. Thus, they lie along the line of muscle action and function as tension recorders, supplying the central nervous system with information about the forces generated by muscles (figure 17.3).

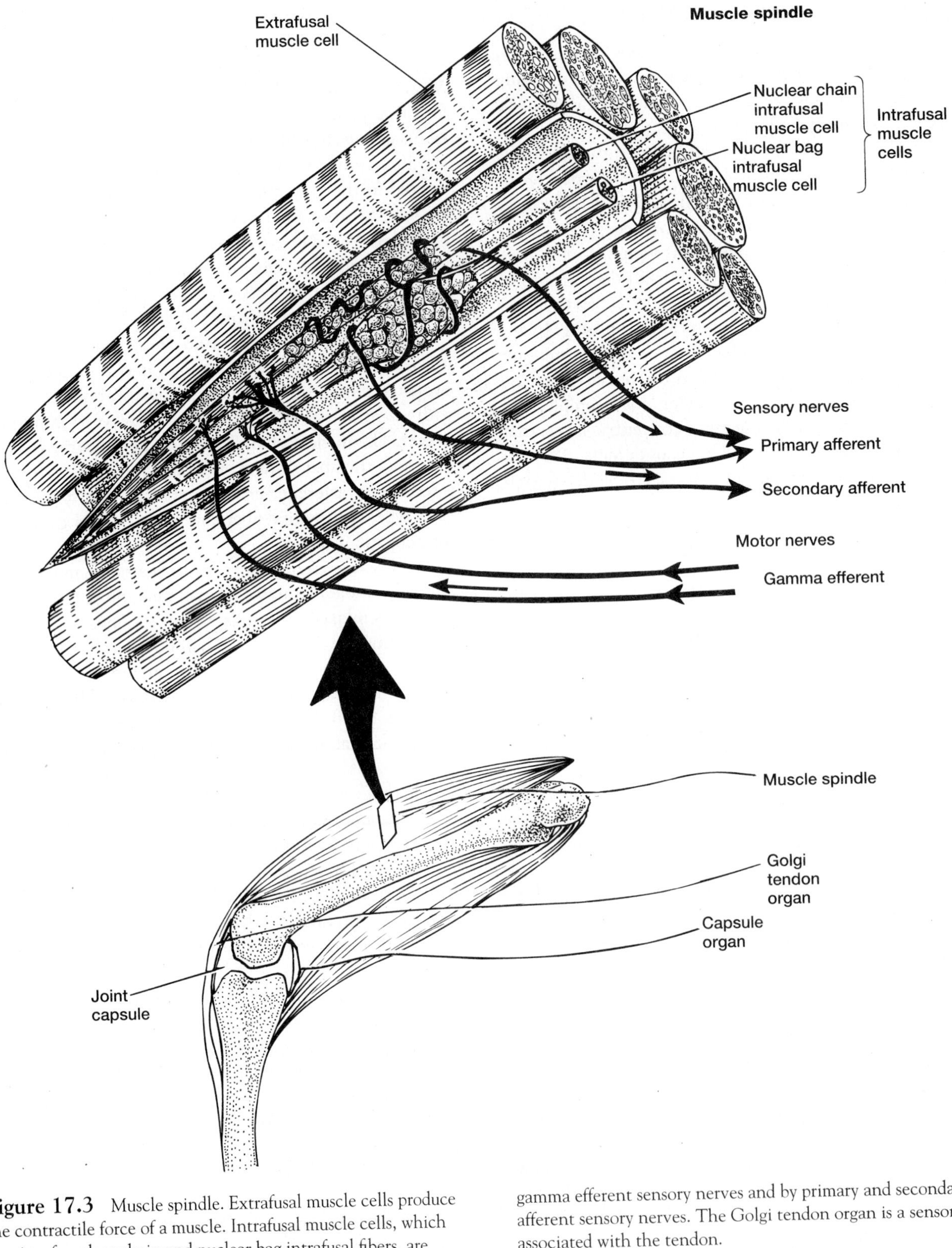


**Figure 17.3** Muscle spindle. Extrafusal muscle cells produce the contractile force of a muscle. Intrafusal muscle cells, which consist of nuclear chain and nuclear bag intrafusal fibers, are modified muscle cells. Intrafusal muscle cells are innervated by gamma efferent sensory nerves and by primary and secondary afferent sensory nerves. The Golgi tendon organ is a sensory fiber associated with the tendon.

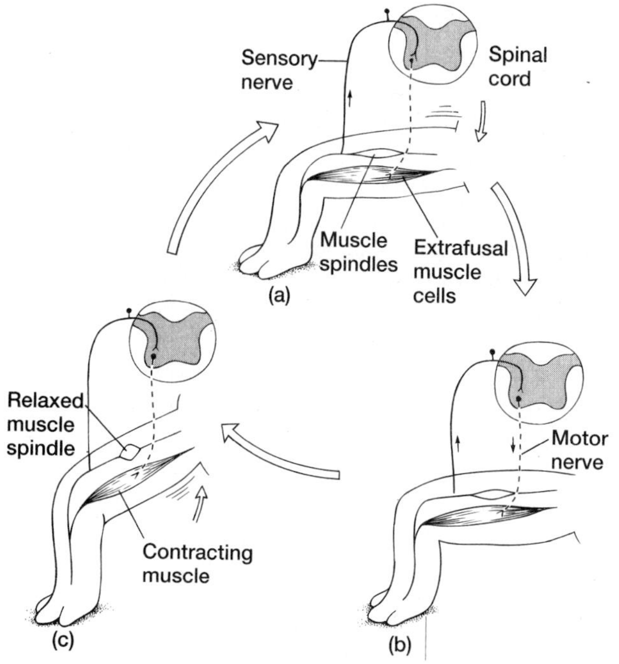


**Figure 17.4** Stretch reflex. (a) When posture changes or a load is placed on an animal's body, muscle spindles are stretched, stimulating associated nerves to generate continuous impulses that travel to the spinal cord (solid line). (b) These impulses travel via afferent fibers to the spinal cord and synapse with motor neurons that conduct these impulses to appropriate extrafusal muscles (dashed line) causing them to contract. (c) Muscle contraction tends to straighten the limb (small open arrow) and relieve the stretch on the muscle spindle. When the muscle spindle relaxes, the associated sensory nerves cease firing and appropriate posture returns.

## Mechanisms of Perceiving Stimuli from General Sensory Receptors

Two theories attempt to explain the relationship between the stimuli that general sensory receptors receive and the perception that the central nervous system produces from them. The **theory of specific nerve energies** proposes that the nerve endings of each sensory receptor are associated exclusively with a specific sense. For instance, stimulation of a Meissner's corpuscle sets up a volley of impulses to the nervous system that, because they come from this type of receptor, are interpreted as tactile stimuli. Impulses from the end-bulb of Krause are interpreted as cold, those from the corpuscle of Ruffini, as hot, and so on.

The alternative concept is the **pattern theory of sensation** through which small complexes of nerve endings are associated with a particular location. Stimulation of the receptors within a specific location makes it possible for different combinations or patterns of sensations to be sent simultaneously to the nervous system, allowing for qualitative differences in interpretation. For example, we recognize that sensations of pain vary in quality and intensity. Some pains are "sharp," whereas others are "dull," or we may sense a "burning" pain. Both theories help to explain how the central nervous system interprets general sensory stimuli.

# Special Sensory Organs

Special sensory organs are usually localized in their distribution, and their responses are restricted to specific stimuli. There are chemical, electromagnetic, mechanical, and electrical stimuli to which sensory organs respond.

## Chemoreceptors

Sensory receptors sensitive to chemical stimuli are **chemoreceptors.** When a chemical contacts an appropriate receptor, it initiates an electrical impulse in the sensory neuron, although just how this occurs is still uncertain. Taste and smell are the most familiar chemoreceptive senses in humans, but this distinction is misleading. The sense of taste recognizes only four basic qualities—salty, sweet, sour, and bitter. What we interpret as the rich "taste" of food is primarily due to the mechanical texture of the food and to the aroma that stimulates our sense of smell. If we are stricken with the flu, which clogs our noses and denies us these stimuli, food loses much of its "taste."

In aquatic vertebrates, the distinction between the sense of taste and smell is even less useful. Some fishes, for example, have chemoreceptors distributed across the surface of their bodies (figure 17.5). Should we say these are used to "smell" or to "taste" the water? Instead of making an arbitrary distinction between taste and smell, we classify chemoreceptors by their locations.

### *Nasal Passages*

The sense of smell, or *olfaction*, involves chemoreceptors usually located in the nasal passages. Anatomically, there are three components of the olfactory circuitry—the olfactory epithelium, the olfactory bulb, and the olfactory tract (figure 17.6).

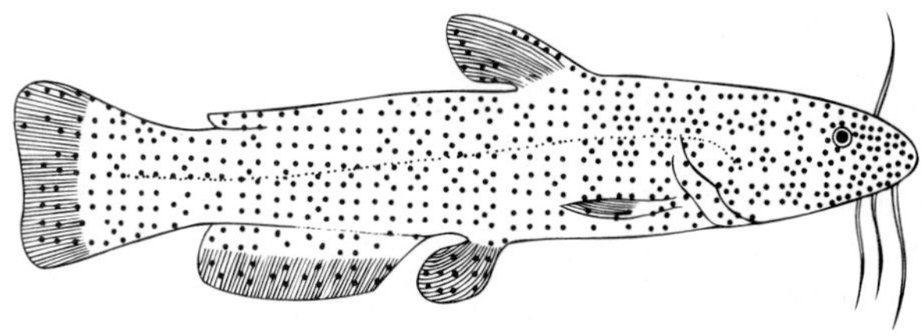

**Figure 17.5** Distribution of chemoreceptors in the catfish. In many fishes, taste buds occur across the surface of the body and on the fins as well. Each dot represents approximately 100 taste buds.

Modifed from Atema, 1971.

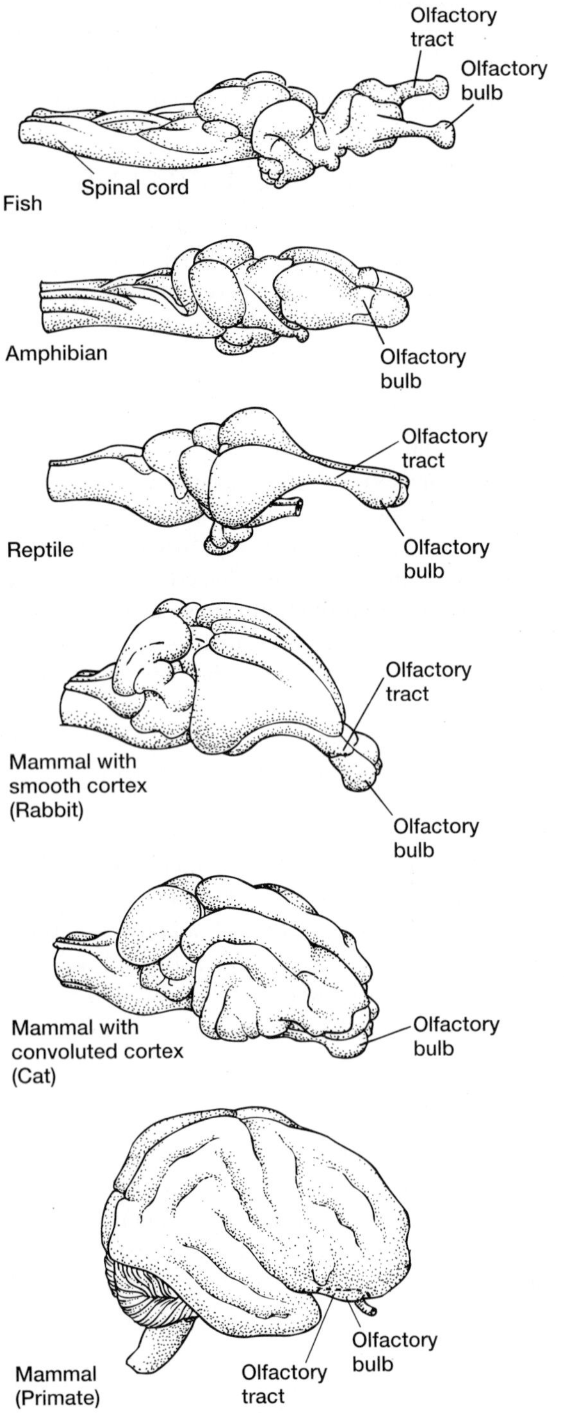


**Figure 17.6** Olfactory bulb and tract. The olfactory tract is an extension of the brain rather than a nerve. The tip of this tract is usually expanded into the olfactory bulb, which receives the short olfactory nerve (not shown) arriving from the olfactory epithelium.

The **olfactory epithelium** is a specialized patch of epithelium within the nasal cavity that contains **basal cells,** which are probably replacement cells, and **sustentacular cells,** which secrete mucus and support the **olfactory sensory cells** (figure 17.7a). The actual chemoreceptor cells in the epithelium are olfactory sensory cells. Each olfactory cell sprouts a tuft of sensory cilia at its apical ends. At its basal end, it sends an axon through the **cribriform plate** into the **olfactory bulb** (figure 17.7b). The term ***olfactory nerve*** is properly applied only to these short axons from the olfactory sensory cells.

Within the enlarged olfactory bulb reside several cell types, the most important of which is the **mitral cell.** Axons of the olfactory sensory cells synapse with mitral cells, which in turn send their long axons, collectively termed the **olfactory tract,** to the rest of the brain (figure 17.7b). Axons in the olfactory tract synapse primarily within the **piriform lobe** and **septum** in the cerebrum before being relayed to other regions of the brain (figure 17.7a). This gives the striatum and limbic systems direct olfactory input.

Olfactory centers (p. 654); limbic system (p. 659)

***Embryology*** The olfactory system begins embryologically as a pair of **olfactory placodes,** thickenings of ectoderm that invaginate dorsally toward the overlying neural tube. The lateral walls of each placode form the respiratory epithelium lining the nasal passages. The central region of the placode forms the olfactory epithelium. Olfactory sensory cells differentiate within this epithelium and sprout axons that grow from the epithelium and through the mesenchyme to reach the forming telencephalon (figure 17.8a). These olfactory fibers induce the telencephalon to produce a swollen outgrowth, the olfactory bulb, that is connected to the rest of the telencephalon by the olfactory tract (figure 17.8b,c). Although neural crest cells may migrate to the vicinity of the differentiating olfactory system, they apparently do not directly form the olfactory sensory cells.

***Phylogeny*** In tetrapods, the nose is associated with breathing, but primitively it arose as an olfactory area. In fact, the projection of olfactory receptors to the medial pallium, one of the earliest regions of the cerebrum, suggests that the olfactory system is quite ancient.

In most fishes, olfactory sensory receptors are recessed in paired, blind-ended pits known as the **nasal sacs** (figures 17.9a and 17.10a). In living cyclostomes and many ostracoderms, these pits were secondarily fused, although the olfactory tracts remained paired. Water carrying chemicals flows in and out of these sacs as the fish swims. However, among some groups of fishes, a one-way flow of water through the nasal sacs is possible. A partial septum divides the sac into incurrent and excurrent apertures. In some lines, these openings become separated (figure 17.9b–d). The excurrent opening may become displaced to the margin of the mouth (e.g., in some crossopterygians) or even open directly into

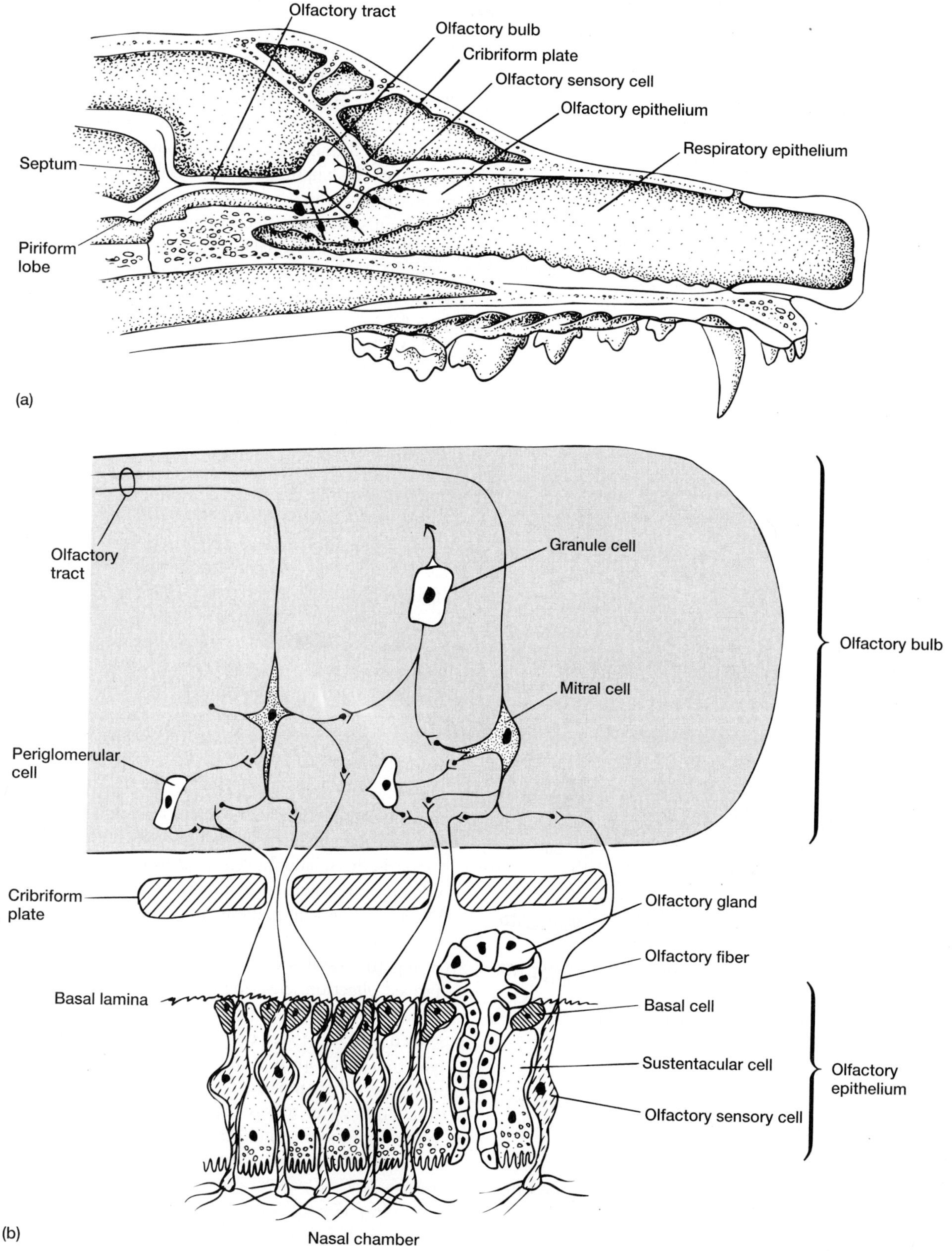


**Figure 17.7** Olfactory epithelium. (a) Nasal passages in mammals are lined with respiratory epithelium. The olfactory epithelium is a small region of this lining that contains specialized neuronal fibers that make contact with neurons of the olfactory tract. These processes relay impulses to the piriform lobe and septal area of the brain. (b) Histology of olfactory epithelium. The olfactory epithelium includes supportive sustentacular cells, basal cells, and olfactory sensory cells. The apical surface of each olfactory cell develops cilia that project into the air passage. Its basal end consists of a nerve fiber that travels through the cribriform plate into the olfactory bulb, where it synapses with periglomerular, mitral, and granule cells. Fibers of the mitral cells constitute the olfactory tract, which goes to the brain.

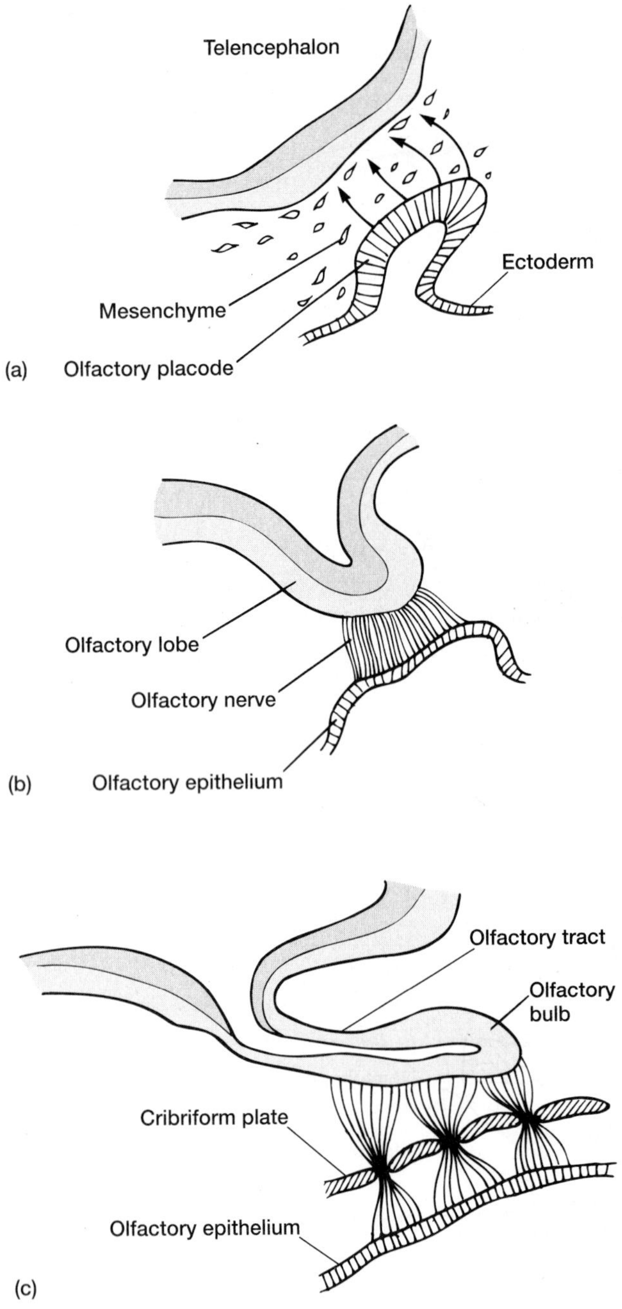


**Figure 17.8** Embryonic development of the olfactory system. (a) Thickening of ectoderm forms the olfactory placode. Cells within it sprout nerve fibers that grow into the nearby telencephalon. (b) These fibers collectively form the olfactory nerve. (c) The outgrowth of the telencephalon that receives the olfactory nerve is the olfactory bulb. The olfactory tract connects the olfactory bulb to the brain.

the buccal cavity (e.g., in holocephalians). The excurrent opening into the mouth has happened several times independently in fish evolution and led to a reevaluation of homologies that is still underway.

**Nasal sac, nasolacrimal duct (p. 245)**

In tetrapods and their immediate fish ancestors, called choanate fishes, a small **external naris** provides access to each nasal passage. The back of the nasal passage opens into the mouth through the **internal naris,** or **choana** (figure 17.10b–e). In amphibians, the nasal sac enlarges between nares, from which a short recess, the **vomeronasal (Jacobson's) organ** projects (figure 17.10c).

In plethodontid salamanders, a pair of sunken grooves connects the front of the mouth with each external naris. These **nasolabial grooves** are thought to deliver chemicals from food in the mouth to the olfactory epithelium (figure 17.11). The groove is unciliated, and fluid apparently moves through it by capillary action.

In reptiles, the nasal sac becomes differentiated into two regions, the anterior **vestibule** that first receives air entering via the external naris, and the posterior **nasal chamber** into which air next flows. In some reptiles, a lateral wall projects into the chamber to form **conchae** or **turbinals,** which are folds that increase the surface area of the respiratory epithelium. Air departs the nasal chamber through the narrow **nasopharyngeal duct** that leads to the internal naris (figure 17.10d). The nasal passages of birds are similar, and their conchae may develop into complicated scrolls.

In mammals, the nasal chamber is large and usually includes extensive turbinals to ensure that the entering air will be warmed and moistened before it flows to the lungs. The olfactory epithelium occupies the posterior wall of the nasal chamber. The remaining lining consists of respiratory epithelium (figure 17.10e).

***Form and Function*** The sense of smell is usually well developed among fishes, but it is one of the secondary senses in birds, bats, and higher primates, including ourselves. Olfactory information is important for fishes who hunt or follow a chemical gradient. In aquatic vertebrates, waterborne chemicals circulate across the olfactory epithelium lining the nasal sacs. Development of one-way flow improved olfaction, ensuring a fairly continuous flow of new water to wash away chemicals that have already been detected and to deliver new ones. Respiratory and olfactory currents are coupled in fishes in whom the excurrent opening enters their mouth. The motions that irrigate the gills also draw water across the olfactory epithelium. By coupling respiratory and olfactory functions, the nasal passages were preadapted for their later role in air breathing in tetrapods.

In tetrapods, air replaces water in chemical transport, although before reaching the sensory receptors, airborne particles are still absorbed into a mucous film covering the olfactory epithelium. Air entering the nostrils must flow past the olfactory epithelium on its way to the lungs; therefore, the olfactory epithelium, can sample chemicals in the airflow. A terrestrial vertebrate may begin to sniff the air when it detects chemicals of special interest. Sniffing is independent of respiration and draws in quick pulses of air to replenish air in the nasal chamber. This increases the

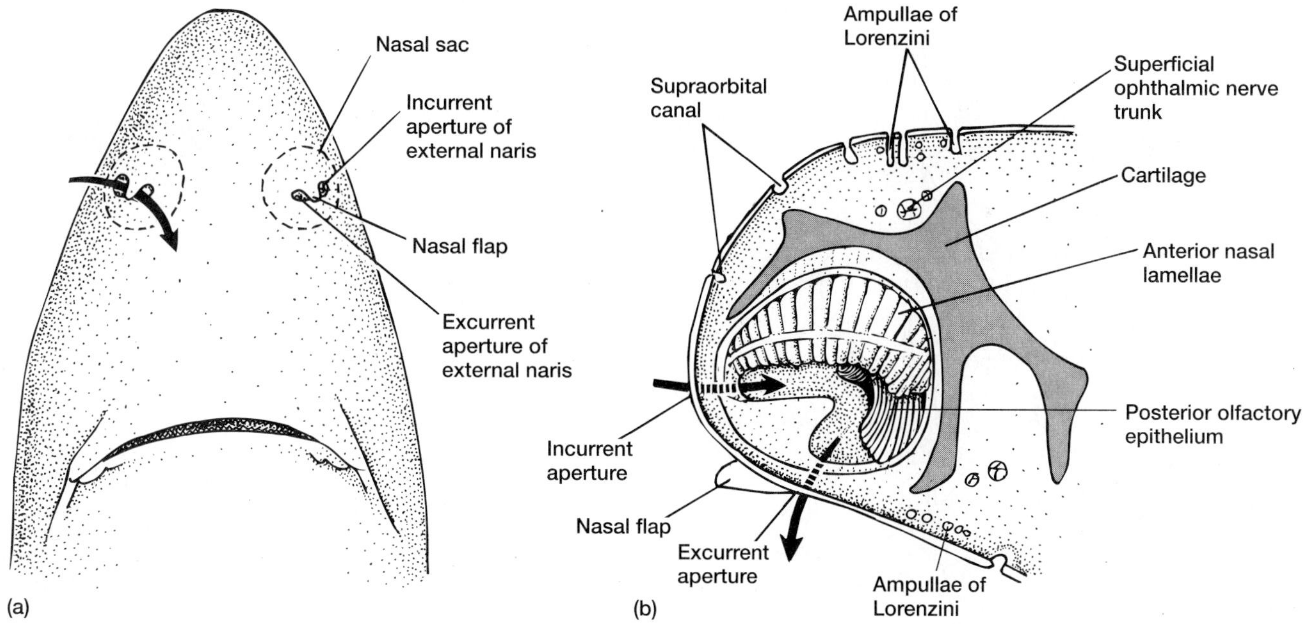


**Figure 17.9** Nasal sacs of a shark. (a) Ventral view of shark head showing the direction water (solid arrow) flows through the nasal sac and across the olfactory epithelium. (b) Cross section of the nasal cavity. (c) Olfactory epithelium showing bipolar sensory cells and associated neurons. (d) Solid arrows indicate the flow of water through the nasal sac. Progressive stages in the establishment of one-way flow across the nasal epithelium in various fishes. *continued.*

turnover of air in the nasal chamber and permits more frequent sampling of environmental odors (figure 17.12).

## *Vomeronasal Area*

Named for the bones that usually house this chemosensory organ, the vomeronasal organ is known only in some tetrapods. It is absent in most turtles, crocodiles, birds, some bats, primates, and aquatic mammals. In amphibians, it is in a recessed area off the main nasal cavity. In reptiles, the vomeronasal organ is a separate pit to which the tongue and oral membranes deliver chemicals. In mammals possessing this organ, it is an isolated area of olfactory membrane within the nasal cavity that is usually connected to the mouth via the **nasopalatine duct** (figure 17.13).

The vomeronasal organ is an accessory olfactory system. It includes basal, sustentacular, and bipolar sensory cells similar to those of the olfactory epithelium. Sensory receptor cells of the vomeronasal organ project into the lumen by means of microvilli, whereas olfactory sensory cells have cilia. The neural circuitry of the vomeronasal system runs parallel to but remains entirely separate from the main olfactory system. Like the main olfactory system, the vomeronasal system can be traced through the limbic system to the hypothalamus and thalamus.

**Thalamus and hypothalamus (p. 652)**

In many vertebrates with a vomeronasal organ, the respiratory airflow carries particles to the organ. But the vomeronasal organ may also establish an association with the mouth, leading some to suggest that it may sense the chemical composition of food in the buccal cavity. However, the vomeronasal also seems to be especially sensitive to chemicals important in social or reproductive behavior. Theoretically, individuals lacking this organ could increase their respiratory ventilation rate to draw more frequent samples of air across the olfactory epithelium, but this would be energetically more expensive and might result in problems with regulation of blood pH. In order for us to recognize the anatomical and functional distinctiveness of the nasal epithelium and the vomeronasal organ, we refer to the chemicals detected as **odors** and **vomodors** and to the processes as **olfaction** and **vomerolfaction,** respectively.

Snakes and lizards extend their tongue from the mouth, sweep air in front of their snout to collect vomodors, and then retract their tongue into the mouth to deliver these chemicals to the vomeronasal organ (figure 17.14a–d). In snakes and possibly in lizards, the returning tongue does not directly enter the lumen of the vomeronasal organ. Instead, the reptile wipes its tongue against the entry ducts to the organ and against small ridges in the lower part of the mouth. These ridges are lifted up to the ducts, apparently adding vomodors they have gathered to those the tongue delivered directly to the entry ducts. In snakes, removal or inactivation of the vomeronasal organ leads to deficits in courtship, pheromonal trailing, and prey detection.

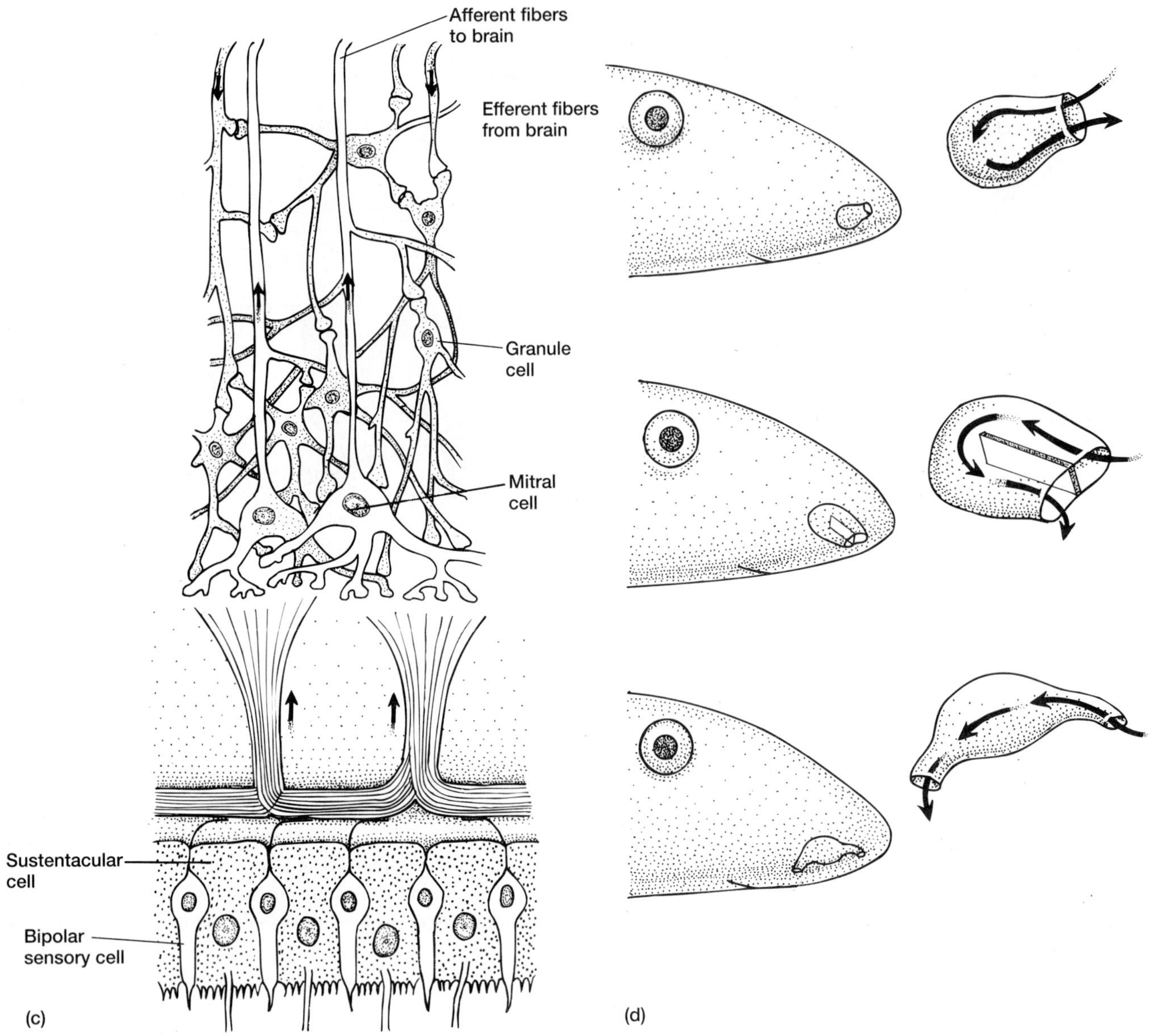


**Figure 17.9** (continued)

## *Mouth*

Taste, like olfaction, centers on the detection of chemical stimuli by chemoreceptors. But the chemoreceptors of taste are **taste buds** located in the mouth. In amphibians, reptiles, and birds, taste buds occur in the mouth and pharynx. Mammalian taste buds tend to be distributed throughout the tongue.

In mammals, three separate cranial nerves deliver sensory information from taste buds to the nervous system—the **facial, vagus,** and **glossopharyngeal nerves** (figure 17.15a). Each taste bud, which pokes through the epithelium via a **taste pore,** is a barrel-shaped collection of 20 or more cells of three types. Basal cells located at the base or periphery of taste buds are thought to be stem cells, that is, cells that replace the other cell types. The life span of taste bud cells is about a week; thus, replacement is a continuous process. Sustentacular cells (dark cells) are supportive and secretory in function. **Gustatory cells** (light cells) are thought to be the primary chemoreceptive cells of the taste bud. Taste bud cells do not have axons. Instead, sensory fibers of the three cranial nerves entwine around these three cell types and establish special synapselike contacts with the gustatory cells (figure 17.15b,c).

Collectively, the taste buds sense chemical stimuli and act as transducers to initiate electrical impulses that travel via the cranial nerves to the central nervous system. Sweet, sour, salty, and bitter substances send different patterns of electrical volleys to the nervous system.

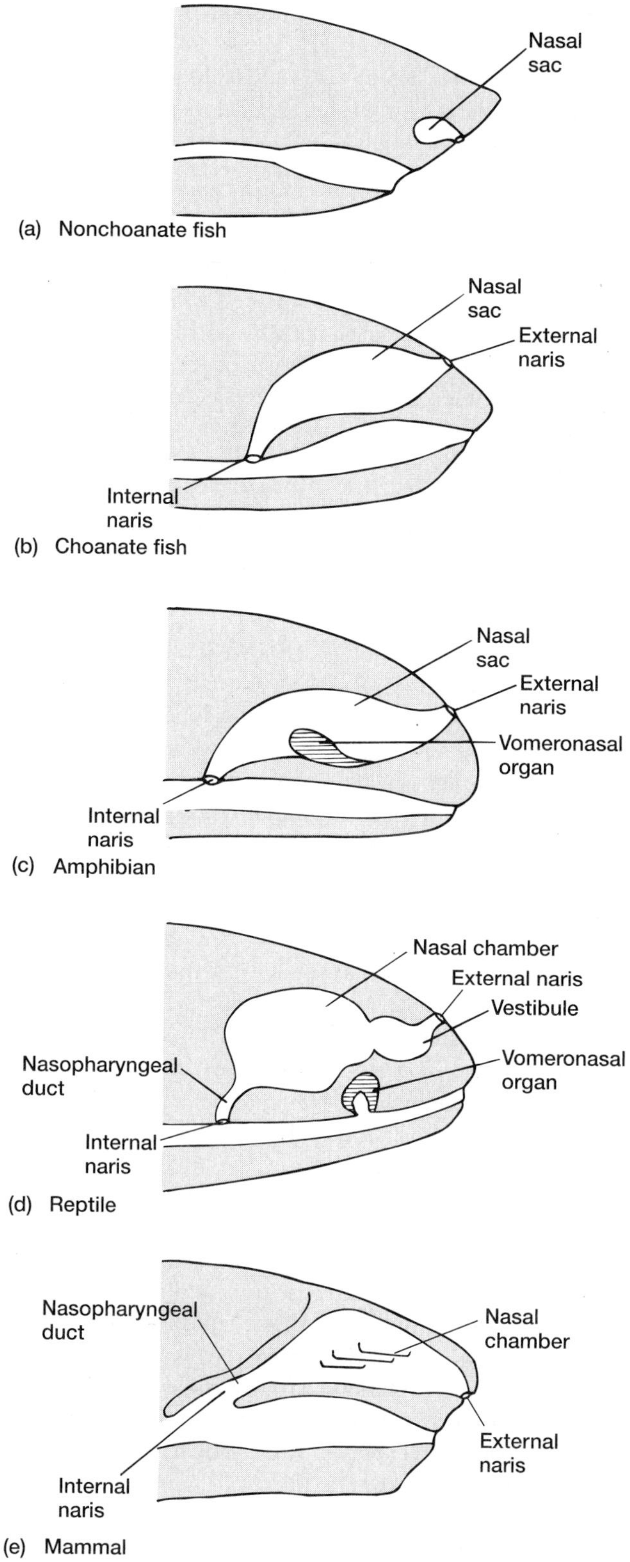


**Figure 17.10** Phylogeny of olfactory organs. Note that the vomeronasal organ is absent in fishes but present in most tetrapods. (a) Nonchoanate fish. (b) Choanate fish. (c) Amphibian. (d) Reptile. (e) Mammal.

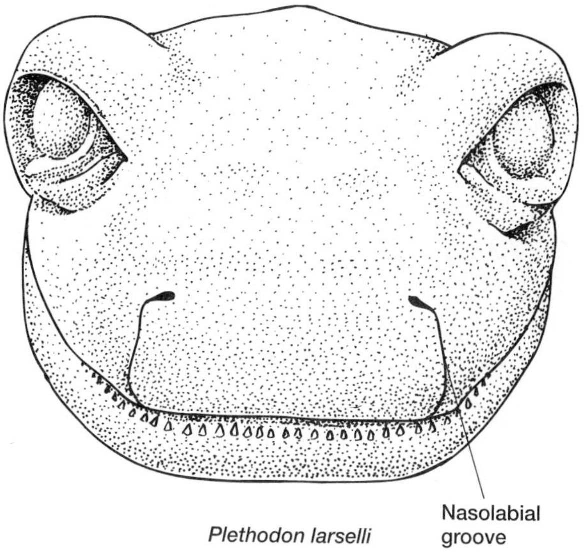


**Figure 17.11** Head of Larch Mountain salamander *Plethodon larselli*. In plethodontid salamanders, a nasolabial groove runs between the mouth and nostrils. It is thought to convey chemicals from mouth to nose.

Based on photographs supplied by J. H. Larsen.

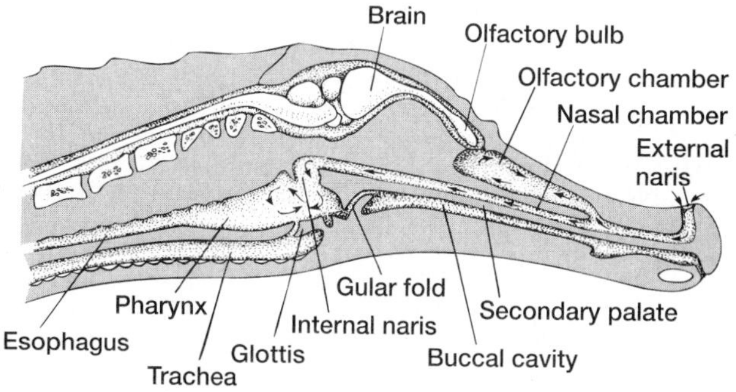


**Figure 17.12** Sniffing by the crocodile. In tetrapods, olfaction often depends on the arival of new chemicals with each respiratory exchange of air to the lungs. Sniffing allows more frequent sampling of air without increasing respiratory rate. The crocodile can close both the glottis and the gular fold, momentarily isolating the nostrils and the mouth. By depressing the floor of the pharynx, fresh air can be drawn just into the olfactory chamber and new chemicals sampled without respiratory ventilation.

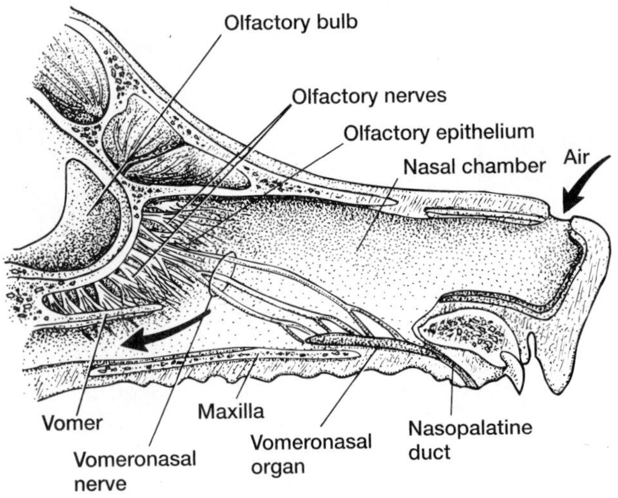


**Figure 17.13** Vomeronasal organ of a dog. The nasopalatine duct passing through the incisive foramen joins the mouth with the nasal chamber. Olfactory nerves travel to a restricted area of the olfactory epithelium.

## Radiation Receptors

Radiation travels in waves. Cosmic radiation has the shortest wavelength and radio waves have the longest. Together with intermediate wavelengths, these constitute the **electromagnetic spectrum.** From the standpoint of an organism, radiation carries information about its intensity, wavelength, and direction.

No organism taps the full range of information available throughout the electromagnetic spectrum. Organisms can perceive only a limited range of wavelengths. Honeybees can see ultraviolet radiation. Some vertebrates (e.g., pit vipers) can detect infrared. Most vertebrates can perceive only a narrow band of electromagnetic radiation between about 380 and 760 nm. This restricted band is called "visible" light, meaning that we can see it (figure 17.16). When we talk of "light," we are being quite provincial because we really mean this very narrow range of electromagnetic radiation out of a much wider spectrum. Similarly, when we speak of the sense of "vision," we are referring to the ability to perceive light within this narrow range.

Vertebrates have evolved a variety of sensory organs that gather electromagnetic radiation. Different regions of the spectrum represent different energies (figure 17.16) and present different levels of stimuli to sensory receptors.

### *Photoreceptors*

**Eye** **Photoreceptors** are sensitive to light. The most obvious and best understood photoreceptor is the eye. The vertebrate eye can focus light on photosensitive cells to form an image of the environment. The capacity to focus light on objects at different distances is called **visual accommodation.**

The nervous system takes advantage of the physical properties of light to interpret the images presented to it. Differences in light intensity are interpreted as contrasts. Within the visible spectrum, different wavelengths are interpreted as different colors.

***Structure of the Eye*** The mammalian eye has three layers—sclera, uvea, and retina (figure 17.17a).

*Sclera.* The outer layer of the eye is the **sclera.** It forms the "white of the eye" and consists of a tough capsule of connective tissue to which the extrinsic ocular muscles attach. Contractions of these muscles rotate the eyeball in its orbit to direct the gaze toward an object of interest. The sclera helps define the shape of the eyeball. In birds, reptiles, and fishes, small plates of bone called **scleral ossicles** are often present to help hold the shape of the sclera. At the front of the eye, the sclera clears to become the transparent **cornea.**

*Uvea.* The middle layer of the eye is the **uvea,** which is composed of three regions. The **choroid** adjacent to the retina is the largest region. Because it is highly vascular, the choroid provides nutritional support to the ocular tissues. The choroid is pigmented. In some nocturnal vertebrates, it includes a special reflective material, the **tapetum lucidum.** Under conditions of dim light, this structure reflects the limited light to restimulate the light-sensitive cells in the retina. The tapetum lucidum produces the "eye shine" of mammals seen at night in car headlights or in flashlight beams.

The second region of the uvea is the **ciliary body,** a tiny circle of smooth muscle around the interior of the eyeball. The **ciliary muscle** controls visual accommodation. It is attached to the flexible **lens** through a circular **suspensory ligament.** Tension on the lens tends to deform it, whereas relaxation allows the lens to restore its shape elastically.

The third region of the uvea is the **iris,** a thin continuation of the uvea across the front of the eyeball. The **pupil** is not a structure but an opening defined by the free edge of the iris. Tiny smooth muscles within the iris act like a diaphragm to reduce or enlarge the size of the pupil and regulate the amount of light that enters the eye.

*Retina.* The innermost layer of the eye is the photosensitive **retina,** which itself is composed of three cell layers. The deepest layer of cells within the retina contain the photoreceptor cells. These include **rods** sensitive to low levels of illumination but not colors, and **cones** sensitive to colors in bright light. The photoreceptor cells synapse with **horizontal** and **bipolar cells.** Proximal to these are layered **amacrine** and **ganglion cells.** This arrangement means that light entering the eye and falling on the retina passes sequentially through ganglion, amacrine, bipolar, and horizontal cells before it reaches the photoreceptive rods and cones (figure 17.17b). The functional significance, if any, of this arrangement is not known.

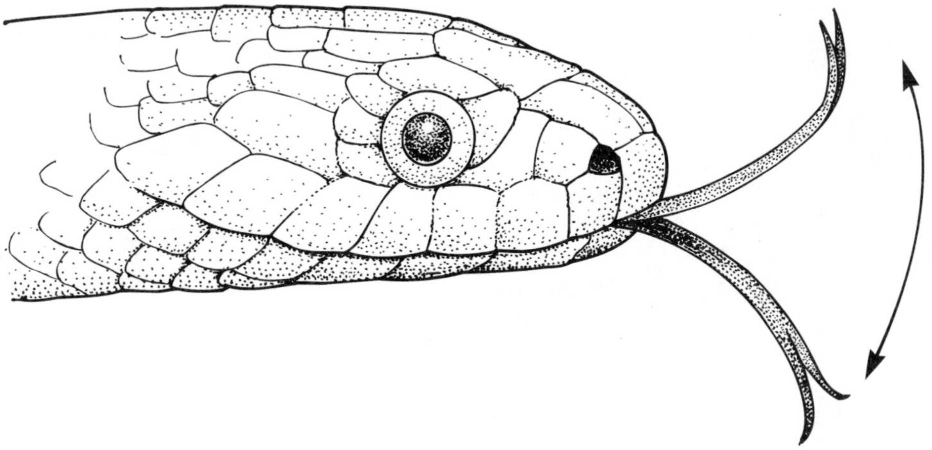


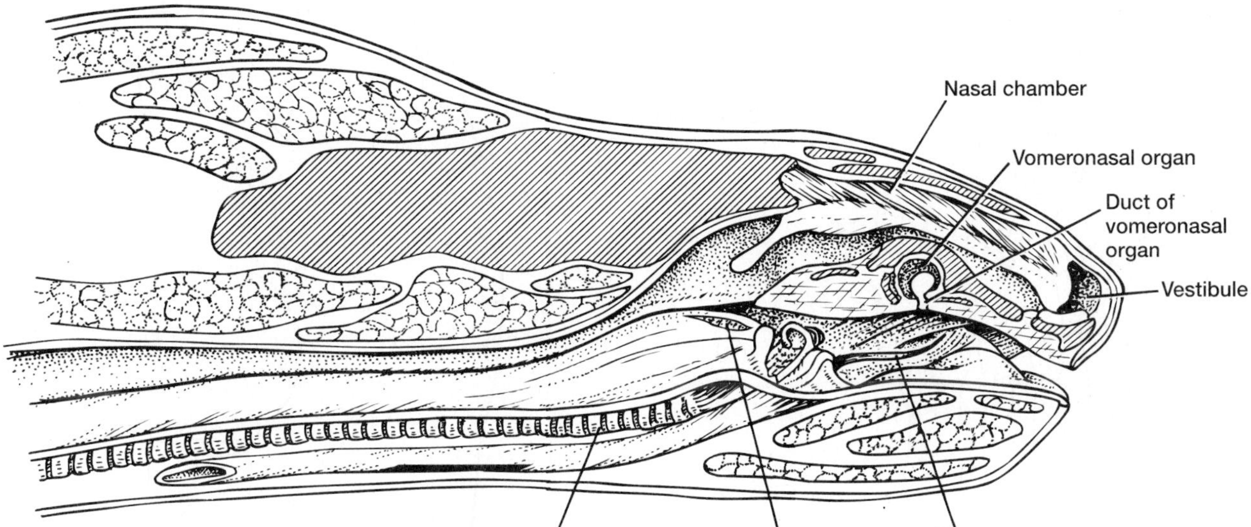


**Figure 17.14** Tongue flicking in snakes. (a) Snakes, like lizards, extend their tongues to sweep air in front of them. The tongue collects and then transports airborne particles into the mouth. Probably along with other oral membranes, the tongue wipes these particles onto the vomeronasal organ on the roof of the mouth. (b) Sagittal section of the head of a boa constrictor. The vomeronasal organ is a blind pocket with a lumen that opens directly into the mouth via a duct. The tip of the retracted tongue projects from its sheath beneath the trachea. (c) Skull and overlying tissue have been cut away to reveal dorsal view of the snake brain. (d) Neuroanatomy of a snake's olfactory organs. The main olfactory bulb receives input from the olfactory epithelium. The accessory olfactory bulb, via a separate tract, receives information from the vomeronasal organ. Vomeronasal and olfactory systems are separate chemoreceptive organs whose fibers travel separately within the olfactory tract. Thereafter, sensory information tends to be brought together in the olfactory cortex of the telencephalon. *continued.*

The retinas of all vertebrates possess rods (able to perceive black and white), but not all vertebrates have cones. Some fishes and amphibians have color vision, many reptiles and all birds possess it, but only a few mammals see in color. Humans and some other higher primates with color vision are exceptions.

In a few vertebrates, the retina is indented, forming a **fovea** (figure 17.17a). The fovea is the point at the back of the eyeball where light converges. Composed entirely of cones, it forms the point of sharpest focus. Although rods are absent from the fovea, they increase peripherally.

Within the eye are three chambers. Two lie in front of the lens, the **anterior chamber** between the iris and the cornea and the small **posterior chamber** between the iris and the lens. The third and largest **vitreal chamber** is located behind the lens. These chambers are filled with a transparent fluid that helps maintain the shape of the eyeball. The anterior and posterior chambers are filled with watery **aqueous humor.** The vitreal chamber contains a thick **vitreous humor,** sometimes called the **vitreous body** because it can be dissected from the eye as a single viscous plug.

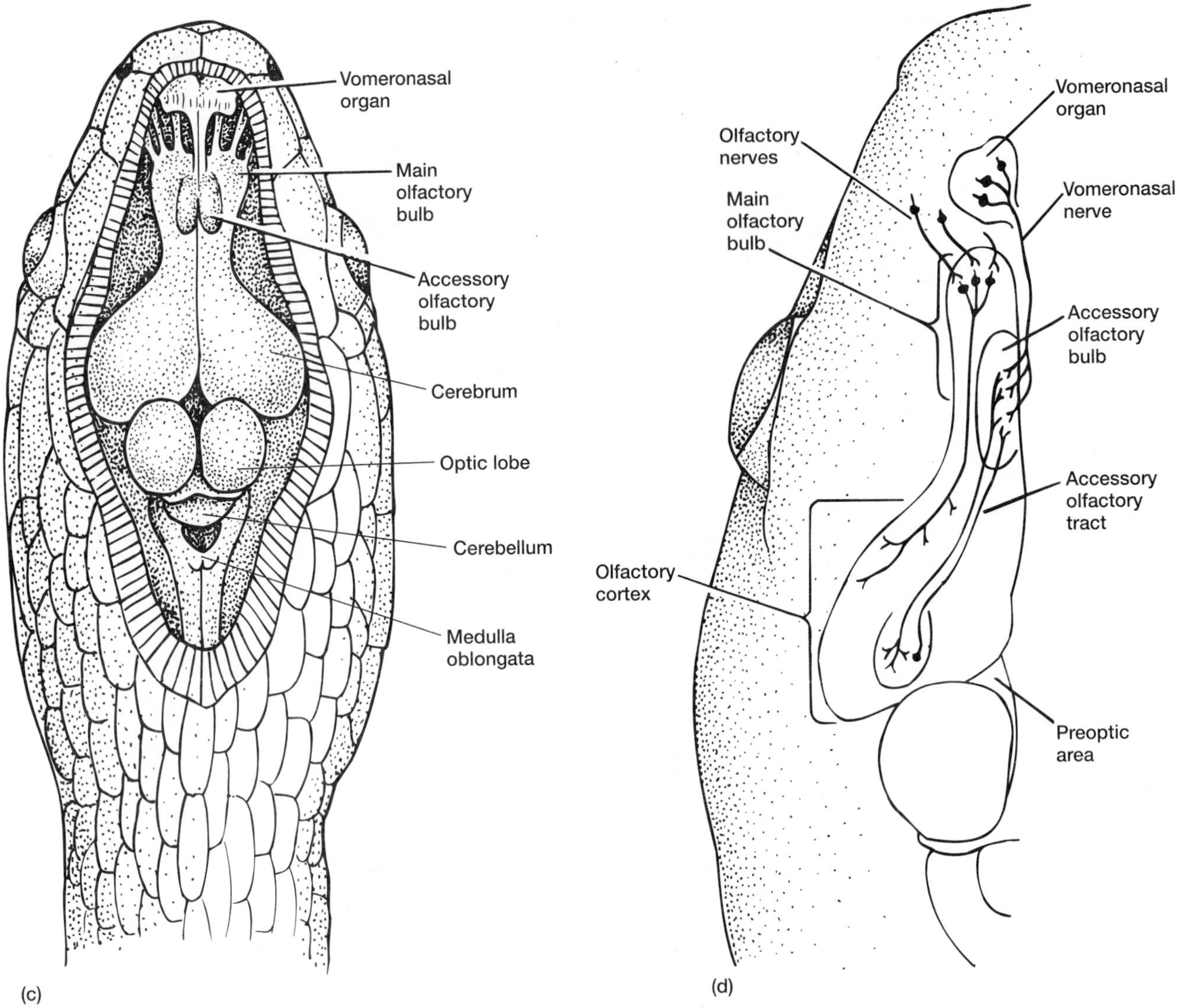


**Figure 17.14** (continued)

***Embryology*** Embryologically, the eye is a composite structure formed from surrounding mesenchyme and the optic placode, a thickened neuroectodermal tissue that gives rise to the optic vesicle. Development of the eye begins with the appearance of paired outgrowths, the **optic vesicles,** from the sides of the future telencephalon (figure 17.18a). As the optic vesicles approach the overlying ectoderm, it thickens to become the optic placode and invaginates to form the **lens primordium** (figure 17.18b). The optic placode pinches off to settle into an indentation, the **optic cup.** Mesenchyme surrounding the developing eyeball condenses to produce the outer coats of the eye (figure 17.18c,d).

Thus, the ectoderm gives rise to the eyelid, cornea, and lens. The mesenchyme forms the choroid and sclera, and the iris and retina develop from the optic cup. The optic vesicle retains its connection with the brain as the **optic stalk,** from which it initially arose. The optic stalk comes to carry the axons of ganglionic cells that project to optic areas in the diencephalon. Although this stalk is actually an extension of the brain and should therefore be termed a tract, in practice it is often termed the **optic "nerve"** and counted as the second cranial nerve.

Cranial nerves (p. 621)

***Phylogeny*** In lampreys, the striated **corneal muscle** of myotomal origin is attached to the spectacle, a clear area of skin over the cornea. Contraction of this muscle tends to draw the spectacle taut and flatten the cornea. This in turn pushes the lens closer to the retina. Accommodation is thus accomplished by deformation of the eyeball from outside. Upon relaxation of the corneal muscle, elasticity of the cornea and vitreous humor return the lens to a resting position (figure 17.19a).

In bony and cartilaginous fishes, the eyeball is often supported by scleral bones or cartilages. The lens is nearly

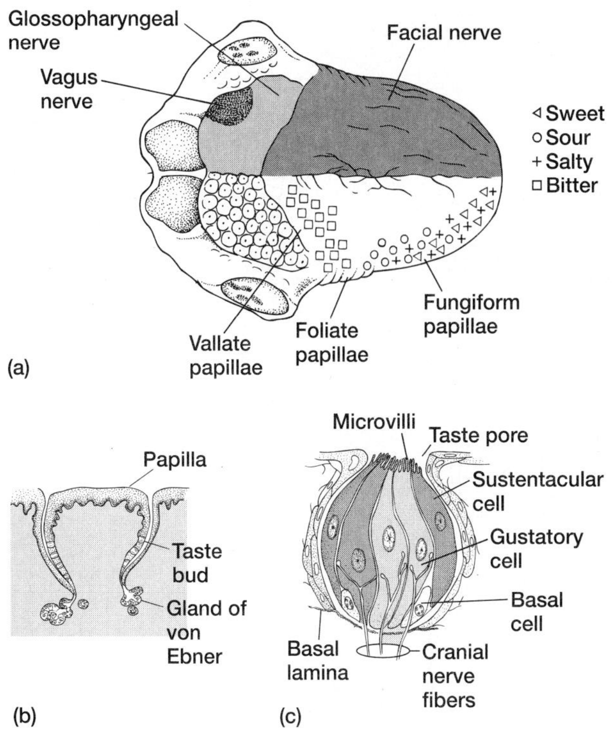


**Figure 17.15** Chemoreceptors of the mouth. (a) Distribution of taste buds on the human tongue. Taste buds detect four basic qualities—sweet, sour, salty, and bitter. Areas supplied by facial, glossopharyngeal, and vagus cranial nerves are indicated. (b) Taste buds reside along the recesses of papillae within the surface epithelium. (c) Each taste bud is made up of several dozen cells, including supportive sustentacular cells, basal cells, and sensory gustatory cells. Microvilli on the apical surfaces of the cells project through the surface epithelium. Afferent nerve fibers are associated with all these cells.

round (figure 17.19b,c). It is held by the suspensory ligament and moved by the **retractor lentis muscle,** which is inserted directly on the lens. To focus the image, the retractor lentis pulls the lens forward in elasmobranchs and backward in teleosts.

The amphibian eye is sometimes reinforced by a cup or ring of scleral cartilage. The lens is nearly round and held by a circular suspensory ligament (figure 17.19d). The retractor lentis muscle of amphibians is inserted on the base of the suspensory ligament instead of directly on the lens. The lens is normally focused on distant objects. In order for amphibians to view near objects, the retractor lentis pulls the lens forward.

Among amniotes, except snakes, the lens changes shape to accommodate the visual image. This usually involves contraction of the ciliary muscle, which may squeeze the lens to change its shape or act through the circular suspensory ligament to stretch the lens and make it flatter. Relaxation of the muscle allows the lens to return, thanks to its resilience, to a more rounded shape. In some reptiles and some birds, but not in mammals, scleral ossicles are present. They are particularly well developed in raptors that reach high rates of aerial speed. A **papillary cone** in reptiles or a **pecten** in birds is projected into the vitreous chamber from the back wall of the eye. These structures are thought to provide supplemental nutritional support for deep ocular tissues (figure 17.19e,f).

***Form and Function*** Water and air present fundamentally different challenges for vision. Water affects light in several ways. It carries dissolved and suspended materials that can block light. In marine waters, the limit of useful vision may be 30 m, but in turbi d freshwater rivers and lakes, the limit often falls to only around 1 to 2 m. The deeper an animal dives, the smaller is the amount of light penetration and the

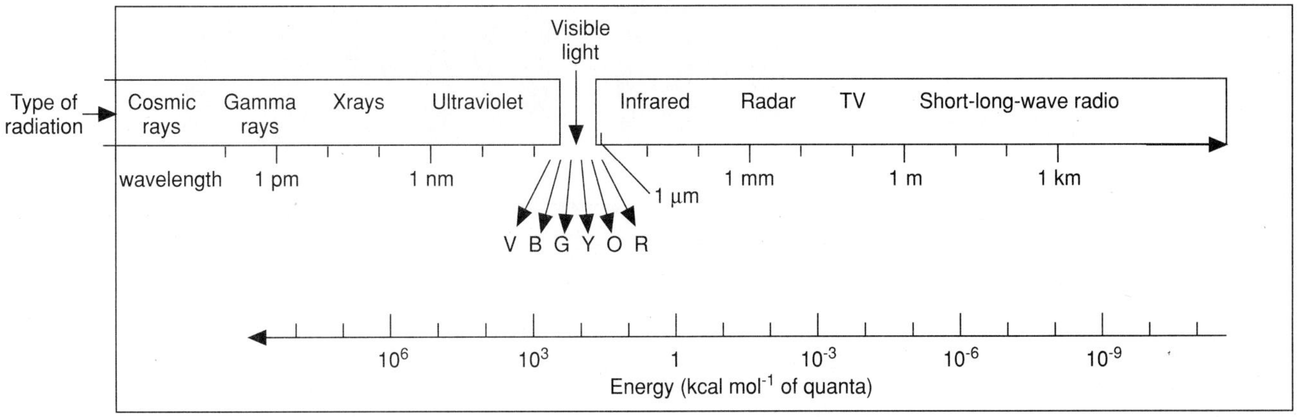


**Figure 17.16** The spectrum of electromagnetic radiation. Between extremely short cosmic rays and long radio waves lies a narrow band of "visible light," to which human eyes are normally sensitive. Wavelength increases to the right. Energy within the electromagnetic radiation increases to the left. Abbreviations: violet (V), blue (B), green (G), yellow (Y), orange (O), red (R).

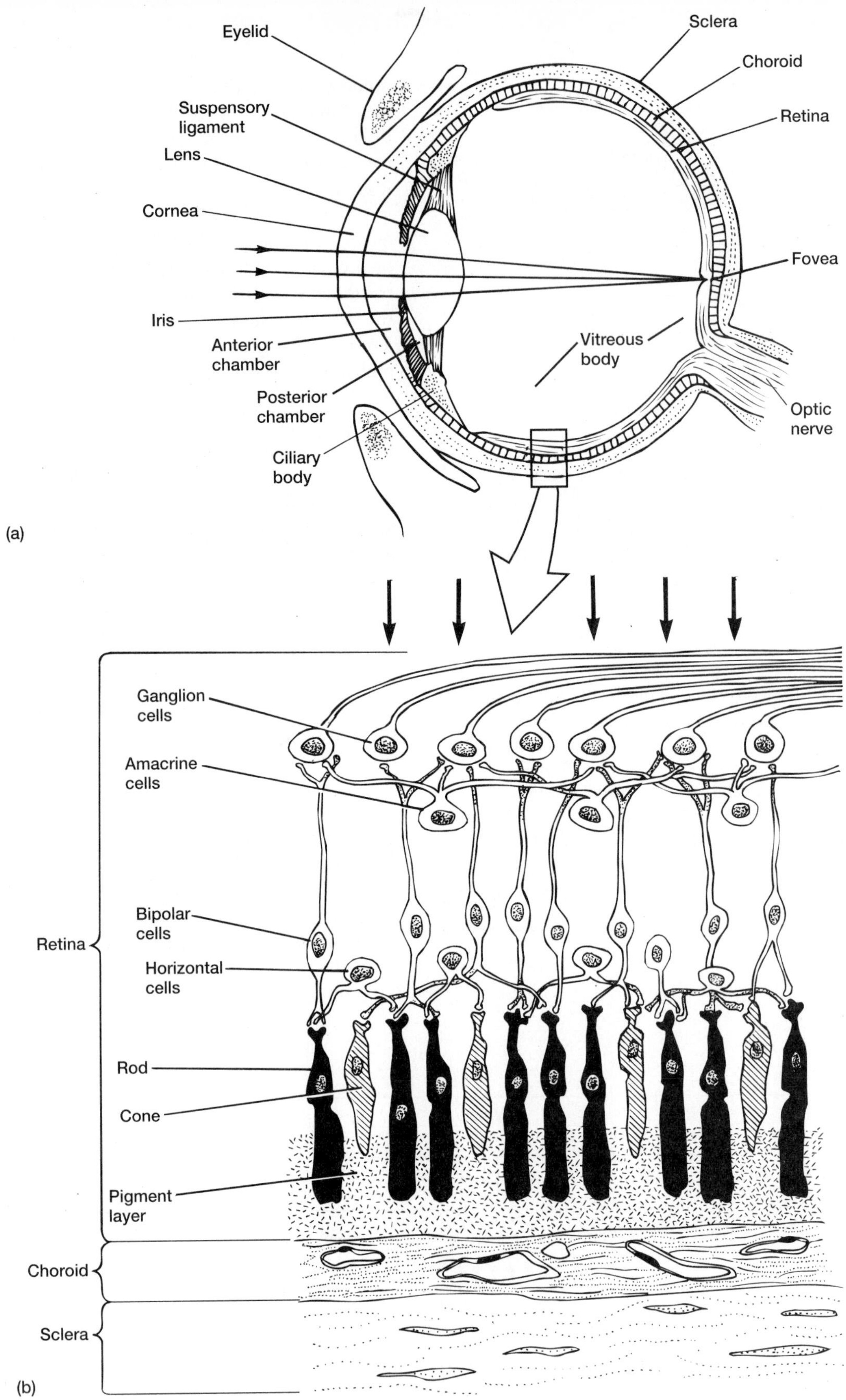


**Figure 17.17** Structure of a higher primate eye. (a) Cross section. (b) Enlarged view of layers in the back wall of the eyeball. Neurons indirectly connect light-sensitive rods and cones to ganglion cells that form the optic nerve. Black arrows indicate the path of entering light.

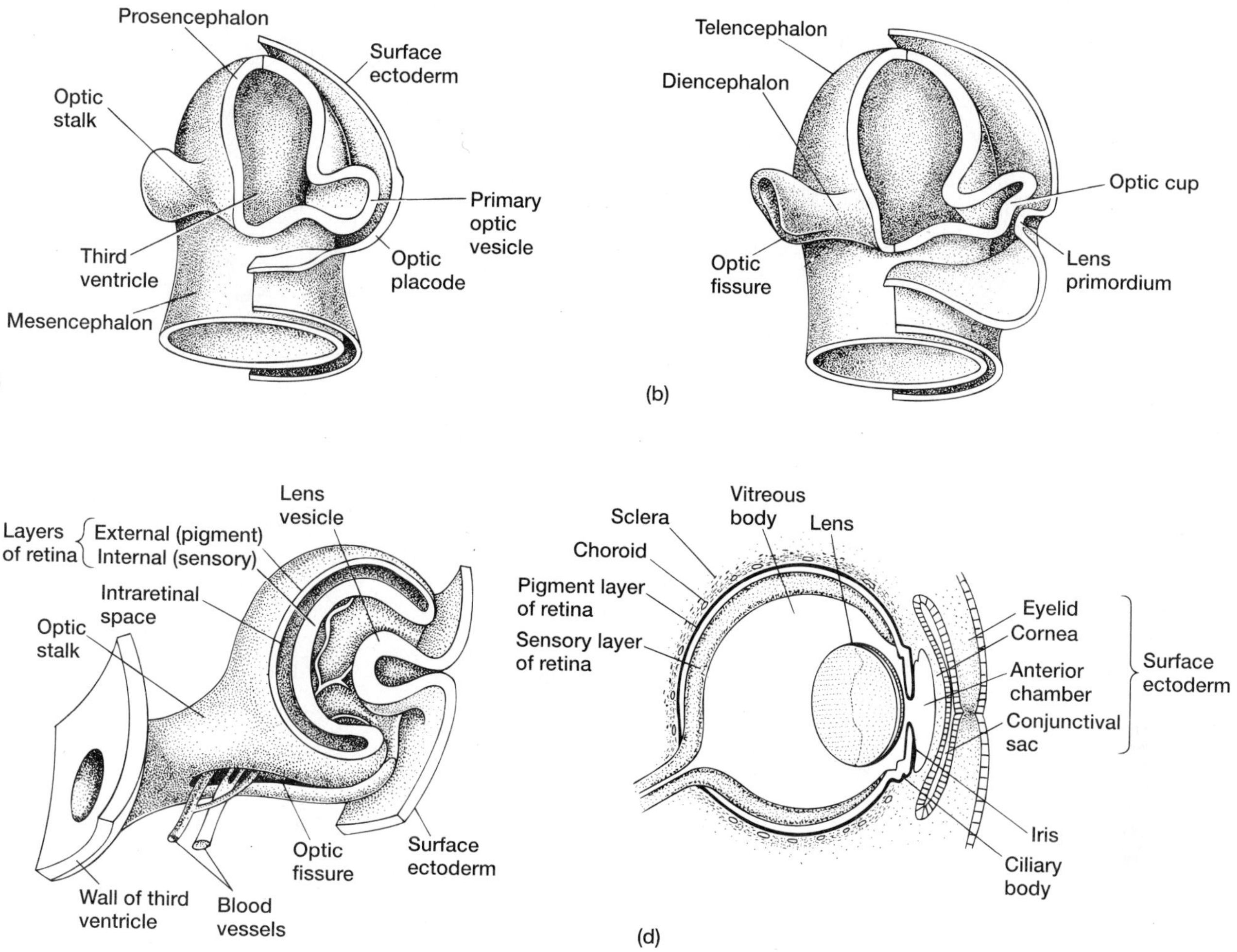


**Figure 17.18** Embryonic development of the vertebrate eye. (a) Early embryo in whom outwardly growing optic vesicles eventually meet the thickened optic placode in the ectoderm. (b) Slightly later, interaction between optic vesicle and optic placode leads to initial differentiation of the lens and what becomes a double-layered secondary optic vesicle or optic cup. (c,d) Successive stages in separation of the lens within the optic vesicle. The eyelid and cornea form, and the wall of the eyeball differentiates into distinct layers. After the lens is pinched off, openings appear within the surface epithelium to delineate the cornea and anterior chamber.

dimmer is the water. Light intensity diminishes selectively with depth. The first wavelength to be absorbed is ultraviolet, and then infrared, red, orange, yellow, green, and finally blue. Virtually no sunlight penetrates below 1,100 m even in clear water. Nevertheless, most fishes at depths greater than 1,100 m possess large and complex eyes, but they do not detect sunlight because light never penetrates to such great depths. Instead, they respond to flashes of bioluminescent light that the fishes themselves produce.

*Visual accommodation.* Water and air also differ in their effects on visual accommodation. To focus an image, light rays must be diverted, or "bent," from their normal parallel lines of travel to converge on the retina. For vertebrates on land, the difference in the **refractive indices** of the eye and air is pronounced, so that light striking the cornea is bent abruptly. As a result, the cornea of terrestrial vertebrates does most of the focusing (figure 17.20a). The lens merely refines the image to bring it into clear focus on the retina. However, in aquatic vertebrates, the cornea contributes very little to focusing. The refractive indices of water and the cornea are nearly equal. Therefore, the lens of fishes refracts most light. The refractive index of the lens is well above that of water thanks to its crystalline structure and considerable thickness (figure 17.20b).

Optics (p. 146)

Focusing in terrestrial vertebrates, except for snakes, depends on changes in lens *shape* (figure 17.21c–e). In contrast to this, focusing in fishes depends on changes in lens *position* within the eye (figure 17.21a,b). It is unclear whether the fish eye at rest is focused on close or distant objects. Different mechanisms of accommodation are

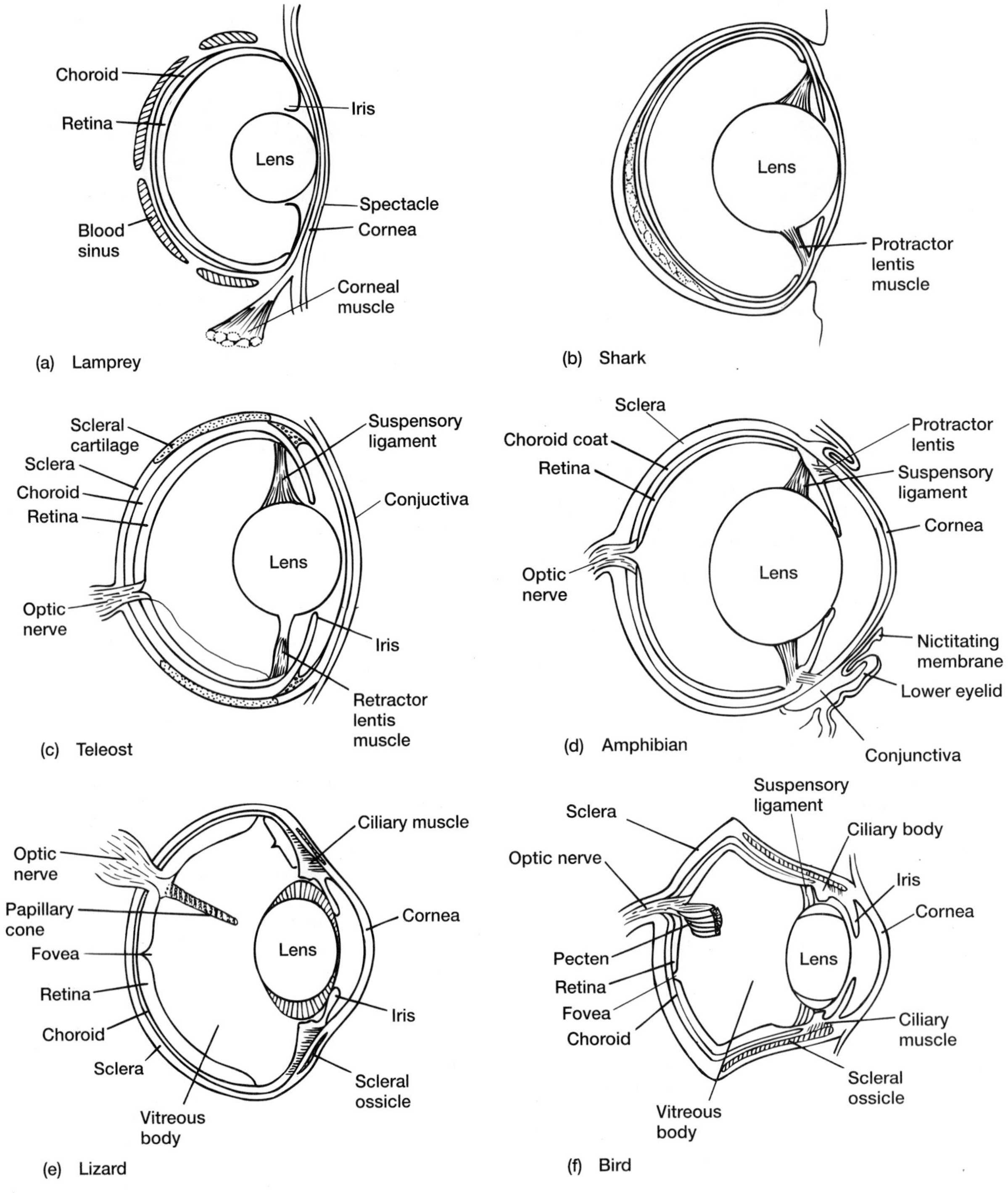


**Figure 17.19** Cross sections of vertebrate eyes. (a) Lamprey. (b) Shark *Squalus*. (c) Teleost. (d) Amphibian. (e) Lizard. (f) Bird.

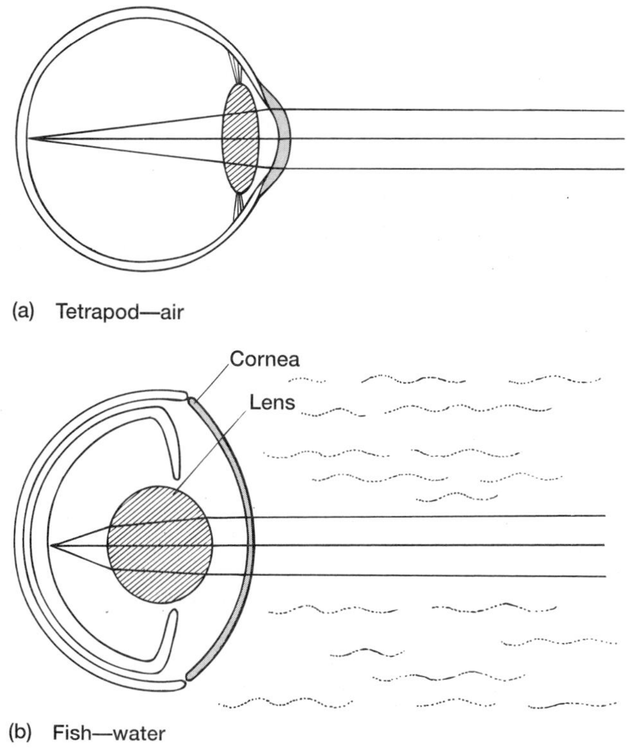


**Figure 17.20** Vision in air and water. (a) Air. Light that passes through air is strongly refracted when it passes through the cornea; therefore, the cornea is primarily responsible for focusing light rays. The lens fine-tunes the focused image. (b) Water. Because the cornea has refractive properties similar to water, incoming light is affected very little when it first enters the eye; therefore, the large lens bears primary responsibility for bringing light rays into focus on the retina.

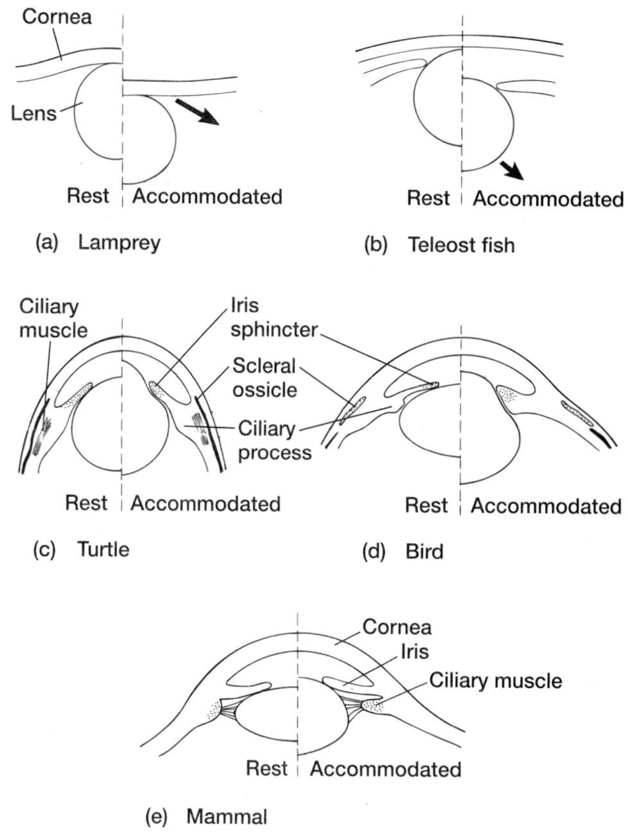


**Figure 17.21** Accommodation in vertebrates. (a) Lamprey. Contraction of the corneal muscle pulls the cornea against the lens, forcing a change in its position and ability to focus incoming light. (b) Teleost. Accommodation in the teleost depends on a change in lens position. (c) Turtle and (d) bird. Accommodation in both involves the iris sphincter muscle squeezing the lens. (e) Mammal. Accommodation is brought about by relaxation of the suspensory ligaments of the eye.

employed by different groups of fishes. When electrically stimulated, the retractor lentis muscles move the lens posteriorly in some fishes, obliquely downward in others, and obliquely upward in still others.

In trout, the lens is round and the retina is ellipsoid, so the two are not concentric (figure 17.22a). This may mean that at rest the near field of view at the center and the far field at the periphery are in focus. During accommodation, the lens is pulled backward. This does not change the focus on peripheral objects much, but it focuses distant objects in the center of the field of view (figure 17.22b).

*Photoreception*. The proportions of rods and cones vary considerably in different animals. In animals active under bright conditions, such as diurnal tetrapods and diurnal shallow-water fishes, both rods and cones are present. Most cones are concentrated near the fovea, and rods predominate at the periphery. Nocturnal animals or those that inhabit dimly lit waters are adapted to poor illumination. In these groups, cones are few or absent, and the retina is composed almost entirely of rods. Photoreceptor cells of these vertebrates possess visual pigments with high light-absorbing capabilities. These pigments are composed of the protein **opsin** conjugated with a light-absorbing derivative of vitamin A.

In humans, there are four visual pigments. One is present in rods, which record the world in black, white, and shades of grey (like a black-and-white television). The other three pigments, which are present in cones, produce a trichromatic system. One cone pigment is sensitive to blue, another to green, and a third to red wavelengths of light.

In diurnal species, single cones tend to synapse with single bipolar cells that synapse with single ganglion cells that project to the central nervous system. This direct one-to-one transfer of impulses is thought to increase the resolution of the retina and hence its acuity (figure 17.23a,b). In nocturnal species, large numbers of photoreceptor cells converge on a small number of interneurons; thus, there is a pooling of information. Acuity decreases, but sensitivity increases. The horizontal cells and perhaps the amacrine cells spread information laterally to help accentuate contrasts.

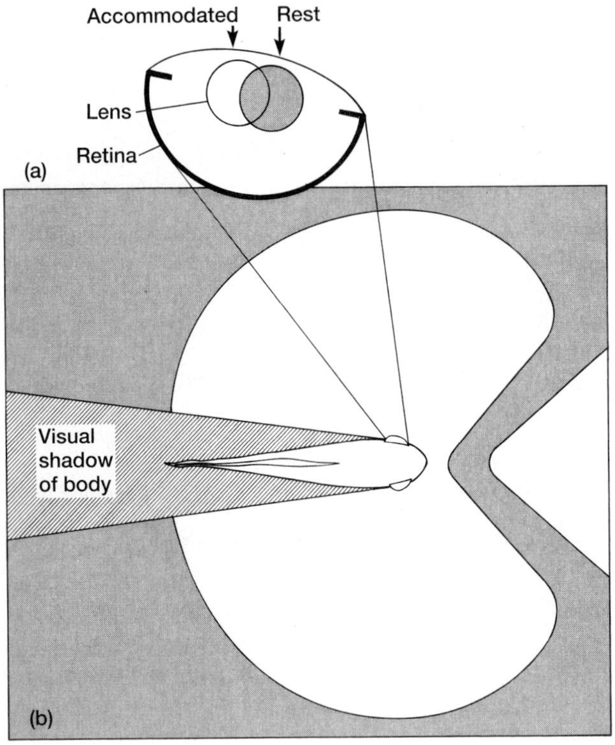


**Figure 17.22** Accommodation in trout. (a) In order for the trout eye to accommodate light, the lens must be pulled posteriorly. Because the curvatures of the lens and the retina differ, it has been proposed that these differing curvatures provide dual focusing. (b) When the eye is at rest, the stippled area, consisting of the V-shaped region immediately in front of the fish and the distant lateral regions, is in focus; the white areas are out of focus. When the retractor muscle pulls the lens posteriorly, the fish can focus on the distant objects in front of it as well as on the adjacent lateral field.

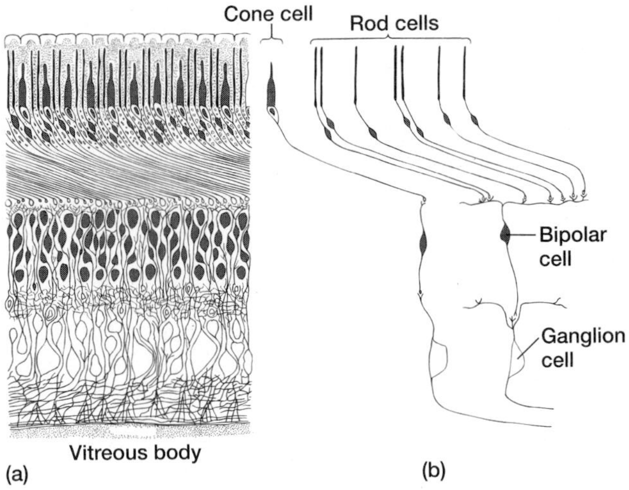


**Figure 17.23** Retinal connections. (a) Diagram of retinal rods, cones, bipolar cells, and ganglion cells. (b) In diurnal species, the one-to-one circuit, in which one cone connects to a single bipolar cell that connects to one ganglion cell, is thought to increase visual acuity, whereas one bipolar cell that receives many rods blends different stimuli but increases sensitivity.

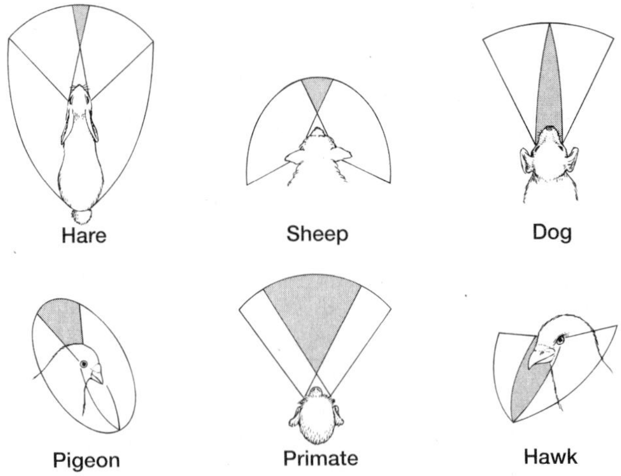


**Figure 17.24** Monocular and binocular vision in birds and mammals. The degree of overlap of visual fields (indicated by shading) varies considerably. Large panoramic fields of view characterize animals susceptible to attack.

*Depth perception*. The position of the eyes on the head represents a trade-off between breadth of the visual field and depth perception. If the eyes are positioned laterally, then each eye scans separate portions of the surrounding world and the total field of view at any moment is extensive. Vision in which the visual fields do not overlap is termed **monocular vision.** Because it allows an individual to see a large portion of its surroundings and to detect potential threats from most directions, it is common in prey animals. Strictly monocular vision in which the visual fields of the two eyes are wholly separate is relatively rare. It occurs in cyclostomes, some sharks, salamanders, penguins, and whales.

The visual field overlaps in animals with **binocular vision** (figure 17.24). Extensive overlap of visual fields characterizes humans. We have as much as 140° of binocular vision, leaving 30° of monocular vision on each side. Binocular vision is important in most classes of vertebrates. Birds have up to 70° of overlap, reptiles up to 45°, and some fishes have as much as 40°. Within the area of overlap, the two visual fields merge into a single image, producing **stereoscopic vision.** The advantage of stereoscopic vision is that it gives a sense of depth perception. If a person with two functioning eyes closes one, she or he will lose much sense of depth.

Depth perception comes from the method of processing visual information. In binocular vision, the visual field seen by each eye is divided. Half the input goes to the same side, and the other half crosses via the **optic chiasma** to the opposite side of the brain. The result is to bring information gathered by both eyes to the same side of the brain. The brain compares the parallax of the two images. **Parallax** is the slightly different view one gets of a distant object from two different points of view. Look at a distant lamppost from one position, and then step a few feet laterally and look at it again. Slightly more of one side can be seen and less of the opposite side. The position of the post relative to background reference points changes as well. This is parallax. The nervous system takes advantage of parallax that results from differences in the positions of the two eyes. Each eye registers a slightly different image due to the distance between the eyes. Although this difference is slight, it is enough for the nervous system to produce a sense of depth from differences in parallax (figure 17.25).

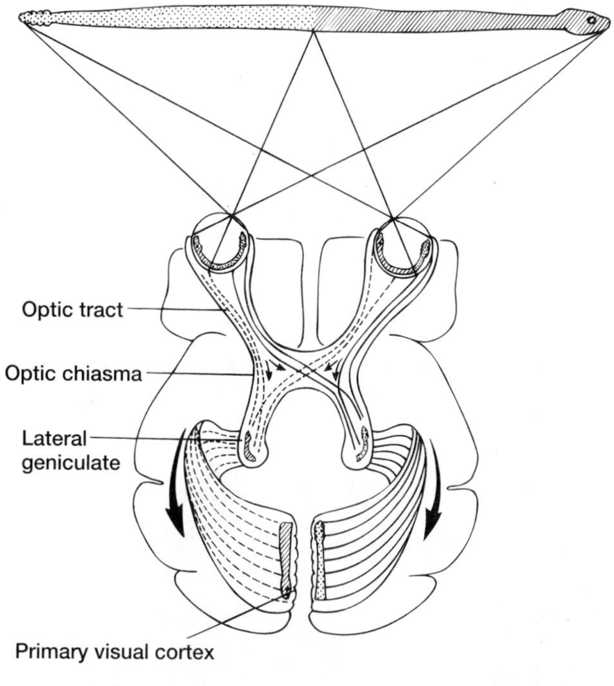


**Figure 17.25** Depth perception. Overlapping fields of view in the eyes collect information about the same object (snake), but some of this information travels to the same side of the brain via the optic chiasma, whereas other information travels to the visual areas on the opposite side of the brain. This permits similar information from the slightly differently placed eyes to be brought together and compared. Such comparisons are thought to be the basis for depth perception. Notice the specific organization of visual information. Information from the left visual field (tail of snake) is received by the right side of each eye. The right visual field (snake head) is received by the left side of each eye. Nerve fibers from the medial retina cross in the optic chiasma. Lateral fibers do not cross. Overall, information from the left visual field (tail) is processed in the primary visual cortex of the right cerebral hemisphere. Information from the right visual field (head) is processed in the left cerebral hemisphere.

Visual accommodation also contributes to depth perception. Even with a single eye, the degree of accommodation required to bring the object into focus can be used to interpret its distance.

*Integration of visual information.* In lower vertebrates, the optic nerve tends to run directly into the midbrain. In amniotes, the axons in the optic tract travel to one of three regions—the **lateral geniculate nuclei** of the thalamus, the tectum of the midbrain, and the pretectal area in the tegmentum of the midbrain.

Most fibers of the optic tract travel to the paired lateral geniculate nuclei of the thalamus. From here, fibers relay information to the primary visual cortex of the cerebrum (figure 17.26). Thalamic cells and photoreceptor cells and some cells of the visual cortex respond to light intensities, but other cells in the visual cortex are more specialized. Some of these respond to visual images in the shape of slits, bars, or edges. Others respond only to moving or to stationary visual edges. Such specialized cells add more information about the size, shape, and movement of the visual image. In other words, the retina responds to light intensity and wavelength, but in the visual cortex, the image gains contour, orientation, and motion. Image formation might be likened to pictures composed of dots of discrete information. As edges, shades, and shapes are sorted, the image progressively emerges (figure 17.27a–c). Furthermore, visual information is thought to reach a conscious level in the visual cortex. When it reaches the cortex, the visual world is consciously perceived.

Before entering the thalamus, some ganglion cells of the optic tract send branches into the midbrain to contact the tectum. In addition to this visual input, neurons of the **superior colliculus** receive information about sound, head position, and feedback from the visual cortex. The tectum, in turn, produces motor output to muscles that rotate the eyes, head, and even the trunk toward the visual stimuli. Humans have two eyes and two visual fields. Mentally, these merge into one field, in part because of the intricately synchronized movements of both eyes as we scan our surroundings. The tectum also sends visual impulses through the **pulvinar nucleus** of the thalamus to the visual cortex. The function of this route of visual input to the cerebral cortex is not known. If the primary visual cortex is damaged or denied direct input from the thalamus, this alternate input from the midbrain may preserve a rudimentary response to visual stimuli.

A few fibers of the optic tract also send input to the **pretectal** area in the tegmentum of the midbrain. A reflex relay from the tegmentum to motor nerves controlling the iris muscles permits pupil size to be adjusted immediately to light intensity.

Thus, visual perception depends on the stimulation of photoreceptor cells in the retina and culminates in the central integrating regions of the brain. It should be emphasized, however, that the processing of visual information begins in the retina itself and continues through the optic

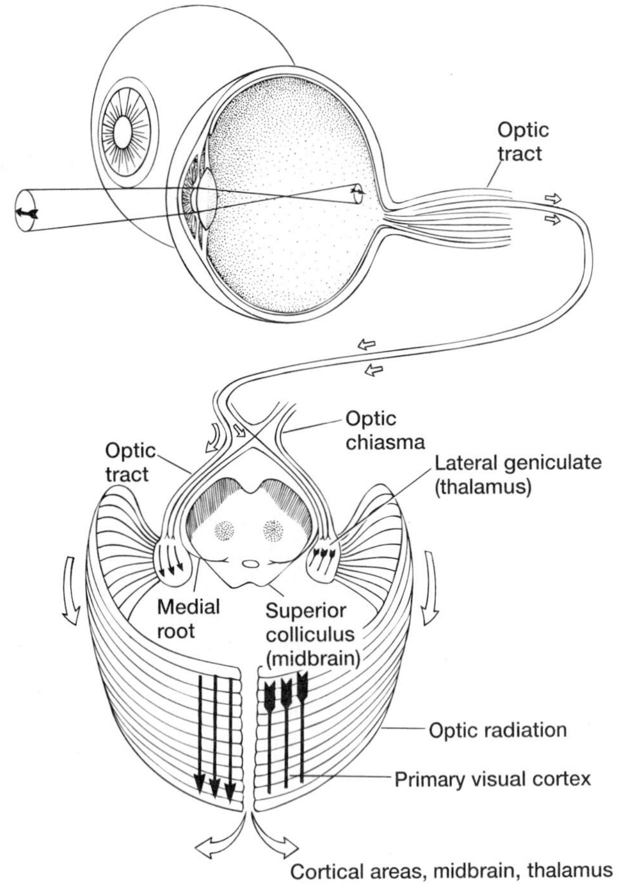


**Figure 17.26** Projections of visual information to the primary visual cortex in primates. The object of focus is a solid arrow. Output from the retina is conveyed via the optic tract to the lateral geniculate nuclei in the thalamus. Selective crossing of some fibers in the optic chiasma results in each lateral geniculate receiving the image (the arrow) from the opposite half of the visual field. The visual image is represented or mapped out in the lateral geniculate as a set of adjacent and highly ordered fields of stimulated neurons. From the lateral geniculate, axons project via an optic radiation to the cerebral hemispheres, where the corresponding parts of the image are now represented as a pile of superimposed fields in the primary visual cortex. This ordering of multiple representations of an image both in the thalamus (lateral geniculate) and in the cerebral hemispheres (primary visual cortex), although not entirely understood, is thought to contribute to neuroanalysis of the binocular image. Notice that the visual pathway does not end in the primary visual cortex but continues to the midbrain and even back to the thalamus. A small branch of the optic tract, the medial root, does not enter the thalamus but instead leads to the superior colliculus (optic tectum). The medial root synapses with motor neurons that control the visual orientation reflex involving eyeball movements, head turning, and rotation of the trunk.

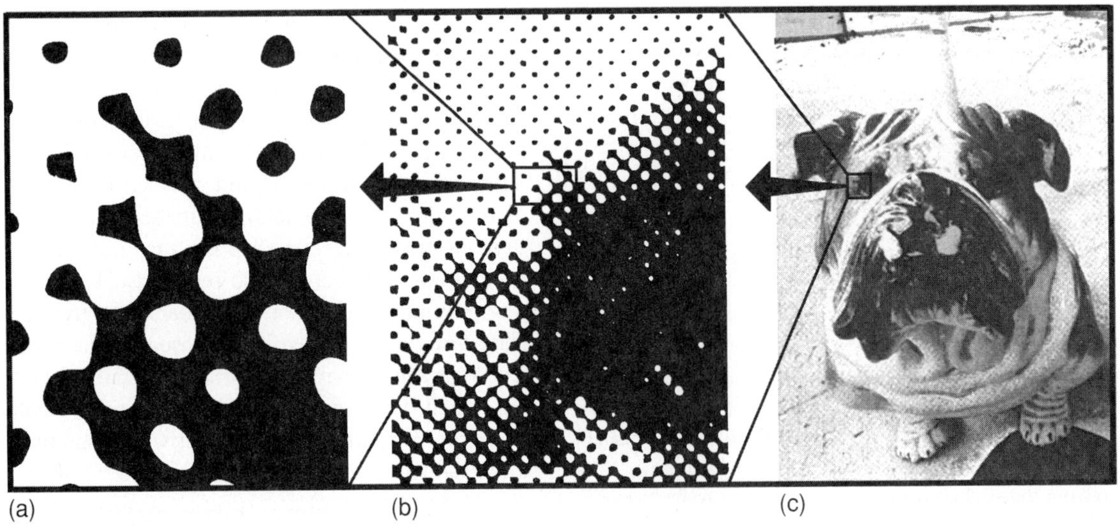


**Figure 17.27** Emergence of visual perception. (a) This photo is an enlargement of a portion of the image to its immediate right. (b) This photo is an enlargement of a portion of the image on the right. (c) The full differentiated image emerges. The visual image in the nervous system begins with processing in the retina, is continued in the thalamus, and is synthesized in the visual cortex. Just as dots merge to form a legible image of a dog on the right, successive processing of visual information in the brain brings the individual stimuli recorded by individual cells together into a synthesized, composite image.

nerve. It has been suggested that as much as 90% of visual information is processed in the retina before it is transmitted to the thalamus.

***Pineal Complex*** In most vertebrates, the roof of the diencephalon, termed the **epithalamus,** produces a single median photoreceptor, the **parietal organ.** However, there is great variation in this organ between groups, and it is often joined by additional adjacent specializations of the epithalamus. To further complicate matters, a phylogenetic change occurs in the function of the parietal organ. It participates in photoreception among lower vertebrates but tends to become an endocrine organ in higher vertebrates.

Endocrine organs (p. 604)

Not surprisingly, a great deal of terminology has grown up around the parietal and adjacent organs of the epithalamus. Let us begin by sorting out this terminology.

*Structure.* Depending on the species, the epithalamus may evaginate to produce up to four structures, each one a discrete organ (figure 17.28a). The most anterior is the **paraphysis,** followed by the **dorsal sac, parietal organ,** and **epiphysis cerebri (epiphysis).** If two or more of these are present together, the collective term **parietal complex** applies.

The functions of the paraphysis and dorsal sac are not well understood, but their structure suggests that they are glandular organs. The epiphysis is sometimes called the **pineal organ** or, if it is largely endocrine in function, the **pineal gland.** Because it is adjacent to the pineal organ, the parietal organ is sometimes called a **parapineal organ,** or a **parietal eye** if it forms a photoreceptive sensory organ. The parietal eye may include a modest cornea, lens, and area of photoreceptive cells that synapse with adjacent ganglion cells forming a nerve that travels to the nervous system (figure 17.28b,c).

*Phylogeny.* A single dorsal parietal foramen through the skull of many ostracoderms testifies to the presence of a parietal organ. Living lampreys possess both an epiphysis and a parietal organ. Both these organs exhibit some capacity for photoreception (figure 17.29a). In elasmobranchs and bony fishes, the epiphysis is prominent, but the parietal organ, if present at all, is only rudimentary (figure 17.29b,c).

Amphibians usually possess both an epiphysis and a parietal organ. In fossil amphibians, the parietal organ forms a distinct parietal eye. Living frogs retain this, but it is absent in salamanders (figure 17.29d,e).

In reptiles, the parietal organ is present. Lizards and *Sphenodon* possess a parietal eye so distinctive that it is often referred to as a **third eye.** The epiphysis is present as well (figure 17.29f).

In birds and mammals, the parietal organ is absent. The epiphysis is present in both classes, but it is exclusively an endocrine organ and is usually referred to as the pineal gland (figure 17.29g,h).

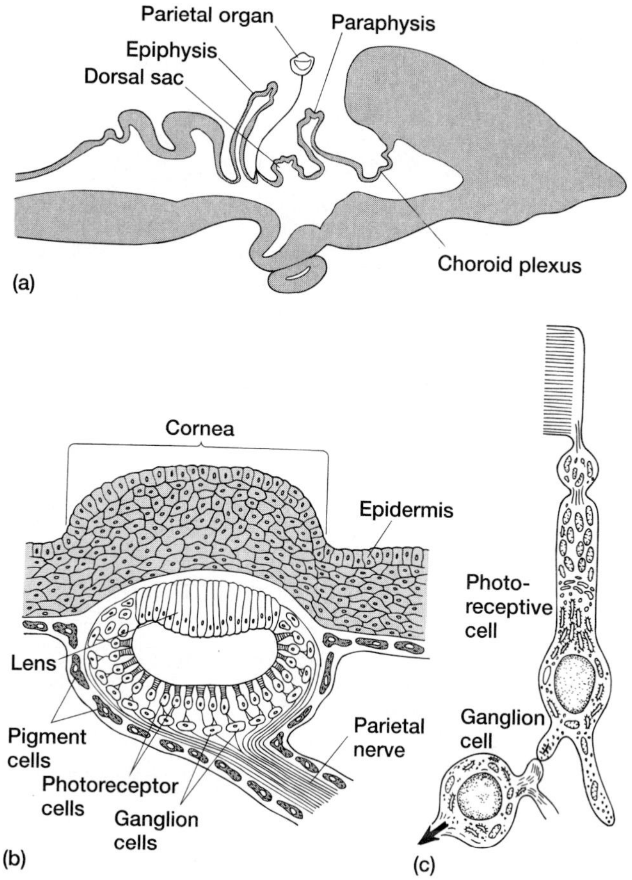


**Figure 17.28** Pineal complex. (a) Sagittal section through the central nervous system of a generalized vertebrate. Up to four evaginations of the roof of the diencephalon may form. (b) Generalized parietal eye. (c) Photoreceptor cell from a parietal organ.

Even within a given class of vertebrates, there is considerable variation in the pineal complex. For instance, the pineal gland is small in owls, shearwaters, opossums, shrews, whales, and bats but large in penguins, emus, sea lions, and seals. In hagfishes, crocodilians, armadillos, dugongs, sloths, and anteaters, the epiphysis is absent entirely. The parietal organ may develop independently of the epiphysis, the two can be closely associated, or, as in some fishes and anurans, they may fuse.

*Form and function.* The primitive pineal complex was a photosensitive organ. Among most cyclostomes, fishes, amphibians, and reptiles, photosensitivity has been experimentally demonstrated in the pineal organ complex. This lends support to the view that this complex is an ancient structure among vertebrates that arose as an accessory sensory system sensitive to photoradiation. Photoreceptive cells from the pineal complex are tall and columnar in shape with specialized apical extension (figure 17.28c). The basal regions synapse with adjacent ganglion cells. These

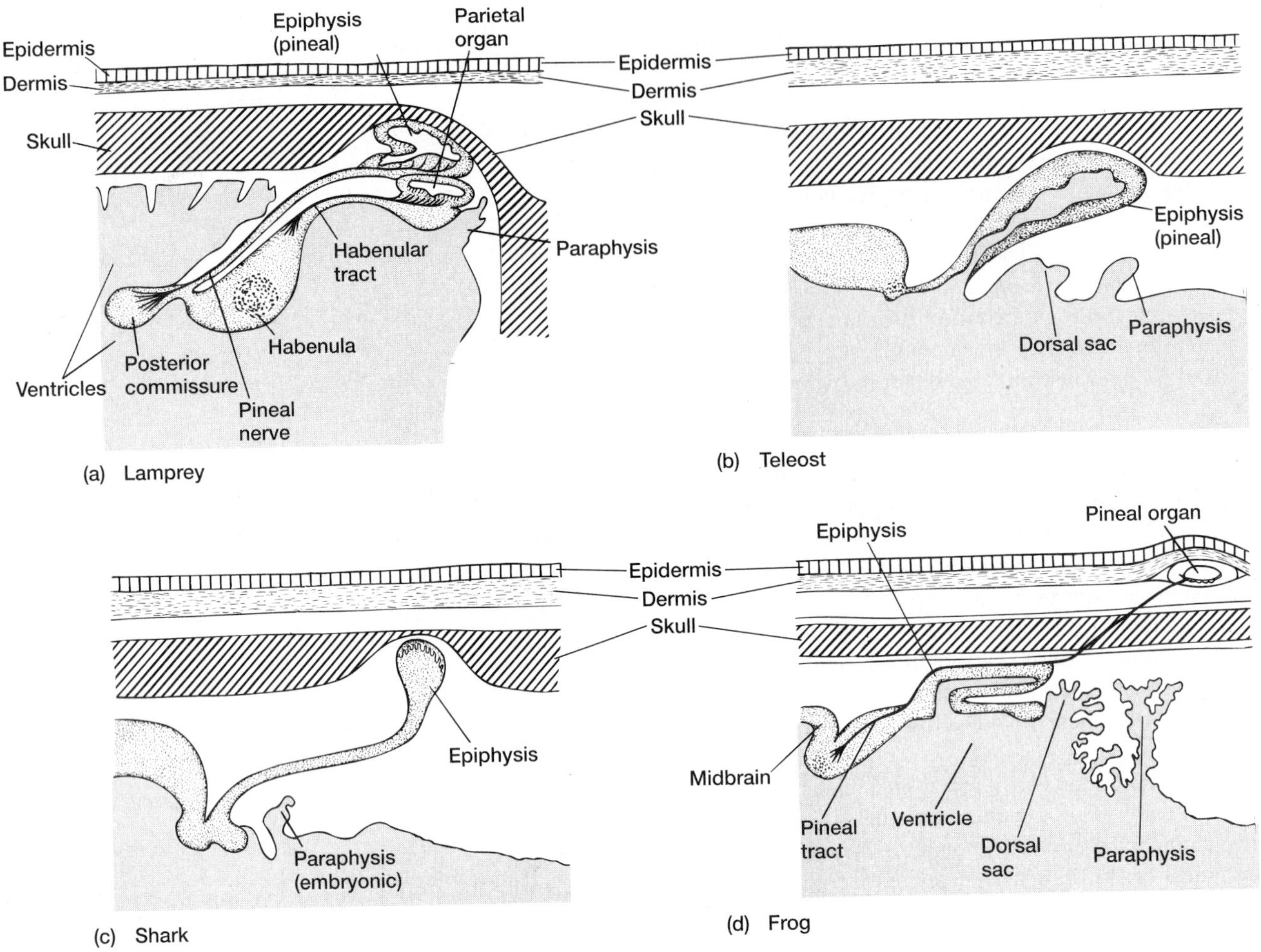


**Figure 17.29** Phylogeny of the pineal complex. (a) Lamprey. (b) Teleost. (c) Shark. (d) Frog. (e) *Necturus*. (f) Lizard. (g) Bird. (h) Mammal.

form the pineal nerve that relays impulses to the **habenular** and other regions of the brain.

The change from photoreception to endocrine secretion occurs in birds and mammals. In some birds, pineal tract fibers have been reported, which suggests a lingering role in photoreception. But, for the most part, the bird epiphysis is glandular and thought to be involved in endocrine secretion. In mammals, pineal fiber projections to the brain are unknown. The mammalian epiphysis is exclusively endocrine in function and composed of secretory cells called **pinealocytes,** which may be modified photoreceptor cells.

## *Infrared Receptors*

Infrared radiation lies just to the right of the visible band of light on the electromagnetic spectrum (figure 17.16). Some vertebrates have special sense organs that respond to infrared radiation. This is especially useful at night when visible light is usually unavailable.

For us to see an object, visible light must strike it and be reflected from it. The natural source of visible light is sunlight. However, infrared light emanates directly from the surface of any object with a temperature above absolute zero, that is, any object warmer than −273°C. Obviously, the sun is well above this temperature, so infrared radiation is included in its spectrum. But all other natural objects are also above this extremely low temperature and give off infrared radiation from their surfaces whether it is day or night. Infrared radiation can be detected by some species of snakes and used to guide their search for prey in the dark.

Because infrared radiation is emitted from warm objects, and objects it shines on are warmed as they absorb it, this type of radiation is sometimes incorrectly termed "heat radiation." Strictly speaking, infrared radiation is a narrow range of wavelengths of electromagnetic radiation, not heat. Nevertheless, this warming effect of infrared radiation on bodies that absorb it is how infrared receptors are stimulated.

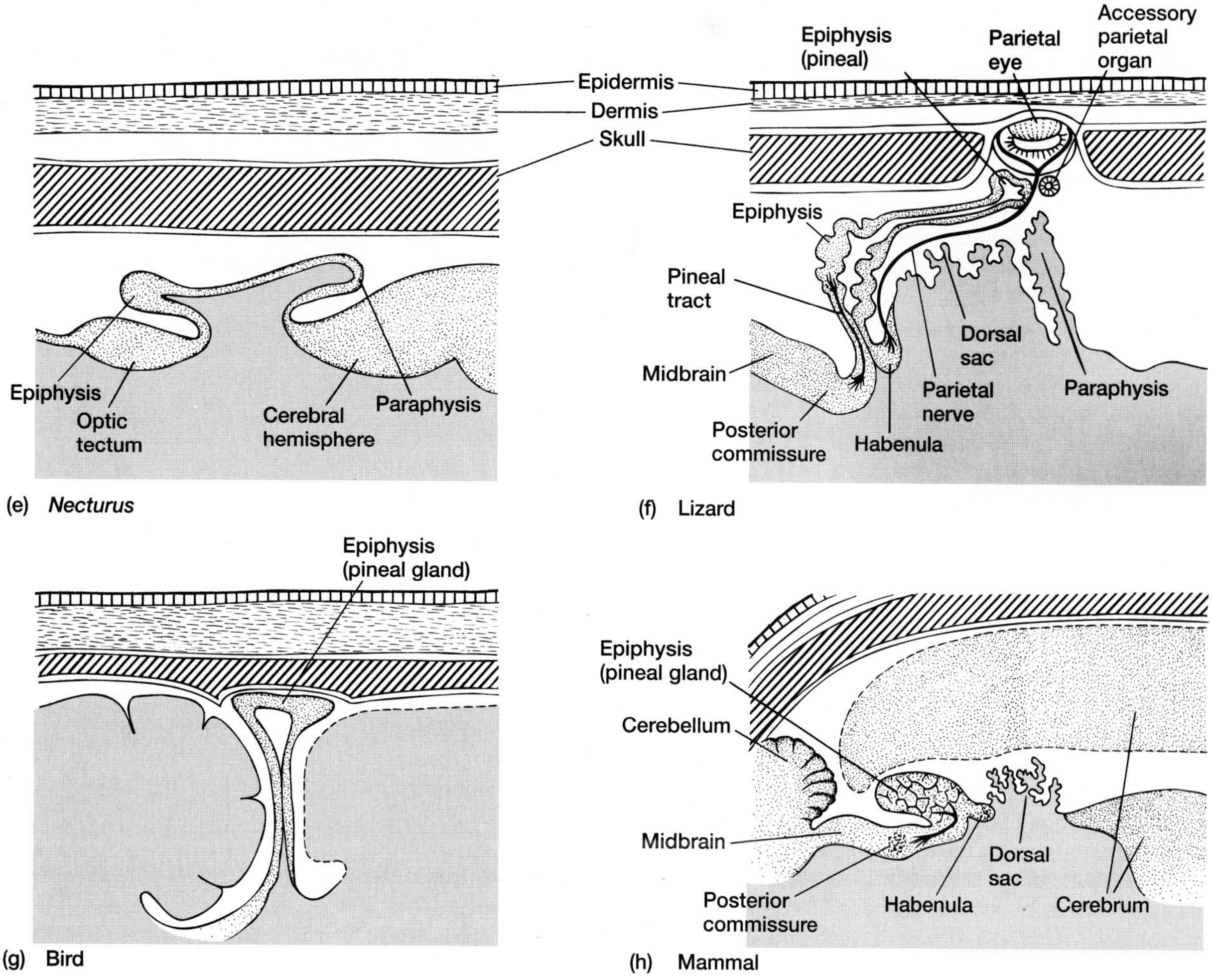


**Figure 17.29** (continued)

Special sense organs containing **infrared receptors (thermoreceptors)** occur in several groups of vertebrates. They are present on the faces of vampire bats that feed on ungulates. Infrared receptors apparently help these bats detect warm blood vessels beneath the thick skin of their prey. The most discrete infrared receptors are found in two groups of snakes, the primitive boas and the advanced pit vipers. In both cases, the sensory receptor is a free nerve ending located in the skin. As the skin absorbs infrared radiation it is warmed; this excites the associated free nerve endings, which transmit this information to the optic tectum of the midbrain.

In boa constrictors, the free nerve endings lie within epidermal scales along the lips (figure 17.30a). In pythons, members of the family Boidae, the nerve endings lie at the bottom of a series of several recessed **labial pits** along the lips (figure 17.30b). The poisonous pit vipers take their name from the presence of a pair of infrared receptors called **facial pits** (loreal pits). Facial pits also are sunken, but they differ from the labial pits of pythons. Sensory nerve endings are suspended in a thin **pit membrane** halfway between the bottom and the top of the pit instead of lying at the bottom of the pit as in pythons (figure 17.30c,d).

When free nerve endings are embedded in an epidermal scale, as in boa constrictors, or associated with tissue at the bottom of a pit, as in pythons, the surrounding tissue may dissipate heat and thus slow the local warming that stimulates the free nerve endings. However, in pit vipers, the nerve endings are warmed rapidly because the free nerve endings are suspended in the thin membrane away from the walls of the pit. This increases their sensitivity to infrared radiation. A pit membrane that is warmed by as little as 0.003°C reaches a threshold sufficient to excite its sensory receptors. For a pit viper, this translates into infrared sensitivity to a mouse about 30 cm away. Pythons and boas can detect mice at distances of about 15 and 7 cm, respectively.

## Box Essay 17.1 Sensory Organs and the Origin of Snakes

Lizards are the closest living relatives of snakes. Snakes may have evolved from lizards, or both groups may share a common ancestor. What is especially intriguing about the origin of snakes is the fact that their derivation from lizards or lizardlike ancestors may have included a fossorial phase in which photoreceptors were reduced in prominence. Following this extended fossorial phase, more modern snakes would have radiated back to aboveground habitats. The photoreceptors that regressed during the fossorial phase were in a sense rebuilt to meet the adaptive requirements of the species moving into diurnally active lifestyles.

The view that snakes went through a fossorial period was championed by Gordon Walls in the early 1940s. It was inspired by his monumental study of the vertebrate eye. Walls noted that the snake eye represented a marked departure from the eyes of all other reptiles, including lizards (box figure 1a). For example, accommodation of the lizard eye occurs through a change in lens shape. In snakes, it occurs by moving the lens forward or back. Typically, lizards have three movable eyelids—upper, lower, and nictitating membrane. Snakes have none. Instead, the snake cornea is covered by the **spectacle,** a transparent derivative of the eyelids that is fixed in position. Scleral ossicles occur in some lizards, but they are absent in all snakes. In the retinas of diurnal lizards, distinct cones and rods are present, but in snakes, the "cones" seem to be modified rods that serve in color perception. Walls noted further differences in circulation, internal structure, and chemical composition that attest to the distinctiveness of the snake eye, not just among reptiles but among all other vertebrates as well. To Walls, the snake eye seemed to be unique.

Walls proposed that these anatomical peculiarities of the snake eye could not be easily explained if snakes evolved from surface-living lizards. Instead, he suggested that snakes evolved from forms in which the eyes were reduced in association with low light conditions. He proposed that snakes went through a burrowing phase. In his view, restructuring of the eye occurred as snakes radiated back to surface-living and diurnal conditions.

Other snake photoreceptors show similar evidence of reconstruction from a regressed reptilian pattern. For example, the parietal organ of many lizards is highly developed into a photoreceptive organ that includes a lens and a photoreceptive layer. The lizard parietal occupies an opening in the skull and resides beneath the skin, where it has direct access to natural light. However, in snakes, the parietal is lost entirely, and only the basal portion of the epiphysis (pineal) is retained. In the remaining ophidian epiphysis, only secretory pinealocytes are present and the pineal gland is beneath the skull (box figure 1b).

The issue of whether the origin of snakes includes a fossorial phase or just a nocturnal phase during which snake ancestors lived under conditions of low light is still being debated. In any case, the special photoreceptors of snakes are certainly a radical departure from those of other reptiles.

**Box figure 1** Comparison of photoreceptors of lizards and snakes. (a) Lizard and snake eyes in cross section. (b) Pineal complexes of a lizard and a snake.

## Mechanoreceptors

Detecting water currents, maintaining balance, and hearing sounds may seem to be different sensory functions. Yet, all are based on **mechanoreceptors,** sensory cells responsive to small changes in mechanical force.

The basic mechanoreceptor is a **hair cell,** an unfortunate term because these cells have nothing to do with hair. The name comes from the microscopic "hairlike" processes at their apical surfaces. These tiny processes include a tight stand of **microvilli** of unequal lengths and a single long **cilium,** which is sometimes called a **kinocilium.** The microvilli are often constricted at their bases and rest on a dense **terminal web** or **cuticular plate.** Each microvillus includes a core of fine microfilaments with molecular crossbridges so that it behaves like a rigid rod. Because they are nonmobile, microvilli are more appropriately called **stereocilia.** A tuft of stereocilia with a kinocilium is a **hair bundle** (figure 17.31a).

Hair cells are transducers that transform mechanical stimuli into electrical signals. Mechanical stimulus of the hair bundle triggers ionic changes in the hair cell. Hair cells are epithelial cells that originate embryologically from surface ectoderm. They lack axons of their own. Instead, each hair cell is embraced by the sensory fibers of neurons sensitive to ionic changes in the hair cell. Through synapses or synapselike contact points between them, electrical excitation is transmitted from the hair cells to their embracing

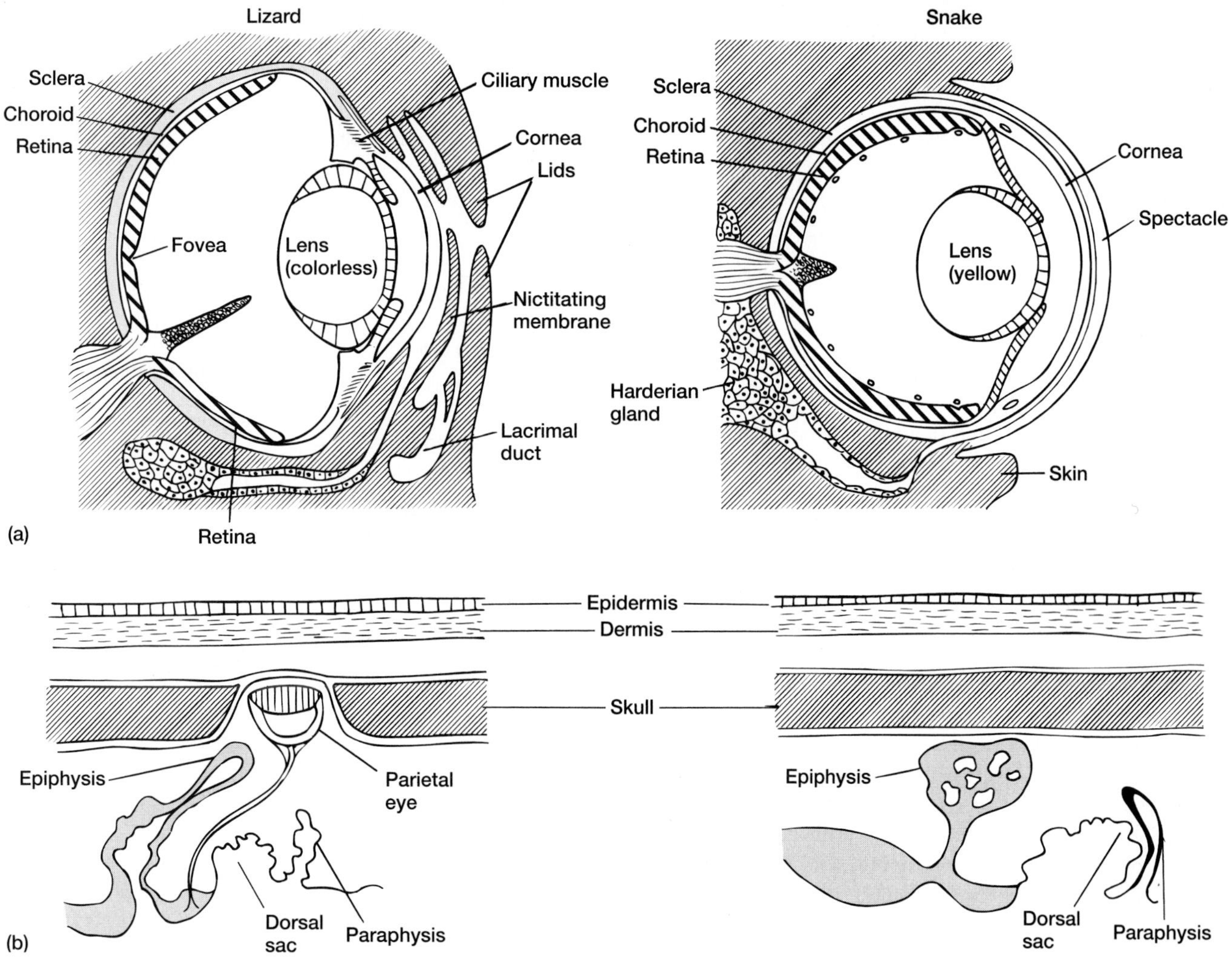


neurons and then to the central nervous system. Hair cells also receive efferent nerves from the central nervous system. Efferent nerves can change the sensitivity of the hair cell or help focus its sensitivity to a restricted range of mechanical frequencies.

Hair cells respond selectively to mechanical stimuli. For example, stimuli applied from one direction will trigger a hair cell, whereas stimuli applied from the opposite direction will not (figure 17.31c). Selectivity is thought to result from the asymmetry of the hair bundle itself, due to the different lengths of its stereocilia.

A **neuromast organ** is a small collection of hair cells, supporting cells, and sensory nerve fibers, composing the most common arrangement of a mechanoreceptor. The projecting hair bundles are usually embedded in a gelatinous cap called the **cupula** (figure 17.31b). The cupula most likely accentuates the mechanical stimulation of the hair cells, thereby increasing their sensitivity. The neuromast organ, or a modification of it, is the fundamental component of all three types of mechanoreceptive systems—the lateral line system that detects water currents, the vestibular apparatus that senses changes in equilibrium, and the auditory system that responds to sound.

## *Lateral Line System*

The **lateral line system** is present within the skins of most cyclostomes, other fishes, and aquatic amphibians, but it is unknown in terrestrial vertebrates, including aquatic birds and mammals. It consists of long recessed grooves, or **lateral**

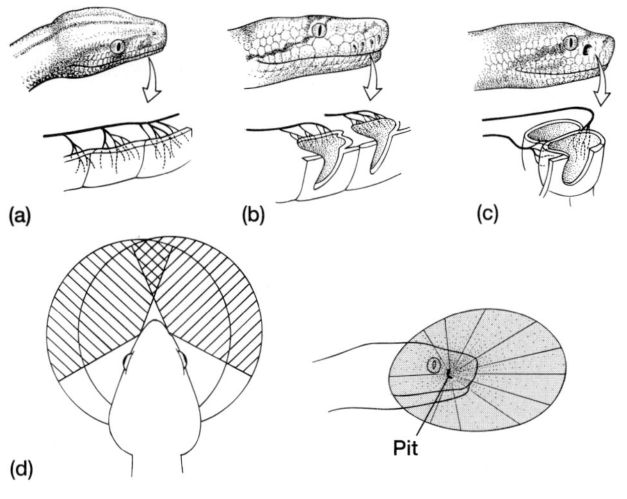


Figure 17.30 Infrared receptors. Sensors receptive to infrared radiation are located within the integument in some boas (a), at the bottom of recessed pits in pythons (b), and on a thin membrane halfway between the opening and bottom of the sensory organ in venomous pit vipers (c). In pit vipers, pit organs scan the shaded cone-shaped area (d) in front and to the side of the infrared receptor.

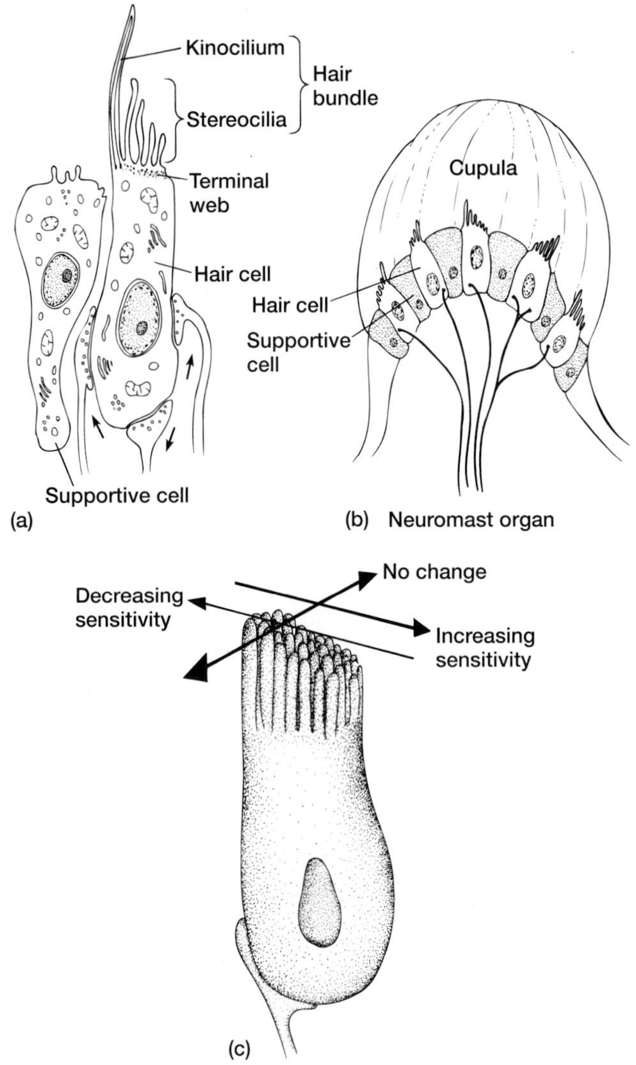


Figure 17.31 Hair cells. (a) From the apical surface of a hair cell projects a hair bundle, composed of stereocilia of unequal length and a single kinocilium. Afferent and efferent nerve fibers are associated with each hair cell. A supportive cell, which is thought not to react to direct mechanical stimulation, is often adjacent to a hair cell. (b) Clumps of hair and supportive cells form a functional unit, the neuromast organ. The cupula is a cap of gelatinous material that fits over the projecting hair bundles. (c) A hair cell responds to direct mechanical stimulation, but its response is selective. Each cell is most responsive to forces from one direction. Mechanical forces in the opposite direction reduce sensitivity. Forces at right angles to the hair cell cause no change. By such selective responses, hair cells indicate the directions of the mechanical forces that impinge upon them.

**line canals,** concentrated on the head and extending along the sides of the body and tail (figure 17.32a,b). Neuromast organs are the sensory receptors of the lateral line system. They can occur separately on the surface of the skin, but they are usually found at the bottoms of the lateral line canals. The canal can be recessed in a valley or sunken and covered by surface skin that has pores through which currents of water flow over the neuromast organs.

The neuromasts respond directly to water currents. Hair cells are oriented with their most sensitive axis parallel to the canal. About half are oriented one direction, the rest in the opposite direction. In the absence of mechanical stimulation, each neuromast generates a continuous series of electrical pulses. Water flowing in one direction stimulates an increase in this discharge rate. If flow is in the opposite direction, the discharge rate falls below its resting rate. If water passes at right angles to these neuromasts, the resting electrical discharge rate is unaffected. This provides information about the animal's direction of movement and about disturbances in the water. Even cave fish that are blind can navigate around obstacles in their environment because of their lateral line systems.

Some fishes use their lateral line canals as a type of "distance touch"; that is, it allows them to detect compression in the water in front of them as they approach a stationary object ahead. In some surface-feeding fishes, the lateral line system detects frequency oscillations produced by insects struggling on the water's surface. If the lateral line nerve is cut or if their grooves are covered, an affected fish loses its ability to navigate or perform schooling behavior.

There is some evidence that the lateral line canals can detect low frequency sounds at least from nearby objects. The uncertainty about its role in hearing is partly the result of the similarity between stimuli. Rapid oscillations of water and low frequency vibrations resulting from sounds traveling through water are mechanically similar. It has also

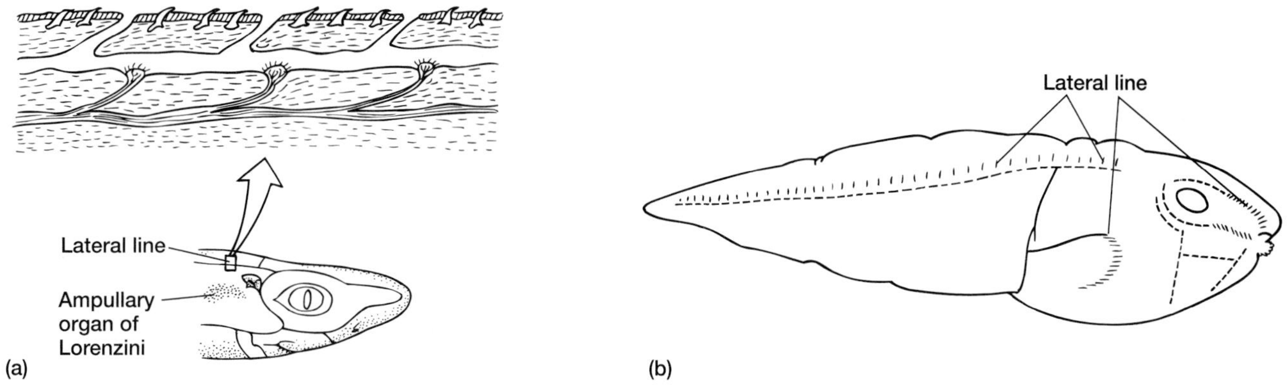


**Figure 17.32** The lateral line system. (a) Section through the skin of a shark showing the sunken lateral line canal opening to the surface via small pores. (b) Distribution of neuromasts on the side of a frog tadpole. The long axis of each neuromast is represented by the orientation of the short bars. Notice how this orientation changes.

been difficult to isolate the lateral line system experimentally from the inner ear, which is also sensitive to sound. It is clear that navigation is the primary role of the lateral line system. Perhaps it detects water disturbances that prey produce, and we should allow for a possible secondary role in hearing as well.

### *Vestibular Apparatus*

The **vestibular apparatus** (membranous labyrinth) is a balancing organ that arises phylogenetically from part of the lateral line system. It is suspended in the **otic capsule** by loose connective tissue. The otic capsule is filled with **endolymph** and surrounded by **perilymph.** Both fluids have a similar consistency to lymph. Embryologically, the vestibular apparatus forms from the **otic placode,** which sinks inward from the surface to produce hair cells, neurons of the otic ganglia, and the vestibular apparatus. In elasmobranchs, the vestibular apparatus maintains continuity with the environment through the endolymphatic duct. In other vertebrates, it is pinched off to form a closed, fluid-filled system of channels.

The vestibular apparatus contains semicircular canals and at least two connecting compartments—the **sacculus** and **utriculus** (figure 17.33a). These are lined by **vestibular epithelium** within which arise neuromast organs that participate in sensing equilibrium and sounds. The three **semicircular canals** are oriented roughly in the three planes of space. Sensory receptors within the semicircular canals are called **cristae** (figure 17.33b). Each crista is an expanded neuromast organ composed of hair cells and cupula. The cristae lie within expanded **ampullae** at the base of each semicircular canal.

The semicircular canals respond to rotation, technically to angular acceleration produced when the head is rotated or turns. When the canals are accelerated, fluid inertia causes the endolymph to lag behind movement of the canal itself. The fluid deflects the cupula, stimulates hair cells, and alters their rhythmic discharge of electrical impulses to the nervous system.

The sensory receptor within the sacculus and the utriculus is the **macula** or the **otolith receptor.** It too is a modified neuromast organ with hair cells and gelatinous cupula, but in addition, tiny calcium carbonate mineral concretions known as **otoconia** are embedded within the surface of the cupula (figure 17.33c).

Maculae respond to changes in orientation within a gravitational field. Moving the head or changing position of the body tilts the maculae and changes their orientation with respect to gravity. Acceleration of the body also causes the maculae to respond. The otoconia are inertial masses that accentuate shearing displacements of the hair bundles in response to linear acceleration or changes in head orientation.

The vestibular apparatus keeps the central nervous system informed as to whether an animal is at rest or in motion and conveys information about its orientation. The maculae sense gravity and linear acceleration, and the cristae respond to angular acceleration. In fishes, the vestibular apparatus is also responsive to sound, although this function does not seem to be well developed. In fishes, reptiles, birds, and especially mammals, the vestibular apparatus produces a specialized region of sound reception, the **lagena** (figure 17.34).

### *Auditory System*

The lagena is involved in hearing. It develops as an enlargement of the sacculus, so it is part of the vestibular apparatus. In terrestrial vertebrates it tends to lengthen, and in most mammals it becomes coiled into the **cochlea.** Within the lagena or cochlea in mammals lies the sensory receptor of sound, the **organ of Corti,** a specialized strip of neuromasts connected to the nervous system via the auditory nerve. The lagena, like the rest of the vestibular apparatus, lies within bone or cartilage of the skull. The ear includes up to

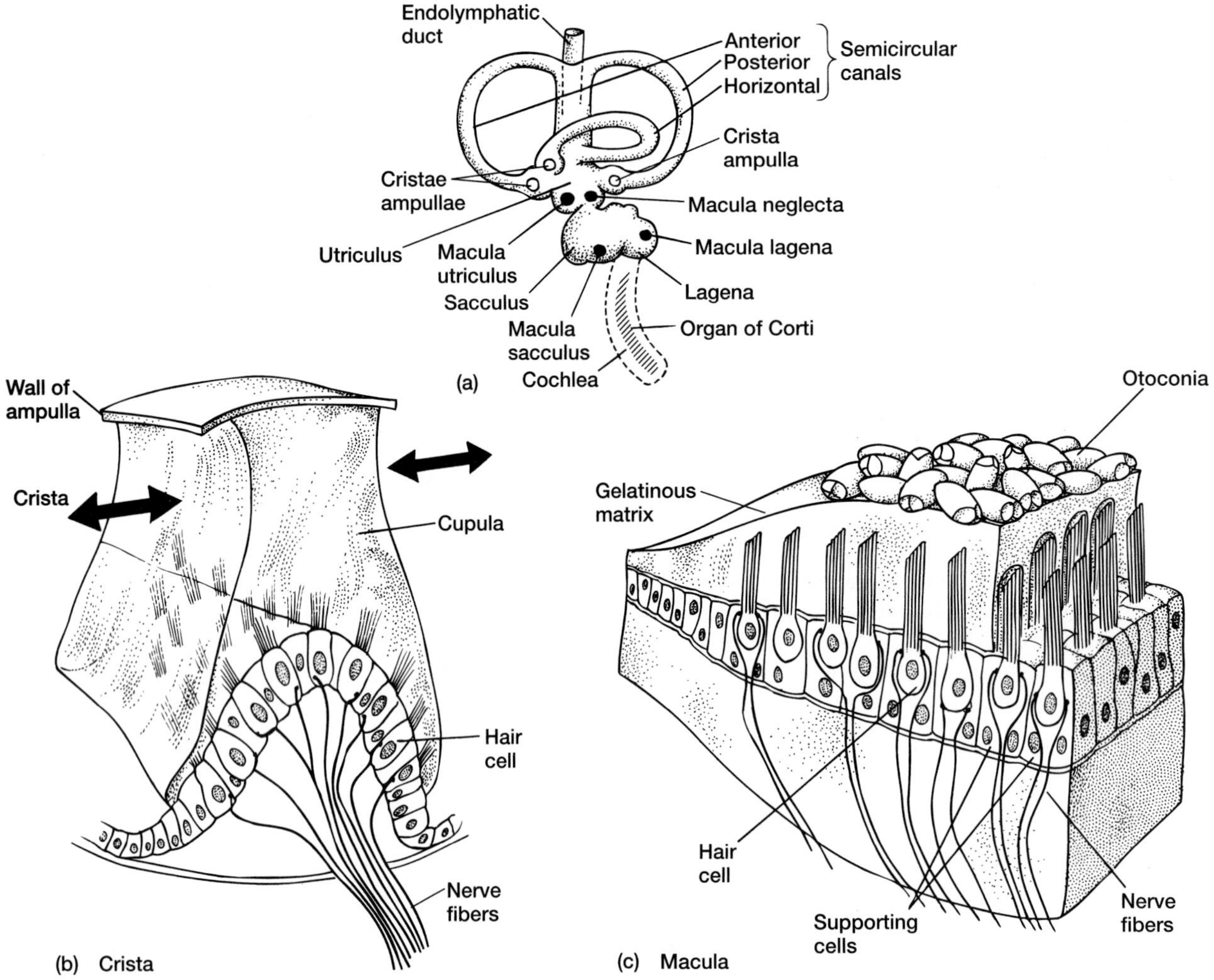


**Figure 17.33** Vestibular apparatus. (a) Generalized vestibular apparatus showing the three semicircular canals and major compartments—utriculus, sacculus, and lagena. (b) The crista is an expanded neuromast organ. One crista resides at the base of each semicircular canal in an enlarged region, the ampulla. The gelatinous cupula extends across the ampula and is attached to the opposite wall. Acceleration of the head (arrows) produces a shearing force of endolymphatic fluid against the cupula, which bends and deforms hair cells embedded within it. (c) The maculae form a neuromast platform containing otoconia. These maculae reside in the three compartments of the vestibular apparatus. They derive their names from these compartments. In some species a fourth macula is present, the macula neglecta.

three adjoining compartments, the external, middle, and inner ears.

*Anatomy of the Ear* The **external ear** is absent in fishes and amphibians. It appears first in reptiles among some lizards and crocodilians, and consists of a short, indented tube, the **external auditory meatus,** that opens to the surface through the **external orifice.** In birds and mammals, the external auditory meatus is elongated. What most people call the "ear" is correctly termed the **pinna.** This external cartilaginous flap surrounding the external orifice is present in most mammals. The irregular shape of the pinna helps to differentiate sounds approaching from different directions and channel them into the external auditory meatus. Just as two eyes with overlapping visual fields give stereoscopic vision, paired ears provide stereophonic hearing.

The **middle ear** consists of three parts, a tympanum, a middle ear cavity or meatus, and one to three tiny bones, or **middle ear ossicles.** The **tympanum,** or tympanic membrane, is present in most tetrapods and first appeared in ancient amphibians. It is a thin taut membrane at the body surface in frogs and a few reptiles, but it is recessed at the bottom of the external auditory meatus in most reptiles as well as birds and mammals.

In primitive tetrapods, the tympanum is stretched across the otic notch. The absence of such a notch in early synapsids and perhaps in several other groups implies the

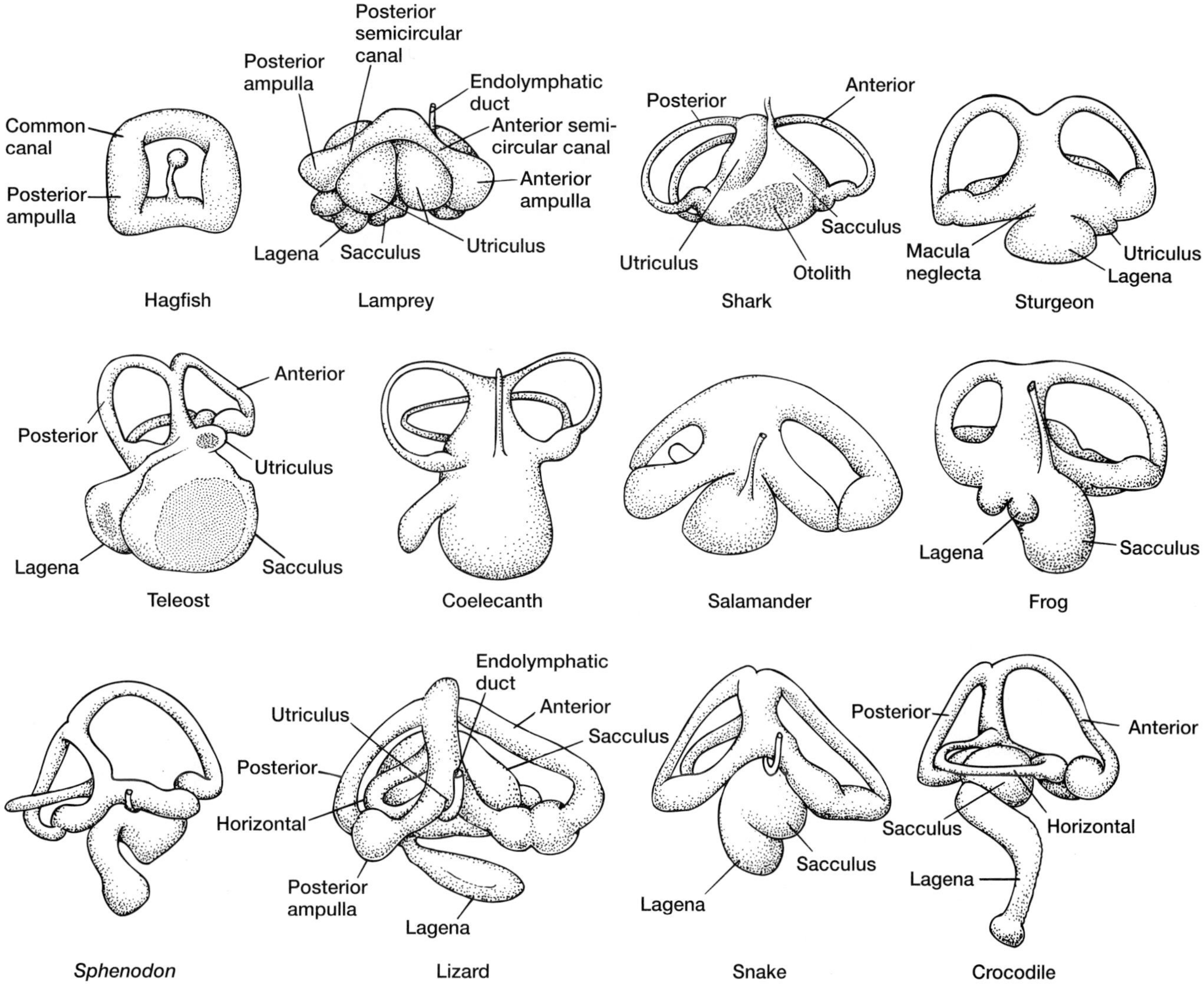


**Figure 17.34** Vertebrate vestibular apparatuses.

absence of the tympanum in ancestors of some modern groups that today have a tympanum. This suggests that the tetrapod tympanum evolved multiple times.

The first pharyngeal pouch enlarges as the tubotympanic recess. Its expanded end forms the **middle ear cavity** (figure 17.35a–e). The remainder of the recess stays open to form the **eustachian tube** that maintains continuity between the middle ear cavity and the pharynx.

The middle ear ossicles lie within the middle ear cavity. Phylogenetically, the most ancient ossicle is the **columella,** which first appeared as a discrete ear ossicle in amphibians. The columella is a small bone that spans the distance from the tympanic membrane to the inner ear. The columella is a much reduced derivative of the fish **hyomandibula,** which functions primarily in jaw suspension in fishes. In some amphibians and in reptiles and birds, the columella is tipped with a cartilaginous extension, the **extracolumella.** This cartilaginous structure, derived from the hyoid arch, rests on the undersurface of the tympanic membrane.

In mammals, there are three middle ear bones. The **stapes** is the reduced reptilian columella. The **incus** and **malleus** are derivatives of the **quadrate** and **articular** bones, respectively. The malleus, incus, and stapes form an articulated chain that spans the middle ear cavity from the tympanic membrane to the inner ear (figure 17.36c).

Evolution of middle ear bones (p. 264)

The **inner ear** includes the vestibular apparatus and the surrounding perilymphatic spaces. As mentioned, in birds and mammals the auditory part of the vestibular apparatus is the tubular lagena. In mammals, the lagena forms a coiled cochlea (figure 17.36b,d). The organ of Corti runs along a central channel suspended within the lagena.

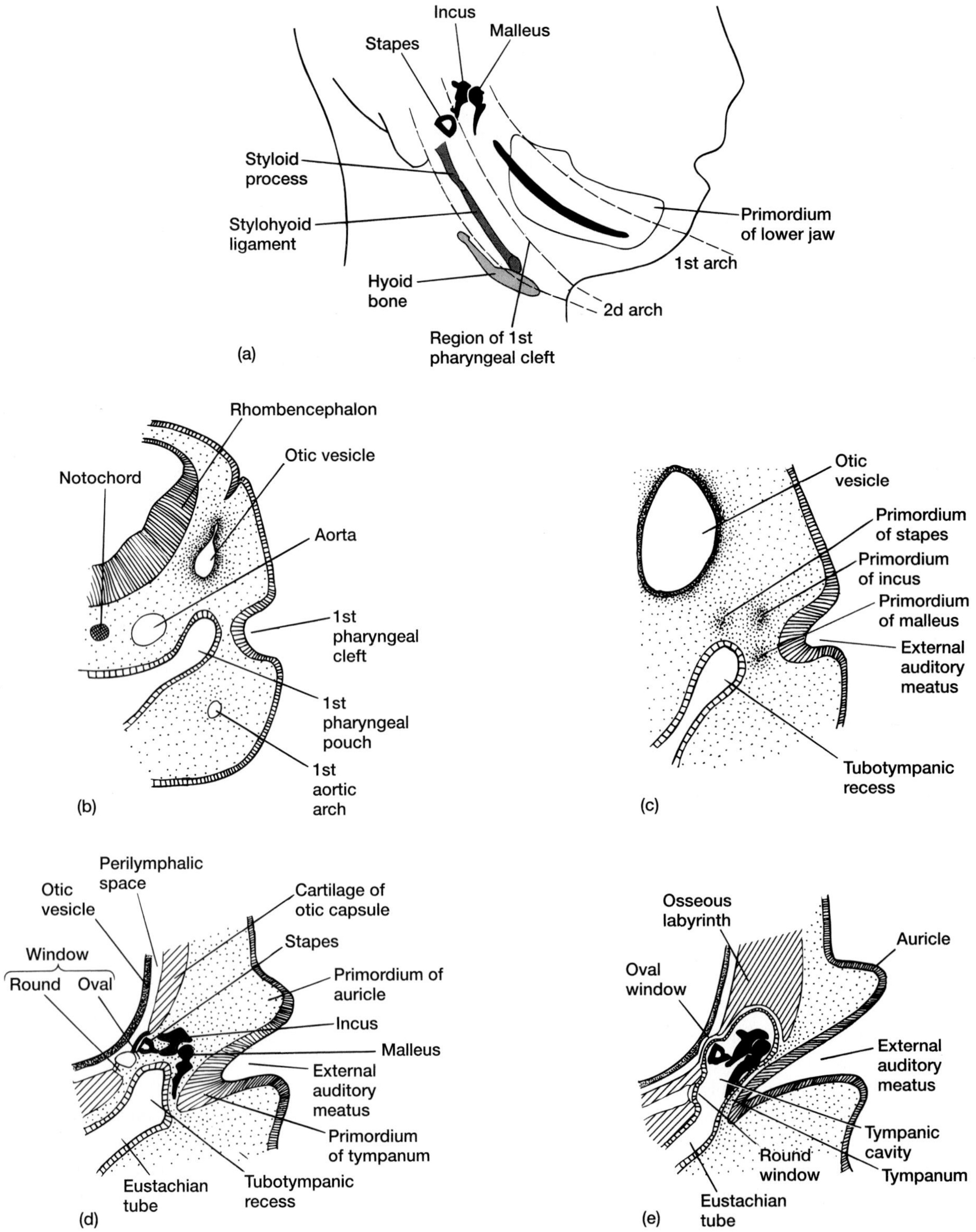


**Figure 17.35** Embryonic formation of the middle ear. (a) Location of the middle ear ossicles relative to derivatives of the splanchnocranium. (b) The surface of the ectoderm thickens, forming an otic placode that sinks beneath the skin and gives rise to the otic vesicle. The otic vesicle moves into the vicinity of the first pharyngeal cleft and pharyngeal pouch. (c) Mesenchyme (indicated by heavy stippling) begins to condense and differentiate into the ear ossicles—the incus, the malleus, and the stapes (d,e).

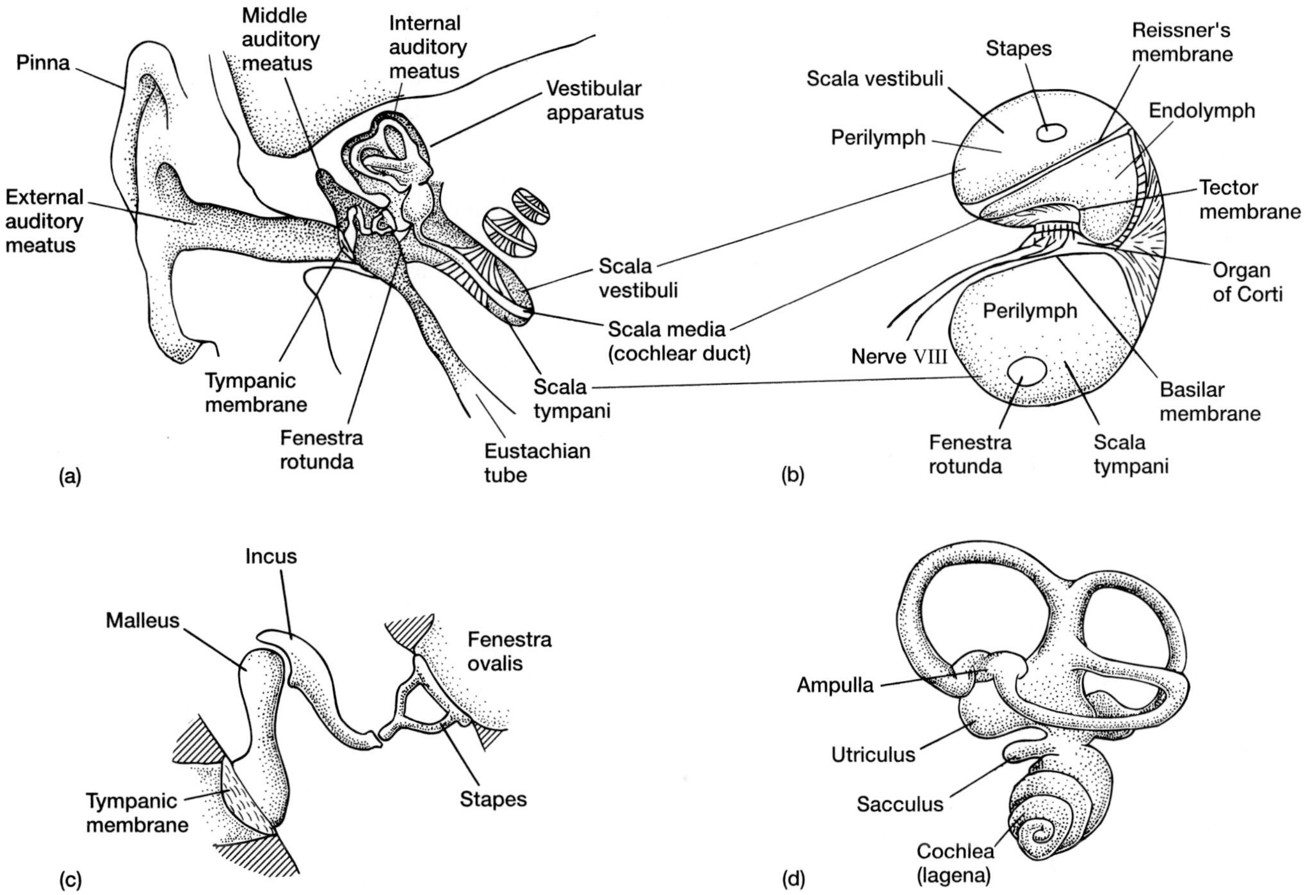


**Figure 17.36** Anatomy of the ear. (a) External, middle, and inner ears. (b) Cross section of the cochlea. (c) Three middle ear ossicles of mammals. (d) Mammalian vestibular apparatus. Note that the lagena is lengthened and coiled to form the cochlea.

Two parallel perilymphatic channels run on either side. Thus, the cochlea consists of three coiled fluid-filled channels. The two perilymphatic channels are the **scala vestibuli** and the **scala tympani,** and the canal between them is usually called the **scala media** (cochlear duct). The **basilar membrane** separates the scala tympani from the scala media, and the organ of Corti vibrates with the basilar membrane in response to sound waves. In many vertebrates, the hair bundles of the organ of Corti are embedded in a firm plate, the **tectorial membrane. Reissner's membrane** is located between the scala vestibuli and scala media (figure 17.36a,b).

Sound enters the inner ear through the **fenestra ovalis,** or oval window (figure 17.36c). One end of the thin columella (or stapes) expands into a pedicle or **footplate** that occupies this window so that sound waves pass from this ear ossicle to the fluid filling the chambers of the inner ear. The **fenestra rotunda,** or round window, at the end of the perilymphatic channels is sealed by a flexible membrane (figure 17.36a,b). The auditory apparatus, containing sensitive hair cells, floats in fluid between the fenestrae ovalis and rotunda. In all vertebrates, sounds conveyed to this fluid vibrate and mechanically stimulate the auditory receptors.

## *Functions of the Ear*

***Fishes*** The inner ear of fishes is very similar to that of tetrapods. There are three semicircular canals (except in cyclostomes, which have two) and three otolithic regions—the sacculus, the utriculus, and the lagena. In addition, a few fishes and a few tetrapods have a **macula neglecta,** a supplementary sensory area near the utriculus.

In fishes, hearing generally involves the sacculus and the lagena, although sometimes one region predominates. The maculae within the sacculus and lagena are the sound receptors. When hair cells of the maculae are set in motion by sound vibrations, they move against the relatively dense and stationary otoliths. Differences in size and shape of the otoconia are thought to lead to slight differences in hair cell stimulation and thus allow different sound frequencies to be detected. Further, the hair cells within the sacculus and the lagena are oriented along perpendicular axes (figure 17.37). Therefore, motion in one direction stimulates one set of hair cells maximally and the other set, minimally. Signals from different directions apparently produce different stimuli that are used by the nervous system to pinpoint the source of a sound.

Fish otoconia usually consist of secreted calcium grains, but in some bottom-dwelling elasmobranchs, sand grains enter through the open endolymphatic duct and settle on the maculae. As mentioned, the maculae of fishes are sensors of gravity and motion as well as of sound.

Sound reaches and stimulates the inner ear via several routes. The tissues of a fish and its environment are primarily composed of water and have similar frequency amplitudes. Thus sound waves pass directly from water to the inner ear and set hair cells in motion. The dense otoconia oscillate at a different frequency from the associated hair cells causing the hair bundles to bend and stimulate these sensory cells.

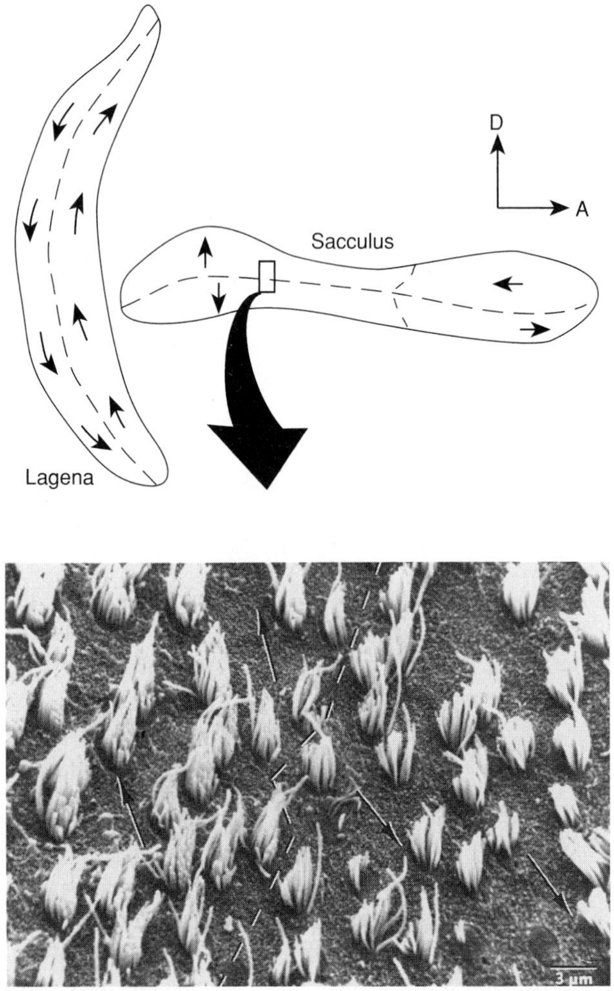


**Figure 17.37** Fish inner ear. Orientations of hair cells within the sacculus and lagena are shown. Arrows indicate the side of the hair bundle on which the kinocilium lies. Dashed lines indicate the divisions between bundles with different orientations. These differences in the placement of hair bundles cause different responses to mechanical vibrations traveling in the surrounding fluid.

Sounds are transmitted to the inner ear of fishes through additional routes. In some species, extensions of the swim bladder come in direct contact with the inner ear. Vibrations picked up by the swim bladder are conveyed directly to the sound-detection apparatus (figure 17.38a). Furthermore, because the air-filled swim bladder and fish's body tissues are very different densities, the light swim bladder may be more responsive to sound and act as a resonator to enhance sound detection. In other species, the swim bladder is connected to the inner ear through a series of tiny bones called **Weberian ossicles** that carry vibrations to sound detectors within the sacculus and lagena (figure 17.38b).

***Tetrapods*** Ears adapted for hearing in an aquatic environment were brought to an air environment during the vertebrate transition from water to land. Most anatomical changes in the tetrapod ear came about to deliver sound energy traveling in air to the small fluid spaces of the inner ear. At issue is a physical problem of **impedance matching** that arises from the different responses water and air have to vibrations. Fluid is thick and air is thin, so the two media differ in the amount of sound energy necessary to vibrate molecules. For a fish in water, sound passes easily from the fluid aquatic environment into the fluid inner ear environment. But for an animal living in an air environment, the responsiveness of inner ear fluid to sound waves differs from the air's responsivenes to sound waves, producing a water/air boundary. Because of its greater viscosity, fluid resists being set in motion by arriving airborne sounds. Consequently, most airborne sounds are reflected away from the ear and go undetected. The structures of the middle ear are largely involved in impedance matching, which means that they gather, concentrate, and deliver airborne sounds to the inner ear at a sufficient level to impart these vibrations to the fluid-filled spaces of the inner ear. This allows the strategically positioned sensory hair cells within the inner ear to respond to oscillations traveling through the surrounding fluid.

Arriving sound waves set the tympanic membrane in motion. It in turn affects the middle ear ossicles. These tiny bones function in three critical ways: (1) they act as a lever system to transmit vibrations to the fenestra ovalis, (2) they transform sound waves in air into sound waves in fluid, and (3) they amplify the sound. The tympanic membrane may be over ten times the area of the fenestra ovalis to which the ossicles transmit sound. By collecting sound over a large area and focusing it on a much smaller area, the tympanic membrane and fenestra ovalis effectively amplify sounds.

The columella or stapes apparently does not act like a piston. Instead it rocks within the fenestra ovalis to disturb the perilymphatic fluid filling the inner ear. This fluid is

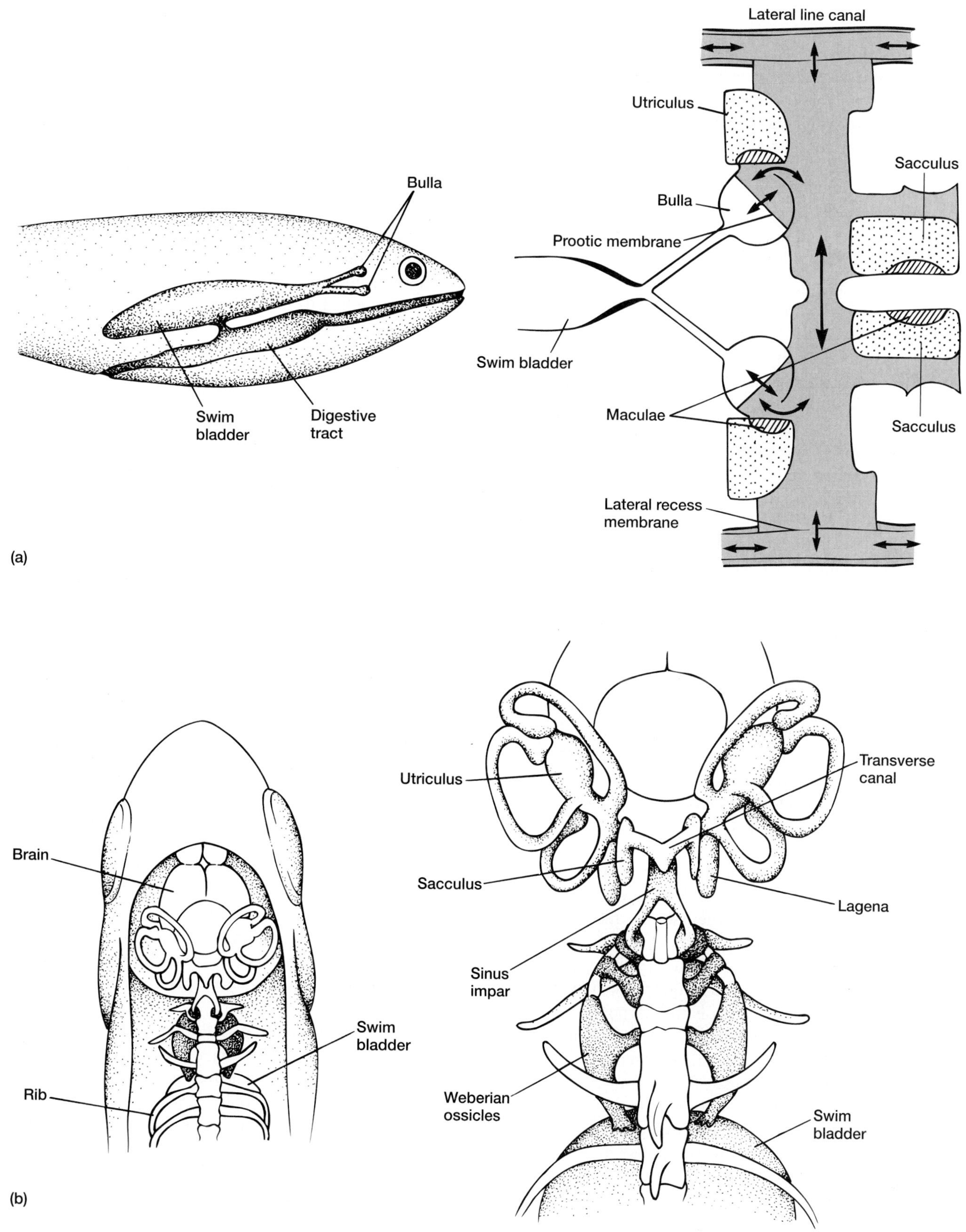


**Figure 17.38** Route of sound transfer to the inner ears of fishes. (a) In some fishes, the swim bladder includes anterior extensions that contact the inner ear. (b) In other fishes, the Weberian ossicles, a tiny series of bones, connect the swim bladder to the inner ear.

incompressible and would resist such compression if it were not for the membrane-covered fenestra rotunda at the end of the perilymphatic channels, which allows some flexibility in the movements of the fluid as sound is transmitted to the inner ear. It has been suggested that the fenestra rotunda helps dampen sounds once they have made a first pass through the cochlea and prevents these waves from ricocheting around the inner ear.

***Amphibians*** There are two auditory receptors in amphibians, the **papilla amphibiorum** (amphibian papilla), unique to amphibians and the **papilla basilaris** (basilar papilla), a possible forerunner of the organ of Corti in later tetrapods. Both are specialized neuromast organs (figure 17.39a).

The best-studied amphibian ears are those of frogs. Most fossil amphibians possess tympanic membranes much like those of frogs. The lagena and utriculus are likely the vestibular receptors, and the sacculus, containing the papillae amphibiorum and basilaris, seems to be the primary site of sound detection. Typically the tympanic membrane in frogs is flush with the surface of the skin. The extracolumella and columella deliver these vibrations in series to the inner ear. The footplate of the columella shares the fenestra ovalis with a small movable bone, the **operculum.** The **opercularis** is a tiny muscle that joins the operculum to the suprascapula of the pectoral girdle. The opercularis is a derivative of the levator scapulae muscle. The tiny **columellaris muscle** is also thought to be a derivative of the levator scapulae and runs from the suprascapula to the columella (figure 17.39a). Both muscles connect the inner ear to the pectoral girdle and, indirectly via the limb, to the ground. In this way, the muscles introduce routes through which **seismic** sound waves may travel from the ground to the inner ear.

Thus, sound reaches the inner ear via the opercular and columellar pathways and sound waves vibrate the fluid in the inner ear, stimulating the auditory receptors. One method frogs use to discriminate sound frequencies apparently takes advantage of these two routes. The papilla amphibiorum responds best to low frequency sounds arriving via the operculum. The papilla basilar responds to higher frequency sounds arriving via the columella. The middle ear muscle enhances this discrimination pattern. Contraction of the opercularis muscle (and relaxation of the columellaris) leaves the columella free to vibrate in the oval window. The reverse situation, contraction of the columellaris and relaxation of the opercularis, immobilizes the columella. In addition, these tiny muscles protect the auditory receptors from violent stimulation by selective contraction. In this way, they might also enhance the inner ear's ability to discriminate different sound frequencies or select sounds arriving along different acoustic pathways.

To localize sounds, frogs might take advantage of another feature of middle ear design. Eustachian tubes join the left and right middle ears through the buccal cavity. Pressure generated by the tympanic membrane responding on one side of the head is transmitted to the buccal cavity, which may act as a resonator before it transmits the sound to the opposite tympanic membrane. Such coupling of tympanic membrane vibrations by a resonator between them means that sound reaching the left and right ears has different acoustic qualities. This difference allows frogs to localize sound sources (figure 17.39b).

In modern salamanders, vibrations transmitted through the columella stimulate the inner ear; however, the tympanic membrane is absent. Instead, the tip of the columella is attached to the squamosal bone through a short **squamosal-columellar ligament.** As in anurans, the footplate of the columella shares the fenestra ovalis with the operculum. Only the opercularis muscle is present in salamanders. In most cases it is derived from the levator scapulae muscle, but in one family, it arises from the cucullaris muscle. Two routes of sound transmission to the salamander inner ear are thus available—one from the squamosal bone to the columella and another from the pectoral girdle to the operculum (figure 17.40a).

Although the routes sound travels to the inner ear are known, the mechanism of reception and the method of processing sound waves in salamanders are not clearly understood. When a salamander is in water, impedance is low and sound reaches the inner ear with little reflected loss. However, lacking a tympanic membrane, the salamander ear would seem to be poorly suited to detection of airborne sounds. Further, the fenestra rotunda is also absent, possibly making the fluids in the inner ear less responsive to pressure vibrations. These observations led to the suggestion that salamanders are deaf in air; however, experimental work refutes this. In fact, salamanders on land respond to seismic and airborne sounds, although their sensitivity to such sounds is less than that of anurans.

Sounds that reach the fenestra ovalis set the fluid of the inner ear in motion. In salamanders, these vibrations ripple across the skull via a channel filled with cerebrospinal fluid in order to reach the inner ear of the opposite side (figure 17.40b). Along this internal route lie the papillae amphibiora that these passing vibrations stimulate.

***Reptiles*** In most reptiles, airborne sounds vibrate the tympanic membrane. This imparts motion to the extracolumella and columella that transmit these vibrations to the inner ear (figure 17.41a,b). The primary sound-sensitive area within the inner ear is the slightly expanded lagena. The primary receptor is the **auditory papilla,** which is stimulated by sound vibrations transmitted to the fluid of the inner ear (figure 17.41c). In addition, various supplemental auditory receptors are present in some species.

Snake ears, unlike those of most reptiles, lack a tympanic membrane. Through a short ligament, the columella is attached at one end to the quadrate bone of the upper jaw, and the opposite end fits into the fenestra ovalis. Despite the popular opinion that snakes are deaf, experi-

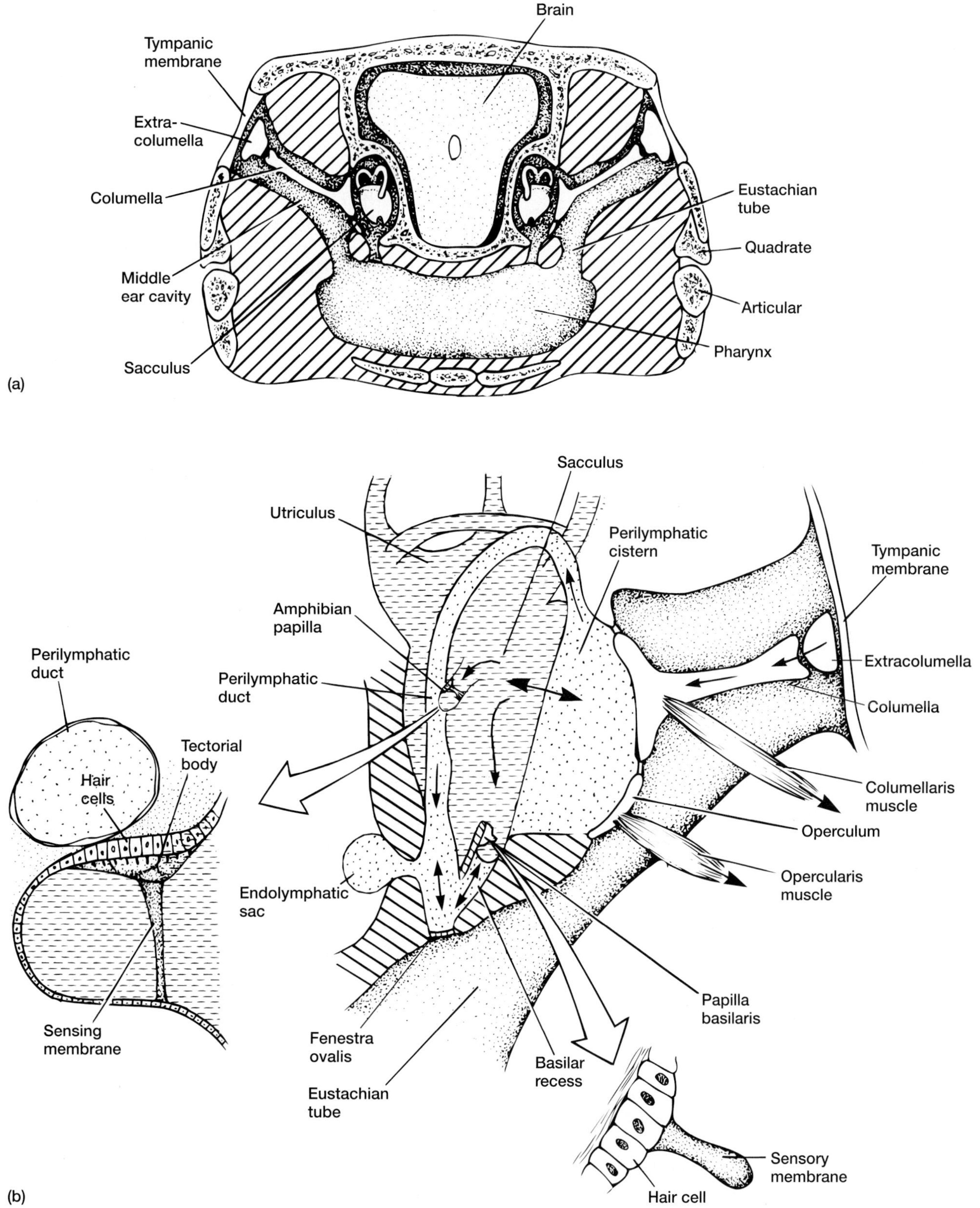


**Figure 17.39** Hearing in frogs. (a) Cross section through the head of a frog. Because the eustachian tubes connect the two ears through the pharynx, a sound that sets one tympanum in motion also affects the ear on the opposite side by producing vibrations in the connecting air passageway. This is thought to allow frogs to localize the source of sounds. (b) Sound arrives at a frog's inner ear via two routes: one involves the tympanum-columella and the other involves the opercularis muscle-operculum. Vibrations arriving by either pathway cause the fluid in the inner ear to vibrate. This vibration stimulates the auditory receptors. The way in which these receptors discriminate sounds is not well understood, but because they seem to be modified neuromast organs, it is thought that they respond selectively to shearing oscillations that arriving vibrations imparted to the fluid of the inner ear.

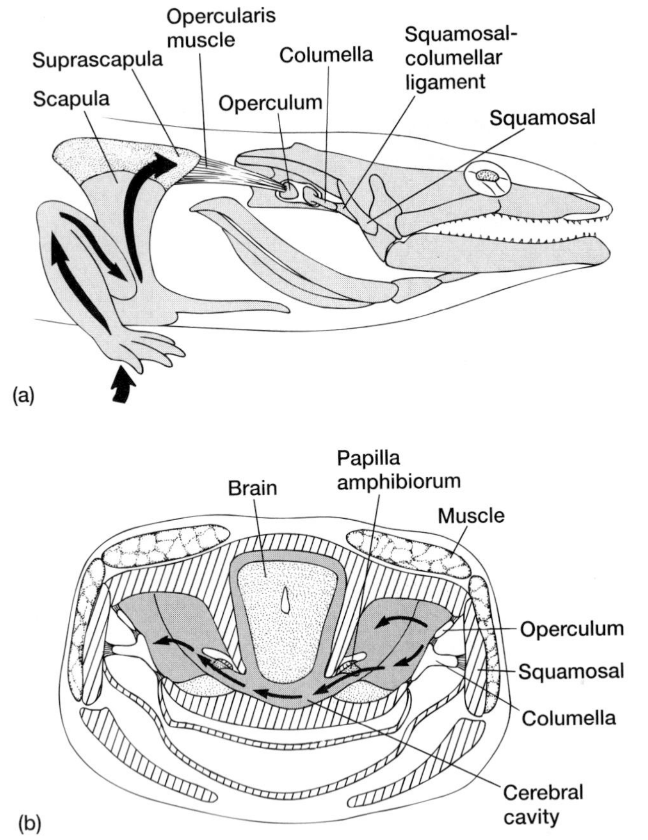


**Figure 17.40** Hearing in salamanders. (a) In many salamanders, sounds reach the inner ear via a squamosal-columella route and via the opercular muscle from the scapula. (b) The two inner ears on opposite sides of the head are connected via a fluid-filled channel that passes through the cerebral cavity. This channel may allow sonic vibrations to spread from one ear to the other (solid arrows).

mental work disproves this. Recordings of electrical activity from the areas of the brain to which auditory nerves travel confirm that the inner ear of snakes is responsive to seismic and airborne sounds, although the range of sensitivity is somewhat restricted.

***Birds*** Because the lagena of the avian inner ear is lengthened in comparison to the reptile's, the stand of hair cells is drawn out into a long strip (figure 17.42a,b). Hair bundles are usually embedded in the tectorial membrane, a continuous sheet that acts to increase shear on these bundles as hair cells are agitated by sound vibrations.

In owls, the tightly packed, raised rim of facial feathers that give owls their "cute" bespectacled look is the functional equivalent of the pinna. This facial ruff of feathers channels sounds into the auditory meatus (figure 17.43a). In an individual owl, the left and right external orifices and their associated auditory meatuses are different in size and shape (figure 17.43b,c), resulting in two ears that produce different acoustic qualities from many directions. The nervous system takes advantage of this to increase the precision with which sources of sound can be pinpointed in the habitat.

***Mammals*** In most mammals, the pinna deflects sounds into the external auditory meatus, where they vibrate the tympanic membrane. All mammals have three ear ossicles that amplify these vibrations and carry them to the fenestra ovalis (figure 17.44a,b). From the fenestra ovalis, vibrations ripple through the fluid in the extensive cochlea. As mentioned, the cochlea is composed of three parallel channels. The middle channel includes the organ of Corti, which consists of an outer and inner row of hair cells. There may be 20,000 to 25,000 of these cells. Hair bundles are embedded in the tectorial membrane (figure 17.44c).

***Discrimination of Different Frequencies*** Sound waves stimulate the neuromast organs (e.g., auditory papilla, organ of Corti), the sites of sound sensation. In some species, the hair bundles protrude directly into the surrounding fluid. When this fluid vibrates in response to sound waves, the bundles are bent by the moving fluid and stimulated. In most higher vertebrates, hair bundles are embedded in the tectorial membrane. Sound waves impart motion to the tectorial membrane, which then deflects the hair bundles. This action stimulates hair cells.

In addition to being able to detect sound waves, auditory receptors can discriminate between different frequencies (tones) of sound. Thus, the inner ear is also a resonance analyzer. Hair cells are tuned into only a narrow range of frequencies. In mammals, differences in the orientation of hair cells in the inner and outer rows of the organ of Corti produce differences in sensitivity in different regions. Sequential grading of the tuned hair cells along the organ of Corti produces tone discrimination over a range of frequencies.

However, there may be more to tone detection than the tuning of hair cells. In mammals, the basilar membrane on which the organ of Corti resides changes gradually in width as the cochlea twists. The basilar membrane may be like a harp in that each section of the membrane may resonate only to a distinctive range of frequencies. If this is the case, tones that enter the cochlea impart the greatest motion to the section of the basilar membrane that corresponds to their frequency; thus, specific tones stimulate specific sections of hair cells. It has been suggested that such differential stimulation of hair cells could contribute to tone discrimination (figure 17.45).

***Evolution of Hearing*** Any organism moving through its environment produces sounds that betray its approach. It is not hard to imagine the adaptive advantages of a sound-detection system that alerts an organism to another's presence. Although the importance of the sense of hearing is easy to understand for audiophiles like ourselves, who have a well-developed ability to discriminate a wide range of

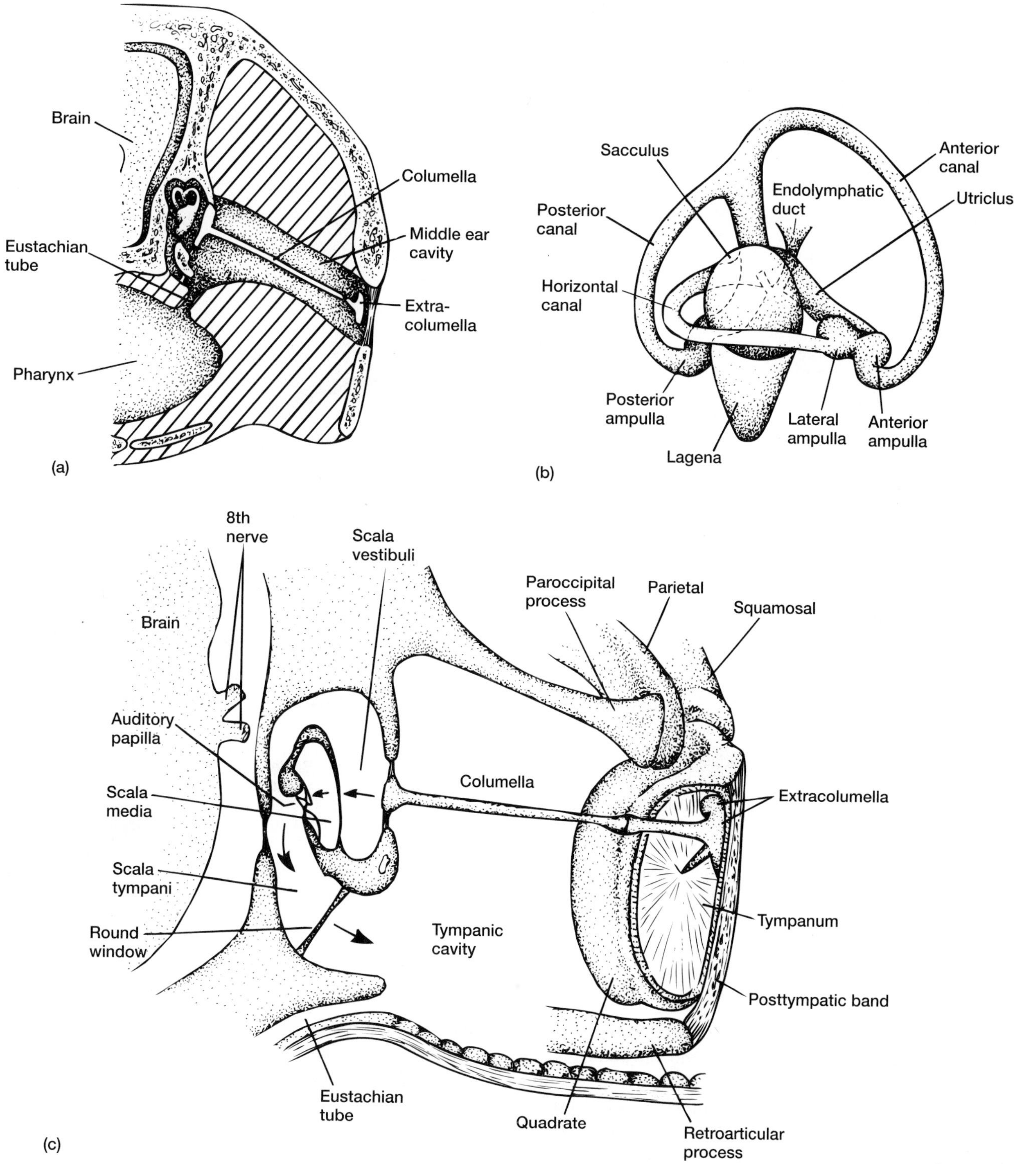


**Figure 17.41** Hearing in reptiles. (a) Cross section through a reptilian head. (b) Vestibular apparatus of a lizard. (c) Section through the ear of an iguana showing the relationship of tympanum, extracolumella, columella, and inner ear.

sounds in our environment, there are still many unanswered questions about the evolution of hearing.

One of the major evolutionary changes in the auditory system occurred during the transition from water to land. This transition involved the appearance or enhancement of the middle and external ears, structures that collect airborne sounds and match them to impedance properties of the inner ear fluid. In addition, the ears of early tetrapods were also able to receive seismic sounds as well. As important as this water-to-land transition was, it has been

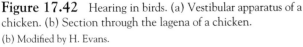

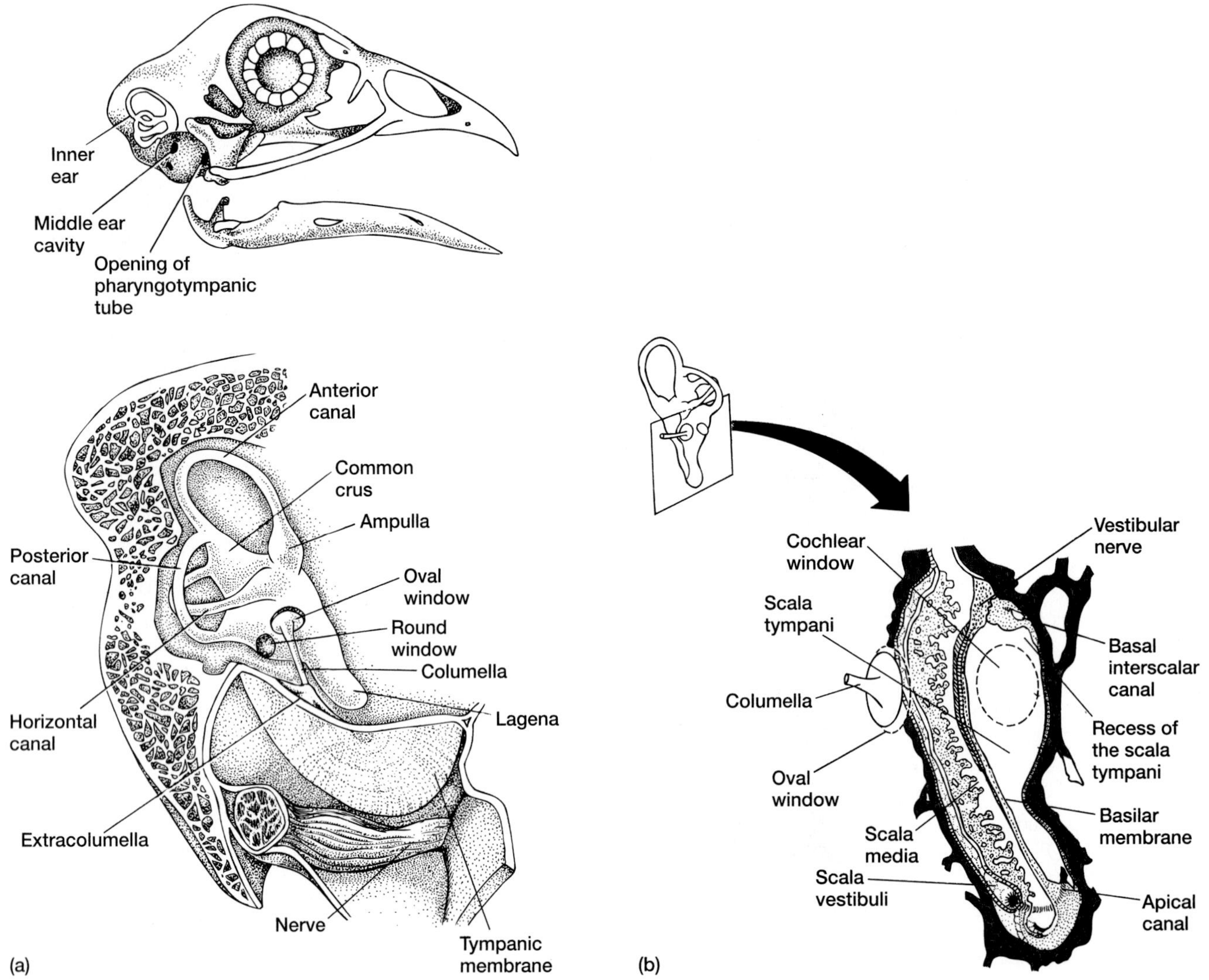


**Figure 17.42** Hearing in birds. (a) Vestibular apparatus of a chicken. (b) Section through the lagena of a chicken.
(b) Modified by H. Evans.

difficult to study directly because there are no living forms close to those of the ancestral labyrinthodont tetrapods.

Early physiologists concentrated on living amphibians. The frog, with its tympanic membrane and sensitivity to airborne sounds, has received much attention. However, it has been suggested recently that the ear of living amphibians may be an independent innovation quite different from the ear that reptiles, birds, and mammals inherited from labyrinthodonts. Some authorities even suggest that living amphibians trace their ancestry to a separate group of primitive fishes rather than sharing an ancestor with amniotes. For our purposes, the point to bear in mind is that the amniote ear may not trace its ancestry through amphibian forms similar to living amphibians.

The inner ear of living amphibians includes two major auditory receptors. The primary auditory receptor is the papilla amphibiorum, a sensory receptor found only in living amphibians. The other is the papilla basilaris that each order of living amphibians seems to have evolved independently. In anurans, the papilla basilaris resides in the sacculus. In salamanders, it is located in the lagena, and in caecilians, in the utriculus. The auditory receptor cells of amphibians are perched on an immobile base attached to the cranium. The hair bundles project into the fluid that sets them in motion. In reptiles, the opposite is the case. Sensory cells move on a flexible basilar membrane, and hair bundles are usually restrained by the tectorial membrane. These differences between amphibians and amniotes, and between various amphibian orders, have made it difficult to construct in detail the early evolution of the tetrapod ear.

Another unresolved problem involves the evolution of the mammalian middle ear ossicles. A good series of fos-

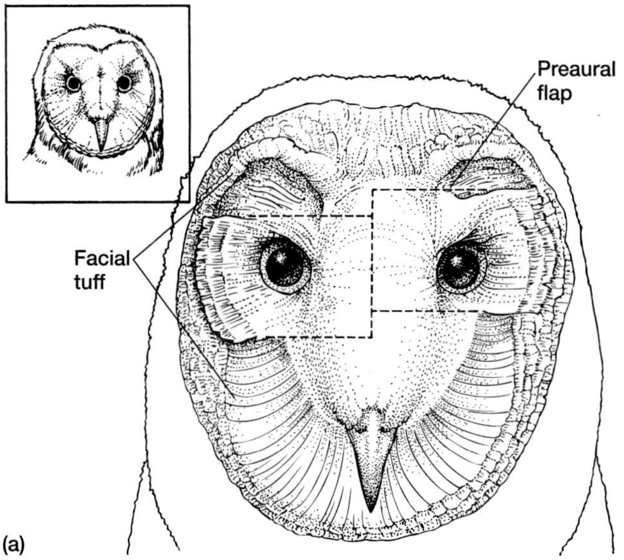


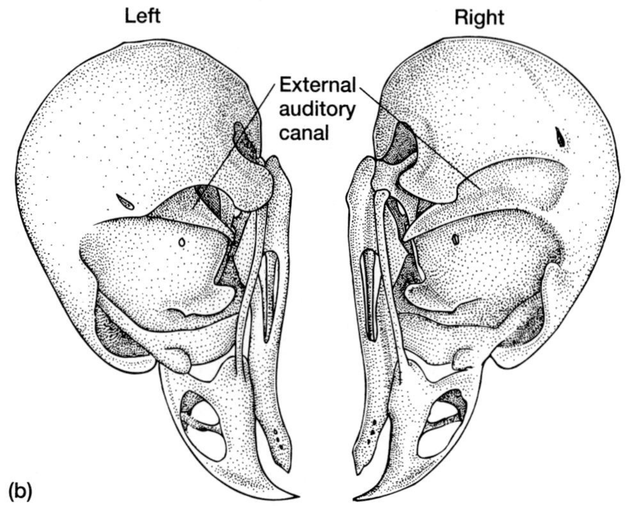


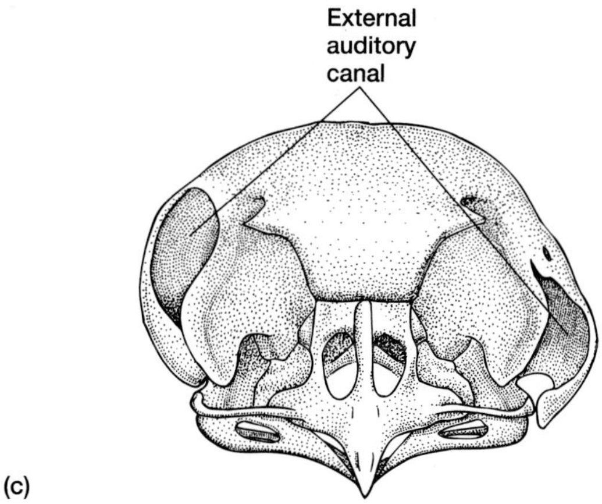


**Figure 17.43** Auditory acuity in owls. (a) Facial disk feathers in normal position (inset) have been removed to reveal sets of auditory feathers. In this barn owl, tightly packed parabolic rims of feathers encircle the face and external orifice of the ear. This facial ruff of feathers, as it is called, collects and directs sounds to the external orifice of the ear. Note the asymmetrical positioning of the preaural flaps of feathers (dashed lines). (b) Left and right sides of the skull of a Tengmalm's owl. There are slight differences in the size of the external auditory canal. (c) Front view showing asymmetry of otic areas in Tengmalm's owl.

sils reveals the anatomical changes that occurred in the transition from reptiles to mammals, but the adaptive advantages favoring such changes are still being debated. The therapsid ancestors of modern mammals tended to walk or run with their bodies off the ground. Some authorities suggest that when therapsids lifted their heads from the ground, seismic sounds could no longer reach the inner ear through their jaws. That left only a long and presumably inefficient route of sound transmission via the limbs. Airborne sounds then became especially important. This may have favored the evolution of a more efficient auditory system. The three middle ear ossicles and the enlarged cochlea allow modern mammals to perceive a greater range of sound frequencies than most other vertebrates experience.

However, the therapsid ear was not the only region of the skull undergoing profound changes. The jaws were changing as well. All dermal bones except the dentary were lost from the lower jaw. These changes were probably related to changes in feeding requirements and in forces that jaw adductors exerted on the mandible. During this transition, some bones that were lost from the lower jaw assumed new roles in hearing. Reduced articular and quadrate bones gave rise to the malleus and incus, respectively. Together with the stapes (derived from the reptilian columella), these three bones now constitute the three middle ear ossicles. Thus, evolution of the mammalian auditory and feeding systems was coupled. This evolutionary pattern has complicated the analysis of each system because it is not yet clear if changes in one system drove the other or if the two evolved under different selective pressures.

**Evolution of middle ear bones (p. 264)**

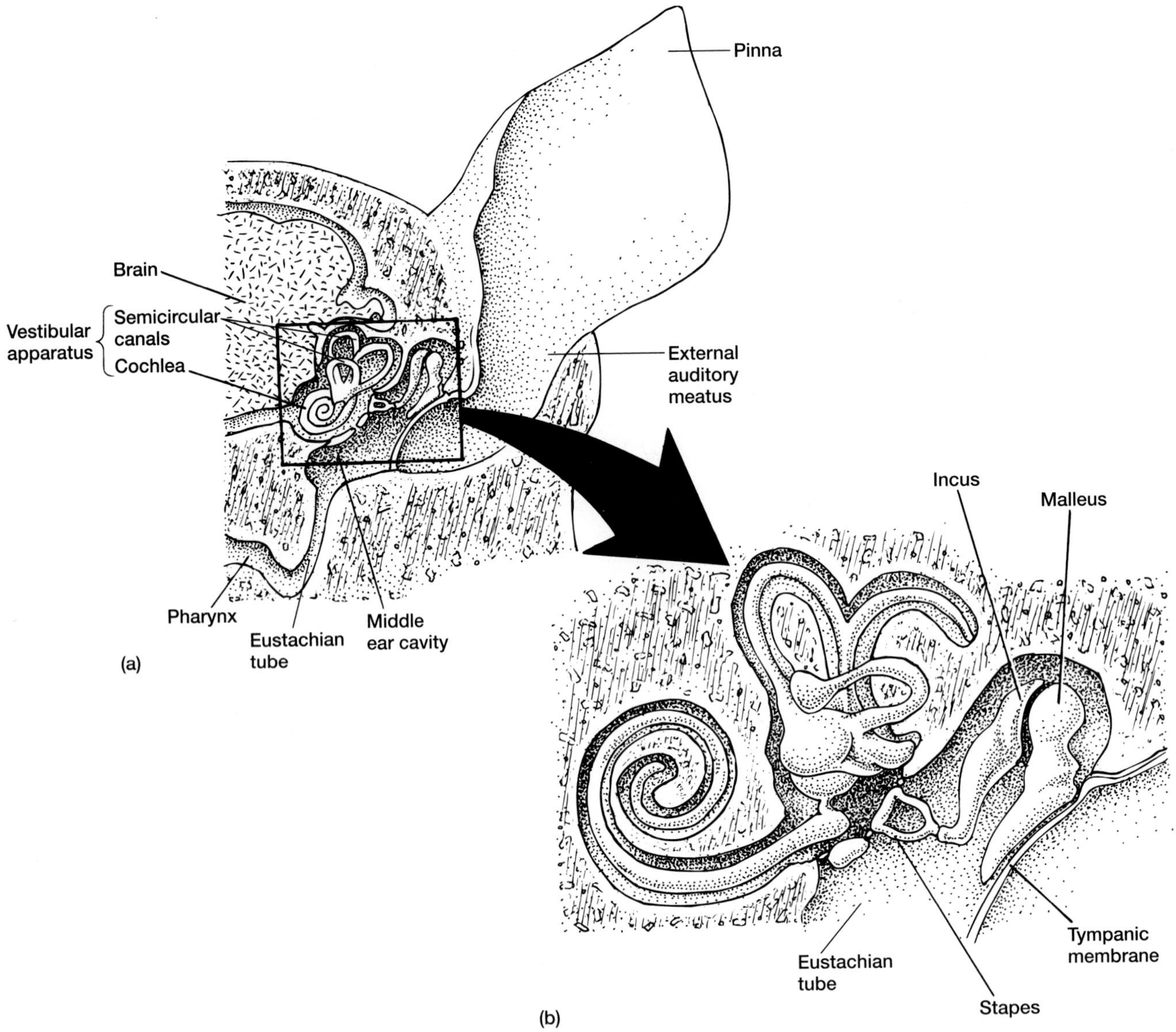


**Figure 17.44** Mammalian ear. (a) Cross section through a mammalian skull. (b) Internal structure of the cochlea. (c) Section through the organ of Corti showing inner and outer rows of hair cells and tectorial membrane in which hair bundles are embedded. Sound waves travel first in the scala vestibuli (solid arrows) before passing at the apex of the cochlea into the scala tympani (dashed arrows).

## Electroreceptors

### *Structure and Phylogeny*

Most fishes, but not tetrapods, possess **electroreceptors,** sensory receptors that are responsive to weak electrical fields. Electroreceptors are modified neuromast organs located in pits within the skin that predominately are concentrated on the fish's head. There are two types of electroreceptors. An **ampullary receptor** contains supportive cells that lie at the bottom of a narrow channel filled with a gelatinous mucopolysaccharide (figure 17.46a). Afferent neurons embrace the receptor cells.

The **tuberous receptor** lies buried under the skin in an invagination beneath a loose layer of epithelial cells (figure 17.46b). This loose layer of epithelial cells may differentiate into **covering cells** over the sensory cells and a superficial set of **plug cells.** This type of electroreceptor is responsive to higher freqencies than the ampullary receptor and is generally adapted to receive electric discharges of the fish's own electric organ. Hence, tuberous receptors are known only in electric fishes so far.

Electroreceptors appeared very early in fish evolution judging from the presence of pits in the dermal bone of acanthodians and some groups of ostracoderm fishes. Among living fishes, electroreceptors are found in all elas-

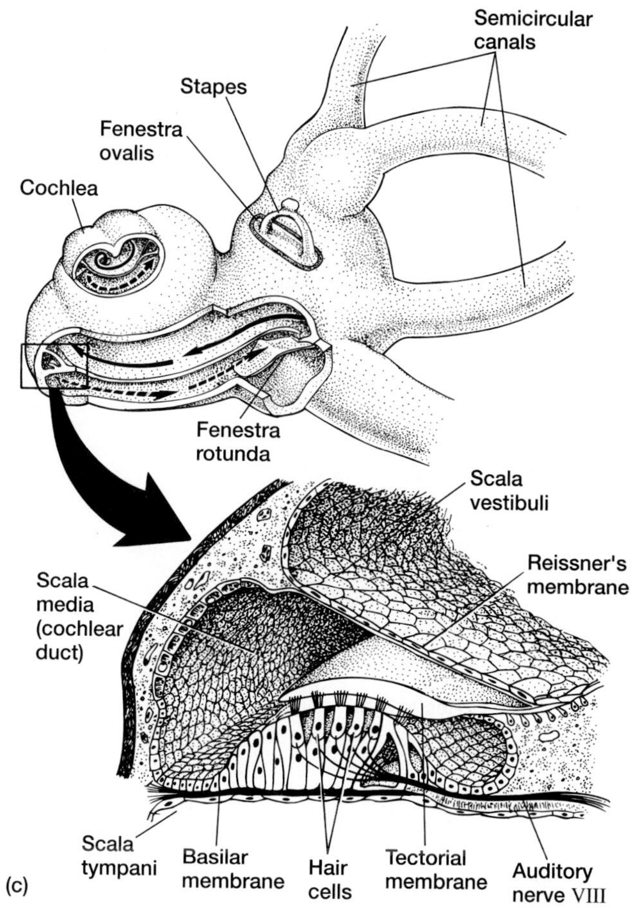


**Figure 17.44** (continued)

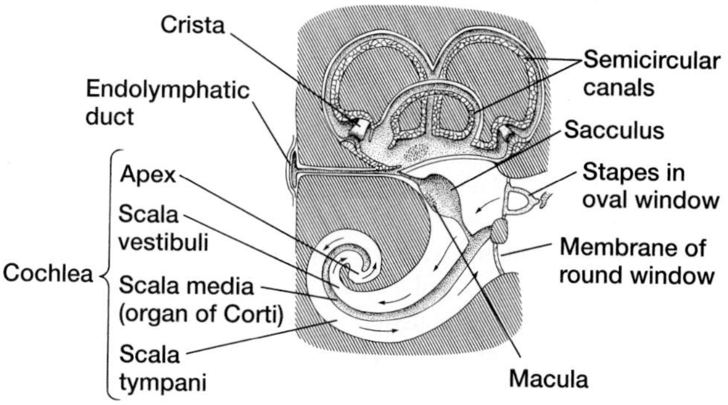


**Figure 17.45** Distribution of sound vibrations through the cochlea. The stapes delivers vibrations at the fenestra ovalis. These vibrations spread through the perilymph within the chamber of the scala vestibuli and around the tip of the cochlea into the connecting chamber of the scala tympani. The cochlear duct, containing the organ of Corti, lies between these two chambers. Passing vibrations are thought to stimulate appropriated sections within this organ. The flexible membrane across the fenestra rotunda serves to dampen sound waves and prevent their rebound back through the cochlea.

mobranchs, catfishes, sturgeons, some teleosts, and the lungfish *Protopterus* (see illustration of electroreceptors of a skate in figure 17.46c).

## *Form and Function*

In electric fishes, specialized blocks of muscle called **electroplaques** form the electric organ. In some, the electric organ can generate a sudden jolt of voltage to stun prey or thwart a predator.

**Electric organs (p. 354)**

In most electric fishes, however, the electric organ produces a mild electrical field around the fish. Electrically conductive and nonconductive objects that enter this field have different effects on the flow of current produced. Living animals, such as other fishes, are relatively salty, making them conductive and causing the lines of the electric field to converge. Rocks are usually nonconductive and cause the current lines to diverge (figure 17.47a). Electroreceptors are sensitive to these distortions in the surrounding electric field. This type of electroreceptor is common in freshwater fishes that live in murky waters or hunt at night. They use this information to navigate and detect prey. The electric fish *Gymnarchus* holds its body rigid to align generating and receiving receptors throughout the body (figure 17.47b).

Patterns of electric organ discharge are also used in intraspecific communication. These patterns change with social circumstances. For example, fishes use electrical communication to recognize the sex and species of other fishes. From these electric signals, they detect threat, submission, and courtship.

Many fishes that do not actively generate their own electrical field nevertheless possess electroreceptors. Elasmobranchs, some teleosts, sturgeons, catfishes, and others are endowed with abundant electroreceptors across their heads, especially concentrated around the mouth. These organs are sensitive to stray electrical fields produced by muscle contractions of prey animals. Sharks, for example, can localize the weak electrical fields produced by the breathing muscles of prey animals buried out of sight in sand or loose sediment (figure 17.47c).

Electroreceptors send spontaneous impulses to the central nervous system at a regular rate. This rate increases or decreases with distortions in the electrical field or with stimuli from stray electrical discharges that prey generate. The mechanisms by which such electrical fields stimulate receptor cells is not clear. Impulses are thought to be transmitted from electroreceptors across synapses to afferent neurons and then to the central nervous system. Most information that electroreceptors gather is transmitted directly to the cerebellum. In fishes that depend on such information, the cerebellum is often enlarged (figure 17.48a,b).

Some electroreceptors respond very fast to stimuli. They are designed to relay information rapidly to the

## Box Essay 17.2 Throwing Light on the Subject

Many fishes carry about their own light source in specialized structures termed **light organs** or **photophores.** The fishes themselves do not produce light, but they carry bioluminescent bacteria in specially designed skin pouches. These pouches of bacteria glow continuously, so to turn off the light, the fish flips them into a black-lined pocket, lifts a shutterlike cover over them, or pulls the pouches into its body.

Freshwater fishes lack light organs, but such organs are present in many marine species. At ocean depths not reached by natural light, many fishes have light organs, but light organs are not restricted to deep-water species. Many shallow-water marine fishes that are presumably active at night also possess light organs.

Light organs are deployed in a variety of roles. By blinking the shutter over their light organs, some species produce characteristic flashes used in signaling members of the same species as part of sexual communication or schooling behavior. During daylight hours, fishes in the upper 1,000 m of the ocean are easily silhouetted from beneath against the light sky above. To camouflage their shapes to predators or prey below, they produce a glow that matches their body color to the light from above. This is accomplished by protruding light organs from gut diverticula. The soft light that is emitted illuminates the fish's ventral surface (box figure 1a). Light organs are also used extensively in feeding. They form lures on the tips of barbels around or in the mouth. Flashlight fish use light organs on their heads to illuminate their prey. Some fishes have carried this a step further. At great depths, the visual pigments of most fishes are sensitive only to blues and greens. However, *Pachystomias* has a red-emitting light organ with red-sensitive retinal pigments that allow it, unlike its prey, to see light. Consequently, they can red-illuminate their unsuspecting prey without alerting it (box figure 1b).

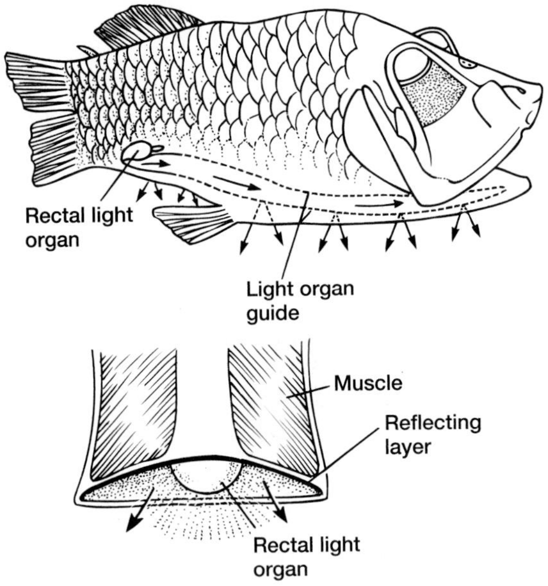


(a) *Opisthoproctus*

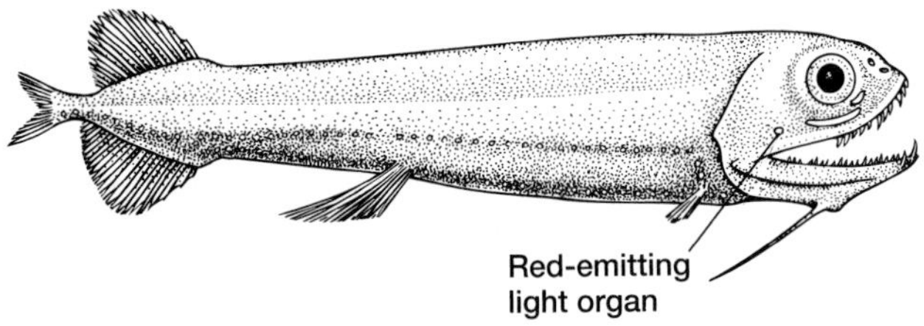


(b) *Pachystomias*

**Box figure 1** Bioluminescent light organs of fishes. (a) The pelagic *Opisthoproctus* has a single bacterial rectal light organ. Radiated light is spread evenly by a long tubular light guide. A reflective layer through the transparent skin of the sole directs the light across the whole ventral surface of the fish. By illuminating the ventral surface, the fish is less silhouetted by downwelling light. Ventral illumination is thought to make this fish more difficult to detect by deeper predators attempting to locate its silhouette above them. A cross-sectional diagram of the tubular light guide projects a view of the rectal light organ at the far end. (b) The deep-ocean fish *Pachystomias* has a light organ near its eye that emits red light to illuminate its prey.

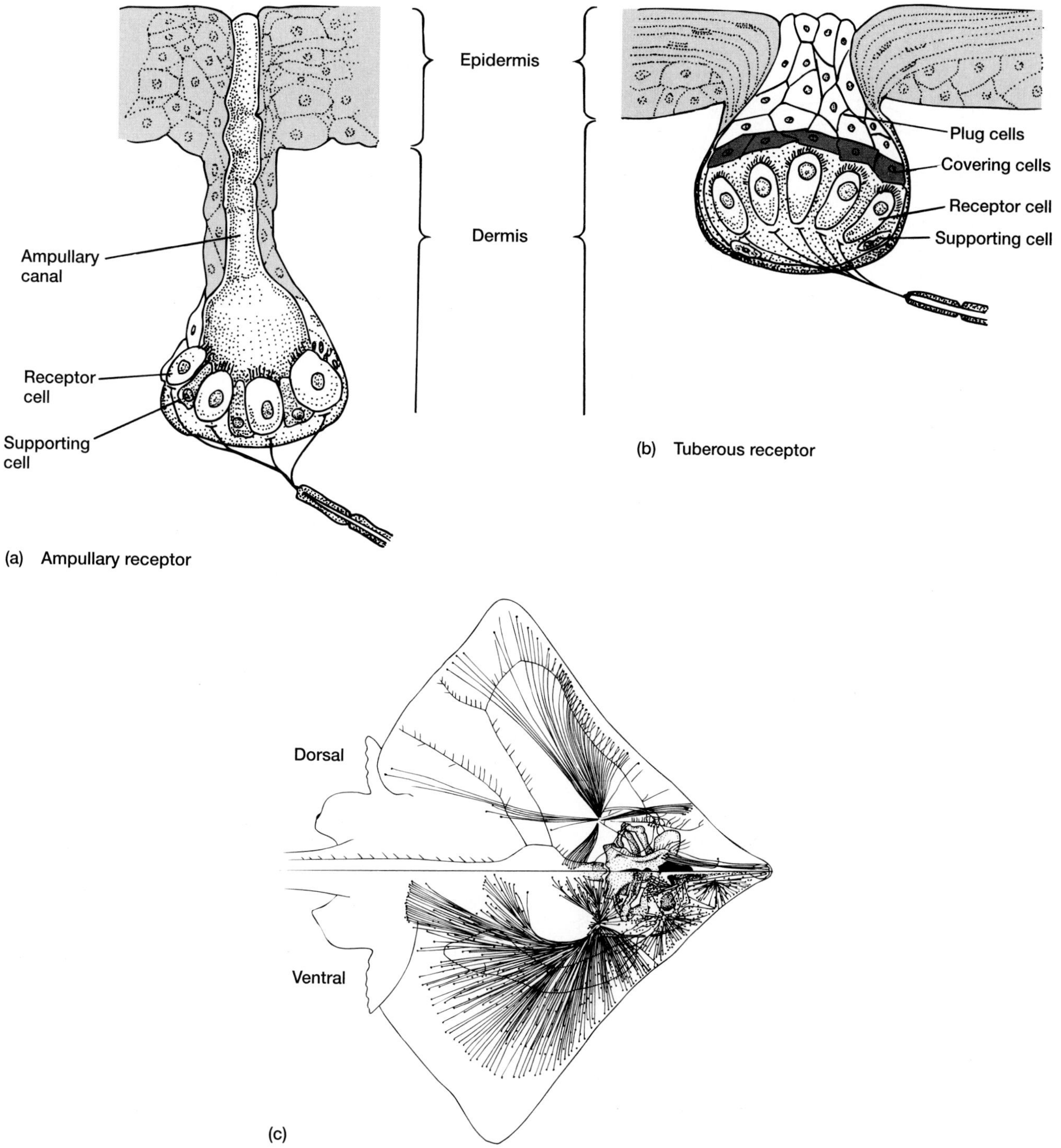


**Figure 17.46** Electroreceptors. (a) Ampullary receptor. In ampullary receptors, the electroreceptor (or receptor) cells lie at the bottom of a deep ampullary canal filled with a mucopolysaccharide. Ampullary receptors are common in fishes that are sensitive to electric energy in the environment. (b)Tuberous receptors. In tuberous receptors, electroreceptor cells form bunches near the body surface within a slight skin depression. They are covered by a loose layer of covering cells and plug cells, both cells being specializations of the epidermis. Tuberous receptors are found only in electric fishes, that is, those capable of producing electric signals. (c) Distribution of electroreceptors (black dots) in the skate *Raja laevis*. Note that distributions across the dorsal and ventral surfaces differ.

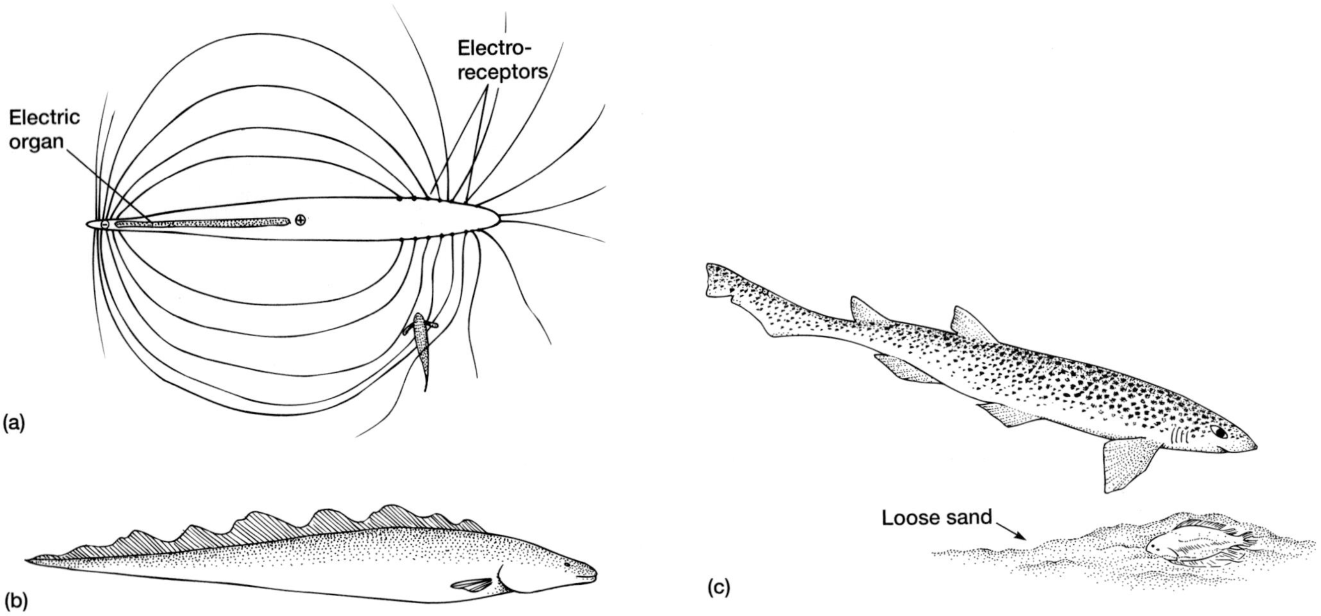


**Figure 17.47** Functions of electroreceptors. (a) Navigation. This electric fish generates its own low electric field. Environmental objects entering the field distort it. Electroreceptors concentrated on the head detect this distortion, and the nervous system interprets it to aid the fish in navigating around such objects. (b) The electric fish *Gymnarchus* holds its body rigid to align generating and receiving receptors throughout the body. Undulating fins produce movement. (c) Prey detection. Electroreceptors around the mouth and head of this shark can detect the low levels of electric discharge given off by active respiratory muscles of its prey. When the prey is buried beneath shallow sediment, these low levels of electric charge can betray its presence.

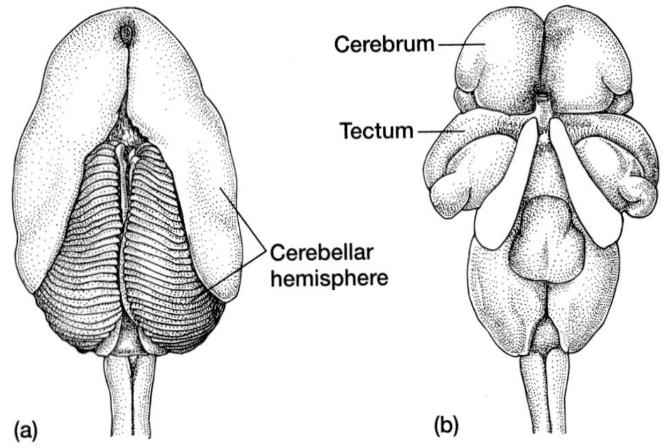


**Figure 17.48** Cerebellum of electric fish. (a) Dorsal view of the brain of a mormyrid fish. Note the large cerebellum. (b) Dorsal view of the same brain with the cerebellar hemispheres removed.

central nervous system. Rapid relay about the electrical discharge patterns of other electric fishes suggests that electroreceptors play a role in communication. Other electroreceptors are sensitive to changes in electrical amplitude. These receptors seem suited to respond to changes in a fish's own electrical field and thus play a role in navigation.

## Additional Special Sensory Organs

Electromagnetic radiation and mechanical and electrical stimuli may not be the only types of information to which vertebrates are sensitive. Sea turtles, for example, seem to use, among other stimuli, the orientation of the Earth's magnetic field to navigate. The Earth's magnetic field runs north/south, and near the poles the lines of the field are increasingly inclined toward the surface of the Earth. Thus, the magnetic field provides information about both direction and latitude. Sea turtles may use this information in an otherwise featureless ocean to reach preferred feeding areas. Years later, they may use magnetic fields, again to guide them back to their breeding grounds. Although experiments have provided evidence of sea turtles using the Earth's magnetic field for navigation, we have not discovered sensory receptors that gather such information.

# Selected References

Allin, E. F., and J. A. Hopson. 1992. Evolution of the auditory system in Synapsida ("mammal-like reptiles" and primitive mammals) as seen in the fossil record. In *The evolutionary biology of hearing*, edited by D. B. Webster, R. Fay, and A. N. Popper. New York: Springer-Verlag, pp. 587–613.

Bell, M. A. 1993. Convergent evolution of nasal structure in sedentary elasmobranchs. *Copeia* 1993(1):144–58.

Burghardt, G. M. 1990. Chemically mediated predation in vertebrates: Diversity, ontogeny, and information. In *Chemical signals in vertebrates*, edited by D. W. Macdonald, D. Müller-Schwarze, and S. E. Natynczuk. Oxford: Oxford University Press, pp. 475–99.

Carr, C. E. 1990. Neuroethology of electric fish. *BioScience* 40:259–67.

Chiszar, D., T. Melcer, R. Lee, C. W. Radcliffe, and D. Duvall. 1990. Chemical cues used by prairie rattlesnakes *(Crotalus viridis)* as they follow the trail of rodent prey. *J. Chem. Ecol.* 16:79–86.

Cooper, W., and G. M. Burghardt. 1990. Vomerolfaction and vomodor. *J. Chem. Ecol.* 16:103–5.

Davson, H. 1980. *Physiology of the eye*. New York: Academic Press.

de Cock Buning, T. 1983. Thermal sensitivity as a specialization for prey capture and feeding in snakes. *Amer. Zool.* 23:363–75.

Dijkgraaf, S. 1963. The function and significance of the lateral line organ. *Biol. Rev.* 38:51–105.

Duvall, D., and D. Chiszar. 1990. Behavioural and chemical ecology of vernal migration and pre- and post-strike predatory activity in prairie rattlesnakes: Field and laboratory experiments. In *Chemical signals in vertebrates*, edited by D. W. Macdonald, D. Müller-Schwarze, and S. E. Natynczuk. Oxford: Oxford University Press, pp. 539–54.

Fay, R. R., and A. N. Popper. 1985. The Octavolateralis system. In *Functional vertebrate morphology*, edited by M. Hildebrand, D. M. Bramble, K. F. Liem, and D. B. Wake. Cambridge, Mass.: Harvard University Press, pp. 291–316.

Halpern, M. 1987. The organization and function of the vomeronasal system. *Ann. Rev. Neurosci.* 10:325–62.

Hopkins, C. D. 1983. Functions and mechanisms in electroreception. In *Fish neurobiology*, edited by R. G. Northcutt, and R. E. Davis. Ann Arbor: University of Michigan Press, pp. 215–59.

Hudspeth, A. J. 1985. The cellular basis of hearing: The biophysics of hair cells. *Science* 230:745–52.

Ketten, D. R. 1990. The marine mammal ear: Specializations for aquatic audition and echolocation. In *The evolutionary biology of hearing*, edited by D. B. Webster, R. Fay, and A. N. Popper. New York: Springer-Verlag, pp. 717–50.

Konishi, M. 1993. Listening with two ears. *Sci. Amer.* 268(4):66–73.

Levine, J. S. 1985. The vertebrate eye. In *Functional vertebrate morphology*, edited by M. Hildebrand, D. M. Bramble, K. F. Liem, and D. B. Wake. Cambridge, Mass.: Harvard University Press, pp. 317–37.

Nakayama, K., and S. Shimojo. 1992. Experiencing and perceiving visual surfaces. *Science* 257:1357–63.

Narims, P. M. 1990. Seismic communication in anuran amphibians. *BioScience* 40:268–76.

Pettigrew, A., S-H. Chung, and M. Anson. 1978. Neurophysiological basis for directional hearing in Amphibia. *Nature* 272:138–42.

Popper, A. N., and S. Coombs. 1980. The morphology and evolution of the ear in Actinopterygian fishes. *Amer. Zool.* 22:311–28.

Radermaker, F., C. Surlemont, P. Sana, M. Chardon, and P. Vandewalle. 1989. Ontogeny of the Weberian apparatus of *Clarias griepinus* (Pisces, Siluriformes). *Can. J. Zool.* 67:2090–97.

Renouf, D. 1989. Sensory function in the harbor seal. *Sci. Amer.* 257(4):90–95.

Rieppel, O. 1988. A review of the origin of snakes. In *Evolutionary biology*, edited by M. K. Hecht, B. Wallace, and G. T. Prance. *Evolutionary Biology* 22:37–130.

Rosowski, J. J. 1990. Hearing in transitional mammals: Predictions from the middle-ear anatomy and hearing capabilities of extant mammals. In *The evolutionary biology of hearing*, edited by D. B. Webster, R. Fay, and A. N. Popper. New York: Springer-Verlag, pp. 615–31.

Schellart, N.A.M., and A. N. Popper. 1990. Functional aspects of the evolution of the auditory system of actinopterygian fish. In *The evolutionary biology of hearing*, edited by D. B. Webster, R. Fay, and A. N. Popper. New York: Springer-Verlag, pp. 295–322.

Schwenk, K. 1988. Comparative morphology of the Lepidosaur tongue and its relevance to squamate phylogeny. In *Phylogenetic relationships of the lizard families*, edited by C. L. Camp, R. Estes, and G. Pregill. Stanford: Stanford University Press, pp. 569–98.

Song, J., and G. Northcutt. 1991. Morphology, distribution and innervation of the lateral-line receptors of the Florida gar, *Lepisosteus platyrhincus*. *Brain Behav. Evol.* 37:10–37.

Tavolga, W. N., A. N. Popper, and R. R. Fay (eds.). 1981. Hearing and sound communication in fishes. New York: Springer-Verlag.

Von der Emde, G., and T. Ringer. 1992. Electrolocation of capacitive objects in four species of pulse-type weakly electric fish. *Ethology* 91:326–38.

Walls, G. L. 1942. *The vertebrate eye and its adaptive radiation*. New York: Hafner.

Wever, E. G. 1978. *The reptile ear*. Princeton: Princeton University Press.

Wever, E. G. 1985. *The amphibian ear*. Princeton: Princeton University Press.

Wu, C. H. 1984. Electric fish and the discovery of animal electricity. *Amer. Sci.* 72:598–607.

Young, B. A. 1990. Is there a direct link between the ophidian tongue and Jacobson's organ? *Amphibia-Reptilia* 11:263–76.

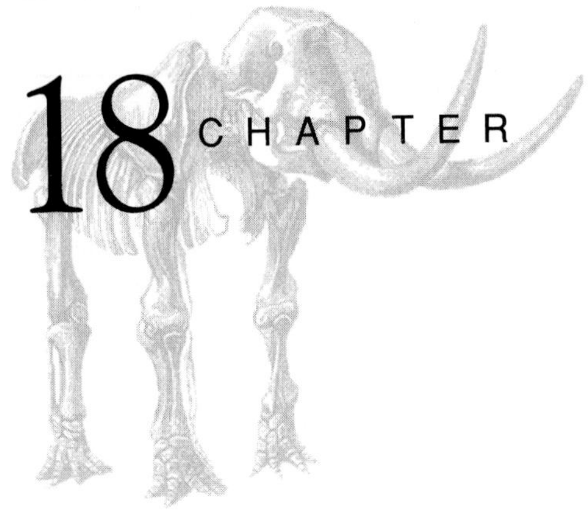

18 CHAPTER

# Conclusions



## Introduction

Morphology holds a central place in the intellectual development of modern biology. However, most persons are unaware of this and think of morphology for its practical applications. For example, in medicine, a knowledge of normal anatomy and function are necessary for physicians to recognize disease states and restore a diseased or injured individual to a healthy condition. Heart valves, kidney machines, artificial limbs, repair of torn ligaments, setting of broken bones, realignment of occlusal surfaces of a tooth row, and so forth are all intended to replace or restore deficient parts to their normal form and function.

As valuable as this may be, morphology is more than just a practical sidekick to medicine. Morphology is also the analysis of animal architecture (figure 18.1a). In addition to just naming parts, morphology delves into the study of why animals are designed the way they are. Certainly architecture, like morphology, has its practical side. Architecture too could be reduced to a mundane and everyday discipline if we saw it only as a means to put a roof over our heads. But for an inquiring mind, it offers much more. The great Gothic cathedrals are more significant and offer much more than just a place to put a congregation (figure 18.1b). To analyze cathedral architecture, we might begin by learning the building's parts—apse, nave, clerestory, entablature, jamb, plinth, spandrel, triforium—but soon we realize that more goes into the architectural design of a cathedral than just the rocks and mortar of its anatomy. The design expresses many things about the people who produced it, the history out of which it came, and the variety of functions it performs. In a similar way, the architectural design of an animal expresses something about the processes that produced it, the history out of which it came, and the functions its parts perform.

When we ask why a particular part of an organism is designed the way it is, we formally approach this question from three integrated analytical perspectives—structure, function, and ecology (figure 18.2). We begin with a description of the architecture of the part, its structure, extend this to include its function, and then look to the ecological setting in which form and function serve. If this is extended through time, we add the evolutionary dimension to our analysis. These three perspectives—form, function, environment—are parallel lines of design analysis. If we seek to understand the evolutionary changes in design, then each of these three becomes an internal critique of the series of morphological changes of interest. Function and environment are not just another set of characters to add to a pot of anatomical characters, but they represent parallel lines of distinct analysis.

This can be illustrated by Gutmann's view of chordate origins. He began with a hypothesized series of anatomical steps from protochordates to chordates, but in parallel with these structural changes, he also included the accompanying functional steps. Mismatches between morphological and functional series are a cross check on each other and hence on the hypothesized course of evolutionary events. For example, an anatomical series that placed pharyngeal slits *before* a notochord would be incongruent with functional requirements. Pharyngeal slits that pierced the pharynx would puncture the hydrostatic skeleton and cause it to fail in its locomotor function. However, if the evolu-

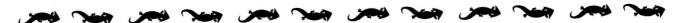

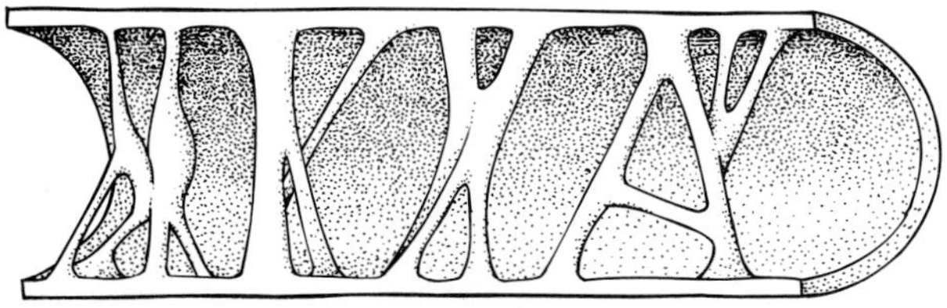

(a) Bird bone

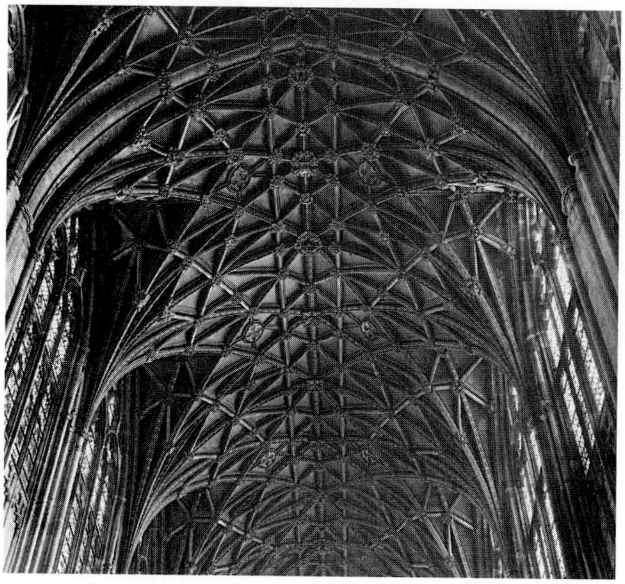

(b) Gloucester Cathedral

**Figure 18.1** Animal architecture. (a) Cross section of a bird bone showing the internal supportive struts. (b) Vault of Gloucester Cathedral (circa 1355) showing the supportive ribs that carry the weight of the ceiling vault to the side wall piers.

tionary order were the other way around, notochord first and pharyngeal slits later, this sequence would not introduce a mismatch between anatomy and functional requirements. Thus, the analysis of design and its evolution must reconcile anatomical and functional events. To be complete, the third part of the analysis should be included, namely, ecology.

Challenges to Gutmann's evolutionary hypothesis can take several forms. One can challenge his anatomical, functional, or ecological analyses or show how the three are incongruent with each other. One can then offer an alternative series of evolutionary steps and test these in a similar way.

Another example of the use of anatomical, functional, and ecological series is in the study of the evolution of birds (figure 18.3). Feathers rose through a series of anatomical modifications starting with scales in ancestral reptiles. Parallel to these anatomical transformations, a series of functional changes that feathers served can be proposed: protection, thermoregulation, leaping, parachuting, gliding, and flight. The ecological settings in which form and function serve might involve transitions from terrestrial to arboreal to aerial life. Attempts to understand the origin of birds center on hypotheses about these parallel changes in form, function, and accompanying ecology. One way to test the vitality of such theories is to examine how compatible the three series are with each other.

Structures evolve not under selection for future roles but for present roles in the environments of the moment, environments in which the organism lives at that particular time. They do not anticipate their future roles. Hypotheses regarding stages in the evolution of birds and the attainment of flight must identify the forms, functions, and ecological relationships that went together at each stage. Each stage represents an adaptive condition brought about by natural selection. Form, function, and ecological setting must be congruent at each stage. As one envisions a series of evolutionary stages, parallel lines of analysis give some idea of how transformations might have occurred in the evolution of birds.

As a practical matter, such an extensive analysis of design is usually too much work for one person to do in one lifetime. This has led in recent years to more interdisciplinary efforts. It is now common to find paleontologists, with a knowledge of fossil anatomy, teamed up with physiologists and ecologists to approach the analysis of design in a fossil series.

Analysis of animal design has taken many forms. Scientists tend to emphasize in their research those techniques that immediately address the problem of analysis design. Descriptive morphologists, interested initially in structural characteristics, give greatest attention to the anatomy itself. Functional morphologists are interested in how parts work and then tend to give their greatest attention to the physiology or biomechanics of the structure. Ecological morphologists seek to examine the interrelationships of an organism in its natural environment to discover how the structures are actually deployed. Evolutionary morphologists usually take advantage of all this information to form hypotheses about the historical development of structures through time.

Scientists from a variety of disciplines have brought a wide range of techniques and philosophical concepts to the study of animal design. At present, new and diverse analytical tools are being incorporated into the study of animal design and expanding it rapidly. Let us look at each of the four perspectives that morphology offers—structure, function, ecology, and evolution—to see what each contributes to the analysis of vertebrate design and what special insights each returns.

## Structural Analysis

Analysis of animal architecture usually begins with a description of the organism or the part of interest. If our interest centers on feeding, we might expect to begin with a careful description of the jaws, teeth, skull, and articulations of cranial structures. Usually we do this in a straightforward matter by dissecting the organism carefully or by

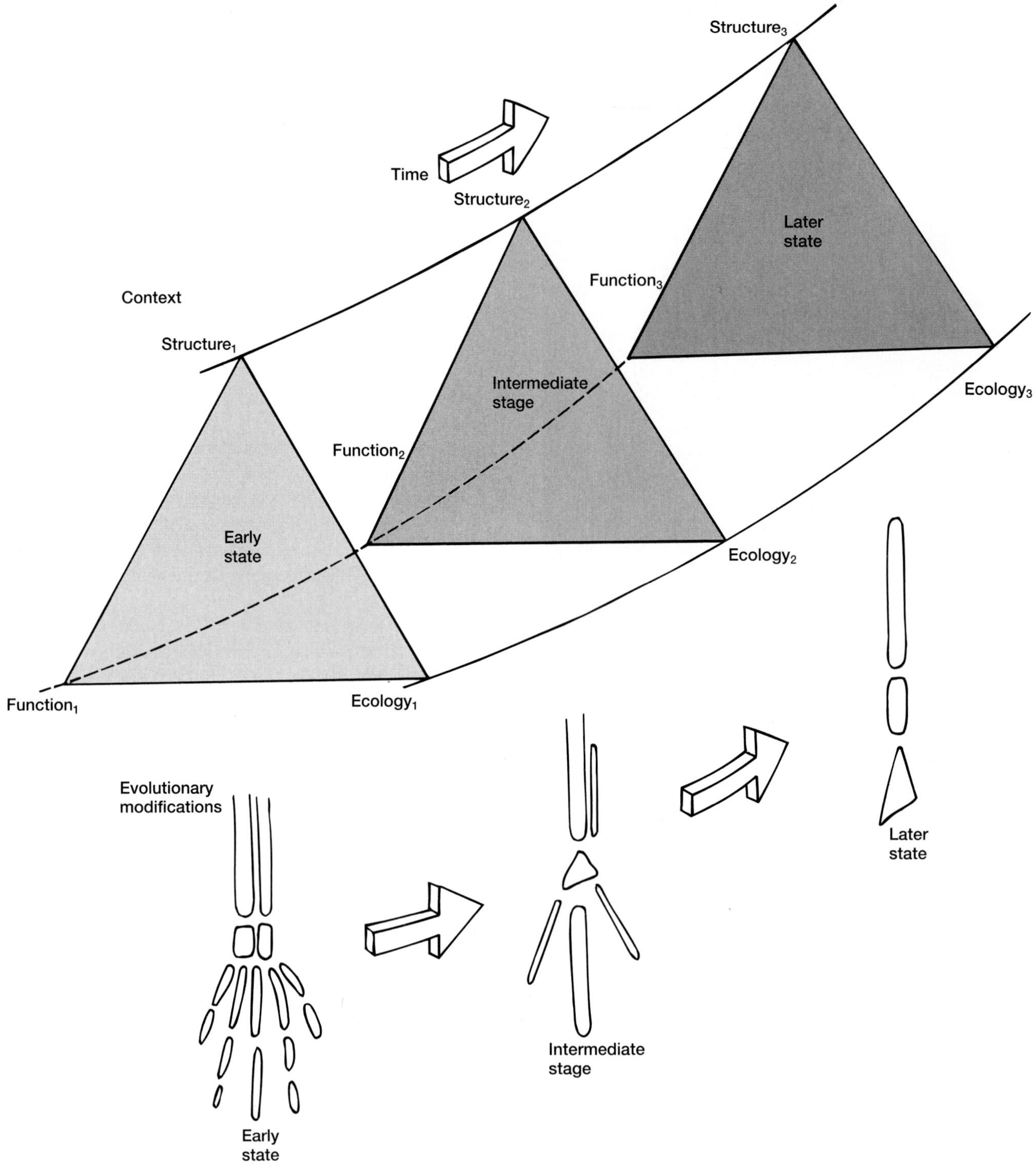


**Figure 18.2** Analysis of design. At any point in time, the particular structure of a part serves a function or functions in a particular ecological setting. Through time, the structure may change along with its function and ecological setting; therefore, analysis of design includes structure, function, and ecology. If followed through time, these change and help explain the evolutionary modifications of overall design.

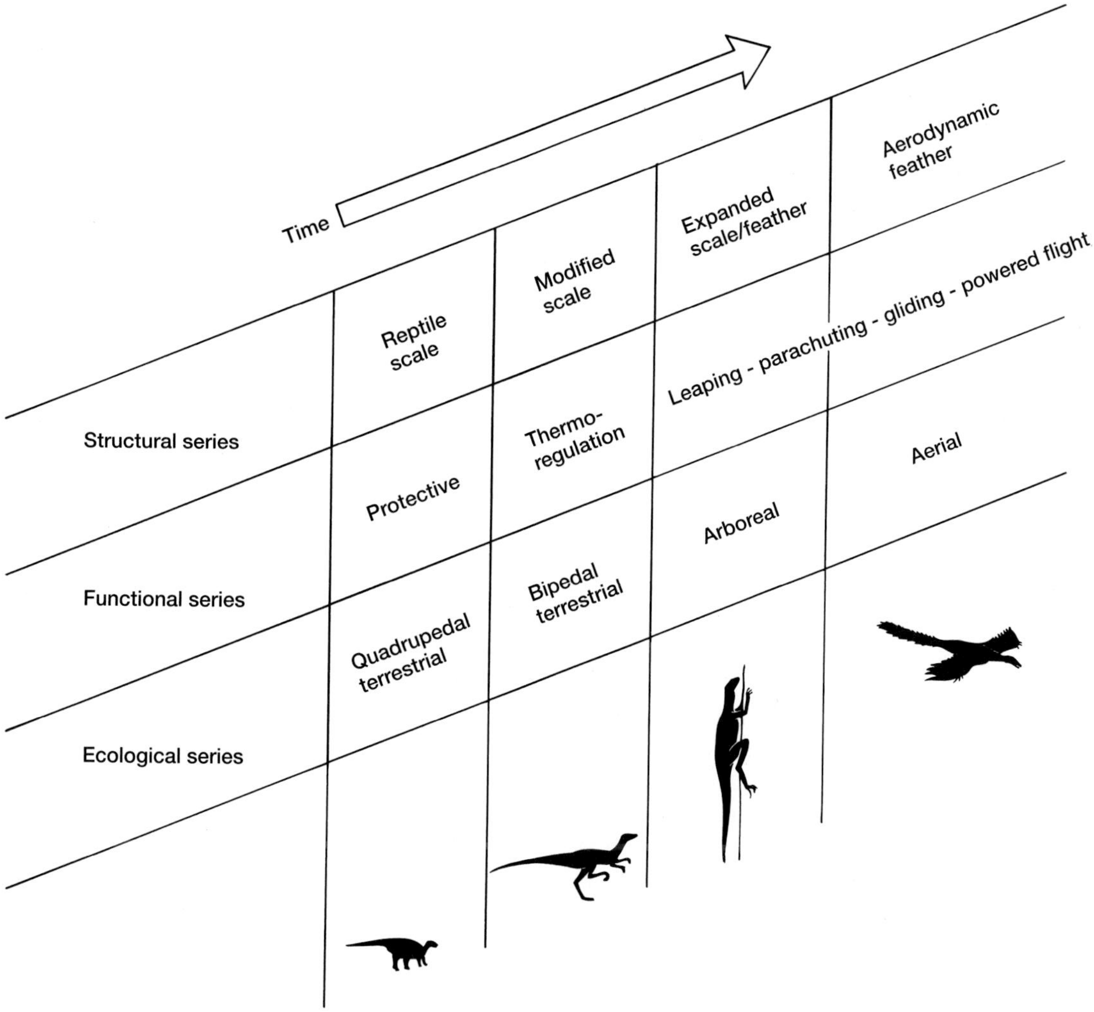


**Figure 18.3** Evolution of flight feathers in birds. Parallel structural, functional, and ecological series are indicated. The characteristic reptile scale affords protection from abrasion and desiccation to the ground-dwelling reptile. The active bipedal reptile takes advantage of a modified scale to thermoregulate its body temperature. Climbing up trees and occasionally leaping onto adjacent branches would make expanded scales advantageous in slowing the descent and in extending the aerial distance traveled. Gliding and powered flight would characterize a derived aerial organism, conditons favoring feathers as aerodynamic structures.

examining small parts under a microscope. From these observations, we produce a description of the anatomy.

As a practical matter, the description of structure begins by making a selection. We choose what level of organization to describe—cell, tissue, organ—and we choose what specifically to highlight. The choice of what level to study is often a matter of personal interest. The choice of what specifically to describe is best determined by considering which structural attributes are significant for the functional properties of the part being considered. Our description must serve the purpose of our study. If we are interested in the lines of muscle action, then points of origin and insertion should be included in our description. If force is of interest, then description of muscle fiber type and perhaps the angle muscle fibers make with the line of action should be added. Descriptive analysis is a distinct endeavor, but it cannot be done in isolation. Knowledge of form and function proceed together. An understanding of function helps us make choices about what to describe. In turn, the resulting description provides detailed information about the elements that we hypothesize to be central to function.

Thus, the description produced is itself a hypothesis about the structure observed. As a hypothesis, the description can be tested. For example, even careful observers might doubt or miss important structural features. This can be especially troublesome with descriptions of fossils. Some claim to see anatomical evidence for hairlike coats on the surface of pterodactyls. Others take exception to this and suggest that these hairlike impressions in the rock surrounding the fossils have another explanation. Similarly, for many years, *Archaeopteryx* was described as having feathers; thus, it was considered to be a bird. Recently this description was challenged, and the fossil feathers were thought to be a hoax. This prompted an anatomical reexamination of the fossils, testing the hypothesis. The challenge proved groundless, and the descriptive hypothesis that *Archaeopteryx* possessed feathers was corroborated.

Even among living forms, new descriptions replace old ones. *Amia* has been a fixture in comparative anatomy courses for decades and studied by generations of students. It was described as lacking a clavicle within the pectoral girdle. That long-standing description has proved inaccurate. Although difficult to find, recent work indicates that this bone is in fact present.

Another approach to forming descriptive hypotheses involves the construction of a hypothetical sequence of events. Glenn Northcutt (1985) took this approach with a study of early vertebrate brains. He could have examined the brains of living hagfishes and lampreys, the most primitive living vertebrates, but he worried that they might have been radically altered during their long evolutionary history. If this were the case, they would not fairly represent a truly primitive condition. Or, he could have examined directly fossils of the earliest vertebrates, but the internal details of ostracoderm brains were not preserved. Instead Northcutt looked for features of the brain shared by hagfishes, lampreys, and gnathostomes. He reasoned that those characteristics common to all these groups must have been present in their common ancestor, and he used this suite of shared characteristics to construct a hypothetical description of the primitive vertebrate brain. This was used as his point of reference when he analyzed subsequent changes in the soft internal anatomy of the vertebrate brain.

Although descriptive morphology is not a glamorous part of science, the importance of description in the analysis of animal architecture cannot be overstated. A description sets forth a hypothesis that is to be analyzed further. If the morphological description is in error, then subsequent analyses of function and ecological role can be sidetracked. Descriptive morphology carries its own implications. Whether or not *Archaeopteryx* had feathers is not a trivial anatomical issue. A description of *Archaeopteryx* with feathers implies quite a different sort of animal than a description without feathers implies. Careful descriptions are the centerpieces of the analysis of animal design.

# Functional Analysis

## How Does It Work?

An analysis of function addresses the question "How does it work?" For some, such a question gives life to descriptive anatomy and these persons are apt to think of themselves as functional morphologists. To study function, functional morphologists borrow the analytical tools from engineering and physics to look at the biomechanics of animal function. Physiology also figures prominently in the functional analysis of many vertebrates systems. With such tools of analysis, the mechanical and physiological relationships between structural elements can be deciphered.

Forces acting on structures when they perform can sometimes be measured directly. For example, Lanyon (1974) glued strain gauges to bones to record fluctuations in forces during loading. Alexander (1974) measured forces that a jumping dog produced on a force platform (figure 18.4). Because reactive forces within the dog must equal those of the force platform, Alexander was able to evaluate the stresses experienced by the dog's hindlimbs.

Swimming in water or flying in air raise problems of fluid mechanics. One approach to the study of the fluid mechanics of fish feeding was done by van Leeuwen (1984). He put tiny polystyrene spheres into water in which fish were feeding and then filmed the ensuing events. During suction feeding, the nearby spheres accelerated along with the captured food. From their pattern of movements, van Leeuwen was able to calculate the velocity of water entering the mouth and map the surrounding area from which it came (figure 18.5a). To study the slow forward flight of birds, Spedding, Rayner, and Pennycuick (1984) floated neutrally buoyant, helium-filled soap bubbles in the path of a flying pigeon. As the bird flew through this cloud of bubbles, photographs recorded the effects of wing beats on the bubbles in the bird's wake (figure 18.5b). The swirling pattern of these bubbles confirmed that the wake is composed of small vortex rings, but oddly enough the forces implied by the observed patterns could not account for the lift produced by the wings. It appeared to be half what would be required to carry the weight of the bird's body, yet the bird could and did fly obviously. This emphasizes the point that biological design is often very subtle, difficult to analyze, and full of surprises to ruffle smug preconceptions.

This type of analysis yields a form-function model that represents the primarily structural and functional elements of the part studied. This form-function model is then tested against observations of the actual part in action. For example, a kinematic linkage model of the jaw of a cottonmouth snake was compared to high-speed films of the actual

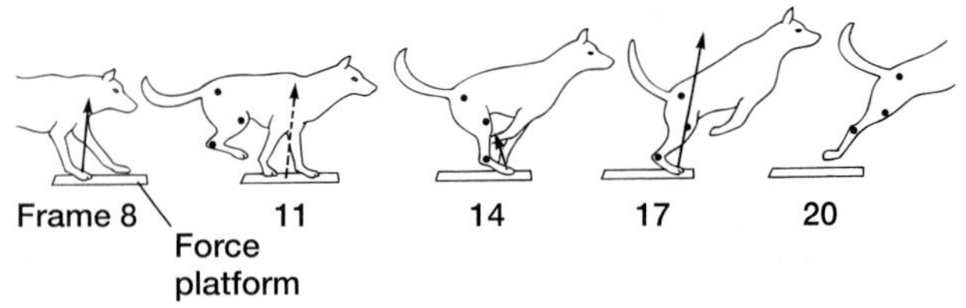


**Figure 18.4** Mechanics of dog jumping. When a force is applied against the force platform, a reactive force (arrow) against the hindlimb of the dog is produced. This can be measured to obtain an indication of the stresses on the structural elements of the skeleton during such a leap.

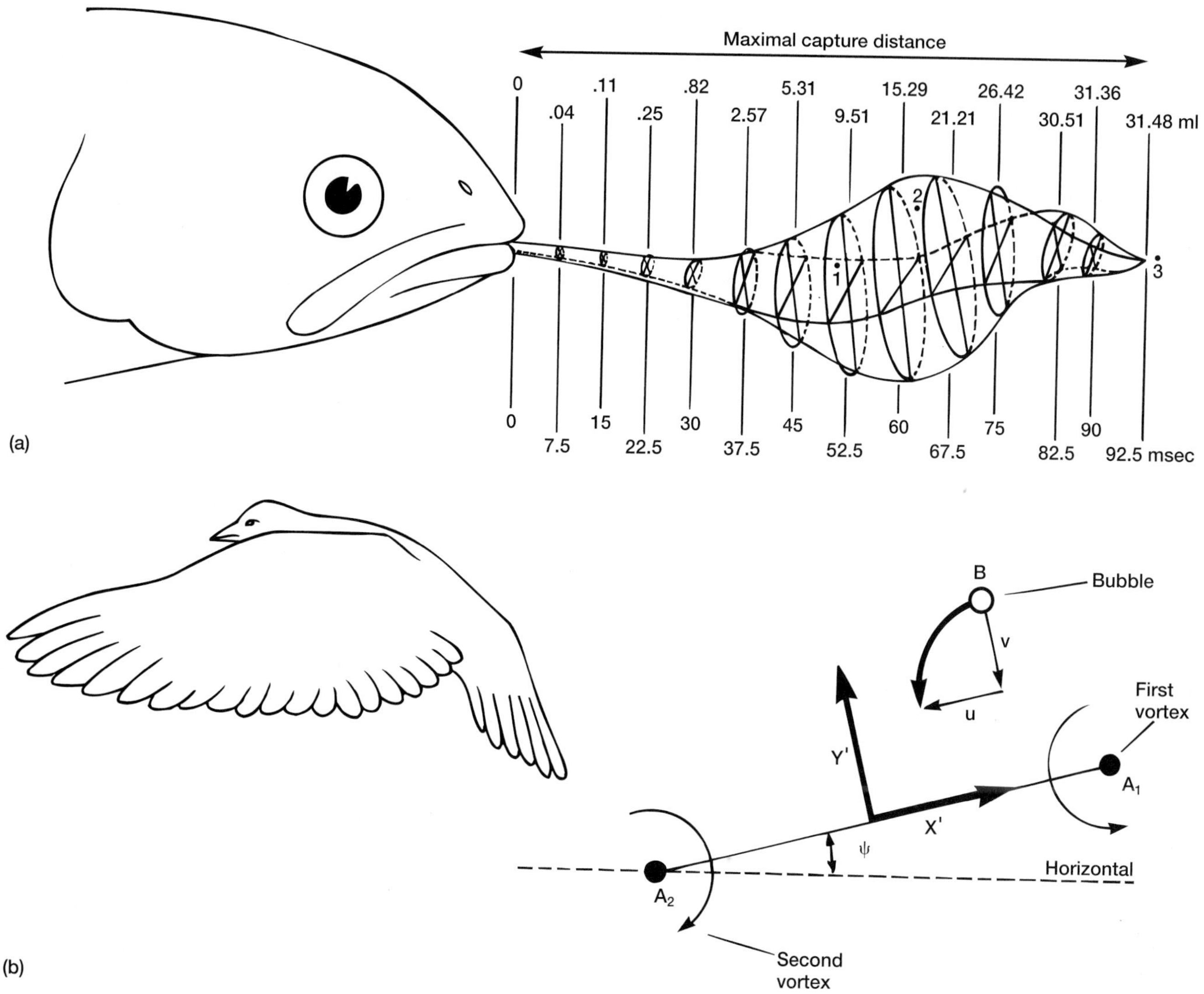


**Figure 18.5** Fluid mechanics. (a) Trout feeding. Small polystyrene spheres floating around the food respond to the suction developed when the trout feeds. Analysis of movements of these spheres reveals the volume of water drawn into the mouth and the shape of the water pulse carrying the food. (b) Slow pigeon flight. Helium-filled bubbles swirl in a vortex as the pigeon flies through them (left). This pattern can be used to calculate the components of motion that wing beats impart and the momentum they deliver (right). $A_1$ and $A_2$ are the cross sections of vortices produced by wing beats. Movements of one selected bubble (B) are shown as $v$ and $u$ components of an axis, X'Y', between vortex centers. Vortex structures in the wake are inclined at an angle, $\psi$, with the horizontal.

Modified from Van Leeuwen, 1984.

strike to see if the model accurately simulated fang rotation during envenomation (figure 18.6a,b). Movement and control of jaw elements in the model matched those of the snake. Lombard and Wake (1976) proposed a form-function model of salamander tongue protrusion and predicted that the rectus cervicis profundus muscle was responsible for retraction of the tongue. When the muscle was cut, the salamander could not retract its tongue, thus corroborating their prediction. Zweers (1982) proposed a form-function model of how pigeons drink. He predicted that the bird's esophagus drew in water by suction and tested this by inserting a fistula into the esophagus to prevent it from developing the negative pressure necessary to produce a suction. However, the pigeon could still drink, falsifying Zweers' form-function model. The pigeon pumped in water rather than sucking it in (Zweers, 1992).

Relationships between form and function can be diagramed to illustrate the mutual influences between units of an organism. Parts that are closely linked form a *functional unit*. The jaws of a shark are structurally united for feeding; the elements of the limbs of a horse, for support and locomotion; the wing bones of a bird, for flight; the penguin's flippers, for swimming. Yet each of these functional units is connected and integrated with other parts of the body as

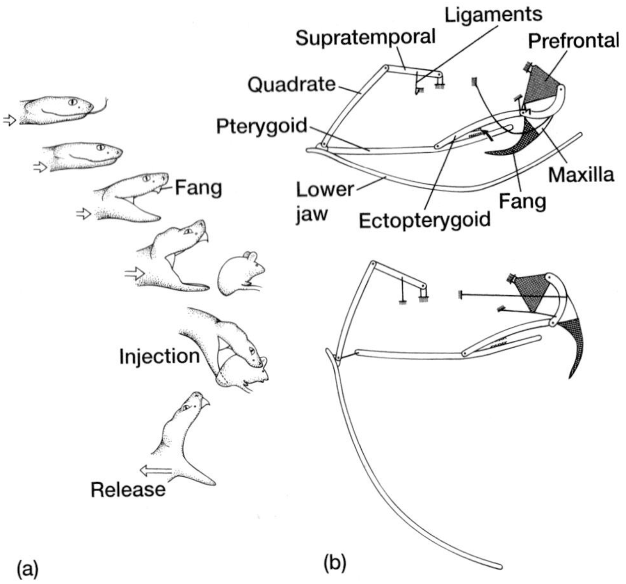


**Figure 18.6** Form-function models. (a) The actual strike of a poisonous snake indicates the sequence of events that result in raising of the fang, injection of the venom, and release of the prey. (b) The kinematic model of the jaws simulates the structural elements and functions of the jaw elements. The accuracy of the form-function model is tested against the actual sequence of events.

well to bring unity to the overall functional performance of the organism. This grid of relationships within and between functional units of an organism places internal constraints on structures. To maintain the functional integrity of a part and to ensure its proper performance, these constraints are necessary. But these constraints also limit the changes possible and thus limit or at least restrict the course of subsequent evolutionary change as well. It is in such a way that the union of form and function are thought to affect likely evolutionary events.

## Functional Coupling, Functional Compromise

The salamander mouth serves both in respiration (lung ventilation) and in feeding (capture prey). Plethodontid salamanders lack lungs altogether and depend entirely on cutaneous respiration. Because this group is without lungs, the mouth no longer participates in lung ventilation. Respiration and feeding are uncoupled in this group. Plethodontid jaws serve feeding almost exclusively and are quite different in design than other salamanders. The hyoid elements, which were previously required for lung ventilation, have been lost. Carrier (1987) has argued that a functional coupling exists between locomotion and respiration in some tetrapods. When a lizard runs, for example, it bends laterally, placing alternate compression on first the left and then the right lung. Air may be pumped back and forth between the lungs, but little new air enters to replenish the spent air (figure 18.7a). By contrast, a galloping mammal first compresses and then expands both its lungs simultaneously. This results in efficient inhalation and exhalation synchronized with flexions of the body (figure 18.7b). Carrier argued that in ancestral tetrapods, as in modern lizards, prolonged locomotion interfered with breathing; this functional coupling constrained subsequent evolution. Ectothermic descendants of ancestral tetrapods became specialized for bursts of activity based on anaerobic metabolism, but such constraints precluded more active modes of locomotion. On the other hand, birds and mammals arose from ancestors that developed morphological changes circumventing these constraints and allowing endothermic tetrapods to adopt prolonged locomotion based in part on aerobic metabolism.

## Multiple Functions

Any structure is likely to have multiple functions, and its design is a compromise between them. The feathers of birds serve flight, but they also insulate, and they can carry colors displayed during courtship. The tail feathers of the peacock function during courtship and may actually hinder flight. Snake jaws snatch prey, but bone articulations must be supple to permit subsequent swallowing of the dispatched animal. Male bighorn sheep use their horns during combat when they butt heads, but the curl of the horns is also used as a visual display to other males. The limbs of a cheetah carry it swiftly in pursuit of prey but are also the instruments used to snag prey when they are within reach.

All functions of a structure should be examined because each function influences design. The vertebrate di-

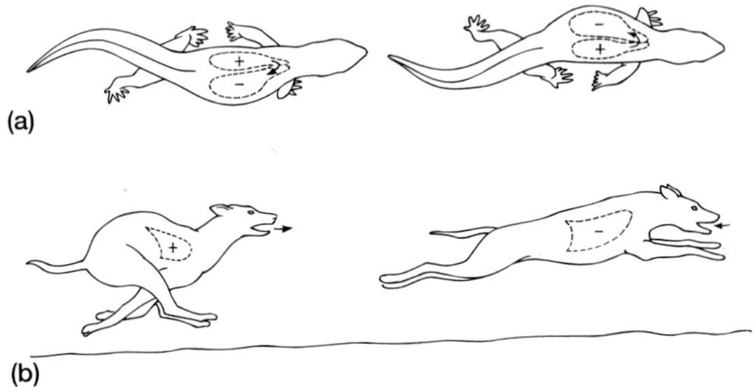


**Figure 18.7** Structural and functional coupling between units. The coupling of lung ventilation and locomotion may either constrain or enhance opportunities for subsequent evolution of alternative designs. (a) Lizard. Lateral bends of the body during locomotion alternatively impinge upon the lungs, interfering with ventilation. (b) Mammal. Dorsoventral bends of the body alternately compress and expand both lungs, serving to aid exhalation and inhalation. Plus and minus symbols indicate positive and negative pressures on the lungs (dashed lines), respectively.

Based on the research of D. Carrier, 1987.

gestive tract plays a central role in digestion, but it also houses lymphoid tissue and is therefore part of the immune system as well. The walls of the cecum, like most of the rest of the digestive tract, carry lymphoid tissue. The vermiform appendix of humans is homologous to this cecum, but it is much reduced. This has led some to suggest that it is without a function, a vestige only. Certainly, the human appendix has lost one major function, cellulose digestion, but it has not lost all function because it still houses lymphoid tissue.

Parts may also have a role to play in what Melvin Moss (1962) has termed a *functional matrix.* The mammalian lung, for example, functions in gas exchange. But, it also supports the rib cage and surrounding tissues. If the lung is removed on one side, the ribs on that side tend to be deformed because they lose the internal "scaffold" the lung provided. The snake lung is another example. In snakes, the long tubular lung runs down the center of the body. Its main function is in respiration. But it also functions to hold and give shape to the snake's body. Snakes lack a sternum to complete the rib cage and, of course, they have no limbs to support the body. Thus, most of the weight of the body rests on the inflated lung (figure 18.8a). The mammalian brain is another example. As the brain grows during development, the outer case of surrounding skull bones conforms to its expanding shape (figure 18.8b). If congenital defects occur during growth and the brain overexpands or underexpands, then the molding of the surrounding bony braincase is affected as well. Thus, lungs, brains, and other parts form matrices around which other elements take shape.

No structure can serve all of its functions equally because functional demands often are contradictory. Thus, compromises in design might be expected. Many marine birds, such as auks, use their wings for flight and for maneuvering underwater in pursuit of fish. In comparison to shorebirds such as gulls, who use their wings very little for swimming, the wings of auks are short and robust. However, in comparison to penguins, who do not use their wings for flight, the auk wing is more slightly built. Between these two extremes, the auk wing is structurally intermediate in design, neither specialized for flight alone nor for swimming alone. Auk wings represent a compromise between the two (figure 18.9).

Compromise in design can occur within one sex when the biological role of that sex poses specific design requirements. The human pelvis is an example. In females, the birth canal is necessarily large to accommodate the passage of the relatively large-headed infant during parturition. However, this widening of the hips to enlarge the birth canal results in a spreading of the limbs well to the side of the midline of the trunk above. Consequently, the limbs are not as well placed beneath the weight of the upper body that they must carry. This results in a tendency for increased asymmetrical loading of the limbs, which increases the chances of overstressing the bones. Hip design in human females seems to represent a

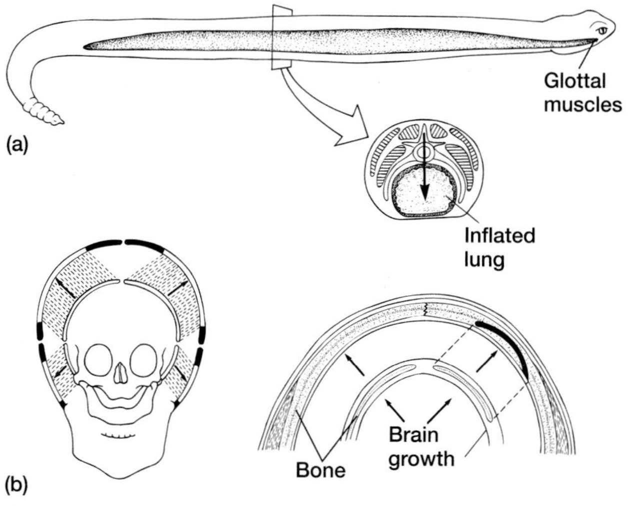


**Figure 18.8** Functional matrices. In addition to serving their primary function, most internal structures support surrounding tissues. (a) Sagittal (top) and cross-sectional (bottom) views of a snake. Small muscles close the glottal opening into the lungs between breaths. The weight of the snake's body rests upon the inflated, elongate lung, which bears the body mass (solid arrow). (b) Frontal (left) and cross-sectional views (right) of developing human skull and brain. The growing brain carries the bony elements outward (arrows). These bones undergo compensatory growth at their margins (shading) to maintain contact and produce the structurally complete braincase.
Based on the research of M. Moss, 1962.

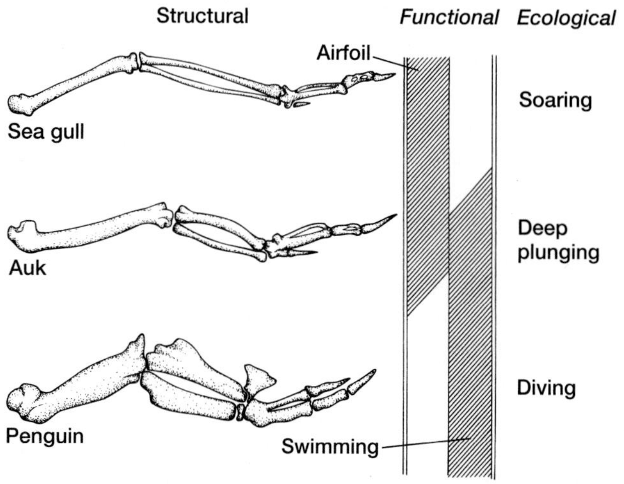


**Figure 18.9** Functional compromise. The wings of the gull serve flight; those of the penguin serve swimming. As a result, the demands on each are quite different, which is reflected in design. The auk, which flies but also uses its wings when diving below the water, is a morphological intermediate. The design of its wing is a compromise between these different mechanical demands.

compromise between the demands of locomotion and the demands of reproduction.

## Performance

If the study of function tells us how a part works, then the study of performance answers the question "How *well* does it work?" One way to rate performance is to compare the structure to an engineering simulation serving the same function. If the engineering model represents the best design that is theoretically possible, it can be considered an *optimum design*. The difference in performance between the actual structure and the ideal model would represent the difference in efficiency.

There are several limitations to such an approach. First, most structures represent a biological compromise among several functions rather than an optimum design for any one function. Thus, a structure might fail to meet expectations of an optimal engineering design but still function quite well. Second, a structure need not be perfect for it to be preserved by natural selection. An organism requires structures that allow it to survive at a frequency equal to or slightly better than its competitors. The parts of an organism thus need not be perfect, only adequate to meet this minimum requirement for survival. Thus, the differences between an actual structure and an optimal one might have no biological significance. To measure performance in a way that is biologically meaningful, the part must be assessed in the environment in which the organism lives. Eventually, what matters most is not how close to a theoretical optimum a structure comes but how well it performs under the ecological conditions in which the part actually serves.

# Ecological Analysis

Living animals are more or less suited to the current environments in which they reside. To complete an analysis of animal architecture, animals must be studied in their current environments to see what biological roles the structures play. Seeing how a part is used helps us understand why it might be designed in the way it is. The long legs of a giraffe increase stride length and, consequently, they would be expected to increase running speed. But an ecological analysis would show instead that the primary survival value of a giraffe's long legs is to lift the body high above the ground so that along with the extended neck the giraffe can reach browse inaccessible to short-legged competitors. Field studies often begin with basic natural history information on migrations, diets, reproductive patterns, and so forth, but they can expand into experimental studies that test ideas about the roles of particular structures in the life of an organism.

# Evolutionary Analysis

## Historical Constraints

Some morphologists study the structure, function, and ecology of an organism and consider these a sufficient basis on which to analyze animal design. However, what this deletes is the history from which the design comes, namely, its evolutionary source. The history behind a structure must be included in the analysis of design, otherwise we cannot address the question of why this particular design has come to characterize this particular organism. Suppose that we analyze the form and function of the tails of a porpoise and an ichthyosaur. We could further relate these to the environment of the porpoise and at least to the likely environment of the ichthyosaur. But, our analysis would not explain why the flukes of the porpoise are *horizontal* and the flukes of the ichthyosaur were *vertical* (figure 18.10a,b). These differences likely result from the different evolutionary backgrounds of these organisms. Porpoises descended from terrestrial mammals in which locomotion included vertical flexions of the vertebral column. Ichthyosaurs descended from terrestrial reptiles that used horizontal undulations of the vertebral column. When porpoises and ichthyosaurs became adapted to aquatic environments, they developed flukes with horizontal and vertical orientations, respec-

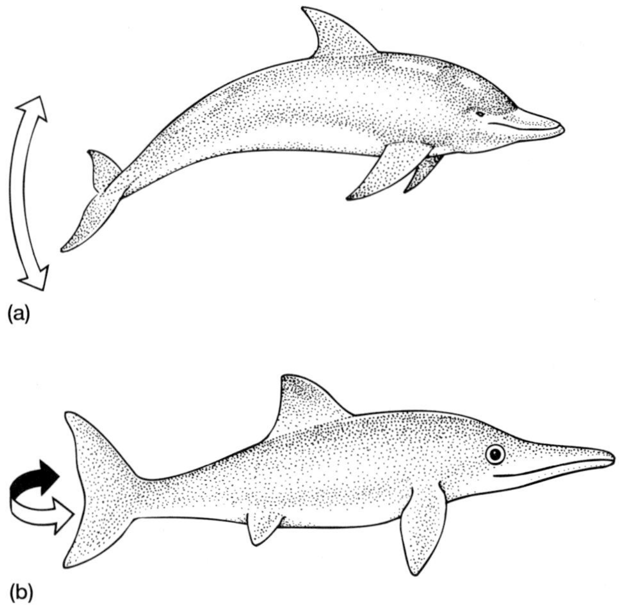


**Figure 18.10** Convergent design. Both porpoises (a) and ichthyosaurs (b) are designed for swimming in similar aquatic habitats. Orientation of the tail flukes is horizontal in porpoises and vertical in ichthyosaurs, however. These differences are probably explained by differences in the evolutionary histories out of which each comes rather than from functional or ecological factors alone.

tively. These orientations took advantage of the preexisting patterns of vertebral column bending. Past history limits future directions of structural and functional change. If we do not include the historical dimension in our analysis of structure, we limit our ability to explain such differences in the designs of organisms.

## Primitive and Advanced

We have already noted that the respective terms *primitive* and *advanced* are used to distinguish species arising early from those that emerged later in a phylogenetic lineage. But, it should be said again that these terms are unfortunate choices because they feed the view that advanced also means better. To replace the term *advanced* with the term *derived* is helpful. Primitive species are those that retain the early features present in the first members of the lineage. Derived species are those with modified characteristics, representing a departure from the primitive condition (figure 18.11).

However, replacing terms will not entirely eliminate the mistaken view that evolving species become progressively better because the misunderstanding lies not with the terms but with the bias most of us bring to the subject of evolution. Many students expect biological evolution to be driven by the same purposes as technological change. Human inventions are usually progressive; they attempt to make life better. Antibiotics improve health. Trains speed travel over the horse and airplanes over the train. Computers replace the slide ruler, which replaced counting on fingers and toes. Certainly there is a price to pay in pollution and expenditure of resources, but most people view these technological changes as progressive improvements. Life gets better.

But this is not the way to look at biological innovation. Mammals are not improvements over reptiles, nor are reptiles improvements over amphibians, nor are amphibians improvements over fishes. Each taxon represents a different but not necessarily a better way of meeting the demands of survival. Each group is equally well adapted to the tasks and lifestyles required for survival. Advanced groups are not better adapted than primitive groups.

This idea of progress is a deep and difficult bias to set aside, even for scientists. In describing the evolution of aortic arches in chapter 12, I pointed out that many scientists succumbed to the mistaken view that aortic arches, incomplete internal septation of the heart, and lungs in lungfishes represented imperfect structures. This presumed mediocrity was excusable in "primitive" animals. Oxygenated and deoxygenated bloodstreams were thought to mix, and this was seen as an unsolved "problem" until advanced groups such as birds and mammals evolved. As one morphologist rejoiced, "the perfect solution" was reached in the advanced avian and mammalian stages, but

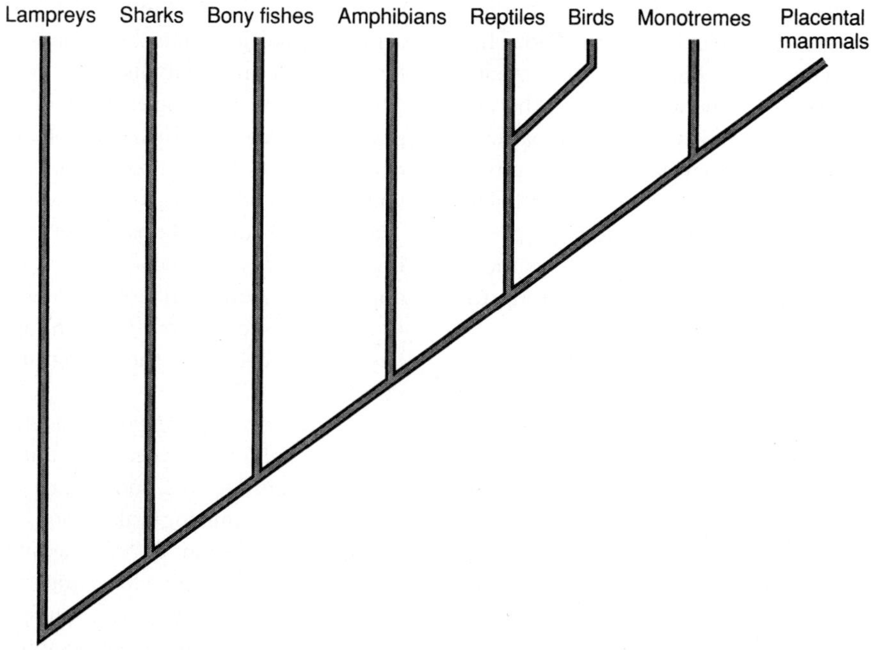


**Figure 18.11** Primitive and derived groups. Lampreys are primitive and mammals are derived or advanced, but this means only that lampreys arose before mammals. The lamprey carries forward much of the early anatomy of ancestral groups; however, it has been around a very long time, and during that interval, it has probably undergone considerable changes in many of its systems. In its own way, the lamprey is highly adapted to its environment and lifestyle, so it would be a mistake to think of it as less well adapted or imperfect in design compared with the mammal.

he conceded that lungfishes and amphibians had made some progress in separation of the two bloodstreams.

Recent research shows that there is in fact little mixing of these bloodstreams, but the point here is more subtle. The mistake is to view lungfishes and amphibians as ineptly designed animals in comparison with birds and mammals. Certainly lungfishes would make poor mammals, but on the other hand, mammals would make poor lungfishes. The design of each must be measured against the environment in which it serves, not against what it might become. Evolution does not look to the future. It is not progressive in the same way that technological changes make life better.

Often students will ask why a primitive fish such as a lamprey continues to survive along with advanced fishes such as trout and tuna. Behind such a question is the assumption that advanced means better, which leads to the mistaken view that the "superior" (i.e., advanced) should have replaced the "imperfect" (i.e., primitive). The notion of "better" does not apply to biological changes. Primitive and advanced species represent *different* ways of surviving *not better* ways of surviving.

## Diversity of Type/Unity of Pattern

The evolution of animal architecture usually proceeds by remodeling, not by new construction. Certainly, novel mutations arise to provide fresh variety, but their effect is usually to modify an existing structure rather than to replace it with something entirely new. As a result, the anatomical specializations that characterize each group are modifications of a common underlying pattern. It is unity of plan and diversity of execution as T. H. Huxley said in 1858. The bird wing is a modified tetrapod forelimb, which is a modified fish fin (figure 18.12a). Similarly, we have seen the diversity of aortic arches derived from a basic six-arch pattern (figure 18.12b). The basic five pharyngeal pouches are the common source for a variety of glands, and the basic tubular digestive tract became diversified into specialized regions in various groups. This remodeling feature of evolution accounts for the structural similarities from one group to the next.

## Mosaic Evolution

Natural selection acts on individual organisms. But different parts of the same organism are under selection pressures of different intensities, so parts change at different rates. For example, in the evolution of horses, the limbs changed considerably between the four- or five-toed ancestors and the single-toed horses of today (figure 18.13a). Teeth and skull changed also but perhaps less radically than the limbs, and relative brain size changed very little. In any evolving lineage, some parts change rapidly, some slowly, some almost not at all. Gavin de Beer (1951) termed such a pattern *mosaic evolution* because of the uneven rates by which parts of an organism undergo modification within a phylogenetic lineage.

When we study evolution of species, we must keep such a mosaic feature of evolution in mind because the selection pressures acting on one part of an organism might be quite different in intensity from those acting on another part. For example, the advanced group of snakes, the Caenophidia, includes a family of mostly nonvenomous species, the Colubridae, and generally two families of highly venomous snakes derived independently from them, the Viperidae (vipers and pit vipers) and the Elapidae (cobras, sea snakes, and their allies). In the evolution of venomous species from nonvenomous ancestors, the jaw apparatus has undergone rather extensive modification, becoming the instrument for delivery of the poisonous toxins produced in a specialized venom gland (figure 18.13b). However, other parts have changed much less dramatically. The vertebrae of venomous snakes are modified slightly from those of nonvenomous ancestors. The basic structure of the scale has hardly changed at all. Because of these different evolutionary rates, our view of the tempo of evolution within a group is likely to be slanted by the particular system we examine.

Failure to recognize the mosaic nature of evolution has led to what might be termed the **missing-link fallacy.** This is the mistaken expectation that evolutionarily intermediate species should in all respects be halfway between the ancestral and the descendant group. For example, *Archaeopteryx* from the mid-Jurassic certainly stands near the transition between reptiles on the one hand and modern birds on the other. Yet, it is not intermediate between these two groups in all respects. It had feathered wings and a furcula like its avian descendants, but it still possessed teeth and a hindlimb like the reptilian ancestors from which it came. A living example of mosaic evolution is the platypus of Australia (figure 18.14). In some ways, it is specialized. It has webbed feet, bears a spike on its hindfeet, and has a broad snout. In some ways evolution has lagged. The pectoral girdle retains elements (e.g., distinct coracoid, interclavicle, and procoracoid) from reptilian ancestors, and reproduction, as in most reptiles, includes a shelled egg. In other ways, evolution has been rapid. The reptilian scale has yielded hair, and the young are nursed at mammary glands.

The prevalent expectation that transitional forms should be halfway between their ancestors and descendants in all respects does not conform to the mosaic character of evolution. The missing-link fallacy has even hindered study of our own evolution. A "missing link" that was intermediate between apes and *Homo sapiens* was envisioned, but none was found. However, because our own evolutionary history, like that of most species, proceeds in a mosaic pattern, we should not expect to find such an intermediate. In fact, the anatomical characteristics that characterize humans evolved at different rates, and modern traits appeared at different times (figure 18.15). Bipedal locomotion

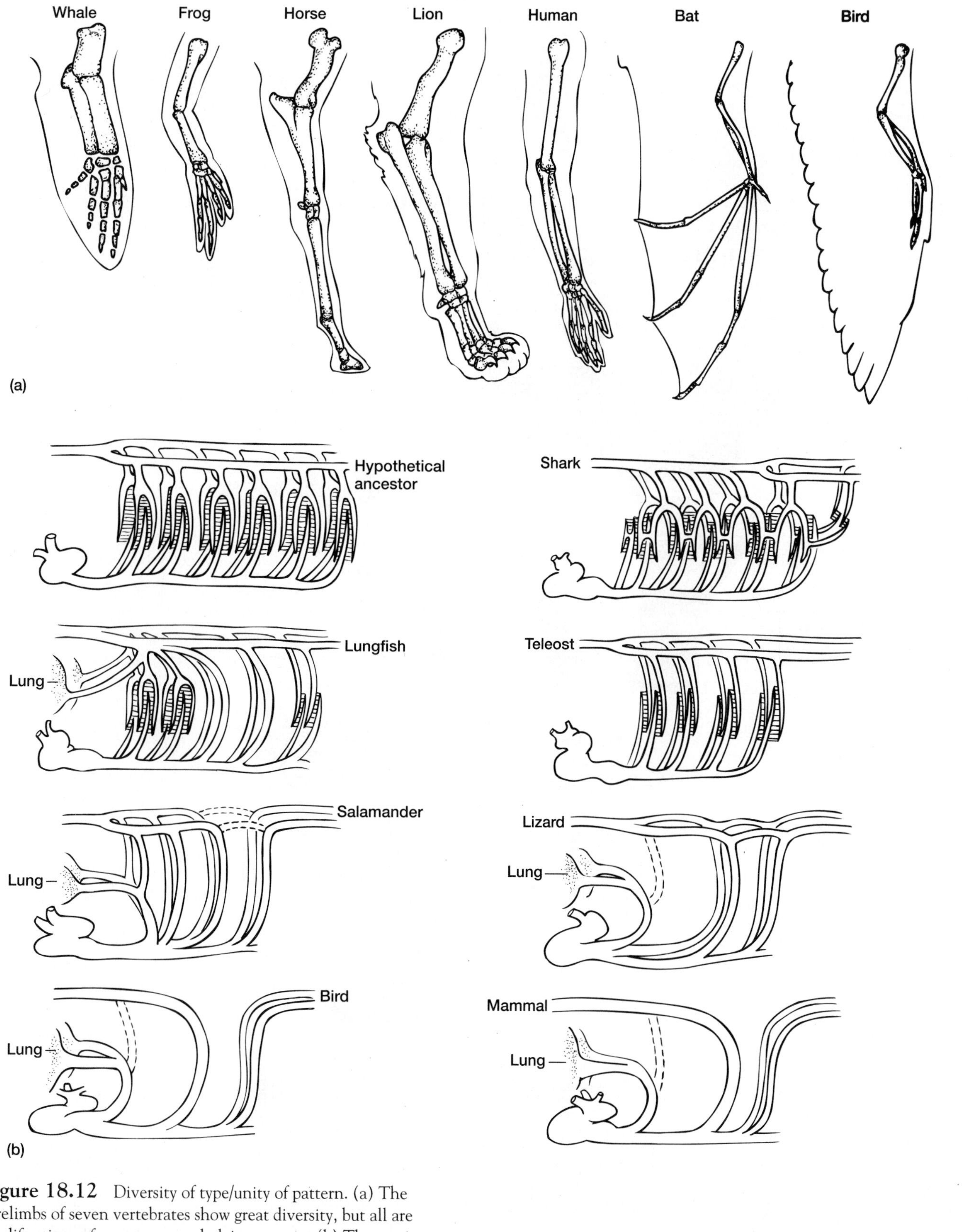


**Figure 18.12** Diversity of type/unity of pattern. (a) The forelimbs of seven vertebrates show great diversity, but all are modifications of a common underlying pattern. (b) The aortic arches of various groups are quite different; however, they seem to be derived from a common six-arch pattern.

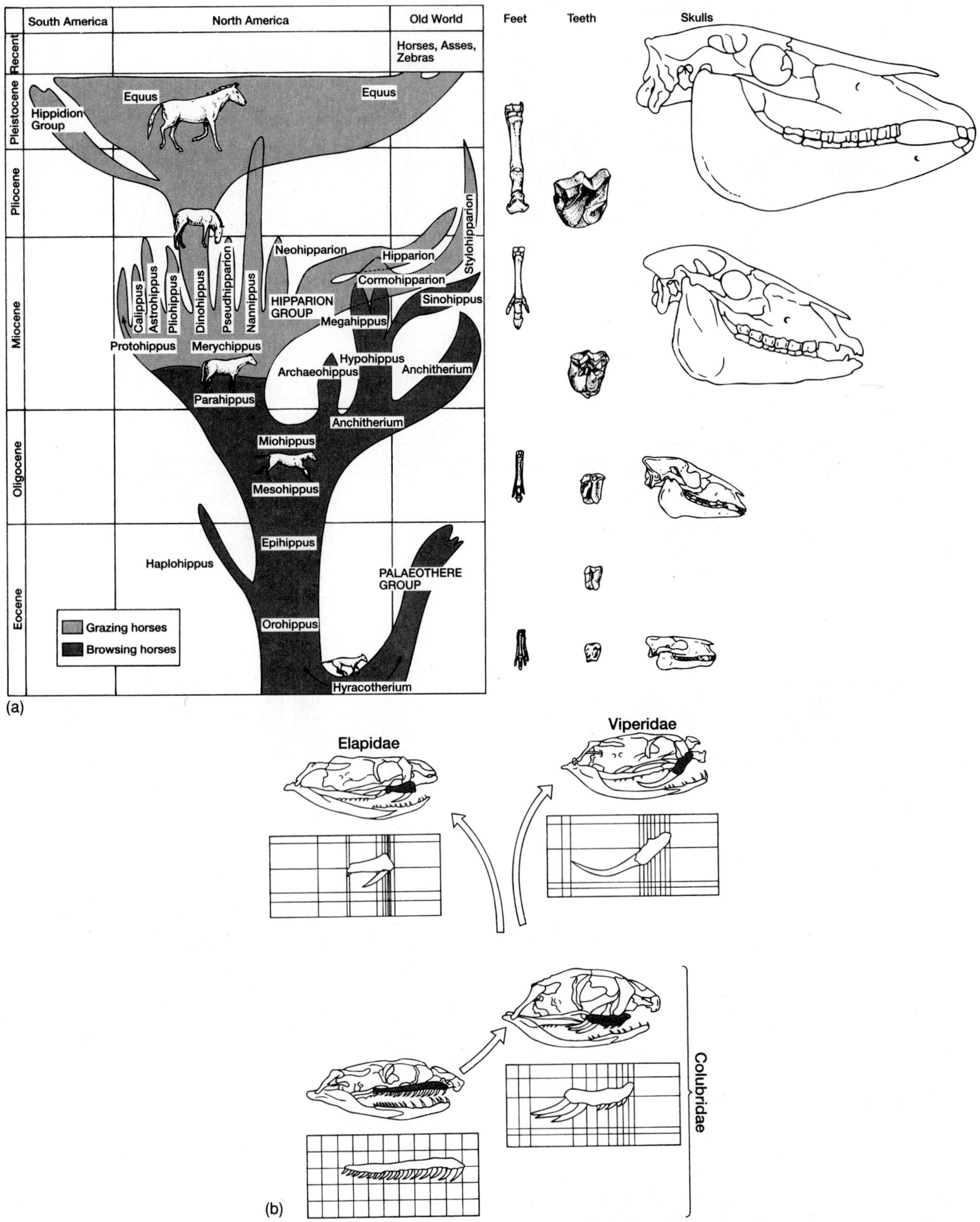


**Figure 18.13** Mosaic evolution. (a) Evolution of the horse has been characterized by relatively rapid changes in foot and tooth structure but little substantial changes in other systems, such as the integument. (b) Evolution of highly venomous snakes has been characterized by relatively rapid and extensive modifications of the maxilla and fang, but the vertebral column and especially the integument changed much less extensively. The maxilla is removed from its position in the skull and enlarged. Transformation grids are used to illustrate its changes within these families of advanced snakes.

evolved early, perhaps occurring in *Australopithecus afarensis*. Grasping hands, opposable thumbs, and a firm clavicle emerged even before this when earlier antecedents to hominids swung through trees; however, large brains, relatively hairless bodies, and speech came much later in our evolution in *Homo habilis*, *H. erectus*, and primitive *H. sapiens*. No single ancestral species possessed all characteristics halfway in transition between ourselves and apes.

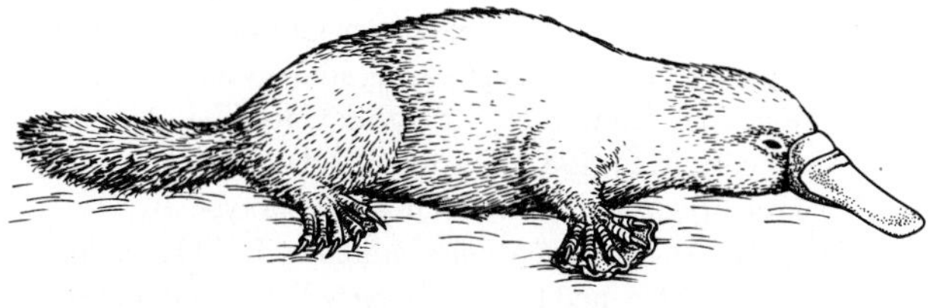

**Figure 18.14** Missing links. The platypus carries some advanced characteristics such as hair and mammary glands, but evolution of its other systems lags. Transitional forms are not necessarily intermediate between ancestors and descendants in all characteristics; instead, they display a mosaic character at different stages of evolutionary modification.

## The Promise of Vertebrate Morphology

The flourishing field of vertebrate morphology is proving to be one of the few disciplines that takes a holistic and comprehensive approach to the study of the individual. It recognizes that there is more to human beings than the molecules of which they are made. The individual is too complex and the effects of molecules that make up the genes too distant from the finished product to account entirely or even mostly for the extraordinary features of design that characterize an individual organism. Because the designs of individuals reflect functional demands, environmental pressures, and constraints of history out of which they evolved, these designs are the result of natural events accessible to discovery and understanding.

Animal architecture holds a mystery that any lively mind must notice. In a congested world in which day-to-day survival predominates, much personal delight is to be found in an intellectual pursuit of questions deeper than the ordinary and everyday. As has been said before, the discipline of vertebrate morphology with the individual at its center holds the promise of helping us to see who we are and what we might become.

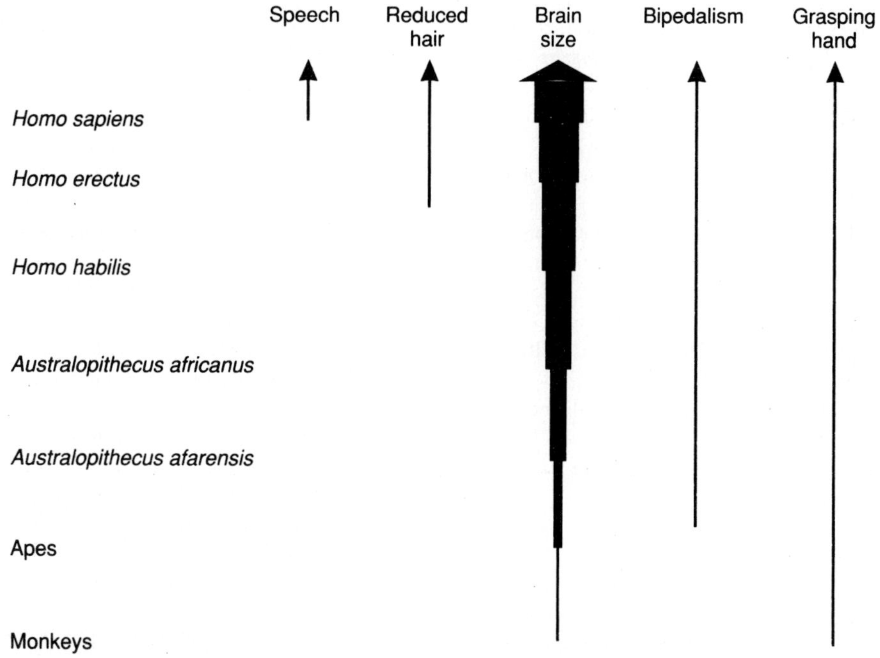


**Figure 18.15** Monkeys to humans. The mosaic character of evolution is evident in hominid evolution. Modern humans possess speech, reduced covering of hair, large brains, bipedal posture, and grasping hands. However, no single ancestral species of *Homo sapiens* possessed all these at equal intermediate states. Grasping hands arose long ago in primates that frequented trees; speech developed recently.

# SELECTED REFERENCES


Alexander, R. McN. 1974. The mechanics of jumping by a dog *(Canis familiaris)*. *J. Zool. (London)* 173:549–73.

Alexander, R. McN. 1987. Bending of cylindrical animals with helical fibres in their skin or cuticle. *J. Theo. Biol.* 124:97–110.

Alexander, R. McN. 1988. The scope and aims of functional and ecological morphology. *Neth. J. Zool.* 38:3–22.

Bock, W. J. 1979. The synthetic explanation of macroevolutionary change—a reductionistic approach. *Bull. Carnegie Mus. Nat. Hist.* 13:20–69.

Bock, W. J. 1988. The nature of explanations in morphology. *Amer. Zool.* 28:205–15.

Bock, W. J. 1989. Principles of biological comparison. *Acta Morphol.* 27:17–32.

Bock, W. J., and G. von Wahlert. 1965. Adaptation and the form-function complex. *Evolution* 19:269–99.

Carrier, D. R. 1987. The evolution of locomotor stamina in tetrapods: Circumventing a mechanical constraint. *Paleobiology* 13:326–41.

Davis, D. D. 1949. Comparative anatomy and the evolution of vertebrates. In *Genetics, paleontology, and evolution*, edited by G. L. Jepsen, E. Mayr, and G. G. Simpson. Princeton: Princeton Universtiy Press, pp. 64–89.

de Beer, G. 1951. *Embryos and ancestors*. Oxford: Oxford University Press.

Denton, E. J. 1970. On the organization of reflecting surfaces in some marine animals. *Phil. Trans. Roy. Soc.* 258B:285–313.

Dullemeijer, P. 1974. *Concepts and approaches in animal morphology*. Assen, Netherlands: Van Gorcum & Comp. B.V.

Feder, M. E., and G. V. Lauder. 1986. Commentary and conclusions. In *Predator-prey relationships*, edited by M. E. Feder and G. V. Lauder. Chicago: University of Chicago Press, pp. 180–89.

Gans, C. 1988. Adaptation and the form-function relation. *Amer. Zool.* 28:681–97.

Herring, S. W. 1988. Introduction: How to do functional morphology. *Amer. Zool.* 28:189–92.

Homberger, D. G. 1988. Models and tests in functional morphology: The significance of description and integration. *Amer. Zool.* 28:217–29.

Huxley, T. H. 1858. On the theory of the vertebrate skull. In *The scientific memoirs of Thomas Henry Huxley*, 4 vols, edited by M. Foster and E. R. Lancaster. London: Macmillan.

Kardong, K. V. 1980. Evolutionary patterns in advanced snakes. *Amer. Zool.* 20:269–82.

Lanyon, L. E. 1974. Experimental support for the trajectorial theory of bone structure. *J. Bone Joint Surg.* 56B:160–66.

Liem, K. F. 1991. A functional approach to the development of the head of teleosts: Implications on constructional morphology and constraints. In *Constructional morphology and evolution*, edited by N. Schmidt-Kittler and K. Vogel. Berlin: Springer-Verlag, pp. 231–49.

Liem, K. F., and D. B. Wake. 1985. Morphology: Current approaches and concepts. In *Functional vertebrate morphology*, edited by M. Hildebrand, D. M. Bramble, K. F. Liem, and D. B. Wake. Cambridge, Mass.: Harvard University Press, pp. 366–77.

Lombard, R. E., and D. B. Wake. 1976. Tongue evolution in the lungless salamanders, family Plethodontidae. I. Introduction, theory and a general model of dynamics. *J. Morph.* 148:265–86.

Moss, M. L. 1962. The functional matrix. In *Vistas in orthodontics*, edited by B. Kraus and R. Reidel. Philadelphia: Lea and Febiger, pp. 85–98.

Norberg, U. M., and J.M.V. Rayner. 1987. Ecological morphology and flight in bats (Mammalia: Chiroptera): Wing adaptations, flight performance, foraging strategy and echolocation. *Phil. Trans. Roy. Soc.* 316B:335–427.

Northcutt, R. G. 1985. The brain and sense organs of the earliest vertebrates: Reconstruction of a morphotype. In *Evolutionary biology of primitive fishes*, edited by R. E. Foreman, A. Gorbman, J. M. Dodd, and R. Olsson. New York: Plenum Press, pp. 81–112.

Radinsky, L. B. 1985. Approaches in evolutionary morphology: A search for patterns. *Ann. Rev. Ecol. Syst.* 16:1–14.

Reif, W-E. 1983. Functional morphology and evolutionary biology. *Paläont. Z.* 57:255–66.

Renous, S., J. Lescure, J-P. Gasc, and V. Bels. 1989. Intervention des membres dans la locomotion et le creusement du nid chez la tortue luth *(Dermochelys coriacea)* (Vandelli, 1961). *Amph.-Rept.* 10:355–69.

Spedding, G. R., J.M.V. Rayner, and C. J. Pennycuick. 1984. Momentum and energy in the wake of a pigeon *(Columba livia)* in slow flight. *J. Exp. Biol.* 111:81–102.

Van Leeuwen, J. L. 1984. A quantitative study of flow in prey capture by rainbow trout, with general consideration of the actinopterygian feeding mechanism. *Trans. Zool. Soc. Lond.* 37:171–227.

Wake, D. B. 1982. Functional and evolutionary morphology. *Persp. Biol. Med.* 25:603–20.

Youm, Y., T. E. Gillespie, A. E. Flatt, and B. L. Sprague. 1978. Kinematic investigation of normal MCP joint. *J. Biomech.* 11:109–18.

Young, B. A. 1993. On the necessity of an archetypal concept in morphology: With special reference to the concepts of "structure" and "homology." *Biol. Phil.* 8:225–48.

Zweers, G. A. 1982. The feeding system of the pigeon *(Columba livia L.)*. *Adv. Anat. Embryol. Cell Biol.* 73:1–111.

Zweers, G. A. 1982. Drinking in the pigeon *(Columba livia)*. *Behavior* 80:274–317.

Zweers, G. A. 1991. Transformation of avian feeding mechanisms: A deductive method. *Acta Biotheor.* 39:15–36.

Zweers, G. A. 1991. Pathways and space for evolution of feeding mechanisms in birds. In *The unity of evolutionary biology*, edited by R. Lolwell and Y. Reseal. Washington, D.C.: Procedure of the International Congress Systematics and Evolutionary Biology, pp. 530–49.

Zweers, G. A. 1992. Behavioural mechanisms of avian drinking. *Net. J. Zool.* 42:60–84.

# APPENDIX A Vector Algebra

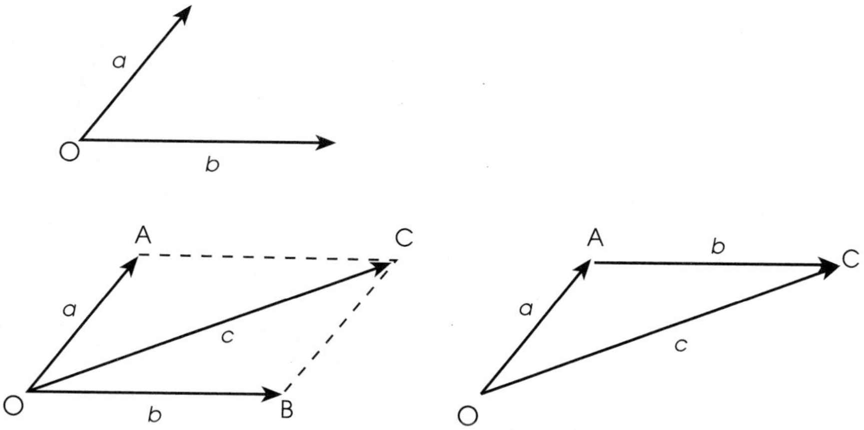


**Figure A.1** Graphic addition of vectors. Two vectors, *a* and *b*, can be added together to give a *resultant*, the single vector *c*, which summarizes the effect of the two separate vectors. One way to determine the resultant is by graphic construction. Graphic addition of vectors can be done by completing a parallelogram (left) or by triangle construction (right).

By using the two vectors as initial sides of a parallelogram, we can determine the opposite sides (dashed lines) to complete the parallelogram. The diagonal, OC, then gives the resultant, *c*.

Construction of a triangle involves adding vectors heads to tails. In this example, vector *b* is moved graphically with its tail positioned at the tip of vector *a*. Although transferred to a new position, its direction and length are, of course, preserved. The connecting distance drawn from the tail of *a* to the head of *b* gives the resultant, *c*.

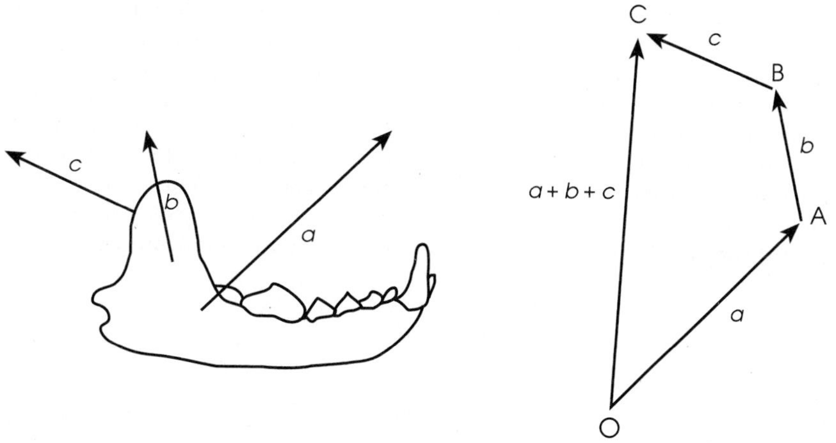


**Figure A.2** Addition of multiple vectors. Several vectors, *a, b, c,* acting simultaneously on the carnivore lower jaw can be added graphically (right) to determine the single resultant, *a + b + c*, of their collective action (left).

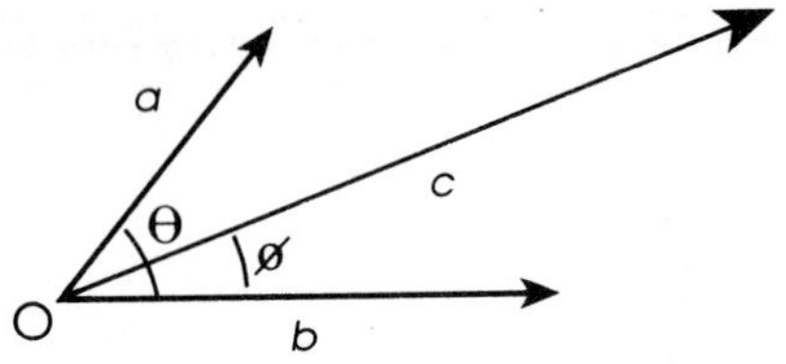


**Figure A.3** Mathematical addition of vectors. Trigonometric addition of the same vectors takes advantage of the law of cosines and sines. If magnitude of the component vectors $a$ and $b$ and the angle between them, $\theta$, are known, then the resultant, $c$, is calculated from the law of cosines:

$$c = \sqrt{a^2 + b^2 + 2ab \cos \theta}.$$

The angle ($\phi$) of the resultant with respect to $b$ is given by the law of sines:

$$\phi = \sin^{-1} (a \sin \theta / c).$$

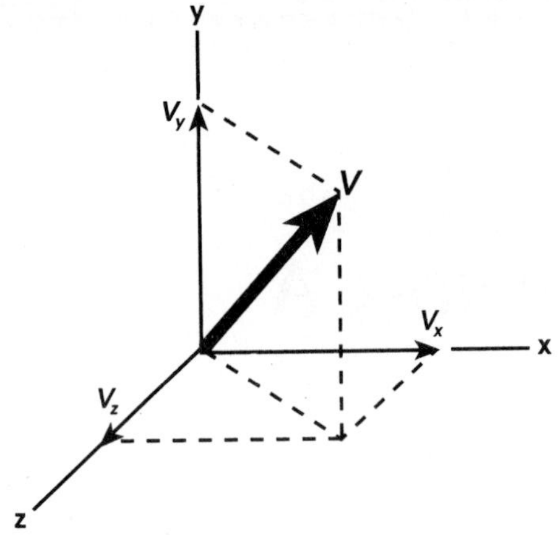


**Figure A.5** Resolution of a vector to multiple axes. So far, vectors have been discussed only in two-dimensional space, but a vector ($V$) can be resolved into three components on the same principle if a third axis, $z$, is used. This then represents the vector ($V$) by three components—$V_x$, $V_y$,$V_z$.

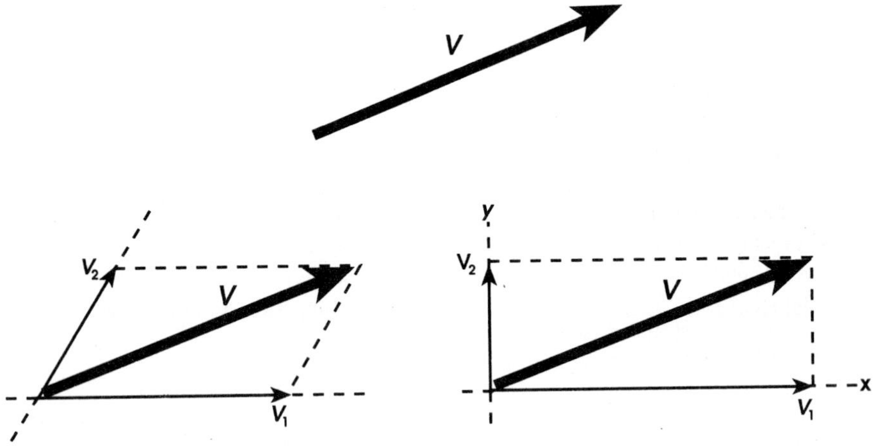


**Figure A.4** Resolution of a vector. By definition, a vector has magnitude and direction. Thus, a car traveling at 55 miles per hour in a northeasterly direction would be a vector quantity. Graphically, a vector is represented by an arrow (**V**) whose length is proportional to its magnitude, and its orientation indicates direction. As discussed in the two preceding figures, adding several vectors gives a single resultant. The reverse process takes a single vector and resolves it into several *vector components*. This process takes advantage of the parallelogram construction. The vector, **V** (left) is projected to each side of the parallelogram to produce its component vectors, $V_1$ and $V_2$. When initially constructing the parallelogram, we are free to choose a slope to the sides that is convenient for our purposes. So long as opposite sides are parallel, projection of the resultant to each side faithfully gives one possible set of component vectors. Usually, orthogonal components are preferred in which all sides of the parallelogram are at right angles (right). This makes it easier to overlay the components on a rectangular Cartesian reference system.

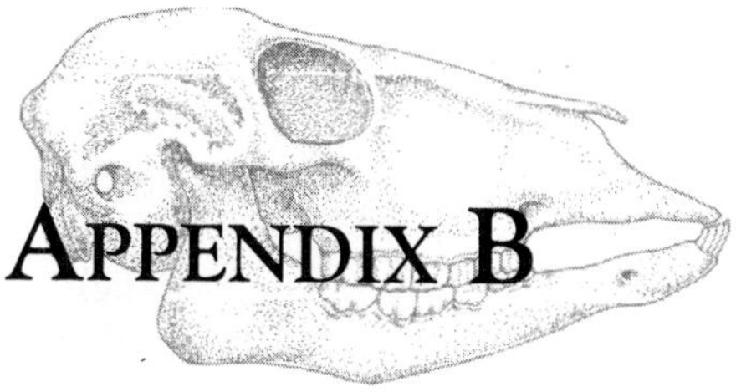

# APPENDIX B

# International System of Units (SI)

The SI system is based on six primary units:

| Quantity | Name of unit | Symbol |
|---|---|---|
| length | meter | m |
| mass | kilogram | kg |
| time | second | s |
| electric current | ampere | A |
| thermodynamic temperature | kelvin | K |
| luminous intensity | candela | cd |

All other SI units are derived from these six basic units. For example, force is kg m $s^{-2}$. Some units are named for the person historically associated with them, for example the newton (N) or watt (W). Some of these derived SI units are:

| Quantity | SI unit | Symbol | SI base units |
|---|---|---|---|
| area | square meter | $m^2$ | |
| volume | cubic meter | $m^3$ | |
| velocity | meter per second | m $s^{-1}$ | |
| density | kilogram per cubic meter | kg $m^{-3}$ | |
| acceleration | meter per second squared | m $s^{-2}$ | |
| frequency | hertz | Hz | 1 Hz = 1 $s^{-1}$ |
| force | newton | N | 1 N = 1 kg m $s^{-2}$ |
| pressure, stress | pascal | Pa | 1 Pa = N $m^{-2}$ |
| energy, work, heat | joule | J | 1 J = 1 N m = 1 kg $m^2$ $s^{-2}$ |
| power | watt | W | 1 W = 1 J $s^{-1}$ 1 kg $m^2$ $s^{-3}$ |
| electric charge | coulomb | C | 1 C = 1 A s |
| electric potential | volt | V | 1 V = 1 W $A^{-1}$ |
| electric resistance | ohm | Ω | 1 Ω = 1 V $A^{-1}$ |

Notice that the units themselves are written in lowercase, but the symbol may be capitalized, for example, newton (N), joule (J), and watt (W). Index notation is recommended. Thus 8 m $s^{-1}$ is preferred rather than 8 m/s.

To indicate multiples, the appropriate prefix is attached to the basic unit. Some examples are:

| Multiple | | Prefix | Symbol | Example |
|---|---|---|---|---|
| 1 000 000 000 | $= 10^{9}$ | giga | G | gigawatt (GW) |
| 1 000 000 | $= 10^{6}$ | mega | M | megawatt (MW) |
| 1 000 | $= 10^{3}$ | kilo | k | kilogram (kg) |
| 100 | $= 10^{2}$ | hecto | h | hectare (ha) |
| 0.1 | $= 10^{-1}$ | deci | d | decimeter (dm) |
| 0.01 | $= 10^{-2}$ | centi | c | centimeter (cm) |
| 0.001 | $= 10^{-3}$ | milli | m | millimeter (mm) |
| 0.000 001 | $= 10^{-6}$ | micro | μ | microsecond (μs) |
| 0.000 000 001 | $= 10^{-9}$ | nano | n | nanometer (nm) |
| 0.000 000 000 001 | $= 10^{-12}$ | pico | p | picosecond (ps) |

| Constants | English | SI |
|---|---|---|
| g-acceleration due to gravity at surface of the earth | 32.17405 ft $s^{-2}$ | 9.80665 $ms^{-2}$ |

## Conversion of Units

**Dimensions**

1 centimeter = 0.3937 inch
1 meter = 39.37 inches
1 meter = 1.0936 yards
1 kilometer = 0.62137 mile
1 square kilometer = 0.386 square mile

1 inch = 2.54 centimeters
1 foot = 30.48 centimeters
1 yard = 0.9144 meter
1 mile = 1.6094 kilometers
1 square mile = 2.59 square kilometers

**Weights**

1 gram = 0.03527 ounce
1 kilogram = 2.2046 pounds
1 ton (English) = 2,000 pounds
1 ton (English) = 1.016 metric ton

1 ounce = 28.35 grams
1 pound = 0.4536 kilogram
1 ton metric (=tonne) = 0.98421 ton (English)

**Volume**

1 cubic centimeter = 0.61 cubic inch
1 liter = 0.2642 gallons

1 cubic inch = 16.39 cubic centimeters
1 gallon = 3.785 liters

1 gallon = 231 cubic inches = .1337 cubic foot

**Force**

1 dyne = $1 \times 10^{-5}$ N
1 pascal = .0014504 psi

1 newton = 100,000 dynes
1 psi = 6894.65 pascals

**SI Conversions**

1 newton = $10^{5}$ dynes
13.55 mm $H_2O$ = 1 mm Hg (at 4°C)

1 dyne = 0.00001 newton

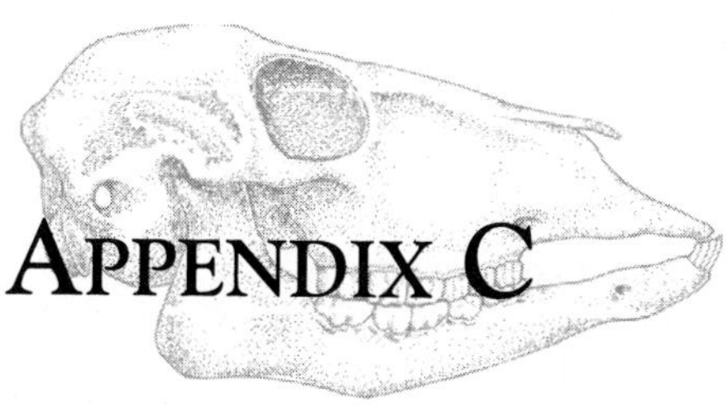

# APPENDIX C

# Common Greek and Latin Combining Forms

| | |
|---|---|
| a-(an) | L., without, not |
| ab- | L., from |
| acinus | L., grape |
| acro- | Gr., extremity |
| ad- | L., to |
| adeno- | Gr., gland |
| adipo | L., fat |
| ala- | L., wing |
| alb- | L., white |
| -algia | Gr., pain |
| alveolus | L., hollow, cavity |
| ambi- | L., both |
| amphi- | Gr., both |
| amyl- | L., starch |
| an- | Gr., without, not |
| ana- | Gr., up |
| ankylo- | Gr., bent |
| ante- | L., before |
| anti- | Gr., against |
| apo- | Gr., from |
| aqua- | L., water |
| archi- | Gr., first |
| areolar | L., small, open space |
| arthro- | Gr., joint |
| -ase | L., enzyme |
| auto- | Gr., self |
| bi- | L., two |
| bio- | Gr., life |
| blast- | Gr., germ, bud |
| bothri- | Gr., pit |
| brachi- | Gr., arm |
| brady- | Gr., slow |
| brevis- | L., short |
| caec- | L., blind |
| -campus | *see* kampos |
| capit- | L., head |
| cata- | Gr., down |
| -cele | Gr., swelling or tumor |
| cer- | Gr., horn |
| cervix | L., neck |
| chlor | Gr., green |
| choan | Gr., funnel |
| chrom- | Gr., color |
| chyl- | Gr., juice |
| cilium | L., eyelid |
| clast- | Gr., to break |
| cleistos | Gr., *see* kleistos |
| coel- | Gr., hollow |
| collum- | L., neck |
| conch- | Gr., a shell |
| corn | L., horn |
| cortico- | L., bark |
| crine (krino) | Gr., secrete, separate |
| cross- | Fr., *see* crusi |
| crusi- | L., cross, ridge |
| crypt- | Gr., hidden |
| cten- | Gr., comb |
| cumulus | L., a heap |
| cyan- | Gr., dark blue |
| cyn- | Gr., dog |
| cyt- | Gr., cell, hollow |
| de- | L., down, from |
| dent- | L., tooth |
| dia- | Gr., thru |
| diplo | Gr., double |
| dis | *see* de |
| dramein | Gr., to run |
| duo- | L., two |
| dura- | L., hard |
| duct- | L., convey |
| dys- | Gr., bad, ill |
| ecto- | Gr., outside |
| ella, -us, -um | L., diminutive |
| emia- | Gr., blood |
| endo- | Gr., within |
| ento- | Gr., within |
| eptero- | Gr., intestine |
| epi- | Gr., on, above |
| ergaster- | Gr., workman |
| erythro- | Gr., red |

| | |
|---|---|
| ex- | L., out |
| extra- | L., outside, beyond |
| eury- | Gr., wide |
| fenestra | L., window |
| fer- | L., to carry, bear |
| flay- | L., yellow |
| follicle | L., little bellows, small bag |
| fossa | L., pit, cavity |
| fug (e) | L., flee |
| gamo- | Gr., marriage |
| gastro | Gr., stomach |
| -gen | Gr., to produce |
| -glia | L., glue |
| glossi- | Gr., tongue |
| glyco- | Gr., sweet, sugar |
| gnath | Gr., jaw |
| -gogue | Gr., leading |
| gon- | Gr., angle, seed |
| graph- | Gr., to write |
| haemo- | Gr., blood |
| halos- | Gr., salt |
| hemi- | Gr., half |
| hepta | Gr., liver |
| hetero- | Gr., different |
| hex- | Gr., six |
| hippo- | Gr., horse |
| histo- | Gr., tissue |
| homo | Gr., same |
| horm- | Gr., to excite |
| hyalo- | Gr., glassy, clear |
| hydro- | Gr., water |
| hymen- | Gr., a membrane |
| hypo- | Gr., under |
| in- | L., not, without |
| in-, en- | L., into |
| inter- | L., between |
| interstitium | L., space between |
| intra | L., within |
| iso- | Gr., equal |
| -itis | L., inflammation |
| kampos | Gr., sea monster |
| kino- | Gr., movable, flexible |
| kleistos | Gr., closed |
| lact | L., milk |
| lucun- | L., pit, lake |
| lamin- | L., layer, plate |
| leio- | Gr., smooth |
| lemma- | Gr., skin |
| leuco- | Gr., white |
| lingua- | L., tongue |
| lipo- | Gr., fat |
| lith- | Gr., stone |
| -logy | Gr., discourse |
| luteus | L., golden yellow |
| -lysis | Gr., divide, destroy |
| macro- | Gr., large |
| macula- | L., spot |
| mal- | L., bad, ill |
| mast- | Gr., breast |
| medi- | L., middle |
| mega- | Gr., large |
| -mere | Gr., a part |
| mes- | Gr., middle |
| meta- | Gr., change, beyond |
| -meter | L., measure |
| micro- | Gr., small |
| mono- | Gr., single |
| morph- | Gr., form |
| morti- | L., death |
| myi- | Gr., fly |
| myo- | Gr., muscle |
| myelo- | Gr., narrow |
| myxo- | Gr., mucus, slime |
| necro- | Gr., dead |
| nemo- | Gr., thread |
| neo- | Gr., new |
| nephto | Gr., kidney |
| neuro- | Gr., nerve |
| nid- | L., nest |
| nigr- | L., black |
| noct- | L., night |
| noto- | Gr., back |
| nuc- | L., nut |
| nud- | L., naked |
| nyssus- | Gr., to prick |
| ocell- | L., small eye |
| oct- | L., sight |
| odont- | Gr., tooth |
| -oid | Gr., like |
| olig- | Gr., few |
| -oma | Gr., tumor |
| omma- | Gr., eye |
| omphalo | Gr., navel |
| oneh- | Gr., barb |
| oo- | Gr., egg |
| opercul- | L., a cover |
| optham- | Gr., eye |
| opisth- | Gr., behind |
| or-, os- | L., mouth |
| orchi- | Gr., testicle |
| ortho- | Gr., straight |
| -ose | L., sugar |
| ossi- | L., bone |
| osteo- | Gr., bone |
| osti- | L., door |
| ot (o) | Gr., ear |
| ov (i) | L., egg |
| oxy- | Gr., sharp |
| pathy- | Gr., thick |
| pataeo- | Gr., ancient |
| pan- | Gr., all |
| par- | L., to beget |

| | |
|---|---|
| para- | Gr., beside |
| pariet- | L., wall |
| path- | Gr., disease |
| ped- | L., foot |
| penicillus | L., paintbrush |
| penta- | Gr., five |
| peri- | Gr., around |
| pertro- | Gr., stone |
| phago- | Gr., to eat |
| phil- | Gr., loving |
| phlebo- | Gr., vein |
| phon- | Gr., voice, sound |
| phot- | Gr., light |
| phyll- | Gr., leaf |
| -physis | Gr., growth |
| phyto | Gr., plant |
| pia- | L., tender |
| pituita | Gr., phlegm, slime |
| plasm- | Gr., formed |
| platy- | Gr., flat |
| pleuro- | Gr., side |
| plexus | L., twine, braid |
| pneumo | Gr., air |
| pnoi | Gr., breath |
| pod (i-o) | Gr., foot |
| poly- | Gr., many |
| pons | L., bridge |
| porta- | L., gate |
| post- | L., after |
| pre- | L., before |
| pro- | Gr., before |
| proct- | Gr., anus |
| protero- | Gr., former, earlier |
| proto- | Gr., first |
| psalter- | Gr., book |
| pseudo- | Gr., false |
| psor- | Gr., itch |
| psych- | Gr., breath, soul |
| psyll- | Gr., flea |
| ptero- | Gr., wing |
| ptyl- | Gr., saliva |
| pubic- | L., flea |
| pulmo- | L., lung |
| pupa- | L., baby |
| pygo- | Gr., rump |
| pyknosis | G., dense mass |
| pyl- | Gr., gate |
| pyo- | Gr., pus |
| pyri- | L., pear |
| ptro | Gr., fire |
| quadr- | L., four |
| rachi- | Gr., spine |
| ram (i) | L., branch |
| re- | L., again |
| rect- | L., straight |
| ren- (i) | L., kidney |
| ret- (e,i) | L., net |
| rhabdo- | Gr., a rod |
| rheo- | Gr., to flow |
| rhino- | Gr., nose |
| rhizo- | Gr., root |
| rhyncho- | Gr., snout |
| rostri- | L., beak |
| rumin- | L., throat |
| saggitta- | L., arrow |
| salpi- | Gr., trumpet |
| sapro- | Gr., putrid |
| sarco- | Gr., flesh |
| scalar- | L., ladder |
| schizo- | Gr., cleft |
| sclera- | Gr., hard |
| sclerosis | L., hardening |
| scoli- | Gr., bent |
| -scope | Gr., see |
| scut- | L., a shield |
| seb- | L., tallow |
| sect- | L., to cut |
| sella- | L., saddle |
| semi- | L., half |
| sept- | L., wall |
| septic- | L., putrid |
| serra- | L., a, was |
| seti- | L., bristle |
| sialo- | Gr., saliva |
| siphon- | Gr., tube |
| siphuncal- | L., small tube |
| soma- | Gr., body |
| somn- | L., sleep |
| sperm- | Gr., seed |
| spheno- | Gr., wedge |
| splanchno- | Gr., viscera |
| squam- | L., scale |
| stat- | L., standing |
| stella- | L., star |
| steno- | Gr., narrow |
| stero- | Gr., solid |
| sterco- | Gr., dung |
| stetho- | Gr., breast |
| stigmo- | Gr., point |
| stoma | Gr., mouth |
| strati- | Gr., layered |
| strepto- | Gr., turned |
| stria- | L., furrowed |
| strongylo- | Gr., round |
| stylo- | L., column, pillar |
| sub- | L., under |
| super- | L., over |
| supra- | L., above |
| sym- | Gr., together |
| syn- | Gr., together |
| syringo- | Gr., pipe |
| tachy- | Gr., rapid |

| | |
|---|---|
| tact- | L., touch |
| taen- | Gr., ribbon |
| tapet | L., carpet |
| tele- | Gr., far, end |
| tenui- | L., thin |
| terato- | Gr., wonder |
| tetra- | Gr., four |
| thalam- | Gr., chamber |
| theco- | Gr., case, covering |
| theli- | Gr., nipple |
| therio- | Gr., breast |
| thermo- | Gr., heat |
| thromb- | Gr., clot |
| thryo- | Gr., door, shield |
| thysan- | Gr., fringe |
| tok- | Gr., birth |
| tomo- | Gr., to cut |
| -tonos | Gr., tone, tension |
| toxo- | Gr., a bow |
| trabecul- | L., a small beam |
| trachel- | Gr., neck |
| trachy- | Gr., rough |
| trema | Gr., hole |
| tremat- | Gr., hole |
| tri- | Gr., three |
| tricho- | Gr., hair |
| troch- | Gr., pulley |
| thrombid | Gr., timid |
| trop- | Gr., turning |
| trophy- | Gr., nutrition |
| trypano- | Gr., an auger |
| tumor- | L., swelling |
| tunic- | L., a garment |
| tympano- | Gr., a drum |
| tyro- | Gr., cheese |
| ultra- | L., beyond |
| unc- | L., hook |
| -uncula | L., little |
| unguli | L., hoof |
| uni- | L., one |
| uro- | Gr., tall |
| uro- | Gr., urine |
| vaso- | L., vessel |
| ven- | L., vein |
| ventra- | L., belly |
| vermi- | L., worm |
| vesicul- | L., blister |
| via- | L., way |
| villi- | L., shaggy |
| vita | L., life |
| vitr- | L., glassy |
| vivi- | L., alive |
| vora- | L., to devour |
| xantho- | Gr., yellow |
| xero- | Gr., dry |
| xylon- | Gr., wood |
| zoo- | Gr., life, animal |
| zygo- | Gr., yoke |
| zym- | Gr., ferment |

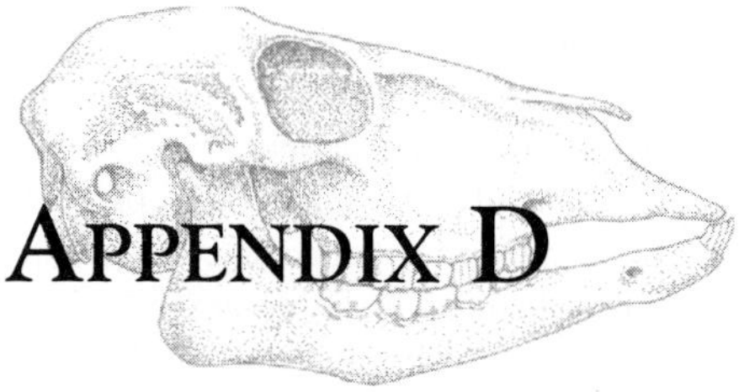

# Appendix D Classification of Chordates

- Phylum Chordata
  - Subphylum Urochordata
  - Subphylum Cephalochordata
  - Subphylum Vertebrata (Craniata)
    - Superclass Agnatha
      - Class Myxini
      - Class Pteraspidomorphi (Diplorhina)
        - Order Heterostraci (Pteraspidiformes)
        - Order Thelodontida
      - Class Cephalaspidomorpha
        - Order Osteostraci
        - Order Galeaspida
        - Order Anapsida
        - Order Petromyzoniformes
    - Superclass Gnathostomata
      - Class Placodermi
        - Order Stensioellida
        - Order Pseudopetalichthyda
        - Order Rhenanida
        - Order Ptyctodontida
        - Order Phyllolepidida
        - Order Petalichthyida
        - Order Acanthothoraci
        - Order Arthrodira
        - Order Antiarchi
      - Class Chondrichthyes
        - Subclass Elasmobranchii
          - Order Cladoselachimorpha
          - Order Xenacanthimorpha
          - Order Selachimorpha—sharks
          - Order Batidoidimorpha—rays (and skates)
        - Subclass Holocephali
      - Class Acanthodii
      - Class Osteichthyes
        - Subclass Actinopterygii
          - Superorder Chondrostei
            - Order Palaeonisciformes
            - Order Acipenseriformes-sturgeons, paddlefish
            - Order Polypteriformes (Cladistia)
          - Superorder Neopterygii
            - Division Ginglymodi
            - Order Lepisosteiformes-gars
            - Division Halecostomi
            - Subdivision Halecomorphi
            - Order Amiiformes-bowfin
            - Subdivision Teleostei
        - Subclass Sarcopterygii
          - Superorder Crossopterygii
            - Division Rhipidistia
            - Order Osteolepiformes
            - Order Porolepiformes
            - Division Coelacanthiformes (Actinistia)
          - Superorder Dipnoi
      - Class Amphibia
        - Subclass Labyrinthodontia
          - Order Ichthyostegalia
          - Order Temnospondyli
          - Order Anthracosauria
            - Suborder Embolomeri
            - Suborder Gephyrostegida
            - Suborder Seymouriamorpha
        - Subclass Lepospondyli
          - Order Aïstopoda
          - Order Nectridea
          - Order Microsauria
          - Order Lysorophia
        - Subclass Lissamphibia
          - Order Gymnophiona (Apoda)
          - Order Urodela (Caudata)
          - Order Salientia (Anura)
      - Class Reptilia
        - Subclass Anapsida
          - Order Cotylosauria
            - Suborder Diadectomorpha
            - Suborder Captorhinomorpha
            - Suborder Procolophonia

Suborder Pareiasauroidea
Suborder Millerosauroidea
Order Mesosauria
Subclass Testudinata
Order Chelonia
Suborder Proganochelydia
Suborder Pleurodira
Suborder Cryptodira
Subclass Synapsida
Order Pelycosauria
Order Therapsida
Suborder Biarmosuchia
Suborder Dinocephalia
Suborder Gorgonopsia
Suborder Cynodontia
Subclass Diapsida
Primitive Diapsids
Order Araeoscelida
Infraclass Lepidosauromorpha
Superorder Lepidosauria
Order Eosuchia
Order Sphenodonta
Order Squamata
Superorder Sauropterygia
Order Nothosauria
Order Plesiosauria
Infraclass Archosauromorpha
Superorder Archosauria
Order Thecodontia
Order Crocodylia
Order Pterosauria
Suborder Rhamphorhynchoidea
Suborder Pterodactyloidea
Order Saurischia
Suborder Theropoda
Suborder Sauropodomorpha
Order Ornithischia
Suborder Ornithopoda
Suborder Pachycephalosauria
Suborder Stegosauria
Suborder Ankylosauria
Suborder Ceratopsia
Infraclass Ichthyopterygia (Ichthyosauria)

Class Aves
Subclass Archaeornithes
Subclass Neornithes
Superorder Odontognathae
Order Hesperornithiformes
Superorder Palaeognathae
Order Tinamiformes-tinamous
Order Struthioniformes-ostriches
Order Rheiformes-rheas
Order Casuariformes-emus, cassowaries
Superorder Neognathae
Order Cuculiformes-cuckoos
Order Falconiformes-falcons, hawks
Order Galliformes-grouse, pheasants
Order Columbiformes-pigeons
Order Psittaciformes-parrots
Order Podicipediformes-grebes
Order Sphenisciformes-penguins
Order Procellariiformes-albatrosses, petrels
Order Pelecaniformes-pelicans, cormorants
Order Anseriformes-ducks, geese
Order Phoenicopteriformes-flamingos
Order Ciconiiformes-herons
Order Gruiformes-rails, cranes
Order Charadriiformes-gulls, plovers
Order Gaviiformes-loons
Order Strigiformes-owls
Order Caprimulgiformes-oilbirds
Order Apodiformes-hummingbirds, swifts
Order Trogoniformes-trogons
Order Coliiformes-mousebirds
Order Coraciiformes-kingfishers
Order Piciformes-woodpeckers, toucans
Order Passeriformes-songbirds

Class Mammalia
Subclass Prototheria
Infraclass Monotremata-monotremes
Infraclass Triconodonta
Infraclass Docodonta
Subclass Allotheria
Order Multituberculata
Subclass Theria
Infraclass Trituberculata
Order Symmetrodonta
Infraclass Metatheria
Order Marsupialia-marsupials
1. Australian groups
2. South/North American groups
Infraclass Eutheria-placentals
Order Xenarthra-anteaters, armadillos
Order Insectivora-shrews, moles
Order Macroscelidea-elephant screws
Order Scandentia-tree screws
Order Chiroptera-bats
Order Dermoptera-colugos
Order Pholidota-pangolins
Order Tubulidentata-aardvark
Order Lagomorpha-pika, rabbits
Order Rodentia
Division Sciurognathi-beavers, mountain beavers, etc.
Suborder Sciuromorpha-squirrels, gophers, chipmunks
Suborder Myomorpha-voles, mice, rats
Division Hystricognathi

Suborder Hystricomorpha-porcupines
Suborder Caviomorpha-cavies
Order Primates
Suborder Prosimii-lemurs, lorises, tarsiers
Suborder Anthropoidea-higher primates
Infraorder Platyrrhini
Family Cebidae-New World monkeys
Family Callithricidae-marmosets
Infraorder Catarrhini
Superfamily Cercopithecoidea
Family Cercopithecidae-Old World monkeys
Superfamily Hominoidea
Family Hylobatidae-gibbons
Family Pongidae-orangutans, chimpanzees, gorillas
Family Hominidae-humans

Order Carnivora
"Fissipeds"
Family Viverridae-mongooses
Family Hyaenidae-hyaenas
Family Felidae-cats
Family Canidae-dogs, wolves, foxes, jackals
Family Ursidae-bears
Family Procyonidae-raccoons
Family Mustelidae-skunks, weasels, otters, badgers
"Pinnipeds"
Family Odobenidae-walrus
Family Otariidae-eared seals, sea lions
Family Phocidae-earless seals
Order Artiodactyla
Suborder Suiformes-pigs, hippos
Suborder Tylopoda-camels, etc.
Family Camelidae-camels, llamas
Suborder Ruminantia-ruminants
Family Cervidae-elk, deer
Family Bovidae-bison, sheep
Family Giraffidae-giraffes
Family Antilocapridae-pronghorns
Order Perissodactyla-horses, tapirs, rhinos
Order Hydracoidea-hyraxes
Order Proboscidea-elephants, mastodons
Order Cetacea
Suborder Odontoceti-toothed whales, dolphins
Suborder Mysticeti-aleen whales
Order Sirenia-sea cows

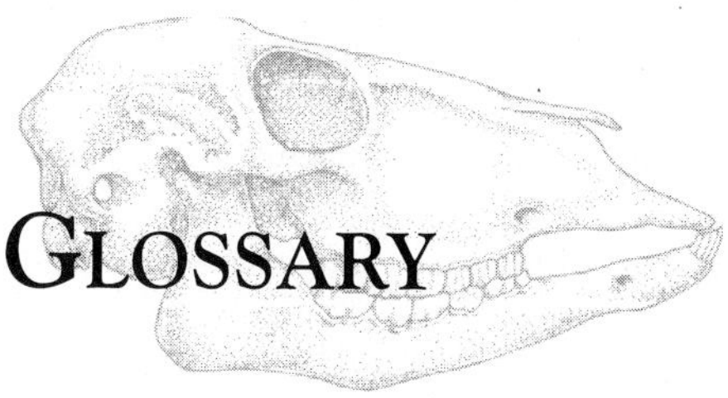

# Glossary

**abomasum** The last of four chambers in the complex ruminant stomach; homologous to the stomach of other vertebrates. See omasum, reticulum, and rumen.

**acceleration** Rate of change of velocity, or how fast velocity is changing.

**accommodation** The eye's ability to bring an object into focus.

**acoustic** Pertaining to hearing or perception of sound.

**action potential** An all-or-none membrane depolarization propagated along a nerve fiber without loss of amplitude.

**activation** Changes in an egg initiated by fertilization that begin cell division.

**acuity** The sharpness or keenness of sensory perception, as in sharp vision and keen hearing.

**adaptation** A phenotypic feature of an individual that contributes to that individual's survival; a feature's form or function and associated biological role with respect to a particular environment.

**adrenergic** Pertaining to nerve fibers that release adrenaline or adrenaline-like neurotransmitter.

**advanced** Referring to an organism or species that is derived from others within its phylogenetic lineage. See derived and compare with primitive.

**aerial locomotion** Active flapping in flight; volant.

**aerobic** Using or requiring oxygen.

**afferent** Refers to the process of bringing to; for example, sensory afferent fibers convey impulses to the central nervous system. Compare with efferent.

**agnathan** A vertebrate lacking jaws.

**air bladder** A gas bladder for respiratory exchange or buoyancy control.

**air capillary** A small compartment within a bird lung in which gas exchange occurs. Compare with alveolus and faveolus.

**airfoil** Any object that produces lift when placed in a moving stream of air (as a bird wing).

**akinetic skull** A skull lacking cranial kinesis, that is, movable joints between skull bones.

**allantois** An extraembryonic extension of the hindgut of amniote embryos that functions in excretion and sometimes in respiration.

**allometry** The study of a change in size or shape of one part correlated with a change in size or shape of another part; this relationship can be followed during ontogeny or phylogeny.

**alpha motor neuron** A nerve cell that innervates extrafusal muscle cells. Compare with gamma motor neuron.

**alveolus** The smallest subdivision of respiratory tissue in mammalian lungs located at the ends of the branching respiratory tree. Compare with air capillary and faveolus.

**amble** A slow gait characterized by the two feet on the same side coming in contact with the ground simultaneously; a slow pace. Compare with pace.

**ammonotelism** Excretion of ammonia directly through the kidneys.

**amnion** A saclike membrane that holds the developing embryo in a compartment of water.

**amniote** A vertebrate whose embryo is wrapped in an amnion.

**amphystyly** Jaw suspension via two major attachments, the hyomandibula and the palatoquadrate.

**amplexus** A mating embrace of male frogs.

**anadromous** Characterizing fishes that hatch in fresh water, mature in salt water, and return to fresh water to breed; for example, salmon. Compare with catadromous.

**anaerobic** Not requiring oxygen.

**analogy** Features of two or more organisms that perform a similar function; common function.

**anamniote** A vertebrate whose embryo lacks an amnion.

**anastomoses** A network of connections between blood vessels.

**angiogenesis** Blood vessel formation.

**angle of attack** The orientation of the edge of a wing as it meets the oncoming airstream.

**angular acceleration** Rate of change of velocity around a point of rotation; rotational acceleration.

**anlage** (pl., anlagen) A primordium or formative embryonic precursor to a later development structure.

**antagonist** A muscle with an action opposite to other muscles. Compare with synergist and fixator.

**antler** A branched, bare bone that grows outward from skull bones on some artiodactyl species; usually grows annually in mature males and is shed during the nonreproductive season. Compare with horn.

**apnea** Temporary cessation of breathing.

**aponeurosis** A broad, flat tendon.

**arboreal locomotion** Movement through trees. Compare with brachiation and scansorial locomotion.

**archetype** The fundamental type or basic underlying blueprint or model on which a definitive animal or animal part is thought to be based.

**archinephric duct** A general term for the urogenital duct; alternative names (wolffian duct) are given it at different embryonic stages (pronephric duct, mesonephric duct, opisthonephric duct) or in different functional roles (vas deferens).

**archipterygial fin** A basic fin type in which the axis (metapterygial stem) runs down the middle of the fin. Compare with metapterygial fin.

**arcualium** An embryonic, cartilaginous anlage to parts of the adult vertebra.

**artery** Blood vessel carrying blood away from the heart; blood carried may be high or low in oxygen tension. Compare with vein.

**aspidospondyly** The condition in which the centra and spines of vertebrae are anatomically separate. Compare with holospondyly.

**aspiration** Drawing in by suction.

**aspondyly** The condition in which centra are absent from vertebrae.

**atmosphere** The weight that a column of air exerts on an object at sea level; 1 atmosphere = 101,000 Pa = 14.7 lb/sq in.

**atrophy** A decrease in size or density. Compare with hypertrophy.

**auditory** Pertaining to the perception of sound.

**autostyly** Jaw suspension in which the jaws articulate directly with the braincase.

**axon** A nerve fiber of a neuron carrying an impulse away from the cell body.

**baculum** A bone within the penis.

**baleen** Keratinized straining plates that arise from the integument in the mouth of some species of whales.

**benthic** Bottom dwelling. Compare with pelagic and planktonic.

**bilateral symmetry** A body in which left and right halves are mirror images of each other.

**biogenetic law** Ernst Haeckel's claim that ontogeny recapitulates (repeats) phylogeny; now discredited.

**biological role** How the form and function of a part perform in an environmental context in order to contribute to the organism's survival. Compare with function.

**biomechanics** The study of how physical forces affect and are incorporated into animal designs.

**bipedal** Walking or running by means of only two hindlegs. Compare with quadrupedal.

**blastocyst** The mammalian blastula.

**blastopore** The opening into the primitive gut formed at gastrulation.

**blastula** The early embryonic stage that follows cleavage and consists of a hollow, fluid-filled ball of cells.

**bolus** Soft mass of food in the mouth or stomach. Compare with chyme.

**bound** A gait in which all four feet strike the ground in unison; a pronk in artiodactyls. Compare with half bound.

**boundary layer** The fluid layer closest to and flowing over the surface of a body.

**brachiation** Arboreal locomotion by means of arm swings and grasping hands, with the body suspended below the underside of branches. Compare with scansorial locomotion.

**brachyodont** Pertaining to teeth with low crowns. Compare with hypsodont.

**bradycardia** Abnormally slow heart rate. Compare with tachycardia.

**braincase** That part of the skull containing the cranial cavities and housing the brain.

**brain stem** The posterior part of the brain comprising the midbrain, pons, and medulla.

**branchial basket** The expanded chordate pharynx that functions in suspension feeding.

**breaking strength** The maximum force a structure reaches just before it fails or breaks.

**budding** A form of asexual reproduction wherein parts separate from the body and then differentiate into a new individual.

**bunodont** Pertaining to teeth with peaked cusps. Compare with lophodont and selenodont.

**bursa** A pouch or sac.

**calcification** The process of calcium deposition in tissue. Compare with ossification.

**cannon bone** Hindlimb bone resulting from fusion of metatarsals III and IV (as in horses).

**canter** A slow gallop.

**capillary** The smallest blood vessel, which is lined only by endothelium.

**carapace** The dorsal, dome-shaped bony part of a turtle shell. Compare with plastron.

**carnassials** Sectorial teeth of carnivores including upper premolars and lower molars.

**catadromous** Characterizing fishes that hatch in salt water, mature in fresh water, and return to salt water to breed; for example, some eels. Compare with anadromous.

**catecholamines** Epinephrine and norepinephrine hormones produced by chromaffin and other tissues.

**caudal** Toward the tail or back end of the body; posterior.

**cecal fermentation** Process by which microorganisms digest food in the ceca of the intestines. See intestinal fermentation.

**cecum** A blind-ended outpocketing from the intestines.

**cementum** Cellular and acellular layers that usually form on the roots of teeth, but in some herbivores, these layers may contribute to the occlusal surface. See enamel and dentin.

**central nervous system** Nervous tissue comprising the brain and spinal cord.

**centrum** The body or base of a vertebra.

**ceratotrichia** Fan-shaped array of keratinized rods internally supporting the elasmobranch fin. Compare with lepidotrichia.

**cheek** The lateral fleshy wall of the mouth, especially in mammals.

**chemoreceptor** A sense organ that responds to chemical molecules. Compare with radiation receptor and mechanoreceptor.

**chiasma** Crossing of fibers.

**choana** The internal naris; the openings of the nasal passage into the mouth.

**cholinergic** Nerve fibers that release the neurotransmitter acetylcholine.

**chondrocranium** That part of the skull formed by endochondral bone or cartilage that underlies and supports the brain; also includes the fused or associated nasal capsules.

**chromaffin tissue** Endocrine tissue and source of catecholamines (e.g., epinephrine); becomes the medulla of the adrenal gland.

**chromatophore** General term for a pigment cell.

**chyme** The liquified bolus of partially digested food after it leaves the stomach and enters the intestine; digesta. Compare with bolus.

**ciliary body** The tiny ring of muscle in the eye that focuses the lens.

**claw** A sharp, curved, laterally compressed nail at the end of a digit; talon.

**cleavage** A rapid series of cell division that follows fertilization and produces a multicellular blastula.

**cleidoic egg** The shelled container in which the fetus is laid, as in reptiles, birds, and primitive mammals. Compare with egg.

**coelom** The fluid-filled body cavity formed within the mesoderm.

**collagen** Protein fibers secreted by connective tissue cells.

**collateral** Accompanying, ancillary, or subordinate.

**colloid** A gelatinous or mucoid substance.

**composite theory** The hypothesis that jaws evolved from several fused anterior branchial arches.

**compressive force** The direction of an applied force that tends to press or squeeze an object together.

**concurrent** Flow of adjacent currents in the same direction.

**contralateral** Occurring on the opposite side of the body. Compare with ipsilateral.

**coprophagy** The eating of feces, a behavior performed usually to process undigested material again; refection.

**copulation** Coitus involving an intromittent organ.

**coracoid** Posterior coracoid; an endochondral bone of the shoulder that first evolved in early synapsids or their immediate ancestors. Compare with procoracoid.

**cornified** Having a layer of keratin; keratinized.

**cortex** The outer portion or rim of an organ.

**corticosteroids** Steroid hormones.

**cosmine** An older term designating a derivative of dentin that covers some fish scales; cosmoid scale.

**countercurrent** Flow of adjacent currents in opposite directions.

**cranial** Toward the head or front end of the body; anterior or rostral.

**cranial kinesis** Movement between the upper jaw and braincase about joints between them; in restricted sense, skulls with a movable joint across the roofing bones. Compare with akinetic skull, prokinesis, mesokinesis, metakinesis.

**cranial nerve** Any nerve entering or leaving the brain. Compare with spinal nerve.

**crista** A mechanoreceptor within the semicircular canals of the vestibular apparatus of the ear; specialized neuromast organ detecting angular acceleration. Compare with macula.

**crop** A baglike expansion of the esophagus.

**cursorial locomotion** Rapid running.

**cutaneous respiration** Gas exchange directly between the blood and the environment via the skin.

**dead space** The volume of used air not expelled upon exhalation. Compare with tidal volume.

**deglutition** The act of swallowing.

**delamination** Splitting of sheets of embryonic tissues into parallel layers.

**dendrite** A nerve fiber of a neuron carrying impulses toward the cell body.

**dendrogram** A branching diagram that represents the relationships or the history of a group of organisms.

**dental formula** Shorthand expression of the characteristic number of each type (incisor, canine, premolar, molar) of upper and lower teeth in a mammalian species.

**dentin** A material that forms the bulk of the tooth and is similar in structure to bone but harder; yellowish in color and composed of inorganic hydroxyapatite crystals and collagen; secreted by odontoblasts of neural crest origin. See enamel and cementum.

**dentition** A set of teeth.

**derived** Denoting an organism or species that evolved late within its phylogenetic lineage; advanced; opposite of primitive.

**dermal papilla** The part of the tooth-forming primordium that is derived from neural crest cells, becomes associated with the enamel organ, and differentiates into odontoblasts that secrete dentin. See enamel organ.

**dermatocranium** That part of the skull formed from dermal bones.

**dermatome** An embryonic skin segment.

**dermis** The skin layer that lies beneath the epidermis and is derived from mesoderm.

**design** The structural and functional organization of a part related to its biological role.

**deuterostome** An animal whose anus forms from or near the embryonic blastopore; the mouth forms at the opposite end of the embryo.

**diffusion** The movement of molecules from an area of high concentration to an area of low concentration; if the movement is random and unaided, it is known as passive diffusion.

**digestion** The mechanical and chemical breakdown of foods into their basic end products, usually simple carbohydrates, proteins, fatty acids, that are absorbed by the bloodstream.

**digitigrade** A foot posture in which the balls of the feet (middle of the digits) support the weight, as in cats and dogs. Compare with plantigrade and unguligrade.

**dikinetic skull** A kinetic skull with two joints passing transversely through the braincase. Compare with monokinetic skull.

**dioecious** Pertaining to female and male gonads in separate individuals. Compare with monoecious.

**diphyodont** A pattern of tooth replacement involving only two sets of teeth, usually milk teeth and permanent teeth.

**diplospondyly** The condition in which a vertebral segment is comprised of two centra. Compare with monospondyly.

**dipleurula** A hypothetical invertebrate larva proposed as the common ancestor of echinoderms and hemichordates.

**discoidal cleavage** Early mitotic divisions restricted to the animal pole; extreme case of meroblastic cleavage. Compare with holoblastic cleavage.

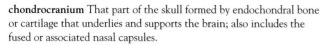

**dissection** The careful exposure of anatomical parts, allowing students to discover and master the extraordinary morphological organization of an animal in order to understand the processes these parts perform and the remarkable evolutionary history out of which they come. Pronounced *dis*-section, as opposed to *di*-section, which means chopping into two halves.

**distal** Toward the free end of an attached part, such as the limb. Compare with proximal.

**diurnal** Active during daylight. Compare with nocturnal.

**dorsal** Toward the back or upper surface of the body; opposite of ventral.

**drag** The force that resists the movement of an object through a fluid; total drag includes parasitic drag and induced drag.

**ecomorphology** The study of the relationship between the form and function of a part and how it is actually used in a natural environmental setting; the basis for determining biological role.

**ectomesenchyme** Loose association of cells derived from neural crest and mesoderm.

**ectotherm** An animal that depends on environmental sources of heat to reach its preferred body temperature. Compare with endotherm.

**edema** Swelling due to collection of fluid in body tissues.

**effector** An organ, such as a muscle or a gland, that responds to nervous stimulation.

**efferent** Refers to the process of carrying away; for example, motor neurons carry impulses away from the central nervous system.

**egg** The haploid cell produced by the female; ovum.

**elastic** In physical terms, the measure of a structure's ability to return to its original size following deformation.

**electric organ** A specialized block of muscles producing electrical fields and often high jolts of voltage.

**electromagnetic radiation** Energy waves through a spectrum that includes radio waves, infrared light, visible light, ultraviolet light, X rays, and gamma rays.

**electromyogram** The electrical record of a muscle contraction.

**electromyography** The study of the pattern of muscle contraction based on detecting its electrical activity.

**electroreceptor** A sensory organ that responds to electrical signals or fields.

**emargination** Large notches in the bony braincase. Compare with fenestra.

**embolomerous vertebra** A dispondylous vertebra in which both centra are separate (aspidospondylous) and of about equal size. Compare with stereospondylous vertebra.

**emulsify** To break up fats into smaller droplets. Compare with digestion.

**enamel** Forms the occlusal cap on most teeth; hardest substance in vertebrate body consisting almost entirely of calcium salts as apatite crystals; secreted by ameloblasts of epidermal origin. See dentin and cementum.

**enamel organ** The part of the tooth-forming primordium that is derived from epidermis, becomes associated with the dermal papilla, and differentiates into the ameloblasts that secrete enamel. See dermal papilla.

**encapsulated sensory receptor** The terminus of a sensory nerve fiber that is wrapped in accessory tissue. Compare with free sensory receptor.

**endochondral bone formation** Embryonic formation of bone preceded by a cartilage precursor that is subsequently ossified; cartilage or replacement bone. Compare with intramembranous bone formation.

**endocrine** Denoting a gland that releases its product directly into blood vessels. Compare with exocrine.

**endocytosis** A phagocytic process in with materials, such as food particles and foreign bacteria, are engulfed by a cell.

**endoskeleton** The supportive or protective framework within the body that lies beneath the integument. Compare with exoskeleton.

**endothelium** The single-celled inner lining of vascular channels. Compare with mesothelium.

**endotherm** An animal capable of maintaining an elevated body temperature with heat produced metabolically from within. Compare with ectotherm.

**enteric nervous system** The network of nerves intrinsic to the digestive system.

**enterocoelom** The body cavity formed within outpocketings of mesoderm. Compare with schizocoelom.

**ependyma** The layer of cells lining the central canal of the chordate spinal cord.

**epiboly** The spreading of surface cells during embryonic gastrulation.

**epidermis** The skin layer over the dermis that is derived from ectoderm.

**epigenetics** The study of developmental events above the level of the genes; embryonic processes not directly arising from the genes that contribute to the developing phenotype.

**epiphysis** 1. The secondary center of ossification on the end of a bone; also refers to the end of a bone. 2. The pineal gland.

**estivation** A prolonged resting state or hibernation during times of heat or drought that is characterized by lowered metabolic levels and breathing rates.

**euryhaline** Having a wide tolerance to salinity differences.

**evolutionary morphology** The study of the relationship between the change in anatomical design through time and the processes responsible for this change.

**excretion** Removal of wastes and excess substances from the body.

**exocrine** Denoting a gland that releases secretions into ducts. Compare with endocrine.

**exocytosis** A process by which the cell releases products.

**exoskeleton** A supportive or protective framework lying on the outside of the body. Compare with endoskeleton.

**extant** Living.

**exteroceptor** A sensory receptor that responds to environmental stimuli. Compare with proprioceptor and interoceptor.

**extinct** Dead.

**extraembryonic** Pertaining to a structure formed by or around the embryo but not retained by or directly contributing to the adult body.

**extrafusal muscle cell** The fiber of striated muscles that actually contributes to the force moving a part. Compare with intrafusal muscle cells and alpha motor neuron.

**extrinsic** Originating outside the part on which it acts. Compare with intrinsic.

**failure** In mechanics, loss of functional integrity and ability to perform; a material may fail but not break. Compare with fracture.

**fascicle** A bundle of muscle fibers defined by a connective tissue coat within a muscle organ.

**fatigue fracture** Reduced breaking strength of an object after prolonged use.

**faveolus** A tiny respiratory compartment within the lung that opens to a central air chamber and results from the subdivisions of the lung lining. Compare with alveolus.

**fenestra** An opening within the bony braincase.

**fermentation** A process in which microorganisms anaerobically extract energy from food in vertebrates by releasing cellulase enzymes that break down plant material.

**fertility** The ability to produce viable eggs or sufficient sperm to propagate offspring; infertility results from nonviable eggs or insufficient sperm. Compare with potency.

**fetus** The embryo at a later stage in development.

**fin** An external plate or membrane that projects from the body of an aquatic animal (as in fishes).

**fixator** A muscle that functions to stabilize a joint. Compare with synergist and antagonist.

**flight** Aerial locomotion accomplished by active flapping of wings. Compare with gliding and parachuting.

**follicle** A small bag that holds cells containing hormones (e.g., thyroid follicle) or one that holds an ovum (e.g., ovarian follicle).

**footfall** Foot contact with the ground during locomotion.

**foramen** A perforation or hole through a tissue wall.

**foramen of ovale** The one-way connection between the right and left atria of an embryonic mammal; closes at birth.

**foramen of Panizza** A connecting vessel between the bases of the left and right aortic arches in crocodilians.

**foregut** Anterior embryonic gut that gives rise to the pharynx, esophagus, stomach, and anterior intestine. Compare with hindgut.

**foregut fermentation** See gastric fermentation.

**formed elements** The cellular components of blood excluding the plasma. Compare with plasma.

**fossorial locomotion** Active removal of soil to produce a burrow; digging.

**fracture** In mechanics, a break or loss of structural integrity; actual separation of material under load. Compare with failure.

**free sensory receptor** The terminus of a sensory nerve fiber that lacks any associated structures. Compare with encapsulated sensory receptor.

**frontal plane** A plane passing from one side of an organism to the other so as to divide the body into dorsal and ventral parts. Compare with transverse plane.

**fulcrum** The point of pivot or the axis of rotation.

**function** How a part performs within an organism. Compare with biological role.

**functional morphology** The study of the relationship between the anatomical design of a structure and the function or functions it performs.

**fusiform** Refers to a narrow shape tapering toward each end.

**gait** The pattern or sequence of foot movements during locomotion.

**gallop** A gait characterized by a high rate of speed and an uneven footfall pattern.

**gametogenesis** The production of mature male and female reproductive cells.

**gamma motor neuron** A nerve cell that innervates intrafusal muscle cells. Compare with alpha motor neuron.

**ganglion** A gathering of nerve cell bodies within the peripheral nervous system.

**ganoin** An older term for a derivative of the enamel that covers some fish scales; ganoid scale.

**gas bladder** A gas-filled bag in fishes derived from the gut. Because the composition of the gases can vary, the term *air* bladder is less appropriate. Compare with swim bladder and respiratory bladder.

**gastralia** Rib-shaped dermal bones located in the abdominal region.

**gastric fermentation** Process in which microorganisms digest food in a specialized stomach; also known as foregut fermentation. See intestinal fermentation.

**gastrocoel** The cavity within the early embryonic gut of the gastrula.

**gastrula** Early embryonic stage during which the basic gut is formed.

**general cranial nerve** A bundle of fibers that detect sensations from the widely distributed viscera. Compare with special cranial nerve.

**genotype** The genetic makeup of an individual.

**gestation** The time that elapses from conception to hatching or birth.

**gill** Aquatic respiratory organ.

**gill slit** Pharyngeal slit associated with a gill.

**gizzard** An especially well muscularized region of the stomach used to grind hard foods.

**gliding** A gradual airborne descent that can be extended by the action of the body and limbs on the relative wind but it is not self-powered. Compare with flight and parachuting.

**glomerulus** 1. A small bed of capillaries associated with the uriniferous tubule. 2. A small cluster of capillaries on the stomochord of hemichordates.

**gnathostome** A vertebrate with jaws.

**graded potential** A nerve impulse proportional to the intensity of the stimulus that produces it and declining thereafter. Compare with action potential.

**gyrus** A swollen ridge on the surface of the brain. Compare with sulcus.

**hair cell** A mechanoreceptive cell with a projecting hair bundle comprised of a kinocilium and several stereocilia.

**half bound** A gait in which the hindfeet contact the ground simultaneously, but the forefeet do not. Compare with bound.

**hemocoel** Blood-filled channels within connective tissue that lack a continuous endothelial lining.

**hemodynamics** The forces and flow patterns of blood circulating within vessels.

**hemopoietic tissue** Blood-forming tissue. Compare with myeloid tissue and lymphoid tissue.

**hepatocyte** A liver cell.

**heterochrony** Within an evolutionary lineage, the change in time at which a characteristic appears in the embryo relative to its appearance in a phylogenetic ancestor; usually concerned with the time of onset of sexual maturity relative to somatic development. Compare with paedomorphosis.

**heterodont** Dentition in which the teeth are different in general appearance throughout the mouth.

**hindgut** Posterior embryonic gut that gives rise to the posterior intestines. Compare with foregut.

**hindgut fermentation** See intestinal fermentation.

**holoblastic cleavage** Early mitotic planes pass entirely through the cleaving embryo. Compare with meroblastic and discoidal cleavage.

**holonephros** A single kidney arising from several regions of the nephric ridge rather than the three types of kidneys (pronephros, mesonephros, metanephros) arising from the nephric ridge.

**holospondyly** The condition in which the centra and spines of vertebrae are anatomically fused into a single bone. Compare with aspidospondyly.

**homeostasis** The constancy of an organism's internal environment.

**homeothermy** The condition of maintaining constant body temperature, without regard to the method.

**homodont** Dentition in which the teeth are similar in general appearance throughout the mouth.

**homology** Features in two or more organisms derived from common ancestors; common ancestry. Compare with serial homology.

**homoplasy** Features in two or more organisms that look alike; similar in appearance.

**hoof** An enlarged cornified plate on the end of an ungulate digit.

**hormone** A chemical messenger that is secreted into the blood by an endocrine organ and affects target tissues.

**horn** An unbranched keratinized sheath with a bony core located on the head; usually occurs in both males and females and are retained year round. Compare with antler.

**hydrofoil** An object that produces lift when placed in a moving stream of water as, for example, the pectoral fin of a shark.

**hydrostatic organ** A structure whose mechanical integrity depends on a fluid-filled core enclosed by walls of connective tissue.

**hydrostatic pressure** Fluid force, as in blood, resulting from heart contraction. Compare with osmotic pressure.

**hyostyly** Jaw suspension primarily through attachment with the hyomandibula.

**hyperplasia** An increase in the number of cells as a result of cell proliferation; usually occurs in response to stress or increased activity. Compare with hypertrophy and metaplasia.

**hyperosmotic** Refers to a solution whose osmotic pressure is greater than the solution that surrounds it; for example, the pressure of the tissue fluid within some fishes is greater that the pressure from the fresh water surrounding them. Compare with hyposmotic and isosmotic.

**hypertrophy** An increase in the size or density of an organ or part, which does not result from cell proliferation. Compare with hyperplasia and atrophy.

**hyposmotic** Refers to a solution lower in osmotic pressure than solution that surrounds it. Compare with hyperosmotic and isosmotic.

**hypoxia** Inadequate levels of oxygen to support metabolic demands.

**hypsodont** Teeth with high crowns. Compare with brachyodont.

**impedance matching** Adjustments of the sound conduction system to address the physical resistance that sound waves meet as they travel from the air to the fluid of the inner ear.

**implantation** The process by which the embryo establishes a viable residence within the wall of the uterus.

**incus** The middle ear bone of mammals derived phylogenetically from the quadrate.

**index fossil** A fossil animal widely distributed geographically but restricted to one rock layer or time horizon; defining species indicator of a stratum.

**induced drag** The resistance to forward travel resulting from the lift produced by an airfoil.

**ingression** A process by which individual surface cells migrate inward to the interior of the embryo.

**insertion** The relatively movable site of attachment of a muscle. Compare with origin.

**integument** The skin covering the body.

**interoceptor** A sensory receptor that responds to internal stimuli. Compare with proprioceptor and exteroceptor.

**interrenal tissues** Endocrine tissue that produces corticosteroids and becomes the cortex of the adrenal gland. Compare with chromaffin tissue.

**interstitial** Pertaining to the fluid-filled space between cells.

**intervertebral body** A pad of cartilage or fibrous connective tissue between articular ends of successive vertebral centra.

**intervertebral disk** A pad of fibrocartilage in the adult mammal that has a gellike core derived from the notochord and is located between articular ends of successive vertebral centra. Compare with intervertebral body.

**intestinal fermentation** Process in which microorganisms digest food in the intestines; also known as hindgut fermentation and cecal fermentation. Compare with gastric fermentation.

**intrafusal muscle cell** The fiber of a striated muscle that is specialized as a sensory organ of proprioception; housed in a muscle spindle. Compare with extrafusal muscle cell, muscle spindle, and gamma motor neurons.

**intramembranous bone formation** Embryonic formation of bone directly from mesenchyme without a cartilage precursor; dermal bone. Compare with endochondral bone formation.

**intratarsal joint** An archosaur ankle in which the line of flexion passes *between* the calcaneus and the astragulus. Compare with mesotarsal joint.

**intrinsic** Belonging entirely to a part; that is, inherent to the part. Compare with extrinsic.

**intromittent organ** The male reproductive organ that delivers sperm into the female reproductive tract; penis or phallus.

**invagination** An indentation or infolding of the surface.

**involution** The turning of embryonic surface cells inward to spread across the interior of the embryo.

**ipsilateral** Occurring on the same side of the body. Compare with contralateral.

**ischemia** Insufficient blood flow to meet the metabolic demands of a tissue.

**isocortex** The cerebral cortex or outer layer of the mammalian cerebrum.

**isolecithal** Pertaining to an egg in which the yolk is evenly distributed throughout the cytoplasm. Compare with telolecithal.

**isometry** Geometric similarity in which proportions remain constant with changes in size.

**isosmotic** Refers to two solutions with equivalent solute levels. Compare with hyperosmotic and hyposmotic.

**jaws** Skeletal elements of bone or cartilage that reinforce the lower borders of the mouth.

**keratin** Fibrous protein.

**keratinization** The process by which the skin forms proteins, especially keratin.

**kinesis** Movement; usually refers to the relative movement of skull bones. See cranial kinesis.

**kinocilium** A modified, rigid cilium of the ear. Compare with microvillus.

**lactation** The release of milk from mammary glands to suckling young.

**lacuna** A small space.

**lamina** A thin sheet, layer, or plate; for example, gill lamella.

**laminar flow** The movement of fluid particles along smooth paths through layers that glide over one another. Compare with turbulent flow.

**larva** An immature (nonreproductive) stage that is morphologically different from the adult.

**lateral** Toward or on the side of the body.

**lecithotrophic** Pertaining to the nutrition that the embryo receives from the yolk of the ovum. Compare with matrotrophic.

**lepidotrichia** A fan-shaped array of ossified or chondrified dermal rods that internally supports the fin of bony fishes. Compare with ceratotrichia.

**lepospondyly** A holospondylous vertebra with a husk-shaped centrum usually pierced by a notochordal canal.

**lever arm** The perpendicular distance from the point at which the force is applied to the point about which a body rotates (moment arm). See moment.

**lift** The force produced by an airfoil perpendicular to its surface.

**lingual feeding** The capture of prey with the tongue.

**load** In mechanics, the forces to which a structure is subjected.

**loop of Henle** A region of the mammalian nephron that includes parts of the proximal and distal tubules (thick limbs) and all the intermediate tubule (thin limb).

**lophodont** Teeth having broad, ridged cusps useful in grinding plant material. Compare with bunodont and selenodont.

**lumen** The space within the core of an organ, especially a tubular organ.

**lymph** Clear fluid carried in lymphatic vessels.

**lymphoid tissue** Blood-forming tissue outside of bone cavities; found, for example, in the spleen and lymph nodes.

**macrolecithal** Pertaining to eggs with large quantities of stored yolk.

**macula** A mechanoreceptor within the vestibular apparatus of the ear; specialized neuromast organ detecting changes in body posture and acceleration. Compare with crista.

**malleus** One of the three middle ear bones in mammals that is phylogenetically derived from the articular.

**mastication** The mechanical breakdown of a large bolus of food into smaller pieces usually with the teeth; chewing of food.

**matrotrophic** Pertaining to the nutrition the embryo receives through the placenta or from uterine secretions. Compare with lecithotrophic.

**meatus** A canal or opening.

**mechanoreceptor** A sense organ that responds to small changes in mechanical force. Compare with chemoreceptor and photoreceptor.

**medulla** The inner portion or core of an organ.

**meroblastic cleavage** Early mitotic planes that do not complete their passage through the embryo before subsequent division planes form. Compare with holoblastic and discoidal cleavage.

**merycism** Remastication together with microbial fermentation of food in nonruminants. Compare with rumination.

**mesenchyme** Loosely associated cells of mesodermal origin.

**mesokinesis** Skull movement via a transverse joint passing through the dermatocranium posterior to the ocular orbit. Compare with prokinesis and metakinesis.

**mesolecithal** Pertaining to eggs with moderate amounts of stored yolk. Compare with microlecithal and macrolecithal.

**mesonephros** A kidney formed of nephric tubules arising in the middle of the nephric ridge; usually a transient embryonic stage that replaces the pronephros, but is itself replaced by the adult opisthonephros or metanephros. Compare with pronephros, opisthonephros, and metanephros.

**mesotarsal joint** An archosaur ankle in which the calcaneous and astragulus fuse and the line of flexion passes between them and the distal tarsals. Compare with intratarsal joint.

**mesothelium** A single-celled lining of body cavities.

**metakinesis** Skull movement via a transverse hinge that lies posterior between the deep neurocranium and outer dermatocranium. Compare with prokinesis and mesokinesis.

**metamorphosis** An abrupt transformation from one anatomically distinct stage (juvenile) to another (adult).

**metanephros** A kidney formed of nephric tubules arising in the posterior region of the nephric ridge and drained by a ureter; usually replaces the embryonic pronephros and mesonephros. Compare with pronephros, mesonephros, and opisthonephros.

**metanephric duct** Ureter; distinct from the pronephric and mesonephric ducts.

**metaplasia** Change of a tissue from one type to another type. Compare with hypertrophy.

**metapterygial fin** Basic fin type in which the axis (metapterygial stem) is located posteriorly in the fin. Compare with archipterygial fin.

**metapterygial stem** The chain of endoskeletal elements within the fish fin that define the major internal supportive axis.

**microcirculation** The capillary beds and arterioles that supply them and venules that drain them.

**microlecithal** Pertaining to eggs that contain small quantities of stored yolk. Compare with mesolecithal and macrolecithal.

**microvillus** A small cytoplasmic projection from a single cell. Compare with villus.

**midsagittal plane** Median parallel plane passing dorsoventrally through the long central axis of the body.

**molariform** A general term describing premolar and molar teeth that appear similar; cheek teeth.

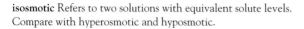

**molt** The shedding of parts or all of the cornified layer of the epidermis; loss of feathers or hair that usually occurs annually; ecdysis.

**moment** The measure of the tendency of a force to rotate a body; the product of force times the perpendicular distance from the point at which the force is applied to the point of rotation (lever arm).

**moment arm** A lever arm.

**monokinetic skull** Skull movement via a single transverse joint passing through the braincase.

**monoecious** Refers to female and male gonads within the same individual; hermaphrodite.

**monospondyly** The condition in which a vertebral segment is comprised of one centrum. Compare with diplospondyly.

**morph** A term referring to the general form or design of an animal; for example, juvenile morph (tadpole) and adult morph (sexually mature stage) of a frog.

**morphological cross section** A plane or cut through the area of a muscle perpendicular to its longitudinal axis at its thickest part. Compare with physiological cross section.

**morphology** The study of anatomy and its significance.

**motor end plate** The neuromuscular junction; specialized ending through which the axon of a motor neuron makes contact with the muscle it innervates.

**motor neuron** A nerve cell carrying impulses to an effector organ. Compare with sensory neuron.

**motor pattern** A defined local pattern of activity produced by muscles that shows little variation when repeated.

**motor unit** One motor neuron and the subset of muscle fibers that it supplies; important in producing graded muscle force.

**mucous gland** An organ secreting a protein-rich mucin that is usually a thick fluid. Compare with serous gland.

**muscle fiber** A muscle cell, that is, the contractile part of a muscle organ.

**muscle organ** Muscle cells together with the noncontractile tissues that support them (connective tissue, blood vessels, nerves).

**muscle spindle** A fusiform bundle in striated muscles that houses specialized sensory receptors known as intrafusal muscle cells.

**myeloid tissue** Blood-forming tissue housed inside bones.

**myoepithelial cell** A cell lining (hence epithelial) a channel or gland and possessing contractile abilities (hence myo-).

**myofibril** A contractile unit of a muscle cell; a chain of repeating sarcomeres composed of myofilaments.

**myofilaments** Thick and thin filaments in the fine structure of muscles composed predominantly of myosin and actin, respectively.

**myomeres** Differentiated segments of a muscle in an adult.

**myotomes** Undifferentiated embryonic blocks of presumptive muscles.

**naris** A nostril.

**natural selection** The process by which organisms with poorly suited features, on average, fare less well in a particular environment and tend to perish, thereby leaving (preserving) those individuals with more favorable adaptations; survival of the fittest.

**neomorph** A new morphological structure in a derived species that has no equivalent evolutionary antecedent.

**neoteny** Paedomorphosis produced by delayed onset of somatic development that is overtaken by normal sexual maturity. Compare with progenesis.

**nephric ridge** The posterior region of the intermediate mesoderm.

**nephridium** A tubular excretory organ.

**nephron** That portion of the uriniferous tubule in which urine is formed; comprised of proximal, intermediate, and distal regions; nephric tubule.

**nephrotome** Segmental forerunner of the nephron in the urinary structure of the early embryo.

**nerve fiber** The cytoplasmic process of a neuron; an axon or a dendrite.

**network** Any structure reticulated or decussated at equal distances with interstices between the intersections.

**neural crest** A paired strip of tissue that separates from the dorsal edges of the neural groove as it forms the neural tube.

**neurocranium** That part of the braincase that contains cavities for the brain and associated sensory capsules (nasal, optic, otic).

**neuroglia** Nonnervous supportive cells of the nervous system.

**neurohormone** A chemical secreted directly into a blood capillary by a neurosecretory neuron at the terminus of its axon.

**neuromast organ** A mechanoreceptive organ comprised of several hair cells, as in the lateral line of the inner ear.

**neuron** A nerve cell.

**neurotransmitter** A chemical released at the synapse of a nerve fiber, usually an axon.

**nerve** A collection of nerve fibers coursing together in the peripheral nervous system.

**nocturnal** Active at night. Compare with diurnal.

**notochord** A long axial rod comprised of a fibrous connective tissue wall around cells and/or a fluid-filled space.

**nucleus** 1. A membrane-bound organelle within the body of a cell. 2. A group of nerve cell bodies within the central nervous system.

**occlusion** The meeting or closure of the upper and lower tooth rows.

**odor** A chemical detected by sensory cells in the nasal epithelium through the olfactory process. Compare with vomodor.

**olfaction** The act of smelling.

**omasum** The third of four chambers in the complex ruminant stomach; a specialization of the esophagus. See abomasum, reticulum, and rumen.

**ontogeny** The course of an individual's development from egg to death.

**operculum** A lid or cover, as over the gills of fishes.

**opisthonephros** The adult kidney formed from the mesonephros and additional tubules from the posterior region of the nephric ridge. Compare with pronephros, mesonephros, and metanephros.

**origin** The relatively fixed site of attachment of a muscle. Compare with insertion.

**osmoregulation** The active maintenance of water and solute balance.

**osmotic pressure** The tendency for fluid solutes to move across a membrane in order to equalize the concentrations of solutes on both sides. Compare with hydrostatic pressure.

**ossification** The process of bone formation; the appearance of bone cells and their surrounding matrix. Compare with calcification.

**osteoderm** A dermal bone located under and supporting an epidermal scale.

**osteon** A highly ordered arrangement of bone cells into concentric rings with bone matrix surrounding a central canal through which blood vessels and nerves run; the Haversian system.

**otoconia** Small calcareous crystals on the maculae of the inner ear; small otoliths.

**otolith** A single calcareous mass in the cupula of hair cells.

**oviduct** A urogenital duct transporting ova and often involved in protection and nourishment of the embryo; müllerian duct.

**oviposition** The act of laying eggs.

**oviparity** The reproductive pattern of egg laying.

**ovulation** The release of the ovum from the ovary.

**pace** A high-speed gait characterized by the two feet on the same side coming in contact with the ground simultaneously; fast amble. Compare with amble.

**paedomorphosis** The retention of general juvenile features of ancestors in the late developmental stages of descendants. See neoteny and progenesis.

**parachuting** An airborne fall slowed by the use of canopylike membranes or body shape that increase drag.

**paradaptation** The concept that some aspects of a feature may not be adaptive or owe their properties to natural selection.

**parallax** The difference in the appearance of an object when it is viewed from two different points.

**parallel muscle** A muscle organ in which all its muscle fibers lie in the same direction and are aligned with its long axis. Compare with pinnate muscle.

**parasagittal plane** A sagittal plane parallel with the midsagittal plane.

**parasitic drag** Resistance to the passage of a body through a fluid as a result of the body's surface friction and adverse backflow in the wake.

**paraxial mesoderm** Paired strips or mesodermal populations forming along the neural tube; in the head it remains as strips of mesoderm called somitomeres, but in the trunk it becomes segmentally arranged as somites.

**parition** The general term for parturition and oviposition.

**partial pressure** The pressure one gas contributes to the total pressure in a mixture of gases.

**parturition** The act of giving birth via viviparity. Compare, oviposition.

**patagium** A stretched fold of skin that forms an airfoil or flight control surface.

**pelagic** Living in open water. Compare with benthic and planktonic.

**pentidactyl** Having five digits per limb; thought to be the basic pattern characteristic of tetrapods but modified by functional demands.

**perfusion** The driving of blood through capillary beds of an organ. Compare with ventilation.

**perichondrium** The sheet of fibrous connective tissue around cartilage.

**perikaryon** The nucleus of a cell and its adjacent cytoplasm, especially applied to nerve cells.

**periosteum** The sheet of fibrous connective tissue around bone.

**peripheral nervous system** The cranial and spinal nerves and their associated ganglia composing that part of the nervous system outside the central nervous system.

**peristalsis** Progressive waves of muscle contractions within the walls of a tubular structure, as within the walls of the digestive tract.

**permissive** Pertaining to hormones that relax insensitive target tissues allowing them to respond to hormonal, neuronal, or environmental stimuli; permitting target tissues to respond.

**pharyngeal slit** An elongated opening in the lateral wall of the pharynx.

**phasic** Recurring cycles of muscle contractions.

**phenotype** The physical and behavioral characteristics of an organism; somatic features. Compare with genotype.

**pheromone** A chemical produced by an organism that affects the behavior of another organism.

**photoreceptor** A radiation sensory receptor responsive to visible light stimuli.

**phylogeny** The course of evolutionary change within a related group of organisms.

**physiological cross section** A plane or cut through the area of all muscle fibers perpendicular to their long axes. Compare with morphological cross section.

**piezoelectricity** Low-level electrical charges arising on the surface of stressed crystals; load stress on bones may produce surface electrical charges.

**pinnate muscle** A muscle organ in which all muscle fibers are aligned obliquely to its line of action. Compare with parallel muscle.

**pituicyte** A nonendocrine cell of the neurohypophysis.

**placenta** A composite organ formed of maternal and fetal tissues through which the embryo is nourished.

**placode** A distinct thickened plate of embryonic ectoderm.

**planktonic** Pertaining to a free-floating microscopic plant or animal that is passively carried about by currents and tides. Compare with benthic and pelagic.

**plantigrade** A foot posture in which the entire sole comes in contact with the ground. Compare with digitigrade and unguligrade.

**plasma** The fluid component of blood without any formed elements.

**plastron** The ventral bony part of turtle shell. Compare with carapace.

**platysma** An unspecialized muscle derived from hyoid arch musculature that spreads as a thin subcutaneous sheet into the neck and over the face.

**plexus** A network of intermingling blood vessels or nerves.

**podocytes** Specialized excretory cells associated with blood capillaries of the kidney.

**polydactyly** An increase in the number of digits over the basic pendidactylous number. Compare with polyphalangy.

**polyphalangy** An increase in the number of phalanges in each digit. Compare with polydactyly.

**polyphyodont** A pattern of continuous tooth replacement. Compare with diphyodont.

**polyspondyly** The condition in which a vertebral segment is comprised of two or more centra.

**portal system** A set of venous vessels beginning and ending in capillary beds or sinuses of the liver.

**potency** The ability of a male to engage in copulation; impotency is an inability to copulate. Compare with fertility.

**power** The amount of work that can be done per unit of time.

**preadaptation** The concept that features possess the necessary form and function to meet the demands of a particular environment before the organism experiences that particular environment. Compare with paradaptation.

**prehension** The rapid grasping and capturing of the prey, usually with the jaws or claws.

**primitive** Denoting an organism or species that appeared early within its phylogenetic lineage; opposite of derived.

**procoracoid** Anterior coracoid (or precoracoid); endochondral bone of the shoulder that first evolved in fishes. Compare with coracoid.

**proctodeum** The embryonic invagination of surface ectoderm that contributes to the hindgut, usually giving rise to the cloaca.

**progenesis** Paedomorphosis produced by precocious onset of sexual maturity in an individual still in the morphologically juvenile stage. Compare with neoteny.

**project** In the nervous system, to transmit neural impulses to.

**prokinesis** Refers to skull movement via a transverse joint that passes through the dermatocranium anterior to the ocular orbit. Compare with mesokinesis and metakinesis.

**pronephros** A kidney formed of nephric tubules arising in the anterior region of the nephric ridge; usually forms only as a transient embryonic structure. Compare with mesonephros, opisthonephros, and metanephros.

**proprioceptor** A specialized interoceptor that responds to limb position, joint angle, and state of muscle contraction. Compare with interoceptor and exteroceptor.

**protandry** Reproduction in which the same individual produces sperm and then later in life produces eggs, but does not do both concurrently.

**protostome** An animal whose mouth forms from or near the embryonic blastopore.

**proximal** Toward the base of an attached part where it joints the body. Compare with distal.

**pterylae** Feather tracks.

**punctuated equilibrium** A description of phylogenetic patterns in which long periods of little or no change are punctuated by short periods of prolific change before returning to a period of little change. Compare with quantum evolution.

**quadrupedal** Walking or running by means of four legs. Compare with bipedal.

**quantum evolution** Adaptive evolutionary change within a lineage characterized by long periods of little change that are suddenly interrupted by short bursts of rapid change. Compare with punctuated equilibrium.

**radial symmetry** A regular arrangement of the body about a central axis.

**radiation receptor** A sensory organ that responds to light and other forms of electromagnetic radiation.

**raptors** Predatory birds that use talons, including hawks, eagles, falcons, and owls.

**rate modulation** The proportionate increase in the contractile force of a muscle as the rate of nerve impulses increases. Compare with motor unit.

**receptor** The end of a nerve fiber that responds to stimuli. Compare with effector.

**recrudescence** Renewal of reproductive interest and readiness of reproductive tracts; usually on a seasonal basis.

**recruitment** The process of initiating contraction of additional muscle cells within a muscle organ during its activity.

**reflex** Involuntary action of effectors mediated by the nervous system.

**refractive index** A measure of the light-bending properties of an object.

**refractory** Not responsive; usually time dependent.

**relative wind** The apparent direction of the airstream across an airfoil; depends on the angle of attack, speed of the airfoil, and so forth.

**release-inhibiting hormone** A hormone that depresses the responsiveness of target tissues.

**releasing hormone** A hormone that initiates target tissue activity. Compare with release-inhibiting hormone.

**renal capsule** The expanded end of the uriniferous tubule that surrounds the vascular glomerulus; ultrafiltrate first forms in the renal capsule; also known as Bowman's capsule.

**respiratory bladder** A gas bladder enriched with capillaries that allow it to function primarily in gas exchange.

**rete** A compact, dense network of capillaries.

**reticulum** The second of four chambers in the complex ruminant stomach; a specialized region of the esophagus. Compare with abomasum, omasum, and rumen.

**rhachitomous vertebra** An aspidospondylous vertebra characteristic of some crossopterygians and early amphibians.

**roe** Fish eggs still within the ovary.

**rumen** The first of four chambers in the complex ruminant stomach; an expanded specialization of the esophagus. Compare with abomasum, omasum, and reticulum.

**ruminant** A placental mammal with a rumen, a specialized expansion of the digestive tract that processes plant material; Ruminantia.

**rumination** Remastication together with microbial fermentation in ruminants.

**sagittal plane** Any plane parallel with the long axis of the animal's body and oriented dorsoventrally.

**sarcolemma** The plasma membrane of a muscle cell.

**sarcomere** A repeating unit of overlapping myofilaments that composes the contractile myofibril of a muscle cell.

**sarcoplasm** The cytoplasm of a muscle cell.

**scaling** Compensatory adjustments in proportion to maintain performance with changes in size.

**scansorial locomotion** Climbing of trees with claws. Compare with brachiation.

**schizocoelom** The body cavity formed by splitting of the mesoderm. Compare with enterocoelom.

**secondary cartilage** Cartilage that forms after initial bone ossification is complete; formed usually in response to mechanical stress, especially on the margins of intramembranous bone.

**secondary plant compounds** Chemicals produced by plants that are toxic or unpalatable to herbivores.

**sectorial teeth** Teeth with opposing sharp ridges specialized for cutting.

**segmentation** A body made up of repeating sections or parts; metamerism.

**seismic** Relating to an earth vibration.

**selection force or pressure** The biological or physical demands arising from the environment that affect the survival of the individual living there.

**selenodont** Teeth with crescent-shaped cusps, as in artiodactyls. Compare with bunodont and lophodont.

**sensory neuron** A nerve cell carrying responses from a sensory organ to the brain or spinal cord. Compare with motor neuron.

**serial homology** Similarity between successively repeated features in the same individual.

**serial theory** The hypothesis that jaws evolved from one of the anterior branchial arches. Compare with composite theory.

**serous gland** An organ secreting a thin, watery fluid.

**sesamoid bone** A bone that develops directly in a tendon, for example, the patella (kneecap).

**sessile** Pertaining to an animal attached to a fixed substrate in its environment.

**sheer force** The direction of an applied force that tends to slide sections of an object across each other.

**sinusoids** Tiny vascular channels that are slightly larger than capillaries and lined or partially lined only by endothelium.

**solenocyte** A single excretory cell with a projecting circle of microvilli around a central flagellum.

**solute** Molecules dissolved in solution.

**somatic** Pertaining to the body, usually to the skeleton, muscles, and skin but not to the viscera.

**somatosensory system** All proprioceptive neurons and neurons receiving stimulation from the skin.

**special cranial nerve** A bundle of fibers that detect stimuli from the local senses—sight, smell, hearing, balance, and lateral line. Compare with general cranial nerve.

**spermatophore** A package of sperm ready for delivery or presentation to the female.

**sphincter** A band of muscle around a tube or opening that functions to constrict or close it.

**spinal nerve** Any nerve entering or departing from the spinal cord. Compare with cranial nerve.

**spiracle** A reduced gill slit that is first in series.

**splanchnocranium** That part of the skull arising first to support the pharyngeal slits and later contributing to the jaws and other structures of the head; branchial arches and derivatives; visceral cranium.

**spontaneous generation** The concept that fully formed organisms arise directly and naturally from inanimate matter.

**stall** Loss of lift due to the onset of turbulent flow across an airfoil.

**stapes** One of the three middle ear bones in mammals that is phylogenetically derived from the columella (hyomandibula).

**stenohaline** Having a narrow tolerance to salinity differences.

**stereocilia** Very long microvilli.

**stereoscopic vision** Denoting the ability to see images in three dimensions.

**stereospondylous vertebra** A monospondylous vertebra in which the single centrum (an intercentrum) is separate (aspidospondylous). Compare with embolomerous vertebra.

**stigmata** An extensively subdivided pharyngeal slit.

**stolon** A rootlike process of ascidians and other invertebrates that may fragment into pieces that asexually grow into more individuals.

**stomodeum** The embryonic invagination of surface ectoderm that contributes to the mouth.

**stratified** Formed of layers.

**stratigraphy** The geology of the origin, composition, and relative chronology of strata.

**stratum** (pl., strata) The geological term for a layer of rocks deposited during about the same geological time.

**strength** The load a structure withstands before failing or breaking.

**streptostyly** The condition in which the quadrate bone is movable relative to the braincase.

**subcortical region** That portion of the telencephalon exclusive of the cerebral cortex.

**subterranean** Pertaining to life lived underground.

**suction feeding** The capture of prey by means of a sudden muscular expansion of the buccal cavity that creates a vacuum to draw in water carrying the prey.

**sulcus** A furrow on the surface of the brain. Compare with gyrus.

**surfactant** A soluble compound that reduces surface tension, as in the lungs.

**suspension feeding** Feeding based on filtering suspended food particles from water; usually involves cilia and secreted mucus; filter feeding or ciliary-mucus feeding.

**swim bladder** A gas bladder functioning primarily in buoyancy control.

**synapse** The region of contact between two neurons or between a neuron and an effector organ.

**syncytium** Multinucleated cytoplasm; an aggregation of cells without cell boundaries.

**synapticules** Cross-linking connections between pharyngeal bars in amphioxus.

**synergist** Two or more muscles cooperating to produce motion in the same direction. Compare with antagonist and fixator.

**tachycardia** Abnormally fast heart rate. Compare with bradycardia.

**tactile** Pertaining to touch.

**talons** Specialized bird claws used in striking or catching live prey.

**target tissue** A group of related cells that respond to a particular hormone.

**telolecithal** Pertaining to eggs in which yolk stores are concentrated at one pole.

**tendon** A noncontractile, fibrous connective tissue band joining a muscle organ to a bone or cartilage. Compare with aponeurosis.

**tensile force** 1. The direction of an applied force that tends to pull apart an object. 2. The force produced by muscle contraction.

**thermoreceptor** A radiation receptor sensitive to infrared energy.

**thermoregulation** The process by which body temperature is established and maintained.

**tidal volume** The total quantity of air inhaled and exhaled in one breath. Compare with dead space.

**tissue fluid** A clear liquid outside of blood or lymphatic vessels that bathes cells.

**tonic fibers** Slow contracting muscle fibers that produce prolonged, sustained contractions with low force. Compare with twitch fibers.

**tonus** Partial muscle contraction with low force when a muscle is in a relaxation state

**tract** A collection of nerve fibers coursing together in the brain or spinal cord. Compare with nerve.

**transverse plane** A plane passing from one side of an organism to the other so as to divide the body into anterior and posterior parts.

**transverse process** A general term for any bony or cartilaginous projection from the centrum or neural arch.

**trophoblast** The outer cellular layer of the mammalian blastocyst.

**trot** A gait characterized by diagonally opposite feet coming in contact with the ground simultaneously.

**turbulent flow** The movement of fluid particles in irregular paths. Compare with laminar flow.

**tusks** Specialized, long teeth protruding from the mouth; elongate incisors (elephants), left upper incisor (narwhal), canines (walruses).

**twitch fibers** Fast contracting muscle fibers whose force may or may not fatigue quickly; phasic fibers. Compare with tonic fibers.

**tympanum** The eardrum or tympanic membrane.

**ungulate** A hoofed placental mammal belonging to the orders Perissodactyla (horses) and Artiodactyla (cattle, deer, pigs).

**unguligrade** A foot posture in which the weight is carried on the tips of the toes (as in horses).

**ureotelism** Excretion of nitrogen in the form of urea.

**ureter** The metanephric duct arising as a ureteric diverticulum and draining the metanephros.

**uricotelism** Excretion of nitrogen in the form of uric acid.

**uriniferous tubule** The functional unit of the kidney comprised of the nephron and collecting tubule.

**vasoconstriction** The narrowing of a blood vessel; usually resulting from smooth muscle contraction. Compare with vasodilation.

**vasodilation** The widening of a blood vessel; may be active or passive enlargement. Compare with vasoconstriction.

**vasoreceptor** Monitors pressure and gas levels in blood passing through the heart and systemic arches.

**vein** A blood vessel carrying blood toward the heart; blood carried may be low or high in oxygen tension. Compare with artery.

**velocity** The rate of change of displacement; how fast a body is traveling in a particular direction.

**ventilation** The active movement of water or air across respiratory exchange surfaces. Compare with perfusion.

**ventral** Toward the belly or bottom of an animal; opposite of dorsal.

**vertebra** One of several bone or cartilage blocks firmly joined into a backbone that defines the major body axis of vertebrates.

**vestibular apparatus** A sensory organ of the inner ear comprised of semicircular canals and associated compartments, such as the sacculus, utriculus, and cochlea (lagena).

**villus** A fingerlike projection of a tissue layer, as in the small intestine. Compare with microvillus.

**viscosity** The resistance of a fluid to flow.

**viviparity** The reproductive pattern of live birth; birth of young not encased in a shell.

**vomeronasal organ** A chemosensory organ present in the nasal chamber or roof of the mouth of some tetrapods.

**vomodor** A chemical detected by sensory cells of the vomeronasal organ. Compare with odor.

**warren** An underground maze of excavated passageways used by animals, usually rabbits.

**wolffian duct** Mesonephric duct.

**zygapophysis** The projection of a neural arch that articulates with the adjacent neural arch.

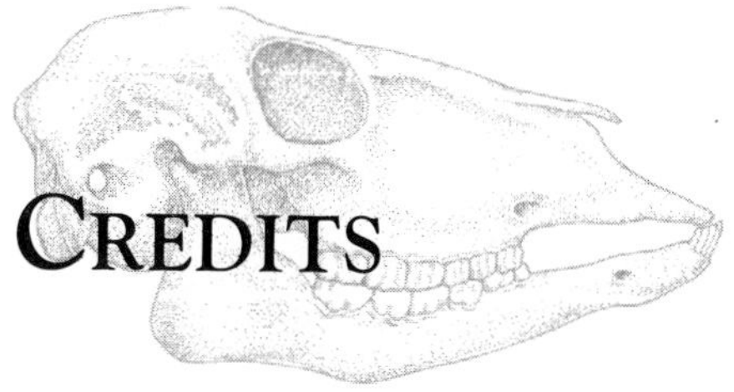

# CREDITS

## Photographs

### Chapter 1

**Figures 1.2:** Neg. 326697, Courtesy Department Library Services. American Museum of Natural History; **1.3:** Historical Pictures/Stock Montage; **1.4:** Museum of Comparative Zoology, Harvard University; **1.5a:** Historical Pictures/Stock Montage; **1.5b:** Courtesy Bibliotheque Centrale MNHN Paris; **1.6:** National Portrait Gallery, London; **1.7:** Historical Pictures/Stock Montage; **1.8:** Historical Pictures/Stock Montage; **1.9a:** Historical Pictures/Stock Montage; **1.23:** © J. Fisher, "Fossil birds and their adaptive radiation" in *The Fossil Record,* The Geological Society of London, 1967; **1.27:** Neg. #336671, Courtesy Department of Library Services, American Museum of Natural History; **1.28:** Courtesy Dr. R. Wild, Staatl. Museum fur Naturkunde, Stuttgart; **1.29:** Dr. John D. Cunningham/Visuals Unlimited; **1.30a:** Neg. #330491 Photo: A. E. Anderson. Courtesy Department Library Services, American Museum of Natural History; **1.30b:** Neg. # 35608 Photo: A. E. Anderson, Courtesy Department Library Services, American Museum of Natural History; **1.31a:** Neg. # 324393 Photo: R. T. Bird, Courtesy Department Library Services, American Museum of Natural History; **1.32a:** Courtesy of Dinosaur National Monument of the National Park Service; **1.32b:** © Dr. David Taylor, Northwest Museum of Natural History, Portland State University, Portland, OR; **1.34:** Illustration of "Chasmosaurus" by Eleanor M. Kish. Reproduced courtesy of the Canadian Museum of Nature, Ottawa, Canada; **1.37b:** Grand Canyon National Park Image #4922; **1.37c:** Zion National Park; **1.37d:** National Park Service; **1.37e:** National Park Service; **1.37f:** © M. W. Williams/WASO-NPS-5786311/National Park Service; **Box figure 1.3-1:** Longmans Green, Inc./Random House.

### Chapter 4

**Figures 4.18a:** Kenneth Kardong; **4.18b:** Kenneth Kardong; **4.39a-b:** Reprinted from Cowin, Hart, Balser, Kohn, "Functional Adaptation in Long Bones: Establishing in Vivo Values for Surface Remodeling Rate Coefficients," *J. Biomechanics,* Vol. 18, No. 9, 1985, with permission from Pergamon Press Ltd., Headington Hill Hall, Oxford OX3 OBW, UK; **4.41b:** Reprinted by permission of the publishers from *Functional Vertebrate Morphology,* edited by M. Hildebrand et al.: The Belknap Press of Harvard University Press, Copyright 1985.

### Chapter 6

**Figure 6.12a:** From *Comparative Vertebrate Histology,* Donald I. Patt & Gail R. Patt.

### Chapter 10

**Figure 10.7b:** From *Comparative Vertebrate Histology,* Donald I. Patt and Gail R. Patt.

### Chapter 11

**Figure 11.27a-b:** Journal of Morphology, 169, 1981, "Ultrastructure of the Lung of the Rattlesnake, *Crotalus viridis oreganus,*" D. L. Luchtel and K. V. Kardong. Reprinted by permission of Wiley-Liss, a division of John Wiley & Sons, Inc.

### Chapter 12

**Figures 12.10a:** Reprinted by permission of McGraw-Hill, Inc. from *Early Embryology of the Chick,* Bradley M. Patten, 1956; **Box figure 12.1-1a:** Reprinted with permission of Macmillan Publishing Company from *Chordate Structure and Function* by Arnold G. Kluge, 1977; **Box figure 12.1-1b:** Reprinted with permission of Macmillan Publishing Company from *Chordate Structure and Function* by Arnold G. Kluge, 1977; **Box figure 12.1-1c:** Reprinted with permission of Macmillan Publishing Company from *Chordate Structure and Function* by Arnold G. Kluge, 1977.

### Chapter 14

**Figure 14.40a:** Courtesy of Dr. Edward J. Zalisko.

### Chapter 16

**Box figure 16.3-1a:** Courtesy of W. H. Freeman & Company.

### Chapter 17

**Figure 17.27:** Courtesy of W. H. Freeman & Company.

### Chapter 18

**Figure 18.1b:** © Royal Commission on the Historical Monuments of England Crown Copyright.

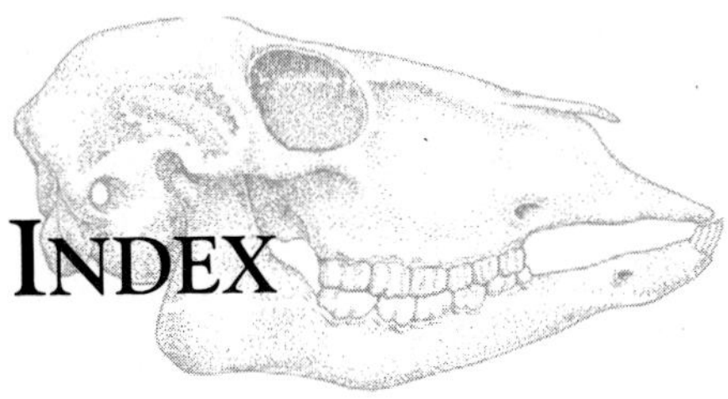

# INDEX